MW00803953

Fundamentals of

Water Treatment Unit Processes

Physical, Chemical, and Biological

Fundamentals of

Water Treatment Unit Processes

Physical, Chemical, and Biological

David Hendricks

CRC Press is an imprint of the
Taylor & Francis Group, an **informa** business

CRC Press
Taylor & Francis Group
6000 Broken Sound Parkway NW, Suite 300
Boca Raton, FL 33487-2742

© 2011 by Taylor and Francis Group, LLC
CRC Press is an imprint of Taylor & Francis Group, an Informa business

No claim to original U.S. Government works

Printed in the United States of America on acid-free paper
10 9 8 7 6 5 4 3 2

International Standard Book Number: 978-1-4200-6191-8 (Hardback)

This book contains information obtained from authentic and highly regarded sources. Reasonable efforts have been made to publish reliable data and information, but the author and publisher cannot assume responsibility for the validity of all materials or the consequences of their use. The authors and publishers have attempted to trace the copyright holders of all material reproduced in this publication and apologize to copyright holders if permission to publish in this form has not been obtained. If any copyright material has not been acknowledged please write and let us know so we may rectify in any future reprint.

Except as permitted under U.S. Copyright Law, no part of this book may be reprinted, reproduced, transmitted, or utilized in any form by any electronic, mechanical, or other means, now known or hereafter invented, including photocopying, microfilming, and recording, or in any information storage or retrieval system, without written permission from the publishers.

For permission to photocopy or use material electronically from this work, please access www.copyright.com (http://www.copyright.com/) or contact the Copyright Clearance Center, Inc. (CCC), 222 Rosewood Drive, Danvers, MA 01923, 978-750-8400. CCC is a not-for-profit organization that provides licenses and registration for a variety of users. For organizations that have been granted a photocopy license by the CCC, a separate system of payment has been arranged.

Trademark Notice: Product or corporate names may be trademarks or registered trademarks, and are used only for identification and explanation without intent to infringe.

Library of Congress Cataloging-in-Publication Data

Hendricks, David W.
Fundamentals of water treatment unit processes : physical, chemical, and biological / David W. Hendricks.
p. cm.
Includes bibliographical references and index.
ISBN 978-1-4200-6191-8 (hardback)
1. Water--Purification. 2. Sewage--Purification. I. Title.

TD430.H457 2011
628.1'62--dc22 2010036943

Visit the Taylor & Francis Web site at
http://www.taylorandfrancis.com

and the CRC Press Web site at
http://www.crcpress.com

Contents

PART I Foundation

PART III Microscopic Particles

PART IV Molecules and Ions

PART V Biological Treatment

Preface

This book is intended primarily as a text for a course in water treatment normally taught to seniors or first-year graduate students. The academic background needed includes the basic undergraduate courses in engineering, that is, mathematics, general chemistry, and fluid mechanics.

The main thrust of the book was to delineate principles that support practice. The "unit processes" approach was the organizing concept. Most of the principles identified are common to any kind of water treatment, for example, drinking water, municipal wastewater, industrial water, industrial wastewater, and hazardous wastewater. The book seeks to identify the strands of theory rather than to keep up with the latest technologies. The underlying idea was that technologies change but principles remain constant.

The chapters are sprinkled with boxes which are explanatory asides. The idea of the boxes was to enlarge the reader's perspective of a topic by including some of the lore and history. Understanding how we arrived at our present state of the art places it in a more logical context. The book intends not only to provide technical proficiency but also to add insight and understanding of the broader aspects of water treatment unit processes. As a note, boxes of living persons were not included without their respective verbal permissions. In writing biography boxes, I did not consider persons who were of my generation or younger, albeit I know many who would warrant such consideration, but perhaps from a younger writer.

The book is organized as follows: Parts I through IV (Chapters 1 through 21) are distillations of *Water Treatment Unit Processes—Physical and Chemical* (CRC Press, Boca Raton, FL, 2006). Part V (Chapters 22 and 23 and on biological treatment) was developed for this book. The intent was to abstract key principles of unit processes with minimal amplification. The 2006 book is more comprehensive, with additional theory and examples of practice. Problems are intended to illustrate principles, but with regard to practice. The SI (kg-m-s) system of units has been used; other metric units have been used as needed. The equivalent U.S. customary units are shown in most tables and figures.

David Hendricks

Acknowledgments

Acknowledgments have been included at the end of most chapters. A number of persons have shown interest in the book and have helped in various ways. Kevin Gertig, manager of water resources and treatment operations, Fort Collins Utilities, has supported the idea of the book and has provided advice and help when needed. Other friends and colleagues in the industry have provided help whenever required.

From the publisher's side, Joseph Clements, editor, Taylor & Francis, CRC Press, Boca Raton, Florida, has nurtured the book during the three years of its development along with Jessica Vakili, production editor. Shayna Murry was the graphic designer for the cover art. Andrea Dale compiled information about the book for dissemination. Robert Sims was the project editor and saw the book through its final stages. Perundevi Dhandapani, project manager, oversaw the copyediting of the text, finding many of the author's oversights and also ways to improve the book's readability. In summary, the manuscript was transformed to a book through the work of the professionals at Taylor & Francis/CRC Press.

Author

David W. Hendricks, received his BS in civil engineering from the University of California, Berkeley, California; his MS in civil engineering from Utah State University, Logan, Utah; and his PhD in sanitary engineering from the University of Iowa, Iowa City, Iowa. He joined Colorado State University, Fort Collins, Colorado, in 1970 and has taught courses in water treatment unit processes and industrial wastes. He has been principal investigator (PI) on some 30 research projects in the field and has over 100 publications to his credit. He has served as a consultant for a range of private firms and government agencies, with various overseas assignments. He is a member of 8 professional organizations, including the American Society of Civil Engineers (fellow), the Association of Environmental Engineering Professors, the American Chemical Society, the American Institute of Chemical Engineers, the International Water Association, the American Water Works Association, and the Rocky Mountain Water Environment Association, and is a board certified environmental engineer (BCEE) by the American Academy of Environmental Engineers.

Downloadable Files

Spreadsheet tables have been included as a means to illustrate applications of design principles that involve computations. They are intended as supplemental material for the reader to modify as may be useful.

The spreadsheets are referenced in the text by a "CD" prefix. For example, "Table CD4.3" in the text indicates that the table is found in spreadsheet form in a file with that designation. In the text, "Table CD4.3" is found as an excerpt from the spreadsheet. The excerpt permits the reader to glean the gist of the spreadsheet contents but without the imperative to access it (except as convenient). The "CD" prefix has no current significance except as a way to designate the respective content as being a computer "file."

Some 60 spreadsheets were generated for this text and are available from the CRC Web site. They are listed in the following "*Contents–Downloadable Files*," along with several figures that are "linked" to their respective spreadsheets. One particular figure file, Figure CD11.17, is an animated walk through of a part of a plant. The files can be accessed through the CRC Web site, at http://www.crcpress.com/product/isbn/9781420061918

Contents—Downloadable Files

PART I Foundation

PART II Particulate Separations

PART III Microscopic Particles

PART IV Molecules and Ions

Part I

Foundation

Taking time to assimilate themes common to all unit processes helps to put order and understanding into learning about water treatment as opposed to a collection of facts and equations. Therefore, Part I describes some of the foundations, or "building blocks," of the field. This *foundation* constitutes the first four chapters.

Unit processes: Chapter 1 describes how water treatment as a topic can be disaggregated to unit processes, principles, and technologies. Examples of treatment trains illustrate a variety of treatment applications. These expand beyond the traditional municipal potable water and wastewater to include tertiary treatment, modified water treatment, industrial process water, and industrial wastewaters. The variety of applications is almost without limit.

Contaminants: Chapter 2 looks at the variety of contaminants found in water. Selected ones must be removed in order to provide for further uses of water either as required by law or as motivated by a private need such as for industrial process water.

Models: In Chapter 3, the idea of a "model" is described in terms of its variety of forms. Models are at the root level of design. We use models in everyday life ranging from mental images to photographs. The designer uses whatever may be available, for example, inspections of existing plants, judgment, rules of thumb, equations, mathematical models, physical models, computer animations, etc. All of these are models and are means to project from the abstract to operation and design.

Reactors: The idea of the "reactor," in Chapter 4, applies to many unit processes. It is the notion that if we pass a dissolved or particulate contaminant through a particular kind of "black-box," changes will occur. The reactor concept is the basis for formulating these changes mathematically. The general idea applies to a variety of unit processes, for example, settling, mixing, deep bed filtration, adsorption in packed columns, ion-exchange, membranes, gas transfer, disinfection, precipitation, oxidation, activated sludge, bioreactors, etc. The concept applies to natural systems where "passive" changes occur, as well as to engineered systems.

1 Water Treatment

The topics covered in this chapter include a review of unit processes, the genesis of water treatment as a technology, a discussion of units, and an overview of how the book is organized.

1.1 WATER TREATMENT IN-A-NUTSHELL

Any water treatment is done in the context of a treatment train, a collection of unit processes. Such unit processes may include screening, sedimentation, flotation, coagulation, filtration, adsorption, ion exchange, gas transfer, oxidation, biological reactions, and disinfection. The aggregation selected, i.e., the treatment train, if applied to full scale with all of the needed appurtenances and engineering to make it function on a continuous basis, is a water treatment plant (WTP). The objective of water treatment, by this treatment train, is to effect a required change in water quality.

The water being treated may be any water, e.g., ambient water used as a source for drinking water, municipal or industrial wastewater, contaminated groundwater, brackish water, seawater, or the product water from any treatment plant. The treatment train employed in a WTP depends, first, upon the source water and the objectives for the effluent water and, second, on other factors, such as capital and operating costs, reliability, ease of operation, traditions, current practices, etc. The relative importance of these other factors depends on the project at hand.

1.1.1 Water Treatment Plants

To bring into existence an operating WTP requires more than a consideration of unit processes. A well-conceived treatment train and the proper design of its unit processes is the heart of any design. But much more is required to support the process design. All plants must have, e.g., stated not in a particular order, various kinds of sampling, metering and monitoring, control of flows with pumps and valves of various types, facilities for receiving and storing chemicals along with their subsequent metering and injection, safety measures regarding chemicals and many other aspects of plant operation, laboratory support, utility tunnels, structural design of various tanks, hydraulic design for various purposes including setting the hydraulic grade line of a plant, etc. In a plant with granular activated carbon (GAC), transport of the treated carbon to and from the reactors must be provided along with provision for regeneration either on-site or off-site. In addition, the storage tanks, usually steel, must be protected from corrosion. The distinction between the design of water treatment processes and a WTP is between principle and implementation. The process principles must be adhered to but they must be provided the means for being implemented, i.e., in terms of a plant. The latter is not excluded in this book, but is not the main focus. The two areas are complementary. Process design can be taught in school. But in putting it all together, experience is most important.

1.1.2 Residuals

As an axiom of water treatment, residuals are always a by-product. They are unwanted, but must be dealt with. For example, in membrane treatment, the "concentrate" water flow is often limiting with respect to the feasibility of an installation. The residuals stream in water treatment includes the sludge from settling basins and the backwash water of filtration. Settling ponds must be provided to decant the water and to dispose of the resulting sludge. In wastewater treatment, a variety of treatment trains may be employed to process the solid streams, with some kind of recycle or disposal being the end result.

Some of these required tasks can be taught and other aspects must be learned by experience. Some introduction is helpful, but experience and visits to installations is essential. Many questions can be answered readily and the problems may be put into perspective by visits to plants complemented by discussions with operators and experienced engineers.

1.2 ORGANIZATION OF WATER TREATMENT KNOWLEDGE

Common themes in organizing water knowledge include

1. Treatment for a particular purpose, such as for drinking water, wastewater, industrial wastewater, and contaminated groundwater
2. Treatment of particular contaminants
3. Unit processes

From about 1880 to 1960, knowledge developed along the lines of the first approach, i.e., drinking water treatment, wastewater treatment, and industrial wastes treatment. Then treatment of hazardous wastes emerged about 1980, employing some of the same unit processes. Desalting of brackish waters and seawater became issues beginning about 1960. Thus, we have had books and persons who think along the lines of drinking water treatment and others who think of municipal wastewater treatment, with commensurate books and journals, and yet another group who specializes in treatment of industrial wastes, etc.

BOX 1.1 CONTEXT AND FIT

A particular type of situation is the *context* for a design and has to do with water quality, treatment objectives, operation capabilities, financing, etc. The *form* is the unit process/technology adopted. The *fit* is the relationship between the *context* and the *form*. An *appropriate* fit is desired. In other words, the process/technology selected must be appropriate for the context. These ideas are central to design.

As an example, a slow sand filter may be appropriate technology for a small community, since materials are available locally, operation is "passive," i.e., not requiring operator skill, and parts are simple and easily repaired locally. On the other hand, slow sand may not be appropriate for a large community as the labor requirement is much more than for a rapid rate plant. For the latter, automation is feasible and parts and skilled labor are likely to be readily available.

In some cases, the paradigm for thinking has been along the lines of treatment for specific contaminants, mostly those that are regulated, or perhaps those that pose an industry problem. For example, halogenated hydrocarbons have been regulated in drinking water in 1978 starting with tri-halomethanes. Nutrients in wastewater have been a treatment focus since the early 1970s as a means to reduce algae blooms in ambient waters. Heavy metals, an industrial waste problem, have also been regulated since the 1970s. There are books as well as articles dedicated to these topics.

In the 1960s, some in the academic community began to adopt the unit operations/unit processes approach in organizing the knowledge of water treatment. This approach was inspired by two books, *Unit Operations of Sanitary Engineering* (Rich, 1961) and *Unit Processes of Sanitary Engineering* (Rich, 1963), and reinforced by persons pursuing graduate degrees who had taken courses in chemical engineering.

In comparing the three approaches, the first is limited in perspective to the problem area at hand (e.g., drinking water treatment, municipal wastewater treatment, industrial wastewater treatment, and hazardous wastes treatment). Screening, e.g., as a bar screen in wastewater treatment is viewed as a means to protect pumps from items that might cause clogging. In water treatment, screening may be a microscreen to remove algae. As a unit process, screening is a means to retain objects and particles for whatever purpose. The difference is in the viewpoint. The first two are by nature empirical and specific, while the third is adaptable to the purpose without the constraints of tradition. With the first two approaches, one lacks the broad perspective that screening may be applied to any problem area.

In this book, we use the unit processes approach. The unit process approach includes operative principles, traditions, practices, empirical methods, technology forms, and the spectrum of applications.

1.3 UNIT PROCESSES

This book presents the topic of water treatment in terms of unit processes, which are discussed in the following chapters. For each unit process considered, principles and practices are explained. The unit processes approach is common to the field of chemical engineering and has been assimilated by the field of environmental engineering.

1.3.1 DEFINITIONS

A water treatment unit process is defined as an engineered system to effect certain intended state changes for the water. Examples include screening, gravity settling, coagulation, flocculation, filtration, gas transfer, ion exchange, adsorption, membrane separations, biological treatment, disinfection, oxidation, and chemical precipitation (Sanks, 1978; Letterman, 1999). These unit processes are the topics of this book. Some of the definitions are as follows:

- *Screening*. The retention of a substance by a screen that has a mesh size smaller than the substance to be retained.
- *Gravity settling*. A particle falling under the influence of gravity is called sedimentation.
- *Coagulation*. The charge neutralization of a negatively charged colloid, usually by chemical means, such as the use of alum or a ferric compound.
- *Flocculation*. A unit process that promotes collisions between particles that attach to each other upon contact, growing in size to increase settling velocity.
- *Filtration*. The convection of a water stream through a porous media with the intent to retain suspended particles within the media.
- *Gas transfer*. The transport of gas between the dissolved phase in water and a gas phase.
- *Ion exchange*. The exchange of benign ions (such as Na^+) bonded to sites within an ion-exchange material (such as a zeolite mineral or a synthetic resin) intended to be displaced by an ion targeted for removal (such as Ca^{2+}) that has a stronger bonding force.
- *Adsorption*. The attachment of a molecule to an adsorption site provided by an internal surface of an adsorbent material. Activated carbon is the best-known adsorbent for an engineered system, although virtually any solid material can provide adsorption sites.
- *Membrane separation*. The four types of membrane processes are as follows:
  1. Microfiltration (removes colloids and bacteria)
  2. Ultrafiltration (removes viruses)
  3. Nano-filtration (removes large molecular weight organic molecules and some ions)
  4. Hyper-filtration, more commonly called "reverse osmosis" (removes molecules and ions)

The membrane processes are different in pore size and in pressures required; defined pores may be lacking, in fact, for the latter two membrane processes:

- *Biological treatment.* A reaction between an organic molecule and a microorganism
- *Disinfection.* The inactivation of microorganisms
- *Oxidation.* The gain of electrons by chemical reaction
- *Precipitation.* The formation of a solid substance from ions in solution

To the extent possible, these unit processes are described in terms of principles. As a rule, however, scientific principles alone do not provide for a "complete" engineering solution.

Engineering has the characteristic that a solution to a problem is the important thing. A problem to be solved cannot wait for a rational explanation. So while the scientific explanation is sought, it is often necessary to make do with some means to make a decision about design. Such methods may involve judgment, a knowledge of scientific principles, modeling, laboratory testing, lore concerning how things have been done in the past, rules of thumb, the use of some kind of calculation method such as a loading rate, associated criteria, etc. Thus, while the book describes what is known as the rationale for a process design, methods used in practice are reviewed also. (The view taken here is that the foregoing provides a means for decision-making. Scientific certainty and accuracy are sought, but the main thing is to provide a basis for achieving a defined level of system performance and doing so economically and with a social "fit.")

A "state" is defined here as the water quality characteristics of a given parcel of water and may include concentrations of suspended solids (including organisms), ions, and molecules; temperature; pH; etc. A treatment "process" is intended to cause a desired change of the state of a volume of water. The idea of a water quality "state" was taken from the field of physical chemistry in which the state of a gas is defined by its temperature, pressure, and volume.

1.3.2 Technologies

For a given unit process, a technology is a means for implementation. For example, a rapid filter is a means to implement a "deep-bed" filtration process. The rapid filter includes the array of appurtenances to make it work, e.g., the filter box depth and area, under-drain system, and backwash system. Rules of thumb, tradition, and manufacturer's standards govern the sizing and characteristics of each component.

In short, a technology is a "package" of design guidelines and components that result in a system that supports a workable process. The manufacturer may provide a complete package or some of all of the supporting components.

1.3.3 Breadth of Unit Processes and Technologies

To illustrate the idea further, Table 1.1 lists some 15 unit processes with samples of associated treatment technologies. The list of associated technologies is not complete, nor does it show the range of variation. The technologies are listed generically. If proprietary technologies were listed, Table 1.1 might be several pages long. All of this is mentioned so that one may gain an appreciation for the breadth of the number of technologies that have been developed.

Table CD1.2 is an excerpt from a matrix with some 700 contaminants listed in rows and 11 basic unit processes listed in columns with selected technologies under each unit process (Champlin and Hendricks, 1993). Columns further to the right list variations. Looking down the columns, the contaminants that may be subject to treatment by a given technology are designated by a code in the intersecting cell based upon the expected percent removal. The matrix expands on what is shown in Table 1.1 but includes a listing of contaminants. Table CD1.2a shows an overall layout of the matrix, i.e., how to navigate, while Table CD1.2b is the large matrix with some 700 rows and 100 columns. The idea of the large matrix is to give an appreciation for the scope of water treatment as a field of practice.

1.3.4 Proprietary Technologies

Proprietary innovation plays a large part in technologies. Several manufacturers offer their own "packages" of technology for the deep-bed filtration process, for example. To illustrate, the Parkson Dynasand® filter is a moving bed filter that takes off the floc-saturated media at the bottom, replenishes the filter bed at the top, and so backwash is not required. The Culligan Multi-Tech® filter is a complete package plant that provides the means for flocculation within a coarse media preceding the main filter and is designed for the complete automation of the filtration process. The Infilco-Dregemont ABW® (automatic backwash) filtration system is a shallow-bed filter comprised of a series of transverse cells 305 mm (1.0 ft) wide. A traveling bridge with a hood to collect the backwash flow and a backwash pump with "shoe" that places the backwash plumbing over the under-drain opening in the finished water channel provides for a short duration backwash of a single cell. The bridge moves then to the next cell and the filter as a whole is never taken out of operation.

The filtration example illustrates the role of proprietary companies in making successful a given process. Essentially, an array of proprietary technologies have been developed for each unit process either to provide support for a generic design such as ancillary equipment or to provide a full operational package technology. Two or three of the unit processes have been developed largely under the impetus of proprietary research. An example is the membrane processes that have evolved commercially since the 1960s and have reached the status of widespread use.

TABLE 1.1
Unit Processes and Associated Technologies

No.	Unit Process	Principle	Treatment Technologies
1	Screening	Retention of objects or particles larger than screen openings	Trash rack Bar screen Fine screen Comminutor Microscreen
2	Chemical precipitation	Solubility product of reaction "product" is very low, e.g., 10^{-10}–10^{-30}; precipitate settles	Lime softening Metals removal
3	Coagulation	Negatively charged colloids are charged neutralized by cation cloud	Rapid mix In-line mixer Static mixer Submerged jet
4	Flocculation	Micro-flocs contact with each other to form settleable flocs. Contacts are induced by controlled turbulence or advection through a sludge blanket	Paddle-wheel flocculator Porous media Solids contact flocculation Turbine flocculator
5	Settling	Gravity force causes fall of particles heavier than water and rise of particles lighter than water	Horizontal flow Up-flow Tube Plate American Petroleum Institute (API) oil separator
6	Flotation	Gas bubbles attach to particles to create buoyant force causing rise	Dissolved air Diffused air
7	Filtration-deep bed	Charge neutralized micro-flocs are advected to "collectors" (usually sand and anthracite) where they "attach" (by van der Waal's forces)	Conventional 760 mm (30 in.) dual media Mixed media Mono media, 1–3 m (3–10 ft) deep Slow sand Proprietary variations
8	Cake filtration	Particles are retained by smaller pores as media is added concurrently (so that the media hydraulic conductivity does not change)	Diatomaceous earth Shapes of septum Plate and frame Candle
9	Membrane filtration	Particles, e.g., ions, organic molecules, viruses, bacteria, cysts, mineral matter, etc., that are larger than the membrane pore size are retained while water and matter smaller than the pore are transported, by advection, under a pressure gradient	Microfiltration Ultrafiltration Nano-filtration Hyper-filtration Shapes of membrane Spiral-wound sheets Hollow fiber Electrodialysis
10	Adsorption	Molecules and particles will adsorb on an "adsorbent," a material that provides "sites" for attachment by van der Waal's forces Ion-exchange is the same except ions attach to sites and force is electrostatic	Granular activated carbon Powdered activated carbon Activated alumina Ion-exchange Zeolite Resin
11	Gas transfer	Concentration gradient for dissolved gas is induced to cause mass transfer Rate of mass transfer is proportional to gradient	Turbine aeration Diffused air Packed towers Steam stripping

TABLE 1.1 (continued)
Unit Processes and Associated Technologies

No.	Unit Process	Principle	Treatment Technologies
12	Oxidation	Electrons are removed from outer shell of substance being oxidized	Chlorine Chlorine dioxide Ozone Ozone-hydrogen peroxide Potassium permanganate Ultraviolet radiation
13	Supercritical oxidation	Pressure and temperature are raised to create supercritical conditions	Wet air oxidation High pressure high temperature oxidation
14	Aerobic biological treatment	Bacterial enzymes permit metabolism of organic molecules with products new bacteria, carbon dioxide, and water	Activated sludge Complete mix Plug flow Aerated lagoon Facultative pond Fixed film reactor Trickling filters (traditional) Bio-filters (forced air) Rotating biological contactor
15	Anaerobic biological treatment	Two stage reaction: acid formers metabolize organic molecules with organic acids as products and methane formers metabolized organic acids to produce methane and carbon dioxide	Anaerobic biological reactor Digester Anaerobic pond Anaerobic filter

TABLE CD1.2
Treatment Technology Matrix (Excerpt from Table CD1.2b)

No.	Water Contaminant	CAS Numbers	Contaminant Type	Empirical Formula
1	Acenaphthene	83-32-9	PAH	$C_{12}H_{10}$
2	Acenaphthylene	208-96-8	PAH	$C_{12}H_8$
3	Acetaldehyde	75-07-0	Aldehyde	C_2H_4O
4	Acetamide	60-35-5	Not found	
5	Acetamide, N[4-[(2-hydroxy-5-methyl . . .)]]	2832-40-8	Not found	
6	Acetone	67-64-1	Ketone	C_3H_6O
7	Acetone cyanohydrin	75-86-5	Nitrile	C_4H_7NO

1.3.5 Status of Unit Processes

The treatment of water became an issue in the first decades of the nineteenth century with drinking water. In this context, James Simpson developed the slow sand filtration technology for London, enumerating design guidelines and support components to have a workable process. Slow sand became an accepted technology for drinking water by the 1870s and was in widespread use, particularly in Europe, by 1890, where it was credited with saving Altoona, Germany, from a cholera epidemic. The treatment of wastewaters became an issue after 1860 in England where chemical precipitation became one of the first unit processes, but did not develop into widespread use. In 1880, the Lawrence Experiment Station was established (Massachusetts, 1953) that later led in the development of several technologies, including trickling filters, and where experimentation inspired Ardern and Lockett to develop activated sludge in England in 1914. Settling, called "subsidence," was well established by 1900; the technology was cheap and was an obvious alternative. By 1885, deep-bed filtration was an innovation that by 1900 had become established in America. In Europe, Klaus Imhoff developed the Imhoff Tank about 1905 and Cameron in England developed the septic tank. How the stage was set for the development of environmental engineering as a field from Ancient times to 1900 was reviewed by Symons (2001).

Ion exchange had been known scientifically by early 1800, and became an established technology for softening by 1924 using zeolites. The use of chlorine, ozone, and ultraviolet radiation as disinfectants were established by 1900. Ozone was adopted widely in Europe, while chlorine became established in America. By the third decade of the twentieth century, gaseous chlorine became the technology of choice (in the United States). Advances in the application of technologies have continued over the following decades. The point is that water treatment technologies have been developing and expanding starting only since 1829 with slow sand. The impetus has been societal issues that were recognized as government responsibilities with advances through research and practice.

1.3.6 Future of Treatment

There is little doubt that technologies will continue to evolve, particularly if the market exists for improved applications of unit processes. Looking at the unit processes, some 15 are listed in Table 1.1; they were identified based upon fundamental principles. Of the 15 identified, and looking at the underlying principles of each, the question would be as follows: Could principles not yet used be applied for separating contaminants from water? While any predictions are uncertain by nature, we could say with safety that any new ones would have to be based on remaining principles that may cause a change of state of a substance.

1.3.7 Energy Expenditure for Treatment

A fundamental axiom of any treatment process is that a separation is involved. The separations involve removing particles from water, ions from water, and molecules from water. In each case the principle is that the entropy of the particles, ions, or molecules must be reduced. By the second law of thermodynamics this can occur only by an input of energy. Any treatment process cannot violate the principle of the second law. In other words, energy must be expended to effect any kind of separation. Then to make any process feasible from an engineering point of view, there must be a compromise between energy cost and the speed of the process (the more irreversible the process, the higher is its velocity, but the higher the energy cost). The second law places an inherent limit on what may be expected.

1.4 TREATMENT TRAINS

The configurations of unit processes that may comprise treatment trains are perhaps as numerous as there are combinations of the unit processes listed in Table 1.1. Those that are most common are (1) for potable water and (2) for wastewater. Figure 1.1 shows schematic sketches of treatment trains for potable water, Figure 1.1a, and for municipal wastewater treatment, Figure 1.1b, respectively. To illustrate the latter, Figure 1.2 shows an aerial photograph of the wastewater treatment plant (WWTP) for the City of Colorado Springs, c. 1972.

The unit processes shown for Figure 1.1a and b are common, but considerable variation is possible. For example, in potable water treatment, Figure 1.1a, plain sedimentation may be omitted if the source is a lake. A variation in the filtration process, called "in-line" filtration, omits flocculation and

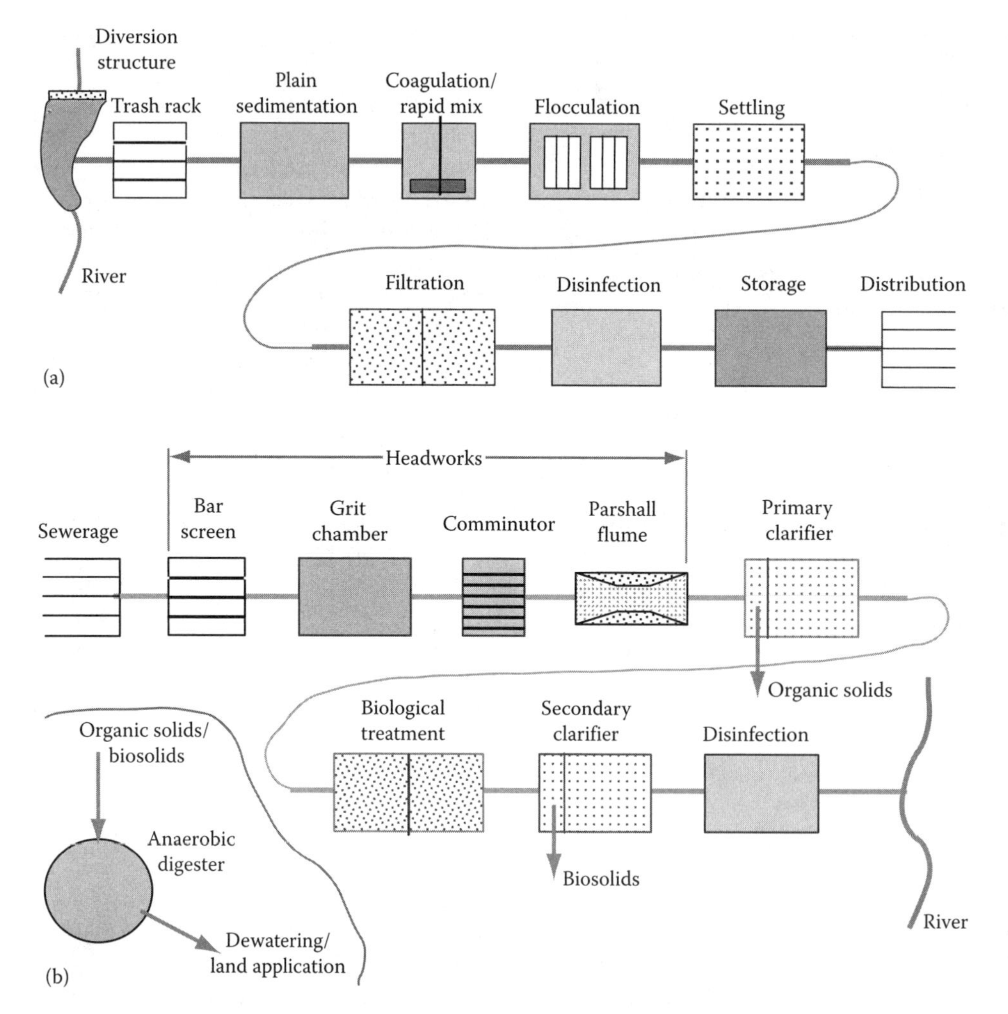

FIGURE 1.1 Treatment trains—schematic drawings: (a) potable water and (b) municipal wastewater.

FIGURE 1.2 Aerial photograph of WWTP, City of Colorado Springs, c. 1972. (Photo courtesy of City of Colorado Springs, Colorado Springs, CO.)

settling. For disinfection, chlorine has been traditional in the United States, while ozone is common in Europe. In municipal wastewater treatment, as in Figure 1.1b, finer bar screens have been used in lieu of comminutors; also, tertiary treatment could be added.

The selection of unit processes depends upon the contaminants present in the source water and the objectives to be achieved by treatment. The selection of technologies depends on contextual factors; costs; preferences of the client; and the engineer's vision, experience, and knowledge. Considerable variation is possible within each treatment train in the selection of specific technologies.

1.4.1 Tertiary Treatment

In municipal wastewater treatment, additional unit processes may include any or all of the following: filtration to reduce particles, precipitation by lime to reduce phosphates, and adsorption with activated carbon to reduce organic molecules. If the goal is to produce potable water, then hyper-filtration (reverse osmosis) may be added along with several other unit processes.

1.4.1.1 Cases

Usually tertiary treatment situations have involved small flows, e.g., 0.044 m^3/s (1.0 mgd). In some cases, such as in providing water for irrigation, sand filtration without coagulants has been used. The Parkson Dynasand moving bed filter has been used in a number of such instances as has the ABW traveling bridge filter of Infilco-Dregemont. These are "package"-type technologies that can be added to any conventional treatment train. These cases illustrate the systems context of most water treatment projects and the role of political factors. Both are inherent in any public project.

1.4.1.1.1 Aspen

The City of Aspen, Colorado, installed an ABW traveling bridge filter, developed for water treatment, to treat effluent from its conventional wastewater treatment train. The effluent discharge permit for discharge to the Roaring Fork River required a "20/20" standard (20 mg/L suspended solids/20 mg/L biochemical oxygen demand). The river has high value as a fishery and for recreational use, and the filters provide a margin of safety.

1.4.1.1.2 Ignacio

Figure 1.3a is a photograph of an ABW system as used for drinking water treatment at Ignacio, Colorado. Figure 1.3b is a cutaway perspective of the ABW system. The basic system is the same for water or wastewater, albeit the media may be different from one use to another. The traditional media is sand with a depth of 305 mm (12 in.). A "cell" is 203 mm (8 in.) wide and its length is across the bed of the filter.

(a)

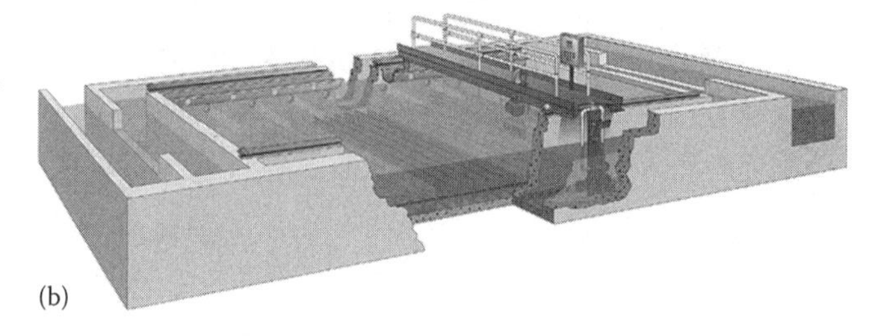

(b)

FIGURE 1.3 ABW (automatic backwash) filtration system. (a) ABW filter installation used for small community. (b) Perspective drawing of ABW filter. (Photo courtesy of Infilco Degremont, Richmond, VA.)

The traveling bridge backwashes each cell individually and moves continuously from one end of the filter bed to the other and back again. The flow through each cell declines as the media clogs and is restored to the clean-bed level after backwash.

1.4.1.1.3 Colorado Springs

Because of a drought during the period 1955–1959, the City of Colorado Springs (Colorado Springs, c. 1972) has had tertiary treatment following traditional treatment since 1960. This initial treatment was filtration only and the water was used for golf course and park irrigation, and was later called the "irrigation circuit"; the treatment capacity was $Q=0.394$ m^3/s (9.0 mgd). In 1970, a second treatment train was added and was called an "industrial circuit," with capacity 0.0876 m^3/s (2.0 mgd). The effluent was used for cooling tower water at a municipal power plant with the cost of treatment about $\$0.07/m^3$ or \$260/mg.

1.4.1.1.3.1 Irrigation Circuit The irrigation circuit had four 3.658 m (12 ft) diameter dual media pressure filters with filtration velocity 36.7 m/h (15 gpm/ft^2). The media was 0.91 m (3 ft) of 1.5 mm sand on the bottom with 1.52 m (5 ft) of 2.8 mm anthracite on top.

1.4.1.1.3.2 Industrial Circuit The treatment train for the industrial circuit was coagulation and settling by means of a "solids-contact" clarifier, re-carbonation with carbon dioxide, filtration with anthracite and sand, and carbon adsorption. The solids-contact unit, 14.63 m (48 ft) diameter, used a lime dose of 300–350 mg/L of CaO, which raised the pH to 11.5. Following the solids contact unit, the pH was lowered to 7.0 by carbon dioxide (from furnace gas used for carbon regeneration) supplemented by sulfuric acid. Filtration using coarse media was the next step in order to provide redundancy in solids removal in the event of upset of the solids-contact clarifier. Carbon adsorption by GAC was the next step, which used two columns in series (Figure 1.4). Each column was 6.096 m (20 ft) diameter and 4.267 m (14 ft) high, packed with 3.048 m (10 ft) of 8×30 mesh GAC with mass 41,864 kg (94,000 lb). For the design flow of 0.0876 m^3/s (2.0 mgd), the hydraulic loading rate was 10.39 m/h (4.25 gpm/ft^2). Removal of COD was 0.50–0.60 kg COD/kg carbon (0.50–0.60 lb COD/lb carbon). The carbon columns were backwashed daily.

FIGURE 1.4 GAC columns for tertiary treatment at Colorado Springs, c. 1972.

1.4.1.1.4 Denver Reuse Demonstration Plant

The Denver Reuse Demonstration Plant, $Q=0.044$ m^3/s (1.0 mgd), was in operation in the period 1985–1991. The source water was effluent from the nearby Denver Metro WWTP (called Denver Metro Water Reclamation Plant). The "reuse plant" treated water that exceeded standards for potable water.

The plant cost was \$20 million with about \$10 million for studies to determine health risks. The treatment train included lime precipitation, filtration, ion exchange, adsorption with activated carbon, ozone oxidation/disinfection, hyper-filtration, and chlorination. Side streams investigated different disinfectants, various membranes such as micro-filtration and ultrafiltration, and ozonation prior to adsorption. Figure 1.5 shows the overall plant schematic (for the "health-effects" treatment train) and three photographs that illustrate the substantial size of the plant.

The plant was "demonstration" size, and was intended to demonstrate the feasibility of direct potable reuse. While the purpose of a pilot plant is to develop design guidelines for a full-scale plant, a demonstration plant is much larger in size and may have several purposes, both technical and political. The former relate to such issues as long-term health effects due to trace amounts of organics, the cost of operation, the manpower required, reliability, and the assessment of unforeseen issues (such as the durability of the lining of the carbon columns). This does not mean that process issues are ignored, but only that the emphasis shifts (as opposed to a pilot plant) to questions that require experience at a full-scale level of operation. The political aspects are equally important. The reuse plant had an attractive, architect-designed, exterior appearance with equally pleasing interior. A water fountain of its product water was located in the lobby, and the plant hosted innumerable tours and published many papers and had high visibility from local to international levels. The program to implement potable water reuse had included the issues of public acceptance, and confidence and need for political support.

1.4.2 Industrial Wastewater Treatment

Regarding industrial wastewater treatment, the treatment technologies are usually the minimum required to meet the regulatory requirements in force. This may range from settling to

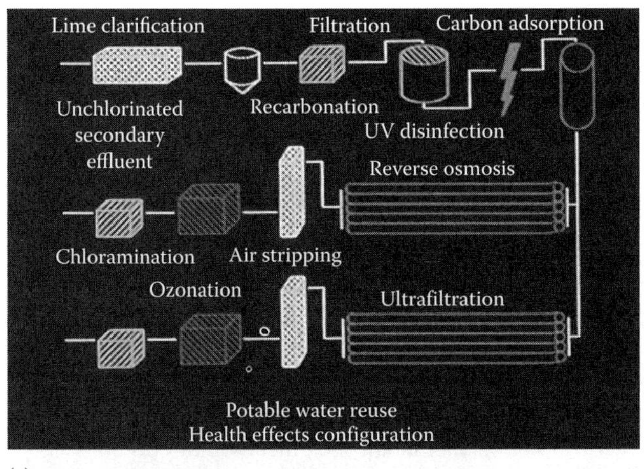

(a)

(b)

(c)

(d)

FIGURE 1.5 Denver Potable Reuse Demonstration Plant, c. 1989. (a) Flow schematic. (b) Lime settling and chemical silos. (c) Pressure filters. (d) Carbon columns. (Courtesy of Denver Water Department, Denver, CO.)

remove soil sediments and land treatment to remove organic molecules for sugar beets, to settling and anaerobic ponds for meat packing, to precipitation of heavy metals for the electronics industry and metals industry. Many industries, depending on policies and plant management, take pride in exceeding legal requirements.

1.4.2.1 Cases

Two cases are reviewed to indicate the variety of circumstances that influence the selection of unit processes that comprise treatment trains for industries. The circumstances of industries are highly variable, and so there is a wide variation in treatment processes.

1.4.2.1.1 Coors Brewery

The Miller-Coors Brewery, formerly Coors Brewery until merger in 2008, in Golden, Colorado, utilizes biological treatment, Figure 1.6, and had a 30/30 (BOD/TSS) effluent discharge permit for the adjacent Clear Creek that flows to the South

FIGURE 1.6 Wastewater treatment at Miller-Coors Brewery, Golden, Colorado. Pure-oxygen-activated sludge reactors are in background and underground. (Courtesy of Miller-Coors Brewery, Golden, CO.)

Platte River about 16 km (10 mi) downstream. The first plant was built in 1954, i.e., 18 years prior to being required by federal/state regulations. In 1981 a new plant, dedicated to the Coors Brewery industrial wastes (i.e., malting, brewing, and packaging) was put on line. (The 1954 plant then served only the City of Golden, along with sanitary wastes from other industrial operations and industrial wastes from the can manufacturing and ceramics facilities.) The new wastewater treatment system (i.e., the 1981 plant) included a coarse screen and grit removal (the latter to remove barley and other grain particles), an equalization basin to mitigate the effects of batch discharges, primary settling, activated sludge treatment using pure oxygen, and secondary settling. The plant was designed with the filtration of secondary settling effluent, using Parkson Dynasand filters with discharge to Clear Creek. In the years since 1981, the plant has been modified to more effectively and more economically achieve its objectives.

In 2001, an anaerobic pretreatment plant was built at site of the 1981 WWTP; the anaerobic plant receives about 60% of the organic loading from the brewery and removes about 65%–85% of the soluble organics, depending upon the hydraulic detention time. (The anaerobic process is sometimes used for high-strength industrial wastes as a means to reduce the loading to an aerobic biological treatment system. In the case of Coors, the BOD varies but about 1800 mg/L may be considered representative.) The Parkson Dynasand filters were taken out of operation in 1999 due to maintenance difficulties caused by adhesion of a cationic polymer to the filter media. The polymer was used in secondary settling, and in the evaluation of the trade-offs was selected in preference to the tertiary filtration. The effluents from the two plants are co-mingled and a single discharge permit then serves both plants.

Some notes on the Coors operation—technical, management, and political—may help to understand better the field of industrial wastewaters:

- First, equalization basins are common to industries subject to batch discharges. The basins are aerated to maintain aerobic conditions and to minimize the deposition of solids. Equalization basins have been advocated for municipal treatment but they have not become assimilated into practice.
- Second, the pure oxygen provides a higher reaction rate and thus smaller aeration basins (i.e., a smaller "footprint," important in the Coors case).
- Third, the filtration process was used originally at Coors to provide a buffer to ensure compliance with the 30/30 permit.
- Fourth, the Miller-Coors Brewery is highly visible because of its size and because of its historical prominence in Colorado. Any noncompliance with environmental standards, particularly on Clear Creek that receives the discharge from Coors, is noted quickly by the news media. The selection of processes and technologies for the treatment train reflects these political factors, i.e., to ensure compliance at a high probability level.
- Fifth, the management philosophy is the most important facet of how an industry deals with its environmental issues. Coors (now Miller-Coors) has had a history of identifying potential issues ahead of the public perception, political movements, and legal mandates, and dealing with them as matters of good business and good citizenship (some call this being "proactive"), which has been the position of many industries, as opposed a perception by some that they all are "foot-dragging."
- Sixth, as a rule, industries want to know the rules and to know that the rules are enforced fairly upon their competitors as well, and to have confidence that there will not be vacillation (i.e., that the rules do not vary from year-to-year as political administrations change).

1.4.2.1.2 Anheuser-Busch Brewery

The Anheuser-Busch Brewery in Fort Collins, Colorado, was put on line about 1986. The brewery is located about 16 km (10 mi) northeast of the City, adjacent to Interstate 25 and in an area that is largely irrigated agriculture. Prior to coming to Fort Collins, an agreement was developed with the City of Fort Collins to utilize its south WWTP, constructed in 1968 and largely mothballed in 1976 after construction of a new plant on the same site, i.e., the North Plant (further modified in 1993). The South Plant was renovated in 1986 to accommodate the brewery wastes and was comprised of primary settling, activated sludge, and secondary settling. A bar screen was located at the plant site. As an alternative to the Fort Collins South Plant, the brewery can discharge to a land treatment site. The site is located about 8 km (5 mi) east of the plant. The site has storage and is set up to spray irrigate the wastewater on the land. The spray irrigation rate does not exceed the infiltration rate of the soil. The system is utilized irrespective of crop growth cycles but is suspended during cold months when freezing may be a problem.

1.4.3 Industrial Process Water Treatment

Industrial water treatment for process water is another significant area. The variation may include industries that can use potable water directly without further treatment, such as for food products, with perhaps dechlorination. The electric energy industry requires mineral-free water for boiler feed, while the electronics industry requires essentially molecular water. The unit processes in each case may require many of those listed in Table 1.1 (and in Tables 1.2a and b).

1.4.4 Hazardous Wastes

Contaminated groundwater is a common context for hazardous wastes. In some instances, the remedial action is to pump the aquifer and pass the flow through a treatment plant, which is called a "pump-and-treat" situation. Organic chemicals are common contaminants and there are thousands of possibilities. Adsorption by GAC is a common unit process/technology. Other situations, such as at mine drainage site, may include heavy metals in which case precipitation is a likely unit process. In some cases, hyper-filtration is added, mainly for redundancy.

1.4.5 Hazardous Wastes: In Situ Treatment

In some cases, in situ treatment is applied. A main technical problem is to get the reactants in contact with one another, i.e., how to "transport" one reactant on the surface to another that is "in situ" is the issue. This may involve an elaborate scheme, in which the "transport" is by convection and diffusion. An example is the practice of punching the ground surface with a grid of holes used to pump water with a "reactant" to the site of adsorbed (on soil) or dissolved (in pore water) contaminants. The "reactant" may be a bacterial species (perhaps genetically modified), chemicals, air, steam, etc. The point is that the "reactor" is the in situ site of the contaminant; actually, it is one of numerous micro-reactors.

1.5 DESIGN

A variety of nontechnical issues are a part of any design. Some are addressed in this section. The technical design itself is not just a computational algorithm, but involves mostly assumptions, judgments, and decisions. An engineering background gives the technical basis for decision-making, e.g., what may be feasible, and a context for continued learning.

1.5.1 Factors: Nontechnical

The end product of the design process is a physical plant. The plant design is designed by a consulting engineering firm with qualified staff. The firm selected is usually one of several who may have competed for the contract. Based upon the design, a contractor is selected, with the engineer acting as the owner's agent (the traditional arrangement). Getting to the point of a completed design involves a host of issues that may involve politics, dealing with neighbors, financing, owner's ideas, operation, esthetics, environmental considerations, water rights, etc. The consulting engineer manages these factors based upon experience and exercise of judgment, as opposed to academic studies.

1.5.1.1 Operation Issues

Design involves a stream of decisions involving "trade-offs" between capital cost and operation. Operation factors may include costs, such as energy and maintenance, the ease of operation, environmental impacts, etc. In addition to such factors, operators nearly always have views that may be important in the design. Some examples include adding a gullet with drain in a pipe gallery for easy hose-down and cleaning; using aluminum hand rails so that painting is not required; venting enclosed spaces and rooms that may be handling chemicals; dehumidifying a pipe gallery; providing attractive lunchrooms and locker rooms with showers; taking into account noise problems associated with pumps; designing rooms with windows; placing signs to accommodate visitors and public tours, a reception area, etc.

1.5.1.2 Managing a Team

In addition to the management of staff, the engineer assumes the *de facto* leadership and management of an informal team that may include the owner and perhaps a representative from the political body, operators, regulatory authorities, equipment manufacturer's representatives, the suppliers of materials required in operation, the contractor, the architect, a landscape architect, a water rights expert, a limnologist, security advisor, accountant, financing expert, etc. Those in the team involved depend upon the size of project and its context. As a rule, these ideas are learned by the engineer "on-the-job." Academic courses tend to focus on technical factors, albeit there are exceptions (see, e.g., Qasim et al., 2006). Excerpts from an article by Lagnese (2000), a former president of the Water Environment Federation (1968–1969), and of the American Academy of Environmental Engineers (1991), who taught a practical design course at the University of North Carolina, provide a firsthand design perspective.

Notes from Joseph F. Lagnese, Jr.

> Design is an arduous, iterative process. Ideas are advanced based upon limited factors and intuition and then evaluated. The initial ideas are then refined and/or eliminated, new ideas proposed, and the process repeated as often as necessary to ultimate completion. Stated in another way, design is inherently a "two steps forward, one step backward" process.... the ultimate products of design are the plans and specifications required to guide the construction of a total facility that is efficient in operating needs and performance capability. Whereas process design theory and fundamentals require mathematical and science capabilities, the other aspects of design rely upon creative and organizational talents and team-play ability, as well as aptitudes in such diverse areas as economics and aesthetics. Design is a knowledge intensive process which requires not only an understanding of many diverse fundamental concepts, but also such practicums as engineering pricing, selection of process equipment and construction materials, architecture, construction techniques and procedures, operational requirements, ergonomics, satisfying relevant regulatory controls, and project financing. As such, design is an optimization process dependent on the appropriate consideration of a broad range of impacting factors. There are few absolutes; most completed designs represent imperfect solutions.

1.5.1.3 Expansion

The provision for expansion may be done by sizing pipes and the procurement of land for a projected ultimate build-out. Not to plan ahead may result in several largely independent plants on one site, each with its own idiosyncrasies of operation, not to mention inordinately higher costs.

1.5.1.4 Esthetics

Esthetic issues are as important as the functional design, and are often overlooked. Does the plant enhance its sight? Does it fit the sight architecturally? Was a landscape architect involved? Are persons working at the plant motivated when they enter the premises? Do the mayor and council view the plant with pride? Are peers pointing to the plant as a showcase for the industry? Esthetic appeal is sometimes discounted by designers and owners. Community pride is nearly always associated with public facilities, not to speak of operator morale.

1.5.1.5 Regulations

There are two aspects to regulations: (1) those that specify effluent requirements, and (2) those that specify design. Compliance with the former is essential. Variance is provided for the latter subject to reasonable rationale or pilot plant results. Design guidelines by state regulatory agencies are sometimes controversial. At the time such guidelines were instituted, which go back to the 1930s and possibly earlier, the art of plant design was not very far along. At that time, most of the expertise regarding design resided among the engineering staff of the state health departments (Ongerth, 1999). As educational institutions developed a cadre of professionals in the field, however, the design capabilities expanded among those in consulting engineering.

1.5.1.6 Institutions

An "institution" is defined here as a "device of society to manage itself" (my interpretation of the term based upon frequent use by a sociologist colleague). Among the institutions in environmental engineering one might include: laws and regulations, the traditions of excellence, practice and lore, education, training, certifications, the licenses to practice, manufacturers, research, government laboratories, the awarding of degrees, construction companies, consulting engineering firms, etc. The respective purposes of these institutions are primarily technical, but they function in a social context.

1.5.1.7 Consulting Engineering

Consulting engineers design systems that provide water supply, treat drinking water, industrial wastes, and municipal wastewaters, and assess the impacts of discharges on receiving water. Any task that contributes to a solution to an environmental problem is within the purview of practice. Consulting engineering has long been regarded as the citadel of technical expertise, ethics, and professionalism. Many consulting engineering firms and engineers within various firms have had careers that match this description.

To give a sense of the nature of consulting engineering and the practitioners, Box 1.2 provides a glimpse through the 70-year career of Harvey F. Ludwig. Box 1.3 describes some of the changes in the character of consulting engineering practice that has evolved since about 1980, i.e., as interpreted by the author. Box 1.4 adds commentary about a "golden" age of environmental engineering that coincided largely with the environmental movement, also as interpreted by the author. These are not the only views of these three areas and are likely to induce alternative or complementary perspectives from others.

BOX 1.2 HARVEY F. LUDWIG ENVIRONMENTAL ENGINEER CONSULTANT

On May 29, 1965, Harvey F. Ludwig (b. 1916) was awarded a doctorate by Clemson University. The citation read, in part,

> engineer, teacher, public servant and world-renown authority on environmental and sanitary engineering, his advancement of the engineering profession in the field of environmental health, his scholarly research, and his outstanding leadership in scientific affairs on a national and international basis.

Harvey F. Ludwig, c. 1968

The citation was at year 27 of what has become a 70-year career of continuing contributions and leadership that has helped shape the modern practice of environmental engineering. Dr. Ludwig obtained his BS degree in 1938 in civil and sanitary engineering from the University of California, with MS in 1941. During World War II Dr. Ludwig was a commissioned officer of the U.S. Public Health Service (USPHS). In 1946, he started a consulting practice and in 1949 became an associate professor at the University of California. In 1951, he became Assistant Chief Engineer, USPHS (under Dr. Mark Hollis). In this position, he presided over the development of the institutions (i.e., laws, federal agencies, programs) that were the foundations for what emerged in final form in the 1970s (and have continued to evolve). At the same time, Dr. Ludwig oversaw research funding at various universities across the United States that fueled research and graduate programs that led to a "flowering" of the field that has continued. In 1956, Dr. Ludwig resigned from the USPHS and started Engineering-Science (ES). His *modus operandi* was to hire mostly MS and PhD students recruited from his network of academic colleagues who would adapt their research knowledge into practice. At that time, hiring engineers with graduate degrees was more unusual than common.

BOX 1.2 (continued) HARVEY F. LUDWIG ENVIRONMENTAL ENGINEER CONSULTANT

ES expanded rapidly with offices at key cities in the United States including a research laboratory and office in Oakland (c. 1956), and later in Washington, District of Columbia (c. 1966) headed by his long-time USPHS associate, Gordon MacCallum, and then in Austin, Texas, started by Dr. Davis Ford (c. 1968). The firm grew rapidly with important projects throughout the United States and started to develop an international clientele. Dr. Ludwig was at this time, in 1969, a "legend" in the field. At the same time, the field was experiencing a "golden age," i.e., research was advancing knowledge, graduate programs were spreading, practice was flourishing, and the public had adopted a widespread environmental ethos that was being translated by politicians into laws and policy. At that time, ES was arguably at the crest of this movement, i.e., one of the most visible of firms in the field and at the forefront of innovation. This was due not only to Dr. Ludwig individually, but to the way he had structured the firm with both depth and breadth of expertise and leadership. In addition, Dr. Ludwig had extensive involvement with professors from throughout the United States.

ES was actually, however, a part of a larger corporate structure. One entity was a construction company that had financial difficulty (due to a low bid on a dam). The "way-out" was a buyout offer in 1968 by Zurn Industries of Erie, Pennsylvania. The new corporate structure did not work out, and Dr. Ludwig left the firm in 1972, setting up his own practice in Washington, District of Columbia (Ludwig 1985). ES was later purchased by its employees and remained prominent in the field through 2004 when its identity was assimilated fully by Parsons, an international construction company.

In 1973, Dr. Ludwig's private practice led to Bangkok where he started a new firm, Seatec International, which has influenced environmental engineering throughout South-East Asia. While there, he has championed the case for adapting environmental standards and designs to match the socioeconomic context of developing countries, i.e., as contrasted to imposing the design approaches of industrialized countries. As of 2008, Dr. Ludwig had some 358 publications ranging from research on coagulation, c. 1941, to strategies for saving the forests in South-East Asia, c. 2005. Four of his papers won awards from organizations such as ASCE, AWWA, WEF. Personal achievement awards have included election to the National Academy of Engineering, in 1969, shortly after its founding; the AEESP Founder's Award; the AAEE Honorary Member award; the 1999 University of California College of Engineering Alumnus of the Year; and various awards in Bangkok.

As to the persona, Dr. Ludwig has been a mentor to countless engineers in practice. He is known by his high standards in writing, in professional practice, and in getting a job done. To quote Professor Donald Anderson, c. 1965, when he headed the Oakland office of ES, "When you work for Harvey Ludwig, . . . ," meaning that much was expected. He was instrumental in founding what is now the American Academy of Environmental Engineers (a certifying organization), c. 1956, and sponsored the founding of the Association of Environmental Engineering and Science Professors (AEESP), c. 1963. In 1966, his firm ES initiated the sponsoring of a "best thesis" cash award within AEESP; his rationale was that the significant cash ($1000 at that time) added prestige to the award. The award has continued under auspices of other firms with inclusion of both master's and doctoral theses.

Dr. Ludwig is known for expressing his candid opinion (on virtually any topic). Consequently, he has both many admirers and a few detractors. He continues an active correspondence, by e-mail, with perhaps a circle of some 20–100 family members, friends, and colleagues. From Bangkok, Dr. Ludwig remains a presence in the field of environmental engineering.

BOX 1.3 GLOBALIZATION OF THE WATER INDUSTRY AND CONSULTING ENGINEERING

In 1974, the water industry in England and Wales was reorganized with 10 regional water authorities with jurisdiction over all water functions, e.g., water supply, wastewater, flood control, and river management. The regional authorities had taken over from some 1600 separate local authorities (Okun, pv, 1977). In 1988, due to a change in political climate, let by the then Prime Minister, Margaret Thatcher, the water industry in England and Wales was again "privatized," with the 10 water authorities remaining. The change had a global significance. Using Thames Water as an example, which was one of the 10 regional water authorities, the organization was freed from the limitations of being a public entity. Over the years following privatization, Thames Water purchased major companies in the water industry and sold its services worldwide in consulting engineering, management, construction, and operation (see http://www.thameswater.co.uk). They joined Vivendi Universal of France in providing "one-stop shopping" for water services. Vivendi Universal was formerly Compagnie Generale des Eaux created in 1853 by imperial decree to provide water to Lyons. The company expanded and diversified from 1980 to 1996,

(*continued*)

BOX 1.3 (continued) GLOBALIZATION OF THE WATER INDUSTRY AND CONSULTING ENGINEERING

becoming involved in music, publishing, TV and films, telecoms, environmental services, etc. The name was changed in 1998 to Vivendi Universal. The environmental services include the divisions of water, waste, energy, and transportation, and were active in over 100 countries. Water customers number about 110 million. The creation of Thames Water and the expansion of Compagnie Generale portended fundamental change in the character of consulting engineering. The "client" became a "customer," something not thinkable until the 1980s.

In the United States, the way had been paved legally for this "new age" institutional form, i.e., a multinational conglomerate. In 1972, the U.S. Justice Department forced ASCE, under consent decree, to change its code of ethics to permit bidding for services and to permit advertising. The upshot was that during the 1980s, bidding for design services became the norm for some clients (not all joined this trend). To compete in the new market place, some of the larger consulting engineering firms formed subsidiary companies to construct and operate plants. The traditional engineer–client relationship was being replaced by "turnkey" projects based on bidding. With this institutional metamorphosis, U.S. engineering firms were then poised to compete globally with Thames Water and Compagnie Generale des Eaux. Professional practice was no longer protected by ethics, tradition, and law. Engineering services were not the same; they were becoming a commodity.

BOX 1.4 GOLDEN AGE OF ENVIRONMENTAL ENGINEERING

Almost every field has a "golden age," perhaps characterized by the nostalgia of those who reminisce. In October 1957, the then Soviet Union launched into orbit the earth satellite *Sputnik*. This event electrified the world and shocked the United States in almost every respect. One could easily view *Sputnik* moving across the skies relative to the field of stars as tangible evidence that the United States was not so advanced as presumed; the view evoked deep emotions. The upshot was introspective questioning of the adequacy of U.S. education, science, mathematics, and engineering. The result was the unprecedented funding of these areas, perhaps epitomized by the 1961 commitment of President Kennedy to "place a man on the moon within this decade." At the same time, an environmental movement was in a period of gestation. In 1962, Rachael Carson's *Silent Spring* was published and set in motion the movement, which had continuing momentum through the 1970s, and was more or less dissipated by the 1980s, perhaps as the political climate was changing; at the same time, the movement was gaining ground in other countries. A series of laws and policies grew out of the movement, however, along with public funding for research, education, and mandates for cleaning up the environment and an associated flourishing of environmental engineering practice. Also, during this period, i.e., in the early 1960s, doctorates became a requisite for academics in engineering and the number of graduate programs expanded beyond the "handful" that had existed up to about 1960. Funding for research was appropriated by the U.S. Congress and the chances for funding was higher in 1960s than in later decades. The result was a science-based understanding of many of the unit processes that was translated toward a more rational practice. Some of the processes that were delineated in scientific terms included the concept of an activated-sludge basin as a "reactor," and the associated materials-balance depiction; reaction rates described in terms of "kinetics"; the adoption of the Michaelis–Menten description of bacterial kinetics; the assimilation of biochemistry in understanding biological reactions and the role of ATP as well as DNA; the assimilation of turbulence theory in understanding coagulation and flocculation; the introduction of chemical equilibrium theory and its application as a basis for understanding processes, e.g., in coagulation, redox reactions, precipitation, acid–base reactions, ion exchange, etc.; the assimilation of physical chemistry as a basis for understanding the role of temperature on reaction rate, the mechanisms of molecular adsorption on activated carbon, the laws for gas transfer, the role of osmotic pressure in membrane desalination, etc.; anaerobic reactions became understood in terms useful for practice. In addition, the "Advanced Waste Treatment Research" program was initiated in 1962 by the Division of Water Supply and Pollution Control, USPHS, which included processes such as adsorption, electrodialysis, distillation, reverse osmosis, ion exchange, etc., with funding for both in-house and extramural projects. The program also contributed to the changing paradigm of the field, i.e., from empiricism to science.

The research during the 1960s was, to a large extent, "unsolicited," meaning that a professor could propose the topic and the approach; the "peer-review" scrutiny was the basis for quality control. This contrasted with the approach after 1972 when EPA was formed and research funding had to "support regulations." The research objectives were formulated by persons within a given agency, which greatly limited continued progress in applying fundamental knowledge to applications. The National Science Foundation, however, remained a traditional unsolicited research organization,

BOX 1.4 (continued) GOLDEN AGE OF ENVIRONMENTAL ENGINEERING

but with not too much funding available in environmental engineering. Some topics remained without "closure" and the rate of advance to the field was slowed and dependent, to a large extent, on what could be gleaned from projects that suited the agencies, particularly EPA. Although the "golden age" has passed, the profession has made the transition from purely empiricism to rational approaches. Some may argue that the pendulum has swung too far, i.e., that we may lack a sense of serving professional practice as the primary guide in engineering education and perhaps even in consulting engineering.

1.6 SUMMARY

As indicated, water treatment covers a great deal of variety. With about 15 unit processes, the combination selected for a treatment train depends on the technical requirements of the treatment task, i.e., the contaminants to be treated and the objectives of the treatment.

For any unit process, a number of technologies exist or may be developed that utilize the principles of the process. Some are generic and some are proprietary. The technologies provide an array of choices that help to tailor a treatment train to the situation at hand.

The intent of the book is to describe theory and practice for each of the unit processes. The specific technologies are described only as useful to illustrate the processes. The technologies represent the variations of principles associated with a particular unit process.

Nontechnical issues play a part in almost every technical decision and are alluded to in order to indicate some of the realities of process design. These factors are learned mostly in the context of experience and are mentioned in this book only to a limited extent so that there is an awareness of their role.

PROBLEMS

1.1 **Unit Processes**
Visit a treatment plant and describe your impressions about the plant and its unit processes. (The idea is to gain some familiarity with a plant and to experience the scale, appearance, functioning, etc. of unit processes. Later, you can compare the variation in technologies between plants.)

1.2 **Unit Processes and Models**
Based upon your recent visit to a WTP, list the treatment processes that you observed. Did you see any physical models? What were the purposes?

1.3 **Sampling and Instrumentation**
Regarding your visit to a water (or wastewater) treatment plant in your vicinity, describe and distinguish between process control and surveillance? What online instruments did you see? What about sampling? Was sampling discussed? What kind of sampling? Was laboratory analysis discussed?

The following problems, Problems 1.4 through 1.14, are intended to illustrate a variety of treatment situations that may be encountered in practice.

1.4 **Treatment Trains**
Suggest treatment trains for a five star hotel located in Cyprus for (a) drinking water, (b) wastewater.

1.5 **Treatment Train**
Describe a traditional treatment train for a municipality that treats surface water.

1.6 **Water Quality**
A small city relies on groundwater. Suggest some water-quality issues that could be present and unit processes that could deal with the issues.

1.7 **Treatment Train for Potable Water**
Describe the treatment train that produces potable water for your locale.

1.8 **Treatment Train at WWTP Site Visit**
Describe the treatment train for wastewater that you visit in your locale.

1.9 **Potable Water Quality Goals and Treatment**
Discuss the future water-quality issues and associated treatment modifications for potable water for the plant in your locale.

1.10 **Water-Quality Goals and Wastewater Treatment**
Discuss the future water-quality issues and associated treatment modifications for wastewater in your locale.

1.11 **Technologies for Unit Processes**
Generate a list of technologies for each of the unit processes identified in your visits to treatment plants (pick two for hand-in).

1.12 **Ambient Water-Quality Processes**
In the ambient environment, list cases in which water-quality changes occur passively. Is knowledge of unit processes applicable to understanding natural systems? Discuss.

1.13 **Passive versus Active Technologies**
Discuss passive technologies versus active (i.e., in which operation is essential to performance). List examples.

1.14 **Groundwater Contaminants**
List 12 contaminants that might be found in a pump-and-treat situation.

1.15 **Form-Fit Context**
Select several kinds of treatment situation that could include any of a variety of possibilities for water-quality profiles and the uses of water. What kind of treatment "fits" would you feel could be appropriate for each hypothetical (or real case from any experiences or knowledge) that you generate.

1.16 **Social Context**
What different social and political contexts would you expect to encounter in the above treatment situations?

Describe how you would respond with respect to treatment technologies.

1.17 Site Visit to WTP

Visit a WTP, and based upon your visit, comment on any aspect of the plant that you found of interest. This could be related to operation, design, the selection of unit processes for the treatment train, the appearance of the influent water, the appearance of the product water, the particular characteristics of any unit process, the control system, the monitoring, the requirements for product water, etc.

1.18 Site Visit to WWTP

Visit a WWTP, and based upon your visit, comment on any aspect of the plant that you found of interest. This could be related to operation, design, the selection of unit processes for the treatment train, the appearance of the influent water, the appearance of the product water, the particular characteristics of any unit process, the control system, the monitoring, the requirements for product water, etc.

ACKNOWLEDGMENTS

The Denver Water Department, Denver, Colorado, is known commonly as Denver Water, which is the citation used in this text. William C. Lauer, American Water Works Association, formerly manager, Denver Potable Water Reuse Plant, filled in knowledge about the plant, its evolution, and its design. Trina McGuire-Collier, manager of Community Relations, Denver Water, facilitated permissions to use brochure material, i.e., Figures 1.5, from the Denver Potable Water Reuse Demonstration Plant.

John Rawlings, Miller-Coors Brewing Company, Golden, Colorado reviewed the section on the Coors WWTP and provided corrections and Figure 1.6. The author is responsible for the use of the material and its accuracy.

Sylvie Roy, communications and marketing manager, Infilco Degremont, Richmond, Virginia, gave permission to use the ABW filter images of Figure 1.3. She also provided additional images to use as needed.

Regarding Colorado Springs, Jim Phillips and later Daryl Gruenwald hosted many class field trips to the plant during the 1970s and provided brochures and plant descriptions from that period that were utilized in this chapter. Since the 1970s, the main plant has been expanded and the tertiary plant has been modified to eliminate the lime clarification and GAC adsorption, with new rapid filters constructed; the 1970s plant was used, however, for the examples in this chapter since it illustrated the points useful for this introductory chapter. Tony Woodrum and Pat McGlothlin in a May 2001 tour of the facility provided an update of the changes that had occurred since the 1970s. Woodrum, Wastewater Operations Superintendent, gave permission (2010) to utilized photographs of the 1970s plant.

Concerning the boxes, the author requested permission from the late Dr. Harvey Ludwig (1916–2010) to use his biography and photograph, and to provide corrections regarding accuracy. Dr. Ludwig commented (October, 2009) that the summary of ES was accurate, subject to a few minor corrections. The biography was done, however, by the author alone based on a long-time personal knowledge going back, in fact, to 1957 and on Dr. Ludwig's autobiography (Ludwig, 1985).

GLOSSARY

Active process: A process that is controlled by actions of the operator.

Box (*n.*): A short, often boxed auxiliary story that is printed alongside a longer article and that typically presents additional, contrasting, or supplemental views.

Constant: A ratio of two or more variables that is characteristic of a group of materials or a system. Examples of the former include the universal gas constant for gases, and the modulus of elasticity for solid materials. Often, a coefficient is called a constant.

Discipline: A family-like grouping of individuals sharing intellectual ancestry and united at any given time by an interest in common or overlapping problems, techniques, and institutions ... Some are happy families, with little controversy over methods and goals. Others are fractured into many research schools, each with a different agenda, each evolving its own traditions of thought and work, and each competing for resources and recognition .... Disciplines not only lend structure and meaning to lives, they also bring order and significance to knowledge. (Excerpts from the Preface of Servos, 1990.)

Engineered process: A unit process that has been designed by an engineer (as opposed to a natural process).

Natural process: An influence within the ambient environment that causes changes in water quality. Examples include dilution due to mixing of streams, dispersion within a stream, heat transfer involving ambient water, microbiological reactions within an ambient water body, etc.

Passive process: A process that occurs largely without operator intervention. Slow sand filtration or trickling filter treatment are examples. All natural unit processes are passive.

Process: An influence that causes change, i.e., a "state" change. As applied to water quality, a process causes change in one or more water-quality characteristics.

Science: Systematized knowledge obtained from observation, study, and experiment in order to determine the nature of that studied. George A. Olah, 1994 Chemistry Nobelist at ACS Symposium as reported in *Chemical & Engineering News*, Vol. 76(35):6, 31 August 1998.

State of water: The quality characteristics of a given water volume, including concentrations of mineral suspended matter, ions, molecules, microorganisms, and such parameters as temperature, pH, specific electrical conductance, etc.

Technology: (1) An anthropogenic device contrived to accomplish a task. The rapid filtration process is encompassed within the filtration technology, which includes all of the appurtenances to make it function to remove suspended particles to a specified concentration level. (2) A collection of devices contrived to accomplish one or more tasks, as in a system. Water treatment technology includes all of the processes and relevant appurtenances to produce potable water.

Technology: A technology is a means to implement a unit process. Any number of technology forms may be devised to embody a unit process. For example, a biofilm reactor may be embodied in several forms, including a traditional trickling filter, a deep-bed trickling filter, a rotating disk reactor, a traditional slow sand filter, a bio-filter for removal of natural organic matter (NOM), etc.

Treatment: Subjecting water to the unit processes of a treatment train.

Treatment train: An aggregation of unit processes.

Unit operation: A term used in chemical engineering to designate a physical change, e.g., pumping, screening, sedimentation, filtration, etc. The term is not used in this book in favor of using a single term, "unit process."

Unit process (chemical engineering): A term used in chemical engineering to designate a chemical change, e.g., oxidation, precipitation, disinfection, and biological treatment. The chemical engineering literature is not unequivocal in the use of the two terms "unit operation" and "unit process," but the definitions given seem to capture the sense of how they are used.

Unit process (this book): As used in this book, the term "unit process" means an engineered effect that causes a "state change." The sense is the same as in chemical engineering except that a "state change" is much broader that being restricted to a "chemical change." A state change may include not only chemical change, but pressure change, temperature change, concentration change, etc. Thus, settling (change in concentration of particles) is a process by this definition as is even pumping (which causes a pressure change).

REFERENCES

Carson, R., *Silent Spring*, Houghton Mifflin Company, Boston, MA, 1962.

Champlin, T. L. and Hendricks, D. W., Treatment train modeling for aqueous contaminants, Volume II, *Matrix of Contaminants and Treatment Technologies*, Environmental Engineering Technical Report 53-2415-93-2 (for U.S. Army Construction Engineering Research Laboratory), Department of Civil Engineering, Colorado State University, Fort Collins, CO, May 1993.

City of Colorado Springs, Two handouts to visitors: The first describes the sewage treatment plant and its history and the second describes the tertiary treatment plant, Department of Public Utilities, City of Colorado Springs, Fort Collins, CO, c. 1972.

Commonwealth of Massachusetts, *Proud Heritage—A Review of the Lawrence Experiment Station Past, Present, and Future*, Commonwealth of Massachusetts, 1953.

Lagnese, J. F., Teaching environmental engineering design—A practitioner's perspective, *Environmental Engineer*, 36(1):8–12, 32, January 2000.

Letterman, R. D., *Water Quality and Treatment*, 5th edn., American Water Works Association, McGraw-Hill, New York, 1999.

Ludwig, H. F., *Adventures in Consulting Engineering*, Seatec International Publications, Bangkok, Thailand, 1985.

Okun, D. A., *Regionalization of Water Management—A Revolution in England and Wales*, Applied Science Publishers, Inc., London, U.K., 1977.

Ongerth, H. J., Personal communication, September 20, 1999. [Henry Ongerth was Chief, Bureau of Sanitary Engineering, State of California, retiring about 1980. He started with the Bureau from the time of his graduation at the University of California in 1936.]

Qasim, S. R., Motley, E. M., and Zhu, G., *Water Works Engineering—Planning, Design & Operation*, Prentice-Hall, New Delhi, India, 2006.

Rich, L. G., *Unit Operations of Sanitary Engineering*, John Wiley & Sons, New York, 1961.

Rich, L. G., *Unit Processes of Sanitary Engineering*, John Wiley & Sons, New York, Rich, 1963.

Sanks, R. L. (Ed.), *Water Treatment Plant Design*, Ann Arbor Science Publishers, Inc., Ann Arbor, MI, 1978.

Servos, J. W., *Physical Chemistry from Ostwald to Pauling: The Making of a Science in America*, Princeton University Press, Princeton, NJ, 1990.

Symons, G. E., The origins of environmental engineering: Prologue to the 20th century, *Journal of the New England Water Works Association*, 115(4):253–287, December 2001.

2 Water Contaminants

Contaminants in water encompass a wide variety of substances. A sampling might include inorganic ions, organic molecules, chemical complexes, mineral particles, microorganisms, and even heat. Larger kinds of contaminants may include oil and scum, natural debris, fish, boards, rags, and whatever may be discarded to the sewer or to ambient waters. Contaminants can number, literally, in the millions. Those that are regulated in the United States number about 4000.

Typical source waters for various purposes include mountain streams, lower reaches of rivers, municipal wastewater, treated municipal wastewater, hazardous waste sites, etc. Each has a typical "profile" of water quality, and some kind of treatment is always required prior to use in order to meet the criteria or standards of that use. Then, after a use of water, and prior to discharge to rivers, lakes, and seas, the product water must meet the standards and/or criteria established for such discharge. The treatment train selected depends upon the particular combination of source water available and product water required. The source water quality available and product water quality required, along with cultural, economic, and operation factors, is the treatment "context." Thousands of such combinations are possible, making each treatment context unique.

2.1 WATER QUALITY: DEFINITIONS

The term *water quality* has to do with the description of given water in terms of its *characteristics*. Characteristics of water quality include temperature; concentrations of various kinds of particles; concentrations of dissolved materials; and parameters such as turbidity, pH, color, conductivity, etc. The term *characteristic* is more inclusive than the term *contaminant* and would include temperature, color, turbidity, conductivity, etc. Two of these categories of characteristics, i.e., particles and dissolved materials, would include thousands of species each. A particular combination, or *set*, of characteristics would comprise a water quality *profile*. Those contaminants that interfere with a particular use may be considered pollutants.

To add further to the definitions that circle about the same idea, the term *parameter* is used frequently. Water quality parameters might include temperature, BOD, pH, specific electrical conductance, UV_{254} absorbance, etc.

With respect to uses of water, the terms *criterion* and *standard* are important. A water quality *criterion* is a contaminant concentration *limit* that, if exceeded, may impair a use or cause a toxic effect in certain animals or plants. As an example, a boron *limit* of 0.5 mg/L is considered appropriate for citrus crops. A criterion could also specify a contaminant or parameter *range*, e.g., $3.3 \leq \text{pH} \leq 10.7$ for trout (McKee and Wolfe, 1963, p. 236). A water quality *standard* is a quasi-legal limit for a contaminant concentration or parameter value, i.e., the value may be referenced in a law but may be either "recommended" or "enforced," depending upon the severity of the effects and the levels that are economically achievable. Usually, there is nothing absolute about the foregoing definitions.

Table 2.1 illustrates a water quality description for a proposed industrial waste discharge. Points of interest in Table 2.1 are (1) some 37 contaminants are listed; (2) concentration limits are shown for each contaminant; (3) two places for discharge—a publicly owned treatment works and a river—are shown, each with its own respective discharge limits; (4) limits are given in terms of the monthly average and the daily maximums; (5) a variety of organic compounds are listed; and (6) a variety of heavy metals are listed. Each treatment situation is different and would have a different list of contaminants and different limits. A similar tabular description, but with different constituents, would apply to a municipal wastewater discharge, a drinking water treatment plant product water, or another industrial waste situation.

2.1.1 CONTAMINANTS

A *contaminant* is defined as a substance that makes another substance impure. As applied to water, any material that is present in the water other than molecular water would be a contaminant. Often, the term has a negative connotation; in other words, we most often refer to an undesired substance as a contaminant. The term's *constituents* and *characteristics* have more neutral connotations and are used here almost interchangeably with the word contaminant. As noted, however, the term *characteristic* is more inclusive and would include such things as temperature and turbidity.

A *pollutant* is a synonym of the word *contaminant* but is more often identified as an *introduced* contaminant from an anthropogenic source. The definition of a pollutant by the US Public Health Service drinking water standards (USPHS, 1962) was

> Pollution, as used in these Standards, means the presence of any foreign substance (organic, inorganic, radiological, or biological) in water, which tends to degrade its quality so as to constitute a hazard or impair the usefulness of the water.

As used later by the regulations pursuant to PL92-500—the 1972 Clean Water Act—the term *pollutant* is associated with an introduced contaminant and implies impaired utility of water.

BOX 2.1 ON WATER QUALITY

In the 1880s, notions of water quality were limited by the knowledge in the two of its science "mother" fields—chemistry and bacteriology. Inorganic constituents were expressed as concentrations of various salts, e.g., calcium sulfate, sodium chloride, etc. Many years would elapse before inorganic chemical water quality would be expressed in terms of cations and anions, e.g., as Ca^{2+}, ${SO_4}^{2-}$, etc. The ion theory had not yet taken hold, being proposed only in 1887 by Svente Arrhenius (1859–1928), professor of chemistry, University of Uppsala, but not accepted until years later. Regarding microbes, the science of microbiology was just being defined, based upon the work of Pasteur in 1861, Lister in 1867, and Koch in 1876 and 1882 (Prescott et al., 2005, p. 8). By about 1882, the science of bacteriology had an identity.

Courses in "sanitary chemistry" that evolved from this background were focused largely on wet chemistry analysis of such constituents as alkalinity, hardness, nitrate, chloride, nitrogen, biochemical oxygen demand (BOD), etc. The Langelier Index, proposed by Professor Wilfred Langelier in 1936 (see Langelier, 1936), brought some degree of rationale from equilibrium chemistry to the problems of deposition of calcium carbonate and corrosion in pipes. The index was applied empirically to handle problems of practice. Much about bacterial growth and enumeration of bacteria was understood by the early twentieth century. Such was, in-a-nutshell, the state of knowledge of water quality about 1950.

By the early1950s, the stage was being set for the modern era. The book *Water Quality Criteria* (McKee, 1952) was published by the State of California, microbiology fundamentals were assimilated in academic studies, and Werner Stumm at Harvard introduced the idea of equilibrium chemistry as a means to model the behavior of natural systems. By the late 1950s, instruments such as atomic absorption, gas chromatography, polarography, fluorescence, TOC analyzers, mass spectrograph, etc. were introduced.

By the 1960s, ideas from chemistry theory, e.g., thermodynamics, kinetics, redox reactions, acid–base reactions, complexation, etc., became assimilated into the nomenclature of aqueous chemistry. Analyses of water could include a complete spectrum of organic compounds along with the traditional ones. Total organic carbon (TOC) was a parameter used to supplement BOD but has yet to supplant the latter. Instrumental methods were displacing wet chemistry and providing the means to analyze for virtually any contaminant, and at microgram per liter levels.

By the 1980s, the idea of water quality had moved well beyond the traditional notions prevalent in the 1950s. Also, the spectrum of contaminants was very broad and might well have included more than 100 in a typical analysis. Because of both perceived health risks and analytical capabilities that included more compounds and at lower levels, the number regulated increased to several thousand contaminants.

2.1.2 State of Water

The *state* of a volume of water, as defined here, refers to its water quality. Characteristics of water quality that may comprise its "state" include temperature; concentrations of various kinds of particles; concentrations of dissolved materials; and parameters such as turbidity, pH, color, conductivity, etc. The idea of "state" (defined usually, in the field of physical chemistry, as pressure, temperature, volume) comes from the field of physical chemistry and is the same as a water quality "profile." Both water quality "state" and water quality "profile" are terms adopted for use in this text.

The term water quality "state" adds the notion that energy is involved. For example, if we reduce the concentration of a substance, as done by a unit process, a state change results and energy is required.

2.1.3 Criteria

A water quality *criterion* refers to a contaminant level, which when not exceeded, will not impair a given beneficial use of water. A great deal of research and deliberation is involved in establishing a criterion for a particular contaminant. Seldom is the result definitive, and considerable uncertainty may be associated with any numerical value determined.

2.1.4 Standards

A criterion becomes the basis for a *standard*, which is a codified criterion. Water quality standards have evolved over the decades of the twentieth century. Usually, standards are *normative* in character, i.e., dependent not only on effects on uses but on economic and cultural factors.

2.1.4.1 Kinds of Water Quality Standards

Water quality standards have been developed for a variety of situations. The first in the United States were in 1914 and applied to drinking water on "common carriers" that crossed interstate boundaries. These standards evolved, incrementally, to the USPHS Drinking Water Standards of 1962. They are useful to review for the following reasons: (1) the standards provide an overview of some of the notions of basic standards for drinking water quality; and (2) the 1962 standards were a starting point for those that have evolved pursuant to PL93-523, the 1974 Safe Drinking Water Act, and its ensuing amendments.

TABLE 2.1
Proposed Discharge Limitations to Illustrate a Variety of Contaminants and Maximum Concentrations for an Industrial Discharge

	Discharge to POTW[a]		Discharge to River[b]	
Parameter	Monthly Average (mg/L)	Daily Maximum (mg/L)	Monthly Average (mg/L)	Daily Maximum (mg/L)
Anthracene	14	36	11	30
Benzene	33	80	19	70
Chlorobenzene	74	197	8	14
Chloroform	62	176	11	24
Ethylbenzene	108	290	17	56
Toluene	18	45	13	41
trans-1,2-Dichloroethene	19	50	11	28
Tetrachloroethene	40	125	11	29
Trichloroethylene	20	53	11	28
1,1,1-Trichloroethane	17	45	11	28
Methylene chloride	63	241	21	46
1,2-Dichlorobenzene	105	410	40	84
1,3-Dichlorobenzene	108	290	16	23
1,4-Dichlorobenzene	73	194	8	14
Bis(2-ethylhexyl) phthalate	72	197	53	144
Naphthalene	14	36	11	30
Nitrobenzene	1704	4878	14	35
Phenol	14	34	8	13
1,2,4-Trichlorobenzene	109	415	35	72
1,1-Dichloroethene	17	46	8	13
1,2-Dichlorpropane	149	605	79	119
2,4-Dimethylphenol	14	36	9	19
4,6-Dinitro-*o*-cresol	59	211	37	63
Di-*n*-butyl phthalate	15	33	14	29
Cadmium	130	130	5	11
Chromium, total	120	230	573	1430
Chromium, hexavalent	60	110	132	266
Copper	110	110	<dL	<dL
Lead	400	400	<dL	<dL
Mercury	5	5	0.2	0.5
Nickel	170	360	872	2054
Zinc	490	490	<dL	<dL
Arsenic	50	50	2	4
BOD5	No limits		45	120
TSS	No limits		183	57
pH (standard units)	6	10	6	9

Source: Adapted from Cooper, A.M. et al., *Environ. Prog.*, 11(1), 18, February 1992.
[a] Pretreatment limitations to POTW (publicly owned treatment works, per PL92-500, the 1972 Clean Water Act).
[b] Direct discharge limitations.

In 1965, PL89-234 required that states develop standards for interstate streams. In 1972, pursuant to PL92-500—the Clean Water Act—an effluent discharge permit was required and was called the National Pollution Discharge Elimination System (NPDES) permit. Drinking water standards were required by PL93-523—the Safe Drinking Water Act—which was the first federal mandate for states to develop drinking water standards. By 1986, some 83 contaminants were regulated. Regulation of toxic pollutants was required under the 1972 Clean Water Act (PL92-500), resulting in the "priority pollutant" list with some 129 contaminants (Keith and Telliard, 1979). In addition, standards have evolved, or have been contemplated, for a variety of other situations such as for water reuse for agriculture, water reuse for drinking water, etc.

2.1.4.2 Normative Standards

There is nothing absolute about water quality standards. Such standards are determined both by the degree of perceived risk that a society is willing to accept and by the knowledge of particular contaminants and their effects, i.e., criteria. It is the *norms* of a society that determine which contaminants are of concern and at what levels. Thus, water quality standards are *normative* in nature.

2.1.4.3 Standards as Targets for Treatment

Water quality standards define the performance requirements for treatment plants. A problem is that such standards have evolved and changed over the decades since their inception, i.e., since the 1962 USPHS drinking water standards, constituting a "moving target," so to speak. Planning a treatment train should, then, build in the flexibility to meet more stringent standards in the future.

The overall "goal" in drinking water treatment has been, since its inception, "to produce a safe, palatable water," as expressed by the American Water Works Association. Traditional objectives of treatment that support the health goal have been to remove disease-causing organisms. Removal of color, odor, and turbidity has supported the goal of a palatable water. The concern about chemicals that could be carcinogenic has led to additional objectives, first to remove trihalomethane precursors (expressed in the 1974 Safe Drinking Water Act, i.e., PL93-523). This objective was expanded to include removal of disinfection by-products (DBPs). Subsequent to this, in later amendments, the SDWA provided the mandate to remove an array of organic chemicals and metals. Thus, while the goals have remained the same, the objectives defined to achieve those goals have expanded to accommodate increasing knowledge about contaminants, lower detection limits and increased accuracy in sampling and analysis, and more encompassing normative standards.

2.1.5 Surrogates

Most measures of water quality involve surrogates, i.e., a quantity that is relatively easy to measure and may be used as an index of the quantity of specific interest. To illustrate, two surrogates—turbidity and coliform bacteria—have been institutionalized in their use. Regarding the latter, there are many enteric pathogens, e.g., hepatitis A virus, ECHO 12 virus, Coxsakie virus, polio virus, *Vibrio cholera* bacterium, *Salmonella typhosi* bacterium, *Shigella dysenteriae* bacterium, *Endamoeba histolytica*, *Giardia lamblia* cysts, *Cryptosporidium parvum* oocysts, etc. By definition, "enteric" means that such organisms may be found in municipal sewage. Therefore, the coliform group of bacteria (more specifically, the fecal coliform subgroup) serves as a surrogate for enteric pathogens in general, and such presence would indicate the presence of pathogens. Turbidity is a general indicator of water quality; a low level of turbidity does not ensure that the water is safe, but a high level is grounds for rejection based on palatability and on the presumption that the water is not safe.

One of the earliest surrogates in wastewater treatment was the 5 day BOD, i.e., BOD5 (or simply BOD, with the 5 day incubation period understood). The BOD is a measure of biodegradable organic matter. The difference between BOD

TABLE 2.2
Examples of Surrogates in Water Treatment

Surrogate	Measurement
Wastewater	
BOD[a]	Organic matter that is subject to biodegradation by biological treatment
SS[b]	Solids that are subject to settling and that will remain in suspension
MLSS[c]	Total suspended matter in an activated sludge reactor
VSS[d]	Index of suspended matter that is organic carbon
Coliforms[e]	Indicator of the presence of pathogens
Potable water	
Turbidity[f]	Suspended matter $\leq 1\ \mu m$
	Index of palatability
Color[g]	Index of palatability
	Index of TOC, organic matter, and fulvic acids
TOC[h]	Collective or group measure of organic matter
TTHMFP[i]	THM precursors
UV_{254}[j]	Index of TOC and TTHMFP
Coliforms	Indicator of the presence of pathogens
MPA[k]	Indicator of whether protozoan cysts could be present in finished water
Particle counts[l]	Indicator of quality of finished water

[a] BOD: Biochemical oxygen demand.
[b] SS: Suspended solids as filtered by filter paper.
[c] MLSS: Mixed liquor suspended solids is a gross parameter of active biomass and requires oven-drying a sample; MLVSS: mixed liquor volatile suspended solids, is considered more accurate measure of active biomass, but requires a placing a solids sample, after oven-drying, in a laboratory furnace at 600°C to combust the organic solids.
[d] VSS: Volatile suspended solids.
[e] Coliforms: Bacteria of the coliform group that ferments lactose. Two subgroups of the coliform group are (1) *Aerobacter aerogenes* and (2) fecal coliforms.
[f] Turbidity: A measure of the light-scattering property of a liquid.
[g] Color: A measure of the color of a water as measured by the cobalt–platinum standard.
[h] TOC: Total organic carbon as measured by converting non-purgable carbon to carbon dioxide, which is measured by infrared absorbance.
[i] TTHMFP: Total trihalomethane formation potential, which is the chlorine consumption of a given water in mg/L over a specified period of time, e.g., 24 or 96 h.
[j] UV_{254}: Absorbance of a water sample by an instrument emitting wavelength 254 nm (in the UV range).
[k] MPA: Microscopic particulate analysis.
[l] Particle counts: Counts of particles in #/mL in the water source being measured.

measurements into and out of a wastewater treatment plant has been a traditional measure of performance. Also, BOD is an index of the impact on the oxygen resources of the ambient receiving water. One factor in favor of adopting such a surrogate is that its use has been institutionalized (e.g., standards have been established, its laboratory protocol is well known, results of monitoring are available for most plants, its interpretation in terms of plant performance or of water quality is clear).

Table 2.2 lists examples of surrogates used in water treatment (both wastewater and potable water). Those that have emerged since about 1980 include TOC, TTHMFP, UV_{254}, MPA, and online particle counts. Some surrogates have a history, e.g., TOC became instrumentally feasible in 1965 through instrumentation developed by Beckman Instruments, Inc., which converted organic carbon to carbon dioxide gas, which then was measured by infrared absorbance. A host of other surrogates could be added. A laundry list, not inclusive, might include TDS (total dissolved solids), EC (specific electrical conductivity), hardness as $CaCO_3$, alkalinity as $CaCO_3$, TKN (total Keldahl nitrogen), MLVSS (mixed liquor volatile suspended solids), etc.

2.2 FEDERAL LAWS

The formal break with the past for the United States with respect to pollution control was in 1965. In 1962, Rachel Carson's book *Silent Spring* (Carson, 1962) precipitated the environmental movement that articulated public concerns about the environment and spawned a host of laws through the 1970s. The Environmental Movement, although historic in itself, was actually a part of a continuum of political events from the Public Heath Movement initiated in the early nineteenth century in England and the Conservation Movement of Gifford Pinchot and Theodore Roosevelt in the early twentieth century.

For engineers, the Environmental Movement resulted in a basic shift in objectives for the design of wastewater treatment plants. From about 1920 when wastewater treatment was underway, most statements of ambient water quality standards were merely to eliminate sludge banks in streams, to prohibit floatables, and to maintain slightly aerobic conditions, i.e., 2 mg/L dissolved oxygen concentration. The idea was to utilize the "assimilative capacity" of a given water body. The main pollutants were BOD and suspended solids (SS).

The new ethic, however, as suggested by the 1965 law and made explicit by the 1972 law, called the Clean Water Act, was to eliminate the discharge of pollutants. In addition to the traditional pollutants, nutrients, i.e., phosphates and nitrates, became a concern along with toxic pollutants. The alga blooms of Lake Erie symbolized the nutrient problem, and the term *eutrophication*, which means nutrient-rich, became a household word. In addition, the 1972 law required a NPDES permit to discharge pollutants. The result was a fundamental transformation in the modus operandi in design objectives for engineers and in the regulation of pollutants.

In the field of drinking water, the U.S. Congress passed the Safe Drinking Water Act, PL93-523, in 1974. This was the first direct involvement of the federal government in drinking water standards—a dramatic break with the past, which was to change the U.S. water industry.

The passage of the Act was precipitated by an Environmental Protection Agency (EPA) report of trihalomethanes (THMs) in New Orleans drinking water and their link to the higher incidence of cancer. The 1986 amendments resulted in another major change, i.e., that 83 substances were listed as drinking water contaminants, with the provision that 25 contaminants were to be added every 3 years. The new ethic of drinking water treatment was articulated by Abel Wolman in a luncheon speech at the 1985 annual conference of the American Water Works Association with the phrase, "if in doubt, take it out." This phrase, coined by the icon in the field, encapsulated a paradigm shift toward articulated objectives in water treatment that had been occurring since the passage of the Safe Drinking Water Act. This shift was similar in character to the shift in water pollution control from the idea of "assimilative capacity" toward the goal of "zero" pollutant discharge (Box 2.2).

BOX 2.2 FEDERAL LAWS ON WATER POLLUTION CONTROL

According to Dworsky (1967), water pollution was a major public concern to the nation's 75 million people at the beginning of the twentieth century. Congress in 1912, however, made the decision to limit its interest to research and technical assistance and to leave to the states the major role in controlling pollution. After the World War II, Congress passed the Water Pollution Control Act of 1948, PL80-845. The Act included financial and technical aid and provision for research and planning. The Act was amended in 1952, 1956, 1961, 1965, and 1966. The 1956 amendment added a phrase, "to establish a national policy for the prevention, control, and abatement of water pollution." The 1965 Act provided for the establishment of water quality standards for interstate streams and other water bodies, giving the states the responsibility, but federal action could be taken in the face of state inaction. The 1972 amendments established the requirement for effluent standards and the National Pollution Discharge Elimination System (NPDES) permit. In 1970, the Environmental Protection Agency (EPA) was established by President Nixon to administer the environmental laws (with some exceptions). The Safe Drinking Water Act was enacted in 1974 by PL93-523 and has had a similar sequence of strengthening amendments.

The federal legislation culminated in the form of the Comprehensive Environmental Response, Compensation, and Liability Act (Superfund) and was enacted in 1980 by PL96-510, spawning an industry on cleanup of hazardous waste sites.

2.2.1 Legal Definitions

As a rule, terms used in federal legislation are defined; many have been assimilated in the general lexicon on water quality. Several are quoted here for reference.

As defined by PL93-523, the Safe Drinking Water Act of December 16, 1974, Section 1401,

> (3) The term *maximum contaminant level (MCL)* means the maximum permissible level of a contaminant in water, which is delivered to any user of a public water system.
>
> (6) The term *contaminant* means any physical, chemical, biological, or radiological substance or matter in water.

As defined by PL92-500, the Clean Water Act of October 18, 1972, Section 502, General Definitions,

> (6) The term *pollutant* means dredged spoil; solid waste; incineration residue; sewage; garbage; sewage sludge; munitions; chemical wastes; biological materials; radioactive materials; heat; wrecked or discarded equipment; rock; sand; cellar dirt; and industrial, municipal, and agricultural waste discharged into water.
>
> (11) The term *effluent limitation* means any restriction established by a state or the administrator on quantities, rates, and concentrations of chemical, physical, biological, and other constituents that are discharged from point sources into navigable waters, the waters of the contiguous zone, or the ocean, including schedules of compliance.
>
> (12) The term *discharge of a pollutant* and the term *discharge of pollutants* means (a) any addition of any pollutant to navigable waters from any point source, (b) any addition of any pollutant to the waters of the contiguous zone or the ocean from any point source other than a vessel or other floating craft.
>
> (13) The term *toxic pollutant* means those pollutants or combinations of pollutants, including disease-causing agents, which after discharge and upon exposure, ingestion, inhalation, or assimilation into any organism, either directly from the environment or indirectly by ingestion through food chains, will, on the basis of information available to the administrator, cause death, disease, behavioral abnormalities, cancer, genetic mutations, physiological malfunctions (including malfunctions in reproduction), or physical deformations in such organisms or their offspring.
>
> (14) The term *point source* means any discernible, confined, and discrete conveyance, including but not limited to any pipe, ditch, channel, tunnel, conduit, well, discrete fissure, container, rolling stock, concentrated animal-feeding operation, or vessel or other floating craft, from which pollutants are or may be discharged.
>
> (17) The term *schedule of compliance* means a schedule of remedial measures, including an enforceable sequence of actions or operations, leading to compliance with an effluent limitation, prohibition, or standard.
>
> (18) The term *industrial user* means those industries identified in the Standard Industrial Classification Manual, Bureau of the Budget, 1967, as amended and supplemented, under the category 'Division D—Manufacturing,' and such other classes of significant waste producers as, by regulation, the administrator deems appropriate.
>
> (19) The term *pollution* means the man-made or man-induced alteration of the chemical, physical, biological, and radiological integrity of water.

2.2.2 Regulations

Federal agencies are charged with the responsibility to implement federal laws. This is done in terms of regulations, which are published in the *Federal Register* as proposed rules, and then after sufficient comment period and hearing if needed, a final rule is published. Finally, the regulation is "codified" in the *Code of Federal Regulations* (CFR). These regulations are "regulatory law" and have the force of law. Parties having an interest in the regulations may comment and influence the outcome as long as the public interest is not compromised. The comments may provide technical data and knowledge that expands the perspective of the agency and provide a forecast of economic impact and other consequences. Other parties may argue for more stringent regulations giving evidence of toxic effects to humans and ecological impacts. Finally, the regulations or their interpretation may be challenged in court (the court ruling is "judicial law").

2.2.3 Priority Pollutants

The Priority Pollutant List originated from Section 307 of PL92-500 (the 1972 Clean Water Act), which required EPA to regulate toxic pollutants. The Priority Pollutant List regulates the discharge of 126 contaminants listed in Section 40, CFR, Part 423, Appendix 2.A. The term "priority pollutant" comes from a lawsuit in 1976 in which the National Resource Defense Fund sued EPA in order to force the regulation of specific toxic compounds. A consent decree resulted from the suit, which forced EPA to accept an additional 65 compounds. The expanded list was incorporated into the 1977 Clean Water Act Amendments, PL95-217, and has become known as the *Priority Pollutant List*. The term "priority pollutant" has become widely used, and so knowledge of its origin helps to understand what is meant.

One outcome was that EPA set numeric limits for 105 water contaminants required by February 1992, listed as follows:

TABLE 2.3
List of Priority Pollutants Regulated by the USEPA

Antimony	Toluene	Hexachlorobutadiene
Arsenic	1,1,1-Trichloroethane	Hexachlorocyclopentadiene
Beryllium	1,1,2-Trichloroethane	Hexachloroethane
Cadmium	Trichloroethylene	Indeno(1,2,3-cd) pyrene
Chromium(III)	Vinyl chloride	Isophorone
Chromium(IV)	2,4-Dichlorophenol	Nitrobenzene
Copper	2-Methyl-4,6-dinitrophenol	*N*-Nitrosodimethylamine
Lead	2,4-Dinitrophenol	*N*-Nitrosodiphenylamine
Mercury	Pentachlorophenol	Phenanthrene
Nickel	Phenol	Pyrene
Selenium	2,4,6-Trichlorophenol	Aldrin
Silver	Acenaphthylene	Alpha-BHC
Thallium	Anthracene	Beta-BHC
Zinc	Benzidine	Gamma-BCH
Cyanide	Benzo[*a*]anthracene	Chlordane
Asbestos	Benzo[*a*]pyrene	4,4′-DDT
2,3,7,8-TCDD (dioxin)	Benzo[*b*]fluoranthene	4,4′-DDE
Acrolein	Benzo[*ghi*]perylene	4,4′-DDD
Acrylonitrile	Benzo[*k*]fluoranthene	Dieldrin
Benzene	Bis(2-chloroethyl) ether	Alpha-endosulfan
Bromoform	Bis(2-ethylhexyl) phthalate	Beta-endosulfan
Carbon tetrachloride	Chrysene	Endosulfan sulfate
Chlorobenzene	Dibenzo[*a,h*]anthracene	Endrin
Chlorodibromomethane	1,2-Dichlorobenzene	Endrin aldehyde
Chloroform	1,3-Dichlorobenzene	Heptachlor
Dichlorobromomethane	1,4-Dichlorobenzene	Heptachlor epoxide
1,2-Dichloroethane	3,3′-Dichlorobenzidine	PCB-1242
1,1-Dichloroethylene	Diethyl phthalate	PCB-1254
1,3-Dichloropropylene	Dimethyl phthalate	PCB-1221
Ethylbenzene	Di-*N*-butyl phthalate	PCB-1232
Methyl bromide	2,4-Dinitrotoluene	PCB-1248
Methyl chloride	1,2-Diphenylhydrazine	PCB-1260
Methylene chloride	Fluoranthene	PCB-1016
1,1,2,2-Tetrachloroethane	Fluorene	Toxaphene
Tetrachloroethylene	Hexachlorobenzene	

Table 2.3 lists the contaminants regulated under Section 307. Of equal importance, the list gives a sense of the variety of metals and organic compounds considered toxic above certain threshold concentration limits and that are subject to treatment. In other words, the idea of a pollutant had been expanded well beyond the traditional notions, prevalent till about 1960, that BOD and SS were the main concerns.

2.3 MATURATION OF WATER QUALITY KNOWLEDGE

From the beginning, before about 1900, knowledge of water contaminants evolved based upon developments that included (1) the mother sciences—chemistry and microbiology, (2) public mandates related to water quality standards, (3) specific knowledge about water contaminants and their ecological and health effects, (4) criteria for various kinds of uses of water, and (5) analysis methods that provide measurement capabilities to nanograms per liter. Thus, the state of knowledge, although not complete, has come a long way from the inception of the modern era of water treatment, i.e., since about 1880 (Box 2.3).

2.3.1 Knowledge of Contaminants

The first formal compilation of substances that comprise water quality was the 1952 book *Water Quality Criteria* by Professor Jack McKee of Cal Tech (and Partner, Camp, Dresser, and McKee), commissioned by the State of California. The book was revised and expanded by Harold Wolfe in 1962. It was an exhaustive treatise on substances that may be found in water and the effects of different concentrations. A similar book called the *EPA Yellow Book* was published in 1973 (USEPA, 1973), which had a wider distribution.

BOX 2.3 THE IONISTS

To give an idea of how recent the notion of water quality has been, the idea of expressing concentrations of mineral substances in terms of ion concentration did not catch on until after about 1900 (and certainly not before 1890). By about 1890, Svante Arrhenius of Uppsala, Jacobus Henricus van't Hoff at Amsterdam, and Wilhelm Ostwald at Leipzig had formulated a comprehensive theory of solutions and became known as the "Ionists" (from Servos, 1990, pp. 13–43).

To put in perspective the ionic theory of solutions, the notion at the time was that a salt in solution retained its identity as an undissociated compound. Thus, compounds in solution were called sulfate of alumina, sulfate of calcium, chloride of sodium, etc., and not as ions, e.g., Ca^{2+}, Cl^-, SO_4^{2-}, etc. The ionic theory was not embraced with enthusiasm as one might expect (e.g., being a conceptual "breakthrough"), but rather with skepticism and doubt. The ionic theory displaced the prevailing notions of solution only gradually, with tradition hanging on for several decades after discovery. The practical significance is that if we are interested, for example, in water quality data c. 1900, we must be prepared to deal with the old expressions.

2.3.2 Measurement Technologies

Prior to 1960, coliforms, turbidity, and pH were the main concerns in drinking water, along with selected anions and cations—mainly those associated with hardness and alkalinity. Water supplies with heavy metals were avoided. Chlorine was an added constituent that required measurement. In wastewater, BOD and SS were the main concerns (after floatables, scum, oil, and settable materials). The coliform density was measured by the most probable number (MPN) method, which was supplemented about 1960 by the membrane filter method. Cations and anions were measured by traditional wet chemistry methods involving titrations to some end point or conversion of a given substance to a colored complex that followed the Lambert–Beer law and could be measured by a spectrophotometer. The Jackson tube measured turbidity; this was a tube with gradations, which was filled with water until the light from a candle at the bottom was extinguished. The reading was stated as Jackson Turbidity Units (JTU).

By the early 1960s, a major technological revolution was underway in instrumental analysis. The TOC analyzer was developed, along with the atomic absorption instrument for metal ions, the gas chromatograph for organic compounds, polarography for specific ion probes, and the mass spectrograph for any compounds. While these instruments became available technologies in the 1960s, they were not used widely in water treatment until years later. These instruments developed further in the 1970s in the degree of sensitivity, and soon measurement in the range of micrograms per liter became feasible (such as in gas chromatography). The ability to measure organic compounds at concentrations in the range of micrograms per liter made possible the development of water quality standards in the same range.

2.4 CATEGORIZATIONS OF CONTAMINANT SPECIES

Contaminant species number in the millions. Such a large number of species must be dealt with in terms of a classification system.

2.4.1 Systems of Categorization

There are several kinds of systems available for organizing water quality knowledge. Examples include

- Alphabetical
- Chemical Abstract System (CAS) number
- Chemical category such as organic, synthetic organic, inorganic, metal, etc.
- Class within a category such as alcohols, aldehydes, phthalates, etc.
- Regulatory lists, e.g., Safe Drinking Water Act Regulated Contaminants, Priority Pollutant List, etc.

Information concerning these contaminants was available in the EPA Register of Lists (USEPA, 1991; Miller, 1993). The Register of Lists (not available by web access in 2010) cross-referenced the chemicals with the statute listing and the EPA office administering. Such listing permitted cross-referencing in terms of CAS number, contaminant type (e.g., polycyclic aromatic hydrocarbon, amine, organic acid, particulate, etc.), empirical formula, contaminant classification (e.g., organic, synthetic organic, volatile organic, inorganic, metal), and the law or regulation under which the contaminant is regulated (e.g., Priority Pollutant List, CWA Section 304, CAA Section 112, RCRA, Sludge Regulations of February, 1993, SARA Section 313, SDWA).

2.4.2 Illustrative System of Contaminant Categorization

As noted, contaminants may be categorized by a variety of systems. Table 2.4 is another illustrative categorization of contaminants. The system shown was devised, for use here, to illustrate the broad scope of contaminant categories, the large number of contaminant species, and the idea that contaminant categorization may be devised to suit the purpose at hand (but is not done arbitrarily). Table 2.5 lists some of the many pathogenic organisms, showing the associated diseases, and is included to give some idea of the microbial hazards of untreated water. Figure 2.1 shows just three organisms that may be found in ambient waters; tens or hundreds of microorganism species are found commonly in a given sample of ambient water, depending on the source. Counts of an individual organism species could range from a few to several thousand.

TABLE 2.4
Contaminant Categorization

Category	Subcategory	Group	Representative Contaminants
Particle	Mineral	Clays	Montmorillonite
			Kaolinite
		Asbestos	Chrysotile
			Amphibole
	Biological	Viruses	Pathogenic (120 types of enteric viruses that cause disease; infectious hepatitis, poliomyelitis, coxsackie, and ECHO are common)
			Nonpathogenic
		Bacteria	Pathogenic
			Nonpathogenic
		Fungi	*Alternaria*, *Aspergillus*, *Cladosporium*, *Penicillium*
		Protozoa cysts	*Giardia lamblia*
			Cryptosporidium parvum
			Entamoeba histolytica
		Plankton	Diatoms
			Green algae
			Blue-green algae
		Invertebrates	Rotifers
			Nematodes
			Crustaceans
			Other arthropods
			Ciliates
		Microscopic materials	Plant debris
			Amorphous debris
			Pollen
Dissolved substances	Inorganic	Cations	Mercury
			Lead
			Nickel
			Copper
		Anions	Chloride
			Sulfate
			Carbonate
			Bicarbonate
			Nitrate
			Nitrite
		Parameters	Alkalinity
			Calcium
			Hardness
			Ammonia ion
			Total dissolved solids (TDS)
			Specific electrical conductance (EC)
	Organic	Chlorinated hydrocarbons	Trihalomethane (THM)
			Chloroform
			Dichlorobromomethane
			Dibromochloromethane
			Bromoform
			Haloacetic acids (HAAs), chloropicrin
			Chloral hydrate
			Haloacetonitriles (HAN)
			Chloropropanes
			Haloketones
			Chlorophenols
			Aldehydes

(*continued*)

TABLE 2.4 (continued)
Contaminant Categorization

Category	Subcategory	Group	Representative Contaminants
		Surrogates	UV_{254}
			THMFP (24 h)
			THMFP (96 h)
		Total organic halides (TOX)	Trichlorophenol
			Chloroacetone
			Chloroethanol
			Chloroacetic acid
		Total organic carbon (TOC)	Trichlorophenol
			Chloroacetone
			Chloroethanol
			Chloroacetic acid
		Total organic carbon (TOC)	
		Natural organic matter (NOM)	
		Synthetic organic compounds	
		Aromatic compounds	Benzene
		Polycyclic aromatic hydrocarbons (PAH)	Alcohols
			Aldehydes
	Dissolved gases		Oxygen
			Nitrogen
			Carbon dioxide
			Ammonia
			Methane
	Characteristics		pH
			Temperature
			Color
			Taste
			Odor
			Floating material
			Esthetic
	Parameters		Specific electrical conductance (EC)
			Turbidity
			Particle counts

TABLE 2.5
Pathogenic Organisms

Category	Species	Disease
120 types of enteric viruses cause disease		
Enteroviruses[a]	Polioviruses (Sobsey, 1975, p. 415) (3)	Poliomyelitis, aseptic meningitis
	Coxsackieviruses A (Sobsey, 1975, p. 415) (23)	Herpangina, aseptic meningitis, exanthem
	Coxsackieviruses B (Sobsey, 1975, p. 415) (6)	Aseptic meningitis, epidemic myalgia, myocarditis, pericarditis
	Echoviruses (Sobsey, 1975, p. 415)	Aseptic meningitis, exanthem, gastroenteritis
Adenoviruses (Sobsey, 1975, p. 415) (31)		Upper respiratory illness, pharyngitis, conjunctivitis
Reoviruses (Sobsey, 1975, p. 415) (3)		Upper respiratory illness, diarrhea, exanthem
Hepatitis A viruses (Sobsey, 1975, p. 415)		Viral hepatitis type A or "infectious hepatitis"
Gastroenteritis viruses		Acute infectious nonbacterial gastroenteritis
	Coliphages	Infect coliform bacterium, *Escherichia coli*

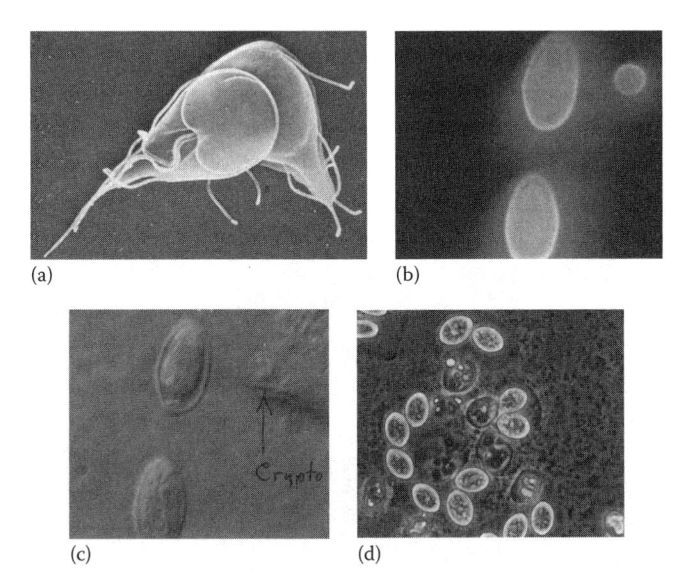

FIGURE 2.1 Examples of microorganisms found in water (the *Giardia* trophozoite is not found in water but occurs in the gut of a warm-blooded animal after excystation). (a) *Giardia* trophozoite, 10 μm width. (Courtesy of Dr. Judith L. Isaac-Renton, Public Health Laboratory Director and Professor, Medical Microbiology, The University of British Columbia, Vancouver, BC.) (b) *Giardia* cyst, 10 μm length, (c) *Cryptosporidium* oocyst, 5 μm width, (d) Oocystis parva, 400×. (Courtesy of Dr. Greg Sturbaum, CH Diagnostics and Consulting Services, Loveland, CO.)

2.5 UTILITY OF WATER QUALITY DATA

In water treatment, uses of water quality data include

1. Preliminary selection of unit processes: Source water quality must be characterized over the time period in which cyclical variation occurs (usually an annual cycle), and product water quality must be specified.
2. Process control: Requires water quality monitoring at selected node points in the treatment train in order to verify proper operation and to diagnose problems.
3. Regulatory compliance: Requires sampling of the effluent to document compliance with standards or criteria.
4. Monitoring of receiving waters: Confirms whether a stream standard has been violated.
5. Laboratory analyses: Data from the laboratories must be organized, understood, and utilized.
6. Data archiving: Data must be archived in a format suitable for long-term storage and retrieval.
7. Supervisory control and data acquisition (SCADA) systems.

Monitoring sensors report to computers for display and use in plant control. Pumps, valves, and motors are controlled.

2.5.1 Contaminants and Water Uses

As stated, contaminants that may be found in water number in the thousands. The utility of water covers a wide variety of uses such as drinking water, industrial water, cooling water, esthetic appreciation, recreation contact, ecological habitats, irrigated agriculture, etc. Each of these categories of water use may be divided further. The objective of water quality criteria is to know the limits of each contaminant level that will not impair the uses intended. The scope of the undertaking may be depicted in a matrix of m uses in columns (perhaps several hundred columns would adequately depict the uses) and of n contaminants in rows (which may number several thousand). The number of possible interactions, i.e., criteria, are several thousand, with size, $m \cdot n$.

Since not all contaminants have a bearing on each use, the problem may be reduced considerably, but still there are hundreds or thousands of criteria that must be delineated. Examples of such criteria include (1) the electronics industry, where particle-free water is required; (2) boiler feed water for electric energy-generating plants, which must be "ion-free" water; (3) growing of citrus crops, where boron levels must be less than 0.5 mg/L, etc. These three examples indicate the nature of the criteria determination issue, e.g., that in-depth knowledge about the contaminant–use interaction must be generated. Most often, criteria are more complex than just a simple number, e.g., temperature affects how fish will survive under low dissolved oxygen conditions, the effect of TDS in irrigated agriculture depends on the plant species and the amount of water applied, etc.

2.6 COMBINATIONS OF QUALITY OF SOURCE WATERS AND PRODUCT WATERS

Source water quality and the required product water quality dictate the design and operation of a treatment train. Each source water is different, and standards for product water vary with the kind of use. Examples of uses include drinking water, discharge of municipal wastewaters to a receiving body, production of particle-free and molecule-free process water for microchip production, etc. The number of combinations is as large as the number of possible water sources times the number of possible water quality standards.

To the extent possible in water treatment where ambient water is used, the water of highest quality is sought as a source. Since high-quality sources have been mostly appropriated or are not available in some regions of countries, secondary sources have been committed to use to an increasingly greater extent. Thus, wastewaters are considered as sources of water, assuming the associated water rights to their use can be established. Also, as unit process technologies develop and expand, the number of treatment options increase, which may facilitate the use of lower-quality waters. The use of membranes, for example, has become increasingly more economical since about the mid-1990s, which has made feasible the use of even seawater as a source.

Table 2.6 indicates the variety of source waters and product water requirements. Other columns indicate examples of

TABLE 2.6
Combinations of Source Waters and Purposes of Treated Water as Contexts for Treatment

Source Water	Examples	Examples of Contaminants to Treat	Treatment	Product Water Purpose	Product Water Quality
Ambient, high-quality water such as streams and lakes	South Platte River in mountains	Turbidity levels <0.5 NTU Algae, rotifers, crustaceans TDS < 50 mg/L	Filtration		
Ambient water degraded by various treated point-source discharges and diffuse inputs	South Platte River below Denver	Turbidity levels <10 NTU $NO_3 \approx$ 40–50 mg/L TDS ≈ 800 mg/L	Membranes		
Ambient water in the lower reaches of rivers degraded by natural hydrologic processes and waste discharges and agricultural runoff	Missouri River in the lower reaches, Mississippi River	Turbidity levels 20–200 NTU $NO_3^- \approx$ 40–50 mg/L TDS ≈ 800 mg/L Hardness 200–300 mg/L			
Groundwater with surface water influence	South Platte alluvium below Denver, used as source water by several riparian communities	Turbidity levels <0.5 NTU $NO_3 \approx$ 40–50 mg/L TDS ≈ 800 mg/L Hardness 200–300 mg/L			
Deep groundwater	Denver aquifer, some 300 m depth	Turbidity levels <0.5 NTU $NO_3 <$ 10 mg/L, TDS < 800 mg/L Hardness 200–300 mg/L Fe^{2+}, Mn			
Municipal wastewater	Denver municipal sewage (with pretreated industrial wastes)	Suspended solids 300 mg/L BOD ≈ 300 mg/L Various pathogenic viruses, bacteria, cysts, worms			
Treated municipal wastewater	Denver Metro Wastewater Reclamation water effluent	Suspended solids 300 mg/L BOD ≈ 30 mg/L Possible low levels of pathogenic viruses, bacteria, cysts, worms			
	Water Factory 21	Viruses, carcinogens	Lime softening Filtration Activated carbon Membranes (hyperfiltration)	Seawater barrier Municipal drinking water	Meets drinking water standards and other criteria—to be established

Treated municipal drinking water	Fort Collins drinking water	Particles TOC TDS	Filtration Activated carbon Membranes (hyperfiltration)	Electronics process water for chip manufacture	Particle-free, TOC $\leq$ DL, TDS $<$ DL
Industrial wastewater	Wastewaters from various industrial plants (steel fabrication, meat packing, metal plating, etc.)	Possible: heavy metals, high BOD	Pretreatment	Discharge to stream	
Industrial wastewater	Wastewaters from various industrial plants (steel fabrication, meat packing, metal plating, etc.)	Possible: heavy metals, toxic organics, high BOD		Discharge to municipal sewer	
Industrial wastewater pretreated	Effluent from HP discharged to municipal sewer	Reduced concentrations of specific contaminants and/or loading to municipal sewer			
Contaminated groundwater, contaminated soils	CERLA hazardous waste sites such as Rocky Mountain Arsenal	Possible: heavy metals, toxic organics (pesticides, herbicides, etc.)			
Source water	Examples	Examples of contaminants to treat	Treatment	Product water purpose	Product water quality
Groundwater not suitable for drinking water	Various groundwater basins such as found in the Intermountain West in the USA	Arsenic	Point of use (under sink or in basement or garage treatment) of ion-exchange or hyperfiltration membrane	Individual household drinking water	Controls on whether drinking water standards are met are not in place
Brackish water	San Francisco Bay	TDS $\approx$ 5000 mg/L	Membranes	Drinking water, industrial uses	
Seawater	Persian Gulf	TDS $\approx$ 35,000 mg/L	Multieffect evaporation distillation	Drinking water—blended with groundwater	Drinking water standards
Connate water	Oil field brines (near Bakersfield, CA)	TDS $\approx$ 38,000 mg/L	Oil removal	Reinjection into oil-bearing aquifer	Connate water with oil removed

source waters, typical contaminants treated, unit processes likely, and the purpose of the product water. As seen in Table 2.6, the variety of treatment situations includes treatment of high-quality surface waters for drinking water, treated municipal wastewater being further treated for agricultural use, industrial cooling water, injection to groundwater, or for drinking water, a contaminated groundwater being renovated by treatment and re-injected into an aquifer, etc. Each requires different criteria or standards and different treatment trains.

Table 2.6, column 3, indicates some of the contaminants to be reduced in concentration (the list is indicative only). Column 4 indicates typical unit processes.

In summary, the source water may be from any source and the product water may be whatever is required to meet the purpose of that water. Water quality must be characterized in each case. As a rule, criteria and standards govern the target of treatment.

PROBLEMS

2.1 Water Quality Profiles

State what you believe may be reasonable estimates for water quality profiles (constituents and concentrations and any important time variations) for several source waters that may be put to some use (for any purpose as listed in the second problem), such as

1. Mountain streams in the Rocky Mountains or in the High Sierras
2. Lower reaches of rivers such as the South Platte, the Missouri, the Ohio, the Sacramento, the Iowa, and the Cedar
3. Lakes such as Lake Superior, Lake Erie, and Lake Tahoe
4. Raw wastewaters to municipal treatment plants
5. Treated wastewaters from municipal treatment plants
6. Tertiary-treated wastewaters from municipal treatment plants
7. Raw wastewaters from industries such as electronics, metal plating, meat packing, brewery, poultry processing, electric energy generation, etc.

Develop case examples for situations that you select.

2.2 Water Quality Criteria

State what you believe may be reasonable expectations for water (constituents and concentrations and any important time variations) for purposes such as

1. Irrigation of citrus; vegetables such as lettuce, sugar beets, etc.
2. Farm uses such as livestock, poultry, etc.
3. Industries such as sugar beet refining, steel manufacturing, manufacture of electronic chips, electric energy generation, poultry processing, dairies, etc.
4. Drinking water such as in New York City, Seattle, Denver, New Orleans, Baghdad, Zurich, Milan, Istanbul, Palermo, etc.
5. Recreation such as swimming pools
6. Protection of saltwater environments such as the Mediterranean, the Caribbean, and the San Francisco Bay
7. Fisheries such as the Blue River near Dillon, Echo Lake in the High Sierras, Lake Michigan, the South Platte near Greeley, San Francisco Bay, the Ohio River, etc.

Develop case examples for situations that you select.

2.3 Source Water Quality and Treatment for Potable Water

Discuss some examples from the literature with respect to water quality profiles of source waters and the degree of treatment needed to meet certain uses that you may select. If you have access to records of treatment plants, then these provide firsthand references and are more "real-world."

2.4 Water Quality Criteria/Standards

Look up criteria and standards for uses that you may select. Pick two categories of uses. Document your sources.

2.5 Organic Carbon over Annual Cycle

Discuss levels of TOC, and color, as they vary over an annual cycle in ambient waters that you may select.

2.6 Particles and Turbidity over Annual Cycle

Discuss levels of particles and turbidity as they vary over an annual cycle in ambient waters that you may select.

2.7 Water Quality Monitoring

Provide some examples of water quality monitoring with respect to

1. Regulatory surveillance
2. Process control
3. Database development

ACKNOWLEDGMENTS

Dr. Judith L. Isaac-Renton, director, Pathology and Laboratory of Medicine, The University of British Columbia, Vancouver, British Columbia, went through her files to locate a *Giardia* trophozoite electron-micrograph, shown in Figure 2.1a, and graciously provided permission for its use. Dr. Greg Sturbaum, president, CH Diagnostics and Consulting Services, Loveland, Colorado, gave permission to use Figures 2.1b,c,d, generated at CSU during the period 1993–1996.

APPENDIX 2.A: ORGANIC CARBON AS A CONTAMINANT

Each contaminant has its own unique story with respect to its occurrence in natural waters and in municipal and industrial discharges, effects on uses of various kinds, the nature and range of treatment for reduction in concentration, and regulatory requirements and their evolution. The story of organic carbon is reviewed here merely to illustrate the depth and range of substantive content that may be extracted from the published material on a given contaminant. The organic carbon story (Box 2.A.1) is more extensive than most due to the health significance of carbonaceous DBPs in drinking water, which started to be understood only since the mid-1970s. In the United States, the subsequent regulations from EPA provided the mandates for the ensuing attention to DBPs. The review here is brief.

BOX 2.A.1 HUMIFICATION

Organic matter is comprised of humic and fulvic acids, indicating that their composition is of the same functional groups that make up lignins and, to a lesser extent, other plant polymers; they have more carboxylic acid functional groups, however, and they are surface active. The components of humus consist of plant polymer segments that have been oxidized to carboxylic acid groups at one or more ends of the segments. In the case of lignin polymers, the unaltered segments are more hydrophobic than the carboxylic groups. A molecule that has both a hydrophobic (nonpolar) part and a hydrophilic (polar) part is called an *amphiphile*.

Humification is a process by which biomass consisting of dead plant and animal remains is converted to humis; this is one of the basic steps of the carbon cycle. The organic compounds that make up plant and animal tissue are thermodynamically unstable in the oxidizing atmosphere of the earth's surface. The tissue is thus converted back to carbon dioxide and water that are catalyzed by enzymes from organisms. Some of the tissue is, however, only partially oxidized, which is the source of the organic compounds that accumulate as humus.

Vascular plants (those with water- and food-conducting tissues) are the dominant group in most terrestrial environments. The tissues of these plants consist of three groups of polymers: (1) cellulose, (2) hemicellulose, and (3) lignin. Lesser quantities of aliphatic polyesters, starches, proteins, phenolic macromolecular species, and lipids are present also.

The degradation of plant polymers involves depolymerization and oxidation reactions that are catalyzed by enzymes. Polysaccharide polymers such as cellulose and hemicellulose usually undergo hydrolytic depolymerization reactions, whereas lignin is degraded mainly by oxidation. Lipids undergo hydrolysis and oxidation. The products from lignin and lipid degradation are, in general, oxidized fragments in which much of the chemical structure of the original polymer is preserved.

In addition to amphiphiles produced by degradation, some of the lignin in wood is present as amphiphilic lignin–carbohydrate complexes. These complexes have number average molecular weights on the order of 6000–8000 Da.

According to Randtke (1988, p. 40), organic contaminants in water may be grouped into three classes:

1. *Natural organic matter* (NOM): Humic substances, microbial exudates, animal wastes, and products of degraded tissue
2. *Synthetic organic chemicals* (SOCs): Pesticides, volatile organic chemicals (VOCs), and other chemicals produced commercially or as waste products of manufacturing
3. *Chemical by-products and additives*: Substances that enter or are formed during treatment or in the distribution system

NOM is the source of color—a traditional contaminant, and is a precursor of DBPs, which are possible carcinogens, an issue since the mid-1970s. In addition, the residual NOM after treatment may serve as substrate for bacterial growths such as in the distribution system. In the United States, the issues of DBPs and SOCs and possible health effects led to the 1974 Safe Drinking Water Act. Enforcement of the Act, by regulations, was stimulated by the capabilities to measure chemical concentrations in micrograms (μg) per liter through developments in instrumental analyses methods. In the case of organic compounds, such instrumental methods were exemplified by advances in gas chromatography and mass spectrography and associated lower costs (see Box 2.1).

2.A.1 Categories of Organics in Water

The organics found in waters are characterized according to various schemes, depending upon the water and the purpose of the characterization. The characterization evolves as new problems become known, and is a function of the purposes of the waters being considered. Wastewaters, for example, have been characterized traditionally in terms of BOD and SS. The BOD measurement is done by a 5 day BOD test and is dependent upon having a properly seeded test bottle. The test measures the organic carbon that is degraded in 5 days by the microbe's species in the seed. The test does not measure TOC, but for a given wastewater, BOD may be proportional to TOC. Only a portion of the organic carbon is biodegradable, and the test measures the portion that is biodegradable within a 5 day period. The test actually calculates BOD based on the measured concentration of dissolved oxygen at the beginning and end of the test under incubation at 20°C for 5 days, taking into account the dilution of the sample with "BOD" water. The BOD water contains a prescribed mixture of nutrients, as specified in *Standard Methods*, and is saturated with oxygen. This test has its origin most probably about 1900.

For natural waters that are sources of municipal water supplies, organic carbon was not of great concern until recent years, i.e., beginning about 1973. NOM occurs in surface waters in concentrations mostly in the range 3–6 mg/L, and as noted, is comprised mostly of humic and fulvic acids. These acids are products of decaying vegetative organic matter that finds its way into ambient waters after rainfall, and of the decay of organic matter within a water body.

NOM causes the water to exhibit "color," measured in terms of "standard color units" (SCU). Color, per se, has no health significance, but it does cause concern as it affects palatability of a drinking water. Since there is no health significance to color, it was assigned the status of a "secondary standard" in the 1973 Safe Drinking Water Act (PL93-523). Color has been considered as a parameter of drinking water quality since about the 1920s, and its reduction has been a traditional objective of water treatment (achieved by coagulation, and as measured by the "jar" test).

In the early 1970s, chlorinated organics were identified as carcinogens (Box 2.A.2) based on reports of higher levels of THMs in New Orleans drinking water, which precipitated the 1974 Safe Drinking Water Act (SDWA)—the first

BOX 2.A.2 DISINFECTION BY-PRODUCTS AS A NATIONAL ISSUE

A review of how disinfection by-products (DBPs) became a national issue was reviewed by James M. Symons (2001a,b), who, in the1970s, was Chief of the Physical and Chemical Contaminants Removal Branch, Drinking Water Research Division, USEPA, Cincinnati. This position provided the vantage point of both perspective and responsibility to provide initiative.

Johannes Rook, a chemist with the Rotterdam Water Works, discovered chloroform in Rotterdam's drinking water in 1971 while looking for sources of taste and odor, based on a "head-space" sampling/analysis technique he developed. Although he took special note of the chloroform as one among a score of micropollutants, there was not any special alarm, especially since the health officer mentioned that chloroform was a constituent of cough syrup and was not know as a toxin.

Also in 1971, Thomas Bellar, a chemist with EPA, was given an assignment to develop an adequate method to measure VOC contaminants in wastewater. The method developed was called by Bellar the "purge and trap" technique, which was an adjunct to gas chromatography. This was an analytical "breakthrough," which opened the door for detecting and measuring organic contaminants at the μg/L level. In measuring contaminants in samples of tap water, Bellar found chloroform but attributed the finding to laboratory contamination. In mid-1973, continuing to find chloroform, he decided to sample other drinking water sources. In sampling at several points in the water treatment train for the City of Cincinnati, he found that DBPs were related to the points of chlorination. For example, chloroform concentrations were (1) Ohio River, 0.9 μg/L; (2) 80 min after chlorination and alum coagulation, 22.1 μg/L; (3) 3 day settled water, 60.8 μg/L; (4) treatment plant settled water following chlorination, 127 μg/L; (5) filter effluent (after powdered activated carbon and filtration), 83.9 μg/L; and (6) finished water after a final chlorination, 94.0 μg/L.

In early 1974, the EPA drinking water group pondered what to do about Bellar's findings. Since the oral lethal dose was 120 mg/L, the issue seemed not an acute problem. In June 1974, however, an article in *Consumer Reports* reviewed the problem of organics in drinking water. The article was based partly on a 1972 report by USEPA and was to have strong influence on future events. The article was an indictment of the water quality of the Lower Mississippi River and stimulated public interest in the problem of organics in drinking water. James Symons returned to Europe in August 1974, and Rook described his theory that NOM, as measured by color, was a precursor to THM formation; his data showed a correlation between TTHM in μg/L and color in Pt–Co units, and other possible precursors were eliminated by experiment and deduction. At that time, the structure of humic substances were not well defined. This meeting with Rook, and his evidence that color was a likely precursor to THMs, had a strong influence on Symons relative to the possible extent of the problem of THM's in drinking water. Further stops in Europe, e.g., at the Swiss Federal Institute of Technology and at Karlsruhe, led to the decision at EPA that the THM issue was indeed important. Then, the National Organics Reconnaissance Survey was started to ascertain how widespread the problem of THMs was in the United States. Following this, in November 1974, the Environmental Defense Fund released an epidemiological study showing that disease rates for persons drinking New Orleans water were higher than those persons living in surrounding communities and drinking water from sources other than the Mississippi River. At the same time, a medical faculty member at the University of New Orleans stated that he had found halogen-substituted organic compounds in the blood of New Orleans residents. He also stated on national television that the water quality (i.e., with respect to chloroform) from the Mississippi River was of considerably better quality than the water produced by the water treatment plant. In December 1974, with an atmosphere that caused a lessening of public confidence in the drinking water industry, the Safe Drinking Water Act (PL93-523) was signed by President Ford.

The national publicity led to pointed criticisms of the drinking water industry. Further, many in the industry could not believe that the trace concentrations of chemicals could be hazardous to public health. (As an editorial aside, the THM issue and the later *Giardia* issue, which emerged in 1978, were combined stimuli that caused a major change in the culture of the U.S. water industry. The drinking water industry became energized and perhaps one might say the "glamour" field, which, during the 1960s, was in wastewater treatment.)

national legislation that would enforce nationwide drinking water standards. Consequently, pursuant to the 1973 SDWA, THM, one of the implicated species of chlorinated hydrocarbons, was regulated in 1978 with a MCL set at 100 μg/L. Subsequent research implicated the whole family of chlorinated hydrocarbons as being possibly carcinogenic, and also a variety of other organic compounds. Thus, in 1986, when the SDWA was reauthorized, the number of regulated contaminants in drinking water was expanded from 25 from the 1962 USPHS-recommended drinking water standards (Anon, 1962), to 83, with the requirement stipulated in the law that 25 new contaminants be added to the list every 3 years. The expanded list included natural organics, synthetic organics, and volatile organics. The change from recommended federal standards to mandated ones constituted a fundamental shift in the treatment of drinking water.

As the issue developed, the whole family of halogens, i.e., chlorine, bromine, and iodine, were implicated. The associated species of halogenated organic compounds, collectively, were termed TOX. Also, measures of THMs were developed. For example, by exposing a sample of organic carbon to chlorine for 24 h, measuring chlorine concentration before and after the test gives a measure called trihalomethane formation potential (THMFP-24 h). The same test extended for 96 h gives THMFP-96 h.

2.A.1.1 Color

Traditionally, dating back to perhaps 1915, the concern with organic matter was color, which was not pleasing esthetically. Other concerns were with its deleterious effects on industrial process waters, its chlorine demand, and interference with coagulation. The USPHS 1962 Drinking Water Standard set the limit as 15 color units on the platinum–cobalt scale, but the AWWA set ≤3 units as a goal for drinking water.

Regarding ambient waters, the color units for snowmelt mountain streams are low, e.g., <5, but may rise to perhaps 50 units during spring runoff. By contrast, for swamp-like waters in the southeastern United States, color units may vary in the range 68–424. Examples of waters in this range include the Suwannee River at Fargo, Georgia; the Florida Everglades 20 miles northwest of Miami; and the Great Dismal Swamp near Norfolk, Virginia (Black and Christman, 1963).

2.A.1.2 Organic Carbon

An index of the organic content of water is TOC, which is a measure of all organic molecules in a water sample (i.e., those that are subject to being converted to carbon dioxide for measurement by infrared absorbance using a TOC analyzer). Figure 2.A.1 shows the ranges of TOC concentrations in seawater, groundwaters, surface waters, and raw and treated wastewaters.

More than 700 specific organic chemicals were identified in various drinking water sources in the United States in 1978 (*FR*43 (28):5759, Feb. 9, 1978). These compounds were from industrial and municipal discharges, urban and rural runoff, natural decomposition of vegetative and animal matter, and chlorination of water and wastewater.

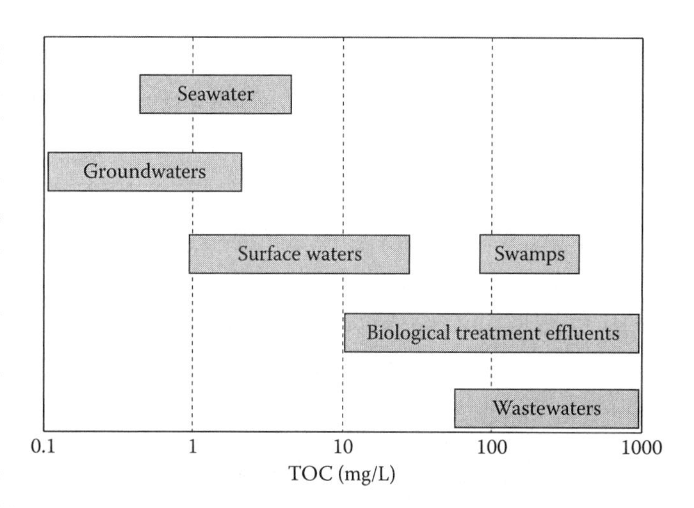

FIGURE 2.A.1 Ranges of TOC for a variety of waters.

At the time of the passage of the SDWA in December 1974, there were more than 12,000 chemical compounds known to be in commercial use, and many new compounds were being added to the list each year. These were called SOCs and included halogenated aliphatic and aromatic hydrocarbons such as carbon tetrachloride, dichloroethane, vinyl chloride, and chlorobenzenes; pesticides such as dieldrin and lindane; aromatics such as benzene, toluene, and styrene; polynuclear aromatics such as fluoranthene; nitrogenous compounds such as aniline and dinitrobenzene; esters such as dibutylphthalate; and many others.

2.A.1.3 UV_{254}

Since the mid-1980s, UV_{254} absorbance has been accepted as a surrogate for TOC. Semmens and Field (1980, p. 477) used UV_{260} (i.e., an ultraviolet light source at 260 nm wavelength), understanding the nature of the relationship. The use of UV converged on the 254 nm wavelength as the 1980s progressed, and was adopted by Hubel and Edzwald (1987) and others.

2.A.1.4 Synthetic Organic Carbon

Synthetic organic carbon compounds number in the tens of thousands and include pesticides and herbicides. This topic is mentioned to indicate its importance, but discussion is beyond the scope of this appendix.

2.A.2 Disinfection By-Products

The most abundant of the DBPs are from reactions with chlorine, with fewer from chloramines and chlorine dioxide, but with six from ozone. Chlorine has been the oxidant investigated most extensively. To illustrate a few structural formulae for some typical DBPs, Table 2.A.1 shows several representative groups. Each compound, i.e., each DBP, could be a part of any total organic halogen (TOX) measure.

Further insight as to the character of the organic carbon present in source waters is seen by the molecular weight

TABLE 2.A.1
Structural Formulae for Selected Disinfection By-Products

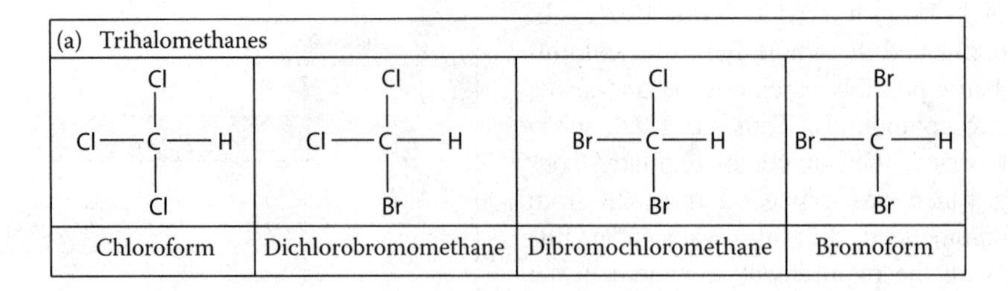

(a) Trihalomethanes			
Cl Cl—C—H Cl	Cl Cl—C—H Br	Cl Br—C—H Br	Br Br—C—H Br
Chloroform	Dichlorobromomethane	Dibromochloromethane	Bromoform

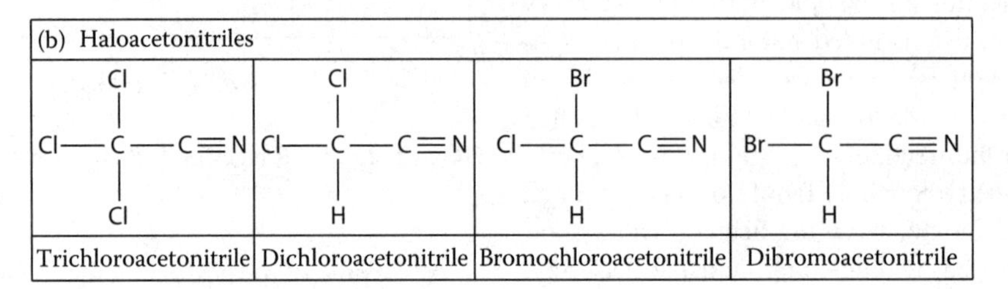

(b) Haloacetonitriles			
Cl Cl—C—C≡N Cl	Cl Cl—C—C≡N H	Br Cl—C—C≡N H	Br Br—C—C≡N H
Trichloroacetonitrile	Dichloroacetonitrile	Bromochloroacetonitrile	Dibromoacetonitrile

(c) Haloketones	
Cl O H Cl—C—C—C—N H H	Cl O H Cl—C—C—C—N Cl H
1,1-Dichloro-propanone	1,1,1-Trichloro-propanone

(d) Miscellaneous		
Cl Cl—C—NO_2 Cl	Cl H Cl—C—C—OH Cl OH	Cl H Cl—C—C—OH Cl OH
Chloroplorin	Chloral hydrate	Cyanogen chloride

(e) Haloacetic acids				
Cl O H—C—C—OH H	Cl O Cl—C—C—OH N	Cl O Cl—C—C—OH Cl	Cl O Cl—C—C—OH Cl	Br O Br—C—C—OH N
Monochloroacetic acid	Dichloroacetic acid	Trichloroacetic acid	Monobromoacetic acid	Dibromoacetic acid

(f) Chlorophenols	(g) Aldehydes	
Cl Cl—(O)—OH Cl	H H—C═O	H H H—C—C═O H
2,4,6 Trichlorophenol	Formaldehyde	Acetaldehyde

Source: Krasner, S.W. et al., *J. Am. Water Works Assoc.*, 74(8), 41, August 1989.

TABLE 2.A.2
Chlorinated Species Detected by Contact between a Filtered Municipal Secondary Clarifier Effluent and 2000 mg/L Chlorine

Compound	Concentration (μg/L)	Compound	Concentration (μg/L)
Chloroform	—	Dichloromethoxytoluene	32
Dibromochloromethane	—	Trichloromethylstyrene (220)	10
Dichlorobutane	27	Trichloroethyl benzene (208)	12
3-Chloro-2-methylbut-1-ene	285	Dichloro-a-methyl benzyl alcohol (190)	10
Chlorocyclohexane (118)	20	Dichloro-bis(ethoxy)benzene (220)	30
Chloroalkyl acetate	—	Dichloro-a-methyl benzyl alcohol (190)	—
o-Dichlorobenzene	10	Trichloro-*N*-methylanisole	—
Tetrachloroacetone	11	Trichloro-a-methyl benzyl alcohol	25
p-Dichlorobenzene	10	Tetrachlorophenol	30
Chloroethylbenzene	21	Trichloro-a-methyl benzyl alcohol	50
Pentachloroacetone	30	Trichlorocumene (222)	—
Hexachloroacetone	30	Tetrachloroethylstyrene	—
Trichlorobenzene	—	Trichlorodimethoxybenzene (240)	—
Dichloroethyl benzene	20	Tetrachloromethoxytoluene (258)	40
Chlorocumene	—	Dichloroaniline derivative (205)	13
N-methyl-trichloroaniline	10	Dichloroaromatic derivative (249)	15
Dichlorotoluene	—	Dichloroacetate derivative (203)	20
Trichlorophenol	—	Trichlorophthalate derivative (296)	—
Chloro-a-methyl benzyl alcohol	—	Tetrachlorophthalate derivative (340)	—

Source: Glaze, W.H. and Henderson IV, J.E., *J. Water Pollut. Control Fed.*, 47(10), 2511, October 1975.
Notes: Parentheses indicates approximate molecular weights.
Sum of concentrations = 786 μg/L; estimated chlorinated organic compounds ≈3000–4000 mg/L.

distributions from a sample of water from the Mississippi River, given as 0–1,000, 48%; 1,000–25,000, 20%; 25,000–100,000, 13%; 100,000–1,000,000, 20%; >1,000,000, 2% (Tate and Fox, 1990, p. 104).

As to the reactions between chloramines and organic carbon, the TOX production is about the same as that resulting from chlorine (Johnson and Jensen, 1986). Further, the health effects of chloramines-treated water are nearly as severe as those of chlorine-treated water. Thus, while the chloramine solves the THM problem, it does not solve the health problem that the THM regulation was intended to address.

2.A.3 Disinfection By-Products in Secondary Effluents

A question pertinent to wastewater treatment is the susceptibility of municipal wastewaters to the formation of DBPs. Such contaminants could constitute a hazard to aquatic life or a problem for downstream drinking water treatment plants. Glaze and Henderson (1975) investigated this issue by "super-chlorinating" (defined for their work as 2000 mg/L chlorine by gas injection, with 60 min contact time) the effluent from the secondary clarifiers at the Denton, Texas, wastewater treatment plant. For reference, 10 mg/L chlorine is a typical dosage of chlorine for wastewaters.

Their gas chromatograms showed over 100 peaks with more than 30 halogenated species detected, identified in Table 2.A.2; most were aromatic derivatives. Some important points are (1) a host of chlorinated organic compounds form when a secondary effluent is subjected to chlorination, and (2) concentrations are in the μg/L range. These compounds formed also using a 10 mg/L chlorine dosage.

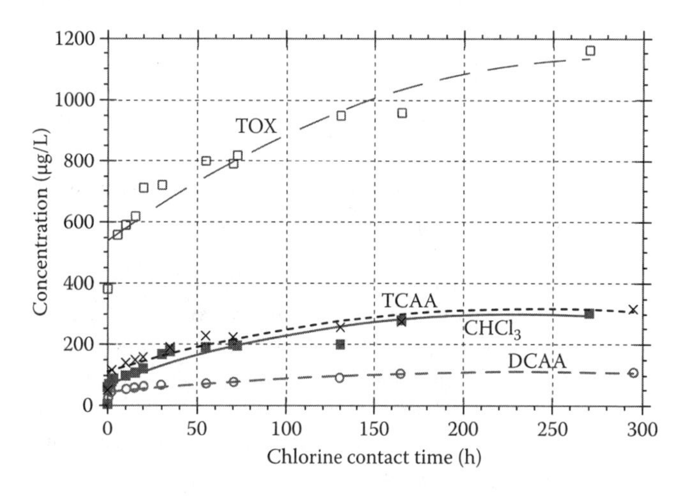

FIGURE 2.A.2 Chlorination by-products of Black Lake fulvic acid as affected by contact time at TOC 4.1 mg/L, pH 7.0, HOCl 20 mg/L. (Plotted from data of Reckhow, D.A. and Singer, P.C., *J. Am. Water Works Assoc.*, 76(4), 151, April 1984.)

2.A.4 Disinfectant Selection

Whenever disinfection occurs, oxidation also occurs. For a "safe" disinfection technology, (1) use a "safe" disinfectant and (2) remove the precursors to oxidation by-products before disinfecting (Trussell, 1992).

Regarding the distribution system, maintenance of a disinfectant residual is important. For factors to consider, include (1) residual stability, (2) residual toxicity, (3) effectiveness in biofilm control, and (4) Oxidation by-product issues (Trussell, 1992).

2.A.5 Other Notes

The formation of chlorinated by-products increases with elapsed time. To illustrate, Figure 2.A.2 shows experimental data from chlorination of fulvic acids (HOCl = 20 mg/L, pH = 7.0, TOC = 4.1 mg/L) (Reckhow and Singer, 1984). The total halide concentration, TOX, was greater than 1100 μg/L after 300 h, with about 600 μg/L after 10 h. Also, as seen, the $CHCl_3$, TCAA, and DCAA concentrations were significant fractions of the TOX concentration.

GLOSSARY

Aromatic compound: A class of molecules with six carbons and six hydrogens in a ring. A key property is their remarkable stability (Aihara, 1992, p. 62), which makes them valuable for many uses, e.g., paint thinner, mothball, gasoline additive, solvents, and as a source for synthetic fibers, resins, and dyes. The carbons are arranged in a closed hexagonal ring—a structure that does not want to react. The ring structure is the distinguishing aspect of the aromatic compounds. Ordinary combustion produces a wide variety of aromatic compounds. Ideally, when hydrocarbon fuels burn, they should form carbon dioxide and water; in reality, complete combustion is rare, and the soot and smoke contain a number of multiring or polycyclic, aromatic compounds, some of which are highly toxic.

Benzene: A ring compound with six carbons and six hydrogens, it is the prototype of a class of molecules known as the aromatic compounds.

Carcinogen: Substance that may cause the development of cancer after exposure at some threshold level or time duration.

Color: Substance in water that causes "color" as measured on the cobalt scale. Two kinds of color are "apparent" color and "true" color. The former is the result of a measurement using a sample of water "as is," and the latter is the same measurement after filtering. Organic color is associated with natural organic matter.

Contaminant: (1) A substance in water other than molecular water. (2) A species added to another species that serves as a matrix. A few molecules of sodium chloride added to a liter of water is a contaminant. (3) To make impure or corrupt by contact or mixture. A contaminant is a substance that is added to a pure substance.

Criterion: A standard, rule, or test on which a judgment or decision can be based.

Disinfection by-products (DBPs): The reactions of various disinfectants, e.g., chlorine, bromine, and ozone, with organic matter—natural organic matter in particular—result in a variety of by-products.

Disinfection by-product precursor: The reactions of various disinfectants, e.g., chlorine, bromine.

Dissolved organic carbon (DOC): Organic carbon passing a 0.45 μm filter (Edzwald, 1993, p. 24).

Fulvic acid: Similar to humic acids, except it is soluble at pH = 1.0 and believed to be in true solution vis-à-vis being colloidal (Randtke, 1988, p. 43). MW < 5000 and can be removed by GAC. Most DBPs are due to reactions with fulvic acid.

Geometric mean (*n.*): *Mathematics* (1) The *n*th root, usually the positive *n*th root, of a product of *n* factors. Parkhurst (1998) states: The geometric mean is the antilog of the mean logarithm of a set of numbers or, equivalently, the *n*th root of the product of *n* numbers. State health departments use this term frequently in referring to concentrations of organisms.

Humic acids: High-MW negatively charged macromolecules having colloidal properties (Edzwald, 1993, p. 24). Also, they are insoluble at pH = 2.0 (Van Benschoten and Edzwald, 1990, p. 1527). MW > 30,000.

Humification: Humification is a process by which biomass consisting of dead plant and animal remains is converted to humis; this is one of the basic steps of the carbon cycle.

Log-normal distribution: The logarithms of the concentrations have a normal distribution. It is common statistical practice to transform such sample concentrations to logs before estimating confidence limits or performing statistical tests such as analysis of variance or *t* tests (Parkhurst, 1998). This term is found frequently in the literature.

Molality: Mass of a solute per unit mass of solvent, i.e., moles(solute)/moles(solvent).

Molarity: Mass of a solute per unit volume of solvent, i.e., moles(solute)/*V*(solvent).

Mole fraction: In a given volume, the amount of a dissolved substance i in moles, i.e., n_i, divided by the summation of the moles of all substances in the solution, n; i.e., mole fraction $i = n_i/n$.

Natural organic matter (NOM): The whole group of natural organic substances of which humic substances are a part.

Normative: Of, relating to, or prescribing a norm or standard: *normative* grammar. Related to the "norms" of a given culture.

NPT: Normal temperature and pressure, defined as 0°C and 101.325 kPa (1.00 atm), used mostly in European publications.

Palatable: (1) Acceptable to the taste (*American Heritage Dictionary*, 1996). (2) In potable water treatment, a drinking water palatability is measured in terms of taste, odor, color, and turbidity.

Particulate organic carbon (POC): Organic carbon passing retained by a 0.45 μm filter. Usually, POC is a small fraction of the TOC compared to DOC and includes bacteria, algae, zooplankton, and organic detritus (Edzwald, 1993, p. 24).

Pollutant: (1) A contaminant level that interferes with, or is perceived to interfere with, a particular use of water. (2) (*n.*) Something that pollutes, especially a waste material that contaminates air, soil, or water. (3) (*n.*) Something that pollutes, especially a harmful chemical or waste material discharged into the water or atmosphere.

Potable: Fit to drink (*American Heritage Dictionary*, 1996).

Specific: A physical quantity divided by its mass; examples include specific heat capacity, specific weight (from Elias, 1997, p. 444).

Standard: An acknowledged measure of comparison for quantitative or qualitative value; a criterion.

State: The "state" of a parcel of water is defined by its physical and chemical characteristics. In physical chemistry, pressure, temperature, and volume are "state" parameters. In a more general sense as applied to unit processes, we would include concentrations of particles and dissolved molecules and ions.

STP: Standard temperature and pressure defined as 20°C and 101.325 kPa (1.00 atm).

Surrogate: A substitute; a quantity that takes the place of another.

Total organic carbon (TOC): All carbon in water when measured by a total organic carbon analyzer is converted to carbon dioxide gas. The carbon dioxide gas is then measured by an infrared wavelength absorbance, calibrated in terms of mg C/L.

TOX: Total organic halides. The principal halides are chlorine, bromine, and iodine (fluorine is also a halogen). If any of these elements react with any organic compound, the product may be termed an organic halide. Thus, the multitude of halogenated organic compounds may be represented collectively by a single parameter, TOX, in lieu of trying to identify each species present and the respective concentrations.

Trihalomethane formation potential (THMFP): The results of a test in which a given water sample is exposed to a known concentration of chlorine and permitted to react over a given time duration, e.g., 24 or 96 h. The loss of chlorine is a measure of THMFP.

Utility: The quality or condition of being useful; usefulness.

UV_{254}: Ultraviolet light wavelength, which refers to absorbance of the 254 nm wavelength by a water sample in a standard cuvette. The 254 nm wavelength has been accepted as a surrogate for TOC.

Water quality: The "state" of a given water volume in terms of concentrations of suspended and dissolved substances and of any other state measures, including temperature and pH.

REFERENCES

Aihara, J., Why aromatic compounds are stable, *Scientific American*, pp. 2–68, March 1992.

American Heritage Dictionary, CD ROM, Softkey International, Cambridge, MA, 1996.

Anon., *Public Health Service Drinking Water Standards, Revised 1962*, Public Health Service Publication No. 956, U.S. Department of Health, Education, and Welfare, Public Health Service, Washington, DC, 1962.

Black, A. P. and Christman, R. F., Stoichiometry of the coagulation of color-causing organic compounds with ferric sulfate, *Journal of the American Water Works Association*, 55(10):1347–1366, October 1963.

Carson, R., *Silent Spring*, Houghton-Mifflin, New York, 1962.

Christman, R. F. and Ghassemi, M., Chemical nature of organic color in water, *Journal of the American Water Works Association*, 58(6):723–741, June 1966.

Cooper, A. M., Torrens, K. D., and Musterman, J. L., On-site evaluation of treatment system requirements to satisfy direct and indirect discharge limits for a complex industrial wastewater: A case study, *Environmental Progress*, 11(1):18–26, February 1992.

Dworsky, L. B., Analysis of federal water pollution control legislation, 1948–1966, *Journal of the American Water Works Association*, 59(6):651–668, June 1967.

Edzwald, J. K., Coagulation in drinking water treatment: Particles, organics, and coagulants, *Water Science and Technology*, 27(11):21–35, 1993.

Elias, H. G., *An Introduction to Polymer Science*, VCH, Weinheim, New York, 1997.

Glaze, W. H. and Henderson IV, J. E., Formation of organochlorine compounds from the chlorination of a municipal secondary effluent, *Journal of the Water Pollution Control Federation*, 47(10):2511–2515, October 1975.

Hubel, R. E. and Edzwald, J. K., Removing trihalomethane precursors by coagulation, *Journal of American Water Works Association*, 79(7):98–196, July 1987.

Johnson, J. D. and Jensen, J. N., THM and TOX formation: Routes, rates, and precursors, *Journal of the American Water Works Association*, 78(4):156–162, April 1986.

Keith, L. H. and Telliard, W. A., Priority pollutants, I-A perspective view, *Environmental Science and Technology*, 13(4):416–423, April 1979.

Krasner, S. W., McGuire, M. J., Jacangelo, J. G., Patania, N. L., Reagan, K. M., and Aieta, E. M., The occurrence of disinfection by-products in US drinking water, *Journal of the American Water Works Association*, 74(8):41–53, August 1989.

Langelier, W. F., The analytical control of anti-corrosion water treatment, *Journal of the American Water Works Association*, 28(10):1500–1521, October 1936.

McKee, J., *Water Quality Criteria*, Resources Agency of California, State Water Quality Control Board, Publication No. 3, Sacramento, CA, 1952.

McKee, J. and Wolf, H., *Water Quality Criteria*, 2nd edn., Resources Agency of California, State Water Quality Control Board, Publication No. 3-A, Sacramento, CA, 1963.

Miller, S., Where all those EPA lists come from, *Environmental Science & Technology*, 27(12):2302–2303, 1993.

Parkhurst, D. F., Arithmetic versus geometric means for environmental concentration data, *Environmental Science & Technology/News*, 32(3), 92A–98A, February 1, 1998.

Prescott, L. M., Harley, J. P., and Klein, D. A., *Microbiology*, 6th edn., McGraw-Hill, New York, 2005.

Randtke, S. J., Organic contaminant removal by coagulation and related process combinations, *Journal of the American Water Works Association*, 80(5):40–56, May 1988.

Rook, J. J., Formation of haloforms during chlorination of natural waters, *Water Treatment Examination*, 23:234–243, 1974.

Safe Drinking Water Act, PL 93-523, 93rd Congress, S. 433, December 16, 1974.

Semmens, M. J. and Field, T. K., Coagulation: Experiences in organics removal, *Journal of the American Water Works Association*, 72(8):476–482, August 1980.

Servos, J. W., *Physical Chemistry from Ostwald to Pauling*, Princeton University Press, Princeton, NJ, 1990.

Sobsey, M. D., Enteric viruses and drinking-water supplies, *Journal of the American Water Works Association*, 67(8):414–418, August 1975.

Symons, J. M., Bellar, T. A., Carswell, J. K., DeMarco, J., Kropp, K. L., Robeck, G. G., Seeger, D. R., Slocum, C. J., Smith, B. L., and Stevens, A. A., National organics reconnaissance survey for halogenated organics, *Journal of the American Water Works Association*, 67(11):634–647, November 1975.

Symons, J. M., The early history of disinfection by-products—A personal chronical (Part I), *Environmental Engineer*, 37(1): 20–26, January 2001a.

Symons, J. M., The early history of disinfection by-products—A personal chronical (Part II), *Environmental Engineer*, 37(1): 7–15, April 2001b.

Tate, C. H. and Fox, A. K., Health and aesthetic aspects of water quality, in Pontius, F.W. (Ed.), *Water Quality and Treatment*, McGraw-Hill, Inc., New York, 1990.

Trussell, R. R., Oxidation by-products complicate disinfectant choices, *Water World News*, January/February, 1992.

United States Public Health Service (USPHS), Drinking Water Standards, 1962.

U.S. Environmental Protection Agency, *Water Quality Criteria, 1972*, EPA R-73-033, USEPA, Washington, DC, March 1973 (also called the *Yellow Book*).

U.S. Federal Register, *FR*43(28):5759, February 9, 1978.

van Benschoten, J. E. and Edzwald, J. K., Chemical aspects of coagulation using aluminum salts—II. Coagulation of fulvic acid using alum and polyaluminum chloride. *Water Research*, 24(12):1527–1535, December 1990.

BIBLIOGRAPHY

Bellar, T. A., Lichtenberg, J. J., and Kroner, R. C., The occurrence of organohalides in chlorinated drinking waters, *Journal of the American Water Works Association*, 66(12):703–706, December 1974.

Black, A. P. and Christman, R. F., Characteristics of colored surface waters, *Journal of the American Water Works Association*, 55(6):753–770, June 1963a.

Black, A. P. and Christman, R. F., Chemical characteristics of fulvic acids, *Journal of the American Water Works Association*, 55(7):897–912, July 1963b.

Christman, R. F., Norwood, D. L., Millington, D. S., and Johnson, D. J., Identity and yields of major halogenated products of aquatic fulvic acid chlorination, *Environmental Science & Technology*, 17(10):625, 1983.

Cotruvo, J. A. and Vogt, C. D., Rationale for water quality standards and goals, in Pontius, F.W. (Ed.), *Water Quality and Treatment*, McGraw-Hill, Inc., New York, 1990.

Edzwald, J. K., Becker, W. C., and Wattier, K. L., Surrogate parameters for monitoring organic matter and THM precursors, *Journal of the American Water Works Association*, 77(4): 122–132, April 1985.

Furman, B., A profile of the United States Public Health Service, 1798–1948, Superintendent of Documents, U.S. Government Printing Office, Washington, DC, 1973.

Hibler, C. P., Analysis of municipal water samples for cysts of *Giardia*, in Wallis, P. M. and Hammond, B. R. (Eds.), *Advances in Giardia Research*, University of Calgary Press, Calgary, Canada, pp. 237–245, 1988.

Kyros, P. N., Legislative history of the Safe Drinking Water Act, *Journal of the American Water Works Association*, 66(10): 566–569, October 1974.

Llao, W. et al., Structural characteristics of aquatic humic material, *Environmental Science & Technology*, 16(7):40, 1982.

McGauhey, P. H., Folklore in water quality standards, *Civil Engineering, ASCE, New York*, 35(6):70–71, June 1965

Reckhow, D. A. and Singer, P. C., The removal of organic halide precursors by preozonation and alum coagulation, *Journal of the American Water Works Association*, 76(4):151, April 1984.

Reckhow, D. A., Singer, P. C., and Trussell, R. R., Ozone as a coagulant aid, in: *Proceedings of the Seminar: Ozonation: Recent Advances and Research Needs, AWWA Annual Conference*, Denver, CO, June 1986.

Rickert, D. A. and Hunter, J. V., Colloidal matter in wastewaters and secondary effluents, *Journal of the Water Pollution Control Federation* 44(1):134–139, January 1972.

Rook, J. J., Haloforms in drinking water, *Journal of the American Water Works Association*, 68(3):168–172, March 1976.

Rook, J. J., Chlorination reactions of fulvic acids in natural waters, *Environmental Science & Technology*, 11(5):478–482, May 1977.

Saville, T., On the nature of color in water, *Journal of the New England Water Works Association*, 31(1):78–123, March 1917.

Singley, J. E., Coagulation and color problems, *Journal of the American Water Works Association*, 62(5):311–314, May 1970.

Symons, J. M., Bellar, T. A., Carswell, J. K., DeMarco, J., Kropp, K. L., Robeck, G. G., Seeger, D. R., Slocum, C. J., Smith, B. L., and Stevens, A. A., National organics reconnaissance survey for halogenated organics, *Journal of the American Water Works Association*, 67(11):634–647, November 1975.

U.S. Environmental Protection Agency, *Quality Criteria for Water*, USEPA, Washington, DC, July 1976 (also called the *Red Book*).

U.S. Environmental Protection Agency, Control of chemical contaminants in drinking water, interim primary drinking water regulations, *Federal Register*, 42(28):5756–5780, February 9, 1978.

U.S. Environmental Protection Agency, Control of chemical contaminants in drinking water, interim primary drinking water regulations, *Federal Register*, 44(231):68624, November 29, 1979.

U.S. Environmental Protection Agency, *Quality Criteria for Water 1986*, EPA 440/5-86-001, USEPA, Washington, DC, May 1, 1986.

U.S. Environmental Protection Agency, Register of Lists, Information Policy Branch (PM-2234), Washington, DC, 1991. Software (ECLIPS Version 1.5), Copyright ©, MicroReg, Inc., Crofton, MD, 1991.

Xie, Y. F., *Disinfection Byproducts in Drinking Water*, Lewis Publishers, CRC Press, Boca Raton, FL, 2004.

3 Models

In education, especially at the undergraduate level, we emphasize problems in which the parameters are well defined. In practice, however, this is not the reality. Knowledge may be incomplete: data on inputs may be lacking, methods of solution may not be well delineated, and even the objectives may be nebulous. All of this is contrary to the common perception of engineering, i.e., that it is deterministic and largely a matter of computation.

This chapter examines some of the approaches for attacking problems. The concept of modeling is a theme common to all. Modeling is an engineering method (Box 3.1).

3.1 UNIT PROCESSES

About 10–15 unit processes comprise the field of water treatment, depending on how they are categorized. Perhaps there are 80–100 technologies developed from them. Table 3.1 lists 13 unit processes and associated technologies. Fundamental principles operative include

- Sieving of particles by screens (ranging from bacteria by membranes to large objects by bar screens)
- Creating conditions for application of a "passive" force on particles (e.g., gravity), or an "active" force (e.g., centrifugal) to cause transport
- Turbulence and diffusion for the transport of particles to cause contacts between reactants
- Charge neutralization
- van der Waals attraction between molecules and a surface (such as activated carbon), or charge attraction (such as between ions and an ion-exchanger material)
- Various chemical reactions such as
  - Redox, acid–base
  - Precipitation
  - Complexation
  - Biochemical
  - Cell synthesis
- Membrane processes involving retention of ions and molecules, i.e., reverse osmosis/nanofiltration

Generally, the unit processes listed in Table 3.1 are the results from a heritage of only since about 1900, albeit the earliest technology was slow sand, with the first installation in 1829 for London. Proprietary innovations have expanded the array of technologies, but most are variations of the unit processes listed in Table 3.1.

3.2 MODELS

A model is a means to represent a portion of a reality. The model is "valid" if the points of the model predict accurately the corresponding points of the system being modeled. As a rule, the system being modeled, i.e., the "prototype," is a full-scale process. Examples include an activated sludge reactor, a biofilm reactor, a plate settler, a slow sand filter, a rapid rate filtration system, a granular activated carbon reactor, an ozone reactor, etc. Natural systems may be modeled also, e.g., water quality of streams and lakes, groundwater, etc., with mathematical models being a primary method of determining the effects of pollutant discharges on such systems.

Table 3.2 lists various forms of models and describes their respective characteristics and positive and negative attributes. The notion of what may comprise a model, Table 3.2, illustrates that a wide range of forms may be encompassed. Thus, a model may include lore, judgment, description, bench testing, pilot plants, demonstration plants, and mathematical models.

3.2.1 Categories of Models

A model is a means to "represent." Thus, a photograph is a model, along with language, a drawing, a painting, a map, a plot, an equation, an array of 0 and 1 digits stored in a computer, or any kind of representation. To be an engineering model, this is necessary but not sufficient. An engineering model, we might assert, must also have utility in projecting from the unknown to the known.

More commonly in engineering, we think of a model in terms of a pilot plant or as a set of coordinated equations in a computer algorithm or in a spreadsheet or even a single equation. Table 3.2 lists some of the more common things that we do in engineering that are really "forms" of models. They qualify as meeting the requirements of a "model," as defined here. Indeed, the various model forms comprise engineering practice, e.g., lore, judgment, extrapolation, bench scale testing, pilot plants, demonstration plants, and mathematical modeling. As an additional note, each model form in Table 3.2 may be thought of as a "black box." In other words, a model as a "black box" accepts a set of "inputs," without regard to how it works, and generates outputs.

3.2.2 The Black Box

The proverbial "black box" has its place as a primary engineering method. Figure 3.1 depicts the concept of the black

BOX 3.1 PHILOSOPHY OF MODELING

Modeling has two themes of logic: inductive and deductive, formalized by Sir Francis Bacon (1561–1626) and René Descartes (1596–1650), respectively. Bacon extolled observation and practical outcomes, while Descartes believed that pure reasoning was the basis for problem solving (Durant, 1926).

The essence of empiricism is observation. Engineering forms include bench scale testing, pilot plants, demonstration plants, evaluations of existing plants, etc. Also included in this category are judgment, lore, and "black box" approaches. Rational models include equations based on a premise leading to an understanding of process mechanisms. Mathematical modeling, scenarios, animations, etc., are modern outcomes. Most problem solving is a blend of empiricism and rationality rather than being exclusively one or the other.

The organization and displays of solutions are important also, as the amount of data generated by physical models or computer models may be overwhelming. Spreadsheets and plots provide a means to organize and present results such that a wide range of conditions can be communicated easily and clearly. Computer animations provide a means to display succinctly and to grasp more easily complex results that could be otherwise difficult to assimilate.

All of these various kinds of models have roles in engineering problem solving. Even when we know little about a problem, some form of model provides a means to identify variables, organize data, test assumptions, generate plausible solutions, and communicate results.

box, illustrating the idea of how the values of dependent variables are generated by maintaining y and z constant while varying x; ϕ and ψ are measured for each level of x at fixed values of y and z. Then y is changed to a new value, and the process is repeated. After all the values of y are explored, z may be changed to a new level, and the foregoing is repeated for each value of z that is to be explored. Suppose that there are 5 levels of x, 8 of y, and 10 of z. Then the number of experiments would be $5 \times 8 \times 10 = 400$. In exploring a hypothetical "surface," a substantial amount of effort is required. An example of the foregoing in more concrete terms is the traditional jar test. Thousands of experiments may be done where a treatment process is being explored, i.e., to determine coagulant dosage, x, and polymer dosage, y, for several seasonal water quality conditions, z.

The "black box" is a device to generate outputs (dependent variables—ϕ and ψ) from selected inputs (independent variables—x, y, z), which may define a useful portion of a functional relationship. Virtually any means to generate outputs from inputs can be considered a "black box." Such means could include judgment, physical models, and mathematical models.

3.2.2.1 Plots

The kind of experimental program outlined above might be called "parametric exploration." Figure 3.2 illustrates the output of ϕ as a function of x and y, with z constant, i.e., $\phi v{\cdot}x$ for $y = y_1, y_2, \ldots, y_n$ and $z = z_1$, where z represents a set of conditions that are maintained constant during the testing. To be more specific, the system being modeled is a rotating drum microscreen. The flow of water through the screen divided by its submerged area is the velocity of water through the screen, v, which is the "dependent" variable, i.e., ϕ. Then v is affected by the independent variables, headloss, h, across the screen, as seen by the curve and the rotational velocity, ω, of the drum, in which h and ω correspond to x and y, respectively. The set of curves of Figure 3.2 is for all other conditions being maintained constant. If, for example, the suspension changes (such as one species of algae instead of another) or the screen size changes, then another set of conditions exists and another set of curves must be generated. Thus, a set of one or more plots, such as seen in Figure 3.2, is the end result of a black box experimental program.

3.2.3 Physical Models

A physical model is a smaller-scale setup of equipment intended to replicate the process being considered. One appeal of a physical model is that variables not anticipated are included passively. The outputs, i.e., dependent variables, thus reflect all independent variables, not just the ones identified.

With the smaller scale, the model is cheaper and easier to operate than a full-scale system. Further, the independent variables can be controlled so that the influence of each on the dependent variables can be investigated. Physical models include bench scale testing, pilot plants, and demonstration scale plants.

3.2.3.1 Bench Scale Testing

Bench scale testing may include jar tests to determine chemical dosages, kinetic coefficients, isotherm constants, and generating relationships between various other kinds of intensive variables. The testing is "one dimensional" in nature, i.e., the intent is to examine the influences of only one or two independent variables (such as screen size) in selected dependent variables (such as effluent concentration).

3.2.3.2 Pilot Plants

One purpose of a pilot plant is to generate functional relationships between dependent and independent variables. The extent to which this is done, i.e., the scope of the experimental program, depends upon the nature of the problem and the budget available.

Another purpose of a pilot plant study may be to determine coefficients of a mathematical model. A mathematical model has greater utility than a set of plots.

Pilot plant experiments will yield, almost without exception, unexpected results that lead to new insights and serendipitous findings. Thus, any plan devised in anticipation of a set of results should have flexibility to incorporate new findings.

TABLE 3.1
Unit Processes and Technologies in Water Treatment

Unit Process	Principle	Technology
1. Screening	Sieving	Bar screens
		Coarse screening
		Microscreening
2. Sedimentation	Gravity force	Plain sedimentation
		Flocculant settling
		Flotation
		Oil separation
		Grit chambers
		Aerated grit chambers
3. Coagulation	Charge neutralization	Rapid mix/coagulants
4. Flocculation	Turbulence	Paddle wheels
		Baffles
5. Chemical precipitation	Equilibrium concentration is exceeded	Softening
		Phosphate removal
		Heavy metal removal
6. Filtration	Adsorption on biofilm	Slow sand
	Adsorption between charge-neutralized particle and collector	Rapid rate
7. Membrane processes	Sieving of micron-size particles	Microfiltration
	Sieving of macromolecules	Ultrafiltration
	Retention of organic molecules	Nanofiltration
	Retention of ions	Hyperfiltration
8. Adsorption	van der Waals attraction	Powdered activated carbon
		Granular activated carbon
	Electrostatic attraction	Ion exchange
		Activated alumna
9. Oxidation	Creating conditions for negative free energy of reaction	Ozone
		Chlorine dioxide
		Supercritical
		Wet air
		Chemical oxidation
10. Gas transfer	Diffusion transport	Oxygen transfer
		Air stripping
11. Biological aerobic treatment	Microbial growth	Activated sludge
		Fixed film reactors
12. Biological anaerobic treatment	Microbial growth	Digestors
		Lagoons
13. Disinfection	Oxidation	Chlorine
		Ozone
		UV

3.2.3.3 Demonstration Plants

A demonstration plant is similar to a pilot plant but is larger in scale. The scale is too large, as a rule, to generate economically the functional relationships between dependent and independent variables. There are many variables that may be difficult to control, e.g., temperature, influent concentration, etc. At the same time, the fact that the demonstration plant operates continuously means that the processes must handle the variations in input variables and exigencies that exist in the "real world." An example may be with the liner used for the steel tanks that define the volume of an activated carbon reactor, which may be acrylic or PVC in a pilot plant. Steel is subject to corrosion, and so a liner (e.g., rubber or fiberglass) is used, which is subject to pinholes or cracks. Many problems of this nature are not identified before the plant is constructed, and so the demonstration scale permits both problem identification and evaluation.

Ostensibly, the demonstration plant should be a "capstone" study for a contemplated full-scale plant. A demonstration plant, however, is large enough to ascertain the impacts of

TABLE 3.2
Forms of Models and Their Characteristics

Model Form	Characteristics	Positive	Negative
Lore	Rules, methods passed by tradition; rationale not necessary	Provides a result that fits with past Familiar	Validity accepted by faith
Judgment	Education and experience, coupled with intuition, provide a basis for decisions	Common A necessary adjunct to any modeling Sometimes the only alternative	Accuracy limited Requires experience
Descriptive	Measurements, impressions, images, etc. used as a basis for transfer to a new design	Inexpensive Based on actual experience Necessary adjunct to any modeling	Qualitative Validity is subjective
Extrapolation	Projection from measured data to new design	Inexpensive Based on actual experience Can evaluate coefficients of mathematical models	Independent variables not controlled Variables not identified may be influential Accuracy limited
Bench testing	Variables isolated to a few and would involve limited kinds of relationships; small in scale	Independent variables can be controlled High accuracy likely	Limited to specific measures
Pilot plant	Complex systems can be simulated with variables controlled	Can maintain constant selected independent variables Can explore the effects of selected independent variables Can develop empirical models Can evaluate coefficients of mathematical models	Requires separate project Generally expensive
Demonstration plant	Emphasis is on maintenance, costs, logistics, operation difficulties	Looks beyond process design to ascertain the roles of dependent variables such as maintenance, costs, logistics, etc.	Expensive and requires time commitment of several years
Mathematical	Independent variables are linked to dependent variables by mathematical relationships	Requires understanding of relationships Experiments can be conducted to explore effects of selected independent variables Complex systems can be evaluated	Expensive to develop coefficients Coefficients may be lacking or inaccurate Validity must be ascertained.
Criteria	Limits are defined by experience, physical modeling, tradition	Simple to apply	May be simplistic, i.e., some key considerations are not included

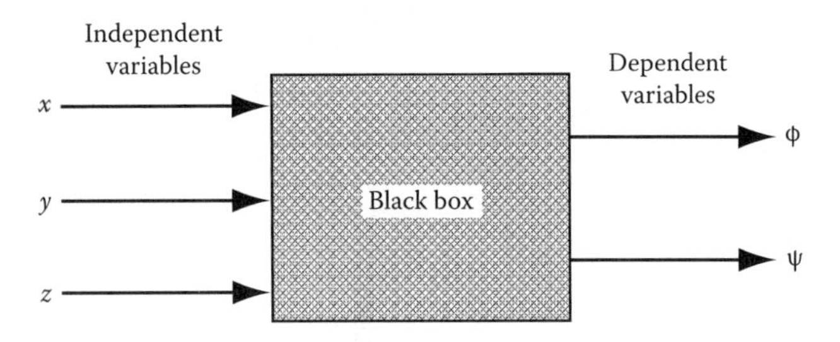

FIGURE 3.1 Black box. x varies, y and z held constant; are measured ϕ and ψ.

operation variables such as storage volumes for chemicals, costs of chemicals, energy, labor, maintenance, etc. In addition, the reliability of the plant can be assessed prior to full scale. Public relations may be another aspect of the demonstration. Examples include the Denver Potable Water Reuse Plant and Water Factory 21 in Orange County, California. These plants have been highly visible and prominent facilities evoking a great deal of public interest as well as political support.

3.2.4 Mathematical Models

A mathematical model epitomizes the deductive approach. The mathematical model starts with a premise. From the premise, we build an "edifice," i.e., the mathematical model. If the premise is not valid, neither is the model.

A system is represented by mathematical relationships that relate dependent variables to independent variables. Usually coefficients or constants are a part of the equations (see Example 5.3).

3.2.5 Computer Models

A computer model is sometimes an extension of a mathematical model, but not necessarily. As an extension of a mathematical model, the computer model may represent a "complex" system depicted by equations, with outputs from one unit comprising inputs to another. The computer model in such a case is a means for "bookkeeping" as variables change in space and time. The steps in organizing the computational scheme are called an "algorithm."

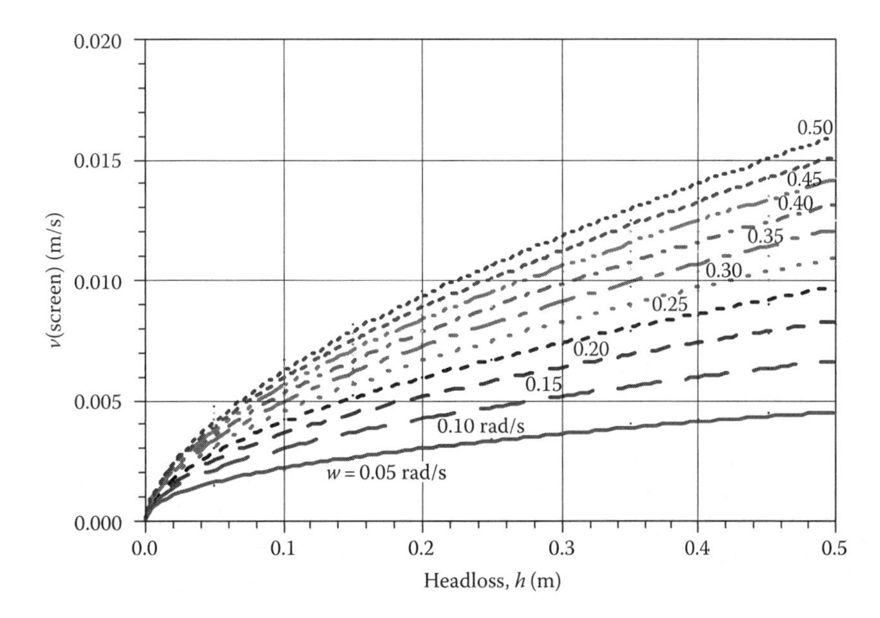

FIGURE 3.2 Plot from hypothetical data generated by "black box," showing ϕ [v(screen)] as a function of $x(h_L)$ and different values of $y(\omega)$ with z (suspended solids concentration, screen mesh size, etc.) held constant.

A computer model may incorporate decision-making steps and thus extend well beyond the concept of strictly mathematical models. Computer models have virtually no limit in the kinds of systems modeled.

3.2.6 Scenarios

For most engineering design problems, there is uncertainty about the inputs to the problem. An approach to this kind of a problem is to consider alternative *sets* of inputs and to calculate the results for each. Each set of inputs, with the results depicted and interpreted, is a "scenario." Usually, the scenario is given a name such as "high-growth," "medium-growth," etc. This takes into account the fact that we do not know the future rate of growth (of population and industry) for which a plant must be designed. The outputs could be the performance of a unit process under different flow conditions, or perhaps the total water output under different demand or plant capacity scenarios. A large number of scenarios could be generated, with the number being limited only by the imagination. For example, a policy change to install water meters would add another demand scenario. Another could be a social trend (in some arid and semiarid urban communities in the United States) away from green grass lawns to natural vegetation, thus changing significantly the summer demand caused by lawn watering. Another could be to examine providing additional finished water storage in lieu of additional treatment plant capacity in order to meet peak summer demands for lawn watering. Any combination of input variables forms the basis for a scenario. Possible water quality changes may be another area for scenarios. The effect of sudden high turbidity levels (due to heavy rainfall) on process performance in a water treatment plant is a common scenario that will be experienced by most water treatment plants. Such effects may require assessment based on judgment of operators as opposed to mathematical models. A plant operating plan may include such scenarios as a systematic means to respond to various exigencies. Another scenario that some plant supervisors are concerned with has to do with acts of terrorism directed at the water supply.

For most engineering problems, the solution is not a single "answer," but a set of results for different assumed conditions. The scenario provides a systematic approach to explore the effects of alternative sets of inputs. Policy, design questions, and operating strategies may be assessed by generation of scenarios (Box 3.2).

3.3 MODELING PROTOCOL

Steps in modeling include

(1) Identify all variables, whether they be independent or dependent (e.g., x, y, z, ψ, ϕ).
  (1.1) Identify independent variables (e.g., x, y, z).
  (1.2) Identify dependent variables (e.g., ψ, ϕ).
(2) Design experiments that include dependent variables as a function of independent variables, e.g., $[\psi, \phi] = f(x, y, z)$.
  (2.1) Hold all independent variables constant except the one whose influence is to be determined, e.g., $[\psi, \phi] = f[x]_{y,z}$.
  (2.2) Generate the respective influences of each independent variable of interest in this manner, e.g., $[\psi, \phi] = f[y]_{x,z}$, $[\psi, \phi] = f[z]_{x,y}$.

BOX 3.2 SCENARIOS

The more sophisticated term for the "What if?" question/answer is the "scenario." An algorithm for a scenario development may be

1. State the problem.
2. Identify the variables that affect the system.
3. Give an identity to the scenario (such as low population growth, high population growth, etc.).
4. Quantify the magnitude of each variable comprising a "set."
5. Select a means to generate "outputs" of interest (a mathematical model, an empirical model, an equation, a judgment, or any "black box" that may be plausible).
6. Generate the scenario outputs.
7. Characterize the outputs in terms of conclusions for the particular scenario.
8. Identify a new scenario, and repeat until the question has been explored to the extent desired.

The idea of the scenario is to permit a more comprehensive exploration of a problem that is done by means of a single set of inputs/answer. Using a spreadsheet, a wide range of scenarios can be applied with easily interpreted outputs. Another combination of inputs comprises another scenario.

Example

1. *Problem*: Pipe flow for different scenarios of corrosion rate
2. *Variables*: Q, f, L, D, h_L
3. *Scenarios*: (a) High rate of f change, (b) low rate of f change
4. *Generation of outputs*: The "black-box" is the Darcy–Weisbach equation
5. *Scenario outputs*: Headloss for each scenario

(3) In analysis, "map" the surface $f(\psi, \phi)$ as a function of x, y, z.

(3.1) To "map" a surface with two dependent variables $f(\psi, \phi)$ and three independent variables (x, y, z), construct plots:

$$\psi = f(x,y)_{z=k1}, \quad \phi = f(x,y)_{z=k1};$$

$$\psi = f(x,y)_{z=k2}, \quad \phi = f(x,y)_{z=k2}$$

(4) Select ranges of interest for the independent variable, and designate on the tables or plots generated.

(5) If the variables measured are "extensive" in nature, consolidate them as either "intensive" variables, or as dimensionless numbers (e.g., Q is extensive, but Q/A is intensive).

(6) Use the relationships generated to extrapolate to full scale.

(6.1) If both the independent and the dependent variables are "intensive" in character, then they may apply directly to the scale-up (and would include, for example, in process design, hydraulic loading rate [HLR], and kinetic coefficients). For example, if all variables are intensive, then the relationships generated for the model may apply also to the prototype (e.g., full scale), i.e.,

$$[\psi = f(x,y)_{z=k1}, \phi = f(x,y)_{z=k1}]_{\text{model}} = [\psi = f(x,y)_{z=k1}, \phi = f(x,y)_{z=k1}]_{\text{prototype}}$$

(6.2) Examples of different variable characteristics

- *Intensive*: Temperature, pressure, HLR (flux density)
- *Properties*: Density, concentration
- *Dimensionless numbers*: Schmidt, Sherwood, Euler, Reynolds, etc.
- *Coefficients*: Isotherm, diffusion, kinetic, weir, orifice, geometric ratios
- *Extensive*: Flow, mass flux, length, velocity, time

A problem with "mapping" an experimental space is that the number of experiments could be in the thousands. Suppose, for example, that ϕ is to be mapped as a function of x, y, and z. To make the idea more tangible, consider the illustration of Figure 3.2 in which v(screen) is plotted as a function of headloss, h_L, and rotational velocity, ω, i.e., $v(\text{screen}) = f(h_L, \omega)$, maintaining the same water quality, temperature, screen size, screen material, etc. To obtain a plot that "maps" the function experimentally, about 10 experiments should be conducted for each v(screen) versus h_L, for a given value for ω. Now, if we have 10 values for ω (e.g., $w = 0.05, 0.10, \ldots, 0.50$ rad/s), the same experiment must be repeated 10 times. The number of experiments to generate the "surface" illustrated in Figure 3.2 is $10 \times 10 = 100$. [Although shown in two dimensions, the plot could be seen as a surface with ω as a third coordinate.] Suppose we have three water quality conditions, such as three levels of algae concentration. For each algae concentration, another set of experiments must be conducted to give 300 experiments to generate three plots similar to the one shown in Figure 3.2. As an example, jar testing, which may examine the effects of different alum dosages and combinations with polymers and concentrations of each along with different raw water conditions, has resulted in some 5000 experiments for some situations.

Rather than conducting some 300 or 3000 experiments, which in some cases may involve a great deal of tedious analytical work and cost, we may apply the idea of "factorial analysis." Consider, for example, looking at selected points in a "space" of independent variables. We move purposefully

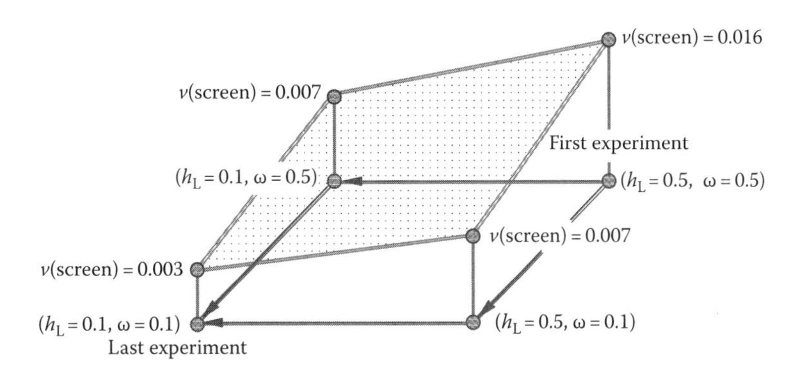

FIGURE 3.3 Illustration of factorial design for hypothetical microscreen experiments.

from one point to another (see, for example, Box et al., 1978; Cochran and Cox, 1992; Hess et al., 1996). For example,

$$\phi_{(1,1,1)} = (x_1, y_1, z_1)\phi_{(1,1,2)} = (x_1, y_1, z_2)$$

$$\phi_{(2,1,1)} = (x_2, y_1, z_1)\phi_{(2,1,2)} = (x_2, y_1, z_2)$$

$$\phi_{(1,2,1)} = (x_1, y_2, z_1)\phi_{(1,2,2)} = (x_1, y_2, z_2)$$

$$\phi_{(2,2,1)} = (x_2, y_2, z_1)\phi_{(2,2,2)} = (x_2, y_2, z_2)$$

So that the idea is more tangible, consider again the case of the microscreen as depicted in Figure 3.2. For the function $v(\text{screen}) = f(h_L, \omega)$, how much can be gained or lost in v(screen) by changing h_L and ω? Suppose the concern in design is that we must have a low screen area, which requires a high v(screen), to minimize capital costs, it means that we are willing to accept the trade-off of higher h_L and ω (resulting in higher operating costs).

Therefore, we can start at the highest permissible values of (h_L, ω) and thus measure the resulting v(screen). Therefore, we do not need to "map" the entire "space" of the $v(\text{screen}) = f(h_L, \omega)$ function. But then, suppose that the v(screen) result is acceptable and the question is to know the effects of decreasing operating costs and which variable, i.e., h_L or ω, will give the most return per unit of change (translated to operating costs). To address this question, let us first lower headloss to $h_L = 0.1$ m (the lowest feasible level). At the same time, try a lower ω to say $\omega = 0.1$ rad/s. Next, let us try lowering both h_L and ω to their minimum values, i.e., $h_L = 0.1$ m and $\omega = 0.1$ rad/s. We may thus explore these effects with only four experiments, not 100 as in "mapping" the v(screen) response surface. These ideas are illustrated in Figure 3.3. Four coordinate points are shown corresponding to the most extreme values of (h_L, ω). Values for v(screen) are shown at each coordinate point and have the approximate magnitudes as seen in Figure 3.2. Note that in this approach, we miss the character of the function $v(\text{screen}) = f(h_L, \omega)$. But, on the other hand, we do not require such knowledge for engineering purposes.

3.3.1 Spreadsheets

Spreadsheets are used routinely for virtually all problems that are quantitative in nature. This was not always the case, as reviewed in Box 3.3. Therefore, most of the problems in this text are intended for spreadsheet software.

The idea of a spreadsheet, as used in this text, is to explore families of solutions based upon certain assumptions for inputs to the problem. This idea is expressed as a "scenario."

BOX 3.3 TECHNOLOGY OF COMPUTATION

In decades past, and up to about 1975, the slide rule was the main instrument of calculation for engineers and scientists. Hand calculation with logarithms was used for precise calculations of large numbers. Hand-operated calculators were developed about 1900 and then became transformed as electronic instruments. In the late 1950s, the computer came on the scene, and Fortran programming made about any kind of modeling feasible, albeit usually with considerable effort. With the advent of personal computers and spreadsheets, the effort needed to program was simplified, and tables and plots could be generated easily.

With personal computer software technologies, families of solutions could be explored based upon parametric programming (changing an independent variable by increments sequentially). Instead of considering a single solution as with the slide rule or reams of data from many pages of a Fortran printout, the spreadsheet technology, since the mid-1980s, has permitted a new approach to problem solving. We can look at the spectrum of inputs that are likely to affect a situation and then examine the associated outputs as either a series of tables or, preferably, as plots. This capability actually makes the solutions intelligible, i.e., in terms of plots, including three-dimensional plots or a series of plots.

The computer software also provides a means to "animate" solutions, i.e., to provide an output that changes with time. In addition to graphical outputs, the solution can show simulations of physical results, e.g., a water surface, a concentration profile, a dispersion effect, etc. This visualization capability also provides a means to comprehend complex solutions to mathematical models.

3.4 UNITS AND DIMENSIONS

Any kind of quantitative work requires units. First, the distinction between units and dimensions should be clarified (Kline, 1965, pp. 8–9). A *unit* is the measure of some physical characteristic of a system. A *dimension* is the characteristic measured. For example, the *length* of an object is a *dimension* and may be measured by a *unit* such as a meter.

In the literature, it is not uncommon to describe a quantity in terms of its dimensions. The quantity *velocity*, for example, has *dimensions* length over time, i.e., L/T. The *units* in the SI system are meters per second (m/s); in general, SI units will be used in this book in lieu of dimensions.

3.4.1 Units

Within this text, SI units are used. To the extent feasible, however, the SI units are accompanied by the equivalent U.S. Customary units. Reasons for favoring the SI system in this book include (1) the SI system has been adopted by all countries (except the United States), (2) it is legal in the United States, (3) engineering societies have adopted the system, (4) it is likely that the United States will adopt the system wholly, and (5) the units are easy to use. Related to (5), the SI system is "coherent," i.e., the product or quotient of two units gives the units of the derived quantity—a quality lacking in the U.S. Customary system.

As a rule, we consider units without much question and in a casual fashion. Some difficulties caused by insufficient attention to units include (1) frequently, in mathematical expressions, units are not stated; (2) dimensional homogeneity in mathematical expressions may be overlooked; (3) conversions between units are prone to mistakes; (4) the procedure used to relate force to mass is especially prone to mistakes; and (5) tables for numerical conversions are often difficult to locate. Table 3.3 summarizes these problems and indicates remedies. While the remedies are quite simple, they are often overlooked.

TABLE 3.3
Problems and Remedies in Use of Units

Task	Problem	Remedy
1. Units not clear	Dimensions not stated or units not stated	Variables are defined with units stated
2. Ensure dimensional homogeneity	Dimensional homogeneity	Use equations that are dimensionally homogeneous
3. Conversions between units	Mistakes	Apply chain rule
4. Force/mass relationship	Not clear on how to do and there are many units	Example problems. Appendix A
5. Numerical conversions	Locating tables	Use Internet

3.4.2 Dimensions

The use of dimensions in lieu of units is not uncommon when presenting mathematical equations. This avoids the problem of selecting a system of units. In this book, units are attached to variables (in lieu of dimensions).

3.5 EXAMPLES OF MODELS

Models can be found in a variety of forms, as noted in Section 3.2.1, e.g., as lore, judgment, physical models, mathematical models, and computer models. To illustrate the broad inclusiveness of the modeling idea, several examples of physical models, i.e., pilot plants, are described here. Even in this category, a wide variety exists.

FIGURE 3.4 Pilot plant with nine unit processes at Engineering Research Center, Colorado State University. (Photograph by Joe Mendoza, Photographic Services, Colorado State University, Fort Collins, CO, 1996.)

FIGURE 3.5 Rapid mix unit (one of three) at pilot plant at the Engineering Research Center, CSU. Force gage is seen mounted at top of basin. Motor rests on a bearing plate for purpose of measuring torque on impeller. (Photograph by Joe Mendoza, Photographic Services, CSU, 1996. Measurement by William F. Clunie, 1996, then graduate student at CSU; presently Technical Manager, Water, AECOM, Wakefield, MA.)

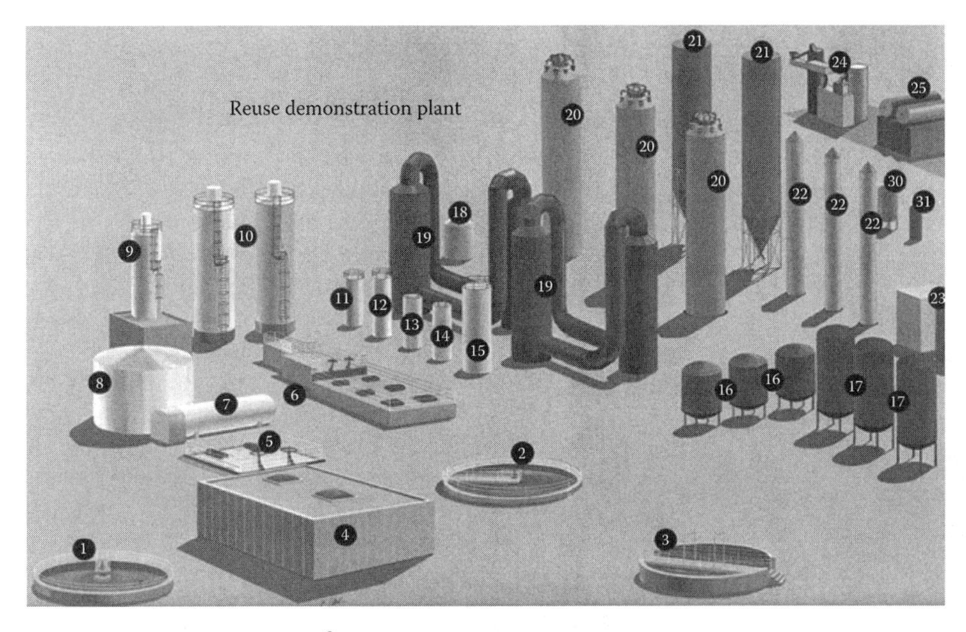

FIGURE 3.6 Denver Water Reuse Plant, 4000 m^3/day (1 mgd) demonstration plant operated 1984–1991 to demonstrate feasibility of treating Denver wastewater to a quality suitable for drinking water. (1) Chemical clarifier No. 1, (2) chemical clarifier No. 2, (3) blackwash equalization basin, (4) recycle pump station, (5) recarbonation basin, (6) flocculation basin, (7) carbon dioxide storage, (8) aluminum sulfate storage, (9) soda ash storage, (10) lime storage, (11) hydrochloric acid storage, (12) sulfuric acid storage, (13) alum storage, (14) muli-purpose coagulant storage, (15) sodium hydroxide storage, (16) filters, (17) clino, (18) brine tank, (19) ARRP towers, (20) first stage carbonation columns, (21) carbon storage, (22) second stage carbonation columns, (23) regenerant clarifier, (24) carbon regeneration furnace, (25) ozone, (26) reverse osmosis system, (27) reverse osmosis pumps, (28) chlorine dioxide, (29) air stripping tower, (30) plant air, (31) instrument air. (Courtesy of Denver Water, Denver, CO, 2010.)

Figure 3.4 shows a pilot plant designed for research, formerly located at Colorado State University. The pilot plant has numerous appurtenances such as sampling taps along the columns, alternative rapid mixes (three were included each with different mixing intensity and detention time, torque measurement for rapid mix impellers, and floc basin paddles), in-line particle counting, turbidity measurement instruments, etc. The pilot plant flow capacity was 76 L/min (20 gpm). Unit processes include rapid mix, flocculation, settling, filtration, ozone reactor, adsorption, ion exchange, air stripping, and membranes.

Figure 3.5 shows one of the three rapid mix units. Each unit was set up with the motor mounted on a bearing plate with a lever arm attached and a force gage at the end for measuring the torque applied to the impeller.

Figure 3.6 shows the Denver Potable Water Reuse Demonstration Plant operated during the period, 1984–1991. The plant was built to demonstrate the feasibility of treating secondary treated wastewater effluent from the Denver Waste Water Reclamation Plant. The plant utilized conventional filtration, adsorption, ion exchange, ozone disinfection, and membranes.

Figure 3.7 shows a pilot plant at the other end of the spectrum, which is a single 51 mm (2 in.) acrylic tube with positive displacement pumps for metering flow and for alum addition. The purpose was to determine alum dosage for a small plant in which jar tests did not yield visible floc.

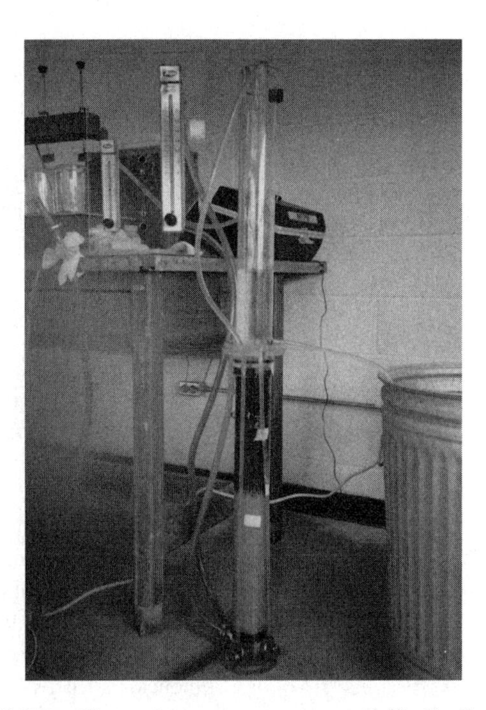

FIGURE 3.7 Filter column set up at a small "in-line" plant to determine coagulant dosage. The water was low turbidity and jar tests did not show a visible floc.

Figure 3.8 illustrates further the variety in physical models. Six 305 mm (12 in.) diameter PVC filters were set up to simulate the slow sand filtration process, operated over a period of 12 months (Bellamy et al., 1985). The first filter was a control filter operated under ambient conditions. The other filters differed from the control filter in one variable that

FIGURE 3.8 Six slow sand filters to evaluate effect of design/operating variables on removals of coliforms and *Giardia* cysts.

was different; e.g., for filter 2, the depth was half of the control filter depth; for filter 3, the temperature was maintained at 4°C; for filter 4, the effective sand size was larger; for filter 5, chlorine was added to minimize growth of organisms before seeding with coliforms and *Giardia*; filter 6 was maintained with nutrients added continuously for 12 months of operation.

3.6 SUMMARY

While ideas and concepts of modeling are reviewed, any practical implementation remains an art, i.e., there are no specific guidelines. A major node point in the development of the field of process design was the publication of the two books by Linvil Rich, i.e., *Unit Operations in Sanitary Engineering*, published in 1961 (Rich, 1961), and *Unit Processes in Sanitary Engineering*, published in 1963 (Rich, 1963). These books were responsible for a paradigm shift toward chemical engineering approaches and a definite departure from empiricism as the modus operandi.

At the same time, mathematical modeling, in the sense of understanding fundamentals, became the quest of academics. In addition to looking for empirical rules and conducting "black box" experiments, the goal was to understand and incorporate "mechanisms" into models, such as mathematical expressions involving kinetics, diffusion, turbulence, boundary layers, adsorption isotherms, equilibrium constants, etc.

The method of problem solving is the same for all engineering problems, simple or complex. The problem must be stated clearly, the objectives stated, and the methods delineated. Regardless of the character of the problem or the method used, this simple procedure is common to all problem solving.

PROBLEMS

3.1 Units and Dimensions

Examine the units used in the reporting forms in a treatment plant. In particular, look at the units associated with the different unit processes. Convert the units to SI.

3.2 Conversions of Units

Demonstrate several conversions of force to mass and vice versa. Use fundamental units for the conversions, not shortcuts. Appendix A provides a guide to units. Examples of problems are listed:

a. Suppose an object shows a reading on a spring scale of 5 lb. Determine its mass in kg. Determine the force it exerts in N.
b. A motor has a power output of 3 hp. Determine its power in watts.

3.3 Scenarios

Consider the design scenarios for a wastewater treatment plant as a whole. Generate several scenarios that have to do with the ability of the plant to meet its effluent discharge permit.

3.4 Spreadsheets

Set up the scenario in 3 on an Excel® spreadsheet, showing the structure for computation. Generate numerical solutions using hypothetical data.

3.5 Models

The Denver Marston Water Treatment Plant treats drinking water drawn from the adjacent Marston Lake (near Quincy Avenue on the south side of Denver). The plant experiences algae blooms that interfere with coagulation and filtration. Suppose that microscreening is a proposed treatment process for removing the algae. A manufacturer has provided a pilot plant that you will use as the basis for a design.

1. Outline an experimental program that you might propose.
2. State dependent variables.
3. Identify the independent variables.
4. Would you do any bench scale testing?
5. Would you visit any microscreen plants?
6. Would mathematical modeling have a place?
7. Describe plots that you would generate from the pilot plant operation.
8. Would you apply mathematical modeling for any aspect of your design? Describe.
9. Describe how you would arrive at a final sizing for a full-scale design.

If you have any ideas for discussing the problem that vary from the above suggested responses, please feel free to bring out the salient points as you feel appropriate.

GLOSSARY

Accuracy: The relationship between a measurement and a "true" value. For example, if a turbidimeter is calibrated and reads the same as, say two or three standards, it may be considered "accurate" (assuming the standard was prepared properly).

Algorithm: A sequence of computer code, such as a Fortran program, that leads to a desired computational outcome.

Animation: A sequence of computer solutions with time such that a motion picture effect is observed.

Bench scale: An experiment set up in a laboratory flask or something equivalent in size. Independent variables are maintained constant while dependent variables are measured.

Black box: A process depiction that may seek a relationship between one or more measured conditions of a given system and one or more output characteristics. A key point is that the internal conditions of the system are not depicted.

Coefficient: A ratio of two or more variables that characterizes the results of a particular experiment or a set of empirical observations about a system. Usually, the equation with the coefficient has one dependent variable on the left side and several independent variables and the coefficient on the right side. An example is the Chezy coefficient in pipe flow, i.e., $v = C(RS)^{1/2}$. A coefficient is the "slack" factor in an empirical equation, i.e., the numerical value that forces equality between variables.

Computer modeling: A computer program that depicts a mathematical model ranging from a few lines of code to a complex system that may take into account contingencies, decisions, time variation, and any other ideas that may simulate a system.

Dependent variable: A variable in a given system that is changed by the influence of an independent variable. The dependent variable is a part of the "effects" caused by a given system.

Dimensionless quantity: A product or ratio of two or more physical quantities that are combined in such a way that the resulting quantity has units of unity (from Elias, 1997, p. 444).

Empirical model: A model that simulates equations that are based upon observation.

Factorial design: In exploring the effect of independent variables, the effect is limited to looking at certain node points, i.e., holding all variables constant and looking at the effect at x_1 and then x_2, then at y_1 and y_2, etc. In other words, one does not "map" the entire solution in two or three, or n dimensions as a continuous function, but only the significant node points of interest.

Goal: A statement of where one "wants to be." For example, to make all streams and lakes in the United States "fishable and swimmable" as stated in the 1972 Clean Water Act (PL92-500) was a "goal" of the legislation. The "goal" of a pilot plant study may be to provide guidance for an economical, effective, and easily operable full-scale treatment process.

Independent variable: A variable in a given system that may influence the changes observed for a dependent variable. The independent variable is a cause of certain effects observed for a system.

Lore: A body of traditions and knowledge on a subject or possessed by a class of people (*Oxford American Dictionary*, Ehrlich et al., 1980).

Model: A means to "map" the magnitudes of dependent variables based upon the variation of selected independent variables.

Normalized: A dependent variable that is a fraction, defined as a quantity divided by perhaps the largest value of a series with the sum of all normalized values equal to 1.0. Usually, a plot is normalized so that the relationship with a dependent variable is applicable to as a general relationship (from Elias, 1997, p. 444). A unit hydrograph is a normalized relationship.

Objective: An objective is a statement of a milestone to be met while striving to meet a goal. Attaining a certain effluent suspended solids level in a plant may be an objective. To generate a headloss versus time curve from a pilot plant study may be an objective. See also *goal*.

Paradigm: An example that serves as pattern or model.

Parameter: (1) A quantity that may have a correlation with the changes observed for a dependent variable. [An independent variable could be called a parameter, but a parameter may not always be called an independent variable.] (2) A water quality attribute. For example, the presence of certain bacteria, the hardness, and the level of sodium are all parameters (Symons et al., 2000).

Parametric variation: Exploration of the effect of independent variables on selected dependent variables done by holding all variables constant except one that is varied between set limits. The process may be repeated, selecting one independent variable at a time to vary.

Pilot plant: A physical model of limited size designed to permit a control of selected independent variables so that the effects on dependent variables may be observed.

Precision: The spread of a set of measurements that may or may not related to a "true" value. In other words, if the standard deviation of a set of measurements is very small, the measurements may be considered to have high "precision." The measurements may not be "accurate." For example, if a turbidity meter is not calibrated, its measurements may not be accurate. The samples may be measured, however, with high precision.

Problem solving: A systematic methodology for arriving at a solution to a problem posed. The classic engineering problem solving goes as follows: state what is known, state what is sought, enumerate a procedure for the solution, and execute the procedure.

Process: An effect that causes change in one or more chemical or physical characteristics of a substance, i.e., a change of state. The effect may be induced, passive, or natural.

Rational model: A model based upon premises and logic, usually mathematical in character.

Scenario: (1) A particular "set" (or configuration) of independent variables that are intended to explore the effect of the set on the dependent variables of interest. (2) Advice given to students: A scenario should be a device to explore a range of possible design outcomes based upon a set of assumptions. Your scenario should be a realistic engineering exploration, but you are limited only by imagination. The spreadsheet offers a means to do this. With the spreadsheet, you hold all variables constant except the one that you wish to explore.

System: A set of components that interact.

Variable: A quantity that is a part of a system.

Water treatment: A set of processes intended to improve the quality of water to prescribed criteria or standards.

REFERENCES

Bellamy, W. D., Hendricks, D. W., and Logsdon, G. S., Slow sand filtration: Influences of selected process variables, *Journal of the American Water Works Association*, 77(12): 62–66, 1985.

Box, G. E. P., Hunter, W. G., and Hunter, J. S., *Statistics for Experimenters*, John Wiley & Sons, Inc., New York, 1978.

Cochran, W. G. and Cox, G. M., *Experimental Designs*, 2nd edn., Wiley Classics Library, New York, 1992.

Durant, W., *The Story of Philosophy*, Simon & Shuster, New York, 1926.

Ehrlich, E., Flexner, S. B., Carruth, G., and Hawkins, J. M. (Compilers), *Oxford American Dictionary*, Oxford University Press, New York, 1980.

Elias, H. G., *An Introduction to Polymer Science*, VCH, Weinheim, Germany, 1997.

Hess, T. F., Chwirka, J. D., and Noble, A. M., Use of response surface modeling in pilot testing for design, *Environmental Technology*, 17:1205–1214, 1996.

Kline, S. J., *Similitude and Approximation Theory*, McGraw-Hill, New York, 1965.

Rich, L., *Unit Operations in Sanitary Engineering*, McGraw-Hill, New York, 1961.

Rich, L., *Unit Processes in Sanitary Engineering*, McGraw-Hill, New York, 1963.

Symons, J. M., Bradley, L. C. Jr., and Cleveland, T. C., *The Drinking Water Dictionary*, American Water Works Association, Denver, CO, 2000.

4 Unit Process Principles

This chapter outlines the spectrum of unit processes and associated technologies that are available for water treatment. The themes of "transport" and "sinks" are described, as is the notion of a "reactor." The materials balance principle is the basis for modeling a reactor. The materials balance principle is the foundation for much of what follows in later chapters, for example, for those that deal with fluidized-bed and packed-bed reactors.

4.1 UNIT PROCESSES

The unit processes for water treatment number only 10–15, depending upon how they are categorized and counted (see, e.g., Section 1.1, Table 1.1, Table 3.1). Under each of the unit processes there are several categories and under the categories there may be numerous technologies. As with contaminants, there are different ways to categorize. The categorization used for this chapter is based upon unit processes.

4.1.1 Spectrum of Unit Processes and Technologies

Table 4.1 (an expansion of Table 3.1) lists unit processes in the left-hand column. As noted, there are only about 15. The respective principles operative for each unit process are indicated in the second column. The third column lists the several categories for a given unit process. The fourth column lists technologies that implement the unit processes; the list is not inclusive. A fifth column, if added, could list the variations for each technology. For example, under settling tanks, there are rectangular tanks, upflow tanks, center feed circular tanks, peripheral feed circular tanks, etc. Also, manufacturers have developed their own variations for implementing each technology.

4.1.2 Matching Unit Process with Contaminant

Some words should be said about the match between contaminants to be treated and the unit processes that could do the job. A tabular array could be developed with contaminants as rows and processes and technologies as columns. The cells of the matrix would indicate the extent to which a given technology could treat the contaminants listed in the rows. Usually, there is more than one unit process that may treat, economically, a given contaminant. In some cases, laboratory testing followed by a pilot plant study is required.

4.1.2.1 Contextual Changes and New Treatment Demands

Since the 1974 Safe Drinking Water Act, the number of regulated contaminants has expanded to include synthetic organics, *Giardia lamblia* cysts, and various disinfection by-products. Not only were new contaminants included, but the traditional contaminants had lower limits. All of this meant that new treatment technologies had to be explored and the traditional ones had to perform to achieve lower effluent concentrations and perform more reliably. For example, to handle disinfection by-products, the idea of "enhanced coagulation" was proposed. While this is merely using higher dosages of alum or ferric coagulant to react with the natural organic matter (NOM), it might be considered as a variation in the established technology. Other technologies that have been explored include granular activated carbon (GAC), biofilters, and nanofiltration membranes. Thus, there is nothing "set in concrete" about which unit processes are most appropriate.

In some cases there may be two or three alternatives in the selection of a unit processes for a treatment train. In the treatment of municipal drinking water, for example, rapid filtration has been traditional in the United States, since about 1910. Other processes that may work include slow sand filtration, diatomaceous earth filtration, and microfiltration. The selection depends upon the "context" (see Box 1.1). In addition to technical considerations such as water quality, effluent or drinking water standards, flow, etc., the context includes such factors as population served, location (e.g., urban or rural), proximity to supplies, operating and maintenance costs, financing capacity of the community, political factors, operator capabilities, etc. The "form," that is, the treatment train selected, should "fit" the context.

4.2 PRINCIPLES

Two themes are common to most unit processes: (1) a "sink" of some sort, and (2) transport to the sink. This section summarizes these ideas. They are amplified further in the unit process chapters.

4.2.1 Sinks

A "sink" is a site within a unit process where something "happens," for example, where a contaminant is removed. To illustrate, a sink may be a surface where a particle settles, an adsorption site within a particle of activated carbon, a surface of granular media to which particles bond, or the multitude of microlocales within a reactor where two reactants collide by turbulence or molecular diffusion and a reaction occurs.

The kinetics of the reaction, that is, its rate, depends upon the rate of transport to the sink or the rate of the reaction. One or the other is rate limiting. Generally, the rate limiting mechanism is the rate of diffusion, which is a transport process, rather than the rate of reaction.

TABLE 4.1
Unit Processes of Water Treatment

Unit Process	Principle	Categories	Technologies/Forms/Examples
Screening	Retention of particles, objects by physical straining	Fine screens	Micro-screens—water tr.
			Micro-screens—ww tr.
		Coarse screens	Drum screens
		Trash racks	Intake cribs
			Bar screen—self-cleaning
			Bar screen—manual cleaning
Sedimentation	Movement of particles through water under gravity force	Grit chambers	Longitudinal grit chamber
			Aerated grit chamber
			Square grit chamber
		Plain sedimentation	Settling tanks—shallow
			Plate settlers
			Tube settlers
		Flocculant settling	Settling tanks—deep
		Flotation	Flotation basin
		Oil separation	API oil–water separator
Coagulation	Charge neutralization of negative charged colloids	Rapid mix	Impeller mixer
			Baffles
			Static mixer
Chemical precipitation	Precipitation of metals	Ca^{2+} removal	Softening
		PO_4^{3-} removal	Tertiary treatment
Flocculation	Providing mechanisms to achieve contacts between particles (diffusion, velocity gradients)	Flocculators	Paddle-wheel
			Baffles
			Impeller
Filtration	Retention of particles on granular media by adsorption	Deep bed	Conventional
			Direct
			In-line
	Physical straining	Diatomaceous earth (also called "pre-coat" filtration)	Plate and frame septum
			Candle septum
Biological filtration	Biological reaction	Slow sand	Filter box with underdrains
		Bioreactors	Filter box with underdrains
Adsorption reactors	Adsorption, i.e., van der Waals bonding to a surface	Granular activate carbon	Fixed-bed reactor
		Powdered activated carbon	Fluidized reactor
		Activated alumina	Fixed-bed reactor
Ion exchange	Electrostatic bonding of ions to a surface	Zeolites	Fixed-bed reactor
		Synthetic resins	Fixed-bed reactor
Gas transfer	Transport of a gas between aqueous dissolved phase and gas phase	Air stripping	Packed towers
			Aerated trays
			Bubble diffusers
		Oxygenation	Diffused aeration
			Surface aerators
Chemical oxidation	Electron loss to an oxidant	Chlorine	Contact basin
		Chlorine dioxide	
		Ozone	
		Ultraviolet radiation	
		UV with ozone	UV tubes in continuous flow
		UV with hydrogen peroxide	Basin
		Supercritical oxidation	
		Wet air oxidation	
Biological treatment—aerobic	Substrate is metabolized under aerobic conditions	Activated sludge	Conventional activated sludge
			Complete mix
			Aerated lagoons
			Sequencing batch reactors
		Biofilm reactors	Trickling filters
			Biofilters
			Rotating biological contactors

TABLE 4.1 (continued)
Unit Processes of Water Treatment

Unit Process	Principle	Categories	Technologies/Forms/Examples
Biological treatment—anaerobic	Substrate is metabolized under anaerobic conditions	Anaerobic digesters	Large tanks—$\theta \approx 30$ days
		Anaerobic reactors	Sealed tank
Disinfection	Interference with organism survival and reproduction		Contact basin
Membrane filtration	Straining small particles	Microfiltration	Tubular—cross flow
	Straining viruses	Ultrafiltration	Ceramic—cross flow
	Straining organic molecules	Nanofiltration	Spiral wound
	Straining ions	Hyperfiltration	Hollow fiber
		Electrodialysis	Electrodialysis

4.2.2 Transport

Transport mechanisms are at two levels: macro and molecular. The macro level includes advection of fluid mass, turbulent diffusion in open channels and pipes, and dispersion in porous media. Molecular diffusion is the other mechanism. As a rule, these mechanisms occur in two or more combinations.

4.2.2.1 Macro Transport: Sedimentation

The settling of a particle through a fluid under the influence of gravity is a transport mechanism. The sink is the surface, that is, the bottom of a tank or an inclined plate, where the particle settles. In flotation, the water surface is the sink.

4.2.2.2 Macro Transport: Advection

At the macro level, advection (sometimes called convection) is the transport of a fluid mass under a pressure gradient. [In open channel flow, gravity provides the pressure gradient.] In other words, advection transports the bulk flow of a fluid.

4.2.2.3 Macro Transport: Turbulent Diffusion

Turbulence is a transport mechanism found in both ambient and engineered environments. It is superimposed on advective transport. Turbulence causes a random transport of molecules and particles; the transport rate is proportional to the intensity of turbulence. The turbulence intensity is dependent upon the rate of energy dissipation and is important in rapid mix/coagulation, flocculation, activated sludge, and other fluidized bed reactors. In some unit processes, such as sedimentation, turbulence is unwanted.

4.2.2.4 Macro Transport: Porous Media Dispersion

In flow through porous media, the advective flow follows "streamlines." At the micro scale, the flow follows a tortuous path due to the irregular pore sizes, which causes commensurate flow velocities that vary randomly about the mean advective velocity. This is porous media dispersion.

To illustrate the idea of dispersion in porous media, consider the depiction of porous media in Figure 4.1 with irregular grain sizes, random packing of media, and random pore sizes. As with turbulent diffusion and molecular diffusion, the velocities are random and are illustrated by the varying random arrow

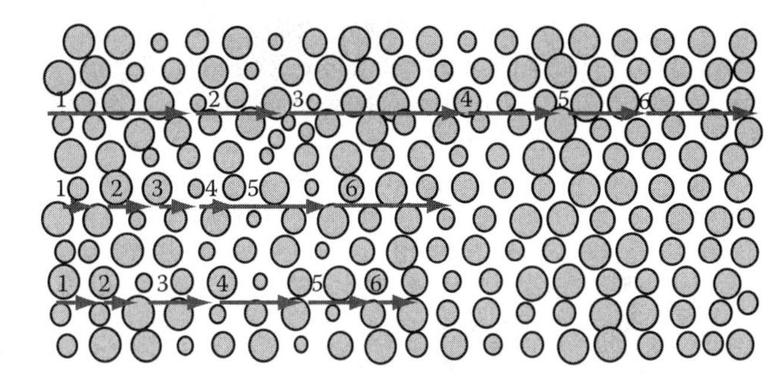

FIGURE 4.1 Illustration of dispersion in flow through porous media (numbers indicate hypothetical sequence of "steps" of tagged molecules).

lengths. Each movement of the advective flow is a "step." Along the top stream path the particular random sequence of six pore velocities shown, and the six "steps", will result in a tagged molecule moving ahead of the others. Along another flow path, say the bottom, the random sequence of pore velocities results in a tagged molecule lagging the others. The average of all pore velocities for a given cross section is the average advective pore velocity for the flow. The actual velocities vary continuously with pore size and also have micro components that vary from the straight lines shown.

The effect of dispersion would be seen if a pseudo vertical line of tagged molecules, say 1000 in number, is placed on the left side of the porous media shown in Figure 4.1. As an initially vertical line of tagged molecules is translated to the right, it becomes a bell-shaped (Gaussian) curve. The bell-shaped curve spreads as the translation progresses. This is the effect of dispersion. In addition, lateral dispersion occurs, and is superposed on the translation. The effect may be observed if tagged molecules are followed from some arbitrary point source on the left side of Figure 4.1 and would be seen as a lateral "spread."

4.2.2.5 Molecular Transport: Diffusion

In any gas or liquid, the molecules comprising the medium, and any contaminants within, are buffeted about randomly by successive incessant collisions with other molecules, called "Brownian motion." This random motion causes a net flux of molecules from a higher to a lower concentration, and is *diffusion*.

To explain diffusion, the distance traveled between collisions is called a "step," and for a single molecule may be depicted, as a *random walk*, that is,

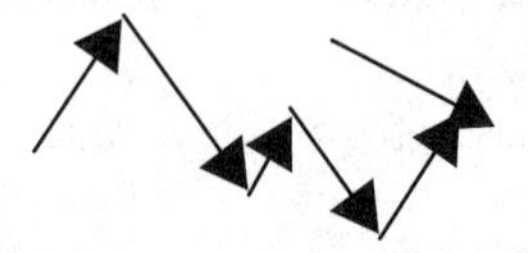

Each "step" is different in length, direction, and velocity (each collision causes a random change in the velocity vector of a given molecule). In a gas, the mean velocity of all molecules increases with temperature, that is, $\text{KE} = mv^{1/2} = kT$ (KE is the kinetic energy of the gas in a given volume, m is the mass of a molecular species, v is the mean velocity of molecules, k is the Boltzman constant, and T is the absolute temperature). Therefore as temperature increases, the mean velocity increases, which means that the number of steps per unit time increases. Thus the rate of diffusion increases.

To illustrate further, if a gas or liquid contains a group of tagged molecules that are confined by some boundary, and if the boundary is removed at a time, $t = 0$, those tagged molecules are free to move to other spaces by random motion. As noted, this random motion is *diffusion*. Also as noted, the *rate of diffusion* is essentially the number of steps per unit time, which depends on temperature.

Diffusion manifests itself as a flux from a higher concentration to a lower concentration. The "flux" of a tagged material is proportional to its concentration gradient, which is Fick's first law (Equations 4.1 or 4.2).

Diffusion is important in gas transfer, adsorption, fixed film reactors, disinfection, and biological reactions. As a rule, diffusion occurs in some combination with advection and/or turbulence, or both. Advection or turbulence brings the contaminant to the vicinity of a reaction site and the final travel may be by diffusion. Turbulence has the effect of bringing reactants into direct contact or it may increase the diffusion concentration gradient.

4.2.2.6 Mathematics of Diffusion, Turbulence, and Dispersion

The three transport mechanisms of diffusion, turbulence, and dispersion are similar in concept and in mathematics. All three are due to random motion that is on the molecular scale for diffusion and on the macroscale for turbulence and dispersion. Random motion is the essence of each. Molecular diffusion is, however, the basis for the discussion that follows; the mathematics is the same, however, for each of the three mechanisms.

If there is a spatial difference in concentration of a contaminant, then any kind of random motion will cause a net transport of material from the more concentrated to the less concentrated regions. Consider, for example, molecules of a certain species (say argon) confined in the proverbial bell jar, as illustrated in Figure 4.2. The molecules within the bell jar are all in a random walk mode at any given time. None can cross the boundary of the bell jar. But let the bell jar be removed at time $t = 0$. At this point the argon molecules are no longer bounded and they are free to move across the

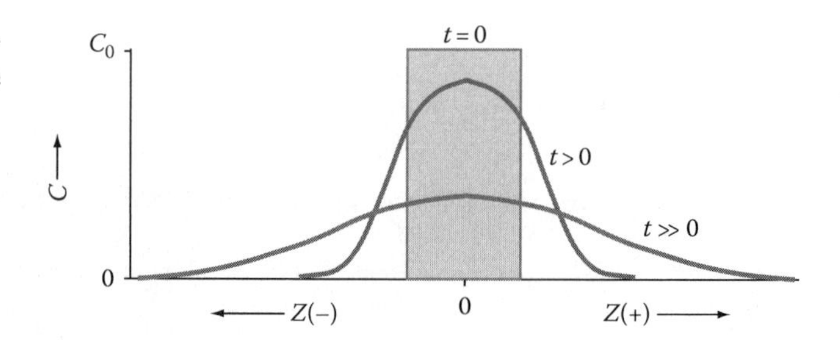

FIGURE 4.2 A bell jar is removed from a tagged gas having concentration profile as indicated at $t = 0$. The gas will diffuse giving continuously flatter "Gaussian" distributions with increases in time. This is due to the Brownian motion of the gas particles.

former boundary. Thus, when the boundary is removed half of the molecules at the former boundary will have velocity vectors toward the external space relative to the former boundary and half will have velocity vectors toward the internal space. Those that find position external to the former boundary at the end of the first step will again have the probability of having the same set of velocity vectors, that is, half will be inward and half outward. In fact all of the argon molecules, even those that started in the interior of the bell jar will have that same probability. Over time, all of the molecules will have dissipated throughout the larger volume (whether that larger volume is a room or the ambient environment). Figure 4.2 shows the concentrations of gas (argon) at time zero and at two other arbitrary times. These profiles, designated $C(Z)_t$, are the result of this random motion. As a note, if a "pulse" input into say an open channel or a pipe should occur, the result observed would be the same as shown for the bell jar. The effect is the same as imposing a translation velocity on the bell jar and will be observed to be similar as seen in Figure 4.2 if the coordinate system takes on the advective velocity of the fluid, that is, the coordinate system is Lagrangian, vis-à-vis Euclidean (i.e., the coordinate system is fixed).

To describe this increasing lateral spread of tagged gas as depicted in Figure 4.2, Fick's law is applicable, that is,

$$\mathbf{j} = -D\nabla C \tag{4.1}$$

in which,
$\mathbf{j}$ is the flux density (kg/s/m^2)
D is the diffusion constant (m^2/s)
∇ is the operator
C is the concentration of a given species (kg/m^3)

Equation 4.1 in one dimension is expressed,

$$\mathbf{j}_z = -D\frac{\partial C}{\partial Z} \tag{4.2}$$

in which,
$\mathbf{j}_z$ is the flux density (kg/s/m^2)
Z is the coordinate (m)

Looking at Figure 4.2, the flux density varies along the curve as the slope, $\partial C/\partial Z$ changes and is highest at the inflection point of each curve (the steepest part of the curve). As time increases, the $C(Z)_t$ curve becomes flatter and therefore the flux density at any Z is less; for example, at time $(t + \Delta t)$ as compared with time, t.

If we apply Equation 4.1 to an infinitesimal volume and let the mass accumulate in that volume, the rate of accumulation is

$$\frac{\partial C}{\partial t} = D\nabla^2 C \tag{4.3}$$

in which t is the elapsed time (s).

Equation 4.3 is Fick's second law. Really it is the mathematical solution to Fick's first law. Its one-dimensional solution is

$$C = \frac{M}{2(\pi Dt)^{1/2}} e\left(-\frac{Z^2}{4D}\right) \tag{4.4}$$

in which M is the mass of argon under the bell jar at $t = 0$ (kg). In other words, Equation 4.4 is the $C(Z, t)$ solution that yields the curves of Figure 4.2.

Fick's first law has broad applicability to many kinds of unit processes and is the key to understanding molecular transport. Examples include gas transfer across a liquid or gas "film," diffusion from the external surface of a particle to the interior as in carbon adsorption, ion exchange, and across organism membranes, diffusion of carbon molecules and oxygen into a biofilm, etc. (Box 4.1).

BOX 4.1 NOTES ON PROBABILITY THEORY AND DIFFUSION AND DISPERSION

The more steps the molecule takes the greater the probability of being found at positions other than Z_0, although Z_0 is still the "most probable" position. Probability theory shows how this result, that is, Equation 4.4 can be arrived at due to a sequence of "coin flipping," which gives the position probability of a particle that takes steps of equal length.

Although illustrated for the random motion of molecules, that is, Brownian motion, as in Figure 4.2, the same arguments apply to the random motion caused in pipe flow or flow through porous media. In a flow situation, the random motion is superimposed on the advective velocity of the flow. Figure 4.3 illustrates the spread that is caused by random motion of turbulence in pipe flow or by the random interstitial velocities of porous media flow. If a pulse (e.g., a salt slug) is injected instantaneously into pipe flow or a stream tube in porous media flow, the salt mass will spread to give a concentration distribution as indicated in Figure 4.2.

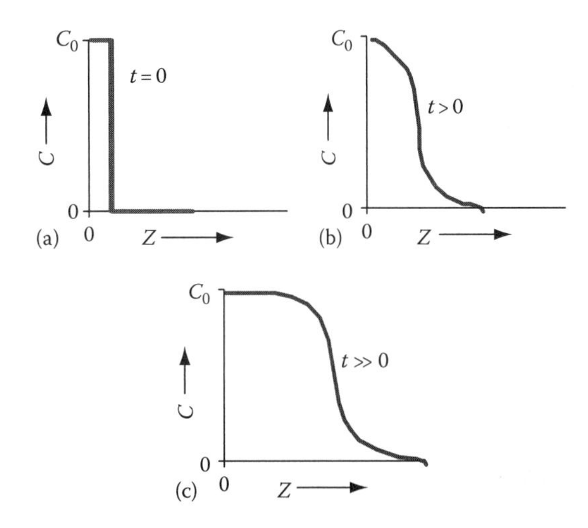

FIGURE 4.3 Effect of dispersion on concentration distance profile—no reaction. (a) Step function at $t = 0$, (b) dispersion curve, $t > 0$, and (c) dispersion curve, $t \gg 0$.

4.2.2.6.1 Frontal Waves

Instead of a pulse input as shown in Figure 4.2, consider the continuous input of a tagged molecule, called a frontal wave, or a "step function," illustrated in Figure 4.3a. The same mechanism of random walk type of motion is operative as described for the pulse input. As the wave progresses with an advection velocity, v, the random walk is superposed. The result is seen, for increasing times, as shown in Figure 4.3b and c, respectively, in the form of progressively flatter S-shaped curves.

The midpoint of the curve in which $C/C_0 = 0.50$ defines the mean travel time, t_0, or the position the step function would occupy had it retained its original shape. This spread about the mean travel time position is called "dispersion" (Beran, 1955; Rifai et al., 1956).

The spread of solute due to dispersion in a moving fluid (with no reaction) is described by the equation

$$\left[\frac{\partial C}{\partial t}\right]_0 = \bar{v}\frac{\partial C}{\partial Z} + D\frac{\partial^2 C}{\partial Z^2} \tag{4.5}$$

in which $\bar{v}$ is the average advective velocity at a given point in a flow field (m/s).

Note that Equation 4.5, a materials balance equation, is the same as Equation 4.3, but with the advection term added. Rifai et al. (1956) have given the solution of this equation as

$$\frac{C}{C_0} = \frac{1}{2}\left[1 \pm \mathrm{erf}\left(\frac{Z - \bar{v}t}{2\sqrt{Dt}}\right)\right] \tag{4.6}$$

The sign is (+) for $Z < \bar{v}t$ and (−) for $Z > \bar{v}t$. Thus a solution for D can be obtained through measurement of a point on the "breakthrough" curve (which is the concentration versus time curve at the exit to a column of porous media). Note that Equation 4.6 describes an S-shaped breakthrough curve and not a bell-shaped curve; the solute here is fed in continuously and not as a slug.

Rifai et al. (1956) presented a method of obtaining D that was done by measuring the whole breakthrough curve. This equation is

$$D = \frac{1}{4\pi}\left[\frac{L}{V_0^2 S_0^2}\right]\bar{v} \tag{4.7}$$

in which

V_0 is the throughput volume at $C/C_0 = 0.50$ (m^3)

S_0 is the slope of the breakthrough curve at $C/C_0 = 0.50$ (m^{-3})

For Equation 4.7, the breakthrough curve is plotted with throughput volume instead of time as the abscissa. This method is in lieu of using Equation 4.6 to solve for D. The throughput volume is given as $V = Q \cdot t$.

Rifai et al. have also shown experimentally that $D/\bar{v}$ is a constant. This is also evident from the analogy with Brownian motion in which it is the number of steps that determines position probability in the diffusion process. Velocity merely determines the number of steps taken per unit of time.

This term $D/\bar{v}$ is a property of the porous media and is the term calculated from the experimental data indicated in Equation 4.7. Thus D is obtained for any value of $\bar{v}$ once $D/\bar{v}$ is evaluated.

4.2.3 Summary

The "sink" is the place where the contaminant is removed from the water stream or is transformed into another species. The manner of reaching the sink depends on the transport mechanism, for example, advection, turbulence, diffusion, or gravity.

4.3 REACTORS

A *reactor* is the place where something "happens," that is, removal of a selected contaminant. Usually, the "place" is within an *engineered* volume where the transport is controlled along with the character of the sinks.

In the ambient environment, the reactor concept is applicable to any arbitrary volume. A volume element of a lake behaves mathematically as a reactor, as does a volume element of a stream, or a volume element within a ground water stream tube.

4.3.1 Examples of Reactors

Under the definition of a reactor stated, the reactor concept is applicable to any kind of unit process. Examples of reactors include: an activated sludge basin, a bed of granular activated carbon, a granular media filter, a rapid mix basin, a flocculation basin, a disinfection basin, etc. Even a sedimentation basin may be considered a reactor under the more liberal definition, that is, as place where something "happens."

4.3.2 Types of Reactors

Two basic reactor types are fluidized volume and fixed bed. Within the former are activated sludge, rapid mix basins, flocculation basins, anaerobic digesters, disinfection basins, ozone contactors, etc. The fixed bed reactors include trickling filters, slow sand filters, rapid rate filters, granular activated carbon beds, etc.

Turbulence characterizes the transport mechanism of a fluidized reactor in which the reactants contact each other by the random motion of turbulence. In the fixed bed reactor, the media is "passive" and the reaction depends upon the other reactant being advected to a site with the final transport step being diffusion (sedimentation may have a role if larger particles are involved).

The foregoing list may be disaggregated further. For example, within the activated sludge category of fluidized-bed reactors, there are complete mix reactors and plug flow reactors. There are also "film" reactors, which are films on a media (this is the fixed bed reactor category). The media may be rocks (trickling filters), granular media (slow sand), and rotating plates (rotating biological contactors). And if there is no flow across the boundaries, then the volume is a *batch reactor*. If there is flow across the boundaries the volume is a *continuous flow reactor*.

From this discussion, a reactor description has several characteristics. We must describe it in terms of whether the contacts within the reactor are caused by the bed being *fluidized* by turbulence, or whether the bed is *fixed*, as in granular media. If the reactor is fluidized it may be *complete mix* and the whole reactor is thereby *homogeneous*; on the other hand, if there is no mixing, the reactor is *plug flow* and nonhomogeneous. Fixed-bed reactors are *nonhomogeneous*. If the reactor is homogeneous, concentrations do not vary spatially; for a nonhomogeneous reactor, concentrations vary spatially.

In terms of its flow, the reactor may be either *batch* or *continuous flow*. The mathematical description of a reactor, that is, mathematical modeling, may have special conditions imposed, depending upon the type of reactor.

4.3.3 Mathematics of Reactors

Reactor mathematics is based upon two principles: (1) materials balance and (2) reaction kinetics. A mathematical depiction of a reactor, so formulated, constitutes a *model*. Such a model is applicable to most of the engineered unit processes (and, as noted, to processes that occur in the ambient environment). A proper kinetic depiction is usually the most difficult part of the model. The materials balance/kinetic model provide a basis for analysis and for rational process design.

The critical aspect of a reactor model is that the materials balance is applicable to a specified volume element. A key point is that the contents of the volume element must be homogeneous, that is, concentration is constant from point to point throughout the volume, regardless of the kind of reactor.

4.3.3.1 Materials Balance: Concept

The basic idea of a reactor model is quite simple and is embodied in Equation 4.8, a word statement of the materials balance principle. Figure 4.4 shows a volume element (two dimensional) which illustrates the terms of Equation 4.8. In this

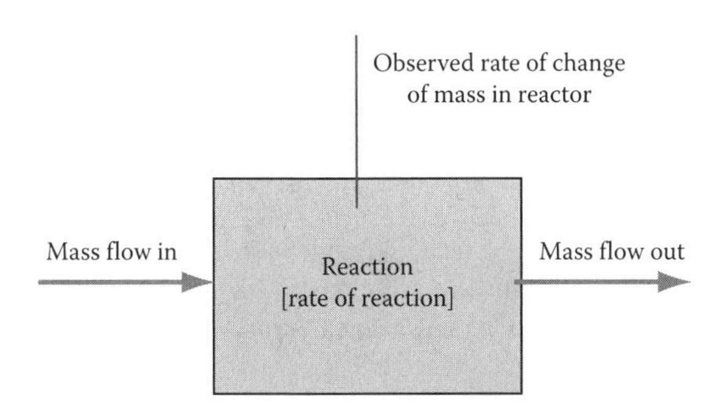

FIGURE 4.4 Materials balance for a homogeneous "complete mix" volume element showing terms in words, as corresponding to Equation 4.8.

case, the volume element is the whole reactor (since the reactor is "complete mix" and therefore the reactor as a whole is homogeneous). The "rate of reaction" is the kinetic term, which depends upon the particular kind of reaction, that is, chemical, biological, coagulation, etc. With this basic concept, reactor modeling is applicable to a wide variety of unit processes. Figure 4.4 is for a *continuous flow* reactor, as is evident by the advection terms. For a "batch" reactor, that is, with no advection in or out, the advection terms would equal zero.

$$\begin{bmatrix}\text{Observed}\\ \text{mass rate}\\ \text{of change}\end{bmatrix} = [\text{Mass flow in}] - [\text{Mass flow out}] - \begin{bmatrix}\text{Rate of}\\ \text{reaction}\end{bmatrix} \quad (4.8)$$

in which

observed mass rate of change = the rate of change of the concentration within a given reactant as observed by measurement (kg reactant A/s)

mass flow in = the product of flow, Q, into the reactor times its concentration, C (kg of A/s)

mass flow out = the product of flow, Q, out of the reactor times its concentration, C (kg of A/s)

rate of reaction = the rate at which A is undergoing change, which is $V{\cdot}\mathrm{d}C/\mathrm{d}t$ (kg of A/s)

4.3.3.2 Comments on Materials Balance

To illustrate the flexibility of Equation 4.8, we may impose different kinds of special conditions, for example:

1. *Steady state operation*: The left side of Equation 4.8 is zero.
2. *Rate of reaction is zero*: The observed rate of change equals the difference between the mass flow in and the mass flow out, such as would occur for a solution of one kind, say saline, replacing another solution, say pure water, in the reactor.
3. *Batch reactor*: There is no mass flow across the boundaries of the reactor, and the observed rate of change is equal to the rate of the reaction.

These conditions are illustrated mathematically in the sections that follow, for example, Section 4.3.4.

4.3.3.3 Materials Balance: Mathematics

The mathematical statement of Equation 4.8 is given for two cases: (1) a complete mix *fluidized* reactor, and (2) a *fixed bed* column reactor, or more commonly called a "*plug-flow*" reactor, in which the homogeneous volume element is an infinitesimal "slice" of the column. The term *fluidized* means that the reactants are being mixed by turbulence, vis-à-vis *fixed bed*, in which case the porous media reactant is fixed in place as in a bed of GAC or a rapid rate filter media bed.

4.3.3.3.1 Complete Mixed Reactor: Finite Volume

Fluidized reactors include complete mix activated sludge, rapid mix coagulation, and any other basin that has a high level of energy input. Most often, this is done by a turbine impeller. The complete mix reactor, sometimes called "completely stirred tank reactor" (CSTR), is homogeneous by definition and therefore the materials balance equation applies to the whole reactor. For the "complete mix" homogeneous volume element of Figure 4.4, Equation 4.8 may be expressed mathematically as Equation 4.9, that is,

$$\left[\frac{\partial(C{\cdot}V)}{\partial t}\right]_{\mathrm{o}} = Q{\cdot}C_{\mathrm{in}} - Q{\cdot}C - \left[\frac{\partial(C{\cdot}V)}{\partial t}\right]_{\mathrm{r}} \quad (4.9)$$

in which

C is the concentration of species of interest in the reactor and leaving reactor (kg/m^3)
V is the volume of reactor (m^3)
C_{in} is the concentration of species in flow into the reactor (kg/m^3)
Q is the flow into and out of the reactor (m^3/s)
t is the elapsed time (s)
subscript "o" indicating "observed" mass rate of change
subscript "r" indicating reaction rate (kinetics)

Figure 4.5 illustrates the same reactor as shown in Figure 4.4, but with the terms depicted in mathematical form (instead of words). The continuous flow, complete mix (and thereby homogeneous) reactor is the general case that embodies the reactor concept. This is the starting point for grasping the mathematical modeling of the other kinds of reactors.

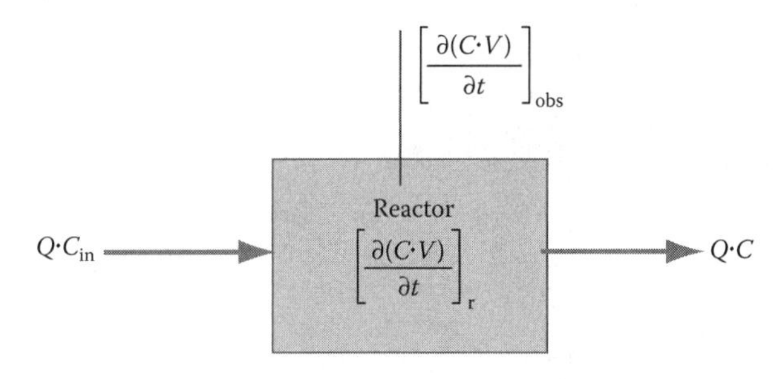

FIGURE 4.5 Materials balance for a homogeneous "complete mix" volume element showing terms in mathematical symbols, as corresponding to Equation 4.9.

4.3.3.3.2 General Applicability of Equation 4.9

First, consider that we may wish to let either Q or C_0, or both at once, vary with time, which is the operating condition for most treatment plants. Such a condition can be imposed merely by writing Equation 4.9 in finite difference form and solving numerically.

In addition, we can impose any special conditions (as noted in Section 4.3.3.1) that we wish, such as for a "batch" reactor, that is, $Q \cdot C_{in} = 0$ and $Q \cdot C = 0$. Or we can impose the "steady state" condition, in which the observed mass rate of change in the reactor, the term on the left side of Equation 4.9, is zero. Finally, we can let the rate of reaction equal zero and solve for the effluent concentration, C, leaving the reactor, such as for a salt solution displacing some other solution from the reactor.

4.3.3.3.3 Packed-Bed Reactor (Column)

Packed-bed reactors include: activated carbon, ion exchange, trickling filters, various biofilm reactors (e.g., slow sand), and rapid rate filters. All packed-bed reactors have concentration changes from the entrance of the reactor to its exit, as Figure 4.6 illustrates. In Figure 4.6, the concentration of the solute entering the reactor is designated, C_0. The concentration declines as a continuous function with distance, Z, along the column, that is, $C(Z)_t$, and may be characterized as the "concentration profile" at a designated time, t.

If the reactor is steady state, such as a trickling filter with no change in Q or C_0 with time, then the profile will remain constant. But if the porous media becomes saturated with time, such as with GAC, an ion-exchange bed, or a rapid rate filter, then the profile will advance downstream with time.

In addition to packed-bed reactors, this same schematic, that is, Figure 4.6, is applicable to a "plug flow" type of fluidized reactor. The latter may include reactors that are long and narrow, such as "conventional" activated sludge. A river would also fit in this category.

4.3.3.3.4 Homogeneity Requirement

As noted previously, the materials balance equation is valid only for a homogeneous volume element. Figure 4.6 shows clearly that the solute concentration within the reactor varies with Z. Therefore, the reactor is not homogeneous. Consequently, the materials balance equation cannot be applied to the reactor as a whole, that is, over the column length. The condition of homogeneity may be true, however, if the volume element is infinitesimal. Figure 4.6 shows such a volume element for a column slice of thickness ΔZ. As ΔZ approaches zero, then the solute approaches homogeneity within the element. Thus, the materials balance equation may be applied to the infinitesimal slice.

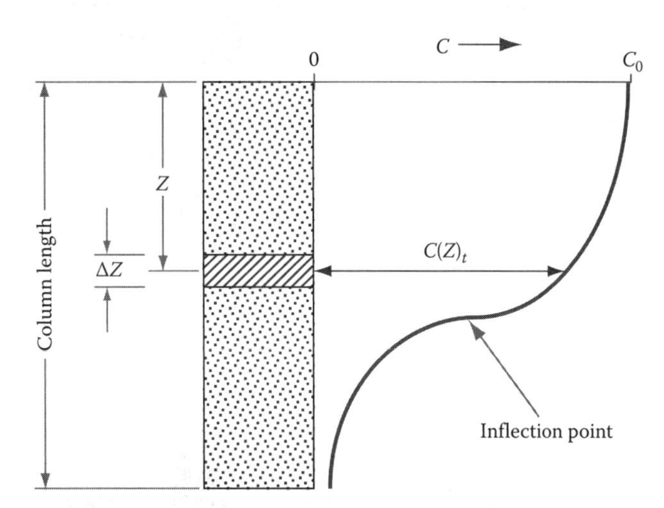

FIGURE 4.6 Concentration profile in a packed-bed reactor.

4.3.3.3.5 Application of Materials Balance Equation

Figure 4.7 shows a materials balance for an infinitesimal column slice of thickness, ΔZ and area, A. The corresponding mathematical statement is given by Equation 4.10. The mass transport terms are advection and dispersion, both illustrated in Figure 4.7.

$$\left[\frac{\partial(C \cdot \Delta Z \cdot A)}{\partial t}\right]_0 = \bar{v} \cdot A \cdot C_{in} - \bar{v} \cdot A \cdot C_{out} + \mathbf{j}_{in}A - \mathbf{j}_{out}A - \left[\frac{\partial(C \cdot \Delta Z \cdot A)}{\partial t}\right]_r \quad (4.10)$$

in which

Z is the distance along column from entrance (m)
$\bar{v}$ is the velocity within column (m/s)
A is the cross-sectional area of column (m^2)
t is the elapsed time from beginning of operation (s)
C_{in} is the concentration of given contaminant into the infinitesimal slice (kg/m^3)
C_{out} is the concentration of given contaminant leaving the infinitesimal slice (kg/m^3)
$\mathbf{j}_{in}$ is the dispersion flux density into element ($kg/m^2/s$)
$\mathbf{j}_{out}$ is the dispersion flux density leaving element ($kg/m^2/s$)

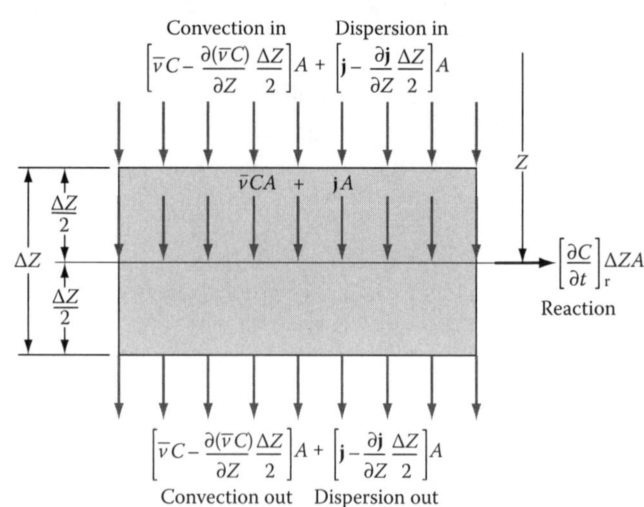

FIGURE 4.7 Materials balance on infinitesimal column slice of thickness, ΔZ, and area, A showing terms for derivation of materials balance equation.

The advection and the dispersion terms are,

advection in

$$\bar{v}C_{\text{in}} = \bar{v}C - \frac{\partial(\bar{v}C)}{\partial Z}\cdot\frac{\Delta Z}{2} \quad \text{(a)}$$

advection out

$$\bar{v}C_{\text{out}} = \bar{v}C - \frac{\partial(\bar{v}C)}{\partial Z}\cdot\frac{\Delta Z}{2} \quad \text{(b)} \tag{4.11}$$

dispersion in

$$\mathbf{j}_{\text{in}} = \mathbf{j} - \frac{\partial \mathbf{j}}{\partial Z}\cdot\frac{\Delta Z}{2} \quad \text{(a)}$$

dispersion out

$$\mathbf{j}_{\text{out}} = \mathbf{j} + \frac{\partial \mathbf{j}}{\partial Z}\cdot\frac{\Delta Z}{2} \quad \text{(b)} \tag{4.12}$$

in which
- C is the concentration of given contaminant at center of the infinitesimal slice (kg/m^3)
- **j** is the dispersion flux density at center of the infinitesimal slice (kg/m^2/s)

In words, the advection flux into the element is the advection flux in the center minus the rate of change of the advection flux with respect to Z times the distance of half the slice, that is, $\Delta Z/2$. The same idea holds for the advection flux leaving the element.

The dispersion flux into the element is the dispersion flux in the center of the element minus the rate of change of the dispersion flux with respect to Z times the distance of half the slice, that is, $\Delta Z/2$. The same idea holds for the dispersion flux leaving the element.

Substituting Equations 4.11 and 4.12 into 4.10, and canceling A,

$$\left[\frac{\partial C}{\partial t}\right]_0 \Delta Z = \left[\bar{v}C - \frac{\partial(\bar{v}C)}{\partial Z}\frac{\Delta Z}{2}\right]_{\text{in}} - \left[\bar{v}C + \frac{\partial(\bar{v}C)}{\partial Z}\frac{\Delta Z}{2}\right]_{\text{out}} + \left[\mathbf{j} - \frac{\partial \mathbf{j}}{\partial Z}\frac{\Delta Z}{2}\right]_{\text{in}} - \left[\mathbf{j} + \frac{\partial \mathbf{j}}{\partial Z}\frac{\Delta Z}{2}\right]_{\text{out}} - \left[\frac{\partial C}{\partial t}\right]_{\text{r}} \Delta Z \tag{4.13}$$

Then, for the condition that $\bar{v}$ is constant, Equation 4.13 simplifies to

$$\left[\frac{\partial C}{\partial t}\right]_0 - \bar{v}\frac{\partial C}{\partial Z} + \frac{\partial \mathbf{j}}{\partial Z} - \left[\frac{\partial C}{\partial t}\right]_{\text{r}} \tag{4.14}$$

The dispersion transport rate is analogous to Fick's law, and is expressed as

$$\mathbf{j} = -D\frac{\partial C}{\partial Z} \tag{4.15}$$

Substituting (4.15) into (4.14),

$$\left[\frac{\partial C}{\partial t}\right]_0 = -\bar{v}\frac{\partial C}{\partial Z} + D\frac{\partial^2 C}{\partial Z^2} - \left[\frac{\partial C}{\partial t}\right]_{\text{r}} \tag{4.16}$$

Equation 4.16 is a "final" form of the materials balance equation for a nonhomogeneous reactor in which the infinitesimal element is a column "slice," as shown in Figure 4.7 (Box 4.2).

BOX 4.2 MATERIALS BALANCE EQUATION FOR PACKED-BED REACTOR

In general, Equation 4.16 must be solved by a finite-difference algorithm using a high-speed computer. The kinetic term is specific to the particular system being modeled, for example, biological, chemical, adsorptive, ion exchange, etc.

Consider granular activated carbon as the media in a packed bed. The reaction term is

$$\left[\frac{\partial C}{\partial t}\right]_{\text{r}} = \rho\frac{(1-P)}{P}\left[\frac{\partial \bar{X}}{\partial t}\right] \tag{4.17}$$

in which
- P is the porosity of porous media
- $\bar{X}$ is the concentration of contaminant in solid phase (kg contaminant/kg adsorbent)
- ρ is the density of adsorbent particles (kg adsorbent/m^3)

Substituting Equation 4.17 in Equation 4.16 gives

$$\left[\frac{\partial C}{\partial t}\right]_0 = -\bar{v}\frac{\partial C}{\partial Z} + D\frac{\partial^2 C}{\partial Z^2} - \rho\frac{1-P}{P}\left[\frac{\partial \bar{X}}{\partial t}\right]_{\text{r}} \tag{4.18}$$

Equation 4.18 cannot be solved mathematically. It must be converted to finite difference form and then solved numerically by computer. Also a kinetic equation is needed for the $[\partial\bar{X}/\partial t]_{\text{r}}$ term. The computational scheme is to divide the column into n slices and to solve the materials balance for each slice starting at the top of the column. In doing this, the output of the ith slice equals the input to the $i+1$ slice and the computation yields C_i for time $t+\Delta t$. Since the variables C_i and $\bar{X}_i$ change with both distance and time, millions of iterations are necessary for a "solution" to Equation 4.18. The form of the solution would be the concentration profile at successive times, that is, $C(Z)_t$. The solution is discussed in more detail in Chapter 15.

In words, Equation 4.16 says: the observed mass rate of change of a given contaminant within the slice equals the net rate of advection for the slice plus the net rate of dispersion for the slice minus the rate of reaction within the slice. As a note, Equation 4.16 is written often in terms of volume rather than time, for example, $[\partial C/\partial V]_0$ instead of $[\partial C/\partial t]_0$ and $[\partial C/\partial V]_{\text{r}}$ instead of $[\partial C/\partial t]_{\text{r}}$. Since the volume of fluid that has passed a certain point is, $V = Qt$; then, $\text{d}V = Q\text{d}t$.

In addition to packed-bed reactors, Equation 4.16 and Figure 4.7 are applicable to fluidized-bed reactors in which the concentration changes with distance. Examples include streams which may receive a wastewater flow, in which concentrations of various contaminants change with distance, a "plug flow" activated sludge reactor, an air-stripping tower, and a chlorination basin.

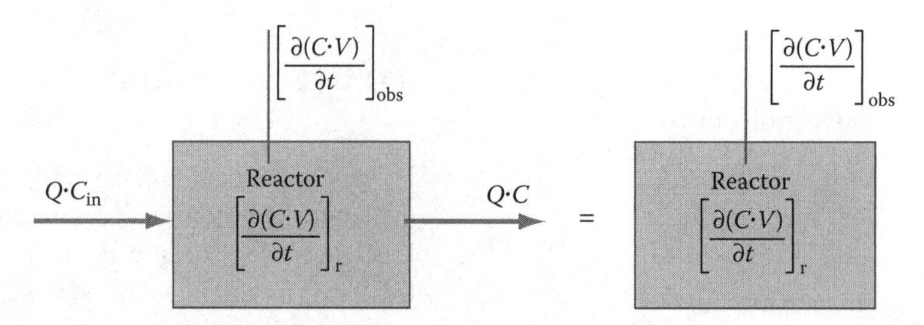

FIGURE 4.8 Materials balance for a batch reactor, that is, for a complete mix reactor when $Q{\cdot}C_0 = 0$ and $Q{\cdot}C = 0$.

4.3.4 Materials Balance: Special Conditions

Several kinds of special conditions can be imposed on Equations 4.8 and 4.9 for finite volume, complete mix reactors. Three of these are (1) batch, (2) steady state, and (3) no reaction. Reviewing these cases illustrates the flexibility of the materials balance principle.

4.3.4.1 Batch Reactor: Complete Mixed

The batch reactor is merely a volume in which there is no mass advected into or out of the reactor. Figure 4.8 depicts this condition. Thus in Equation 4.9, $QC_0 = 0$ and $QC = 0$, and so Equation 4.9 has these operations imposed, that is,

$$\left[\frac{\partial(C\cdot V)}{\partial t}\right]_0 \overset{0}{\cancel{Q}}\cdot C_{\text{in}} - \overset{0}{\cancel{Q}}\cdot C - \left[\frac{\partial(C\cdot V)}{\partial t}\right]_{\text{r}} \tag{4.19}$$

to become Equation 4.20, that is,

$$\left[\frac{\partial(C\cdot V)}{\partial t}\right]_0 = \left[\frac{\partial(C\cdot V)}{\partial t}\right]_{\text{r}} \tag{4.20}$$

In other words, with no advection of mass across the boundaries of the reactor, the "observed" rate of change of mass of a given contaminant within the reactor equals the rate at which the reaction is taking place. If we measure the observed rate of change of mass, that is, the left side of Equation 4.20, this is the rate of reaction, that is, the right side of Equation 4.20. To measure the observed rate of change of C in an experiment, we would measure C at specified times, t, and then plot C versus t and fit an equation to the curve. The derivative, that is, the slope of the curve, is the mass rate of change (assuming the reactor volume is maintained constant). If we go further in the analysis of the curve, we may be able to fit a rate equation to the curve, such as a first-order kinetic equation. Such an experiment would then yield a kinetic equation for the right side of Equation 4.20.

4.3.4.2 Steady State Reactor: Complete Mixed

As noted, Equation 4.9 is general with no restrictions (for a complete mix reactor). But if we impose the condition that the "mass flow in" and the "mass flow out" do not change with time, then the "observed rate of change" of C will be zero. Figure 4.9 illustrates this condition. Note that the observed rate of change of $(C{\cdot}V)$ is zero.

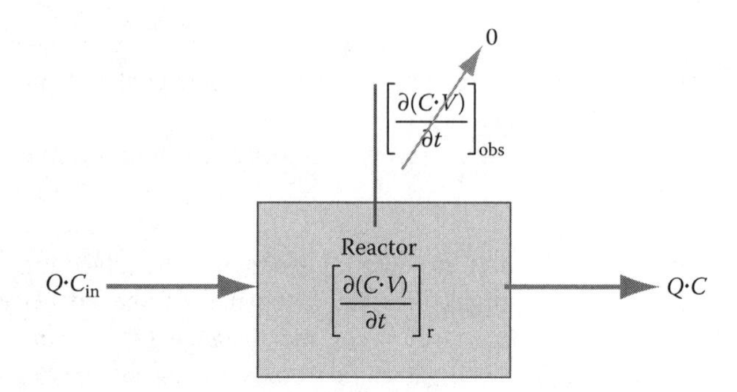

FIGURE 4.9 Steady state materials balance for a batch reactor, that is, for a complete mix reactor when the observed mass rate of change of C in the reactor is zero.

Equation 4.9 with the steady state condition imposed is,

$$\overset{0}{\cancel{\left[\frac{\partial(C\cdot V)}{\partial t}\right]_0}} = Q\cdot C_{\text{in}} - Q\cdot C - \left[\frac{\partial(C\cdot V)}{\partial t}\right]_{\text{r}} \tag{4.21}$$

to become

$$0 = Q\cdot C_{\text{in}} - Q\cdot C - \left[\frac{\partial(C\cdot V)}{\partial t}\right]_{\text{r}} \tag{4.22}$$

Equation 4.22 says in words that the mass rate of reaction equals the mass rate of flux (influent mass rate) to the reactor minus the mass rate of flux (effluent mass rate) from the reactor. As noted, Figure 4.9 illustrates this concept. Equation 4.22 may be the basis for a mathematical solution for C. Since we assume constant V, and since the detention time $\theta = V/Q$, then Equation 4.22 may be stated as

$$C_{\text{in}} - C = \left[\frac{\mathrm{d}C}{\mathrm{d}t}\right]_{\text{r}} \cdot \theta \tag{4.23}$$

in which θ is the reactor detention time (s).

Equation 4.23 adds additional insight into the materials balance concept, that is, that the amount of contaminant reacted in terms of concentration, that is, $(C_{\text{in}} - C)$, equals the rate of reaction times the detention time, θ, in the reactor. By the same token, C is the concentration of contaminant leaving the reactor. From Equation 4.23, C decreases as the rate of reaction increases, or as the detention time (the reaction

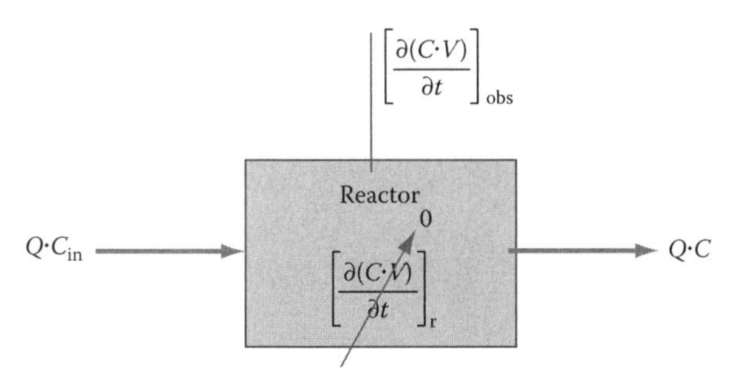

FIGURE 4.10 Steady state materials balance for a batch reactor, that is, for a complete mix reactor when the observed rate of reaction is zero.

time) increases. All that remains in this case is to apply a kinetic model for the dC/dt term. A "first-order" kinetic model is most common, that is, in which $[dC/dt]_r = kC$, that is, the rate of reaction is proportional to the concentration of the reactant being changed.

4.3.4.3 Zero Reaction: Complete Mixed

For the case of zero reaction rate, such as for sodium chloride salt or a nonreactive tracer such as rhodamine-B or pontacyl pink dye, the reaction term in Equation 4.9 goes to zero. Figure 4.10 illustrates this condition and Equation 4.24 shows the materials balance equation with the "no reaction" condition imposed, giving Equation 4.25. Note that in Equation 4.24, the volume, V is constant, along with Q, and $V/Q = \theta$.

$$\left[\frac{\partial(C\cdot V)}{\partial t}\right]_0 = Q\cdot C_{in} - Q\cdot C - \overset{0}{\left[\frac{\partial(C\cdot V)}{\partial t}\right]_r} \qquad (4.24)$$

Divide by Q, and since the rate of reaction is assumed zero, and as C increases, dC/dt is negative:

$$\left[-\frac{dC}{dt}\right]_0 \theta = (C_{in} - C) \qquad (4.25)$$

Separating variables in Equation 4.25 and applying limits of integration,

$$\int_{C_i}^{C} \frac{-dC}{(C_{in} - C)} = \frac{1}{\theta}\int_0^t dt \qquad (4.26)$$

gives after integration,

$$\ln\left[\frac{C_{in} - C}{C_{in} - C_i}\right] = -\frac{t}{\theta} \qquad (4.27)$$

or,

$$\left[\frac{C_{in} - C}{C_{in} - C_i}\right] = e^{[-(t/\theta)]} \qquad (4.28)$$

Figure 4.11 is a plot of Equation 4.28, showing the change in the ratio, $[(C_{in} - C)/(C_{in} - C_i)]$ as a function of t/θ. The plot

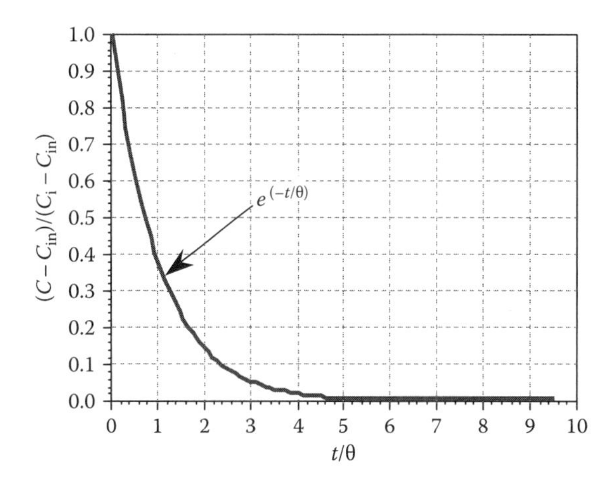

FIGURE 4.11 Constant flow displacement of solution in complete mix reactors plotted by Equation 4.28.

shows the change in the dimensionless concentration in the reactor (and thus leaving it) as a function of the dimensionless time, t/θ. Thus, the dimensionless concentration in the reactor has changed from 1.0 to 0.37 within one detention time, that is, $t/\theta = 1$, and to 0.14 when $t/\theta = 2$, and to 0.05 when $t/\theta = 3$. Thus by three detention times, the reactor is almost purged by the incoming substance. Note that Equation 4.28 is generally applicable, regardless of the initial conditions.

4.3.4.3.1 Initial Conditions

Application of Equation 4.28 requires that the initial conditions be stated. In general some initial concentration in the reactor, C_i, is displaced by the inflow with another concentration, C_{in}. The respective values of C_i and C_{in} make no difference in the application.

To further illustrate the application of Equation 4.28, consider two initial conditions: (1) the initial concentration in the reactor is, $C_i = 100$ mg/L and is displaced by a flow with $C_{in} = 1000$ mg/L, and (2) the initial concentration in the reactor is, $C_i = 1000$ mg/L and is displaced by a flow with $C_{in} = 100$ mg/L. Figure 4.12a and b are plots obtained after inserting these two sets of initial conditions, respectively, into Equation 4.28. The results are another form of Figure 4.11 and show, for example, that when $t/\theta = 1$, then $C = 669$ mg/L in Figure 4.12a and 431 mg/L in Figure 4.12b. Note that the effect of letting $C_i = 100$ mg/L has an effect on each result in that this value is the starting point in (a) and the asymptote in (b).

4.3.4.4 Nonsteady State Reactor

If we express Equation 4.9 in finite difference form, then nonsteady state conditions can be handled. This means that we can let Q and C_0 (either or both) vary with time. The starting point is Equation 4.9

$$\left[\frac{\partial(C\cdot V)}{\partial t}\right]_0 = Q\cdot C_{in} - Q\cdot C - \left[\frac{\partial(C\cdot V)}{\partial t}\right]_r \qquad (4.9)$$

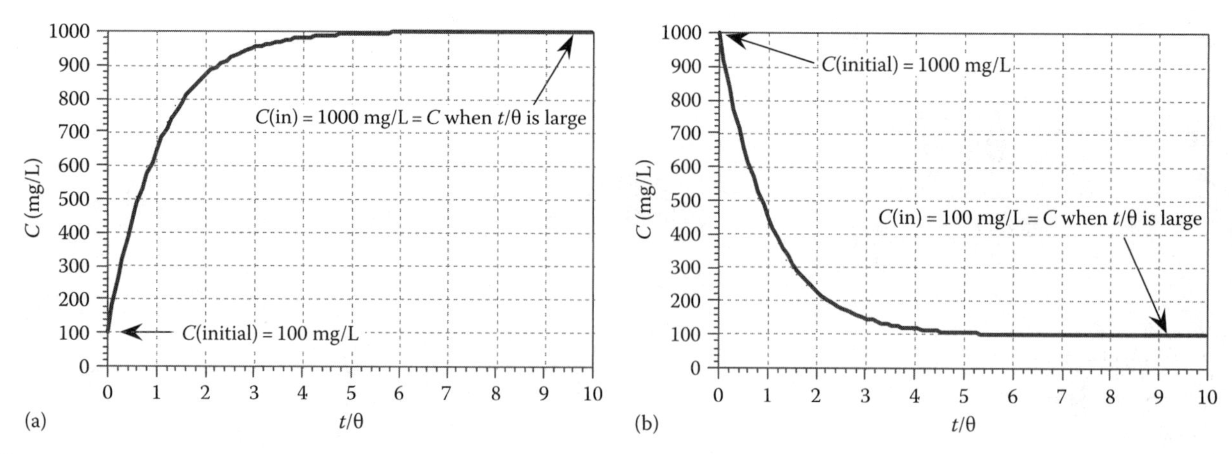

FIGURE 4.12 Constant flow displacement of solution in complete mix reactors plotted by Equation 4.28 under two boundary conditions. (a) Salt flow into complete mix reactor displaces dilute solution. (b) Dilute solution flow into complete mix reactor displaces salt solution.

Expanding the left side of Equation 4.9 gives

$$\left[\frac{dC}{dt}\right]_0 V + \left[\frac{dV}{dt}\right]_0 C = Q \cdot C_{in} - Q \cdot C - \left[\frac{dC}{dt}\right]_r V \tag{4.29}$$

If the reactor is not at constant volume, such as a waste stabilization pond in a temperate climate that fills in the winter and is drawn down in the summer, then Equation 4.29 applies and should not be simplified by assuming constant volume. The expansion of Equation 4.9 to Equation 4.29 illustrates its generality. Now, if we let the volume be constant, then

$$\left[\frac{dC}{dt}\right]_0 V + \left[\frac{dV}{dt}\right]_0^{\nearrow 0} C = Q \cdot C_{in} - Q \cdot C - \left[\frac{dC}{dt}\right]_r V \tag{4.30}$$

which gives the constant volume form

$$\left[\frac{dC}{dt}\right]_0 V = Q \cdot C_{in} - Q \cdot C - \left[\frac{dC}{dt}\right]_r V \tag{4.31}$$

Expressing the differential on the left side in finite difference form

$$\left[\frac{\Delta C}{\Delta t}\right]_0 V = Q \cdot C_{in} - Q \cdot C - \left[\frac{dC}{dt}\right]_r V \tag{4.32}$$

Dividing by V gives

$$\left[\frac{\Delta C}{\Delta t}\right]_0 = \frac{Q}{V}[C_{in} - C] - \left[\frac{dC}{dt}\right]_r \tag{4.33}$$

Note that the reaction term is not expressed in finite difference form because a kinetic model must be inserted. These terms may be constant or vary with time or they may be functionally dependent on other variables. [The kinetic models are specific to the kind of situation being modeled.]

Now, since $\Delta C = C_{t+\Delta t} - C_t$, Equation 4.33 may be expressed as

$$C_{t+\Delta t} = C_t + \left\{\frac{Q_t}{V} \cdot (C_{in,t} - C_t) - \left[\frac{dC}{dt}\right]_{r,t}\right\} \Delta t \tag{4.34}$$

Equation 4.34 is general for the complete mix materials balance equation, that is, Equation 4.9, except that constant volume is assumed. For example, the flow, Q, may change with time and C_{in} may change with time. If these functions are known, then C as a function of t may be calculated, as per Equation 4.34. Figure 4.12a and b illustrates the results, that is, C, of such calculations for pulse durations, measured by t/θ, of 0.4 and 0.9, respectively. The two pulses are shown, along with the calculated C_t curves. A mathematical solution is not feasible for such conditions and therefore the approach is by the finite difference technique, as per Equation 4.34.

4.3.4.5 Spreadsheet Method to Solve Finite Difference Form of Mass Balance Equation

In applying Equation 4.34, a spreadsheet is required, with columns set up to incorporate the respective variables. Table CD4.2(a) is such a spreadsheet, but without the reaction term, that is, let $[dC/dt]_r = 0$. The reaction term requires a kinetic equation (see, e.g., Section 4.4.1).

The spreadsheet file, Table CD4.2(a), provides the means to impose any input conditions of interest (such as unsteady flow or time varying concentration). The file has two parts: (1) the trial-and-error selection of a Δt needed for the finite difference model, Equation 4.34, that is, as in Table CD4.2(a) on the left side, and (2) application of the Δt to the finite difference model in spreadsheet form, that is, as in Table CD4.2(b), on the right side. The Δt value

TABLE CD4.2

(a) Comparison between Finite Difference and Mathematical Solutions for Continuous Input of Salt Starting at $t = 0$

Finite Difference Solution

Δt (s)	t (s)	V (m^3)	Q (m^3/s)	$\theta = V/Q$ (s)	t/θ	$C_{in,t}$ (kg/m^3)	C_t (kg/m^3)	$C_{t+\Delta t}$ (kg/m^3)
0.01	0.00	1000	1000	1	0.00	1000	100.000000	109.000000
	0.01				0.01	1000	109.000000	117.910000
	0.02				0.02	1000	117.910000	126.730900

Mathematical Solution

e($-t/\theta$)	($C_{in} - C_{t=0}$) (kg/m^3)	($C_{in} - C$) (kg/m^3)	C (kg/m^3)	Diff. (kg/m^3)
1.00000	900	900.000000	100.000000	0.000000
0.99005	900	891.044850	108.955150	0.044850
0.98020	900	882.178806	117.821194	0.088806

Notes: (1) Δt is by trial and error.
(2) $t = t + \Delta t$
(3) V is a design input.
(4) Q is the flow through the reactor.
(5) q is the detention time.
(6) C_{in} is the specified salt concentration flowing into reactor.
(7) C_t is the reactor concentration at time, $t - \Delta t$.
(8) $C_{t+\Delta t}$ is the reactor concentration at time, $t + \Delta t$, calculated by finite difference equation.
(9) Mathematical solution—dimensionless.
(10) The quantity, $(C_{in} - C_t)$ is calculated as $e^{-t/\theta} \cdot (C_{in} - C_{t=0})$.
(11) $C_t = C_{in,t} - (C_{in,t} - C_t)$.
(12) Difference $= C_t$(numerical) $- C_t$(mathematical).

TABLE CD4.2

(b) Solution Finite Difference Mass Balance Equation—Continuous Input ($0.1 \leq t/\theta \leq 0.5$) of Salt[a]

Δt (s)	t (s)	V (m^3)	Q (m^3/s)	$\theta = V/Q$ (s)	t/θ	$C_{in,t}$ (kg/m^3)	C_t (kg/m^3)	$C_{t+\Delta t}$ (kg/m^3)
0.01	0.00	1000	1000	1	0.00	100	100.000	100.000
	0.01				0.01	100	100.000	100.000
	0.02				0.02	100	100.000	100.000

Notes: (1) Δt is from "calibration" with mathematical solution $t + \Delta t$.
(2) $t = t + \Delta t$.
(3) V is a design input.
(4) Q is the flow that must be processed.
(5) θ is the detention time.
(6) C_{in} is the specified salt concentration flowing into reactor.
(7) C_t is the reactor concentration at time, $t - \Delta t$.
(8) $C_{t+\Delta t}$ is the reactor concentration at time, $t + \Delta t$, calculated by finite difference equation.

[a] This table is a copy of Table CD4.2 (a) with changes to reflect the pulse loading.

is selected when the finite difference solution approaches the mathematical solution, that is, when the "difference" between the two solutions is small and the finite difference solution is "stable," meaning that the solution starts to "converge" as Δt decreases.

Table CD4.3 is an excerpt from Table CD4.2(b) showing selected columns and enough rows to illustrate that a pulse loading of salt occurs. Figure 4.13a is a plot of the output, C_t from Table CD4.2(b) for a pulse loading for the time period, $0.1 \leq t/\theta \leq 0.4$. Figure 4.13b is a plot of the same thing, but for a pulse loading for the time period, $0.1 \leq t/\theta \leq 1$.

Example 4.1 Pulse Concentration Input to Complete Mix Reactor:

1. *Statement*: Determine C versus t for a complete mix reactor with the rate of reaction equal to zero, with constant flow and for $C_{in} = 1000$ mg/L as a "pulse" function for the period, $0.1 \leq t/\theta \leq 0.5$.
2. *Solution scheme*: Impose mathematical conditions for the problem as stated, that is, $[dC/dt]_r = 0$, for which Equation 4.34 becomes

$$C_{t+\Delta t} = C_t + \left\{\frac{Q_t}{V} \cdot (C_{in,t} - C_t)\right\}\Delta t \quad \text{(Ex4.4.1)}$$

TABLE CD4.3
Solution Finite Difference Equation—Pulse Input ($0.1 \leq t/\theta \leq 0.5$) of Salt[a]

Δt (s)	t (s)	V (m³)	Q (m³/s)	$\theta = V/Q$ (s)	t/θ	$C_{in,t}$ (kg/m³)	C_t (kg/m³)	$C_{t+\Delta t}$ (kg/m³)
0.01	0.00	1000	1000	1	0.00	100	100.000	100.000
	0.01				0.01	100	100.000	100.000
	0.02				0.02	100	100.000	100.000
	0.03				0.03	100	100.000	100.000
intentional discontinuity in spreadsheet								
	0.46				0.46	1000	373.228	379.496
	0.47				0.47	1000	379.496	385.701
	0.48				0.48	1000	385.701	391.844
	0.49				0.49	1000	391.844	397.925
	0.50				0.50	100	397.925	394.946
intentional discontinuity in spreadsheet								

Notes: (1) Δt is from "calibration" with mathematical solution.
(2) $t = t + \Delta t$.
(3) V is a design input.
(4) Q is the flow through the reactor.
(5) θ is the detention time.
(6) C_{in} is the specified salt concentration flowing into reactor.
(7) C_t is the reactor concentration at, $t - \Delta t$.
(8) $C_{t+\Delta t}$ is the reactor concentration at time, $t + \Delta t$, calculated by finite difference.

[a] Printout of an excerpt from Table CD4.2 (b) for condition of "pulse" loading.

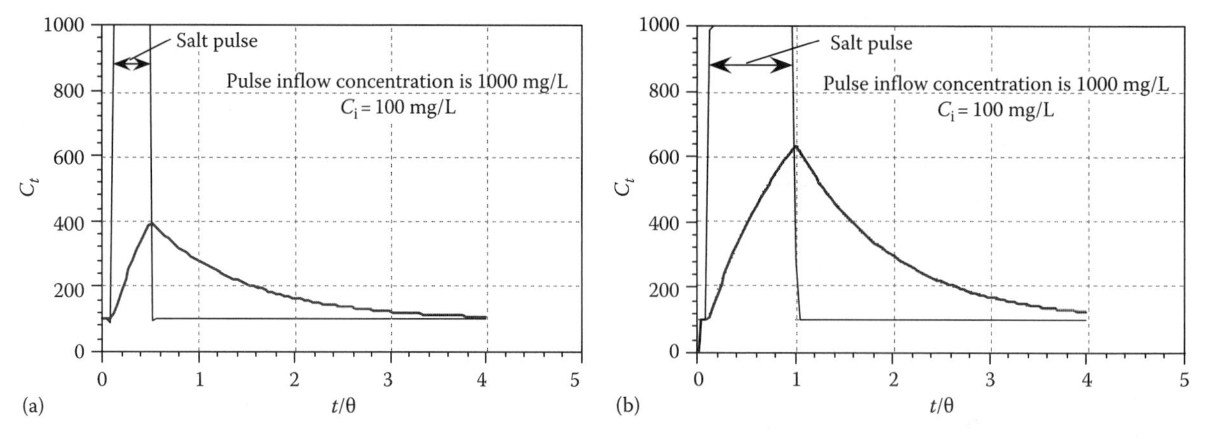

FIGURE 4.13 Pulse flow displacement of solution in complete mix reactors plotted by Equation 4.34. (a) Pulse duration: $t/\theta = 0.4$. (b) Pulse duration: $t/\theta = 0.9$.

3. *Set up spreadsheet solution*: The first step is to open the spreadsheet file, Table CD4.2(a). The second step is to make a copy of the file to solve the problem at hand. The copy with the required pulse loading imposed is shown in the file as Table CD4.2(b) (copied side-by-side with Table CD4.2(a)). Table 4.3 is an excerpt of Table CD4.2(b). Figure 4.3 is a plot of the output, that is, t versus Ct from Table CD4.2(b), seen also as in the copy, Table CD4.3.
4. *Discussion*: To determine Δt in Equation E.4.1, a trial and error solution is required such that the solution does not "blow up," that is, become unstable. Usually, picking successively smaller Δt values will soon result an appropriate value. But then if a smaller Δt is used than is required, the computing time becomes higher than necessary.

For Equation 4.34, the boundary conditions may be imposed such that solutions from the finite difference model and the mathematical model, Equation 4.28, are about the same, for example, suppose a continuous input of a salt solution at $C_{in} = 1000$ mg/L displaces a $C(\text{reactor}) = 100$ mg/L solution in the complete mix reactor. Since the mathematical solution is exact, the effect of choosing different Δt's can be seen by comparing C_t values for each solution. In the spreadsheet file, Table CD4.2(a), this may be evaluated by a column, "Diff," that shows the difference between the two solutions. A suitable Δt has been selected when the "Diff" column is within acceptable limits, for example, 2%–3%.

4.3.4.6 Utility of Finite Difference Equation and Tracer Tests

In practice, the C versus t curves are determined most often by tracer tests since often the geometric configurations are more complex than simple complete mix reactor. These tests involve adding a "slug" of brine solution (or a fluorescent dye such as rhodamine-B or ponticyl pink) at some point of injection. An example is to evaluate the effect of dispersion in a clear well in a drinking water treatment plant so that the effect on the "contact-time" for a disinfectant can be evaluated. Samples are taken downstream at the point of interest, such as at the outflow of the basin (a conductivity sensor with output to a computer provides a continuous trace of the C versus t relationship). Another example includes what to expect in sampling after seeding a pilot plant with microorganisms. Such results will help determine when the seed arrives and when it has full effect.

4.4 KINETIC MODELS

Every reactor situation, unless the reaction term is zero, requires a kinetic model. The topic of kinetics encompasses a body of knowledge; kinetics is most often the limiting factor in reactor modeling. In most cases, we lack sufficient data on the kinetic coefficients. To establish a broad enough database of kinetic coefficients, however, is probably beyond the scope of being feasible since the number of situations is almost without limit. Thus, where a kinetic model is needed, the effort is usually ad hoc to fit the needs of the case at hand.

Despite the aforementioned caveats, there are quite a few things that we can say about kinetics. Some of the fundamentals are reviewed.

4.4.1 First-Order Kinetics

The most widely used kinetic model is "first-order" that says simply that the rate of a given reaction for a disappearance of a constituent, C, is proportional to the negative of the concentration at any instant and is expressed mathematically as

$$\frac{\mathrm{d}C}{\mathrm{d}t} = -kC \tag{4.35}$$

in which

C is the concentration of a given constituent (kg/m^3)
t is the elapsed time from a given concentration, C_0 (s)
C_0 is the concentration of a given constituent at time, $t = 0$ (kg/m^3)
k is the kinetic rate constant (s^{-1})

Separation of the variables and integration gives

$$\ln \frac{C}{C_0} = -kt \tag{4.36}$$

Figure 4.14 illustrates the type of plot resulting. The slope is the kinetic constant, k.

To ascertain whether the reaction rate for a given reaction is first order, C versus t data are generated using a laboratory flask, that is, a batch reactor, and samples are withdrawn for analyses at intervals needed to delineate the relationship. Conditions must be specified and controlled. If the ln C versus t relationship is a straight line, then the reaction is first-order. This is determined by a plot of log C versus t. If the data plot is within an acceptable statistical significance, then the reaction may be accepted as being first order and the slope of the plot is the kinetic constant divided by 2.303, that is, k/2.3.

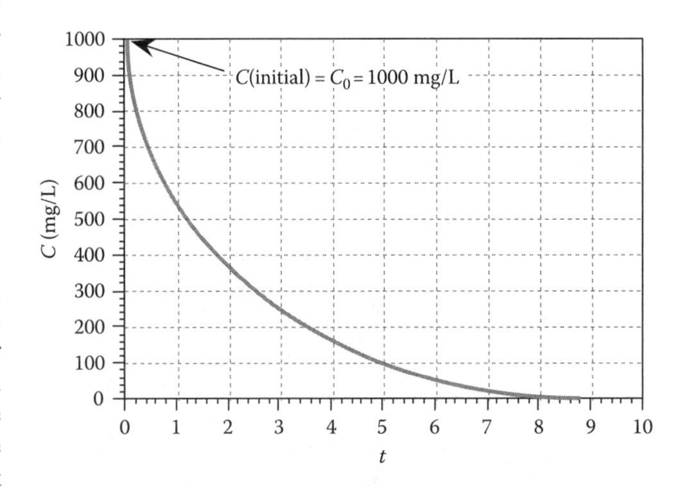

FIGURE 4.14 Illustration of a first-order kinetic relation as described by Equation 4.36.

First-order kinetics is applicable to a wide range of situations. Examples include disinfection, chemical precipitation, biological growth and degradation, chemical oxidation, etc. If the rate of transport governs, vis-à-vis the rate of reaction, that is, for diffusion into a spherical porous solid, the relation is

$$\left[\frac{dC}{dt}\right]_r = -D\frac{dC}{dr} \tag{4.37}$$

in which
D is the diffusion coefficient (m/s)
R is the radius of particle (m)

4.4.2 Second-Order Kinetics

The other widely used reaction rate description is second-order kinetics. The rate for a second-order reaction is proportional to the square of the concentration of a given reactant or the product of two different reactants, that is,

$$\frac{dC}{dt} = -kC^2 \tag{4.38}$$

$$\frac{dC}{dt} = -kC \cdot A \tag{4.39}$$

in which A = concentration of constituent A (kg/m^3) (Box 4.3).

4.4.3 Examples of Kinetic Equations

Examples of kinetic equations are in paragraphs that follow. They are intended to indicate the nature of the $[dC/dt]_r$ term in the materials balance equation and its variety of forms.

4.4.3.1 Example: Gas Transfer

The rate of gas transfer is diffusion limited and occurs across two pseudo films: (1) a gas film, and (2) a liquid film. For the liquid film, the rate is approximated

$$\frac{dC}{dt} - K_La(C^* - C) \tag{4.40}$$

> **BOX 4.3 UNITS IN CHEMISTRY**
>
> In chemistry, concentrations are expressed in equivalents per liter and moles per liter in addition to mg/L or the SI equivalent kg/m^3. The convention is that a straight bracket means moles per liter, that is, $[A]$ means concentration of A in moles per liter, for example, $[Mg^{2+}]$ = 0.02 mol/L. The brackets, $\{A\}$ refer to activity, in which $\{A\} = \gamma_A[A]$, and γ_A = activity coefficient for A. For dilute solutions $\gamma_A \rightarrow 1$ and thus $[A] \approx \{A\}$; therefore we can use concentrations for most situations in this field. An exception would be for seawater.

and

$$K_La = \frac{D}{\delta}aV \tag{4.41}$$

in which
C is the concentration of dissolved gas in the bulk of solution (kg/m^3)
C^* is the concentration of dissolved gas in solution at the gas–liquid interface in equilibrium with the gas phase (kg/m^3)
K_La is the mass transfer coefficient (s^{-1})
D is the diffusion constant for gas dissolved in water (m^2/s)
δ is the thickness of pseudo film (m)
a is the area of interface (m^2)
V is the volume of reactor (m^3)

4.4.3.2 Example: Biological Degradation of Substrate

The kinetics for a substrate degradation in a biological reaction is tied to the rate of synthesis of cells, that is,

$$\frac{dS}{dt} = -\frac{1}{Y}\frac{dX}{dt} \tag{4.42}$$

The substrate may be any compound that limits the rate of reaction, or a mix of compounds. The rate of cell synthesis is a first-order kinetic equation with respect to cells, that is,

$$\frac{dX}{dt} = \mu X \tag{4.43}$$

The kinetic rate constant, μ, is based upon the degree of saturation of microbial enzymes with a substrate, that is,

$$\mu = \hat{\mu}\frac{S}{K_m + S} \tag{4.44}$$

in which
S is the substrate concentration (kg substrate/m^3)
X is the concentration of cells (kg cells/m^3)
Y is the stoichiometric constant (kg cells synthesized/kg substrate degraded)
μ is the kinetic rate constant (s^{-1})
$\hat{\mu}$ is the maximum kinetic rate constant (s^{-1})
K_m is the half saturation constant (kg substrate/m^3)

As may be evident, the kinetic description for substrate degradation is tied to a sequence of relationships, as indicated. A key point is that substrate degradation occurs only with concurrent cell synthesis.

4.4.3.3 Example: Trickling Filter

The substrate, for example, biochemical oxygen demand (BOD), for a trickling filter varies from the top to the bottom of the filter bed, with $S = S_0$ at $Z = 0$. By definition, the filter bed is nonhomogeneous. The materials balance for an infinitesimal slice of the filter bed is

$$\left[\frac{\partial(S\cdot\Delta Z\cdot A)}{\partial t}\right]_{\mathrm{o}} = \bar{v}\cdot A\cdot S_{\mathrm{in}} - \bar{v}\cdot A\cdot S_{\mathrm{out}} - \left[\frac{\partial(S\cdot\Delta Z\cdot A)}{\partial t}\right]_{\mathrm{r}} \tag{4.45}$$

Dividing by $\Delta Z\cdot A$ and following the derivation of Equation 4.14, and substituting the kinetics of Equations 4.41, 4.43, and 4.44,

$$\left[\frac{\partial S}{\partial t}\right]_{\mathrm{o}} = \bar{v}\cdot\frac{\partial S}{\partial Z} - \frac{1}{Y}\cdot\hat{\mu}\frac{S}{K_{\mathrm{m}}+S}\cdot X(\mathrm{rock})\cdot\sigma \tag{4.46}$$

in which

- $X(\mathrm{rock})$ is the mass of active cells per unit area of rock surface (kg active cells/m^2 rock surface)
- σ is the specific surface area of rock (m^2 rock/m^3 bulk volume of media

Equation 4.45 must be solved numerically by transforming it to finite difference form, that is,

$$\frac{S_{t+\Delta t,i}-S}{\Delta t} = \bar{v}\cdot\frac{S_{i+l,t}-S_{i-l,t}}{2\cdot\Delta Z} - \frac{1}{Y}\cdot\hat{\mu}\frac{S_{i,t}}{K_{\mathrm{m}}+S_{i,t}}\cdot X\cdot\sigma \tag{4.47}$$

in which

- i is the slice number from top of column
- Δt is the time increment for numerical solution (s)
- ΔZ is the distance increment for numerical solution (m)
- $S_{i,t}$ is the substrate concentration at slice i and time t (kg substrate/m^3 pore volume)

Most of the terms, other than $S_{i,t}$, in the kinetic part of Equation 4.47 would be "lumped," for example, let $K = (1/Y)\cdot\hat{\mu}\cdot X\cdot\sigma\cdot[1/(K_{\mathrm{m}}+S_{i,t})]$ and determined by empirical means, for example, using a pilot plant or even a full-scale plant. The $S_{i,t}$ term that is "lumped" into the coefficient, must be imputed (to give the best fit). The idea would be to "fit" the pseudo kinetic coefficient to the model. The model is near-rationale and is appropriate for engineering, albeit the "lumped" coefficient limits its applicability to the range for which it was determined.

PROBLEMS

4.1 Introduction to the Materials Balance Relation

Describe a checking account or a savings account in terms that are parallel to the materials balance equation.

4.2 Calibration of Finite Difference Equation

Using Table CD4.2 (an Excel file), explore different Δt values to determine the effect on the "Diff" column. Is a lower Δt value necessary or desirable? [Keep in mind that any solution is an approximation to the real situation and cannot be exact in any case because of factors not incorporated into the model.] To explore this question, the total time elapsed should be large enough such that the simulation is realistic and trends are evident. This should be done graphically, that is, "Diff" versus t for different Δt, with the spreadsheet outputs providing the documentation.

4.3 Complete Mix Reactor

(a) Assume a pulse loading of salt in a complete mix reactor. Let the pulse loading be for the period, $0.2 \leq t/\theta \leq 0.5$. Using the setup shown in Table CD4.3 (an Excel file) which demonstrates how to employ Equation 4.24, generate the associated C versus t relationship and show graphically. Let the period of simulation be $0 \leq t/\theta \leq 5$.

(b) For the problem setup developed for (a), let the flow vary with time over the period of simulation (or any part of that period that demonstrates a point that you may have in mind). The variation can be whatever you wish. [If you wish to relate to a WWTP, as an example of a real situation, a sinusoidal variation would approximate a 24 h variation in flow to the plant. For a complete mix activated sludge tank, $\theta \approx 6$ h for the average daily flow.]

4.4 Pulse Loading of a Nonreactive Material in a Complete Mix Reactor

Based upon results from a pulse load to a rapid mix reactor using a salt solution, determine when a polymer injected into a rapid mix basin (as a step function) becomes effective.

4.5 Pulse Loading of a Nonreactive Material in a Complete Mix Reactor

An industrial waste coming into an activated sludge basin has a pulse input of a toxin. Determine the concentration-time effect of the toxin on the activated sludge basin. Assume a 6 h detention time for the basin and make any other reasonable assumptions necessary for the model.

4.6 Mathematical Solution of C versus t for Salt Loading to Complete Mix Reactor

Suppose for a complete mix reactor with zero rate of reaction, as depicted in Figure 4.10, is $C_{t=0} = 0$ mg/L, $C_{\mathrm{in}} = 1000$ mg/L at $t = 0$, and the inflow of salt is a "step" function. Show the plot of C versus t/θ. [The associated C versus t/θ tabular data should be shown also, as generated by a spreadsheet.]

4.7 Mathematical Solution of C versus t for Reactive Substance in Complete Mix Reactor

Suppose for a complete mix reactor with a specified rate of reaction is $C_{t=0} = 0$ mg/L, $C_{\mathrm{in}} = 1000$ mg/L at $t = 0$, and the inflow of a degradable organic material is a "step" function. Show the plot of C versus t/θ. [The associated C versus t/θ tabular data should be shown also, as generated by a spreadsheet.] Assume that the rate of reaction is, say, $0.2\cdot Q\cdot C_{\mathrm{in}}$ and that $\theta = 6$ h.

4.8 Trickling Filter Model

Set up on a spreadsheet the solution for a trickling filter finite difference model as depicted in Equation 4.50. For the finite difference solution, Equation 4.47, let constants be

$\hat{\mu} = 0.2\ h^{-1}$
$K_m = 6$ mg/L
$Y = 0.31$ kg cells synthesized/kg substrate degraded
$= 31.4\ m^2$ media surface area/m^3 bulk volume
$\bar{X} = 1.0$ kg cells/m^2 media surface area
porosity, $P = 0.40$.

Let design variables be:
depth of the filter, $D = 2.0$ m
$HLR = Q/A = 5.0\ gpm/ft^2 = 3.395 \cdot 10^{-3}$ m/s and recall, $\bar{v} = HLR/P$

Operating conditions are:
$S_0 = 200$ mg BOD/L, and
$S(2.0\ m) = 30$ mg BOD/L

[Note that $S(2.0)$ is a nomenclature adopted here that means the effluent concentration is at 2.0 m depth in the filter.]

(a) Assume steady state conditions and determine a "calibration" coefficient, K for the model. The value for K is determined by trial and error and is the value selected when the calculated $S(2.0)$ equals 30 mg/L (the assumed actual effluent concentration).
(b) Assume a flow variation over a 24 h period (use sinusoidal variation if you wish), and determine the $S(2.0\ m)$ versus t.

4.9 Global Atmosphere as a Reactor
Suppose that a gas (any gas) is emitted by various surface sources. Let the gas be nonreactive in the atmosphere, but will dissolve in water (in accordance with Henry's law). If the atmosphere has a uniform concentration of the gas at $x\ kg/m^3$, estimate the number of years for the atmosphere to purge itself to a concentration 0.05 fraction of the original concentration. [Hint: The rate of contact with water surfaces is probably rate limiting.] As a second aspect of the problem, consider the complexity of the movement of air masses and the assumption of homogeneity (i.e., complete mix is only a first approximation).

4.10 Global Atmosphere as a Reactor with Carbon Dioxide
Suppose that the gas in the previous problem is carbon dioxide. Develop a rough global model of a carbon balance.

GLOSSARY

Advection: The transport of molecules or mass by a fluid flow with such flow responding to a pressure gradient.

Appropriate fit: A "form" that relates to a given "context" in an appropriate manner.

Biofilm: Bacteria film that adheres to a surface.

Contactor: A reactor in which reaction occurs.

Context: The situation or setting that comprises the basis for a design.

Convection: The movement of a fluid under a pressure gradient (see *advection*). [In this text, convection was used initially and then was replaced with "advection"; in splitting hairs to decide, the scales were tipped in favor of advection; convection seemed to be more associated with heat.]

CSTR: (1) A common acronym for "completely stirred tank reactor." (2) A "complete-mix" reactor is also a CSTR.

Detention time: The residence time of a fluid in a volume of some kind in which the hypothetical "plug" flow occurs.

Diffusion: Molecular transport by random motion of molecules. Fick's first law describes the transport rate.

Dispersion: The "spread" of a substance in flow. In porous media flow and in turbulent flow, the mechanism is a statistical variation of velocity about the mean advective velocity. In flow through basins, "short circuiting" causes a portion of the flow to reach the end of the basin sooner than the "plug-flow" velocity. At the same time, a portion of the flow will exchange mass with dead zones of the basin which will cause a long tail on a dispersion curve. A dye test or a salt test is used commonly as a means to evaluate dispersion. Intuitively, if 100 molecules are transported to a point in front of or behind the mean flow due to a statistical mechanism the same statistics are applicable to those 100 molecules in the new position. In the next "step" some of the 100 will be behind the average and some will be ahead.

Fick's first law: The mathematical relation in which flux density of a given substance is proportional to the concentration gradient.

Finite difference equation: The expression of a differential equation in terms of infinitesimal differences, that is, ΔC, Δt, ΔZ,

Fit: The relationship between a "form" and its "context."

Flow net: Analysis method for "irrotational flow" and is applicable to turbulent flow where viscosity is not a major influence and to ground water flow for cases in which $\mathbf{R} \leq 1$ (which is usual). The flow net is comprised of streamlines and potential lines (or surfaces) and is unique for a particular set of "boundary conditions."

Flux: Rate of transport of mass (or heat) across a boundary (kg/s)

Flux density: Rate of transport of mass (or heat) across a boundary per unit area ($kg/s/m^2$)

Form: The kind of design synthesized that may or may not "fit" the "context" at hand.

Gaussian: The probability curve that is "normal" in shape, that is, bell-shaped; it may be the outcome of measurements of a given random variable.

Homogeneous reactor volume: The part of a reactor in which the contaminant of interest, for example, A has a constant concentration from point to point within the volume. It is important to designate because the materials balance relation is valid only for a homogeneous volume. A "complete-mix" activated sludge reactor could approach this homogeneity. A plug-flow reactor does not, by definition. The concentration of A within a packed-bed reactor, for example, an adsorption column, changes continuously along its length. An infinitesimal slice of the reactor may be assumed homogeneous and thus the materials balance may apply in differential form.

Infinitesimal volume: The small hypothetical volume used as the basis for a mathematical formulation.

Iteration: The solution of a finite difference equation within a one-step change, such as a Δt.

Kinetics: A description of the rate of a chemical reaction. Usually, the description is in terms of the rate of disappearance of a *reactant* or, alternatively, the rate of appearance of a *product*.

Kinetics, first order: The rate description which is proportional to the concentration at any given instant.

Kinetics, second order: The rate description which is proportional to the concentration squared at any given instant or to the concentrations of two constituents.

Materials balance: For a given complete-mix reactor, the observed mass rate of change of a substance within a reactor equals the mass flow in minus the mass flow out plus or minus the mass rate of reaction within the reactor summed for all reactions.

Model: A mimic of a full-scale system in which a one-to-one correspondence is sought between the system mimicked and the model. Models may be mathematical or physical.

Numerical solution: A solution to a differential equation expressed in finite difference form.

Organism, *Cryptosporidium parvum* oocyst: The survival form of the protozoan in the phylum Apicomplexa that are referred to as coccidia that affect humans. The organism is round and about 5 μm in diameter. Dubey et al. (1990) list 21 species.

Organism, *Cyclospora*: (1) A family of protozoan organisms in the subclass Coccidia, distributed worldwide. (2) *Cyclospora cayatensis*: A coccidian parasite distributed worldwide related to *Cryptosporidium parvum* about 9–10 μm in size; organism infects immunocompetent and immunosuppressed children and adults (Symons et al. 2000, p. 100, 101).

Organism, *Giardia lamblia* cyst: The survival form of a protozoan infectious to humans and is about 5×10 μm in size and is common to sewage and should be assumed present in any ambient water.

Organism, *Microspora*: (1) Phylum in subkingdom *Protozoa*: They are small microsporans, 3–6 μm, and are obligatory intracellular parasites (Prescott et al. 1993, p. 552, 556). (2) Microsporidia. Small, unicellular, obligate intracellular, spore-forming (spores 1–4.5 μm) protozoan parasites that are widely distributed in nature and include more than 100 genera and about 1000 species. They are pathogens of insects, fish, birds, and mammals, including humans. Species found in humans include *Encephalitozoon cuniculi*; *E. Hellem*; *E. intestinalis*; *Enterocytozoon bieneusi*; *Nosema conneri*; *Pleistophora*; *Trachipleistophora hominis*; and *Vittaforma corneum* (Symons et al. 2000, p. 264).

Organism, *Microsporidium*: A "catch-all" genus name for microsporidia that have not yet been classified; see also *microsporidia* (Symons et al. 2000, p. 264).

Plug flow: (1) Advective flow in which there is no turbulent diffusion or dispersion. (2) In plug flow, an element of thickness, ΔZ, and area, A, may be advected downstream with no transfer of mass across the boundaries and the mass balance equation is that the observed rate of change of C within the element is the rate of reaction that is occurring. (3) Alternatively, a volume element may be fixed in space with mass transfer occurring across the boundaries; a finite difference solution is required.

Porous media: In water treatment, a packed bed of granular media. A ground water aquifer is also porous media.

Potential: A term used in the field of ground water that is the sum of pressure head and gravity head, that is, $\phi = p/g + h$.

Potential line: A locus of points that describes a surface of constant pressure in a flow field.

Product: In any chemical reaction the substances synthesized from the reaction are called "products" (see also *reactant*).

Reactant: In any chemical reaction the substances utilized and transformed by the reaction are called "reactants" (see also *product*).

Reaction: A transformation of substance from one "state" to another "state"; boiling water would be a reaction in this rather broad definition. In a stricter sense, a reaction is the transformation of substances, that is, reactants, to substances, that is, "products," such that the products are a molecularly different chemical species than the reactants.

Reaction, adsorption: A reaction in which an adsorbent is one of the reactants.

Reactor: A volume in which a reaction occurs, or in which there is some other kind of treatment.

Reactor, batch: A volume in which the flow across its boundaries is zero.

Reactor, biological: A reactor in which one of the reactants is an organism, usually a bacteria. Also, a product may be bacteria.

Reactor, column: A reactor which is not homogeneous from top to bottom. Packed-bed reactors, such as a granular activated carbon or a trickling filter, are examples.

Reactor, complete mix: A reactor that is mixed in such a way that the contents are homogeneous throughout the volume and therefore the materials balance equation is applicable.

Reactor, continuous flow: A reactor volume in which mass crosses its boundaries continuously.

Reactor, homogeneous: A reactor that is mixed in such a way that the contents are spatially non-varying and are thus homogeneous throughout the volume and therefore the materials balance equation is applicable.

Reactor, nonsteady state: A reactor in which the mass flow in or out or both varies with time.

Reactor, packed bed: A reactor packed with some kind of media such as granular activated carbon, sand, or rocks.

Reactor, pond: A volume that may vary from one design to another, usually several hectares in area with depth varying from 1 m to perhaps 10 m, depending on the kind of pond.

Reactor, steady state: A reactor in which the mass rate of flux in and/or out is constant.

Reactor, trickling filter: A kind of biofilm reactor in which a depth of rocks, usually about 80–120 mm in size, are in a "column" form.

Sink: A site where a reaction occurs or a surface where particles are removed from the reactor. The term, "sink" is not unusual in the literature; as used here, however, it has a broader connotation that while not contrary to normal usage, it is used in a more explanatory sense.

Streamline: The locus of points that may define boundaries that confine equal flows and in which there is no advective flow across. Streamlines respond to the geometrical configurations that shape the boundary.

Tracer: A nonreactive substance that can be used to obtain concentration time curves at some point downstream in an advective flow. Salt, that is, the chloride ion of sodium chloride, is a suitable tracer since the chloride ion is nonreactive. Rhodamine-WT is another. A tracer may be used to develop isoquants of constant concentration in two- or three-dimensional flow situations.

Tracer test: The injection of a tracer at some upstream location as a "slug" or as a "step" function with the ensuing measurement downstream by a set of grab samples or by an in situ sensor.

Transport: The movement of a constituent (or a fluid mass) from one set of spatial coordinates to another.

Transport, advection: The transport of a substance by advective flow.

Transport, diffusion: The transport of a substance by the random motion of molecules.

Transport, dispersion: As a part of the dispersion that occurs in fluid flow, molecules, say dye, will be in a part of the fluid mass that has moved ahead of the mean flow, or similarly, with respect to the fluid mass retained. See also *dispersion*.

Transport, gas transfer: The transport of a gas across a gas–water interface.

Transport, sedimentation: The movement of particles under the influence of gravity to a "sink."

Transport, turbulence: The random motion of a fluid caused by energy dissipation.

Turbulence: Random motion of fluid mass from one momentum neighborhood to another causing a fluid shear, which becomes a work that is dissipated as heat.

Wave, frontal: The tracer test result caused by a "step" function input of a tracer.

Wave, pulse: The tracer test result caused by a "pulse" function input of a tracer.

REFERENCES

Beran, M. J., Dispersion of soluble matter in slowly moving fluids, Microfilm copy of unpublished PhD dissertation, Harvard University, Cambridge, MA, 1955.

Champlin, T. L. and Hendricks, D. W., Treatment train modeling for aqueous contaminants, in: *Matrix of Contaminants and Treatment Technologies*, Vol. II, Environmental Engineering Technical Report 53-2415-93-2, Department of Civil Engineering, Colorado State University, Fort Collins, CO, May 1993.

Prescott, L. M., Harley, J. P., and Klein, D. A., *Microbiology*, 2nd edn., Wm. C. Brown Publishers, Dubuque, IA, 1993.

Rifai, M. N. E., Kaufman, W. J., and Todd, D. K., Dispersion phenomena in laminar flow through porous media, Report No. 2, I.E.R. Series 90, Sanitary Engineering Research Laboratory, University of California, Berkley, CA, 1956.

Symons, J. M., Bradley, L. C. Jr., and Cleveland, T. C., *The Drinking Water Dictionary*, American Water Works Association, Denver, CO, 2000.

Part II

Particulate Separations

Screening and gravity settling are the topics of Part II. The former includes intake cribs, bar screens, and fine screens. The latter includes settling basins, flocculent settling, plate settlers, oil separators, flotation, and grit chambers. These treatment technologies may be employed to remove a wide variety of particle types that occur in different categories of water (e.g., rivers and lakes for drinking water and various kinds of wastewaters).

Particles found in water include clay, silt, sand, algae, bacteria, viruses, chemically coagulated colloids, chemical precipitates, biological flocs, organic solids, etc. Solid objects include refuse items and natural debris, that is, objects larger than those that would be considered particles. All such particles may be removed by either of these processes by applying the appropriate technology. As a note, centrifuging belongs with gravity settling, that is, the force acting on a given particle occurs within a force field.

5 Screening

Screening is the retention of particles either by a grid or longitudinal bars with openings smaller than the particles to be removed. Screens have a wide variety of forms and range from microscreens, with openings as small as 1 μm, to trash racks. The design of screens is an art that may involve selection of materials, structural calculations, mechanical appurtenances, hydraulics, cleaning, conveyance of screening wastes, disposal of screenings, and provision for maintenance. Usually, the design is done by equipment manufacturers who provide a "package," i.e., a screening "subsystem" that has a "fit" within the treatment train. The role of the designer is to select the appropriate screening equipment from a manufacturer. A knowledge of screens is necessary in order to do this properly, i.e., work with the manufacturer's representative and other technical personnel who may in some cases be from the manufacturer's headquarters.

While screening is an essential part of almost any treatment train, it is not, as a rule, an area that sparks enthusiasm. The success of a screen is not only in terms of its effectiveness in removing intended particles but in the support functions. The latter include cleaning of the screen; appurtenances for collection and removal of solids; having materials that are durable, strong, and noncorrosive; and providing a system that can be maintained easily. All of this is a part of screening technology that continues to be developed by proprietary companies. Consequently, the lore of screening technology is found primarily in the catalogs of manufacturers and is associated with some of the experienced personnel.

5.1 THEORY OF SCREENING

Screening is a method of separating particles according to size alone. The objective of the screen is to accept a "feed" containing a mixture of particles of various sizes and separate it into two fractions, an "underflow" that is passed through the screen and an "overflow" that is rejected by the screen. An ideal screen would sharply separate the feed mixture in such a way that the smallest particle in the overflow would be just larger than the largest particle in the underflow. Such an ideal separation defines a cut diameter, d_c, which marks the point of separation between fractions (McCabe and Smith, 1956, 1967, 1976, 1993).

The probability of passage of a particle through a screen depends upon the fraction of the openings relative to the gross surface area, on the ratio of the diameter of the particle to the width of an opening in the screen, and on the number of contacts per unit of flow between the particle and screen surface, and on the shear force caused by the fluid velocity relative to a resisting particle.

A means to size a screen is in term of its "capacity," i.e., the limit of the mass of material fed per unit time per unit area of the screen. Capacities may be expressed in kilograms of overflow material per hour per square meter. Screens must be rated by data from operation with the particulate matter of interest. Examples of numerical values for capacities are 0.045–0.18 metric tons/h/m^2/mm mesh size (0.05 and 0.2 tons/h/ft^2/mm mesh size) for grizzlies and 0.18–0.73 metric tons/h/m^2/mm mesh size (0.2 and 0.8 tons/h/ft^2/mm mesh size) for vibrating screens. Whether the capacity of a screen is exceeded or not depends upon the rate of feed to the unit.

Regarding names, a "grizzly" is a grid of parallel metal bars set at an incline and is stationary. A gyrating screen has a significant amplitude and a low frequency. A vibrating screen has a much higher frequency, e.g., 1800–3000 vibrations per minute, and much smaller amplitude than a gyrating screen. Usual vibrating screen installations are woven or mesh screen, inclined at a small angle to the horizontal.

5.2 TYPES OF SCREENS

For any screening task a variety of configurations exist. One type, for example, is the traveling screen consisting of wire mesh panels attached to a belt system that operates in a vertical path. The flow of water passes through the screen and debris is retained on the upstream side, which is removed by a jet of water after the panel rises on the belt to the air portion of its travel. Other configurations include bar screens, drum screens, disk screens, and microscreens (Pankratz, 1988).

Purposes are also varied. Bar screens in wastewater treatment or coarse screens in drinking-water treatment are used to protect equipment such as pumps. Trash racks in intakes for drinking-water treatment or industrial cooling water are designed to keep out debris that could clog pipes and pumps, cause nuisances, and interfere with treatment. Other screens are designed to exclude fish and crustaceans. At the other end of the screen spectrum are microscreens, which actually provide a form of treatment in that the intent is to remove small particles, such as algae and biological flocs.

5.2.1 Bar Screens

Bar Screens are found in the "headworks" of every wastewater treatment plant and are made of steel bars with openings perhaps 10–30 mm (0.4–1.2 in.). The purpose of a bar screen is to protect equipment vis-à-vis treatment (albeit treatment must occur). The bar screen should exclude large objects and rags which could clog intake pipes, flow measuring devices, fine screens, or pumps. The openings of the bar screen should

TABLE 5.1
Bar Screen Openings

Type of Bar Screen	Opening (mm)	Opening (in.)	Material Screened
Trash racks in rivers for water treatment plant intakes	80–160	3–6	Logs, timbers, stumps
Bar screen ahead of raw sewage pumps and grit chambers	50–150	2–6	Large objects, rags
Bar screen ahead of other devices or processes	20–50	0.75–2	
Comminuting	6–20	0.25–0.75	Clinging materials
Bar screen mechanically cleaned	25	1	Large objects

(a)

(b)

FIGURE 5.2 Manually cleaned bar screens for 0.44 m^3/s (1.0 mgd) plant flow. (a) Bar screen is at turn in channel; grate is for screenings. (b) Bar screen is in channel at right; grate is for screenings.

be smaller than the size of the smallest pipe or pump. The common openings of several bar screen applications are given in Table 5.1. At least two units should be installed; both to account for the variation in loading and so that one unit may be out of operation for maintenance.

5.2.1.1 Cleaning

Bar screens must be kept clean either by hand or by mechanical means. If clogging becomes too severe a backup will occur, such as in an open channel. The rapid increase in flow after hand cleaning may result in an undesirable surge. Because of this, and the need for additional labor with manually cleaned screens, mechanically cleaned bar screens are preferred for most sewage treatment plants. Figure 5.1 shows a mechanically cleaned bar screen installation.

5.2.1.2 Manually Cleaned Bar Screens

In some situations a manually cleaned bar screen may be selected. Such situations could include small plants, especially those in remote locations where obtaining parts could be a problem or if skilled mechanics are not available. In other situations, such as in some countries, economic factors could favor manual cleaning. This would include the desire to maintain funds internally and to employ local labor. Figure 5.2 is a photograph of an installation for a small plant, i.e., $Q \approx 0.44$ m^3/s (1.0 mgd).

FIGURE 5.1 Mechanically cleaned bar screen Fort Collins Wastewater Treatment Plant (WWTP). (Photo courtesy of Uma Wirutskulshai, a former student at Colorado State University, Fort Collins, CO, 1996.)

5.2.1.3 Screenings

The quantity of screenings collected in a sewage treatment plant varies with the size of bar screen openings. Figure 5.3 is based upon data from 133 plants. The approximate range of practice is shown. The finer screens may be favored if the comminutor is omitted from the headworks.

5.2.1.4 Bar Size

The selection of bar size depends on the width and depth of the screen channel, and on conditions expected in the normal

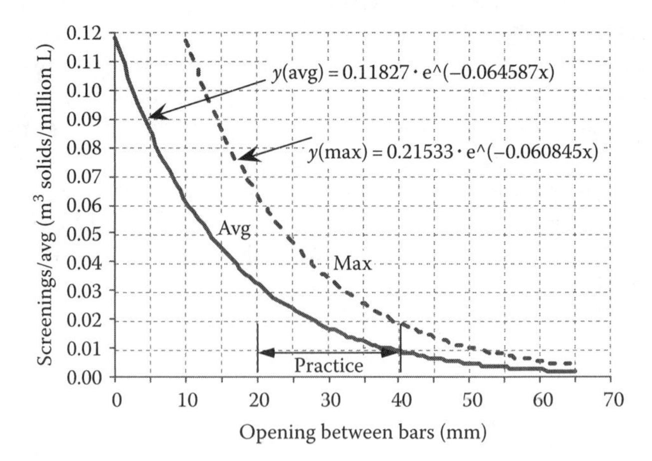

FIGURE 5.3 Quantity of screenings—in a sewage treatment plant for different size bar screen openings. (From Rexnord, 1955, Data Sheet No. 315-2.307; now 2010, WSG & Solutions, Montgomeryville, PA. With permission.)

operation of the screen. According to Rex Chainbelt (1955), later Rexnord, then Envirex, it is standard practice to use 8 mm × 80 mm (5/16 in. × 2 in.) bars for bars up to 1.83 m (6 ft) in length or 95 mm × 64 mm (3/8 in. × 2 1/2 in.) bars for bars up to 3.66 m (12 ft) in length. The freeboard allowance should be at least 9 in. for the bar protrusion above maximum sewage level.

5.2.1.5 Hydraulic Design

Velocities through a sewage treatment plant bar screen rack in a clean condition should not exceed 0.61 m/s (2 ft/s) for normal sewage flows or 0.91 m/s (3 ft/s) for storm water flows (Rexnord, 1955). A key concern is to minimize the carry through of retained material. These velocities are the basis for determining the gross channel area (see Example 5.1). In addition, the headloss through the bar screen must be calculated to determine whether or not appreciable backup will occur in the sewer. Equation 5.1 is recommended to calculate the headloss through a bar screen (Rexnord, 1955). Example 5.1 illustrates the calculation procedure and Table CD5.2 translates the procedure into a spreadsheet algorithm.

$$h_L = \frac{1}{0.7} \cdot \frac{v_s^2 - v_c^2}{2g} \tag{5.1}$$

where
- h_L is the headloss across bar screen
- v_s is the velocity through the rack (m/s), (ft/s)
- v_c is the velocity in channel above rack (m/s), (ft/s)
- g is the acceleration due to gravity (9.81 m/s^2), (32.2 ft/s^2)

Example 5.1 Bar Screen Design (Adapted from Rex Chainbelt, 1955)

1. Assume

$$Q(\text{max sewage flow}) = 0.176\,\text{m}^3/\text{s} \;[=4\,\text{mgd} = 4 \cdot 1.547 = 6.2\,\text{ft}^3/\text{s}]$$

$$Q(\text{max storm flow}) = 0.308\,\text{m}^3/\text{s} \;[=7\text{mgd} = 7 \cdot 1.547 = 10.8\,\text{ft}^3/\text{s}]$$

2. Net area, A_n, of bar rack for sewage flow condition is

$$A_n(\text{max sewage flow}) = \frac{Q(\text{max sewage flow})}{V_s(\text{max sewage flow})} = \frac{0.176\,\text{m}^3/\text{s}}{0.61\,\text{m}^3/\text{s}} \left[= \frac{6.2\,\text{ft}^3/\text{s}}{2.0\,\text{ft/s}}\right] = 0.29\,\text{m}^2 \;[=3.1\,\text{ft}^2]$$

3. Net area, A_n, for storm flow condition is

$$A_n(\text{max storm flow}) = \frac{Q(\text{max storm flow})}{V_s(\text{max storm flow})} = \frac{0.308\,\text{m}^3/\text{s}}{0.91\,\text{m/s}} \left[= \frac{10.8\,\text{ft}^3/\text{s}}{3.0\,\text{ft/s}}\right] = 0.34\,\text{m}^2 \;[=3.6\,\text{ft}^2]$$

TABLE CD5.2
Bar Screen Design Based on Hydraulic Criteria

Criteria

v(sewage) Š = 0.61	m/s
v(storm) Š = 0.91	m/s

Given

Q = 4.0	mgd
Bars 51	mm deep
25	mm opening
8	mm bar width
Ratio(free area) = 0.76	

Q(sewage)		Q(storm)		A (net/sew)	A (net/storm)	A (net/sew)	A (gross)	V (screen)	V (channel)		h_L (1/2 Plug) (m)
(mgd)	(m^3/s)	(mgd)	(m^3/s)	(m^2)	(m^2)	(m^2)	(m^2)	(m/s)	(m/s)	h_L (m)	
1.00	0.044	4.0	0.175	0.07	0.19	0.19	0.25	0.91	0.17	0.058	0.239
2.00	0.088	5.0	0.219	0.14	0.24	0.24	0.32	0.91	0.28	0.055	0.236
4.00	0.175	7.0	0.307	0.29	0.34	0.34	0.44	0.91	0.39	0.049	0.230
8.00	0.350	10.0	0.438	0.57	0.48	0.57	0.76	0.61	0.46	0.012	0.093

A(net/sew) = Q/v(sewage)   Channel area Q/A(gross)
A(net/storm) = Q/v(storm)   Max vel. $= (v_s^2 - v_c^2)/2g/.7$
Select larger A

The larger of the two net areas should be used, i.e., for the case at hand,

$$A_n(\text{max storm flow}) > A_n(\text{max sewage flow}),$$

and therefore,

$$A_n = 0.34\,\text{m}^2\ [=3.6\,\text{ft}^2]$$

4. The gross area is calculated as follows:
First, select a rack consisting of 51 mm × 8 mm (2 in. × 5/16 in.) bars space to provide a clear opening of 25 mm (1 in.).
The ratio of free area to gross area is

$$\frac{25\,\text{mm}}{25\,\text{mm} + 8\,\text{mm}} = 0.76\ \left[= \frac{1\,\text{in.}}{1\,\text{in.} + (5/16)\,\text{in.}}\right]$$

Therefore, the gross projected area of the bar rack, A_G, is

$$\begin{aligned} A_G &= \frac{A_n}{0.762} \\ &= \frac{0.34\,\text{m}^2}{0.76}\ \left[= \frac{3.6\,\text{ft}^2}{0.76}\right] \\ &= 0.45\,\text{m}^2\ [=4.74\,\text{ft}^2] \end{aligned}$$

5. Velocity, v_c, in channel

$$\begin{aligned} v_c &= \frac{Q(\text{max storm flow})}{A_G} \\ &= \frac{0.308\,\text{m}^3/\text{s}}{0.45\,\text{m}^2}\ \left[= \frac{10.8\,\text{ft}^3/\text{s}}{4.74\,\text{ft}^2}\right] \\ &= 0.68\,\text{m/s}\ [=2.3\,\text{ft/s}] \end{aligned}$$

6. Headloss, h_L, through the rack

$$\begin{aligned} h_L &= \frac{1}{0.7} \cdot \frac{v_s^2 - v_c^2}{2g} \\ &= \frac{1}{0.7} \cdot \frac{(0.91\,\text{m/s})^2 - (0.68\,\text{m/s})^2}{2 \cdot 9.806\,\text{m/s}^2} \\ &\left[= \frac{1}{0.7} \cdot \frac{(3.0\,\text{ft/s})^2 - (2.3\,\text{ft/s})^2}{2 \cdot 32.2\,\text{ft/s}^2}\right] \\ &= 0.026\,\text{m}\ [=0.082\,\text{ft}] \end{aligned}$$

7. Headloss through half-plugged rack
If the net area of the screen is cut in half, the velocity must double. The headloss under the half-plugged condition is

$$\begin{aligned} h_L &= \frac{1}{0.7} \cdot \frac{v_s^2 - v_c^2}{2g} \\ &= \frac{1}{0.7} \cdot \frac{(2 \cdot 0.91\,\text{m/s})^2 - (0.68\,\text{m/s})^2}{2 \cdot 9.806\,\text{m/s}^2} \\ &\left[= \frac{1}{0.7} \cdot \frac{(2 \cdot 3.0\,\text{ft/s})^2 - (2.3\,\text{ft/s})^2}{2 \cdot 32.2\,\text{ft/s}^2}\right] \\ &= 0.21\,\text{m}\ [=0.68\,\text{ft}] \end{aligned}$$

This amount of headloss is not desirable because of surges caused by removing the material on the rack, and also because of resulting slower velocities in the sewer, which permits the deposition of settleable material. Therefore, the cleaning cycle should assure that such conditions do not occur.

5.3 COMMINUTORS

A comminutor is a bar screen with a cutting device aligned with the bar. The blades shred the retained material, usually stringy items, allowing it to pass. In most installations the screens are circular in shape, although a rack may be used. A typical comminutor is shown in Figure 5.4. The unit is located usually after the grit chamber, as a part of the "headworks" of a wastewater treatment plant. Any installation should be designed for the maintenance or repair of one or more units. A bypass channel, usually with a manually cleaned bar rack, is recommended to handle contingencies.

5.3.1 Design

Ordinarily, a manufacturer's catalog is used to select a comminutor. The particular sizing depends on the flow capacity for a given unit and how many units are desired for a given headworks flow. The number of units equals maximum plant flow divided by flow capacity for given unit. The flow capacity depends upon the diameter of the unit and the water levels desired for operation, the latter being determined by rating curves.

FIGURE 5.4 Photograph of comminutor installation in a small sewage treatment plant, i.e., 0.44 m^3/s (1.0 mgd).

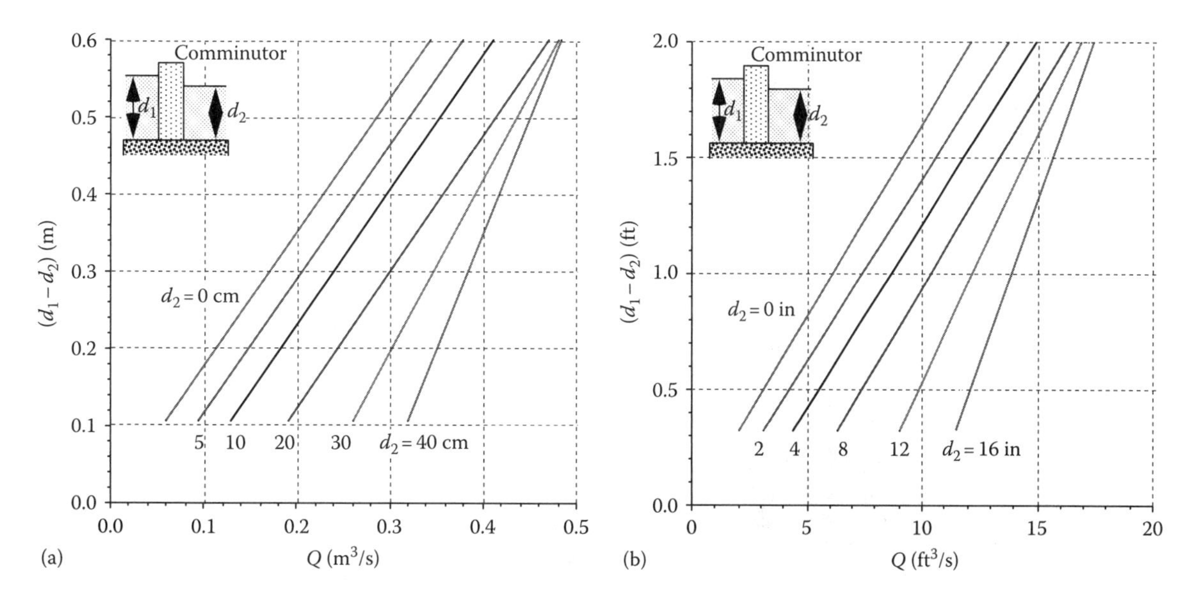

FIGURE 5.5 Rating curves for 16 in. Lyco comminutor. (a) Metric units, (b) U.S. customary units. (Courtesy of Lyco Systems, Inc., Williamsport, PA, c. 1969).

Figure 5.5 is a set of rating curves for an installation similar to that of Figure 5.4. For the curves shown, h_1 is the upstream water depth from the comminutor and h_2 is the downstream water depth. For a given flow, Q, the $(h_2 - h_1)$ value can be determined from Figure 5.5. The downstream depth, h_2, is controlled by the structures downstream, such as a Parshall flume. From this depth, h_2, the upstream depth, h_1 can be calculated from the $(h_2 - h_1)$ value from the rating curve. This permits the hydraulic profile to be established (i.e., for the part of the hydraulic profile affected by the comminutor) for any flow scenario of interest.

The rating curves of Figure 5.5 are for illustration as the specific curves depend upon the size of the unit, usually expressed as diameter, and on the characteristics from the manufacturer. Basically, the flow through the comminutor is the velocity, as given by the orifice equation, times the area of the slot; the total flow is the sum of the flows through each slot. If the comminutor is submerged fully, the flow would be simply, $Q = C(2g)^{0.5}(h_2 - h_1)^{0.5} \cdot A(\text{slots})$. If the comminutor is partially submerged then the calculation is more complex, but still follows the same principle.

Example 5.2 Comminutor Design

1. Problem: For a sewage flow of 7 mgd (10.8 ft^3/s), design a comminutor system. For this problem, in order to illustrate the procedure, flow variation, i.e., the diurnal change seen in municipal wastewater treatment plants will not be considered.

2. Approach: Select a comminutor for which headloss relationships are provided by the manufacturer. Lyco Systems provided such data in their catalog for several sizes of comminutors (Anon, 1969). Figure 5.5 is one such headloss curve for a 16 in. diameter unit. Entering the curve with $Q = 10.8$ ft^3/s, gives $(h_1 - h_2) = 0.65$ ft for $h_2 = 12$ in. Or, $h_1 = 20$ in.

5.4 FINE SCREENS

While bar screens and comminutors are designed to protect equipment only, fine screens are for the purpose of treatment. Fine screens are defined as screens with size openings 12 mm (0.5 in.) or less (Pankratz, 1988, p. 167) and may be used for both drinking-water and wastewater treatment applications. The latter includes primary treatment in lieu of clarifiers and solids recovery in industrial process streams. The configurations for fine screens include rectangular elements mounted on a vertically oriented belt, rotary screens, disk screens, and static screens (Pankratz, 1988, pp. 167–181).

5.4.1 Drum Screens and Disk Screens

The rotary screens represent one of the advances in screening technology wastewater. The fabric is wedge-wire stainless steel with slot openings 0.25–2.5 mm (0.010–0.10 in.). The screens are both externally fed and internally fed and have been used to replace clarifiers with a BOD reduction of about 35% (Pankratz, 1988, p. 168). With internally fed cylinders the solids leave the cylinder after a tumbling action at the bottom, in a dewatered condition. Figure 5.6 is a photograph that illustrates the operation of a drum screen at a wastewater treatment plant.

Disk screens have been used at the intakes of municipal water supplies in order to exclude debris and fish. For low turbidity waters, such screening along with chlorination has been the only treatment; such cases were not uncommon in decades before about 1990.

5.4.2 Wedge-Wire Static Screens

A wedge-wire screen is one with a wire made of V-shaped strands with length perpendicular to the flow. They have

(a)

(b)

FIGURE 5.6 Drum screen, internally fed, used in lieu of primary clarifier. (a) Side view of drum. (b) Perspective view of drum. (Courtesy of Centennial Water and Sanitation District, Highlands Ranch, CO.)

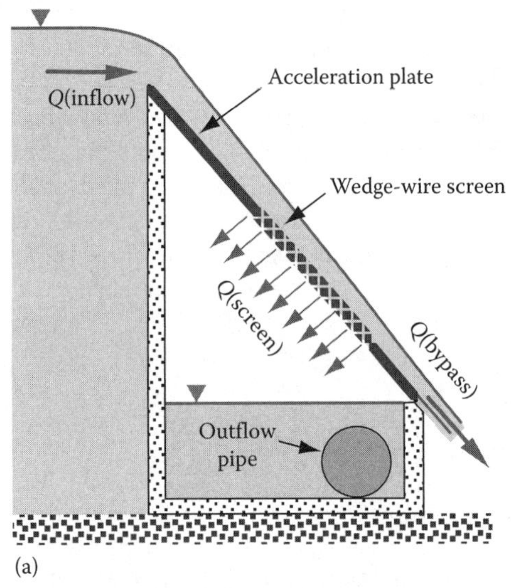

(a)

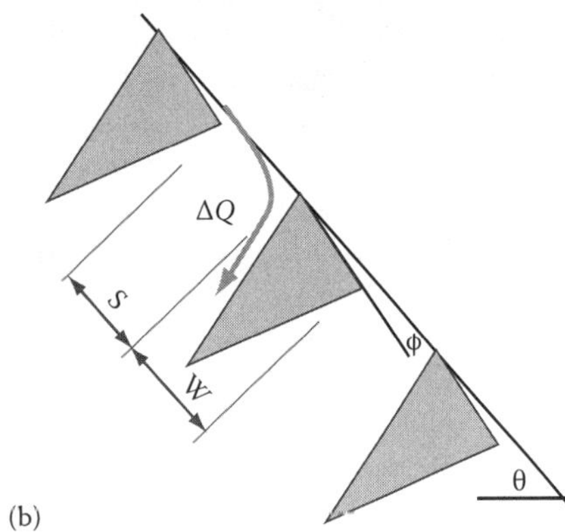

(b)

FIGURE 5.7 Wedge screen schematics. (a) Cross section showing setup, (b) cross section showing detail of wire and Coanda effect flow. (Adapted from Wahl, T.L., *J. Hydraul. Eng. Am. Soc. Civil Eng.*, 127(6), 2001, available from http://www.usbr.gov/pmts/hydraulics_lab/ twahl/coanda/, 16 pp.)

been used in mining since 1955, and for many years in food processing and wastewater treatment (Wahl and Einhellig, 2000, p. 2, Wahl, 2001, p. 1). An evolution in the design, introduced by patent in 1983, changed the orientation of the individual wire strands such that the wire surface was "tilted" downstream at a slight angle; at the same time, the functioning of the screen was changed (as explained subsequently).

Figure 5.7a is a side-view schematic drawing as an installation might be configured and Figure 5.7b is a side-view detail of the wire and flow and shows the "tilt" angle, ϕ, of the individual wires. Figure 5.8 is a photograph of an installation at Empire, Colorado, with a population of about 500, located at its intake in a diversion structure at Mad Creek that provides water to its slow sand filter.

The merit of the wedge-wire screen is that it is "self-cleaning," i.e., there is no accumulation of matter on the surface. The high velocity of water across the screen, v(bypass), transports most debris away from the slot openings to exit the screen; therefore, it is necessary that $Q(\text{bypass}) \gg 0$. The screens may be flat or concave-down with a radius of curvature 3–4 m. The hydraulic characteristics of the screens have been investigated by the U.S. Bureau of Reclamation in Denver (see Wahl, 1995, 2001; Wahl and Einhellig, 2001).

To expand on the self-cleaning feature of the screen, the tangential flow or the water and debris across the face of the screen causes about 90% removal of the debris that is 50% of the screen wire spacing, i.e., its slot width. For example, a 0.5 mm wire spacing will remove 90% of all matter larger than 0.25 mm. This feature is further enhanced by the fact that particles and debris have mass and momentum as they flow down the face of the screen and tend to continue in the direction that they are moving. The abrupt change in the fluid flow direction caused by the shearing action of the wires enhances the self-clean feature of the screen (Weir, 2002).

FIGURE 5.8 Wedge-wire screen installation (Hydroscreen™) at Mad Creek for slow sand filter at Empire, Colorado (persons in photo are Robert Weir, consultant and Julie Holmes, Town of Empire. With permission). $A(\text{screen}) = 1.07\ \text{m}^2$ for $Q(\text{screen}) \approx 0.145\ \text{m}^3/\text{s}$ (3.3 mgd).

5.4.2.1 Mathematical Relationships

From Figure 5.7a, the screen is installed on an inclined plane (about 60° recommended) just below an "acceleration plate." The flow balance, seen from the drawing, is

$$Q(\text{inflow}) = Q(\text{screen}) + Q(\text{bypass}) \tag{5.2}$$

and the flow through the screen is

$$Q(\text{screen}) = v(\text{screen}) \cdot A(\text{screen}) \tag{5.3}$$

where
- Q(inflow) is the inflow of raw water to screen (m^3/s)
- Q(screen) is the raw water flow passing through the screen (m^3/s)
- Q(bypass) is the raw water flow passing over the screen surface and leaving (m^3/s)
- v(screen) is the apparent velocity of raw water through the screen (m/s)
- A(screen) is the total area screen surface (m^2)

Other variables, seen in Figure 5.7b, are: the width of the wire, w, the slot opening, s, the angle of inclination of the screen, θ, and the inclination of the plane of the top wire surface with respect to the plane of the screen, ϕ; the flow through a single slot opening is ΔQ. To clarify the definition of A(screen), if the number of wire elements is n(wire elements) and if the width of the screen is w(screen), then $A(\text{screen}) = w(\text{screen}) \cdot n(\text{wire elements}) \cdot (s + w)$.

5.4.2.2 Theory

As stated by Wahl (1995, p. 2), each V-shaped wire is tilted at an angle, $\phi \approx 5°$, giving a tilt so that the upstream edge is offset to the flow, as shown in Figure 5.7b. A thin layer of the flow is thus sheared off at the bottom, which means, at the same time, that there is no boundary layer and hence no friction. The mechanical shearing action of the leading edge of each of the tilted wires is enhanced by the Coanda effect (after Henri-Marie Coanda who observed the phenomenon in 1910), which is the tendency of a fluid jet to remain, attached to a solid boundary.

Due to this effect, which is prevalent at supercritical velocities, the flow remains attached to the top surface of a given upstream wire and is directed to hit the face of the next downstream wire (Wahl and Einhellig, 2000, p. 3). For subcritical velocity, v(slot) is calculated by the orifice equation, i.e., is proportional to the square root of the depth of water above the slot. Thus, a portion of the flow is directed down through the slot opening of width, s. The incremental discharge, ΔQ, through each opening is a function of the flow velocity and the thickness of the sheared water layer. The velocity over the screen depends, in turn, on the elevation drop from the crest of the acceleration plate to the screen.

The variables that affect Q(screen) are (Wahl and Einhellig, 2000)

$$\Delta q(\text{screen}) = \mathbf{F}, s, w, \phi, \theta, H, \text{ screen arc radius} \tag{5.4}$$

$$= Cs[2gH]^{0.5} \tag{5.5}$$

where
- $\mathbf{F}$ is the Froude number (dimensionless)
- H is the specific energy at a given slot location (m)
- C is the discharge coefficient (dimensionless)
- q(screen) is the flow per unit width of screen ($m^3/s/m$)

The discharge coefficient (two are lumped here to simplify the discussion) is dependent on **F** and the geometric variables, with relationships given by Wahl (2001) from experimental data and computations. Since **F** changes along the length of the screen, Q(screen) must be computed slot by slot and summed.

A less accurate but easier-to-apply approach to determine v(screen) is to use a relationship from empirical data as given by Wahl (1995, p. 5) for specific conditions, i.e., for an arc screen, $w = 1.52$ mm, $s = 1.0$ mm, i.e.,

$$v(\text{screen}) = a + b \cdot q \tag{5.6}$$

where
- a = intercept of experimental curve (m/s) = 0.71 m/s for data of Wahl (1995, p. 4)
- b = slope of experimental curve (dimensionless) = −1.83 for data of Wahl (1995, p. 5)
- q = specific flow to screen ($m^3/s/m$ screen width)

Equation 5.6 provides an estimate of v(screen) for the stated conditions. For a screen with $s = 0.5$ mm, v(screen) is reduced about 18% (Wahl, 2001, p. 13). [Wahl's data were given in terms of q (flow per unit width of screen) for an arc screen of length 0.457 m (1.5 ft); v(screen) was calculated as $v(\text{screen}) = q/w(\text{screen})$.]

5.4.2.3 Design

Table 5.3 summarizes data from Wahl (2001) as may be useful for an initial estimation of design variables, such as sizing the screen, setting the angle, θ, determining the total head drop, and in selecting a fabric. The steep angle, θ, serves two purposes: (1) to cause supercritical velocity, i.e., $\mathbf{F} \gg 1$, and (2) to, in turn, have a velocity high enough to ensure that the screen is self-cleaning. The maximum practical screen dimension (one piece) is 2.4 × 5.5 m (8 × 18 ft); screens are usually fabricated, however, in smaller sections and bolted together (Hydroscreen, 2002). Also the screens are usually designed to accommodate the required flow and existing conditions of available head and installation footprint size.

TABLE 5.3
Typical Design Data for Wedge-Wire Screens

	Dimensions	Low	Nominal	High
v(screen)[a]	m/s	0.25		0.53
Velocity down screen face	m/s		2–3	
F		2		30
Total head drop across structure	M		1.2–1.5	
w	mm		1.52	
s[b]	mm	0.3	1.0	13
ϕ	°	3	5	6
θ	°		60	

Sources: Wahl, T.L., Hydraulic testing of static self-cleaning inclined screens, in: *1st International Conference on Water Resources Engineering*, American Society of Civil Engineering, San Antonio, TX, August 14–18, 1995, available from Web site: http://www.usbr.gov/pmts/hydraulics_lab/twahl/coanda/; Wahl, T.L., *J. Hydraul. Eng. Am. Soc. Civil Eng*., 127(6), June, 2001, available from Web site: http://www.usbr.gov/pmts/hydraulics_lab/twahl/coanda/, 16 pp.

[a] Calculated by upper and lower limits of experimental plot of Wahl (1995) from which Equation 5.6 was derived.

[b] Higher and lower values from Hydroscreen (2002).

5.5 MICROSCREENS

Microscreens are a special category of fine screens which have fabric openings of microns size. The removal mechanism is straining based upon the size of opening in the fabric. But like many screens, the retained material which forms a mat functions to strain particles, perhaps smaller than the microscreen openings.

Table 5.4 lists some examples of microfabrics which are commercially available. A variety of metals and plastic are used to make the fabrics; carbon steel is common (Burns and Roe, 1971).

TABLE 5.4
Microscreens—Sizes and Manufacturers

Opening Size (Microns)	Density of Mesh (No./in.^{2})	Density of Mesh (No./cm^2)	Manufacturer
23	144,000	22,320	Crane Co., King of Prussia, Pennsylvania
25			Walker Equipment Co., Chicago, Illinois
35	80,000	12,400	Crane Co., King of Prussia, Pennsylvania
35	120,000	18,600	Zurn Industries, Inc., Erie, Pennsylvania
40			Walker Equipment Co., Chicago, Illinois
60	58,500	9,067	Crane Co., King of Prussia, Pennsylvania

Source: Burns and Roe, Inc., *Process Design Manual for Suspended Solids Removal*, US Environmental Protection Agency, Washington, DC, October, 1971.

5.5.1 Equipment and Installation

A proprietary microstrainer is constructed with a fabric covering a steel drum support frame, which rotates. In operation, raw water enters the interior of the drum and passes through the fabric with a loss of head; a weir at the end of the basin on the outer side maintains the effluent-side water level. The influent-side water level may rise to the level required in order that the flow will pass, i.e., so that there is sufficient headloss. As the unit rotates, trapping suspended matter from the feed stream, the fabric is backwashed by a jet above the drum on the outer side. The weave and shape of the individual fabric wires permit the water from the jets to penetrate and detach the solids mat, which forms on the inside of the screen.

5.5.2 Applications

Microscreens have been in use since about the early 1950s (Burns and Roe, 1971). Frequent applications have been for algae removal in both water treatment and in effluent from wastewater stabilization ponds. In conventional wastewater treatment microscreens have been used following secondary treatment, especially to help ensure that effluent standards are met.

5.5.3 Performance

The effectiveness of a microscreen in removal of solids depends upon the mesh size and on the material being removed that forms the filter mat. Removals of suspended solids and BOD are given in Table 5.5 for tertiary treatment applications for two mesh sizes (Burns and Roe, 1971). The BOD removals shown are associated with the suspended solids removal. The mat of previously trapped solids provides a finer filtration or straining capability; undoubtedly this accounts for the high suspended-solids removal. Table 5.5 is indicative of performance for tertiary treatment. Removals, for a given mesh size and suspension, can be assessed by bench scale tests.

In addition to effectiveness in removal of particles, performance is based also on operating factors such as hydraulic loading rate (HLR), requirements for cleaning, and maintenance factors. An evaluation of such performance factors as cleaning and maintenance must be done by means of records examination of full-scale installations. For HLR determination, pilot plant testing is recommended at the site of the installation.

5.5.4 Operation

Some operating problems are, depending on the application, screen clogging by slimes, iron or magnesium buildups, and perhaps oil and grease (Burns and Roe, 1971). Units must be taken out of service on a regular basis for cleaning when clogging occurs. Cleaning may be done by a chlorine solution for slimes, and acid solution for iron or magnesium,

TABLE 5.5
Microscreener Performance in Tertiary Treatment

Location	Screen Size (μm)	Plant Size (mgd)	Plant Size (m^3/day)	SS (% rem.)	BOD (% rem.)	Backwash (%)	Manufact.
Luton[a]	35	3.6	13,680	55	30	3	Crane
Bracknell[a]	35	7.2	27,320	66	32	NA	Crane
Harpendon[a]	35	0.3	1,140	80	NA	NA	Crane
Brampton[b]	23	0.1	380	57	54	NA	Crane
Chicago	23	2.0	7,600	71	74	3	Crane
Lebanon, OH	23	Pilot		89	81	5	Crane
Lebanon, OH	35	Pilot		73	73	5	Crane

Sources: Burns and Roe, Inc., *Process Design Manual for Suspended Solids Removal*, US Environmental Protection Agency, Washington, DC, October, pp. 8–11, 1971.

[a] Luton, Bracknell, Harpendon are in England.

[b] Brampton is in Ontario, Canada.

and hot water and/or steam for oil and grease, but with limitations based upon the screen material.

Recommended headloss through a microscreening unit is about 300–450 mm (12–18 in.). Headloss may be reduced by increasing the rate of drum rotation and by increasing the pressure and flow of the backwashing jets, i.e., maintaining a cleaner screen. Backwashing jets usually require 1%–5% of the throughput flow. Manifold pressure depends upon the nozzle flow desired and nozzle design, but 140–400 kPa (20–60 psi) are indicative.

5.5.5 Sizing

Both solids loading and hydraulic loading determine the size of unit required. Criteria are sparse in the literature, but a maximum solids loading was given by Burns and Roe (1971) as 4.3 kg/day/m^2 (0.88 lb/day/ft^2) for an activated sludge secondary effluent. HLR depends upon several variables, such as the particles being removed, solids loading, rotational velocity, and mesh size.

Table 5.6 lists microscreen capacity ranges for different drum diameters and widths. Sizes for matching motors for drums and backwash pumps are also given. Specific sizing is available from manufacturer's literature.

TABLE 5.6
Microscreen Sizes and Motors, as Related to Capacities

Drive Sizes (mm) Diameter	Drive Sizes (mm) Width	Motors (W) Drive	Motors (W) Wash Pump	Drum Capacity (m^3/s)
(a) Metric				
1524	305	373	746	0.0044–0.022
1524	914	560	2238	0.0132–0.066
2286	1524	1492	3730	0.035–0.176
3048	3048	3730	5595	0.132–0.44

Drive Sizes (ft) Diameter	Drive Sizes (ft) Width	Motors (bhp) Drive	Motors (bhp) Wash Pump	Drum Capacity (mgd)
(b) U.S. Customary[a]				
5.0	1.0	0.50	1.0	0.1–0.5
5.0	3.0	0.75	3.0	0.3–1.5
7.5	5.0	2.00	5.0	0.8–4.0
10.0	10.0	5.00	7.5	3.0–10.0

[a] Burns and Roe, Inc. (1971, pp. 8–11).

5.5.6 Operating Data

Table CD5.7 shows an excerpt of operating data from 33 microscreen installations in the United States (Envirex, 1985). Table CD5.7 describes, for each installation, the size of the units, the application (i.e., coarse removal, water treatment, polishing lagoon effluent, polishing secondary effluent), the flow, the material, and the size of microscreen openings.

The data were compiled further on an Excel spreadsheet, Table CD5.7, and then calculations were made for each installation to estimate the respective HLRs for each installation. From the HLRs the coefficients, K, were calculated for the equation, $\mathrm{HLR} = [K \cdot \omega \cdot h_L]^{1/2}$ (see Example 5.3) using assumed values of h_L and ω (e.g., $h_L = 0.3048$ m and $\omega = 1.05$ rad/s, which are representative of practice). The outcome is shown as frequency plots for HLR and K coefficients, seen in Figure 5.9a and b, respectively.

The data for the different mesh sizes plot on the same frequency curves in Figure 5.9, except that the data for the 1 μm mesh size was markedly lower. The plots showed 1 log HLR variation for $10 \le P \le 90\%$, i.e., $0.001 \le \mathrm{HLR} \le 0.01$ for the 6, 21, 74 μm screens and about 0.3 log, i.e., $0.0001 \le \mathrm{HLR} \le 0.0003$ $m^3/s/m^2$ for the 1 μm screen. In other words, the HLRs as shown in Figure 5.11a represent the range of practice for the statistical sample.

A more useful parameter is K in that this coefficient permits the application of the equation, $\mathrm{HLR} = [K \cdot \omega \cdot h_L]^{1/2}$, (see Example 5.3) as a mathematical model. In other words, knowing K, one can explore the effect of ω and of h_L on HLR.

TABLE CD5.7
Microscreen Coefficient and Subsequent Use of Coefficient for Design (Data from Envirex, 1985)—Excerpt Showing 24 Out of 40 Columns and 5 Out of 39 Rows

Start of spreadsheet (columns a to f):

Plant Data

Plant	Purpose	No Units	Material	Size (μm)	Size (mesh)
Marin City, CA	Coarse removal	2	Polyester		60
Berthoud, CO	Water treatment (algae)	1	Polyester	35	
Sterling, CO	Polishing lagoon effluent	5	Polyester	1	
Ft. Meyers, FL	Polishing secondary effluent	2	Polyester	21	
	discontinuity between rows 4 to 37				
Superior, WI	Polishing secondary effluent	2	Polyester		50

continuation of table (columns g to p):

Calculation of HLR for Different Plants

Q (mgd)	*Q* (gal/min)	*Q* (m^3/s)	*D* (ft)	*D* (m)	*L* (ft)	*L* (m)	*A*(gross screen) (ft^2)	*A*(gross screen) (m^2)	Submergence[a] (%)
10	6945	0.440	8	2.44	6	1.83	150.8	14.0	70.0
15	10417	0.660	10	3.05	10	3.05	314.2	29.2	68.3
0.7	486	0.031	12	3.66	16	4.88	603.2	56.0	70.4
1.8	1250	0.079	10	3.05	10	3.05	314.2	29.2	68.3
				discontinuity between rows 4 to 37					
6.5	4514	0.286	6	1.83	8	2.44	151	14.0	70.4
									Calculated[a]

continuation of table (columns ag to an):

Design of Microscreen of Horsetooth Reservoir Water

Scenario Screen Size	K^b	h_L^c (m)	w^c (rad/s)	HLR[a] ($m^3/m^2/s$)	Q^c (mgd)	Q^c (m^3/s)	*A*(net)[a] (m^2)
6	0.00036	0.30	0.10	0.00339	7	0.31	90.8
	0.00036	0.30	0.21	0.00479	7	0.31	64.3
6	0.00036	0.61	0.10	0.00479	7	0.31	64.2
			discontinuity between rows 4 to 37				
	0.00036	2.13	0.10	0.00897	7	0.31	34.3
	Selected[b]	Assumed[c]					

[a] Calculated.
[b] Selected.
[c] Assumed.

While Figure 5.9 provides data from practice, the distribution covers about 1 log for HLR and about 1.5 log for *K*, which is too much variation to be useful for design. Therefore, obtaining data from a full-scale plant that is similar to the one being considered for design in character of water to be treated is one approach to determine a more accurate estimate *K* or HLR. The ultimate solution is a pilot plant which provides a means to investigate all variables comprehensively.

5.5.7 Microscreen Model

Figure 5.10 is a schematic of a microscreen cross section, intended to depict some of the process variables. The steps in developing a mathematical model for a microscreen are enumerated below as Example 5.3. The idea of a mathematical model for microscreening may seem "far-fetched" at first glance and it was considered here, from the standpoint of an

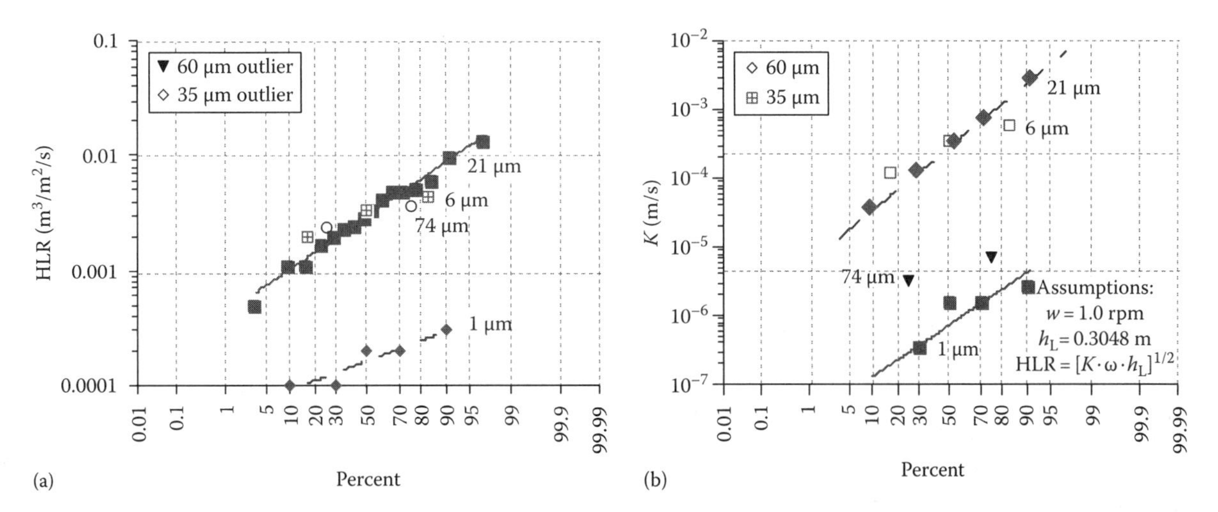

FIGURE 5.9 Microscreen parameter frequencies by screen size. (a) HLR frequency for $h_L = 0.1$ rad/s. (b) Coefficient, K, frequency for $h_L = 0.3$ m, $\omega = 0.1$ rad/s. (From Envirex, Envirex Data Sheet 315-3.201, pp. 1–3, 1982. With permission.)

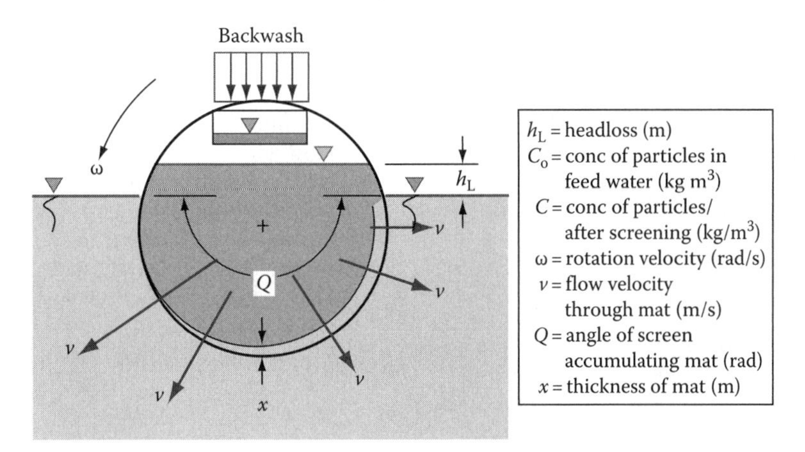

FIGURE 5.10 Microscreen cross-section schematic showing variables of interest in mathematical model development.

exercise, as a challenge. The idea is that a systematic approach to problem analysis is applicable to virtually any kind of process. Example 5.3 was intended to illustrate this tenet using the microscreen for illustration.

Example 5.3 Development of Mathematical Model for Microscreen

Step 1: *State purpose of model*
A theory in screening, in general, is lacking. The problem is seen largely as the application of a technology to practice. A mathematical depiction of microscreen performance could serve to aid design and operation by better understanding the mechanisms of microscreen performance, thus the role of variables.

Step 2: *State objectives*

1. Explore the utility of mathematical relations obtained.
2. Determine headloss across the screen and mat as a function of rotational velocity, suspended solids loading, hydraulic loading, degree of cleaning, etc.

Step 3: *Identify variables, aided by diagram* (Figure 5.10)

1. Dependent variables:
   C = concentration of suspended particles leaving screen (kg/m^3)
   C_r = concentration of suspended particles removed by screen (kg/m^3)
   X = thickness of deposited mat of suspended matter at any Θ (m)
   X_M = thickness of deposited mat of suspended matter at Θ_M (m)
   h_L = headloss across screen and mat (m)
2. Independent variables:
   L = length of microscreen (m)
   C_o = concentration of suspended particles in raw water (kg/m^3)
   Θ = angle from initial outside water line to any location on the screen (rad)
   Θ_M = angle from initial outside water line to final outside water line on screen (rad)
   ω = rotational velocity, omega, of screen (rad/s)
   Q = flow to screen (m^3/s)

ρ = specific mass of suspended matter as deposited on screen (kg/m^3)
HLR = hydraulic loading rate (m^3/m^2/s)
M = mass of suspended matter deposited on screen at any given time (kg)
V = volume of suspended matter deposited on screen at any given time (m^3)
v = velocity of flow of water through screen (m/s)
A = area of screen (m^2)
r = radius of screen (m)
t = time for rotation of screen from initial water line (s)
k(screen) = coefficient of hydraulic conductivity for screen (m/s)
k(mat) = coefficient of hydraulic conductivity for mat of suspended matter (m/s)

Step 4: *State materials balance for screen*

Rate of mass retention on screen
= (mass rate of suspended solids to screen)
− (mass rate of suspended solids leaving screen) (Ex5.3.1)

Step 5: *Express materials balance mathematically*

$$\frac{dM}{dt} = QC_o - QC \qquad \text{(Ex5.3.2)}$$

Substitute: $M = V \cdot \rho$

$$\frac{d(V\rho)}{dt} = QC_o - QC \qquad \text{(Ex5.3.3)}$$

Substitute: $V = A \cdot X$

$$\frac{d(XA\rho)}{dt} = QC_o - QC \qquad \text{(Ex5.3.4)}$$

Note that $v = Q/A$

$$\frac{d(X)}{dt}\rho = vC_r \qquad \text{(Ex5.3.5)}$$

Step 6: *Apply Darcy's law for flow across screen*

$$v = k(\text{screen}) \cdot \frac{h_L(\text{screen})}{X(\text{screen})} \qquad \text{(Ex5.3.6)}$$

Step 7: *Again apply Darcy's law, this time for flow across mat*

$$v = k(\text{mat}) \cdot \frac{h_L(\text{mat})}{X(\text{mat})} \qquad \text{(Ex5.3.7)}$$

Note: By continuity, the velocity across the screen equals the velocity across the mat, thus Equations Ex5.3.6 and Ex5.3.7 are equal.

Also, for later reference recall,

$$h_L = h_L(\text{screen}) + h_L(\text{mat}) \qquad \text{(Ex5.3.8)}$$

Since, for most of the screening duration, $h_L(\text{mat}) \geq\geq h_L(\text{screen})$, then we can neglect h_L(screen) in Equation Ex5.3.8 to give

$$h_L \approx h_L(\text{mat}) \qquad \text{(Ex5.3.9)}$$

Step 8: *Substitute Equation Ex5.3.7 in Equation Ex5.3.5*

$$\frac{dX(\text{mat})}{dt}\rho = \left[k(\text{mat})\frac{h_L(\text{mat})}{X(\text{mat})}\right]C_r \qquad \text{(Ex5.3.10)}$$

Step 9: *Separate the variables and integrate*

$$\int X(\text{mat})dX(\text{mat}) = \frac{k(\text{mat})h_L(\text{mat})C_r}{\rho\omega}\int_0^{\Theta_M} d\Theta \qquad \text{(Ex5.3.11)}$$

giving a "final" equation:

$$\frac{1}{2}X(\text{mat})^2 = \frac{k(\text{mat})h_L(\text{mat})C_r}{\rho\omega}\Theta_M \qquad \text{(Ex5.3.12)}$$

To simplify, substitute Equation Ex5.3.9 for h_L(mat) to give

$$\frac{1}{2}X(\text{mat})^2 = \frac{k(\text{mat})h_L C_r}{\rho\omega}\Theta_M \qquad \text{(Ex5.3.13)}$$

Step 10: *Illustrate graphically selected relations in Equation Ex5.3.13*

Discussion: Figure 5.11 says that the mat builds up quickly and then declines toward an asymptote. Since Θ is fixed within narrow limits a rapid buildup of mat is inevitable. At the same time, the mat thickness declines exponentially with ω and then declines toward a lower asymptote. In other words, a slight increase in ω can do much to reduce the mat thickness. The mat thickness is that at Θ_M.

Step 11: *Find relation for Q*

$$Q = \int^A v\, dA \qquad \text{(Ex5.3.14)}$$

$$Q = \int^A \left[k(\text{mat})\frac{h_L(\text{mat})}{X(\text{mat})}\right] dA \qquad \text{(Ex5.3.15)}$$

$$Q = \int_0^{\Theta_M} k(\text{mat})\frac{h_L(\text{mat})}{\left[((2k(\text{mat})h_L(\text{mat})C_r)/(\rho\omega))\Theta\right]} Lr\, d\Theta \qquad \text{(Ex5.3.16)}$$

$$Q = \left[\frac{2\rho\omega k(\text{mat})h_L(\text{mat})\Theta_M}{C_r}\right]^{1/2} Lr \qquad \text{(Ex5.3.17)}$$

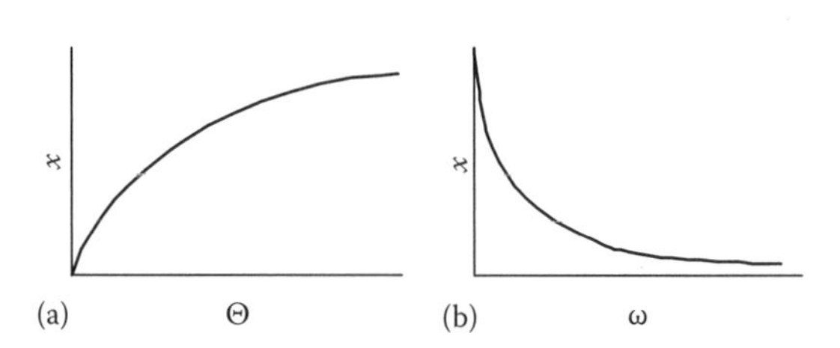

FIGURE 5.11 Mat thickness, calculated as function of submergence angle and rotational velocity. (a) Relation x versus Θ. (b) Relation x versus ω.

HLR is by definition:

$$\text{HLR} = \frac{Q}{L\Theta_M r} \quad \text{(Ex5.3.18)}$$

Combining (Ex5.3.17) and (Ex5.3.18) gives

$$\text{HLR} = \left[\frac{2\rho\omega k(\text{mat})h_L(\text{mat})}{\Theta_M C_r}\right]^{0.5} \quad \text{(Ex5.3.19)}$$

Again, to simplify, substitute Equation Ex5.3.13 for h_L(mat) to give

$$\text{HLR} = \left[\frac{2\rho\omega k(\text{mat})h_L}{\Theta_M C_r}\right]^{0.5} \quad \text{(Ex5.3.20)}$$

5.5.7.1 Interpretation of Model Results

Equation Ex5.3.20 says that the HLR for design is a function of the variables listed in the equation to the 1/2 power. The two design-independent variables are h_L and Θ_M, while ω is an operating variable, and ρ, k(mat) and C_r are "passive" variables, i.e., those that are ambient or are otherwise not controlled. The mass loading rate per unit area (MLR), which is essentially the same as the mass removal rate, since $C \approx 0$, is obtained by multiplying both sides of Equation Ex5.3.20 by C_r. The resulting equation is merely a derivative of Equation Ex5.3.20 and is not shown.

The "passive" variables in Equation Ex5.3.20 may be consolidated into a single coefficient, K, which must be determined by pilot plant testing, or from data obtained from a full-scale plant. The variable, Θ_M, is also consolidated in the coefficient. The result is an equation that has more utility, i.e.,

$$\text{HLR} = K\omega^{0.5}h_L^{0.5} \quad \text{(Ex5.3.21)}$$

Figure 5.12a and b shows plots of HLR vs. h_L and HLR vs. ω, respectively for $K = 3.6 \cdot 10^{-4}$ m/s. Figure 5.12a shows that increasing headloss toward the higher end of practice can yield significantly higher HLR. Figure 5.12b shows that providing for a several-fold increase in ω can permit significantly higher HLR, which would therefore be an important variable in operation.

By the same token, since $A = Q/v$, the plots, Figure 5.13a and b, respectively, are derivatives of Figure 5.12a and b, respectively. Figure 5.13a shows that the required area for a 0.31 m^3/s (7 mgd) flow declines exponentially with an increase in headloss applied to the design. Also, if ω is applied

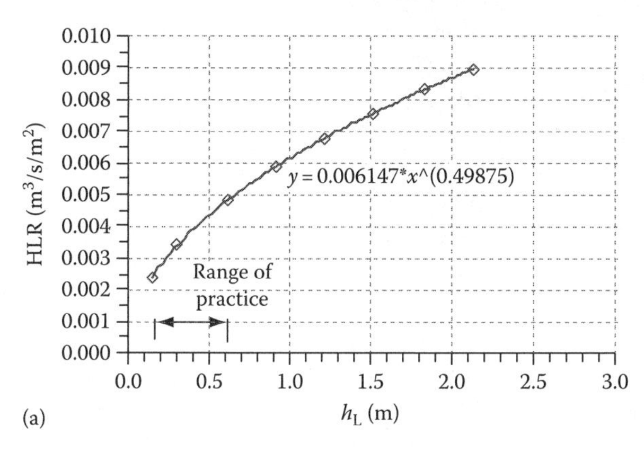

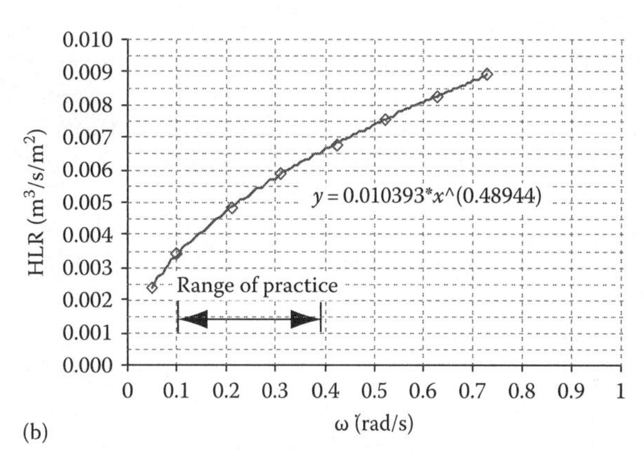

FIGURE 5.12 Microscreen calculated HLRs for K (50%). (a) HLR versus h_L for $h_L = 0.3$ m. (b) HLR versus ω for $\omega = 0.1$ rad/s.

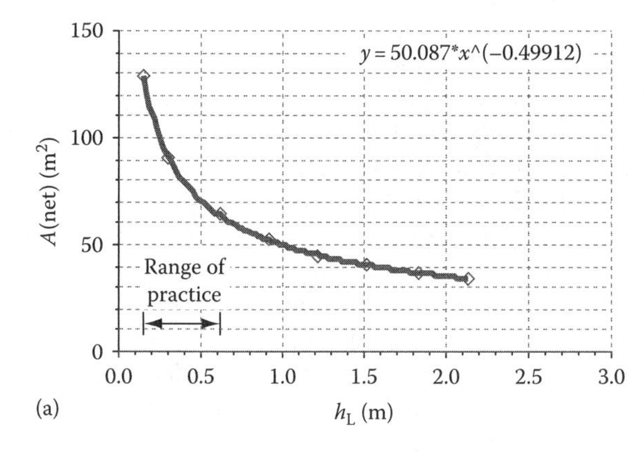

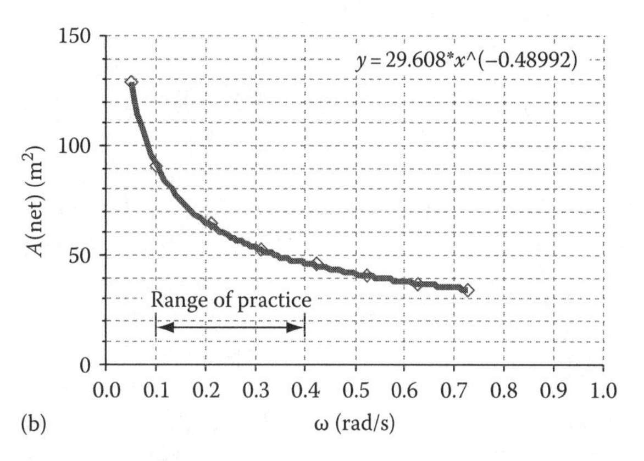

FIGURE 5.13 Microscreen calculated net screen area for K (50%). (a) A(net) versus h_L for $h_L = 0.3$ m. (b) A(net) versus ω for $\omega = 0.1$ rad/s.

to design instead of operation, the same dividends can be realized. As a note, plots of this nature do not substitute for knowledge gain from pilot plant tests. They provide guidance for what to explore in pilot plant studies. For example, once the appropriate K is determined by pilot plant studies, the mathematical model may be applied as indicated here. Of course, some spot checking of the mathematical model would be advisable.

PROBLEMS

Bar Screens

5.1 Traditional Approach to Bar Screen Sizing
For a bar screen design for a municipal wastewater treatment plant, let Q(max sewage flow) = 0.396 m^3/s (9.0 mgd) and let Q(max storm flow) = 0.44 m^3/s (10.0 mgd).

5.2 Spreadsheet for Traditional Approach to Bar Screen Sizing
Design a spreadsheet to accomplish all of the design tasks illustrated in Example 5.1 for any flow. Assume values for Q(max sewage flow) and Q(max storm flow) and bar sizes and bar spacing. Apply the criteria for velocity through the screen. The spreadsheet should provide a design that meets the criteria stated. Also, calculate headlosses.

5.3 Scenarios on Spreadsheet for Traditional Approach to Bar Screen Sizing
Modify your spreadsheet to address different scenarios of operation. In other words, the design is fixed. Therefore, apply the spreadsheet in #2 to explore the effects of different scenarios of flow. These might include some unexpected storm flows, or, by contrast, very low sanitary flows (to simulate a draught, for example, such as the one in California in the 1980s).

5.4 Half-Clogged Bar Screen Added to Traditional Approach to Sizing
For a sewage flow, Q = 7 mgd, size a bar screen system. Provide a drawing of your design showing approximate dimensions of the bar screen and channel. Suppose the screen becomes "half-clogged." Show on your spreadsheet how this affects your solution. Describe problems caused by this condition.

Assumptions for "baseline" scenario:

$$\text{Flows are: } Q(\text{avg. sewage flow}) = 7\,\text{mgd}$$
$$Q(\text{max. storm flow}) = 12\,\text{mgd}$$

5.5 Case Study on Cleaning Frequency for Bar Screen
The Marcy Gulch WWTP in Colorado has a 6 mm (1/4 in.) bar screen (Parkson, Inc.). In this case, the screen in a moving screen that makes an incremental movement up after a period of screenings accumulation, which may be perhaps 2–3 min. How would you determine the frequency of screen renewal, if you were able to travel to the site and take measurements? What would be your criteria? Determine for reference the headloss for the clean screen.

5.6 Comminutors
For a sewage flow, Q = 0.308 m^3/s (7 mgd), size a comminutor. Provide a drawing of your design showing approximate dimensions of the comminutor and channel. Suppose the screen becomes "half-clogged." Show on your spreadsheet how this affects your solution. Describe problems caused by this condition.

5.7 Hydraulic Profile for Headworks
Show the hydraulic profile for the headworks of a wastewater treatment plant.

5.8 Microscreen Design with Incomplete Data
Reference: Excel® spreadsheet file for microscreen design, Table CD5.7. The left side of Table CD5.7 provides data on the headloss coefficient, K; these data were generated for the purpose of generating frequency of occurrence of K's, plotted in Figure 5.13. The right side provides an algorithm for design, i.e., sizing for A(net), based upon different "scenarios." For the above context

(a) Select a microscreen mesh size (or opening size in μm).
(b) Explore the effects of uncertainty regarding the coefficient, K, with respect to the effect on headloss.
(c) Suppose Q is increased from 0.308 m^3/s (7 mgd) to 0.616 m^3/s (14 mgd) for an existing microscreen. Determine the associated headloss.

Assumptions for "baseline" scenario: Flow, Q = 0.308 m^3/s (7 mgd), ω = 6.28 rad/min.

5.9 Microscreen Modeling
The Denver Marston Water Treatment Plant treats drinking water drawn from the adjacent Marston Lake (near Quincy Avenue on the south side of Denver). The plant experiences algae blooms that interfere with coagulation and filtration. Suppose that microscreening is a proposed treatment process for removing the algae. A manufacturer has provided a pilot plant which you will use as the basis for a design. For this context, or a similar one with which you are familiar, (a) Outline an experimental program that you might propose. (b) State dependent variables. (c) Identify the independent variables. (d) Would you do any bench scale testing? (e) Would you visit any microscreen plants? (f) Would mathematical modeling have a place? (g) Describe plots that you would generate from the pilot plant operation. (h) Would you apply mathematical modeling for any aspect of your design? (i) Describe how you would arrive at a final sizing for a full-scale design.

5.10 Variables and Scenarios in Microscreen Design
As a choice in a design exercise, the 50% frequency would be a reasonable choice for input to a design spreadsheet, which would then be the basis for exploration of design outcomes using different input "scenarios" (combinations of independent variables). The spreadsheet should include several such "scenarios," comprising different flows, i.e., Q, and other uncertainties concerning the design, e.g., substance to be removed. Each situation is unique, however, and testing

is recommended to select a suitable fabric and to determine HLR and operating conditions, such as rotation velocity and backwash velocity (in terms of pressure) and flow. Enumerate the variables and uncertainties and generate scenarios that provide a reasonable exploration that relates to a design.

5.11 Microscreen Appurtenances
Size the appurtenances to a microscreen design, such as motor power, backwash pump power, flow capacity for backwash, weir design for effluent flow, etc. [This will be handled most easily if manufacturer's catalogs are available.]

ACKNOWLEDGMENTS

The comminutor rating curves, Figure 5.5, were from a catalog (c. 1969) of Lyco Equipment, Williamsport, Pennsylvania. In trying to trace the lineage of the company since 1969, in order to seek permission to use the figure, Ranvir Singh, director of research and development for Lyco during the period 1967–1969, and presently with the Office of Surface Mining, Denver, Colorado, confirmed that the company had gone out of business about 1971. This meant, of course, that the question was moot regarding permission. Singh was coauthor of the Lyco catalog and stated that the catalog data were generated from their own laboratory testing or by contracts with university laboratories.

The section on wedge-wire hydro-screens was added after lectures and demonstrations of their performance by Robert K. Weir, presently a consulting water engineer in Denver, Colorado (he retired from the Denver Water Department in 1999 as the deputy director of operations and maintenance). Weir also designed, fabricated, and installed the screen at Empire, Colorado, which provided a field demonstration of how well the screens have worked to eliminate debris from the slow sand filters both during summer and in winter with ice and freezing conditions prevalent. The Town of Empire also provided hospitality to visit its plant for drinking-water treatment on numerous occasions.

The photograph of mechanically cleaned bar-screen, Figure 5.1, was taken by Uma Wirutskulshai, a former graduate student at Colorado State University, (1996); (currently (2010) a doctoral student at the Asian Institute of Technology (AIT), Bangkok, Thailand). In addition, Steve Comstock, formerly supervisor, Fort Collins WWTP, granted permission to use the photograph. Comstock hosted my class on numerous occasions during which photographs were often taken with his permission.

The late Paul Grundeman, Marcy Gulch WWTP, Centennial Water and Sanitation District, Highlands Ranch, Colorado, provided hospitality to visit the plant to obtain several photographs, including the drum screen installation. Dr. John Hendrick, general manager of the district gave permission to use the photographs taken at the plant for this book.

Permission to use the data for construction of Figure 5.9 was given by Michael Quick, Siemens Envirex Division, Waukesha, Wisconsin Quick noted that the manufacturing of microscreens was discontinued by Envirex in the early 1990s.

GLOSSARY

Bar screen: A screen that has steel bars typically $25 < \text{opening} < 50$ mm (1–2 in.) able to withstand the impact of large objects. Bar screens are an integral part of the headworks of every municipal WWTP. See also *comminutor*.

Coanda effect: A phenomenon in which a high-velocity flow tends to adhere to a surface—observed by Henri-Marie Coanda in 1910.

Comminutor: A screen with blades that cut material that has attached, e.g., stringy matter. The comminutor is located usually after the bar screen and prior to pumping (if required). In the 1990s, the trend was to omit comminutors and use, instead, a bar screen with $5 < \text{opening} < 12$ mm (1/4–1/2 in.).

Disk screen: A screen that is disk shaped and that rotates with material collected being removed from the disk area that leaves the intake flow.

Drum screen: A screen with fabric that covers a drum structure that rotates on an axis, with partial submergence; water flow is from inside to outside. As the drum area leaves the water, the material collected is removed by water jet to be carried away in a flume structure.

Fabric: Material that covers the screen structure. The term is applicable to fine screens and other screens that have a mesh material used to exclude particles.

Intake crib: A screen structure at the bottom of a lake intended to exclude fish and whatever other larger material might occur in the vicinity of the intake.

Manually cleaned bar screen: A bar screen in which the debris that accumulates is removed manually, usually by a pitchfork-like tool.

Microscreen: A screen with opening $1 < \text{opening} < 60$ μm. Such screens came into use about the 1950s with the intent to exclude filter-clogging algae that were a particular problem for eutrophic lake water sources. In the 1980s, a slot type of microscreen was used in wastewater treatment, eliminating primary settling.

Trash rack: A large opening bar screen, e.g., $25 < \text{opening} < 75$ mm (1–3 in.), intended to exclude large debris such as branches as found in streams.

Wedge-wire: A screen with V-shaped screen wire with flat portion parallel to the flow or offset at a slight angle to the main flow velocity vector. Opening is about 1 mm.

REFERENCES

Anon, *Catalog of Water Treatment Equipment*, Lyco Engineering, Inc., Williamsport, PA, c. 1969.

Burns and Roe, Inc., *Process Design Manual for Suspended Solids Removal*, U.S. Environmental Protection Agency, Washington, DC, October 1971.

Envirex, Rex Products, Envirex–A Rexnord Company, Waukesha, WI, 1982.

Hydroscreen, http://www.hydroscreen.com, 2002.

McCabe, W. L. and Smith, J. C., *Unit Operations of Chemical Engineering*, 1st edn., McGraw-Hill, New York, 1956.

McCabe, W. L. and Smith, J. C., *Unit Operations of Chemical Engineering*, 2nd edn., McGraw-Hill, New York, 1967.

McCabe, W. L. and Smith, J. C., *Unit Operations of Chemical Engineering*, 3rd edn., McGraw-Hill, New York, 1976.

McCabe, W. L., Smith, J. C., and Harriott, P., *Unit Operations of Chemical Engineering*, 5th edn., McGraw-Hill, New York, 1993.

Pankratz, T. M., *Screening Equipment Handbook (for Industrial and Municipal Water and Wastewater Treatment)*, Technomic Publishing Company, Lancaster, PA, 1988.

Rex Chainbelt, Product Manual, Sanitation Equipment and Process Equipment Division, Rex Chainbelt, Inc., Milwaukee, WI, 1955.

Wahl, T. L., Hydraulic testing of static self-cleaning inclined screens, in: *First International Conference on Water Resources Engineering*, American Society of Civil Engineering, San Antonio, TX, August 14–18, 1995. [Available from website: http://www.usbr.gov/pmts/hydraulics_lab/twahl/coanda/]

Wahl, T. L., Hydraulic performance of Coanda-Effect screens, *Journal of Hydraulic Engineering, American Society of Civil Engineering*, 127(6), June 2001. [Available from website: http://www.usbr.gov/pmts/hydraulics_lab/twahl/coanda/, 16 pp.]

Wahl, T. L. and Einhellig, R. F., Laboratory testing and numerical modeling of Coanda-effect screens, in: *Joint Conference on Water Resources Engineering and Water Resources Planning and Management*, American Society of Civil Engineering, Minneapolis, MN, July 30–August 2, 2000. [Available from website: http://www.usbr.gov/pmts/hydraulics_lab/twahl/coanda/]

6 Sedimentation

Gravity settling, or "sedimentation," is used at several points in both water and wastewater treatment trains. The difference in each application is the nature of the suspension to be settled. Table 6.1 describes these suspensions, where they occur in treatment, and the respective kinds of settling units.

6.1 KEY NOTIONS IN DESIGN

Basic themes in the design of sedimentation basins are (1) the suspension characteristics, and (2) basin hydraulics. These two themes are the basis for theory and practice (Camp, 1946).

The suspensions in Table 6.1 can be classified in settling as discrete settling, flocculent settling, hindered settling, and compression settling. Each suspension has its own characteristic settling behavior, described in Sections 6.2.3 and 6.4.

The second theme of the basin design is hydraulics. A settling particle is subject to the vagaries of water flow, i.e., the patterns of current, the superimposed eddies, and the microscale turbulence. Such effects are not predictable (except by CFD computer technology as noted in Box 6.1) and so the concept of the *ideal basin* (Section 6.3.3) has become the point of departure in depicting basin hydraulics.

6.2 PARTICLE SETTLING

The settling velocity of a single particle is the starting point of settling theory, leading to the concept of the ideal settling basin. As stated previously, the notion of the "ideal" basin is the reference for understanding the behavior of real systems.

6.2.1 Particle Settling Principles

Figure 6.1 is a free body diagram for a falling particle in dynamic equilibrium. Dynamic equilibrium means the drag force, F_D, equal the propulsion force, W_B. The propulsion force for a falling particle is its buoyant weight, i.e.,

$$W_B = V(\gamma_s - \gamma_f) \tag{6.1}$$

$$= V(\rho_s - \rho_f)g \tag{6.2}$$

where
- W_B is the buoyant weight of particle in fluid medium (N) or (lb)
- V is the volume of water displaced by particle (m^3) or (ft^3)
- γ_s is the specific weight of the particle (N/m^3) or (lb/ft^3)
- γ_f is the specific weight of fluid medium (N/m^3) or (lb/ft^3)
- ρ_s is the specific mass of particle (kg/m^3) or ($slugs/ft^3$)
- ρ_f is the specific mass of fluid medium (kg/m^3) or ($slugs/ft^3$)
- g is the acceleration of gravity (9.81 m/s^2) or (32.2 ft/s^2)

Drag forces, F_D, act on any object, e.g., a particle moving through a fluid. In turbulent flow, drag forces are caused by (1) boundary shear (skin friction) and (2) unequal pressure distribution around the object (form drag). In laminar flow, drag is due to the viscous shear forces distributed through the fluid. The general expression for the drag force due to fluid motion is

$$F_D = C_D \rho A \frac{v_s^2}{2} \tag{6.3}$$

where
- F_D is the drag force on particle (N) or (lb)
- C_D is the drag coefficient
- A is the projected area of particle normal to the direction of flow (m^2) or (ft^2)
- ρ is the density of fluid (kg/m^3) or ($slugs/ft^3$)
- v_s is the velocity of particle (m/s) or (ft/s)

For dynamic equilibrium, per Figure 6.1, the drag force developed equals the propulsion force, i.e.,

$$F_B = F_D \tag{6.4}$$

Substituting Equations 6.2 and 6.3 into Equation 6.4 gives

$$V(\rho_s - \rho_f)g = C_D A \rho \frac{v_s^2}{2} \tag{6.5}$$

6.2.2 Stokes' Law

The drag coefficient of Equation 6.5, C_D, is functionally related to Reynolds number, **R**. For the special case when $\mathbf{R} \leq 1$, i.e., the laminar flow range, $C_D = 24/\mathbf{R}$. Now, recall that $\mathbf{R} = \rho v_s d/\mu$ and substitute in Equation 6.5, to give,

$$V(\rho_s - \rho_f)g = \frac{24}{\rho v d/\mu} A \rho \frac{v_s^2}{2} \tag{6.6}$$

where
- d is the diameter of particle (m) or (ft)
- μ is the dynamic viscosity of fluid ($N\ s/m^2$) or ($lb\ s/ft^2$)

TABLE 6.1
Suspensions in Water and Wastewater Treatment

Suspension	Occurrence	Settling Unit
Mineral particles	Raw water supply	Plain sedimentation
Oil	Refinery wastes	Separators
Floc particles with bubbles attached	Air flotation	Flotation thickening
Organic particles	Raw sewage	Primary settling
Biological floc	Biologically treated sewage	Final settling
Chemical floc	Chemically treated water/sewage	Flocculent settling
Sludge's	Settled chemical and biological flocs and settled organics	Thickening compartments of settling basins

BOX 6.1 THE COMPLETE MATHEMATICAL MODEL OF FLUID FLOW

As seen in fluid mechanics texts, the mathematical description of any fluid flow is provided by the classical Navier–Stokes equations. They were formulated by Navier, Cauchy, and Poisson, early in the nineteenth century, and by Saint-Venant and Stokes in the mid-nineteenth century. Equations are named as a matter of custom after the first and last of these investigators (Rouse, 1959, p. 208), hence Navier–Stokes. The Navier–Stokes equation was merely an expansion, in differential form, of Newton's second law, i.e., the familiar, $F = ma$. In the expansion, expressed as a differential equation, the left-hand side included all of the forces that act on an infinitesimal volume of fluid, e.g., pressure, gravity, viscous, surface energy. The *ma* side is the dynamic response to the forces. The Navier–Stokes expression has several dependent variables and so has defied solution, i.e., until the advent of the computer, which provided the means for a numerical solution, done in the 1960s by Fortran programming in the 1990s by "computational fluid dynamics" (CFD) software.

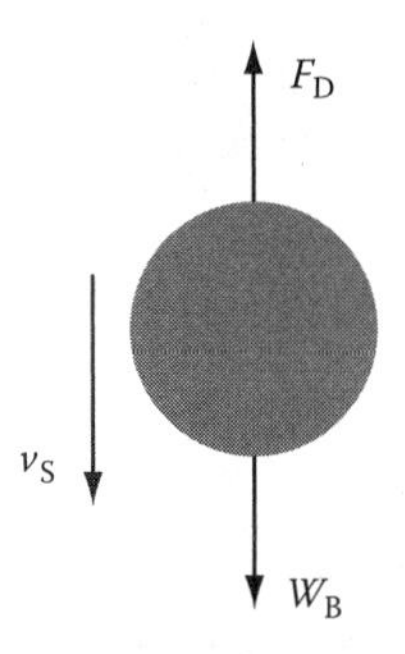

FIGURE 6.1 Forces acting on a falling particle.

After substituting the projected area for a sphere, i.e., $A = \pi d^2/4$, and the volume, $V = \pi d^3/6$, Equation 6.6 becomes Stokes' law:

$$v_s = \frac{1}{18}\frac{g}{v}(SG_s - SG_f)d^2 \quad (6.7)$$

where
v is the kinematic viscosity of fluid (m^2/s) or (ft^2/s)
SG_s is the specific gravity of particle
SG_f is the specific gravity of fluid

While Stokes' law is useful, its important to keep in mind that its merely an equality of forces for the special case in which $\mathbf{R} \leq 1$, i.e., the viscous range for which $C_D = 24/\mathbf{R}$. Example 6.1 illustrates numerical calculations; Table CDEx6.1 is set up as an algorithm for computations. A relationship for C_D for spheres that includes the range, $1 \leq \mathbf{R} < 10^5$ was given by Fair et al. (1968, p. 25-3), i.e., $C_D = 24/\mathbf{R} + 3/\mathbf{R}^{0.5} + 0.34$.

Example 6.1 Application of Stokes' Law

Illustrate the application of Stokes' law for different situations of fall of a quartz sand particle in water.

1. *Fall velocity*: Calculate from Equation 6.7 the fall velocity, v_s, for a quartz sand particle of 0.1 mm with equivalent diameter at 20°C (68°F),

$$v_s = \frac{1}{18}\frac{1}{\mu}(\gamma_s - \gamma_f)d^2 \quad \text{(Ex6.1.1)(6.7)}$$

$$v_s = \frac{1}{18}\left(\frac{1}{1.002 \cdot 10^{-3}\ \text{Ns/m}^2}\right)(998.21\ \text{kg/m}^3)$$
$$\times (9.81\ \text{m/s}^2)(2.65 - 1.00)(0.1 \cdot 10^{-3}\ \text{m})^2 \quad \text{(Ex6.1.2)}$$
$$= 0.0090\ \text{m/s} = 0.90\ \text{cm/s} \quad \text{(Ex6.1.3)}$$

Note that $g_s = r_s \cdot g$, and $g_f = r_f \cdot g$, where g_s is the specific weight of particle (N/m^3) and g_f is the specific weight of fluid (N/m^3).

2. *Largest particle diameter at 10°C*: Calculate the largest diameter quartz sand particle (SG = 2.65) for which Stokes' law is applicable at 0°C, i.e., $\mathbf{R} = 1$.

   a. *Trial-and-error solution*: The solution is by trial and error and involves the following steps: (1) assume a value for *d*; (2) using the assumed *d*, calculate v_s from Stokes' law; and (3) from the calculated v_s and assumed *d*, calculate **R**. Since Stokes' law is valid for $\mathbf{R} \leq 1.0$, the largest *d* is for $\mathbf{R} = 1.0$. Therefore, if the assumed *d*, gives $\mathbf{R} < 1.0$, increase *d* for the next trial; if the assumed *d*, gives $\mathbf{R} > 1.0$, then decrease *d* for the next trial.
   b. *Spreadsheet*: The easiest way to execute the foregoing algorithm is by means of a spreadsheet (Table CDEx6.1). Several trials are shown, i.e., $d = 0.010$, 0.050, 0.10, 0.15, ..., 0.124 mm, with the last being the size that meets the criterion $\mathbf{R} = 1.00$.

TABLE CDEx6.1
Determination of Maximum Particle Size for Stokes' Law to be Applicable

(a) Temperature coefficients (Table QR.4)

	M0	M1	M2	M3	M4	M5
μ(water)	0.0017802356694	−5.6132434302E−05	1.0031470384E-06	−7.5406393887E−09		
ρ(water)	999.84	6.82560E−02	−9.14380E−03	1.02950E−04	−1.18880E−06	7.15150E−09

Formula: $y = M0 + M1 \cdot T + M2 \cdot T^2 + M3 \cdot T^3 + M4 \cdot T^4$

(b) Fixed data

g = 9.80665	m/s^2	(Table QR.1)
SG_s = 2.65	dimensionless	
SG_f = 1.00	dimensionless	

(c) Calculation of v_s and R for assumed T and d

T (°C)	μ(water) (N s/m^2)	ρ(water) (kg/m^3)	D (m)	v_s (m/s)	v_s (mm/s)	R
10	0.00131	999.700	0.000010	6.85E−05	0.069	0.001
10	0.00131	999.700	0.000050	1.71E−03	1.71	0.065
10	0.00131	999.700	0.000100	6.85E−03	6.85	0.522
10	0.00131	999.700	0.000150	1.54E−02	15.4	1.76
Final trial in next row involved changing d until **R** = 1.00						
10	0.00131	999.700	0.000124	1.05E−02	10.5	1.00
Assume	Formula	Formula	Assume	Stokes' law	Convert	Calculate
				$v_s = (1/18)(\rho g/\mu)(SG_s - SG_f)d^2$		$\mathbf{R} = \rho v_s d/\mu$

3. *Largest particle diameter at 20°C, 0°C*: Calculate the largest diameter sand particle (SG = 2.65) for which Stokes' law is applicable at 20°C, 0°C, i.e., $\mathbf{R} \leq 1$. An excerpt from the same spreadsheet, used in 2, shows that $d(20) = 0.104$ mm and $d(0) = 0.152$ mm.

(c) Calculation of v_s and R for assumed T and d

T (°C)	μ(water) (N s/m^2)	ρ(water) (kg/m^3)	d (m)	v_s (m/s)	v_s (mm/s)	R
20	0.00100	998.204	0.000104	0.00965457	9.655	1.000
0	0.00178	999.840	0.000152	0.011695382	11.70	1.000

4. *Discussion*: Example 6.1 illustrates how the numerical data are applied to various calculations using Stokes' law. Note that the dynamic viscosity and the kg units must be related by Newton's second law relation. Parts 2 and 3 show that Stokes' law is not applicable to particles larger than 0.15 mm at 0°C and 0.10 mm at 20°C. Note also that the shape of the particle is not a factor in the Reynolds number calculation as long as a characteristic dimension of the particle is used.

6.2.3 Suspensions

The four types of settling behaviors are discrete, flocculent, hindered, and compression, called Type I, Type II, Type III, and Type IV, respectively (Katz et al., 1962a; WPCF, 1985, p. 3), illustrated in Figure 6.2. As seen in Figure 6.2, the fall velocity of the particles of Type I is constant, e.g., distance settled divided by time is a straight line. For the Type II suspension, the particles grow in size and, thus, their fall velocities increase as they fall in the suspension. Finally, as the particles approach the bottom of the basin, their concentrations, for both Type I and Type II suspensions, increase to such extent that the particles interfere with each other and the settling is "hindered," i.e., the suspension becomes Type III. Table 6.2 gives some of the characteristics of suspensions as found in the various settling units.

Regarding Type IV settling, the solids are not in suspension but are supported by particles below; consolidation is the phenomenon (Camp, 1937). Figure 6.2 provides an overview of settling characteristics and design guidelines for various settling situations.

6.2.3.1 Type I: Discrete Particle Suspensions

Type I suspensions consist of discrete particles that settle in accordance with Stokes' law. Grit removal in municipal wastewater treatment, and silt removal by plain sedimentation in drinking water treatment are examples, albeit turbulence is ever present and probably is a dominant influence in most situations.

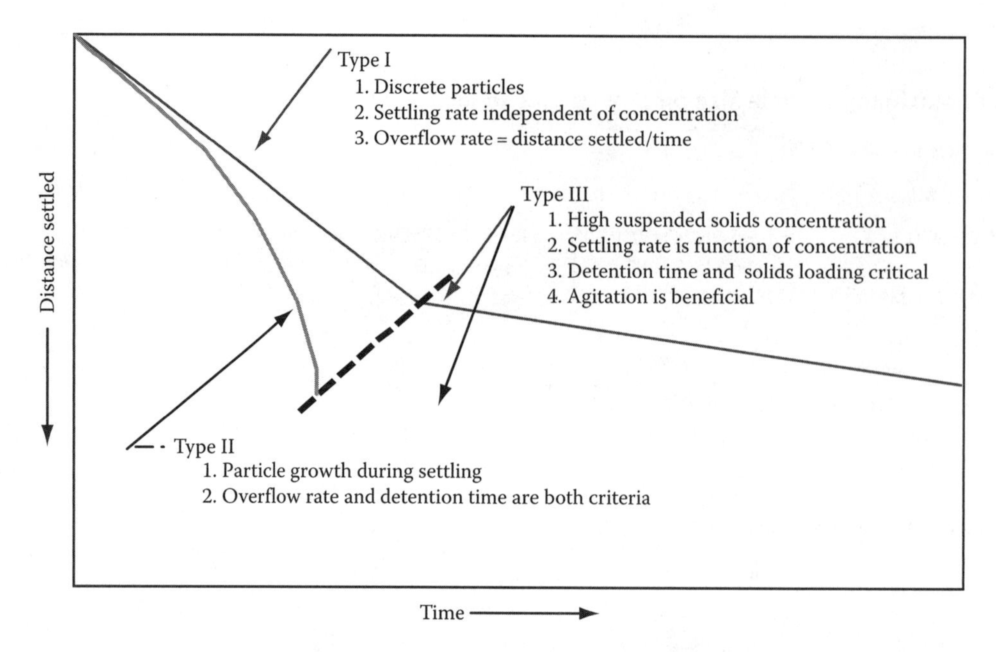

FIGURE 6.2 Settling velocities of three categories of suspensions. (Adapted from Katz, W.J. et al., *Concepts of Sedimentation Applied to Design*, Part 1, *Water and Sewage Works*, 162, April 1962a.)

TABLE 6.2
Characteristics of Suspensions and Overflow Rates for Various Categories of Settling Units in Water and Sewage Treatment

Settling Unit	Particle Composition	Sus. Type	Particle SG	Part. Size (mm)	v_o (mm/s)	v_o (gpd/ft^2)	θ (h)
Grit chamber	Sand, seeds, coffee grounds[a]	I[a]	1.2–2.65	≤0.2	23		
Primary settling	Irregular particles, mostly organics	II[b]	1.0–1.2	<5	≤0.3	≤600[c]	1–2[d]
Final settling	Biological floc	III[a]		≈1.0[a]	≤0.4	≤800[c]	2–3[e]
Plain sedimentation	Silica, clay, silt mineral particles	I	2.65	<1	<100		
Flocculent settling of chem. precipitates	Aluminum flocs	II	1.18	<3	≤0.5	≤1000[f]	≥2[f]
	Iron flocs	II	1.34	<3	≤0.5	≤1000[f]	≥2[f]
	$CaCO_3$	II	2.7 cr 1.2 fl	<3	0.5 (hin)	≤1000[f]	

Note: Sus., suspension; SG, specific gravity; v_o, overflow velocity, i.e., $v_o = Q/V$(plan); θ, detention time, i.e., $\theta = Q/A$(basin); cr, crystal; fl, floc; hin, hindered; *A*(plan), plan area of basin; *V*(basin), volume of basin.

[a] Camp (1946).

[b] References indicate that primary settling is Type II but lack data to corroborate.

[c] Ten States Standards (1968) for plants having *Q*(avg. daily) ≤ 1 mgd.

[d] ASCE-WPCF (1959).

[e] Ten States Standards (1968) for plants having *Q*(avg. daily) ≥ 1 mgd.

[f] EPA (1971).

6.2.3.2 Type II: Flocculent Suspensions

Type II flocculent suspensions are those composed of particles that tend to grow in size. Such particles, as they collide with others, agglomerate, and fall at increasingly higher velocity, i.e., in accordance with Stokes' law, Equation 6.7, velocity increases as the square of the diameter). Examples of Type II suspensions include activated sludge floc and chemical flocs of various sorts, e.g., metal hydroxides.

6.2.3.3 Type III: Hindered Settling

In Type III suspensions, a clearly delineated solid–liquid interface forms. Figure 6.3 is a photograph of such an interface at

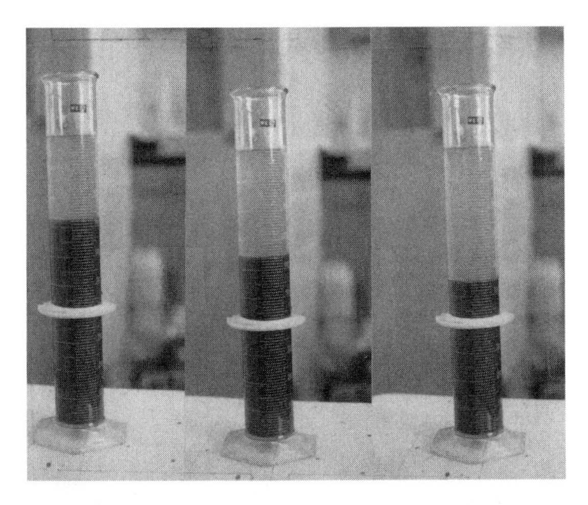

FIGURE 6.3 Zone settling—activated sludge from an aeration tank.

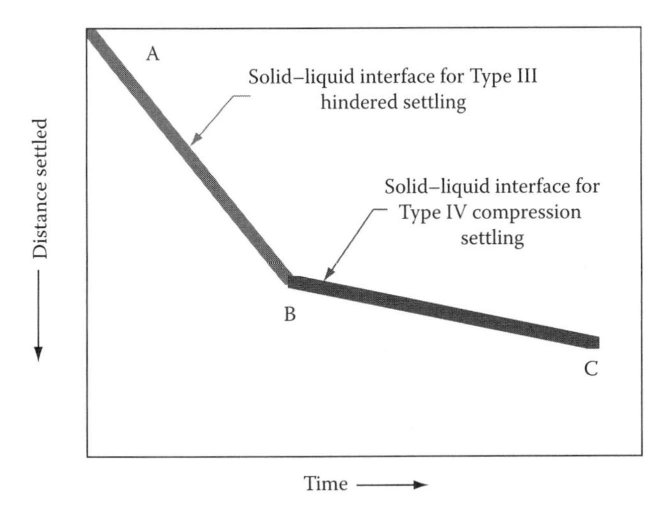

FIGURE 6.4 Solid–liquid interface for Type III and Type IV suspensions. (Adapted from Rich, L.G., *Unit Operations of Sanitary Engineering*, John Wiley & Sons, Inc., New York, 1961.)

three times after starting with a mixed suspension. The initial rate of subsidence of this interface may be used as a design-settling velocity and may be determined using a 1000 mL cylinder, as illustrated in Figure 6.3, and timing its rate of subsidence. The results are illustrated by curve A–B in Figure 6.4, which is a Type III settling curve.

6.2.3.4 Type IV: Compression Settling

Type III suspensions, after settling to a certain point, reach a density where the particles support each other, and further settling must be by compression. The Type III suspension thus changes to a Type IV suspension characterized by shallower interface elevation versus time slope, illustrated by the B–C portion of the curve in Figure 6.4. The compression results from the weight of particles above any given level, literally "squeezing" the water from the pore volume.

For the Type III suspensions, gentle stirring may increase the settling rate. To illustrate, Figure 6.5 (Katz et al., 1962a) shows settling for "gentle stirring" (top curve) compared with "no mixing" (bottom curve).

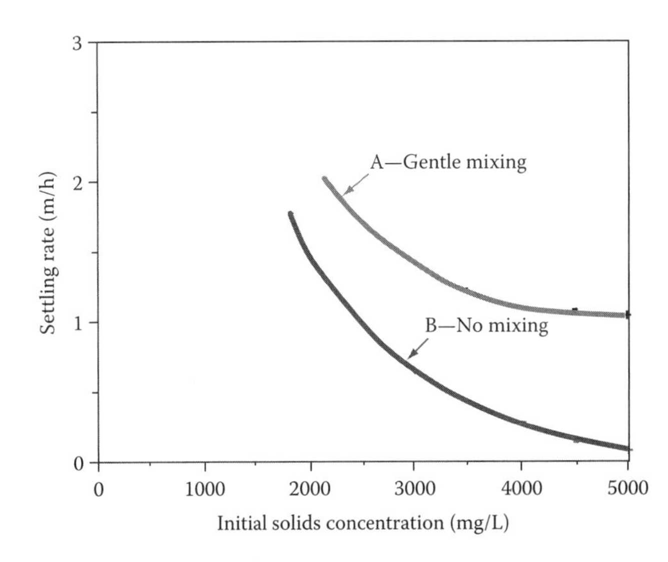

FIGURE 6.5 Rate of interface decline for Type III suspensions. (Adapted from Katz, W.J. et al., *Concepts of Sedimentation Applied to Design*, Part 1, *Water and Sewage Works*, 162, April 1962a.)

6.3 SETTLING BASINS

Two of the classical figures in sanitary engineering are Allen Hazen and Thomas R. Camp, whose careers are outlined in Box 6.2. They were instrumental in shaping the tenets of settling basin design, as reviewed in this chapter.

6.3.1 The Ideal Basin

The theory of settling basins starts with T. R. Camp's notion of the "ideal" basin (Camp, 1946). The roots were, however, in a paper by Allen Hazen published in a 1904 paper by ASCE (Hazen, 1904). Camp's 1946 paper was the first comprehensive theory for sedimentation basin design; key notions are described in the sections that follow.

6.3.1.1 Camp's Conditions for the Ideal Basin

A hypothetical settling tank in which settling takes place in exactly the same manner as in a quiescent settling container of the same depth is called an "ideal basin" (Camp, 1946). The ideal basin has four zones, as illustrated in Figure 6.6, which are characterized by (1) an inlet zone in which the suspension is dispersed uniformly over the cross section, (2) the settling zone in which all settling takes place, (3) an outlet zone in which the clarified liquid is collected and directed to an outlet conduit, and (4) a sludge zone at the bottom.

6.3.1.2 Overflow Velocity

As in the ideal basin, consider a column located at the inlet of the settling basin with the particles distributed uniformly top to bottom. Now, move the column horizontally to the right at velocity, v_H. The path of the settling particles will be the vector sum of the horizontal velocity, v_H, of the column, and the settling velocity of the particles, v_s. Hence, a settling test may be used to predict the concentration of particles at any point in the settling zone of an ideal basin.

BOX 6.2 HAZEN AND CAMP

Allen Hazen (1869–1930) was one of the early sanitary engineers who started at the Lawrence Experiment Station in Massachusetts, c. 1890–1892. He was a prolific contributor to the advancement of the field with well-known books, such as *The Filtration of Public Water Supplies* (John Wiley, London, U.K., 1913), and numerous papers. His 1904 paper marked the beginning of the modern sedimentation theory with the notion that basins should be shallow. In this paper, in Proposition 14, he describes what, essentially, was the plate-settler concept that came into practice about 1970. Hazen went into consulting in 1895, forming the firm Noyes & Hazen, which became Hazen & Whipple in 1904, Hazen, Whipple & Fuller in 1915, and Malcolm Pirnie in 1930; the latter name remains current (from Malcolm Pirnie, Inc., c. 1995).

Allen Hazen. (Courtesy of Malcolm Pirnie, White Plains, NY.)

In 1916 Thomas R. Camp received his BS degree from Texas A&M; he served in the U.S. Army 1917–1919, and in 1925 was awarded an MSCE degree from MIT. In 1932, he was appointed associate professor of Sanitary Engineering at MIT. At MIT, he started research in the field of sedimentation. At that time, Hazen's ideas had been discussed but not assimilated; and so practice had remained largely empirical. In general, Camp's papers were "classics" (ASCE, 1973) characterized by identifying principles and demonstrations on how to apply them to practice. In 1947, Camp resigned his professorship at MIT to start the firm, Camp Dresser and McKee.

Thomas Camp. (Courtesy of Camp Dresser and McKee, Boston, MA.)

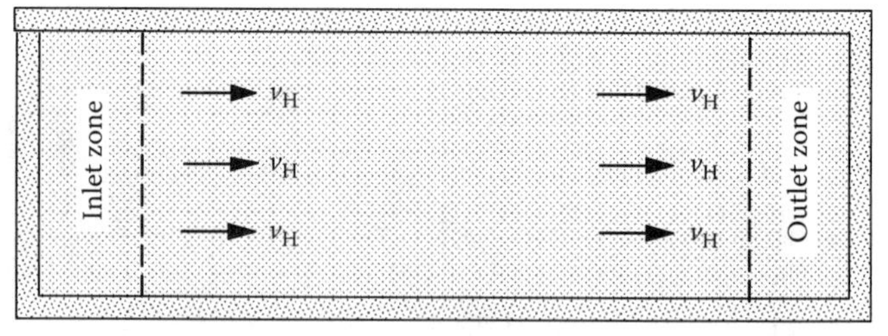

Top view

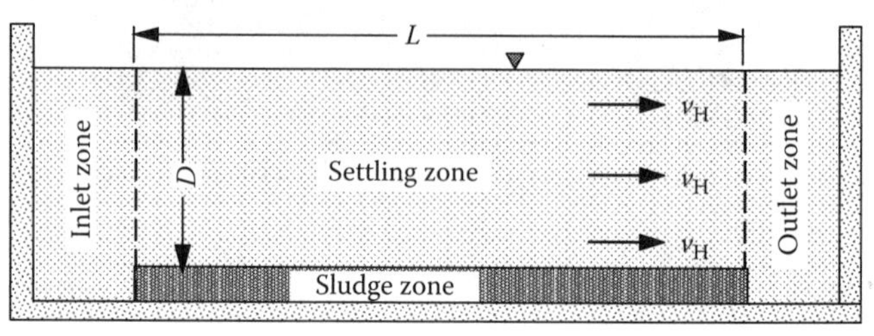

Side view

FIGURE 6.6 Ideal rectangular settling basin. (From Camp, T.R., *Trans. ASCE*, III, 895, 1946.)

Figure 6.7 shows that those particles being translated at velocity, v_H, starting at the top of the water surface and falling at velocity, $v_s = v_o$, will reach the bottom at depth, D, within a basin length, L. Thus, by similar triangles

$$\frac{v_o}{v_H} = \frac{D}{L} \quad (6.8)$$

where
v_o is the overflow velocity (m/s)
v_H is the horizontal velocity (m/s)
D is the depth of settling basin (m)
L is the length of settling basin (m)

Moving v_H to the right side and recalling, $v_H = Q/wD$, gives

$$v_o = \frac{D}{L} \cdot v_H \quad (6.9)$$

$$= \frac{D}{L} \cdot \frac{Q}{wD} \quad (6.10)$$

Therefore, the *overflow velocity*, v_o, is

$$v_o \equiv \frac{Q}{wL} \quad (6.11)$$

where
w is the width of the basin (m)
Q is the flow through basin (m^3/s)

The term, "overflow velocity," v_o, is also termed, "surface overflow rate" (SOR). These terms are used interchangeably

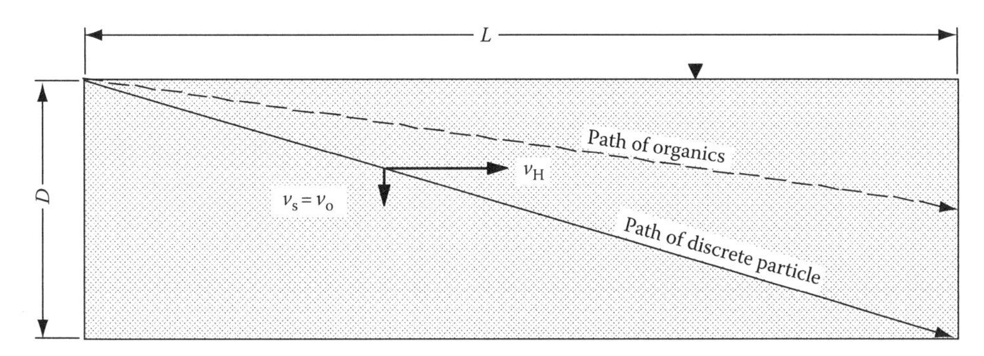

FIGURE 6.7 Overflow velocity as related to particle trajectory for discrete particles.

except that v_o may be associated with a "design"-settling velocity of particles, while SOR is more often associated with a regulatory limit.

6.3.1.3 Significance of Overflow Velocity

Camp's idea of overflow velocity was a fundamental innovation, in that basin depth, D, was deleted from the design relationship and, at the same time, detention time was shown to have no bearing on removal. In other words, overflow velocity, v_o, is the basis for sizing a basin. Thus, a settling basin can be as shallow as desired. The caveat is that the horizontal velocity, v_H, should not cause scour, i.e., resuspension of settled particles, that is unacceptable. With such limit defined for v_H, the product, wD, is determined, i.e., by Equation 6.12:

$$v_H = \frac{Q}{wD} \tag{6.12}$$

The ratios w/D and w/L are based upon empirical guidelines and practical considerations. For example, if v_o is given as a criterion, then by Equation 6.11, wL is fixed, and if w/L is given by guidelines, L may be calculated (and then D).

6.3.1.4 Insignificance of Detention Time

As a note, the relationship between detention time and depth is as follows: $\theta = V/Q = A \cdot D/Q = D/v_o$; therefore, $v_o = D/\theta$. Thus, if v_o is "set" (based on particle-size-to-be-removed or an empirical guideline), any change in θ requires a proportionate change in D. Example 6.2 illustrates how to apply the foregoing principles.

Example 6.2 Dimensions of an Ideal Basin

Given
An ideal settling basin that removes clay.

Required
That size of the basin.

Solution

a. *Fall velocity, v_o.*
Table 6.9 gives fall velocities for various particles. For clay, $v_o = 0.12$ m/min (3.0 gpm/ft^2).

b. *Spreadsheet*
Set up a spreadsheet that applies the foregoing principles, e.g., Equations 6.11 and 6.12. (1) Assume a flow, e.g., $Q = 0.05$ m^3/s (1.14 mgd) for a very small plant serving a population of about 10,000; (2) obtain a value for v_o for clay particles from Table 6.9; (3) calculate the quantity (wL); (4) assume several trial-and-error values for v_H, e.g., ≤ 0.30 m/s, such that scour is not likely; (5) calculate the corresponding value for (wD); (6) assume values for D and calculate, w, i.e., $w = (wD)/D$; and (7) finally, calculate, L, i.e., $L = (wL)/w$.

Discussion
Table Ex6.2 illustrates how to determine the basin dimensions, D, w, and L with a few trials based upon the idea of a selected overflow velocity, v_o, and a limit for horizontal velocity, v_H. As noted, all the particles with settling velocity, v_s, that are less than the overflow velocity, v_o, i.e., $v_s < v_o$, will be removed. The table could be set up by assuming w/L as an alternative to assuming D (or, for that matter, w or L). A spreadsheet provides a more convenient means for doing such calculations.

6.3.1.5 Partial Removals for Particles with Fall Velocities, $v_s < v_o$

Any suspension has a distribution of particle sizes, some with $v_s < v_o$, and some with $v_s > v_o$. All particles having $v_s \geq v_o$, will be removed completely. Particles having $v_s < v_o$ will be removed partially.

To determine the partial removal of particles with $v_s < v_o$, consider the fraction of particles having a specific fall velocity v_1, where $v_1 < v_o$. Figure 6.8 shows the resultant velocity vector, $\vec{v}_1 + \vec{v}_H$, for particles of this size. Letting this vector intercept the bottom of the basin at its far right, and then extrapolating the vector back to the entrance of the basin shows that it intercepts the vertical plane on the left side at depth d_1. Thus, all particles with fall velocity v_1 that enter the basin at $d \leq d_1$ will be removed. Therefore, the proportion of particles, r_1, of fall velocity, v_1 that will be removed must be (Camp, 1946)

$$r_1 = \frac{d_1}{D} \Delta P_1 \tag{6.13}$$

TABLE Ex6.2
Dimensions of Ideal Basin with Illustration of Trade-Offs

Q		v_o		wL	v_H	wD	D	w	L
(m³/s)	(mgd)	(m/s)	(gpm/ft²)	(m²)	(m/s)	(m²)	(m)	(m)	(m)
0.05	1.14	0.00204	3.00	24.55	0.30	0.17	0.30	0.56	44.18
0.05	1.14	0.00204	3.00	24.55	0.30	0.17	0.10	1.67	14.73
0.05	1.14	0.00204	3.00	24.55	0.10	0.50	0.30	1.67	14.73
0.05	1.14	0.00204	3.00	24.55	0.05	1.00	0.30	3.33	7.36
Given		Table 6.9 for clay		$v_o = Q/wL$ (Equation 6.11)	Assumed	$wD = Q/v_H$ (Equation 6.12)	Assume D	$W = (wD)/D$	$L = (wL)/w$

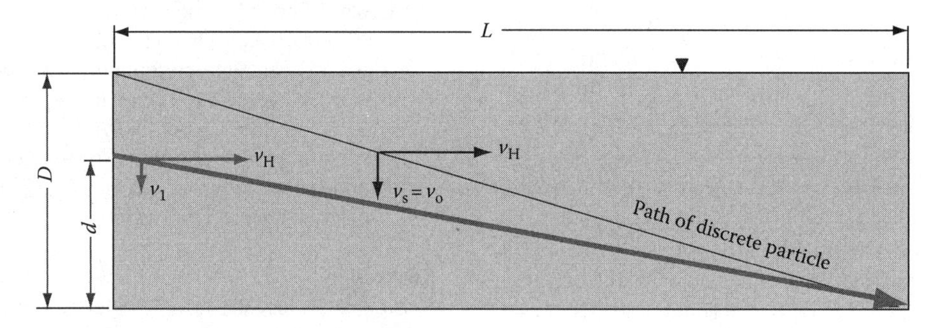

FIGURE 6.8 Removal of particles having fall velocity, v_1, where $v_1 < v_o$.

where
- r_1 is the fraction of particles having velocity v_1 that will be removed (fraction)
- d_1 is the depth at entrance of the basin of intercept of vector for particles with fall velocity, v_1, extrapolated from the far right intercept with the basin floor (m)
- ΔP_1 is the portion of particles having diameter, d_1, expressed as a decimal fraction

A visual rationale for Equation 6.13 is then evident in Figure 6.8. To extend this idea, replace d_1/D in Equation 6.13 by v_1/v_o, i.e., Equation 6.17 in Box 6.3, to give

$$r_1 = \frac{v_1}{v_o}\Delta P_1 \qquad (6.14)$$

where v_1 is the fall velocity of particles having equivalent diameter, d_1 (m/s).

In other words, for a particular group of particles with fall velocity v_1, the proportion of particles removed is given by Equation 6.14. This notion was carried further by Camp to provide a rationale for summing these partial removals for all particles, $v_s < v_o$, of a suspension (see Section 6.4).

BOX 6.3 CONVERSION OF DEPTH RATIO TO VELOCITY RATIO

To convert Equation 6.13 to a more useful form, i.e., Equation 6.17, consider two similar triangles in Figure 6.8, i.e.,

$$\frac{d_1}{L} = \frac{v_1}{v_H} \qquad (6.15)$$

and

$$\frac{v_o}{v_H} = \frac{D}{L} \qquad (6.8)$$

Now divide Equation 6.15 by Equation 6.8, i.e.,

$$\frac{d_1/L = v_1/v_H}{D/L = v_o/v_H} \qquad (6.16)$$

to give

$$\frac{d_1}{D} = \frac{v_1}{v_o} \qquad (6.17)$$

6.4 CHARACTERIZING SUSPENSIONS

Camp, in his 1946 paper, not only revived the notion of the ideal settling basin of Allen Hazen and gave us the basic principle that the overflow velocity was the critical parameter in basin design, but also recognized that a suspension of discrete particles is not uniform in size. In addition, he provided a means to characterize a suspension by a settling test and to calculate removals for a heterogeneous suspension.

6.4.1 Characteristics of Discrete Particle Suspensions and Removal Analysis

The particles size distribution for any suspension may be illustrated by a histogram, such as Figure 6.9a. The sizes are represented by the associated fall velocities, e.g., v_1, v_2, v_3, v_4, ..., v_s, and are indicated by the vertical velocity vectors. Figure 6.9b shows the resultant velocity vectors for these same particles in a horizontal flow settling basin. A common horizontal velocity, v_H, is added as a vector to the respective fall velocity vectors, e.g., v_1, v_2, v_3, $v_4, \ldots, v_s$, to give the associated resultants. Plotted as a continuous function, the histogram of Figure 6.9a becomes a frequency distribution curve, Figure 6.10a. Figure 6.10b is a *cumulative* particle size distribution, or the proportion, P, of particles having a fall velocity, v_s or smaller and is obtained from the frequency distribution curve. In Figure 6.10b, P_o is the fraction of particles having settling velocity v_o or less. Equation 6.14 is illustrated graphically as the ΔP portion of the plot of Figure 6.10b, which is considered further in Section 6.4.3.

6.4.2 Graphic Depiction of Size Fraction Removed

In Figure 6.11, the particles with fall velocities, v_1, v_2, v_3, v_4, etc., are distributed uniformly over the cross section at the basin inlet. For each particle in Figure 6.11, the resultant velocity vector, $\vec{v}_{Ri}$, is the vector sum of fall velocity, v_{si}, and v_H, i.e., $\vec{v}_{Ri} = \vec{v}_{si} + \vec{v}_H$, also illustrated in Figure 6.9b.

To illustrate how this vector addition applies to particles entering the basin, consider particles of size "3" in which, $\vec{v}_{R3} = \vec{v}_3 + \vec{v}_H$. As seen by the resultant velocity vectors for particles of size 3, all size 3s that enter the basin at depth d_3 and below will be removed. The computation of the removal, v_3, for this particle size range is therefore $r_3 = (d_3/D) \cdot \Delta P_3$, i.e., Equation 6.13, which is equivalent to $r_3 = (v_3/v_o) \cdot \Delta P_3$, i.e., Equation 6.14.

By the same token, we can see that *all* particles of size "2" will be removed. The resultant velocity vector, for the size 2 particles, is $\vec{v}_{R2} = \vec{v}_2 + \vec{v}_H$. That all size 2 particles will be removed is verified by visual inspection of Figure 6.11; we

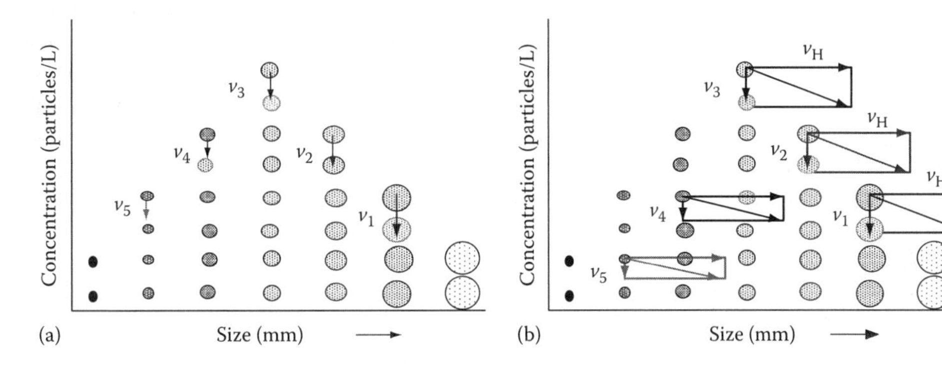

FIGURE 6.9 Particle size distributions in suspension in horizontal-flow basin: (a) distribution of particles of different sizes and associated fall velocities, v_s, and (b) vector addition of particle velocities, i.e., $\vec{v}_s + \vec{v}_H = \vec{v}_R$.

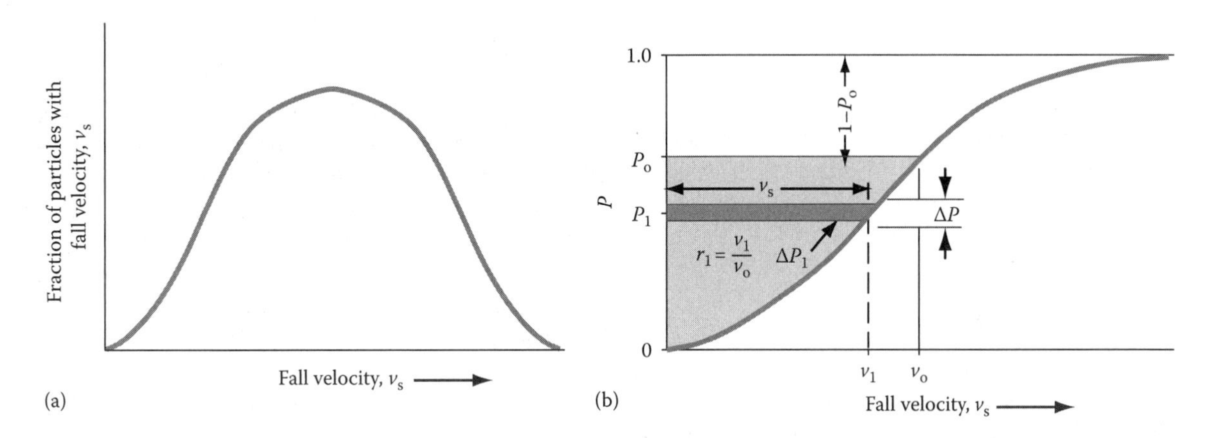

FIGURE 6.10 Distribution of particle fall velocities by two kinds of plots: (a) distribution of fall velocities for different particles and (b) cumulative distribution of fall velocities for different particles.

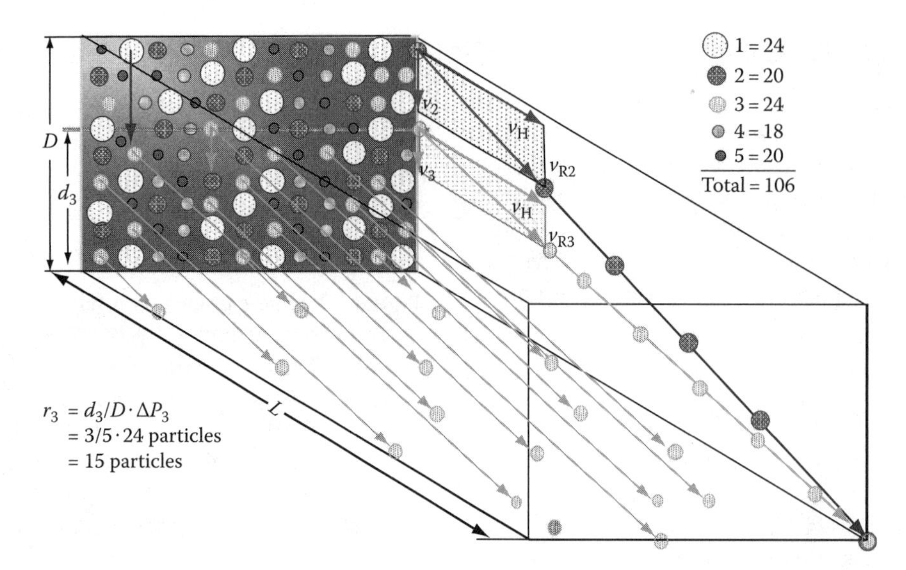

FIGURE 6.11 Ideal settling basin showing settling paths of the particles of the "3" size fraction. All size "3" particles are removed below level d_3. Overflow velocity, v_o, equals fall velocity for size "2" particles, i.e., v_2.

can see that a size 2 particle located at the top of the entrance cross section will just reach the bottom of the basin after distance, L. By definition then, $v_2 = v_o$, i.e., the "overflow velocity."

The main idea of Figure 6.11 is to provide a means to better visualize the idea of a specific particle fraction, e.g., ΔP_i, and the proportion removed, e.g., $r_3 = (d_3/D) \cdot \Delta P_3$ (Example 6.3).

Example 6.3 Illustrate the Fraction Removals for a Heterogeneous Suspension

Given

Suppose there are 24 particles having fall velocity, v_3, i.e., $\Delta P_3 = 24$ particles (in Figure 6.10, this is the count of size 3 particles). Also suppose that $d_3/D = 3/5$ (measuring these distances in Figure 6.10).

Solution

Based upon the mathematics noted previously, the number of d_3 particles removed is $r_3 = (d_3/D) \cdot \Delta P_3 = 3/5 \cdot 24 = 15$.

Discussion

Counting the size "3" particles confirms that there are 15 particles coming into the basin that are at depth $\leq d_3$. Also, to reinforce the idea that fall velocities may be used for the same calculation, since $d_3/D = v_3/v_o$, and since fall velocities are used to characterize a suspension, we can also use the expression, $r_3 = (v_3/v_o) \cdot \Delta P_3$ in lieu of d_3/D.

6.4.3 Mathematics of Removal

Figure 6.10b indicates the method of calculation of the total removal, R. For a given overflow velocity, v_o, all particles where $v_s \geq v_o$, or the $(1 - P_o)$ fraction, will be 100% removed. Particles with $v_s < v_o$ will be partially removed. Regarding these particles, i.e., $v_s < v_o$, consider, as in Figure 6.10b, a particle size designated "1," which has an associated size fraction, ΔP_1. The proportion of those particles removed, $r_1 = (v_1/v_o) \cdot \Delta P_1$, as seen in Figure 6.10b. Figure 6.11 illustrates the concept intuitively. The removal of all particles where $v_s < v_o$ is the sum of the infinitesimal removals, r_i, i.e., $1/v_o \int_0^{P_o} v_s \, dP$, which is the shaded area shown in Figure 6.10b divided by v_o. The total removal, R, is the sum of the integral and the quantity, $(1 - P_o)$, and is stated as

$$R = (1 - P_o) + \int_0^{P_o} \frac{v_s}{v_o} dP \quad (6.18)$$

where

- R is the total fraction of suspension removed in the ideal settling basin
- v_o is the overflow velocity (m/s)
- v_s is the fall velocity of any particle in suspension (m/s)
- P is the fraction of suspension associated with any v_s
- P_o is the fraction of suspension associated with overflow velocity, v_o

To apply Equation 6.18, the integral must be evaluated graphically. The problems are for two cases: (1) R is specified and v_o must be determined, and (2) v_o is specified and R must be determined. For Case 1, different v_o values are assumed and trial-and-error calculations are done until the right-hand side of Equation 6.18 satisfies the specified R. For Case 2, the determination is a straightforward graphical integration and the value of R is whatever is calculated. Figure 6.12 shows results of a settling column analysis of a hypothetical discrete particle suspension (Camp, 1946), as illustrated in Figure 6.10b. For any assumed overflow velocity, e.g., $v_o = 0.11$ cm/s, with an associated $P_o = 0.94$, all particles with $v_s \geq 0.11$ cm/s are

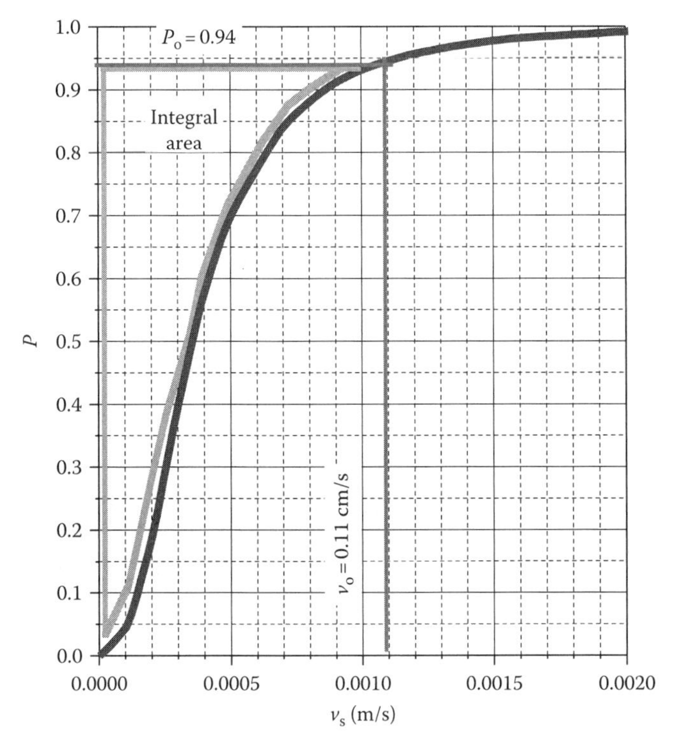

FIGURE 6.12 Distribution of settling velocities. Each square is (0.01 cm/s) · 0.05 fraction = 0.0005 cm/s/square. (From Camp, T.R., *Am. Soc. Civil Eng.*, III, 895, 1946.)

removed, e.g., $R = (1 - 0.94) = 0.06$. Particles with $v_s < v_o$ are partially removed as indicated by the integral part of Equation 6.18, which is on the left side of the curve. The evaluation procedure is by graphical integration as outlined by Camp (1946). For the plot, the technique starts with sizing a square, i.e., (0.01 cm/s) · 0.05 fraction = 0.0005 cm/s square; therefore, count the squares above the curve to the desired v_o, then multiply by 0.0005 cm/s/square, and divide by v_o to integrate the integral graphically. Example 6.4 illustrates this method.

6.4.4 Up-Flow Basins: A Special Case

As a matter of interest, up-flow basins are used sometimes, particularly as retrofits since they are not subject to "short circuiting" (see Section 6.8.1). For an up-flow basin, the total removal is $(1 - P_o)$ since only particles having $v_s \geq v_o$ are removed. Particles with $v_s \leq v_o$ are not removed, i.e., the integral portion of Equation 6.18 is zero, and the equation reduces to $R = (1 - P_o)$.

6.4.5 The Role of Ideal Settling Basin Theory

The foregoing theory applies to a suspension of discrete particles with a distribution of sizes settling in an ideal basin. The ideal settling basin theory has application to practice but cannot deal with the hydraulic problems (mainly turbulence and short circuiting) of real basins.

Conclusions from Camp's ideal settling theory applicable to practice are

1. The term "overflow velocity," sometimes called "surface overflow rate" (SOR), has been established in the vernacular in practice and as a way of thinking about settling basins.
2. The fraction of a suspension removed is a function of the "overflow velocity," v_o. Since, $v_o = Q/A$ (plan), the plan area of the basin is the focus.
3. Depth, and by corollary detention time, has no role in the removal of a discrete particle suspension.
4. The settling path of a particle having fall velocity v_s is a straight line from its position at the inlet end of the settling zone. The iso-concentration lines for a given suspension are straight lines from the inlet end of the basin.
5. Long narrow basins can minimize the effects of short circuiting.

For flocculent suspensions, removal depends upon both overflow velocity and detention time. This is the topic of Section 6.5.

Example 6.4 Removal of Type I Suspension by Ideal Settling

Problem

Figure 6.12 shows percent remaining versus overflow velocity, i.e., P versus v_o, for a discrete particle suspension. Assume an overflow velocity, $v_o = 0.11$ cm/s. Determine removal R for the suspension.

Method

Apply Equation 6.18, i.e.

$$R = (1 - P_o) + \frac{1}{v_o}\int v_s dP \qquad \text{(Ex6.4.1)(6.18)}$$

From Equation 6.18, evaluate the function, $R = f(v_o)$. This can be done by integrating the integral graphically and adding the associated $(1 - P_o)$ value.

Solution

From Figure 6.12, for the overflow velocity, $v_o = 0.11$ cm/s, $P_o = 0.94$. The graphical integration is done as indicated in Figure 6.10b. Each square is: [0.01 cm/s · 0.05 fraction] = 0.0005 cm/s/square. Counting the squares above the curve of Figure 6.12 between the limits 0 and 0.11 cm/s, with corresponding $P_o = 0.84$, gives about 73 squares. Thus, 73 squares · 0.0005 cm/s/square = 0.0365 cm/s, the value of the integral. Next, dividing by $v_o = 0.11$ cm/s, gives, (0.0365 cm/s)/(0.11 cm/s) = 0.33 fraction of suspension removed between the limits, $0 < v_s < 0.11$ cm/s. The total removal, $R = (1 - 0.94) + 0.33 = 0.39$.

Comment

As seen by the flatter curvature of Figure 6.12 as v_o increases, the counting error increases. Thus, the count of 73 squares for $v_o = 0.11$ cm/s is perhaps ±1 square.

6.5 FLOCCULENT SUSPENSIONS (TYPE II)

Flocculation occurs when particles collide and attach to each other thus increasing in size. As such, particles collide during settling and grow in size as they settle at an increasingly higher rate, i.e., proportional to the diameter squared, per Stokes' law, Equation 6.6.

Figure 6.13 shows the paths of two particles, designated "1" and "2" in a horizontal flow-settling basin, with horizontal velocity v_H. For each particle, the initial fall velocities are v_1 and v_2, respectively. As the particles grow in size along their mean trajectories, their fall velocities increase correspondingly, causing a changing resultant velocity vector and consequently curved paths.

Inspecting Figure 6.13 shows that size 1 particles will be 100% removed. For the size 2 particles starting at position d_2 above the bottom, the particles reaches the bottom in length, L. This means that all size 2 particles *below* d_2 will be removed and those above will leave the basin. The removal of the size 2 particles is thus $(d_2/D) \cdot \Delta P_2$.

Looking further at Figure 6.13, suppose that 60% of the particles have fall velocities higher than "1," then the trajectory of "1," starting at 0 depth, is also the 60% removal line. Now suppose 70% of the particles have fall velocities higher than "2," then the trajectory of particle "2" at 0 depth is also the 70% removal line. As another means of looking at the trajectories of different particles, consider conducting a settling test in a quiescent cylinder and moving the column along the length of the basin at velocity, v_H. The concepts are the same as for discrete particle settling, except that flocculent particles settle with increasing velocity.

6.5.1 Settling Test for a Flocculent Suspension

Results from a hypothetical settling test of a flocculent suspension are shown in Table 6.3; the column depth is 3.35 m (11 ft). The data are concentrations and can be converted to percent removal by subtracting the values given from 100. The percent removals, as calculated from Table 6.3, are recorded in Figure 6.14 using the respective coordinate positions, i.e., depth is the ordinate, with origin at the top, and time is the abscissa. From the plotted numerical data, iso-percent removal lines were drawn, which are the same as the trajectories of particles of different sizes, as noted previously.

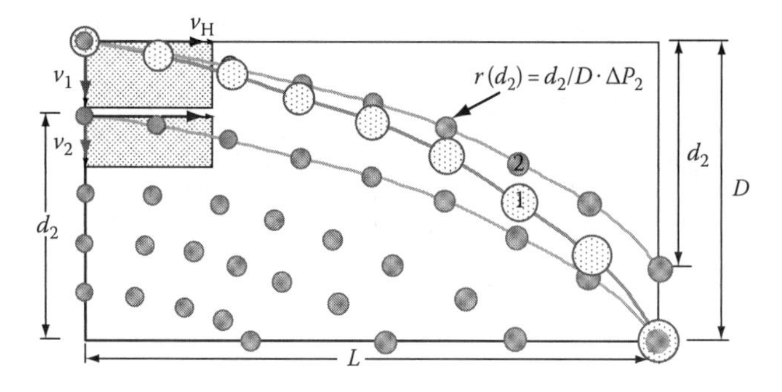

FIGURE 6.13 Illustration of flocculent settling for size 1 and size 2 particles.

TABLE 6.3
Results of Settling Column Test for Type II Suspension

Liquid Depth		Suspended Solids Concentration (mg/L) Time (min)							
(m)	(ft)	0	10	30	50	80	110	140	170
0.61	2.0	100	84	60	23	0	0	0	0
1.22	4.0	100	91	72	50	22	0	0	0
1.83	6.0	100	93	79	64	37	13	0	0
2.44	8.0	100	95	83	70	47	24	7	0
3.05	10.0	100	95	85	73	52	33	15	0

Note: Data are hypothetical.

6.5.2 Determining Percent Removals

Several points are pertinent to Figure 6.14:

1. The iso-percent removal lines represent the settling paths of particular particle sizes, as discussed with respect to Figure 6.13.
2. Consider the 30% line. At any point along the 30% removal line, 30% of the suspension will have been removed.
3. At any point along the 30% line, the average settling velocity can be calculated, and the detention time is a coordinate.

To illustrate the calculation of settling velocities, the third point, consider the coordinate point (3.00 m, 54 min) [(9.84 ft, 54 min)]. The average velocity, v_s, of particles removed was $v_s = 3.00$ m/54 min = 0.055 m/min = 0.00092 m/s ($v_s = 9.84$ ft/54 min = 0.18 ft/min). All particles having a settling velocity equal to or greater than $v_s = 0.055$ m/min (0.18 ft/min) will be removed.

Particles with lower velocities than v_s will be removed in proportion to d/D (or v_s/v_o); to illustrate, consider the 30%–40% removal increment at D(30–40 average) = 2.4 m, $t = 54$ min. The average settling velocity is $v_s = 2.4$ m/54 min = 0.044 m/min ($v_s = 7.87$ ft/54 min = 0.146 ft/min). Therefore, the increment of removal is 0.044 m/min/0.055 m/min) × 10% (or 0.15/0.18 × 10%) to equal 8%. The next 10% increment is handled in the same way. At this point, it should be noted that because the times are the same, the depth ratio is the same as the velocities ratio. From this, a more general algebraic expression can be obtained for percent removal from a horizontal flow basin (see also WPCF, 1985, p. 11):

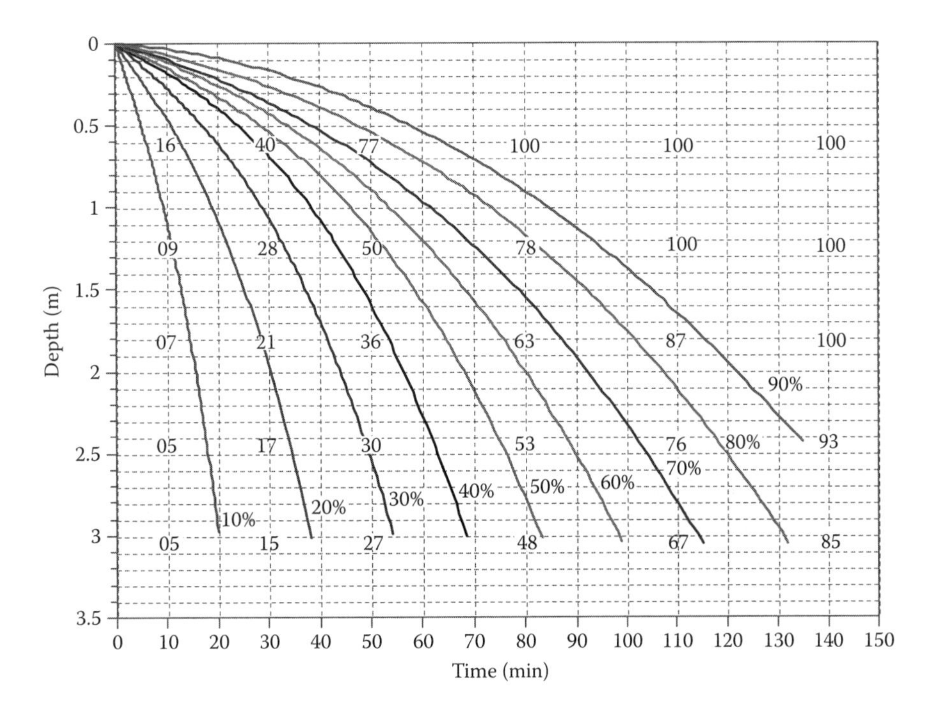

FIGURE 6.14 Percent removals in flocculent settling, plotted from Table 6.3.

$$R(d_o, T) = P_o + \frac{d_n}{d_o}\Delta P_{o,n} + \frac{d_{n-1}}{d_o}\Delta P_{n,n-1} + \frac{d_{n-2}}{d_o}\Delta P_{n-1,n-2} + \qquad (6.19)$$

where

n is the number of increments of equal percent removals from P_o to 0.

P_o is the iso-percent removal line intercepting the depth, d_o, at a given time, T (min)

d_o is the depth of the level at which calculation is started (m)

$d_n, d_{n-1}, \ldots$ is the median depths between each succeeding two iso-percent removal lines, e.g., P_{n-1} and P_{n-2} (m)

Example 6.5 Calculation of Percent Removal at One Coordinate Point for Flocculent Settling

Problem statement

Calculate the percent removal at depth 2.44 m after detention time $T = 50$ min.

Method

Figure 6.14 shows the plot for the results of the settling column test given in Table 6.3. To obtain the total percent removal at depth 2.44 m and 50 min detention time, i.e., R (2.44 m, 50 min), follow the steps given below:

1. Draw a horizontal line at 2.44 m (8.0 ft) depth.
2. Draw a vertical line at $T = 50$ min.
3. Calculate removals by Equation 6.19 as illustrated, i.e.,

$$R(2.44\text{ m}, 50\text{ min}) = 30\% + \frac{1.98\text{ m}}{2.44\text{ m}}\cdot 10\% + \frac{1.37\text{ m}}{2.44\text{ m}}\cdot 10\% + \frac{1.07\text{ m}}{2.44\text{ m}}\cdot 10\% + \frac{0.85\text{ m}}{2.44\text{ m}}\cdot 10\% + \frac{0.67\text{ m}}{2.44\text{ m}}\cdot 10\% + \frac{0.46\text{ m}}{2.44\text{ m}}\cdot 10\% + \frac{0.15\text{ m}}{2.44\text{ m}}\cdot 10\% = 57\%$$

The corresponding calculation in U.S. customary units is

$$R(8.0\text{ ft}, 50\text{ min}) = 30\% + \frac{6.5\text{ ft}}{8.0\text{ ft}}\cdot 10\% + \frac{4.5\text{ ft}}{8.0\text{ ft}}\cdot 10\% + \frac{3.5\text{ ft}}{8.0\text{ ft}}\cdot 10\% + \frac{2.8\text{ ft}}{8.0\text{ ft}}\cdot 10\% + \frac{2.2\text{ ft}}{8.0\text{ ft}}\cdot 10\% + \frac{1.5\text{ ft}}{8.0\text{ ft}}\cdot 10\% + \frac{0.5\text{ ft}}{8.0\text{ ft}}\cdot 10\% = 57\%$$

Discussion

Note that R is the total percent removal, i.e., includes partial removals for particles in which $v_s < v_o$.

6.6 HINDERED AND COMPRESSION SETTLING (TYPE III AND TYPE IV SUSPENSIONS)

Hindered settling (Type III) and compression settling (Type IV) occur in the final clarifiers that follow activated sludge reactors. The analysis of the settling characteristics of such a suspension is important to the performance of final settling basins and is reviewed in this section in several steps.

6.6.1 Settling Velocity as Affected by Solids Concentration

As illustrated in Figure 6.2, particles in a Type III suspension settle with a distinct interface, leaving a clear supernatant above the top of the sludge blanket. As the particles approach the bottom and accumulate, a transition occurs to a Type IV suspension, characterized by compression and decreasing velocity.

6.6.1.1 Settling Tests

To illustrate settling behavior, Figure 6.15 shows results from a settling test for a suspension taken from an activated sludge reactor done in a 1000 mL graduate, such as seen in Figure 6.2. As seen, the rate of subsidence is high initially and then declines continuously; at $t \approx 10$ min, the Type III suspension makes the transition to Type IV and the subsidence rate declines further.

6.6.1.2 Characterizing Settling Velocity

The interfacial settling velocity, v_i, of a Type III/Type IV suspension has been characterized mathematically by Vesilind (1968a) as declining exponentially with X_i, i.e.,

$$v_i = v_I e^{-bX_i} \tag{6.20}$$

where

- v_i is the settling velocity of suspension water–solids interface at depth layer, "i" (m/h)
- v_I is the intercept in the semilog plot of experimental data (m/h)
- b is the slope · 2.303 in the semilog plot of experimental data (L/mg)
- X_i is the solid concentration in the final clarifier at any given level (mg/L)

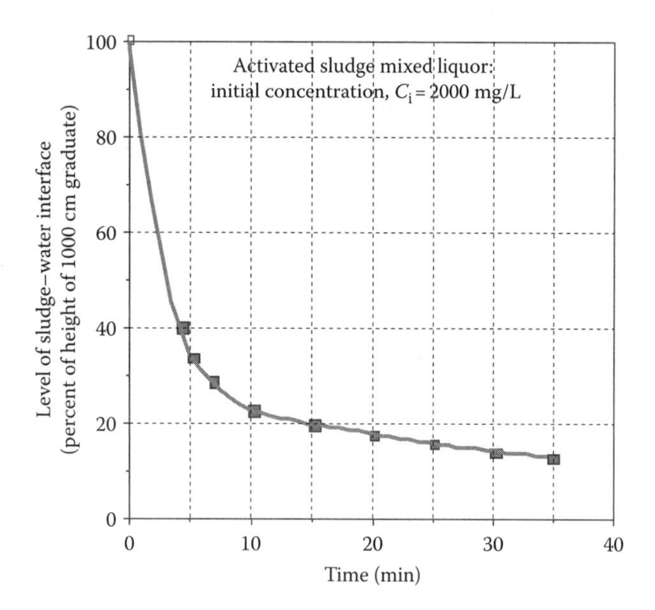

FIGURE 6.15 Settling and compression of activated sludge. (Adapted from Camp, T.R., *Trans. ASCE*, III, 895, 1946.)

TABLE 6.4
Coefficients for Vesilind Equation for Different Suspensions

Floc	v_I (m/h)	b (L/mg)	Reference
Activated sludge[a]	18.6	0.00076	Wahlberg et al. (1993)
Activated sludge	7.62	0.00024	Watts et al. (1996)
Activated sludge	6.37	0.00092	Dick (1970)
Activated sludge	7.80	$0.148 + 0.00210 \cdot \text{SVI}$[b]	Daigger and Roper (1985)
Activated sludge	$v_I = 15.3 - 0.0615$ SVI (stirred)	See footnote[c]	Wahlberg and Keinath (1988)
$Al(OH)_3$	6.62	0.00122	Bhargava and Rajagopal (1993, p. 463) using plots of their data
$Fe(OH)_3$	2.77	0.00099	
$CaCO_3$	2.08	$1.24 \cdot 10^{-5}$	
Bentonite	2.13	$5.36 \cdot 10^{-5}$	

[a] 20 determinations; $\sigma(v_I) = 5.3$, $\sigma(b) = 0.00026$ from San Jose, California. (σ is standard deviation from mean).
[b] Best fit for $36 \leq \text{SVI} \leq 402$ mL/g; 236 settling velocities were measured for two Milwaukee plants.
[c] $b = 0.426 - 0.00384 \cdot \text{SVI (stirred)} + 0.000\,0543 \cdot \text{SVI (stirred)}^2$, in which, SVI(stirred) = volume of 1 g of dry sludge with slow stirring.

The v_I and b constants are obtained from a set of (v_i, X_i) data points, generated from tests conducted using different concentrations for a given suspension. The tests should be conducted using a cylinder about the same depth as the settling basin (Dick, 1970).

Table 6.4 gives v_i and b coefficients of Equation 6.20 for activated sludge suspensions, chemical flocs, and bentonite. The data illustrate the variation in the coefficients and indicate they are unique for a given suspension. The coefficients are affected also by stirring (Vesilind, 1968a). The b coefficient has been related to the sludge volume index (SVI; see glossary) (Daigger and Roper, 1985; Wahlberg and Keinath, 1988). The settling velocity is also affected by the diameter of the cylinder used for the test (Vesilind, 1968b), with a 914 mm (36 in.) cylinder used for reference.

6.6.2 Final Settling as Affected by Limiting Flux Density

A final clarifier with Type III/Type IV settling behavior, as characterized by Equation 6.20, has a limiting settling flux density of suspended solids, which may govern sizing of the plan area. Overflow velocity is the other criterion. The larger of the two plan areas, i.e., A(plan), governs selection. The procedure for determining the limiting flux density, **j**(limit), is outlined here.

6.6.2.1 Activated Sludge

A "conventional" activated sludge system consists of a reactor and a final settling basin, which are illustrated schematically in Figure 6.16. The flows of various water suspensions are shown by arrows; terms are defined in the drawing. The recycle of a thickened sludge, i.e., bacterial cells, back to the reactor, serves as a part of the reaction between the cells at concentration, X, in the reactor and biodegradable wastes at concentration, S. The product of the rate of recycle flow, R, and the cell concentration, X_r, in the thickened underflow determines, in part, the concentration, X, in the reactor (see Section 23.2.2 and Equation 23.14). In principle, the higher the value of X, the faster the reaction rate; a usual achievable target is $X \approx 2000$–2500 mg/L. As seen in Figure 6.16, final settling is an integral part of an activated sludge system.

6.6.2.2 Final Settling Basin Processes

As depicted in Figure 6.16 the final settling basin has two zones: (1) settling and (2) thickening. First, the incoming cells from the reactor, comprising a Type III suspension, must be separated from the water to produce a clarified effluent that meets regulatory requirements, e.g., $X_e \leq 30$ mg/L. Second, the cells that have settled, comprising a Type IV suspension, must be "thickened" (Dick and Ewing, 1967). The thickened solid leaving the basin at concentration, X_r, is often called the "underflow," i.e., $(R + W)$. The value of X_r affects R, X, and W. As a rule, $6{,}000 < X_r < 10{,}000$ mg/L; $X_r \rightarrow 10{,}000$ mg/L is a goal.

6.6.2.3 Mass Balance Relations

Referring to Figure 6.16, several mass balance relations for the final settling basin can be derived, which are useful for later reference. First, the mass flux of solids, J(solids), into the basin is

$$J(\text{solids})_{\text{in}} = (Q + R) \cdot X \tag{6.21}$$

where

$J(\text{solids})_{\text{in}}$ is the mass flux of solids into the final settling basin (kg solids/s)
Q is the flow of water to the system (m^3/s)
R is the return flow of water back to the reactor with solids suspension (m^3/s)
X is the concentration of solids in the reactor and those entering the final settling basin (kg solids/m^3)

The solids flux leaving the basin in the underflow is

$$J(\text{solids})_{\text{out}} = (R + W) \cdot X_r \tag{6.22}$$

where

$J(\text{solids})_{\text{out}}$ is the mass flux of solids into the final settling basin in underflow (kg solids/s)
W is the waste flow of water-solids suspension leaving the system (m^3/s)
X_r is the concentration of solids in the underflow leaving the final settling basin (kg solids/m^3)

Equating (6.21) and (6.22) is a mass balance on the final settling basin, as seen in (23.14). The mass balance shows the relationship between the variables, illustrating the importance of final settling to achieve a high X_t and thus a higher X in the reactor.

Actually, $J(\text{solids})_{\text{inflow}} = J(\text{solids})_{\text{underflow}} + (Q - W)X_e$, but since $X_e \leq 30$ mg/L by regulation, the product $(Q - W)$ X_e is considered negligible; therefore, we let $J(\text{solids})_{\text{inflow}} \approx J(\text{solids})_{\text{underflow}}$, so the two may be used interchangeably, depending on the purpose of a relation.

An initial plan area for a basin, A(plan), is calculated from the expression for overflow velocity, i.e., Equation 6.11:

$$(Q - W) = \text{SOR} \cdot A(\text{plan})_{\text{SOR}} \tag{6.23}$$

This value for A(plan) is used as a first trial for the calculation of the "bulk" velocity of water, u, as it falls in the basin, i.e.,

$$(R + W) = u \cdot A(\text{plan})_{\text{underflow}} \tag{6.24}$$

where

u is the bulk velocity of water falling vertically through the basin (m/s)
A(plan) is the plan area of the basin (m^2)

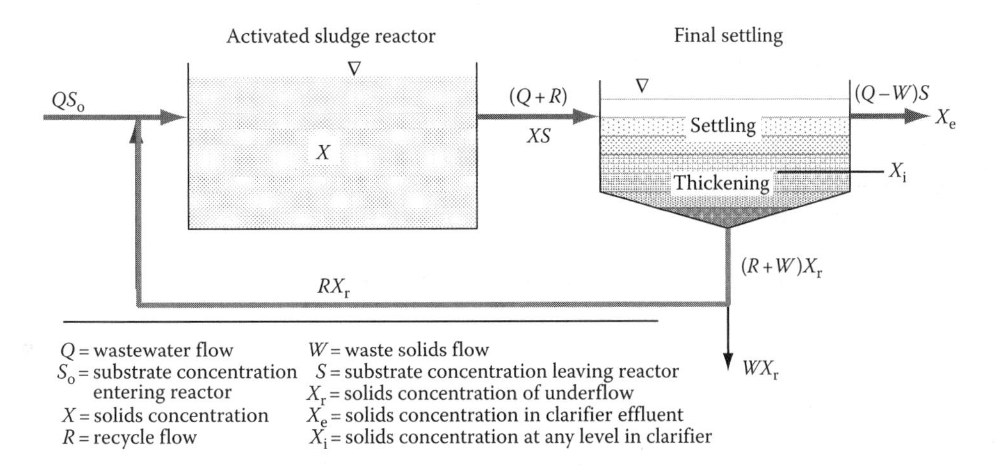

FIGURE 6.16 Schematic diagram of activated sludge system illustrating role of final clarifier.

$A(\text{plan})_{SOR}$ is the plan area of the basin as calculated from SOR (m^2)
$A(\text{plan})_{underflow}$ is the plan area of the basin (m^2)

As a first trial, u is calculated letting $A(\text{plan})_{SOR} = A(\text{plan})_{underflow}$. Equations 6.23 and 6.24 show that for the calculations, the inflow is split into two parts, i.e., overflow $(Q - W)$ and underflow $(R + W)$. In other words, these are the flows associated with $A(\text{plan})_{SOR}$ and $A(\text{plan})_{underflow}$, respectively. Another point is that the value of SOR is from sources such as regulations or the literature, or possibly a settling test.

An identity with Equation 6.24, which gives an alternative for calculating u, is obtained by multiplying both sides of Equation 6.24 by X_r, i.e.,

$$J(\text{solids}) = (R + W) \cdot X_r \quad (6.25)$$

then dividing both sides by A(plan) to give

$$\mathbf{j}(\text{solids}) = uX_r \quad (6.26)$$

where
J(solids) is the total solids flux falling across a horizontal plan of the basin (kg/s)
A(plan) is the actual plan area of the basin based on (m^2)

Thus, Equation 6.26 is an alternative for calculating u, i.e., based on a solids mass balance as opposed to a water balance.

6.6.2.4 Limiting Flux Density

An important issue in the final settling is that the solids flux density, **j**(total), is limited by a "bottleneck" level of X_i, at which a limit, **j**(limit), occurs. The total solids flux falling across a horizontal plane of the basin is

$$J(\text{solids}) = \mathbf{j}(\text{total}) \cdot A(\text{plan}) \quad (6.27)$$

where
J(solids) is the total solids flux falling across a horizontal plan of the basin (kg/s)
j(total) is the total solids flux density falling across a horizontal plan of the basin in the thickening zone (kg solids/s/m^2)
A(plan) is the actual plan area of the basin based on (m^2)

The bottleneck is due to the gradation in solids concentration, X_i, with depth, i.e., as depicted by Figure 6.16. The area of the clarifier, A(plan), must be sufficiently large to accommodate the limiting flux density, **j**(limit), calculated as

$$(R + W) \cdot X_r = \mathbf{j}(\text{limit}) \cdot A(\text{plan})_{\mathbf{j}(\text{limit})} \quad (6.28)$$

where
j(limit) is the limit in total solids flux density falling across a horizontal plan of the basin in the thickening zone (kg solids/s/m^2)
$A(\text{plan})_{\mathbf{j}(\text{limit})}$ is the calculated plan area of the basin based on **j**(limit) (m^2)

If it happens that $A(\text{plan}) < A(\text{plan})_{\mathbf{j}(\text{limit})}$, the excess, i.e., the difference, **j**(total) − **j**(limit), will accumulate above the critical zone and will leave in the clarifier effluent. Thus, the plan area, A(plan), of the final settling basin must be the larger of the two values, i.e., $A(\text{plan})_{SOR}$ or $A(\text{plan})_{\mathbf{j}(\text{limit})}$. As will be seen, the solution for u and, thus, **j**(limit) and, therefore, $A(\text{plan})_{\mathbf{j}(\text{limit})}$ becomes a trial-and-error procedure.

The limiting flux density theory was described in a 1970 ASCE paper by Richard I. Dick (1970), then an assistant professor at the University of Illinois and later a faculty member at Cornell University. The theory has been adopted in practice along with considerations of what is known as "bulking-sludge," which is sludge that does not flocculate well. Bulking sludge will affect the settling curve, i.e., v_i versus X_i, and thus will affect the limiting flux density.

As a note for reference, $J(\text{waste}) = W \cdot X_r$, which equals the mass rate of synthesis of cells in the reactor, i.e., $[dX/dt] \cdot V$, where V is the volume of the reactor. Thus, the performance of the reactor and the final settling basin are linked.

6.6.2.5 Limiting Flux Density: Evaluation Procedure

Figure 6.17 shows four plots that encapsulate the procedural steps to determine **j**(limit). The plots may be incorporated in a spreadsheet algorithm to facilitate the calculation of $A(\text{plan})_{\mathbf{j}(\text{limit})}$.

First, Figure 6.17a shows how the settling velocity of the sludge water interface, v_i, varies with solids concentration, X_i. The plot may be approximated by Equation 6.20 if v_i and b are determined from a least-squares fit to data, or taken from Table 6.4, i.e., if data are not available.

Second, Figure 6.17b is a plot of the product $(v_i \cdot X_i)$, which is **j**(settling) versus X_i, i.e., **j**(settling) $= (v_i \cdot X_i)$. Due to the mathematical character of the Figure 6.17a plot, the exponential decline of v_i with X_i, the flux density, **j**(settling) in Figure 6.17b reaches a peak and then declines with X_i.

Third, Figure 6.17c shows the solids flux density, **j**(bulk), which is the product, (uX_i) versus X_i. As stated, the term u is the "bulk" velocity (also called the drawdown velocity due to advection), which is caused by the "underflow" from the basin, $(R + W)$.

Finally, Figure 6.17d shows the plot, total flux density, **j**(total) versus X_i, where **j**(total) is the sum of **j**(settling) in Figure 6.17b and **j**(bulk) in Figure 6.17c, i.e.,

$$\mathbf{j}(\text{total}) = \mathbf{j}(\text{settling}) + \mathbf{j}(\text{bulk}) \quad (6.29)$$

$$= (v_i \cdot X_i) + uX_i \quad (6.30)$$

where
j(settling) is the flux density of solids due to settling (kg/s/m^2)
j(bulk) is the flux density of solids due to drawdown caused by underflow (kg/s/m^2)

The salient part of Figure 6.17d is that **j**(total) shows a trough, which is the "limiting-flux-density," and is designated, **j**(limit). This is the key design parameter for sizing the basin;

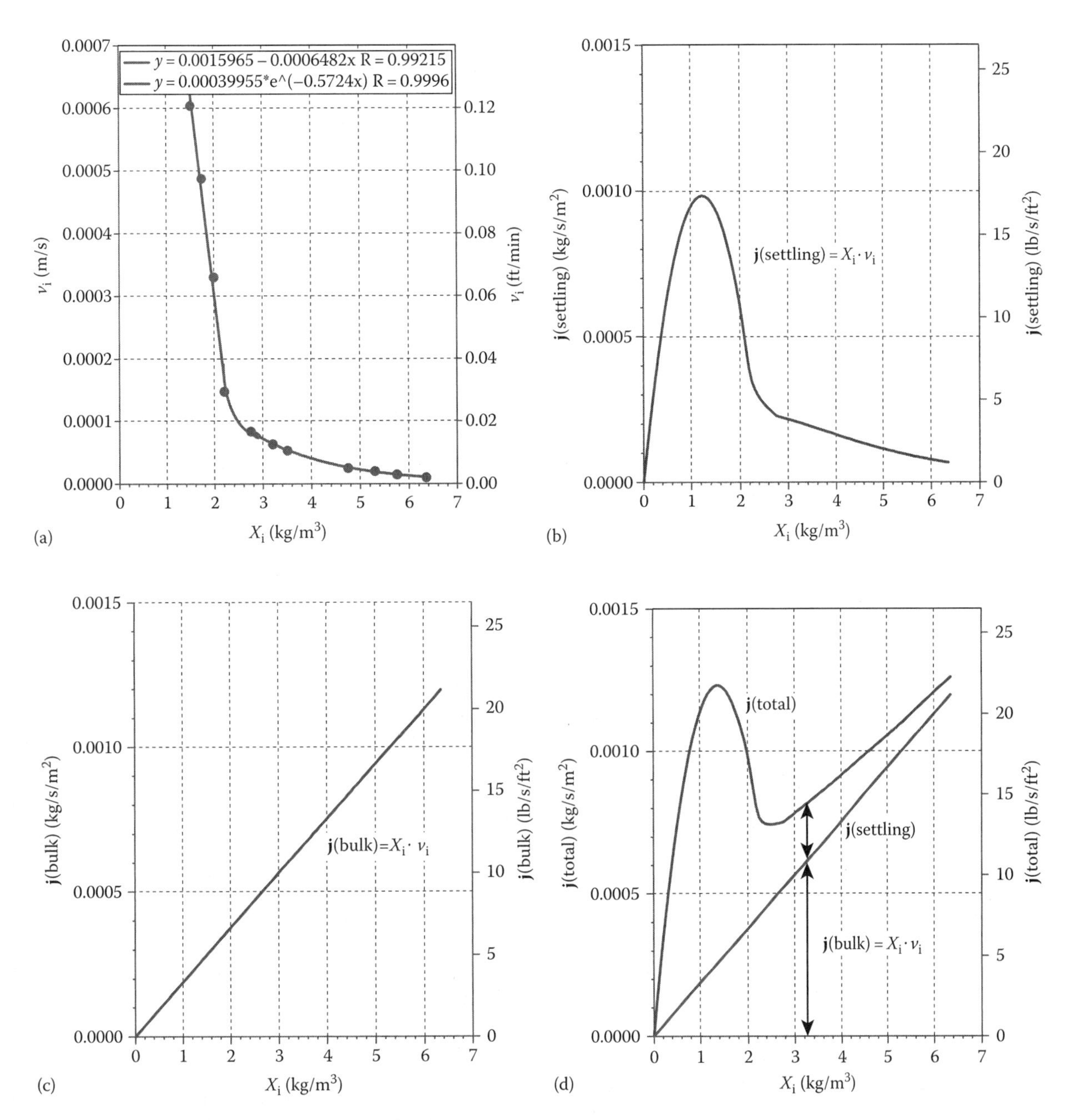

FIGURE 6.17 Solids flux in final clarifier for wastewater treatment: (a) initial settling velocity v, X_i, (b) solids flux density by settling, (c) interface fall due to bulk flow, and (d) total flux density by settling and bulk. (Adapted from Dick, R.I., *J. Sanit. Eng. Div., ASCE*, 96 (SA2), 423, April 1970.)

the level in the sludge blanket where the associated X_i occurs is the "bottleneck." Thus, a design requirement is that $A(\text{plan})_{\mathbf{j}(\text{limit})} \geq J(\text{solids})/\mathbf{j}(\text{limit})$. If $A(\text{plan}) < A(\text{plan})_{\mathbf{j}(\text{limit})}$, the excess solids will leave the clarifier, i.e., $J(\text{excess}) = [J(\text{solids})/A(\text{plan}) - J(\text{solids})/A(\text{plan})_{\mathbf{j}(\text{limit})}] \cdot A(\text{plan})$.

6.6.2.6 Example of Limiting Flux Density Using Plots

Example 6.6 illustrates how to apply the foregoing ideas in terms of numerical data. Table CD6.5 incorporates the equations that generate plots similar to those of Figure 6.17; in addition, it also computes the revised values of X and X_r. A new operating value of R may be tried as a revised input, for which a new value of **j**(limit) is obtained for the linked plots. The foregoing is an introduction to principles involved in taking into account the "solids flux theory." Techniques for evaluating the effects of R/Q and solids loading on X and X_r have been given by Keinath (1985, 1990) and Hermanowicz (1998).

Example 6.6 Sludge Thickening (Adapted from Dick, 1970)

Given

Activated sludge system with $Q = 0.044$ m³/s (1.0 mgd), $X = 2000$ mg/L, $X_r = 6000$ mg/L.

TABLE CD6.5
Materials Balance Calculations for Area of a Final Settling Basin (Dick, 1970)—Excerpt from Spreadsheet

Q (m^3/s)	A (m^2)	X (mg/L)	X_r (mg/L)	R (m^3/s)	u (m/s)	X_i (mg/L)	v_i (m/h)	j(settling) (kg/day/m^2)	j(drawdown) (kg/day/m^2)	j(total) (kg/day/m^2)
0.0438	116.1	2000	6000	0.022	0.00019	0	0.00	0.00	0.00	0.00
						50	21.95	26.33	0.81	27.15
						500	4.39	52.67	8.15	60.82
discontinuity in table; see spreadsheet for full table with explanations of terms and linked plots										
						4500	0.09	10.07	73.33	83.41
						5150	0.05	6.78	83.93	90.71
						5700	0.04	5.00	92.89	97.89

Required

Determine area of final clarifier, A(plan), based on sludge thickening requirements.

Method

Apply the foregoing theory from Dick (1970)

Solution

(a) Determine R.

Equating (6.27) and (6.28), permits calculation of R. Assume $W = aR$, and let $a = 0$ for this illustration. Thus,

$$(Q + R)X = (R + W)X_r$$
$$(0.044 + R) \cdot 2000 = (R + 0) \cdot 6000$$
$$R = 0.0219 \text{ m}^3/\text{s (0.5 mgd)}$$

(b) Calculate J(solids),

$$J(\text{solids}) = (Q + R)X$$
$$= (0.044 \text{ m}^3/\text{s} + 0.0219 \text{ m}^3/\text{s}) \cdot 2000 \text{ mg/L}$$
$$= 0.132 \text{ kg/s (0.290 lb/s)}$$
$$= 11{,}405 \text{ kg/day (25,148 lb/day)}$$

The same result is obtained using Equation 6.28.

(c) Obtain **j**(limit) from Figure 6.17d, i.e., **j**(limit) $\approx$ 0.00075 kg/s/m^2 (65 kg/day/m^2/13.6 lb/day/ft^2)

(d) Apply Equation 6.19,

$$J(\text{total}) \leq \mathbf{j}(\text{limit}) \cdot A(\mathbf{j}\text{limit})$$
$$11{,}405 \text{ kg/day} \leq 65 \text{ kg/day/m}^2 \cdot A(\mathbf{j}\text{limit})$$

or, in SI units,

$$0.1314 \text{ kg/s} \leq 0.00075 \text{ kg/s/m}^2 \cdot A(\mathbf{j}\text{limit})$$
$$A(\mathbf{j}\text{limit}) = 175 \text{ m}^2 \text{ (1850 ft}^2\text{)}$$

(e) Calculate plan area based on surface overflow velocity, SOR, i.e., A(SOR), by Equation 6.23,

$$(Q - W) = \text{SOR} \cdot A(\text{SOR})$$
$$(0.044 - 0.000) \text{ m}^3/\text{s} = 0.00038 \text{ m/s} \cdot A(\text{SOR}) \text{ 175 m}^2$$
$$= 116 \text{ m}^2 \text{ (1250 ft}^2\text{)}$$

Guidelines suggest SOR $\leq$ 800 gal/day/ft^2, which is the same as SOR $\leq$ 0.00038 m/s in SI units.

Comments

Of the two areas calculated for A(plan), i.e., A(SOR) = 116 m^2 and A(**j** limit) = 175 m^2, the largest is selected. This means that another iteration is required, since u is affected per Equation 6.24, $(R + W) = u \cdot A$(plan), which changes **j**(bulk) and thus, **j**(limit), with X and X_r changing also, e.g., the latter changing per Equation 6.26, **j**(total) $= u \cdot X_r$.

The clarifier cannot handle a flux density higher than **j**(limit), e.g., 0.00075 kg/s/m^2 (65 kg/day/m^2/13.6 lb/day/ft^2) as seen as the low point in the plot, i.e., Figure 6.17d. As Dick (1970, p. 430) stated, it is not physically possible for **j**(total)>65 kg/day/m^2 (13.6 lb/day/ft^2) to pass through a solids layer in which $X_i \approx 2500$ mg/L. If **j**(total) > **j**(limit), the clarifier cannot handle the load as the excess solids flux cannot reach the bottom and will appear as effluent overflow. The effect of insufficient solids handling capacity may not be recognized and instead may be viewed as poor settling. The latter may be true also, but the two effects must be distinguished.

Table CD6.5 is a spreadsheet developed from the above procedure and provides for modifying any of the primary input variables, such as Q, R, SOR, and perhaps X, X_r. The spreadsheet can also incorporate different experimental data, such as in Figure 6.17a. The last two columns of the spreadsheet give the new values, $X = 1.316$ kg/m^3, $X_r = 3.948$ kg/m^3, which deviate from the goals, $X = 2.000$ kg/m^3, $X_r = 6.000$ kg/m^3. As Dick (1970, p. 430) stated, alternate operating conditions cannot produce the desired values of X and X_r. For example, letting $(R/Q) = 0.70$ will increase u, and thus **j**(bulk), with a higher value of **j**(limit), with $X = 1.485$ kg/m^3, $X_r = 3.606$ kg/m^3, i.e., the gain is not significant. If a design is pending, the solution is iterative using different values of SOR and R. Other techniques are reviewed by Dick (1970, p. 431); see also Wahlberg and Keinath (1988), Keinath (1990), and Hermanowicz (1998).

6.7 HYDRAULICS OF SETTLING BASINS

Settling in real basins is affected by flow patterns that deviate from the ideal basin. Dispersion dye or salt tests permit evaluation of such deviations. Alternatively, flow patterns may be depicted by CFD, see Box 6.1, which is based upon hydraulic theory and executed by computer simulation. Review of these topics helps to understand behavior of real settling basins, albeit it does not permit predictions of performance.

6.7.1 Flow Patterns and Short Circuiting

Real flow is characterized by "short circuiting" and "dead zones," both illustrated in Figure 6.18. The configuration of inlet "source" and outlet "sink" are exaggerated. The term "short circuiting" means that a portion of the flow entering the basin reaches the basin exit at time, t_i, much more quickly than the detention time, q; i.e., $t_i \ll q$ (where t_i is the time for initial appearance of tagged molecules after being added to the flow at the entrance to a given basin). At the same time, the dead zones result in a portion of the flow leaving the basin at time $t \gg q$. This spread is called hydraulic "dispersion" and can be evaluated by inserting a dye or other tracer, e.g., brine or chloride ion concentrate, in the inflow to the basin.

As implied by Figure 6.18, the design of inlet and outlet are important in determining the flow patterns within a basin. While it is not possible to avoid short-circuiting it can be minimized through attention to design of inlet and outlet. A long narrow basin has a smaller proportion of its volume taken up by dead zones, and the inlet and outlet flow distortions are proportionately less, and is favored for both theoretical and practical reasons.

6.7.2 Density Currents

The causes of density currents may be (1) a cold source water enters a basin of warmer water; (2) a warm source water enters a basin of colder water; and (3) a concentrated suspension, e.g., activated sludge "mixed liquor," enters a final settling basin. As an example, a cold-water density current results in a "plunging" flow with a warmer-water dead zone on top. Even small differences in temperature, e.g., 0.3°C, between the influent water and the water in the basin may cause a density flow as may a turbidity differential of 50 NTU (Kawamura, 1996, p. 133). Density currents may override all other kinds of short-circuiting.

6.7.3 Dispersion Tests Using a Tracer

The traditional means to assess the hydraulic characteristics of a basin is to perform a dye dispersion test (Camp, 1946). To conduct the test, a "tracer" is injected in the influent flow; its concentration in the basin effluent is then measured with time. A suitable tracer may be any substance that does not react or degrade and that may be detected at low concentrations. Such tracers include Rhodamine WT a fluorescent dye (which is detectable at very low concentrations with a fluorometer), a brine solution (measuring conductivity), chloride ion, fluoride ion, and lithium chloride (Crawford, 1994, p. 25). To conduct a test, a highly concentrated batch is poured into the inflow to the basin (in a gallon jug, a barrel, etc.), with size depending on the scale involved; the mass input should be such that the tracer can be detected in the effluent.

6.7.3.1 Results of Dispersion Tests

The dispersion of the fluid may be measured by the parameters, t_i/θ and t_A/θ (t_i is the time of initial appearance of the tracer; t_A is the time to the center of gravity of the area under the dispersion curve; and q is the detention time of the basin, a computed term, in which, $V(\text{basin}) = Q \cdot q$). Small values of t_i/θ indicate significant short-circuiting; the term "short-circuiting" is relative so there is no threshold value of t_i/θ.

Figure 6.19 illustrates the variation that occurs for diverse types of basins. Basins A and F are theoretical extremes; A is a "complete-mix" basin while F is "plug-flow" for an ideal basin; they are included to permit comparisons with real basins. For example, if the dispersion parameters for a given real basin approach those of "A," it would likely be rejected as a design. On the other hand, the closer the parameters approach those of "F," the better the performance. Thus, the curves, B, C, and D in Figure 6.19 illustrate progressively "better" types of settling tanks as measured by how far they deviate from F, the ideal basin.

Table 6.6 summarizes the dimensions and dispersion parameters, t_i/θ and t_A/θ, for basins B, C, and D. In reviewing Figures 6.19 and 6.6, it is evident that B, the radial flow basin, has dispersion characteristics that deviate the most from F, the "ideal basin" with C and D becoming progressively closer. Of the three real basins, the long narrow shape is closest to the ideal, which was corroborated by Langelier

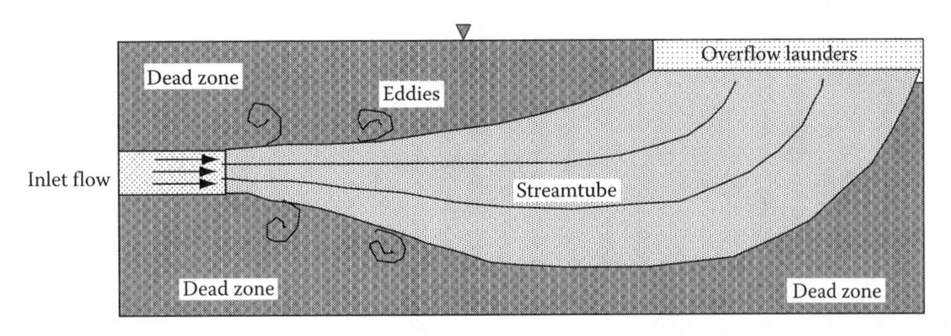

FIGURE 6.18 Illustration of short-circuiting for submerged jet flow to overflow launders showing streamlines enclosing equal flows and dead zones; eddies "peel" from main flow.

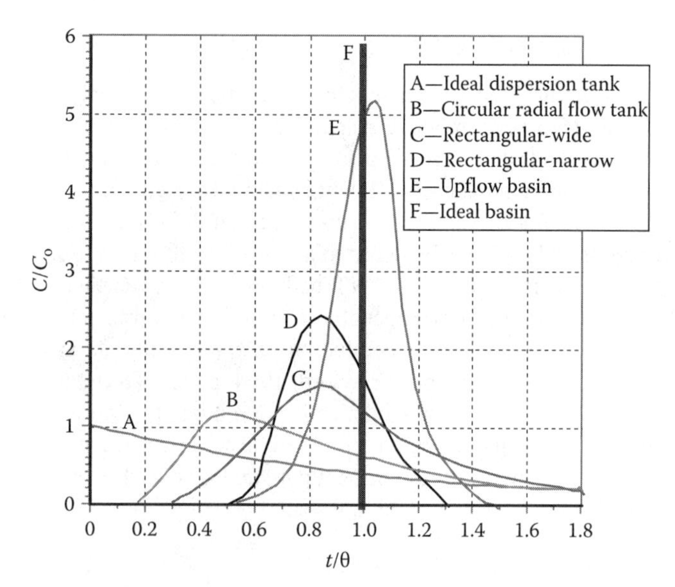

FIGURE 6.19 Examples of dye dispersion curves for six basins. (Adapted from Camp, T.R., *Trans. ASCE*, III, 895, 1946.)

(1930) in a study of the performance of such a basin—a 2 m wide, 1 m deep, 2650 m long tunnel used to settle a flocculent suspension. As noted previously, the up-flow basin circumvents some of the issues of horizontal-flow basins, as seen by its dispersion parameters.

6.7.4 Computational Fluid Dynamics

The advent of CFD using commercial software, c. 1995, has provided a means to model the intricacies of fluid flow in a given basin and to predict the dispersion curve (see also Box 6.1). The use of such software requires knowledge of hydraulic theory, skill, and, in accordance with the adage of any mathematical modeling, valid inputs.

6.8 DESIGN PRACTICE

Empirical guidelines have been the basis for a variety of settling basin designs. Information is provided in this section to give perspective on the range of practice and to illustrate designs.

6.8.1 Categories of Basins

Basins have been categorized by a variety of defining characteristics including

1. Direction of flow, i.e., horizontal flow and up-flow
2. Shape, e.g., long and narrow, square, circular
3. Position in the treatment train
   a. Plain sedimentation (to settle non-colloidal mineral sediments in ambient water)
   b. Flocculent settling (for chemical floc to remove mineral colloids from ambient water)
   c. Primary settling in wastewater treatment (for settleable organic matter)
   d. Secondary settling (for biological floc)
   e. Grit chambers (which are a part of the headworks in wastewater treatment, Chapter 7)
4. Tube settlers and plate settlers (Section 6.10)

TABLE 6.6
Hydraulic Characteristics of Typical Tanks from Figure 6.19 Curves

Curve	Type of Tank	w (m)	D (m)	L (m)	v_H (m/min)	t_i/θ	t_A/θ
A	Complete mix					0	0.693
B	Circular—radial flow		4.3		0.34	0.14	0.831
C	Rectangular—wide	41.1	5.6	100	0.95	0.30	0.925
D	Rectangular—narrow	4.9	4.3	83	1.58	0.52	0.903
E	Up-flow basin					0.5	0.95
F	Ideal basin—plug flow					1.0	1.0

Note: Notes pertaining to Table 6.6 and Figure 6.19 (except for *E*) are from Camp (1946), Dimensions are from basins with measurements by Camp. Basin dimensions are w, width; D, depth; L, length, v_H, horizontal velocity; t_i, time for initial appearance of tracer; t_A, time to center of gravity of area under dispersion curve; θ, detention time, i.e., q(basin) = V(basin)/Q. Blank cells indicate data were not available.

Basin	Comments
A	Complete mix is a theoretical limit in settling basin design; settling cannot occur and curve is shown for reference
B	Primary settling tank 200 ft in diameter (proposed for the Detroit Sewage Treatment Works) $\theta = 90$ min
C	Sedimentation basins at the Springwells Filtration Plant, Detroit
D	Primary settling tanks constructed for the Detroit Sewage Treatment Works (in preference to the circular tanks characterized by curve B)
E	From Burns and Roe (1971); the up-flow basin is included for reference
F	Theoretical, i.e., "plug-flow" shown for reference as a theoretical limit

5. Solids contact basins in which the coagulated water passes through a blanket of solids, i.e., a chemical floc
6. Flotation basins (Chapter 8)
7. Innovative technologies
   a. Actiflow® process is an example in which sand is attached by polymers to the flocculent material. The higher specific gravity material incorporated into the floc results in a markedly higher gravity force and thus higher fall velocity.
8. Suspension category:
   a. Discrete particle
   b. Flocculent particle
   c. Hindered
   d. Compression
   e. Oil–water separations

With these different kinds of settling basins, the same fundamentals apply. To reiterate, these are (1) characteristics of the suspension and (2) the hydraulics of the basin.

6.8.2 Examples of Designs

Design practice encompasses a wide range of basin types, geometry, suspensions, and parameter values. Data on primary wastewater basins are given for reference (as adapted from ASCE-WPCF (1959); used and cited by Nemerow (1971) and Nemerow and Agardy (1998)). In addition, guidelines are compiled from various sources for design of basins for the four suspension types.

6.8.2.1 Horizontal Flow

The traditional design is the horizontal flow rectangular basin, conceptually like the ideal basin as illustrated in Figure 6.8 for a real basin, however, appurtenances must be added for inlet water distribution, for outlet flow, and for sludge removal. Figure 6.20 is a sketch that depicts a real basin showing the inlet, outlet, solids scraping equipment, solids storage, and settling zone. A hydraulic roll characterizes the settling zone, i.e., the associated velocity vectors are horizontal along the bottom, up toward the overflow launders, and then along the top and back toward the entrance.

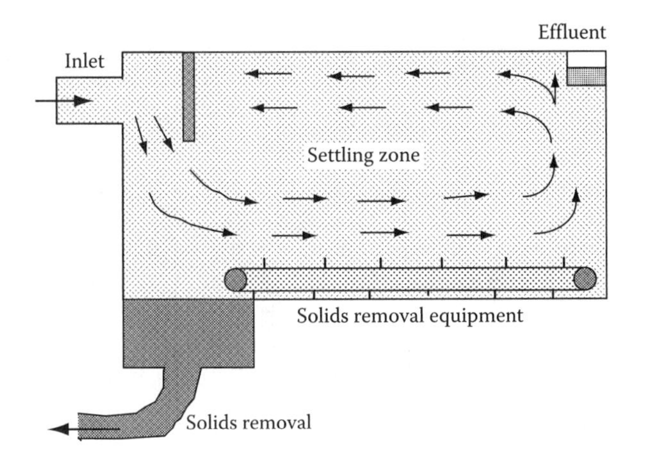

FIGURE 6.20 Conventional rectangular settling basin illustrating hydraulic roll. (Adapted from Katz, W.J. et al., *Concepts of Sedimentation Applied to Design*, Part 2, *Water and Sewage Works*, 169, May 1962b.)

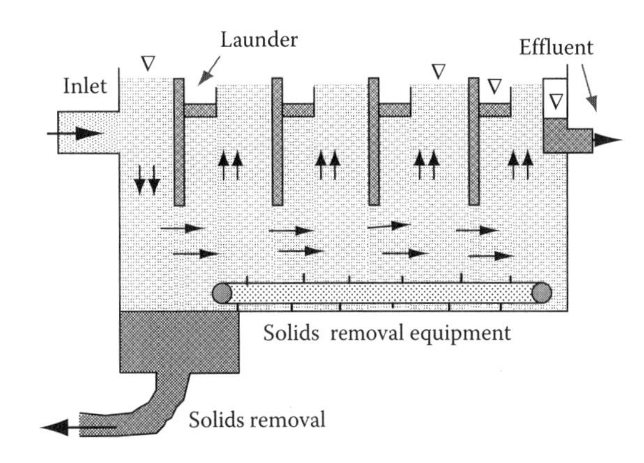

FIGURE 6.21 Up-flow sedimentation basin. (Adapted from Katz, W.J. et al., *Concepts of Sedimentation Applied to Design*, Part 2, *Water and Sewage Works*, 169, May 1962b.)

6.8.2.2 Up-Flow

Figure 6.21 illustrates the idea of an up-flow basin. The plan area is covered with launders, which forces the flow vertically upward, over the weirs. The basin is fitted with the same appurtenances as the horizontal flow basin. The upward velocity, v_o, is set as the overflow velocity and so there is no partial removal of particles in which $v_s < v_o$. The up-flow basin circumvents many of the hydraulic problems of horizontal-flow basins.

6.8.2.3 Data from Real Basins

Tables 6.7 and 6.8 show flows, dimensions, θ and v_o, suspended solids, and BOD data for rectangular and circular primary settling basins, respectively. Comparing these characteristics shows a wide range for each of the two basin types. Flow per basin is as low as 0.0033 m^3/s (0.07 mgd) and as high as 4.3 m^3/s (12 mgd) for rectangular basins. Depths varied 2.4–4.6 m (8–15 ft), while lengths were up to 82 m (270 ft), with widths 3.7–36 m (12–117 ft) and θ ranged 1.0–5.1 h. Overflow velocity ranged 0.76–3.75 m/h (450–2210 gpd/ft^2). Inspecting removals, the 67% suspended solids removal and 33% BOD removal for primary settling basins, the rule of thumb (for removals) from lore, appears to be corroborated. Despite the problems inherent to circular basins, they appear to perform about the same as the rectangular basins, as seen by comparing the removals in Tables 6.7 and 6.8.

6.8.3 Guidelines and Criteria for Design

Design parameters for settling basins include detention time, θ, overflow velocity, v_o, solids loading rate, horizontal velocity, v_H, weir loading rate, Q/L(weir), and depth (WPCF, 1985, p. 4). Also, w/L, and L/D should be included. As seen, practice encompasses a wide variation in the foregoing parameters as do the guidelines abstracted from the literature and summarized in Tables 6.9 through 6.11.

TABLE 6.7
Rectangular Primary Settling Tank Data

Plant Location	Q		Tanks	L		w		D		L/w	L/D	θ	v_o		Weir	Flow	SS_{in}	R(SS)	BOD_{in}	R(BOD)
	(m^3/s)	(mgd)	(#)	(m)	(ft)	(m)	(ft)	(m)	(ft)			(h)	(m/h)	(gpd/ft^2)	(m^3/h/m)	(gpd/ft)	(mg/L)	(%)	(mg/L)	(%)
Hartford, Connecticut	1.06	24.3	8	30.5	100	20.7	68	2.74	9	1.5	11.4	3.53	0.764	450	29.39	56,800	173	61	240	42
Detroit, Michigan	18.31	418	8	82.3	270	35.7	117	3.96	13	2.3	20.8	1.41	2.801	1650	211.11	408,000	184	44	153	39
Racine, Wisconsin	0.74	17	4	42.7	140	12.2	40	3.05	10	3.5	13.3	2.48	1.290	760	55.10	106,500	149	67	133	48
NYC (Bowery Day)	1.80	41	3	37.8	124	15.8	52	3.66	12	2.5	10.3	0.98	3.752	2210	146.95	284,000	152	39	169	22
NYC (Talkmans Is.)	1.36	31	3	37.8	124	15.2	50	3.66	12	2.5	10.7	1.25	2.835	1670	111.24	215,000	137	55	128	39
Fort Wayne, Indiana	0.82	18.7	3	30.5	100	10.1	33	3.96	13	3.3	7.7	1.25	3.208	1890	48.90	94,500	409	61	231	34
Kenosha, Wisconsin	0.56	12.77	4	40.2	132	9.8	32	3.05	10	4.1	12.7	2.49	1.282	755	51.74	100,000	138	48	102	48
Jackson, Michigan	0.40	9.17	3	20.4	67	9.4	31	3.05	10	2.2	6.7	1.22	2.495	1470	61.06	118,000	193	16	134	22
Hammond, Indiana	0.91	20.7	6	36.6	120	4.9	16	3.96	13	7.5	9	1.32	3.056	1800	12.42	24,000	273	30	206	25
NYC (26th Ward)	1.80	41	4	49.4	162	20.4	67	3.66	12	2.4	13.5	2.16	1.579	930	18.37	35,500	139	31	127	28
NYC (Hunts Point)	4.16	95	4	51.2	168	33.2	109	3.66	12	1.5	14	1.70	2.207	1300	50.19	97,000	140	48	113	30
Abington, Pennsylvania	0.05	1.24	2	15.2	50	4.3	14	3.05	10	3.6	5	2.02	1.451	855	22.99	44,430	237	39	98	29
Portsmouth, Virginia	0.32	7.36	0.4	30.5	100	4.6	15	3.05	10	6.5	10	1.49	2.037	1200	23.80	46,000	153	63	185	45
Canton, Ohio	0.74	17	3	37.8	124	9.8	32	3.35	11	3.9	11.7	1.33	2.428	1430	110.73	214,000	577	40	253	33
Niles, Michigan	0.10	2.3	6	22.9	75	4.3	14	2.74	9	5.4	8.3	1.86	0.615	362	14.07	27,200	250	69	106	57
Dallas, Texas	0.85	19.4	2	54.9	180	15.2	50	3.66	12	3.6	15	2.00	1.833	1080	12.42	24,000	358	66	256	41
Richmond, Virginia	0.27	6.1	4	29.0	95	4.9	16	4.27	14	5.9	6.5	2.64	1.686	993	12.94	25,000	159	40	133	23
Lansing, Michigan	0.72	16.45	16	26.8	88	4.9	16	3.05	10	5.5	8.7	2.45	1.248	735	12.26	23,700	445	76	201	68
Waterbury, Connecticut	0.61	13.94	3	64.6	212	10.1	33	3.05	10	6.4	21.2	2.71	1.120	660	7.50	14,500	144	54	166	33
Oklahoma City, Oklahoma	0.23	5.19	3	25.9	85	10.1	33	3.05	10	2.5	8.5	2.91	1.051	619	10.56	20,400	242	50	228	31
Tampa, Florida	0.54	12.3	4	51.8	170	12.2	40	3.96	13	4.2	13.1	5.12	0.772	455	8.95	17,300	215	69	183	37
Roanoke, Virginia	0.34	7.76	2	36.6	120	9.8	32	3.05	10	3.8	11.4	1.87	1.715	1010	62.09	120,000	230	67	190	51
East Hartford, Connecticut	0.07	1.5	2	38.1	125	9.8	32	2.44	8	3.9	16.7	7.18	0.317	187	6.47	12,500	212	54	242	50

Sources: Adapted from ASCE-WPCF, *Sewage Treatment Plant Design*, ASCE Manual of Practice No. 36, WPCF Manual of Practice No. 8, American Society of Civil Engineers and Water Pollution Control Federation, Headquarters of the Society, New York, 1959; Nemerow, N.L., *Liquid Wastes of Industry, Theories, Practices, and Treatment*, Addison-Wesley, Reading, MA, 1971; Nemerow, N.L. and Agardy, F.J., *Strategies of Industrial and Hazardous Waste Management*, Van Nostrand Reinhold, New York, 1998.

TABLE 6.8
Circular Primary Tanks Performance Data

Location	Q (m³/s)	Q (mgd)	Tanks (#)	Dia (m)	Dia (ft)	D (m)	D (ft)	θ (h)	v_o (m/h)	v_o (gpd/ft²)	Suspended Solids SS_{in} mg/L	Suspended Solids SS_{out} mg/L	Suspended Solids Rem (%)	BOD BOD_{in} mg/L	BOD BOD_{out} mg/L	BOD Rem (%)	Sludge Solids (%)	Sludge Vol. (%)
Washington, District of Columbia	5.97	136.3	12	32.3	106	4.3	14	1.88	2.292	1350	163	83	49	173	120	30.5	8.1	67
Winnipeg, Manitoba	1.00	22.8	2	35.1	115	3.7	12	1.98	1.867	1100	348	159	55	310	231	25.5	9.0	70
Battle Creek, Michigan	0.22	4.92	2	24.4	80	3.0	10	3.66	0.832	490	282	85	70	264	174	34.1	5.5	82
Buffalo, New York	5.91	135	4	48.8	160	4.6	15	1.6	2.869	1690	209	114	46	138	107	22.5	5.8	59
Albuquerque, New Mexico	0.22	5.0	1	24.4	80	3.7	12.2	2.21	1.689	995	254	91	61	282	ISO	44.5	3.9	81
Yakima, Washington	0.42	9.5	4	27.4	90	2.7	9	4.32	0.633	373	110	23	74	175	92	50	7.0	74
Appleton, Wisconsin	0.21	4.8	2	21.3	70	3.0	10	2.9	1.058	623	276	63	77	284	141	50	5.6	58
Baltimore, Maryland	3.92	89.5	3	51.8	170	3.7	12	1.64	2.309	1360	214	83	61	281	20t	27.5	3.9	83
Springfield, Ohio	0.65	14.8	2	27.4	90	3.0	10	1.55	1.969	1160	166	63	62	90	43	52		
Mansfield, Ohio	0.13	3.0	1	19.8	65	3.7	12	2.38	1.536	905	208	87	58	227	139	38.8	4.2	76
Cedar Rapids, Iowa	0.18	4.21	1	21.3	70	3.5	11.5	1.95	1.782	1050	354	132	63	383	291	24	5.5	81
Austin, Texas	0.25	5.64	1	22.9	75	3.7	12	1.69	2.164	1275	263	95	64	285	152	46.3	4.0	83
Denver, Colorado	2.02	46	4	42.7	140	3.0	9.7	2.34	1.273	750	187	44	77	212	108	49	5.4	76
Ypsilanti, Michigan	0.07	1.66	2	12.2	40	2.7	9	2.5	1.120	660	I236	87	62	14′	95	33	8.2	71
Monroe, Louisiana	0.19	4.3	2	25.9	85	2.3	7.5	3.55	0.643	379	329	75	77	135	73	46	5.2	68

Sources: Adapted from ASCE-WPCF, *Sewage Treatment Plant Design*, ASCE Manual of Practice No. 36, WPCF Manual of Practice No. 8, American Society of Civil Engineers and Water Pollution Control Federation, Headquarters of the Society, New York, 1959; Nemerow, N.L., *Liquid Wastes of Industry, Theories, Practices, and Treatment*, Addison-Wesley, Reading, MA, 1971; Nemerow, N.L. and Agardy, F.J., *Strategies of Industrial and Hazardous Waste Management*, Van Nostrand Reinhold, New York, 1998.

TABLE 6.9
Suggested Surface Overflow Velocities for Discrete Particles

Particle	Size (mm)	SG	v_o (m^3/min/m^2)	v_o (gpm/ft^2)
Sand	1.0	2.65	60	144
Silt	0.2	2.65	1.27	30
Clay	0.04	2.65	0.12	3
Alum floc	1–4	1.001	0.012–0.055	0.3–1.3
Lime floc	1–3	1.002	0.025–0.072	0.6–1.7

Source: Adapted from Kawamura, S., *Integrated Design of Water Treatment Facilities*, John Wiley & Sons, New York, 1991, p. 132.

The most common guidelines are those of state regulatory agencies; most have been adopted from the standards of the Great Lakes—Upper Mississippi Board of State Sanitary Engineers (1968, 1987), usually called the "Ten States Standards" (see http://www.hes.org). Other references include manuals of practice by ASCE-WPCF (1959), AWWA-ASCE (1998), and WPCF (1985).

The specific guidelines reviewed here were organized in terms of the four types of suspensions. Subcategories include potable water treatment and wastewater treatment, whether the suspension is a chemical flocs (alum, iron, lime) or a biological floc, and type of basin, e.g., horizontal flow, up-flow, rectangular, circular.

6.8.3.1 Discrete Particle Suspensions: Type I

Included in discrete particle suspensions are plain sedimentation in potable water treatment and grit chambers in municipal wastewater treatment (design of grit chambers is the topic of Chapter 7). Type I suspensions are found in many industrial process streams as well. Design parameters for Type I suspensions include overflow velocity, v_o, horizontal velocity, v_H, weir loading rate, w/L ratio, and L/D ratio. In general, the basins should be as shallow as possible with depth being governed by the wD product required to prevent scour, i.e., $v_H = Q/(wD)$. Definitions are: w is the width of basin, L is the length of basin, D is the depth of basin, v_H is the average horizontal velocity of water within basin, and Q is the flow of water through basin. Weir loadings should be the same as other categories of basins. Table 6.9 (Kawamura, 1991, p. 132) lists several surface loading rates for different suspensions.

6.8.3.2 Flocculent Suspensions: Type II

Flocculent suspensions include chemical flocs and biological flocs. Design parameters are detention time, θ; overflow velocity, v_o; horizontal velocity, v_H; weir loading rate; depth; w/L ratio; and L/D ratio (WPCF, 1985). Primary settling (of municipal wastewaters) is considered Type II (Camp, 1953; WPCF, 1985, p. 10). In potable water treatment, the turbidity of settling basin effluent, i.e., after flocculation, should be ≤ 2 NTU (Kawamura, 1996, p. 131). Table 6.10 is a summary of parameter criteria for basins settling Type II suspensions, with sources indicated by footnotes.

6.8.3.3 Flocculent Suspensions–Hindered Settling: Type III

In hindered settling, the particles (either discrete or flocculent) interfere with one another as they settle. Flocculent suspensions are most common and include both chemical and biological floc. Usually, a suspension enters a basin as Type II and becomes Type III as it settles, and then becomes Type IV at the bottom. Design parameters include detention time, θ; overflow velocity, v_o; solids loading rate; horizontal velocity, v_H; weir loading rate; depth; w/L ratio; and L/D ratio (WPCF, 1985, p. 4). Table 6.11 provides guidelines for these parameters from sources indicated. The "no-data" cells are included to indicate that information is lacking.

6.8.3.4 Compression Settling: Type IV

As stated in Section 6.2.3.4, the Type IV suspension develops as the Type III suspension increases in concentration near the bottom of the clarifier such that the particles support one another and "consolidation" occurs. Usually, the suspension is stored in a bottom compartment. Chemical floc may be stored for longer periods than biological floc, which will result in further consolidation. Biological floc is reactive and may be stored only for a few hours, e.g., 2–6 h, depending on the rate of production of gas (a mixture of methane and carbon dioxide), since the gas causes the sludge to float. With respect to the effect of anaerobic conditions on the viability of activated sludge, Dick (1976, p. 638) stated that cells could be stored for ≤ 24 h without affecting the reactor performance.

Type IV settling occurs at the bottom of primary settling tanks used for municipal wastewater treatment. Usually, the tanks are constructed with a recessed volume incorporated in the bottom. The dimensions are about 3–4 m (10–12 ft) diameter and 1 m (3 ft) deep. The sludge is scraped toward this volume where it is stored and thickened. The "thickening" is by means vertical structural posts comprising a part of the scraping structure (called a "picket fence thickener"). The posts move slowly through the solids displacing water (and trapped gases) and allow consolidation. Without the picket fence thickener, the sludge may be perhaps 3% solids but near 5% with its use (with about 6% as an upper limit). After thickening, the sludge is pumped to a digester. The higher solids concentration permits longer solids residence time in the digester; since the mass flow, i.e., WX_r, does not change, the volumetric flow is less.

6.9 REAL BASINS

Figures 6.22 and 6.23 are schematic drawings of a rectangular basin and a circular basin, respectively, and illustrate the appurtenances used in practice. The appurtenances include an inlet design, overflow launders, effluent pipe, baffles, a flight of scrappers, and a sludge holding pocket. To illustrate

TABLE 6.10
Design Guidelines or Basins Settling Type II Suspensions

Criteria	Units	Primary WW Treatment[a]	Water Treatment	Up-Flow Water Treatment	Solids Contact
v_o	$m^3/min/m^2$	<0.028[b]	0.0006–0.0017[c]	0.0009–0.0013[c]	0.0014–0.0021[c]
		0.005–0.028 [d,a]	0.022–0.026 alum floc[g]		
		0.012–0.046[e,f]			
	gpm/ft^2	<0.69[b]	0.015–0.04	0.5–0.75[c]	0.8–1.2[c]
		0.14–0.69[d,a]	0.53–0.65 alum floc[g]		
		0.31–1.12[e,f]			
	gpm/ft^2		0.34–1.0[c]	0.5–0.75[c]	0.8–1.2[c]
v_H	m/min	Shield's equation	0.9[h]	NA	
	ft/min	<2[i]	3		
θ	h	1–4[j]	1.5–3[c]	1–3[c]	1–2[c]
		1–2[i]			
L/w		=20[k]	>1/5	NA	
Weir loading	$m^3/min/m$	0.10–0.25[l]	0.18[c]	0.12[c]	0.12–0.25[c]
		0.87–0.25[m]			
	gpm/ft	2.11–6.34[l]	<15[c]	10[c]	10–20[c]
		6.94–20.1[m]			
Depth (D)	m	1–5[n]; >2.1[o]	3–5[c]	3–5[c]	
	ft	3–15[n]; >7[o]	10–16[c]	10–16[c]	
		7–12[i]			

[a] Primary settlers are sized on overflow rate alone.
[b] Does not distinguish between rectangular and circular (WPCF, 1985).
[c] Kawamura and Lang (1986) for chemical floc in water treatment plants.
[d] Camp (1953).
[e] Chemical sedimentation 0.31–1.12 gpm/ft^2 for iron and polymer coagulated primary sewage, generally $v_o = 0.69$ gpm/ft^2, $T = 2$ h.
[f] The upper limit rate to lime floc which is used most often tertiary treatment; lime floc settles 1–4 times greater than alum flocs.
[g] Surface loading rates 0.52–0.63 $m^3/min/m^2$ (0.53–0.65 gpm/ft^2) were recommended (for an alum floc suspension).
[h] Shield's equation is recommended.
[i] European practice uses $v_H < 0.61$ m/min (2 ft/min) for rectangular basins, with D/L 1/10 to 1/30, with $w = 6.1$–9.1 m (20–30 ft) and depths 1.5–3.0 m (5–10 ft). In lieu of specifying v_H, American practice uses depths 2.1–3.7 m (7–12 ft) with $\theta = 1$–2 h.
[j] Value dependent on flow and suspension to be settled; data from primary clarifiers at Denver showed marked reduction in suspended solids removal at higher overflow velocities.
[k] Camp (1953) suggests that $L/w = 20$, such that inlet and outlet zones are about 10% of tank length with settling zone about 90%.
[l] Weir loading rates for primary settling tanks in wastewater treatment; circular tanks fall in this range for flows (WPCF, 1985). According to some authorities, weir loading is not as important as weir placement and tank design.
[m] Weir rates 7.64–20.1 gpm/ft depending on nature of chemical solids.
[n] Deeper clarifiers are recommended but the question of depth is debated.
[o] Burns and Roe (1971).

further, Figure 6.24 is a photograph showing a rectangular settling basin at the City of Fort Collins Wastewater Treatment Plant, c. 1996.

6.9.1 Inlet Design

The goal of the inlet zone is to distribute the flow such that the entire cross-sectional area of the basin is utilized. A slotted baffle device, as shown in Figure 6.25, is often used. The openings, e.g., slots or orifices, must be distributed over the cross-sectional area; the higher the headloss across the baffle, the more uniform the flow distribution. At the same time, the associated jet velocity is higher, which is not desired, and so there is a "trade-off." For ready reference in setting up a spreadsheet model, the flow across any orifice (or slot) is $Q(\text{orifice}) = C \cdot A(\text{orifice})(2g\Delta h)^{0.5}$. The headloss, i.e., Δh, is equal for all orifices regardless of the orifice depth. Also, $v(\text{orifice}) = Q(\text{orifice})/A(\text{orifice})$.

Inlet diffuser guidelines are (Kawamura, 1996, p. 132) (1) a distribution plate, or diffuser wall, with ports should be used to distribute the flow; (2) the ports should be distributed uniformly across the diffuser wall; (3) the maximum number of ports should be provided so that dead zones between ports are minimized; (4) the headloss across the diffuser wall should be 0.3–0.9 mm to equalize the flow distribution with minimal floc breakage; (5) the size of the ports should be uniform and large enough, i.e., 75–150 mm, to avoid clogging by algae and other matter; (6) the ports should be spaced 250–400 mm

TABLE 6.11
Design Guidelines for Basins with Hindered Settling

Criteria	Units	Rectangular[a]	Up-Flow	Solids Contact	Sludge Thickening
v_o	$m^3/min/m^2$	0.0028–0.034[b] 0.028–0.085[c] 0.0226[d]	No data	No data	No data
	gpm/ft^2	0.069–0.83[b] 0.69–2.08[c] <0.55[d]	No data	No data	No data
v_H	m/min ft/min	No data	NA	NA	NA
θ	h	2–3[e]	No data	No data	No data
L/w		No data	NA	No data	No data
Weir loading	$m^3/min/m$	0.086–0.258[f] 0.283–0.424[g]	No data	No data	No data
	gpm/ft	6.94–20.8[f] 6.94–10.4[g]	x	x	x
D	m ft	2.4–4.6[h] 8–15[h,d]	No data	No data	No data
Solids loading	$kg/day/m^2$ $lb/day/ft^2$	245[i] 50	No data	No data	No data

[a] Secondary settler design based on overflow rate and solids loading, i.e., kg dry solids/day/m^2 (lb dry solids/day/ft^2), see Dick (1970). Solids loading can be obtained from lab settling tests. Areas based on settling and solids loading are compared and the larger is used.
[b] Recommended $v_o \sim 0.023$ $m^3/min/m^2$ (0.55 gpm/ft^2); temperatures of 0°C decreases v_s by a factor of about 1.75 compared with 20°C (WPCF, 1985).
[c] Final settling should be designed on the basis of overflow velocity rather than detention time, with allowances for the depth of sludge blanket for activated sludge final settling (Camp 1953).
[d] Secondary settlers are quite different than primary settlers due to the amount and nature of the solids. Ten States Standards $v_o = 0.55$ gpm/ft^2 and $D > 8$ ft.
[e] Burns and Roe (1971).
[f] Most sources indicate that weir placement is more important than weir loading; for center feed tanks, optimum weir location is about 0.67–0.75 of the radial distance from the center (WPCF, 1985).
[g] Weir loading (p. 7–8) is given as 0.28 $m^3/min/m$ for <0.044 m^3/s (<6.94 gpm/ft for <1 mgd) and 0.42 $m^3/min/m$ (<10.4 gpm/ft) for larger plants for secondary settling.
[h] Deeper basins are recommended for final settling of activated sludge; for large final clarifiers, the average depth was 4.6 m (15 ft) with range 3.7–6.2 m (12–20 ft); studies showed that effluent concentration was lower as depth increased; other factors being equal (WPCF, 1985).
[i] Maximum allowed by Ten States Standards 1978 edition. Recommended: 49 $kg/day/m^2$ at SVI = 300–290 $kg/day/m^2$ at SVI = 100.

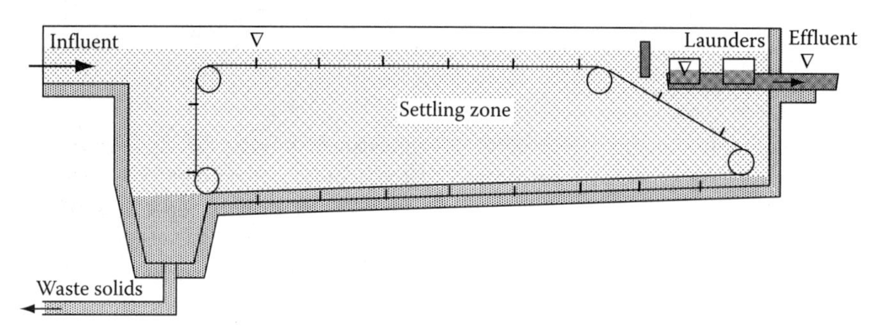

FIGURE 6.22 Rectangular settling basin schematic.

to provide structural strength for the diffuser wall; and (7) the diffuser wall should be located 1.8–2.1 m downstream of the inlet pipes. One and, preferably, two intermediate diffuser walls were recommended. Two guide walls extending from the inlet walls a short distance, perhaps 20% of the basin length, and parallel to the side walls were found to stabilize flow.

For flocculent settling that follows paddle wheel flocculation, the two processes may be integrated as a single basin. A slotted wood baffle is recommended before the settling,

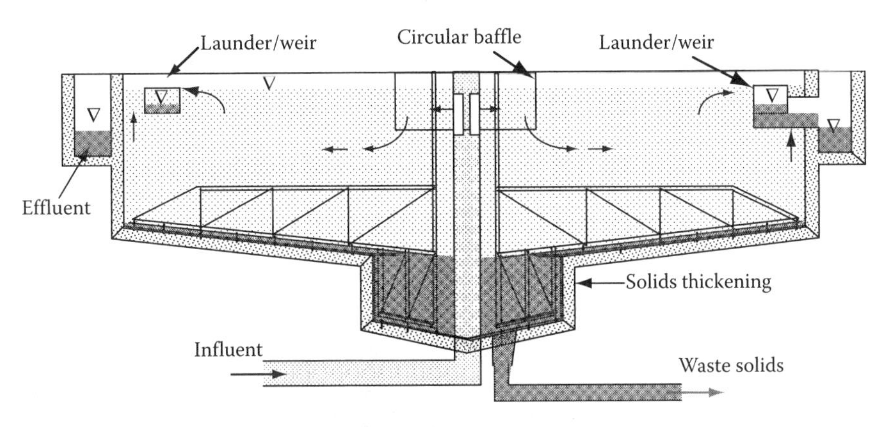

FIGURE 6.23 Circular settling basin schematic.

FIGURE 6.24 Photograph of a rectangular primary settling basin (FCWWTP). Photo Courtesy of Fort Collins Utilities, Fort Collins, CO.)

FIGURE 6.26 Photograph of overflow launders for rectangular primary settling basin.

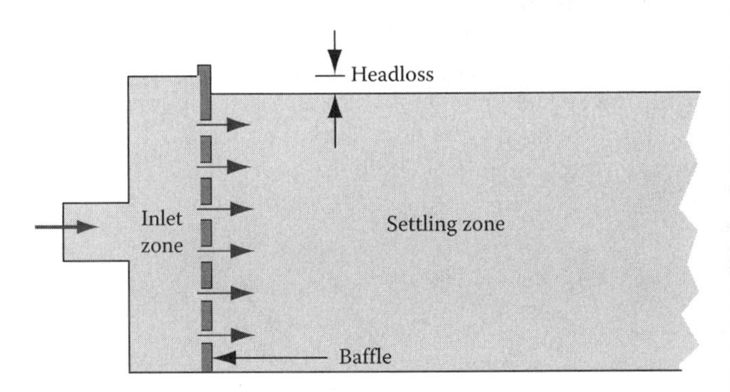

FIGURE 6.25 Slotted baffle inlet. (Adapted from Katz, W.J. et al., *Concepts of Sedimentation Applied to Design*, Part 2, *Water and Sewage Works*, 169, May 1962b.)

FIGURE 6.27 Photograph of weir plates for overflow launders.

however, in order to reduce extraneous currents. Although plate settlers or tube settlers (see Section 6.10) provide uniform flow, measures to minimize extraneous currents is recommended.

6.9.2 Outlet Design

The outlet design consists usually of overflow weirs, arranged to minimize short-circuiting. Thus, collection launders should be uniform and perpendicular to the flow direction, i.e., across the tank. In practice, however, the launders are often found parallel to the length of the tank, e.g., as in Figure 6.26. A grid layout is another recommended configuration. A common regulatory agency's weir-loading limit is 264 m^3/day/m of launder length (21,260 gpd/ft) (Kawamura, 1996, p. 132). The weir plates have multiple V-notches, as illustrated in Figure 6.27. The plates are made adjustable so that leveling is feasible.

6.9.3 Summary Notes for Practical Design

Camp (1946, 1953) departed from the conventional wisdom in a number of areas. His ideas are worth summarizing: first, because they reinforce basic concepts of basin design and, second, because of his credibility regarding settling theory and practice. Camp advocated long narrow basins to reduce the effects of hydraulic factors. For final settling, Camp suggested shallower basins. In deep tanks, he noted, the activated sludge plunges along the bottom of the basin and then upturns at the end (see also Kawamura (1996) and Esler (1998)). With a shallow rectangular basin, integrated with the activated sludge reactor, the flow is distributed better initially and the suspension passes through the settling phases, i.e., Type II, then Type III further, and finally Type IV at the bottom. Concerning effluent launders, Camp (1953) suggested placing them concurrent with flow and across (forming a grid) and into the basin in order to draw the flow from the clarified zone and away from the sludge upturn-back roll. Finally, Camp suggested withdrawing the sludge at the effluent end.

BOX 6.4 ON PLATE SETTLERS

Two of the Swedish companies that market plate settlers are Purac and Parkson. Originally there was one company, Axel Johnson, which was family owned. In the 1950s, however, Purac was formed by some of the family members and the two companies each have established identities in the same field of business and both produce plate settlers. At the same time, Waterlink Technologies was a part of Nordic Water Products AB, formerly an Axel Johnson company (Waterlink Technologies, 1997), which produced a plate-settler system similar to the Purac system after the Purac patent expired. According to the Waterlink annual report (from a January 10, 2008 google search on plate settlers), Waterlink was formed in December 1995 as a holding company to consolidate various individual companies to form a comprehensive company that could offer a complete array of services, including a "design-build" service. According to the report, they acquired Purac in 1998. Their annual report listed a number of companies that comprised the overall company, perhaps on the same order of size as U.S. Filter, Infilco-Dregemont, Vivendi, and others.

6.10 PLATE SETTLERS AND TUBE SETTLERS

Inclined plate and tube settlers have evolved over the years. The concept started with Hazen in 1904 as horizontal "tray settlers" and then was advanced further by Camp (1946, 1953) and implemented in Sweden in the 1950s (Fischerström, 1955). Sludge removal was the problem, however, since the plates were horizontal. The answer was found by tilting the plates so that, after some amount of accumulation of mass, the solids would slide from the surface by gravity. Tube settlers are in the same category, with respect to principle, as plate settlers. The inclined plate settlers and tube settlers circumvented the issues of short-circuiting, dead zones, and turbulence.

6.10.1 Plate Settlers

There are three types of inclined plate settlers: (1) up-flow, (2) down-flow, and (3) cross-flow. Up-flow plate settlers are most common. The removal principles are the same for tube settlers.

6.10.1.1 Particle Path: Analysis

For an inclined surface settler, i.e., plate settler or tube settler, the particles take paths that are the vector sum of v_s and v_P, i.e., v_R

$$\vec{v}_s + \vec{v}_P = \vec{v}_R \tag{6.31}$$

where
- $\vec{v}_s$ is the fall velocity of any particle (m/s)
- $\vec{v}_P$ is the advection velocity of water flow between plates of settler (m/s)
- $\vec{v}_R$ is the vector sum of fall velocity of particle and advection velocity of water (m/s)

Figure 6.28 illustrates the application of Equation 6.31 for the special condition in which $v_s = v_o$.

To simplify the analysis, the velocity profile between the plates is assumed to be uniform, which actually is parabolic (as occurs in the viscous flow range, i.e., for $R < 1000$); the velocity vector, $\vec{v}_P$, is considered the mean velocity. Based on

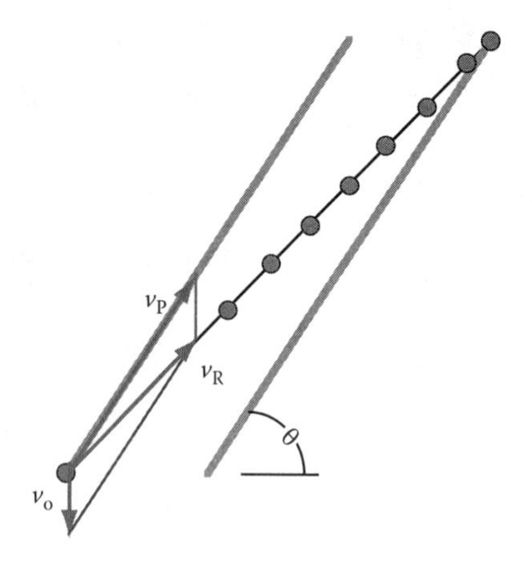

FIGURE 6.28 Velocity vectors within plate.

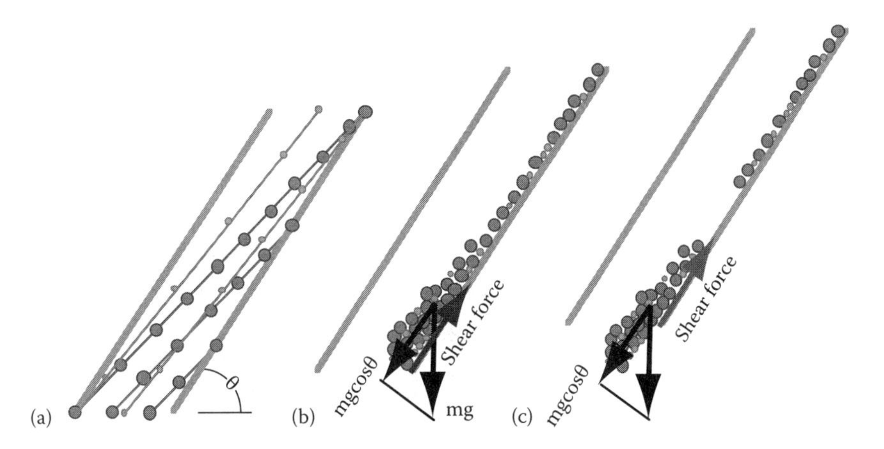

FIGURE 6.29 Three phases of particle removal by plate settler: (a) paths of two particle sizes, (b) particles accumulating on lower plate surface, and (c) Particle mass weight component exceeds shear force.

this assumption, particles for which $v_s \geq v_o$ will strike the lower plate and thus will be removed from the suspension.

6.10.1.2 Sludge Removal

At some angle, θ, the mass of accumulated solids will slide down the plate (or tube), with $\theta = 55°$ being adopted for plates and $q = 60°$ for tubes. The movement occurs when the weight component of the sludge mass in the direction of the plate exceeds the shear resistance of the solids mass. Figure 6.29 illustrates the sequence of particle removal from the suspension. To elaborate, Figure 6.29a depicts the paths of particles toward the surface of the lower plate. Figure 6.29b shows the accumulation of particles. Figure 6.29c shows the breaking loose of a given mass of sludge, i.e., m(sludge mass)$g \cos \theta \geq$ shear-resistance-force-of-sludge-mass. Such sludge masses break off randomly across the surface of any given plate and slide down the plate and fall as clumps, perhaps 0.5–2 cm in size. This is seen in the photograph (Figure 6.30). The plates are spaced at 51 mm (2 in.) and set in a tank 2438 mm × 1219 mm × 1219 mm (4 ft × 4 ft × 8 ft). Sludge masses are visible in different stages of accumulation across the area of any given plate. Several clumps are seen falling below the plates.

FIGURE 6.30 Side view of plate settlers with sludge accumulation; CSU Water Treatment Research Pilot Plant. (Photo courtesy of Joe Mendoza, Colorado State University, Fort Collins, CO, 1995.)

6.10.1.3 Plate Settler Systems

Plate settlers were introduced into the market about 1970 (Yao, 1973) under the trade names, Lamella Separator™ (Parkson Corporation) and GEWE™ Lamella Sedimentation System (Purac Corporation, www.lackebywater.se/inc/pdf/en_purac_gewe.pdf). The SuperSettler™ of WaterLink Technologies was introduced later and is an adaptation of the Purac GEWE Plate Settling System.

Figure 6.31 shows a perspective drawing of a GEWE plate settler system of Purac Corporation. Of particular interest is the influent flow, which is constrained by channels on each side and in the center so that density currents have no opportunity to form. The inflow then enters the plate cells from the side of the influent channel at the lower part of the plate. The water entering the plate cells from each side then turns up and flows to V-notch effluent weirs on each side of a given cell where the collected effluent flow leaves the system to enter the next unit process (e.g., deep bed filters in water treatment). The plate cells extend above the effluent weirs, which hydraulically separate each cell. The sludge, as it accumulates as a mass on the plate surfaces, slides down and falls to the sludge-collection zone where it is removed by a hydraulic vacuum to a sludge hopper and then is pumped to a sludge-holding pond.

For the effluent, v-notched weirs are located on each side of any given cell. The weirs are individually adjustable so that each weir takes its share of flow, i.e. for each cell, $Q(\text{plate})_{\text{cell }1} = Q(\text{plate})_{\text{cell }2} = Q(\text{plate})_{\text{cell }n}$. Similarly, for each weir, $Q(\text{weir}) = Q(\text{plate})/2$. The plates rest on holders attached to the outside channels at an inclination of 55°. For large installations, the plates are installed in basins designed with the needed brackets and appurtenances while for smaller installations, package systems are available.

The foregoing description applies only to the GEWE system, and will vary from one proprietary system to another. To illustrate further, Kamp (1989, p. 210) provided a description for a GEWE system installed for the 1981 WRK III WTP

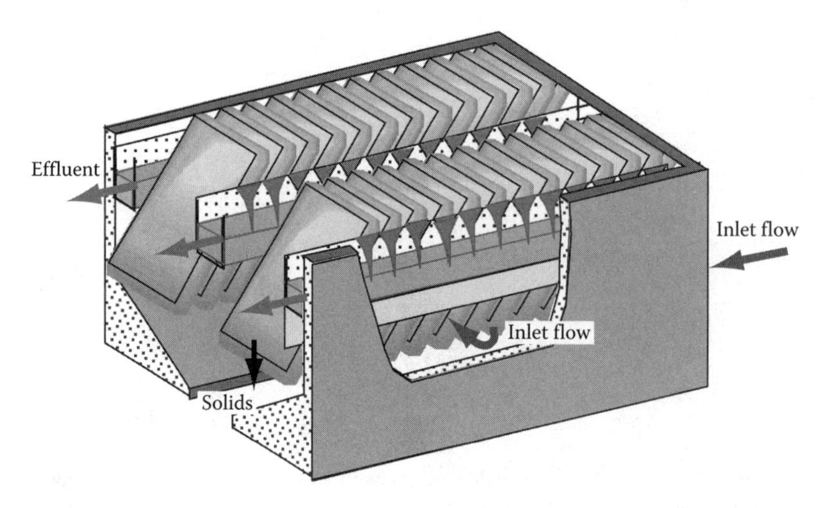

FIGURE 6.31 Perspective of GEWE plate system. (Adapted from Hane-Weijman, H., Lamella separation—old technology in a modern concept, Unpublished paper for short course Water Treatment Plant Design at Colorado State University, in June, 1991, Purac Engineering, Inc., Wilmington, DE, 1991.)

at Enkhuizen, Holland with $Q(\text{max}) = 15{,}000\ \text{m}^3/\text{h}$ or 4.17 m^3/s (95 mgd). The dimensions were $w(\text{plate}) = 1.2$ m (3.9 ft), $L(\text{plate}) = 3$ m (9.8 ft), $\theta = 55°$, $d = 80$ mm (3.2 in.); the plates were of stainless steel.

Plate settlers have been used for a variety of municipal and industrial installations. For Purac®, as an example, some 400 installations were listed worldwide (Hane-Weijman, 1991) that included municipal water treatment, municipal wastewater treatment, and industrial wastewater treatment. The listings included such diverse plants such as a 4.17 m^3/s (95 mgd) waterworks at Enkhuizen, Holland, and at a cement factory in Japan with 0.0020 m^3/s (32 gpm).

6.10.1.4 Sizes of Units

The sizes of plate systems depend upon the manufacturer, but typical dimensions are as follows: $d = 80$ mm (3.2 in.), $L = 3$ m (9.8 ft), with $w = 1220$ mm (4.0 ft), and $\theta = 55°$. Plate spacing, d, depends upon the concentration of the suspension and varies from 51 to 102 mm (2–4 in.) (Hane-Weijman, 1991). For the Purac system, the effective plate length is about 2 m (6.6 ft) since the flow enters from the side about 1 m (3.2 ft) from the bottom. In general, $\mathbf{R} < 100$ for these systems.

6.10.1.5 Surface Overflow Rates

The SOR is the flow per unit area of plate as projected on the horizontal plane and is an identity with v_o, as seen in Figure 6.28. The SOR is comparable to the same measure as used for horizontal flow basins. The calculation for a plate settler is, $v_o = \text{SOR} \equiv Q(\text{plate})/[L(\text{plate})\cos q \cdot w(\text{plate})]$, where $Q(\text{plate}) = Q(\text{plate assembly})/[n(\text{plates-in-assembly}) - 1]$.

For plate settlers, surface loading rates for several installations are given in Table 6.12. The SOR values are seen to vary by a factor of nearly 2:1. For comparison, iron and alum floc at 10°C settle at about 50 mm/min, i.e., 72 m/day (1.2 gal/min/ft^2) (Yao, 1973, p. 624). For further reference, the effluent turbidities from the plate settlers at Fort Collins are ≤ 0.1 NTU.

TABLE 6.12
Surface Overflow Rates—Plate Settler Installations

Location	Flow (m^3/s)	Flow (mgd)	Surface Overflow Rate ($\text{m}^3/\text{m}^2/\text{day}$)	Surface Overflow Rate (gpm/ft^2)
Andijk WRK III, Holland	4.16	95	38.7	0.66
Markis I, Yugoslavia	2.19	50	36.4	0.62
Ringssoverket, Sweden	1.18	27	24.1	0.41
Jezero, Yugoslavia	1.31	30	36.4	0.62
Fort Collins, Colorado	0.88	20	19.9	0.34
Colorado Springs, Colorado	3.94	90	22.3	0.38

Source: Adapted from Hane-Weijman, H., Lamella separation—old technology in a modern concept, Unpublished paper for short course Water Treatment Plant Design at Colorado State University, in June, 1991, Purac Engineering, Inc., Wilmington, DE, 1991.

Note: $\text{SOR} = A(\text{plan}) = Q(\text{plate})/[L(\text{plate}) \cdot \cos\theta \cdot w(\text{plate})]$.

6.10.1.6 Theory

Three types of plate settlers, as noted, are (1) up-flow, (2) down-flow, and (3) cross-flow (Kamp, 1989; Hane-Weijman, 1991) with particle trajectories illustrated by the vector diagrams of Figures 6.32, 6.33, and 6.34, respectively. From these diagrams, the mathematical relationships between v_o and v_P, and d and L can be determined.

6.10.1.6.1 Mathematics

The mathematical relationships, based on similar triangles between the velocity vectors (v_o, v_P) and the plate geometry (d, L, θ), are summarized below:

1. For up-flow, from Figure 6.32

$$\frac{v_o}{v_P} = \frac{d/\cos\theta}{L + d/(\cos\theta\sin\theta)} \tag{6.32}$$

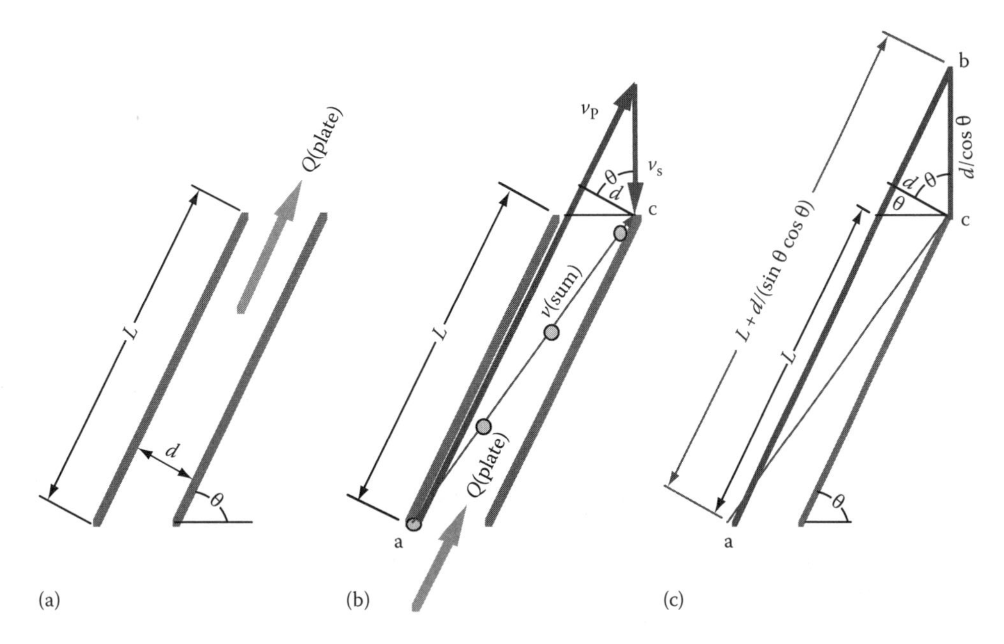

FIGURE 6.32 Up-flow settler: (a) flow direction, (b) velocity vectors, and (c) geometric similarity.

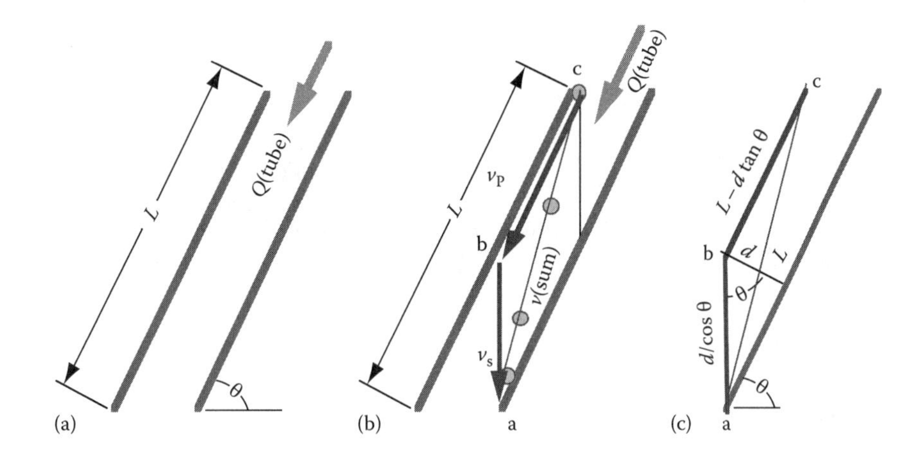

FIGURE 6.33 Down-flow settler: (a) flow direction, (b) velocity vectors, and (c) geometric similarity.

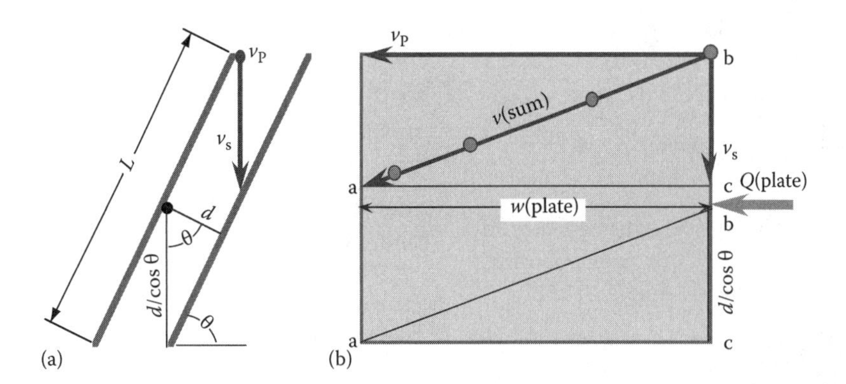

FIGURE 6.34 Cross-flow settler: (a) flow direction and (b) velocity vectors and geometric similarity.

2. For down-flow, from Figure 6.33

$$\frac{v_o}{v_P} = \frac{d/\cos\theta}{(L - d\tan\theta)} \tag{6.33}$$

3. For cross-flow, from Figure 6.34

$$\frac{v_o}{v_P} = \frac{d/\cos\theta}{w(\text{plate})} \tag{6.34}$$

where
- v_P is the mean velocity within plate cell (m/s)
- v_o is the fall velocity of smallest particles to be removed completely (m/s)
- L is the length of plate (m)
- d is the distance of plate separation (m)
- w(plate) is the width of plate assembly (m)
- Q(plate) is the flow per cell between plates (m^3/s)

6.10.1.6.2 Effective Plate Area for Up-Flow Plate Settlers

To arrive at the effective plate area, the pattern of Section 6.3.3.1 for conventional basins is followed, i.e.,

1. Modify Equation 6.32 to place cos θ in the denominator:

$$\frac{v_o}{v_P} = \frac{d}{L + \dfrac{d}{\cos\theta\sin\theta}\cos\theta} \tag{6.35}$$

$$= \frac{d}{L\cos\theta + \dfrac{d}{\sin\theta}} \tag{6.36}$$

2. Now recall

$$v_p = \frac{Q(\text{plate})}{w(\text{plate})\cdot d} \tag{6.37}$$

3. Substitute Equation 6.37 in Equation 6.36:

$$v_o = \frac{Q(\text{plate})}{w(\text{plate})}\cdot d\frac{d}{L\cos\theta + \dfrac{d}{\sin\theta}} \tag{6.38}$$

$$= \frac{Q(\text{plate})}{L\cos\theta + \dfrac{d}{\sin\theta}\cdot w(\text{plate})} \tag{6.39}$$

4. The effective area for settling is the denominator. If we let $\theta = 60°$, then $d/\sin\theta = 1.16d$, which is small relative to L, can be neglected to give

$$v_o \approx \frac{Q(\text{plate})}{L\cos\theta\cdot w(\text{plate})} \tag{6.40}$$

5. In other words, the effective area of the plate is the projection of its area to the horizontal plane. The surface overflow velocity, SOR, then is (neglecting the 1.16d factor)

$$v_o \approx \frac{Q(\text{plate})}{A(\text{projection-of-plate-area-on-horizontal-plane})} \tag{6.41}$$

The equality is true for up-flow, down-flow, or horizontal-flow plate settler systems, or for tube settlers (usually operated as up-flow).

6.10.1.6.3 Relationship between SOR, HLR(cross section), and HLR(horiz plane)

Three definitions of unit flows for plate settlers are

1. Surface overflow rate: The SOR is the flow between two plates divided by the area of a single plate projected on the horizontal plane:

$$\text{SOR} \equiv v_o \equiv \frac{Q(\text{plate})}{L(\text{plate})\cos\theta\cdot w(\text{plate})} \tag{6.42}$$

$$= v_P\left[\frac{d(\text{plate})/\cos\theta}{L(\text{plate})}\right] \tag{6.43}$$

2. Hydraulic loading rate based on plate cross section. HLR(cross section) is defined as the flow between two plates, Q(plate), per unit of cross-sectional area normal to the velocity vector v_P:

$$\text{HLR(cross-section)} \equiv v_P \equiv \frac{Q(\text{plate})}{d(\text{plate})\cdot w(\text{plate})} \tag{6.44}$$

3. Hydraulic loading rate based on basin area projected to horizontal plane, HLR(horiz-plane),

$$\text{HLR(horiz-plane)} \equiv v_P\sin\theta$$

$$\equiv \frac{Q(\text{plate})}{[d(\text{plate})/\sin\theta]\cdot w(\text{plate})} \tag{6.45}$$

The first, i.e., SOR, appears most frequently in the literature and was adopted here. The other two are given for reference. If SOR is used, its meaning is usually understood, i.e., as defined in Equation 6.42. On the other hand, hydraulic-loading-rate (HLR) is often used without being defined; the term should be defined, i.e., in terms of cross section normal to the plate length axis or to the plate separation distance projected to the horizontal plane. Conversions between the expressions are seen in Equations 6.43 through 6.45.

Willis (1978, p. 334), in an article on tube-settlers, draws attention to the issue of clearly defining the hydraulic loading rate and suggests as a criterion, HLR(horiz-plane) ≤ 2.5 gpm/ft^2-face-area-of-tubes, with "face-area" defined as the area of the tubes intersecting the horizontal plane. For reference, the Colorado Department of Health (1987, p. 32) gives as a criterion HLR(horiz plane) ≤ 0.102 m^3/min/m^2 (2.5 gpm/ft^2) and defines the term as the total tube or face area of the plane containing the open end of all tubes. An equivalent interpretation is that the HLR(horiz plane) is the flow per basin divided by the plan area of the basin.

Example 6.7 Floc Particle Sizes Removed by Plate Settlers

Given

Consider the Fort Collins Colorado Water Treatment Plant for which $v_o \equiv$ SOR = 19.9 m/day (Table 6.12). Plate dimensions are (Kamp, 1989), i.e., dimensions, L(plate) = 3.0 m, w(plate) = 1.2 m, d = 80 mm, and $\theta = 55°$; also, the flow enters each plate from the sides 1.0 m from the bottom. Assume $T = 20°$C.

Required

a. *Floc size*: Determine the smallest size of floc particle, d(floc), removed 100% by the plate settlers, assuming Stokes' law applies.
b. *Laminar flow check*: Check to determine whether the flow between the plates is laminar, i.e., $\mathbf{R} \leq 100$.

Solution

a. Calculate d(floc)

Overflow velocity, SOR, is given as,

$$v_o = \text{SOR} = 19.9 \text{ m/day} = 2.30 \cdot 10^{-4} \text{ m/s}$$

Particles having fall velocity, $v_s \geq v_o$, will be removed 100%; particles having $v_s < v_o$ will be partially removed. Therefore, particles for which $v_s = v_o$ are the smallest that will be 100% removed for a given SOR and for a particular plate design, which is depicted by Figure 6.28.

If the flow between the plates is laminar, i.e., $\mathbf{R}$(plate flow) ≤ 1000, the particle size associated with a particle fall velocity, $v_s = v_o$, is given by Stokes' law, Equation 6.7. By definition, Stokes' law applies to a particle falling in the range, $\mathbf{R}$(particle) ≤ 1. For such Reynolds number range, turbulence is not a factor. Also, the parabolic velocity distribution between the plates is neglected as a factor in the particle trajectory.

To apply Stokes' law, assume the specific gravity, SG(alum floc) = 1.05. For the conditions stated, application of Stokes' law to yield d(floc) is

$$v_s = \frac{1}{18}\frac{g}{\nu}(SG_s - SG_f)d^2 \quad \text{(Ex6.7.1) (6.7)}$$

$$2.30 \cdot 10^{-4} \text{ m/s} = \frac{1}{18} \cdot \frac{9.81 \text{ m/s}^2}{1.004 \cdot 10^{-6} \text{ m}^2/\text{s}} \cdot (1.05 - 1.00) \cdot d(\text{floc})^2 \quad \text{(Ex6.7.2)}$$

$$d(\text{floc}) = 0.092 \text{ mm} \quad \text{(Ex6.7.3)}$$

$$\approx 0.1 \text{ mm} \quad \text{(Ex6.7.4)}$$

In other words, d(floc) ≥ 0.1 mm will be 100% removed and d(floc) < 0.1 mm will be partially removed.

b. Determine $\mathbf{R}$

First, apply Equation 6.36 to calculate v_P:

$$\frac{v_o}{v_P} = \frac{d(\text{plate})}{(L(\text{plate}) \cos\theta + d(\text{plate})/\sin\theta)} \quad \text{(Ex6.7.5) (6.36)}$$

$$\frac{2.30 \cdot 10^{-4} \text{ m/s}}{v_P} = \frac{0.080 \text{ m}}{[2.0 \text{ m} \cdot \cos 55° + (0.080 \text{ m}/\sin 55°)]} \quad \text{(Ex6.7.6)}$$

$$v_P = 0.00358 \text{ m/s} \quad \text{(Ex6.7.7)}$$

Calculate $\mathbf{R}$ at $T = 20°$C:

$$\mathbf{R} = \frac{\rho(T = 20°\text{C}) \cdot v_P \cdot d(\text{plate})}{\mu(T = 20°\text{C})} \quad \text{(Ex6.7.8)}$$

$$= \frac{(998 \text{ kg/m}^3)(0.00358 \text{ m/s})(0.08 \text{ m})}{(0.001002 \text{ Ns/m}^2)} \quad \text{(Ex6.7.9)}$$

$$= 285 \quad \text{(Ex6.7.10)}$$

Discussion

Laminar flow in a pipe requires, $\mathbf{R} \leq 1000$; the same criterion is assumed to apply for flow between two flat plates. Thus, since $\mathbf{R}$(plate) = 285, the flow between the plates is laminar. The limitation in the accuracy of the calculation is the assumed value of floc-specific gravity, i.e., SG(floc) = 1.05. Also, as noted, the parabolic velocity profile between the plates was neglected. A spreadsheet may be set up to calculate d(floc) to test the sensitivity of the value to SG(floc).

6.10.2 Tube Settlers

The tube settler system was developed by Neptune Microfloc, Inc. and appeared on the scene during the mid-1960s; their experimental program started in 1964 (Hansen and Culp, 1967; Culp et al., 1968). The two versions of tubes were (1) hexagonal tubes, each tube 760 mm (30 in.) long and 51 mm (2 in.) across the flats and $\theta = 5°$, and (2) square tubes, each tube 610 mm long and 51 mm (2 in.) square, and $\theta = 60°$. The hexagonal tube type was designed for the settling of floc prior to filtration with cleaning of accumulated solids by rapid draining timed to occur simultaneously with filter backwash.

The square tube type was designed for the self-cleaning of solids as they accumulate. For the square tubes inclined at 60°, the bundle was fabricated of PVC plastic with tubes arranged,

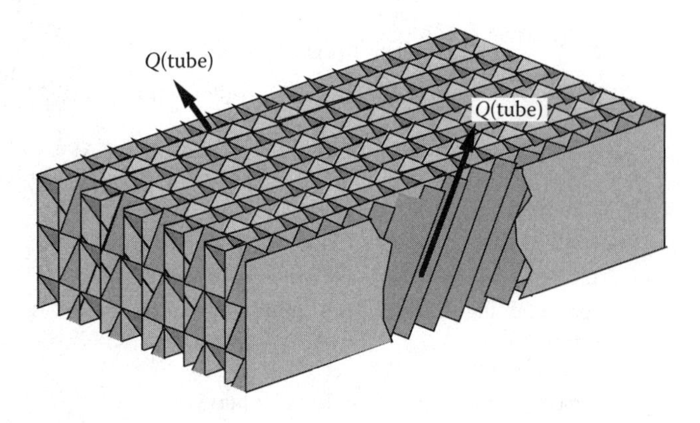

FIGURE 6.35 Bundle of square tubes. (Adapted from Culp, G. et al., *J. Am. Water Works Assoc.*, 60, 681, June 1968; Culp, G.L. and Conley, W., High-rate sedimentation with the tube clarifier concept, in: Gloyna, E. and Eckenfelder, W. W. (Eds.), *Water Quality Improvement by Physical and Chemical Processes*, Water Symposium No. 3., University of Texas Press, Austin, TX, 1970.)

as shown in Figure 6.35. These bundles may be used to retrofit conventional settling basins, round or square, or may be installed in new basins designed specifically for tube settlers. Due to the fabrication arrangement with the rows of tubes having 60°/120° angles alternating, the module acts as a beam and can support its own weight with supports necessary only at each end. The square tubes were made of extruded ABS plastic with rows separated by thin sheets of PVC (Culp and Conley, 1970, p. 149). Installations have included both potable water treatment and industrial applications, including separation of oil. The 60° tubes are steep enough that most sludges will slide along the bottom surface, bearing a given tube, after a sufficient mass has accumulated.

PROBLEMS

6.1 Basin Sizing

Given

Suppose that $Q = 0.05$ m^3/s (1.14 mgd). Also, let $v_o = 0.00038$ m/s (800 gal/day/ft^2). A long narrow basin is preferred.

Required

(a) By spreadsheet, determine the plan area required and its distribution between width and length. Keep in mind that a long narrow basin is desired.

(b) Show how the different w/L ratios are affected by depth, D, i.e., $v_H = Q/(wD)$. Is the solution unique or are there many solutions?

6.2 Sedimentation Theory

Given

Flocculent particles of alum may vary perhaps $1 < d(\text{floc}) < 5$ mm after flocculation. Let SG = 1.05.

Required

(a) Determine the range of fall velocities of the particles, assuming Stokes' law applies.

(b) For a $d(\text{floc}) = 5$ mm particle, compare the Stokes' law result with a drag coefficient, C_D, in the turbulent range.

6.3 Plain Sedimentation—Iso-Percent Removals for Discrete Particle Suspension

Given/Required

For a discrete particle suspension, explain how the iso-concentration lines are determined. Extend this to a depth versus time plot that shows iso-concentration lines.

6.4 Total Removal of Particles for Hypothetical Basin

Given

Referring to Figure 6.11, consider the particles shown at the entrance to basin as representative of the suspension.

Required

Calculate the removal, R, based upon Equation 6.18 for the basin as shown.

Hint: Use a straight-edge and scale as needed in utilizing Figure 6.11.

6.5 Rectangular Sedimentation Basin for Discrete Particles

Given

A discrete particle suspension is described in Figure 6.12.

Required

Design a rectangular sedimentation basin to remove 75% of the suspension.

Hint: This problem is similar to Problem 6.1 except (1) the determination of a specified R (fraction of suspension removed) involves a trial-and-error procedure, and (2) scour is an additional consideration.

6.6 Horizontal Flow Basin

Given/Required

Using the protocol of Problem 6.5, size a horizontal flow sedimentation basin assuming that the suspension fall velocity distribution is as in Figure 6.12. Use any flow that you wish.

6.7 Using Suspension Test Results to Design an Ideal Basin

Given/Required

Determine the overflow velocity required for a 0.70 fraction removal of the suspension characterized by Figure 6.12. Determine the dimensions of a horizontal-flow basin for this result. Use Shield's equation to determine maximum horizontal velocity. The proportion of suspension scoured is removed from the net deposit.

6.8 Spreadsheet for Basin Sizing

Given/Required

Set up a spreadsheet to accomplish a basin sizing for different scenarios, with calculation of corresponding removals.

6.9 Sizing Horizontal Flow Basin—Two Approaches

Given/Required

Examine the sizing of a horizontal flow sedimentation basin recognizing two basic approaches:

a. Assume velocities (overflow and then horizontal flow).

b. Alternatively, assume basin dimensions that result in the calculation of v_o and v_H.

6.10 Histograms for Removal

Given/Required

The tabular data of Tables 6.7 and 6.8 are for primary settling basins. Plot histograms of percent removal of suspended solids for (a) rectangular basins and (b) circular basins. Discuss your findings.

6.11 Removals of Suspensions as Function of Independent Variables

Given/Required

For the basins of Tables 6.7 and 6.8, plot percent removal of suspended solids *v.* v_o for (a) rectangular basins and (b) circular basins. Discuss your findings. Do the same using detention time as the independent variable. Try a three-dimensional plot that combines v_o and θ as independent variables.

6.12 Basin Design for Flocculent Settling

Given/Required

Design a rectangular sedimentation basin to remove 70% of the suspension described in Figure 6.14 (and with removals compiled in Figure 6.19 (Hendricks, 2006) reproduced below).

6.13 Basin Design for Flocculent Suspension

Given/Required

For the suspension of Figure 6.14, select basin dimensions for $Q = 0.044$ m^3/s (1.0 mgd); let $R = 80\%$. Extend the spreadsheet to Table 6.7 to include the determination of w and L. Discuss your outcome for v_H and the basin proportions.

6.14 Basin Performance for Flocculent Suspension if Q Increases

Given/Required

Determine the performance of the basin in terms of R and effect on scour if Q is increased by a factor of two.

6.15 Linking Activated-Sludge Reactor Variables to Secondary Clarifier Performance (Requires Additional Study)

Given/Required

Based on the paper of Hermanowicz (1998) in which plots were developed for the relationship, $X = f(R/Q, X_r, u, \text{SOR}, \text{SVI})$, (a) develop the same thing by means of a spreadsheet, i.e., the output should be X as a function of X_r for given values of R/Q, u, and SVI, (b) duplicate the graphical outputs as illustrated in Figure 7 of his paper. Comment on the significance of this integration of variables. Note that $\text{SOR} = (Q - W)/A(\text{plan})$, Equation 6.23, and $u = (R + W)/A(\text{plan})$, Equation 6.24. Definitions are $X = \text{MLSS}$ (mixed liquor suspended solids) and R is calculated from assumed, X, X_r.

6.16 Application of the Vesilind Equation (6.6.1.2 Settling Tests)

Given/Required

Estimate v_I and b for the plot of Figure 6.17a, which is from the data of Dick (1970). Compare with the data of Watts et al. (1996) in which $v_I = 7.62$ m/h and $b = 0.00024$ L/mg $= 0.24$/g (Table 6.4).

6.17 Application of Flux Theory to Design (6.7.1.2 Sludge Thickening)

Given/Required

a. From Figure 6.17a, obtain the associated plot for the flux density, **j**(settling) versus X_i, as in Figure 6.17b.
b. Assume data for underflow and estimate **j**(bulk).
c. For $Q = 0.044$ m^3/s (1.0 mgd), determine the area required for a secondary clarifier.

6.18 Application of Flux Theory to Secondary Clarifier Operation

Given/Required

For the final clarifier design of Example 6.6, explore the effect of modifying operating conditions. The spreadsheet, Table CD6.5 is recommended. Modification may be useful. Some of the questions that may be explored include: (a) suppose the underflow solids concentration is say, $X_r = 4000$ mg/L, (b) $X_r = 8000$ mg/L, (c) suppose the operator chooses to increase R, and (d) suppose the future solids loading to the final clarifier is increased by say 30% (caused by increased organic loading to the reactor and consequent greater production of bio-solids). Can the clarifier handle this increased load? Will solids overflow? Describe in terms of theory.

6.19 Application of Flux Theory to Performance Evaluation

Given/Required

Suppose, in the Example of 6.6, a clarifier of diameter 12.2 m (40 ft) already existed. Let $R = 0.0219$ m^3/s (0.5 mgd). Determine the effect on underflow concentration, X_r, and X. Describe the performance consequences of this design.

6.20 Application of Flux Theory to Effect of Changing Operating Variables

Given/Required

Suppose, in the Example of 6.6, a clarifier of diameter 12.2 m (40 ft) already existed. As an operating variable, R can be increased. Describe the scenario quantitatively and descriptively. As a second consideration, evaluate the effect on the reactor (detention time is the direct effect).

Hint: Consider the effect of R on **j**(bulk) and the associated materials balance.

6.21 Hindered Settling

Given

A typical concentration of activated sludge solids (mixed liquor suspended solids, MLSS) is about 2000 mg/L. After settling, the suspension concentration, X_r, is about 10,000 mg/L.

Required

Calculate the size for an ideal final settling basin for an activated sludge suspension. Compare your size with a real basin. Use real data for flow to the basin from a nearby wastewater-treatment plant.

6.22 Plate Settler Design

Given

A sedimentation basin is comprised of plate settlers.

Assume

- L(plate) = 100 cm long.
- d(plates) = 6.0 cm.
- Suspension has characteristics shown in Figure 6.18.

Required

Determine the effect on performance if the plates are 150 cm long (instead of 100 cm).

6.23 Performance of Plate Settlers in Holland

Given

Data from Table 6.12 for the Andijik WRK III WTP in Holland is given as an excerpt from the table, i.e.,

	Flow		Surface Loading Rate, SOR	
Location	(m^3/s)	(mgd)	(m^3/m^2/day)	(gpm/ft^2)
Andijk WRK III, Holland	4.16	95	38.7	0.66

Required

Estimate the smallest particle size, d(particle), that may be removed by the plant.

6.24 Reynolds Number Calculation

Given/Required

Calculate the Reynolds number for the flow in the plate cells at the Fort Collins WTP (see Example 6.7).

6.25 Spreadsheet for Plate-Settler Performance Evaluation—Discrete Particles

Given

Suppose the particles to be removed by a plate settler system are mineral sediments with size distribution as shown by Figure 6.12.

Required

Show by plot obtained by spreadsheet the relationship between plan area of plates and the percent removal of the suspension.

6.26 * Spreadsheet for Plate Settler Performance Evaluation—Flocculent Particles

(*Note*: This is a * problem, i.e., more difficult)

Given

Suppose the particles to be removed by a plate-settle-rsystem comprise a flocculent suspension characterized by Figure 6.14 (and Figure 6.18 of Problem 6.22).

Required

Show by plot obtained by spreadsheet the relationship between plan area of plates and the percent removal of the suspension. Provide your own rationale on how to take care of the time factor. Estimate the detention time in the plate settlers.

6.27 Design for Andijik WRK III

Given/Required

For the Andijik WRK III WTP in Holland, determine the plan area required showing plate arrangement and hydraulic flows.

6.28 Particle Sizes Removed by Plate Settlers at Andijik WRK III

Given/Required

For the Andijik WRK III WTP in Holland, estimate the smallest size of particles that will be removed completely. Determine the plan area required showing plate arrangement and hydraulic flows.

6.29 Particle Sizes Removed by Plate Settlers at CSU WTRPP

Given

The following data pertain to the plate settler system of a water-treatment pilot plant, c. 1996, at Colorado State University:

- w(plate) = 1.22 m (4.0 ft)
- L(plate) = 0.61 m (2.0 ft)
- d = 64 mm (2.5 in.)
- Q = 76 L/min (20 gpm)
- n(cells) = 25
- Temperature varies, 2°C $\leq T \leq$ 12°C

Required

Estimate the smallest size of alum floc particles that should be removed.

ACKNOWLEDGMENTS

Judy Berkun, manager special projects, Malcolm Pirnie, Inc., White Plains, New York, provided the photograph of Allen Hazen. Marlene Hobel, vice president, Corporate Communications, Camp Dresser & McKee (now CDM), Boston, Massachusetts, provided the photograph of Thomas Camp. The author appreciates their help.

Steve Comstock, supervisor, Fort Collins WWTP, granted permission to use the photographs of the sedimentation basin at the Fort Collins plant. Comstock and staff often hosted student field trips and individual visits to the plant.

GLOSSARY

Center feed clarifier: A settling basin with flow from center ports and discharge around peripheral overflow weirs.

Clarifier: See *settling basin*.

Dead zone: A region of a flow field in which there is little participation in the main flow and the exchange with the main flow is small.

Detention time: Basin volume divided by flow, i.e., $T = V(\text{basin})/Q$.

Discrete particle: A particle with fixed size.

Dispersion: The "spread" of a substance during a flow due to currents, turbulence, or molecular diffusion.

Dynamic viscosity: Defined as the coefficient, μ, in the relation, $t = \mu(dv/dy)$, where μ = dynamic viscosity (Ns/m^2), t = shear stress in laminar fluid flow (N/m^2), (dv/dy) = velocity gradient (s^{-1}).

Final clarifier: The final settling basin in biological treatment for the separation of biological floc suspension before discharge of the clarified water, often called secondary settling basin.

Floc: A particle comprised of aggregation of other like particles. Examples include biological floc, alum floc, and iron floc.

Flocculent settling: Settling of flocculent particles, i.e., those that grow in size.

Flow: Defined as volume per unit time that passes a given plane and often is called "flow rate."

Flow net: Configuration of streamlines and potential lines that define a flow field between a given source and sink and boundaries.

Flux: Volume, mass, or particle numbers transferred per unit time across a given area that is normal to the velocity vector of the volume, mass, or particles undergoing transport. The particles may be discrete, molecules, ions, or photons. The term is more general, however, and may apply to a field (gravitational, electric, or magnetic) or to energy.

Flux density: Flux per unit area—the area being normal to the velocity vector. Hydraulic loading rate and solids loading rate are examples.

Fractal: Ill-defined geometry of something, e.g., a particle. A floc particle is an example of a fractal.

Hydraulic loading rate (HLR): By definition, HLR = Q(plate)/A(cross section), in which A(cross section) is the area normal to the orientation of the flow direction of the individual plates or tubes.

Hindered settling: Settling in which the flow field of the settling particles are mutually affected by the others.

Horizontal flow basin: A traditional basin in which flow enters one end of a basin and leaves the other, leaving as a rule by means of overflow weirs.

Ideal basin: As defined by Camp (1946), an ideal basin has uniform velocity in all parts of the settling zone, and thus each molecule of water remains in the basin for a time equal to its detention time, the suspended particles are uniformly distributed at the inlet, and a particle that reaches the bottom is removed from the suspension.

Ideal flow: Flow that is frictionless and for which a flow net is applicable.

Inlet zone: The zone of turbulence dissipation at the entrance to a horizontal flow settling basin, which is not effective in settling.

Kinematic viscosity: Calculated as the dynamic viscosity of a fluid divided by its density, i.e., $n = \mu/r$, where n = kinematic viscosity (m^2/s), r = density of fluid (kg/m^3). Units conversion is (m^2/s) = (N s/m^2) · (m^3/kg) · (kg m/N s^2). Example calculation for T = 20°C: $n = \mu/r = (1.005 \cdot 10^{-3}$ Ns/m^2)/(998.2040 kg/m^3) = $1.007 \cdot 10^{-6}$ m^2/s. See also *dynamic viscosity*.

Loading rate: Although used frequently, this term should be used with full designation such as hydraulic loading rate, surface loading rate, or solids loading rate with units specified.

Outlet zone: The zone of a horizontal flow settling basin that is dedicated to the outlet flow and is not counted as being effective in settling.

Overflow launders: The weirs of a settling basin over which the effluent flow is discharged.

Overflow rate (OR): See *surface overflow rate* or *surface overflow velocity*, v_o.

Overflow velocity (v_o): See *surface overflow rate* or *surface overflow velocity*, v_o.

Plain sedimentation: A settling process in water treatment intended to settle mineral sediments in natural waters, usually rivers. If used, plain sedimentation would be after intake screening and helps to reduce the load on other processes. The process may be most efficacious during spring runoff, but may be required during the entire year for those rivers that are inherently sediment laden.

Plate settler: An array of plates inclined at an angle of 55° or 60°. The suspension flows between the plates and by gravity is carried to the surface of the lower plate. As the solids build up on the lower plate, the gravitational vector parallel to the plate at some point exceeds the shear resistance and the mass slides down to a solids collection zone. This usually occurs in a random fashion as "patches" of solids may be observed sliding down the plate at random locations.

Primary clarifier: The first settling basin in wastewater treatment that separates settleable solids before biological treatment.

Relationship between HLR and HLR(horiz plane): The HLR(horiz plane) is the flow, Q(plate), divided by the cross-sectional area of an individual plate, or tube, projected on the horizontal plane, i.e., A(horiz plane). For a plate, the A(horiz plane) = w(plate) · $d/\sin\theta$. Therefore, HLR(horiz plane) = HLR · $\sin\theta$. The HLR(horiz plane) has been used by regulatory agencies (Colorado Health Department, 1987, p. 32) and by those in the industry using tube settlers.

Rim-flow clarifier: A clarifier in which the influent flow is from a peripheral feed inlet.

Sedimentation: Synonymous with *settling*.

Settling: Unit process for separation of a suspension by gravity settling of particles.

Settling basin: As defined by Camp, a settling basin has four zones: inlet, settling, outlet, and sludge collection.

Settling zone: The "effective" zone of a clarifier in which settling occurs; the term was derived from the "ideal" basin.

Shield's equation: An empirical mathematical relationship that relates particle characteristics to the critical channel velocity for scour.

Short circuiting: The main flow within a basin is direct from inlet to outlet with little participation by the intended volume.

Sink: The outlet of a flux.

Sludge volume index (SVI): The volume (using a graduated 1000 mL cylinder) of 1 g of sludge after 30 min settling. The measurement usually is applied to a sample of activated sludge from an aeration basin to provide and index of settling behavior. SVI = 100 is the same as a sludge concentration of 10,000 mg/L, i.e., 1000 mg/0.10 L, and is considered a desired target for settled sludge, and thus for X_r (see also Dick and Vesilind, 1969). Abbreviated SVI.

Solids contact clarifier: Other names are reactor clarifier and sludge blanket clarifiers. These are circular basins to remove colloidal particles or to create a chemical floc in which chemical mixing is accomplished in a rapid mix unit in the center of the unit. The chemically treated water is forced to flow through a solids blanket and upward and over weirs at the surface of the basin. Most systems are proprietary and are sold as a system. They are used to remove colloidal solids and in chemical reactions, such as softening.

Solids loading rate: Solids flux per unit of area, e.g., $kg/min/m^2$.

Source: The source of a flux.

Stokes equation: The relationship for laminar flow conditions that relates fall velocity of a particle to the shear resistance.

Stream tube: The volume between two streamlines such that the flow within each streamtube is the same.

Streamline: Locus of points of a velocity vector in a flow field. Any given streamline is perpendicular to the potential gradient at any point. An array of streamlines and potential lines done according to certain rules is a "flow net." In a general sense, streamlines are the same as flux lines.

Surface overflow rate (SOR): Flow divided by plan area, i.e., $SOR = Q/(wL)$, also called surface loading rate, overflow velocity, and overflow rate. Although SOR and v_o may be used interchangeably, the sense of the two are different. The SOR is also an "intensive" variable and is the hydraulic loading applied to the plan area. A complementary idea is the surface overflow velocity, v_o (surface overflow velocity). For plate settlers, SOR (plate settlers) $= Q/\Sigma n(\text{plates}) \cdot w(\text{plate}) \cdot L(\text{plate}) \cdot \cos\theta$; in other words, it is the flow divided by the total plate area projected on to the horizontal plane.

Surface overflow velocity (v_o): Flow divided by plan area, i.e., $v_o = Q/(wL)$, or the rate at which the water surface will hypothetically fall in transport from a basin entrance to its exit. When compared with the fall velocity of particles, it must be true then that all particles will be removed for which, $v_s \geq v_o$, for the ideal basin.

Tracer: A chemical that dissolves in water and has negligible effect on density, which can be detected easily and accurately at low concentrations. In cases in which the dispersion patterns were to be observed, such tracers as potassium permanganate and methylene blue dyes have been used. Later, fluorescein was used, but as it decayed in light, it was replaced by Rhodamine-B and Rhodamine-WT. The latter can be detected at very low concentrations by means of a fluorometer. Sodium chloride is often used since the chloride ion can be detected easily by titration or by a specific ion electrode, or a conductivity meter may be useful when source waters are low specific electrical conductivity. Specific ion electrodes have also been used to detect fluoride ion. Online instruments can be used in place of sampling.

Tracer test: The procedure of injecting a tracer, such as a dye or salt, into a flow with measurement of tracer concentrations in the effluent so that a dispersion curve can be constructed. Two methods of conducting the test are the step dose and the slug dose, with the latter being most common and easiest to conduct.

Tube settler: An array of square tubes, usually 25 mm (1 in.) square, inclined at 60°, which functions the same as plate settlers. The suspension settles to the lower surface of the tubes and is removed by sliding to a removal zone below the tubes.

Type I settling: Discrete particle settling.

Type II settling: Flocculent particle settling.

Type III settling: Settling in which there is mutual interference between settling particles; an interface forms.

Type IV settling: Settling that occurs by the displacement of water due to the weight of the mass of particles above a given level.

Up-flow basin: A settling basin with weirs arranged across the entire basin. Flow is forced upward to be removed by the weirs.

Weir loading rate: The flow per unit length of weir, e.g., m^3/day flow/m weir length.

REFERENCES

ASCE, *Civil Engineering Classics, Outstanding Papers of Thomas R. Camp*, American Society of Civil Engineers, New York, 1973.

ASCE-WPCF, *Sewage Treatment Plant Design*, ASCE Manual of Practice No. 36, WPCF Manual of Practice No. 8, American Society of Civil Engineers and Water Pollution Control Federation, Headquarters of the Society, New York, 1959.

AWWA-ASCE, *Water Treatment Plant Design*, 3rd edn., McGraw-Hill, New York, 1998.

Bhargava, D. S. and Rajagopal, K., Differentiation between transition zone and compression in zone settling, *Water Research*, 27(3):457–463, 1993.

Burns and Roe, Inc., *Process Design Manual for Suspended Solids Removal*, Environmental Protection Agency, Washington, DC, October 1971.

Camp, T. R., A study of the rational design of settling tanks, *Sewage Works Journal*, 8:742–758, 1936.

Camp, T. R., Discussion of paper, Slade, J. J. Jr., Sedimentation in quiescent and turbulent basins, *Transactions of the ASCE*, 102:306–314, 1937.

Camp, T. R., Sedimentation and the design of settling tanks, *Transactions, of the ASCE*, III:895–958, 1946.

Camp, T. R., Studies of sedimentation basin design, *Sewage and Industrial Wastes*, 25:1–12, 1953.

Colorado Department of Health, Drinking Water Design Criteria, Colorado Department of Health, Water Quality Control Division, Denver, Colorado, September 1987.

Crawford, S., Plant-scale tracer tests, *Rumbles, Rocky Mountain Section AWWA-WEA Newsletter*, July 1994.

Culp, G. L. and Conley, W., High-rate sedimentation with the tube clarifier concept, in: Gloyna, E. and Eckenfelder, W. W. (Eds.), *Water Quality Improvement by Physical and Chemical Processes*, Water Symposium No. 3., University of Texas Press, Austin, TX, 1970.

Culp, G., Hansen, S., and Richardson, G., High rate sedimentation in water treatment works, *Journal of the American Water Works Association*, 60(6):681–698, June 1968.

Daigger, G. T. and Roper, R. E., The relationship between SVI and activated sludge settling characteristics, *Journal of Water Pollution Control Federation*, 57(8):859–866, 1985.

Dick, R. I., Role of activated sludge final settling tanks, *Journal of Sanitary Engineering Division, ASCE*, 96(SA2):423–436, April 1970.

Dick, R. I., Folklore in the design of final settling tanks, *Journal of Water Pollution Control Federation*, 48(4):633–644, April 1976.

Dick, R. I. and Ewing, B. B., Evaluation of activated sludge thickening theories, *Journal of Sanitary Engineering Division, ASCE*, 93(SA4):9–29, August 1967.

Dick, R. I. and Vesilind, P. A. The sludge volume index—What is it?, *Journal of Water Pollution Control Federation*, 41 (7):1285–1291, July 1969.

Esler, J. K., Optimizing clarifier performance—An outline, in: *Rocky Mountain Section WEA Preconference Workshop*, Snowmass Village, CO, September 13, 1998.

Fair, G. M., Geyer, J. C., and Okun, D. A., *Water and Wastewater Engineering*, Volume 2. *Water Purifications and Wastewater Treatment and Disposal*, John Wiley & Sons, New York, 1968.

Fischerström, C. N. H., Sedimentation in rectangular basins, *Proceedings Sanitary Engineering Division, ASCE*, 81, Separate No. 687:1–29, 1955.

Great Lakes—Upper Mississippi River Board of State Sanitary Engineers, *Recommended Standards for Sewage Works*, Health Education Service, Box 7283, Albany, NY, 1968.

Great Lakes—Upper Mississippi River Board of State Engineers, *Recommended Standards for Water Works*, Health Education Service, Albany, NY, 1987.

Hane-Weijman, H., Lamella separation—Old technology in a modern concept, Unpublished paper for short course Water Treatment Plant Design at Colorado State University, in June, 1991, Purac Engineering, Inc., Wilmington, DE, 1991.

Hansen, S. P. and Culp, G. L., Applying shallow depth sedimentation theory, *Journal of the American Water Works Association*, 59(9):1134–1148, September 1967.

Hazen, A., On sedimentation, *Transactions of the ASCE*, 53:45, 1904.

Hermanowicz, S. W., Secondary clarification of activated sludge: Development of operating diagrams, *Water Environment Research*, 70(1):10–13, 1998.

Kamp, P. C., Research, design, and operating experience—WRK III treatment plant, Enkhuizen, Holland, paper without publication identification, c. 1989.

Katz, W. J., Geinopolos, A., and Mancini, J. L., *Concepts of Sedimentation Applied to Design*, Part 1, *Water and Sewage Works*, 162–165, April 1962a.

Katz, W. J., Geinopolos, A., and Mancini, J. L., *Concepts of Sedimentation Applied to Design*, Part 2, *Water and Sewage Works*, 169–171, May 1962b.

Katz, W. J., Geinopolos, A., and Mancini, J. L., *Concepts of Sedimentation Applied to Design*, Part 3, *Water and Sewage Works*, 257–259, July 1962c.

Kawamura, S., Hydraulic scale model simulation of the sedimentation process, *Journal of the American Water Works Association*, 73(7):372–379, December 1986.

Kawamura, S., *Integrated Design of Water Treatment Facilities*, John Wiley & Sons, New York, 1991.

Kawamura, S., Optimization of basic water treatment processes—Design and operation: Sedimentation and filtration, *Aqua*, 45(3):130–142, 1996.

Keinath, T. M., Operational dynamics and control of secondary clarifiers, *Journal of Water Pollution Control Federation*, 57(7):770–776, 1985.

Keinath, T. M., Diagram for designing and operating secondary clarifiers according to the thickening criterion, *Research Journal of the Water Pollution Control Federation*, 62(3):254–258, 1990.

Langelier. W. F., Shallow sedimentation basins, *Journal of the American Water Works Association*, 23(11):1484–1489, 1930.

Malcolm Pirnie, Inc., *The First Century, 1895–1995*, Malcolm Pirnie, Inc., White Plains, NY, c. 1995

Nemerow, N. L., *Liquid Wastes of Industry, Theories, Practices, and Treatment*, Addison-Wesley, Reading, MA, 1971.

Nemerow, N. L. and Agardy, F. J., *Strategies of Industrial and Hazardous Waste Management*, Van Nostrand Reinhold, New York, 1998.

Purac, *Eleven Hundred Plants in Forty Counties*, Purac Corporation, Wilmington, DE, c. 1991

Rich, L. G., *Unit Operations of Sanitary Engineering*, John Wiley & Sons, Inc., New York, 1961.

Rouse, H. (Ed.), *Advanced Mechanics of Fluids*, John Wiley & Sons, Inc., New York, 1959.

Vesilind, P. A., The influence of stirring in the thickening of biological sludge, PhD Dissertation, University of North Carolina, Chapel Hill, NC, 1968a (cited in Daigger and Roper (1985), Wahlberg and Keinath (1988), etc.).

Vesilind, P. A., "Discussion of Dick, R. I. and Ewing, B. B., Evaluation of activated sludge thickening theories, *Journal of Sanitary Engineering Division, ASCE*, 93(SA4):9–29, August 1967," *Journal of Sanitary Engineering Division, ASCE*, 94 (SA1):185–191, February 1968b.

Wahlberg, E. J. and Keinath, T. M., Development of settling flux curves using SVI, *Journal of Water Pollution Control Federation*, 60:2095–2100, 1988.

Wahlberg, E. J., Stahl, J. F., Chen, C., and Augustus, M., Field application of the Clarifier Research Technical Committee's Protocol for evaluating secondary clarifier performance: Rectangular, cocurrent sludge removal clarifier, in: Paper presented at *66th Annual Conference and Exposition, Water Environment*, Anaheim, CA, October 4, 1993.

Waterlink Technologies, Inclined Plate Settlers, Waterlink Brochure WL-609-5M-6/97-MCA, Nordic Water Products, Inc., Nunäshamn, Sweden, 1997.

Watts, R. W., Svoronos, S. A., and Koopman, B., One-dimensional modeling of secondary clarifiers using a concentration and feed velocity-dependent dispersion coefficient, *Water Research*, 30(9):2112–2124, 1996.

Willis, R. M., Tubular settlers—A technical review, *Journal of the American Water Works Association*, 70(6):331–335, June 1978.

WPCF, Clarifier design, in: *Manual of Practice FD-8*, Water Pollution Control Federation, Washington, DC, 1985.

Yao, K. M., Design of high rate settlers, *Journal of the Environmental Engineering Division, ASCE*, 99(EE5):621–637, October 1973.

7 Grit Chambers

This chapter summarizes the horizontal-flow grit removal process (Camp, 1942). Aerated grit chambers are included, an innovation on the scene since the 1950s. Proprietary grit removal processes are not included, in keeping with the principle-oriented theme.

Grit removal is a sedimentation process and the chapter could have been included in Chapter 6, but was felt to warrant being dealt with separately in order to focus on grit as a special case.

7.1 GRIT

Not much has been published on grit removal as compared to other unit processes in water treatment. Perhaps the topic has lacked "glamour," or perhaps there was not much to be said after Camp's 1942 paper (Camp, 1942), and became an established process (Box 7.1). Ancillary technologies, that is, to make the process work, include scrapers, lift devices, and grit washing.

Purposes of grit removal: The two purposes of grit removal are: (1) to protect pump and other metal surfaces from an excessive rate of abrasion wear, and (2) to prevent nuisance conditions.

Nuisance conditions: Two categories of nuisance include:

1. Grit accumulations in digesters, and in pipelines and channels. Grit in digesters causes sluggish circulation and accumulates at the bottom. Grit in pipes, for example, from primary clarifiers to digesters, may cause clogging; such a problem is not uncommon and has been reported frequently by operators.
2. Excessive organics in grit-to-be-disposed (Camp, 1942). The "quality" of grit refers to its organic fraction; the approximate range is $0.05 \leq$ organic-fraction and ≤ 0.50 g-organics/g-inert suspended-solids. The lower fraction is considered a relatively "clean" grit. The higher value is not an uncommon value for some grit removal operations and would be considered unsuitable for land disposal due to its putrescible nature.

Objectives of design: The design of a grit chamber requires two processes be operative: (1) settling of grit, and (2) scour of organics. As recommended by Camp (1942), the settled grit should have a diameter of 0.2 mm and larger, with specific gravity of 2.65. Camp asserted that most of the grit that causes problems has larger size. Removal of $\geq$0.2 mm grit with $\leq$0.05 organic fraction has been accepted as objectives for practice. To accomplish scour of organics, the horizontal velocity of the grit chamber should be adequate to suspend deposited organics but not the grit particles larger than 0.2 mm, that is, $0.23 \leq v_H \leq 0.38$ m/s ($0.75 \leq v_H \leq 1.25$ ft/s).

Removal principle: Grit removal is a sedimentation process, that is, the force of gravity acting on particles. Alternatively, the force acting could be "generated" by centrifuging, in which case proprietary equipment is involved.

7.2 HORIZONTAL FLOW GRIT CHAMBERS

The classic grit removal process is the horizontal flow grit chamber.

7.2.1 Theory

Grit removal is a sedimentation process, with principles delineated in Chapter 6. The essential elements are summarized in this section. Scour of organics is an added factor.

7.2.1.1 Ideal Basin

The theoretical design of a grit chamber is based upon Camp's concept of the ideal settling basin (1946), outlined in Section 6.3.3, and described first in his 1942 article on grit chambers. Key assumptions of ideal settling theory, for a horizontal flow basin, are: (1) the horizontal flow velocity is uniform over depth, (2) a discrete particle subsides at a constant fall velocity. Combining these two velocities will define the path of a subsiding grit particle, shown in Figures 7.6 through 7.8. The path shown is for a 0.2 mm particle of grit to be 100% removed. The corresponding fall velocity, calculated by Stoke's law, for a 0.2 mm particle with specific gravity 2.65 is 21 mm/s (0.069 ft/s).

The function of a grit chamber is to remove silica sand particles of $\geq$0.2 mm, while passing putrescible organic particles. Since organic particles have a slower fall velocity than grit, the subsidence path has an angle with the horizontal much less than the grit; thus, as indicated in Figure 7.1, most organic particles will be carried out with the flow.

At the entrance to the grit chamber, however, organic particles as well as other particles are distributed uniformly over depth. Therefore, the path of some of these organic particles will intercept the bottom of the grit chamber, as indicated in Figure 7.2.

7.2.1.2 Scour

If the grit chamber is to function effectively, any deposited organics should be resuspended. This is accomplished by maintaining a sufficient scour velocity which will remove

BOX 7.1 CAMP AND GRIT CHAMBER PRACTICE

Camp's 1942 paper on grit chamber design provided definitive design guidelines. Like most of his papers, this one also was comprehensive, that is, characterized by theory, criteria for practice, and application examples. This chapter was constructed mostly from Camp's 1942 paper.

Other than Camp's paper, the literature on grit has been sparse, with only a few papers appearing over the decades; for example, Morales and Reinhardt (1984) who investigated the performance of grit chambers in several plants; Londong (1989), who developed a rationale based on pilot plant studies; Hirano et al. (1998), who reported on the problems of grit overwhelming the San Francisco WWTP; and a series of four papers by Wilson et al. (2007a–d), who revisited some of the premises of grit characteristics and removal. Principles are constant, however, and so Camp's work has remained the primary reference.

the lighter organics from the bed while leaving the grit. Thus, control of horizontal velocity is necessary.

The ideal horizontal velocity, v_H, is 0.3 m/s (1.0 ft/s) for all flows (from minimum to maximum). Accepted operating limits are 0.23–0.38 (0.75–1.25 ft/s). The 0.3 m/s (1.0 ft/s) criteria has been well established since the 1909 experiments of N.A. Brown in Rochester (reported by Metcalf and Eddy, 1916) who related grit size, horizontal flow velocity, and percent of grit retained for each size. Camp (1946) in analyzing the mathematical relations and experimental work of A. Shields formulated an operational equation to describe the critical mean channel velocity, v_c, for incipient motion of particles of size, d, and specific gravity, SG, given here as Equation 7.1,

$$v_c = \left[\frac{8\beta}{f} \cdot g(\text{SG} - 1)d\right]^{0.5} \tag{7.1}$$

where
- v_c is the critical mean horizontal velocity to start incipient motion of particles (m/s)
- d is the size of particle (m)
- SG is the specific gravity of particle
- β is the some unknown function of d/d_b (where d_b is the boundary film thickness) which can be assumed a constant (see Figure 11 in Camp, 1946), varying from 0.04 for smooth sand beds to 0.10 for a bed of sand ripples
- f is the Darcy–Weisbach friction factor, taken as 0.03
- g is the acceleration of gravity 9.81 m/s^2 (32.2 ft/s^2)

The equation will work for any units desired as long as g is in those same units.

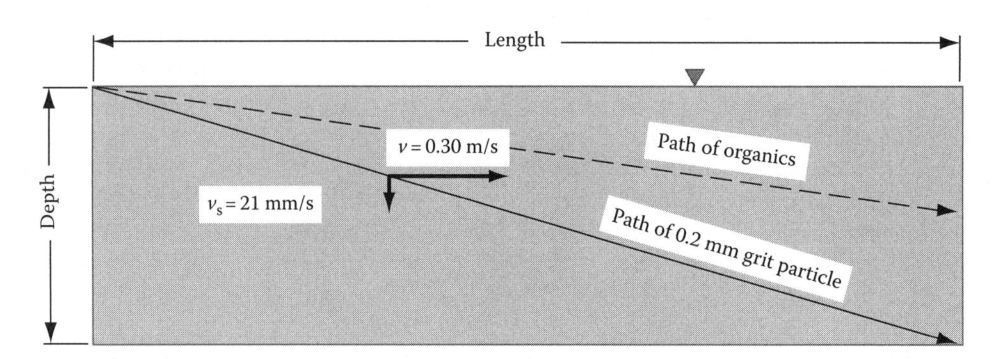

FIGURE 7.1 Path of subsidence of the smallest particle of grit to be completely removed. (Adapted from Rex Chainbelt, Grit Collectors, Product Manual, Sanitation Equipment, Conveyor and Process Equipment Division, Rex Chainbelt, Inc., Milwaukee, WI, Data Sheets 315-4.001-315-4.531, 1965.)

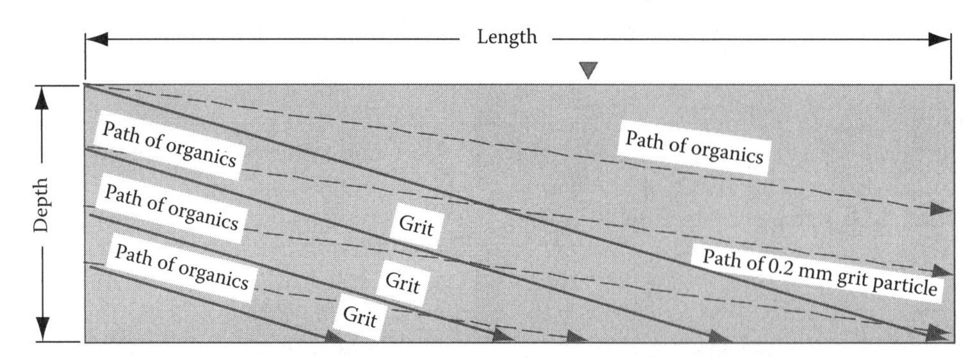

FIGURE 7.2 Path of lighter organics relative to grit. (Adapted from Rex Chainbelt, Grit Collectors, Product Manual, Sanitation Equipment, Conveyor and Process Equipment Division, Rex Chainbelt, Inc., Milwaukee, WI, Data Sheets 315-4.001-315-4.531, 1965.)

TABLE CD7.1
Solution for Example 7.1—Shield's Equation for Scour in Grit Chamber

v_c (m/s)	β	f	g (m/s²)	SG	d (m)	v_c (m/s)
(a) Scour velocity for 0.2 mm quartz particle:						
	0.06	0.03	9.81	2.65	0.0002	0.23
(b) Diameter of quartz particle scoured at 0.3 m/s (1 ft/s) and sensitivity to constants, β and *f*:						
0.30	0.06	0.03	9.81	2.65	0.0004	
0.30	0.08	0.023	9.81	2.65	0.0002	
(c) Size of organic size matter scoured at SG = 1.2 and 1.05:						
0.30	0.06	0.03	9.81	1.2	0.003	
0.30	0.06	0.03	9.81	1.05	0.012	

Source: Londong, J., *Water Sci. Technol.*, 21, 13, 1989.

Example 7.1 Scour Velocity in Grit Chamber (Adapted from Camp, 1946)

Problem Statement

(a) Estimate the highest velocity in a grit chamber to avoid scour of sand particles 0.2 mm is size.
(b) Determine the sand size that will be started in incipient motion if the velocity is 0.3 m/s (1.0 ft/s).
(c) Determine the size of organic matter (assume SG = 1.2 and then 1.05) that will be scoured.
(d) Look at any other variations that may be of interest.

Solution

The most expedient means to solve (a), (b), and (c) is by means of a spreadsheet, that is, Table CD7.1, which is based on the Shield's equation.

Comments

The Shield's equation is used to estimate design limits for channel velocity for a horizontal flow grit chamber. With respect to task (d), once set up, as in Table CD7.1, the algorithm can be used in a sensitivity analysis to assess the effect of errors, in choosing β and f and the effects of varying v_c on particle sizes that may be scoured, etc.

7.2.2 Horizontal Velocity Control

As noted in Section 7.2.1.2, the importance of maintaining a constant horizontal flow velocity of 0.3 m/s (1.0 ft/s) has long been understood. The use of two or more parallel channels will accomplish this; a number of such examples are given in Metcalf and Eddy (1916). In other words, as flow increases above a certain "set-point," the second grit chamber will come into operation. Practice, however, has favored a single channel having some method for velocity control to maintain constant v_H. Since flow, Q, is the product of v_H and A, the cross-sectional area, it follows that A must be a linear function of Q if v_H is held constant, that is,

$$Q = v_H \cdot A \tag{7.2}$$

where $v_H = 0.3$ m/s (1.0 ft/s)—a constant. This mathematical condition can be accomplished by several methods, described in the following sections.

7.2.2.1 Proportional Weir

For channels with vertical walls, the width, w, is constant, and therefore the depth must be a linear function of flow, Q. A proportional weir, shown in Figure 7.3, has this characteristic, that is, $Q = kd$. The proportional weir may be dimensioned by Equations 7.3 through 7.5. Terms are defined in Figure 7.3.

To expand further, a proportional weir is a plate that fits across the channel (Figure 7.3). Below the weir, a free

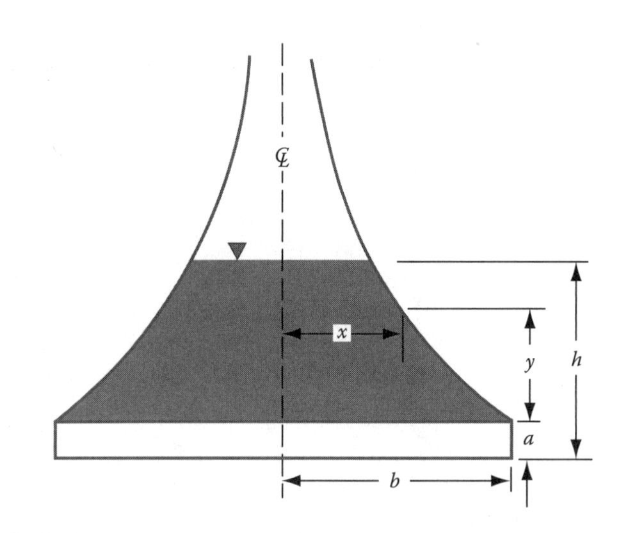

FIGURE 7.3 Proportional weir—used with grit chambers to maintain constant velocity.

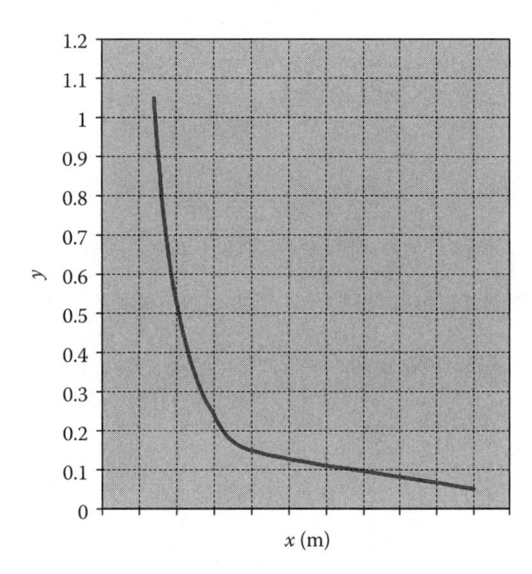

FIGURE 7.4 Proportional weir geometry.

fall is required, which results in a larger headloss than with a Parshall flume.

$$x = w\left[1 - \frac{2}{\pi}\tan^{-1}\left(\frac{y}{a}\right)^{0.5}\right] \tag{7.3}$$

The flow for the half-weir section is calculated (Babbitt, 1940, p. 379; ASCE-WPCF, 1959, p. 69; ASCE-WPCF, 1977, p. 141):

$$Q(\text{half-section}) = K\left[h + \frac{2}{3}a\right] \tag{7.4}$$

$$K = Ca^{0.5}b(2g)^{0.5} \tag{7.5}$$

where

- x is the horizontal distance from vertical center line to point on curved edge of weir (m)
- y is the vertical distance from the base of the curve, at height a, to same point on curved edge of weir (m)
- b is the width of half of the rectangular base part of weir (m)
- a is the height of rectangular base part of weir (m)
- Q (half-section) is the flow through half of the weir section, that is, either the half to the right or the half to the left of the vertical center-line (m^3/s)
- K is the coefficient for weir (m^2/s)
- h is the depth of water from bottom of rectangular base to water surface (m)
- C is the discharge coefficient, which is taken as 1.0 in the literature (dimensionless)
- g is the acceleration of gravity (9.81 m/s^2)

Table CD7.2 is a spreadsheet with associated plots that illustrate the application of Equations 7.3 through 7.5. Figure 7.5 shows the weir shape plotted from the calculated coordinate, x. Figure 7.5 shows the plot of calculated flows for assumed values of h (which confirms the linear relationship between flow and depth).

7.2.2.2 Parshall Flume

A Parshall flume is a frequent choice for a grit chamber control section. Figure 7.6 is a photograph of an installation in Colorado; the grit chamber is located upstream. The merits of a Parshall flume are its low headloss and its use as a standard technology for flow measurement. As a caveat, however, the velocity through the grit chamber is not constant with depth (as it is with the proportional weir).

Background: The Parshall flume was developed by Ralph M. Parshall at Colorado State University (CSU) during the late 1920s (Parshall, 1926); publications on the topic were published

TABLE CD7.2
Spreadsheet Showing Calculation of Proportional Weir Sizing and Flow

w (m)	a (m)	y (m)	x (m)	H (m)	K (m^2/s)	Q (m^3/s)
1.00	0.050	0.0	1.00000	0.1	0.614	0.020
		0.1	0.39183	0.2		0.082
		0.2	0.29517	0.3		0.143
		0.3	0.24675	0.4		0.205
		0.4	0.21635	0.5		0.266
		0.5	0.19498	0.6		0.328
		0.6	0.17891	0.7		0.389
		0.7	0.16626	0.8		0.450
		0.8	0.15596	0.9		0.512
		0.9	0.14737	1.0		0.573
		1.0	0.14005	1.1		0.635
Assumed	Assumed	Assumed	$x = w[1-(2/\pi)\tan^{-1}(y/a)\text{^}0.5]$			$Q = K[-a/3]$
				$H = y + a$	$K = C(a\text{^}0.5)\ w(2g)\text{^}0.5$	

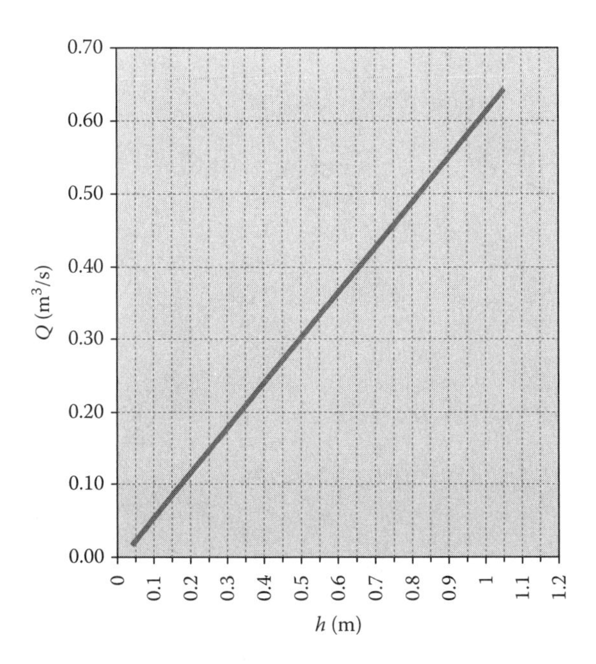

FIGURE 7.5 Flow vs. depth, proportional.

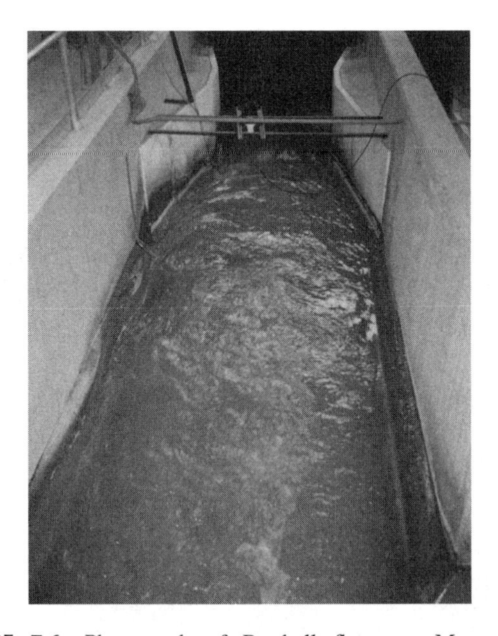

FIGURE 7.6 Photograph of Parshall flume at Marcy Gulch WWTP, Colorado, c. 2003. (Courtesy of Centennial Water and Sanitation District, Highlands Ranch, CO.)

in bulletins of the CSU Agricultural Experiment Station and the U.S. Department of Agriculture. The report of Skogerboe et al. (1967) was the first systematic calibration of the Parshall flume coefficients since the original work of Parshall in the late 1920s and the early 1930s and was used as the primary reference for this section.

Water level tie: Figure 7.7 is a perspective drawing showing a grit chamber and a Parshall flume together as a "system." As indicated, the water level at the grit chamber exit and the flume inlet are at the same elevation. To achieve such a water level "tie," the relationship between the depth of water in the grit chamber, d, and the depth in the flume, H_a, is, $d = H_a + \Delta Z$, in which ΔZ is the difference between the two depths. The elevation of the floor of the flume is "set" by this relationship. The continuity between the water levels defines the "hydraulic profile."

7.2.2.2.1 Free Flow and Submerged Flow

If the flow exceeds, what is called the "free-flow" condition, a "submerged-flow" condition exists. The "free-flow" condition means that a single measurement, H_a (which is upstream of the "throat"), is sufficient to measure flow. If the flow is "submerged," an additional measurement, H_b (which is downstream of the "throat"), is necessary. The water level downstream from the flume box should be low enough such that a backup does not occur, that is, resulting in "submerged-flow" condition. The "throat" is the narrow section of the flume and is defined in terms of its width dimension.

Free flow: For the "free-flow" condition, which means that "super-critical" velocity occurs in the throat section, the expression for flow is in terms of H_a,

$$Q = CH_a^n \tag{7.6}$$

where

Q is the flow through a given Parshall flume (m^3/s)
C is the coefficient, specific to throat width, taken from Table CD7.3
H_a is the water depth measured at a section two-third of the length of the entrance section upstream from the start of the flume "throat" (m)
n is the exponent for flume equation, specific to throat width, taken from Table CD7.3

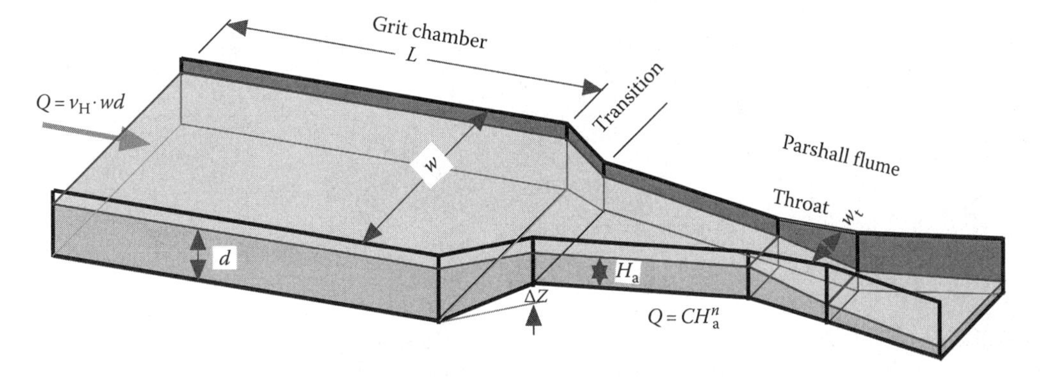

FIGURE 7.7 Parshall flume perspective showing grit chamber.

TABLE CD7.3
Free Flow Ranges, Coefficients, and Exponents for Parshall Flumes of Various Throat Widths[a]

Q (ft^3/s)		Q (m^3/s)		w_t[b]		C			
Min	Max	Min	Max	U.S. Cust.	Metric (m)	U.S. Cust.	Metric	n	S_t[c]
0.01	0.02	0.00028	0.00057	1 in.	0.0254	0.338	0.0605	1.55	0.56
0.02	0.4	0.00057	0.0113	2 in.	0.051	0.676	0.085	1.55	0.61
0.03	0.6	0.00085	0.017	3 in.	0.076	0.992	0.178	1.55	0.64
0.05	2.9	0.0014	0.0821	6 in.	0.152	2.06	0.382	1.58	0.55
0.1	5.1	0.00283	0.1444	9 in.	0.229	3.07	0.536	1.53	0.63
0.4	16	0.0113	0.4531	12 in.	0.305	4.00	0.69	1.52	0.62
0.5	24	0.0142	0.6797	18 in.	0.457	6.00	1.06	1.54	0.64
0.7	33	0.0198	0.9346	24 in.	0.61	8.00	1.43	1.55	0.66
0.8	41	0.0227	1.1611	30 in.	0.762	10.00	1.80	1.555	0.67
1.00	50	0.0283	1.416	3.0 ft	0.914	12.00	2.17	1.56	0.68
1.3	68	0.0368	1.9258	4.0 ft	1.219	16.00	2.93	1.57	0.70
2.2	86	0.0623	2.435	5.0 ft	1.524	20.00	3.70	1.58	0.72
2.6	104	0.0736	2.945	6.0 ft	1.829	24.00	4.50	1.59	0.74
4.1	121	0.116	3.427	7.0 ft	2.134	28.00	5.32	1.60	0.76
4.6	140	0.130	3.965	8.0 ft	2.438	32.00	6.08	1.60	0.78
6.0	200	0.170	5.664	10.0 ft	3.048	40.13	7.52	1.59	0.8

Source: Adapted from Skogerboe, G.V. et al., Parshall Flumes, Report WG31-3, OWRR Project No. B-006-Utah, Utah Water Research Laboratory, Utah State University, Logan, UT, March, 1967.

[a] Metric units were calculated from Skogerboe et al. (1967) data in this table.

[b] Throat width of flume.

[c] S_t is called the "transition submergence"; the condition occurs approximately when $H_b/H_a > S_t$.

Because the flow in the throat section is "supercritical," any flow disturbance downstream cannot have any effect upstream. Therefore, the Parshall flume operating in the free-flow condition is a suitable "control" section for the headworks of a wastewater treatment plant (Figure 7.8). At the same time, only a single depth measurement, H_a, is necessary for flow measurement, as seen by Equation 7.6. In design, the downstream hydraulic profile should be set such that the free-flow condition is maintained. In other words, the downstream water surfaces should be low enough such that the flume does not experience a "submerged" condition.

Flume selection: Parshall flumes are sized by the "throat" width, w_{throat}, which is the narrowest part of the construction looking at a plan view. For a given maximum flow expected through the headworks during operation for which free-flow performance is desired, the flume selected should have a free-flow capacity that is greater, that is,

$$Q(\max) \leq Q(\max\ \text{free flow for given}\ w_{throat}) \quad (7.7)$$

where

Q(max) is the maximum flow expected for normal operation of the plant (m^3/s)

Q(max free flow for given w_{throat}) is the free-flow capacity of flume for a given throat size, w_{throat}, as given in Table 7.3 (m^3/s)

The flume is "selected" based upon the maximum flow expected. Table CD7.3 gives the throat widths, w_{throat}, for

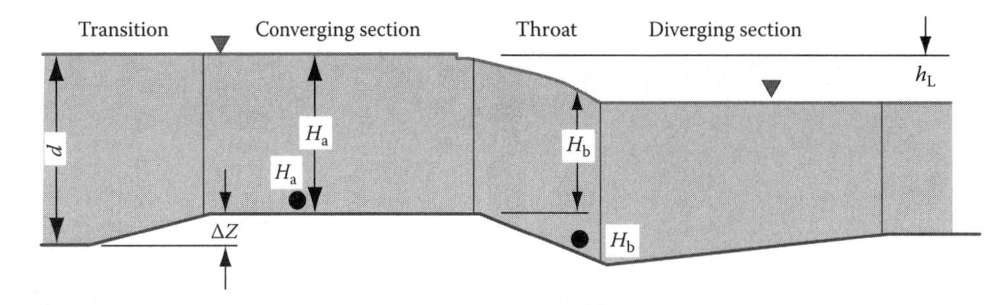

FIGURE 7.8 Installation of Parshall Flume to operate under free-flow conditions, showing definitions used in design. (From Skogerboe, G.V. et al., Parshall Flumes, Report WG31-3, OWRR Project No. B-006-Utah, Utah Water Research Laboratory, Utah State University, Logan, UT, March, 1967.)

different maximum flows, along with the associated coefficients, C and n, and the S_t criteria. Example 7.2 illustrates a selection of flume size.

Example 7.2 Selection of Parshall Flume

Problem statement

Select a flume for a flow, $Q(\text{max}) = 0.35\ \text{m}^3/\text{s}$ ($12.36\ \text{ft}^3/\text{s}$ or 8.0 mgd).

Solution

Referring to Table CD7.3, a flume with $w_{\text{throat}} = 0.305$ m (12 in.) has a "free-flow" capacity, $Q(\text{max}) = 0.45\ \text{m}^3/\text{s}$ ($16\ \text{ft}^3/\text{s}$). The associated coefficients are: $C = 0.69$ in SI units (4.00 in. for ft^3 and s units) and $n = 1.52$.

Discussion

The next larger flume size, that is, $w_{\text{throat}} = 0.457$ m (18 in), has a capacity, $Q(\text{max}) = 0.68\ \text{m}^3/\text{s}$ ($24\ \text{ft}^3/\text{s}$), which is more than required. Also depth differences for the flow variation would be appreciably less. Therefore, the selection should be $w_{\text{throat}} = 0.305$ m (12 in.).

Submerged flow: Submerged flow is caused by a downstream backup of the flow such that the super-critical velocity in the throat of the flume no longer exists. The S_t values in Table 7.3 are the criteria, for respective throat widths, that determine the point of transition to the "submerged" flow condition. When $H_b/H_a = S_t$ the flow begins to become "unstable" and as the ratio H_b/H_a increases, that is, $H_b/H_a > S_t$, the submerged-flow condition becomes established.

Hydraulic profile: The important question for a Parshall flume is the maximum level of H_b for which incipient submergence occurs. This can be determined by the relation

$$\frac{H_b}{H_a} = S_t \tag{7.8}$$

where S_t is the maximum ratio of H_b/H_a for incipient submergence from Table 7.3.

For a given Q, H_a may be calculated by Equation 7.6 and then H_b from Equation 7.8. The latter is the maximum level of H_b that can be tolerated before the onset of the submerged-flow condition. If $H_b(\text{measured}) \gg H_b(\text{calculated})$, then the submerged condition exists.

Example 7.3 Determination of Maximum Level of H_b for Incipient Submergence

Problem

For a 305 mm (12 in.) Parshall flume, determine the maximum level of H_b, that is, $H_b(\text{max})$ for the maximum-flow condition, $Q(\text{max})$.

Solution

Extract from Table 7.3 the row for the 305 mm flume, as shown in Table CD7.4a. Set up Table CD7.4b to calculate $H_a(\text{max})$ by Equation 7.6 and from this result, calculate $H_b(\text{max})$ by Equation 7.8. The result was $H_b(\text{max}) = 470$ mm.

Discussion

If the tailwater level should rise such that $H_b \geq H_b(\text{max})$, then submerged conditions will occur. If Q has lesser value, the same procedure is followed, that is, H_a is calculated by Equation 7.6, and H_b is calculated by Equation 7.8. If the tailwater level rises to cause a larger value of H_b than is calculated by Equation 7.8, then the submerged condition will occur.

Construction data: The Parshall flume must be constructed in accordance with certain dimensions for each throat size, w_{throat}. Figure 7.9 designates the dimensions for any Parshall flume. The dimensions for each flume size are given in Table CD7.5a and b for metric and U.S. Customary units, respectively. As seen in Figure 7.9, the converging flow section is level. The throat section has a downward slope (where $H/E = 9/24$) and then an upward slope (where $(H-K)/F = 1/26$). Parshall flumes up to a certain size may be prefabricated and installed in a prepared channel. Proprietary prefabricated units are available up to a certain sizes. For units constructed in place, concrete is used (commonly).

TABLE CD7.4
Hydraulic Profile Calculation

(a) Extract of data from Table CD7.3 for 0.305 m flume

Q (ft^3/s)		Q (m^3/s)		w_t[b]		C		
Min	Max	Min	Max	U.S. Cust.	Metric (m)	Metric	n	S_t[c]
0.4	16	0.0113	0.4531	12 in.	0.305	0.69	1.52	0.62

(b) Calculation of H_b for maximum flow

Q(max) (m^3/s)	C Metric	n	H_a(max) (m)	H_b(max) (m)
0.4531	0.69	1.52	0.758	0.470
Table 7.3			$Q = CH_a^n$	$H_b/H_a = S_t$

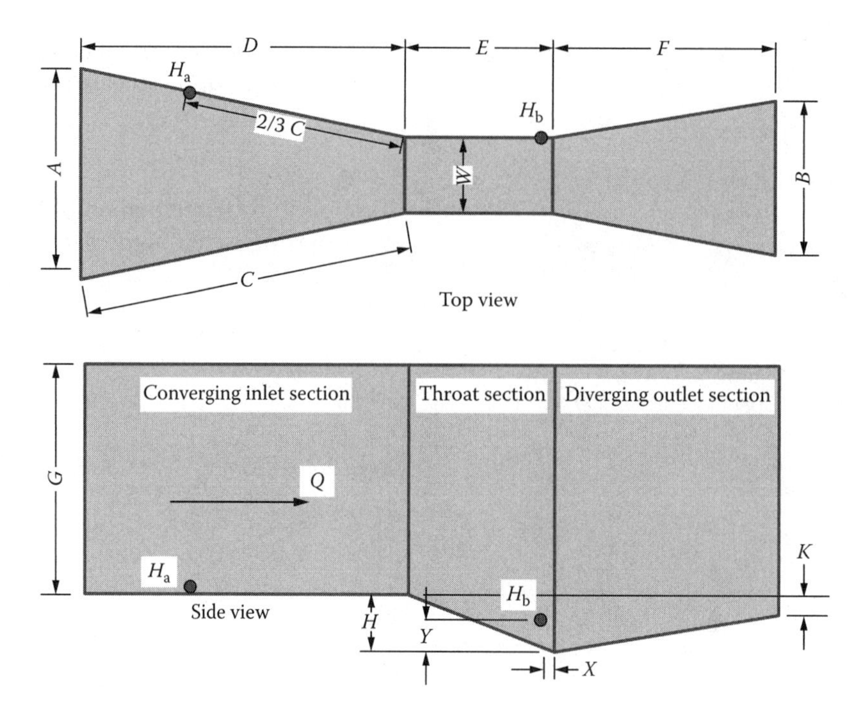

FIGURE 7.9 Parshall flume, views defining sections, dimensions, and points of depth measurement. (Adapted from Skogerboe, G.V. et al., Parshall Flumes, Report WG31-3, OWRR Project No. B-006-Utah, Utah Water Research Laboratory, Utah State University, Logan, UT, March, 1967.)

Deviations from the dimensions given imply differences in the coefficients, C and n.

7.2.2.3 Rectangular Section

The rectangular section for a grit chamber is easier to construct than a parabolic or trapezoidal section and has been favored in most plant headworks. Most often in practice a Parshall flume is used as a control, albeit the match is not compatible mathematically (as it is with a parabolic section) to yield a constant v_H. Because of this mismatch, the sizing of the grit chamber must serve to minimize the term $[v_H - v_H(\text{criterion})]$ for the different depths of operation. The general procedure is to select the smallest Parshall flume that will pass $Q(\text{max})$ so that the depth differential is the greatest attainable. As a note, a modified flume with horizontal bottom was proposed by Walker et al. (1973), which was intended as a control section for a rectangular grit chamber.

7.2.2.3.1 Parshall Flume Selection and Rectangular Section Grit Chamber

The steps in the selection of a Parshall flume and a rectangular section grit chamber are enumerated as follows:

1. Determine maximum flow, $Q(\text{max})$, expected for normal operation.
2. Select a Parshall flume from Table 7.3 for the smallest w_{throat} that will pass $Q(\text{max})$.
3. For $Q(\text{max})$, calculate H_a by Equation 7.6.
4. Let $v_H = 0.305$ m/s (Camp's velocity criterion).
5. Calculate $w(\text{grit chamber}) \cdot d(\text{grit chamber})$ from the continuity equation.

$$Q = A(\text{grit chamber}) \cdot v_H \qquad (7.9)$$

$$= [w(\text{grit chamber}) \cdot d(\text{grit chamber})] \cdot v_H \qquad (7.10)$$

With Q being stated and v_H being specified by criterion, the product $[w(\text{grit chamber}) \cdot d(\text{grit chamber})]$ is calculated.

6. Assume values of $w(\text{grit chamber})$ and calculate $d(\text{grit chamber})$.
7. Set the floor of the Parshall flume by the relation seen in Figure 7.8, that is,

$$d(\text{grit chamber}) = H_a + \Delta Z \qquad (7.11)$$

where

$d(\text{grit chamber})$ is the depth of water in grit chamber (m)

ΔZ is the difference in elevation between grit chamber floor and Parshall flume floor (m)

The "tie" between the grit chamber and the Parshall flume is the water surface, as seen in Figure 7.8. In other words, the Parshall flume floor is placed at an elevation differential, ΔZ, either above or below the grit chamber floor.

8. Determine v_o for a 0.2 mm diameter sand particle (by Stoke's law, $v_o(\text{SG} = 2.65) \approx 30$ mm/s).
9. Knowing v_o, calculate the plan area for the grit chamber, that is,

$$Q = A(\text{plan}) \cdot v_o \qquad (7.12)$$

$$= [w(\text{grit chamber}) \cdot L(\text{grit chamber})] \cdot v_o \qquad (7.13)$$

TABLE CD7.5

Dimensions and Capacities for Parshall Flumes

(a) Metric units

Throat Width (m)	Free Flow Capacities (m^3/s)		Dimensions (m)											
			A	B	C	2/3C	D	E	F	G	H	K	X	Y
0.025	0.00028	0.0006	0.17	0.093	0.365	0.242	0.356	0.076	0.204	0.152	0.029	0.019	0.008	0.013
0.051	0.00057	0.0113	0.21	0.135	0.414	0.029	0.406	0.114	0.253	0.204	0.043	0.022	0.016	0.025
0.076	0.00085	0.017	0.26	0.178	0.466	0.305	0.457	0.152	0.305	0.381	0.057	0.076	0.025	0.038
0.152	0.0014	0.0821	0.39	0.394	0.621	0.415	0.61	0.305	0.61	0.457	0.114	0.076	0.051	0.076
0.229	0.00283	0.1444	0.58	0.381	0.878	0.588	0.863	0.305	0.457	0.61	0.114	0.076	0.051	0.076
0.305	0.0113	0.4531	0.84	0.61	1.372	0.914	1.344	0.61	0.914	0.914	0.229	0.076	0.051	0.076
0.457	0.0142	0.6797	1.02	0.762	1.448	0.966	1.42	0.61	0.914	0.914	0.229	0.076	0.051	0.076
0.61	0.0198	0.9346	1.21	0.914	1.524	1.015	1.497	0.61	0.914	0.914	0.229	0.076	0.051	0.076
0.762	0.0227	1.1611	1.39	1.067	1.631	1.085	1.6	0.61	0.914	0.914	0.229	0.076	0.051	0.076
0.914	0.0283	1.416	1.68	1.219	1.676	1.119	1.646	0.61	0.914	0.914	0.229	0.076	0.051	0.076
1.219	0.0368	1.9258	1.94	1.524	1.829	1.219	1.792	0.61	0.914	0.914	0.229	0.076	0.051	0.076
1.524	0.0623	2.435	2.30	1.823	1.981	1.295	1.945	0.61	0.914	0.914	0.229	0.076	0.051	0.076
1.829	0.0736	2.945	2.67	2.134	2.134	1.423	2.091	0.61	0.914	0.914	0.229	0.076	0.051	0.076
2.134	0.116	3.427	3.03	2.438	2.286	1.524	2.24	0.61	0.914	0.914	0.229	0.076	0.051	0.076
2.438	0.130	3.965	3.40	2.743	2.438	1.60	2.39	0.61	0.914	0.914	0.229	0.076	0.051	0.076
3.048	0.170	5.664	4.76	3.658	4.349	1.83	4.27	0.914	1.829	1.219	0.343	0.152	0.305	0.229

(b) U.S. Customary units

Throat Width (ft)	Free Flow Capacities (ft^3/s)		Dimensions (ft)											
			A	B	C	2/3C	D	E	F	G	H	K	X	Y
0.083	0.01	0.02	0.55	0.305	1.19	0.794	1.167	0.25	0.67	0.5	0.094	0.062	0.026	0.042
0.166	0.02	0.4	0.70	0.442	1.36	0.906	1.333	0.375	0.83	0.67	0.14	0.073	0.052	0.083
0.25	0.03	0.6	0.85	0.583	1.53	1.02	1.50	0.50	1.00	1.25	0.188	0.25	0.083	0.125
0.5	0.05	2.9	1.29	1.292	2.036	1.36	2.00	1.00	2.00	1.50	0.375	0.25	0.167	0.25
0.75	0.1	5.1	1.89	1.25	2.88	1.93	2.83	1.00	1.50	2.00	0.375	0.25	0.167	0.25
1	0.4	16	2.77	2.00	4.50	3.00	4.41	2.00	3.00	3.00	0.75	0.25	0.167	0.25
1.5	0.5	24	3.36	2.50	4.75	3.17	4.66	2.00	3.00	3.00	0.75	0.25	0.167	0.25
2	0.7	33	3.96	3.00	5.00	3.33	4.91	2.00	3.00	3.00	0.75	0.25	0.167	0.25
2.5	0.8	41	4.56	3.50	5.35	3.56	5.25	2.00	3.00	3.00	0.75	0.25	0.167	0.25
3	1.00	50	5.16	4.00	5.50	3.67	5.40	2.00	3.00	3.00	0.75	0.25	0.167	0.25
4	1.3	68	6.35	5.00	6.00	4.00	5.88	2.00	3.00	3.00	0.75	0.25	0.167	0.25
5	2.2	86	7.55	6.00	6.50	4.25	6.38	2.00	3.00	3.00	0.75	0.25	0.167	0.25
6	2.6	104	8.75	7.00	7.00	4.67	6.86	2.00	3.00	3.00	0.75	0.25	0.167	0.25
7	4.1	121	9.95	8.00	7.50	5.00	7.35	2.00	3.00	3.00	0.75	0.25	0.167	0.25
8	4.6	140	11.1	9.00	8.00	5.25	7.84	2.00	3.00	3.00	0.75	0.25	0.167	0.25
10	6.0	200	15.6	12.0	14.27	6.00	14.0	3.00	6.00	4.00	1.125	0.50	1.00	0.75

Source: Adapted from Skogerboe, G.V. et al., Parshall Flumes, Report WG31-3, OWRR Project No. B-006-Utah, Utah Water Research Laboratory, Utah State University, Logan, UT, March, 1967.

Example 7.4 Parshall Flume Selection and Rectangular Grit Chamber Sizing

Statement

A maximum expected flow through a treatment plant is $Q(\max) = 0.765\ m^3/s$ ($27.0\ ft^3/s$). Let $Q(\min) = 0.425\ m^3/s$ ($15.0\ ft^3/s$).

Required

Select a Parshall flume and size a compatible grit chamber.

Solution

From Table 7.3, a 0.61 m (24 in.) flume has a capacity of $0.934\ m^3/s$ ($33.0\ ft^3/s$) and gives coefficients $C = 1.43$ SI (8 US) and $n = 1.55$. Determining H_a from Equation 7.6

gives $H_a = 0.76$ m (2.50 ft). From Table 7.5, the width of the flume at entrance is 1.206 m (3 ft–11 1/2 in.).

1. *By trial-and-error determine a suitable width, w, for grit chamber*: For each trial, a series of steps are required. The idea is to first calculate the depth, *d*, of the grit chamber, based upon an assumed width at *Q*(max). We can also let $v_H(\text{max}) = 0.38$ m/s (1.25 ft/s), the maximum permissible scour velocity. Knowing *d*, the depth of the grit chamber, the next step is to "set" the elevation of the flume floor, which is located a distance $H_a(\text{max})$ below the water surface at *Q*(max); keep in mind that the grit chamber water surface must be contiguous with that of the Parshall flume at all depths. Note that the flow at *Q*(max) is what sets the floor of the grit chamber. Next, v_H at *Q*(min) may be calculated. If $v_H < 0.23$ m/s (0.75 ft/s), an increase in *w* will result in a higher v_H at *Q*(min).

Trial 1 Assume the grit chamber has a width of 1.22 m (4 ft 0 in.).

1.1 Depth of water, *d*, in grit chamber at *Q*(max),

$$d = \frac{Q(\max)}{w \cdot v_H}$$

$$= \frac{0.765\,\text{m}^3/\text{s}}{1.219\,\text{m} \cdot 0.381\,\text{m}}$$

$$= 1.65\,\text{m}\ (5.40\,\text{ft})$$

1.2 Setting the Floor of the Flume

Figure 7.8 shows how the floor of the flume is to be set relative to the grit chamber. Thus, with the upstream water surface as the control, the floor of the flume is set at $H_a(\text{max}) = 0.67$ m (2.19 ft) below the water level. At the same time, the floor of the flume is 0.98 m (3.21 ft), that is, $\Delta Z = d(\text{max}) - H_a = 1.65 - 0.67 = 0.98$ m $(5.40 - 2.19 = 3.21$ ft) above the floor of the grit chamber.

1.3 Check v_H at *Q*(min)

- Check horizontal velocity, v_H, in grit chamber when $Q(\text{min}) = 0.425\ \text{m}^3/\text{s}$ (15 ft^3/s).
- From Equation 7.6, $Q(\text{min}) = 0.425\ \text{m}^3/\text{s} = 1.43 H_a^{1.55}$; $H_a(\text{min}) = 0.457$ m.

And, in U.S. Customary units, $Q(\text{min}) = 15.0\ \text{ft}^3/\text{s} = 8.0 H_a(\text{min})^{1.55}$; $H_a(\text{min}) = 1.50$ ft.

- Depth in grit chamber: $d(\text{min}) = H_a(\text{min}) + \Delta Z = 0.457 + 0.98 = 1.44$ m.

In U.S. Customary units, $d(\text{min}) = 1.50 + 3.21 = 4.71$ ft.

- Therefore, $v_H = (0.425\ \text{m}^3/\text{s})/(1.219\ \text{m} \cdot 1.44\ \text{m}) = 0.24$ m/s; which is close enough to the 0.23 lower limit.

In U.S. Customary units, $v_H = (15\ \text{ft}^3/\text{s})/(4.0\ \text{ft} \cdot 4.71\ \text{ft}) = 0.80$ ft/s.

- Although $v_H(\text{min}) = 0.24$ m/s (0.80 ft/s) approximates, closely enough, to $v_H(\text{min}) = 0.23$ m/s (0.75 ft/s), further trials might yield a more satisfactory result.

Trial 2 Assume the grit chamber preceding has a width of 2.74 m (9′0)″.

2.1 Depth of water, *d*, in grit chamber at *Q*(max),

$$d = \frac{Q(\max)}{w \cdot v_H}$$

$$= \frac{0.765\,\text{m}^3/\text{s}}{2.74\ \text{m} \cdot 0.381\ \text{m}}$$

$$= 0.73\ \text{m}\ (2.40\,\text{ft})$$

2.2 Setting the Floor of the Flume

Figure 7.8 shows how the floor of the flume is to be set relative to the grit chamber. The floor of the grit chamber is set at $d(\text{max}) = 0.73$ m (2.40 ft) below the water level. The floor of the flume is $H_a(\text{max}) = 0.67$ m (2.19 ft). At the same time, the floor of the flume is, $\Delta Z = d(\text{max}) - H_a = 0.73 - 0.67 = 0.06$ m $(2.40 - 2.19 = 0.21$ ft) above the floor of the grit chamber.

2.3 Check v_H at *Q*(min)

- Check horizontal velocity, v_H in grit chamber when $Q(\text{min}) = 0.425\ \text{m}^3/\text{s}$ (15 ft^3/s).
- From Equation 7.6, $Q(\text{min}) = 0.425\ \text{m}^3/\text{s} = 1.43 H_a^{1.55}$; $H_a(\text{min}) = 0.457$ m.

$$Q(\min) = 15.0\,\text{ft}^3/\text{s} = 8.0 H_a(\min)^{1.55};$$
$$H_a(\min) = 1.50\,\text{ft}.$$

- Depth in grit chamber: $d(\text{min}) = H_a + \Delta Z = 0.457 + 0.06 = 0.52$ m $(1.50 + 0.21 = 1.71$ ft).
- Therefore, $v_H(\text{min}) = Q(\text{min}) = (0.425\ \text{m}^3/\text{s})/(2.74\ \text{m} \cdot 0.52\ \text{m}) = 0.30$ m/s and in U.S. Customary units, $(15\ \text{ft}^3/\text{s})/(1.71 \cdot 9.0) = 0.97$ ft/s).
- The value of $v_H(\text{min})$ approaches the $v_H = 0.30$ m/s (1.0 ft/s) ideal.

2. *Selection*: A spreadsheet, following the foregoing protocol, would facilitate further trials that could factor in other considerations, for example, achieving a long and narrow grit chamber while, at the same time, maintaining $0.23 \le v_H \le 0.38$ m/s $(0.75 \le v_H \le 1.25$ ft/s). Of the two widths investigated, the $w = 2.74$ m (9.0 ft), more closely approximates the ideal than $w = 1.22$ m (4.0 ft).

3. *Length of grit chamber*: The settling velocity for a 0.2 mm quartz sand particle $(\text{SG} = 2.65,\ T = 20°\text{C})$ is $v_o \approx 25$ mm/s (0.08 ft/s) (ASCE-WPCF, 1977, p. 143). Also, from preceding calculations, let *w*(grit chamber) = 2.74 m (9.0 ft). Then from the relation,

$$Q(\text{grit chamber}) = w(\text{grit chamber}) \cdot L(\text{grit chamber}) \cdot v_o$$

$$0.425\ \text{m}^3/\text{s} = (2.74\ \text{m}) \cdot L(\text{grit chamber}) \cdot (0.025\ \text{m/s})$$

$$L(\text{grit chamber}) = 6.20\ \text{m}\ (20\ \text{ft})$$

The length to width ratio is *L*(grit chamber)/*w*(grit chamber) = 6.20 m/2.74 m ≈ 2.2, which is marginally acceptable. Further trials would be in order to search for a higher *L*/*w* ratio.

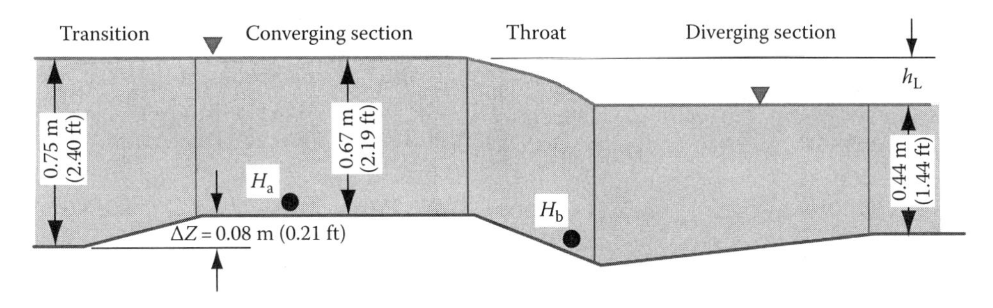

FIGURE 7.10 Installation of 0.61 m (24 in.) Parshall flume to operate under free-flow conditions, with 2.74 m (9.0 ft) wide grit chamber, flowing at $Q(\max) = 0.76\ \mathrm{m^3/s}$ (27.0 cfs).

4. *Submergence limit*: For the 0.61 m (24 in.) Parshall flume, the transition submergence is $S_t = 0.66$ (Table 7.3), where $H_b/H_a = S_t$. Therefore, at $Q(\max)$, $H_b(\max) = H_a(\max) \cdot S_t = 0.67\ \mathrm{m} \cdot 0.66 = 0.44\ \mathrm{m}$ (1.44 ft). If the tailwater elevation increases such that $H_b = H_b(\max)$, the hydraulic condition is called transition-submergence. Further increase in H_b results, eventually, in a "submerged" condition, that is, H_a increases and the "free-flow" hydraulic condition ceases. This means that measurement of H_a is no longer sufficient, by itself, to calculate Q. For proper functioning of the grit chamber the free-flow condition should be maintained, although Q may be calculated for the submerged condition if it is not excessive.

5. *Hydraulic profile*: For the 0.61 m (24 in.) Parshall flume, the transition submergence, $S_t = 0.66$ (Table 7.3), permits calculation of the maximum downstream water depth, which is $H_a = 0.44$ m (1.44 ft). The tailwater depth should be maintained lower, however, in order to ensure a certainty that submergence will not occur. As a note, the corresponding headloss across the flume is $h_L = H_a - H_b = 0.67 - 0.44 = 0.23$ m (0.75 ft) (Figure 7.10).

Spreadsheet algorithm: A spreadsheet algorithm, as illustrated by Table CD7.6, can facilitate the grit

TABLE CD7.6
Design of Rectangular Grit Chamber with Parshall Flume as Control

(a) Metric units

Flows			Parshall Flume Coefficients				Grit Chamber: Design for Q(max)				Grit Chamber: Depth, v for Q(min)		
Q(avg) (m³/s)	Q(max) (m³/s)	Q(min) (m³/s)	C	n	S_t	H_a(max) (m)	w (m)	v(max) (m/s)	d(max) (m)	ΔZ (m)	H_a(min) (m)	d(min) (m)	v(min) (m/s)
0.44	0.66	0.22	1.06	1.54	0.64	0.73	2.13	0.30	1.01	0.28	0.36	0.64	0.16
						0.73	2.13	0.38	0.81	0.08	0.36	0.43	0.24
						0.73	2.44	0.30	0.88	0.15	0.36	0.51	0.18
						0.73	2.44	0.37	0.74	0.00	0.36	0.36	0.25
						0.73	2.74	0.30	0.79	0.05	0.36	0.41	0.19
						0.73	2.74	0.34	0.71	−0.02	0.36	0.34	0.23

(b) U.S. Customary units

Q(avg) (ft³/s)	Q(max) (ft³/s)	Q(min) (ft³/s)	C	n	S_t	H_a(max) (ft)	w (ft)	v(max) (ft/s)	d(max) (ft)	ΔZ (ft)	H_a(min) (ft)	d(min) (ft)	v(min) (ft/s)
15.47	23.21	7.74	6.00	1.54	0.64	2.41	7.00	1.00	3.32	0.91	1.18	2.09	0.53
						2.41	7.00	1.25	2.65	0.25	1.18	1.42	0.78
						2.41	8.00	1.00	2.90	0.49	1.18	1.67	0.58
						2.41	8.00	1.20	2.42	0.01	1.18	1.19	0.81
						2.41	9.00	1.00	2.58	0.17	1.18	1.35	0.64
						2.41	9.00	1.10	2.34	−0.06	1.18	1.12	0.77

Q(avg) was assumed
$Q(\max) = 1.5 \cdot Q(\text{avg})$
$Q(\min) = 0.5 \cdot Q(\text{avg})$

C, n, S_t for 18 in. flume
H_a calculated:
$Q(\max) = CH_a^n$

w was assumed
$v(\max)$ was assumed
$d(\max) = Q(\max)/w \cdot v(\max)$
$\Delta Z = d(\max) - H_a(\max)$

H_a calculated:
$Q(\min) = CH_a^n$
$d(\min) = H_a(\min) + \Delta Z$
$v(\min) = Q(\min)/b \cdot d(\min)$

chamber-Parshall flume design by permitting exploration of many alternatives. Example 7.6 illustrates the use of such a spreadsheet, that is, Table CD7.6. The formulae used in the cells and the calculation procedure are documented in the sections below the four categories of calculations.

Example 7.5 Design of Grit Chamber/Parshall Flume by Spreadsheet

Problem
Explore the design of a Parshall flume and grit chamber combination for a 0.44 m^3/s (10 mgd) average daily flow.

Solution
Set up the protocol for calculation on a computer spreadsheet, Table CD7.6. The first category of the spreadsheet is flow and the full range is entered. Next, knowing Q(max), a 0.457 mm (1.50 ft) flume is selected. From these data the H_a(max) depth is calculated by formula, that is, $Q = CH_a^{\ n}$. The grit chamber calculations start with assumptions for channel width, w, and with assumption of channel velocity at maximum flow, $v_H(\text{max}) \leq 0.38$ m/s (1.25 ft/s). From these data, the channel depth, d(max), is calculated, along with ΔZ, that is, $d = H_a + \Delta Z$. The next concern is to check the channel velocity for Q(min) to ascertain whether $v_H(\text{min}) \geq 0.75$ ft/s. The selection is indicated by the blocked out area, based upon the v_H criteria being met, that is, $0.23 \leq v_H \leq 0.38$ m/s ($0.75 \leq v_H \leq 1.25$ ft/s).

Tables CD7.6(a) and (b) (metric units and U.S. Customary units, respectively) illustrate the foregoing spreadsheet description. Formulae for cells are indicated at the bottom.

7.2.2.4 Parabolic Section

The problem of maintaining constant v_H in the grit chamber is achieved by a parabolic section, matched with a Parshall flume control section. The mathematical characteristics of a parabolic section are reviewed here.

7.2.2.4.1 Mathematics of a Parabolic Section

The coordinates for a parabola, oriented with axis in the y-direction and vertex at (0,0), were given by Griffin (1936, p. 293) as

$$x^2 = 4py \tag{7.14}$$

where

x is the x-coordinate of the parabolic shape (m)
y is the y-coordinate of the parabolic shape (m)
p is the mathematical constant characteristic of the parabola shape, for example, narrow or wide (m)

The "spread" of the parabola depends on the value of p, for example, when $y = p$, then $x = \pm 2p$. The area of the parabola at any y is given by integration of Equation 7.14,

$$A(\text{parabola}) = \frac{8}{3}\sqrt{p} \cdot y^{3/2} \tag{7.15}$$

7.2.2.4.2 Calculation of p

If the bottom of a parabolic grit chamber is the same as the floor of a Parshall flume, v_H will remain constant, that is, $v_H = 0.3$ m/s (1.0 ft/s) (Camp, 1942). The match is obtained by solving for p of the parabola, obtained by forcing the equality of flows between the grit chamber and the Parshall flume, for example,

$$Q(\text{parabolic grit chamber section}) = Q(\text{Parshall flume}) \tag{7.16}$$

Now substitute Equations 7.15 and 7.6, respectively, after multiplying A(parabola) by v_H,

$$v_H \cdot \frac{8}{3}\sqrt{p} \cdot y^{3/2} = \text{C}H_a^n \tag{7.17}$$

Now by letting $v_H = 0.305$ m/s (i.e., Camp's velocity criterion) and assuming flow, Q, to yield H_a by Equation 7.6, and if $y = (\text{set})H_a$, then p can be calculated. The "throat" width of the flume selected should be as narrow as possible to accommodate the maximum flow, that is, Q(max). As a caveat, the width of the flume at maximum depth could be too large to be practical.

7.2.2.4.3 Spreadsheet for Parshall-Flume with Parabolic Section

Table CD7.7 illustrates a protocol for sizing a Parshall-flume-control with a parabolic-section-grit-chamber in terms of a spreadsheet. The spreadsheet is formatted in three parts: (1) selection of Parshall flume, (2) determination of p, and (3) calculation of the coordinates for the parabolic section. Figure CD7.11 is a plot of the resulting parabolic half-section, which is linked to the spreadsheet. Example 7.6 illustrates the design protocol.

Example 7.6 Design of Parabolic Grit Chamber Section

Given
Let the expected maximum flow for a headworks of a wastewater treatment plant, be $Q(\text{max}) = 0.3067$ m^3/s (7 mgd).

Required
Select a Parshall flume for the flow stated and determine the associated parabolic grit chamber section.

Solution
Table CD7.7 is a spreadsheet that follows the foregoing protocol. Part (a) shows that a 305 mm (12 in.) Parshall flume has a capacity of 0.453 m^3/s (10.3 mgd), which is greater than Q(max). [The 229 mm (9 in.) flume lacks sufficient capacity and the 457 mm (18 in.) flume has too much capacity.] The coefficients C and n are given in Part (a). Determination of p is given in Part (b) by applying Equation 7.17 for any Q. Knowing p permits calculation of the coordinates of a matching parabolic section as seen in Part (c); Figure CD7.11 shows the half-parabolic section plotted by the coordinates given.

TABLE CD7.7
Calculated Parabolic Grit Chamber Section for Selected Parshall Flume

Q		Parshall Flume				Parabolic Grit Chamber Section					
		Flume			H_a	v_H	y	p	X	A(parab)	v_H
(mgd)	(m^3/s)	(m)	C	n	(m)	(m/s)	(m)	(m)	(m)	(m^2)	(m/s)
(a) Selection of Parshall flume											
	0.453	0.305	0.69	1.52	0.758						
(b) Determination of p for parabolic grit chamber section											
1.00	0.044	0.305	0.69	1.52	0.163	0.305	0.163	0.6693			
(c) Calculation of H_a, x, A(parab), and verification of v_H for grit chamber											
0.00	0				0		0.000		0.000	0.000	0
0.20	0.0088				0.0566		0.057		0.389	0.029	0.299
0.50	0.0219				0.1033		0.103		0.526	0.072	0.302
1.00	0.0438				0.163		0.163		0.661	0.144	0.305
2.00	0.0876				0.2572		0.257		0.830	0.285	0.308
3.00	0.1314				0.3359		0.336		0.948	0.425	0.309
4.00	0.1752				0.4059		0.406		1.042	0.564	0.311
6.00	0.2628				0.53		0.530		1.191	0.842	0.312
7.00	0.3067				0.5865		0.587		1.253	0.980	0.313
8.00	0.3505				0.6404		0.640		1.309	1.118	0.313
10.00	0.4381				0.7416		0.742		1.409	1.393	0.314

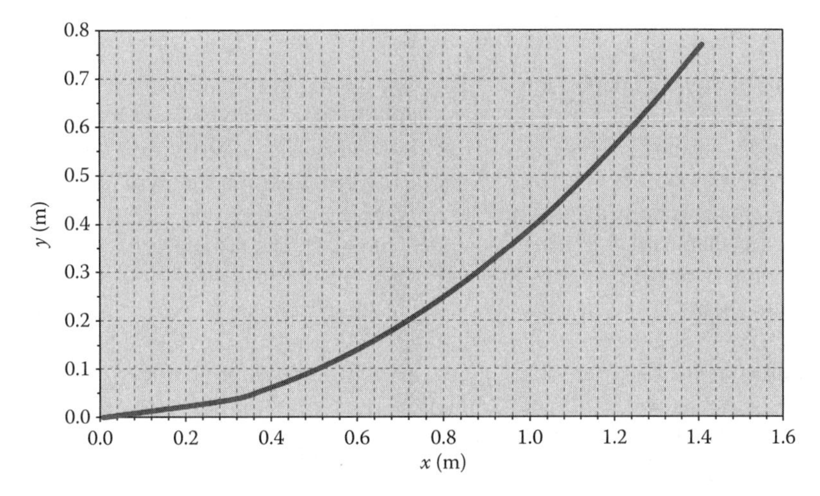

FIGURE CD7.11 Calculation of parabolic section.

Comments
As seen, $0.296 \leq v_H \leq 0.314$ m/s, corroborates the design, that is, $v_H \rightarrow 0.305$ m/s (1.0 ft/s). Also, w(top-of-parabolic-section at 0.44 m/s $\approx$ 2.81 m (9.2 ft) for $Q = 0.44$ m/s (10 mgd). The width of the grit chamber would be considered within an acceptable limit.

7.2.3 Practice—Horizontal Flow Grit Chambers

As with any sedimentation basin, the hydraulic behavior of a grit chamber deviates from the "ideal" model of Camp (1942, p. 31). Such deviations are due to turbulence at entrance and exit, nonuniformity of velocity with depth, etc. (see Section 6.91). In addition to hydraulic factors, equipment is pertinent to proper functioning of grit chambers. Grit loading is another performance factor and varies with the characteristics of the sewer system, that is, sanitary or combined, soils, and kinds of inputs to the system. Some of these factors are reviewed in this section, along with data from actual designs, including sizing and performance.

7.2.3.1 Design and Performance—Examples

Table 7.8 gives data on grit chamber sizing for installations that illustrate the classic long and narrow plan area with shallow depth that conforms most closely to theory. Other

TABLE 7.8
Grit Chamber Dimensions for Different Cities

Location	No. Channels	Dimensions (m)			Dimensions (ft)		
		Depth	Length	Width	Depth	Length	Width
South Milwaukee, WI	1	0.99	12.95	1.22	3.25	42.50	4.00
Ephrata, WA	1	0.76	7.62	0.91	2.50	25.00	3.00
Vicksburg, MS	1	1.22	14.33	1.52	4.00	47.00	5.00
Watertown, NY	2	1.22	6.10	0.48	4.00	20.00	1.58
Toledo, OH	2	1.52	13.41	1.22	5.00	44.00	4.00
Portsmouth, VA	3	1.45	17.60	1.83	4.75	57.75	6.00
Oswego, NY	1	0.30	6.00	0.51	1.00	19.67	1.67
Mt. Pleasant, IA	1	0.91	10.67	0.91	3.00	35.00	3.00
Freeport, IL	2	1.30	13.72	1.22	4.25	45.00	4.00
Salisbury, NC	2	0.76	12.19	1.22	2.50	40.00	4.00
Libertyville, IL	1	1.07	9.14	0.76	3.50	30.00	2.50
Waukesha, WI	1	1.30	16.25	1.62	4.25	53.33	5.33
Roxboro, NC	1	0.86	11.28	1.37	2.83	37.00	4.50
Meriden, CT	2	0.59	12.19	1.98	1.92	40.00	6.50
South Cobb County, GA	2	1.68	9.75	1.22	5.50	32.00	4.00
Brigantine, NJ	1	0.46	6.10	0.91	1.50	20.00	3.00
Eufala, OK	1	0.43	5.18	0.61	1.42	17.00	2.00
Portsmouth, VA	3	1.45	18.90	1.83	4.75	62.00	6.00
Middlesex, NJ	1	1.98	12.22	1.37	6.50	40.08	4.50
Bordentown, NJ	1	1.07	10.67	0.76	3.50	35.00	2.50

Source: Adapted from Rexnord, Grit chambers data sheet, in: *Rexnord Product Manual*, Rexnord, Inc., Waukesha, WI, 1980.

shapes have been used in practice, for example, square plan area with shallow depth.

Table 7.9 provides data on the character of the sewer system (sanitary or combined), plant flow, grit accumulation, and grit quality. From these data it is seen that average grit production may range from less than 0.0075 L grit/m^3 water (1.0 ft^3 grit/mg water) to more than 0.15 L grit/m^3 water (20 ft^3 grit/mg water). There seems no consistently higher rate of grit accumulation from combined sewers as opposed to those that are sanitary, but for combined sewers, surges in grit quantity do occur and can overwhelm the system. The volatile solids content of grit varies—with 20% to 50% not unusual; 5% or less of volatile solids is a goal. Rex Chainbelt (1965) reported an average of 8.5%.

Table 7.10 shows sieve analyses of grit taken from four operating plants and illustrates the variable composition of removed grit. On grit sizes the Kenosha (Wisconsin) plant removed particles primarily 0.2 mm and larger. The other three plants removed particles smaller than 0.2 mm.

7.2.3.2 Removal Equipment

Grit collects on the bottom of the chamber and is scraped by a flight of blades. The scraper speed is about 3.0 m/min (10 ft/min). Usually, the grit enters a hopper for collection; removal is by a chain-and-bucket lift or a screw conveyor. After washing, the grit drops into a truck for transport to a landfill.

Figure 7.12 shows the general layout of a proprietary aerated grit chamber with associated appurtenances labeled by the numbers shown. Flow enters from the right and leaves on the left side with level maintained by a rectangular weir. An air header is along the concrete wall (to cause a "roll"). The chain-and-bucket system scrapes the grit to the head of the grit chamber and then transfers the material to a screw conveyor. The same system is available as a horizontal-flow grit chamber, in which case a proportional weir (instead of a rectangular weir) is used to maintain constant horizontal velocity, v_H, at different flows, Q; also, there is no airflow header pipe (9) or wooden baffle (3).

7.3 AERATED GRIT CHAMBERS

Figure 7.13 illustrates the basic elements of an aerated grit chamber, showing the approximately square cross section with fillets in the corners, a diffuser that emits air bubbles, a baffle to confine the stream of air bubbles, a grit removal zone at the bottom, and a spiral circulation. Grit particles and organic particles are depicted as suspended; a portion of the removed particles are shown at the bottom.

The air bubbles rise with a velocity in accordance with Stoke's law, inducing a drag on the water and associated velocity, v_T. By adjusting airflow, Q_a, the operator can control v_T, and thus the mass fraction of organic matter in the grit. The horizontal velocity, v_H, is $v_H = Q/A$(cross section); the vector sum, $v_H + v_T = v_R$, defines a spiral. As with horizontal flow grit chambers, the ideal scour velocity, that is, v_T in the case of aerated grit chambers, should be $v_T \approx 0.3$ m/s

TABLE 7.9
Grit Quantities and Quality for Grit Chambers in Selected U.S. Cities

City	Percent Sewers Comb.	Q(design)		Q(1954)		Area Served		Year	Grit Quantity (ft^3/mg)			Grit Quantity (L Grit/ML Flow)			Grit Quality	
		(mgd)	(ML/day)	(mgd)	(m^3/day)	(mi^2)	(km^2)		max	min	avg	max	min	avg	Moisture (%)	VSS (%)
Bridgeport, CN[a,b]	100	14	53	5.1	19	3.1	8	1953	3.0	0.8	1.3	22.4	6.0	9.4	53.1	48.4
Cranston, RI	0	5.5	21	2.7	10	6.7	17	1951	4.3	1.9	2.0	32.2	13.8	15.0		
Dallas, TX	0	18	68	19.4	73	148	383	1954	5.5	0.1	2.4	41.1	0.7	18.3		
Detroit, MI[a]	9	450	1703	444	1681			1950–1953	6.5	2.4	4.0	48.6	18.0	29.9	46.0	40.0
Duluth, MN[d]	0	12	45	13.6	51			1953	1.6	0.4	0.8	12.0	3.0	6.0		
Fairfield, CT	0	4	15	1.4	5	6	16	1954	3.9	1.7	3.0	29.2	12.7	22.4		
Lansing, MI	98	20	76	16.5	62			1953	8.6	3.6	5.5	64.3	26.9	41.1		
Madison, WI[c] Metro	0	18	68	11.1	42			1953	3.5	1.5	3.0	26.2	11.2	22.4	65.0	50.0
Marshalltown, IA	0	4.0	15	3.1	12	6.2	16	1953	4.7	2.6	3.4	35.2	19.5	25.4		
Paul, MN[c]	90	134	507	134	507	110	285	1950–1954	24.1	0.1	5.2	180.3	0.7	38.9	13.4	9.5
Oshkosh, WI[c]	90	6.0	23	5.2	20	7.4	19	1954	2.7	0.5	1.0	20.2	3.7	7.5	50.0	43.6
Port Huron, MI	95	13.5	51	9	34	7	18	1953	15.0	0.9	3.5	112.2	6.7	26.2		
Portsmouth, VA	0	9.7	37	7.3	28			1954	0.9	0.1	0.4	6.7	0.7	2.9		
Stamford, CT[e]	0	10	38	6.3	24	4.5	12	1954	6.6	1.8	2.1	49.4	13.5	15.7	33.0	20.0
Tampa, FL[f] (raw)	0	36	136	12.3	47	19	49	1954	1.0	0.4	0.6	7.5	3.0	4.7		11.6
-washed																1.0
Uniontown, PA	100	3	11	2.3	9			1954	21.0	2.7	10.5	157.1	20.2	78.6		
Washington, DC[g,h]	50	175.0	662	163.8	620			1939–1953[h]	7.5	0.6	1.9	56.1	4.5	13.8	33.0	23.0

Source: ASCE-WPCF, *ASCE Manual of Engineering Practice No. 36 and the WPCF Manual of Practice No. 8*, American Society of Civil Engineers, New York, 1959.

[a] Water spray grit washing.

[b] Rubber and cereal wastes.

[c] Grit washing by screw conveyer.

[d] Grit washing by reciprocating rake.

[e] Grit washing by organic return pump.

[f] Grit washer.

[g] Grit washing by spiral conveyer.

[h] 6 year avg.

TABLE 7.10
Sieve Analysis of Particles Collected in Grit Chambers

Sieve Size (U.S. Series)	Sieve Open (mm)	Percentage Retained			
		Green Bay	Kenosha	Tampa	St. Paul
4	4.76				1–7
8	2.38				5–20
10	2.08	3.7	12		
20	0.84	9.1			2–53
40	0.42	19.8	70		
50	0.30	29.6		2.3	20–67
65	0.21	51.7			
80	0.18		95		
100	0.15	78.2		59.3	97–99.9
200	0.07	96.1		99.5	

Source: Adapted from ASCE-WPCF, *ASCE Manual of Engineering Practice No. 36 and the WPCF Manual of Practice No. 8*, American Society of Civil Engineers, New York, 1959.

(1.0 ft/s). Also, as with horizontal flow grit chambers, the goal is to remove grit particles, $d(\text{grit}) \geq 0.2$ mm, for SG ≈ 2.65 with organic-mass-fraction ≤ 0.05.

7.3.1 Principles of Aerated Grit Chamber Operation

Theoretically, the fraction of grit removed is proportional to the number of "rolls," which favors a long-narrow shape, that is, with more "rolls" per unit length.

7.3.2 Theory of Aerated Grit Chambers

Figure 7.14 shows the velocity vectors of a grit particle in the separation zone, that is, the velocity, v_T, of the circulation, and the settling velocity, v_S. The resultant velocity is v_R. Particles that enter the separation zone will be removed if the resultant velocity vector, v_R, intersects the bottom, that is, the top plane of the grit collector. Each rotation provides another "pass" across the separation zone, and thus the opportunity for further settling to the collection zone. A third vector, v_H, transports the particle in direction of the flow, that is, normal to the plane of the paper. Thus, $\mathbf{v}_R(\text{particle}) = \mathbf{v}_S + \mathbf{v}_H + \mathbf{v}_T$. The particle is transported with the flow but with the settling velocity vector, $\mathbf{v}_S$, superimposed on the fluid motion.

7.3.2.1 Calculation of Grit Removal

Figure 7.15 illustrates the spiral path of the circulation for a section of the grit chamber with length, ΔL. Particles that start at a point A will be advanced to point B over one rotation.

During each "rotation" of the water mass with entrained grit particles, a fraction, P, of the grit will enter the "separation zone" and will be removed. After the "pass," the turbulence will redistribute the remaining grit particles. During the next pass, the same fraction, P, of remaining particles will be removed. To illustrate, suppose $P = 0.2$ and let the suspension contain 10 particles at point A of Figure 7.15. Then two that is of the particles will be removed and eight particles will remain in suspension at point B and circulated again, with the same proportion of particles removed over the second "pass."

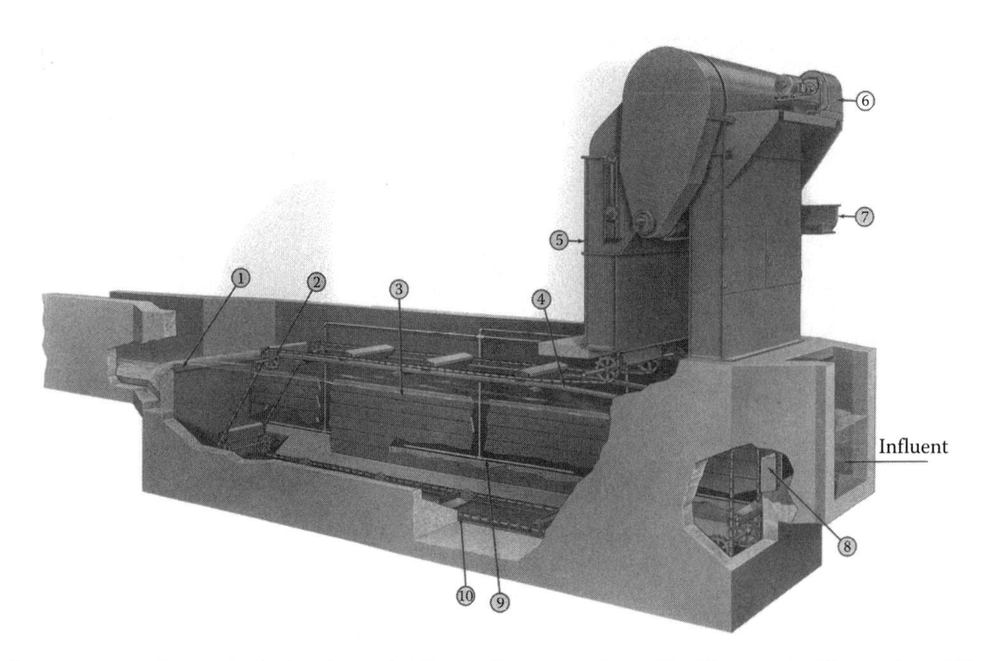

FIGURE 7.12 General layout of aerated grit chamber and collector. Key to numbers: (1) effluent weir; (2) sprockets; (3) circulation baffle, wooden; (4) guided chain support; (5) housing; (6) motorized drive; (7) screw conveyor for discharge of grit; (8) inlet baffle for distribution of flow around baffles; (9) air inlet pipe and headers, perforated pipe is adequate; (10) chain bucket collector mechanism. (Courtesy of Siemens Envirex products WSG & Solutions, Montgomeryville, PA.)

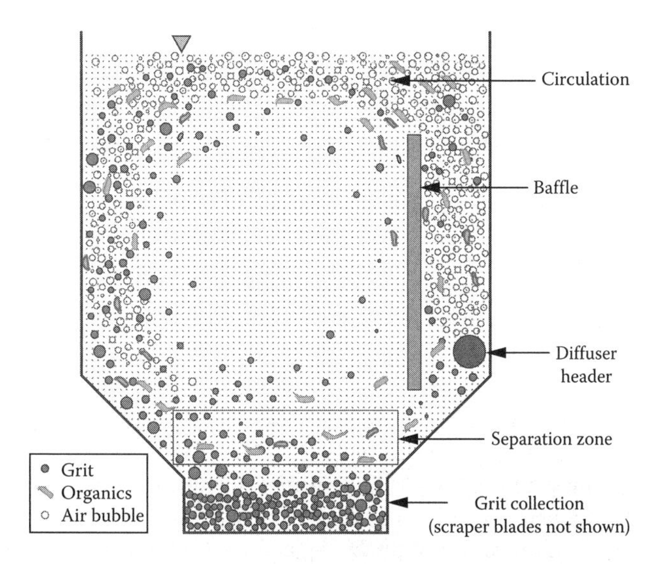

FIGURE 7.13 Elements of an aerated grit chamber. (Adapted from USFilter Envirex Products Brochure, c. 1980 now WSG & Solutions. With permission).

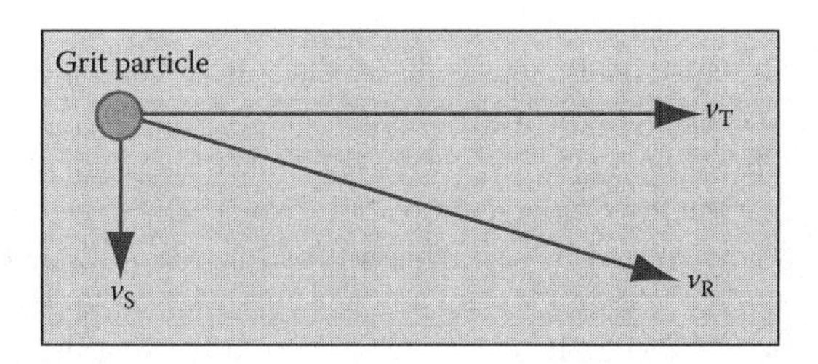

FIGURE 7.14 Schematic of separation zone in aerated grit chamber (drawing is in vertical plane). The resultant vector for a given particle is $v_R(\text{particle}) = v_S + v_H + v_T$.

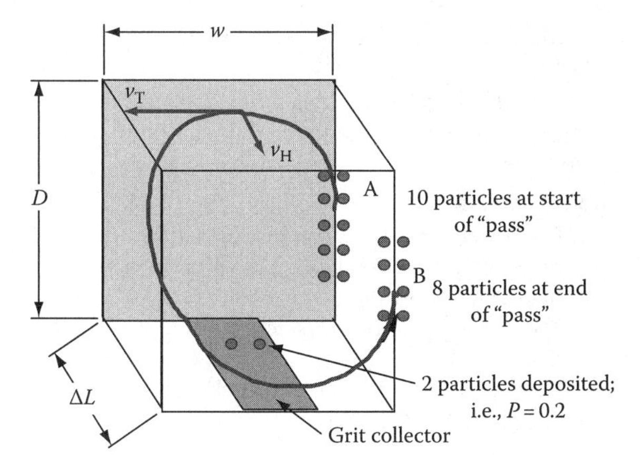

FIGURE 7.15 Spiral rotation of grit particles showing axial velocity vector, v_H, and tangential velocity vector, v_T; the particles also settle with velocity, v_S. Ten particles start at A, two are deposited in the grit collector, and eight appear at the corresponding position, B. The length of the spiral from A to B is DL.

The grit remaining after the first compartment is designated $R(n = 1)$ and is, $(1 - P)$, that is,

$$R(n = 1) = (1 - P) \tag{7.18}$$

The grit remaining after the second compartment, $n = 2$, designated as $R(n = 2)$, will be $(1 - P)(1 - P)$, that is,

$$R(n = 2) = (1 - P)(1 - P) \tag{7.19}$$

$$= (1 - P)^2 \tag{7.20}$$

Thus, the general equation for fraction of grit remaining after n compartments is,

$$R = (1 - P)^n \tag{7.21}$$

where
R is the fraction of grit remaining after n passes
P is the fraction of grit removed after one pass
n is the number of "rotation compartments" in the length of the grit chamber, or the number of rotations of the water mass in length, L

7.3.2.2 Calculation of Spiral Length, *DL*

Figure 7.16 depicts an "unfolded" spiral (top view), where a particle starts at A and ends at B, as in Figure 7.15. The length of advance, along the axis of the grit chamber, due to a rotation is DL over a rotation distance is approximately, πD. The circulation velocity, v_R, has components v_T and v_H. From Figure 7.16, the two triangles, shown by different shades, are similar, that is,

$$\frac{\Delta L}{\pi D} = \frac{v_H}{v_T} \tag{7.22}$$

where
ΔL is the length of a "compartment" of the grit chamber, defined as the axial length of the grit chamber for one rotation of the spiral flow (m)
D is the depth of water in the grit chamber, that is, from the water surface to the top plane of the grit collector (m)
v_H is the velocity component of spiral flow in the axial direction (m/s); $\approx Q/A$(cross section)
v_T is the velocity component of spiral flow tangent to the spiral circulation path in the plane of the grit chamber cross section (m/s); ≈ 0.3 m/s

7.3.2.3 Empirical Guidelines

From empirical guidelines: $w/D \approx 0.8$; $3 \le D \le 5$ m (10–15 ft); scour velocity is, $0.2 \le v_T \le 0.3$ m/s (0.75–1.0 ft/s); and from continuity, $v_H = Q/(wD)$. Thus, ΔL can be calculated by Equation 7.22.

7.3.2.4 *n* Determination

The length of the grit chamber can be calculated as

$$L = n \cdot \Delta L \tag{7.23}$$

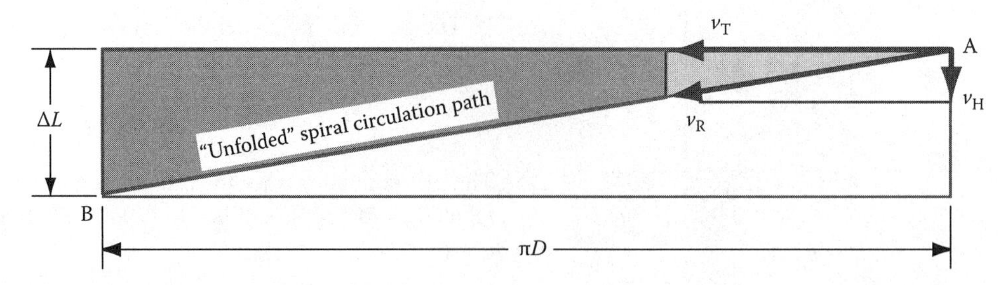

FIGURE 7.16 Spiral circulation path is shown "unfolded" to illustrate relations, that is, $\Delta L/\pi D = v_H/v_T$ (by similar triangles).

Three methods to estimate n are:

1. *Operating grit chamber*: For a design similar to one being contemplated, suppose R is measured (by laborious effort) along with L. Then ΔL may be calculated by Equation 7.22, that is, $DL/\pi D = v_H/v_T$, and n by Equation 7.23, that is, $L = n \cdot \Delta L$. From this value of n, P can be calculated by Equation 7.21, that is, $R = (1 - P)^n$. The grit chamber to be designed would have a similar P, and therefore with a specified performance, n can be calculated again using Equation 7.21.
2. *Impute a P*: A second approach is to assume a P, say $P = 0.2$, specify an R, and calculate n from Equation 7.21, that is, $R = (1 - P)^n$. This is best done by a spreadsheet in which, for a given R, different P's can be assumed and systematically changed. Then another R is assumed and the process is repeated. The value of n is determined for each (R, P) pair by Equation 7.21. From these calculations, the uncertainties can be estimated and the variation in n can be assessed. Then ΔL and L can be estimated by Equation 7.22, that is, $DL/\pi D = v_H/v_T$, and Equation 7.23, that is, $L = n \cdot \Delta L$, respectively.
3. *Set L and calculate n*: Select a value for L and calculate n from Equation 7.23, that is, $L = n \cdot \Delta L$, with ΔL from Equation 7.22, that is, $DL/\pi D = v_H/v_T$. From the n so calculated, and an assumed R, calculate P by Equation 7.21, that is, $R = (1 - P)^n$. Use this P to determine a new n for any specified R, and thus calculate L by Equation 7.23, that is, $L = n \cdot \Delta L$.

7.3.2.5 Algorithm for Calculations

The forgoing procedure may be summarized as an "algorithm." The algorithm can be applied best as a spreadsheet procedure:

1. *Determine w and D*: Assume both w and D from empirical guidelines, for example, let $w = D \approx 3$–5 m.
2. *Determine v_H*: Knowing Q and having assumed w and D, calculate v_H (i.e., $v_H = Q/(wD)$.
3. *Determine v_T*: From Camp's criterion for scour velocity for organic matter and for deposit of 0.2 mm grit particles and from $0.2 \leq v_T \leq 0.3$ m/s (0.75–1.0 ft/s) (Camp, 1942; Londong, 1989). The value of v_T may be controlled by adjusting the airflow, that is, Q(air).
4. *Determine ΔL*: With D, v_H, and v_T determined, calculate ΔL by Equation 7.22, that is, $DL/\pi D = v_H/v_T$.
5. *Determine n*: Estimate or assume an n (or a series on n values in a spreadsheet) by one of the three methods summarized in Section 7.3.2.4.
6. *Determine L*: Calculate L by Equation 7.23, that is, $L = n \cdot \Delta L$. This is an estimate and should be calculated on a spreadsheet so that the variation in the estimate of L can be taken into account. After review of the spreadsheet, L is selected as a "decision" based on judgment.
7. *Operation—Effect of Increased Flow on R*: Suppose that the flow for a given plant is increased. From this a new ΔL can be calculated from Equation 7.22, that is, $DL/\pi D = v_H/v_T$ and a new n follows from Equation 7.23, that is, $L = n \cdot \Delta L$. Therefore, using the P as determined previously, the new R can be estimated by Equation 7.21, that is, $R = (1 - P)^n$. In other words, once a design is determined, then the effects of different flow "scenarios" can be assessed (again, by spreadsheet).

Example 7.7 Aerated Grit Chamber Design

This example problem illustrates the application of the equations developed in the rational design of an aerated grit chamber.

Problem statement

(a) For an aerated grit chamber with flow, $Q = 0.75$ m^3/s (17.1 mgd), sketch a cross-section view and a plan view, showing dimensions.
(b) Show the air diffuser system and estimate the airflow.
(c) Document the criteria that you choose.
(d) Suggest alternative criteria.

Criteria: The design criteria could be from either Morales and Reinhart (1984) or Londong (1987, 1989). The criteria from the former are based upon practice as distilled from their study of grit chambers at five plants. The criteria from

TABLE 7.11
Criteria for Design of Aerated Grit Chamber

Variable	Magnitude	Units	Reference
θ	20	min	Londong (1989)
w/D	0.8	m/m	
$w \cdot D$	1–7	m^2	
v_H	0.20	m/s	
v(axial)	0.10	m/s	
L	15–60	m	
d(diffusers)	$0.7 \cdot D$	m	
Slope of fillets	45	Degrees	

Source: Adapted from Londong, J., *Water Sci. Technol.*, 21, 13, 1989.

Londong (1989) are based on a pilot plant study and have a rationale. The criteria of Londong will be used and are summarized in Table 7.11.

Also, the equations of Londong (1989) for airflow, Q_a, will be used, that is,

$$Q(\text{air, coarse bubble}, v_T = 0.2\,\text{m/s}) = [0.07 + 0.76 \ln d(\text{diffuser})]^{-1.33} \quad (7.24)$$

$$Q(\text{air, fine bubble}, v_T = 0.2\,\text{m/s}) = [0.63 + 0.52 \ln d(\text{diffuser})]^{-0.62} \quad (7.25)$$

where

Q_a is the air flow per unit volume of basin to maintain $v_T = 0.2$ m/s (m^3 air/h/m^3 grit chamber)

d(air header) is the depth of diffusers below water surface (m)

For calculation, set up a spreadsheet or refer to Table CD7.14 for airflow calculations.

Determine sizing: Criteria in Table 7.11 will be applied to determine w, D, L, and d(air header). Figure 7.17 shows the configuration of the geometry of the basin and summarizes the dimensions resulting from the calculations that are enumerated as follows:

Determine w, D:

$$A = w \cdot D - 0.5(w/2)^{0.5}$$
$$w/D = 0.8$$

Try $A = 8\ \text{m}^2$:

$$8.0 = w \cdot (0.8w) - 0.5(w/2)^{0.5}$$
$$w = 3.44\,\text{m} \approx 3.50\,\text{m}$$
$$D = 0.8 \cdot 3.50\,\text{m} = 2.80\,\text{m}$$

Determine L:

$$w \cdot D \cdot L = Q \cdot \theta$$
$$3.50\,\text{m} \cdot 2.80\,\text{m} \cdot L = 0.75\,\text{m}^2/\text{s} \cdot (10\ \text{min} \cdot 60\,\text{s/min})$$
$$L \approx 46\,\text{m}$$

Air header placement:

$$d(\text{air header}) = 0.70 \cdot D = 0.70 \cdot 2.80\,\text{m} = 2.00\,\text{m}$$

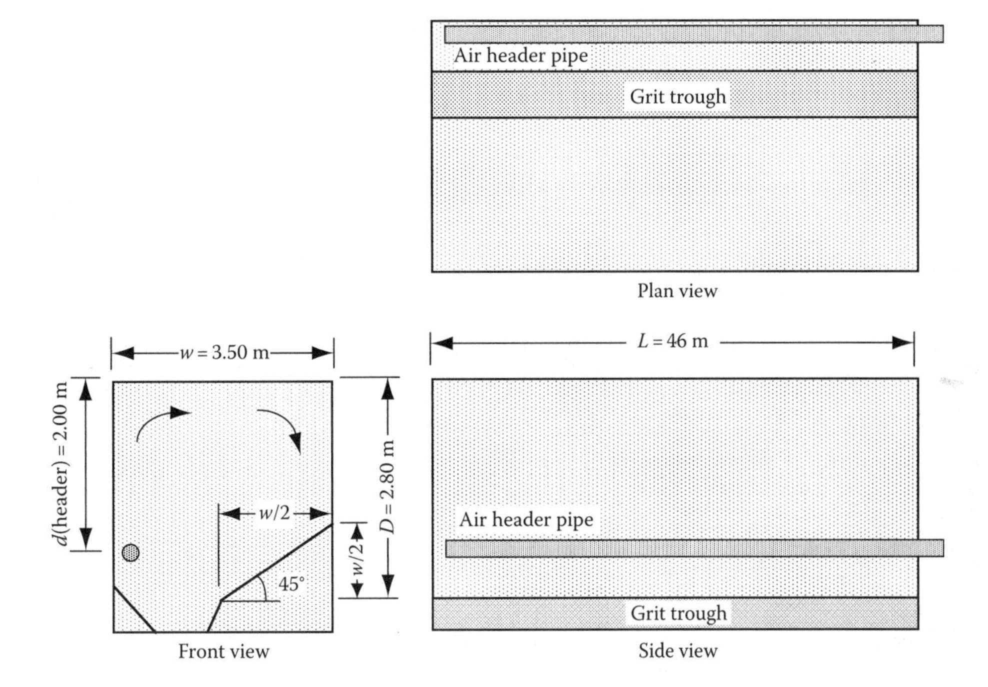

FIGURE 7.17 Grit chamber showing dimensions as calculated. (Adapted from Londong, J., *Water Sci. Technol.*, 21, 13, 1989.)

Determine airflow: Select coarse bubble diffuser because it is less likely to clog and will require less maintenance.

$$Q(\text{air, coarse bubble}, v_H = 0.2\ \text{m/s})$$
$$= [0.07 + 0.76 \ln d(\text{air header})]^{-1.33}$$
$$= [0.07 + 0.76 \ln 2.00]^{-1.33}$$
$$= 2.00\ \text{N m}^3/\text{m}^3\ \text{h}$$
$$Q(\text{air-for-basin}) = (2.00\ \text{N m}^3/\text{m}^3\ \text{h}) \cdot (8.0\ \text{m}^2 \cdot 46\ \text{m})$$
$$= 736\ \text{N m}^3/\text{h}$$
$$= 12.3\ \text{N m}^3/\text{min}$$

(Note, the N in front of m^3 means "normal" temperature and pressure, that is, 20°C, 1.0 atm pressure.)

Alternate criteria: Morales and Reinhart (1984) give

$$\theta = 2\text{–}5\ \text{min}$$
$$d(\text{particles removed}) \geq 0.21\ \text{mm}$$
$$R(0.21\ \text{mm particles}) \geq 95\%$$

These criteria are based upon a study of empirical data without regard to rationale. The detention time recommended by Londong, that is, about 20 min, is much longer than the findings of Morales and Reinhart (1984) for American practice, that is, 2–5 min. The longer detention time, if in terms of a longer grit chamber, will permit more "compartments," and thus "rotations," over the separation zone, and hence higher removals.

Concerning airflow, $4.6 \leq Q(\text{air-per-unit-length}) \leq 12.4$ L/s/m of length, or $3 \leq Q(\text{air-per-unit-length}) \leq 8\ \text{ft}^3/\text{min/ft}$ (ASCE-WPCF, 1977, p. 137). The range for Q(air-for-basin) calculates as

$$Q(\text{air-for-basin-min}) = Q(\text{air-per-unit-length}) \cdot L(\text{basin})$$
$$= (4.6\ \text{L/s/m of length}) \cdot 46\ \text{m}$$
$$= 212\ \text{L air/s}$$
$$= 3.5\ \text{N m}^3\ \text{air/min}$$

The range is thus $3.5 \leq Q(\text{air-for-basin}) \leq 9.5\ \text{N m}^3\ \text{air/min}$, which is less than the value calculated based on the Londong (1989) equation.

7.3.3 Practice: Aerated Grit Chambers

Guidelines for aerated grit chamber practice are sparse, and most designs rely on empirical rules-of-thumb. This section reviews one paper that illustrates the diversity found in operating plant designs.

7.3.3.1 Guidelines from Five Designs

A paper by Morales and Reinhart (1984) summarized the designs of five aerated grit chambers located at different plants in Atlanta, Georgia; associated performance data are seen in Table 7.12. The grit chambers differed in shape and other characteristics, for example, whether inlet and outlet

TABLE 7.12
Comparative Design and Performance Data for Five Aerated Grit Chambers[a]

Parameter	Units	RMC[a]	UC	IC	SR	FR
Design						
Width	(m)	3.7	7.6	5.5	5.5	8.5
Length	(m)	27.2	7.6	6.0	4.3	8.5
Depth	(m)	4	3.4	2.3	2.9	3.7
q	(min)	4.6	3.8	3.6	5.6	7.8
Surface velocity	(m/s)	0.21–0.61	0–0.61	0–0.54	0.24–0.61	0.09–0.67
Bottom velocity	(m/s)	0.03–0.24	0.03–0.24	0–0.40	0.15–0.36	0.03–0.15
Inlet baffle		No	No	No	Yes	Yes
Outlet baffle		No	Yes	Yes	Yes	Yes
Flow direction		Axial	Cross	Cross	Cross	Cross
Aeration device		perf.baf.	discfuser	discfuser	discfuser	Air lift *p*
Air header placement		Side	Center	Side	Side	Center
Grit removal equip.		bucket	grit *p*	Convey	Bucket	Air lift *p*
Grit elevator		screw	screw	tubular	bucket	Air lift p
Grit washing		No	Yes	No	No	Yes
Performance						
Grit removal, wet	(mL^3 grit /m^3 water)	10.9	7	39	17	34
Total solids	(g dry solid/g moist grit)	0.79	0.70	0.65	0.28	0.53
Volatile solids of grit	(g dry VS/g dry grit)	0.08	0.22	0.13	0.38	0.26
Grit removal	(fraction ≥ 0.2 mm)	0.95	0.82	0.98	0.95	0.99

Source: Morales, L. and Reinhart, D., J. *Water Pollut. Control Fed.*, 56(4):337–343, 1984.
Note: Conversion of grit volume is: 100 mL grit/m^3 water = 13.5 ft^3 grit/mg water.
[a] Plants are RMC-RM Clayton, UC-Utoy Creek, IC-Intrenchment Creek, SR-South River, FR-Flint River (Atlanta, Georgia).

baffles were used, air header placement, grit removal equipment, and whether grit washing was used.

The conclusion was that the RMC plant (see Table 7.12) was an "optimal" design in terms of operation and performance. The dimensions were length, 27.2 m; width, 3.7 m; and depth, 4.0 m. The plant had no inlet or outlet baffles, which were deemed not essential due to its long-narrow geometry. The aeration header was a perforated pipe (i.e., resulting in coarse bubbles and nonclogging operation) with a parallel longitudinal baffle that aided the hydraulic roll. Regarding performance, the grit removal was 0.95 fraction for particles $\geq$0.2 mm, the volatile solids fraction of the grit removed was 0.08 fraction, and the solids fraction was 0.79 (i.e., only 0.21 fraction water). The tracer test for the RMC plant showed an inverted "V-shape," that is, close to the plug-flow ideal. The theoretical detention time was 3.7 min with actual measured (based on the tracer curve) at 4.6 min. The bottom roll velocities were $0.03 \leq v_T \leq 0.24$ m/s.

While some three of the other grit chambers, that is, IC, SR, FR (see Table 7.12), had high grit removal fractions, the organic fractions were high. Operational differences for these plants were considered as contributing to the higher levels of grit removal, for example, lower overflow velocities. In some cases, grit had deposited on the bottom (i.e., not in the collection zone). Also, tracer tests showed accentuated short-circuiting with long tails (the latter indicating dead zones). Detention times varied 3.6–7.8 min and bottom velocities were generally 0.2–0.4 m/s.

Two general deficiencies of the grit-chamber designs included: (1) airflows could not be measured, that is, no flow meters; (2) the airflows could be adjusted manually, but with difficulty. The characteristics of a well-designed grit chamber included not only effective performance but a unit that is relatively trouble-free and had operating flexibility. Grit that has low organic matter was deemed important so that grit handling and land disposal may be done easily, that is, with minimum nuisance and in accordance with regulations. In summary, the RMC grit chamber had the closer-to-ideal dimensions (based on theory) and the overall best performance.

7.3.3.2 Summary of Guidelines

The following paragraphs summarize some of the important design criteria for aerated grit chambers. As noted previously, most are empirical.

Grit quantity: Table 7.12 shows a range of grit collected ranging 7–39 mL grit/m^3 water (0.9–5 ft^3/mg), as measured "wet." For comparison, Tchobanoglous and Burton (1991, p. 462) give 4–200 mL grit/m^3 water (0.5–27 ft^3/mg) for aerated grit chambers.

Detention time: About 3–5 min for detention time seems to be entrenched in the lore of American practice. However, Table 7.12 shows a range 3.6–7.8 min, and for comparison, Tchobanoglous and Burton (1991, p. 462) give 2–5 min for peak flow. By contrast, Londong (1989) recommended, based on a modeling study, $\theta > 10$ min for storm water flow and $\theta > 20$ min for dry weather flow.

Shape and size: Aerated grit chambers have been constructed with a variety of shapes as seen in Table 7.12. To promote the spiral flow, a rectangular cross section is recommended with corner fillets in the corners to reduce the hydraulic dead zones. Londong (1989) recommends a width to depth ratio of 0.8, that is, $w/D = 0.8$.

Air diffuser placement: The header placement of the air diffusers line is almost always parallel to the direction of the flow and near one of the sidewalls, the deviations of Table 7.12 notwithstanding. Londong (1989) recommended that the air header be placed at a depth, $0.7 \cdot D$, below the water surface.

Types of air diffusers: Most diffusers are coarse-bubble, which are less likely to clog. Fine bubbles, however, exert a higher amount of drag on the water mass. Fine bubble diffusers require a higher pressure drop than a coarse bubble diffuser for a given airflow. Whatever the diffuser type, the airflow versus pressure drop (across the diffuser) should be determined, that is, Q_a vs. ΔP(diffuser) so that the compressor can be sized accurately.

Airflow control and measurement: The airflow is controlled by a valve to the header pipe located just after the compressor. The airflow, Q(air), should be measured so that performance can be related to airflow. An orifice meter or a Venturi meter with differential manometer are instruments common for flow measurement.

Required airflow: The required airflow, Q(air), may be determined: (1) by a pilot plant study, (2) by an experimental use of a full-scale grit chamber, or (3) by data from practice (i.e., operating plants). The airflow "capacity" should be in excess of the maximum airflow needed. In a pilot study, the top and bottom velocities of the roll should be measured for each airflow. The compressor power required is based on an equation for an "adiabatic" compression, with inputs, Q(air) and p_2 (where p_2 is the pressure at the discharge side of the compressor).

Equations for experimental system: Equations for airflow for coarse bubble and fine bubble diffusers were obtained by Londong (1989) for a particular experimental system and are given here as Equations 7.24 and 7.23, respectively. The equations give airflows that result in tangential velocities of 0.2 m/s, that is, $v_T = 0.2$ m/s and give airflow as a function of the depth of the air header:

$$Q(\text{air, coarse bubble}, v_T = 0.2\,\text{m/s}) = [0.07 + 0.76 \ln d(\text{air header})]^{-1.33} \tag{7.24}$$

$$Q(\text{air, fine bubble}, v_T = 0.2\,\text{m/s}) = [0.63 + 0.52 \ln d(\text{air header})]^{-0.62} \tag{7.25}$$

where

Q is the air flow necessary to maintain a circulation velocity of 0.2 m/s, that is, $v_T = 0.2$ m/s (m^3/s)

d is the depth of submergence of air diffusers (m)

Table CD7.14, Parts 1–3 provides a calculation algorithm for Q(air) for different diffuser depths for each of the two diffuser types.

7.3.3.3 Pressure in Header Pipe

The *absolute* pressure in the header pipe is the sum of the atmospheric pressure plus the pressure at the depth of submergence of the diffuser plus the pressure loss across the diffuser, the latter being a function of Q_a(diffuser). The pressure at outlet of the blower equals the pressure in header pipe at the diffuser, plus the pressure losses in the pipes and bends, plus the pressure loss across the airflow control valve. The relations are best understood by sketching the "pneumatic" grade-line from the ambient air pressure in the water above the tank to the ambient air pressure at the blower intake. The mathematical relation is

$$p(\text{atmosphere}) + \Delta p(\text{compressor}) - \Delta p(\text{pipe friction}) - \Delta p(\text{pipe losses}) - \Delta p(\text{valve}) = p(\text{atmosphere}) + \Delta p(\text{water depth}) + \Delta p(\text{diffuser}) \quad (7.26)$$

where

p(atmosphere) is the barometric pressure or standard pressure at a given elevation (kPa)

Δp(compressor) is the pressure gain across the compressor (kPa)

Δp(pipe friction) is the pressure loss within the pipe due to friction (kPa)

$\Delta p(\text{pipe friction}) = fL/d\rho v^2/2$, $f = 0.012$ for smooth pipe, L is the length of pipe (m), d is the diameter of pipe (m), ρ is the density of air at temperature and pressure (kg/m^3), v is the mean velocity of air (m/s)

Δp(pipe losses) is the pressure losses due to bends and appurtenances in pipeline (kPa)

k is the loss coefficient for appurtenance (a value of 0.1 is common)

Δp(valve) is the pressure loss across the air flow control valve (kPa)

$\Delta p(\text{valve}) = k\rho v^2/2$

k is the coefficient that depends upon type of valve and degree of closure and takes up the "slack" in Equation 7.26

Δp(water depth) is the pressure due to water depth, h (kPa), which is calculated as the specific weight of water, γ(water), times the diffuser depth, h(diffuser)

Δp(diffuser) is the pressure loss across diffuser (kPa)

$\Delta p(\text{diffuser}) = Q(\text{diffuser}) = CA(\text{diffuser})[\Delta p(\text{diffuser})]^{0.5}$

C is the orifice coefficient (use 0.62 as a default value)

The "pneumatic" grade-line is an extension of the "hydraulic" grade-line concept, which depicts water pressure in terms of "head" of water (meters or feet). The "pneumatic" grade-line does the same thing but depicts absolute air pressure (kPa or psi); alternatively, a manometer could be tapped into a pipe to measure the *relative* pressure of the gas as mm of water.

7.3.3.4 Blower Power

The power required of a blower is given by the equation for an adiabatic compression, shown in Table 7.13 for both SI units and U.S. Customary units. The associated definitions in both systems of units are given below the respective equations. A calculation algorithm is set up as Parts 3 and 4 of Table CD7.14.

Table CD7.14 illustrates in spreadsheet format how to determine the airflow Q(air) and the power required by a

TABLE 7.13
Power Required for Adiabatic Compression of Air

SI Units

$$P = \frac{Q_a\rho(\text{air}) \cdot (R(\text{univ})/\text{MW}(\text{air})) \cdot T(\text{air})}{(k-1)/k}\left[\left(\frac{P_2}{p_i}\right)^{(k-1)/k} - 1\right] \quad (7.27a)$$

Definitions

P = power output of compressor (Nm/s)

Q = flow of air (m^3/s)

ρ = mass density of air (kg/m^3)

R = universal gas constant = $8.314510\ \text{N} \cdot \text{m} \cdot \text{K}^{-1}\ \text{mol}^{-1}$

MW(air) = 28.9 g/mol = 0.028964 kg/mol

T(air) = absolute temperature of air (K) = 273.15+°C

p_2 = absolute pressure on outlet side of compressor (N/m^2)

p_i = absolute pressure on inlet side of compressor (N/m^2)

k = ratio of heat capacity at constant pressure, C_p, to the heat capacity at constant volume, C_v (i.e., 1.395 and is dimensionless)

U.S. Customary Units

$$P = \frac{Q_a\rho(\text{air}) \cdot (R(\text{air}) \cdot T(\text{air})}{(k-1)/k}\left[\left(\frac{P_2}{p_i}\right)^{(k-1)/k} - 1\right] \quad (7.27b)$$

P = power output of compressor (ft lb/s)

Q = flow of air (ft^3/s)

ρ = mass density of air (lbm/ft^3)

$R(\text{air})$ = gas constant for air $\left(\text{i.e., } 53.3\ \frac{\text{ft} \cdot \text{lb}_f}{\text{lb}_f \cdot {}^\circ\text{R}}\right)$

$$R(\text{air}) = \frac{R_u}{M(\text{air})} = \frac{\left(1544\ \frac{\text{ft} \cdot \text{lb}_f}{\text{lb}_f - \text{mol}\ {}^\circ\text{R}}\right)}{\left(29\ \frac{\text{lb}_f}{\text{lb}_f - \text{mol}}\right)}$$

T(air) = absolute temperature of air (°R) = 459.6+°F

p_2 = absolute pressure on outlet side of compressor (lb/ft^2)

p_i = absolute pressure on inlet side of compressor (lb/ft^2)

k = ratio of heat capacity at constant pressure, C_p, to the heat capacity at constant volume, C_v (i.e., 1.395 and is dimensionless)

TABLE CD7.14
Airflow and Power Calculations for Aerated Grit Chamber

1 Coefficients

Coeff.	Coarse	Fine
	Diffuser Type	
a =	0.07	0.63
b =	0.76	0.52
c =	−1.33	−0.62

2 Constants

Constant	Value	Units
R =	8.314510	J K^{-1}mol^{-1}
MW(air) =	0.028964	kg/mol
k =	1.395	

3 Air Flow Calculations / **4 Compressor Power Required**

d(diff) (m)	*Q*(air, co) (m^3/m/h)	*Q*(air, *f*) (m^3/m/h)	*V*(Gr Ch) (m^3)	*Q*(air) (m^3/s)	Elev. (m)	*p*(atm) (Pa)	*T* °C	*r*(air) (mol/m^3)	*r*(air) (kg/m^3)	p_2 (Pa)	*P* (kW)	(hp)
0.90	>1000	1.409	375	0.1	0	101325	20	41.57	1.204	303975	21	27.5
1.00	34.357	1.332	375	0.1	0	101325	20	41.57	1.204	303975	19	25.9
1.50	3.645	1.113	375	0.1	0	101325	20	41.57	1.204	303975	16	21.7
2.00	1.987	1.006	375	0.1	0	101325	20	41.57	1.204	303975	15	19.6
2.50	1.425	0.939	375	0.1	0	101325	20	41.57	1.204	303975	14	18.3
3.00	1.142	0.893	375	0.1	0	101325	20	41.57	1.204	303975	13	17.4

5 Notes on columns and equations

Dimensions are m^3 air flow (at normal temperature and pressure) per hour per m^3 of grit chamber volume

$Q(\text{air}) = [a + b \ln(d)]^c$ (from Londong, 1989)

$Q(\text{air}) = Q(\text{air, fine}) \cdot V(\text{Gr Ch})$

Assumed

Diffuser $p + Dp$(losses)

d(diff) = depth of diffuser

$V(\text{Gr Ch}) = w \cdot D \cdot L$

Q(air, co) = air flow for coarse bubble diffuser

Assumed

Q(air, *f*) = air flow through fine bubble diffuser

Alternatively, $Q(\text{air}) = 3\text{–}8$ ft^3/min/ft length (ASCE, 1977) $= 0.0046\text{–}0.0124$ m^3/s/m length of tank

Use barometric pressure or default value adjusted for elevation, that is, $p(\text{atm}) = 101{,}325 * 10^{-0.00005456*Z}$

$r(\text{air}) = p(\text{atm})/RT$

$r(\text{air}) = p(\text{mol/m}^3) * \text{MW(air)}/1000$

Power required by compressor is for an adiabatic compression and is calculated by relations in Table 7.13

compressor for an adiabatic compression for different depths of diffuser submergence. The spreadsheet (on a CD) also has the same table in U.S. Customary units.

The pressure, p_2, on the discharge of the compressor was assumed as $p_2 = 3 \cdot p(\text{atm})$, which was an arbitrary assumption for the purpose of illustrating the spreadsheet functioning. The value of p_2 should be calculated by Equation 7.26, which requires another linked spreadsheet and involves utilizing submergence depth, for example, $Dp(\text{submergence}) = r(\text{water}) \cdot g \cdot z(\text{submergence})$, determining pipe sizes, and calculations such as pressure losses due to pipe friction, for example, by the Darcy–Weisbach equation, and minor losses for valves, a flow meter, etc.

PROBLEMS

Given Conditions

For all problems, let $Q(\text{avg}) = 0.22$ m^3/s (5 mgd), $Q(\text{min}) = 0.30 \cdot Q(\text{avg})$ and $Q(\text{max}) = 3 \cdot Q(\text{avg})$, unless otherwise specified. Use a spreadsheet for each problem unless advised otherwise.

7.1 Rectangular Grit Chamber with Proportional Weir Control

Given

Assume flows: $Q(\text{avg}) = 0.22$ m^3/s (5 mgd), $Q(\text{min}) = 0.30 \cdot Q(\text{avg})$, and $Q(\text{max}) = 3 \cdot Q(\text{avg})$.

Required

Design a rectangular grit chamber with a proportional weir as control.

7.2 Rectangular Grit Chamber with Parshall Flume Control

Given

Flows: $Q(\text{avg}) = 0.22$ m^3/s (5 mgd), $Q(\text{min}) = 0.30 \cdot Q(\text{avg})$, and $Q(\text{max}) = 3 \cdot Q(\text{avg})$.

Required

Design a rectangular grit chamber with a Parshall flume as control.

7.3 Parabolic Grit Chamber with Parshall Flume Control

Given/Required

Design a parabolic section grit chamber with a Parshall flume as control.

Solution

Flows: $Q(\text{avg}) = 0.22\ \text{m}^3/\text{s}$, $Q(\text{max}) = 0.66\ \text{m}^3/\text{s}$, $Q(\text{min}) = 0.066\ \text{m}^3/\text{s}$. Table CDprob7.3 and Table CD7.7 contains the algorithm for calculation of the Q(max) and Q(min) depths, H_a(max) and H_a(min), in the "a" and "b" sections, respectively. The "c" section shows the calculations for the parabolic cross section and the v_H values for different depths, "y," and flows (columns "a" and "b"; as seen, the v_H values calculated are all about 0.31 m/s (1.0 ft/s). The formulae are given at the bottom of the table. The plot of the parabolic cross section at the bottom of the table is linked to the table values of (x, y).

7.4 Trapezoidal Grit Chamber with Parshall Flume Control

Given

Flows: $Q(\text{avg}) = 0.22\ \text{m}^3/\text{s}$, $Q(\text{max}) = 0.66\ \text{m}^3/\text{s}$, and $Q(\text{min}) = 0.066\ \text{m}^3/\text{s}$.

Required

Design a trapezoidal section grit chamber with a Parshall flume as control, approximating this from the parabolic section in Problem 7.3.

7.5 Aerated Grit Chamber—Rational

Given

$Q(\text{avg}) = 0.22\ \text{m}^3/\text{s}$, $Q(\text{max}) = 0.66\ \text{m}^3/\text{s}$, and $Q(\text{min}) = 0.066\ \text{m}^3/\text{s}$.

Required

Design an aerated grit chamber using guidelines from Londong (1989).

7.6 Aerated Grit Chamber—Empirical

Given

$Q(\text{avg}) = 0.22\ \text{m}^3/\text{s}$, $Q(\text{max}) = 0.66\ \text{m}^3/\text{s}$, $Q(\text{min}) = 0.066\ \text{m}^3/\text{s}$.

Required

Design an aerated grit chamber using empirical guidelines.

7.7 Model to Estimate Performance

Given

Flows: $Q(\text{avg}) = 0.22\ \text{m}^3/\text{s}$, $Q(\text{max}) = 0.66\ \text{m}^3/\text{s}$, and $Q(\text{min}) = 0.066\ \text{m}^3/\text{s}$.

Required

Estimate by Equation 7.21 the difference in performance of any grit chamber designed for the range of flow specified.

7.8 Model to Estimate Aerated Grit Chamber Performance

Given

The Fort Collins WWTP (1976 North Plant at Drake Road and Cache La Poudre River) had (before a plant expansion, c. 1991) an aerated grit chamber that was about 5 m wide · 5 m deep · 10 m long in dimensions for a flow of about 6 mgd. Performance is hypothetical, but assume that $C_o \approx 1000$ mg/L for grit particles about 0.2 mm in size and that $C \approx 100$ mg/L (for 0.2 mm grit leaving the grit chamber). Assume the new grit chamber must handle flows of about 15 mgd. Flows: $Q(\text{avg}) = 0.22\ \text{m}^3/\text{s}$, $Q(\text{max}) = 0.66\ \text{m}^3/\text{s}$, $Q(\text{min}) = 0.066\ \text{m}^3/\text{s}$.

Required

Using the 1976 grit chamber performance and sizing as a guide, determine a sizing for a new one, along with determining airflow and compressor horsepower. Using your spreadsheet, explore alternatives that may be considered. Also determine Q(air) and P(compressor).

7.9 Grit Quantity

Given

Flows: $Q(\text{avg}) = 0.22\ \text{m}^3/\text{s}$ 5.0 mgd), $Q(\text{max}) = 0.66\ \text{m}^3/\text{s}$ (15 mgd), and $Q(\text{min}) = 0.066\ \text{m}^3/\text{s}$ (1.5 mgd).

Required

Estimate the volume rate of grit to be expected at a WWTP in your locale for

1. Daily dry weather flow.
2. Wet weather flow.
3. What size (volume) of container would you set below your grit conveyor belt and how frequently would you expect to transport the grit to a disposal site?
4. Determine the disposal site for the plant. What regulations pertain?

7.10 Pressure, p_3, at Compressor Outlet—For Aerated Grit Chamber

This problem involves an extensive spreadsheet; in lieu of setting up the spreadsheet, an alternative is to review the problem in principle, referring to the spreadsheet already developed.

Given

1. *Flows*: $Q(\text{avg}) = 0.22\ \text{m}^3/\text{s}$ (5.0 mgd), $Q(\text{max}) = 0.66\ \text{m}^3/\text{s}$ (15 mgd), and $Q(\text{min}) = 0.066\ \text{m}^3/\text{s}$ (1.5 mgd).
2. *Assumptions*: Assume the distance equals 50 m (164 ft) between the compressor and the diffuser header pipe in the grit chamber. The pipe size should be a part of the spreadsheet so that you can examine pipe losses for different pipe sizes. For sizing the grit chamber, let $q = 20$ min, with V(grit chamber) based on Q(max). For the cross-section, let width = depth = 5.0 m. Also, let Equation 7.24 be the means to estimate the air flow for the aerated grit chamber, that is, $Q(\text{air, coarse bubble}) = [0.07 + 0.76 \cdot \ln(d)]^{-1.33}$, where d is the depth of the diffuser header pipe below the water surface of the grit chamber; let $d = 3.50$ m.

Required

1. *Compressor outlet pressure*: Set up a spreadsheet to determine pressure, p_3, in outlet pipe from the compressor.
2. *Compressor power*: Determine the power required by the compressor for an adiabatic compression.

7.11 Compressor Power for an Aerated Grit Chamber

Given

A municipal WWTP has flows,

Q(max) = 1.06 m^3/s (24 mgd), which is the hydraulic capacity for the plant

Q(avg) = 0.53 m^3/s (12 mgd), which was used as the basis for design

Q(min) = 0.26 m^3/s (6.0 mgd)

Two aerated grit chambers are installed, each with dimensions, $w = 4.6$ m, $D = 4.6$ m, $L = 7.6$ m (15 ft · 15 ft · 25 ft). Assume both aerated grit chambers are in operation at the same time, that is, half the incoming plant flow goes to each grit chamber.

Three positive displacement compressors, each 5.6 kw (7.5 hp), were installed. The air flow capacity for one compressor is Q(air total for one compressor) = 0.050 m^3/s NTP, that is, normal temperature and pressure (105 scfm) at discharge pressure, p_2(gage) = 46.2 kPa (6.7 psi). Two compressors are operated at once (one for each of the two grit chambers); the third compressor is standby and is rotated with the other two. Keep in mind that the given flows pertain to two parallel grit chambers.

Required

1. For the conditions given, calculate the power required for an adiabatic compression, that is, P(adiabatic).
2. Compare P(adiabatic) with the installed motor power and calculate the overall wire-to-compressor-output efficiency.

7.12 Volume of an Aerated Grit Chamber—Installed and Calculated

Given

A municipal WWTP has flows as follows:

Q(max) = 1.06 m^3/s (24 mgd), which is the hydraulic capacity for the plant

Q(avg) = 0.53 m^3/s (12 mgd), which was used as the basis for design

Q(min) = 0.26 m^3/s (6.0 mgd)

Two aerated grit chambers are installed, each with dimensions, $w = 4.6$ m, $D = 4.6$ m, $L = 7.6$ m (15 ft · 15 ft · 25 ft). Assume both aerated grit chambers are in operation at the same time, that is, half the incoming plant flow goes to each grit chamber. Q(min for one grit chamber) = 0.13 m^3/s (3.0 mgd).

Required

Compare the size of the grit chamber as constructed with empirical guidelines.

ACKNOWLEDGMENTS

Permission to use Figures 7.12 and 7.13 was granted by Tom Quimby, director of marketing, WSG & Solutions, Montgomeryville, Pennsylvania (www.wsgandsolutions.com).

As a note, Rex Chainbelt is a historic company, which became Rexnord, then Envirex, and later became a division of Siemens Water Technologies Corp. In 2007, the headworks business (bar screens and grit collectors) was sold by Siemens to become WSG & Solutions, Montgomeryville, Pennsylvania (see www.wsgandsolutions.com/). The latter company designs, manufactures, and provides a complete product line of headworks equipment.

GLOSSARY

Adiabatic: Without heat transfer; an adiabatic compression means that the commensurate equation is used to estimate power requirement.

Aerated grit chamber: Grit chamber with diffused air along one side to cause a rotation of the fluid mass.

Blower: For air compression, a centrifugal pump that builds up significant pressure.

Collection zone: A rectangular inset at the bottom of the grit chamber that collects and stores grit to be scraped or transported by a screw to a hopper at the head end of the grit chamber.

Control section: A particular section following the grit chamber used to control the depth velocity relation in the grit chamber; a necessary condition is that for the control section, $\mathbf{F} > 1$.

Critical velocity: The velocity at which the Froude number, $\mathbf{F}$, is $\mathbf{F} > 1$.

Detention time: The mean residence time of a fluid in a given volume; mathematically, $\theta = V/Q$.

Diffuser: Device with holes spread out over a plate or membrane through which air flows.

Free flow: Refers to flow condition for a Parshall flume in which the velocity is higher than "critical," that is, $\mathbf{F} > 1$.

Grit: Technically, grit is defined as particles with diameter, $d \geq 0.2$ mm with SG = 2.65. Any particles that settle readily in a grit chamber, however, could be considered grit.

Grit chamber: A particular settling basin technology intended to settle grit particles and scour organic particles that may have settled.

Header: For the case of a grit chamber, a manifold pipe that distributes flow uniformly through diffusers.

Hydraulic grade line (HGL): Graphical depiction of pressure profile (usually in feet of water or meters of water) of a pressurized pipeline with associated components, for example, a reservoir, a pump, pipeline, bends, valves, and perhaps a terminal reservoir. Common in civil engineering.

Organics: With regard to grit chambers, organic matter entrained in the raw wastewater entering the grit chamber.

Pneumatic: Refers to a gas characteristic, usually air, for example, pressure, velocity.

Pneumatic grade line (PGL): Graphical depiction of pressure profile from air intake, through a compressor, through the pipes, finally to the header pipe and through the diffusers, terminating at the water surface above the diffusers. The air bubbles emitted from the diffusers might be considered the terminal point as opposed to the water surface since the pressure in the air bubbles equals the depth of submergence of the diffusers. This description applies to the aerated grit chamber application, but may be generalized. The graphical depiction would be in terms of pressure (absolute pressure would be the clearest, as opposed to relative pressure), whereas the HGL is in terms of meters (or feet) of water. Normally, the pressure units would be as kilopascals (kPa) or "psi" in U.S. Customary units. The term PGL was suggested by Professor Robert M. Meroney, Professor of Civil and Environmental Engineering, Colorado State University during a conversation, c. 2000, when the topic of HGL was brought up and the author questioned him about a corresponding relation for air, since the HGL is common in hydraulic depictions.

Pressure: Dimensions are force per unit area. SI units are Newtons per square meter, or pascals. Air pressure at sea level, $p_a = 101,325$ Pa $= 101.325$ kPa (14.7 psi). Conversion to feet of water is obtained from the relation, $p = z \cdot g_w$, for example, z(feet of water) $= p_a/g_w = (14.7 \text{ lb/in}^2 \cdot 144 \text{ in.}^2/\text{ft}^2)/62.4 \text{ lb/ft}^3 = 33.9 \text{ ft} = 10.34$ m.

Pressure (absolute): Pressure of a fluid with reference to zero. Usually the local atmospheric pressure would be added to a measured gage pressure. A mercury barometer with an evacuated tube is an accepted standard method of measuring local atmospheric pressure (in mm of mercury).

Pressure (relative): Pressure with reference to the local atmosphere, also called "gage" pressure. A Bourdon gage is used commonly, in which the pressure on the outside of the elastic coil is atmospheric and the inside of the gage contains the fluid in which the pressure is measured. A manometer or piezometer would also measure relative pressure.

Proportional weir: A special weir plate that has narrowing width toward the top and is designed to give a constant velocity in the horizontal flow rectangular grit chamber.

Separation zone: A hypothetical area near the bottom of an aerated grit chamber that will capture any grit particle that enters the zone.

Shield's equation: Empirical mathematical relation that relates the mean horizontal velocity for incipient scour to the properties of the particles in question.

Submerged flow: A flow condition for a Parshall flume in which the tailwater below the flume is increased in depth such that the upstream depth H_a is affected. This means that the depth of flow in thc throat of the flume exceeds "critical-depth." The submerged-flow condition occurs when $H_b/H_a \gg S_t$; see also "transition submergence." Determination of flow then requires measurement of both H_a and H_b.

Super critical: An hydraulic term that may be defined by depth or velocity with reference to the "critical" depth or velocity, respectively. If the velocity in a given channel exceeds "critical," it is termed "super critical"; another characteristic of super-critical flow is that a "wave" cannot travel upstream. As a given channel becomes narrower, at some point the depth is forced through "critical." Alternatively, as the slope of the channel increases, the velocity passes through "critical." The foregoing is a cursory description of this topic; more thorough explanations are given in most fluid mechanics or hydraulics texts).

Throat: The mid section of a Parshall flume that is used to characterize the flume; for example, a 0.30 m (12 in.) Parshall flume is one that has a throat width of that dimension.

Transition-submergence: The condition that occurs when the tailwater depth, as measured by H_b, is high enough such that an additional increase may cause a "submerged-flow" condition. The criterion for incipient submergence is that $H_b/H_a = S_t$. The tailwater depth is measured by H_b.

REFERENCES

ASCE-WPCF, Sewage treatment plant design, in: *ASCE Manual of Engineering Practice No. 36 and the WPCF Manual of Practice No. 8*, American Society of Civil Engineers, New York, 1959.

ASCE-WPCF, Wastewater treatment plant design, in: *ASCE Manual of Engineering Practice No. 36 and the WPCF Manual of Practice No. 8*, American Society of Civil Engineers, New York, 1977.

Babbitt, H. E., *Sewerage and Sewage Treatment*, 5th edn., John Wiley & Sons, New York, 1940.

Camp, T. R., Grit chamber design, *Sewage Works Journal*, 14: 368–381, 1942.

Camp, T. R., Sedimentation and the design of settling tanks, Transactions of the ASCE, III: 895–958, 1946.

Griffin, F. L., *An Introduction to Mathematical Analysis*, revised edition, Houghton-Mifflin Co., New York, 1936.

Hirano, R., Pitt, P., Chen, R., and Skelley, E., Grit Overload—Oceanside Plant in San Francisco, California overcomes grit accumulation problems, *Water Environment Technology*, 10(11):55–58, November, 1998.

Londong, J., Beitrag Zur Bemessung Beluufteter Sandfänge Unter Besonderer Berücksichtigung der Gleichzeitigen Nutzung Als Adsorptionsstufe, 94, Technische Hochschule Aachen, ISSN 0342-6068, Aachen, 1987.

Londong, J., Dimensioning of aerated grit chambers and use as a highly loaded activated sludge process, *Water Science and Technology*, 21:13–22, 1989.

Metcalf, L. and Eddy, H. P., *American Sewerage Practice* (Volume III, Disposal of Sewage), McGraw-Hill, New York, 1916.

Morales, L. and Reinhart, D., Full-scale evaluation of aerated grit chambers, *Journal of the Water Pollution Control Federation*, 56(4):337–343, April, 1984.

Parshall, R. L., The improved Venturi flume, *Transactions ASCE*, 89:841–880, 1926.

Rex Chainbelt, Grit Collectors, Product Manual, Sanitation Equipment, Conveyor and Process Equipment Division, Rex Chainbelt, Inc., Milwaukee, WI, Data Sheets 315-4.001-315-4.531, 1965.

Rexnord, Grit chambers data sheet, in: *Rexnord Product Manual*, Rexnord, Inc., Waukesha, WI, 1980.

Skogerboe, G. V., Hyatt, M. L., England, J. D., and Johnson, J. R., Parshall Flumes, Report WG31-3, OWRR Project No. B-006-Utah, Utah Water Research Laboratory, Utah State University, Logan, UT, March, 1967.

Tchobanoglous, G. and Burton, F. L., for Metcalf & Eddy, Inc., *Wastewater Engineering—Treatment, Disposal, and Reuse*, McGraw-Hill, Inc., New York, 1991.

Walker, W. R., Skogerboe, G. V., and Bennett, R. S., Flow-measuring flume for wastewater treatment plants, *Journal of the Water Pollution Control Federation*, 45(3):542–551, March, 1973.

Wilson, G., Tchobanoglous, G., and Griffiths, J., The Nitty Gritty – Grit sampling and analysis, *Water Environment and Technology, Water Environment Federation*, 19(7):64–68, July, 2007a.

Wilson, G., Tchobanoglous, G., and Griffiths, J., The Nitty Gritty – Peak flows and light grit, *Water Environment and Technology, Water Environment Federation*, 19(8):72–75, August, 2007b.

Wilson, G., Tchobanoglous, G., and Griffiths, J., The Nitty Gritty – Designing a grit removal system, *Water Environment and Technology, Water Environment Federation*, 19(9):115–118, September, 2007c.

Wilson, G., Tchobanoglous, G., and Griffiths, J., The Nitty Gritty – Grit washing system design, *Water Environment and Technology, Water Environment Federation*, 19(10):81–84, October, 2007d.

8 Flotation

The flotation process involves (1) generation of air bubbles, (2) contact between the air bubbles and the particles to be removed, (3) flotation of particles by the buoyant force created, and (4) removal by skimming. In modern practice, flotation utilizes dissolved air as a source of bubbles and is called "dissolved air flotation," with the common acronym "DAF."

Examples of particles to be floated include algae; chemical precipitates; coagulant flocs such as alum or ferric, perhaps strengthened with a polymer; and biological flocs. The objective may be either to separate solids and water or to "thicken" the solids, e.g., to raise the solids concentration from say 1% to perhaps 4% (such as in the case of activated sludge).

8.1 DEVELOPMENT OF FLOTATION

The flotation process was developed about 1875 for ore separations in the mining industry, and in the 1950s it was adopted for treatment of industrial wastewaters. During the decades since, applications have been extended to include thickening of activated sludge, removal of algae from oxidation pond effluents, and in drinking-water clarification in lieu of gravity settling.

8.1.1 Beginning Design Practice

Some of the same empirical design parameters as are current were seen in the 1950s literature. For example, Eckenfelder et al. (1958, p. 257) proposed guidelines for the air-to-solids (A/S). Then, Vrablik (1960) gave guidelines for overflow rate and saturator pressure (Section 8.4). In other words, those active in that time evidently had insight into critical process parameters.

Design theory evolved further during the 1990s based on research by Edzwald, Fukushi, and Haarhoff and their coworkers (Edzwald, 1995; Edzwald and Walsh, 1992; Fukushi et al., 1998; Haarhoff and Rykaart, 1995; Haarhoff and van Vuuren, 1995) directed mostly toward water treatment. Principles delineated were based on saturator pressure, bubble–particle contact probability, and rise-velocities of bubble–particle agglomerates.

8.1.2 Water and Wastewater Applications

Kalinske (1958, p. 228) listed 11 applications of flotation in industrial wastewater treatment. Only one application was mentioned in municipal wastewater treatment, i.e., thickening of activated sludge, which was not established as a technology.

Some of the specific applications of DAF for industrial wastewater treatment have included: canneries in which organic suspended solids are removed; chemicals in which recoveries of fine particles are required such as colloidal metals, calcium sulfate, and metal hydroxides; finely divided coal; scale and oil in steel mill wastewater; solids and fatty acids in laundry wastes; grease from meat product wastewaters; free or emulsified oil from refineries; pulp and paper wastewaters; and recovery of fats and oils in soap manufacturing (Vrablik, 1960).

8.2 DAF SYSTEM DESCRIPTION

The DAF "process" comprises a system, i.e., a coordinated collection of components that results in achieving its objectives. The objectives depend, of course, on the application, but the common thread of all is separation of water and solids by flotation.

8.2.1 Synopsis of DAF Process

Figure 8.1 shows details of a DAF system, e.g., rapid mix, flocculators, air saturator, and the flotation basin. The descriptions that follow refer to Figure 8.1.

8.2.1.1 Coagulation

Coagulant chemicals, e.g., alum or ferric ion, are added in the rapid mix, where particles to be removed are charge neutralized, resulting in "microflocs."

8.2.1.2 Flocculation

The "micro-flocs" formed grow in size in the flocculation basins to form "flocs," preferably to a size of 10–50 μm but not larger than 50 μm. The size of the flocs is controlled by the coagulant dose, flocculation turbulence intensity, i.e., G in s^{-1}, and detention time, $q \approx 10$ min. Recommended G values are $G \approx 70\ s^{-1}$ for alum or ferric coagulation, and $G \geq 30\ s^{-1}$ for polyaluminum chloride. For wastewater, suggested limits were $60 \leq G \leq 80\ s^{-1}$ (Ødegaard, 1995).

8.2.1.3 Contact Zone

After flocculation, the raw-water flow enters the "contact zone" of the flotation basin where the floc particles contact precipitated gas bubbles in the recycle flow from the saturator. The recycle flow is emitted just below the contact zone in jets from a bank of nozzles attached to the manifold.

8.2.1.4 Saturator

The saturator is a tank where "gas transfer" occurs, i.e., the water gains dissolved oxygen and nitrogen (and minor gases such as argon and carbon dioxide) from air under pressure. The air flows upward from beneath a "packing" material and the gas transfer occurs in the "recycle" water flowing down.

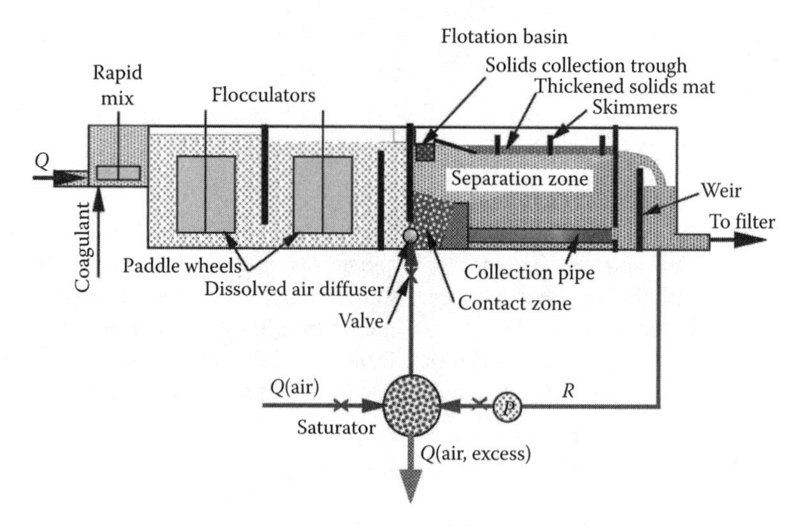

FIGURE 8.1 Side-view schematic of DAF system. (Adapted from Valade, M.T. et al., *J. Am. Water Works Assoc.*, 88(12), 36, December, 1996.)

The concentration of each gas is proportional to its partial pressure in the tank (the total pressure ranges 4–8 atm gage) and is sometimes termed "super-saturated." Typically, the recycle flow, R, ranges 5%–15% of Q, the influent flow. The water flow bifurcates at each piece of packing, thus creating a large water surface area within the voids.

8.2.1.5 Gas Precipitation

The "super-saturated" water flows into a manifold, just under the "contact zone" of the flotation basin and through a bank of nozzles. The sudden pressure drop across the nozzle "throat" causes a portion of the dissolved gases to "precipitate" as bubbles, typically they are 20–50 μm median diameter; the mass quantity of gas precipitated is termed "released" gas (also called "excess" gas). The effect is the same as opening a bottle of carbonated beverage (see Chapter 18 and Section H.3.1).

8.2.1.6 Bubble–Floc Agglomerate

The small bubbles of precipitated gas rise within the "contact zone" with a fraction "attaching" to floc particles. The bubbles create buoyancy and cause the bubble–floc agglomerate to rise. Their rise velocity in the separation zone depends upon the number of air bubbles that are attached per floc particle (or more accurately the volume of air bubbles attached relative to the mass of the floc particle).

8.2.1.7 Float Layer

The bubble–floc agglomerate, as it reaches the surface, forms a "float layer." The float layer is removed by skimmer blades that push the float over a "beach" and into a trough. The solids flow is about 2%–4% solids and consumes about 1%–2% of the total flow, Q.

8.2.1.8 Clarified Water

The partially clarified water leaves at the bottom of the separation zone by means of a set of perforated collection pipes to enter a head box, with the water surface controlled by means of a weir. The overflow from this box enters another head box control before entering a filtration system.

8.2.1.9 Further Processing

The solids stream leaving by the float layer is subject to additional processing, such as further thickening, e.g., by centrifuging, belt filter-press, etc. The water stream also may be processed further, depending on the context. For example, filtration would be expected if the context is drinking-water treatment.

8.3 PRINCIPLES OF DAF FLOTATION

A DAF system has four process phases (Edzwald, 1995):

1. Gas transfer across the air–water interface (from the gas phase to the dissolved phase) in a "saturator" tank.
2. Gas, i.e., air, in the dissolved state "precipitates" to form gas bubbles.
3. Transport of the gas bubbles to solid particles to achieve "contact" and then "attachment" (which occurs in the "contact zone" of the flotation basin).
4. Flotation of the bubble–particle agglomerate in the "separation zone" of the flotation basin.

These four phases are illustrated in Figure 8.1. Principles that underlie them are reviewed here.

8.3.1 Gas Transfer

The "mass transfer" of gases in air to the aqueous phase occurs in the "saturator." The several facets of this process phase are summarized in this section. Chapter 18 reviews "gas transfer," and Appendix H reviews equilibrium between the gas phase and the aqueous phase.

8.3.1.1 Henry's Law

The gas phase–aqueous phase equilibrium is given by Henry's law, which states merely that the equilibrium concentration of a gas, "A," in the aqueous phase is proportional to the partial pressure of gas "A" at the gas–water interface, i.e.,

$$C(\text{gas A}) = H(\text{gas A}) \cdot P(\text{gas A}) \quad (8.1)$$

in which

C(gas A) is the concentration of dissolved gas A in aqueous solution (kg gas A/m^3 water)
H(gas A) is the Henry's constant for a given gas, e.g., "A" (kg gas A/m^3 water/kPa gas A)
P(gas A) is the partial pressure of gas A above gas–water interface (kPa gas A)
A is the designation for a particular species of a gas, e.g., O_2 or N_2

Table H.5 gives Henry's constants for different gases with temperature as a variable. In addition, coefficients, i.e., "A" and "B," are given at the top of Table H.5 for "best-fit" empirical equations for H(gas A) vs. temperature, i.e., $H(\text{gas A}) = A \cdot \exp(B \cdot T°\text{C})$. As a historical note, Henry's law was mentioned by Masterson and Pratt (1958, p. 233) and by Eckenfelder et al. (1958, p. 251). In other words, application of Henry's law to flotation has been established since the 1950s.

The partial pressure of "gas A" is implied in Dalton's law, i.e., it states that the total pressure in a mixture of gases is the sum of the respective partial pressures, expressed for a single gas as

$$P(\text{gas A}) = X(\text{gas A}) \cdot P(n \text{ gases}) \quad (8.2)$$

in which

X(gas A) is the mole fraction of gas A in gas phase (mol gas A/m^3/sum of mol n gases/m^3)
P(n gases) is the total pressure of all gases above gas–water interface (kPa n gases)
n is the number of species of gases in a given volume

Then combining Equations 8.1 and 8.2,

$$C(\text{gas A}) = H(\text{gas A}) \cdot X(\text{gas A}) \cdot P(n \text{ gases}) \quad (8.3)$$

Equation 8.3 is the operational equation. If it happens that $P(n = 1)$, i.e., a "pure" gas A is involved, then it follows that $X(\text{gas A}) = 1.0$. Examples 8.1 and 8.2 illustrate the application of Henry's law, which is straightforward in accordance with Equation 8.3. Example 8.3 shows the calculation of a "pseudo" Henry's constant for air, H(air), which again is straightforward, albeit involving mole fraction weighted average of the respective Henry's constant.

Example 8.1 Calculation of Dissolved Oxygen Concentration at Sea Level by Henry's Law at 20°C

The problem refers to water at equilibrium with oxygen at sea level at 20°C based on 101.325 kPa total atmospheric pressure and demonstrates the application of Henry's law.

Given
$H(O_2) = 0.0004383$ (kg dissolved oxygen/m^3 water/kPa oxygen) at 20°C
$X(O_2) = 0.209$ mol O_2/mole air (Table B.7)
P(atmosphere) = 101.325 kPa (stated)

Required
Calculate the concentration of dissolved oxygen at equilibrium by Henry's law.

Solution
Substitute given data, for oxygen, in Equation 8.3,

$$C(O_2) = H(O_2) \cdot X(O_2) \cdot P(\text{total}) = 0.0004383\ (\text{kg dissolved } O_2/m^3\text{water/kPa } O_2) \quad (\text{Ex8.1.1})$$

- 0.209 mol O_2/moles all gases in air
- 101.325 kPa total pressure
  = 0.00928 kg dissolved O_2/m^3 water (9.3 mg/L)

Discussion
As stated, the foregoing simple calculation demonstrates the application of Henry's law. Two ideas are involved. First, in accordance with Henry's law, each gas dissolves in water in proportion to its partial pressure. Second, Dalton's law states that each gas in a mixture of gases exerts a partial pressure in accordance with its respective mole fraction, e.g., $X(O_2)$. In addition, note that the units are delineated fully, which helps in applying Henry's law.

Example 8.2 Calculation of Dissolved Nitrogen Concentration at Sea Level by Henry's Law at 20°C

The problem refers to water at equilibrium with air at sea level at 20°C based on 101.325 kPa total atmospheric pressure.

Given
$H(N_2) = 0.0001875$ (kg dissolved nitrogen/m^3 water/kPa nitrogen)[a] at 20°C
$X(N_2) = 0.78084$ moles N_2/mole air (Table B.7)
P(atmosphere) = 101.325 kPa (stated)

Required
Calculate the concentration of dissolved nitrogen at equilibrium by Henry's law.

Solution
Substitute given data, for nitrogen, in Equation 8.3,

$$C(N_2) = H(N_2) \cdot X(N_2) \cdot P(\text{total}) = 0.0001875\ (\text{kg dissolved } N_2/m^3\ \text{water/kPa air})$$

- 0.78084 moles N_2/moles all gases in air
- 101.325 kPa
  = 0.01483 kg dissolved N_2/m^3 water (14.8 mg/L)

Discussion
The relevant points are stated in Example 8.1.

Example 8.3 Determine an Equivalent "H(air, 20°C)"

Problem
Determine a Henry's constant for air

Given
$H(O_2) = 0.0004383$ (kg O_2 dissolved/m^3 water/kPa O_2)—Table H.5, 20°C
$X(O_2) = 0.209$ mol O_2/moles all gases in air—Table B.7
$H(N_2) = 0.0001876$ (kg N_2 dissolved/m^3 water/kPa N_2)—Table H.5, 20°C
$X(N_2) = 0.781$ mol N_2/moles all gases in air—Table B.7

Solution

1.
$$\begin{aligned} C(O_2 + N_2) &= H(O_2) \cdot X(O_2) \cdot P(\text{total}) + H(N_2) \cdot X(N_2) \cdot P(\text{total}) \\ &= [0.0004383 \cdot 0.209 + 0.0001876 \cdot 0.781] \cdot 101.325\ \text{kPa} \\ &= [9.16 \cdot 10^{-5} + 14.65 \cdot 10^{-5}] \cdot 101.325\ \text{kPa} \\ &= 0.0241\ \text{kg}\ (O_2 + N_2)/\text{m}^3\ \text{water} \end{aligned} \qquad \text{(Ex8.3.1)}$$

2. Substitute in Equation 8.3,

$$C(\text{air}) = H(\text{air}) \cdot X(\text{air}) \cdot P(\text{total})$$
$$0.0241\ \text{kg}\ (O_2 + N_2)/\text{m}^3\ \text{water} = H(\text{air}) \cdot X(\text{air}) \cdot P(\text{total})$$
$$0.0241\ \text{kg}\ (O_2 + N_2)/\text{m}^3\ \text{water} = H(\text{air}) \cdot 1.00 \cdot 101.325\ \text{kPa}$$
$$H(O_2 + N_2) \approx 0.000238\ (\text{kg}\ O_2 + N_2\ \text{dissolved/m}^3\ \text{water/kPa air}) \qquad \text{(Ex8.3.2)}$$

Discussion
The result is for 20°C. Such an equivalent Henry's constant for air is an "artifice" that some may prefer as opposed to dealing with the individual gases. In dissolving, however, the gases act individually, i.e., oxygen and nitrogen mostly (which comprise 0.99032 mol fraction of air). The reason for considering "air" as a dissolved gas, i.e., in lieu of oxygen and nitrogen (and other minor gases) independently, is that the literature refers frequently to "dissolved air." The minor gases include CO_2, Ar, Ne, He, Kr, Xe, CH_4, and H_2. As a matter of interest, the sum of the mole fractions of O_2, N_2, CO_2, Ar equals, 0.9999700, i.e.,

$$\begin{aligned} [X(O_2) &= 0.2094760 + X(N_2) = 0.7808400 + X(CO_2) \\ &= 0.0003140 + X(\text{Ar})] = 0.0093400] \\ &= 0.9999700. \end{aligned}$$

The equivalent Henry's coefficient, if CO_2 and Ar are included, is (from Table CD8.3)

$$H(O_2 + N_2 + CO_2 + \text{Ar}) = 0.0002561\ (\text{kg}\ O_2 + N_2 + CO_2 + \text{Ar})\ \text{dissolved/m}^3\ \text{water/kPa air})$$

The other gases, i.e., Ne, He, Kr, Xe, CH_4, and H_2, comprise only 0.0000300 mol fraction.

As to dealing with a range of temperatures for "H(air)," Table CD8.3, Part (a) gives that capability by substituting different values of temperature in "cell B3" to generate a range of results for H(air), plotting H(air) vs. T, and then applying the coefficients, e.g., "a" and "b" in a best-fit equation.

8.3.1.2 Application of Henry's Law to Saturator

The "saturator" is a tank where oxygen and nitrogen (and minor gases) in air under high pressure are transferred to water in accordance with Henry's law. The gas-transfer rate, i.e., from gas to aqueous solution, is governed by the pressure in the saturator and the water surface area that has contact with the gases (i.e., air, which is a mixture). The dissolved gas concentration leaving the saturator, C(saturator), is always less than the equilibrium level as stated by Henry's law. The ratio of these two values is the "efficiency-factor," f, which is defined (Edzwald, 1995, p. 7) by

$$f = \frac{C(\text{saturator, gas A})}{H(\text{gas A}) \cdot X(\text{gas A}) \cdot P(n\ \text{gases})} \qquad (8.4)$$

in which
f is the ratio of gas concentration leaving saturator to gas concentration by Henry's law
C(saturator, gas A) is the concentration of dissolved gas "A" in water leaving the saturator, which is the same as C(saturator), the collective term (kg dissolved gas A/m^3 water)

Equation 8.3 thus has a modified form for a saturator, i.e.,

$$C(\text{saturator, gas A}) = f \cdot H(\text{gas A}) \cdot X(\text{gas A}) \cdot P(n\ \text{gases}) \qquad (8.5)$$

For packed-bed saturators, $f \approx 0.9$, and for unpacked saturators, $f \approx 0.7$ (Edzwald, 1995). The value for f depends upon the packing, the hydraulic loading rate, and the saturator depth. The gas A may be any gas, e.g., O_2, N_2, Ar, CO_2, or in terms of practice, "air," which is the aggregate of the component gases. For the usual case of "air" as the gas, $X(\text{gas A}) = 1.0$ (which is the approximate sum of the major component gases), i.e., $X(N_2) = 0.78084$, $X(O_2) = 0.209476$, $X(\text{Ar}) = 0.00934$, $X(CO_2) = 0.000314$; the sum of the mole fractions for these four gases out of the 12 listed in Table H.1 is, $X(N_2, O_2, CO_2, \text{Ar}) = 0.99997$.

The saturator capital cost is about 12% of the cost of a DAF plant and about 50% of the operating cost (Haarhoff and Rykaart, 1995). Therefore, a higher f will provide a cost saving in operation that may justify the added capital cost.

8.3.1.3 Saturator

Henry's law, modified for saturator application, is the basis for calculating C(saturator), i.e.,

$$C(\text{saturator}) = f \cdot H(\text{air}) \cdot P(\text{saturator}) \qquad (8.6)$$

where

$H(\text{air})$ is the Henry's constant for air (kg dissolved air/m^3 water/kPa air); a value for $H(\text{air})$ may be approximated as a weighted average for $H(O_2)$ and $H(N_2)$, i.e., $H(O_2 + N_2) \approx 0.000238$ (kg $O_2 + N_2$ in gas phase/m^3 water/kPa air)

$P(\text{saturator})$ is the pressure of gas air in saturator tank, i.e., at gas–water interface (kPa)

The saturator pressure should be sufficient to provide the required value of $C(\text{saturator})$, i.e., in order to provide the bubbles needed to float the particles entering the contact zone.

8.3.1.4 Gas Concentration at Nozzle Depth

The concentrations of dissolved gases collectively, i.e., air, at the depth of nozzle submergence, C_a, is also calculated by Henry's law, i.e.,

$$C_a = H(\text{air}) \cdot \left[P(\text{atm-at-}Z\text{-elev}) + \frac{D(\text{nozzle})}{10.33\ \text{m}} \cdot 101.325\right] \quad (8.7)$$

where

$H(\text{air})$ is the Henry's constant for air (kg dissolved air/m^3 water/kPa air); a value for $H(\text{air})$ may be approximated as a weighted average for $H(O_2)$ and $H(N_2)$, i.e., $H(O_2 + N_2) \approx 0.000238$ (kg $O_2 + N_2$ in gas phase/m^3 water/kPa air)

$P(\text{atm})$ is the pressure of air at elevation of flotation tank (kPa)

$D(\text{nozzle})$ is the depth of nozzle and manifold (m)

10.33 m is the depth of water having pressure of 1.0 atm or 101.325 kPa

The value of C_a is also the concentration at which gas precipitation occurs and is the maximum concentration of gas in solution.

8.3.1.5 Saturator Mass Balance

Figure 8.2 shows the flows of water and air entering and leaving the saturator tank. As seen, the recycle flow, R, enters the system at a dissolved air concentration, C_a, and is pressurized by a pump to the required saturator pressure, usually in the range of $400 \leq P(\text{saturator}) \leq 800$ kPa gage (4–8 atm gage), and leaves at a concentration, $C(\text{saturator})_e$, calculated as per Equation 8.5. The airflow, $Q_a(\text{STP})$, enters the system and is compressed to the saturator pressure; the airflow is "dead-ended," i.e., no airflow leaves the system, and thus the air is supplied at the rate of dissolution. The change in mass flow of dissolved air through the saturator, i.e., $R[C(\text{saturator}) - C_a]$, equals the rate of gas dissolution from the airflow, i.e., $Q(\text{STP}) \cdot r(\text{STP})$. Although the airflow is shown as entering the saturator through a compressor, a traditional means to supply air is by an intake on the suction side of the pump (see, e.g., Kalinske, 1958, p. 225), which requires the pump to be set at an elevation that results in a negative gage pressure at the pump inlet.

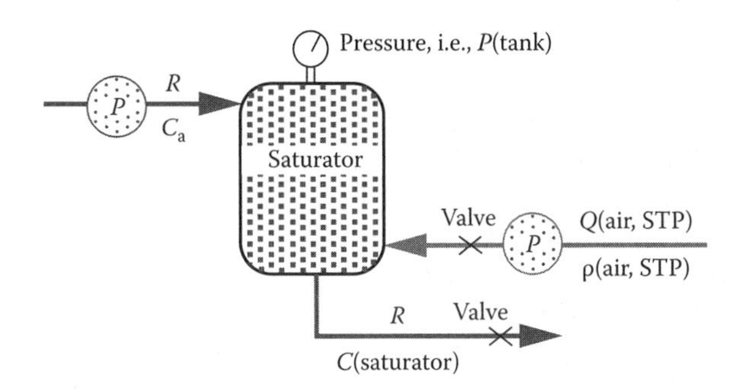

FIGURE 8.2 Schematic showing materials balance of gases for saturator.

8.3.1.6 Saturator Packing

Figure 8.3 shows (a) an "unpacked" saturator, and (b) a "packed" saturator. In each case, the purpose is to generate a high air–water interfacial area so that the oxygen and nitrogen (comprising 0.99 fraction of air) and the minor gases may transfer from the gas state to the dissolved state.

For both the unpacked and packed saturators, the water flows through a distribution plate at the top; the water at the bottom of the saturator is "ponded." With the unpacked saturator, i.e., Figure 8.3a, the gas transfer occurs as the water falls, and by the entrained air in the flow that plunges into the "pool" at the bottom. With the packed-bed saturator, i.e., Figure 8.3b, the gas transfer occurs mostly within the packing interstices due to the large air–water surface area created by the bifurcation in flow at each packing object comprising the bed. A valve located after the saturator can control the flow, R, and the level of the "ponding." A water manometer, shown in each diagram, can monitor the level of the "ponded" water (the air pressure operates on both sides of the manometer and so the differential is the water depth, as shown). Empirical guidelines for saturator design and operation are (Edzwald, 1995, p. 16)

Pressure: $400 \leq P(\text{sat}) \leq 800$ kPa gage
Hydraulic loading rate: $50 \leq \text{HLR} \leq 80$ m/h
Depth of packing material: $0.8 \leq Z_D \leq 1.2$ m

8.3.1.7 Hydraulic Grade Line

Figure 8.4 illustrates the "relative" hydraulic grade line (HGL) for the recirculation flow, R; "relative" means the HGL (locus of points of pressure head plus elevation head) is located with respect to atmospheric pressure (i.e., gage pressure). The important points are (1) the saturator pump must develop the "head" necessary to pressurize the tank to a specified level, and (2) the major headloss occurs across the nozzles in the "contact zone." The flow through the packing is "unsaturated" and so the headloss across the packing is due only to the loss in elevation through the depth of the bed. The nozzles are shown pointed up only to illustrate their shape schematically; their direction, as installed, would be horizontal.

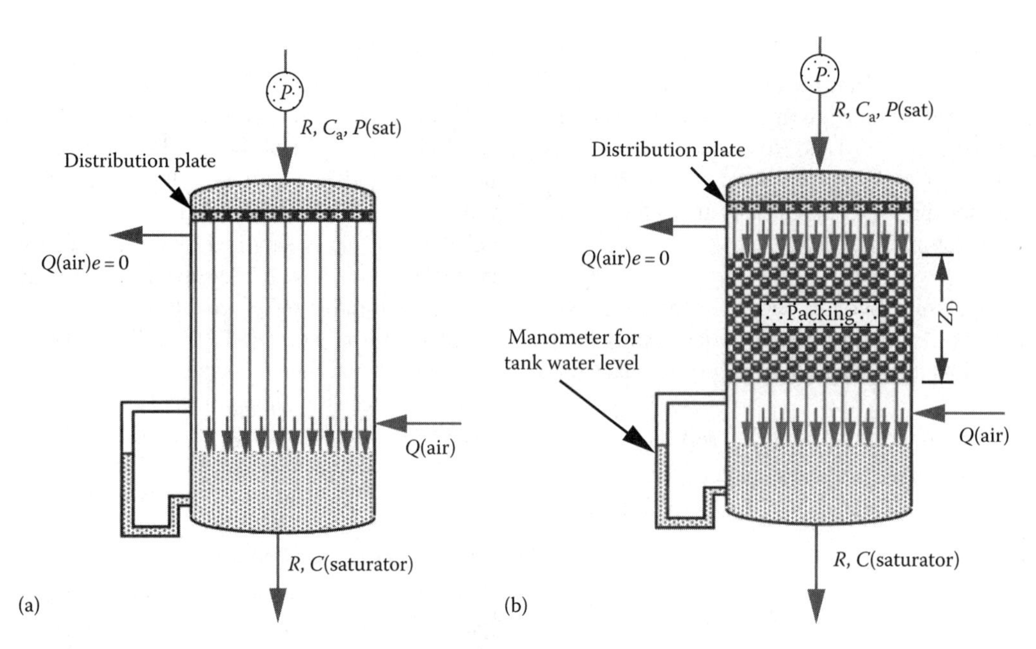

FIGURE 8.3 Sketches of two types of saturators: (a) unpacked saturator and (b) packed saturator.

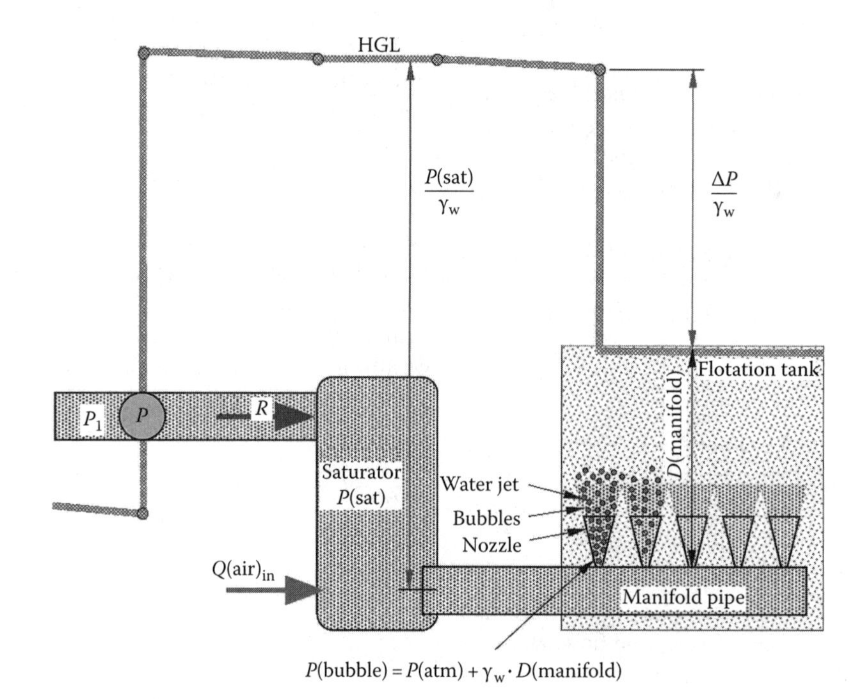

FIGURE 8.4 Schematic drawing showing HGL as indicator of pressure changes for gas dissolution and for gas precipitation. *Notes:* (1) scale is distorted to emphasize important characteristics, (2) HGL location is relative.

8.3.2 Gas Precipitation

The occurrence of bubbles and their characteristics are central to the flotation process. This section reviews bubble formation, their size, size distribution, and number concentration.

8.3.2.1 Bubbles

The spontaneous occurrence of gas bubbles in water is called "gas precipitation" (Section H.3). They occur if a dissolved gas is "supersaturated" under pressure, and then if the pressure is reduced quickly. There are many examples. A common one is the appearance of gas bubbles when a bottle of carbonated beverage is opened. Boiling water is another example (gas bubbles appear when the vapor pressure of the water equals atmospheric pressure). Bubbles may sometimes be observed around algae blooms, which presumably are pure oxygen. Often bubbles (nitrogen, carbon dioxide, methane) can be seen breaking the surface of a primary clarifier; also, they may be dislodged from low-elevation lake or river muds during summer months.

A dissolved gas, e.g., N_2, will precipitate if its concentration exceeds that which would be in equilibrium with a *pseudo-pure-N_2-gas*, which is at the pressure of the water at the particular point in question. For a flotation system, using nitrogen gas to illustrate, the idea may be expressed as follows:

$$C(\mathrm{N_2}, \text{nozzle contraction}) > H(\mathrm{N_2}) \cdot P(\textit{pseudo-pure-N}_2\textit{-gas} \text{ at depth of } \textit{flotation tank-at-nozzle}) \quad (8.8)$$

in which

$C(N_2$, nozzle contraction) is the actual concentration of dissolved N_2 in the nozzle contraction, i.e., just before the throat and immediately preceding bubble formation that occurs in the nozzle expansion (mg dissolved N_2/L water) or (kg dissolved N_2/m^3 water)

$H(N_2)$ is the Henry's coefficient for N_2 at a specified temperature as given in Table H.5 (19.01 mg dissolved N_2/L water/atm N_2) or (0.0001876 kg dissolved N_2/m^3 water/kPa N_2)

$$P(\textit{pseudo-pure-N}_2\textit{-gas} \text{ at depth of flotation tank-at-nozzle}) = P(\text{atm}) + \gamma_w \cdot D(\text{flotation-tank-depth}) \text{ (kPa)}$$

P(atm) is the pressure of atmosphere at elevation of water surface (kPa); for reference, P(atm at sea level) = 101.325 kPa

D(flotation-tank-depth) is the depth of nozzles in flotation tank (m)

γ_w is the specific weight of water (N/m^3)

$\gamma_w = r_w \cdot g$, r_w is the mass density of water (998.2063 kg/m^3 at 20°C, Table B.9) and g is the acceleration of gravity (9.806650 m/s^2)

In other words, if the left side of Equation 8.8 exceeds the quantity on the right side, gas precipitation occurs. The left side is the actual dissolved gas concentration from the saturator while the right side is that calculated by Henry's law and is the maximum dissolved gas concentration that can exist. The difference between the left side and the right side is the "released gas" (also called "excess gas") that becomes bubbles. The dissolved gas concentration in the "contact zone," and at the nozzle elevation (after gas precipitation), is at "equilibrium" with the gas bubbles, and is designated, C_a, where, by Henry's law, $C_a = H(N_2) \cdot P(\textit{pure-N}_2\textit{-gas}$ at depth of flotation tank-at-nozzle).

It is not known whether the bubbles formed are pure gas of a single species, e.g., N_2, or a mixture, e.g., N_2 and O_2. Based on theory, it is assumed here that a bubble of pure gas will form initially, but the gas–water interface formed would facilitate another gas species diffusing into the bubble. Gas bubbles collected from primary clarifier sludge were found (by mass-spectrograph analyses) to comprise a mixture of gases, e.g., N_2, CH_4, and CO_2 (Hendricks, 1966). Most likely, the mixture of gases observed in bubbles is due to collisions.

Example 8.4 Saturation Concentration

Given

The nozzles of a flotation tank are located at a depth of 4.0 m at a sea-level location, $T = 20°C$.

Required

Determine the critical concentration at which gas precipitation will occur.

Solution

Apply the principle that the concentration of a gas cannot exceed that which would be at pseudo equilibrium with the local pressure of the water being considered, i.e.,

$$\begin{aligned}\lim C(\text{dissolved } \mathrm{O_2}) &= H(\mathrm{O_2}) \cdot X(\mathrm{O_2}) \cdot P(\text{Nozzles}) \\ &= (43.39\,\text{mg/L/atm}) \\ &\quad \cdot (1.0\,\text{mol } \mathrm{O_2}/\text{mol pure } \mathrm{O_2}) \\ &\quad \cdot [1.0\text{ atm} + (4.0\text{ m}/10.33\text{ m})\text{ atm}] \\ &= 56.0\,\text{mg/L}\end{aligned}$$

Comments

The concentration of dissolved oxygen must exceed 56 mg/L in order for gas precipitation to occur.

Example 8.5 Saturation Concentration—Bubbles Adjacent to Algae Mats

Given

Bubbles are sometimes observed adjacent to "mats" of floating algae sometimes found in summer in quiescent natural waters. Assume this occurs at sea level and at or near the water surface.

Required

Determine the maximum concentration of dissolved oxygen in the vicinity of the algae.

Solution

The gas will precipitate at concentration.

$$\begin{aligned}\lim C(\text{dissolved } \mathrm{O_2}) &= H(\mathrm{O_2}) \cdot X(\mathrm{O_2}) \cdot P(\text{atmosphere}) \\ &= (43.39\,\text{mg/L/atm}) \\ &\quad \cdot (1.0\,\text{mol } \mathrm{O_2}/\text{mol pure } \mathrm{O_2}) \\ &\quad \cdot [1.0\text{ atm} + (0.0\text{ m}/10.33\text{ m})\text{ atm}] \\ &= 43.4\,\text{mg/L}\end{aligned}$$

Comments

The dissolved oxygen (a reaction product of photosynthesis) concentration cannot exceed 43.4 mg/L since gas precipitation will preclude higher levels.

8.3.2.2 Bubble Size

Bubble size decreases with increasing saturator pressure and with increasing flow, as illustrated by the experimental data of Figure 8.5. In the "recommended practice" range, i.e., 400–600 kPa gage, the bubbles were in an acceptable size range, i.e., 40–60 μm. The pressure change was across a needle valve with three experimental flows.

The merits of the small bubble size include (de Rijk et al., 1994, p. 467) (1) small bubbles attach to floc more easily; (2) the floc–bubble collision probability is proportional to the bubble number concentration, not bubble size; and (3) small bubbles have a lower rise velocity giving a longer residence time in the contact zone.

Bubble sizes larger than 150 μm diameter are called "macro-bubbles" (Rykaart and Haarhoff, 1995). They impair the flotation process by hindering interaction between floc and microbubbles; also, they rise rapidly, causing a turbulent wake, and they are likely to disrupt the float layer.

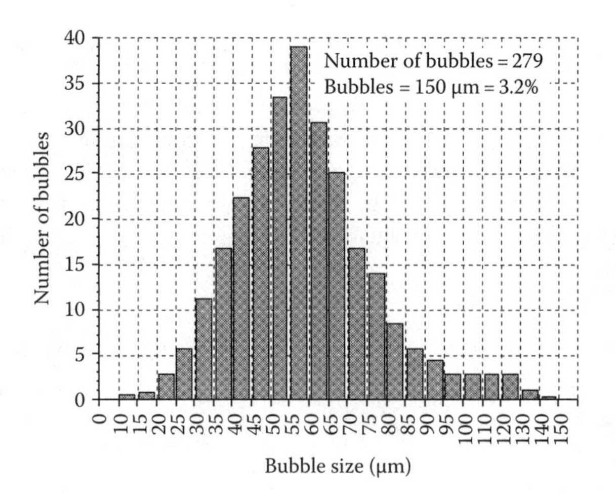

FIGURE 8.6 Frequency distribution of bubble sizes from photomicrograph. (Adapted from Haarhoff, J., Factors Influencing Bubble Formation in Dissolved Air Flotation, *International Association of Water Quality Yearbook*, 1994–95, 1995.)

8.3.2.3 Bubble Size Distribution

Typically, bubble sizes follow a Gaussian distribution, as shown in Figure 8.6 (Haarhoff, 1995); the median size shown is 54 μm. About 40 μm is considered, however, a reasonable goal (Edzwald, 1995).

8.3.2.4 Bubble Numbers

Using a mean bubble diameter of 60 μm, P(saturator) = 490 kPa, at $T = 15°C$, and r(recycle-ratio) = 0.1, Fukushi et al. (1998, p. 79) calculated 58,600 bubbles/mL, which was confirmed by counting bubbles on microscopic video photographs. This observation provides a reference for an achievable bubble number concentration.

8.3.2.5 Nozzle Design

It is in the nozzle where the pressure change occurs; the change should be within a short distance rather than over a gradual transition (Rykaart and Haarhoff, 1995). The time for the pressure drop across a nozzle is the orifice length divided by the flow velocity and should be less time than that required to precipitate the gas from the solution, estimated to be about 1.7 ms. The bubbles form at nucleation centers and commence immediately after the pressure reduction. Larger bubbles will form, however, by coalescence of smaller ones due to turbulence-caused contacts. If the nozzle is directed to an obstruction, such as a plate, the bubbles will be broken up to smaller sizes by the energy of impact.

Experiments by Rykaart and Haarhoff (1995) found median bubble sizes as small as 39 μm with a pressure of 500 kPa gage with a plate 5 mm distance from the orifice and a jet velocity of about 15 m/s. Without a plate obstruction, or with the plate ≥10 μm distance, the bubble size was about 62 μm for pressures of both 200 and 500 kPa.

A tapered outlet, shown in Figure 8.7, as designed by Rykaart and Haarhoff (1995) for their experiments, limited eddy formation. This resulted in median bubble diameters of 61 and 30 μm for 200 and 500 kPa gage, respectively. The fraction of macro-bubbles was only about 0.01.

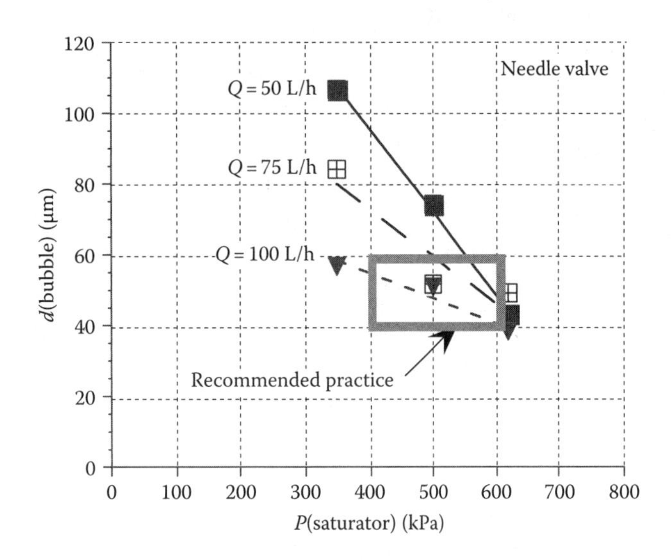

FIGURE 8.5 Bubble sizes vs. saturator gage pressure. (From de Rijk, S.E. et al., *Water Res.*, 28(2), 465, 1994.)

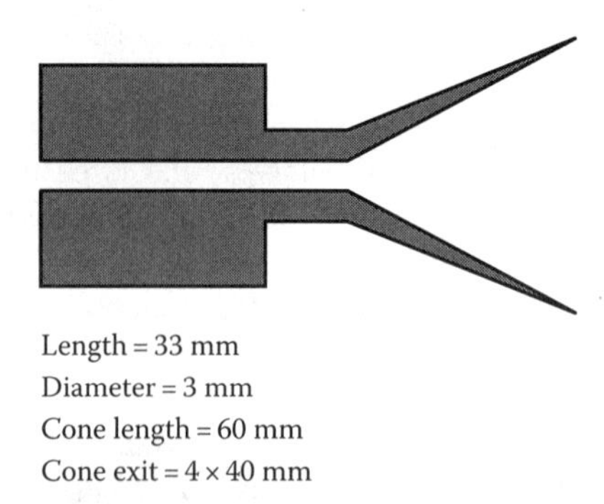

FIGURE 8.7 Tapered orifice design. (Adapted from Rykaart, E.M. and Haarhoff, J., *Water Sci. Technol.*, 31(3–4), 25, 1995.)

8.3.3 Contact Zone

As to mechanism, in the contact zone there are two steps. First, the bubbles and particles must contact; the higher the concentration of bubbles, the higher the probability of contact. Second, the bubbles that contact must attach; the fraction attaching is in the range of 0.3–0.4, depending on the portion of the surface already occupied by bubbles, and other factors (Matsui et al., 1998). After attachment, the particle–bubble agglomerates become buoyant and rise. The "rise" occurs in the "separation" zone.

Figure 8.8 depicts the contact zone and the adjacent separation zone. Floc particles enter the contact zone where a fraction of the bubbles attach. The bubble–particle agglomerates then rise in the separation zone and form a float layer at the water surface, which is moved by skimmer blades to a trough for removal.

Delivery of the dissolved gas is through a "manifold," a pipe across the width of the contact zone, indicated in Figure 8.8. The diameter should be large enough such that the friction headloss is negligible. From the manifold, the flow R is distributed through a bank of nozzles; the major change in pressure between the saturator and the contact zone is across the "throat" of each nozzle.

In the "contact zone," the bubbles released in the flow, R, from the saturator are dispersed into the flow of floc particles, Q. The "transport" of the bubbles occurs first by the random motion due to turbulence, with a fraction making "contact" with the floc. At the same time, the bubbles rise with a fraction making contact with the floc by "interception." Once contact is made, a fraction of the particles "attach." The contact zone is where these two phases of the project occur.

8.3.3.1 Floc–Bubble Transport and Attachment

Figure 8.9 illustrates two mechanisms of bubble transport and attachment to solid particles. Figure 8.9a shows a gas bubble being "transported" to make contact with a solid particle.

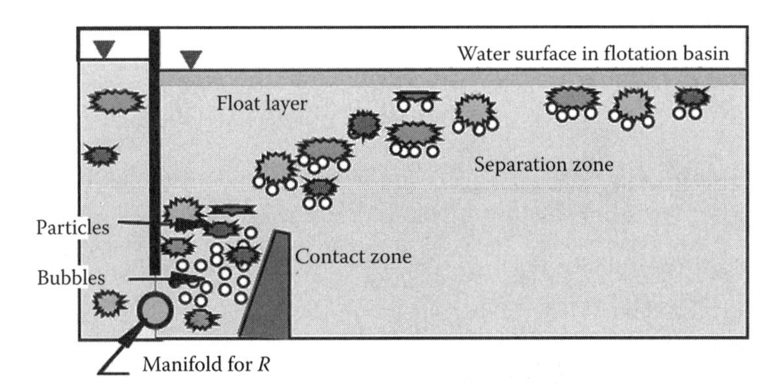

FIGURE 8.8 Schematic of flotation basin showing contact zone and separation zone.

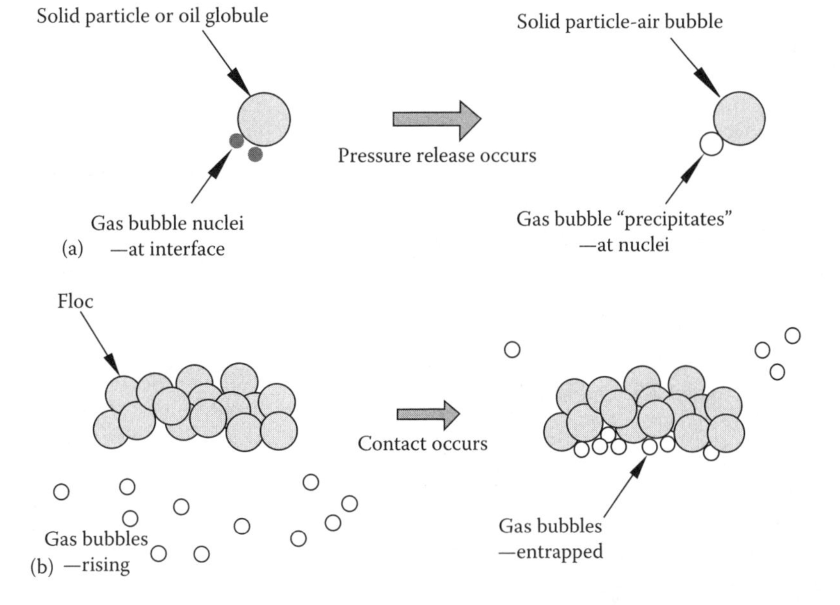

FIGURE 8.9 Mechanisms of flotation: (a) gas bubble–particle transport, then contact and attachment; (b) entrapment by floc structure of rising gas bubbles. (Adapted from Nemerow, N.L., *Liquid Wastes of Industry, Theories, Practices, and Treatment*, Addison-Wesley, Reading, MA, 1971.)

The transport may be by turbulence, diffusion, interception, sedimentation, etc. The most likely transport mechanism is probably interception as the bubbles rise, aided by turbulence.

After contact, the bubble may "attach" to a solid particle, usually a "floc." The "floc" is a particle created by coagulation (Chapter 9) and flocculation (Chapter 11), which precede the "contact zone." Since the bubbles are negatively charged, the floc particle must be positively charged. Their size is controlled by coagulant dose, flocculation intensity, and detention time with a goal of 10–30 μm, but not larger than 50 μm (Edzwald, 1995, p. 20). As indicated, the floc particle is not large relative to the bubble.

Figure 8.9b illustrates the mechanism of entrapment, which is more likely if the floc particles are larger in size, i.e., $d_p > 100$ μm. Of the two bubble–particle interactions, the first is felt to predominate in water treatment.

8.3.3.2 Bubble–Particle Contact

The rate of particle–bubble adhesions is proportional to the respective concentrations of bubbles and particles, and other factors, as described in Equation 8.9 (Edzwald, 1995, p. 9):

$$\frac{dN_p}{dt} = -\alpha_{pb}\eta_T A_b v_b N_b N_p \qquad (8.9)$$

in which

- N_p is the particle number concentration (# particles/m^3)
- N_b is the bubble concentration in contact zone (# particles/m^3)
- t is the elapsed time (s)
- α_{pb} is the adhesion efficiency, i.e., ratio of particle–bubble adhesions to particle–bubble contacts (dimensionless)
- η_T is the transport function is the ratio of the number of particle–bubble contacts to the number of bubbles being transported to the vicinity of a given particle (dimensionless)
- A_b is the projected area of bubble (m^2)
- v_b is the rise velocity of bubble relative to water as calculated by Stoke's law (m/s)

As stated in a previous paragraph, the transport function, η_T, is influenced mostly by turbulent diffusion and interception; sedimentation and molecular diffusion have smaller effects. All have been evaluated quantitatively by theoretical equations (see Edzwald, 1995, p. 9).

On adhesion efficiency, a_{pb} decreases as the bubbles attach, taking up more area. Concerning attachment, floc particles are a matrix of positively charged coagulant hydroxides and negatively charged suspended solids (Fukushi et al., 1998, p. 80). A bubble, which is negatively charged (−100 zeta-potential for oxygen and −150 mV for precipitated air bubbles at pH 7), must attach at a positively charged site. Values of α_{pb} were estimated at 0.35 for 2.5 mg/L alum and $\alpha_{pb} = 0.40$ for 5.0 mg/L alum (Fukushi et al., 1998). The particle–bubble attachment can be increased further by adding polymers (de Rijk et al., 1994, p. 467).

Another result of the negative charges on the bubbles is that they repel one another. Thus, the bubbles maintain their spacing to give a uniform blanket of rising bubbles increasing the probability of particle–bubble contacts, i.e., by "interception" (a component of the transport function). In forming larger bubbles, the repulsive energy must be overcome, e.g., by turbulence.

8.3.3.3 Parameter Values

Table 8.1 gives size ranges for bubbles and particles, concentrations of bubbles and particles, and other values. The particle diameter, d_p, and particle density, N_p, are controlled for a given water by the coagulant concentration, polymer usage, and flocculation turbulence intensity. As noted, the bubble diameter, d_b, is dependent upon the saturator pressure. The bubble number concentration, N_b, is dependent upon the "released" (or "excess") dissolved gas concentration, i.e., that available to form bubbles.

8.3.4 Separation Zone

The particles rise in the "separation zone," i.e., as illustrated in Figure 8.8. The "overflow velocity," v_o (Section 6.3.1.2), is the basis for determining the plan area, i.e., $v_o = (Q + R)/A$ (plan). In principle, the value of v_o is based upon a characteristic rise velocity of the particle–bubble agglomerate.

8.3.4.1 Rise Velocity of Bubbles

For reference, the rise velocity of a bubble may be calculated by Stoke's law, Equation 6.8, applied to a bubble, i.e.,

$$v_b = \frac{g d_b^2 (\rho_w - \rho_b)}{18\mu} \qquad (8.10)$$

in which

- d_b is the diameter of bubble (m)
- v_b is the rise velocity of bubble (m/s)
- ρ_w is the density of water (998.2063 kg/m^3 at 20°C, Table B.9)
- ρ_b is the density of air bubble (1.2038 kg/m^3 at 1.00 atm pressure and 20°C)
- g is the acceleration of gravity (9.8066 m/s^2)
- μ is the viscosity of water at stated water temperature ($1.002 \cdot 10^{-3}$ NS/m^2 at 20°C)

8.3.4.2 Rise Velocity of Particle–Bubble

The buoyant force on a particle–bubble agglomerate equals the weight of the volume of water displaced, i.e., Archimedes principle (Section 6.2.2). The associated rise velocity may be calculated by Stoke's law (Section 6.2.2), derived in the steps outlined (Edzwald, 1995, p. 13) as follows:

1. Determine the density for a particle–bubble agglomerate:

$$\rho_{pb} = \frac{\left[\rho_p d_p^3 + B\left(\rho_b d_b^3\right)\right]}{\left[d_p^3 + B d_b^3\right]} \qquad (8.11)$$

TABLE 8.1
Parameter Values in Flotation

Parameter	Definition	Value
α_{pb}	Attachment coefficient	$\alpha_{pb} \rightarrow 1.0$ with effective coagulation
d_p	Diameter of floc particle	$10 < d_b < 100$ μm, with median 40 μm; A strong floc of size range $10 \leq d_p \leq 30$ μm is a goal of flocculation (Edzwald, 1995, p. 20), which may be controlled by a low alum dosage and flocculation intensity e.g., $G \approx 70\ s^{-1}$, and duration, e.g., 5–10 min.
d_b	Diameter of bubble	$40 < d_p < 80$ μm, with median 50 μm; a size range $20 \leq d_b \leq 40$ μm is a goal
N_b	Bubble density	$10^3 < N_b < 2.4 \cdot 10^5$ bubbles/mL
N_p	Particle density	$10^5 < N_p$ (reaction zone) $< 10^5$ particles/mL
$B = N_b/N_p$	Ratio of bubbles to particles	$N_b \geq 10 \cdot N_p$; $B = N_b/N_p$ ratio
ρ_p	Mass density of particles	1010 kg/m^3
ρ_w	Mass density of water	998.2063 kg/m^3 at 20°C, Table B.9
ρ_{air}	Mass density of air at STP	1.2038 kg/m^3 at 20°C, Table B.7 The mass density of air may be calculated by the ideal gas law, i.e., $PV = n$RT. Rearranging gives the molar density, i.e., $n/V = P$/RT. Mass density is r(air) = (P/RT) · MW(gas)/1000 = 101,325 Pa/(8.31451 Nm/K mol · 293.15 K) · (28.9641 g/mol/1000 kg/mol); r(mass) = 1.204 kg air/m^3 gas.
g	Acceleration of gravity	9.8066 m/s^2, Appendix QR
μ	Dynamic viscosity of water	$1.002 \cdot 10^{-3}$ Nm/s^2, Table B.9

References: Rows 1–7 from Edzwald (1995, p. 12); Rows 8–11 from Appendix B.
STP is an acronym for "standard temperature and pressure."

in which

ρ_{pb} is the density of the particle–bubble agglomerate (kg/m^3)
ρ_p is the density of particles (1010 kg/m^3); from Edzwald (1995, p. 14)
d_p is the diameter of floc particle (m)
B is the number of attached bubbles (bubbles per particle)

2. Determine the equivalent spherical diameter, d_{pb}:

$$d_{pb} = \left[d_p^3 + Bd_b^3\right]^{1/3} \qquad (8.12)$$

in which d_{pb} is the diameter of the particle–bubble agglomerate (m)

3. The rise velocity, v_{pb}, by Stoke's law is

$$v_{pb} = \frac{g(\rho_w - \rho_{pb})d_{pb}^2}{18\mu} \qquad (8.13)$$

v_{pb} is the velocity of particle–bubble agglomerate (m/s).

Table CD8.2 gives results of computations to obtain v_{pb} for various particle diameters, d_p, for 1, 2, and 10 bubbles attached per particle based upon Equations 8.11 through 8.13,

TABLE CD8.2
Particle Rise Velocities as Function of Number of Bubbles Attached, B[a,b]

	B_n ($n = 1$)			B_n ($n = 2$)			B_n ($n = 10$)		
d_p (μm)	ρ_{pb} (g/mL)	d_{pb} (μm)	v_{pb} (m/h)	ρ_{pb} (g/mL)	d_{pb} (μm)	v_{pb} (m/h)	ρ_{pb} (g/mL)	d_{pb} (μm)	v_{pb} (m/h)
10	0.02	40	3.1	0.01	50	5.0	0.003	86	14.5
20	0.11	42	3.0	0.06	51	4.9	0.01	87	14.5
50	0.67	57	2.1	0.20	63	3.9	0.17	92	13.7
100	0.95	102	1.0	0.90	104	2.2	0.62	118	10.4
200	1.01	200	0	0.99	201	0.3	0.94	205	5.2
500	1.01	500	0	1.01	500	0	1.01	501	0

[a] From Table 3, Edzwald, 1995 (reconstructed by spreadsheet computations, Table CD8.2)
[b] Computations are based on Equations (8.11)), (8.12), and (8.13) where
$d_b = 40$ μm
$d_p = 50$ μm
$T = 20$°C, and initial
$\rho_p = 1.01$ g/cm^3 [= 1010 kg/m^3]

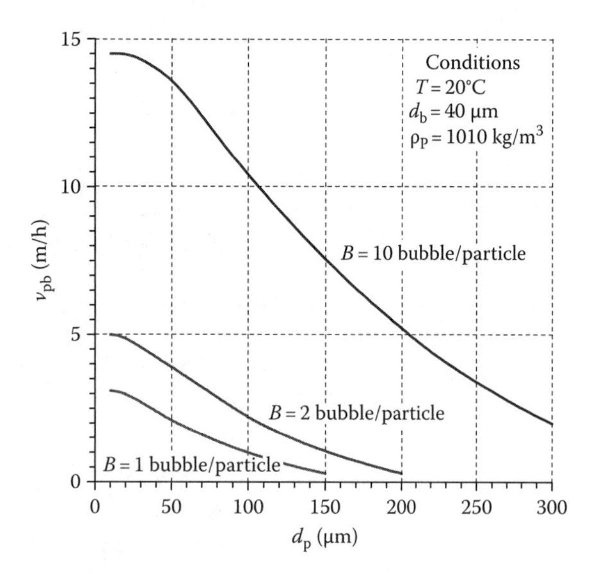

FIGURE 8.10 Rise velocity of particle–bubble agglomerate as a function of d_P for different values of B. (From Edzwald, J.K., *Water Sci. Technol.*, 31(3–4), 1, 1995.)

respectively. Figure 8.10 shows the results in graphical form. Theoretically, the v_{pb} values can be used as a guide to estimate overflow velocity for the separation zone. From the curves, v_{pb} is highest for $d_P < 50$ μm and $B_n \geq 10$ bubbles/particle. The floc-size, d_P, may be controlled by coagulant dose, and flocculation intensity and duration.

8.3.4.3 Bubble–Particle Ratio

On the bubble–particle ratio, $B = N_b/N_p \approx 12$, should ensure more bubbles than necessary to result in high v_{pb} (Edzwald, 1995, p. 14). Assuming a floc particle concentration, $N_p \approx 10^4$ particles/mL and $N_b/N_p \approx 12$, then $N_b \approx 1.2 \cdot 10^5$ bubbles/mL, where N_b is the bubble number concentration.

8.3.4.4 Concentration Expressions

Important concentration definitions (Edzwald, 1995, p. 9) concerning air supplied as precipitated gas include Φ_b, the volume concentration of dissolved gas; N_b, the bubble number concentration; and C_r, the mass concentration of bubbles precipitated from solution in the contact zone, also termed "released air." The definitions are

1. Bubble volume concentration Φ_b,

$$\Phi_b = \frac{C_r}{\rho(\text{air})} \tag{8.14}$$

in which
- Φ_b is the volume concentration of gas bubbles, e.g., air, "precipitated" from solution (m^3 gas/m^3 water)
- ρ(air) is the mass density of air or dry air (it is 1.204 kg dry air/m^3 water at 20°C)
- C_r is the mass concentration of the air bubbles that must be generated to satisfy the demand by the particle number concentration in order to float the particles; also termed, the air "released" by gas precipitation (kg gas/m^3 water)

2. Bubble number concentration N_b,

$$N_b = \frac{\Phi_b}{\pi d_b^3/6} \tag{8.15}$$

in which
- N_b is the bubble number concentration (# bubbles/m^3 water)
- d_b is the mean diameter of air bubbles (cm)

After substituting (8.14) in (8.15),

$$N_b = \frac{6C_r}{\rho(\text{air}) \cdot \pi d_b^3} \tag{8.16}$$

The required bubble number concentration, N_b, is the product of the particle concentration times the bubbles required per particle, i.e.,

$$N_b = N_p \cdot B \tag{8.17}$$

in which B is the bubble numbers per particle (# bubbles/particle).

After substituting (8.17) in (8.16) and solving for C_r,

$$C_r = N_p \cdot B \cdot \rho(\text{air}) \cdot (\pi d_b^3/6) \tag{8.18}$$

The value of C_r determines the value of C(saturator), the two being related by mass balance about the flotation tank, as outlined in the section following. Once C(saturator) is determined, P(saturator) may be calculated by Henry's law. Other variables are described in Equation 8.19 and Table CD8.3. Example 8.6 illustrates calculation of C_r based on Equation 8.18.

Example 8.6 Calculation of Mass Density of Air Bubbles, C_r

Given

Let $N_p = 1.2 \cdot 10^4$ particles/mL $= 1.7 \cdot 10^{10}$ particles/m^3; $B = 10$ bubbles/particle;

$$\begin{aligned} r(\text{air}) &= (P/RT) \cdot MW(\text{gas})/1000 \\ &= 101325 \text{ Pa}/(8.31451 \text{ Nm/K mol} \cdot 293.15 \text{ K}) \\ &\quad \cdot (28.9641 \text{ g/mol}/1000 \text{ kg/mol}); \\ &= 1.204 \text{ kg air/m}_3 \text{ gas}; \\ d(\text{mean})_b &= 40 \cdot 10^{-6} \text{ m/bubble.} \end{aligned}$$

Required

Mass density of air bubbles, C_r, i.e., "released air."

TABLE CD8.3

Calculation of Required Saturator Pressure to Float Solids for Stated Conditions by Mass Balance

(a) Gas law data and calculations to obtain density of air for standard conditions and Henry's constant for air at any given temperature

P(atm) = 101325	Pa		k = 1.4	
R = 8.31451	Nm/K mol			
T = 20	C =	293.15		K

Gas	MW (g/mol)	X(gas) (mol fraction)	ρ(molar) (mols/m³)	ρ(mass) (kg/m³)	A	B	H(gas i) (mg/L/atm)	H(gas i) (kg/m³/atm)	H(gas i) (kg/m³/kPa)			
N_2	28.0134	0.7808400	32.46027	0.90932	27.593	−0.01710	19.60					
O_2	31.9988	0.2094760	8.70812	0.27865	64.750	−0.01862	44.62					
CO_2	28.0104	0.0003140	0.01305	0.00037	3129.9	−0.02955	1733.25					
Ar	39.948	0.0093400	0.38827	0.01551	64.750	−0.01862	44.62					
Other gases		0.00002715								oxygen	47.8	10.0
Air	28.9641	0.9999971	41.57084	1.20406			25.61	0.02561	0.0002528	nitrogen	21.0	16.4
Other gases are:	Table H.1		ρ(molar) = n/V = P/RT		Table H.7	Table H.7	H = A · exp(B · T°C)					
Ne, He, I, Xe,	See also Table B.7	0.99032		ρmass = ρ(molar) · MW(gas)/1000			A and B for argon is an estimate based on			Air	26.4	26.4
CH_4, H_2	Equivalent MW(air) based on composition of air using weighted average of MW and X values for different gases.						web information that argon solubility is about the same as oxygen					

(b) Vapor pressure

T (°C)	P(vapor) (kPa)	ρ(vapor) (kg vapor/m³ gas)
20	2.33847	0.000017
20	2.33847	0.000017
Assumed	P(water vapor) = M0 +M1 · T +M2 · T^2 +M3 · T^3+M4 · T^4 +M5 · T^5+M6 · T^6 i.e., Equation H.27	ρ(mass) = (P/RT) · MW(gas)/1000

(*continued*)

TABLE CD8.3 (continued)
Calculation of Required Saturator Pressure to Float Solids for Stated Conditions by Mass Balance

(c) Calculation of air concentration to float solids

ρ(air) (kg air/m³ gas)	d_b (m/bubble)	Mass(bubble) (kg air/bubble)	N_p (#/mL)	(#/m³)	B (bubbles/particle)	N_b (bubbles/m³)	ϕ_b (m³ air/m³ water)	C_r (kg air/m³ water)
1.20406	4.E-05	4.03486E-14	12000	1.2.E+10	10	1.2E+11	0.0040	0.0048
1.20406	4.E-05	4.03486E-14	17000	1.7.E+10	10	1.7E+11	0.0057	0.0069
Density of air calculated from PV = nRT ρ(molar) = n/V = P/RT ρ(mass) = ρ(molar) · MW (gas)/1000 = (P/RT) · MW(gas)/ 1000-not corrected for Water-vapor pressure in air bubbles	Diameter of bubble—assumed	Mass(bubble) = ρ(air) · V(bubble)	N_p is the particle number concentration. Edzwald (1995, p. 13) gave a range of 10^3–10^5 particles/mL The value used was based on matching calculations of Edzwald (1995, p. 9)		B is the number of bubbles generated per particle and is based on Figure 8.10 (Edzwald, 1995, p. 14) showing rise velocity of bubble–particle agglomerate as function of B and dp	Nb is the "bubble-number concentration" $N_b = N_p \cdot B$ $1.6*10^{11}$-ok	$\phi_b = N_b \cdot (\pi d_b^3/6)$ 0.006—ok ϕ_b was termed "air-bubble-volume concentration" by Edzwald (1995, p. 8)	$C_r = Nb \cdot \pi d_b^3/6)$ $\cdot$ ρ(air) 0.007—ok C_r was termed, "air released" by Edzwald (1995, p. 8) and is the airbubble '-mass concentration *check: this is only 6.9 mg/L "required," as calculated to float the floc at B = 10

(d) Gas concentration, C_a, by Henry's law for 20 °C

Z(elevation) (m)	D(nozzles) (m)	P(atmosphere) (kPa)	P(nozzle depth) (kPa)	$H(O_2$, T °C) (mg O_2/L water/atm O_2)	$H(N_2$, T °C) (mg N_2/L water/atm N_2)	H(air, 20 C) (kg air/m³ water/kPa air)	C_a (kg air/m³ water)	C_o (kg air/m³ water)
0	1.00	98.987	108.569	44.62	19.60	0.000243	0.0264	0.024
1585	3.00	80.693	104.127	44.62	19.60	0.000243	0.0253	0.024
Elevation above sea level	Depth of manifold/nozzles below water surface	Atmospheric pressure as a function of Z, i.e., P(atmosphere) = 101325 · 10^ (−0.00005456 · Z) '-P(water vapor). Note correction for water-vapor pressure in using Henry's law	Absolute pressure at depth of nozzles, i.e., P(nozzle depth) = P(atm) · (1+D/10.33) where 10.33 is the water depth equal to one atmosphere of pressure	Calculated as from best fit of empirical data, i.e., $H(O_2)$ = $A(O_2)$ · exp($B(O_2)$ · T°C) Data reference is Table H.6	Calculated as from best fit of empirical data, i.e., $H(N_2)$ = $A(N_2)$ · exp($B(N_2)$ · T°C) Data reference is Table H.6	H(air) calculated as weighted molar fraction for $H(O_2)$ + $H(N_2)$, i.e., H(air) ≈ $X(O_2) \cdot H(O_2)$ + X $(N_2) \cdot H(N_2)$- as in Example 8.4	Saturation concentration of air with respect to pressure in air bubbles at depth of nozzles, I.e., C_s = H (air) · P(nozzle depth)	Concentration of air entering system (assumed to be saturated with respect to atmosphere at elevation of basin), i.e., C_o = H (air) · P(atm)

First Row Inputs					Second Row Inputs				
(e) Calculation of C(saturator)			(f) P(saturator)		(e) Calculation of C(saturator)			(f) P(saturator)	
r	K	C(saturator) (kg air/m^3 water)	f	P(saturator) (kPa)	r	k	C(saturator) (kg air/m^3 water)	f	P(saturator) (kPa)
0.05	0.002	0.175	0.90	798	0.05	0.00	0.19	0.70	1141
0.10	0.002	0.103	0.90	470	0.10	0.00	0.11	0.70	665
0.15	0.002	0.079	0.90	361	0.15	0.00	0.09	0.70	507
0.20	0.002	0.067	0.90	307	0.20	0.00	0.07	0.70	427
0.25	0.002	0.060	0.90	274	0.25	0.00	0.06	0.70	380
0.30	0.002	0.055	0.90	252	0.30	0.00	0.06	0.70	348
0.35	0.002	0.052	0.90	236	0.35	0.00	0.06	0.70	325
0.40	0.002	0.049	0.90	225	0.40	0.00	0.05	0.70	308
0.45	0.002	0.047	0.90	216	0.45	0.00	0.05	0.70	295
0.50	0.002	0.046	0.90	208	0.50	0.00	0.05	0.70	284
0.55	0.002	0.044	0.90	202	0.55	0.00	0.05	0.70	276
0.60	0.002	0.043	0.90	197	0.60	0.00	0.05	0.70	268
0.65	0.002	0.042	0.90	193	0.65	0.00	0.04	0.70	262
0.70	0.002	0.042	0.90	190	0.70	0.00	0.04	0.70	257
0.75	0.002	0.041	0.90	186	0.75	0.00	0.04	0.70	253
0.80	0.002	0.040	0.90	184	0.80	0.00	0.04	0.70	249
0.90	0.002	0.039	0.90	179	0.90	0.00	0.04	0.70	242
1.00	0.002	0.038	0.90	176	1.00	0.00	0.04	0.70	237
r is the recycle ratio, i.e., r = R/Q, and is an assumed value	k = C_a–C_o	C(saturator) is the dissolved air concentration leaving the saturator in the recycle flow, R	$C_r = \frac{r[C(\text{saturator}) - C_a] - (C_a - C_o)}{(1+r)}$ f is the saturator efficiency, i.e., C(saturator) = f · H(air) · P(saturator)		r is the recycle ratio, i.e., r = R/Q and is an assumed value	k = C_a–C_o	$C_r = \frac{r[C(\text{saturator}) - C_a] - (C_a - C_o)}{(1+r)}$ C(saturator) is the dissolved air concentration leaving the saturator in the recycle flow, R	f is the saturator efficiency, i.e., (mass flow of dissolved air leaving saturator)/(mass flow of air into saturator)	C(saturator) = f · H(air) · P(saturator)

Solution
The calculation is by Equation 8.18

$$C_r = N_p \cdot B \cdot r(\text{air}) \cdot (\pi d_b{}^3/6) \quad (8.18)$$
$$= (1.2 \cdot 10^{10}\ \text{particles/m}^3\ \text{water})$$
$$\cdot (1.204\,\text{kg air/m}^3\ \text{gas} \cdot (10\ \text{bubbles/particle})$$
$$\cdot \pi \cdot (40 \cdot 10^{-6}\ \text{m/bubble})^3/6$$
$$= 0.0048\,\text{kg air/m}^3\ \text{water}$$

Comments
A spreadsheet would facilitate computations.

8.3.5 Materials Balance for Dissolved Gas in Flotation Basin

Figure 8.11 shows a schematic drawing of a DAF basin and saturator. The associated materials balance for dissolved air in the flotation basin is formulated.

8.3.5.1 Mass Balance for Flotation Basin

Figure 8.11 depicts a flotation basin with recycle, showing mass flows in and out for the boundary shown (i.e., that excludes the saturator),

$$QC_o + RC(\text{saturator}) = (Q+R)C_r + (Q+R)C_a \quad (8.19)$$

Dividing by Q and rearranging, gives

$$C_r = \frac{r[C(\text{saturator}) - C_a] - (C_a - C_o)}{(1+r)} \quad (8.20)$$

in which
Q is the flow of water into system (m^3/s)
R is the flow of recycle water through saturator (m^3/s)
r is the ratio, Q/R (dimensionless)
C_o is the mass concentration of dissolved gas coming into system in flow, Q (kg gas/m^3)

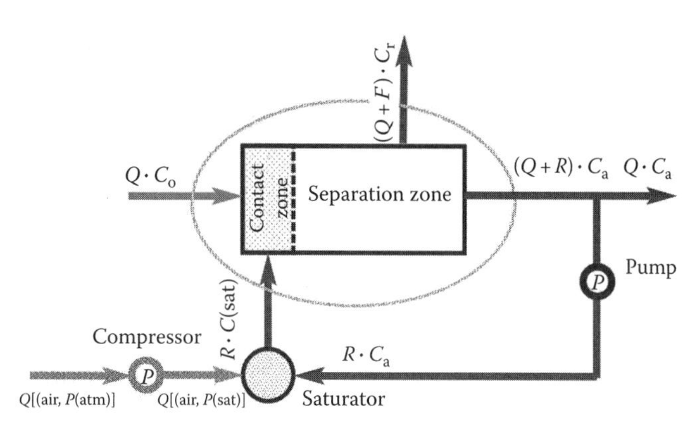

FIGURE 8.11 Materials balance for DAF system with recycle. (Adapted from Edzwald, J.K., *Water Sci. Technol.*, 31(3–4), 1, 1995.)

$C(\text{saturator})$ is the mass concentration of dissolved gas leaving saturator in recycle flow, R, then flowing through the manifold and nozzles; the "excess" concentrations of the gases are precipitated in the expansion part of the nozzle and then enter the contact zone of the flotation basin (kg gas/m^3 water) as bubbles
C_a is the mass concentration of dissolved gas leaving contact zone and then the separation zone, being transported in the flow, $(Q+R)$, which also leaves the basin (kg gas/m^3)
C_r is the mass concentration of dissolved gas precipitated as bubbles in the expansion part of the nozzles, which then enter the contact zone, after which they rise as bubbles and bubble–particle agglomerates in the separation zone, being transported from the contact zone in the combined flow, $(Q+R)$ (kg gas/m^3)

Comments on values of terms in Equation 8.20

- r is a parameter with values assumed in order to calculate effect on $C(\text{saturator})$; typically, the range is $0.05 \le r \le 0.15$ (Edzwald, 1995, p. 9).
- C_o is the concentration of dissolved air entering the flotation tank. In water treatment, C_o, may be taken as the saturation concentration with respect to atmospheric pressure at the elevation above sea level of the flotation tank, e.g., $C_o = H(\text{air}) \cdot P(\text{atm})$; in wastewater treatment, however, C_o would be zero.
- C_a is the dissolved gas concentration leaving the tank and is the saturation concentration with respect to atmospheric pressure plus the pressure at the depth of the manifold/nozzles, e.g., $C_a = H(\text{air}) \cdot P(\text{atm}) \cdot [1.0 + (\text{nozzle depth}/10.33\ \text{m})]$. Note that 10.33 m is the depth of water that exerts a pressure of 101.325 kPa, i.e., 1 atm pressure at sea level. The effect of the nozzle-depth can probably be ignored since a portion of the dissolved gases will be lost due to mass transfer to the atmosphere; based on this assumption, $C_a \rightarrow C_o$, and so, $C_a \approx C_o$. At the same time, C_a is the criterion for gas precipitation in a flotation tank; any higher values of C_a in solution will result in gas precipitation and so C_a is the gas concentration limit.
- C_r is the "released air" required to float the solid and is determined by a calculation procedure described in the previous section, i.e., by Equation 8.20.

Typically, $3.5 \cdot 10^{-3} \le C_r \le 10 \cdot 10^{-3}$ kg air/m^3 water (Edzwald, 1995, p. 9)

- $C(\text{saturator})$ is the concentration of dissolved air in the flow, R, leaving the saturator and is calculated by Equation 8.20. Once $C(\text{saturator})$ is determined, $P(\text{saturator})$ may be calculated by Henry's law, as described previously, i.e., Equation 8.5.

8.3.5.2 Mass Balance Calculations by Spreadsheet

Table CD8.3 shows the sequence of calculations that permit determination of P(saturator) for various r. The algorithm utilizes the materials balance equation as basis for calculating the required C_r. The spreadsheet is set up to permit the exploration of "what-if" scenarios, with respect to temperature, elevation, N_p, B, and r. Note that the best-fit polynomial equations for the effect of temperature on $H(O_2)$ and $H(N_2)$, from Table H.5, was applied to approximate H(air, T).

Example 8.7 Rational Design

The purpose of this example is to illustrate a design algorithm based on theory, as described in the previous sections. A similar algorithm is used in Table CD8.3.

Given

$Q = 0.0876\ m^3/s$ (2.0 mgd)	Recycle ratio, $r = 0.10$
$d_P = 50\ \mu m$ (average)	Saturator efficiency, $f = 0.9$
$d_b = 40$	Elevation = sea level, i.e., $Z = 0.00$ m
$N_P = 1.2 \cdot 10^4$ particles/mL	$T = 20°C$
$B = 10$ bubbles/particle	

Required

A(basin), P(sat)

Solution

1. *Determine average rise velocity of particle–bubble combination*:
   v_{pb}(10 bubbles/particle, $d_p = 50\ \mu m$) = 13.7 m/h (Table CD8.2)
2. *Determine area of separation zone, i.e., A*(basin)
   - Enter Figure 8.10 with arguments, $d_p = 50\ \mu m$, $B = 10$ bubbles/particle to obtain, $v_o = 13.7$ m/h = 0.0038 m/s
   - Calculate A(basin):

$$v_o = \frac{Q}{A(\text{basin})}$$

$$0.0038\ \text{m/s} = \frac{0.0876\ \text{m}^3/\text{s}}{A(\text{basin})}$$

$$A(\text{basin}) = 23.0\ \text{m}^2$$

3. *Proportion length to width; let $L/w \approx 10{:}1$, i.e.,*

$$w(\text{basin}) = 2.0\ \text{m}$$
$$L(\text{basin}) = 12.0\ \text{m}$$

4. *Depth*
   From experience, depth is about 3–5 m; let $D = 3.0$ m.
5. *Detention time, θ*:

$$V = 3\ \text{m} \cdot 24.0\ \text{m}^2 = 72.0\ \text{m}^3$$

$$\theta = V/Q = 72.0\ \text{m}^3/0.0876\ \text{m}^3/\text{s} = 822\ \text{s} = 14\ \text{min}$$

6. *Determine C_r*:
   (1) Calculate N_b for $N_b/N_p = B = 10$,

$$\frac{N_b}{1.2 \cdot 10^4\ \text{particles/mL}} = 10$$

$$N_b = 1.2 \cdot 10^5\ \text{bubbles/mL}$$

   (2) Recalling $F_b = C_r/r(\text{air})$, calculate C_r from (8.16),

$$N_b = \frac{\Phi_b}{\pi d_b^3/6} \tag{8.16}$$

$$1.2 \cdot 10^{11}\ \frac{\text{bubbles}}{\text{m}^3\ \text{water}} = \frac{C_r/(1.204\ \text{kg air/m}^3\ \text{air})}{[\pi(40 \cdot 10^{-6}\ \text{m})^3/6]/\text{bubble}}$$

$$C_r = 4.785 \cdot 10^{-3}\ \text{kg air/m}^3\ \text{water}$$

7. *Determine C*(saturator):
   From the mass balance Equation 8.20, with arguments, $C_r = 4.785 \cdot 10^{-3}$ kg/m^3 and $r = 0.10$, C(saturator) is calculated,

$$C_r = \frac{r[C(\text{saturator}) - C_a] - (C_a - C_o)}{(1 + r)}$$

$$4.785 \cdot 10^{-3} = \frac{0.10 \cdot [C(\text{saturator}) - 0.031] - (0.031 - 0.024)}{(1 + 0.10)}$$

$$5.26 \cdot 10^{-3} = 0.10 \cdot [C(\text{saturator}) - 0.031] - (0.031 - 0.024)$$

$$0.0526 = [C(\text{saturator}) - 0.031] - 0.07$$

$$0.0526 = C(\text{saturator}) - 0.101$$

$$C(\text{saturator}) = 0.154\ \text{kg air/m}^3\ \text{water}$$

8. *Note that C_a and C_o in Equation 8.20 are calculated as follows*:
   Assume D(nozzles) = 3.0 m; therefore, C_a is calculated by Henry's law, i.e.,

$$C_a = H(\text{air}) \cdot [P(\text{atm}) + (D(\text{nozzle})/10.33\ \text{m}) \cdot (101.325\ \text{kPa})]$$

$$C_a = (0.000238\ \text{kg air/m}^3\ \text{water/kPa air}) \cdot [(101.325 - 2.338)\ \text{kPa} + (3.00\ \text{m}/10.33\ \text{m}) \cdot (101.325\ \text{kPa})]$$
$$= 0.031\ \text{kg air/m}^3\ \text{water}$$

   Also, calculate C_o by Henry's law, i.e.,

$$C_o = H(\text{air}) \cdot [P(\text{atm}) - \text{vapor pressure}]$$
$$= (0.000238\ \text{kg air/m}^3\ \text{water/kPa air}) \cdot [(101.325 - 2.338)\ \text{kPa}]$$
$$= 0.0236 = 0.3056\ \text{kg air/m}^3\ \text{water} \tag{Ex8.7.1}$$

9. *The saturator pressure, P*(saturator) *is also calculated by Henry's law, i.e.,*

$$C(\text{saturator}) = f \cdot H(\text{air}) \cdot P(\text{saturator})$$

$$0.154\ \text{kg air/m}^3\ \text{water} = 0.9 \cdot (0.000238\ \text{kg air/m}^3\ \text{water/kPa air}) \cdot P(\text{saturator})$$

$$P(\text{saturator}) = 719\ \text{kPa absolute}$$
$$= 618\ \text{kPa gage}$$

Comment

The value for $v_o = 13.7$ m/h compares with the range for practice, $0.05 \leq v_o \leq 100$ m/h (Table 8.2). The value for P(saturator) = 618 kPa gage is at the upper end of the range for practice, $300 \leq P(\text{sat}) \leq 600$ kPa gage (Table 8.4).

TABLE 8.4
Criteria from Practice in Five Countries for Dissolved Air Flotation

Parameter	Range[a]	Typical[a]	South Africa[b]	Finland	The Netherlands	UK	Scandinavia
Flocculation							
Alum dose (mg/L)	5–30	20					
G (s^{-1})	10–150	70	50–120				
Time (min)	5–15	10	4–15	20–127	8–16	20–29	28–44
Separation zone							
SOR[c] (m/h)	5–15	8	5–11	2.5–8	9–26		
Time (min)	5–15	10				11–18[d]	
Depth (m)	1.0–3.2	2.4	2.5–3.0				
Freeboard (m)	0.1–0.4	0.3					
Bubble size, d_p (μm)	10–120	40–50					
Recycle ratio (%)	6–30	6–12	6–10	6–42	6–15	6–10	10
Unpacked saturator							
P(sat) (kPa gage)			400–600			400–500[d]	460–550
HLR (m/h)			20–60				
Time (s)			20–60				
Efficiency, f		0.90					
Packed saturator							
P(sat) (kPa gage)	350–620	485	300–600			400–500[d]	
HLR (m/h)			50–80				
Packing depth (m)			0.8–1.2				
Efficiency, f		0.90					
Float layer solids							
Percent solids	0.2–6	3					

Source: Adapted from Edzwald, J.K., *Water Sci. Technol.*, 31(3–4), 16, 1995. With permission.

[a] Design parameters are from Edzwald and Walsh (1992, p. 2) who give the "range" and "typical" values from practice as compiled from three European sources.

[b] Recommended minimum and maximums by Haarhoff and van Vuuren (1995) based on review of 14 plants in South Africa used for drinking clarification; 12 other plants which included sludge thickening were included in the survey, but guidelines were not included in this abstract.

[c] SOR $= (Q + R)/A$(flotation zone); A(flotation zone) is the plan area of the flotation zone.

[d] From Edzwald (1996, p. 16).

Note that the spreadsheet, Table CD8.3, shows that if D(nozzles) = 0.0 m, where $C_a = C_o$, then P(saturator) = 353 kPa absolute. In other words, with a higher value for C_a, there is a lower concentration of "excess" dissolved gases available for "release" as bubbles. Thus, placing the nozzles at low depth of submergence translates to lower saturator pressure and lower operating cost. The spreadsheet, Table CD8.3, facilitates iterations for such variables.

8.4 PRACTICE

Flotation practice has evolved from industrial wastewater applications in the 1950s to include solids thickening in the 1960s and then water treatment applications in the 1980s. Then, during the 1990s, theory evolved to supplement empirical guidelines.

8.4.1 Design Criteria

Design and operating criteria include recycle ratio, saturator pressure, saturator depth, saturator packing, coagulant dosage, flocculation intensity, overflow rate, detention time, and air-to-solids ratio. Most criteria have been established by practice, with theory providing rationale. A pilot plant study may refine empirical guidelines and provide more certainty to design and operation. If a proprietary package plant is used, guidelines are incorporated, as a rule, in a manufacturer's recommendation.

8.4.1.1 Flotation in Water Treatment

Table 8.4 summarizes a range of design criteria in water treatment practice from five countries. The parameters cover flocculation, the contact zone, the separation zone, the recycle ratio, and both packed and unpacked saturators.

Concerning the separation zone, empirical guidelines from Vrablik (1960) were: overflow rate 3.7–6.0 m/h, which compares with 5–11 m/h for South Africa, 2.5–8 m/h for Finland, and 9–16 m/h for the Netherlands in Table 8.8 for the 1990s. Vrablik indicated saturator pressures in the range of $3 \leq P(\text{saturator}) \leq 8$ kPa gage, depending on the recycle ratio, r, with $0.0 \leq r \leq 0.5$ ($r = 0$ indicates that the full flow, Q, passed through the saturator).

8.4.1.2 Flotation for Sludge Thickening

In thickening, the solids loading rate is more important than SOR (Haarhoff and van Vuuren, 1995, p. 209). Guidelines for solids loading are (1) without coagulation, 2.0–6.0 kg/m^2/h; and (2) with coagulants, 6.0–12.0 kg/m^2/h.

8.4.1.3 Air-to-Solids Ratio

The air-to-solids ratio (i.e., A/S) is an empirical parameter defined as the ratio of the mass fluxes of air and solids, respectively, where A is the mass flux of air from the saturator, and S is the mass flux of solids entering the "contact zone," i.e.,

$$A = R \cdot C(\text{saturator})(\text{kg air in flow from saturator/s})$$
$$S = Q \cdot C(\text{floc})(\text{kg solids/s})$$

Thus, $A/S = R \cdot C(\text{saturator})/Q \cdot C(\text{floc})$ with units (kg air/kg solids). As stated in Section 8.1.2, the parameter was used first by Eckenfelder et al. (1958, p. 257) who gave limits as, $0.03 \leq A/S \leq 0.10$ kg air/kg solids. They also presented empirical plots showing that increasing A/S results in increasing percent solids in the float layer. Matsui et al. (1998, p. 16) reviewed the use of the A/S parameter by Eckenfelder et al., and recommended $A/S \gg 0.01$, which ensures an excess of air bubbles relative to the volume needed to float the floc particles. The A/S ratio is similar in concept to the bubble-to-particle ratio.

8.4.2 Pilot Plants

A pilot plant can address design issues such as flocculation turbulence intensity, basin sizing, and saturator design. Operation questions involve selection of coagulants and dosages, including polymers, sensitivity to flow variation, airflow required, and recycle flow.

8.4.2.1 Pilot Plant Study

Figure 8.12 shows sketches of equipment for subunit processes, i.e., rapid mix, flocculation, and saturator, respectively, indicating some of the variables of the flotation process. Table 8.5 lists dependent and independent variables for the respective unit processes. The dependent variables include those that are the outcome of change in the independent variables. A dependent variable for one unit process may serve as an independent variable for another.

Looking at the variables in Table 8.5, it is clear that a large number of functional relationships could be involved. To be economical, a study would have to be limited to only the dependent/independent variables relevant to the design and

TABLE 8.5
List of Variables for Pilot Plant Study

	Variables	
Unit Process	**Dependent**	**Independent**
Rapid mix	Zeta potential	θ, G
		Metal coagulant selection
		Coagulant dosage
		Polymer selection
		Polymer dosage
Flocculation	Floc size	N(compartments)
	Floc strength	θ
	$C(\text{particles})_{\text{out}}$	P/V
Saturator	$C(O_2)_e$	R, P(sat), HLR, K_La
		Packing type
		L(packing)
		Pump sizing (head, flow)
		Compressor sizing
Flotation	$C(\text{particles})_e$	Floc size
		Floc strength
		$C(\text{particles})_{\text{in}}$
		$C(O_2)_{\text{sat}}$
		R
		Q
		v_o
		Tank design

Notes: (1) In lieu of "air," dissolved oxygen may serve as a surrogate since its concentration may be measured easily by instrument of by the "Winkler" method (see *Standard Methods*).

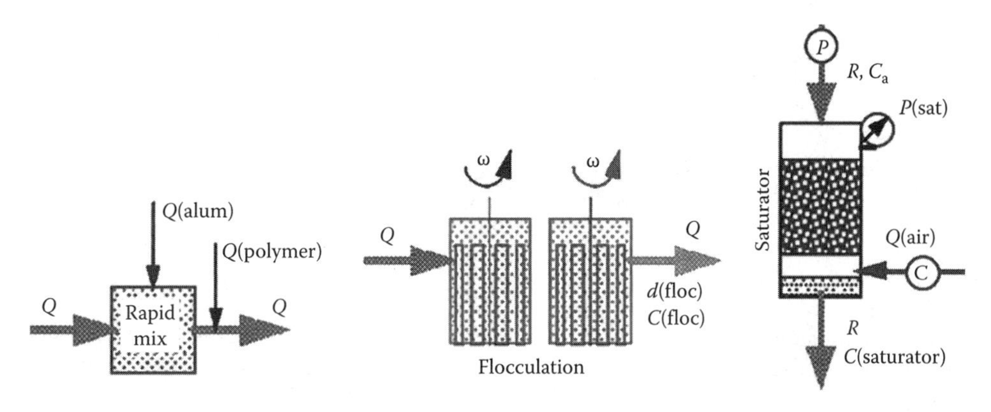

FIGURE 8.12 Sketch showing variables for pilot plant study.

operation, and to the time and money available. As a rule, a study requires a proposal that describes the purpose (why); the scope in terms of objectives (what), work to be done, estimated time, and budget; and the personnel to be involved. A party who has budget authority must commit to the study. Example 8.8 suggests some considerations on how a pilot plant study might be conducted.

Example 8.8 Pilot Plant Study for Design of an Air Flotation System

Outline a pilot plant study to design a DAF system to remove a flocculent suspension. Identify both relevant independent variables and relevant dependent variables to be measured. Indicate how you would use the results in sizing the dissolved air portion of the system and in sizing the basin.

Solution

The dependent variables must be identified first, and then the independent variables. Before going ahead, however, how a flotation basin functions will be reviewed.

1. *Identify dependent variables*: The flotation unit must remove a portion of the flocculent suspension. The primary dependent variable then will be the concentration of the suspension in the effluent, *C*.
2. *Identify independent variables*: The dependent variables include
   2.1 Flotation variables
   - Q, flow of water into the system (m^3/s)
   - C_o, influent suspension concentration (mg/L)
   - [gas i], concentration of gas i, such as oxygen, i.e., $[O_2]$, in (mg gas/L)
   - C_o, influent suspension concentration (mg/L)
   - $[gas]/C_o$, ratio of dissolved gas concentration to influent solids concentration; oxygen would be the most convenient gas to measure as an indicator (mg ss/L/mg gas/L)
   - v_o, overflow rate ($m^3/m^2/s$)
   - θ, detention time in basin (s)
   - w, L, d, width, length, and depth of flotation basin, respectively (m)

   2.2 Variables that affect dissolved gas concentrations, i.e., [gas]:
   - Q_a, flow of air into the pressurized gas transfer vessel (gr/s)
   - P, pressure of water (and the gas bubbles) at diffusers (atm)

   The fundamental independent variables include [Co/[gas] and overflow velocity, v_o. Detention time, θ, is not a factor (except Z_D is a primary variable).
3. *Pilot plant study*: Variables to be measured in a pilot plant study include those listed in 2, with $[gas]/C_o$ and v_o being calculated. The pilot plant study would seek to determine C as a function of $[gas]/C_o$ and v_o. The effect of [Co/[gas] would be examined first. Then when the "optimum" $[gas]/C_o$ is established, that value would be used and held constant as the effects of v_o are investigated. Since the pilot plant will have fixed dimensions, the only way to vary v_o is to change Q (holding other variables constant). If the depth can be changed, then the effect of θ can be investigated also (θ is not a relevant variable, however, since the reactor is not homogeneous). The dissolved air portion of the design can be investigated by varying pressure and airflow and Z_D, or *L*(saturator), to investigate their respective effects on dissolved gas concentration, such as $[O_2]$.
4. *Plots*: The relationships expected include C as a function of $[gas]/C_o$, and of the hydraulic variable, vo. For the dissolved air design, $[O_2]$ will be plotted as a function of P, Q_a. From these relationships the basin can be sized and the dissolved air system can be sized (which would include water pump to give the needed pressure, the air compressor, and the vessel size).

Discussion

Designing a pilot plant study is an art since many judgments must be made and experience helps to converge on the key issues. Any pilot plant study could become a full research study with useful design knowledge being generated. Budget and time are always limited, however, and so the study should develop priorities.

8.4.3 Case: Birmingham

The Frankley Water Treatment Plant provides water to the City of Birmingham, in the United Kingdom and is operated by Severn Trent Water (data abstracted from Schofield, 1996). The plant capacity is 450,000 m^3/day, serving a population of 1.15 million persons. Investigations started in 1987 to look at treatment options related to a redevelopment plan to provide new treatment facilities to ensure compliance with a European Community Drinking Water Directive. Based on pilot plant studies, ferric sulfate was selected as the coagulant for the removal of color and organics at a pH of about 5.5. The surface loading rate was set at $v_o \leq 11$ m/h. The maximum flotation area was set at 100 m^2, with maximum width, $w \leq 8.5$ m, with $d \geq 1.5$ m. With $A(\text{plan}) = 100$ m^2, and $v_o = 11$ m/h, $Q(\text{flotation cell}) = 26{,}400$ m^3/day, requiring 17 cells (which was set at 20 cells for the design).

The flow scheme started with two raw water streams passing through static mixers, where lime and/or carbon dioxide are added for pH control. Ferric sulfate is added just upstream from a measuring weir for each inlet channel, with the coagulant mixing provided by the turbulence below the weir. The flow is then distributed to the 20 DAF cells. Each DAF unit has a flocculation basin comprising three compartments each with vertical variable speed paddle wheels. The detention time was 30 min, and with $25 < G \leq 80$ s^{-1}. The flocculated water is transported to the contact zone of the DAF basin located midway in length such that the cell is divided into two half-cells where the flow is directed upward to the contact zone. In the contact zone, the saturated water at 400 kPa pressure is injected through a series of needle valves.

A surface float is removed periodically by raising the water level a cell and then scrapping the float to a collecting trough where it is removed as "sludge." The sludge is further dewatered to 25% solids by means of a filter press so that it is acceptable for landfill disposal. The final phase in commissioning the new facilities was 12 months of performance testing for optimization of operation.

8.4.4 Equipment

Figure 8.1 indicates major components of the DAF technology. Supporting appurtenances are not shown and are reviewed here. First, the recycle line requires flow measurement and a valve for flow control, as well as a pump for pressurizing the saturator. Dissolved oxygen (DO) probes should be located in the influent and effluent streams, with DO serving as a surrogate for dissolved air. Dissolved nitrogen may be assumed to be present with concentration calculated by the product, $H(N_2) \cdot X(N_2) \cdot P(\text{saturator})$. The airflow to the saturator also requires flow measurement and a valve for flow control and a compressor.

The DAF tank requires a nozzle so that the gas precipitation occurs before mixing with the feed flow. The feed flow requires rapid-mix/coagulation and flocculation prior to entering the contact zone. Once the floc–bubble attachments occur and the agglomerates float to the surface, the float so formed must be removed by a skimmer blade assembly that transfers the solids to a collection trough. The effluent flow is collected near the bottom of the tank with a portion becoming recycle flow.

Usually, for small systems, designs provide a complete "package" with the necessary appurtenances. Table 8.6 gives dimensions of several package plants from one manufacturer that illustrates tank sizes for different flows. Figure 8.13a and b shows photographs of the saturator tank and the pump, respectively, for a proprietary DAF unit used for waste-activated sludge thickening prior to anaerobic digestion.

TABLE 8.6
Dimensions of Package Plants

Model	Design (m^3/s)	Flow (mgd)	*L* (m)	*L* (ft)	*W* (m)	*W* (ft)	*D* (m)	*D* (ft)
LFT-250	0.011	0.25	6.83	22.42	2.34	7.67	2.90	9.5
LFT-500	0.022	0.50	9.55	31.33	2.95	9.67	2.90	9.5
LFT-750	0.033	0.75	12.27	40.25	3.25	10.67	2.90	9.5
LFT-1000	0.044	1.0	13.64	44.75	3.56	11.67	3.35	11.0
LFT-1500	0.066	1.5	18.95	62.17	3.86	12.67	3.51	11.5
LFT-2000	0.088	2.0	21.72	71.25	4.32	14.17	3.51	11.5

Source: Adapted from Leopold, *Dissolved Air Flotation*, brochure, The F. B. Leopold Company, Inc. Zelienople, PA, 8pp., 1996 (revised 2001, 2003).

(a)

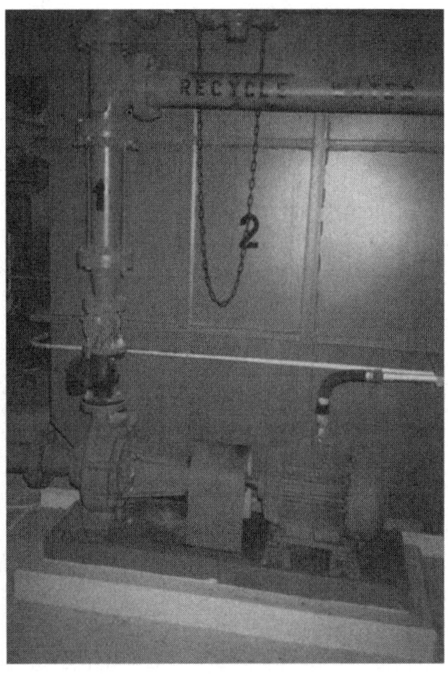

(b)

FIGURE 8.13 Photographs illustrating DAF components at Marcy Gulch Wastewater Treatment Plant: (a) saturator for a DAF sludge-thickening unit; (b) pump for pressurizing saturator for DAF sludge-thickening unit. (Courtesy of Centennial Water and Sanitation District, Highlands Ranch, CO.)

PROBLEMS

8.1 History

Problem

Summarize the difference between the state-of-the-art of flotation c. 1959 and in the 1990s (or other dates if you wish). Cite one or two papers for each period.

8.2 System Description

Problem

Summarize the flotation process, identifying key phases.

8.3 Dissolved Oxygen Concentration in Saturator by Henry's Law

The problem refers to a saturator; air is the gas, with 700 kPa total pressure imposed. Let $f = 1.0$ for the purpose of the calculation. Reference is Section 8.3.1.

Given

$T = 20°C$
$H(O_2) = 0.000428$ (kg oxygen dissolved/m^3 water/kPa oxygen) at 20°C
$X(O_2) = 0.209$ mol O_2/mole air (Table B.7, Composition of Air)
$P(\text{total}) = 700$ kPa (stated)

Required

Calculate the concentration of dissolved oxygen at equilibrium by Henry's law.

8.4 Dissolved Nitrogen Concentration in Saturator by Henry's law

The problem refers to a saturator; air is the gas, with 700 kPa total pressure imposed. Let $f = 1.0$ for the purpose of the calculation.

Given

$T = 20°C$
$H(N_2) = 19.01$ (mg nitrogen dissolved/L water/atm nitrogen)
$= 0.0190$ (kg nitrogen dissolved/m^3 water/atm nitrogen)
$= 0.0001875$ (kg nitrogen dissolved/m^3 water/kPa nitrogen)
$X(N_2) = 0.78084$ mol N_2/mole air
$P(\text{total}) = 700$ kPa

Required

Calculate the concentration of dissolved N_2 at equilibrium by Henry's law.

8.5 Dissolved Air Concentration in Saturator by Henry's Law

Given

Let $T = 20°C$, $P(\text{saturator}) = 700$ kPa, which results in $C(O_2) = 0.0628$ kg dissolved O_2/m^3 water and $C(N_2) = 0.10248$ kg dissolved N_2/m^3 water. Let $f = 1.0$ for the purpose of the calculation.

Required

Calculate the concentration of "air" leaving the saturator.

8.6 Dissolved Air Concentration in Saturator Using *H*(air)

Let the context of the problem be a saturator, with air as the gas, and 700 kPa total pressure imposed.

Given

$$T = 20°C$$
$$H(\text{air}) = H(O_2) \cdot X(O_2) + H(N_2) \cdot X(N_2)$$
$$= 43.49\ \text{mg}O_2/\text{L}/\text{atm} \cdot 0.209 + 19.01\ \text{mg}\ N_2/\text{L}/\text{atm} \cdot 0.78084$$
$$= 23.93\ \text{mg air}/\text{L}/\text{atm}$$
$$= 0.0002362\ (\text{kg air dissolved}/\text{m}^3\ \text{water}/\text{kPa air})$$
$$X(\text{air}) = 1.00\ \text{mol air}/\text{mol air}$$
$$P(\text{total}) = 700\ \text{kPa}$$

Required

Calculate the concentration of dissolved air at equilibrium by Henry's law.

8.7 Condition for Gas Precipitation

Given

Consider a flotation basin at sea level and $T = 20°C$,

- Atmospheric pressure, $P(\text{atm}) = 101.325$ kPa (1.00 atm)
- Let nozzle depth below water surface, $d(\text{noz}) = 3.0$ m (9.84 ft)
- Let $f \approx 0.9$ (saturator efficiency)

Required

Determine the minimum saturator pressure for gas precipitation in a flotation basin.

8.8 Separation Zone: Rise Velocity, v_{pb}, and SOR

Reference is Section 8.3.4.

Given

$B = 10$ bubbles/particle
$d_b = 40$ μm
$d_p = 50$ μm
$T = 20°C$
$\rho_p = 1.01$ g/cm^3 [= 1010 kg/m^3]

Required

Calculate the of rise velocity, v_{pb}, of a particle–bubble agglomerate for conditions stated.

8.9 Separation Zone: *A/S* ratio from Bubble-Particle Ratio

Reference is Section 8.3.4.

Given

$d_p = 50$ μm	$r = 0.10$
$N_p = 17000$ #/mL	$C(\text{saturator}) = 0.136$ kg air/m^3 water
$r(\text{floc}) = 1010$ kg solids/m^3 water	Table CD8.3

Required

Calculate the air-to-solids ratio, i.e., A/S.

8.10 Mass Balance and Released Air Required

Reference is Section 8.3.5 and Table CD8.3.

Given

$T = 20°C$	$N_p = 12{,}000$ #/mL
$d_b = 40$ μm	$B = 10$ bubbles/particle

Required

Calculate C_r, the "air released" or excess air required to float the floc particles.

8.11 Mass Balance and Maximum Air Concentration at Nozzles

Reference is Section 8.3.5 and Table CD8.3.

Given

$T = 20°C$
D(nozzles) = 40 μm

Required

Calculate C_a, the maximum "air" concentration at nozzle level and which leaves the flotation basin in the flows, Q and R, respectively (assuming no losses in dissolved gases due to interface mass-transfer exchanges).

8.12 Mass Balance and Recycle Ratio

Reference is Section 8.3.5 and Table CD8.3.

Given

The input variables for Table CD8.3 are

$T = 20°C$	$N_p = 12{,}000$ #/mL	$f = 0.90$
$d_b = 40$ μm	$B = 10$ bubbles/particle	
D(nozzles) = 40 μm	$r = 0.10$	

Calculation results from Table CD8.3 that go into the mass balance relation are

$C_r = 0.0048$ kg air/m^3 water
$C_a = 0.024$ kg air/m^3 water
$C_o = 0.0311$ kg air/m^3 water

Required

Calculate P(saturator) from Table CD8.3.

8.13 Mass Balance and Role of Temperature

The effect of temperature on P(saturator) and other dependent variables is explored through application of the spreadsheet, Table CD8.3.

Given

The input variables for Table CD8.3 are

(1) Independent variables (input to Table CD8.3)

Variable	Units	Magnitude of Variable		
T	(°C)	20	2	30
Z (elevation)	(m)	0	0	0
D(nozzles)	(m)	3.0	3.0	3.0
d_b	(μm)	40	40	40
d_p	(μm)	50	50	50
N_p	(#/mL)	12,000	12,000	12,000
B	(bubbles/particle)	10	10	10
r	(dimensionless)	0.10	0.10	0.10

Note that the independent variables other than temperature are the same.

Required

Examine the effect of temperature on N_b, C_r, C_a, C_o, P(saturator).

8.14 Mass Balance and Role of Elevation

The effect of elevation above sea level on P(saturator) and other dependent variables is explored through application of the spreadsheet, Table CD8.3. Reference is Section 8.3.5.

Given

The input variables for Table CD8.3 are shown with variation in Z.

(1) Independent variables (input to Table CD8.3)

Variable	Units	Magnitude of Variable		
T	(°C)	20	20	20
Z (elevation)	(m)	0	1,586	3,000
D(nozzles)	(m)	3.0	3.0	3.0
d_b	(μm)	40	40	40
d_p	(μm)	50	50	50
N_p	(#/mL)	12,000	12,000	12,000
B	(bubbles/particle)	10	10	10
r	(dimensionless)	0.10	0.10	0.10

Note that the independent variables other than elevation, Z, are the same.

Required

Examine the effect of elevation on N_b, C_r, C_a, C_o, P(saturator).

8.15 Sizing Plan Area of Basin by Empirical Criteria

Reference is Section 8.4.1 and Table 8.4.

Given

$Q = 0.044$ m^3/s (1.0 mgd)
$r = 0.10$
$2.5 \leq \text{SOR} \leq 26$ m/h (Table 8.4)

Required

A(flotation basin)

8.16 Sizing a System by Table of Equipment Sizes

Reference is Section 8.4.4 and Tables 8.6 and 8.4.

Given

Assume $Q = 0.044$ m^3/s (1.0 mgd).

Required

Select a package plant for the flow given.

ACKNOWLEDGMENTS

The photographs in Figure 8.13 were taken by the author and used by permission from the late Paul Grundeman, supervisor, Marcy Gulch Wastewater Treatment Plant, Highlands Ranch, Colorado. Dr. John Hendrick, General Manager of the district gave permission to use the photographs taken at the plant for this book (2010).

GLOSSARY

Air: A mixture of gases; nitrogen and oxygen comprise 0.9903 mole fraction of air. The other gases are Ar, CO_2, Ne, He, Kr, Xe, CH_3, and H_2, O_3, and Rn. Adding the mole fractions of argon and carbon dioxide to those of oxygen and nitrogen give 0.9999971. The molecular weights and mole fractions are given along with Henry's constants, where available.

	N_2	O_2	Ar	CO_2	Ne
MW	28.0134	31.9988	39.948	44.0098	20.1797
***X*(gas)**	0.78084	0.209476	0.00934	$3.14 \cdot 10^{-4}$	$1.818 \cdot 10^{-5}$
***H*(20°C)**	19.01	43.39		1688	
	He	**Kr**	**Xe**	**CH_3**	**H_2**
MW	4.0026	83.80	131.29	16.0428	2.01588
***X*(gas)**	$5.24 \cdot 10^{-6}$	$1.14 \cdot 10^{-6}$	$8.7 \cdot 10^{-8}$	$2 \cdot 10^{-6}$	$5 \cdot 10^{-7}$
***H*(20°C)**				23.18	1.603

Notes: (1) Data from Table H.1. (2) *H*(20°C) is Henry's constant at 20°C with units (mg gas i/L water/atm gas i).

Attachment: Bonding of one particle to another by an adhesion force. For two particles that have opposite charges, e.g., a negatively charged bubble and a positively charged floc particle (i.e., after coagulation), the force is electrostatic. For particles without charge, bonding is by van der Waal's force.

Attachment coefficient, α: Ratio of particle–bubble attachments to particle–bubble contacts.

Bifurcation: Division of a fluid flow.

Contact zone: Zone in flotation basin just after bubble formation and as the particle stream enters in the flow Q to provide the opportunity for bubble–particle contacts. The term "contact zone" was adopted by a consensus of persons discussing the terminology at the *1994 International Joint Specialized Conference on Flotation Processes in Water and Sludge Treatment*, Orlando, Florida, April, 1994. The term "contact zone" was favored over "reaction zone" at the conference; the definitions are the same.

DAF: Acronym for dissolved air flotation.

Detention time: Defined as the volume of a basin divided by flow, i.e., $\theta = V/Q$.

Diffuser: A device with many small orifices for flow of a fluid for the purpose of mixing with another fluid.

Diffusion: Transport of material by random motion, such as molecular diffusion caused by the thermal motion of molecules. Turbulent diffusion is due to the random motion of turbulence.

Dissolved air flotation: Flotation process in which the air bubble source is dissolved air that "precipitates" in a zone of lower pressure where contacts are made with the solid particles.

Film thickness: For any motion of a fluid across a surface, a velocity gradient exists with zero at the surface to a finite value in the bulk of flow; the distance is called the "boundary layer." If such a fluid motion exists across a concentration gradient, say in water, then the distance from the gas–water interface where the concentration could be say that in equilibrium with a gas, to a point in the bulk of the solution where the concentration gradient has become near zero is called the "film thickness."

Float layer: The collected air–solid-particle agglomerates that have risen to the water surface.

Flotation: Unit process in which air bubbles are brought into contact with solid particles to cause the latter to become buoyant and rise to the surface, where they are skimmed by blades into a collection trough. The solid particles may be pretreated by means of coagulation and flocculation to form a floc.

Gas precipitation: Formation of gas bubbles due to gas concentration in water exceeding the equilibrium concentration for the local pressure of the bubbles.

Gas saturation: Concentration of gas, species i, in water that is in equilibrium with the partial pressure of gas i in the gas phase across a common gas–water interface.

Henry's constant: An equilibrium constant defined here as the ratio of concentration of gas i dissolved in water to the partial pressure of i in the gas phase. Magnitude depends upon the units for aqueous phase concentration and gas pressure. Also, some definitions are the reciprocal of others.

Henry's law: Concentration of gas species i in water is proportional to partial pressure of i in the gas phase. See also Appendix H.

Hydraulic grade line (HGL): Locus of points of pressure head (in meters), i.e., pressure divided by specific weight of water ($r_w g$).

Interfacial area: Surface area of air–water interface.

$K_L a$: Mass-transfer coefficient that says the rate of transfer of a gas between phases is proportional to the concentration difference between an equilibrium level and that existing at a given time.

Mole fraction: Ratio of molecules of gas *i* in a given volume to the total number of molecules of all species with number of molecules expressed as moles.

Nozzle: Convergence, then divergence of a pipe.

Packing: A material or plate system within a saturator that is intended to cause bifurcations of the water stream to create a substantial interfacial area of water. The packing may be a granular media or a system of plates staggered one above the other or "rings" of various sorts. An airflow occurs counter-current within the pores of the packing material so as to come into contact with the water. The water flow through the packing material must be "unsaturated" so as to maintain a high water surface area (which is the intent of the packing material).

Partial pressure: According to Dalton's law, the partial pressure of a gas *i* is proportional to its mole fraction.

Phenolphthalein: An indicator that changes color from pink at $pH < 8.3$ to colorless when $pH \geq 8.3$, as in titration with N/50 sodium hydroxide solution to determine dissolved carbon dioxide concentration (Sawyer and McCarty, 1967, p. 67).

Reaction zone: Zone in flotation basin just after bubble formation and as the particle stream enters in the flow Q to provide the opportunity for bubble–particle contacts. Same as "contact zone," which was the term adopted by a consensus of persons discussing the terminology at the 1994 *International Joint Specialized Conference on Flotation Processes in Water and Sludge Treatment*, Orlando, Florida, April, 1994.

Recycle: The portion of the flow that leaves the flotation basin that is returned to the saturator to pick up dissolved air and then return to the flotation tank.

Recycle ratio: The fraction, *r*, of the flow, *Q*, that is recycled, i.e., *R*, to pass through the saturator, i.e., R/Q.

Saturator: A tank under pressure that has some means of creating an interfacial area between air and water. One method is to use a "packing." Another is to create a high surface area of air bubbles within the saturator.

Separation zone: The zone in the flotation basin after the contact zone for particle–bubble agglomerates to rise to the surface.

Skimmer: Horizontal blades that traverse the surface of the flotation basin and are translated in the longitudinal direction by means of a chain drive.

Transport coefficient: The ratio of the number of bubbles contacting floc particles to the number of bubbles in a given volume within the contact zone.

Unpacked saturator: Saturator without packing. The water enters the saturator at the top and falls through air at high pressure to a pool of water at the bottom of the tank. The entering water may impinge against a splash plate to increase the interfacial area as it falls thought the air. Air bubbles are entrained as the water falls into the water pool below and is believed by some to be where most of the gas transfer occurs.

Vacuum flotation: A technology that imposes a negative pressure on the surface of the flotation tank. Gas precipitation may occur if the incoming flow of water to be treated is "saturated" with respect to the local atmosphere of the raw water and would be more successful at low elevation.

REFERENCES

de Rijk, S. E., Jaap, H. J. M., van der Graaf, J. M., and den Blanken, J. G., Bubble size in flotation thickening, *Water Research*, 28(2):465–473, 1994.

Eckenfelder, W. W., Jr., Rooney, T. F., Burger, T. B., and Gruspier, J. T., Chapter 2–10, in: McCabe, J., and Eckenfelder, W. W., Jr., *Biological Treatment of Sewage and Industrial Wastes* (Volume II, Anaerobic Digestion and Solids-Liquid Separation), Reinhold Publishing Corporation, New York, 1958.

Edzwald, J. K., Principles and applications of dissolved air flotation, *Water Science and Technology*, 31(3–4):1–23, 1995.

Edzwald, J. K. and Walsh, J. P., *Dissolved Air Flotation: Laboratory and Pilot Plant Investigations*, American Water Works Association Research Foundation, Denver, CO, 1992.

Fukushi, K., Matsui, Y., and Tambo, N., Dissolved air flotation: Experiments and kinetic analysis, *Aqua*, 47(2):76–86, 1998.

Haarhoff, J., Factors influencing bubble formation in dissolved air flotation, International Association of Water Quality Yearbook, 1994–95, 1995.

Haarhoff, J. and Rykaart, E. M., Rational design of packed saturators, *Water Science and Technology*, 31(3–4):179–190, 1995.

Haarhoff, J. and van Vuuren, L., Design parameters for dissolved air flotation in South Africa, *Water Science and Technology*, 31(3–4):203–212, 1995.

Hendricks, D. W., Oxidative Liquor Return to the Eimco Clari-Thickener, The Eimco Corporation, Internal Report, Salt Lake City, UT, 1966.

Kalinske, A. A., Flotation in waste treatment, Chapter 2–7, in: McCabe, J. and Eckenfelder, W. W., Jr., *Biological Treatment of Sewage and Industrial Wastes* (Volume II, Anaerobic Digestion and Solids-Liquid Separation), Reinhold Publishing Corporation, New York, 1958.

Leopold, *Dissolved Air Flotation*, brochure, The F.B. Leopold Company, Inc., Zelienople, PA, 8pp., 1996 (revised 2001, 2003).

Masterson, E. M. and Pratt, J. W., Application of pressure flotation principles to process equipment design, Chapter 2–8, in: McCabe, J. and Eckenfelder, W. W., Jr., *Biological Treatment of Sewage and Industrial Wastes* (Volume II, Anaerobic Digestion and Solids-Liquid Separation), Reinhold Publishing Corporation, New York, 1958.

Matsui, Y., Fukushi, K., and Tambo, N., Modeling, simulation, and operational parameters of dissolved air flotation, *Aqua*, 47(1): 9–20, 1998.

Nemerow, N. L., *Liquid Wastes of Industry, Theories, Practices, and Treatment*, Addison-Wesley, Reading, MA, 1971.

Ødegaard, H., Optimization of flocculation/flotation in chemical wastewater treatment, *Water Science and Technology*, 31(3–4):73–82, 1995.

Rykaart, E. M. and Haarhoff, J., Behavior of air injection nozzles in dissolved air flotation, *Water Science and Technology*, 31(3–4):25–35, 1995.

Sawyer, C. N. and McCarty, P. L., *Chemistry for Sanitary Engineers*, McGraw-Hill, New York, 1967.

Schofield, T., Design and Operation of the World's Largest Dissolved Air Flotation Water Treatment Plant, *Yearbook*, 1995–96, International Association on Water Quality, 1996.

Valade, M. T., Edzwald, J. K., Tobiason, J. E., Dahlquist, J., Hedberg, T., and Amato, T., Particle removal by flotation and filtration: pretreatment effects, *Journal of the American Water Works Association*, 88(12):35–47, December, 1996.

Vrablik, E. R., Fundamental principles of dissolved air flotation, in: *Proceedings of the 14th Industrial Waste Conference*, May 5–7, 1959, Purdue University, West Lafayette, IN, Extension Series No. 104, Engineering Bulletin, v. 44, No. 5, pp. 743–779, May, 1960.

Part III

Microscopic Particles

Microscopic particles include those that do not settle readily and must be removed by means other than settling. The sizes of such particles are $<100\,\mu m$. Number concentrations may range from 1,000 #/mL to perhaps 100,000 #/mL.

A "quasiparticulate" material is natural organic matter (NOM), which is macromolecular and behaves like colloidal matter. Practice has evolved since about 1900 to remove NOM by coagulation.

The traditional means to remove small particles (and NOM) is by coagulation as a first step. Subsequent steps depend on the treatment train. In "conventional" filtration, the subsequent steps are mixing (Chapter 10), flocculation (Chapter 11), settling (Chapter 6), and filtration (Chapter 12). In flotation (Chapter 8), the sequential steps are coagulation, flocculation, flotation, and filtration. For "in-line" filtration, the steps are coagulation and filtration.

9 Coagulation

The term *coagulation* refers to the process of inducing contacts between a chemical and colloidal particle to effect a reaction. The reaction product is called here, a *microfloc*. Engineering the process has two phases: (1) choosing the proper chemicals, the proper dosages, and the proper pH to achieve a microfloc product, and (2) causing contacts between the coagulant chemicals and the colloidal particles. The latter is the topic of Chapter 10. Inducing the growth of microflocs to form flocs is the topic of Chapter 11. The effectiveness of the coagulation process determines the effectiveness of the subsequent processes, for example, flocculation–flotation; flocculation–sedimentation; flocculation–settling–filtration, or filtration.

9.1 COAGULATION IN-A-NUTSHELL

Coagulation is chemical in character. Simply stated, coagulation is the reaction between a chemical and particles to form a "microfloc." Actually, a body of knowledge is involved; key elements of the topic are summarized in this section.

9.1.1 Defining Coagulation

An array of terms are involved in the coagulation process and are explained here. The glossary includes a more extended list.

9.1.1.1 Particles to Be Removed

Ambient waters contain a wide variety of particles in the general categories: mineral (e.g., clays), biological (viruses, bacteria, algae, protozoan cysts, etc.), and organic matter (e.g., natural organic matter (NOM)). These particles range in size from nanometers to perhaps 200–300 μm.

9.1.1.2 Coagulation

Trivalent metal ions, that is, Al^{3+} and Fe^{3+}, are the traditional coagulants used in water treatment practice. They react with water to form "hydrolysis products" and/or precipitate, which react in turn with negatively charged particles to form what is called here, "microflocs." This process is termed, "coagulation."

9.1.1.3 Microflocs

Small discrete particles that leave the rapid-mix are called here, "microflocs." In the case of coagulation with Al^{3+}, the objective is that the reaction products of Al^{3+} and water combine with the microscopic particles as found in the ambient water being treated.

9.1.1.4 Rapid-Mix

Mixing (Chapter 10) is the means to cause contacts between the coagulant chemicals and the particles to be removed in the raw water. The mixing unit in coagulation is termed "rapid-mix." Several design forms are found in practice, for example, "back-mix" reactor, "flash-mix," baffles, elbows, etc.

9.1.1.5 Flocculation

The collision between microflocs to form larger aggregated particles is called "flocculation" (Chapter 11) and the resulting particles are called "flocs."

9.1.1.6 Themes of Coagulation Theory

The three themes that have emerged in coagulation theory are as follows: (1) a sequence of reactions between aluminum and water, (2) adsorption destabilization with positively charged polynuclear aluminum species, and (3) "sweep-floc" based on positively charged aluminum hydroxide.

9.1.2 Coagulation Practice

The many aspects of coagulation practice include

- Selection of type of coagulant (generally alum or a ferric salt)
- Selection of a coagulant aid, that is, a polymer (which may or may not be used)
- Determination of dosage of each chemical
- Design of rapid-mix
- Evaluation of coagulation effectiveness

The preceding enumeration has related, traditionally, to particle removal, with ancillary focus on removing color, a surrogate for NOM. In the mid-1970s NOM was identified as a precursor to halogenated organics, the latter being considered carcinogenic. NOM became a major focus of coagulation practice along with particles.

9.1.2.1 Dosage

Coagulation is the most critical phase of the depth filtration process (Chapter 12) and coagulant dosage is the determining factor (along with pH). Generally, dosage is determined by a jar test. The term, "enhanced-coagulation" refers to the addition of an alum or ferric dosage sufficient to remove NOM as well as particles.

9.1.2.2 Coagulation Effectiveness

Coagulation effectiveness has been measured traditionally by removal of turbidity. Color removal is another traditional objective in drinking water treatment, but for esthetic reasons. Since about 1980, however, color has became a health issue as well, since it is caused mostly by NOM.

9.2 PARTICLES IN AMBIENT WATERS

Coagulation theory is based on the mineral colloid "model." The main characteristics of the mineral colloid model are (1) a negative charge with a diffuse double layer, and (2) a particle size ≤ 10 μm. In practice, however, a variety of kinds of particles, both mineral and biological, occur in ambient waters and must be removed. There are perhaps thousands of different species of microscopic organisms found commonly in water supplies. The general groups include: viruses, bacteria, cysts, algae, spores, rotifers, fecal debris from rodents, etc. Other particles may include organic debris such as parts of animal fecal matter, and organic molecules such as humic and FAs, that is, NOM, which are essentially macromolecules.

9.2.1 Particle Variety

Figure 9.1 indicates the different kinds of particles found in ambient surface waters (i.e., in rivers and lakes). Size ranges are shown and compared with several common references, for example, visible to the unaided eye, laboratory microscopic, and the electron microscope. As seen, colloids are those particles nominally ≤ 10 μm, viruses are 10–800 Å, bacteria are about 1 μm, algae are 1–50 μm, protozoa 5–100 μm. The coagulation process is likely to be effective for most of these particles (Singley et al., 1971, p. 99).

9.2.2 Particle Characteristics

Table 9.1 lists some of particles that may be coagulated and gives sizes, "zeta potentials," and representative counts. As seen, except for chrysotile, the particles have negative zeta potentials; the chrysotile is listed only to indicate that a positive charge, albeit the exception, may occur. The particles listed include viruses, bacteria, algae, protozoan cysts, plant debris, nematode eggs, and other kinds of particles having biological origin and reflect the character of the water.

9.2.2.1 Colloids

A colloid is a discrete particle that remains in suspension, for example, does not settle in water. Clay is a typical colloid, as seen in Table 9.1, and remains in suspension due to its small size, that is, ≤ 10 μm, and negative charge, the latter causing mutual repulsion between particles. Such a suspension is called a "sol" and is termed, "stable." An objective of coagulation is to "destabilize" the suspension, that is, reduce the magnitude of the colloid negative charge, which is done in traditional water treatment practice by trivalent cations, for example, Al^{3+} or Fe^{3+}.

9.2.2.2 Microscopic Particles

In addition to mineral colloids, ambient water has a variety of particle categories, mostly biological, that include: viruses,

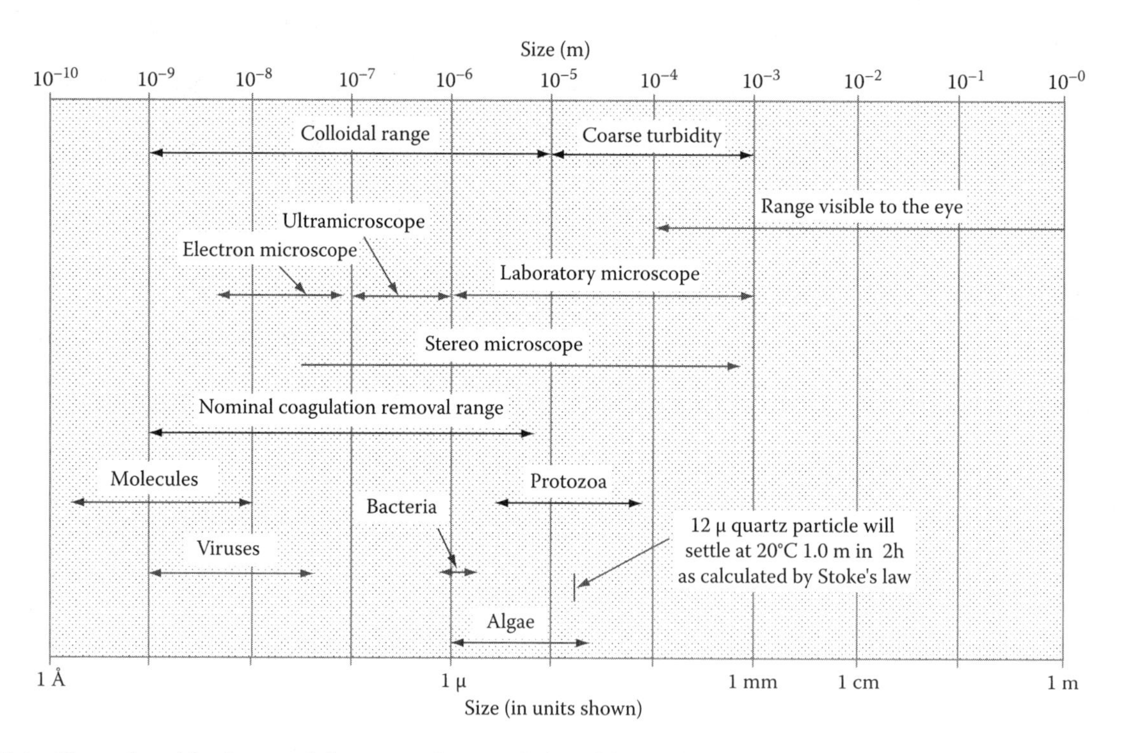

FIGURE 9.1 Sizes of particles in coagulation removal range. (Adapted from Riddick, T.M., *J. Am. Water Works Assoc.*, 53(8), 1007, August 1961; Amirtharajah, A. and O'Melia, C.R., Coagulation processes: Destabilization, mixing, and flocculation, in Pontius, F.W., Ed., *Water Quality and Treatment*, 4th edn., McGraw-Hill, New York, 1990, Chapter 6.)

TABLE 9.1
Particles Found in Ambient Water in North America and Their Characteristics

Category	Group/Name	Size (μm)	Zeta Pot. (mV)	Counts (#/L)
Mineral	Clays[a] (kaolinite, montmorillonite)	0.001–1[b]	−15 to −20[c]	
	Asbestos fibers—amphibole	<2[d]	Negative[d]	10^6–10^8 fibers/L[e]
	Asbestos fibers—chrysotile	<2[d]	Positive[d]	10^6–10^9 fibers/L[e]
Biological	Viruses	0.01–0.1[b]		10^4–10^{10}
	Bacteria	0.3–10[f]	−50[g]	10^3–10^9
	Bacillus subtilis spores		−18[h]	10^3–10^5
	In raw municipal sewage			≥200
	Giardia lamblia cysts	10[i]	−40[j], −14[h]	0.1–20
	In raw municipal sewage			≥200
	Cryptosporidium parvum oocysts	4–5[j,k]	−35[j], −7[h]	0.1–20
	Algae	30–50[k]	−3 to −30[l]	10–10^9[m]
	Diatoms	3–70[g]	−3 to −30[l]	10–10^5[m]
	Parasite eggs	10–50[k]		0.01[n]
	Nematodes eggs, free living	10[k]		0.01[n]
	Nematodes eggs, parasitic	10[k,n]		0.001[n]
	Coccidia	3–10[n]		0.01[n]
	Pollen	30–150[n]		1[n]
	Ciliates	20–800[n]		1[n]
	Crustaceans/eggs	100[n]		1[n]
	Insects	100–1000[n]		1[n]
	Flagellates	5–80[n]		1[n]
	Free-living amoebae	30–30[n]		0.1[n]
	Biological concentrate	mixture	−13.5[c]	10,000[n]
Other particles	Amorphous debris, small	1–5[j,n]		10,000[n]
	Amorphous debris, large	5–500[j,n]		100[n]
	Fecal debris, rodent	50–1000[n]		1[l]
	Organic colloids[o]	1–100		
	Color (organic macro-mol.)		Negative[p]	
	Activated carbon	1–1000	−35[j]	

Source: Adapted from Hendricks et al. *Manual of Design for Slow Sand Filtration*, AWWA Research Foundation and American Water Works Association, Denver, CO, 1991.

[a] Clays include montmorillonite, kaolinite, and illite, to name a few groups.
[b] Tate and Trussel (1978).
[c] Al-Jadhai and Hendricks (1989) measured −16.4 mV for water from the Cache La Poudre River having turbidity 0.25 NTU and particle counts 600,063 particles/10 mL using a 70 μm aperture.
[d] O'Melia (1979, p. 2). Note that chrysotile is the only particle listed with positive charge. The statements from O'Melia and later from Schleppenbach, c. 1983, p. 22 on the positive charge were not documented with references and are included to indicate the possibility that some particles may have positive charge.
[e] Logsdon (1979, p. 24 for amphibole, p. 43 for chrysotile).
[f] Beard and Tanaka (1977).
[g] Bean et al. (1964); range in zeta potential for bacteria was −35 to −70; bacteria species were *Balantidium coli*, *Bacillus subtilis*, and Proteus.
[h] Fox and Lytle (1996); they detected no discernible effect of pH for *Bacillus subtilis* spores, for *G. lamblia* cysts, or *C. parvum* oocysts over for $4 < pH < 11$.
[i] Jakubowski and Hoff (1979).
[j] Ongerth and Pecoraro (1996), data are for pH = 7.0. Formalin fixed cysts had $z = 17$ at pH = 7.1.
[k] Hibler, C. P., Personal communication (August 23, 1990); note that the list of biological particles cited in footnotes "k" and "n" were as compiled by Hibler for microscopic examination of cartridge filter samples obtained by water treatment plant operators and sent to Dr. Hibler at Colorado State University.
[l] Cushen (1996).
[m] Dr. Paul Kugrens, Professor of Biology, Colorado State University, personal communication, February 15, 2001.
[n] Hancock, C. M., Personal communication (February 14 and 19, 2001); concentrations are approximate "averages" based on experience in microscopic examination of perhaps thousands of samples from ambient waters and will vary with different waters and seasons.
[o] Safe Drinking Water Committee (1977) lists but gives no data.
[p] Black (1948).

bacteria, cysts, algae, spores, rotifers, fecal debris from rodents, etc. Most of these particles also have negative charges, as measured by "zeta potential," and are removed coincidentally with mineral colloids.

9.2.2.3 Natural Organic Matter and Color

Natural organic matter (NOM) is comprised of humic and FAs, found in almost all surface waters, and are due to breakdown of vegetation. Humic acids (HA) have molecular weights ranging from several hundred to several thousand and are comprised primarily of aromatic compounds (i.e., with benzene-rings in various forms). FAs are similar in structure but have lower molecular weights. A property of NOM is "color;" its removal has been an ancillary objective of coagulation since about 1900. About 1975, however, NOM was found to be a "disinfection by-product" (DBP) precursor (Symons et al., 1975); after being determined carcinogenic, became a removal priority. NOM has a negative charge and is removed partially by coagulation, for example, in the range 10%–90%. Concentrations of NOM in surface waters range from about 1–50 mg/L as dissolved organic carbon (DOC), with a median about 4 mg/L. The molecular weight of most organic color and humic substances is in the range, $700 < MW < 200{,}000$ (O'Melia et al., 1979, p. 590).

9.2.2.4 Total Organic Carbon

A measure of NOM is total organic carbon (TOC), which has two components, that is, particulate organic carbon (POC) DOC) In ambient waters, the DOC portion dominates, for example, during spring runoff of a mountain stream, measurements were, $TOC = 9.6$ mg/L; $DOC \approx 9.4$ mg/L; $POC \approx 0.2$ mg/L (Carlson and Gregory, 2000, p. 559).

9.2.2.5 Turbidity

Light scattering, that is, "nephelometry," is a property of colloidal suspensions. Thus, if a light source is directed into such a suspension the light will "scatter." A detector located at a right angle to the light source will show a reading that is approximately proportional to the concentration of colloidal matter (for a given suspension). An instrument arranged in such fashion is called a turbidimeter. A turbidity unit is a measure obtained by a turbidimeter from a standardized suspension, called a nephelometric turbidity unit (NTU). Natural waters have a wide range of turbidities, ranging from as low as 0.2 NTU for mountain streams in the winter to several hundred NTU for a muddy river. Values vary seasonally and with rainfall events as well; for example, a mountain stream may have turbidities of perhaps 50 NTU during spring runoff.

9.2.2.6 Particle Counts

Particle counting technology evolved from about 1970 for discrete samples to continuous "on-line" particle counting during the late 1980s becoming common in water treatment during the 1990s. These instruments can measure particle concentrations in selected size ranges, for example, >2 μm, >5 μm, etc., as well as total counts, that is, all particle sizes larger than a given lower limit size, for example, 2 μm. Particle counting instruments require regular maintenance, for example, calibration and cleaning.

9.2.3 Turbidity and Particle Counts in Ambient Waters and Finished Waters

Turbidity and particle counts vary from place to place and seasonally. Despite the wide variation in source waters, the variation in finished waters is remarkedly narrow.

9.2.3.1 Spatial Variation in Source Waters Compared with Plant Effluents

Table 9.2 lists raw and filtered water turbidity and particle counts for 10 water treatment plants in the United States. As seen in Table 9.2, raw water turbidities vary from as low as 0.2 NTU for Lake Mead near Las Vegas to as high as 49 for Winnetka, Illinois. Particle counts vary from 3,200 #/mL for Lake Mead near Las Vegas to 500,000 #/mL for Winnetka, Illinois. Filter effluent turbidities in Table 9.2 vary from 0.01 to 0.07 NTU with effluent particle counts as low as 41–720 #/mL. The data show a representative range in turbidity levels and corresponding particle counts for both raw water and filtered water.

9.2.3.2 Seasonal Variation

The water quality characteristics of every ambient water show a characteristic profile with respect to time. Average monthly turbidities and alkalinities are shown in Figure 9.2a and b, respectively, for the Cache La Poudre River, a mountain stream in Colorado. These profiles illustrate the variability that occurs for most water quality constituents over an annual cycle which must be considered in water treatment.

9.3 CHEMISTRY

Coagulation is based on chemistry. In the early decades, c. 1900–1930, and through the 1980s, practice was based largely on empiricism. Theory began to evolve, c. 1920 and was assimilated during each subsequent decade through the 1990s.

9.3.1 Chemistry of Coagulation: Evolution of Theory and Practice

Chemical treatment of municipal wastewaters, for example, using lime or "sulfate of alumina," was well established in Great Britain in 1870 (Metcalf and Eddy, 1916, p. 5). In 1885, due to difficulties in filtration of Mississippi River water for New Orleans, alum was tried which led to the modern practice of rapid filtration (as opposed to slow sand). Since the ionic theory of solution was yet to be formulated, that is, by Svante Arrhenius in his 1884 doctoral thesis, the early practice had little theoretical basis. Later, in reviewing 1920s research, Willcomb (1932, p. 1418) was aware of the role of pH on alum coagulation.

TABLE 9.2
Turbidity and Particle Counts for Various Water Treatment Plants

Location	Plant	Turbidity Influent (NTU)	Turbidity Effluent (NTU)	Particle Influent (#/mL)	Counts Effluent (#/mL)
Tuscaloosa, Alabama	Ed Love WTP	1.2	0.07	17,000	540
Glendale, Ariz	Cholla WTP	4.4	0.04	69,000	590
Contra Costa, California[a]	R. D. Bollman WTP	9.0	0.05	179,000	290
Loveland, Colorado	Chasteens Grove	1.5	0.03	11,000	120
Winnetka, Illinois	Winnetka WTP	49.0	0.01	500,000	500
Diluth, Minnesota	Lakewood WTP	0.4	0.02	4,000	41
Merrifield, Virgina[b]	Corbalis WTP	10.0	0.03	51,000	210
Los Angeles, California[c]	LA Aqueduct WTP	3.6	0.07	55,000	630
Las Vegas, Nevada[d]	A. M. Smith WTP	0.2	0.06	3,200	290
East Bay MUD, California[e]	Orinda WTP	0.4	0.06	8,800	720

Source: Adapted from Cleasby, J.L. et al., *Design and Operation Guidelines for Optimization of the High-Rate Filtration Process: Plant Survey Results*, AWWA Research Foundation, Denver, CO, September 1989, pp. 73–74.

[a] Source water is from the Sacramento—San Joaquin River Delta; peak raw water turbidity 80 NTU, DOC $\leq$ 11 mg/L, seasonally high plankton populations up to 50,000 organisms/mL.

[b] Source is Potomac River with peak turbidity 180 NTU and peak color 100 apparent color units.

[c] Source is Owens River, carried by pipeline to Los Angeles.

[d] Source is Lake Mead.

[e] Source is Calaveras River, Sierra Nevada mountains.

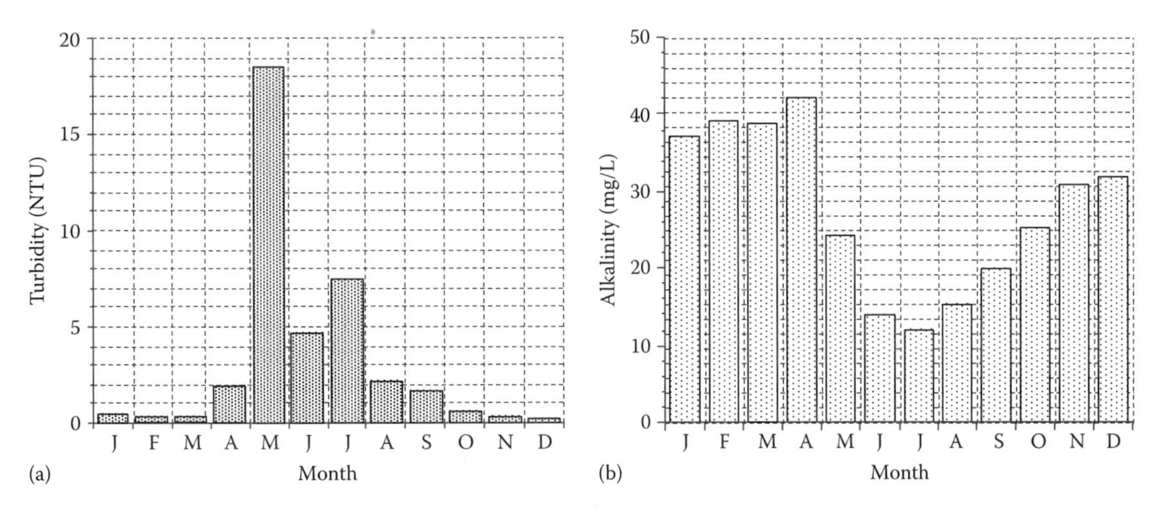

FIGURE 9.2 Profiles of turbidity and alkalinity for Cache La Poudre River, 1985. (a) Turbidity and (b) alkalinity. (Courtesy of Dr. Keith Elmund, City of Fort Collins, CO.)

The work of these investigators may be summed up as follows: when a polyvalent sulfate solution is added to a water two kinds of floc may be formed depending upon the resultant pH of the solution. From pH 4.0 up to 5.0 the strong coagulating power of the trivalent aluminum ion acting on negatively charged colloidal color is in evidence, forming what is called color floc. The second kind of floc known as alum floc approaches completion of precipitation at pH 5.4 and is complete at pH 8.5.

So, by about 1900, the groundwork had been laid in both practice and theory. In other words, the innovations to come were a part of building on the work of others, not spontaneous discoveries.

9.3.1.1 Key Innovations

One of the major developments of the 1920s was paddle-wheel flocculation technology and along with this, the "jar-test" (Langelier, 1921, 1925; Box 9.3). The jar-test became available as a commercial apparatus, c. 1934, and soon became the main tool in practice for assessing effective coagulation. Langelier continued research in the field; for example, delineating the "zones" of coagulation (Langelier and Ludwig, 1949), which, at the same time, advanced theory. Box 9.1 outlines some of the history of these developments.

BOX 9.1 EARLY STUDIES ON COAGULATION

In 1921, Professor Wilfred Langelier published a landmark paper, "Coagulation of water with alum by prolonged agitation." The paper was related to the design of a new filtration plant at Sacramento by Professor Charles Gilman Hyde and, through the use of an in-house fabricated "stirring-device," established such innovations as the "jar-test" and paddle-wheel flocculation. In addition, his paper recognized the importance of rapid-mix, water characteristics (e.g., turbidity, color, alkalinity, colloids) and their effect on coagulant dosage, and laboratory studies. The issues of coagulant dosage, coagulant mixing, and flocculation were addressed through a systematic experimental program.

In 1942, Harvey Ludwig (see Box 1.2) presented a Master of Science thesis on coagulation of turbid water, building upon Langelier's 1921 paper. The ensuing paper (Langelier and Ludwig, 1949), with publication deferred because of the war, assimilated principles of colloid chemistry from the well-known soil scientist, Professor Hans Jenny (see Ludwig, 1985, pp. 2–12). The paper outlined the idea of double layer suppression, the concept of "replaceable," or adsorbed ions in the double layer, with the increasing order of attraction given as, Na^+, K^+, Mg^{2+}, Ca^{2+}, H^+, Al^{3+}, Fe^{3+}, the notion of "zeta potential" as the measure of negative repulsive force, the idea of charge neutralization as the key to effective coagulation, and the role of a hydroxide precipitate (which they characterized as a "binder" material).

The paper introduced the terms *peri-kinetic* flocculation and *ortho-kinetic* flocculation to mean, respectively, aggregation of flocs as permitted by neutralization of zeta potential of colloids and aggregation of flocs as caused by enmeshment with other flocs and colloids (Langelier and Ludwig, 1949, p. 165; see also *ortho-kinetic* in the glossary). The basis for the Langelier and Ludwig paper was several thousand tests employing a variety of coagulant chemicals at different dosages, different pHs and different alkalinities, with 12 different soils (the latter being used as the basis for synthesizing 21 artificial raw waters used for the experiments). Conclusions from the study are encapsulated by one of their many residual turbidity versus coagulant dosage curves, which was a major underpinning of theory until extended by Black (1967), Stumm, and O'Melia (1968) and then by Amirtharajah and Mills (1982).

Wilfred F. Langelier c. 1940, Professor of Sanitary Engineering, University of California (from Langelier, 1982)

Harvey Ludwig c. 1945, graduate student under Langelier (used with permission by Dr. Ludwig)

9.3.1.2 Color

Color was recognized early as colloidal and with negative charge (Saville, 1917). Coagulation of color was considered by Black (1934, p. 1714), who noted that in the acid pH range positive ions form and may react with negatively charged color colloids. In the higher pH range, he noted that aluminum and iron form "microcrystals," which are seen as gelatinous flocs. The latter have positive charge, and can neutralize and precipitate negatively charged color colloids.

9.3.1.3 Modern Theory

The modern notion of coagulation was outlined by Moffett (1968, p. 1256), and in a preliminary form by Black (1948, p. 143), as

1. Hydrolysis of water reacts with Al^{3+} to multi-nuclear hydrolysis species
2. Adsorption of the hydrolysis species at the solid–solution interface of the colloid resulting in the formation of a "microfloc"
3. Aggregation of the microflocs as "flocs" occurs due van der Waals forces

Coagulation theory, involving colloids, color, coagulants, and the coagulation process, evolved over decades. Two of the key players in the formative decades of coagulation chemistry, A.P. Black and Wilfred Langelier, were academic chemists in the water treatment industry and are a part of the lore of the field. Many others who contributed to modern theory are cited (and many others were not cited due to limitations in the scope of this text).

9.3.2 Coagulation Reactions

The reactions of metal ions, for example, Al^{3+} and Fe^{3+}, with water result in a variety of products, with the species dependent on pH, dosage, ionic strength, alkalinity, and perhaps other factors. The reactions are complex and are only summarized here.

9.3.2.1 Metal Ion Reactions with Water

At low pH, for example, $4 < pH < 6$, when Al^{3+} or Fe^{3+} react with water, the reaction products are "complexes" with water. At higher pH levels, for example, $6 < pH < 10$, and especially at higher dosages, metal hydroxide is the major product. The two categories of reactions are

1. Complexes,

$$Al_2(SO_4)_3 + nH_2O \rightarrow 2Al \cdot (OH)_n(H_2O)_n{}^{n+} + H^+ + 3SO_4{}^{2-} \quad (9.1)$$

2. Metal ion precipitate,

$$Al_2(SO_4)_3 + 6H_2O \rightarrow 2Al(OH)_3 + 6H^+ + 3SO_4{}^{2-} \quad (9.2)$$

where
$Al_2(SO_4)_3$ is the aluminum sulfate, that is, alum
$Al \cdot (OH)_n(H_2O)_n{}^{n+}$ is the hydrated aluminum complex with water (variable charge and variable waters of hydration)
$Al(OH)_3$ is the aluminum hydroxide precipitate (waters of hydration not shown)

9.3.2.2 Two Coagulation Mechanisms

The two coagulation zones in Al^{3+} and Fe^{3+} coagulation are as follows: (1) charge neutralization for $pH < 6$; and (2) sweep floc for $pH > 6$. The reaction products for the two zones are "hydrolysis-products" and $Al(OH)_3$ precipitate, respectively (Amirtharajah and Mills, 1982).

9.3.2.2.1 Charge-Neutralization ($pH < 6$)

In Equation 9.1, the Al product is actually an array of "polynuclear" Al cations, also called aluminum "hydrolysis" products, or "complexes," and occur in the pH range, $pH \leq 6$. The complexes are positively charged, have only a few milliseconds of life, and react, in turn, with negatively charged microscopic particles. The reaction is termed "charge-neutralization," and is depicted as

$$\text{colloid}^{n-} + Al \cdot (OH)_n(H_2O)_n{}^{m+} \rightarrow \text{colloid} \cdot [Al \cdot (OH)_n(H_2O)_n]^{(m-n)+} \quad (9.3)$$

where
colloid^{n-} is the negatively charged colloid with n negative charges
$\text{colloid} \cdot [Al \cdot (OH)_n(H_2O)_n]^{(m-n)+}$ is the colloid and metal ion complex

The positively charged complexes, on the left side in Equation 9.3, become incorporated within the diffuse "double layer" of a negatively charged particle, which results in the reaction product, that is, a colloid–metal ion complex, on the right side, called here a "microfloc." The van der Waals attractive forces between the resulting "microflocs" then dominate, permitting them to attach to one another, forming a precipitate. The job of rapid-mix is to induce collisions between the reactants on the left side of Equation 9.3 to form such microflocs. Since the life of the complexes is very short, that is, in terms of milliseconds, a very large fraction of the collisions between reactants must occur before leaving the "rapid-mix" (Chapter 10).

9.3.2.2.2 Sweep Floc ($pH > 6$)

The "sweep-floc" reaction (Amirtharajah and Mills, 1982) is between aluminum hydroxide precipitate, Equation 9.2, and colloids in suspension and occurs at $pH > 6$,

$$\text{colloid}^- + Al(OH)_3 \rightarrow \text{colloid} \cdot [Al(OH)_3] \quad (9.4)$$

In sweep-floc, the positively charged $Al(OH)_3$ precipitate contacts the negatively charged colloids through random contacts, for example, through turbulence in "rapid-mix," resulting in particle attachment and enmeshment. The aluminum hydroxide floc is amorphous in nature with large surface area, for example, 159–234 m^2/g for ferric floc (Randtke, 1988, p. 41), which also facilitates particle enmeshment (Matijevic, 1967, p. 337). The removal of the suspended microscopic particles is, in general, proportional to the floc surface area.

Figure 9.3 shows two photomicrograph examples of alum floc enmeshing diatoms. Both figures (Figure 9.3a and b) illustrate the amorphous but differing character of an aluminum hydroxide floc particle enmeshing diatoms. The particles are not free to disengage, that is, they appear to involve bonding between the particles and the alum floc.

9.3.2.3 NOM Removal by Metal Coagulatants

Independent variables that affect NOM removal include pH and coagulant dosage, with dosage being proportional to the humic concentration, that is, it is stoichiometric (Black and Willems, 1961, p. 592; O'Melia et al., 1979, p. 594; Dempsey et al., 1984). The "zones" of coagulation for NOM were delineated by Edwards and Amirtharajah (1985), which extended work done for turbidity removal by Amirtharajah and Mills (1984).

(a)

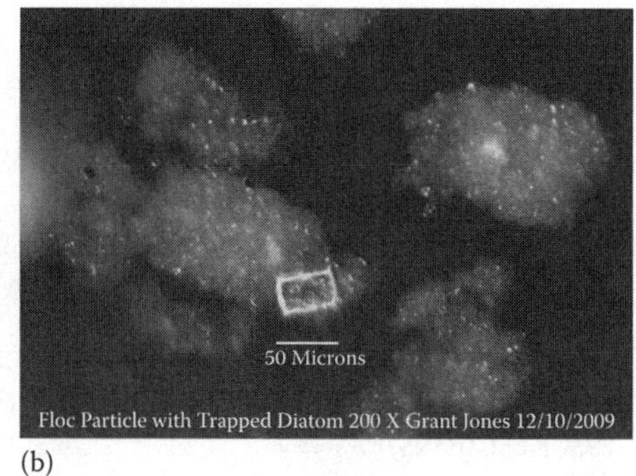

(b)

FIGURE 9.3 Alum floc—two examples. (a) Amorphous alum floc with enmeshed particles and a 100 μm diatom; (b) Discrete alum flocs with enmeshed particles and a 50 μm diatom. (Courtesy of Grant Williamson-Jones, WTP, City of Fort Collins, CO.)

9.3.2.3.1 Alum Reacting with Color

Figure 9.4a illustrates the effect of alum dosage on color, and electrophoretic mobility (EM) from top to bottom, respectively. The top plot shows that color declines sharply with alum dosage, then increases; the bottom plot shows that EM increases sharply with alum dosage to an asymptote. Figure 9.4b illustrates the effect of pH on color and EM from top to bottom, respectively, at the optimum dosage of 120 mg/L; also that $pH \approx 4.5$ is optimum.

9.3.2.3.2 Coagulation Zones for Color

Effective coagulation zones for fulvic acids (FA) were defined on p[Al] versus pH diagrams (Edwards and Amirtharajah, 1985), which were similar to the p[Al] versus pH diagram for turbidity (Amirtharajah and Mills, 1982), that is, as in Section 9.5.3.4. The coagulation zones are described as follows: (1) at $4 < pH < 6$, the negatively charged HA sols adsorb highly charged polynuclear cationic aluminum hydrolysis species, and (2) at $6 \leq pH \leq 8$, with dosage $>2 \times 10^{-4}$ M Al, aluminum hydroxide precipitate occurs and removal is by adsorption (Edzwald et al., 1977, p. 990; Dempsey et al., 1984; Edwards and Amirtharajah, 1985, p. 51).

9.3.2.3.3 Adsorption of NOM by Aluminum Hydroxide

Regarding adsorption, the uptake of organic matter follows a Langmuir isotherm (Chapter 15) such as shown in Figure 9.5

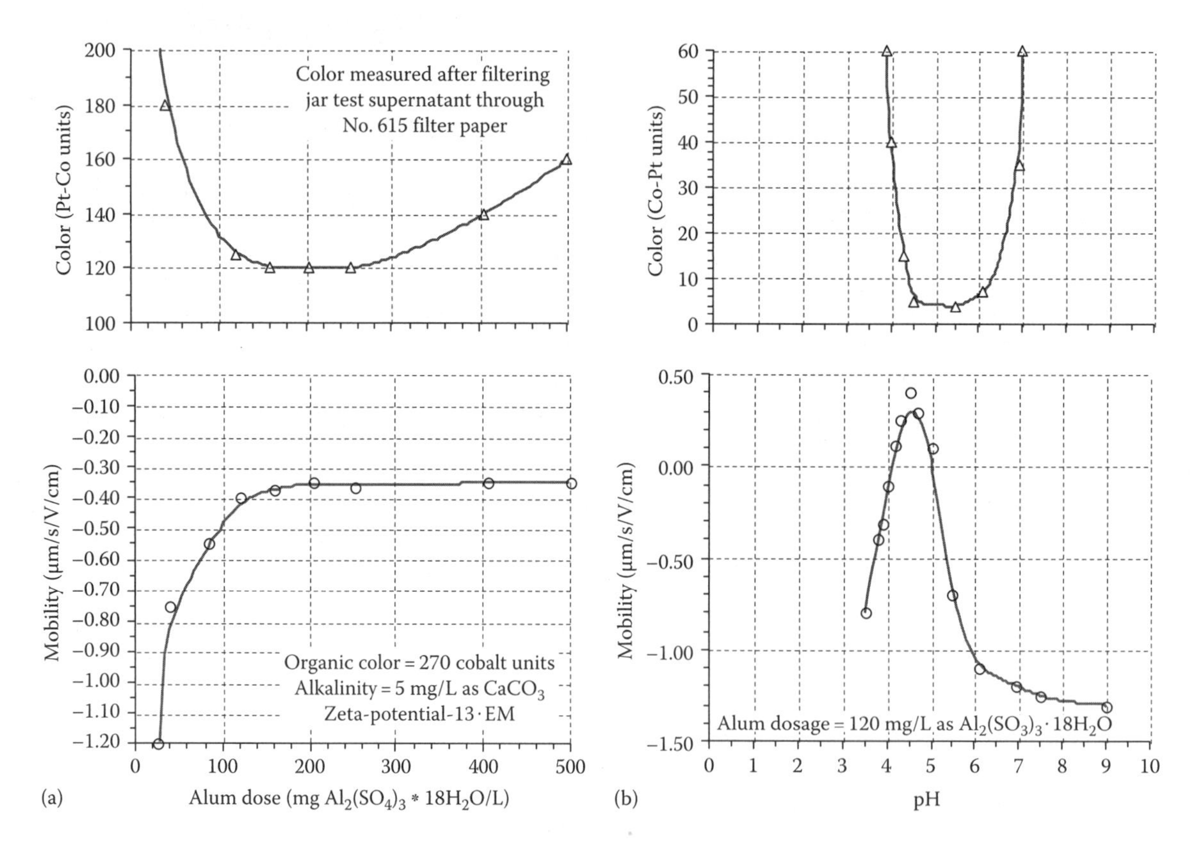

FIGURE 9.4 Role of alum dosage and pH on EM and residual color (a) Effect of alum dosage (b) Effect of pH at constant alum dosage. (From Black, A.P. and Willems, D.G., *J. Am. Water Works Assoc.*, 53(5), 592, May 1961. With permission.)

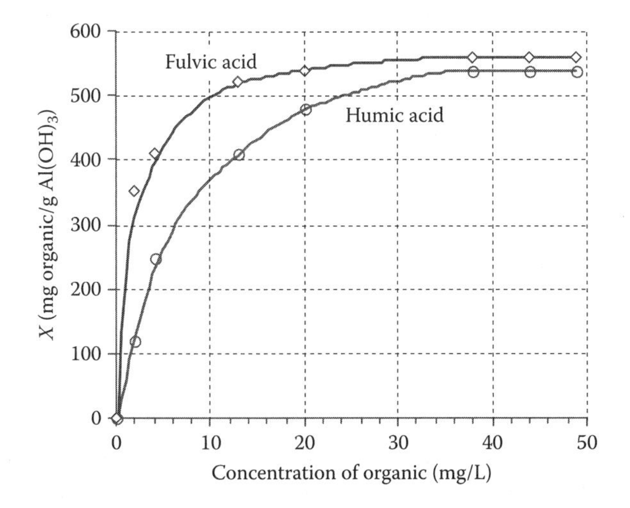

FIGURE 9.5 Isotherm showing equilibrium concentrations of organics in solution and adsorbed to $Al(OH)_3$. (Adapted from Mazet, M. et al., *Water Res.*, 24(12), 1509, December 1990.)

for HA and FA) in equilibrium with preformed aluminum hydroxide precipitate. Maximum capacities of the aluminum hydroxide precipitate for HA and FA were about 540 mg HA/g $Al(OH)_3$ and 560 mg FA/g $Al(OH)_3$. About 0.8 fraction of the uptake occurred within 30 min with stirring at 50 rpm.

9.3.2.3.4 Summary for NOM Removal

Chemical reactions that relate to NOM removal (Edzwald, 1993, p. 29) include

1. Under neutral or acidic pH conditions, humic and FA organic ligands may complex with Al, creating a demand for Al that must be satisfied before precipitation of $Al(OH)_3$ can occur.
2. Regarding Al hydrolysis products, the organic ligands compete with OH^-.
3. NOM adsorbs on amorphous $Al(OH)_3$ precipitate.

9.3.2.4 Organics in Wastewaters

Wastewaters include municipal and industrial sources, which include a wide variety of organic substances. Municipal wastewaters include fats, proteins, carbohydrates, urea, etc. Industrial wastewaters may include variety of contaminants, including synthetic organics, depending on the industry.

9.3.2.4.1 Composition of Municipal Wastewater

Table 9.3 shows a distribution of organic matter sizes in a typical raw municipal wastewater and compared with the same fractions for effluent after secondary settling (provided here for reference). As seen, organic matter is given as volatile solids (VS), TOC, and chemical oxygen demand (COD), which are different laboratory measures of organic matter in water.

9.3.2.4.2 Chemical Treatment

Effluent from primary settled municipal wastewater can be rendered clear to the eye by alum coagulation, which can be demonstrated readily by a laboratory jar test. As noted (Section 9.3.1), lime and alum have long been recognized as agents that precipitate nonsettleable solids in water. In tertiary treatment of municipal wastewaters, that is, effluent from secondary clarifiers, lime is used frequently as a coagulant (e.g., City of Colorado Springs, Rawhide Power Plant, Wellington, Colorado).

9.3.2.5 Coagulation of Synthetic Organics

Regulated organics included some 615 contaminants (Champlin and Hendricks, 1993). In a spiking experiment

TABLE 9.3
Distribution of Organic Matter in Wastewater and Secondary Effluent

		VS		TOC		COD	
Substance	**Fraction**[a]	**(mg/L)**	**(%)**	**(mg/L)**	**(%)**	**(mg/L)**	**(%)**
Wastewater	Soluble	116	48	46	42	168	40
	Colloidal	23	9	12	11	43	10
	Supra-colloidal	43	18	22	20	87	21
	Settleable	59	25	29	27	120	29
	Total	*241*		*109*		*418*	
Secondary effluent[b]	Soluble	62	67	16.5	69	46	74
	Colloidal	6	7	1.5	6	3	5
	Supra-colloidal	24	26	6	25	13	21
	Settleable	0	0	0	0	0	0
	Total	*92*		*24*		*62*	

Source: Adapted from Rickert, D.A. and Hunter, J.V., *J. Water Pollut. Control Fed.*, 44(1), 135, January 1972.

[a] *Definitions*: Settleable, >100 μm; supra-colloidal, 1–100 μm; colloidal, 10^{-3}–1 μm; soluble, <1 μm.

[b] Treatment was conventional activated sludge.

with three synthetic organics (trichloroethylene, toluene, naphthalene), removals were only 5% for each by conventional filtration (Carlson et al., 1993). In a review of alum coagulation of synthetic organics percent removals were: DDT, 84–95; dieldrin, 10–55; endrin, 0–35; aldrin, 10; toxaphene, 0; parathion, 0; malathion, 0; 2,4-D herbicide, 0; rotenone insecticide, 0 (O'Melia et al., 1979, p. 598). In general, removal of synthetic organic carbon (SOCs) by coagulation is low or uncertain in effectiveness.

9.4 DOUBLE LAYER THEORY

The negative surface charge of a colloidal particle (along with small size) causes its special properties, that is, mutual repulsion of particles and lack of settling. From this, the "double layer" theory has evolved, which is reviewed in this section.

9.4.1 Double Layer Description

Figure 9.6 depicts, in a simple fashion, the three parts of the double layer theory. First, Figure 9.6a shows a negatively charged particle, for example, a clay colloid, which is the starting point of the theory. Second, Figure 9.6b shows the same particle with a "bound layer" of positive counterions, often called the Stern layer. Third, Figure 9.6c illustrates the "ion cloud" of mostly positive ions, often called the Gouy–Chapman layer that surrounds the particle.

9.4.1.1 Beginning

The names of several persons involved in colloid science from 1879 to 1948 are associated with the double layer theory. In 1879, H. von Helmholtz postulated an initial double layer theory, that is, that a negatively charged thin film of bound-water was attached to a surface, which was countered by a thicker layer of water molecules with positive charges. This was before the 1884 ion theory of Arrhenius and so the idea that the positive charges were "ions" was in the future (Sonon et al., 2001).

9.4.1.2 Surface Charge

Solids, such as clays, silica, hydrous metal oxides, pulp fibers, bacteria, take on a net electrostatic charge at the solid–water interface, which may be either negative or positive (usually its negative). Five possible causes of the charge at a solid surface include: differential ion solubility; ionization of surface groups; isomorphous ion substitution; specific ion adsorption; anisotropic crystals (Myers, 1991, p. 71). In each case, a charged chemical group is involved. To illustrate, assume that the surface group is COOH. In solution, dissociation occurs, that is, $COOH \rightarrow COO^- + H^+$. Therefore, the COO^- group that remains gives the surface a negative charge but the system retains overall neutrality, that is, with H^+ going into the solution.

9.4.1.3 Gouy–Chapman Model

The Gouy–Chapman model described a diffuse cloud of mobile positive point charges surrounding a negatively charged colloidal particle. The model, developed by G. Gouy in 1910 and D.L. Chapman in 1913, permitted calculation of distributions of both counterions and electric potential (O'Melia, 1969). The weakness of the theory when proposed was that it did not recognize that the point charges were positive ions (Sonon et al., 2001).

9.4.1.4 Fixed Layer

In 1924, Otto Stern by delineated a "fixed layer" of a colloid as including positive ions that were bound to a negatively charged particle. These two parts, the fixed layer and the ion cloud, were termed the "double layer" and comprised the beginning of modern theory of colloidal behavior (O'Melia, 1969; Myers, 1991, pp. 69–85, 423; Sonon et al., 2001).

9.4.1.5 Effect of Ionic Strength of Solution

The counterion cloud surrounding the colloid reacts differently depending upon the ionic strength of the solution in which the colloid is immersed. Figure 9.7 extends the ideas of Figure 9.6. Figure 9.7a through c depicts the Stern layer and ion cloud for the three illustrative conditions: (1) dilute solution, (2) concentrated ionic solution, and (3) mostly trivalent ions.

The second sequence, Figures 9.7d through f illustrates the electrostatic potentials (ψ potentials) that correspond to the double layer depictions in Figure 9.7a through c, respectively. In Figure 9.7d, the potential curve associated with a dilute

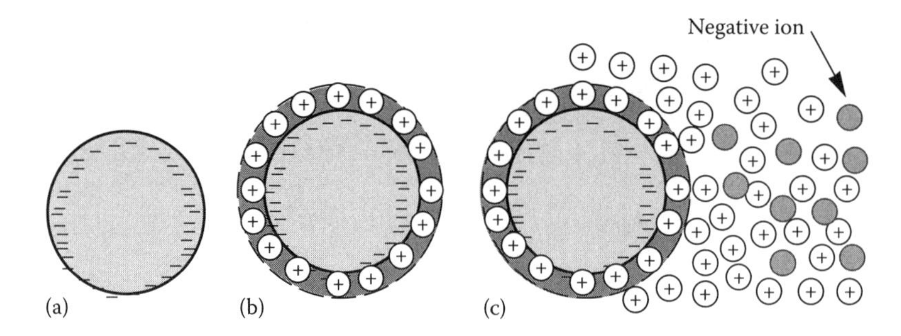

FIGURE 9.6 Colloidal particle showing: (a) negatively charged colloid surface; (b) colloid with positive counterions comprising fixed (Stern) layer; and (c) diffuse (Gouy–Chapman) layer. (Courtesy of Zeta-Meter, Inc., Long Island City, New York, 1988.)

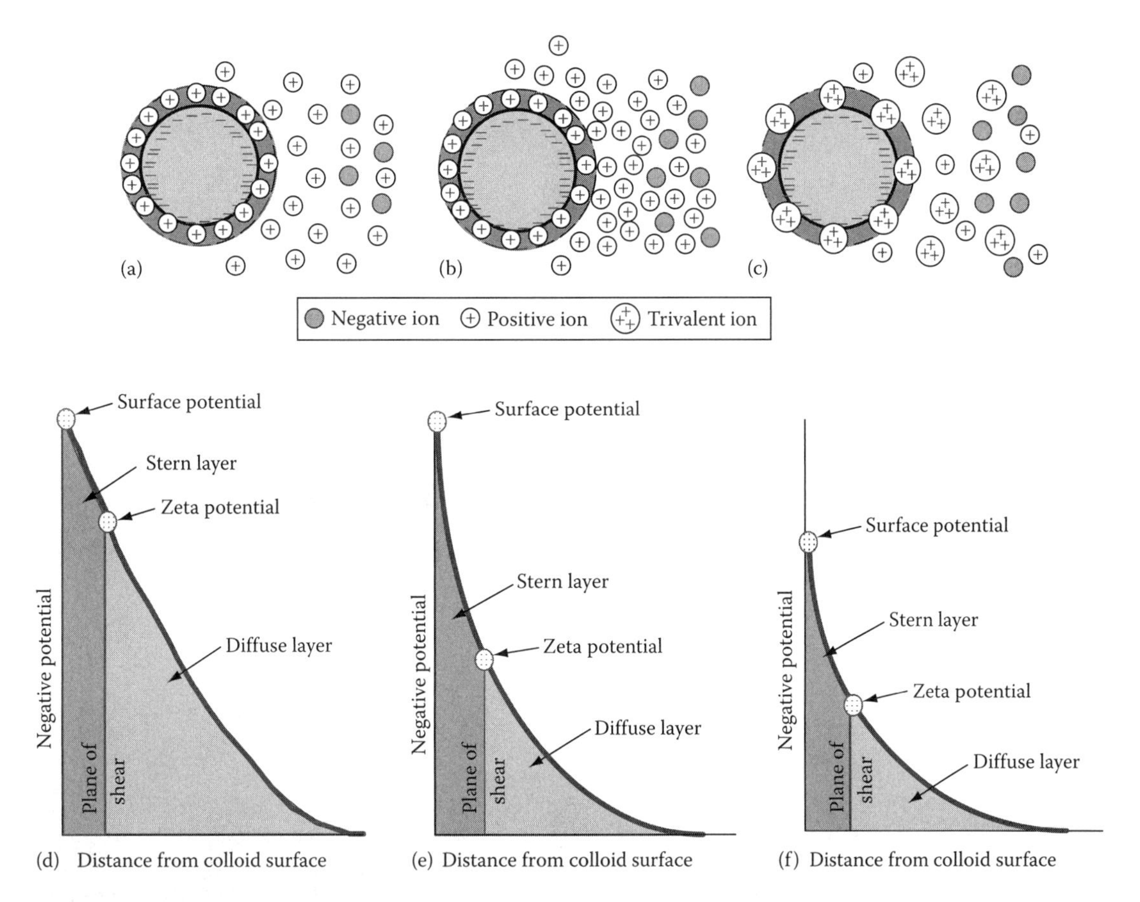

FIGURE 9.7 Negative potential of colloid, as affected by ionic strength, (a) and (b), and by charge neutralization, (c). Potential curve is "compressed" in (e) and (f), reducing negative zeta potential. (a) Stern layer and diffuse counterions in dilute ionic solution; (b) Stern layer and diffuse counterions in concentrated ionic solution; (c) Stern layer and trivalent counterions reducing surface potential; (d) electrostatic potential in dilute ionic solution; (e) electrostatic potential in concentrated ionic solution; and (f) electrostatic potential after addition of trivalent cationic coagulant. (Adapted from Singley, J.E. et al., *J. Am. Water Works Assoc.*, 63(2), 100, February 1971.)

solution is high. In Figure 9.7e, the curve is "compressed" due to a higher concentration of ions. In Figure 9.7f, the trivalent positive ions have two effects: (1) they lower the surface potential within the Stern layer, and (2) the lower the potential distribution within the ion cloud.

9.4.1.6 Electrostatic Potentials

Designations are given for two specific electrostatic potentials seen in Figure 9.7d through f. The anchor point for the potential curve is at the particle surface and is the "surface potential," sometime called the *Nernst* potential. If the charged particle moves relative to the water under the influence of an electric field, the Stern layer moves with it. The "shear plane" demarks the separation between the fixed ions and the ion cloud. The electrostatic potential at the shear plane is called the "zeta potential" and is a reference in coagulation theory. An objective in coagulation is to increase the zeta potential; for example, from perhaps −20 to −5 mV.

9.4.1.7 DLVO Theory

The Gouy–Chapman model as developed further is called the DLVO theory, after Derjaguin and Landau in a 1945 paper and Verwery and Overbeek in a 1948 paper. The DLVO theory combines the van der Waals intermolecular attractive forces with the electrostatic repulsive forces. The "net" potential due to addition of these two forces determines the strength of the colloid interactions.

Figures 9.8a through c show the same electrostatic curves as seen in Figure 9.7d through f, respectively, but adds the van der Waals potential curves, which are depicted in the lower half of each drawing. Combining the electrostatic and van der Waals curves in Figures 9.8a through c, gives the respective "net" potentials shown as the bold-line curves. The net potential curve has a characteristic shape, which starts positive near the particle surface, rises rapidly, and then falls. The net-potential curve at and near the particle surface is characterized as an "energy trap," that is, once two particles penetrate this zone attraction occurs and the two particles bond. The

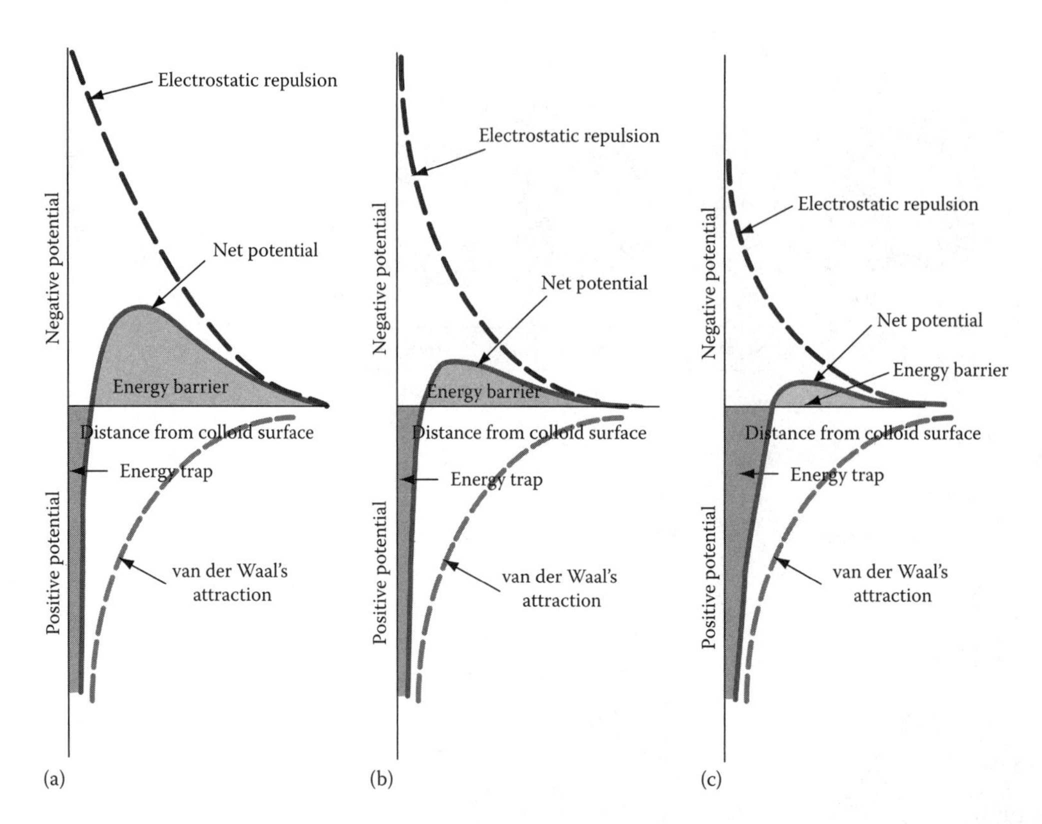

FIGURE 9.8 Net potential of colloid, as affected by ionic strength, (a) and (b), and by charge neutralization with polynuclear reaction products as in (c). Energy barrier is reduced in (b) and (c), by higher ionic strength of solution and by charge neutralization, respectively. (a) Net potential in dilute ionic solution; (b) net potential in concentrated ionic solution; and (c) net potential after addition of trivalent cationic ion coagulant. (Adapted from O'Melia, C.R., *Public Works*, 100, 90, May 1969; Weber, W.J., Jr., *Physicochemical Processes for Water Quality Control*, Wiley-Interscience, New York, 1972, p. 65; Gregory, J., *Particles in Water*, IWA Publishing, CRC/Taylor & Francis, Boca Raton, FL, 2006, p. 79.)

"hump" shown is an "energy-barrier," that is, another particle must have an energy level higher than the hump in order to penetrate to the energy-trap and bond.

The height of the potential barrier determines whether the particles may come together, which is requisite to "destabilize" the suspension, which is "effective" coagulation. In Figure 9.8a, the potential barrier is too high to permit contacts. Thus, an objective of coagulation is to lower this potential barrier. Figure 9.8b and c show lower potential barriers, caused by the two conditions, that is, higher ionic solution concentration and adding trivalent ions, respectively. Coagulation is achieved in practice by adding trivalent cations. Actually, however, several polynuclear hydrolysis species, discussed subsequently, are operative in lieu of the trivalent cations. The energy level of Brownian motion (that is, thermal energy) must be sufficient to overcome the energy barrier hump and to reach the attractive zone of the "energy trap."

A full review of the DVLO theory was given by Gregory (2006, pp. 63–92) who also provides quantitative theoretical calculations (for the curves such as in Figure 9.8) in terms of V_T/kT (where V_T is the net-potential curve). For example, for a 1:1 electrolyte solution concentration of 100 mM/L (0.01 mol/L), Figure 4.8 (p. 82) shows an energy barrier, that is, $V_T/kT \approx 35$, which gives $V_T \approx$ 90 kJ/mol, that is,

$$V_T \approx 35 \cdot kT \cdot N(\text{Avogadro}) \approx 35 \cdot (1.38 \cdot 10^{-23}\ \text{J/K molecule}) \cdot 300\ \text{K} \cdot (6.022\ \text{molecules/mol}).$$

This compares with the average thermal energy of particles of $3kT/2$ (Gregory, 2006, p. 79), which gives KE(particle) $\approx$ 3 kJ/mol, that is, much lower than the energy barrier. As the energy barrier is lowered by means of higher electrolyte concentration or by trivalent ions, the V_T energy curve is lowered with an energy barrier to near zero or negative. For the "energy-trap," that is, at separation distance between particles of about 0.5 nm, V_T("energy-trap")$/kT \approx -50$; thus, V_T("energy-trap") $\approx -50 \approx -2.1 \cdot 10^{-19}$ J/molecule $\approx$ -125 kJ/mol. This value for V_T("energy-trap") may be on the high side, since by comparison, Stumm and Morgan (1996, p. 517) give 10–40 kJ/mol. Also, by comparison, the bonding energy of a covalent bond is about 100–300kT (Myers, 1991, p. 39).

9.5 TRIVALENT METAL IONS: REACTIONS WITH WATER

The only trivalent coagulant chemicals are Al^{3+} and Fe^{3+}, with Al^{3+} being the most common in practice. As noted, these ions react with water and form "complexes"; the resultant

TABLE 9.4
Molecular Weights of Alum and Ferric Forms

Atomic Weights		Aluminum Forms		Ferric Forms	
Element	**Atomic Weight**	**Compound**	**MW**	**Compound**	**MW**
Al	26.982	$Al_2(SO_4)_3 \cdot 14.3H_2O$[a]	599.536	$Fe_2(SO_4)_3 \cdot 4.5H_2O$[b]	480.872
Fe	55.845	$Al_2(SO_4)_3 \cdot 14H_2O$	594.136	$Fe_2(SO_4)_3$[b]	399.876
S	32.066	$Al_2(SO_4)_3 \cdot 18H_2O$	666.132	Fe_2O_3	159.687
O	15.999	$Al_2(SO_4)_3$	342.15	Fe_2	119.690
H	1.000	Al_2O_3	101.961	$FeCl_3$[c]	162.203
O	15.999	Al_2	53.964	$FeCl_3 \cdot 6H_2O$[c]	270.197
Cl	35.4527				

[a] The number of waters of hydration for aluminum sulfate is sometimes given as 14.3; the number recommended by C. Lind (General Chemical) was 14, that is, without the decimal.
[b] The number of waters of hydration for ferric sulfate is 4.5, which is an approximate value (AWWA Standard B406-97, AWWA, 1997). The hydrated form dissolves readily; the anhydrous form does not. The liquid form contains about 50% of the dry form by weight.
[c] Ferric chloride is available in liquid form; waters of hydration are not included in the molecular weight (AWWA Standard B407-98). The AWWA Standard is for the liquid form only. The solid form is available in two forms: hexahydrate and anhydrous.

species depends on the pH and concentration (that is, ionic strength) of the solution.

9.5.1 Aluminum and Ferric Ions

The trivalent ions, aluminum, Al^{3+}, and ferric iron, Fe^{3+}, have similar reactions with water. The difference between the two is mostly in the equilibrium constants, that is, not in the kind of complexes formed.

9.5.1.1 Waters of Hydration

Knowledge of waters of hydration is needed to calculate the molecular weight of a given coagulant and to designate the coagulant used. In manufacturing the aluminum salt is hydrated as $Al_2(SO_4)_3 \cdot 14H_2O$, which is commercial alum. The reagent grade alum is hydrated as $Al_2(SO_4)_3 \cdot 18\ H_2O$. Ferric sulfate is hydrated as $Fe_2(SO_4)_3 \cdot 4.5\ H_2O$ (AWWA Standard B406-97, AWWA, 1997). Solid ferric chloride may be hydrated as $Fe_2Cl_3 \cdot 6H_2O$, but may be anhydrous; in liquid form it is not hydrated (AWWA Standard B407-98, AWWA, 1998a).

9.5.1.2 Expressing Concentrations

The concentration of aluminum or ferric ion may be expressed in a variety of ways, for example, Al^{3+}, $Al_2(SO_4)_3$, $Al_2(SO_4)_3 \cdot 14H_2O$, $Al_2(SO_4)_3 \cdot 18H_2O$. For alum, the expression used most frequently is $Al_2(SO_4)_3 \cdot 14H_2O$, and is understood as a rule. Reagent grade alum, on the other hand, which is used often in research, is always expressed as $Al_2(SO_4)_3 \cdot 18H_2O$. In research, concentrations are often expressed as "mol/L," along with the mass concentration in "mg salt species/L." The important thing is to qualify the concentration in the terms of the expression used. Example 9.1 illustrates a calculation protocol for conversion between forms.

Table 9.4 lists chemical formulae for compounds of each of the two trivalent ions, that is, Al^{3+}, and Fe^{3+}, respectively. The associated molecular weights are listed which permits conversion from one form to another. The atomic weights are listed on the left for convenience.

Example 9.1 Conversion of Concentration Expressions

Problem

Suppose that 100 mg/L alum as $Al_2(SO_4)_3 \cdot 14H_2O$ is a required coagulant dosage. Determine equivalent expressions in terms of mg $Al_2(SO_4)_3$/L and mg Al^{3+}/L. Table 9.4 provides ready reference for atomic weights and molecular weights, respectively.

Solution

The ratio of the molecular weights is the basis for conversions. Thus,

1. To convert from $Al_2(SO_4)_3 \cdot 14H_2O$ as mg/L to mol/L,

$$\frac{100\ \text{mg}\ Al_2(SO_4)_3 \cdot 14H_2O}{\text{L solution}} \cdot \frac{Al_2(SO_4)_3 \cdot 14H_2O\ \text{mol}}{594{,}136\ \text{mg}\ Al_2(SO_4)_3 \cdot 14H_2O}$$
$$= \frac{0.0001683\ \text{mol}\ Al_2(SO_4)_3 \cdot 14H_2O}{\text{L solution}}$$

2. Regardless of the form of expression number of moles is the same. Therefore, in terms of $Al_2(SO_4)_3$,

$$\frac{0.0001683\ \text{mol}\ Al_2(SO_4)_3}{\text{L solution}} \cdot \frac{342{,}150\ \text{mg}\ Al_2(SO_4)_3}{\text{mol}\ Al_2(SO_4)_3}$$
$$= \frac{57.58\ \text{mg}\ Al_2(SO_4)_3}{\text{L solution}}$$

3. In the case of Al^{3+}, however, there are two atoms/molecule, that is,

$$\frac{0.0001683 \text{ mol } Al_2(SO_4)_3}{\text{L solution}} \cdot \frac{2 \cdot 26982 \text{ mg Al}}{\text{mol } Al_2} = \frac{9.08 \text{ mg Al}}{\text{L solution}}$$

Discussion

An important point, sometimes overlooked, is that there are 2 mol Al^{3+} per mol $Al_2(SO_4)_3 \cdot 14H_2O$, which is seen in the factor "2" in (3). The conversion procedure is relevant in understanding coagulation diagrams which usually have two scales, that is, mols Al^{3+}/L and mol $Al_2(SO_4)_3 \cdot 14H_2O$/L, for example, 100 mg $Al_2(SO_4)_3 \cdot 14H_2O$/L = 0.0003366 mol Al^{3+}/L = $3.34 \cdot 10^{-4}$ mol Al^{3+}/L (since there are 2 mol Al^{3+} per mol $Al_2(SO_4)_3 \cdot 14H_2O$). Also, for later reference, concentration may be expressed as p[Al^{3+}], that is, p[Al^{3+}] = −log[Al^{3+}] = −log[$3.34 \cdot 10^{-4}$ mol Al^{3+}/L] = 3.47.

9.5.1.3 Liquid Alum

Since its introduction in the 1950s, liquid alum has become more widely used than the hydrated solid crystal form. The reason is convenience and cost. Costs have been reduced due to the dissemination of distribution centers in the United States which has lowered the haul distance. Handling of a solid requires storage, metering, mixing, and cleaning. Liquid alum, on the other hand, is delivered by truck (or rail) to a storage tank and then is metered directly into the rapid-mix. If delivered at a specific gravity of 1.335, the corresponding alum concentration as $Al_2(SO_4)_3 \cdot 14H_2O$ is 647 mg/L. For reference, the equivalent concentration expressed as Al^{3+} is $(54/594) \cdot 647 = 58.8$ mg/L.

9.5.2 Alkalinity

Part of the lore of coagulation practice has been that alkalinity (Box 9.2) is necessary for coagulation. Both the traditional view and the modern view are described for reference.

BOX 9.2 ALKALINITY DEFINED

Alkalinity is defined as the sum of HCO_3^-, CO_3^{2-}, and OH^-. By convention, alkalinity is expressed in terms of the $CaCO_3$ equivalent with molecular weight 100. To illustrate, let the HCO_3^- concentration be say 48 mg/L as HCO_3^-, which is $(100/61) \cdot 48 = 78.7$ mg/L as $CaCO_3$. The total alkalinity then is the sum of the concentrations of all three, expressed as $CaCO_3$. The $CaCO_3$ expression is also used for Ca^{2+} and Mg^{2+}, the sum of which is defined as "hardness."

9.5.2.1 Role of Alkalinity as a Buffer

For many years alum chemistry was described in terms of its reaction with alkalinity. The well-known "classical" reaction between alum and the alkalinity in water is (Black, 1948, p. 142),

$$Al_2(SO_4)_3 + 3Ca(HCO_3)_2 + 6H_2O \rightarrow 3CaSO_4 + 2Al(OH)_3 + 6H_2CO_3 \quad (9.5)$$

After omitting the "spectator" ions, Ca^{2+} and SO_4^{2-}, the reaction becomes

$$2Al^{3+} + 6HCO_3^- + 6HOH \rightarrow +2Al(OH)_3 + 6H_2CO_3 \quad (9.6)$$

An equivalent depiction focuses on the idea of Al^{3+} reacting with H_2O to form $Al(OH)_3$, releasing H^+ as in Equation 9.7. Removing the H^+, by the HCO_3^- buffer as in Equation 9.8, drives the reaction to the right, forming H_2CO_3. Summing the two equations, that is, Equation 9.9, the result is the same as Equation 9.6, but the emphasis is on HCO_3^- as a buffer.

$$Al^{3+} + 3H_2O \rightarrow Al(OH)_3 + 3H^+ \quad (9.7)$$

$$3HCO_3^- + 3H^+ \rightarrow 3H_2CO_3 \quad (9.8)$$

$$Al^{3+} + 3HCO_3^- + 3H_2O \rightarrow Al(OH)_3 + 3H_2CO_3 \quad (9.9)$$

The reactions are stoichiometric, meaning that the $Al(OH)_3$ precipitate will be produced in proportion to the availability of HCO_3^-. Regarding carbonate equilibria, HCO_3^- is predominant in the range $4.35 < pH < 10.33$ which are the respective pK_a's that separate HCO_3^- from H_2CO_3 and CO_3^{2-}, respectively.

9.5.2.2 Effect of Alkalinity on Demand for Alum

That "alum-demand" is proportional to alkalinity is seen in the experimental plot, Figure 9.9a, which is consistent with Equation 9.9. Alum-demand was defined as the "critical coagulant concentration" (CCC) to achieve a zeta potential of −5 mV for the suspension being treated, which was the zeta potential that corresponded to minimum settled water turbidity. Figure 9.9b shows that as the alkalinity increases the settled water turbidity declines toward an asymptote and that the residual pH hovers near neutral (Tseng et al., 2000).

9.5.2.3 Effect of Alum on pH

Alum and ferric iron act as "Bronsted acids," which means that they may donate a proton (H^+ ion) to the solution, thus depressing the pH. The effect is marked for a low alkalinity water (i.e., a water that has little buffer capacity), as demonstrated in Figure 9.10. Figure 9.10 shows a pH depression from pH = 5.0 at 40 mg/L alum to pH = 4.2 at 120 mg/L alum. As seen, alkalinity will react with the H^+ to maintain pH levels. Without alkalinity as a buffer, the H^+ generated acts to depress the pH. In another example, for pilot plant

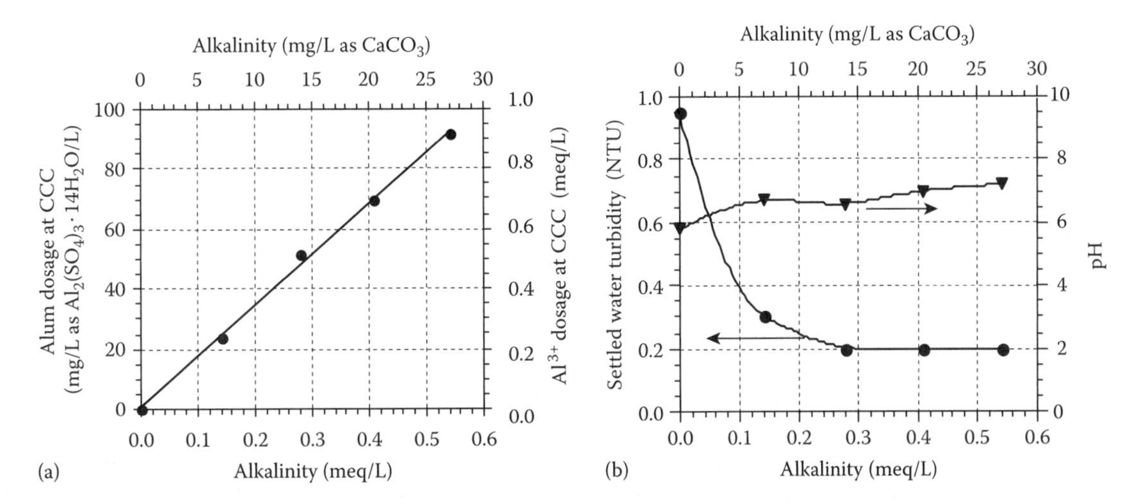

FIGURE 9.9 Results of experiments to assess the role of alkalinity for two low turbidity, low alkalinity snowmelt waters characterized by turbidities ≤ 0.5 NTU and alkalinities 30–50 mg/L as $CaCO_3$: (a) Critical alum dosage (CCC) at minimum turbidity versus alkalinity added and (b) settled water turbidity and pH versus alkalinity added. (Adapted from Tseng, T. et al., *J. Am. Water Works Assoc.*, 92(6), 48, June 2000.)

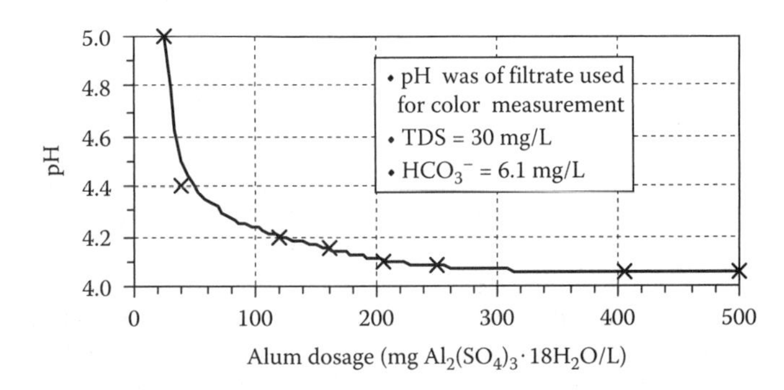

FIGURE 9.10 pH as affected by alum dosage. (Adapted from Black, A.P. and Willems, D.G., *J. Am. Water Works Assoc.*, 53(5), 593, May 1961.)

experiments over a 12 month period, the raw water varied $7.0 \leq pH \leq 7.7$, $27 \leq$ alkalinity ≤ 30 mg $CaCO_3$/L; with alum dose 26 mg $Al_2(SO_4)_3 \cdot 14H_2O$/L, the pH just after the rapid-mix unit was depressed to the range $5.8 \leq pH \leq 6.3$ (Hendricks et al., 2000, p. 16).

9.5.3 Reactions between Alum/Ferric Iron and Water

When alum or ferric iron is added to water, the reaction product is a *complex* with six water *ligands* (i.e., the six waters each share a *coordinated bond* with the *central metal ion*). The formula, $Al(H_2O)_6^{3+}$, was postulated from empirical and theoretical considerations and later, in 1972, confirmed by "nuclear-magnetic-resonance" spectra (Nordstrom and May, 1996, p. 45).

9.5.3.1 Beginning

The idea that ferric iron or alum occur as free trivalent ion began to be questioned by about 1960 when A.P. Black (1960) recognized that they may be present only at $pH < 3.0$ and $pH < 4.5$, respectively. Black noted also that when either alum or a ferric salt is added to water *hydrolysis* occurs. The hydrolysis products were more effective than trivalent ions in reducing the zeta potential of particles (Black, 1960; Black and Willems, 1961, p. 597). At the same time, Packham (1962) recognized the role of aluminum hydroxide precipitate, that is, that clay particles are "enmeshed" in the "mass of flocculating aluminum hydroxide" (later called "sweep floc").

9.5.3.2 Sequential Hydrolysis Reactions

Aluminum or ferric iron salts react with water forming a wide array of complexes or precipitate, depending on pH. To illustrate, a postulated sequence of hydrolysis reactions for Al^{3+}, that is, involving the "splitting" of the water molecule is (Stumm and Morgan, 1962; Amirtharajah and Mills, 1982, p. 210),

$$Al(H_2O)_6^{3+} + H_2O \rightarrow Al(H_2O)_5(OH)^{2+} + H_3O^+ \quad (Al.1)$$

$$Al(H_2O)_5(OH)^{2+} + H_2O \rightarrow Al(H_2O)_4(OH)_2^+ + H_3O^+ \quad (Al.2)$$

$$Al(H_2O)_4(OH)_2^+ + H_2O \rightarrow Al(H_2O)_3(OH)_3(s) + H_3O^+ \quad (Al.3)$$

$$Al(H_2O)_3(OH)_3(s) + H_2O \rightarrow Al(H_2O)_2(OH)_4^- + H_3O^+ \quad (Al.4)$$

$$Al(H_2O)_2(OH)_4^- + H_2O \rightarrow Al(H_2O)(OH)_5^{2-} + H_3O^+ \quad (Al.5)$$

$$Al(H_2O)(OH)_5^{2-} + H_2O \rightarrow Al(OH)_6^{3-} + H_3O^+ \quad (Al.6)$$

The foregoing are "proton-transfer" reactions and are essentially instantaneous, limited in rate only by mixing and diffusion (Nordstrom and May, 1996, p. 44). The hydration waters are "ligands" (see glossary). A similar reaction sequence occurs if Fe^{3+} is added to water.

9.5.3.2.1 Metal Ion Polymers

In addition to the products shown in Equations Al.1 through Al.6, examples of others include: $Al_2(OH)_{34}{}^{5+}$; $Al_8(OH)_{20}{}^{4+}$; $Al_6(OH)_{15}{}^{3+}$; $Al_7(OH)_{17}{}^{4+}$ (Amirtharajah and Mills, 1982, p. 211). These are "hydroxo metal complexes" that readily adsorb on surfaces, and, at the same time, are polymers, that is, include repeating units. These products result from other hydrolysis reactions, with about 15 identified by various authors, for example, Stumm and Morgan (1962); Stumm and O'Melia (1968); O'Melia (1979); and Amirtharajah and Mills (1982).

At "equilibrium" in the sequence of hydrolysis reactions, a distribution of aluminum-complex species results. The distribution depends on the concentration of the Al^{3+} (or Fe^{3+}) added and the resulting pH. The distribution of species may be calculated by writing the equilibrium statements for each reaction, the mass balance equation, and imposing the condition of electroneutrality (a topic in water chemistry).

9.5.3.2.2 Brevity in Writing Equations

The correct depiction of the hydrolysis reactions, from the standpoint of accepted theory, is as indicated in Equations Al.1 through Al.6, that is, with the water ligands. Often, for brevity in equation writing, however, the water ligands are omitted. Thus, repeating the first two equations, that is, Equations Al.1 and Al.2, without the water ligands gives,

$$Al^{3+} + H_2O \rightarrow AlOH^{2+} + H^+ \quad (9.10)$$

$$AlOH^{2+} + H_2O \rightarrow Al(OH)_2{}^+ + H^+ \quad (9.11)$$

The discussion here favors retaining the water ligands in the equations since they are primary participants in the reactions.

9.5.3.2.3 Aluminum/Ferric Iron Hydroxide Precipitate

The theory of coagulation focused first on the double layer theory. The second stage of theory delineated the chemistry of metal ion hydrolysis. In practice, however, $pH > 6.0$ (usually), and so aluminum hydroxide forms as an amorphous precipitate and is the predominant species, enmeshing colloids (Stumm and O'Melia, 1968, p. 523), which is called "sweep-floc" (O'Melia, 1979; Amirtharajah and Mills, 1982).

9.5.3.3 Species Equilibrium

As with any reaction equation, those of Equations Al.1 through Al.6, may be expressed as equilibrium equations; then, taking negative logs of each side gives p(concentration) versus pH. Table 9.5 includes Equations Al.1 and Al.4 and three others as deemed important in coagulation (Amirtharajah and Mills, 1982) in terms of the reaction, equilibrium statement, and pC versus pH, respectively. Figure 9.11 illustrates the associated equilibrium lines constructed from the respective log-form of the equilibrium equations.

The equilibrium lines of Figure 9.11 are obtained as indicated from the respective rows in Table 9.5, that is, (1) write the reaction expression, (2) write the associated equilibrium statement, and (3) take the logs of each side of the equilibrium statement, then multiply by "−1" and write the equation in terms of "p," and (4) plot p[concentration-of-a-given species] versus pH. Alternatively, plot [concentration-of-a-given species] on a log-scale versus pH, in which [concentration-of-a-given species] $= 10^{-pC}$.

Example 9.2 illustrates the method of developing the logarithmic expressions, such as given in Table 9.5, and from this constructing an equilibrium diagram. Example 9.3

TABLE 9.5
Equilibrium Relations for Hydrolysis Reactions[a]

Reaction	Equilibrium Statement	Logarithmic Form
$Al^{3+} + 3H_2O \Leftrightarrow Al(OH)_3(s) + 3H^+$	$K_o = \frac{[H^+]^3}{[Al^{3+}]} = 10^{+10.4}$	$p[Al^{3+}] = 3pH + pK_o$ $pK_o = -10.4$
$Al^{3+} + H_2O \Leftrightarrow Al(OH)^{2+} + H^+$	$K_{1,1} = \frac{[Al(OH)^{2+}][H^+]}{[Al^{3+}]} = 10^{-5.55}$	$\log K_{1,1} = \log[Al(OH)^{2+}] + \log[H^+] - \log[Al^{3+}]$ $-\log K_{1,1} = -\log[Al(OH)^{2+}] - \log[H^+] + \log[Al^{3+}]$ $pK_{1,1} = p[Al(OH)^{2+}] + p[H^+] - p[Al^{3+}]$ $p[Al(OH)^{2+}] - p[Al^{3+}] = -pH + pK_{1,1}$ $p[Al(OH)^{2+}] - p[Al^{3+}] = -pH + pK_{1,1}$ $pK_{1,1} = +5.55$
$Al(OH)_3(s) + 2H^+ => AlOH^{2+} + 2H_2O$	$K_5 = [AlOH^{2+}]/[H^+]^2 = 10^{-4.85}$	$p[AlOH^{2+}] = 2pH - pK_5$ $pK_5 = +4.85$
$Al(OH)_3(s) + H_2O \Leftrightarrow Al(OH)_4{}^- + H^+$	$K_4 = [Al(OH)_4{}^-][H^+] = 10^{-12.35}$	$p[Al(OH)_4{}^-] = -pH + pK_4$; $pK_4 = 12.35$
$8Al^{3+} + 20H_2O \Leftrightarrow Al_8(OH)_{20}{}^{4+} + 20H^+$	$K_{8,20} = \frac{[Al_8(OH)_{20}{}^{4+}][H^+]^{20}}{[Al^{3+}]^8} = 10^{-68.7}$	$p[Al_8(OH)_{20}{}^{4+}] = 4pH + pK_5$ $pK_5 = -14.5$

Source: Amirtharajah, A. and Mills, K.M., *J. Am. Water Works Assoc.*, 74(4), 210, April 1982.

[a] The water ligands are omitted for brevity. For reference, however, the second and fourth equations with the water ligands included are seen as Equations Al.1 and Al.4, respectively. Documentation that showed the water ligands for the first, third, and fifth equations was not found.

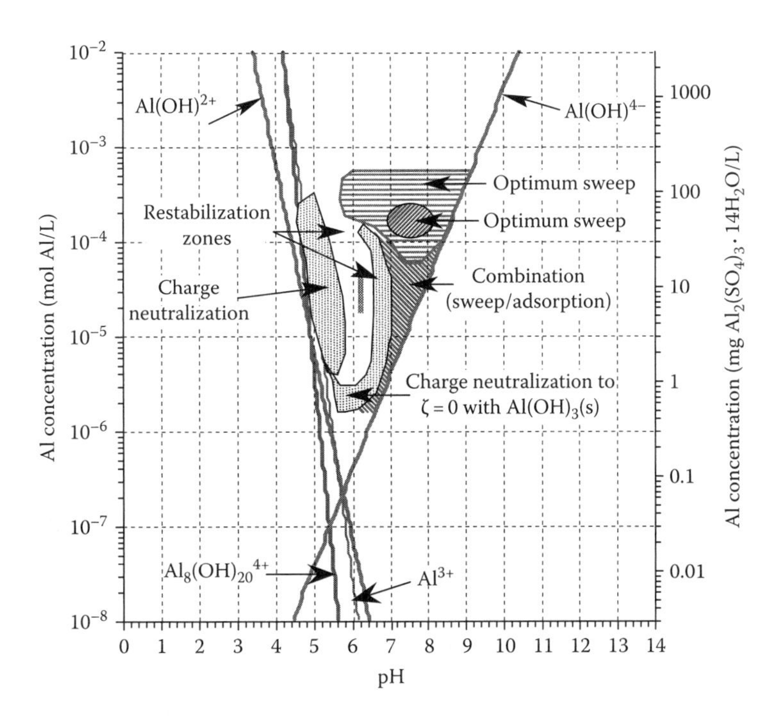

FIGURE 9.11 Zones of coagulation as affected by alum dosage and pH. [Multiply left y scale in mol Al/L by (594,000/2) to get concentration in mg $Al_2(SO_4)_3 \cdot 14H_2O$/L.] (From Amirtharajah, A. and Mills, K.M., *J. Am. Water Works Assoc.*, 74(4), 210, April 1982.)

demonstrates how to convert between concentration in mg/L and mol/L, which is provided for convenience (often chemistry fundamentals are not at the fingertips of those who work in the area infrequently).

Tables CD9.6 and CD9.7 demonstrate further, that is, in specific detail, the construction of equilibrium diagrams for aluminum and ferric additions to water, respectively. The solution steps are enumerated along with the linked plot showing the associated equilibrium lines.

Example 9.2 Illustrate Construction of an Equilibrium Line for Coagulation Diagram

Given

Let the reaction equation be the equilibrium between aluminum hydroxide precipitate, $Al(OH)_3$ (am), and the aluminate ion, $Al(OH)_4^-$.

Solution

The reaction equation is, from Table 9.5, Equation Al.4,

$$Al(OH)_3(s) + H_2O \rightarrow Al(OH)_4^- + H^+ \quad \text{(Al.4)}$$

The equilibrium statement is

$$K_4 = [Al(OH)_4^-]\,[H^+] = 10^{-12.35} \quad \text{(Ex9.2.1)}$$

The logarithmic form is

$$\log\,[Al(OH)_4^-] + \log\,[H^+] = \log K_4 = -12.35 \quad \text{(Ex9.2.2)}$$

Multiplying each side by "−,"

$$-\log\,[Al(OH)_4^-] - \log\,[H^+] = -\log K_4 = +12.35 \quad \text{(Ex9.2.3)}$$

In p-form,

$$p[Al(OH)_4^-] + pH = pK_4 = +12.35 \quad \text{(Ex9.2.4)}$$

For plotting, the form is

$$p[Al(OH)_4^-] = -pH + pK_4, \quad pK_4 = +12.35 \quad \text{(Ex9.2.5)}$$

or

$$p[Al(OH)_4^-] = -pH + 12.35 \quad \text{(Ex9.2.6)}$$

From Equation Ex9.2.3, the equilibrium line that gives the $Al(OH)_3$ solid precipitate and $[Al(OH)_4^-]$ relationship versus pH may be plotted, for example,

- pH = 1.0, $p[Al(OH)_4^-] = 11.35$, or $\log[Al(OH)_4^-] = -11.35$, or $[Al(OH)_4^-] = 10^{-11.35}$ mol/L = $4.47 \cdot 10^{-12}$ mol/L)
- pH = 5.0, $p[Al(OH)_4^-] = 7.35$, or $\log[Al(OH)_4^-] = -7.35$, or $[Al(OH)_4^-] = 10^{-7.35}$ mol/L = $4.47 \cdot 10^{-8}$ mol/L)
- pH = 11, $p[Al(OH)_4^-] = 1.35$, or $\log[Al(OH)_4^-] = -1.35$, or $[Al(OH)_4^-] = 10^{-1..35}$ mol/L = $4.47 \cdot 10^{-2}$ mol/L)

pH = 10, $p[Al(OH)_4^-] = 2.35$, or $\log[Al(OH)_4^-] = -2.35$, or $[Al(OH)_4^-] = 10^{-2.35}$ mol/L = $4.47 \cdot 10^{-3}$ mol/L).

TABLE CD9.6
Distribution of Aluminum Ion Hydrolysis Species with Varying pH

1 Reactions				2 Equilibrium Statements	3 Logarithmic Forms	
	Equilibrium Constant					
Reaction	Arithmetic	Log K	Reference	(a) Basic Form		pK
(1) $Al^{3+} + 3H_2O \rightarrow Al(OH)_3\ (s) + 3H^+$	$K_1 = 10^\wedge +8.5$	8.5	Morel & Hering, p. 283	$K_1 = \frac{[H^+]^3}{[Al^{3+}]}$	$-\log[Al^{3+}] = -3\log[H^+] + \log K_1$ $p[Al^{3+}] = 3pH - pK_1$	8.5
(2) $Al^{3+} + 2H_2O \rightarrow Al(OH)_2{}^+ + 2H+$	$K_2 = 10^\wedge +9.3$	9.3	Morel & Hering, p. 283			
(3) $Al^{3+} + H_2O \rightarrow AlOH^{2+} + H^+$	$K_3 = 10^\wedge + 4.97$	4.97	Morel & Hering, p. 283			
To calculate $Al(OH)_2{}^+$, subtract (1) from (2) to obtain,						
(4) $Al(OH)_3\ (s) + H^+ \rightarrow Al(OH)_2{}^+ + H_2O$	$K_4 = 10^\wedge + 0.8$	0.8	Calculated	$K_4 = \frac{[Al(OH)_2^+]}{[H^+]}$	$-\log[Al(OH)_2{}^+] = -\log\{H^+] - \log K_4$ $p[Al(OH)_2{}^+] = pH + pK_4$	0.8
To calculate $AlOH^{2+}$, subtract (3) from (1) to obtain,						
(5) $Al(OH)_3\ (s) + 2H^+ \rightarrow AlOH^{2+} + 2H_2O$	$K_5 = 10^\wedge + 3.53$	3.53	Calculated	$K_5 = \frac{[AlOH^{2+}]}{[H^+]^2}$	$-\log[AlOH^{2+}] = -2\log\{H^+] - \log K_5$ $p[AlOH^{2+}] = 2pH + pK_5$	3.53
(6) $Al(OH)_3\ (s) + H_2O \rightarrow Al(OH)_4{}^- + H^+$	$K_4 = 10^\wedge - 14.5$	14.5	Stumm & Morgan, p. 275	$K_4 = [Al(OH)_4{}^-][H+]$	$\log[Al(OH)_4{}^-] = -pH + \log K_4$	14.5
(7) $8Al^{3+} + 20H_2O \rightarrow Al_8(OH)_{20}{}^{4+} + 20H^+$	$K_8 = 10^\wedge - 68.7$	−68.7		$K_8 = \frac{[Al_8(OH)_{20}{}^{4+}][H^+]^{20}}{[Al^{3+}]^8}$	$\log[Al_8(OH)_{20}{}^{4+}] = -4pH + \log K_8 - 8\log K_5$	−14.5

4 Calculation of Data for pC vs. pH Diagram

pH	(1) $p[Al^{3+}]$	(4) $p[Al(OH)_2{}^+]$	(5) $p[AlOH^{2+}]$	(6) $p[Al(OH)_4{}^-]$	(7) $p[Al_8(OH)_{20}{}^{4+}]$
0	−8.5	0.8	−3.53	14.5	−14.5
1	−5.5	1.8	−1.53	13.5	−10.5
2	−2.5	2.8	0.47	12.5	−6.5
3	0.5	3.8	2.47	11.5	−2.5
4	3.5	4.8	4.47	10.5	1.5
5	6.5	5.8	6.47	9.5	5.5
6	9.5	6.8	8.47	8.5	9.5
7	12.5	7.8	10.47	7.5	13.5
8	15.5	8.8	12.47	6.5	17.5
9	18.5	9.8	14.47	5.5	21.5
10	21.5	10.8	16.47	4.5	25.5
11	24.5	11.8	18.47	3.5	29.5
12	27.5	12.8	20.47	2.5	33.5
13	30.5	13.8	22.47	1.5	37.5
14	33.5	14.8	24.47	0.5	41.5

5 Plot[a]

pH
1 2 3 4 5 6 7 8 9 10 11 12 13 14
pC
0 1 2 3 4 5 6 7 8 9 10 11 12 13 14
$Al_8(OH)_{20}{}^{4+}$
$Al(OH)_4{}^-$
$Al(OH)_3$
$Al(OH)_2$
$Al(OH)_2{}^+$
Al^{3+}

TABLE CD9.7
Determining the Distribution of Ferric Iron Hydrolysis Species with Varying pH

1 Reactions			2 Equilibrium Statements	
Reaction	Equilibrium Constant	Table 9.7 Equation	(a) Basic Form	(b) Rearranging
3 Taking logarithms of each expression				
4 Calculation of data for log concentration diagram				
5 Plot				

Note: Only table headings are shown in text; the spreadsheet may be downloaded.

Discussion
Figure 9.11 shows a plot of $[Al(OH)_4^-]$ versus pH, along with the equilibrium lines for the other species listed in Table 9.5. Table CD9.6 demonstrates, in spreadsheet form, the construction of the same diagram in the form, $p[Al(OH)_4^-]$ versus pH. With respect to nomenclature, recall that the brackets, "[−]," represent concentration in mol/L.

Example 9.3 Illustrate Conversion of Concentration in mol/L to mg/L for Alum

Given
Assume that hydrated aluminum sulfate is added to water at 50 mg $Al_2(SO_4)_3 \cdot 14H_2O$

Required
Determine the concentration of $[Al_2(SO_4)_3 \cdot 14H_2O]$ in mol/L

Solution
1. $MW[Al_2(SO_4)_3 \cdot 14H_2O] = 594$ g/mol (Table 9.4)
2. Calculate molar concentration,

$$[Al_2(SO_4)_3 \cdot 14H_2O] = 50\frac{mg\,Al_2(SO_4)_3 \cdot 14H_2O}{L} \cdot \frac{g}{1000\,mg} \cdot \frac{mol\,Al_2(SO_4)_3 \cdot 14H_2O}{594\,g\,Al_2(SO_4)_3 \cdot 14H_2O} = 0.000084\frac{mol\,Al_2(SO_4)_3 \cdot 14H_2O}{L}$$

On the log scale, this concentration reads, log $[0.000084] = -4.076$, or $[Al_2(SO_4)_3 \cdot 14H_2O] = 10^{-4.076}$ mol/L.

Discussion
The conversion factor is seen to be 594,000. Thus to convert from mol/L to mg/L, multiply mol/L by 594,000 mg $Al_2(SO_4)_3 \cdot 14H_2O$/mol $Al_2(SO_4)_3 \cdot 14H_2O$. As noted in Example 9.1, however, caution should be exercised when Al only is expressed in mol/L as there are two Al atoms in an $Al_2(SO_4)_3 \cdot 14H_2O$ molecule. Thus, $10^{-4.075}$ mol $Al_2(SO_4)_3 \cdot 14H_2O/L \cdot 2$ mol Al^{3+}/mol $Al_2(SO_4)_3 \cdot 14\,H_2O = 2 \cdot 10^{-4.075}$ mol $Al^{3+}/L = 1.68 \cdot 10^{-4}$ mol Al^{3+}/L.

9.5.3.4 Coagulation Zones

Table 9.5 lists five selected reaction equations for alum as a coagulant (Amirtharajah and Mills, 1982). The respective equilibrium statements and their logarithmic forms (as "p" concentration which is the negative log of concentration in mol/L) are also shown. Figure 9.11 shows the corresponding equilibrium diagram as "species-concentration versus pH"; concentrations are in "mol Al/L" on the left-hand scale and in "mg $Al_2(SO_4)_3 \cdot 14H_2O)$/L" on the right-hand scale. Each line in the diagram depicts the respective equilibrium equation of Table 9.5.

The "coagulation diagram," as it is known, was assimilated into practice during the 1980s after its description by Amirtharajah and Mills (1982), building on equilibrium theory (Stumm and O'Melia, 1968). In the diagram, pH is the "master-variable," shown on the *x*-axis, with alum species concentrations on the *y*-axis. In the diagram, several "zones" of coagulation are identified, for example, charge-neutralization, sweep-floc, combination, and restabilization. Regarding the precipitate, the pH controls the effect as follows:

- $pH < 7.0$: "re-stabilization" is likely for the positively charged precipitate
- $7.0 < pH < 8.0$: at an alum dose of about 30 mg $Al_2(SO_4)_3 \cdot 14H_2O)$/L the charge is also positive, which will react with negatively charged colloids
- $pH > 8.0$ the precipitate is weakly negative

In practice, the diagram provides guidance on what to expect with different pH and coagulant dosage combinations, and on the zones to seek or avoid. Since "every water is different," a common expression, the zones should be confirmed by jar testing and/or pilot plant.

9.5.3.5 Spreadsheet Construction of Coagulation Diagrams

Table CD9.6 outlines in a spreadsheet the construction of the alum coagulation diagram, adopting the equations given in Table 9.5. The logic of the construction of the diagram is displayed stepwise along the columns of the spreadsheet. The starting point is to write the respective reaction equations and the associated equilibrium constants. The next step is to write the equations for equilibrium, and then their conversions to their logarithmic "p" forms. Using the equations, that is, pC versus pH, the tabular outputs of concentrations of the different species as a function of pH are calculated. From the tabular output, the pC versus pH diagram is obtained as an embedded plot (located below the tabular output). The pC

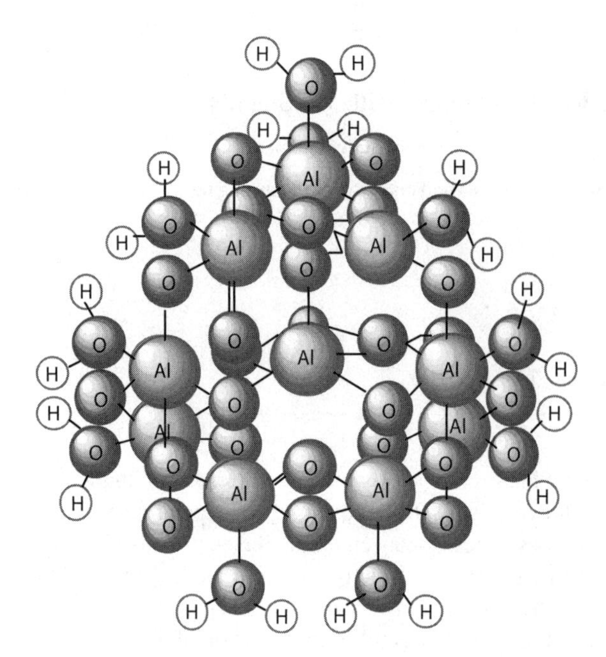

FIGURE 9.12 Ball-and-stick representation of $AlO_4Al_{12}(OH)_{24}(H_2O)_{12}^{7+}$. (Adapted from Bertsch, P.M. and Parker, D.R., Aqueous polynuclear aluminum species, in *The Environmental Chemistry of Aluminum*, 2nd edn., Sposito, G., Ed., CRC-Lewis Publishers, Boca Raton, FL, 1996, Chap. 4, Figure 2, p. 124.)

form may be modified to moles per liter by the conversion, $[mol/L] = 10^{-pC}$. Table CD9.7 is the corresponding ferric iron coagulation diagram.

9.5.3.6 Polynuclear Species

Figure 9.12 depicts a "ball-and-stick" model of an Al-hydrated complex (Bertsch and Parker, 1996, p. 124) and is indicative of the wide variety of such complexes. As seen, the water molecules are shown attached to Al central atoms with hydrogen atoms vulnerable to detachment. Of the variety of aluminum hydrolysis species, with a few given in Table 9.5, those that have convincing experimental support include: $Al_2(OH)_2^{4+}$, $Al_2(OH)_5^{+}$, $Al_3(OH)_8^{+}$, $Al_3(OH)_4^{5+}$, $Al_8(OH)_{20}(H_2O)_5^{4+}$, $Al_6(OH)_{12}(H_2O)_{12}^{6+}$, $Al_{54}(OH)_{144}(H_2O)_{36}^{18+}$, and $Al_{13}O_4(OH)_{24}(H_2O)_{12}^{7+}$ (Bertsch and Parker, 1996, p. 122).

9.5.3.7 Summary of Alum Speciation

The aqueous equilibrium chemistry of aluminum may be explained in terms of the following species (Dempsey et al., 1984):

- Three polymeric species: $Al_2(OH)_2^{4+}$, $Al_3(OH)_4^{5+}$, and $Al_{13}O_4(OH)_{24}^{7+}$
- Five monomers: Al^{3+}, $AlOH^{2+}$, $Al(OH)_2^{+}$, $Al(OH)_3$, and $Al(OH)_4^{-}$
- A solid precipitate, $Al(OH)_3(s)$

Regarding the alum chemistry of sweep-floc coagulation, two views are given by Bache et al. (1999, p. 210). The first is that the positive-charged Al monomers (see preceding paragraph), the effective species, form within ≤1 s before being transformed into an amorphous aluminum hydroxide precipitate with neutral charge. Therefore, the aluminum hydroxide coagulant must be dispersed uniformly, by effective rapid mix, within the first second of introduction. The second view is that positive-charged polynuclear species (see Section 9.5.3.6) such as Al13 are also formed, and are effective in coagulation; but such species are more stable and have a longer life and so the rapid dispersal of alum is not so critical. They emphasize, however, that mixing is important whatever the alum chemistry. As a note, the amorphous aluminum hydroxide precipitate, if allowed to age, ends up eventually as gibbsite and bayerite, the most stable forms (Marshall, 1964, p. 148). A conclusion is that alum chemistry is complex. A comprehensive review of alum chemistry is provided by Gregory (2006, pp. 121–153), who also gives assessments of the state of knowledge on the various facets of the topic.

9.6 SYNTHETIC ALUMINUM POLYMERS

About 1985, a new commercial product, synthetic polyaluminum chloride (PACl), appeared on the market; a similar product, poly-aluminum sulfate (PAS), appeared soon after.

9.6.1 Characteristics of PACl

The utility of PACl is that the synthesis process produces preformed charged polymers, for example, $Al_{13}O_4(OH)_{24}^{7+}$, as well as Al^{3+}, $Al(OH)^{2+}$, and $Al(OH)_4^{-}$ (Pernitsky and Edzwald, 2000). Such species retain their identity over time and over a broad pH range and are effective over the range of water treatment temperatures (Edzwald, 1993, p. 27). Both PACl and PAS are available in liquid form and are easy to use, for example, can be metered in liquid form.

9.6.1.1 Description of PACl

The manufactured PACl product is a pale yellow liquid, with properties: $1.12 < SG \leq 1.28$, $4.3 \leq$ viscosity ≤ 5.2 cp; $2.0 \leq pH \leq 2.6$ (PPG Industries, c. 2000). The product is available in tank trucks, 2000 lb plastic containers, and 55 gal plastic drums. The liquid is slightly corrosive and should be stored in tanks lined with materials such as epoxy, rubber, PVC, FRP, etc. The feed pump should be of material that will withstand acids The product is stable for several months if the temperature is maintained from −10°C to −40°C (15°F −96°F).

9.6.1.2 Electrophoretic Mobility: Comparing Alum and PACl

Figure 9.13a and b show EM of alum and PACl, respectively, as a function of pH in deionized water, that is, near zero particle concentration and negligible ion concentrations. In both plots, EM decreases as pH increases. Comparing, the two plots, that is, Figure 9.13a and b, the overall higher EM values for PACl is evident. Figure 9.13a shows that for alum EM is affected by temperature, with values lower at 25°C than at 4°C. Figure 9.13b shows that for PACl the temperature has no discernible effect.

The qualities indicated (high EM at all pH, no discernable temperature effect, and positive EM values even in deionized

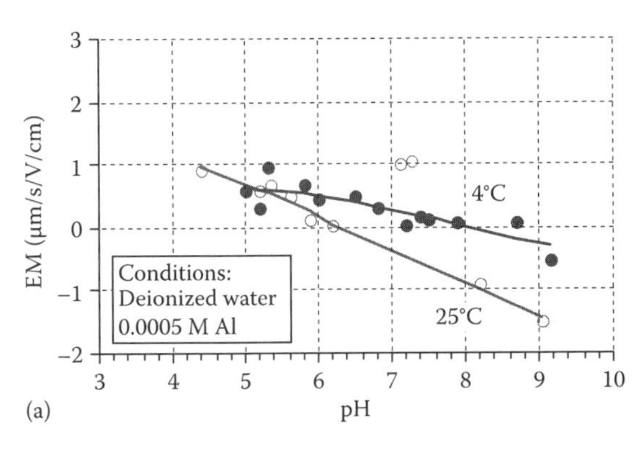

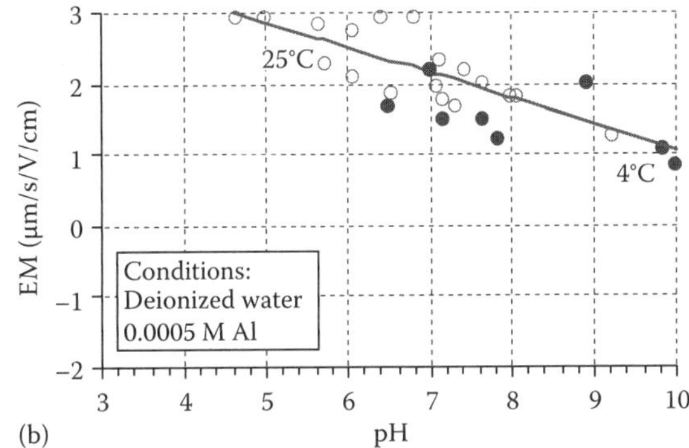

FIGURE 9.13 Electrophoretic mobility versus pH for alum and PACl (a) alum and (b) poly-aluminum chloride (PACl). (Adapted from van Benschoten, J.E. and Edzwald, J.K., *Water Res.*, 24(12), 1524, December 1990.)

water) have made PACl an attractive coagulant. In addition, with PACl there is certainty that the preformed hydrolysis products are the ones desired, that is, those that result in effective coagulation.

9.7 ZETA POTENTIAL, CHARGE DENSITY, AND STREAMING CURRENT POTENTIAL

The idea of measuring charges on particles goes to the heart of coagulation theory and is appealing because it provides a rationale for determining coagulant dose. Three approaches to measuring this charge (or a surrogate), are as follows: (1) EM/zeta potential, (2) colloid titration, and (3) streaming current.

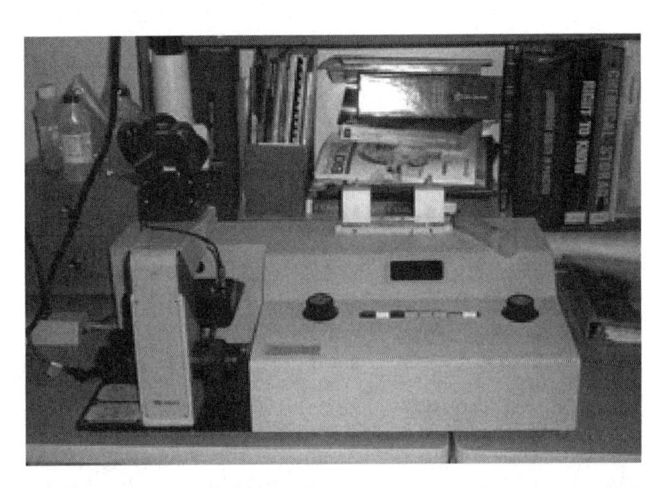

FIGURE 9.14 Photograph of a zeta potential measurement apparatus. (Courtesy of Fort Collins Utilities, City of Fort Collins, CO.)

9.7.1 Basic Notions of Electrophoretic Mobility

When placed in an electric field, negatively charged colloidal particles in water move toward the positive. The speed at which the particles move is determined by the applied voltage gradient, the charge on the particle, and the viscosity of water. The ratio of the velocity of the particle to the voltage gradient is termed the "electrophoretic mobility" (EM) (Pilipovich et al., 1958; Black, 1960; Riddick, 1960; Black and Hanna, 1961; O'Melia, 1969); units are μm/s/V/cm. The movement of charged particles in an electric field is called "electrophoresis."

When the colloid moves in the electric field, some but not all of the counterions in the ion cloud around the particle move with it. A plane of shear is developed in the diffuse layer. The electric potential in volts from the plane of shear to the bulk of the solution is the "zeta potential," designated by the symbol, ζ, which is a measure of the particle charge causing the motion. The magnitude of the zeta potential is calculated from measurements of EM. Figure 9.14 is a photograph of a zeta potential measurement apparatus, c. 2000.

At the "isoelectric point," the zeta potential is zero (Pilipovich et al., 1958, p. 1470). The isoelectric point can be determined by plotting the zeta potential versus coagulant dose or zeta potential versus pH. In theory, the zeta potential should be zero when the proper coagulant dosage is added. Thus, zeta potential should be a surrogate for proper dosage. Two concerns are (O'Melia, 1969) as follows: (1) there is no single value of zeta potential at which aggregation will always be produced, and (2) for a particular water and a specific coagulant, it is possible that an unique value of zeta potential may be observed whenever aggregation is achieved, and must be determined experimentally. In practice, these points are resolved in terms of establishing a "set-point," defined as the measured zeta potential at which settled water turbidity is minimum or when filter water turbidity is minimum. Usually the set point occurs at $-5 < \zeta < +5$ mV.

9.7.2 Mathematical Relations for Electrophoresis

The mathematical expressions for electrophoresis are several and are given here for reference. Examples illustrate their application.

9.7.2.1 Electrophoresis

The measurement of EM is done in a cell which contains two plates to which a voltage is applied and which permits the measurement of the particle velocity (by microscopic technique) in the electric field. The mathematical expression is

$$\text{EM} = \frac{v}{(\delta V/\delta x)} \tag{9.12}$$

where

EM is the EM (μm/s/V/cm)
v is the velocity of particle in electric field (cm/s)
δV is the voltage drop across electrode plates (V)
δx is the distance of separation between electrode plates (m)

The measurements are taken on individual particles that are visible and tracked visually in the cell with time noted for traveling a set distance. Enough results are obtained to delineate a histogram and the mean is taken as the result. With the early instruments the measurements and calculations were laborious involving the use of a stopwatch, thermometer, with associated calculations. Example 9.4 illustrates how to calculate EM from basic cell measurements. With the advent of solid state electronics in the 1970s, laser optics in the 1980s, combined with personal computers in the 1980s, most measurements are done automatically and data are compiled by computer with only a minimal amount of labor required.

9.7.2.2 Zeta Potential

The zeta potential, ζ, is calculated from EM measured data (Matijevic, 1967). The Helmholtz–Smoluchowski equation is common and is given for reference. Another relation by Hunter (1981) is dimensionally homogeneous and also computes the correct value of ζ, and is favored here.

9.7.2.2.1 Helmholtz–Smoluchowski Equation

The Helmholtz–Smoluchowski equation is (Pilipovich et al., 1958, p. 1474; Riddick, 1961, p. 1021; Black and Smith, 1962, p. 925),

$$\zeta = \frac{4\pi\mu}{D} \cdot \text{EM} \tag{9.13}$$

where

ζ is the zeta potential (mV)
μ is the viscosity of water medium (N s/m^2)

- For water $\mu(25°\text{C}) = 0.89 \cdot 10^{-3}$ N s/m^2
- In the literature, viscosity is given often in poises. The definition is: poise = g/cm/s = dyne · s/cm^2
- The conversion is $0.89 \cdot 10^{-2}$ poises at 25°C = $0.89 \cdot 10^{-3}$ N s/m^2

D = dielectric constant for medium (dimensionless)
= 78.36 the dielectric constant for water at 25°C (Lide, 1996, pp. 6–18)

9.7.2.2.2 Hunter Equation

A modification of the Helmholtz–Smoluchowski and Debye–Hückel equations for the EM to ζ conversion was given by Hunter (1981, pp. 61, 359), which is

$$\zeta = \frac{\mu}{\varepsilon_o D} \cdot \text{EM} \tag{9.14}$$

where ε_o = permittivity in a vacuum (F/m) = 8.854 187 817 F/m (Lide, 1996, back cover).

The difference is that the permittivity, that is, ε (in which $\varepsilon = \varepsilon_o D$), replaces the dielectric constant alone and the π and numerical terms cancel. The equation is dimensionally homogeneous and gives a conversion factor, that is, 12.9 reported by several investigators (next section).

9.7.2.2.3 Empirical Relation

According to Black and Willems (1961, p. 592), Hall (1965, p. 198) the group $(4\pi\mu/D)$ has the numerical value, $(4\pi\mu/D) = 13$, giving

$$\zeta \approx 13 \cdot \text{EM} \tag{9.15}$$

in which ζ units are mV and EM units are μm/s/V/cm.

9.7.2.2.4 Examples of EM and Zeta Potential Calculations

Example 9.4 illustrates the calculation of EM from measured data. Example 9.5 shows the following: (a) in Equation 9.13 the group, $(4\pi\mu/D) \neq 13$; (b) Equation 9.13 is not dimensionally homogeneous; (c) Equation 9.14 is dimensionally homogeneous and the group, $(\mu/\varepsilon_o D) = 12.9$. Example 9.6 illustrates the application of Equation 9.15.

Example 9.4 Calculation of Electrophoretic Mobility (Black and Smith, 1962, p. 934)

This example illustrates how to convert basic measurements to an EM value.

Given

Black and Smith (1962, p. 934) obtained the following data using a Briggs cell:

- d(field) = 49 μm = $49 \cdot 10^{-6}$ m
- A(cell) = 0.100 cm × 1.73 cm = 0.001 m · 0.0173 m = $1.73 \cdot 10^{-5}$ m^2
- t(avg) = 7.8 s (average)
- $i = 4.0 \cdot 10^{-4}$ A
- R = 4450 ohm-cm = 44.50 ohm-m

Required

Calculate EM, based on measurements using the Briggs cell.

Solution

Equation Ex9.4.1, from Black and Smith (1962, p. 934), may be used to calculate EM (with the Black and Smith units converted to SI). Insertion of data from the Briggs cell gives

$$\text{EM} = \frac{d(\text{field})A(\text{cell})}{t(\text{particle})iR}$$

$$= \frac{49 \cdot 10^{-6}\ \text{m} \cdot (1.00 \cdot 10^{-3}\ \text{m} \cdot 1.73 \cdot 10^{-2}\ \text{m})}{7.8\,\text{s} \cdot 4.0 \cdot 10^{-4}\ \text{amps} \cdot 44.5\,\text{ohm-m}}$$

$$= 0.061 \cdot 10^{-7} \frac{\text{m}^2}{\text{s} \cdot \text{amps} \cdot \text{ohms}}$$

Since amps · ohms = volts, and converting from SI to traditional units

$$= 0.061 \cdot 10^{-7} \frac{m^2}{s \cdot V} \cdot \frac{10^6\ \mu m}{m} \cdot \frac{10^2\ cm}{m}$$

$$= 0.61 \cdot 10^{-2}\ \frac{\mu m/s}{V/cm}$$

Discussion
The main idea of this example was to illustrate how electrophoretic data are converted to EM. In modern instruments, all of this is done by software.

Example 9.5 Derivation of Factor to Convert EM (μm/s/V/cm) to ζ in mV

The problem of conversion of EM to ζ is confounding in that the older equation, that is, Equation 9.13 is not dimensionally homogeneous and does not yield the correct result numerically. The modification by Hunter, however, is dimensionally homogeneous and does give the correct conversion factor numerically.

Given
Equations 9.13 and 9.14 give equation to convert EM to zeta potential. A more expedient conversion is given as Equation 9.15, which is valid for 25°C (the viscosity of water, a term in both equations, is a function of temperature). The factor in Equation 9.15 is actually 12.9, and is rounded off to 13 and converts EM in (μm/s/V/cm) to ζ in mV. Such an equation is useful since electrokinetic data are given often in EM. Therefore, merely multiplying EM by 13 gives ζ in mV.

Required
Derive the factor, 12.9 in Equation 9.15.

Solution

1. The three referenced data are
   a. D = the dielectric constant for water at 25°C (dimensionless, i.e., no units) = 78.36 from Lide (1996, pp. 6–18)
   b. μ (25°C) = viscosity of water at 25°C (N s/m^2) = $0.89 \cdot 10^{-3}$ N s/m^2
   c. ε_o = permittivity of vacuum (F/m) = $9.854187817 \cdot 10^{-12}$ F/m (Lide, 1996, back cover)
2. Start with Equation 9.13, which is attributed to Helmholtz–Smoluchowski and Debye–Hückel, that is,

$$\zeta = \frac{4\pi\mu}{D} \cdot \text{EM} \tag{9.13}$$

   Calculation of the collection of terms: $4\pi\mu/D = 4\pi \cdot (0.89 \cdot 10^{-3}\ \text{N s/m}^2)/78.36 = 1.43 \cdot 10^{-4}$ N s/m^2. The number sought is 13. The discrepancy is not reconcilable in either numerical value or units.

3. Apply the Equation 9.14 from Hunter

$$\zeta = \frac{\mu}{\varepsilon_o D} \cdot \text{EM} \tag{9.14}$$

$$= \frac{4\pi(0.895 \cdot 10^{-3}\ \text{N s/m}^2)}{(8.854 \cdot 10^{-12}\ \text{F/m}) \cdot (78.36)} \cdot \text{EM}$$

$$= 1.29 \cdot 10^6 \frac{\text{N s}}{\text{F m}} \cdot \text{EM}$$

Recall that EM was given in traditional units, that is, μm/s/V/cm; now convert to SI

$$\zeta = 1.29 \cdot 10^6\ \frac{\text{Ns}}{\text{F m}} \cdot \left(\text{EM}\ \frac{\mu m \cdot cm}{V \cdot s} \cdot \frac{m}{10^6\ \mu m} \cdot \frac{m}{10^2\ cm}\right)$$

$$= 0.0129 \cdot \frac{\text{Nm}}{\text{F V}} \cdot \text{EM}$$

Now recall: Nm = Ws and F = A s/V

$$\zeta = 0.0129 \cdot \frac{\text{Ws}}{\frac{\text{A s}}{\text{V}}\text{V}} \cdot \text{EM}$$

After canceling units

$$= 0.0129 \cdot \frac{\text{W}}{\text{A}} \cdot \text{EM}$$

Now recall, W = V · A, to give

$$\zeta = 0.0129 \cdot \text{V} \cdot \frac{\text{EM}}{D}$$

and to convert to mV,

$$\zeta = 0.0129 \cdot \text{V} \cdot \frac{1000\,\text{mV}}{\text{V}} \cdot \text{EM}$$

$$= 12.9 \cdot \text{EM (mV)}$$

Discussion
The outcome starting with Equation 9.14 shows that the equation is homogeneous with respect to units and yields the accepted empirical factor for the EM to ζ conversion, that is, 12.9.

Example 9.6 Calculation of Zeta Potential from EM

Given
Let an EM measurement be −1.20 μm/s/V/cm at 25°C.

Required
Calculate zeta potential

Solution
As noted, Black and Willems (1961, p. 592), gave a rule of thumb to convert EM in (μm/s/V/cm) to ζ in mV as, ζ = 13 · EM (also confirmed in Example 9.5). Applying this rule of thumb to the problem at hand

$$\zeta = 13 \cdot \text{EM} \tag{9.15}$$

$$\approx 13 \cdot (-1.20\ \mu m/s/V/cm)$$

$$= -15.6\,\text{mV}$$

Discussion

The factor 12.9, rounded to 13, is an accepted empirical conversion factor to convert EM (μm/s/V/cm) to ζ (mV).

9.7.3 Measured Zeta Potentials

In general, and as indicated in Table 9.1, zeta potentials of most kinds of particles are negative, varying from −3 to −50 mV. In addition to variation between particle categories, however, zeta potentials are affected by pH, alum dosage, and by ionic strength of solution. The increase in zeta potential caused by alum dosage is, of course, a desired result of coagulation (assuming the zeta potential of a given particle is negative).

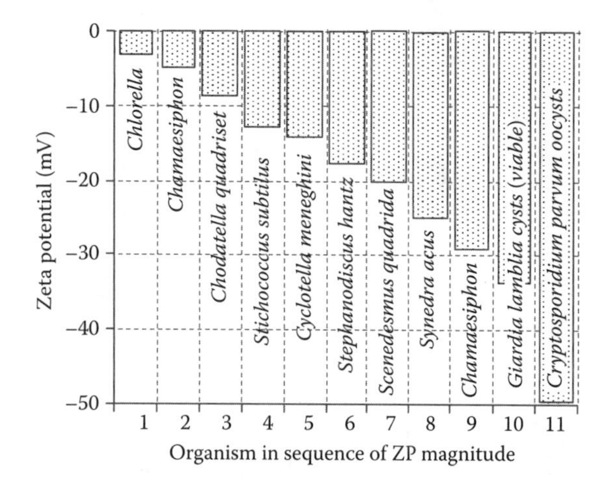

FIGURE 9.15 Zeta potentials of algae in log-growth phase and viable cysts/oocysts at pH = 7.0. (Adapted from Cushen, A.D., Zeta-potentials of selected algae, Cryptosporidium oocysts, and Giardia cysts, MS Thesis, Department of Civil Engineering, Colorado State University, Fort Collins, CO, 1996.)

9.7.3.1 Typical Zeta Potentials

Figure 9.15 is a bar chart showing zeta potentials for 11 organisms measured at pH = 7.0. Algae sizes ranged 2–70 μm, with shapes spherical, oval, rod, needle, oval with spines, etc. An important point is that zeta potentials varied between organisms. The log removals by filtration were found to increase with increasingly negative zeta potentials, for example, log R(chorella) ≈ 2.0 with $3.0 \leq \log R \leq 3.5$ for the other organisms (Hendricks et al., 2000, p. 130, 2005, p. 1628).

The effect of pH on zeta potential for kaolinite clay (particle sizes 0.6–2 μm) is shown in Figure 9.16a for two ionic strengths, that is, $1 \cdot 10^{-2}$ and $5 \cdot 10^{-2}$ mol/L, for the top and bottom curves, respectively. For each curve, as the pH increased zeta potential decreased. Figure 9.16b shows that pH influences the zeta potentials for viable *Giardia lamblia* cysts and viable *Cryptosporidium parvum* oocysts; as seen, each organism showed a unique response of zeta potential to pH variation. As a note, the best fit curve for *Giardia* cysts is merely one interpretation, since any two points could be outliers; data by Ongerth and Pecoraro (1996) showed ζ values declining (more negative) with increasing pH, that is, (pH = 3.5, ζ = −9 mV) declining to (pH = 7.2, ζ = −42 mV), that is, the trend was consistent with the first three data points.

9.7.3.1.1 *Effect of Alum Dosage*

Electrophoretic mobility as a function of alum dosage was determined experimentally by Pilipovich et al. (1958, p. 1478) for several clays. Figure 9.17 shows their plot for their "illite 35" clay, which was representative of other clays used, for example, montmorillonite and kaolinite. The data show that the EM increased with alum dosage and then leveled off, presumably as the cation exchange capacity of the clay (31.4 μe/L) was satisfied. The alum dosage needed for good coagulation was determined to be about 27 μe/L, that is, near

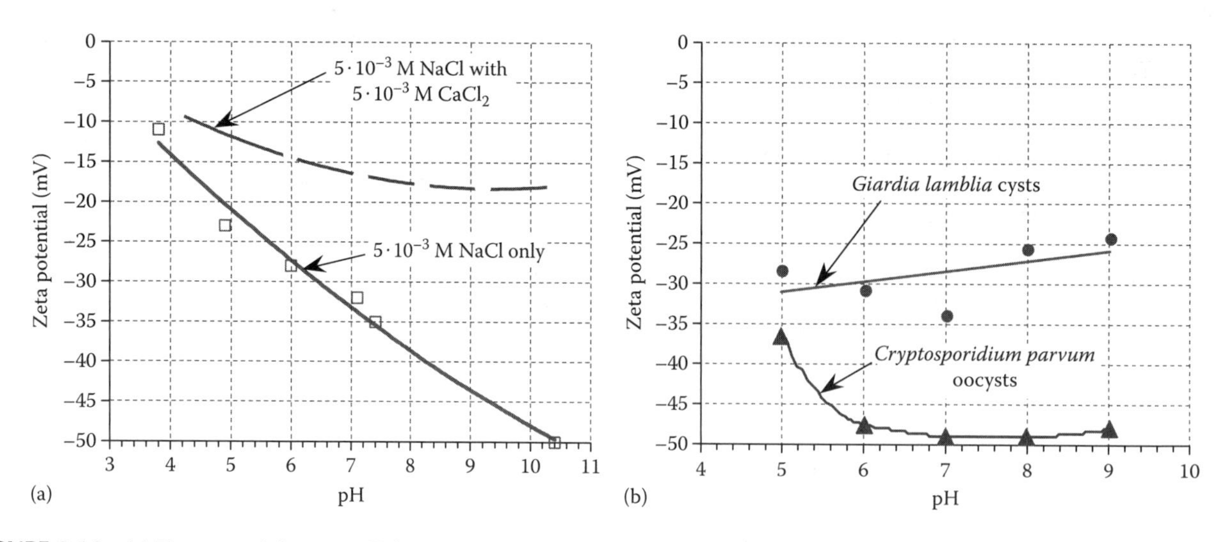

FIGURE 9.16 (a) Zeta potential versus pH for kaolinite for two chemical environments. (Adapted from Loganathan, P. and Maier, W.J., *J. Am. Water Works Assoc.*, 67(6), 340, February 1975.). (b) Zeta potentials versus pH for viable *Giardia lamblia* cysts and *Cryptosporidium parvum* oocysts. (Adapted from Cushen, A.D., Zeta-potentials of selected algae, Cryptosporidium oocysts, and Giardia cysts, MS Thesis, Department of Civil Engineering, Colorado State University, Fort Collins, CO, 1996, p. 4-3.)

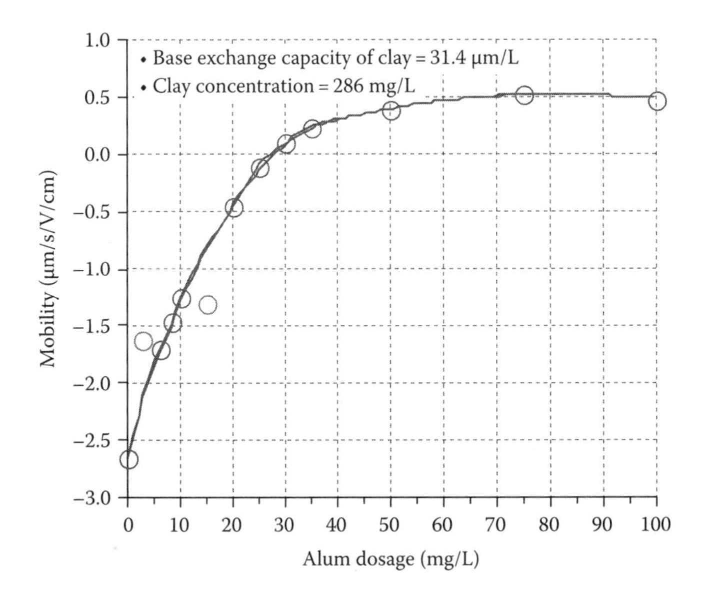

FIGURE 9.17 Electrophoretic mobility as affected by alum dosage for Illite 35 clay. (Adapted from Pilipovich, J.B. et al., *J. Am. Water Works Assoc.*, 50(11), 1478, November 1958.)

the asymptote. The investigation by Pilipovich et al. (1958) helped to launch the incorporation of zeta potential as a variable in coagulation practice.

9.7.4 Colloid Titration

While EM measures the rate of movement of single particles under the influence of an electric field, colloid titration measures the colloid charge directly by titration (Jorden, 1996). The colloid titration measure is for aggregate charge of the suspension as a whole (in meq/L), vis-à-vis EM which is for individual particles (Kawamura et al., 1967). This is an important difference in that a substantial portion of the aggregate charge is contained in particles that are smaller in size than can be seen by the EM technique (which requires visual tracking of individual particles). On the other hand, for several systems tested by Kawamura et al. (1967) both EM and colloid charge increased with increasing alum dosage, with trends being near-parallel to one another. The isoelectric point was coincident for each. Also, the CCC alum dosage for color removal was coincident with the dosage for attaining the isoelcctric point. In each case, the CCC alum dosage for turbidity removal was less than that required to attain the isoelectric point (and for color removal). In other words, no particular advantage was seen by the use of colloid titration over EM.

9.7.5 Streaming Current Monitor

The streaming current technique for determining coagulant dosage is a variation of the principle applicable in zeta potential measurement. A current instead of a potential is measured. An instrument was proposed by W. F. Gerdes in 1966 (Smith and Somerset, n.d.) based upon the discovery that the walls of a capillary quickly take on the charge characteristics of the colloidal particles or other charge-influencing species in the fluid.

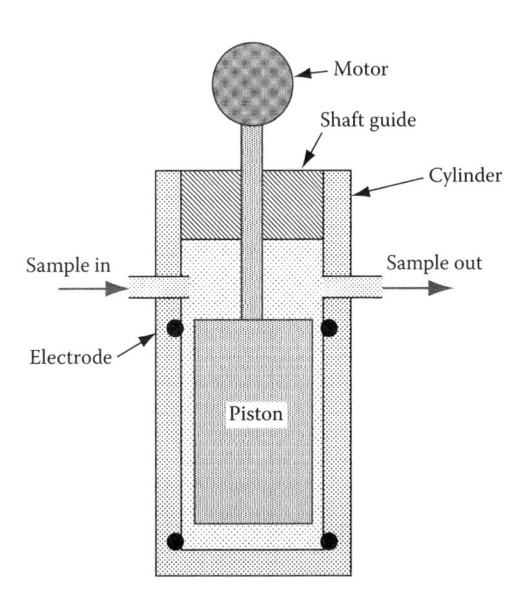

FIGURE 9.18 Streaming current monitor cell schematic. (Adapted from Bryant, R.L., *Waterworld News*, 1(3), 18, May/June 1985.)

Figure 9.18 shows the elements of a streaming current monitor (SCM) instrument. The instrument cell is the annular space between a piston and a cylinder. The piston has a four-cycle reciprocating motion. The water sample with the charged particles is displaced within the annular space as the piston reciprocates. The basis for the instrument is that the charged colloidal particles are presumed immobilized, that is, adsorbed on the surfaces of both the piston and the cylinder. As the water in the annular space is forced past the stationary colloidal particles, the motion will physically shear the counterions from the particle, which will generate a *streaming current* (Cardile et al., 1982; Amirtharajah and O'Melia, 1990; Peterson, 1992, pp. 3, 6).

In using the instrument, a "set point" must be determined. This is done by relating streaming current potential to jar test residual turbidities or to effluent turbidities from pilot filters (or the full-scale plant) plotted as a function of coagulant dosage.

The streaming current is related to zeta potential by the relationship (Smith and Somerset, n.d.)

$$i = \frac{\zeta D \Delta P r^2}{4\mu L} \tag{9.16}$$

where
- i is the streaming current (A)
- ΔP is the pressure drop across cell (Pa)
- r is the radius of diaphragm (m)
- L is the length of diaphragm (m)

Advantages of a streaming current detector over an electrophoresis instrument are: (1) the SCM is set up online (that is, continuous monitoring occurs), and (2) a coagulant metering pump can be set up to provide coagulant to the raw water such that the dosage satisfies a SCM "set point." The streaming current of coagulated particles is measured, based on a 4–20 ma output signal, which is sent to a process controller.

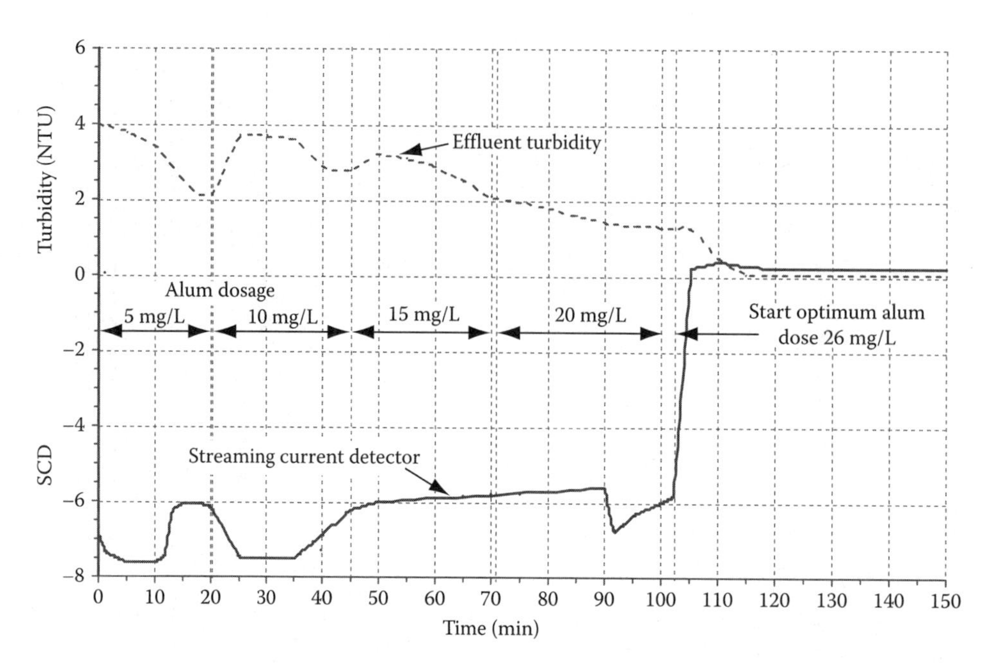

FIGURE 9.19 Responses of continuous effluent turbidity response and streaming current potential to alum dosage for in-line mono media (anthracite) filtration, Water Treatment Research Pilot Plant at ERC, CSU. (From Hendricks, D.W. et al., *Biological Particle Surrogates for Filtration Performance Evaluation*, AWWA Research Foundation, Denver, CO, 2000, p. 73. With permission.)

The controller may be set up to pace the strokes per minute (and/or stroke length) of a positive displacement chemical metering pump to maintain an established set point (Veal, 1988). Since the continuous sample is taken after rapid-mix, the adjustment in coagulant dosage is immediate, that is, not having to wait for the time lag between a coagulant change and the associated effluent turbidity result from filters or from a laboratory jar test.

By the year 1985, the SCM technology had been adopted by 150 water treatment plants in the United States and abroad (Bryant, 1985). Some of the factors important to successful use of SCM includes: proper location of sample intake, routine maintenance and cleaning of the sensor, proper sample delivery to the cell, standard calibration procedure, protocol to establish a set point for optimized dosage, and supervisory control and data acquisition (SCADA) recording to permit examination of trends (Kramer and Harger, 2001).

Figure 9.19 compares effluent turbidity (top curve as measured by a continuous reading Hach 1720D® turbidimeter) with corresponding SCM measurements (bottom curve as detected by a 4–20 ma output), each for increasing increments of alum dose for a 76 L/min (20 gpm) in-line filtration pilot plant. The diversion point of the sidestream to the SCM instrument was located just after the rapid-mix. As seen, the turbidity dropped to about 0.05 NTU at alum dosage 26 mg $Al_2(SO_4)_3 \cdot 14H_2O$/L; the corresponding SCM was about +0.5 units, which would be the "set point" for this particular water and filtration mode.

In all experiments, the alum addition was based on achieving the +0.5 SCM units, which was more expedient than waiting for the effluent turbidity. The lag time between the addition of 26 mg/L alum and the SCM response of about +5 mV was about 3 min, while the lag time to an effluent turbidity of 0.05 NTU was about 13 min (compared with 12 min for a salt tracer test). The times, that is, 3 and 13 min, can be seen in Figure 9.19 for the bottom solid line for SCM and for the top dotted line, respectively. In other words, the results illustrate the practical utility of SCM.

9.8 PHYSICAL MODELS

Two kinds of physical models (Chapter 3) are the jar test and the pilot plant. The jar test can be a basis for massive screening of coagulants and dosages. A pilot plant can take into account a larger number of variables, however, and thus can simulate the full-scale process. The pilot plant may be used most efficiently in conjunction with jar testing.

9.8.1 Jar Tests

The jar test is a means to estimate the coagulant dosage and the effective pH regions, and to explore the use of polymers with respect to type and dosage. The jar test apparatus consists of a set of six square-shaped jars, about 2000 mL in size, which are used in conjunction with a gang stirrer. The apparatus permits all six jars to be controlled simultaneously, for example, start and stop and rotational speed of paddles. Figure 9.20a shows a basic jar test apparatus as manufactured by Phipps & Bird, Inc., c. 1970. The jar test apparatus permits control of rotational speed; the setup shown has in-house fabricated 2000 mL square jars and a constant-temperature water bath as set up for a research project in 1984. Figure 9.20b shows Phipps & Bird®, c. 2000 "top-of-the-line" jar test apparatus, which has a programmable speed control (to permit change from high speed rapid-mix simulation to low speed flocculation simulation to zero rpm for settling).

(a)

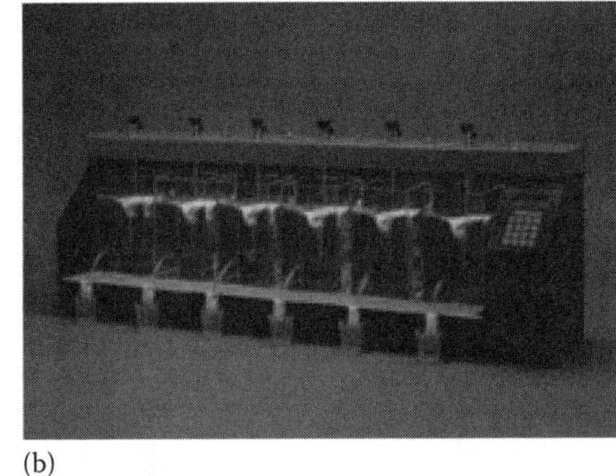
(b)

FIGURE 9.20 Jar test apparatus manufactured by Phipps & Bird, Inc. (a) Jar test apparatus with water bath; used for research, c. 1970 and (b) programmable jar tester, Model PB-900™, c. 2000. (Courtesy of Phipps & Bird.)

Typically, the jar test consists of filling the six jars with 1000 mL each of the sample water. Different dosages of coagulant are added to the respective jars to span the expected optimum dosage range. The coagulants should be added simultaneously after the rapid-mix is started. The sequence normally is 3 min rapid mix at 100 rpm and about 15–20 min slow mix to simulate flocculation followed by 30 min settling. The turbidities of the supernatants are measured and plotted as a function of coagulant dosage. After plotting settled-water turbidity versus coagulant dose, another test is done with smaller dosage increments, for example, to confirm an "optimum" dose, and to obtain increased resolution. In addition, alkalinity may be added at an optimum coagulant dosage, then various polymers, etc. In one case (known to the author), some 5000 jar tests were done by the consulting engineer retained for plant modification (Box 9.3).

BOX 9.3 ORIGIN OF THE JAR TEST

A jar-test apparatus was developed and used by Professor Wilfred Langelier (Langelier, 1921) who stated:

> The apparatus used to determine the ideal conditions of agitation consists of a wooded frame on which are mounted four rotating paddles or propellers, each geared to a horizontal shaft driven by a small electric motor. The water is held in cylindrical glass jars 5 in. in diameter and 10 in. deep. A later modification of the device is designed with friction disks which permit the operation of each paddle at a speed independent of the others. The same effect of variable agitation can be produced in the original apparatus by using paddles of different sizes. These simple devices have proved extremely useful in studying coagulation phenomena. Similar apparatus could with profit be included in the laboratory equipment of all water purification plants using coagulation.

The context was his work with Professor Charles Gilman Hyde in the design of a water treatment plant at Sacramento, California (Langelier, 1982). The issue that Langelier was addressing was coagulation. At that time, with the modern version of rapid filtration technology being put into practice only about a decade earlier, the need for initial mixing of alum with water was recognized as important, but the idea of forming flocs with velocity gradients was contrary to the conventional wisdom; rather, quiescent settling was the method. One outcome of his studies with the jar-test apparatus was the technology of paddle-wheel flocculation, first implemented at the Sacramento WTP. Further, by means of the apparatus, he was able to study not only the effect of rotational velocity mixing on coagulation and flocculation, but the effects of pH, dosage, alkalinity, etc. on the character of the floc produced. Black and Harris (1969, p. 49) confirmed the foregoing account, stating that the test was first used by W. F. Langelier and Charles Gilman Hyde in 1918 to determine the proper treatment for the water of Sacramento. The laboratory studies, they stated, led to the first large-scale application of "mechanical agitation," that is, paddle-wheel flocculation (Chapter 11), which was at the Sacramento WTP in 1921, as recommended by Langelier.

In a later study, A. P. Black with two colleagues, Owen Rice and Edward Bartow (Black and Rice, 1933), used six "battery jars" of 3.5 L capacity for their coagulation experiments, with 2.0 L water used for each jar. The standard jar-test apparatus as an "off-the-shelf" item was developed and marketed by Phipps & Bird Inc. and was available by 1935.

9.8.2 Bench Scale Filters

The standard jar test does not work well for some water, for example, low turbidity (e.g., <0.5 NTU), low alkalinity (e.g., <50 mg/L as $CaCO_3$) snowmelt waters found in the Rocky Mountains, the Sierra Nevada Mountains, the Cascades, and other places. While microflocs form, as evidenced by the increase in turbidity after coagulation with alum in a jar test, a settleable floc does not develop as in higher turbidity waters. For such situations, "in-line" filtration is appropriate

(Al-Ani et al., 1986), which is rapid-mix and filtration. A bench scale filter was developed for such cases to perform a more rapid screening of chemicals and dosages as compared to the use of a pilot plant (Brink et al., 1988). The filters used were six, 51 mm (2 in.) diameter lucite or clear PVC columns about 762 mm (30 in.) long, packed with media to be used in the filtration process (or 0.9 mm anthracite or 0.5 mm sand). The jar test may proceed to the point of rapid-mix completion at which point a portion of the contents of the jar are poured through the media at a slow rate. As with the standard jar test, the effluent turbidities from the respective filters should be plotted against the dosage of coagulant. The procedure was adapted as a proprietary apparatus, with an integrated jar-tester, c. 2000, by Phipps & Bird, Inc.

9.8.3 Pilot Plants

A pilot plant is a physical model of the system as a whole, that is, rapid-mix, flocculation and settling (if used), and rapid filtration, and therefore is the most accurate means to assess the effects of coagulation. The jar test and the pilot plant are complementary tools for operation of the filtration process, that is, the jar test for initial screening and the pilot plant for final assessment and "fine tuning."

9.8.3.1 Independent Variables

The key independent variable in coagulation is coagulant dosage, although others may be important, such as pH and alkalinity. In addition, polymer selection and dosage may be assessed as a primary coagulant, as a coagulant aid, as a flocculent, or as a filter aid. Also, mixing variables (turbulence intensity, circulation pattern, detention time) are relevant to coagulation performance. At a Colorado State University (CSU) pilot plant (Figures 3.4 and 3.5), the mixing variables were taken into account to some extent by inclusion of three alternative rapid-mix basins (different detention times and each with variable impeller speed), any one of which could be included in the treatment system by opening and closing the appropriate ball valves.

9.8.3.2 Dependent Variables

Coagulation determines the potential for effective flocculation, settling efficiency, filtration effluent turbidity, rate of headloss increase, and the mass flux of solids as waste. All of these effects may be evaluated by means of a pilot plant. If filtration is "conventional," this may be associated with effective flocculation (e.g., large, dense, tough floc), which in turn leads to efficient settling. The next in priority is to minimize the waste solids flux while not diminishing the other gains. Finally, another goal of filtration is to minimize TOC, which may require "enhanced" coagulation, that is, higher dosage.

9.8.3.3 Pilot Plant Design

The pilot plant should be constructed to replicate the processes of the full-scale plant. Its scale should be large enough such that scale effects are not an important concern, for example, $Q \geq 38$ L/min (10 gpm), albeit there is no specific flow required. The method is the same as with any experimental program: the effect of one variable is investigated while the others are maintained constant. Other guidelines are given in Chapter 12.

9.9 POLYMERS

Synthetic organic polymers were considered in the 1950s (see Johnson, 1956) and after about 1960 were used increasingly in drinking water treatment, industrial wastewater treatment, and for conditioning of chemical and biological sludges. In 1967, the first synthetic cationic polymer was accepted by the USPHS for use in drinking water treatment (AWWA B451-87); over one thousand polymer products were approved by the USEPA through 1985 (Dentel et al., 1986). They are used by virtually every water and wastewater utility and industrial treatment facility.

Usually polymers are selected based on manufacturer's recommendation, word-of-mouth, jar-test data, or pilot plant results. Important characteristics include: molecular weight, charge density, and chemical structure of the monomer. Although these characteristics have not been cataloged, manufacturers may provide limited information about their polymers, for example, whether it is cationic, anionic, or nonionic, and a category of molecular weight.

The selection of a polymer, its dosage, point of addition, mixing intensity, etc. remains a trial and error as opposed to a rational procedure (Ghosh et al., 1985). The tools for empirical testing are the jar test and a pilot plant (Dentel et al., 1986). Dependent variables in coagulation include: floc size and shear strength. In filtration the dependent variables are rate of headloss increase and effluent turbidity and whether the filter media becomes clogged.

9.9.1 Definitions

A *polymer* is defined as a repeating *monomer*; if formed by a single monomer, it is called a *homopolymer*. If two or more monomers are involved, the product is called a *copolymer*. A polymer that has ionized sites along its length is called a *polyelectrolyte*. With respect to charge, a polymer without charge is *nonionic*; with positive charged sites its *cationic*; with negatively charged sites its *anionic*. A polyelectrolyte may function as a *primary coagulant*, that is, if it is added to the rapid-mix instead of alum or ferric iron (as a rule, not done due to cost, among other considerations). As used here, the term *polymer* is used without necessarily specifying whether it is a *polyelectrolyte*; but it is usually implicit that a polyelectrolyte is the context.

If a polyelectrolyte is added along with alum or ferric iron to the rapid-mix and if the polyelectrolyte provides sites for microflocs (or floc particles), it functions as a *coagulant aid*; *bridging* between microflocs. A coagulant aid may help to tailor size, settling characteristics, and shear strength of the resulting floc. The purpose is to improve settling or filtration performance and/or to reduce the dosage of alum or ferric iron (with perhaps less sludge and associated lower cost).

A *flocculant* is a polymer added just before flocculation and functions the same as a coagulant aid, that is, by providing attachment sites for microflocs (or flocs). In practice, the terms *coagulant aid* and a *flocculent* are likely to be used interchangeably.

A *filter aid* is a polyelectrolyte that functions by attaching to filter grains, which then provide attachment sites for coagulated particles. The result is higher filtration efficiency (due to increasing the transport coefficient and perhaps by increasing the attachment coefficient, described in Chapter 12). If used improperly, a filter aid could also "gum-up" the filter media; also, if used in excess, there is potential for "mudball" formation. A filter aid is likely to be added in the pipe or channel just before the filters.

Black (1960) calculated that a polyelectrolyte having a molecular weight of 100,000 at a dosage of 0.2 mg/L would have 120 trillion active chains per liter of water. Such a value gives some appreciation of the capacity of a polymer to react with a particle of opposite charge.

9.9.2 Characteristics of Polymers

A few characteristics of polymers include molecular weight, structure, commercial form (e.g., solid, liquid, emulsion), charge concentration, and specific gravity. As a rule, proprietary companies provide only limited information, for example, the name given to a particular product.

9.9.2.1 Charge Concentration

The negative charge of NOM may be neutralized stoichiometrically by positive charged polymers with polymer dosage in terms of μeq positive charge/L (Edzwald et al., 1987). Example 9.7 illustrates the calculation of charge concentration provided by a polymer.

9.9.2.2 Specific Gravity

The specific gravity for liquid polymers varies $1.0 \leq SG \leq 1.2$ (Chamberlain, 1981, p. 246).

Example 9.7 Calculate Charge Concentration for the Cationic Polymer Magnifloc 572C

Given

Positive charge density of Magnifloc 572C = 3.1 meq/g neat solution at pH = 7.0; fraction "active" polymer ≈ 0.50 in neat solution (Dentel, 1988). Assume SG(Magnifloc 572C) ≈ 1.05.

Required

Concentration of positive charge in meq/L polymer.

Solution

$$\begin{aligned} C(\text{positive-charge}) &= (3.1 \text{ meq charge/g neat solution}) \\ &\quad \cdot (1.05 \text{ g neat solution/L neat solution}) \\ &= 3.3 \text{ meq/L neat solution} \end{aligned}$$

Discussion

1. *Chemical formula*. Magnifloc 572C is an "Epi/DMA" polymer (i.e., an "epichlorohydrin/dimethylamine quaternary amine" and has about 0.50 fraction "active" polymer per unit mass of liquid polymer as supplied (Dentel and Gucciardi, 1989); it is also called a "polyquaternary amine" Edzwald (1985, p. 171).
2. *Charge density*. The charge density of Magnifloc 572C (an Epi/DMA polymer) was given as 7.95 meq/g active polymer, with no effect of pH in the range $4.0 \leq \text{pH} \leq 8.0$, and 0.508 g "active" polymer/g neat polymer solution (Edzwald, 1985, pp. 173, 175; Edzwald et al., 1987, p. 172). For these data and with a dosage of neat solution given as 9.5 mg polymer product/L raw water,

$$\begin{aligned} C(\text{positive charge}) &= (9.5 \text{ mg neat solution / L raw water}) \\ &\quad \cdot (0.508 \text{ g active polymer / g neat solution}) \\ &\quad \cdot (\text{g}/1000 \text{ mg}) \\ &\quad \cdot (7.95 \text{ meq charge / g activepolymer}) \\ &\quad \cdot (1000\ \mu\text{eq / meq}) \\ &= 38\ \mu\text{eq/L raw water} \end{aligned}$$

This concentration of positive charge, that is, 38 μeq/L raw water, was the "optimum" concentration to reduce color from about 135 Pt-Co units to about 15 at 4 mg/L. The corresponding initial HA concentration was 10 mg/L with associated DOC concentration 4.3 mg/L.

9.9.3 Polymers in Wastewater Treatment

In addition to applications in treatment of drinking waters, polymers have found application in conditioning of wastewater sludges, that is, in thickening and dewatering. As with other topics, a body of knowledge is involved, which is reviewed here briefly.

9.9.3.1 Sludge Conditioning

Two kinds of sludge conditioning are thickening and dewatering. The purpose of each is to reduce the amount of water for improving the functioning of subsequent processes and reducing the cost of transport, respectively. The operative mechanism in treatment of sludges is "inter-particle bridging," in which a polymer molecule attaches to sites on several particles; important factors are as follows: mixing, $Gt \approx 10{,}000$; dosage, about 5 mg/L; and molecular weight, $MW \geq 10^6$ (Novak, 1983).

9.9.3.1.1 Thickening

A sludge that has more water than is desired for a subsequent unit process may be "thickened" to reduce the water content. Examples may include flotation of waste activated sludge before anaerobic digestion (which results in less water in the reactor increasing the detention time), and thickening of primary sludge prior to anaerobic digestion (again to reduce the water content and increase of detention time). Industrial wastes treatment often utilizes flotation for thickening of both organic and inorganic wastes.

9.9.3.1.2 Dewatering

The purpose of a polymer in sludge dewatering is to increase the solids concentration (as in thickening). The benefits of increased solids concentration has to do with the reducing the costs of transport to a disposal site, for example, land spreading. For example, if the use of a polymer results in 24% solids instead of 20%, the savings would be to transport 4 kg less water per hundred kg of sludge. Sludges that are dewatered include those from water treatment (e.g., settling, filter backwash), anaerobic digestion, and aerobic digestion. Dewatering methods include centrifuging, belt press processing, filter press processing, vacuum filtration. Other means of dewatering include drying beds, freezing, heating, etc.

9.9.4 Structure of Polymers

Several configurations are possible in polymer structure, for example, linear, branched, dendritic. Then, to function as a polyelectrolyte a polymer must have incorporated *functional groups* that may dissociate to leave charged sites. Because of their high molecular weights, ranging from tens of thousands to tens of millions, polymers are considered *macromolecules*.

To illustrate the idea of a polyelectrolyte, Figure 9.21a depicts a *linear* polymer with attached COOH *groups*. If the pH is increased, consuming the H^+, the residual charges of the groups are negative, and the polymer is a *polyelectrolyte*, that is, as in Figure 9.21b. Because the negative charges repel one another, the polymer becomes "stretched."

9.9.4.1 Functional Groups

Functional groups are a part of the structure of polyelectrolytes; they have charged sites with a mobile counterion. Some of the common functional groups are listed in Table 9.8. Anionic groups include carboxylate and sulfonate, while the most common cationic group is quaternary amine (see *amines* in glossary). The structure of an amine is similar to NH_3 but is protonated to give NH_4^+, but with R groups instead of H attached to the nitrogen. The "R" groups are any hydrocarbon, for example, the methyl group, $-CH_3$. Some of the common structures that include quaternary amine are listed in the lower row of Table 9.8, for example, DMAEM-MCQ, DADMAC, and Mannich, respectively.

9.9.4.2 Monomers

Some of the monomers that comprise a polymer structure are categorized in Table 9.9 as cationic, anionic, and nonionic. Common cationic polymers are DADMAC and DMA. Common anionic polymers include sodium polystyrene sulfonate and acrylate. The most common nonionic polymers are acrylamide and epichlorohydrin (epi). Such monomers may be synthesized as homopolymer or copolymers, with the myriad of possible configurations.

TABLE 9.8
Functional Groups with Charge

Anionic

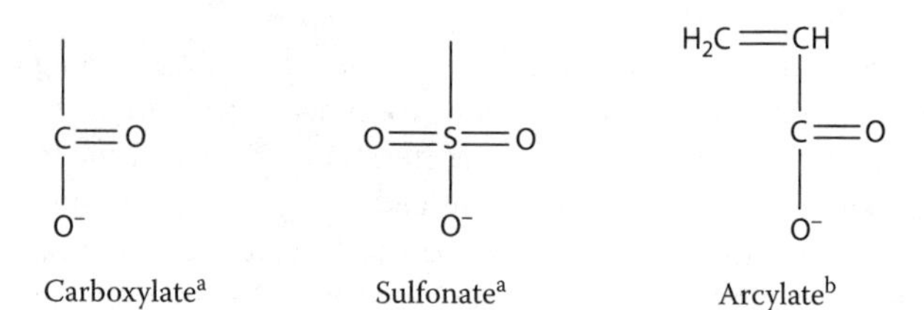

Carboxylate[a] Sulfonate[a] Arcylate[b]

Cationic

R
R—N⁺—R
R

Quaternary amine–examples of structures

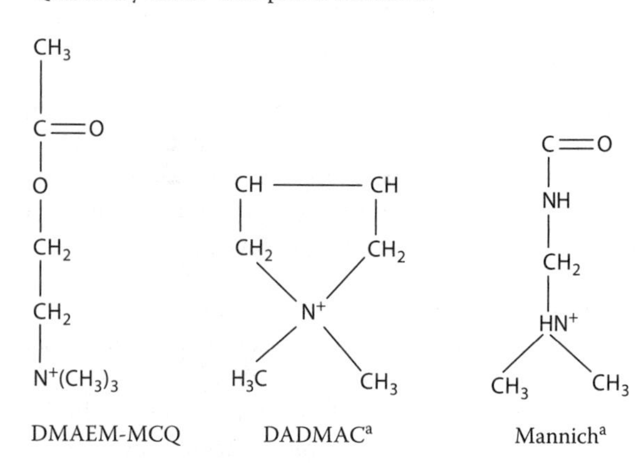

DMAEM-MCQ DADMAC[a] Mannich[a]

[a] May (1988).
[b] Rose (1988).

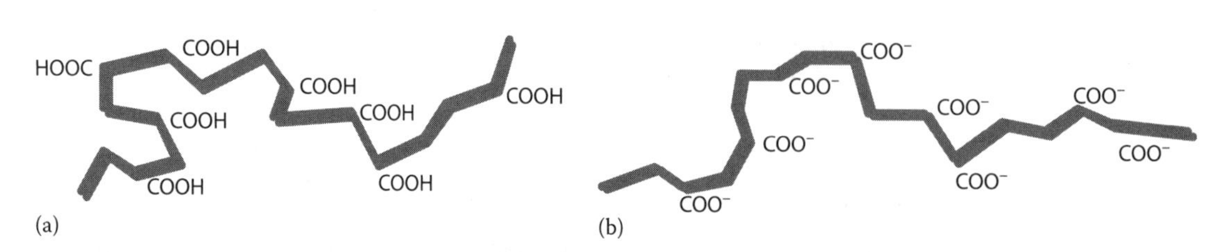

FIGURE 9.21 Polymers to illustrate the effect of charge on shape. (a) Polymer without charge. (b) Polymer with charged sites. (Adapted from Black, A.P., *J. Am. Water Works Assoc.*, 52(4), 493, April 1960.)

TABLE 9.9
Examples of Monomers That May Be Used to Synthesize Polymers

Cationic	Anionic	Non-ionic
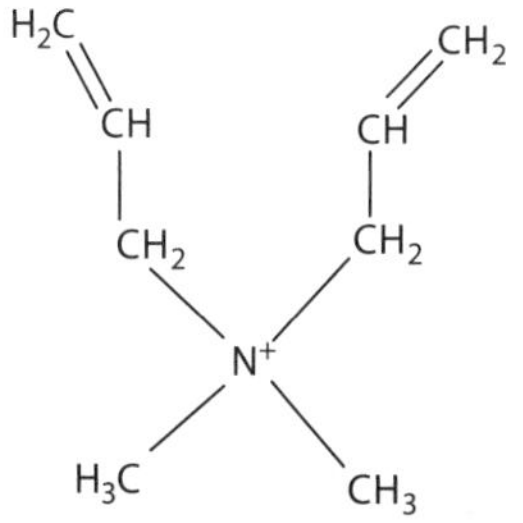	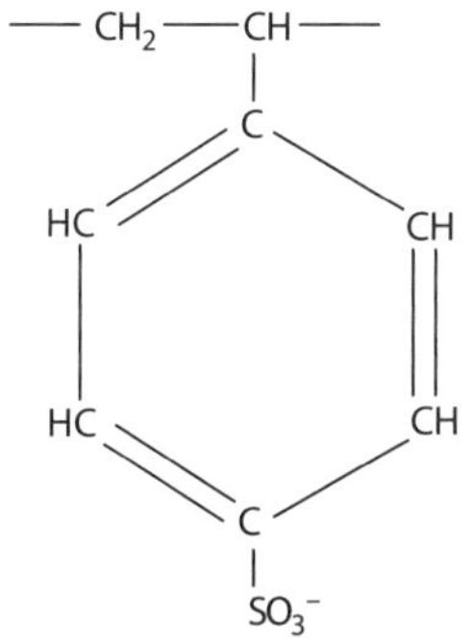	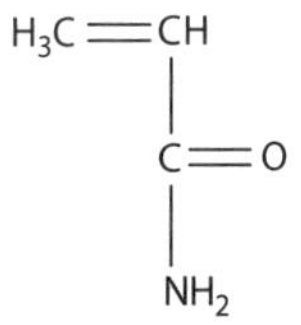
Polydiallyldimethyl-ammonium chloride PDADMAC (Cat-Floc)	Sodium polystyrene sulfonate	Acrylamide
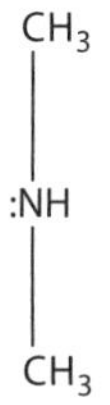	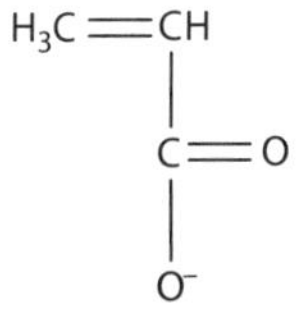	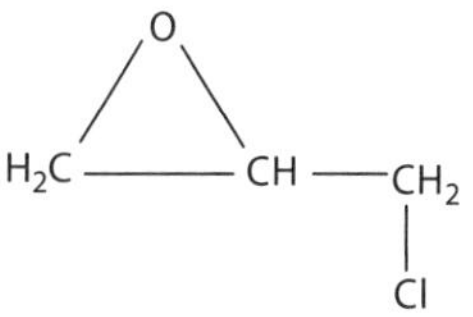
Dimethylamine (DMA)	Acrylate	Epichlorohydrin (epi)

9.9.4.3 Polymers

As noted, polymers are composed of a sequence of monomers, and so their naming follows their composition. To give an idea of the variation in polymer structures Tables 9.10 through 9.12 provide examples of cationic, anionic, and nonionic polymers, respectively. In addition to their structures on the left, the right-hand column indicates (for a given polymer and as data were available) the number of manufacturer's, the number of products, the molecular weight, and the charge density. In polymer nomenclature, the "degree of polymerization" is given by n in the formula (Ravve, 1995, p. 1).

1. *Cationic polymers*: Common cationic polymers are listed in Table 9.10.
2. *Anionic polymers*: The anionic polymers are generally of high molecular weight. Common anionic polymers are listed in Table 9.11.
3. *Nonionic polymers*: The nonionic polymers are mostly polyacrylamide, depicted in Table 9.12, and have high molecular weight.

9.9.5 Selection of Polymers

Thousands of polymers exist as commercial products. As noted, the approach to polymer selection has been based largely upon recommendations of manufacturer's representative and then jar testing for screening and/or confirmation. Scientific guidelines have not been developed for practice, partly because polymers are patented, and their basic characteristics are not provided sufficiently for application of any such guidelines, and partly because a trial and error process would be involved in any case. While both are true, guidelines based upon the character of the polymer, its charge density, its molecular weight, the proposed application, and quantitative knowledge (water characteristics, other coagulants used, point of application, concentration, and settled water turbidity with and without the polymer, filtered water turbidity with and without the polymer, length of filter run with and without the polymer) can aid in selection.

9.9.5.1 Polymer Screening

Screening of a group of selected polymers may utilize jar testing (see Section 9.8.1), which provides the flexibility to test the effects of one independent variable at a time holding all others constant. Dependent variables may include floc size and settling velocity and settled water turbidity (which are all related).

Pilot plant testing, that is, after preliminary screening by jar tests, is useful for a "final screening" of the coagulation process chemicals, including polymers. Dependent variables measured may include: effluent turbidity (or particle counts), rate of headloss increase, ripening time, and clean bed headloss change after repeated filtration cycles. The independent variables to be explored are the same as for the jar test (Mangravite, 1983; Ghosh et al., 1985).

TABLE 9.10
Cationic Polymer Structures and Characteristics

Polymer	Data
(a) Polydiallyldimethylammonium (PDADMAC)	131 Manufacturers, 1216 products[a] Commercial example: Catfloc T[b] Molecular weight: $0.5 \cdot 10^6$ Charge density: 6 meq/g active polymer[c]
(b) Acrylamide-DADMAC	Molecular weight: 10^4–10^5[c] Charge density: 8 meq/g active polymer[d]
Primary amine, Secondary amine, Tertiary amine (c) Polyamines	104 Manufacturers, 749 products[a] Molecular weight: $15 \cdot 10^6$ Charge density: high
(d) epi/DMA polyquaternary amine[f]	131 Manufacturers, 1216 products[a] MW $\approx 0.67 \cdot 10^5$ Charge density≈7.95 meq/g solid Commercial example: Magnifloc 572C[d] MW = 63,000 g/mol Charge density: 3.1 meq/g neat at pH = 7.0 American Cyanamid[e]
(e) Mannich	

[a] National Sanitation Foundation, www.NSF.com, November 27, 2000.
[b] Luttinger (1981).
[c] Edzwald (1993, p. 27).
[d] Hubel and Edzwald (1987, p. 99).
[e] Dentel and Gucciardi (1989, p. 25, Table 2).
[f] Edzwald (1985, p. 172).

TABLE 9.11
Anionic Polymer Structures and Characteristics

Polymer	Data
$-[CH_2-CH(C(=O)NH_2)]_x-[CH_2-CH(C(=O)O^-)]_y-$ (a) Acrylamide/acrylate (AcAm/Ac)	
$-[CH(C(=O)NH_2)-CH(C(=O)NH_2)-CH(C(=O)O^-)-CH(C(=O)NH_2)]_n-$ (b) Polyacrylamide chain[b]	113 manufacturers, 1335 products[a] O– is carboxyl group, affected by pH Commercial example: Magnifloc 835A[b] Molecular weight: $15 \cdot 10^6$ Charge density: high Commercial example: Magnifloc 820A[b] Molecular weight: $6 \cdot 10^6$ Charge density: intermediate Commercial example: Magnifloc 837A[b] Molecular weight: $15 \cdot 10^6$ Charge density: low
$[-CH_2-CH(C_6H_4SO_3^-)-]_n^{n-}$ (c) Polystyrene sulfonate	Commercial example: Purifloc A-21 (Dow)[b] Molecular weight: $15 \cdot 10^6$ Charge density: very high

[a] National Sanitation Foundation, www.NSF.com, November 27, 2000.
[b] Luttinger (1981).

TABLE 9.12
Nonionic Polymer Structures and Characteristics

Polymer	Characteristics
$-[CH_2-CH(C(=O)NH_2)-]_n$ Polyacrylamide[a,b]	Commercial name: Magnifloc 905N[a] 113 manufacturers, 1335 products[b] Molecular weight: $15 \cdot 10^6$ Charge density: none Commercial name: Purifloc N-17 Molecular weight: $5–10 \cdot 10^6$ Charge density: none

[a] Luttinger (1981).
[b] May (1988).

9.9.5.2 Polymer Packaging

Polymers may be purchased as solid, for example, powder, beads, flakes, granular, solution or emulsion. Dry polymers may absorb moisture and stick to equipment and may dissolve slowly, for example, 15–60 min. Solution polymers are preferred by many because they are convenient to use since metering is simple, as is cleanup. High molecular weight polymers are rarely sold as water solutions, with most being available as powders or as emulsions (Mangravite, 1983, p. 5). Emulsion polymers consist of a polymer and water suspended in mineral oil, with polymer content 20%–30% by weight and water content 0%–50% by weight. Such products must be agitated prior to use. The first dilution must be carried out within a specified concentration range, typically about 1 percent product, to ensure complete dissolution of polymer (Mangravite, 1983, p. 5). Solid polymers require feeding equipment first to wet the particles and then to agitate them in solution for a specified period of time, with subsequent transfer to a holding tank. The solid form contains 75%–95% polymer, with the remainder being moisture or salts such as sodium carbonate or sodium sulfate.

9.9.5.3 Specification Sheets

Each manufactured product has a specification sheet that states such items as principal uses, typical properties (physical form, ionic nature, density, viscosity, charge, freezing point), preparation and feeding, typical dosages, materials for storage and feed, handling and storage, safety information, shipping information. These are available from manufacturer's representatives, brochures, or Web sites. Some manufacturers state the kind of structure used, for example, polyacrylamide, polyamine, etc.

9.9.5.4 Prepared Batches

Prepared batches of polymer are used normally within 24–48 h to ensure use before loss of activity. Stock solutions are made usually with 0.1%–0.5% (1000–5000 mg/L), as a compromise between storage volume, batch life, and viscosity. Polymers are considered more effective when fed as dilute solutions because they are easier to disperse (i.e., to distribute uniformly). Typically, the feed strength is between 0.01% and 0.05% (100–1000 mg polymer/L solution), but should be based on the recommendations of the manufacturer.

9.9.5.5 Feed of Polymer

In feeding low/medium molecular weight polymers, the neat solution, as provided by the manufacturer, may be pumped to a dilution line, preferably just prior to a static mixer, and then pumped (by metering pump) to the point of application. Alternatively, the neat solution may be pumped to intermediate batch mixing and then to the point of application. The concentration at the point of application may be 1%–5% (10,000–50,000 mg polymer/L solution). Mixing at the point of application is the next step, which is critical to effective coagulation. Example 9.8 illustrates how to calculate the metering rate of a polymer, given the concentration desired and the raw water flow. An emulsion polymer requires intermediate mixing before being metered into the raw water flow. Example 9.9 shows how such a polymer is prepared prior to metering.

9.9.5.6 Concentration: Convention (Adapted from AWWA B453-96)

In feeding low/medium molecular weight polymers, the polymer may be fed as neat solution. In designating the concentration of a polymer, three possible forms are as follows: (1) as-sold, (2) total solids, and (3) active polymer (AWWA, 1996, p. 11).

9.9.5.6.1 As Sold

The "as sold" basis means as it comes out of the container. For example, a 10 mg/L solution means that 10 mg of product from the container is associated with 1000 mL of raw water, that is, 10 mg product/L raw water.

9.9.5.6.2 Total Solids

The "total solids" basis means that the fraction of solids in the product is determined. For example, suppose the fraction of total solids in the product is 0.40, and a 10 mg total solids/L solution is desired. Since 400,000 mg solids is associated with 1000 g product, then 10 mg solids is associated with x g product. Then, $x = [10^6/(0.4 \cdot 10^6)] \cdot 10 = 25$ mg product. Therefore, 25 mg product is metered into 1000 mL raw water.

9.9.5.6.3 Active Polymer

The "active polymer" basis means that the fraction of solids in the product is determined. For example, suppose the active polymer is 0.50 and the total solids in the product is 0.40, and a 10 mg active polymer/L solution is desired. Since $0.50 \cdot 400{,}000 = 200{,}000$ mg active polymer is associated with 1000 g product, then 10 mg solids is associated with x g product. Then, $x = [10^6/(0.2 \cdot 10^6)] \cdot 10 = 50$ mg product. Therefore, 50 mg product is metered into 1000 mL raw water.

9.9.5.6.4 Discussion

The "as sold" basis is most common convention for stating concentration since it is straightforward and easy to determine. In expressing concentration, the basis should be specified, for example, 10 mg product/L raw water.

Example 9.8 Metering a Neat Polymer Solution

Given

As an example of a concentration calculation, suppose that the flow of raw water is 1.0 m^3/s (22.8 mgd) and that the concentration of polymer is to be 0.40 mg/L on an "as sold" basis. The polymer density (as supplied) = 1.064 kg polymer emulsion/L emulsion.

Required

Determine the rate of feed flow of neat polymer solution.

Solution

1. *Specification sheet data*. Suppose that the polymer is Clarifloc A-210P. The specification sheet dated 1991 (polydyneinc.com) states the polymer has a medium charge, is a polyacrylamide, is in emulsion form, and is approved by the National Sanitation Foundation (NSF) for clarification of potable water at dosages ≤ 1.0 mg/L, with density = 8.88 lb/gal (1.064 kg/L).
2. *Calculation*

Mass flux of polymer in raw water flow

$$= Q \cdot C(\text{polymer})$$
$$= 1000\,\text{L/s} \cdot 0.40\,\text{mg/L}$$
$$= 400\,\text{mg polymer/s}$$

Mass flux of polymer as neat emulsion

$= 400\ \text{mg polymer/s}$

$= Q(\text{neat}) \cdot C(\text{neat})$

$= Q(\text{neat}) \cdot 1{,}064{,}000\ \text{mg polymer/L}$

$$Q(\text{neat}) = 0.000376\ \text{L/s} = 0.376\ \text{mL/s} = 22.5\ \text{mL/min}$$

Discussion

The example is given to illustrate how the polymer concentration is calculated on the "as supplied" mass of polymer as it comes from the barrel. A second point is that the flow of polymer from the barrel, Q(neat), is very low, for example, 0.000376 L/s, which must be mixed with a flow of 1000 L/s, that is, a factor of 10^6. The problem is to adequately mix the polymer with the raw water flow so that particle-polymer contacts are made (mixing is the topic of Chapter 10).

Example 9.9 Design for Emulsion Polymer Storage, Dilution/Mixing

Given

The polymer Clarifloc A-210P, an emulsion polymer (see Example 9.8), has been selected for water treatment with a flow, $Q = 1000$ L/s.

Required

Apply the manufacturer's recommendation for feeding the polymer into the raw water flow and translate this into a design.

Solution

1. *Excerpts from Clarifloc A-210P Data Sheet*
   - Polymer density (as supplied) = 1.064 kg/L.
   - Suggested in-plant storage life is 6 months in unopened drums.
   - Bulk tanks should be mixed by periodically recirculating the contents bottom to top. Bulk tanks can also be fitted with an agitator type mixer that reaches the bottom 2 ft of the tank. Drums should be mixed very well before first use.
   - In most cases, the product should not be applied neat.
   - One method for dilution is adding the neat polymer into the vortex of a mixing tank at a concentration between 0.25% and 1.0% polymer (0.5% is recommended) by weight.
   - The best-feed systems use initial high energy mixing (>1000 rpm) for <30 s to achieve good dispersion followed by low energy mixing (<400 rpm) for a longer time, that is, $10 < t < 30$ min.
   - Polymer solution should be aged for 15–60 min. Solution shelf life is 8–16 h.
2. *Storage*. Based on a uniform raw water flow of 1000 L/s and polymer concentration 0.4 mg/L, the mass amount required for 6 month's polymer storage is

Mass polymer (6 months storage)

$= 1000\ \text{L/s} \cdot 0.4\ \text{mg/L} \cdot 3600\ \text{s/h} \cdot 24\ \text{h/day} \cdot 180\ \text{day}$

$= 6221\ \text{kg polymer}$

$$V(\text{barrel}) = 55.0\ \text{gal} = 208.2\ \text{L}$$

$$\text{Mass(barrel)} = 208.2\ \text{L} \cdot 1.064\ \text{kg/L} = 222\ \text{kg/barrel}$$

N(barrels) = Mass polymer(6 month storage)/Mass(barrel)

$= 6221\ \text{kg polymer}/222\ \text{kg polymer/barrel}$

$= 28\ \text{barrels}$

For delivery in 275 gal "tote" tanks with $V = 275$ gal $= 1041$ L,

$$\text{Mass(tote tank)} = 1041\ \text{L} \cdot 1.064\ \text{kg/L} = 1108\ \text{kg/tote tank}$$

N(tote tank) = Mass polymer × (6 month storage)/Mass(tote tank)

$= 6221\ \text{kg polymer}/1108\ \text{kg polymer/barrel}$

$= 5.6$ tote tanks per 6 month period

3. *Initial dilution/mixing*. Dilute to the recommended concentration, that is, 0.5% by weight. The initial high-energy mixing tank should have a detention time of 30 s. Since Q (neat) = 0.376 mL/s (Example 9.8),

$$M(\text{mass flux neat}) = 0.376\ \text{mL/s} \cdot \text{L}/1000\ \text{mL} \cdot 1.064\ \text{kg/L} = 0.00040\ \text{kg/s}$$

Concentration(diluted polymer)

$= 0.005\ \text{kg polymer/kg water}$

$= 5000\ \text{mg polymer/L water}$

Since mass flux polymer = 400 mg/s

and

$$\text{Mass flux polymer} = Q(\text{first dilution}) \cdot C(\text{diluted polymer})$$

$$400\ \text{mg polymer/s} = Q(\text{first dilution}) \cdot 5000\ \text{mg polymer/L water}$$

$$Q(\text{first dilution}) = 0.08\ \text{L/s} = 80\ \text{mL/s}$$

$$V(\text{tank, first dilution}) = Q(\text{first dilution}) \cdot t(\text{tank}) = 0.080\ \text{L/s} \cdot 30\ \text{s} = 2.4\ \text{L}$$

The tank requires mixing at about 1000 rpm, which can be done using a small mixer set-up; probably a 100 W motor would work.

3. *Second mixing tank*. For the second mixing tank, the detention time recommended is 10–30 min. Let t (second tank) = 30 min. Then,

$$V(\text{second tank}) = Q(\text{first dilution}) \cdot t(\text{second tank})$$
$$= 0.080\,\text{L/s} \cdot 1800\,\text{s}$$
$$= 144\,\text{L}$$

From the Clarifloc A-210P data sheet, "polymer solution should be aged for 15–60 min for best results. Solution shelf life is 8–16 h." The 30 min detention time in the second tank takes care of the aging requirement and consumes the polymer before the 8–16 h shelf life has occurred.

Discussion

Since the tote tanks provide for continuous mixing of the emulsion and would involve less handling, tote tank delivery would be preferred. To comply with the 6-month shelf life, the delivery schedule should be about three tanks every 6 months. As a second point, the flow of diluted polymer, Q(first dilution) = 0.080 L/s, compares with Q = 1000 L/s, a factor of about 10^4, compared with about 10^6 for the neat solution. Thus, the mixing requirement remains a major issue, but is less severe than if the neat emulsion was used.

PROBLEMS

9.1 Shulze-Hardy Rule

Given/required

Describe the logic of the Shulze-Hardy rule.

9.2 Theories of Coagulation

Given/required

Coagulation theory has evolved to include such ideas as ion exchange, double layer, and sweep-floc. Describe briefly some of the key aspects of the evolution of the current theory.

9.3 Potential Barrier

Given/required

Describe methods to reduce the magnitude of the potential barrier in the double layer theory.

9.4 Bonding Energy between Particles—from DVLO Theory

Given/required

Determine the order of magnitude of the bonding energy between two particles, expressed in kJ/mol. (1) Charge-neutralization occurs and (2) DVLO theory applies.

Hint: Section 9.4.1.7 may serve as a starting point.

9.5 Equilibrium Constants—Aluminum

Given/required

Generate the lines of the alum concentration versus pH diagram by means of a spreadsheet.

9.6 Equilibrium Constants—Ferric Iron

Given/required

Generate the lines of the ferric iron concentration versus pH diagram by means of a spreadsheet.

9.7 Conversion—moles/L to mg/L

Given/required

The problems are stated as (a) and (b). The solutions are straightforward and are freshman chemistry type conversions (mentioned so that the problems are seen as not too involved). Example 9.3 illustrates the conversion procedure.

9.8 Metering Liquid Alum

Given/required

In treating the low turbidity in the water of a mountain stream (e.g., the Cache La Poudre River in Colorado) the lowest effective alum concentration was determined to be 26 mg $Al_2(SO_4)_3 \cdot 14H_2O$/L (from a pilot plant study). Assume the flow of raw water to the plant is Q(raw water) = 0.76 m^3/min (200 gpm), which is a small plant.

9.9 Alum Manufacture

Given/required

Describe the manufacture of alum. Describe sources, manufacturing steps, annual production in the United States, how it is packaged and delivered, the distribution infrastructure, and costs.

9.10 Historical Persons

Given/required

Outline the contributions of A. P. Black to knowledge of coagulation chemistry. W. F. Langlier. W Stumm, and former students.

9.11 Alum in History

Given/required

Outline coagulation practice in water treatment from antiquity.

9.12 Wastewater Coagulation

Given/required

What could be expected if coagulation is used as a unit process in treatment of raw wastewater in reduction of BOD (or COD or TOC) and suspended solids? Verify with laboratory jar tests.

9.13 Tertiary Wastewater Coagulation

Given/required

What could be expected if lime coagulation is used as a unit process in treatment of secondary treated wastewater in reduction of BOD (or COD or TOC) and suspended solids? Verify with laboratory jar tests. You may wish to consider other coagulants, for example, ferric iron or alum.

9.14 Selection of Coagulants

Given/required

How would you go about selection of coagulation chemicals for a new water treatment plant installation? Outline important considerations.

9.15 Issues in Wastewater Coagulation

Given/required

Outline the issues for a program to determine if coagulation should be used in treating a raw wastewater.

Describe the nature and extent of each of these phases (i.e. what you would hope to learn). Enumerate "measures of effectiveness" relating to technical effectiveness of the coagulation process and cost criteria. As a note, it is not likely, for most situations, that the chemical treatment would be used in lieu of biological treatment. The problem is intended as an exercise in approaching a new problem as may be found in practice.

9.16 Models in Coagulation

Given/required

Discuss the role of models in understanding coagulation (e.g., descriptive, empirical, mathematical).

9.17 Metering Solid Polymers

For this kind of problem, the manufacturer's specifications sheet should serve for reference to provide density of dry polymer, storage conditions and time, strength and viscosity of first-stage polymer solution in first stage mixing, mixer (impeller type, size, mounting angle in basin, motor power), strength of second stage storage, strength of polymer feed solution, etc. In addition, the manufacturer may have package systems for mixing and feeding of the solid polymer, including pump type and sizing.

Given

- A solid polymer is to be prepared for metering into a raw water flow for a water treatment plant.
- Q(plant) = 1.00 m^3/s (23 mgd).
- The polymer is 90% active polymer.
- The concentration of polymer required for treatment is 0.5 mg/L polymer mass as provided by the manufacturer (as distinguished from active polymer mass).

Required

Determine the following:

1. The volume of storage required for the polymer for a 60 day supply
2. The rate of solids feed to a first stage mixing unit
3. The volume and motor size for the first stage mixing unit
4. The volume and motor size for the second stage mixing unit
5. The rate of pumping from the second stage mixing unit to the raw water flow
6. Recommend a particular metering pump
7. The quality control set-up (i.e., a 1000 mL cylinder set up on-line) to verify the metering rate
8. Aspects of handling that are relevant, for example, labeling, materials for pipelines, storage vessels, pumps, etc. (see AWWA Standards)

9.18 Metering Liquid Polymers

Given

A liquid polymer is to be prepared for metering into a raw water flow, $Q = 1.00$ m^3/s (23 mgd), for a water treatment plant. The concentration of polymer required for treatment is 0.5 mg/L polymer mass as provided by the manufacturer (as distinguished from active polymer mass).

Required

Determine the following:

1. The volume of storage required for the polymer for a 60 day supply
2. The rate of feed to a first stage batch mixing unit
3. The volume and motor size for the first stage mixing unit
4. The rate of pumping from the first stage mixing unit to the raw water flow
5. Recommend a particular metering pump
6. The quality control set-up (i.e., a 1000 mL cylinder set up online) to verify the metering rate
7. Aspects of handling that are relevant, for example, labeling, materials for pipelines, storage vessels, pumps, etc.

9.19 Mixing and Metering Liquid Polymers

Given

A liquid polymer is to be prepared for metering into a raw water flow, $Q = 1.00$ m^3/s (23 mgd), for a water treatment plant. The concentration of polymer required for treatment is 0.5 mg/L polymer. Assume that an alternative metering system is planned, which is to blend the polymer with a dilution water prior to injection into the raw water.

Required

Design a system to feed the polymer from polymer storage to the injection jets into the raw water. Assume the injection is to follow rapid-mix.

Solution

Example 9.9 addresses some of the issues of mixing of an emulsion, but which pertains, by and large, to any mixing problem. For a neat polymer solution, the solution may be pumped directly from the storage tank and blended into a flow of dilution water; a static mixer could be used. While there is no known rule for dilution, about 100/1, that is, dilution flow to neat polymer flow, could be used. The next step is to get the diluted polymer flow mixed into the raw water flow after the rapid-mix. This may be done within a pipe with flow in the turbulent flow regime and by a diffuser, with four to eight jets directed upstream. The jets might be placed in a ring manifold or perhaps a cross manifold that covers the cross section of the pipe.

9.20 Handling Polymers

Given

Two kinds of polymers are being considered for use in a water treatment plant: (1) polyacrylamide, and (2) epi/DMA polyamine.

Hint: The AWWA Standards are first sources of information on such questions.

Required

For one of these polymers, specify considerations for handling, such as follows:

1. Materials used in storage, pipes, pumps, etc. (In general, bulk storage of polymer solutions utilizes fiberglass tanks, stainless steel, and lined tanks of rubber, glass, or fiberglass)
2. Labeling
3. Training of personnel
4. Laboratory testing
5. Possible health concerns
6. Other issues

ACKNOWLEDGMENTS

Kevin Gertig, water resources & treatment operations manager (formerly supervisor, Fort Collins Water Treatment Plant, City of Fort Collins, Colorado) was available for consultation at any time and provided photographs as needed from the extensive collection at the plant and made available the library or the plant. Grant Williamson-Jones, City of Fort Collins, Colorado, provided the two photo micro-graphs of alum floc, Figure 9.3, from his collection.

Figure 9.19 from a report of the Water Research Foundation (WaterRF), Denver, Colorado, was reproduced with permission provided by Adam Lang, publishing manager.

Jarid Kling, presently Equipment Diagnostics Center lead, Alyeska Pipeline Service Company, took my water treatment courses in 1998–1999; his homework on coagulant chemical—pH equilibria spreadsheets was utilized as Table CD9.7 which he graciously gave permission to use again (2010), i.e., for the present text. His work on this topic overcame some difficulties in constructing a correct set of equations and associated linked plots.

GLOSSARY

Acidity: Refers to the loss of a proton, that is, by HA (Streitwieser and Heathcock, 1985, p. 60). An example reaction is

$$HA = H^+ + A^-$$

conjugate acid conjugate base

Adsorption destabilization: Assimilation of hydrolysis products of a metal ion, generally Al^{3+} or Fe^{3+}, into the diffuse double layer of a colloidal particle to effect in a reduction of negative repulsive force between colloids so that van der Waals attractive forces may dominate when the particles overcome the "potential barrier." Same as charge neutralization.

Alkalinity: Defined as the sum of its three forms, HCO_3^-, CO_3^{2+}, OH^- and usually expressed as $CaCO_3$.

Alum: Aluminum sulfate. See Appendix F for description.

Amines: Organic relatives of ammonia, which are derived by replacing one, two, or all three hydrogens of ammonia with organic groups (Hart, 1991, p. 305). For convenience, amines are classified as primary, secondary, or tertiary, depending on whether one, two, or three organic groups are attached to the nitrogen, that is,

H—N̈—H, with H on N	R—N̈—H, with H on N	R—N̈—R, with H on N	R—N̈—R, with R on N
Ammonia	Primary amine	Secondary amine	Tertiary amine

The R groups may be identical to or different from one another. In secondary or tertiary amines, the nitrogen may be part of a ring. In quaternary ammonium salts, all four hydrogens of the ammonium ion are replaced by organic groups; the result is

R—N⁺—R, with R above and R below N

Amphoteric: Substances, such as metal oxides, can develop either negative or positive surface charges depending on pH (see also isoelectric point and point of zero charge).

Anion: Ion with negative charge, for example, Cl^-, HCO_3^-,. . . .

Anionic polymer: Polymer that has a negatively charged group; the charge manifests itself when an associated cation dissociates from the group. Common cations include: H^+, Na^+, K^+, Ca^{2+}, etc.

Aquometal ion: Central metal ion with waters of hydration bonded as ligands, for example, $Al(H_2O)_6^{3+}$.

Basicity: Refers to the gain of a proton, that is, by A^- (Streitwieser and Heathcock, 1985, p. 62).

Benzene: The compound, C_6H_6, with a ring structure. Benzene is the parent hydrocarbon for a whole family of organic compounds. As a group, the benzene-like compounds were called aromatic because many of them have characteristic aromas (Streitwieser and Heathcock, 1985, p. 562).

Boltzman constant: Same as the ideal gas constant per molecule ($k = 1.380658 \cdot 10^{-23}$ J/K), that is, instead of per mol. Note that $R = 8.314510$ J/K mol and that $N(\text{Avogadro}) = 6.022 \cdot 10^{23}$ molecules/mol. Therefore,

$$k = \frac{R}{N}(\text{Avagadro}) = \frac{8.314510\,\text{J/Kmol}}{6.022 \cdot 10^{23}\,\text{molecules/mol}}$$
$$= 1.380689 \cdot 10^{-23}\,\text{J/(K} \cdot \text{molecule)}$$

The main point is that the units are really J/K/molecule, as opposed to J/K, which usually is understood in the literature, but is not apparent to persons not trying to reconcile units given in some aspects of coagulation theory, that is, bonding energy, which is given sometimes in terms of kT; to obtain a magnitude in terms of energy per mole; the units work out if we multiply kT (J/K/molecule) by Avogadro's number (molecules/mol), which gives, J/mol.

Boltzman equation: A gas occupying a given volume has an overall average energy per mole; the gas molecules actually have a distribution of energies as defined by the Boltzman equation, that is, $N/N_o = e^{-(E/kT)}$.

Bridging: Generally, the term applies to the condition in which single polymer molecules attach to two or more suspended particles.

Briggs cell: A cell of specified dimensions that provides for a specified flow through with a specified voltage between plates (see Black and Smith, 1962 who describe an operational cell as referenced in a 1940 article by D.R. Briggs).

Bronsted acid: A substance that may contribute a proton to solution or cause a proton to be contributed to solution. A water ligand bound to a central metal ion may contribute a proton to solution. The complex is thus acting as a Bronsted acid.

Brownian motion: Random motion of molecules in a gas due to thermal energy, measured by the parameter, kT in gases (in which k is the Boltzman constant and T is the absolute temperature). Note that $k = R/N_o$, where R is the universal gas constant and N_o is Avogadro's number, that is, the number of molecules in a mole.

Carboxyl group: A chemical group, $-COOH$, which may be attached to a larger carbon-based molecule, for example, a polymer. A characteristic of the group is its ionization, which may leave a charged site, that is, at $pH > 4$, $-COOH + H_2O \rightarrow -COO^- + H_3O^+$.

Cartridge filter: Fiber-wound filter about 30 cm long and 12 cm diameter and hollow in center, used commonly since the 1980s for evaluation of filter performance or assessment of organisms in natural water bodies. The results are expressed as numbers of organisms of different species and particles retained per liter of water that has passed through the filter. The sampling/analysis procedure and results is designated, 'microscopic particulate analysis' (MPA). Figure 13.2 shows two cartridge filters as used in the evaluation of particle removal for a slow sand filter. The cartridge is placed in a pressurized holder designed to permit water flow from the outside to the hollow center. Recommended flow through the cartridge is 4 L/min (1 gpm), albeit flows of perhaps 40 L/min (10 gpm) have been used. After removal, the cartridge is placed in a 1 gal size Zip-loc® plastic bag and placed on ice for transport to a laboratory for MPA analysis. See microscopic particulate analysis.

Cation: Ion with positive charge, for example, Na^+, Ca^{2+}, Al^{3+}, Fe^{3+}, ...

Cationic polymer: Polymer with attached ionic groups that are positive.

Central ion: Usually a metal ion that coordinates with a ligand.

Charge neutralize: Attachment of a positive charged polymeric metal ion species or cationic polymer to a negatively charged colloid.

Chelate: Complex involving multidentate ligand. See *ligand*.
- Monodentate ligand—Ligands that attach at only one point, such as H_2O, OH^-, Cl^-, and CN^-.
- Multidentate ligand—Ligands that attach at two or more sites.

Chelate: Polydentate ligands can produce a chelate (Greek for claw), a complex in which a ligand forms a ring that includes the metal ion (Shriver and Atkins, 1999, p. 220).

Chelating agent: A multidentate ligand.

Clay: Soil fraction characterized by small size, for example, ≤ 2 μm, negative charge, and plasticity when mixed with water (Mitchell, 1993, pp. 18–30). Some aluminosilicate clays listed are as follows: kaolinite, illite, vermiculite, smectite (group includes montmorillonite), and chlorite (Hemingway and Sposito, 1996, p. 87). Clays may have cations on the layered surfaces that may be exchanged easily with other cations in solution and are called "exchangeable" cations. The amount of exchangeable cations expressed in milleequivalents per 100 g of dry clay is called the "cation exchange capacity" (CEC), sometimes called the "base exchange capacity." Montmorillonite has a $CEC \approx 70$ meq/100 g. Also, montmorillonite is characterized by interlayer swelling. Illites, on the other hand, are distinguished by the lack of interlayer swelling. For kaolinite, $1 < CEC < 10$ meq/100 g (van Olphen, 1977, pp. 57–76).

Coagulant: (1) A substance, usually a trivalent cation, which may combine directly with colloids to form "microflocs;" more likely however, the hydrolysis products of Al^{3+} (or Fe^{3+}) reacting with water combine with suspended colloids and charged particles. (2) Any chemical that destabilizes a sol suspension (adapted from O'Melia, 1978, p. 241).

Coagulant aid: A substance that may be added to rapid-mix to improve the results of coagulation. The most common is a polymer. In cases of low turbidity water, a clay, such as bentonite has been used to aid coagulation.

Coagulation: (1) Coagulation comes from the Latin word, *coagulare*—to be driven together and is brought about by a reduction of the repulsive potential of the electrical double layer in accordance with the DLVO theory causing particle destabilization (Black, 1967, p. 277; Stumm and O'Melia, 1968). (2) The chemical process of reducing the zeta potential and destabilization of particles (Moffett, 1968, p. 1256). (3) The chemical reduction of repulsive forces between particles such that the van der Waals forces become dominant so that particles will stick when they collide and form aggregates called "microflocs." The process remains an "art" (Gregory, 1975, p. 61). (4) Concerning kinetics, the coagulation reactions are nearly instantaneous and the only time required for their completion is what is necessary for dispersing the chemicals throughout the water (Conley and Evers, 1968).

Coagulation/flocculation: Moffett (1968, p. 1256) noted that the terms coagulation and flocculation are ambiguous, as used in the water field. This was affirmed by O'Melia (1970), who stated that coagulation and flocculation have no commonly agreed upon definitions. Opinion seems to have converged, however, on the following definitions:

1. Coagulation refers to the chemical alteration of a colloid such that aggregation between destabilized colloids and coagulant chemicals *can* occur upon mutual contact
2. Flocculation is the process of aggregating coagulant chemicals and colloids by contact induced by some transport mechanism, for example, Brownian motion or turbulence.

Colloid: Small particles that do not settle due to an electric surface charge, usually negative. The characteristic size range is 10^{-9} to 10^{-5} m (10–100 Å or 10 μm). Sennett and Olivier (1965, p. 34) suggest an upper size limit of 1 μm for purposes of classifying a "sol," but point out that larger particles may exhibit characteristics of a colloid.

Colloidal system: A "classical" colloidal system is a *sol*, that is, dispersions of small solid particles in a liquid medium. These small particles are larger than the molecular size. In a colloidal system there are two phases: (1) the *dispersed phase*, a solid, liquid, or gas that is finely divided and dispersed uniformly throughout a second substance which is the *dispersion medium* or *continuous phase*, and (2) the *continuous* phase may be a solid, liquid, or gas (Myers, 1991, pp. 187–195). An *emulsion* is a liquid dispersed in another liquid. A second class of colloids is where the aggregates of molecules that may be simultaneously a molecular solution and a true colloidal system (FA and HA would be in this category). Yet another class of colloids is the *lyophilic* colloids which are solutions but in which the solute molecules, that is, polymers, are much larger than those of the solvent.

Color: Substance in water that causes "color" as measured on the cobalt scale. Two kinds of color are "apparent" color and "true" color. The former is the result of a measurement using a sample of water "as-is" and the latter is the same measurement after filtering. Organic color is associated with NOM.

Complex: In aqueous solution, free metal ions are *complexed* with water. The metal ions are said to be *hydrated*. The interaction of these hydrated metal ions with acids and bases is a *ligand* exchange reaction that is commonly called *hydrolysis* (or protolysis). A complex is the same as a coordination compound. To summarize,

$$\text{ligand} + \text{central metal ion} \rightarrow \text{complex}$$

Coordination compound: Has one or more central atoms or central ions, usually metals, with a number of ligands attached. Same as a *complex*.

Copolymer: If the polymer molecule is formed from more than one type of repeating chemical unit, it is called a copolymer (Singley et al., 1971).

Coprecipitation: Contamination of a precipitate by an impurity that is otherwise soluble under the conditions of precipitation (Randtke, 1988, p. 41).

Coulomb: Quantity of electric charge. Charge is related to current by the relation, $q = i \cdot t$, in which q = quantity of charge (C), i = current (A), and t = time (s). For reference, 1.00 As = 1.00 C.

Coulombic interaction: Attraction or repulsion between charged particles in accordance with Coulomb's law, that is,

$$F = \frac{q_1 q_2}{4\pi D \varepsilon_o r^2} \tag{G9.1}$$

where

F is the force between q_1 and q_2 (N)
q_1 is the charge on particle "1" (coulombs)
q_2 is the charge on particle "2" (coulombs)
$\varepsilon_o = 8.854 \cdot 10^{-12}$ ($C^2/(mJ)$
D is the dielectric constant (dimensionless)
r is the separation distance between particles "1" and "2" (m)

Thus such forces are found only in systems containing charged species. The strengths of the interactions equal or exceed those of the covalent bonds (Myers, 1991, p. 39).

Covalent bond: When two atoms bind to form a typical nonionic molecule, the forces involved in bond formation are termed, covalent. The characteristic of such a bond are shared electrons between two or more atoms. The covalent bonds are short range, that is, they act over a bond distance of 0.1–0.2 nm. The energies of normal covalent bonds range from 150 to 900 kJ/mol (100–300 kT), and generally decrease in strength as the bond length increases (Myers, 1991, p. 39).

Critical coagulant concentration (CCC): The minimum concentration of coagulant above which destabilization will take place (Stumm and O'Melia, 1968, p. 516).

Cross-linking: Incorporation of divinylbenzene into the polymerization of styrene results in two vinyl groups participating in two separate chains, producing a three dimensional network. Polystyrene, for example, is soluble in many solvents but with 0.1% divinylbenzene the polymer no longer dissolves but only swells (Streitwieser and Heathcock, 1985, p. 1113).

Crystal lattice: A solid that has a regular geometric arrangement of atoms in space that determine its properties.

Dielectric coefficient: The field between a pair of oppositely charged parallel plates and the induced charges on the surfaces of a dielectric adjacent to the plates. An insulator, or dielectric, is a substance within which there are a few or no charged particles free to move under the influence of an electric field. Representative values of the coefficient, D, are D(vacuum) = 1, D(glass) = 5–10, D(rubber) = 2.5–35, D(water) = 81, D(air) = 1.00059 (Sears, 1947, p. 177). See also *permittivity*.

Diffuse layer: Ion cloud of counterions having charge opposite the surface charge of the colloid; also called the "Gouy" layer.

Disinfection by-product (DBP): Products formed due to the reaction between a disinfectant and organic carbon.

Dispersed phase: Two disperse phases important in water systems include a gas in a liquid and particles in water (Stumm and O'Melia, 1968, p. 532; Myers, 1991, p. 191).

Dispersion: A colloidal dispersion is the solid and liquid phases together. A system of fine particles is called a colloidal dispersion, or a sol (Sennett and Olivier, 1965, p. 33).

Dispersion forces: The same as London-van der Waals forces and sometimes called dispersion forces because the electron oscillations involved are also responsible for the dispersion of light; the term is distinguished from dispersions of particles (Gregory, 2006, p. 67).

Dissolved organic carbon (DOC): Organic carbon passing through a 0.45 μm filter (Randtke, 1988, p. 43).

DLVO theory: A theory of colloid stability synthesized by four investigators in colloid science, Derjaguin–Landau–Verwey–Overbeek that integrates electrical repulsion between particles with van der Waals attraction.

Double layer: A negatively charged colloid in water has both a "fixed layer" of positive ions bound to the particle (as described by Helmholz and Stern) and an "ion cloud" that emanates from the fixed layer (as described by Gouy and Chapman). These two parts, the fixed layer and the ion cloud, were termed the "double layer." The diffuse layer is characterized by a gradual change in potential, due to the thermal energy of the ions (Sennett and Olivier, 1965, p. 36).

Electric field: A potential gradient. Also, a charged particle at a point, A, causes an electric field (Sears, 1947, p. 17).

Electrokinetic: When two phases exist, that is, a solid and a liquid, each surface is likely to have a charge. When one phase moves past the second phase, four effects are possible, depending on the way the motion is induced: (1) electrophoresis, (2) electroosmosis, (3) streaming potential, (4) sedimentation potential (Hunter, 1981, p. 2).

Electron volt: Work done to an electron in moving it through an electric field under a potential difference of 1 V. 1.0 eV = $1.60217733 \cdot 10^{-19}$ C; also, 1.0 eV = $1.6022 \cdot 10^{-19}$ J. [C = 1.0 A · 1.0 s]

Electrophoresis: The movement of a charged particle in an electric field. A charged particle moves toward a plate of opposite charge at a velocity proportional to the potential gradient in V/cm.

Electrophoretic mobility (EM): The electrophoretic mobility, EM, is the particle velocity per unit of field strength with SI units, m/s/V/m (Pilipovich et al., 1958, p. 1469; Gregory, 1975, p. 65, 67).

Emulsion: Suspension of one liquid in another, for example, globules of fat in water (Saville, 1917, p. 80).

Enhanced coagulation: The addition of alum (or ferric ion) in excess of that required for removal of colloidal particles to the extent that a substantial amount of NOM is also removed. The term has been around since about 1984 in general usage, being introduced by Kavanaugh (1978) as "modified" coagulation.

Enhanced coagulation end point: Dosage of coagulant and/or pH value which, when achieved, no longer produces significant TOC reduction (Pizzi and Rodgers, 2000, p. 4).

Equilibrium: As pertaining to a chemical reaction, the products and reactants have concentrations that do not change with time.

Equilibrium constant: The ratio of product concentrations in mol/L to reactant concentrations in mol/L, the stoichiometric coefficients are exponents. The equilibrium constant is designated, K_a. The negative log of K_a is designated, pK_a, that is, $-\log K_a = pK_a$. As an example of the use of pK_a, for carbonate equilibria, HCO_3^- is predominant in the range, $4.35 < pH < 10.33$, which are the respective pK_a that separate HCO_3^- from H_2CO_3 and CO_3^{2-}, respectively.

Equilibrium diagram: A diagram that delineates equilibrium concentration relationships between a pair of species in a chemical reaction.

Faraday: Amount of electric charge required to liberate 1 g equivalent of material at an electrolytic cell 1 Faraday = 96,514 C

Ferric iron salt: The ferric salts used in water treatment include $FeCl_3$ and $Fe_2(SO_4)_3$; usually, they are hydrated with 6 and 4.5 waters, respectively.

Floc: An aggregation of microflocs induced usually by hydraulic turbulence causing a growth in size such that settling occurs readily. Size may range 0.1–5 mm.

Flocculation: Flocculation is the physical process of bringing the coagulated particles into contact to promote floc formation. The term, flocculation comes from the Latin word, *flocculus*—a small tuft of wool or a loosely fibrous structure (Black, 1967, p. 277).

Flocculent: A chemical added to produce growth in size of destabilized particles or precipitate. Usually flocculents are commercial polymers introduced into the flow prior to mechanical flocculation.

Fractal: An object or substance of irregular shape that may be difficult to define in terms of dimensions.

Fulvic acid: Similar to HAs except it is soluble at $pH = 1.0$ and believed to be in true solution, vis-à-vis being colloidal (Randtke, 1988, p. 43).

Functional groups: An ionic group attached to a bonding site on a polymer molecule that ionizes. Examples of such groups include: carboxyl ($-COOH$); amino ($-NH_2$); and sulfonic acids. Carboxyl groups ionize at $pH > 4$ (i.e., $-COOH + H_2O \rightarrow -COO- + H_3O+$) and the amino group binds a proton at $pH < 10$ (i.e., $-NH_2 + H_3O+ \rightarrow -NH_3+ + H_2O$).

Gouy–Chapman double layer: An accounting for the charge distribution in an electrical double layer, giving the ψ_o distribution with distance from the interface. An exact treatment of the spherical double layer is possible only by numerical techniques (see Loeb, 1961).

Gram-equivalent weight: Atomic or molecular weight of an ion divided by its valency, for example, $Cu^{2+} = 63.57/2 = 31.79$.

Hydration: The "state" of a free metal ion that has been complexed with water. In aqueous solution, all free metal cations are complexed with water, that is, are hydrated (Stumm and Morgan, 1996, p. 258).

Humic acid: An organic acid insoluble at $pH = 1.0$.

Humic substances: Typically, humic substances are divided into the more soluble FAs and the less soluble HAs. The humic molecules are chemically complex; they are part aromatic and part aliphatic (Dempsey, 1989, p. 2).

Hydrolysis: Dissecting the word hydrolysis, it refers to "lysis," a breaking apart, of something through the action of water. In aquatic chemistry the "hydrolysis of metal ions" is defined as a *lysis* of *water itself* by the metal ion, not vice versa, for example, $Mg^{2+} + H_2O \rightarrow MgOH^+ + H^+$ (Gregory, 2006, p. 123).

Illite: Illites are distinguished by the lack of interlayer swelling. See also *clay*.

Ionogenic group: Functional group attached to a surface that has dissociable ion, for example, H^+, Na^+, etc.

Ionic strength: The definition of ionic strength is (Alberty and Silby, 1992, p. 246):

$$I = \frac{1}{2}\sum_i m_i z_i^2 = \frac{1}{2}\left(m_1 z_1^2 + m_2 z_2^2 + \cdots\right) \quad \text{(G9.2)}$$

where

I is the ionic strength (mol/kg solvent)

m_i is the molal concentration of an ion, i, in solution (mol i/kg solvent)

z_i is the valence of ion

The summation pertains to all ions in solution, that is, all positive ions and all negative ions.

Isoelectric point: When a particle does not migrate toward either electrode in an electric field, it is said to be at its isoelectric point (Pilipovich et al., 1958).

Isotherm: Equilibrium relationship between an adsorbate (such as an organic compound) and an adsorbent (such as activated carbon). In the case of NOM, the main interest relates to the capacity of a metal hydroxide precipitate, for example, $Al(OH)_3$ as the adsorbent. Two types of mathematical expressions for isotherms are the Langmuir and the Freundlich (see Chapter 14).

Jar test: A setup of six 1.5 L square jars each filled with 1.0 L of test water and arranged under a gang paddle stirring apparatus. A common use is to determine coagulant dosage by applying a different dosage to each jar with about 2 min rapid-mix at 300 rpm followed by 10–15 min slow mix at about 10–20 rpm, followed by about 20 min settling. The supernatant is sampled for turbidity or other characteristic of interest.

Kinetics: Rate of reaction. Reaction orders are zero, first, and second. A second-order reaction is defined mathematically, $dC/dt = -kC$.

Lennard-Jones 6-12 potential: As molecules approach one another at molecular distances, that is, nanometers, another force, the Born repulsion, becomes significant and is given as $w(\text{Born}) = B/r^{12}$. This, combined with the London force gives the net interaction, that is, $w(r) = B/r^{12} - C_L/r^6$, which is called, commonly, the Lennard-Jones 6-12 potential. The Born potential is repulsive and shows an exponential decline with r; the London force is attractive and shows an exponential increase. Adding the two functions, results in a "potential well," which is a zone of adhesion.

Ligand: Ions or molecules that are attached to a central atom or ion as a part of a complex. The central species is an electron acceptor and the ligand is an electron donor.

Log *R*: Log removal of a given constituent, defined,

$$\text{Log}\,R = \log\left[\frac{C(\text{effluent})}{C(\text{influent})}\right]$$

In other words, suppose $C(\text{effluent}) = 0.01$ cysts/L and $C(\text{influent}) = 10$ cysts/L. Then,

$$\text{Log}\,R = \log\left[\frac{0.01\ \text{cysts/L}}{10\ \text{cysts/L}}\right] = -3.0$$

The expression is, commonly, "3-log removal."

London forces: A portion of the van der Waals interaction forces, also called "dispersion" interaction. The potential energy of molecule separation is inversely proportional to the sixth power of their separation (Shriver and Atkins, 1999, p. 55).

Lyophilic colloid: The attraction between the colloid and the dispersing medium (e.g., water) is large (Black, 1948, p. 143).

Lyophobic colloid: The attraction between the colloid and the dispersing medium (e.g., water) is small (Black, 1948, p. 143).

Masking agent: Ligand that reduces the concentration of the hydrated metal ion to a point at which the metal ion does not significantly participate in a metal ion reaction because a stable complex has been formed (Freiser, 1996).

Micelle: Molecular aggregates (Myers, 1991, p. 301), which is brief definition.

Microfloc: The term "microfloc" seems to have originated with A.P. Black (1948, p. 143), who stated: "When the positively charged aluminum or ferric ions have neutralized a considerable portion, perhaps most of the negatively charged colloidal particles of color or turbidity, the resulting particles may be called *microflocs* for the purpose of this discussion since they are beyond the limits of visibility and are far too small to settle under the influence of gravity. If subject to flocculation the microflocs will grow in size to form settleable flocs." Pilipovich et al. (1958, p. 1468) adopted this term and its definition by Black as did Moffet (1968, p. 1263), among others.

Microscopic particulate analysis (MPA): Microscopic examination of particles retained by a cartridge filter, which may include counting the species present. (Hibler, 1988; Hancock et al. 1996). The examination does not include bacteria or viruses. See *cartridge filter*.

Molecular weight: The sum of atomic masses of all atoms comprising a molecule.

Mono-disperse: All particles in a colloidal system are approximately the same size, that is, they have a narrow size distribution (Myers, 1991, p. 192).

Monomer: The individual repeating units that make up a polymer are called monomers (Singley et al., 1971).

Montmorillonite: A type of clay noted by interlayer swelling with a CEC $\approx$ 80–150 meq/100 g. The external specific surface area $\approx$ 50–120 m^2/g and the secondary specific surface area $\approx$ 840 m^2/g; the latter is exposed by an expanded lattice (Mitchell, 1993, p. 31). See also *clay*.

Natural organic matter: Organic molecules that occur in ambient waters as a result of natural processes. A portion of NOMis FA and HA. See also Chapter 2.

Neat solution: Without dilution. For example, liquid alum as provided by the manufacturer is a "neat" solution. The same is true for a liquid polymer.

Nephelometeric turbidity unit (NTU): A unit of turbidity based on the light scattering principle as calibrated by a standard. A formazin suspension is a common standard, made up in accordance with prescribed protocol.

Nernst potential: The potential at the surface of a colloid.

Nonionic polymer: A polymer without ionizable groups and thus zero charge.

Orthokinetic flocculation: (1) Langelier and Ludwig (1949, p. 165) introduced the terms *perikinetic* flocculation and *orthokinetic* flocculation, taken from the field of soil chemistry, to mean, respectively, aggregation of flocs as permitted by neutralization of zeta potential of colloids and aggregation of flocs as caused by enmeshment with other flocs and colloids. (2) Argaman and Kaufman (1968) gave definitions based on the type of energy associated with the reaction: (a) perikinetic coagulation is by contacts between reactants being due to Brownian motion, and (b) orthokinetic coagulation is due to turbulence.

Particle: Defined here as any contiguous matter that may range in size from a few angstroms, such as macromolecules or viruses to perhaps 1000 μm. Examples include viruses, bacteria, protozoan cysts, alga cells, rotifers, and mineral particles such as colloids and grains of sand.

Particle count: The results of a particle count instrument, which may be by total count and counts by particle size ranges. The instrument may be batch or online. Online instruments have become increasingly common since about 1990.

Particle destabilization: The process of reducing the repulsive forces between colloidal particles such that when contacts occur the particles will attach to each other.

Particulate organic carbon (POC): Organic carbon passing retained by a 0.45 μm filter. Usually, POC is a small fraction of the TOC compared to DOC (Edzwald, 1993, p. 24).

Permittivity: When electric charges are immersed in a dielectric medium (like water) the strength of the electric field is significantly reduced because the molecular dipoles tend to align themselves in such a way as to cancel part of the field. The capacity of a substance to affect the electric field strength is measured by its permittivity. The reference is the permittivity of a vacuum, ε_o. The dielectric constant, D, is the correction for the medium. Thus, $\varepsilon(\text{water}) = \varepsilon_o D(\text{water})$ (Hunter, 1981, p. 349).

Perikinetic coagulation: (1) The first phase of coagulation in which the zeta potential of the colloid is reversed, neutralized, or reduced to a point where the London-van der Waals forces become predominant and primary coagulation takes place (Black and Willems, 1961, p. 599). (2) The process of contacting between particles caused by Brownian diffusion. [peri is from the Greek, meaning around, near; kinetic is from the Greek *kinein*, to move.] See also *orthokinetic coagulation*.

Physicochemical treatment: Term referring to chemical precipitation and settling, activated carbon adsorption, filtration, etc. (Weber et al., 1970; Burns and Shell, 1971). The term came to use, c.1967, and related mostly to wastewater treatment; the idea was to propose an alternative to biological treatment.

pK_a: Defined, p$K_a = -\log K_a$. See also, *equilibrium constant*.

Polyaluminum chloride: Alum that has been reacted with OH^- under special conditions forms a polynuclear complex that remains stable over a wide range of pH and temperature conditions.

Polydisperse: Particles in a colloidal system are of different sizes, that is, they have a broad size distribution (Myers, 1991, p. 192).

Potential barrier: The potential field of a charged particle varies, that is, increasing and then decreasing, with radial distance from the edge of a colloidal particle. The "hump" that characterizes the variation is called the "potential barrier."

Polymer: A repeating chain of structural groups of atoms, that is, monomers. A polymer may contain more than one type of subunit.

Polymer: A collection of monomers bonded by chemical forces. The polymer may have more than one monomer.

Polyelectrolyte: A polymer having ionizable groups or charged sites.

Precipitate: A solid compound formed in a solution when the solubility product has been exceeded.

Restabilization: As coagulant is added to a suspension, settled water turbidity decreases at some critical coagulant (CCC) dosage. As dosage is increased further, "charge reversal" occurs, which is "restabilization." Further addition of coagulant may result in precipitation of the metal hydroxide with associated turbidity increase.

Salt: A compound that dissociates in solution into anions and cations.

Shulze-Hardy Rule: The rule states that coagulation efficiency of coagulation increases with metal ion valence in the ratio of 1:100:1000 as the charge of the counterion increases in the ratio of 1:2:3, for example, for Na^+, Ca^{2+}, and Al^{3+}, respectively (Sawyer and McCarty, 1967; O'Melia, 1969, p. 89, p. 221). The rule is quoted frequently and fits with the double layer theory concept, except that Gregory (2006, p. 82), mentions that while the theoretical dependence is $1/z^6$, it is experimentally $1/z^3$.

Sol: (1) A colloidal suspension. (2) A dispersion composed of particles smaller than 1 μm is considered a sol; a dispersion of particles larger in size is considered a suspension. Emulsions and suspensions may exhibit colloidal properties and may be so treated (Sennett and Olivier, 1965, p. 33).

Specific ultraviolet absorbance (SUVA): SUVA = 100 · [UV254 (cm)]/[DOC (mg/L)]

Units are m^1/mg/L (Edzwald, 1993, p. 21; Edwards, 1997, p. 80). To obtain SUVA, UV_{254} is plotted against different experimental values of DOC (for a given water); the slope of the plotted line is defined as SUVA. See Chapter 2.

Stable suspension: A suspension of charged colloidal particles, that is, a sol, which has not been "coagulated."

Stability constant: Equilibrium constant for a complex reaction, with complex as product.

Stern layer: The fixed layer of positive ions associated with a negatively charged colloid, usually only a few angstroms in thickness, called the "fixed" layer or the "Stern" layer.

Stoichiometric: Refers to the fact that chemical reactions occur with chemicals combining in "definite" proportions, that is, definite mole ratios.

Streaming potential: (1) The electric field generated when a liquid is forced to flow past a stationary charged surface (Myers, 1991, p. 82). (2) When a solution is forced through a porous plug or tube of material which acquires charge in contact with solution, a streaming potential is set up (Gregory, 1975, p. 65). An instrument that measures this is a streaming current detector.

Supersaturation: A product concentration that exceeds that which may exist in equilibrium with other constituents.

Surface potential: The potential, ψ_o, at the surface of a colloid, also called the Nernst potential.

Thermal energy: The usual measure of thermal energy is given as kT. At 25°C, $kT = 25.7$ mV (the kT value was from Adamson, 1967, p. 212). The kT term is used often an energy reference, with kT being in the denominator of a dimensionless ratio.

Total organic carbon (TOC): Organic carbon as measured by a TOC analyzer. See Chapter 2.

Trivalent positive ion: An ion that has a valence of +3, for example, A^{3+}, Fe^{3+}.

Turbidimeter: An instrument with a light source and a detector that measures the turbidity of a water. The instrument must be calibrated with a standard. Formizon has been used as an accepted standard. Turbidimeters may be bench instruments or "online."

Turbidity: A measure of the light scattering property of a water by an instrument called a turbidimeter.

Van der Waals interaction: Attractive forces between atoms and molecules were postulated earlier than 1870 by van der Waals to explain the nonideality of real gases. The three types of forces are due to: (1) orientation, (2) induction, and (3) dispersion (Gregory, 1975, p. 77).

Zeolites: Silicates with open 3D structures with "cages" of molecular dimensions containing water and cations (van Olphen, 1977, p. 64).

Zero point of charge (ZPC): (1) The concentration of an additive, for example, an ion, to a colloidal suspension at which the particles are rendered neutral is called the zero point of charge. If the recharging ions are H+ or OH− and the process is followed by a change in pH, the pH at which the particles do not move in an electric field is usually referred to as the isoelectric point (IEP) (Matijvic, 1967, p. 331).

Zeta potential: Electric potential of a particle at the plane of shear; at the plane of shear, $\psi_o = \zeta$ (Pilipovich et al., 1958, p. 1469).

REFERENCES

Al-Ani, M. Y., McElroy, J. M., Hendricks, D. W., Hibler, C. D., and Logsdon, G. S., Removal of giardia cysts and other substances from low turbidity, low temperature water by rapid rate filtration, *Journal of the American Water Works Association,* 78 (5):66–73, May 1986.

Al-Jadhai, I. S. and Hendricks, D. W., Removal mechanisms in rapid filtration of low turbidity water, Environmental Engineering Technical Report, Department of Civil Engineering, Colorado State University, Fort Collins, CO, 1989.

Adamson, A. W., *Physical Chemistry of Surfaces*, 2nd edn., Interscience Publisher, John Wiley & Sons, New York, 1967.

Alberty, R. A. and Silbey, R. J., *Physical Chemistry*, John Wiley & Sons, New York, 1992.

Amirtharajah, A. and Mills, K. M., Rapid mix design for mechanisms of alum coagulation, *Journal of the American Water Works Association*, 74(4):210–216, April 1982.

Amirtharajah, A. and O'Melia, C. R., Coagulation processes: Destabilization, mixing, and flocculation, in: Pontius, F.W. (Ed.), *Water Quality and Treatment*, 4th edn., Chapter 6, McGraw-Hill, New York, 1990.

AWWA, AWWA Standard for Polyacrylamide, ANSI/AWWA B453-96, American Water Works Association, Denver, CO, 1996.

AWWA, AWWA Standard for Ferric Sulfate ANSI/AWWA B406-97, American Water Works Association, Denver, CO, 1997.

AWWA, AWWA Standard for Liquid Polyaluminum Chloride ANSI/AWWA B408-98, American Water Works Association, Denver, CO, 1998.

Bache, D. H., Johnson, C., Papavasilopoulos, E., Rasool, E., and McGilligan, F. J., Sweep coagulation: Structures, mechanisms, and practice, *Aqua—Journal of Water Supply Research and Technology*, 48(5):201–210, 1999.

Bean, E. L., Campbell, S. J., and Anspach, F. R., Zeta potential measurements in the control of coagulation chemical doses, *Journal of American Water Works Association*, 56(2):214–227, February 1964.

Beard, J. D. and Tanaka, T. S., A comparison of particle counting and nephelometry, *Journal of American Water Works Association*, 69(10):533–538, October 1977.

Bertsch, P. M. and Parker, D. R., Aqueous polynuclear aluminum species, in: Sposito, G. (Ed.), *The Environmental Chemistry of Aluminum*, 2nd edn., Chapter 4, CRC-Lewis Publishers, Boca Raton, FL, 1996.

Black, A. P. and Rice, O., Formation of floc by aluminum sulfate, *Industrial and Engineering Chemistry*, 25(7):811–815, July 1933.

Black, A. P., Coagulation with iron compounds, *Journal of the American Water Works Association*, 26(11):1713–1718, November 1934.

Black, A. P., Chemistry of water treatment—Part I. Coagulation, *Water and Sewage Works*, 95:142–144, April 1948.

Black, A. P., Basic mechanisms of coagulation, *Journal of the American Water Works Association*, 52(4):493–504, April 1960.

Black, A. P. and Hannah, S. A., Electrophoretic studies of turbidity removal by coagulation with aluminum sulfate, *Journal of the American Water Works Association*, 53(4): 438–452, April 1961.

Black, A. P. and Willems, D. G., Electrophoretic studies of coagulation removal of organic color, *Journal of the American Water Works Association*, 53(5):589–604, May 1961.

Black, A. P. and Smith, A. L., Determination of the mobility of colloid particles by microelectrophoresis, *Journal of the American Water Works Association*, 54(5):926–934, August 1962.

Black, A. P., Electrokinetic Characteristics of Hydrous Oxides of Aluminum and Iron, in Faust, S. D. and Hunter, J. V. (Eds.), *Principles and Applications of Water Chemistry (Proceedings of the Fourth Rudolf's Research Conference,* Rutgers University), John Wiley & Sons, New York, pp. 274–300, 1967.

Black, A. P. and Harris, R. H., New Dimensions for the Old Jar Test, *Water and Wastes Engineering,* 6(12):49–51, December, 1969.

Brink, D. R., Choi, S., Al-Ani, M., and Hendricks, D. W., Bench-scale evaluation of coagulants for low turbidity water, *Journal of the American Water Works Association*, 80(4):199–204, April 1988.

Bryant, R. L., Precise coagulant control using streaming current measurement, *Waterworld News*, 1(3):18–19, May/June 1985.

Burns, D. E. and Shell, G. L. Physical-chemical treatment of a municipal wastewater using powdered activated carbon, in: *44th Annual WPCF Conference*, San Francisco, CA, October 3–8, 1971.

Cardile, R., Clark, S., and Patzelt, R., Controlling clarification using a streaming current detector (SCD), in: *43rd Annual Meeting International Water Conference*, Pittsburgh, PA, October 25–27, 1982.

Carlson, K. H., Champlin, T. L., Hendricks, D. W., Sullivan, L. P., Walters, R. W., and Zibell, T. G., Treatment train modeling for aqueous contaminants, Volume I Summary Report, Environmental Engineering Technical Report No. 53-2415-93-1, Department of Civil Engineering, Colorado State University, Fort Collins, CO, December 1993.

Carlson, K. H. and Gregory, D., Optimizing water treatment with two-stage coagulation, *Journal of Environmental Engineering Division, American Society of Civil Engineers*, 126(EE6):556–561, June 2000.

Chamberlain, R. J., Polyelectrolyte makeup and handling, in: Schwoyer, W. L. K. (Ed.), *Polyelectrolytes for Water and Wastewater Treatment,* Chapter 8, CRC Press, Inc., Boca Raton, FL, 1981.

Champlin, T. L. and Hendricks, D. W., Treatment train modeling for aqueous contaminants, Volume II Matrix of Contaminants and Treatment Technologies, Environmental Engineering Technical Report No. 53-2415-93-2, Department of Civil Engineering, Colorado State University, Fort Collins, CO, May 1993.

Cleasby, J. L., Dharmarajah, A. H., Sindt, G. L., and Baumann, E. R., *Design and Operation Guidelines for Optimization of the High-Rate Filtration Process: Plant Survey Results*, AWWA Research Foundation, Denver, CO, September 1989.

Conley, W. R. and Evers, R. H., Coagulation control, *Journal of the American Water Works Association*, 60:2:165–174, February 1968.

Cushen, A. D., Zeta-potentials of selected algae, *Cryptosporidium oocysts*, and *Giardia cysts*, MS Thesis, Department of Civil Engineering, Colorado State University, Fort Collins, CO, 1996.

Dempsey, B. A., Ganho, R. M., and O'Melia, C. R., The coagulation of humic substances by means of aluminum salts, *Journal of the American Water Works Association*, 76(4):141–150, April 1984.

Dempsey, B. A., Bench-scale production, characterization, and application of polyaluminum sulfate, Research Report, American Water Works Association Research Foundation, Denver, CO, 1989.

Dentel, S. K., Gucciardi, B. M., Bober, T. A., Shetty, P. V., and Resta, J. J., *Procedures Manual for Polymer Selection in Water Treatment Plants*, AWWA Research Foundation, Denver, CO, October, 1986.

Dentel, S. K., Application of the precipitation—Charge neutralization model of coagulation, *Environmental Science and Technology*, 22(7):825–832, July 1988.

Dentel, S. K. and Gucciardi, B. M., Practical methods for characterizing organic polyelectrolytes used in water treatment, in: *Proceedings of Annual Conference of American Water Works Association*, Los Angeles, CA, June 1989, American Water Works Association, Denver, CO, 1989.

Edzwald, J. K., Coagulation in drinking water treatment: Particles, organics and coagulants, *Water Science and Technology*, 27(11):21–35, 1993.

Edwards, M., DOC removal during enhanced coagulation, *Journal of the American Water Works Association*, 89(5):78, May 1997.

Edwards, G. A. and Amirtharajah, A., Removing color caused by humic acid, *Journal of the American Water Works Association*, 77(3):50–57, March 1985.

Edzwald, J. K., Cationic polyelectrolytes in water treatment, in: *Proceedings of the Engineering Foundation Conference on Flocculation, Sedimentation, and Consolidation*, Sea Islands, GA, January 27–February 1, 1985, pp. 171–180, AIChE, 1985.

Edzwald, J. K., Haff, J. D., and Boak, J. W., Polymer coagulation of humic acid water, *Journal of Environmental Engineering Division, American Society of Civil Engineers*, 103(EE6):989–1000, December 1977.

Edzwald, J. K., Becker, W. C., and Tambini, S. J., Organics, polymers, and performance in direct filtration, *Journal of Environmental Engineering Division*, American *Society of Civil Engineers*, 113(EE1):167–185, February 1987.

Fox, K. R. and Lytle, D. A., Zeta potential and particle counting for optimizing pathogen removal, in: *Proceedings of AWWA National Conference at Toronto*, American Water Works Association, Denver, CO, June 1996.

Freiser, H., Lecture notes in analytical chemistry, Department of Chemistry, University of Arizona, Freiser Equilograph Co., Tucson, AZ, 1996.

Ghosh, M. M., Cox, C. D., and Prakash, T. M., Polyelectrolyte selection for water treatment, *Journal of the American Water Works Association*, 77(3): 67–73, March 1985.

Gregory, J., Interfacial phenomena, in: Ives, K. J. (Ed.), *The Scientific Basis of Filtration* (*Proceedings of NATO Advanced Study Institute*, Cambridge, U.K.), Noordhoff International Publishing, Leyden, the Netherlands, 1975.

Gregory, J., *Particles in Water*, IWA Publishing, CRC/Taylor & Francis, Boca Raton, FL, 2006.

Hall, E. S., The zeta potential of aluminum hydroxide in relation to water treatment coagulation, *Journal of Applied Chemistry*, 15:197–205, May 1965.

Hancock, C. M., Personal communication, February 14 and 19, 2001.

Hancock, C. M., Ward, J. V., Hancock, K. W., Klonicki, P. T., and Sturbaum, G. D., Assessing plant performance using MPA, *Journal of the American Water Works Association*, 88(12):24–34, December 1996.

Hart, H., *Organic Chemistry*, 8th edn., Houghton Mifflin Company, Boston, MA, 1991.

Hemingway, B. S. and Sposito, G., Inorganic aluminum-bearing solid phases, in: Sposito, G. (Ed.), *The Environmental Chemistry of Aluminum*, 2nd edn., Chapter 3, CRC-Lewis Publishers, Boca Raton, FL, 1996.

Hendricks, D. W. (Ed.), *Manual of Design for Slow Sand Filtration*, AWWA Research Foundation and American Water Works Association, Denver, CO, 1991.

Hendricks, D. W., Clunie, W. F., Anderson, W. L, Hirsch, J., Evans, B., McCourt, B., Wendling, P., Nordby, G., Sobsey, M. D., and Hunt, D. J., *Biological Particle Surrogates for Filtration Performance Evaluation*, AWWA Research Foundation, Denver, CO, 2000.

Hendricks, D. W., Clunie, W. F., Sturbaum, G. D., Klein, D. A., Champlin, T. L., Kugrens, P., Hirsch, J. et al., Filtration removals of microorganisms and particles, *Journal of the Environmental Engineering Division, American Society of Chemical Engineers*, 131(12):1621–1632, 2005.

Hibler, C. P., Analysis of municipal water samples for cysts of *Giardia*, in: Wallis, P. M. and Hammond, B. R. (Eds.), *Advances in Giardia Research*, pp. 237–245, University of Calgary Press, Calgary, Canada, 1988.

Hibler, C. P., Personal communication, August 23, 1990.

Hubel, R. E. and Edzwald, J. K., Removing trihalomethane precursors by coagulation, *Journal of the American Water Works Association*, 79(7):98–106, July 1987.

Hunter, R. J., *Zeta Potential in Colloid Science*, Academic Press, New York, 1981.

Jakubowski, W. and Hoff, J. C., Waterborne transmission of giardia, EPA/600/9-79-001, US Environmental Protection Agency, Cincinnati, OH, 1979.

Johnson, C. E., Polyelectrolytes as coagulants and coagulant aids, *Industrial and Engineering Chemistry*, 48:1080–1083, June 1956.

Jorden, R. M., Colloid charge titration—Its measurement, problems and promise, in: *Seventh International Gothenburg Symposium*, Edinburgh, Scotland, September 1996.

Kavanaugh, M. C., Modified coagulation for the removal of trihalomethane precursors, *Journal of the American Water Works Association,* 70(11):613–620, November, 1978.

Kawamura, S., Hanna, G. P., Jr., and Schumate, K. S., Application of colloid titration technique to flocculation control, *Journal of the American Water Works Association*, 59(8):1003–1013, August 1967.

Kling, J., Calculation of distribution of ferric hydrolysis species with pH from equilibrium constants, Homework Problem in CE541, Department of Civil Engineering, Colorado State University, Fort Collins, CO, January 20, 1999.

Kramer, L. and Harger, J., The good, the bad & the successful—Streaming current monitor used to optimize coagulant dosages, *Water World*, 17(3):10, 22, March 2001.

Kugrens, P., Personal communication, January 25, 2001, February 13, 2001.

Langelier, W. F., Coagulation of water with alum by prolonged agitation, *Engineering News Record*, 86:924–928, 1921.

Langelier, W. F., Water treatment and laboratory control, *Journal of American Water Works Association*, 14(4):476–480, April 1925.

Langelier, W. F., Teaching, research, and consulting in water purification and sewage treatment, UC Berkeley 1916–1955, Interview conducted by Malca Chall in 1970, Regional Oral History Office, Copy No. 7, The Bancroft Library, University of California, Berkeley, CA, 1982.

Langelier, W. F. and Ludwig, H. F., Mechanism of flocculation in the clarification of turbid waters, *Journal of the American Water Works Association*, 41(2):163–181, February 1949.

Lide, D. R. (Ed.), *Handbook of Chemistry and Physics*, 77th edn., CRC Press, Boca Raton, FL, 1996.

Loeb, A. L., Overbeek, J. T. G., and Wiersema, P. H., *The Electrical Double Layer around a Spherical Colloid Particle*, The MIT Press, Cambridge, MA, 1961.

Loganathan, P. and Maier, W. J., Some surface chemical aspects in turbidity removal by sand filtration, *Journal of the American Water Works Association*, 67(6):336–342, February 1975.

Logsdon, G. S., Water filtration for asbestos fiber removal, EPA-600/2-79-206 Municipal Environmental Research Laboratory, US Environmental Protection Agency, Cincinnati, OH, December 1979.

Ludwig, H. F., *Adventures in Consulting Engineering (Story of H.F. Ludwig),* Seatec International Publications, Bangkok, Thailand, 1985.

Luttinger, L. B., The use of polyelectrolytes in filtration processes, in: Schwoyer, W.L. K. (Ed.), *Polyelectrolytes for Water and Wastewater Treatment,* CRC Press, Inc., Boca Raton, FL, 1981.

Mangravite, F. J., Synthesis and properties of polymers used in water treatment, paper in Seminar 3: *Use of Organic Polyelectrolytes in Water Treatment*, Research Division, *AWWA National Conference*, Las Vegas, NV, June 5, 1983.

Marshall, C. E., *The Physical Chemistry and Minerology of Soils, Volume 1: Soil Materials,* John Wiley & Sons, New York, 1964.

Matijevic, E., Charge reversal of lyophobic colloids, in: Faust, S. D. and Hunter, J. V. (Eds.), *Principles and Applications of Water Chemistry* (*Proceedings of the Fourth Rudolf's Research Conference*, Rutgers University, Newark, NJ), John Wiley & Sons, New York, 1967.

May, L. M., Organic polymers used in coagulation/flocculation, in: *Nalco Water Treatment Chemicals Application Seminar*, Denver, CO, January 14, 1988.

Mazet, M., Angbo, L., and Serpaud, B., Adsorption of humic acids onto preformed aluminum hydroxide flocs, *Water Research*, 24(12):1509–1518, December 1990.

Metcalf, L. and Eddy, H. P., *American Sewerage Practice, Volume III, Disposal of Sewage,* McGraw-Hill, New York, 1916.

Mitchell, J. K., *Fundamentals of Soil Behavior*, 2nd edn., John Wiley & Sons, New York, 1993.

Moffett, J. W., The chemistry of high rate water treatment, *Journal of the American Water Works Association*, 60(11):1255–1271, November 1968.

Myers, D., *Surfaces, Interfaces, and Colloids*, VCH Publishers, Inc., New York, 1991.

National Sanitation Foundation, *Chemicals*, www.NSF.org, November 2000.

Nordstrom, D. K. and May, H. M., Aqueous equilibrium data for mononuclear aluminum species, in: Sposito, G. (Ed.), *The Environmental Chemistry of Aluminum*, 2nd edn., Chapter 2, CRC-Lewis Publishers, Boca Raton, FL, 1996.

Novak, J. T., Polymers in sludge conditioning, Seminar paper, Sunday seminar 3: Use of organic polyelectrolytes in water treatment, Research Division, AWWA National Conference, Las Vegas, NV, 1983.

O'Melia, C. R., A review of the coagulation process, *Public Works*, 100:87–98, May 1969.

O'Melia, C. R., Coagulation in water and wastewater treatment, in: Gloyna, E. F., and Eckenfelder, W. W. Jr. (Eds.), *Water Quality Improvement by Physical and Chemical Processes*, University of Texas Press, Austin, TX, 1970.

O'Melia, C. R., Coagulation in wastewater treatment, in Ives, K. J. (Ed.), *The Scientific Basis of Coagulation*, Sijthoff & Noordhoff, Alphen aan den Rijn, the Netherlands, 1978.

O'Melia, C. R., Coagulation, Typed paper, April 12, 1979.

O'Melia, C. R., Coagulation, unpublished paper, April 12, 1979.

Ongerth, J. and Pecoraro, J. P., Electrophoretic mobility of *Cryptosporidium* oocysts and *Giardia* cysts, *Journal of the Environmental Engineering Division,* ASCE, 122(3):228–231, March, 1996.

Packham, R. F., The coagulation process II. Effect of pH on the precipitation of aluminum hydroxide, *Journal of Applied Chemistry*, 12:564–568, December 1962.

Pernitsky, D. J. and Edzwald, J. K., Polyaluminum chloride—Chemistry and selection, *Proceedings of 2000 Annual Conference at Denver* (CDROM), American Water Works Association, Denver, CO, 2000.

Peterson, T. D., Streaming current monitor controls coagulant addition, *Opflow*, *American Water Works Association*, 18(5):3, 6, May 1992.

Pilipovich, J. B., Black, A. P., Eidness, F. A., and Stearns, T. W., Electrophoretic studies of water coagulation, *Journal of the American Water Works Association*, 50(11):1467–1482, November 1958.

Pizzi, N. and Rodgers, M., Testing your enhanced coagulation endpoint, *Opflow*, *American Water Works Association*, 26(2):1–5, February 2000.

PPG Industries (One PPG Place, Pittsburgh, PA) for its product, PRODEFLOC AC100S.

Raave, A., Principles of Polymer Chemistry, Plenum Press, New York, 1995.

Randtke, S. J., Organic contaminant removal by coagulation and related process combinations, *Journal of the American Water Works Association*, 80(5):40–56, May 1988.

Rickert, D. A. and Hunter, J. V., Colloidal matter in wastewaters and secondary effluents, *Journal of Water Pollution Control Federation*, 44(1):134–139, January 1972.

Riddick, T. M., Discussion of Black, A. P., Basic mechanisms of coagulation, *Journal of the American Water Works Association*, 52(4):493–504, April 1960.

Riddick, T. M., Zeta potential and its application to difficult waters, *Journal of the American Water Works Association*, 53 (8):1007–1030, August 1961.

Rose, G. R., Coagulation/flocculation in water, wastewater and sludge treatment, in: *Nalco Water Treatment Chemicals Application Seminar*, Denver, CO, January 14, 1988.

Saville, T., On the nature of color in water, *Journal of the New England Water Works Association*, 31(1):78–123, March 1917.

Sawyer, C. N. and McCarty, P. L., *Chemistry for Sanitary Engineers*, McGraw-Hill, New York, 1967.

Schleppenbach, F. X. (Diluth Water and Gas Department), Water Filtration at Diluth, Agreement S804221 Between EPA and City of Diluth, Municipal Environmental Research Laboratory, US Environmental Protection Agency, Cincinnati, OH (copy available was not bound and with no date), c. 1983.

Sears, F. W., *Principles of Physics II—Electricity and Magnetism*, Addison-Wesley Press, Inc., Cambridge, MA, 1947.

Sennett, P. and Olivier, J. P., The interface symposium—1, *Industrial and Engineering Chemistry*, 57(8):33–50, August 1965.

Shriver, D. F. and Atkins, P. W., *Inorganic Chemistry*, 3rd edn., W. H. Freeman and Co., New York, 1999.

Singley, J. E., Birkner, F., Chen, C., Cohen, J. M., Ockershausen, R. W., Shull, K. E., Weber, W. J., Jr., Matijevic, E., and Packham, R. F., State of the art of coagulation—Mechanisms and stoichiometry, *Journal of the American Water Works Association*, 63(2):99–108, February 1971.

Smith, C. V., Jr. and Somerset, I. J., Streaming current technique for optimum coagulant dose (paper provided by Milton Roy; no date or journal citation).

Sonon, L. S., Chappell, M. A., and Evangelou, V. P., A brief history of soil chemistry, *Division S-2 Soil Chemistry Newsletter, Soil Science of America*, 10(1), 2001, excerpt taken from Web site, http://ces.ca.uky.edu/s2/Newsletter/feb-2001.htm, on September 14, 2008.

Streitwieser, A., Jr. and Heathcock, C. H., *Introduction to Organic Chemistry*, 3rd edn., Macmillan Publishing Company, New York, 1985.

Stumm, W. and Morgan, J. J., Chemical aspects of coagulation, *Journal of the American Water Works Association*, 64(8): 971–993, August 1962.

Stumm, W. and Morgan, J. J., *Aquatic Chemistry*, 3rd edn., John Wiley & Sons, Inc., New York, 1996.

Stumm, W. and O'Melia, C. R., Stoichiometry of coagulation, *Journal of the American Water Works Association*, 70(5):514–539, May 1968.

Symons, J. M., Bellar, T. A., Carswell, J. K., DeMarco, J., Kropp, K. L., Robeck, G. G., Seeger, D. R., Slocum, C. J., Smith, B. L., and Stevens, A. A., National organics reconnaissance survey for halogenated organics, *Journal of the American Water Works Association*, 67(11):634–647, November 1975.

Tate, C. H. and Trussell, R. R., The use of particle counting in developing plant design criteria, *Journal of the American Water Works Association,* 70(12):691–698, 1978.

Tseng, T., Segal, B. D., and Edwards, M., Increasing alkalinity to reduce turbidity, *Journal of the American Water Works Association*, 92(6):44–54, June 2000.

van Benschoten, J. E. and Edzwald, J. K., Chemical aspects of coagulation using aluminum salts—I. Hydrolytic reactions of alum and polyaluminum chloride, *Water Research*, 24 (12):1519–1526, December 1990.

van Olphen, H., *Clay Colloid Chemistry*, 2nd edn., John Wiley & Sons, New York, 1977.

Veal, C. R., Jr., Streaming current monitor controls coagulation, *Opflow, American Water Works Association*, 14(8):4–5, August 1988.

Weber, W. J., Jr., Hopkins, C. B., and Bloom, R., Jr., Physiochemical treatment of wastewater, *Journal of Water Pollution Control Federation*, 42(1):83–99, January 1970.

Weber, W. J., Jr., *Physicochemical Processes for Water Quality Control*, Wiley-Interscience, New York, 1972.

Willcomb, J., Floc formation and mixing basin practice, *Journal of the American Water Works Association*, 24(9):1416–1441, September 1932.

Zeta-Meter, Inc., *Everything You Want to Know about Coagulation & Flocculation*, Zeta-Meter, Inc., Long Island City, New York, 1988.

10 Mixing

The mixing unit process is found throughout treatment plants. Examples include rapid mix (also called "initial mixing"), flocculation, disinfection; activated sludge, the anaerobic process, and gas dissolution. Usually, the mixing is not visible since the equipment may be located under floors, in pipes, in tanks, or behind walls.

For most unit processes, mixing is a critical supporting step causing (1) contacts between reactants, (2) creation of interfacial area, or (3) reduction of the "film" thickness (to maximize the diffusion gradient across an interface). In design, the objective is to ensure, to the extent feasible, that mixing is not a rate-limiting factor.

Design questions include the selection of type of mixing unit, its size, and power input. For a given mixing application, a certain type of mixing technology may be more appropriate than another. The intent of this chapter is to cover mixing principles and practice.

10.1 DEFINITIONS AND APPLICATIONS

The world of mixing includes an array of technologies, any one of which may be employed for one of the numerous applications. The principles involved, i.e., advection, turbulence, diffusion (Box 10.1), are applicable also to the variety of mixing situations that occur in the ambient environment, e.g., re-aeration in streams, chemical and biological reactions in streams and lakes, and dispersion in streams.

10.1.1 Definitions

The term "mixing" seems self-explanatory. Several ancillary terms are used, however, which are near-synonyms.

10.1.1.1 Mixing

Mixing is defined as inducing the random distribution of two or more initially separate phases through one another (McCabe et al., 1993, p. 235). Some commonplace examples are: in the kitchen, e.g., tossing a mixed salad with dressing, blending chocolate chips or nuts into cookie dough; in the bar, e.g., mixing drinks; or in the garage, e.g., mixing paint.

10.1.1.2 Near-Synonyms

The following terms are found in the literature without fine distinctions. Nevertheless, each has its own connotation.

10.1.1.2.1 Blending

In blending, one substance may lose its identity within another. A criterion may be when the concentration of a material in a solution is some arbitrary fraction of an ultimate value, e.g., 0.99.

10.1.1.2.2 Dispersion

The dictionary defines dispersion as "scatter," which results in mixing. In flow through a porous medium, the mechanism is usually random velocity variation along any flow path. The same mechanism applies to turbulent diffusion (e.g., in the case of open channel flow or pipe flow). Examples include the spread of a solute in porous media flow (both longitudinally and laterally), the mixing of two rivers (lateral dispersion), the mixing of pipeline discharge into a stream, mixing by a submerged jet (lateral and longitudinal), and peeling-off of eddies from main circulation currents in a mixing basin.

10.1.1.2.3 Agitation

Agitation involves the induced motion of a material (McCabe et al., 1993, p. 235), i.e., a disturbance of the status quo. The chemical engineering literature seems to use this term preferentially.

10.1.2 Application Categories

Processing pairs applicable to water treatment include (1) liquid–solid, (2) liquid–gas, (3) two immiscible liquids, (4) two miscible liquids, and (5) fluid motion. Descriptions of each are given with reference to an impeller–tank. Jet mixers, static mixers, baffled basins, etc., are other technologies that could be applicable.

10.1.2.1 Liquid–Solid

A "liquid–solid" pair includes particles that dissolve in water. For example, a solid chemical in pellet or powder form must be suspended and "agitated" to achieve mass transfer from the solid phase to the dissolved phase.

10.1.2.2 Liquid–Gas

A gas flow emerging from a tube below an impeller is "dispersed" throughout the liquid volume first by shear to create bubbles and then by pumping and shear along the advected stream to spread them; the diffusion of gas from the bubbles to the water is the final step.

10.1.2.3 Immiscible Liquids

High molecular weight polymers are often dissolved in an oil base. Prior to metering into a water flow, the emulsion must be dispersed by means of a high-shear zone, i.e., an impeller, within a small flow of water, which is then metered into the main flow.

BOX 10.1 MIXING IN-A-NUTSHELL

Mixing can be described in terms of a simple task such as stirring sugar into a cup of coffee or tea. The sugar granules, poured into a cup, come to rest on the bottom and dissolve. If the solution is quiescent, the dissolved sugar molecules diffuse, in accordance with Fick's law, forming a concentration gradient from the bottom of the cup to the top. Over hours of time the gradient approaches uniformity.

As everyone knows, however, stirring causes the sugar granules to be "dispersed" rapidly throughout the cup. Two things occur in stirring, e.g., by a spoon: (1) the dissolved sugar circulates throughout the cup, i.e., is "advected" by the "large-eddies" from the spoon, and (2) a "wake eddy" is induced and the sugar granules and molecules "blend" with the water molecules.

In summary, three transport mechanisms are operative, i.e., advection, turbulence, and diffusion. They apply to virtually all mixing situations, e.g., gas transfer, chemical dissolution, polymer blending, coagulation, disinfection, and oxidation reactions.

10.1.2.4 Miscible Liquids

The term "blending" is sometimes given to one miscible liquid being dispersed, i.e., "blended" into another. Metering liquid alum into a water flow is an example.

10.1.2.5 Fluid Motion

When the required fluid motion is achieved, the other parts of the process will also be satisfied (Oldshue, 1983, p. 5). For an impeller–basin system, the characteristics of fluid motion include the pumping capacity of the impeller, the flow patterns within the basin, and the shear zones and their extent and intensities.

10.1.2.6 Pumping and Shear

The impeller energy is divided into pumping and shear. The proportion of each depends upon the impeller–tank system. Whether advection flow is preferred at the expense of shear or vice versa depends upon the application (Oldshue, 1983, p. 8).

10.1.2.7 Examples

Applications of mixing in water treatment include rapid mix after the addition of chemicals; disinfection after the addition of chlorine (or other oxidant); the dispersion of mixed liquor return activated sludge into the reactor flow; the dispersion of an immiscible liquid, e.g., a polymer in oil in a water concentrate prior to metering into raw water; the dispersion of a polymer for sludge thickening; the dispersion of a gas through water as small bubbles. In the anaerobic process, advection mixing predominates, while for the activated sludge process the shear rate is higher.

10.1.3 Mixing as Rate Limiting

Mixing may control the reaction rate for anaerobic reactors, aerobic reactors, coagulation in water treatment, oxygen dissolution, gas stripping, etc. For diffusion-limited processes, the rate of the creation of surface area governs the total area available as a mass transfer interface. At the same time, with fresh area being created, a steep concentration gradient (small δ) is maintained between the concentration in the bulk of solution, C_o, and the interface concentration, C_i. In other words, mixing is important in paving the way for other processes to function effectively. In such cases, mixing determines the rate of mass transfer to some asymptote limit, where higher mixing energy will have no further effect.

10.2 HISTORY OF MIXING

Mixing has been an empirical practice since the early days of water treatment, i.e., about 1900 with theory getting its start in the 1940s with the work of Camp and Stein (1943). This section covers the evolution of theory and practice.

10.2.1 Drinking Water Treatment

Mixing in drinking water treatment is important in coagulation (i.e., initial mixing or rapid mix), in flocculation (Chapter 11), and in disinfection. Also, mixing is required to dissolve solid chemicals (e.g., lime), to disperse polymer emulsions, to dissolve solid polymers, to blend a liquid chemical (e. g., fluoride), to dissolve a gas (e.g., CO_2 for pH reduction), etc.

10.2.1.1 Initial Mixing

Willcomb (1932, p. 1427) stated that none of the plants built during the period 1900–1911 under his direction had mixing. Rather, the gravity plants had their alum introduced just before the raw-water meter, and the pumping plants had their alum dose added to the pump suction. By the time of his article in 1932, rapid mix as a unit process was becoming established. Alternative mixing methods mentioned included aeration, the current of a flume, the turbulence associated with a valve, and the hydraulic jump.

The importance of mixing was recognized by Hansen (1936) who noted that mixing of coagulants with raw water was essential, followed by flocculation (he also cited the importance of Langelier's work in Sacramento in 1919, Langelier, 1921). Despite the trend toward recognizing the importance of initial mixing, Babbitt and Doland (1939, p. 529) did not wholly endorse such a position with the following view:

> It is probable that rapid mixing devices for the somewhat violent initial mixing of the chemicals with the raw water will not come into general use, because of their cost, both for construction and for operation, is not justified by the slight increase in efficiency of operation. Where low lift pumps are available, they may serve the purpose without added expense.

FIGURE 10.1 One of two flocculation-settling basins. c. 1920. City of Fort Collins Water Treatment Plant No. 1, adjacent to Cache La Poudre River, about 10 miles from the city center. This system was later supplanted by upgrades.

An example of initial mixing followed by a flocculation–sedimentation basin, c. 1920, is seen in Figure 10.1. A channel, not shown, led into the basin at the far end with alum feed directly into the channel. In other words, rapid mix occurred in the channel. The large quiescent basins were continuous flow but they were taken out of operation, one at a time, as sediment accumulated. Hydraulic detention time was 24–48 h.

By 1961, Skeat (1961, p. 504) recognized that chemicals "should be rapidly and evenly distributed throughout the mass of water being treated." It is usual practice, he stated, to introduce the chemical at some point of high turbulence with mixing times 30–60 s. A drop in head of 0.23–0.46 m (9–18 in.) was recommended for a weir or other in-channel method. For a mixing basin, he recommended 3–6 $kw/m^3/s$ (0.25–0.5 hp/1000 gpm). Possible basin designs included an impeller in a draft tube with recirculation and pumping from a basin to return the flow as a submerged jet to impinge against a flat plate with the alum solution introduced adjacent to the core zone of the jet. Figure 10.2a and b illustrates the respective designs. Both are "back-mix" reactors, i.e., in each case, recirculation occurs through the reaction zone, the high-turbulence zone. Also, in each case, the chemical feed is injected into the part of the flow that will become a part of the high-turbulence zone.

10.2.1.2 Gas Dissolution

Gas transfer as a unit process in drinking water treatment has included the dissolution of carbon dioxide for pH adjustment (prior to coagulation); the removal of high concentrations of dissolved gases (nitrogen and oxygen may precipitate from solution if present in excess); and the dissolution of chlorine, chlorine dioxide, or ozone (usually for disinfection). In most cases for dissolution, a diffuser may be placed in a side stream of water without much regard for formal design. In the case of chlorine gas dissolution, the engineer's role has been largely to specify equipment recommended by a selected manufacturer, with the technologies having been developed by the latter.

10.2.2 Wastewater Treatment

In wastewater treatment, mixing applications have included dispersing, within a reactor basin, clarified wastewater, return-activated sludge, and dissolving oxygen. Usually, design has been by empirical guidelines. The dissolution of chlorine gas or ozone has followed the approaches mentioned in Section 10.2.1.

In anaerobic digestion, the role of mixing was not discovered until the late 1950s. The story was that an equipment

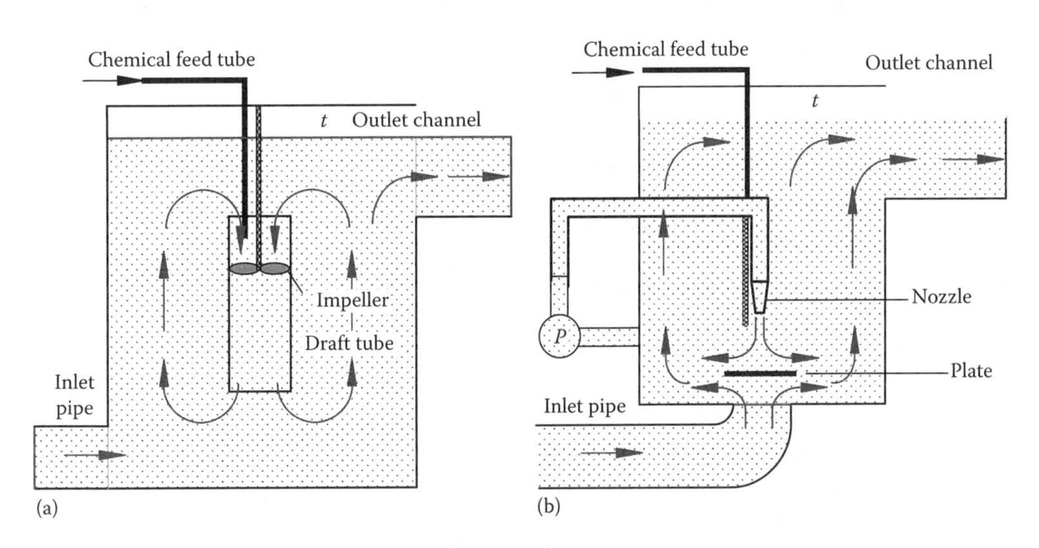

FIGURE 10.2 Mixing basin designs. (a) Impeller in draft tube. (b) Submerged jet. (Adapted from Skeat, W. O. (Ed.), *Manual of British Water Engineering Practice*, 3rd edn., Published for the Institution of Water Engineers, W. Heffer & Sons Ltd., Cambridge, U.K., pp. 504, 505, 1961.)

manufacturer (Chicago Pump) found that the return of methane gas caused a higher reaction rate. Initially, the explanation was that the methane acted as a catalyst. Soon after, the mixing induced by the methane gas bubbles was determined to be the cause. Thereafter, mixing was incorporated into the design of anaerobic digesters. Prior to this, the digesters were permitted to merely "sit" quiescently, except for the fill or draw phases.

10.2.3 Evolution of Mixing Theory

In 1918, Smoluchowski published a mathematical expression for collision frequency between particles in a suspension for laminar flow; this was the basis for subsequent developments in theory. Later, Camp and Stein (1943) applied the Smoluchowski concept to turbulent mixing.

As related to impeller–basin systems, the results of the first systematic experimental studies were by Rushton and his associates in the late 1940s, with results published in the 1950s. They expressed their results in terms of dimensionless numbers, e.g., the "power number" vs. the Reynolds number.

The theory of turbulence is the basis for another thread of theory that supports a further understanding of mixing, with modern ideas beginning to crystallize also about 1950 (see Batchelor, 1953, for example). Then, starting about 1960, with the advent of modern computers, the Navier–Stokes equation, considered a theoretical abstraction for decades, became amenable to solution by finite difference techniques. By the 1990s, these techniques were developed into commercial software technologies for use with "work-station" computers. By the end of that decade, such software found its way into engineering practice and was used to address various "what if?" design scenarios in water treatment.

10.2.3.1 Development of Collision Frequency Mathematics

The mathematical expression for the rate of collision in a turbulent flow field was developed from the equations of Smoluchowski. The enumeration of steps, presented here, is from Argaman and Kaufman (1968, p. 5).

Step 1 *Brownian Motion Equation by Smoluchowski (1916)* The first equation of von Smoluchowski (1916) was for the condition that collisions were due to Brownian motion (called "perikinetic" motion) calculated as the diffusion flux of particles in the radial direction around a single stationary particle,

$$N(\text{diffusion})_{ij} = 4\pi D_{ij}(r_i + r_j)^3 n_i n_j \tag{10.1}$$

where

$N(\text{diffusion})_{ij}$ is the number of contacts per unit time per unit volume between i and j particles due to diffusion flux (collisions/m^3/s)
D_{ij} is the combined diffusion coefficient, $D_i + D_j$ (m^2/s)
r_i is the radius of particle i (m)
r_j is the radius of particle j (m)
n_i is the number concentration of particles of radius r_i (#/m^3)
n_j is the number concentration of particles of radius r_j (#/m^3)

Step 2 *Laminar Motion Equation by Smoluchowski (1918)* The second equation of von Smoluchowski (1918) was for collisions between particles induced by fluid motion (called "orthokinetic" motion). For laminar flow the collision frequency was given as

$$N(\text{laminar})_{ij} = (4/3)(r_i + r_j)^3 n_i n_j (\mathrm{d}v/\mathrm{d}y) \tag{10.2}$$

where

$N(\text{laminar})_{ij}$ is the number of contacts per unit time per unit volume between i and j particles due to laminar fluid motion (collisions/m^3/s)
$\mathrm{d}v/\mathrm{d}y$ is the velocity gradient due to laminar flow (s^{-1})

Step 3 *The Next Step by Camp and Stein (1943)* Camp and Stein, according to Argaman and Kaufman (1968, p. 5), noted that turbulent conditions exist in most cases of flocculation and adapted Smoluchowski laminar flow equation to the case of turbulent fluid motion,

$$N(\text{turbulent})_{ij} = (4/3)(r_i + r_j)^3 n_i n_j G \tag{10.3}$$

where

$N(\text{turbulent})_{ij}$ is the number of contacts per unit time per unit volume between i and j particles due to laminar fluid motion (collisions/m^3/s)
G is the velocity gradient averaged over volume, V, of reactor (s^{-1})

The velocity gradient, G, was defined by Camp and Stein (1943) as

$$G \equiv \frac{\mathrm{d}v}{\mathrm{d}y} \tag{10.4}$$

$$= \left(\frac{P}{\mu V}\right)^{0.5} \tag{10.5}$$

where

v is the velocity at a point in space (m/s)
y is the coordinate normal to the velocity vector $\mathbf{v}$ (m)
P is the power dissipated by fluid motion, e.g., viscosity or turbulence (W)
μ is the dynamic viscosity of fluid (N s/m^2)
V is the volume in which power dissipation occurs (m^3)

The derivation of Equation 10.5 is given in Section 10.2.3.2 with further elaboration in Box 11.1. The flaw in the relation was that in turbulent flow, the viscosity term, μ, is small relative to inertia term (see, for example, Cleasby, 1984, pp. 876, 877). Nevertheless, the G parameter was widely accepted and is embedded in practice (Amirtharajah et al., 2001, p. 162), its limitations notwithstanding.

10.2.3.2 Derivation of G

The starting point in the derivation of Camp and Stein's G is the well-known relation between shear and velocity gradient, given here in one dimension, i.e.,

$$\tau = \mu \frac{dv}{dn} \tag{10.6}$$

where
- τ is the overall shear stress (N/m^2)
- μ is the dynamic viscosity of fluid (N s/m^2)
- v is the local velocity of fluid (m/s)
- n is the coordinate normal to velocity vector and coincident with velocity gradient, $\nabla\vec{v}$ (m)

Multiplying both sides of Equation 10.6 by dv/dn, and since $t(dv/dn) = dP/dV$ (recalling that shear is force per unit area and power is force times velocity),

$$\frac{dP}{dV} = \mu\left(\frac{dv}{dn}\right)^2 \tag{10.7}$$

In terms of a finite volume, the average power expended is

$$\frac{P}{V} = \mu\left(\frac{dv}{dn}\right)^2 \tag{10.8}$$

Since Equation 10.8 is the same as Equations 10.4 and 10.5 combined the derivation is established.

Dividing both sides of Equation 10.8 by the fluid density, r, gives a variation in Equation 10.8 from Saffman and Turner (1956), which is power expended per unit mass, i.e.,

$$\varepsilon_m \frac{P}{rV} = n\left(\frac{dv}{dn}\right)^2 \tag{10.9}$$

where
- ε_m is the work of shear per unit mass per unit time (N · m/s/kg)
- ρ is the density of fluid (kg/m^3)
- v is the kinematic viscosity of fluid (m^2/s)
- $n \equiv \mu/r$

The ε_m parameter and Equation 10.9 are given because they are used often in the literature, i.e., as opposed to P/V. The relation between G and ε_m is, $G = [\varepsilon_m/n]^{0.5}$.

10.2.3.3 Modifying Camp and Stein's G

The most obvious issue with the Camp and Stein G is that the dynamic viscosity of the fluid is incorporated in Equation 10.5, i.e., $G = [P/\mu V]^{0.5}$ (Cleasby, 1984, p. 875). Another issue identified by Clark (1985, p. 759) was that a root-mean-square average velocity gradient is not representative of the complex structure of a turbulent flow regime. Regarding the viscosity issue, Cleasby (1984, p. 876) proposed, for fluid shear term,

$$t = (\mu + h)\frac{dv}{dn} \tag{10.10}$$

and

$$\mathbf{e}_v \equiv \frac{h}{r} \tag{10.11}$$

where
- h is the eddy viscosity of the fluid for turbulent conditions, i.e., analogous to μ (N s/m^2)
- $\mathbf{e}_v$ is the kinematic eddy viscosity for turbulent conditions, i.e., analogous to n (m^2/s)

The eddy viscosity, h, is a characteristic of the fluid flow but it has the advantage of being analogous to the molecular viscosity, μ; but dividing by r gives another parameter, $\mathbf{e}_v$, where $\mathbf{e}_v = l(v'^2)^{1/2}$, which depends only on the eddy size, l, and the root mean square of the velocity deviations from the mean, i.e., $(v'^2)^{1/2}$ (Rouse, 1946, p. 178). The eddy viscosity, h, is dominant in turbulent flow, i.e., at high **R**; on the other hand, the dynamic viscosity, μ, is a property of the fluid and is dominant in laminar flow, i.e., for low **R**, e.g., $\mathbf{R} \leq 2000$. Eddy viscosity is important in understanding the role of turbulence in collisions between particles, but cannot be evaluated quantitatively (as contrasted with μ the dynamic viscosity).

Cleasby (1984, p. 894) recommended that (1) G is valid only for particles smaller than the Kolmogorov microscale, which is not typical, but may be applicable, or rapid mixing of short duration in which the initial phases of aggregation of particles smaller than microscale occur; (2) otherwise, the parameter $e_m^{2/3}$ is more appropriate for practice because the turbulent eddies are larger than the Kolmogorov microscale, and is of independent temperature.

10.2.3.4 Empirical Parameters

Empirical parameters include P/Q, P/V, G, q, and ε_m; Table 10.1 gives some representative guidelines from the literature. The G parameter remains in water treatment practice largely because of its history, because guidelines are available, and it has not been supplanted. The water treatment research community has criticized the G criterion increasingly since the work of Argaman and Kaufman (1968, 1970), who noted that G calculates average energy dissipation over the volume of the whole reactor and does not take into account that the velocity field of a reactor is nonuniform. The P/V parameter is used most frequently in chemical engineering and is used also in water treatment, often in lieu of G.

10.2.3.5 G and θ

The effect of G and θ on settled water turbidity was determined for various pH and alum dosages that covered the adsorption–destabilization and sweep-floc zones of the alum coagulation diagram (Amirtharajah and Mills, 1982, pp. 214–216). The span of $G\theta$ covered was $16{,}000 \leq G\theta \leq 18{,}000$, with $G \approx 300, 1{,}000, 16{,}000$ s^{-1}, and $\theta = 60, 20, 1$ s, respectively. For the adsorption–destabilization zone, their findings showed markedly lower settled water turbidities at

TABLE 10.1
Parameter Guidelines for Mixing

Parameter					
P/Q		P/V		G	θ
(kW/mL/d)	(hp/mgd)	(kW/m^3)	(hp/ft^3)	(s^{-1})	(s)
Basin–impeller mixing					
$0.05 \le P/V \le 0.2$[a]	$0.25 \le P/V \le 1.0$[a]			300[a]	$10 \le \theta \le 30$[a]
				$500 \le G \le 1000$[b]	120[b]
				10,000[b]	120[b]
				$600 \le G \le 1000$[c]	$10 \le \theta \le 60$[c]
		$0.2 \le P/V \le 2$[d]	$0.0076 \le P/V \le 0.076$[d]	$200 \le G \le 2000$[d]	
In-pipe impeller mixing					
0.02–0.53[e]	0.08–2.67[e]				
In-pipe jet mixing					
0.02–0.09[f]	0.11–0.45[f]				

[a] ASCE-AWWA-CSSE (1969, p. 72).
[b] Gemmel in AWWA (1972, p. 128).
[c] AWWA-ASCE (1998. p. 92) in the 3rd edition of *Water Treatment Plant Design*.
[d] Myers et al. (1999, p. 35) suggested for high-energy impeller mixing of low viscosity liquids (low value) and emulsification and gas dispersion (high value).
[e] Kawamura (2000, p. 309); for 12 in-pipe impeller-flash-mix installations; P/Q(average) $\approx$ 0.18 kw/mL/day (0.93 hp/mgd).
[f] Kawamura (2000, p. 309); for 12 in-pipe jet-mix installations; P/Q(average) $\approx$ 0.03 kw/mL/day (0.17 hp/mgd).

$G = 16{,}000\ \text{s}^{-1}$. For the sweep-floc zone, their findings showed the same settled water turbidities for $300 \le G \le 16{,}000\ \text{s}^{-1}$. For the restabilization zones, G made no difference.

The ostensible guidelines are: (1) $G \gg 1000\ \text{s}^{-1}$, i.e., $G \approx 16{,}000\ \text{s}^{-1}$, is imperative for adsorption–destabilization; (2) a back-mix reactor is suitable for the sweep-floc zone but not for adsorption–destabilization; (3) a high-energy blender is essential for the adsorption–destabilization zone and will not harm the other zones. A key point is that if uncertainty exists as to whether the zone is sweep-floc, then to err on the side of high energy will not harm, and possibly help. Exceptions are suggested, however, in the literature (see for example, Amirtharajah and Mills, 1982) and so pilot studies are recommended.

10.2.4 Technologies of Mixing

Mixing technologies include a hydraulic jump, a series of elbows, over-and-under baffles, end-around baffles, submerged jets, a grid across a pipe, static mixers, a pump impeller, and stirred mixing basins. Figure 10.3 illustrates six of these technologies.

In coagulation the prevalent mixing technologies from about 1910–1930 included the hydraulic jump, end-around baffles, over-and-under baffles, an open channel, pipe turbulence, and a pump impeller. Diffusion alone in quiescent settling was another technology for coagulation, illustrated in Figure 10.1.

The mixing basin with an impeller seems to have been instituted about the 1930s; it did not supplant the other technologies but was added to the repertoire. By the 1970s the mixing basin was questioned by some because of the problem of variable residence times. In the 1980s, static mixers and in-line mixers (an impeller in a pipe) were added, with the latter being used most extensively. By the 1990s, jet mixing and wake turbulence came to be favored increasingly as the industry became more cognizant of the arguments calling attention to the limitations of impeller–basins and the advantages of in-line and jet mixers (Vrale and Jorden, 1971; Stenquist and Kaufman, 1972; Chao and Stone, 1979; Jorden, 2001) as related to the variable residence times inherent with back-mix reactors.

10.3 THEORY OF MIXING

Mixing theory involves fluid transport, which has three forms: (1) advection, (2) turbulent diffusion, and (3) molecular diffusion. The first two may be modeled by the classical Navier–Stokes equation, expressed in finite difference form with execution by computer algorithm. Molecular diffusion is modeled by Fick's first law, another classical equation which, as a rule, is approximated.

Another approach to understand mixing is physical modeling. Data obtained may be summarized in terms of dimensionless numbers. Mathematical modeling and physical modeling help to understand mixing but neither is sufficient for design. The fallback is empirical guidelines and experience. As with engineering in general, empiricism and theory are complementary.

10.3.1 Transport Mechanisms

Mixing is transport of one substance through another, e.g., dispersing a coagulant, chemical, or a dissolved gas in water. As stated, the three forms of transport are (1) advection, (2) turbulence, and (3) diffusion, depicted in

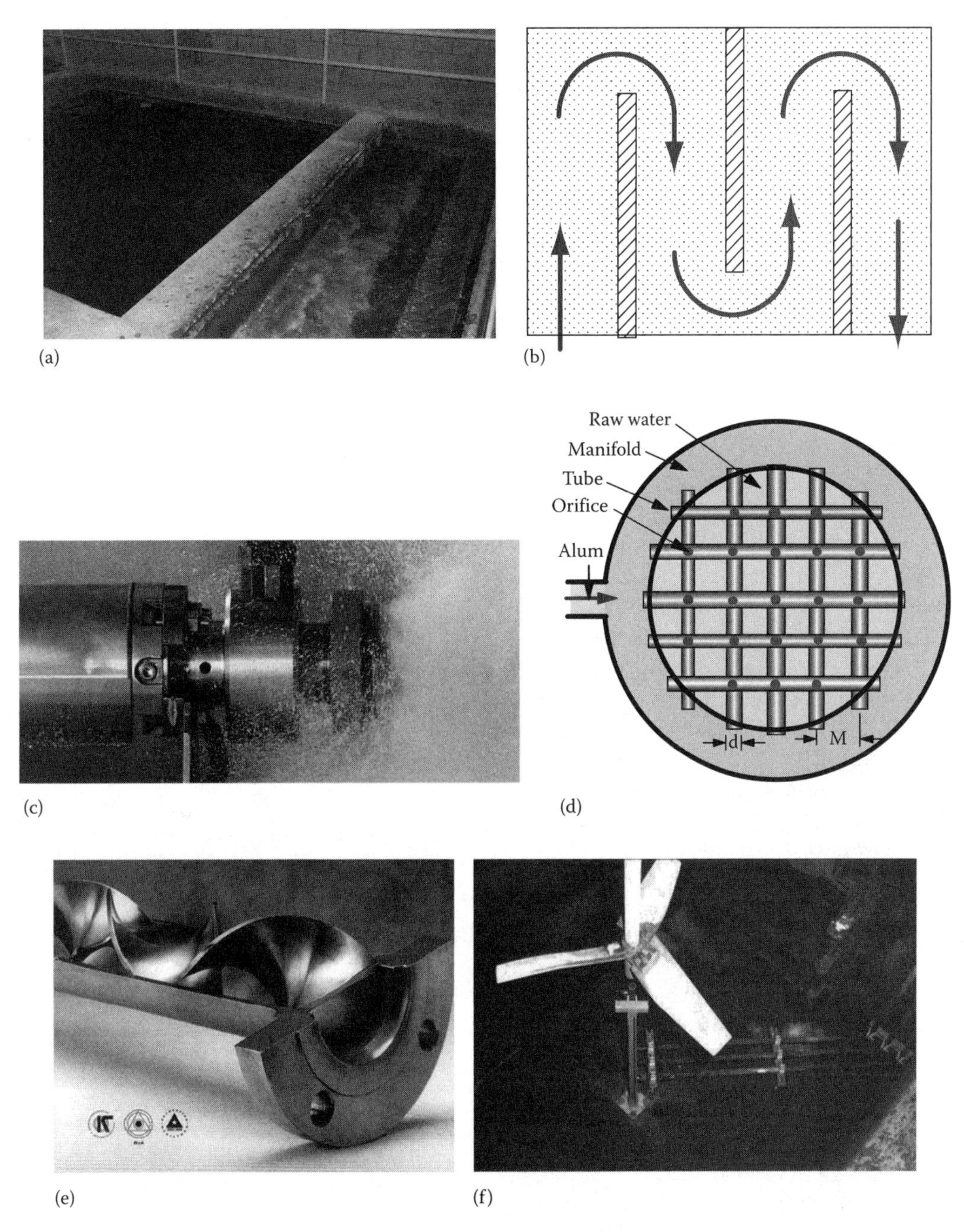

FIGURE 10.3 Examples of mixing technologies. (a) Hydraulic jump with alum distribution. (Courtesy of Gertig, K., personal communication, March 2002). (b) Baffles—end-around. (c) Submersible chemical induction unit (Water Champ®, Siemens Water Technologies, 2010). (d) Grid across pipe. (From Stenquist, R.J. and Kaufman, W.J., Initial Mixing in Coagulation Processes, Report to Environmental Protection Agency, EPA-R2-72-052, Office of Research and Monitoring, Washington, D.C., 1972. With permission.) (e) Static mixer. (Chemineer–Kenics, Static Mixing Technology, Dimension Sheet, Style KMS, CA812, Ref. Bulletin Series 800, http://www.chemineer.com/images/pdf/bulletin_800.pdf, 2010.) (f) Stirred basin. (Courtesy of Gertig, K., personal communication, March 2002.)

Figure 10.4a through c, respectively, and are reviewed in the sections that follow.

10.3.1.1 Advection

Advection is the movement of a fluid mass and is illustrated in Figure 10.4a for a hypothetical two-dimensional case. The inertia of the incoming flow, depicted by the arrow at the bottom left of the sketch, causes a "large" eddy as indicated by the circular motion.

As illustrated, advection functions to move a mass of one fluid through another. The scale is perhaps decimeters (dm) or meters (m), depending on the velocity of the initial jet and the tank size. While such an advective stream retains its essential identity, as shown in Figure 10.4a, some secondary patterns are generated from the main stream due to fluid shear. The main flow loses momentum flux, i.e., $Q(\text{jet}) \cdot r(\text{water}) \cdot v(\text{jet})$, as its primary flow, $Q(\text{jet})$, and its initial velocity, $v(\text{jet})$, are reduced due to fluid shear stress.

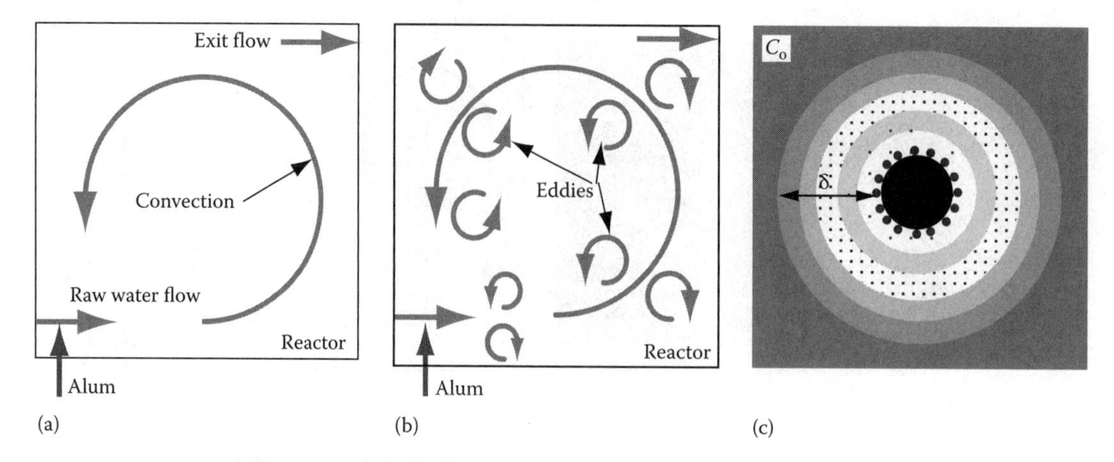

FIGURE 10.4 Depiction of basic transport mechanisms of mixing. (a) Advection: m-dm. (b) Turbulence: dm-mm. (c) Diffusion: μm.

The shear stress causes eddies, actually vortex tubes, to peel-off from the main flow (which is turbulence).

10.3.1.2 Turbulence

Velocity gradients cause fluid shear and the generation of eddies, which is "turbulence." Figure 10.4b shows eddies "peeling-off" from the large eddy with energy input from the incoming mass of flow depicted in the lower left corner. The eddies become "vortex tubes" whose shear causes further shedding of vortices, resulting in an "energy cascade." Ultimately, as the vortex "tails" become smaller, viscous forces predominate and the energy is dissipated as heat.

The effects of turbulence in mixing are (1) to disperse (or blend) one constituent within the mass of another; (2) to cause collisions between macro-reactants, i.e., ≥ 1 μm (between floc particles, for example); (3) to bring particles $\ll 1$ μm (including molecular substances) into the "diffusion proximity" of another molecular substance or surface; (4) to increase a diffusion concentration gradient; (5) to create new interfacial surface area; and (6) to maintain a suspension of particles that otherwise would settle. The average intensity of turbulence, G, i.e., velocity gradient, dv/dy (root mean square of velocity gradient), for a given volume determines the rate of these processes. At the same time, the scale of turbulence, i.e., the size of the eddies, λ, controls the propensity of particles to be transported. How these factors, i.e., G and λ, relate to mixing is the main topic of this section.

10.3.1.2.1 Views of Turbulence

Two views of turbulence theory are given here: (1) classical theory which is statistical in nature, and (2) the "deterministic" view which is in terms of vortices. The literature in the field is extensive (see, for example, books by Batchelor, 1953; Tennekes and Lumley, 1999; Frisch, 1995; Wilcox, 1997) with the latter view, i.e., vortices, becoming prevalent.

- *Classical theory*: The classical theory of turbulence is purely statistical and is characterized by random velocities with position and time, illustrated in Figure 10.5. Also shown are some common terms used to

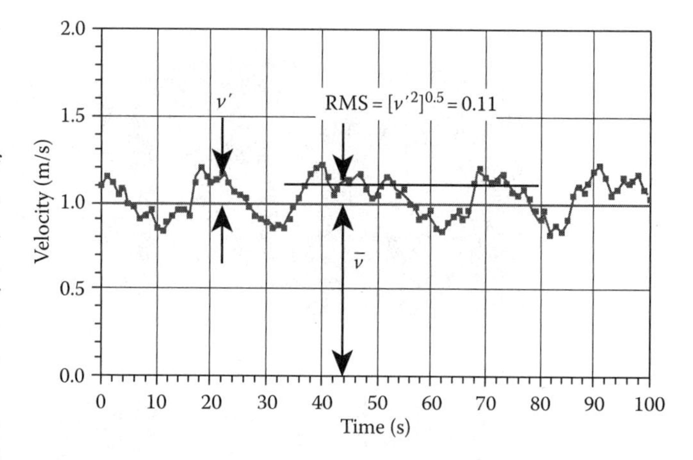

FIGURE 10.5 Typical trace of velocity variation about mean. (Adapted from Rouse, H., *Elementary Mechanics of Fluids*, John Wiley & Sons, New York, p. 178, 1946.)

characterize turbulence, e.g., the mean velocity of advection, $\bar{v}$ (m/s), and the root mean square of the instantaneous velocity deviation, $\sqrt{\overline{(v')^2}}$, i.e., RMS (m/s). The theory had its origins with Ludwig Prandtl, c. 1925, who proposed the ideas of mixing length and eddy viscosity (Rouse, 1946, p. 178). According to Batchelor (1953, p. 8), however, modern notions of turbulence originated, c. 1935, with publications of Goeffrey I. Taylor, a professor of mathematics at Cambridge (and a consultant on the Manhattan Project). Taylor had a background in aeronautics dating to wind tunnel studies during WWI.
- *Vortex tubes*: Through the 1960s and 1970s, the notion of an eddy developed toward the idea of a vortex, then to "vortex stretching," to become a "vortex tube." The terms "eddy" and "vortex," are used interchangeably but the latter has a three-dimensional connotation, which fits the notion of a "vortex tube." With a constant energy source, vortices are created continuously, resulting in a "tangle" of vortex tubes filling the volume of a reactor (Hanson and Cleasby, 1990;

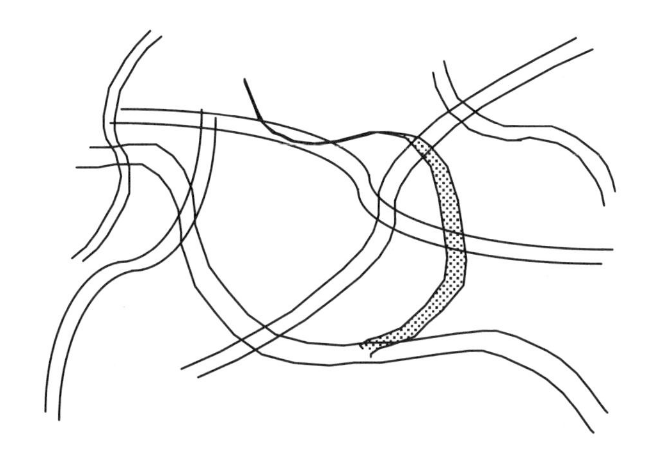

FIGURE 10.6 Space-filling vortex tubes. (Adapted from Hanson, A. T. and Cleasby, J.L., *J. Am. Water Works Assoc.*, 82(11), 56–73, 1990.)

Meroney, 2001; Bienkiewicz, 2001). Figure 10.6 is a depiction of such an array of vortex tubes having a spectrum of diameters, lengths, orientations, and energy levels. The tubes also interfere with one another, causing a given tube to combine with another and perhaps breaking (Bienkiewicz, 2001). In addition, a given vortex tube sheds smaller daughter vortices, losing some of its energy. Eventually, the vortices become small enough that viscous forces predominate and the kinetic energy of the vortex dissipates as heat that occurs in the "viscous sub-range" of the "energy spectrum" (Section 10.3.1.2.3). The vortex tangle is thus a part of a dynamic system and has a deterministic character in that it is due to the forces and acceleration on infinitesimal particles, as described by the Navier–Stokes equation. The continuous advection of vortex tubes past any given point with changing orientation is a plausible explanation for the Eulerian (i.e., fixed coordinates) observation of turbulence as illustrated in Figure 10.5. Figure 10.6 illustrates the alternate "Lagrangian" view, i.e., a relative coordinate system in which the observer moves with the motion of a given parcel.

10.3.1.2.2 Turbulence Generation

As stated, turbulence is associated with velocity gradients (i.e., fluid shear). To generate turbulence for mixing, boundary discontinuities give the highest velocity gradients, i.e., high shear. Form separation and submerged jets are two approaches for causing such high-shear zones (Rushton and Oldshue, 1953b, p. 268) and are illustrated in Figure 10.7a and b, respectively. In each case, the discontinuity causes eddies; the shear (velocity gradient) between a given eddy and the surrounding fluid results in an "eddy cascade."

The turbulence generated by the separation seen in (a) is termed "wake turbulence" and causes "wake mixing" (Stenquist and Kaufman, 1972, p. 7). In (b) there is no formal designation but the term, "jet mixing," is descriptive. These two themes, i.e., wake mixing and jet mixing, permeate virtually all mixing technologies.

- *Wake turbulence*: Wake turbulence is created by devices such as a flat-blade radial-flow impeller, an orifice plate in a pipeline, a sudden pipe expansion, etc. Figure 10.8 illustrates the process of eddy

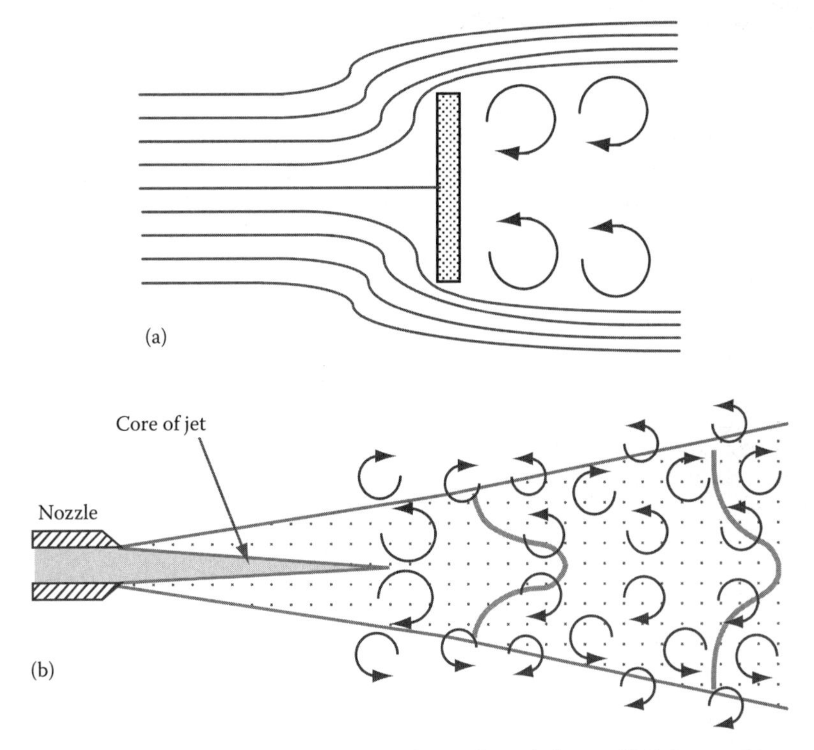

FIGURE 10.7 Types of boundary discontinuities. (a) Flat plate causing wake turbulence. (b) Submerged jet generating eddies.

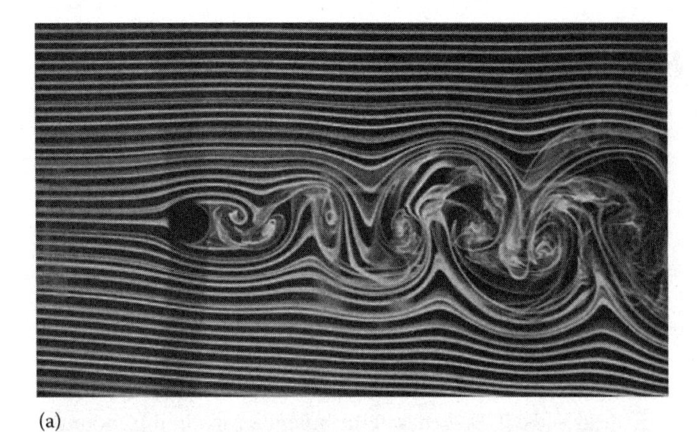

(a)

(b)

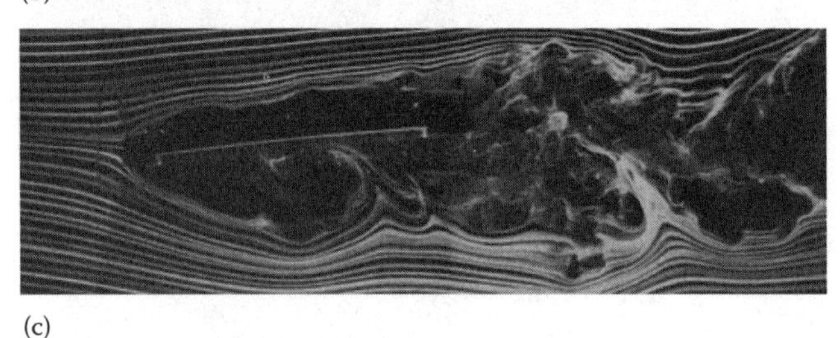

(c)

FIGURE 10.8 Examples of eddy formation due to different disturbances; $R \approx 10^5$. (Courtesy of Professor B. Bienkiewicz, Colorado State University, Fort Collins, CO, from wind tunnel experiments). (a) Cylinder. (b) Multiple airfoils. (c) Flat plate.

formation due to flow past three shapes: (a) a cylinder, (b) multiple airfoils, and (c) an inclined flat plate. For each of the three cases, $\mathbf{R} \approx 10^5$. Figure 10.8a shows an initial eddy about the size of the cylinder and just behind it the eddy grows in size as it is advected downstream. In Figure 10.8b multiple airfoils cause the same effect but downstream the adjacent eddies impinge against each other. In Figure 10.8c the inclined flat plate causes separation on the underside with intense mixing. As noted, the fate of the eddies is an "energy-cascade" with an eventual dissipation of the energy as viscous forces and heat (Hanson and Cleasby, 1990; Hanson, 2001).

- *Submerged jet*: Figure 10.9 illustrates the general characteristics of a submerged jet as it expands into the surrounding ambient environment. As seen, the velocity profile flattens with increasing x-coordinate. Also, as illustrated, the velocity gradient $\mathrm{d}v/\mathrm{d}r$ changes with r at a given x, which means that eddies are created accordingly and define the mixing zone (seen as a conical expansion of the jet from the nozzle).

Two notes are relevant at this point: (1) A submerged jet is also seen in a hydraulic jump, in the plunging nappe from a weir, and in the flow from an impeller (the jet of an axial flow impeller is parallel to the axis of the impeller and is in the radial direction in a radial-flow impeller). (2) In coagulation with alum, the flow must be distributed in the core of the jet, before emergence from the orifice, so that the very small flows of neat alum are carried into the turbulence zone with its subsequent mixing (which occurs as lateral and axial dispersion).

10.3.1.2.3 Energy Spectrum

For a given mixing volume with a continuous energy input, such as an impeller, a submerged jet, or flow across a disturbance, turbulence occurs as vortex tubes in a spectrum of sizes that fill the volume in a "tangle," as described

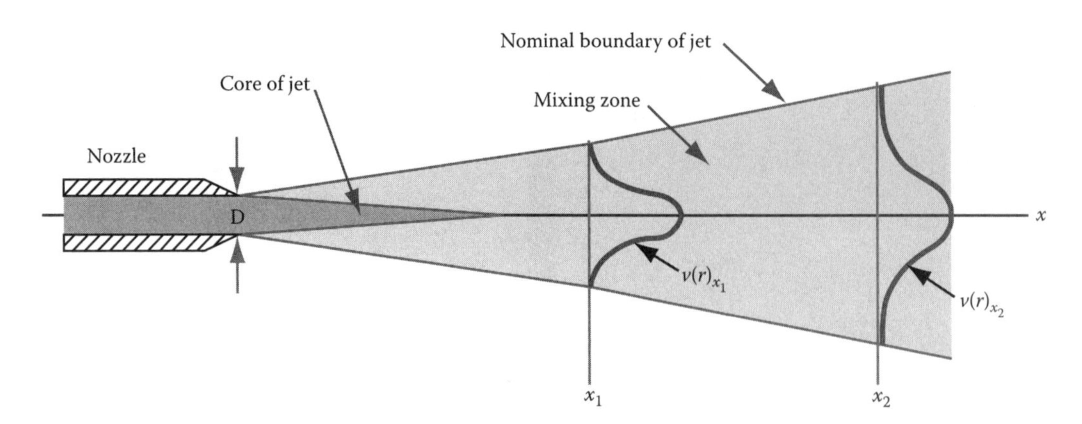

FIGURE 10.9 Mixing zone in terms of a submerged jet, diameter (D) and velocity profiles as indicated. (Adapted from Albertson, M.L. et al., *Am. Soc. Civil Eng. Trans.*, 115, 657, 1950.)

previously (i.e., as in Figure 10.6). Vortex sizes range from the dimension of the disturbance to the "Kolmogorov micro-scale" (Box 10.2); which is micrometers in size. For eddies having this micrometer-scale size, their energy is dissipated as viscous shear. Each vortex tube has a different amount of kinetic energy and the distribution of energy with size is called the "energy spectrum" (Stenquist and Kaufman, 1972, p. 11). In the parlance of theoreticians, the reciprocal of the eddy size in meters is called the "wave number." The total kinetic energy per unit mass of fluid is (Batchelor, 1953, p. 36)

$$\overline{\frac{1}{2}u_i(x)u_i(x)} = \int_0^\infty E(\kappa)\mathrm{d}\kappa \qquad (10.12)$$

and

$$\kappa \equiv \frac{1}{\lambda} \qquad (10.13)$$

where
- u_i is the velocity of eddy i (m/s)
- κ is the wave number (1/m)
- λ is the diameter of vortex tube (m)
- $E(\kappa)$ is the energy as function of wave length, i.e., the "energy spectrum" (N m^2/kg)
- $\overline{\frac{1}{2}u_i(x)u_i(x)}$ is the mean kinetic energy of all eddies collectively (m^3/s^2), which is same as (N m^2/kg)

BOX 10.2 KOLMOGOROV'S CONTRIBUTION TO TURBULENCE THEORY

According to A. N. Kolmogorov's 1941 paper, there is a range of high frequencies where the turbulence is statistically in equilibrium and uniquely determined by the energy dissipation rate, $E(\kappa)$, and the kinematic viscosity, ν (Batchelor, 1953, p. 115; Argaman and Kaufman, 1968, p. 32). The $E(\kappa)_2$ curve of Figure 10.10 is called the "universal equilibrium range" of wave numbers. Kolmogorov postulated this "sub-range" in 1941 referring to it as the theory of small eddies (Batchelor, 1953). Initially, his paper received little attention due to the war and the fact that the Russian journal where it was published was not readily accessible to the researchers in other countries. Later, as Kolmogorov's work was assimilated, Batchelor (1953, pp. 103–168) considered that the idea of an equilibrium range of the energy spectrum, and the theory that was built upon it, constituted the most important development during the 1940s decade. Wilcox (1997, p. 9) gives a further description of Kolmogorov's universal equilibrium theory:

We begin by noting that the cascade process present in all turbulent flows involves a transfer of turbulence kinetic energy (per unit mass), $E(\kappa)$, from larger eddies to smaller eddies. Dissipation of kinetic energy to heat through the action of molecular viscosity occurs at the scale of the smallest eddies. Because small-scale motion tends to occur on a short timescale, we can assume that such motion is independent of the relatively slow dynamics of the large eddies and of the mean flow. Hence, the smaller eddies should be in a state where the rate of receiving energy from the large eddies is very nearly equal to the rate at which the smallest eddies dissipate the energy to heat. This is one of the premises of Kolmogorov's (1941) universal equilibrium theory.

These two sub-ranges of eddies (Figure 10.10) are also known collectively as the "universal equilibrium range," i.e., where the rate of energy received by the small eddies in the viscous sub-range equals the rate of energy received from the large eddies in the inertial sub-range.

Figure 10.10 illustrates the two kinds of $E(\kappa)$ distributions: (1) the inertial sub-range, $E(\kappa)_1$, and (2) the viscous sub-range, $E(\kappa)_2$ (Hanson and Cleasby, 1990). The largest vortices (i.e., those about the size of the disturbance) are in the "inertial" sub-range and contain most of the kinetic energy of the system and essentially "drive" the ensuing "energy cascade" toward their ultimate fate in the viscous sub-range. When the viscous forces dominate, the vortex is no longer a "free vortex," which means that the mechanism driving the energy cascade is no longer present and the lower scale of turbulence has been reached. Once the vortices reach this length scale, $\kappa > \kappa^*$ in Figure 10.10, their energy is quickly dissipated as viscous shear and consequently, heat (Hanson and Cleasby, 1990). As a note, the author asked Professor Robert Meroney to comment on the $E(\kappa)$ units; based on his three-paragraph reply (September 15, 2009), the issue is much more

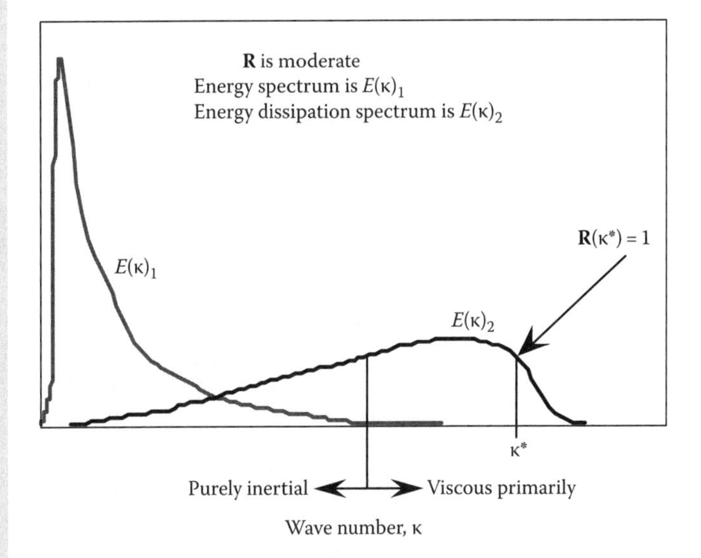

FIGURE 10.10 Typical Kolmogorov energy spectrum. (Adapted from Hanson, A.T. and Cleasby, J.L., *J. Am. Water Works Assoc.*, 82 (11), 56, 1990.)

complex than indicated here. But the short answer is that while he confirmed that the N m/kg units were correct (and are consistent with the development shown), other alternative units may be preferred by those in the field.

- *Kolmogorov microscale*: A particular wave number indicated on curve $\varepsilon(\kappa)_2$ is κ^*, called the "Kolmogorov microscale." A characteristic is that in the range $\kappa = \kappa^*$, then $\mathbf{R} = 1$; in other words, the inertia forces equals the viscous forces. For $\kappa > \kappa^*$, the energy function drops sharply with the energy being dissipated to heat. The Kolmogorov microscale is the practical limit of eddy size. At such size scales, diffusion, the final transport step, becomes dominant. The length scale, λ^*, and the vortex velocity, v^*, may be calculated as (Batchelor (1953, p. 1115; Argaman and Kaufman, 1968, p. 32; Frisch, 1995, p. 91; Clark, 1996, p. 207; Logan, 1999, p. 196),

$$\lambda^* = \left(\frac{\nu^3}{\varepsilon}\right)^{1/4}, \quad (10.14)$$

and

$$v^* = (n \cdot e)^{1/4} \quad (10.15)$$

where
λ is the size of eddy at Kolmogorov microscale (m)
n is the kinematic viscosity (m^2/s)
e is the rate of energy dissipation (N m/s/kg)
v is the peripheral velocity of vortex at Kolmogorov microscale (m/s)

The Reynolds number for the Kolmogorov microscale, $\mathbf{R}^*$, has a value of unity, i.e., $\mathbf{R}^* = 1$ and is calculated (Argaman and Kaufman, 1968, p. 32),

$$R^* = \frac{v^*\lambda^*}{\nu} \quad (10.16)$$

- *Role of large eddies*: Regarding the big slow eddies, they interact very weakly with the remainder of the turbulence and preserve their energy virtually intact (Batchelor, 1953, p. 91). The energy spectrum of very small wave numbers (the larger eddies) suffers very little modulation during the whole of the decay process; the reason is that most of the energy of the $E(\kappa)_1$ inertial sub-range is associated with smaller wave numbers, as seen in Figure 10.10. On the other hand, the energy in higher wave numbers (the smaller eddies) of the spectrum, $E(\kappa)_2$, i.e., in the viscous sub-range, is rapidly dissipated by viscosity. From the foregoing, it follows that ultimately the big eddies will supply most of the energy of the turbulence (Batchelor, 1953, p. 92). The practical outcome is that the advective motion of large eddies is essential to distribute a given species to different neighborhoods of the fluid bulk where smaller eddies, i.e., vortices, can do the job of final mixing.

  As related to Figure 10.10, as G increases, the spectra distribution extends to higher frequencies (Argaman and Kaufman, 1968, pp. 84–94, 1970). These results are in accordance with the theory which predicts the relationship between the power input and the Kolmogorov microscale of turbulence, i.e., Equation 10.14, $\lambda^* = (\nu^3/\boldsymbol{\varepsilon}_m)^{1/4}$, resulting in decreasing λ^* with increasing $\boldsymbol{\varepsilon}_m$ (or G); therefore, the $E(\kappa)_2$ curve shifts to the right at higher G values. Also, the area under the $E(\kappa)_1$ curve is larger with increasing G, since more energy must be distributed over the $E(\kappa)_2$ energy spectrum.

- *Particle transport by eddies*: Only those eddies within certain size ranges will influence particle transport (Argaman and Kaufman, 1968, p. 19). Some of the ideas underlying particle transport mechanisms are illustrated by Figure 10.11a through c, respectively, i.e.,
  1. In Figure 10.11a, i.e., where $\lambda(\text{eddy}) \gg d(\text{separation})$, the eddy may transport particles, but the scale is such that the particles are entrained to move with the eddy rather than to be effective in collisions with other particles.
  2. In Figure 10.11b, i.e., where $\lambda(\text{eddy}) < d(\text{particle})$, the eddy is too small to transport the particle. Eddies that are smaller than the particle size, do not contribute substantially to turbulent diffusion (p. 19).
  3. In Figure 10.11c, i.e., where $\lambda(\text{eddy}) \gg d(\text{particle})$ and $\lambda(\text{eddy}) < d(\text{separation})$, the eddy is of such size as to cause effective particle transport and collisions with other particles. In other words, the eddy-induced motion of two particles relative to one another will be governed by eddies that are larger than their size and smaller than their separation.

  To summarize, from Hanson and Cleasby (1990), the process of vortex stretching will create myriad localized shear fields, which drive the process of contacts between primary particles and coagulants (and also between like particles as in flocculation).

- *Design implications*: Eddies sizes about the size of the primary particles (or slightly larger) are most effective in causing collisions, as illustrated in Figure 10.11c. Thus the source of turbulence should not be too large (e.g., in rapid mix, the blade width of a radial-flow impeller would be narrow rather than wide). A large

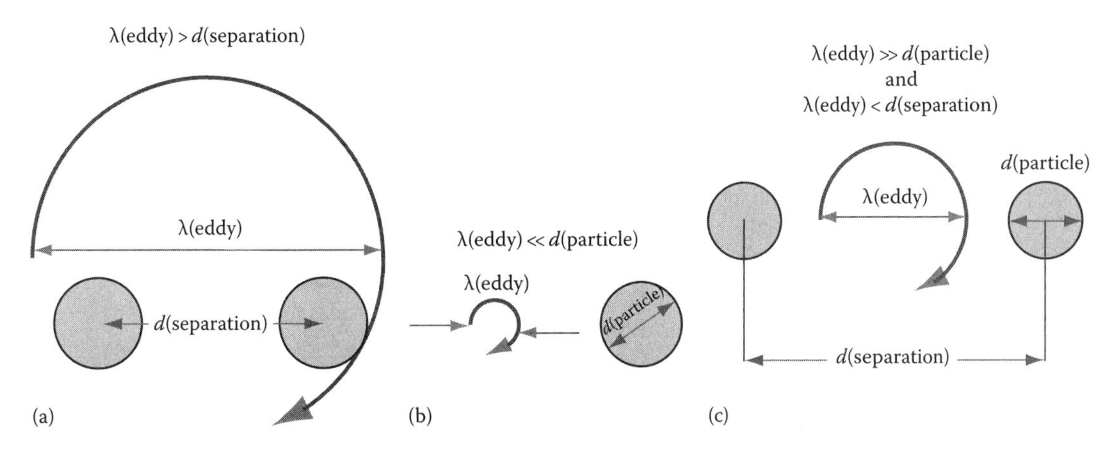

FIGURE 10.11 Eddy sizes relative to effective mixing. (a) Eddy too large. (b) Eddy too small. (c) Effective eddy. (Adapted from Argaman, Y. and Kaufman, W.J., Turbulence in orthokinetic flocculation, SERL Report No. 68-5, Sanitary Engineering Research Laboratory, University of California, Berkeley, CA, 1968.)

form will generate large inertial eddies with an energy spectrum $E(\kappa)_1$ shifted to the left, i.e., with small wave numbers as illustrated by Figure 10.11a. But a smaller form drag will result in a shift to the right, which distributes more energy toward the higher wave numbers. At the same time, the higher the energy input, the lower the Kolmogorov microscale, λ^*. Forms that generate suitable turbulence include a grid across a pipe. For example, Stenquist and Kaufman (1972) showed that mixing was more effective with a grid rather than with a back-mix reactor. The grid size may be designed to be compatible with the scale of turbulence desired. The same is true for flocculation; the paddle blades should be not too large so the $E(\kappa)_1$ energy does not have to cascade from very large eddy sizes to the $\varepsilon(\kappa)_2$ energy distribution.

In adsorption–destabilization, much of the process occurs at the small length scales. The more energy added to the water in a localized region, the smaller the microscale of turbulence. Therefore, the higher the energy input per unit mass, the more the energy spectrum is shifted to the right and the more effective is the mixing for small primary particles. On the other hand, if the particles are larger in size, e.g., $d(\text{particle}) \gg 1\ \mu\text{m}$, the energy input may have some upper limit for effective mixing, i.e., such that $d(\text{eddy}) > d(\text{particles})$, so that Case 3, Figure 10.11c prevails.

Example 10.1 Shear of Floc Particles with Respect to Kolmogorov Microscale (from Hanson and Cleasby 1990)

Given

Two cases are summarized in the following table: (1) floc is smaller than the Kolmogorov microscale, and (2) floc is larger than the microscale.

Required

For each case, determine whether the floc will rupture.

Solution

The tabular summary indicates the outcome based upon the foregoing guidelines, i.e., as outlined in Figure 10.11.

Case	Case Description	G (s^{-1})	λ^{*a} (μm)	d(floc) (μm)	Outcome
1	Floc smaller than microscale	10	316	250	Floc will not break since floc is smaller than smallest turbulent eddies
2	Floc larger than microscale	30	200	250	Floc will interact with turbulent flow field and rupture

[a] Calculated by Equation 10.14.

Discussion

For Case #1, the particles are smaller than the Kolmogorov microscale, i.e., $d(\text{floc}) < \lambda^*$. A particle in this size range is like a mosquito inside a car; the mosquito does not know when the car turns a corner. And so it is for a particle inside a vortex tube; the particle with size, $d(\text{particle}) < \lambda^*$, does not feel the inertial interactions of the vortex (simile from discussions with Professor Adrian Hanson, September 10, 2001); the vortex does not cause mixing.

For Case #2, the particles are larger than the Kolmogorov microscale, i.e., $d(\text{floc}) > \lambda^*$. A particle in this size range is subject to the shear field generated by the smallest turbulence (but does not feel the turbulence of the larger vortices).

10.3.1.2.4 *Diffusing Substances, Diffusion Rate, and Sinks*

Figure 10.12 illustrates in (a), (b), and (c), respectively, substances that diffuse, a graphic depiction of mathematical

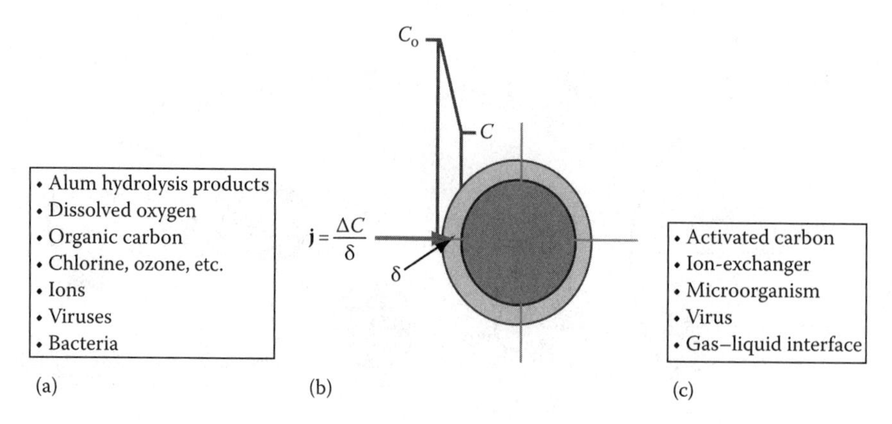

FIGURE 10.12 Reactions requiring diffusion, facilitated by turbulence. (a) Diffusing substance. (b) Diffusion to surface. (c) Surface material.

terms, and the kinds of "sinks" that take up molecules or particles that diffuse. As seen in Figure 10.12a, diffusing substances include aluminum hydrolysis products, dissolved oxygen, organic carbon, a disinfectant (such as chlorine or ozone), ions, etc. The reacting sites, or "sinks," include discrete particulate surfaces such as activated carbon, an ion-exchanger (resin or green sand), a microorganism, a virus, a gas–liquid interface, etc. As in Figure 10.12b, the flux density, **j**, to a sink is governed by Fick's first law, i.e., $\mathbf{j} = \nabla DC)$, where ∇ is the operator for the total differential, D is the diffusion constant, and C is the solution concentration at a given point in space, and is approximated as, $\mathbf{j} = -D(C_o - C)/\delta$, with terms illustrated in Figure 10.12b.

The role of turbulence is threefold: (1) to reduce the film thickness, δ; (2) to maintain the concentration at near the level of the bulk of solution, C_o; and (3) to increase the total surface area across which diffusion may occur, i.e., ΣA_i, which is relevant where creation of an interface is involved (such as causing smaller bubbles in gas transfer). With respect to the (1) and (2), reducing δ and maintaining C_o increases the concentration gradient, which, of course, increases **j**.

10.3.1.3 Transport Regime

The parameters that define the relative roles of turbulence-transport and diffusion transport are delineated here. From this, we can see the approximate ranges of G and particle sizes that define each respective regime, i.e., turbulence and diffusion.

10.3.1.3.1 Perikinetic Transport

The frequency of particle contacts caused by diffusion (*perikinetic* transport) was given by O'Melia (1970) as

$$J_{pk} = \frac{dN}{dt} = \frac{-4\alpha kT}{3\mu}N^2 \qquad (10.17)$$

where

J_{pk} is the rate of change of total particles in suspension by perikinetic transport (particles/m^3/s)
N is the total concentration of particles in suspension at time, t (particles/m^3)
t is the elapsed time since initial measurement of N (s)
α is the collision efficiency factor (fraction of collisions producing aggregates/collision)
k is the Boltzman's constant (10^{-23} J/K)
T is the absolute temperature of suspension (K)
μ is the dynamic viscosity of the water suspension (Pa · s or N s/m^2)

10.3.1.3.2 Orthokinetic Transport

The frequency of particle contacts caused by turbulence (*orthokinetic* transport) was given also by O'Melia (1970) as

$$J_{ok} = \frac{dN}{dt} = \frac{-2\alpha G d^3 N^2}{3} \qquad (10.18)$$

where

J_{ok} is the rate of change of total particles in suspension by orthokinetic transport (particles/m^3/s)
d is the diameter of colloidal particles (m)

10.3.1.3.3 Ratio of Turbulent Transport to Diffusion Transport

The ratio of contact rate by turbulence (orthokinetic transport) to such rate by diffusion (perikinetic transport), J_{ok}/J_{pk}, is, dividing Equation 10.18 by Equation 10.17,

$$\frac{J_{ok}}{J_{pk}} = \frac{\mu G d^3}{2kT} \qquad (10.19)$$

Figure 10.13 was plotted from Equation 10.19, and shows how the ratio J_{ok}/J_{pk} is influenced by G for particle diameters of $d = 0.1$, 1, and 10 μm, respectively

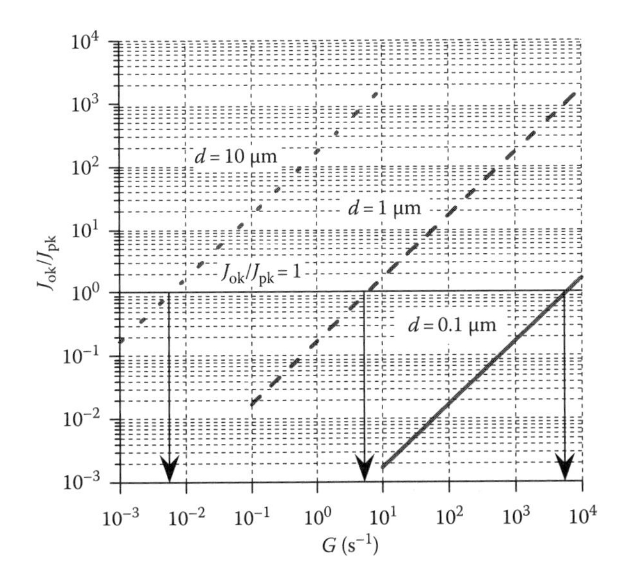

FIGURE 10.13 Plot of Equation 10.19 for 25°C showing how the J_{ok}/J_{pk} ratio is affected by G and particle diameter, d. Perikinetic transport is dominant to the left of the vertical lines with arrows and orthokinetic transport is dominant to the right.

(O'Melia, 1970). As implied, orthokinetic transport is dominant for ratios $J_{ok}/J_{pk} > 1$. The plot shows that at small particle diameter, such as 0.1 μm, perikinetic transport is clearly dominant until $G \approx 7000\ \text{s}^{-1}$, which is based on the criterion $J_{ok}/J_{pk} > 1$. For a particle diameter of 1 μm, orthokinetic transport is dominant at $G > 7$. For a particle diameter of 10 μm, turbulence, i.e., orthokinetic transport, is clearly dominant. Turbulent transport is necessary even for small particles, however, so that small particles are dispersed to the proximity of the resistance film to maintain a constant bulk concentration, C_o, at the edge of the film.

10.3.2 Navier–Stokes Equation

The Navier–Stokes equation, along with continuity, is considered the definitive description of fluid flows, including turbulence (see, for example, Wilcox, 1997). Interest in the equation was mostly academic until the advent of mainframe computers, c. 1958, which made it amenable to solution by writing it in finite difference form and solved in terms of Fortran programming (c. 1960). By the 1990s, the solution protocol had evolved into software called computational fluid mechanics (CFD).

10.3.2.1 Mathematics of Navier–Stokes Equation

Newton's second law applied to an infinitesimal fluid element equates the sum of the various forces acting on the element to the inertia force, ma. The resulting differential equation is the proverbial Navier–Stokes equation (Munson et al., 1998, p. 360) (Box 10.3), given here without elasticity or surface energy terms.

For the x-direction, $\Sigma\vec{\mathrm{F}}_\mathbf{x} = m\mathbf{a}_\mathbf{x}$ is in expanded form for an infinitesimal fluid element,

> **BOX 10.3 NAVIER AND STOKES**
>
> A certain amount of mystique is associated with the Navier–Stokes equation. A brief account of its development (from Rouse and Ince, 1957, pp. 91, 92, 104, 193–198) may help to gain some familiarity. The equation was developed almost in its present form by Louis Marie Henri Navier (1785–1836). A graduate of Ponts et Chaussées, he taught there and at Ecole Polytechnique. In an 1822 paper to the *Académie des Sciences*, he built on an earlier work by Leonhard Euler (1707–1783), i.e., that body forces plus pressure gradient equals local acceleration plus advective acceleration. Navier, however, added a hypothetical molecular attraction, embodied in a term, ε. Then, in 1845, Sir George Gabriel Stokes (1819–1903), in a paper before the Royal Society, replaced the coefficient, ε, with the dynamic viscosity, μ, giving the equation its current form. Stokes was the first person after Newton to hold the Lucasian professorship at Cambridge.

$$-\frac{\partial p}{\partial x} + \rho g_x + \mu\left(\frac{\partial^2 u}{\partial x^2} + \frac{\partial^2 u}{\partial y^2} + \frac{\partial^2 u}{\partial z^2}\right) = \rho\left(\frac{\partial u}{\partial t} + u\frac{\partial u}{\partial x} + v\frac{\partial u}{\partial y} + w\frac{\partial u}{\partial z}\right) \quad (10.20a)$$

For the y-direction, $\Sigma\vec{\mathrm{F}}_\mathbf{y} = m\mathbf{a}_\mathbf{y}$ is

$$-\frac{\partial p}{\partial y} + \rho g_y + \mu\left(\frac{\partial^2 v}{\partial x^2} + \frac{\partial^2 v}{\partial y^2} + \frac{\partial^2 v}{\partial z^2}\right) = \rho\left(\frac{\partial v}{\partial t} + u\frac{\partial v}{\partial x} + v\frac{\partial v}{\partial y} + w\frac{\partial v}{\partial z}\right) \quad (10.20b)$$

For the z-direction, $\Sigma\vec{\mathrm{F}}_\mathbf{z} = m\mathbf{a}_\mathbf{z}$ is

$$-\frac{\partial p}{\partial z} + \rho g_z + \mu\left(\frac{\partial^2 w}{\partial x^2} + \frac{\partial^2 w}{\partial y^2} + \frac{\partial^2 w}{\partial z^2}\right) = \rho\left(\frac{\partial w}{\partial t} + u\frac{\partial w}{\partial x} + v\frac{\partial w}{\partial y} + w\frac{\partial w}{\partial z}\right) \quad (10.20c)$$

In vector notation the preceding equations are, in consolidated form,

$$-\nabla p + \rho\mathbf{g} + \mu\nabla^2\mathbf{v} = \rho\left(\frac{\partial v}{\partial t} + \mathbf{v}\cdot\nabla\mathbf{v}\right) \quad (10.21)$$

where
- p is the pressure (N/m^2)
- $\mathbf{g}$ is the acceleration of gravity (m/s^2)
- x, y, z are the coordinate directions (m)
- u, v, w are the velocities of fluid in x, y, z directions, respectively (m/s)

∇ is the vector operator, a shorthand notation for the three partial differential equations of Equation 10.20 (no dimensions)
$\mathbf{v}$ is the velocity vector for infinitesimal mass (m/s)

The terms on the left side are due to pressure gradient, gravity, and viscous shear, respectively. On the right side are the inertia terms due to unsteady flow acceleration and advective acceleration, respectively. In many situations, e.g., steady and uniform pipe flow, the terms on the right side equal zero. The dimensionless numbers, e.g., Euler, Froude, Reynolds, are empirical forms of this equation for the special conditions that only one force kind is dominant in the acceleration, e.g., pressure, gravity, viscous shear. In mixing, advective acceleration is present throughout the volume due to both curvature of flow and changes in velocity from point to point. For steady-state flow the unsteady term is zero.

The Navier–Stokes equation, when combined with the conservation of mass equation, i.e., $\nabla \cdot \mathbf{v} = 0$, provides a complete mathematical description of the flow of incompressible Newtonian fluids. There are two equations and two unknowns, i.e., $\mathbf{v}$ and p (the velocity field and the pressure field), respectively. The Navier–Stokes equation, however, is a "nonlinear, second order, partial differential equation" that is not amenable to mathematical solution (except for special cases with simplifying assumptions). This was changed in the 1960s by expressing the equations in finite difference form; Fortran algorithms for numerical solutions evolved, enabled by the advent of high-speed computers. Mainframe computing started about 1958 in a few university computer centers; the latter were present in most research universities by about 1960. During the 1960s the advances were rapid in mainframe computing and by about 1970, some initial animations were done, a characteristic of CFD that developed during the 1990s.

The approach to creating a numerical solution to a differential equation is illustrated in Section 4.3 for several kinds of reactors. The same approach is followed in creating a numerical solution to the Navier–Stokes equation. To reiterate from Section 4.3, the computational scheme is to first divide the volume being modeled into infinitesimal "cells" (numbering in the thousands or millions, depending on the problem), which may be depicted as a "mesh." Next, the "boundary conditions" (velocities, pressures, state conditions at the inflow and outflow boundaries at time $t = 0$) for the problem are applied. Any variety of "boundary conditions" may be incorporated to give different solutions. In mixing, for example, the solutions may depict the effects of different shapes and sizes of impellers, different configurations of tank geometry, different points of addition of a coagulant, etc.

10.3.2.2 Computational Fluid Dynamics

By the mid-1990s, the Fortran solution algorithms for the Navier–Stokes equation had evolved into commercial computer software (e.g., Fluent™, 2001) for computer "workstations"; the application of such software was termed "computational fluid dynamics," abbreviated, "CFD." By 2000 the procedure was becoming a technology for engineering practice, albeit special expertise was required.

Once a CFD model has been defined for a given mixing situation, variables may be changed to investigate a variety of "what if?" scenarios. For example, the rotational velocity of the impeller may be changed, the flow through the reactor may be varied, the chemical reactions/kinetics may be changed, the impeller diameter may be increased or decreased, baffles may be changed in size, or the type of impeller may be changed. The same is true if the mixing technology is an "in-pipe" technology, such as a submerged jet or a grid.

For steady-state conditions, a solution may be depicted graphically; for solutions involving changes with time, animation can depict the outputs. Some of the outputs include velocity fields, pressure fields, concentration fields, etc., either steady state or in terms of time-varying animation.

Specific applications of CFD have included a variety of hydraulic problems common to water and wastewater treatment, e.g., design of disinfection basins to minimize short-circuiting, evaluation of the effect of density currents in settling basins, evaluation of inlet and outlet designs for sedimentation basins, and for mixing. Figure 10.14 illustrates the results of CFD animation of vortices generated by a radial-flow "Rushton-type" impeller system showing perspective, top, and side views, respectively, of vortices at one instant of time. The simulation was a "large eddy simulation" and used a "Fluent 5" code and involved 763,000 cells with $0.01 \leq \Delta t \leq 0.05$ s with a Sun "Ultra 60" dual processor (Bakker et al., 2000). The point in mentioning these particulars about the computer is that such simulations were formerly relegated to large mainframes, i.e., "super-computers"; the desktop machines had become powerful enough to handle the computation.

The merit of CFD is that the solutions involve few simplifications; virtually all factors, e.g., geometry, fluid properties, functional dependencies, may be incorporated in the solution. In other words, the method involves the application of science as contrasted to the traditional empirical approach. As a caveat, however, CFD solutions should be verified by field measurements or physical modeling. The application of CFD, however, is not routine and requires special expertise.

10.3.3 Similitude

The principles of similitude in mixing were developed by J.H. Rushton and his associates (Box 10.4) in the 1950s. Several kinds of similitude scale-up are pertinent to mixing, e.g., geometric, dynamic, kinematic, detention time, power per unit volume (P/V), G, and impeller tip speed (Oldshue, 1983, p. 197). The issue is that they are not simultaneously compatible with one another, and so their application is limited. Nevertheless, a knowledge of similitude principles aids design, e.g., determining volume of basin, dimensions, impeller type, impeller rotational speed, and impeller power. Also, as noted by Oldshue and others, modeling can aid in design in providing an understanding of what not to do or how to modify a proposed system.

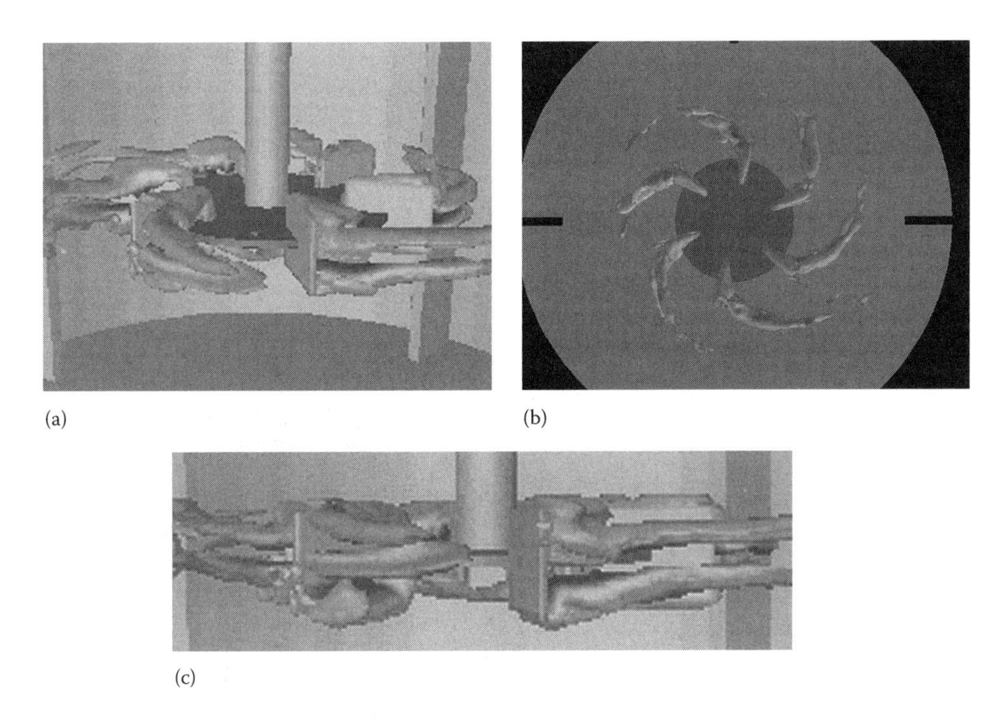

(a) (b) (c)

FIGURE 10.14 Example of mixing animation by CFD for a Rushton system. (a) Perspective. (b) Top view. (c) Side view. (From Bakker, A. et al., *The Use of Large Eddy Simulation to Study Stirred Vessel Hydrodynamics*, The On-line CFM Book, www.bakker.org/cfm, updated 2010 (downloaded as an Adobe Acrobat pdf file). With permission.)

> **BOX 10.4 RUSHTON'S ROLE IN MIXING**
>
> The idea of similitude in mixing started in 1950 when J. H. Rushton and his associates published the initial results of an experimental program conducted over the previous several years at the Mixing Equipment Company (Rushton et al., 1950a,b). The experiments numbered several thousand and involved various diameters of tanks and impellers, different impeller types, varying baffles sizes and numbers, various power levels, and different fluid properties such as viscosity and density. The plots generated were in terms of dimensionless variables, e.g., the power number vs. Reynolds number for constant geometry. This initial work was followed by a series of articles on mixing similitude and how to scale up from models and limitations of scale-up (Rushton 1951, 1952a,b, 1954; Rushton and Oldshue, 1953a,b). These works represented the first systematic studies of the effects of different variables on mixing regime and power required. They remain primary references on the topic. Later state-of-the-art books on the topic by Oldshue (1983) and Amirtharajah and Tambo (1991) lead to the conclusion that mixing design remains an "art."

10.3.3.1 Dimensionless Numbers

The dimensionless numbers involving inertia are really ratios of the two sides of Newton's second law, i.e., $F = ma$, in which F is the sum of the different forces acting on a fluid element. If all but one force can be neglected then the ratio, ma/F, is a particular dimensionless number with the designation of F, depending upon which force is dominant. The forces that pertain to mixing include viscous, pressure, and gravity; the respective dimensionless numbers are Reynolds, **R**; Euler, **E**; and Froude, **F**. These numbers are defined in Table 10.2 in terms of a rotating impeller; as a note, the power number is a form of the Euler number.

TABLE 10.2
Dimensionless Numbers in Mixing (see Appendix G)

Name	Group	Equation
Reynolds number	$\mathbf{R} = \dfrac{\rho n D^2}{\mu}$	(10.22)
Power number	$\mathbf{P} = \dfrac{P}{\rho n^3 D^5}$	(10.23)
Froude number	$\mathbf{F} = \dfrac{n^2 D}{g}$	(10.24)

where

- **R** is the Reynolds number: ratio inertia force/viscous force
- ρ is the density of fluid (kg/m^3)
- n is the rotational velocity of impeller (rev/s)
- D is the diameter of impeller (m)
- μ is the dynamic viscosity (N s/m^2)
- **P** is the power number: ratio drag force/inertia force
- P is the power applied to impeller (N m/s)
- **F** is the Froude number: inertia force/gravity force
- g is the gravitation constant (9.806650 m/s^2)

Note: In Equation 10.23 for the power number the units include N (Newtons) on top, in the power term, P, and kg (kilograms) on the bottom (in the density term). To reconcile this discrepancy, substitute the definition, $\text{N} = \text{kgm/s}^2$, in the numerator to eliminate N which results in cancellation of all units.

10.3.3.2 Variables of Impeller–Basin Mixing

Rushton et al. (1950a) grouped mixing variables as geometric, dynamic, and fluid properties. The functional relationship between dependent and independent variables was in terms of dimensionless groups, i.e.,

$$\mathbf{P} = K_{\text{d}}\mathbf{R}^m\mathbf{F}^n\left(\frac{T}{D}\right)^t\left(\frac{H}{D}\right)^h\left(\frac{C}{D}\right)^c\left(\frac{S}{D}\right)^s\left(\frac{L}{D}\right)^l \times \left(\frac{W}{D}\right)^w\left(\frac{J}{D}\right)^j\left(\frac{B}{R}\right)^b \tag{10.25}$$

where

K_{d} is the constant, disaggregated from geometric variables (dimensionless)
P is the Power number
R is the Reynolds number
F is the Froude number (pronounced, fr-ud)
D is the impeller diameter (m)
T is the tank diameter (m)
H is the depth of water in tank (m)
C is the distance of center of impeller from bottom of tank (m)
S is the pitch of impeller (marine type) (m)
L is the length of impeller blade (flat blade type) (m)
W is the height of impeller (flat blade type) (m)
J is the baffle width (m)
B is the number of blades on impeller
R is the number of baffles

The exponents are empirical.

As seen in Equation 10.25, the reference dimension for the length ratios is the impeller diameter, D, also called D(impeller). A given exponent indicates the quantitative effect of a particular ratio. Rushton et al. (1950a,b) actually obtained such exponents by means of several thousand experiments for a wide variety of impeller types and basin geometries.

10.3.3.2.1 Simplifying Special Cases

If the geometry of an experimental system remains fixed, then Equation 10.25 simplifies to

$$\mathbf{P} = K\mathbf{R}^m\mathbf{F}^n \tag{10.26}$$

The presence of a vortex indicates that gravity forces influence **P**, and thus, $n > 0$. But with baffles or if the impeller axis is off-center with the basin or is tilted at an angle to the vertical, then a vortex will not form; thus, $n = 0$ and Equation 10.26 becomes

$$\mathbf{P} = K\mathbf{R}^m \tag{10.27}$$

As evident in Equation 10.27, **P** is a function of **R** (for the condition of no vortex); Rushton (1950, p. 401) gives empirical data to define m. When $m = 0$, i.e., the viscous effect is not dominant and Equation 10.27 becomes

$$\mathbf{P} = K \tag{10.28}$$

This is the level part of the **P** vs. **R** experimental curve and occurs where inertia forces are dominant, i.e., high **R**; thus the geometry of the system controls.

10.3.3.3 Experimental Plots

Figure 10.15 is a $\mathbf{P}/\mathbf{F}^n$ vs. **R** plot from Rushton et al. (1950) for a radial-flow impeller–basin; proportions are as given in the plot. The plot as published showed several hundred experimental plotted points (Box 10.5). Flow regimes illustrated are viscous, transition, gravity, and inertial; characteristics include the following:

- For $\mathbf{R} \leq 10$, the slope of the log $\mathbf{P}/\mathbf{F}^n$ vs. log **R** curve is "-1" and viscous forces govern.
- For a basin with no baffles, a "swirl" vortex formation is observed as **R** increases due to higher rotational velocity, which is the bottom curve; Equation 10.26 applies.
- For $10 \leq \mathbf{R} \leq 300$ all data points fall on the single curve and viscous forces still dominate; Equation 10.27 applies.
- For tanks with baffles, and for $\mathbf{R} \geq 10{,}000$, the system geometry is the sole influence and Equation 10.28 applies. To illustrate, for the plot of Figure 10.15, as J/T (baffle width to tank diameter) increases, K increases (all other geometric variable are constant). By the same token, for a given baffle proportion, the K value will change with impeller shape, blade width, and blade length.

10.3.3.3.1 Power Function Curves for Different Impellers and Tanks

All impeller–tank combinations have characteristic curves similar to those illustrated in Figure 10.15. Rushton et al. (1950, p. 472) have provided **P** and **R** data for $\mathbf{R} = 5$, 200, and 100,000, which permit an approximate sketch of such characteristic curves for a given system. Table 10.3 provides examples of such data for four impeller shapes with geometry ratios as given in the columns. The data show, in particular, the effects of the J/T ratio and impeller type on **P** at $\mathbf{R} \geq 10^5$.

10.3.3.4 Scale-Up by Fluid Similitude

To utilize one of the $\mathbf{P}/\mathbf{F}^n$ vs. **R** curves generated by Rushton et al. (1950a,b), e.g., Figure 10.15 or Table 10.3, the system being considered must be geometrically similar to the one for which the respective curve was generated, i.e., a "Rushton" impeller–basin. An approach is to find a point on the level part of the curve, i.e., **P**, at $\mathbf{R} \geq 10^4$, which yields reasonable

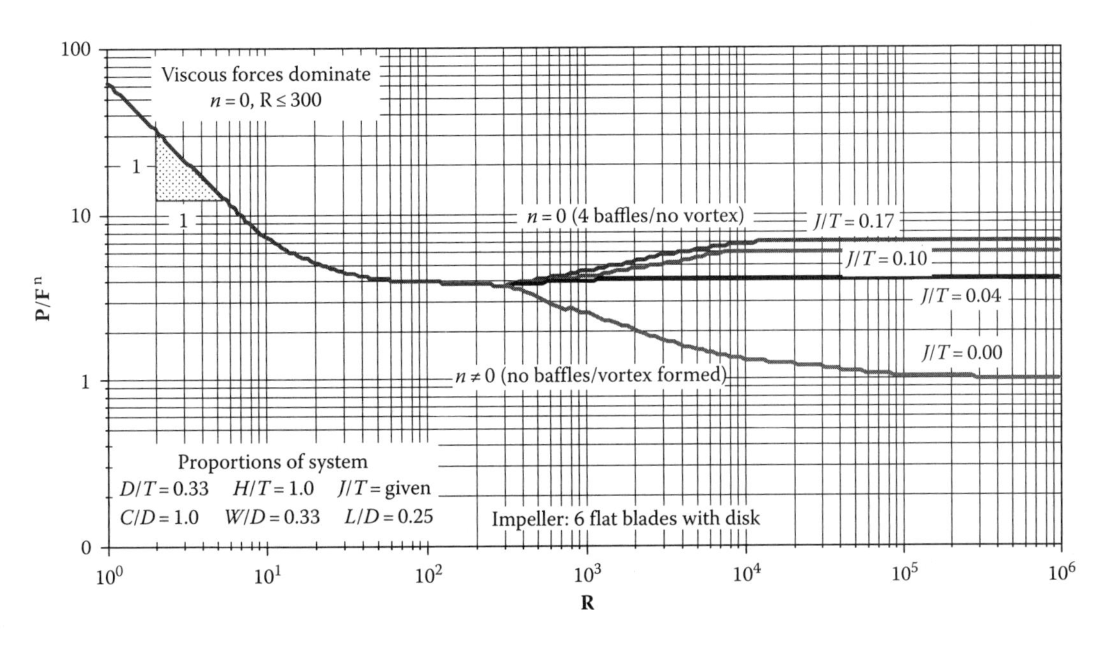

FIGURE 10.15 Characteristic plot for power function for a given radial-flow impeller and geometric proportions as given (Rushton et al. 1950, p. 468); for the "Rushton" system, $J/D = 0.10$.

BOX 10.5 RUSHTON'S EXPERIMENTS

The work of Rushton et al. (1950a,b) was based on using impellers and tanks of different sizes to generate a series of **P** vs. **R** curves, e.g., $76 \leq D$(impeller) ≤ 1220 mm ($3 \leq D \leq 48$ in.) and $216 \leq T \leq 2438$ mm ($8.5 \leq T \leq 96$ in.). They also used different fluids with viscosities ranging $0.001 \leq \mu \leq 40$ N s/m^2 ($1 \leq \mu \leq 40{,}000$ cp), which included water, kerosene–carbon tetrachloride mixtures, lubricating oil, linseed oil, corn syrup solutions. Densities varied $955 \leq \rho \leq 1442$ kg/m^3 ($59.6 \leq r \leq 90$ lb/ft^3). For reference, μ(water, 20°C) ≈ 0.001 N s/m$^2 = 0.010$ poises $= 1$ cp and ρ(water, 20°C) $= 998.2$ kg/m^3; also for reference, the viscosity of carbon tetrachloride is about the same as water.

The studies by Rushton and his associates were classic and have remained useful for reference (see, for example, McCabe et al., 1993) and as a basis for designing modeling studies. Reference to the "Rushton" impeller (which means the Rushton impeller–basin system) is common in the literature on impeller mixing. The system is described in the glossary.

operating values, e.g., P, P/V (or G); this involves changing **R** parametrically. An algorithm is enumerated as follows:

1. Scale up geometrically in terms of θ, selecting a value based on practice, e.g., 10 s, and calculate the dimensions of the system from the relation, $V(\text{basin}) = Q\theta$.
2. Select values of **R** (in the turbulent range), parametrically, and calculate n values.
3. From **P** for the system selected, calculate P.
4. Calculate P/V (or G if preferred) for different **R** (or n).
5. Select a P/V (or G) based on practice.
6. For the P/V (or G) selected, P is calculated and n is unique (calculated from mathematical definitions of **P** and **R**, respectively).

Example 10.2 illustrates the algorithm, which is applied in Table CD10.4; for each row in the table a different **R** was selected with the calculated values for n, P, P/V, and G shown in the different columns. As indicated, the row is selected that meets the criterion set for P/V or G. The approach used in the spreadsheet was to maintain the geometry of a "Rushton basin," with $\theta = 10$ s, vary **R**, then look at the effect on G, to give $\mathbf{R} = 8 \cdot 10^5 \rightarrow G \approx 1000$ s^{-1}. The associated impeller speed and power are $n = 138$ rpm, $P = 5.2$ kW (7 hp), and $P/V \approx 1.12$ kW/m^3. A test of the results is to confirm experimentally that the "$C(t)/C_o$ vs. t/θ" relationship yields $C(t)/C_o \approx 0.99$ at $t \leq \theta$. Table CD10.4 shows that n and P are highly sensitive to **R**.

Example 10.2 Imposing Similitude for Design

Given

A Rushton impeller–basin (six blades) is to be designed based upon the characteristic **P** vs. **R** curve of Figure 10.15. The detention time is $\theta = 10$ s and $Q_p = 0.438$ m^3/s (10 mgd). Assume operation is in the turbulent range, i.e., $\mathbf{R} \geq 10^4$.

TABLE 10.3
Characterizing Data for *P* vs. *R* Curves

									P		
Impeller	***D/T***	***H/T***	***C/D***	***J/T***	***S/D***	**Blades**	***L/D***	***R*** = **5**	***R*** = **200**	***R*** = $\mathbf{10^5}$
Marine—3 blade	0.33	1.0	1.0	0	1.02	3		8.3	0.75	0.22
	0.33	1.0	1.0	0.083	1.03	3		8.3	0.77	0.26
Propeller—off center	0.33	1.0	1.0	0	1.03	3		8.3	0.77	0.32
Flat blade w/disk	0.33	1.0	1.0	0		6	4.0	14.2	3.7	1.15
	0.33	1.0	1.0	0.083		4	4.0	14.2	3.7	4.3
	0.33	1.0	1.0	0.083		6	4.0	14.2	3.7	6.0
	0.33	1.0	1.0	0.083		12	4.0	14.2	3.7	9.9
Paddle—2 blades	0.33	1.0	1.0	0.083		2	3	7.3	1.60	1.63

Source: Rushton, J.H. et al., *Chem. Eng. Prog*., 46 (9), 472, 1950b.
S/D is the pitch of the impeller and is the ratio of linear advance of a pseudo stream of water to the diameter.
The column under "$\mathbf{R} = 10^5$" gives *K* values for Equation 10.28, i.e., $\mathbf{P} = K$.

Required

Determine the basin size, the impeller diameter, *D*(impeller), rotational velocity, *n*, and power required by the impeller.

Solution

1. From Figure 10.15, in the turbulent range, $\mathbf{P} = K \approx 6.0$ for six blades and $J/T = 0.10$.
2. Set up a spreadsheet as shown in Table CD10.4, with groups of columns for fluid properties, basin sizing by detention time, impeller speed determined by **R**, power from the equation for **P**, and empirical guidelines, *P*/*V*, *P*/*Q*, and *G*.
3. The application of the similitude equations, Equations 10.22 and 10.23, yield *n* and *P*, respectively and further calculations yield *P*/*V*, *P*/*Q*, and *G*. As a first try, row 1, shows that *n*, *P*, ..., *G*, etc., are unreasonably low. If **R** is selected as an independent variable, and changed parametrically in successive rows (all other variables constant), the effect on *n*, *P*, ..., *G* may be seen.
4. Using *G* as a criterion, select $G \approx 1000\ s^{-1}$. The row in Table CD10.4 that meets $G \approx 1000\ s^{-1}$ is shown shaded. The associated design values are $V = 4.38\ m^3$, *D*(impeller) = 591 mm, *n* = 138 rpm, *P* = 5.2 kW = 7 hp (*V* and *D*(impeller) are the same for all rows).

Discussion

1. Looking at Table CD10.4, the dependent variables, *n*, *P*, ..., *G*, are highly sensitive to **R**. Therefore, as seen, for $G \approx 1000\ s^{-1}$, the associated $\mathbf{R} \approx 8 \cdot 10^5$ for $\theta = 10$ s.
2. The application of the similitude equations is easier said than done. The dependent variables in the similitude equations are highly sensitive to the variables *n* and *D*(impeller); therefore, as seen in Table CD10.4, unreasonably low or high values of *P*, ..., and *G* will result if **R** is not selected in an appropriate range.
3. Another point, critical in applying similitude, is that any model used would have unreasonable high values of *P*, ..., *G*, if the same value of **R** is used. This point is a limitation in the application of similitude principles. Thus, the similitude principle cannot be imposed between model and prototype using only water as a common fluid; the model must have a different fluid if the $\mathbf{R} = 8 \cdot 10^5$ is to be imposed on the model.
4. Table CD10.4 shows that as **R** is varied along the **P** vs. **R** curve for $\mathbf{R} \geq 10^4$ a constant **P** is imposed as illustrated in Figure 10.15. Between model and prototype, the relationship "slides" along the horizontal **P** vs. **R** line but cannot be at the same "point" (at least not with reasonable "*n*" and *P* values). In other words, only two of the three similitude conditions may be achieved, e.g., geometric and dynamic, i.e., **P**, but not, at the same time, kinematic, i.e., **R**.
5. The merit in the application of the Rushton modeling results, i.e., Figure 10.15, is that the impeller power and rotational velocity may be estimated; note, however, the paragraph following that abstracts comments from Oldshue (1983, p. 197).

10.3.3.5 Scale-Up Dilemma

A number of mixing parameters may seem appropriate to be maintained constant for scale-up. A list may include those dimensionless ones delineated by Rushton et al. (1950a,b), i.e., geometric, kinematic, dynamic similarity (scale, **R**, and **P**, respectively). Others may include detention time, θ; *P*/*V*; *P*/*Q*; flow number, ***Q***; blend number, nt_r; tip speed, *nD*; etc. (see Oldshue, 1983, pp. 194–197). In practice empirical parameters are adopted for sizing a basin, e.g., *q* and *G*, or

TABLE CD10.4
Calculations of *n*, *P*, *P*/*V*, *P*/*Q*, and *G* based on R and P

1 Fluid Properties				2 Basin Sizing						3 Impeller Speed			4 Power to Impeller			5 Empirical Criteria		
T_m (°C)	$\mu(water)_m$ (N s/m^2)	ρ_m (kg/m^3)	υ_m (m^2/s)	θ_p (s)	Q_p (mgd)	(m^3/s)	V_p (m^3)	T_p (m)	D_p (m)	R	n (rev/s)	p (rpm)	P	P (watts)	p (hp)	P/V (kW/m^3)	P/Q (hp/mgd)	G (s^{-1})
20	0.000998294	997.548	1.001E−06	10	10	0.438	4.381	1.772	0.591	10000	0.029	1.7	6.0	0.01	0.00	0.000	0.000	1.52
20	0.000998294	997.548	1.001E−06	10	10	0.438	4.381	1.772	0.591	50000	0.143	8.6	6.0	1.27	0.00	0.000	0.000	17
20	0.000998294	997.548	1.001E−06	10	10	0.438	4.381	1.772	0.591	100000	0.287	17	6.0	10.15	0.01	0.002	0.001	48
20	0.000998294	997.548	1.001E−06	10	10	0.438	4.381	1.772	0.591	500000	1.433	86	6.0	1269	1.70	0.290	0.170	539
20	0.000998294	997.548	1.001E−06	10	10	0.438	4.381	1.772	0.591	700000	2.007	120	6.0	3483	5	0.795	0.467	892
20	0.000998294	997.548	1.001E−06	10	10	0.438	4.381	1.772	0.591	800000	2.294	138	6.0	5199	7	1.187	0.697	1090
20	0.000998294	997.548	1.001E−06	10	10	0.438	4.381	1.772	0.591	1000000	2.867	172	6.0	10153	14	2.318	1.361	1524

Note: Impeller: flat blade w/disk—6 blades.

P/V are common. The dilemma is that similitude for all of the dimensionless parameters cannot occur simultaneously. Certain ones may be selected for scale-up, i.e., remain constant; all or most of the others will change. This incongruity may be seen by substituting algebraic expressions for one parameter into another, as illustrated for the power number in the following paragraph.

10.3.3.5.1 *Inherent Incongruity of Scale-Up—Illustration*

Scale-up from model to prototype requires geometric, kinematic, and dynamic similarity. Geometric similarity may be obtained by setting $\theta_m = \theta_p$, i.e., with $V_m/Q_m = V_p/Q_p$. Kinematic similarity is based on setting $\mathbf{R}_m = \mathbf{R}_p$, and dynamic similarity is obtained by setting $\mathbf{P}_m = \mathbf{P}_p$. respectively. The rule is, however, that two parameters may be maintained constant between model and prototype, but not three. To illustrate, if L_m/L_p and $\mathbf{R}$ are to be maintained constant, substitute $n = \mathbf{R}\mu/(\rho D^2)$ for the "n" in the power number, $\mathbf{P}$, which gives

$$\frac{\mathbf{P}_p}{\mathbf{P}_m} = \left(\frac{r_m}{r_p}\right)^2 \cdot \left(\frac{\mu_p}{\mu_m}\right)^3 \cdot \left(\frac{\mathbf{D}_m}{\mathbf{D}_p}\right) \quad (10.62)$$

In other words, if $r_m = r_p$ and if $\mu_p = \mu_m$, then it is evident that for the power ratio, $\mathbf{P}_p/\mathbf{P}_m \neq \mathbf{D}_m/\mathbf{D}_p$. The only means to achieve dynamic similarity is for fluid properties, i.e., r and μ, to vary; thus, the scale-up of a model using water is not feasible. In other words, the selection of parameters for scale-up and how to scale up remains an art (see, for example, Oldshue, 1983, p. 197).

10.3.3.5.2 *Further Notes on Scale-Up—From Oldshue (1983, p. 197)*

The most common guideline for scale-up is to maintain geometric similarity. A major point is that trying to maintain constant selected parameters may add false confidence to a scale-up and could lead to a failure of the mixer to achieve a required process result. Some may avoid scale-up as being too risky and the process result too uncertain; a more useful role of models, as opposed to direct scale-up is to study, in an empirical fashion, the effect of each of the parameters on the process being considered.

10.3.4 Injection of Coagulant Chemicals

Dispersing a coagulant solution, e.g., a neat solution of liquid alum, into a raw-water flow is a major task in mixing for water treatment, often overlooked. The two issues are (1) the disparity of flows, and (2) the reaction time for Al^{3+} (or Fe^{3+}) hydrolysis to occur. To utilize alum (or ferric) for charge neutralization, the dispersion of coagulants throughout the raw-water flow must occur within <1 s and ideally <0.1 s (Clear Corp., 2001). Jet mixers, in-line mixers, or static mixers are the technologies of choice to approach resolving these two issues.

10.3.4.1 Disparity of Flows

As noted, the problem in mixing of a coagulant flow with the raw-water flow is that a large disparity exists between the two flows. For example, suppose $Q(\text{raw water}) = 3785$ m^3/day (1.0 mgd) $= 0.044$ $m^3/s = 44{,}000$ mL/s; let alum concentration, $C(\text{alum}) = 10$ mg/L, and from Appendix F, $C(\text{neat alum}) = 647{,}000$ mg/L. Then, by mass balance as illustrated by Example 10.3, $Q(\text{alum, neat}) = 0.68$ mL/s. As seen by calculation, the ratio of the two flows is $Q(\text{raw water})/Q(\text{alum, neat}) = 44{,}000$ mL/s/0.68 mL/s $= 65{,}000/1$. Kawamura (2000, p. 307) gives 50,000 as a nominal value and mentions that many design engineers are not cognizant of this high ratio. That these two flows must be mixed is the major challenge.

10.3.4.2 Advection of Neat Alum

The approach in dispersing one milliliter neat alum solution in 65,000 mL of raw water is twofold: (1) the entire flow of raw water must enter turbulence zones created for mixing, and (2) the flow of alum must be distributed uniformly by advection into the same zones of turbulence to comingle with the raw water. The methods of creating turbulence zones within a pipe were reviewed in Section 10.3.1.2, and involved submerged jets or disturbances such as an orifice or a pipe constriction (wake turbulence). To disperse a neat alum solution by advection, a ring manifold within the raw-water pipe discharging the alum under high pressure, through six orifices, was used by Vrale and Jorden (1972). Another method is to feed the alum just before the nozzle of an orifice that creates one or more high-velocity submerged jets.

10.3.4.2.1 *Injection of Neat Alum around an Impeller*

The velocity gradients are steepest at the outflow (i.e., at the tip) from a radial-flow impeller, and are less steep for the axial-flow impeller. Therefore, the radial-flow impeller (i.e., as opposed to a marine impeller) is favored for blending, which is the case for rapid mix in coagulation, or for dispersing a polymer emulsion. A side stream of neat alum solution will mix most rapidly by injection within this high-shear zone, i.e., at the tip of the impeller, as illustrated in Figure 10.3c. Injection of alum just upstream of the shear zones of two or more "in-line" impellers would be more effective than a single injection point using one impeller. In other words, the idea is to try to distribute the alum over the raw-water flow to the extent feasible.

Example 10.3 Mass Flow of Neat Alum

Given

Assume that the concentration of alum required for coagulation is 10 mg $Al_2(SO_4)_3 \cdot 14H_2O$/L raw water. Let $Q(\text{raw water}) = 3785$ m^3/day (1.0 mgd). Also, let $d(\text{pipe}) = 500$ mm; therefore, $v(\text{pipe}) = 5.1$ m/s. The alum is to be injected into the core zone of a jet mixer that disperses the alum into raw-water flow in the pipe.

Required

1. Determine the mass flow of neat alum required.
2. Determine the volume flow of neat alum.

Solution

1. The mass flow of alum, *J*(alum), required is concentration in the raw water times the flow of raw water, i.e.,

$$\begin{aligned} J(\text{alum}) &= C(\text{alum}) \cdot Q(\text{raw water}) \\ &= 10 \text{ mg } Al_2(SO_4)_3 \cdot 14H_2O/L \\ &\quad \cdot 3785 \text{ m}^3/\text{day} \cdot (1000 \text{ L/m}^3) \\ &= 37{,}850 \text{ g/day} \\ &= 37.85 \text{ kg/day} \end{aligned}$$

2. First, obtain the concentration of neat alum,

$$C(\text{neat alum}) = 647 \text{ g } Al_2(SO_4)_3 \cdot 14H_2O/L \text{ (Figure F.4).}$$

Next, determine the volume flow of neat alum,

$$\begin{aligned} J(\text{alum}) &= C(\text{neat alum}) \cdot Q(\text{neat alum}) \\ 37{,}850 \text{ g/day} &= 647 \text{ g } Al_2(SO_4)_3 \cdot 14H_2O/L \cdot Q(\text{neat alum}) \\ Q(\text{neat alum}) &= 58.5 \text{ L/day} \\ &= 40.6 \text{ mL/min} \\ &= 0.68 \text{ mL/s} \end{aligned}$$

Discussion

1. *Pumping alum*: The flow of neat alum is done by a positive displacement metering pump. The pump should be fitted with a "snubber" to mitigate the pulse flow characteristic of a positive displacement pump. An alternative system should be in place to account for a possible failure of the pump and/or clogging of the line. The alum feed line should be set up for cleaning while the other is in use, e.g., by hot water.
2. *Mixing ratio*: The raw-water flow is, $Q(\text{raw water}) = 3785 \text{ m}^3/\text{day} = 0.0438 \text{ m}^3/\text{s} = 43{,}800 \text{ mL/s}$. The ratio: $Q(\text{raw water})/Q(\text{neat alum}) = 43{,}800 \text{ mL/s}/0.68 \text{ mL/s} = 64{,}423 \approx 64{,}000$ mL raw water/mL neat alum.
3. *Alternative mixing ratio*: An approach to reduce the mixing ratio is to dilute the neat alum solution just prior to injection, e.g., such that $C(\text{alum-solution}) \approx 65 \text{ g } Al_2(SO_4)_3 \cdot 14H_2O/L$ and thus, $Q(\text{alum solution}) \approx 6.4$ mL/s. The ratio is still very high. The use of neat alum is recommended.

10.4 MIXING TECHNOLOGIES

A variety of mixing technologies have been developed; three major categories are (1) impeller–tank systems, (2) jet mixers, (3) static mixers. Table 10.5 lists some of the respective technologies; any of those listed may have a "fit" in practice, depending on the "context."

TABLE 10.5
Mixing Technologies

	Category of Mixer		
	Impeller	**Jet**	**Static**
Technology	Open basin	Nozzles	Orifice
	Draft tube in basin	Orifice	Pipe constriction
	In-line mixer	Hydraulic jump	Air bubbles
	Pump	Parshall flume	Bifurcation vanes
		Baffles—over-and-under	Elbows in sequence
		Baffles—end-around	Flow obstruction
		Weir	

Example 10.4 Selection of Mixing Technology

Given

The engineer for a small community, e.g., 2000 persons, must select a rapid mix for alum coagulation.

Required

Suggest a "passive" technology.

Solution

Table 10.5 lists a selection of technologies. If a centrifugal pump is used just prior to the plant, the coagulant could be metered-in on the suction side. If the plant is located lower than a reservoir, a nozzle or orifice could work. If water pressure is not available, a weir, a Parshall flume, or end-around baffles would be most reliable.

Discussion

Capital cost, reliability, minimal operation skills, and low operating cost are factors that relate to selection. The alum feed is subject to clogging and provision should be made for cleaning; an alternate feed should be provided.

10.4.1 Impeller Mixing

Impeller mixing has been the most frequent technology used in water treatment. This section reviews theory for both the traditional "back-mix" reactor and the later "in-line" mixers.

10.4.1.1 Reactors—Back-Mix and In-Line

Figure 10.16a shows an impeller in a tank, commonly called a "back-mix reactor" (Levenspiel, 1972). As seen, the flow recirculates and thus may make multiple passes through the high-shear zone (i.e., the mixing zone); the number of passes depends upon the impeller pumping rate relative to the raw-water flow. Figure 10.16b shows an "in-line" mixer, which is

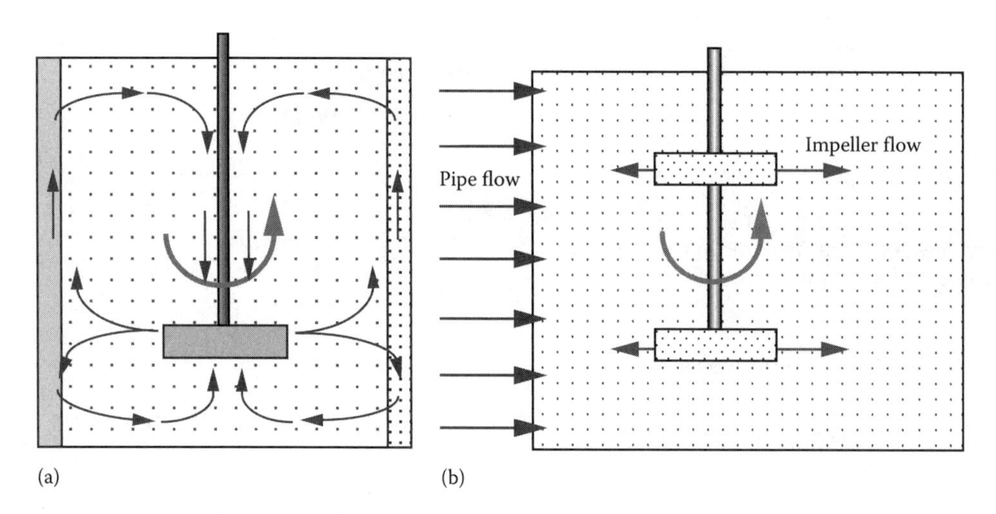

FIGURE 10.16 Two modes of impeller mixing. (a) Back-mix reactor. (b) In-line mixing.

characterized by once-through advective flow. The two impellers illustrated can capture a larger proportion of the flow than one impeller. A number of configurations have been used in practice, but the key idea is to force a large proportion of the flow through the high-turbulence zones.

Figure 10.17 shows photographs of each, i.e., Figure 10.17a is a back-mix reactor, and Figure 10.17b is an in-line mixer system. Within these two categories, a variety of configurations are found in practice with variations based on vessel shape, entry and exit locations, impeller type, number of impellers on the shaft, impeller location, etc.

10.4.1.1.1 Back-Mix Reactors

A "back-mix" reactor is characterized by the return of circulated water through the impeller multiple times, which requires

$$\frac{Q(\text{impeller})}{Q(\text{raw water})} \gg 1 \qquad (10.29)$$

where

Q(raw water) is the flow of raw water (m^3/s)
Q(impeller) is the pseudo flow of water pumped by impeller, i.e., is not measurable (m^3/s)

By the continuity principle,

$$\begin{aligned} V(\text{basin}) &= Q(\text{impeller}) \cdot \theta(\text{impeller}) \\ &= Q(\text{raw water}) \cdot \theta(\text{raw water}) \end{aligned} \qquad (10.30)$$

Rearranging,

$$\frac{Q(\text{impeller})}{Q(\text{raw water})} = \frac{\theta(\text{raw water})}{\theta(\text{impeller})} \qquad (10.31)$$

where

θ(raw water) = detention time of raw water in basin (s)
q(impeller) = pseudo average time for one circulation of water due to impeller pumping, i.e., is not measurable (s)

The number of passes through the impeller is

$$n(\text{passes-through-impeller}) = \frac{Q(\text{impeller})}{Q(\text{raw water})} \qquad (10.32)$$

If, for example, n(passes-through-impeller) = 5, fraction blending ≈ 0.99 (McCabe et al., 1993, p. 258). The corresponding "pseudo" detention time for the pumped circulating flow is $\theta(\text{impeller}) \approx (1/5) \cdot \theta(\text{raw water})$; at the same time n(passes-through-impeller) is a "pseudo" value.

10.4.1.2 Circulation Criterion for 0.99 Blending in a Back-Mix Reactor

As noted, if the number of circulations, i.e., $n(\text{circulations}) \geq 5$ per raw-water detention time, then the blend fraction is ≥ 0.99 (McCabe et al., 1993, p. 258). Mathematically, the criterion is expressed,

$$\frac{C(t_{5R})}{C_o} \geq 0.99 \qquad (10.33)$$

where

$C(t_{5R})$ is the effluent concentration of a substance, e.g., alum, at time, t_{5R} (s)
t_{5R} is the elapsed time since the start of a continuous flow of substance to be mixed that results in five circulations due to impeller pumping in not more than one raw-water detention time (s)
C_o is the theoretical calculated concentration of a substance, e.g., alum, which occurs when $t/q \gg 1.0$, where t is elapsed time since start of substance flow (kg/m^3)

The value of C_o is calculated by a mass balance for the alum (or a tracer, e.g., a dye, conductivity, Cl^-), i.e.,

$$Q(\text{raw water}) \cdot C_o = Q(\text{neat alum}) \cdot C(\text{neat alum}) \qquad (10.34)$$

(a)

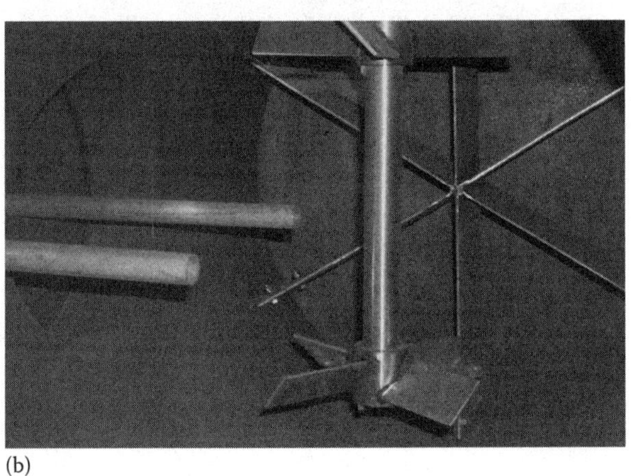

(b)

FIGURE 10.17 Two kinds of impeller mixer systems. (Courtesy of City of Fort Collins Utilities, Fort Collins, CO, 2010). (a) Back-mix reactor. (b) In-line mixer.

where

Q(neat alum) is the flow of neat alum into reactor (m^3/s)

C(neat alum) is the concentration of neat alum solution as provided by manufacturer (647 kg neat alum/m^3 neat alum solution)

10.4.1.2.1 Blend Number

The product "n(impeller) $\cdot$ t_{5R}" is a constant for a given system (McCabe et al., 1993, p. 258), i.e.,

$$n(\text{impeller}) \cdot t_{5R} = K(0.99 \text{ blend}) \tag{10.35}$$

where

n(impeller) is the rotational velocity of impeller (rev/s)

$t_{5R} \equiv$ time for five circulations through impeller (s) $= 5 \cdot \theta(\text{impeller})$

K(0.99 blend) is the mixing constant for 0.99 fraction blend valid for turbulent zone (dimensionless)

The "blend number," i.e., "n(impeller) $\cdot$ t_{5R}," is designated here as K(0.99 blend) and is constant in the turbulent range, i.e., high **R**. The blend numbers for four systems are given in Table 10.6 and illustrate range in values, e.g., $36 \leq K$(0.99 blend, turbines) ≥ 60; and $120 \leq K$(0.99 blend, propellers) ≥ 550. The utility of the blend number is that from it the rotational velocity of the impeller, n(impeller), may be estimated.

To illustrate how the blend number may be used, assume the system is a turbine in a baffled tank, in which K(0.99 blend) as in Table 10.6, and assume that θ(raw water) = 10 s.

First, from Equation 10.35,

$$n(\text{impeller}) \cdot t_{5R} = K(0.99 \text{ blend})$$

and since we let

$$t_{5R} = 5 \cdot \theta(\text{impeller}) = \theta(\text{raw water}),$$

then,

$$n(\text{impeller}) \cdot \theta(\text{raw water}) = K(0.99 \text{ blend})$$

Next, substitute numerical values, i.e., θ(raw water) = 10 s, and from Table 10.6, for a turbine in a baffled, tank, K(0.99 blend) = 36, to give

$$n(\text{impeller}) \cdot (10\sigma) = 36$$

Thus, n(impeller) = 3.6 rev/s.

TABLE 10.6
Values of "Blend Number," K(0.99 Blend)

Impeller Type	Conditions	D/T[b]	H/T[b]	**R**[c]	K(0.99 Blend)
Propeller		1/6	1.0	≥20,000	550
Propeller		1/3	1.0	≥20,000	120
Turbine[a]		1/3	1.0	≥4,000	60
Turbine	Baffled	1/3	1.0	≥2,000	36

Source: Adapted from McCabe, W.L. et al., *Unit Operations of Chemical Engineering*, 5th edn., McGraw-Hill, New York, p. 259, 1993.

[a] A turbine is a radial-flow impeller.

[b] D, diameter of impeller; T, diameter of tank; H, depth of water; **R**, Reynolds number.

[c] Threshold value of Reynolds number at which K(0.99 blend) becomes constant.

10.4.1.2.2 Derivation of Blend Number

While the "blend number" appears to be empirical, a rationale may be seen, as given in the account following.

From Equation 10.35, i.e., $t_{5R} \equiv 5 \cdot \theta(\text{impeller})$, substitute $q(\text{impeller}) = V(\text{reactor})/Q(\text{impeller})$,

where $V(\text{reactor}) = \pi T^2 H/4$ and $Q(\text{impeller}) = QnD(\text{impeller})^3$, then collect the numerical terms, let $H = T$, and move n to the left side, to give, $n(\text{impeller}) \cdot t_{5R} = (35.34/\boldsymbol{Q})$. Then, from Table 10.6,

$K(0.99 \text{ blend, Rushton system}) \approx 36$,

which gives $\boldsymbol{Q} \approx 0.98$, which is larger than the accepted range $0.54 \leq \boldsymbol{Q} \leq 0.88$ for the Rushton 6-blade impeller. While $\boldsymbol{Q} \approx 0.98$ is higher than "hoped-for," the rationale to obtain it is logical and the discrepancy is not large.

10.4.1.3 Time Ratio, *t*/*Q*(reactor), to Attain 0.99 Blending—Experimental Procedure (a)

As noted, the ratio $C(t)/C_o \approx 0.99$ is a criterion that may be adopted to define when adequate blending occurs. For a given rotational speed, i.e., $w(\text{impeller})$, the curve $C(t)/C_o$ vs. t/q (raw water) may be defined experimentally by adding a tracer as a step function and then sampling the reactor effluent. If the rotational speed is changed, another curve may be generated. A series of such curves may be generated for rotational speeds between the limits, $0 \leq w(\text{impeller}) \leq w(\text{max})$. A "direct-current" motor is required in order to vary $w(\text{impeller})$. The experimental method may be used, however, to evaluate any kind of mixing system, e.g., static mixer, back-mix reactor, in-line mixer, etc. For the static mixer, the question is to determine the number of elements for 0.99 blend.

10.4.1.4 Impeller Speed, *w*(impeller), to Attain 0.99 Blending—Experimental Procedure (b)

For each $w(\text{impeller})$ value, the particular value of t/q(raw water) that occurs when $C(t)/C_o \approx 0.99$ permits a second plot, i.e., $[t/q(\text{raw water})]_{C(t)C_o \approx 0.99}$ vs. $w(\text{impeller})$. The second plot should have enough points to define the curve, e.g., for $[t/q(\text{raw water})]_{C(t)C_o \approx 0.99} \gg 1.0$. Entering the plot at $[t/q(\text{raw water})]_{C(t)C_o \approx 0.99} \approx 1.0$ yields the sought value of impeller speed, i.e., $[w(\text{impeller})]_{C(t)C_o \approx 0.99}$. In other words, this is the impeller speed that results in 0.99-fraction blend for one raw-water detention time.

10.4.1.5 Complete-Mix Reactors

The defining characteristic of a "complete-mix" reactor is that upon the addition of a finite mass of a substance "A" to the reactor, a homogeneous concentration results instantaneously. For a real mixer, such an instantaneous distribution is hypothetical and may be approached but never attained.

10.4.1.5.1 Mathematics of Complete-Mix

The mathematics of a complete-mix reactor is reviewed in Section 4.3.3, e.g., as described by Equations 4.24 through 4.28. Equation 4.28 is repeated as Equation 10.36, i.e.,

$$\left[\frac{C_{\text{in}} - C}{C_{\text{in}} - C_{\text{i}}}\right] = e^{[-t/\theta]} \tag{10.36}$$

where

C_{in} is the concentration of substance A in flow entering reactor at any time, $t \geq 0$ (kg/m^3)

C_{i} is the initial concentration of substance A in reactor, i.e., at $t = 0$ (kg/m^3)

C is the concentration of substance A in reactor at any time, t (kg/m^3)

t is the elapsed time after start of substance A in flow to reactor (s)

$\theta(\text{reactor})$ is the detention time of reactor, i.e., $\theta(\text{reactor}) = Q/V$ (s)

To illustrate, the behavior of Equation 10.36, let $C_{\text{in}} = 0$ at $t > 0$, which gives $C/C_{\text{i}} = e^{(-t/\theta(\text{reactor}))}$. For this case, let pure water enter the reactor at $t > 0$; thus, at $t = 0$, $C = C_{\text{i}}$. Table CD10.7 shows calculated values C/C_{i} vs. (t/θ) and Figure CD10.18 is a plot linked to the calculated relationship. Other columns of Table CD10.7 are for illustration and show as a function of t: concentration of A, residual mass of A, and mass of A that has exited; the initial concentration, C_{i}, the detention time, $\theta(\text{reactor})$, and the flow, Q, were assumed, as given in the spreadsheet. The calculation formulae are shown at the bottom of each respective column. Note that by definition, for a "complete-mix" reactor the concentration does not vary spatially at any given instant, i.e., it is

TABLE CD10.7
Complete-Mix Reactor Calculations of Residual Concentrations and Mass Remaining

$\theta = 10$ s $\quad V = 1.000$ m^3
$C_{\text{i}} = 1000$ kg/m^3 $\quad Q = 0.100$ m^3/s

t/q	C/C_{i}	t (s)	C (kg/m^3)	Mass R (kg)	Mass Exited (kg)
0.0	1.00	0.00	1000	1000	0.00
0.1	0.90	1.00	905	905	95
0.2	0.82	2.00	819	819	181
0.3	0.74	3.00	741	741	259
0.4	0.67	4.00	670	670	330
0.5	0.61	5.00	607	607	393
0.6	0.55	6.00	549	549	451
0.7	0.50	7.00	497	497	503
0.8	0.45	8.00	449	449	551
0.9	0.41	9.00	407	407	593
1.0	0.37	10.00	368	368	632
1.5	0.22	15.00	223	223	777
2.0	0.14	20.00	135	135	865
3.0	0.05	30.00	50	50	950
4.0	0.02	40.00	18	18	982
5.0	0.01	50.00	7	7	993
	$=\exp(-t/q)$	$=(t/q)\cdot q$	$=(C/C_i)\cdot C_i$	$=C\cdot V(\text{reactor})$	$=[C_i - C]\cdot V(\text{reactor})$

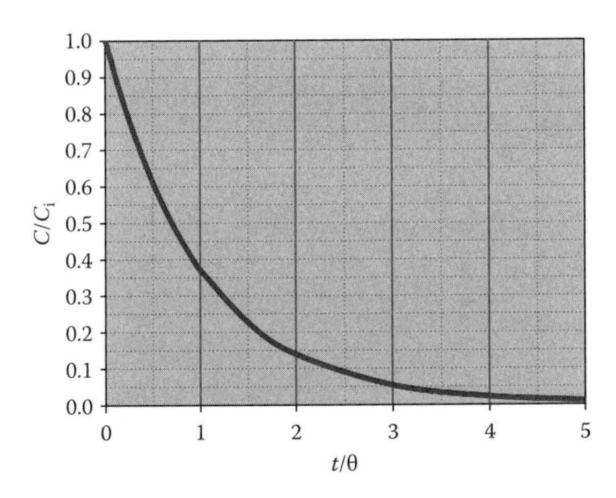

FIGURE CD10.18 Residual concentration fraction as a function of t/θ.

"homogeneous"; thus the effluent concentration is the same as that in the tank.

For different initial condition, e.g., the case in which a tracer, e.g., a salt, replaces a water of concentration, C_i, entering the reactor at time, $t < 0$; let the concentration of the salt solution entering the reactor at time, $t \geq 0$ be C_{in}, and Equation 10.36 applies directly. To illustrate, at time, $t = 0$, $C = C_i$, and therefore at time, $t = 0$, $[(C_{in} - C_i)/C_{in} - C_i)] = e^{[-0/\theta]} = 1.0$; for $t > 0$, the function declines exponentially with time toward zero, i.e., $(C_{in} - C) \rightarrow 0$ as $t \gg 0$, as in Figure CD10.18.

Example 10.5 Residence Times in a Complete-Mix Reactor

Given

Calculated residence times in the reactor as given by Equation 10.36, Table CD10.7, and Figure CD10.18.

Required

Discuss the characteristic residence times for a complete-mix reactor.

Solution

From Figure CD10.18, at $t/\theta = 1.0$, 0.63 fraction of the flow volume coming in during the period, θ, will have passed through the basin. At the same time 0.37 fraction of that same volume remains in the basin. Other residence times are if $t/\theta \approx 0.2$, only 0.18 fraction will have passed through the basin and 0.82 fraction remains. On the other hand, if $t/q \approx 3$, then 0.95 fraction will have passed through the basin and only 0.05 fraction remains.

Discussion

An inherent limitation of the back-mix reactor is that, as a complete-mix basin, the flow has various residence times. A "way out" of this limitation, not perfect, is to achieve the required mixing in time, $t \ll \theta$. This can be done by providing sufficient power to the impeller such that the five passes through the high-shear zone occur at $t/\theta \ll 1.0$. For example, if $t/\theta \approx 0.2$, only 0.18 fraction will have passed through the basin. In other words, if the pumping rate is high enough, such that the five passes occur within $t/\theta \approx 0.2$, then a large portion of the reactor contents, i.e., 0.82 fraction, will have received the required mixing. This does not mean that 0.18 fraction will not be treated; some lesser fraction will have inadequate treatment. A portion of the 0.18 fraction will have made passes through the high-shear zone, the amount being not determinate (unless one wishes to go through a detailed analysis). At the same time, a significant portion will remain in the basin well over the time needed for complete treatment and will be "overtreated," i.e., the microflocs formed will consume alum since those formed will remain in the reactor for various residence times (Stenquist and Kaufman, 1972, p. 7; Jorden, 2001; Kawamura, 2000). This is an inherent flaw in the "complete-mix" reactor. Such a flaw is true in theory, as illustrated in Figure CD10.18, and is no less true in practice, albeit in practice, "complete-mix," i.e., instantaneous distribution, is not achievable, but may be approached.

Example 10.6 Rapid-Mix Design

Given

$Q = 1.000\ m^3/s$; $T = 20°C$; $\theta \approx 1$ s; the rapid-mix system is a Rushton radial-flow impeller and basin.

Required

Basin size; rotational velocity of impeller, n; power required by impeller

Solution

a. Sizing of basin and impeller

$$V(\text{basin}) = Q \cdot \theta = 1.000\ m^3/s \cdot 1\ s = 1.000\ m^3$$
$$V(\text{basin}) = \pi T^2 H = 1.000\ m^3$$

And since $T = H$ for a Rushton basin,

$$T = H = 0.683\ m$$
$$D = 0.33T = 0.33 \cdot 0.683\ m = 0.228\ m$$

b. Rotational velocity, n

For an average of five circulations within one raw-water detention time, θ(raw water),

$$t_{5R} \approx \theta(\text{raw water}) = 1\ s$$

From Table 10.6, the blend number, $nt_{5R} = 36$,

$$nt_{5R} = 36,$$
$$n \cdot 1.0\ s = 36;$$
$$n = 36\ \text{rev/s}\ (2{,}160\ \text{rpm})$$

c. Power to impeller, P

$$\mathbf{P} = P/(\rho n^3 D^5)$$
$$6.0 = P/[998\ kg/m^3 \cdot (36\ \text{rev/s})^3 \cdot (0.228\ m)^5]$$
$$P = 172\ kW\ (231\ hp)$$

Discussion

Obviously, the result is not practical. Some practical criterion, e.g., P/V or G, will give a more reasonable result for power. A smaller impeller would reduce the power, i.e., if the impeller diameter is 0.228 m/2, then $P \approx 5.4$ kW (7.2 hp). Also, reducing n has a large effect on P. Such changes affect **P**, the blend number, nt_{5R}, and **R**; so the problem could be explored further (most conveniently) using a spreadsheet, e.g., Table CD10.4. Note that in trying to achieve similarity for the parameter, nt_{5R}, adds an additional parameter, which reinforces the idea that multiple points of similarity are simply not feasible. These parameters indicate "goals," which warrant cognizance but are not simultaneously achievable. As noted by Oldshue, they provide guidance for a modeling study or, indeed, by evaluation of a full-scale system.

10.4.2 Impellers and Tanks

Impellers and tanks go together as a "system." The reference system is the "Rushton" design, which is referred to frequently in the literature (see glossary).

10.4.2.1 Impeller Variety

Figure 10.19 shows a sample of four impeller types. Figure 10.19a and b are the two basic ones: a marine impeller and a radial-flow impeller, respectively. Between these two types, there is a large variety, e.g., perhaps hundreds, with many developed as proprietary products (Oldshue, 1983, p. 2). Two examples of others are: mixed flow, Figure 10.19c, and curved-blade radial flow, Figure 10.19d.

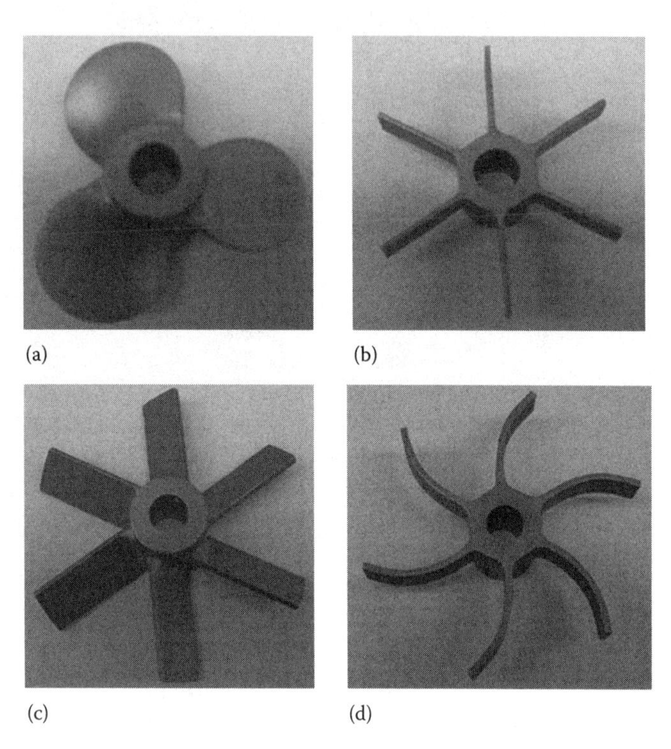

FIGURE 10.19 Sample of impeller types (photograph of model impellers from a set). (a) Marine impeller. (b) Radial-flow impeller. (c) Mixed-flow impeller. (d) Curved-blade radial-flow impeller.

10.4.2.2 Impeller Characteristics

Figure 10.20a and b illustrates the characteristic velocity profiles for axial-flow and radial-flow impellers, respectively. The maximum shear is located at the radial distance from the centerline of the jet at which dv/dy is maximum, which occurs, in each case, at the inflection point of the Gaussian curve. Comparing the two velocity profiles, it is seen that dv/dy is higher for the radial-flow impeller, i.e., higher shear rate.

10.4.2.2.1 Axial-Flow Impellers

For an axial-flow impeller, the velocity of the fluid being pumped is, by definition, parallel to the axis of the impeller. The "classic" one is the marine impeller, seen in Figure 10.20a, which has an increasing blade angle from blade tip to hub, and a constant pitch and a constant shear rate across its diameter. The marine impeller is the starting point in considering the variety of axial-flow impellers.

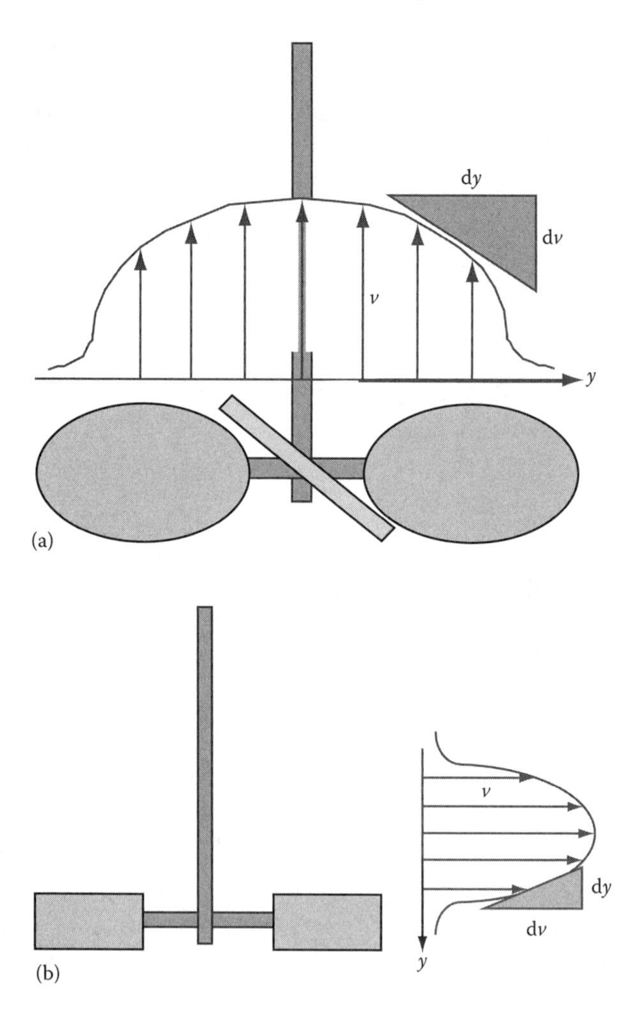

FIGURE 10.20 Velocity profiles for two categories of impellers: Axial flow and radial flow. (a) Marine impeller. (b) Radial-flow impeller. (Adapted from Oldshue, J.Y. and Trussell, R.R., Design of impellers for mixing, in: Amirtharajah, A., Clark, M.M., and Trussell, R.R. (Eds.), *Mixing in Coagulation and Flocculation*, American Water Works Research Foundation, Denver, CO, p. 311, 1991.)

10.4.2.2.2 Radial Flow

The motion induced by a radial-flow impeller, Figure 10.20b, is characterized by "high shear" and "low flow" and has three main components:

1. A radial flow of high velocity, which creates a high-shear zone just at the tip of the impeller
2. The flat blade of the impeller causes a separation effect, i.e., wake turbulence" on the trailing side
3. The radial-flow jet sets up advection currents that move up and down the walls of a tank, recirculating to the center. These currents also experience shear and thus eddies peel off. Eventually, all of the energy is dissipated as turbulence.

10.4.2.3 Impeller Pumping

In general, the purpose of mixing is to disperse chemicals so that the intended reactions such as coagulation, disinfection, oxidation, etc., can occur through increasing the probability of contacts. The objectives are (1) to create turbulence, i.e., a shear zone; and (2) to provide advection through the shear zone.

10.4.2.3.1 Flow Patterns

Flow patterns are determined by the configuration of the impeller and the confining boundary. For a "back-mix" reactor, the primary flow pattern is circulation. For a "flow-through" reactor, the flow pattern must be designed for a single pass through the turbulence zone.

10.4.2.3.2 Pumping Rate

The pumping rate for a radial-flow impeller is proportional to its tip speed, i.e., nD(impeller), and the area swept, i.e., $D(\text{impeller})^2$, to give

$$Q(\text{impeller}) = \mathbf{Q}nD(\text{impeller})^3 \qquad (10.37)$$

where

Q(impeller) is the flow of water pumped by impeller (m^3/s)
n is the rotational velocity of impeller (rev/s)
D(impeller) is the diameter of impeller (m)
$\mathbf{Q}$ is the empirical constant, i.e., the "flow number" (dimensionless)

10.4.2.3.3 Flow Number

The constant, $\mathbf{Q}$, in Equation 10.37 is called the "flow number," a dimensionless number, that characterizes the pumping capacity of an impeller (McCabe et al., 1993, p. 244). Table 10.8 provides flow numbers for several impeller types for $\mathbf{R} \geq 1000$; the basin proportions are assumed the same as given for a "Rushton basin" (see glossary).

Example 10.7 Pumping Capacity from Q

Given

A six-blade radial-flow impeller is installed in a Rushton-type basin. $Q(\text{raw water}) = 1.0\ m^3/s$.

Required

Determine the pumping capacity of a suitable impeller, i.e., Q(impeller) and its rotational speed, n.

Solution

1. Let $q(\text{raw water}) = 1.0$ s
2. $V(\text{basin}) = Q(\text{raw water}) \cdot q(\text{raw water}) = (1.0\ m^3/s) \cdot (1.0\ s) = 1.0\ m^3$
3. For a Rushton system: T = diameter of tank (m); H(water) = depth of water in tank (m);

$$T = H(\text{water}) \quad \text{and} \quad D(\text{impeller})/T = 0.33.$$

4. Basin dimensions: $[\pi T^2/4] \cdot H(\text{water}) = 1.0\ m^3$

$$T = H(\text{water}) = 1.08\ \text{m}$$
$$D(\text{impeller})/T = 0.33$$
$$D(\text{impeller}) = 0.33 \cdot 1.08\ \text{m} = 0.36\ \text{m}$$

5. Flow number, $\mathbf{Q}$,
   From Table 10.8, let $\mathbf{Q} = 0.70$
6. Impeller pumping, Q(impeller)
   Let $Q(\text{impeller})/Q(\text{raw water}) \approx 5$

$$Q(\text{impeller}) = 5 \cdot 1.0\ m^3/s = 5\ m^3/s$$

7. Impeller rotational speed, n,

$$Q(\text{impeller}) = \mathbf{Q}nD(\text{impeller})^3 \qquad (10.37)$$
$$5 \cdot (1.0\ m^3/s) = 0.70 \cdot n \cdot (0.36)^3$$
$$n = 153\ \text{rev/s}$$

Discussion

With $q(\text{raw water}) = 1.0$ s, the impeller pumping rate was selected as five times the raw-water flow, which was a "guess" on the high side that should give high probability, i.e., 0.99 fraction, that almost all raw-water particles will have contact with coagulant as they pass through the reactor. The other selections are also arbitrary,

TABLE 10.8
Flow Numbers for Representative Impellers

Impeller Type	Q	Reference
Rushton—four blades	0.70	Oldshue (1983, p. 169)
Rushton—six blades	$0.54 \leq \mathbf{Q} \leq 0.88$	Oldshue (1983, p. 169)
Marine-square pitch	0.5	McCabe et al. (1993, p. 244)
Turbine—four blades, 45°	0.87	McCabe et al. (1993, p. 244)

e.g., q(raw water) = 1.0 s, as well as $\mathbf{Q}$. With these assumptions, the calculated result, i.e., $n = 153$ rev/s, is too high from a practical standpoint, e.g., impeller power is proportional to n to the third power. A spreadsheet setup would be useful to examine the trade-offs, e.g., q(raw water) and thus D(impeller), Q(impeller)/Q(raw water), and n. The use of the power number, $\mathbf{P} = P/(rn^3D^5)$ and the plot for the Rushton system power number, $\mathbf{P}$, provide a basis for estimating the power required for the system and thus the reasonableness of any design trials, e.g., for n and D(impeller). As noted, changes in the geometric proportions of the system, i.e., deviations from the Rushton design, result in changes to Q and $\mathbf{P}$, which are not necessarily available in the literature.

10.4.2.3.4 Power Dissipation

The power transferred to an impeller is the product of the torque applied and its rotational velocity, i.e.,

$$P = \mathbf{T} \cdot \omega \tag{10.38}$$

where
P is the power dissipated by the impeller (W)
$\mathbf{T}$ is the torque exerted by impeller due to fluid drag (N m)
ω is the rotational velocity (rad/s)

$$\omega \text{ (rad/s)} = n \text{ (rev/s)} \cdot 2\pi$$

The experimental procedure to measure $\mathbf{T}$ and w is described in Box 10.6.

10.4.2.3.5 Shear

The power imparted to the water by an impeller causes a pressure (head) increase if confined by a casing. In the absence of a casing, the impeller energy is distributed between advection and turbulence, aggregated here as H(shear), since the advective flow sheds eddies and ends up as turbulence. The power dissipated is thus (Myers et al., 1990, p. 35),

$$P = Q(\text{impeller})\gamma_w H(\text{shear}) \tag{10.39}$$

where
γ_w is the specific weight of water, i.e., $\gamma_w = \rho_w g$ (N/m^3)
H(shear) is the pseudo head due to impeller rotation dissipated as shear (m)

Equating the power terms of Equation 10.23, i.e., $\mathbf{P} = P/(n^3D^5\rho)$, Section 10.3.3.1, and Equation 10.39, i.e., $P = Q(\text{impeller})\gamma_w H$, and substituting Equation 10.37, Q (impeller) $= \mathbf{Q}nD^3$, for Q, gives (Meyers et al., 1990, p. 35),

$$H(\text{shear}) = \left(\frac{P}{Qg}\right) \cdot n^2 D(\text{impeller})^2 \tag{10.40}$$

where $\mathbf{P}$ is the power number for impeller–tank system (dimensionless).

BOX 10.6 MEASUREMENT OF IMPELLER TORQUE

For an experimental impeller mixing system, fabricated by a machinist, the torque, $\mathbf{T}$, imparted to an impeller may be measured if the drive motor is mounted on a bearing plate (such as a "lazy susan" from a hardware store) above the basin. A lever rod is attached to the motor and rotates with the motor; the motor as set up is free to rotate. A force gage is then mounted on the top of the basin with a hook to restrain the lever rod (Figure 3.5). When an impeller is attached to the shaft from the motor, the associated force may be measured. The measured force in Newtons (or lb) times the lever arm distance in m (or ft) is the torque in N · m (lb · ft) generated by the motor to overcome the drag resistance of the impeller. If a luminescent tape strip is attached to the shaft, its rotational velocity, n, may be measured by a strobe. Since $n \cdot 2\pi = \omega$, the power, P, dissipated by the impeller is, $P = \mathbf{T} \cdot (n \cdot 2\pi)$. The power dissipated by friction may be measured by removing the impeller and taking the same measurements. Ordinarily, the measured friction torque should approach zero. To convert from kilograms-force to Newtons, N = kg-force · 9.80665; i.e., a kilogram of force (2.2 lb force) is that force exerted by gravity on a mass of 1 kg.

In other words, the energy dissipated as shear, i.e., turbulence, is proportional to the impeller speed squared, n^2, and impeller diameter squared, D(impeller)^2.

10.4.2.3.6 Shear/Flow Ratio

For most mixing situations, high shear (i.e., turbulence) is required, along with some level of advection. To obtain an expression for the shear-to-flow ratio, divide Equation 10.40 by Equation 10.37 to give (Myers et al., 1999, p. 35),

$$\frac{H(\text{shear})}{Q(\text{impeller})} = \left[\frac{\mathbf{P}}{gQ^2}\right] \cdot \left(\frac{n}{D}\right)$$
$$\frac{H(\text{shear})}{Q(\text{impeller})} = \left[\frac{\mathbf{P}}{g\mathbf{Q}^2}\right] \cdot \left(\frac{n}{D}\right) \tag{10.41}$$

Equation 10.41 shows that for a given system, i.e., for unique values of $\mathbf{P}$ and $\mathbf{Q}$, the shear-to-flow ratio is proportional to the n/D ratio, i.e., higher values of n and smaller values of D(impeller). Illustrative values are (Myers et al., 1999, p. 36).

Shear	Pumping	n (rpm)	D (m)
High	Low	3500	0.1
Low	High	100	1.0

In other words, the distribution between shear and pumping flow may be controlled by the n/D ratio. Also, as seen, the ratio is also affected by the system constants, i.e., the "power-number," **P**, and the "flow-number," ***Q***.

10.4.2.3.7 P/V Ratio

In the final analysis, the empirical "P/V" is the only quantitative guideline available; typically, the power dissipated per unit volume varies $0.2 \leq P/V \leq 2$ kW/m^3 water ($0.0076 \leq P/V \leq 0.076$ hp/ft^3) (Myers et al., 1999, p. 35); these values correspond to $200 \leq G \leq 2000$ s^{-1} at 20°C. As indicated, the values cover a wide range, albeit the upper limits may be much higher; there are no definitive numerical values.

Example 10.8 In-Line Pipe Mixing

Given

Let Q(raw water) = 1.0 m^3/s (22.8 mgd); let d(pipe) = 300 mm as an initial trial.

Required

Provide an impeller and alum injection design for an in-line mixing system.

Solution

The challenge for in-line mixing is to disperse the coagulant uniformly within the raw-water flow so that a large fraction of this "sub-mixture" is exposed to high turbulence and thus approaches complete mixing, i.e., blend fraction ≈ 0.99.

1. *Selection of pipe diameter*
   First trial: Assume the pipe diameter is 300 mm; v(pipe) ≈ Q(raw water)/A(pipe) = (1.0 m^3/s)/[$\pi \cdot 0.300^2/4$] = 14 m/s. Let d(impeller) ≈ 0.1 m; assume the turbulence zone is x(turbulence) ≈ 0.5 m. Thus, q(turbulence) ≈ x(turbulence)/v(pipe) = (0.5 m)/(14 m/s) = 0.04 s. About 0.5 s would be better (for the adsorption–destabilization zone of coagulation).

   Second trial: Assume the pipe diameter is 1000 mm; v(pipe) ≈ Q(raw water)/A(pipe) = (1.0 m^3/s)/[$\pi \cdot 1.000^2/4$] = 1.3 m/s. Let d(impeller) ≈ 0.1 m; again, assume the turbulence zone is x(turbulence) ≈ 0.5 m. Thus, q(turbulence) ≈ x(turbulence)/v(pipe) = (0.5 m)/(1.3 m/s) = 0.4 s, which is close enough to 0.5 s.
2. *Impeller design*
   Select d(impeller) ≈ 0.1 m, which is arbitrary but within the guidelines suggested by Myers et al. (1999, p. 36). Also assume that the turbulence zone x(turbulence) ≈ 0.5 m as in the previous paragraphs. The turbulence should extend throughout the pipe section; therefore, select about six impellers with two vertical shafts. Locate the shafts each about 200 m from the centerline of the pipe. Locate the impellers, for each shaft, one in the center of the pipe and one each 150 mm above and below the horizontal centerline, respectively. The intent of the design is to fill the pipe section with turbulence to the extent feasible.
3. *Motors for impellers*
   The motors should be direct current with variable speed control, with power sufficient to give $n \leq 3500$ rpm. This may be determined best by trial, i.e., mounting two or more candidate motors. It may be that $n \leq 3500$ rpm is an overkill, but the rationale is that for such a large capital expenditure as a water treatment plant, testing is warranted in order to reduce the uncertainty of the design. Also, note that a high shear-to-flow ratio is desired, which requires a high n/D ratio.
4. *Alum distribution*
   Let the alum be distributed by a stainless steel tube ring, about 400 mm diameter with tube about 50 mm diameter. The ring should be secured by cross beams such that the vibration of the ring is not an issue. The tube should be located about two impeller diameters distance upstream from the center of the impeller array. The alum should be emitted downstream through six orifices, with the orifices located just opposite the respective impellers. The pressure should be sufficient such that the alum flow is the same through each orifice. A pressure sensor should be located near the intake to the ring-manifold; the alum feed should be by positive displacement pump with an air chamber or other device to reduce pressure pulses.
5. *Repair and maintenance*
   The distribution ring and impellers assembly should be located in a flanged section so that removal can be accomplished with replacement by a straight pipe section. The distribution tube and orifices should be connected to a hot-water/chemical cleaning solution for routine maintenance. The pipe section with alum distribution and impellers should have an observation window with internal lighting. After cleaning, a dye (e.g., Rhodamine-B) may be injected to visually confirm that the orifices are functioning.
6. *Redundancy*
   The alum feed and mixer assembly should be located with an identical assembly located in another section of pipe such that the plant can continue operation without disruption while one assembly is being maintained or repaired.
7. *Testing*
   The alum distribution and mixing setup should be tested with a brine solution or Rhodamine-B dye solution to determine the impeller mixing speed and whether the configurations should be changed. The tracer concentration may be measured on the effluent side of the mixer system after a "step-input" of tracer.

Discussion

The design procedure suggested indicates the uncertainty of the mixing state of the art. A testing procedure should produce greater certainty of mixing outcome, but with a higher-than-usual budget required. The same considerations should be given to any mixing system, e.g., a jet-mixer system or an in-line static-mixer system.

10.4.2.4 Tanks

The tank and impeller are a "system"; therefore the geometry of each and other characteristics must be specified. The ensuing pressure and velocity fields are unique to a particular system.

10.4.2.4.1 Geometry

Mixing tanks are usually circular or square in plan view. The proportions for a circular tank (Oldshue, 1983, p. 12; McCabe et al. 1993, p. 241) are $H/T = 1$ (water depth, H, to tank diameter, T). A recommended impeller diameter is $0.5T <$ D(impeller) $< 0.8T$ (McCabe et al., 1993). If $H/T = 1$, the optimum impeller placement for blending is at $0.5H$.

10.4.2.4.2 Baffles

A circular or square tank with impeller shaft in the center and oriented vertically and without baffles will cause the whole fluid mass to "swirl," creating a vortex. With baffles, the rotation of the fluid mass will be reduced and the vortex may be eliminated. The general guideline baffle width, J, for a tank with six baffles, is, J, $T/12 \leq J \leq T/10$ (Oldshue, 1983, p. 17). If the baffles do not reach the bottom of the tank, then solids accumulation in the tank bottom may be reduced. For a square tank, the same sizing guidelines apply except only four baffles are used. Another approach is to set the shaft of the impeller at some location other than the center, or by tilting the axis of the pump at some angle to the vertical.

10.4.2.4.3 Draft Tubes

Figure 10.2a is a schematic drawing of a draft tube system. As a rule, a draft tube is used with a marine impeller and helps to control the direction and velocity of the flow (McCabe et al., 1993, p. 241).

10.4.2.5 Rushton System

A particular impeller–basin combination used by Rushton et al. (1950a,b) to develop similitude relations was later designated a "Rushton basin," which became a quasi-standard for comparing with other mixing systems. Its proportions are given in the glossary. The Rushton system is used often as a default design since its performance has been well established, e.g., its **P** vs. **R** curve, as seen in Figure 10.15, was based on literally hundreds of data points and the system has been used often in practice.

The Rushton basin is scaled to fit the situation at hand and is based on the volume of the basin, i.e., $V = Q\theta$, with detention time, θ, specified to be the same for model and prototype. With Q given, and since $H = T$, then $T(\pi T^2/4) = V$, yielding the dimensions H and T. Since $D(\text{impeller}) = T/3$, then the other dimensions, may be calculated. As noted, however, in Section 10.3.3.5, geometric scale-up does not result in both dynamic and kinematic similarity.

10.4.2.6 In-Line Mixers

An "in-line" mixer (one of several types of "flash" mixers) consists of an impeller located in a pipe; Figure 10.16b is an example. As seen, the impeller is open to the raw-water flow of the pipe. A key feature is that there is not a confining geometry and the raw-water flow passes across the impellers just once. The flow pattern is that of pipe flow, i.e., straight lines parallel to the pipe, distorted by the superimposed circulation pattern of the impeller.

As with back-mix reactors, numerous configurations exist for in-line mixing. Figure 10.16b, illustrates a case in which two "mixed-flow" impellers (300 mm, 12 in. diameter) each cause flow both outward and toward the center of a 910 mm (36 in.) diameter pipe; the two tubes entering from the side are for a neat alum solution and/or a polymer, respectively. The advective flow for the top impeller is down and is up for the bottom impeller. The two advective flows meet halfway between the two impellers with a resulting turbulence and in which the two chemicals are mixed with the raw-water flow. Further mixing occurs by hydraulic dispersion within the pipe.

Table 10.9 lists some 12 "flash-mix" installations compiled by Kawamura (2000, p. 309) that give plant flow, number of

TABLE 10.9
Impeller–Basin Flash-Mix Installations

Plant	Place	State	Flow (mgd)	Flow (mL/day)	No. Units	Power/Unit (hp)	Power/Unit (kw)	Total Power (hp)	Total Power (kw)	Power/Unit of Flow (hp/mgd)	Power/Unit of Flow (kw/mL day)
Behner	Pasadena	CA	7.5	28	1	10	7	10	7	1.33	0.26
Davenport	Davenport	IA	15	57	2	7.5	6	15	11	1.00	0.20
Badger	San Diguito	CA	27	102	2	10	7	20	15	0.74	0.15
Stanton	Stanton	DE	30	114	4	20	15	80	60	2.67	0.53
LartonWTP	Fairfax	VA	40	151	2	25	19	50	37	1.25	0.25
Helix	Helix	CA	67	254	1	50	37	50	37	0.75	0.15
Jersey	Jersey City	NJ	80	303	3	15	11	45	34	0.56	0.11
Bridgeport	Bridgeport	CT	100	379	6	7.5	6	45	34	0.45	0.09
La Mesa	Manila		200	757	3	40	30	120	89	0.60	0.12
Foothills	Denver	CO	250	946	4	5	4	20	15	0.08	0.02
Aqueduct	Los Angeles	CA	600	2271	8	100	75	800	597	1.33	0.26
Guarau	Sao Paulo		750	2839	2	150	112	300	224	0.40	0.08
Average										0.93	0.18

Source: Kawamura, S., *Aqua—J. Int. Water Assoc.*, 49(6), 309, 2000.

units, power of each, and power per unit of flow. The compilation is indicative of the range in designs for flash-mix installations and is given for reference. For comparison, 0.05–0.2 kw/mLd (0.25–1 hp/mgd) was recommended by ASCE-AWWA (1997) as reported by Kawamura (2000, p. 308).

10.4.3 Jet Mixers

A "jet" is a high-velocity flow of a given fluid. A "submerged jet" discharges into a fluid medium and is characterized by significant penetration, with its energy being dissipated by turbulence. A jet mixer comprises one or more nozzles that emit a high-velocity flow into a water medium, as illustrated in Figure 10.7b.

Applications of jet mixing (common to other kinds of mixing as well) include coagulation, disinfection, gas bubble dispersion, mixing within anaerobic reactors, etc. The jet-mixing technology has developed since about 1980 (see, for example, initial papers by Kawamura, 1976, p. 332; Amirtharajah, 1979, p. 137; Chao and Stone, 1979) and seems to have become a favored method of "flash-mixing," especially during the 1990s, with about 40 installations counted to Year 2000 in the United States and Australia (Kawamura, 2000, p. 307). The advantages of jet mixing include simplicity such as no moving parts (except for a pump), low capital cost, reliability, and effectiveness. A disadvantage is the possible clogging of nozzles.

There are many kinds of jet-mixer configurations, e.g., radial flow of several jets toward the center of a pipe, radial flow from a center ring, a single jet pointed upstream in the pipe, and so on Figure 10.21 illustrates the second. Regarding design, the jets emerge from nozzles attached to a pressurized manifold with velocity given as $v = C(2g\Delta h)^{0.5}$, where Δh is the headloss between the manifold and the ambient fluid (the pressure drop is $\Delta p = (\rho g)\Delta h$). The manifold is pressurized by a pump or, in some cases, available head from a reservoir. As a rule of thumb in coagulation mixing, the total jet flow is about 10% of the raw-water flow. The coagulant may be added at a point where its dispersion into the whole of the raw-water flow is most likely, such as in the manifold pipe.

10.4.3.1 Flash Mixing by Submerged Jets

The purpose of "flash-mixing" is to disperse coagulants, e.g., alum, within 1–2 s throughout the raw-water flow. The submerged jet has the two key elements needed to accomplish this task, i.e., advection and turbulence. Jet mixing and wake mixing are the only technologies that may approach this mixing ideal. Guidelines for jet design were given initially by Chao and Stone (1979) with later amplification by Kawamura (1991, 2000). In jet mixing, the alum is injected into the carrier water just prior to its dispersion in the main flow of raw water or sometimes at the point of emergence of the jet flow.

Data compiled by Kawamura (2000, p. 309) are shown in Table 10.10 for 12 installations giving flow, number of units, power per unit, total power, and power per unit of flow. The power data are for that expended by electric motors rather than by the jets. The actual power imparted to the jets may be determined by placing a flow meter and a pressure gage in the jet pipeline and another pressure gage in the pipeline, i.e., $P(\text{jet}) = Q \cdot (\rho g) \cdot \Delta h$; using a motor efficiency of 0.7 and allowing for headloss in the jet pipeline, the power consumed at by jet-induced turbulence would be perhaps 0.7 fraction of the values given. Comparing values for power expended per unit of flow in Tables 10.9 and 10.10, lower values are seen for the jet mixing (these are motor power in each case).

10.4.3.1.1 Design Components

Figure 10.21 illustrates, schematically, the components of a jet-diffuser system, i.e., the raw-water pipeline, the circulation pipe for the jet, the jet diffuser (comprising a set of nozzles), and the coagulant feed. The raw-water flow is shown as being terminated by an overflow weir. The water for the jet flow is taken upstream from the jet diffuser with a pump, which pressurizes the jet pipe; appurtenances for operation include a throttling valve, a flow meter, and a pressure gage (the throttling valve would be fully open, ordinarily). The jet diffuser comprises a set of nozzles that are arranged around the jet pipe; the nozzles may be placed in a larger manifold. The alum is fed as a neat solution near the end of the jet pipe, pressurized by a positive displacement metering pump.

Table 10.11 categorizes the four component categories of a jet-mixer system mentioned previously, listing also some of the

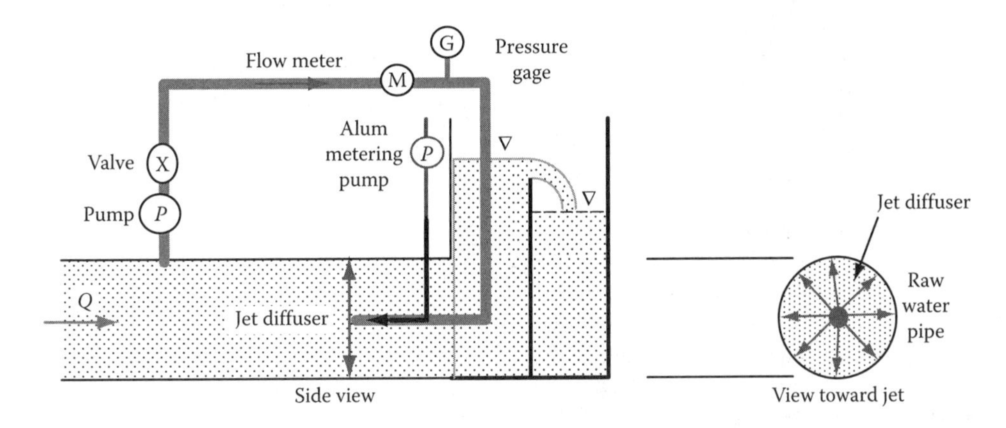

FIGURE 10.21 Schematic drawing of jet-mixing system.

TABLE 10.10
Jet-Mix Installations

Plant	Place	State	Flow (mgd)	Flow (mL/day)	No. Units	Power/Unit (hp)	Power/Unit (kw)	Total Power (hp)	Total Power (kw)	Power/Unit of Flow (hp/mgd)	Power/Unit of Flow (kw/mL day)
Royer Nesbit	Cucamonga	CA	4.5	17	1	1	1	1	1	0.22	0.04
Waterman	Fairfield	CA	15	57	1	3	2	3	2	0.20	0.04
Southeast	Salt Lake City	UT	20	76	1	3	2	3	2	0.15	0.03
UTE	Grand Junction	CO	35	132	1	3	2	3	2	0.09	0.02
Wemlinger	Aurora	CO	40	151	1	5	4	5	4	0.13	0.02
Otay	San Diego	CA	40	151	1	5	4	5	4	0.13	0.02
Bollman	Contra Costa	CA	80	303	1	10	7	10	7	0.13	0.02
Anstay Hill	Adelaide	Aus.	85	322	1	10	7	10	7	0.12	0.02
Santa Teresa	Santa Clara	CA	100	379	2	8	6	15	11	0.15	0.03
Val Vista	Phoenix	AZ	140	530	1	15	11	15	11	0.11	0.02
Jordan Valley	Salt Lake City	UT	190	719	1	30	22	30	22	0.16	0.03
Henry Mills	MWD-S.Ca.	CA	330	1249	3	50	37	150	112	0.45	0.09
Average										0.17	0.03

Source: Adapted from Kawamura, S., *Aqua—J. Int. Water Assoc.*, 49(6), 309, 2000.
Note: For installations where the number of units exceeds one, the assumption is that the flow is separated into the number of streams that equal the number of jet-mixing units.

TABLE 10.11
Components in Design of Jet-Mixing System

Raw-Water Pipeline	Jet Pipeline	Nozzles	Coagulant Feed
Diameter	Metering pump	Number	Point of feed
Flow	Flow and head	Flow and head	Orientation of flow
Flow variation	Total flow	Flow—each	Feed velocity
Velocity of water	Diameter	Diameter	Flow
Pressure at jet	Appurtenances	Velocity of core	Pump selection
	Valves-actuated	Orientation	
	Valves-control	Upstream or downstream	
	Flow meter	Radial from peripheral	
	Pressure gage	Radial from center	
		Spray angle	
		Diffusion plate (or not)	

specific aspects of design, with most seen in Figure 10.21. The items in each category indicate decisions and calculations that should be considered in jet-mixer design. Example 10.9 provides an algorithm for incorporating most of the considerations listed in Table 10.11. The location of the alum feed is one of the considerations that could be perplexing. The alum feed point seen in Figure 10.21 is logical in that the neat alum is mixed with the jet flow just prior to its dispersion into the raw-water flow.

10.4.3.1.2 Nozzles

A variety of nozzles are available for a wide range of industrial purposes (see, for example, Catalog 70 from Spraying Systems Co., 2010), which may serve in a jet-mixing system. Numerous "full-jet" types with different capacities are available with conical spray cone angles, $0 \leq \theta \leq 120°$. Another type suitable for jet mixing has a flat spray pattern with angles of coverage, $0 \leq \theta \leq 120°$ (Spraying Systems Co. VeeJet® nozzles, catalog 70, pp. C3–C11). The nozzles are available in different materials including stainless steel and may be screwed into a manifold. Flow vs. pressure data are given for each nozzle. Data from one of the nozzles, i.e., H-15, capacity size #15280, (15° angle and 28.0 gpm at 40 psi reference pressure), with orifice diameter given as "0.391" were plotted and are shown in Figure 10.22, with scales in both SI and U.S. customary units (see Spraying Systems Co. Catalog 70, p. B27 for nozzle shape and pp. B28–29 for data). The same data were used in the spreadsheet, Table CD10.12. The "nominal" operating range shown is intended to show the expected limits of operation. The data, plotted on log–log scales, show a straight-line relationship giving the coefficients, $C(\text{SI}) = 0.927$ $(C(\text{US}) = 0.943)$ and $n(\text{SI}) = 0.510$ $(n(\text{US}) = 0.505)$; flow may be calculated from the standard orifice equation, i.e.,

$$Q = CA(\text{jet})\sqrt{2gh^n} \qquad (10.42)$$

where

C is the discharge coefficient (dimensionless)
A(jet) is the area of orifice or nozzle (m^2)
n is the exponent (dimensionless)
h is the head of water applied to orifice (m)

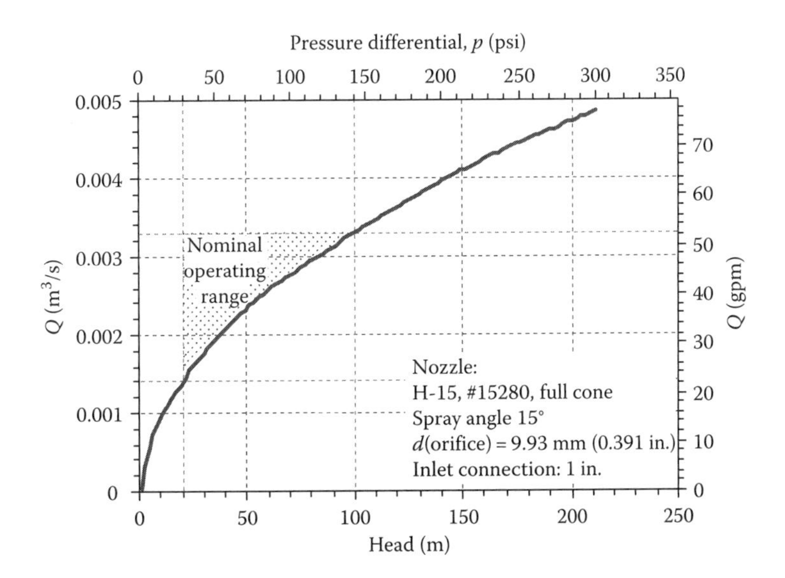

FIGURE 10.22 Flow vs. head for nozzle. (Data from Spraying Systems Co., Wheaton, IL, Catalog 70, p. B28, 2010.)

For a sharp edge orifice, $C = 0.611$; also for any orifice, $n = 0.5$, which conforms to value given in Catalog 70, p. B28 for their nozzles. For a nozzle, the edge is not sharp and differs for each model and so the discharge coefficient must be determined by a log–log plot such as shown. The exponent, n, may be obtained from the same plot. The $\sqrt{2g}$ could be assimilated in a coefficient but is shown separate here (so that more variables can be discerned).

10.4.3.1.3 Design Algorithm

The design of a jet-mixing system is based first on knowing the relationship between flow and head, i.e., knowing the discharge coefficient in Equation 10.42 and/or a plot such as given in Figure 10.22 (or data from a catalog). The merit of an equation for the flow–head relationship is that it is amenable to use in a spreadsheet algorithm for design. The velocity of the jet may be calculated from knowing the orifice diameter. The velocity head from the jet is dissipated as turbulence (with a portion, not known, going to heat directly). The power required should be based on the head for the nozzle (not just the velocity head, since there is a loss of head through the nozzle). The wire-to-water power, i.e., the power required by the pump motor, may be calculated knowing the efficiency of the pump (in its operating range) and the motor efficiency. The cost of operation may be obtained from the local cost of energy per kilowatt-hour (kW-h). For initial estimates a pump efficiency of 0.7 and a motor efficiency of 0.7 may be used.

Example 10.9 enumerates a design algorithm for a jet-mixer system. The spreadsheet, Table CD10.12, shows for increasing manifold pressure: the jet flow, jet velocity, assumed pipe flow, jet flow as a fraction of pipe flow, the power dissipated by the jet, the distance, Δz for the intersection of the jet trajectory with the pipe wall, the mixing time, the hypothetical mixing volume, G, $G\theta$, pump power imparted to water, energy use per month (based on assumed wire-to-water efficiency, monthly cost (based on an assumed cost of electric energy), and power dissipated per unit of flow (energy consumed per unit of flow is about twice that calculated in Table CD10.12).

Example 10.9 Design of Jet-Mixer System

Given

Raw water is delivered to a WTP by a 1067 mm (42 in.) pipeline from Horsetooth Reservoir with water surface elevation 61 m (200 ft) above the plant. Let $T = 20°C$ for working purposes.

Required

Design a jet-mixer system for coagulation.

Solution

1. *General design*
   Select Spraying Systems, nozzle H-15, #15280 (as in the plot shown in Figure 10.23)
   The jets have a full cone spray with cone angle 15°
   Let n(nozzles) = 24, intended to cover the cross-sectional area of the raw-water pipeline
   Let the jets be oriented radial direction outward from a center manifold
   Let the coagulant (alum) be injected into the manifold flow just prior to the plane of the jet circle.
2. *Calculations*
   Set up a spreadsheet such as developed for the problem as Table CD10.12. The spreadsheet develops an algorithm to calculate for increasing manifold pressure: jet flow, jet velocity, assumed pipe flow, jet flow as a fraction of raw-water flow, the power dissipated by the jet, the distance, Δz for the intersection of the jet trajectory with the pipe wall, the mixing time, the hypothetical mixing volume, G, $G\theta$, pump power imparted to water, energy use per month (based on assumed wire-to-water efficiency), monthly cost (based on an assumed cost of electric energy), and power dissipated per unit of flow (energy consumed per unit of flow is about twice that calculated in Table CD10.12). Table CD10.12b shows advective trajectories for two jet flows. Figure CD10.23 is a linked plot, from trajectories restated in Table CD10.12f, showing how the trajectory is modified by jet velocity.

TABLE CD10.12

Calculations of Jet Flow, Jet Velocity, Pipe Velocity, Power Dissipated, *G* Values, *Gθ* Values, Trajectory from Jet, etc. (Nozzle: 1H-15280, 1″ Orifice Diameter; from Spraying Systems Catalog 70, p. B28, 2010)

(a) Given Data

$g(SI) = 9.80665$ m/s^2 $C(SI) = 0.927$
$g(US) = 32.1756187$ ft/s^2 $C(US) = 0.943$

(b) Calculations of Jet Flow, Jet Velocity, Velocity Head Dissipated

$k(SI) = 0.50949$
$k(US) = 0.50481$

Δp				d(orifice)		Q(jet)		v_o(jet)		$v_o(\text{jet})^2/2g$		P(vel. head)		n(Jets)	P(vel. head)		Q(all jets)	
(psi)	(kPa)	(m)	(ft)	(in)	(m)	(ft^3/s)	(m^3/s)	(ft/s)	(m/s)	(ft/s)^2	(m/s)^2	(hp)	(kw)	(#)	(hp)	(kw)	(ft^3/s)	(m^3/s)
0	0	0	0	0.391	0.00993	0	0	0	0									
10	68.95	7.03	23.06543	0.391	0.00993	0.03075	0.00086	36.88	11.09	21.14	6.27	0.074	0.053	24	1.770	1.267	0.246	0.007
20	137.9	14.06	46.13086	0.391	0.00993	0.04364	0.00122	52.33	15.78	42.56	12.70	0.211	0.152	24	5.057	3.656	0.349	0.010
30	206.84	21.09	69.19629	0.391	0.00993	0.05355	0.00150	64.22	19.41	64.09	19.20	0.389	0.283	24	9.346	6.795	0.428	0.012
40	275.79	28.12	92.26172	0.391	0.00993	0.06192	0.00174	74.26	22.47	85.69	25.74	0.602	0.439	24	14.448	10.547	0.495	0.014
60	413.68	42.18	138.3926	0.391	0.00993	0.0760	0.00214	91.13	27.63	129.04	38.91	1.112	0.817	24	26.699	19.601	0.608	0.017
80	551.58	56.25	184.5563	0.391	0.00993	0.08787	0.00248	105.38	31.99	172.56	52.18	1.720	1.268	24	41.288	30.435	0.703	0.020
86.7	597.78	60.96	200	0.391	0.00993	0.09151	0.00258	109.74	33.33	187.15	56.63	1.943	1.434	24	46.631	34.415	0.732	0.021
100	689.48	70.31	230.6871	0.391	0.00993	0.09834	0.00278	117.94	35.84	216.16	65.50	2.412	1.783	24	57.884	42.802	0.787	0.022
150	1034.21	105.46	346.0143	0.391	0.00993	0.12068	0.00341	144.73	44.07	325.49	99.00	4.457	3.314	24	106.957	79.540	0.965	0.027
200	1378.95	140.61	461.3414	0.391	0.00993	0.13954	0.00395	167.35	51.02	435.18	132.72	6.890	5.144	24	165.350	123.463	1.116	0.032
300	2068.42	210.92	692.0285	0.391	0.00993	0.17124	0.00486	205.36	62.73	655.34	200.63	12.732	9.561	24	305.561	229.458	1.370	0.039

Pressure differential given

Spraying systems catalog data for nozzle H-15, p. 94

$Q(\text{jet}) = C \cdot (2g)^{0.5} \cdot h^k$

$v_{x=0}(\text{jet}) = Q(\text{jet})/A(\text{jet})$

Velocity head dissipated

$P = Q \cdot (rg) \cdot (v_o(\text{jet})^2/2g)$

Number of jets in plane

Total power dissipated by all jets

$n(\text{jets}) \cdot P(\text{vel. head})$

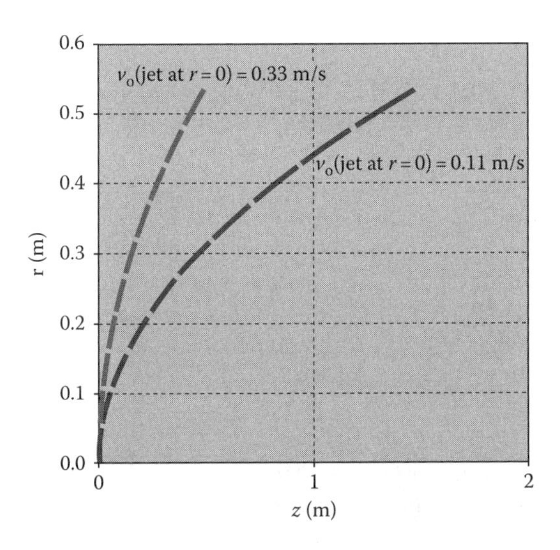

FIGURE CD10.23 Trajectories of two jets.

Discussion

The spreadsheet, Table CD10.12, provides a means to investigate the effect of design variables on jet flow as a fraction of raw-water flow, G, power dissipated by turbulence, motor power required, monthly cost of energy, and power consumed per unit of flow. The latter are higher than given in Table 10.10 for 12 operating plants. This does not mean overdesign in Table CD10.12, but rather that the jet system has a higher amount of turbulence dissipated.

Table CD10.12b calculates two jet trajectories, which give a basis for estimating a hypothetical mixing volume. Table CD10.12f is a repeat of (b) to delineate the plotting coordinates for Figure CD10.23.

In addition to the center jets described in the design of the jet system of Table CD10.12, a set of peripheral jets (in the pipe wall and directed inward in the radial direction) should be placed at a distance where the advective trajectory of the center jets intersect the pipe wall. The second set of jets would provide for the penetration of jet turbulence to the center of the raw-water pipe, thus filling a void from the center feed jets. A third set of peripheral jets may be added, to increase the $G\theta$ product. The third set could be operated after testing to ascertain their effect.

10.4.3.1.4 Proprietary Systems

Jet mixing seems to have limited proprietary "packages." One such system is the Water Champ® from Siemens Water Technologies, which demonstrates the two key elements of submerged jet mixing, i.e., advection and turbulent diffusion. Figure 10.24a shows the Water Champ™ and the jet emanating from it; Figure 10.24b is a cross-section drawing showing its key features: (1) draft tube, (2) a small impeller at the end of the draft tube, (3) an opening in the side of the draft tube for water intake and for chemical intake. The impeller is a small diameter axial-flow type that rotates at 3450 rpm and forms a small diameter, high-velocity flow, i.e., a jet. Sizes and rated power for units range from 4 to 6 in. and 2–20 hp, respectively. The discharge from the unit behaves as a submerged jet, i.e., with centerline

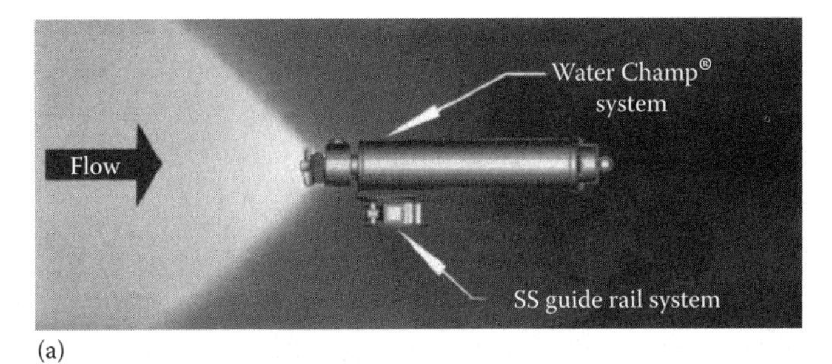

(a)

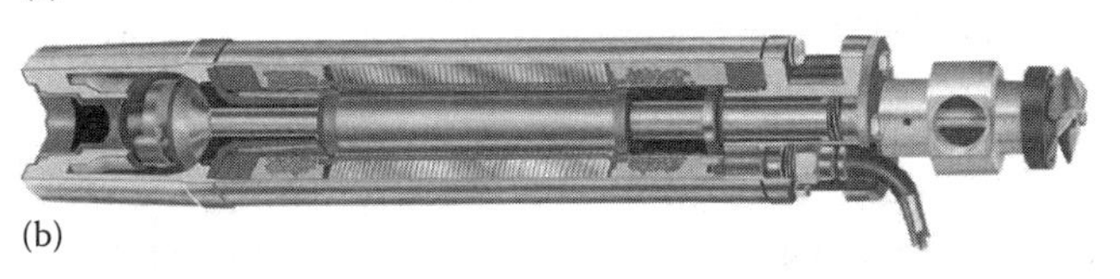

(b)

FIGURE 10.24 Water Champ® chemical induction system. (a) Photograph of zone of influence jet diffusion cone. (b) Cross section. US Filter Siemens, US Filter Stanco Brochure ST-WC-BR-0601, US Filter Siemens, 2010.

velocity decreasing with distance and expanding as cone. The induction of the chemical through the draft tube and the shear of the impeller cause thorough mixing between chemical and jet water. Once beyond the impeller the jet behaves as any other jet, i.e., with (1) advective flow which transports the "mixed chemical" into the raw-water flow, (2) concurrent shear which forms a cone of turbulent diffusion resulting in the mixing of the "mixed chemical" and the raw water, shown as the "zone-of-influence". Visual assessment of the jet diffusion, as in Figure 10.24a, indicates that the jet mixing would encompass essentially all of the raw-water flow (assuming the latter was well defined as in a pipe flow). Applications of the Water Champ™ include flash mix for coagulation, disinfection, odor control, pH control, gas dissolution, etc. Some of the specific chemicals amenable to the mixing by the unit, as listed by the manufacturer, are given by unit process category, as follows.

Process	Chemical
Coagulation	Aluminum sulfate
	Ferric chloride
Gas dissolution	Chlorine
	Ozone
	Oxygen
	Carbon dioxide
	Sulfur dioxide
Chemical feed	Calcium hypochlorite,
	Sodium hypochlorite
	Hydrogen peroxide
	Potassium permanganate
	Anhydrous ammonia
	Soda ash
	Hydrochloric acid, sulfuric acid
	Lime slurry
	Sodium thiosulfate
	Sodium sulfite

To illustrate specific data for a flash mix coagulation installation, Bob Reed, Superintendent Soldier Canyon WTP (which serves the Fort Collins, Colorado environs) provided the following information (Reed, February 8, 2002). The design flow is 0.220 m^3/s (5 mgd); mixing is by a 4 in. 3.73 kW (5 hp) Water Champ™ with shaft entering a 1067 mm (42 in.) raw-water pipeline from the side, i.e., the jet flow is from near the pipe wall directed toward the center. The unit has an external motor with the coagulant feed entering an induction feed tube inside the raw-water pipeline. The coagulant was also fed in front of the propeller, as an experiment, instead of through the induction tube. The result was higher alum dosage for the former feed configuration. The plant also has two jet-mixer systems, each with center feed, providing redundancy to facilitate maintenance.

10.4.4 Static Mixers

The term "static mixer" usually means a proprietary pipe insert consisting of a number of twisted blades, e.g., Komax™, or Kenics™, that cause a sequence of bifurcations. The static mixer is a "passive" technology, i.e., there is no operator involvement. Generically, a static mixer could mean any stationary "form" that induces wake turbulence (Section 10.3.1.2).

10.4.4.1 General Principles

Elbows, baffles, and other flow obstructions, create wake turbulence in which headloss for a single unit is

$$h_L = K(\text{unit})\frac{v^2}{2g} \quad (10.43)$$

where

h_L is the headloss across unit (m)
K(unit) is the loss coefficient for a single unit (dimensionless)
v is the average velocity of flow within unit (m/s)
g is the acceleration of gravity (9.806 650 m/s^2)

The coefficient, K(unit), may be available from published data or may be determined by head measurements, e.g., by piezometers, at the entrance and exit of an installation. The total headloss for successive units is the sum of the headloss for a single unit, i.e., $\Sigma h_L(n \text{ units}) = n(\text{units}) \cdot h_L(\text{unit})$. Mixing is by advection (in changing the direction of the flow) and the turbulence associated with each unit (caused by separation type form drag). For any of the various kinds of static mixers, the power dissipation for a single unit is

$$P(\text{unit}) = Q(\text{unit}) \cdot \gamma(\text{water}) \cdot h_L(\text{unit}) \quad (10.44)$$

where

P(unit) is the power dissipated by turbulence across a given unit, e.g., an elbow (watts)
Q(unit) is the flow of water in across a given static mixer unit (m^3/s)
γ(water) is the specific weight of water (N/m^3)

10.4.4.2 Baffles

Another "passive" kind of mixing device is a system of baffles placed in the way of the gravity flow in an open channel; two kinds are (1) over-and-under, and (2) end-around. The first use of baffles for a large-scale water treatment plant was at New Orleans in 1909 (Willcomb, 1932, p. 1431). Of some 30 plants built during the 1920s, about one-third used over-and-under baffles and one-third used end-around (Willcomb, 1932, p. 1431). Their use was common until supplanted by a trend toward impeller–basin systems beginning probably about 1940.

10.4.4.2.1 Over-and-Under

An over-and-under baffle system comprises a combination of weirs and sluices (most likely, for rapid mix, a single weir or sluice unless several chemicals are added, e.g., lime, alum, polymer in succession). As with the other mixing systems, design is based upon hydraulic principles. The water entering the first compartment rises to a level sufficient to overflow the first weir plate. The coagulant is added to the overflow, and is distributed over the width by means of several orifices from a manifold parallel to the weir crest. The water falls into the pool, below. The depth of the pool below should be as small as feasible, but is determined by the opening of the sluice (if used).

10.4.4.2.2 End-Around Baffle Systems

In an end-around basin, the raw water enters the first chamber and flows around successive bends, giving a "cascade" to the exit (see, for example, Harhoff, 1998); Figure 10.25 shows a top view. Alum may be added by an orifice manifold oriented parallel to the baffles, positioned just downstream from the first baffle slot and just above the flow. If the baffle system is designed for disinfection, the same principles apply. The power dissipated around a given bend is given by the relation, $P = Q \cdot (\rho g) \cdot \Delta h(\text{end})$. Therefore, as with a weir, the power dissipation for a single "end-around" is proportional to the water surface drop in elevation, i.e., h(end). The end slot must be narrow enough to cause a backup that forces the flow through "critical," which also requires a downstream drop in the floor elevation. The "specific discharge" diagram (see, for example, Rouse, 1946, p. 136) is the basis for the calculations. A series of channels produces a "cascade" to the final channel which may be designed as a "pool" to provide the headwater for an outlet pipeline or it could be an open channel to the next unit process, e.g., flocculation.

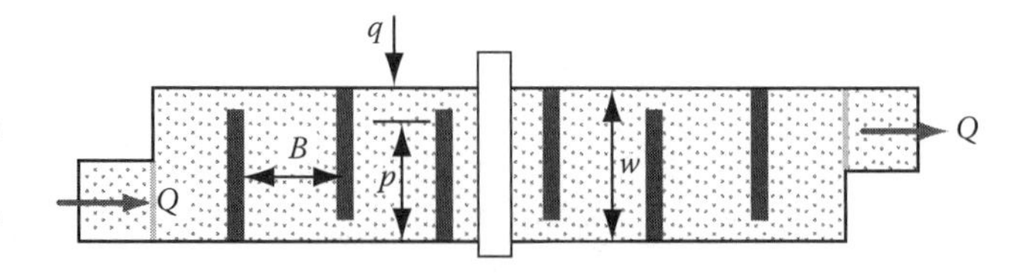

FIGURE 10.25 End-around baffled mixing basin.

10.4.4.3 Static Mixers

Static mixers comprise proprietary "elements" that are inserted in a pipe to cause a sequence of flow bifurcations. In water treatment, static mixers may be used for the same purposes as for other mixers, e.g., flash mixing for coagulation, pH control, blending a powdered activated carbon with raw water, chlorine or ozone dissolution, blending a flocculent aid with coagulated water prior to flocculation, blending of nutrients with raw wastewater prior to biological treatment, and blending a polymer with dilution water prior to use as a coagulant aid (Mutsakis and Rader, 1986).

10.4.4.3.1 Mixing with Static Mixers

A variety of proprietary forms have been fabricated for in-pipe static mixers. Figure 10.26 shows a Kenics™ static mixer with elements configured as helical blades placed in a 51 mm (2 in.) tube; the ridges of successive blades are orthogonal. A flow bifurcation occurs at the beginning of each element; the twist in the blade causes fluid rotation first clockwise and then counterclockwise at the edge of the next element.

Figure 10.27 illustrates conceptually the idea of mixing based on five elements. Assume that two fluids, a "dark" and a "light," each bifurcate at the edge of each element but retain their respective identities, i.e., do not mix. Assume that at the edge of the first element, the dark fluid enters the top half and the light fluid enters the lower half. At the edge of the second element, half of the dark goes to the left and half goes to the right; at the same time, half of the light goes to the left and half goes to the right. In this manner, the second element has four discrete fluid identities, i.e., two dark and two light. The same kind of split occurs at the entrance to the third element, i.e., the four tagged fluid elements become eight. At the entrance to the fourth element, the 8-tagged fluid elements (4 dark and 4 light) become 16, i.e., 2^4. Finally, at the beginning of the fifth element, the 16 tagged fluid elements become 32 (16 dark and 16 light), i.e., 2^5. With a sixth element, the 32 tagged elements become 64 (32 dark and 32 light), i.e., 2^6. At this point, any two of the different tinted fluid elements are not distinguishable. If the flow should be turbulent, and we may assume that in most cases the Reynolds number is high enough such that turbulent mixing occurs within each element, mixing would be virtually complete at the end of the second or third element.

FIGURE 10.26 Static mixer in 25 mm line for 76 L/min (20 gpm) pilot plant (1995 dwh).

10.4.4.3.2 Headloss

The headloss from a static mixer is given by the same relationship as for pipe friction (Amirtharajah et al., 2001, p. 37), i.e.,

$$h_{\mathrm{L}}(\text{static mixer}) = \mathbf{f}(\text{static mixer})\frac{L(\text{static mixer})}{d(\text{pipe})}\frac{v(\text{pipe})^2}{2g} \tag{10.45}$$

where

h_{L}(static mixer) is the headloss across static-mixer elements (m)
f(static mixer) is the loss coefficient for static mixer (dimensionless)
L(static mixer) is the length of static mixer (m)
d(pipe) is the diameter of pipe for used for static-mixer elements (m)
v(pipe) is the calculated velocity of water in pipe in which static-mixer elements are located (m/s)
g is the acceleration of gravity (9.806650 m/s^2)

Figure 10.28 is a plot of **f**(static mixer) vs. **R**; the plot includes data for three proprietary static mixers, i.e., the Chemineer-Kenics KMS helical (Chemineer, Dayton, OH, 2002); the Sulzer ChemTech SMV (Koch-Glitsch, Wichita, Kansas; Koch, 2002); and the TAH spiral (TAH Industries, Robbinsville, New Jersey), where $\mathbf{R} = \rho \cdot v(\text{pipe}) \cdot d(\text{pipe})/\mu$; pipe diameters were $25 < d(\text{pipe}) < 51$ mm (1–2 in.), and $2 \leq$ (# elements) ≤ 6. The data points for the three static mixers (as obtained from Amirtharajah et al. 2001, pp. 37, 61–63) plotted coincident. The "**f**" loss factor in Figure 10.28, i.e., $1.5 \geq \mathbf{f}(\text{static mixer}) \geq 1.0$ for $2000 \leq \mathbf{R} \leq 30{,}000$, compares with $\mathbf{f}(\text{pipe}) \approx 0.06$ for $\mathbf{R} \geq 2000$, which is 15–20 times lower, but is consistent with a standard pipe-friction plot.

10.4.4.3.3 Design Criteria

The work of Amirtharajah et al. (2001) considered **R**, G, h_{L}, #elements, θ, and $G\theta$ but did not find a basis to recommend

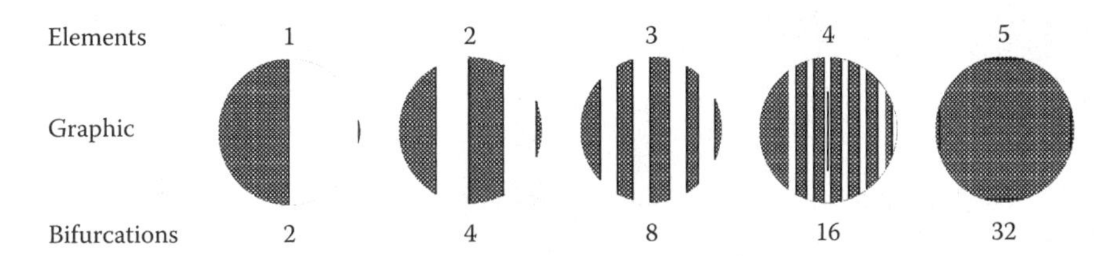

FIGURE 10.27 Illustration of effect of bifurcations. (Adapted from Avalosse, T. and Crochet, M.J., *AIChE J.*, 43(3), 589, 1997b.)

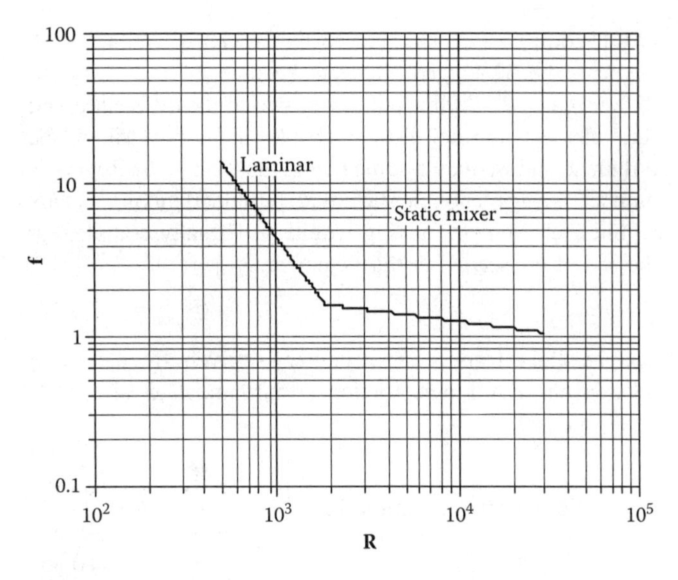

FIGURE 10.28 Static mixer friction loss coefficient as a function of **R**. (Adapted from Amirtharajah, A. et al., Mixing in Coagulation and Flocculation, Report 90841, American Water Works Research Foundation, Denver, CO, pp. 37, 61–63, 2001.)

design criteria for any of the foregoing parameters. They found that the number of elements made little difference for (#elements) > 2; using four or six elements did not, in general, result in lower effluent dosages or lower filter effluent turbidities. Amirtharajah et al. (2001, p. 37) also compared static mixers and back-mix reactors with respect to alum dosages and filter effluent turbidities and found, in general, lower alum dosages and lower effluent turbidities using a static mixer.

Table 10.13 illustrates the nature of data available from catalogs, e.g., giving length of units for different pipe diameters (≤610 mm or 24 in.) and for different model numbers (given as the number of elements). Units are available, however, as large as 1219 mm (48 in.).

Example 10.10 Static-Mixer Design

Given

Let $Q = 0.696\ m^3/s$ (20 mgd).

Required

Design a mixing system for alum coagulation.

Solution

1. Set up a spreadsheet as shown in Table CD10.14 to calculate R, h_L, G.
2. Add rows to explore the effects of pipe diameter, *d*(pipe), number of elements of static mixer selected, and variation in *Q*.
3. Develop "scenario" sets (each row is a scenario) to examine each of the foregoing independent variables:
   a. Effect of *d*(pipe). Select *d*(pipe) = 762 mm (30 in.). Smaller pipes result in much higher headloss and larger pipes give much lower *G* values.

TABLE 10.13
Length of Static-Mixer Elements for Different Pipe Diameters and Number of Elements

Pipe Diameter (mm)	(a) Length (mm[a]) KMS Model 2	3	4	6	12	18	24
13			114	156	292	432	568
19			140	203	384	568	754
25			178	257	492	727	962
38	152	210	270	391	740	1108	1476
51	191	267	333	498	965	1429	1908
64	222	318	410	594	1153	1708	2280
76	270	387	498	727	1416	2115	2813
102	346	502	648	948	1832	2775	3680
152	511	762	994	1454	2870	4277	5671
203	664	968	1295	1924	3800		
254	841	1232	1626	2435	4820		
305	994	1451	1959	2896	5213		
356	921	1327	1759	2648	5213		
406	921	1327	1759	2648	5213		
457	1124	1645	2216	3258	6483		
508	1124	1645	2216	3258	6483		
610	1327	1949	2623	3867	7703		

Pipe Diameter (in.)	(b) Length (in.)[b] KMS Model 2	3	4	6	12	18	24
0.50			4.50	6.1	11.5	17	22.4
0.75			5.50	8.0	15.1	22.4	29.7
1.0			7.00	10.1	19.4	28.6	37.9
1.5	6	8.2	10.6	15.4	29.1	43.6	58.1
2.0	7.5	10.5	13.1	19.6	38.0	56.2	75.1
2.5	8.8	12.5	16.1	23.4	45.4	67.2	90
3.0	10.6	15.2	19.6	29	55.8	83.2	111
4.0	13.6	19.8	25.5	37	72.1	109	145
6	20.1	30.0	39	57	113	168	223
8	26.1	38.1	51	76	150		
10	33.1	48.5	64	96	190		
12	39.1	57.1	77	114	205		
14	36.2	52.2	69	104	205		
16	36.2	52.2	69	104	205		
18	44.2	64.7	87	128	255		
20	44.2	64.8	87	128	255		
24	52.2	76.8	103	152	303		

Source: Kenics KMS models, 2002. (*Note:* Kenics is now, 2010, Chemineer-Kenics.)

[a] Metric dimensions calculated from units in (b).

[b] Dimensions from Kenics (2002, p. 20).

TABLE CD10.14

Example of Headloss, *P*, *G* Calculations for Static Mixer

$g = 9.80665\ m/s^2$

T	μ(water)	ρ(water)	Q		d(pipe)		v(pipe)			Elements	L(mixer)	h_L	P	V(mixer)	P/V	G
(°C)	($N\ s/m^2$)	(kg/m^3)	(mgd)	(m^3/s)	(in.)	(m)	(m/s)	R	f	(#)	(m)	(m)	(W)	(m^3)	(kW/m^3)	(s^{-1})
(a) Effect of d(pipe) and selection																
20	0.001312	999.7	20	0.696	18	0.457	4.2	1476889	1.0	4	1.96	3.922	26761	0.321	83.30	7968
20	0.001312	999.7	20	0.696	24	0.610	2.4	1107667	1.0	4	2.623	1.248	8513	0.766	11.12	2911
20	0.001312	999.7	20	0.696	30	0.762	1.5	886134	1.0	4	3.26	0.508	3468	1.487	2.33	1333
20	0.001312	999.7	20	0.696	36	0.914	1.1	738445	1.0	4	3.91	0.245	1673	2.570	0.65	704
20	0.001312	999.7	20	0.696	42	1.067	0.8	632953	1.0	4	4.57	0.132	903	4.081	0.22	411
20	0.001312	999.7	20	0.696	48	1.219	0.6	553834	1.0	4	5.22	0.078	529	6.092	0.09	257
(b) Effect of # elements																
20	0.001312	999.7	20	0.696	30	0.762	1.5	886134	1.0	2	1.63	0.254	1734	0.744	2.33	1333
20	0.001312	999.7	20	0.696	30	0.762	1.5	886134	1.0	3	2.45	0.381	2601	1.115	2.33	1333
20	0.001312	999.7	20	0.696	30	0.762	1.5	886134	1.0	4	3.26	0.508	3468	1.487	2.33	1333
20	0.001312	999.7	20	0.696	30	0.762	1.5	886134	1.0	6	4.89	0.762	5202	2.231	2.33	1333
(c) Effect of flow variation																
20	0.001312	999.7	10	0.348	30	0.762	0.8	443067	1.0	4	3.26	0.127	434	1.487	0.29	471
20	0.001312	999.7	20	0.696	30	0.762	1.5	886134	1.0	4	3.26	0.508	3468	1.487	2.33	1333
20	0.001312	999.7	30	1.044	30	0.762	2.3	1329200	1.0	4	3.26	1.144	11705	1.487	7.87	2449
20	0.001312	999.7	40	1.392	30	0.762	3.1	1772267	1.0	4	3.26	2.033	27746	1.487	18.66	3771
Given	Calculated	Calculated	Given		Trial		$v = Q/A$	$\mathbf{R} = \rho v d/\mu$	Amirtharajah et al. (2001)			$h_L = f(L/D)(v^2/2g)$		$V = A \cdot L$		
										Trial	Calculated based on units of Kenics (2001)					
													$P = Q \cdot (rg) \cdot h_L$			

b. Effect of #elements. Now examine the effect different numbers of elements. A four-element static mixer has only 0.508 m of headloss with 3.468 kW (4.6 hp) and $G\theta \approx 2850$ ($\theta = 2.1$ s).
c. Effect of flow variation. The third scenario set is to explore effect of Q on h_L, P, and G. The G value is on the low side at 0.348 m^3/s (10 mgd), i.e., $G \approx 471$ s^{-1}. For $Q = 1.044$ m^3/s (30 mgd), $P = 11{,}705$ kW (15.6 hp) which would probably require a booster pump (unless the raw-water source has sufficient head).

Discussion
As seen by the spreadsheet exploration, there is no single definitive design (in terms of *d*(pipe) and #elements). The **R**-values are seen to be in the turbulent range, i.e., $\mathbf{R} \gg 2000$. The design is for illustrative purposes only. A manufacturer's catalog should be consulted to obtain *L*(mixer) and to confirm the value of **f**.

10.5 SUMMARY

Mixing has evolved from its first uses, c. 1900, toward being recognized as important, c. 1950, toward having both empirical and science foundations, c. 2000 (i.e., with CFD maturing as a science). There is a general awareness of its importance in numerous applications, e.g., coagulation, polymer mixing, dissolution of solid chemicals, gas dissolution, disinfection, aerobic bacterial reactions, and anaerobic bacterial reactions. Mixing may be the rate-limiting factor in most of these applications, i.e., the more intense the mixing, the more complete the reaction, up to the point that diffusion is rate controlling.

On understanding, the three transport mechanisms, linked as a sequence of steps, are (1) advection, (2) turbulence, and (3) diffusion. The first two mechanisms must be developed in any design to either promote direct contacts between reactants or to bring a given substance into the diffusion proximity of a surface, i.e., a "sink."

With an improved understanding of turbulence, as comprising perhaps a tangle of vortex tubes, and of the role of eddy size and particle size in particle transport, more sophisticated notions of the role of turbulence have evolved since the 1950s. While these ideas help to understand how turbulence affects particle transport and mixing, they have not been reduced to the format of design guidelines.

For design, the old standby of Camp and Stein, G, has not been discarded. Alternatively, P/V, of which G is a derivative, has been used by tradition. A limitation of either of these parameters is that there are not unequivocal criteria. Recommendations have vacillated for some 40 years.

While model studies are invaluable, the main value is to better understand the role of independent variables in the process being considered. Scale-up remains an art. This is evident from the number of parameters that may be operative, e.g., geometric scale and detention time, tip speed, Reynolds number, Froude number, and Power number. There is not compatibility in scale-up for more than two.

Regarding technologies, the three categories are (1) impellers, (2) jets, and (3) static. Within each category a number of specific technologies are listed. The "back-mix" reactor, a traditional technology within the first category, has inherent limitations in that each parcel of fluid entering the reactor has a different residence time (evident from a simple reactor analysis). The submerged jet (i.e., a high-velocity flow from a nozzle that disperses throughout the raw-water flow) within the second category has been favored since the 1990s in coagulation. Submerged jets have been used also in anaerobic reactors. The proprietary "static mixer" has found increasing use in various applications. Both of these have, to some extent, displaced the back-mix reactor. Some of the jet-mixer technologies, e.g., weirs and baffles, are "passive" in nature and may have appeal in places where mechanical systems may be "not appropriate" (such as in small systems). The "bottom-line" is that design of any mixing system remains an "art," i.e., there is not an "algorithm," or protocol, that leads to an unequivocal design.

PROBLEMS

10.1 Examples of G and θ Parameters in Practice

Given
Locate a nearby water treatment plant that has a rapid-mix basin, i.e., with an impeller. If it is not feasible to use data from an operating plant, make assumptions about the system. Assume the plant capacity is 0.876 m^3/s (20 mgd) and has two trains of 0.438 m^3/s (10 mgd) capacity each and that the system is a "Rushton" type (see glossary).

Required
For the foregoing rapid mix, calculate (a) G, (b) θ, and (c), $G\theta$. Do this for a full range of flow conditions, and summarize in tabular form, and compare with values for comparable plants (using data from the literature). What range in operator control is possible for G?

10.2 Turbidity

Given/Required
What is the turbidity of the input water and of the output water for a rapid-mix unit in a water treatment plant in your vicinity?

10.3 Chemical Feed

Given/Required
Describe the chemical feed methods into the rapid mix. Is the chemical injection a point source to the rapid mix? Is it a ring of point sources? Is the coagulant fed neat or diluted? If diluted, what is the pH of the feed solution? What should be the pH?

10.4 Thermal Energy

Given/Required
What is the order of magnitude of thermal energy, in kcal/mol, associated with diffusion (perikinetic) mixing?

10.5 Turbulent Energy

Given/Required

What is the order of magnitude of kinetic energy dissipated by turbulence associated with jet (orthokinetic) mixing? Assume the jet discharges through a nozzle from a reservoir with water surface 10.0 m above the centerline of the nozzle.

10.6 Role of G in Mixing

Given

Let d(particle) $\approx 1\ \mu$m and let $G \approx 1000\ \text{s}^{-1}$.

Required

Estimate the ratio of particle contacts by turbulence and diffusion. Reference is Section 10.3.1.4 and Figure 10.12.

10.7 Rapid-Mix Basin Design

Given

Rapid-mix basin in which, Q(plant-flow) $= 0.876\ \text{m}^3/\text{s}$ (20 mgd) divided into two treatment trains.

Required

Determine the following:

(a) Dimensions and geometry of rapid-mix basin
(b) Power of the mixing motor to be installed
(c) Size, type, and setting of impeller, and speed
(d) Chemicals required, sequence and rate of feed
(e) pH of rapid-mix basin
(f) Recommend rapid-mix equipment from available catalogs

10.8 Flow Variability

Given/Required

Address the effect of flow variability on design parameters, e.g., q, t_{5R}/q, G, P/Q, for a "Rushton"-type mixing system.

10.9 Laboratory Exercise on Scale-Up

Given/Required

This problem is intended to provide a pseudo "hands-on" laboratory experience. The scale-up will be inconclusive with conflicts seen between the various approaches, e.g., geometric, detention time, G, Reynolds, number, and power number. The exercise should be approached looking for a "half-full" glass, vis-à-vis one that is "half empty."

10.10 Back-Mix Reactor

Given

For a complete mix reactor, let $Q = 0.100\ \text{m}^3/\text{s}$, $V = 5.0\ \text{m}^3$.

Required

(a) Plot the C vs. t curve for a substance, A, fed continuously into the reactor. (b) Assume that the substance A is a neat alum solution. Describe what happens in the reactor with respect to the alum–particle reactions.

Hint: (a) Use the spreadsheet, Table CD10.7/CDprob10.10 (see CRC website for excel spreadsheets; Table CD10.7 and CDprob10.10 are identical), to calculate the curve, i.e., with the plot linked to the table. (b) Focus on the raw-water particles that will have varying residence times.

10.11 Complete-Mix Reactor

Given

A complete-mix reactor is used for coagulation.

Required

Discuss the problem of coagulation with a complete-mix reactor.

10.12 Imposing Similitude for Design*

Given

A Rushton impeller–basin (six blades) is to be designed based upon the characteristic **P** vs. **R** curve of Figure 10.14. The detention time is $\theta = 10$ s and $Q_p = 0.438\ \text{m}^3/\text{s}$ (10 mgd). Assume operation is in the turbulent range, i.e., $\mathbf{R} \geq 10^4$.

Required

Determine the basin size, the impeller diameter, D(impeller), rotational velocity, n, and power required by the impeller. Reference is Section 10.3.3.4.

*Indicates a more difficult problem.

10.13 Time for Five Passes through Turbulent Zone and 0.99 Blend

Given

Complete-mix coagulation basin has flow $Q = 0.100\ \text{m}^3/\text{s}$ and $\theta = 10$ s.

Required

(a) Determine the detention time, q(basin), and impeller speed required so that 0.99 fraction mixing occurs such that 0.90 remains in the basin. (b) Suppose $\theta = 20$ s. Determine n(pump impeller speed) for 0.99 fraction blend.

10.14 Power Required to Achieve Five Passes Through Turbulent Zone

Given

The problem statement is the same as in problem 10.13, i.e., a complete-mix Rushton-type coagulation basin has flow $Q = 0.100\ \text{m}^3/\text{s}$ and $\theta = 10$ s.

Required

(a) Determine the power required for problem 10.13 (a). The "solution for problem 10.13 (a) is given here for reference: as seen in Table CD10.7, the time if we accept the criterion that 0.90 fraction of the fluid should be subjected to five passes through the high-turbulence zone of the impeller, which will achieve 0.99 fraction blend (Section 10.4.1.2), then the associated $t/\theta = 0.10$, and if $\theta = 10$ s, as given, then, $t_{5R} = 1$ s. The task is to then calculate, $n = K/t$(five passes). Assuming we have a Rushton design, and that $\mathbf{R} > 1000$, then from Table 10.6, the "blend number," $n \cdot t_{5R} = 36$. Thus, $n \cdot (1\ \text{s}) = 36$, and $n = 36$ rps.

(b) Determine the power required for #10.13 (b). The "solution for problem 10.13 (b) is given for reference: If $\theta = 20$ s, $t/\theta = 0.10$, and thus, t_{5R}(required) $= 2$ s. Again, from Table 10.6, the "blend number," $n \cdot t_{5R} = 36$. Thus, $n \cdot (2\ \text{s}) = 36$, and $n \approx 20$ rps. In other words, with more time in the reactor, the pumping rate can be reduced.

Hint: (a) Determine from Equation 10.37 the time for $C/C_i = 0.90$, which is $t/\theta = 0.10$. This is the time t = set t(five passes). Then calculate, $n = K/t$(five passes).

10.15 Q(impeller) for Five Passes

Given
Back-mix reactor with Rushton impeller and let Q(basin flow) = 0.1 m^3/s and θ(basin flow) = 10 s.

Required
Determine Q(impeller).

10.16 Calculation of n for 0.99 Blending

Given
Select a Rushton impeller–basin system and let $n = 10$ rps (as a first trial to be adjusted after calculation of power, etc.) Assume, Q(raw-water) = 1.000 m^3/s.

Required
Calculate t_{5R} for five passes through the impeller and determine the detention time, θ(raw water). Reference is Section 10.4.1.2.

10.17 Volume of Back-Mix Reactor

Given
Assume that the raw-water flow for coagulation is $3.785 \cdot 10^6$ L/day (1.0 mgd). Assume that the coagulation chemistry regime is adsorption–destabilization.

Required
Determine the volume of a back-mix reactor. Reference is Sections 10.4.1.2 and 10.4.1.5.

10.18 Geometric Similitude

Given
Consider a Rushton impeller–basin system with four impeller blades. The system is to operate at $\mathbf{R} \geq 10^4$. Let $\theta = 10$ s and $Q_m = 0.076$ m^3/min (20 gpm), $Q_p = 0.438$ m^3/s (10 mgd).

Required
(a) Calculate V_m and V_p. (b) Determine the dimensions of each basin.

10.19 Imposing Similitude for Design

Given
A new impeller using a Rushton-type basin ($J/T = 0.10$) is to be tested by means of a model; for most tests, $\theta = 10$ s and $Q_p = 0.00126$ m^3/s (20 gpm). A variable speed direct current motor has been installed with maximum power 2 kW; assume that the efficiency is 0.7 and is constant over all rotational velocities. The model has been fitted with a bearing plate and scale to read torque; the rotational velocity is measured by a strobe. The maximum rotational velocity of the impeller is n(max) = 3600 rpm. For working purposes, assume that the curve follows the standard viscous flow range curve and that for the level part, $\mathbf{P} = K = 0.5$. The calculations will have to be revised after a final K is determined. A heat exchanger is used to cool circulated fluids.

Required
Design an experimental program such that a **P** vs. **R** curve is generated. The curve will be used to design prototype systems of various sizes.

10.20 Calculate Flow of Liquid Alum and Ratio of Alum to Raw-Water Flows

Given
For a raw-water flow, let Q(raw water) = 1.00 m^3/s (22.8 mgd) and let C(alum) = 26 mg as $Al_2(SO_4)_3 \cdot 14\ H_2O$/L.

Required
Calculate the flow of liquid alum required as a neat solution. Reference is Section 10.3.4.2.

10.21 Injection of Neat Alum

Given
For a raw-water flow, let Q(raw water) = 1.00 m^3/s (22.8 mgd) and let C(alum) = 26 mg as $Al_2(SO_4)_3 \cdot 14\ H_2O$/L. Neat alum solution, i.e., "liquid-alum," is to be injected into the core zone of a jet mixer that disperses the raw-water flow in a 500 mm (20 in.) pipe.

Required
Determine the nozzle size for a single jet of neat alum. Reference is Section 10.3.4.2.

10.22 Exploration of Mariotte Siphon for Small-System Alum Feed

Given
The engineer for a small community, e.g., 2000 persons, must select for alum-feed system for coagulation using a rapid mix. For reference, Example 10.4 provides a context.

Required
Consider whether the use of a Mariotte siphon (see Glossary) might be considered in lieu of a metering pump for alum feed (for a possible "passive" alum-feed system).

10.23 Design of Jet-Mixer System

Given
Raw water is delivered to a WTP by a 1067 mm (42 in.) pipeline from Horsetooth Reservoir with water surface elevation 61 m (200 ft) above the plant. Let $T = 20$°C for working purposes.

Required
State how the operation should be adjusted to account for the flow variation. How would you design the system to provide for the operating flexibility?

10.24 Design of Second and Third Jet Rings

Given
Raw water is delivered to a WTP by a 1067 mm (42 in.) pipeline from Horsetooth Reservoir with water surface elevation 61 m (200 ft) above the plant. Over the annual cycle, $2 \leq T \leq 16$°C. Example data

and calculations for a jet system are given in the spreadsheet, Table CD10.12. Second and third rings of jets are to be designed for installation in planes 150 and 300 mm downstream of the first ring; the flow from each of the jet rings is radially toward the center of the pipe.

Required

Design two jet rings.

10.25 Jet-Mixer System Test Program

Given

Raw water is delivered to a WTP by a 1067 mm (42 in.) pipeline from Horsetooth Reservoir with water surface elevation 61 m (200 ft) above the plant. Over the annual cycle, $2 \leq T \leq 16°C$. Second and third rings of jets are installed in planes 150 and 300 mm downstream of the first ring; the flow from each of the jet rings is radially toward the center of the pipe. If used, these rings would require higher operating expense. The jets were installed because the philosophy of the plant management was to provide for uncertainty in operating effectiveness if only small added capital expenses are involved.

Required

(a) Outline a test program to determine optimum jet flow for a jet-mixer system.
(b) Describe alternative feed points for alum and how you would test for which is most effective.
(c) Outline a test program to determine whether the downstream jet rings should be placed in operation.
(d) Indicate, in your opinion, the effect of temperature variation.

10.26 Rationale for Design of Jet-Mixer System

Given

Table 10.10 from Kawamura (2000) shows that most of the operating plants have much lower power per unit of flow values than seen in Table CD10.12 (for right columns).

Required

Discuss reasons for the differences. Would you design based upon the Kawamura data as a guide or would you apply the rationale of Example 10.9 and the associated computational algorithm of Table CD10.12?

10.27 Static Mixer for Alum Addition

Given

A Kenics KMS Model 6 helical static mixer is to be used for alum mixing. Assume $Q = 1.0$ m^3/s and let d(pipe) = 457 mm (18 in.). From Table 10.13, for d(pipe) = 457 mm (18 in.), L(static mixer) = 3258 mm (128 in.). Assume liquid alum is to be fed; alum concentration = 25 mg $Al_2(SO_4)_3 \cdot 14H_2O$.

Required

Determine the headloss across the six-element static mixer and the P/V value. Reference is Section 10.4.4.3.

10.28 Design of End-Around Baffle Mixing System

Given

An end-around mixing basin is considered for alum coagulation. Let $Q = 0.044$ m^3/s (1.0 mgd), $w = 0.330$ m, slot = 0.200 m.

Required

Determine the power dissipated around the first bend and G. Also determine, y_1, y_2, B(channel), V(channel).

10.29 Over-and-Under Baffle System Design for Initial Mixing

Given

Let $Q \approx 0.088$ m^3/s (2 mgd) with variation, $0.044 \leq Q \leq 0.132$ m^3/s ($1 \leq Q \leq 3$ mgd) Also let $T = 20°C$ and assume C(alum) ≈ 20 mg/L as $Al_2(SO_4)_3 \cdot 14H_2O$ and is fed as liquid alum by means of a manifold placed above the water surface in the plane of the weir plate. Let the weir plates be adjustable (such as by wooden slats each about 10 cm (4 in. height); also assume the sluice gate can be raised or lowered as needed. A reference is Harhoff (1998).

Required

Determine the baffle settings for the two weirs and the sluice for the minimum, average, and maximum flows such that $G_1 \approx 1000$ s^{-1} for each flow. Show in a drawing, the water surface profiles. Table CDprob10.29 may be used to facilitate calculations.

10.30 Static-Mixer Design

Given

Let $Q = 0.348$ m^3/s (10 mgd) for flow through a static mixer.

Required

(a) Recommend a raw-water pipe diameter.
(b) Recommend a pipe diameter for a static mixer in the raw-water line (possibly resulting in a smaller pipe size).
(c) Calculate the associated G and repeat with as many trials as needed to arrive at a G value that seems most appropriate.
(d) Calculate the monthly energy cost for each alternative.

10.31 Design of Elbow Assembly Rapid Mix

Given

Flow of water is 76 L/min (20 gpm) into a pilot plant with a 51 mm (2 in.) PVC raw-water pipe. An elbow assembly is proposed as a rapid mix. A coagulant is to be added at the entrance to the assembly.

Required

Determine the size, d(pipe), for each elbow and the number, n(elbows), in the sequence.

ACKNOWLEDGMENTS

Professor Bogusz J. Beinkiewicz, Department of Civil Engineering, Colorado State University, spent several sessions with me to review the topic of turbulence with particular attention to Kolmogorov's microscale, and to provide advice on several books that were consulted on the topic. Professor Robert Meroney, Department of Civil Engineering, Colorado State University, also gave advice about turbulence and clarified points concerning "computational fluid dynamics", i.e., cfd.

Professor Adrian Hanson, Department of Civil Engineering, New Mexico State University, shared his knowledge on turbulence and on Kolmogorov's microscale in several sessions in Jackson, Wyoming, Fort Collins, and Angel Fire, New Mexico, over a 2 year period.

Kevin Gertig, Water Resources and Treatment Operations Manager, City of Fort Collins Utilities, provided photographs of mixing systems used at the Fort Collins WTP, and many other photographs, some of which were used in this chapter. In addition he was always available for discussions regarding his experiences with mixing systems. Bob Reed, superintendent, and Steve Hall, operator, Soldier Canyon WTP, Fort Collins, shared their experiences with jet mixers and provided access to Soldier Canyon WTP to examine their installations and spent considerable time in reviewing drawings of their jet-mixer systems.

Tim Rice, Municipal Treatment, Inc., Lakewood, Colorado, arranged for the use of Figures 10.3c and 10.24a and b with Glen P. Sundstrom, product promotion manager, Siemens Water Technologies Corp., Rockford, Illinois, who granted permission to use the figures and updated information on Siemens products. Sundstrom also provided high-resolution photographs of the Water Champ mixing system for use in the text.

All references to Spraying Systems Co. products and literature found in this book were provided with the permission of Spraying Systems Co., Wheaton, Illinois. Scott Monson, Monson & Associates, Olathe, Kansas, regional representative for Spraying Systems Co. (2010), arranged for permission to use catalog material (http://www.spray.com/misc/legal.asp).

Terry Naughton, director of order processes, Chemineer, Dayton, Ohio re-granted permission to use the materials as obtained from the Chemineer Website, and references to the Chemineer-Kenics™ brand name. The original contact with Naughton was through Jim Martinson, Falcon Supply, Niwot, Colorado.

GLOSSARY

Advection: (1) Transport of a fluid mass; the transport generally is in response to a potential gradient as in pipe flow, an open channel, an ocean current, etc. (2) The transfer of a property of the atmosphere, such as heat, cold, or humidity, by the horizontal movement of an air mass; the horizontal movement of water, as in an ocean current. (3) Latin: advecti, advectin, *act of conveying*; from advectus, past participle of advehere, *to carry*. (2) and (3) From *Merriam-Webster Online*: http://www.merriam-webster.com on November 27, 2001). See also "convection." Some difference of opinion relates to whether "advection" or "convection" should be used in describing bulk movement of fluids; "advection" has been adopted here and seems to be favored in the literature and by those with whom I have discussed the matter.

Agitation: Induced motion of a material in a specified fashion, usually a circulatory pattern within a vessel (McCabe et al., 1993, p. 235). Purposes of agitation, they say (p. 236), are: to suspend solid particles, to blend miscible liquids, to disperse a gas through a liquid in the form of small bubbles, dispersing a second liquid, immiscible with the first, to from an emulsion of suspension of fine drops. See also, "mixing" and "blending."

Axial flow: Advective flow along the axis of an impeller.

Back-mix reactor: (1) Mixing between the fluid elements which have been present in the reactor for different lengths of time (Stenquist and Kaufman, 1972, p. 7). The consequence is, they stated, that reactants entering the mixer may "react" with previously formed "reaction" products rather than with the constituents of the water stream of concern. (2) See "CSTR," "recycle reactor," or "mixed reactor" (Levenspiel, 1972, p. 97). (3) The connotation is that flow returns "back" through the reaction zone, either by returning a discrete flow or by the circulation caused by an impeller in a closed tank. The raw water once blended with coagulant chemicals, for example, is recycled back through the reaction zone (see also Kawamura, 1991, p. 74). (4) Amirtharajah (1979, p. 133) designates a "mechanical mixer" as a back-mix reactor.

Baffle: In mixing, a flat blade attached to the side of a tank, oriented vertically.

Basin: A defined volume usually associated with some purpose such as storage, settling, reaction, etc. A tank is volume having geometric proportions, which may or may not be true for a basin.

Blender, in-line: Mixer placed in a pipeline.

Blending: Mixing to achieve molecular solution of one species with another, e.g., of a saline solution with a water solution, or of a tracer dye solution with a water solution. A criterion for achievement of sufficient blend may be set arbitrarily. For batch blending this could be at the time when the concentration is 0.999 times the concentration at mixing at time equals infinity (the latter could be defined as the time when further change in concentration is not detectable). If the blending is to be achieved for a continuous flow reactor, e.g., with an impeller in a tank, with a relatively large flow of raw water and a very small flow of coagulant, such as a neat liquid alum solution, the blending may be considered to be achieved when the concentration of alum is 0.999 times the theoretical concentration of mixing when the two proportions are combined. The efficiency of the impeller–tank system may be evaluated by means of a dye or salt solution being injected at the desired location in the system and under the desired conditions (raw-water flow and rotational velocity of the impeller) with measurements taken from the effluent stream. If the blending is not efficient, the concentration of the outflow may be much less than the potential. Once the conditions are attained to meet the blending criterion, additional mixing may be necessary if reaction is involved.

Coagulation: (1) The driving together of colloidal particles by chemical forces (ASCE-AWWA, 1969, p. 65). (2) Colloid destabilization due to reduction of zeta-potential. (3) Destabilization of colloidal particles through the addition of coagulant to water (Stenquist and Kaufman, 1972, p. 34).

Complete-mix reactor: Continuous flow reactor, in which the reactor contents are homogeneous and are the same as the effluent concentration. Mathematically, the distribution of residence times is declining exponential. If two or more reactors are arranged in series the effluent distribution curve, i.e., $C(t)$, becomes S-shaped, with the steepness of the S-curve increasing as the number of reactors increases. See also *CSTR*.

Computational fluid dynamics: The term computational fluid dynamics, abbreviated, "CFD," refers solution of the Navier–Stokes equation (along with the continuity equation) for a given set of boundary conditions by means of a finite element computation utilizing a computer. The solution is displayed, as a rule, in terms of a velocity field or a pressure field or as a flow net (depending on conditions) or by means of color coding. If the solution is unsteady, it may be displayed in animated format.

CSTR: Constant flow stirred tank reactor; reactor contents are uniform throughout; also called "mixed reactor" and "back-mix reactor" (Levenspiel, 1999, p. 91). Also, the reactor is called a "complete-mix" reactor. The essential feature is that the contents of the reactor are homogeneous at any instant in time and the effluent concentration is the same as the reactor. See also "complete-mix" reactor.

Dead zone: The residence time of a portion of the fluid is much greater than the average, i.e., $t_p \gg \theta$, where, t_p is the residence time of any given parcel of fluid. See also, "short-circuiting."

Disperse: (1) Exchange of mass between adjacent coordinates in fluid flow. (2) To scatter,.. or to send in different directions (*Oxford American Dictionary*, Oxford University Press, New York, 1980).

Dispersion: (1) A mixture, (2) see glossary, Chapter 4.

Draft tube: A tube with diameter about the same as the associated impeller, positioned with axis coincident with the impeller that directs advective flow into the eye of the impeller. The impeller may be positioned above, within, or below the draft tube.

Eddy: A circular motion within a fluid induced by a surface discontinuity, e.g., a flat plate normal to the velocity vector, or a turbulent shear, i.e., velocity gradient, or a submerged jet. A radial-flow impeller in a mixing basin induces eddies by the submerged jet (the flow from the impeller), by the eddies induced by the flat blade as it rotates, and by the general circulatory motion induced by the pumping action (advective flow). The terms, "eddy" and "vortex" are used interchangeably (Hanson, 1989). The modern concept of turbulence is, however, in terms of vortices. See also *vortex*.

Emulsion: A dispersion in which one liquid is not miscible with water.

Eulerian coordinates: (1) A coordinate system in which the observer is stationary and the mathematics of motion are with respect to a stationary reference. (2) The analysis of occurrences at a fixed point is usually called the Eulerian method (Rouse and Ince, 1957, p. 107).

ε: Defined, $\varepsilon = (P/V\rho)$, which is energy dissipated by turbulence per unit mass of fluid.

Field: The loci of a given vector or scalar quantity, e.g., concentration, pressure, velocity, acceleration, force, gravitation, magnetic that is usually represented graphically in space as a result of computations or experimental measurements. If the field is unchanging with time, then it is steady state; if changing with time, then the field is unsteady.

Flash mix: Short residence time CSTR utilizing an impeller with high rotational velocity to produce homogeneous mixtures through advective dispersion and turbulent mixing (Stenquist and Kaufman, 1972, p. 7). See also *rapid mix*.

Flash mixing: Mixing of coagulant chemicals with raw water; a short detention time is implied, e.g., ≤ 1 s. Usually, a flash mixer is the same as an "in-line" mixer, i.e., an impeller in a pipe, but may be any device that distributes a coagulant and generates turbulence (e.g., a hydraulic jump, baffles, submerged jets, an impeller–basin system) As a rule, the effectiveness of the mixing depends on the intensity of turbulence, a short detention time, high turbulence form of initial mixing, i.e., rapid mixing.

Flocculation: (1) Refers to the assimilation of coagulated particles into floc particles (ASCE-AWWA, 1969, p. 65). (2) The purpose of flocculation is to remove particles in the 1 μm size range and place them in the ≥ 10 μm size range so that the inertial and gravitational forces will transport the particles (Hanson and Cleasby, 1990, p. 68). (3) Collision and aggregation of destabilized particles into relatively large aggregates known as flocs. The term is applied to the hydrodynamics of aggregation and floc formation (Stenquist and Kaufman, 1972, p. 34).

Flow number: The flow number is defined, $\mathbf{Q} = Q(\text{impeller})/(nD^3)$.

G: Defined, $G = (P/V\mu)^{1/2}$, which has the dimensions of velocity gradient, e.g., s^{-1}. An alternative equation which is equivalent to $G = (P/V\mu)^{1/2}$ and used frequently by researchers is $G = (\varepsilon/\nu)^{1/2}$.

Impeller: A device on a shaft intended to cause motion of a fluid. Two broad categories of impellers are axial flow and radial flow.

Initial mixing: Injection of one or more chemicals (e.g., alum, a base, an acid, a polymer individually or in sequence) into a flow of raw water with mixing by turbulence. Synonymous with flash mixing, rapid mixing. The term has been used sometimes in lieu of the term "rapid mix" (Stenquist and Kaufman, 1972, p. 7; Letterman et al. 1973, p. 716; Chao and Stone, 1979, p. 570, Kawamura, 2000, p. 307).

In-line mixers: An in-line mixer is placed in the raw-water pipe and has become used increasingly since the early 1970s.

Jet: A high-velocity fluid flow that has significant penetration into a fluid medium, the latter being of relatively large flow (for in-line mixing) or of large mass (for mixing in a tank).

Jet Mixer: A nozzle, or a collection of nozzles, that emits high-velocity jets for mixing. The resulting submerged jet entrains the substance to be mixed within the subsequent zone of turbulence.

Kolmogorov microscale: The smallest range of eddies has been termed the "universal equilibrium range. It has been further divided into a low eddy size region, the viscous dissipation sub-range, and a larger size region, the inertial convection sub-range. The two sub-ranges are divided by the Kolmogorov microscale (Cleasby, 1984, p. 878).

Lagrangian coordinates: (1) A coordinate system in which the observer moves with the advection of a particle and describes the motion of particles relative to the moving frame of reference. (2) Analysis of the fate of a particular particle [Named after Joseph Louis Lagrange (1736–1813). Lagrange was born in Turin of French ancestry and in 1776 took Euler's professorship at Berlin; he also succeeded Euler as the world's leading mathematician (from Rouse and Ince, 1957, p. 107).]

Marine impeller: An axial flow impeller with three large blades with constant pitch from tip to axis.

Mariotte siphon: A closed container with a submerged air tube, which maintains constant head as a liquid surface lowers in elevation, which maintains a constant flow from the container. Usually, the system is used in laboratory systems which can be sealed easily, e. g., using a 19 L (5 gal) carboy. Any engineered system for alum feed should be tested first using a carboy with a section of tubing sized to give the desired flow; in addition, clogging and cleaning frequency would be concerns.

Mixed-flow reactor: See CSTR.

Mixed reactor: As given by Levenspiel (1972, p. 97), an "ideal" reactor in which the contents are stirred and homogeneous; thus, the exit flow has the same composition as the fluid within the reactor and is also called a "mixed flow reactor," or a "backmix" reactor, or a "CFSTR" (continuous flow stirred tank reactor). See also *CSTR*.

Mixing: Random distribution, into and through one another, of two or more initially separate phases (McCabe et al., 1993, p. 235). See also, *agitation* and *blending*.

Mixing length: When a fluid mass with velocity, v, is displaced due to turbulent motion in the transverse direction, e.g., from y_1 to y_2, with a change in velocity Δv (the velocity fluctuation in the x-direction; the distance $(y_2 - y_1)$ is the mixing length, l_m.

Model: A representation of some portion of reality (see Chapter 4). With respect to a mixing impeller–basin system, a model is a smaller version of the larger system that it is intended to simulate.

Navier–Stokes Equation: Mathematically, the Navier–Stokes equation, an expansion of Newton's second law, is expressed as a partial differential equation (see, for example, Rouse (Ed.), 1959, p. 207; Munson et al., 1998, p. 361) that has the form, in vector notation,

$$-\nabla p + \rho \mathbf{g} + \mu \nabla^2 \mathbf{v} = \rho\left(\frac{\partial \mathbf{v}}{\partial t} + \mathbf{v} \cdot \nabla \mathbf{v}\right) \quad (10.21)$$

where

p is the pressure (N/m^2)
$\mathbf{g}$ is the acceleration of gravity (m/s^2)
$x, y, z =$ are coordinate directions (m)
u, v, w are velocities of fluid in x, y, z directions, respectively (m/s)
$\mathbf{v}$ is the velocity vector for infinitesimal mass (m/s)

The equation expresses all of the variables that influence fluid motion, such as acceleration, pressure, viscosity, surface energy, elastic energy, etc. Its solution provides a full depiction of a flow field or an acceleration field or a pressure field, etc. Analytical solutions are not feasible but finite difference solutions started about the decade of the 1970s. Since the mid-1990s, commercial software has been available to provide solutions which, nevertheless, still require knowledgeable persons to formulate.

Newtonian fluid: A fluid which conforms to the relationship that shear is proportional to velocity gradient, i.e., $\tau = \mu dv/dy$.

Orthokinetic: (1) Collision of suspended particles due to their motion as induced by fluid motion (Langelier and Ludwig, 1949, p. 165; Argaman and Kaufman, 1968, p. 5). The latter stated that Smoluchowski was believed to have selected the term "ortho" to describe the ordered nature of shear flocculation in contrast to the disordered nature of Brownian or perikinetic flocculation. They use the term as referring to any form of flocculation caused by fluid motion. (2) Greek: Rectangular, upright (*Aa Dictionary*, Apple, Inc., 2007); the rationale of Smoluchowski could have been that the collision rate in laminar flow is normal to the velocity vector, i.e., the velocity gradient is dv/dy.

Paddle: Flat-blade impeller with surface normal to flow-inducing radial flow. Sometimes, blades are pitched. Otherwise, there is no axial flow. They rotate at slow to moderate speeds, e.g., 20–150 rpm, with axis located in the center of a vessel. The diameter is about 0.5–0.8 times the diameter of the vessel with width 0.17–0.1 times the impeller diameter. Baffles are necessary to prevent rotation of the fluid mass and the development of a vortex (McCabe et al., 1993, p. 237).

Perikinetic: (1) Collision of suspended particles due to their motion as induced by Brownian motion (Langelier and Ludwig, 1949, p. 165; Argaman and Kaufman, 1968, p. 5). (2) Greek: prefix; around, about (*Aa Dictionary*, Apple, Inc., 2007).

Plug-flow reactor: An "ideal" reactor characterized by orderly flow with no element overtaking or mixing with any other ahead or behind; the necessary and sufficient condition is for the residence time in the reactor to be the same for all elements of the fluid (Levenspiel, 1972, p. 97). For a "plug-flow" reactor, the average residence time is $\theta = V(\text{reactor})/Q$.

Power number: A dimensionless number, defined as, $\mathbf{P} = P/(\rho n^3 D^5)$. The power number is an identity with the drag coefficient.

Pressure field: The description of pressures at points in space for a given system at a given time.

Primary particles: The particles that are the target of coagulation as contrasted with the product particles formed as a result of coagulation.

Propeller: Axial flow, high speed, 1150–1750 rpm or 400–800 rpm for larger sizes, impeller that causes currents in an axial direction that continue until deflected by a floor or wall. The flow plume is helical and entrains adjacent stagnant fluid. Because of the persistence of the currents, propellers are effective for large vessels. A standard, three-blade marine impeller is most common; size rarely exceeds 457 mm (18 in.) regardless of the size of the vessel. In a deep tank, two or more impellers may be mounted on the same shaft. Shear is high at the surface of the impeller (McCabe et al., 1993, p. 237).

Prototype: The term prototype means a full-scale version of a system that was previously simulated by a model. The terms "model" and "prototype" go together.

Pumping number: Same as flow number, **Q**.

Rapid mix: Mixing of coagulant chemicals, usually associated with high-intensity turbulence generated, as a rule, by an impeller–basin system. Synonymous with flash mixing, initial mixing. Related definitions are: (1) A mixer in water treatment intended to promote the coagulation process. (2) The function is to disperse the applied coagulant or other chemicals throughout the water to be treated.

Recycle reactor: The product stream from a plug-flow reactor is recycled back to the entrance (Levenspiel, 1999, p. 136). As the recycle flow increases, the behavior of a plug flow reactor approaches a CSTR. Recycle is a means to obtain various degrees of "backmixing" with a plug-flow reactor (p. 136).

Rushton: The "Rushton" impeller and mixing basin system became a standard reference for mixing systems. The proportions of such a basin are given by McCabe et al. (1993, p. 242) and Rushton et al. (1950a,b). Table G10.1 gives nomenclature and proportions for a Rushton basin.

TABLE G10.1
Proportions for Rushton Basin

Proportion	Fraction	Proportion	Fraction
Tank shape is round	Round	Baffles =	4
$D(\text{impeller})/T$ =	0.33	$W(\text{blade})/D(\text{impeller})$ =	0.20
C/T =	0.33	$L(\text{blade})/D(\text{impeller})$ =	0.25
$H(\text{water})/T$ =	1.0	$J(\text{baffle})/H(\text{tank})$ =	0.083–0.10

Nomenclature
D(impeller) is the diameter of impeller (m)
T is the diameter of tank (m)
C is the distance from floor of tank to mid-point of impeller blade (m)
D(water) is the depth of water in tank (m)
H(tank) is the depth of water in tank (m)
W(blade) is the height of impeller blade, i.e., vertical dimension (m)
L(blade) is the length of blade, i.e., horizontal dimension (m)
J(baffle) is the width of baffle (m)

Scale-up: The idea of a model is to predict some portion of the behavior of the prototype and, at the same time, to try to scale up in terms of geometry, turbulence, power required, etc.

Shear: Hydraulic shear is given as $\tau = \mu dv/dy$; units are N/m^2.

Shear rate: Shear divided by viscosity, i.e., $\tau/\mu = dv/dy$; units are s^{-1}.

Short-circuiting: The residence time of a portion of the fluid is much less than the average, i.e., $t_p \ll \theta$, where, t_p is the residence time of any given parcel of fluid. See also, *dead-zone*.

Similarity: The idea of similarity applies to modeling as a primary application. Three kinds of similarity are as follows (Rushton, 1952, p. 34):

1. *Geometric similarity*: Two systems of different size have the same ratio of length for all corresponding boundaries and positions.

2. *Kinematic similarity*: Paths of fluid motion are geometrically similar and the ratios of velocities at corresponding points are equal to the ratios at other corresponding points.
3. *Dynamic similarity*: The ratio of masses and forces at corresponding points are equal to the ratios and other corresponding points, and geometric and kinematic similarity exist.

Other parameters of similarity include detention time, θ; power per unit volume, P/V; average velocity gradient, G; tip velocity, nD; etc.

Static mixer: Ordinarily a static mixer is considered as one of the proprietary elements placed in a pipe that has blades causing a flow bifurcation and helical twist to the next element where another bifurcation occurs. In a broader sense, a static mixer may be considered any obstruction placed in a flow that causes bifurcations or turbulence, e.g., wake turbulence, or a jet.

Submerged jet: A high-velocity fluid flow that enters a large mass of stagnant fluid such that a characteristic velocity profile with associated turbulence zone is created (see Albertson et al., 1950). In practice the jet may enter a larger flow of fluid (as opposed to a stagnant fluid mass) and the velocity profile and turbulence zone may deviate somewhat from those for the stagnant fluid mass.

Tank: A defined volume having defined geometric proportions. See *basin*.

Tracer: Dye or salt injected into a flow stream for the purpose of downstream detection to discern the shape of $C(t)$ curve.

Turbine: The term, turbine, is not defined explicitly, although it seems to be used in chemical engineering as a radial-flow-type impeller. In this sense, turbines have a variety of shapes, e.g., flat blade, six vanes normal to flow, flat blades mounted on a plate, curved blades with surface normal to flow, flat blades with surface at an angle to plane of rotation. Flow is both radial and tangential to the impeller circle. The currents persist throughout the vessel and are effective in penetrating what would otherwise be stagnant zones. High shear occurs in the vicinity of the impeller. Baffles are necessary to impede fluid rotation (McCabe et al., 1993, p. 237). [Axial flow impellers seem to be excluded from the definition of a turbine by McCabe et al.]

Turbulence: Random velocities of fluid superimposed on the general advective velocity vector.

Turbulence, homogeneous: Random velocities whose average properties are independent of position in the fluid (Batchelor, 1953, p. 1). Homogeneous turbulence is difficult to attain in laboratory situations but may be approached by a grid. Usually, real-world turbulence is more complex than homogeneous turbulence (p. 2).

Turbulence intensity: In general, the higher the amount of energy dissipated as random fluid motion with a given volume, the higher the turbulence intensity. Also, the higher the **R**, the higher the velocity gradient as G, the higher the turbulence intensity. A mathematical definition is: $I = \sqrt{\overline{(v')^2}}/v$ (see Section 10.3.1.2).

Turbulence, isotropic: (1) The turbulence characteristics are not dependent on the direction of the axes of reference. Isotropic turbulence can exist only when homogeneous turbulence already exists (Batchelor, 1953, p. 3). (2) Isotropic turbulence occurs if its statistical features do not change with rotation or reflection of a set of coordinate axes; all three components of the root mean square velocity variations must be equal, based on a rectangular coordinate system; in isotropic turbulence there is no shear stress and no gradients of mean velocity (Stenquist and Kaufman, 1972, p. 12).

Velocity field: (1) The description of velocity vectors in space for a given system at a given time. (2) Velocity field: The configuration of velocity vectors for a given set of boundaries. Thus the geometry of a basin and the impeller act as a system to result in a particular velocity field. Any changes in such a system will result in a different velocity field.

Vortex: (1) Curvature of surface in vessel caused by tangential velocity component induced by the rotation of impeller in center. At high impeller velocities, the vortex may reach the impeller. (2) Same as eddy (see eddy). Generally, a vortex is a rotational motion of fluid that decreases in diameter with length. (3) A vortex tube, i.e., an "eddy."

Vortex stretching: A vortex tube that extends in length; its diameter decreases as its length increases.

Wake: The whole region of nonzero vorticity on the downstream side of a body in an otherwise uniform stream of fluid (Batchelor, 1967, p. 348).

Wave number: A mathematically defined characteristic of turbulence given the symbol, κ, and defined as $\kappa = 1/\lambda$, in which λ is the length scale of the turbulence.

REFERENCES

Albertson, M. L., Dai, Y. B., Jensen, R. A., and Rouse, H., Diffusion of submerged jets, Paper 2409, *American Society of Civil Engineers Transactions*, 115:639–697, 1950.

Amirtharajah, A., Design of rapid mix units, in: Sanks, R. L. (Ed.), *Water Treatment Plant Design*, Ann Arbor Science Publishers, Ann Arbor, MI, 1979.

Amirtharajah, A. and Mills, K. M., Rapid-mix design for mechanisms of alum coagulation, *Journal of the American Water Works Association*, 74(4):210–216, April 1982.

Amirtharajah, A. and Tambo, N., Mixing in water treatment, in: Amirtharajah, A., Clark, M. M., and Trussell, R. R. (Eds.),

Static Mixers for Coagulation and Disinfection, American Water Works Research Foundation, Denver, CO, 1991.

Amirtharajah, A., Jones, S. C., Skeens, B. M., Heindel, H. L., Li, W., Hardy, S. A., and Latimer, R., Mixing in coagulation and flocculation, Report 90841, American Water Works Research Foundation, Denver, CO, 2001.

Argaman, Y. and Kaufman, W. J., Turbulence in orthokinetic flocculation, SERL Report No. 68-5, Sanitary Engineering Research Laboratory, University of California, Berkeley, CA, 1968.

ASCE-AWWA, *Water Treatment Plant Design*, American Water Works Association, New York, 1969.

Avalosse, T. and Crochet, M. J., Finite-element simulation of mixing: 1. Two-dimensional flow in periodic geometry, *AIChE Journal*, 43(3):577–587, March 1997a.

Avalosse, T. and Crochet, M. J., Finite-element simulation of mixing: 2. Three-dimensional flow through a kenics mixer, *AIChE Journal*, 43(3):588–597, March 1997b.

AWWA-ASCE, American Water Works Association, American Society of Civil Engineers, *Water Treatment Plant Design*, 3rd edn., McGraw-Hill, New York, 1998.

Babbitt, H. E. and Doland, J. J., *Water Supply Engineering*, McGraw-Hill, New York, 1939.

Bakker, A., The colorful fluid mixing gallery, http://www.bakker.org/cfm, April 2010.

Bakker, A., Oshinowo, L. M., and Marshall, E. M., *The Use of Large Eddy Simulation to Study Stirred Vessel Hydrodynamics*, The On-line CFM Book, http://www.bakker.org/cfm, updated August 22, 2000 (downloaded as an Adobe Acrobat pdf file).

Baron, T. and Alexander, L. G., Momentum, mass, and heat transfer in free jets, *Chemical Engineering Progress*, 47(4):181–185, April 1951.

Batchelor, G. K., *The Theory of Homogeneous Turbulence*, Cambridge University Press, Cambridge, U.K., 1953.

Batchelor, G. K., *An Introduction to Fluid Dynamics*, Cambridge University Press, Cambridge, U.K., 1967.

Bienkiewicz, B. J., personal communications, August–December, 2001.

Camp, T. R. and Stein, P. C., Velocity gradients and internal work in fluid motion, *Journal of the Boston Society of Civil Engineers*, 30:219–237, October 1943.

Chao, J.-L. and Stone, B. G., Initial mixing by jet injection blending, *Journal of the American Water Works Association*, 71 (10):570–573, October 1979.

Chemineer-Kenics, Static Mixing Technology, Dimension Sheet, Style KMS, CA812, Ref. Bulletin Series 800, http://www.chemineer.com/images/pdf/bulletin_800.pdf, 2010).

Clark, M. M., Critique of Camp and Stein's RMS velocity gradient, *Journal of the Environmental Engineering Division, ASCE*, 111(EE6):741–754, December 1985.

Clark, M. M., Mixing: Should We Really Ignore It?, Seminar, Association of Professors in Environmental Engineering, Vancouver, B.C., June 19, 1992.

Clark, M. M., *Transport Modeling for Environmental Engineers and Scientists*, John Wiley & Sons, New York, 1996.

Clark, M. M., Srivastava, R. M., Lang, J. S., Trussell, R. R., McCollum, L. J., Bailey, D., Christie, J. D., and Stolarik, G., *Selection and Design of Mixing Processes for Coagulation*, American Water Works Association Research Foundation Report, American Water Works Association, Denver, CO, 1994.

Cleasby, J. L., Is velocity gradient a valid turbulent flocculation parameter? *Journal of the Environmental Engineering Division, ASCE*, 110(EE5):875–897, 1984.

Fluent, Inc., Welcome to fluent online, http://www.fluent.com, June, 2001.

Frisch, U., *Turbulence—The Legacy of A.N. Kolmogorov*, Cambridge University Press, Cambridge, U.K., 1995.

Gemmell, R. S., Mixing and sedimentation, Chapter 4, in: *Water Quality and Treatment*, 3rd edn., McGraw-Hill, New York, 1971.

Gertig, K., personal communication, March 2002.

Guyon, E., Nadal, J., and Pomeau, Y., *Disorder and Mixing—Advection, Diffusion and Reaction in Random Materials and Processes*, NATO ASI Series, Vol. 152, Kluwer Academic Publishers, London, U. K., 1988.

Hansen, P., Treatment and purification of water progress and developments during 1936, *Journal of the American Water Works Association*, 29(10):1443, 1936.

Hanson, A. T., The effect of water temperature and reactor geometry on turbulent flocculation. Doctoral dissertation, Iowa State University, Ames, IA, 1989.

Hanson, A. T., personal communication, Angel Fire, NM, September 11–12, 2001.

Hanson, A. T. and Cleasby, J. L., The effects of temperature on turbulent flocculation: Fluid dynamics and chemistry, *Journal of the American Water Works Association*, 82(11):56–73, November 1990.

Hanson, A. T. and Srivastava, R. M., personal communication, Fort Collins, CO, April 13, 2001.

Harhoff, J., Design of around-the-end hydraulic flocculators, *Aqua*, 47(3):142–152, 1998.

Jorden, R., Rapid Mixing of Alum/Ferric for Efficient Coagulation In Drinking Water Treatment, Technical Application Note C.01, ClearCorp, Longmont, CO, 2001.

Kawamura, S., Considerations on improving flocculation, *Journal of the American Water Works Association*, 68(6):328–336, June 1976.

Kawamura, S., *Integrated Design of Water Treatment Facilities*, John Wiley & Sons, New York, 1991.

Kawamura, S., Optimization of basic water treatment processes – design and operation: Coagulation and flocculation, *Aqua—Journal of International Water Association*, 45(1):35–47, February 1996.

Kawamura, S., Initial (flash) mixing by water jet diffusion, *Aqua—Journal of International Water Association*, 49(6):307–319, December 2000.

Langelier, W. F., Coagulation of water with alum by prolonged agitation, *Engineering News Record*, 86:924, 1921.

Langelier, W. F. and Ludwig, H. F., Mechanism of Flocculation in the Clarification of Turbid Waters, *Journal of the American Water Works Association*, 41(2):163–181, February 1949.

Letterman, R. D., Quon, J. E., and Gemmell, R. S., Influence of rapid-mix parameters on flocculation, *Journal of the American Water Works Association*, 65(11):716–722, November 1973.

Levenspiel, O., *Chemical Reaction Engineering*, 2nd edn., John Wiley & Sons, New York, 1972.

Levenspiel, O., *Chemical Reaction Engineering*, 3rd edn., John Wiley & Sons, New York, 1999.

Logan, B. E., *Environmental Transport Processes*, John Wiley & Sons, New York, 1999.

McCabe, W. L., Smith, J. C., and P. Harriott, *Unit Operations of Chemical Engineering*, 5th edn., McGraw-Hill, New York, 1993.

Meroney, R. N., personal communication, August 18, 2001.

Munson, B. R., Young, D. F., and Okishi, T. H., *Fundamentals of Fluid Mechanics*, John Wiley & Sons, New York, 1998.

Mutsakis, M. and Rader, R., Static mixers bring benefits to water/wastewater operations, *Water Engineering and Management*, 133(11):30–34, November 1986.

Myers, K. J., Reeder, M. F., Ryan, D., and Daly, G., Get a fix on high-shear mixing, *Chemical Engineering Progress*, 95 (11):33–42, November 1999.

Oldshue, J. Y., *Fluid Mixing Technology*, McGraw-Hill, New York, 1983.

Oldshue, J. Y. and Trussell, R. R., Design of impellers for mixing, in: Amirtharajah, A., Clark, M. M., and Trussell, R. R. (Eds.), *Mixing in Coagulation and Flocculation*, American Water Works Research Foundation, Denver, CO, 1991.

O'Melia, C. R., Coagulation in Water and Wastewater Treatment, in Gloyna, E. F. and W. W. Eckenfelder, Jr., Water Quality Improvement by Physical and Chemical Processes, University of Texas Press, Austin, 1970.

Reed, R., Supervisor, Soldier Canyon WTP, personal communication, February 8, 2002.

Rouse, H., *Elementary Mechanics of Fluids*, John Wiley & Sons, New York, 1946.

Rouse, H., and Ince, S., *History of Hydraulics*, Iowa Institute of Hydraulic Research, State University of Iowa, Iowa City, IA, 1957.

Rouse, H., (Ed.), Appel, D. W., Hubbard, P. G., Landweber, L., Laursen, E. M., McKnown, Rouse, H., J. S., Siao, T. T., Toch, A., and Yih, C. S., *Advanced Mechanics of Fluids*, John Wiley & Sons, New York, 1959.

Rushton, J. H., The use of pilot plant mixing, *Chemical Engineering Progress*, 47(9):485–488, September 1951.

Rushton, J. H., Applications of fluid mechanics and similitude to scale-up problems — Part I, *Chemical Engineering Progress*, 48(1):33–38, January 1952a.

Rushton, J. H., Applications of fluid mechanics and similitude to scale-up problems — Part II, *Chemical Engineering Progress*, 48(2):95–102, February 1952b.

Rushton, J. H., How to make use of recent mixing developments, *Chemical Engineering Progress*, 50(12):587–589, December 1954.

Rushton, J. H. and Oldshue, J. Y., Mixing—Present theory and practice, *Chemical Engineering Progress*, 49(4):161–168, April 1953a.

Rushton, J. H. and Oldshue, J. Y., Mixing—Present theory and practice, *Chemical Engineering Progress*, 49(5):267–275, May 1953b.

Rushton, J. H., Costich, E. W., and Everett, H. J., Power characteristics of mixing impellers—Part I, *Chemical Engineering Progress*, 46(8):395–404, August 1950a.

Rushton, J. H., Costich, E. W., and Everett, H. J., Power characteristics of mixing impellers—Part II, *Chemical Engineering Progress*, 46 (9):467–476, September 1950b.

Saffman, P. G. and Turner, J. S., On the collision of drops in turbulent clouds, *Journal of Fluid Mechanics*, 1(Part 1): 16–30, 1956. Siemens Water Technologies Corp., Water Champ® Chemical Induction Unit 4 Submersible F Series Model No. SWC3F, Product Sheet, www.siemens.com/stranco, 2010. Note: The siemens web site is extensive and the foregoing provides links to many Water Champ® models as well as other kinds of technologies for water treatment.

Skeat, W. O. (Ed.), *Manual of British Water Engineering Practice*, 3rd edn., Published for the Institution of Water Engineers, W. Heffer & Sons Ltd., Cambridge, U.K., 1961.

Spraying Systems Co., Industrial Spray Products, Catalog 60A, Spraying Systems Co., Wheaton, IL, 2010 (see also http://www.spray.com).

Spraying Systems, A guide to optimizing in tank agitation and mixing using educators, Industrial Spray Products, Bulletin 635, Catalog 70, Spraying Systems Co., Wheaton, IL, http://service.spray.com/web/register/view_lit.asp?code = B635, 2010.

Stenquist, R. J. and Kaufman, W. J., Initial mixing in coagulation processes, Report to Environmental Protection Agency, EPA-R2-72-052, Office of Research and Monitoring, Washington, D.C., 1972.

Sulzer ChemTech, SMV Static Mixers, Mixing and Reaction Technology, Sclzer ChemTech, Ltd., Winterthur, Switzerland, www.sulzerchemtech.com, 2010.

Tennekes, H. and Lumely, J. L., A First Course in Turbulence, MIT Press, Cambridge, MA, 1999.

Vrale, L. and Jorden, R. M., Rapid mixing in water, *Journal of the American Water Works Association*, 63(1):52–58, January 1971.

Wilcox, D. C., *Basic Fluid Mechanics*, DCW Industries Inc., La Cañada, CA, 1997.

Willcomb, J., Floc formation and mixing basin practice, *Journal of the American Water Works Association*, 24(9):1416–1441, September 1932.

BIBLIOGRAPHY

The papers cited here are often cited in the literature but without complete description of title and source. These papers have had a prominent role in the history of the mixing and are listed for the convenience of interested persons.

Kolmogoroff, A. N., The local structure of turbulence in incompressible viscous fluid for very large Reynolds numbers, *Comptes Rendus de l'Académie des Sciences*, U.R.S.S., 30:301, 1941a (from Batchelor, 1951, p. 192).

Kolmogoroff, A. N., On degeneration of isotropic turbulence in an incompressible viscous liquid, *Comptes Rendus de l'Académie des Sciences*, U.R.S.S., 31:538, 1941b (from Batchelor, 1951, p. 192).

Kolmogoroff, A. N., Dissipation of energy in locally isotropic turbulence, *Comptes Rendus de l'Académie des Sciences*, U.R.S. S., 32:16, 1941c (from Batchelor, 1951, p. 192).

Kolmogoroff, A. N., On the disintegration of drops in turbulent flow, *Doklady Akademii Nauk S.S.S.R.*, 66:825, 1949 (from Batchelor, 1951, p. 192).

von Smoluchowski, M., Drei Vortrage uber Diffusion, Brownsche Molekular Bewegung und Koagulation von Kolloidteilchen, *Physik. Z.*, 17:557, 1916 (from Argaman and Kaufman, 1968, p. 161).

von Smoluchowski, M., Versuch einer mathematischen Theorie der Kaogulationskinetick kolloider Lösungen, *Zeitschrift für physikalische Chemie*, 92:155, 1918 (from Argaman and Kaufman, 1968, p. 161).

11 Flocculation

Flocculation is the unit process in which larger particles are formed from smaller particles due to collisions between them. The collisions occur by the "transport" mechanisms of velocity gradients (induced by laminar flow or turbulence), Brownian motion, or a sequence of the two.

In water treatment, the starting point is the rapid mix where "microflocs" form from coagulation chemicals and which may incorporate "primary" particles, e.g., mineral turbidity, microorganisms, and other "microscopic particulates." The next step is flocculation; the objective is to cause collisions such that the microflocs grow in size to become "floc" particles. If the treatment train is conventional filtration, the flocculation objective is to produce large settleable flocs. In flotation, on the other hand, the objective is to form a floc of small size, e.g., ≤ 50 μm, that rises readily when small gas bubbles attach.

The common flocculation technology is the paddle wheel. Other technologies include impeller basins, baffles, and proprietary innovations. These various technologies provide the "transport" mechanisms that induce floc collisions.

11.1 DEFINITIONS

The flocculation terms used here were based on usage in the literature. In a few cases, the literature is ambivalent and therefore some liberties were taken in gleaning what was believed the most coherent definition that fitted with past usage. The glossary provides a more inclusive listing.

11.1.1 FLOC

Floc is an agglomeration of "primary particles" combined with "microflocs" (see Sections 9.1, 9.1.1, and Glossary of Chapter 10); the latter are usually of diameter, $d \leq 1$ μm. The flocs grow in size by collisions with one another, i.e., by "flocculation." Size may vary from "pinpoint," i.e., $5 \leq d \leq 10$ μm, to perhaps 1–5 mm depending on the turbulence intensity. A general description is that a floc is an aggregate with a complex random structure with low average density (Meakin, 1989, p. 250).

11.1.1.1 Biological Floc

Biological floc is an agglomeration of the diverse microorganisms as found in a biological reactor. As it occurs in an activated sludge reactor, the biological floc is termed "mixed liquor suspended solids" (MLSS). The character of the floc is important in settling, i.e., as in "final settling." Filamentous bacteria, for example, result in floc that does not settle well.

11.1.1.2 Chemical Floc

Floc formed from a chemical substance as the primary particles is *chemical floc*. The most common ones are those formed from alum, ferric ion, lime, etc.

11.1.1.3 Primary Particles

The smaller particles that comprise floc are called *primary particles*; generally the size is $d \leq 1$ μm. In the case of chemical flocs, the primary particles are *microflocs* (Ham and Christman, 1969, p. 483); see also Chapter 9. In the case of bioflocculation, the primary particles are individual microbes. Also, primary particles may include viruses, bacteria, cysts, other microscopic organisms, and turbidity.

11.1.2 FLOCCULATION

The term *flocculation* is the process of causing collisions between

- Primary particles
- Primary particles and floc
- Floc particles and other floc

The purpose of flocculation is to cause growth in particle size such that the floc will settle *or* be amenable to flotation. Effective flocculation requires that the colliding particles adhere to one another.

11.1.2.1 Orthokinetic Flocculation

Collisions induced by turbulence is called *orthokinetic* flocculation (Langelier and Ludwig, 1949, p.165; Argaman and Kaufman, 1968, p. 5). The collision process is most effective with eddy sizes that are approximately the same size as the floc particles (Casson and Lawler, 1990, p. 68). Large eddies are not effective except to maintain the particles in suspension.

11.1.2.2 Perikinetic Flocculation

Collisions induced by Brownian motion is called *perikinetic* flocculation. The terms "perikinetic flocculation" and "orthokinetic flocculation" were used by Langelier and Ludwig (1949, p. 165), as adopted from the field of soil chemistry. Later Argaman and Kaufman (1968, p. 5) used the terms and credited their first use to Smoluchowski in his 1916 and 1918 papers.

11.1.2.3 Flocculent

A polymer added prior to flocculation to add a "toughness" to the floc is called a *flocculent*. Note that here we say that a chemical added before rapid mix is a "coagulant" and a chemical added before flocculation is a "flocculent"; often the literature does not make this distinction, calling both "flocculents."

11.2 APPLICATIONS

Flocculation is applied to a variety of applications in water treatment. Common ones are mentioned here.

11.2.1 Conventional Filtration

Flocculation is a part of the treatment train in conventional filtration, i.e., rapid mix, flocculation, settling, filtration. The objective is to cause floc growth to a size such that a high settling velocity results.

11.2.2 Direct Filtration

Flocculation is a part of the treatment train in direct filtration, i.e., rapid mix, flocculation, filtration. The objective is to create a floc size that will penetrate the filter media, such as a pinpoint floc. In general, the floc size should be less than what is visible, e.g., $\leq 30\ \mu m$, as in the case of water from Deer Creek Reservoir in Utah for which $G\theta \approx 42{,}000$ was satisfactory (Treweek, 1979, p. 100). Higher $G\theta$ values resulted in larger settleable flocs; the latter were not suitable for direct filtration since they did not penetrate the filter bed.

11.2.3 Flotation

In flotation, flocculation follows coagulation. The floc size is important; Edzwald (1995, p. 12) suggested that $10 \leq d(\text{floc}) \leq 100\ \mu m$, with median size 40 μm, the smaller sizes being more desired. There are no specific guidelines for G and detention time, but Edzwald (1995, p. 16) found that for South African practice, the range was $50 \leq G \leq 120\ s^{-1}$ and $4 \leq \theta \leq 15$ min.

11.2.4 Activated Sludge Floc Settling

The floc that forms in an activated sludge reactor should settle readily. Practice has found that such floc needs no induced flocculation and is ready to settle when it reaches the final settling basin, depending on the design of the basin (Section 6.6.7). Filamentous floc, on the other hand, does not settle readily, and is an issue in operation.

11.2.5 Softening

In "softening," i.e., the removal of calcium and magnesium hardness by lime precipitation, a solid precipitate is formed of $Ca(OH)_2$ or $Mg(OH)_2$. The precipitate grows in size to form a floc that will settle.

11.2.6 Tertiary Treatment

Removal of orthophosphate from secondary municipal wastewater effluent is formed by precipitation with Ca^{2+}. The precipitate, $Ca_3(PO_4)_2$, is the basis for a floc that grows in size for subsequent settling.

11.3 HISTORY

In *The Quest for Pure Water*, M.N. Baker (1949, p. 299) noted that coagulation as an aid for the clarification of household water supplies has been practiced since ancient times. At large industrial plants, coagulation was established sometime before 1830 (p. 304). Then in 1885, the Hyatt brothers, in developing their proprietary "mechanical" filters, found that the addition of alum improved performance; this was the beginning of coagulation and the subsequent flocculation in water treatment practice.

11.3.1 Practice

The role of alum in causing removal of colloidal particles was documented in 1869 by a commission in the Netherlands. They described a "flocculent precipitate" which took up the turbidity of water and left it perfectly clear (Baker, 1949, p. 308).

11.3.1.1 Quiescent Basins

Initially, alum was used in fill-and-draw basins prior to filtration. Such a basin can be seen at the Fort Collins Water Treatment Plant #1 in the Cache La Poudre River Canyon (see Figure 10.1). Alum was added to the raw water at a "chemical house"; the treated water then flowed by open channel to an open basin that provided about 24 h detention time. There was no designed "agitation"; the coagulation occurred passively in the open channel and the flocculation was in the basin itself. Brownian motion and differential settling were the mechanisms for contacts. The basin was built probably between 1910 and 1925 (the filters were constructed about 1910). In some cases, in the 1900s, coagulation and flocculation were the only treatments with the latter occurring in "settling reservoirs." Examples included Omaha, Nebraska in 1889; Chester, Pennsylvania in 1901; Kansas City, Missouri in 1902; Nashville, Tennessee in 1908 (Baker, 1949, p. 311). In such cases, flocculation was a "passive" process.

11.3.1.2 Langelier's Paddle Wheels

In 1916, at the invitation of Professor Charles Gilman Hyde, Professor Wilfred F. Langelier moved from the Illinois State Water Survey as a staff chemist to the University of California, Department of Civil Engineering. Professor Hyde was active as a consultant and in 1919 he asked Professor Langelier to review preliminary drawings for a water filtration plant for the City of Sacramento and suggest changes, with particular attention to coagulation with alum prior to filtration (Chall, 1970, p. 23). Prior to this time, Professor Langelier had observed that unsatisfactory performance of small filtration plants was traceable usually to poor coagulation. On one occasion, for example, he had collected a sample of clear filter effluent and observed clouding after stirring with a pencil. Also, alum sludge was found in the water mains in large quantities, and "it was obvious that either insufficient mixing or insufficient time had elapsed between the addition of coagulant and the passage of water through the sand." Sedimentation of the "coagulum" prior to filtration was a recognized part of the treatment train. Initially, the Sacramento plant had ample sedimentation detention time but did not have provision for thorough mixing of the alum after its addition. This led to dramatic breakthroughs in water treatment, described in Langelier's words in an oral history interview by Chall (1982, pp. 24–26):

> It immediately occurred to me that I could demonstrate in our laboratory the advantages of thorough mixing in effecting good coagulation. I proposed this to Professor Hyde who readily agreed with my suggestion. I devised an apparatus for the demonstration. It consisted of six in-line, one liter, clear glass jars, each provided with a slowly revolving paddle. The paddles were driven by a motorized shaft equipped with mitre gears.
>
> The first test runs were made with muddy water from the Sacramento River, and from the beginning our results were strikingly successful and far beyond our expectations. The multi-jar feature made it possible to observe the effect of any one variable; for example, the coagulant dosage, mixing time period, or rotational speed on the effectiveness of floc formation and subsequent clarification through sedimentation.
>
> In these early tests, the outstanding observation was that in the absence of prolonged agitation or stirring beyond that required to instantly diffuse the added coagulant chemical solution, visible coagulation did not occur for several hours and clarification often required an overnight settling period. With prolonged stirring, tiny flocs began to form within a few minutes and as time progressed up to about 10 min, the individual floc particles continued to grow in size and ultimately became widely separated. When the stirring was stopped, the flocs settled within a few minutes leaving a clear water above. After many repeated tests, it was concluded that adequate clarification with least chemical for a given water normally required a mixing period of about 10 to 20 min.
>
> Using synthetic waters, we noted a moderate increase in the required coagulant with increased turbidity, but more significantly, and much to our surprise, we noted that the alum demand increased more directly with the bicarbonate alkalinity of the water. This intrigued us very much and we tried increasing the bicarbonate content of the water by adding small increments of bicarbonate of soda.
>
> In all these tests, we noted that optimum flocculation occurred when a definite fraction of the total alkalinity had been neutralized. Without going into the chemical theory too deeply, this indicated that optimum flocculation was occurring at a constant hydrogen ion concentration or pH, slightly above true neutrality. We had arrived at an important conclusion in a roundabout way.
>
> The first few series of the jar tests proved very convincing and he (Professor Hyde) immediately arranged for a small, continuous flow pilot installation in Sacramento for the following summer. My recollection is that the Sacramento city filtration plant, the first to use prolonged mechanical agitation to induce flocculation, was placed in service in January, 1924.

The research and the ensuing design for the Sacramento plant (Langelier, 1921), established the guidelines for flocculation practice. The outcomes of this work were as follows:

1. The jar test was invented as a means to explore the role of coagulation variables.
2. The concept of flocculation was discovered.
3. The role of gentle mixing to promote floc growth was established.
4. Paddle-wheel flocculators were developed.
5. An on-site pilot plant study complemented the laboratory work.
6. The studies led to the first full-scale plant to implement paddle-wheel flocculators, i.e., at the new water filtration plant at Sacramento, California, put on line in January, 1924.
7. The importance of pH and alkalinity was discovered and principles were established that pH control was an important part of the coagulation process.
8. Langelier recognized the complexity of the coagulation–flocculation process and that the jar test apparatus and a pilot plant were the means to isolate the variables.

Figure 11.1 shows the design of the Sacramento WTP as modified by Langelier in 1919 to include paddle-wheel flocculators (but, as seen, did not include rapid mix). Langelier also delineated the details of design, specifying that the velocities of the water should be up to about 2.0 ft/s (as induced by the paddle wheels) and flocculation detention times should be about 30 min. Langelier also compared end-around baffling with the paddle-wheel design, in a separate pilot plant constructed for such purpose, and concluded that the paddle-wheel method was the most desirable because of greater flexibility in operation.

11.3.1.3 Design Guidelines

To illustrate the evolution of practice, Leopold (1934, p. 1070) described the "mechanical agitation" used for alum floc formation in a 11.4 ML/day (3 mgd) plant at Winnetka, Illinois treating water from Lake Michigan. The plant was expanded to 22.8 ML/day (6 mgd) in 1932 with six flocculation basins with combined detention period of 30 min. The mixer assembly was a 1.5 kW (2 hp), 5 speed motor, a speed reducer, one set of bevel gears, and a vertical drive paddle shaft. Each agitator had one paddle with a total surface area of 0.61 m^2 (6.6 ft^2) each or 125 m^3 basin volume/m^2 paddle surface area (409 ft^3/ft^2). The tip velocity of the paddle wheel was variable, e.g., $0.24 \leq v(\text{tip}) \leq 0.74$ m/s ($0.78 \leq v(\text{tip}) \leq 2.41$ ft/s) and θ (all floc basins) = 30 min. Camp (1955) adopted the first two criteria in his recommendations for design of flocculation basins; paddle-wheel area, however, was a fraction of the cross-sectional area of the basin, i.e., 0.15–0.20 to avoid rotation of the water mass (p. 11) instead of as a ratio with respect to the basin volume. In other words, in advancing the concept of flocculation to practice-specific guidelines were the key, and evolved with experience.

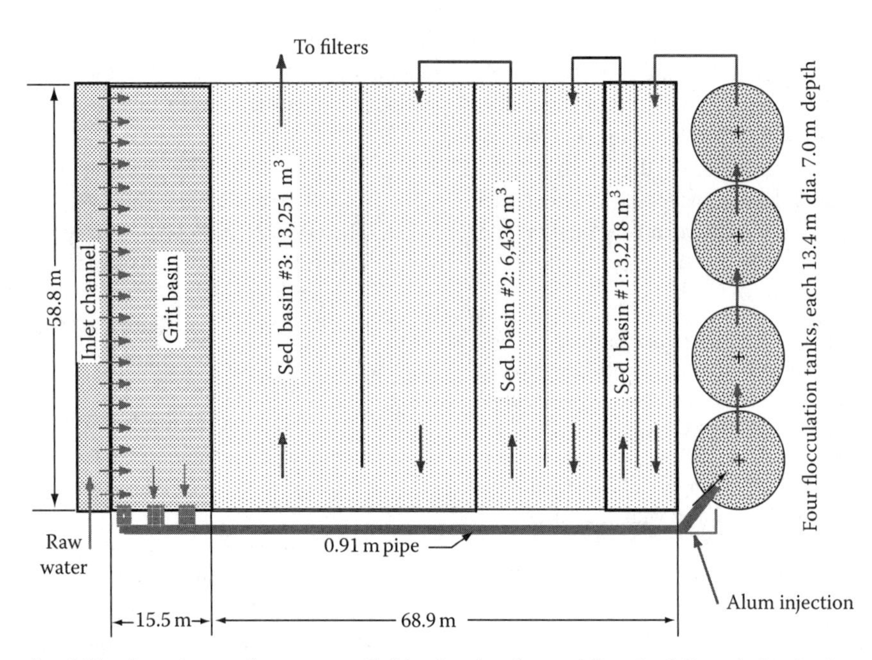

FIGURE 11.1 Schematic of filtration plant at Sacramento, California, showing paddle-wheel flocculation basins and sedimentation basin (Langelier, 1921). Q(design) = 2.2 m^3/s (50 mgd); q(flocculation basins) ≈ 30 min; q(sedimentation basins) ≈ 63 h.

11.3.1.4 Flocculation Practice, c. 1940

Table 11.1 gives data from Camp (1955) for 14 flocculation basins, c. 1940. He computed tip velocity, v; detention time, θ; headloss, h_L, for baffled systems (or power consumed, P/Q, for paddle-wheel basins); power per unit volume, P/V; G; and $G\theta$. Of the 14 plants, most had end-around baffles; only four were paddle-wheel basins. The baffled basins were not satisfactory according to Camp (1955, p. 5) because the gradients were excessive at the bends and deficient in the straight portions. Camp's 1955 paper was the basis for rational design of floc basins, adopted widely in practice.

11.3.2 Evolution of Theory

Three "milestones" in flocculation theory were (1) the equations of Smoluchowski for "perikinetic" and "orthokinetic," fluid motion in 1916 and 1918, respectively; (2) the observations of Langelier, c. 1919, on the effect of slow stirring on the formation of settleable floc; and (3) the work of Camp in formulating the basis for the theory and practice in 1943 (Camp and Stein, 1943) and in 1955 (Camp, 1955), respectively.

11.3.2.1 Langelier

As applied to water treatment practice, Langelier's flocculation experiments as related to the Sacramento plant, c. 1919, was the basis for an initial descriptive theory. Langelier recognized the role of stirring in causing the growth of floc particles and used his jar-test apparatus as the means for investigating the effects of other independent variables. Colloid chemistry entered the picture, c. 1941, through an MS thesis by H. Ludwig (Ludwig, 1942) under Langelier; the basis was Professor Hans Jenny's course on the topic.

11.3.2.2 Smoluchowski's Collision Equations

The equations of Smoluchowski are referenced frequently in the literature as the starting point for discussion of kinetic theory of particle collisions. The Smoluchowski equation on orthokinetic motion was referenced by Harris et al. (1966, p. 96) who credited its introduction into water treatment theory to Camp and Stein (1943, Equation 23). They derived the equation from fundamentals, however, and did not cite Smoluchowski, i.e., probably they were not aware of his papers. The enumeration of Smoluchowski's equations here is from Argaman and Kaufman (1968, p. 5), which were described previously in Section 10.2.3.1; the numbering is retained.

11.3.2.2.1 Perikinetic Motion

The first equation of Smoluchowski (1916) was for the condition that collisions were due to Brownian motion (called "perikinetic" motion) calculated as the diffusion flux of particles in the radial direction around a single stationary particle:

$$N(\text{diffusion})_{ij} = 4\pi D_{ij}(r_i + r_j)^3 n_i n_j \qquad (10.1)$$

where

$N(\text{diffusion})_{ij}$ is the number of contacts per unit time per unit volume between i and j particles due to diffusion flux (collisions/m^3/s)
D_{ij} is the combined diffusion coefficient, $D_i + D_j$ (m^2/s)
r_i is the radius of particle i (m)
r_j is the radius of particle j (m)
n_i is the number concentration of particles of radius r_i (#/m^3)
n_j is the number concentration of particles of radius r_j (#/m^3)

TABLE 11.1
Data on Flocculation Basin Designs in Water Treatment Plants, c. 1940

Plant	Date Built[a]	Q^{b} (m^3/s)	Q^{b} (mgd)	Type of Agitation	v_H[c] (m/s)	v_H[c] (ft/s)	Q (min)	h_L (m)	h_L (ft)	h_L (kW)	h_L (hp)	P^{d} ($kW/m^3/s$)	P^{d} (hp/mgd)	P/V (kW/m^3)	P/V ($ft \cdot Lb/s/ft^3$)	G (s^{-1})	$G\theta^{e}$
Baltimore	1928	4.906	112	End-around baffles	0.396	1.3	30	0.91	3					0.005	0.104	61	111,000
Cleveland	1918	6.133	140	Over-and-under baffles	0.122	0.4	39	0.12	0.4					0.001	0.011	20	46,800
Denver	1924	2.628	60	End-around baffles	0.091	0.3	20	0.15	0.5					0.001	0.026	31	37,200
Detroit	1931	11.916	272	End-around baffles	0.122	0.4	17	0.15	0.5					0.001	0.030	33	33,600
Flint	1924	1.227	28	Over-and-under baffles	0.183	0.6	35	0.37	1.2					0.002	0.036	36	75,600
Fort Worth	1923	0.876	20	End-around baffles	0.152	0.5	40	0.30	1					0.001	0.026	31	74,400
Grand Rapids	1924	1.752	40	End-around baffles	0.305	1.0	45	0.30	1					0.001	0.023	29	78,000
Kansas City, Kansas	1927	1.095	25	End-around baffles	0.335	1.1	10	0.12	0.4					0.002	0.041	39	23,400
Knoxville	1927	0.657	15	Paddle wheel	0.457	1.5	20			4.5	6	6.8	0.4	0.006	0.120	66	79,200
Miami	1926	0.876	20	Paddle wheel	0.305	1.0	20			2.2	3	2.6	0.15	0.002	0.045	40	48,000
New Orleans	1928	3.154	72	End-around baffles	0.244	0.8	60	0.30	1					0.001	0.017	25	90,000
Oakland	1927	0.526	12	Paddle wheel	0.518	1.7	20			3.0	4	5.6	0.33	0.005	0.098	60	72,000
Oklahoma City	1923	0.701	16	Over-and-under baffles	0.183	0.6	30	0.61	2					0.003	0.071	51	92,000
Tampa	1926	0.613	14	Paddle wheel	0.244	0.8	100			6.0	8	9.7	0.57	0.002	0.034	35	210,000

Source: Data from Camp, Water treatment plant design, in: *Manual of Engineering Practice No. 19*, ASCE, 1940, p. 29, 1955, Table 2.

[a] Date of construction.

[b] Q is plant capacity in mgd.

[c] v_H is horizontal velocity in basin.

[d] P/V is power dissipated per unit volume.

[e] $G\theta$ is parameter in which θ is measured in seconds.

11.3.2.2.2 Orthokinetic Motion

The second equation of Smoluchowski (1918) was for collisions between particles induced by fluid motion (called "orthokinetic" motion). For laminar flow, the collision frequency was given as

$$N(\text{laminar})_{ij} = \left(\frac{4}{3}\right)(r_i + r_j)^3 n_i n_j \left(\frac{dv}{dy}\right) \tag{10.2}$$

where
- $N(\text{laminar})_{ij}$ is the number of contacts per unit time per unit volume between i and j particles due to laminar fluid motion (collisions/m^3/s)
- dv/dy is the velocity gradient due to laminar flow (s^{-1})

11.3.2.3 Camp's *G*

Camp and Stein (1943) derived an expression for velocity gradient called G (Camp and Stein, 1943), which is based on the forces acting on a fluid element:

$$G \equiv \sqrt{\frac{P}{\mu V}} \tag{10.5}$$

where
- G is the average velocity gradient for basin (s^{-1})
- P is the power applied to paddle wheel (N m/s)
- μ is the dynamic viscosity of water (N s/m^2)
- V is the volume of flocculation basin (m^3)

Later, in a 1955 paper (Camp, 1955), Camp delineated how to apply G to flocculation for diffused aeration, paddle wheels, and reciprocating blades. The procedure that he described remains current. Following Langelier's suggestion for tapered flocculation, Camp suggested higher G values, i.e., $G \approx 70$ s^{-1} in the first compartment, tapering to $G \approx 20$ s^{-1} in the third or fourth, based upon data that he compiled from operating plants, as described in Table 11.1.

11.4 THEORY OF FLOCCULATION

The theory of flocculation has to do with the rate of aggregation of flocs. The starting point has been, nearly always, the classic Smoluchowski equations in 1916 and 1918 for Brownian motion, i.e., perikinetic, and fluid motion, i.e., orthokinetic, respectively (see Sections 10.2.3.1 and 11.3.2.2).

11.4.1 KINETICS

For flocculation, the kinetic model given by O'Melia (1978), an adaptation of the Smoluchowski model (Section 10.2.3.1), incorporates terms that can be expanded to explain the rate of aggregation of flocs. This section deals with those factors.

11.4.1.1 Frequency of Particle Collisions

The general equation for collision frequency is (O'Melia, 1978, p. 227) given by

$$J(i,j) = \gamma(i,j) \cdot n_i n_j m, \tag{11.1}$$

where
- $J(i,j)$ is the frequency of collision per unit volume of suspension between i species and j species (i–j collisions/m^3/s)
- $\gamma(i,j)$ is the kinetic rate constant (m^3/s)
- n_i is the concentration of particles of species i (particles i/m^3)
- n_j is the concentration of particles of species j (particles j/m^3)

In other words, the frequency of collisions is proportional to the concentrations of particles that are subject to interaction, i.e., as used in the Smoluchowski models. The frequency coefficient, $\gamma(i,j)$, however, incorporates considerable complexity; basic tenants are reviewed in paragraphs that follow.

11.4.1.1.1 The Kinetic Rate Constant

The kinetic rate constant, $\gamma(i,j)$, is the product

$$\gamma(i,j) = \alpha(i,j) \cdot k(i,j) \tag{11.2}$$

where
- $\alpha(i,j)$ is the collision frequency factor, i.e., fraction of particle collisions that result in collisions for curvilinear model as compared to the rectilinear model (collisions that occur by curvilinear model/collisions that occur by rectilinear model)
- $k(i,j)$ is the rate constant for rectilinear model (m^3/s)

11.4.1.1.2 The α Factor

The traditional form of Equation 11.2 is for $\alpha(i,j) = 1$, which describes the "rectilinear" model of collision frequency (Ives, 1978, p. 45). The $\alpha(i,j)$ term is a "correction" factor to take into account that the "i" particle path is not rectilinear, but curvilinear, and also as the particles come into the proximity of one another they are subject to "short-range" forces, e.g., van der Waals (Han and Lawler, 1992, p. 83). The $\alpha(i,j)$ term is defined as the ratio

$$\alpha(i,j) = \frac{x_c^2}{(d_i + d_j)^2} \tag{11.3}$$

where
- x_c is the "critical" diameter just outside the trajectory "shadow" of particle i beyond which particle j will not collide with particle i (m)
- d_i is the diameter of particle i (m)
- d_j is the diameter of particle j (m)

The values of $\alpha(i,j)$ are given additional resolution by Han and Lawler (1992) in separate equations for Brownian motion, shear, and settling, respectively.

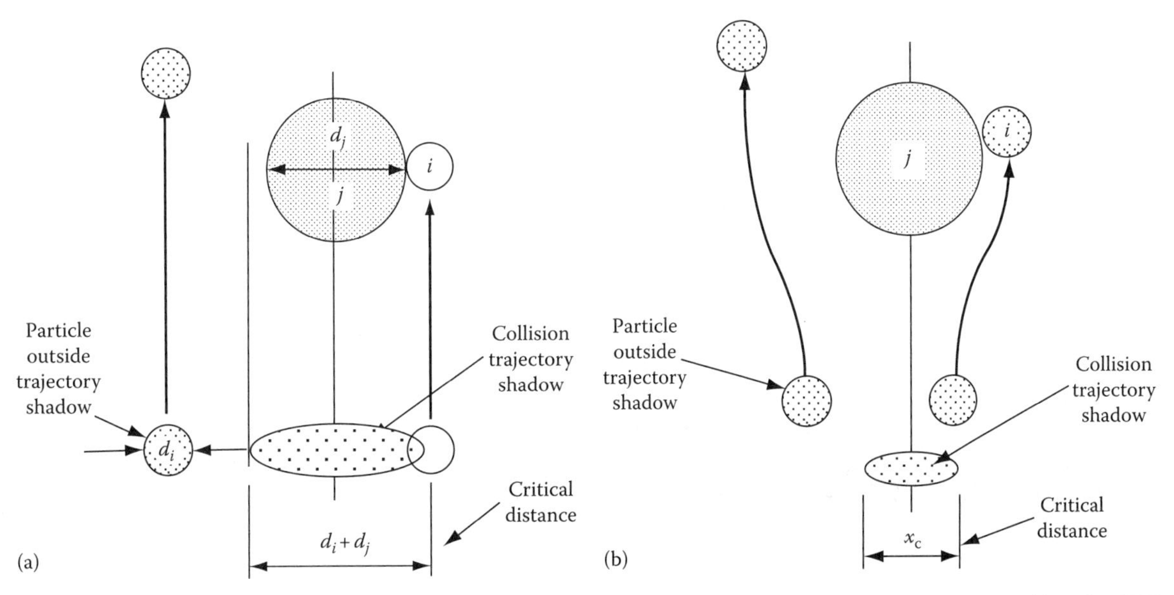

FIGURE 11.2 Collision trajectory models. (a) Rectilinear model. (b) Curvilinear model. (Adapted from Han, M. and Lawler, D.F., *J. Am. Water Works Assoc.*, 84(10), 80, 1992.)

11.4.1.1.3 Rectilinear Model

To expand on the distinction between the rectilinear and the curvilinear models, the rectilinear model is traditional and is based upon the assumption that particles follow a straight-line path and interparticle forces are not considered. Figure 11.2a depicts the rectilinear model. The "collision trajectory shadow" is the pseudo area that has a diameter, $(d_i + d_j)$; the i particles within this area traveling in a straight line toward the particle j will strike the j particle. The i particles outside the "trajectory shadow" will miss the larger j particle.

11.4.1.1.4 Curvilinear Model

Figure 11.2b depicts the idea of the curvilinear model. As seen, the trajectory of a given i particle curves around the j particle. If an i particle originates within the "collision trajectory shadow," a collision will occur with the j particle. For either the rectilinear model or the curvilinear model, any i particle outside the "critical diameter" will not strike the j particle.

11.4.1.1.5 Relation of Rectilinear and Curvilinear Models to $\alpha(i, j)$

As seen in Figure 11.2a, $\alpha(i, j) = 1$ for the rectilinear model, i.e., the area shadow of the particles is the same as the "collision trajectory shadow." Also, Equation 11.1 reduces to the traditional form, i.e., $\gamma(i, j) = k(i, j)$. Next, as seen in Figure 11.2b, a depiction of the curvilinear model, the area of the "collision trajectory shadow" is smaller than the area shadow of the i particles; therefore, for the curvilinear model, $\alpha(i, j) < 1$, which is a "correction factor" for the rectilinear collision frequency model. Ives (1978, p. 45) considered the curvilinear effect to be small. The issue was revisited by Han and Lawler (1992), however, who found that significant corrections to the rectilinear model were warranted; they delineated the corrections for particle transport due to the Brownian diffusion, shear, and settling, i.e., $\alpha_B(i, j)$, $\alpha_{SH}(i, j)$, $\alpha_S(i, j)$, respectively.

11.4.1.1.6 Rate Coefficients, $k_B(i, j)$, for Rectilinear Model

The rate coefficients applied traditionally to the rectilinear collision model are in three categories, Brownian diffusion, shear, and differential settling (O'Melia, 1978, p. 227), respectively, i.e.,

$$k_B(i,j) = \frac{2}{3}\frac{kT_{abs}}{\mu}\frac{(d_i + d_j)^2}{d_i d_j} \tag{11.4}$$

$$k_{SH}(i,j) = \frac{(d_i + d_j)^3}{6}G \tag{11.5}$$

$$k_S(i,j) = \frac{\pi g(SG_p - 1)}{72\nu}(d_i + d_j)^3(d_i - d_j) \tag{11.6}$$

where

$k_B(i,j)$ is the rate constant due to Brownian transport for rectilinear model (m^3/s)
$k_{SH}(i,j)$ is the rate constant due to velocity gradient transport for rectilinear model (m^3/s)
$k_S(i,j)$ is the rate constant due to differential settling for rectilinear model (m^3/s)
k is the Boltzmann's constant $= 1.38 \cdot 10^{-23}$ J/K
T_{abs} is the absolute temperature (K)
μ is the absolute viscosity of water at temperature, T (N s/m^2)
G is the velocity gradient (s^{-1})

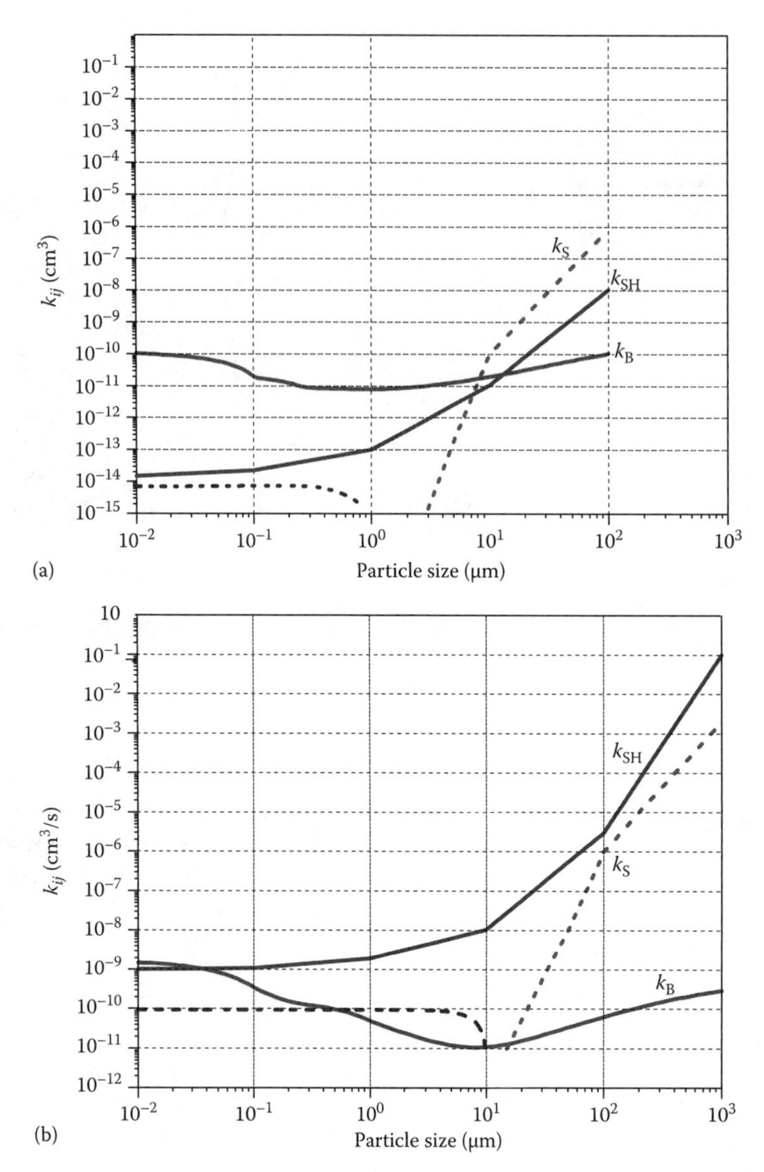

FIGURE 11.3 Effect of target particle size on collision rate constants for conditions indicated. (a) Settling basins. (b) Flocculation basins. (Adapted from O'Melia, C.R., Coagulation in wastewater treatment, in: Ives, K.J. (Ed.), *The Scientific Basis of Flocculation*, Sijthoff & Noordhoff International Publishers, Alphen aan den Rijn, the Netherlands, p. 228, 1978.)

g is the acceleration of gravity = 9.806650 m/s^2
SG_p is the specific gravity of solid particles (dimensionless)
ν is the kinematic viscosity of water at temperature, T (m^2/s)

Figure 11.3a and b shows how the three rate coefficients, i.e., $k_B(i,j)$, $k_{SH}(i,j)$, $k_S(i,j)$ vary with particle diameter for the conditions of a settling tank and a flocculation basin, respectively. As seen in Figure 11.3a, Brownian motion is dominant for d(particle) < 10 μm but for d(particle) > 10 μm, settling becomes dominant. For a flocculation basin, Figure 11.3b shows that shear is dominant for all particle sizes, but settling is equally important for d(particle) > 100 μm. Also, as noted in Section 10.3.1.2, e.g., Figure 10.11, for "mixing" and reaffirmed for flocculation, shear is effective in causing collisions only for eddies that are about the same size as the particles (Casson and Lawler, 1990).

	Conditions	
Parameter	**Settling Tank**	**Flocculation Tank**
Tank	Settling	Flocculation
Reference	O'Melia (1978, p. 228)	
Figure	11.3a	11.3b
d_i (μm)	1	10
d_j (μm)	10^{-2}–10^2	10^{-2}–10^3
T (°C)	5	20
SG	1.02	1.02
G (s^{-1})	0.1	10
Subscript	**Motion**	**Equation Reference**
B	Brownian motion	11.4
SH	Fluid shear	11.5
S	Settling	11.6

11.4.1.1.7 Temperature Effect

In one of the first studies of the influence of temperature on coagulation–flocculation settling, Leopold (1934, p. 1072) examined settling in laboratory jar tests after flocculation at temperatures 2.2°C, 7.2°C, 12.8°C, 18.3°C, and 23.9°C (36°F, 45°F, 55°F, 65°F, 75°F) and found no difference. Hanson and Cleasby (1990) confirmed this finding and cited others who did as well, e.g., Velz (1934), Camp et al. (1940), Morris and Knocke (1984), Cleasby (1984). These findings are plausible if the conditions were turbulent as explained by Cleasby (1984) and Hanson and Cleasby (1990). In quiescent settling, in accordance with Stoke's law, viscosity is a part of the mathematical relationship; thus temperature has an effect.

11.4.1.2 Rate of Formation of New Particles, *k*

An extension of Equation 11.1 is

$$\frac{dn_k}{dt} = \frac{1}{2}\sum_{i+j=k}\gamma(i,j)n_i n_j - n_k\sum_{\text{all } i}\gamma(i,k)n_i \quad (11.7)$$

where n_k is the concentration of particles of size k (particles k/m^3).

The left side of Equation 11.7 is the rate of change of size k particles. The first summation on the right side is the rate at which k particles are formed due to collisions between i and j particles. The relation under the summation means that the sum of the i and j volumes equals the k volume. The factor 1/2 is applied so that the collisions are not counted twice, i.e., once for the i particles and once for the j particles. The second summation is the loss of size k particles by flocculation to become a larger size. Equation 11.7 is a traditional relationship given by Ives (1978, pp. 41, 47), Casson and Lawler (1990, p. 55), and Han and Lawler (1992, p. 80).

11.4.2 Nature of Flocs and Flocculation

The flocculation process starts with microflocs, which may collide with primary particles, e.g., mineral turbidity, viruses, bacteria, protozoan cysts, etc., with a fraction being assimilated. The aggregates formed may, in turn, adhere to one another upon subsequent collisions. How these flocs are formed and their characteristics are reviewed in this section.

11.4.2.1 Characteristics of Flocs

The characteristics of flocs, the aggregation product of the flocculation process are important in the subsequent processes of settling and filtration or flotation. Floc size, density, shape, fractal appearance, age, shear resistance, settling velocity, and shear resistance, are reviewed in this section.

11.4.2.1.1 Size

With regard to size, flocs are referred to often as "microflocs," "pinpoint," "intermediate," and large. These sizes are, respectively, microfloc, <10 μm; pinpoint, 10–50 μm; intermediate, 50–100 μm; and large, >100 μm (p. 25). Within a floc basin, flocs of several mm, e.g., 0.5–3 mm may be seen (e.g., by a light beam) in the third basin, i.e., with the very slow paddle wheels. Within each size group, the distribution is usually "normal."

Figure 11.4a shows the average floc length, $L(\text{floc})_{\text{avg}}$, as affected by G for three alum concentrations, with the

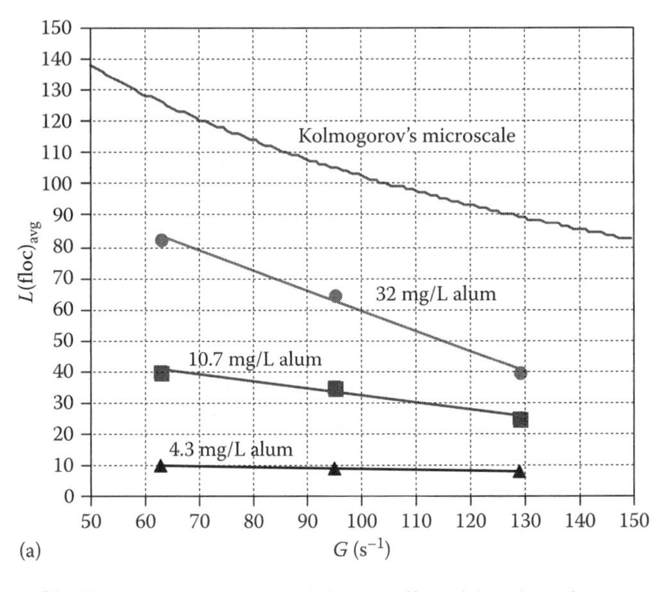

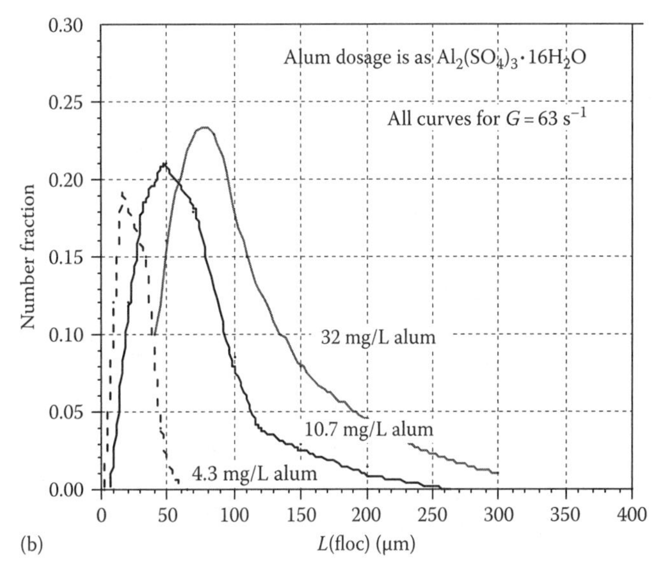

FIGURE 11.4 Floc characteristics as affected by alum dosage and velocity gradients. (Adapted from Spicer, P.T. and Pratsinis, S.E., *Water Res.*, 30(5), 1051, 1996a. (a) Floc size versus G for different alum dosages. (b) Floc size distributions for different alum dosages at $G = 63\ \text{s}^{-1}$.

Kolmogorov microscale shown for reference. Figure 11.4b shows the floc size distributions for the same three alum dosages, all for $G \approx 63\ s^{-1}$ (Spicer and Pratsinis, 1996a, p. 1051). With increasing G, the floc size distribution shifts to smaller diameters (Spicer and Pratsinis, 1996b). Relevant observations from the two plots are as follows:

- The lowest alum dosage at 4.3 mg/L is for adsorption destabilization. The average floc size is lowest in this zone, i.e., $L(\text{floc})_{avg} \approx 10\ \mu m$ for $63 \leq G \leq 129\ s^{-1}$. The associated floc size distribution, seen in Figure 11.4b, is narrowest for this zone.
- For an medium alum dosage at 10.7 mg/L, the zone is mixed. The average floc size is intermediate and is affected moderately by G. The associated floc size distribution seen in Figure 11.4b is broader.
- For the highest alum dosage at 32 mg/L is for the sweep-floc zone. The average floc size is highest in this zone and is affected strongly by G, i.e., $L(\text{floc})_{avg}$, declines sharply as G increases. The associated floc size distribution seen in Figure 11.4b is broadest for this zone.
- Also, since the smallest eddy size is approximately the same as λ, i.e., Kolmogorov's microscale, $\lambda^* = (\nu^3/\varepsilon)^{1/4}$ (see Equation 10.14), the shear field determines the size characteristics of the resulting flocs at high floc concentrations (Spicer and Pratsinis, 1996a, p. 1051), and so $L(\text{floc})_{avg}$ declines approximately the same as λ.

Other observations (Spicer and Pratsinis, 1996a, p. 1051) were as follows:

- Higher alum concentration produces stronger as well as larger flocs. Steady state size is attained more quickly at the higher alum dosage, e.g., in 5 min at $G \approx 63\ s^{-1}$.
- As the flocs grow in size, they become more porous.
- As shear rate increases, the higher rate of breakage/regrowth and restructuring produces more compact floc structures.

11.4.2.1.2 Density

The inner particle pore space in a fractal aggregate is considered to come from two sources: (1) the fractal nature, which is related to the fundamental aggregation process, and (2) the packing effects, as related to the shapes of the primary particles and the aggregates. Therefore, regarding the latter, the "packing-factor," $\zeta = 1$, which means that there is no pore space resulting from the shape effects of the aggregates. There is still pore space, however, due to the fractal nature of the aggregate, represented by the fractal parameter, D_F (Lee et al., 2000, p. 1990).

The buoyant density of a floc particle, as related to size, is (Reed and Mery, 1986, p. 75; Gregory 1989, p. 216) given by

$$\rho_f - \rho_w = kA^{-\alpha} \tag{11.8}$$

where

ρ_f is the mass density of floc (kg/m^3)
ρ_w is the mass density of water (kg/m^3)
k is the empirical coefficient, given as 0.349 for d(floc) < 1.5 mm
A is the projected area of floc (m^2)
α is the empirical exponent, given as 0.338 based on floc geometry

Calculation of floc specific gravities for different assumed floc sizes based on Equation 11.8 is shown in Table CD11.2. The specific gravities in Table CD11.2 are just slightly higher than the experimental results of Lagvankar and Gemmell (1968, p. 1044) for $d(\text{floc}) \leq 1.5\ \mu m$; for $d(\text{floc}) > 1.5\ \mu m$ their experimental results showed only a slightly declining specific gravity and Equation 11.8 gave increasingly lower SGs.

As the floc particles grow in size, their surface geometry becomes more complex, i.e., a "fractal," resulting in progressively poorer fit to Equation 11.8 and lower floc density (Reed and Mery, 1986, p. 75). Observations of floc particles showed that small particles were dense spheres while the larger particles were loose agglomerations of the denser small particles.

TABLE CD11.2
Calculation of Floc Specific Gravity by Equation 11.9

d(floc)[a] (mm)	d(floc) (m)	A (m^2)	k	α	$(\rho_f - \rho_w)$ (kg/m^3)	SG(floc) (Calc.)	SG(floc)[b] (Exp.)
0.5	0.0005	1.96E−07	0.349	0.338	64.53	1.065	1.048
1	0.001	7.85E−07	0.349	0.338	40.39	1.040	1.037
2	0.002	3.14E−06	0.349	0.338	25.28	1.025	1.030
5	0.005	1.96E−05	0.349	0.338	13.61	1.014	1.029

[a] d(floc) values were assumed; calculations were by spreadsheet.
[b] Experimental results of Lagvankar and Gemmell (1968, p. 1044).

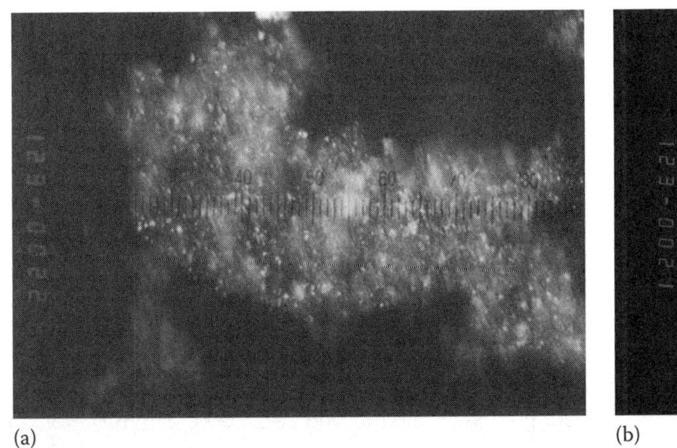

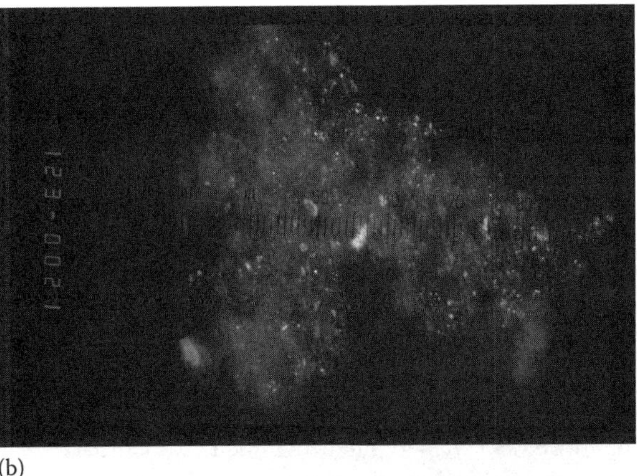

(a) (b)

FIGURE 11.5 Floc particles; one small division on the scale shown = 10 μm. (a) Alum dose 20 mg/L, *d*(floc) = 845 μm, pH = 5.6, $G = 28{,}200\ s^{-1}$, (p. 122). (b) Alum dose 30 mg/L, *d*(floc) = 731 μm, pH = 6.9, $G = 10{,}100\ s^{-1}$, (p. 66). (From Khan, Z., Floc characteristics as affected by coagulation and as affecting filtration of low turbidity water, Doctoral dissertation, Department of Civil Engineering, Colorado State University, Fort Collins, CO, 1993. With permission.)

11.4.2.1.3 Floc Volume Concentration

The floc volume concentration is defined as the volume of floc particles per unit volume of suspension, i.e., mL floc particles/mL water suspension. Values of floc volume concentration ranged from 300 (high) to 50 mL floc/10^6 mL water suspension (low) for coagulation turbulence intensity values of $100 < G < 1000\ s^{-1}$ (Camp, 1968, p. 247).

11.4.2.1.4 Shape

The shape of a floc particle was reported to vary with the kind of primary particle involved in coagulation. Gorczyca and Gahczarczyk (1992, p. 8) reported significant differences in floc shape and size depending upon the primary particles. Of four mineral suspensions, montmorillonite floc was the most irregular and calcite flocs were the most regular, with illite and silt being intermediate.

11.4.2.1.5 Fractal Appearance

Figure 11.5 shows two samples of alum floc formed under different conditions of alum dosage, pH (coagulated water), and *G*. Photographs from a number of experiments (Khan, 1993) showed no characteristic size, shape, or appearance that related to the conditions. Most of the flocs seen were "fractal" in appearance, i.e., irregular shape, similar to the examples.

11.4.2.1.6 Fractal Geometry

Floc is irregular in shape, i.e., amorphous; such a particle is called, by definition, a *fractal*. The traditional assumption of floc development is that the volume of a given floc particle is the sum of its individual units, such as depicted in Figure 11.6a. This is called the Euclidean model of "coalesced spheres" (see also Han and Lawler, 1992, p. 81). An alternative model, called the "coalesced fractal sphere," is based upon the idea of a floc particle as a "fractal" (Jiang and Logan, 1991; Lee et al., 2000, p. 1989) and is depicted in Figure 11.6b. The

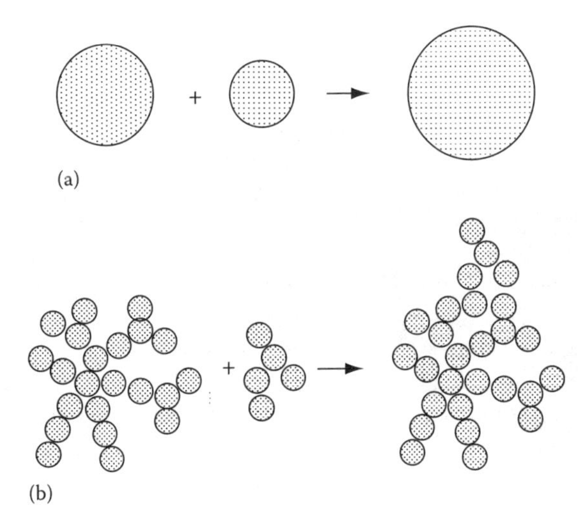

FIGURE 11.6 Models of floc formation. (a) Model of "coalesced spheres." (b) Coalesced fractal sphere. (Adapted from Lee, D.G. et al., *Water Res.*, 34(7), 1987, 2000.)

volume of the new particle formed is assumed to be the sum of the volumes of the particles that have collided and attached to each other. The inner pore space depends upon both the packing arrangement of the individual particles, i.e., ζ, and the fractal nature of the particle formed (Lee et al., 2000, p. 1990). The number of spherical monomers that can be packed into a larger sphere is given as (Gmachowski, 1995, p. 1815; Lee et al., 2000, p. 1990):

$$N = \zeta\left(\frac{d}{d_o}\right)^{D_F} \tag{11.9}$$

where

N is the number of spherical monomers of diameter, d_o (#)
d is the diameter of fractal, characterized as a sphere (m)
d_o is the diameter of spherical monomer (m)

ζ is the packing factor, indicating how the monomers are packed (dimensionless)

D_F is the parameter that characterizes the fractal dimension of aggregates with respect to its three-dimensional geometry; see Box 11.1 (dimensionless)

BOX 11.1 FRACTAL DIMENSION, D_F

The fractal dimension, D_F, seen as the exponent in Equation 11.9 has become a common parameter among those who wish to pursue the idea of floc geometry in terms of fractal theory, as developed largely since the about the mid-1980s. Gregory (1989, p. 215) explained this parameter as follows: A solid three-dimensional body has a mass, which depends on the third power of some characteristic length (such as the diameter of a sphere), so that a log–log plot of mass against size should give a straight line with a slope of three. When such plots are made for aggregates, however, lower slopes are found, with non-integer values. The slope of the line is known as the fractal dimension, D_F. In three-dimensional space, D_F may take values between 1 and 3, the lower value representing a linear aggregate and the upper one an aggregate of uniform density or porosity. Generally intermediate values are found, and the lower the fractal dimension, the more "open" or "stringy" the aggregate structure. The earliest attempts at computer simulation of aggregation were based on the random addition of single particles to growing clusters, which gives $D_F = 2.75$, indicating a fairly compact aggregate structure. An alternative model is for cluster–cluster aggregation which is more like real flocculation which leads to a much lower value of D_F, i.e., $D_F = 1.75$, indicating a fairly "open" structure.

For the Euclidian model, $D_F = 3$ and $\zeta = \pi/(3\sqrt{2}) = 0.7405$; the latter applies for close-cluster packing. The value of ζ depends also on the shapes of the monomers, i.e., if other than spherical, which is likely. The usual assumption for the traditional model, however, as depicted in Figure 11.6a, is that $\zeta = 1$, which means that there is no pore space due to packing effects (which, of course, cannot be true). The diameter, d, of the fractal aggregate is a pseudo dimension, since a fractal, by definition, is difficult to characterize. Dimensions that could serve include (1) the "hydraulic" diameter (based upon the fall velocity with diameter calculated by Stoke's law) and (2) the diameter calculated from the radius of gyration (Lee et al., 2000, p. 1990; Chakraborti et al., 2000, p. 3969). For fractals, i.e., $D_F < 3$, the lower values represent large, highly branched, and loosely bound structures (Chakraborti et al., 2000, p. 3969). For reference, values given for coagulation of minerals with 4.5 mg/L alum and 1 mg/L polymer (Purifloc A-23, Dow Chemical) were $D_F(\text{illite}) \approx 1.49$, $D_F(\text{montmorillonite}) \approx 1.79$, $D_F(\text{calcite}) \approx 1.65$, and $D_F(\text{silt}) \approx 1.37$.

Table 11.3 gives D_F values with descriptions of suspensions for lake water and a montmorillonite suspension after alum coagulation by charge neutralization, and "sweep floc," respectively (Chakraborti et al., 2000, p. 3969). The initial suspension was without coagulant. The charge neutralization stage was defined by the coagulant dosage required to give a floc zeta potential for a minimum settled water turbidity before restabilization (and higher turbidity). For lake water, this was for zeta potential ≈ -1 mV and for the montmorillonite suspension, zeta potential ≈ -15 mV (with alum dosages 3 and 2 mg/L, respectively). The sweep-floc stage was defined as the minimum alum dose that resulted in a settled water turbidity ≈ 1 ntu (14 mg/L for lake water and 20 mg/L for the montmorillonite suspension). As seen in Table 11.3, the fractal dimension parameter decreases with increasing fractal size, indicating a looser, more spread-out structure as corroborated by in situ photographs. Also, as described, the larger aggregates are more irregular in structure, with the primary particles for the sweep floc being surrounded

TABLE 11.3
Descriptions of Floc at Three Stages of Coagulation

Suspension	Coagulation Stage	D_F	Description of Suspension
Lake water	Initial suspension	2.93 ± 0.20	Heterodisperse
	Charge neutralization	2.57 ± 0.20	Small flocs, irregular in shape
	Sweep floc	2.12 ± 0.50	Large aggregates of many primary particles surrounded by gel-like alum floc
Montmorillonite	Initial suspension	2.71 ± 0.20	Heterodisperse
	Charge neutralization	2.51 ± 0.20	
	Sweep floc	2.39 ± 0.30	

Source: Adapted from Chakraborti, R.K. et al., *Environ. Sci. Technol.*, 34(18), 3969, 2000.

by a "gel-like" alum floc. For comparison, Bellouti et al. (1997, p. 1230) found for anaerobic flocs, D_F(avg) = 1.84 (for measurements of 54 particles).

11.4.2.1.7 Aging

Aging is the irreversible change of texture and floc structure from the moment flocs are completely formed (François, 1987a, p. 523). Thus, the character of aluminum hydroxide floc changes with time. Its solubility product is given with respect to the reaction

$$\mathrm{Al(OH)_3} \downarrow + \mathrm{H_2O} \Leftrightarrow \mathrm{Al(OH)_4}^- + \mathrm{H}^+ \qquad K(\text{fresh}) = 1 \cdot 10^{-13} \tag{11.10}$$

After 12 days, K(aged) = $1.1 \cdot 10^{-14}$. During the continuing polymerization, the metal ions are linked with an increasing number of hydroxide groups, which decreases the pH (Francois, 1987a, p. 524). In addition to monomeric complexes, e.g., $Al(H_2O)_6^{3-}$, $Al(OH)(H_2O)_5^{2+}$, $Al(OH)(H_2O)_4^{+}$, $Al(OH)_3$, $Al(OH)_4^-$, $Al(OH)_5^{2-}$, some unstable polynuclear structures with 20–400 atoms per structure are formed during hydrolysis of alum coagulant. The polynuclear structures grow with time and are finally converted into microcrystalline particles. Depending on the OH^-:Al^{3+} ratio, different aged products are formed. As a first step, an amorphous boehmite gel is always formed. For OH^-:$Al^{3+} > 2$–2.75, crystallization occurs during aging. For OH^-:$Al^{3+} > 3$–3.3, crystals of gibbsite, bayerite, nordstrandite, or a mixture are formed. Clay particles increase the rate of formation of microcrystals (François, 1987a, p. 524).

The electrophoretic mobility (EM) remains constant during the first 8–12 h and afterward decreases sharply. During the first period, the EM is due to amorphous aluminum hydroxide flocs which enmesh the destabilized particles (François, 1987a, p. 527). When the coagulant dosage was not optimized, the clay particles still had a negative charge. When an optimized dose was used, the destabilized clay particles had a small, positive charge (François, 1987a, p. 528).

11.4.2.1.8 Activated Sludge

Activated sludge floc has a filamentous network to which clusters of primary particles cling (Parker et al. 1972, p. 88). The floc grows in size as the particles collide with one another, becoming settleable as they enter the final settling basin.

11.4.2.1.9 Settling Velocity: Biological Floc

The settling velocity of activated sludge flocs was determined experimentally by Li and Ganczarczyk (1989, p. 1386). A best fit of their experimental plot conformed to the empirical relationship,

$$v(\text{floc}) = 0.37 + 1.25d(\text{floc}) \tag{11.11}$$

where,

v(floc) is the settling velocity of activated sludge floc (mm/s)

d(floc) is the longest dimension of floc, as a fractal (mm)

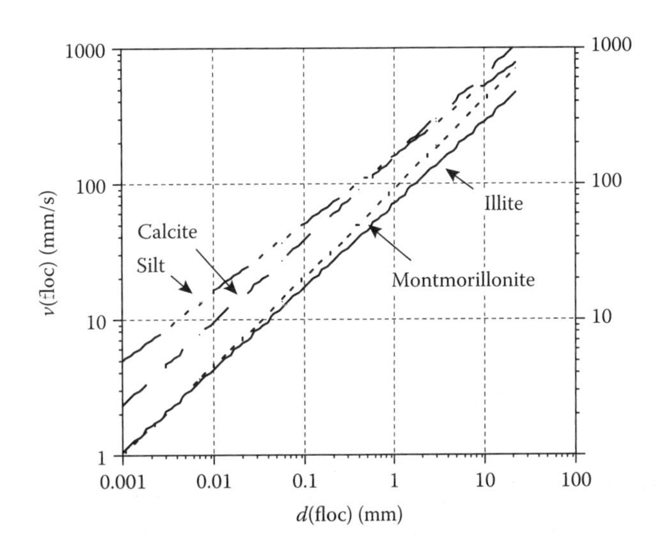

FIGURE 11.7 Settling velocity of different alum flocs as a function of equivalent circular diameter for floc fractal. (Adapted from Parker, D.S. et al., *J. Environ. Eng. Division, ASCE*, 98(SA1), 89, 1972.)

11.4.2.1.10 Settling Velocity: Aluminum Hydroxide Floc

Figure 11.7 shows experimentally obtained settling velocities for aluminum hydroxide flocs based on four kinds of suspensions of primary particles, i.e., illite, montmorillonite, silt, and calcite (experimental data points not shown). All four plots show the same trend, i.e., linear on a log–log plot, i.e., $y = ax^b$, but differ in intercepts and exponents. The data may be useful in the design of settling basins but the settling velocities in Figure 11.7 are generally higher than those used in practice and so may result in a basin plan area that is smaller than determined by guidelines (see Sections 6.8.3 and Table 6.10). For illite, the equation for the lower plot line is, $y = 71 \cdot x^{0.61}$. For comparison, Stoke's law is $v_s = (g/18n) \cdot (SG_s - SG_f)d^2$, which shows the disparity between exponents, e.g., 0.61 versus 2.

In some cases, flocs of low density are desired, e.g., as in flotation. But, if settling follows flocculation, a large floc is desired, preferably of higher density. In-line filtration, on the other hand requires a pinpoint floc, i.e., to penetrate the filter. For these reasons, an improved understanding of floc, in terms of knowing the fundamental factors that affect its characteristics, will help tailor a floc to fit a specified purpose (Gregory, 1989, p. 217).

11.4.2.1.11 Shear Resistance

The magnitude of the mean shear stresses during coagulation, e.g., $G \geq 12{,}500\ s^{-1}$, is only 11 N/m^2, which is equivalent to a shear energy of $2.0 \cdot 10^{-7}$ kJ/mol of water ($4.8 \cdot 10^{-8}$ kcal/mol). By comparison, the chemical bond energy of weak van der Waals forces or hydrogen bonds are only about 21 kJ/mol (5 kcal/mol), which is about 10^8 times the rupture forces. Thus, a ferric oxide crystal is not susceptible to rupture by hydraulic shear (foregoing from Camp, 1968, pp. 252, 255). Note that the SI system favors joules over calories, but the "kcal" is common in the literature prior to about 1980.

11.4.2.2 Floc Breakup

The net effect of the flocculation process is the aggregation rate minus breakup rate (Parker et al., 1972, p. 79). Two kinds of breakup are (1) surface erosion of primary particles and (2) deformation and floc fragmentation (Parker et al., 1972, p. 81; Gregory, 1989, p. 221). Surface shearing of primary particles is a function of the scale of turbulence (Parker et al., 1972, p. 82). Eddies that are sufficiently large to fully entrain the floc produce zero relative velocity and no surface shear. Eddies much smaller than the floc result in little surface shear. Eddies of a scale approximating the floc diameter would impart the maximum relative velocity and maximum surface shear (see Figure 10.11). Therefore, there will be a maximum stable floc size in which the maximum stress imposed upon the floc will just equal the surface shear strength (Parker et al., 1972, pp. 82, 85). A force balance on a floc particle equates viscous drag and acceleration relative to the fluid with buoyancy and gravity.

In Figure 11.8a, the structural backbone is the filamentous network. Such filaments rupture by tensile failure. The essence of this structure represented in Figure 11.8b shows two spherical clusters connected by a strand of filaments. For the floc to rupture, two eddies must act on the floc entraining each end and causing stress in opposite directions on the clusters, thereby causing tensile stress on the filamentous strand (Parker et al., 1972, p. 89). If the floc is "stable," the tensile strength of the filamentous strands is greater than the tensile force induced, i.e., the strand is not broken. The maximum size stable floc is that size where the tensile strength of the filament stands equals the eddy-induced tensile stress (Parker et al., 1972, p. 89).

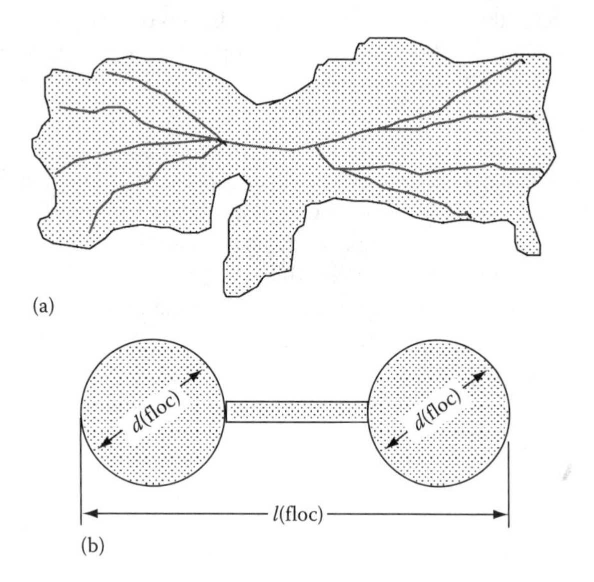

FIGURE 11.8 Depictions of activated sludge floc with filaments. (a) Characterization of actual floc. (b) Structural characterization of floc. (Adapted from Gorczyca, B. and Gahczarczyk, J., *The AWWA Annual Conference*, Vancouver, BC, p. 5, 1992.)

Figure 11.9a shows empirical data that relates average floc size to velocity gradient for both alum and ferric floc. The sizes are compared with Kolmogorov's microscale; the latter is generally smaller than the floc sizes, except as G increases. By comparison, floc sizes are smaller in the study summarized in Figure 11.4 by Spicer and Pratsinis (1996a, p. 1051). Figure 11.9b shows the maximum length of activated sludge floc observed for different velocity gradients for three activated sludges, one with sludge age markedly less than the other two. As seen, the biological floc size is generally larger than the chemical floc.

In general, and as seen in Figure 11.9, the relationship between $d(\text{floc})$ and G plotted on a log–log scale is linear (François, 1987b, p. 1024), i.e.,

$$d(\text{floc}) = KG^{-\gamma} \tag{11.12}$$

where
$d(\text{floc})$ is the floc diameter (m)
K is the empirical constant
γ is the empirical exponent

A theoretical analysis by some investigators have shown that $\gamma = 2$ for the inertial range of turbulence and $\gamma = 1$ for the viscous range (François, 1987b, p. 1024). Flocs with dimensions of the order of magnitude of the turbulence scale for inertial advection will be ruptured into large fragments. Those subjected to viscous forces will be ruptured by erosion of small particles from the floc surface (François, 1987b, p. 1029). The empirical coefficient, γ, is an index of floc strength against hydraulic shear. Some values from different experimenters are seen in Table 11.4 for different G values and for different coagulants and chemical conditions.

11.4.2.3 Bioflocculation

The aggregation of microbes is called "bioflocculation." Such aggregation, in the form of settleable flocs, is an essential part of the activated sludge process.

The "glue" that binds the bacteria cells to form an aggregate is thought to be exocellular biopolymers produced by the bacteria. They can be attached to the cell as a capsule or excreted into the surrounding medium as a slime (Higgins and Novak, 1997, p. 479). Exocellular polysaccharides have been thought to play the major role in flocculation but Higgins and Novak (1997, p. 479) have indicated that exocellular proteins, i.e., lectin in particular, may have an important role as well, with divalent cations being necessary to provide bridging between negatively charged sites within the biopolymer. The model they proposed is illustrated in Figure 11.10; it

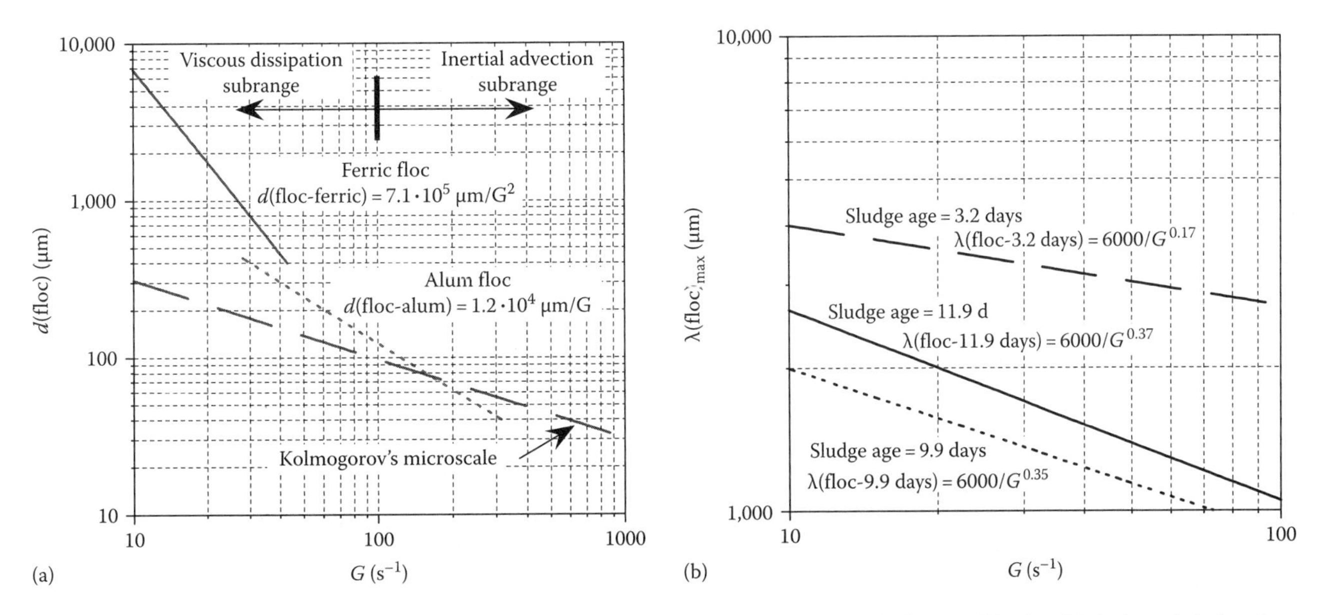

FIGURE 11.9 Stable floc size as affected by G (data points not shown). (a) Alum and ferric floc—stable size (b) Activated sludge floc—maximum length. (Adapted from Parker, D.S. et al., J. *Environ. Eng. Division, ASCE*, 98(SA1), 92, 1972.)

TABLE 11.4
Experimental Data for γ and *G*, Equation 11.12

Investigator	γ	G (s^{-1})	Type of Floc
Argaman and Kaufman (1970)	1	30–120	Aluminum hydroxide + kaolinite in distilled water
Camp (1968)	0.74	180–1000	Aluminum hydroxide + kaolinite + NaCl
Hoppe et al. (1977)	1.18 ± 0.09	75–250	Polyacrylamide flocs + kaolinite + $CaCl_2$
Lagvankar and Gemmell (1968)	0.67	10–40	Iron hydroxide flocs + kaolinite in distilled water
Leentvaar and Rehbun (1983)	0.59	20–150	Iron hydroxide flocs + domestic wastewater
	1.1	20–150	Iron hydroxide flocs + tap water
Parker et al. (1971)	0.36	10–100	Activated sludge
Stevenson (1972)	0.80	3.8–40	Iron hydroxide flocs + ground water
François (1987b, p. 1025)	0.3–0.5	34–1398	Aluminum hydroxide + kaolinite

Source: François, R.J., *Water Res.*, 21(9), 1023, 1987b.

- Argaman, Y. and Kaufman, W. J., Turbulence and Flocculation, *Journal of the Sanitary Engineering Division*, American Society of Civil Engineers, 96(SA2):223–241, 1970.
- Camp, T. R., Floc volume concentration, *Journal of the American Water Works Association*, 60(6):656–673, June, 1968.
- Hoppe, H., Tröger. W., and Winkler, F., Das Kinetische Modell der Orthokinetischen Phase der Flocking am Beispiel Einer Flockulierten Kaolinsuspension, *Wiss. Z. Tech. Hochsch. Chem. Leuna-Merseb.*, 19:399–408, 1977.
- Lagvankar, A. L. and Gemmell, R. S., A size-density relationship for flocs, *Journal of the American Water Works Association*, 59(9):1040–1046, September, 1968.
- Leentvaar, J. and Rehbun, M., Strength of ferric hydroxide flocs, *Water Research*, 17:895–902, 1983.
- Parker, D. S., Kaufman, W. J., and Jenkins, D., Physical conditioning of activated sludge floc, *Journal Water Pollution Control Federation*, 43:1817–1833, 1971.
- Stevenson, D. G., Flocculation and floc behavior, *Journal of the Institution of Water Engineers*, 3:155–169, 1972.
- François, R. J., Strength of aluminum hydroxide flocs, *Water Research*, 21(9):1023–1030, 1987b.

involves a lectin-like protein attaching to the bacteria on one side and a polysaccharide chain on the other, with cations providing bridging between polysaccharide chains, between lectins and lectins, and between lectins and polysaccharides. The divalent cations associate with the negatively charged groups on the molecules mentioned.

11.4.3 Flocculents

A polymer (Box 11.2) is sometimes added as a *flocculent*, subsequent to rapid mix, in order to create *bridging* between floc particles as they develop within the flocculation basin; the purpose is to strengthen the floc and to facilitate growth to a

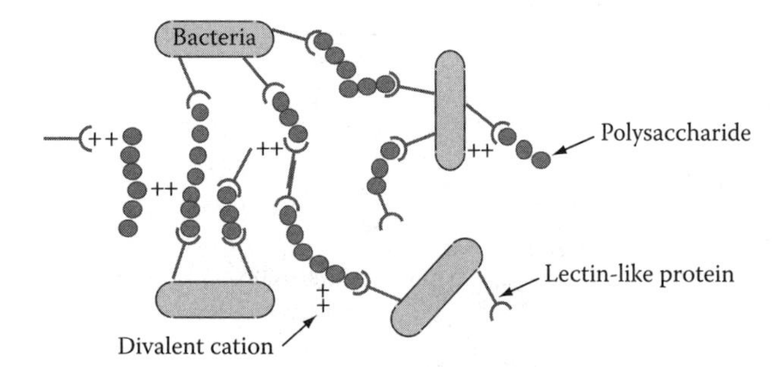

FIGURE 11.10 Depiction of role of lectins, along with polysaccharides and divalent metals, in aggregation of bacteria. (Adapted from Higgens, M.J. and Novak, J.T., *J. Environ. Eng. Division, ASCE*, 123(EED5), 484, 1997.)

BOX 11.2 POLYMERS

A review of the topic of polymers is given by Gregory (1987) and abstracted as follows (see also, Chapter 9, Section 9.9).

- Synthetic polymers for coagulation/flocculation have been available since the early 1950s (p. 164).
- Most commercial products are based on polyacrylamide, since they can give polymer of high MW, i.e., MW $\approx 2 \cdot 10^7$ (p. 165).
- Polyacrylamide is nominally nonionic in character.
- Controlled hydrolysis gives polyacrylamides with different degrees of anionic character and charge density (p. 165).
- Cationic polyelectrolytes based on polyacrylamide are prepared by copolymerization of acrylamide with a suitable cationic monomer, e.g., dimethylaminoethyl acrylate or methacrylate, which are quaternized after polymerization. This is the most convenient method of preparing cationic polymers of very high MW. The proportion of cationic monomer determines the charge density (p. 165).

larger mean size. As a rule, an anionic polymer is used for this purpose.

11.4.4 Design Principles for Paddle-Wheel Flocculators

The most-used technology for creating random transport motion to cause collisions between floc particles is the paddle wheel. This section outlines the theory and protocol for paddle-wheel design, as outlined in a classic paper by Camp (1955).

FIGURE 11.11 Paddle-wheel flocculators installed in a basin.

Figure 11.11 is a photograph that illustrates the layout of paddle wheels in a compartment. Typically, there are three compartments in series; in some cases there are four. In the case shown, the flow is transverse from under the baffle wall to the left, then across the compartment, and under the baffle wall on the right. In other cases, the flow schematic is under-and-over, etc. Yet another flow schematic is serpentine along the axis of the paddle wheels, then end-around to the next compartment.

As seen, a paddle wheel consists of a number of blades attached at different radial distances; its overall diameter is typically 1–2 m (3–6 ft). The paddles are oriented such that the surface of each is normal to the direction of its motion. The rotational velocity of the paddle wheel combined with the blade area is designed to produce a specified G value for the volume of the particular paddle-wheel compartment, i.e., as in Equation 10.5. The mathematical relation that expresses Equation 10.5 in terms of paddle-wheel dimensions are derived here.

Figure 11.12a is a sketch of a paddle-wheel flocculator showing the basic layout of a set of paddles on a large rotating frame. The inset in Figure 11.12b shows a single blade traveling at velocity v_b with the drag force, F_D, induced by the motion, which must be overcome by the torque exerted by the shaft on the arms and paddles. These figures are the basis for the mathematical development, which follows, relating power expended to the drag forces on the rotating paddles.

11.4.4.1 Derivation of Camp's Equation for Paddle-Wheel Flocculation

Consider the drag force on a flat blade moving through water at velocity, v_b, such as shown in Figure 11.12b, which is (see Box 11.3 for drag coefficient discussion),

$$F_D = C_D \frac{\rho_w}{2} v_b^2 A_b \tag{11.13}$$

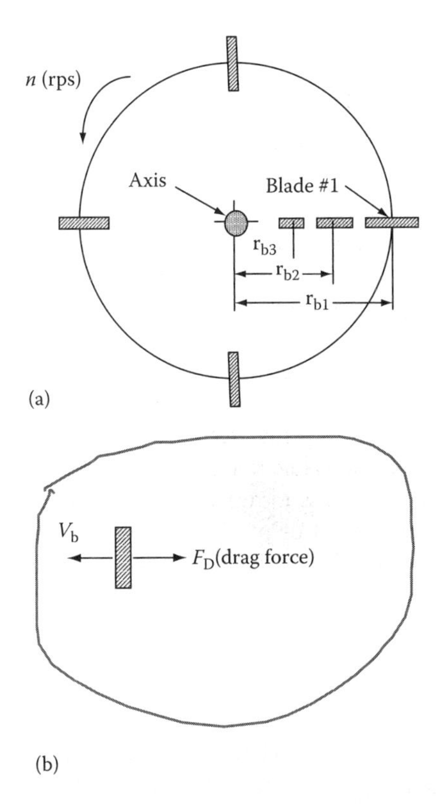

FIGURE 11.12 Sketch of paddle-wheel flocculator showing terms used in mathematical development of design equations. (a) Side view sketch of paddle wheel. (b) Inset diagram showing drag force on a blade with velocity v_b.

BOX 11.3 DRAG COEFFICIENT

For a flat plate with surface normal to the velocity, $C_D = 1.8$ when $\mathbf{R} > 10^3$ (Rouse, 1946, p. 247; Figure D.1, Appendix D). Thus, a paddle-wheel blade would have this drag coefficient. For $1 < \mathbf{R} < 10^3$, the C_D versus R experimental curves should be used (see, for example Rouse, 1946, p. 247). The following drag coefficients are given for $\mathbf{R} \geq 10^3$ by Rouse (1966, p. 249) for flat plates of different proportions: $L/w = 1$, $C_D = 1.16$; $L/w = 5$, $C_D = 1.20$; $L/w = 20$, $C_D = 1.50$; $L/w = \text{infinity}$, $C_D = 1.90$. For $\mathbf{R} \leq 1$, Stoke's law applies and $C_D = 24/\mathbf{R}$.

where
- F_D is the drag force due to motion of blade (N)
- C_D is the drag coefficient for a flat plate (dimensionless)
- ρ_w is the density of water (kg/m^3)
- v_b is the velocity of paddle (m/s)
- A_b is the area of paddle (m^2)

The power, P, expended by the blade is the drag force, F_D, times its velocity, v_b, i.e.,

$$P(\text{blade}) = C_D \frac{\rho}{2} v_b^3 A_b \qquad (11.14)$$

where P(blade) is the power dissipated by blade (N m/s).

Now if the blade is connected to a rotating paddle-wheel arm with rotational velocity, ω rad/s, the velocity of the blade can be expressed as $v_b = r_b \cdot \omega$, in which r_b is the radial distance to the blade. Inserting this in Equation 11.14 gives

$$P(\text{blade}) = C_D \frac{\rho_w}{2} \omega^3 r_b^3 A_b \qquad (11.15)$$

where
- r_b is the distance to center of blade from shaft (m)
- ω is the rotational velocity of shaft (rad/s)

Let there be a set of n(blades) per paddle-wheel arm and N(arms) per paddle wheel. In practice, $2 \leq N(\text{arms}) \leq 4$. Using the subscript, i, to designate any blade, the power expended by all blades per paddle wheel is

$$P(\text{paddle wheel}) = C_D \frac{\rho_\omega}{2} \omega^3 N(\text{arms}) \sum_1^{n(\text{blades})} r_{bi}^3 A_{bi} \qquad (11.16)$$

Since the whole water mass is subject to being set in motion by the paddle wheel, a slippage factor k is inserted into Equation 11.16 to give

$$P(\text{paddle wheel}) = \frac{1}{2} C_D \rho (1-k)^3 \omega^3 N(\text{arms}) \times \sum_1^{n(\text{blades})} r_{bi}^3 A_{bi} \qquad (11.17)$$

where
- N(arms) is the number of arms, with blades that comprise the paddle wheel (m)
- k is the slippage factor which is ratio of rotational velocity of water mass to rotational velocity of paddle wheel, $k = 0.24$–0.32

To explain the $(1-k)\omega$ term, let the relative velocity of the blade past the water be designated $v_{b/w}$. The velocity of the rotating water mass is v_w. Therefore, $v_{b/w} = v_b - v_w$. If $v_w = kv_b$, then, $v_{b/w} = v_b - kv_b = v_b(1-k)$. Since $v_b = r_b\omega$, the grouping in Equation 11.17 follows.

The effect of the radial spokes can be calculated in the same fashion, only an integral expression is involved. If the spoke is a flat blade shape, its integral is

$$1/4 \cdot w(\text{blade}) \cdot r_o^4 \qquad (11.18)$$

and must be added to the summation, multiplied by the number of spokes. Thus, the complete equation is

$$P(\text{paddle wheel}) = \frac{1}{2} C_D \rho (1-k)^3 \omega^3 \times \left[N(\text{arms}) \sum_1^{n(\text{blades})} r_{bi}^3 A_{bi} + \frac{N(\text{spokes})}{4} w(\text{blade}) \cdot r_o^4 \right] \qquad (11.19)$$

11.4.4.2 *P*(paddle-wheel) with Units

Equation 11.19 may be solved by inserting the appropriate numerical data and units for each term. Table 11.5 delineates the conversions for both SI units and U.S. Customary units, respectively.

11.5 DESIGN

The design procedure for paddle-wheel flocculation basins was established by Camp (1955) and remains largely the same as set forth in his guidelines. For baffled basins, the procedure has remained empirical as established in the 1930s, but with more recent update of guidelines (e.g., Haarhoff, 1998).

11.5.1 Design Procedure from Camp

Not much had been done with G in the first few years after being introduced by Camp and Stein (1943). Camp's 1955 paper brought G into the picture, however, for quantitative design, showing how to compute G for different kinds of flocculation technologies, e.g., paddle-wheel basins, baffled basins, diffused air, and reciprocating blades. For each technology, Camp showed how to compute G from fundamentals, e.g., Equation 11.19 for paddle-wheel flocculators. The 1955 paper also characterizes Camp's work in that he (1) showed how to apply fundamentals, (2) developed practical criteria, e.g., for G and $G\theta$, and (3) gave practical guidelines, e.g., for tip velocity, area of paddles, tapered flocculation, etc. Thus, Camp had delineated a design protocol for flocculation.

11.5.1.1 Camp's Criteria

Camp (1955) recommended upper limits of G for a flocculation basin ranging from 74 s^{-1} for the first compartment to 20 s^{-1} for the third compartment. The upper limit of $G = 20\ s^{-1}$ in the third compartment was to minimize floc breakup. The criterion for the total number of collisions, $G\theta$ had a wide range, i.e., $23{,}000 \leq G\theta \leq 210{,}000$.

The values adopted for G and $G\theta$ criteria represented limits of design practice for 20 operating plants, as discussed in Section 11.3.1.4, with data given in Table 11.1. As seen in Table 11.1 (c. 1918–1931), 14 of the plants used baffle flocculation; only 4 used paddle wheels. After Camp's paper, paddle wheels were used in most designs.

For flotation, G values are higher, e.g., $G \approx 70\ s^{-1}$. Only one or two compartments are used since a smaller floc, e.g., $d(\text{floc}) \approx 10\ \mu m$, is desired.

11.5.1.2 Camp's Guidelines

Some guidelines for flocculation basin design, abstracted from Camp (1955), are enumerated as follows:

1. The turbulence intensity, G, should be "tapered" along the length of the basin such that for the first compartment, $G \approx 70$–$80\ s^{-1}$ and for the last compartment $G \approx 10$–$20\ s^{-1}$. In an installation designed by Camp in Cambridge, Massachusetts (Camp, 1955), the values of G in the first three compartments of the flocculation basin were, respectively, 5.6, 3.8, and 1.9 times the G value in the fourth compartment.
2. As seen in Equations 11.13 and 11.19, any one of several independent variables may be imposed to change G, i.e., V(compartment), n, or A(paddles).
3. The design should provide for variation in rotational velocity, n, by the operator. At Cambridge, for example, $2.0 \leq n(\text{compartment 1}) \leq 5.2$ rpm, and $1.1 \leq n(\text{compartment 4}) \leq 2.9$.
4. The paddle wheels may be oriented with the axes either normal to the flow direction or parallel to it.
5. Paddle area for any one paddle wheel should range 10%–25% of the cross-sectional area of the basin.
6. Stators are advisable to mitigate the tendency for the whole water mass to rotate with the paddles.
7. Peripheral speed of paddles may range from 0.1 to 1.0 m/s (0.3–3.0 ft/s).
8. The total basin detention period, θ, may range from $30 \leq \theta \leq 60$ min.
9. The $G\theta$ parameter should be distributed as uniformly as possible among the different flocculation compartments.
10. Where there is conflict in the above guidelines with the results of calculations based upon the G and $G\theta$ criteria, the latter should prevail.

Some additional guidelines, also from Camp (1955), except as noted, are as follows:

- If the width of a paddle is too wide, the water in front is carried along by the velocity of the paddle (p. 10); mixing does not occur to any extent.
- The blades of a paddle wheel should be relatively narrow and more in number (as opposed to wider and fewer in number). This is in accordance with mixing theory and experimental findings, e.g., that the scale of the turbulence should be about the same as the size of the floc desired (Section 10.3.1.2, Figure 10.4, Figure 10.11; Section 11.4.3.5 and Figure 11.9). This would favor smaller width blades, and larger number, with larger blades in each successive compartment (so that the scale of the turbulence increases with floc size). A caveat is that in the third compartment the flow regime is likely to be laminar.
- Without stator blades, $0.15 \leq A(\text{blades}) \leq 0.20$ times cross-sectional area to prevent rolling water. If $A(\text{blades}) \geq 0.25$, major rotation will occur.
- The P/V dissipative function is an average over the basin volume; local variation, i.e., from point to point, may be considerable.
- Since $k = 0$ at $t = 0$, the startup power greatly exceeds the equilibrium power, so the paddles must be brought to equilibrium speed slowly.
- Figure 11.13 shows power versus rpm for the paddle-wheel flocculator at Cambridge, Massachusetts, as determined by Camp (1955), showing an exponential increase in power with rotational velocity.

TABLE 11.5
Delineation of Paddle-Wheel Energy Dissipation Equations

SI Units		U.S. Customary Units	
1. Mass to force conversion—for reference,			
$\rho_w\ (\mathrm{kg/m^3}) = \dfrac{\gamma\ (\mathrm{N/m^3})\cdot \mathrm{s^2}}{9.81\ \mathrm{m}} = \dfrac{9790\ (\mathrm{N/m^3})\cdot \mathrm{s^2}}{9.81\ \mathrm{m}} = 998\ (\mathrm{N\cdot s^2/m^4})$	(11.20a)	$\rho_w(\mathrm{slugs/ft^3}) = \dfrac{\gamma\ (\mathrm{lb_f/ft^3})\cdot \mathrm{s^2}}{32.2\ \mathrm{ft}} = \dfrac{62.4\ (\mathrm{lb_f/ft^3})\cdot \mathrm{s^2}}{32.2\ \mathrm{ft}} = 1.94\ (\mathrm{lb_f\cdot s^2/ft^4})$	(11.20b)
2. Dimensions inserted in equation,			
$P(\text{paddle wheel}) = \dfrac{1}{2}C_D\left(\rho_w\ \dfrac{\mathrm{kg}}{\mathrm{m^3}}\right)(1-k)^3\left(\omega^3\ \dfrac{\mathrm{rad^3}}{\mathrm{s^3}}\right)N(\text{arms}) \times \sum^{n(\text{blades})} (r_i^3\ \mathrm{m^3})(A_i\ \mathrm{m^2})$	(11.21a)	$P(\text{paddle wheel}) = \dfrac{1}{2}C_D\left(\rho_w\ \dfrac{\mathrm{slugs}}{\mathrm{ft^3}}\right)(1-k)^3\left(\omega^3\ \dfrac{\mathrm{rad^3}}{\mathrm{s^3}}\right)N(\text{arms}) \times \sum^{n(\text{blades})} (r_i^3\ \mathrm{ft^3})(A_i\ \mathrm{ft^2})$	(11.21b)
3. Density of water is inserted,			
$= \dfrac{1}{2}C_D\left(998\ \dfrac{\mathrm{kg}}{\mathrm{m^3}}\right)(1-k)^3\left(\omega^3\ \dfrac{\mathrm{rad^3}}{\mathrm{s^3}}\right)N(\text{arms}) \sum^{n(\text{blades})} (r_i^3\ \mathrm{m^3})(A_i\ \mathrm{m^2})$	(11.22a)	$= \dfrac{1}{2}C_D\left(1.94\ \dfrac{\mathrm{lb_f\cdot s^2}}{\mathrm{ft^4}}\right)(1-k)^3\left(\omega^3\ \dfrac{\mathrm{rad^3}}{\mathrm{s^3}}\right)N(\text{arms}) \sum^{n(\text{blades})} (r_i^3\ \mathrm{ft^3})(A_i\ \mathrm{ft^2})$	(11.22b)
4. Dimensions are consolidated,			
$= \dfrac{1}{2}C_D\left[998\ \dfrac{\mathrm{N\cdot s^2}}{\mathrm{m^4}}\ \dfrac{1}{\mathrm{s^3}}\ \mathrm{m^3 m^2}\right](1-k)^3\omega^3 N(\text{arms}) \sum^{n(\text{blades})} r_i^3 A_i$	(11.23a)	$= \dfrac{1}{2}C_D\left(1.94\ \dfrac{\mathrm{lb_f\cdot s^2}}{\mathrm{ft^4}}\ \dfrac{\mathrm{rad^3}}{\mathrm{s^3}}\ \mathrm{ft^3 ft^2}\right)(1-k)^3\omega^3 N(\text{arms}) \sum^{n(\text{blades})} r_i^3 A_i$	(11.23b)
5. Final result for expression in SI/U.S. Customary units,			
$= \dfrac{1}{2}C_D\left[998\mathrm{N}\ \dfrac{\mathrm{m}}{\mathrm{s}}\right](1-k)^3\omega^3 N(\text{arms}) \sum^{n(\text{blades})} r_i^3 A_i$	(11.24a)	$= \dfrac{1}{2}C_D\left(1.94\ \dfrac{\mathrm{lb_f ft}}{\mathrm{s}}\right)(1-k)^3\omega^3 N(\text{arms}) \sum^{n(\text{blades})} r_i^3 A_i$	(11.24b)
6. Expressing in n rev/s gives,			
$= \dfrac{1}{2}C_D\left[998\mathrm{N}\ \dfrac{\mathrm{m}}{\mathrm{s}}\right](1-k)^3(2\pi n)^3 N(\text{arms}) \sum^{n(\text{blades})} r_i^3 A_i$	(11.25a)	$= \dfrac{1}{2}C_D\left(1.94\ \dfrac{\mathrm{lb_f ft}}{\mathrm{s}}\right)(1-k)^3(2\pi n)^3 N(\text{arms}) \sum^{n(\text{blades})} r_i^3 A_i$	(11.25b)
7. Consolidating numerical terms,			
$= 124C_D\left[998\mathrm{N}\ \dfrac{\mathrm{m}}{\mathrm{s}}\right](1-k)^3 n^3 N(\text{arms}) \sum^{n(\text{blades})} r_i^3 A_i$	(11.26a)	$= 240C_D\left(\dfrac{\mathrm{lb_f ft}}{\mathrm{s}}\right)(1-k)^3 n^3 N(\text{arms}) \sum^{n(\text{blades})} r_i^3 A_i$	(11.26b)

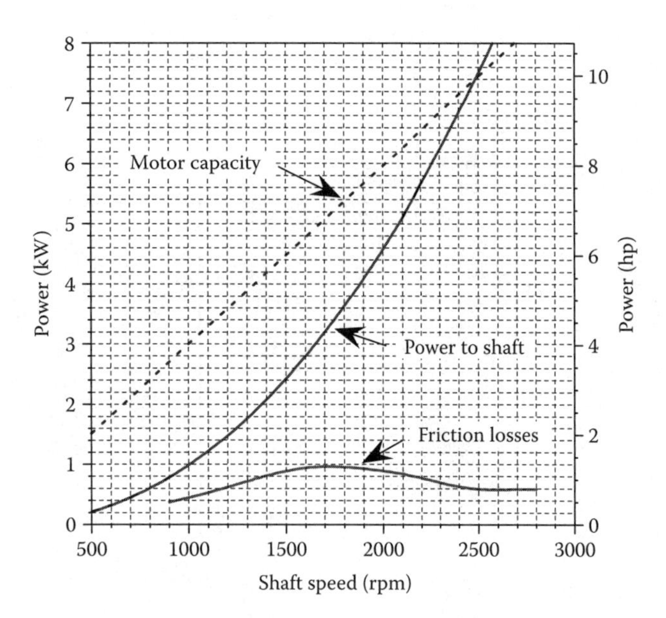

FIGURE 11.13 Paddle-wheel motor test results at Cambridge. (Adapted from Camp, T.R., *Trans. Am. Soc. Civil Eng.*, 120, 13, 1955.)

- Figure 11.13 shows about 0.7 kW (1 hp), is lost through the transmission, stuffing box, and bearings for motor speeds $500 \le n(\text{motor shaft}) \le 2600$ rpm. The motor power was 7.5 kW (10 hp) for each shaft. The rotational speed of the motor shafts were reduced by gears and adjustable to give paddle speeds within the ranges $1.1 \le n(\text{low-speed paddle wheel}) \le 2.9$ rpm and $2.0 \le n(\text{high-speed paddle wheel}) \le 5.2$ rpm.
- Camp (1955) found that $0.24 \le k \le 0.32$ for the Cambridge flocculators.
- The sedimentation basin should be contiguous with the floc basin, i.e., no pipes between, in order to minimize floc breakup.

11.5.1.3 Spreadsheet Algorithm

Table 11.6 is a spreadsheet solution for the application of Camp's equation (11.19), which also incorporates his recommended guidelines. The spreadsheet is set up with formulae that incorporate the radial distance to the center of each blade and the width of each blade, respectively, for the first blade (which may be modified as desired). The criterion for detention time, G, and other parameters are incorporated into the spreadsheet. Once the flow is stated, the design and operating results are outputs providing calculated values of G, P, P/V, rpm, v_b, and $G \cdot \theta$. In the event the parameter outputs are outside the ranges of Camp's guidelines, the design parameters may be modified by trial and error. The spreadsheet is structured to accommodate modifications.

As constructed, Table 11.6 shows the effect of three temperatures, i.e., 0°C, 20°C, 40°C (32°F, 68°F, 104°F) on P/V, G, and the power required for the particular design being explored. Of particular interest, the power required for the design shown (diameter = 3.9 m (12.8 ft) and $G = 60\ \text{s}^{-1}$) is 1.323 kW (1.8 hp). Using a motor efficiency of 0.67, the motor power would be 1.97 kW (2.6 hp), which corresponds to similar installations in practice (albeit the diameter used in this example is larger than used in most installations in practice).

11.5.2 Model Flocculation Basin

Figure 11.14 shows a flocculation basin with three compartments with paddle wheels, each with vertical shaft. The motor for the first compartment was mounted on a bearing plate with lever arm attached to the motor frame and with a force gage attached at the end of the lever arm. This arrangement permitted measurement of the force exerted by the lever arm and calculation of torque. The rotational velocity for a given motor controller setting could be measured simply by counting the rotations with a stopwatch.

11.5.2.1 Calculations

Table CD11.7 shows data obtained from the first compartment of the flocculation basin shown in Figure 11.14 along with calculated values of torque on the shaft, shaft power, paddle wheel G, power number, **P**, and Reynolds number, **R**. Formulae in the cells are torque = force gage reading in kg · 9.81 · length of lever arm

power to shaft = torque · (rpm/60)
$G = P/\mu V(\text{comp})$
$\mathbf{P} = P/(n^3 \cdot D^5 \cdot \rho)$
$\mathbf{R} = \omega D^2 \rho/\mu$

The viscosity is calculated using an empirical intercept-slope formula (as opposed to the polynomial formula given in Table CDQR.5).

11.5.2.2 Plots

From the data of Table CD11.7, plots were generated and are shown in Figure 11.15a through d, for G versus rpm, G versus **R**, **P** versus **R**, power-to-shaft versus rpm, respectively. The plots show the range of practice, i.e., $10 \le G \le 100\ \text{s}^{-1}$, as shaded areas. Features of each plot are enumerated.

- Figure 11.15a shows how G is affected by rotational velocity of the shaft, given as rpm at two temperatures. For the first compartment assuming $G \approx 100\ \text{s}^{-1}$, about 13 rpm is the upper limit for 10°C; about 10 rpm is the upper limit for 22°C.
- Figure 11.15b is the same as (a) except that Reynolds number is used as the abscissa, which takes viscosity out of the picture; in other words, data points for the two temperatures line up along the same curve.
- Figure 11.15c shows the power number, **P** versus Reynolds number, **R**. The curve has the same classic shape as defined by the various mixing impellers

TABLE 11.6
Generic Design of Flocculation Basin Using Spreadsheet

1 Fixed Data

$Q = 6$ mgd	$C_d = 1.8$
$= 4167$ gpm	$\rho_w = 998$ kg/m^3
$= 0.263$ m^3/s	$k = 0.24$
	$k = 0.60 + 0.007 \cdot$ rpm

2 Viscosity Formula Coefficients

Limits	Intercept	Slope
$0 \le T \le 20$	0.00179	−0.0126
$20 < T \le 50$	0.00153	−0.0093

3 Computations

Conditions					Basin Sizing						
Q (m^3/s)	(mgd)	Temperature (°C)	μ (N · s/m^2)	θ (min)	V (m^3)	Comp (desig.)	N(com) (No. of)	V(com) (m^3)	Depth (m)	L(com) (m)	W(basin) (m)
0.2628	6.0	0	0.00179	40	630.8	1	3	210.3	4	4	13.1
0.2628	6.0	0	0.00179	40	630.8	2	3	210.3	4	4	13.1
0.2628	6.0	0	0.00179	40	630.8	3	3	210.3	4	4	13.1
0.2628	6.0	20	0.001	40	630.8	1	3	210.3	4	4	13.1
0.2628	6.0	20	0.001	40	630.8	2	3	210.3	4	4	13.1
0.2628	6.0	20	0.001	40	630.8	3	3	210.3	4	4	13.1
0.2628	6.0	40	0.00065	40	630.8	1	3	210.3	4	4	13.1
0.2628	6.0	40	0.00065	40	630.8	2	3	210.3	4	4	13.1
0.2628	6.0	40	0.00065	40	630.8	3	3	210.3	4	4	13.1
Given		Given	Formula	Assumed	$V = Q \cdot \theta$	Designated compartment	No. of compartments	V/N(com)	Assumed	Equals depth	$V/D \cdot L$

Paddle Wheel Design

Blade Length, Width						Arms							
L(blade) (m)	$w(b)_1$ (m)	$w(b)_2$ (m)	$w(b)_3$ (m)	$w(b)_4$ (m)	$w(b)_5$ (m)	N(arms)/paddle	Wheel Dia (m)	$r(b)_1$ (m)	$r(b)_2$ (m)	$r(b)_3$ (m)	$r(b)_4$ (m)	$r(b)_5$ (m)	$\Sigma A_i \cdot r(b)_i$ (m^5)
13.04	0.15	0.09	0.06	0.05	0.03	4	3.9	1.88	1.41	0.99	0.56	0.13	17.04
13.04	0.15	0.09	0.06	0.05	0.03	4	3.9	1.88	1.41	0.99	0.56	0.13	12.90
13.04	0.15	0.09	0.06	0.05	0.03	4	3.9	1.88	1.41	0.99	0.56	0.13	17.04
13.04	0.15	0.09	0.06	0.05	0.03	4	3.9	1.88	1.41	0.99	0.56	0.13	17.04
13.04	0.15	0.09	0.06	0.05	0.03	4	3.9	1.88	1.41	0.99	0.56	0.13	17.04
13.04	0.15	0.09	0.06	0.05	0.03	4	3.9	1.88	1.41	0.99	0.56	0.13	17.04
13.04	0.15	0.09	0.06	0.05	0.03	4	3.9	1.88	1.41	0.99	0.56	0.13	17.04
13.04	0.15	0.09	0.06	0.05	0.03	4	3.9	1.88	1.41	0.99	0.56	0.13	17.04
13.04	0.15	0.09	0.06	0.05	0.03	4	3.9	1.88	1.41	0.99	0.56	0.13	17.04
W(basin)-0.3	Assume	$0.6 \cdot w(b)_1$	$0.4 \cdot w(b)_1$	$0.33 \cdot wb1$	$0.2 \cdot w(b)_1$	Assume	$D - 2 \cdot 0.05$	$(D/2) - 0.5 \cdot w_1$	$0.75 \cdot r_1$	$0.53 \cdot r_1$	$0.30 \cdot r_1$	$0.07 \cdot r_1$	$\Sigma A_i \cdot r(b)_i$

(continued)

TABLE 11.6 (continued)
Generic Design of Flocculation Basin Using Spreadsheet

Operation						Criteria			
G (s^{-1})	P/V ($N \cdot m/s/m^3$)	P ($N \cdot m/s$)	ω_1 (rad/s)	n (rpm)	k	ω_2 (rad/s)	v_b (m/s)	$G \cdot \theta$	$\Sigma G \cdot \theta$
60	6.43	1353	0.37	3.53	0.08	0.31	0.69	48,000	
40	2.86	601	0.31	2.96	0.08	0.26	0.58	32,000	
15	0.40	85	0.15	1.40	0.07	0.12	0.28	12,000	92,000
60	3.60	757	0.30	2.91	0.08	0.25	0.57	48,000	
40	1.60	336	0.23	2.22	0.08	0.19	0.44	32,000	
15	0.23	47	0.12	1.16	0.07	0.10	0.23	12,000	92,000
60	2.35	494	0.26	2.52	0.08	0.22	0.50	48,000	
40	1.04	220	0.20	1.93	0.07	0.17	0.38	32,000	
15	0.15	31	0.10	1.00	0.07	0.09	0.20	12,000	92,000
Assumed	$P/V = \mu G^2$	$P = (P/V) \cdot V$					$0.1 < v_b < 0.9$		$23{,}000 < \Sigma GT < 210{,}000$

$\omega \cdot 60/2\pi$ 0.06 + 0.007 rpm

$\omega_1 = [(P/V)*V/((1/2)*Cd*r_w*(1 - k)\Sigma A_i Rbi^\wedge 3]^\wedge 0.333$

ω_2 = second calculation

4 Summary

Temperature (°C)	G (s^{-1})	(rpm)	P ($N \cdot m/s$)
0	60	3.53	1353
0	40	2.96	601
0	15	1.40	85
20	60	2.91	757
20	40	2.22	336
20	15	1.16	47
40	60	2.52	494
40	40	1.93	220
40	15	1.00	31

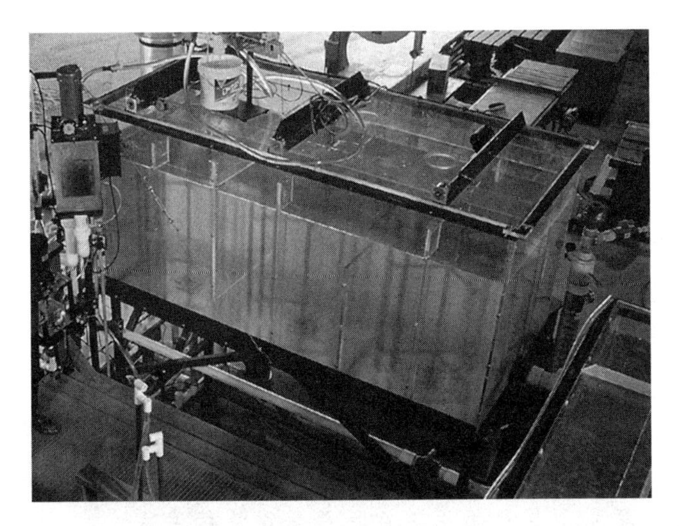

FIGURE 11.14 Flocculation basin for 76 L/min (20 gpm) pilot plant at Engineering Research Center, Colorado State University, Fort Collins, CO.

given in Rushton's work (see Section 10.3.3.3). The power number is affected by Reynolds number as a straight-line relationship in the range, $\mathbf{R} < 10^5$, and for $\mathbf{R} > 10^5$, $\mathbf{P} \approx 4.1$. If $G \approx 100\ s^{-1}$, as in the first compartment, $\mathbf{R} \approx 10^5$, which is in the turbulent range. If $G \approx 15\ s^{-1}$, as in the third compartment, $\mathbf{R} \approx 20{,}000$, which is in the laminar range and Camp's equation (10.5) $G = [P/\mu V]^{0.5}$ applies, i.e., with μ in the denominator.

- Figure 11.15d shows shaft power versus rotational velocity of the shaft. As seen, the curve shows an exponential rise in power with rpm with exponent = 2.26. The power required for a given shaft rotational velocity can be obtained. For $n = 12$ rpm, for example, $P \approx 10$ W (0.013 hp).
- The plots shown in Figure 11.15 apply only for the system tested. The nature of the relationships shown, however, and their general shapes should apply to any other system.

11.5.2.3 Slip Factor

Table CD11.8 shows the application of Equation 11.19 to the data of Table CD11.7 to give k, i.e., the "slip factor," values for different rotational velocities. Figure 11.16 is a plot of the data showing k versus rpm as a linear relation, i.e.,

$$k = 0.074 + 0.007 \cdot \text{rpm} \tag{11.27}$$

The intercept (0.074) and slope (0.007) may differ from one system to another, but the Equation 11.27 form should be true regardless of the system (such as model or full scale). If a slip factor is selected for a design that lacks empirical data, then Camp's slip factor, i.e., $0.24 < k < 0.32$ gives a reasonable estimate; for reference, the k values in Table CD11.8 are within the same range.

11.5.3 Plant Design

While unit processes are the foci of any plant design, a host of other considerations are necessary to support any process functioning. The layout of the overall plan showing all unit processes as they are integrated as a system is the first task. Other tasks required include the sizing of pipes, motors, meters, etc.; selection of materials for pipes, paddle wheels, walls; locations of pipes and valves; and methods of adjustment for paddle-wheel rotational velocity. How all of this fits together is shown in drawings with specifications giving supplemental details. A modern adjunct to traditional drawings is an animation derived from the drawings by means of software.

Figure CD11.17a and b are excerpts from "walk-through" animations illustrating the design for the Floc-Sed basin 2000 addition to the Fort Collins WTP. The animation starts with a view seen upon entering the building. The plant addition has two identical parallel treatment trains, each with a four-compartment flocculation basin; the flow leaves the fourth basin and then to an assembly of Lamella plate settlers on the east side of the building. The first view is from the northwest corner looking south along the west side of the building and along the first compartments of both trains. Walking south, a left turn is made heading east between the two trains. The plate settler basins are encountered at the end of the floc basin. Walking to the east side of the building and then south along the plate settler basin, the tops of the plates are seen to the right. Walking around the assembly gives a more detailed view of the third compartment of the flocculation basin. The serpentine path from one compartment to another is clearly visible at this point. Also, the detail of the number of arms for each paddle wheel may be seen, i.e., compartments #1 and #2 paddles each have three arms while compartments #3 and #4 each have two. The separation walls between each compartment are only to channel the water flow and are made of redwood. Figure CD11.17b (animation) starts at the same place but the walk leads down the stairs to the lower level where a pipe gallery and paddle-wheel motors are seen; the motors are larger in size as the walk moves from compartment #4 toward compartment #1. Turning the corner, the main pipe gallery is seen with large pipes that deliver coagulated water to each of the floc basins.

As stated previously, design walk-through animations are software derivatives of the traditional engineering drawings as done by drawing software (such as AutoCad™). They provide a means to visualize the project as constructed, which permits inspection and perhaps modifications. The animation permits "seeing" in places difficult to visualize by drawings alone. For example, at about 0.45 completion of (b), a recessed space with sludge drain pipes is seen, which is difficult to visualize from the traditional drawings. The pseudo walk through lets the designer determine whether, for example, pipes are crossing paths of one another at some point, whether the overall layout is reasonable, and whether the plant seems operable. These same points are of interest to the persons in operation.

TABLE CD11.7

Floc Basin Paddle-Wheel Data for Different Motor Controller Settings and Associated Calculations of Torque, Power Expended, G, P, and R for First Compartment of a 76 L/min (20 gpm) Pilot Plant

1 Fixed Data

Q = 0.0288 mgd
= 20 gpm
= 0.0013 m^3/s
C_d = 1.8

ρ_w = 998 kg/m^3
k = 0.24
Lev Arm = 0.46 m

Motor Pacific Scientific #xxxxx, 1.0 hp

2 Viscosity Formula Coefficients

Limits	Intercept	Slope
$0 \leq T \leq 20$	0.00179	−0.01256
$20 < T \leq 50$	0.00153	−0.00926

3 Volume

V(comp) = 0.719 m^3
θ(comp) = 9.50 min

4 Compartment

L(comp) = 820
W(comp) = 1070
d(water) = 914
B(baffles) = 130

5 Paddle Wheel

D(paddle) = 762
H(blades) = 940
w(blades) = 25.4

6 Blades

r_1 = 152
r_2 = 216
r_3 = 305
r_4 = 368
r_o = 381
(Meas. in mm)

7 Variable Data

Date	Controller Setting	RPM Meas. (rpm)	Force Gage (kgf)	Torque (N · m)	Power (watts)	Temp. (°C)	Viscosity (N · s/m^2)	G (s^{-1})	Power No.	R
12/21/1994	10	1.82	0.25	1.13	0.22	22	0.000958	18	30.2	18,341
	18	4	0.38	1.72	0.72	22	0.000958	32	9.5	40,310
	20	6.67	0.66	2.99	2.09	22	0.000958	55	5.9	67,217
	25	8.57	0.95	4.31	3.86	22	0.000958	75	5.2	86,364
	30	10.43	1.35	6.12	6.68	22	0.000958	98	5	105,108
	35	12	1.7	7.7	9.68	22	0.000958	118	4.7	120,929
	40	14.12	2	9.06	13.4	22	0.000958	139	4	142,294
	45	15	2.3	10.42	16.37	22	0.000958	154	4.1	151,162
	45	14.54	2.1	9.52	14.49	22	0.000958	145	4	146,526
	50	16	2.8	12.69	21.26	22	0.000958	176	4.4	161,239
	55	19.2	3.5	15.86	31.89	22	0.000958	215	3.8	193,487
	65	20.7	4.25	19.26	41.75	22	0.000958	246	4	208,603
	70	20	4.1	18.58	38.92	22	0.000958	238	4.1	201,549
	85	22.2	4.5	20.39	47.41	22	0.000958	262	3.7	223,719
	100	22.2	4.7	21.3	49.52	22	0.000958	268	3.8	223,719
11/28/1995	2	6.25	0.23	1.04	0.68	10	0.001338	27	2.4	45,107
	3	10.5	1.18	5.35	5.88	10	0.001338	78	4.3	75,780
	4	13	1.72	7.8	10.61	10	0.001338	105	4.1	93,823
	5	16.5	2.72	12.33	21.3	10	0.001338	149	4	119,083

Notes: (1) Data for 12/21/94 were from system without flowing water and so the water temperature had equilibrated with the surroundings, i.e., to 22°C. (2) Data for 12/28/94 were from system with flowing water and so the water temperature was the same as the incoming water, i.e., 10°C.

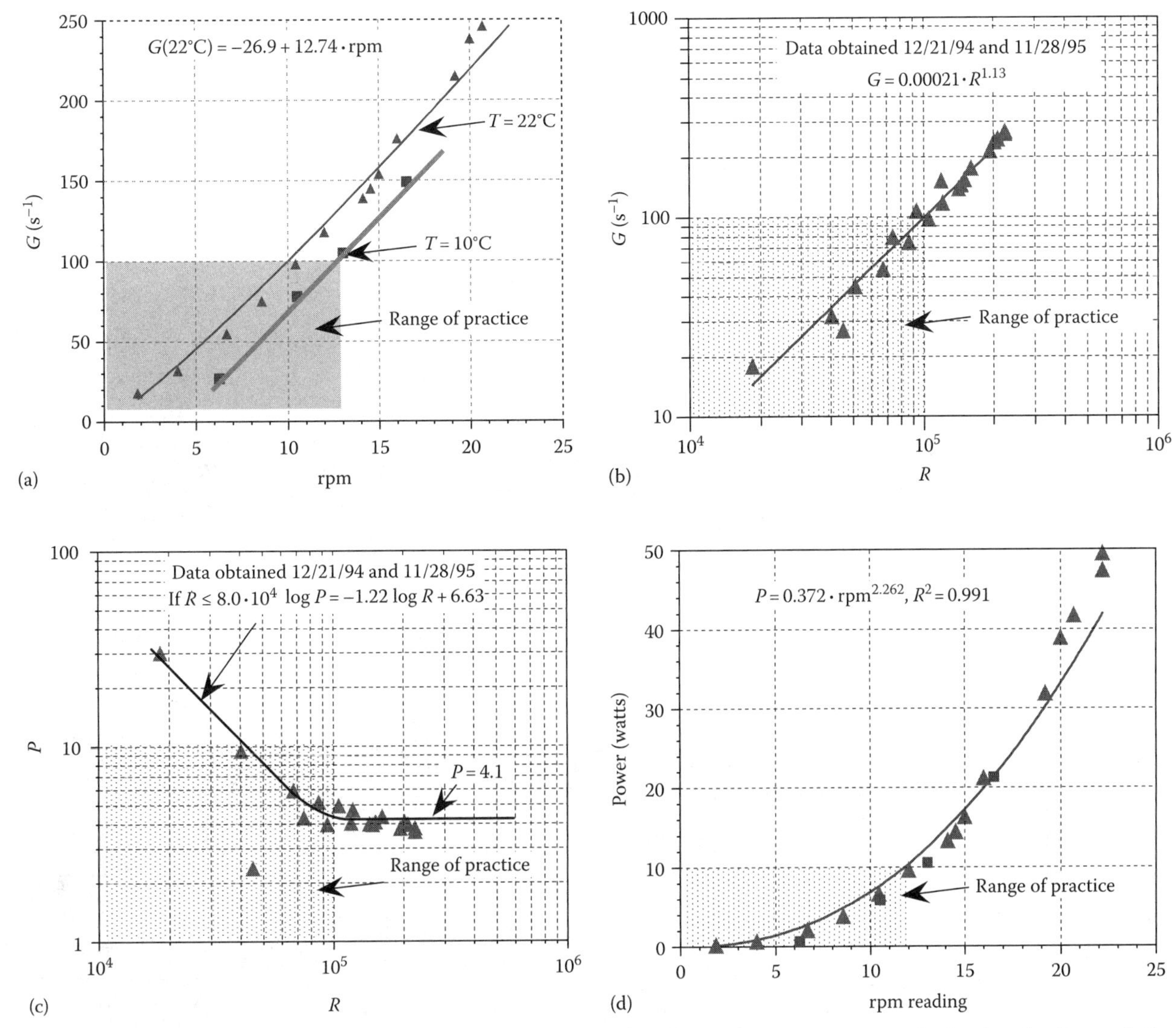

FIGURE 11.15 Plots generated from first compartment of flocculation basin of 76 L/min (20 gpm) located at Engineering Research Center, Colorado State University, Fort Collins, CO. The range of practice, shown as shaded area, is assumed $10 \le G \le 100$. (a) G versus n, (b) G versus R, (c) P versus R, and (d) P versus n.

11.5.4 Other Technologies

Turbines, baffles, sludge-blanket clarifiers, and contact flocculation are among the other technologies that have evolved as flocculation technologies. The design procedures are empirical and are reviewed in this section.

11.5.4.1 Turbines

Turbine flocculators utilize a mixing impeller, e.g., the Rushton type. Usually, their diameter is large and the rotational velocity is low. The criterion, G, is the basis for design, e.g., $20 < G < 90\ \text{s}^{-1}$, in which G is the average for the basin. The "con" side of a turbine flocculator is that the G at the tip of the impeller is several times the average G.

11.5.4.2 Baffles

Baffled basins, as noted, were among the first technologies for flocculation. They remain an important option because of their passive character, i.e., no moving parts and no operator decisions, and at the same time, they are simple. For these reasons they remain favored in many kinds of contexts, e.g., rural areas and developing countries (Bhargava and Ojha, 1993, p. 465).

The design approach for a baffled basin is the same, in principle, as for use of a baffled basin for rapid mix (see, for example Hendricks, 2006, Section 10.4.3.3). The headloss dissipated is, as for any form drag (Bhargava and Ojha, 1993, p. 466; Swamee, 1996, p. 1046; Haarhoff, 1998, p. 145),

$$h_L(\text{slot}) = K\frac{v(\text{slot})^2}{2g} \qquad (11.28)$$

where

$h_L(\text{slot})$ is the headloss due to turbulence induced by slot (m)
K is the loss coefficient for slot (dimensionless)
$v(\text{slot})$ is the velocity between baffle edge and wall of basin (m)

TABLE CD11.8
Calculation of Camp's Slip Coefficient, *k*, Using Data from Pilot Plant

1 Fixed Data

$Q = 0.0288$ mgd
$= 20$ gpm
$= 0.0013\ m^3/s$
$C_d = 1.8$
$\rho_w = 998\ kg/m^3$
$k = 0.24$
Lev Arm $= 0.46$ m
Motor Pacific Scientific #xxxxx, 1.0 hp

2 Viscosity Formula Coefficients

Limits	Intercept	Slope
$0 \leq T \leq 20$	0.00179	−0.01256
$20 < T \leq 50$	0.00153	−0.00926

3 Volume

V(comp) = 0.719 m^3
T(comp) = 9.50 min

4 Compartment

L(comp) = 820
W(comp) = 1070
d(water) = 914
B(baffles) = 130

5 Paddle Wheel

D(paddle) = 762
H(blades) = 940
w(blades) = 25.4

6 Blades

$r_1 = 152$
$r_2 = 216$
$r_3 = 305$
$r_4 = 368$
$r_o = 381$
(Meas. in mm)
N(arms) = 4

7 Variable Data

Date	RPM (rev/min)	ω (rad/s)	*P* N · m/s	$1/2C_d\,\rho\,\omega^3$	*N*(arms) · $\Sigma A_i \cdot r_{bi}^3$ (m^5)	Integral $8 \cdot w(bl) \cdot r_o^4$	Sum	*k*
12/21/1994	1.82	0.19	0.22	6	0.00829	0.00107	0.06	−0.56
	4.00	0.42	0.72	66	0.00829	0.00107	0.62	−0.05
	6.67	0.70	2.09	306	0.00829	0.00107	2.9	0.10
	8.57	0.90	3.86	649	0.00829	0.00107	6.1	0.14
	10.43	1.09	6.68	1170	0.00829	0.00107	11.0	0.15
	12.00	1.26	9.68	1782	0.00829	0.00107	16.7	0.17
	14.12	1.48	13.40	2904	0.00829	0.00107	27.2	0.21
	15.00	1.57	16.37	3481	0.00829	0.00107	32.6	0.21
	14.54	1.52	14.49	3171	0.00829	0.00107	29.7	0.21
	16.00	1.68	21.26	4225	0.00829	0.00107	39.6	0.19
	19.20	2.01	31.89	7301	0.00829	0.00107	68.4	0.22
	20.70	2.17	41.75	9149	0.00829	0.00107	85.7	0.21
	20.00	2.09	38.92	8252	0.00829	0.00107	77.3	0.20
	22.20	2.32	47.41	11286	0.00829	0.00107	105.7	0.23
	22.20	2.32	49.52	11286	0.00829	0.00107	105.7	0.22
Date of data	RPM = rev/Dt	ω = 2šRPM/60	P = Tω	Camp's #1	Camp's #2	Camp #3 (added)	Camp's term	*k* = 1 − (*P*/Sum)^33

Note: $R > 10^5$ for the shortest radial distance blade and the slowest velocity; therefore $C_d = 1.8$ for all conditions.

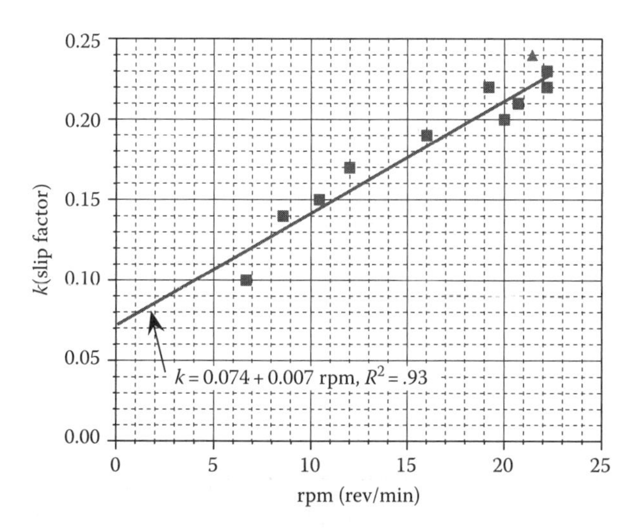

FIGURE 11.16 Slip factor plot generated from data from first compartment of flocculation basin of 76 L/min (20 gpm) pilot plant located at Engineering Research Center, Colorado State University, Fort Collins, CO.

(for an over-and-under baffle, v(baffle) is the velocity between bottom edge of the baffle and floor of basin)
g is the acceleration of gravity (9.806650 m^2/s)

The power dissipated is

$$P(\text{slot}) = Q(\text{slot}) \cdot \rho\text{water} \cdot g \cdot h_L(\text{slot}) \qquad (11.29)$$

where
P(slot) is the power dissipated by flow around a single baffle (W)
Q(slot) is the flow through baffled basin (and around any single baffle or through "slot") (m^3/s)
ρwater is the density of water (kg/m^3)

The velocity gradient, G, is the same as defined previously, i.e., Equation 10.5, but with P(slot) as defined as in Equation 10.43, i.e.,

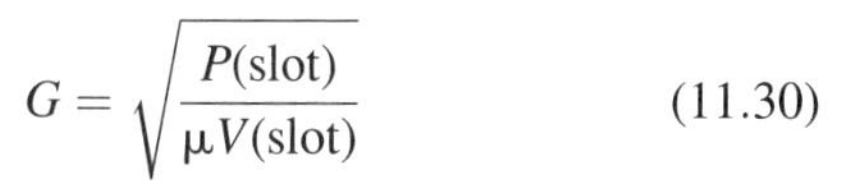

$$G = \sqrt{\frac{P(\text{slot})}{\mu V(\text{slot})}} \qquad (11.30)$$

where
G is the velocity gradient for baffle slot (s^{-1})
V(slot) is the volume of water through which h_L(slot) is dissipated (m^3)

11.5.4.2.1 End-Around Flocculation Basins

Figure 11.18a and b shows a top view and side view, respectively, for a hypothetical system. Dimensions are indicated as proportions with respect to the channel width, B (see Haarhoff, 1998, pp. 142–152). Following the method of Haarhoff, the outcome of a design should be the channel width, B (the "master variable"), the number of channels, n(channels), the proportions, p, q, w, and r, and the headloss through the slot, h_L(slot). Tables 11.10 and 11.11 summarize data from 12 plants, as compiled by Haarhoff (1998, pp. 143, 144). Facilities B and C in Table 11.9 and Facility P6 in Table 11.10 show data for a sequence of three stages for tapered flocculation. The two tables indicate a range of practice for the sample of 12 facilities.

Design guidelines for end-around baffles are summarized in Table 11.11. The goal of the design is to have an even, continuous energy loss along the successive channels (Haarhoff, 1998, p. 144).

11.5.4.2.2 Tapered Flocculation

The basin may be designed for tapered flocculation by adjusting the slot width, i.e., increasing the slot width in the downstream direction such that the velocity is reduced sufficient to result in a lower headloss, as calculated by Equation 11.28 that conforms to the G specified.

11.5.4.2.3 Flow Variation

The downstream control may be used to maintain the same velocity gradients throughout all channels in the basin. For decreased flows, the depth may be decreased proportionately

(a)

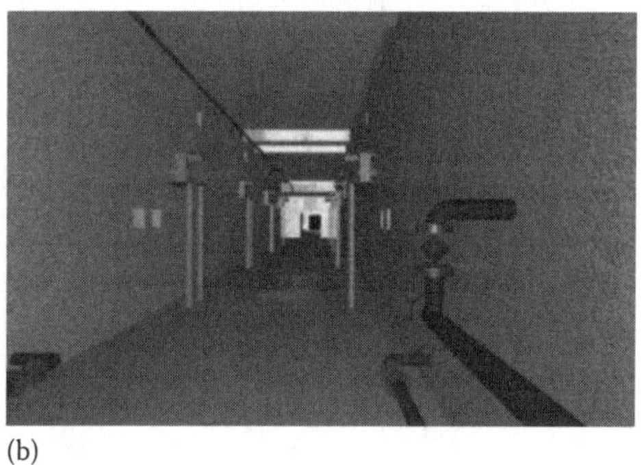
(b)

FIGURE CD11.17 Flocculation and sedimentation finished designs (animated walk through of Fort Collins Water Treatment Plant, 2000 Addition. (a) Animation 1 Flocculation Basin—Plate Settlers Walk-Through upstairs—PAK1B AVI (excerpt shows flocculation basins). (b) Animation II Flocculation Basin—Plate Settlers Walk through downstairs—PAK1A AVI (excerpt shows corridor between basins; paddle-wheel motors are visible on the walls). (Courtesy of Kevin Gertig, Water Resources & Treatment Operations Supervisor, FCWTFP and Kevin Heffernan, CH2M Hill, Inc., Denver, CO.)

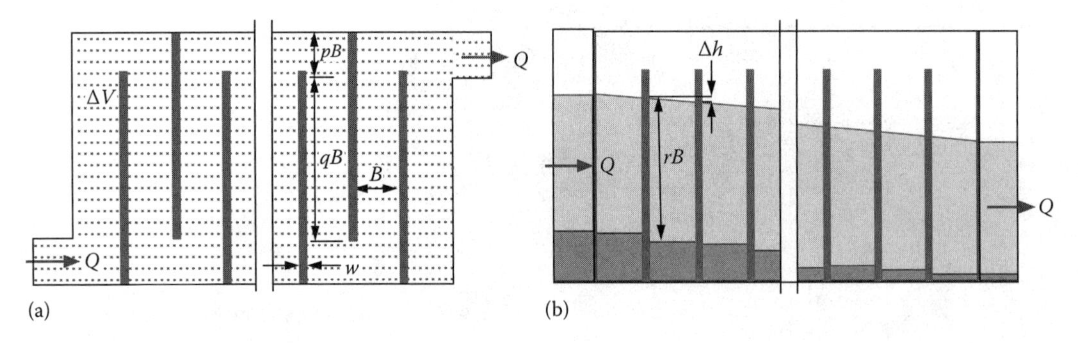

FIGURE 11.18 Side baffled mixing basin. (a) Top view. (b) Side view.

TABLE 11.9
Design Data from Literature as Compiled by Haarhoff (1998, p. 143)

	Facility Designation									
		B			C					
		Flocculation Stage			Flocculation Stage					
Variable	**A**	**#1**	**#2**	**#3**	**#1**	**#2**	**#3**	**D**	**E**	**F**
Q (m^3/s)	0.283	0.500	0.500	0.500	0.116	0.116	0.016	0.010	0.100	1.00
G (s^{-1})	40	70	35	20	50	35	25	50	50	50
Flocculation time (s)	1800	500	500	500	420	420	420	1000	1000	1000
n(channels)	20	22	18	14	23	17	14	18	18	18
Channel width, B (m)	0.763	2.590	3.170	4.070	0.450	0.62	0.770	0.288	0.944	2.86
Slot width ratio, p	0.33	0.21	0.31	0.36	1.51	1.50	1.56	0.52	0.50	0.52
Baffle overlap ratio, r	26.3	0.82	0.09	−0.18	9.11	6.06	4.49	44.1	12.8	3.50
Water depth ratio, r	3.20	0.77	0.63	0.49	2.22	1.61	1.30	0.52	0.50	0.52
Channel velocity (m/s)	0.152	0.097	0.079	0.061	2.257	0.187	0.150	0.233	0.225	0.23
Time per turn (s)	95	24	29	38	19	26	32	59	59	59

TABLE 11.10
Full-Scale Baffled Flocculators in South Africa

	Facility Designation							
						P6		
						Flocculation Stage		
Variable	**P1**	**P2**	**P3**	**P4**	**P5**	**1**	**2**	**3**
Q (m^3/s)	1.389	0.058	0.231	0.174	0.695	0.174	0.174	0.174
G (s^{-1})								
Flocculation time (s)	630	279	576	619	331	76	244	115
n(channels)	19	29	43	26	17	6	17	7
Channel width, B (m)	2.000	0.370	0.900	1.175	1.400	0.680	0.870	1.174
Slot width ratio, p	1.000	1.000	1.000	1.000	0.98	1.000	1.000	1.000
Baffle overlap ratio, r	2.70	1.51	0.89	1.33	1.88	3.74	2.48	1.32
Water depth ratio, r	1.20	2.82	1.39	0.74	1.24	1.18	0.82	0.56
Channel velocity (m/s)	0.289	0.150	0.205	0.169	0.287	0.318	0.278	0.227
Time per turn (s)	33	10	13	24	19	13	14	18

Source: Haarhoff, J., *Aqua*, 47(3), 144, 1998.

TABLE 11.11
Design Guidelines for End-Around Flocculators

Parameters				Limits		
Name	**Symbol**	**Units**	**Condition**	**Lower**	**Median**	**Upper**
No. of channels	*n*(channels)					
Channel width	*B*	m		0.45		
Water depth ratio	*r*			0.5		2.0
Slot width	*p*			0.5		1.5
Overlap	*q*			0.9		3.7
Wall thickness	*w*(timber)	mm		20		40
Slope of floor	*S*(floor)	m/m	Constant water depth			
Velocity	*v*(channel)	m/s		0.15		0.4
Time in channel	θ(channel)	s		19	35	95
Headloss	h_L(slot)	m	if *p* =	then		*K*(slot) =
			1.0	3.2		3.5
			1.5	2.5		4.0
			<1.0	1.5		1.5

Source: Adapted from Haarhoff, J., *Aqua*, 47(3), 142, 1998.
Explanation of the variables as abstracted from Haarhoff (1998, p. 145) are as follows:

- *n*(channels) = number of channels. The value of *n*(channels) is assumed based upon practice.
- *B* = channel width. The minimum value, i.e., $B \geq 0.45$ m, is set to facilitate cleaning; there is no firm guideline for an upper limit.
- *r* = water:depth ratio with respect to *B*, i.e., *D* = water depth = *rB*. The channel depth is a function of flow, *Q*, and the dimensions of a control section, e.g., a channel constriction or weir. The channel flow is "nonuniform" and to calculate exact depth along the length of each channel is tedious and not useful since the change is likely to be small. Therefore, the depth can be determined from the control section depth, with changes determined by the Manning equation (assuming a flat bottom and uniform flow, the slope obtained is a pseudo slope for the water surface); the depth for the section above adds the headloss through the slot.
- *p* = slot width ratio with respect to *B*, i.e., slot width = *pB*. As seen in Tables 11.10 and 11.11, *p* has a wide range with particular values based on the practice of a particular region or consultant. The objective, as stated, is that the energy dissipation is continuous. If *p* is too small, the flow would be forced through critical and the energy would be dissipated as a series of cascades, rather than continuous.
- *q* = overlap ratio with respect to *B*, i.e., overlap = *qB*. If *q* is too high, discrete energy losses dominate; if too short, the flow will not be forced through a full 180° turn and will meander instead. The overlap is the distance from the end of one baffle end to the next.
- *w* = baffle thickness. The baffle thickness has negligible effect on design equations and is considered only from the standpoint of structural rigidity and material that resists deterioration. To provide for adjustment, if needed, decay-resistant timber is recommended.
- $S(\text{floor}) = \{\Sigma h_L(\text{slot}) + \Sigma h_L(\text{friction})\}/[n(\text{channels}) \cdot L(\text{channel})]$. The slope of the floor is sufficient to maintain constant water depth from the bottom of the last channel to the top of the first channel. Calculation is as given by the foregoing equation; the friction loss is usually small and is neglected. As noted, the starting point is the downstream water level which is determined by the control section and the flow, *Q*. The floor of the downstream channel is set by the depth desired for the first channel (which should be the same for each subsequent upstream channel by letting the increase in floor elevation equal the headloss between the same two points).
- *v*(channel) = velocity of water in channel. Minimum velocity should be great enough to prevent deposits, recommended as 0.15–0.3 m/s (Bhargava and Ojha, 1993, p. 468).

so the same velocity and thus the same headloss is maintained through each slot. In other words, contrary to popular opinion, operator control for baffled basins is feasible (Haarhoff, 1998, p. 149). A spreadsheet facilitates the calculations.

11.5.4.2.4 Optimization by CFD

Design of baffled flocculators was studied experimentally and by CFD by Haarhoff and van der Walt (2001, pp. 149–159) to optimize the ratios *p*, *q*, and *r* in terms of maintaining a uniform *G* throughout the flocculator and to minimize back mixing. For $G = 50\ \text{s}^{-1}$, optimum values were $0.9 \leq p \leq 1.1$, $4 \leq q \leq 5$, $1 \leq r \leq 3$; the optimum ratios would change for different *G* values. The variation in *G* was not too sensitive to *r*, with *p*, the slot width ratio, being the most important. The overlap ratio, *q*, was important to permit the flow to become uniform after a bend. The highest *G* values were immediately after the plane of the baffle at the slot location and downstream about the distance of the slot width (as would be expected, eddies are generated by 180° change in direction at the slot location).

11.6 PROPRIETARY TECHNOLOGIES

As with other processes, proprietary technologies for flocculation have been available since the 1930s with the advent of solids contact units. Most of the technologies are sized by manufacturer's recommendations. Rationales that explain their behavior have followed in some cases, e.g., solids contact units. Design is most often based on the manufacturer's recommendations.

11.6.1 Turbine Flocculators

The use of axial flow and turbine impellers of the Rushton type (Sections 10.3.3.4, 10.4.2.5, and Glossary of Chapter 10) for flocculation was investigated by Walker (1968), then president, Walker Process Equipment, who used a 1676 mm (66 in.) Rushton impeller in a 12.2 m (40 ft) diameter by 4.6 m (15 ft) tank with 1.52 m (5 ft) depth of submergence. The circulation patterns for the axial flow impeller was up and around with return flow to the impeller eye. While Walker was striving for uniform turbulence, the tests showed that the large eddies that predominated in the reactor volume were not efficient in flocculation, i.e., consistent with later theory.

11.6.2 Solids Contact Units

Solids contact units, sometimes called "sludge-blanket clarifiers," contain rapid mix, flocculation, and settling in one unit; Figure 11.19 is a schematic diagram. As seen, the raw water flows into a rapid mix basin with coagulant flow being injected into the rapid mix. The flow from the rapid mix passes through a flocculation zone and then is forced through the sludge blanket with upflow through the clarification zone and into radial overflow troughs, which flow into a gullet around the periphery of the clarifier. The sludge blanket is fluidized floc particles and is maintained at a designated level by means of sludge wasting from the sludge pocket. A "picket fence" thickener of vertical bars attached to a rotating shaft helps to increase the density of the sludge before it leaves the sludge pocket. Each manufacturer has its own variation of the kind of system shown in Figure 11.19. Applications include drinking water treatment and chemical treatment of secondary-treated wastewaters. The latter may include lime as the chemical for phosphate reduction.

The solids contact technology was developed by equipment manufacturers and the units have been in use since the 1930s. They are widely used, especially in small plant situations e.g., $\leq 0.876\ m^3/s$ (20 mgd) plant capacity. They are viewed with disfavor by some (Burns and Roe, 1971, pp. 4–6) because of lack of control of each of the unit processes. An additional disadvantage is that the flow of water is likely to be distributed nonuniformly below the sludge blanket. The flow of water will seek its own path of least resistance through the sludge blanket, thus creating a channel of higher velocity flow leaving much of the sludge blanket an inert mass. The advantage in their use is that three unit processes, i.e., rapid mix, flocculation, sedimentation, are combined in one unit with consequent smaller size than the three unit processes as separate units.

11.6.2.1 Principles

The processes in a solids contact unit start in the rapid mix where the coagulant chemicals and the raw water are mixed and collide with floc particles pumped into the reactor from the sludge blanket, as shown in Figure 11.19. The flow then enters the flocculator where further collisions between particles occur due to reduced turbulence in the flow. The flocculated particles then enter the sludge blanket and presumably flow upward through the blanket to collide with previously formed floc particles, thus growing in size and settling more readily.

As to the size of floc particles in the floc blanket, Tambo and Hozumi (1979, p. 441) mention $0.3 \leq d(\text{floc})_{max} \leq 0.5$ mm (300–500 μm) with incoming microfloc sizes $5 \leq d(\text{microfloc}) \leq 10$ μm. The contacts are between particles with these size differences. Equation 11.1 is applicable to the collision frequency within the floc blanket. The decline in concentration of microflocs with distance in the blanket is (Tambo and Hozumi, 1979, p. 446) given by

$$\frac{C}{C_o} = e^{-kz} \qquad (11.31)$$

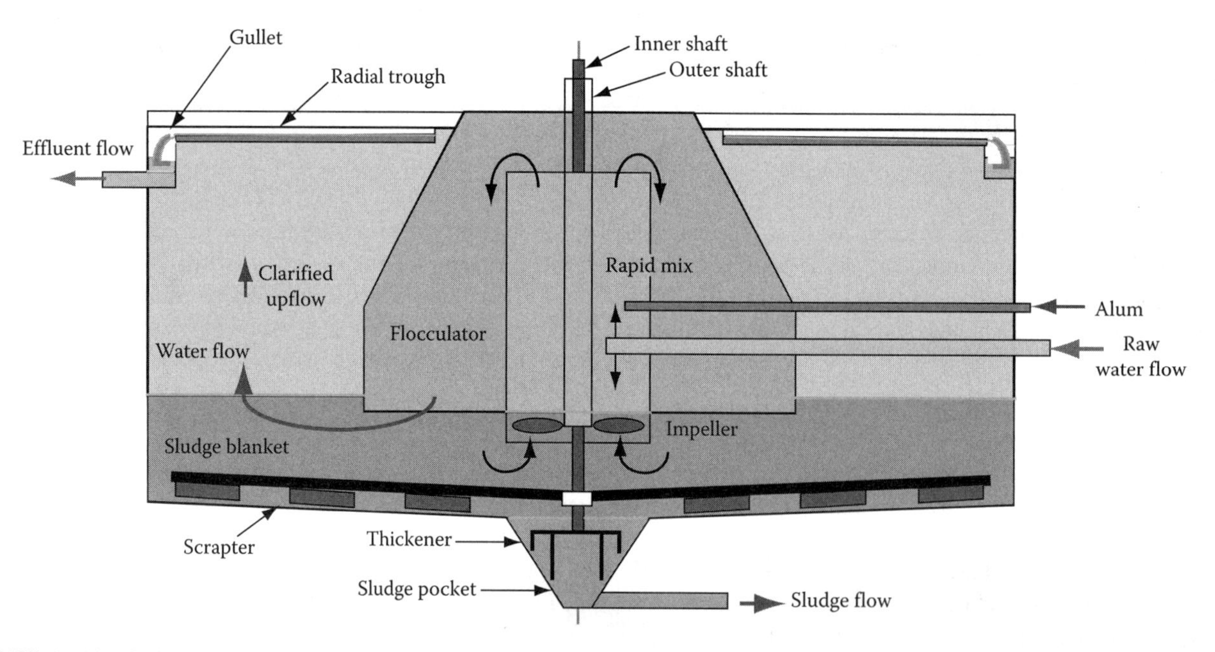

FIGURE 11.19 Solids contact clarifier—generic schematic (EPA Manual, 1971).

where

C is the concentration of primary particles (microflocs) at a distance, z within the floc blanket ($\#/m^3$)
C_o is the concentration of primary particles (microflocs) entering the floc blanket ($\#/m^3$)
k is the coefficient of collision frequency (m^{-1})
z is the distance from the bottom of the floc blanket (m)
v is the velocity through blanket, defined as Q/A (m/s) where Q is the flow through blanket (m^3/s) A is the area of blanket as defined by its confining dimensions (m^2)

Tambo and Hozumi (1979, p. 446) found that $k = 4.37, 28.8$ m for $d = 1$ μm, and 2 μm, respectively, for $v = 0.03$ m/min.

11.6.2.2 Design Practice, Equipment, Operation

Design for solids contact units is largely a matter of equipment selection, although some design latitude may be worked in based on overflow rates and detention time. Empirical guidelines, based upon available equipment, is the usual basis for design, complemented by experience with actual installations. Overflow rates of 1220–2440 m/h (500–1000 gpm/ft^2) are given for alum or ferric treatment and 2928–4148 m/h (1200–1700 gpm/ft^2) for lime treatment (Burns and Roe, 1971, pp. 6–7).

The key operating variable related to floc formation is the level of the sludge blanket. The operator must monitor the blanket depth to keep it deep enough that good flocculation occurs, but shallow enough that floc will not be carried out of the basin. How often and how much sludge is to be wasted is readily determined by the operator after sufficient experience with a given system.

11.6.3 Super-Pulsators™

The super-pulsator is a proprietary unit of Ondeo-Dergemont (formerly Infilco Dregemont), which functions as a solids contact basin. Its shape is rectangular, namely, circular. Following the flow path, a coagulant chemical, e.g., alum, is added to the raw water pipeline which enters a vacuum chamber and then to a distribution manifold (a pressure conduit) from which perforated distribution laterals distribute water uniformly below parallel-plate separators (which function to hold the sludge blanket). The sludge between the plates must be agitated to give a homogeneous sludge concentration. The sludge blanket is agitated by means of a nonsteady flow created by the head of water in the vacuum chamber provided when a vent valve above is opened. In effect, a hydraulic "pulse" occurs to agitate and expand the suspended solids contained between the plates. A unique feature is that short-circuiting of flow cannot occur, i.e., the entire sludge blanket functions in the solids contact coagulation process.

11.6.4 Culligan Multi-Tech™

The Culligan Multi-Tech™ system consists of a bed of coarse media followed by a filter media, each in separate pressure vessels, called the "contact flocculator" and "depth filter," respectively (Section 12.7.2.2). Coagulant is injected on the suction side of a centrifugal pump that pressurizes the system. Tank diameters range from 508 mm (20 in.) for a 30 gpm model to 1372 mm (54 in.) for a 220 gpm model.

The Culligan Multi-Tech™ system was developed during the period 1976–1979 responding to a need that some 60,000 small water systems would fall under regulations emanating from PL93-523, i.e., the 1974 Safe Drinking Water Act. Its genesis was an idea from Professor E. Robert Baumann, a consultant with Culligan for a "contact flocculator" comprised of a bed of gravel which produced "controlled turbulence," and hence flocculation, followed by a filter bed.

11.7 SUMMARY

The flocculation process evolved largely as an empirical practice since being delineated by the 1919 work of Langelier in Sacramento for the first paddle-wheel units. Design principles were articulated by Camp in a 1955 paper that were adopted almost universally in practice. Turbulence theory was adopted by Kaufman and Argaman in the late 1960s to lay the foundation for flocculation kinetics, which was developed further in the 1980s and 1990s.

PROBLEMS

11.1 Flocculation Discoveries of Langelier at Sacramento

Given

Excerpts of Professor Wilfred Langelier's oral history interview with Malca Chall (1982) given in Section 11.3.1.2.

Required

Identify discoveries of Langelier, c. 1921, as related to the design of the Sacramento WTP that were important in developing knowledge of coagulation and/or flocculation or that affected practice.

11.2 Distinction between Coagulation and Flocculation

Given

Excerpts of Professor Wilfred Langelier's interview with Malca Chall (1982).

Required

How does Langelier deal with the distinction between "coagulation" and "flocculation?"

11.3 Settling Velocity–Biological Floc

Given

An activated sludge floc has an average "longest dimension," d(floc-avg) ≈ 1.0 mm. Assume a standard deviation for the floc size distribution is 0.12 mm, which means about 0.15 fraction of the floc has a size smaller than (1.0–0.12 mm), i.e., d(lower std. dev.) ≈ 0.88 mm, and, by the same token (for clarification), d(higher std. dev.) ≈ 1.12 mm.

Required

Calculate the overflow velocity of a tank that will remove all particles larger than $d(\text{floc}) \approx 0.88$ mm. Reference is Section 11.4.2.1.

11.4 Examples of G and θ Parameters in Practice

Given/Required

For a nearby water treatment plant employing paddle wheel flocculators, calculate (a) G, (b) θ, and (c) $G\theta$. Do this for a full range of flow conditions, and summarize in tabular form, and compare with values for comparable plants in Table 11.1. What range in operator control is possible for G.

11.5 Floc Description

Given/Required

Describe the floc formed at the different stages of the floc basin. Alternatively, describe the floc formed during a jar test at different times of rapid mix and flocculation.

11.6 Air Bubbles for Flocculation

Given

A basin is underlain by a grid of diffusers spaced at 300 mm. Let the basin be 3.0 m in depth.

Required

Calculate G as affected by the flow of air, $Q(\text{air})$. Arrive at a design for a flocculation system.

Hint: $F_D = C_D A \gamma v^2/2g$, $P = F_D v$, and $V(\text{basin}) = Q\theta$. For C_D, use the relation given by Fair, et al. (1968, p. 25-3), Section 6.2.2, for the laminar range through the transitional range, i.e., $C_D = 24/\mathbf{R} + 3/\mathbf{R}^{0.5} + 0.34$.

11.7 Approaching Uniform Turbulence With Paddle-Wheel Flocculator

Given

A paddle-wheel flocculator is to be designed.

Required

The paddle wheel should be designed with blade widths and spacing such as to approach uniform turbulence.

Hint: A spreadsheet may be set up dividing the paddle wheel into concentric volumes that are equal. The sizing of the blades and their spacing should result in equal G values for each concentric volume.

11.8 Design of Paddle-Wheel Flocculation System—Power Required

Given

A flocculation basin is to be designed for $Q = 0.263$ m^3/s (6.0 mgd) at $T = 20°C$ with other data the same as that of Table 11.6 except that the paddle-wheel diameter is to be 3.0 m instead of 3.9 m.

Required

Calculate the power, P, required by the paddle wheel.

Hint: Apply a trial-and-error solution using Table 11.6.

11.9 Design of Paddle-Wheel Flocculation System—Design Algorithm

Given

A flocculation basin is to be designed for $Q = 0.263$ m^3/s (6.0 mgd) at $T = 20°C$ with other data the same as that of Table 11.6 except that the paddle-wheel diameter is to be 3.0 m instead of 3.9 m.

Required

Set up a design algorithm for four compartments.

Hint: Apply a trial-and-error solution using Table 11.6.

11.10 Design of Paddle-Wheel Flocculation System for a Flotation System

Given

A two-compartment flocculation basin is to be designed for $Q = 0.263$ m^3/s (6.0 mgd) at $T = 20°C$ for a flotation system.

Required

Set up a design algorithm with appropriate criteria for G.

Hint: Apply a trial-and-error solution using Table 11.10.

11.11 Utilization of a P versus R Plot

Given

Figure CD11.14 is a photograph of a floc basin, which is a part of a water treatment pilot plant at Colorado State University. To give an idea of size, the floc basin was constructed of four 1219 mm × 2438 mm (4 ft × 8 ft) acrylic sheets to form the top, bottom, and two sides with 1219 mm × 1219 mm (4 ft × 4 ft) sheets forming the ends. The paddle wheels were oriented with vertical shafts, each with a direct current motor with adjustable speed. The paddle wheels were 762 mm (30 in.) diameter × 940 mm (37 in.) long. The blades were 25.4 mm (1 in.) wide. The first compartment had five blades on each of four arms, spaced at radial distances given in Table CD11.7; the third compartment had only two blades per arm. For the first compartment, the motor was mounted on a 150 mm "lazy Susan" ball bearing plate (designed for use with rotating shelves for a kitchen cabinet). A 460 mm brass rod was attached to the motor with end restrained by a hook attached to a force gage. Without restraint, the motor would rotate freely without turning the paddle-wheel shaft. The force of the lever arm at distance 460 mm from the shaft was measured by the force gage. The product of the force times the lever arm distance was the torque exerted on the paddle wheel. The torque times the rotational velocity in radians per second was the power dissipated by the blades of the paddle wheel due to form drag. This power varied with rotational velocity. Table CD11.7 shows data collected on two separate occasions and the associated calculations. For different rotational velocities, calculations were made for power dissipated, P; turbulence intensity, G; power number, $\mathbf{P}$, and Reynolds number, $\mathbf{R}$. The related plots are shown in Figure 11.15: (a) G versus n; (b) G versus $\mathbf{R}$; (c) $\mathbf{P}$ versus R; and (d) $\mathbf{P}$ versus n. The range of practice, selected as $G \leq 100$ s^{-1}, is indicated in each plot.

Required

Discuss the significance of the plots.

11.12 Characteristics of Full-Scale Paddle Wheel by Mathematical Model

Given

A flocculation basin is to be designed for $Q = 0.263$ m^3/s (6.0 mgd) at $T = 20°C$.

Required

(a) Calculate the power, P, required by the paddle wheel using Table 11.6 such that $G = 60\ s^{-1}$.

(b) Generate plots as in Figure 11.15a through d.

Reference

Table 11.6, i.e., file "11.6FlocBDes.082105.xls," is copied and renamed, as file "TableCDprob11.12/11.6FlocBDes.082105.xls."

11.13 Algorithm for End-Around Flocculation Basin

Given

Assume flow, $Q = 1.0\ m^3/s$.

Required

Set up a spreadsheet algorithm to design an end-around flocculator basin. Use headings to identify variables with rows for the assumed numerical values with calculated values across. Show a design sketch, i.e., both plan and profile as well as the design spreadsheet.

11.14 Plot of Power and G versus rpm for Paddle-Wheel Flocculator

Given

Data of Table CD11.7 give measurements for the pilot scale flocculation basin of Figure 11.14. For this problem, consider the third basin, Basin #3, which is the same as Basin #1, except that there are only two blades per arm. Radial distances to the center of each blade are: $r_1 = 241$ mm, $r_2 = 368$ mm.

Required

Determine the power versus rpm curve and the G versus rpm curve using a spreadsheet and show the results on an associated plot.

ACKNOWLEDGMENTS

Kevin Heffernan, PE, principal project manger, CH2M HILL, INC., Denver office, provided the three-dimensional animated drawing files, seen as Figures CD11.17a and b. The animations were included by permission from both the City of Fort Collins and CH2M HILL, INC. Heffernan was design manager and resident engineer during the construction of the new flocculation basins and building, i.e., as illustrated in Figure CD11.17, during the period 1998–2000.

Kevin Gertig, water resources and treatment operations manager, City of Fort Collins, was superintendent of the Fort Collins Water Treatment Facility during the aforementioned flocculation capacity expansion. As with other projects, Gertig was intimately involved with the flocculation system expansion and passed-on his knowledge freely as related to its design, construction, and operation, which helped in formulating this chapter.

APPENDIX 11.A: DERIVATION OF CAMP AND STEIN G FOR THREE-DIMENSIONAL CUBE

TABLE 11.A.1
Development of Camp and Stein G for Three-Dimensional Infinitesimal Cube

Term	One Dimension	Three Dimensions[a]
Velocity gradient	$\frac{dv}{dn}$	$\left(\frac{\partial u}{\partial y}+\frac{\partial v}{\partial x}\right)+\left(\frac{\partial u}{\partial z}+\frac{\partial w}{\partial x}\right)+\left(\frac{\partial v}{\partial z}+\frac{\partial w}{\partial y}\right)$
Total shear	$\tau = \mu\frac{dv}{dn}$	$\tau = \mu\left[\left(\frac{\partial u}{\partial y}+\frac{\partial v}{\partial x}\right)+\left(\frac{\partial u}{\partial z}+\frac{\partial w}{\partial x}\right)+\left(\frac{\partial v}{\partial z}+\frac{\partial w}{\partial y}\right)\right]$
Work of shear/unit time	$\tau\cdot\frac{dv}{dn} = \mu\frac{dv}{dn}\cdot\frac{dv}{dn}$	$\tau\frac{dv}{dn} = \mu\left[\left(\frac{\partial u}{\partial y}+\frac{\partial v}{\partial x}\right)^2+\left(\frac{\partial u}{\partial z}+\frac{\partial w}{\partial x}\right)^2+\left(\frac{\partial v}{\partial z}+\frac{\partial w}{\partial y}\right)^2\right]$
Work of shear/unit time	$\frac{P}{V} = \mu\left(\frac{dv}{dn}\right)^2$	$\Phi = \mu\left[\left(\frac{\partial u}{\partial y}+\frac{\partial v}{\partial x}\right)^2+\left(\frac{\partial u}{\partial z}+\frac{\partial w}{\partial x}\right)^2+\left(\frac{\partial v}{\partial z}+\frac{\partial w}{\partial y}\right)^2\right]$
G defined	$G \equiv \frac{dv}{dn}$	$G \equiv \left[\left(\frac{\partial u}{\partial y}+\frac{\partial v}{\partial x}\right)^2+\left(\frac{\partial u}{\partial z}+\frac{\partial w}{\partial x}\right)^2+\left(\frac{\partial v}{\partial z}+\frac{\partial w}{\partial y}\right)^2\right]^{1/2}$
G in practical terms	$G \equiv \frac{dv}{dn} = \left(\frac{P}{\mu V}\right)^{1/2}$	$G = \left[\frac{\Phi}{\mu}\right]^{1/2}$

Source: Abstracted from Camp, T.R. and Stein, P.C., *J. Boston Soc. Civil Eng.*, October, 1943.

[a] In addition to G, another defined term introduced by Camp and Stein (1943) was

$$\Phi \equiv \frac{P}{V} \tag{11.A.1}$$

where Φ is the work of shear per unit volume per unit time (N $m/m^3/s$).

The term has not been adopted in this work in order to minimize the use of defined terms; it is given here for reference.

GLOSSARY

Adhesion: Attachment of floc to filter grains (Hannah et al., 1967, p. 844).

Aging: The irreversible change of texture and floc structure from the moment flocs are completely formed, i.e., when they have their maximum dimension (François, 1987a, p. 523).

Anisotropic turbulence: Anisotropic turbulence is characterized by unequal strain rates, i.e., velocity gradients, with respect to direction. According to Tennekes and Lumley (1972, p. 262), small eddies exhibit "local isotropy," where any sense of direction is lost with turbulence being increasingly scrambled at small scales. The range of wave numbers exhibiting local isotropy is called the *equilibrium range*.

Bioflocculation: The aggregation of microbes is called "bioflocculation"; usually, the objective is to attain a larger mass that is settleable.

Blowdown: Continuous periodic removal of a solids stream so that there is no buildup of solids but a concentration maintained at some specified level. The rate of generation of solids must equal the flux of solids leaving the system by means of "blowdown."

Bond: Forces of adhesion or cohesion between particles and another surface or between particles and particles, respectively (Hannah et al., 1967, p. 844).

Bridging: 1 Reactive functional groups located along long-chain polymers form chemical-type bonds at reactive sites on the surface of particles to be agglomerated, thereby forming interparticle bridges (Ham and Christman, 1969, p. 482).

Brownian motion: Random thermal motion of molecules (Hannah et al., 1967, p. 845). Such motion may cause buffeting of small particles, i.e., ≤ 1 μm but is generally not sufficient to cause flocculation within time frames important in water treatment.

Carryover: Refers to floc in suspension leaving the floc blanket and carried with the flow to the overflow weirs leaving the basin.

Coagulation: The process of chemical reaction of a coagulant in water that requires intense mixing to distribute the coagulating agent uniformly throughout the water so that it makes contact with the suspended particles before the reaction is completed; particle concentrations in water are nominally $\geq 10^7$ #/mL (Hudson and Wolfner, 1967, p. 1257).

Cohesion: Attachment of floc to other floc particles (Hannah et al., 1967, p. 844).

Contact flocculation: (1) Contact between primary particles and floc particles causing a growth in the floc particles (see Tambo and Hozumi, 1979, p. 441). (2) Larger suspended flocs adsorb incoming primary particles on their surfaces (Tambo and Hozumi, 1979, p. 448). (3) A water clarification process whereby the water is applied directly to the filter without prior clarification by sedimentation, and which is designed to bring about coagulation, flocculation, and solids separation directly in the filter bed (Shea et al., 1971, p. 41).

Density: See *floc density*.

Eddy: Localized circular motion of fluid.

Electrostatic forces: Forces that cause repulsion or attraction between like charge and unlike charged particles, respectively (Hannah et al., 1967, p. 845). Insoluble hydrous oxide formed from alum or iron salts may have either a positive or negative charge depending on pH.

Floc: (1) An aggregate of particles created by collisions from smaller particles that adhere to one another by bonding forces likely to be van der Waal's. The shape, usually, is ill-defined, i.e., is amorphous, or fractal. (2) The aggregate of destabilized particles known as micelles (Argaman, 1971, p. 775).

Floc density: The density of a floc, ρ_F, is the mass of particles and includes water divided by the envelope volume. The "effective" density, ρ_E, is this value minus the density of water, i.e., $\rho_E = \rho_F - \rho_W$ (Gregory, 1989, p. 216).

Floc specific gravity: The specific gravity of a floc is the density of floc particles relative to the density of water, i.e., $SG(floc) = \rho_F/\rho_W$.

Floc strength: Resistance of floc to shear stress as induced by hydraulic velocity gradients (Hannah et al., 1967, p. 843).

Floc̨ulent: (1) The term flocculent is used by many, e.g., Halvorson and Panzer (1980, p. 489), to mean the chemical agents that create *microflocs* from colloidal particles. (2) The term *flocculent* is used more commonly in practice to designate a chemical, i.e., a polymer, added to the beginning of the flocculation process (the inflow to a flocculation basin) intended to cause interparticle bridging to aid in floc growth and toughness (shear resistance).

Flocculation: (1) The grouping and compacting of coagulated particles into larger assemblages called "floc" particles. Flocculation time should be, ordinarily $30 \leq \theta \leq 60$ min (Hudson and Wolfner, 1967, p. 1257). [Tambo and Watanabe (1979, p. 429) define flocculation as including what is called here, coagulation. This seems not uncommon and is mentioned here so that the reader is aware of this alternative, more inclusive definition.] (2) A primary water treatment process that changes the size distribution of particles from a large number of small particles to a small number of large particles for removal in later processes (Casson and Lawler, 1990, p. 54). (3) The term is used by some researchers in a more inclusive sense to mean charge neutralization and the subsequent agglomeration (see, for example, Gregory and Guibai, 1991, pp. 3–4).

Flocculent: (1) A chemical, e.g., a polymer, that contributes to the aggregation and growth of floc particles; the mechanism would be "bridging," i.e., two or more floc particles are attached to a molecular strand of polymer. (2) A coagulant chemical, e.g., alum or ferric ion is sometimes called a "flocculent" (this definition is mentioned only because of its use in the literature and is not defined unequivocally; rather, its definition is implied).

Fractal: (1) A structure of formed aggregates (Gmachowski, 1995, p. 1815). (2) See Chapter 9. (3) The concepts of fractal geometry provide a mathematical framework for description of the structure of irregular flocs (Spicer and Pratsinis, 1996a, p.1052).

Fractal dimension: (1) An exponent in the relation, $N = \zeta(d/d_o)^{D_F}$ that characterizes the aggregate mass of fractal (dimensionless). (2) A solid three-dimensional body has a mass which depends on the third power of some characteristic length (such as the diameter of a sphere), so that a log–log plot of mass against size should give a straight line with a slope of 3. When such plots are made for aggregates, lower slopes are found, with non-integer values. The slope of the line is known as the fractal dimension, D_F. In three-dimensional space, D_F may take values between 1 and 3, the lower value representing a linear aggregate and the upper one an aggregate of uniform density or porosity. Generally, intermediate values are found, and the lower the fractal dimension, the more "open" or "stringy" the aggregate structure (Gregory, 1989, p. 215).

***G*:** A term defined by Camp and Stein (1943) as $G = dv/dy = [P/(\mu V)]^{0.5}$.

Heterodisperse: A suspension having a distribution of particle sizes.

Hydrogen bond: Considered having a minor role in intraparticle bonding because of their short range; energies are 3–10 kcal/mol (Hannah et al., 1967, p. 846). Considered, however, to be a major force in attachment of flexible polymer chains to floc particles to permit bridging and agglomeration.

Isotropic turbulence: Isotropic turbulence is characterized by equal strain rates, i.e., velocity gradients, with respect to direction. According to Tennekes and Lumley (1972, p. 262), small eddies exhibit "local isotropy," where any sense of direction is lost with turbulence being increasingly scrambled at small scales. The range of wave numbers exhibiting local isotropy is called the *equilibrium range* (see also *anisotropic turbulence*).

Kolmogorov's microscale: (1) The Kolmogorov's microscale of turbulence is a particular eddy size, calculated by the fluid viscosity and the energy dissipation (François, 1987, p. 1023), i.e., $\lambda(\text{Kolmogorov}) = (\nu^3/\varepsilon)^{1/4}$. (2) Based on the universal equilibrium theory, the Kolmogorov's microscales of length, time, and velocity are defined as the smallest scales of fully turbulent motion in a turbulent flow field (Casson and Lawler, 1990, p. 55) (see also Chapter 10).

Kolmogorov's universal equilibrium theory: Relates to the net rate of change in the energy contained in the small-scale motions to the energy dissipation rate of these motions (Casson and Lawler, 1990, p. 55).

Monodisperse: A suspension having but one particle size.

Orthokinetic flocculation: (1) Refers to flocculation induced by a velocity gradient (Argaman and Kaufman, 1968, p. 5; Ives, 1978, p. 39). (2) See Chapter 9.

Perikinetic flocculation: (1) Refers to flocculation induced by a Brownian motion (Argaman and Kaufman, 1968, p. 5; Ives, 1978, p. 39). (2) See Chapter 9.

Primary particles: (1) Particles at the commencement of flocculation in which all particles are considered to be covered with coagulant and have diameters nearly equal (Tambo and Watanabe, 1979, p. 430). (2) Destabilized particles ready for aggregation (Argaman, 1971, p. 775). (3) The particles to be removed from the raw water, e.g., turbidity, microorganisms (viruses, bacteria, cysts, oocysts, algae, and other microscopic particulates).

Root mean square velocity gradient: The root mean square (rms) of the velocity fluctuation, $\overline{u'}$, is defined,

$$\overline{u'} \equiv \left(\overline{u'^2}\right)^{1/2}$$

In a turbulent flow field, the velocity at any point fluctuates randomly; this fluctuation is dealt with statistically as the rms.

Scale of turbulence: Distance across which the velocity of an eddy changes (François, 1987b, p. 1023). Energy content of an eddy depends upon the scale of turbulence; the large eddies contain most of the energy of the system and to not dissipate energy.

Sludge blanket clarifier: See *solids contact unit*.

Solids contact unit: (1) A tank that has a center well used for coagulation, with outer portion maintained as a "sludge blanket." The microflocs resulting from coagulation are forced up through the blanket, perhaps 0.7 m (2 ft) depth and fluidized, of larger floc particles; during their flow through the blanket, these primary particles make contact during with these previously formed floc particles. The solids blanket is maintained at a desired depth by waste flow of sludge from the bottom. The effluent flow leaves the clarifier by peripheral weirs. For those floc particles that are suspended in the flow leaving the solids blanket, they may fall back to the sludge blanket. (2) Proprietary devices that combine rapid mixing, flocculation, and sedimentation in one unit. These units provide separate coagulation and flocculation zones and are designed to cause contact between newly formed floc and settled solids.

Specific gravity: See *floc specific gravity*.

Technology: An anthropogenic device used to perform some function.

Toughness: Refers to the resistance of floc to fragmentation by shear (see *floc strength*).

Universal equilibrium range: The range of the scales of turbulence that dissipate energy (François, 1987, p. 1023). The upper zone, the zone in which energy is dissipated by inertia, is separated from the viscous zone by Kolmogorov's microscale of turbulence, λ(Kolmogorov). The eddies that transmit energy and those which dissipate energy are not the same and they are independent of each other (François, 1987b, p. 1023).

van der Waals forces: Molecular forces resulting from the interaction of induced dipoles; bond energies are 1–2 kcal/mol (Hannah et al., 1967, p. 845).

Vortex: Same as eddy.

Vortex tube: Localized circular motion of a fluid that extends over a distance into the fluid, but may have curvature.

REFERENCES

Argaman, Y. A., Pilot plant studies of flocculation, *Journal of the American Water Works Association*, 63(12):775–777, December 1971.

American Water Works Association, American Society of Civil Engineers, *Water Treatment Plant Design*, 3rd edn., McGraw-Hill, New York, 1990.

Amirtharajah, A. and Tambo, N., Mixing in water treatment, in: Amirtharajah, A., Clark, M. M., and Trussell, R. R. (Eds.), *Mixing in Coagulation and Flocculation*, American Water Works Association Research Foundation, Denver, CO, 1991.

Argaman, Y. and Kaufman, W. J., Turbulence In Orthokinetic Flocculation, SERL Report No. 68-5, Sanitary Engineering Research Laboratory, University of California, Berkeley, CA, 1968.

Argaman, Y. and Kaufman, W. J., Turbulence and flocculation, *Journal of the Sanitary Engineering Division, American Society of Civil Engineers*, 96(SA2):223–241, 1970.

Baker, M. N., *The Quest for Pure Water*, American Water Works Association, New York, 1949.

Bellouti, M., Alves, M. M., Novais, J. M., and Mota, M., Flocs vs Granules: Differentiation by Fractal Dimension, Research Note, *Water Research*, 31(5):1227–1231, 1997.

Bhargava, D. S. and Ojha, C. S. P., Models for design of flocculating baffled channels, *Water Research*, 27(3):465–475, 1993.

Burns and Roe, Inc., Process Design Manual for Suspended Solids Removal, Technology Transfer Program, United States Environmental Protection Agency, October 1971.

Camp, T. R., Flocculation and flocculation basins, *Transactions, American Society of Civil Engineers*, 120:1–16, 1955.

Camp, T. R., Water treatment plant design, in: Manual of Engineering Practice No. 19, ASCE, 1940, p. 29, Table 2, 1955.

Camp, T. R., Floc volume concentration, *Journal of the American Water Works Association*, 60(6):656–673, June 1968.

Camp, T. R. and P. C. Stein, Velocity gradients and internal work in fluid motion, *Journal of the Boston Society of Civil Engineers*, October 1943.

Camp, T. R., Root, D. A., and Bhoota, B. V., Effects of temperature on rate of floc formation, *Journal of the American Water Works Association*, 32(11):1913–1927, November, 1940.

Casson, L. W. and Lawler, D. F., Flocculation in turbulent flow: Measurement and modeling of particle size distributions, *Journal of the American Water Works Association*, 82(8): 54–68, August, 1990.

Chakraborti, R. K., Atkinson, J. F., and Van Benschoten, J. E., Characterization of alum floc by image analysis, *Environmental Science and Technology*, 34(18):3969–3976, September 15, 2000.

Chall, M. and Langelier, W. F., Teaching, research, and consulting in water purification and sewage treatment, UC Berkeley 1916–1955, Interview conducted by Malca Chall in 1970, Regional Oral History Office, Copy No. 7, The Bancroft Library, University of California, Berkeley, CA, 1982.

Clark, M. M., A critique of camp and Stein's RMS velocity gradient, *Journal of the Environmental Engineering Division, ASCE*, 111(EE6):741–753, December 1985.

Cleasby, J. L., Is velocity gradient a valid turbulent flocculation parameter, *Journal of the Environmental Engineering Division, ASCE*, 110(EE5):875–897, October 1984.

Edzwald, J. K., Principles and applications of dissolved air flotation, *Water Science and Technology*, 31(3–4):1–23, 1995.

François, R. J., Ageing of aluminum hydroxide flocs, *Water Research*, 21(5):523–531, 1987a.

François, R. J., Strength of aluminum hydroxide flocs, *Water Research*, 21(9):1023–1030, 1987b.

François, R. J. and Van Haute, A. A., Structure of hydroxide flocs, *Water Research*, 19(10):1249–1254, 1985.

Gmachowski, L., Mechanism of shear aggregation, *Water Research*, 29(8):1815–1820, 1995.

Gorczyca, B. and Gahczarczyk, J., Influence of the nature of the turbidity on some properties of alum coagulation flocs, in: *The AWWA Annual Conference*, Vancouver, BC, June 18–22, 1992.

Gregory, J., Flocculation by polymers and polyelectrolytes, in: *Solid/Liquid Dispersions*, Academic Press, Inc., London, U.K., 1987.

Gregory, J., Fundamentals of flocculation, *CRC Critical Reviews in Environmental Control*, 19(3):185–230, CRC Press, Boca Raton, FL, 1989.

Gregory, J. and Guibai, L., Effects of dosing and mixing conditions on polymer flocculation of concentrated suspensions, *Chemical Engineering Communication*, 108:3–21, 1991.

Haarhoff, J., Design of around-the-end hydraulic flocculators, *Aqua*, 47(3):142–152, 1998.

Haarhoff, J. and van der Walt, J. J., Toward optimal design of around-the-end hydraulic flocculators, *Aqua*, 50(3):149–152, 2001.

Halvorson, F. and Panzer, H. P., Flocculating agents, in: Grayson, M. (Ed.), *Kirk-Othmer Encyclopedia of Chemical Technology*, Volume 10, John Wiley & Sons, Inc., New York, 1980.

Ham, R. K. and Christman, R. F., Agglomerate size changes in coagulation, *Journal of the Sanitary Engineering Division, ASCE*, 95(SA3):481–502, June 1969.

Han, M. and Lawler, D. F., The (relative) insignificance of G in flocculation, *Journal of the American Water Works Association*, 84(10):79–91, October 1992.

Hannah, S. A., Cohen, J. M., and Robeck, G. G., Measurement of floc strength by particle counting, *Journal American Water Works Association*, 59(7):843–858, July 1967.

Hanson, A. T. and Cleasby, J. L., The effects of temperature on turbulent flocculation: Fluid dynamics and chemistry, *Journal of the American Water Works Association*, 82(11):56–72, November 1990.

Harris, H. S., Y., Kaufman, W. J., and Krone, R. B., Orthokinetic flocculation in water purification, *Journal of the Sanitary Engineering Division, American Society of Civil Engineers*, 92(SA6):95–111, 1966.

Hendricks, D. W., *Water Treatment Unit Processes—Physical and Chemical*, CRC Press/Taylor & Francis Group, Boca Raton, FL, 2006.

Higgens, M. J. and Novak, J. T., Characterization of exocellular protein and its role in bioflocculation, *Journal of the Environmental Engineering Division, ASCE*, 123(EED5):479–485, May 1997.

Hudson, H. E. and Wolfner, J. P., Design of mixing and flocculation basins, *Journal of the American Water Works Association*, 59(10):1257–1267, October 1967.

Infilco Degremont, Inc., Superpulsator Clarifier, DB-585 (brochure), Richmond, VA, December 1990.

Infilco Degremont, Inc., Superpulsator Clarifier, Brochure DB-585, ONDEO Dergemont, Inc., Richmond, VA, October, 2001 (brochure downloaded as Adobe Acrobat file on June 24, 2002).

Ives, K. J., Theory of operation of sludge blanket clarifiers, *Proceedings, Institute of Civil Engineers*, 39(2):243–263, 1978.

Ives, K. J., Rate Theories, in: Ives, K. J. (Ed.), *The Scientific Basis of Flocculation*, Sijthoff & Noordhoff, The Netherlands, 1978.

Jiang, Q. and Logan, B. E., Fractal dimensions of aggregates determined from steady-state size distributions, *Environmental Science and Technology*, 25:2031–2038, 1991.

Khan, Z., Floc characteristics as affected by coagulation and as affecting filtration of low turbidity water, Doctoral dissertation, Department of Civil Engineering, Colorado State University, Fort Collins, CO, 1993.

Lagvankar, A. L. and Gemmell, R. S., A size-density relationship for flocs, *Journal of the American Water Works Association*, 59(9):1040–1046, September 1968.

Langelier, W., Coagulation of water with alum by prolonged agitation, *Engineering News Record*, 86(22):924–928, June, 1921.

Langelier, W. F. and Ludwig, H. F., Mechanism of flocculation in the clarification of turbid waters, *Journal of the American Water Works Association*, 41(2):163–181, February 1949.

Lawler, D. F., Izurieta, E., and Kao, C., Changes in particle size distributions in batch flocculation, *Journal of the American Water Works Association*, 75(12):604–612, December 1983.

Lee, D. G., Bonner, J. S., Garton, L. S., Ernest, A. N. S., and Autenrieth, R. L., Modeling coagulation kinetics incorporating fractal theories: A fractal rectilinear approach, *Water Research*, 34(7):1987–2000, 2000.

Leopold, C., Mechanical agitation and alum floc formation, *Journal of the American Water Works Association*, 26(8):1070–1084, August, 1934.

Letterman, R. D., Quon, J. E., and Gemmell, R. S., Influence of rapid-mix parameters on flocculation, *Journal of the American Water Works Association*, 65(11):716–722, 1973.

Li, D. and Ganczarczyk, J., Fractal geometry of particle aggregates generated in water and wastewater treatment processes, *Environmental Science and Technology*, 23(11):1385–1389, 1989.

Ludwig, H. F., Properties of the dispersed phase as the controlling factor in the flocculation of turbid water, MS Thesis, University of California, Berkeley, CA, 1942.

Meakin, P., Fractal aggregates, *Advances in Colloid and Interface Science*, 28:249–331, Elsevier Science Publishers, Amsterdam, the Netherlands, 1989.

Morris, J. K. and Knocke, W. R., Temperature effects on the use of metal-ion coagulants for water treatment, *Journal of the American Water Works Association*, 76(3):74–79, March, 1984.

O'Melia, C. R., Coagulation in wastewater treatment, in: Ives, K. J. (Ed.), *The Scientific Basis of Flocculation*, Sijthoff & Noordhoff International Publishers, Alphen aan den Rijn, the Netherlands, pp. 219–268, 1978.

Parker, D. S., Kaufman, W. J., and Jenkins, D., Floc breakup in turbulent flocculation processes, *Journal of the Environmental Engineering Division, ASCE*, 98(SA1):79–99, February 1972.

Reed, G. D. and Mery, P. C., Influence of floc size distribution on clarification, *Journal of the American Water Works Association*, 78(8):75–80, August 1986.

Rouse, H., *Elementary Mechanics of Fluids*, John Wiley & Sons, New York, 1946.

Shea, T. G., Gates, W. E., and Argaman, Y. A., Experimental evaluation of operating variables in contact flocculation, *Journal of the American Water Works Association*, 62(1): 41–48, January 1971.

Spicer, P. T. and Pratsinis, S. E., Shear-induced flocculation: The evolution of floc structure and the shape of the size distribution at steady state, *Water Research*, 30(5):1049–1056, 1996a.

Spicer, P. T. and Pratsinis, S. E., Coagulation and fragmentation: Universal steady-state particle-size distribution, *Journal of the American Institute of Chemical Engineers*, 42(6):1612–1620, June 1996b.

Stenquist, R. J. and Kaufman, W. J., Initial Mixing in Coagulation Processes, Report to Environmental Protection Agency, EPA-R2-72-052, Office of Research and Monitoring, Washington DC, 1972.

Swamee, P. K., Design of flocculating baffled channel, technical note, *Journal of the Environmental Engineering Division, ASCE*, 122(EED11):1046–1048, November 1996.

Tambo, N. and Hozumi, H., Physical aspect of flocculation process—II. Contact flocculation, *Water Research*, 13(5):441–448, 1979.

Tambo, N. and Watanabe, Y., Physical aspect of flocculation process—I. Fundamental treatise, *Water Research*, 13 (5):429–439, 1979.

Tennekes, H. and Lumley, J. L., *A First Course in Turbulence*, MIT Press, Cambridge, MA, 1972.

Treweek, G. P., Optimization of flocculation time prior to direct filtration, *Journal of the American Water Works Association*, 71(2):96–101, February 1979.

Velz, C. J., Influence of temperature on coagulation, *Civil Engineering*, 4(7):345–349, July 1934.

von Smoluchowski, M., Versuch einer Mathematischen Theorie der Koagulations-Kinetik Kolloider Losungen, *Z. Physik. Chem.*, 92:129, 1917 (from citation in Lee et al., 2000).

Walker, J. D., High-energy flocculation units, *Journal of the American Water Works Association*, 60(11):1271–1279, November 1968.

12 Rapid Filtration

The common filtration technology since about 1900 in the United States has been "rapid filtration." As we have gained more understanding the "process," as distinct from the "technology," has been called "depth" filtration.

The topics of this chapter include (1) a description of rapid filtration, (2) a review of the theory, (3) elements of practice, (4) a description of operation, (5) the use of pilot plants, and (6) an introduction to proprietary systems.

12.1 DESCRIPTION OF RAPID FILTRATION

As a technology, rapid filtration evolved from a variety of proprietary innovations in the 1880s and as an empirical practice from about 1900 through perhaps the 1980s. By the mid-1980s process theory began to influence design and operation. The "process," called "depth filtration," has to do with the removal mechanisms of particulates being "transported" to the granular media comprising the filter bed and "attaching" to the surfaces, that is, being removed from the fluid flow.

The modern design of a depth (or rapid) filter system has two phases: (1) process design, and (2) design of the technology support system. Process design, in the modern sense, requires knowledge of theory coupled with a pilot plant study. The technology support system requires knowledge of the practices that have been developed since the first reinforced concrete rapid filters were put on line after 1901 (according to Fuller (1928) the first was in 1901 at Little Falls, New Jersey). The technology support has to do with hydraulic design, for example, getting coagulated water to the filter, collecting filtered water, backwashing, collecting the backwash water, and disposal of waste solids (see also Qasim et al., 2006, pp. 355–422).

12.1.1 Filtration Technology

The filtration process and the supporting components to make it work is called here, "filtration technology." The "process," that is, "depth filtration," occurs within the media bed. The supporting components include an underdrain system, pipes, valves, controls, and various instruments. Numerous variations exist which are both generic and proprietary.

12.1.1.1 In-a-Nutshell

The media bed is where the "process" part of rapid filtration occurs, that is, it is where the changes occur. Floc particles in suspension (following coagulation-mixing and possibly flocculation-settling) are removed within the media, resulting in a lower effluent concentration of suspension. The resulting filter effluent should have low turbidity, for example, ≤ 0.10 nephelometric turbidity units (NTU).

12.1.1.1.1 Filtration Processes

Two kinds of filtration processes are as follows: (1) straining, and (2) depth filtration. Straining is the retention of particle that are too large to pass through the pores of the media. Depth filtration is the retention of floc particles that "attach" to the media grains or to floc that has previously attached to media grains. Such attachment is due to bonding by electrostatic and surface forces.

12.1.1.1.2 Clogging and Headloss

Straining causes an exponential headloss increase with time, which is not desired. With depth filtration, however, the time rate of headloss increase is linear, which is desired. As a filter run progresses, the media "stores" the attached floc, which encroaches on the void spaces, and is termed "clogging." A goal of the filter design is that the clogging progresses through the bed depth, that is, not just within the top layer of media. Media design is important to permit approaching this goal, along with control of floc size. As clogging occurs, the headloss across the bed increases.

12.1.1.1.3 Length-of-Run

At some point, the clogging causes a designated "terminal headloss" to be reached. At the same time, the effluent particle concentration may reach a designated "breakthrough" limit, as measured by turbidity or particle counting. Whichever occurs first determines the "length-of-run." The breakthrough limit should occur first, however, which means setting an adequate depth of the filter box.

12.1.1.1.4 Transport

Removal of floc particles within the filter bed is a two step sequence: (1) transport and (2) attachment. The transport step is the movement of a suspended particle through the water to a media grain. Mechanisms include (1) diffusion, (2) gravity settling, and (3) interception. The ratio of particles striking a media grain to those approaching is the transport coefficient, η. The value of η is determined by media size, the floc particle size, temperature, filtration velocity, etc. (Yao et al., 1971).

12.1.1.1.5 Attachment

Attachment is the bonding between the particle and the media grain, which is affected by the zeta potentials of the media and the floc particles. Depth filtration is thus transport from the

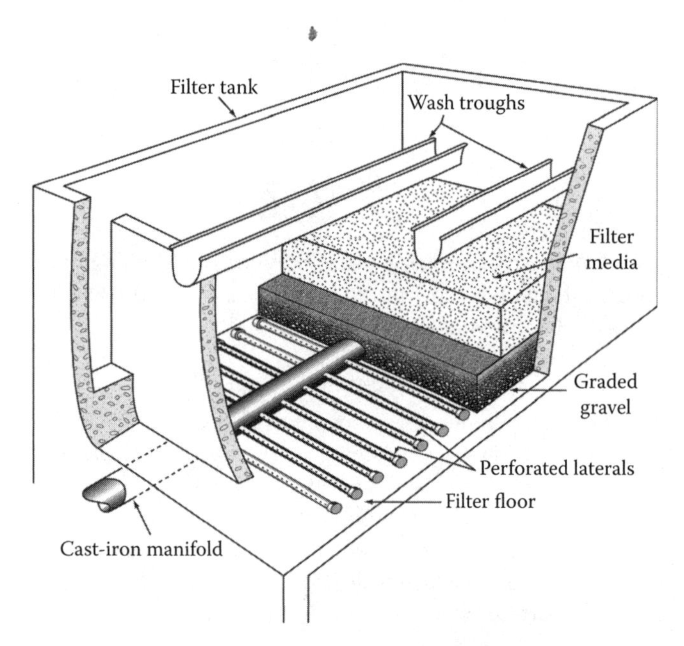

FIGURE 12.1 Cut-away perspective of rapid filter. (Adapted from McNamee, R.L. et al., *J. Am. Water Works Assoc.*, 49(7), 795, 1956.

suspension to the media and then removal by attachment. The ratio of particle attaching to those striking is the attachment coefficient, α. If coagulation is not effective, then $\alpha \ll 1.0$ and if coagulation is effective, then $\alpha \to 1.0$.

12.1.1.2 Support Components

Figure 12.1 is a cut-away perspective drawing of a rapid filter. The bed of granular media (sand or anthracite) is where the particles are removed, that is, where the "depth filtration process" occurs. Components that support the process include (1) the filter box, (2) the granular media bed, (3) the gravel support, (4) the under-drain system, and (5) overflow launders. Components not shown include pretreatment, for example, coagulation, flocculation, settling, the pipe gallery with pipes and associated valves for raw water flow, backwash flow, and wastewater flow. Control of valves and collection of data generated by online instruments has been largely by SCADA (supervisory control and data acquisition) systems since the mid to late 1980s. Ideally, each filter should be instrumented with online turbidimeters, online particle counters, piezometers or pressure sensors in the headwater and under-drain system, and flow measurement to or from the filter bed.

The functioning of the system involves (1) filtration and (2) backwash. The filtration process occurs in the granular media bed with collection of filtered water by the under-drain system. The water entering the filter bed must be coagulated with proper dosages of alum, which may include polymer. Backwash is a reverse flow through the under-drain system and upward through the granular media with sufficient velocity to "fluidize" the bed and dislodge the solid deposits by hydraulic shear and scrubbing, and to remove them from the bed. Ancillary features in backwash may include a surface wash and/or an air scour; one or both is essential to cleaning the media adequately. The wastewater from the backwash is removed by overflow launders, that is, troughs. After the backwash is completed, the filtration cycle is started again. Provision for filter-to-waste is recommended, which may be used during the filter "ripening," that is, at the start of the filter run.

The system also includes provision for treated water storage, both to provide detention time for disinfection and to account for diurnal fluctuation in demand, and perhaps help meet days of peak demand. Backwash water may be stored separately and elevated, which avoids the risk of a cross connection with treated water storage. The wastewater is conveyed to a solids storage pond where the decanted water is returned to the head of the plant.

12.1.1.3 Filtration Mode

The particular process steps prior to depth filtration, determine the "mode" of filtration, that is, inline, direct, or conventional, illustrated in Figure 12.2a through c, respectively. When the term "rapid filtration" is used, the mode used should be specified but "conventional" is the default mode.

12.1.1.3.1 Inline Filtration

The "inline" mode consists of rapid-mix/coagulation followed by filtration. This mode is used most often with low turbidity waters.

12.1.1.3.2 Direct Filtration

The "direct" filtration mode is rapid-mix/coagulation followed by flocculation, followed by filtration (Logsdon, 1978). Direct filtration is used only infrequently.

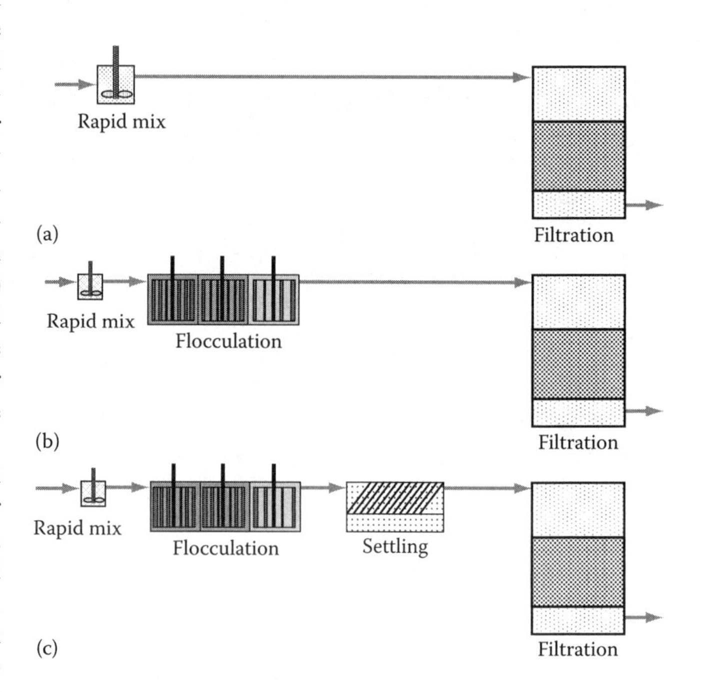

FIGURE 12.2 Three filtration modes, in-line, direct, and conventional. (a) In-line filtration. (b) Direct filtration. (c) Conventional filtration.

12.1.1.3.3 Conventional Filtration

The "conventional" filtration mode is coagulation in a rapid mix followed by flocculation, followed by settling, followed by filtration. Conventional filtration is the most common mode and is essential when floc load must be reduced prior to filtration.

12.1.2 Applications

Rapid filtration may be applied for any water treatment task that involves removal of particles. Municipal wastewaters may be treated for such purposes as irrigation, boiler feed water, recharge of groundwater, etc. Industrial water supplies and industrial wastewaters may be treated by rapid filtration, with variations that depend upon the purpose, unique to each industry.

12.1.3 Variations

In addition to the three "modes" of rapid filtration, the technology has many other variations. Some of the generic alternatives include (1) deep-bed mono media or dual media, (2) constant flow or declining flow, (3) effluent control or rising headwater, (4) pressurized versus gravity, air scour versus surface wash or both, (5) gravel under-drain system or proprietary, etc. Proprietary "package" systems may include combinations of the foregoing and, most likely, a unique feature.

12.2 DEVELOPMENT OF RAPID FILTRATION

The rapid filtration technology evolved from proprietary systems as proposed in the 1880s and took a generic form during the period 1910–1920. Further innovations were in the 1960s, for example, with media alternatives, (e.g., from sand to dual media of anthracite and sand), and higher hydraulic loading rates, for example, from 0.082 to 0.204 m/min (2.0–5.0 gpm/ft^2). By the mid-1980s, theory, coupled with experiments with pilot plants, began to provide guidance for design with deep beds of mono media (e.g., 2.0 m) and high filtration rates (e.g., 0.55 m/min or 13.5 gpm/ft^2). By the early 1990s, pilot plants became established as essential for design and as permanent installations for operation. Pilot plants were not new, however, having been used by James Simpson in developing the slow sand technology for London and later at the Lawrence Experiment Station in Massachusetts, c. 1890.

12.2.1 Development of Rapid Filtration

During the initial period of American water treatment practice, about 1870–1900, the so-called "English" filtration method, that is, slow sand, was emulated. At the same time, during the early 1880s, experiments were underway with "mechanical filters," so named because the method of cleaning was mechanical rather than by manual labor as in slow sand (Baker, 1948, p. 179). The mechanical devices included jets of water applied on or just below the surface, a reverse flow wash of the sand bed, and revolving sand agitators which loosened the media from top to bottom.

The impetus for the development the rapid filtration technology for municipal drinking water seems to have been the fact that the water in many American rivers was too turbid for slow sand.

12.2.1.1 Hyatt Filter

According to Baker (1948, p. 183), the genesis of the modern rapid filter was the Hyatt mechanical filter. This filter was the result of an 1880 patent by Patrick Clark for a sand bed that could be cleaned by downward jets of water. In December, 1880, Clark, John W. Hyatt, and Albert Westervelt incorporated as the Newark Filtering Co. to build and market the filter. John Hyatt used Clark's patents and added his own innovations that included a closed tank (a pressure filter) and a common header pipe to several filters for both raw water and backwash water. He obtained a patent in 1881 for what would be the prototype for the rapid filtration concept. The idea was to design a filter that could be cleaned by "mechanical" means.

At the same time, Col. L.H. Gardner, Superintendent of the New Orleans Water Co., was experimenting with coagulation and was convinced that it was more effective than slow sand filtration (called at that time, simply, "filtration") for dealing with muddy water. Isaiah Smith Hyatt, older brother of John, was on the scene in New Orleans as sales agent for the Newark Filtering Co., trying to clarify Mississippi River water for a New Orleans industrial plant. Col. Gardner suggested using a coagulant, which was done in conjunction with the filtration and was successful. On February 19, 1884, Isaiah Hyatt obtained a patent for simultaneous coagulation filtration, having had used "perchloride of iron" for the filtration of the Mississippi River waters. Also, Baker reported that by 1889, John Hyatt had developed a filter that incorporated a backwash to expand the filter bed in what was essentially the genesis of the modern rapid rate filter.

12.2.1.2 Warren Filter

Another proprietary filter was developed by John E. Warren, agent of the S.D. Warren & Co. paper mills, Cumberland, Maryland, who in 1884, planned and constructed a 45,000 m^3/day (12 mgd) filter plant for the mills, illustrated in Figure 12.3. These filters were gravity type, contained in wooden tanks. An adjunct to the filter was an alum dosing apparatus, patented in 1890 by Professor Henry Carmichael of Malden, Massachusetts (Baker, 1948, p. 199). As with the Hyatt filters, several installations were completed for municipal water supply systems. A coagulant was not used, however, until the fourth plant was completed in 1892 at Macon, Georgia.

12.2.1.3 Other Proprietary Filters

After the mid-1880s, several more proprietary filters emerged on the scene. One prominent company was the National

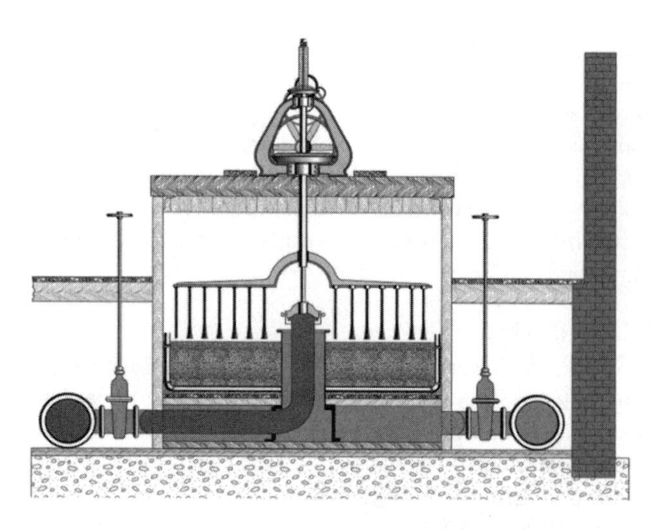

FIGURE 12.3 Warren gravity filter with revolving rakes forced into media to aid backwash, c. 1889. (Adapted from *The Quest for Pure Water*, 1948, p. 198, used by permission, American Water Works Association. Copyright © 1948, American Water Works Association; redrawn by Shane Tribolet, Colorado State University, Fort Collins, CO.)

Water Purifying Co. of New York City, incorporated in 1886 to promote a filter patented by William Deutsch, a former salesman for the Newark Filtering Co. (Baker, 1948, p. 199), shown in Figure 12.4. Albert Leeds, Professor of Chemistry, Stevens Institute of Technology transferred rights in his water aeration patents to the company and became its technical advisor. An 1896 patent was for an air and water wash; the air wash was described in terms similar to modern use. Three other Deutsch patents were in 1900; they were for sedimentation and coagulation based on Fuller's 1896 Louisville experiments which showed the necessity to clarify highly turbid waters by sedimentation before filtration.

The National Water Purifying Co. undertook, according to Baker (1948, p. 205), one of the boldest and most disastrous attempts ever made to filter the water supply of a city. Despite the advice of Professor Leeds and others against the venture, the company contracted in 1891 to supply a constant supply

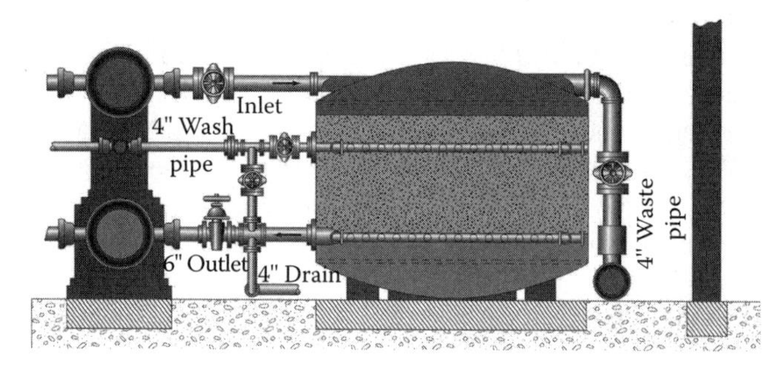

FIGURE 12.4 National pressure filter with double backwash jet, c. 1891 (Adapted from *The Quest for Pure Water* p. 200 and re-drawn by Shane Tribolet, Colorado State University, Fort Collins, CO, by permission. Copyright © 1968, American Water Works Association.) (As a matter of historical interest, the drawing was published orginally by Eng. News, 25:127, 1891.)

of clear water to New Orleans from the Mississippi River as the water source. The General Superintendent of the water company, George Earl, in fact had advised the company to construct 1 filter, instead of the 30 for the full plant and find out first what could be accomplished. The plant was put in operation in 1893 and, after tests, the New Orleans Water Company refused to accept the plant on grounds of nonfulfillment of guarantee, which was upheld in the courts and cost the company $134,500, after which the company became bankrupt. At the same time, the company also was being litigated against by the Newark Filtering Co. for infringement of patents. Professor Leeds, reported on the disaster in 1896 before a meeting of the American Water Works Association.

The Jewell filter was also well known. Omar Jewell and his sons Ira and William took out 50 patents during the period 1888–1900. The first filter was for boiler feed water and was constructed in 1885 in Chicago, financed by James B. Clow & Sons (which became one of the recognized companies in the water works industry). A "rate of flow controller" was patented in 1897 by William Jewell. By 1896, 21 plants had been completed using Jewell filters.

In 1892, four companies merged to form the New York Filter Manufacturing Co. Many such mergers occurred during the period leading to 1900, along with law suits for patent infringements. By 1900, the New York Continental Jewell Filtration Co. stood nearly alone in the field and was heir to scores of patents (Baker, 1948, p. 226) most of which had expired. By 1909, the company had completed 360 plants.

12.2.1.4 Fuller's Experiments

The capstone events that marked the transition to modern filtration practice, amid all of the ferment among proprietary filtration companies, were the experiments conducted by George W. Fuller at Louisville in 1895–1897 and then at Cincinnati in 1898–1899 (Box 12.1). Prior experiments in the 1880s at Louisville demonstrated that slow sand filters could not cope with the turbid waters of the Ohio River (Baker, 1948, p. 228). The 1895–1897 experiments by George W. Fuller at Louisville were focused on evaluating the proprietary filters of four companies. Each company was to operate its own filters at their own expense using 11 L/s (0.25 mgd or 174 gpm) units.

Regarding the experiments, Fuller (1898, p. 3) stated as follows:

Sulfate of alumina (or alum) was added to the river water, as it entered the devices in quantities varying with the character of the water. By combining with lime naturally dissolved in the river water the sulfate of alumina formed a white, gelatinous, solid compound, called hydrate of alumina. This latter compound gradually coagulated the suspended matter in the river water, in a manner similar to the well-known action of white of egg when added to turbid coffee. In the settling basins, where the river water first entered, this coagulation progressed so that, as the water left the settling basins and entered the sand layer, the river water had lost some of the

BOX 12.1 FULLER

George Warren Fuller and Allen Hazen were "alumni" of the Lawrence Experiment Station in Massachusetts. For the Louisville experiments, Fuller led a team that included Robert Spurr Weston and George A. Johnson, who also became prominent in the field. Fuller, a legend in the field, was president of the American Water Works Association in 1923 and is known also by an award in his name. He had a successful consulting practice and was candid in his assessments of a client's needs. An example was his review of his engagement by New York City to draw up plans for a filtration system. In a paper (Fuller, 1914), he criticized the decision by the City not to proceed. Appendix 12.A reviews the New York case.

In his report, *The Purification of the Ohio River Water at Louisville, Kentucky*, D. Van Nostrand, New York, 1898, Fuller's remarks demonstrate that even at the time he understood the complexities of the filtration process and that every situation is unique.

While much careful attention has been given to the art of water purification for more than 60 years, the general solution of the problem on a practical basis for large cities is, however, far from satisfactory or complete at its present stage of development. This is due partly to varying effects of the adopted processes with different natural waters, partly to the lack of a widely practical and scientific understanding of the influence of a number of factors of the processes themselves, and partly to the great cost involved in the construction of adequate filtration works.

In his report, he reviewed slow sand filtration starting in 1829 in London. Based on this review, he referred to the slow sand filters as "English" filters which was done to distinguish them from the "American" filters which were those that had alum addition and backwash to remove accumulated material, which as Fuller noted, were the distinguishing features of the latter.

FIGURE 12.5 George Warren Fuller, c1930. (Courtesy of American Water Works Association.)

mud suspended in it, and the mud and clay which it did contain were formed into flakes of sufficient size to allow a very rapid flow of water through the sand layer, with satisfactory results. The claim that this method of water purification was more economical for the Ohio River water than those practiced in Europe was based on the assertion that comparatively small amounts of sulfate of alumina permitted a very great reduction in the necessary area of filtering surface.

Several points are important: (1) the alum addition depended on the raw water quality, (2) lime was considered a part of coagulation (alkalinity was recognized later as the important constituent), (3) the term "coagulation" was used in the current sense, (4) settling of coagulated water prior to filtration, with "passive" flocculation, was recognized as necessary to reduce the solids load to the filters, (5) less surface area was required as compared with slow sand.

By 1898, Fuller (1898, p. 10) estimated that 100 rapid rate filter systems were in place. He pointed out that slow sand filtration never got a firm start in the United States, first because water-borne epidemics were not in evidence, as in Europe, and by the time that filtration may have been contemplated, the rapid filtration technology was being developed. Fuller noted also (p. 10) that alum addition (patented in 1884) was a distinguishing feature of rapid filtration, as was the development of a successful backwash operation. Alum addition to coagulate water had been practiced, he stated, in various ways for many centuries, with its description in the scientific literature starting about 1830.

Fuller was the President of the American Water Works Association in 1923. Figure 12.5 is a photograph of George Warren Fuller. He was a prominent personality for four decades and is given credit for launching the modern practice of rapid filtration.

12.2.2 Emergence of Filtration Practice

By about 1900, a generic practice in rapid filtration began to emerge. First, patents of proprietary systems were expiring, which would make less clouded the development of a generic technology and practice. Second, the private companies had expended much of their working capital in litigation and competition (Fuller, 1933, p. 1571). The first generic plant, designed by Fuller, was the Little Falls plant of the East New Jersey water company, placed in service in September 1902.

The Little Falls plant did not, however, lead to the sudden emergence of a generic technology. Evidence of this is seen in Fuller's description of his filtration system design for New York (see Appendix 12.A). He explained to a critic, Mr. Alexander Potter, that he recommended slow sand in his 1907 report because only two rapid filtration systems had

TABLE 12.1
Use of Filters in the United States

Year	Population Total	Urban	Population with Filtered Water[b] Rapid	Slow Sand	Total	Percent
1870	39,800,000[c]		0	0	0	0
1880	50,200,000[c]	13,300,000[a]	0	30,000	30,000	0.2
1890	62,980,000[c]	21,400,000[a]	275,000	35,000	310,000	1.4
1900	76,212,000[c]	29,500,000[a]	1,500,000	360,000	1,860,000	6.3
1904	82,619,000[c]	32,700,000[c]	2,600,000	560,000	3,160,000	9.7
1924	112,900,000[c]	60,200,000[ac]	18,610,000[d]	5,054,000[d]	23,664,000[d]	percent

Source: Adapted from Turneaure, F.E. and Russell, H.L., *Public Water Supplies*, John Wiley & Sons, New York, 1913.

[a] Total urban population in the United States in towns above 2500 from Turneaure and Russell (1913).

[b] From Turneaure and Russell (1913).

[c] Census data rounded; interpolated for 1904 and 1924.

[d] The 1924 line was from Gillespie (1925, p. 124) and was for plants of size $\geq$3785 m^3/day ($\geq$1.0 mgd).

been built that were of fair size and he did not feel that the "art had attained sufficient standing" to warrant its use. On the other hand, in the years between his 1912 report, he felt that as sufficient progress had been made he felt no hesitation in recommending rapid filtration (Fuller, 1914, pp. 454–465). A point to be extracted from this is that we might say that rapid filtration technology had its gestation from about 1880–1885, development as a proprietary technology 1885–1898, its development as a generic technology 1898–1912, and its maturation as a practice 1912–1960, with the knowledge from research applied after 1960.

12.2.2.1 State of the Art, 1890 and 1990

Fuller's conclusion that plain sedimentation and then settling of coagulated water were necessary prior to filtration was a "breakthrough" for practice that made rapid filtration applicable to virtually the full range of raw water conditions. The subsequent practice that evolved was the sequence: *plain sedimentation-coagulation-settling-filtration*, which was notable because it lacked explicit turbulent flocculation. At that time, the role of flocculation was not understood well. Flocculation occurred in the settling basin, but by Brownian motion. This required 24–48 h detention time. The floc that formed settled to bottom of the basin and was removed after a sufficient amount of accumulation. The "conventional" filtration mode matured to its modern state in 1922 when Professor Wilfred Langelier introduced the idea of paddle-wheel flocculation (Chapter 11) at the plant being designed for Sacramento, California. From this, the process sequence: *coagulation-flocculation-settling-filtration* evolved, which is known today as "conventional" filtration.

12.2.2.2 Growth of Waterworks Industry

During the period, 1850–1896, the number of waterworks in the United States grew from 83 to 3196 (Turneaure and Russell, 1913). In 1870, none of the population had treated water. In 1880, out of the 13,300,000 urban population, 30,000 had treated water. Table 12.1 summarizes data of 1924, which shows a steady growth of treated water provided to the urban population of the United States, with rapid filtration growing faster than slow sand. The 1924 installed capacities were: 15,321,680 m^3/day (4048 mgd) among 587 plants for rapid filters and 3,467,060 m^3/day (916 mgd) among 47 plants for slow sand (Gillespie, 1925, p. 124). Most of the plants in Gillespie's survey were constructed during the years 1910–1924.

12.2.3 Progress in Filtration Practice

Filtration practice through about the 1950s evolved largely from the work of Allen Hazen, George Warren Fuller, Wilfred Langelier, John Baylis, and others. Baylis (1937, p. 1011) described some of the issues of the period from 1915, the start of his career in filtration, to 1937, for example, maintaining effective coagulation, the lack of laboratory control, education of city officials and those in operation, under-drain design, and the problem of adequate backwash to clean the media adequately which was associated with controlling mudballs and surface cracks. Some of the innovations were the submarine light, about 1918, which permitted operators to judge effluent turbidity in the clear well and the surface wash developed by Baylis after noting that most of the clogging occurred in the top 150 mm (6 in.).

The practice, as it evolved, was codified by state regulations which specified both process design and hydraulic design. By the time of the 1950s, practice codification included media size (about 0.45 mm sand), bed depth (600–760 mm or 24–30 in.), hydraulic loading rate (4.9 m/h or 2 gpm/ft^2), and support components such as the under-drain system, overflow launders, rate-of-flow-controller, backwash storage, etc. The accepted filtration mode was conventional filtration, that is, coagulation, flocculation, settling, filtration.

12.2.3.1 Dual Media

Although dual media of anthracite and sand is usually associated with the 1960s, Hansen (1936) mentions "renewed" interest in the use of crushed anthracite as a filtering medium (see also McNamee et al., 1956, p. 805, who mention that the use of anthracite dated back to 1914 in Cumberland, Maryland, and was in current use in 26 some states). He mentions that for a 10 year period, crushed anthracite on top of sand was used at the Marston plant in Denver. Its use was favored partly because mudball and cracking problems were less. The widespread adoption of anthracite and sand as a dual media was in the early 1960s when promulgated by Walter Conley, who, at the same time, introduced mixed media (anthracite, sand, and garnet) filter beds. By the early 1970s, dual media was common in practice. Mixed media was a proprietary product of Neptune Microfloc® and became widely used.

12.2.3.2 Breaking the HLR Barrier

Fair (1963, p. 820) refers to the filtration velocity and bed depth in the following statement:

> Among the practices from which the profession was eventually liberated by clear-thinking operators and imaginative designers were a slavish adherence to 30-in. beds of non-uniform sand and constant rates of filtration of 2 gpm/sq ft of bed surface.

McNamee et al. (1956, p. 793) shed more light on 2.0 gpm/ft^2 rate noting that in the early years of rapid filtration, sizing of filter beds was based on the filtration rate of 1.4 mm/s or 5.0 m/h (2.0 gpm/ft^2), which was selected because high quality water was associated with this rate. They noted, however, that most plants operated at peak hourly rates at perhaps up to 3.5 mm/s or 12.5 m/h (5.0 gpm/ft^2). Baylis was, most probably, the person who "broke" the 5.0 m/h (2.0 gpm/ft^2) filtration rate barrier. He described (Baylis, 1956) the operating experience from 1948 to 1955 in which he compared performance of 10 of 80 filters with filtration rates varying from 5.0 m/h (2.0 gpm/ft^2) to 12.5 m/h (5.0 gpm/ft^2). His conclusion was that with the higher filtration rates there was no deterioration of filtrate quality and that the productivity per filter was higher with the higher rates. In commenting on the work of Baylis, Hudson (1956, p. 1146) noted that filtration rates of 10.0 m/h (4.0 gpm/ft^2) and even 25.0 m/h (10.0 gpm/ft^2) are possible without deterioration of water quality provided proper grain size and depth of filter medium are selected. This concept of filter bed design was perhaps 25 years ahead of notions that prevailed in practice.

In the 1960s, this higher rate was advocated for mixed media proprietary filters as an average rate and by 1980 the 3.5 mm/s or 12.5 m/h (5.0 gpm/ft^2) filtration rate was accepted widely in practice. State regulations moved toward this higher rate by about the mid-1980s. Practice developed toward even higher rates, for example, 9.1 mm/s or 32.5 m/h (13.3 gpm/ft^2) at the Los Angeles Aqueduct Plant (Kawamura, 1996).

12.2.3.3 Alternative Modes of Filtration

Conventional filtration (coagulation-flocculation-settling-filtration) has been the standard for practice from the time of Fuller's experiments—keeping in mind that the distinction between coagulation and flocculation was not well delineated until about 1920 with Langelier's design of paddle wheel flocculators. The compelling rationale was that the filtration system could then handle higher seasonal increases in surface water turbidities.

In 1968, Conley and Evers, working with the low turbidity (1-2 NTU) waters of the Columbia River, advocated the idea of coagulation-filtration (Conley and Evers, 1968). This process, that is, coagulation-filtration, is called *inline* filtration, a term suggested by Cleasby (1984) in his American Society of Civil Engineers (ASCE) Simon Freese Lecture at Boulder, Colorado.

12.2.4 Modern Filtration Practice

Modern practice is characterized by its focus on the *process*, that is, understanding what happens within the filter bed, coupled with excursions from traditional guidelines of past decades. This change has been "enabled" by theory with stimuli from federal mandates regarding turbidity and *Giardia* cysts. The "tool" has been the pilot plant, used widely in both design and operation (Logsdon, 1982).

12.2.4.1 The Federal Role

Filtration practice languished with small incremental improvements over the decades until 1974 when the Safe Drinking Water Act (SDWA) was passed, the first direct foray into federal regulation of drinking water. The turbidity standard adopted by the states to this time was mostly the 5 Jackson Candle turbidity units (JTU) based upon the 1962 Drinking Water Standards for Interstate Carriers (USPHS, 1962, p. 6). The 1974 SDWA, however, mandated a 1 NTU standard for drinking water, which provided the impetus for the industry to reassess its practices. Another impetus were the regulations, called the "Filtration Rule," published in the June 29, 1989 *Federal Register* (*FR*54:124:27486) which promulgated a 0.5 NTU standard (effective June 1993). The goal of many in the industry is a 0.1 NTU standard and some plants have set that as in internal standard. In addition, the "Filtration Rule" required an overall 3 log reduction in *Giardia* cysts, administered by giving so many "credits" for filtration and so many for disinfection that complied with criteria established by the "Rule." For example, with conventional filtration, a 2 log credit was given, and for disinfection by chlorination a 1 log credit was given. The Filtration Rule was a guideline, but did not ensure complete safety of product water. For example, during the April 1993 Milwaukee outbreak of 403,000 cases of cryptosporidiosis (estimated) the plant was in compliance with standards.

12.2.4.2 Modern Practice

Two influences changed the character of practice during the 1980s: (1) theory became assimilated into practice and (2) pilot plants became the basis for design and an aid to operation. Theory provided the rationale for the excursions from standard designs and pilot plants provided the empirical confirmation.

12.3 THEORY

Experimental work aimed at discovering filtration mechanisms started with Eliassen (1941) reporting on doctoral research completed at MIT in 1935, followed by Stein (1940), Stanley (1955), and Ives (1961,1962). Filtration theory, in the mathematical modeling sense, started with a paper by Iwasaki (1937) (Section 12.3.3.1). Ives (1962) coupled his experimental work with mathematical modeling, building on Iwasaki's work.

12.3.1 Quest of Theory

Goals of filtration theory are as follows: (1) to describe variables that affect particle removal, (2) to develop mathematical models that describe filtration behavior, and (3) to explain the mechanisms of particle removal in depth filtration. Issues include the rate of clogging, the rate of headloss increase, effluent particle counts with time, backwash effectiveness, etc.

12.3.1.1 Dependent Functions in Filtration

In filtration, the suspended solids concentration and the deposited solids both change with depth and with time. At the same time, the local hydraulic gradient changes with distance and time due to the continuing depositing of solids along the depth of the filter column. The respective functional relations may be expressed as

$$C(Z, t) \tag{12.1}$$

$$\sigma(Z, t) \tag{12.2}$$

$$i(Z, t) \tag{12.3}$$

where

- C is the concentration of suspended matter of a given species (kg suspended solids/m^3 of water)
- σ is the specific solids deposit on media (kg of solids/m^3 of total volume of filter)
- i is the local hydraulic gradient (m headloss/m of filter depth)
- Z is the coordinate distance in vertical direction (m)
- t is the elapsed time since introduction of suspended matter to filter medium (s)

The three functional relations are aspects of the same phenomenon, that is, the loss of a portion of the suspended solids from suspension within the pores of the filter bed, and their subsequent deposit on the grains of the filter medium, and the "clogging" of the medium that causes increase in hydraulic gradient.

12.3.1.2 Definitions

The $C(Z)_t$ curve is called here the "wave front"; the $C(t)_Z$ curve is called the "breakthrough" curve; the two curves are related mathematically. The $s(Z)_t$ curve is the "clogging-front." The hydraulic gradient profile, $i(Z)_t$ is related to the clogging-front, which is also related to the wave front.

12.3.2 Process Description

Several experimental investigations have provided data that describe the filtration process in terms of $C(Z, t)$ results.

12.3.2.1 Experimental $C(Z)_t$ Results of Eliassen

Figure 12.6a shows experimental $C(Z)_t$ curves, that is, "wave fronts" from Eliassen (1941); Figure 12.6b is the first part of the associated $C(t)_Z$ curve, that is, the "breakthrough" curve, that is, $C(t < 120\text{ h})_{Z=60\text{ cm}}$. The Eliassen curves characterize filtration behavior, that is,

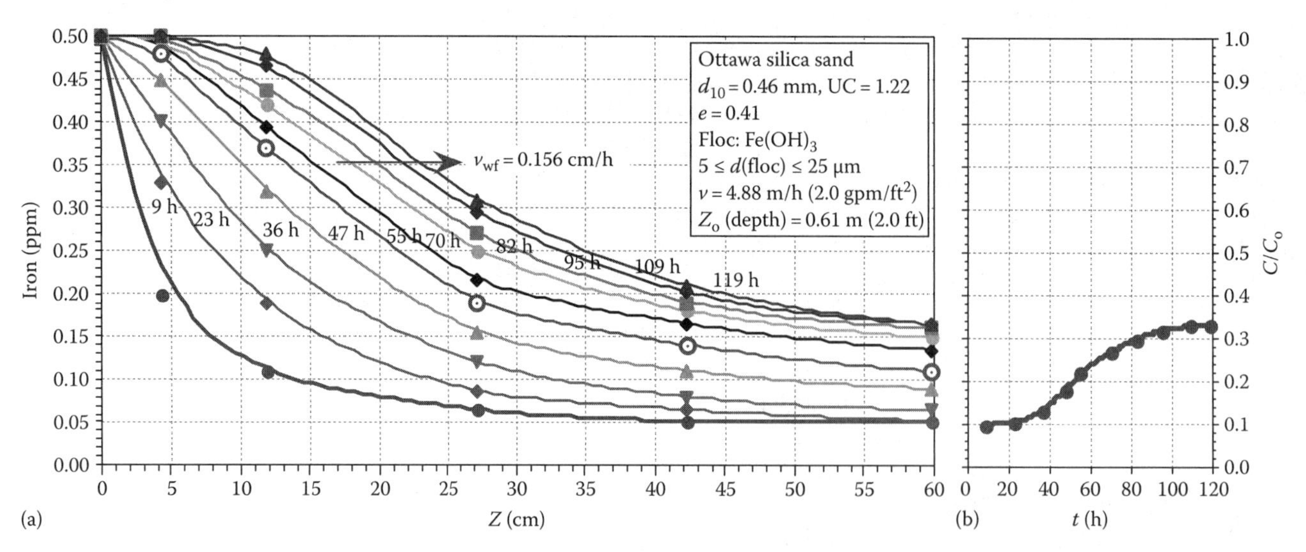

FIGURE 12.6 Experimental $C(Z, t)$ curves. (a) $C(Z)_t$ curves. (b) $C(t)_{Z=Z_o}$ curve. (Adapted from Eliassen, R., *J. Am. Water Works Assoc.*, 33(5), 936, 1941.)

1. The $C(Z)_{t\approx 0}$ curve has exponential decline with distance (per Iwasaki's Equation 12.12).
2. The $C(Z)_{t>0}$ curves may show a steeper exponential decline due to "ripening."
3. When the upper layer becomes "saturated" with solids, the $C(Z)_{t\gg 0}$ curve takes a steady state shape.
4. At about $t=55$ h the steady state wave front translates at a velocity, $v_{wf}\approx 0.156$ cm/h.
5. The $C(t)_{Z=Z_o}$ curve is defined as the wave front that emerges from the bottom of the filter at $Z=Z_o$.

12.3.2.2 Experimental $C(Z)_t$ Results of Ives

Figure 12.7 shows another set of $C(Z)_t$ curves, obtained by Ives (1962) through tagging algae with Cs-137 (with measurements by a scintillation counter). As with the Eliassen experiments, the first profile in Ives' results at $t=20$ min shows the suspended solids concentration declining exponentially with distance. Later, "ripening" sets in and the curve becomes steeper, for example, as seen in the profiles at $t=80$ and $t=160$ min. As the solids continue to deposit, the upper part of the filter becomes "saturated" with deposited solids. This causes the corresponding suspended solids concentration to be the same as the input, that is, C_o.

As this "saturated zone" of deposits develops, the $C(Z)_t$ curve, which is a distance profile at a specified time, that is, the "wave front," takes a steady state shape and translates downstream. Its velocity depends on both the solids flux into the column and the capacity of the medium to accumulate solids, for example, Equation 12.9. The wave front at $t=440$ min is the approximate start of the "steady state" shape. The wave front then advanced downstream without change in shape, as seen in the profiles at $t=680$ and $t=1440$ min; its velocity, $v_{wf}\approx 0.90$ cm/h.

12.3.2.3 $C(Z,t)$ in Three Dimensions

Figure 12.8 illustrates the $C(Z)t$ curves of Figure 12.7 in three dimensions, that is, as $C(Z,t)$. In the plot, concentration, C(ppm), is the vertical axis; distance, Z (cm), is the first horizontal axis on the left; and time, t (minutes), is the second horizontal axis on the right. The $C(Z,t)$ plot was the result of model simulation by Ives (1962) and gives an overall perspective to the same results seen in the $C(Z)_t$ plot.

12.3.2.4 Mass Transfer Similarities between Adsorption and Filtration

The filtration process falls within a class of packed bed reactor problems involving both materials balance and similar kinetics, for example, granular activated carbon and ion exchange (Vagliasindi and Hendricks, 1992; Adin and Rebhun, 1977; see also Sections 12.3.3.6 and 12.3.4.1).

12.3.2.5 Relation between the $C(Z)_t$ Wave Front and the $C(t)_{Z=Z_o}$ Breakthrough Curve

The translation of the $C(Z)_t$ wave front with its continuous emergence at the bottom of a filter bed defines the $C(t)_{Z=Z_o}$ "breakthrough" curve. The two curves are related mathematically by the chain rule, that is,

$$\frac{\partial C}{\partial t}=\frac{\partial C}{\partial Z}\frac{\partial Z}{\partial t} \qquad (12.4)$$

and since $\partial Z/\partial t=v$(wave front)

$$\frac{\partial C}{\partial t}=v(\text{wave front})\cdot\frac{\partial C}{\partial Z} \qquad (12.5)$$

where

t is the elapsed time from start of filter run (s)
Z is the distance from top of filter bed to a point on the wave front (m)
v(wave front) is the velocity of wave front (m/s)

12.3.2.5.1 Graphical Depiction of Chain Rule

Figure 12.9 depicts Equation 12.1 in the form, $C(Z)t_1$, $C(Z)t_2$, $C(Z)t_n$, with horizontal scales, Z and t. As seen, a succession of wave fronts are emerging from the bottom of the filter bed

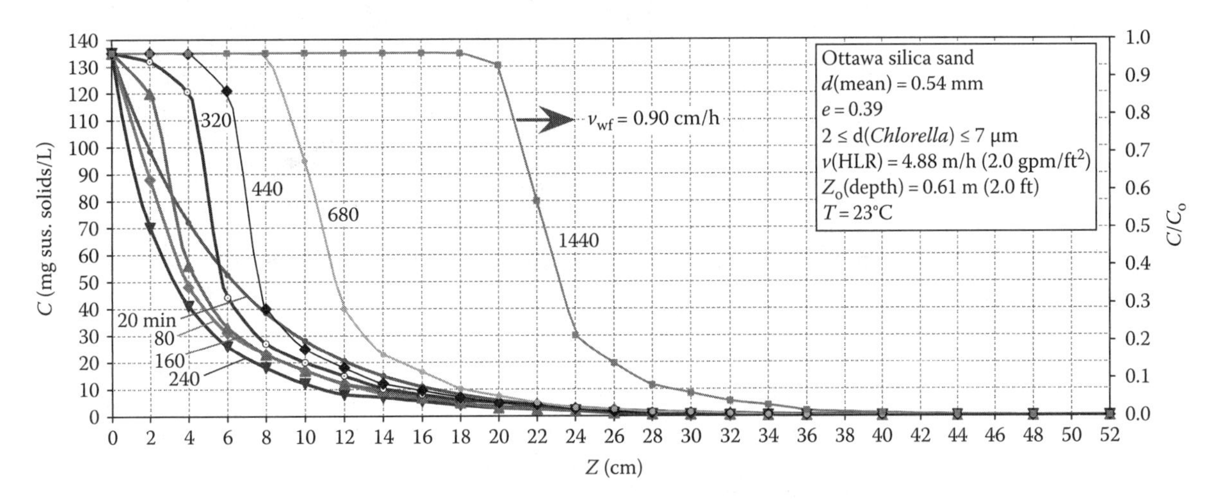

FIGURE 12.7 Measured data from experiments with radioactive algae. (Adapted from Ives, K.J., *Trans. ASCE*, 127(Part III), 382, 1962.)

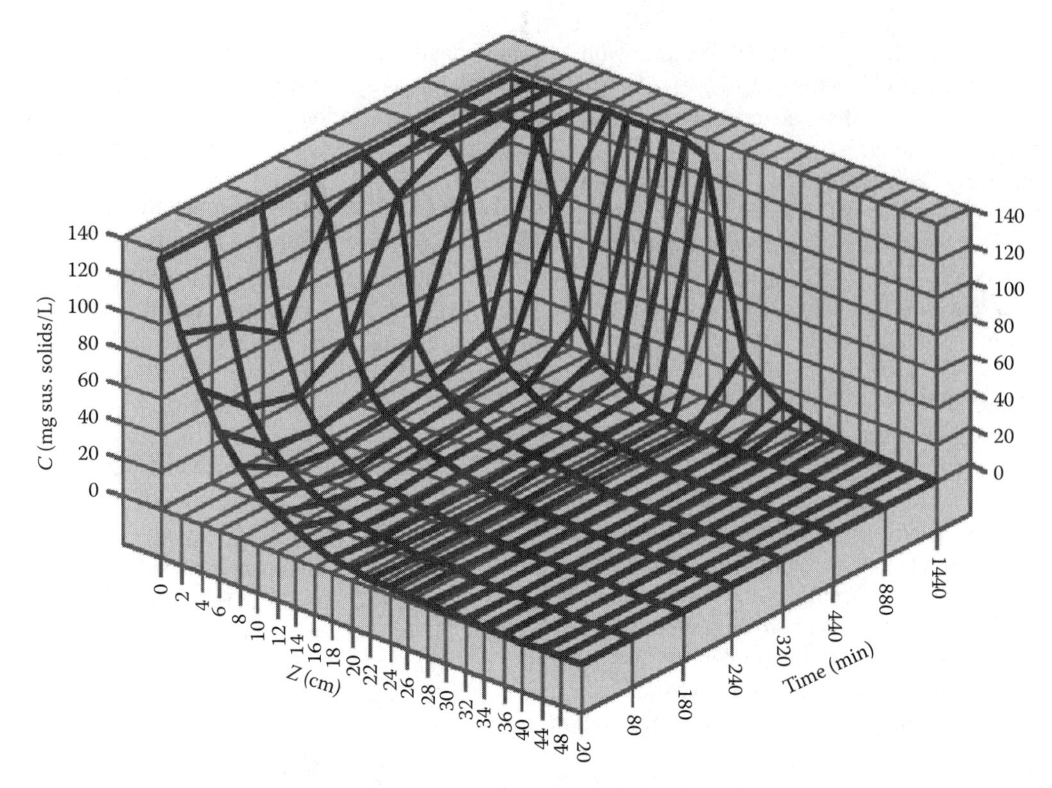

FIGURE 12.8 $C(Z, t)$ plot of computed output of Ives' model. (Adapted from Ives, K.J., *Trans. ASCE*, 127(Part III), 372, 1962.)

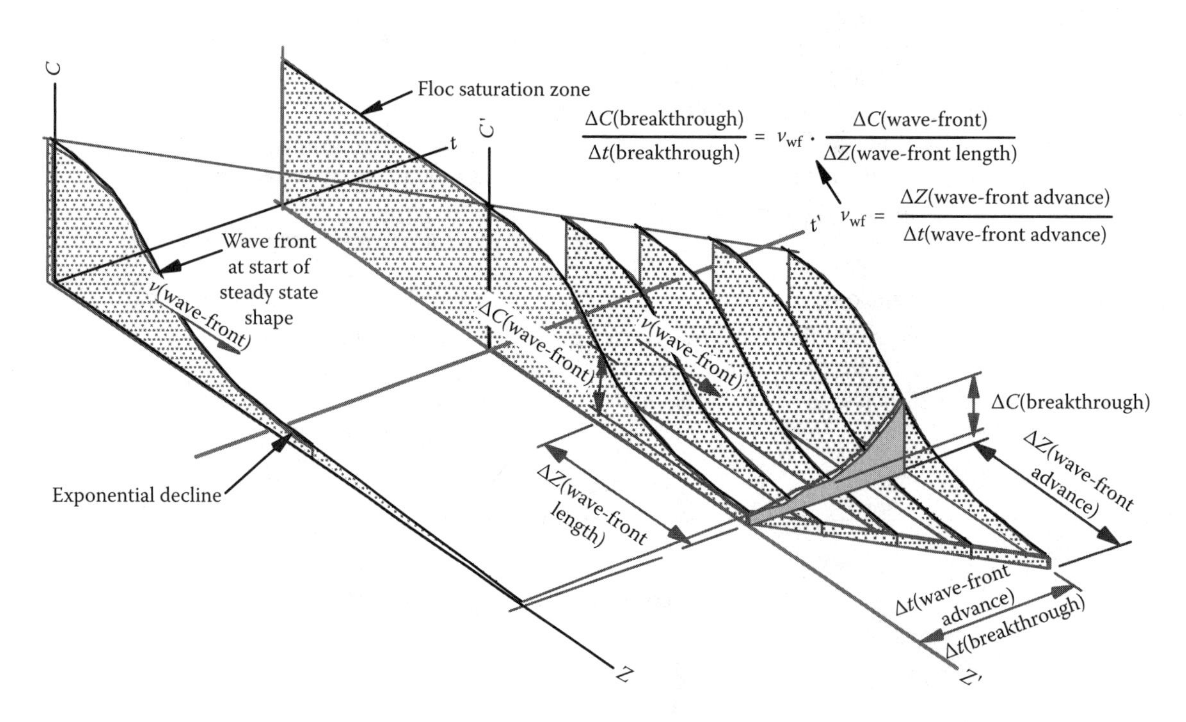

FIGURE 12.9 Emergence of the wave front at bottom of filter column and formation of breakthrough curve showing the associated mathematical relationships.

as time increases. Also, as seen in Figure 12.9 (and in Figures 12.8 through 12.10) the wave front is steady state in shape and it translates along the depth of the column. The wave front translation is due to increasing solids accumulation within the medium. As the wave front emerges from the bottom of the filter column, it defines the "breakthrough" curve, which is seen as the $C(t)_Z$ curve perpendicular to the Z-axis.

In depicting Equation 12.5 consider the wave front for the increment of concentration, ΔC(wave front) with

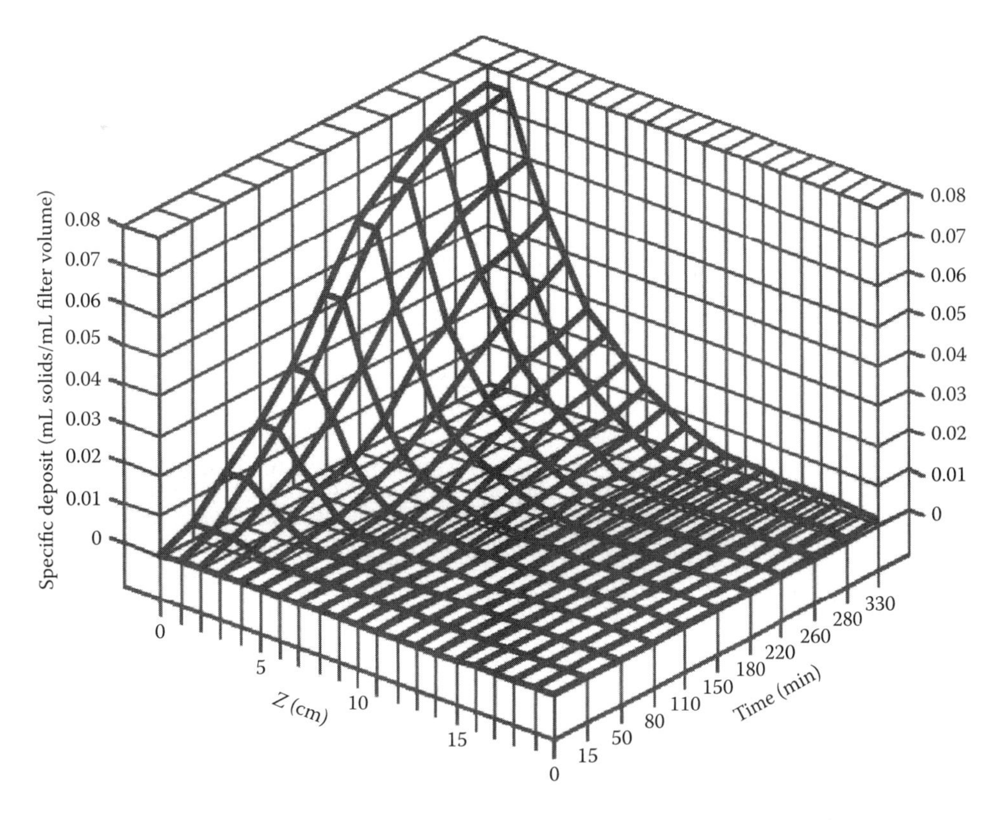

FIGURE 12.10 $\sigma(Z, t)$ plot of measured data from radioactive algae. (Adapted from Ives, K.J., *Trans. ASCE*, 127(Part III), 378, 1962.)

corresponding length, ΔZ, seen in Figure 12.9. This same concentration increment is seen as ΔC(breakthrough) when it emerges from the bottom of the filter after time, Δt (looking to the right along the t-axis). The terms $\Delta C/\Delta Z$ and $\Delta C/\Delta t$ are seen depicted on the wave front and breakthrough curve, respectively. The velocity of the wave front, v_{wf}, is the slope, $\Delta Z/\Delta t$, also seen graphically (the triangle in the Z–t plane). Since Figure 12.9 shows ΔC(wave front) = ΔC(breakthrough), then the depiction shows also that the velocity of the wave front, v_{wf}, times its Δt increment equals its associated ΔZ distance. In other words, given the $C(Z)_t$ curve and v_{wf}, the $C(t)_Z$ curve can be "mapped."

12.3.2.6 Specific Solids Deposit, $\sigma(Z, t)$

Regarding the specific solids deposit, the form of Equation 12.2 depicted more commonly is $\sigma(Z)_t$, that is, the distance profile of specific solids concentration at a given time. Ives (1962) measured the $\sigma(Z)_t$ profiles at various times for a pilot scale filter using Cs-137 tagged algae and a scintillation counter; Figure 12.10 is a three-dimensional plot of the data. The plot shows, as expected, zero deposits along the filter profile at $t=0$. At $t>0$, the profile has an exponential decline with distance, but changes in shape with time as the mass of deposit increases. At $t=290$ min, the deposit at $Z=0$–1 cm becomes "saturated" and no further "net" deposit occurs; this flattened part of the curve, that is, at $Z=0$–1 cm, $t \gg 0$, is called here the "saturated zone" of the filter. In the saturated zone, solids deposit and encroach into the pore volume; deposition and shear occur at the same rate. As the deposits continue to distribute their mass approximately as illustrated in Figure 12.10, the saturated zone migrates with increasing depth within the filter bed. The profile eventually attains a "steady state" shape and translates approximately the same as the solids in suspension.

12.3.2.7 Clogging Front

As the top of the filter bed becomes "saturated" with solids, the "clogging front" translates downstream with a constant (unchanging) shape as with a fully developed wave front (Herzig et al., 1970). This idea is illustrated further in Figure 12.11, which shows the saturated zone and the clogging wave front just ahead. The clogging front (i.e., $\sigma(Z)_t$) and the concentration wave front for suspended solids (i.e., $C(Z)_t$) have similar distributions. In other words, the emergence of the clogging front from the bottom of the filter bed is the same as the suspended solids wave front emergence. Stanley (1955, p. 592) confirmed this experimentally with iron hydroxide floc.

12.3.2.7.1 Velocity of Clogging Front

The materials balance principle, applied to a column as a whole, is the basis for calculating the velocity of the clogging front (Stanley, 1955, p. 592; Tien and Payatakes, 1979, p. 741), that is,

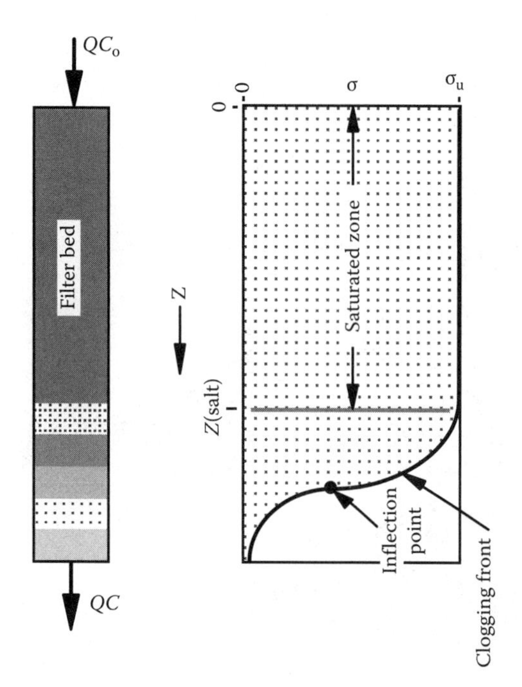

FIGURE 12.11 Wave front in relation to saturated zone.

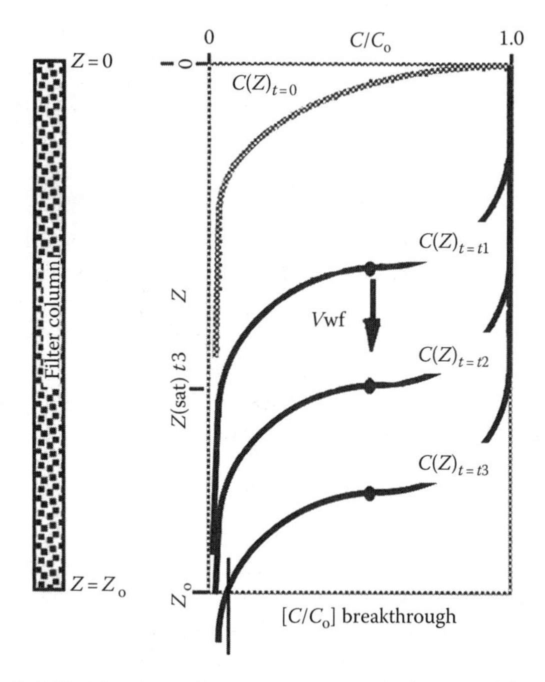

FIGURE 12.12 Wave front movement and advance of the saturated zone.

mass of suspended solids input to filter columnin time t(sat)
= mass of solids in saturated zone
+ mass of solids associated with clogging front

(12.6)

The concept of Equation 12.6 is illustrated in Figures 12.11 and 12.12. The corresponding equation in mathematical terms are:

$$QC_o t(\text{sat}) = \sigma_u A Z(\text{sat}) + \int \sigma \, dZ \tag{12.7}$$

where

t(sat) is the time associated with a given position of the saturated zone, that is, Z(sat) (s)
Z(sat) is the distance along the column to the end of the saturated zone (m)
σ_u is the maximum level of solids that may exist in the filter media (kg solids/m^3 filter bed)

The velocity of the clogging front is

$$v_{wf} = \frac{Z(\text{sat})}{t(\text{sat})} \tag{12.8}$$

Substituting Equation 12.8 in 12.7 and dropping the integral term gives

$$v_{wf} \approx \frac{vC_o}{\sigma_u} \tag{12.9}$$

where

v_{wf} is the velocity of the wave front (m/s)
v is the velocity of water, that is, $v = Q/A$ (m/s)
C_o is the concentration of suspended solids entering the filter column (kg solids/m^3 water)
σ_u is the ultimate capacity of the porous medium to hold solids (kg solids/m^3 filter bed)

The importance of the integral term depends upon the ratio of solids in the saturated zone to the solids associated with the clogging front. Therefore, the longer the saturated zone, the more accurate is the approximation of Equation 12.9. It is likely that traditional filter beds of only 76 cm (30 in.) deep would not have sufficient length of saturated zone that would permit accurate application of Equation 12.9, but it should work well for longer columns of mono-media. The filter bed should be deep enough to permit a significant length of "saturated zone" to develop.

12.3.2.8 Local Hydraulic Gradient, *i*(*Z*, *t*)

The "local" hydraulic gradient, that is, $i(Z, t)$, reflects the magnitude of the "local" specific deposit, that is, $\sigma(Z, t)$, which "clogs" the pores causing higher velocities and therefore increased hydraulic gradient. The "total" headloss, that is, the headloss between the top and the bottom of the filter, is the measure of practical interest. How this total headloss changes with time determines the "length-of-run" (the concentration breakthrough should occur first).

Figure 12.13a shows headloss versus Z at different times for the filter column used by Ives (1962); the slope of any of the curves at any point is the local hydraulic gradient, that is, $i(Z, t)$. As seen by comparing Figure 12.13a with Figure 12.10, the local hydraulic gradient mirrors the specific solids deposit at any (Z, t). The dotted diagonal line in Figure 12.13a is the "clogging front," defined as the point where the hydraulic gradient approximates that of the "clean-bed," that is, $i(Z, t = 0)$. The slopes of the curves at $Z \geq Z$(clogging front) are parallel to the clean-bed headloss curve.

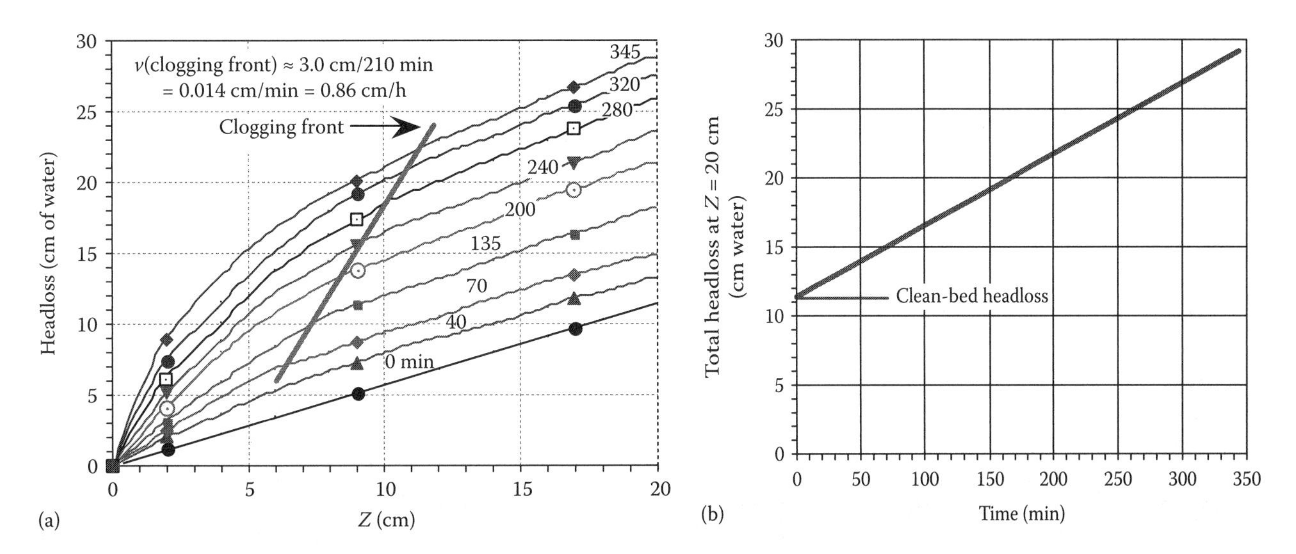

FIGURE 12.13 Measured data from experiments with radioactive algae. (a) $h_L(Z)_t$ plot. (b) $h_L(t)_{Z=20\text{cm}}$ plot. (Adapted from Ives, K.J., *Trans. ASCE*, 127(Part III), 384, 1962.)

The rate of progress of the clogging front (the slope of the diagonal line) is about the same as the velocity of the wave front, that is, 0.84 cm/h for the clogging front and 0.90 cm/h for the wave front, as seen by the results of Ives, Figures 12.15 and 12.9, respectively. The velocity of the clogging front (or wave front), v_{wf}, times the desired length of run, t(run time), gives the length of the saturated zone at run termination, that is, (L(sat zone) $= v_{wf} \cdot t$(run time); adding the length of the wave front, L_{wf}, gives the length of column. In other words, Equation 12.10 is demonstrated further.

Figure 12.13b illustrates a linear increase in *total* headloss versus time at depth, $Z = 20$ cm. The question is of interest in practice and the linear increase with time is confirmed by the data of Ives. The plot data were obtained from Figure 12.13a.

Figure 12.14a shows plot of $h_L(Z)_t$ as measured by Adin et al. (1979); the rate of total headloss increase is linear (9.0 cm/h), confirming further the linear characteristic of the h_L(total) versus time function. Figure 12.14b was derived from Figure 12.14a by plotting the Z position of the clogging front versus time. Its slope is the velocity of the clogging front, which is the same as the velocity of the wave front, for example, $v_{wf} \approx 2.8$ cm/h. This approach to evaluate v_{wf} is feasible by instrumentation of a pilot plant with piezometers spaced at intervals at say 100 mm. Knowing v_{wf} and L_{wf} permits calculation of the length of run.

12.3.2.9 Rational Design

A derivative of the foregoing discussion is that with a means to determine v_{wf}, then only L_{wf} needs to be estimated in order to design a filter column. The L_{wf} term may be estimated by side port sampling of a pilot plant filter column after enough time has elapsed for a steady state wave front shape to develop, or from the "breakthrough" curve (as outlined in Section 12.3.2.5). A definition for L_{wf} is arbitrary but may be defined as the distance, $0.05 \leq C/C_o \leq 0.95$. Examples of L_{wf} from Figures 12.7 and 12.10 are 16 and 50 cm, respectively.

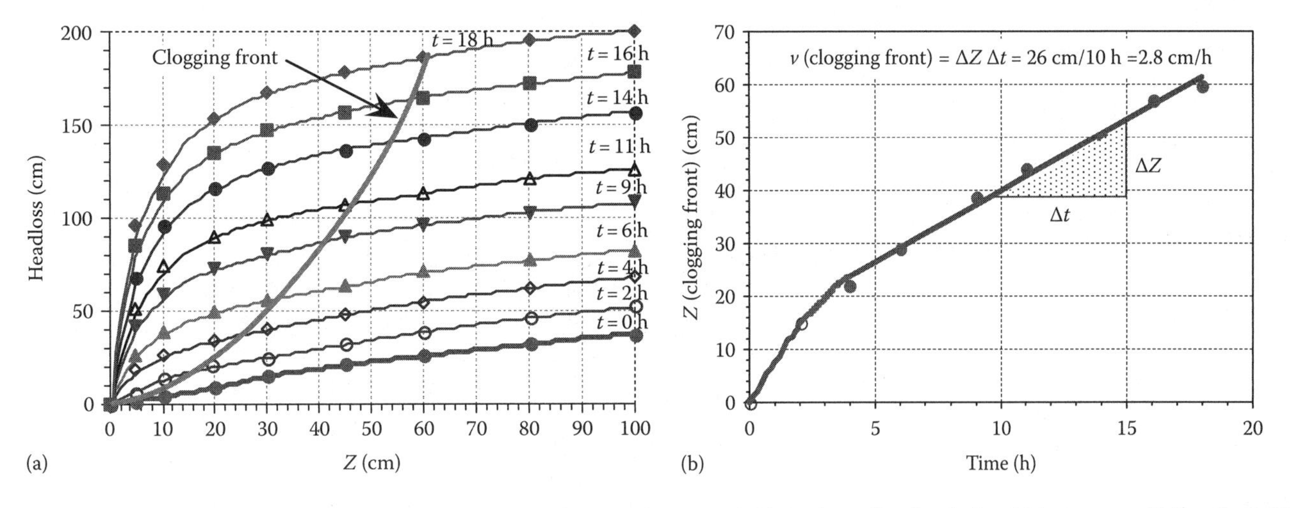

FIGURE 12.14 Measured data from experiments. Clean-bed headloss: $i = 0.31$ cm/cm; d(grain size) $= 1.21$ mm, $v = 20.0$ m/h (8.20 gpm/ft^2). The rate of total headloss increase is 9.0 cm/h. (a) $h_L(Z)_t$ plot showing clogging front. (b) Movement of clogging front with time. (From Adin, A. et al., *J. Am. Water Works Assoc.*, 71(1), 20, 1979.)

A rational design of a filter may be based on the length of run desired times the velocity of the wave front plus the length of the wave front. The corresponding equation is

$$L(\text{column}) = v_{wf} \cdot t(\text{run time}) + L_{wf} \tag{12.10}$$

where

L(column) is the length of filter column (m)
v_{wf} is the velocity of wave front (m/s)
t(run time) is the elapsed time from start of run to end of run (s)
L_w is the length of wave front (m)

Example 12.1 Calculation of Depth of Filter Bed

Problem

Calculate the length of filter column, L(filter bed), required based on the experiments of Ives (1962) for a run length, $t(\text{run}) = 30$ h.

Solution

1. Data for Equation 12.10.
   From Figure 12.7, $v_{wf} = 0.90$ cm/h and $L_{wf} \approx$ 16 cm.
2. Apply Equation 12.10,

$$L(\text{column}) \approx v_{wf} \cdot t(\text{run time}) + L_{wf}$$
$$= 0.90\,\text{cm/h} \cdot 30\,\text{h} + 16\,\text{cm} = 43\,\text{cm}$$

Discussion

Obtaining v_{wf} and L_{wf} must be based on a pilot plant study. Methods to determine v_{wf} may be based on $C(Z)t$, or on $\sigma(Z)t$. The latter is the only practical approach and can be determined by fitting the column with piezometers and plotting headloss profiles at different times as in Figures 12.13a and 12.16b. The magnitude of v_{wf} will vary with v, C_o and σ_u; the latter depends on the media size, for example, d_{10} and UF. The length of the wave front, L_{wf}, must be defined, for example, $0.05 \le C/C_o \le 0.95$, is a reasonable definition. To measure the $C(Z)_t$ curve is a means to obtain an approximate determination of L_{wf}. To measure the breakthrough curve with online instruments, either turbidity or particle counting, gives the $C(t)_{Z=Z_o}$ curve. From this curve

$$\Delta t\left(0.05 \le \frac{C}{C_o} \le 0.95\right) \cdot v_{wf} \approx L_{wf}\left(0.05 \le \frac{C}{C_o} \le 0.95\right).$$

12.3.2.10 Total Headloss and Components of Headloss

The three headloss components, illustrated in Figure 12.15, are

1. Clean-bed headloss
2. Clogging headloss
3. Straining headloss

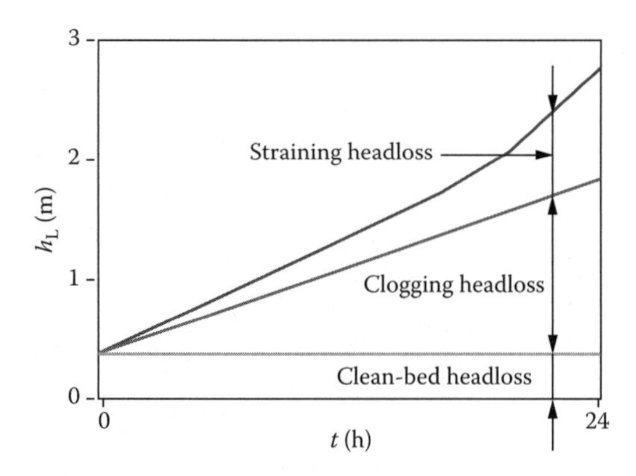

FIGURE 12.15 Headloss development with time in filter bed, showing the three components. (Adapted from Ives, K.J. Mathematical models of deep bed filtration, in Ives, K.J. (Ed.), *The Scientific Basis of Filtration*, Proceedings of NATO Advanced Study Institute, Cambridge, U.K., Noordhoff International Publishing, The Netherlands, 220, 1975c. With permission.)

As seen in Figure 12.15, the clean-bed headloss is the initial headloss, that is, at $t = 0$, which does not change with time. The clogging headloss causes a linear headloss increase with time, and is the middle curve. Finally, the straining headloss increases exponentially with time, and is the upper line. A goal in filtration is that the straining headloss, due to its exponential increase with time, should be negligible. The straining effect may be controlled by either increasing the media size or decreasing the floc size, or some combination. A pilot plant is the only practical means to assess such effects.

12.3.2.11 Characteristics of $C(t)_Z$ for a Filter Cycle

Figure 12.16 shows $C(t)_Z$ curves for effluent particles and turbidity for a pilot filter from the start of a filtration cycle. Three phases characteristic of any filtration cycle, identified in Figure 12.16, are

1. Chemical conditioning (also called "ripening")
2. Steady state
3. Breakthrough

Questions relate to

1. The factors affecting ripening and its duration.
2. The lowest turbidity and particle count numbers attainable for the "steady state" phase.
3. The time when the respective breakthrough curves cross the maximum particle count or maximum turbidity permitted, indicating the end of the run.
4. The total headloss versus time curve and the time when the maximum headloss permitted occurs.

As noted, the third and fourth questions determine the "length of run," based on either time to maximum permitted particle counts or turbidity or time to maximum headloss.

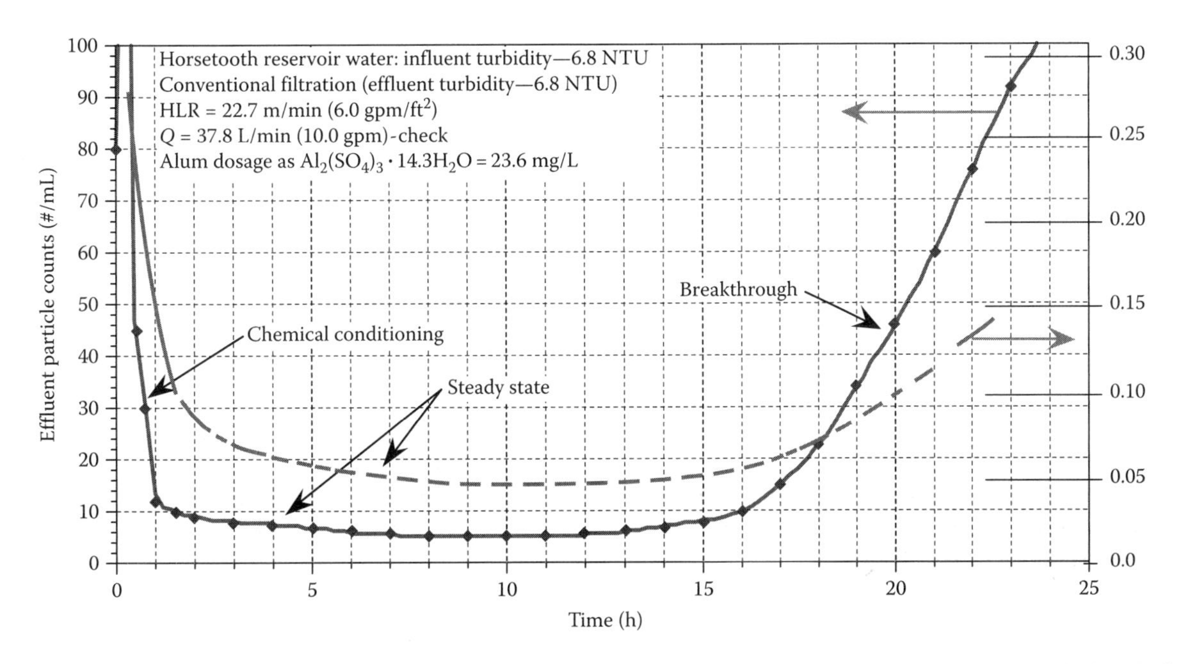

FIGURE 12.16 Effluent particle counts and turbidity from pilot filter. (Courtesy of Marinelli, F. and Carlson, K., Colorado State University, Fort Collins, CO, Run A3C, May 6, 1999.)

12.3.2.11.1 Chemical Conditioning (Ripening)

Filter "ripening" has two causes:

1. Hydraulic dispersion
2. The need for "chemical conditioning" of the filter medium

Hydraulic dispersion (see Sections 4.2.2.4 and 4.2.2.6) is the displacement of remnant water in the headwater and the pore water by the coagulated water. A tracer test, such as with a salt solution, can evaluate the dispersion effect. The ripening time cannot be less than the hydraulic dispersion time; 15 min is representative of the latter for the pore water (Mosher and Hendricks, 1986).

For most cases, the ripening period is perhaps 30–120 min. which includes dispersion. The reasoning is that the attractive force between floc particles and a bare filter grain is much less than that between two floc particles (see, e.g., Amirtharajah, 1985, for example, α(floc-grain) $\ll$ 1.0, whereas α(floc-floc · grain) $\rightarrow$ 1.0. Therefore, the media grains must be coated partially with floc particles for attachment to be effective; filtration is merely an extension of flocculation in this view (O'Melia, 1985).

12.3.2.11.2 Start of Filter Run Cycle

Figure 12.17 delineates further the phases of the filtration cycle and the effect of remnant water on the quality of filter effluent water. Remnant water is defined as backwash water that remains in the filter box after backwash and includes water in the under-drain system, pore water, and head water. The initial few minutes of the start-up (Amirtharajah and Wetstein, 1980, p. 518) is characterized by

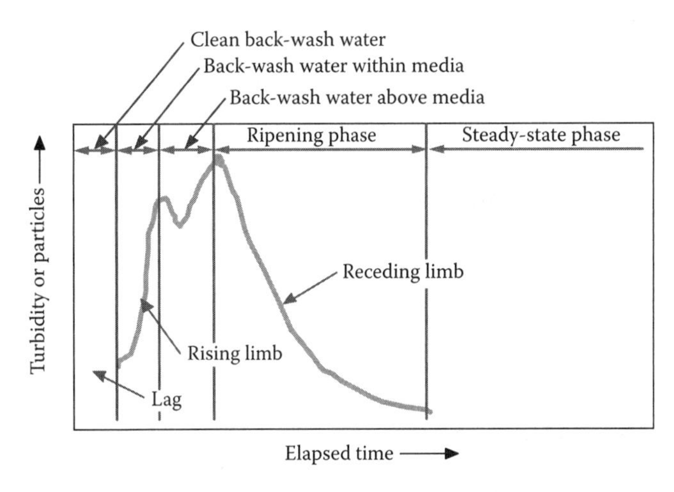

FIGURE 12.17 Detail of effluent quality changes during filter startup. (Adapted from Amirtharajah, A. and Wetstein, D.P., *J. Am. Water Works Assoc.*, 72(9), 519, 1980.)

1. A "lag" in which the water is of high quality, attributed to low turbidity remnant water in the under-drain system
2. A rising limb of increasing turbidity due first to remnant water in the filter pores giving a first peak and then to remnant water in the headwater (water above the filter bed) giving the second peak
3. A receding limb, that is, "ripening" (or chemical conditioning)

Usually the two peaks are seen within the first 5–15 min, but will vary depending upon the system design. The water at the highest level in the filter box has the most particulates and thus is responsible for the second peak

(Amirtharajah, 1985). Hydraulic dispersion also plays a role in each of these events.

12.3.3 Mathematical Modeling

The starting point for mathematical modeling was a paper by Iwasaki (1937) which described the basic equations of the filtration process. The work languished, however, until the 1950s when Ives commenced his research on filtration (see Section 12.3.2.2).

Two other papers were notable in setting direction. First, O'Melia and Stumm (1967) delineated fundamental factors affecting particle collector attachment, α, that is, in terms of double-layer interactions and van der Waals forces, and the role of associated chemical factors. Second, Yao et al. (1971) described the transport coefficient, η. These studies were the basis for more sophisticated mathematical models developed during the 1970s and 1980s, reviewed in this section.

12.3.3.1 Iwasaki's Equations

In 1936, Iwasaki published, "Some Notes on Sand Filtration," which proposed three equations to describe the removal of particles within a sand bed:

$$0 = -\bar{v}\frac{\partial C}{\partial Z} + \frac{1}{\varepsilon}\frac{\partial \sigma}{\partial t} \tag{12.11}$$

$$\frac{\partial C}{\partial Z} = -lC \tag{12.12}$$

and

$$\lambda = \lambda_0 + c\sigma \tag{12.13}$$

where
- C is the mass concentration of suspended material in water entering filter bed (kg suspended matter/m^3 water)
- Z is the depth from surface of filter (m)
- λ is the "local" filter coefficient at any depth and time (m^{-1})
- λ_0 is the filter coefficient for clean-bed (m^{-1})
- λ is the filter coefficient for clean-bed after deposits of solids (m^{-1})
- t is the elapsed time since start of filtration (s)
- σ is the mass of suspended material retained per unit of filter volume (kg suspended matter/m^3 filter bed)
- ε is the porosity of media, that is, volume pores/volume of filter bed (m^3 of void volume/m^3 of filter bed volume)
- c is the coefficient (m^2 filter bed/kg suspended matter)
- $\bar{v}$ is the interstitial filtration velocity, that is, $\bar{v} = Q/(A\varepsilon)$ (m/s)
- Q is the flow of water into the column (m^3/s)
- A is the cross sectional area of column (m^2)

The Iwasaki equations have been accepted almost universally as the starting point for modern theory. The equations were modified here in the units of the σ term to convert it to a mass concentration in lieu of volume concentration, the latter being the form given by Iwasaki.

1. *Materials balance*: Equation 12.11 states that the rate of accumulation of particles within an infinitesimal slice of the filter bed, at depth, Z, equals the net rate of advection of particles to and from the slice. As a note, the relationship is for a "steady state" condition, which means that the left side of Equation 12.11 is zero. The suspended solids concentration changes with time, however, at any given Z, as seen in Figures 12.8 through 12.10 and so a mass accumulation term, for example, $V\varepsilon[\partial C/\partial Z]_{observed}$, should have been included in the left side of Equation 12.11. This point is rectified in Equations 12.22 through 12.30, Section 12.3.3.6.
2. *Kinetics*: Equation 12.12 states that the rate of removal of particles, with respect to filter depth, $\partial C/\partial Z$, is proportional to the concentration, C. The coefficient, λ, is a measure of the probability of removal of a particle per unit of filter bed depth. The relation describes the beginning of the filtration cycle, that is, when the filter is "clean," but is not valid as the filtration cycle progresses due to "clogging" of pores with solids removed, which changes λ. The integrated form of Equation 12.12 is

$$\frac{C}{C_o} = e^{-lZ} \tag{12.14}$$

with the equivalent natural log form

$$\ln\left(\frac{C}{C_o}\right) = -lZ \tag{12.15}$$

and the equivalent common log form

$$\log\left(\frac{C}{C_o}\right) = -\left(\frac{\lambda}{2.3}\right)\cdot Z \tag{12.16}$$

Figure 12.18 illustrates the forms of Equations 12.14 and 12.16, respectively. For reference, Figure 12.18a shows a hypothetical filter column. Examining first Equation 12.14, its form is an exponential decline with distance, Z, seen in Figure 12.18b for time, $t = 0$. Its equivalent logarithmic form, that is, Equation 12.16, is a straight line as seen in Figure 12.18c. The filter coefficient is the slope in natural log (i.e., "ln") cycles per unit of length. To illustrate the mathematics, suppose $C(Z = 0) = 10{,}000$ particles/mL. At the distance for 1 ln cycle decline, the concentration is 1000 particles/mL, then after another such distance, the concentration is 100 particles/mL, and so on. The slope of a common log plot is $\lambda/2.3$, that is, 1 log cycle/m = (λ/2.3); therefore, $l = 2.3 \cdot$ (1 log cycle/ Z m); or, $Z = (2.3/1$ log cycle)/(l log cycle/m), which is the distance for 1 log-cycle of decline in concentration. The coefficient, l, on the other hand is the number of cycles of concentration decline per unit of

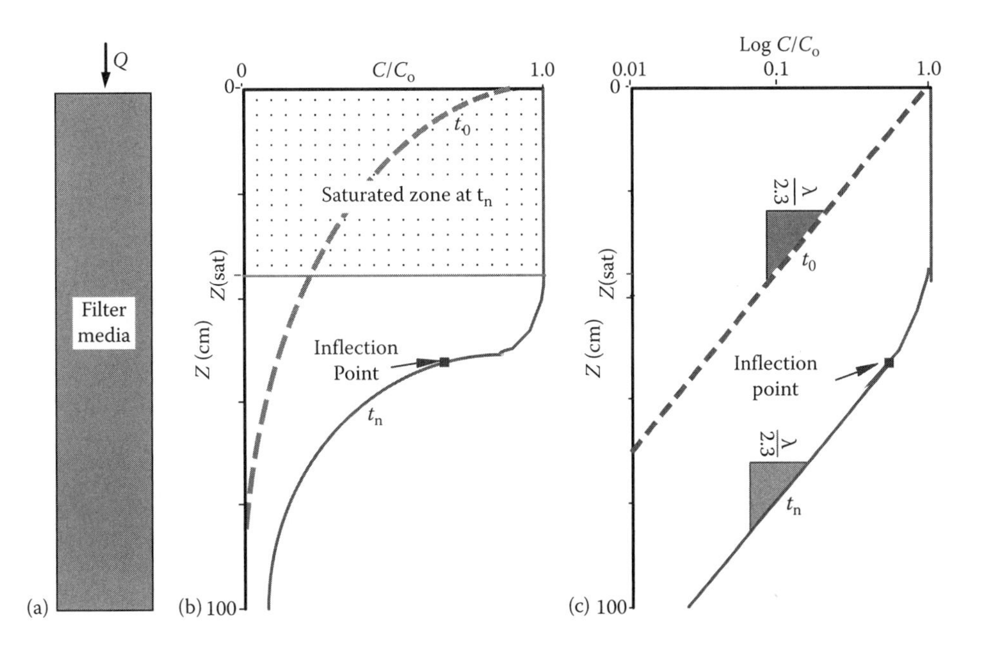

FIGURE 12.18 Iwasaki's kinetic equation illustrated. (a) Column. (b) $C(Z)_t$ plots, arithmetic scale. (c) $C(Z)_t$ plots, semi-log scale.

depth of the filter bed, for example, ln cycles per meter. Note that this discussion belabors the point a bit, especially to one adept in basic mathematics.

3. *Filter coefficient*: For reference, the magnitudes of the clean-bed filter coefficient, λ_o, from several sources are compiled in Table 12.2. As seen, λ_o varies with the kinds of particles entering the filter bed, the kind and size of filter media, and the filtration velocity. Values range $0.2 \leq l_o \leq 20$ ln-cycles/m (or $0.087 \leq l_o/2.3 \leq 8.7$ log-cycles/m); for reference, Appendix C.3.3 reviews the natural-ln to common-log conversion. To summarize, the steeper the $\log(C/C_o)$ versus Z curve, the higher the value of $l_o/2.3$ and the more effective the filtration process. The trends of $\lambda_o/2.3$ with particle size, and filtration velocity are seen in the polystyrene data,

TABLE 12.2
Filter Coefficients as Affected by Variables

	Particles		Filter Medium		v		
Source[a]	Material	Size (μm)	Material	Size (mm)	(mm/s)	(gpm/ft²)	λ_o (m^{-1})
Maroudas	Polystyrene-angular	65	Glass spheres	2.00	6.30	7.64	2.6
					25.00	30.34	0.7
					88.50	107.39	0.2
		90			25.00	30.34	1.2
					100.00	121.34	0.4
		125			22.80	27.67	2.1
					83.50	101.32	1.3
Ives	Algae	5	Sand	0.54	1.40	1.70	14.0
				0.70	1.40	1.70	14.7
Eliassen	Ferric hydroxide	6–20	Sand	0.46	1.40	1.70	20.0
Fox/Cleasby	Ferric hydroxide	4–25	Sand	0.70	2.70	3.28	10.0
					1.40	1.70	0.6
Khan[b]	Aluminum hydroxide	633	Anthracite	$d_{10} = 1.10$	4.12	5.00	4.8
		597		UC = 1.6	4.12	5.00	4.1
		399			4.12	5.00	3.4

Source: Adapted from Herzig, J.P. et al., *Ind. Eng. Chem.*, 62(5), 22, 1970.
[a] Sources, except for Khan, are listed by Herzig et al. (1970).
[b] Khan (1993).

whereas the hydroxide floc data with sand and anthracite are more useful as reference for practical application. Note that Equation 12.13, that is, $l = l_o + cs$, is included first for historical reasons since it was included in Iwasaki's paper, and second it acknowledges that the filter coefficient changes as the media becomes "clogged" with solids. The c coefficient is affected, however, by a number of variables and cannot be quantified easily.

12.3.3.2 Filter Coefficient

O'Melia and Stumm (1967) recognized that the filtration process has two steps: (1) *transport* (affected by physical factors) and (2) *attachment* (affected by chemical factors). A particle to be removed must reach a sand grain, that is, a "collector," and then it must *attach*. These steps explain then, in simple terms, the process of depth filtration. Later, Yao et al. (1971) disaggregated the Iwasaki filter coefficient (see Equation 12.14) mathematically in terms of these two steps, that is, transport and attachment, that is,

$$l = \left(\frac{3}{2}\right) \cdot \left[\frac{1-\varepsilon}{d}\right] \cdot \alpha\eta \qquad (12.17)$$

where

- ε is the filter bed porosity (m^3 of void volume/m^3 of filter bed volume)
- d is the grain diameter (m)
- α is the number of contacts which produce a particle collector adhesion divided by the number of particle collector collisions, called *attachment efficiency* (number of particle collector attachments/number of particle collector collisions)
- η is the rate at which particles strike a collector divided by the rate at which particles flow toward the collector, called the *transport efficiency* (particle collector collisions/particle flux associated with a given collector)

The first group of terms, that is, $(3/2) \cdot (1-\varepsilon)/d$, is the grain surface area per unit volume of filter bed, that is, the "specific surface" (Ives and Sholji, 1965, p. 3) with the 3/2 referring to a spherical particle shape. The transport and attachment coefficients, that is, α and η, respectively, are delineated further in the sections that follow.

12.3.3.3 Transport Coefficient

The transport step involves getting a coagulated particle to a *collector* (a term used often by theoreticians in referring to a grain of the filter medium). The three transport mechanisms are interception, diffusion, sedimentation (Yao et al., 1971). Figure 12.19 illustrates the path of a single particle for each mechanism. The diffusion and sedimentation mechanisms cause particles to cross streamlines and thus be transported to the proximity where attachment with a media grain could occur. With interception, the particle follows the streamline and may "brush" a collector (for streamlines that pass within one-half particle diameter distance from a collector). Two other mechanisms are inertia and shear, which are considered not important.

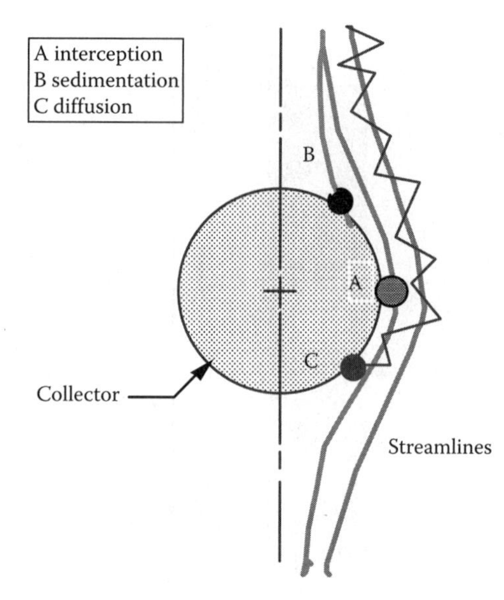

FIGURE 12.19 Transport mechanisms. (Adapted from Yao, K.M. et al., *Environ. Sci. Technol.*, 11(5), 1106, 1971.)

1. *Interception*: The transport of a coagulated particle to a spherical collector by advection along a streamline is interception. The particle at A in Figure 12.19 illustrates.
2. *Diffusion*: Random motion due to thermal energy is superimposed upon the advective motion within the filter media, as defined by a given streamline. The particle at C in Figure 12.19 illustrates.

   The contact frequency between particles and collectors depends upon the number of random "steps" per unit time, which is proportional to temperature (i.e., from molecular theory of gases, $1/2mv^2 = kT$, which is not perfectly transferred to particles in liquids). If N = number of steps/s, then the number of steps per unit length along a streamline is $N/\bar{v}$. Therefore, the lower the interstitial velocity, the more steps per unit distance, which in turn increases the probability of particle-filter grain contact by diffusion. Thermally induced random motion can be observed microscopically, for example, for *Staphylococcus aureus* bacteria, which is about 1 μm in size (Hendricks et al., 1970, p. 19).
3. *Sedimentation*: The third major transport mechanism, sedimentation, is described mathematically by Stoke's law. Adding the gravitational velocity vector (as defined by Stoke's law) to the advective velocity vector, which is tangent to a streamline at any given point, gives a resultant particle trajectory that incorporates the influence of gravity. The particle at point B in Figure 12.19 illustrates how the particle trajectory is modified from its advective path along a streamline to a path influenced by gravity.

12.3.3.3.1 Collisions within Depth of Filter Media

As is evident in Figure 12.19, some of the particles will have contact with a given media grain and some will not, depending upon the proximity of the particle to a grain and the magnitude of attachment forces. If a particle is not in a streamline proximity for making contact with one collector, the particle has another chance at each new collector level. The probability is quite high that a contact will be made at some level within the filter bed, depending upon velocity, grain diameter, and other variables involved in transport efficiency. At the same time, since we are dealing with a probability phenomenon, some particles, will escape collision with a collector and leave the filter. The number escaping depends on the filter coefficient and media depth. Habibian and O'Melia (1975) refer to this as "contact opportunities"; in other words, the larger values of η result in a higher rate of contact opportunities. If η is low, then a deeper filter bed can compensate to give the total number of contact opportunities needed to meet a specified filter effluent concentration.

Example 12.2 Calculation of Effluent Concentration from Iwasaki's Equation

Given

Suppose 5200 particles per mL enter a filter bed with length 1.00 m. The conditions are as given in Table 12.2, Eliassen's data with $\lambda = 0.20\ \text{cm}^{-1} = 20$ log cycles/m.

Required

Calculate the concentration of particles leaving the filter bed.

Solution

1. Apply Equation 12.16,

$$\log\left(\frac{C}{C_o}\right) = -\left(\frac{\lambda}{2.3}\right) \cdot Z$$

2. Substitute numerical data,

$$\begin{aligned}\log\left[\frac{C}{C_o}\right] &= -\left(\frac{\lambda}{2.3}\right)Z \\ &= -(20 \ln \text{cycles/m}) \cdot (1/2.3) \cdot 1.00\,\text{m} \\ &= -8.7 \text{ log cycles}\end{aligned}$$

$$\begin{aligned}\frac{C}{C_o} &= 10^{-8.7} \\ C &= 5200 \text{ particles/mL} \cdot 10^{-8.7} \\ &= 0.000\ 010\ 4 \text{ particles/mL}\end{aligned}$$

Discussion

Suppose the coefficient from Khan (1993, as given in Table 12.2) is used, that is, $\lambda = 0.048\ \text{cm}^{-1} \approx 4.8$ ln cycles/m. Then,

$$\begin{aligned}\log\left[\frac{C}{C_o}\right] \log\left[\frac{C}{C_o}\right] &= -\left(\frac{\lambda}{2.3}\right)Z \\ &= -(4.8 \ln \text{cycles/m}) \cdot (1/2.3) \cdot 1.00\,\text{m} \\ &= -2.1 \text{ log cycles}\end{aligned}$$

$$\begin{aligned}\frac{C}{C_o} &= 10^{-2.1} \\ C &= 5200 \text{ particles/mL} \cdot 10^{-2.1} \\ &= 41 \text{ particles/mL}\end{aligned}$$

The difference between the two filter coefficients, that is, $\lambda = 20$ and $4.8\ \text{m}^{-1}$, respectively, results in a difference of 10^7 in order of magnitude between the effluent concentrations.

12.3.3.3.2 Particle Transport Equations

The respective influences of interception, diffusion, and gravity, are given in terms of the component coefficients, η_I, η_D, and η_G, defined mathematically by Yao et al. (1971), as

$$\eta_I = \frac{3}{2}\left(\frac{d_p}{d_c}\right)^2 \quad (12.18)$$

$$\eta_D = 0.9\left(\frac{kT}{\mu d_p d_c v_o}\right)^{2/3} \quad (12.19)$$

$$\eta_G = \frac{\left[(\rho_P - \rho_w) g d_P^2\right]}{18 \mu v_o} \quad (12.20)$$

where

- η_I is the collision frequency coefficient due to interception
- d_P is the diameter of particle (m)
- d_c is the diameter of collector; same as grain diameter, d (m)
- η_D is the collision frequency coefficient due to diffusion
- k is the Boltzmann constant ($1.38 \cdot 10^{-23}$ J/K/molecule)
- T is the absolute temperature (K)
- v_o is the interstitial velocity of water (m/s)
- μ is the dynamic viscosity (Newton $\cdot$ s/m^2)
- η_G is the collision frequency coefficient due to sedimentation
- ρ_P is the density of suspended particle (kg/m^3)
- ρ_w is the density of water (kg/m^3)
- g is the acceleration of gravity (m/s^2)

The overall coefficient, η, is the sum of the three components,

$$\eta = \eta_I + \eta_D + \eta_G \quad (12.21)$$

where η is the overall transport coefficient, defined as ratio of particles striking a collector to the particle flux approaching (dimensionless).

The above equations, that is, 12.18 through 12.21 identify the independent variables that affect particle-collector contacts and permit calculation of η.

Table CD12.3 is a spreadsheet that calculates the transport coefficients η_I, η_D, η_G, and η as defined in Equations 12.18

TABLE CD12.3
Excerpt—Calculation of Transport Coefficients and Filter Coefficients by Excel Spreadsheet

Fixed Data

Viscosity formula:

$$m = 0.001787 - 5.61324 \cdot 10^{-5} \cdot T + 1.0031 \cdot 10^{-6} \cdot T^2 - 7.541 \cdot 10^{-9} \cdot T^3$$

Boltzmann $= 1.39E{-}23$ kg · m²/s²

Variable	Value	Units	Range
$e =$	0.35		0.35–0.55
$d =$	0.45	mm	0.30–0.5
$d_p =$	0.1	μ	0.01–500
$T =$	25	°C	0–30
$v_o =$	5	gpm/ft²	0.5–15
$r =$	1.05	g/cm³	1.01–1.20

Data					Diffusion					
e	*d* (mm)	*d* (m)	*a*	3(1 − e)/2*d* (m⁻¹)	*k* (kg · m²/s²)	*T* (°C)	*T*(abs) (°K)	μ kg/(m s)	d_p (m)	HLR (gpm/ft²)
0.35	0.45	4.50E−04	1.00	2.17E + 03	1.39E−23	25	298	8.93E−04	1.00E−07	5.0
0.35	0.45	4.50E−04	1.00	2.17E + 03	1.39E−23	25	298	8.93E−04	3.00E−07	5.0
0.35	0.45	4.50E−04	1.00	2.17E + 03	1.39E−23	25	298	8.93E−04	5.00E−07	5.0
0.35	0.45	4.50E−04	1.00	2.17E + 03	1.39E−23	25	298	8.93E−04	7.00E−07	5.0
0.35	0.45	4.50E−04	1.00	2.17E + 03	1.39E−23	25	298	8.93E−04	1.00E−06	5.0

through 12.21, respectively, as a function of particle diameter and showing also the assumed values for other variables (which may be changed to explore their respective effects). In addition, the filter coefficient, λ, is calculated from Equation 12.17. From the calculated values in Table CD12.3, Figure 12.20a depicts graphically the relationships for η_I, η_D, η_G, and η as function of d_p. Figure 12.20b shows how the filter coefficient, λ, varies with interstitial filtration velocity, v_o, for two filter grain diameters, *d*. As seen, the transport coefficient is a minimum at $d_p \approx 1$ μm. The filter coefficient is seen to decline with increasing v_o. From these plots one can understand better the trade-offs between the effects of operative variables of filtration. For example, a floc size of perhaps 10 μm would provide a high enough transport coefficient (and filter coefficient), but probably not so high that clogging would be a problem. As to the effects of the filter coefficient, Figure 12.20b would indicate that a larger filter grain diameter would result in a smaller value of λ, and thus a longer filter column is necessary to achieve the same removal efficiency. At the same time, the run length would be longer simply because the filter bed pores are larger and can store more floc.

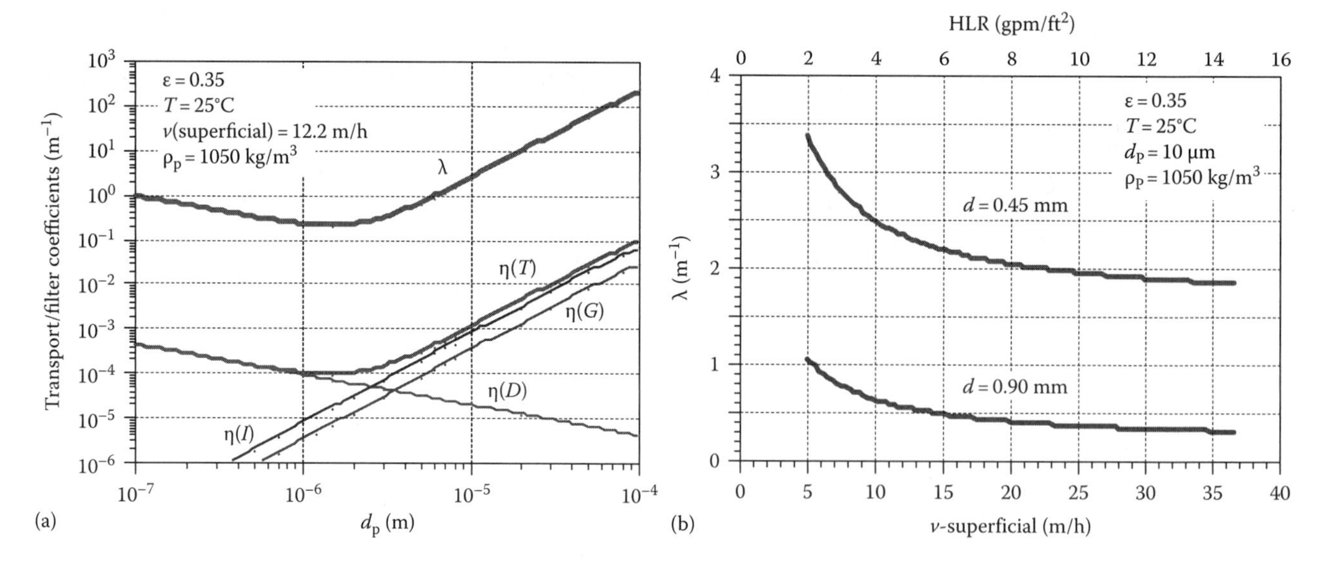

FIGURE 12.20 Calculations from spreadsheet (Table CD12.3) for exploring effects of independent variables on transport coefficients and filter coefficient. (a) Filter coefficient and transport coefficients affected by particle diameter. (b) Filter coefficient as a function of filtration velocity.

Example 12.3 Set Up a Spreadsheet to Calculate η and λ and the Associated C/C_o

Problem description

The transport coefficient has three components, that is, interception, diffusion, and sedimentation, each affected by independent variables, as given in Equations 12.18 through 12.20. The effect of each independent variable on the component transport coefficients and the total transport coefficient can be seen most easily by setting up the equations in a spreadsheet with a linked plot to the results. The effect of particle size is demonstrated in this example.

Given

Assume default values for the independent variables of Equations 12.18, 12.21, and 12.20 as given in Table CD12.3.

Required

A spreadsheet is needed in order to do a sensitivity analysis to estimate the effect of any independent variable.

Solution

Table CD12.3 shows a spreadsheet formatted to calculate dependent variables with associated plots, as seen in Figure 12.20a and b; similar figures are embedded in the spreadsheet, but without numbers. Change selected variables, such as particle size to estimate the effect on λ. Particle size is a key variable and can be explored with all other variables held constant.

Discussion

The Excel® spreadsheet is set up to explore the effect of any of the independent variables on η and λ. In the case shown, d_p is changed over the range that may be found in practice and the effect on the dependent variables, η and λ can be seen in the linked plot. The effect of any of the other variables may be explored in the same fashion. The plot shows a minimum value in the transport coefficient, η (and at the same time the filter coefficient, λ), at $d_p \approx 1$–2 μm, which seems to be a consensus among theoreticians (see Logan et al., 1995). This conclusion was confirmed experimentally by Habibian and O'Melia (1975, p. 578) who found $C/C_o \approx 0.25$, 0.18, and 0.03 for $d_p = 1.0$, 0.1, 7.6 μm, respectively, in filtration with optimum concentrations of cationic polymer.

12.3.3.4 Attachment Coefficient

The second part of the filtration process is attachment. Once a particle collector contact occurs, the particle will either attach or not attach. The ratio of particles attaching to the number the particle collector contacts is called the attachment coefficient, α. In general, whether attachment occurs depends upon how effectively the particle was *charge-neutralized* by coagulation chemicals to reduce its zeta potential. The goal of coagulation is that $\alpha \to 1.0$ for the preponderance of particles in the water being treated.

12.3.3.4.1 Collector Zeta Potential

Ives and others have reported that sand and anthracite have zeta potentials of about −20 mV. Examples of other zeta potentials for different materials and pH levels are shown in Table 12.4.

TABLE 12.4
Zeta Potentials for Filter Media

Material	pH	Zeta Potential (mV)
Ottawa sand (sieved and rinsed)	4.0	−30
	5.5	−68
	7.0	−100
Ottawa sand—coated with Al(OH)3	5.5	+80
	8.8	−30
Glass beads—washed (chromic acid/acetone/dist. water)	5.5	−140

Source: Adapted from Truesdale et al., *J. Environ. Eng. Div., ASCE*, 124(12), 1220, 1998.

12.3.3.4.2 Factors Affecting Attachment

O'Melia and Stumm (1967) stated that particle adhesion to sand grains is promoted at low pH (resulting in higher zeta potentials as seen in Table 12.4 for Ottawa sand). Attachment to previously deposited particles is highest at the iso electric point (the iso electric point is the pH at which the zeta potential is zero and by interpolation is at pH = 7.9 for the data of Table 12.4). O'Melia and Ali (1978) suggested further that filtration in "ripened" filter media is merely an extension of flocculation in that floc particles attach to floc-coated collectors.

12.3.3.4.3 Forces in Attachment

O'Melia and Stumm (1967) proposed that the forces between a suspended particle and a filter grain were the sum of the van der Waals attractive forces and the coulombic repulsion. As in coagulation, suppression of the double layer by positive ions reduces the energy barrier of the electrostatic repulsive field, for both the particle and the collector, and allows the van der Waals attractive force (which is not affected by chemical factors) to become predominant, resulting in attachment, illustrated experimentally in Figure 12.21. As shown, as the $[Ca^{2+}]$ concentration increases, a increases; finally, when $[Ca^{2+}] =$ 0.1 mol/L, then a $\to$ 1.0. This confirms the role of chemical influence on a, which was attributed to double layer suppression. The values of z for the 4 μm latex particles change from $z \approx -70$ mV at $[Ca^{2+}] \approx 0$ mol/L, to $z \approx -8$ mV at $[Ca^{2+}]$ ≈ 0.1 mol/L. For the glass beads, the values of z change from $z \approx -40$ mV at $[Ca^{2+}] \approx -0$ mol/L (pH ≈ 7.0, $[Na^+] \approx 0.01$ mol/L), to $z \approx 0$ mV at $[Ca^{2+}] \approx 0.1$ mol/L.

12.3.3.5 Effect of Attachment Efficiency on Filter Ripening

Tobiason and O'Melia (1988) showed the effect of the attachment efficiency, α, on the filter ripening phase of the $C(t)_Z$ curve. To help explain filter ripening, they introduced another term, α_p, which is the attachment efficiency between particles and particles attached to collectors, with α being the attachment efficiency between particles and the collector. In

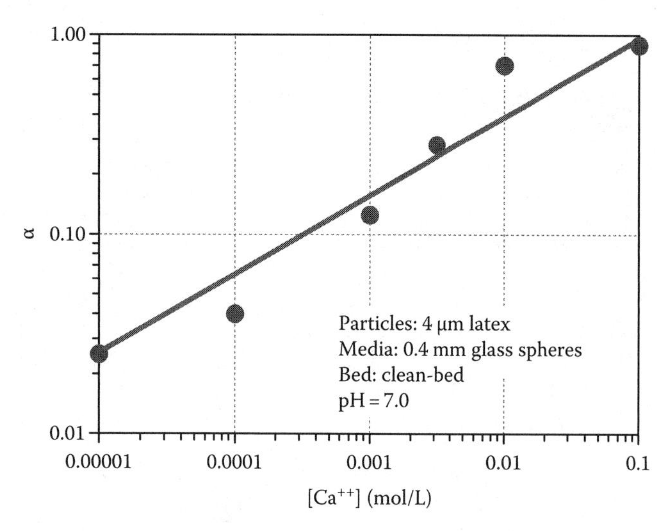

FIGURE 12.21 Attachment efficiency versus ion concentration. (Adapted from Tobiason, J.E. and O'Melia, C.R., *J. Am. Water Works Assoc.*, 80(12), 61, 1988.)

ripening, if $\alpha \ll 1$ and $\alpha_p = 1$, the C versus Z profile declines at a rate that reflects the exact value of a; the lower the value of a, the longer the time for the "bare" collectors to be covered with particles. As this coverage occurs, however, $\alpha_p \approx 1$ and the curve steepens.

12.3.3.6 Derivation of Materials Balance Expression

About all mathematical models of the filtration process start with the materials balance principle (see Section 4.3.3.1). As stated previously in Section 4.3.3.3, the principle applies to a homogeneous volume. For a "packed bed" reactor, in which a filter column is (as is a granular activated carbon column or an ion exchange column), only an infinitesimal element is homogeneous (since concentration changes with depth). Such an element is illustrated in Figure 12.22. Therefore, the resulting expression of the materials balance equation is in differential form.

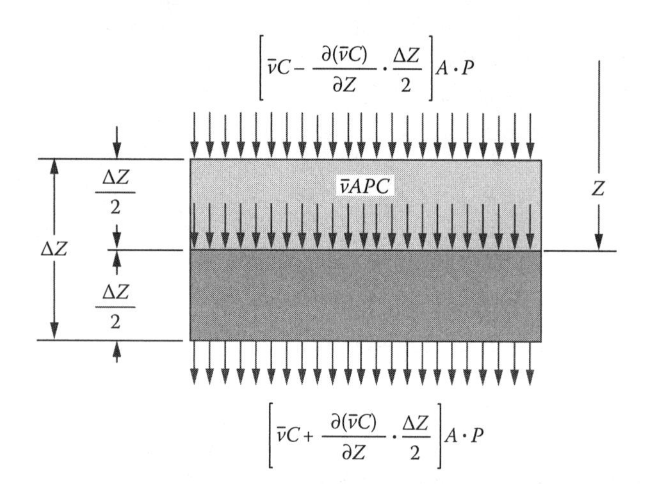

FIGURE 12.22 Materials balance for slice.

Consider in Figure 12.22 the filter bed element of area A and thickness ΔZ. From this element, a mathematical statement of the materials balance principle is

$$\left[\frac{\partial C}{\partial t}\right]_o \Delta Z A\varepsilon = [\bar{v}C_{in}A\varepsilon - \bar{v}C_{out}A\varepsilon] + [\mathbf{j}_{in}A\varepsilon - \mathbf{j}_{out}A\varepsilon] + \left[\frac{\partial C}{\partial t}\right]_r \Delta Z A\varepsilon \quad (12.22)$$

where

$[\partial C/\partial t]_o$ is the rate of change of suspended floc concentration as observed in the infinitesimal volume element (kg suspended floc in element/m^3 volume)
$[\partial C/\partial t]_r$ is the rate of change of suspended floc concentration in pore water of the volume element due to deposit on filter media (kg suspended floc in element/m^3 volume of element)
$\mathbf{j}_{in}$ is the dispersion flux density to infinitesimal volume element (kg suspended floc/m^2/s)
$\mathbf{j}_{out}$ is the dispersion flux density out from infinitesimal volume element (kg suspended floc/m^2/s)
A is the cross section area of column (m^2)
$\bar{v}$ is the interstitial flow velocity, that is, Q/Ae (m/s)

In words, Equation 12.22 says merely that the

[observed rate of change of suspended floc within the infinitesimal element pore volume]
= [advection flux of floc in − advection flux of floc out]
+ [dispersion flux of floc in − dispersion flux of floc out]
+ [rate of change of floc deposit on filter media] (12.23)

The dispersion flux is neglected, as a rule, for packed bed reactors, since its magnitude is not large for such cases. In examining the remaining individual terms, consider first the advection flux. The advection fluxes both in and out are as depicted in Figure 12.22 and expressed, respectively, as

advection flux in, advection flux out,

$$\bar{v}C_{in}A\varepsilon = \bar{v}CA\varepsilon - \frac{\partial C}{\partial Z}\frac{\Delta Z}{2}A\varepsilon \quad \bar{v}C_{out}A\varepsilon = \bar{v}CA\varepsilon + \frac{\partial C}{\partial Z}\frac{\Delta Z}{2}A\varepsilon \quad (12.24)$$

Substituting 12.24 in 12.22,

$$\left[\frac{\partial C}{\partial t}\right]_o \Delta Z A\varepsilon = \left[\bar{v}CA\varepsilon - \bar{v}\frac{\partial C}{\partial Z}\frac{\Delta Z}{2}A\varepsilon\right]\left[\bar{v}CA\varepsilon + \bar{v}\frac{\partial C}{\partial Z}\frac{\Delta Z}{2}A\varepsilon\right] + [\mathbf{j}_{in}A\varepsilon - \mathbf{j}_{out}A\varepsilon] + \left[\frac{\partial C}{\partial t}\right]_r \Delta Z A\varepsilon \quad (12.25)$$

Simplifying, the $\bar{v}\cdot CAe$ terms drop out, the $\Delta ZA\varepsilon$ terms cancel, and dispersion is neglected:

$$\left[\frac{\partial C}{\partial t}\right]_{\text{o}} = -\bar{v}\frac{\partial C}{\partial Z} + \left[\frac{\partial C}{\partial t}\right]_{\text{r}} \tag{12.26}$$

Equation 12.26 describes what occurs in an infinitesimal slice of the reactor, that is, the "the observed rate of change of suspension in the slice" equals the "net rate of advection" to and from the slice plus the "rate of concentration change in the slice due to uptake to the solid phase."

12.3.3.6.1 Kinetics

The "reaction" term in Equation 12.26 represents the rate of depletion of solids from the suspension that is deposited on the filter media, that is,

$$-\left[\frac{\partial C}{\partial t}\right]_{\text{r}}\Delta ZA\varepsilon = \left[\frac{\partial \sigma}{\partial t}\right]\Delta ZA \tag{12.27}$$

where $[\partial s/\partial t]$ is the rate of increase of solids deposit on filter media (kg suspended solids deposited/m^3 of bed volume).

Note that in Equation 12.27, the left side is in terms of pore water concentration which requires multiplication by ε, while the right side is in terms of solids concentration for the filter as a whole. Equation 12.27 simplifies to

$$-\left[\frac{\partial C}{\partial t}\right]_{\text{r}}\varepsilon = \left[\frac{\partial \sigma}{\partial t}\right] \tag{12.28}$$

Substituting Equation 12.28 in Equation 12.26 gives

$$\left[\frac{\partial C}{\partial t}\right]_{\text{o}} = -\bar{v}\frac{\partial C}{\partial Z} + \frac{1}{\varepsilon}\left[\frac{\partial \sigma}{\partial t}\right] \tag{12.29}$$

12.3.3.6.2 Discussion

The materials balance expression as given in Equation 12.29 is a common starting point for modeling of the filtration process. The equation says merely that the observed rate of change of suspended solids concentration within the pore volume of an infinitesimal column slice equals the net advection rate minus the uptake rate of solids by adsorption on collectors. Note that the gradient, $\partial C/\partial Z$ is usually negative, that is, concentration decreases as Z increases.

To relate back to the Iwasaki materials balance equation, Equation 12.11, assumes that the left side of Equation 12.29 is zero, that is, $[\partial C/\partial t]_{\text{o}} = 0$. Others, such as Ives have done this also. Later, Horner et al. (1986) called attention to the fact that the left side, that is, $[\partial C/\partial t]_{\text{o}}$, term had been neglected in filtration modeling over the decades since Iwasaki's work.

12.3.3.6.3 Finite Difference Form of Materials Balance Equation

Equation 12.29 cannot be solved analytically. A finite difference form is required, which can be obtained by replacing the infinitesimal designation (i.e., partial differential), ∂, with finite the symbol, Δ, that is,

$$\left[\frac{\Delta C_Z}{\Delta t}\right]_{\text{o}} = -\bar{v}\frac{\Delta C_t}{\Delta Z} + \frac{1}{\varepsilon}\left[\frac{\Delta \sigma}{\Delta t}\right] \tag{12.30}$$

which can be expressed as

$$\left[\frac{C_{t+\Delta t,Z} - C_{t,Z}}{\Delta t}\right]_{\text{o}} = -\bar{v}\frac{(C_{Z+\Delta Z,t} - C_{Z,t})}{\Delta Z}\frac{1}{\varepsilon}\left[\frac{\Delta \sigma}{\Delta t}\right]_{Z,t} \tag{12.31}$$

where

- $C_{t+\Delta t,Z} - C_{t,Z} = \Delta C_Z$, the change in concentration of the interstitial suspension between time t and time $t + \Delta t$, at a given Z (kg solids in suspension/m^3 suspension)
- $C_{Z+\Delta Z,t} - C_{Z,t} = \Delta C_t$, the change in concentration of the interstitial suspension between slice Z and slice $Z + \Delta Z$, at a given time, t (kg solids in suspension/m^3 suspension)

A second equation is required for the term representing solids uptake rate. Adin and Rebhun (1977) developed such an equation (reviewed in the next section).

12.3.4 Synthesis of a Model

Modeling has, in general, followed Ives approach which has been to consider the effect of solids deposits on the filter coefficient, $\ddot{\lambda}$ and to compute the entire curve from the Iwasaki kinetic relation, with the correction for λ. The approach given here is that of Adin and Rebhun (1977) which provides an expression of the solids uptake rate, which, in turn, is inserted into the materials balance relation, Equation 12.29.

12.3.4.1 Solids Uptake Rate

Adin and Rebhun (1977) have noted that the filtration process falls within a class of packed bed reactor problems (e.g., filters, granular activated carbon, ion exchange) involving materials balance and kinetics. They formulated a kinetics expression as a second order relation for uptake with a scour term for solids depletion as

$$\frac{\partial s}{\partial t} = [k_1\bar{v}C(F - s)] - [k_2 si] \tag{12.32}$$

where

- k_1 is the accumulation coefficient (m^2 water/kg suspended matter)
- k_2 is the detachment coefficient (m^3 bed volume/ kg suspended matter/s)

F is the theoretical filter capacity, or amount of retained material per unit of bed volume which could clog the pores completely (kg suspended matter/m^3 bed volume)
i is the hydraulic gradient (m headloss per m of filter bed)

The first term on the right side is an accumulation term and depends on the particle flux density, $\bar{v}C$, and on the available capacity for solids deposit at any instant, $(F - \sigma)$. The term is a "second-order" kinetic expression, that is, it is proportional to both the advective flux density and the capacity to hold floc, $(F - \sigma)$. The second term is the expression for the rate of detachment which is proportional to the concentration of solids previously attached, σ, and the hydraulic gradient, i, a surrogate for the shear. At some point, based on the pore space occupied by adsorbed floc, the rate of detachment due to shear equals the rate of accumulation (at a given Z and t). Observations with an endoscope by Ives (1989, p. 864) confirmed the detachment of particles, that is, as proposed by Professor D.M. Mintz (1966), with redeposit lower in the filter and confirmed also by Cleasby (1969); Tien and Payatakes (1979, p. 755); Saatci and Halilsoy (1987).

Returning to Equation 12.32, recall Darcy's law, that is, $\bar{v} = (-K\rho_w g/\varepsilon\mu)i$, keeping in mind that i is a negative quantity, and substitute for i, to give

$$\frac{\partial\sigma}{\partial t} = k_1\bar{v}C(F - \sigma) - k_2\sigma\left[\frac{\bar{v}\varepsilon\mu}{K\rho_w g}\right] \quad (12.33)$$

The hydraulic conductivity term is reduced from its clean-bed value, K_o, by the ratio of solids deposit, σ, to capacity, F (Adin and Rebhun, 1977),

$$K = K_o\left[1 - \left(\frac{\sigma}{F}\right)^{0.5}\right]^3 \quad (12.34)$$

where
K is the intrinsic permeability of porous media as clogged with solids (m^2)
ρ_w is the mass density of water at a given temperature (kg/m^3)
μ is the viscosity of water at a given temperature (N s/m^2)

Now substitute Equation 12.34 in Equation 12.33, to give, after grouping terms,

$$\frac{\partial\sigma}{\partial t} = k_1\bar{v}C(F - \sigma) - k_2\bar{v}\left(\frac{\varepsilon\mu}{\rho_w g}\right)\frac{\sigma}{K_o\left(1 - \sqrt{\sigma/F}\right)^3} \quad (12.35)$$

12.3.4.1.1 Finite Difference Form of Solids Uptake Rate

Equation 12.35 applies to a particular depth, Z, and time, t. Applying these subscripts and expressing the left side as a finite difference,

$$\left[\frac{\Delta\sigma}{\Delta t}\right]_{Z,t} = k_1\bar{v}C_{Z,t}(F - \sigma_{Z,t}) - k_2\bar{v}\left(\frac{\varepsilon\mu}{\rho_w g}\right) \times \frac{\sigma_{Z,t}}{K_o\left(1 - \sqrt{\sigma_{Z,t}/F}\right)^3} \quad (12.36)$$

which becomes

$$\frac{\sigma_{Z,t+\Delta t} - \sigma_{Z,t}}{\Delta t} = k_1\bar{v}C_{Z,t}(F - \sigma_{Z,t}) - k_2\bar{v}\left(\frac{\varepsilon\mu}{\rho_w g}\right) \times \frac{\sigma_{Z,t}}{K_o\left(1 - \sqrt{\sigma_{Z,t}/F}\right)^3} \quad (12.37)$$

where $(\sigma_{t+\Delta t,Z} - \sigma_{t,Z})$ is the change in concentration of solids deposited on the collectors between time t and time $t + \Delta t$, at a given Z (kg solids in deposited/m^3 filter bed).

Equation 12.37 is the basis for calculating the $\sigma(Z, t)$, that is, the accumulation of solids deposits within the filter bed. Finally, Equation 12.37 when substituted in Equation 12.31 permits calculation of $C(Z, t)$.

12.3.4.1.2 Computational Protocol

Equation 12.37 may be used in conjunction with the materials balance equation, Equation 12.31, which permits calculation of $C_{t+\Delta t,Z}$. The first step is to calculate $C_{Z,t=0}$ by the Iwasaki relations. These values of C are the initial input values for all Z. The constants, k_1, k_2, F, and K_o are determined as outlined by Adin and Rebhun (1977). The second step is to calculate $\Delta\sigma/\Delta t$ as given by Equation 12.36, which then goes into Equation 12.31. At the same time, $\sigma_{t+\Delta t,Z}$ is calculated by Equation 12.37.

12.3.4.2 Conditions at Equilibrium

The equilibrium condition within the filter is defined as the zone where the rate of attachment is equal to the rate of detachment, thus, $\partial\sigma/\partial t = 0$; by definition, this is the "saturated zone," that is, $Z \leq Z(\text{sat})$. For such zone, $\partial C/\partial Z = 0$, $C = C_o$, $\sigma = \sigma_u$, and there is no net change in the suspension concentration, or $[\partial C/\partial t]_o = 0$ (see also Ives, 1982, p. 4). Therefore, for Equation 12.35, the left side equals zero, to give, after canceling $\bar{v}$s,

$$0 = k_1 C_o(F - \sigma_u) - k_2\left(\frac{\varepsilon\mu}{\rho_w g}\right)\frac{\sigma_u}{K_o\left(1 - \sqrt{\sigma_u/F}\right)^3} \quad (12.38)$$

where
C_o is the concentration of suspended solids entering the filter bed (kg suspended matter/m^3 water volume)
σ_u is the operational storage capacity for suspended solids within the pores of filter (kg suspended matter/m^3 bed volume)

In other words, at equilibrium, the rate of attachment equals the rate of detachment.

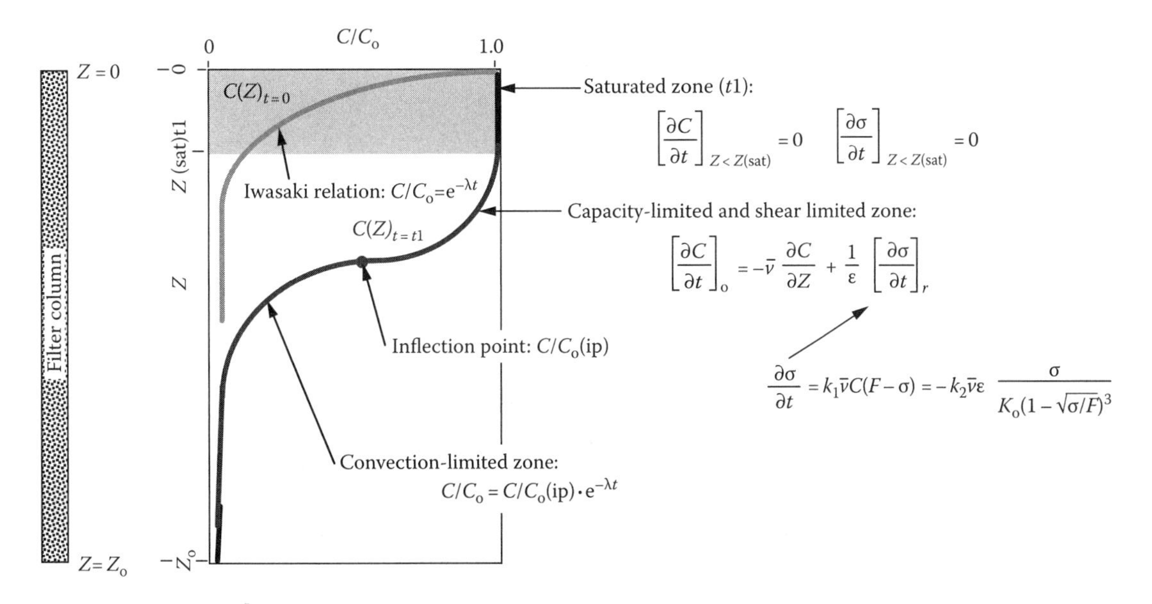

FIGURE 12.23 Zones of wave front.

12.3.4.3 Zones of Wave Front

Figure 12.23 shows the zones of the wave front as described by Equation 12.29 combined with Equation 12.35.

12.3.4.3.1 Saturated Zone

The saturated zone is seen as the region in which both $\partial C/\partial Z = 0$ and $\partial \sigma/\partial Z = 0$ and by the same token, $\partial C/\partial t = 0$ and $\partial \sigma/\partial t = 0$.

12.3.4.3.2 Uptake-Limited and Shear Zone

The uptake-limited and shear zone is between the saturated zone and the inflection point. In this zone, the specific deposit takes up void space and so there is less capacity to take additional deposit, which affects the rate of uptake. Also, the specific deposit, as it intrudes into the void space, causes higher interstitial velocities, which causes an increased rate of shear. Both of these effects are seen in the first and second parts of Equation 12.46, respectively.

12.3.4.3.3 Advection-Limited Zone

From the inflection point forward, that is, at all $Z \geq Z_{ip}$, the specific deposit is not sufficient to limit uptake of solids or to cause significant increase in shear. Therefore, the rate of uptake of solids is limited only by the rate of advection to a given slice. In this zone, since $\sigma \approx 0$, Equation 12.29 when combined with Equation 12.35 approaches the Iwasaki equation.

12.3.5 Summary

The Adin and Rebhun (1977) approach, as outlined above, is a coherent model in that it has rational components and it accounts for the three zones of the depth filtration process. By inclusion of the $[\partial C/\partial C]_{obs}$ term, the model can also account for the advance of the wave front with time. Figure 12.23 summarizes the key ideas of the Adin and Rebhun (1977) approach modified to include the $[\partial C/\partial C]_{obs}$ term. Inspection of each zone of the wave front and the associated equations helps to understand the depth filtration process.

12.4 DESIGN

Design has two parts: process design and design of subsystem support. In the process design the objectives are two: (1) to ensure an economical "length-of-run" and (2) to maximize the "net-water-production." The "length-of-run" is defined when the "breakthrough concentration," such as particle counts or turbidity, exceeds some criterion, or when terminal headloss occurs. Ideally, breakthrough occurs just before terminal headloss. The "net-water-production" is defined as the total water production per unit area of filter minus requirements for backwash, filter-to-waste, and other support functions and is related not only to length-of-run, but also to other factors, for example, superficial filtration velocity (the same as HLR).

To achieve a design that meets the process objectives, a pilot plant study is advisable. One reason is that "every water is different," a cliché in the industry, but true. The output of a process design is a sizing of the process components, for example, the filter bed area, media selection, media depth, and an estimate of terminal headloss. These factors, then "drive" the rest of the design, that is, the subsystems that support the process design. Such subsystems include, very broadly, getting the coagulated water to the filters, transporting treated water, backwashing filters, and processing wastewaters. Figure 12.24 illustrates major subsystem groups, which include

1. The influent flow system to the filter box gullet with open/close valve
2. The filter box with media, under-drain, wash-water trough, gullet
3. The effluent discharge with flow measurement and modulating valve
4. The backwash water waste line from filter gullet, with open/close valve

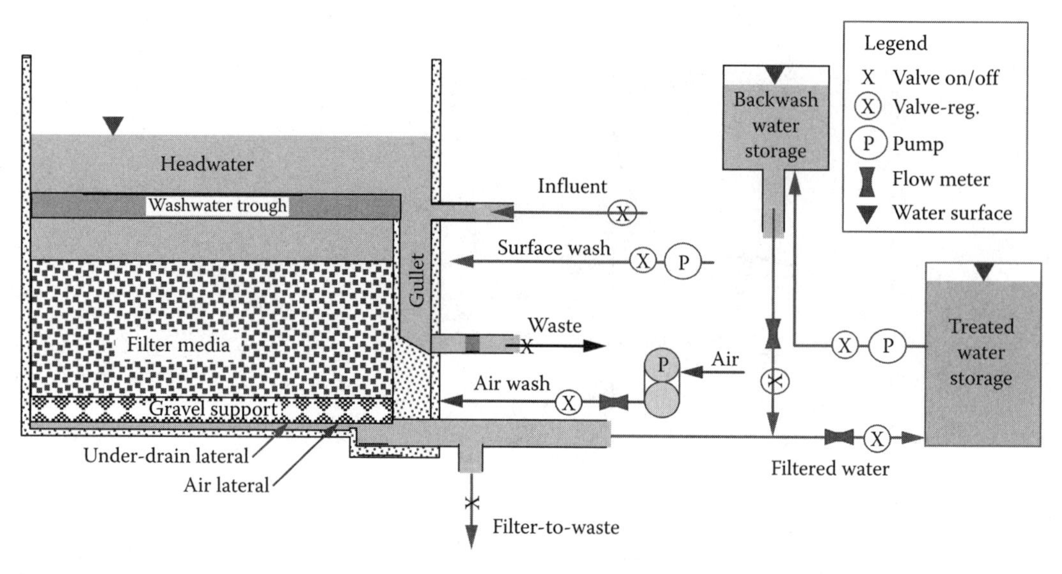

FIGURE 12.24 Filter showing subsystems, for example, air scour, backwash, effluent discharge, valves.

5. Backwash flow from storage with pump, flow measurement, modulating valve, and open/close valve
6. Air scour supply with compressor, flow measurement, regulating valve, open/close valve, and air header in under-drain system
7. Instrumentation that provides for flow regulation, opening and closing valves, and for reporting flows for effluent, backwash, and air scour, and for reporting water level in filter, headloss across filter from headwater to effluent pipe, water level in backwash storage, pressure of air supply in under-drain header, pressure of backwash water in header pipe
8. Online turbidity and particle counting instruments with data to SCADA system

12.4.1 External Parameters

A variety of nontechnical issues are a part of any design. Some are addressed in this section. The technical design involves many assumptions, judgments, and decisions.

12.4.1.1 Design Decisions

Some of the questions of process design must be settled before specific components are selected or sized and include: filtration mode (e.g., inline or conventional), filtration media (e.g., mono-media or dual media), depth and size of media, type of media, and filtration velocity. Once the process questions are determined, then the subsystem issues may be resolved and include filtration hydraulics (effluent rate controlled, declining rate, increasing water level in filter box), type of backwash (conventional backwash only, air-wash only, or air-wash and surface-wash, etc.), method of backwash (elevated backwash or pumped backwash), type of under-drain system (generic or proprietary), etc.

Once these decisions are made, then the details can be determined, for example, filter area, and number of filters, depth of filter box, size of influent channel, sizes of pipes for effluent flow, backwash sizing, air-wash sizing, tailwater elevation, size of clear-well, control valve locations and means of actuation, sampling points, SCADA system design, etc.

12.4.1.2 Cost

The capital cost of the filtration part of a water treatment plant (WTP) may range 15%–45% of the total cost (Letterman, 1980, p. 280); the higher end of the range is more likely. The distribution percentages among components of a filter building in one case noted by Letterman (1980) was given as foundations, 26; structure, 31; filter media, 4; wash troughs, 3; filter bottoms, 9; piping, 11; energy utilities, 12; surface wash, 4. The annual cost is the amortized capital cost plus operating costs. Design, for example, filtration velocity, filtration mode, backwash volume, etc., affects both categories. To illustrate, as the filtration velocity increases capital cost declines while operating cost increases, resulting in a minimum at some point, for example, at 24 m/h (10 gpm/ft^2) in an example by Letterman (1980, p. 288). While estimates of costs are necessary, an inherent uncertainty is usually associated with the assumptions.

12.4.2 Components of Filter Design

Process design is the theoretical aspect of filter design. Other considerations include filter layout, pipe gallery, under-drains, backwash system, etc. This section summarizes some of the practices in providing for these components.

12.4.2.1 Layout of Filters

The layout of the filters must be integrated with the other parts of the plant and with the site. Filtration involves repeating units and so their layout usually is linear with one bank of filters on one side of a pipe gallery and an identical bank on the other side. Figure 12.25 shows plan and profile schematic drawings for filters in Bellingham, Washington, a 75,700 m^3/day (20 mgd) plant. The plan shows three filter bays on each side of the operating floor, with two filters per bay.

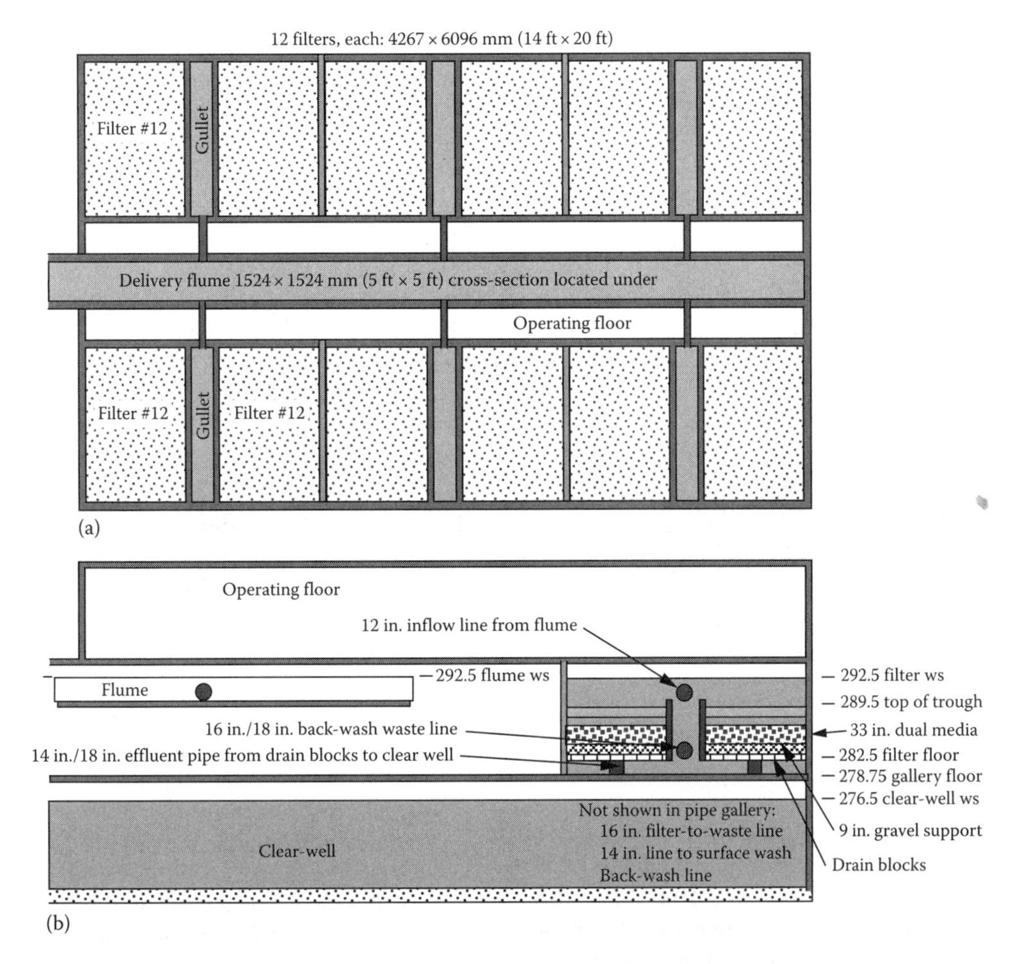

FIGURE 12.25 (a) Plan view of filter and (b) profile view of operating floor, flume, pipe gallery, filter section, and clear well—75,700 m^3/day (20 mgd) rated capacity illustrating layout and approximated sizes of pipes. (Adapted from plant drawings, City of Bellingham, WA.)

12.4.2.2 Hydraulic Modes of Filtration

Three of the hydraulic schemes of filtration are as follows: (1) constant rate and constant headwater level, (2) constant rate-rising headwater, and (3) declining rate. Constant rate is most common and is characterized by a "rate-of-flow-controller," for example, a valve that is in the near-closed condition at the start of a filter run and is open at the end. Constant rate, rising head is characterized by inflow control and constant flow; the filter effluent valve is fully open and water level in the filter box rises as the filter clogs to account for the increasing headloss; Figure 12.26 is an example of an installation. Declining rate is characterized by the filter effluent valves being continuously fully open for all filters, which have a common headwater elevation and a common tailwater elevation. As a filter is put back in operation after backwash, its filtration velocity is highest and declines as the filter clogs.

FIGURE 12.26 Constant rate–rising head filter.

12.4.2.3 Water Distribution

Figure 12.27 shows the "gullet," without water, for the Bellingham, Washington, plant (reference is also to Figure 12.25). A shallow flume from the rapid-mix basin distributes coagulated water to each of three "bays" (two filters per bay) on each side, that is, to six bays total. The flume is below the operating floor and runs the length of the three bays. A 250 mm (10 in.) pipe distributes water to each of the six gullets (between the two filters in a filter bay). The inlet pipe is at the top of the gullet and is not visible in the photograph. The outlet pipe for backwash water is at the bottom of the gullet (also not visible due to the workman in the gullet). The workman provides a size perspective for the system. In filter operation, the water from the inlet flume fills the gullet and flows into the backwash troughs and spills into the residual headwater in the filter box. The backwash troughs distribute the

FIGURE 12.27 Gullet in filter bay serves two filters, dissipating velocity and receiving backwash water. (City of Bellingham, WA.)

coagulated water and thus reduce the impact on the headwater; a cushion of headwater dissipates the energy from the falling water and must be deep enough, for example, 18 in. (450 mm) such that the media is not disrupted. When a filter run is terminated, the water is drained from the gullet and backwash water overflows into each backwash water trough, shown on each side of the gullet, and into the gullet, leaving the gullet through the pipe at the bottom. When completed, the cycle is repeated.

12.4.2.4 Media

Three kinds of media designs are as follows: (1) dual media of anthracite and sand, (2) tri-media of anthracite, sand, and garnet, and (3) deep bed coarse mono media, usually of anthracite. The media design is usually based on arbitrary decisions, tradition, or a standard approach and unless pilot plant studies are done the answer in most instances will be dual media (Monk, 1987). A typical design for dual media is depth(anthracite) = 610 mm (24 in.), depth(sand) = 254 mm (10 in.); d_{10}(anthracite) = 0.90 mm, uniformity coefficient (UC)(anthracite) = 1.5, and d_{10}(sand) = 0.45 mm, UC(sand) = 1.5. In designing filters with layers of different media, each layer should fluidize equally; otherwise loss of media or a dirty bottom layer may be a consequence (Kawamura, 1996, 1999, p. 80). As a design for a tri-media filter, Kawamura (1991, p. 215) recommended:

				Depth	
Medium	**SG**	**d_{10} (mm)**	**UC**	**(m)**	**(ft)**
Anthracite	1.4	0.9–1.4	1.4–1.7	0.45	1.5
Sand	2.65	0.45–0.65	1.4–1.7	0.30	1.0
Garnet	4.2	0.25–0.3	1.2–1.5	0.08	0.3

Table 12.5 shows some media designs that illustrate the range of practice for drinking water treatment. Plant capacities and filtration rates are given also. Concerning the UC, a value of about 1.5 is used frequently in practice, with the ideal being 1.0; lower values provide higher void volume for floc storage and longer filter runs (Kawamura, 1999, p. 81). The coarse media deep bed filter has found increasing favor because of longer filter runs and has been used with higher filtration velocities.

12.4.2.5 Pipe Gallery

Figure 12.28 shows a pipe gallery with header pipes on each side serving adjacent filters. The pipes are color coded and labeled. Table 12.6 lists the categories of influent and effluent flumes and pipes that serve a bank of filters.

1. *Manifold pipes*: Those pipes that deliver or receive water to or from several pipe "laterals" are, by definition, "manifold" pipes ("header" pipe). The water flow exiting each lateral pipe or orifice must be approximately the same (achieved by using a large header pipe so that the pressure loss is small). The under-drain pipe distribution system below the filter media is comprised of a manifold pipe with adjacent laterals. The under-drain system receives filtered water and distributes backwash water.

   Most pipes found in a pipe gallery are large, such as 300–900 mm (12.36 in.), especially if they serve as manifolds. A manifold pipe in the pipe gallery may serve to collect filtered water from several filters simultaneously. A backwash pipe in the filter gallery, on the other hand, serves only one filter at a time and therefore does not function hydraulically as a manifold pipe. Pipes are usually sized by specifying velocities. Kawamura (1991, p. 220) gives maximum velocities for both channels and pipes as follows:

Conduit	**Ordinary (m/s)**	**Filters (ft/s)**	**Self-Back (m/s)**	**Wash Filters (ft/s)**
Influent channel	0.61	2	0.61	2
Inlet valve	0.91	3	0.15	0.5
Forebay channel	0.15	0.5	0.15	0.5
Effluent valve	1.5	5	0.61	2
Effluent channel	1.5	5	0.61	2
Backwash main	3.0	10	0.91	3
Backwash valve	2.4	8	1.5	5
Surface-wash line	2.4	8	2.4	8
Wash-waste main	2.4	8	2.4	8
Wash-waste valve	2.4	8	2.4	8
Filter-to-waste valve	5.2	17	5.2	17
Inlet to filter under-drain lateral	1.4	4.5	1.4	4.5

The use of velocities to size a conduit provides a starting point. Using such velocities, hydraulic grade

TABLE 12.5
Filter Designs Illustrative of Practice

Plant	Plant Type	Q (m³/day)	Q (mgday)	HLR (mm/s)	HLR (gpm/ft²)	Media	d_{10} (mm)	UC	Depth (mm)	Depth (in.)	Headloss (m)	Headloss (ft)
Bellingham, Washington[a]	In-line	$73 \cdot 10^3$	19.3	2.72	4.0	Anthracite	1.10		467	18	4.86	16
						Sand	0.53		305	12		
						Garnet	0.35		76	3		
									838	33		
Oakland[b]	In-line	$662 \cdot 10^3$	175	3.40	5.0	Anthracite	0.9	1.5	457	18	1.83	6
Orinda Plant						Sand	0.5	1.6	305	12		
						Garnet			0	0		
									762	30		
Los Angeles[b]	Direct	$2270 \cdot 10^3$	600	9.03	13.3	Anthracite	1.5	1.5	1829	72	2.43	8
Aqueduct Plant						Sand				0		
						Garnet				0		
									1829	72		
Tuscaloosa[b]	Conventional	$57 \cdot 10^3$	15	2.72	4.0	Anthracite	1	1.7	419	16.5	1.83	6
Ed E. Love Plant						Sand	0.55	1.8	229	9		
						Garnet	0.28	2.2	114	4.5		
									762	30		
Corvallis[b]	Conventional	$80 \cdot 10^3$	21	5.09	7.5	Anthracite	1.1	1.3	533	21	2.43	8
H. D. Taylor Plant						Sand	0.4	1.3	305	12		
						Garnet				0		
									838	33		
Las Vegas[b]	Conventional	$1550 \cdot 10^3$	400	3.40	5.0	Anthracite	1.1	1.7	483	19	2.43	8
A. M. Smith Plant						Sand	0.5	1.8	229	9		
						Garnet	0.28	2.2	38	1.5		
									750	30		

[a] Hendricks et al. (2000, p. 198); total media depth 838 mm (33 ft); depth of each layer was estimated based on photograph of display section.
[b] Cleasby et al. (1989), Appendix B of report.

FIGURE 12.28 Photograph in pipe gallery.

lines may be ascertained by a spreadsheet with a plot linked to the output. This permits iterations based usually on minimizing headloss. Metal pipes are subject to corrosion or deposits with size and friction factor changing accordingly over time. Pipes that have been in service under similar conditions may be examined to estimate changes.

2. *Filtered water*: The drawing of Figure 12.29 shows a cross section of a flume for filtered water; the flume is along the length of a bank of filters with effluent weir at one end controlling depth. For such a method of collecting filtered water, the flume should be covered and should have provision for chlorine addition to control microbial films. Usually, the filtered water is collected by means of a common header pipe located in the pipe gallery. The open channel is an alternative depiction, indicating that a head difference must be provided between the headwater and the tailwater.
3. *Backwash header pipe*: A backwash header pipe, sized to serve one filter at a time, runs the length of the filter gallery with actuated valve connections to the under-drain system of each filter. If the storage is from an elevated reservoir, a rate-of-flow controller may be provided in the line in order to set the desired flow. If the storage is not elevated, a variable-speed motor and pump may be used to control the flow (Monk, 1987). Flow variations in water flow for a simultaneous air-water backwash include (1) gradual start, (2) backwash at rate prescribed, (3) adjustment

TABLE 12.6
Descriptions of Various Flumes and Pipes Associated with Water Transport to or from a Filter

Category	Description
Influent flows	
Coagulated source water	Flume from previous treatment distributes coagulated source water to each filter by means of pipe from bottom of flume to filter gullet
	Note: If the flume is placed high enough such that a slight free fall occurs from the pipe to the filter gullet then the filter may be operated either as a constant rate filter (headwater is at full height level to start with effluent flow regulating effluent flow) or as declining rate (headwater seeks its own level and effluent valve is open fully) Alternatively, a header pipe in the pipe gallery (instead of a flume) may deliver coagulated source water to the filter
Backwash supply	Header pipe along pipe gallery with lateral pipe to individual filter under-drains
Surface wash	Header pipe along pipe gallery with lateral pipe to individual filter surface wash
	Note: The surface wash may be a fixed grid or rotary
Air wash	Header pipe along pipe gallery with lateral pipe to individual filter air distribution manifold system which distributes air uniformly under filter media
Effluent flows	
Filtered water	Lateral pipe from under-drain system to filtered water header pipe in pipe gallery
	Note: Alternative pipe may be from under-drain system of a given filter to clear well
Backwash wastewater	Gullet within filter box collects backwash water, which drains into a wastewater pipe
	Note: A terminus of the pipe may be a wastewater flume in the pipe gallery. The discharge should be below the highest water level so that an air gap exists
Filter-to-waste	Lateral pipe from under-drain system to filtered water header pipe in pipe gallery has a "T" connection to a waste pipe that discharges into a waste-water flume located along the side of the pipe gallery
	Note: The discharge pipe should exit above the highest water level so that an air gap Exists

in flow to restratify media if dual media is used with air-wash, (4) gradual decline in flow to zero over a period of 30–60 s.

4. *Backwash wastewater*: For the backwash wastewater, a flume is shown in Figure 12.38, vis-à-vis a manifold pipe. An important point in the scheme shown is that a cross-connections between filtered water and unfiltered water is not possible as the air gaps are imposed, such as between the backwash effluent pipe and the wastewater flume below the floor of the pipe gallery.
5. *Surface-wash*: Another manifold must be provided for the surface-wash, if used. The pipe diameter is sized such that the headloss is not large.

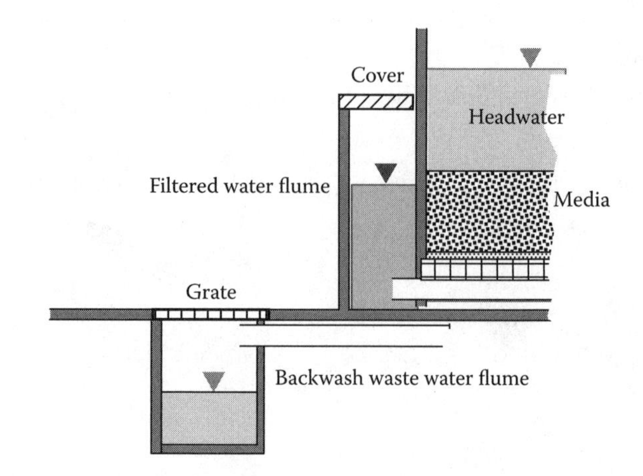

FIGURE 12.29 Effluent flume adjacent to filter to control tailwater elevation and backwash flume.

6. *Air-wash*: If air-wash is used, another manifold pipe is located in the pipe gallery, also sized such that the pressure loss is not large. As noted, provision for an air-wash is recommended.
7. *Flumes versus manifold pipes*: If flumes are used for influent flow, filtered water, and wastewater, the pipe gallery will contain manifold pipes only for backwash flow, surface-wash, and air-wash. The piping scheme is simple under such an arrangement. Another advantage is that the possibility of cross connections between filtered water and nonfiltered water is minimized.
8. *Valves*: Each of the four phases of filtration (filtration, backwash, filter-to-waste, air-wash), requires an open path for water flow, with the others closed. This involves a number of valves. Butterfly valves are the most common for switching between on and off modes. Most are actuated by air pressure. A butterfly valve is used sometimes to regulate flow. A problem with flow regulation with a butterfly valve is that the major part of the flow reduction occurs only as the valve approaches closure. Water hammer is a result of sudden closure. Manually operated gate valves are desired at some locations.

12.4.2.6 Clear-Well

Three functions of a clear-well design are as follows: (1) to provide a water surface elevation equal to the top of the filter media (if the clear-well provides the tailwater elevation), (2) to provide adequate detention time for disinfection, and (3) to provide a portion of the treated-water-storage. The total volume of treated-water-storage should be sufficient to handle the variation in water demand over a 24 h cycle so that the water production rate from the filters does not have to vary sharply.

12.4.2.7 Control Systems

In former times, each individual filter was operated through a console for that filter located on the operating floor. The console was fitted with valves that controlled air flow to open and close valves and switches for any pumps and motors. The end of a filter run could be initiated by headloss or time. A manual override was possible. For the plant as a whole, a wall was fitted with various switches and gauges for control. During the 1980s, computer control became more common, that is, SCADA systems. The open/closed position of actuated valves and sensor data were input to an "interface board," which in turn provided data that may be interpreted by computer software. Motors were controlled by micro-relays on the same board. The valves and motors could be controlled by an operator at the computer or through "set-points" with value monitored by the software.

12.4.3 Filter Box

Area and depth are the main issues of filter box design. Both are based largely on practice, except the pilot plant data that may provide the basis for exploring alternatives, for example, higher than normal filtration velocities, deep media bed, backwash alternatives, etc.

12.4.3.1 Filtration Rate

The filtration rate selected, that is, superficial filtration velocity, Q/A, may be based either on practice or on a pilot plant study. The traditional HLR was 4.88 m/h (2 gpm/ft^2) from about 1900 to about 1950. By 1970, the norm was 12.2 m/h (5 gpm/ft^2), which is current, albeit there is no absolute upper limit. As noted, the Los Angeles Aqueduct Plant at Sylmar, placed in operation in 1987, demonstrated $v \approx 32.3$ m/h (13.3 gpm/ft^2).

12.4.3.2 Area of Filters

A first estimate of total filter area is determined by the selected filtration rate, that is,

$$A(\text{total filter area}) = \frac{Q(\text{peak day})}{v(\text{selected})} \tag{12.39}$$

where
- A(total filter area) is the calculated area of all filter beds to be constructed (m^2)
- Q(peak day) is the average flow for the estimated peak day during the design period (m^3/s)
- v(selected) is the superficial filtration velocity selected by practice or pilot plant (m^3/m^2/s)

The filter bed area requires a uniform backwash, which is the major limitation for bed area (as filter bed area increases the difficulty of achieving a uniform backwash rate increases due to the need for longer lateral pipes and thus increased headloss). A rule of thumb for American practice is that the area of a single filter bed should be less than 93 m^2 (1000 ft^2). Usually, filters are much smaller; for example, the ones in Figure 12.25 are each 26 m^2 (280 ft^2).

12.4.3.3 Net Water Production

An expression to calculate total filter area that takes into account the time used for backwash and the water used for backwash was given by Letterman (1977, 1982), that is,

$$Q(\text{design}) = A(\text{filter}) \cdot [\text{number of filter runsper unit time} \cdot \text{net water production per unit area of bed per run}] \tag{12.40}$$

The number of filter runs per day is

$$N\left(\frac{\text{filter runs}}{d}\right) = \left(\frac{1440 \text{ min}/d}{(\text{WP}/v) + \text{TB}}\right) \tag{12.41}$$

and the "net water production" per filter run per unit area of filter bed is

$$\text{NWP} = \text{WP} - \text{VB} \tag{12.42}$$

Therefore, Equation 12.40 is

$$Q(\text{design}) = A(\text{filter}) \cdot \left[N\left(\frac{\text{filter runs}}{d}\right) \cdot \text{NWP}\right] \tag{12.43}$$

$$= A(\text{filter}) \cdot \left[\frac{1440 \text{ min}/d}{(\text{WP}/v + \text{TB}) \cdot (\text{WP} - \text{VB})}\right] \tag{12.44}$$

where
- Q(design) is the design flow to a given filter (m^3/day)
- A(filter) is the plan area of filter bed (m^2)
- WP is the volume of water production per unit area of filter bed per filter run (m^3 water produced/m^2 filter plan area/filter run)
- v is the filtration rate (m^3/m^2/min)
- TB is the filter down time per backwash(min downtime/filter run)
- VB is the volume of water required per unit of filter bed area per backwash(m^3 backwash water/m^2 filter bed area/filter run)
- NWP is the net water production unit area of filter bed per filter run (m^3 water produced/unit area of filter bed/run)

Values given by Letterman (1977, p. 24) for some parameters are as follows: $10 \le \text{TB} \le 30$ min, $2.037 \le \text{VB} \le 20.37$ m^3 backwash water/m^2 filter bed area ($50 \le \text{VB} \le 500$ gal/ft^2).

12.4.3.4 Depth of Filter Box

Guidelines from Kawamura (1991, p. 210) for the overall depth of the filter box are as follows:

Type of Filter	Depth (m)	Range (ft)	Depth (m)	Average (ft)
Ordinary gravity	3.7–6.1	12.20	5.2	17
Self-backwash	5.5–7.6	18–25	6.7	22

Looking at the filter box in more detail, its depth depends on the under-drain system (e.g., size of under-drain plenum or support media), media depth, and headloss desired. Headloss varies in the range, 2–3 m (6–9 ft), but there is no limit theoretically (except that turbidity breakthrough will occur eventually). Plenum systems can range 610–915 mm (24–36 in.) in depth (Monk, 1987). If self backwashing is used, a deeper filter box is required, as described by Monk (1987).

Example 12.4 Under-Drain Design

Problem statement

A filter bed area of 120 m^2 (1300 ft^2) is proposed and a generic under-drain system is to be used, that is, header and laterals with orifices. The filtration rate proposed is HLR ≈ 22.0 m/h (9 gpm/ft^2) and the backwash rate determined by pilot tests is 61 m/h (25 gpm/ft^2).

Required

Estimate the difference in orifice flow between the two extremes of the under-drain system, that is, at the first orifice of the first lateral and the last orifice of the last lateral.

Solution

The solution depends on the size of header, laterals, and orifices. Table CDD.2 is set up with an algorithm to calculate the kind of results as illustrated in Figure 12.30. The first trial might use a header pipe that is say 381 mm (15 in.) with laterals say 203 mm (8 in.), and orifices say 6 mm (1/4 in.), with spacing say 203 mm (8 in.). After comparing flows at the first orifice in the first lateral and the last orifice in the last lateral, the difference should be no more than say 5% as a criterion. Pipe sizes and orifices may be revised, depending on results.

Example 12.5 Depth of Filter Box

Problem statement

A filtration process will utilize a media depth of 3000 mm (118 in.) of anthracite with $d_{10} = 2$ mm, UC = 1.5. Assume the backwash system is generic with laterals and gravel support. An air wash system is also installed, but requires no additional depth. Assume that the tailwater weir is at the same elevation as the top of the media and that the headloss permitted is 3048 mm (10 ft) and that the freeboard is 610 mm (24 in.).

Required

Estimate the depth of the filter box.

Solution

The depth of filter box is

$$\begin{aligned} D(\text{filter box}) &= D(\text{gravel support}) + D(\text{media}) \\ &\quad + \text{headloss} + \text{freeboard} \\ &= 381\text{ mm} + 3000\text{ mm} + 3048\text{ mm} + 610\text{ mm} \\ &\quad (=15\text{ in.} + 118\text{ in.} + 120\text{ in.} + 24\text{ in.}) \\ &= 7039\text{ mm } (23\text{ ft}-1\text{ in.}) \end{aligned}$$

12.4.4 Backwash

The backwash system has several components: backwash flow distribution, disposal of wastewater, surface wash, and air wash. These components are described in this section.

12.4.4.1 Manifold Principles

Figure 12.30 shows a hydraulic profile for a generic under-drain system; the illustration is for the backwash mode. Table CDD.2 provides a spreadsheet calculation algorithm to calculate the pressure surface for any given manifold

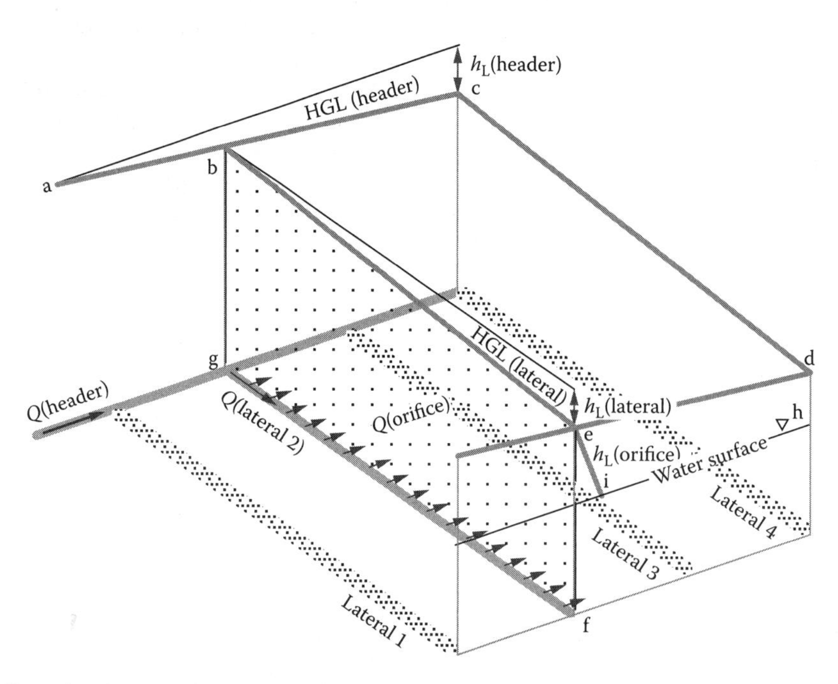

FIGURE 12.30 Three-dimensional perspective drawing of backwash system.

design. The spreadsheet may be used for design of a generic under-drain system. Any combination of design variables may be selected to assess the effect on orifice flows and the pressure surface for the manifold system as a whole.

The objective of manifold design is to have uniform orifice flow over the whole manifold system. The maximum orifice flow occurs at the first lateral and the first orifice while the minimum orifice flow is at the last lateral and the last orifice. The goal of manifold design is to minimize the difference between these two extremes in orifice flow but with the constraints that the sizes of the header, laterals, orifices and the number of laterals, and orifices should be within practical guidelines. As a corollary to objective of having near-uniform orifice flows is that the pressure surface should have a minimum difference between the first and last orifice.

12.4.4.1.1 Hydraulic Grade Line for Backwash System

Figure 12.30, a three-dimensional perspective drawing for backwash, applies also to the air wash and illustrates the idea of relative pressure losses. For air, with pressure dimensions applied to the Bernoulli relation, we may think of a "pneumatic" grade line (PGL, which has pressure as the energy dimension, that is, FL/L^3) instead of an hydraulic grade line (HGL, which has length as the energy dimension, that is, FL/F).

The Darcy-Weisbach equation is applicable; for gases the form is $D_p = f \cdot (L/D) \cdot r(\text{gas}) \cdot (v^2/2)$. Equation D.23, $Q(\text{orifice}) = A(\text{orifice}) C_D [2g\Delta h]^{0.5}$, is for orifice flow. Table CDD.2 may be adapted as a spreadsheet solution to calculate the power of an adiabatic compression required for the compressor (default compressor and motor efficiencies for 0.70 for each, respectively, are incorporated in the spreadsheet). The orifice coefficient of 0.61 is also a default value, as is, **f**, the pipe friction coefficient. The pressure for an air wash system is determined by the water depth at the orifice, the Δp(orifice) for the required air flow per orifice, and the friction and minor losses through the laterals, manifold, and other pipes, which gives the pressure, p_2 for the compressor; p_1 pressure is the ambient air pressure. The air flow, Q(air for one filter) combined with the ratio, p_2/p_1, and inlet temperature, permits calculation of the compressor power.

Example 12.6 Compressor Flow and Pressure for Air Wash

Given

Suppose a filter bed is 4.27 × 6.10 m (14 × 20 ft) plan area and that the air wash capacity in terms of surface loading should be 1.524 m^3/m^2/min (STP) (5 ft^3/ft^2/min). Let temperature be 25°C and let atmospheric pressure be sea level, 101,325 kPa. Let the depth of water in the filter box be say 3.0 m (9.84 ft), that is, from the orifices to the crests of the backwash water troughs. Let the temperature be say 20°C and assume an altitude of 1585 m (5200 ft).

Required

Compressor flow capacity and discharge pressure and power.

Analysis

The first task is to determine the air flow, that is, Q(air wash), required by the system. The second task is to calculate the pressure required at the outlet of the compressor. This is done as depicted in Figure 12.30. First the pressure in the bubbles as they emerge from any given orifice is the static pressure of the water, that is, $p(\text{bubble}) = \rho_w g h(\text{water})$ plus atmospheric pressure. Then the pressure losses due to pipe friction must be calculated, for example, by Equation D.32, along with any other losses such as bends. This gives, by the Bernoulli relation, the pressure at the compressor exit. Knowing the absolute pressure at the compressor exit, the power can be calculated by the equation for an adiabatic compression, that is, Equation D.75.

Solution

1. Determine the air flow, Q(air wash),

$$\begin{aligned} Q(\text{air-wash}) &= \text{Loading(air-wash)} \cdot A(\text{filter}) \\ &= 1.52\ \text{m}^3/\text{min (STP)} \cdot (4.267\ \text{m} \cdot 6.096\ \text{m}) \\ &= 39.52\ \text{m}^3/\text{min (1396 standard ft}^3/\text{min)} \end{aligned}$$

2. Determine the pressure, p_2. which is the friction loss and other losses in the pipe system plus the water pressure at the depth of submergence, that is,

$$p_2 = \Delta p(\text{friction}) + \rho w g h(\text{water})$$

where

$$\Delta p(\text{friction}) = f\left(\frac{L}{D}\right)\rho(\text{air})\left(\frac{v^2}{2g}\right) \qquad (D.43)$$

3. Calculate the theoretical compressor power for an adiabatic compression by Equation D.72, that is,

$$P = Q(\text{air}) \cdot p_1 \cdot (\kappa/(\kappa-1)\{(p_2/p_1)^{[(\kappa-1)/\kappa]} - 1\} \qquad (D.59)$$

4. Assume the compressor efficiency is say 0.7 and the motor efficiency is 0.7 to calculate actual power required by the compressor and the power required by an electric motor.

Comments

The value of "P" can be calculated most expediently by means of the spreadsheet, Table CDD.5.

12.4.4.2 Types of Backwash Systems

The three kinds of backwash systems are (Monk, 1987) as follows: (1) direct pumping, (2) pump and reservoir, and (3) self backwashing. In direct pumping, filtered water is pumped from the clear-well and pressurizes the under-drain system. In the "pump-and-reservoir" type, filtered water is pumped from the clear-well to separate reservoir, usually elevated. A self backwashing system utilizes the head from a tailwater overflow with an associated intermediate reservoir before the clear-well. This intermediate reservoir has an overflow weir to the clear-well that has high enough crest elevation to backwash a filter. The weir crest must be higher enough than the top of the media to fluidize the media to the extent required. Flow from the other filters must exceed the backwash flow as the weir must continue to overflow during the backwash. Such a system has the advantage of simplicity, that is, no pumps and no elevated storage, but the filter box must be deeper to provide for the higher tailwater elevation. Kawamura (1999, p. 79) recommends that at least four and preferably six filters should feed into the intermediate backwash reservoir.

12.4.4.3 Backwash Volume

For an hydraulic backwash, the rate and duration are key questions. Amirtharajah (1985) derived the relation

$$\ln C = \ln k - \left(\frac{vt}{d}\right) \tag{12.45}$$

where

C is the concentration of particles in backwash water (kg/m^3)
k is the coefficient (kg/m^3)
v is the superficial velocity of backwash, that is, Q(backwash)/A(filter bed) (m/s)
t is the elapsed time since start of backwash(s)
d is the representative diameter of collectors, that is, media grains (m)

Figure 12.31 shows the relationship for experimental data (data not shown) with $r = 0.93$ for 16 runs with expanded bed porosities ranging from 0.55 to 0.78 (Amirtharajah, 1985). For a given grain diameter, d, the exponent is the volume of backwash water per unit area of filter bed and is the product vt, that is, the backwash superficial velocity time elapsed time.

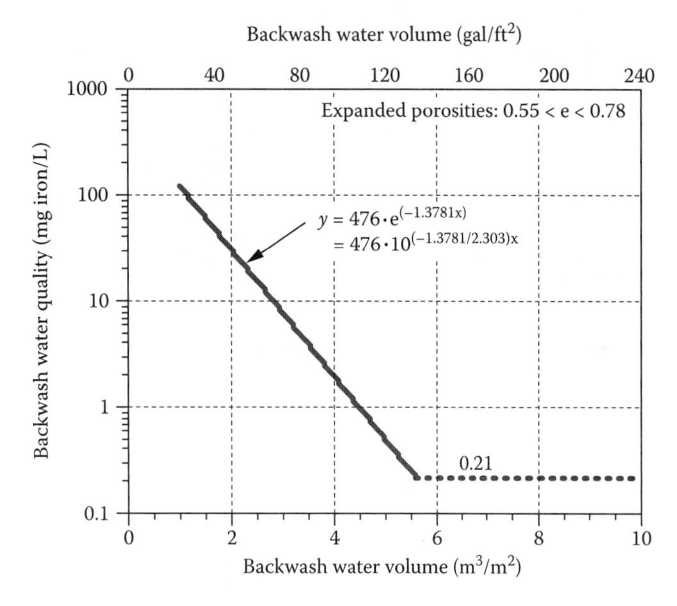

FIGURE 12.31 Backwash water quality as function of backwash water volume. (From Amirtharajah, A., *Water Res. (J. Int. Water Qual. Assoc.)*, 19(5), 587, 1985.)

In other words, a high backwash rate for a short duration yields the same result as a moderate backwash rate for a longer duration; the volume of backwash water is key. For the data shown, a backwash volume of greater than about 6.0 m^3/m^2 showed no improvement in water quality of the backwash water, leveling off at about 0.21 mg iron/L at volumes $\geq$6.0 m^3/m^2. The coefficients k and (vt/d), the intercept and slope, respectively of Equation 12.45 should be determined by pilot plant for the situation at hand.

Example 12.7 Storage Volume for Backwash Water

Given

Suppose a filter bed is 4267 × 6.096 m (14 × 20 ft) plan area and that the backwash HLR is 1.018 m/min (25 gpm/ft^2).

Required

Storage volume for 30 min backwash.

Solution

$$\begin{aligned} V(\text{backwash}) &= \text{HLR(backwash)} \cdot A(\text{filter}) \cdot \Delta t(\text{backwash}) \\ &= 1.018\,\text{m/min} \cdot (4.267\,\text{m} \cdot 6.096\,\text{m}) \cdot 30\,\text{min} \\ &= 795\,\text{m}^3\ (210{,}000\,\text{gal}) \end{aligned}$$

If cubic, the size is about 9.3 × 9.3 × 9.3 m (30.5 × 30.5 × 30.5 ft)

If cylindrical, the size is about 10 m high, 10 m diameter.

Comments

If HLR(filtration) = 0.122 m/min (3.0 gpm/ft^2), the run duration is 20 h, the water production during a filter run is

$$\begin{aligned} V(\text{production}) &= \text{HLR(filtration)} \cdot A(\text{filter}) \cdot \Delta t(\text{filter-run}) \\ &= 0.122\,\text{m/min} \cdot (4.267\,\text{m} \cdot 6.096\,\text{m}) \cdot (20 \cdot 60\,\text{min}) \\ &= 3{,}808\,\text{m}^3\ (1{,}008{,}000\,\text{gal}) \end{aligned}$$

If the backwash volume used in a normal backwash is about 400 m^3, the percentage of water production used for backwash is about 400/3808 = 0.105, or 10%.

Example 12.8 Volume of Backwash Water

Given

Figure 12.31 represents a backwash water quality versus backwash volume relationship for a contemplated plant with 4.27 m × 6.10 m (14 ft × 20 ft) filters. The two filters occupy a "bay" with common gullet for backwash water and water from the coagulated water flume enters the bay to serve both filters simultaneously.

Required

Estimate the backwash storage volume required.

Solution

1. Service Need The backwash water volume must be sufficient to serve both filters in one bay in a sequential backwash, that is, one filter and then the other backwashed.

2. Volume Criterion for Backwash Water Assume that the 6 m^3/m^2 criterion applies (from Amirtharajah, 1985, illustrated in Figure 12.31). Therefore,

$$V(\text{backwashwater}) = V(\text{loading}) \cdot A(\text{filter}) \cdot 2\ \text{filters/bay}$$
$$= 6.0\,\text{m}^3/\text{m}^2 \cdot (4.27 \cdot 6.10\,\text{m}^2)/\text{filter} \cdot 2\ \text{filters/bay}$$
$$= 312\,\text{m}^3/\text{bay}\ (82{,}580\ \text{gal/bay})$$

Discussion

Backwash is usually done on a time basis. The use of a backwash turbidimeter would provide a process that has a more focused approach to backwash duration. A plot of backwash turbidity versus backwash water volume per unit area of filter bed, similar to Figure 12.31, would be useful for a given installation. Such a plot should be generated for each kind of raw water filtration season and for different pretreatment conditions. To estimate the fraction of water produced used in backwash, assume HLR (filtration) = 12.2 m/h (5.0 gal/min/ft^2), and assume $t(\text{run}) \approx 24$ h. $V(\text{water produced/run}) \approx (12.2\ \text{m/h}) \cdot (24\ \text{h}) \cdot (4.27\ \text{m} \cdot 6.10\ \text{m}) = 7626\ \text{m}^3$ (2.015 mg). The backwash water volume is $312/7626 \approx 0.04$ fraction of the water produced, which is slightly less than the "rule of thumb" value of about 5%.

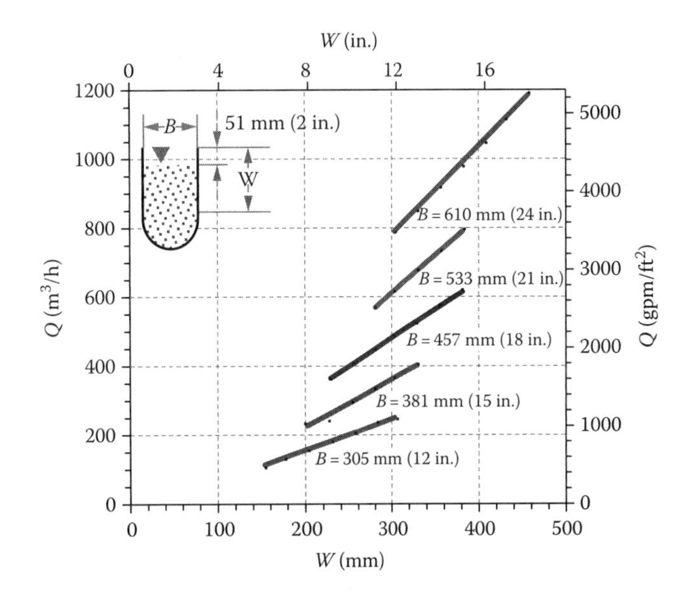

FIGURE 12.32 Flow capacities for washwater troughs. (Adapted from ITT Water & Wastewater Leopold Inc., Fiberglass Wash-Water Troughs, Product Data Sheet FRP-200; used by permission from ITT Leopold, Zelienople, PA.)

12.4.4.4 Backwash Water Troughs

The functions of wash water troughs are as follows: (1) to collect and convey backwash water to a gullet and then through pipes or channels and pipes to a storage pond, and (2) to distribute coagulated raw water over the filter bed so that there are no localized high velocities (albeit the filter is started with backwash remnant water above the media). Some issues in design are (1) spacing, (2) cross-section dimensions, (3) distance above media, and (4) whether to adopt proprietary methods of reducing media loss.

Regarding spacing, the distance selected is arbitrary and may vary 2–3 m (6–9 ft). The distance depends also on the flow capacity of the backwash water troughs. For example, the maximum backwash water flow, $Q(\text{max})$, is divided by the flow per trough, $Q(\text{trough})$. Therefore, the number of troughs is $n(\text{troughs}) = Q(\text{max})/Q(\text{trough})$, where $n(\text{troughs})$ is rounded to the nearest whole number. For a rectangular filter of dimensions $w(\text{filter}) \cdot L(\text{filter})$, the spacing is then, $w(\text{troughs}) \approx L(\text{filter})/n(\text{troughs})$. When set up on a spreadsheet, the procedure may be repeated until satisfactory spacing and trough size are obtained.

The distance above the media depends upon the backwash practice. In the United States, the bed expansion is 20%–50% and the vertical distance from the media surface to the crest of the wash-water troughs is 0.7–1.0 m (Cleasby, 1992). In the United Kingdom, with air first and water second, with bed expansion <10%, the distance is 0.1–0.2 m. In Europe, with air-water first and water second, with bed expansion about "zero" percent, the distance is 0.5 m. Kawamura (1999, p. 85) recommended about 1.8 m (6 ft) for a concurrent air-water backwash so that the grains of media do not reach the wash-water troughs.

In selecting wash-water troughs, the semicircular bottoms minimize trapping of air and foam (Kawamura, 1991, p. 199). A filter bed width of ≤6 m (20 ft) permits the use of "off-the-shelf" proprietary type of troughs of fiberglass or other synthetic material.

On sizing the troughs, Figure 12.32 gives flow capacities for different sizes for a proprietary fiberglass trough. The troughs should have the hydraulic capacity for the highest backwash flow. The trough discharge into a gullet should provide an adequate free-fall so that there is no backup. The gullet itself requires the same consideration with respect to hydraulic capacity. Kawamura (1990, p. 202) gives a formula that relates flow and depth for a flat bottom rectangular channel as

$$h_o = \left(\frac{Q}{CB}\right)^{0.667} \tag{12.46}$$

where

h_o is the depth of water at the upstream end of channel (m)
Q is the flow in channel (m^3/s)
B is the width of channel (m)
C is the coefficient

- In SI units, $C = 1.38$
- In U.S. Customary units, $C = 2.49$ and other units are ft and s

12.4.4.5 Under-Drain Systems

The under-drain system has two functions: (1) to collect filtered water while retaining the media, and (2) to distribute backwash water uniformly over the area of the filter bed. Another requirement is that these functions remain intact and do not deteriorate during decades of operation, that is, they remain reliable.

Since the under-drain system is not accessible, being beneath the media, the system cannot be inspected and so there must be a confidence that the orifices will not clog or corrode. It is expected that the media will remain in place for at least a 20–30 year period with satisfactory performance from the under-drain system. The attributes of an "ideal" under-drain system are outlined below, along with the "practical" aspects. The gravel support is also described.

1. *The ideal*: The perfect under-drain technology would have the following attributes:
   - Backwash water is distributed uniformly over the whole filter area.
   - Filtered water is collected uniformly over the whole area.
   - Orifices are not susceptible to clogging, either by precipitation or by particulates.
   - Air and water wash can be done simultaneously.
   - Orifices for air and water flows are spaced close enough that a there are no "dead-zones" for either flow.
   - The media can rest on a porous plate and thus eliminate the need for a gravel support (which has potential for displacement and occupies 300–450 mm (12.18 in.) of depth in the filter box).
2. *The practical*: A perfect under-drain system has not been achieved as yet, but some may have approached the ideal through proprietary innovations. Some realities of under-drain design are as follows:
   - Headers and laterals will have some friction headloss and so the differential pressure across orifices will be different between those at the beginning of the first lateral and the end of the last lateral. The larger the conduits the smaller the friction headloss as calculated by the Darcy-Weisbach equation, that is, $h_L = f(L/d)(v^2/2g)$. A maldistribution in orifice flow of $\leq 10\%$ is considered acceptable (Beverly, 1995), while the Leopold (1999a,b,c) states $\leq 5\%$ is achieved with their system. Table CDD.2 is a spreadsheet model that computes pressures and orifice flows for any assumed sizes of the header pipe, laterals, and orifices.
   - The requirement for uniform distribution of flow mandates a significant headloss across the orifices, which implies a smaller orifice diameter (or larger orifice size with increased spacing).
   - To minimize dead zones, the Leopold system uses a 50 mm (2 in.) spacing between orifices over the floor area (Leopold, 1999a,b,c). At the same time, the closer the spacing, the smaller the orifices. For example the Tetra U® block technology has orifices of 6.35 mm (0.25 in.).
   - To minimize surface chemical changes, a ceramic or plastic material is preferred to a metal.
   - A porous plate is available that retains media and passes flows of both water and air; headloss is about 150 mm (6 in.) at a backwash rate of 50 m/h (20 gpm/ft^2) (Leopold, Product Data Sheet IMS-100, 1995) at 13°C (56°F).
   - For proprietary slotted under-drain nozzles that rest in the media, they must have openings smaller than the smallest size of the media. Nozzles with slotted areas are available with widths 0.25–9.0 mm (Monk, 1987). An important concern is that the orifice area should be small enough such that significant headloss occurs, (i.e., the headloss across any given orifice should be about the same as every other orifice). This cannot be true if the orifice headloss is small compared to header and lateral headlosses; the orifice headloss equation is, $Q(\text{orifice}) = C(\text{orifice})A\ (\text{orifice})\ [2g\Delta h(\text{orifice})]^{0.5}$, which shows that Q (orifice) is affected by $\Delta h(\text{orifice})$.]

Most under-drain systems are available as proprietary equipment. The classic generic design is to use perforated pipe laterals with a coarse gravel layer covering the laterals with graded gravel above. Laterals for air may be added so that a simultaneous air–water backwash may be used.

Retrofits of existing filter beds often use plastic blocks with a porous plate above, which eliminates the need for the gravel layer. Such systems permit a greater depth of media and may provide for simultaneous air–water backwash. Gravel support layers are subject to upset, and with the development of suitable porous plates the latter are increasingly favored. Thus, the long quest to eliminate the gravel support, with its inherent risks, is finding suitable alternatives.

3. *Sizing guidelines*: For a generic system, that is, with header, laterals, orifices, and gravel support, the following guidelines for sizing an under-drain system (ASCE et al., 1969, p. 136):

$$\frac{\text{sum orifice area}}{\text{bed area}} = \frac{0.0015}{1} - \frac{0.005}{1} \tag{12.47}$$

$$\frac{\text{sum lateral area}}{\text{sum area of orifices served}} = \frac{2}{1} - \frac{4}{1} \tag{12.48}$$

$$\frac{\text{manifold area}}{\text{sum lateral area}} = \frac{1.5}{1} - \frac{3}{1} \tag{12.49}$$

Other recommendations by Cleasby (1991) taken from Weber (1972) are as follows: $6 \leq d(\text{orifice}) \leq 13$ mm (1/4–1/2 in.); spacing of orifices: $76 \leq x(\text{orifice spacing}) \leq 300$ mm (3–12 in.); $76 \leq x(\text{lateral spacing}) \leq 300$ mm (3–12 in.); $L(\text{lateral})/d(\text{lateral}) < 60$.

The guidelines have been used in practice as a simple means to size the under-drain system. They may be used as a starting point for the spreadsheet simulation of an under-drain system as given in

Table CDD.2, Appendix D, which is illustrated graphically in Figure 12.30.

4. *Gravel support*: The gravel support, if a generic under-drain system is used, is graded from coarse at bottom to fine gravel at the interface with the media. For the top layer of gravel, garnet, with SG ≈ 4.6, is recommended to reduce movement potential during backwash (Monk, 1987). Recommendations by Cleasby (1991) are

Medium-to-gravel interface:	d_{10}(interface gravel)/d_{10} (medium) ≈ 4
Layer-to-layer fine:	d_{10}(layer below)/d_{10}(layer above) ≈ 2
Layer-to-layer coarse:	d_{10}(layer below)/d_{10}(layer above) ≈ 4
Gravel-to-orifice:	d_{10}(layer above)/orifice size ≈ 2 or 3
Depth of layers:	d(layer) ≥ 70 mm (3 in.)

Specific sizes and depths of gravel layers for a generic gravel support are given as follows (Kawamura, 1991, p. 218):

	Sieve Size					
	Passing Size		Retaining Size		Depth	
Layer	(mm)	(in.)	(mm)	(in.)	(mm)	(in.)
1	13	1/2	19	3/4	100–150	4–6
2	19	3/4	12	1/2	75	3
3	13	1/2	6	1/4	75	3
4	6	1/4	3	#6	75	3
5	3	#6	1.7	#12	75	3
				Total	16–18	400–450

5. *Proprietary systems*: Proprietary systems are of two types: (1) laterals that are composed of rectangular channel blocks with perforations in the floor upon which rests the graded gravel bed and (2) orifices that have direct contact with the media with the channel within a length of blocks forming the lateral. The blocks may have two compartments, one for water and the other for air. It is important that the air be purged from the system during backwash.

The Leopold Type S™ under-drain, shown in Figure 12.33, provides for simultaneous air and water backwash. The structure is polyethylene with plastic porous plate (IMS™ cap) about 25 mm (1 in.) thick. The porous plate causes little pressure loss for either air or water, with only 115–140 mm (4.5–5.5 in.) water at a backwash velocity of 0.81 m/min (20 gpm/ft^2). The Leopold Type S™ and Type SL™ under-drain blocks will accommodate an air flow range of 0.30–1.52 m^3/m^2/min at STP (1–5 scfm/ft^2) (Leopold, Brochure FIL-100, 1999c).

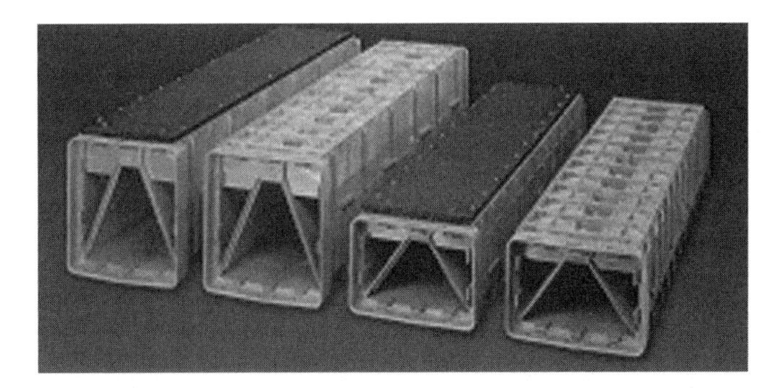

FIGURE 12.33 Leopold Type S® under-drain with IMS® (Integral Media Support) porous plate cap suitable for simultaneous air and water backwash. (From Leopold, Leopold underdrain, Brochure UNN-100, F. B. Leopold Company, Zelienople, PA, 1999. With permission.)

12.4.4.6 Bed Fluidization

The filter bed must be partially or wholly fluidized in order to clean the media. This section describes the criteria for minimum backwash velocity for bed fluidization and the relationship between the expanded bed and the backwash velocity, that is, for v(backwash) > v_{mf}.

12.4.4.6.1 Description

In a static bed of media, saturated with water, the grain-to-grain pressure is due to the buoyant weight of the grains. This pressure is reduced during backwash due to the upward drag forces on the grains. At some point, as v (superficial velocity) increases, the grain-to-grain pressure becomes zero, which is the point of incipient fluidization; the associated backwash superficial velocity and incipient hydraulic gradient are designated, v_{mf} and i_{mf}, respectively. Any higher backwash velocity, that is, $v > v_{mf}$, will fluidize the bed. At the same time, the hydraulic gradient will not increase higher than i_{mf}, although $v > v_{mf}$ during backwash (Amirtharajah and Cleasby, 1972, p. 55). The height of the bed will rise, however, with each increment of "v."

12.4.4.6.2 Headloss versus Backwash Velocity—Experimental

Figure 12.34 illustrates an experimental relationship between headloss through the bed, Δh, and v(backwash). As seen, the relationship is initially linear, that is, in accordance with Darcy's law, but as v(backwash) increases, a transition starts. At some point, Δh remains constant as v(backwash) increases and the bed is "fluidized." The value of v(backwash) at the start of bed fluidization is designated, v_{mf}, also designated for clarity, $v(\text{backwash})_{mf}$, which is the superficial velocity of water through the bed.

12.4.4.6.3 Calculation of v_{mf}

An empirical relationship (Amirtharajah and Cleasby, 1972; Hewitt and Amirtharajah, 1984) that describes when incipient fluidization occurs is

$$v_{mf} = \frac{3.2193 \cdot 10^{-11}(d_{60})^{1.82}\left[\left(\gamma_w^2(\text{SG(medium)} - 1)\right)\right]^{0.94}}{\mu^{0.88}} \tag{12.50}$$

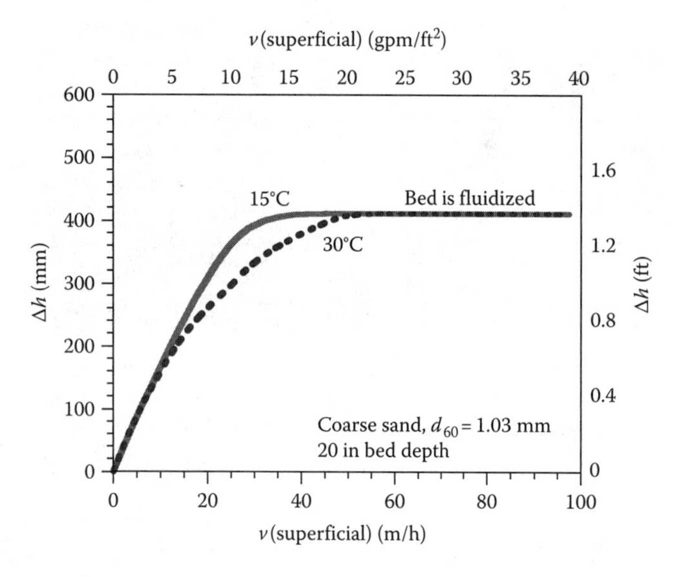

FIGURE 12.34 Headloss versus backwash velocity. (Adapted from Amirtharajah, A. and Cleasby, J.L., *J. Am. Water Works Assoc.*, 64(1), 56, 1972.)

where

v_{mf} is the backwash velocity for incipient fluidization (m/s)
d_{60} is the media size that is 60% finer than (mm)
SG(medium) is the specific gravity of medium
μ is the dynamic viscosity of water (N s/m^2)

Converted to U.S. Customary units, the coefficient is, $3.81 \cdot 10^{-3}$ (instead of $3.22 \cdot 10^{-11}$); the exponents are the same for both systems of units.

12.4.4.6.4 Bed Expansion Calculation Protocol

Equation 12.50 for v_{mf} is valid for $\mathbf{R} \leq 10$. For $10 < \mathbf{R} < 300$, a correction factor K_R is applied to v_{mf}, that is, v_{mf}(corrected) = $K_R \cdot v_{mf}$, and is calculated as

$$K_R = 1.775\mathbf{R_{mf}}^{-0.272} \quad (12.51)$$

where

K_R is the correction factor for $10 < \mathbf{R} < 300$, applied to v_{mf}
$\mathbf{R_{mf}}$ is the Reynolds number calculated for whatever conditions exist when $\mathbf{R} > 10$

and

$$\mathbf{R} = \frac{\rho v_{mf} d_{60}}{\mu} \quad (12.52)$$

The second part of the protocol is to determine the bed expansion for $v \geq v_{mf}$, which may be done by two empirical relations (Amirtharajah and Cleasby, 1972; Cleasby and Fan, 1981, p. 460), that is,

$$\varepsilon^{4.7}\mathbf{Ga} = 18\mathbf{R} + 2.7\mathbf{R}^{1.687} \quad (12.53)$$

and

$$\mathbf{Ga} = \frac{d_{30}^3 r_w^2 (\mathrm{SG_s} - 1)g}{\mu^2} \quad (12.54a)$$

$$\mathbf{Ga} = \frac{d_{60}^3 \rho_w^2 (\mathrm{SG_s} - 1)g}{\mu^2} \quad (12.54b)$$

where **Ga** is the Galileo number (dimensionless).

The bed expansion is

$$\frac{(h - h_o)}{h_o} = \frac{(e - e_o)}{(1 - e)} \quad (12.55)$$

The calculation protocol that applies to Equations 12.50 through 12.55 is embedded in Table CD12.7, which was formatted to calculate the expanded bed depth as a function of assumed superficial backwash velocities. In other words, the calculation protocol assumes a v, which permits calculation of **R** by Equation 12.52. At the same time **Ga** is calculated by Equation 12.54. Then $\varepsilon^{4.7}$ is calculated from Equation 12.53. The minimum fluidization velocity, v_{mf}, is calculated by Equation 12.50 with correction by Equation 12.51 incorporated (if $10 < \mathbf{R} < 300$). Calculation of v_{mf} is necessary since Equations 12.52 through 12.54 are valid only for $v > v_{mf}$. Table CD12.7 may be used to calculate bed expansion versus an assumed backwash velocity for assumed conditions, for example, temperature, specific gravity of medium, d_{10}, etc.

Amirtharajah and Cleasby (1972) showed comparisons between the expanded bed depth calculation, that is, h, by

TABLE CD12.7

Excerpt—Bed Expansion as Function of Backwash Superficial Velocity and Calculation of Minimum Fluidization Velocity with Correction (Equations from Amirtharajah and Cleasby, 1972)

(a) Calculation of bed expansion by Wen/Yu Equation from Amirtharajah and Cleasby, 1972

g = 9.80665

Data Concerning Media and Conditions

T (°C)	μ (Ns/m²)	d_{10} (mm)	d_{60} (mm)	e_o	d_o (mm)	SG(media)	r_w (kg/m³)	v (m/h)	v (m/s)	v (gpm/ft²)	Ga	R	$e^{4.7}$	e
20	0.001005264	0.43	0.69	0.412	513	2.648	998.37	2.0	0.00056	0.81833	5237	0.38	0.0014	0.25
20	0.001005264	0.43	0.69	0.412	513	2.648	998.37	4.0	0.00111	1.63666	5237	0.76	0.0029	0.29
22	0.000957291	0.50	0.689	0.407	457	2.655	998.37	6.0	0.00167	2.45499	5770	1.20	0.0044	0.31
22	0.000957291	0.50	0.689	0.407	457	2.655	998.37	8.0	0.00222	3.27332	5770	1.60	0.0060	0.34

TABLE 12.8
Media Used to Compare Experimental and Calculated Bed Expansions

Media	d_{10} (mm)	d_{60} (mm)	ε_o	h_o (mm)	SG(media)
Sand A	0.43	0.69	0.412	513	2.648
Coarse sand B	0.62	1.029	0.410	513	2.653
Coarse fraction	0.75	1.145	0.422	345	2.653
Fine fraction	0.55	0.74	0.400	168	2.653
Corning sand C	0.50	0.736	0.407	457	2.655
Coal D	1.00	1.321	0.502	498	1.680

Source: Amirtharajah, A. and Cleasby, J.L., *J. Am. Water Works Assoc.*, 64(1), 52–59, 1972.

Equations 12.52 through 12.54 and experimental data for four media is described in Table 12.8. The calculated results agreed with the experimental results of bed expansion within about 2%–5%. For the anthracite, however, the equations predict h values that are perhaps 10%–15% less than the measured h values for backwash velocities up to 98 m/h (40 gpm/ft^2).

Example 12.9 Bed Expansion as a Function of Fluidization Velocities

Given

Table 12.8 provides the relevant data for calculation of bed expansion as a function of backwash velocity; Table Ex12.9.1 is an excerpt from the associated spreadsheet, Table CD12.7a.

Required

Applying Equations 12.52 through 12.55, estimate the bed expansions at different backwash velocities for the Corning Sand C in Table 12.8.

Solution

1. Insert the data for Corning sand C data of Table 12.8 in the spreadsheet, Table CD12.7b.

Media	d_{10} (mm)	d_{60} (mm)	ε_o	h_o (mm)	SG
Corning Sand C	0.50	0.74	0.407	457	2.655

Regarding nomenclature, let d = diameter of filter grain, for example, d_{10} and d_{60}; h_o = initial bed depth before expansion, h = bed depth after expansion.

2. Calculate h and $(h - h_o)/h_o$ for different superficial velocities by Table CD12.7a. Excerpts from the table are given as Table Ex12.9.1.
3. Calculation of minimum fluidization velocity, v_{mf}. From Table CD12.7a,

$$v_{mf} = 20.1 \text{ m/h } (8.2 \text{ gpm/ft}^2)$$

4. Figure 12.35 is a plot of bed expansion shown as $(h - h_o)/h$ versus backwash velocity, v.

Discussion

Table Ex12.9.1 and Figure 12.35 show excerpts of calculation results of the spreadsheet calculations for the Corning C sand (from Corning, Iowa). The calculation used a weighted average for d, that is, d_{eq} as recommended by Amirtharajah and Cleasby (1972), instead of d_{60} as shown

TABLE Ex12.9.1
Excerpts from Table CD12.7a Showing Assumed Superficial Velocities and Calculated[a] Expansion, $(h - h_o)/h_o$

v			h		
(m/h)	(m/s)	(gpm/ft^2)	(mm)	(in.)	$(h - h_o)/h_o$
18.0	0.00500	7.36498	460	18.1	0.01
20.0	0.00556	8.18331	469	18.4	0.03
25.0	0.00694	10.22913	490	19.3	0.07
30.0	0.00833	12.27496	510	20.1	0.12
35.0	0.00972	14.32079	529	20.8	0.16
40.0	0.01111	16.36661	548	21.6	0.20
45.0	0.01250	18.41244	567	22.3	0.24
50.0	0.01389	20.45827	585	23.0	0.28
60.0	0.01667	24.54992	623	24.5	0.36

[a] Calculation in table excerpted from spreadsheet Table CD12.7a for corning sand C (Table 12.8); the spreadsheet is available on the CRC website.

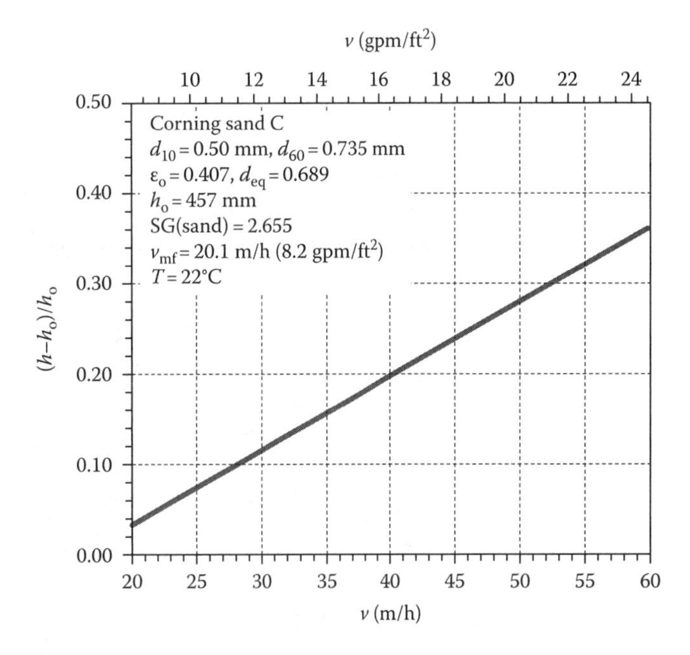

FIGURE 12.35 Plot of calculated bed expansion versus v(backwash).

in the protocol above, which compared with experimental data within about 0.1%–5%. Note that $d_{eq} < d_{60}$ results in lower values for $(h - h_o)/h$ by perhaps 5%–10%. For an estimate of bed expansion, the d_{60} value is probably adequate.

12.4.4.6.5 Comparison between Measured and Calculated Fluidization Velocities

Table 12.9 compares measured and calculated v_{mf} values for different sieve sizes of sand and anthracite for two temperatures. As seen, the comparisons show the same trends with discrepancies varying about 3%–10%. The procedure is shown at the bottom of the table and is similar to that shown in Table CD12.7, but modified.

TABLE 12.9
Minimum Fluidization Velocities for Filter Media

	Minimum Fluidization Velocities, v_{mf}			
	(25°C)		(40°C)	
U.S. Sieve Range	Measured (m/h)	Predicted (m/h)	Measured (m/h)	Predicted (m/h)
Sand				
10–12	79.2	87.5	93.6	97.6
14–16	57.6	57.2	57.6	67.3
18–20	32.4	34.2	43.2	42.8
30–35	18.0	13.3	19.8	18.0
Anthracite				
5–6	97.2	99.4	100.8	105.5
6–7	86.4	85.3	93.6	92.5
7–8	72.0	72.4	82.8	79.9
12–14	36.0	38.2	50.4	45.7

Source: Adapted from Cleasby, J.L., Backwash and underdrain considerations, unpublished paper for short course at Colorado State University on design of filtration systems, June, 1991. With permission.

Note: The "predicted" fluidization velocities were obtained from the procedure of Wen and Yu as described by Cleasby (1991).

$$\mathbf{R}_{mf} = [33.7^2 + 0.0408\mathbf{Ga}]^{0.5} - 33.7 \tag{12.56}$$

$$\mathbf{R}_{mf} = \frac{d_{90} v_{mf} \rho}{\mu} \tag{12.57}$$

$$\mathbf{Ga} = \frac{d_{90}^3 \rho(\rho_s - \rho) g}{\mu^2} \tag{12.58}$$

where
$\mathbf{R}_{mf}$ is the Reynolds number at minimum fluidization
$\mathbf{Ga}$ is the Galileo number
v_{mf} is the minimum fluidization velocity, that is, Q(backwash)/A(filter)

To determine v_{mf}, solve for **Ga** by Equation 12.58, then solve for $\mathbf{R}_{mf}$ by Equation 12.49, and then v_{mf} by Equation 12.57. To obtain a backwash rate, multiply v_{mf} by 1.3 as a factor of uncertainty.

12.4.4.7 Surface-Wash

Surface-wash involves high velocity jets of water that impact the media surface and penetrate into the filter bed, that is, to a depth of about 1.2 m (4 ft). With the advent of dual media, a second jet is often installed at the interface between the sand and anthracite, which operates during bed expansion. Usually surface-wash is done before backwash and continues simultaneously with backwash for a short duration. The jets may be rotating or fixed grid; the fixed type is recommended because of lack of moving parts (Kawamura, 1999, p. 82).

Recommended surface rates are given in Table 12.10. As seen, there is some difference between the recommendations of Kawamura and Cleasby that merely illustrates that the guidelines are not absolute. Higher pressure allows cushion for uncertainty.

While air-wash has been the trend over the past two decades in the United States, Kawamura (1996) suggests

TABLE 12.10
Surface Wash Velocities and Pressures[a]

		Velocities		Pressure	
Type	Source	(m/h)	(gpm/ft^2)	(kPa)	(psi)
Fixed nozzle	Kawamura	7.2–9.6	3–4	55–83	8–12
	Cleasby	9.6–14.4	4–6	344–689	50–100
Rotating arm (single arm)	Kawamura	1.2–1.8	0.5–0.7	489–690	70–100
	Cleasby	2.4	1.0	344–689	50–100
Rotating arm (dual arms)	Kawamura	3.0–3.6	1.3–1.5	500–600	80–100

Sources: Kawamura, S., *Integrated Design of Water Treatment Facilities*, John Wiley & Sons, Inc., New York, 1991, p. 213; Cleasby, J.L., Backwash and underdrain considerations, unpublished paper for short course at Colorado State University on design of filtration systems, June, 1991.

[a] Velocities are total surface wash flow divided by filter bed area.

Nozzles should be placed about 25 mm (1 in.) above un-expanded bed.

TABLE 12.11
Backwash Rates in the U.S. and Europe Practice

Location	Sand Size (mm)	Sequence	Air Velocity (m/min)	Water Velocity (m/h)
United States	0.5–1.2	Air first	0.9–1.5	
		Water second		36–54
	Dual media		Same	Same
United Kingdom	0.6–1.2	Air first	0.3–0.5	
		Water second		12.6–18
	Dual media	Water second	0.5	
Europe	1–2	Air + water first	0.9–1.5	12.6–18
		Water second		Same or double
	Dual media	Air + water first	≤1.5	
		Water second		Expand bed 0.1–0.2
	2–4	Air + water first	1.8–2.4	14.4–18
		Water second		Same or double

Source: Adapted from Cleasby, J.L., Backwash and underdrain considerations, unpublished paper for short course at Colorado State University on design of filtration systems, June, 1991.

both air-wash and surface-wash. For surface-wash with backwash, the procedure is to draw the water level to about 100 mm (4 in.) above the media bed and then begin the surface-wash. The sequence and time recommended are as follows: surface wash, 0–4 min; backwash, 2–6 min (Kawamura, 1999, p. 84).

12.4.4.8 Air-wash

Air-wash is accomplished by a separate manifold system with orifices that deliver air to the filter media. As in surface-wash, the air-wash causes particle-to-particle contacts, which is felt necessary for effective cleaning. Three kinds of air scour are (Amirtharajah and Trusler, 1982) are as follows: (1) channels of air run through the bed and the air has little effect on the media not in direct contact with the channels, (2) simultaneous water and air washes in which discrete bubbles, in moving upward, cause turbulence throughout the bed along with a high rate of particle contacts and improved cleaning, (3) a "subfluidization" transition range between the two foregoing conditions that oscillates between channels and bubbles which is typical of European practice (Hewitt and Amirtharajah, 1984, p. 592).

Two types of air scouring are as follows: sequential and concurrent and both perform well when the systems are properly designed and operated (Kawamura, 1999, p. 83). As noted in Section 12.4.4.9, however, the most effective backwash procedure is a concurrent air–water backwash. Empirical guidelines for air-wash were recommended by as 0.75–0.90 $m^3/m^2/min$ (2.5–3.0 $scf/ft^2/min$) for "ordinary" filter beds and 0.9–1.2 $m^3/m^2/min$ (3–4 $scf/ft^2/min$) for "deep" filter beds (Kawamura, 1991, p. 216).

In the layout of an air system, the orifices should all be at the same level so that there is the same external water pressure, and hence the same air flow from each orifice (Monk, 1987). An air scour will cause a dual or tri-media bed to become mixed and so the bed must be fluidized with water alone following the air–water backwash so that the media may again be stratified. Table 12.11 summarizes backwash practices in the United States, the United Kingdom, and Europe with associated air and water velocities.

A separate manifold/lateral pipe/orifice system is required for the air-wash, with the laterals being placed in the support gravel (see Monk, 1987). The design of an air-wash system is, in principle, the same as the hydraulic design for the water backwash. In other words, the header pipes should be large so that the pressure losses are minor and with large diameter laterals for the same reason but smaller in diameter than the header pipe because if there are n laterals, the flow is only $1/n$th the flow of the header pipe. The major pressure loss should be across the orifices so that whatever losses there are in the headers and laterals there are a small proportion of the pressure loss across any orifice. As in the distribution of water, an objective is that the air flow from the furthest orifice from the header entrance should be $\leq 5\%$ of the flow at the header entrance.

12.4.4.9 Air–Water Concurrent Backwash

A simultaneous air–water backwash, done at subfluidization velocities and followed by water alone, is the most effective procedure for media cleaning. Figure 12.36 provides a quick guide to velocities for both air and water and those recommended for practice by General Filter Company (Cleasby, 1991). The sequence of water and air is as follows: (1) slow water backwash at 8–15 m/h (3.5–6.1 gpm/ft^2) for time $0 \leq t \leq$ 5 min with air scour at 0.014–0.10 m/h (0.5–3.5 cfm/ft^2) for time $0 \leq t \leq 6$ min; (2) stop the air scour and purge the air from the media using water backwash at 29 m/h (12 gpm/ft^2) for time $7 \leq t \leq 13$ min, with increase in water flow (optional) to 45 m/h (18 gpm/ft^2) for time $13 \leq t \leq 15$ min. The procedure is to draw down the water level to about 100 mm (4 in.) above the media bed. The slow backwash should be stopped before the water level rises to 150 mm (6 in.) below the trough weir.

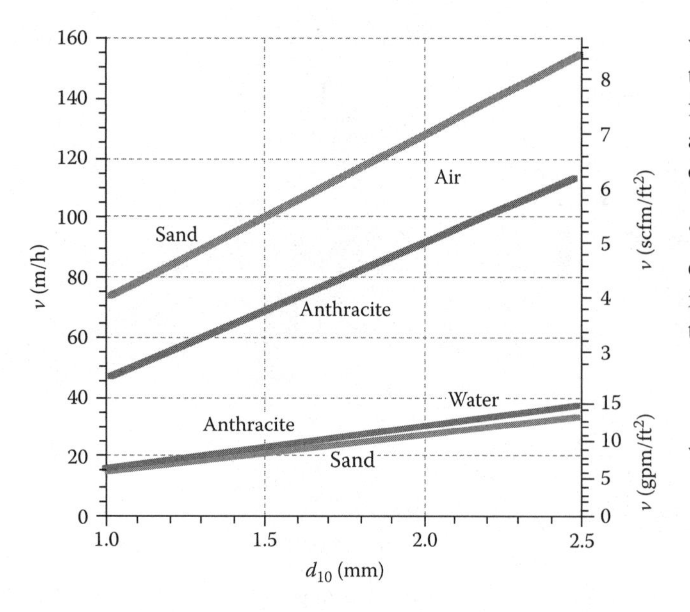

FIGURE 12.36 Velocities for simultaneous air-water backwash. (Adapted from Cleasby, J.L., Backwash and underdrain considerations, unpublished paper for short course at Colorado State University on design of filtration systems, June, 1991.)

12.4.4.10 Collapse Pulsing

The "collapse pulsing" mode of media cleaning was considered to be most effective based on visual and film observations of bubble-media behavior. Based on observations, "collapse pulsing" occurs at certain combinations of v/v_{mf} and v(air) in the subfluidization range of backwash (Amirtharajah, 1984; Hewitt and Amirtharajah, 1984). This mode is characterized by the following description. The air flow moves up through the media via air channels, eventually forming air pockets. As more air flows into a given pocket, another channel forms above the pocket and air flows up to form yet another pocket while the first pocket collapses. The first air pocket collapses due to sand being thrown up into the pocket from below due to air flow below. Within this dynamic system, the media grains slip and slide against each other resulting in abrasion and detachment of particles (Hewitt and Amirtharajah, 1984, p. 592).

The particular combinations of v/v_{mf} and v(air) in which "collapse pulsing" is observed is a "locus-of-points" defined empirically, such as seen in Figure 12.37 for three sands, that is, $d_{10} = 0.46$, 0.64, and 0.88, respectively. The best fit relationship, based on observations, is

$$\frac{v}{v_{mf}} = 0.49 - 0.119 \cdot v(\text{air}) \tag{12.59}$$

where

v is the backwash water velocity (m^3 water/m^2 filter bed area/s)

v(air) is the air flow per unit of filter bed area (m^3 air/m^2 filter bed area/s)

Referring to Figure 12.37, for air–water flow combinations below-the-line, air channeling predominates; this region is not effective. On-the-line, the air–water backwash is effective because air transfers momentum to the sand grains, which, in turn, is dissipated both by shear and by random collisions with other sand grains. In addition, the upward backwash velocity reduces the grain-to-grain pressure that is sufficient to facilitate displacement by the air bubbles and permits relative movement between grains. For operating points above-the-line, the media-water matrix begins to behave as if fluidized and the cleaning effectiveness is reduced and media loss may be appreciable (Hewitt and Amirtharajah, 1984). Media loss may be minimized by (1) operating along-the-line, (2) using larger media, for example, $d_{10}(\text{sand}) \geq 0.88$ mm, (3) locate the crest of the backwash troughs at least 760 mm (30 in.) above the media surface.

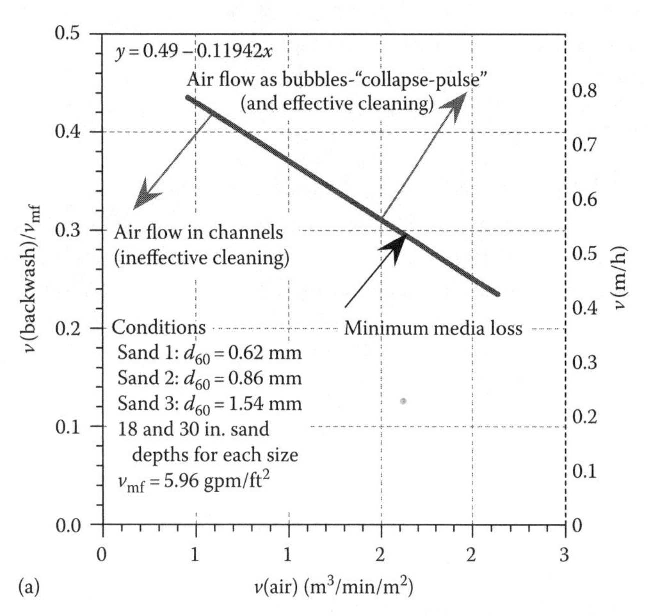

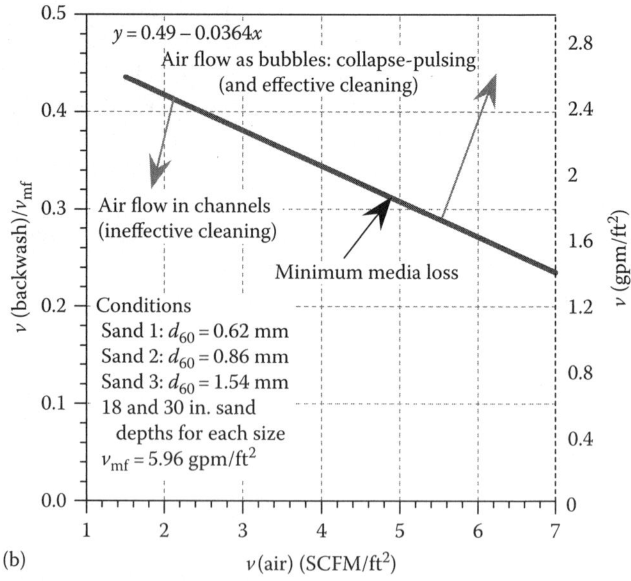

FIGURE 12.37 Air–water combinations for collapse pulsing in backwash. (a) Metric units. (b) U.S. Customary units. (Adapted from Hewitt, S.R. and Amirtharajah, A., *J. Environ. Eng., ASCE*, 110(3), 601, 1984.)

Example 12.10 Enumerate Protocol for Air–Water Backwash in "Collapse Pulsing" Mode

Given

A filter bed is sized 4.27 m × 6.10 m (14 ft × 20 ft) and uses an air–water backwash in accordance with the procedures outlined by Amirtharajah and Trusler (1982). The backwash troughs weir crests are 760 mm (30 in.) above the media. Assume the media is sand and $d_{10} = 0.46$ mm, depth 760 mm (30 in.).

Required

Determine: (1) the backwash rates for air and water, (2) determine a recommended backwash sequence, (3) estimate the volume of backwash water used, and (4) estimate the compressor power for the air wash.

Solution

1. *Backwash rates*: From Figure 12.37a, select a coordinate pair with low backwash rate and higher air rate, as recommended by Amirtharajah (1984), for example,

$$\frac{v}{v_{mf}} \approx 0.30;$$

$Q(\text{air}) \approx 1.7\ \text{m}^3/\text{min/m}^2$ (5.6 standard $\text{ft}^3/\text{min/ft}^2$)

Calculate HLR (or "v"),

$$\begin{aligned} \text{HLR} &\approx 0.30 \cdot v_{mf} \\ &= 0.30 \cdot 14.54\ \text{m/h} \\ &= 4.36\ \text{m/h}\ (1.8\ \text{gpm/ft}^2) \\ &= 73\ \text{mm/min} \end{aligned} \quad (\text{Ex}12.10.1)$$

Calculate Q,

$$\begin{aligned} Q &= \text{HLR} \cdot A(\text{filter}) \\ &= 4.36\,\text{m}^3/\text{h/m}^2 \cdot (4.27 \cdot 6.10\,\text{m}^2) \\ &= 112.76\,\text{m}^3/\text{h} \\ &= 1.88\,\text{m}^3/\text{min} \end{aligned}$$

Calculate Q(air),

$$\begin{aligned} Q(\text{air}) &\approx Q(\text{air}) \cdot A(\text{filter}) \\ &= 1.7\,\text{m}^3/\text{h/m}^2 \cdot (4.27 \cdot 6.10\,\text{m}^2) \\ &= 44.3\,\text{m}^3/\text{min}\ (1.8\,\text{ft}^3/\text{min}) \end{aligned}$$

2. *Backwash sequence (Amirtharajah, 1984)*:
   a. Lower water level in filter to media surface
   b. Begin water backwash at 1.88 m^3/min (500 gal/min)
   c. When the bed is flooded by 80–160 mm (3–6 in.) of water, which requires about 3 min from top of media, begin introducing air slowly to an air flow, $Q(\text{air}) = 44.3\ \text{m}^3/\text{min}$ (1.8 standard ft^3/min)
   d. Time to reach water level 160 mm (6 in.) from the weir crest, that is, 760 $\text{mm}^{-2} \cdot 160$ mm = 440 mm (30 − 12 = 28 in.), is

$$\begin{aligned} t(\text{rise of water}) &= \frac{440\,\text{mm}}{73\,\text{mm/min}} \\ &= 6\,\text{min} \end{aligned}$$

   e. Terminate air when the water level is about 160 mm (6 in.) from the weir crest, that is, after 6 min
   f. Increase the water backwash rate to cause about 0.20 bed expansion for say 10 min, calculated by Equations 12.52 through 12.71 as incorporated in Table CD12.7
   g. Terminate the backwash slowly so that the porosity of the bed approaches a minimal level (see Trussell and Chang, 1999; Trussell et al. 1999)
3. *Backwash volume*:

Backwash water volume		
Volume to start of air-wash:	1.88 m^3/min · 3 min = 5.64 m^3	(200 ft^3 or 1500 gal)
Volume during air-wash:	1.88 m^3/min · 6 min = 11.28 m^3	(400 ft^3 or 3000 gal)
Volume during backwash:	1.88 m^3/min · 6 min = 11.28 m^3	(400 ft^3 or 3000 gal)
Total volume:	1.88 m^3/min · 6 min = 28.20 m^3	(1000 ft^3 or 7500 gal)

4. *Compressor power*:
   From (1)

$$Q(\text{air}) = 44.3\ \text{m}^3/\text{min}\ (1.8\,\text{ft}^3/\text{min})$$

Table CDD.5 calculates P, given the data inputs, Q_a, p_2, p_1, that is,

$$P = Q_a p_1 (\kappa/(\kappa - 1))[(p_2/p_1)^{(\kappa-1)}/\kappa - 1] \quad (\text{D}.75)$$

Comments

The calculated power, P, must be increased by an efficiency factor, for example, 0.67 to calculate the compressor power. The power required by the electric motor is the compressor power increased by another efficiency factor, for example, 0.67. Calculate P using the spreadsheet; p_1(absolute pressure at compressor intake) and p_2(absolute pressure on discharge side of compressor). To determine p_1, the elevation, and atmospheric pressure is used as a rule (or subtract losses if the intake pipe is long or if there are obstructions). To determine p_2, start with absolute atmospheric pressure at the filter, and calculate water depth, bubble pressure, orifice pressure loss, pipe friction losses and any other losses, back to the compressor intake; p_2 must be high enough to overcome these losses. All of this can be seen most easily in terms of a pneumatic-grade-line (similar to an hydraulic-grade-line, i.e., HGL).

12.5 OPERATION

Operation has many facets, such as performing the functions of the filter cycle, monitoring performance, ensuring that equipment and instruments function properly, maintaining records, relating to the public, providing security, managing staff, working with consultants, reporting to regulatory agencies, interacting with city officials, and anticipating and minimizing problems whether looking ahead to new regulations or responding to unexpected exigencies in ambient water quality. This section reviews only the operation functions that relate to the depth filtration process.

A sampling of the ubiquitous issues includes mud-balls, air-binding of media, variable ambient water quality, unexpected water quality events, ripening duration, backwash duration, bacterial films on filter walls, and localized high backwash velocities.

12.5.1 Filter Operating Cycle

Figure 12.38 illustrates four phases of operation:

1. Filter-to-waste
2. Filtration
3. Draining filter for backwash
4. Backwash

The transitions from one phase to the next involves opening and closing the proper valves to direct the flow of water through the appropriate pipes and channels. In modern plants, the tasks of opening and closing valves and measuring flows are accomplished by a SCADA system.

12.5.2 Filtration Hydraulics

For $\mathbf{R} \leq 1$, the hydraulics of filters follows Darcy's law, $v = -(k/\mu) \cdot i$, albeit at high filtration rates ($\mathbf{R} > 1$ and $\mathbf{R} \gg 1$) the "Forcheimer" relation, Equation E.3, $i = a_F \cdot v + b_F \cdot v^2$, applies. For a filter in clean-bed condition, the application of Darcy's law is straightforward and simple, that is, headloss is linear with distance. As the filter bed clogs with solids, however, its intrinsic permeability changes with depth and so the hydraulic gradient changes commensurately.

12.5.2.1 Clean-Bed Headloss

The headloss at the start of the filtration cycle is called the "clean-bed" headloss. The headloss depends on the superficial velocity, v (i.e., flow to filter bed divided by area of filter bed, sometimes called hydraulic loading rate); the media (which determines the clean-bed intrinsic permeability, k); headloss, Δh; bed depth, ΔZ; and the temperature, T (which affects viscosity, μ); that is, the variables in Darcy's law, $v = (kr_w g/\mu) \cdot (dh/dZ)$,

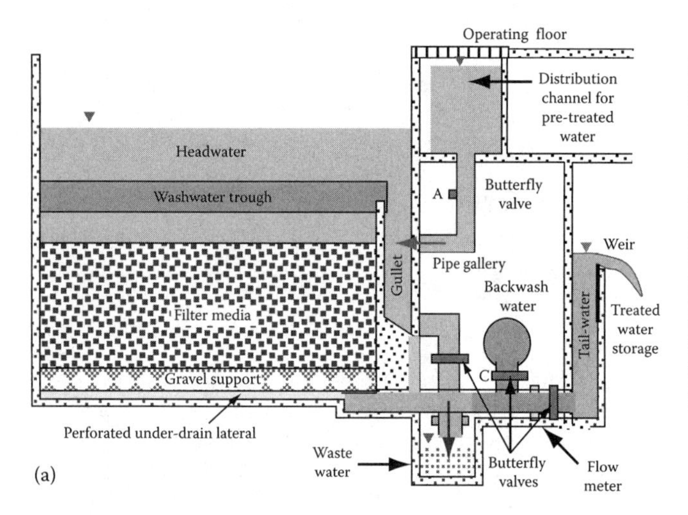

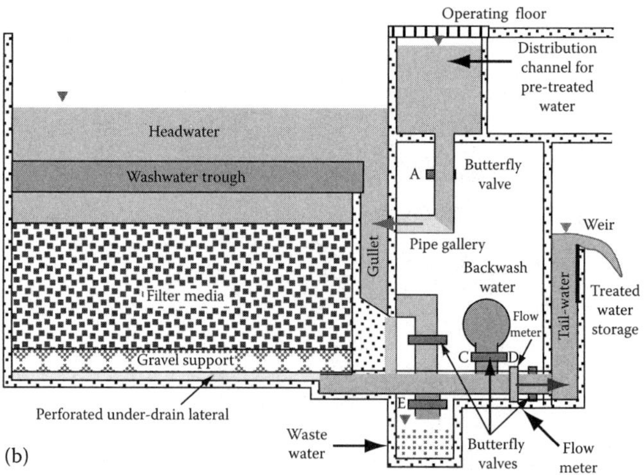

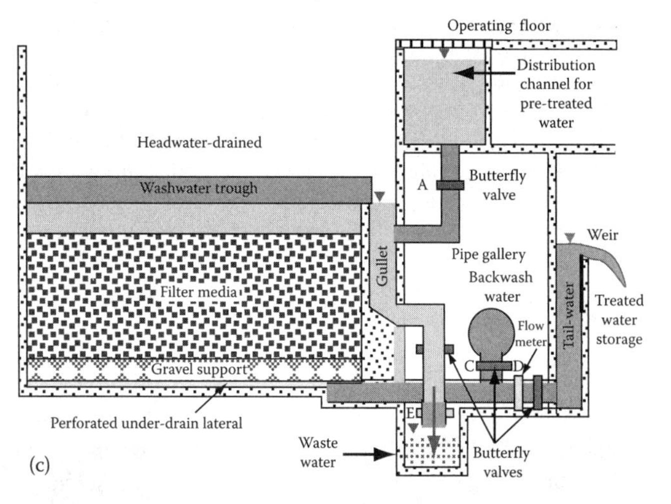

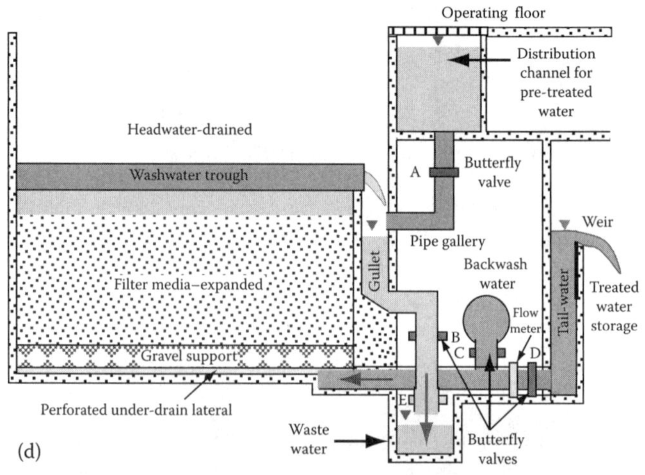

FIGURE 12.38 Four phases of filter operation. (a) Filter-to-waste. (b) Filtering. (c) Draining headwater. (d) Backwash.

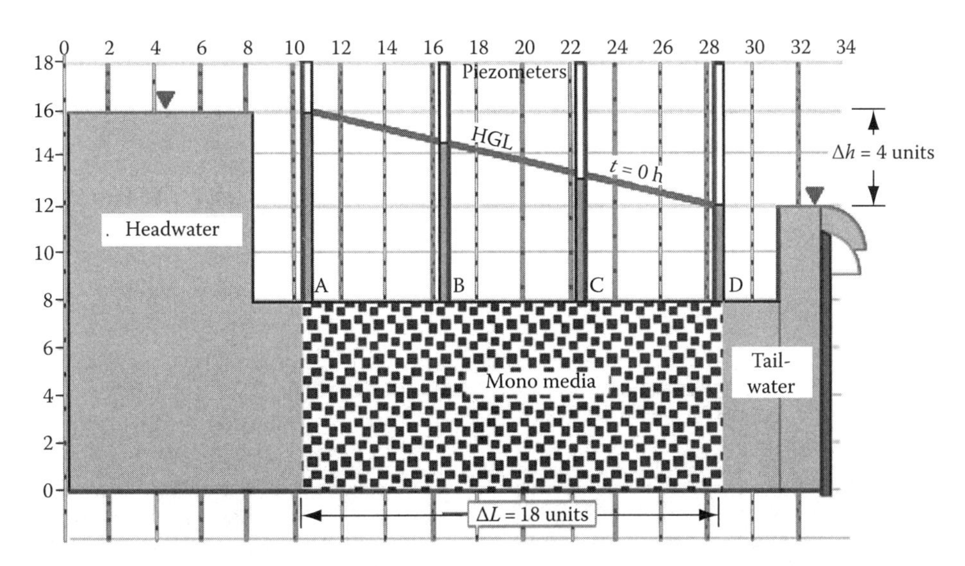

FIGURE 12.39 Piezometric head along length of clean-bed (Equation E.5).

Equation E.5. A typical value for the initial headloss might be 30 cm (12 in.). Figure 12.39 illustrates the linear hydraulic gradient for a clean-bed, oriented horizontally.

12.5.2.2 Progression of Headloss with Filter Run

Figure 12.40a is essentially the same as Figure 12.39, but the bed is oriented vertically and it shows the progress of changes in hydraulic gradient as the bed "clogs," that is, during a filter-run. At $t = 0$, that is, the "clean-bed" condition, the hydraulic grade line (HGL) is linear; as in Figure 12.39. As the filter run progresses, however, the bed clogs with solids, starting in the upper levels, and the HGL changes commensurately. The clogging effect is illustrated at $t = 2$, 4, and 6 h, with the HGL declining exponentially with vertical depth and progressing

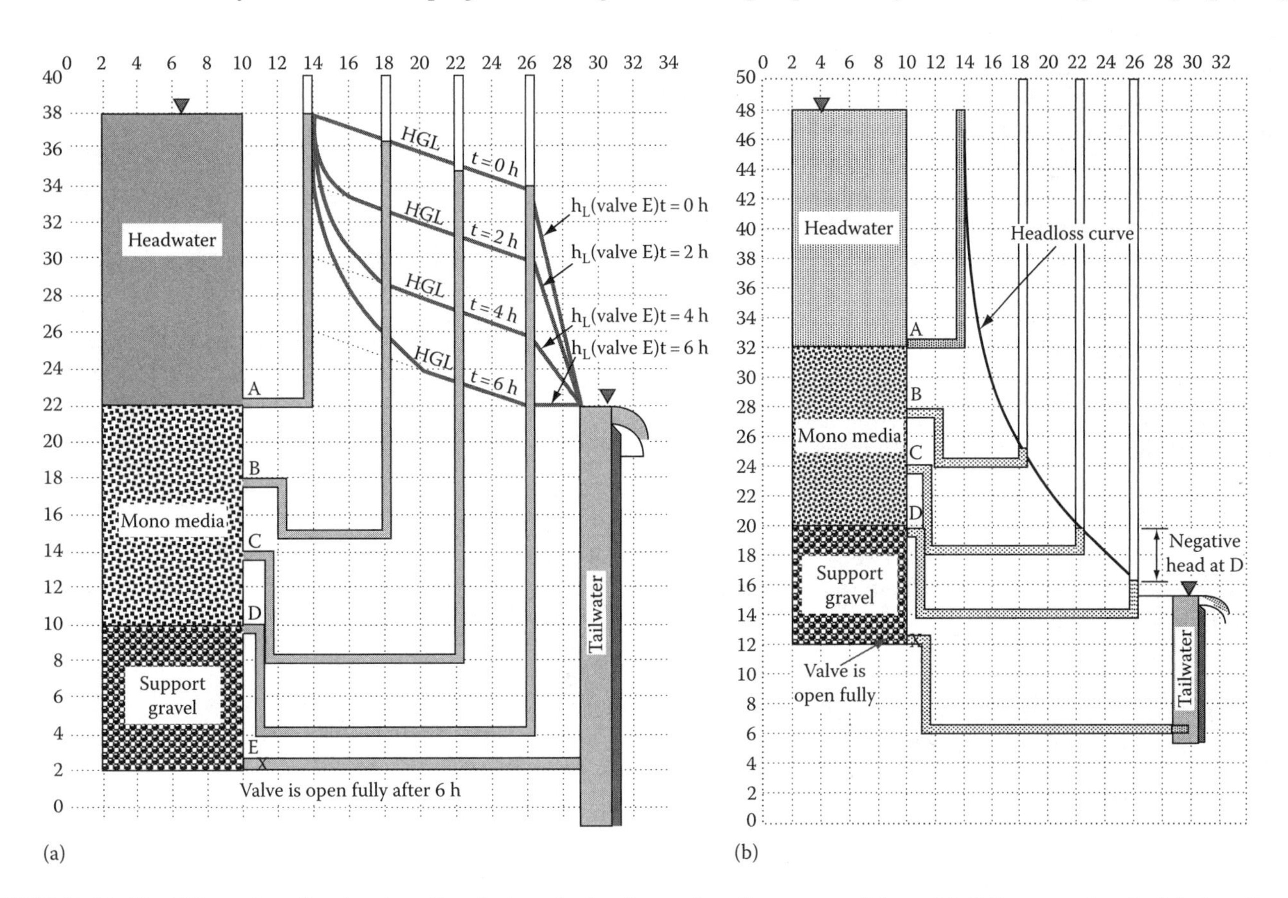

FIGURE 12.40 Effect of media clogging filter bed on hydraulic gradient for two conditions. (a) Tailwater at level of bed surface. (b) Tailwater below level of bed surface (Equation E.6). (Reprinted from Mackay, D. J., *Opflow*, 14(11), 1, 1988b. With permission. Copyright © 1988, American Water Works Association.)

with time. The "clogging front" is seen to move downward with time and is detected by the beginning of the linear part of the headloss versus distance plot. The illustration indicates that "terminal" headloss occurs at $t \approx 6$ h. Valve "E" is a rate-of-flow-controller and is opened only a slight amount at the start of the run, that is, at $t = 0$, but is opened fully at $t = 6$ h. The total headloss available is distributed between the media-bed and Valve E. In the illustration, the tailwater elevation is the same as the surface of the media-bed.

12.5.2.3 Negative Pressure

Figure 12.40b shows the same filter bed as seen in Figure 12.40a but the tailwater is below the level of the media-bed surface. The HGL is shown only for terminal headloss; for this condition, the HGL elevations are below the media taps, for example, taps, B, C, D, and E. Therefore, negative heads (or pressures, i.e., $p = gh$) occur in the media at these respective tap elevations. The "head" (or pressure) at a given tap equals the difference between its HGL elevation and its tap elevation. Consider Tap "D," for example: let Elev (HGL-D) = 8.00 m and Elev(tap-at-D) = 10.00 m; then Head(D) = 8.00 − 10.00 = −2.00 m. In other words the pressure within the filter bed at Tap D is negative, for example, −2.00/10.33 = −0.19 atm.

12.5.2.4 Air Binding

Gas precipitation, when it occurs in filters, causes "air binding" (Fair and Geyer, 1961, p. 699; see also, Section 12.5.2.3), which occurs when gas bubbles occupy volume within the filter bed. In such a case, the headloss increases inordinately and at the same time causes higher than average interstitial velocities. The effects of air binding may be observed during backwash as "boils" of large air bubbles breaking the water surface. Such precipitated air may disrupt a gravel support.

Gas precipitation can be avoided by positioning the weir crest of the tailwater at the same level as the top of the media-bed, or not too far below (see also Monk, 1987). This works unless the gas concentrations exceed what would exist at equilibrium with the atmospheric conditions at hand, that is, when "supersaturated."

Example 12.11 Evaluation of Whether Gas Will Precipitate in Filter

Given

Let elevation of a filter be 1524 m (5000 ft). The tailwater elevation is lower than the surface of the filter bed as shown in Figure 12.40b. A piezometer tap "D" has a HGL level-2.0 below the level of the tap.

Required

Determine whether gas precipitation will occur.

Solution

1. Partial pressure of oxygen and nitrogen in atmosphere is as follows:
   $P(O_2$, atmosphere, Table H.1 or B.7) = 0.209476 (mol fraction O_2)
   $P(N_2$, atmosphere, Table H.1 or B.7) = 0.78084 (mol fraction N_2)
2. Atmospheric pressure at 1524 m (5000 ft) from Figure H.2 is, P(atm, 1524 m) ≈ 0.85 atm
3. In terms of water pressure, 0.85 atm · 10.33 m water/atm ≈ 8.78 m water (absolute pressure).
4. Absolute pressure at Tap D is,
   H(D-absolute) = 8.78 m − 2.0 m = 6.78 m water ≈ 0.65 atm.
5. Saturation concentrations of pure gases at 1.0 atm are as follows:
   C(1.0 atm pure O_2, 20°C) = 43.39 mg/L
   C(1.0 atm pure N_2, 20°C) = 19.01 mg/L
6. Saturation concentrations of gases at their respective partial pressures in the atmosphere at 1524 m elevation are as follows:
   $C(O_2$, 20°C, 0.85 atm absolute) = 7.73 mg/L (i.e., 43.39 · 0.209 · 0.85)
   $C(N_2$, 20°C, 0.85 atm absolute) = 12.62 mg/L (i.e., 19.01 · 0.781 · 0.85)
7. Saturation concentrations of gases at elevation −2.0 m water pressure at 1524 m elevation, which is 0.65 atm absolute pressure, is their respective partial pressures in the atmosphere at 1524 m elevation are:
   $C(O_2$, 20°C, 0.65 atm absolute) = 5.91 mg/L (i.e., 43.39 · 0.209 · 0.65)
   $C(N_2$, 1524 m, 20°C, 0.65 atm absolute) = 9.65 mg/L (i.e., 19.01 · 0.781 · 0.65)
8. Therefore, since
   7.73 > 5.91 mg O_2/L, oxygen gas will precipitate.
   12.62 > 9.65 mg N_2/L, nitrogen gas will precipitate.

Discussion

A not uncommon design has been to locate the clear-well below the filter bed bottom, with a pipe from the underdrains discharging into the clear-well. Thus when the rate-of-flow-controller valve is open all the way the HGL will drop below the media elevation (as illustrated in Figure 12.40) which is, by definition, a negative pressure, which may cause gas precipitation. As another issue, if the water is "supersaturated" on entering the filter bed, for example, due to air bubbles being entrained in a pipeline that drops in elevation, or due to algae photosynthesis, gas precipitation could occur even if the tailwater elevation is at the same level as the filter bed surface. To avoid gas precipitation, the gas must be removed before entering the filter bed (see Chapter 18 and Appendix H).

12.5.3 Backwash

At the end of a filter run, the filter is backwashed to remove the attached floc from the media grains. Traditional backwash involves bed fluidization. Ancillary steps may include surface-wash or air scour, or both (see also Logsdon, 2008, pp. 115–145). An inadequate backwash is likely to result in "mudball" formation.

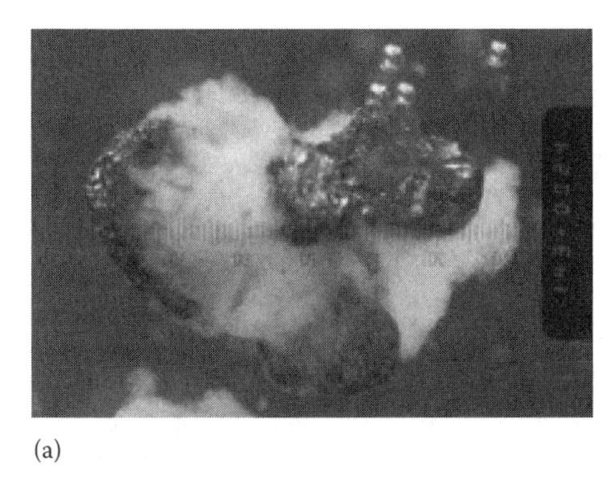

(a)

(b)

FIGURE 12.41 Mudballs from filters: beginning and fully formed. (a) A microscopic agglomerate of alum floc and anthracite. (Courtesy of Grant Williamson-Jones, City of Fort Collins Utilities, Font Collins CO.) (b) Fully formed mudball. (Reprinted from Mackay, D.J., *Opflow*, 14(11), 1, 1988b. With permission. Copyright © 1988, American Water Works Association.)

12.5.3.1 Mudballs and Surface Cracks

Figure 12.41a shows a small agglomerate of floc and media, which is the beginning stage of a mudball formation. Figure 12.41b shows an example of a fully formed mudball. If not disrupted by cleaning, the small agglomerate of floc and media, seen in Figure 12.41a will grow in size during continued filter cycles with the formation of mudballs, as in Figure 12.41b. Baylis (1937, p. 1020) described the problem.

The data presented showed that much of the filter bed trouble originated from an accumulation of compacted coagulated material not removed by the washing system, and to coatings on the sand grains. Shrinkage or settlement of the beds while in service usually is caused by a soft coating around the sand grains which is not removed by backwashing the filters. This shrinkage allows cracks to open along the sidewalls and occasionally other parts of the bed.

The most effective control is to clean the media adequately through backwash with surface-wash or air-wash. Surface-wash is especially effective in breaking up compacted surface layer, thought to be a precursor to mudball formation. The use of polymers may also predispose the media to "stickiness," and development of mudballs. If mudballs build up to a great extent the media may require replacement.

12.5.3.2 Floc-to-Grain Bonding

The bonding of an aluminum or iron floc to a grain of sand or anthracite is due to four forces: (1) van der Waals, (2) electrical double-layer, (3) Born repulsion, and (4) structural (Raveendran and Amirtharajah, 1995). The hydraulic shear and grain-to-grain shear and impact forces developed by backwash must overcome these bonding forces in order to dislodge attached particles.

12.5.3.3 Practice

In backwash, the media expands and the grains are maintained apart due to the nature of the pressure field associated with an array of particles. Therefore, grain-to-grain contacts do not occur and fluid shear is the only removal mechanism (Hewitt and Amirtharajah, 1984). Both are necessary, however, for effective cleaning. Therefore, an adjunct to bed fluidization, that is, either air-wash or surface-wash, is required.

12.5.3.4 Operating Protocol

An operating protocol recommended by Amirtharajah (1982) for both effective cleaning and minimizing media loss is as follows:

1. Lower water level in filter to media surface.
2. Begin water backwash.
3. Introduce air slowly after bed is flooded by 80–160 mm (3–6 in.) of water, using low water velocity and high air loading rate which also prolongs the time to reach the weir crest of the wash-water trough.
4. Terminate air when the water level is about 160 mm (6 in.) from the weir crest.
5. Increase the water backwash rate to cause about 0.20 bed expansion.

The procedure requires considerable operator attention. A SCADA system can be programmed to perform these functions.

12.6 PILOT PLANTS

For reference, Section 3.5 describes the general issues of pilot plant experimentation. Examples of permanent pilot plants used as an adjunct to operation have been at Fort Collins, Colorado (1988) and Bellingham, Washington (1993), respectively. The Fort Collins pilot plant was set up with two conventional treatment trains, with 37 L/min (10 gpm) flow each, which can be changed to in-line or direct modes by valve adjustment with flow directed to any one or all of three filter columns, each with a different media design. The pilot plant has been used to address design questions as changes are contemplated and for a continuing array of operational questions. The Bellingham pilot plant has a flow of 37 L/min (5 gpm) per train, with three in-line treatment trains. Figure 12.42 shows one of the 305 mm (12 in.) square filter columns. The pilot plant was installed to address process design questions, anticipating plant expansion as population increases. One question was whether increased filtration velocity, as a means of handling increasing future demands,

FIGURE 12.42 One of three pilot plant filters at Bellingham WTP; rapid-mix precedes filters. (City of Bellingham, WA.)

would cause a reduction in water quality and to determine the effect on length of run. Other questions were whether chemical dosages could be reduced as an economic question and to confirm the capability of the system to remove various organisms.

The theoretical theme of a pilot plant, as in a full-scale plant, is that chemical factors determine the attachment efficiency and that physical factors determine the transport efficiency. The former is related to operation and the latter to design, for example, coagulant dosage, polymer selection, pH, etc., and media depth, media size, HLR, etc., respectively. In addition, the pilot plant study may determine the rate of headloss increase, which is affected by the size of floc and the pore sizes, that is, both chemical and physical factors.

In experiment design, all independent variables, for example, source water, pH, coagulant dosage, HLR, media design, etc., are maintained constant. The selected dependent variables, for example, effluent turbidity, headloss, are then measured during the course of the run. For the next several runs, one independent variable is changed, for example, HLR, and its effect on turbidity and headloss is determined. The single treatment train may be split, if desired, to two or three filters each with a different media; this permits comparing the effluent turbidity versus HLR relation for three media designs (i.e., physical variables). On the other hand, if two treatment trains are used, that is, each with its own rapid-mix, identical filtration columns, etc., then the effect of chemical variables may be explored. The same thing may be done with only a single treatment train by a sequential set of experiments, which increases the time for an experimental program. Conducting the experiments in parallel reduces the time, but requires more manpower.

12.6.1 Equipment

The pilot plant schematic of Figure 12.43 shows some of the equipment required and its configuration to investigate chemical factors, for example, rapid-mix conditions may vary or chemicals may vary in dosages, polymers, sequence, etc. If physical conditions, for example, in filter design, are to be investigated, a common pretreated water header would feed one to four filters. The functions of a pilot plant which are the same as a full-scale plant, for example, metering flow, measuring selected influent and effluent characteristics, providing for chemical additions, sampling taps, backwash, etc., are also indicated in Figure 12.43. If the pilot plant set-up is duplicated, for example, two, three, or even four trains, the experimental program may progress faster. Too many treatment trains, on the other hand, could load the laboratory and perhaps the ability to process and assimilate data.

12.6.1.1 Contaminant Injection

Quite often in pilot plant work, there is interest in testing removals of specific contaminants, for example, MS-2 virus, total coliform bacteria, certain algae, *Giardia lamblia* cysts, *Cryptosporidium parvum* oocysts, etc. (Hendricks et al., 2005; Al-Ani et al., 1986). Such tests provide a confirmation of the log removals of such organisms, which may be important for various reasons, and may be done by injecting a flow of contaminants by a metering pump.

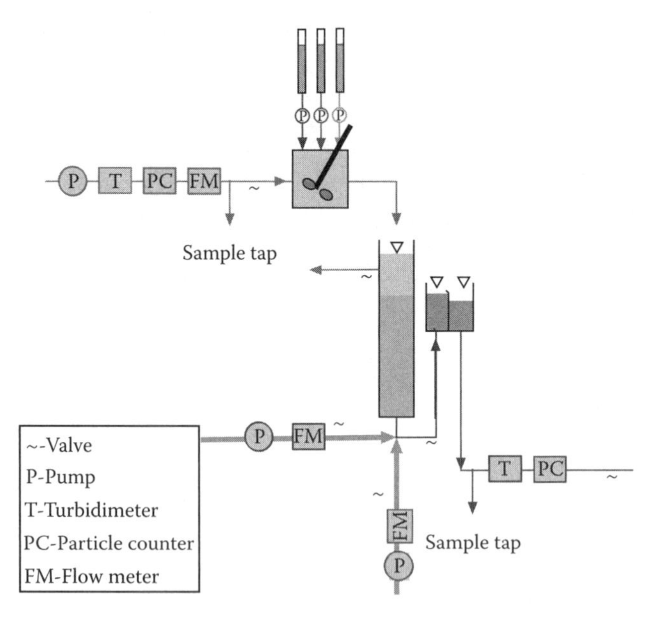

FIGURE 12.43 Pilot plant set-up to investigate chemical factors of design.

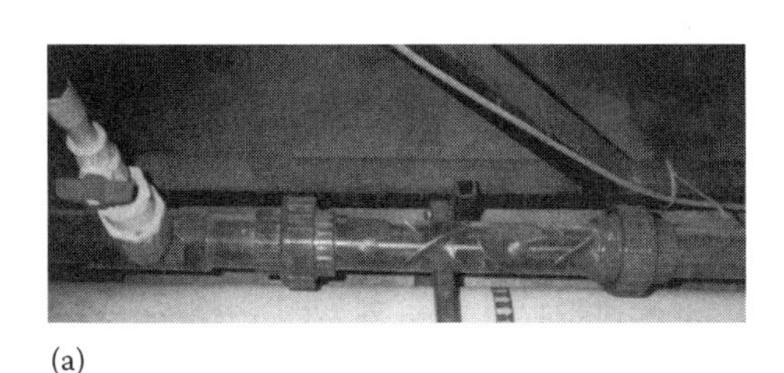
(a)

(b)

FIGURE 12.44 Contaminant injection and sampling in 51 mm (2 in.) PVC influent pipe for 76 L/min (20 gpm) pilot plant, Engineering Research Center, Colorado State University, Fort Collins, CO. (a) Contaminant injection point upstream from static mixer. (b) Contaminant sampling points for grab sampling and continuous flow sampling (cartridge filter sample), respectively.

The concentration of the contaminants mixed with inflow should be adequate to detect numbers on the effluent side (e.g., without the use of detection limits). For viruses and bacteria, grab sampling is recommended for both influent and effluent at say 15 min intervals. For algae, cysts, and oocysts, cartridge filters with 1 μm size may be used to collect perhaps 200 L on the influent sample and perhaps 1000 L of effluent sample, that is, large enough to obtain countable numbers.

Figure 12.44a shows an injection point for test organisms (or other contaminants of interest) located just upstream from a static mixer. Figure 12.44b shows sampling points downstream from the static mixer after the contaminants have been mixed with the raw water inflow.

Example 12.12 Metering Contaminants

Given

Suppose the detection limit of *Giardia* cysts is say 0.01 cyst/L. Removals are about 3.5-log; assume 4-log for a conservative estimate. The pilot plant flow is 76 L/min (20 gpm) for a 610 mm (24 in.) square filter column.

Required

Number of cysts in a 2000 mL volume, to be metered into the flow over 120 min.

Solution

1. Let $C(\text{effluent}) = 0.01$ cyst/L
2. If $\log R = \log(C/C_o) = -4.0$, then,

$$\log(0.01/C_o) = -4.0, \text{which gives,}$$
$$C_o = 100 \text{ cysts/L}$$

3. $Q(\text{contaminant metering}) = V(\text{contaminant})/t(\text{metering time})$

$$= 2000\,\text{mL}/120\,\text{min}$$
$$= 16.67\,\text{mL/min}$$

4. $Q(\text{contaminant metering}) \cdot C(\text{contaminant metering}) = Q(\text{pp flow}) \cdot C_o$

$$(0.01667\,\text{L/min}) \cdot C(\text{contaminant metering}) = (76\,\text{L/min}) \cdot (100\,\text{cysts/L})$$
$$C(\text{contaminant metering}) = 455{,}908\,\text{cysts/L}$$

5. $N(\text{cysts}) = Q(\text{contaminant metering}) \cdot C(\text{contaminant metering}) \cdot t(\text{metering time})$

$$= 0.01667\,\text{L/min}) \cdot 455{,}908\,\text{cysts/L} \cdot 120\,\text{min}$$
$$= 911{,}998\,\text{cysts}$$
$$\approx 1.0\,\text{million cysts}$$

Comments

Cysts up to perhaps 100 million may be obtained from various sources. Typical numbers for injection in a large pilot plant are around 10 million. As a caution, injections of organisms in the vicinity of operating plants may be not advisable unless precautionary measures are taken for proper isolation of the pilot plant and for disposal of solids produced and filter effluent. The solids will be contaminated and the filter effluent should be presumed to be contaminated.

12.6.1.2 Filter Column

The filter column is a simple tube, but a number of considerations enter into its design. Common sizes for filter columns are 51 mm (2 in.) diameter, 102 mm (4 in.) diameter, 152 mm (6 in.) diameter. For a permanent installation, a 305 mm (12 in.) square size is common. To construct the column, clear PVC is favored because it is relatively cheap, durable, and is easy to glue and cut without a machinist. Piezometers taps, 6.3 mm (1/4 in.) in size, may be spaced at 100 mm on centers along the column length to ascertain the hydraulic gradient. The approximate length of the column should be the sum of length of bottom fittings to porous plate + depth of media + increment of bed expansion during backwash + distance above expanded media for backwash effluent + length of top fittings + additional head desired such that filter runs are terminated by effluent concentration, not by headloss limitation. The backwash flow may exit the column about 300 mm (12 in.) above the expanded media, or whatever is necessary to avoid loss of media. The coagulated water flow should enter the column with low velocity such that the medium is not disturbed. The flow is metered and controlled on the influent side. The effluent flow should leave the system by means of an overflow weir with crest about the same elevation as the top of the media. Ball valves are the cheapest and most functional for on-off control.

12.6.1.3 Pilot Plant System

Regarding pilot plant flow, there are no firm rules. A flow for a single treatment train, or filter, in the range, $20 < Q < 80$ L/min ($5 < Q < 20$ gal/min) is manageable in terms of having flows large enough that coagulant flows may be metered and measured accurately and yet the logistic demands, for example, for chemicals, contaminants-to-be-injected, etc., are not dominant factors.

12.6.1.4 Data Handling

Data procurement and processing requires an organized approach. If attention is given to this phase of pilot plant work the data generated can be processed in terms of final plots and tables and archived in a form that permits easy retrieval. To minimize mistakes and to facilitate data processing, metric units are preferred. Conversions to U.S. Customary units can be done easily by spreadsheet for any final results. A 24 h clock also reduces confusion, recorded to the minute, for example, 2145 h. Data should be recorded on forms designed for the project at hand and transferred to a spreadsheet daily. Generally, a separate line should be provided for each clock time that a data set is recorded.

12.7 WASTEWATER FILTRATION

Not too much has been formalized about guidelines for filtration of wastewaters, albeit by the late 1980s, the practice became fairly widespread. Proprietary systems have assumed a prominent role, sometimes with a prior pilot plant study and other times a unit has been placed online with the idea of working out operational procedures in the course of water production.

Typical treatment trains for wastewater filtration given by Tchobanoglous and Eliassen (1970) are as follows:

1. Chemical treatment of raw sewage, then filtration, followed by further treatment
2. Chemical treatment of secondary effluent, then filtration, followed by further treatment
3. Filtration of secondary effluent with or without further treatment

Further treatment could include other unit processes, for example , adsorption, ion exchange, oxidation, disinfection, etc. The third treatment train is the focus of this section, that is, filtration of biological floc. The first two, are primarily for removal of nutrients.

12.7.1 Background

Filtration of wastewaters has to do with making the water suitable for some further use, for example, cooling water, irrigation of golf courses, irrigation of certain crops, industrial water, and even as a precursor to further treatment that could include drinking water. Rapid filtration may both reduce the overall suspended solids loading to an ambient water and attenuate fluctuations.

The first study for filtering secondary wastewater effluent, illustrated in treatment train (3), was by Tchobanoglous and Eliassen (1970). Their pilot filter was set up with piezometers and sampling taps along its depth and also had one side that could be removed to examine deposits of solids. Uniform sand, that is, UC $\approx$ 1.0 with "equivalent diameters" of 0.49, 0.68, and 0.98 mm were used as media to filter secondary effluent with suspended solids concentrations 5–18 mg/L. Particles sizes were bimodal at about 5 μm and about 90 μm average size. Zeta potentials were about −20 mV as a mean. The suspended solids were reduced about exponentially within the first 50 mm (2 in.) of bed depth with decline being unchanging with depth at $Z > 50$ mm for all three sand sizes and for three filtration velocities, that is, 4.88, 14.2, 24.4 m/h (2.0, 5.8, 10.0 gpm/ft^2). Removal for the 0.49 mm sand at 4.88 m/h (2.0 gpm/ft^2) was only 0.40 fraction at depths $Z > 50$ mm, which was the highest and was lowest, with removal about 0.2 for the 0.98 mm sand at 24.4 m/h (10.0 gpm/ft^2). They did not observe a moving wave front as seen in filtration of metal flocs and determined that the removal mechanism was straining in the top layer. As one indication that the removal mechanism was straining was that the headloss curves for each media increased with time with the shape of a power function.

12.7.2 Forms of Practice

Filtration may be added to a biological treatment train for wastewaters, or it may follow chemical treatment. Designs have included denitrification as well by adding methanol to induce a biological reaction in the filter. The procedure involves a short backwash (3 min) at 4–8 h intervals to remove nitrogen gas.

12.7.2.1 As a Unit Process within a Water Treatment Train

A common designation for treatment of wastewater following secondary biological treatment is "tertiary treatment." In some cases, a conventional water treatment train is employed to follow secondary wastewater treatment. The main idea is to remove particulates. Particular problems that are characteristic of wastewater as a source water include (1) biofilms may appear more luxuriant than in ambient source waters and (2) the sludge is more putrescent. In treatment, this means that cleaning must be frequent enough to control growths. Hosing, followed by a disinfectant is usual when dealing with wastewaters that have organic matter as a predominant characteristic.

12.7.2.2 As a Stand-Alone Process Following Biological Treatment

Filtration alone following biological treatment has been practiced in two forms: (1) as cake filtration and (2) as depth filtration. If the particles to be removed are smaller than the pores of the media, straining occurs and the particles are retained on the media surface, forming a "cake." Further removal is by straining by the cake, which increases in thickness as particles are further retained and accumulated. Because the headloss increases rapidly frequent backwash is required.

TABLE 12.12
Performance Data for Three Filters at Ames, Iowa

Parameter	Parameter Units	Filter Influent	Filter Effluent		
			Dual Media	Tri-Media	Coarse Sand
BOD	(mg/L)	30.38	12.68	12.99	14.46
Soluble BOD	(mg/L)	9.67	7.21	7.27	7.78
Suspended solids	(mg/L)	34.08	7.05	6.82	9.46
TOC	(mg/L)	19.86	12.02	12.77	12.99
SOC	(mg/L)	13.41	12.00	11.83	12.98
Turbidity	(FTU)	17.60	4.80	6.78	4.66
	Media Characteristics				
	Anthracite	d_{10} (mm)/UC	1.03/1.57	1.03/1.57	
		Depth in mm (in.)	381 (15 in.)	381 (15 in.)	
	Sand	d_{10} (mm)	0.49/1.41	0.49/1.41	2.0/1.52
		Depth in mm (in.)	229 (9 in.)	229 (9 in.)	1168 (46 in.)
	Garnet	d_{10} (mm)		0.27/1.55	
		Depth in mm (in.)		76 (3 in.)	

Source: Adapted from Cleasby, J.L. and Lorence, J.C., *J. Environ. Eng. Div., Proc. Am. Soc. Civil Eng.*, 104(EE4), 759, 1978.

Notes: BOD, biochemical oxygen demand (5 day is understood unless subscripted otherwise). Soluble BOD = BOD from filtrate of filter paper. Suspended solids = solids retained on filter paper after oven drying. TOC, total organic carbon. SOC, soluble organic carbon. Turbidity measured by light scattering instrument calibrated by formazin standard (formazin is a chemical made commercially available for this purpose). Filter influent is from secondary clarifier that follows trickling filter treatment from Ames, Iowa WWTP. Filter effluent is from each of three filters operated in parallel at 7.8 m/h (3.2 gpm/ft^2).

In filtration of wastewater by depth filtration, the media selection, its depth, and backwashing are all important. A summary of performance of three filter designs is given in Table 12.12. The available headloss for each filter was 1.83 m (6 ft) and length of runs were about 12 h for dual media and tri-media filters and about 24 h for the coarse sand filter. In comparing the three filter designs, the coarse sand had only slightly higher effluent suspended solids than the dual media and mixed media filters, with effluent turbidity levels being about the same. The $d_{10} = 2.0$ mm size for the coarse sand was also found to be most appropriate in terms of floc capture per unit of headloss. The most effective backwash procedure was subfluidization of the bed coupled with air-wash.

12.8 PROPRIETARY EQUIPMENT

In every generic unit process various kinds of ancillary items of equipment are required for the process to function, for example, pipes and valves, surface-wash, under-drain blocks, media support, etc. At the same time, proprietary firms have produced "package" water treatment systems. The discussion here is intended to indicate the kinds of products available.

12.8.1 Ancillary Equipment

The support equipment includes surface-wash nozzles, pipes and valves, media, backwash systems, air scour systems, under-drain systems, control systems, and instrumentation. The latter may include water level measurement, pressure measurement, flow measurement, and online turbidity and online particle counting (or whatever else may be of interest, with fewer and fewer limitations). Equipment catalogs and internet web sites and local (or regional) representatives are sources of information.

12.8.2 Package Filtration Systems

A package plant is a small unit that has all of the components required to facilitate operation as a system. For small systems, such as for populations up to 5000, package plants are often used rather than a generic design. For fewer than 1000 persons, a package plant would be strongly favored over a generic design. A variety of proprietary pilot plants are on the market and a few examples are described here.

12.8.2.1 Deep Bed Filtration—Parkson DynaSand®

The Parkson DynaSand Filter has been installed for municipal water treatment, industrial water treatment, industrial wastewater treatment, and treatment of municipal wastewater. The Dyna Sand® filter is a continuous backwash upflow, deep bed granular media filter. The filter media is cleaned continuously by recycling the sand internally by means of an air lift pipe and sand washer with the clean sand redistributed on the top of the bed. Units were installed at the Coors/Golden, Colorado, wastewater treatment plant (WWWTP) for final polishing of wastewater effluent before discharge to the adjacent Clear Creek. Bed depths of 1016–2032 mm (40–80 in.) are available and sizes from 0.914 to 3.43 m (3 ft-0 in.–11 ft-3 in.) diameter.

FIGURE 12.45 Culligan Multi-Tech® filtration system as set up at Colorado State University, 1985, for evaluation of removals of *Giardia* cysts. (Courtesy of ABW® Infilco-Degremont, Richmond, VA.)

FIGURE 12.46 Automatic backwash (ABW® Infilco-Dregemont) at Ignacio, Colorado.

Concrete modules have been constructed up to 4.65 m^2 (50 ft^2). Parkson listed some 300 installations for industrial water and wastewater treatment (Parkson DS1, 1991).

12.8.2.2 Deep Bed Filtration—Culligan Multi-Tech®

The Culligan Multi-Tech® came about from the collaboration with Professor E. Robert Baumann (Iowa State University) and John Scanlan (Culligan) in the early 1980s. The system was intended for small municipal water systems as a package technology, made feasible by modern SCADA systems (the Culligan had a form of this by the mid-1980s). Figure 12.45 is a photograph of a system, c. 1985, which was effective in reducing turbidity to low levels and in removing Giardia lamblia cysts (Horn et al., 1986; Horn et al., 1988). The key components are: a centrifugal pump that pressurizes a system, a first stage filter of coarse media, and a second stage filter of graded media, finer than in the first stage. The coagulant is injected into the first stage. Backwash may be programmed on time, on headloss, or turbidity breakthrough. The coagulant dosage must be determined by the operator.

12.8.2.3 Shallow Bed Filtration—ABW®

The automatic back wash filtration system (ABW®) was developed in the 1950s by Hardinge and was later acquired by Environmental Elements, Inc. of Baltimore, later assimilated by Infilco-Dregremont, Inc. (http://www.degremont-technologies.com/dgtech.php?article390). The filter has had skeptics due to its shallow depth and ostensibly short duration backwash. Despite this the system has been adopted rather widely for both municipal drinking water and for filtration of secondary effluent from biological wastewater treatment. The bed may be 30 m (100 ft) long × 6.1 m (20 in.) wide and is divided into cells about 305 mm (12 in.) wide with plastic partitions. The headwater is the same for all cells and thus the filtration scheme is "declining rate." The under-drain for a given cell collects water from that cell only with conveyance to a filtered water flume on one side that runs the length of the filter giving a common tailwater elevation for all cells. Backwash is accomplished by a traveling bridge that has a backwash pump and a "shoe" that fits over the under-drain block for a particular cell during the time of backwash. The backwash water source is the flume of finished water. The bridge moves continuously from one cell to the next and from one end of the bed to the other and back. Figure 12.46 shows an ABW® system as installed for municipal water treatment at Ignacio, Colorado. The system has been used for filtration of secondary treated wastewater. An example is at Aspen, Colorado, where such a filter is used following secondary biological treatment and before discharge of the effluent into the Roaring Fork River.

12.8.2.4 Package Filtration—EPD Wearnes USA®

The EPD Wearnes USA® filter system started as a swimming pool filter in the 1980s (under its predecessor company Environmental Products Division, Hoffinger Industries) and then, in the early 1990s, after further development, found a market in small drinking water system applications. The filter system is two stage, each stage in a cylindrical tank with horizontal axes. The two tanks are filled with media of garnet to a depth of 305 mm (12 in.) with the first stage coarser than the second. For the first unit, $d_{10} = 0.27$ mm, UC = 1.6, and for the second unit $d_{10} = 0.18$ mm, UC = 1.3. The backwash rate was 37 m/h (15 gpm/ft^2) without loss of media. The system is automated and is backwashed based on pressure, time, or effluent turbidity. Coagulant feed and turbidity and particle counting instrumentation were a part of the "package," along with a SCADA system.

12.8.3 Evaluation of Products

Information on virtually any product is available through various sources, for example, manufacturer's brochures, Web sites, and manufacturer's representatives. Trade shows are the best place to obtain information. Demonstrations are set-up, as a rule, and knowledgeable representatives from the company are always present to explain and consult. As a rule, proprietary systems are available as skid-mounted pilot plant units that may be shipped to any location. In most cases, the units are ready to connect to raw water or wastewater pipes or hose.

PROBLEMS

12.1 Theory of Depth Filtration

Given/Required

Discuss the relationship between the clogging rate in a filter and the transport efficiency.

12.2 Practical Design of a Rapid Filter

Given/Required

Design the operational components of a rapid filter. Include filter box sizing, media selection and depth, surface-wash system, backwash storage, gravel support and tile under-drain system, wash-water trough and gullets, pipe sizing, valves, control devices and instrumentation. Draw appropriate diagrams showing arrangement of all components.

12.3 Backwash Protocol

Given/Required

Outline the sequence of steps for backwash, specifying duration of each step.

12.4 Pilot Plant for Filtration of Wastewater Effluent

Given/Required

A physicochemical treatment scheme is proposed for a municipal waste effluent. Most likely, this approach would be considered only for a small flow, for example, a camp in a remote area, for example, $\leq$500 persons with water use estimated at 400 L/person/day giving about 200 m^3/day (50,000 gpd). Most likely, the solids would be transported from the site.

(a) Outline a pilot plant program for process design of the filtration step.
(b) What kinds of design results will the pilot plant studies give?
(c) How will data be analyzed in terms of plots?
(d) Show a data sheet which might be used.
(e) Sketch the experimental setup.
(f) Consider how you would handle the solids.

12.5 Filter Bed Hydraulics

Given/Required

Figure 12.39 shows a filter cross section along with a hypothetical hydraulic grade line (HGL) after some hours of operation. Use the sketch as the basis for the following tasks.

(a) Sketch the HGL at the start of a filtration run.
(b) Looking at the HGL as drawn, indicate where most of the solids have accumulated.
(c) Show the negative head at point C.
(d) Show the headloss at the valve, point E, at the beginning and end of the filter run.

12.6 Optimizing Rapid Filter Design by Pilot Plant

Given/Required

(a) Sketch a hypothetical h_L versus time curve obtained by pilot plant experiment along with a hypothetical concentration breakthrough curve.
(b) Illustrate how the filter box depth may be determined by extrapolating the h_L versus time curve to the time at which concentration breakthrough occurs.

12.7 Negative Pressure in Depth Filtration

Given

A typical filter design showing headwater and tailwater (the latter being the filtered water) was given by McNamee et al. (1956, pp. 798, 799). The media is sand, 27 in. depth with 12 in. graded gravel with Wheeler bottom with a plenum perhaps 18 in. deep. The filtered water reservoir is below the filters with the maximum water level 12.0 ft below the headwater above the filter bed. A rate-of-flow controller maintains a constant effluent flow from the filter. The pipe from the filter bottom to the tailwater is "closed," that is, no air gap.

Required

Show the hydraulic grade line (HGL) from the headwater to the tailwater (a) at the start of the filter run, and (b) at the end of the run with the rate-of-flow controller valve wide open.

12.8 Headloss for Rapid Filter Backwash

Given

Suppose a "self-backwashing system" is to be designed as part of a rapid-rate filtration system. The filter box is described as follows:

$$\begin{aligned}D(\text{filterbox}) &= D(\text{under-drains} + \text{gravel support})\\ &\quad + D(\text{media}) + \text{max. headloss} + \text{freeboard}\\ &= 600\,\text{mm} + 2000\,\text{mm} + 3000\,\text{mm} + 600\,\text{mm}\\ &= 24\,\text{in.} + 66\,\text{in.} + 120\,\text{in.} + 24\,\text{in.}\\ &= 6200\,\text{mm}\ (244\,\text{in. or } 20\,\text{ft})\end{aligned}$$

The foregoing dimensions are approximations for guidance; they may be modified as desired. The bed is anthracite with $d_{10} = 1.0$ mm, UC = 1.3 and SG = 1.68 (as used by Amirtharajah and Cleasby, 1972, p. 58).

Required

Determine the depth of filter box required in order to implement the self-backwashing feature of the design. Assume the HLR(filtration-mode) $\approx$ 24 m/h (10 gpm/ft^2).

Hint: Assume, as a first try, and if needed, that the backwash velocity is 61 m/h (25 gpm/ft^2). For filter headloss, the intrinsic permeability, *k*, may be estimated from Figure E.2. Table CDE.2 provides a means to calculate *k* from *K*. An alternate approach is to utilize Table CDE.4, based on the Forchheimer equation. As seen, the problem is essentially three problems: (a) to show the filter box, in a drawing, (b) to select the headloss advisable for normal operation, and (c) to estimate the headloss required for the backwash.

12.9 Headloss to Expand Bed of A Rapid Filter

Given

A bed of sand has $d = 400$ mm and $\varepsilon \approx 0.40$.

Required

Calculate Δh for incipient fluidization.

12.10 Scenario Explorations by Mathematical Modeling Using Spreadsheet

Given

From a pilot plant study, the filter coefficient, λ, was determined as, $\lambda = 0.06\ \text{cm}^{-1}$

Other conditions.

- Media is anthracite, $d_{10} = 1.1$ mm
- Porosity, $P = 0.40$
- HLR = 5 gpm/ft^2
- Temperature, $T = 20$°C
- ρ(floc) ≈ 1.05 g/cm^3
- d(floc) can range from 1 to 300 μm, depending on coagulant dosage, rapid mix, flocculation factors (flocculation may be bypassed if desired)

Required

Explore design by mathematical modeling using the theory outlined. An objective is about 2–3 log removal as measured by turbidity and/or particles.

(a) Explore the feasibility of a deep filter bed, such as 3–4 m.
(b) Explore the feasibility of a high HLR.
(c) Ascertain the effect of floc particle size on performance and your recommendation for the resulting design.

Show plots as appropriate.

ACKNOWLEDGMENTS

Kevin Gertig, supervisor, Fort Collins Water Treatment Plant, was available for consultation at any time and clarified points regarding plant operation. Grant Williamson-Jones, City of Fort Collins, Colorado, provided the photo micrograph of floc-media grains.

The City of Bellingham, Public Works Department, Ted Carlson, director, granted permission (2010) to use photographs and other material from their water treatment facility, including Figures 12.25, 12.27, and 12.41. The department also granted permission to use water quality data from the plant for Problems 2.5 and 2.6 of the Solutions Manual.

Carol Sosak, marketing coordinator, ITT Water & Wastewater Leopold, Inc., F Zelienople, Pennsylvania, granted permission to use graphics taken from ITT Leopold materials.

Sherry Morrison, senior administrative assistant, Publishing Group, American Water Works Association, Denver arranged for permission to use Figures 12.3 through 12.5, and 12.41b. Figure 12.5 was from the files of Kurt Keeley, data base manager, American Water Works Association, Denver, Colorado.

Figure 12.46 was provided by Sylvie Roy, communications and marketing manager, Infilco Degremont, Richmond, Virginia, and was used with permission.

APPENDIX 12.A: FILTRATION IN NEW YORK

A distinguishing aspect of civil engineering projects is the role of political factors, in the sense that various kinds of values are represented and must come to bear in the decision making. The history of the New York water supply is an interesting case study (see Gibson, 1982, p. 25) that illustrates this idea.

The drama started with Aaron Burr, who as a state assemblyman, wrote a bill, "An Act for supplying the City of New York with a pure and wholesome water." Foreseeing a surplus of funds, Burr organized the Manhattan Company, which was then given the charter in 1799 by the legislature to supply New York City with wholesome water. The Manhattan Company would use the surplus capital from the water works financing to start a new bank to compete with the Federalist's banks associated with Alexander Hamilton. Instead of bringing in outside water, presumably from the Bronx River, the company sunk more wells into the polluted aquifer. The bank became the Chase-Manhattan and later, c. 2005, Chase.

The Manhattan Company essentially tied the hands of any progress toward a satisfactory water supply and so nothing was done until after 1830. Finally, in October 1842, water from the Croton River was delivered to New York with a maximum flow of 95 mgd (this system was later called "Old Croton"). Then in 1893, the "New Croton" was completed with an aqueduct capacity of 302 mgd. When added to the 28 mgd Bronx River conduit, completed in 1885, the total conduit capacity was 425 mgd (Wegmann, 1896, p. iii).

George Warren Fuller entered the picture on May 23, 1906, when the Board of Estimate engaged Fuller and Rudolph Hering to investigate the Croton water supply and to prepare plans for filter construction (Fuller, 1914, p. 152). Their recommendations, in an October 30, 1907 report, was to build 42 acres of slow sand filters at Jerome Park (the site of a major reservoir for the Croton water), superposed on a filtered-water reservoir. According to Fuller (p. 153),

> This project was pigeonholed, very likely because the panic of that year and the difficulty of selling bonds made public improvements of this character impractical.

Then on May 18, 1911, an appropriation of $8,690,000 was made by the Board for construction of the Jerome Park Filters. Plans for construction were made under the direction of Fuller as consulting engineer and the chief engineer and division engineer for the City. The procedure involved comparing the relative merits and costs of slow sand and "mechanical" filters, based on a filtered water flow of 320 mgd. In a report of May 21, 1912, Fuller recommended mechanical filters,

> I am firmly convinced, therefore, of the soundness of the conclusion that it is best to build mechanical filters for the purification of the Croton water supply.

The plans and estimates with data were submitted to a five member board of experts appointed by the Commissioner (of the Department of Water Supply, Gas, and Electricity) that included Allen Hazen and George A. Johnson. They stated that the "the mechanical filters shown by the plans are in general accordance with modern practice." The plans for the mechanical filter plant were completed June 25, 1912. The plans included a mixing chamber, five settling basins with six hours detention time, a sixth basin to settle wash water, a coagulant addition at "Gate House 8" through a rubber pipe grid, from two 5 in. coagulant supply pipes, 80 filters divided into four series of 20 each, with 10 on each side of the operating gallery. The net area of the filter beds was 2.7 acres, giving a filtration velocity of 1.89 gpm (calculated for a design flow of 320 mgd). Sand size specified was $0.60 \leq d_{10} \leq 0.70$. After filtration the water was to be treated with "hypochlorite of lime." A three story building for a laboratory was to be connected to the south end of the four filter galleries. The filtered water reservoir had a net area of 53.55 acres with capacity of 356 mg. The plans provided for alum storage and lime or soda storage, alum solution tanks, lime suspension tanks, and hypochlorite solution tanks. The cost estimate was $5,916,700, with nine bids received ranging from a low 13% less to 14% higher. The Board of Estimate responded, after their review and suggestions for minor changes with the remark,

> In general, specifications are in our judgment admirable in form and arrangement and exceptionally free from ambiguity, and we can see no reason why they should not be approved by the Committee.

At the time the plans were prepared opposition appeared which criticized the site selected, and the method of filtration and the matter was referred to an advisory commission of five engineers (four were consulting engineers with one, Mr. John H. Frazee being Assistant Engineer, Department of Finance. The majority report addressed all issues including the quality of the Croton water, the site at Jerome Park, the design, the costs, the overall economics (related to users not having to filter their own water, which was common especially for hotels), and concluded the that the system should be built. Their summary statement (Fuller, 1914, p. 170), sheds some light on the general feeling about the water filtration, that is,

> It is of the opinion of this commission that the filtration of the Croton water supply is not imperative, but that it is highly advisable to filter the water at this time both with a view to improving the physical quality and as a safeguard against potential danger, and we so recommend with the exception of one member dissenting.

At the time, the "danger" was measured in terms of number of typhoid deaths per 100,000 persons, which was 10 in the year 1912, and was lower than say Detroit with 18 or Washington, District of Columbia with 22, or New Orleans with 14. [The current similarity might be in terms of cases of giardiasis or cryptospiridousis.] The dissenting member, Mr. Frazee stated,

> I am of the opinion that it would be as satisfactorily efficient at this time, and far more economical, to strike at the sources of trouble, some of which were recognized as existing ten years ago, but toward the abatement of which nothing has been done.
>
> The advancement of sanitary and medical science in recent years in the development of this treatment (hypochlorite application) and the innocculation against typhoid, tend strongly to discount the conjectural possibility of future outbreaks pending delay and particularly so if the pollution sources known to exist in part and to be indicated are abated.

Mr. Frazee's dissent continued noting that the problems of water quality were mainly esthetic and that the water at times exceeded standards set by experts. He further called for what was later called a "protected" watershed. He calculated the financial savings of delay at $31,000,000 (Fuller, 1914, p. 171); Fuller took issue with the premises on Mr. Frazee's calculations were based. On May 22, 1913, a Mr. Mitchell, who became mayor, cast the vote that killed the project. His rationale was that hypochlorite would make the water safe and that aeration would remove 50%–75% of the tastes and odors and color and turbidity.

Fuller argued for filtration with both rationale and passion (Fuller, 1914, p. 136):

> ...And it is noteworthy that such filtration is considered necessary, not because any tests indicate a danger of infection, but because a surface supply is always liable to pollution from unexpected sources.

and (Fuller, 1914, p. 174),

> The various phases of this matter have been clearly before the city officials for many years. It has never been represented as absolutely imperative. Neither is it imperative for the city to clean its streets daily, to maintain good smooth pavements, or to build a $15,000,000 office building. But it is certainly desirable that a city of the rank of ours should have a pride in doing these things; and it should have a pride in drinking a water for which it need not apologize as reasonably safe, at most times, and fairly decent looking at most times.
>
> Real safety we are not getting. Limited by the need of adding only enough hypochlorite not to spoil the taste, and with an organic content in the water varying from day to day, it is not physically possible to vary the hypochlorite dose so as to oxidize both the organic matter and the bacterial content and leave no objectionable excess. And conditions may arise at any time when the hypochlorite treatment will not be effective in preventing the transmission of disease.

After the decision not to proceed, Fuller stated (Fuller, 1914, p. 174),

> It is rather curious that a ten year long attempt to give the citizens of New York a drinking water supply of a quality such as is elsewhere demanded should have finally proven abortive. And even more curious that the same officials who had voted to give New York a proper drinking water should later with the same information before them reverse themselves, that a Board of Alderman, who but a short time before were considering a vote of censure on the executive department for their slowness in providing filtered water should a few months later repeal their ordinance and condemn the project; that the newspapers which had just stopped printing attacks on the city officers for their negligence in furnishing bad water, should write to oppose the so-called steal when they were about to receive their demands.

Fuller's paper provoked a great deal of discussion. Mr. Alexander Potter, took issue with Fuller in a discussion (Fuller, 1914, pp. 456–463) noting that Fuller had reversed his position in method of filtration as he recommended slow sand in his 1907 report and then mechanical filtration in his 1912 report. Also, he quoted extensive expert witness testimony by both Fuller and George A. Johnson in which they supported the practice of treatment with hypochlorite as sufficient for Jersey City. Mr. Potter also pointed out several other contradictions in Fuller's position.

Fuller's response was that more was known at the present about hypochlorite than when the Jersey City testimony was given and, in his words (Fuller, 1914, p. 464),

> When the time comes, as it will in the course of years, that the Croton water is again largely polluted and contains much organic content and possibly greater turbidity, the effect of the hypochlorite treatment will be far less, and if it should happen that at that time the Croton water be polluted by typhoid bacilli, it is quite within the bounds of probability that a typhoid epidemic, perhaps of small and perhaps of large extent, may result. Such a danger is sufficient to warrant the expenditure of 25 cents per capita per year to avoid.

Concerning the change in his position on slow sand versus mechanical filtration between 1907 and 1912, Fuller stated (Fuller, 1914, p. 464),

> Mr. Potter raises the question of the change in the type of filters recommended between the years 1907 and 1914. The reasons are to be found in the development of the art of mechanical filtration during these seven years. The writer feels in no danger of being accused of having been a foe to mechanical filtration. At the time of the early investigations into the Croton water two fair-sized mechanical filters of modern type had been built, namely, the Little Falls plant and the Hackensack plant. While these plants had been entirely successful, the writer did not feel that the art had attained sufficient standing at the time of his 1907 report to warrant the adoption of mechanical filtration for this particular water. In the interval since 1907 modern mechanical filtration has reached such position that the writer feels no hesitation in recommending it for any water whatever where the desired results could be obtained by this method at a less total annual cost that with sand filters.

A major point in the above rebuttal by Fuller was that he did not feel secure that state-of-the-art of rapid filtration practice was sufficiently advanced in 1907 to warrant recommending it, whereas he did feel comfortable with the practice of the technology by 1912.

Regarding the quality of the Croton water, Fuller's paper contained both tables of water quality data and descriptions, for example, (Fuller, 1914, p. 169),

> As at present delivered to the consumer, the Croton water would be characterized by the water analyst as noticeably colored and slightly turbid, at times quite turbid; at other times containing numbers of microscopic organisms with an odor persistently vegetable and occasionally aromatic, grassy or even fishy; reasonably soft; a good boiler water and generally satisfactory for industrial purposes; ordinarily safe but at times sufficiently polluted to indicate the possible danger of infection from water borne diseases.

The story of the New York water supply does not end with Fuller. In a 1988 paper, Abel Wolman reviewed the status of the filtration versus protected watershed debate (Wolman, 1988). At that time, systems that did not filter included New York, Boston, Seattle, San Francisco, Portland, and Rochester; Los Angeles had started to filter its Owens River supply in the early 1980s. He recounted his tenure on a 1951 panel that produced a report, "Future Water Sources of the City of New York." The panel recommended,

> Regardless of considerations as to additional water supply... the matter of improving the protection of all supplies through the construction of and operation of a modern filtration plant be given immediate consideration and that preliminary plans and estimates of cost of such a project be provided within the next few years.

That further consideration was given to filtering the Croton water supply is seen in a report by Fulton and Hazen (1979) in which they reported to their clients on the outcome of pilot testing at the Jerome Park Reservoir in order to recommend a treatment train. Pilot testing of diatomaceous earth filtration was recommended and in the late 1980s a large pilot plant (1 mgd) was constructed. This was in anticipation of a 100 mgd full scale diatomite plant. Later, in the early 1990s, the idea of a full-scale diatomite plant was apparently abandoned.

In 1993, an "Expert Panel" (Okun, 1993) published a report sponsored by the US Environmental Protection Agency regarding the risks associated with the New York water supply. The issue was in compliance with the Surface Water Treatment Rule (*FR*, June 29, 1989). Although the Croton water was to be filtered the report states, about 30 years will be needed to bring the filtration system on line. The Panel concluded "that New York should not be granted an avoidance from filtration." They stated further "that New York City should be obligated to adopt a firm time schedule for pilot plant studies, preliminary design, site studies, environmental impact assessments, final design, and the letting of a construction contract. The time period for completing these phases of the work should be no more than six years."

The City of New York was planning to filter their Croton water supply, which would have brought a resolution to the approximately 100 years of concern on the question of filtration of the Croton water. The issue may have been resolved as far as the City of New York was concerned when the discovery was made about 1998 that ultraviolet (UV) light could kill *Cryptosporidium* oocysts; subsequent reports were that a UV installation was underway. One reason for not moving ahead on filtration was the very high cost involved.

GLOSSARY

Advection: The mass flow of a quantity that occurs with the bulk velocity. Same as convection.

Air binding: Gas, that is, oxygen and nitrogen, that displaces volume within the filter bed causing blockage of pores and thus higher headloss is called "air binding."

Air-wash: An adjunct to backwash to help dislodge attached floc by impacts between filter grains. Protocols vary but an investigation by Amirtharajah (1985) showed that a "collapse pulsing" procedure at subfluidization of the filter bed was most effective.

American filter: During the early days of the rapid filtration technology, the prevailing filtration technology in Europe was slow sand. Since rapid filtration was an American invention, the filters were called "American filters."

Attachment: The bonding between a particle and a filter grain (which may be called a "collector," depending on the context). The bonding is due to surface forces principally van der Waals attractive forces after reduction in electrostatic repulsive forces. Chemical factors affect the amount of reduction of the electrostatic repulsive forces.

Attachment coefficient: The ratio of particles striking a filter grain to those that attach.

Backwash: The reverse flow through a filter, that is, through the under-drain system, upward through the media, with waste floc removed by overflow launders. The bed of granular media is "fluidized" by the backwash.

Bed expansion: The bed expansion is usually a percentage referenced to the "at rest" bed, that is, bed expansion $= (h - h_o)/h_o$. In theory, the expanded bed porosity is also used as a parameter of bed expansion. The relationship is $(h - h_o)/h_o = (\varepsilon - \varepsilon_o)/(1 - \varepsilon)$, in which h_o and ε_o refer to bed depth and porosity before expansion and h and ε refer to bed depth and porosity after expansion.

Blinding: A deep bed filter that retains suspended particles by straining may cause an exponential increase in headloss, which is then said to be blinded.

Boucher's law: The relationship, $\Delta p = k_1 \exp(k_2 V)$, is Boucher's law. It was discovered in observing that the pressure drop across a steel mesh screen with 35 μm pore opening increased exponentially with volume of suspension passed. Applied to the surface deposition on a deep bed filter, the Boucher equation describes the "blinding" of the filter (see Ives, 1975b, p. 186).

Breakthrough: The third and final phase of a filtration cycle (excluding backwash) is called "breakthrough." It is the rising leg of the $C(t)_z$ curve and occurs when the filter is near exhaustion with respect to its capacity to store floc. Actually, this phase is merely the observation of the "wave front" as it emerges from the bottom of the filter bed.

Breakthrough curve: The $C(t)_z$ curve in its entirety may be sometimes called the breakthrough curve. More commonly, breakthrough curve refers to the final phase of the $C(t)_{Z=Z_o}$ curve in the filtration cycle in which the curve begins to rise steeply.

Cake filtration: The slurry, that is, the body feed is the filtering agent which accumulates and forms a cake. Filtration of small particles may occur in the depth of the diatomite layer. (Ives, 1975a, p. 1) The mechanism is primarily straining.

Chemical conditioning: The first phase of the filtration cycle in which the concentration declines with time to some acceptable level, that is, $C \leq C(\text{limit})$, which is maintained for the second phase of the filtration cycle called the "steady state" phase. The duration of the chemical conditioning phase may be say 15 min to about 2 h and depends, most likely, on the initial value of the attachment coefficient, α. As floc particles attach to the media, they will serve as collectors, presumably with higher α than with the bare media.

Clogging front: A term used by Adin and Rebhun (1977) as a designation for the $\sigma(Z)t$ profile. The wave front may be a measurable surrogate for the clogging front. Another surrogate that indicates the approximate end of the clogging front is the headloss versus Z profile at a given time; the intersection of the curve with the clean-bed headloss versus Z curve, a straight line, indicates the end of the clogging front.

Coagulant aid: The connotation is that a polymer is used in addition to the metal coagulant. The intent is that microflocs may bind to sites on a polymer which may occur partially during coagulation.

Collector: A granular media surface that exhibits surface forces that may cause particle attachment. A single grain of granular media, for example, sand or anthracite or garnet, is called a "collector." As the term implies, the particles-to-be-removed "attach" to collector surfaces and are removed.

Collector: May refer to a porous medium grain that provides a surface for particle adhesion. The term may be used in the context of a particular model of a porous medium. Some choices in geometry include capillary tubes, constricted cell, sphere, or a sphere within cell of Happel (1958).

Collision: A "collision" between a particle and a filter grain occurs when the particle trajectory brings the particle into the proximity of the force field of the filter grain where it may be either repulsed or attracted. A "collision" may also be defined as a "contact."

Contact: See *collision*.

Conventional filtration: The "conventional" filtration mode is rapid-mix/coagulation followed by flocculation, followed by settling, followed by filtration. Conventional filtration is the most common mode and is essential when floc load must be reduced prior to filtration.

d_{10} size: In a sieve analysis of a filter media, the d_{10} size is the size of which 10% of the media is smaller by weight. The d_{60} size is also noted and sometimes the d_{90} size is used. Usually when one refers to a media size of say 0.9 mm, the implicit reference is to the d_{10} size.

Declining-rate filtration: A means to distribute water to a group of filters is to apply the coagulated water to all without control valves. The water levels are the same for each filter of the group. At the same time the effluent flow from the under-drain system of each filter is not restricted by a control valve. Thus the headloss across each filter is constant and is the same at the start of the filter run as at the end of the run. The filtration velocity declines, however, as the filter clogs.

Deep bed filtration: See *depth filtration*.

Deposit: The particles-to-be-removed attach to the collectors and are termed deposits. Also we may call these deposits, in aggregate, "solids" as distinguished from the same material as comprising the suspension.

Depth filtration: In depth filtration, the suspended particles penetrate into the porous medium and attach to "collectors," that is, filter grains, at different depths. The process involves "transport" of particles to a collector (grain of granular media) and then "attachment" on the collector surface. Ives (1975a, p. 1) and Tien and Payatakes (1979, p. 733) state simply that depth filtration is filtration through a deep bed of granular media. The particles interact with each other and with the filter media in a manner involving colloid and interfacial forces at various depths in the filter bed. The term "deep bed" filter is used often in chemical engineering to distinguish the process from "cake" filtration.

Detachment: Ives (1975b, p. 199) states that experimental evidence indicates that increasing the flow in a deep bed filter, when deposited particles are present in the pores, leads to detachment of some of these particles causing locally increased suspension concentration. Mints of the USSR believes that such detachment occurs even at constant flow because the deposits cause local increases in interstitial velocity with consequent increase in shear. Rupture of a part of the deposit causes a particle or aggregate of particles to be detached and entrained in the flow; thus it behaves like any other particle and is subject to attachment at some distance downstream. Such deposits also cause a higher local pressure gradient, which is another manifestation of the higher shear stresses. The exact detachment point depends upon the shear stress of the attached floc, which is indeterminate.

Deterministic: The sense of usage is that a variable outcome is due to predictable factors. A "deterministic" model has the characteristics that a given dependent variable may be a function of certain independent variables. In a given model if the dependent and independent variables are identified and if the mathematical functions are proposed, the model is "deterministic." If the model is tested experimentally and if the dependent variables behave as predicted by changes in selected independent variables, then the model is "verified." The antithesis of "deterministic" is "stochastic"; ultimately, most stochastic variables have deterministic characteristics if they are understood well.

Diffusion: The random motion of particles due to thermal energy, that is, Brownian motion, may bring particles into proximity with a collector. For particles larger than 1 μm diameter, the mean-free-path of particles is at the most one or two particle diameters and so diffusion is not important. An important parameter is the coefficient, B, in the Stokes-Einstein equation, $B = kT/(3\pi\mu d_c)$. The ratio, Brownian velocity/Advective velocity $= B/(d_c v)$, which is $1/\mathbf{P}$ ($\mathbf{P}$ is the Peclet number)

where

B is the Stokes-Einstein diffusion coefficient (m^2/s)

d_c is the diameter of granular media particle (m)

In water filtration, $10–8 < 1/\mathbf{P} < 0.5 \cdot 10^{-5}$.

Direct filtration: The "direct" filtration mode is rapid-mix/coagulation followed by flocculation, followed by filtration.

Electro-osmosis: Flow of liquid through a porous plug (or tube) under the influence of an applied electric field (Gregory, 1975, p. 65).

Electrophoresis: Migration of charged particles in an electric field (Gregory, 1975, p. 65).

Electrophoretic mobility: Velocity of a particle in an electric field per unit of field strength, that is, $U = V_e/E = \varepsilon\zeta/\eta$; and, a quick conversion between ζ an U is, $\zeta \approx 12.8U$ (Gregory, 1975, p. 65).

English filter: During the early days of the rapid filtration technology, the prevailing filtration technology in Europe was slow sand, which had its inception in London in 1929. To distinguish slow sand filtration from the rapid filters being developed in the United

States, the former were called "English filters," and the latter, "American filters."

Field: The idea of a "field" refers to the distribution of some quantity throughout a given geometric bounded volume or area. The quantity could be a force (electrical, gravity, centrifugal), velocity, acceleration, potential (e.g., hydraulic head), concentration of suspended solids or dissolved solids, etc. A change in geometry changes the configuration of the field, but a change in input level has no effect. A flow net is a combination of a potential field and streamlines; a velocity field and a convective acceleration field are each derivatives of the streamline configuration.

Filter aid: Polymer added prior to filtration. The premise is that one part of the polymer will attach to the media with strands extending into the flow which, in turn, may provide attachment sites for coagulated particles.

Filter coefficient: The coefficient, λ, in the Iwasaki relation, $\partial C/\partial Z = \lambda C$.

The filter coefficient can be related to the several transport mechanisms, shown as dimensionless numbers, that is, (Ives, 1975b, p. 196)

$$\lambda d(\text{grain}) = K(d/D)^{a}(1/\mathbf{P})^{b}\mathbf{S}^{c}(1/\mathbf{R})^{d}$$

This shows a minimum at $d(\text{particle}) \approx 1$ μm.

Filter medium: The granular material that comprises a filter bed is called a filter medium. The plural is "media."

Filtration: Filtration is a unit process for achieving a separation between a fluid and its suspended matter by passage of the suspension through a porous medium.

Filtration efficiency: A common definition is simply the ratio of particle concentration leaving the filter, C, divided by the particle concentration entering the filter, C_o, i.e., the ratio, C/C_o.

Filtration mode: The filtration "mode" as used here refers to whether the treatment train is in-line, direct, or conventional, illustrated in Figure 12.1a,b, and c, respectively. These terms are defined subsequently. When the term "rapid filtration" is used, one of these modes must be specified.

Filtration process: See *filtration*.

Filtration technology: The filtration process and the supporting components to make it work is called here, "filtration technology." The "process" occurs within the media bed. The supporting components include: an under-drain system, pipes, valves, controls, and various instruments. Numerous variations exist which are both generic and proprietary.

Filtration theory: The quest of filtration theory is to provide a means to understand how the three *dependent* functions in filtration, that is, the effluent turbidity/particle concentration versus time, the solids concentration depth profile, and the rate of clogging headloss with time, is affected by the most important of the *independent* variables.

Flocculant: The implication is that a polymer is added after coagulation and preceding flocculation. The term is not strictly defined except that the use of a polymer is implied. Some persons may use the term interchangeably with coagulant aid. Others may intend that a flocculent in intended to aid in binding microflocs and floc particles to become larger and "tougher" floc particles.

Fluidize: Water flow with velocity vertical and up will at a certain point cause a bed of granular media to be maintained in suspension caused by the drag on the particles; when this occurs the bed is "fluidized."

Gravity: The effect of gravity on a particle trajectory may cause deviation from the advective transport along its streamline to cause impingement against a collector surface. The effect becomes more significant as particle size approaches about 100 μm. The idea was suggested by Allen Hazen who described the pores of a slow sand filter as miniature settling basins. (see Ives, 1975b, p. 189). In typical rapid filter with $v = 2$ mm/s, $P = 0.4$, $T = 20°\text{C}$, $v_i = 5$ mm/s and applying Stoke's law to a 10 μm clay particle, $v_s = 0.1$ mm/s, which is 2% effect. Near the surface of a granular media the local velocities may be near 0.2 mm/s indicating that particles close to the surface may be transported to the collector surface. Experiments have shown, however, that particles deposit on the top of the granular media giving a dome effect on top.

Hamaker constant: A constant associated with calculation of intermolecular forces (see, for example, Myers (1991, p. 65). A more complete explanation was given by Gregory (1975, pp. 78–79). In 1937, H.C. Hamaker published a paper that extended the concept of van der Waals forces to include the energy of interaction between finite particles of various shapes, separated by distance, d. His mathematical relations for the bonding energy between two flat plates and two spheres were, respectively, $U(\text{flat plates}) = -A_{12}/12\pi d^2$ and $U(\text{two spheres}) = -A_{12}a_1a_2/6d(a_1 + a_2)$, in which A_{12} is the Hamaker constant for materials 1 and 2 in a vacuum. For a sphere and a flat surface, the model used in filtration for a particle-grain surface attachment is $U(\text{sphere/flat plate}) = -A_{12}a_1/6d$. Gregory (p. 79) extended the idea to the case in which water is a medium. The magnitude of the Hamaker constant for materials in water is about $0.1\text{–}10 \cdot 10^{-20}$ J (Gregory, 1975, p. 81), with the exact magnitude depending on the materials.

Hamaker constant: Hamaker showed how interactions between molecules could be integrated to give the energy of interaction between particles of various shapes. The results for parallel flat plates and for spheres are of special interest and are given below. These expressions are based on the assumption of complete additivity of intermolecular interactions and contain a constant A12, which depends only on physical properties of the interacting materials 1 and 2. This is known as the Hamaker constant (Gregory, 1975, p. 78).

Happel collector: Refers to a particular geometry that represents a single collector of a granular media filter bed used in mathematical modeling. The geometry selected by Happel in 1958 and used by other modelers, for example, Tien, is a spherical granular media particle enveloped by a spherical volume of water that has the same volume as the pore volume for that particle.

Headloss: The difference in head across the filter bed is the headloss. The gravel support and the underdrain system may be included. *Note*: The term is found in the literature frequently as two words, i.e., "headloss", and less-frequently as "headloss"; the latter is used here, however, since its used throughout the text. Its combination represents the well-known concept regarding loss of energy head due to pipe friction, pipe bends, porous media flow, etc.

Heterodisperse suspension: Suspended particles are heterogeneous in size (Darby et al., 1992).

In-line filtration: The "in-line" mode of filtration is comprised of rapid-mix/coagulation followed by filtration. This mode is used most often with low turbidity waters.

Inertia: Streamlines approaching a filter pore converge as the flow passes through. If the particles have sufficient inertia they maintain a trajectory that may cause the particle to impinge against a collector. In water filtration, inertia is not a significant transport mechanism, but in air filtration it has a major effect.

Interception: When a particle is transported by a streamline that passes within a distance $d/2$ from a collector (d being the diameter of the particle), the particle will brush the collector; the contact is called interception. The probability of an interception contact is proportional to $d(\text{particle})/d(\text{pore})$. When $d(\text{particle})/d(\text{pore}) \Rightarrow 1$, straining becomes dominant. In filtration of water and wastewater, $2 \cdot 10^{-4} < d(\text{particle})/d(\text{pore}) < 1 \cdot 10^{-1}$, with higher values causing straining (from Ives, 1975b, p. 188).

Iwasaki equations: Three equations proposed by Iwasaki in his 1937 paper described, in part, the filtration process and were (1) a kinetic equation, describing the rate of change of concentration with distance, (2) a materials balance equation that accounts for the loss of particles from the fluid suspension as the gain to an attached phase, and (3) a statement that postulates that the filter coefficient, λ, changes over time as the filter pores clog.

Limiting trajectory: A trajectory that separates the trajectories that intercept the collector from those that do not. To amplify, the limiting trajectory approach calculates the trajectory of the particle that includes all other particles that would strike a collector at the end of a unit cell. Particles outside the limiting trajectory would not strike collectors within a given collector pore volume. The proportion of particles that would strike the collector, that is, η, can then be calculated.

log Removal

$$\log \mathbf{R} = \log C_o - \log C$$
$$= \log \frac{C_o}{C}$$

where log **R** is the common log of the removal ratio.

The relationship between the $\%R$ and log **R** is

$$\%R = 100 - 10^{-\log \mathbf{R}}$$

Manometer: A tube that penetrates the wall with a fluid on the other side that has a U-shape for reading the difference in pressure between the tap point and some other point, that is, a "differential manometer." If the other end is open to the atmosphere, the manometer is a special type, called a piezometer. The manometer fluid may be any fluid that is different from the fluid whose pressure is being measured so that an interface is visible. The U-tube is fitted with a scale so that the vertical elevations of each interface can be read, which must also be related to the elevations of each pressure tap.

Mass flow diagram: A plot of the total volume of flow versus time is a mass flow diagram. The two conditions are (1) raw water flow and filter plant flow versus time and (2) demand flow and filter plant flow versus time.

1. *Raw water storage*: If the diagram is used to estimate raw water storage for a water treatment plant that has a steady flow, the latter is a straight line. The maximum difference between the line representing the total water supply volume and the straight line representing the plant flow is the raw water storage required. The time scale on the raw water side usually is in years so that years of drought conditions can be incorporated; the more years included, the better the estimate of storage required. The time scale could also be for a year, to reflect seasonal variation, or daily to reflect daily variation.
2. *Treated water storage*: Similarly, on the demand side, the maximum difference between the total demand volume, over a 24 h period (which is a varying line with some sharp increases), and the total water produced by the plant at a constant rate (which is a straight line) is the volume of treated water storage required. On the treated water side, the time scale usually is 24 h and reflects the

variation in demand over a 24 h period. See also Rippl diagram in various texts.

Mechanical filtration: During the early years of the rapid filtration technology, for example, from about 1880 to perhaps 1910, the technology was called "mechanical filtration."

Monodisperse suspension: All suspended particles have the same size (Tobiason and Vigneswaran, 1994, p. 335).

Mudball: A mudball is the result of adhesion between media particles caused by inadequate removal of coagulant chemicals. According to Mackay (1988a,b), mudballs are formed when grains of filter media are not cleaned thoroughly; the sticky residue causes the grains to clump together. The size of a mudball may be perhaps 25–100 mm (1–4 in.).

Negative pressure: When the hydraulic grade line in a pipe or porous medium drops below the level of a given point in the pipe or porous medium, a negative pressure results. By definition, the vertical distance between a point in a conduit and the hydraulic grade line is the pressure head.

Net water production: Net water production per filter run per unit area of filter bed is defined as

$$NWP = WP - VB$$

where

NWP is the net water production unit area of filter bed per filter run (m^3 water produced/unit area of filter bed/run)

WP is the volume of water production per unit area of filter bed per filter run (m^3 water produced/m^2 filter plan area/filter run)

VB is the volume of water required per unit of filter bed area per backwash (m^3 backwash water/m^2 filter bed area/filter run)

See also unit filter run volume (UVRV), which is the same neglects the VB term.

Orthokinetic flocculation: Refers to velocity gradient flocculation (see Ives, 1970, p. 206).

Packed bed reactor: Both the rapid and the slow sand filtration processes are done in a bed of packed granular media. Also, a bed of granular activated carbon and an ion exchange column are both "packed bed" reactors.

Particle: Any particle in suspension in the water applied to the filter is termed here a "particle." As applied to rapid filtration, most of the ambient particles would be either charge neutralized with alum species attached or they might be enmeshed in hydroxide precipitates and these would then be the particles-to-be-removed.

Particle counter: An instrument that measures particle number and sizes in the sample provided. Batch particles counters are for a given sample and "in-line" particle counters read the data at specified intervals for a continuous sample.

Piezometer: A tube that penetrates the wall of a pipe with a fluid on the other side will transmit the fluid pressure at the point of penetration causing the fluid, generally water, to rise to an elevation above the penetration point such that the pressure due to the fluid in the tube equals the pressure in the fluid at the point of penetration. A piezometer is a special kind of manometer characterized by an open end to the atmosphere.

Pilot plant: A pilot plant is a physical model that emulates the full-scale filtration process. They may range from coagulant injection into a pipe/elbow mixing and a simple tube filled with media to an elaborate SCADA run system with alternative treatment trains.

Plenum: (1) A condition, space, or enclosure in which air or other gas is at a pressure greater than that of the outside atmosphere. (2) The condition of being full; fullness. (3) A space completely filled with matter. American Heritage Dictionary, Softkey International, 1995.

Porosity: Defined as the ratio of void volume to bed volume.

Proprietary filters: Manufactured and sold by on particular firm, usually under a patent. Examples of proprietary filters are:

- Hardinge "automatic backwash" (ABS) filter (comprised of filtration cells about 305 mm (12 in.) wide that are individually backwashed one at a time by an apparatus that moves along the filter bed) marketed presently by Infilco-Dregemont.
- Parkson Dynasand™ filter which is a moving bed upflow filter with a sand removed continuously from the bottom, washed, and returned to the bed at the top.
- Culligan Multi-Tech™ is a pressurized package system with "contact" flocculation.
- Environmental Products Division High Rate Filtration System™, a pressurized cylinder oriented horizontally.

Frequently, the proprietary filters are sold as "package" systems, that is, coagulant feed, pumps, SCADA, etc. are provided such that the filter is ready to operate. In some cases, the filters are prefabricated and others may require construction in accordance with specifications of the manufacturer.

Quality assessment (QA): A process of measuring and evaluating quality. Often this assessment involves evaluating such information as control charts, replicate measurements, spiked samples, and standard reference materials (Symons et al., 2000, p. 384).

Quality assurance: (1) An overall system of management functions designed to provide assurance that a specified level of quality is being obtained. It can be thought of as being composed of quality control (QC) and quality assessment (QA). (2) The management of products, services, and production or delivery processes to ensure the attainment of operational performance, product, or both in keeping with quality requirements (Symons et al., 2000, p. 384).

Quality control (QC): A system of functions carried out at a technical level for the purpose of maintaining and documenting quality. It includes such features as personnel training, standard operating procedures, and instrument calibrations (Symons et al., 2000, p. 384).

Rapid filtration: Rapid filtration is defined here as a *technology* that provides the means to implement the "depth filtration" process. The technology includes the under-drain system, the backwash system, the wastewater system, the instruments, and the control system to permit the depth filtration process to work; coagulation is the most important part of the process.

Rate-of-flow controller: A valve in the effluent pipe from a filter under-drain system that regulates flow at a given set-point by opening or closing the valve, usually in small increments. The valve has associated with it a flow meter and that provides the signal for valve adjustment. At the start of the filter run, with the water level in the filter box at full depth, the valve is near-closed and takes up most of the headloss of the system. As the filter clogs, the valve opens further, reducing the valve headloss. As the run approaches termination, the valve opens fully, at which point all of the headloss occurs in the media and none in the valve. For a declining rate filter scheme the rate-of-flow controller valve is not needed, nor is it needed for a constant flow, rising water level scheme. In every case a flow meter should be installed in the pipe from the under-drain system for each individual filter.

Regime: A prevailing pattern of behavior or the pattern of a regulated system. Adapted from *American Heritage Dictionary*.

Remnant water: Remnant water is defined as backwash water that remains in the filter box after backwash and includes water in the under-drain system, pore water, and headwater.

Ripening: The first stage of a filtration cycle in which the effluent turbidity or particle concentration declines with time toward a "steady state" level, say 0.05 NTU or say 10 particles/mL.

Saturated zone: At the influent end of the filter (the top in most cases), the influent flux of floc particles is received. As these particles adhere to the media grains and previously deposited particles, they occupy void space causing a higher interstitial velocity with a proportionately increasing rate of shear. As the rate of shear increases, the rate of detachment of particles at some point equals the rate of attachment. At this point the filter bed is "saturated."

SCADA: An acronym for "supervisory control and data acquisition." Refers to data acquisition by electronic sensors for such quantities as pressure, temperature, specific electric conductivity, pH, turbidity, particle counts, dissolved oxygen, specific ions, flow meters, etc. Signals are 4–20 mA and are proportional to the sensor reading and are transmitted to an analog board. The analog board transmits the signal to a computer where it is interpreted in terms of a calibration to read the actual value, for example, 7.5 mg/L dissolved oxygen. Control is by actuated valves that generally have on-off functions but may, alternatively, regulate water flow. The computer sends a signal to a micro-dac board which contains a micro-relay. For an open valve the signal requires the relay to permit current to be transmitted to the valve to perform the function intended.

Single collector: A term that arises often in discussions of modeling is the "single collector." The term "single collector" refers to a single constricted cell, Happel's sphere-in-cell, etc., that by their nature exclude the presence of neighboring grains. The effect is that the viscous flow field, for example, as well as other fields of a complex system, vis-à-vis a single collector, are not included in the model.

Slow sand filtration: Slow sand filtration is characterized by low hydraulic loading rate, with consequent large filter area and the raw water without coagulation is applied to the filter. Headloss increase is due to a surface deposit on the top of the filter, called the *schmutzdecke*. At terminal headloss, the bed is dewatered to just below the bed surface and the *schmutzdecke* is removed by scraping. Slow sand filtration is characterized by both straining and depth filtration. Straining occurs as ambient particles deposit on the surface of the filter bed, forming a schmutzdecke. Depth filtration occurs as a portion of the ambient particles penetrate the filter bed. If the filter has been in operation for some months, the granular particles will have developed biofilms which then serve as collectors. At the risk of being simplistic, we might say that in rapid filtration we condition the particles (by coagulation) and in slow sand filtration we condition the collectors (by hoping for or promoting growth of biofilms). Slow sand is a distinct technology and is the topic of Chapter 13.

Specific solids deposit: The mass of solids deposit per unit volume of filter bed, that is, pore volume plus solid volume; units are kg solids per m^3 filter volume.

Standard temperature pressure: The most common standards for temperature and pressure are 25°C and 101 325 kPa (1.00 atm), respectively; this condition is designated STP, meaning standard temperature and pressure. In U.S. Customary units, the values are 68°F and 1.00 atm. In Europe, the common term used is, "normal" temperature and pressure (NTP).

Steady state: Influent flow and water quality are not changing with time. The term may refer to characteristics of the filtration process, as well, such as the "steady state" portion of the $C(t)_{Z=Z_o}$ curve, or that the shape of the $C(Z)_t$ curve has attained a "steady state" shape. The sense of the term is that a phenomenon or characteristic does not change with time.

Stochastic: The random occurrence of the magnitude of a variable would be considered as "stochastic." A "random variable" in a statistical sense has the characteristic of being "stochastic." The occurrence of Giardia lamblia cysts in a river, as detected by sampling, is "random," and thus is "stochastic." The opposite of stochastic is "deterministic."

Straining: A particle in suspension that is larger than the interstitial pores of a medium is removed by straining. As $d(\text{particle})/d(\text{pore}) \rightarrow 1$, interception becomes straining, which is the limiting condition of interception (Ives, 1975b, p. 194).

Streaming potential: The potential established when a solution is forced through a porous plug (or tube) of material which acquires a charge in contact with the solution (Gregory, 1975, p. 65).

Subfluidization: Backwash velocity, v, in which $v \leq v_{mf}$, in which v_{mf} is the minimum velocity for incipient fluidization.

Surface wash: An adjunct to backwash. High pressure jets with nozzles about 25 mm (1 in.) above the media are directed into the media before the backwash and during the backwash as the bed expands.

Technology: (1) The scientific study of mechanical arts and applied sciences (as engineering). (2) These subjects and their practical application in industry, etc. (*Oxford American Dictionary*, Oxford University Press, New York, 1980).

Theory: (1) A set if ideas formulated by reasoning from known facts to explain something; 4 A statement of the principles on which a subject is based (Oxford American Dictionary, Oxford University Press, New York, 1980).

Transport: A suspended particle is "transported" to the proximity of a "collector" (a media grain) where it may come under the influence of surface forces.

Transport coefficient: The ratio of particles striking a filter grain to those approaching.

Turbidimeter: An instrument that measures particle turbidity in the sample provided. Bench turbidimeters are for a given sample and "in-line" turbidimeters read the data at specified intervals for a continuous sample. See also "turbidity."

Uniformity coefficient: The uniformity coefficient, U, is defined as the ratio d_{10}/d_{60}.

Unit filter run volume (UFRV): Volume of water filtered per unit of area per run, which is the product of filtration velocity and run time. Units are m^3 water/m^2 filter bed area/filter run (gal/ft^2/run). Reference is Kawamura (1999, p. 80).

van der Waals force: About 1875, van der Waals postulated the presence of intermolecular forces to explain the fact that the behavior of real gases deviated from the ideal gas law at high pressures. Three types of intermolecular forces were postulated by others: (1) orientation effects due to two molecules having permanent dipole moments, (2) induction effect due to one molecule having a permanent dipole moment inducing a dipole in a neighboring molecule and (3) dispersion effect due not to a permanent dipole of either molecule but to fluctuating dipoles due to the random motion of electrons within a molecule. The van der Waals forces are likely to be the major cause of attraction and bonding between say a destabilized floc particle and a mineral filter grain, or a floc coated mineral grain (Adapted from Myers, 1941, pp. 41–67).

Wave front: The concentration versus distance profile at any given time, that is, $C(z)_t$. The concentration may be any parameter; usually turbidity or particle counts would be measured. See also, "clogging front."

Zeta potential: The potential at the boundary between the fixed and mobile phases of the double layer associated with a colloid particle, that is, at the "slipping plane," which is just outside the Stern layer (Gregory, 1975, p. 65).

REFERENCES

Adin, A. and Rebhun, M., A model to predict concentration and headloss profiles, *Journal of the American Water Works Association*, 69(8):444–453, August 1977.

Adin, A., Baumann, E. R., and Cleasby, J. L., The application of filtration theory to pilot-plant design, *Journal of the American Water Works Association*, 71(1):17–27, January 1979.

Al-Ani, M. Y., Hendricks, D. W., Logsdson, G. S., and Hibler, C. P., Removing *Giardia* cysts from low turbidity waters by rapid rate filtration, *Journal of the American Water Works Association*, 78(5):66–73, May 1986.

Amirtharajah, A., Fundamentals and theory of air scour, *Journal of the Environmental Engineering Division*, 110(EE3):573–590, June 1984.

Amirtharajah, A., The interface between filtration and backwashing, *Water Research (Journal of the International Water Quality Association)*, 19(5):581–588, May 1985.

Amirtharajah, A. and Cleasby, J. L., Predicting expansion of filters during backwash, *Journal of the American Water Works Association*, 64(1):52–59, January 1972.

Amirtharajah, A. and Wetstein, D. P., Initial degradation of effluent quality during filtration, *Journal of the American Water Works Association*, 72(9):518–524, September 1980.

Baker, M. N., *The Quest for Pure Water*, American Water Works Association, New York, 1948.

Baylis, J. R., Experiences in filtration, *Journal of the American Water Works Association*, 29(7):1010–1048, 1937.

Baylis, J. R., Seven years of high rate filtration, *Journal of the American Water Works Association*, 48(5):585–596, 1956.

Beverly, R. P., Gravity-Filter Design and Maintenance 101, Waterworld Review, Pennwell Publishing Company, Tulsa, OK, January/February 1995. (Re-printed by Leopold Water and Wastewater Products).

Cleasby, J. L., Approaches to a filterability index for granular filters, *Journal of the American Water Works Association*, 61(8): 372–381, August 1969.

Cleasby, J. L., Is velocity gradient a valid turbulent flocculation parameter, *Journal of the Environmental Engineering Division, ASCE*, 110(EE5):875–897, October 1984.

Cleasby, J. L., Declining rate filtration, *Fluid/Particle Separation Journal*, 2(1):1–4, March 1989.

Cleasby, J. L., Backwash and underdrain considerations, unpublished paper for short course at Colorado State University on design of filtration systems, June 1991.

Cleasby, J. L. and Fan, K., Predicting fluidization and expansion of filter media, *Journal of the Environmental Engineering Division*, Proceedings American Society of Civil Engineers, 107 (EE3):455–471, June 1981.

Cleasby, J. L. and Lorence, J. C., Effectiveness of backwashing for wastewater filters, *Journal of the Environmental Engineering Division*, Proceedings American Society of Civil Engineers, 104(EE4):749–765, August 1978.

Cleasby, J. L., Dharmarajah, A. H., Sindt, G. L., and Baumann, E. R., Design and Operation Guidelines for Optimization of the High-Rate Filtration Process: Plant Survey Results, AWWA Research Foundation and American Water Works Foundation, 1989.

Conley, W. R. and Evers, R. H., Coagulation control, *Journal of the American Water Works Association*, 60(2):165–174, February 1968.

Darby, J. L., Attanasio, R. E., and Lawler, D. F., Filtration of heterodisperse suspensions: Modeling of particle removal and headloss, *Water Research*, 26(6):711–726, June 1992.

Eliassen, R., Clogging of rapid sand filters, *Journal of the American Water Works Association*, 33(5):926–942, May 1941.

Fair, G. M., Fifty years of progress in water purification, 1913–63, *Journal of the American Water Works Association*, 55(7):813–823, July 1963.

Fair, G. M. and Geyer, J. C., *Water Supply and Wastewater Disposal*, John Wiley & Sons, New York, 1961.

Federal Register, Filteration Rule, Vol. 54(124):27486, June 29, 1989.

Fuller, G. W., *The Purification of the Ohio River Water at Louisville Kentucky*, D. Van Nostrand Co., New York, 1898.

Fuller, G. W., The Croton water supply: Its quality and purification, *Journal of the American Water Works Association*, 1(1):135–187, 1914 (with discussion by Alexander Potter, pp. 456–463 and response by George W. Fuller, pp. 463–465).

Fuller, G. W., Historic review of the development of sanitary engineering in the United States during the past 150 years—Water works, *Transactions ASCE*, 92:1209–1224, 1928.

Fuller, G. W., Progress in water purification, *Journal of the American Water Works Association*, 25(11):1566–1576, 1933.

Fulton, G. P. and Hazen, R., Report on Pilot Water Treatment Plant at Jerome Park Reservoir—Croton Water System, Contract No. HED-486 between Metcalf & Eddy of New York, Inc. and Hazen and Sawyer and the City of New York, Bureau of Water Supply, Department of Environmental Protection, June 1979.

Gibson, J. I., *Centennial, American Water Works Association*, 1881–1981, Denver, CO, 1982.

Gillespie, C. G., Filtration Plant Census of 1924, *Journal of the American Water Works Association*, 14:123–142, 1925.

Ginn, T. A. Jr., Amirtharajah, A., and Karr, P. R., Effects of particle detachment in granular media filtration, *Journal of the American Water Works Association*, 82(2):66–76, 1992.

Gregory, J., Interfacial phenomena, in: Ives, K. J. (Ed.), *The Scientific Basis of Filtration*, Proceedings of NATO Advanced Study Institute, Cambridge, U.K., Noordhoff International Publishing, the Netherlands, 1975.

Habibian, M. and O'Melia, C. R., Particles, polymers, and performance in filtration, *Journal of the Environmental Engineering Division*, 101(4):567–583, December 1975.

Hansen, P., Treatment and purification of water progress and development during 1936, *Journal of the American Water Works Association*, 29(10):1443–1466, 1936.

Happel, J., Viscous flow in multiparticle systems: Slow motion of fluids relative to beds of spherical particles, *AIChE Journal*, 4(2):197–201, 1958.

Hendricks, D. W., Post, F. J., Khairnar, D. R., and Jurinak, J. J., Bacterial Adsorption on Soils—Thermodynamics, Report PRWG-62-1, Utah Water Research Laboratory, Utah State University, Logan, UT, July 1970.

Hendricks, D. W., Clunie, W. F., Anderson, W. L., Hirsch, J., Evans, B., McCourt, B., Monaghan-Welding, P., Nordby, G., Sobsey, M. D., and Hunt, J. D., Biological Particle Surrogates for Filtration Performance Evaluation, Report for Project #181, AWWA Research Foundation, Denver, CO, May 2000.

Hendricks, D. W., Clunie, W. F., Sturbaum, G. D., Klein, D. A., Champlin, T. L., Kugrens, P., Hirsch, J., McCourt, B., Nordby, G. R., Sobsey, M. D., Hunt, D. J., and Allen, M. J., Filtration removals of microorganisms and particles, *Journal of the Environmental Engineering Division*, ASCE, 131(12): 1621–1632, December 2005.

Herzig, J. P., Leclerc, D. M., and LeGoff, P., Flow of suspensions through porous media—Application to deep filtration, *Industrial & Engineering Chemistry*, 62(5):8–35, 1970.

Hewitt, S. R. and Amirtharajah, A., Air dynamics through filter media during air scour, *Journal of Environmental Engineering*, ASCE, 110(3):591–606, June 1984.

Horn, J. B. and Hendricks, D. W. , Removal of *Giardia* Cysts and Other Particles from Low Turbidity Waters Using the Culligan Multi-Tech™ Filtration System, Environmental Engineering Technical Report No. 86-296530-1, Department of Civil Engineering, Colorado State University, Fort Collins, CO, August 1986.

Horn, J. B., Hendricks, D. W., Scanlan, J. M., Rozelle, L. T., and Trnka, W. C., Removing *Giardia* Cysts and Other Particles from Low Turbidity Waters Using Dual-Stage Filtration, *Journal of the American Water Works Association*, 80(2):68–77, 1988.

Horner, R. M. W., Jarvis, R. J., and Mackie, R. I., Deep bed filtration: A new look at the basic equations, *Water Research*, 20(2):215–220, February 1986.

Hudson, H. E. Jr., Factors affecting filtration rates, *Journal of the American Water Works Association*, 48(9):1139–1153, 1956.

Ives, K. J., Filtration using radioactive algae, *Journal of the Sanitary Engineering Division*, ASCE, 87(SA3):23–37, May 1961.

Ives, K. J., Filtration using radioactive algae, *Transactions, ASCE*, 127(Part III):372–390, 1962.

Ives, K. J., Rapid filtration – Review paper, *Water Research*, 4:201–233, 1970.

Ives, K. J., Mathematical models of deep bed filtration, in: Ives, K. J. (Ed.), *The Scientific Basis of Filtration*, Proceedings of NATO Advanced Study Institute, Cambridge, U.K., Noordhoff International Publishing, the Netherlands, 1975a.

Ives, K. J., Capture mechanisms in filtration, in: Ives, K. J. (Ed.), *The Scientific Basis of Filtration*, Proceedings of NATO Advanced Study Institute, Cambridge, U.K., Noordhoff International Publishing, the Netherlands, 1975b.

Ives, K. J., Mathematical models of deep bed filtration, in: Ives, K. J. (Ed.), *The Scientific Basis of Filtration*, Proceedings of NATO Advanced Study Institute, Cambridge, U.K., Noordhoff International Publishing, the Netherlands, 1975c.

Ives, K. J., Fundamental of filtration, in: Weller, R. and Jansem, J. (Eds.), *Proceedings of International Symposium on Water Filtration*, Antwerp, Belgium, May 1982.

Ives, K. J., Filtration with endoscopes, *Water Research (Journal of the International Water Quality Association)*, 23(7):861–866, 1989.

Ives, K. J. and Sholji, I., Research on variables affecting filtration, *Journal of the Sanitary Engineering Division*, 91(SA4):1–18, August 1965.

Iwasaki, T., Some notes on sand filtration, *Journal of the American Water Works Association*, 29, 1591, December 1937.

Kawamura, S., *Integrated Design of Water Treatment Facilities*, John Wiley & Sons, Inc., New York, 1991.

Kawamura, S., Optimization of basic water treatment processes – design and operation: Sedimentation and filtration, *Aqua, Journal the International Water Association*, 45(3):130–142, 1996.

Kawamura, S., Design and operation of high-rate filters, *Journal of the American Water Works Association*, 91(12):77–90, December 1999.

Khan, Z., Floc characteristics as affected by coagulation and as affecting filtration of low turbidity waters, PhD Dissertation, Department of Civil Engineering, Colorado State University, Fort Collins, CO, 1993.

Leopold, General Sales Catalog, Leopold Water and Wastewater Products, The F.B. Leopold Company, Inc., Zelienople, PA, c. 1999a.

Leopold, Leopold Fiberglass Wash-Water Troughs, Product Data Sheet FRP-200, The F.B. Leopold Company, Inc., Zelienople, PA, 1999b.

Leopold, Leopold underdrain, Brochure UNN-100, The F.B. Leopold Company, Inc., Zelienople, PA, 1999c.

Letterman, R. D., Filtration II—Fundamental Considerations, *Proceedings of the Nineteenth Annual Public Water Supply Engineers Conference*, pp. 23–37, Champaign, IL, April 5–7, 1977.

Letterman, R. D., Economic analysis of granular-bed filtration, *Journal of the Environmental Engineering Division*, 106 (EE2):279–291, April 1980.

Letterman, R. D., Effect of pre-treatment on filter performance and system economics, *Proceedings of the Water Filtration Symposium*, Sponsored by the European Federation of Chemical Engineering, pp. 4.1–4.11, Antwerp, Belgium, April 1982.

Logan, B. E., Jewett, D. G., Arnold, R. G., Bouwer, E. J., and O'Melia, C. R., Clarification of clean-bed filtration models, *Journal of the Environmental Engineering Division*, 121(12):869–873, December 1995.

Logsdon, G. S., Direct filtration – Past, present, future, *Civil Engineering*, 48(7):68–73, July 1978.

Logsdon, G. S., Pilot plant studies—From study planning to project implementation, *Proceedings of Seminar on Design of Pilot Plant Studies*, AWWA Seminar at Miami Beach, FL, Publication No. 20164, American Water Works Association, Denver, CO, May 16, 1982.

Logsdon, G. S., *Water Filtration Practices*, American Water Works Association, Denver, CO, 2008.

Mackay, D. J., Backwashing ensures effective filtration, *Opflow*, 14 (10):4–5, October 1988a.

Mackay, D. J., Filter media requires scheduled maintenance, *Opflow*, 14(11):1, 4, November 1988b.

McNamee, R. L., Baylis, J. R., Borchardt, J. A., Cosens, K. W., Hazen, R., Hill, K. V., Howard, N. J., Kenmir, R. C., and Schram, W. B., Filtration (AWWA Committee 8913 P), Filtration – Revision of water quality and treatment, Chapter 11, *Journal of the American Water Works Association*, 49(7):787–817, July 1956.

Mintz, D. M., *Modern Theory of Filtration, Proceedings*, Vol. II, *Seventh Congress*, International Water Supply Association, Special Subject No. 10, Barcelona, Spain, pp. 225–241, October 3–7, 1966.

Monk, R. D. G., Design options for water filtration, *Journal of the American Water Works Association*, 79(9):93–106, September 1987.

Mosher, R. R. and Hendricks, D. W., Rapid rate filtration of low turbidity water using field scale filters, *Journal of the American Water Works Association*, 78(12):42–51, December 1986.

Myers, D., *Surfaces, Interfaces, and Colloids*, VCH Publishers, Inc., New York, 1991.

Okun, D. A., Report of the Expert Panel on New York City's Water Supply, U.S. Environmental Protection Agency, March 24, 1993.

O'Melia, C. R., Particles, pretreatment, and performance in water filtration, *Journal of the Environmental Engineering*, 111(6): 874–890, 1985.

O'Melia, C. R. and Ali, W., The role of retained particles in deep bed filtration, *Progress in Water Technology*, 10(5/6):167–182, 1978.

O'Melia, C. R. and Stumm, W., Theory of water filtration, *Journal of the American Water Works Association*, 59(11):1397–1412, November 1967.

Parkson DS1, Parkson Corporation, Dyna Sand® Filter, Case Study: Apollo Beach WWTP Re-Uses DynaSand® Filter Effluent, Two page performance description, Parkson Corporation, Fort Lauderdate, FL, 1991.

Qasim, S. R., Motley, E. M., and Zhu, G., *Water Works Engineering: Planning, Design & Operation*, Prentice-Hall of India, New Delhi, India, 2006.

Raveendran, P. and Amirtharajah, A., Role of short-range forces in particle detachment during filter backwashing, *Journal of the Environmental Engineering Division*, 121(12):860–868, December 1995.

Saatchi, A. M. and Halilsoy, M., A new solution of the deep bed filter equations, *The Chemical Engineering Journal*, 34:147–150, 1987.

Stanley, D. R., Sand filtration studies with radiotracers, *Proceedings, ASCE*, 81(592):1–23, 1955.

Stein, P. C., A study of the theory of rapid filtration of water through sand, ScD Thesis, Department of Civil and Sanitary Engineering, Massachusetts Institute of Technology, Cambridge, MA, 1940.

Symons, J. M., Bradley, L. C. Jr., and Cleveland, T. C., *The Drinking Water Dictionary*, American Water Works Association, Denver, CO, 2000.

Tchobanoglous, G. and Eliassen, R., Filtration of treated sewage effluent, *Journal of the Sanitary Engineering Division*, ASCE, 96(2):243–265, 1970.

Tien, C. and Payatakes, A. C., Advances in deep bed filtration, *AIChE Journal*, 25(5):737–759, September 1979.

Tobiason, J. E. and O'Melia, C. R., Physicochemical aspects of particle removal in depth filtration, *Journal of the American Water Works Association*, 80(12):54–64, December 1988.

Tobiason, J. E. and Vigneswaran, B., Evaluation of a modified model for deep bed filtration, *Water Research*, 28(2):335–342, February 1994.

Truesdail, S. E., Westermann-Clark, G. G., and Shah, D. O., Apparatus for streaming potential measurements on granular filter media, *Journal of the Environmental Engineering Division, ASCE*, 124(12):1220–1232, December, 1998.

Trussell, R. R. and Chang, M., Review of flow through porous media as applied to headloss in water filters, *Journal of the Environ-*

mental Engineering Division, 125(11):998–1005, November 1999.

Trussell, R. R., Chang, M., Lang, J. S., and Hodges, W. E. Jr., Estimating the porosity of full-scale anthracite filter, *Journal of the American Water Works Association*, 91(12):54–63, November 1999.

Turneaure, F. E. and Russell, H. L., *Public Water Supplies*, John Wiley & Sons, New York, 1913.

Vagliasindi, F. and Hendricks, D. W., Wave front behavior in adsorption reactors, *Journal of the Environmental Engineering Division, ASCE*, 118(4):530–550, July/August 1992.

Weber, W. J., *Physicochemical Processes for Water Quality Control*, Wiley-Interscience, New York, 1972.

Wegmann, E., *The Water-Supply of the City of New York, 1658–1895*, John Wiley & Sons, New York, 1896.

Wolman, A., Filtration – The issue, in: *Water Supply Symposium*, New York State Department of Health, Albany, NY, September 19, 1988.

Yao, K. M., Habibian, M. T., and O'Melia, C. R., Water and waste filtration: Concepts and applications, *Environmental Science and Technology*, 11(5):1105–1112, 1971.

13 Slow Sand Filtration

As a "process," slow sand might be classified as "biofiltration." It has its own identity as a "technology" and a "practice" that is reviewed here briefly, along with theory (Box 13.1).

13.1 DESCRIPTION

A slow sand filter is merely a bed of sand, confined within a box, with appurtenances to deliver and remove water. Figure 13.1 illustrates the main elements of the process and technology.

13.1.1 Slow Sand Technology

Slow sand consists of a filter box with appurtenances that measure and control flows for the operation, for example, inflow, filtered water drainage, backfilling after scraping, headwater drainage, etc. All of this is to support a biological depth-filtration process, which is within the sand bed, coupled with a straining process, by a layer of material deposited, and perhaps with organism growth. The straining occurs on top of the sand bed, called the *schmutzdecke*. Slow sand filtration may be characterized as a *passive* process in that it occurs without operator intervention.

13.1.1.1 Filter Box and Appurtenances

Usually, the sand bed is 1.0–1.2 m (3–4 ft), but there is no rigid rule. As seen in Figure 13.1, the sand bed is supported by a graded gravel support, usually 0.3 m (1.0 ft), contained within a concrete "box." On the floor of the box, and within the gravel support, are the underdrains to remove the filtered water. The filtered water then enters the tailwater whose elevation is controlled by a weir plate on the effluent side of the filter. The flow of raw water to the filter is controlled by a valve on the influent side, with flow meter preceding. The energy of the incoming raw water must be dissipated so that the sand bed is not eroded, which would cause higher local velocity through the sand bed. In other words, the energy of the portion of the diffusion cone (illustrated in Figure 13.1) as it intercepts the surface of the sand bed should not be sufficient to lift the sand grains. This can be done by providing about 300–600 mm (12–24 in.) minimum depth of water at the start of the filter run as a "rule of thumb," and by distributing the flow around the filter box.

13.1.1.2 Sand Bed

The sand bed must be biologically "mature" to be effective in filtration. This means that the sand grains must have developed a biological film, that is, a community of microorganisms adhering to their surfaces, to which particles in the ambient water may "stick" once a collision occurs. The removal mechanism is "depth-filtration" (described in Sections 12.3 and 13.2.1.2).

13.1.1.3 *Schmutzdecke*

A characteristic of slow sand is the formation of a surface deposit called the *schmutzdecke*, a German word meaning "sludge blanket," or "dirty layer," both apt terms for the phenomenon observed, also illustrated in Figure 13.1. As the *schmutzdecke* develops, whether by deposit of material or growth of organisms, or both, the headloss increases. A major task of the operation is to remove the *schmutzdecke* (see Section 12.2.1.1) when the headloss exceeds some established criterion, usually $\geq$2.0 m (6 ft).

13.1.1.4 Design Approach

To make the process work, design guidelines should be followed and appurtenances must be added. The design goal is to provide *passive* operation. In other words, the design engineer must endeavor to ensure that proper plant operation will be self-evident and self-implementing to the operator through the life of the filter. If the design engineer achieves this goal, then slow sand filtration can be an *appropriate* technology for small communities, where operators may have other responsibilities.

13.1.2 Attributes

Slow sand is effective as a process and is economical in certain contexts, especially for small communities. Being simple in design and construction and *passive* in operation adds to its appeal.

13.1.2.1 Selection Criteria

Preferred source water turbidity levels are $<$10 nephelometric turbidity units (NTU). As an upper limit, 30–50 NTU has been mentioned as a rule of thumb, but the limit is more a matter of engineering and judgment than an absolute level. Factors include: effluent turbidity, length of run, and whether pretreatment may be acceptable. Color and volatile organics are other considerations; if the raw water concentrations are high, then a pilot plant study may be advisable.

13.1.2.2 Effectiveness

The slow sand filtration process is expected to remove biological particles such as cysts, oocysts, algae, bacteria, viruses, parasite eggs, nematode eggs, and amorphous organic debris at 2-log to 4-log levels when the filter bed is biologically *mature* (or "ripened"). The foregoing applies to biological particles present in the influent flow. There are other organisms that may find ecological niches within the *schmutzdecke* or within the filter bed. Those organisms that adhere to the sand grains within the filter bed comprise the biofilm that gives the sand bed "maturity" (also called a "ripened" sand bed).

BOX 13.1 ORGANIZING DILEMMA

Slow sand has a history: it is the first successful technology for municipal drinking water treatment. The 1829 London filter (Sections 13.1.4.1 and 13.1.4.2) was the foundation for present practice. Its design by James Simpson was based on studies of filtration efforts that were not successful and was refined by a pilot plant study. Therefore, because of its history that fits the themes of the book, for example, principles, modeling, pilot plants, practice, and, the fact that it is one of the important technologies that is "on-the-shelf" for use, and is appropriate for many situations, and perhaps because of sentiments about slow sand, the decision was to allocate a chapter to the topic.

To assess the overall performance of a slow sand filter, cartridge filter sampling is useful. Figure 13.2 shows cartridge filters after sampling the influent raw water and the filter effluent, respectively, for the slow sand filter at Empire, Colorado. The visual inspection gives an impression of the overall effectiveness of the filtration process. If the influent cartridge filter is black in color after 100–200 L of water throughput and the effluent cartridge remains white (or off-white), then high removals may be expected of all organisms, for example, algae, cysts, bacteria, viruses, etc. Heterotrophic plate counts may be higher in the effluent, however, due to the growths within the biofilm. Also, while high turbidity removals are expected, this may not always be the case, for example, for particles ≤ 1 μm (Bellamy, 1984; Bellamy et al., 1985a,b).

To provide more quantitative assessment of slow sand effectiveness, Figure 13.3 shows influent and effluent turbidities and influent and effluent coliforms in (a) and (b), respectively; the data were compiled by Sims and Slezak (1991) from results of a national performance survey of slow sand filtration plants. The plots are indicative of the quality of raw water sources used for slow sand and of the effectiveness of slow sand in reductions of turbidity and coliforms, respectively, for example, as seen in the plots, log R(turbidity) ≈ 0.8, log R(coliforms) ≈ 2.

13.1.2.3 Economy

The economic attractiveness of slow sand depends upon the context. The appeal in current times is mostly for small communities. Communities having populations of 1000 to 2000 persons, or even 5000 persons, should not be too large for slow sand to be "appropriate." At larger plants, however, the labor costs of processing sand (i.e., cleaning) will be greater than the cost of operation for rapid filtration. The point of this crossover will depend upon circumstances. The City of Salem, Oregon, however, with a population of 107,000 (serving 135,000) uses slow sand filtration, as does West Hartford, Connecticut with a population of 300,000 served; some others are listed by Slezak and Sims (1984). In Germany, slow sand is not uncommon and London has 47 ha (116 ac) of slow sand filters. Slow sand has been adopted in recent times in developing countries (van Dijk and Oomen, 1978; Kerkhoven, 1979; Komolrit et al., 1979; Alagarsamy and Gandhirajan, 1981; Paramasivam et al., 1981; van Markenlaan, 1981), and in Puerto Rico (Gaya, 1992).

13.1.2.4 Labor

Most of the labor is in scraping the sand bed, removing the sand from the box, washing, moving sand to and from storage, and rebuilding the sand bed. The frequency of scraping has been about monthly at Empire, Colorado (population 450) and required about 30–60 min for two persons for one filter bed having an area of 76.5 m^2 or 825 ft^2 (Seelaus et al., 1986, p. 4). The sand was deposited on the ground outside the filter, however, which deferred the labor of storing the sand for later washing and resanding. The plant required a daily visit for flow adjustment, water level measurements, turbidity measurements, and recording of data. By contrast, the Denver Kassler plant (151,000 m^3/day or 40 mgd) employed 20 persons for continuous activity in scraping, washing,

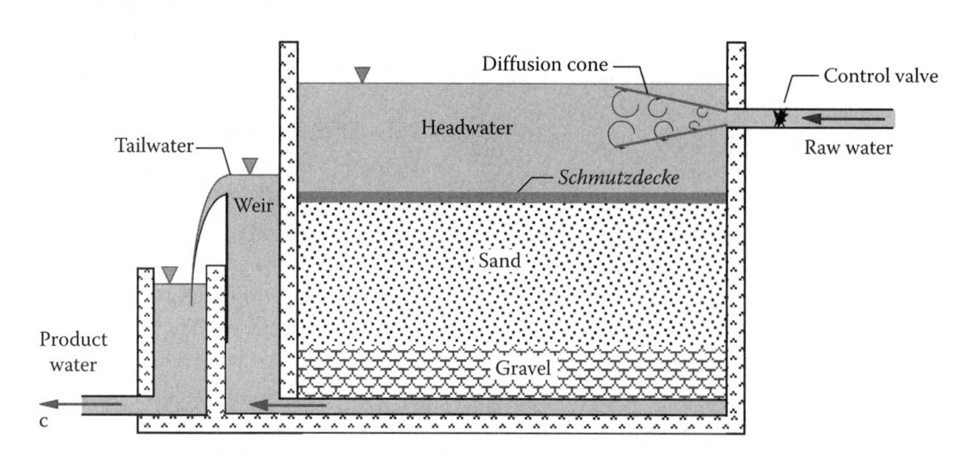

FIGURE 13.1 Slow sand filter schematic cross section. (Adapted from Hendricks, D.W. (Ed.), *Manual of Design for Slow Sand Filtration*, AWWA Research Foundation and American Water Works Association, Denver, CO, p. 2, 1991.)

FIGURE 13.2 Comparisons between influent and effluent cartridge filters (1 μm pore size) for slow sand filters at Empire, Colorado. (Adapted from Hendricks, D.W. et al., Filtration of *Giardia* cysts and other particles under treatment plant conditions, Research report on water treatment and operations, AWWA Research Foundation, Denver, CO, February, 1988; Hendricks, D.W. (Ed.), Manual of Design for Slow Sand Filtration, AWWA Research Foundation and American Water Works Association, Denver, CO, p. 31, 1991.)

storage placement, and rebuilding the sand bed for a filter area of 2.47 ha (10.5 ac) (Beer and Dice, 1982; Hendricks et al., 1991, p. 92) with filter runs of 3–6 months.

13.1.2.5 Materials

The materials for a slow sand filter are mostly sand, gravel, and concrete, which, for many installations, may be obtained locally. A local contractor may be available also. All of this makes slow sand attractive economically since most of the construction funds may be spent within the community. Some appurtenances also may be obtained locally. For example, an orifice plate may be used for flow measurement, with fabrication by a local machinist. Valves and underdrain pipe laterals must be ordered from vendors.

13.1.2.6 Contextual Factors

Slow sand has special appeal for small communities, regardless of country. The reasons include: (1) being a passive process, skilled operation is not required; (2) keeping capital and operating costs local helps the economy of the community; (3) requiring more labor, for larger installations, employs more persons, which may be a societal objective; and (4) the slow sand process is effective in removal of organisms. For such reasons, the International Reference Center (IRC) in Delft, the Netherlands, has done much to promote slow sand in India, SE Asia, and Latin America (Kerkhoven, 1979; van Dijk and Oomen, 1978; van Markenlaan, 1981).

13.1.3 History

By 1800, the notion of filtration to purify drinking water had evolved in several cities in Great Britain (Baker, 1948, pp. 64–115). The filters of that period were designed for downward, upward, or horizontal flow, with the latter being prevalent. All of these filters "failed," since there was no means for in-place cleaning. Two cleaning methods that eventually evolved were: (1) backwash and (2) scraping the surface. The former was patented in 1791 and the technology became known as "mechanical" filtration. The patent was utilized in 1827 by Robert Thom to engineer the first installation for municipal use at Geenock, Scotland (p. 91); his design was aided by observing clogging of existing filters. James Simpson also observed existing filters (p. 93) but took a different path. He noted that the detritus removed was at or just below the surface of the sand and so concluded that the

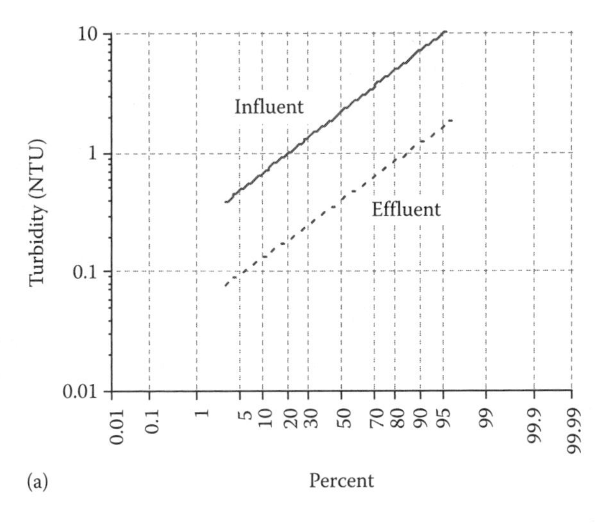

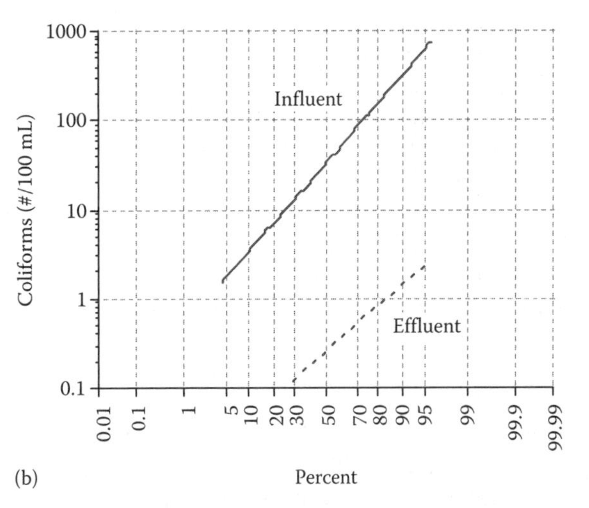

FIGURE 13.3 Comparisons between influent and effluent for slow sand filtration from a national survey. (a) Turbidity, (b) coliforms. (Adapted from Sims, R.C. and Slezak, L., Present practice of slow sand filtration in the United States, in Logsdon, G.S. (Ed.), *Slow Sand Filtration Manual of Practice*, American Society of Civil Engineers, New York, 1991.)

best method to clean it was to scrape the surface layer, wash it, and replace it at intervals.

13.1.3.1 James Simpson and the Start of Slow Sand

On January 14, 1829, a 1 ac slow sand filter was put in operation at Chelsea on the north bank of the River Thames (Baker, 1948, pp. 99–113). The Chelsea slow sand filter was the first successful technology for municipal filtration. Its design became the basis for developing a practice of slow sand filtration that spread to other countries in Europe and then to the United States. The 1 ac filter produced 8,250–11,300 m^3/day (2.25–3 mgd), with associated $0.09 \leq HLR \leq 0.12$ m/h ($2.25 \leq HLR \leq 3$ mgad).

James Simpson, engineer for the Chelsea Water Works Company, had worked out the design of the Chelsea filter based on knowledge of various filtration attempts and on operation of a pilot filter. At the age of 24, James Simpson (1799–1869) was appointed engineer for the Chelsea Water Works Company and at 26 was elected to the Institution of Civil Engineers.

13.1.3.2 Evolution of Practice

From its start in London, the slow sand technology spread to continental Europe and, finally, in 1872, came to America. The basic technology has remained unchanged from the pilot filters of James Simpson.

13.1.3.2.1 London

With the experience of the Chelsea filter, the health benefits of filtered water became apparent, and in 1839 the city's five commercial suppliers began filtering their water. Then, in 1852, the London city government required filtration prior to public sale, and later established the Thames Conservancy Board to regulate potable water quality (Hazen, 1913). By 1894, five successive increases in area had occurred for the Chelsea filters and the total surface area was 47 ha (116 ac), producing 890,000 m^3/day (234 mgd). In 1977, the filter area was 72 ha (178 ac), with zero-coagulation rapid filters or microstrainers preceding the slow sand filters (Poynter and Slade, 1977, p. 75). The slow sand filters numbered eight in 1988 treating 2 million m^3/day (530 mgd) (Rachwal et al., 1988, p. 331). This "double filtration," that is, zero-coagulation rapid filters followed by slow sand, became prevalent over much of western Europe.

13.1.3.2.2 Continental Europe

From the example of the Chelsea filters in London, continental Europe began filtering public water supplies by the 1850s, with installations and dates as follows: Berlin, 1856; Altona, 1860; Zurich, 1884; Hamburg, 1893; Budapest, 1894 (Hazen, 1913). The health benefit of slow sand filtration was demonstrated dramatically by the 1892 cholera epidemic in port city of Hamburg (population 640,000) which caused 8605 deaths (Cosgrove, 1909, p. 112). By contrast, the adjacent City of Altona (population 143,000) had only 323 deaths. The water intake for Hamburg was from the Elbe River upstream from the combined sewage discharge of the two cities, but the intake was under the influence of the floodtide and so was subject to some degree of sewage transport to the intake. The intake for Altona's water supply was also from the Elbe River, downstream from Hamburg and downstream from the combined sewage discharge. Hamburg had no treatment at the time of the outbreak while Altona had slow sand filtration. As noted, Altona had installed slow sand filtration in 1860 and Hamburg followed in 1893, just after the cholera outbreak. The two cities were a part of a contiguous metropolitan area and the only difference was that the water supply systems were separate. The case became a classic in demonstrating the value of filtration, to wit (Cosgrove, 1909, p. 111):

> To use Professor Kochs own words: 'Chlorera in Hamburg went right up to the boundary of Altona and then stopped. In one street, which for a long way forms the boundary, there was cholera on the Hamburg side, whereas on the Altona side was free from it, and yet there was only one detectable difference, and one only, between the two adjacent areas– they had different water surfaces.'

13.1.3.2.3 Worldwide

While Europe, including Great Britain, was where slow sand grew to its greatest extent, there was some diffusion to the United States, Canada, South America, and Asia. Table 13.1 summarizes the installed capacity of slow sand filtration, worldwide in 1900 as compiled by Allen Hazen in his 1913 book, *The Filtration of Public Water Supplies*. The installed capacity of continental Europe was about the same as Great Britain, with not too much in South America and Asia and only a modest amount in North America.

The slow sand technology has been disseminated to less developed countries, that is, in Asia and Africa by the Slow Sand Project of the International Reference Centre for Community Water Supply and Sanitation (IRC). Raman et al. (1981, p. 44) reported that in the Haryana State, India, about 100 slow sand plants were in operation for both urban and village supplies. In Thailand, the influence of IRC had resulted in numerous village scale slow sand plants. Symposiums on the topic (Graham and Collins, 1988, 1996) have helped to update the body of knowledge and applications among persons active in the field. The U.S. Environmental Protection Agency started to revive slow sand under the guidance of Dr. Gary Logsdon, who along with several others, has continued to infuse the technology into practice (see, for example, Logsdon, 2008).

13.1.3.2.4 America

The first slow sand filter in the United States was at Poughkeepsie, New York (1870 population 20,000), and began operation on December 1, 1872 (Baker, 1948, pp. 148–158). The engineer identified with the project was James P. Kirkwood (1807–1877), perhaps the most eminent sanitary engineer of the time (and the second president of the American Society of Civil Engineers). The source water for the filter was the Hudson River, deemed by the chief engineer for the city, J. B. G. Rand, as a polluted source; also the Hudson was

TABLE 13.1
Summary of Cities Using Slow Sand Filters in 1900, World Wide

Place	Population	Area of Filters (ha)	Area of Filters (ac)	Filters (#)	Average Daily Use (mL/day)[a]	Average Daily Use (mgd)
United States	259,774	7.00	17.31	45	101.7	26.87
British Columbia	16,841	0.33	0.82	3	6.8	1.80
South America	500,000	1.68	4.15	3		
Holland	1,414,021	9.20	22.75	47	119.2	31.48
Great Britain	10,100,738	65.48	161.80	161	1448.6	382.73
Germany	4,639,080	42.99	106.22	185	443.3	117.13
Other European Countries	2,984,839	14.06	34.74	88	336.3	88.84
Asia	1,397,000	2.71	6.69	23		
Totals	21,411,293	143.45	354.48	555	2455.9	648.85

[a] mL = million liters.

under tidal influence at low flows. Both Rand and Kirkwood recommended other sources that were free of known pollution but were overruled by the city's water commissioners in favor of the Hudson (p. 149). Kirkwood resigned as the city's consulting engineer on December 31, 1872. There were two filters each with an area of 1,300 m^2 (14,000 ft^2); the overall depth was 1,829 mm (72 in.) with 610 mm (24 in.) sand and 457 mm (18 in.) graded gravel. In 1886, a trough 61 m (200 ft) long was used to wash the sand and in 1896 jet washers were used. Allen Hazen (1869–1930) entered the picture in 1913 and advised on several modifications, including using an upland source (Baker, 1948, p. 138).

By about 1910, the rapid filtration technology had developed to the point that it was favored, almost without question, over slow sand. The upshot of this favoring of rapid filtration over slow sand was that by 1940, the United States had about 100 slow sand filters and about 2275 rapid filters (Baker, 1948, p. 148). Some 39 slow sand plants were built from the late 1960s through 1988 with 11 proposed plants (Logsdon and Fox, 1988). Of the 39 plants, 23 had capacities of ≤ 1000 m^3/day. In 1994, there were about 225 slow sand plants, with about 25–30 utilities evaluating slow sand in 1996 (Brink and Parks, 1996, p. 14). One of the issues with slow sand was that it was not suitable for some of the turbid waters of the mid-west, for example, the Ohio and the Mississippi.

13.2 SLOW SAND AS A PROCESS

Removal mechanisms and headloss are the two process concerns that relate to design and operation. The former is reviewed in terms of straining and depth filtration. The latter is understood best in terms of Darcy's law.

13.2.1 Removal Mechanisms

Both straining and depth filtration are operative in slow sand filtration. Neither are understood definitively but a review of what is known helps to understand slow sand removal mechanisms.

13.2.1.1 *Schmutzdecke* and Its Role in Straining

Retention of particles by smaller openings, through which flow occurs, is *straining*. When particles in the raw water are retained on the surface of the sand bed, then the pore openings are partially blocked, resulting in yet smaller openings with capacity to retain still smaller particles. The retained particles form a layer that has been termed the *schmutzdecke*, a German word which translates literally as "dirty skin." The word, used first in Germany, was adopted in British practice by 1902 Baker (1948, p. 124). The *schmutzedecke* is defined here as "a layer of matter, both deposited and synthesized, on the top of the filter bed which causes headloss disproportionate to its thickness."

13.2.1.1.1 Schmutzdecke: Further Notes

Schmutzdecke, as traditionally described, is a gelatinous zoogeal mass of living and dead microorganisms (Babbitt and Doland, 1939, p. 536; Huisman and Wood, 1974, p. 20). Such a classic *schmutzdecke* was reported for slow sand pilot filters at Portsmouth, New Hampshire, and at Ashland, New Hampshire, that is, as being comprised of gelatinous organic matter (Collins, 1990).

As compiled by experiences in various locations, the character of the *schmutzdecke* varies widely. At Empire, Colorado, for example, the surface deposit was a light, easily suspended, inert black carbonaceous deposit, comprised of light flakes about 1 mm in diameter. Figure 13.4 is a photograph showing the contrast between the scraped sand and the *schmutzdecke* on the right, comprised of the carbonaceous forest detritus. Figure 13.2 illustrates further showing two 1 μm cartridge filters for influent and effluent, respectively, for the Empire slow sand filter, each with a throughput of about 100 L when the photograph was taken. The raw water turbidity in Mad Creek, the raw water source for Empire, seldom exceeded 0.5 NTU and was not associated with the deposit on the filters. The black color was due to the carbonaceous forest detritus as deposited on the surface of the slow sand filters.

FIGURE 13.4 Scraped sand bed at Empire, Colorado, contrasted with *schmutzdecke* after 30 days operation, comprised of carbonaceous deposit. (Adapted from Hendricks, D. W. (Ed.), *Manual of Design for Slow Sand Filtration*, AWWA Research Foundation and American Water Works Association, Denver, CO, p. 13, 1991.)

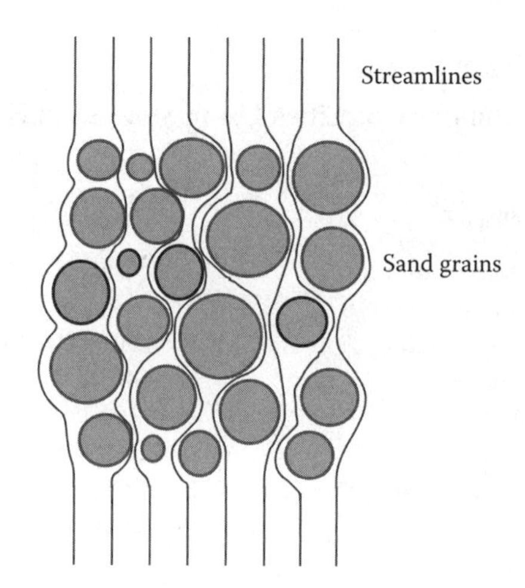

FIGURE 13.5 Streamlines within a sand bed. (Adapted from Hendricks, D. W. (Ed.), *Manual of Design for Slow Sand Filtration*, AWWA Research Foundation and American Water Works Association, Denver, CO, p. 9, 1991.)

Schuler et al. (1991) describe similar observations in filtering a water in Pennsylvania having turbidities 0.1–5.8 NTU. The *schmutzdecke* was "tightly packed and unattached to the sand." In pilot filters at Colorado State University (Bellamy et al., 1985a,b), a well-defined *schmutzdecke* was not visible, but the headloss increased with time consistent with the development of a *schmutzdecke*. Scraping the surface resulted in recovery of the clean-bed headloss.

Whatever the character of the *schmutzdecke*, a deposit of some sort always occurs in every slow sand filter and causes headloss to increase. Removing the *schmutzdecke* by scraping will cause the headloss to recover to the "clean-bed" level (plus some incremental headloss due to deposits or biofilm development within the sand bed).

13.2.1.2 Depth Filtration

Within the sand bed, ambient raw water particles (viruses, bacteria, cysts, mineral turbidity, etc.) that are not removed by the *schmutzdecke* have some probability of being transported to a sand grain surface during its passage through the interstices. If a biofilm has developed on the grains comprising the sand bed, such particles may *attach* (and thus be removed). Such removal within the sand bed is, by definition, *depth filtration*. The two facets of depth filtration are (1) transport and (2) attachment (Section 12.3.3).

13.2.1.2.1 Interstitial Flow

To better visualize the transport step of depth filtration (Iwasaki, 1937), Figure 13.5 depicts a packed bed of sand grains with associated streamline configuration assumed by a flow of water from top to bottom. As seen, within a packed bed with many sand grains, the streamlines have a tortuous configuration. The stream tubes bifurcate and rejoin and bifurcate again at random points. This continuous bifurcation creates opportunity for collisions between particles and sand grains. The *probability* of an impingement within a given distance of travel depends upon the size of the sand grains, the interstitial velocity, and temperature. Figure 12.19 provides another perspective of the same idea from Yao et al. (1971). The smaller the sand grains, the higher the probability of an impingement; there are simply more bifurcations per unit distance for an interstitial stream with smaller sand. Also, the lower the interstitial velocity, the higher the probability of impingement; lower velocity permits more "steps" of random motion by diffusion per unit distance and more time for an impingement to result from gravity acting on a particle and altering its trajectory. By the same token, higher temperature gives more random motion "steps" per unit time (for small particles) than lower temperature, and hence there is a higher probability of impingement (Section 12.3.3.3). Removals by interception are not affected, however, by velocity (in the laminar flow regime). The three transport mechanisms, interception, sedimentation, and diffusion, are discussed subsequently.

As implied in Figure 13.5, a particle within the interstitial stream will most likely, at some point during its path, impinge upon a sand grain (due to one of the three transport mechanisms). Whether it attaches or not depends, in the case of biofiltration, on whether a biofilm exists on the sand grain surface. Plain particles may attach to bare sand grains in some cases, depending on inter-particle forces. Evidently such attachments occur, especially for microorganisms, since biofilms do develop.

13.2.1.2.2 Attachment Coefficient and the Role of Biofilm

Unless attachment occurs, there is no removal. The fraction of particles that attach, relative to the number of collisions, is by definition, the coefficient, α. Research suggests that biofilm development on the sand grains provides an adsorptive surface for such attachment. Another idea is that extracellular enzymes will coagulate some biological particles to permit attachment (i.e., the enzyme alters the zeta-potential of the particle to permit attachment). If so, these particles become the biofilm. Once attachment has occurred, the biofilm may metabolize biological particles and organic contaminants.

Slow sand has been termed, in fact, a "biological filter," with the biofilm described as: "a teeming mass of micro-organisms, bacteria, bacteriophages, predatory organisms such as rotifers and protozoa, all feeding on the adsorbed impurities and upon each other" (Huisman and Wood, 1974). Protozoa and Rotifera were found to be the dominant interstitial microfauna removing bacteria within the sand bed (Lloyd, 1973). Core sampling of the Hampton and Ashford Common filters in England showed aerobic bacteria, flagellates, ciliates, rotifers, flatworms, gastrotrichs, nematoda, annelida, and arthropoda (Duncan, 1988, p. 168). Bacterial densities were 10^9–10^{10}/mL sand; protozoans (flagellates, ciliates [e.g., *Vorticella*], amoeba) numbered in the thousands per mL. The *Vorticella* were considered the most efficient "filter-feeders" of suspended particles.

The rate and extent of biofilm development increases both with nutrient concentration and temperature (Bellamy et al., 1985a,b; Barrett, 1989; Bryck et al., 1987a,b). Slow sand filters located in a nutrient-limited situation may be expected to have 2-log coliform removals after biofilm *maturity* while filters using nutrient rich waters may expect to have 3-log coliform removals (Bellamy et al., 1985a,b) and even 4-log removals (Barrett, 1989). Straining within the sand bed is not likely to be a major removal mechanism since removals are not significant until a biofilm has been developed. To illustrate, removals were zero in filter beds that were chlorinated to disinfect and then de-chlorinated and purged before seeding with coliforms (Bellamy et al., 1985a,b).

13.2.1.2.3 Clean Sand Bed Removals

Removals of organisms by clean-beds of sand have been variable. As noted previously (Bellamy et al., 1985a,b), chlorinated and purged sand beds were found to have virtually zero removals of organisms. But in a classic study of virus removals for various conditions, Roebeck et al. (1962, p. 1280) found that removals of attenuated poliovirus varied with hydraulic loading rate (HLR) (i.e., superficial filtration velocity), as shown in Figure 13.6. Further, the fraction removed was higher for the $d_{10} = 0.28$ mm UC = 1.4 sand than the 0.78 mm UC = 1 sand. As seen, removals were about 0.98 for HLRs less than the slow sand range, that is, $0.04 \leq \text{HLR} \leq 0.4$ m/h and decreased to ≤ 0.1 for the range of rapid filtration.

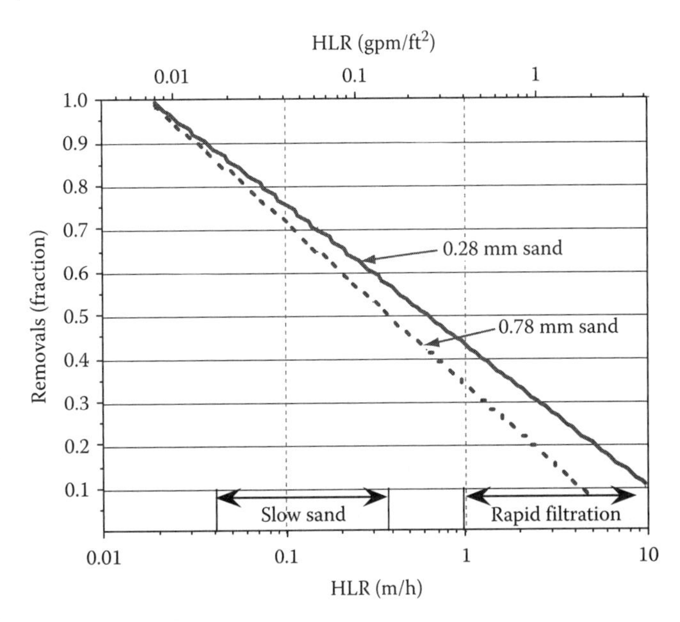

FIGURE 13.6 Virus (attenuated polio) removals for clean sand bed, 0.610 m (2.0 ft), as affected by HLR. (Adapted from Robeck, G.G. et al., *J. Am. Water Works Assoc.*, 54(10), 1280, October, 1962.)

13.2.2 Hydraulics

Headloss within a slow sand filter is caused by flow through the *schmutzdecke* and the sand bed. As the filter is operated, the *schmutzdecke* develops and its hydraulic resistance increases, causing most of the headloss. Removing the *schmutzdecke*, for example, by hand-scraping, will permit the headloss to recover to near clean-bed level. The "clean-bed" headloss of the sand bed is perhaps 100–200 mm, depending upon the HLR, the temperature, and the sand bed media characteristics. The well-known Darcy's law integrates these variables.

13.2.2.1 Darcy's Law

Headloss through a porous medium is described by Darcy's law (Sections E.2.1 and E.2.2), which, in finite difference form, is

$$v = -K\frac{\Delta h}{\Delta Z} \tag{13.1}$$

where

- v is the superficial velocity, also called hydraulic loading rate, that is, HLR $= Q/A$ (m/s)
- h is the hydraulic head at any point along a bed of porous medium (m)
- Δh is the difference in hydraulic head between any two points along a bed of porous medium, which may be expressed as headloss, h_L (m)
- Z is the flow distance though porous medium (m)
- K is the hydraulic conductivity of porous medium (m/s)

Figure E.5 illustrates the Δh and ΔZ terms where the hydraulic gradient is shown for an horizontally oriented sand bed. Figure 12.39 illustrates the same thing but with the sand bed oriented vertically; the piezometers show the hydraulic gradient.

13.2.2.2 Intrinsic Hydraulic Conductivity

The hydraulic conductivity, K, of a porous medium is dependent on the intrinsic hydraulic conductivity, k (Section E.3.1), the water density, ρ_w, and the water viscosity, μ, that is,

$$K = k\frac{\rho_w g}{\mu} \tag{13.2/E.4}$$

TABLE CD13.2/CDE.2
Conversion between K and k Including Headloss Calculation from k

(a) K–k

	$g = 9.807$				Enter K to Calculate k					
Media Name	d_{10} (mm)	d_{60} (mm)	d_{50} (mm)	UC	K (m/d)	K (m/s)	T (°C)	μ (N-s/m^2)	ρ_w (kg/m^3)	k (m^2)
Sand	0.50			1.5	2.42E+02	2.80E−03	3	0.00162	999.965	4.622E−10
Anthracite	0.91			1.5	1.26E+03	1.46E−02	3	0.00162	999.965	2.419E−09
Flatiron masonry	0.24			2.7	3.77E+01	4.37E−04	3	0.00162	999.965	7.215E−11
Flatiron masonry	0.24			2.7	4.08E+01	4.72E−04	3	0.00162	999.965	7.804E−11

(b) k–K

	$g = 9.807$				Enter k to Calculate K					
Media Name	d_{10} (mm)	d_{60} (mm)	d_{50} (mm)	UC	k (m^2)	T (°C)	μ (N-s/m^2)	ρ_w (kg/m^3)	K (m/s)	K (m/d)
Sand	0.50			1.5	4.62E−10	3	0.00162	999.965	2.80E−03	2.4162E+02
Anthracite	0.91			1.5	2.42E−09	3	0.00162	999.965	1.46E−02	1.2644E+03
Flatiron masonry	0.24			2.7	7.21E−11	3	0.00162	999.965	4.37E−04	3.7717E+01
Flatiron masonry	0.24			2.7	7.80E−11	3	0.00162	999.965	4.72E−04	4.0800E+01

Note: $\mu(\text{water}) = 0.00178024 - 5.61324 \cdot 10^{-05}T + 1.003 \cdot 10^{-06}T^2 - 7.541 \cdot 10^{-09}T^3$.
(water) $\rho = 999.84 + 0.068256 \cdot T - 0.009144 \cdot T^2 + 0.000102957T^3 - 1.1888 \cdot 10^{-06}T^4 + 7.1515 \cdot 10^{-09}T^5$.

where

- k is the intrinsic hydraulic conductivity, also called intrinsic permeability (m^2)
- ρ_w is the mass density of water at a given temperature (kg/m^3)
- μ is the dynamic viscosity of water at given temperature (N-s/m^2)

The intrinsic hydraulic conductivity, k, is a characteristic of the porous medium while the permeability, K, incorporates fluid properties. Table CD13.2/CDE.2(a)/(b) provides computations for conversions between k and K. Note that a hydraulic conductivity test, for example, as in Section E.4.1, yields, K and from this k may be calculated. Once k is determined, then K may be calculated for any other condition, for example, different temperatures, for use in Darcy's law. Table CD13.2/CDE.2(a) gives a few examples of k and K values for different media.

Combining Darcy's law, Equation 13.1, with the relation for K, Equation 13.2/E.4, gives the Darcy equation in terms of intrinsic hydraulic conductivity, k, that is,

$$v = -k\frac{\rho g}{\mu}\frac{\Delta h}{\Delta z} \tag{13.3}$$

Equation 13.3 has utility when k is given, such as in a table of values. Table CDE.2 and Figures E.2, E.3, and E.4 (Appendix E) provide such data.

The intrinsic hydraulic conductivity, k, of the clean sand bed is a function of the sand size, sand size distribution, and the aggregation (i.e., the extent to which the sand packs grain to grain, vis-à-vis voids being formed by particle bridging). The hydraulic conductivity, k, cannot be predicted with accuracy and must be based on measurements. Table 13.3 shows data from a lab test, several pilot filters (305 mm diameter) and several full-scale slow sand filters. All results are within one order of magnitude, that is, with k ranging from $2.55 \cdot 10^{-11}$ to $3.07 \cdot 10^{-10}$ m^2 for $d_{10} = 0.13$ mm and $d_{10} = 0.92$ mm, respectively.

In terms of its utility, k may serve to monitor whether clogging of the filter bed is occurring and whether a sand being considered for an installation has a k that falls within an expected range. In addition, one may calculate the clean-bed headloss under different, HLR, sand bed depth, and temperature scenarios. Example 13.1 illustrates the utility of having k data in estimating the clean-bed headloss and shows how Darcy's law may be applied.

Example 13.1 Darcy's Law Calculation

Calculate the clean-bed headloss for the slow sand filter at Empire, Colorado, having bed depth 1.22 m, at HLR = 0.2 m/h, and $T = 15$°C.

1. Apply Darcy's Law

Darcy's law in the form of Equation 13.3 and using k as given in Table 13.3, and μ (15°) from Table B.9:

$$v = -k\frac{\rho g}{\mu}\frac{\Delta h}{\Delta Z} \tag{13.3}$$

$$0.2\left(\frac{\text{m}}{\text{h}}\right) \cdot \left(\frac{\text{h}}{3600\ \text{s}}\right) = 7.03 \cdot 10^{-11}(\text{m}^2) \cdot \frac{999.102\left(\frac{\text{kg}}{\text{m}^3}\right) \cdot 9.807\left(\frac{\text{m}}{\text{s}^2}\right)}{1.139 \cdot 10^{-3}\left(\frac{\text{N-s}}{\text{m}^2}\right)} \cdot \frac{h_L}{1.22\ (\text{m})}$$

TABLE 13.3
Hydraulic Conductivities for Slow Sand Filters

Installation	d_{10}[a] (mm)	UC[b]	Method	Diameter (mm)	K (25°C)[g]		k[h] (m^2)
					(m/h)	(m/s)	
Empire[c]	0.21	2.67	Full-scale		2.78	7.72E.04	7.03E.11
Empire			Lab column	50	2.0	5.56E.04	5.06E.11
100 Mile House[d]	0.25	3.5	Full-scale		2.04	5.67E.04	5.16E.11
CSU pilot plants[e]							
Phase I							
Filter #1	0.27	1.63	Pilot column	305	1.46	4.06E.04	3.69E.11
Filter #2	0.27	1.63	Pilot column	305	2.28	6.33E.04	5.76E.11
Filter #3	0.27	1.63	Pilot column	305	3.31	9.19E.04	8.37E.11
Phase II							
Filters #2,3, 4, 6	0.29	1.53	Pilot column	305	3.56	9.89E.04	9.00E.11
Filter #5	0.62	1.59	Pilot column	305	10	2.78E.03	2.53E.10
Phase III							
Filter #5	0.13	1.6	Pilot column	305	1.01	2.81E.04	2.55E.11
CU pilot plants[f]							
Sand #1	0.22	2.5	Pilot column	305	2.74	7.61E.04	6.93E.11
Sand #2	0.92	2.28	Pilot column	305	11.05	3.07E.03	2.79E.10
Sand #3	0.2	4.15	Pilot column	305	2.54	7.06E.04	6.42E.11

Source: Hendricks, D.W. (Ed.), Manual of Design for Slow Sand Filtration, AWWA Research Foundation and American Water Works Association, Denver, CO, p. 19, 1991.

Notes: r (25°C) = 997.048 kg/m^3. μ (25°C) = 0.00089 N-s/m^2. g = 9.80665 m/s^2.

[a] The term d_{10} is defined as the sand size of which 10% is finer, based on a sieve analysis.

[b] The term UC is called the uniformity coefficient and is defined as the ratio of the d_{60} size to the d_{10} size.

[c] For Empire, full-scale: headloss, $h_L = 0.15$ m; sand bed depth, $\Delta Z = 1.22$ m; HLR = 0.22 m/h at 8°C (Seelaus et al., 1986, p. 8, 11). From Darcy's law, Equation 13.1, $K = 1.79$ m/h = $5 \cdot 10^{-4}$ m/s. Measurement of 1.79 m/h at 8°C converts to 2.78 m/h at 25°C (Seelaus et al., 1988).

[d] For 100 Mile House, K is calculated for 25°C, that is, $\mu = 0.90 \cdot 10^{-3}$ N-s/m^2 (Bryck et al., 1987a,b).

[e] For the CSU filters, calculations of k and K were based upon graphical headloss data (Bellamy et al., 1985a,b) and interpolation was accurate about ±30% and so the calculations for k and K will reflect this uncertainty.

[f] CU pilot plant data compiled from Barrett (1989).

[g] K, hydraulic conductivity, was calculated from measured data inserted in Darcy's law, that is, $v = -K(\Delta h/\Delta z)$.

[h] k, intrinsic hydraulic conductivity (permeability), was calculated from relation, $K = k(\rho g/\mu)$.

2. Solve for h_L,

$$h_L = 0.112 \text{ m } (15°\text{C})$$

3. At 0°C,

$$h_L = 0.175 \text{ m } (0°\text{C})$$

4. Discussion

The terminal headloss permitted for the Empire filter is about 1.95 m. The example shows that the clean sand bed accounts for only 0.112–0.175 m of this total. Thus, at the end of the run, the *schmutzdecke* accounts for most of the headloss, that is, about 0.91 fraction. Pilot plant modeling is most important in order to know the hydraulic character of the *schmutzdecke* and how it changes over the annual cycle. Understanding the hydraulic theory, as outlined above, helps in interpreting pilot plant behavior and in anticipating the behavior of the full-scale filter.

13.2.2.3 Hydraulic Profile and Headloss

Figure 13.7 shows the hydraulic profile across a sand bed after development of a *schmutzdecke*, as measured by piezometers. Piezometers are installed in the walls of the filter box, approximately as shown. The largest headloss occurs across the *schmutzdecke*, measured by piezometers A–B. The headloss across the sand bed is measured by piezometers B–C–D. As illustrated, the headloss across the sand bed is small relative to that across the *schmutzdecke*. Monitoring the piezometer water levels is useful in operation, for example, knowing when to scrape, to dewater, and to backfill.

13.3 DESIGN

Slow sand filter design was established largely by James Simpson for the 1829 London filters with guidelines given by Allen Hazen in his 1913 book. In over 125 years, the design of slow sand filters has changed little from the London

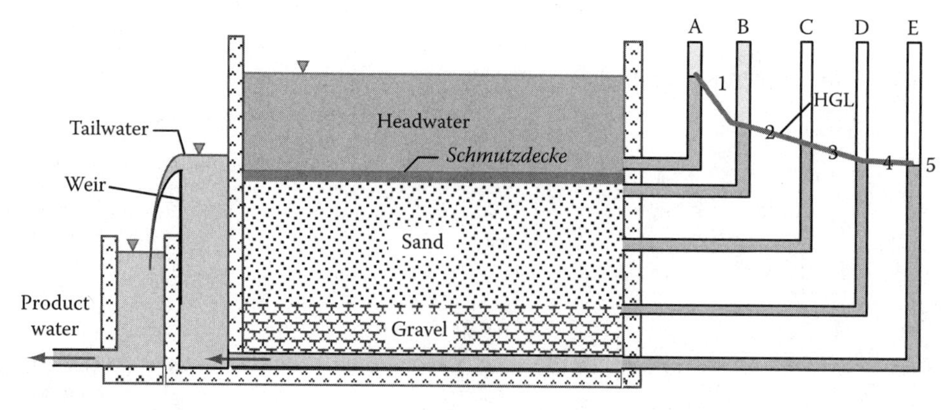

FIGURE 13.7 Hydraulic head profile across sand bed with developed *schmutzdecke*. (Adapted from Hendricks, D. W. (Ed.), *Manual of Design for Slow Sand Filtration*, AWWA Research Foundation and American Water Works Association, Denver, CO, p. 18, 1991.)

filters (van de Vloed, 1955, p. 565). Design parameters established included: HLR, d_{10}, and uniformity coefficient (UC) values for the sand, sand bed depth, and gravel support specifications.

13.3.1 Filter Box

Design issues of the filter box include area, number of cells, layout, depth, structural design, and water tightness. Table 13.4 summarizes data from three slow sand filters constructed during the period 1985–1990, giving the design population, total bed area, number of cells, bed area per cell, time out of operation for scraping (see footnote), time required for scraping, and number of persons involved in scraping.

13.3.1.1 Hydraulic Loading Rate and Area

The range filtration velocities are $0.04 \leq \text{HLR} \leq 0.4$ m/h ($1.0 \leq \text{HLR} \leq 10$ mgad; $0.016 \leq \text{HLR} \leq 0.16$ gpm/ft^2). By comparison for rapid rate the range is $5 \leq \text{HLR (rapid rate)} \leq 32$ m/h [$2 \leq \text{HLR (rapid rate)} \leq 13$ gpm/ft^2]; the ratio of areas is nominally a factor of 100. In principle, the HLRs selected determines the filter area needed, that is, HLR = Q(filters)/A(filters). The issue, however, involves such exigencies as peak day flow, peak hour, filtered water storage, and number of cells. The ideal protocol is (1) select HLR (peak day); (2) determine Q (peak day); (3) determine volume of filtered water storage for peak day, considering hourly variations and peak hour; (4) determine corresponding constant flow for peak day; and (5) using Q (peak day, constant flow) and HLR (peak day), calculate area required. The treated water storage is determined by a cumulative demand versus time plot over a 24 h period along with a filtered water flow that is constant (and a straight line in the plot), called a "mass-flow" plot. Therefore, once the filter bed is sized, the HLR will vary as operating conditions change, for example, based on variation in demand, and taking filters in and out of operation.

TABLE 13.4
Data on Slow Sand Filters

Place/Design Population	Total Bed Area (m^2)	Cells (#)	A(cell) (m^2/cell)	Time for Scraping (h)[a]	Persons in Scraping (#)
Empire/1000[b]	153	2	76.6	2	2
100 Mile H./2300[c]	774	3	258	4	3
Moricetown/900[d]	180	2	90	2	2

Source: Hendricks, D.W. (Ed.), Manual of Design for Slow Sand Filtration, AWWA Research Foundation and American Water Works Association, Denver, CO, p. 98, 1991.

[a] The "downtime" for a filter includes the time to drain the headwater, dewater the sand bed about 30–50 mm (1–2 in) below the sand bed surface, scrape and remove the sand, backfill from the bottom, and refill the headwater. The downtime for the three filters above is about 24 h for each.

[b] Population was 450 in 1984. Plant was designed for 0.946 million L/day (250,000 gpd) for projected population of 1000 persons, with HLR of 0.26 m/h (6.5 mgad) for a total bed area 153 m^2 (1650 ft^2). Flow demand in 1985 ranged from 0.189 to 2.000 million L/day (50,000–530 gpd).

[c] Population was 1925 in 1987. Plant was designed for a peak flow of 7.26 million L/day, to serve a population of 2300 (Bryck et al., 1987a,b).

[d] Plant design for peak day flow of 0.922 million L/day to serve a projected 1996 population of 900; the plant can be expanded to serve 1240 persons, projected for 2006 (Dayton & Knight, Ltd., 1989).

13.3.1.2 Number of Cells

Every slow sand filter should have two or more cells so that when one is out of service for scraping or other reasons, one filter bed can continue producing sufficient water for the community. Whether more than two filter beds are used depends upon the time required for scraping and capital costs. As discussed above, the number of cells directly affects how the system operates and performs as a system.

Example 13.2 Number of Cells Comprising a Filter Bed

Using the data from Empire, calculate an upper limit area for a single cell for a slow sand filter.

1 Local conditions—assumptions

Assume that the filter bed can be out of operation for 16 h, and that three persons are available for scraping. Also assume that 10 h are needed to drain the headwater (the time required at Empire), and 4 h are needed to put the filter back into operation, leaving 2 h available for scraping. Also assume scraping rate = 38 m^2/person/h (the rate at Empire).

2 Calculate maximum area of a cell

$$A_{cell} = (\text{scraping rate in m}^2/\text{person/h}) \cdot (\text{No. of persons}) \cdot (\text{Hours allotted to scraping})$$

$$A_{cell} = 38\,\text{m}^2/\text{person/h} \cdot 2\ \text{persons} \cdot 2\,\text{h} = 152\,\text{m}^2$$

in which A_{cell} is the maximum area of a single cell based upon allotted downtime of a single cell, crew size, and rate of scraping per crew member (m^2).

3 Number of cells, using Empire data

$$\text{No. of cells} = A\ (\text{total filter area})/A_{cell} = 153\,\text{m}^2/152\,\text{m}^2/\text{cell} \approx \text{one cell}$$

which means that Empire does not have enough total area to warrant concern about the HLR criterion.

4 Discussion

By comparison, the Empire filter has a total bed area of 153 m^2, with two cells giving 76.6 m^2/cell and, as noted, is removed from operation for only about 16 h. By comparison, the 100 Mile House plant has 774 m^2 total bed area, with three cells, giving 258 m^2/cell. The total downtime to scrape one of the three cells at 100 Mile House is about 24 h. The calculations illustrate the factors relevant in determining the upper limit in area for a single cell. The magnitudes of the factors will depend upon local circumstances.

13.3.1.3 Layout

An important part of the design is the layout of the filter area. Filter layout determines the plumbing configuration, the economy of the filter box construction, and whether future plant expansion will be feasible. Figure 13.8 is a schematic of the filter box and plumbing layout for the slow sand filter plant at 100 Mile House, British Columbia. For comparison, Figure 13.11 shows another layout, that is, for Moricetown, British Columbia, which is also logical and simple with respect to operation. The plumbing layout is located in a pipe gallery along the front ends of the filter boxes, making access easy for both operation and maintenance. At 100 Mile House, the pipe gallery is a part of the operations building that has laboratory, office, chlorine room, chlorine contact tank, and a clearwell for the treated water pumps. In addition, the filter boxes at 100 Mile House have common walls, each having the capacity to withstand the hydraulic pressure when the adjacent filter box is drained. By contrast, the slow sand filter layout at Empire, Colorado was constrained by the site, adjacent to Mad Creek, which required rock excavation.

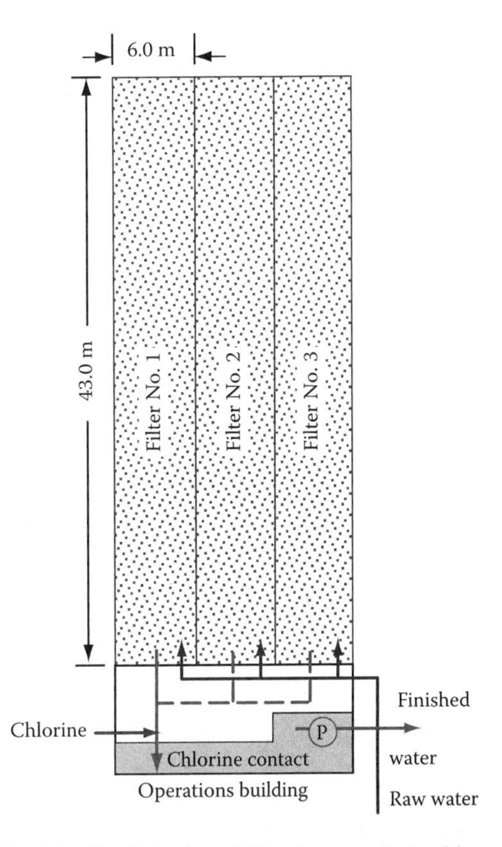

FIGURE 13.8 Configuration of filter boxes and plumbing at slow sand filter plant, 100 Mile House, British Columbia. (Adapted from Hendricks, D.W. (Ed.), *Manual of Design for Slow Sand Filtration*, AWWA Research Foundation and American Water Works Association, Denver, CO, p. 100, 1991.)

13.3.1.4 Depth of Box

The depth of the filter box can be stated as

$$\begin{aligned}\text{Depth of filter box} &= \text{depth of gravel support } (0.61\,\text{m}) \\ &+ \text{depth of filter media } (1.0 - 1.5\,\text{m}) \\ &+ \text{maximum depth of water } (2.0 - 3.0\,\text{m}) \\ &+ \text{freeboard depth } (0.3\,\text{m}) \\ \text{Total} &\approx 3.9 - 5.4\,\text{m } (13.18\,\text{ft})\end{aligned}$$

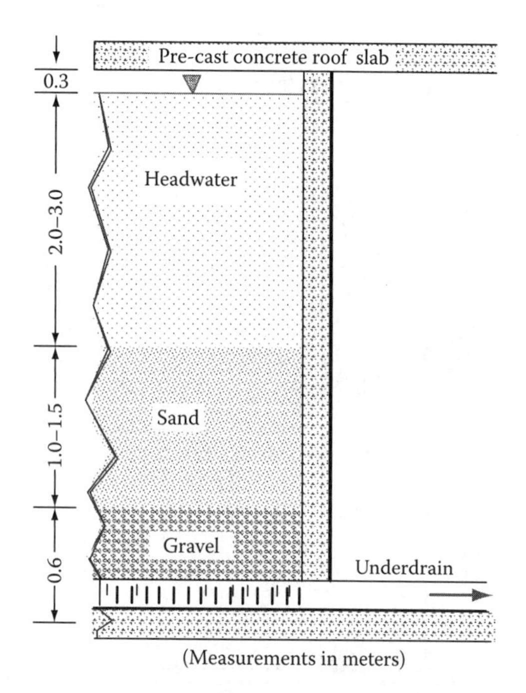

FIGURE 13.9 Cross section of slow sand filter showing filter box with gravel support, sand, headwater, freeboard, and underdrain. (Adapted from Hendricks, D.W. (Ed.), *Manual of Design for Slow Sand Filtration*, AWWA Research Foundation and American Water Works Association, Denver, CO, p. 102, 1991.)

A most important concern is that a person of about 1.83 m (6.0 ft) can stand comfortably when scraping the sand bed and not touch the underside of roof, or roof truss, when the filter bed is at maximum height. As noted, the depths of the headwater and of the initial sand bed are limited by economic considerations rather than absolute criteria. The underdrain pipes are placed on the bottom of the filter floor and the gravel is placed around the pipes and so the pipe sizes are not measured in computing depth of filter box.

Figure 13.9 is a cross section drawing of a slow sand filter, showing underdrain, gravel support, sand, headwater, and freeboard. At 100 Mile House, British Columbia, the total distance from roof slab to concrete floor is 3.80 m (12.5 ft), with gravel 0.83 m (2.7 ft), sand 1.05 m (3.4 ft), maximum headwater depth 1.80 m, and freeboard 0.12 m. At Empire, Colorado, the depth of the filter box is 3.66 m (12.0 ft), with gravel 0.61 m (2.0 ft), sand 1.22 m (4.0 ft), maximum headwater depth 1.52 m (5.0 ft), freeboard 0.31 m (1.0 ft) (Arix, 1984). The respective depths of the headwater and of the sand bed at 100 Mile House and at Empire are representative of practice.

13.3.1.5 Structural Design

The structural design of the filter box depends upon the hydraulic pressure on the inside, and upon the soil pressure on the outside. The hydraulic pressure at any water depth is

$$p = \gamma_w h \qquad (13.4)$$

where

p is the pressure at depth h (Pa) or (lb/ft^2)
h is the depth below water surface (m) or (ft)
γ_w is the specific weight of water (9990 N/m^3) or (62.4 lb/ft^3)

The force on a wall due to the hydraulic pressure is calculated by applying the concept of the pressure prism:

$$F = (\gamma_w h/2) \cdot A(\text{wall}) \qquad (13.5)$$

where

F is the force on wall (N or lb)
A(wall) is the area of wall (m^2 or ft^2)

Example 13.3 illustrates the application of the foregoing equation.

Example 13.3 Pressure and Force

Calculate the pressure at the bottom of the filter at Empire and the force per unit width of wall.

1 Data for Empire filter

Maximum depth of water for Empire slow sand filter is 3.65 m (12.0 ft).

2 Pressure at depth 3.65 m (12.0 ft)

$$\begin{aligned} p\,(3.65\,\text{m}) &= g_w h\,(3.65\,\text{m}) \\ &= 9{,}990\,\text{N/m}^3 \cdot 3.65\,\text{m} \\ &= 36{,}463\,\text{N/m}^2 \\ &= 36.46\,\text{kPa} \\ &= 750\,\text{lb/ft}^2 \end{aligned}$$

3 Force on wall

$$\begin{aligned} F &= (g_w h/2) \cdot A \\ &= (9{,}990\,\text{N/m}^2 \cdot 3.65\,\text{m}/2) \cdot (3.65\,\text{m} \cdot 1.0\,\text{m}) \\ &= 66{,}546\,\text{N (for a 1 m vertical section)} \\ &= 4{,}492\,\text{lb (for a 1 ft vertical section)} \end{aligned}$$

4 Discussion

The wall should be designed to withstand the force indicated above. In addition, the bending moment at the base of the wall should be calculated. The design should be done by an engineer knowledgeable in the field of structural design. The example is intended to give an idea of the horizontal force on a typical wall of a slow sand filter. Soil pressure should be determined for the local conditions by a geotechnical engineer. In addition, the bottom floor and the roof require structural design.

13.3.2 Hydraulics

A number of design decisions are driven by hydraulic analysis. Major hydraulic functions are (1) to backfill the sand bed and headwater through the underdrain system after scraping;

(2) to distribute the raw water without erosion of the sand bed; (3) to collect water uniformly from the filter; (4) to drain the headwater for sand bed scraping; (5) to install an overflow weir below the level of the top of the filter box; (6) to measure the flow to the filter; (7) to control the flow through the filter; (8) to measure headloss through the filter bed; (9) to provide for a variety of plumbing needs, such as filter-to-waste, drains, directing flows, filling the dry bed from the bottom, etc.; and (10) to avoid negative pressures within the sand bed. Most of these functions may be performed by the raw water distribution manifold/valve system or the drainage system.

13.3.2.1 Backfilling after Scraping

After scraping, the dewatered filter must be backfilled with filtered water through the underdrain system. Backfilling can be accomplished easily with the valve configuration shown in Figure 13.10. Normal operation is illustrated in Figure 13.10a, which shows that the valve connecting the two filters is closed. Figure 13.10b shows backfilling from Filter 1 to Filter 2, which assumes Filter 2 was the one scraped and needs to have about 300 mm (12 in.) water depth above the sand bed surface. Thus, the water level in Filter 1 must be higher than the intended water level in Filter 2. The valve connecting the filters must be open, with the filtered water line valves closed for both filters. The treated water storage should be sufficient to satisfy demand during the time Filters 1 and 2 are not in operation. The backfill can be done using elevated storage, if available.

The importance of starting a filter run with about 300–600 mm (12–24 in.) water depth over the sand bed is illustrated in the erosion of the sand bed of the Empire filters, as seen in Figure 13.11. The Empire filter (Box 13.2) did not have provision for connecting the two filters. Thus, the backfilling occurred from the raw water inflow to the top of the filters; the water spilling onto the sand bed caused the erosion. The Empire system has two filter beds and both have the same

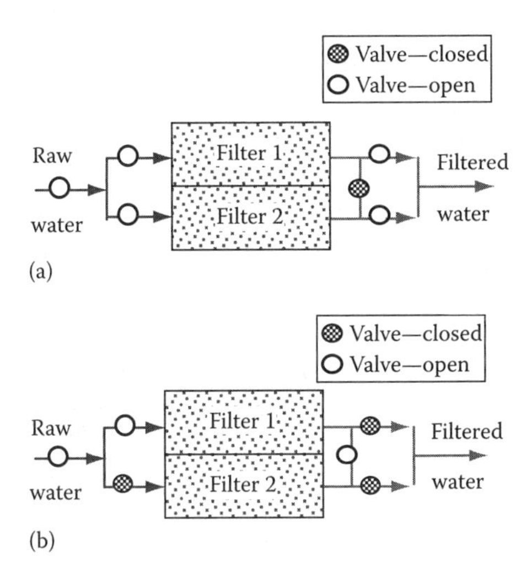

FIGURE 13.10 Schematic of how backfill is accomplished from one filter to another. (a) Normal operation: two filters, (b) backfill mode: Filter 1 to Filter 2.

FIGURE 13.11 Sand bed erosion at Empire, Colorado. (Adapted from Hendricks, D.W. (Ed.), *Manual of Design for Slow Sand Filtration*, AWWA Research Foundation and American Water Works Association, Denver, CO, p. 69, 1991.)

BOX 13.2 POSTMORTEMS IN DESIGN

The Empire filter had a several design deficiencies that were evident in hindsight, and some in foresight (Seelaus et al., 1988). Issues included no interconnection between filters to permit backfilling of the "just-scraped" filter, lack of a headwater dewatering pipe, filter bed erosion due to raw water inflow at startup, no piezometers, overflow pipe was subject to freezing thus permitting the filter box to overflow, tailwater level not adjustable, no storage provided for scraped sand, valves underground with adjustment only by long rods, no flow measurement, no trash-rack, no pipe gallery, etc. A pipe gallery would have alleviated other design problems and facilitated operation. Some issues were due to lack of recent experiences with slow sand but an inadequate budget was the main limitation. Despite the issues, the slow sand filter has functioned well and served the community since its completion in 1984. Re-sanding became an issue, however, in 2004, and in lieu of washing and reinstalling the sand bed, a new source of sand had to be found.

The context was that the town of about 500 persons was under a mandate from the state health department in 1983 to provide filtration. At the same time, a nationwide recession affected the local economy and the demand for metals causing layoffs at a nearby mine. The funding for the project was dependent entirely on grants arranged for by the consulting engineer and the state health department. Slow sand was selected for various reasons, but partly because of its "passive" character. To complicate matters, the site required removal of considerable rock, which used more than half of the fixed budget.

(*continued*)

BOX 13.2 (continued) POSTMORTEMS IN DESIGN

An absolute limit on the budget and unexpected factors shaped the kind of project that resulted. Regardless of benefits to operation that might accrue in adding some of the amenities noted, the town simply did not have the capacity to increase the funding, that is, in addition to the grant monies provided. Such is the nature of virtually all engineering projects, that is, there are always social, political, and economic factors that comprise the context of a given project and add complexity. Even with adequate funding, social and political factors are likely to rise to the surface. Rarely do technical factors alone govern the course of a project.

pattern of erosion, below their respective raw water inflow pipes, that is, a depression about 300 mm (12 in.) deep and diameter about 600 mm (24 in.).

13.3.2.2 Air Binding

Backfilling a just-scraped filter through the underdrain system displaces air in the top 100–200 mm (4–8 in.) of the filter bed (after it has been dewatered for scraping). If the filter box is filled from above, the air in the pores of the top of the sand bed will be trapped, that is, "air-binding" occurs. This causes disruption of the downward flow of water and possibly "boils" of air will emerge from the sand bed. Air binding may also occur by "gas precipitation" (see Sections 12.5.2.3, 12.5.2.4, and Appendix H.3).

13.3.2.3 Distribution of Raw Water Inflow Kinetic Energy

Although the most likely cause of sand bed erosion is not to have sufficient depth of water over the sand when the filtering cycle is started again after scraping, the problem can be mitigated further by reducing the kinetic energy of the raw water discharge. This may be accomplished by having multiple discharge points and larger orifices to reduce the discharge velocity. Probably, 10 discharge points distributed around the periphery of the sand bed is adequate and not expensive. Figure 13.12 illustrates how the raw water influent flow may be distributed around the filter box. At the same time, the depth of water above the top of the filter bed should be about 0.5 m (1.6 ft).

13.3.2.4 Drainage System

The drainage system has several functions: collection of filtered water; collection of overflow from the filter box (to prevent overflow of the filter-box walls); fast drawdown of the headwater to just above the level of the sand bed; provision for filter-to-waste; and backfill of filtered water from one filter and through the underdrains and upward through a just-scraped filter bed.

Figure 13.13 is a schematic representation for the piping layout for the slow sand filter at Moricetown, British Columbia. All of the foregoing drainage functions may be discerned by a study of the layout. The functions may be performed by opening or closing the appropriate valves.

13.3.2.5 Underdrain Manifold Design

Figure 13.14a shows the underdrain pipe layout, before the gravel was installed, for one of the three slow sand filters at 100 Mile House, British Columbia; floor dimensions were 43 m (141 ft) × 6 m (19.7 ft). Figure 13.14b is a photograph of the slotted pipe used for the underdrains. The slotted pipes were 152 mm (6 in.) diameter SDR 26 PVC with 131 slots/m/row, with three rows around the diameter of the pipe. Each slot was 1 mm (0.039 in.) wide and 2.5 cm (1 in.) long. The underdrain pipes were spaced at 2 m (6.6 ft). For reference, Huisman (1978, p. 161) recommended laterals with an 80 mm inside diameter with spacing of about 1.5 m, and holes 10 mm at the underside, 5 holes/m.

As noted, the underdrain system is a *manifold*; an empirical guideline for a manifold design is that the headloss across the system points at which flow is distributed should be large compared to the headloss within the manifold header pipe. The idea is that the pressure within the header pipe or within any lateral, at each of the points of distribution, should be about equal (see Sections 12.4.4.6 and Appendix D.2.5).

13.3.2.6 Depth of Sand

Although Hazen (1913) suggested 0.67–1.3 m (2–4 ft) as the range of bed depths, about 1 m (3 ft) has become traditional. There is no reason, however, to limit the bed to this depth. The Empire filter, for example, had a bed depth of 1.22 m

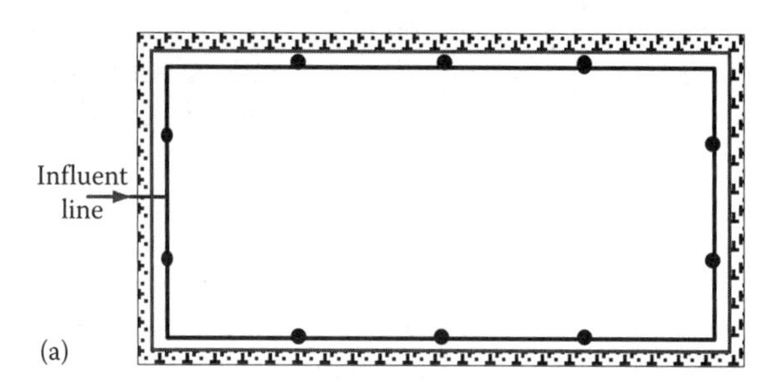

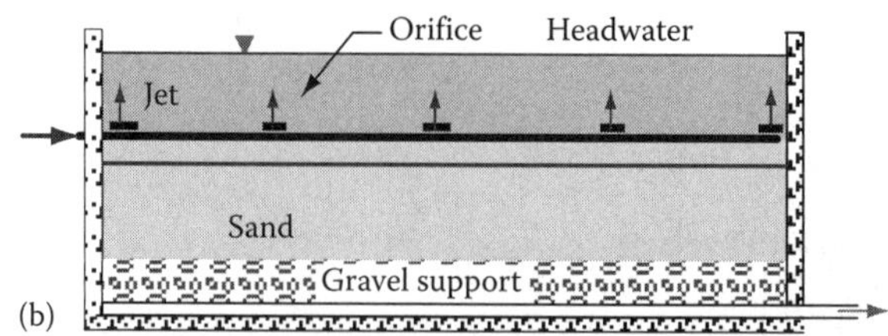

FIGURE 13.12 Distribution of flow around filter box to reduce sand bed erosion: (a) plan view, (b) profile view. (Adapted from Hendricks, D.W. (Ed.), Manual of Design for Slow Sand Filtration, AWWA Research Foundation and American Water Works Association, Denver, CO, p. 70, 1991.)

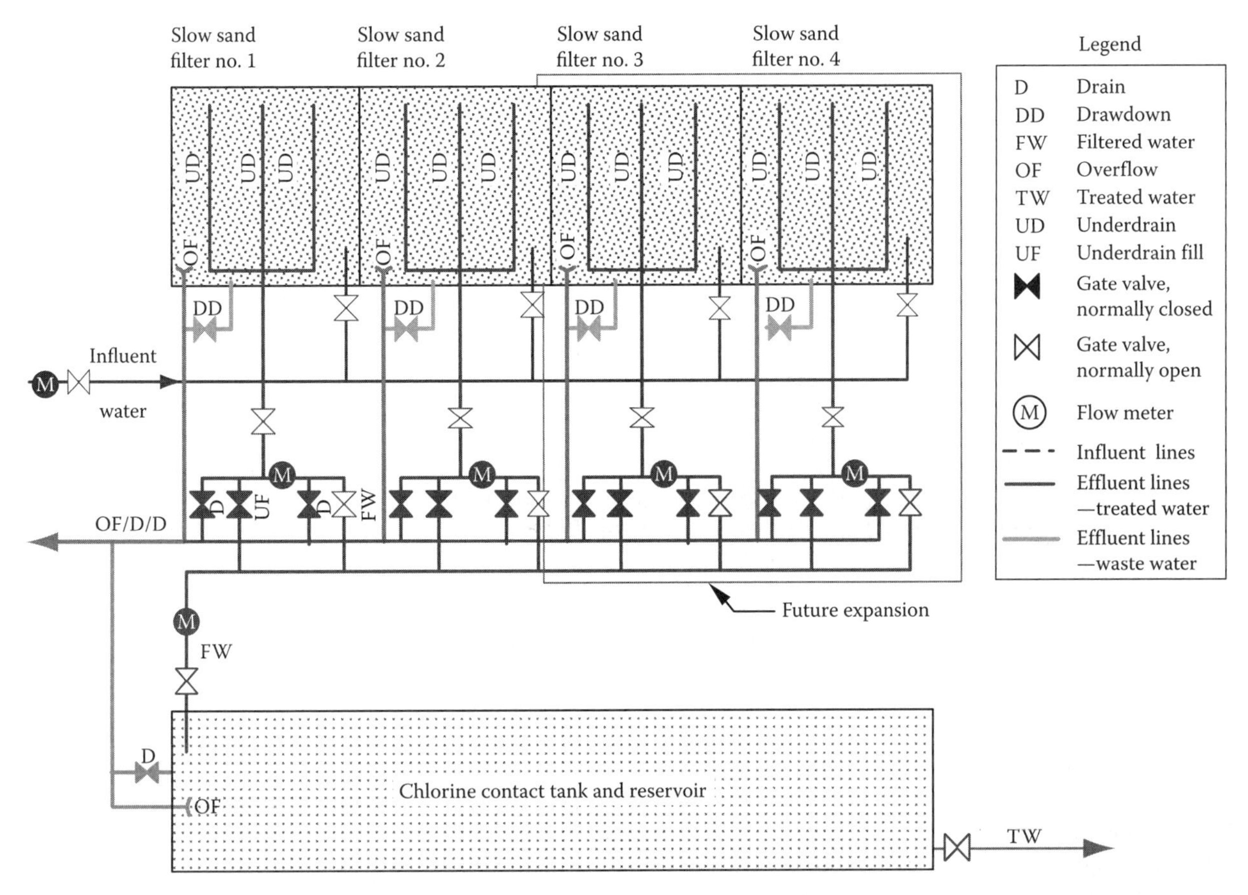

FIGURE 13.13 Flow configuration for slow sand filters at Moricetown, British Columbia. (Adapted from Construction drawing by Dayton & Knight, Vancouver, BC, 1988 and used in Hendricks, D.W. (Ed.), *Manual of Design for Slow Sand Filtration*, AWWA Research Foundation and American Water Works Association, Denver, CO, p. 68, 1991.)

FIGURE 13.14 Perforated underdrain pipe as used at 100 Mile House, British Columbia. (a) Underdrain pipes as placed on floor, (b) close-up of underdrain pipe. (Courtesy of Dayton & Knight, Vancouver, BC.)

(4 ft). The result was to prolong the bed life by about 5 years (Seelaus et al., 1988, p. 24) as compared with the bed life of a 1.0 m (3 ft) bed depth (based on a scraping rate of about 3–4 mm/month).

13.3.2.7 Sand Size

Huisman and Wood (1974, p. 53) recommended $0.15 \leq d_{10} \leq 0.35$ mm and $1.5 \leq \text{UC} \leq 2.0$. The UC is low so that the sand will have sufficient porosity. Their recommended upper limit was UC < 3. Bellamy et al. (1985a,b) found that "biological maturity" was the key issue in removal effectiveness for sands with d_{10} sand sizes of 0.13, 0.29, and 0.62 mm, for example, with average coliform removals of 99.4, 98.6, and 96.0, respectively, over a 10 month period. In other experimental work, Barrett (1989) showed that large sand size of $d_{10} = 0.92$ mm gave removals of $3 \leq \log R \leq 4$ over a 2.5 month

TABLE 13.5
Sand Sizes for Selected Slow Sand Filters Completed in the United States and Canada during the Period 1965–1989

Installation	d_{10} (mm)	UC	Comments
Empire, Colorado	0.21	2.67	\$21/metric ton[a,b] (\$19/U.S. ton)
100 Mile House, British Columbia	0.2–0.3	3.3	\$11.60[c,d]/m^3
Moricetown, British Columbia	0.15–0.35	2.0–2.5	\$34[e]/m^3
CSU Pilot Plant	0.28	1.46	Muscatine: \$128/metric ton

Source: Hendricks, D.W. (Ed.), Manual of Design for Slow Sand Filtration, AWWA Research Foundation and American Water Works Association, Denver, CO, p. 117, 1991.

[a] 1 metric ton (mT) = 1.1 U.S. tons (short, or 2000 lb).

[b] Seelaus et al. (1986, p. 38).

[c] U.S. dollars; conversion was made from Canadian dollars using December 1984 exchange rate of \$0.7577 Canadian = \$1.00 U.S. Significant figures shown are due to applying conversion from Canadian dollars to U.S. dollars, and does not connote precision.

[d] Calculated from following data: 1000 m^3 sand plus 560 m^3 gravel for \$24,000 Canadian plus \$10,000 for installation (personal communication, Jack Bryck, Sept. 20, 1990).

[e] Calculated from following data: 250 m^3 sand \$11,250 Canadian (Contract Bid Document 168.21.2, Dayton & Knight, Vancouver, British Columbia, 1988).

sampling period, after a ripening time of about two weeks with nutrient-enriched feed water.

The UC is important in that UC ≤ 2.0 ensures that the pores are open enough so that clogging does not become an issue. To consider a sand that has UC > 3 may be necessary, however, in many cases for economic reasons. The use of local sand as with the case at Empire, in lieu of a sand that meets specifications strictly, may save much money and keep the funds local. For example, the cost for the local sand used for the Empire filter was only \$21/metric ton versus \$128/metric ton for Muscatine sand that met specifications. Table 13.5 gives d_{10} and UC data for four examples. As seen, the d_{10} sizes are within recommended range while the UC values are at the higher end.

13.3.2.8 Gravel Support

The gravel support is aptly named because its function is to support the sand bed and to facilitate uniform drainage from the overlying sand. To accomplish both purposes, the gravel support must be graded, with finer material at the top and coarser at the bottom. The size of gravel in each layer, the respective depths, and the headloss are topics that follow.

13.3.2.8.1 Size

The top layer of the gravel support should not permit migration of sand from the sand bed, nor should the gravel of any layer find its way to a lower level. The bottom layer should not permit entry of gravel to the underdrain orifices. Huisman and Wood (1974) gave rules for design of gravel support layers as

1. d_{90} (given layer)/d_{10} (given layer) ≤ 1.4
2. d_{10} (lower layer)/d_{10} (upper layer) ≤ 4
3. d_{10} (top layer)/d_{15} (sand) ≥ 4
4. d_{10} (top layer)/d_{85} (sand) ≤ 4
5. d_{10} (bottom layer) $\geq 2d$ (drain orifice diameter)

13.3.2.8.2 Depth of Gravel Layers

Another rule from Huisman and Wood is that the thickness of each gravel layer should be greater than three times the diameter of the largest stones. Table 13.6 provides media sizes and depths of gravel support designs at three installations: Empire, Colorado; 100 Mile House, British Columbia; and Moricetown, British Columbia. The Empire design was based on the recommendations of the Great Lakes Upper Mississippi River Board (1987).

TABLE 13.6
Examples of Gravel Support Designs at Three Slow Sand Installations

Place	Layer	Size Range (d_{10}–d_{90}) (mm)	Depth (mm)	Depth (in.)
Empire, Colorado	Top	3–6	50	2
	Second	6–13	100	4
	Third	13.19	100	4
	Fourth	19–38	130	5
	Bottom	38–64	230	9
100 Mile House, British Columbia	Top	3–6	150	6
	Second	9–14	150	6
	Bottom	20–63	300	12
Moricetown, British Columbia	Top	2.5–3	150	6
	Second	10–15	150	6
	Bottom	40–60	300	12

Source: Hendricks, D.W. (Ed.), Manual of Design for Slow Sand Filtration, AWWA Research Foundation and American Water Works Association, Denver, CO, p. 13, 1991.

13.3.2.8.3 Headloss

Huisman and Wood calculated the total headloss across the gravel support as 1.37 mm applying Darcy's law to gravel layers of sizes 0.7–36 mm and assuming HLR = 0.5 m/h (12.5 mgad). In other words, the headloss through the gravel support is not significant.

13.3.3 Support Systems

For a process to work as a technology, various kinds of support systems are necessary. These include instruments for flow measurements, a pipe gallery, a means to access the filters for scraping, etc.

13.3.3.1 Flow Measurements

The locations of flow measurement instruments are shown in Figure 13.15. The flow meters shown include (1) an orifice meter on the influent side for the whole plant, (2) orifice meters on the influent side for the individual filters, and (3) the total flow meter on the exit side for the whole plant. Alternatives include Venturi meters, magnetic flow meters, and propeller meters. The influent flow meter is used to adjust the flow to the plant using a gate valve just downstream. The individual filters should have flow measurement capability to make sure each filter receives the same flow. The volumetric flow meter on the filtered waterside can provide data for records on water usage by the community.

13.3.3.2 Piezometers

Piezometers measure hydraulic head and may show visually the hydraulic gradient from the headwater, through the filter bed, and to the tailwater. Piezometers will also show the level of the water in the filter, for example, when dewatering for scraping and when backfilling to restart operation after scraping. The piezometers can be clear plastic tubes (e.g., clear PVC) connected to points along a vertical line. Those within the sand bed (at maximum sand bed depth) should be spaced at about 200 mm vertical distance. Figure 13.16 is a photograph of the piezometers in slow sand filters at the village of 100 Mile House, British Columbia. The tubes are large enough in diameter to be read easily and, in fact, have floats to facilitate reading.

FIGURE 13.16 One of three piezometers installed for slow sand filters at Village of 100 Mile House, British Columbia. Jack Bryck, PE, is pointing to level of water in piezometer.

13.3.3.3 Turbidimeters

If electricity is available, an online turbidimeter on the effluent side is advisable in order to detect any anomalies in performance. Otherwise, a sample may be taken daily by an operator in the process of inspecting the facility and obtaining other data. Sampling taps should be provided for each filter.

13.3.3.4 Flow Control

Flow control to the overall filter should be on the influent side by means of a gate valve located downstream from the metering device, as indicated in Figure 13.17. The flow should be steady over a 24 h period, using the treated water storage to provide for varying hourly demand over the daily cycle.

13.3.3.5 Tailwater Control

To control the tailwater, a vertical overflow pipe, that is, a circular weir, may be placed with lip elevation of about 300–600 mm (12–24 in.) above the top of the sand bed, which provides a means to dissipate the kinetic energy of the raw water inflow and provides a positive pressure in the filter bed. The overflow from the weir is captured by an outer cylinder

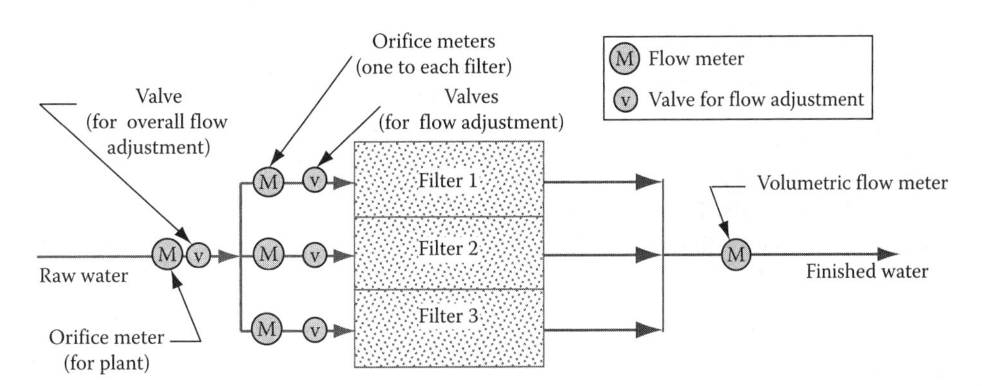

FIGURE 13.15 Flow meters for slow sand filter. (Hendricks, D.W. (Ed.), *Manual of Design for Slow Sand Filtration*, AWWA Research Foundation and American Water Works Association, Denver, CO, p. 82, 1991.)

(a)

(b)

FIGURE 13.17 Design features of slow sand filter at 100 Mile House, British Columbia. Brian Walker, and Jack Bryck, Dayton & Knight, c. 1985. (a) Pipe gallery showing one of three pilot filters, (b) ship doors give access to filter beds. (Courtesy of Dayton & Knight, Vancouver, BC.)

with an outlet pipe to treated water storage. The water level in the filter box rises as the run progresses. A piezometer tapped into the headwater permits observation of the water level in the filter box. Alternatively, a weir plate can serve to control the tailwater level. If the weir elevations are adjustable, for example, by a telescoping pipe or an adjustable weir plate, then the system is more flexible with respect to utilizing additional water head as the sand bed is lowered over the years due to scraping.

13.3.3.6 Pipe Gallery

A pipe gallery is a necessary adjunct to a slow sand filter. Figure 13.17a is a photograph showing the pipe gallery for the slow sand filter at 100 Mile House, British Columbia. All lines are seen as simple and uncluttered, with valves easy to operate and maintain. The operator should be provided easy access to all valves, meters, and the piezometers within the pipe gallery, without having to stoop or to bend in difficult positions. The pipe systems include: raw water (influent), finished water (effluent), headwater drainpipes, backfill from finished water from operating filter to a drained filter, and filter-to-waste. The pipes should be color coded and labeled so that the function of each pipe, valve, and meter is self-evident.

13.3.3.7 Access to Filters

The filter beds must have easy access to operators for both inspection during operation and removing sand after scraping and to resand the bed when the time arrives. The access for scraping and resanding should be adequate so that the operators can complete the tasks in a normal working fashion. The sand removal and placement should not require abnormal body positions that could result in injury. At 100 Mile House, British Columbia, access to the filters is by means of ships doors, which are shown in the photograph, Figure 13.17b.

13.3.3.8 Plumbing Functions

In addition to the above hydraulic functions related to making slow sand filtration work as a *process*, a number of ancillary needs should be provided for, such as filter-to-waste, valves to direct flows, and drains of various sorts. Figures 13.13 and 13.15 show a portion of the plumbing configuration needed; the drawings are simplified. Filter-to-waste is imperative when a plant is placed in operation at startup or when the sand bed is replaced. During this beginning period, the sand bed will shed its fines, causing short-term higher turbidity levels. Figure 13.13 shows the piping to drain the filter, which may be used also for filter-to-waste. In addition, all basins and floors should be provided with drains and gutters to facilitate cleaning and removal of unwanted water.

13.3.3.9 Hydraulic Profile

The starting point for any hydraulic analysis is the hydraulic profile. Such a profile should be drawn so that pipe sizes can be determined (for headloss calculations) and the filter box elevation can be set subject to the constraints of source water elevation and filtered water storage elevations.

13.3.3.10 Headroom

Adequate headroom should be provided so that persons scraping the sand bed can assume normal posture. The distance between the top of the sand bed and the roof should be $\geq$2.0 m. Since the amount of headloss to be provided should be $\geq$2.0 m, there is no conflict.

13.3.3.11 Designing to Avoid Freezing

In the northern latitudes, freezing temperatures must be considered. Two approaches are (1) to accept the presence of an ice block in winter and to slope the interior walls to handle the thrust of the expansion, and (2) to prevent the occurrence of an ice block. To accept the presence of ice is not good policy, however, as the nuisance and problems of ice are worth avoiding. Scraping with even a thin crust of ice will cause an inordinate increase in labor requirement, since the ice must be removed in order to scrape.

The Kassler slow sand filter, an outdoor filter operated from 1906 to 1985 by the City of Denver, operated routinely in the winter months with a floating ice block usually 0.3–0.6 m (1–2 ft) in thickness. The filter area, 2.47 ha (10.5 ac), was too extensive to provide a cover. The important point in operation was that the ice block did not touch the surface of the sand bed. The horizontal force caused by expansion of the ice block was deflected by the sloped sidewalls of the filter. The sidewalls were 150 mm (6 in.) concrete under earth berms

(a)

(b)

FIGURE 13.18 Slow sand filters at two installations in British Columbia showing earth embankments used for insulation; about 0.3 m (1 ft) earth covers slab roof in each case. (a) 100 Mile House, British Columbia, (b) Moricetown, British Columbia. (Courtesy of Dayton & Knight, Ltd., Vancouver, BC.)

sloped 1:2; there was no structural difficulty with ice. Scraping was timed for just before the winter season, so that the filter run could extend to spring when the ice block was expected to melt.

To prevent the occurrence of an ice block, the filter must be covered. The slow sand filters placed in operation at Empire, Colorado; 100 Mile House, British Columbia; and at Moricetown, British Columbia were all covered. Figure 13.18a is a photograph of the slow sand filter at 100 Mile House, British Columbia, which has a flat roof of precast concrete and earth sidewalls; the operations building is shown in the foreground. The pipe gallery is below a floor grate in the building and treated water storage is below the concrete floor. With the insulation provided by the earth sidewalls, the 100 Mile House filter had no auxiliary heat and has had only a thin skin of ice on the surface during operation since November, 1985. The lack of ice problem was helped also by the use of a small pump which maintains circulation at the headwater surface.

Figure 13.18b shows the earth insulation at Moricetown, British Columbia, which is placed along the sides and has a depth of about 0.3 m (1 ft) on the slab roof. The top of the filter box has a port to the operations room to permit heat to advect above the headwater. Figure 13.19 shows the Empire installation; ice formation was avoided by propane heaters during the first years of operation. When the propane was exhausted during one season, a thin ice cake formed over part of the water surface which resulted in an inordinately more difficult task in scraping.

FIGURE 13.19 Slow sand filter at Empire, Colorado adjacent to Mad Creek, raw water source.

13.3.3.12 Sand Recovery System

The arguments for onsite sand recovery are (1) sand will be at hand for resanding; (2) the cost of resanding will be minimal; (3) the sand will have been washed and ready to use; (4) the sand will not be a nuisance because of possible indiscriminate discard; and (5) in any new sand acquisition, some degree of uncertainty exists concerning whether proper attention will be given to specifying a proper sand or whether it can be obtained at a reasonable price. Jordan (1920, p. 13) emphasized the importance of sand recovery, to wit: "The fundamental proposition is that sand handling is the key to (successful) operation of the slow sand filtration plant." In other words, sand recovery facilitates a sustainable operation.

Elements of a sand washing system include (1) a storage bin for dirty sand; (2) a flume that carries the dirty sand to a settling box; (3) a settling box that provides for overflow of the dirty water and settling of the clean sand; (4) removal of the dirty water to an approved land site; and (5) a sluice from the settling box to a bin for washed sand, which provides for drainage. A front-end loader may be useful for moving large volumes of sand.

13.4 PILOT PLANT STUDIES

A pilot plant may address the questions of (1) headloss versus time for different seasons, (2) effluent turbidity for different influent turbidity levels (due to storms or seasonal changes), and (3) log removals of organisms (e.g., coliforms, *Giardia* cysts, *Cryptosporidium* oocysts). Other concerns relate to design variables such as the effect of HLR on the foregoing, the effect of a shift in HLR, both higher and lower, due to taking an adjacent filter out of operation or being returned to operation, the time for sand bed "ripening," that is, to reach "maturity," and the effect of sand size (d_{10} and UC) on the dependent variables. The first question is sufficient, by itself, to warrant a pilot plant study. The phases of a pilot plant study include study plan, pilot plant construction, execution, data handling, data interpretation, and application of study results to design. In some cases, these tasks may warrant a deliberate plan; in other cases, the study can get underway within 2–3 days, depending on circumstances.

13.4.1 Pilot Plant Construction

Essentially, the pilot plant is merely a cylinder (or pipe), which holds the gravel support, the sand, and the headwater. The media cylinder should be long enough to hold the gravel support, the sand bed, and the headwater. This means that the cylinder should be about 4 m (13 ft) high. The diameter is not critical from a process standpoint, but a 305 mm (12 ft) diameter is easier to use than is a smaller diameter. For cylinder material, a SC200 PVC pipe works well as its walls are easy to tap. Piezometers should be located in the headwater, along the sand bed, and within the gravel support layer. The headloss through the gravel support layer and the effluent piping will be negligible compared with headloss in the sand bed. The piezometers should be clear plastic with a diameter $\geq$10–20 mm (0.5–0.8 in.) to minimize capillary effect. The measurement of flow should be done volumetrically using a 1000 mL cylinder and stopwatch, complemented by a rotometer to ascertain flow at any time.

13.4.2 Case Study

A classic pilot plant study by Dayton & Knight, Ltd. involved rapid filtration, diatomaceous earth, and slow sand for the village of 100 Mile House, British Columbia. Slow sand was selected for the full-scale because of anticipated lower operating costs, although the capital cost was slightly higher than rapid filtration and diatomaceous earth. A review of their report (Dayton & Knight, 1983) reveals how a slow sand pilot plant study can be conducted at minimal cost to answer the most critical questions: length of run and effluent turbidity.

13.4.2.1 Context

In 1981, the village of 100 Mile House had 60 confirmed cases of giardiasis attributed to the unfiltered, chlorinated water supply from Bridge Creek. After initial consideration of rapid rate filtration and cost estimate in 1982, the consulting firm (Dayton & Knight, Ltd.) examined also diatomaceous earth and slow sand. Subsequently, they recommended to the village that a pilot plant study be conducted to compare the three basic filtration technologies. Study data would be used to select the technology while effectively removing *Giardia* cysts. Such a study was conducted during the months of July–October, 1983. The slow sand portion of the study is reviewed here.

13.4.2.2 Pilot Plant Setup

Figure 13.20a and b is a photograph and section drawing, respectively, of the 1983 pilot plant setup at 100 Mile House, British Columbia. The photograph illustrates the simplicity of the setup, with only an influent pipeline of raw water, a filtration tube of precast concrete pipe 1050 mm (42 in.) diameter, and a 100 mm (4.0 in.) PVC drain pipe. The section drawing shows the details. As seen, the sand bed was supported by three layers of graded gravel (top layer 100 mm with $d_{10} = 0.6$ mm, middle layer 100 mm with $d_{10} = 5$ mm, bottom layer 250 mm with $d_{10} = 15$ mm).

(a)

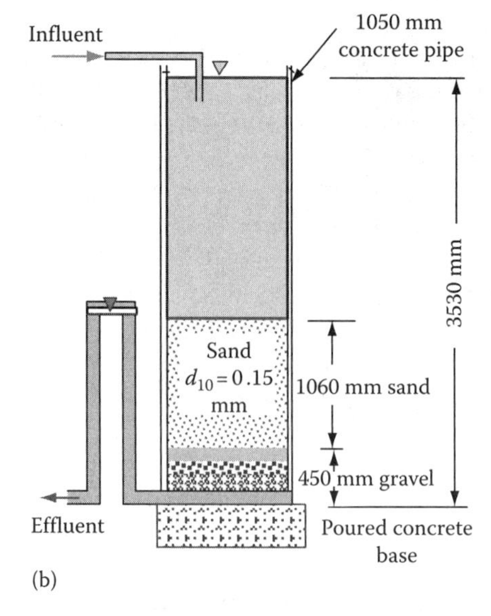

(b)

FIGURE 13.20 Pilot plant setup at 100 Mile House, British Columbia. (a) Photograph, (b) sketch of setup. (Courtesy of Dayton & Knight, Ltd., Pilot water treatment program, Village of 100 Mile House, Report for Client, Dayton & Knight, Ltd., Consulting Engineers, West Vancouver, British Columbia, December, 1983.)

The 100 mm diameter PVC drainpipe was perforated in the portion that drained the bottom gravel layer, and the bend served as a weir with a 20–30 mm (1 in.) hole drilled in the top of the bend to ensure an air break for the tailwater control. The headwater level was measured by a scale attached to the inside wall above the sand bed and to the top of the pipe. A drain valve was attached to the lower part of the effluent pipe to drain the column. This type of pilot plant setup is easy to construct, and is inexpensive since it uses materials at hand.

13.4.2.3 Results

Figure 13.21a and b show headloss and influent/effluent turbidities, respectively, for the first cycle of operation before scraping. As seen in Figure 13.21, the maximum headloss possible, based upon the height of the column, was 2000 mm. Figure 13.21a shows that the run time with 2000 mm headloss would be about 33 days (extrapolating the curve). This run time was deemed acceptable. The second critical question was whether an effluent turbidity of <0.5 NTU could be achieved. Figure 13.21b answers that question, showing effluent turbidities in the range <0.1–0.5 and for influent turbidities 1–4 NTU.

13.4.2.4 Discussion

Several lessons can be extracted from the 100 Mile House pilot plant study. First, the pilot plant setup was improvised on the spot, keeping expenses low and without the need to wait for supplies. Second, local persons were trained to obtain the needed measurements of headwater elevation and turbidity. Third, the pilot plant study was simple, with limited scope, designed to answer only the critical questions of run time and effluent turbidity. Fourth, the duration of the study was only 4 months and included the most adverse raw water quality condition, that is, algae growth.

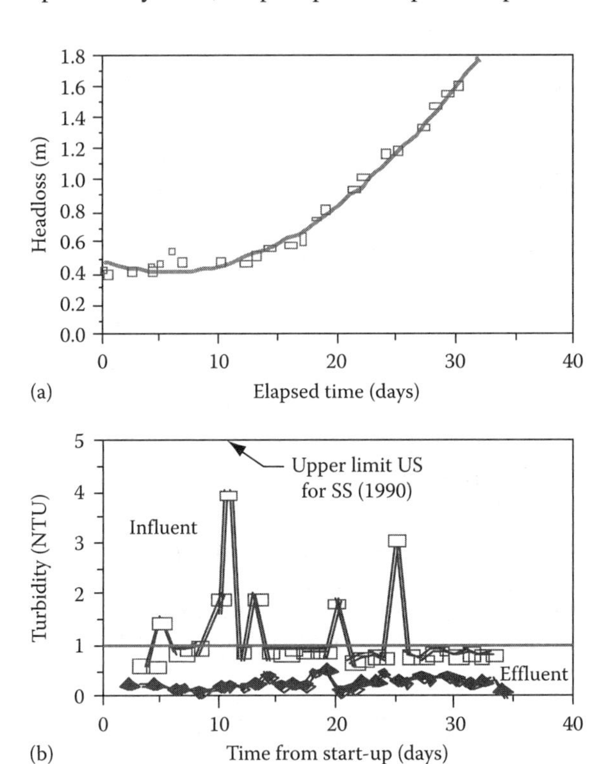

FIGURE 13.21 Data from pilot plant at 100 Mile House, British Columbia, July 18–August 17, 1983. (a) Headloss versus time for first filtration cycle, (b) turbidity versus time for influent and effluent water for first filtration cycle. (Courtesy of Dayton & Knight, Ltd., Pilot water treatment program, Village of 100 Mile House, Report for Client, Dayton & Knight, Ltd., Consulting Engineers, West Vancouver, BC, December, 1983.)

13.5 OPERATION

A major appeal of slow sand filtration is its simplicity in operation. Nevertheless, certain tasks must be performed. The initial task is plant start-up. The routine tasks include scraping, handling of used sand, monitoring, and maintenance.

13.5.1 Plant Start-Up

Following the completion of construction, the plant requires a break in period before the production of potable water can begin. The first task is to fill the bed from the bottom with raw water. The second task is to operate in the filter-to-waste mode until the water produced has acceptable quality. During this period of conditioning, the meters should be calibrated, the laboratory should be set up, and the monitoring routines and data forms developed. Also, it is assumed that the filter bed will have attained biological "maturity" during a nominal period of operation, for example, 2–4 weeks depending on temperature and other conditions.

13.5.2 Operating Tasks

The tasks involved in operating a slow sand filter pertain largely to the filter media and filter bed. These tasks include scraping the sand surface, washing and storing of sand, and rebuilding the filter bed.

13.5.2.1 Scraping

Slow sand filters should be scraped when the headwater level rises to the overflow. The following process is recommended for scraping the filters:

1. Drain the headwater to just above the sand bed.
2. Slowly drain the water level to several cm below the surface of the sand.
3. Scrape the top 5–30 mm of sand.
4. Remove the scraped sand from the filter box.
5. Place the scraped sand in a storage bin.

Asphalt rakes were used for scraping the sand at the 100 Mile House facility in British Columbia and at Empire, Colorado. The filter should be put back in operation as soon as possible so that the biofilms can recover.

The time required for scraping a slow sand filter (in person-hours) depends largely on two factors: (1) the depth of sand removed, and (2) the method used to convey the dirty

FIGURE 13.22 Scraping operation at Empire, Colorado. (Hendricks, D. W. (Ed.), *Manual of Design for Slow Sand Filtration*, AWWA Research Foundation and American Water Works Association, Denver, CO, p. 181, 1991.)

sand from the filter. Letterman and Cullen (1985) reported a typical time requirement of 5 person-hours per 93 m^2 (1000 ft^2) when 25 mm (1.0 in.) of sand was removed using shovels and a hydraulic conveyance to remove the dirty sand from the filter. At the Empire facility, where sand is removed in buckets, scraping was performed at a rate of 19 m^2/person/h (205 ft^2/person/h). Figure 13.22 is a photograph of the scraping operation at Empire, Colorado. The tool shown is an "asphalt rake," which is used to scrape the black *schmutzdecke* into windrows. The operators have learned to scrape about 5 mm depth, removing only the thin deposit on the sand surface. After scraping the sand into the windrows, the sand is shoveled into 20 L (5 gal) buckets and carried out the door.

The frequency of scraping varies for different installations. The scraping frequency at Empire, for example, was about 30 days, with raw water turbidity only 0.5 NTU. In a study of seven slow sand installations in New York, Letterman and Cullen (1985) found scraping frequencies varied from 1 to 7 months. Raw water turbidities were $\leq$3.0 NTU with about 8 NTU for one site.

13.5.2.2 Rebuilding the Sand Bed

On minimum bed depth before rebuilding the sand bed, Bellamy et al. (1985a,b) found excellent coliform removals through filter beds of 0.5 m (20 in.). This was also a minimum bed depth recommended by Visscher et al. (1987). Huisman and Wood (1974) describe the rebuilding process and recommended setting aside the portion of the filter bed that has been retained in place, placing the new sand on the bottom, that is, just above the gravel support, with the old sand serving as the top layer of the reconstituted sand bed.

13.5.3 Monitoring and Reporting

Certain operational and water quality parameters must be regularly monitored and reported. These parameters include headloss and such water quality characteristics as turbidity, disinfectant, and concentrations of certain biological and chemical contaminants. Cartridge filter samples may be taken perhaps every 3 months to ascertain process effectiveness.

PROBLEMS

13.1 Transport and Filter Coefficients for Viruses and *Giardia* Cysts

Given

Two sands, California dune ($d_{10} = 0.28$ mm, UC ≈ 1.4) and Muscatine ($d_{10} = 0.78$ mm, UC ≈ 1.0), were used by Roebeck et al. (1962, p. 1277) in filtration experiments with 0.610 m (2.0 ft) sand columns using attenuated polio with results as shown in Figure 13.6.

Required

Calculate the transport coefficient, η, and the filter coefficient, λ, for attenuated poliovirus and *Giardia* cysts, respectively, and plot against HLR.

Hint: Set up a spreadsheet, that is, Table CD12.3, and apply Equations 12.15 through 12.17. For reference, see Sections 12.3.3 and 13.2.1.2 and Figure 13.6.

13.2 Filtration Removals of Viruses and *Giardia* Cysts

Given

Two sands, California dune ($d_{10} = 0.28$ mm, UC ≈ 1.4) and Muscatine ($d_{10} = 0.78$ mm, UC ≈ 1.0), were used by Roebeck et al. (1962, p. 1277) in filtration experiments with 0.610 m (2.0 ft) sand using attenuated polio with results as shown in Figure 13.6. See Section 13.2.1.2.

Required

1. Calculate, by Iwasaki's equation, Equation 12.14, the filter coefficient, λ, for attenuated polio for the two sands as a function of HLR and plot the results, that is, λ versus HLR.
2. Second, using the η values as calculated in a spreadsheet, that is, Table CDProb13.2, calculate α by Equation 12.15.
3. Discuss: Are the results reasonable? Should there be a "calibration" for η or α?

Hint: Set up a spreadsheet, that is, Table CD12.3, and apply Equations 12.15 through 12.17. For reference, see Section 12.3.3 and 13.2.1.2 and Figure 13.6.

13.3 Effect of Additional Bed Depth on Initial Headloss

Given

Suppose a sand bed has an intrinsic hydraulic conductivity, $k = 7.2 \cdot 10^{-11}$ m^2 (Flatiron masonry) as given in Table CD13.2 and also in Table 13.3. See Sections 13.2.2.1 and 13.2.2.2.

Required

1. Determine the initial headloss, that is, at $t = 0$, for bed depths of 1.0 m (3.3 ft) and 1.3 m (4.3 ft).
2. Determine the same for 0°C and 20°C.

Hint: See Sections 13.2.2.1, and 13.2.2.2.

13.4 Effect of Additional Bed Depth on Life of Sand Bed (13.3.2.6)

Given

Let alternative initial bed depths of a sand bed be 1.0 m (3.3 ft), and 1.3 m (4.3 ft). The sand bed scraping frequency is about 1.0 month. The amount of sand scraped is about 10 mm each scraping. The sand bed is to be resanded when the depth is about 300 mm (1.0 ft). See Section 13.3.2.6.

Required

Estimate the life of the sand bed before scraping in each case.

13.5 Depth of Sand Bed (13.3.2.6)

Given

The depth of the sand bed is being considered for a new slow sand installation. Assume the scraping rate of sand is 5 mm/month.

Required

Assume a convenient d_{10} and UC. Then evaluate the pros and cons of different bed depths. For each depth considered, include such concerns as

1. Life of the sand bed before rebuilding
2. Clean-bed headloss
3. Depth of the filter box
4. Estimated cost of the different filter box depths, etc.

13.6 Operation of a Slow Sand Filter

Given

The piping layout for the slow sand filter at Moricetown, British Columbia, is shown schematically in Figure 13.13.

Required

Make several copies of Figure 13.13 and show by colored highlighter type of felt-tip pen how the various functions of the drainage system are executed.

13.7 Effluent Weir Design

Given

Consider the slow sand filter at Empire, Colorado. Suppose that each of the two filter beds has a vertical pipe telescoping overflow weir. The overall flow average to both filters is taken as $Q(\text{Empire, avg}) = 946\ \text{m}^3/\text{day}$ (250.000 gpd). The maximum flow is taken as three times the average flow.

Required

Determine the size of a sleeve pipe overflow weir capable of handling the maximum flow.

13.8 Sand Bed Erosion

Given

Data given by Seelaus et al. (1986, p. 4) are

$$Q(\text{avg}) \approx 946\,\text{m}^3/\text{day}\ (250{,}000\,\text{gpd}) = 0.010949\,\text{m}^3/\text{s}$$

$$\text{HLR} \approx 0.26\,\text{m/h}\ (6.5\,\text{mgad or } 0.10\,\text{gpm/ft}^2)$$

$$A(\text{total of filter beds}) = 153\,\text{m}^2(1650\,\text{ft}^2)$$

$$\text{Number of filters} = 2$$

$$d(\text{raw water influent pipe}) = 152.4\,\text{mm}\ (6\,\text{in.})$$

Required

1. Estimate the kinetic energy of the raw water discharge jets for estimated maximum flow at Empire discharging at single points in each of the two filter boxes.
2. Estimate the kinetic energy associated with free-fall of water to the sand bed.
3. Discuss the effect of the water fall to the sand bed with no headwater and with about 500 mm of headwater.

13.9 Design of a Sand Recovery System

Given

The scraped sand at Empire has been discarded by empting the buckets just outside the door at the top of the filter box. See Section 13.3.3.12.

Required

Design a sand recovery system for the Empire slow sand filter. Consider storage of dirty sand, washing system, and storage of clean sand (if any).

13.10 Design of a Pilot Plant Study

Given

Mad Creek, a mountain stream, has been used for decades by the town of Empire, Colorado for its drinking water. A slow sand filter is to be constructed to provide the town with filtered water.

Required

Design a pilot plant study plan for Empire. A small community in your region may be selected if you wish (in lieu of Empire). Set up an Excel data sheet for collection and processing of data.

13.11 Critique of a Slow Sand Filter Design

Given

Assume that you have visited a slow sand plant in your region.

Required

Critique the design with respect to a goal to achieve a "passive" operation. [If you have not visited a plant, let the "required" be points for discussion.] Examples of points to consider are

1. Pipe gallery. Examine the valve system for the plant. Where are they located? How does the operator open and close valves?
2. Effluent overflow. Is the tailwater weir elevation at the proper level? Describe the control for the tailwater elevation. Was there evidence of gas "boils" or did the operator comment on this?
3. Flow measurement. Describe the flow measurement. Do you have recommendations?
4. Scraping. (a)What is the nature of the *schmutzdecke*? (b) What is the frequency of scrapping? (c) How long does it take to dewater the headwater bed? (d) What is the duration required to scrape the sand bed? (e) How much sand is removed? (f) Can water be admitted to the sand bed from the underdrain system?

5. Sand recovery. What is done with the sand after it is taken through the door? Recommend what should be done?
6. Depth of sand bed. How many years will the sand bed last until the sand bed must be rebuilt?
7. Headwater. State the elevation of the headwater above the bottom of the filter box? About how much operating headloss is available? Could there be more? Or should the headloss available be less?
8. Describe the water quality provided by the slow sand filter?
9. Describe other points that come to mind after your visit. For example, is there provision for overflow in case the operator does not scrape the filter bed before overflow? Was there erosion of the sand around the inflow? Were there multiple points of inflow discharge?

ACKNOWLEDGMENTS

This chapter drew, to a large extent, from Hendricks, D.W. (Ed.), *Manual of Design for Slow Sand Filtration*, a project of the Water Research Foundation (formerly Awwa Research Foundation) and American Water Works Association, Denver, Colorado, 1991. Coauthors of the manual were Joy M. Barrett (director of training and technical services, Rural Community Assistance Partnership, RCAP); Jack Bryck (Dayton & Knight, Ltd., Consulting Engineers, Vancouver, British Columbia, presently with Malcolm Pirnie Consulting Engineers), M. Robin Collins (University of New Hampshire), Brain A. Janonis (presently director of utilities, City of Fort Collins), and Gary S. Logsdon (retired from the Environmental Protection Agency, Cincinnati, Ohio, and later retired from Black and Veatch Consulting Engineers, Cincinnati, Ohio). The project manager was Martin J. Allen (WRF).

The author gratefully acknowledges the permission from the Water Research Foundation for the use of figures and tables from the manual, which included Figures 13.1, 13.3 through 13.5, 13.7 through 13.9, 13.11 through 13.13, 13.15, and 13.22, and Tables 13.3, 13.5, and 13.6. Sherry Morrison, senior administrative assistant, Publishing Group, American Water Works Association, Denver, arranged for permission to use these figures and tables.

Jack Bryck, PE, presently senior associate, Malcolm Pirnie, Phoenix, formerly with Dayton & Knight, Ltd., gave permission to use Figure 13.16 and 13.17b, i.e., each of which includes his photograph. Brian Walker, formerly principal and owner, Dayton & Knight, West Vancouver, British Columbia, gave permission to use Figure 13.17a and b, each of which includes his photograph. In addition Jack Lee, PEng, Vice President, Operations, Dayton & Knight kindly gave permission to use Figures 13.13, 13.14, 13.16 through 13.18, 13.20, 13.21; Lee also gave permission to use Figure 14.16 which shows a diatomaceous earth pilot plant (photograph and cross-section drawing) at 100 Mile House, British Columbia, c. 1983.

GLOSSARY

Air binding: The presence of gas bubbles in the sand bed, causing increased headloss and possible rupture of the *schmutzdecke* on their release.

Appurtenances: Accessories, such as meters.

Atmosphere: Unit of pressure. 1 atm = 14.7 lb/in.^{2} = 101.325 kPa = 760 mmHg.

Attachment coefficient, α: Ratio of particles attaching to collector surface/particle striking collector surface.

Backfilling: Saturation of the sand bed by filling slowly from the bottom upward, to displace air in the bed.

Biologically mature: See *mature*.

Biologically mature: This term was used by Bellamy et al. (1984) to mean that the sand bed had developed sand grain biofilms to its near-maximum extent. The extent of biofilm development depends upon the influent nutrient concentration. See also *mature*.

Bleeding: Allowing faucets to drip in order to avoid freezing.

Breakthrough: Passage of contaminants through the sand bed.

cfu, colony forming unit: A measure of bacteria concentration.

Chlorine demand: The chlorine fed into the water that reacts with oxidizable impurities and may, therefore, not be available for disinfection, reported in units of mg/L.

Chlorine residual: The concentration of chlorine, in mg/L, remaining in the water after the chlorine demand has been met.

Cholera: Waterborne disease caused by *Vibrio cholerae*.

Clean-bed headloss: Headloss amount due to sand bed and gravel support prior to formation of *schmutzdecke* or biofilm development within the sand bed.

Clearwell: A facility for the storage of treated water.

Coliform: A gram-negative, nonsporing, facultative rod that ferments lactose with gas formation within 48 h at 35°C (Prescott et al., 1993, p. G6). See also *enteric bacteria*. The coliform group includes *E. coli*, *Enterobacter aerogenes*, and *Klebsiella pneumoniae* (p. 839).

Coliform removal, coliform removal efficiency: The extent of retention of coliform bacteria by a filter (see also Unger and Collins, 2008).

Constant head box: A device which controls flow rate to the filter(s) by maintaining a constant head of liquid over fixed orifices; used in pilot plant studies.

Control building: A facility that houses meters, valves, laboratory instruments, office, etc.

***Cryptosporidium parvum*:** A pathogenic protozoan which causes enteritis and/or severe diarrhea; the oocyst is resistant to chlorine disinfection.

CT: The product of residual disinfectant concentration (C) in mg/L and the corresponding disinfectant contact time (t) in min.

Cumulative flow demand: The total volume of flow calculated over a given time period; the cumulative flow volume is calculated by summing the product of the flow rate and the duration of that flow rate over the time period of interest.

d_{10}: The size of the sieve opening through which 10% of the sand will just pass; also called the "effective size."

Design flow: The maximum daily flow for a projected population.

Detritus: Nonliving organic matter.

Dewatering: Draining the water level in the filter box to below the sand bed surface. The "headwater," that is, the water above the sand bed surface, should be drained first by pipe outlets above the sand bed.

Diatomaceous earth: High-silica skeletal remains of algae, used as a filtration material.

Drawdown: Drainage of the headwater.

Effective size: See d_{10}.

Energy grade line, EGL: A graphic representation of the total energy head, which shows the rate at which energy decreases; the EGL always drops downward in the direction of flow unless there is an energy input from a pump.

Enteric bacteria: (1) Members of the family *enterobacteriaceae* (gram negative, straight rods, etc.). (2) Bacteria that live in the intestinal tract. (Prescott et al., 1993, p. 435, p. G9). Included are the genera, *Escherichia*, *Enterobacter*, *Proteus*, *Salmonella*, and *Shigella* (pp. 435–438). *Salmonella typhi* causes typhoid fever and gastroenteritis; *Shigella*, bacillary dysentery; some strains of *E. coli* cause gastroenteritis (p. 438).

Enterovirus: Genus of viruses of the human GI tract (of the family Picornaviridae); polio is an example (Prescott et al., 1993, p. A29).

Fecal coliform: Coliforms derived from the intestinal tract of warm-blooded animals, which can grow at 44.5°C.

Filter harrowing: Disturbing the filter media surface with a large implement consisting of a series of teeth or disklike blades.

Filter mats: Nonwoven synthetic fabrics placed on the surface of the filter media to facilitate cleaning.

Filter-to-waste: The process of disposing of tailwater (filter effluent) while the filter product water is of unacceptable quality.

Fines: The smallest particles in unwashed sand.

Gas precipitation: A phenomenon whereby dissolved gas comes out of solution due to supersaturation.

***Giardia*:** Name often used in conversation for the pathogenic organism, *Giardia lamblia*. Actually, there are several species of *Giardia*.

***Giardia lamblia*:** A protozoan pathogenic to humans which causes severe diarrhea. The organism is ingested in cyst form. Excystation occurs in the stomach.

***Giardia muris*:** A protozoan that infects mice and thought to be not pathogenic to humans.

***Giardiasis*:** Disease caused by protozoan *Giardia lamblia*.

Gravel support: Coarse media on which the filter media is placed, and which surrounds and covers the underdrains.

Headloss: Loss of media permeability; increased flow resistance.

Headwater: The raw water in the filter box directly above the filter bed; also called supernatant water.

Hydraulic conductivity: Permeability, reported as length/time.

Hydraulic grade line, HGL: The piezometric head line, that is, a graphic representation of what would be the free surface if one could exist, and the same conditions of flow were maintained; if the velocity head is constant, the drop in the hydraulic grade line between any two points is the value of the loss of head between those two points.

Hydraulic loading rate: Volumetric flow rate divided by filter surface area, resulting in units of length/time.

Log removal, log R: Defined: $\log \mathbf{R} \equiv \log N_{\text{in}} - \log N_{\text{out}}$ in which N_{in} is the concentration in the influent and N_{out} is the concentration in the effluent of whatever constituent species is being measured. Conversion to $\%\mathbf{R}$ is: $\%\mathbf{R} = 1 - 10^{-\log \mathbf{R}}$.

Mature: The state of a filter when coliform removal has reached its optimum level. Also, a "ripened" filter.

Mature filter bed: Maximum development of biofilm within filter bed for given conditions.

Newton: Unit of force. 1 N = 0.2248 lb force.

NPDES: National Pollution Discharge Elimination System (an acronym introduced with PL-500).

NPDOC: Non-purgable dissolved organic carbon.

Package plant: A commercially available prefabricated filter.

Pascal: Unit of pressure. 1 Pascal = 1 Newton/m^2 = 0.000145 lb/ft^2.

Peak flow: A community's maximum daily water demand.

Percent removal, %R: Defined: $(N_{\text{in}} - N_{\text{out}})/N_{\text{in}} \cdot 100$, see also *log removal*.

Performance capacity: The flow rate above which the performance of the filter (effluent turbidity, rate of headloss development, etc.) no longer satisfies regulatory or community requirements.

PFU: A measure of the number of cell-infectious viruses capable of forming plaques per unit volume.

Pilot plant: A small-scale replica of a proposed or existing full-scale facility, useful in determining, at relatively low expense, the feasibility of the full-scale plant in achieving the desired finished water quality given the raw water characteristics.

Poise: Unit of dynamic viscosity. 1 poise = 0.10 N · s/m^2.

Poliovirus: The poliovirus is of the family Picornaviridae (MW ≈ 2.5 · 106, diameter 22–30 nm), genera Enterovirus (viruses of the gastrointestinal tract), (Prescott et al., 1993, p. A29). The shape of the poliovirus is spherical (Ibid., p. 356) with a protein shell surrounding its nucleic acid, called a "capsid" (Ibid., p. 355).

Poliovirus, attenuated: A poliovirus inactivated, unable to cause infection.

Preozonation: Oxidation of the raw water prior to filtration.

Reservoir: The filter box zone above the filter media; the location of the headwater or supernatant water.

Ripening: (1) The process whereby a diverse biological community develops within a filter bed. (2) The sense of the term is that a biofilm develops on the sand surfaces that may then serve as for "attachment" of microorganisms that are transported to the surface.

Roughing filter: A pretreatment consisting of a series of chambers of coarse media which serve to reduce raw water turbidity.

Run length: The period of time between filter startup and terminal headloss.

Sand bed: The filter media.

***Schmutzdecke*:** A German word which translates literally as "dirty layer," and adopted early in American practice. The *schmutzedecke* is defined here as a layer of material deposited on the top of the filter bed which causes headloss disproportionate to its thickness. As stated by van de Vloed (1955, p. 568) the term was used by Piefke in 1880 (there is no further citation available), which is useful to date the recognition of the phenomenon as well as the German involvement.

Scour: Disturbance of the filter media, usually caused by high-velocity discharge of water into the filter.

Scraping: Removing the *schmutzdecke* from the surface of the filter bed by manual raking. The depth of removal ranges from about 5 mm in the case of Empire to perhaps 10–20 mm in other installations.

Sedimentation: Settling.

Small community: There is no strict definition but probably a small community would have a population of $\leq$5000, but more likely $\leq$1000 and it is likely that one person has several jobs, for example, operation of plants for drinking water and wastewater plants, water mains, street maintenance, animal control, etc. Usually public enterprises are limited by money to a greater extent than in larger communities.

Supernatant water: Headwater; the raw water in the reservoir.

SWTR: Surface water treatment rule. A federal regulation adopted by the USEPA pursuant to the 1994 Safe Drinking Water Act (PL93-523). In general, the SWTR requires a three-log inactivation of *Giardia lamblia* cysts and a four-log removal of viruses. The regulation requires filtration of surface waters, with exceptions. As applied to slow sand the turbidity limit was set at 1 NTU. (USEPA, 1989; Pontius, 1990, p. 41).

Tailwater: The filtered water that emerges from the filter underdrain system and flows over a weir. The elevation of the weir, along with the headwater elevation, controls the pressure gradient in the filter bed.

THMFP: Trihalomethane formation potential.

Transport coefficient, η: Ratio of particles striking collector surfaces in a given unit area in a plane normal to the flow in a given time increment to particles approaching collectors that have crossed the same area and plane within the same time increment. See Chapter 12.

Turbidity: Cloudiness of the water due to small particles.

Underdrain: A system of perforated pipes in the gravel support that serves to collect filtered water and channel it out of the filter box to an overflow weir.

Uniformity coefficient, UC, d_{60}/d_{10}: The ratio of the sieve size through which 60% of the sand will pass to the size through which 10% will pass.

UVA: Ultraviolet absorbance. Term relates to instrumental method to measure dissolved organic carbon.

Weir: A plate or other device that serves as for overflow of water and which may function to control the water elevation and to measure the flow.

REFERENCES

Alagarsamy, S. R. and Gandhirajan, M., Package and water treatment plants for rural and isolated communities, *Journal of Indian Water Works Association*, 13(1):73–80, 1981.

Arix, A Professional Corporation, Project manual for water system improvements—Filter plant, Town of Empire, CO, Project No. 83206.00, February 1984.

Babbitt, H. E. and Doland, J. J., *Water Supply Engineering*, 3rd edn., McGraw-Hill, Inc., New York, 1939.

Baker, M. N., *The Quest for Pure Water*, The American Water Works Association, New York, 1948.

Barrett, J. M., Improvement of slow sand filtration of warm water by using coarse sand, PhD dissertation, Department of Civil, Environmental, and Architectural Engineering, University of Colorado, Denver, CO, 1989.

Beer, C. R. and Dice, J. C., Denver's slow-sand filters, in: *Research News*, No. 38, AWWA Research Foundation, Denver, CO, November 1982.

Bellamy, W. D., Slow sand filtration of *Giardia Lamblia* and other substances, PhD dissertation, Department of Civil Engineering, Colorado State University, Fort Collins, CO, 1984.

Bellamy, W. D., G. P. Silverman, and D. W. Hendricks, Filtration of *Giardia* Cysts and Other Substances, Volume 2: Slow Sand Filtration. Project Summary, Report No. EPA-600/S2-85/026, Water Engineering Research Laboratory, USEPA, Cincinnati, OH, May 1985a (NTIS Report No. PB85-191633/AS).

Bellamy, W. D., Silverman, G. P., and Hendricks, D. W., Removing *Giardia* cysts with slow sand filtration, *Journal of the American Water Works Association*, 77:52–60, February 1985b.

Brink, D. R. and Parks, S., Update on slow sand/advanced biological filtration research, in: Graham, N. J. D. (Ed.), *Slow Sand Filtration: Recent Developments in Water Treatment Technology*, John Wiley & Sons, Chichester, England, 1996.

Bryck, J., *Giardia* removal by slow sand filtration—Pilot to full scale, in: *Proceedings Sunday Seminar on Coagulation and Filtration: Pilot to Full Scale*, Annual Conference of the American Water Works Association, Kansas City, MO, June 14, 1987.

Bryck, J., Personal communication, September 20, 1990.

Bryck, J., Walker, B., and Hendricks, D. W., *Slow Sand Filtration at 100 Mile House, British Columbia*, Supply and Services Canada Contract ISV84-00286, Dayton & Knight, Ltd., Consulting Engineers, West Vancouver, BC, June 1987.

Collins, M. R., personal communication, August 16, 1990.

Cosgrove, J. J., *History of Sanitation*, Standard Sanitary Mfg. Co., Pittsburgh, PA, 1909.

Dayton & Knight, Ltd., Pilot water treatment program, Village of 100 Mile House, Report for client, Dayton & Knight, Ltd., Consulting Engineers, West Vancouver, BC, December 1983.

Dayton & Knight, Ltd., Moricetown band water treatment plant, Contract No.168.21.2, (contract documents and plans), Dayton & Knight, Ltd., Consulting Engineers, West Vancouver, BC, May 1988.

Dayton & Knight, Ltd., Fact sheet on Moricetown band water treatment plant, Dayton & Knight, Ltd., Consulting Engineers, West Vancouver, 1989.

Duncan, A., The ecology of slow sand filters, in: *Slow Sand Filtration: Recent Developments in Water Treatment Technology*, Graham, N. J. D. (Ed.), Ellis Horwood Ltd., Chichester, UK, 1988.

Gaya & Associados and Metcalf & Eddy de Puerto Rico, Draft preliminary design report, El Portal Del Yunque Caribbean National Forest On-Site Water Supply and Distribution System, August 20, 1992.

Graham, N. J. D. (Ed.), *Slow Sand Filtration: Recent Developments in Water Treatment Technology*, John Wiley & Sons, Chichester, England, 1996.

Graham, N. J. D. and Collins, R. (Eds.), *Advances in Slow Sand and Alternative Biological Filtration*, John Wiley & Sons, New York, 1988.

Great Lakes Upper Mississippi River Board of State Public Health and Environmental Managers, "Recommended standards for water works," Health Research Inc., Health Education Services Division, Albany, NY, 1987.

Hazen, A., *The Filtration of Public Water Supplies*, 3rd edn., John Wiley, New York, 1913.

Hendricks, D. W. (Ed.), *Manual of Design for Slow Sand Filtration*, AWWA Research Foundation and American Water Works Association, Denver, CO, 1991.

Hendricks, D. W., Seelaus, T., Saterdal, R., Alexander, B., Jones, G., Gertig, K., Jones, F., Blair, J., Mosher, R., Bauman, J., and Hibler, C., Filtration of *Giardia* cysts and other particles under treatment plant conditions, Research report on water treatment and operations, AWWA Research Foundation, Denver, CO, February 1988.

Huisman, L., Developments of village-scale slow sand filtration, *Progress in Water Technology* (India), 11(1):159–165, 1978.

Huisman, L. and Wood, W. E., *Slow Sand Filtration*, World Health Organization, Geneva, Switzerland, 1974.

Iwasaki, T., Some notes on sand filtration, *Journal of the American Water Works Association*, 29:1591, December 1937.

Jordan, H. E., Fifteen years filtration practice in Indianapolis, in: *Proceedings of the Thirteenth Annual Convention of the Indiana Sanitary and Water Supply Association*, Indianapolis, IN, pp. 1–67, 1920.

Kerkhoven, P., Third world tests for sand filters, *World Water*, September 1979.

Komolrit, K., Chainarong, L., and Buaseemuang, S., Results of a slow sand filtration programme in Thailand, *Aqua, Thailand*, 4:12–17, 19, 21, 1979.

Letterman, R. D. and Cullen, T. R. Jr., Slow sand filtration maintenance: Costs and effects on water quality, *EPA/600/S2-85/056*, Water Engineering Research Laboratory, USEPA, Cincinnati, OH, August 1985.

Lloyd, B., The Construction of a sand profile sampler; its use in the study of the Vorticella populations and the general interstitial microfauna of slow sand filters, *Water Research*, 7(7): 963–973, July 1973.

Logsdon, G. S., *Water Filtration Practices*, American Water Works Association, Denver, CO, 2008.

Logsdon, G. S. and Fox, K., Slow sand filtration in the United States, in: Graham, N.J.D. (Ed.), *Slow Sand Filtration: Recent Developments in Water Treatment Technology*, Ellis Horwood Ltd. and John Wiley & Sons, New York, 1988.

Paramasivam, R., Mhaisalkar, V. A., and Berthouex, P. M., Slow sand filter design and construction in developing countries, *Journal of the American Water Works Association*, 73 (4):178–185, April 1981.

Pontius, F. W., Complying with the new drinking water quality regulations, *Journal of the American Water Works Association*, 82(2):32–52, February 1990.

Poynter, S. F. B. and Slade, J. S., The removal of viruses by slow sand filtration, *Progress in Water Technology*, 9(1):75–88, January 1977.

Prescott, L. M., Harley, J. P., and Klein, D. A., *Microbiology*, 2nd ed., Wm. C. Brown Publishers, Dubuque, IA, 1993.

Rachwal, A. J., Bauer, M. J., and West, J. T., Advanced techniques for upgrading large scale slow sand filters, in: Graham, N. J. D. (Ed.), *Slow Sand Filtration: Recent Developments in Water Treatment Technology*, Ellis Horwood, Ltd., Chichester, U.K. and John Wiley, New York, 1988.

Raman, R., Singh, G., and Nagpal, J. L., Slow sand test-beds help refine design, *World Water* (International Reference Centre for Community Water Supply and Sanitation, The Netherlands), pp. 44–45, December 1981.

Robeck, G. G., Clarke, N. A., and Dostal, K. A., Effectiveness of water treatment processes in virus removal, *Journal of the American Water Works Association*, 54(10):1275–1291, October 1962.

Schuler, P. F., Ghosh, M. M., and Gopalan, P., Slow sand and diatomaceous earth filtration of cysts and other particulates, *Water Research*, 25(8):995–1005, August 1991.

Seelaus, T. D., Hendricks, D. W., and Janonis, B. J., Design and operation of a slow sand filter, *Journal of the American Water Works Association*, 78(12):35–41, December 1986.

Seelaus, T., Hendricks, D. W., and Janonis, B. J., Slow sand filtration at Empire, Colorado, Vol. 1 in Hendricks, D. W. et al. (Eds.), *Filtration of Giardia Cysts and Other Particles under Treatment Plant Conditions*, Research Report, AWWA Research Foundation, Denver, CO, Feb., 1988.

Sims, R. C. and Slezak, L., Present practice of slow sand filtration in the United States, in Logsdon, G.S. (Ed.), *Slow Sand Filtration Manual of Practice*, American Society of Civil Engineers, New York, 1991.

Slezak, L. A. and Sims, R. C., The application and effectiveness of slow sand filtration in the United States, *Journal of the American Water Works Association*, 76(12):38–43, 1984.

USEPA (US Environmental Protection Agency), National primary drinking water regulations: Filtration, disinfection: Turbidity, *Giardia lamblia*, viruses, *Legionella*, and heterotrophic bacteria; final rule, *Federal Register*, 54(124):27486–27541, June 29, 1989.

van Dijk, J. C. and Oomen, J. H. C. M., Slow sand filtration for community water supplies in developing countries, A design and construction manual, Technical Paper No. 11, International Reference Center for Community Water Supply and Sanitation, The Hague, The Netherlands, 1978.

van Markenlaan, J. C., Slow sand filtration for community water supplies in developing countries, A report of an *International Appraisal Meeting* held in Nagpur, India, September 15–19, 1980, *Bulletin Series No. 16*, International Reference Center for Community Water Supply and Sanitation, The Hague, The Netherlands, March 1981.

van de Vloed, A., Comparison between slow sand and rapid filters, in: *Proceedings Third Congress*, July 18–23, London, International Water Supply Association, 1955.

Visscher, J. T., Paramasivam, R., Raman, A., and Heijnen, H. A., Slow sand filtration for community water supply, planning, design, construction, operation, and maintenance, Technical Paper No. 24, International Reference Center for Community Water Supply and Sanitation, The Hague, The Netherlands, 1987.

Yao, K. M., Habibian, M. T., and O'Melia, C. R., Water and waste filtration: Concepts and applications, *Environmental Science and Technology*, 11(5):1105, 1971.

14 Cake Filtration

As with all unit processes, "cake" filtration has its own nomenclature and technology. Also, as with other unit processes, it has its own "subculture," defined largely, in this case, by the industry that manufactures the filter media. The latter relates to how the field evolved from the discovery of the geologic deposits of fossil diatoms in Germany through its application to drinking-water treatment. Finally, as with all unit processes, the design of a system is circumscribed, to a large extent, by its past. Understanding these themes, that is, the past, the technology, and the process principles, gives rationale for the design.

14.1 DESCRIPTION

Figure 14.1 encapsulates the key aspects of the cake filtration process. As illustrated, a "cake" filter (also called a "pre-coat" filter) has a media, usually diatomaceous earth, deposited on a "septum," comprising a stainless steel fabric. The mesh size of the fabric must be small enough to retain the media. The circulation of a specified mass of slurry through the septum transfers the media to a "pre-coat" deposit on the septum, illustrated as the portion of the filter cake just to the left of the septum in Figure 14.1. After the pre-coat forms, the filtration process can begin. The raw water is introduced, along with a "body feed" injected into the raw-water flow to give a specified concentration, usually of the same material as the pre-coat. At this point the filter run begins. The body-feed deposits on the pre-coat, building up the filter "cake," illustrated in Figure 14.1 as the portion of the filter cake just to the left of the pre-coat. When the headloss reaches a specified design limit, or when the cake thickness reaches a limit as defined by the spacing between septum leaves, whichever occurs first, then the cake is removed and the cycle is repeated. As also illustrated in Figure 14.1, the suspended matter deposits within the matrix of the cake that has been deposited. Without the body feed, the suspended matter would deposit as a layer, forming its own "cake" on the pre-coat, resulting in a rapid rate of headloss increase. Not shown in Figure 14.1 is the structural support for the septum that must be provided in order to withstand the pressure differential between the two sides of the filter cake.

14.1.1 CAKE FILTRATION IN-A-NUTSHELL

Cake filtration has its own nomenclature, operating procedures, and process characteristics. These are described briefly to provide an orientation for the chapter.

14.1.1.1 Applications

Applications of diatomite filtration have included potable-water treatment, tertiary treatment of wastewater, treatment of backwash water, treatment of industrial process water, industrial wastewater treatment, clarification of swimming pool water, and filtration of aquarium water. Although this chapter is built around the application to potable water treatment, the principles apply to any of the applications mentioned.

14.1.1.2 Definitions

Most of the terms here are requisite to further learning about diatomite filtration. The "glossary" expands on the terminology given.

14.1.1.2.1 Filter Aids

Filter media have been called, in the industry, a "filter aid." The most common medium is diatomaceous earth, called "diatomite" by those in the industry. Sometimes it is called "DE." Another filter aid, less common, is a volcanic ash, manufactured under the name, perlite®. Yet another is cellulose, made from wood pulp.

14.1.1.2.2 Septum

The filter media are retained on a fabric, called a "septum." The most common fabric is stainless steel, but other materials include carbon steel, titanium, hasteloy, polypropylene, etc. The mesh opening most often cited is 60 μm. The septum must be supported by a structural grid of several layers to withstand the pressure differential between the two sides, for example, 103–280 kPa (15–40 psi), depending upon the system and to minimize flex that could cause cracks in the pre-coat.

14.1.1.2.3 Pre-Coat

An initial deposit of filter media on the septum is called a pre-coat. The pre-coat has three primary functions: (1) to limit the passage of particles at the start of a filtration cycle ("immediate clarity"), (2) protect the septum from "fouling", and (3) aid in cake release at the cycle termination. The rule of thumb for the deposit is 0.5–1.0 kg/m^2 (0.1–0.2 lb/ft^2). A mid-range, that is, 0.75–1.0 kg/m^2 (0.15 lb/ft^2) was applied in studies that adopted DE for portable field water treatment units for military use (Lowe et al., 1944). The higher number, that is, 1.0 kg/m^2 (0.2 lb/ft^2), was recommended to alleviate the passage of cyst-size radioactive beads (Logsdon et al., 1981, p. 113) and *Giardia murus* cysts (Logsdon and Lippy, 1982, p. 655). The thickness of the deposit is about 3–5 mm (1/8 in.).

14.1.1.2.4 Body Feed

The filter media are metered into the raw-water flow, which add to the pre-coat deposit, building up the thickness of the filter cake. The media concentration is established by trial such that the rate of headloss increase is linear, that is, it does not increase exponentially, as would be the case without

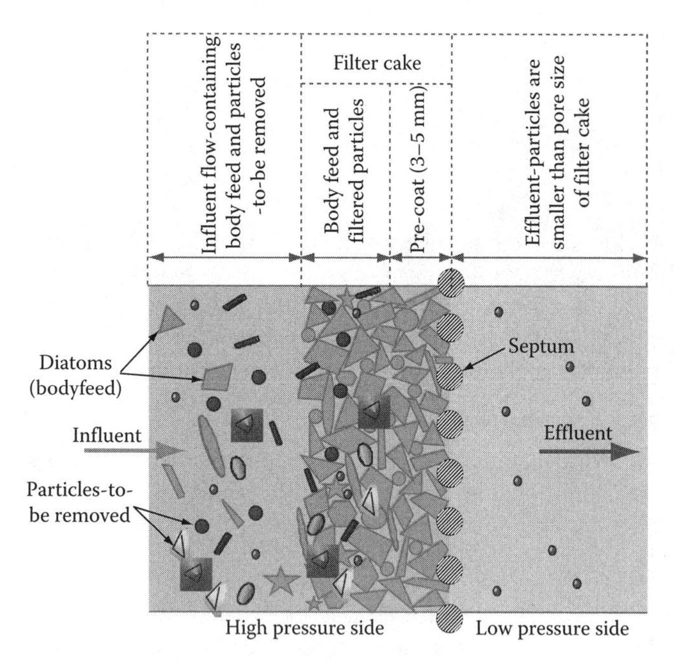

FIGURE 14.1 Septum, pre-coat, body-feed cake, particles-to-be-removed, water flow. (Adapted from McIndoe, R.W., *Water and Wastes Engineering*, 50, 1969a.)

body feed. A body-feed concentration of about 25 mg/L, after mixing with the raw water, would be representative.

14.1.1.2.5 Filter Cake

The accumulation of body feed on the septum, added to the pre-coat, builds in thickness as a filter cake. The thickness of the cake depends on the headloss permitted and the clearance needed between filter cakes deposited on adjacent septa. Having adequate space between the cakes minimizes the probability of the cake being dislodged by hydraulic shear.

14.1.1.3 Phases of Operation

Figure 14.2 illustrates the three phases of DE operation. That is, pre-coat, body feed, and cleaning. These three phases are explained in the paragraphs of this section.

14.1.1.3.1 Pre-Coat

In the pre-coat phase, the septum is prepared for operation with a deposit of DE as illustrated in Figure 14.2a. As shown, a pre-coat tank has a "charge" of DE adequate for a specified deposit, for example, 0.5–1.0 kg/m^2 (0.1–0.2 lb/ft^2). The DE charge in the pre-coat tank is maintained in suspension as a slurry and is pumped, in a closed loop, through the septum. Circulation is continued until the slurry becomes clear. The pre-coat is distributed uniformly since any nonuniform areas of deposit carry a higher velocity and thus a higher DE mass flux density to the deficient area.

During the circulation, a bridging of media particles occurs such that they are retained on the septum fabric. The openings of the fabric may be larger than the DE particles but not so large as to not permit bridging and retention.

The purpose of the pre-coat is to protect the septum from the particles to be removed and to support the filter cake. The pre-coat also acts as an initial filtering media, which is effective at the start of the filter run (McIndoe, 1969a, p. 51).

14.1.1.3.2 Body Feed

Upon completion of the pre-coat, body feed is started, as illustrated in Figure 14.2b. The body-feed tank is charged with a mass of DE sufficient for operation for the expected number of hours (i.e., before the headloss reaches a designated maximum). The slurry concentration should be high enough that the tank size is not excessively large and low enough that the DE may be maintained in suspension. As in the pre-coat tank, the DE in the body-feed tank is maintained as a slurry by means of a mixer (not high speed).

The transition from the pre-coat phase to the body-feed phase must be "smooth," that is, without pressure shock to the pre-coat that has just been deposited. To attain the smooth transition, the raw-water valve is opened slowly while the valve from the pre-coat line is closed slowly; at the same time, the body feed is started and the effluent valve from the tank-septum assembly is opened slowly. A filter-to-waste period is advisable until the effluent turbidity reaches the objective level.

14.1.1.3.3 Cleaning

Figure 14.2c illustrates the cleaning cycle. As illustrated, clean water is returned to the pressurized tank in the form of (1) high pressure jets that impinge on the cake that has formed on each septum surface, and/or (2) as a backflow across the septum surfaces. Drainage is at the bottom of the tank through pipes large enough that permit an easy flow of spent DE slurry and removed particles. The pressurized tank that contains the septum assembly should be at atmospheric pressure during the cleaning phase through an open valve at the top of the tank. The high pressure jet method may be facilitated, in a leaf system, by a provision to rotate the leaf elements such that the entire septum area is subject to the cleansing action of the jet. Other variations in cleaning are provided by equipment manufacturers, for example, "dry-cake" discharge.

14.1.1.4 Process Description

The filter media provides a rigid porous structure that passes the water being treated, and retains particles. The body feed entrains and "embeds" the particles, that is, blocks their movement, leaving the pore channels substantially intact within the filter cake for the flow of fluid (Cummins, 1942, p. 403). The correct body-feed concentration occurs when the DE structure dominates the pore structure rather than the particles being removed. If the particles being removed dominate the pores, then additional diatomite is needed. Without body feed, the particles to be removed will be retained on the surface of the pre-coat and will "blind off" the filter, that is, causing an inordinately high rate of headloss increase.

14.1.1.5 DE Selection

In water treatment, the general rule is to select the coarsest DE grade that provides the requisite effluent quality. If the run length is too short, then DE grade is probably not a good fit

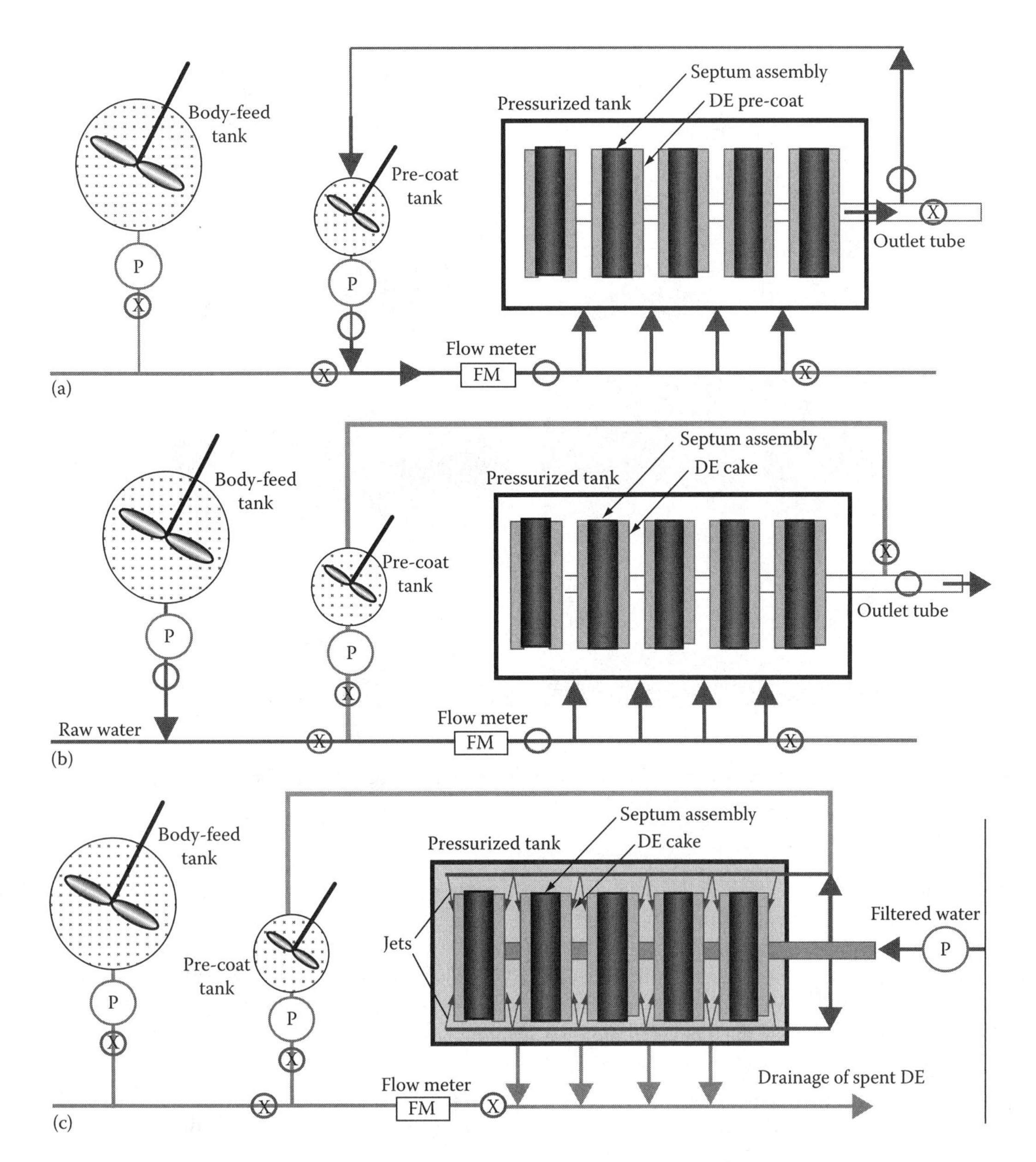

FIGURE 14.2 Cycles of diatomite filtering operation: (a) pre-coat, (b) body feed, and (c) cleaning.

for the situation. The run length is determined by the rate of headloss increase and the selected pressure differential limit.

14.1.2 Media

The media used in cake filtration are manufactured products. Selection of a product must be matched to the removal task at hand.

14.1.2.1 Kinds of Media

Three kinds of media are used in cake filtration: (1) diatomite; (2) volcanic ash, for example, Perlite®; and (3) cellulose. The most common media is diatomite and is the basis for the discussions in this chapter.

14.1.2.1.1 Diatomaceous Earth: Fossil Diatoms

Diatomite is the manufactured form of a product obtained from a sedimentary rock that comprises fossil diatoms. The fossilized diatom is hydrous silica, that is, $SiO_2 \cdot nH_2O$, in composition (Kadey, 1975, p. 605). Figure 14.3a shows a mixture of diatoms that illustrates the wide variety in shapes and sizes. The scale is 50 μm. Most diatoms are a few microns to over 100 microns in size, the latter applying to rod-shaped diatoms. The species of diatoms number 12,000–16,000 (Kadey, 1975, p. 609). At Celite's Lompoc mine, the ore comprises seawater species. Figure 14.3b through d shows three examples. At the Nevada deposits of Eagle-Picher the species are freshwater.

14.1.2.2 Sources of Media

The filter media, that is, diatomite and perlite®, are mined as minerals in "rock" form with about 40% moisture and then processed to form desired grades of filter media. Deposits of DE are located in many locations over the world. In the United States, the largest deposit is at Lompoc, California,

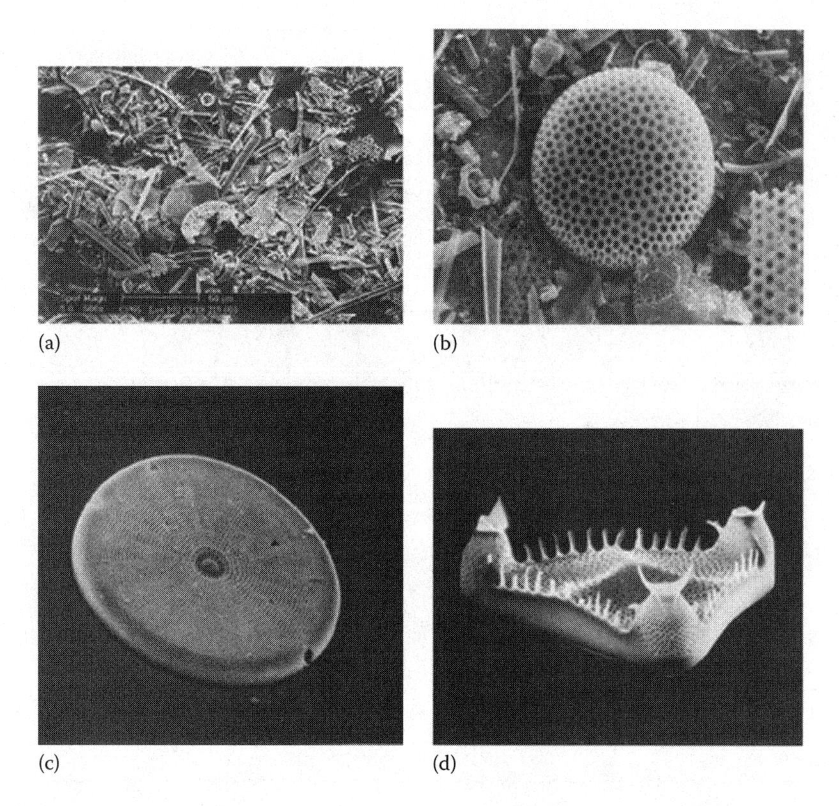

FIGURE 14.3 Examples of the diversity of diatoms found in deposits: (a) mixture (scale is 50 μm), (b) no identification, (c) *Arachnoidiscus orantus*, and (d) no identification. (Courtesy of Celite/World Minerals Inc., Lompoc, CA.)

which has been worked since about 1922. Other deposits are near Lovelock, Nevada and at Clark, Nevada (near Reno). The largest and most uniform deposits are at Lompoc, which is about 300 m (1000 ft) thick. The Celite/World Minerals Inc. mines the deposits; the Dicalite Co. also produces from the deposit. Most of the deposits in the United States are mined by open pit quarries. Figure 14.4a is an aerial view of the Lompoc mining operation, giving some idea of the extent of the deposit. Figure 14.4b illustrates the ore excavation. Figure 14.5 is a photograph of ore, that is, crude diatomite. The Lompoc deposits are over 15 million years in age and are the result of a high rate of diatom production in the adjacent ancient ocean, involving more than 10,000 species of diatoms (McIndoe, 1969a, p. 50).

14.1.2.3 Manufacturing of Media

At the Lompoc site, crude diatomite is mined and processed into various powders, called Celite®, for example, Celite 503®, Celite 545®. Because the diatoms are what give the diatomite its uniqueness, the milling is done with care to preserve this basic structure (Kadey, 1975, p. 620). Without further processing, the diatomite is a "natural" product. Further adjustment of particle size is done by "calcining," which is heating the powder to incipient fusion in large

FIGURE 14.4 Mining operation at Lompoc, California: (a) aerial view and (b) loader digging into ore body. (Courtesy of George Christoferson, Celite/World Minerals Inc., Lompoc, CA, December, 2002.)

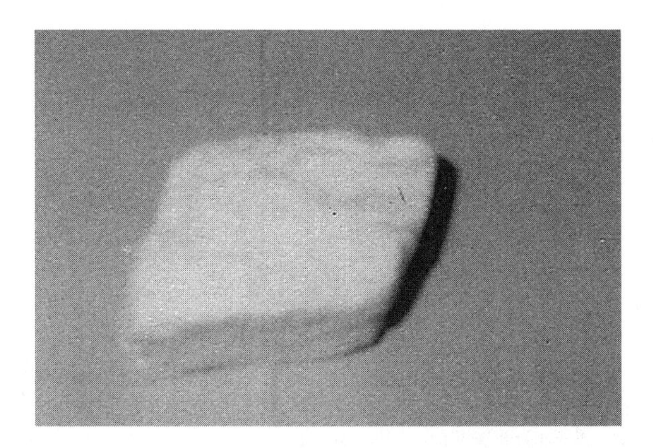

FIGURE 14.5 Ore "chunk" as obtained from mining operation (ore from Ray McIndoe, c. 1984).

rotary kilns, followed by further milling and classifying; the resulting powder is light pink in color. A "flux calcined" diatomite is manufactured by adding a "flux," that is, soda ash, before the calcining step, which gives still coarser grades of DE, which are white in color (Kadey, 1975, p. 620).

With respect to Celite®, the finest grade is Filter-Cel®, a "natural" DE, which has been selectively quarried, dried, milled, and air classified. Three calcined grades of DE include Celite 505®, Standard Super-Cel®, and Celite 512® (Johns Manville Brochure FA-304 6-80, p. 7). The flux-calcined grades include Hyflo Super-Cel®, Celite 501®, Celite 503®, Celite 535®, Celite 545®, Celite 540®, and Celite 560® (McIndoe, 1969a, p. 52). The flux-calcined products have a more open pore structure within the bulk of the material (i.e., not within the individual particles).

The finished product is packaged in 23 kg (50 lb) bags or shipped bulk in trucks or box cars; bulk loading is pneumatic. The average cost per ton was \$55/metric ton (\$50/U.S. ton), 1960–1962; in 1972, the cost was \$72/metric ton (\$65/U.S. ton). The total world production in 1972 was 2.2 million metric tons (2 million U.S. tons) (Kadey, 1975, p. 625).

14.1.2.4 Characteristics of Media

The properties of diatomite are based in part on its method of production, that is, whether it is a natural, calcined, or flux-calcined material. The properties of interest include color, particle size and distribution, pore size and distribution, permeability, specific gravity, and dry bulk densities. Manufacturers designate their respective products by "grade," for example, Celite-560™. Table 14.1 lists selected properties of different DE grades as designated by two manufacturers, that is, Celite® (Celite/World Minerals Inc., Lompoc, California) and Celatom® (Eagle-Picher, Inc., Reno, Nevada). The grades of diatomite are listed in an ascending order of particle size, which corresponds to increasing pore size and increasing permeability. As seen, the median particle sizes are in the range of $7 \leq d(\text{diatoms})_{50} \leq 100$ μm and median pore sizes are proportionately smaller, that is, in the range of $1.5 \leq d(\text{pore})_{50} \leq 22$ μm. Figure 14.6 shows size distributions for several grades of Celite®, which show that for each grade a distribution exists, which results in the void volume being filled to a large extent. The grades are differentiated by their $d(50)$ sizes.

> **BOX 14.1 INTERNATIONAL DIATOMITE PRODUCER'S ASSOCIATION**
>
> The International Diatomite Producer's Association (IDPA) was formed in 1988 to provide a central office to promote the use of diatomite. The IDPA represents all diatomite producers who have elected to belong. The association provides information about diatomite and may act as a conduit from interested parties to the diatomite producers regarding the availability of the product, its characteristics, and cost.

14.1.3 Attributes

Some of the merits of DE filtration include compact size, low operation and maintenance costs, simplicity of operation, and a system that can be shut down quickly and started quickly (Spenser and Collins, 1995, p. 72). Examples are listed which illustrate the range of contextual situations.

- Meeting summer peaking demand. *Example*: The DE plant at Vacaville, California is operated 8 h per day only during the summer.
- Small footprint. *Example*: To treat the Croton water supply for New York, a diatomite plant was proposed to be located in Jerome Park; Delaware would utilize a smaller land area than rapid filtration.
- Cost competitive. *Example*: For Georgetown, Colorado a cost comparison between diatomite filtration and a rapid filtration pilot plant in 1986 indicated lower annual costs using diatomite (Wirsig, 1986).
- Automated operation. *Example*: Hearst Castle in California has a varying demand for drinking water over the annual cycle; two DE plants, rated at 113 L/min (30 gpm) and 155 L/min (41 gpm), serve the demand pattern. The plant is operated by PLC (programmable logic controller), that is, pre-coat circulates until turbidity is <0.5 NTU, then the unit switches to produce treated water, the unit stops at 241 kPa (35 psi), and goes into the backwash mode (Rogers, 1997, p. 8).
- Filtration is needed to follow another process. *Example*: The water source for Lompoc, California is a hard groundwater (782 mg/L as $CaCO_3$), which is treated by lime soda. The effluent turbidity from this process is about 2 NTU. The DE plant that follows the softening treatment produces a finished water with about 0.1 NTU (Rogers, 1997, p. 7).
- Cysts in ambient source waters are a particular issue. *Example*: DE has demonstrated high removals, that is, $4 < \log R < 6$ for cysts of *Endameba histolytica*,

TABLE 14.1
Properties of Diatomaceous Earth Filtration Media

Manufacturer[a]								
Celite/World Minerals Inc.					**Eagle-Picher Minerals, Inc.**			
Grade	**Color[b]**	**Particle $d(50)^b$ (μm)**	**Pores $d(50)^b$ (μm)**	**Perm[b] (Darcys)[d]**	**Grade**	**Color**	**Part. $d(50)^c$ (μm)**	**Perm[c] (Darcys)[d]**
Filter Cel	Grey-natural	7.5	1.5	0.07				
Celite 577	Pink-C	12	2.5	0.16				
Celite 505	Pink-C		2.0	0.06	Celatom FP-2	Buff/pink	12.8	0.100
Standard Super-Cel	Pink-C	14	3.5	0.25	Celatom FP-4	Buff/pink -FC	15	0.300
Celite 512	Pink-C	16	5.0	0.50	Celatom FW-6	Light pink-FC	18	0.480
Hyflo Super-Cel	White-FC	22	7.0	1.10	Celatom FW-12	Light pink-FC	24	0.800
Celite 501	White-FC	24	9	1.3	Celatom FW-18	White-FC	31	1.70
Celite 503	White-FC	23	10	1.9	Celatom FW-20	White-FC	33	2.10
Celite 535	White-FC	34	13	3.0	Celatom FW-50	White-FC	42	3.50
Celite 545	White-FC	36	17	4.0	Celatom FW-60	White-FC	48	5.0
Celite 550	White-FC			7.4				
					Celatom FW-80	White-FC	77	10.0
Celite 560	White-FC	106	22	30				

[a] All grades are registered trademarks of the companies indicated.

[b] Particle $d(50)$, pore $d(50)$, and perm (permeability); data set from http://www.diatomite.com, Nov. 2002.

[c] Particle $d(50)$, and perm (permeability) from Eagle-Picher Minerals, Inc., Celatom Technical Data Sheet (2003); see www.epminerals.com/celatomgrade.html for current data on grades and properties.

[d] A permeability of 1 darcy permits a flow of 1 mL of liquid of 1 centipoise viscosity through 1 cm^2 of porous solid 1 cm thick under a differential pressure of 1 atm in 1 s (Bell, 1962, p. 1247). To convert to "intrinsic permeability," "k," multiply the value given in darcys by $0.987 \cdot 10^{-12}$ m^2.

Other properties: SG(DE) ≈ 2.3 (McIndoe, 1983, p. 6). Dry bulk densities for each grade are: Filter-Cel® = 112; C-577® = 128; C-505® = 128; Standard Super-Cel® = 128; C-512® = 128; Hyflo Super-Cel® = 144; C-501® = 152; C-503® = 152; C-535® = 192; C-545® = 192; C-550® = 288; C-560® = 313 kg/m^3 (from "Johns-Manville Celite filter aids for maximum clarity at lowest cost"). Specific surface areas of representative Celite® grades are: Filter-Cel® 30,000 cm^2/g; Standard Super-Cel® 20,000 cm^2/g; Hyflo Super-Cel® 10,000 cm^2/g; Celite-503® 8,000 cm^2/g; Celite-535® 7,000 cm^2/g; Celite-545® 6,000 cm^2/g (from Cummins, 1942, p. 405); divide by 10 to obtain m^2/kg.

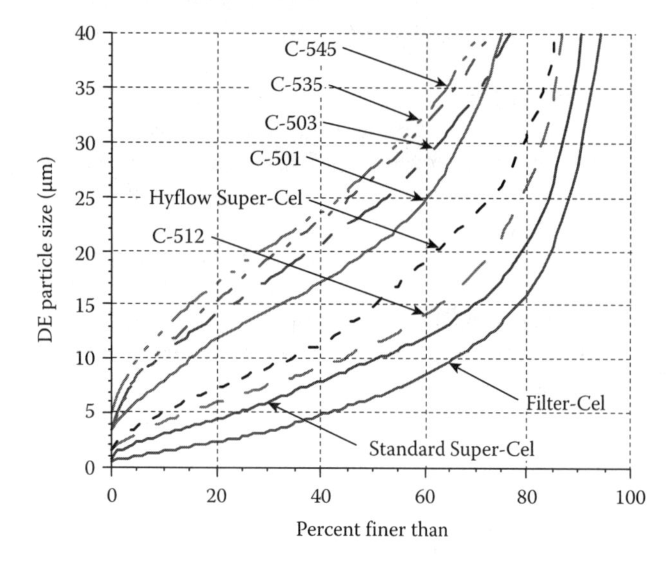

FIGURE 14.6 Particle size distribution of several Celite™ grades. (From Johns-Manville, *Maximum Clarity at Lowest Cost*, Celite Filter Aids, FA-84A, 5-81, no date but estimated at 1981 from designation 5-81; Table 1, Celite/World Minerals Inc. With permission.)

Giardia lamblia, and *Cryptospordium parvum*. Black Hawk, Colorado is a 4.3 mL/day (1.15 mgd) DE plant (Rogers, 1997) that treats water from a low turbidity mountain stream, which addresses this issue; the plant produces effluent turbidities generally <0.5 NTU with raw water generally <1 NTU.

14.1.4 History

Diatomite filtration was used as early as 1893 to clarify beet sugar solutions, and then from about 1900 by the brewery industry to clarify beer (Cummins, 1975, pp. 1–15). An article by Cummins (1942) describes DE filtration with reference to applications in the sugar industry; there is no reference to applications involving water treatment. Since DE is an inert substance, there is no taste or color imparted by its use for filtration.

14.1.4.1 1940s' Military Use of DE Filtration

During World War II, the U.S. Army was seeking a more suitable technology for the field production of drinking water

BOX 14.2 RAY MCINDOE

A person who was a major force for diatomite filtration during the 1970s and the 1980s was Ray McIndoe, Marketing Manager, Johns-Manville Corp. McIndoe was associated with a diatomite filtration primer (McIndoe, 1969b) and was the driving force in its 1988 update. He was always ready to go the "extra mile" in helping to facilitate the adoption of diatomite, whether at a small plant serving 200 persons or in being called out of retirement, c. 1989, to supervise the operation of a 1.0 mgd pilot plant for New York City. Alan Wirsig, an associate of McIndoe at Johns-Manville, remained in the field as a consulting engineer after his 1982 retirement at Johns-Manville. They are shown while operating a pilot plant setup at a field site in northern Colorado, c. 1985 (McIndoe and Wirsig, left to right).

than the sand filtration then available. Effective removal of *E. hystolytica* cysts (15 μm) was a specific objective (Lowe et al., 1944). In the Pacific and India–Burma–China theaters of war, for example, *E. Hystolytica* was endemic in the populations and the strains were particularly virulent (p. 16). Thus, there was a strong impetus for an adequate treatment technology and the need was considered immediate. The other objective was portability (Black and Spaulding, 1944; Lowe et al., 1944).

Relative to the foregoing background, in 1943 (April–August) the U.S. Army, in collaboration with the U.S. Public Health Service, conducted a series of experiments to test both their U.S. Army Portable Water Purification Unit Model 1940 (pressurized sand filter with alum and soda-ash feed) and eight systems that utilized diatomaceous earth. The test units were seeded with a large numbers of cysts, for example, hundreds of thousands or millions. The outcome of the experiments was that all grades of DE tested, even Celite 545, resulted in passage of zero cysts, albeit the latter was less effective in removal of turbidity. In addition, a technology was at hand, for example, diatomite manufacturing, septum designs, tank, and operating protocol.

14.1.4.2 1950s′ Adaptation of DE for Municipal Use

Based upon the success of the U.S. Army in adopting diatomite for drinking water for field use, a plant was constructed in 1949 as a standby supply for Gasport, New York, a community with a population of about 800 (Baumann et al., 1965). Then during the period 1950–1960, nine DE plants were installed along the Saginaw–Midland pipeline in Michigan. The smallest plant had one filter with a septum area of 4.18 m^2 (45 ft^2) serving 120 persons and the largest had nine filters with a total area of 141.4 m^2 (1522 ft^2) serving 5000 persons (Vander Velde and Crumley, 1962, p. 1495). By 1969 there were some 130 municipal installations with sizes of $38 < Q(\text{plant}) \leq 38{,}000$ m^3/day ($0.010 < Q(\text{plant}) \leq 10$ mgd) (McIndoe, 1969a, p. 50). By 1983 some 170 plants had been built, most in the 3,800–38,000 m^3/day (1–10 mgd) category, with the largest being 76,000 m^3/day (20 mgd) at San Gabriel, California (McIndoe, 1983, p. 1) and with 47 plants in New York.

14.1.4.2.1 New York City

In 1979, two consulting firms engaged by the City of New York (i.e., Metcalf & Eddy and Hazen and Sawyer) proposed a $610 \cdot 10^3$ m^3/day (160 mgd) ozone-DE plant for treatment of the Croton water supply (Bryant and Brailey, 1980). An 11,355 m^3/day (3 mgd) pilot plant was to have been built first to provide design criteria for the larger plant. As a note, in 1980, $610 \cdot 10^3$ m^3/day (160 mgd) was the average daily use of water from the Croton watershed, which had a "safe yield" of $910 \cdot 10^3$ m^3/day (240 mgd). The drainage area was 2400 km^2 (380 mi^2) on the eastern side of the Hudson River in Dutchess, Putnam, and Westchester counties, with a permanent population in 1980 of 100,000. Turbidity averaged 1–2 NTU with color about 15 units. The site of the proposed DE plant was at the 35 ha (85 ac) Jerome Park Reservoir, in the Bronx, near the northern border of New York City, which is the terminus for the aqueducts of the Croton system. Bulk deliveries of DE would be removed from railroad cars, slurried, and pumped to day tanks from a site approximately 900 m (3000 ft) from the plant. Pretreatment would be bar screens and traveling fine screens. The water would then be pumped through variable speed pumps to 48, 2 m (80 in.) diameter, 140 m^2 (1500 ft^2) septum area DE filters arranged in 12 clusters with HLR(max) ≈ 6.12 m/h (2.5 gpm/ft^2). The relatively small size of the proposed system and the prospect of reclaiming filter media were major motivations for the selection of DE (Spenser and Collins, 1995, p. 72).

The planned 11,355 m^3/day (3 mgd) ozone-DE pilot plant was constructed and operated during the period c. 1988–1990 to determine criteria for a full-scale 1100 ML/day (300 mgd) plant. The pilot plant effluent produced a finished water turbidity of <0.1 NTU and color <2 units (Parmelee, 1990, p. 1). To filter the Croton water supply, the DE process was recommended for several reasons: (1) diatomite filtration equipment had a relatively small "footprint," which was important in the location selected, that is, the densely developed Bronx; (2) the operational flexibility, that is, to use the Croton supply only in times of drought or high demand;

(3) effluent turbidities have been <0.1 NTU consistently; (4) at 3.7 m/h (1.5 gpm/ft^2), the filter runs have been 4–5 days (before reaching terminal headloss; (5) the spent DE could be reclaimed; and (6) waste DE, since there is no chemical coagulant added, can be disposed of in a land fill. About 1990, the engineering firms had some questions about the selection of diatomite filtration by their predecessors. In the meantime, New York had been feeling pressure, mostly from EPA, to filter their Croton water supply. In 1998, however, the discovery that UV254 would render *C. parvum* oocysts nonviable made moot the question of filtration. Thus, UV disinfection was the technology selected.

Example 14.1 Comparison of Footprints for DE and Rapid Filtration

Given

The full-scale diatomite filtration system for the Croton water supply was to have had 48, 2032 mm (80 in.) diameter, DE filters, each with 140 m^2 (1500 ft^2) septum area, arranged in 12 clusters with HLR(max) ≈ 6.12 m/h (2.5 gpm/ft^2).

Required

Estimate the "footprint" area of the collection of pressure vessels and compare with an estimate of the "footprint" for rapid filters.

Solution

1 Diatomite system

1.1 One tank: The footprint for one upright DE filter is

$$A(\text{tank}) = \pi d(\text{tank})^2/4$$
$$= \pi 2.032^2/4$$
$$= 3.243 \text{ m}^2$$

1.2 A(cluster): Let the tanks be spaced with 1.000 m between tanks, to give 5.000 m^2 for a cluster, that is, $A(\text{cluster}) \approx 25.0 \text{ m}^2$.

1.3 Total area: Let the spacing between clusters be 2.000 m. If the clusters are arranged 4 × 3, the layout is then 14 m × 10 m, that is, $A(\text{all tanks}) = 140 \text{ m}^2$.

1.4 Capacity: The total flow for the full-scale plant is

$$Q(\text{plant}) = A(\text{septum area/tank}) \cdot N(\text{tanks}) \cdot \text{HLR}$$
$$= 140 \text{ m}^2/\text{tank} \cdot 48 \text{ tanks} \cdot 6.1 \text{ m/h}$$
$$= 40{,}992 \text{ m}^3/\text{h}$$
$$= 983{,}800 \text{ m}^3/\text{day (260 mgd)}$$

2 Rapid filtration

2.1 HLR: Assume that HLR = 12.2 m/h (5 gpm/ft^2)

2.2 Plant flow: Q(plant) = 983,800 m^3/day (260 mgd)

2.3 A(filters)

$$A(\text{filters}) = Q(\text{plant})/\text{HLR}$$
$$= (983{,}800 \text{ m}^3/\text{day})/(24 \text{ h/day})/12.2 \text{ m/h}$$
$$= 336 \text{ m}^2$$

Discussion

Comparing areas,

$$A(\text{diatomite tank collection}) = 140 \text{ m}^3$$
$$A(\text{rapid filters}) = 336 \text{ m}^3$$

The area of the diatomite tank collection does not include the pre-coat and body-feed tanks, pumps, piping, diatomite storage, waste storage, etc. The area for the rapid filters is for the filtration bed area only; gullets, pipe gallery, chemical storage, etc., are not included. The diatomite filtration system would require less area for appurtenances than would the rapid filtration system.

14.1.4.3 Research

Just after the initial research for drinking-water applications in 1943, the U.S. Army Engineering Research and Development Laboratories (ERDL) at Fort Belvoir, Virginia, supported contracts to universities and equipment manufacturers to develop further the diatomite filtration process. Then, in 1979, the Environmental Protection Agency, the legislative mandate to develop enforceable standards for drinking-water quality, reopened the diatomite filtration research.

14.1.4.3.1 Process Research 1950–1970s

Academic research started in 1946 (Baumann, 1957) at the University of Illinois under the direction of Harold E. Babbitt and E. Robert Baumann with funding from ERDL. Based on their report titles (Baumann et al., 1965), the research included such topics as filter septa, pre-coat, hydraulics, hydraulic loading rates, compression and effects on permeability, removals of *E. histolytica*, design and operation factors, effect of body feed, effect of prior coagulation on filterability, and economics. This research laid the foundation for municipal water treatment practice that was developing at the same time; it was the basis for the 1954 doctoral thesis of Baumann at Illinois. In 1953 Baumann was appointed associate professor at Iowa State University (where he remained until 1991 when he retired) and continued research in diatomite filtration with colleagues and graduate students on topics such as vacuum systems, theory of diatomite filtration, and economic optimization of design, and design criteria (Box 14.3).

14.1.4.3.2 Research for EPA Regulatory Support

Just as with slow sand, diatomite had a resurgence in interest about 1979 due to the fact that giardiasis had become recognized as a major waterborne disease. Dr. Gary Logsdon, EPA, Cincinnati, was familiar with the 1943 publications concerning the U.S. Army research on the filtration of *E. histolytica* cysts and the adoption of diatomite filtration by the military. Based on the promise of diatomite filtration for the removal of *Giardia* cysts, some initial research was done at the EPA labs in Cincinnati along with work at the University of Washington, followed by research sponsored by EPA at Colorado State University (CSU), facilitated by personnel at Johns-Manville, Denver (Ray McIndoe and AlanWirsig). The CSU research established that all grades of DE removed *G. lamblia* cysts and that turbidity removal and bacteria removals depended on grade, and that alum or polymer additions to DE body feed resulted in high removals for the latter, even for coarse grades of DE (Lange, 1983; Lange et al., 1986).

BOX 14.3 ROBERT BAUMANN

Professor E. Robert Baumann (b. 1921) is listed in *Who's Who in America*, 56th edition (2002) as "environmental engineering educator." Regarding diatomite filtration, he has had a sustained role in the field starting in 1946 at the University of Illinois and continuing until about 1979 when he wrote a definitive chapter (Baumann 1979) on the topic in *Water Treatment Plant Design* (Sanks, 1979). His work has delineated most of the guidelines that are established that have made possible the technology as we know it today. He moved to Iowa State University in 1953 after completing his doctorate at Illinois, retiring as Distinguished Professor Emeritus in 1991, becoming a full-time consultant. His work in diatomite filtration has probably been eclipsed by accomplishments in research, teaching, writing textbooks, consulting, and in leadership positions in several professional societies. Numerous awards have attested to his prominence in both water treatment and wastewater treatment, and teaching. His photograph is c. 1960.

14.1.4.3.3 Further Research

The CSU research was followed by a microscopic analysis of removal mechanisms by Harris Walton (1988) of Johns-Manville; further work was done on alum and cationic additives by Schuler and Ghosh (1990). Work on removals of *C. parvum* oocysts was sponsored by IDPA which found log R(oocysts) $\gg 4$ (Ongerth and Hutton, 1997, 2001).

14.2 CAKE FILTRATION PROCESS

In DE filtration, it is axiomatic that the grade selected must be smaller in pore size than the size range of the particles to be removed. On the other hand, a grade that is fine enough to remove particles in all categories of interest, especially fine turbidity, viruses, and bacteria will, more than likely, cause a higher-than-desired rate of headloss increase. Thus, there is a trade-off between low effluent particle concentrations and a higher rate-of-headloss increase (McIndoe, 1969a, p. 52).

14.2.1 Particle Removal Effectiveness

Suspended particles that may be removed by diatomite filtration include iron and manganese oxides, residual solids from lime-soda softening, algae, bacteria, cysts and oocysts, and amorphous organic matter (McIndoe, 1969b, p. 49). The lower limit of particle size that may be removed by the finest grade of diatomite is perhaps 0.1–1 μm (McIndoe, 1969b, p. 50). The use of alum or polymer-coated diatomite adds to the removal effectiveness, for example, mineral colloids, viruses, bacteria, but results in a higher rate of headloss. Alternatively, for episodes of higher turbidity, a finer grade of DE may be kept on hand (Logsdon, 2008, p. 245).

14.2.1.1 Turbidity and Bacteria

Figure 14.7 shows removals of turbidity, standard plate count (SPC) bacteria, and coliforms, plotted as a function median particle size (showing also the corresponding grade of diatomite). The plots were from experimental work by Lange et al. (1984) with data generated by seeding a 0.093 m^2 (1.0 ft^2) septum area diatomite pilot plant with HLR = 2.44 m/h (1.0 gpm/ft^2). As seen, removals were about 3-log for the

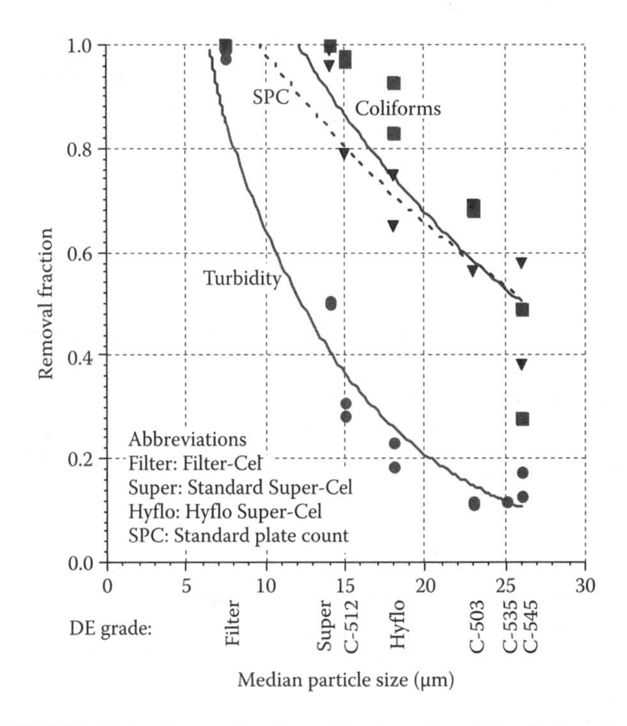

FIGURE 14.7 Removals of particles as affected by DE grade. (Adapted from Lange, K.P. et al., Removal of *Gairdia Lamblia* cysts and other substances, *Diatomaceous Earth Filtration*, Vol. 1, Municipal Environmental Research Laboratory, United State Environmental Protection Agency, EPA-600/S2-84-114 (available from NTIS Order No. 84-212 703), September 1984, pp. 66–67.)

finest grade of diatomite, that is, Filter-Cel®, but then were only 1-log for the coarsest grade, that is, Celite 545®.

In another case, that is, another raw water, using both Hyflo Super-Cel and Celatom FW-12 (which are comparable grades as seen in Table 14.1), turbidity levels were reduced from about 0.9 NTU in the raw water to <0.10 NTU in the filter effluent for HLR = 2.44 m/h (1 gpm/ft^2), body feed = 5 mg/L, and pre-coat = 1.0 kg/m^2 (0.20 lb/ft^2) (Ongerth and Hutton, 2001).

14.2.1.2 Particle Counts

Using Celite Hyflo Super-Cel® and Celatom FW-12®, particle counts were reduced from about 4100–4500 #/mL in the raw water to 105–250 #/mL in the filter effluent for HLR = 2.44 m/h (1 gpm/ft^2), body feed = 5 mg/L, and pre-coat = 1.0 kg/m^2 (0.20 lb/ft^2) (Ongerth and Hutton, 2001). As with turbidity, particle count removals are dependent upon the particular water being filtered.

14.2.1.3 Iron and Manganese

Representative concentrations of soluble iron and manganese in ambient waters are perhaps 4–8 mg/L Fe^{2+} and 0.3–1.5 mg/L Mn^{2+} (Coogan, 1962, p. 1509). The oxide forms, for example, Fe_2O_3 and MnO_2, are particulate, which may be removed by DE filtration. Iron may be reduced to ≤0.2 mg/L (Coogan, 1962, p. 1514); data are not available for MnO_2.

14.2.1.4 Asbestiform Fibers

Two kinds of asbestiform fibers are found in Lake Superior near Diluth at levels up to 17 million fibers/L for total fibers. Amphibole fibers were about 10^4–10^5 fibers/L and chrysotile fibers numbered 10^5 fibers/L. Removals by DE were 100% and perhaps 70%–80%, respectively, achieved using either C-512 or Hyflo Super-Cel with alum coating (Baumann, 1975, p. 12).

14.2.1.5 Biological Particles

Removals of biological particles depend on both their size and the grade of diatomite. Removals of cysts, oocyst, bacteria, and viruses have been determined in various investigations since about 1972 (with the first being the assessment of removals of *E. histolytica* cysts by the U.S. Army in 1944 (see Section 14.1.5.1).

14.2.1.5.1 *Giardia* Cysts

Figure 14.8 is an electron photomicrograph of cysts of *G. lamblia*; as is well known, the cysts are about 5 μm wide by 10 μm long, confirmed in the photograph. Also, the cysts are remarkably consistent in size and shape. Based on numerous tests, seeding a 0.09 m^2 (1.0 ft^2) septum area pilot plant generally with $\geq 1 \cdot 10^6$ cysts, none were detected in the effluent; based on calculated detection limits, log $R \geq 3$ even for the coarsest grade of diatomite, that is C-545®, and for 2.44 ≤ HLR ≤ 9.76 m/h (1.0 ≤ HLR ≤ 4.0 gpm/ft^2) (Lange et al., 1984, 1986).

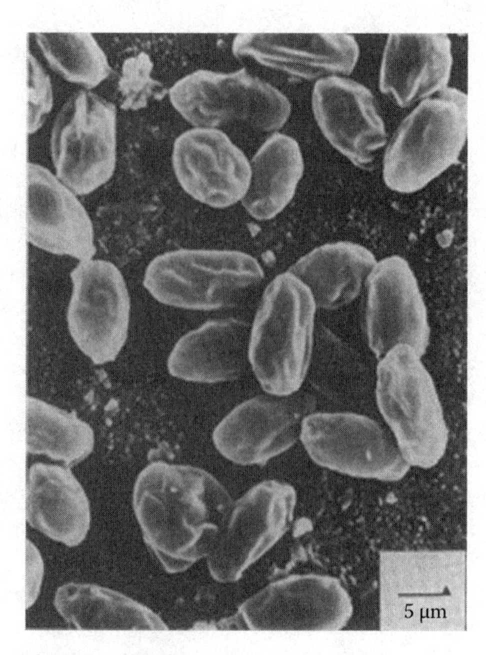

FIGURE 14.8 Cysts of *G. lamblia*. (From Walton, H.G., Diatomite filtration: Why it removes *Giardia* from water, in Wallis, P.M., and Hammond, B.R. (Eds.), in *Advances in Giardia Research*, University of Calgary Press, Calgary, Alberta, Canada, 1988. With permission; Courtesy of Celite® Corp., Lompoc, CA.)

14.2.1.5.2 *Cryptosporidum oocysts*

In a study of removals of *Cryptosporidium* oocyts using Walton test filters (a bench-scale flow-through apparatus with a septum area of 15 cm^2), Ongerth and Hutton (1997) found

Celite 512®	Celatom FW-6®	$6.03 \leq \log R(Cryptosporidium) \leq 6.53$
Celite Hyflo Super-Cel®	Celatom FW-12®	$5.79 \leq \log R(Cryptosporidium) \leq 6.12$
Celite 535®	Celatom FW-50®	$3.60 \leq \log R(Cryptosporidium) \leq 4.51$

By comparison, for rapid filtration, removals are on the order of $3.0 < \log R(Cryptosporidium) < 4.0$.

14.2.1.5.3 Bacteria

Removals of bacteria (*E. coli*) were dependent on grade and whether alum or polymer was added (Schular and Ghosh, 1990, p. 71). Removals were Celite-512® = 3-log; Celite-503® = <1-log; Celite-545® = ≤1-log. The addition of alum increased removals for all grades >1-log; addition of anionic polymer increased removals to near 2-log.

14.2.1.5.4 Viruses

Virus removal by DE is effective only with alum (or ferric ion) or a polymer (Malina et al., 1972). T-2 bacteriophage was removed 100% by 2% alum coating of Hyflo B. Then, a virulent strain, Mahoney Type I polio virus, was found to be removed 100% by Hyflo, based on pretreatment with 0.14 mg/L Dow C-31 polymer or with 0.5 mg C-31 per gram of Hyflo. Operating conditions were pre-coat

0.73 kg/m^2 septum (0.15 lb/ft^2), body feed 50 mg/L, and HLR 2.44 m/h (1.0 gpm/ft^2).

14.2.2 Removal Mechanisms

Straining is the primary removal mechanism in diatomite filtration; the definition of "straining" is expanded (see Glossary), however, to include a particle being "embedded" within the media. If alum or a polymer are added to coat the media, then the particles to be removed may also attach to the media, that is, "adsorb."

14.2.2.1 Straining and Embedding

Figure 14.9 is a scanning electron photomicrograph (SEM) at 2000× showing *Giardia* cysts embedded in C-545® as deposited on a septum (obtained by Harris Walton, 1988, then at Manville Corp., Denver, Colorado). The cysts are distinguished from the media by their oblong form. The cysts are embedded in the media not able to progress further. To be more specific, the hydraulic shear force on any given cyst would be resisted by an equal and opposite force of the diatomite particles on the cyst. If, however, the hydraulic shear forces cause movement of the cyst, progress will be limited since some fraction of the pores will be smaller than the cysts. Embedding might be considered as a subset of straining, that is, the particles strained are larger than the openings.

14.2.2.2 The Role of Body Feed

The idea of body feed is to prevent the particles to be removed from accumulating within the pores and causing blockage.

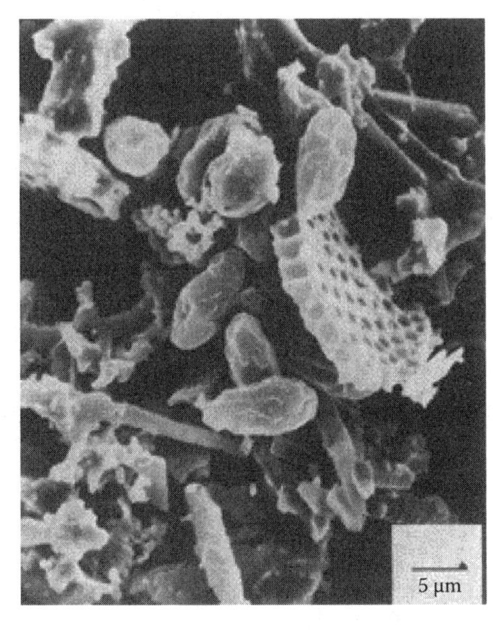

FIGURE 14.9 C-545 diatomite with *Giardia* cysts interspersed. (From Walton, H.G., Diatomite filtration: Why it removes *Giardia* from water, in Wallis, P. M., and Hammond, B. R. (Eds.), *Advances in Giardia Research*, University of Calgary Press, Calgary, Alberta, Canada, 1988, Figure 8. With permission.)

Insufficient body feed causes an exponential increase in the rate of headloss increase. If the body feed is sufficient, then enough unblocked pores are provided such that the rate of headloss is proportional to the thickness of the cake, that is, the rate of headloss increase is linear. The particles to be removed are embedded within the cake and, as noted, are prevented from movement unless the particles to be removed are smaller than the smallest fraction of pore sizes. Viruses (10–400 nm), for example, are small enough to move within the C-545 cake, but not *Giardia* cysts (10 μm). Also, bacteria (1 μm) are small enough to move within the pore structure of C-545, but not within the pores of the smallest grades of DE, for example, Filter-Cel™, which has an average pore size of about 1.5 μm (Logsdon and Lippy, 1982, p. 655).

14.2.2.3 Adsorption

The zeta potentials of the DE particles were measured in distilled water as $-47 \leq \zeta \leq -24$ mV (Oulman and Baumann, 1964, p. 920). Adding a trivalent ion, that is, Al^{3+} or Fe^{3+}, will cause a coating on the DE particle surface, with consequent charge reversal, that is, a positive zeta potential in the range of $+22 \leq \zeta \leq +30$ mV using about 0.05 g Al^{3+}/g DE (similar data were obtained by Schuler and Ghosh, 1990, p. 68). A cationic polymer coating will result in $\zeta \approx +40$ mV. The coating may be by mixing the alum or polymer with the pre-coat and the body feed prior to application. A zeta potential, $\zeta \approx +60$ mV, was achieved with C(Purifloc™601) ≈ 0.001 g Purifloc™601/g C-545, which enabled the adsorption of clay particles in accordance with a Langmuir isotherm relation (Baumann and Oulman, 1970b, p. 689). In other words, the DE media, having negative charge, is likely to repel negatively charged particles. By charge neutralizing the media, however, negatively charged particles, for example. *Giardia* cysts, may be attracted.

14.2.2.4 Comparisons between Filtration Processes

At this point it is useful to review the three main filtration processes and their respective filtration mechanisms: (1) depth filtration works by the particles to be removed penetrating the filter media and "attaching" to the grains of media; (2) slow sand filtration works by the straining of the particles to be removed by previously deposited matter that has been retained on the surface of the media bed and by depth filtration for those particles that penetrate the media bed attaching to biofilms on sand grains; and (3) cake filtration works by straining/embedment *within* the media cake, that is, the particles to be removed are enmeshed within the matrix of media grains and blocked from movement by those grains.

14.2.3 Hydraulics

Flow of water through the filter cake and associated pumps, pipes, and tanks that comprise a diatomite filtration system is hydraulic in nature. The sections that follow review the hydraulic principles relevant to diatomite filtration.

14.2.3.1 Hydraulics of Cake Filtration

The water flow in filtration follows Poiseuille's law, which rearranged results in Darcy's law (Hoffing and Lockhart, 1951). The adaptation to diatomite filtration was taken from the work of Baumann et al., (1962).

14.2.3.1.1 Darcy's Law

Flow of water through a porous medium cake is in accordance with Darcy's law (Equation E.2), that is,

$$v = -k\frac{\rho_w g}{\mu}\frac{dh}{dz} \tag{14.1}$$

where

v is the filtration velocity (the same as hydraulic loading rate, Q/A) (m/s)
k is the intrinsic permeability of medium (m^2)
ρ_w is the mass density of water (kg/m^3)
g is the acceleration of gravity (9.806 650 m/s^2)
μ is the viscosity of water at a given temperature (N-s/m^2)
h is the hydraulic head, that is, $p/\gamma + z$ (m)
z is the coordinate in direction of velocity vector (m)

Applying Equation 14.1 to the pre-coat and the filter cake, Baumann et al. (1962, p. 1114), developed the equations,

$$v = -k(\text{pre-coat})\frac{\rho_w g}{\mu}\frac{h_L(\text{pre-coat})}{\Delta z(\text{pre-coat})} \tag{14.2}$$

$$v = -k(\text{cake})\frac{\rho_w g}{\mu}\frac{h_L(\text{cake})}{\Delta z(\text{cake})} \tag{14.3}$$

where

k(pre-coat) is the intrinsic permeability of pre-coat (m^2)
k(cake) is the intrinsic permeability of body-feed cake (m^2)
h_L(pre-coat) is the headloss across pre-coat (m)
h_L(cake) is the headloss across body-feed cake (m)
Δz(pre-coat) is the thickness of pre-coat (m)
Δz(cake) is the thickness of body-feed cake (m)

Rearranging each equation in terms of the h_L term and then adding $h_L = h_L(\text{pre-coat}) + h_L(\text{cake})$,

$$h_L = v\frac{\mu}{\rho_w g}\left[\frac{\Delta z(\text{pre-coat})}{k(\text{pre-coat})} + \frac{\Delta z(\text{cake})}{k(\text{cake})}\right] \tag{14.4}$$

Now, Δz's for the pre-coat or the body-feed cake are, respectively,

$$\Delta z(\text{pre-coat}) = \frac{W(\text{pre-coat})}{\rho(\text{pre-coat})} \tag{14.5}$$

$$\Delta z(\text{cake}) = \frac{W(\text{cake})}{\rho(\text{cake})} \tag{14.6}$$

where

W(pre-coat) is the mass loading application of pre-coat per unit area of septum (kg pre-coat/m^2 septum)
W(cake) is the mass loading application of body feed per unit area of septum (kg body feed/m^2 septum)
ρ(pre-coat) is the bulk density of pre-coat (kg pre-coat/m^3 pre-coat)
ρ(cake) is the bulk density of body-feed cake (kg body-feed cake/m^3 body-feed cake)

As a last relation, the mass loading of body feed per unit area of septum application is

$$W(\text{cake}) = C(\text{body feed}) \cdot v \cdot t \tag{14.7}$$

where

C(body feed) is the concentration of body feed in raw-water flow (kg body feed/m^2 water)
t is the elapsed time since start of body feed (s)

Substituting Equation 14.7 in Equation 14.6 and then inserting Equations 14.5 and 14.6 in Equation 14.4 gives

$$h_L = v\frac{\mu}{\rho_w g}\left[\frac{W(\text{pre-coat})}{k(\text{pre-coat}) \cdot \rho(\text{pre-coat})} + \frac{C(\text{body feed}) \cdot v \cdot t}{k(\text{cake}) \cdot \rho(\text{cake})}\right] \tag{14.8}$$

which is Equation 14.12 as given by Baumann et al. (1962, p. 1114) and verified in over 500 experiments.

Equation 14.8 yields the following conclusions:

- The pre-coat headloss is proportional to the pre-coat diatomite load on the septum and the filtration velocity.
- The headloss caused by the body feed is proportional to the body-feed concentration and the elapsed time since the start of the body feed, and the square of the filtration velocity.

14.2.3.1.2 Calculations

Table CD14.2 is a spreadsheet for the calculation of headloss versus time in accordance with Equation 14.8; the final columns are assumed time, t, and the resulting headloss, h_L, respectively. Figure CD14.10 is a plot of headloss versus time calculations from the spreadsheet for five grades of diatomite; also, the figure is embedded within the spreadsheet. Figure CD14.10 shows that the rate of headloss increase varies markedly with the grade of DE. A "nominal" limit to the headloss is shown as the horizontal line at 300 kPa (44 psi).

Example 14.2 Calculation of Headloss by Equation 14.8

Given

Assume Hyflo Super-Cel® is being considered for the treatment of a reservoir water. The pertinent data are

TABLE CD14.2

Pressure Loss as a Function of Hydraulic Variables as Calculated by Equation 14.8

		Coefficients for Temperature Dependence					
Variable	**Units**	***M0***	***M1***	***M2***	***M3***	***M4***	***M5***
μ(water)	$(N \cdot s/m^2)$	0.001787	−5.61E−05	1.00E−06	−7.54E−09		
ρ(water)	(kg/m^3)	999.84	0.068256	−0.0091438	0.00010295	−1.188E−06	7.15E−09
G	(m/s^2)	9.806 650					

Formula: $y = M0 + M1 \cdot T + M2 \cdot T^2 + M3 \cdot T^3 + M4 \cdot T^4 + M5 \cdot T^5 + M6 \cdot T^6$

TABLE CD14-*

Headloss Calculations as Function of Hydraulic Variables (Calculated by Equation 14-*)

	DE Properties					Water				Conditions						
				k(cake)						HLR			*C*(body feed)		*W*(pre-coat)	
Grade of DE	**ρ(Dry Density) (kg/m^3)**	**ρ (pre-coat) (kg/m^3)**	**ρ(body feed) (kg/m^3)**	**(Darcys)**	**(m^2)**	**T (°C)**	**ρ_w (kg/m^3)**	**μ $(N\text{-}s/m^2)$**	**$\mu/\rho_w(g)$**	**(gpm/ft^2)**	**(m/h)**	**(m/s)**	**(mg/L)**	**(kg/m^3)**	**(lb/ft^2)**	**(kg/m^2)**
Hyflo Super-Cel	144	144	144	1.10	1.08570E−12	10	1.00E+03	0.00132	1.32E−07	1.50	3.66	0.001408	25	0.025	0.15	0.73
Reference data																
Filter Cel	112	112	112	0.07	6.90900E−14											
Celite 577	128	128	128	0.16	1.57920E−13											
Celite 505	128	128	128	0.06	5.92200E−14											
Standard Super-Cel	128	128	128	0.25	2.46750E−13											
Celite 512	128	128	128	0.50	4.93500E−13											
Hyflo Super-Cel	144	144	144	1.10	1.08570E−12											
Celite 501	152	152	152	1.3	1.28310E−12											
Celite 503	152	152	152	1.9	1.87530E−12											
Celite 535	192	192	192	3.0	2.96100E−12											
Celite 545	192	192	192	4.0	3.94800E−12											
Celite 550	228	228	228	7.4	7.30380E−12											
Celite 560	312	312	312	25.0	2.46750E−11											

$$h_L = v \frac{\mu}{\rho_w g} \left[\frac{W(\text{pre-coat})}{k(\text{pre-coat}) \cdot \rho(\text{pre-coat})} + \frac{C(\text{body feed}) \cdot v \cdot t}{k(\text{cake}) \cdot \rho(\text{cake})} \right]$$

(continued)

TABLE CD14-* (continued)
Headloss Calculations as Function of Hydraulic Variables (Calculated by Equation 14-*)

Elapsed Time Intermediate Calculations								
t			*W*(body feed)			ΔP		
(min)	(h)	(day)	(kg/m^2)	First Group	Second Group	(m Water)	(kPa)	(psi)
0	0.00	0.00	0.0000	4.694E+09	0.0000000	0.872	8.5	1.24
800	13.33	0.56	1.2200	4.694E+09	7.803E+09	2.320	22.8	3.30
1,600	26.67	1.11	2.4400	4.694E+09	1.561E+10	3.769	37.0	5.36
2,400	40.00	1.67	3.6600	4.694E+09	2.341E+10	5.218	51.2	7.42
3,200	53.33	2.22	4.8800	4.694E+09	3.121E+10	6.666	65.4	9.48
4,000	66.67	2.78	6.1000	4.694E+09	3.902E+10	8.115	79.6	11.55
4,800	80.00	3.33	7.3200	4.694E+09	4.682E+10	9.564	93.8	13.61
5,600	93.33	3.89	8.5400	4.694E+09	5.462E+10	11.013	108.0	15.67
6,400	106.67	4.44	9.7600	4.694E+09	6.243E+10	12.461	122.2	17.73
7,200	120.00	5.00	10.9800	4.694E+09	7.023E+10	13.910	136.4	19.79
8,000	133.33	5.56	12.2000	4.694E+09	7.803E+10	15.359	150.6	21.85
8,800	146.67	6.11	13.4200	4.694E+09	8.584E+10	16.808	164.8	23.91
9,600	160.00	6.67	14.6400	4.694E+09	9.364E+10	18.256	179.0	25.97
10,400	173.33	7.22	15.8600	4.694E+09	1.014E+11	19.705	193.2	28.03
11,200	186.67	7.78	17.0800	4.694E+09	1.092E+11	21.154	207.4	30.09
12,000	200.00	8.33	18.3000	4.694E+09	1.171E+11	22.603	221.6	32.16
12,800	213.33	8.89	19.5200	4.694E+09	1.249E+11	24.051	235.9	34.22
13,600	226.67	9.44	20.7400	4.694E+09	1.327E+11	25.500	250.1	36.28
14,400	240.00	10.00	21.9600	4.694E+09	1.405E+11	26.949	264.3	38.34

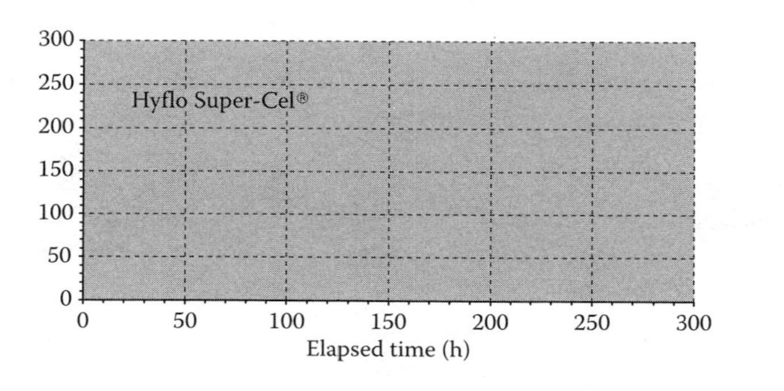

Figure CD14-* Headloss versus time for precoat plus body feed for Hyflo-Super-Cel®.

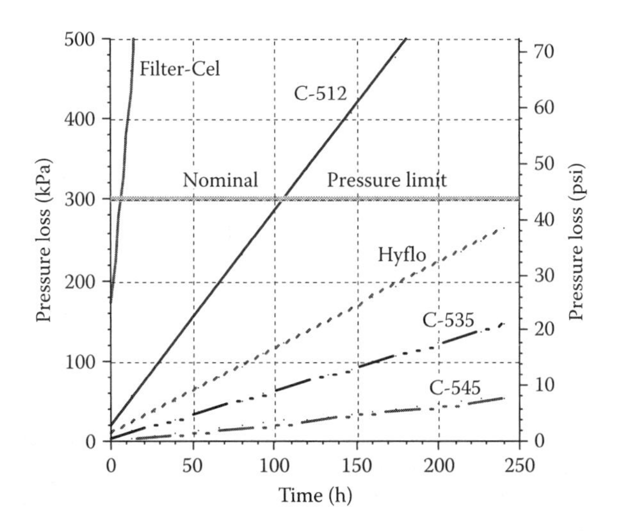

FIGURE CD14.10 Headloss versus time plots for five grades of diatomite as calculated by Table CD14.2.

$T = 11°C$,
Turbidity = 4 NTU
HLR = 3.66 m/h (1.5 gpm/ft^2)
$\rho_w = 998$ kg/m^3
$g = 9.81$ m/s^2
ρ(pre-coat) = 144 kg/m^3
ρ(cake) = 144 kg/m^3
C(body feed) = 25 mg/L

Required

(a) Calculate, from Equation 14.8, a plot of h_L versus t.
(b) Show the effect of doubling the HLR to HLR = 7.32 m/h.
(c) Describe the effect of increasing C(body feed).

Solution

The steps for solution are enumerated below.

1. Equation: The calculation of headloss is given by the equation from Baumann et al. (1962):

$$h_L = v\frac{\mu}{\rho_w g}\left[\frac{W(\text{pre-coat})}{k(\text{pre-coat})\cdot\rho(\text{pre-coat})} + \frac{C(\text{body feed})\cdot v\cdot t}{k(\text{cake})\cdot\rho(\text{cake})}\right] \quad (14.8)$$

The mass loading of body feed per unit area of septum, W(cake), is

$$W(\text{cake}) = C(\text{body feed})\cdot v\cdot t \quad (14.7)$$

2. Spreadsheet: Table CD14.2 is set up to calculate headloss in accordance with Equation 14.8. For the spreadsheet, let ρ(cake) = ρ(pre-coat) = dry density. For example from Table 14.1, ρ(Hyflo dry density) = 144 kg/m^3.

Discussion

The spreadsheet, Table CD14.2, may be used to explore the effect of any of the independent variables on the headloss versus time outcome. The independent variables include: Grade of DE, HLR, temperature, and W(pre-coat).

14.3 DESIGN

The first task of design is to select a particular diatomite technology, for example, pressure or vacuum, septum type, orientation, etc. The sizing of equipment is next, including specification of appurtenances, for example, tanks, pumps, pipes, etc., and the control system.

14.3.1 Diatomite Technologies

A variety of equipment types is available from manufacturers that relate to septum housing, septum type, and the method of cleaning. To illustrate the kind of filter housing/septum assembly available, Figure 14.11 shows a pressure filter with retractable housing, which permits a dry-cake discharge. The same catalog source has a wide variety of tank–septum

FIGURE 14.11 Pressure tank/septum assembly. (Courtesy of Durco Filters by Ascension Industries, North Tonawanda, NY.)

assemblies, for example, wet-cake discharge, vertical oriented tank-vertical septum leaves, vertical tank-horizontal leaves, as well as different kinds of septum materials and construction. The sizes of single tanks may be in the range of $914 \leq d(\text{tank}) \leq 1829$ mm (36–72 in.) with $6 \leq N(\text{leaves}) \leq 34$ (Durco Filters by Ascension Industries, 2009). A system may be purchased as a "package" from a manufacturer or the tank–septum system is purchased with components designed and obtained separately. For smaller systems, a skid-mounted "package" system may be more practical and for larger systems, individual design may be more appropriate.

14.3.1.1 Equipment

As noted previously, a variety of septum shapes have been provided by manufacturers along with different kinds of tanks and approaches to septum cleaning. A system design always involves ≥ 2 tanks, each with its own set of components. In other words, the system design is modular, that is, the same design is repeated for each tank. This means also that there is no theoretical limit to the size of an installation.

14.3.1.1.1 Pressure Filters

A pressure filter is characterized by positive pressures in the tank, such as shown in Figure 14.11. Figure 14.12 illustrates the pressure changes through the system in terms of the hydraulic grade line (HGL). As seen, the HGL is sloped down (i.e., negative slope) at the beginning to show friction headloss in the influent pipe. At the pump, the pipe and, consequently, the tank, is pressurized. On the effluent side, a tailwater overflow weir depicts the terminal head. The sharp drop in the HGL on the right side of the tank is the headloss across the filter tank (more specifically across the filter cake). Alternatively, if a reservoir is located on the influent side, the pump may be omitted, provided the reservoir has enough head, for example, 25–30 m (80–100 ft).

14.3.1.1.2 Vacuum Filters

A vacuum filter may have the same tank–septum configuration as the pressure filter except that the tank is open to the atmosphere and the pump is on the effluent side of the filters, creating a negative pressure within the septum.

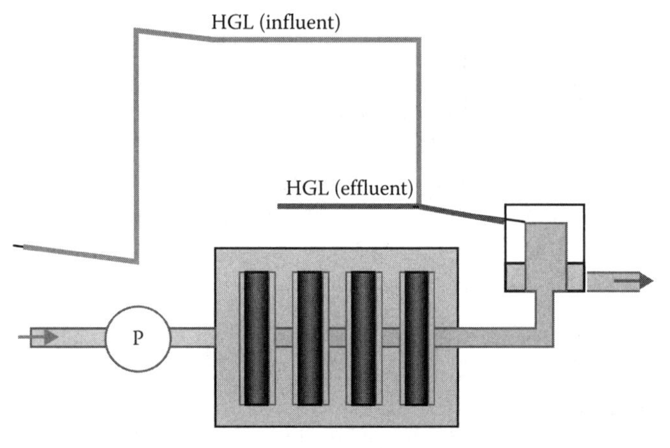

FIGURE 14.12 HGLS for DE pressure filter.

14.3.1.1.3 Septum

The most common septum shape is the vertical leaf. As seen in Figure 14.11, the vertical leaf filter is a disk with septum on both sides with vertical orientation. To permit the buildup of cake and to provide for the uniform circulation of diatomite slurry, without cake erosion, the discs should be spaced sufficiently far apart.

The septa have been made of a variety of materials, for example, stainless steel wire cloth, synthetic fabrics, and porous stone. The stainless steel wire cloth fits most of the criteria for a suitable septum material. Required openings in one direction are ≤ 0.13 mm (0.005 in.) (Bell, 1962, p. 1245). A sieve size of U.S. Standard 60, however, has an opening of 0.25 mm (a Tyler 60 sieve has the same opening), which was used at Vacaville. Generally, the septum openings are larger than the coarsest grades of DE. Bridging over the openings occurs as the pre-coat is circulated; once the bridging starts, the pre-coat is formed.

While the purpose of the septum is to retain pre-coat, it requires support by a structure that can withstand the pressure differential, for example, 207–345 kPa (30–50 psi) without significant flex. The septum material should also have negligible hydraulic resistance, which also provides for backflow hydraulic flushing. The design of the septum assembly also requires provision for easy drainage.

14.3.1.1.4 Sizes

Tank–septum assemblies may be oriented with axis horizontal or vertical and so their "footprint" area may vary with orientation. But to give an indication of size, data are excerpted from a Durco catalog (Durco Filters by Ascension Industries, North Tonawanda, NY, 2009) given in Table 14.3. Tank lengths are not given but may be calculated from the leaf spacing, which is center to center. The length of the 914 mm tank with 10 leaves (third row) and 102 mm spacing is $L(914\text{ mm, }10\text{ leaves}) = 102\text{ mm/leaf} \cdot 10\text{ leaves} + 180\text{ mm} = 1200\text{ mm}$ (47 in.). The length of the 1829 mm tank with 34 leaves (last row) and 76 mm spacing is $L(1829\text{ mm, }34\text{ leaves}) = 76\text{ mm/leaf} \cdot 34\text{ leaves} + 216\text{ mm} = 2800\text{ mm}$ (110 in.). [The Durco catalog has a variety of forms of tank–septum systems, e.g., horizontal tank with wet discharge, vertical oriented tanks with vertical leaves, etc.]

14.3.1.1.5 Filter Cleaning

There are several methods of removing the cake from the filter septa. The most common are (McIndoe, 1969a, p. 53) (1) dry discharge, in which the filter leaves are pulled out of the tank and the cake may be removed as a mud by mechanical scrapping or as a slurry by manual sluicing; (2) wet discharge with jet sluicing, which causes a peeling of the cake from the septum leaves, which is then flushed out of the drain; and (3) wet discharge with reversible flow backwash, in which the spent cake can be dislodged and flushed from the tank.

TABLE 14.3
Tank-Septum Data for Dry Cake Discharge (Durco[a] Catalog, 2009)

Septum Area		Tank Dia.		Leaves	Leaf Spacing[b]		Tank Volume		Weight	
(m^2)	(ft^2)	(mm)	(in)	(#)	(mm)	(in)	(m^3)	(gal)	(kg)	(lb)
4.35	46.8	914	36	6	102	4	0.79	210	1021	2250
5.79	62.4	914	36	8	75	3	0.79	210	1066	2350
7.24	78.0	914	36	10	102	4	1.05	278	1134	2500
9.42	101.4	914	36	13	76	3	1.05	278	1202	2650
11.14	120	1219	48	8	102	4	1.74	461	1860	4100
15.33	165	1219	48	11	76	3	1.74	461	1942	4280
18.11	195	1219	48	13	102	4	2.21	584	1987	4380
20.90	225	1219	48	15	76	3	2.21	584	2064	4550
28.20	303.6	1524	60	12	102	4	3.44	909	2130	4645
32.90	354.2	1524	60	14	76	3	3.44	909	2155	4750
37.60	404.8	1524	60	16	102	4	4.16	1100	2220	4895
47.00	506	1524	60	20	76	3	4.16	1100	2291	5050
56.93	612.8	1829	72	16	102	4	5.68	1500	2405	5300
78.27	842.6	1829	72	22	76	3	5.87	1550	2643	5825
92.50	995.8	1829	72	26	76	3	6.43	1700	2813	6200
120.97	1302.2	1829	72	34	76	3	7.95	2100	3176	7000

[a] Durco Filters by Ascension Industries, North Tonawanda, NY, 2009.
[b] Spacing is distance center-to-center between leaves. The cake thickness is assumed to be 38–95 mm leaf spacing.

14.3.1.2 System Components

Those components of a diatomite filtration system that make it work as a technology include the filter housing, septum assembly, tanks for pre-coat and body feed, metering pumps, flow meters, pressure gages, pipes, valves, control system software, etc.

14.3.1.2.1 Tanks for Pre-Coat and DE Body Feed

The volume sizing for both pre-coat tank and the body-feed tank is $V(\text{tank}) \cdot C(\text{slurry}) = \text{Mass DE}$. The pre-coat tank is sized merely to circulate a given mass charge at the highest concentration that can be circulated while maintaining a uniform suspension of DE. The body-feed tank is sized based on both the concentration of the slurry that is feasible and a desired duration for the tank to provide a body feed, which depends upon *C*(body feed) to the filters. The relation is, as a mathematical expression,

$$V(\text{tank})_{BF} \cdot C(\text{slurry})_{BF} = \text{Mass DE}_{BF} = C(\text{body feed}) \cdot Q(\text{septum tank}) \cdot t(\text{tank}) \quad (14.9)$$

where

$V(\text{tank})_{BF}$ is the volume of body-feed tank (m^3)
$C(\text{slurry})_{BF}$ is the concentration of diatomite slurry in body-feed tank (kg/m^3)
Mass DE_{BF} is the mass charge of diatomite to body-feed tank (kg)
C(body feed) is the concentration of body feed to septum assembly (kg/m^3)
Q(septum tank) is the flow of water septum assembly in a given filter housing (m^3/s)
t(tank) is the time duration for volume of slurry in body-feed tank to last (s)

Calculation of tank size, $V(\text{tank})_{BF}$, depends upon the magnitudes of the variables in Equation 14.9. Generally, operation will be served best if $t(\text{tank}) \geq 48$; since *C*(body feed) and *Q*(septum tank) are established prior to this stage of design, then Mass DE_{BF}, that is, the mass charge to the tank, is directly proportional to *t*(tank).

The slurry concentration is not so low that the tank volume is excessive, nor so high that the slurry is difficult to maintain as a uniform suspension. Some examples of tank concentrations are provided in Table 14.4, with data and calculated concentrations given in tabular form.

As seen, the pre-coat slurry concentrations are about an order of magnitude higher than the body-feed slurry concentrations, that is, about 120–240 kg/m^3 for the former and 10–24 kg/m^3 for the latter. In each case, the slurries should be agitated to maintain them in suspension; mixers with rotational speed 40–60 rpm are recommended as high-speed mixers may degrade the media particles (McIndoe et al., 1988, p. 21).

14.3.1.2.2 Pipes

The main concern in the sizing of slurry pipes is to avoid the settling of DE due to low velocity. On the other hand, if velocities are too high, solids may accumulate and compact around bends with consequent clogging (Bell, 1962, p. 1250).

TABLE 14.4
Examples of Pre-Coat and Body-Feed Masses, Volumes, Concentrations

Plant	Plant Capacity (mL/day)	Plant Capacity (mgd)	Pre-Coat Mass (kg)	Pre-Coat Volume (L)	Pre-Coat $C(\text{slurry})_{PC}$ (kg/m^3)	Body Feed Mass (kg)	Body Feed Volume (L)	Body Feed $C(\text{slurry})_{BF}$ (kg/m^3)
Las Virgines	56.78	15	136.4	946	144			
Hearst Castle						9.09	378	24.0
Mills	75.70	20	159	1325	120			
Saratoga	15.14	4				11.36	1,250	9.1
Rawlins	32.55	8.6	273	1136	240	136	11,355	12.0

The velocity range for a DE slurry is $1.0 \leq v(\text{pipe}) \leq 1.83$ m/s ($3.5 \leq v(\text{pipe}) \leq 6$ ft/s).

14.3.1.2.3 Pumps

For pre-coat, a centrifugal pump may be used since a simple circulation is all that is required; in addition, the water flow from a centrifugal pump may be stopped by a valve without building up excess pressure. The duration of the pre-coat step is about 15 min. To use one volume change per min for the slurry tank gives a pump size, $Q(\text{pre-coat}) = V(\text{pre-coat})/(1\ \text{min})$. The headloss will be due only to the plumbing configuration initially (which probably is negligible) but will rise to the pre-coat headloss at the end of the cycle, for example, 10–20 kPa (2–4 psi), depending on the grade of diaomite. The power required will be determined as

$$P(\text{pre-coat pump}) = Q(\text{pre-coat}) \cdot \gamma_w \cdot h_L(\text{pre-coat}) \tag{14.10}$$

where

$P(\text{pre-coat pump})$ is the power required to circulate pre-coat (watts)
γ_w is the specific weight of water (9808 N/m^3)
$h_L(\text{pre-coat})$ is the headloss of pre-coat circulation flow when pre-coat is deposited (m water)

For body feed, a positive displacement pump is recommended, for example, a positive displacement rotary progressive cavity pump (e.g., a Moyno®) will avoid pressure pulses. The motor should be direct current in order to provide for variable body-feed flow.

For the raw-water pump, a rotary screw (progressive cavity) pump may be used if the flows are not too high, since the flow is constant for a given motor rotation velocity. A direct current motor permits the raw-water flow to be adjusted from one run to the next.

14.3.1.2.4 Instruments

Flow meters, pressure gages, temperature sensors, turbidity meters, and perhaps on-line particle counters are required for operation. Each instrument category is available as an analog sensor (i.e., 4–20 mA) that can be attached to an interface point, for example, Microdac®, with recording by computer software (the signal is calibrated by an instrument reading and the instrument is calibrated to a standard at a specified frequency).

14.3.1.2.5 Control System

To control the flows, actuated valves have been available since the 1980s along with pumps controlled by micro-relays (Mirliss, 2002, p. 14). The actuated valves and the pumps may be set to operate based on signal levels, that is, pressure, time, turbidity, etc., that may be set by the operator using the software, for example, Wonderware®.

14.3.1.3 Layout

Figure 14.13 is an illustrative layout showing a module of four DE tanks. The point is that through proper pipe layout and the use of an adequate number of ball valves, about any configuration is possible. Whatever the configuration, its logic should be self-evident. The system should have ≥two modules for redundancy with each module having ≥2 tanks. For small systems, however, a module may be one tank, with a minimum of two modules. The backwash system is not shown in Figure 14.13, but is depicted in Figure 14.2c.

The floor area needed is the sum of the areas used by each module (e.g., two tanks per module, four tanks per module, etc.). Other kinds of functions requiring floor area include

- A dock area for receiving bagged diatomite and other materials
- A storage area for bagged diatomite
- A transfer area for breaking bags and transfer of diatomite to a slurry tank (the transfer area should be designed to alleviate any problems of DE dust)
- A bin, for storage of spent DE
- Outside storage for weeks or months accumulation of spent DE
- A laboratory for bench instruments and any wet chemical analyses deemed advisable and for calibrations of on-line instruments
- An office for administrative tasks (computer for reports to regulatory agency, storing data, etc.)

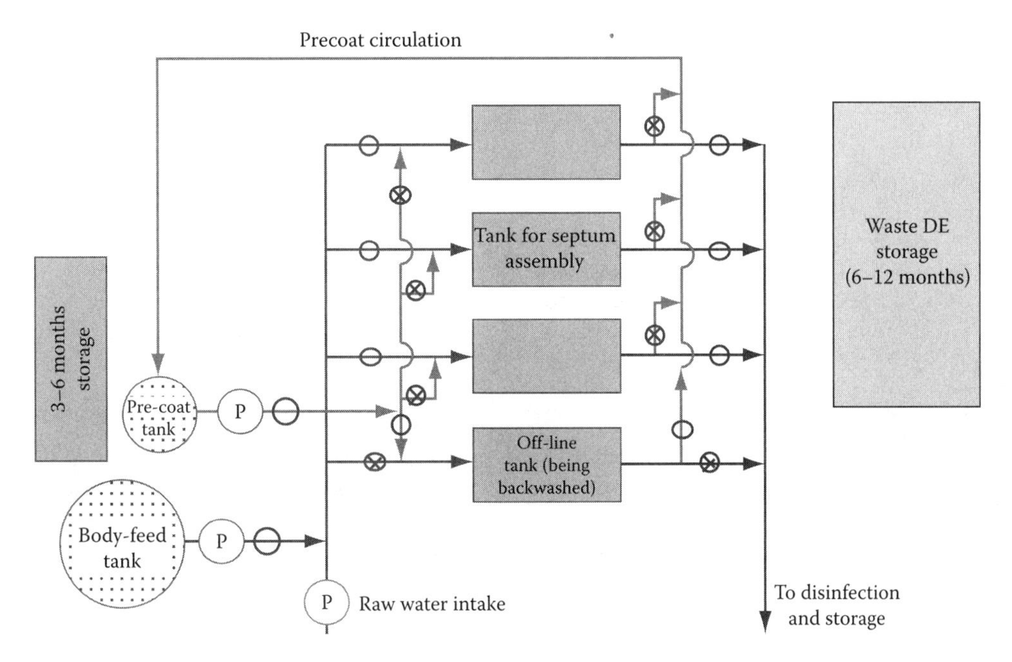

FIGURE 14.13 Schematic layout of a DE system, indicating elements of a system.

- Conference room for discussions with consultants, supervisory persons, regulatory persons, etc.
- Lunch room for workers
- Locker rooms, showers, restrooms for men and women

A transfer area should be designed to minimize the problems of lifting bags of diatomite. For example, a portable conveyor belt or an electric fork lift may be used to move bags to near the point where bag opening occurs.

Easy cleaning should be provided for, with floor drains to permit the hosing of equipment or floor. Gutters may be provided along walls to convey wastewater to a drain. All tanks and pipes should have low points for easy drainage. Pipes should have tees for flushing any material that could have accumulated and for the drainage of flushed matter.

14.3.2 Design Parameters

Design criteria for any diatomite plant must be built around the nature of the turbidity in the water supply and the requirements for finished water quality (Bell, 1962, p. 1254). In this section variables are identified and some guidelines are given for magnitudes.

14.3.2.1 Variables

The process variables in the diatomite process may be grouped as follows:

1. Dependent
   a. Effluent concentration of selected contaminants, for example, turbidity, particle counts, cysts, bacteria, viruses
   b. Rate of headloss increase, for example, $\Delta p/\Delta t$
2. Independent
   a. Raw-water quality, for example, turbidity, particle counts, and occurrence of cysts, bacteria, and viruses, and temperature
   b. Grade of diatomite
   c. Alum or polymer-coated diatomite
   d. HLR
   e. W(pre-coat)
   f. C(body feed)
   g. Selected pressure limit

A general rule for the selection of DE grade is to use the coarsest grade that meets water-quality objectives (LaFrenz and Baumann, 1962, p. 851). The coarsest grade results, in turn, in the lowest rate of headloss increase for the designated water-quality objectives. Alum or polymer additions are means to further control the effluent quality, that is, by introducing adsorption as a removal mechanism (with the trade-off being higher $\Delta p/\Delta t$). For reference, length of run = (selected pressure limit)/$\Delta p/\Delta t$. The selected pressure limit determines the tank design (for a pressure tank) and the diameter of septum permitted and its associated structural support.

14.3.2.2 Guidelines and Criteria

Usually designs are based on guidelines from manuals (e.g., Logsdon, 2008) or regulatory criteria (Ten States Standards). These are starting points in any design and deviations are usually based either on a pilot plant study or experience.

14.3.2.2.1 HLR and Septum Area

The determination of septum area is by HLR, that is,

$$Q(\text{plant}) = \text{HLR} \cdot A(\text{septum}) \tag{14.11}$$

where

Q(plant) is the flow of water to be treated (m^3/s)
HLR is the filtration velocity (m/s)
A(septum) is the total septum area for plant, that is, for all tanks (m^2)

In general, HLR = 2.44 m/h (1.0 gpm/ft^2), but higher values are not uncommon, as seen in Table CD14.5. For particles that are embedded, that is, cysts, the HLR has little effect on removals; for those that are mobile, for example, bacteria in larger grades of diatomite, the HLR has a moderate effect (Lange et al., 1986, p. 82).

14.3.2.2.2 Tank Pressure

For tank pressures, there are no firm upper limits but 300 kPa (45 psi) is seen most often. The actual limit would depend on the tank design and whether the septum can take the pressure without excessive flex, which could cause the cake to crack.

14.3.3 Design Examples

A literature survey search coupled with plant visits was conducted by Rogers (1997) who provided data for 12 diatomite plants. His findings are summarized in this section.

14.3.3.1 Data from 12 Plants

Table CD14.5 is a tabular compilation of data from the 12 DE plants showing various kinds of design data including plant name, plant capacity, equipment manufacturer, type of tank, septum area per tank, type of septum, HLR, diatomite manufacturer and grade, body-feed rate, and performance in terms of effluent turbidity, run length, and terminal pressure loss. Such data are useful for reference in designing a plant.

14.3.3.2 Plant Descriptions

The footnotes of Table CD14.5 add narrative description to the tabular data for each of the 12 plants. As seen (1997), the plants served a wide variety of purposes, for example, to filter water already given treatment but stored in a reservoir (Las Virgines) to running only 8–10 h/day to meet peak demand in the summer (Vacaville). The well-established HLR criterion, that is, HLR = 2.44 m/h (1.0 gpm/ft^2) is seen to be exceeded in several plants, for example, $\gg$2.44 m/h (1.0 gpm/ft^2) is not uncommon in practice. A septum size d(septum) = 1626 mm (64 in.) is found in one plant, that is, Saratoga.

General comments on the 12 plants (numbers in parentheses refer to plant in table) and notes on pre-coat-to-body-feed transition

(1) Las Virgines. Pre-coat-to-body-feed transition: pre-coat 136.4 kg (300 lb) suspended in 946 L water is recirculated until effluent turbidity <0.5 NTU; then start body feed; then open effluent valve. Operators feel that the type of DE is the most important factor in operating the plant and is based on septum mesh size; freshwater diatoms are used due to ease of backwashing. The plant draws water from an impoundment reservoir that has been treated by MWD (Metropolitan Water District); the plant is used only for summer peaking.
(2) Lompoc. Pre-coat-to-body-feed transition is the same as (1), that is, pre-coat mass is suspended in a given volume of water, which is then recirculated by pumping until effluent turbidity <0.5 NTU; then start body feed; then open effluent valve. Pre-coat mass is 102.3 kg (225 lb) per filter tank. Typical influent turbidities are 0.3–0.4 NTU until after softening and then go up to 2.0 NTU. Celite 535 is an alternate pre-coat/body feed.
(3) Hearst Castle China Hill. Pre-coat-to-body-feed transition is the same as (1), that is, pre-coat mass is suspended in a given volume of water, which is then recirculated by pumping until effluent turbidity <0.5 NTU; then start body feed; then open effluent valve. Body-feed tank charge is 9.09 kg per 3785 L (20 lb/100 gal) = 2402 mg/L to raw water. Amorphous "blobs" have been found in effluent in winter.
(4) Hearst Castle #2. Pre-coat-to-body-feed transition is the same as (1), that is, pre-coat mass is suspended in a given volume of water, which is then recirculated by pumping until effluent turbidity <0.5 NTU; then start body feed; then open effluent valve. Body-feed tank charge is 9.09 kg per 3785 L (20 lb/100 gal) = 2402 mg/L to raw water. Amorphous "blobs" found in effluent in winter.
(5) Sandhill. Pre-coat-to-body-feed transition: pre-coat recirculated 15 min and pneumatic influent and effluent valves opened slowly and body feed started. HLR = 7.32 m/h (3 gpm/ft^2) granted based on particle count data; body-feed rate based on influent turbidity which varies daily and seasonally. Septum leaves rotate for cleaning.
(6) Glendale. The equipment specified for the Glendale plant was by Westfall, but the design engineers permitted the contractor to substitute Sepramatic, a manufacturer of swimming pool filters. Neither the design firm nor the manufacturer had worked with potable water. Once started, the equipment did not integrate with the SCADA system. Other issues included pinholes in the septum; filter operation was designed to stop at 381 mmHg (0.50 atm), but operators noted that the vacuum did not increase, so the filter run was then based on time (2 weeks). Body-feed concentration in slurry tank is 10,000 mg/L; neither rate nor concentration is known in raw water.
(7) Mills. Pre-coat-to-body-feed transition: recirculate pre-coat (159 kg in 1325 L, that is, 350 lb DE in 350 gal water per filter tank) 25 min before filter comes on line; valves opening is controlled by PLC. Alum is added to body-feed tank giving good results. Alum added to the body feed at a ratio: 2.0 kg alum as $Al_2(SO_4)_3 \cdot 14H_2O$ per kg DE and 1.0 kg soda

TABLE CD14.5
Description of 12 DE Plants and Operating Protocols

	Plant				Equipment										
					Tanks										
			Capacity					Flow per Tank		Septum	Area/Tank			HLR	
	Name	State	(mL/day)	(mgd)	Manufacturer	Type	(#)	(L/min)	(gpm)	(#)	(m^2)	(ft^2)	Type	(m/h)	(gpm/ft^2)
1	Las Virgines	California	56.78	15	Westfall	Vacuum	10	4883	1290	*	119.86	1290	Leaf	2.44	1.00
2	Lompoc	California	23.47	6.2	Westfall	Vacuum	3	5526	1460	*	133.78	1440	Leaf	2.44	1.00
3	Hearst Castle, China Hill	California		*	Schneider	Pressure	1	151	40		3.71	40	Leaf	2.44	1.00
4	Hearst Castle, Plant #2	California		*	Schneider	Pressure	1	227	60	*	5.57	60	Leaf	2.44	1.00
5	Sand Hill	California	88.19	23.3	U.S. Filter	Pressure	4	6813	1800	*	55.74	600	Leaf	7.32	3.00
					U.S. Filter	Pressure	4	8516	2250	*	69.68	750	Leaf	7.32	3.00
6	Glendale	California	6.25	1.65	Sepramatic	Vacuum	*	*	*	52	126.35	1360	Leaf	*	*
7	Mills	Wyoming	75.70	20	Industrial	Pressure	2	4542	1200	*	111.48	1200	Leaf	2.44	1.00
8	Saratoga	Wyoming	15.14	4	US Filter	Pressure	1	5678	1500	*	179.86	1936	Leaf	1.88	0.77
							1	4921	1300	*	147.16	1584	Leaf	2.15	0.82
9	Black Hawk	Colorado	4.35	1.15	R. P. Adams	Pressure	2	1514	400	*	22.11	238	Tube	4.10	1.68
10	Georgetown	Colorado	3.79	1	Dreco	Pressure	2	1325	350	*	25.55	275	Tube	3.10	1.27
11	Rawlins	Wyoming	32.55	8.6	Durco	Pressure	3	5678	1500	*	185.81	2000	Leaf	1.85	0.75
					Industrial	Pressure	1	5678	1500	*	185.81	2000	Leaf	1.85	*
12	Vacaville	California	45.42	12		Pressure	8	*	*	*	*	*	Tube	3.71	1.52

(continued)

TABLE CD14.5 (continued)
Description of 12 DE Plants and Operating Protocols

Filter Aid		Pre-Coat		Body Feed	Performance				
					Turbidity			Δp	
Manufacturer	Grade	(kg/m²)	(lb/ft²)	(mg/L)	Raw (NTU)	Effluent (NTU)	Run (day)	(kPa)	(psi)
Eagle-Picher	FW-20 (DE)	0.11	0.23	10.8	0.6–0.9	0.1	≤14	54	7.9
Grefco	4200 (DE)	0.08	0.16	15–20	0.3–0.5	0.08–0.16	≤10	57	8.3
Eagle-Picher	FP1W(DE) and FP2 (DE)	0.10	0.2		2–8	0.08–0.5	≤4	241	35
Eagle-Picher	FP1W(DE) and FP2 (DE)	0.10	0.2		2–8	0.08–0.5	≤4	241	35
Celite	Hyflo Super-Cel (DE)		*	*	1–8	0.04–0.10	4–14	290	42
			*	*					
Celite	SW10 (C)/C-545 (DE)	0.09	0.18	*	2.0	<0.1	14	0	0
Harborlite	1500 S (P)	0.14	0.29	12	3–10	0.3	*		*
Harborlite	700, 900, 1500 (P)	0.05	0.1	63	3–200	0.1–0.5	3.3	483	70
		0.06	0.13	63					
Celite	C-535 and Aquacel 1:1 (DE)	0.10	0.21	4–20	0.3–0.6	0.1–0.9	12	207	30
Eagle-Picher	FW12 (DE)/ FW20 (DE)	0.26	0.54	40	1–100	0.08–0.1	3	186	27
Harborlite	900 W (P)	0.04	0.075	12.6	0.2	0.02–0.6	14	276	40
Celite	Hyflo Super-Cel (DE)	0.09	0.18	25–40	5–12	0.3	0.4	96	14

Source: Complied from Nolte and Associates, 1997a, Tables 1 and 2, pp. 3&4. With permission.
Note: DE: Diatomaceous earth; P: perlite; C: Cellulose (Provided by Pavlakovich, 2010).

ash, that is, Na_2CO_3 in about 10 L water (1.0 lb alum, 1.0 lb DE and 1.0 lb soda ash in 1 gal water). Effluent turbidities were reported to be lower for a given coarse grade of DE without a decrease in run time, which was reported not to decrease.

(8) Saratoga. Pre-coat-to-body-feed transition: pre-coat recirculated 30 min, then valves open automatically for raw water, body feed, and effluent. When influent turbidities are high plant has difficulty in maintaining effluent turbidities <1 NTU; operators claim blending of DE grades (at 2:3:3) lowers effluent turbidities. Body-feed suspension by adding 11.36 kg perlite to 1250 L tank (25 lb in 330 gal) gives 9100 mg/L body-feed concentration. Leaf septum is 1626 mm (64 in.) round. Primary filter has 44 vertically oriented leaves; the secondary filter has 36. Pressure differential at start-up is about 27.6 kPa (4 psi).

(9) Blackhawk. Pre-coat-to-body-feed transition: recirculate pre-coat 15 min, then start raw-water pumps, body-feed pumps, open influent valve, open filter-to-waste valve until effluent turbidity is 0.6–0.8 NTU (typically 45–60 min). Stones plug up after extended use. Well water may be blended with surface water to maintain influent turbidities <0.9 NTU using two basins adjacent to the plant (one pond for each source). Tube septa are each 108 mm (4.25 in) diameter × 1219 mm (48 in.) in length, giving 0.1116 m^2/tube (1.20 ft^2/tube); there are 349 tubes/tank giving 25.5 m^2 (275 ft^2) septum area per tank (which is less than the calculated number).

(10) Georgetown. Pre-coat-to-body-feed transition: recirculate pre-coat 18 min, then filter is brought on line. The mixing of DE in body-feed tank is approximate. Filter tank/septum manufacturer is Dreco, Model No. AWP6-42-275.

(11) Rawlins. Pre-coat-to-body-feed transition: water is recirculated for 20 min at 1200 gpm after which influent and effluent valves are opened. Pre-coat cake is at 0.37 kg/m^2 (0.075 lb/ft^2), to give 1.6 mm thickness. Pre-coat slurry is about 272.7 kg (600 lb) perlite in 1136 L (300 gal) water and recirculates about 20 min at a rate of 4542 L/min (1200 gpm). Body feed is mixed in a 11,355 L (3,000 gal) tank with perlite added at the rate of 0.454 kg perlite/37.85 L (1 lb/10 gal) water to give a slurry concentration of 12,000 mg/L and 12.6 mg/L in a raw-water flow of 10.825 mL/day (2.86 mgd) for a slurry feed rate such that the slurry lasts 48 h. Septum are 1.981–2.234 m (6.5–7.0 ft) diameter, depending on manufacturer. The Durco filters have 39 leaves each filter and the Industrial has 36 leaves. Water supply is from wells and a spring; each source can be brought directly to the plant or discharge to a holding pond. Wells are high in sodium and when used are blended with the spring water.

(12) Vacaville. The Vacaville plant treats water from Lake Berryessa, delivered by the Putah South Canal. The plant is operated in the summer and fall months when demands are higher and raw-water turbidities are low. The septum is tubular stainless steel, 60 mesh. During operating seasons, the plant operates 8–10 h to meet peak diurnal demands and terminal headloss is never reached. The plant is not operated when turbidity ≥15 NTU. Pre-coat is applied by circulating the DE from the suction side of the pump basin through the septum and back to the basin until all of the pre-coat is deposited (about 15 min). A mass of polymer (Nalco Ulitron 8157) is fed to the raw-water screen area over a 15 min period and body feed is started at the same time while water from the filters continues to be circulated. When filtered water turbidity is <0.40 NTU, the filters go on line (10–30 min after the start of body feed). A second mass of polymer is added just after the filters go on line. One hour after going on line a third mass of polymer is added. Body feed is paced with buildup of differential pressure; when terminal headloss is reached, all body feed will have been applied; the run is terminated by operators when peak demand is met, typically in 8–10 h.

14.4 OPERATION

The operation involves ordering diatomite, maintaining equipment and instruments, monitoring instruments, recording and archiving data from instruments, reporting on operation, maintaining order and cleanliness, storage and disposal of waste diatomite, etc.

14.4.1 Operating Protocol

Pre-coat and body feed are the important factors in the diatomite filtration process. Executing this involves charging tanks with needed masses, metering flows, and operating pumps. Much of this may be automated, leaving the operator with the role of monitoring operations (after the software is programmed with values needed for operation).

14.4.1.1 Pre-Coat Deposit

Guidelines for pre-coat deposit have ranged at $0.50 \leq L(\text{DE}) \leq 1.00$ kg/m^2 (0.10–0.20 lb/ft^2) (Baumann et al., 1962, p. 1114). Since about 1980, the tendency has been toward the higher end of the range.

14.4.1.1.1 Pre-Coat Tank Charge

The pre-coat tank is filled with a given amount of water and with a given mass charge of diatomite that results in the specified pre-coat area density. Table 14.4 gives examples of pre-coat slurry concentrations (and associated mass charges and water volumes for several plants). The mass charge is given as

$$\text{Mass(pre-coat)} = L(\text{DE}) \cdot A(\text{septum}) \cdot N(\text{septa}) \quad (14.12)$$

$$= C(\text{pre-coat slurry}) \cdot V(\text{pre-coat suspension}) \quad (14.13)$$

where
- Mass(pre-coat) is the mass of diatomite required to provide pre-coat charge to a tank of module of several tanks (kg)
- L(DE) is the mass loading of DE on septum (kg DE/m^2 septum area)
- A(septum) is the septum area for a single element, for example, two sides for a leaf septum (m^2)
- N(septa) is the number of septa in a given assembly to be pre-coated (#)
- C(pre-coat slurry) is the concentration of pre-coat slurry (kg/m^3)
- V(pre-coat suspension) is the volume of pre-coat water for diatomite slurry (m^3)

The quantity to be calculated is V(pre-coat suspension), with C(pre-coat slurry) assumed based upon experiments or upon reported practice, for example, Table 14.4 in which $120 \leq C$(pre-coat slurry) ≤ 240 kg/m^3.

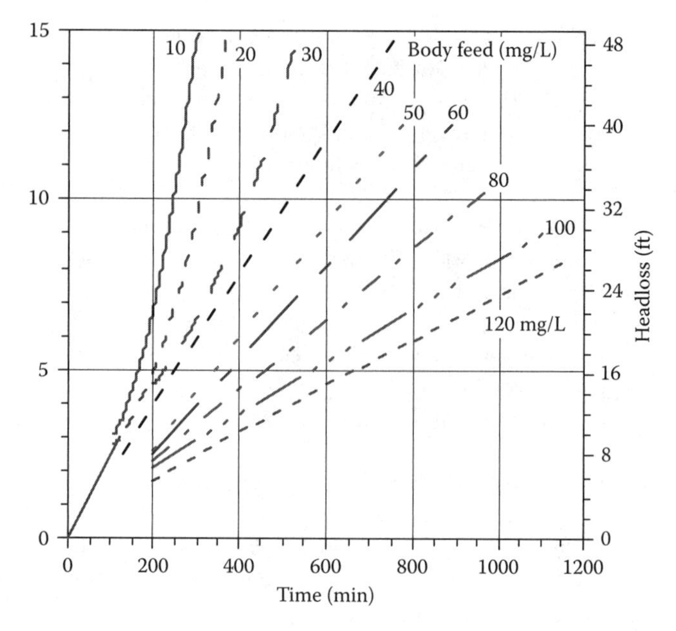

FIGURE 14.14 Headloss increase with time. (Adapted from Baumann, E.R. et al., *J. Am. Water Works*, 70(9), 1109, 1962, p. 111 for 8 mg/L Fe at v = 2.44 m/h, 20°C.)

14.4.1.2 Body Feed

Selection of DE grade and determination of concentration and mixing of slurry are the tasks associated with body feed. Usually the grade and concentration are determined as a first estimate during the design phase and refined during operation.

14.4.1.2.1 Selection of Grade

The grade of body feed should be the coarsest that results in the effluent quality meeting treatment objectives. This may be determined by bench-scale tests. For water treatment, the coarsest grades are usually those with $18 \leq d(50) \leq 24$ μm. If alum or polymer is required to meet the treatment objectives, then a coarser grade of diatomite may be used; but the rate of headloss increase is likely to be higher than desired.

14.4.1.2.2 Body-Feed Concentration

The body-feed concentration depends upon the sizes and kinds of particles to be removed and upon the grade of DE. Concentrations in the range of 20–40 mg/L are usual. Ideally, the concentration to be used should be based on a pilot plant study for the particular conditions at hand. Otherwise, the concentration may be adjusted at full scale, starting on the high side and working down to lower concentration.

14.4.1.2.3 Body-Feed Adjustment: Based on Linear Rate of Headloss Increase

Figure 14.14 illustrates the effect of body-feed concentration on the respective headloss versus time curves. As seen, for body-feed concentrations of 10, 20, and 30 mg/L, the h_L versus time curves increase exponentially with time, but at C(body feed) ≥ 40 mg/L, the relationships are linear. The body feed to be used is the minimum concentration that gives a linear h_L versus time curve, that is, 40 mg/L for the case shown. The curves shown are unique for the conditions given.

The exponential h_L versus time curves at lower DE concentrations are due to the particles to be removed dominating the cake pores, effectively "blinding" the cake.

14.4.1.2.4 Body-Feed Charge

The body-feed tank is filled with a given amount of water and then given a mass of diatomite that results in the specified slurry concentration. Table 14.4 gives slurry concentrations in the range of $12 \leq C$(body-feed slurry) ≤ 24 kg/m^3. Equation 14.10 indicates the variables involved; t(tank) is the associated time that the slurry volume will last. The mass charge is calculated as the right-hand side of Equation 14.10.

14.4.1.3 Valve and Pump Operation

Valves and pumps should be operated by a PLC (programmable logic controller) or by desktop computer and control software. The entire operating protocol may be programmed such that "hands-on" control by the operator is not significant in terms of labor required. Rather, the operating labor involves, for the most-part, adding the pre-coat and body-feed charges, monitoring instrument read-outs, servicing equipment as required, reporting to regulatory authorities, etc.

14.4.2 Monitoring

Monitoring involves whether the magnitudes of variables are within specified limits and whether higher-than-expected rates of change occur. In either case, the computer may take prescribed actions should deviations be outside the ranges, for example, shut down the module, or initiate other action such as a telephone call to an operator. The other function is to provide a documentation of performance to regulatory authorities.

14.4.2.1 Flow versus Time

The flow should be recorded for each module of tanks during a given run. Inspection should be in terms of the magnitude of the flow, that is, whether it is within the range specified, and whether the flow is constant. The software may be programmed to detect and take a prescribed action should unexpected deviation from normal operating range occur.

14.4.2.2 Headloss versus Time

Headloss versus time varies with the grade of diatomite and water, that is, kind of particles to be removed and their concentrations, pre-coat rate, body-feed concentration, and HLR. Whether the headloss versus time relationship is linear should be documented by computer software. Cycle lengths also depend on terminal pressure permitted for the particular installation. For 12 operating plants, as reviewed in Table CD14.5, $3 < \text{run-length} < 14$ days for diatomite grades with $18 < d(50) < 24$ μm and with $54 < h_L(\text{terminal}) < 483$ kPa ($8 < h_L(\text{terminal}) < 70$ psi).

14.4.2.3 Turbidity versus Time

Raw-water turbidity should be monitored for the whole plant for effluent turbidity from each tank, preferably by on-line turbidimeters. Raw-water turbidity should be plotted for the annual cycle in order to characterize the turbidity changes. The monitoring of effluent turbidity is essential for plant operation and is required for reporting. A spike in effluent turbidity or a "step" increase is reason to terminate the run.

14.4.2.4 Criteria for Run Termination

Run termination may be based on turbidity exceeding given criteria, for example 0.5 NTU, or headloss exceeding a given limit, for example, 345 kPa (50 psi). The former may be imposed by regulations and the latter is determined by equipment pressure limits or based on economic analysis. In addition, electric power and flow must be monitored to ascertain that interruptions have not occurred. Any sudden change in flow or loss of power is the basis for run termination. Such changes are likely to cause loss of integrity of the filter cake; most likely some or all of the cake will fall from the septum.

14.4.3 Cleaning and Start-Up

Cleaning, pre-coat, and body feed have their respective protocols, described in Figure 14.1, which is useful for reference. Cleaning and precoat are described in this section.

14.4.3.1 Protocol

The run termination is initiated by closing the effluent valve, then the body-feed pump may be shut down along with the raw-water pump (unless the latter serves other modules). Valves should be closed to isolate the tank being cleaned with the appropriate backwash valves (air vent, drain, jet for septa) opened while the backwash pump is started.

14.4.3.2 Start-Up

After backwash the septum tank is isolated for pre-coat circulation. The pre-coat pump is started with the effluent valve closed until the slurry displaces the air (i.e., the air vent valve remains open until the tank is filled), at which time the effluent valve is opened to permit the circulation of pre-coat. After the pre-coat is deposited on the septum area (as determined by turbidity $\leq$0.5 NTU), the transition is initiated to operation with body feed; the first step is to close the pre-coat circulation valve (after a tee from the tank effluent line). The body feed is then started, then the raw-water pump, followed by the opening of the tank effluent valve.

14.4.4 Disposal of Waste Diatomite

Wastes are generated and must be dealt with as a part of any design and of any operation. Things are easier in the subsequent operation if the issue, in terms of waste storage and waste disposal, has been dealt with by the design engineer in collaboration with the operating personnel.

14.4.4.1 Waste Storage

The waste is removed from the tank by slurry during backwash. The slurry may be transported to a sump, small in volume and designed to avoid the settling of diatomite, with the slurry being pumped subsequently to a bin with provision for drainage and storage. As noted, the waste diatomite may be stored for several months. The bin should be designed such that a front-end loader can easily pick up the waste diatomite and load it on a truck.

14.4.4.2 Waste Disposal

The options for waste disposal are few. Regulatory restrictions limit the manner of disposal for most solid wastes. Landfill disposal is the most common method of disposal, since diatomite is an inert substance and is contaminated only by the suspended matter removed from the raw water. The most hazardous of the matter removed are cysts, which are subject to eventual decay. Determining an acceptable disposal option is a task that involves ferreting out from among those generated by usually more than one person, including discussions with operation personnel, the owner, and regulatory authorities.

14.5 PILOT PLANT STUDIES

As in other unit processes, a DE pilot plant study is useful in evaluating its suitability for a particular application, in selecting parameter values, and in recommending operating procedures. It is likely that a pilot plant study would involve comparing diatomite filtration with alternatives, for example, with respect to treatment effectiveness, unit cost of water produced, etc.

14.5.1 Questions for a Pilot Plant Study

The questions for a pilot plant study relate mostly to process design. The main task is to select a diatomite grade that is effective in attaining treatment objectives and that results in the longest filter run. The answer depends on raw-water characteristics, which may vary seasonally. Other questions may relate to the effect of increasing HLR on the economics of operation and on effluent quality. The headloss versus time curves are of interest for each HLR so that the effect of increasing HLR can be anticipated. The pilot study can also provide guidelines for pre-coat and body feed and whether to add alum or a polymer for short periods, for example, in responding to a turbidity event such as after a rainfall. In addition to its utility in design, a pilot plant can serve as a continuing tool for operation and may be retained on-site usually for a nominal cost.

14.5.1.1 Functional Relationships

The questions for a pilot plant study may be formulated in terms of the operative variables, for example, as given in Section 14.3.2.1. In such a case, the dependent variables are some function of the independent variables, that is,

$$\begin{bmatrix} \bullet\ \text{Effluent concentrations of selected contaminants} \\ \bullet\ \text{Rate of headloss increase} \end{bmatrix} = f\begin{bmatrix} \bullet\ \text{Raw water quality} \\ \bullet\ \text{Grade of diatomite} \\ \bullet\ \text{Alum/polymer concentration} \\ \bullet\ \text{HLR} \\ \bullet\ W(\text{precoat}) \\ \bullet\ C(\text{bodyfeed}) \end{bmatrix} \quad (14.14)$$

Empirical relationships generated from a pilot plant based on Equation 14.14 are necessarily limited by time and budget (a well-known truism), and so the study must be pared down from including all of these variables. In paring down, knowledge of research and practice can serve as a guide, for example, diatomite grades for water treatment generally have a size range of $18 < d(50) < 24$; smaller causes excessive headloss and larger does not, as a rule, remove sufficient turbidity. Also $2.44 \leq \text{HLR} \leq 4.88$ m/h ($1.0 \leq \text{HLR} \leq 2$ gpm/ft^2), with little effect on water quality if higher values are used. Whether to use HLR $\gg 4.88$ m/h (2 gpm/ft^2) in sizing units involves trade-offs between capital and operating costs (Box 14.4).

14.5.2 Cases

Two diatomite filtration pilot plant cases may illustrate the diversity of situations. Every case has a different aspect.

> **BOX 14.4 PILOT PLANT STUDIES**
>
> Pilot plant studies are common in the field of diatomite filtration. Skid-mounted units are available, as a rule, from manufacturers of equipment or diatomite producers. In the case of cylindrical elements, however, a pilot plant must be fabricated. A pilot plant study can either confirm that a contemplated design will perform as intended or possibly save funds in construction.

14.5.2.1 SR Ranch, Colorado

The SR Ranch in Colorado had been using a 38 L/min (10 gpm) diatomite filtration system starting about 1970. The raw-water source was the Boyd Hanson Feeder Canal, near Loveland, Colorado, which was snow-melt water. The turbidity range was 2.3–6.7 NTU during the period of testing, which was typical during most of the year. Excursions to 30 NTU may occur, however, during spring runoff. Cysts of *G. lamblia* have been found in the canal. The system was operated about 8 h/day with storage used to meet the off-line demand of about 30–50 persons. The equipment was a U.S. Army Erdlator system that had been obtained as military surplus. The unit had supplied field troops with drinking water from surface supplies. Since the system supplied ≥ 25 persons, it was classified as a public water system under the 1974 U.S. Safe Drinking Water Act and therefore was required to meet drinking-water standards. For diatomite filtration, the turbidity standard was ≤ 1 NTU set by EPA and adopted by the Colorado Health Department. At the time of the investigation in 1984, the pressure tank was found to have a few pinpoint holes due to corrosion and a high rate of headloss increase. The maximum pressure differential across the filter cake was $\Delta P(\text{cake}) \approx 207$ kPa (30 psi). The management was interested in updating the system and in operating parameters being defined for meeting drinking-water standards. Since the SR Ranch had a limited budget, an exploratory gratis study was undertaken by Mr. Ray McIndoe, then with Johns Manville in Denver and Mr. Alan Wirsig, a consultant recently retired from Johns Manville.

14.5.2.1.1 Pilot Plant

The pilot plant used to investigate the issues at SR Ranch was a Johns-Manville 0.091 m^2 (1.0 ft^2) stainless steel unit shown in Figure 14.15a. Figure 14.15b shows the same pressure vessel/septum assembly with some of the adjunct instrumentation required to perform the pilot plant study. Such instruments include a raw-water screw-type metering pump, a raw-water flow meter, a pre-coat tank and pump for circulation, a body-feed tank and metering pump, pressure gage for the pressure vessel, air vent for the pressure vessel, a constant head overflow for the effluent, injection port in the raw-water line, and sampling ports in the raw-water line and the effluent line. The setup was moved from a laboratory at CSU to the field site adjacent to the Boyd Hansen Feeder canal, the water source, requiring only a few hours for the move. The setup was essentially the same as required for a full-scale plant.

Effluent turbidities ≈ 0.5 NTU for C512 and Hyflo and 0.7 NTU for Celatom FW20, with $\Delta P/\Delta t \approx 5.5$ kPa/h (0.8 psi/h) for run length $\approx$ 207 kPa/5.5 kPa/h = 38 h, which is much longer than the needed 8 h. Effluent turbidity was 3.2 NTU for the coarser grade, C545, which of course was not satisfactory. Adding alum as a coating to the diatomite reduced markedly the effluent turbidities for each of the grades, as expected, without appreciable $\Delta P/\Delta t$ penalty. While simplicity of operation is a main goal in any diatomite system, the alum performance provided knowledge that there is a "way out" should the need for lower effluent turbidities be necessary.

(a)

(b)

FIGURE 14.15 Stainless steel pilot plant provided by Johns-Manville (now Celite®) c. 1985. (a) Pilot plant pressure vessel and (b) pilot plant with instrumentation.

14.5.2.2 100 Mile House, British Columbia

In 1981, the Village of 100 Mile House, 1984 population of about 2075, experienced an epidemic of giardiasis; their water supply from Bridge Creek, an outlet from Horse Lake with turbidity <2 NTU and not filtered, was implicated since beaver and muskrat inhabited the watershed. The peak-day demand was estimated as 3.30 mL/day (0.87 mgd) for 1984 and 6.67 mL/day (1.76 mgd) for 2002 with projected population 4200. Minimum demands for 1983 were about 0.2–0.3 of the peak demand with an average day demand of about half of the peak day. In 1983, the Village engaged their consulting engineer, Dayton & Knight, Vancouver, British Columbia, to recommend a course of action. Subsequently, Dayton & Knight embarked on a pilot plant study that compared filtration alternatives, that is, rapid filtration package plants (both gravity and pressurized), slow sand, and diatomite. Based upon estimates of capital costs, cost of labor, cost of energy, etc., the annual costs were compared. The pilot plant studies provided a basis for estimating energy and labor costs, and generated data on effluent turbidities.

14.5.2.2.1 Pilot Plant

Figure 14.16 shows the pilot plant located at 100 Mile House, British Columbia. Figure 14.16a shows the pilot plant with pre-coat tank, body-feed tank and pumps, all skid mounted, as set up at the field site. Figure 14.16b shows the pressure vessel/septum assembly. As seen, each septum is a "leaf" with two sides; the interior has a structure to support the septum mesh in order to withstand the pressure with minimum deflection.

14.5.2.2.2 Results and Discussion

In the pilot plant study, the consultant for the village examined rapid filtration (a package plant), slow sand, and

(a)

(b)

FIGURE 14.16 Diatomite pilot plant at 100 Mile House, British Columbia. Unit is Type 122, Industrial Filter Mfg. Co., Cicero, Illinois, with four stainless steel leaves, septum area 0.93 m^2 (10 ft^2). (a) Field setup showing appurtenances and (b) pressure vessel and septum assembly. (Courtesy of Dayton & Knight, Ltd., Pilot water treatment program, Village of 100 Mile House, Report for Client, Dayton & Knight, Ltd., Consulting Engineers, West Vancouver, British Columbia, December 1983.)

diatomite, and did an economic comparison. The diatomite pilot plant produced effluent turbidities 0.20–0.28 NTU with raw-water turbidity generally <0.8 NTU using Hyflo Super-Cel and body-feed concentration 20 mg/L giving run length of 80 h at terminal pressure 276 kPa (40 psi). Calculation gives the rate of headloss increase as 3.45 kPa/h (0.5 psi/h). Based on the projected 2002 peak-day demand, the required septum area $= 6.67 \cdot 10^3$ m^3/day/24 h/day/(2.44 m/h) = 114 m^2 (1226 ft^2). The estimated annual cost of electric energy for 1992 was \$5100, based on 275 kPa (40 psi) terminal pressure and pumping power of 15 kW (20 hp). The 1992 projected use of diatomite was 31,000 kg (68,000 lb) with cost \$15,320 (at the 1983 price of \$49/kg or \$0.22/lb). Based on a density of 1440 kg/m^3 (90 lb/ft^3), the estimated annual volume of diatomite waste was about 21 m^3 (27 yd^3); the cost of landfill disposal was estimated as \$417. While the unit costs of water production (i.e., operating costs and amortized capital cost) were comparable for all three technologies, the technology selected was slow sand based largely on the fact that it was a "passive" technology and appropriate for the situation (see Section 13.1.1.4).

PROBLEMS

14.1 Support for Septum

Given

Assume that the pressure differential between the two sides of the filter cake is 280 kPa (40 psi). Assume that the system consists of a pressurized tank with leaf septa 1000 mm (40 in.) diameter.

Required

Calculate the force on one side of the leaf septum. Also depict, by means of a diagram, the pressure profile. Comment on the likelihood of flex in the septum as the pressure differential increases.

14.2 Footprint for the NYC Demonstration Plant

Given

The demonstration plant at Jerome Park operated to filter the Croton water supply for New York was sized to filter 11.4 mL/day (3 mgd) at HLR = 6.1 m/h (2.5 gpm/ft^2) (see Section 14.1.4.2).

Required

Design a layout for a horizontal tank with leaf septa with surfaces in the vertical plane.

14.3 Logistics of Diatomite

Given

A 3.785 mL/day (1.0 mgd) diatomite filtration plant at Georgetown, Colorado takes water from an adjacent mountain creek. The body-feed rate of diatomite is 40 mg/L.

Required

Estimate the monthly mass requirement of diatomite in terms of kg/month, and number of 22.7 kg (50 lb) bags/month. About how much storage is needed for a month's supply of diatomite. Should the storage be larger to reduce the number of deliveries per year? How much diatomite should be on hand at any given time, for example, just prior to a delivery?

14.4 Calibration of Baumann et al. Equation

Given

Initial headloss data and the rate of headloss increase data from pilot plant experiments are given in tabular form in the tabular data that follow for four grades of diatomite (Nolte, 1997b,c). The dry densities and permeabilities of each grade are given in Table 14.1 (the permeabilities in "Darcys" are converted to "m^2" in Table CD14.2). The septa were 60 mesh "candles" 28.58 mm (1–1/8 in.) × 895 mm (35 in.) long as used in a pilot plant.

	Nolte Data (kPa)	Nolte Data (kPa/h)
C-512	6.9	8.3
Hyflo Super-Cel	4.8	7.6
C-503	3.5	6.9
C-545	3.4	9.0

Required

Try to "calibrate" the Baumann et al. equation using the Nolte data on the initial headloss and the rate of headloss increase.

14.5 Volume of Body-Feed Tank

Given

Let the Q(raw water) = 0.044 m^3/s (1.00 mgd) and let C(body feed) = 25 mg/L. The duration of time, t(tank), for the volume of the tank slurry to last is to match the length of run for the filter, which is about 48 h, typically.

Required

Estimate the size of body-feed tank.

14.6 Storage of Waste Diatomite

Given

A 37,850 m^3/day (10 mgd) diatomite plant has a body feed of 25 mg/L.

Required

(a) Design a storage area for the waste diatomite.
(b) Determine the options for the disposal of waste diatomite in your area. Indicate regulations that may facilitate or limit the disposal options.

ACKNOWLEDGMENTS

Brian Walker, principal and owner, Dayton & Knight, Ltd., Vancouver, British Columbia, gave permission (2002) to use material from their 1983 report; permission was renewed (2010) by Jack Lee, PEng, vice president, Operations, Dayton & Knight. Jack Bryck, then with Dayton & Knight, presently (2010) with Malcolm Pirnie, was the engineer on site.

The survey data that comprise Table CD14.5 were obtained from a 1997 report by Nolte and Associates, Inc., Sacremanto, California, and was used by permission (2002, and renewed in 2010). Their client was the City of Vacaville and involved a pilot plant study as well as the survey; the three reports for the city are listed in references. Todd Rogers, PE, then with Nolte, obtained the data from operating plants; see also an interim report by Rogers (1997).

Walter Pavlakovich, technical services and marketing manager, Celite Corp/World Minerals Inc. (a division of Imerys Corp.), Lompoc, California, provided photographs and data that were used as cited. Pavlakovich reviewed the manuscript and made several corrections, which improved accuracy, and provided material on cellulose and perlite, which was added to the Glossary. George Christoferson (retired from Celite Corp/World Minerals Inc.) provided the photographs of the mining operations (Figure 14.4 used in the 2006 edition and in this revision).

O. Alan Wirsig, formerly with Johns-Manville and later principal of Alan Wirsig Engineering, gave permission to use correspondence material from his work with communities in Colorado and to use his image in Box 14.2. The late Ray McIndoe, formerly with Johns-Manville, gave permission, c. 2002, to use his image in Box 14.2 and to summarize his career in diatomite filtration.

Ascension Industries (Durco Filters by Ascension), North Tonawanda, New York, gave permission to use a photograph of one of their units, downloaded from their website, www.asmfab.com, and material from one of their catalogs as cited in Table 14.4. Bruce Hindle, director of filtration services, provided updated references for Durco filter products.

GLOSSARY

Adsorption: As related to filtration, the particles to be removed become attached to the media grains (see Section 12.3.3.4).

Air bump: A pulse of air pressure to dislodge the filter cake from the septum at the end of the filtration cycle.

Asbestiform fibers: A term for an asbestos fiber found in L. Superior near Diluth at levels to 17 million fibers/L for total fibers. Other kinds of fibers include amphibole and chrysotile.

Blinding: Particle concentration at a surface causing inordinate increase in headloss.

Bridging: Joining of two or more particles by arching over individual openings in the filter septum or between the individual filter elements.

Body feed: Continuous flux of DE into the raw-water flow that results in particles to be removed becoming embedded in the filter cake and a linear rate of headloss increase.

Cake: A layer of filter media on the septum that develops due to body feed, after the pre-coat but the pre-coat and the body-feed cake are referred to collectively as the "cake" or the "filter cake."

Cake filtration: Filtration through the layer of filter media, that is, filter cake, that has developed on the septum.

Calcined: (1) To heat (as inorganic materials) to a high temperature but without fusing in order to effect useful physical and chemical changes such as (a) to convert to a powder or to a friable state by heating, (b) to heat in order to drive off volatile matter (as carbon dioxide from limestone, ores, or concentrates, or chemically combined water from clay) (http://www.unabridged.merriam-webster). (2) The process used in manufacturing diatomaceous earth by melting and fusing particles.

Celite: Registered trademark, 1915. In 1917, the corporate name was changed from Kieselguhr of America to the Celite Products Co. The term Celite was applied not only to the company but was used as a name for the DE of the Lompoc deposit; this latter usage was abandoned, however, as it was recognized as being restrictive (Cummins, 1975, No. 16, p. 10).

Cellulose: (1) Fibrous material of vegetable origin (Celite/World Minerals Brochure FA-84A, 2010). (2) Cellulose fibers bridge rapidly on coarse septum to form a firmer pre-coat. Additionally, cellulose forms stronger cakes less likely to be disturbed by pressure changes. Filtration fibers are made from both hardwood and softwood pulp in both bleached and unbleached grades.

Table 14.G.1 lists important properties of Fiber-Cel® Cellulose (Manufactured by International Fiber Corp, Tonawanda, New York and sold by Celite/World Minerals Inc.) in descending order of particle size, which corresponds to decreasing permeability.

TABLE 14.G.1
Typical Properties of Fiber-Cel Cellulose Filter Aid

Grade	Fiber Length (μm)	Permeability (Darcy's)	Bulk Density (lb/ft^3)	Through 200 Mesh (%)
SW5	120	20	7	15
SW10	120	16	8	20
BH	100	9	9	35
BH	65	6.5	12	40
BH	40	3.5	16	70
BH	35	2.5	18	75
BH	30	2	18	85

Source: Courtesy of Walter Pavlakovich, Celite Corp/World Minerals Inc. (a division of Imerys Corp.), Lompoc, CA.

Cycle: Filtration interval, length of time filter operates before cleaning.

Darcy: A permeability of 1 darcy permits a flow of 1 mL of liquid of 1 centipoise viscosity through 1 cm^2 of porous solid 1 cm thick under a differential pressure of 1 atm in 1 s (Bell, 1962, p. 1247). Darcy $= 0.987 \cdot 10^{-12}$ m^2. See also Equation E.5 and Appendix E—Glossary.

Detection limit: The minimum concentration of an organism or chemical that can be detected by the given sampling protocol and particular analysis procedure.

Diatom: (1) Unicellular photosynthetic microscopic organism characterized by its silica shell. The products of the photosynthesis include a fatty oil, which has a fishy odor, and is a nutritious food for microscopic marine animals (Cummins, 1975, No. 2, p. 2.]. (2) Diatom populations in the ocean vary with location and season but average perhaps 7–8 · $10^9/m^2$. Diatoms have the capacity to extract silica, which has a solubility of up to 40 mg/L, as silicic acid and create a silica shell. Siliceous sediments are an end result. (3) "Beginning with curved outlines, we find among them perfect circles of all sizes, others passing into oval, elliptical, crescent, serpentine, sigmoid, and other curved contours with variations almost beyond number." (Cummins, 1975, No. 17, p. 1). (4) Diatoms are classified in the Kingdom *Protista*, Division *Chrysophyta*, which has about 6000 species (Prescott et al., 1993, p. 536). The various species of algae are grouped into seven "divisions," of which five are under the Kingdom *Protista*, and two are under the Kingdom *Plantae*.

Diatomaceous earth: Same as diatomite. Suggested as a term in 1860 as a substitute for kieselguhr (Cummins, 1975, No. 16, p. 2).

Diatomite: (1) A light friable siliceous material resembling chalk that is derived chiefly from the remains of diatoms and is used as a filter aid, adsorbent, filler (as for paints and plastics), and abrasive, and for thermal insulation—also calleddiatomaceous earth. See also kieselguhr (http://www.merriam-webster.com/). (2) Diatomites are essentially SiO_2, that is, about 90% and contain minor amounts of aluminum, iron, calcium, magnesium, and traces of other elements. Al_2O_3 varies from 2%–5% (Cummins, 1975, No. 15, p. 12). (3) Preferred term, introduced about 1880, for the rock composed of diatom residues (Cummins, 1975, No. 16, p. 3).

Differential pressure: Pressure differential across the filter cake, pre-coat, septum, and filter leaf, usually expressed as ΔP.

Dynamite: Explosive containing nitroglycerine with a solid substance as an absorbent. Powdered diatomite was established in 1863 by Alfred Nobel as a substance that stabilized the nitroglycerine and made it relatively safe for use.

Embed: See *straining*.

Feed: A mixture of particles and fluid that is introduced into the filter. Terms used synonymously include influent, incoming slurry, and raw water.

Filter aid: (1) Term introduced in 1914 referring to DE powders when employed for filtration and clarification of liquids. Adopted by the Celite Co. shortly after (Cummins, 1975, No. 16, p. 3). (2) Diatomaceous earth, volcanic ash, or other material that may form a filter cake. [The term "filter aid" is also used in rapid filtration by some to indicate a coagulant.]

Filter Cel: Natural milled DE for use in filtration (Cummins, 1975, No. 16, p. 10); trademark granted in 1915.

Filtration: The process by which particles are separated from a fluid by passing the fluid through a permeable material.

Flow: The rate of flow of a given fluid across a given area whose surface is normal to the velocity vectors of the flow. Also called "flow rate," "rate of flow," etc.

Flux: A flow of something; for example, a fluid, a suspension, electrons, a magnetic field, etc.

Hyflo Super-Cel: Trademark granted in 1925 (Cummins, 1975, No. 16, p. 10).

Kieselguhr: A term first applied in 1808 to earth from the Isle de France and later became a applied generally to include paste-like materials occurring in bogs and other deposits because of its supposed resemblance to a fermenting mass, that is, kiesel (silica) and guhrer (to ferment). At that time, while the silica nature of the materials was known, the fact that it was primarily diatom residues was unknown. Later in 1836, the nature of kieselguhr as fossil diatoms was established. The discovery that kieselguhr was diatoms then stimulated a search for deposits. German deposits near Hanover (Luneburger) were the most important until those of Lompoc reached full-scale operation in 1922. [Foregoing from Cummins, 1975, No. 1, pp. 1–6.]

Leaf: A plate-and-frame type of filter element that includes the support structure and septum.

Lompoc: Town in California and the site of one of the major diatomite deposits in the United States and worldwide. Discovery was in 1888 (Cummins, 1975, No. 3, p. 7); full commercial development was achieved by 1922. The deposits are a mixture of >300 species and are of marine origin.

Mesh: Number of openings in a lineal inch of wire cloth.

Micron: A metric unit of length; 10^{-6} m.

Particle size: The distribution obtained from a particle count grouped by specific micron sizes. Usually, particle size is expressed as some characteristic of the distribution obtained, for example, d10 (i.e., 10% finer than the size stated), d50, etc.

Perlite: A mediam used in the pre-coat filtration process. Perlite ore is formed from a volcanic magma flow of pure alumina silicate glass deposited onto the surface of the earth where the molten mass cools and subsequently hydrates water. When perlite ore is heated to 1600°F–2400°F, it becomes molten glass, and the water of hydration within each granule is released as expanded water vapor. Accomplished rapidly and under carefully controlled conditions, this glass liquefaction/water vaporization event results in the virtually instantaneous formation of partially fractured, low-density, multicellular particles. Additional milling and classification operations result in a variety of grades with well-defined porosity and density to meet the needs of a wide variety of

FIGURE 14.G.1 SEM micrograph of perlite filter aid illustrating its unique morphology; the product was manufactured by Harborlite/ World Minerals Inc. (Photo courtesy of Walter Pavlakovich, World Minerals Inc./Imerys Corp., Lompoc, CA.)

TABLE 14.G.2
Perlite Filter Aid

Grade	Color	Particle d(50) (µm)	Permeability (Darcy's)
Harborlite® 500	White	32.5	0.5
Harborlite® 600	White	33.2	0.8
Harborlite® 635	White	33.4	1.2
Harborlite® 700	White	35.6	1.3
Harborlite® 800	White	36.5	1.7
Harborlite® 900	White	37.4	1.9
Harborlite® 1500	White	40.3	2.3
Harborlite® 1800	White	41.8	2.8
Harborlite® 1900S	White	42.5	3.5
Harborlite® 2000	White	43.1	7.0

Source: Courtesy of Walter Pavlakovich, Celite Corp/ World Minerals Inc. (a division of Imerys Corp.), Lompoc, CA.

filtration applications. Figure 14.G.1 is a SEM photomicrograph illustrating the morphology and size of perlite as a manufactured product. Table 14.G.2 lists size and permeability data for different commercial grades of perlite.

Permeability: The property of the filter medium that permits a fluid to pass through under the influence of differential pressure. The ability of a material to permit a substance to pass through it.

Porosity: Ratio of volume of voids to total volume of media plus voids.

Pre-coat: Initial deposit of media formed by circulation of a given mass of charge through the septum area of a tank.

Pre-coat filtration: Same as *cake filtration*.

Sensitivity analysis: Changing the magnitude of one independent variable in a mathematical model yields an associated change in the dependent variable. To examine the incremental change of a dependent variable caused by a given incremental change to a selected independent variable is a sensitivity analysis. The sensitivity analysis may be repeated for as many of the independent variables that are of interest.

Septum: A screen or porous material used to support the filter media.

Straining: Retention of particles by a screen or porous media in which the pore openings are smaller than the particles to be removed. A variation of this mechanism, that is, physical retention, is what some call being "embedded." In diatomite filtration, within the body feed, the particles-to-be-removed, if they are larger than the adjacent pores within the porous media, are "embedded" by the particles around.

Voids: The pore space in a filter medium.

REFERENCES

Baumann, E. R., Diatomite filters for municipal installations, *Journal of the American Water Works Association*, 49(2):174–186, February 1957.

Baumann, E. R., Diatomite filters for asbestiform fiber removal from water, *American Water Works Association Conference Proceedings*, Denver, CO, Paper No. 10-2c, 1975.

Baumann, E. R., Pre-coat filtration, in R. L. Sanks (Ed.), *Water Treatment Plant Design*, pp. 313–371, Ann Arbor Science Publishers, Ann Arbor, MI, 1979.

Baumann, E. R. and Oulman, C. S., Polyelectrolyte coatings for filter media, 1070 Society Gold Medal Award Paper, *Proceedings of the Filtration Society*, *Filtration and Separation*, Elliccot, MD, 7(6):682–690, November/December 1970.

Baumann, E. R., Cleasby, J. L., and LaFrenz, R. L., A theory of diatomite filtration, *Journal of the American Water Works*, 70(9):1109–1119, September 1962.

Baumann, E. R. et al., Diatomite filters for municipal use, task group report, *Journal of the American Water Works Association*, 57(2):157–180, 1965.

Bell, G. R., Design criteria for diatomite filters, *Journal of the American Water Works Association*, 54(10):1241–1256, October 1962.

Black, H. H. and Spaulding, C. H., Diatomite water filtration developed for field troops, *Journal of the American Water Works Association*, 36(11):1208–1221, November 1944.

Bryant, E. A. and Brailey, D., Large-scale ozone-DE filtration: An industry first, *Journal of the American Water Works Association*, 88(11):604–611, November 1980.

Celite/World Minerals Brochure FA-84A, World Minerals, Inc., www.worldminerals.com, Celite Corp., Lompoc, CA, May 2010.

Coogan, G. J., Diatomite filtration for removal of iron and manganese, *Journal of the American Water Works Association*, 54(12):1507–1517, December 1962.

Cummins, A. B., Clarifying efficiency of diatomaceous filter aids, *Industrial and Engineering Chemistry*, 34(4):403–411, April 1942.

Cummins, A. B., *Terra Diatomacea*, Johns-Manville Corp., Denver, CO, 1975 (date estimated).

Dayton & Knight, Ltd., Pilot water treatment program, Village of 100 Mile House, Report for Client, Dayton & Knight, Ltd., Consulting Engineers, West Vancouver, British Columbia, December 1983.

Durco, Durco pressure leaf filters, Bulletin EF-21, Filtration Systems Division, The Duriron Co., Inc., Angola, NY, 1982.

Eagle-Picher, *Celatom Diatomite Filter Aids*, Brochure, Eagle-Picher Industries, Inc., Cincinnati, OH, 1970.

Eagle-Picher Minerals, Inc., Diatomite, perlite, cellulose filter aids, http://www.epminerals.com, Reno, NV, January 10, 2003.

Hoffing, E. H. and Lockhart, F. J., Resistance to filtration, *Chemical Engineering Progress*, 47(1):3–10, January 1951.

Johns-Manville, *The Filtration of Water*, Celite Filter Aids, New York, 1964.

Johns-Manville, *Maximum Clarity at Lowest Cost*, Celite Filter Aids, New York, FA-84A, 5-81, no date but estimated at 1981 from designation 5-81.

Kadey, F. L. Jr., Diatomite, in Lefond, S. J. (Ed.), *Industrial Minerals and Rocks*, 4th edn., pp. 605–635, American Institute of Mining, Metallurgical and Petroleum Engineers, New York, 1975.

LaFrenz, R. L. and Baumann, E. R., Optimums in diatomite filtration, *Journal of the American Water Works Association*, 70(7):847–851, July 1962.

Lange, K. P., Removal of *Gairdia lamblia* cysts and other substances by diatomaceous earth filtration, MS thesis, Department of Civii Engineering, Colorado State University, Fort Collins, CO, 1983.

Lange, K. P., Bellamy, W. D., and Hendricks, D. W., Removal of *Gairdia lamblia* cysts and other substances, *Diatomaceous Earth Filtration*, Vol. 1, Municipal Environmental Research Laboratory, United State Environmental Protection Agency, EPA-600/S2-84-114 (available from NTIS Order No. 84-212 703), September 1984.

Lange, K. P., Bellamy, W. D., Hendricks, D. W., and Logsdon, G. S., Diatomaceous earth filtration of *Giardia* cysts and other substances, *Journal of the American Water Works Association*, 78(1):76–84, January 1986.

Logsdon, G. S., Water filtration practices—Including slow sand filters and precoat filtration, American Water Works Association, Denver, CO, 2008.

Logsdon, G. S. and Lippy, E. C., The role of filtration in preventing waterborne disease, *Journal of the American Water Works Association*, 74(12):649–655, December 1982.

Logsdon, G. S., Symons, J. M., Hoye, R. L. Jr., and Arozarena, M. M., Alternative filtration methods for removal of *Giardia* cysts and cyst models, *Research and Technology*, 89(2):111–118, February 1981.

Lowe, H. N., Brady, F. J., Jones, M. F., Wright, W. H., and Black, H. H., Efficiency of standard army water purification equipment and of diatomite filters in removing cysts of *Endamoeba Histolytica* from water, Project WSS 346, The Engineer Board, Fort Belvoir, VA and/or The Chief of Engineers, U.S. Army, Washington, DC, July 3, 1944.

Malina, J. F., Moore, B. D., and Marshall, J. L., Poliovirus removal by diatomaceous earth filtration, Civil Engineering Department, The University of Texas, Austin, TX, June 1972.

McIndoe, R. W., Diatomaceous earth filtration for water supplies—Part 1, *Water and Wastes Engineering*, 6:50–53, October 1969a.

McIndoe, R. W., Diatomaceous earth filtration for water supplies—Part 2, *Water and Wastes Engineering*, 6:48–52, November 1969b.

McIndoe, R. W., Diatomite filtration—An old/new process, Paper presented at *Meeting of New York Section*, American Water Works Association, Liberty, NY, September 1983.

McIndoe, R. W., Logsdon, G. S., Ris, J. L., and Wirsig, A., *Pre-coat Filtration*, 1 edn., AWWA Manual M30, American Water Works Association, Denver, CO, 1988.

Metcalf and Eddy and Hazan and Sawyer. Report on pilot water treatment plant at Jerome Park Reservoir, Croton Water System. City of New York, HED-486, 1979.

Mirliss, M. J., References to DE outdated, *Opflow*, 28(8):14, August 2002.

Nolte and Associates, Inc. (Tom Mingee, Principal-in-Charge; Report Author: Todd B. Rogers), City of Vacaville water treatment plant filter optimization study technical memorandum, Task 2—DE plant survey, Draft, Nolte and Associates, Inc., Sacramento, CA, August 1997a.

Nolte and Associates, Inc. (Tom Mingee, Principal-in-Charge), City of Vacaville water treatment plant filter optimization study, Volume 1—Report, Nolte and Associates, Inc., Sacramento, CA, December 1997b.

Nolte and Associates, Inc. (Tom Mingee, Principal-in-Charge), City of Vacaville water treatment plant filter optimization study, Volume 2—Appendices, Nolte and Associates, Inc., Sacramento, CA, December 1997c.

Ongerth, J. E. and Hutton, P. E., DE filtration to remove *Cryptosporidium*, *Journal of the American Water Works Association,* 89(12):39–46, December 1997.

Ongerth, J. E. and Hutton, P. E., Testing of diatomaceous earth filtration for removal of *Cryptosporidium oocysts*, *Journal of the American Water Works Association,* 93(12):54–63, December 2001.

Oulman, C. S. and Baumann, E. R., Streaming potential in diatomite filtration of water, *Journal of the American Water Works Association*, 72(7):915–930, July 1964.

Parmelee, M. A., New York City prepares to meet SWTR exemption criteria, *AWWA Mainstream*, 34(6):1, 8, 1990.

Prescott, L. M., Harley, J. P., and Klein, D. A., *Microbiology*, 2nd edn., Wm. C. Brown Publishers, Dubuque, IA, 1993.

Rogers, T. B., Technical memorandum: Task 2—DE plant survey, City of Vacaville water treatment plant optimization study, Draft, Nolte and Associates, Sacramento, CA, August 1997.

Sanks, R. L. (Ed.), *Water Treatment Plant Design*, Ann Arbor Science Publishers, Ann Arbor, MI, 1979.

Schuler, P. F. and Ghosh, M. M., Diatomaceous earth filtration of cysts and other particulates using chemical additives, *Journal of the American Water Works Association*, 82(12):67–75, December 1990.

Spencer, C. M. and Collins, M. R., Improving precursor removal, *Journal of the American Water Works Association*, 87(12): 71–82, December, 1995.

Vander Velde, T. L., Crumley, C. C., and Moore, G. W., Experiences with municipal diatomite filters in Michigan and New York, *Journal of the American Water Works Association*, 54(12):1493–1506, December 1962.

Walton, H. G., Diatomite filtration: Why it removes *Giardia* from water, in Wallis, P. M. and Hammond, B. R. (Eds.), in *Advances in Giardia Research*, University of Calgary Press, Calgary, Alberta, Canada, 1988.

Wirsig, A., Letters to client G. using water from Clear Creek, CO, Alan Wirsig Engineering, Littleton, CO, January 11, 1986, February 5, 1986, August 4, 1986.

World Minerals, Inc., http://www.worldminerals.com, Celite Corp., Lompoc, CA, May 2010.

Part IV

Molecules and Ions

Any ambient water is a *solution* comprised of dissolved molecules and ions. As a rule, the molecular species are carbon compounds, which number in the millions. The common cations are such species as Ca^{2+}, Na^{+}, K^{+}, and Mg^{2+}. The ones that give trouble are Fe^{2+}, Mn^{2+}, and the various heavy metals such as Cr^{6+}, Cd^{2+}, Pb^{2+}, etc. Common anions are Cl^{-}, ${SO_4}^{2-}$, ${HCO_3}^{-}$, to name a few. Those that have biological nuisance effects include ${PO_4}^{3-}$ and ${NO_3}^{-}$. Dissolved gases are also present in any ambient water and include CO_2, H_2S, and O_2; some gases may require addition or removal (Chapter 18—Gas Transfer), depending on the context.

Methods to treat organic compounds include adsorption, gas transfer, or oxidation. For inorganic compounds, ion-exchange will remove both cations and anions; oxidation may be used to change the valence state of some; precipitation works as well, depending on the species.

A sure method of removing any ion or molecule or particle is by a membrane. There are several kinds of membranes characterized by their pore size. Some will remove particles such as colloids and bacteria and may be used in lieu of deep-bed filtration—the process is called *microfiltration*. Next in line is *ultrafiltration*, which has small-enough pore sizes to remove viruses and perhaps some large molecules. *Nanofiltration* will remove most organic molecules and some ions, such as Ca^{2+}. Reverse osmosis (also called *hyperfiltration*) will remove virtually all ions since the size of the membrane structure is in angstroms.

Disinfection is addressed in Chapter 19, which includes the inactivation of pathogenic organisms by chlorine gas, ozone, chlorine dioxide, and ultraviolet radiation.

For some substances in water, several treatment choices may be available. The process train selected depends upon economics and other factors, such as operating costs versus capital costs, ease of operation, esthetics, context, and site-specific considerations.

15 Adsorption

Adsorption is the attachment of molecules or particles to a surface. The surface may be a part of any solid matter, but some are more effective than others. Molecules that adsorb are largely organic, and include both natural and synthetic. Particles include viruses, bacteria, and others such as cysts and algae. Ions do not adsorb but attach by electrostatic attraction to sites within ion-exchangers or soils—the topic of Chapter 16.

Adsorption occurs extensively in the natural environment. Random contacts between molecules and particles occur throughout the hydrologic cycle and in many kinds of aquatic systems. For engineered adsorption systems, the context for "contacts" is a reactor and the solid is usually activated carbon.

Applications of adsorption include drinking water treatment, tertiary treatment of wastewaters, treatment of high-purity industrial process waters, pretreatment of industrial wastewaters prior to discharge to municipal sewer systems, pump-and-treat groundwater treatment, etc. The solid used most extensively is granular activated carbon.

15.1 DESCRIPTION

Definitions, kinds of adsorbents, adsorbates, applications, and history provide an introduction to the adsorption process. These topics help to lay the groundwork for the subsequent sections on theory, pilot plant studies, design, and operation.

15.1.1 Adsorption In-a-Nutshell

This section provides a perspective of the adsorption process and includes definitions, process description, operation, and performance measures.

15.1.1.1 Definitions

The adsorption process has its own vernacular. To understand adsorption, the definitions help to set the stage for subsequent explanations.

Adsorption: Adsorption is the bonding of two particles. The particle providing the bonding sites is called the adsorbent; its scale is usually "macro," i.e., >100 μm. The particle bonding to the site is the adsorbate; its scale is usually "micro," i.e., most are probably ≪10 μm. The bonding forces are usually "physical" in that there is not a change in chemical structure.

Absorption: Absorption is another phenomenon. According to Weber (1972), it "is a process in which the molecules or atoms of one phase interpenetrate nearly uniformly among those of another phase to form a *solution* with the second phase."

Sorption: A term that includes both adsorption and absorption is "sorption." The term seems to be used most frequently in chemical engineering; it is included only for reference.

Adsorbate: One of the particles, or the material being concentrated, is called the *adsorbate*; the other, or the adsorbing phase, is called the *adsorbent*. An adsorbate may be any particle which has attraction for an adsorbent. Molecules, macromolecules, viruses, bacteria, colloidal particles, etc., are examples.

Adsorbent: An adsorbent may be any solid material which provides bonding "sites." Soils, biological floc, chemical precipitates, macromolecules, wood, char, activated alumina, activated carbon—in fact almost any solid substance—can act as an adsorbent. The best-known and most widely used commercial adsorbent is activated carbon.

Activated carbon: Carbon that has been processed, i.e., "activated," to increase the surface area within its pores (to perhaps 1000 m^2/g). Various kinds of activated carbon are manufactured with the specific properties depending on the source material, the temperature of activation, the oxidizing substance used, such as air, steam, etc., and other factors.

GAC: Activated carbon with average size perhaps $d(50) \approx 0.5$ mm and used to "pack" a "column," resulting in a "packed bed." The "feel" within the fingers is like medium sand.

PAC: Activated carbon with average size $d(50) \approx 0.1$ mm and used in water treatment in the form of a slurry added before rapid filtration. The appearance is as a dust. Some give the size as less than U.S. mesh size 50, i.e., <0.3 mm.

Isotherm: A relation that describes the solid-phase adsorbate concentration versus the aqueous-phase concentration at a constant temperature. Common isotherms include the Langmuir, the Freundlich, and the Brunauer–Emmett–Teller (BET).

Reactor: As applied to adsorption, a confined volume, engineered to effect the adsorption reaction to an extent specified. The most common adsorption reactor is a column that has been "packed" with an adsorbent (usually GAC).

Complete-mix reactor: A reactor in which the adsorbate is homogeneous with respect to any spatial dimension. The homogeneity is usually accomplished by a mixer. For a continuous-flow reactor, any entering solute is distributed "instantaneously," by definition.

Fluidized-bed reactor: A reactor in which the adsorbent, e.g., GAC, is suspended by fluid drag due to upflow and/or turbulence.

Batch reactor: A defined-volume reactor, e.g., a tank, in which there is zero influent flow and zero outflow during a period of reaction.

Continuous-flow reactor: A reactor in which the influent flow >0 and the effluent flow >0. Usually, such reactors are either complete-mix or a column. The former are more common for powdered activated carbon; the latter are more common for granular activated carbon.

Column: A packed-bed reactor that is shaped as a column and is a continuous-flow. The reactor column is nonhomogeneous with respect to adsorbate concentration with depth, *Z*.

Breakthrough curve: A term given to the characteristic S-shaped concentration versus time curve that is seen in the effluent of a packed-bed column as the column nears exhaustion, i.e., as it approaches "saturation" near the bottom of the column. The mathematical statement is $C(t)_{Z=Z_{max}}$.

Wave front: The concentration, *C*, versus distance, *Z*, curve at a given time, i.e., $C(Z)_t$ within a column, or packed-bed reactor. The emergence of the wave front (i.e., its translation) at the bottom of the column defines the "breakthrough" curve. What is called here the "wave front" is also called, synonymously, the "mass-transfer zone" (see also Neethling and Culp, 1990, p. 132, Snoeyink, 1990, p. 795). Johnson et al. (1964, p. 9) refer to the phenomenon of a wave front (of dissolved organic matter) moving downward through the column and that complete exhaustion is not accomplished until the total bed length has had this "front" pass through it.

15.1.1.2 Process Description

The adsorption process is the assimilation of adsorbate molecules or particles by the surface of a solid, i.e., the adsorbent. This process may be depicted by the reaction,

$$X + A \rightarrow A \cdot X \tag{15.1}$$

where

- X is the concentration of adsorbent, X, in aqueous suspension (kg/m^3)
- A is the concentration of adsorbate, A, in aqueous solution (kg/m^3)
- $A \cdot X$ is the concentration of A in adsorbed state on or within X (kg A/kg X/m^3 solution)

The adsorbate may be any dissolved molecular substance or particles in suspension. Usually, for a designed system, the adsorbent is activated carbon.

15.1.1.3 Operation

Regardless of the kind of system, an adsorbent, e.g., GAC, will eventually become "saturated," i.e., exhausted, by continuous exposure to a given adsorbate. Therefore, to utilize the GAC effectively, several columns, e.g., 2 or 3, may be arranged serially. An additional column may serve as standby. The first column will become saturated first and then must be removed from operation. At same time, the standby column will be placed in operation as operating column #3, which ensures that the breakthrough curve is perpetually within column #2, or at least not extending to the end of column #3.

15.1.1.4 Performance Measures

Several measures of performance have been used. The "effectiveness" of the process is whether the effluent concentration is less than a target value, C_b (the breakthrough value) i.e., $C_e \leq C_b$. Another measure of effectiveness is the volume of water treated when $C_e \approx C_b$ occurs, i.e., V_b. A measure of efficiency is the carbon usage rate (CUR); CUR = (mass of carbon involved when $C_e \approx C_b$ occurs)/V_b (Reed et al., 1996). A parameter used in research is "bed volumes" treated, in which a bed volume is defined as the pore volume in the GAC reactor bed, i.e., pore volume of reactor bed = V(reactor bed) $\cdot P$ (in which P = porosity). The relation between "bed volumes" treated and time of operation, t, is "bed volumes treated" = $Q \cdot t/(V(\text{reactor bed}) \cdot P)$.

15.1.2 Adsorbents

The range of manufactured adsorbents is not large; the most common is activated carbon. Therefore, it is described more extensively.

15.1.2.1 Kinds of Adsorbents

Almost any solid substance can serve as an adsorbent. In nature, rocks in mountain streams may provide a surface for biofilms; sand may serve the same function in the percolation of water through soil to groundwater. Clay or silt may provide adsorption sites for bacteria (and probably viruses). In water treatment, biological or chemical flocs may provide adsorption sites for organic molecules.

15.1.2.1.1 Activated Carbon

The most common adsorbent used for engineered adsorption systems is activated carbon, which can adsorb a wide spectrum of organic compounds. Its large surface area, e.g., 1000 m^2/g, provides the highest adsorption capacity of any material. Its mean pore diameter is about 3–6 μm (McGuire and Suffet, 1978, p. 623).

15.1.2.1.2 Synthetic Resins

Several commercial resins (i.e., "macroreticular" resins, meaning that they resemble a net in internal appearance or structure) are available, e.g., XAD-1®, XAD-2®, XAD-4®, XAD-7®, XAD-8®, etc. (Kunin, 1980) with surface areas 100, 300, 725, 450, and 160 m^2/g, respectively, with mean pore diameters 200, 100, 48, 85, and 150 Å, respectively. The first three are nonpolar adsorbents and XAD-7®, XAD-8® are polar. The pore size distribution may be quite large with XAD-7® having a range of 30–1800 Å. These resins are polymeric, i.e., they are pure polymers without ion-active groups, and depend upon their large surface areas for their adsorption capacities. The XAD-8® resin (Rohm and Haas, Philadelphia, Pennsylvania) has been used to concentrate natural organic matter (NOM) from natural waters at low pH, e.g., by addition of acid prior to the column. The NOM is eluted by adding a concentrated base.

15.1.2.1.3 Activated Alumina

Activated alumina is amphoteric, i.e., it has a positive charge at pH < pHpzc and a negative charge at pH > pHpzc (pHpzc is the pH at which the charge is zero). The removal of ions is complex and occurs by chemisorption, forming a surface complex via covalent bond, or by ion-exchange (Fleming, 1986, pp. 159–160). Activated alumina is an oxide or hydroxide of aluminum. It is one of the few options for the removal of fluoride ion. It is classified here as an ion-exchange material rather than an adsorbent and, therefore, is discussed more thoroughly in Chapter 16.

15.1.2.1.4 Aluminum Hydroxide Floc

Natural organic matter (NOM) has been found to adsorb on aluminum hydroxide floc (see Section 9.3.2.3) in accordance with an isotherm relation. While the removal of NOM during coagulation has become a goal in drinking water treatment, the context is that of a passive removal vis-à-vis an engineered reactor. In other words, NOM removal occurs, but coagulation is the primary objective; NOM removal is increased in accordance with a stoichiometric addition of alum (called "enhanced coagulation," Section 9.1.2.1).

15.1.2.1.5 Soil Organic Carbon

The organic fraction in soils is largely amorphous matter and (Marshall, 1964, p. 158). Its adsorptive character is difficult to determine.

15.1.2.1.6 Soil Minerals

Clays, e.g., montmorillonite, illite, kaolinite, function as ion-exchangers (see Chapter 16) and are noted for small size and large specific surface areas, e.g.,

	Particle Size[a] (μm)	Specific Surface Area[a] (m^2/g)
Kaolinite	0.3–3	10–20
Illite	0.1–2	80–100
Montmorillonite	0.1–1	800

[a] Lambe and Whitman (1969).

As seen, nominal sizes are 1 μm, with specific surface areas 10–800 m^2/g; also, the clays tend to be "platy" in shape. The large surface areas for these clays, i.e., even 10–20 m^2/g for kaolinite, render them amenable to adsorption as well as ion-exchange.

15.1.2.2 Sources of Activated Carbon

Almost any carbonaceous matter can serve as raw material in the manufacture of activated carbon. Some of these materials include wood, coal, peat, lignin, nut shells, lignite, bone, blood, petroleum residues, etc. The characteristics of activated carbon are derived from both the nature of the raw material used and the conditions of manufacture. For example, coal as a source results in a hard and dense product; but no two coal seams are alike (Greenbank and Spotts, 2003).

15.1.2.3 Manufacturing of Activated Carbon

The preparation of an activated carbon from a given raw material generally involves three main steps (West, 1971, p. 6): (1) grinding, (2) carbonization, and (3) activation. The grinding produces a specified particle size. Carbonization or pyrolysis, the second step, is the heating of the raw material in an inert atmosphere to drive off volatile substances and increase the proportion of pure carbon. The process temperature is about 800°C; the product is called a "calcinate." The final step is "activation," an oxidation reaction, which burns residual hydrocarbons at temperatures 800°C–950°C. The oxidizing gas may be steam, carbon dioxide, or oxygen. This reaction increases the pore volume producing a network of microscopic pores and creates active functional groups on the internal surfaces. To give an idea of the effect of carbonization, in one example involving a Wyoming coal, the percent of fixed carbon increased from about 45% in the coal to about 90% in the calcinate product.

The skeleton of calcinate is composed of plates of graphite created from the starting material. The size and degree of imperfection of the plates (i.e., spacing, angle, size of gaps between neighboring plates, and the general structural order) is related to the chemical building blocks that comprised the starting material and the heat of activation. Two characteristics of interest are (1) the surface area of the resulting carbon, and (2) kinetics. Surface area is determined by the voids between plates and the pore structure. Kinetics is determined by the size of the pores and their nature (e.g., their tortuosity).

15.1.2.4 Characteristics of GAC

Activated carbon is manufactured as powdered or granular. Granular carbons are those larger than US Sieve Series No. 50, i.e., $d(50) \approx 0.3$ mm, while powdered carbons are smaller. Granular carbons are used in packed beds or fluidized beds while powdered carbons are used in slurries or in complete-mix reactors. Table 15.1 compares data for four manufactured granular activated carbons, illustrating the variation in properties; each manufacturer produces a range of products.

15.1.2.4.1 Physical Properties

The properties of activated carbon which are of interest include hardness, resistance to abrasion, density, and size. The values of a few representative carbons are given in Table 15.1. Standard tests (USEPA, 1973) have been devised as measures of these properties and are summarized in the following paragraphs.

Hardness number: Hardness is a quality of the retention of original surfaces. It is measured by subjecting 50.0 g of

TABLE 15.1
Properties of Four Commercial Activated Carbons

Properties/Specifications	ICI America Hydodarco 3000	Calgon Filtrasorb 300	Westvaco Nuchar WV-L	Witco 517
Physical properties				
Surface area, BET (m^2/g)[a]	600–650	950–1050	1000	1050
Apparent density (kg/L)	0.43	0.48	0.48	0.48
Particle density (g/mL)	1.4–1.5	1.3–1.4	1.4	0.92
Effective size (mm)	0.8–0.9	0.8–0.9	0.85–1.05	0.85
Uniformity coefficient	1.7	≤1.9	≤1.8	1.4
Specifications				
Sieve Size (U.S. Std. Series)[b]	8 × 30	8 × 30	8 × 30	12 × 30
Iodine No.	650	900	950	1000
Ash (%)	No data	8	7.5	0.5
Moist. as packed (max%)	No data	2	2	1

Source: USEPA, *Process Design Manual for Carbon Adsorption,* Technology Transfer Division, U.S. Environmental Protection Agency, Cincinnati, OH, pp. 2–5, 1973.

[a] The surface area of a substance is measured by a standard test using nitrogen gas based on the BET multilayer isotherm from a paper by Branauer et al. (1938).

[b] Size nomenclature: Assume a mesh size, e.g., 8 × 30. The first number, i.e., "8," means that most (95%) of the granular material is smaller than the U.S. sieve #8 (2.38 mm opening); the second number, i.e., "30," means that most (95%) of the granular material is larger than the U.S. sieve #30 (0.59 mm opening).

carbon of 6 × 8 mesh size to the action of steel balls and noting the weight of carbon retained on the No. 8 sieve. Hardness No = weight of material retained on sieve × 2.

Abrasion number: Abrasion is the resistance to the degradation of material by mechanical action. In the test, 100 g of carbon is subjected to the action of steel balls. Sieve analyses, from which mean particle diameters are calculated, are done before and after the abrasive action.

Apparent density: The apparent density, also called bulk density, is measured in grams of carbon per mL of total volume occupied in place of carbon. The "in place" carbon volume includes the sum of the volumes of the carbon particles and the volume of the voids between particles.

Particle density: The density of the carbon particle itself in gm/mL is particle density.

Particle size distribution: This is obtained by sieve analysis of the activated carbon. The results are plotted as sieve opening in mm on the ordinate versus cumulative percent passing on the abscissa.

Mean particle diameter: The mean particle diameter, D_m, is based upon measurements from a sieve analysis. It is calculated as an integral of the weight of particle size *i* times the particle diameter of size *i* divided by the total weight of sample; a sieve analysis of a sample is used for the calculation.

Effective size: That sieve opening in millimeters which passes 10% of the total sample is the effective size, designated, d_{10}.

Uniformity coefficient: The uniformity coefficient is the ratio d_{60}/d_{10}.

Pore volume of particles: This term refers to the total volume of pores within carbon particles and may range 0.40–1.1 mL/g (West, 1971).

Porosity of carbon particles: This is the ratio of pore volume to particle volume, which may range 0.4–0.76 (West, 1971).

Porosity of bulk carbon: The ratio of the volume of intraparticle voids to total bulk volume is defined as porosity, $P = V_v/V$, where P = porosity; V_v = void volume; $V = V_v +$ V(solids).

15.1.2.4.2 Index Numbers

Another empirical approach to adsorption capacity utilizes what is called here, "index numbers," i.e., iodine number, molasses number, methylene blue number, etc., The numbers are obtained by prescribed procedures for contacting selected adsorbates such as iodine, molasses, methylene blue, alizarin red, alkyl-benzene sulfate (ABS) number, etc., with a selected adsorbent, usually activated carbon. Definitions for several are given in the Glossary.

15.1.2.4.3 Internal Structure

Two characteristics of activated carbon that define adsorption capacity and diffusion rate are surface area and structure. Surface area refers to the area of the internal pores. Structure refers to internal pore volume, openness of pores, pore size distribution, and size of pores. A large surface area provides more sites for adsorption, while structure determines the size of molecules which can be adsorbed and the capabilities of the activated carbon to permit the migration of the adsorbates within the pores.

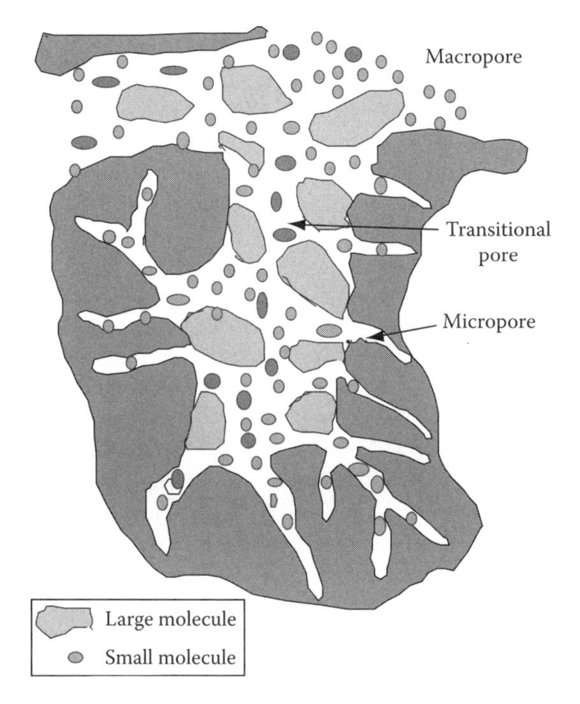

FIGURE 15.1 Sketch of internal pore structure of activated carbon. (Adapted from Pittsburgh Activated Carbon, *Descriptive Brochure*, Pittsburgh Activated Carbon, Pittsburgh, PA, p. 7, 1968.)

Figure 15.1 is a sketch depicting schematically the internal structure of activated carbon pores. Some of the macropores open directly to the external surface of the GAC particle, while the transitional pores connect the macropores and the micropores. As indicated in the sketch, the small molecules have access to the micropores while the larger molecules may gain entrance only to the transitional pores. Thus, the capacity of the GAC for the large molecules is not as high as for the small molecules (on a moles adsorbate per gram of GAC basis). In terms of surface area, the micropores provide about 0.9 of the total.

Figure 15.2 shows a sequence of four photomicrograph of an activated carbon particle at magnifications from 20× to 50,000× for (a) and (d), respectively, and show the actual pores. Scales are indicated on each photograph with (a) at 1000 (i.e., 1 μm), with (d) showing 3000 Å (i.e., 3 μm). Some of the features of each photomicrograph are given below.

Figure	Comments
15.2a	Overall view of particle showing irregular surface
15.2b	A close-up of the surface shows macropore openings into carbon
15.2c	Going closer, the irregular crystalline structure becomes evident, along with pores
15.2d	At 50,000×, the irregular microcrystalline platy structure becomes evident

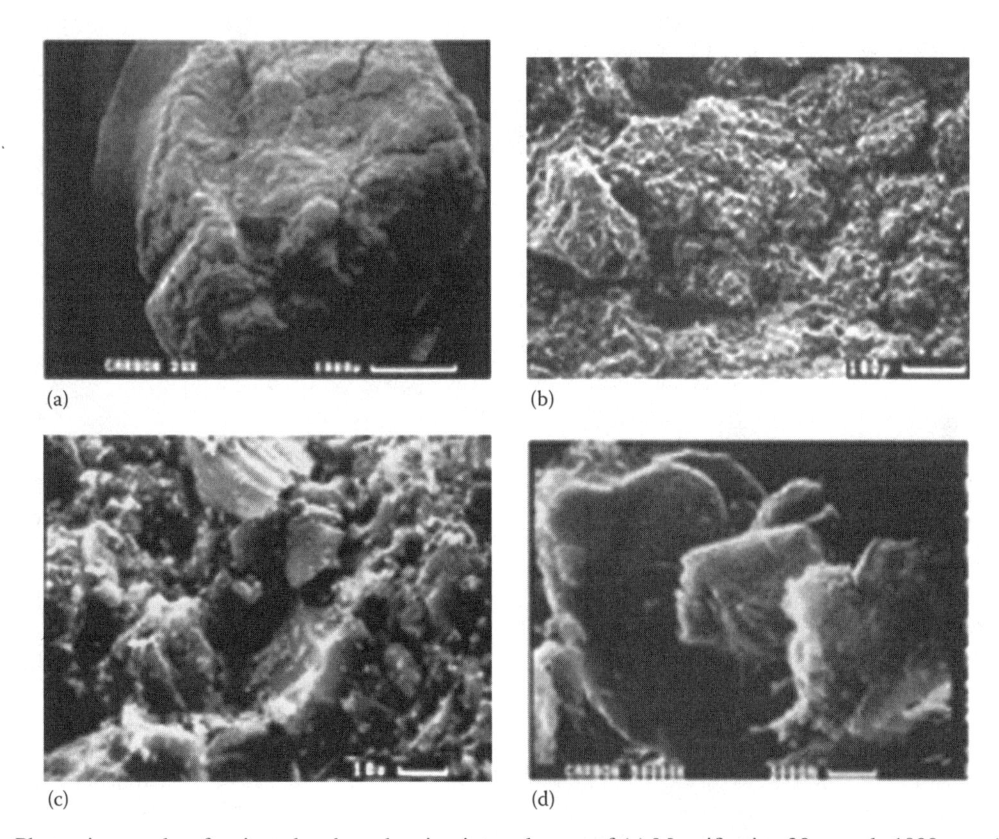

FIGURE 15.2 Photomicrographs of activated carbon showing internal pores of (a) Magnification 20×; scale 1000 μm. (b) Magnification 100×; scale 100 μm. (c) Magnification 1000×; scale 10 μm. (d) Magnification 50,000×; scale 3000 Å. (Courtesy of Brooks, D.R., Calgon Activated Carbon Co., Pittsburgh, PA, 2003.)

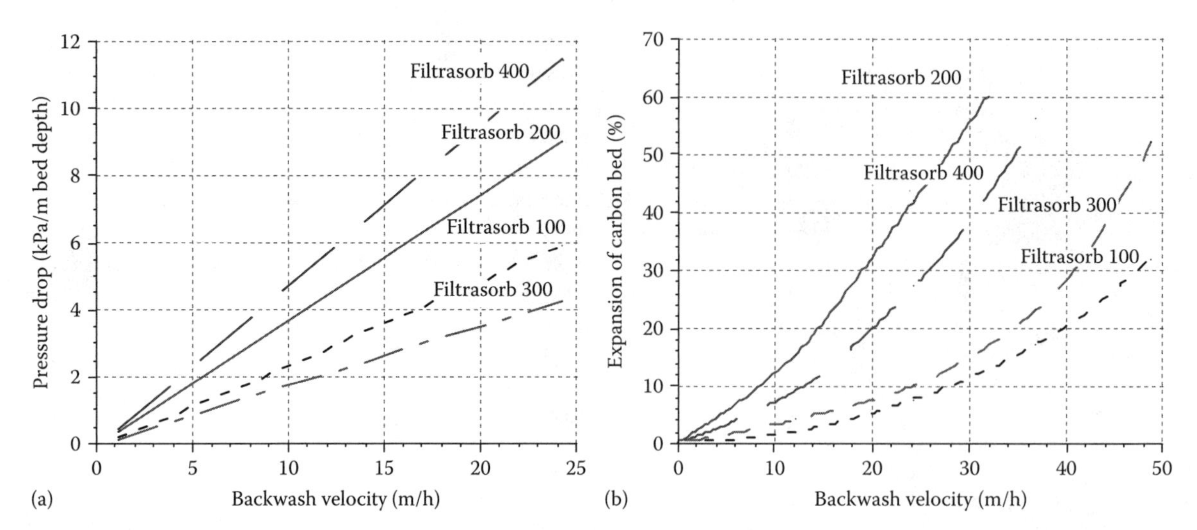

FIGURE 15.3 Effects of backwashing of four sizes of GAC packed beds. (a) Hydraulic gradient. (b) Expansion of bed. (Adapted from Calgon Bulletins 20-3 and 20-4, 1966.)

15.1.2.4.4 Pore Size

The pore sizes vary with the carbon, i.e., raw material and production process. A size classification is given as macropores ≥100 Å radius and micropores ≤100 Å radius (West, 1971).

15.1.2.4.5 Microscopic Structure

Looking at the carbon in more detail, the apparently amorphous carbon consists of "flat plates" in which the carbon atoms are arranged in a hexagonal lattice, i.e., "crystallites" (Hassler, 1963, p. 186, 1974, p. 185). Each microcrystallite is a stack of graphite planes. The channels through the graphite regions and interstices between microcrystallites are *macropores*, whereas the fissures within and parallel to the graphite planes are the *micropores*.

Surface area: The surface area is measured based on the BET isotherm using nitrogen gas as the adsorbent. Surface area may range from 100 to 2000 m^2/g, but generally it is in the neighborhood of 1000 m^2/g (West, 1971, p. 6).

15.1.2.4.6 Hydraulics of Packed Beds

The hydraulic characteristics of packed beds of carbon must be known for design. Of interest are the hydraulic gradient versus velocity through the packed bed (HLR) and the percent expansion of carbon bed versus HLR. Figure 15.3a and b gives these relationships graphically for four commercial granular activated carbons. Example 15.1 illustrates the application of the plots.

Example 15.1 Pressure Loss through a Column

Given

A GAC column is packed with Filtrasorb® 400 and is to be operated at HLR = 12.2 m/h (5.0 gpm/ft^2) in the downflow mode; the temperature is 10°C and the column depth is L(reactor) = 10.0 m.

Required

Determine the headloss across the packed bed.

Solution

1. Assume $T = 20°C$ for the results of Figure 15.4 and calculate k

   From Figure 15.4, for HLR = 12.2 m/h, $th \approx 5.0$ kPa/m.
2. Calculate k from the Darcy Equation E.5, i.e.,

$$v = -k\left(\frac{\rho g}{\mu}\right)\cdot\frac{\Delta h}{\Delta Z}$$

3. Utilize the spreadsheet, Table CDEx15.1(c), for $T = 20°C$, i.e., the first row, to obtain k, i.e.,

$$k = 6.78\cdot 10^{-10}\ \mathrm{m}^2$$

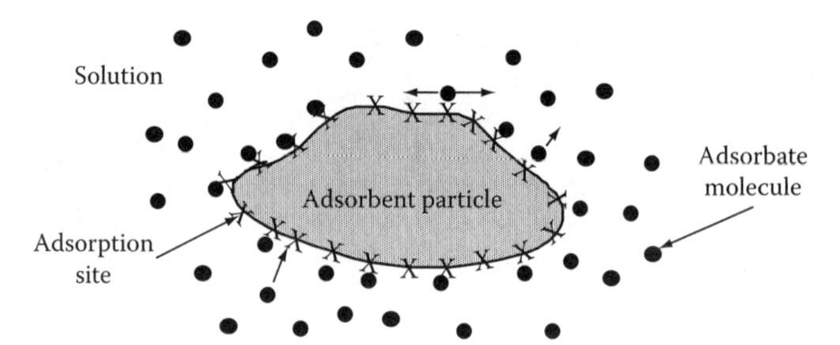

FIGURE 15.4 Adsorption as a statistical phenomenon. (From Hendricks, D.W. and Kuratti, L.C., Kinetics Part I—A laboratory investigation of six proposed rate laws using batch reactors, Utah Water Research Laboratory, Utah State University, Report PRWG66-1, p. 4, July 1973. With permission.)

TABLE CDEx15.1
Determination of Headloss in GAC Column Given k

(a) Temperature coefficients

	M0	M1	M2	M3	M4	M5
μ(water)	0.0017802356694	−5.6132434302E-05	1.0031470384E-06	−7.5406393887E-09		
ρ(water)	999.84	6.82560E-02	−9.14380E-03	1.02950E-04	−1.18880E-06	7.15150E-09

Formula; $y = M0 + M1 \cdot T + M2 \cdot T^2 + M3 \cdot T^3 + M4 \cdot T^4$

(b) Fixed data

$g = 9.80665 \text{ m/s}^2$

(c) Calculation of headloss in GAC column

T	μ(water)	ρ(water)		HLR		L(reactor)	k	K	Δh		d_{10}	
(°C)	(N s/m²)	(kg/m³)	GAC	(m/h)	(m/s)	(m)	(m²)	(m/h)	(m)	(kPa)	(mm)	R
20	0.00100	998.204	Filtrasorb 400	12.20	0.00339	10.00	1.00E-09	3.53E+01	3.46	33.90	0.65	2.2
10	0.00131	999.700	Filtrasorb 400	12.20	0.00339	10.00	1.00E-09	2.69E+01	4.53	44.46	0.65	1.7

4. Calculate Δh(col) for $T = 10°C$, from the second row in the spreadsheet, to give, for L(reactor) = 10.0 m,

$$\Delta h(\text{col}, T = 10°C) = 6.69 \text{ m } (65.6 \text{ kPa})$$

5. Compare Δh(col, $T = 20°C$)
   From the first row in Table CDEx15.1(c), for $T = 20°C$, L(reactor) = 10.0 m,

$$\Delta h(\text{col}, T = 20°C) = 5.10 \text{ m } (50.0 \text{ kPa})$$

6. Check the Reynolds Number, **R**, for $T = 20°C$,
   From http://www.calgoncarbon.com/carbon_products/documents/Filtrasorb400.pdf,
   $0.55 \leq d_{10} \leq 0.75$ mm; therefore, let, $d_{10} \approx 0.65$ mm, to give, $\mathbf{R} = \rho v d_{10}/\mu \approx 2.2$

Discussion

First, since the Darcy relationship only starts to deviate from linearity at **R** = 1, the Darcy assumption gives a reasonable estimate of headloss, facilitated by Table CDEx15.1. The Forcheimer equation, Section E.7, Equation E.3, would give a more accurate estimate. An on-site pilot plant is always preferred, however, to a calculation.

15.1.2.5 Shipping Data

For one manufacturer, carbons are packaged in 60 lb bags (net). They are available on trailer-truck pallets, 38 bags per pallet, 16 pallets per standard 40 ft trailer. Rail car pallets are 42 bags per pallet, 24 pallets per standard 50 ft car. Bulk shipment by truck is 30,000 lb. minimum; by rail 50,000 lb minimum (Calgon Bulletin 20-1, 1966). Reactor design may be based, in part, on shipping data.

15.1.3 Adsorbates

Adsorbates include mostly organic compounds. Macroscopic particles, e.g., bacteria and viruses, may adsorb, but "passively" (as opposed to being the intended "target" material to be removed).

15.1.3.1 Organic Compounds

Organic compounds are virtually countless in number with some 4-million synthetic compounds listed by the ACS registry as of 1977. Some 33,000 chemicals were in common use and some 1260 were found in ambient waters (Dobbs and Cohen, 1980, p. 1), with >1000 being regulated. The latter estimate was based upon a survey of regulated substances; some 4000 were identified (Champlin and Hendricks, 1993). Among those regulated are the aromatic compounds that include pesticides, herbicides, surfactants, and phenol (Cookson, 1978, p. 264). Examples of organic compounds found in drinking water, i.e., after treatment, in a survey involving 11 cities, include (Clark and Lykins, 1989, p. 4)

Benzene	Hexachlorobenzene	Chlorobenzene
carbon tetrachloride	Lindane	Chloromethylether
bis(2-chloroethyl)ether	PCB	Dibromochloromethane
Chloroform	Tetrachloroethylene	1,3-Dichlorobenzene
1,2-dichloroethane	Trichloroethylene	Dichloromethane
Dieldrin	Vinyl chloride	Methylene chloride
DDT, DDE	Bromodichloromethane	Vinylidene chloride
Heptachlor		

TABLE 15.2
Organic Compounds Amenable to Adsorption by GAC

Class	Examples
Aromatic solvents	Benzene, toluene, xylene
Polynuclear aromatics	Naphthalene, biphenyl
Chlorinated aromatics	Chloro benzene, PCB's, endrin, toxaphene, DDT
Phenolics	Phenol, cresol, resorcinol, nitrophenols, chlorophenols, alkyl phenols
Aromatic amines and high MW aliphatic amines	Aniline, toluene, diamine
Surfactants	Alkyl benzene sulfonates
Soluble organic dyes	Methylene blue, textile dyes
Fuels	Gasoline, kerosene, oil
Chlorinated solvents	Carbon tetrachloride, perchloroethylene
Aliphatic and aromatic acids	Tar acids, benzoic acids
Pesticides/herbicides	2.4-D, atrazine, simazine, aldicarb, alachlor, carbofuran

Source: Adapted from Groeber, M.M., Granular Activated Carbon Treatment, Engineering Bulletin EPA-540/2-91/024, Office of Research and Development, U.S. Environmental Protection Laboratory, Cincinnati, OH, October, 1991.

15.1.3.1.1 Adsorbable Categories of Compounds

Compounds not amenable to adsorption are those that have low MW and high polarity (Groeber, 1991, p. 2); examples include low MW amines, nitorsamines, glycols, and certain ethers. The readily adsorbed compounds include pesticides, polynuclear aromatic hydrocarbons, phthalates, phenolics, and substituted benzenes (Dobbs and Cohen, 1980, p. 3). Examples of absorbable substances within different categories of organic compounds are given in Table 15.2.

15.1.3.2 Natural Organic Matter

Natural organic matter (NOM) is quantified, in general, by total organic carbon (TOC), and if filtered, the TOC measure is dissolved organic matter (DOC) (see Appendix 2.A). Humic and fluvic acids comprise 0.4–0.9 fraction of DOC (Karanfil et al., 1996a, p. 2187). Adsorption of humic matter is governed largely by molecular size with lower MW fractions being removed preferentially.

In treatment trains for drinking water, GAC reactors are located, as a rule, after coagulation and filtration. The latter processes reduce ambient DOC concentrations 0.1–0.9 fraction, with about 0.3 removal expected. Generally, the higher MW compounds are removed by coagulation/filtration removals; thus the influent to the GAC reactors is mostly the lower MW compounds (Karanfil et al., 1996b). If synthetic organic compounds (SOCs) are to be treated at the same time, the "preloading" of DOCs will preempt the removal of the SOCs (Carter and Weber, 1994; Müller et al., 1996).

15.1.4 Applications

For engineered systems, activated carbon is the adsorbent of choice in most situations in which organic carbon is an issue. Such situations include the following: (1) NOM occurs in ambient waters (see Appendix 2.A) and reacts with chlorine (and other halogens). (2) Organic carbon residuals in municipal wastewaters may impede reuse unless removed. (3) Synthetic organic carbon (SOC) compounds discharged into ambient freshwaters are a health hazard in potable water and must be removed. (4) In some cases, e.g., the Love Canal, SOCs are a health hazard with respect to various forms of exposure possibilities. (5) Industrial process waters often have exacting specifications, which require removal of any organics. (6) Industrial wastewaters may need pretreatment prior to discharge into municipal sewerage systems in order to meet standards for acceptance to a municipal wastewater treatment plant; if discharged to an ambient water, effluent discharge standards must be met.

15.1.5 History

The use of wood chars and bone chars in medicine dates to 1550 BC (Hassler, 1974). Their use in modern industry started in the early years of the nineteenth century to purify sugar. The development of modern activated carbons is dated to patents in 1900. The first commercial production (Kornegay, 1979, p. 1) was in 1913 by the Industrial Chemical Company (later, in 1979, the company became the Chemical Division, Westvaco Corporation). The first municipal use of activated carbon was at the Hackensack Water Company, Milford, New Jersey in 1930 when PAC was used for taste and odor control. By 1943, 1200 water treatment plants were using PAC (Baker, 1948, p. 454). PAC was favored over GAC because of the relatively low cost for storage and feed equipment and the flexibility in operation.

As a unit process, GAC adsorption emerged in 1960 as a part of the *Advanced Water Treatment Research Program* (AWTR) of the U.S. Public Health Service (Anon., 1962). The groundwork for this program was the survey of organics in ambient surface waters in the late 1950s that used the CCE (carbon chloroform extract) method of sampling and analysis. Based on the findings of this survey, the AWTR program was initiated in 1960 and authorized by Congress in the 1961 amendments to the Federal Water Pollution Control Act, i.e., Public Law 87-88. The mandate was to search for new technologies that could treat secondary effluents such that the product was suitable for further uses (Morris and Weber, 1962, p. iii).

The AWTR program sponsored extramural research in activated carbon adsorption, which included pilot plant studies, demonstration scale plants, and one full-scale plant (South Lake Tahoe) built during the 1960s. The program paved the way for the use of granular carbon beds as a unit process in a "tertiary-treatment" train. Tertiary treatment became the means to use effluent streams from municipal wastewater treatment plants for uses such as industrial cooling water, irrigation of public parks, and other uses as may be dictated by water rights or market factors.

The issue of trace organics (found in the finished water of New Orleans in 1974) was the stimulus for the expanded use of GAC in drinking water. The motivation was to control carcinogenic substances, not just taste and odor compounds. To document the extent of carcinogenic substances, the USEPA conducted a survey of drinking water, i.e., the National Organics Reconnaissance Survey reported in 1975 (Symons et al., 1975). The surveys led to the 1977 Interim Primary Drinking Water Regulations giving impetus to the use of activated carbon in controlling organics and a 1978 proposed regulation published in the *Federal Register* that required activated carbon treatment under certain circumstances. The proposed regulation was highly controversial and was withdrawn (Weber, 1984, p. 906).

15.1.5.1 Lore

Concerning the design of GAC reactors, two criteria emerged from the AWTR program, e.g., $4.9 \leq \text{HLR} \leq 24$ m/h (2–10 gpm/ft^2) and the use of "empty-bed contact time (EBCT). Both criteria have become established as a basis for sizing reactors. By definition,

$$\text{HLR} \equiv \frac{Q}{A} \tag{15.2}$$

where

HLR is the hydraulic loading rate, also called "superficial velocity" (m/s)
Q is the flow into packed bed (m^3/s)
A is the cross-sectional area of packed bed (m^2)

By definition, EBCT = V(packed bed)/Q. Dividing the numerator and denominator on the right side by A(packed bed) gives

$$\text{EBCT} = \frac{L(\text{packed-bed})}{\text{HLR}} \tag{15.3}$$

where

EBCT is the detention time in portion of reactor packed with media for empty condition (s)
L(packed bed) is the length of packed-bed portion of reactor (m)
V(packed bed) is the volume of packed-bed portion of reactor without the media, i.e., the "empty" bed (m^3)

Recommended criteria for HLR and EBCT (Culp et al., 1978, p. 185) are $4.9 \leq \text{HLR} \leq 24$ m/h (2–10 gpm/ft^2), and $10 \leq \text{EBCT} \leq 50$ min, respectively. The EBCT has been accepted in practice as a parameter for sizing a GAC reactor and was considered a most important design parameter in sizing a reactor (Hager and Flentj, 1965, p. 1445), with HLR having little effect (Culp and Culp, 1974, p. 236). As seen in Equation 15.3, for a given HLR, if EBCT changes by some factor, L(packed bed) must change by the same factor. The relationship is illustrated in Example 15.2.

Example 15.2 Relation between EBCT and L(packed bed)

Given
Data from the Pomona, California pilot plant (Section 15.3.2.1), which had four GAC reactors in series, are EBCT = 37 min; HLR = 18.6 m/h (7.6 gpm/ft^2).

Required
Calculate L(packed bed)

Solution
Apply Equation 15.3, i.e.,

$$37 \text{ min} = L(\text{packed bed})/(18.6 \text{ m/h} \cdot \text{h}/60 \text{ min}),$$

$$L(\text{packed bed}) = 11.60 \text{ m}$$

Discussion
A packed bed is not homogeneous and so the idea of "contact time" has no bearing on the performance of the reactor. The value of L(packed bed) must, however, be longer than the "reaction zone"; also the longer the L(packed bed), the longer the run time, i.e., before "breakthrough" and column exhaustion. Thus L(packed bed) is related conceptually to reactor performance (see also Section 15.2.3.1).

15.1.5.2 Science

The "science" of adsorption as a water treatment process evolved from the AWTR program. From the research sponsored, several key ideas developed, which are given below.

1. Isotherms were established as a methodology to express capacity of activated carbon
2. Kinetics became a means to express the rate of uptake of the adsorbate by the GAC
3. Modeling by pilot plants became a means to take out the uncertainty in process design
4. Mathematical modeling, based on the materials balance principle and kinetics, became a means to better understand reactor design and operation

15.1.5.3 Practice

As a "technology" GAC treatment has emerged only since about 1960, based on stimuli from the USPHS Advanced Water Treatment Program (Anon., 1962). Examples of tertiary wastewater treatment have included plants at Pomona, California, c. 1966; South Lake Tahoe, c.1966; Colorado Springs, c. 1968; Orange County Water District, California, c. 1974; Denver, c. 1982. Water treatment examples have included Nitro, West Virginia, c. 1963; Cincinnati, Ohio, c. 1992; Commerce City, Colorado, c. 1989. These and other examples are reviewed in Section 15.4.3.

15.2 ADSORPTION PROCESS THEORY

The two key elements of adsorption theory are (1) the adsorption reaction, and (2) kinetics. Reactor modeling (Section 4.3.3, Section 15.2.3) in terms of a materials balance relation is the basis for assimilating these elements into a theory for design and operation.

15.2.1 EQUILIBRIUM

The first step in modeling adsorption is to depict equilibrium. Complementary aspects of equilibrium include the reaction, statistical equilibrium, the Langmuir isotherm, the Freundlich isotherm.

15.2.1.1 Reaction

The adsorption reaction may be depicted,

$$\mathrm{C} + \mathrm{X} \leftrightarrow \mathrm{C} \cdot \mathrm{X} \tag{15.4}$$

or rewritten as

$$\mathrm{C} + \mathrm{X} \leftrightarrow \overline{\mathrm{X}} \tag{15.5}$$

where
- C is the adsorbate concentration ($\mathrm{kg/m^3}$)
- X is the unoccupied adsorbate sites per unit of adsorbent material per unit of volume (kg unoccupied adsorbate capacity/kg adsorbent/$\mathrm{m^3}$ solution)
- C · X is the concentration of sites occupied by adsorbate molecules (kg adsorbate adsorbed/kg adsorbent/$\mathrm{m^3}$ solution)

$$\overline{\mathrm{X}} = \text{same as } \mathrm{C} \cdot \mathrm{X}$$

Figure 15.5 depicts the equilibrium between adsorbate molecules in solution and the adsorbent surface. The adsorbate particles are shown (1) in solution, (2) adsorbed on a bonding site, (3) being desorbed from a site, and (4) oscillating about a site due to thermal energy.

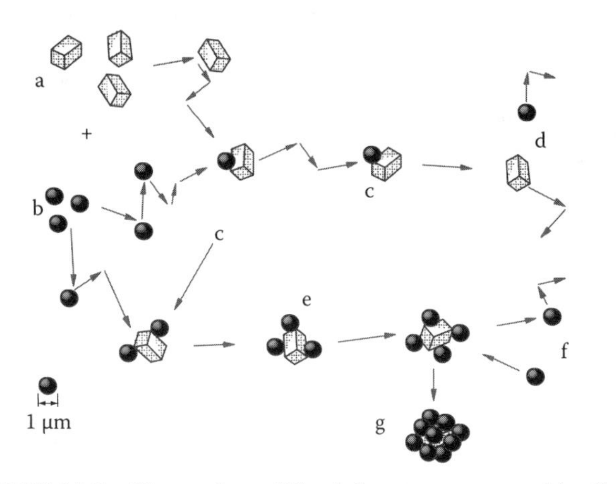

FIGURE 15.5 Observations of *Staphylococcus aureus* and kaolinite clay particles in "wet-slide" at 1000×. (From Hendricks, D.W. et al., Bacterial adsorption on soils—thermodynamics, Utah Water Research Laboratory, Utah State University, Report PRWG62-1, p. 19, July 1970.)

15.2.1.1.1 Equilibrium Statement

The equilibrium statement for Equation 15.4, or Equation 15.5, can be expressed in the usual form for any reaction. From Equation 15.4, the equilibrium statement is

$$K_{\mathrm{eq}} = \frac{[\mathrm{C} \cdot \mathrm{X}]^*}{[\mathrm{C}]^*[\mathrm{X}]^*} \tag{15.6}$$

where
- K_{eq} is the equilibrium constant for adsorption reaction
- $[\mathrm{C}]^*$ is the concentration of adsorbate, C, at equilibrium (kg adsorbate/$\mathrm{m^3}$ of solution)
- $[\mathrm{X}]^*$ is the concentration of unoccupied adsorption sites (number of sites/kg adsorbate)
- $[\mathrm{C} \cdot \mathrm{X}]^*$ is the concentration of adsorbate adsorbed on adsorbent (kg adsorbate/kg adsorbent/$\mathrm{m^3}$ solution)

15.2.1.1.2 Statistical Concept of Equilibrium

At equilibrium, the sites are occupied in proportion to the intensity of competition among the adsorbate particles for the adsorption sites. In other words, the higher the concentration of adsorbate, the higher the proportion of occupied sites. From this point of view, adsorption is a statistical phenomenon. At equilibrium, the adsorption and desorption rates are equal. Those molecules that desorb are those that have sufficient thermal energy, exhibited as vibration about the site, to break the bond of attachment. If the temperature rises, a higher fraction of molecules have enough energy to break the attachment bond and hence the fraction of adsorbent coverage is less than at a lower temperature.

15.2.1.1.3 Microscopic Observations of Statistical Character of Equilibrium

The depiction of Figure 15.5 was observed under a "wet slide," i.e., under a 1000× microscope, as sketched in Figure 15.8. The adsorbent was kaolinite clay of nominal size about 1 μm; the adsorbate was *Staphylococcus aureus* bacteria. As seen under microscope, both the clay particles and the spherical bacteria could be observed being buffeted randomly, with collisions and attachments occurring also randomly. Both adsorption and desorption could be observed, with a certain fraction of bacteria observed to be attached, which was confirmed by experiment in terms of a Langmuir isotherm (Hendricks et al., 1970, p. 19; Hendricks et al. 1979).

15.2.1.2 Langmuir Isotherm

The Langmuir isotherm equation can be obtained from Equation 15.6 by reworking the nomenclature, i.e., let, $\alpha = K_{\mathrm{eq}}$; $X^* = [\mathrm{C} \cdot \mathrm{X}]^*$; $C^* = [\mathrm{C}]^*$; $X^* = [\mathrm{X}]^*$. Then substitute in Equation 15.6, i.e.,

$$\alpha = \frac{X^*}{C^* X^*} \tag{15.7}$$

Another relation is that the total number of adsorption sites, X_m, equals the sum of unoccupied sites, X, plus occupied sites, $\overline{X}$, i.e., at any time,

$$X_m = X + \overline{X} \tag{15.8}$$

and at equilibrium,

$$X_m = X^* + \overline{X}^* \tag{15.9}$$

where X_m is the total number of available adsorption sites per unit mass of adsorbent (# sites/kg adsorbent/m^3 solution).

When Equation 15.9 is inserted in Equation 15.7 the same equilibrium statement, with the new nomenclature, is

$$\alpha = \frac{\overline{X}^*}{C^*(X_m - \overline{X}^*)} \tag{15.10}$$

Rearranging (15.10) gives

$$\frac{\overline{X}^*}{X_m} = \frac{\alpha C^*}{1 + \alpha C^*} \tag{15.11}$$

Equation 15.11 is the usual expression for the Langmuir isotherm. The equation is well known and can be derived also from a statistical thermodynamics approach.

15.2.1.2.1 Plots

Figure 15.6a depicts Equation 15.11 graphically, i.e., solid-phase equilibrium concentration, $\overline{X}$, versus aqueous-phase equilibrium concentration, C^*; X_m is the maximum solid-phase concentration for the particular adsorbate used. The plotted points are experimental data obtained from 4 L stirred flasks operated for a time duration such that the aqueous-phase concentration of adsorbate did not change with time, i.e., after several days or weeks.

15.2.1.2.2 Determining Constants/Linearized Form

Equation 15.11 can be arranged to give a linearized form,

$$\frac{C^*}{\overline{X}^*} = \frac{1}{\alpha X_m} + \frac{1}{X_m} C^* \tag{15.12}$$

The same data used in Figure 15.6a are shown replotted in Figure 15.6b. This is a straight-line plot whose intercept is $1/\alpha X_m$ and whose slope is $1/X_m$. Because the relationship is linear, a best fit can be obtained by regression analysis. Actually Figure 15.6b is plotted first, before Figure 15.6a, so that α and X_m can be used in plotting the latter curve.

15.2.1.2.3 Equilibrium Experiments

Each of the experimental points in Figure 15.6 represents a single equilibrium experiment. Twenty-eight of these experiments were used to define the isotherm that related the equilibrium between Rhodamine-B dye used as the adsorbate and Dowex 50 (a cation exchange resin) used as the adsorbent at 20°C. The experimental procedure involved the following: (1) placing 4000 mL of distilled water in each of 28 flasks, with each flask having a different initial concentration of dye; (2) placing the flasks in a water bath held at 20°C (using a heat-exchanger that recirculates the water in the bath) and setting up a continuous stirrer for each flask; (3) at time $t = 0$, adding 1.00 g of Dowex 50 resin to each flask; (4) measuring liquid-phase concentrations of Rhodamine-B until no further change was noted. Dowex 50 resin was used as an adsorbent because of the high uptake of Rhodamine-B, and that the resin was a light-beige color which stained red in proportion to its uptake to the solid phase. Since the study cited for this illustration was to try to discern a kinetic relation for adsorption uptake and involved a great many experimental

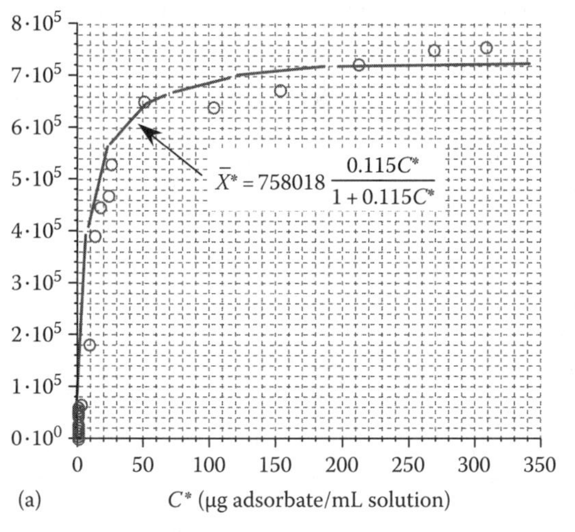

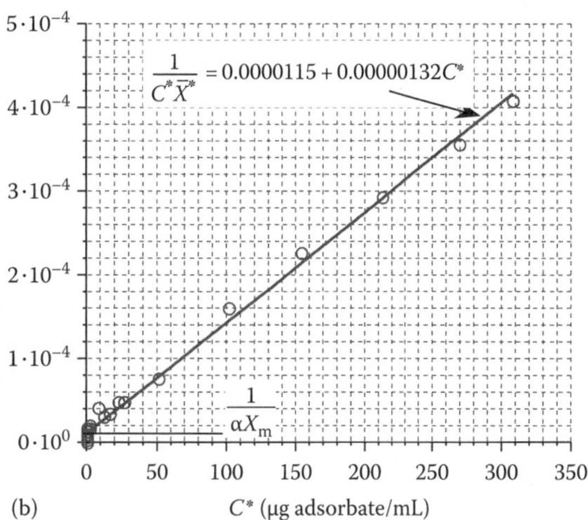

FIGURE 15.6 Langmuir isotherm fit to experimental equilibrium data; $T = 20°C$; adsorbate Rhodamine-B dye; adsorbent Dowex 50 resin H+; $\overline{X}^* = 758018[0.115C^*/(1 + 0.115C^*)]$. (a) Arithmetic Langmuir isotherm. (b) Linearized Langmuir equation. (Adapted from Hendricks, D.W. and Kuratti, L.C., Kinetics Part I—A laboratory investigation of six proposed rate laws using batch reactors, Utah Water Research Laboratory, Utah State University, Report PRWG66-1, p. 58, July 59, 1973.)

measurements, an easily measurable concentration of adsorbate was the main concern.

The adsorbate depletion curve and the associated adsorbent uptake curves are described by the equations,

$$(C_0 - C)/C_0 = 10\wedge(-k_C/2.3) \cdot t \quad (15.13)$$

$$(\overline{X}^* - \overline{X})/\overline{X}^* = 10\wedge(-k_X/2.3) \cdot t \quad (15.14)$$

In logarithmic form the two equations are

$$\log(C_0 - C) = \log C_0 - (k_C/2.3) \cdot t \quad (15.15)$$

$$\log(\overline{X}^* - \overline{X} = \log \overline{X}^* - (k_X/2.3) \cdot t \quad (15.16)$$

where

C_0 is the initial concentration of adsorbate (kg/m^3)
k_C is the rate coefficient for depletion of adsorbate from solution (s^{-1})
k_X is the rate coefficient for uptake of adsorbate from solution to adsorbent (s^{-1})

Example 15.3 Analysis of Adsorption Uptake Data to Determine Equilibrium Point

Given

Figure 15.7a shows the depletion of Rhodamine-B dye from solution; Figure 15.7b shows the associated uptake of by Dowex 50 resin, calculated from the depletion data.

Required

Analyze the data to obtain an isotherm coordinate point, i.e., $(C^*, \overline{X}^*)$, for Run #7.

Solution

The asymptotes of Figure 15.7a and b give C^* and $\overline{X}^*$ respectively, i.e.,

$C^* \approx 0.004$ μg Rh-B/mL solution
$\overline{X}^* \approx 117{,}200$ μg Rh-G/g Dowex 50

Discussion

The isotherm of Figure 15.6 was defined by 28 equilibrium points; each point was determined by an adsorbate depletion curve and a calculated uptake curve such as illustrated for Run #7. A more common method of determining an isotherm is to pulverize the adsorbate and to use a smaller flask and a water bath with a shaker mechanism.

Example 15.4 Analysis of Adsorption Uptake Data to Determine Kinetic Coefficients

Given

Figure 15.7a shows the depletion of Rhodamine-B dye from solution; Figure 15.7b shows the associated uptake of by Dowex 50 resin, calculated from the depletion data.

Required

Analyze the data to obtain kinetic coefficients that define the adsorbate depletion and uptake curves for Run #7.

Solution

A regression analysis of the logarithmic forms of the data for the two curves, i.e., Equations 15.15 and 15.16, gives the kinetic coefficients, i.e.,

$$\text{slope}(C \text{ depletion}) \approx -0.0343 \quad \text{and}$$

$$\text{slope}(\overline{X} \text{ uptake}) \approx 0.0356$$

Multiplying both by 2.303 gives

$$k_C - 0.079 \text{ h}^{-1} \quad \text{and} \quad k_X = -0.082 \text{ h}^{-1}.$$

The intercepts are

$\overline{X}$(intercept) = 113,418 μg Rh-B/g resin and
C(intercept) = 14.12 μg Rh-B/L.

Discussion

Knowing the kinetic coefficients permits a more in-depth kinetic analysis of empirical data, useful in kinetics research. The rates of uptake, i.e., $d\overline{X}/dt$, may be obtained by taking the derivative of Equation 15.14.

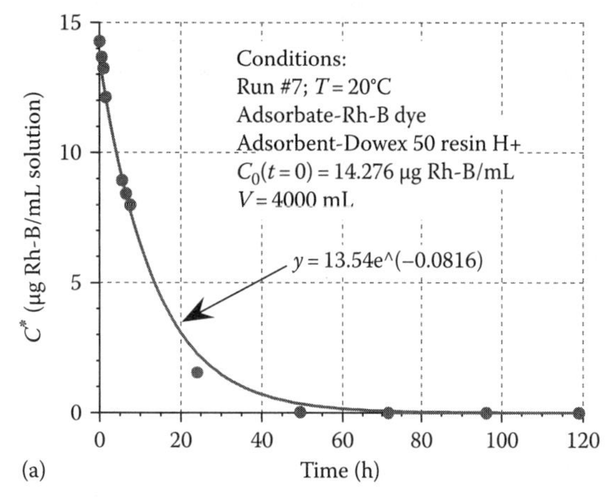

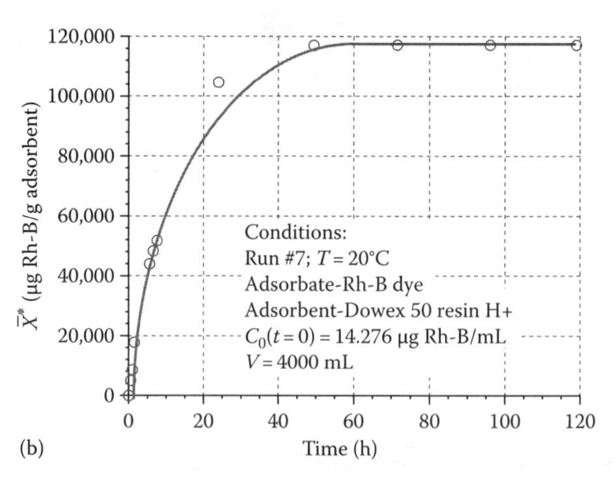

FIGURE 15.7 Experimental data for batch-reactor kinetics yielding isotherm pair $(C^*, \overline{X}^*)$. (a) Depletion of adsorbate. (b) Uptake of adsorbate. (Adapted from Hendricks, D.W. and Kuratti, L.C., Kinetics Part I—A laboratory investigation of six proposed rate laws using batch reactors, Utah Water Research Laboratory, Utah State University, Report PRWG66-1, p. 56, July 1973.)

15.2.1.2.4 Thermodynamics

The effect of temperature on adsorption equilibrium is given by the van't Hoff equation (Alberty and Silbey, 1992, p. 160),

$$\alpha = C_{\text{vh}} e^{-\Delta H^\circ/RT} \tag{15.17}$$

or, in terms of base 10 exponent,

$$\alpha = C_{\text{vh}} 10^{-\Delta H^\circ/2.3RT} \tag{15.18}$$

where

- ΔH° is the standard state enthalpy of reaction (kilocalories per mole),
- R is the gas constant (1.98 kcal per mole per degree),
- T_A is the temperature (K). The temperature symbol is used, nearly always, in conjunction with units; thus in T (K) means temperature in Kelvin, which is synonymous with absolute temperature. T (°C) means temperature in degree Celsius.
- C_{vh} is the constant, dimensionless

Figure 15.8 shows two experiment plots of α versus $1/T_A$. Figure 15.8a data are for Rhodamine-B dye and Dowex 50 resin while Figure 15.8b data are for *Staphylococcus aureus* and Mendon silt loam. The point is that the van't Hoff relation is applicable to a wide variety of adsorption phenomenon. The negative slope means that ΔH° is positive. In terms of the Langmuir constant, α, higher temperatures increase the value of α. For the plots shown, especially Figure 15.8b, the plot should be considered in terms of the van't Hoff relation being applicable, but the ΔH° would be interpreted as an empirical coefficient (since the gas constant, R, would not apply to bacteria).

As related to competitive adsorption, the compounds with higher net energy of adsorption will displace those with lower values, with the latter appearing in the effluent of a GAC column, i.e., as a chromatographic effect (McGuire and Suffet, 1980, p. 102).

15.2.1.3 Freundlich Isotherm

The Freundlich isotherm is used commonly, i.e., used more frequently than the Langmuir, to describe adsorbate–adsorbent equilibrium (see, for example, Snoeyink et al., 1969, Culp et al., 1978, p. 178) and is given,

$$\overline{X}^* = KC^{*1/n} \tag{15.19}$$

where

- K is the coefficient unique to a particular adsorbate–adsorbent equilibrium
- n is the exponent unique to a particular adsorbate–adsorbent equilibrium

The logarithmic form is

$$\log \overline{X}^* = \log K + \frac{i}{n} \log C^* \tag{15.20}$$

As implied by the form of the equations, the experimental data are plotted on log–log paper. The slope and the intercept of the best-fit straight line give $1/n$ and K, respectively. The slope, $1/n$, is the "intensity" of adsorption while K is an index of the "capacity" of the adsorbent and is the intercept (Adamson, 1967, p. 401; Adamson and Gast, 1997, p. 393).

15.2.1.3.1 Plots

Figure 15.9a and b shows Freundlich isotherm plots in terms of Equations 15.19 and 15.20, respectively. The plots are for

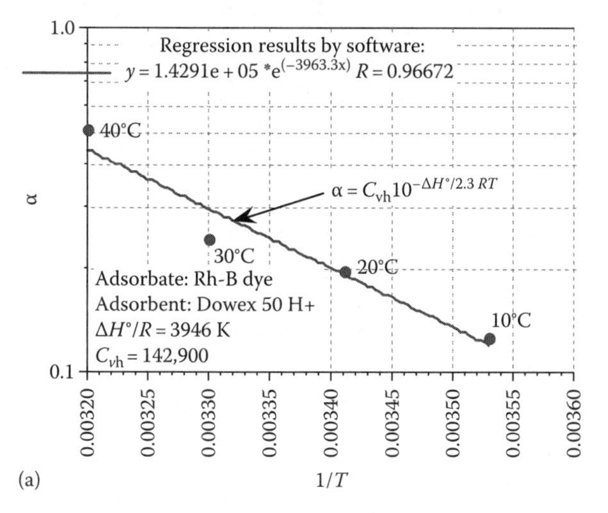

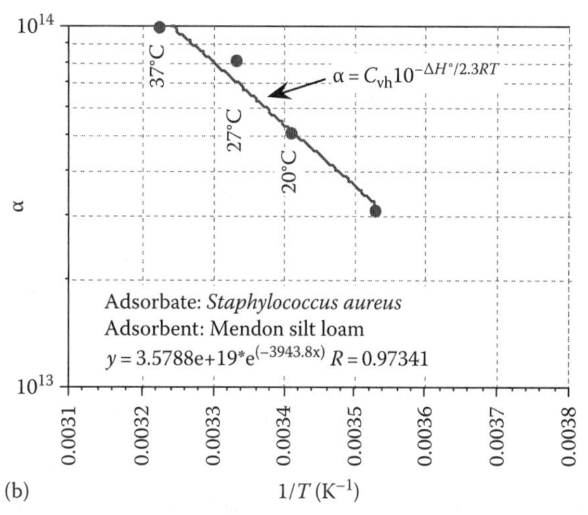

FIGURE 15.8 Langmuir constant data plotted in terms of van't Hoff equation, $\alpha = C_{\text{vh}} e^{(\Delta H^\circ/RT)}$. (a) Dye and resin. (From Hendricks, D.W. and Kuratti, L.C., Kinetics Part I—A laboratory investigation of six proposed rate laws using batch reactors, Utah Water Research Laboratory, Utah State University, Report PRWG66-1, p. 20, July 1973.) (b) Bacteria and soil. (From Hendricks, D.W. et al., *Water, Air, Soil Pollut.*, 12, 226, 1979.)

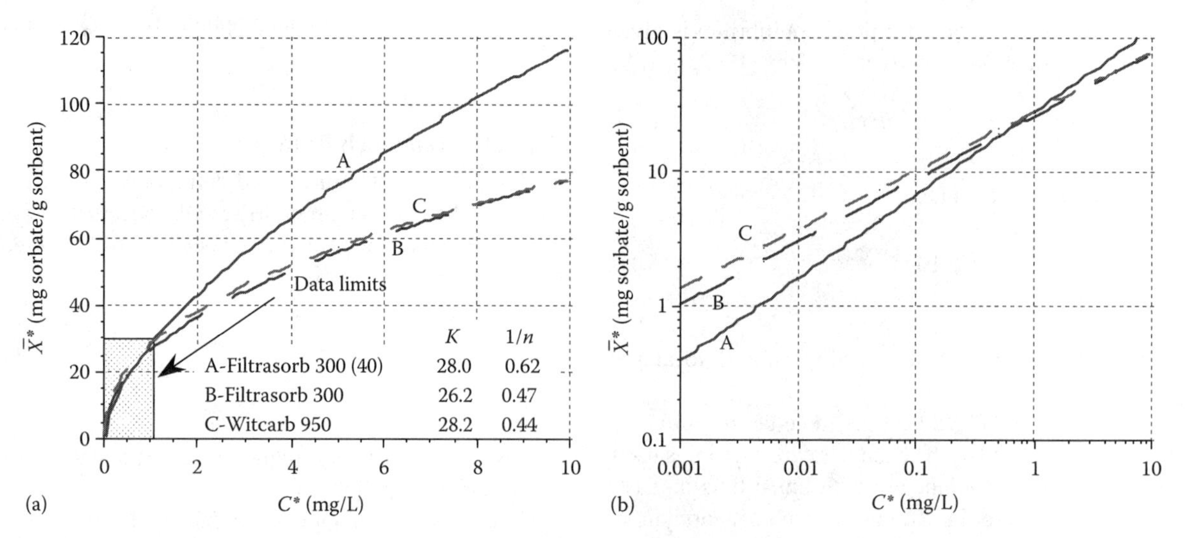

FIGURE 15.9 Freundlich isotherm plots for trichloroethylene (TCE). (From Love, O.T. et al., Treatment of volatile organic compounds in drinking water, Report EPA-600/8-83-019, Municipal Environmental Research Laboratory, U.S. Environmental Protection Laboratory, Cincinnati, OH, May 1983.). (a) Arithmetic. (b) Log–log.

TCE with three GAC adsorbents, with the Freundlich coefficients from Figure 15.3. Comparing the two plots:

- Most of the log–log plot is seen in only a small part of the arithmetic scale plot, e.g., for $C^* < 1$; this range is seen in Figure 15.9a as the shaded area. All of the experimental data are given, in fact, for $C^* < 1$; the curves for $C^* \geq 1$ are extrapolated in order to see the character of the Freundlich equation.
- Figure 15.9a shows a curved shape, while the log–log plots, i.e., Figure 15.9b are straight lines, as expected.
- The arithmetic-scale plots continue to rise, as of course do the log–log plots, i.e., there is no leveling off toward a maximum value of $\bar{X}^*$ as in a Langmuir plot.

15.2.1.3.2 Published Coefficients

Table 15.3 provides Freundlich coefficients, K and $1/n$ for volatile organic compounds along with several adsorbents. The K and $1/n$ data illustrate the variation for different organic compounds and also the effect of different activated carbons for the same compound. Also of interest, the different structural forms of carbon compounds, e.g., the *cis* versus the *trans* have different coefficients. The main point is that each particular adsorbate–adsorbent combination is unique with respect to its adsorption behavior.

15.2.1.3.3 Freundlich versus Langmuir

The Freundlich isotherm seems to be favored in practice, since only a $\bar{X}^*$ versus C^* relationship is sought. While it is considered strictly empirical by some, this concept has been refuted by others (see, for example, Adamson, 1967, p. 401). The Langmuir isotherm has a rational basis and may be derived from a reaction equilibrium statement (see Section 15.2.1.2) or from statistical thermodynamics (Hill, 1960). For that reason, it has more academic appeal and it has practical utility as well, given enough data to define the relationship. The Freundlich isotherm, on the other hand, may be applied for cases of limited data, i.e., only a few data points are required to define the relationship.

15.2.1.4 General Isotherm

Figure 15.10 shows a "general" isotherm from superimposed plots from a number of individual isotherms obtained from the literature (McGuire and Suffet 1980, p. 108). The shaded area, i.e., the "general" isotherm, envelopes most of the individual isotherms. Those compounds that are better adsorbed are nearer the envelope boundary and those less well adsorbed are significantly lower. Urea, for example, is not well adsorbed and is seen as having low equilibrium concentrations on the solid phase, even at very high solution equilibrium concentrations. At the same time, urea has a low net energy of adsorption and so the lower isotherm is expected. The "general" isotherm provides guidance as to the upper limits of adsorption potential of GAC and, for a mixture of solutes, may indicate those compounds more favorably adsorbed.

15.2.1.5 Multicomponent Equilibria

Mixtures are the usual reality in adsorption. The equilibrium models that deal with mixtures are, however, more complex than those for single component adsorption (see, for example, Radke and Prausnitz, 1972).

Another kind of competitive effect is "pre-loading" of organic matter. Pre-loading occurs when the wave front of the more weakly adsorbing background organic moves ahead of the target compounds, causing reductions in equilibrium capacity and rates for target compounds in the downstream sections of the reactor (Carter and Weber, 1994, p. 614).

TABLE 15.3
Freundlich Isotherm Coefficients

Adsorbate	Structure	Adsorbent	K	$1/n$
Trichloroethylene	C_2HCl_3	Wittcarb 950	28.2	0.44
		Filtrasorb 300	26.2	0.47
		Filtrasorb 300 (40)	28.0	0.62
Tetrachloroethylene	C_2Cl_4	Wittcarb 950	84	0.4
		Filtrasorb 300 (40)	51	0.6
Cis-1,2 dichloroethylene	$C_2H_2Cl_2$	Wittcarb 950	8.4	0.5
		Filtrasorb 300 (40)	6.5	0.7
Trans-1,2 dichloroethylene	$C_2H_2Cl_2$	Filtrasorb 300 (40)	3.1	0.5
1,1 Dichloroethylene	$C_2H_2Cl_2$	Filtrasorb 300 (40)	4.9	0.5
Vinyl chloride	C_2H_3Cl			
1,1, 1-Trichloroethane	$C_2H_3Cl_3$	Wittcarb 950	9.4	0.5
		Filtrasorb 300 (40)	2.5	0.3
1,1, 1-Trichloroethane	$C_2H_4Cl_2$	Wittcarb 950	5.7	0.5
		Filtrasorb 300 (40)	3.6	0.8
Carbon tetrachloride	CCl_4	Norit	28.5	0.8
		Nuchar WV-G	25.8	0.7
		Filtrasorb 400	38.1	0.7
		Hydrodarco 1030	14.2	0.7
		Filtrasorb 300 (40)	11.1	0.8
Methylene chloride	CH_2Cl_2	Filtrasorb 300 (40)	1.3	1.2
		Filtrasorb 400 (47)	1.6	0.7
Benzene	C_6H_6	Norit	49.3	0.6
		Nuchar WV-G	29.5	0.4
		Filtrasorb 400	16.6	0.4
		Hydrodarco 1030	14.2	0.4
		Filtrasorb 300 (40)	1.0	1.6
Chlorobenzene	C_6H_5Cl	Filtrasorb 300 (40)	91	0.99
1,2-Dichlorobenzene	$C_6H_4Cl_2$	Filtrasorb 300 (40)	129	0.4
1,3-Dichlorobenzene	$C_6H_4Cl_2$	Filtrasorb 300 (40)	118	0.4
1,4-Dichlorobenzene	$C_6H_4Cl_2$	Filtrasorb 400 (62)	226	0.4
		Filtrasorb 300 (40)	121	0.5
1,2,4-Trichlorobenzene	$C_6H_3Cl_3$	Filtrasorb 300 (40)	157	0.31

Source: Love, O.T. et al., Treatment of volatile organic compounds in drinking water, Report EPA-600/8-83-019, Municipal Environmental Research Laboratory, U.S. Environmental Protection Laboratory, Cincinnati, OH, May 1983.

15.2.2 Kinetics

Kinetics refers to the rate of a reaction. As related to adsorption, the rate is controlled by (1) advection from the ambient solution to the vicinity of the adsorbent particle; (2) liquid-phase diffusion, i.e., from the solution to the external surface of the adsorbent particle (across a *pseudo* "film"); and (3) solid-phase diffusion, i.e., from the external surface of the particle to sites within the particle.

15.2.2.1 Graphical Depiction

Figure 15.11 depicts the overall concentration gradient, from the bulk of solution, with adsorbate concentration, C_0, to the interior of a particle, at two times, i.e., $t_1 \approx 0$ and $t_2 \gg 0$. As illustrated, C_0 declines to C_i, at the water–particle interface. The associated solid-phase concentration at the particle surface, i.e., $\overline{X}^*$, assumes equilibrium with the interface concentration, i.e., C_i, which is designated, $\overline{X}^*(C_i)$. Since C_i is not measurable, C_0 is taken as a surrogate for C_i, i.e., $C_i = C_0$. As seen in Figure 15.11, the adsorbate concentration gradient at $t_1 \approx 0$ is steep in both the aqueous phase and in the solid phase. Then, at $t_2 \gg 0$ the solid-phase profile becomes markedly less steep and the aqueous-phase profile adjusts accordingly such that the diffusion rates are equal for the two phases (which is always the case). Most often, the solid-phase diffusion is rate controlling. The diffusion rate is proportional to the concentration gradient, which declines as the particle adsorption "sites" become increasingly occupied by adsorbate molecules, i.e., as the particle approaches "saturation" with respect to the isotherm (Box 15.1).

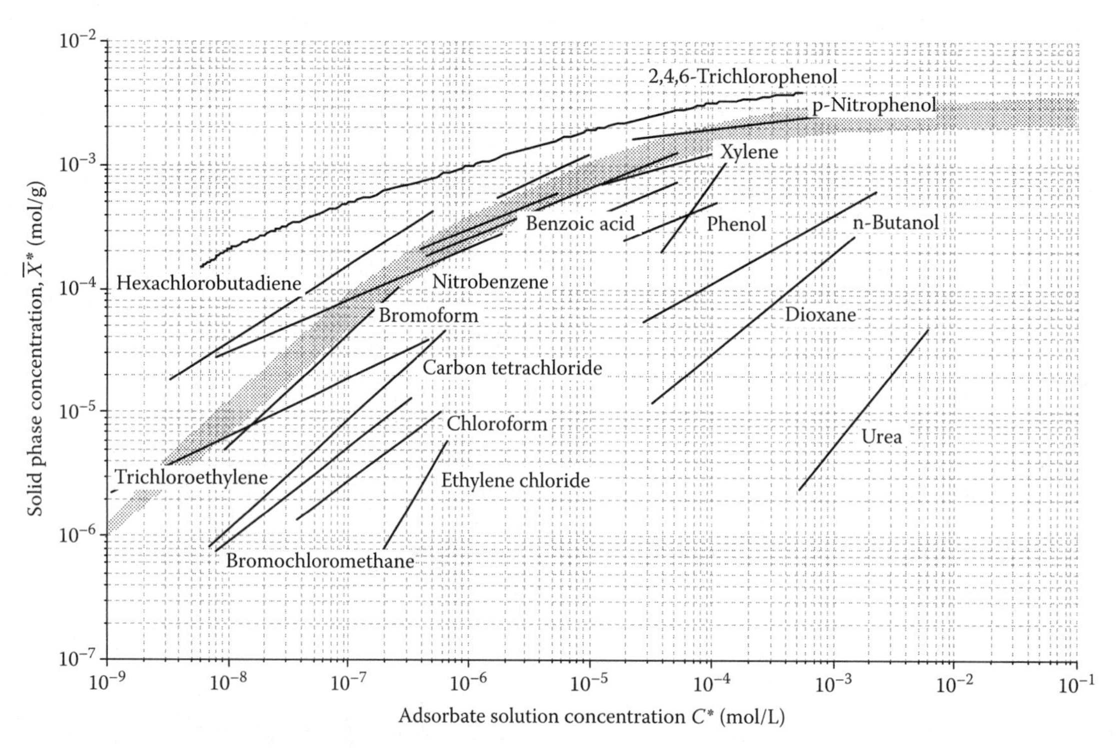

FIGURE 15.10 General isotherm synthesized from specific isotherms. (Adapted from McGuire, M.J. and Suffet, I.H., The calculated net adsorption energy concept, Chap. 4, in: Suffet, I. H. and McGuire, M. J. (Eds.), *Activated Carbon Adsorption of Organics from the Aqueous Phase*, Vol. 1, Ann Arbor Science Publishers, Ann Arbor, MI, p. 108, 1980. With permission; see also SDWC (Safe Drinking Water Committee), *Drinking Water and Health*, Vol. 2, National Academy Press, Washington, DC, p. 262, 1980.)

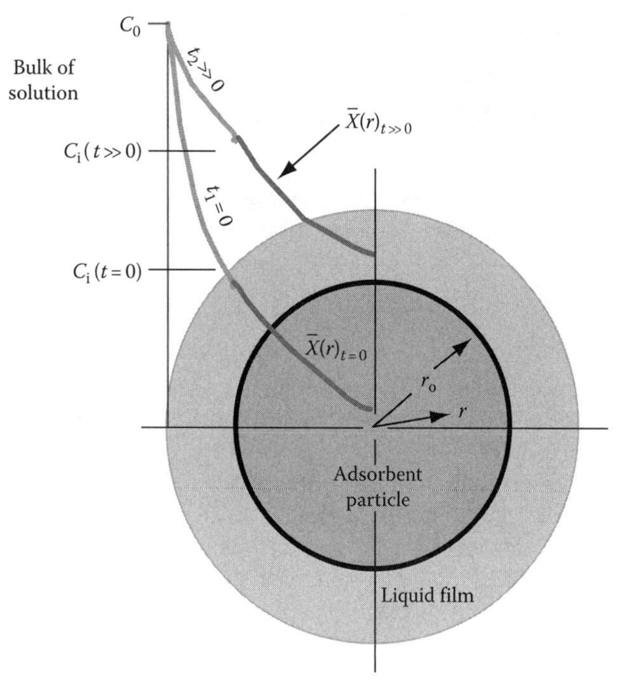

FIGURE 15.11 Concentration gradients across aqueous film and in solid phase from bulk of solution.

BOX 15.1 NOMENCLATURE ON C_0

The term, C_0, is most often used as a "starting concentration," e.g., with respect to distance or time, for various kinds of situations. In the case of diffusion to and within a particle, C_0 is designated here the adsorbate concentration in the "bulk-of-solution," e.g., as in Figure 15.11. In the case of a "packed bed," the starting concentration at the top of the column is also designated C_0, in which case the particle usually is not considered and so the distinction in terms is not an issue. For a particle in a packed bed, however, the "local" concentration at some depth, Z, is designated C, or $C(Z)$, which is the same as the "bulk-of-solution" concentration, C_0, in Figure 15.11. The context in the use of C_0 gives the distinction. But, if the two models, e.g., packed-bed and particle, are "linked," as in a computer algorithm, the distinction is necessary. The same ideas hold true for a "batch reactor."

15.2.2.2 Rate of Uptake: Theoretical

Kinetic theory is based on diffusion, as described by Fick's law. The modeling here is based on solid-phase diffusion, albeit the aqueous-phase diffusion could be rate limiting, especially before the solid-phase accumulates much adsorbate.

15.2.2.2.1 Fick's First Law

Fick's first law of molecular diffusion is applicable for either liquid or solid phase. Its mathematical statement in three-dimensional coordinates, i.e., Equation 4.1 is

$$\mathbf{j} = -D\nabla C \tag{15.21}$$

where
- $\mathbf{j}$ is the flux density of diffusing adsorbate (kg adsorbate/s/m^2)
- D is the diffusion coefficient for migrating adsorbate in liquid or solid (m^2/s)
- C is the concentration of adsorbate (kg/m^3 solution)
- ∇ is the vector operator, i.e., $\partial/x + \partial/y + \partial/z$, in rectangular coordinates

For diffusion into a particle, the vector operator would be in terms of spherical coordinates, which is more complex.

15.2.2.2.2 Kinetics in Solid Phase

A mass balance equation for a spherical particle was given by Weber and Van Vliet (1980, p. 30), i.e.,

$$\frac{\partial \overline{X}}{\partial t} = D_s \frac{\rho(\text{solid})}{r^2} \frac{\partial}{\partial r}\left[r^2 \frac{\partial \overline{X}}{\partial r}\right] \tag{15.22}$$

where
- $\overline{X}$ is the solid phase adsorbate concentration, at radius r, and at time, t (kg adsorbate/kg adsorbent)
- t is the elapsed time from a reference (s)
- D_s is the diffusion coefficient for adsorbate in adsorbent particle (m^2/s)
- ρ(solid) is the density of adsorbent particle (kg adsorbent/m^3 solid particle)
- r is the radial distance from center of a solid particle (m)

The $\partial\overline{X}/\partial t$ term is the average rate of concentration change within the particle, at time, t. The principle in deriving Equation 15.22 is that an adsorbate mass balance for any infinitesimal volume is the mass rate of change within the volume equals the mass flux-in minus the mass flux-out. For a particle mass flux-in is the diffusion flux-in, i.e., $\mathbf{j} = D_s\partial\overline{X}/\partial r$, Equation 15.21. The Weber and Van Vliet derivation was modified here to neglect pore water and the $D_s(\overline{X})$ function. The materials-balance principle was reviewed in Chapter 4 (see, for example, Section 4.3.3). The method of solution, i.e., illustrating the mass balance principle and diffusion kinetics, was illustrated for a cylindrical particle by Weber and Rumer (1965) and summarized in Hendricks (2006, p. 860). A similar method was used by Abdelrasool (1992) for spherical particles, applying a finite difference form of Equation (15.22) which was "fit" to empirical laboratory results for $d\overline{X}/dt$ to calculate the solid-phase diffusion coefficient. D_s.

15.2.2.3 Empirical Rate Equation

An empirical rate equation is in terms of the rate coefficient, $\overline{D}_3$, the adsorbate concentration in solution at any time, C, and the "saturation-deficit" for the particle as a whole at any time, i.e.,

$$\frac{d\overline{X}}{dt} = \overline{D}_3 C(\overline{X}^* - \overline{X}) \tag{15.23}$$

where
- $\overline{X}$ is the solid-phase adsorbate concentration at any time, t, averaged for the whole particle (kg adsorbate/kg adsorbent)
- $\overline{X}^*$ is the solid-phase adsorbate concentration at any time, t, averaged for the whole particle but in equilibrium with the aqueous-phase concentration, C_0, as calculated by an isotherm relation, e.g., Langmuir or Freundlich isotherm (kg adsorbate/kg adsorbent)
- $\overline{D}_3$ is the kinetic coefficient for Equation 15.23 [m^3/(s $\cdot$ kg)]
- C is the concentration of adsorbate in ambient solution (kg/m^3)

The coefficient, $\overline{D}_3$, was found to vary with the adsorbate-solution concentration, C, and the degree of particle saturation, $\overline{X}/\overline{X}^*$. This complicates the matter, but it does not negate the concept. To delineate the function $\overline{D}_3(C, \overline{X}/\overline{X}^*)$ a considerable amount of laboratory data must be generated and analyzed, e.g., by Fortran (see Hendricks and Kuratti, 1973, 1982).

15.2.3 Reactor Theory for Packed Beds

For a packed-bed column, the adsorbate concentration varies along the length of the "reaction zone," expressed as $C(Z)_t$. Thus, the homogeneity assumption is valid only for an infinitesimal slice, ΔZ. For such a slice, the mass balance must be in terms of a differential equation, e.g., Equation 4.18, which is reproduced here as Equation 15.24.

15.2.3.1 Mathematics

Figure 15.12 illustrates schematically a $C(Z)_t$ curve for the "reaction zone" of a packed-bed reactor column. For the column slice, ΔZ, the mass balance differential equation (see Hiester and Vermeulen, 1952, which is probably the classic paper on this topic) is

$$\left[\frac{\partial C}{\partial t}\right]_0 = \overline{v}\frac{\partial C}{\partial Z} + D\frac{\partial^2 C}{\partial Z^2} - \rho\frac{1-P}{P}\frac{\partial \overline{X}}{\partial t} \tag{15.24}$$

where
- C is the concentration of adsorbate species in liquid phase (kg/m^3)
- t is the time from a convenient reference point, such as initial introduction of adsorbate (s)
- Z is the distance from a reference point, such as the top of the packed bed (m)

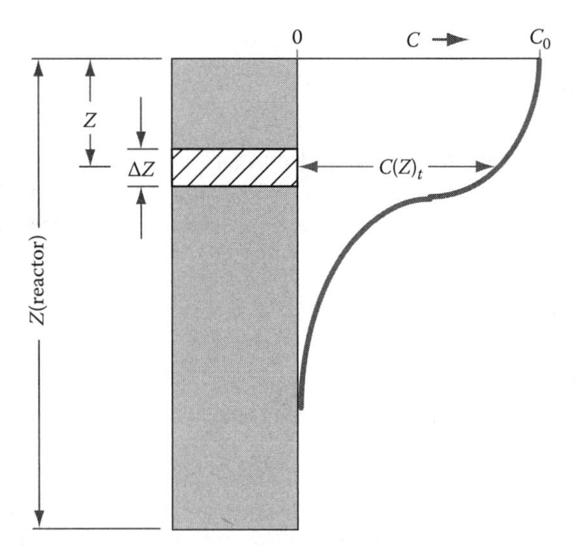

FIGURE 15.12 Concentration profile in a packed-bed reactor.

$\bar{v}$ is the interstitial velocity, i.e., within the media pores, $\bar{v} = Q/(AP)$ (m/s)
D is the coefficient of dispersion for the porous media at interstitial velocity, $\bar{v}$ (m^2/s)
ρ is the dry density of the individual granular particles comprising the porous media (kg/m^3)
P is the porosity of the porous media (m^3 voids/m^3 voids + solids)

Equation 15.24 is the basic mathematical formulation of a mass balance for a packed-bed reactor (see Section 4.3.3.3 for derivation and Box 4.3 for discussion of its application). The formulation is done commonly in terms of "bed volumes," i.e., $V = Q \cdot t$ (see Keinath and Weber, 1968; Weber and Smith, 1987). It is applicable to any "slice" along the length of the column. It says that the rate of change of adsorbate concentration within the slice depends upon the adsorbate transport flux into the slice by advection and dispersion, minus that carried out by the same transport mechanisms, minus the rate of uptake to the solid phase. The verbal statement of mass balance is

$$\begin{bmatrix}\text{net rate of} \\ \text{change of adsorbate} \\ \text{concentration}\end{bmatrix} = [\text{net conversation rate}] \\ +[\text{net dispersion rate}] \\ -[\text{adsorption rate}] \qquad (15.25)$$

15.2.3.1.1 Terms in the Materials Balance Equation

The various terms in the mass balance equation requiring experimental determination are ρ, the particle density; P, the porosity of the packed bed; D, the dispersion coefficient; and $\partial\bar{X}/\partial t$, the rate of uptake to the solid phase. For reference, the dispersion, D, varies with $\bar{v}$; the parameter, $D/\bar{v}$, is constant for a given porous media, however, and may be determined by Equation 4.7 using data from a breakthrough curve, e.g., for Cl^-. For the Dowex 50 resin, $D/\bar{v} \approx 0.8$ cm. For the kinetic term, $\partial\bar{X}/\partial t$, isotherms must be determined along with kinetic coefficients. In addition, the conditions imposed for the simulation must be specified, which includes C_0, the adsorbate influent concentration; and, $\bar{v}$, the interstitial velocity ($\bar{v} = Q/AP$, in which Q is the flow into the column, and A is the cross-sectional area of the column) (Box 15.2).

BOX 15.2 REVIEW OF SALIENT POINTS ABOUT PACKED-BED REACTORS

1. *Nomenclature*. The curve, $C(Z)_t$, Figure 15.12, is the "adsorbate concentration profile"; sometimes, for brevity, it is called the "wave front," or, alternatively, the "mass-transfer zone."
2. *Mass Balance Applies Only to a Homogeneous Volume*. As stated in Sections 4.3.2, 4.3.3, 4.3.3.1, and 15.2.3.1, the mass balance principle applies only to a homogeneous volume. In the case of a packed-bed column, the homogeneity condition is valid only for an infinitesimal volume element of thickness, ΔZ, as seen in Figure 15.12.
3. *Derivation*. Section 4.3.3.3 gives the derivation of Equation 15.24.
4. *Repetition*. Figure 15.12 is the same as Figure 4.6 and Equation 15.24 is the same as Equation 4.18 with derivation in Section 4.3.3.3. The repetition here is for convenience.
5. *General Applicability*. The mass balance principle in differential form applies to any infinitesimal volume element, such as in modeling various situations in nature, e.g., in groundwater flow in which adsorption occurs, and in bodies of water (e.g., lakes and estuaries) in which biological reactions occur and concentrations vary spatially. For a cubic element, more terms are involved, which are needed to account for inflows and outflows across the six faces of the element.
6. *One-Dimensional Form*. The one-dimensional differential equation form may apply to a filter, or an ion-exchange column, as well as to an adsorption column. Also, the equation may apply to a "plug-flow" activated sludge basin or a stream (in which case the infinitesimal element may be permitted to move at its advective velocity). It applies also to a fluidized-bed adsorption reactor.

15.2.3.2 Advection Kinetics

As long as adsorbate is delivered to an adsorbent particle faster than its rate of uptake, particle kinetics, as represented by Equation 15.23 controls the rate of adsorption. If, on the other hand, the advection rate of adsorbate to a layer (of adsorbent particles) is less than can be taken up by the solid-phase, then the advection rate to the adsorbent particle governs, which has been designated "advection kinetics" (see Hendricks 1973, 1980). These two mechanisms, i.e., particle kinetics and advection kinetics, are distinguished symbolically by the terms, $[\partial\overline{X}/\partial t]_P$ and $[\partial\overline{X}/\partial t]_A$, respectively. Stated mathematically, when

$$\left[\frac{\partial\overline{X}}{\partial t}\right]_A < \left[\frac{\partial\overline{X}}{\partial t}\right]_P \tag{15.26}$$

then advection transport to the particle is rate controlling. The numerical solution to Equation 15.24 involves a test of this criterion at each slice, $1 \leq i \leq n$, in the column and for each Δt time pass in the iteration.

15.2.3.2.1 Advection Kinetics Model

An expression for advective transport (i.e., advection kinetics) can be derived starting with Equation 15.24. The premise is that the uptake rate capacity by the particle, i.e., $[\partial\overline{X}/\partial t]_P$, exceeds the transport rate to the particle by advection and dispersion, i.e., $[\partial\overline{X}/\partial t]_A$.

When the adsorption rate is limited by the transport of adsorbate to an adsorbent particle (by advection and dispersion) none of the adsorbate molecules making contact with the external surface of an adsorbent particle is "rejected." For this condition, the concentration profile in the zone where advective-dispersion advection kinetics governs, is approximately "steady state" and at any Z in this zone, the observed $[dC/dt]_0 \approx 0$. Thus Equation 15.24 is

$$\left[\frac{\partial C}{\partial t}\right]_0 = 0 = \overline{v}\frac{\partial C}{\partial Z} + D\frac{\partial^2 C}{\partial Z^2} - \rho\frac{1-P}{P}\frac{\partial\overline{X}}{\partial t} \tag{15.27}$$

Solving for $\partial\overline{X}/\partial t$ gives

$$\left[\frac{\partial\overline{X}}{\partial t}\right]_A = \rho\frac{1-P}{P}\left[-\overline{v}\frac{\partial C}{\partial Z} + D\frac{\partial^2 C}{\partial Z^2}\right] \tag{15.28}$$

Thus the uptake rate $[\partial\overline{X}/\partial t]_A$ is dependent solely upon the rate of transport of adsorbate by advection, $(\overline{v}\partial C/\partial Z)$, and dispersion, $(D\partial^2 C/\partial Z^2)$.

The task now is to find a $C(Z)_t$ function which will allow a solution of Equation 15.28. To deduce a $C(Z)_t$ relationship for advection kinetics, consider a porous medium with a given flow velocity. An adsorbate molecule will have a probability of say 0.50 of making a collision with an adsorbent particle within a certain distance of travel (call it the "half distance" if it is desired to see the analogy with radioactive disintegration with time). Now if we consider 100 particles in the fluid stream initially, 50 will remain after one half distance, 25 after the next half distance, and so on. This suggests a decay equation of the form,

$$C(Z') = C_0' e^{-\lambda Z'} \tag{15.29}$$

where $C_0' = C(Z_0)$, and Z_0 is the "balance point" where $(\partial X/\partial t)_P = (\partial X/\partial t)_A$; also the distance Z' is $Z' = Z - Z_0$. The term λ is an experimentally determined coefficient, which depends upon $\overline{v}$; the function is unique for a given porous media. Taking first and second derivatives of Equation 15.29,

$$\frac{\partial C}{\partial Z} = \lambda C_0' e^{-\lambda Z'} \tag{15.30}$$

$$= -\lambda C(Z') \tag{15.31}$$

and

$$\frac{\partial^2 C}{\partial Z^2} = \lambda^2 C(Z') \tag{15.32}$$

Substituting (15.31) and (15.32) into Equation 15.28 gives

$$\left[\frac{\partial\overline{X}}{\partial t}\right]_A = \frac{1}{\rho}\frac{P}{1-P}(\overline{v} + D\lambda)\lambda C(Z') \tag{15.33}$$

Substituting (15.30) in (15.33) gives

$$\left[\frac{\partial\overline{X}}{\partial t}\right]_A = \frac{1}{\rho}\frac{P}{1-P}(\overline{v} + D\lambda)\lambda C_0' e^{-\lambda Z'} \tag{15.34}$$

Equation 15.33 or 15.34 has an interesting physical significance. The terms $\overline{v}$, D, and $C(Z)$ are indicative of the transport rate at the position Z, but the term λ tells how many collisions will occur per unit distance. In other words, the expression given in (15.33) or (15.34) gives the probability that within the time period Δt the solid-phase adsorbent concentration at Z will have increased by $\Delta\overline{X}$. At the same time, in the zone where convective dispersion transport governs, i.e., $Z > Z_0$, the concentration profile, $C(Z')_t$, may be computed by Equation 15.29. The transition point, i.e., where kinetics changes from particle kinetics being rate controlling to where advection/dispersion is rate controlling is the inflection point in the concentration profile, i.e., the "wave front."

15.2.3.2.2 Model Delineation

In a packed bed of granular adsorbent, Equation 15.30 is the "mathematical model" which depicts the adsorbate concentration profile in the column to the right of the inflection point, i.e., as seen in Figure 15.13. To the left of the inflection point, particle kinetics governs and Equation 15.23 is the basis for computation of $C(Z, t)$. The $C(Z)_t$ curve in the advection–dispersion kinetics zone is constant in shape. But then Z_0' moves downstream as $\overline{X}(Z=0)$ increases (shifting control to particle kinetics). At the same time, the inflection point, i.e., where $[\partial\overline{X}/\partial t]_P = [\partial\overline{X}/\partial t]_A$, starts to move downstream for $t > 0$. At some point in time, i.e., $t \gg 0$, the $C(Z)_t$ profile assumes a steady-state shape, such as in Figure 15.13, and "translates" downstream.

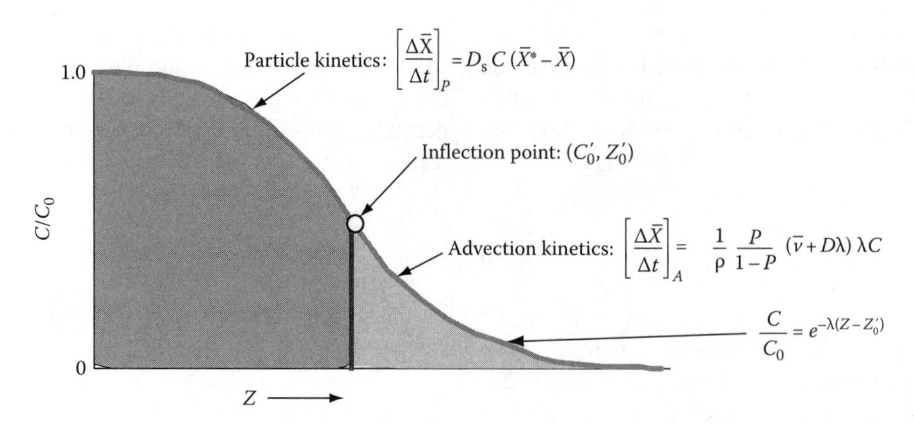

FIGURE 15.13 Illustration of model components showing zones of influence of governing kinetic equations. (Adapted from Vagliasindi, F. and Hendricks, D.W., *J. Environ. Eng. Div.*, ASCE, 118(4), 532, 1992.)

15.2.3.2.3 Determination of λ

The probability coefficient, λ, can be determined only by experimental data to yield a plot of log $C(Z')_t$ versus Z', which is best done at low adsorbate concentration for the likely condition that $[\partial\bar{X}/\partial t]_P \gg [\partial\bar{X}/\partial t]_A$. The plot is a straight line in the zone $Z > Z_0$, whose slope is λ/2.3 per Equation 15.46 and its logarithmic form, Equation 15.35,

$$\log C(Z') = \log C_0' - \frac{\lambda}{2.3}Z' \quad (15.35)$$

15.2.3.3 Simulation Modeling

Equation 15.24 may be solved numerically by means of a computational algorithm (Keinath, 1975; Vagliasindi and Hendricks, 1992). Several million to several tens of millions of iterations may be involved.

15.2.3.3.1 Numerical Solution of Mass Balance Equation

Solutions for $C(Z,t)$ and $\bar{X}(Z,t)$ are arrived at by means of a numerical scheme executed by a computer algorithm, e.g., Fortran. The algorithm is straightforward, relying on repetition. First, Equation 15.24 is rewritten in finite-difference form, i.e.,

$$\frac{\Delta C}{\Delta t} = -\bar{v}\frac{\Delta C}{\Delta Z} + D\frac{\Delta\cdot\Delta C}{\Delta Z^2} - \rho\frac{1-P}{P}\frac{\Delta\bar{X}}{\Delta t} \quad (15.36)$$

To solve Equation 15.36, the column is divided into slices of thickness, ΔZ. Then, using the central-difference method from numerical analysis, Equation 15.36 is applied to a slice "*i*"; its restatement is

$$\frac{(C_i)_{t+\Delta t} - (C_i)_t}{\Delta t} = -\bar{v}\frac{C_{i+1}+C_{i-1}}{2\Delta Z} + D\frac{C_{i+1}-2C_i+C_{i-1}}{\Delta Z^2} - \rho\frac{1-P}{P}\left[\frac{\Delta\bar{X}}{\Delta t}\right]_{i,t} \quad (15.37)$$

Rearranging Equation 15.37 to solve for the new concentration in slice "*i*" after a time increment Δt gives

$$(C_i)_{t+\Delta t} = (C_i)_t + \left[-\bar{v}\frac{C_{i+1}+C_{i-1}}{2\Delta Z} + D\frac{C_{i+1}-2C_i+C_{i-1}}{\Delta Z^2} - \rho\frac{1-P}{P}\left[\frac{\Delta\bar{X}}{\Delta t}\right]_{i,t}\right]\Delta t \quad (15.38)$$

Equation 15.38 allows computation of $C(Z,t)$ where Z is calculated as $i\cdot\Delta Z$ and t is calculated as $t_2 = t_1 + \Delta t$. A printout for time, t, gives columns with slice, "*i*," distance, "*Z*," and the computed, C_i. In other words, the columns provide $C(Z)_t$ output, or the "wave front."

The printout interval is for whatever is convenient to illustrate the changes with time, e.g., 1.0 h, 10 h, depending on the rate of change. Equation 15.38 is applicable to either particle kinetics or advection kinetics, whichever applies. Advection kinetics is applicable when $[(\partial\bar{X}/\partial t]_P]_{i,t} \geq [[(\partial\bar{X}/\partial t]_A]_{i,t}$.

15.2.3.3.2 Solid-Phase Concentration

To obtain adsorbate concentration in the solid phase, i.e., $\bar{X}(Z,t)$ the calculation for the concentration profile, i.e., $\bar{X}(Z)_t$, is

$$(\bar{X}_i)_{t+\Delta t} = (\bar{X}_i)_t + \left[\frac{d\bar{X}}{dt}\right]_{i,t}\cdot\Delta t \quad (15.39)$$

The adsorption uptake term, $d\bar{X}/dt$, must be determined for two cases: $[(\partial\bar{X}/\partial t]_P]_{i,t}$, and $[(\partial\bar{X}/\partial t]_A]_{i,t}$ for each slice *i*, and for each time iteration. The computer program will test each of these equations by Equation 15.26 to determine which is smallest; the smallest governs. Then, from the computation of Equation 15.39, $\bar{X}(Z)_t$ is obtained.

15.2.3.3.3 Computer Algorithm

Figure 15.14a depicts an adsorption reactor column showing column slices, $1 \geq i \leq n$, each of thickness, ΔZ. The shading represents adsorbate concentration in the solid phase as being proportional to the shade of grey, discussed in the next section. The slices are designated, "*i*," e.g., $i = 1$, $1 = 2$, $i-1$, i, $i+1$, $i = n$. The mass balance differential equation applies to any slice, *i*. The computation protocol is to start at $t = 0$ and at $i = 1$ with $C_{i=1} = C_0$ with $C(i)_{t=0} = C_0' e^{-\lambda\cdot i\Delta Z}$, $1 \leq i \leq n$

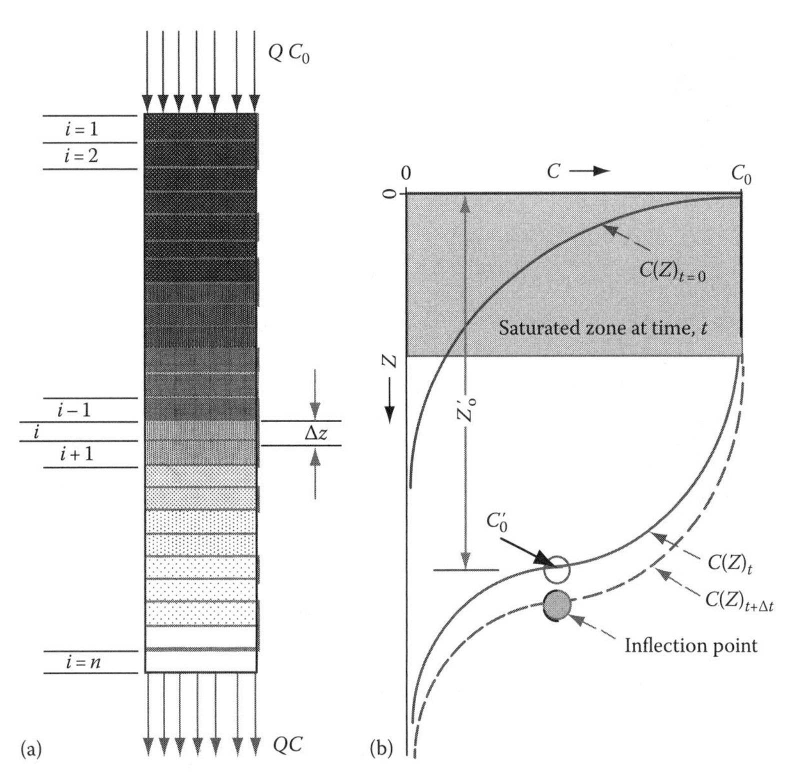

FIGURE 15.14 Elements of adsorption column and associated adsorbate concentration profiles. (a) Slices of adsorption column. (b) Adsorbate concentration profiles. (Adapted from Vagliasindi, F. and Hendricks, D.W., *J. Environ. Eng. Div.*, ASCE, 118(4), 532, 1992.)

(in which $C_0' = C_0$). Then, Equation 15.38 is applied to each slice, i, from $1 \leq i \leq n$, calculating $C(i)_{t+\Delta t}$. At the same time, $\overline{X}(i)$, $1 \leq i \leq n$, is calculated by Equation 15.39.

At time, $t = t + \Delta t$, the computation is repeated, i.e., as a "DO loop" (in Fortran), letting $C(i)_t = C(i)_{t+\Delta t}$ and $\overline{X}(i)_t = \overline{X}(i)_{t+\Delta t}$ at each successive iteration. To illustrate the amount of computation involved, for a simulation time, $t \approx 150$ h, about 2–3 h of Cray computer time (c. 1990) was required ($\Delta Z = 0.50$ cm, $\Delta t = 0.001$ min, $Z = 60$ cm) for the case of Rhodamine-B and Dowex 50. The associated number of iterations was about 20–30 million (Vagliasindi and Hendricks, 1992).

15.2.3.4 Characteristics of Output Curves

The "solutions" to Equation 15.24 are the output curves, $C(Z, t)$, in the form, $C(Z)_t$, the "wave front," or alternatively, $C(t)_{Z=Z_{max}}$, the breakthrough curve. These curves have characteristics that are reviewed in the following sections.

15.2.3.4.1 Concentration Profile, $C(Z)_t$

Figure 15.14b depicts adsorbate concentration profiles at $C(Z)_{t=0}$, $C(Z)_t$, and $C(Z)_{t+\Delta t}$, as depicted in Figure 5.21b. Salient points from the profiles include the following:

- At $t = 0$, the "initial" profile is that $C(Z)_{t=0} = C_0' e^{-\lambda Z}$; in other words, the assumed kinetic condition at $t = 0$ is that $[(\partial\overline{X}/\partial t)_P]_{i,t=0} \geq [(\partial\overline{X}/\partial t)_A]_{i,t=0}$ for $1 \leq i \leq n$.
- The saturated zone is depicted as shaded. This zone moves downstream as t increases. At all points within the saturated zone, $C(Z) \rightarrow C_0$. The rate of migration of the saturated zone depends upon the adsorbent capacity as determined by the isotherm and the feed concentration of adsorbate, C_0. Saturation must be *defined*, e.g., that $\overline{X} \geq 0.99 \cdot \overline{X}^*(C_0)$, the latter term meaning that the adsorbent is saturated in accordance with the isotherm relation for the argument, $C^* = C_0$.
- For the wave front, the $C(Z)_t$ profile changes rapidly; it is the zone where "mass transfer" occurs.
- Each computation iteration results in an infinitesimally different $C(Z)_t$ profile, i.e., a new $C(Z)_{t+\Delta t}$ profile is depicted for time $t + \Delta t$. Once the profile achieves a steady-state shape, it advances downstream at a constant velocity, maintaining the same shape (see Vagliasindi and Hendricks, 1992).

Also, as seen by Equation 15.24, and assuming $[dC/dt]_0 \approx 0$, the steeper the $\partial C/\partial Z$ curve, the higher the rate of uptake to the solid phase, $d\overline{X}/dt$. By the same token, the steeper the $\partial C/\partial Z$ curve, the higher the net rate of advection (and dispersion) to a given column slice. This steepest part of the $C(Z)_t$ curve occurs, by definition, at the inflection point, which is also the point at $[(\partial\overline{X}/\partial t)]_{i,t} = [(\partial\overline{X}/\partial t)_A]_{i,t}$ and is designated as C_0', which occurs at distance Z_0', as seen in Figure 15.14.

15.2.3.4.2 Examples of Concentration profile, $C(Z)_t$

Figure 15.15 shows $C(Z)_t$ profiles generated by computer solution of Equation 15.24, i.e., Equation 15.36 in finite-difference form. Both Figure 15.15a and b shows $C(Z)_t$

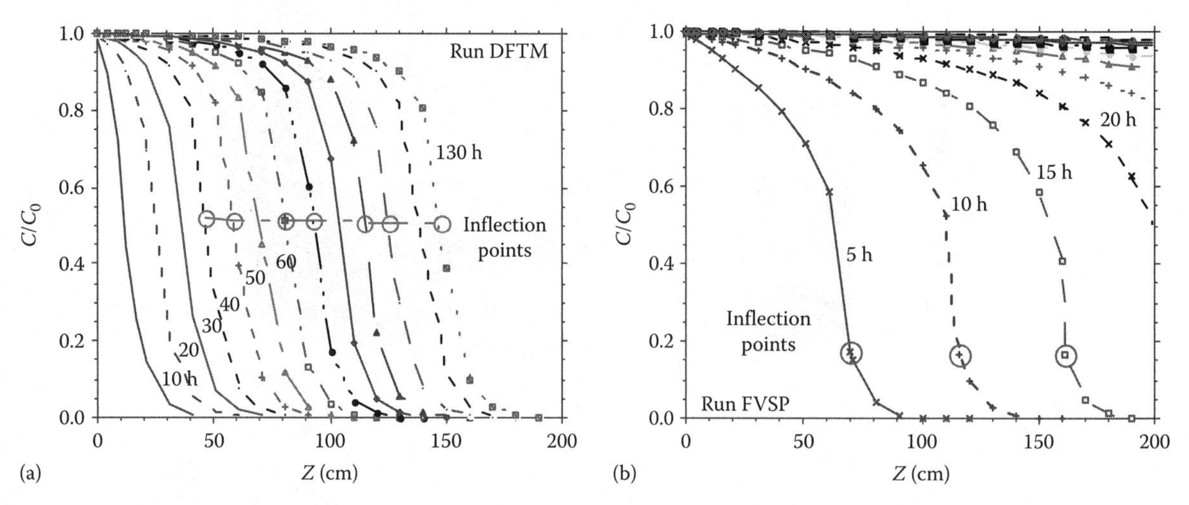

FIGURE 15.15 Computer-generated adsorbate concentration profiles. (a) Scale is 200: $C_0 = 200$, HLR = 56.1 cm/min $Q = 640$. (b) Scale is 200: $C_0 = 2000$, HLR = 56.1 cm/min $Q = 640$. (Adapted from Vagliasindi, F. and Hendricks, D.W., *J. Environ. Eng. Div.*, ASCE, 118(4), 535, 536, 1992.)

profiles at 10 and 5 h increments, respectively, i.e., for $10 \leq t \leq 130$ h and $5 \leq t \leq 60$ h. The initial profiles are in a "developing" phase and after some time attain a "steady-state" form and translate downstream at constant velocity. The translation velocity of the wave front, v_{wf}, depends upon the isotherm capacity of the adsorbent as defined by the influent adsorbate concentration and the product HLR $\cdot\, C_0$ (see Section 15.2.3.1). The inflection points are designated by circles, which remain at the same point for all steady-state wave fronts. The $C/C_0(Z)_{t=0}$ curve starts out with a shape calculated by Equation 15.29, i.e., exponential decay, since at $t = 0$, $Z_0 = 0$, and $[\partial\overline{X}/\partial t]_P > [\partial\overline{X}/\partial t]_A$.

Comparing Figure 15.15a and b, shows the effect of different influent concentrations, i.e., $C_0 = 200$ μg/mL and $C_0 = 2000$ μg/mL, respectively. As seen, for $C_0 = 2000$ μg/mL the profiles assume a more elongated form and the translation velocity is evidently higher than for $C_0 = 200$ μg/mL. A similar effect would be seen holding constant $C_0 = 200$ μg/mL and increasing HLR.

15.2.3.4.3 Breakthrough Curve, $C(t)_Z$

A *breakthrough curve* is the description of the adsorbate concentration with time, i.e., $C(t)_{Z=Z_{max}}$, as detected in the water flowing out of the column at $Z = Z_{max}$. As the column becomes "saturated," the effluent concentration will increase and eventually will approach the feed concentration, C_0, i.e., $C(t \gg 0)_{Z=Z_{max}} \rightarrow C_0$. In other words, the column becomes exhausted as the adsorbent becomes saturated with adsorbate.

Figure 15.16a shows $C(Z)_t$ profiles for Run #QUPR and the associated "breakthrough," i.e., $C(t)_{Z=Z_{max}}$, is seen in Figure 15.16b (Vagliasindi and Hendricks, 1992) The

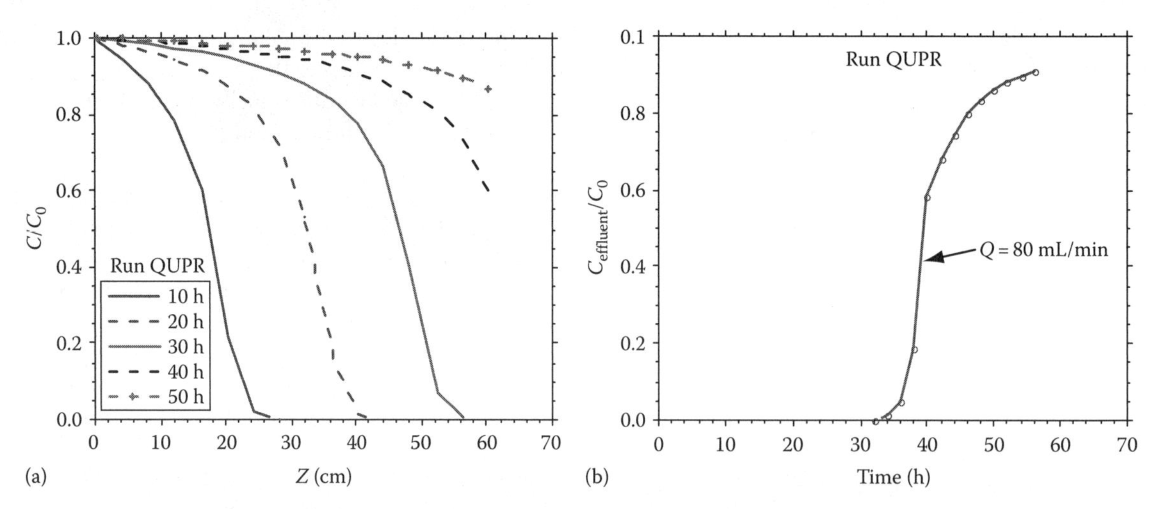

FIGURE 15.16 Advance of wave front for Run QUPR, i.e., (a), and ensuing breakthrough curve (b). (a) Wave-front advance for Run QUPR ($C_0 = 2000$, HLR = 7.0 cm/min, $Q = 80$ mL/min, $\Delta t = 0.0001$ min, $\Delta Z = 0.05$ cm). (b) Breakthrough curve for Run QUPR at Z(max) = 60 cm ($C_0 = 2000$, HLR = 7.0 cm/min, and $Q = 80$ mL/min).

breakthrough curve for Run #QUPR is seen to emerge at about $t \approx 32$ h with midpoint at about $t \approx 39$ h.

The breakthrough curve may, in fact, be "mapped" from the wave front, i.e., by the chain rule (see any text on differential equations), i.e.,

$$\frac{dC}{dt} = \frac{\partial C}{\partial Z} \cdot \frac{\partial Z}{\partial t} \tag{15.40}$$

and, for a given position on the wave front,

$$v_{wf} = \frac{\partial Z}{\partial t} \tag{15.41}$$

where v_{wf} is the velocity of wave front (m/s).

Therefore,

$$\frac{dC}{dt} = v_{wf} \cdot \frac{\partial C}{\partial Z} \tag{15.42}$$

Equation 15.42 shows the relationship between the breakthrough curve, $C(t)_{Z=Z_{max}}$ and the wave front, $C(Z)_t$; from this the breakthrough curve for Run #QUPR, shown in Figure 15.16b, may be mapped from the wave front, shown in Figure 15.16a. In other words, the emergence of the wave front from the column defines the breakthrough curve (see, for example, Vagliasindi and Hendricks, 1992).

15.2.3.4.4 Solid-Phase Adsorbate Concentration Profile, $\overline{X}(Z)_t$

Uptake to the solid phase is given as

$$\overline{X}_{Z,t+\Delta t} = \overline{X}_{Z,t} + \left[\frac{d\overline{X}}{dt}\right]_{Z,t} \cdot \Delta t \tag{15.43}$$

which is the same as Equation 15.39. As the adsorbent becomes "saturated" with adsorbent at the higher levels of the column, the solid-phase "wave front" translates downstream in the same fashion as the $C(Z)_t$ profile. Figure 15.17 shows the computed solid phase $\overline{X}(Z)_t$ profile for Run #DFTM, which plots synoptically with the $C(Z)_t$ profile of Figure 15.15a, thus establishing that the latter reflects the former and may be used to monitor the occurrence of saturation of the adsorbent media (Box 15.3).

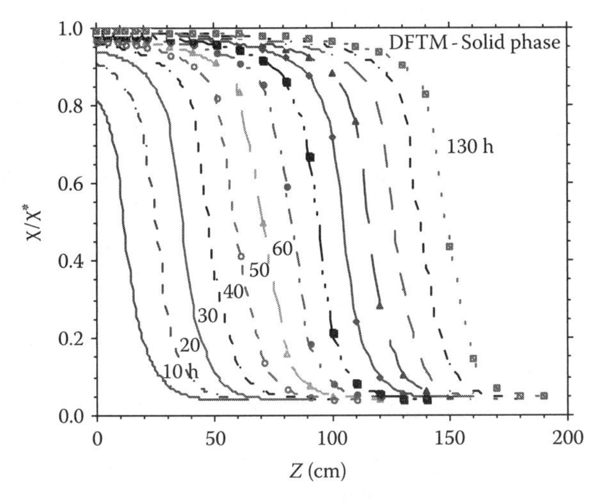

FIGURE 15.17 Solid-phase adsorbate concentration profile ($C_0 = 200$, HLR = 56 cm/min, $A = 11.43$ cm^2).

> **BOX 15.3 MODELING: MATHEMATICAL VERSUS PHYSICAL**
>
> Mathematical modeling of packed-bed reactors has not been applied often (i.e., directly to design) for several reasons: (1) the variables in the model—especially the kinetic ones—require extensive laboratory testing to determine; (2) the model must be in the form of a computer algorithm to execute, e.g., Fortran; (3) a high-speed computer is required to run the program; (4) the model is complex. The main limiting factor, however, is the extensive laboratory effort required to generate the needed kinetic coefficients and then to analyze the results. A final limitation, that warrants special mention, is that the general model described is for a single adsorbate, whereas most real situations involve a mixture of adsorbates.
>
> A physical model, i.e., a pilot plant, essentially "short circuits" the difficulties of mathematical modeling by empirically "integrating" the effects of independent variables in terms of $C(Z)_t$ profiles or, alternatively, of "breakthrough" curves, i.e., $C(t)_{Z=Z_{max}}$. In addition, the physical model may incorporate variables that were not anticipated in mathematical modeling. Biological growths are an example of the latter. i.e., the extent and the effect of biological growth. Competitive effects, e.g., of two or more adsorbates or background organic matter, are integrated as well. Two scales of physical models are (1) pilot plant, and (2) demonstration (discussed in subsequent sections). Due to slow particle uptake rate in some situations, physical modeling could require several weeks or months of operation.
>
> While not practical for direct application in design or operation, a mathematical model has other kinds of utility: (1) scenarios may be imposed to examine the trends caused by different variables; (2) the model helps to plan pilot plant experiments.

15.2.4 Rational Design

Two approaches to rational design, i.e., in terms of sizing an adsorbate reactor column, are (1) to solve the mass balance mathematical model, and (2) to size the column based on the wave-front velocity, v_{wf}. The main design outcomes sought are the length of the wave front and the position of the wave front at any given time. When the breakthrough concentration reaches the end of the reactor column, i.e., at time, t(breakthrough), this is the duration of the run. Subtracting from this L(wave front) gives the length of the saturated zone at the end of the run, i.e., $L(\text{sat}) = L(\text{packed bed}) - L(\text{wave front})$.

15.2.4.1 Quick-and-Dirty Mass Balance

The solid-phase wave front, $\overline{X}(Z)_t$, advances at the same velocity as the aqueous-phase wave front, $C(Z)_t$ (Section 15.2.3.4). This provides a basis for estimating the length of run from only isotherm and flow data.

15.2.4.1.1 Velocity of Concentration Profiles by Materials Balance

The materials-balance principle, applied to a packed-bed reactor column, with an effluent concentration of near-zero, may be stated,

$$V_t \cdot C_0 = [\rho A(1-P)]L(\text{sat})\overline{X}^*(C_0) + [\rho A(1-P)] \times \int_{L(\text{sat})}^{Z} \overline{X}(Z)\text{d}Z \quad (15.44)$$

where

- V_t is the volume of adsorbate solution that has flowed into the column after elapsed time, t (m^3)
- L(sat) is the length of saturated zone of column, defined in terms of $\overline{X}/\overline{X}^*(C_0)$, e.g., $1.0 \geq \overline{X}/\overline{X}^*(C_0) \geq 0.95$; the second limit is arbitrary.
- $\overline{X}^*(C_0)$ is the solid-phase adsorbate concentration at saturation based on feed concentration, C_0 (kg adsorbate/kg adsorbent)
- $\overline{X}(Z)$ is the solid-phase adsorbate concentration as it varies with Z in the reaction zone of the column (kg adsorbate/kg adsorbent)

Equation 15.44 states that the adsorbate mass lost from the aqueous phase equals that taken up by the solid phase. The first term on the right is the mass of adsorbate in the saturated zone of the solid phase of length, $Z = L(\text{sat})$, while the integral gives the mass of adsorbate in the mass-transfer zone, i.e., L_{wf}. As the column continues operation, the length of the saturated zone, i.e., L(sat), increases and, therefore, the mass of adsorbate in the mass-transfer zone of the column, as depicted by the integral, is less, relative to the mass of adsorbate in the saturated zone. Therefore, for a long column and after a long time of operation, the mass transfer to the solid phase can be approximated by neglecting the integral in Equation 15.44, i.e., $L(\text{sat}) \gg L_{\text{wf}}$, to give

$$V_t \cdot C_0 \approx [\rho A(1-P)]L(\text{sat})\overline{X}^*(C_0) \quad (15.45)$$

The velocities of the wave fronts for both solid phase and the aqueous phase are the same and may be approximated by the rate of advance of the saturated zone, i.e.,

$$v_{\text{wf}} \approx \frac{L(\text{sat})}{t} \quad (15.46)$$

where

- v_{wf} is the velocity of wave front, both solid phase and aqueous phase (m/s)
- t is the elapsed time from start of run (s)

A second relation is to express V_t in terms of HLR, A, and t, i.e.,

$$V_t = \text{HLR} \cdot A \cdot t \quad (15.47)$$

Now combining Equations 15.46 and 15.47 with Equation 15.45 gives an expression for the velocity of the wave front, i.e.,

$$v_{\text{wf}} \approx \frac{\text{HLR} \cdot C_0}{\overline{X}^*(C_0) \cdot \rho \cdot (1-P)} \quad (15.48)$$

Thus, when the $L(\text{sat}) \gg L_{\text{wf}}$, the wave-front velocity, v_{wf}, may be approximated by Equation 15.48.

15.2.4.1.2 Length of Packed-Bed Reactor

The length of the reactor bed at exhaustion, L(reactor), is the sum of the length of the saturated zone, L(sat), plus the length of the wave front, L_{wf}, i.e.,

$$L(\text{reactor}) = L(\text{sat}) + L_{\text{wf}} \quad (15.49)$$

where

- L(reactor) is the length of packed-bed portion of the reactor column (m)
- L_{wf} is the length of wave front, defined in terms of C/C_0, i.e., $0.01 \leq C/C_0 \leq 0.99$ (m)

As with L(sat), the definition of L_{wf} is arbitrary and may be taken as the distance for which $0.05 \leq C/C_0 \leq 0.95$. Substituting Equation 15.46 for L(sat) in Equation 15.49 gives

$$L(\text{reactor}) \approx v_{\text{wf}} \cdot t + L_{\text{wf}} \quad (15.50)$$

where v_{wf} can be calculated by Equation 15.48. If the saturated zone is long, relative to the wave-front length, then $L_{\text{wf}} \ll L(\text{sat})$, and Equation 15.49 simplifies to give $L(\text{reactor}) \approx L(\text{sat})$. The term, L_{wf}, may be determined by a pilot plant study, with sample taps at 50 mm increments along the length of the column. Both computer simulations and pilot plant measurements using Rhodamine-B dye and Dowex-50 resin and GAC have indicated that $L_{\text{wf}} \leq 1.0$ m. To illustrate the significance of the foregoing relations, if a column has a packed- bed of say 4.0 m (either a continuous bed or several in series) about 1.0 m may be allocated to the wave front (lacking specific data from a pilot plant).

Example 15.5 Sizing a Packed-Bed Reactor from Isotherm

Given

Benzene, a "priority pollutant" is to be removed from a groundwater to be used for drinking water. $Q = 3785$ m^3/day (1.0 mgd) and C_0(benzene) = 2.5 mg/L.

Required

Determine the area and length of a reactor for 30 days of operation.

Solution

1. Freundlich isotherm coefficients.
   From Table 15.3, for Filtrasorb 300, K(benzene) = 1.0 and $1/n$(benzene) = 1.60.
2. Calculate $\overline{X}(C_0)$, from Freundlich isotherm relation.

$$\overline{X}^* = KC_0^{1/n}$$
$$= 1.0 \cdot (2.5 \text{ mg/L})^{\wedge 1.60}$$
$$= 4.3 \text{ mg benzene/g carbon}$$
$$= 0.0043 \text{ kg/m}^3$$

3. Assume HLR = 12.2 m/h = 0.0034 m/s (5.0 gal/min/ft^2)
4. Assume constants,

$$\rho(\text{GAC}) \approx 1.40 \text{ g/mL}$$
$$P = 0.40$$

5. Calculate v_{wf} from Equation 15.48,

$$v_{wf} \approx \frac{\text{HLR} \cdot C_0}{\overline{X}^*(C_0) \cdot \rho \cdot (1 - P)} \quad (15.48)$$
$$= (0.0034 \text{ m/s}) \cdot (0.0025 \text{ kg/m}^3)/[(0.0043 \text{ kg B/kg C}) \cdot (1400 \text{ kg/m}^3) \cdot (1 - 0.35)]$$
$$= 2.17 \cdot 10^{-6} \text{ m/s}$$
$$= 0.19 \text{ m/day}$$

Then, assume $L_{wf} \approx 1.0$ m and calculate L(reactor) by Equation 15.50, with v_{wf} by Equation 15.48.

$$L(\text{reactor}) \approx v_{wf} \cdot t + L_{wf}) \quad (15.50)$$
$$= 0.19 \text{ m/day} \cdot 365 \text{ day} + 1.0 \text{ m}$$
$$= 6.6 \text{ m}$$

Discussion

The method of Equation 15.48 provides a more expedient means to estimate the wave-front velocity (as compared to a pilot plant).

15.2.4.1.3 Past Studies

Several literature sources refer to the "wave front" concept in terms of v_{wf} and L_{wf}, e.g., USEPA (1971, pp. 4–9), Zogorski and Faust (1978, pp. 753–776), Mullins et al. (1980, pp. 273–307). These corroborate what has been described in previous sections, e.g., Section 15.2.4.1. For example, according to USEPA (1971), as the carbon becomes saturated, the zone of adsorption moves downward with only a gradual increase in effluent adsorbate concentration. Then, when the leading part of the wave front reaches the end of the column, the adsorbate concentration increases rapidly until it equals the influent concentration (i.e., the breakthrough curve develops).

Zogorski and Faust (1978, pp. 753–776) discuss the length of the mass-transfer zone, i.e., L_{wf}, and its velocity, v_{wf}, as basic parameters characterizing column behavior. They give the methodology for determining both L_{wf} and v_{wf}: (1) pilot scale experiments are conducted in which several columns of different lengths are set up; (2) for each column, "breakthrough times" are plotted against "bed depth"; (3) the slope of the plot is the reciprocal of v_{wf} and the intercept is L_{wf}. They determined v_{wf} and L_{wf} for different GAC particle sizes, adsorbent species, influent concentrations, and HLR values (described in the next section).

Mullins et al. (1980, p. 278) obtained $0.0061 \leq v_{wf} \leq 0.012$ m/day ($0.02 \leq v_{wf} \leq 0.04$ ft/day) and $0.2 \leq L_{wf} \leq 0.3$ m for three GACs adsorbing chloroform in Louisville tap water at HLR ≈ 10.2 m/h (4.2 gpm/ft^2). For a 1.0 m-long GAC column they found that a column could be operated 6–12 weeks before breakthrough, i.e., 1.0 m/v_{wf}. They noted that the small diameter adsorbent particles have steeper wave fronts, i.e., a smaller length mass-transfer zone than larger particles. This is due to the faster diffusion per unit of GAC mass for the smaller particles.

15.2.4.2 Empirical Data for L_{wf} and v_{wf}

Table 15.4 gives L_{wf} and v_{wf} as determined for different experimental conditions (from Zogorski and Faust, 1978, p. 762). The effects of media size, adsorbate species, C_0, and HLR are shown. The data show that one might expect $0.11 \leq L_{wf} \leq 0.49$ m and $0.02 \leq v_{wf} \leq 2.2$ m/day for a wide range of operating conditions. These ranges may serve as default values for preliminary designs.

15.2.4.3 Theoretical Results for L_{wf} and v_{wf}

The results of Table 15.5 were generated by a Fortran model that solved Equation 15.24 by finite difference, i.e., using Equation 15.39, for conditions stated (see headings and footnotes). The important points are the following: (1) both L_{wf} and v_{wf} are proportional to C_0; (2) both L_{wf} and v_{wf} are proportional to HLR; (3) the effects of HLR and C_0 on L_{wf} and v_{wf} were as expected, as seen by Equation 15.48; (4) the values of L_{wf} and v_{wf} are in the same order of magnitude as seen in Table 15.4 (which may be fortuitous).

15.2.5 Problems

The first limitation of the foregoing theory is that competition is likely if two or more adsorbates are present. A second is that bacterial growths are probable and cause deviations in the velocity of wave fronts and the shapes of breakthrough curves for any given adsorbate.

15.2.5.1 Competition between Adsorbents

For a given adsorbent, different adsorbates have different affinities to adsorb, as reflected in the isotherm coefficients,

TABLE 15.4
Length and Velocity of Wave Front for Different Conditions—from Pilot Plant Data

	Conditions				Results	
Adsorbate	HLR (m/h)	C_0 (mmol/L)	Mesh	d(particle) (mm)	L_{wf} (m)	v_{wf} (m/day)
Effect of media size[a]						
2.4-Dichlorophenol	6.1	5.6	12 × 16	1.2–1.7	0.49	0.53
—			16 × 10	1.2–2.0	0.34	0.55
			20 × 25	0.7–0.8	0.19	0.55
			25 × 30	0.6–0.7	0.12	0.50
			30 × 40	0.4–0.6	0.11	0.58
Effect of adsorbate species[a]						
Phenol	6.1	5.5	20 × 25	0.7–0.8	0.074	0.79
4-Methoxyphenol					0.095	0.70
4-Nitrophenol					0.16	0.60
2,4-Dichlorophenol					0.15	0.53
Effect of C_0[a]						
2.4-Dichlorophenol	6.1	0.18	20 × 25	0.7–0.8	0.11	0.024
		1.06			0.11	0.12
		5.5			0.15	0.53
		5.6			0.16	0.53
Effect of HLR[a]						
2.4-Dichlorophenol	4.9	5.5	20 × 25	0.7–0.8	0.15	0.43
	9.3				0.24	0.86
	17.1				0.36	1.61
	23.7				0.47	2.21

Source: Adapted from Zogorski, J.S. and Faust, S.D., Operational parameters for optimum removal of phenolic compounds from polluted waters by columns of activated carbon, Chap. 20, in: Cheremisinoff, P.N. and Ellerbusch, F. (Eds.), *Carbon Adsorption Handbook*, Ann Arbor Science Publishers, Inc., Ann Arbor, MI, p. 762, 1978.

[a] Adsorbent was Columbia LCK GAC; temperatures were 27°C, 25°C, 25°C, and 27°C, respectively.

α and $1/n$ for the Langmuir and Freundlich isotherms, respectively. The upshot is that some adsorbates will adsorb preferentially (DiGiano et al., 1980, Frick et al., 1980, Fritz et al., 1980; Crittenden and Weber, 1978a,b,c). For example, if the isotherm constant, α, is higher for adsorbate A than adsorbate B, the wave front of A will lag that of B. The one with the higher α has the stronger attraction for the adsorbent.

15.2.5.2 Chromatographic Effect

The effect of competitive adsorption is sometimes illustrated in terms of a "chromatographic" effect, illustrated in Figure 15.18a for competition between *p*-nitrophenol and *p*-chlorophenol (Fritz et al., 1980, p. 205). The *p*-nitrophenol has stronger bonding with the GAC and shows a "classic" breakthrough curve. The *p*-chlorophenol has weaker bonding and is displaced by the *p*-nitrophenol, causing C/C_0 (*p*-chlorophenol) $\gg 1.0$; i.e., the *p*-chlorophenol has higher-than-feed concentration in the breakthrough curve. The displacement of one compound by the other causing different wave fronts and associated velocities constitute the "chromatographic effect" and is an expected result of a solute mixture fed into an adsorbent column.

Figure 15.18b shows breakthrough curves for four compounds in a mixture of equal-molar-feed concentrations. The 1,4-dioxane was adsorbed until the carbon column was exhausted (with respect to this particular adsorbate). Nitromethane, methyl ethyl ketone, and *n*-butanol continued to be adsorbed after the 1,4-dioxane had broken through, indicating that they are more successful competitors for adsorption sites than 1,4-dioxane (McGuire and Suffet, 1978, p. 626).

15.2.5.2.1 TOC as a Surrogate

The chromatographic effect occurs in full-scale GAC columns where the influent water may contain a variety of organic compounds, all with different degrees of affinity for the carbon surface. In most cases, however, an aggregate measure, e.g., TOC, is used in place of trying to assess each individual component, which masks the chromatographic effect.

15.2.5.3 Bacterial Colonization

An effect not anticipated initially in GAC technology for water treatment (i.e., about the mid-1960s) was the colonization by bacteria and a subsequent removal of

TABLE 15.5
Length and Velocity of Wave Front for Different Conditions by Computer Simulation

Conditions				Results	
Adsorbate	HLR (m/h)	C_0 (μg/L)	Run	L_{wf} (m)	v_{wf} (m/day)
Effect of C_0					
Rhodamine-B	4.2	200	TEBV	0.20	0.037
dye		500	PYSO	0.28	0.090
		1000	QGUR	0.40	0.170
		2000	QUPR	1.00	0.360
Effect of HLR					
Rhodamine-B	4.2	500	PYSO	0.28	0.090
dye	8.4		DFTP	0.45	0.180
	33.7		DFUH	1.30	0.68

Source: Adapted from Vagliasindi, Wave front behavior in adsorption reactors, MS Thesis, Department of Civil Engineering, Colorado State University, Fort Collins, CO, p. 74, 1991.

Adsorbent was Dowex 50 cation exchange resin. $\bar{X}^* = 726{,}434$; 745,060; 751,483; 754,736 μg Rhodamine-b/g Dowex 50 resin for $C_0 = 200$, 500, 1000, 2000 μg Rh-B/L, respectively.

some compounds by bacterial metabolism. In fact, the presence of a metabolizing compound is likely to promote such colonization. While some view the bacterial colonization as a positive effect, the bacteria may cause higher impedance to the diffusion of adsorbate molecules to adsorption sites.

15.3 LABORATORY AND PILOT PLANT STUDIES

The most important laboratory data to be generated are isotherms. A pilot plant study can add more specific information about the behavior of a given system.

15.3.1 Questions for a Laboratory/Pilot Plant Study

Some of the questions for a pilot plant investigation include type of carbon, bed depth, hydraulic loading rate, hydraulic conductivity of the packed bed, backwash head needed, pretreatment advisable, etc. The foregoing list comprises independent variables which may be imposed on the column, one at a time, to assess the effect. The dependent variables of interest include the wave front, $C(Z)_t$ and the breakthrough curve, $C(t)_{Z=Z_{max}}$.

15.3.1.1 Isotherm Determination

As described in Section 15.2.1.2 the experimental determination of an isotherm is the basis for the prediction of wavefront velocity, v_{wf}, as described by Equation 15.48, Section 15.2.4.1. As noted, most waters contain more than one soluble organic compound and so the isotherm for a specific compound in a particular mixture would be unique. If TOC reduction is an objective, then a *pseudo* isotherm, as measured by TOC, may be used as an estimate of adsorbent capacity.

15.3.1.2 Determine v(wave front)

The velocity of the wave front, v_{wf}, may be determined by (1) successive direct measurement from sample taps along a

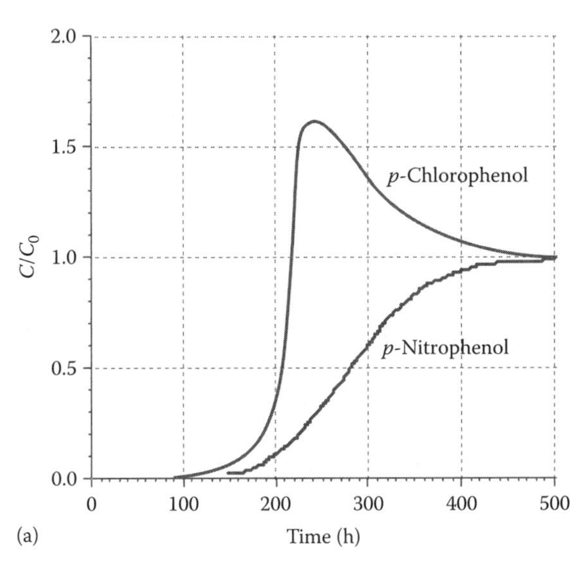

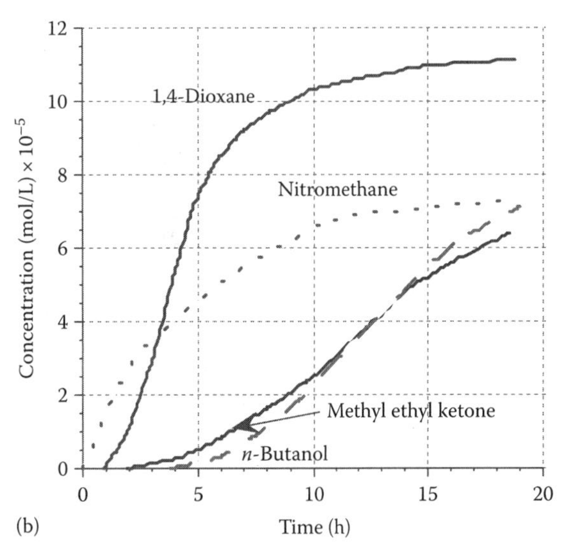

FIGURE 15.18 Examples of breakthrough curves that illustrate chromatographic effect for multi-solute adsorbate systems (data points not shown). (a) Bi-solute adsorbate system. (From Fritz, W. et al., Competitive adsorption of dissolved organics on activated carbon, Chap. 9, in: Suffet, I.H. and McGuire, M.J. (Eds.), *Activated Carbon Adsorption of Organics from the Aqueous Phase*, Vol. 1, Ann Arbor Science Publishers, Ann Arbor, MI, p. 205, 1980.) (b) Four-solute adsorbate system. (From McGuire, M.J. and Suffet, I.H., The calculated net adsorption energy concept, Chap. 4, in: Suffet, I.H. and McGuire, M.J. (Eds.), *Activated Carbon Adsorption of Organics from the Aqueous Phase*, Vol. 1, Ann Arbor Science Publishers, Ann Arbor, MI, p. 625, 1980.)

column, (2) measurement of the breakthrough curve, and (3) calculation using Equation 15.48.

15.3.1.3 *L*(wave front)

The length of the wave front, L_{wf}, may be determined by direct measurement from sampling taps. Alternatively, L_{wf} may be estimated by calculation from the breakthrough curve measurement using Equation 15.42.

15.3.1.4 Breakthrough Curve

The breakthrough curve is the most common measurement obtained from pilot plant studies. As noted in the foregoing paragraphs, v_{wf} and L_{wf} may be determined from the measurement of a breakthrough curve.

15.3.1.5 Rate of Headloss Increase

The measurement of hydraulic gradient may be obtained by piezometer measurements along the length of the column. The simultaneous measurement of Q and calculation of HLR and temperature permits determination of intrinsic permeability, k, per Equation E.5. Whether k increases with time is of interest also.

15.3.1.6 Backwash Velocity

Backwash velocity requires empirical determination by noting the HLR when the bed becomes suspended. As the HLR increases, the hydraulic gradient increases at the same time; when the bed is suspended, the hydraulic gradient does not increase with further increase in HLR (see Section 12.4.4.6).

15.3.1.7 Assess Competitive Effects of Different Adsorbates

As noted in Section 15.2.5.2, competitive effects must be measured by a pilot plant study. This may be a tedious exercise, especially if a number of organic compounds are involved. If knowledge of such compounds is needed, then a gas chromatograph "scan" of the influent water, just to reveal the peaks, will give an idea of how many compounds are in the water. This may be followed by the identification of compounds, and perhaps quantitative analyses.

15.3.1.8 Discover Effects of Unanticipated Problems

Usually, in the operation of any kind of system, unanticipated behavior may occur, or in some cases, problems may develop. The discovery of such behavior or problems may be sufficient justification for the operation of a pilot plant. Some of the more common influences on column behavior include temperature changes, variation in influent concentration, the development of biological growths, corrosion, slurry transport issues, etc. Each of these influences may be investigated by means of laboratory-scale and/or pilot-scale studies.

15.3.1.9 Fabrication

A column 2–3 m in length and about 51 mm (2 in.) diameter will suffice, with sampling taps spaced at 50 mm. The sampling taps may be 1 mm brass tubes inserted in rubber stoppers (or a glued-and-machined fitting) with a non-adsorbing gauze over each of the tubes to prevent adsorbent particles from entering. The column may be acrylic plastic (Plexiglas), or clear PVC (which is not transparent). Fittings can be attached easily to PVC (using a cement) while a machinist may be needed to work with the acrylic. Piezometers should be installed, e.g., at 100 mm, mounted on a board with scale attached.

15.3.2 Demonstration-Scale Plants

As noted in Chapter 3, a demonstration-scale plant addresses questions of process design, but is large enough to provide additional data on such issues as labor requirements, logistics of GAC supplies, storage, regeneration of carbon, energy needs, estimation of costs, and unforeseen problems. Two examples are the Pomona, California plant, c. 1967 and later the Denver Water Reuse Plant, c. 1982.

15.3.2.1 Pomona

One of the best known activated carbon demonstration-scale pilot plants in the 1960s was a 1,136 m^3/day (0.3 mgd) installation at the 17,033 m^3/day (4.5 mgd) Pomona Water Reclamation Plant operated by the Federal Water Quality Administration (which became USEPA in 1971) and the Sanitation Districts of Los Angeles County (Parkhurst et al., 1967). Initially, sponsorship was a part of the USPHS AWTR program to explore new technologies for wastewater treatment and help to develop those that were most promising. The design, along with reactor dimensions, is shown in Figure 15.19. The internal diameter of each reactor was 1.829 m (6.0 ft) with GAC depth of (2.9 m) (9.5 ft), giving a carbon mass of 3019 kg (6650 lb), supported on a double perforated stainless steel plate screen. Activated carbon is corrosive to steel and so the tanks were lined with tar epoxy. The column operated at

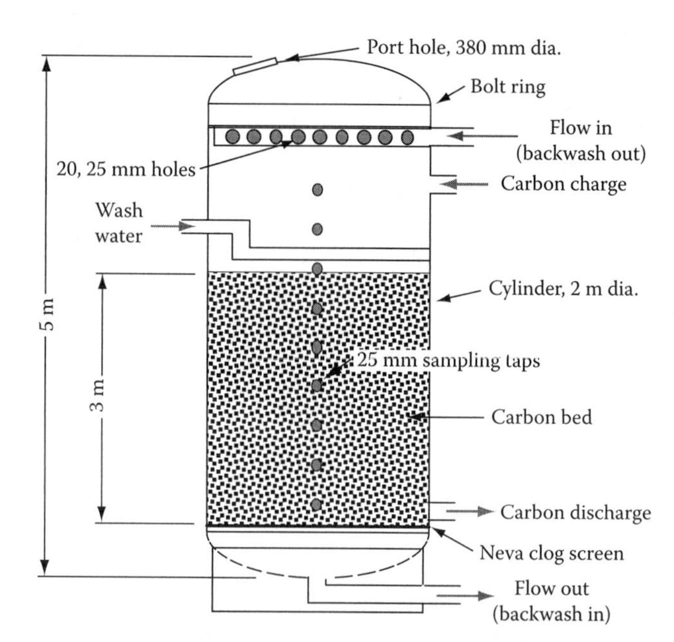

FIGURE 15.19 Carbon reactor design, Pomona pilot plant. (Adapted from Parkhurst, J.D. et al., *J. Water Pollut. Control Federation*, 39(10, Part 2), R69, October 1967.)

12.6 L/s (200 gpm) or at a hydraulic loading rate of 17.1 m/h (7 gpm/ft^2). Backwashing rate was 12.62 L/s (200 gpm) maximum, which allowed for 35% bed expansion, with a total flow of about 18,925 L (5,000 gal) per backwash. The lead reactor in the series was backwashed 30–45 min. each day.

The plant had five carbon bed reactors, with four operated in series, and in the downflow mode. One was operated alone for greater experimental flexibility. The reactors were rotated in position designations, i.e., A, B, C, D; the A designation was the lead reactor. The position change was accomplished by changing valve open/close positions in the plumbing. The reactor with the freshly regenerated carbon always came on line in the D position, while the reactor taken out of service for regeneration always came from the A position.

The influent COD varied $25 \leq \text{COD} \leq 39$ mg/L and showed a random pattern of variation. Carbon (as COD) remaining after column D varied $2 \leq \text{COD(effluent)} \leq 13$ mg/L, with nominal values ≈5 mg COD/L. About 76 million L (20 mg) of water could be treated before the product water COD reached 12 mg/L. In terms of carbon capacity, the adsorption of organics was about 0.60 kg COD/kg carbon (60 lb COD/100 lb of carbon).

15.3.2.2 Denver Reuse Plant

The 3785 m^3/day (1.0 mgd) Denver Potable Water Reuse Demonstration Plant ("Denver Reuse Plant") was conceived in the 1960s as a means to reuse Denver's "foreign" water, i.e., domestic wastewater that was treated by the Denver Wastewater Reclamation Plant (Denver Metro). The motivation was the 1957 Blue River Decree, which implied that Denver should utilize its imported water before seeking additional supplies from the "west-slope." Colorado water law permits use and reuse of imported water, i.e., from basins outside the South Platte River basin. Water appropriated for use within the basin may be used only once by each appropriator.

The means to reuse the foreign water, at that time, was as drinking water. Thus, a pilot plant study was initiated in 1968 under contract with the University of Colorado; the pilot plant (capacity 19 L/min or 5 gpm) was located at Denver Metro with secondary effluent as feed water. A variety of unit processes were investigated, including biological treatment, which continued until 1974. Based on the pilot plant study, construction of a demonstration-scale plant was started in 1981 and placed in operation initially about 1984, being put on-line in 1985 (after the "shakedown" operation and working out "bugs," e.g., corrosion control). The construction cost was about \$20 million with \$7 million from USEPA. The site for the Reuse Plant was just across the South Platte River from "Denver Metro"; the flow of secondary effluent taken for the Reuse Plant was $Q(\text{GAC}) = 3785$ m^3/day (1.0 mgd).

The 3785 m^3/day (1.0 mgd) main treatment train was lime coagulation, flocculation, settling, filtration, ion-exchange, and first-stage GAC, with 0.9 fraction of the flow going to industrial use with a 380 m^3/day (0.1 mgd) sidestream to ozone, second-stage GAC, reverse osmosis, air stripping (to remove CO_2 and VOCs), and disinfection. The plant had the capability to investigate different treatment train sequences and to examine proposed unit processes or technology variations by means of sidestreams.

Objectives were (1) to investigate the health effects of its effluent on laboratory animals, (2) to estimate the operating costs of a full-scale plant, (3) to discover any exigencies in operation (i.e., unanticipated issues), and (4) to provide a basis for the design of a full-scale plant (if built). The plant completed its plan of study in December 1991 and was taken out of service at that time.

Figure 15.20 shows the GAC portion of the Denver Reuse Plant. Figure 15.20a is a drawing, showing ammonia recovery and removal process (ARRP) towers (#19) and three larger reactors (#20), on the left; two storage vessels for regenerated GAC (#21); and three sidestream reactors on the right (#22); with a regeneration furnace on the far right (#24). Figure 15.20b is a photograph showing the three 1.22 m (4 ft) diameter sidestream reactors toward the center with the regenerated GAC storage vessels on the right.

Table 15.6 summarizes design and operating data for the GAC system. For the three first-stage GAC columns, $Q(\text{design}) = 3785$ m^3/day (1.0 mgd), giving HLR(design) = 13.4 m/h (5.5 gpm/ft^2). For the three second-stage reactors, $Q(\text{design}) = 378.5$ m^3/day (0.1 mgd). The "first-stage" GAC columns took the full plant flow, i.e., $Q(\text{plant}) = 0.0425$ m^3/s (0.97 mgd). The "second-stage" GAC columns, in series with

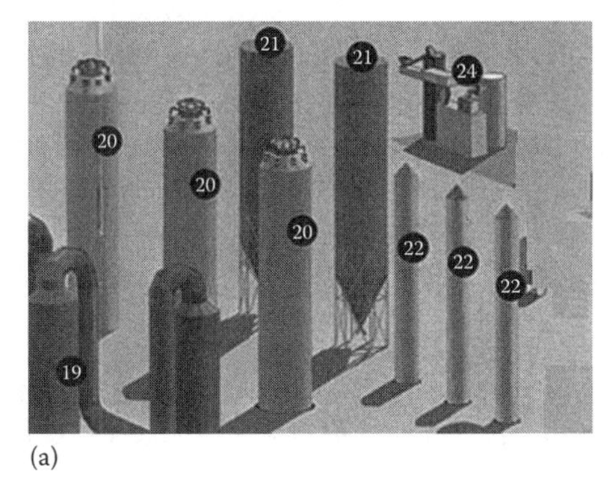

(a)

(b)

FIGURE 15.20 GAC reactors at Denver Potable Water Reuse Plant. (a) Schematic of GAC reactors. With permission. (b) Photograph of GAC reactors. (Courtesy of Denver Water, c. 1986.)

TABLE 15.6
Design and Operation Parameters for GAC at Denver Reuse Plant[a,b]

	First Stage		Second Stage	
Parameter	**(SI Units)**	**(U.S. Units)**	**(SI Units)**	**(U.S. Units)**
GAC	Filtrasorb 300	Virgin GAC	Filtrasorb 300	Regenerated
Mode	Downflow		Downflow	
Problems	Bio-mat at top		(no bio-mat)	
Mass(GAC)			5,136 kg	11,300 lb
V(GAC)	90.6 m^3	3200 ft^3	9.06 m^3	340 ft^3
Q	0.0425 m^3/s	0.97 mgd	0.0038 m^3/s	0.082 mgd
HLR	14.4 m/h	5.9 gpm/ft^2	11.7 m/h	4.5 gpm/ft^2
A(calculated)	10.61 m^2	114.17 ft^2	1.176 m^2	12.654 ft^2
d(column)[c]	3.68 m	12.06 ft	1.22 m	4.0 ft
H(packed-bed)[c]	8.54 m	28.0 ft	8.25 m	27 ft
ΔP(packed-bed)	60–75 kPa	9–11 psi	45 kPa	7 psi
EBCT	34 min		42 min	
Q(backwash)	0.126 m^3/s	2000 gpm	0.013 m^3/s	200 gpm
HLR(backwash)	44 m/h	18 gpm/ft^2	39 m/h	15.9 gpm/ft^2
θ(backwash)	15 min		15 min	
TOC(influent)	19 mg/L to plant			
TOC(effluent)			1.2 mg/L 12/88	from GAC
			5 mg/L, 09/89	from GAC
TOX(influent)	84 μg/L to plant			
TOX(effluent)			35 μg/L	from GAC

[a] Willis (1989).
[b] Lauer (1998).
[c] Calculated from *V*(GAC), *Q*, HLR data.

the first stage, took about 0.10 fraction of the plant flow, and followed ozone oxidation. The purpose of the oxidation was to break down large organic molecules to render them more amenable to adsorption. The first-stage GAC used virgin Filtrasorb 300®, while the second-stage utilized in-house regenerated carbon. For the first stage, the entire flow was passed through one of the three columns. The second column was maintained in standby position, filled with virgin carbon, and the third column was empty. For the second-stage GAC, the same pattern was used.

A variety of organic compounds were present in the influent from the secondary clarifiers at Denver Metro as seen by a gas chromatogram (Lauer et al., 1985, p. 54), which showed about 135 peaks. Regarding performance, the influent TOC concentrations were about 19 mg TOC/L with second-stage GAC effluent ranging 1–5 mg TOC/L. Operation was generally in the downflow mode but the upflow mode was an option.

15.4 DESIGN

The main idea in process design is to size the reactor, i.e., total area of all columns in parallel, and the total length of all columns operated in series. An economical reactor length, i.e., *L*(reactor), is sought. The longer the packed bed, the higher the fraction that is saturated, before breakthrough, and the longer the run. On the other hand, the saturated zone essentially "stores" the spent GAC. This leads to a dilemma in that saturated GAC does not adsorb contaminants. Adsorption occurs only along the wave front, i.e., the mass-transfer zone (MTZ). In lieu of a long column, a series of columns, e.g., as at Pomona (Section 15.3.2.1) may be used with the first one removed from the operation after it is saturated. A fresh column is then placed in the last place.

15.4.1 Design Variables

Table 15.7 shows groupings of variables in both dependent and independent categories. Within each category, the variables are identified with ranges and typical values given with descriptions given for nonquantitative variables.

15.4.1.1 Independent Process Variables

The independent variables are seen in Table 15.7 as the second group, which includes adsorbent properties, reactor design, and operation. These variables are those that affect the wave-front/breakthrough curve, i.e., length, L_{wf}, velocity, v_{wf}, and *t*(breakthrough), respectively.

15.4.1.1.1 Adsorbent Properties

Table 15.7 lists adsorbent variables as adsorbent type, shape, size (mesh limits, d_{50} or d_{10} and UC), porosity, particle density, apparent density, surface area, molasses number, iodine number, abrasion number, dispersion coefficient, intrinsic permeability.

TABLE 15.7
Design Variables and Magnitudes for a Packed-Bed Reactor

	Variable		Magnitude		
Variable Group	**Basic**	**Derivative**	**SI Units**	**Range**	**Typical**
Dependent					
Column performance	$C(Z, t)$		(kg/m^3)		
Wave front	$C(Z)_t$		(kg/m^3)		
		v_{wf}	(m/s)	$f(\overline{X}^*, HLR, C_0)$[a]	0.3 m/day[a]
		L_{wf}	(m)	$f(HLR, C_0)$[a]	0.8[a]
Breakthrough curve	$C(t)_{Z=Z_{max}}$		(kg/m^3)		
		t(breakthrough)	(s)		150 d[b]
Independent					
Adsorbent (GAC)					
Carbon type				Commercial[c]	
Shape				All angular	Crushed
Size					
mesh			Mesh	0.5–4 mm[d]	8 × 30 mesh
d_{50}	d_{50}		mm	1.2–1.7[e]	1.5[e]
d_{10}			mm	0.8–1.1[e]	0.8[e]
UC				1.4–1.9[e]	1.7[e]
Porosity	P		m^3 voids/m^3 bed	0.3–0.6	0.5
Particle density	ρ_s		kg sol/m^3 sol	0.92–1.5[e]	1.3[e]
Apparent density		$\rho(\text{apparent}) = \rho_s(1 - P)$	kg solids/m^3 bed	430–480[e]	480[e]
Surface area			m^2/g	300–2500	1000[e]
Molasses number			min		230[d]
Iodine number			min	600–1000[e]	900[e]
Abrasion number			min	70–85[e]	70[e]
Dispersion coeff.	$D/\overline{v}$			0.75–0.80[f]	0.75[f]
Intrinsic permeability	k		(m^2)		
		Headloss	(m)		
Bed expansion headloss			(m)		
Reactor design					
HLR	HLR		(m/s)	0.081–0.41[e]	0.20[e]
Area		A	(m^2)	$A = Q/HLR$	
Length of bed	L(bed)		(m)	3.0–9.0	9.0
Contact time		EBCT = L/HLR	(min)	10–50	30
Reactor type					Pressure steel Gravity concrete
Lining					epoxy, rubber
Operation					
Adsorbate species				1000's	NOM, TCE, etc.
Influent concentration	C_0		μg/L	20–1000	20 μg TCE/L
			mg/L	1–5	2 mg NOM/L
Isotherm capacity	$\overline{X}^*$ (NOM)		kg NOM/kg C		0.5
	$\overline{X}^*$ (CH_2Cl_2)		mg/g C		1.0[g]
	$\overline{X}^*$ (1,2,4 $C_6H_3Cl_3$)		mg/g C		100[g]
Isotherm coefficients	K			1–273[g]	30[g] CCl_4
	$1/n$			0.3–1.6[g]	0.8[g] CCl_4
Headloss rate	$\Delta p/\Delta L$		(m water/m)[h]	0.05–0.2[h]	0.1[h]
Backwash rate	HLR(backwash)		(m/s)	0.61–0.81[e]	0.61[e]
Bed expansion			percent	10–50	20
Flow mode	Upflow Downflow				

(*continued*)

TABLE 15.7 (continued)
Design Variables and Magnitudes for a Packed-Bed Reactor

	Variable		Magnitude		
Variable Group	Basic	Derivative	SI Units	Range	Typical
Carbon use rate					
Tertiary			(kg GAC/m^3)	0.12–0.23	
			(lb GAC/mg)[i]	200–400	
Physical chemical tr.			(kg PAC/m^3)	0.29 = 1.04	
			(lb carb/mg)	500–1800	

[a] Vagliasindi and Hendricks (1992).
[b] Clark and Lykins (1989, p. 282); EBCT = 15 min → L(column) ≈ EBCT · HLR ≈ 0.25 h · 12.2 m/h ≈ 3.1 m (10 ft) → $v_{wf} \approx 0.0211$ m/day.
[c] Selected from commercial designations given by manufacturers; e.g., Filtrasorb (Calgon); Norit,; etc. Some 142 companies distribute activated carbon are listed in a web search (February 2003). Norit Americas, Inc. mentions that 150 types are manufactured by their company (norit.com).
[d] Norit Americas, Inc., Atlanta, GA, Norit 1240 specification for size is 5% is retained by U.S. sieve 10 mesh (2.00 mm) and 0.5% passes U.S. sieve 40 mesh (0.42 mm); the specification that follows would be, therefore, 10 × 40. [See norit.com]
[e] USEPA (1973, pp. 2–5).
[f] Hendricks (1973) for Filtrasorb 200.
[g] Love e al. (1983).
[h] USEPA (1973); units are: (meters water headloss/meter GAC bed depth).
[i] (lb car/mg) = pounds carbon per million gallons water processed.

All GAC is crushed and, therefore, angular in shape. Size is measured by sieve analysis (U.S. standard sieves are the most common). A sieve designation such as 8 × 30, means that about all of a given sample passes (except perhaps <5%) the #8 sieve and about all is retained by a #30 sieve (except perhaps <5%).

Porosity depends upon the uniformity coefficient, UC, the angularity, and the packing. For spheres, values may range P(rhombic packing) = 0.26, and P(face centered) = 0.48 m^3 voids/m^3 packed bed (Section E.4.2). From Section E.4.2, $0.35 \leq P(\text{sand}) \leq 0.45$, $1.23 < \text{UC(sand)} < 1.31$; $0.46 \leq P(\text{anthracite}) \leq 0.58$, $1.24 < \text{UC(anthracite)} < 1.33$. Particle density, ρ_s, is the mass of carbon per unit volume of particle with range, $1300 < \rho_s < 1500$ kg carbon/m^3 carbon particle. Apparent density is the mass of carbon per unit volume of packed bed with range, $430 < \rho(\text{apparent}) < 480$ kg carbon/m^3 packed bed. Equation 15.51 gives the relation between ρ(apparent) and ρ_s, i.e.,

$$\rho(\text{apparent}) = \rho_s(1 - P) \qquad (15.51)$$

The derivation is based on defining $\rho(\text{apparent}) = W_s/V = \rho_s V_s/V = \rho_s(V - V_v)/V = \rho_s(1 - P)$, in which W_s = mass of solids (m^3), V = total volume of packed bed (m^3), V_s = volume of solids (m^3), and V_v = volume of voids (m^3); also recall, $P \equiv V_v/V$.

Indices of adsorption capacity (see Table 15.7 and Glossary) include surface area (as measured by BET isotherm using nitrogen gas), molasses number, iodine number, etc. The surface area is a measure of the total surface area of all internal pores that are accessible to N_2 gas. Molasses number is an index of the larger pore sizes while the iodine number is an index of the smaller pore sizes.

The dispersion coefficient is a measure of the "spread" of a solute in passage through a porous medium due to statistical variation in velocity (see Section 4.2.2.4). Representative values of the $D/\bar{v}$ are given in Table 15.7. Representative intrinsic permeability values are $k \approx 2.8 \cdot 10^{-9}$ m^2 for Witco 12 × 30 and $k \approx 7.6 \cdot 10^{-10}$ m^2 for Filtrasorb 400 12 × 40.

15.4.1.1.2 Design Variables

The second group of independent variables in Table 15.7 includes those that affect design: HLR, bed area, length of reactor bed, contact time, reactor type. The hydraulic loading rate, i.e., the apparent velocity, is selected based on practice, with HLR ≈ 12.2 m/h (5.0 gpm/ft^2) being representative. Higher values may be used, which will result in a longer wave front, i.e., L_{wf}, and a higher wave-front velocity, v_{wf}, which translates to a shorter run (see also Section 15.2.3.4).

15.4.1.1.3 Operating Variables

The operating variables in Table 15.7 include those that affect operation: adsorbate species, influent concentration, C_0, adsorbent capacity as a function of C_0, (in terms of isotherm coefficients), hydraulic gradient (i.e., headloss per unit length), backwash rate, flow mode (i.e., upflow or downflow), and carbon use rate for the given application (in terms of kg carbon used per m^3 of water treated). The carbon use rate is a common parameter of operation. The GAC usage may be monitored in terms of the kg GAC used per m^3 water treated (lb GAC used/mg water treated).

15.4.1.2 Guidelines and Criteria

There are no institutionalized (e.g., Ten States Standards) guidelines for the design of GAC reactors. The pertinent process variables for sizing are HLR and L(bed). The latter, i.e., L(bed), depends on the duration of run desired, i.e., $t(\text{run}) = L(\text{bed})/v_{wf}$. The general ranges, usual in practice, are $4.9 \leq \text{HLR} \leq 24.6$ m/h ($2.0 \leq \text{HLR} \leq 10$ gpm/ft^2) and $3.0 \leq L(\text{bed}) \leq 9.0$ m ($10 \leq L(\text{bed}) \leq 30$ ft), respectively (USEPA, 1973).

15.4.2 Design Protocol

An objective of process design is to size the reactor columns, i.e., to determine the total packed-bed cross-section area and the number of reactors in parallel; the total length of packed bed, i.e., L(reactor) and the associated number of reactors in series. Ancillary objectives include: to select a length of run, i.e., t(run); to determine the layout of the system, whether the reactors are upflow or downflow; and to estimate the cost of the virgin carbon for the packed bed. Incorporation of a design algorithm into a spreadsheet facilitates exploration of alternative scenarios.

15.4.2.1 Spreadsheet Layout

Table CD15.8 is a spreadsheet that incorporates a design algorithm for a GAC reactor column. Inputs are required at several points along the sequence of steps and are categorized as "laboratory," "given," "assumed," and "calculated" and are explained in the following paragraphs.

15.4.2.1.1 Laboratory Data

The laboratory generated data include the following:

- Isotherm constants for the given adsorbate–adsorbent pair, e.g., K and $1/n$ for Freundlich isotherm
- Particle density, ρ_s, and porosity, P (or, alternatively apparent density, ρ(apparent))

15.4.2.1.2 Given Data

The given data describe the circumstances of the situation and include the following:

- Adsorbate species, e.g., TCE, chloroform, benzene, etc. The calculation protocol, i.e., Table CD15.8, applies only to a single adsorbate.
- Influent adsorbate concentration, C_0. The value of C_0 may vary, depending upon the situation. For the purposes of design, a non-varying value of C_0 is used, as a rule.
- Again, the flow to the plant, Q, is whatever occurs, but may also be set at a certain level, depending upon the circumstances.
- Unit cost of GAC. The unit cost of the GAC is usually based on a bid price and depends upon quantity. For reference, a bulk price range for virgin activated carbon was given by USEPA (2000) as \$1.54–2.64/kg (\$0.70–1.20/Lb).
- Unit cost of GAC regeneration. For large installations, on-site regeneration is usually recommended. For smaller installations, the carbon may be transported for regeneration by special tanker truck.

15.4.2.1.3 Assumed Data

The assumed data may take on any values desired but usually are constrained by practical guidelines such as given in Table 15.7.

- Adsorbent, e.g., Norit, Fitrasorb 300, Wittcarb 950, etc. The GAC is selected based upon the pore structure, cost, and results of isotherm tests or upon given data from the literature.
- Hydraulic loading rate, HLR. The HLR is assumed as a means to size the total bed area, i.e., $A(\text{bed}) = Q/\text{HLR}$.
- Length of saturated zone of reactor, L(sat). The value of L(sat) is assumed as a basis for calculating t(run). If t(run) is very long, e.g., $t(\text{run} > 300$ days, then a shorter value for L(sat) may be used. Alternatively, t(run) may be assumed and L(sat) may be calculated.
- Length of wave front, L_{wf}, Ideally, L_{wf} would be obtained from pilot plant data. If assumed, then an estimate of $L_{wf} \approx 1.0$ m is probably adequate, especially if L(sat) is large and if t(run) is long, which would compensate for uncertainty.
- Number of columns, n(col), to be operated in parallel. The value of n(col) operated in parallel depends upon such factors as the maximum bed area desired per column, redundancy, i.e., $n(\text{col}) \geq 2$, economy, etc.

15.4.2.1.4 Calculated Results

The calculated values, as given in the spreadsheet, are listed below.

- The adsorbent capacity to adsorb adsorbate, $\overline{X}^*(C_0)$, i.e., Equation 15.19, $\overline{X}^* = KC_0^{1/n}$. The Freundlich relation is given because values for the coefficients appear in the literature most frequently.
- Velocity of wave front, v_{wf}, Equation 15.48, $v_{wf} = \text{HLR} \cdot C_0/[\overline{X}^* \cdot \rho(1-P)]$.
- Length of run, i.e., time duration until breakthrough occurs, t(run), Equation 15.46, i.e., $t(\text{run}) = L(\text{sat})/v_{wf}$.
- Length of packed-bed portion of reactor, i.e., L(reactor), Equation 15.50, $L(\text{reactor}) = L(\text{sat}) + L_{wf}$.
- The total area of the packed bed for all columns, A(total), Equation 15.2, $A(\text{total}) = Q/\text{HLR}$.
- The area of a single column, A(col), in a collection of packed beds that are operated in parallel, $A(\text{col}) = Q/n(\text{col})$.
- Diameter of one of several parallel columns, D(col), of area, A(col), i.e., $D(\text{col}) = (4/\pi) \cdot A(\text{col})$.
- Volume of GAC packed bed, V(bed), that includes all beds in system, $V(\text{beds}) = L(\text{reactor}) \cdot A(\text{total})$.
- Mass of GAC in all packed beds of system, M(carbon), i.e., $M(\text{carbon}) = \rho(\text{apparent}) \cdot V(\text{beds})$.
- Cost of virgin GAC, i.e., $\text{Cost(GAC)} = \text{unit cost} \cdot M(\text{carbon})$.

TABLE CD15.8
Design Protocol for GAC Packed-Bed Reactor

				C_0		$\overline{X}^*$		ρ			ρ(app)		HLR		
Adsorbent	Adsorbate	K	$1/n$	(mg/L)	(kg/m^3)	(mg ate/gC)	(kg ate/kg C)	(gC/mLC)	(kgC/m^{3}C)	P	(kgC/m^{3}bed)	(gpm/ft^2)	(m/h)	(m/s)	(m/s)
Wittcarb 950	TCE	28.2	0.44	0.10	0.00010	10.24	0.01024	1.50	1500	0.55	675	5.0	12.2	0.003388889	0.000000049
Wittcarb 950	TCE	28.2	0.44	0.10	0.00010	10.24	0.01024	1.50	1500	0.55	675	5.0	12.2	0.003388889	0.000000049
Wittcarb 950	TCE	28.2	0.44	0.10	0.00010	10.24	0.01024	1.50	1500	0.55	675	5.0	12.2	0.003388889	0.000000049
Filtrasorb 300		26.2	0.47	0.10	0.00010	8.88	0.00888	1.50	1500	0.55	675	5.0	12.2	0.003388889	0.000000057
Norit	CCl_4	28.5	0.8	0.10	0.00010	4.52	0.00452	1.50	1500	0.55	675	5.0	12.2	0.003388889	0.000000111
Nuchar WV-G		25.8	0.8	0.10	0.00010	4.09	0.00409	1.50	1500	0.55	675	5.0	12.2	0.003388889	0.000000123
Hydrodarco 1030		14.2	0.7	0.10	0.00010	2.83	0.00283	1.50	1500	0.55	675	5.0	12.2	0.003388889	0.000000177
Hydrodarco 1030	Benzene	10	1.6	0.10	0.00010	0.25	0.00025	1.50	1500	0.55	675	5.0	12.2	0.003388889	0.000001999
Dowex 50	Rh-B	758,018	0.115	2000	2.000	754.737	0.75474	1.30	1300	0.34	858	1.7	4.197	0.001165778	0.000003601
		(μg/g)	(mL/μg)	(μg/mL)											
				Feed concentration Given		Calculated $\overline{X}^*(C_0) = KC_0^{1/n}$		Particle density Given		Porosity = ρ(1−P) Given		Given			Wave-front velocity

15.4.2.2 Spreadsheet Scenarios

The spreadsheet, i.e., Table CD15.8 may be used to explore different design "scenarios," i.e., the results from imposing a set of design conditions. For example, one may explore the effect of different selections of HLR, L(sat) on t(run) and M(carbon) used per unit volume of water treated for per unit of time (e.g., monthly carbon use). If additional columns are added, the annual cost of carbon may be assessed for these different conditions. In addition, the effects of variations in operating conditions may be explored, e.g., Q and C_0.

15.4.3 Design Examples

Several full-scale applications of GAC are described which illustrate the variety of designs found in practice. The examples range from taste-and-odor control to SOC removals.

15.4.3.1 Examples of Sites

Table 15.9 shows a spectrum of contamination sources for various organic compounds. The examples of sources include truck spills, rail car spills, chemical spills, on-site storage tanks, landfill sites, gasoline spills and tank leakage, chemical by-products, manufacturing residues, and chemical landfills. The tabular summary is indicative of the kinds of treatment situations and the kinds of compounds that may be found at different sites. A wide variation is seen in influent concentrations (in mg/L); in some cases these are high, relative to what is found in say groundwaters (usually reported in μg/L). Carbon use rates varied 0.05–1.6 kg GAC used/m^3 of water treated (0.4–13 lb/1000 gal) with typical values of about 0.1 kg/m^3 (1 lb/1000 gal).

15.4.3.2 GAC for Taste-and-Odor Control

The use of PAC for taste-and-odor control for drinking water started in the late 1920s. The use of GAC for this purpose started in the 1960s; two cases illustrate such application.

15.4.3.2.1 Buckingham, England

Most of the activated carbon applications for drinking water prior to the 1960s were for taste-and-odor control. GAC was used for such purpose in Buckingham, England and found to be cheaper than PAC for a 7570 m^3/day (2 mgd) plant constructed in 1960. The reactors were 2.438 m (8 ft) diameter, 0.914 m (3 ft) length, eight in number, rubber lined, packed with 18×60 mesh GAC, and operated in parallel at 12.2 m/h (5 gpm/ft^2). Backwash was once per week and the GAC lasted 4 years before being replaced (Hager, 1969).

15.4.3.2.2 Goleta Water District, California

The Goleta Water District obtained water from Lake Cachuma treated by filtration and GAC in a plant constructed in 1962 and enlarged in 1964 to 9462 m^3/day (2.5 mgd). The raw water was pre-chlorinated and filtered through two diatomaceous earth pressure filters and the through four GAC reactors, 3.658 m (12 ft) diameter, 2.438 m (8 ft) bed depth, with 14 × 60 mesh (U.S. Sieve Series) GAC, operated in parallel and under pressure at 14.6 m/h (6 gpm/ft^2). The steel GAC vessels were protected from corrosion with a baked epoxy coating. Backwashing removed entrapped air and turbidity. The usual threshold odor number of the influent was 20 with a maximum 70, with effluent numbers approaching zero. Bed life was about 3.5 years (Hager, 1969).

15.4.3.3 Chemicals in Drinking Water Sources

Rivers in the United States and in other countries worldwide are conduits for waste discharges, runoff, seepages, etc., which accumulate downstream. GAC was recognized in the early 1960s as a means to reduce the chemical concentrations in drinking water. Three cases illustrate how GAC was applied.

15.4.3.3.1 Nitro Plant, West Virginia Water Company

The Nitro WTP was the first to utilize GAC for treatment of drinking water. Pilot plant studies were conducted starting in 1962 (Dostal et al., 1965).

v_{wf}						Q									
(m/h)	(m/day)	L(sat) (m)	t(run) (d)	L_{wf} (m)	L(reactor) (m)	(mgd)	(m³/s)	A(total) (m²)	n(col) (#)	A(col)	D(bed) (m)	V(bed) (m³)	M(carbon) (kg)	Unit Cost ($/kg C)	$(carbon) ($)
0.0001765	0.0042	10	2360	1.00	11.00	2.00	0.0876	25.85	1.00	25.85	5.74	284	191964	0.66	126,696
0.0001765	0.0042	10	2360	1.00	11.00	2.00	0.0876	25.85	2.00	12.93	4.06	284			
0.0001765	0.0042	10	2360	1.00	11.00	2.00	0.0876	25.85	4.00	6.46	2.87	284			
0.0002036	0.0049	10	2047	1.00	11.00	2.00	0.0876	25.85	4.00	6.46	2.87	284			
0.0004001	0.0096	10	1041	1.00	11.00	2.00	0.0876	25.85	1.00	25.85	5.74	284	191964	0.66	126,696
0.0004420	0.0106	10	943	1.00	11.00	2.00	0.0876	25.85	1.00	25.85	5.74	284			
0.0006379	0.0153	10	653	1.00	11.00	2.00	0.0876	25.85	1.00	25.85	5.74	284			
0.0071954	0.1727	10	58	1.00	11.00	2.00	0.0876	25.85	1.00	25.85	5.74	284	191964	0.66	126,696
0.0129618	0.3111	10	32	1.00	11.00	2.00	0.0876	25.85	1.00	25.85	5.74	284	244008	0.66	161,045
		Given	$t = L(\text{sat})/v_{wf}$		$L = L(\text{sat}) + L_{wf}$			$A = Q/\text{HLR}$		$A(\text{col}) = A(\text{total})/n(\text{col})$				Given	Cost = Unit Cost
$v_{wf} = \text{HLR} \cdot C_o/(X^*(C_o) \cdot \rho \cdot (1-P)$				Estimated		Given			Given			$V(\text{bed}) = L(\text{reactor}) \cdot A(\text{total})$			$\cdot M(\text{carbon})$
													$M(\text{carbon}) = V(\text{bed}) \cdot \rho(\text{app})$		

Context: The Nitro WTP is on the Kanawha River, surrounded by a number of chemical industry plants. The issue was to provide protection for the drinking water of Nitro for which the River served as the source.

GAC treatment: In 1966, the sand in the filter beds was replaced by GAC, giving a treatment train of aeration, chemical clarification, disinfection, GAC adsorption/filtration. The 10 filters were concrete boxes with pipe-lateral under-drains in graded gravel which supported 0.76 m (2.5 ft) GAC with $2.44 \leq \text{HLR} \leq 4.88$ m/h ($1 \leq \text{HLR} \leq 2$ gpm/ft²). Backwash was every 4 days. Reactivation was by a multiple-hearth furnace. Turbidity was removed to 0.05–0.15 NTU with threshold odor number being reduced from 20 to 80 in filter/GAC influent to ≤ 3 in the effluent. The carbon chloroform extract (CCE) was reduced from 200 as influent to the filter/GAC to <40 in the effluent (Hager, 1969).

Micro-pollutants—River Elbe: The River Elbe (1100 km in length) has DOC concentrations ≈6 mg/L, caused mostly by runoff. The pesticide atrazine is a typical micro-pollutant with average concentrations ≈0.14 μg/L, which exceeds the drinking water guideline 0.1 μg/L. The waterworks along the river use bank filtration as a first treatment step in which biological processes decrease the concentration of DOC. The residual DOC adsorbs more strongly on GAC, which decreases the removal efficiency of micro-pollutants, e.g., atrazine (Foregoing from Müller et al., 1996).

15.4.3.3.2 Cincinnati Municipal Plant

In 1992, GAC treatment was added to conventional drinking water treatment for Cincinnati, Ohio. The first objective was to reduce overall organic carbon levels, which were about 1.5 mg/L TOC before GAC treatment to about 0.2–0.6 mg/L after GAC treatment. A second objective was to reduce the hazard of SOCs, due to the large number of industrial discharges and due to spills that have been known to occur. GAC pilot plant studies were initiated in 1977.

Background: The raw water source for Cincinnati's California WTP is the Ohio River, with intake some 745 km (463 mi) downstream from the headwaters at the confluence of the Allegheny and Monongahela Rivers. Six major tributaries contributed about 270 point source discharges. The streams are subject also to agricultural runoff and accidental spills. For example, an incident in 1977 involved a discharge of some 64 metric tons (70 U.S. tons) of carbon tetrachloride into the Kanawha River (upstream) and subsequently was found in intakes for drinking water along the Ohio River. The river is also a major transportation artery for coal, grain, petroleum products, etc. The raw water source for the plant, the Ohio River, has ambient TOC ≈ 3 mg/L with about half of that removed after coagulation and filtration. A gas chromatogram (with flame ionization detection) showed, however, about 30 peaks with some 15 synthetic compounds enumerated in order of detection frequency, e.g., chloroform, benzene, toluene, dichloromethane, 1,2-dichlorobenzene, and so on (Westerhoff and Miller, 1986, p. 148). In GAC pilot studies, the number of peaks found in the effluent was only four (also present in a "blank"). Some 200 organic contaminants had been found in trace amounts in the Ohio River (AWWA Mainstream, 1992).

15.4.3.3.3 California WTP

The California WTP had a design capacity of 833,000 m³/day (220 mgd), with measured flow about 454 mL/day (120 mgd) in winter and 503 mL/day (133 mgd) in summer; the treatment train was "conventional." The GAC addition to the plant was the largest worldwide (c. 2003), serving about 1 million persons, with Q(max daily) = 470 mL/day (175 mgd). Reactors numbered 12, with 7–11 operated in parallel; the others were on standby or out of service for reactivation. The elements of the GAC system included (1) GAC reactors, (2) storage and transport, (3) regeneration, and (4) plant controls.

GAC reactors: The reactors were to be gravity, rectangular in shape, with the box of reinforced concrete. The reactors were

TABLE 15.9
GAC Performance for Different Kinds of Contaminant Sources and Organic Contaminants

		Concentration	
Source	**Contaminant**	**Influent (mg/L)**	**Effluent (μg/L)**
Truck spill	Methylene chloride	21	<1.0
	1,1,1-Trichloroethane	25	<1.0
Rail car spills	Phenol	63	<1.0
	Orthochlorophenol	100	<1.0
	Vinylidine chloride	4	<10
	Ethyl acrylate	200	<1.0
	Chloroform	0.02	<1.0
Chemical spills	Chloroform	3	<1.0
	Carbon tetrachloride	135	<1.0
	Trichloroethylene	3	<1.0
	Tetrachloroethylene	70	<1.0
	Dichloroethyl ether	1.1	<1.0
	Dichloroisopropyl ether	0.8	<1.0
	Benzene	0.4	<1.0
	DBCP	3	<1.0
	1,1,1-Trichloroethane	0.4	<10
	Trichlorotrifloroethane	6	<10
	Cis 1,2-dichloroethylene	0.005	<1.0
On-site storage tanks	*Cis* 1,2-dichloroethylene	0.5	<1.0
	Tetrachloroethylene	7	<1.0
	Methylene chloride	1.5	<100
	Chloroform	0.5	<100
	Trichloroethylene	8	<1.0
	Isopropyl alcohol	0.2	<10
	Acetone	0.1	<10
	1,1,1-Trichloroethane	12	<5.0
	1,2-Dichloroethane	0.5	<1.0
	Xylene	8	<1.0
Landfill site	TOC	20	<5000
	Chloroform	1.4	<1.0
	Carbon tetrachloride	1	<1.0
Gasoline spills, tank leakage	Benzene	11	
	Toluene	7	<100
	Xylene	10	
	Methyl *t*-butyl ether	0.03	<5.0
	Di-isopropyl ether	0.04	<1.0
	Trichloroethylene	0.06	<1.0
Chemical by-products	Di-isopropyl methyl phosphonate	1.2	<50
	Dichloropentadiene	0.5	<10
Manufacturing residues	DDT	0.004	<0.5
	TOC	9	
	1,3-Dichloropropene	0.01	<1.0
Chemical landfill	1,1,1-Trichloroethane	0.08	<1.0
	1,1-Dichloroethylene	0.01	0.005

Source: Adapted from Groeber, M.M., Granular Activated Carbon Treatment, Engineering Bulletin EPA-540/2-91/024, Office of Research and Development, U.S. Environmental Protection Laboratory, Cincinnati, OH, October, 1991.

open at the top, downflow, with stainless steel under-drains, i.e., a header pipe the full length of the vessel with horizontal laterals of wedge-wire screen (slots 0.25 mm (0.010 in.). Removal of spent GAC was to be as slurry, achieved by fluidizing the bed, with provision for removal both at midlevel and at the bottom. For storage prior to regeneration, six tanks each with capacity 480 m^3 (17,000 ft^3) were to be located adjacent to the furnace room. Slurry pipe velocities were designed $0.9 \leq v(\text{pipe}) \leq 1.5$ m/s ($3 \leq v(\text{pipe}) \leq 5$ ft/s); stainless steel piping for slurry transport was to be fabricated to facilitate convenient removal for maintenance.

Design criteria: The design criteria were EBCT $\approx$ 15 min; L(reactor bed) = 3.4 m (11 ft); HLR = 13.4 m/h (5.5 gpm/ft^2); mass GAC used (average over the year) $\leq$ 22,700 kg/day (50,000 lb/day); and GAC sieve size 12 × 40, i.e., $0.55 \leq d(\text{particle}) \leq 0.75$ mm (Westerhoff and Miller, 1986, p. 149). Other data were V(GAC per reactor) = 600 m^3 (21,000 ft^3), mass(GAC per reactor) = 250,000–300,000 kg (560,000–660,000 lb); for backwash, $0.02 \leq$ bed expansion $\leq$ 0.30 fraction of bed depth with HLR(backwash) $\leq$ 26.8 m/h (11 gpm/ft^2).

Reactivation of GAC: The carbon was reactivated on-site by two multiple-hearth furnaces, each with capacity 18,144 kg/day (40,000 lb/day), which calculates to be 7–8 days for the reactivation of one GAC reactor. The volume loss is about 9% for each reactivation. Removal of TOC was found effective through six cycles of reactivation. Pore sizes were reduced with each reactivation, with loss of micropores, with larger pore sizes increasing. Apparent density of the virgin GAC was 0.501 g/mL increasing to 0.525 g/mL after the carbon was saturated with TOC. After six cycles of reactivation, the density was only 0.407 g/mL (The foregoing description was from Moore et al., 2003.)

15.4.3.4 Pump and Treat

The term "pump and treat," refers to pumping from a contaminated water body, usually groundwater, and then passing the water through a treatment facility, usually, GAC. A number of such cases emerged as a result of CERCLA (Comprehensive Environmental Response, Compensation, and Liability Act) in 1980, preceded by the famous case of the Love Canal. Another case was the Rocky Mountain Arsenal near Denver, Colorado. Both are reviewed.

15.4.3.4.1 Love Canal

The environmental movement of the 1960s was still a popular cause and a viable political force through the 1970s. This was the context for the public health issues of chemical toxins in the environment which were exemplified by the Love Canal and was a stimulus for CERCLA. The Love Canal case illustrates not only an application of treatment by GAC but is indicative of the unique history of most such cases.

Background: The Love Canal Hazardous Waste Site in the City of Niagara Falls, New York, has been an infamous example of chemical contamination. The name comes from

William T. Love, an entrepreneur who, in 1892, saw a canal as a means for ships to bypass Niagara Falls or, alternatively, to generate electric energy. Because of loss of financial backing, only 1 mi of the canal was dug, leaving a body of water 60 ft wide and 3000 ft long. By 1920, the land was sold and afterward became a municipal and chemical disposal site. From 1942 through 1953, the Love Canal Landfill was used principally by Hooker Chemical, one of the many chemical plants located along the Niagara River. During this period, nearly 19,000 metric tons (21,000 U.S. tons) of "toxic chemicals" were dumped at the site. In 1953, with the landfill at maximum capacity, Hooker filled the site with layers of soil.

Human exposures: The Niagara Falls Board of Education then purchased the Love Canal land from Hooker Chemical for $1. Included in the deed transfer was a "warning" of the chemical wastes buried on the property and a disclaimer absolving Hooker of any further liability. Single-family housing soon surrounded the Love Canal site, and, as the population grew, the 99th Street School was built directly on the former landfill. At the time, homeowners were not warned or provided information of potential hazards associated with the former landfill site. According to residents who lived in the area, from the late 1950s through the early 1970s repeated complaints of odors and "substances" surfacing in their yards brought City officials to visit the neighborhood. By 1978, the Love Canal neighborhood included approximately 800 private, single-family homes, 240 low-income apartments, and the 99th Street Elementary School—located near the center of the landfill. On August 7, 1978, President Jimmy Carter declared the Love Canal area a federal emergency, which provided funds to relocate the 239 families living in the first two rows of homes encircling the landfill. By 1980, U.S. government scientists had identified 248 individual chemicals in the Love Canal waste site. (The foregoing paragraphs were abstracted from http://ublib.buffalo.edu/libraries/ projects/lovecanal/.)

Chemical conditions: An analysis of composite samples from 14 shallow wells before treatment was started which gave the following general results: pH $\approx$ 5.6; TOC $\approx$ 4,300 mg/L; SOC $\approx$ 4,200 mg/L; COD $\approx$ 11,500 mg/L; oil/grease $\approx$ 90 mg/L; suspended solids $\approx$ 200 mg/L; dissolved solids $\approx$ 15,700 mg/L; chloride $\approx$ 9,500 mg/L; sulfate $\approx$ 240 mg/L; sodium $\approx$ 1,000 mg/L; calcium $\approx$ 2,500 mg/L; iron $\approx$ 330 mg/L; mercury < 0.0005 mg/L; lead $\approx$ 0.4 mg/L. An equilibrium study of $\bar{X}^*$(SOC) versus C^*(SOC) generated a well-defined isotherm, Figure 15.21, which showed, for example, that at C^*(SOC) $\approx$ 4200 mg/L, $\bar{X}^*$(SOC) $\approx$ 200 mg SOC/g GAC.

GAC pilot plant: A GAC pilot system was used to assess treatment parameters, consisting of two columns in series, each 1.2 m (4 ft) long. The first-stage column was removed from operation when the breakthrough occurred in the effluent from the second-stage column (simulating full-scale operation). The effluent TOC was about 260 mg/L. Figure 15.21 shows an isotherm generated for the aggregate SOCs at the site. The pilot study also indicated the rate of carbon exhaustion as 34 kg GAC/m^3 water treated (284 lb/1000 gal).

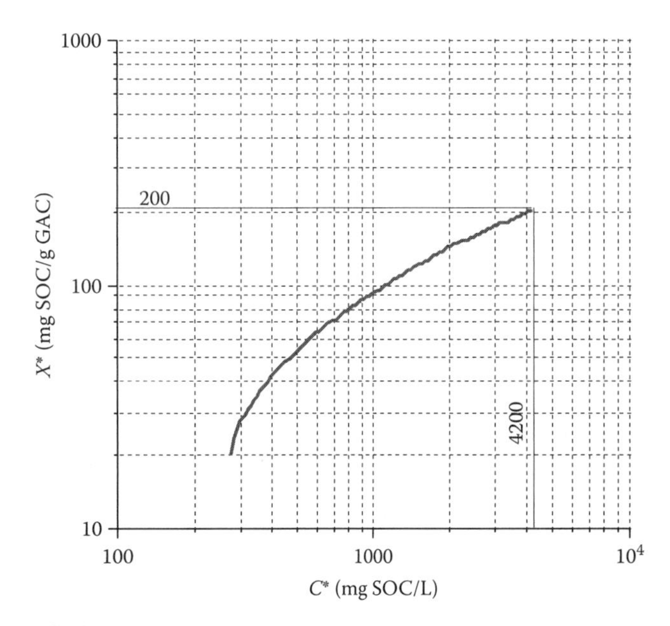

FIGURE 15.21 Isotherm for aggregate of synthetic organic compounds (SOC) at Love Canal. (Adapted from McDougall, W.J. et al., *J. Water Pollut. Control Federation*, 52(12), 2916, December 1980.)

A temporary treatment system was set up and included caustic soda addition, clarification/sludge disposal, filtration, GAC adsorption, and discharge to sanitary sewer. The GAC treatment included two skid-mounted pressure reactors, each holding 9,072 kg (20,000 lb) of GAC with $Q \approx 662$ L/min (175 gpm). For this temporary full-scale system, TOC(influent) $\approx$ 800 mg/L and TOC(effluent) $\approx$ 80 mg/L. The Love Canal site was 6.47 ha in area and was essentially a "bathtub" in that it was underlain by bedrock and there was no flow of groundwater through the site. Thus, it was a classic "pump and treat" situation (foregoing from McDougall et al., 1980).

GAC performance and observations: Table 15.10 lists the priority pollutant compounds (all 29 were found) with concentrations indicated before and after GAC treatment. Important points to note include (1) the data pertain to a mixture; (2) an isotherm, Figure 15.21, was defined from the SOC mixture; (3) influent concentrations were high for most compounds; (4) GAC treatment reduced most effluent concentrations to "not detectable" (ND); and (5) a "chromatography effect" (one or more compounds being displaced by others) was not seen.

15.4.3.4.2 Rocky Mountain Arsenal, Colorado

The Rocky Mountain Arsenal in Colorado, near Denver, was a 9.66 km × 8.66 km (6 mi × 6 mi) area of land located on the NE side of Denver and about 5 km east of the South Platte River, which flows in a northerly direction. Groundwater flow through a shallow aquifer is north-west, toward the South Platte.

Background: The site was developed during WWII to manufacture munitions; nerve gas was the most prominent. The facility was taken over, by lease arrangement beginning in

TABLE 15.10
Love Canal Leachate Treatment of Priority Pollutants by GAC

	Leachate	
Compound	Raw (GAC Influent) (μg/L)	GAC Effluent (μg/L)
2,4,6-Trichlorophenol	84	<10
2,4-Dichlorophenol	5,100	ND
Phenol	2,400	<10
1,2,3-Trichlorobenzene	870	ND
Hexachlorobenzene	110	ND
2-Chloronaphthalene	510	ND
1,2-Dichlorobenzene	1,300	ND
1,3 and 1,4-Dichlorobenzene	960	ND
Hexachlorobutadiene	1,500	ND
Anthracene and phenanthrene	29	ND
Benzene	28,000	<10
Carbon tetrachloride	61,000	<10
Chlorobenzene	50,000	12
1,2-Dichloroethane	52	ND
1,1,1-Trichloroethane	23	ND
1,1-Dichloroethane	66	ND
1,1,2-Trichloroethane	780	<10
1,1,2,2-Tetrachloroethane	80,000	<10
Chloroform	44,000	<10
1,1-Dichloroethylene	16	ND
1,2-*Trans*-dichloroethylene	3,200	<10
1.2-Dichloropropane	130	ND
Ethylbenzene	590	<10
Methylene chloride	140	46
Methyl chloride	370	ND
Chlorodibromomethane	29	ND
Tetrachloroethylene	44,000	12
Toluene	25,000	<10
Trichloroethylene	5,000	ND

Source: Adapted from Groeber, M.M., Granular Activated Carbon Treatment, Engineering Bulletin EPA-540/2-91/024, Office of Research and Development, U.S. Environmental Protection Laboratory, Cincinnati, OH, October, 1991.

1946 by Shell Chemical to manufacture agricultural chemicals, which was continued until 1982. Government use continued with the site being used for disposal of munitions. Contamination of groundwater was recognized in the late 1950s and by about 1970 became a public issue and decontamination of munitions were started and by 1982 full environmental decontamination was underway, e.g., surface water basins, buildings, groundwater, etc. The aquifer was a source for drinking water for the adjacent community of Commerce City, on the west boundary of the Arsenal (http://www.epa.gov/region08/superfund/sites/co/rma). Some of the contaminants found included diisopropyl methyl phosphonate (DIMP), associated with nerve gas production and dicyclopentadiene (DCPD), associated with manufacture of pesticides. Cease and desist orders were issued by the Colorado Department of Health in 1975.

Clay barrier: As a part of the cleanup strategy, a 1 m (3 ft) wide 457 m (1500 ft) bentonite clay barrier (extending into an impervious layer of clay shale bedrock) was constructed to block a contaminant plume. Six wells were located on the upstream side of the barrier to pump the contaminated water to a GAC treatment plant with injection on the downstream side through 12 recharge wells.

GAC treatment: The treatment train consisted of a sump for the storage of contaminated water, filtration, and GAC treatment. Two GAC reactors contained a total of 9,091 kg (20,000 lb) of GAC. Flow through the system was 0.632 m^3/min (167 gpm). The removals of DIMP and DCPD were 0.99 fraction of influent concentrations. The GAC system was operated by contract with Calgon Corporation for an annual service charge of $125,000, which included exchange of carbon.

Water source: The community of Commerce City (and environs), served by the South Adams Water and Sanitation District, obtained its water supply from the same shallow aquifer. The community obtained federal funding from the U.S. Army Corps of Engineers to construct a GAC plant which was placed on-line in 1989.

Wildlife refuge: The former arsenal site was converted to a wildlife refuge, which was the final resolution after some 40 years as a local and national political issue.

15.4.3.5 Tertiary Treatment

The idea of tertiary treatment emerged in the 1960s as a postsecondary treatment method to make waters discharged from WWTPs suitable for further use or to protect sensitive waters. Three such cases illustrate the applications of tertiary treatment that have involved GAC: South Lake Tahoe, the Orange County Water District (OCWD), and Colorado Springs.

15.4.3.5.1 South Lake Tahoe

Lake Tahoe on the California–Nevada state line is a pristine mountain lake that emerged in the 1950s from a few resorts and second homes around the shoreline toward an urbanizing region. The problem was recognized and the fastest-growing community, South Lake Tahoe was designated for a tertiary-treatment plant. The effluent from the plant, although of very high quality, was pumped from the Tahoe basin, which was the final resolution of the issue, c. 1966.

Tertiary plant and GAC: The South Tahoe, California tertiary-treatment plant had a design capacity of 28,388 m^3/day (7.5 mgd), with HLR $\leq$ 0.33 $m^3/min/m^2$ (8 gpm/ft^2) (USEPA, 1973). The plant had eight steel columns, diameter = 3.7 m (12 ft) and height = 7.3 m (24 ft). The columns had conical tops and bottoms and an effective carbon depth of 4.3 m (14 ft). Each column held 22–25 tons of 8 × 30 mesh size of Calgon 300 granular carbon. The eight columns were

TABLE 15.11
Representative Performance Data, South Tahoe Tertiary-Treatment Plant

Quality Parameter	Raw Wastewater	Secondary Effluent	Separation Bed Effluent	Carbon Column Effluent
BOD (mg/L)	200–400	20–1	<1	<1
COD (mg/L)	400–600	80–160	20–60	1–25
Total organic carbon (mg/L)	—	—	8–18	1–6
ABS (mg/L)	2.0–4.0	0.4–2.9	0.4–2.9	<0.01–0.50
PO_4 (mg/L as PO_4)	—	25–30	0.1–1	0.1–1
Color (units)	—	—	10–30	<5
Turbidity (units)	—	30–70	<0.5–3	<0.5–1
Nitrogen (mg/L as organic N)	10–15	4–6	2–4	1–2
Ammonia N (mg/L as N)	25–35	25–32	25–32	25–32
NO_3 and NO_2 (mg/L as N)	0	0	0	0
Unchlorinated:				
Coliforms (MPN/100 mL)	—	2,400,000	9,300	11,000
Fecal coliforms (MPN/100 mL)	—	150,000	930	930
Viruses	—	—	Negative	Negative
Chlorinated				
Coliforms (MPN/100 mL)	—	—	8.6	<2.1
Fecal coliforms (MPN/100 mL)	—	—	<2.1	<2.1

Source: Adapted from Slechta, A.F. and Culp, G.L., *J. Water Pollut Control Federation*, 39(5), 788, May, 1967.

operated in parallel using the countercurrent mode of operation. The flow of water was upward and the carbon flow was downward. As a note, exhausted carbon is heavier than fresh carbon or fresh regenerated carbon. Culp and Roderick (1966) reported that performance results in operation of the full-scale plant were much like the pilot plant. Table 15.11 summarizes some of the operating data for the plant, showing the effect of the carbon columns in the right-hand column. As seen, the BOD was reduced to <1 mg/L with $1 \leq \text{COD} \leq 25$ mg/L and $1 \leq \text{TOC} \leq 6$ mg/L. The objective was to protect the pristine waters of Lake Tahoe.

15.4.3.5.2 Colorado Springs Tertiary Plant

The tertiary plant at Colorado Springs was a joint venture between the City of Colorado Springs and the Environmental Protection Agency. The plant was brought on line in 1970 with a total cost of about $1 million. The plant had two circuits: (1) a 9 mgd design capacity irrigation circuit, providing dual media filtration, and (2) a 2.0 mgd industrial circuit consisting of lime treatment, re-carbonation, filtration, and carbon adsorption.

GAC reactors: Three carbon columns of downflow type were used for the plant; two were operated in series while the third column was down for regeneration. The carbon depth was 3048 mm (10 ft) of 8×30 mesh granular carbon in a tank with diameter 6096 mm (20 ft), which amounts to 186 ft^3 GAC per column, giving EBCT $\approx$ 17 min/column and with calculated HLR $\approx$ 10.4 m/h (4.24 gpm/ft^2). The removals of COD were $0.50 \leq$ Removal ≤ 0.60 kg COD/kg carbon. The columns were backwashed twice a week with air and water at 0.0152–0.0203 m^3 air/min/m^2 area (3–4 ft^3 air/min/ft^2 area) and 49 m/h (20 gpm/ft^2), respectively. Backwash was about 30 min duration (City of Colorado Springs, c. 1974). These reactors were taken out of operation in the 1980s and replaced with rapid filters (no coagulation) to provide water for power plant cooling and irrigation.

Post-GAC: The GAC treatment was considered not required, c. 1998, with the reactors replaced by the filtration of secondary effluent.

15.4.3.5.3 Water Factory 21

A 56,780 m^3/day (15 mgd) capacity tertiary-treatment plant called "Water Factory 21" was put on line in 1974 by the Orange County Water District, California (USEPA, 1973). The objective was to protect groundwater from seawater intrusion, using a line of injection wells to replenish the aquifer. The portion of the flow diverted through the GAC reactors was 34,100 m^3/day (9 mgd).

GAC Reactors: The plant was constructed with 17 carbon columns, each 3.658 m (12 ft) diameter, with a 7.32 m (24 ft) straight wall and EBCT = 30 min. Each of the columns contained (2700 ft^3) of 8×30 mesh carbon and the hydraulic loading at 34,100 m^3/day (9 mgd) was 14.2 m/h (5.8 gpm/ft^2) with EBCT $\approx$ 30 min, depending on how many columns are operated at the same time. The columns were operated in parallel and the countercurrent mode of operation was used (alternatively, the downflow mode could be used). Carbon could be added continuously or intermittently. For the latter mode of operation, the usual practice was to withdraw and replace 5%–10% of the carbon at one time. The removal of TOC by the GAC was about 0.7 fraction. The water flow of the reactor column was up from the bottom while the flow of GAC was down in a countercurrent fashion.

Regeneration: The GAC is removed at the bottom and is transported in slurry form in a 51 mm (2 in.) line to a holding and drainage tank in preparation for regeneration by a multiple-hearth furnace. The furnace capacity was 5455 kg GAC/day with recovery about 0.93 of the feed carbon rate (USEPA, 1973, http://www.ocwd.com, February, 2003).

15.5 OPERATION AND COSTS

Although operation is always important for any treatment process, only a paragraph is given here, albeit the topic is extensive. Cost information is provided in order to provide a perspective of the topic.

15.5.1 Operation Characteristics

To a large extent, operation is "passive" in that the adsorption process requires only monitoring for flows, headloss and breakthrough, etc. Other measurements may be worked into the routine, however, such as a weekly sampling of influent and effluent for GC/MS (gas chromatograph/mass spectrograph) analysis, sensor calibration, etc. Monitoring is required, usually through a SCADA system, which also aids in the compilation of reports internal review and for regulatory agencies. Usually, a GAC plant fits into a system, which has its own unique requirements for operation, e.g., water levels of service reservoirs, pump operation, flow monitoring.

15.5.2 Costs

Cost data are given, as a rule, for the year reported. Updates may be done by the ENR (*Engineering News Record*) Cost Indices (see Section C.6 and Table C.3).

15.5.2.1 South Lake Tahoe

Table 15.12 provides operating and capital cost data for GAC treatment at the South Tahoe plant (USEPA, 1973, pp. 5–14). The operating cost includes on-site regeneration of GAC. Capital costs were adjusted to USEPA, STP index 127.0, amortized at 5% for 25 years. For comparison, the cost of GAC treatment at Water Factory 21 (OCWD, 2003) was $0.05/m^3 ($0.18/1000 gal). The ENR CCI ratio (see Appendix C.6 and Table C.3) between 1973 and 2001 is 3.4; multiplying the 1973 $0.007/m^3 cost for S. Lake Tahoe by this factor gives $0.025/m^3 ($0.09/1000 gal), which is about half of the 2001 cost for Water Factory 21 (OCWD).

TABLE 15.12
Cost of GAC Treatment at S. Lake Tahoe[a]

Plant	Cost Category	Unit Costs[b,c] ($/m^3)	($/1000 gal)
S. Lake Tahoe[a]	Operating	0.0029	0.011
	Capital	0.0042	0.016
	Total (USEPA, 1973, pp. 5–14)	0.0071	0.027

[a] USEPA (1973).
[b] In 1973, the ENR BCI(1973) = 1138, and the ENR CCI(1973) = 1895 (ENR, 2002, p. 78).
[c] In 2001, the ENR BCI(2001) = 3574, and the ENR CCI(2001) = 6334 (ENR, 2002, p. 79).

15.5.2.2 Virgin GAC

The cost of virgin granular activated carbon in 1973 was about $0.73/kg ($0.33/lb). Regeneration costs in cents per pound started at about $0.66/kg ($0.30/lb) (USEPA, 1973, pp. 5–9, 5–15) for low capacity furnaces, but declined to about $0.22 /kg ($0.10/lb) for regeneration systems having the capacity of >900 kg/day (2,000 lb/day) (USEPA, 1973, pp. 5–16).

15.5.2.3 Regeneration

Regeneration is the process of recovering the adsorption capacity of the exhausted activated carbon, i.e., GAC, removed from reactor packed beds. The regeneration process is accomplished usually by a thermal excitation of adsorbed molecules.

15.5.2.3.1 Regeneration Process

Thermal regeneration has three process steps (McGuire and Suffet, 1978, p. 624):

1. Drying at 100°C, where the moisture and adsorbed volatile organics are thermally desorbed.
2. Pyrolysis of the other adsorbed organics at 650°C–750°C, with the char product being deposited in the pores.
3. Activation of the surface by burning the char from the pores at 870°C–980°C in the presence of steam with the oxygen fraction of the reactivation atmosphere being controlled.

Variables that could affect the character of the regenerated carbon include pyrolysis temperature, oxidation temperatures, oxidant gas (steam or CO_2), oxidant flow rate, and extent of oxidation (Canon et al., 1992). A previous step in the treatment train, such as iron or aluminum coagulation, lime precipitation, etc., could also affect the regenerated GAC.

15.5.2.3.2 Multiple-Hearth Furnace

As an example, the South Tahoe plant had a six-hearth furnace. The burners were located on hearths 4 and 6; temperatures on the hearths were No. 1, 426°C; No. 2, 538°C; No. 3, 704°C; No. 4, 916°C; No. 5, 871°C; No. 6, 916°C. The total fuel requirement for the regeneration system was about 9868 kJ/kg (4250 btu/lb) of carbon regenerated, i.e., for furnace heat and steam.

15.5.2.3.3 Effects of Regeneration

A decrease in micropore volume occurs with each regeneration (West, 1971; Cannon et al., 1993); thus, the carbon does not return to its virgin state. The molasses number of the regenerated carbon is about the same as that of the virgin carbon. The iodine number is lower, as a rule, indicating fewer pores in the 5–14 Å radius range. The net loss of absorption capacity is up to 13% on the first regeneration cycle; this loss diminishes with further cycles; the carbon loss for each successive cycle ranges from 5% to 10%.

15.5.2.3.4 Cost of Regeneration

Generally, on-site reactivation is economical only for a carbon use rate >900 kg/day (2000 lb/day); otherwise, off-site reactivation may be more economical (Groeber, 1991, p. 4); others give the threshold rate >4500 kg/day (10,000 lb/day) (Stenzel, 1993, p. 42). Usually, off-site reactivation is done on a contract basis, with spent carbon taken away by truck and regenerated carbon delivered at the same time. For such a case, a reactor column may serve as storage for spent GAC. Typical cost is \$1.75–2.20/kg (\$0.80–1.00/lb) for virgin carbon and \$1.30–1.75/lb for regenerated carbon (Stenzel, 1993, p. 42).

PROBLEMS

15.1 Mass of GAC from Laboratory Isotherm Plots

Given

Consider the experimental isotherm of Figure 15.6 as being applicable for the design of an adsorption reactor. Let the influent concentration of Rhodamine-B dye be whatever you wish to select (it should be in the range of your isotherm curve). Assume the adsorbate is Dowex-50 resin, as in the figure.

Required

(a) For assumed levels of Rhodamine-B dye, in the μg/L range, calculate the amount (i.e., mass in kg) of Dowex-50 resin required to treat the water for 30 days. Assume the flow of water to be treated is 11 mgd.

(b) Suppose the hydraulic loading rate for the reactor is 5 gpm/ft^2. Determine the size of the reactor required. If the depth is quite short, select a longer time so that the depth of the reactor is at least 3–5 m.

(c) Suppose that the Rhodamine-B dye levels are doubled. Calculate the amount of carbon required.

15.2 Mass of GAC for Two Adsorbate Influent Concentrations

Given

Assume TCE is to be removed by GAC. Assume the flow of water to be treated is 0.482 m^3/day (11 mgd) and 12.2 m/h (HLR = 5.0 gpm/ft^2).

Required

Determine the mass of GAC required for two concentrations of TCE (let $C_2 = 2 \cdot C_1$.

Solution (gist of the problem)

1. Capacity of carbon for two influent concentrations
   - Consider TCE on Filtrasorb 300(40) (Love et al., 1983, p. 12)
     $K = 28$
     $1/n = 0.62$
   - Let $C_1 = 0.01$ mg/L $\rightarrow \bar{X}^* = KC^{1/n} = 28 \cdot 0.01^{0.62} = 1.61$ mg TCE/g carbon
   - Let $C_2 = 0.02$ mg/L $\rightarrow = 28 \cdot 0.02^{0.62} = 2.47$ mg TCE/g carbon
2. Mass of carbon needed:
   mass of adsorbate to be removed = mass of adsorbate adsorbed by adsorbent

$$Q \cdot C_0 \cdot \Delta t = V(\text{carbon bed}) \cdot (1 - P) \cdot \rho \cdot \bar{X}^*$$

3. The ratio of carbon required for C_2 is $(2.47/1.61) \cdot$ carbon required for C_1

15.3 Mass of GAC from Published Isotherm Plots

Given

Obtain isotherm data for two or more organic compounds that are classified as "contaminants." Suppose that these compounds are to be removed by activated carbon adsorption.

Required

(a) For assumed levels of contaminants, in the μg/L range, calculate the amount of carbon required to treat the water for 30 days. Assume the flow of water to be treated is 11 mgd.

(b) Suppose that the contaminant levels are doubled. Calculate the amount of carbon required.

15.4 Scenarios for Velocity of Wave Front Based on Isotherms

Given

Table CD15.A.1 has Fruendlich constants, K, n, for a list of organic compounds, with some listed in Table CDprob15.3. For the problem, use the following data as necessary:

- Adsorbent is granular activated carbon
- $\rho(\text{carbon}) \approx 1.4$ g/cm^3
- Porosity, $P = 0.40$
- HLR = 5 gpm/ft^2
- Temperature, $T = 20$°C
- Assume the carbon bed should remain in operation for say 3–6 months before it is exhausted. If this duration results in a very small length of bed then make it a reasonable length and calculate the time of operation before exhaustion.
- Most carbon beds in practice for water treatment vary in their length dimension within a fairly narrow range. For example, at the Klein Water Treatment Facility the beds are 3–4 m deep. The Denver Reuse Plant columns were perhaps 10 m depth.
- The GAC may be any brand for which isotherm data are available.

Required

(a) Select two different adsorbates for your problem.
(b) Pick a reasonable input concentration.
(c) Prepare a spreadsheet for your "solution" that is suitable for exploring various "scenarios."
(d) Pick two or more "scenarios" to explore using your spreadsheet.
(e) The spreadsheet should have as an "output" the velocity of the wave front. Another output should be the length of the column required (depth of carbon bed).
(f) Your scenarios could be anything that relates to design (e.g., different input adsorbate concentrations, different HLRs, different times to exhaustion, different column lengths, or different temperatures).
(g) Equation 15.48 is a means to determine the length of the carbon bed.

Hints

- Most carbon beds in practice for water treatment vary in their length dimension from about 3–4 m (e.g., at the Klein Water Treatment Facility) to perhaps 10 m (Denver Reuse Plant).
- The GAC may be any brand for which isotherm data are available.
- Your scenarios could be anything that relates to design (e.g., different input adsorbate concentrations, different HLRs, different times to exhaustion, different column lengths, or different temperatures)
- Equations 15.71, 15.72, and 15.73 provide a means to determine the length of the carbon bed.

15.5 Mass Balance as Design Principle

Given

HLR = 10.2 m/h (4.2 gpm/ft^2), adsorbate chloroform in Louisville tap water, L(packed bed) = 10 m, $6 \leq t(\text{breakthrough}) \leq 12$ weeks, $0.2 \leq L_{wf} \leq 0.3$ m, $0.0061 \leq v_{wf} \leq 0.012$ m/day. See Section 15.2.4.20.

Required

Estimate a plausible influent concentration of chloroform in the tap water.

Solution

From Equation 15.48, estimate the ratio $C_0/\overline{X}^*(C_0)$. Determine whether there is an unique point on the Freundlich isotherm for chloroform that satisfies the ratio calculated; if so, determine the associated values of C_0 and $\overline{X}^*(C_0)$.

Discussion

The "back calculation" to determine C_0 and $\overline{X}^*(C_0)$ provides a means to corroborate empirical data.

15.6 Mass Balance as Design Principle (See Section 15.2.4.3)

Given

- HLR = 6.1 m/h (2.5 gpm/ft^2),
- $C_0 \approx 5.6$ mmol/L,
- Adsorbate 2,4-dichlorophenol,
- Adsorbent Columbia LCK, mesh 25 × 30,
- L(packed-bed) = 10 m,
- $L_{wf} \approx 0.12$ m,
- $v_{wf} \leq 0.50$ m/day

Required

(a) Calculate v_{wf} by Equation 15.48 and compare with the value given in Figure 15.4.
(b) Estimate t(breakthrough).

Solution

(a) Formula for 2,4-dichlorophenol is $C_6H_4OCl_2$; MW (2,4-dichlorophenol) = 163.0 g/mol. Therefore [2,4-dichlorophenol] = 5.6 mmol/L · 163 g/mol · mol/1000 mmol · 1000 mg/g = 913 mg/L. From Table CD15.A.1, the Freundlich coefficients for Filtrasorb 300® are $K = 157$ and $1/n = 0.15$ (estimates since the GAC's are different). Now calculate v_{wf} from Equation 15.48.
(b) Calculate t(breakthrough) from Equation 15.45.

Discussion

Since the isotherm for 2,4-dichlorophenol and Columbia LCK is not available readily, that for 2,4-dichlorophenol and Filtrasorb 300®, Table CD15.A.1 is used as a surrogate. This gives an approximation for $\overline{X}^*$.

15.7 Rate of GAC Exhaustion (Love Canal Example)

Given

Assume Filtrasorb 300® GAC is used for a packed-bed reactor to remove organics from water pumped from Love Canal groundwater. Assume also that 2,4,6 trichlorophenol at concentration 84 μg/L, pH = 6, is to be removed (Table 15.10). Assume HLR ≈ 12.2 m/h (5.0 gpm/ft^2) and ignore, for this problem, the issue of competitive effects.

Required

Estimate the GAC exhaustion in terms of the velocity of the wave front, i.e., v_{wf}.

15.8 Effect of TCE Concentration on Run Time in GAC Reactor (Section 15.2.3.1)

Given

Let C(TCE) = 20 μg/L in groundwater used as a source for GAC treatment. GAC = Filtrasorb 300®.

Required

Determine the capacity of the GAC for TCE at C(TCE) = 20 μg/L, 5, μg/L, 40 μg/L, 100 μg/L. Plot the capacity in μg TCE/g GAC. Discuss the effect of higher concentrations of TCE on the capacity of the GAC to adsorb TCE.

Solution

Table CD15.A.1 gives Freundlich isotherm coefficients for 141 different synthetic organic compounds for Filtrasorb 300®; i.e., $K(\text{TCE}) = 28$, $1/n(\text{TCE}) = 0.62$. Apply Equation 15.19 to solve for $\overline{X}^*$ at different concentrations of TCE.

Discussion

The idea of the problem is to indicate the effect of increasing adsorbate concentration on the capacity of GAC to adsorb. The point is that doubling the adsorbate concentration requires, as a rule, less than doubling of the GAC concentration.

15.9 Estimated Run Time for GAC Reactor at Klein WTF

Given

Let $C(\text{TCE}) = 20$ μg/L in groundwater used as a source for GAC treatment. $Q(\text{reactor}) = 0.022$ m^3/s (250 gpm); d(reactor) 3.048 m (10.0 ft); h(packed bed) = 2.44 m (8.0 ft); V(packed bed) = 20.2 m^3 (714 ft^3); mass(GAC) = 9,100 kg (20,000 lb). GAC = Filtrasorb 300®.

Required

Determine the run time for the reactor to attain complete saturation as the first column in a series of two. Compare the result with the 2.5 year run time for the column as experienced by the Klein WTP, Commerce City, Colorado.

15.10 Determination of Headloss and Backwash Velocity

Given

A GAC column is packed with Filtrasorb 200®. Hydraulic characteristics are as given in Figure 15.3a and b. Let the column be 9 m in depth, i.e., $L(\text{reactor}) = 9.0$ m.

Required

Determine the headloss for the column and the HLR required for bed expansion.

ACKNOWLEDGMENTS

Staff members of the Klein Water Treatment Facility (Klein WTF), South Adams Water and Sanitation District, provided the information on the plant operation. Roger Dirrim, water operations and maintenance supervisor (2004) provided the design and operation data for the plant and Charlene Seedle, water quality lab supervisor (2004) provided data on operation of the plant and on use of the GC-MS instrument. Jim Jones, general manager of the district (2010) and manager at the Klein WTF (2004), has provided continuing hospitality for student field trips and delegations from other countries from the time the plant started functioning in 1989 and through the decade of the 1990s. The Klein WTF was named after Jean Klein, long time board president and board member, who was instrumental in bringing the facility into reality. The plant was designed by Harley Bryant, PE, Black and Veatch, Denver (BSCE, Colorado State University, 1972).

Daniel R. Brooks, national sales manager (2004), Calgon Carbon Corporation, Pittsburgh, Pennsylvania, provided the photo-micrographs of granular activate carbon used in Figure 15.2. Zachary R. Navarro, Calgon Sales, kindly granted permission (2010) to use the images for this current edition.

William C. Lauer, American Water Works Association, formerly manager, Denver Potable Water Reuse Plant, Denver Water, filled in knowledge about the plant, its evolution, and its design.

Trina McGuire and Sabrina Hall, Denver Water, facilitated permissions (2010) to use brochure material from the Denver Potable Water Reuse Demonstration Plant.

APPENDIX 15.A: FREUNDLICH ISOTHERM COEFFICIENTS

Table CD15.A.1 is a compilation of Freundlich isotherm coefficients, i.e., K and $1/n$ for some 141 organic compounds from experimental data of Dobbs and Cohen (1980). A few data are provided for different pH values, e.g., pH = 3, 7, 9. The compounds were selected based upon the following criteria: (1) annual production, (2) concentrations that cause adverse environmental impacts, (3) probability of occurrence in wastewater, (4) persistence in the water environment, (5) solubility. During the course of their study, the Occupational Safety and Health Administration's list of regulated carcinogens was developed, as was the list of "priority pollutants" (some 29) named in the consent decree of the USEPA. Both these lists are included, i.e., in addition to those selected by the authors. Only one adsorbent, Filtrasorb 300® (Calgon Carbon Corporation 2003b) was used for all experiments.

Although other Freundlich coefficients are available (in the literature), most of these were generated on an *ad hoc* basis. The Dobbs and Cohen work is the most extensive and comprehensive experimental work on isotherms. Concerning other work, Love et al. (1983) generated Freundlich coefficients for some 17 volatile organic compounds (VOC's) with compilation in Table 15.3. Where there is overlap with the Dobbs and Cohen coefficients, the K and $1/n$ values are within about 0.01–0.05 fraction of each other.

GLOSSARY

Abrasion number: (1) An index of the resistance of particles to physical degradation. (2) Final mean particle diameter/original mean particle diameter as determined by an abrasion test involving sieve analysis (USEPA, 1973, p. B-20).

Absorb: The assimilation of matter within the pore structure of a material.

Activated alumina: A porous, highly adsorptive alumina; made usually by heating alumina hydrates and used chiefly in drying gases and liquids (http://unabridged.merriam-webster.com) [see Chapter 16].

Activated carbon: Plant based material that has been processed to leave the basic structure, resulting in a large internal surface area, e.g., perhaps 1000 m^2/g carbon.

TABLE CD15.A.1

Freundlich Isotherm Coefficients for 141 Synthetic Organic Compounds—Adsorbent was Filtrasorb300®[a] (Dobbs and Cohen, 1980)[b]

Compound	K	1/n
Acenapthene	190	0.36
Acenaphthalene	115	0.37
Acetone cyanohydrin	*not*	*ads.*
Acetophenone	74	0.44
2-Acetylaminofluorene	318	0.12
Acridine orange	180	0.29
Acridine yellow	210	0.14
Acrolein	1.2	0.65
Acrylonitrile	1.4	0.51
Adenine	38	0.38
Adipic acid	20	0.47
Aldrin	651	0.92
4-Aminobiphenyl	200	0.26
Anethole	300	0.42
o-Inisidine	50	0.34
Anthracene	376	0.70
Benzene	1	1.60
Benzidine dihydrochloride	220	0.37
Benzoic acid pH 7	76	1.80
pH 3	51	0.42
3,4-Benzofluoranthene	57	0.37
Benzo(k)fluoranthene	181	0.57
Benzo(ghi)perylene	10.7	0.37
Benzo(a)pyrene	33.6	0.44
Benzothiazole	120	0.27
Alpha-BHC	303	0.43
Beta-BHC	220	0.49
Gamma-BHC (lindane)	256	0.49
Bromoform	19.6	0.52
4-Bromophenyl phenyl ether	144	0.68
5-Bromouracil	44	0.47
Butylamine	*not*	*ads.*
Butylbenzyl phthalate	1520	1.26
N-Butyl phthalate	220	0.45
Carbon tetrachloride	11.1	0.83
Chlorobenzene	91	0.99
Chlordane	245	0.38
Chlorethane	0.59	0.95
Bis(2-chloroethoxy)methane	11	0.65
Bis(2-chloroethyl)ether	0.086	1.84
2-Chloroethyl vinyl ether	3.9	0.80
Chloroform	2.6	0.73
Bis(2-chloroisopropyl)ether	24	0.57
Parachlorometa cresol	122	0.29
2-Chloronaphthalene	280	0.46
1-Chloro-2-nitrobenzene	130	0.46
2-Chlorophenol	51	0.41
4-Chlorophenyl phenyl ether	111	0.26
5-Chlorouracil	25	0.58
Choline chloride	*not*	*ads.*
Cyclohexanone	6.2	0.75
Cyclohexylamine	*not*	*ads.*
4.6-Dinitro-*o*-cresol pH 3	237	0.32
pH 5	169	0.35
pH 9	43	0.9
2.4-Dinitrophenol pH 3	160	0.37
pH 7	33	0.61
pH 9	41	0.25
2.4-Dinitrotoluene	146	0.41
2.6-Dinitrotoluene	145	0.32
Diphenylamine	120	0.31
1,1-Diphenylhydrazine	135	0.16
1.2-Diphenylhydrazine	16,000	2.00
Alpha-endosulfan	194	0.5
Beta-endosulfan	615	0.83
Endosultan sulfate	686	0.81
Endrin	666	0.80
Ethanol	*not*	*ads.*
Ethylbenzene	53	0.79
Ethylenediamine	*not*	*ads.*
Ethylenediamenetetraaceticacid-EDTA	0.86	1.5
Bis(2-ethylhexyl) phthalate	11,300	1.50
Fluoranthene	664	0.61
Fluorene	330	0.28
5-Fluorouracil	5.5	1.00
Guanine pH = 3	75	0.48
pH 7–9	120	0.40
Heptachlor	1220	0.95
Heptachlor epoxide	1038	0.70
Hexachlorobenzene	450	0.60
Hexachlorobutadiene	258	0.45
Hexachlorocyclopentadiene	370	0.17
Hexachloroethane	97	0.38
Hexamethylenediamine	*not*	*ads.*
Hydroquinone	90	0.25
Isophorone	32	0.39
Methylene chloride	1.3	1.16
4,4′-Methylene-bis(2-chloroaniline)	190	0.64
Morpholine	*not*	*ads.*
Naphthalene	132	0.42
Alpha-naphthol	180	0.32
Beta-naphthol	200	0.26
Alpha-naphthylamine	140	0.25
Beta-naphthylamine	150	0.30
p-Nitroaniline	140	0.27
Nitrobenzene	68	0.43
4-Nitrobiphenyl	370	0.27
2-Nitrophenol pH 3	101	0.26
pH 5.5	99	0.34
pH 9	85	0.39
4-Nitrophenol pH 3	80	0.17
pH 5.4	76	0.25
pH 9	71	0.28
N-Nitrosodiphenylamine	220	0.37

TABLE CD15.A.1 (continued)
Freundlich Isotherm Coefficients for 141 Synthetic Organic Compounds—Adsorbent was Filtrasorb300®[a] (Dobbs and Cohen, 1980)[b]

Compound	*K*	1/*n*
Cytosine pH 7–9	1.1	1.60
DOE	232	0.37
DDT	322	0.50
Dibenzo(a,h)anthracene	69	0.75
Dibromochloromethane	4.8	0.34
1,2-Di bromo-3-chloropropane	53	0.47
1,2-Dichlorobenzene	129	0.43
1,3-Dichlorobenzene	118	0.45
I,4-Dichlorobenzene	121	0.47
3,3-Dichlorobenzidine	300	0.20
Dichlorobromomethane	7.9	0.61
1,1-Dichloroethane	1.8	0.53
1,2-Dichloroethane	3.6	0.83
1,2-*trans*-Dichloroethene	3.1	0.51
1,1-Dichloroethene	4.9	0.54
2,4-Dichlorophenol	157	0.15
1,2-Dichloropropane	5.9	0.60
1,2-Dichloropropene	8.2	0.46
Dieldrin	606	0.51
Diethylene glycol	*not*	*ads.*
Diethyl phthalate	110	0.27
4-Dimethylaminoazobenzene	249	0.24
N-Dimethylnitrosoamine	$6.8 \cdot 10^{-5}$	6.60
2,41-Dimethylphenol pH 3	78	0.44
pH 5.8	70	0.44
pH 9	108	0.33
Dimethylphenylcarbinol pH 3	110	0.60
pH 7	210	0.34
Dimethyl phthalate	97	0.41
N-Nitrosodi-*n*-propylamine	24	0.26
p-Nonylphenol pH 3	53	1.04
pH 7	250	0.37
pH 9	150	0.27
PCB 1221	242	0.70
PCB 1232	630	0.73
Pentachlorophenol pH 3	260	0.39
pH 7	150	0.42
pH 9	100	0.41
Phenanthrene	215	0.44
Phenol	21	0.54
Phenylmercuric acetate pH 3–7	270	0.44
pH 9	130	0.54
Styrene	120	0.56
1.1,2,2-Tetrachloroethane10	11	0.37
Tetrachloroethene (tetrachloroethylene)	50.8	0.56
1,2,3,4-Tetrahydronaphthalene	74	0.81
Thymine	27	0.51
Toluene	26	0.44
1.2.4-Trichlorobenzene	157	0.31
1,1,1-Trichloroethane	2.5	0.34
1,1,2-Trichloroethane	5.8	0.60
Trichloroethene (trichloroethylene)	28	0.62
Trichlorofluoromethane	5.6	0.24
2,4,6-Trichlorophenol pH 3	219	0.29
pH 6	155	0.40
pH 9	130	0.39
Triethanolamine	*not*	*ads.*
Uracil	11	0.63
p-Xylene	85	0.19

[a] Filtrasorb 300® is a product of Calgon Corporation, Pittsburgh, Pennsylvania.
[b] A compilation of the data by Westates Carbon Co. (1985) facilitated the present compilation.

Activation: The first step in manufacture is dehydration of the raw material; the second is carbonization; the third is activation. The dehydration and carbonization is by slow heating in the absence of air. Excess water, including structural water must be driven from the organic material. Carbonization converts the organic material to primary carbon, which is a mixture of ash (inert organics), tars, amorphous carbon, and crystalline carbon. During carbonization, some decomposition products or tars will be deposited in the pores but will be removed during activation. Activation is a two-phase process: (1) burn-off of amorphous decomposition products, i.e., tars, and (2) enlargement of pores in the carbonized material. Burn-off frees the pore openings increasing the number of pores and activation enlarges the pores. Steam at 750°C–950°C burns off the decomposition products exposing pore openings for subsequent enlargement (USEPA, 1973, pp. 2–2).

Advection: The movement of particles that are suspended or dissolved in a fluid by bulk flow of the solvent, e.g., under a hydraulic gradient.

Adsorb: The bonding of adsorbate to an adsorbent surface.

Adsorbate: (1) Particles that bond to sites provided by an adsorbent. (2) Reactant in adsorption reaction. (3) Adsorbate particles are molecules normally but may include viruses or bacteria if they bond to sites.

Adsorbent: (1) Solid surface that provides sites for adsorbate particles to bond. (2) Reactant in adsorption reaction. (3) The most common adsorbent in the adsorption treatment process is activated carbon.

Adsorber: Name sometimes given to a GAC reactor column.

Adsorption: (1) Uptake of an adsorbate by an adsorbent. (2) The bonding of adsorbate molecules (or particles) to "sites" within the pore structure of an adsorbent, e.g., GAC. (3) The concentration and collection of contaminants at the surface of a solid. The solid or "adsorbent" can then be removed from the liquid by simple mechanical means, taking associated impurities along with it (Anon., 1962, p. 17).

Adsorption reaction: A reaction between particles and a surface.

Adsorption zone: The linear distance within an adsorption reactor column that encompasses most of the uptake between adsorbate by an adsorbent.

Alumina: The oxide of aluminum Al_2O_3 that occurs native as corundum and in hydrated forms, that is made usually from bauxite, in various forms (as a white powder obtained by calcination or a hard crystalline substance resembling natural corundum obtained by heating calcined aluminum oxide almost to the fusion point), and that is used chiefly as a source of metallic aluminum, as an abrasive and refractory, as a catalyst and catalyst carrier, and as an adsorbent as in drying gases and liquids and in chromatography; see also aluminum hydroxide (http://unabridged.merriam-webster.com).

Aluminum hydroxide: Any of several white gelatinous or crystalline hydrates $Al_2O_3{\cdot}nH_2O$ of alumina found in nature, especially in bauxite, or obtained as precipitates by treating solutions of aluminum salts with hydroxides. Hydrated alumina; especially the trihydrate $Al_2O_3 \cdot 3H_2O$ or $Al(OH)_3$ of alumina, regarded as acting both as a weak base and as a weak acid, that occurs as gibbsite and is used chiefly in ceramics, in pigments, and as a reinforcing agent for rubber (http://unabridged.merriam-webster.com).

Apparent density: The mass weight of dry adsorbent per unit volume of a packed bed as found *in situ* in a column (USEPA, 1973, p. B-23). Related to particle density, ρ_s by relation, $\rho(\text{apparent}) = \rho_s(1 - P)$.

Aromatic compound: (1) Characterized by the presence of at least one benzene ring; included are monocyclic, bicyclic, and polycyclic hydrocarbons and their derivatives as benzene, toluene, naphthalene, phenol, aniline, salicylic acid (http://www.merriam-webster.com/).

AWTR: Acronym for the "Advanced Water Treatment Research Program" of the Division of Water Supply and Pollution Control, U.S. Public Health Service, which was ongoing from 1960 to 1967.

Batch reactor: A reactor in which there is no flow through, but mixing is usually provided.

Bed volumes: The ratio of throughput volume for an adsorption column to the pore volume, i.e., bed volumes = throughput volume/($P \cdot V$(reactor bed), Bed volumes is a parameter often used in lieu of time as the abscissa for the breakthrough curve. The relation is "bed volumes treated" = $Q \cdot t/(V(\text{reactor bed}) \cdot P)$.

BET isotherm: The multilayer isotherm model proposed by Brunauer, Emmett, and Teller (Brunauer et al., 1938).

BOM: Background organic matter.

Bonding: Usually refers to the forces that cause attachment of a molecule or particle at an adsorption "site."

Bonding site: The surface of an adsorbent may have a potential morphology that provides for a stronger bonding force at a particular point or localized area than adjacent points or areas, e.g., the potential field varies spatially over the surface with some areas having a more favorable potential for bonding with a molecule or particle than other areas.

Breakthrough: (1) The time of the first appearance of an adsorbent in the effluent of a GAC reactor column. While the first appearance may be the first detectable concentration of an adsorbent, it is more likely to be with respect to a defined threshold concentration. (2) The time of the first appearance of a given contaminant that exceeds a defined treatment objective.

Breakthrough curve: The concentration versus time curve as it emerges from the end of a reactor column, i.e., $C(t)_{Z=Z_{max}}$. Usually, the curve has an S-shape.

Brownian motion: The random motion of molecules due to thermal energy. Velocities of molecules are higher at higher temperatures, i.e., $1/2mv^2 = kT$.

Carbon chloroform extract (CCE): A collective index of volatile organic carbon concentration in water. A sample of river or lake water is pumped through a small GAC volume; some or all of the VOCs present will adsorb on the carbon. A portion of the VOCs that have adsorbed may be eluted by chloroform and analyzed. Thus, very low concentrations in the ambient water may be concentrated on the carbon, which brings the VOC concentrations to detectable levels.

Column: (1) A reactor in which there is ordinarily upward or downward flow of adsorbate. (2) The connotation associated with the term, "column," is that the length is longer than the diameter.

Competitive adsorption: Two or more adsorbates in a solution will compete for adsorption sites; the adsorbent will favor the adsorbate species with strongest bonding interaction.

Concentration profile: The concentration versus distance, i.e., $C(Z)_t$, along the depth of an adsorption reactor.

Contact time: (1) A hypothetical time of contact between adsorbate and adsorbent within a column reactor. Since the adsorbate concentration varies along the reaction zone of the reactor, the term lacks a rationale. (2) The same as EBCT.

Contactor: A reactor containing an adsorbent, usually activated carbon.

Continuous-flow reactor: A defined volume in which flow enters and leaves where a planned reaction occurs.

Countercurrent: The operation of an adsorbent reactor in which the flow of water is upward and the flow of carbon is downward. The carbon is added at the top of a column and withdrawn from the bottom.

Crystal: (1) A body that is formed by the solidification of a chemical element, a compound, or a mixture and has a regularly repeating internal arrangement of its atoms and often external plane faces (meriamwebster.com). (2) Any solid material in which the component atoms are arranged in a definite pattern and whose surface regularity reflects its internal symmetry (Britannica.com).

Desorption: The movement of adsorbed molecules into solution based on reduced adsorbate concentration.

Diffuse (verb): The transport of molecules spatially from higher concentration to lower concentration.

Diffusion: The transport of molecules spatially from higher concentration to lower concentration. The transport flux density is described by Fick's first law of diffusion.

Dispersion: The transport of molecules based on random pore velocities.

DOC: Dissolved organic carbon.

Dowex 50: Resin manufactured by Dow Chemical Co.

Eductor: A device with no moving parts used to force activated-carbon water slurry to flow through a pipe (USEPA, 1973, p. A-3).

Empty-bed contact time: (1) The volume of reactor filled adsorbent divided by flow, i.e., EBCT $= V(\text{adsorbent})/Q$. Dividing both numerator and denominator by area of bed, A, gives EBCT $= L(\text{reactor})/\text{HLR}$. (2) An empirical design parameter used to size an adsorption reactor. The higher the EBCT, the deeper the reactor bed and the longer the time to breakthrough (Clark and Lykins, 1989, p. 35).

Equilibrium: (1) That adsorbate concentration on the adsorbent that is in equilibrium with the adsorbate concentration in solution. (2) The adsorbed adsorbate concentration at which the rate of adsorption equals the rate of desorption.

Exhausted: See *saturated*.

Fick's first law: Mathematical relation, $\mathbf{j} = -\nabla D$ (in three dimensions); $\mathbf{j} = -\partial C/\partial X$ (one dimension).

Film thickness: In diffusion from the bulk of solution to an adsorbent particle surface, the concentration changes within a short distance from Co to a concentration, Cs, at the particle surface. The distance in which the concentration change occurs is called the film thickness. The change, although in accordance with Ficks first law, is considered linear.

Filtration velocity: Mean interstitial velocity, i.e., $\bar{v} = \text{HLRP}$.

Fixed bed: The adsorption reactor bed is fixed in place, i.e., it remains in place until saturated with adsorbate and then is removed from service (USEPA, 1973, p. A-3)

Fluidized bed: Bed of granular media that has been expanded but with steady HLR. In an upflow packed bed, headloss increases with HLR. At some point the headloss versus HLR relation becomes flat, i.e., the headloss reaches a limit and does not increase with further HLR increases. At this point the bed is fluidized. A fluidized bed has constant headloss, permits continuous feed (i.e., the unit is not taken out of service for backwash or batch removal of the GAC), and avoids collection of particulates and purges air bubbles (eliminating air binding). At the same time, the saturated particles of GAC have a higher density and thus are classified, sinking to the bottom. The saturated adsorbent particles are always at the bottom and may be removed on a continuous basis and the less saturated particles are at the top and provide for the polishing of the effluent (that exits from the top of the reactor).

Flux: The mass flow of a given substance, such as water, a particular ion, a particular compound, etc.

Freundlich isotherm: Empirical relationship, i.e., $\bar{X}^* = KC^{*1/n}$, between the solid-phase equilibrium concentration of a given adsorbate and the aqueous-phase equilibrium concentration.

Hydraulic loading rate: Defined, HLR $= Q/A$.

Gas chromatograph: Analytical instrument developed in the late 1940s to identify and measure concentrations of organic compounds. The instruments became widely available by the early 1960s. A gas chromatograph has an injection point, a "column," and a detector. The column is usually a long tube of small diameter, perhaps 60 m as shown in Figure G15.1b, and packed with an absorbent film that may adsorb volatile organic compounds. The three kinds of detectors are thermal, electron-capture, and flame ionization.

GC-MS: A mass spectrograph detects different compounds or elements by means of mass differences, separated by centrifugal force. The GC-MS combines the two instruments. Figure G15.1a, left side, shows such an instrument.

GAC: Acronym for "granular activate carbon." Mean particle size is 1.2–1.6 mm (Anon., 1986).

Hardness number: (1) An index of the resistance of carbon particles to degradation by a steel ball abrasion test, involving also a sieve analysis. (2) The mass weight of material retained on a sieve assembly after loss due to an abrasion test.

Homogeneous reactor: A reactor in which the concentration of reactants is non-varying spatially within the volume; a "complete-mix" reactor would approach such spatial homogeneity. A packed-bed-column reactor is at the opposite end of the spectrum of reactor types in that the concentration of the reacting substance, e.g., the adsorbate, varies, within the mass-transfer zone, along the length of the reactor.

Inflection point: At the point along the mass-transfer zone of a column reactor, the S-shaped $C(Z)_t$ curve where $d^2C/dZ^2 = 0$.

Intra-particle: A term that refers to the pores of a particle of granular activated carbon (or other substance).

Iodine number: Defined as the mg of iodine adsorbed by 1.00 g carbon when the iodine concentration of the residual filtrate is 0.02 N (USEPA, 1973, p. B-2).

Isotherm: (1) The locus of point of equilibrium between the solid-phase equilibrium concentration of a given adsorbate and the aqueous-phase equilibrium

(a)

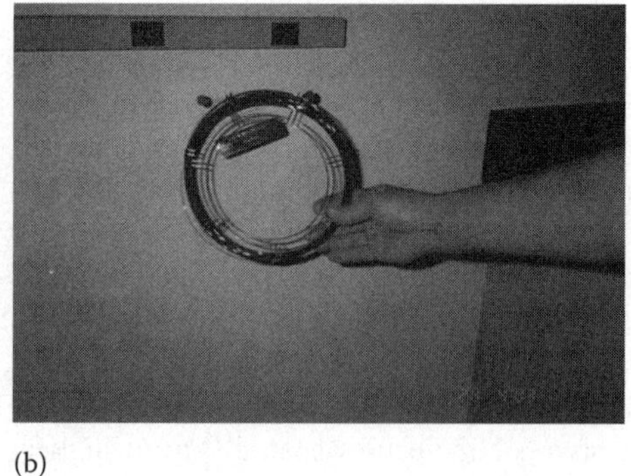

(b)

FIGURE G15.1 GC-MS instrument for volatile organic compounds. (a) Left-to-right: GC-MS, purge-and-trap unit, auto-sampler, computer. (b) Column for GC, 60 m length, 0.32 mm i.d., 1.8 μm silicon film. (Courtesy of Klein Water Treatment Facility, Commerce City, CO, 2003, with cooperation of Charlene Seedle.)

concentration at a given temperature. (2) The adsorption reaction is always between particular adsorbate and a particular adsorbent, resulting in a particular equilibrium constant. The equilibrium equation may be manipulated to give a mathematical relationship called an isotherm whose constants are unique for the particular adsorbate–adsorbent reaction. The relation is called an "isotherm" because it is valid for only a particular temperature (i.e., the equilibrium constant, the isotherm constants, and the associated graphical depiction). [The temperature effect is described by the well-known van't Hoff equation (see Section 15.2.1.2).]

Kinetics: A term that refers to the rate of change of a given reactant with respect to time. With reference to adsorption, kinetics refers to the rate of uptake of adsorbate by the adsorbent.

Kinetic theory: The most common kinetic models are (adapted from Carter and Weber, 1994, p. 614) (1) the pore-surface diffusion model and its two major derivatives: (a) the pore diffusion model, and (b) the homogeneous surface diffusion model. These models are described as follows:

***Pore-surface diffusion model*:** The pore-surface diffusion model incorporates a mathematical description of the major physicochemical mechanisms involved in the adsorption of an adsorbate from the flowing solution and into the adsorbate particles. These mechanisms include axial flow (i.e., advection within the reactor), hydrodynamic dispersion, local equilibrium at the particle surface, mass-transfer resistance across the hydrodynamic boundary layer (known as the "film") surrounding the particle, and intra-particle diffusion along the pore surfaces and through the pore solution within the particle. The surface–diffusion coefficient is designated, D_s.

***Film diffusion model*:** The film diffusion model depicts diffusion across the boundary layer in terms of the concentration differential between the bulk of solution and at the surface of the particle. The mass transfer coefficient is designated, k_f.

***Pore diffusion model*:** The pore diffusion model neglects the mechanism of surface diffusion. The pore–diffusion coefficient is designated, D_p.

***Homogeneous surface diffusion model*:** The homogeneous surface diffusion model is identical to the pore–surface diffusion model except that intra-particle pore diffusion is considered negligible compared to surface diffusion. The surface–diffusion coefficient is designated, D_s.

[See also Weber and DiGiano (1996) for further explanations.]

Langmuir isotherm: An equilibrium relationship between adsorbate and adsorbent as defined for a given temperature.

Linearized Langmuir isotherm: An algebraic rearrangement of the Langmuir relation can yield a linear relationship between variables that is useful in determining the constants, α and $\overline{X}_m$, from experimental data of (C^*, $\overline{X}^*$) pairs. At least three variations of such relationships are possible.

London dispersion forces: (1) A subset of van der Waals forces defined by F. London in 1930. "The force is always attractive and arises from fluctuating electron clouds in all atoms that appear as oscillating dipoles created by the positive nucleus and negative electrons." (Adamson and Gast, 1997, p. 228). (2) "London developed a quantum mechanical perturbation theory according to which the continuous motion of electrons in atoms and molecules effect rapidly fluctuating temporary dipole and quadrupole moments. Such fluctuating dipole and quadrupole moments in foreign molecules or atoms approaching a solid surface can perturb the electron distribution of surface molecules to induce temporary dipoles and quadrupoles therein, and conversely. This leads to extensive permutation of induced and temporary dipole-dipole, dipole-quadrupole and quadrupole-quadrupole attractive interactions, constituting the London dispersion forces of physical adsorption." (Weber and Van Vliet, 1980, p. 17).

Macromolecule: A molecule, usually a polymer, that may have $10^4 < MW < 10^6$.

Macropore: Pores in activated carbon larger than 1000 A, i.e., 1 μm (USEPA, 1973, p. A-4).

Macroreticular: See *reticular*.

Mass-transfer zone: The portion of the wave front in a GAC column in which mass transfer is occurring; in other words, the adsorbate is being taken up, i.e., adsorbed, by the adsorbent in the MTZ. The shorter the length of the zone, the higher is the rate of uptake.

Materials-balance equation: A mathematical relationship that accounts for a given substance in terms of a given volume that says, basically, that mass flux-in minus mass flux-out equals mass created or depleted.

Mathematical model: A mathematical equation or a set of linked equations that is intended to simulate the behavior of certain variables of a physical system.

Mesh size: Mesh size refers to the number of open spaces in a square inch of screen through which particles pass (Spotts, 2003). For a 20 × 50 mesh size, the 20 refers to the largest size that the particles pass, e.g., 0.03 fraction may be larger than a 20 mesh screen. The 50 refers to the smallest size that the particle can pass, e.g., 0.01 fraction may pass the 50 mesh screen (Spotts, 2003).

The sizes of particles are measured by obtaining a sample, e.g., 100 g, and passing through a nest of sieves, from larger to smaller (sizes of very small particles are determined by settling tests). The sieve number, e.g., #8, #12, ..., #300, etc., is in sequence, larger size to smaller, respectively. Most particles of a given kind, e.g., sand, anthracite, activated carbon, occur as a distribution of sizes. Conventions in designating sizes vary with the purpose. In filtration, d_{10} and d_{60}, i.e., 10% passes the d_{10} size and 60% passes the d_{60} size, respectively. In activated carbon, the size is given in terms of sieve size passing all material and sieve size retaining all material of the sample. For example, 8 × 30 means that the sample passes the #8 sieve and is retained by the #30 sieve. Common GAC sizes include 8 × 30, 8 × 30, 12 × 40, etc. Two common sieve standards are the U.S. sieve series and the Tyler sieve series, with the United States being most common (Spotts, 2003).

Methylene blue number: Defined as the mg methylene blue adsorbed by 1.0 g carbon in equilibrium with a solution of methylene blue having a concentration of 1.0 mg/L (USEPA, 1973, p. B-12).

Micropore: Pores in activated carbon smaller than 1000 A, i.e., 1 μm (USEPA, 1973, p. A-4).

Model: A representation of a portion of reality, e.g., a drawing, a photograph, a mathematical model, etc. See Chapter 3.

Molasses number: An index of the adsorption capacity of a given carbon as compared to a standard. Referring to one of two tests, molasses number $= K \cdot B/A$, where K is the molasses value of standard carbon, B is the optical density of filtrate from standard carbon, A is the optical density of filtrate from test carbon (USEPA, 1973, p. B-6).

Numerical solution: A computer solution of a finite-difference form of a differential equation that relies on small increments of change of a variable to calculate a new solution, e.g., solution at $t_2 =$ solution at $t_1 + \Delta t$ solution increment.

PAC: Powdered activated carbon; size range is less than U.S. mesh size 50, i.e., <0.3 mm. Others give <0.1 mm as the mean particle diameter of PAC.

Particle density: The mass weight of adsorbent divided by volume of adsorbent particles.

Particle kinetics: Refers to the rate of change of adsorbate in solution due to diffusion within the pores of the GAC particles.

Particle size: Usually refers to sizes of two screens, either in the U.S. Sieve Series or the Tyler Series between which the bulk of the carbon sample falls. For example, 8 × 30 means most of the carbon passes a No. 8 screen but is retained on a No. 30 screen (USEPA, 1973, p. A-5).

Partition: The relationship between two phases of a given substance, e.g., the solution phase and the adsorbed phase of an adsorbate.

Phenomenological: (1) An observable fact or event, e.g., an item of experience or reality. (2) appearance of things as contrasted with their true being; (3) a fact or event susceptible to scientific description and explanation (adapted from http://www.merriam-webster.com/).

Physical forces: Physical adsorption is due to van der Waals forces, which comprise London dispersion forces and classical electrostatic forces.

Physical–chemical treatment (PCT): A treatment sequence in which physical and chemical processes are used to the exclusion of biological processes (USEPA, 1973, p. A-5). A PCT treatment scheme has been considered as a substitute for conventional biological wastewater treatment and was an idea in vogue in the late 1960s and early 1970s. The concept probably had its origin in the USPHS Advanced Wastewater Treatment Program (AWTR, c. 1961–1968). A PCT scheme following an existing biological treatment train is called "tertiary" treatment.

Pore size distribution: A measure of the pore structure of activated carbon which gives the relationship between pore size and the volume of pores smaller or larger than that size. Pore size distributions in the micropore range are determined from nitrogen adsorption isotherms while distributions in the macropore range are measured by means of a mercury penetrometer. Micropore distributions can be related to the adsorptive capacities of different molecular weight substances. Macropore distributions may be correlated with rates of adsorption (USEPA, 1973, p. A-5).

Porosity (*P*): Defined as the volume of voids (pores) divided by total volume, i.e., volume of solids plus volume of voids.

Powdered activated carbon: See *PAC*.

Profile: As used here, the adsorbate concentration distance along the direction of flow in an adsorbent packed bed.

Pulsed-flow reactor: An upflow reactor bed in which GAC is added at the top either continuously or intermittently, i.e., as a "pulse," and removed continuously at the bottom.

Random walk: A model of diffusion in which each "step" of a molecule or particle is a random vector, i.e., with respect to direction and distance traveled, before a collision that causes another vector reaction.

Reactor: A defined volume in which a reaction may occur. The reactor may be "batch" or "continuous flow."

Real density: The density of the skeleton of a carbon particle and usually comes close to that for graphite (USEPA, 1973, p. A-6).

Refractory: Contaminants that are not removed by conventional treatment.

Regeneration: The restoration of a portion of the adsorption capacity of virgin carbon. About 0.9 restoration is common. Usually, the regeneration is by thermal means, i.e., at 900°C–954°C (1650°F – 1750°F).

Reticular: Resembling a net in appearance or structure (http://www.merriam-webster.com/).

Safe Drinking Water Act: PL93-523 passed by U.S. Congress in 1974. The legislative basis for regulations and guidelines promulgated by the U.S. Environmental Protection Agency that have resulted in the definition of contaminants and the specifications of MCLs (maximum contaminant levels) for drinking water and MCL goals.

Saturated: (1) When the carbon is at equilibrium with the influent solution adsorbate concentration, i.e., C_0, the carbon is "saturated" with respect to the adsorbate. (2) Same as "exhausted."

Saturation: The maximum amount of adsorbate that an adsorbent may adsorb per unit mass for a given adsorbate concentration. The "maximum" is actually the equilibrium amount as defined by an isotherm.

Scenario: (1) The construction of a set of imposed conditions, hypothetical, for which possible outcomes are generated by means of a model of some sort. The model may range from ones imputing outcomes based on one's imagination to the results of quantitative computations based upon a computer model. A mathematical model is useful particularly when the interactions among variables are complex. Computer simulations provide a means of keeping track of these changes. (2) A sequence of events especially when imagined; especially, an account or synopsis of a possible course of action or events.

Simulation: A modeling process in which the behavior of a prototype system is "simulated" by another system. An electric analog is a good example, in which voltage may simulate hydraulic head and current simulates flow of water in a pipe.

Sorption: A term that is inclusive of both adsorption and absorption.

Stochastic: A random or probabilistic occurrence. The occurrence of *Giardia* cysts in a sampling of a stream may be "stochastic."

SOC: Acronym for synthetic organic compound.

Structure: Structure determines the size of molecules which can be adsorbed and the capabilities of the activated carbon to permit the migration of the adsorbates within the pores (as measured by a diffusion coefficient).

Superficial velocity: Same as hydraulic loading rate, HLR, i.e., Q/A.

Surface area: Defined as the surface area per unit mass of carbon. Usually, the surface area is determined from the nitrogen adsorption isotherm by the Brunauer–Emmett–Teller method, i.e., BET method. Usually surface area is expressed as m^2 surface area/g carbon (USEPA, 1973, p. A-6).

Tertiary treatment: A term that came into use in the 1960s, probably as an outgrowth of the USPHS AWTR program in which research was conducted to develop treatment processes that could be added to biological treatment to result in an effluent water quality that met some more stringent criterion, such as for further use of the water (Morris and Weber, 1962, p. 7).

TOC: Total organic carbon. According to Anon. (USPHS, 1965, p. 89) the Dow Chemical Company developed a dry oxidation system for organic compounds in water that employed a hot tube furnace and an infrared CO_2 analyzer. The system was acquired by Beckman Instruments by 1968.

TOX: Total organic halogens.

Transport kinetics: As used here, the term "transport kinetics" refers to the advection of adsorbate to the diffusion proximity to an adsorbent particle.

Tyler sieve series: A sieve series used to provide a size distribution of granular media.

U.S. standard sieve series: A sieve series used to provide a size distribution of granular media. Sieve sizes and openings are given in the table following for both the Tyler and the U.S. series (U.S. series data from Culp and Culp, 1971, p. 151; Tyler series equivalencies from Spotts, 2003).

Sieve Series		Sieve Opening	
Tyler	U.S.	(mm)	(in.)
4	3.5	5.660	0.223
	4	4.760	0.187
	5	4.000	0.157
6	6	3.360	0.132
	7	2.830	0.111
8	8	2.380	0.0937
	10	2.000	0.0787
10	12	1.680	0.0661
12	14	1.410	0.0555
14	16	1.190	0.0469
	18	1.000	0.0394
20	20	0.840	0.0331
24	25	0.710	0.0280
	30	0.590	0.0232
32	35	0.500	0.0197
35	40	0.420	0.0165
42	45	0.350	0.0138
48	50	0.297	0.0117
60	60	0.250	0.010
65	70	0.212	0.008
80	80	0.180	0.007
100	100	0.150	0.006

Wave front: The portion of the concentration profile, i.e., $C(Z)_t$ curve, which is the reaction zone of the reactor. Same as "mass-transfer zone."

REFERENCES

Abdelrasool, F., Kinetics of adsorption, PhD thesis, Department of Civil Engineering, Colorado State University, Fort Collins, CO, 1992.

Adamson, A. W., *Physical Chemistry of Surfaces*, 2nd edn., Interscience Publishers, New York, 1967.

Adamson, A. W. and Gast, A. P., *Physical Chemistry of Surfaces*, 6th edn., Interscience Publishers, New York, 1997.

Alberty, R. A. and Silbey, R. J., *Physical Chemistry*, 1st edn., John Wiley & Sons, New York, 1992.

AWTR: Advanced Waste Treatment Research Program, Robert A. Taft Sanitary Engineering Center, Division of Water Supply and Pollution Control, U.S. Public Health Service, Cincinnati, OH. As related to activated carbon adsorption, the research in this program produced the reports listed below during the period 1961–1968.

- Anon., Summary Report: June 1960–December 1961, Advanced Waste Treatment Research-1, May, 1962.
- Anon., Summary Report: Jan 1962–June 1964, Advanced Waste Treatment Research, AWTR-14, April, 1965.
- Anon., Summary Report: July 1964–July 1967, Advanced Waste Treatment Research, AWTR-19, 1968.
- Johnson, R. L., Lowes, F. J. Jr., Smith, R. M., and Powers, T. J., Evaluation of the Use of Activated Carbons in Treatment of Wastewater, AWTR-11, May, 1964.
- Joyce, R. S. and Sukenik, V. A., Feasibility of Granular Activated-Carbon Adsorption for Wastewater Renovation 2, AWTR-10, October, 1965.
- Morris, J. C. and Weber, W. J. Jr., Preliminary Appraisal of Advanced Waste Treatment Processes, Technical Report W62-24, September, 1962.
- Morris, J. C. and Weber, W. J. Jr., Adsorption of Biochemically Resistant Materials from Solution 1, AWTR-9, May, 1964.
- Morris, J. C. and Weber, W. J. Jr., Adsorption of Biochemically Resistant Materials from Solution 2, AWTR-16, March, 1966.
- Williamson, J. N., Heit, A. H., and Calmon, C., Evaluation of Various Adsorbents and Coagulants for Wastewater Renovation, AWTR-12, June, 1964.

[Topics of other reports in the series included: AWTR-3 wet oxidation and incineration; AWTR-4 freezing; AWTR-5 foaming; AWTR-6 distillation costs; AWTR-7 distillation; AWTR-8 ultimate disposal; AWTR-13 electrochemical treatment; AWTR-17 market projections; AWTR-18 electrodialysis.]

Baker, M. N., *The Quest for Pure Water*, American Water Works Association, New York, 1948.

Brooks, D. R., Personal communication, Calgon Activated Carbon Co., Pittsburgh, PA, 2003.

Brunauer, S., Emmett, P. H., and Teller, E., *Journal of the American Chemical Society*, 60, 309, 1938.

Calgon Carbon Corporation, Contaminated groundwater cleaned by granular carbon system, http://www.calgoncarbon.com/articles, February 17, 2003a.

Calgon, Filtrasorb® 300 & 400, Product Bulletin, PB-1042-08/98, Calgon Carbon Corporation, Pittsburgh, PA, http://www.calgoncarbon.com, 2003b.

Cannon, F. S., Snoeyink, V. L., Lee, R. G., Dagois, G., and DeWolfe, J. R., Effect of calcium in field-spent GAC's on pore development during regeneration, *Journal of the American Water Works Association*, 85(3):76–89, March, 1993.

Champlin, T. L. and Hendricks, D. W., Treatment train modeling for aqueous contaminants, Volume II, Matrix of contaminants and treatment technologies, Environmental Engineering Technical Report 53-2415-93-2, Department of Civil Engineering, Colorado State University, Fort Collins, CO, May, 1993.

Carter, M. C. and Weber, W. J. Jr., Modeling adsorption of TCE by activated carbon preloaded by background organic matter, *Environmental Science and Technology*, 28(4):616–623, 1994.

Charlie, W. A., Personal communication, May 5, 2003.

Cheremisinoff, P. N. and Ellerbusch, F., *Carbon Adsorption Handbook*, Ann Arbor Science Publishers, Inc., Ann Arbor, MI, 1978.

Clark, R. M. and Lykins, B. W., *Granular Activated Carbon—Design, Operation and Cost*, Lewis Publishers, Chelsea, MI, 1989.

Colorado Springs, City of, Tertiary-Treatment Plant (handout paper for plant visitors), Colorado Springs, CO, c. 1974.

Cookson, J. T., Adsorption mechanisms: The chemistry of organic adsorption on activated carbon, Chap. 7, in: Cheremisinoff, P. N. and Ellerbusch, F. (Eds.), *Carbon Adsorption Handbook*, Ann Arbor Science Publishers, Inc., Ann Arbor, MI, 1978.

Cookson, J. T. and North, W. J., Adsorption of viruses on activated carbon, *Environmental Science and Technology*, 1(1):46–52, January, 1967.

Crittenden, J. C. and Weber, W. J. Jr., Predictive model for design of fixed-bed adsorbers: Parameter estimation and model development, *Journal of the Environmental Engineering*

Division, American Society of Civil Engineers, 104 (EE2):185–197, 1978a.

Crittenden, J. C. and Weber, W. J. Jr., Predictive model for design of fixed-bed adsorbers: Single component model verification, *Journal of the Environmental Engineering Division,* American Society of Civil Engineers, 104(EE3):433–443, June, 1978b.

Crittenden, J. C. and Weber, W. J. Jr., Model for design of multicomponent adsorption systems, *Journal of the Environmental Engineering Division,* American Society of Civil Engineers, 104(EE6):1175–1195, 1978c.

Culp, R. L. and Culp, G. L., *Advanced Wastewater Treatment*, Van Nostrand Reinhold, New York, 1971.

Culp, G. L. and Culp, R. L., *New Concepts in Water Purification*, Van Nostrand Reinhold, New York, 1974.

Culp, R. L. and Roderick, R. E., The Lake Tahoe reclamation plant, *Journal of the Water Pollution Control Federation*, 38(2):147–155, February, 1966.

Culp, R. L., Wesner, G. M., and Culp, G. L., *Handbook of Advanced Wastewater Treatment,* 2nd edn., Van Nostrand Reinhold, New York, 1978.

Denver Water, Water reuse demonstration plant, brochure for visitors, Denver Board of Water Commissioners, Denver, CO, c. 1986.

DiGiano, F. A., Baldauf, G., Frick, B., and Sontheimer, H., Simplifying the description of competitive adsorption for practical application in water treatment, Chap. 10, in: Suffet, I. H. and McGuire, M. J. (Eds.), *Activated Carbon Adsorption of Organics from the Aqueous Phase*, Vol. 1, Ann Arbor Science Publishers, Ann Arbor, MI, 1980.

Dobbs, R. A. and Cohen, J. M., Carbon adsorption isotherms for toxic organics, Report EPA-600/8-80-023, Municipal Environmental Research Laboratory, U.S. Environmental Protection Laboratory, Cincinnati, OH, April, 1980.

Dostal, K. A., Pierson, R. C., Hager, D. G., and Robeck, G. G., Carbon bed design criteria study at Nitro, *W.Va., JAWWA*, 57(5):663–674, 1965.

Fleming, H. L., Application of aluminas in water treatment, *Environmental Progress (*AIChE), 5(3):159–166, August 1986.

Frick, B., Bartz, R., Sontheimer, H., and DiGiano, F. A., Predicting the competitive adsorption effects in granular activated carbon filters, Chap. 11, in: Suffet, I. H. and McGuire, M. J. (Eds.), *Activated Carbon Adsorption of Organics from the Aqueous Phase*, Vol. 1, Ann Arbor Science Publishers, Ann Arbor, MI, 1980.

Fritz, W., Merk, W., Schlünder, and Sontheimer, H., Competitive adsorption of dissolved organics on activated carbon, Chap. 9, in: Suffet, I. H. and McGuire, M. J. (Eds.), *Activated Carbon Adsorption of Organics from the Aqueous Phase*, Vol. 1, Ann Arbor Science Publishers, Ann Arbor, MI, 1980.

Greenbank, M. and Spotts, S., Six criteria for coal-based carbon, Coalification series, Water Technology, Calgon Carbon Corporation, www.calgoncarbon.com/articles/Six_Criteria, 02/17/203, 2003.

Groeber, M. M., Granular Activated Carbon Treatment, Engineering Bulletin EPA-540/2-91/024, Office of Research and Development, U.S. Environmental Protection Laboratory, Cincinnati, OH, October, 1991.

Hager, D. G., Adsorption and filtration with granular activated carbon, *Water and Wastes Engineering*, 6:39–43, August, 1969.

Hager, D. G. and Flentje, M. E., Removal of organic contaminants by granular carbon- filtration, JAWWA, 57(11):1440–1450, November, 1965.

Hassler, J. W., *Activated Carbon*, Chemical Publishing Company, Inc., New York, 1963.

Hassler, J. W., *Purification with Activated Carbon*, Chemical Publishing Company, Inc., New York, 1974.

Hendricks, D. W., Sorption Kinetics Part II-Simulation and Testing of Sorption in a Packed Bed Reactor, Utah Water Research Laboratory, Utah State University, Report PRWG66-2, July 1973.

Hendricks, D. W., Advection limited sorption kinetics in packed-beds, *Journal of the Environmental Engineering Division*, ASCE, 106(EE4):727–739, August, 1980.

Hendricks, D. W., *Water Treatment Unit Processes – Physical and Chemical*, CRC Press and Taylor-Francis, Boca Raton, FL, 2006.

Hendricks, D. W. and Kuratti, L. C., Kinetics Part I—A laboratory investigation of six proposed rate laws using batch reactors, Utah Water Research Laboratory, Utah State University, Report PRWG66-1, July 1973.

Hendricks, D. W. and Kuratti, L. G., Derivation of an empirical sorption rate equation by analysis of experimental data, *Water Research*, 16:829–837, June, 1982.

Hendricks, D. W., Post, F. W., Khairnar, D. R., and Jurinak, J. J., Bacterial adsorption on soils—Thermodynamics, Utah Water Research Laboratory, Utah State University, Report PRWG62-1, July 1970.

Hendricks, D. W., Post, F. W., and Khairnar, D. R., Adsorption of bacteria on soils: Experiments, thermodynamic rationale, and application, *Water, Air, and Soil Pollution*, 12:219–232, 1979.

Hiester, N. K. and Vermeulen, T., Saturation performance of ion-exchange and adsorption columns, *Chemical Engineering Progress*, 48(10):505–516, October, 1952.

Hill, T. L., *Introduction to Statistical Thermodynamics*, Addison-Wesley, Reading, MA, 1960.

Karanfil, T., Kilduff, J. E., Schlautman, M. A., and Weber, Walter, J. Jr., Adsorption of organic macromolecules by granular activated carbon. 1. Influence of molecular properties under anoxic solution conditions, *Environmental Science and Technology*, 30(7):2187–2194, 1996a.

Karanfil, T., Schlautman, M. A., Kilduff, J. E., and Weber, Walter, J. Jr., Adsorption of organic macromolecules by granular activated carbon. 1. Influence of dissolved oxygen, *Environmental Science and Technology*, 30(7):2194–2201, 1996b.

Keinath, T. M., Modeling and simulation of the performance of adsorption contactors, Chap. 1, in: Keinath, T. M. and Wanielista, M. P. (Eds.), *Mathematical Modeling for Water Pollution Control Processes*, Ann Arbor Science Publishers, Inc., Ann Arbor, MI, 1975.

Keinath, T. M. and Weber, W. J. Jr., A predictive model for the design of fluid bed adsorbers, *Journal of the Water Pollution Control Federation*, 40(5):741–765, May, 1968.

Kornegay, B. H., Control of synthetic organic chemicals by activated carbon – theory, application, and regeneration alternatives, in: *Seminar on Control of Organic Chemical Contaminants in Drinking Water*, U.S. Environmental Protection Agency, February, 1979.

Kunin, R., Porous Polymers As Adsorbents–A review of Current Practice, *Amber-Hi-Lites*, Rohm and Haas Company, Philadelphia, PA, No. 163, Winter, 1980.

Lambe, T. W. and Whitman, R. V., *Soil Mechanics*, John Wiley & Sons, New York, 1969.

Lauer, W. C., The demonstration of direct potable water reuse: Denver's Landmark Project, in: Asano, T. (Ed.), *Wastewater Reclamation and Reuse*, Technomic Publishing Co., Inc., Lancaster, PA, 1998.

Lauer, W. C., Rogers, S. E., and Ray, J. M., The current status of Denver's Potable Water Reuse Project, *Journal of the American Water Works Association*, 77(7):52–59, July, 1985.

Love, O. T., Miltner, R. J., Eilers, R. G., and Frank-Leist, C. A., Treatment of volatile organic compounds in drinking water, Report EPA-600/8-83-019, Municipal Environmental Research Laboratory, U.S. Environmental Protection Laboratory, Cincinnati, OH, May, 1983.

Marshall, C. E., *The Physical Chemistry and Mineralogy of Soils*, John Wiley & Sons, Inc., New York, 1964.

McDougall, W. J., Fusco, R. A., and O'Brien, R. P., Containment and Treatment of the Love Canal Landfill leachate, *Journal of the Water Pollution Control Federation*, 52(12):2914–2924, December, 1980.

McGuire, M. J. and Suffet, I. H., Adsorption of organics from domestic water supplies, *Journal of the American Water Works Association*, 70(11):621–636, November, 1978.

McGuire, M. J. and Suffet, I. H., The calculated net adsorption energy concept, Chap. 4, in: Suffet, I. H. and McGuire, M. J. (Eds.), *Activated Carbon Adsorption of Organics from the Aqueous Phase*, Vol. 1, Ann Arbor Science Publishers, Ann Arbor, MI, 1980.

Moore, B. C., Cannon, F. S., Metz, D. H., and DeMarco, J., GAC pore structure in Cincinnati during full-scale treatment/reactivation, *Journal of the American Water Works Association*, 95(2):103–112, February, 2003.

Müller, U., Hess, F., and Worch, E., Impact of organic matter adsorbability on micropollutant removal by activated carbon, *Aqua*, 45(6):273–280, December, 1996.

Mullins, R. L. Jr., Zogorski, J. S., Hubbs, S. A., and Allgeier, G. D., The effect of several brands of granular activated carbon for the removal of trihalomethanes from drinking water, Chap. 14, in: Suffet, I. H. and McGuire, M. J. (Eds.), *Activated Carbon Adsorption of Organics from the Aqueous Phase*, Vol. 1, Ann Arbor Science Publishers, Ann Arbor, MI, 1980.

Neethling, J. B. and Culp, G. L., Capital and operating costs of GAC facilities, in: *Sunday Seminar on Engineering Considerations for GAC Treatment Facilities, AWWA National Conference*, Cincinnati, OH, June, 17, 1990.

OCWD, WF21 Costs, Orange County Water District, Fullerton, CA (http://www.ocwd.com, February, 2003).

Parkhurst, J. D., Dryden, F. D., McDermontt, G. N., and English, J., Pomona activated carbon pilot plant, *Journal of the Water Pollution Control Federation*, 39(10, Part 2): R69–R81, October, 1967.

Pittsburgh Activated Carbon, *Descriptive Brochure*, 11pp., Pittsburgh Activated Carbon, Pittsburgh, PA, 1968.

Radke, C. J. and Prausnitz, J. M., Thermodynamics of multisolute adsorption from dilute liquid solutions, *Journal of the American Institute of Chemical Engineers*, 18(4):761–768, July, 1972.

Reed, B. E., Jamil, M., and Thomas, B., Effect of pH, empty bed contact time and hydraulic loading rate on lead removal by granular activated carbon columns, *Water Environment Research*, 68(5):877–882, July/August, 1996.

SDWC (Safe Drinking Water Committee), *Drinking Water and Health*, Vol. 2, National Academy Press, Washington, DC, 1980.

Slechta, A. F. and Culp, G. L., Water reclamation studies at the South Tahoe PUD, *Journal of the Water Pollution Control Federation*, 39(5):787–814, May, 1967.

Snoeyink, V. L., Adsorption of organic compounds, in: Pontius, F. W. (Ed.), *Water Quality and Treatment*, pp. 781–867, McGraw-Hill, New York, 1990.

Snoeyink, V. L., Weber, W. J. Jr., and Mark, H. B. Jr., Sorption of phenol and nitrophenol by active carbon, *Environmental Science and Technology*, 2(10):918–926, October, 1969.

Spotts, S. D., Understanding carbon mesh size, *Water Technology*, March, 1993.

Stenzel, M. H., Remove organics by activated carbon adsorption, *Chemical Engineering Progress*, 89(19):36–43, April, 1993.

Symons, J. M., Bellar, T. A., Carswell, J. K., DeMarco, J., Kropp, K. L., Robeck, G. G., Seeger, D. R., Slocum, C. J., Smith, B. L., and Stevens, A. A., National organics reconnaissance survey for halogenated organics, *Journal of the American Water Works Association*, 67(11): 634–647, November, 1975.

USEPA, *Process Design Manual for Carbon Adsorption,* Technology Transfer Division, U.S. Environmental Protection Agency, Cincinnati, OH, 1971.

USEPA, *Process Design Manual for Carbon Adsorption,* Technology Transfer Division, US Environmental Protection Agency, Cincinnati, OH, 1973.

USEPA, Watewater technology fact sheet—granular activated carbon absorption and regeneration, US Environmental Protection Agency, EPA 832-F-00-017, Office of Water, Washington, DC, September, 2000.

Vagliasindi, Wave front behavior in adsorption reactors, MS Thesis, Department of Civil Engineering, Colorado State University, Fort Collins, CO, 1991.

Vagliasindi, F. and Hendricks, D. W., Wave front behavior in adsorption reactors, *Journal of the Environmental Engineering Division*, ASCE, 118(4):530–550, July/August, 1992.

Weber, W. J. Jr., *Physicochemical Processes for Water Quality Control*, Wiley-Interscience, New York, 1972.

Weber, W. J. Jr., Evolution of a technology, *Journal of the Environmental Engineering Division,* American Society of Civil Engineers, 110(EE5):899–917, October, 1984.

Weber, W. J. Jr. and Rumer, R. R., Intraparticle transport of sulfonated alkylbemaenes in a porous solid: Diffusion with nonlinear adsorption, *Water Resources Research,* 1(3):361–373, Fall Quarter, 1965.

Weber, W. J. Jr. and Smith, E. H., Simulation and design models for adsorption processes, *Environmental Science and Technology*, 21(11):1040–1049, November, 1987.

Weber, W. J. Jr. and Van Vliet, B. M., Fundamental concepts for application of activated carbon in water and wastewater treatment, Chap. 1, in: Suffet, I. H. and McGuire, M. J. (Eds.), *Activated Carbon Adsorption of Organics from the Aqueous Phase*, Vol. 1, Ann Arbor Science Publishers, Ann Arbor, MI, 1980.

Weber, W. J. Jr. and DiGiano, F. A., *Process Dynamics in Environmental Systems*, John Wiley & Sons, New York, 1996.

West, R. E., Effect of porous structure on carbon activation, Water Pollution Control Research Series, 17020 DDC 06/71, U.S. Environmental Protection Agency, U.S. Government Printing Office, Washington, DC, 94pp., 1971.

Westerhoff, G. P. and Miller, R., Design of the GAC Treatment Facility at Cincinnati, *Journal of the American Water Works Association*, 78(4):146–155, April, 1986.

Willis, B., Denver's Potable Water Reuse Demonstration Facility, CE 541 paper, Colorado State University, Fort Collins, CO, May 2, 1989.

Zogorski, J. S. and Faust, S. D., Operational parameters for optimum removal of phenolic compounds from polluted waters by columns of activated carbon, Chap. 20, in: Cheremisinoff, P. N. and Ellerbusch, F. (Eds.), *Carbon Adsorption Handbook*, Ann Arbor Science Publishers, Inc., Ann Arbor, MI, 1978.

FURTHER READINGS

Additional depth concerning transport processes and reactor theory may be found in several books some of which are listed. The principles reviewed are applicable to a wide variety of applications, including adsorption.

Clark, M. M., *Transport Modeling for Environmental Engineers and Scientists*, John Wiley & Sons, New York, 1996.

Logan, B. E., *Environmental Transport Processes*, John Wiley & Sons, New York, 1999.

Weber, W. J. Jr. and DiGiano, F. A., *Process Dynamics in Environmental Systems*, John Wiley & Sons, New York, 1996.

Weber, W. J. Jr., *Environmental Systems and Processes*, John Wiley & Sons, New York, 2001.

16 Ion-Exchange

Ion-exchange has three applications in water treatment: (1) softening, (2) specific ion removal, and (3) demineralization. The ion-exchange technology is another that is "on the shelf," so to speak, to be used as needed for particular situations, i.e., rather than for routine application.

As a science, the ion-exchange phenomenon has been known since about 1850. Its use for softening in water treatment practice began about 1924 (Powell, 1929; Behrman, 1934). Synthetic ion-exchangers were developed about 1935 (Dow Chemical, 1964, p. v).

16.1 DESCRIPTION

Definitions are the first step in understanding ion-exchange and in providing an orientation. Other aspects involve becoming familiar with ion-exchanger media and its characteristics, applications, and history.

16.1.1 ION-EXCHANGE IN-A-NUTSHELL

In theory, ion-exchange is about the same as "adsorption," described in Chapter 15. It differs in that the species attaching to the solid are ionic, rather than molecular. Also, the solid ion-exchanger has *charged* sites within its interior, rather than sites that exhibit molecular forces of attraction. Another difference is that each has its own history of development, applications, literature, and practices.

16.1.1.1 Definitions

A few definitions help to get started. The "glossary" has a more extensive list of terms.

16.1.1.1.1 Ion-Exchange

Ion-exchange is a reaction in which an ionic species, A^+, in solution is exchanged with another ion species, B^+, attached to a solid phase, called an "ion-exchanger," designated as X. The reaction can be depicted by

$$X \cdot B + A^+ \rightarrow X \cdot A + B^+ \tag{16.1}$$

The ion-exchanger has "sites" on a macromolecular or crystalline framework, for resins and minerals, respectively, to which the ions are attached. The number of bonding sites per unit of ion-exchanger is usually expressed as milliequivalents ion-exchange sites per gram of ion-exchanger, i.e., meq/g. The attachment force is electrostatic, having a bonding energy, $\Delta H^\circ \approx 2$ kcal/mol (Helferrich, 1962, p. 8).

16.1.1.1.2 Ion-Exchanger

An ion-exchanger can be any solid substance having an array of ion-exchange sites. The ion-exchanger material can be a *cation-exchanger* as in Equation 16.1, or an *anion-exchanger*, which bonds anions. Materials which have both kinds of sites are called *amphoteric*.

16.1.1.1.3 CounterIons

The ions which exchange are called *counterions*, in that they are opposite to the charge of the sites of the ion-exchanger.

16.1.1.1.4 Co-Ions

Ions in solution having the same charge as the ion-exchanger framework are called *co-ions*.

16.1.1.2 Process Description

The ion-exchanger is a solid material that provides internal sites for the bonding of ions. The "target" ion species to be removed, e.g., Ca^{2+}, the counterion, has a stronger bonding energy because of its double charge than the ion being replaced, e.g., Na^+. Common ion-exchangers include the zeolites and synthetic resins. Usually, the reactor is a packed-bed of ion-exchanger and is operated the same as a packed-bed of GAC. The inflow to the packed-bed is an aqueous solution that includes the ions to be removed. A portion of the target ions, and most likely other counterions, are removed after transport by advection and diffusion to exchange sites, with the column effluent being largely free of the target ions.

Ion-exchange follows the same principles that apply to adsorption, e.g., the reaction equations are similar; a Langmuir isotherm describes equilibrium, as does an equilibrium constant; the van't Hoff equation is applicable (that relates the equilibrium constant to temperature); and Fick's law of diffusion applies to kinetics (in both aqueous and solid phases). Also reactor mathematics, describing the "wave front" and the "breakthrough curve," are the same for both, and so, in principle, reactor design should follow the same logic. The difference is that the adsorbate is molecular in the case of adsorption and ions in the case of ion-exchange, with bonding forces being van der Waal's and electrostatic attraction, respectively.

16.1.1.3 Phases of Operation

Ion-exchange has an operating cycle similar to adsorption. First, the ion-exchanger bed becomes saturated or "exhausted" with respect to the "target" ion, e.g., Ca^{2+}. Then, the recharge part of the cycle is started, which consists of flooding the bed with the recharge ion, e.g., Na^+, in the form of a concentrated brine solution, which displaces the adsorbed Ca^{2+} and conditions the ion-exchanger for another cycle of Ca^{2+} removal. To better utilize the ion-exchanger bed, two columns may be operated in series: when exhausted, the first column is removed for recharge; the second column is place at the beginning; a third standby column that has been recharged is placed in the second position.

16.1.2 History

The field of ion-exchange has its origins in science in the modern sense of conducting systematic inquiry to discover knowledge. Practice is built on this knowledge (Rohm and Haas, 1989).

16.1.2.1 Science

The ion-exchange principle was discovered by H.S. Thompson, an English agricultural chemist, who in 1848 noted that in treating a soil with ammonium sulfate or ammonium carbonate, most of the ammonium was absorbed and Ca^{2+} (then identified as "lime" since the ion theory had yet to be developed) was released (Kunin, 1958). Thompson reported this finding to J. Thomas Way, who followed up, during the period 1850–1854, with a systematic study. Way found that: (1) the exchange of calcium and ammoniums ions noted by Thompson was verified; (2) the exchange of ions in soils involved equivalent quantities; (3) certain ions were exchanged more readily than others; (4) the extent of exchange increased with concentration, reaching a leveling-off value; (5) aluminum silicates present in soils were responsible for the exchange; (6) exchange materials could be synthesized from soluble silicates and alum; and (7) exchange of ions differed from physical adsorption. These principles remain inviolate. It is not clear how all of this was explained without the ionic theory of solutions, i.e., as proposed by Arrhenius, van't Hoff, and Ostwald in the 1880s (Servos, 1990, pp. 35–45).

16.1.2.1.1 Zeolites

In 1858, Eichhorn established the principle of reversibility of the ion-exchange reaction. He showed also that natural zeolites acted as ion-exchangers, which comprised hydrated double silicates (Behrman, 1925, Applebaum, 1925). According to Behrman (1925) this early work laid the theoretical foundation for ion-exchange practice, which was not to commence until the early part of the twentieth century with the work of German chemist, Robert Gans. Gans developed aluminum silicates as "synthetic zeolites," which had a higher exchange capacity than the natural ones, and established their utility in treating sugar solutions to replace K^+ ion with Ca^{2+} ion to increase the yield of crystallizable sugar, and for softening waters (Kunin, 1958).

16.1.2.1.2 Resins

In 1935, two British scientists, B.A. Adams and B.L. Holmes found that synthetic resins had ion-exchange properties. They made the first ion-exchange resins and learned how to add different ionic groups (Dorfner, 1972). They showed that stable high-capacity cation-exchange resins could be prepared by adding sulfonic acid groups, and that anion-exchange resins could be prepared by adding polyamine groups. Later, strong-base quartenary amines were developed, i.e., $–N^+–$, $–N^+(CH_3)_3$, etc. Still later, i.e., before 1962, other strong base groups came into being and included, $–P^+–$, and $>S^+–$. As a note anion-exchangers did not exist prior to the development of synthetic resins (foregoing from Helfferich, 1962, p. 47).

16.1.2.1.3 Cross-Linked Polystyrene Resins

The "Adams and Holmes" patents were purchased by IG-Farbenindustrie A.G., where the development and production of ion-exchanger resins have continued since 1936. In 1945, G. F. d'Alelio, in the United States, patented his synthesis process of incorporating sulfonic acid groups into cross-linked polystyrene resins. This development formed the basis for the modern industry in ion-exchange resins. The polystyrene anion-exchange resins were developed in 1949. Further development was directed toward synthesizing resins with specific ion-exchange properties. Presently, a wide variety of such resins are commercially available. Kunin (1983) viewed the development of anion-exchange resins as the final breakthrough that paved the way for the widespread development of the ion-exchange technology.

16.1.2.1.4 Maturing of Ion-Exchange Technology

Liberti and Helfferich (1983, p. v) and Millar (1983a, p. 2) viewed the 1950s as the "Golden Age" in the development of synthetic resins for ion-exchange. During this period, the technology of ion-exchange was developed and the foundations were laid for modern theory. By 1982, they viewed ion-exchange as a "mature" technology. As an index of activity, worldwide ion-exchange production in thousands of cubic meters was: 1967: 60; 1970: 70; 1974: 138; 1977: 125; and in 1981: 140 (Millar, 1983a, p. 4). Of the amount sold in 1981, >120,000 m^3 resin was polystyrene strong-acid or strong-base gel resins in bead form.

16.1.3 Applications

Ion-exchange has been applied for softening in water treatment, ammonia removal in wastewater treatment, demineralization for various industrial purposes, and specific ion removal in hazardous wastes.

16.1.3.1 Municipal Use

Softening of water is the removal of "hardness" ions (that consume soap), which are mostly Ca^{2+} and Mg^{2+} (see Chapter 21). Water-softening practice began in 1905 with the first municipal softening plant which used precipitation. In 1906, Professor Robert Gans, Director of Chemistry of the German Geological Survey, obtained a patent for softening water by "base-exchange" (AWWA, 1951). The first applications were for hotels, apartments, laundries, and boiler feed water (Applebaum, 1925). The Permutit Co., which was related to the well-known German company, I.G. Farber, had the patents of Gans and had forged ahead in developing the ion-exchange technology for water treatment.

In 1920, the discovery of the New Jersey greensands (glauconite) made ion-exchange economically feasible for municipal water softening (AWWA, 1951). The greensands, (which are not zeolites) had a higher rate of exchange than the synthetic zeolites. The first zeolite plant was completed in 1924, with 10 plants in 1930, 45 in 1935, 110 in 1938, 199 in 1941, and in 1945 there were 238 plants serving 4.4 million persons in the United States. By 1950 about 150 municipal systems had been installed and about 30,000 ion-exchange systems were used by the industry (AWWA, 1951). Proprietary soft water service got underway in 1937 at Rockford, Illinois and then in Wheaton, Illinois, followed by Hagerstown, Maryland in 1938. By 1944, there were 200 central service regenerating plants serving 310 communities.

16.1.3.2 Removals of Specific Ions

The list of specific ions that may be removed by ion-exchange includes: ammonia, various heavy metals, radioactive ions, boron, nitrates, fluoride. Nitrate pollution is not uncommon in shallow aquifers and heavy metals are a concern at hazardous wastes sites. Excess fluoride occurs in some drinking water sources. Boron may be a residual in seawater deionization.

The perfectly uniform pore size of a given zeolite aids its selectivity for certain ions, e.g., Cs^{2+}, Sr^{2+}, and heavy metals such as lead, i.e., Pb^{2+}. Clinoptilolite, a mineral in the zeolite family, is useful in concentrating radioactive ions or certain heavy metals for containment and controlled disposal (Vaughan, 1988, p. 27). Ammonia removal from sewage effluents is another application of clinoptilolite (Vaughan, 1988, p. 27). The attenuation of ammonia peaks in treated sewage effluent, e.g.,

$$C(\text{column influent, } NH_4{}^+) \approx 16 \text{ mg/L}$$

to

$$C(\text{column effluent, } NH_4{}^+) \approx 1 \text{ mg/L},$$

was demonstrated in laboratory studies using clinoptilolite packed-bed columns (Beler Baykal et al., 1996, 1997).

16.1.3.3 Deionization

Deionization has many industrial applications, e.g., boiler feed water, microchip production, analytical laboratories, and process water for various purposes. The process involves a cation-exchanger charged with H^+ and an anion-exchanger charged with OH^- and may be done sequentially through each bed or through a "mixed-bed"; the cations and anions are exchanged for H^+ and OH^-, respectively.

For reference, the specific electric conductance (see glossary) of deionized water is typically about 0.05 μS/cm, which is about the same as distilled water. Typical values for ambient waters are: Lake Tahoe, 97 μS/cm; Lake Mead, 850 μS/cm; Atlantic Ocean, 43,000 μS/cm.

16.1.4 Media

An "ion-exchanger" is a solid substance having an open framework, which may be either crystalline or macromolecular, depending upon the ion-exchanger material. Because the framework is open the exchange sites on it are accessible by diffusion to any ions in a surrounding aqueous solution that are smaller than the pores within the solid. As a rule, mineral ion-exchangers are granular, i.e., irregular, while resins are beads, with sizes generally about 0.5–1 mm.

TABLE 16.1
Minerals Having Ion-Exchange Properties

Class	Name	Chemical Formula
Zeolites	Analeite	$Na[Si_2AlO_6]_2 \cdot HOH$
	Chabezite	$(Ca,Na)\ [Si_2AlO_6]_2 \cdot 6HOH$
	Harmotome	$(K,Ba)\ [Si_5Al_2O_{14}] \cdot 5HOH$
	Heulandite	$Ca[Si_3AlO_8] \cdot 5HOH$
	Natrolite	$Na_2[Si_3Al_2O_{10}] \cdot 2HOH$
	Sadalite	$Na_8Al_6\ Si_6O_{24}C_{12}$
	Siliceous zeolite	$xNa_{20} \cdot Al_2O_3 \cdot zSiO_2 \cdot aHOH$
Clays	Montmorilonite	$Al_2[Si_4O_{10}(OH)_2] \cdot yHOH$
	Illite	$(K_2H_3O)\ (Al,Mg,Fe)_2(Si,Al)_4\ O_{10}[(OH)_2, (H_2O)]$
	Kaolinite	$Al_2O_3 \cdot 2SiO_2 \cdot 2H_2O$
	Vermiculite	$(Mg,Fe^{2+},Al)_3\ (Al_2Si)_4O_{10}(OH)_2 \cdot 4H_2O$

16.1.4.1 Mineral Ion-Exchangers

Most natural ion-exchange minerals are crystalline aluminosilicates with cation-exchange properties (Helferrich, 1962). Table 16.1 lists seven minerals of the zeolite group and three clays. The zeolites have a lattice structure which is open with channels connecting. The lattice has a negative charge for each aluminum atom. The charge is balanced by cations which are free to move within the lattice. The clays in Table 16.1, on the other hand, have a loose layer structure and can swell, increasing the interlayer distance.

16.1.4.2 Clays

A clay is a mineral particle characterized by its small size, i.e., ≤1 μm. A variety of mineral types comprise the clays; common ones include montmorillonite and kaolinite. Another, also with ion-exchange properties, is vermiculite; cadmium uptake was shown to follow the Freundlich isotherm for $0.2 < C(\text{Cd}) \leq 2$ mg Cd/L, i.e., $\overline{X}^*(\text{Cd}^{2+}) = 1.82 \cdot C(\text{Cd}^{2+})^{0.948}$, in which $\overline{X}^*(\text{Cd}) = (\text{mg Cd}^{2+}/\text{g vermiculite})$ and $C(\text{Cd}^{2+}) = (\text{mg Cd}^{2+}/\text{L solution})$ (Das and Bandyopadhyay, 1993, p. 2).

16.1.4.3 Zeolites

By the early twentieth century, chemists were trying to synthesize zeolite crystals and invented structures not found in nature (Kerr, 1989, p. 100). Some 40 natural zeolites and over 70 synthetic zeolites have been discovered (Kato, 1995, p. 7). These numbers have increased, however, since research on zeolites has been continuous (see INZA, 2010; IZA, 2010). For example, 194 zeolite "Framework Types," each designated by a code of three capital letters, e.g., ABW, ACO, AEI, . . . , YUG, ZON, are seen within a matrix format (IZA, 2010). Each code within the matrix is linked to 3D wire-frame drawing and other depictions that describe its characteristic "framework." Within a given framework, numerous "materials" may be included. For example, the Framework Type

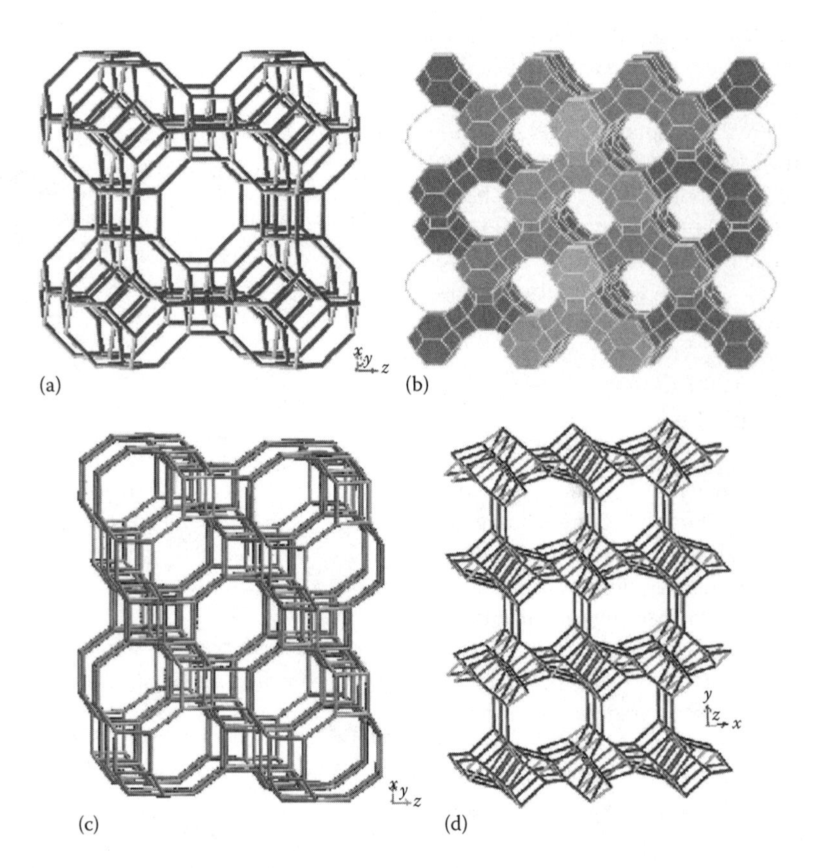

FIGURE 16.1 Examples of zeolites illustrating different Framework Types with codes in parenthesis. (a) Linde Type A (LTA)–synthetic; (b) Faujasite (FAU)–mineral; (c) Chabazite (CHA)–mineral; (d) Clinoptilolite (HEU)–mineral. (From Baerlocher, Ch. and McCusker, L. B., Database of zeolite structures, http://www.iza-structure.org/databases, June 2003. With permission from International Zeolite Association.)

ABW has some 22 "materials," e.g., [Be-As-O]-ABW; [Be-As-O]-ABW; [Be-P-O]-ABW; [Ga-Si-O]-ABW; ..., |Na|[Zn-P-O]-ABW (Baerlocher et al., 2007, p. 14; IZA, 2010). The convention is that the vertical bold bars encompass the "guest" species and the brackets encompass the "host" species (McCusker et al., 2001, p. 388).

Examples: Figure 16.1 illustrates four of the 194 zeolite Framework Types. The first, Linde Type A, under the "LTA" Framework Type is synthetic. The others, i.e., Faujasite-FAU, Chabazite-CHA, Clinoptilolite-HEU, are minerals, i.e., found in nature. Table 16.2 lists some of the zeolites mentioned in the literature and further illustrates the IZA classification system; see for example, footnotes "d" and "e" that pertain to FAU. As illustrated, Linde Type X and Linde Type Y (both synthetic zeolites) and Faujasite (a mineral zeolite), have the FAU Framework Type.

IZA classification system: To further explicate in terms of the IZA system, a zeolite Framework Type, e.g., "FAU" (IZA, 2010), Figure 16.1 and Table 16.1, respectively, describes only the "connectivity" of the framework, while a zeolite structure is that of a specific material, i.e., for a given framework there are often many zeolite structures. The hierarchy within the IZA web-site is: IZA (2010)/Zeolite Framework

TABLE 16.2
Examples of Zeolite Materials and Respective IZA Framework Type Codes

Mineral[a]	Synthetic
Sodalite[b]/SOD	ZSM-5[b]/MFI
Clinoptilolite (Heulandite)/HEU	UZM-4[c]/MAZ
Chabazite/CHA	UZM-5[b]/UFI
Analcime/ANA	Linde Type A[a]/LTA
Faujasite[d]/FAU[e]	Linde Type X[b,d]/FAU[e]
	Linde Type Y[b,d]/FAU[e]

Source: Adapted from Helfferich, F., *Ion-Exchange*, McGraw-Hill, New York, p. 29, 47, 1962.

[a] Helfferich, 1962, p. 10;
[b] Kerr, 1989, pp. 102, 104;
[c] Jacoby, 2003, p. 38;
[d] Faujasite is a rare zeolite mineral. The synthetic versions, used commercially, are known as Linde Type X, and Linde Type Y, respectively; they differ in their Si/Al ratios, i.e., for Linde Type X: ($1.0 \leq$ Si/Al ≤ 1.5); and for Linde Type Y: ($1.5 \leq$ Si/Al ≤ 3.0).
[e] A number of zeolites have the FAU Framework Type, including Faujasite, Linde Type X, Linde Type Y, and about 20 others. Others not listed in Table 16.1.

Types/ FAU/ Related Materials. The IZA web-site has links to numerous animated 3-D images, which give a perspective of any of the respective Framework Types in their database.

The publisher's abstract of the book by Baerlocher et al. (2007) describes further the idea of a Zeolite Framework Type:

Each time a new zeolite framework structure is reported, it is examined by the Structure Commission of the International Zeolite Association (IZA-SC), and if it is found to be unique and to conform to the IZA-SC's definition of a zeolite, it is assigned a three-letter framework type code. This code is part of the official IUPAC (International Union of Pure and Applied Chemistry) nomenclature for microporous materials. The *Atlas of Zeolite Framework Types* is essentially a compilation of data for each of these confirmed framework types. These data include a stereo drawing showing the framework connectivity, features that characterize the idealized framework structure, a list of materials with this framework type, information on the type material that was used to establish the framework type, and stereo drawings of the pore openings of the type material.

As may be evident, the field of zeolite structures and chemical compositions is specialized. For water treatment applications, usually related to removals of specific ions, particular zeolites may be recommended based on past use. The IZA web-site gives access to in-depth structural knowledge of any zeolite.

16.1.4.3.1 Characteristics

The zeolites are soft minerals and therefore have low resistance to abrasion. They have internal cage-like structures with surface areas up to several hundred square meters per gram and cation-exchange capacities up to several equivalents per kilogram (Haggerty and Bowman, 1994, p. 452).

16.1.4.3.2 Pores

A zeolite has a unique crystal structure with a center channel of uniform size. Zeolite pores range in size, 2.5–8 Å ($2.5 \cdot 10^{-10}$–$8.0 \cdot 10^{-10}$ m) diameter, depending on the structure (Kerr, 1989, p. 100; Kato, 1995, p. 7). The pores may be: (1) three-dimensional, i.e., with channels that intersect from three directions such as Type A synthetic; (2) two-dimensional, i.e., with two intersecting channels such as ZSM-5 synthetic; and (3) one-dimensional, i.e., resembling a pack of straws (Vaughan, 1988, p. 27).

16.1.4.3.3 Structure

The basic building block in all zeolites is a tetrahedral structure of an aluminum or silicon atom surrounded by four oxygen atoms; each tetrahedron is connected with others through shared oxygen atoms to form a framework. The tetrahedral building blocks can arrange themselves in varied combinations resulting in different framework geometries, i.e., different crystal structures (Vaughan, 1988, p. 25; Kerr, 1989, p. 100). Different combinations of the same secondary building unit may give many kinds of distinctive zeolite structures; theoretically, thousands are possible (Vaughan, 1988, p. 25).

16.1.4.3.4 Charge

The proportion of silicon to aluminum has little effect on the overall structure of a zeolite. But whether the zeolite may function as an adsorbent or as an ion-exchanger is affected. The two extremes are: (1) a zeolite with the same number of silicon and aluminum atoms (always, it must be true that: # aluminum atoms $\leq$# silicon atoms in a crystal), and (2) a zeolite with nearly all silicon atoms. The first case results in a net negative charge in the pores; free ions are needed to balance the charge and most often these are Na^+, K^+, Ca^{2+}, Mg^{2+}. In the second case, there is no net charge and the structure is hydrophobic and will attract neutral molecules based on van der Waal's forces (Kato, 1995, p. 8).

16.1.4.3.5 Clinoptilolite

Clinoptilolite is the most abundant naturally occurring zeolite with cation-exchange capacity, CEC = 150 meq/100 g (Marshall, 1964, p. 120). The channel structure has a large cavity size measuring 4.4 × 7.2 Å (Haggerty and Bowman, 1994, p. 452). The chemical composition was determined as $[Na_2O]_{0.92}[K_2O]_{0.78}[Fe_2O_3]_{0.27}[MgO]_{0.21}[Al_2O_3]_{1.93}[SiO_2]_{9.70}$ (Kesraoul-Ouki, 1993, p. 1115). Figure 16.2 is a scanning electron microscope (SEM) photomicrograph of a Clinoptilolite surface, showing its irregularity.

16.1.4.3.6 Greensands

One of the best known of the natural ion-exchangers is glauconite, which is a sodium aluminosilicate (Powell, 1954), commonly called "greensand (which is classed by mineralogists as a mica clay, not a zeolite)." Zeo-Dur® and Inversand® have been trade names for this material. The particles are greenish-black with exchange capacities of 6.18–8.70 kg/m^3 (2700–3800 g/ft^3). Shallow deposits in New Jersey were reported as being almost inexhaustible (Babbitt and Doland, 1949, p. 513).

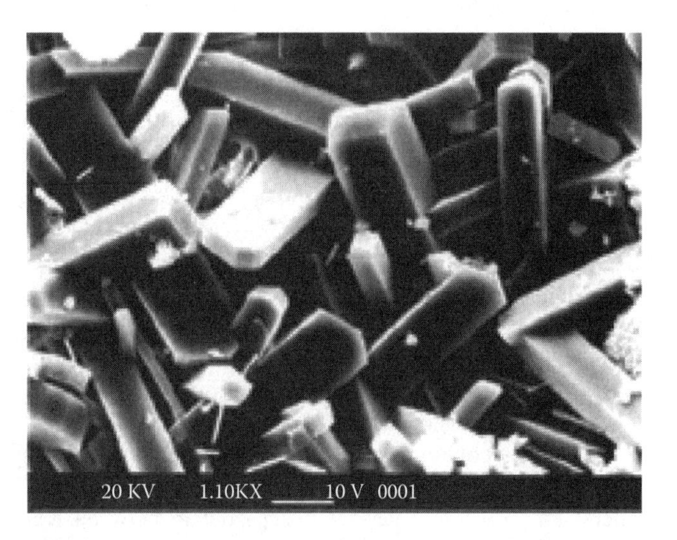

FIGURE 16.2 SEM photomicrograph of clinoptilolite surface. (Reprinted from INZA, International Natural Zeolite Association web-site, http://inza.edu/pics_crystals.php, 2010. With permission from Steve Chipera.)

The glauoconites are: SiO_2, 50%; iron oxides and alumina, 29%; CaO, 2%; MgO, 3%; K_2O, 8%; and water, 8% (Collins, 1937). Grains of natural glauconite can be broken with the fingernail; thus they are processed to harden. The processing of glauconite from the raw material involves the following:

1. Washing to remove mud, fine grains, quartz sand (amount of material removed may be 50% of original) giving a product size of 0.2–6.2 mm
2. Scouring with caustic to dissolve less stable grains and to increase exchangeable sodium
3. Sodium silicate and weak-acid treatment to provide a hard silica envelope
4. Marketing under a trade name

16.1.4.3.7 Synthetic Zeolites

The synthetic zeolites are either precipitated gel types or fusion types. The former are obtained by precipitating a reaction from solutions of sodium silicate and sodium sulfate or sodium aluminate. The dried gel is broken into sizes needed for softening. The fusion types are obtained by heating mixtures of minerals, such as sodium carbonate, kaolin, and feldspar, to the melting point. After cooling, the material is crushed and sized (Collins, 1937). Commercial zeolites are usually in the form of pellets, granules or beads, and powders. Glues and binders are used to form these shapes.

16.1.4.4 Synthetic Resins

A resin is a solid or liquid organic polymer (*Apple Dictionary*, 2009). A model of a synthetic ion-exchange resin suggested by Robert Kunin (1983, p. 46) is that of a plate of congealed spaghetti. The strands of pasta may represent the polymer chains and the points of contact between strands are the cross-linked points. The ion-active groups are located along the strands, i.e., polymer chains. The mass as whole is a solid, made so by the strands adhering to one another, but with a porous character. Microscopically, the mass is heterogeneous, i.e., the cross-linking is not uniform, nor are the ion-active groups spaced uniformly. Externally, i.e., macroscopically, the resin bead appears homogeneous. By contrast, a zeolite has an ordered structure, i.e., a crystal, with regular pore sizes and spacing.

16.1.4.4.1 Cross-Linking

The cross-linked polymer network is the framework for the ion-exchanger. The degree of cross-linking can be adjusted by controlling the DVB (di-vinyl benzene) proportion of the reactants, i.e., the molar ratio of DVB to styrene. The degree of cross-linking is expressed quantitatively as the "nominal DVB content" and is defined as "the mole percent of di-vinyl benzene in the polymerization mixture." For example, 10% DVB means one DVB molecule per nine styrene molecules. General purpose ion-exchanger resins contain between 8% and 12% DVB but resins with as little as 0.25% DVB and as much as 25% DVB have been prepared (Helfferrich, 1962).

The degree of cross-linking determines the pore size of the network and the swelling propensity of the resin. For a highly cross-linked resin the mesh size (within the polymers) is only a few angstroms but for one that is not strongly cross-linked the mesh size may be 100 Å after swelling. The cross-linking also serves to make the polymer framework insoluble in water (Millar, 1983b, p. 23); without the cross-linking, the resin would be soluble.

16.1.4.4.2 Framework of Resins

The most commonly synthesized ion-exchange resins are made of styrene monomers, which are polymerized, and cross-linked with di-vinyl benzene (DVD). Fixed ionic groups are added, which contain the charge sites. Figure 16.3 delineates the building of such an ion-exchange resin (adapted from Dow Chemical, 1964, p. 3).

Figure 16.3a shows the benzene molecule, i.e., the "benzene-ring," which is one of the building blocks of a styrene monomer. Figure 16.3b shows the benzene molecule combined with a vinyl group, i.e., $H_2C{=}CH_2$, to give a styrene monomer (vinyl benzene). Figure 16.3c illustrates the styrene monomers combined to form a styrene polymer. The styrene polymer is a linear, two-dimensional structure. The cross-linking is by di-vinyl benzene (DVD), shown shaded in Figure 16.3d. The next step involves combining the DVB (shaded) with polystyrene, illustrated in Figure 16.3e. The copolymerization of the linear polymer with the DVD, giving the cross-linking, ties the linear chains together, and yields an insoluble three-dimensional structure. As noted, the DVD proportion can be varied. Figure 16.3f shows a portion of the final molecule, with a sulfonic acid group, which provides the exchange site when attached to a styrene molecule. The H attached to the SO_3^- is really H^+ and is the exchangeable ion.

Figure 16.4 illustrates the more complete cross-linked polystyrene showing SO_3^- functional groups attached to the styrene molecules and with DVB linkages between the styrenes that have no functional groups. As seen, the system may be extended indefinitely. With each DVB cross-linkage, many additional styrene/functional groups may be added, to which yet other DVD groups may be added. As indicated, the DVB has the role of cross-linking the styrene molecules, which is the basis for building a complex macromolecule of high molecular weight.

A variety of functional groups may be attached to the styrene molecule. Sulfonate is the most common for strong-acid ion-exchangers and is used for the illustrations in both Figures 16.3 and 16.4, with H^+ as the counterion.

16.1.4.4.3 Functional Ionic Groups

After the desired cross-linked matrix is formed, the remaining task is to add the desired fixed ionic groups. This may be done by substitution of these groups in the benzene ring during or after polymerization, or by starting with monomers that carry ionic groups. The ionic groups are attached only to the "para" position on the benzene ring of the styrene and not to the DVB. The point is illustrated in Figure 16.3f. The ion-exchange behavior of a resin is determined by the particular species of fixed ionic group attached to the benzene ring. Table 16.3 shows ionic groups commonly used in commercial ion-exchange resins. The number of groups per 100 benzene

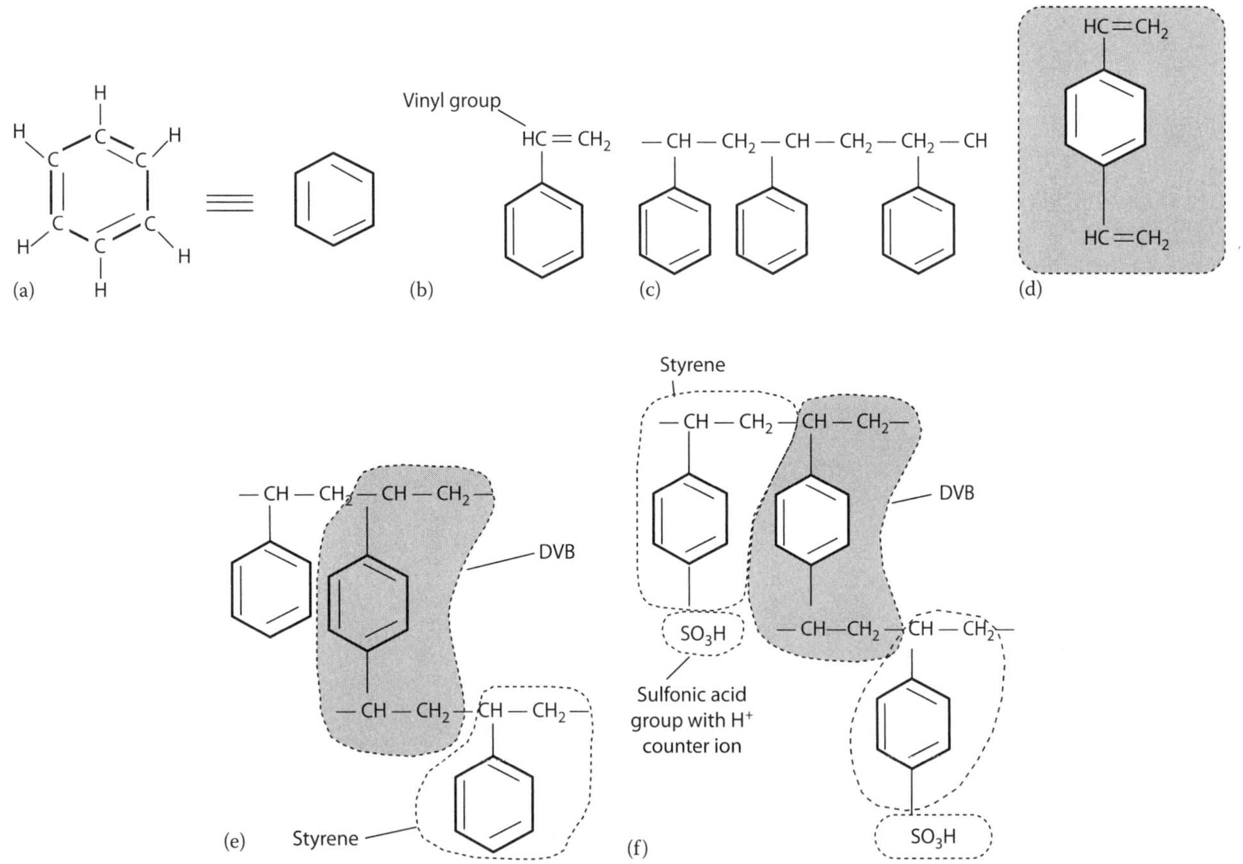

FIGURE 16.3 Chemical components of an ion-exchange resin and final product. (a) Benzene ring; MW = 78; (b) styrene monomer (vinyl benzene); MW = 106; (c) styrene polymer (polystyrene); MW = n (106); (d) di-vinyl benzene (DVB); MW = 134; (e) cross-linked polystyrene; (f) cross-linked polystyrene with sulfonic acid groups.

rings determines ion-exchange capacity. The maximum is to have one functional group attached to each benzene ring, not occupied by a DVD, giving in turn maximum ion-exchange capacity. According to Kunin (1983, p. 47), polymer chemists have the capability to introduce almost any functional group (ion-active group) onto many cross-linked polymer structures.

16.1.4.4.4 Strong-Acid/Strong-Base Groups

Ion-exchange resins are classified as "strong-acid" and "weak-acid" for the cation-exchangers, and "strong-base" and "weak-base" for the anion-exchangers. Table 16.3 summarizes some of the species of fixed ionic groups associated with each category. The dissociated behavior of each of the groups in Table 16.3 can be explained in terms of their respective pK's, or affinities for H^+ or OH^-. For example, strong-acid groups have little affinity for H^+ and remain ionized even at low pH. Thus, for strong-acid groups the ion-exchange site is open and available over a wide range of pH. By the same token, strong-base groups, such as quartenary amine, e.g., $-N^+-$, and $-N^+(CH_3)_3$, remain ionized even at high pH.

16.1.4.4.5 Weak-Acid/Weak-Base Groups

By contrast, weak-acid groups, such as –COO–, are ionized only at high pH. At low pH they combine with H^+, forming undissociated –COOH–, and thus do not have fixed charges. In similar fashion, weak-base groups such as $-NH_3^+$ lose a proton, forming uncharged $-NH_2$, when the pH is high. In other words, the ion-exchange capacities of weak-acid and weak-base ion-exchangers are pH dependent. These, i.e., weak-acid groups or weak-base groups, have limited use in water treatment (Dow Chemical, 1964).

16.1.4.4.6 Chelating Resins

A chelating resin (e.g., Dowex A-1) contains chemical groupings similar to chelating compounds, e.g., EDTA, which are attached to a cross-linked matrix. These compounds tightly bond certain metal species, e.g., heavy metals, iron, copper, zinc (Dow, 1964, pp. 10, 33). After removal of the ion the resin must be disposed.

16.1.4.4.7 Commercial Designations

A number of companies provide synthetic resins having different properties. The Rohm and Haas Company (Philadelphia, Pennsylvania), for example, lists some 27 different ion-exchange resins; most are designated Amberlite, e.g., Amberlite IR-120 Plus®, which is a sulfonated polystyrene cation-exchange resin (16.50 mesh, density = 51 lb/ft^3 or 816 g/L in H^+ form, 0.45 fraction moisture, effective size = 0.50 mm, UC = 1.6, $0.35 \leq P \leq 0.40$, as given by Rohm and Haas, 1987, p. 27).

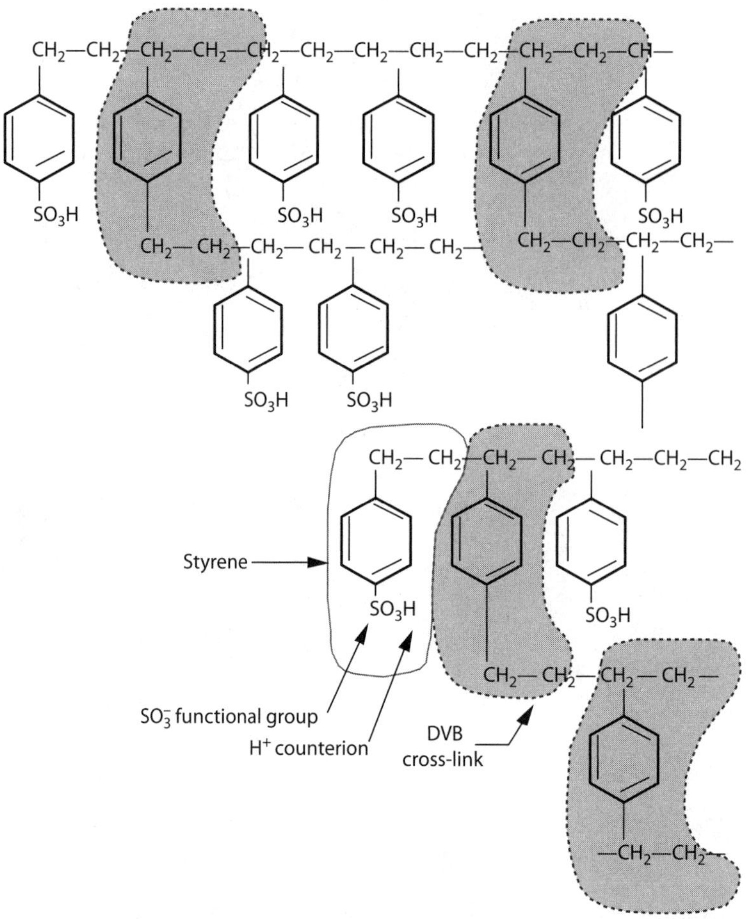

FIGURE 16.4 Macromolecular matrix of strong-acid ion-exchange resin showing polystyrene, DVB cross-linking, sulfonic acid functional groups, and H^+ counterions.

TABLE 16.3
Fixed Ionic Groups Used to Formulate Four Major Categories of Ion-Exchanger Resins

Cation-Exchangers		Anion-Exchangers	
Strong-Acid $4 \leq pH \leq 14$	**Weak-Acid** $6 \leq pH \leq 14$	**Strong-Base** $1 \leq pH \leq 2$	**Weak-Base** $0 \leq pH \leq 7$
SO_3^2	$-COO^-$–	$-\overset{\vert}{\underset{\vert}{N}}{}^+-$	$>NH_2^+$
		$-N^+(CH_3)_3$	$-NH_3^+$
	PO_3^{2-}		
	$-O^-$	$-\overset{\vert}{\underset{\vert}{P}}{}^+-$	$\equiv N^+$
	$-PO_2H^-$	$>S^+-$	$BR\text{-}N(CH_3)_3$
	AsO_3^{2-}		BR-N-OH
	$-SeO_3^-$		$-CH_2NH_3^+-$

Source: Helfferich, F., *Ion-Exchange*, McGraw-Hill, New York, 1962, pp. 29, 47.

Notes: (1) $-SO_3^-$ ion-exchangers include Amberlite IR-120; Dowex 50; Nalcite HCR; Permutit Q; Duolite C-20, C-25; Lewatit S-100 (Helfferich, 1962, p. 574). (2) Two most common groups for strong-base resins are $-\overset{\vert}{\underset{\vert}{N}}{}^+-$, and $-N^+(CH_3)_3$. (3) Strong-base Type I quaternary ammonium anion-exchangers include; Nalcite SBR; Duolite A-42; Dowex 1; Amberlite IRA.400; Permutit S-1 (Helfferich, 1964, p. 578). (4) Strong-base Type II quaternary ammonium anion-exchangers include Nalcite SAR; Duolite A-40; Dowex 2; Amberlite IRA.410; Permutit S-2; preferred for Cl^-, HCO_3^- (Helfferich, 1964, p. 579). (5) Maximum ion-exchange capacity for $-SO_3^-$ group attached to one polystyrene monomer with formula weight 184.2 is $\approx$5.43 meq/g; with 8% DVD cross-linking, CEC = 5.35 meq/g (Helfferich, 1964, p. 74). (6) $-COO^-$ $K = 10^{-5}$–10^{-7}; selectivity for Ca^{2+}, Mg^{2+}; AsO_3^{2-} has affinity for V^+.

The nomenclature used by Dow Chemical (1964, p. 4) is:

Type	DVD %	Mesh Size	Ionic Form
Dowex 50	X8	20–50	Na

The type is Dowex 50® which has 8% DVD cross-linking (meaning 8 molecules of DVD combine with 92 molecules of styrene to form the resin material) and mesh size 20–50 (meaning most of the resin passes mesh 20 and is retained by mesh 50). The ion-active group is sulfonic acid, i.e., $-SO_3^-$, which is strongly acidic. The ionic form of the particular resin in the example is Na^+, meaning the strongly acidic $-SO_3^-$, groups are combined with Na^+ counterions. The Na^+ ions are the exchangeable ions.

An example of a strong-base anion-exchanger is Dowex Marathon A, Type I, styrene-DVB gel, quaternary amine, with total exchange capacity 1.3 eq/L (28.4 kg/ft^3 as $CaCO_3$); water content is 50%–60%, UC = 1.1, mean particle size = 575 ± 50 μm, total swelling ($Cl^- \rightarrow OH^-$) = 20%, particle density = 1.08, shipping weight = 670 g/L (42 lb/ft^3); pressure drop is 0.5 bar/m bed depth at HLR = 36 m/h (14.8 gpm/ft^2), $T = 25°C$ (from http://www.dow.com/ liquidseps/pc/ jump/dowex/ dm_a.htm).

16.1.4.4.8 Mesh Sizes

Size equivalents of the U.S. Standard mesh sizes are (Dow Chemical, 1964, p. 15):

Mesh Range	Screen Analysis	Particle Diameter (mm)
16–20	Wet	1.168–0.84
20–50	Wet	0.84–0.297
50–100	Dry	0.297–0.149
100–200	Dry	0.149–0.074
200–400	Dry	0.074–0.038

As seen, the larger sizes are sieved wet, while the smaller sizes are sieved dry.

16.1.4.4.9 Moisture

The fraction moisture, fM, is defined, $fM = [\rho(\text{moist}) - \rho(\text{dry})]/\rho(\text{dry})$; from this the relation between moist density, ρ_M, and dry density, ρ_D, is, $\rho_M = \rho_D(1 + fM)$.

16.1.4.4.10 Density: Moist and Dry Particle

The density of resins are 800–848 g/L (50–53 lb/ft^3) as shipped, with moisture content reported as 45% in the Na^+ form and 53% in the H^+ form (ranges for Dowex 50 and Amberlite IR-120). For reference, a laboratory determination (Abdelrasool, 1992) of seven samples of Dow AGW50-X8 cation-exchange resin, H^+ form for of moist density (as packed) and dry density after oven drying were $\rho(\text{moist}) \approx 0.82$ (g/mL) and $\rho(\text{dry}) \approx 0.52$ (g/mL), with $fM \approx 0.58$.

16.1.4.4.11 Density: Bulk and Particle

The relation between the dry bulk density and the dry particle density is

$$\rho_b = (1 - P)\rho_s \tag{16.2}$$

in which
- ρ_b is the bulk density of ion-exchanger in packed-bed (kg dry resin/m^3 bulk volume of resin in packed-bed)
- P is the porosity of packed-bed (volume of voids/volume of packed-bed)
- ρ_s is the density of dry solid ion-exchanger (kg dry resin/m^3 solid resin)

16.1.4.4.12 Swelling

A resin swells as it takes up moisture. The degree of swelling depends upon the particular resin and the degree of cross-linkage, i.e., percent DVD. As an example, Dowex 50 H^+ resin has a ratio of swollen bead diameter/dry bead diameter ≈2.0 at 2% DVD to 1.4 at 10% DVD.

16.1.4.5 Aluminas

The removal of fluoride ion is also associated with alumina, which is selective for other ions, e.g., metals, sulfates, nitrates. The fluoride-removal cycle involves (Fleming, 1986, p. 162): (1) filtration with influent adjustment of pH to pH ≈ 5.5; (2) backwash with treated water; (3) regeneration with 1% NaOH; (4) rinse with treated water; (5) neutralization with 0.5% H_2SO_4. Commercial names for aluminas include: Alcoa W-100 with selectivity for certain cations, and exchange capacity of 0.75 meq/g; Alcoa F-1 (replaced by Alocoa DD-1), used for the removal of arsenic (trivalent and pentavalent). Alcoa F-1 has been used for removals of orthophosphates, e.g., $H_2PO_4^-$, HPO_4^{2-}, PO_4^{3-}; pyrophosphates, e.g., $P_2O_7^{4-}$; tripolyphosphate, e.g., $P_3O_{10}^{3-}$; and hexametaphosphate, e.g., $(PO_3)_6^{3-}$.

As to physical characteristics, the aluminas resist oxidation, do not swell, and have high crushing strength, abrasion resistance, and low chemical solubility. Other characteristics, from a product data sheet, gave: Al_2O_3 = 92.2%; moisture = 5% maximum; bulk density 674.31 kg/m^3 (42 lb/ft^3); surface area = 290 m^2/g (Whittaker, Clark, Daniels, South Plainfield, NJ, Activated Alumina #4204, CAS#1344-28-1). Alcoa® F-1 activated alumina was reported as being available in several mesh sizes, e.g., 1/4 × 8, 8 × 14, 14 × 28, 28 × 48, 48 × 100, −100, −325, and was packed as sling bags, paper bags, and drums (Alcoa, 1989).

16.1.4.5.1 Activated Alumina

Activated alumina is formed by dehydrating amorphous aluminum hydroxide, i.e., $Al(OH)_3$. It is a mixture of amorphous and crystalline aluminum oxide of approximate composition, Al_2O_3. The adsorption process involves surface complexation and exchange of hydroxide ions for the contaminants. The selectivity sequence is

$$OH^- > H_2AsO_4^- > F^- > SO_4^{2-} > HCO_3^- > Cl^- > NO_3^- \text{ (Chwirka et al., 2000).}$$

Typically, the granules for packed-beds are 28–48 mesh (0.3–0.6 mm). The site charges are related to pH as follows (Carlson and Thomson, 2001):

At $pH < pH_{zpc}$, the sites are charged as OH_2^+
At $pH = pH_{ZPC}$, the sites are charged as OH
At $pH > pH_{ZPC}$, the sites are charged as O^-
For reference, $pH_{ZPC} = 8.2–9.1$, where pH_{ZPC} is the pH at zero-point charge

Regarding the background on the development of practice, Savinelli and Black (1958, p. 34) reviewed the use of activated alumina for fluoride removal in a full-scale plant constructed in 1952 at Bartlett, Texas using a 14.15 m^3 (500 ft^3) bed. They cited C.S. Boruff as the person who initiated the development of activated alumina, who in 1934 described regeneration with NaOH, followed by neutralization with HCl. A 1936 patent by H.V. Churchill mentioned specifically the use of activated alumina (as contrasted with aluminum oxide). The alumina was "activated" by heating it to 400°C–500°C in the presence of alkali metal ions; the heating was the only distinction seen in the literature between activated alumina and alumina.

The steps in fluoride removal (Rubel and Woosley, 1978, p. 45; Rubel, 1984) were summarized as: (1) the treatment mode should be at $5.0 < pH < 6.0$, adjusted by sulfuric acid addition to the raw water; (2) backwash at HLR (backwash) ≈ 18 m/h (7.4 gpm/ft^2) for about 10 min to remove small particulates and break up any packing of the bed; (3) regeneration which has several steps and consists of up-flow rinse with raw water at 12.2 m/h (5.0 gpm/ft^2), drainage to top of bed, down-flow with 1% by weight of NaOH at 6.12 m/h (2.5 gpm/ft^2). Each step is about 30 min duration. The foregoing was based on a bed depth of 1.50 m (5.0 ft). A fourth step is neutralization with H_2SO_4. For the treated water, adjustment of pH to pH ≈ 7.5 should be done by blending or addition of NaOH. Meenakshi and Maheshwari (2006) reported fluoride concentrations >10 mg/L at a survey of some 17 locations in India. Removal by activated alumina were on the order of 0.75 fraction, with as high as 0.90 fraction; inflow concentrations were 4–9 mg/L and required a pH range, $5 < pH \leq 6$. At $pH > 7$ silicate and hydroxide compete strongly for exchange sites and at $pH < 5$ the activated alumina tends to dissolve into solution. The regeneration interval was every 4–6 months.

16.2 ION-EXCHANGE THEORY

Ion-exchange theory involves ion affinity, capacity, equilibria, kinetics, and hydraulics. The associated principles are similar to adsorption.

16.2.1 Capacity of Media

For a displacing ion, the fraction of sites occupied by that ion depends upon the concentration in the solution from which the ion is being removed, which may be seen in terms of an isotherm. For a given influent concentration of counterion, e.g., "A^+," C_o (A^+), to a packed-bed, the quantity adsorbed is $\overline{X}^*$ ("A^+").

16.2.1.1 Expressions of Capacity

Helfferich (1962) lists seven expressions of ion-exchanger capacity and mentions that confusion often is the result. The "mass capacity" is recommended, and is defined as "milliequivalents of exchange capacity per gram of dry resin," or meq/g dry resin. A holdover expression from the early days of softening practice is "kilograins as $CaCO_3/ft^3$." Conversions between various units of capacity are given in Appendix 16.A.

16.2.1.2 Upper Limit of Capacity

The upper limit of ion-exchanger "capacity" is the number of sites per unit mass of dry media, designated as X_M, expressed usually in milliequivalents of sites per gram of dry media. For a zeolite, X_M must be determined empirically. For resin, X_M may be calculated as illustrated in Example 16.1.

Example 16.1 Theoretical Exchange Capacity Calculation (Adapted from Helfferich, 1964, p. 74)

Given
Assume the resin is sulfonated polystyrene with 8% DVB.

Required
Calculate the theoretical exchange capacity in meq/g dry resin.

Solution
The steps in the calculation are enumerated as follows:

1. The sulfonated styrene consists of the monomers	2. The DVB consists of the monomers
–CH–CH$_2$– (benzene ring) $SO_3 \cdot H^+$	–CH–CH$_2$– (benzene ring) –CH–CH$_2$–
For each monomer, MW ($C_8H_8O_3S$) = 186.2	For each monomer, MW ($C_{10}H_{10}$) = 130

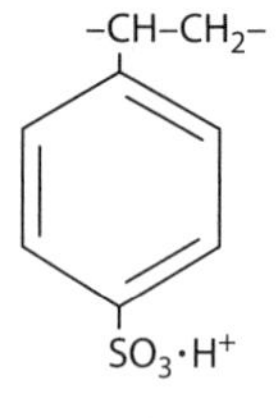

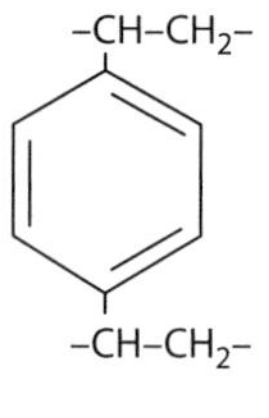

3. With 8% DVB, there are 8 DVB units for every 92 sulfonated styrene units, giving a total MW = $92 \cdot 186.2 + 8 \cdot 130 = 17{,}130 + 1{,}040 = 18{,}170$ for 100 units.
4. For the 92 units of sulfonated styrene there are also 92 gram-equivalents of H^+, which are the same as 92 gram-equivalents of exchange capacity. In other words a sulfonic acid group may occupy a place on the styrene that does not have a DVD monomer attached.

5. The theoretical exchange capacity of the 100 units is 92 equivalents (or 92,000 meq).
6. The theoretical exchange capacity per gram of dry resin is: 92,000 meq/18,170 g = 5.12 meq exchange capacity/g dry resin. In other words, the molecular weight of the 92 sulfonated styrene units plus 8 DVB units is 18,170 g.

Discussion

As can be seen by the calculation, the higher the DVB fraction the lower the ion-exchange capacity. This is because the fraction of sulfonated styrene decreases as DVB increases. According to Helfferich, the above method of calculation agrees well with experimental determinations. The weight capacities of other resins are similar in order of magnitude. Capacities will be in the range of 2.5–4.0 meq/g, however, for polystyrene anion-exchangers. For anion-exchangers, not every benzene ring carries one ionic group, as it may with cation-exchangers. The exchange capacity for Dowex 50 resin in the H^+ form was given as 5.0 meq/g dry resin.

16.2.2 Equilibria

The term equilibria refers to an ion-exchange reaction, such as Equation 16.1, and to the final concentrations of all species both in solution and attached to ion-exchange sites. In other words, an ion-exchange reaction is essentially the same as any chemical reaction.

16.2.2.1 General Reaction and Equilibrium Equations

A general ion-exchange reaction is depicted by Equation 16.1. The equilibrium relation is given by (16.3), which assumes that the ions are monovalent and that the activity coefficients equal 1.0,

$$X\cdot \mathrm{B} + \mathrm{A}^+ \leftrightarrow X\cdot \mathrm{A} + \mathrm{B}^+ \tag{16.1}$$

$$K = \frac{[X\cdot \mathrm{A}][\mathrm{B}^+]}{[X\cdot \mathrm{B}][\mathrm{A}^+]} \tag{16.3}$$

in which

K is the equilibrium constant (dimensionless)
$[X\cdot \mathrm{A}]$ is the concentration of ion-exchanger sites holding A^+ ions (meq/g ion-exchanger)
$[X\cdot \mathrm{B}]$ is the concentration of ion-exchanger sites holding B^+ ions (meq/g ion-exchanger)
$[\mathrm{A}^+]$ is the concentration of A^+ ions (mol/L solution)
$[\mathrm{B}^+]$ is the concentration of B^+ ions (mol/L solution)

16.2.2.2 Isotherm Expression of Equilibrium

A more convenient method of addressing the question of ion-exchange equilibrium is in terms of an isotherm, as in adsorption. To obtain an isotherm relation from Equation 16.3, substitute for $[X\cdot \mathrm{B}]$ from the materials balance relation,

$$X_\mathrm{M} = [X\cdot \mathrm{A}] + [X\cdot \mathrm{B}] \tag{16.4}$$

in which X_M is the capacity of ion-exchanger (meq ions/g ion-exchanger) and rearrange to obtain Equation 16.5.

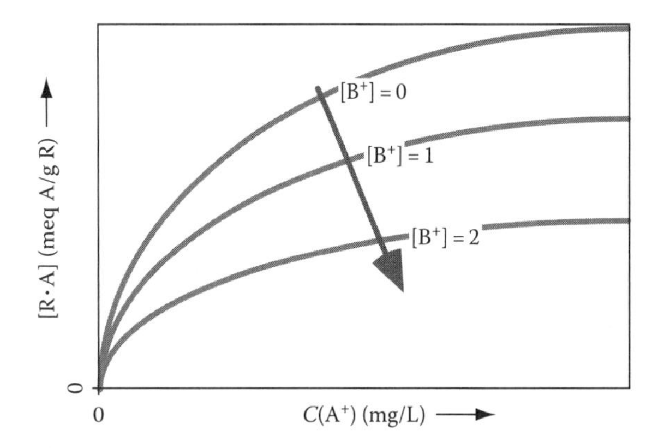

FIGURE 16.5 Plot of Equation 16.6 isotherms showing effect of different concentrations of B^+, a competing cation.

$$\frac{[X\cdot \mathrm{A}]}{X_\mathrm{M}} = \frac{K\cdot[\mathrm{A}^+]}{[\mathrm{B}^+] + K\cdot[\mathrm{A}^+]} \tag{16.5}$$

Now dividing both numerator and denominator of the right side by $[\mathrm{B}^+]$, and let $\alpha_\mathrm{B} = K/[\mathrm{B}^+]$, to give

$$\frac{[X\cdot \mathrm{A}]}{X_\mathrm{M}} = \frac{\alpha_\mathrm{B}\cdot[\mathrm{A}^+]}{1 + \alpha_\mathrm{B}\cdot[\mathrm{A}^+]} \tag{16.6}$$

Equation 16.6 is the same form as the Langmuir isotherm relation. Figure 16.5 is a plot of Equation 16.6 for different levels of $[\mathrm{B}^+]$. The graphical depiction of Figure 16.5 is useful first because the concept of equilibria is displayed in terms easy to understand. Second, it illustrates that ion-exchange for this special case and adsorption, have a common form of equilibrium expression.

16.2.2.3 Selectivity of Counterions

Factors that influence selectivity of one ion in preference to another include: counterion valence, ionic solvation and swelling pressure, sieve action, ion-pair bonding, electrostatic attraction, London interactions, complex formation in solution, and precipitate formation (Helfferich, 1962). In general, the higher the valence of a counterion, the stronger is the attraction; the smaller the ion the stronger the attraction. The order of ion selectivity based upon electrostatic attraction is, for cations and anions, respectively, is given below:

Cations

$$\mathrm{Fe}^{3+} > \mathrm{Al}^{3+} > \mathrm{Pb}^{2+} > \mathrm{Ba}^{2+} > \mathrm{Sr}^{2+} > \mathrm{Cd}^{2+} > \mathrm{Zn}^{2+} > \mathrm{Cu}^{2+} > \mathrm{Fe}^{2+} > \mathrm{Mn}^{2+} > \mathrm{Ca}^{2+} > \mathrm{Mg}^{2+} > \mathrm{K}^+ > \mathrm{NH_4}^+ > \mathrm{Na}^+ > \mathrm{H}^+ > \mathrm{Li}^+$$

Anions

$$\mathrm{CrO}^{2-} > \mathrm{SO_4}^{2-} > \mathrm{SO_3}^{2-} > \mathrm{HPO_4}^{3-} > \mathrm{CNS}^- > \mathrm{CNO}^- > \mathrm{NO_3}^- > \mathrm{NO_2}^- > \mathrm{Br}^- > \mathrm{Cl}^- > \mathrm{CN}^- > \mathrm{HCO_3}^- > \mathrm{HSiO_3}^- > \mathrm{OH}^- > \mathrm{F}^-$$

16.2.3 Kinetics

The theory of ion-exchange kinetics is virtually the same as the theory of adsorption kinetics. The three basic steps in each case are: (1) film diffusion, (2) particle diffusion, and (3) attachment. For adsorption, the substances of interest are molecules or small particles, and for ion-exchange they are ions. A difference is that for the ion-exchanger, mass transport of ions in one direction, e.g., out of the particle, must be balanced by mass transport in the other direction, e.g., into the particle, which preserves electroneutrality.

16.2.3.1 Rate-Determining Step

As with adsorption, the rate-determining step is either film diffusion or particle diffusion. Figure 16.6 illustrates successive $C(r)t$ curves for the two cases, respectively. In Figure 16.6a the rate of particle diffusion is high, i.e., with no substantial concentration gradient within the particle, causing film diffusion to be rate controlling. Such a condition occurs when the particle structure is open as with low degree of cross-linking and small particle size. In Figure 16.6b particle diffusion is rate controlling. Here film diffusion is faster and the needed film concentration gradient is slight to achieve the needed flux into the particle (to equal the particle flux).

16.2.3.2 Fick's First Law

Whether the diffusion of an ion species, say "A," is rate limiting in the liquid phase or in the solid phase, the random motion of ions and molecules causes a flux which is proportional to the concentration gradient (see also Sections 4.2.2.6). This is described by Fick's First Law, i.e.,

$$j_A = -D_A \text{ grad } C_A \tag{16.7}$$

in which

- j_A is the flux density of species A (mol A/cm²/s)
- D_A is the diffusion coefficient (cm²/s)
- C_A is the concentration of A in (mol A/cm³)
- grad is the mathematical term for "gradient," i.e., concentration gradient in the case at hand

The application of Equation 16.7 is difficult because usually, D_A is not constant; it varies with temperature and with concentration. Then for the solid phase, its magnitude is unique for a given ion-exchanger. Thus, as stated by Helfferich, Equation 16.7 is essentially a definition of D_A.

16.2.3.2.1 Particle Diffusion

Equation 16.7 applies to particle diffusion as noted. For an ion species A^+, within an ion-exchanger, it has the form,

$$\bar{j}_A = -\bar{D}_A \text{ grad } \bar{C}_A \tag{16.8}$$

in which

- $\bar{j}_A$ = flux density of ion species A^+ (mol A^+/cm²/s)
- $\bar{D}_A$ = diffusion coefficient within ion-exchanger (cm²/s)
- $\bar{C}_A$ = concentration of A^+ in ion-exchanger (mol A^+/cm³)

The bars refer to the interior of the ion-exchanger.

To evaluate the changes within the ion-exchanger as a function of time, Fick's second law is applicable, i.e.,

$$\frac{\partial \bar{C}_A}{\partial t} = -\text{div } \bar{j}_A \tag{16.9}$$

div = divergence

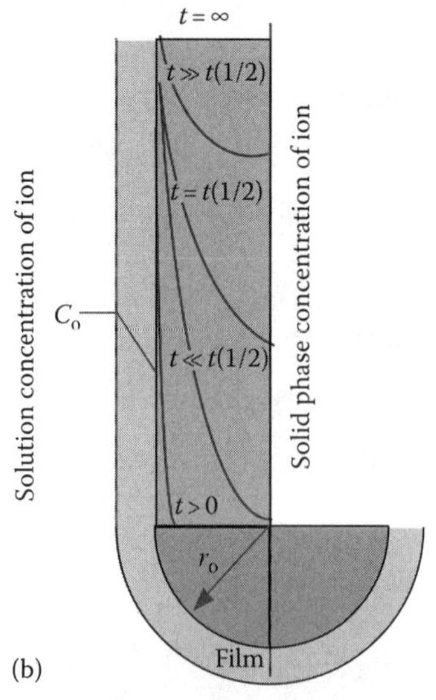

FIGURE 16.6 Concentration profiles, $C(r)t$, of diffusing ions illustrating rate-controlling mechanisms. (a) Film diffusion rate controlling; (b) particle diffusion rate controlling. (Adapted from Helfferich, F., *Ion-Exchange*, McGraw-Hill, New York, 1962, p. 254.)

Fick's second law is a materials balance relationship. Assume the ion-exchanger particle has spherical geometry and $\overline{D}_A$ is constant, then combining (16.8) and (16.9) gives

$$\frac{\partial \overline{C}_A}{\partial t} = \bar{D}_A \left[\frac{\partial \overline{C}_A}{\partial r^2} + \frac{2\partial \overline{C}_A}{r \partial r} \right] \tag{16.10}$$

16.2.3.2.2 Film Diffusion Control

The flux across the film around the ion-exchanger particle is

$$j_A = D_A \frac{\Delta C_A}{\delta} \tag{16.11}$$

in which

D_A is the diffusion coefficient of A^+ in the liquid phase (cm^2/s)

ΔC_A is the concentration difference across film, i.e., from the bulk of solution to the ion-exchanger external surface (mg A^+/mL)

δ is the film thickness (cm)

16.2.3.2.3 Diffusion Coefficients

The diffusion coefficient, D_A, of a species A^+ within the ion-exchanger depends on the size, valence, and chemical nature of species A^+, the degree of swelling, mesh width, charge density, chemical nature of the solid phase matrix, composition of the pore liquid, and temperatures. Figure 16.7 illustrates the influence of three of the above factors, e.g., degree of cross-linking, species, and temperature, on D_A, the self-diffusion coefficient. The plots illustrate the effects of cross-linking, ion valence, and temperature on D_A and indicate the approximate magnitudes of $\overline{D}_A$.

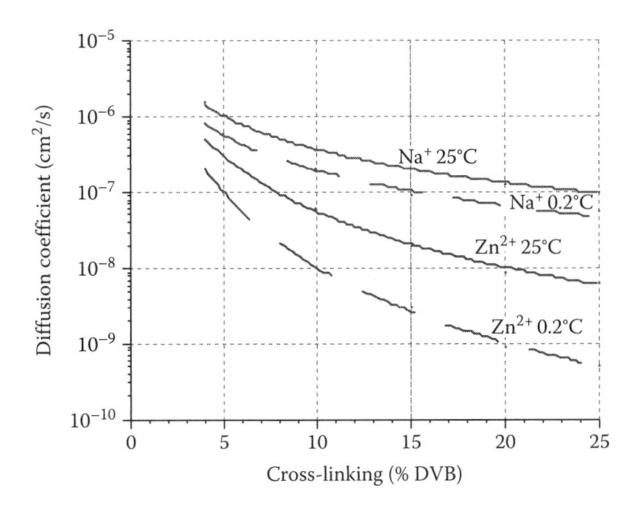

FIGURE 16.7 Influence of three factors on self-diffusion coefficient, $\overline{D}_A$, in particle: (1) degree of cross-linking, (2) species, and (3) temperature. Resin is sulfonated styrene cation-exchanger. (Adapted from Helfferich, F., *Ion-Exchange*, McGraw-Hill, New York, 1962, p. 305.)

16.3 DESIGN

Design involves not only selection of an ion-exchanger, but sizing of the support components with suitable materials. Much of this information is available in manufacturers' design manuals in hard copy (e.g., Dow Chemical, 1964; Rohm and Haas, 1987) or in Adobe Acrobat® format for a Web site (e.g., http://www.purolite.com/Re1Id/33637/ISvars/default/Home.htm, 1999).

16.3.1 Selection of Ion-Exchangers

The particular ion-exchanger used depends upon the purpose (e.g., softening, demineralization, ammonia removal from wastewaters, nitrate removal, boron removal, heavy-metal removal), properties desired (e.g., high exchange capacity, minimal swelling, no attrition), and cost. In addition, site specific factors may be overriding (e.g., availability of natural zeolites, cost of re-generant chemicals, disposal issues).

16.3.1.1 Resins

For most water treatment applications, a polystyrene with DVB cross-linking, strong-acid cation-exchange resin is recommended, having SO_3^- as the functional group. For the removal of anions, as in demineralization, a strong-base anion-exchanger having a quaternary amine functional group is commonly used.

The suitability of the resin selected should be confirmed by laboratory tests and, preferably, by pilot plant testing. The laboratory testing can provide data on fundamental parameters such as dry density, size, ion-exchange capacity, selectivity, etc. Pilot plant testing can determine the concentration profiles, i.e., the shapes of the wave fronts for various ions, run time before breakthrough, hydraulic loading rates, bed porosity, backwash rate, quantity of regenerate needed, etc.

Manufacturers' Web sites and catalogs provide a wealth of data on their respective products. For example, the Rohm and Haas Engineering Manual (1987) has plots on pressure drop per unit of bed length versus HLR, bed expansion, shipping weight, effective size, UC, density, void volume, matrix type, functional group, recommended HLR, and exchange capacity. Similar data are available from other manufacturers with the most current data available from Web sites (http://www.rohmhaas.com/ and http://www.purolite.com/RelId/33637/ISvars/default/Home.htm).

16.3.1.2 Zeolites

Zeolites may be selected over resins mostly because their cost is lower, especially for natural zeolites. A disadvantage is lower exchange capacities with the capacities of natural zeolites being lower than synthetic zeolites. The selection also depends upon the purpose, e.g., for ammonia removal, in which case clinopthiolite has been favored. A particular pore size maybe desired, for which a certain zeolite may be favored. Zeolites can serve only for the exchange of cations.

16.3.1.3 Range of Ion-Exchangers and Properties

Examples of data that characterize ion-exchanger groups are given in Table 16.4, e.g., strong-acid, strong-base, weak-base

TABLE 16.4
Examples of Ion-Exchanger Groups and Associated Properties

Name	Group	Cat[c]	DVB (%)	Capacity[d] meq/g	Capacity[d] meq/mL	d_{10} (mm)	UC	Size (Mesh)	ρ (g/L)	*P*	Moisture (%)
Polystyrene	$-SO_3^-$	SA	8	5.35							0
Amberlite IR-120[a]	$-SO_3^{-a}$	SA	8	5.0	1.9 wet	0.50	1.6	16–50	816	0.40	44–48
Dowex 50W[b]		SA	8	5.0	1.7 wet			20–50	787	0.40	53
Amberlite IRA-400	$-N(alkyl)_3^+$	SB	8	2.6	1.2						42–48
Dowex Marathon A[b]	Quaternary amine	SB	Gel	2.4		0.58	1.1		670		50–60
Amberlite IR-45[a]	Amino groups	WB		5	2						37–45
Amberlite IRA 743[a]	Methylglucamine				4–7 mg B[e]	0.7	1.6		700		
Clinoptilolite				0.15[f]							
Glauconite				0.23[g]							
Activated alumina				0.0046	mg F-						

Notes: Density is g moist resin/L packed-bed. Moisture is percent on dry weight basis. All resins listed are polystyrene matrix. All resins listed are in bead form.

[a] Rohm and Haas (1987, p. 27) Philadelphia, PA.

[b] Dow Chemical, Midland, MI (1964, p. 74); Marathon A from www.dow.com/liquidseps/pc/jump/dowex/

[c] Categories are: SA, strong acid; WA, weak acid; SB, strong base; WB, weak base.

[d] Capacity: (1) meq/g dry resin; (2) meq/mL packed-bed.

[e] Boron selective; capacity is as mg B/mL packed-bed.

[f] Marshall (1964, p. 120).

[g] Babbitt and Doland (1949, p. 513).

(an example of a weak-acid type is missing), chelating, zeolite, and activated alumina. Available data are given for two zeolites and activated alumina. The purpose of the table is to provide an idea of the variety of ion-exchangers and values of characteristic parameters.

16.3.2 System Design

An ion-exchange system has three subsystems: (1) pretreatment, (2) reaction, and (3) regeneration. Figure 16.8 illustrates, showing the components of each subsystem.

16.3.2.1 Pretreatment

The ion-exchange bed should not function as a filter, or as a biological reactor. Thus suspended matter and dissolved organics, if present, should be removed prior to ion-exchange. Ion-exchange is one of the final steps in any treatment scheme.

16.3.2.2 Reactor Cycle

A basic ion-exchange reactor system comprises three reactor units, of which two are operated simultaneously while the third is being recharged or is on standby. Figure 16.9 illustrates the operating scheme. With two reactors the second position reactor, i.e., "B," is switched to the first position, after "A" is taken out of service for recharge, a step which lets the "B" reactor approach 100% saturation. The operating sequence in Figure 16.9a through c is, respectively, A-B, B-C, and C-A.

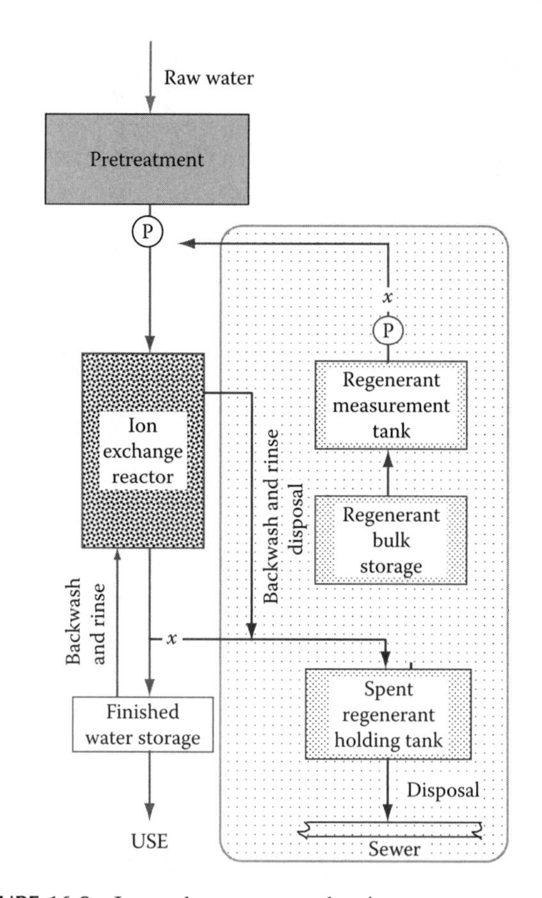

FIGURE 16.8 Ion-exchange system showing components.

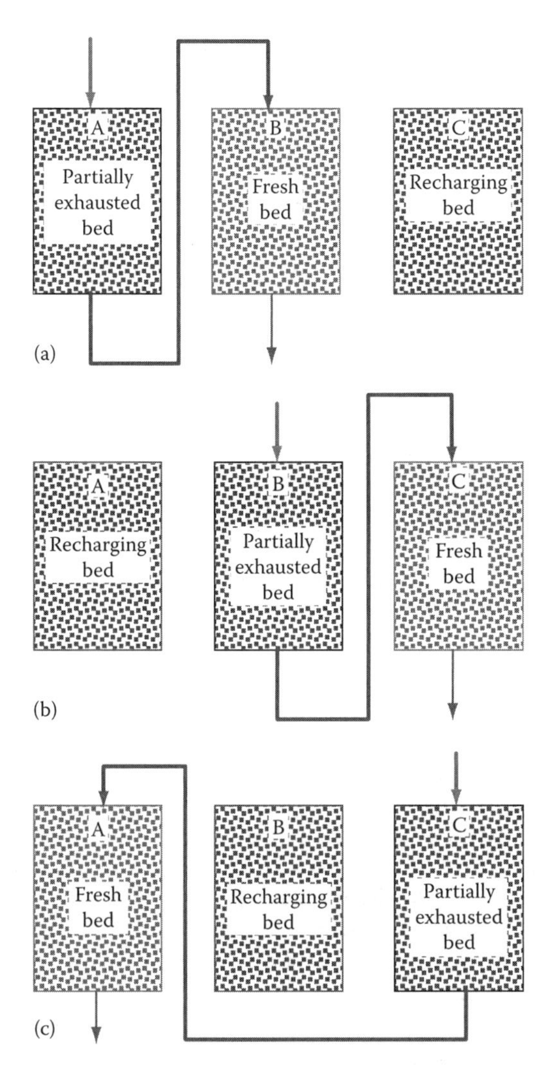

FIGURE 16.9 Operating scheme for a three-reactor system. (a) Start of cycle: "A" partially exhausted, "B" is fresh bed, "C" is recharging. (b) Second phase of cycle: "B" is partially exhausted, "C" is fresh bed, "A" is recharging. (c) Third phase of cycle: "C" is partially exhausted, "A" is fresh bed, "B" is recharging.

16.3.2.3 Regeneration

Figure 16.8 shows the basic components of the regeneration portion of the ion-exchange system. The components of the regeneration system include: (1) finished water storage for backwash and rinse, (2) pumps and piping for backwash and rinse, (3) holding tank for spent regenerate and backwash and rinse water, (4) sewer for disposal, (5) tank or tanks for bulk regenerate storage, and (6) regenerate measurement tanks. These units also have pumps and associated plumbing. The finished water storage should provide for about 10–15 min backwash at 50% bed expansion; the HLR depends upon the specific gravity and size of ion-exchanger material; as an example, for Amberlite IR-120 resin, 50% bed expansion is achieved at HLR $\approx$ 15 m/h (6 gpm/ft^2) at 22°C (Rohm and Haas, 1987, p. 27). Following regeneration additional the regenerate is drained and with flushing by finished water. The design should provide for drainage and rinse.

16.3.3 Reactor Design

The reactor is the heart of the ion-exchange system. As noted in Section 16.2, the theory is similar to any packed-bed, e.g., adsorption or filtration (see also Section 4.2.2.6).

The reactor design requires knowledge of ion-exchange capacity and an understanding of kinetics as related to wave fronts (Section 15.2.3), e.g., the steeper the wave front the faster the rate of uptake. Practically, the design requires determination of the hydraulic loading rate, and sizing of the reactor volume (Section 15.2.4). Both are described in the following.

16.3.3.1 Summary of Design Data

A hydraulic loading rate (HLR) criterion is the basis for determining the cross-sectional area. The reactor volume is based on the duration of run desired and the capacity of the ion-exchanger (in terms of the isotherm for the influent concentration of the target ion to be removed). The bed depth results from these calculations. Pressure loss is another design issue, and depends upon both the media and its depth.

16.3.3.1.1 Hydraulic Loading Rate

Examples of recommended HLR values (Rohm and Haas, 1987, pp. 27, 33) are HLR (operation) $\approx$ 2.5–5.0 m/h (1.0–2.0 gpm/ft^2). The hydraulic loading rate selected yields the total cross-sectional area for all columns in aggregate, i.e.,

$$A = \frac{Q}{\text{HLR}} \tag{16.12}$$

which "sizes" this part of the column. After determining the volume, the length of the packed-bed may be calculated, i.e., $L(\text{packed-bed}) \cdot A(\text{cross section}) = V(\text{packed-bed})$.

16.3.3.1.2 Reactor Volume

The calculation of the mass (or volume) of an ion-exchanger is the same as given for adsorption, i.e., as in Section 15.2.4. The corresponding mass balance equation (the mass flow of an ion, "A" into the reactor for a time duration, t, equals the mass of "A" that is retained by the ion-exchanger) is

$$Q \cdot C_o \cdot t = M(\text{dry ion} - \text{exchanger}) \cdot \bar{X}^*(\text{A}) \tag{16.13}$$

in which

Q is the flow of water to be treated, into a given reactor column (m^3/s)

C_o is the influent concentration of ions reactor column, to be removed (kg A^+/m^3 solution)

t is the desired duration of run before exhausting (s)

M (dry resin) is the mass of dry resin to be used in reactor bed (kg)

$\bar{X}^*(\text{A})$ is the isotherm capacity of ion-exchanger for species "A^+," to be removed at specified temperature and equilibrium concentration, C^*, in which $C^* = C_o$ and for a given concentration of a competing ion, e.g., "B^{+}" (eq A^+/kg dry ion-exchanger)

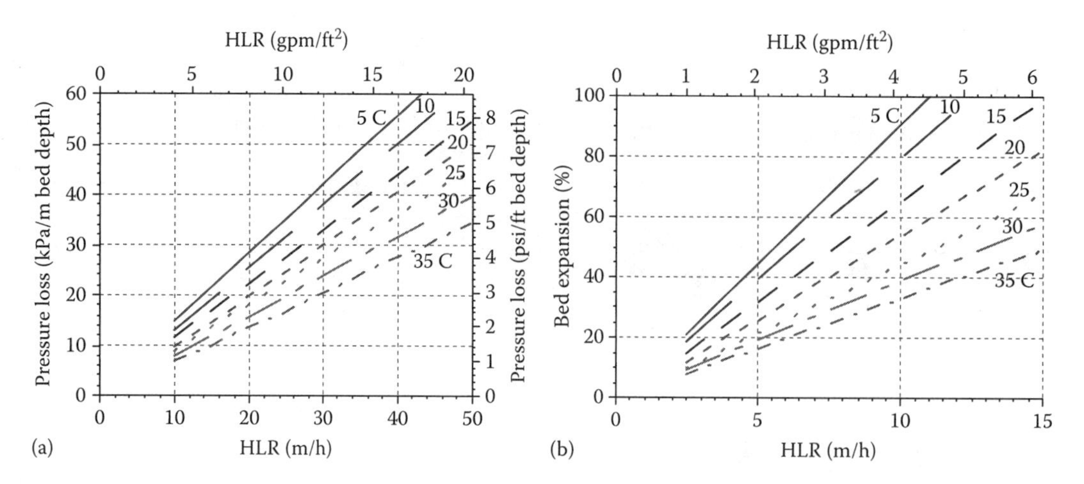

FIGURE 16.10 Example of hydraulic data (Purolite Engineering Manual, www.purolite.com/library, 1999) A-200 is strong-base polystyrene DVB gel Type II, quaternary ammonium, density = 680–715 g/L (42.5–44.5 lb/ft^3); size 0.3–1.2 mm; Cl^- form goes to OH^- with 15% swelling; SG = 1.08 wet; exchange capacity = 1.3 eq/kg resin. (From Purolite, *Purolite Engineering Manual—Puropack packed-Bed Technology*, The Purolite Company, Bala Cynwyd, PA, 1999, downloaded as Adobe Acrobat file from www.purolite.com/library. With permission.) (a) Pressure loss versus HLR and temperature (Puropack resins, p. 147); (b) Expansion of bed versus HLR and temperature (Puropack A-200. p. 159).

An alternative form of Equation 16.54 is in terms of volume, i.e.,

$$Q \cdot C_o \cdot t = V(\text{bulk}) \cdot \bar{X}^*(\text{bulk A}) \qquad (16.14)$$

in which

- V (bulk) is the bulk volume of moist resin used in reactor bed (m^3)
- $\bar{X}^*$(bulk A) is the volumetric isotherm capacity of ion-exchanger for species "A^+," to be removed at specified temperature and equilibrium concentration, C^*, in which $C^* = C_o$ and for a given concentration of a competing ion, e.g., "B^{+}" (eq A^+/m^3 bulk moist ion-exchanger)

Equation 16.55 may be more useful in practice since the capacity data are often given in terms of ion removed per unit of bulk volume. The determination of reactor volume requires a decision on how long it is desired that the reactor be operated before saturation. Note that $\bar{X}^*(\text{bulk A}) = \rho(\text{bulk}) \cdot \bar{X}^*(\text{A})$.

16.3.3.1.3 Bed Depth

Once the volume of the reactor is determined, based on Equation 16.14, i.e., V(bulk), the length of the reactor bed, L(reactor), may be calculated as follows:

$$L(\text{reactor}) = \frac{V(\text{bulk})}{A(\text{cross-section})} \qquad (16.15)$$

The results should be assessed in terms of what seems reasonable. For example, if the resulting volume is distributed as a short, wide reactor, the dimensions should be adjusted. If the reactor length is too long, resulting in a very high headloss, then the length should be reduced. A spreadsheet is a convenient means to explore alternatives.

16.3.3.1.4 Pressure Loss and Bed Expansion

In general, the pressure loss through an ion-exchanger bed is the same as through any packed-bed, i.e., conforming to Darcy's law (Appendix E.3). Figure 16.10a shows $\Delta P/\Delta L$ data versus HLR for Puropack® resin at different temperatures and Figure 16.10b shows bed expansion versus HLR. The relationships are specific for these particular ion-exchangers and are shown for illustration. For other ion-exchangers manufacturers' data or laboratory data should be obtained. The pressure loss may be estimated for a given bed depth and temperature from plots such as Figure 16.10a. The bed expansion may be specified, e.g., 50%; the associated HLR may then be determined from a relation such as Figure 16.10b.

16.3.3.2 Pilot Plant Studies

A pilot plant study may be important for a large installation. For smaller plants, e.g., serving a population of ≤2000, the plant may be sized based on criteria as recommended by manufacturers, e.g., using HLR ≤ 4 m/h (2 gpm/ft^2) and with volume based upon published data for exchange capacity and headloss from published data. As the size of the plant increases, however, a pilot plant study becomes more desirable in order to resolve uncertainties and to optimize design and operation.

The approach to a pilot plant study and the physical setup is similar to that for adsorption (Chapter 15). Questions may relate to issues of a particular water, e.g., the potential for fouling, length of run, concentration of regenerate chemicals, and HLR for regenerate flow.

16.4 OPERATION

The operation may be automated to a large extent; alarms may be set in the event that data are outside the ranges stipulated. The operator will need to pay attention to computer monitor screens, pressure gages that are located on the equipment and obtain samples periodically.

16.4.1 Operating Cycle

The operating cycle comprises production and regeneration. Regeneration has three steps: backwash, elution, and rinse. The run may be terminated based on effluent concentration, time of operation, or volume of water treated, or headloss.

16.4.1.1 Production

Figure 16.11 shows what happens during the production cycle of an ion-exchange bed in which Ca^{2+} is being removed. The Ca^{2+} wave front starts out as in (a), and moves through the bed as in (b); when it reaches the end as in (c) then "breakthrough" occurs. When the breakthrough concentration exceeds a specified criterion, C_b, then the bed is termed "exhausted" or "saturated"). The bed is then taken out of service and is regenerated by chemical elution, such as with Na^+. Alternatively, the nearly exhausted bed may be placed in the lead position in a two-bed reactor series in order to achieve complete exhaustion and, therefore, complete utilization of the bed.

16.4.1.2 Regeneration

The first step in regeneration is to backwash the resin to remove debris. While the water introduced into an ion-exchange bed should be free of particles, accumulations may occur. The backwash rate should be enough to expand the bed by about 30%–50%.

After the bed is cleaned of particles, it is regenerated by a reverse flow of concentrated regenerate chemical. A measure of the proportion of the regenerate chemical that is transferred from the concentrate to sites on the resin is called the efficiency of the regenerate, E (regenerate), and is defined, as the fraction of sites occupied by the regenerating counterion, e.g., Na^+. For softening, $45\% < E(\text{regenerate}) < 70\%$, applying 48.7–162 kg/m^3 (3–10 lb NaCl/ft^3) resin (Weber, 1972). The third step in regeneration is rinse. This is done with finished water and should be for enough bed volumes such that the effects of hydraulic dispersion are taken into account.

16.4.1.3 Disposal

How to dispose off spent regenerate water is a major concern for any ion-exchange system. Disposal to a municipal WWTP requires permission. Discharge to a stream requires a discharge permit. Ordinarily, groundwater should not be used as a sink. If the purpose of the ion-exchange is specific ion removal and if such ions are toxic or pose an environmental hazard then concentration and solidification may be recommended.

16.5 CASE STUDIES

A case study may be useful to illustrate the principles of ion-exchange and to indicate the variety of applications. The case study selected is but one of many situations in which ion-exchange may be applicable.

16.5.1 Nitrate Removal at Glendale, Arizona

Nitrate and arsenic are common in some groundwaters used for drinking water; the problem is prevalent particularly in the southwestern part of the United States, e.g., in New Mexico and Arizona. A mobile ion-exchange pilot plant has been operated for many years, i.e., since the late 1970s, by Professor Dennis Clifford, University of Houston, with sponsorship, in part, by EPA Cincinnati and has been used in many investigations relative to the applicability of ion-exchange for some of these situations (e.g., Clifford and Liu, 1993).

One such situation was at Glendale, Arizona, where the mobile pilot plant was used to examine the applicability of ion-exchange for nitrate removal (Clifford et al., 1985). The

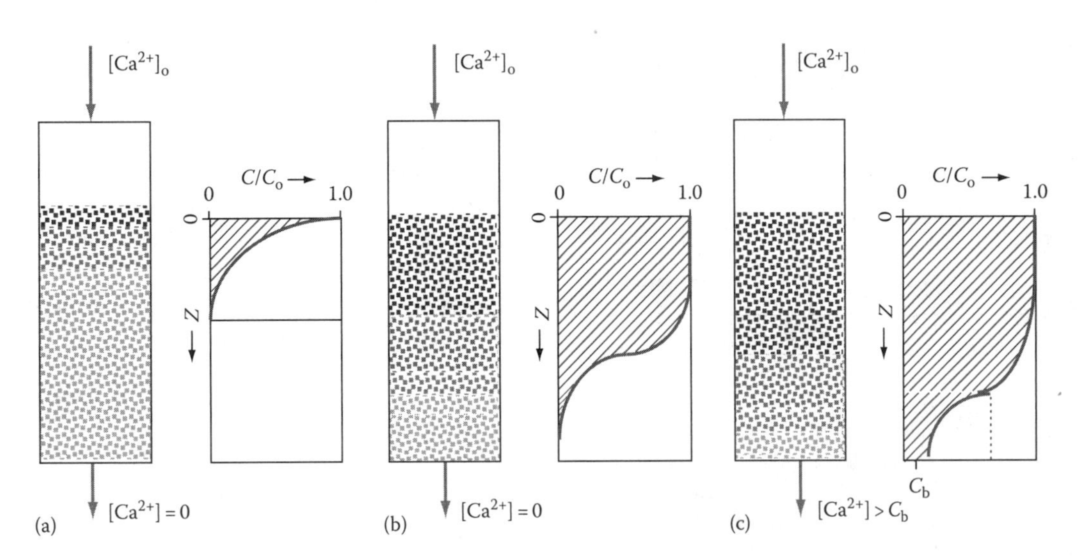

FIGURE 16.11 Ca^{2+} concentration profiles within ion-exchange bed at phases of cycle. (a) Beginning of cycle; (b) partial exhaustion; (c) exhaustion.

ion-exchange runs were done using a 203 mm (8 in.) diameter column with bed depth 750 mm (29.5 in.) containing 24.3 L (0.86 ft^3) of Cl^- charged anion-exchange resin. Neither the particular strong-base resin used, nor the empty bed contact time (EBCT) had significant effect on the bed volumes (BV)-to-nitrate breakthrough, which typically was 400 BV. For example, HLR $\approx$ 4.8 m/h (1.95 (gpm/ft^2) and HLR $\approx$ 11.1 m/h (4.54 (gpm/ft^2) with EBCT = 9.44, 4.05 min, respectively, both had about 400 BVs-to-nitrate breakthrough.

The raw water at Glendale had concentrations: NO_3^-–N = 19–25 mg/L; Cl^- = 123 mg/L; SO_4^{2-} = 42 mg/L; HCO_3^- = 124 mg/L; TDS = 532 mg/L; SiO_2 = 23 mg/L; pH = 8.0. The treated water was blended with raw water to produce a product water with NO_3^-–N < 10 mg/L. For dilute feed waters, i.e., <10 N mg NO_3^-–N/L, the resins showed the preference order: $SO_4^{2-} > NO_3^- > Cl^- > HCO_3^-$. This preference order results in the breakthrough order of appearance: $HCO_3^- > Cl^- > NO_3^- SO_4^{2-}$ During regeneration by NaCl, SO_4^{2-} appeared first during the elution, then NO_3^-. For more concentrated feed waters, i.e., >10 N mg NO_3^-–N/L, the resins showed the preference order: $NO_3^- > SO_4^{2-} > Cl^- > HCO_3^-$. The anion-exchange capacities for four resins were 1.39, 1.33, 1.33, 1.17 meq anion/mL resin bed. An increase in sulfate concentration resulted in a decrease in BVs-to-nitrate breakthrough, e.g., if SO_4^{2-} concentration increased from 42 to 140 mg/L, the BVs-to-NO_3^- breakthrough decreased from 400 to 240.

Figure 16.12 shows breakthrough curves for Cl^-, which is being displaced from the strong-base resin; HCO_3^- is seen to be the first ionic group to appear with essentially little uptake; NO_3^- appears after about 18 h or 220 BV; SO_4^{2-} is the last to appear after 25 h or 280 BV. The HCO_3^- levels off at about 130 mg/L, which is about the same as the influent concentration. The NO_3^- levels off also at about the influent concentration, while the SO_4^{2-} rises to about 140 mg/L, which is considerably higher than the influent level of 42 mg/L.

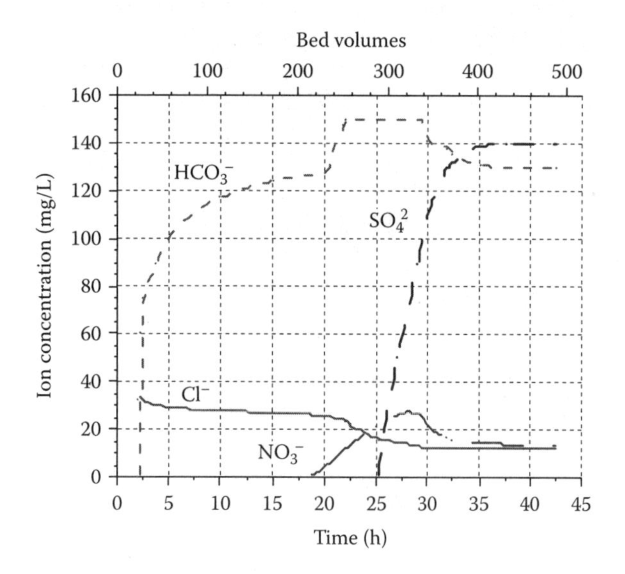

FIGURE 16.12 Breakthrough curves from strong-acid ion-exchangers used for removal of nitrates from groundwater at Glendale, Arizona. (Adapted from Clifford, D. et al., Salt conservation, selectivity reversal and breakthrough detection in ion-exchange for nitrate removal, in Liberti, L., and Millar, J. R. (Eds.), *Fundamentals and Applications of Ion-exchange*, NATO ASI Series, Martinus Nijhoff Publishers, the Hague, the Netherlands, 1985, p. 110.)

PROBLEMS

16.1 Sources of Commercial Ion-Exchangers

Given

Journal articles, journal cards, web access.

Required

Determine three sources of ion-exchangers in the categories: (a) natural zeolites, (b) synthetic zeolites, (c) activated alumina, (d) synthetic resins.

16.2 Properties of Commercial Ion-Exchangers

Given

Web access, manufacturers' brochures.

Required

Examine manufacturers' literature from manuals, brochures, and Web sites and list quantitatively some of the properties of comparable ion-exchangers in tabular format, e.g., properties in columns with a given set of properties in a row. What properties are of interest? Are the units given in forms that are useful for engineering purposes?

16.3 Mass of Clinoptilolite Zeolite to Treat a Mine Wastewater

Given

The lead concentration of wastewater from a mine is 250 mg/L. The water is to be treated with clinoptilolite zeolite. Let Q(wastewater) $\approx$ 1900 m^3/d (0.50 mgd). Assume the ion-exchange capacity $\approx$200 mg Pb^{2+}/g clinoptilolite when $C(Pb^{2+}) \approx$ 250 mg/L.

Summary of conditions:

- C(Pb) = 250 mg/L
- Adsorbent = clinoptilolite
- Q = 1900 m^3/day (0.50 mgd)
- Ion-exchange capacity $\approx$ 200 mg Pb^{2+}/g clinoptilolite when $C(Pb^{2+}) \approx$ 250 mg/L
- t(run) = 30 day

Required

Calculate the mass of clinoptilolite required to remove the lead for a 30-day "run," assuming the column is saturated 100%. If the number turns out to be not reasonable, suggest a run duration such that the mass required is reasonable.

Mass (clinoptilolite).

16.4 Exchange Capacity of Strong-Acid Cation-Exchanger

Given

A strong-acid cation-exchanger of polystyrene matrix with 12% DVB cross-linking and with all benzene rings sulfonated.

Required

Determine the exchange capacity in meq/g ion-exchanger.

16.5 Exchange Capacity of Strong-Base Anion-Exchanger

Given

A strong-base anion-exchanger of polystyrene matrix with 12% DVB cross-linking and with all benzene rings having attached a fixed ionic group, $-N^{+}(CH_3)_3$.

Required

Determine the exchange capacity in meq/g ion-exchanger.

16.6 Desalination

Given

A resort on an island in the Aegean Sea must desalinate seawater for drinking water and other potable water uses. The resort will have 1000 guest rooms and a staff of 500. A mixed-bed ion-exchange system is being considered.

Required

Size the ion-exchange system.

16.7 Volume of Ion-Exchanger for Hardness Removal

Given

Groundwater that serves a supply for a community, population 5000, is considered "hard," with Ca^{2+} concentration of about 200 mg/L as $CaCO_3$. Alkalinity concentration is similar.

Required

Determine the volume of ion-exchanger required for a 30-day operation before recharge.

16.8 Exchange Capacity (ExC) Conversion

Given

(a) Amberlite IR-120 exchange capacity is given as ExC (kg as $CaCO_3/ft^3$ packed-bed) = 25 after acid regeneration at 160 g acid/L resin (packed-bed). The concentration of H_2SO_4 recommended was 10%. (b) For sodium cycle operation, the recommended concentration NaCl = 10%, with HLR ≈ 2.4 m/h (1.0 gpm/ft^2); with recommended application 400 g NaCl/L resin. For these conditions, ExC (kg as $CaCO_3/ft^3$ packed-bed) = 34.

Assume $P = 0.39$ and that fM = 0.40 (if needed). (All values from Rohm and Haas (1987), IR-120 data sheet.)

Required

Calculate ExC (meq/g dry solids) and ExC (g as $CaCO_3$/L packed-bed) for each regeneration condition. *Hint*: Use Table CD16.A.2 for calculations.

ACKNOWLEDGMENTS

Steve Chipera, presently (2010) Senior Geologist/XRD Specialist, Chesapeake Energy Corporation, Oklahoma, City, Oklahoma, facilitated permission to use Figure 16.2, a SEM image of a clinoptilolite surface, which was generated by he and coworkers while employed at the Los Alamos National Laboratory, and which was available at the International Natural Zeolite Association web site (INZA, 2010).

Dr. Lynne B. McCusker, Laboratory of Crystallography, Zurich gave permission to use IZA images cited and helped to clarify the associated information about zeolites. The author is responsible for its interpretation and use. The web site of the International Zeolite Association (IZA, 2010) has a site-menu that gives access to a wealth of information about IZA and about zeolites.

The distinction should be noted between the International Zeolite Association (IZA) and the International Natural Zeolite Association (INZA). The former would include all zeolites while the latter includes only those found in nature. The respective web sites provide information on each organization.

APPENDIX 16.A: ION-EXCHANGE CONVERSIONS

A variety of units both in the literature and in manufacturers' data sheets have been used to express ion-exchange capacity and the density of ion-exchange materials. Certain units may be preferred by some but different ones may be more useful to others. Often, conversions are needed. Therefore, for convenience, two spreadsheet conversion tables are provided, i.e., Tables CD16.A.1 and CD16.A.2, for density (ρ) and exchange capacity (ExC), respectively. The units seen in a given row are converted to those of a column by multiplying the value given by the factor in the cell.

The numerical data used are seen in each of the cells by clicking on the cell of interest; the conversion formulae are given in the footnotes. The principle used for each conversion is the "chain-rule," i.e., to provide for cancellations of units until the desired result is achieved.

Notes on spreadsheets for unit conversions

Table CD16.A.1 converts densities from any units seen in rows to units in the columns. Each row in the left-hand column gives densities in different units. The columns to the right each have densities in the same sequence. The coordinates of the matrix are defined as (row, column), which identifies a matrix cell and which contains the respective conversion formula. For example, matrix cell (1, 2) gives the formula for the conversion from dry solid density, ρ_{Ds} to moist solid density, ρ_{Ms}; the formula is $\rho_{Ms} = \rho_{Ds}\,(1 + fM) = \rho_{Ms}$. Each of the conversions is identified in the "notes" below Table CD16.A.1 and the formula for the respective cell is shown. The numerical value for the resulting conversion is given in the cell. To illustrate (albeit cumbersome),

1.00 g dry solid/mL solid) · [1.40 (g moist solid/ mL solid)/(g dry solid/mL solid)] = 1.40 g moist solid/ mL solid.

All of the input values for Table CD16.A.1 are "dummy" values, given in the second column from the left. The applicable input should be inserted in place of the dummy value. The result will be a calculation for all other units for that parameter along the respective matrix row. As seen, an input for units in a given row may be converted to the units seen in a column. Table CD16.A.2

TABLE CD16.A.1
Conversions of Density of Particles (Inputs Shown Are Default Values)

Bed porosity, P = 0.39 mL voids/mL packed-bed
Particle moisture fraction, fM = 0.40 g moisture/g solid
ρ_D = 1.00 g dry solid/mL packed-bed
ρ_M = 1.00 g moist solid/mL packed-bed

	Value of Parameter (to Left)	Mass per Unit Volume of Dry Solid	1 ρ_{Ds} (g Dry Solid/mL Solid)	2 ρ_{ms} (g Moist Solid/mL Solid)	Density (Mass/Vol Packed-Bed)	3 $\rho_{D\text{-Bulk}}$ (g Dry Solid/mL Packed-Bed)	4 $\rho_{Ms\text{-Bulk}}$ (g Moist Solid/mL Packed-Bed)	5 $\rho_{Ms\text{-Bulk}}$ (g Moist Solid/L Packed-Bed)	6 $\rho_{Ms\text{-Bulk}}$ (lb Moist Solid/ft^3 Packed-Bed)	
Mass per unit volume of dry solid										
ρ_{Ds} (g dry solid/mL solid).	1.00		x	1.40		0.60	1.20	1200	7.63	1
ρ_{ms} (g moist solid/mL solid)	1.00		0.50	x		0.31	0.6	600	38.15	2
Density (mass/vol packed-bed)										
$\rho_{D\text{-bulk}}$ (g dry solid/mL packed-bed)	1.00		1.67	3.33		x	2	2000	127.2	3
$\rho_{Ms\text{-bulk}}$ (g moist solid/mL packed-bed)	1.00		0.83	1.67		0.50	x	1000	63.6	4
$\rho_{Ms\text{-bulk}}$ (g moist solid/L packed-bed)	1.00		0.00	0.00		0.00	0	x	0.1	5
$\rho_{Ms\text{-bulk}}$ (lb moist solid/ft^3 packed-bed)	1.00		0.013	0.03		0.01	0.02	16	x	6

Notes:

Task (Row, Column)	Conversion Formula
(1, 2) Dry solid to moist solid	$\rho_{Ds}\,(1 + fM) = \rho_{Ms}$
(1, 3) Dry solid to dry packed-bed	$\rho_{D\text{-bulk}} = \rho_{DS}\,(1 - P)$
(1, 4) Dry solid to moist packed-bed	$\rho_{M\text{-bulk}} = \rho_{DS}\,(1 - P)\cdot(1 + fM)$
(1, 5) Dry solid to moist packed-bed	$\rho_{M\text{-bulk}} = \rho_{DS}\,(1 - P)\cdot(1 + fM) * 1000$
(1, 6) Dry solid to moist packed-bed (lb/ft^3)	$\rho_{M\text{-bulk}} = \rho_{DS}\,(1 - P)\cdot(1 + fM)\cdot(28317\text{ mL/ft}^3)\cdot(\text{lb}/445.36\text{ g})$
(2, 1) Moist solid to dry solid	$\rho_{Ds} = \rho_{Ms}/(1 + fM)$
(2, 3) Moist solid to dry packed-bed	$\rho_{D\text{-bulk}} = (\rho_{Ms}/(1 + fM))\cdot(1 - P)$
(2, 4) Moist solid to moist packed-bed	$\rho_{M\text{-bulk}} = \rho_{MS}\,(1 - P)$
(2, 5) Moist solid to moist packed-bed (L)	$\rho_{M\text{-bulk}} = \rho_{MS}\,(1 - P) * 1000$
(2, 6) Moist solid to moist packed-bed (lb/ft^3)	$\rho_{M\text{-bulk}} = \rho_{MS}\,(1 - P)\cdot(28317\text{ mL/ft}^3)\cdot(\text{lb}/445.36\text{ g})$
(3, 4) Dry bulk to moist bulk	$\rho_{M\text{-bulk}} = \rho_{D\text{-bulk}}\,(1 + fM)$
(3, 5) Dry bulk to moist bulk (L)	$\rho_{M\text{-bulk}} = \rho_{D\text{-bulk}}\,(1 + fM)\cdot 1000$
(3, 6) Dry bulk to moist bulk (lb/ft^3)	$\rho_{M\text{-bulk}} = \rho_{D\text{-bulk}}\,(1 + fM)\cdot(28317\text{ mL/ft}^3)\cdot(\text{lb}/445.36\text{ g})$
(4, 5) Moist bulk (g/mL) to moist bulk (g/L)	$\rho_{M\text{-bulk}}\,(\text{g/L}) = \rho_{M\text{-bulk}}\,(\text{g/mL})\cdot 1000$
(4, 6) Moist bulk to moist bulk (lb/ft^3)	$\rho_{M\text{-bulk}}\,(\text{lb/ft}^3) = \rho_{M\text{-bulk}}\,(\text{g/mL})\cdot(28317\text{ mL/ft}^3)\cdot(\text{lb}/445.36\text{ g})$
(5, 6) Moist bulk (L) to moist bulk (lb/ft^3)	$\rho_{M\text{-bulk}}\,(\text{lb/ft}^3) = \rho_{M\text{-bulk}}\,(\text{g/L})\cdot(28.317\text{ mL/ft}^3)\cdot(\text{lb}/445.36\text{ g})$

TABLE CD16.A.2

Conversions of Ion-Exchange Capacity

Bed porosity, P = 0.37

Particle moisture fraction, fM = 0.45

ρ(bulk) = 0.55 (g dry solid/mL packed-bed)

Exchange Capacity	Value of Parameter (to Left)	1 ExC (meq/g Dry Solids)	2 ExC (meq/g Moist Solids)	3 ExC (meq/mL Packed-Bed)	4 ExC (eq/ft^3 Packed-Bed)	5 ExC (kg as $CaCO_3$/ft^3 Packed-Bed)	6 ExC (g as $CaCO_3$/L Packed-Bed)	
ExC (meq/g dry solids)	1.00	x	0.69	0.55	15.6	12.02	28	1
ExC (meq/g moist solids)	1.00	1.45	x	0.80	22.6	17.4	39.9	2
ExC (meq/mL packed-bed)	1.00	1.82	1.25	x	28.32	21.85	50	3
ExC (eq/ft^3 packed-bed)	1.00	0.064	0.044	0.035	x	0.772	1.77	4
ExC (kg as $CaCO_3$/ft^3 packed-bed)	1.00	0.08	0.06	0.05	1.30	x	2.3	5
ExC (g as $CaCO_3$/L packed-bed)	1.00	0.04	0.025	0.020	0.566	0.437	x	6

Notes:

Task (Row, Column)	Conversion Formula
(1, 2) meq/g dry solids to meq/g moist solids	ExC (moist) = ExC (dry)/(1 + fM)
(1, 3) meq/g dry solids to meq/mL packed-bed	ExC (meq/mL bulk) = C (meq/g dry solid) · ρ (bulk)(g dry solid/mL bulk)
(1, 4) meq/g dry solids to eq/ft^3 packed-bed	ExC (eq/ft^3 bulk) = C (meq/g dry solid) · r (bulk)(g dry solid/mL bulk) · (28317 mL/ft^3) · (eq/1000 meq)
(1, 5) meq/g dry solids to kg as $CaCO_3$/ft^3 packed-bed	ExC (kg as $CaCO_3$/ft^3 packed-bed) = ExC (meq/g dry solid) · r (bulk)(g dry solid/mL bulk) · (28317 mL/ft^3) · (eq/1000 meq) · (50 g $CaCO_3$/eq wt) · (g/0.064799 g) · (kg/1000 g)
(1, 6) meq/g dry solids to g as $CaCO_3$/L packed-bed	ExC (kg as $CaCO_3$/ft^3 packed-bed) = ExC (meq/g dry solid) · r (bulk)(g dry solid/mL bulk) · (28317 mL/ft^3) · (eq/1000 meq) · (50 g $CaCO_3$/eq wt)
(2, 3) meq/g moist solids to meq/mL packed-bed	ExC (meq/mL bulk) = ExC (meq/g moist solids) · (1 + fM) · r (g dry solid/mL bulk)
(2, 4) meq/g moist solids to eq/ft^3 packed-bed	ExC (eq/ft^3 bulk) = ExC (meq/g moist solids) · (1 + fM) · r (g dry solids/mL bulk) · (28317 mL/ft^3 bulk) · (eq/1000 meq)
(2, 5) meq/g moist solids to kg as $CaCO_3$/ft^3 packed-bed	ExC (kg as $CaCO_3$/ft^3 packed-bed) = ExC (meq/g moist solids) · (1 + fM) · r (g dry solids/mL bulk) · (28317 mL/ft^3 bulk) · (eq/1000 meq) * (50 g $CaCO_3$/eq) · (g/0.064799) · (kg/1000 g)
(2, 6) meq/g moist solids to g as $CaCO_3$/L packed-bed	ExC (g as $CaCO_3$/L packed-bed) = ExC (meq/g moist solids) · (1 + fM) · r (g dry solids/mL bulk) · (eq/1000 meq) * (50 g $CaCO_3$/eq) · (1000 mL/L)
(3, 4) meq/mL packed-bed to eq/ft^3 packed-bed	ExC (eq/ft^3 bulk) = ExC (meq/mL packed bed) · (eq/1000 meq) · (28317 mL/ft^3)
(3, 5) meq/mL packed-bed to kg as $CaCO_3$/ft^3 packed-bed	ExC (kg as $CaCO_3$/ft^3 packed-bed) = ExC (meq/mL packed-bed) · (eq/1000 meq) · (50 g $CaCO_3$/eq) · (g/0.064799) · (kg/1000 g) · (28317 mL/ft^3)
(3, 6) meq/mL packed-bed to g as $CaCO_3$/L packed-bed	ExC (g as $CaCO_3$/L packed-bed) = ExC (meq/mL packed-bed) · (eq/ 1000 meq) · (50 g $CaCO_3$/eq) · (1000 mL/L)
(4, 5) eq/ft^3 packed-bed to kg as $CaCO_3$/ft^3 packed-bed	ExC (kg as $CaCO_3$/ft^3 packed-bed) = ExC (eq/ft^3 packed-bed) · (50 g $CaCO_3$/eq) · (g/0.064799) · (kg/1000 g)
(4, 6) eq/ft^3 packed-bed to g as $CaCO_3$/L packed-bed	ExC (g as $CaCO_3$/L packed-bed) = ExC (eq/ft^3 packed-bed) · (50 g as $CaCO_3$/eq) * (ft^3/28.317 L)
(5, 6) kg as $CaCO_3$/ft^3 packed-bed to g as $CaCO_3$/L packed-bed	ExC (g as $CaCO_3$/L packed-bed) = ExC (kg as $CaCO_3$/ft^3 packed-bed) · (1000 g/kg) * (0.064799 g/g) · (ft^3/28.317 L)

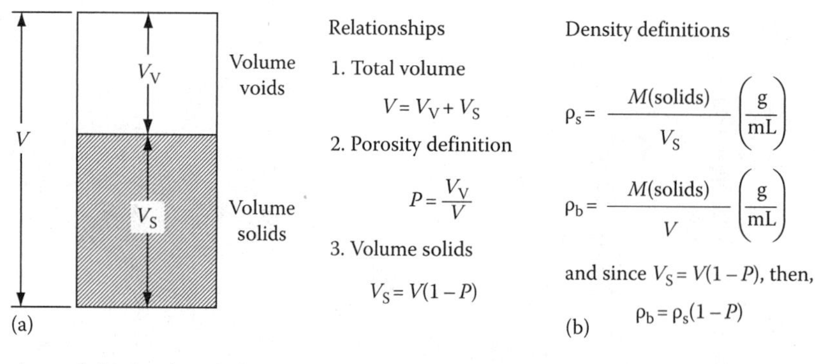

FIGURE 16.A.1 (a) Porosity and (b) density relations for a packed-bed.

is structured the same way as Table CD16.A.1, except that Table CD16.A.2 gives conversions of exchange capacity (ExC) units.

Example 16.A.1 illustrates a density conversion and Example 16.A.2 illustrates the conversion protocol for ion-exchange capacity. For reference, Figure 16.A.1 shows the basic definitions for packed-bed porosity, void volume, particle density, and bulk density.

Example 16.A.1 Density Conversion

Given

Amberlite IR-120 density is given as ρ_{Mb}(IR120) = 51 lb moist solids/ft^3 packed-bed. Assume $P = 0.39$. Assume also that fM = 0.40. [All values from Rohm and Haas (1987), IR-120 data sheet.]

Required

Calculate ρ_{Ms}(IR120).

Solution

Inserting the argument, ρ_{Mb}(IR120) = 51 lb moist solids/ft^3 packed-bed, into Table CD16.A.1 gives

$$\rho_{Ms} = 1.34 \text{ g moist solid/mL solid}$$

Discussion

The density of the dry solid is seen also, i.e., $\rho_{DS} = 0.67$ g dry solid/mL solid, also with values for other units.

Example 16.A.2 Convert from C (Kilograins as $CaCO_3$/ft^3 Packed-Bed) to C (meq/mL Packed-Bed):

$$C\left(\frac{\text{meq}}{\text{mL packed bed}}\right) = C\left(\frac{\text{kgr as CaCO}_3}{\text{ft}^3\text{ packed-bed}}\right) \cdot \left(\frac{1{,}000 \text{ g}}{\text{kg}}\right) \cdot \left(\frac{0.0648 \text{ g}}{\text{g}}\right) \cdot \left(\frac{\text{eq}}{50 \text{ g CaCO}_3}\right) \cdot \left(\frac{1{,}000 \text{ meq}}{\text{eq}}\right) \cdot \left(\frac{\text{ft}^3}{28{,}317 \text{ mL}}\right)$$

$$= 0.048 \text{ meq exchange capacity/mL packed-bed}$$

Discussion

The conversion factor, 0.0458, is seen in Table CD16.A.2, cell (5, 3).

GLOSSARY

Activated alumina: (1) A porous, highly adsorptive alumina made usually by heating alumina hydrates and used chiefly in drying gases and liquids (http://unabridged.merriam-webster.com). (2) Formed by dehydrating amorphous aluminum hydroxide, i.e., $Al(OH)_3$ Typically, the granules are 28–48 mesh (0.3–0.6 mm). At pH < pH_{zpc}, the sites are charged as OH_2^+; at pH = pH_{zpc}, the sites are charged as OH; at pH > pH_{zpc}, the sites are charged as O^- (Carlson and Thomson, 2001). (3) The material is highly porous with a surface area of 50–300 m^2/g. Fluoride reacts with the alumina in a ligand exchange, i.e., not adsorption, exchanging OH^- for F^- in an acid solution. The hydroxide ion regenerates the alumina (Clifford, 1999, p. 9.7, 9.8).

Alumina: (1) The oxide of aluminum Al_2O_3 that occurs native as corundum and in hydrated forms, that is made usually from bauxite, in various forms (as a white powder obtained by calcination or a hard crystalline substance resembling natural corundum obtained by heating calcined aluminum oxide almost to the fusion point), and that is used chiefly as a source of metallic aluminum, as an abrasive and refractory, as a catalyst and catalyst carrier, and as an adsorbent as in drying gases and liquids and in chromatography; see also aluminum hydroxide (http://unabridged.merriam-webster.com). (2) A generic name for the oxides and hydroxides of aluminum. The phase chemistry is complex with five thermodynamically stable phases, plus a large number of metastable and transition forms (Fleming, 1986, p. 159).

Aluminum hydroxide: Any of several white gelatinous or crystalline hydrates $Al_2O_3 \cdot nH_2O$ of alumina found in nature, especially in bauxite, or obtained as precipitates by treating solutions of aluminum salts with hydroxides. Hydrated alumina; especially the trihydrate $Al_2O_3 \cdot 3H_2O$ or $Al(OH)_3$ of alumina, regarded as acting both as a weak base and as a weak acid, that occurs as gibbsite and is used chiefly in ceramics, in pigments, and as a reinforcing agent for rubber (http://unabridged.merriam-webster.com).

Amine: A class of organic compounds that are structurally similar to ammonia; an amine contains a nitrogen atom that is covalent bonded to one or more carbon atoms and that has an unshared pair of electrons. An amine is a weak base, i.e., does not have a strong affinity for H^+. Examples of amines are:

$$CH_3-\ddot{N}(H)-H \qquad CH_3-\ddot{N}(CH_3)-H \qquad CH_3-\ddot{N}(CH_3)-CH_3$$

The reaction with a strong acid drives the reaction toward an ionic form, e.g., $CH_3\ddot{N}H_2 + H{-}\ddot{Cl}\!: \rightarrow CH_3\overset{+}{N}H_3 + :\ddot{Cl}:^-$ (Fessenden and Fessenden, 1994, p. 27). The amine group, $CH_3\overset{+}{N}H_3$, and $>NH_2^+$ are weak-base amino groups used for anion-exchangers. Strong-base quaternary amines include $-\overset{|}{\underset{|}{N^+}}-$, $-N^+(CH_3)_3$, etc. (Helfferich, 1962, p. 47).

Amphoteric: Materials which have both kinds of sites, i.e., both positive and negative, are called *amphoteric*.

Anion: Atom of an element or complex in solution that has a negative charge, e.g., Cl^-, F^-, SO_4^2.

Benzene ring: Hexagonal ring structure with carbon atom at each vertex. The carbons share electrons with each other and with other atoms.

Bond: Three kinds of bonds are: (1) ionic, (2) covalent, and (3) metallic (Silberberg, 1996, p. 327).

In ionic bonding electrostatic attraction occurs between positive and negative ions (or groups). Covalent bonding is most important between nonmetal atoms. Each nonmetal atom holds onto its own electrons tightly and tends to gain other electrons as well. The attraction of each nucleus for the valence electrons of the other draws the atoms together. A shared valence electron pair is localized in that it spends most of its time between two atoms, linking them in a covalent bond. The Lewis dot system maybe used to depict the number of valence electrons; the element symbol represents the nucleus and the dots around the four sides represent the number of valence electrons, e.g., for nitrogen, the depiction is: $\cdot\dot{N}\!:$ which may also be depicted, $\cdot\overset{|}{\underset{|}{N}}$ (for typesetting, it is a long dash on the left side, vertical dashes on top and bottom in center, nothing on right).

A double bond is two bonding pairs, i.e., four electrons shared between two atoms, which occurs in ethylene: $H_2C{=}CH_2$; the double bond is represented by two parallel dashes. A triple bond is represented by three parallel dashes, such as with nitrogen, which has five valence electrons.

Carboxyl group: The group, $-CO_2H$, also written, $-COOH$, or showing the bonds is, $-\overset{O}{\overset{\|}{C}}OH$. Compounds containing a carboxyl group are weak acids and are called carboxylic acids; acetic acid, i.e., CH_3CO_2H, is an example. A carboxyl group in a carboxylic acid donates a hydrogen ion to a base and in doing so the carboxyl group becomes a carboxylate-ion, which is an anion. Carboxylic acid compounds weak acids and so acid-base reactions do not proceed to completion. To drive the reaction to completion, a base stronger than water, i.e., NaOH, must be used. To illustrate, the reaction with water is, $CH_3\overset{O}{\overset{\|}{C}}-O-H + OH^- \rightarrow CH_3\overset{O}{\overset{\|}{C}}-O^- + HOH$ (Fessenden and Fessenden, 1994, p. 15).

Cation: Atom of an element or complex in solution that has a negative charge, e.g., Na^+, Ca^{2+}, Mg^{2+}.

Chelate: Chemical group characterized by a covalent bond.

Chelating resin: Resin with chemical grouping similar to a conventional chelating compound, e.g., EDTA, but attached to a matrix for gross insolubility (Dow, 1964, p. 10). These compounds attach ions to be removed by a coordinating-type bond.

Clinoptilolite: Widely abundant mineral of the zeolite group that has a crystal structure.

Co-ion: Those ions in solution having the same charge as the ion-exchanger framework are called *co-ions*.

Complex: Same as chelate.

Copolymer: A mixture of two polymers, e.g., styrene and di-vinyl benzene.

Counter ion: The ions which exchange are called *counter-ions*, in that they are opposite in charge to the charge of the sites of the ion-exchanger.

Covalent bond: A bond between two atoms that results due to the sharing of a pair of electrons. Bonding energies, ΔH, may be 88, 163, and 230 kcal/mol, for a single bond (e.g., $CH_3{-}CH_3$), a double bond (e.g., $CH_2{=}CH_2$), and a triple bond (e.g., $CH{\equiv}CH$), respectively (Fessenden and Fessenden, 1994, p. 15).

Cross-linked resins: A benzene ring may attach to a vinyl groups of two polystyrene molecules, resulting in di-vinyl benzene (DVB). This configuration of molecules is called "cross-linking." As an example, a 4% cross-linkage is made with beads composed of 4% DVB and 96% styrene based on molecular composition. The percent cross-linkage is indicated by an "*X*" number, i.e., *X*8 means the copolymer is constituted of 8% DVB, i.e., 8 molecules of DVB per 92 molecules of styrene. The DVB contributes the third dimension to the polymer network and makes it insoluble (Dow Chemical Company, 1964, p. 17).

Crystal: An ordered solid chemical structure having a regularly repeating internal arrangement of its atoms.

Demineralization: Removal of anions and cations from solution by exchange of an anion for OH^-, and of a cation for H^+ by means of reactions with a cation-exchanger and an anion-exchanger, respectively.

For example, Na^+ may attach to the cation-exchanger site for H^+ and Cl^- may attach to the anion-exchanger site for OH^- to result in water as a product. Usually, the process is in a "mixed-bed," but it could be done in separate beds.

Electric conductance: See specific electric conductance.

Equilibria: The term equilibria refers to an ion-exchange reaction, and to the final concentrations of all species both in solution and attached to ion-exchange sites.

Equilibrium constant: The constant that indicates the ratios of reactants and products with reference to its associated stoichiometric equation; valences are the exponents.

Equivalent weight: Atomic weight divided by valence of ion. For example, $MW(Ca^{2+}) = 40.078$ g/mol and $EqWt(Ca^{2+}) = MW(Ca^{2+})/2 = 40.078$ g/mol/2 valence = 20.039 g/equivalent weight. An "equivalent" weight is abbreviated, "eq."

Exchange capacity: Sites for ion-exchange per unit mass of ion-exchanger, expressed usually in milliequivalents of ions per gram of ion-exchanger, i.e., meq/g.

Free radical: An atom such as H•, or group of atoms such as H_3C•, that contains an unpaired electron, also called simply, a "radical." Radicals are usually electrically neutral. Because a radical contains an atom with an incomplete octet, most radicals are unstable and therefore are highly reactive. A radical is represented by a single dot in its formula, e.g., $H\ddot{\underset{..}{O}}$• is represented as HO• (Fessenden and Fessenden, 1994, p. 14).

Functional group: A complex with an exchangeable ion that attaches to the benzene ring of a polystyrene resin; an example is the sulfonate group, i.e., $-SO_3^{2-}$ Also called the "ion-active group."

Gel-type ion-exchange resin: A gel-type ion-exchange resin may be modeled conceptually as a plate of congealed spaghetti. The strands, i.e., polymers are entangled but there are randomly distributed points of adhesion between the strands. The structure is open, porous, and heterogeneous, but externally appears as a homogeneous solid mass (foregoing from Kunin, 1983, p. 45). Generally, the pore size of a gel-type resin is ≤ 30 Å (Kunin, 1979, p. 1).

Gibbs free energy: A state function, commonly defined as, $G = H - TS$, (G is Gibbs free energy, H is enthalpy, and S is the entropy of a system). Equilibrium is attained when the free energy of the products of a reaction equals the free energy of the reactants. The change in free energy is independent of the path. The term, "free-energy," was coined by Hermann von Helmholtz in 1882 to describe that portion of the energy of a reaction that was not bound in the form of heat but could be freely converted to other forms of energy. A similar quantity, the "chemical potential," had been independently defined by Willard Gibbs, professor at Yale University in a paper published in 1878 in the *Transactions of the Connecticut Academy of Arts and Sciences* (Servos, 1990, p. 352). The paper was recognized by the "big-three" founders of the discipline of physical chemistry, i.e., Wilhelm Ostwald, University of Liepzig; Jacobus Henricus van't Hoff of the University of Amsterdam; Svante Arrhenius, of the Hogskola in Stockholm (Servos, 1990). The G was assigned in the 1970s in recognition of Gibbs' contribution.

Glauconite: (1) Ferrous alumnosilicates containing exchangeable potassium, with rather dense and rigid crystal structure; cation-exchange occurs at the crystal surfaces (Helfferich, 1964, p. 11). (2) A mineral under the mica group within the subclass, "Phyllosilicate" (from Greek, *phyllon*, leaf or sheet silicate). The taxonomy is (http://en.wikipedia.org/wiki/Silicate_minerals):

Phyllosilicate subclass (the sheet structures)/
Mica group/glauconite – $(K,Na)(Al,Mg,Fe)_2$ $(Si,Al)_4O_{10}(OH)_2$

Groups under the Phyllosilicate subclass include: the Serpentine group; the Clay mineral group; the Mica group; and the Chlorite group. (3) For reference to explicate further the taxonomy, another sub-class (parallel to "Phyllosilicates") is "Tectosilicates," or "framework silicates," which are characterized by a three-dimensional framework. With the exception of the quartz-group, the Tectosilicates are "aluminosilicates," and include the Feldspar group, the Feldspathoid group, a Petalite, the Scapolite group, the Analcime group, and the Zeolite group 3 A misconception is that glauconite is a "zeolite," which it is not; as stated, glauconite and zeolite fall under different mineral subclasses. See *zeolite*.

Greensand: (1) A natural zeolite made of the mineral glauconite, a greenish-black material found in the form of kidney-bean-shaped granules (Babbitt and Doland, 1949, p. 513). The material was one of the earliest recognized ion-exchange materials, being applied to water treatment about 1925 and was mined from deposits in New Jersey. (2) The term "zeolite" is often applied to greensand; it is a misnomer, dating to the early days of use (such as by Babbitt and Doland, 1949). (3) Greensands have been used in iron and manganese removal by oxidation of Fe^{+2} or Mn^{+2} with $KMnO_4$, which then form solid precipitates, removed by the greensand.

Hardness: An indication of the soap-consuming capacity of a water, which was one of the early definitions. The causative ions include Ca^{2+}, Mg^{2+}, Fe^{2+}, Mn^{2+}, Zn^{2+}; Ca^{2+}, and Mg^{2+} are the most common. Hardness is usually expressed in terms of mg $CaCO_3$/L.

Hardness conversion: Suppose an analysis of a hard water gives Ca^{2+} concentration = 128 mg Ca^{2+}/L and Mg^{2+}/L concentration = 32 mg Mg^{2+}/L. These concentrations in equivalents per liter are: 88 mg Ca^{2+}/L/(40.078 mq/2/meq) = 4.391 meq Ca^{2+}/L and 16 mg Mg^{2+}/L/(24.305 mq/2/meq) = 1.317 meq Ca^{2+}/L, giving total hardness = 5.708 meq/L.

To convert to hardness expressed as $CaCO_3$, $MW(CaCO_3) = 100.087$ g/mol or 50.044 g/equivalent. Therefore, 5.708 meq hardness/L · 50.044 mg $CaCO_3$/meq = 285 mg hardness as $CaCO_3$/L, which is a "hard" water. (As a note, expressing hardness in terms of $CaCO_3$ is probably used because of tradition and because it has been used so extensively that the relative concentrations in terms of $CaCO_3$ are understood with respect to relative impacts. Hardness as equivalents per liter would be a more direct and a more rational expression.)

Heavy metal: A metal that has a "higher" molecular weight such as lead, cadmium, mercury, etc.; usually these metals are considered toxic at low concentrations, e.g., μg/L.

Ion-active group: See *functional group*.

Ion-exchange: (1) The process of exchanging an ion in solution for one of the same charge that was attached to a "site" within an ion-exchanger. Changing Ca^{2+} in solution for Na^+ in the ion-exchanger is an example. The foregoing is a traditional softening application. In years since the 1980s, USEPA drinking water regulations have been increasingly important, and engineers have sought ion-exchangers that have selective properties, e.g., to remove arsenic, fluoride, boron, nitrate. (2) A separation process which utilizes functional or active sites on so-called inert matrices (Calmon, 1985, p. 2).

Ion-exchanger: An ion-exchanger can be any solid substance having an array of ion-exchange sites. Usually the substances-providing sites are either crystalline or macromolecules. A material, e.g., a zeolite or other crystal structure, a synthetic resin, or other material, that may have exchange sites within its structure. The sites are accessible by diffusion from an external solution. Ions that are smaller than the pore size may be excluded from the space.

Isotherm: The relationship between the solid phase concentration of a given ion and its aqueous phase concentration at a specified temperature.

Leakage: The appearance in the effluent of an ion-exchange column of ions which should be removed. Leakage may be due to unfavorable equilibria or incomplete regeneration (Dow, 1964, p. 73).

Macro-porous: A generic term that refers to resins with pores ≤200 Å (Kunin, 1979, p. 1). On the other hand, the macro-porous resins may also refer generically to those resins that have large pore sizes, which are, at the same time, of the macro-reticular type with pore sizes 50–1,000,000 Å (Kunin, 1979, pp. 2–3).

Macro-reticular ion-exchange resin: A highly porous resin prepared by adding an organic solvent for the resin monomers, but a poor solvent for the polymer, to the polymerization mixture. As polymerization progresses, the solvent is squeezed out by the growing copolymer regions. Spherical beads obtained with pores several hundred angstrom units provide easy access to the interior of the beads (Helfferich, 1964, p. 61). The term was coined by Robert Kunin and his associates at Rohm and Haas (Kunin, 1983, p. 48) to distinguish the foregoing pore structure (network) from the gel-type (intertwined polymer strands with cross-linking to provide adhesions between strands). The macro-reticular structure results in pore sizes 50–1,000,000 Å, which, therefore, provide for large molecules to gain access to the interior sites. The surface areas are 7–600 m^2/g dry resin (Kunin, 1979, p. 2).

Microporous: Zeolite is classed as a microporous solid because of its regular repeating array of pores within crystal units connected by angstrom-sized channels (see Van Tassel et al., 1994, p. 925). As related to a resin, the term microporous refers to pores ≤20 A; gel-type resins are in this category (Kunin, 1979, p. 1).

Milliequivalents: The number of equivalent weights divided by 1000. Usually, the expression is applied to the expression of the concentration of a given ion in meq/L.

Mixed-bed: A mixed-bed contains a cation-exchanger charged with H^+ and an anion-exchanger charged with OH^-; both are synthetic resins, in bead form. In the removal cycle, the various cations, e.g., Na^+, Ca^{2+}, Mg^{2+}, exchange with hydrogen and the various anions, e.g., Cl^-, HCO_3^-, SO_4^{2-}, NO_3^-, exchange with OH^-, with water being the product.

Monomer: A unit of a polymer. Styrene is a monomer that comprises polystyrene.

pH: Defined as the negative log of the hydrogen ion concentration. For water, $[H^+][OH^-] = 10^{-14}$ and pH = 14-pOH^-.

p*K*: The negative log of the equilibrium constant for an acid-base reaction. To illustrate, consider the dissociation of acetic acid: HA => $H^+ + A^-$. The equilibrium statement is, $K_a = [H^+][A^-]/[HA]$. Take logs of both sides, $\log K_a = \log[H^+] + \log[A^-] - \log[HA]$. Multiply both sides by (−1) to get, $-\log K_a = -\log[H^+] - \log[A^-] + \log[HA]$. Then, $pK_a = pH + p[A^-] - p[HA]$, which defines pK_a for the dissociation of acetic acid. For acetic acid, pK_a(acetic acid) = 4.7.

Polymer: A repeating monomer unit, e.g., polystyrene.

Poly-electrolyte: A repeating monomer unit, e.g., polystyrene that has ion-functional groups, e.g., sulfonic acid, attached to a monomer.

Pore (for an ion-exchanger resin): Spaces of distances between polymer chains and cross-links. The sizes depend on the degree of cross-linking (Kunin, 1983, p. 48).

Quaternary: Etymology: Middle English, from Latin quaternarius, adjective, consisting of four each, from quaterni four each (from quater four times) + -arius –ary. 1 a : a group of four. (http://unabridged.merriam-webster.com)

Resin: (1) An ion-exchange resin is a special type of synthetic polymer, i.e., macromolecule or poly-electrolyte, which is insoluble and may be visualized as an elastic three-dimensional network to which ion-active groups are attached (Dow Chemical, 1964, p. 3). (2) Ion-exchange resins are poly-electrolyte gels. Cation-exchangers in H^+ form and anion-exchangers in OH^- form may be considered as insoluble acids and bases, respectively (Helfferich, 1962, p. 81).

Reticular: From Latin reticulum little net. (1) Resembling a net in appearance or structure. (2) Resembling a net in operation or effect. (http://unabridged.merriam-webster.com)

Sieman: A measure of specific electrical conductivity (same as "mhos," which is the reciprocal of "ohms"). The standardized measurement involves immersing a standard cell in water and reading, at specified voltage, the current between two platinum electrode plates, each 1.0 cm × 1.0 cm, separated by 1.0 cm. Often the units $\mu S/cm$ are used in lieu of the SI units, S/m. The conversion is, $(S/m) \cdot (m/cm \cdot 10^2) \cdot (\mu S \cdot 10^6/S) = \mu S \cdot 10^4/cm$. In other words, multiply (S/m) by 10^4 to get ($\mu S/cm$). For example, EC(seawater) ≈ 5.5 S/m = 55,000 $\mu S/cm$. See also specific electric conductance.

Softening: Removal of hardness-causing ions; as a rule, the predominant one is calcium with magnesium being the second in concentration.

Specific electric conductance: A measure of the electrical conductivity of water, usually given in Siemans or micro-Siemans. Examples are: de-ionized water, 0.05 $\mu S/cm$; distilled water, 0.05 $\mu S/cm$; Lake Tahoe, 97 $\mu S/cm$; Lake Mead, 850 $\mu S/cm$; Atlantic Ocean, 43,000 $\mu S/cm$. See also, *Sieman*.

Stoichiometric: Refers most often to a chemical reaction in which molecular ratios are defined in accordance with a "balanced" equation.

Strong-acid ion-exchanger: A cation-exchange resin in which complete ionization occurs; for example, when the functional group is sulfonate, i.e., $-SO_3^-$, H^+ dissociates readily from its attached site. A solution equivalent would be sulfuric acid or hydrochloric acid.

Strong-base ion-exchanger: An anion-exchange resin in which complete ionization occurs; for example, when the functional group is a quaternary amine, e.g., $-N^+-$, or $-N^+(CH_3)_3$, OH^- dissociates readily from its attached site. A solution equivalent would be sodium hydroxide.

Styrene monomer: A vinyl group attached to a benzene ring. Polystyrene, with di-vinyl benzene cross-linking, is a common constituent of synthetic resins.

Sulfonic acid functional groups: The complex $-SO_3^-$ which forms a bond with the benzene ring of a polystyrene molecule and which has an exchangeable site for a cation.

Valence: The ionic charge of an atom in the solution phase [The cation, e.g., Na^+, loses an electron and the anion, e.g., Cl^-, gains an electron].

Vinyl group: The compound, $H_2C{=}CH_2$, which is attached to the benzene ring in a styrene polymer and which provides the basis for attachment of additional monomers. Also, vinyl may be attached to the lower position of the benzene ring to provide cross-linkage to other styrene polymers.

Weak-acid ion-exchanger: A cation-exchange resin in which incomplete ionization occurs; for example, when the functional group is carboxylate, i.e., $-COO^-$; H^+ does not dissociate readily from its attached site, except at high pH; thus they are not functional at low pH. A solution equivalent would be carbonic acid, H_2CO_3 acid or acetic acid, HAc.

Weak-base ion-exchanger: An anion-exchange resin in which incomplete ionization occurs; e.g., when the functional group is $>NH_2^+$, OH^- does not dissociate readily from its attached site, except at low pH; thus they are not functional at high pH.

Wet capacity: Exchange capacity per unit volume of swollen resin (Dow, 1964, p. 10).

Zeolite: (1) Crystalline aluminosilicates composed of connected cage-like structures which are large enough to house a small number of guest molecules, generally <20 angstroms, (see Van Tassel et al., 1994, p. 925). (2) Zeolites are the aluminosilicate members of the family of microporous solids known as "molecular sieves." The term molecular sieve refers to a particular property of these materials, i.e., the ability to selectively sort molecules based primarily on a size exclusion process. This is due to a very regular pore structure of molecular dimensions. The maximum size of the molecular or ionic species that can enter the pores of a zeolite is controlled by the dimensions of the channels. (3) The zeolite taxonomy is

silicates/tectosilicate subclass (the framework silicates)/zeolite group/
- Analcime family
- Chabazite family
- Gismondine family
- Harmotome family
- Heulandite family (includes Clinoptiloltie)
- Natrolite family
- Stilbite family

(http://www.galleries.com/minerals/silicate/tectosil.htm).

(4) As noted in the text, zeolite minerals occur (by definition) in nature and include some 54 structures, e.g., amicite, analcime, barerite, bellbergite, bikitaite, goggsite, brewsterite, chabizite, clinoptilolite, cowlesite, etc. (http://en.wikipedia.org/wiki/ zeolite). The mineral zeolites are distinct from the synthetic zeolites (see text). Some mineral structures have been created by synthesis, however, which are given different names under the same zeolite framework type (illustrated in Table 16.1).

REFERENCES

Abdelrasool, F., Kinetics of adsorption, PhD Thesis, Department of Civil Engineering, Colorado State University, Fort Collins, CO, 1992.

Alcoa, Product Data Sheet F-1 Granular Activated Alumina, SEP 921, Separations Technology Division, Alcoa, Warrendale, PA, 1989.

Applebaum, S. B., Characteristics and properties of zeolites for water softening, *Journal of the American Water Works Association*, 13(2):213–220, February 1925.

AWWA, *Water Quality and Treatment*, 2nd edn., the American Water Works Association, Inc. New York, 1951.

Babbitt, H. E. and Doland, J. J., *Water Supply Engineering*, McGraw-Hill Book, New York, 1949.

Baerlocher, Ch. and McCusker, L. B., Database of zeolite structures, http://www.iza-structure.org/databases, March 2010.

Baerlocher, Ch., McCusker, L. B., and Olson, D. H., *Atlas of Zeolite Framework Types*, 6th revised edition, Elsevier, Amsterdam, the Netherlands, 2007.

Behrman, A. S., Early history of zeolites, *Journal of the American Water Works Association*, 13(2):221–231, February 1925.

Behrman, A. S., Progress in municipal zeolite water softening, *Journal of the American Water Works Association*, 26(5):618–628, May 1934.

Beler Baykal, B. and Guven, D. A., Performance of clinoptilolite alone and in combination with sand filters for the removal of ammonia peaks from domestic wastewater, *Water Science and Technology*, 35(7):47–54, 1997.

Beler Baykal, B., Oldenburg, M., and Sekoulov, I., The use of ion-exchange in ammonia removal under constant and variable loads, *Environmental Technology*, 17:717–726, 1996.

Calmon, C., Ion-exchange towards the twenty first century, in Liberti, L., and Millar, J. R. (Eds.), *Fundamentals and Applications of Ion-exchange*, NATO ASI Series, Martinus Nijhoff Publishers, the Hague, the Netherlands, 1985.

Carlson, K. and Thomson, B., *Activated Alumina for Arsenic Removal, Power Point® presentation, Rocky Mountain Section AWWA, WEA Annual Conference*, Angel Fire, NM, September 2001.

Chwirka, J. D., Thomson, B. M., and Stomp III, J. M., Removing arsenic from groundwater, *Journal of the American Water Works Association*, 92(3):79–88, March 2000.

Clifford, D. A., Ion-exchange and inorganic adsorption, in Letterman, R. D. (Ed.), *Water Quality and Treatment*, Chap. 9, 5th edn., McGraw-Hill Inc., New York, 1999.

Clifford, D. and Liu, X., Ion-exchange for nitrate removal, *Journal of the American Water Works Association*, 85(4):135–143, April 1993.

Clifford, D., Horng, L., and Lin, C., Salt conservation, selectivity reversal and breakthrough detection in ion-exchange for nitrate removal, in Liberti, L., and Millar, J. R. (Eds.), *Fundamentals and Applications of Ion-exchange*, NATO ASI Series, Martinus Nijhoff Publishers, the Hague, the Netherlands, 1985.

Collins, L. F., A study of contemporary zeolites, *Journal of the American Water Works Association*, 29(10):1472–1514, October 1937.

Das, N. C. and Bandyopadhyay, M., Rational design of vermiculite column adsorber for removal of cadmium, printed paper, Department of Civil Engineering, Indian Institute of Technology, Kharagpur, India, 1993.

Dorfner, K., *Ion-Exchangers—Properties and Applications*, Ann Arbor Science Publishers, Ann Arbor, MI, 1972.

Dow Chemical Company, *Dowex: Ion-Exchange*, The Dow Chemical Company, Midland, MI, 1964.

Fessenden, R. J. and Fessenden, J. S., *Organic Chemistry*, 5th edn., Brooks/Cole Publishing Co., Pacific Grove, CA, 1994.

Fleming, H. L., Application of aluminas in water treatment, *Environmental Progress* (*AIChE*), 5(3):159–166, August 1986.

Haggerty, G. M. and Bowman, R. S., Sorption of chromate and other inorganic anions by organo-zeolite, *Environmental Science and Technology*, 28(3):452–458, March 1994.

Helfferich, F., *Ion-Exchange*, McGraw-Hill, New York, 1962.

INZA, International Natural Zeolite Association web-site, http://inza.nmt.edu/pics_crystals.php, 2010.

IZA, International Zeolite Association web-site, http://www.iza-online.org/, 2010.

Jacoby, M., Custom zeolites, *Chemical and Engineering News*, 81(20):38, May 19, 2003.

Kato, D. M., Promising applications for water treatment, *California Engineer*, 73(4):7–11, May 1995.

Kerr, G. T., Synthetic zeolites, *Scientific American*, 261(1):100–105, July 1989.

Kesraoul-Ouki, S., Cheeseman, C., and Perry, R., Effects of conditioning and treatment of chabazite and clinoptilolite prior to lead and cadmium removal, *Environmental Science and Technology*, 27(6):1108–1116, June 1993.

Kunin, R., *Ion-Exchange Resins*, 2nd edn., John Wiley & Sons, New York, 1958.

Kunin, R., Two decades of macroreticular ion-exchange resins, Amber-hi-lites, Rohm and Haas Company, No. 184, Spring, 1979.

Kunin, R., The nature and properties of acrylic exchange resins, in Liberti, L. and Helfferich, F. G. (Eds.), *Mass Transfer and Kinetics of Ion-Exchange, Proceedings of A NATO Advanced Study Institute*, Martinus Nijhoff Publishers, the Hague, the Netherlands, 1983.

Liberti, L. and Helfferich, F. G. (Eds.), *Mass Transfer and Kinetics of Ion-Exchange, Proceedings of A NATO Advanced Study Institute*, Martinus Nijhoff Publishers, the Hague, the Netherlands, 1983.

Liberti, L. and Millar, J. R. (Eds.), *Fundamentals and Applications of Ion-Exchange*, NATO ASI Series, Martinus Nijhoff Publishers, the Hague, the Netherlands, 1985.

Marshall, C. E., *The Physical Chemistry and Mineralogy of Soils*, John Wiley & Sons, Inc., New York, 1964.

McCusker, L. B., Liebau, F., and Engelhardt, G., Nomenclature of structural and compositional characteristics of ordered microporous and mesoporous materials with inorganic hosts, *Pure and Applied Chemistry*, 73(2):381–394, 2001.

Meenakshi, R. C. and Maheshwari, R. C., Fluoride in drinking water and its removal, *Journal of Hazardous Materials*, B137:456–463, 2006.

Millar, J., On the synthesis of ion-exchange resins, in Liberti, L. and Helfferich, F. G. (Eds.), *Mass Transfer and Kinetics of Ion-Exchange, Proceedings of A NATO Advanced Study Institute*, Martinus Nijhoff Publishers, the Hague, the Netherlands, 1983a.

Millar, J., On the structure of ion-exchange resins, in Liberti, L. and Helfferich, F. G. (Eds.), *Mass Transfer and Kinetics of Ion-Exchange, Proceedings of A NATO Advanced Study Institute*, Martinus Nijhoff Publishers, the Hague, the Netherlands, 1983b.

Powell, S. T., Trends in zeolite softening, *Journal of the American Water Works Association*, 29:1722–1728, 1929.

Powell, S. T., *Water Conditioning for Industry*, McGraw-Hill, New York, 1954.

Purolite, *Purolite Engineering Manual—Puropack Packed-Bed Technology*, The Purolite Company, Bala Cynwyd, PA, 1999 (available 2010 as Adobe Acrobat file from http://www.purolite.com/customized/uploads/pdfs/PuroPack%20Packed%20Beds.pdf).

Rohm and Haas Company, *Engineering Manual for the Amberlite® Ion-Exchange Resins*, Rohm and Haas Company, Philadelphia, PA, 1987 (binder: updated periodically).

Rohm and Haas Company, 50 Years, Ion-exchange Resin Technology, 1939–1989, Amber Hi-Lites, No. 185, November 1989.

Rubel, F. Jr. 1984. *Design Manual: Removal of Fluoride from Drinking Water Supplies by Activated Alumina*, U.S. Environmental Protection Agency, Municipal Environmental Research Laboratory, Cincinnati, OH, EPA-600/2-84-134 (NTIS Order No. PB85-113991), 1984.

Rubel, F. Jr. and Woosley, R. D., The removal of excess fluoride from drinking water by activated alumina, *Journal of the American Water Works Association*, 71(1):45–48, January 1979.

Savinelli, E. A. and Black, A. P., Defluoridation of water with activated alumina, *Journal of the American Water Works Association*, 50(1):33–44, January, 1958.

Servos, J. W., *Physical Chemistry from Ostwald to Pauling*, Princeton University Press, Princeton, NJ, 1990.

Silberberg, M., *Chemistry*, Mosby Publishing Co., St. Louis, MO, 1996.

Van Tassel, P. R., Davis, H. T., and McCormick, A. V., New lattice model for adsorption of small molecules in zeolite micropores, *AIChE Journal*, 40(6):925–934, June 1994.

Vaughan, D. E. W., The synthesis and manufacture of zeolites, *Chemical Engineering Progress*, 84(2):25–31, February 1988.

Weber, W. J., *Physico Chemical Treatment*, John Wiley & Sons, New York, 1972.

BIBLIOGRAPHY

Abrams, I. M., The Duolite® phenolic resins: their properties and applications, Amber-hi-lites, Rohm and Haas Company, No. 184, Winter, 1988.

Anderson, R. E., Toxic metals decontamination. Does ion-exchange fit? Amber-hi-lites, Rohm and Haas Company, No. 181, Summer, 1987.

Baer, C., Meier, W. M., and Olson, D. H., *Atlas of Zeolite Framework Types*, Fifth revised edition, Copyright International Zeolite Association, Published on behalf of the Structure Commission of the International Zeolite Association, Elsevier, Amsterdam, 2001, ISBN 0-444-50701-9. [Downloadable in 2010 as a pdf file from the International Zeolite Association Web site, http://www.iza-structure.org/databases/books/Atlas_5ed.pdf].

Berg, L., Kindschy, E. O., Reveal, W. S., and Saner, H. A., Alumina activated with anhydrous hydrogen fluoride, *Chemical Engineering Progress*, 47(9):469–471, September 1951.

Clifford, D. A., Ion-exchange and inorganic adsorption, in Letterman, R.D. (Ed.), *Water Quality and Treatment*, 5th edn.,, Chap. 9, McGraw-Hill, Inc., New York, 1999.

Kunin, R. and Myers, R. J., *Ion-Exchange Resins*, John Wiley & Sons, Inc., New York, 1951.

Meyers, R. J. and Easters, J. W., Synthetic resin ion-exchangers for water purification, *Industrial Engineering Chemistry*, 33, 1203, 1941.

Nachod, F. C., *Ion-exchange*, Academic Press, New York, 1949.

Sheerman, J. and Vermeulen, T. (Eds.), *Adsorption and Ion-Exchange—Progress and Future Prospects*, AIChE Symposium Series, No. 233, Vol. 80, American Institute of Chemical Engineers, New York, 1984.

17 Membrane Processes

Membrane treatment of water contaminants has evolved as a technology since about 1960. An array of membrane applications has emerged, ranging from basic water treatment involving particle removal to desalting of seawater that removes ions. The practice of membrane treatment has become not only feasible as a technology but economical as well.

17.1 DESCRIPTION

A membrane installation involves a "package" of components. First, the membranes must be selected and sized. Other aspects of the design include involve pretreatment, pump selection, instrumentation, cleaning chemicals, ancillary basins, etc. (Box 17.1).

17.1.1 MEMBRANES IN-A-NUTSHELL

Two categories of membrane technologies are (1) those that reject primarily particles and (2) those that reject primarily molecules and ions. Microfiltration (MF) and ultrafiltration (UF) are for particle removal. Nanofiltration (NF) and reverse osmosis (RO) are for removals of molecules and ions.

The two membrane categories are fundamentally different as to how water moves through the membrane material; technologies used, for example, hollow fiber (mostly) or spiral-wound sheets (mostly), kinds of contaminants removed, contaminant purging from pressure vessel, and cleaning method. The processes have common features in that they both involve synthetic membrane materials, are "pressure-driven" processes, have modular designs, and have common support organizations, for example, AMTA, AWWA, IWA, and IDA. The IDA, however, focuses on desalination; technologies available for desalination include RO, electrodialysis (ED), and distillation (which includes two to three technologies). A committee chaired by Zander (2008) reviewed the topic of desalination from history, global growth, economies, etc., providing a perspective on the topic, including the use of membranes. Table 17.1 lists some of the distinctions between the two categories (Box 17.2).

17.1.1.1 Analysis: Flow Balance Principle

A key principle of analysis is that flow-in and flow-out of a membrane pressure vessel must be accounted for, that is, $Q_F = Q_P + Q_C$, in which Q_F = feed flow, Q_P = permeate flow, and Q_C = concentrate flow, also called "reject" flow (all in m^3/s). Figure 17.1a is a schematic depiction of this principle for a cross-flow membrane.

For a hollow-fiber membrane, the flow is "dead end," that is, there is no cross-flow. But there is a periodic flush of concentrate by an air-bump or a hydraulic backflow around the fibers (not through the fibers). Figure 17.1b depicts an intermittent period (i.e., 20–30 s) of backwash flow of permeate into the membrane housing as Q_{bw}, a dotted light arrow, and a simultaneous flow, Q_{bw} leaving (but laden with suspended solids, i.e., "concentrate"). Since Q_{bw} comes from permeate storage (not shown), the net effect is that Q_{bw} is all that is necessary to depict the flow balance, that is, the flow balance is still $Q_F = Q_P + Q_C$. In this case, Q_C is the pulse volume divided by the time interval between backwash pulses and therefore is a *pseudo*-flow and is used only to depict the overall flow balance. As a rule, Q_{bw} occurs for a time duration $\Delta t \approx 20$–30 s and at intervals 20–50 min, depending on the manufacturer. In other words, $Q_C(\text{pseudo-flow}) = Q_{bw} \times \Delta t(\text{pulse})/\Delta t(\text{backwash interval})$, which may seem overdone but the idea is that the nomenclature for the hollow fiber be consistent with the spiral-wound membrane.

Figure 17.2a is a photograph of a 203 mm (8 in.)-diameter spiral-wound membrane element and Figure 17.2b shows a hollow-fiber membrane element with a portion of the pressure vessel cut away. The 203 mm (8 in.)-diameter spiral-wound membrane element has a total surface area of about 33.9 m^2. The hollow-fiber element has perhaps thousands of fibers per vessel, which may be about 2 m (6 ft) long. The spiral-wound membrane is used mostly for NF/RO applications and the hollow-fiber membrane is used mostly for MF/UF applications.

17.1.1.2 Definitions

Several key definitions help to explain the membrane process. Examples of definitions include concentrate flow, feed flow, membrane, membrane element, spiral-wound membrane element, hollow-fiber membrane element, membrane process, membrane technology, flow balance, module, pressure vessel, stage, pass, rack, skid, permeate flow, reject flow, solute rejection, silt-density index, etc. These definitions, and others, are given in the glossary.

17.1.1.3 Acronyms for Membrane Materials and Membranes

Acronyms are used commonly to designate different membrane materials. Some include polysulfone (PS), polyvinylidenefluoride (PVDF), polyacrylonitrile (PAN), polyamines (PAs). For membranes, microfiltration is MF, ultrafiltration is UF, nanofiltration is NF, and reverse osmosis is RO (RO has been called also "hyperfiltration").

17.1.1.4 Process Description

The transport of water through a membrane material under a pressure gradient with retention of contaminants is the

BOX 17.1 MEMBRANE KNOWLEDGE BASE

Membrane technology is underpinned by a network of scientists, engineers, consulting firms, equipment manufacturer's, manufacturer's representatives, academics, operators, industries, and municipalities using the technology, etc. The topic is discussed at meetings, seminars, journals, Web sites, etc., which result in the exchange of ideas and advancement and dissemination of knowledge. The larger parent areas of knowledge include environmental engineering, mechanical engineering, polymer science, the water treatment industry, etc. The foregoing description is not, of course, unique to membranes; other water treatment unit processes have their own respective knowledge bases.

BOX 17.2 MICRO/ULTRA AND NANO/RO

Another organizing dilemma is whether to consider the two categories of membranes as separate or to integrate them. They are integrated for the purposes of this chapter. Categories also could be based on source water, application, membrane material, shape of membrane element (e.g., spiral-wound, hollow tube, and hollow fiber), configuration of a membrane assembly, driving force, etc.

"membrane process" (see also Van der Bruggen et al., 2003). For MF/UF, particles are the contaminants to be retained, for example, mineral, cysts, bacteria, and viruses. For NF/UF, molecules and ions are the contaminants to be retained. In general, MF/UF retains particles presumably by "screening." The mechanism is not clear for NF/RO (see Section 17.3.2).

17.1.1.5 Membrane Technology

The membrane technology is the means to make the membrane process "work." This has involved finding suitable membrane materials, packaging, and designing the ancillary components of a system. Thus, membranes may be categorized with respect to pore size (e.g., MF, UF, NF, and RO), chemical composition of membrane material (e.g., cellulose acetate [CA] and polyamide), the technique for packaging the membrane (e.g., spiral wound, hollow fiber, and plate-and-frame), etc. Membrane technologies

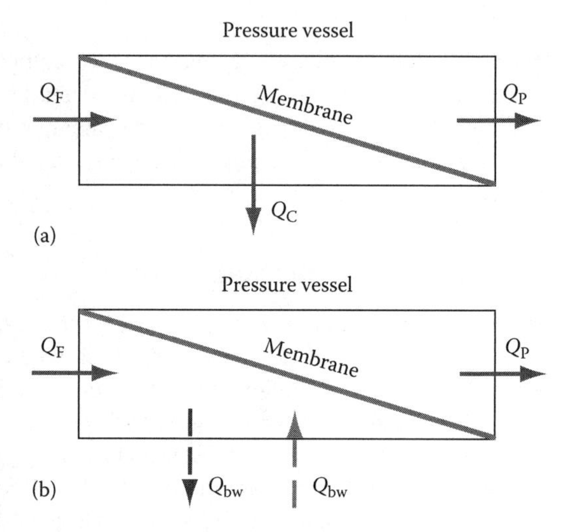

FIGURE 17.1 Definition sketch showing feed flow, Q_F, permeate flow across membrane, Q_P, and concentrate flow, Q_C. (a) Cross-flow (spiral-wound membrane or hollow tube): Q_C is steady. (b) Dead end (hollow fiber): Q_{bw}, is intermittent, for example, at intervals, 20–50 min, $\Delta t \approx$ 20–30 s.

TABLE 17.1
Distinctions between Micro/Ultra and Nano/RO Membrane Processes

	Membrane Category	
Issue	**Micro/Ultra**	**Nano/RO**
1. How water moves through membrane	Pores	Molecular structure
2. Membrane fabrication/packaging	Hollow fiber mostly	Spiral wound mostly, but also hollow fiber
3. Mechanism of contaminant rejection	Screening (particles/molecules)	Rejection of ions (ill-defined mechanism)
4. Flow	Dead-end flow (with hollow fibers)	Cross-flow
5. Contaminant (concentrate) removal	Concentrate accumulates then is flushed out	Concentrate is removed continuously by cross-flow
6. Δp	Low	High

Notes: (1) Ceramic tubes have been developed for cross-flow microfiltration. (2) AMTA is the American Membrane Technology Association; AWWA is the American Water Works Association; and IDA is the International Desalting Association. Other organizations are involved with membranes also. (3) How water moves through an RO membrane seems to be by passage through the molecular structure, vis-à-vis through a discrete pore. (4) Hollow fibers are available for RO. (5) The mechanisms of solute rejection by RO membranes have not been elucidated with certainty. (6) The term "concentrate" has been used most frequently by the membrane industry, although the terms "retentate" and "reject" have been used and still appear. The term "concentrate" refers to the matter retained and entrained in the flow leaving the membrane pressure vessel that has not passed through the membrane.

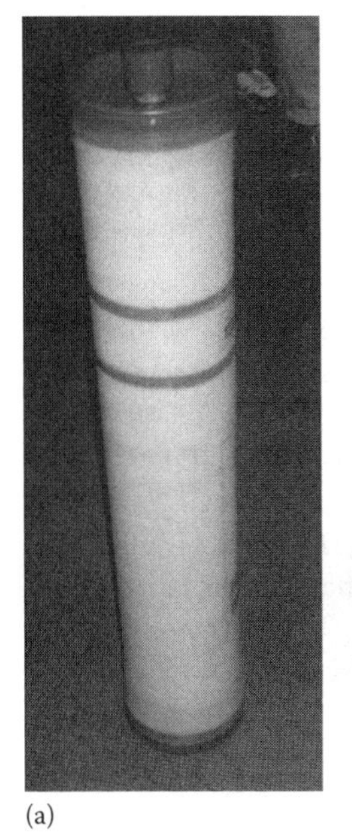

(a) (b)

FIGURE 17.2 Two kinds of membrane elements: spiral wound and hollow fiber. (a) Spiral-wound RO membrane element, 203 mm (8 in.) diameter 1016 mm (40 in.) long. (Courtesy of Rodney Evans, James (Ed) Burke, Brighton, CO.) (b) Hollow-fiber membrane element (Photo courtesy of Tory Champlin; element is Siemens Memcor.)

FIGURE 17.3 Portion of a reverse-osmosis membrane racks at Brighton, CO, c. 1997. (Courtesy of City of Brighton, CO.)

may be configured for almost any combination of choices within these categories.

17.1.1.6 Racks

Membrane systems are usually arranged with a "rack" of membranes that may be expandable by adding additional racks. Figure 17.3 shows a portion of one rack of a three-rack installation at Brighton, CO as placed in operation in 1993; another two racks were added by 2003. A rack contains six columns and nine rows of pressure vessels. Each pressure vessel contains six membrane "elements" arranged linearly in a series. Each membrane element is 203 mm (8 in.) diameter × 1016 mm (40 in.) long. The three-rack plant was designed for $Q_P \approx 15{,}000\ m^3/day$ (4.0 mgd), with Q(blended) $\approx 26{,}500\ m^3/day$ (7 mgd) serving a population of 14,000 (with 2003 population ≈ 24,000).

17.1.1.7 Treatment Train

The components of a treatment train depend on the kind of system. For all membrane systems, the feed flow should not have constituents that cause run lengths that are shorter in duration than expected, for example, in terms of months for most membranes. Thus, some kind of pretreatment for particle removal is necessary for virtually all surface waters, and perhaps some groundwaters. For NF or RO, a cartridge type of filter usually precedes the membrane process, that is, to remove particles that could be present. Other causes of clogging, for example, bacterial growth and precipitation, must also be addressed by pretreatment.

Permeate treatment may include pH adjustment (in the case of RO), perhaps blending, and disinfection. The system must also include secure storage for cleaning chemicals, for example, acid and detergent.

17.1.1.8 Operation

Operation involves monitoring flow and pressure primarily, along with specified water quality constituents. Pressure is monitored both along the series line of membrane elements and across the membranes, that is, the feed-flow pressure to the permeate pressure. The pressure across the membrane will increase over time as fouling occurs. In some cases, for example, at Brighton, the pressure loss was very little, for example, $\Delta P/\Delta t \approx 1\text{–}2$ kPa/year.

Membrane rupture, while not probable, has severe consequences if it occurs (it's more probable in a hollow-fiber membrane system). An index of membrane integrity would be some indicator that will not pass through a membrane but through a small hole. For an RO membrane, the indicator may be an ion found in the feed water. An indicator for an MF membrane may be a bacterium, for example, coliform, which is easy to measure. Membrane integrity testing is an important part of operation.

Also, cartridge filters, if used for pretreatment, must be changed when the pressure drop becomes larger than some designated limit, for example, about 100 kPa (1 atm). Pumps that pressurize the membranes must be monitored and maintained. Cleaning of the membranes is done with periodic flushing, for example, at intervals ranging from several weeks to several months, depending on how long in operation before the flux decline exceeds about 10% (or how long for the pressure increase to exceed 10%). Usually, this time should be >30 day for a membrane process to be acceptable. Chemicals may include a detergent, citric acid, a specialized proprietary chemical, or some combination in sequence. Hollow fiber membranes may be cleaned by a pulse of water every 0.5–1 min, with more complete chemical cleaning after several weeks.

17.1.2 Global Capacity

Figure 17.4 shows the worldwide capacity for RO in 1970–1984 and for MF/UF in 1980–2002. As seen, the RO-installed capacity started in about 1980 to grow as a viable technology while MF/UF got underway about 1990. In terms of RO capacity in 2005, the numbers were as following

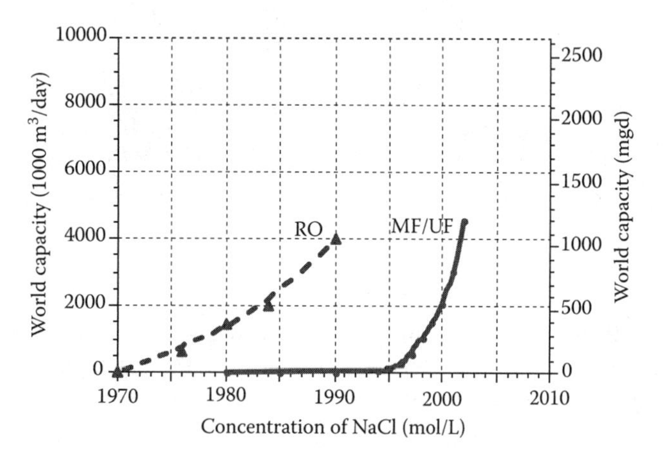

FIGURE 17.4 Global membrane capacity by year (Porter (1990) for RO; Adham et al. (2003) for MF/UF drinking water). (Mickley and Hamilton, 1991, p. 437 for RO 1990 point.)

(Hanft, 2005): United States, 16.3×10^6 m^3/day (4300 mgd); Europe, 2.8×10^6 m^3/day (750 mgd); Asia/Pacific, 1.9×10^6 m^3/day (500 mgd); Middle East/Africa, 4.2×10^6 m^3/day (1100 mgd); and Caribbean/Latin America, 0.5×10^6 m^3/day (140 mgd). The total, that is, global, 2005 RO capacity was therefore 25.7 million m^3/day (6790 mgd). This number compares with a global, 2005 desalting capacity of 47 million m^3/day (12,400 mgd), spread among 12,300 projects in 155 countries (Abstract, Global Water Intelligence (GWI), 2006). In the twenty-first IDA survey (GWI, 2008), the installed desalination capacity was given as 52 million m^3/day (13,800 mgd), with fractions distributed among the various technologies, that is, 0.04 ED, 0.09 MED (multieffect distillation), 0.27 MSF (multistage flash evaporation), and 0.59 RO. Feed-water fractions were 0.05 wastewater, 0.19 brackish water, 0.08 river water, and 0.62 seawater. Consumption fractions were 0.01 tourism, 0.23 industrial, 0.02 irrigation, 0.01 military, 0.67 municipal, and 0.06 power. The 10 countries with the largest capacities were 0.17 Saudi Arabia, 0.13 UAE, 0.13 United States, 0.08 Spain, 0.05 Kuwait, 0.04 Algeria, 0.04 China, 0.04 Qatar, 0.02 Japan, and 0.02 Australia.

As to NF plants, the first ones started operation in 1976 with capacities of 38–380 m^3/day (0.01–0.1 mgd). By 1993, there were numerous plants with capacities in excess of 60,600 m^3/day (16 mgd) (Wiesner, 1993). On the other hand, very few UF plants have been installed for potable water treatment. Probably, the number in 1993 was less than 100, worldwide, with capacities averaging 38 m^3/day (0.01 mgd) (Wiesner, 1993).

17.1.3 Membrane Types

Table 17.2 lists the four common membrane types along with their applications, pore sizes, and operating pressures.

TABLE 17.2
Membrane Types, Applications, Pore Sizes, and Operating Pressures

Membrane Type	Application (Removal)	Pore Size (μm)	Flux Density ($m^3/m^2/h$)	Flux Density (gpm/ft^2)	Operating Pressure[a] (kPa)[b]	Operating Pressure[a] (psi)
Microfiltration	Particles	0.1–2	167–1670	68–684	35–210	5–30
Ultrafiltration	Particles	0.01–0.1	42–500	17.205	105–415	15–60
Nanofiltration	NOM/Ca^{2+}	0.001–0.01[c]	25–42	10–17	550–830	80–120
Reverse osmosis	Desalting	0.0001–0.001[c] 5–20 Å	5–33	2–14	1,380–13,600	200–2,000

Source: Adapted from Thompson, M.A., The role of membrane filtration in drinking water treatment, Seminar S5—Engineering considerations for design of membrane filtration facilities for drinking water supply, in *Proceedings of Annual Conference American Water Works Association*, Denver, CO, June 18, 1992.

[a] For reference, the energy required for desalination by RO was given by Furukawa (2006) as 1.8–2.4 kWh/m^3 or 6.8–8.9 kWh/1000 gal, which at \$0.08US/kWh calculates to about \$0.14–9.19/m^3 (\$0.54–0.72/1000 gal); in other words, these numbers are for the energy dissipated in overcoming osmotic pressure and the added transmembrane pressure. The numbers depend on the added transmembrane pressure (i.e., in addition to the osmotic pressure) and the permeate flux density. The numbers do not include capital costs or other kinds of operating costs.

[b] Weisner (1993, p. 4) gives pressure ranges as 30–300, 50–700, 350–1000, and 800–8000, respectively.

[c] The idea of "pore size" is a convenience for NF and RO; the flux most likely permeates through the molecular structure of the membrane. For reference, Mulder (1991) gives 0.1–10 μm for MF (p. 114) and <2 nm for RO (p. 217).

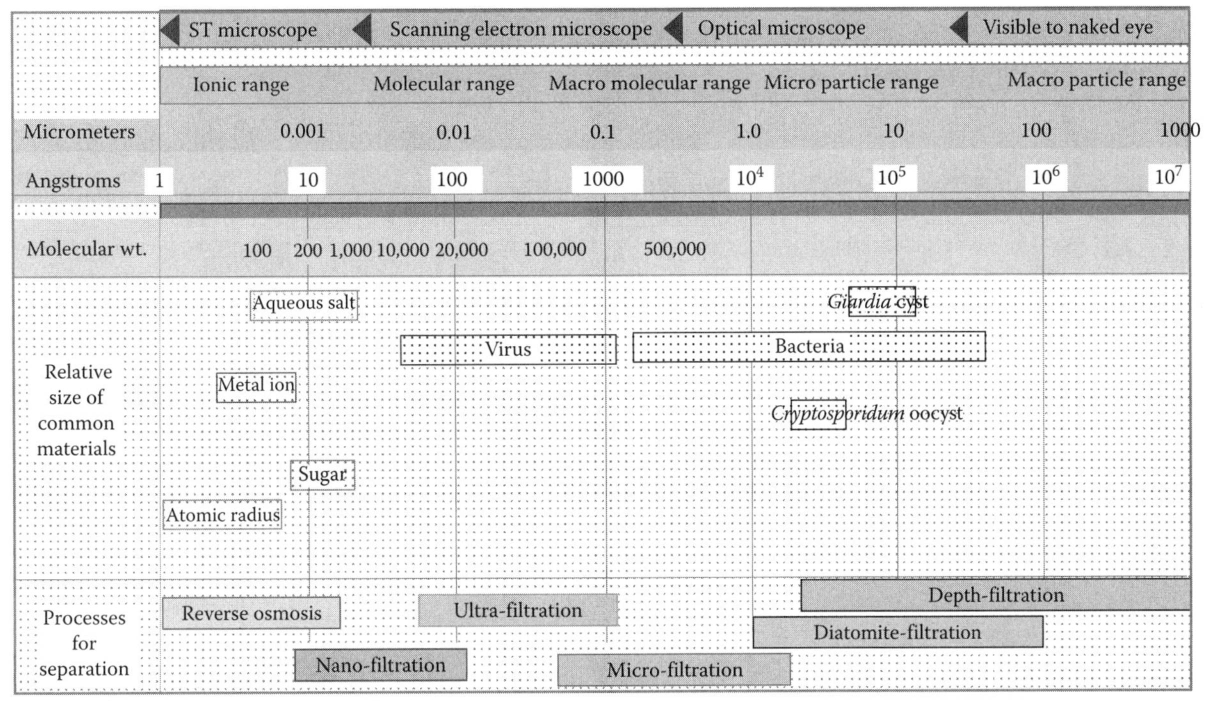

FIGURE 17.5 Membrane types compared with scales for pore sizes and particles. (Adapted from chart by General Electric Company, *The Filtration Spectrum*, General Electric Company, Minnetonka, MN, 1993. With permission.)

As seen, the different membrane types encompass a range of pore sizes and operating pressures. The MF membranes remove particles >1 μm and have the lowest operating pressure, while the RO membranes may remove ions and have the highest operating pressures. Figure 17.5 shows membrane pore sizes against a seven-log scale and also shows size ranges for ions, molecules, viruses, and bacteria.

17.1.4 Membrane Materials

Table 17.3 lists membrane materials and their applications; the listing also illustrates the wide variety of materials. Of these materials, CA, polyamide, PS, polyvinylidene, and polytetrafluorethylene are the most common (Cheryan, 1986). All of the materials listed are synthetic organic polymers. CA was one of the first membrane materials developed and may tolerate ≤2 mg/L of chlorine. Most polymer-based membranes are, however, unable to treat chlorinated water. Inorganic ceramic membranes (not listed) include various metal oxides, for example, aluminum (γ-Al_2O_3, α-Al_2O_3), silicon (SiO_2), cerium (CeO_2), titanium (TiO_2), and zirconium (ZrO_2), which are proprietary and are used mostly for MF.

17.1.5 Membrane Structure

Membranes are categorized also by pore geometries into either microporous or asymmetric membranes. Figure 17.6 illustrates the further breakdown of these categories (from Champlin, 1996, p. 8).

17.1.5.1 Microporous Membranes

The left branch of Figure 17.6 shows a microporous membrane, which is either isotropic (uniform pore size) or anisotropic (pores increase in size with depth from the active filtration surface). MF membranes are microporous in structure.

17.1.5.2 Asymmetric Membranes

The right branch of Figure 17.6 shows asymmetric membranes that are characterized by a thin-film layer on the surface of a microporous support. As shown, the latter may be either isotropic or anisotropic. UF, NF, and RO are usually asymmetric membranes.

17.1.5.2.1 Composite Membranes

Another type of asymmetric membrane is the composite membrane. While the standard asymmetric membrane is manufactured in a one-step process (annealing of a microporous membrane), the composite membrane requires a two-step procedure. The first step is the preparation of a suitable microporous support. The second step is the preparation of the dense-film layer (selective layer) and its lamination to the surface of the microporous support. Figure 17.7a illustrates the lamination step. Figure 17.7b shows an example of an asymmetric composite membrane cross section as seen by a scanning electron microscope. A PS surface layer is seen as a distinct layer on top of a polyester microporous support. The film layer controls the kinds of substances removed.

TABLE 17.3
Organic Membrane Materials and Types of Membranes Where Used

Material	Microfiltration	Ultrafiltration	Reverse Osmosis
Cellulosic polymers			
Cellulose acetate (CA)	X	X	X
Cellulose triacetate	X	X	X
CA/triacetate blend			X
Cellulose ester (mixed)	X		
Cellulose nitrate	X		
Cellulose (regenerated)	X	X	
PAN		X	
Polyvinylchloride	X		
Polyvinylchloride copolymer	X	X	
Polyamide (aromatic)	X	X	X
Polysulfone	X	X	
Polybenzimidiazole			X
Polybenzimidiazolone			X
Polycarbonate (track-etch)	X		
Polyester (track-etch)	X		
Polyether			
Polyimide		X	X
Polypropylene	X		
Polyelectrolyte complex		X	
Polytetraflouoethylene	X		
PVDF	X	X	
Polyacrylic acid + zirconium oxide (skin layer of dynamic membrane)		X	X
Polyethleneimine + toluene diisocyanate (skin of thin-film composite)			X
Polyurea			

Sources: Cheryan, M., *Ultrafiltration Handbook*, Technomic Publishing Company, Inc., Lancaster, PA, 1986; Wiesner, M. R., An Overview of ΔP Membrane Processes, Presented to Association of Environmental Engineering Professors, San Antonio, Texas, June 7, 1993.

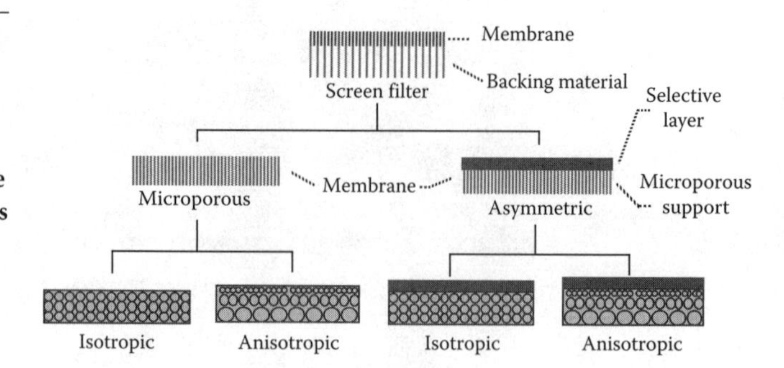

FIGURE 17.6 Different categories of membrane screen filters. (From Champlin, T.L., Membrane Filtration (handout prepared for Operators Annual Short Course), University of Colorado, Boulder, CO, p. 8, March 26, 1996. With permission.)

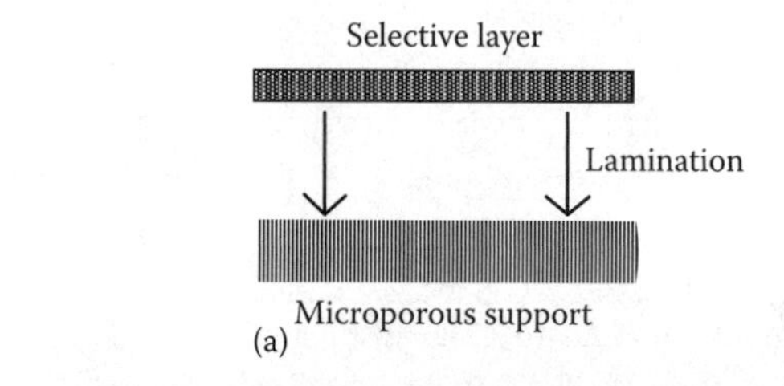

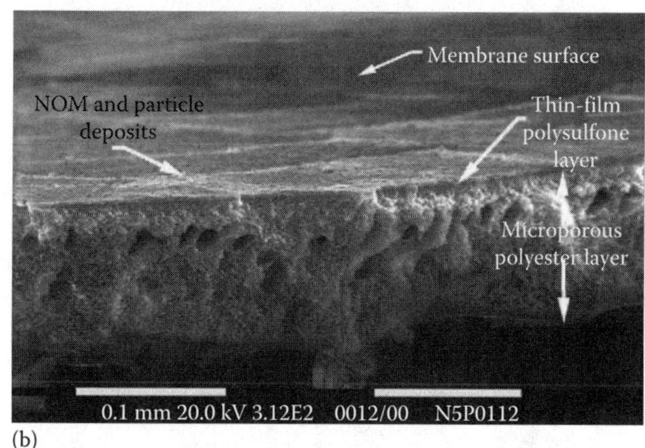

FIGURE 17.7 Composite asymmetric membrane. (a) Illustration of lamination. (b) Nanofiltration composite membrane. (SEM by Champlin, T.L., Natural organic matter and particle fouling of spiral-wound nanofiltration membrane elements, Doctoral thesis, Colorado State University, Fort Collins, CO, 1998, Figure 7.29. With permission.)

17.1.6 Manufacturing

This section provides a general description of the manufacturing process. Knowledge of specific manufacturing is proprietary since each manufacturer has its own unique membrane materials and methods.

17.1.6.1 Flat Sheets

Flat sheets are generally manufactured on a casting machine (Porter, 1990). A polymer solution is spread by a casting knife onto a polyester or polyethylene support paper, which is supplied continuously from a roller. Then, the cast polymer film is fed to a precipitation bath, where the actual membrane is formed. After a certain amount of time in the rinse bath, where solvents are removed, the membrane is collected as a flat sheet on a take-up roller. The membrane is produced as a continuous sheet up to 2 m wide. From this raw material, the desired membrane element is manufactured.

17.1.6.2 Tubes

The three types of membrane tubes are hollow fiber, capillary, and tubular; each is defined primarily by its outer diameter (Koros, 1995):

Hollow fiber	≤0.9 mm od
Capillaries	0.5–5 mm od
Tubular	5–15 mm od

17.1.6.2.1 Hollow Fiber and Capillary

Capillary and hollow fiber membranes are manufactured generally by a wet-spinning process using a spinneret with a double-bore nozzle. The polymer solution is fed to the outer bore of the nozzle. The central fluid can be either a precipitant or an inert gas, depending on the intended location of the active filtration surface, either on the inside of the tube or the outside of the tube, respectively. For a "bore-feed" hollow fiber (the filtration of water is from the inside to the outside of the tube), a precipitant fluid is fed continuously through the center bore. This allows the precipitation of the polymer to occur from the inside out, producing a thin film on the inside of the tube membrane. In the opposite case where the filtration of water is "shell feed" (from the outside to the inside), an inert gas is fed through the central bore and the fiber is spun into a precipitation bath. Hollow-fiber membranes are used for MF, UF, and RO.

17.1.6.2.2 Tubular

Tubular membranes are prepared either by an ultrasonic welding process using a flat membrane sheet or by directly casting on a porous support tube using a conical casting device pulled through a porous tube. Since the outer diameter of the casting cone is slightly smaller than the inner diameter of the tube, a thin polymer solution film is formed on the inside of the tube, which is converted into membrane by immersing the entire tube into a precipitation bath. Typical applications include MF and UF membranes.

17.1.7 Packaging

Membranes are packaged into a variety of modules: (1) plate-and-frame, (2) spiral wound, (3) tubular, and (4) hollow fiber modules. Tubular filters are used mostly for MF.

17.1.7.1 Plate-and-Frame Modules

Plate-and-frame modules are used in industrial applications primarily; they are constructed using a repetitive stacking of feed-flow spacers, membranes, and porous support plates, clamped together between two end plates. Feed-flow spacers allow for the feed water to come in contact with the membrane surface and provide for a cross-flow pattern. Porous plates support the membrane and channel the permeate out of the filter. MF, UF, and RO membranes have been used in plate-and-frame modules. The modules have low area per unit volume of pressure vessel.

17.1.7.2 Spiral-Wound Membrane Modules

Figure 17.8 illustrates the configuration of a spiral-wound module showing several kinds of detail. The eight views illustrate the construction and assembly: (a) layout of the sheets showing two sheets with permeate spacer with flow collected by the center tube; (b) same assembly but with feed-water spacers added on each side of membranes and with assembly rolled (several such assemblies are rolled on the tube); (c) feed-water mesh spacer showing fibers; (d) path taken by water, that is, over-and-under alternate fibers of mesh (showing also how deposits form with respect to mesh geometry); (e) membrane element with membrane rolls unfurled on left with rolled membrane on right; (f) end caps that distribute water for 203 mm (8 in.) and 102 mm (4 in.) membrane elements; (g) element inserted partially into pressure vessel; and (h) side view of RO membrane rack at Brighton for 203 mm (8 in.) membranes, six elements per module.

As seen, spiral-wound membrane modules use a stacking arrangement with sheets of feed-flow spacers, membranes, and permeate-flow spacers that are rolled around an inner tube. The inner tube collects the permeate flow by perforations along its length that are between glue lines of the attached membrane sheets. The permeate flow is spiral, while the feed flow is parallel to the tube and distributed over the cross section of the spacers. If the membrane sheets are unfolded, the feed flow would be seen as flowing between them and parallel to the axis and through the spacers, also called, "cross-flow." Figure 17.8c and d illustrates how the feed flow manages to flow over-and-under the mesh-net of a given spacer. The feed-flow, Q(feed-flow) minus the permeate flow, Q(permeate), is the concentrate flow, Q(concentrate), which leaves the tube at the end opposite the feed-flow, which is the water-mass balance statement for the membrane element. The pressure gradient of the feed flow (along the tube length) is controlled by a valve at the end of the membrane. The permeate flux is proportional to the pressure differential, Δp, between the feed flow and the permeate tube. Since there is also a pressure gradient along the feed flow, the Δp varies along the length of the tube and the permeate flux varies in proportion to Δp (along the tube length).

Spiral-wound membranes are used for UF, NF, and RO. They are manufactured in three sizes: (1) 51 mm (2 in.) diameter × 600 mm (24 in.) long, (2) 102 mm (4 in.) diameter × 203 mm (40 in.) long, and (3) 203 mm (8 in.) diameter × 1016 mm (40 in.) long. The areas are 1.115 m^2 (12 ft^2), 7.897 m^2 (85 ft^2), and 33.90 m^2 (365 ft^2), respectively (see also http://www.hydranautics.com). An even larger membrane, that is, 406 mm (16 in.) diameter × 1016 mm (40 in.) long has been assessed as more economical than smaller sizes (Furukawa, 2006; Yun et al., 2006).

17.1.7.3 Hollow-Fiber Modules

Hollow-fiber membranes are long thin flexible tubes perhaps 1–2 m (3–6 ft) in length with id ≈ 1 mm, od ≈ 2–3 mm. The fibers may be configured as "bore feed" or "shell

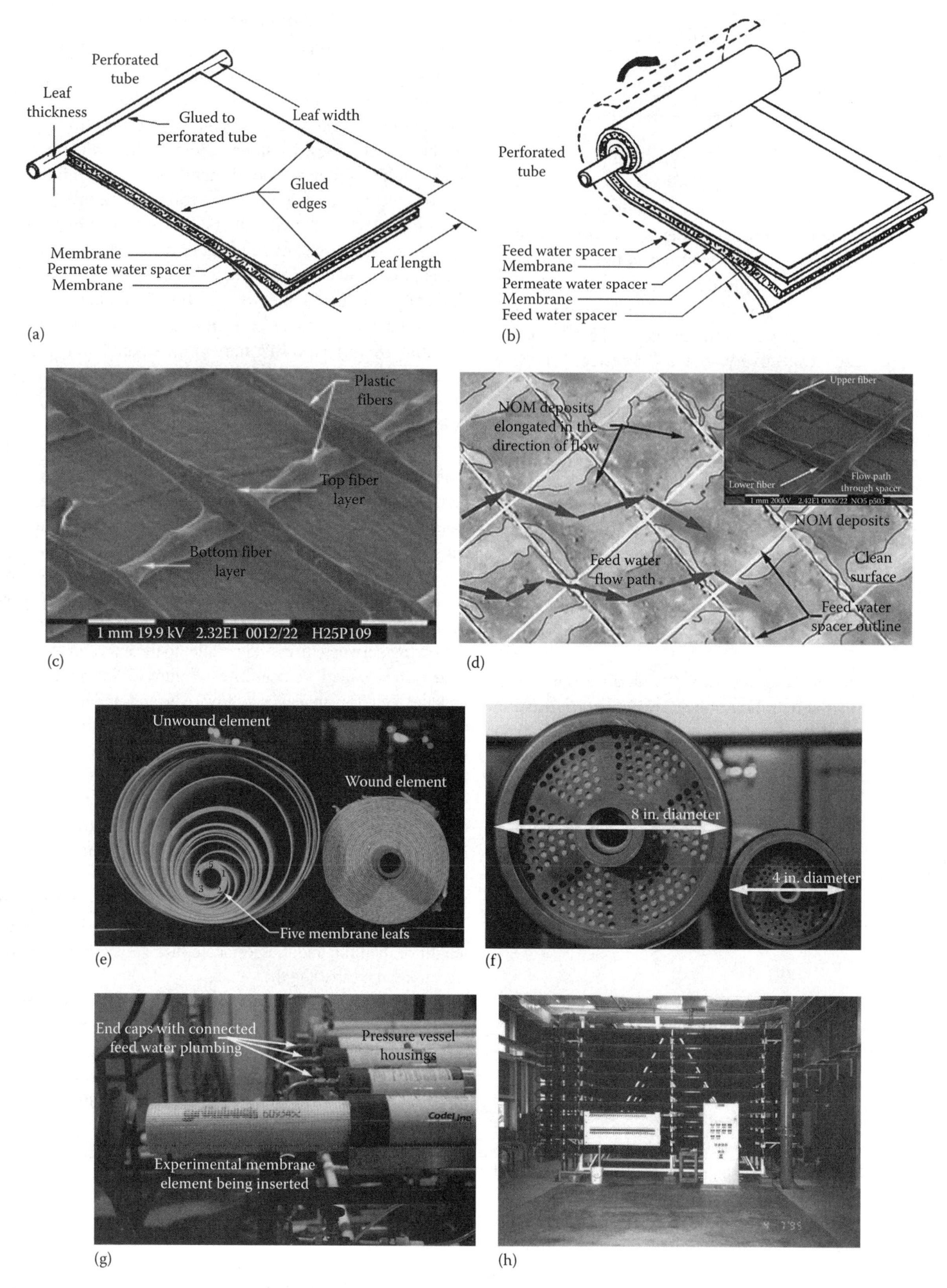

FIGURE 17.8 Spiral-wound membrane details: (a) Membrane sheets showing spacers, (b) sheet assembly being rolled on tube, (c) SEM closeup of spacer mesh, (d) flow path over/under spacer mesh, (e) spiral unfurled and rolled and taped, (f) end caps and permeate center tubes, (g) element and pressure vessel, and (h) side view of RO rack at Brighton, CO. (Photographs (a)–(g) Reproduced from Champlin, T.L., Natural organic matter and particle fouling of spiral-wound nanofiltration membrane elements, Doctoral thesis, Colorado State University, Fort Collins, CO, 1998. With permission; (h) Courtesy of City of Brighton, CO.)

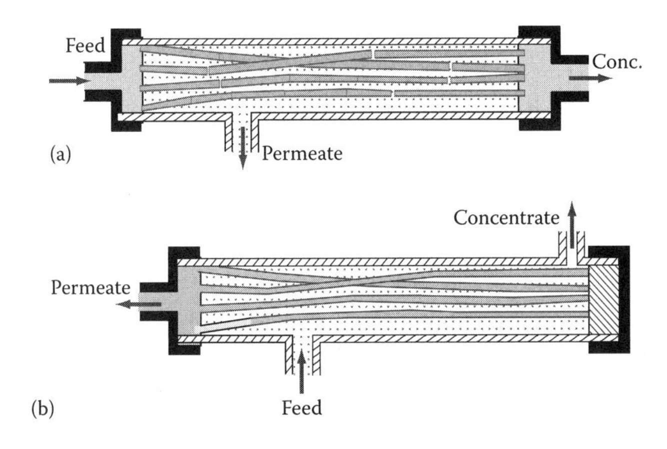

FIGURE 17.9 Schematics of capillary/hollow-fiber membrane modules. (Adapted from Mulder, M., *Basic Principles of Membrane Technology*, Kluwer Academic Publishers, Dordrecht, the Netherlands, p. 113, 1991.)

feed," as illustrated in Figure 17.9a and b, respectively. Also, as illustrated, a "bundle" of fibers may comprise an "element" and is fitted in a pressure vessel, which is a "module" (as defined here). A "bundle" of fibers in six MF or UF plants varied in number from 6,000 to 20,000 (Hugaboom et al., 2003). The hollow-fiber membranes have the highest area/unit volume ratio of any of the kinds of membrane packaging, for example, $\leq$30,000 m^2/m^3 (Baker, 2000, p. 472). Most modules are in pressure vessels but some, for example, Zenon Seaweed® hollow-fiber membranes are placed in open basins to provide both biological treatment (by a biofilm on the individual hollow fibers) and solids separation and are free of a pressure vessel. In the latter case, the membranes are usually operated in a "vacuum" mode (since the pressure in the bore is negative with respect to the atmosphere).

17.1.7.4 Flow within Membrane Element

The feed flow for a spiral-flow membrane element is "cross-flow" as discussed in Section 17.1.7.2, and for a hollow fiber is "dead end." The "bore-feed" hollow fiber has used a periodic "air-bump" to effect cleaning. The "shell-feed" hollow fiber uses a periodic flow of permeate water to flush accumulated solids.

17.1.7.4.1 Cross-Flow

Continuous flow across the surface of a permeable membrane surface is called "cross-flow" filtration, illustrated in Figure 17.10. The feed-flow velocity vector is parallel to the membrane surface (and the axis of the membrane element), and the permeate velocity vector is normal to the surface. Figure 17.10 shows hypothetical particles, some passing across the membrane surface and some retained on the surface, for example, by adsorption. At equilibrium, the particle concentration that adheres to the surface is constant, that is, the rate of deposit equals the rate of shear. Such deposits

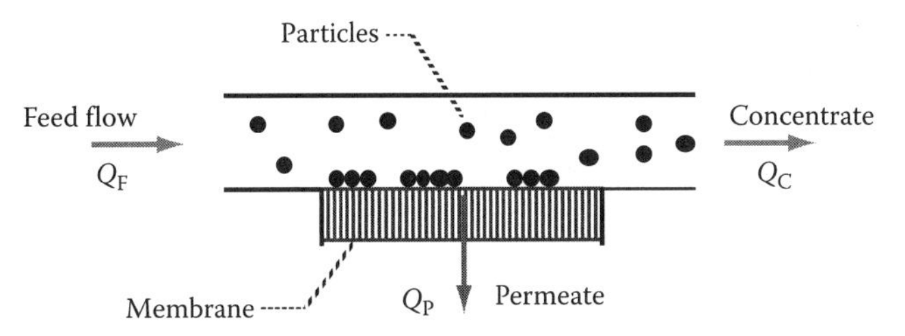

FIGURE 17.10 Cross-flow membrane schematic showing equilibrium between particle deposit and shear.

constitute "fouling" that causes the permeate flux density to decline with time.

17.1.7.4.2 Through-Flow (Dead End)

In through-flow filtration, the entire raw water flow is filtered by the media. With respect to particles, the filter is "dead end"; those that are retained build up on the surface, causing increasing headloss with time. Solids in the water may either remain on the surface, collect in the pores of the filtering media, or pass-through into the filtrate, depending on the solids and the type of membrane (e.g., MF, UF, NF, and RO). At some point in the operation, the filtering media must be purged of solids. For shell-feed hollow-fiber membranes, the solids are purged by a backflow of permeate through the pressure vessel. For bore-feed hollow-fiber membranes, an "air-bump," that is, a pulse of compressed air is used to purge solids. For the clean-in-place protocol, there may be a backflow of cleaning solution under pressure gradient through the membrane.

17.1.7.5 Ratings

Membranes have a pore-size distribution, as illustrated in Figure 17.11. For the case of microporous membranes, the distribution curve is the widest for the anisotropic membrane and narrowest for isotropic membranes. Microporous membranes (for MF/UF) are manufactured to

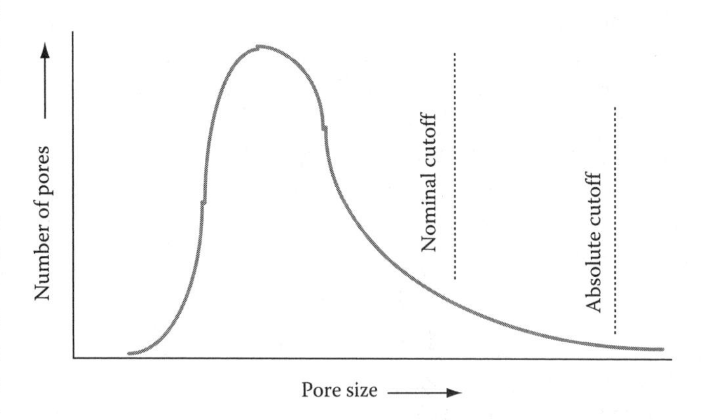

FIGURE 17.11 Hypothetical membrane pore-size distribution illustrating "nominal" and "absolute" cutoff ratings (Mulder, 1991, p. 113).

reject all particles above a certain size. These membranes are given an "absolute" rejection rating based on the largest pore size.

Asymmetric membranes (for NF/RO) are usually given a "nominal" rejection rating; the pore size of the "selective" layer governs the shape of the distribution curve, with narrow curves for both anisotropic and isotropic membranes. Associated with the distribution curve for a given membrane are membrane ratings, that is, absolute cutoff and nominal cutoff, also illustrated in Figure 17.11. The term "absolute cutoff" means that *all* particles of a given size and larger will be "rejected" by the membrane. The "nominal cutoff" rating means that only a portion of the particles (albeit about 98%–99% usually) is rejected. Membrane manufacturers generally state nominal rejections based on experimental data from ASTM tests (ASTM Method D4194-85, ASTM, 1987), which use a standard solution, e.g., a salt, an organic compound, or particles, to test for the fraction of the test substance rejected. As an example, if a standard fructose solution was tested, with MW = 180 daltons, with an observed test rejection of 98%, such a result be the basis for comparing performances of different membranes. An alternative designation for the nominal rejection rating is molecular weight cutoff (MWCO) of solute molecules, which is applicable to NF and RO membranes.

Cheryan (1986, p. 57) shows a distribution for a PS UF membrane similar to Figure 17.11 with low range about 10 Å and with the high about 150 Å with mean about 60 Å. As a matter of interest, he determined that the pore density was about 4×10^{11} pores/cm^2 of membrane surface area. He calculated the flux density of water through a similar membrane by the equation of Poiseuille, that is, $\mathbf{j}$ (cm/s) $= \varepsilon d_p^2 \Delta p / 32\delta\mu$ (Rouse, 1946, p. 158), in which ε = porosity, d_p = mean pore diameter, Δp = pressure differential, and μ = dynamic viscosity of water. For $N = 3 \times 10^9$ pores/cm^2 (from a photomicrograph) and $d_p = 175$ Å, the product, $N \times \pi d_p^2/4 = 7.216 \times 10^{-3}$ cm^2 pore openings/cm membrane surface ≈ 0.0072 fraction of membrane surface is occupied by pores $= \varepsilon$. The membrane skin thickness, $\delta = 0.2$ μm and μ (20°C) $= 10^{-2}$ g/cm/s. His calculations (cgs units) gave $\mathbf{j} = 3.45 \times 10^{-3}$ cm/s $= 124$ L/m^2/h. For comparison, experiment gave 80 L/m^2/h. The Poiseuille model is for straight circular cylinders; by contrast, the pores of a membrane are quite complex. The calculation demonstrates, however, that general notion of the Poiseuille model is likely to be applicable. For the NF/UF membranes, "pores" as such may not exist and so the water permeates by a pressure gradient, but apparently through the molecular structure of the membrane.

17.1.7.6 Variations in Manufacturer's Products

Table 17.4 lists variations in type, pore size, and materials for MF and UF membrane elements representative of those manufactured in 1993. The membranes listed are subject to change and the intent is to provide an indication of the variation that exists. As seen, the membrane types include hollow fiber (shell feed and bore feed), spiral wound, tubular, and flat.

TABLE 17.4
Representative Pore Sizes and MWCO's for Various Membranes

Type	Effective Pore Diameter[a] (μm)	Material[a]
Hollow fiber (shell feed)	0.2	Polypropylene
Tubular	0.2–0.8	Alumina
	MWCO (Da)[a]	
Hollow fiber, tubular	6,000–13,000	Acrylonitrile, glass
Tubular, spiral wound	8,000	Polyamide
	5,000	Polysulfone
	18,000	PVDF
Hollow fiber	1,000–300,000	Polysulfone
Tubular, spiral wound	300 Da–0.1 μm	PTFE (teflon), polysulfone, cellulosic, thin-film composites
Tubular	0.6 μm	Alumina
Spiral wound, flat (DDS)	2,000–500,000	Thin-film composite, polysulfone
Spiral wound	10,000–100,000	Polysulfone
Hollow fiber	100,000	Cellulosic
Spiral wound	1,000 Da–0.1 μm	Cellulosic, polysulfones, VF, polypropylene
Tubular	100,000	PDF
Hollow fiber	100,000	Polysulfone
Tubular	0.02–0.05 μm	Zirconia
Hollow fiber (bore feed)	600–800	Polysulfone
Hollow fiber (shell feed)	400–600	Polysulfone
Tubular	100,000	PVDF

Source: Adapted from Wiesner, M. R., An overview of ΔP membrane processes, in: *Association of Environmental Engineering Professors*, San Antonio, TX, p. 14, June 7, 1993.

[a] A dalton is a unit of mass equal to 1/12 the mass of a carbon-12 atom, that is, about the same as a hydrogen atom. The atomic radius of carbon is given as 0.77 Å (see glossary, Dalton); therefore, to infer that a dalton is about 0.1–1 Å is probably a reasonable inference as to the size associated with a dalton. For a more definitive reference, Silberberg (1996, p. 51) gives the nominal size of an atom as about 10^{-10} m, while a nucleus size is about 10^{-14} m. Also, for reference, 1 μm = 10^{-6} m and 1 Å = 10^{-9} m.

17.1.8 Applications

Initially, in the 1960s, membranes were considered for desalting of seawater. Since the 1990s, the scope of membrane filtration has been expanded to include removals of organisms (cysts, oocysts, bacteria, and viruses), organic compounds, and ions (e.g., selected ones, such as nitrates, boron, etc., or all that are in the feed water). The purposes have included: drinking water treatment, industrial process water, ultrapure industrial water for electronics, reuse of wastewaters, and separation of solids in wastewater treatment by immersed membranes.

17.1.8.1 Particle Removals

Since the early 1990s, MF membranes have been used increasingly for drinking water treatment, often in lieu of the more traditional rapid filtration. MF will remove most of the suspended particles found in ambient waters including such protozoan cysts as *Giardia lamblia* cysts and *Cryptosporidium parvum* oocysts. The removals are absolute. In addition, as seen in Figure 17.5, MF will remove most bacteria. Also, as seen in Figure 17.5, UF will remove all bacteria (e.g., *Salmonella typhi* and *Vibrio cholerae*) and many viruses. Whatever virus species UF does not remove, NF will definitely remove (e.g., hepatitis, polio).

17.1.8.2 Removal of Organics

Most organic molecules from natural organic matter (NOM), described in Appendix 2.A, are removed by NF at perhaps the 0.8–0.9 fraction (see, e.g., Amy et al., 1990; Blau et al., 1992; Champlin and Hendricks, 1994). Because of the lower Δps required, NF is favored over RO for cases in which NOM is the primary target. Some synthetic organics may be removed by NF, but to a lesser extent. RO, on the other hand, removes essentially all organics, that is, NOM or synthetic.

17.1.8.3 Removal of Cations and Anions

With some exceptions, RO is used almost exclusively for removals of cations and anions. Exceptions are for alkalinity and Ca^{2+}, Mg^{2+}, and perhaps a few others, which may be removed about 0.7–0.8 fraction by NF. RO has been used increasingly since the 1980s for seawater desalination (as opposed to distillation) and also for a number of specific applications such as for nitrate removal and to lower the total dissolved solids (TDSs) levels for borderline supplies. Examples of cations and anions removed by RO are (1) cations: Na^+, Ca^{2+}, Mg^{2+}, K^+, Fe^{2+}, Mn^{2+}, Al^{3+}, NH_4, Cu^{2+}, Ni^{2+}, Zn^{2+}, Sr^{2+}, Cd^{2+}, Ag^{2+}, and Hg^{2+} and (2) anions (with names given): chloride-Cl^-, sulfate-$SO_4{}^{2-}$, nitrate-$NO_3{}^-$, fluoride-F^-, silicate-$SiO_2{}^{2-}$, phosphate-$PO_4{}^{3-}$, bromide-Br^-, borate-$B_4O_7{}^2$, chromate-$CrO_4{}^{2-}$, cyanide-CN^-, sulfite-$SO_3{}^{2-}$, thiosulfate-$S_2O_3{}^{2-}$, and ferrocyanide-$Fe(CN)_6{}^{3-}$. As a rule, the larger ions are rejected more completely than the smaller ones.

17.1.9 Pros and Cons

As with other processes, membranes have their own unique advantages/disadvantages. Some of both categories are enumerated as follows:

17.1.9.1 Advantages

- *Replace conventional treatment*—MF and UF are capable of removing particles and pathogenic organisms for which rapid filtration has been the most common treatment technology.
- *Modular design*—As requirement for water increases, additional membrane racks can be added to existing facilities (assuming that the requisite pipe sizes have been provided in the original design).
- *Less operator attention*—Membrane treatment facilities normally require fewer operators and less operator attention as compared with conventional treatment facilities, which has been less true since the late 1980s as rapid filtration plants were being automated.
- *Effective removal*—Membranes are capable of reducing levels of disinfection by-products, organic chemicals, inorganics, and microorganisms, depending on the type of membrane, for example, whether MF, UF, NF, and RO.

17.1.9.2 Disadvantages

- *Costs*—Operation costs include membrane element replacement, pretreatment and posttreatment costs (e.g., fouling inhibitors), cleaning chemicals, and high pressures. In addition, membranes must be replaced after so many years of use.
- *Pretreatment*—Pretreatment may include particle removal using either conventional filtration or membranes of larger pore size (e.g., cartridge filters or MF or UF prior to NF or RO). Inhibitors and antiscalents are also added to prevent biological fouling and inorganic precipitation, respectively.
- *Posttreatment*—NF and RO membranes can effectively "soften" water, resulting in corrosive waters. To counter the corrosive effect, water alkalinity levels are normally increased by the adding a caustic (e.g., lime).
- *Disposal of concentrate*—Concentrate from membrane filtration systems contain all of the "reject" substances from the "feed" water, for example, particles, organics, inorganics, and microorganisms. Concentrate disposal is a special problem for many situations (especially cases in which it's not feasible to discharge to the sanitary sewer for treatment by the wastewater treatment facility.
- *Cleaning of membranes*—When water flux declines become significant, membrane elements require cleaning. Membrane modules are taken off-line and cleaned using stored permeate water and chemical solutions.

17.2 HISTORY

Membrane science started about 1748 when animal membranes were used to conduct osmosis experiments (Presswood, 1981). The technology started about 1939 in Germany for filtering and culturing bacteria for rapid enumeration that became widely adopted by about 1960.

17.2.1 Membranes in Science

Osmosis was the main phenomenon associated with membranes through the nineteenth century that culminated with van't Hoff formulating the law of osmotic pressure.

17.2.1.1 Beginnings

Asymmetry of pores in animal membranes was reported in 1845. Also, in 1845, nitrocellulose, a generic name for the nitration products of cellulose, was synthesized accidentally, which paved the way for synthetic membranes. Collodion, a term for such cellulosic polymers, is a solution of nitrocellulose in an ether–alcohol mixture or an acetic acid–acetone mixture, which when poured over a flat surface was found to form a thin film. Pore structure was controlled by the time allowed for the evaporation of solvents. By 1872, collodion membranes were formed in sheets. At this time, membranes were viewed as objects of curiosity but with little practical value.

17.2.1.2 The Development Period

In 1918, Zsigmondy and Bachmann at the University of Göttingen patented a graded series of membranes and used cellulose and esters of cellulose in a method proposed for commercial production of membrane filters. The membranes had conical pores, as with Brown's membranes. The Zsigmondy method led to the developmental period of membrane technology.

Before World War II, membranes were used mostly to remove microorganisms and particles from liquids, in diffusion studies, and for sizing macromolecules. The art of using the membrane filters for growing bacterial colonies on the surface was to come later. This idea was developed by the Germans in World War II as a means to determine more rapidly the bacterial quality of their drinking water. Mueller and others developed the membrane filtering and culturing process, the one that is commonly used today, to quickly assess drinking water quality. This was developed further in 1951 by Goetz at the California Institute of Technology, which led to the widespread adoption of the membrane filter method for enumeration of bacteria, that is, by the Millipore Corporation.

17.2.1.3 Modern Period

RO membranes were invented independently by Reid and Brenton at University of Florida and by Sourirajan at UCLA in 1958. Later Leob working with Sourirajan invented a film casting technique for the production of asymmetric CA membranes that serves as the basis for modern membrane manufacturing (Saurirajan, 1981; Loeb, 1981). Following this, that is, during the 1960s and 1970s, a sequence of materials were used and included cellulose triacetate, polyvinyl chloride, nylon, polycarbonate, polyamide, and polysulfone (see also, Turbak, 1981a,b).

17.2.2 Membranes in Water Treatment Practice

The major breakthrough that led to modern membrane water treatment practice was the work of Sourirajan and Leob who discovered how to make anisotropic membranes and then learned how to cast a membrane sheet. So it was about 1962 when several companies were borne from embryo start-ups along the Los Angeles–San Diego axis (from discussion with D. Furukawa, c. 1996) with the goal of seawater desalination. As a note, David H. Furukawa was inducted into the American Membrane Technology Association (AMTA) "Hall-of-Fame" in 2009, honouring his many decades of innovation and leadership in the field. By 1968, there was hope of desalination by RO, but the technology was expensive (Anon., 1968, pp. 60–64). By about 1975, however, membranes were operational for the Orange County Water District Plant (Water Factory 21) as the final unit process in a treatment train for treating the wastewaters for recharging groundwater.

17.3 THEORY

Issues of theory relate to solute or particle rejection and permeate flux density. These are the "dependent" variables. The independent variables depend, in general, on the kind of membrane, operating factors, and feed-flow water quality.

17.3.1 Performance Variables

The key variables of performance are effluent concentrations (of constituents of interest), flux density vs. time (and thus "run-length"), transmembrane pressure, and costs (capital and operating). Table 17.5 lists these as "dependent" variables and shows in the right-hand column the associated "independent" variables.

17.3.2 Solute/Particle Rejection

Membranes are, technically speaking, "screen" filters (as opposed to "depth" filters). Screen filters, by definition, retain material that is smaller than the opening of the screen. Screen filters have defined pore sizes and can therefore be given a

TABLE 17.5
Variables in Membrane Process

Dependent Variables	Independent Variables
Concentrations in effluent	Concentrations in influent
	Membrane pore size
Product flow, Q_P	Feed flow, Q_F, Q_C
Flux density, $\mathbf{j_P} = Q_P/A$ (membrane)	Concentrate flow, Q_C
	$R = Q_P/Q_F$
	Pressure differential, Δp, across membrane
	Solute chemical composition
	Solute MW
	pH
	Membrane used
Run length	Particle kind and composition
	SDI
	Scaling potential
	Biological fouling potential
Disposal cost	Site
Cost	Pretreatment
	Membrane used
	Reject disposal
	Source water

quantitative rating in terms of size of particles rejected, for example, MWCO for RO membranes.

Membrane performance is judged primarily by two criteria: (1) rejection fraction of intended solutes and (2) permeate flux density across the membrane and its rate of decline due to fouling. These "dependent" variables are affected by a number of "independent" variables, which include

1. Membrane pore size
2. Membrane composition
3. Membrane configuration
4. Feed-water quality
5. Changes in membrane properties due to induced stresses
6. Concentration polarization
7. Operational parameters

17.3.3 Models Describing Water and Solute Flux through Membranes

Several models have been proposed explaining the transfer of water and rejection of solutes by membranes. For microporous membranes, that is, MF and UF, most agree that water moves across the membrane by advective transport and that the "rejected" suspended particles are physically screened, that is, removed at the surface of the membrane (Cheryan, 1986; AWWA, 1992). The permeate flow follows the Poiseuille law, and is called the "pore-flow" model (Baker, 2000, p. 18).

For the asymmetric membranes, that is, NF and RO, the explanation is not so clear. A number of models have been proposed such as: diffusion and advection, solution–diffusion, pore flow, frictional models based on irreversible thermodynamics, and empirical correlations (Pham et al., 1985). Sourirajan (1970) in the late 1950s suggested a preferential sorption/capillary model where water was preferentially sorbed to the surface of the membrane while solutes experience preferential repulsion. This behavior was dependent on the both chemical nature of the membrane surface and the size of the pores. This sorption of water and repulsion of solutes resulted in the formation of a multimolecular layer of pure water at the membrane–solution interface (Sourirajan, 1970). This layer of pure water then flows through the pores of the membrane under pressure. According to this model, separation of solutes from the water occurs near the interface of the solution and multimolecular layer of pure water. The chemical nature of the membrane provides the conditions for sorption of water and repulsion of solutes, but there is not a direct sieving of solutes.

Later, work by Reid and Breton (1959) suggested that water and solute transfer across CA membranes was due to hydrogen bonding and diffusion. Ions and molecules in the water that could form hydrogen bond with the membrane would be transported across the membrane by alignment-type diffusion. Hole-type diffusion transport was explained for ions and molecules unable to form hydrogen bond with the membrane (Sourirajan, 1970).

The differences in transport models conceived by Sourirajan (1970) and Reid and Breton (1959) during the developmental stages of membrane filtration later divided researchers into two groups: (1) those that believe water and solute transport across asymmetric membranes is due to advection/diffusion transport, and (2) those that believe in a solution–diffusion transport model. Baker (2000, p. 8) stated that the latter, that is, the solution–diffusion model, has been accepted since about 1980. The pores of an RO membrane are on the order of magnitude of 3–5 Å and oscillate in size due to molecular vibration within the membrane polymer. As a rule of thumb, the transition from the solution–diffusion model to the pore-flow model is in the range of 5–10 Å (Baker, 2000, p. 19).

The Poiseuille equation (Equation D.11), used in fluid mechanics to describe laminar flow through pipes, is used commonly to model advective transport through membrane pores where the pressure differential on either side of the membrane drives the transport (Weber, 1972; Duranceau et al., 1992; Amjad, 1993; Laisure et al., 1993). That water flux is proportional to the Δp across the membrane, and is inversely proportional to viscosity, is well established by experimental evidence. What is not clear is that since an RO membrane has a homogeneous molecular structure that does not show discernable "pores," the laminar flow model does not seem plausible (in the sense that the pipe flow model may be valid, which assumes zero velocity at the pipe wall and a parabolic velocity distribution).

17.3.4 Basic Notions for a Cross-Flow Membrane Element

A "cross-flow" membrane is by definition one in which the "feed flow" between two membrane sheets passes through the element and leaves it as "concentrate-flow." The cross-flow is diminished by the permeate flow along the axis of the membrane. Also, the cross-flow results in a shear along the membrane surface, which may erode some deposits.

17.3.4.1 Flow Balance

Figure 17.12 (the same as Figure 17.1a) shows flows in and out of a cross-flow type of membrane element, depicting a flow balance, that is,

$$Q_F = Q_P + Q_C \tag{17.1}$$

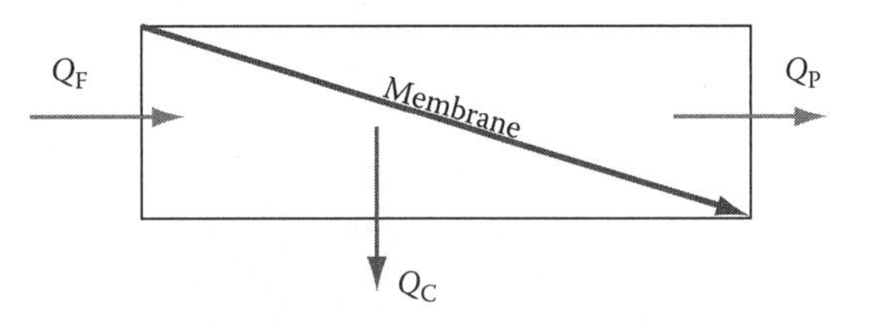

FIGURE 17.12 Flow balance definition sketch for a cross-flow type membrane element.

where
Q_F is the feed flow (m^3/s)
Q_P is the permeate flow (m^3/s)
Q_C is the concentrate flow (m^3/s)

The flow balance is the basis for further definitions. The feed flow, Q_F, is the influent flow to a given membrane element after whatever pretreatment occurs. The permeate flow, Q_P, is what passes through the membrane, and the concentrate flow, Q_C, leaves the membrane pressure vessel (but not through the membrane).

17.3.4.2 Mass Balance and Pressures

Figure 17.13 defines the flows, pressures, and solute concentrations for a cross-flow membrane element. The system is the same as depicted in Figure 17.12 except that concentrations and pressures are shown also.

The terms in Figure 17.13 are defined:

Q_F = flow of feed water (m^3/s)
Q_C = flow of concentrate water (m^3/s)
Q_P = flow of permeate water (m^3/s)
C_F = solute concentration in feed water (kg/m^3)
C_C = concentration of solute in concentrate water (kg/m^3)
C_P = concentration of solute in permeate water (kg/m^3)
C_m = concentration of solute at the surface of membrane (kg/m^3)
P_F = pressure of feed water (N/m^2)
P_C = pressure of the concentrate water (kg/m^2)
P_P = pressure of the permeate water (kg/m^2)
$\mathbf{j_w}$ = flux density of water through membrane ($m^3/s/m^2$)
$\mathbf{j_s}$ = flux density of a given solute through membrane ($kg/s/m^2$)

Based on Figure 17.13, the mass balance equation is

$$Q_F C_F = Q_P C_P + Q_C C_C \tag{17.2}$$

Other definitions seen often in the literature are

$$R \equiv \frac{Q_P}{Q_F} \tag{17.3}$$

$$R^\circ \equiv \frac{C_F - C_P}{C_F} \tag{17.4}$$

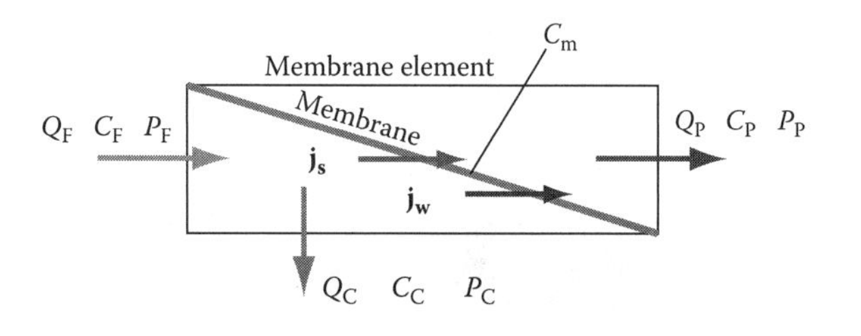

FIGURE 17.13 Mass balance and pressures definition sketch for a cross-flow membrane element.

where
R is the water recovery (dimensionless)
R° is the solute rejection (dimensionless)

17.3.4.3 Water Flux Density

Water flux density through a membrane is defined:

$$\mathbf{j_w} \equiv \frac{Q_P}{A} \tag{17.5}$$

where A = surface area of the membrane (m^2).

17.3.4.4 Solute Mass Flux

Solute mass flux density is defined as the mass of a solute passing through a unit area of membrane per unit of time, that is,

$$\mathbf{j_S} = \frac{Q_P \cdot C_P}{A} \tag{17.6}$$

where
$\mathbf{j_S}$ is the mass flux of solutes through the membrane ($kg/s/m^2$)
A is the total area of membrane in filtering module (m^2)
Q_P is the flow of the permeate water ($m^3/m^2/s$)
C_P is the solute concentration in the permeate water (kg/m^3)

17.3.4.5 Transmembrane Pressure

The "transmembrane pressure," ΔP, is defined as the difference in pressure between the two sides of the membrane. For a spiral-wound membrane, however, pressure loss occurs between the entrance to the membrane element and the exit. Therefore, to obtain ΔP, these two pressures are averaged and the permeate pressure is subtracted, that is,

$$\Delta P = \left(\frac{P_F + P_C}{2}\right) - P_P \tag{17.7}$$

where
ΔP is the transmembrane pressure (kPa)
P_F is the pressure of the feed water (kPa)
P_C is the pressure of the concentrate water (kPa)
P_P is the pressure of the permeate water (kPa)

The permeate flow should be controlled by an overflow weir or some means to maintain a positive pressure within the membrane element and thus reduce the incidence of gas precipitation (bubble formation).

17.3.5 Poiseuille Law

The equation of Poiseuille for laminar flow-in pipes (Rouse, 1946, p. 158) is commonly adopted to model water flux

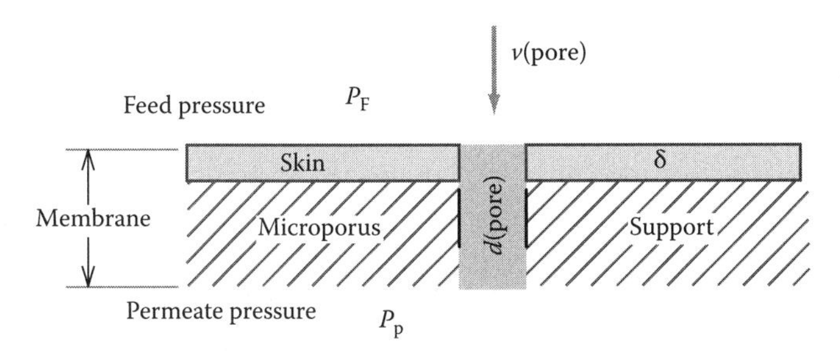

FIGURE 17.14 Membrane cross section with reference to Poiseuille equation. (Adapted from Cheryan, M., *Ultrafiltration Handbook*, Technomic Publishing Company, Inc., Lancaster, PA, 1986, Figure 4.3. With permission.)

through a microporous membrane. Figure 17.14 illustrates the flow through a membrane pore with pressures on each side of the membrane indicated, with $\Delta P = P_F - P_P$.

Referring to Figure 17.14, the Poiseuille equation adapted for membrane pore flow is (see also, Equation D.10),

$$\frac{\Delta P}{\delta} = \frac{32\mu}{d(\text{pore})^2} v(\text{pore}) \tag{17.8}$$

where
- ΔP is the pressure difference across membrane (N/m^2)
- δ is the membrane thickness (m)
- μ is the dynamic viscosity of water (N × s/m^2)
- v(pore) is the velocity through pores (m/s)
- d(pore) is the diameter of pore (m)

Equation 17.8 can be rearranged in terms of pore velocity, that is,

$$v(\text{pore}) = \frac{d(\text{pore})^2}{32\delta} \frac{\Delta P}{\mu} \tag{17.9}$$

Thus, the velocity through the pore is proportional to the diameter squared of the pore and the pressure difference across the membrane, and is inversely proportional to the viscosity of the water and the thickness of the membrane.

To calculate the flux, **j**, through a membrane, v(pore) can be combined with the continuity equation, that is,

$$\mathbf{j}_w = v(\text{pore}) \cdot a(\text{pore}) \cdot n(\text{pores}) \tag{17.10}$$

$$= \left[\frac{d(\text{pore})^2}{32\delta} \frac{\Delta P}{\mu}\right] \cdot \left[\frac{\pi d^2}{4}\right] \cdot n(\text{pores}) \tag{17.11}$$

$$= \left[\frac{\pi}{128}\right] \cdot \left[\frac{n(\text{pores}) d(\text{pore})^4}{\delta}\right] \cdot \left[\frac{\Delta P}{\mu}\right] \tag{17.12}$$

$$= K(\text{membrane}) \cdot \left[\frac{\Delta P}{\mu}\right] \tag{17.13}$$

Combining Equations 17.12 and 17.13, gives for K(membrane),

$$K(\text{membrane}) = \left[\frac{\pi}{128}\right] \cdot \left[\frac{n(\text{pores}) d(\text{pore})^4}{\delta}\right] \tag{17.14}$$

where
- $\mathbf{j}_w$ is the flux density of water (m^3 water/m^2/s)
- a(pore) is the area of pore (m^2)
- n(pores) is the pore density or number of pores per unit area of membrane (pores/m^2)
- K(membrane) is the coefficient characteristic of the membrane (m)

Equation 17.14 groups the independent variables related to the membrane characteristics, that is, n(pores), d(pore), and δ, and operating variables, that is, ΔP and μ (albeit μ is a function of water temperature and is not controlled in operation).

Equation 17.12 is useful in separating the membrane variables from the operating variables. For example, with constant ΔP, the flux density may vary as temperature changes (see Table B.9, which gives the effect of temperature on viscosity of water). As related to membrane characteristics, Equation 17.12 shows that flux density is affected by pore diameter to the fourth power, that is, d(pore)^4. Thus, the pore-size distribution is important. Suppose, for example, that $d(\text{pore})_1 = 1$, $d(\text{pore})_2 = 2$, and $n(\text{pore})_1 = n(\text{pore})_2$; the flow through pore 2 would be 16 times the flow through pore. The larger pores will therefore transport a disproportionately higher fraction of the water flux through the membrane. If the membrane has a distribution of pore sizes, the small pores will transport relatively low flows.

Equation 17.14 incorporates the fundamental membrane variables into a coefficient, K(membrane), which is an "intrinsic" mass transfer coefficient and must be evaluated empirically by measurements of $\mathbf{j}_w$ and ΔP and μ. The coefficient, K(membrane) applies to a "clean" membrane. Membrane "fouling" will occur, however, due to development of a layer on top of the membrane, or by deposits within the membrane pores. For those membranes that are open to flow, the "cross-flow" velocity affects the rate of deposition. Fouling will cause K(membrane) to increase and so in this sense K(membrane) is an empirical coefficient. Its value when new, based on the "clean-membrane" flux density and flow at a given temperature, may be designated the "clean-membrane" mass transfer coefficient and designated, K^o(membrane).

BOX 17.3 CONVERSIONS OF K_W

To convert K_w from units of (L/day/m^2/kPa) to (day^{-2}), multiply (L/day/m^2/kPa) by the factor, 0.00981. Both sets of units are often seen in the literature, along with SI units. Ordinarily, unit conversions do not warrant a special mention; this particular conversion is an exception, as may be seen by doing the conversion in either direction, that is, since it could cause a mild frustration.

17.3.6 Osmosis

Figure 17.15a through c, depicts a solution (left side) and pure water (right side) for the initial condition, the occurrence of osmosis, and the imposition of pressure on the solution side to cause RO, respectively. These three figures capture, in-a-nutshell, the key ideas of osmosis.

17.3.6.1 Osmotic Pressure

Osmosis is a natural phenomenon that occurs when two solutions having different solute concentrations come in contact with a semipermeable membrane, that is, a membrane permeable only to water. To explain in more detail, Figure 17.15a depicts the initial condition when the solution is placed on the left side of the membrane and pure water is placed on the right side. Figure 17.15b depicts the water flow from the pure water side to the solution side. The water level on the solution side will rise to a point equivalent to the osmotic pressure, π, of the concentrated solution. This is the osmosis process.

When the water level rises on the solution side such that the osmotic pressure is achieved, the water fluxes in both directions across the membrane are equal. One explanation is that the pure water side on the right has no blockage of pores due to the dissolved substances and so there is an uninhibited diffusion flux of water from the right side to the left side. On the left side, pores are blocked in proportion to the mole fraction of solute and so the diffusion flux from left to right is less. As the water level rises, the water flux from left to right increases due to the pressure differential until the two fluxes are equal.

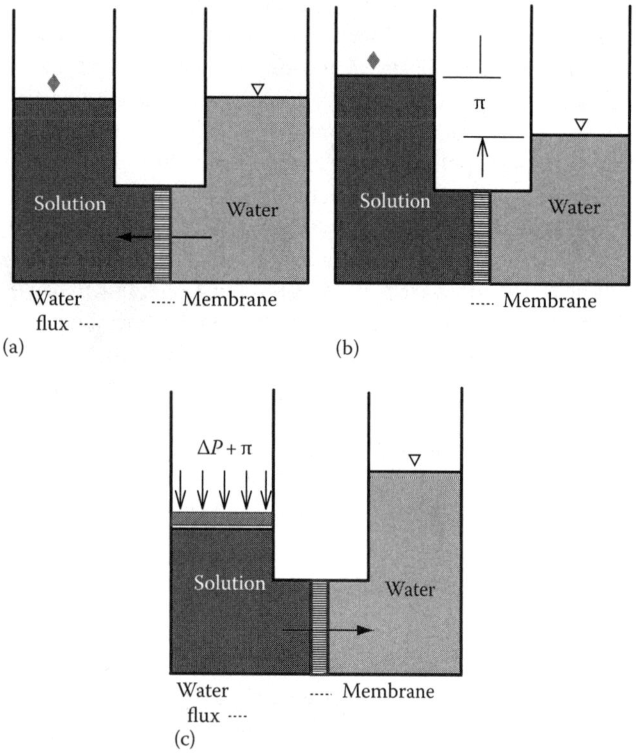

17.3.6.1.1 Calculation of Osmotic Pressure

The van't Hoff law for dilute solutions permits calculation of the osmotic pressure as a function of solute concentration, that is,

$$\pi = C_B RT \qquad (17.15)$$

where

π is the osmotic pressure (kPa)
C_B is the concentration of solute "B" (mol/m^3)
R is the universal gas constant (8.314510 N-m/g-mol K)
T is the temperature of solution (K)

The van't Hoff relation may be derived starting with an equilibrium statement that equates the partial molar free energy of the pure water solution with the water in solution (see, e.g., Mulder, 1991, p. 201). The equation gives a reasonably accurate estimate of osmotic pressure. Figure 17.16 shows the results of calculations of π for sodium chloride and gives references for brackish water and seawater; concentration ranges are about 2,000–5,000 and 35,000–39,000 mg/L, respectively. The calculations are less accurate as concentration increases.

17.3.6.2 Reverse Osmosis

Figure 17.15c depicts RO. Any additional pressure on the left side, that is, the solution side, that is higher than the osmotic pressure will cause an increase in flux density from the solution side to the pure water side. The increase is proportional to the excess pressure, that is, net driving pressure ($\Delta P - \pi$). The process of causing this net flow across the membrane from the solution side to the pure water side has been termed "reverse osmosis."

Solution
Water
Water flux
Membrane
(a)
Solution
π
Water
Membrane
(b)
$\Delta P + \pi$
Solution
Water
Water flux
Membrane
(c)

FIGURE 17.15 Membrane between solution and pure water. (a) Initial conditions, (b) osmosis, and (c) reverse osmosis.

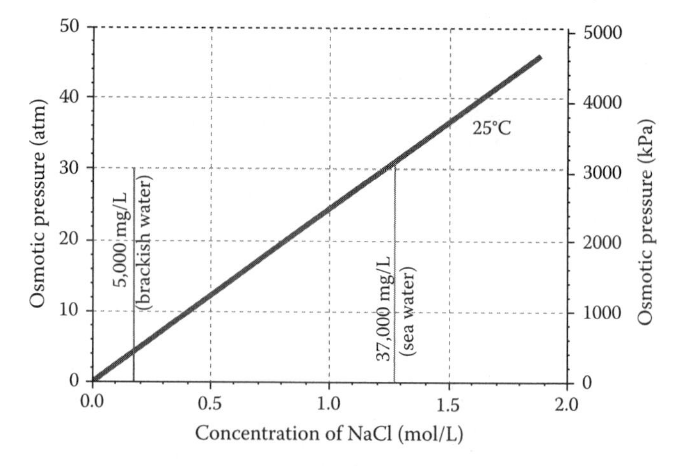

FIGURE 17.16 Osmotic pressure as a function of molar concentration according to van't Hoff's law.

Example 17.1 Calculation of Osmotic Pressure

Given

A solution of $MgCl_2$ has a concentration of 1000 mg/L. Let the temperature of the solution be 25°C.

Required

Calculate the osmotic pressure.

Solution

A spreadsheet, Table CD17.6, shows the sequence of calculations. Starting with the compound, the molecular weight is determined (based on data in Table B.1). The molar concentration of the compound itself is given. But when dissolved, the compound dissociates into ions, giving three ions per molecule. Therefore, the multiplier is "3" to give a total molar concentration given in the next column. Next, the van't Hoff equation, Equation 17.15, is applied to give the osmotic pressure, π, in kPa (using $R = 8314.510$ Pa-L/g-mol K). The van't Hoff equation must be divided by 1000 since the units are kPa (i.e., not Pa). Divide by 101.325 kPa/atm to obtain the osmotic pressure in atmospheres.

Discussion

The spreadsheet shows the van't Hoff calculation for several compounds and may be used for other calculations that may be of interest. Consider, for example, water with TDS $\approx$ 500 mg/L. Let NaCl serve as a surrogate. Calculation by the spreadsheet, Table CD17.6, gives $\pi = 0.42$ atm (43 kPa or 6 psi). In other words, the membrane flux is proportional to the transmembrane minus π.

17.3.6.3 Effect of Membrane Pressure on Water Flux Density

RO forces a solvent from the solution side, as depicted in Figure 17.15c, to the pure solvent side by application of a pressure higher than the osmotic pressure. The flux is proportional to the pressure difference between the two sides of the membrane minus the osmotic pressure, that is, the net driving pressure $(\Delta P - \pi)$, to give,

$$\mathrm{j_w} = \frac{K(\text{membrane})}{\mu}(\Delta P - \pi) \tag{17.16}$$

where

ΔP is the pressure difference between the two sides of membrane (kPa)

π is the osmotic pressure of solute side of membrane (kPa)

The equation applies to any membrane that is not permeable to salt. Technically, the term π should be $\Delta\pi$, where $\Delta\pi = \pi(\text{salt side}) - \pi(\text{pure water side})$. But since, as a rule,

TABLE CD17.6

Calculation of Osmotic Pressure by van't Hoff Law of Dilute Solutions (Equation 17.15)

R = 0.0820578 L-atm/g-mol K

R = 8.314510 Pa-m^3/g-mol K

R = 8.314510 N-m/g-mol K

R = 8314.510 Pa-L/g-mol K (used in calculations)

Compound	MW (g/mol)	C_B (mg/L)	(g/L)	(mol/L)	(ions/mol)	(mol/L)	T (C)	π (kPa)	π (atm)
NaCl	58.442	37,000	37.00	0.633	2	1.266	25	3139	30.977
NaCl	58.442	5,000	5.00	0.086	2	0.171	25	424	4.186
NaCl	58.442	1,000	1.00	0.017	2	0.034	25	85	0.837
$NaHCO_3$	83.991	1,000	1.00	0.012	2	0.024	25	59	0.583
Na_2SO_4	142.022	1,000	1.00	0.007	3	0.021	25	52	0.517
$MgSO_4$	120.347	1,000	1.00	0.008	2	0.017	25	41	0.407
$MgCl_2$	95.2104	1,000	1.00	0.011	3	0.032	25	78	0.771
$CaCl_2$	110.9834	1,000	1.00	0.009	3	0.027	25	67	0.661
Sucrose	342.241	1,000	1.00	0.00292	1	0.003	25	7	0.071
Dextrose	180	1,000	1.00	0.00556	1	0.006	25	14	0.136

Source: For gas constant: Lide, D.R., *CRC Handbook of Chemistry and Physics*, 77th edn., CRC Press, Boca Raton, FL, 1996.

Sucrose is $C_{12}H_{22}O_{11}$

Compound	MW
Ca	40.078
Cl	35.4527
C	12.01100
H	1.00794
Mg	24.305
Na	22.989768
O	15.994
S	32.066

π(pure water side) ≈ 0, we can simplify the equation, that is, $\Delta\pi = \pi$(salt side) $- \pi$(pure water side) $= \pi$(salt side) $- 0$. Then simply let $\Delta\pi = \pi$(salt side) $= \pi$ in order to simplify the nomenclature.

17.3.7 Electrodialysis

The term, *dialysis* is the transfer of solute molecules across a membrane by diffusion from a concentrated solution to a dilute solution. The term *electrodialysis* refers to the transfer of ions across membranes due to the influence of an electric field (Helferrich, 1962, p. 397). An electric field is applied between two electrodes to mobilize ions to move in the direction of the oppositely charged plate across ion-selective membranes. The basic innovation goes back to the 1930s (Lonsdale, 1982), when desalting was demonstrated in a three-compartment cell. The multicompartment cell, the configuration of current technology, came in 1940 and is illustrated in Figure 17.17. As seen, the positive electrode plate is on the left and the negative plate is on the right. Therefore, the anions (represented by Cl^-) are attracted toward the left plate and the cations (represented by Na^+) toward the right plate. The membranes are selective, passing anions or cations, respectively. For the system depicted, the center cell becomes depleted of both anions and cations, and desalting occurs. ED has been applied most often to desalination of brackish waters.

Referring to Figure 17.17 again, a "cell-pair" is comprised of a concentrating cell and an ion depleting cell; the cell includes "spacers" that separate the membranes and permit flow to occur. In practice, a membrane "stack" is composed of some several hundred cell pairs (Meller, 1984, p. 13). The system must be set up with a manifold system that collects the demineralized water and the saline water in separate flows; the anode and cathode streams are minor and are waste also.

The anode cell and the cathode cell are special cases. What happens at each electrode is depicted also in Figure 17.17. As seen, an oxidation reaction occurs at the anode, that is, Cl^- loses electrons to form chlorine gas. Also water is dissociated to form H^+ and O_2 with a loss of electrons. The anode is a sink for these electrons. At the cathode, a reduction reaction occurs; the cathode gives off electrons producing OH^- and H_2 gas.

17.3.7.1 Applications

Most ED systems are for desalinization of brackish waters. More than 1000 ED plants have existed around the world (Lonsdale, 1982), with 310 having capacities more than 95 m^3/day (25,000 gpd). The combined desalting capacity was about 272,000 m^3/day (72 mgd). The ED technology has competed with RO for brackish water desalinization since the first plant in 1969 (Reahl, 2006, p. 1). Initially, c. 1969, ED was the only viable nondistillation type desalination technology. The 2008 share of worldwide capacity was about 0.04 fraction (Section 17.1.2), having being reduced proportionately as RO membranes became cheaper.

17.3.8 Fouling

A reduction in membrane flux density due to foreign material is called fouling. Almost all substances in water have the potential to foul membranes and include (1) particles, (2) organic compounds, (3) inorganic compounds, and (4) biota that grow on the membrane surface. Particles foul membranes by either collecting inside membrane pores or by blocking pores due to surface deposition. Organics, inorganics, and biota may adsorb to membrane surfaces and pores. The nature and extent of fouling is influenced by the chemical and biological nature of the water, the chemical composition of the membrane, solute–solute type interactions, and membrane–solute type interactions.

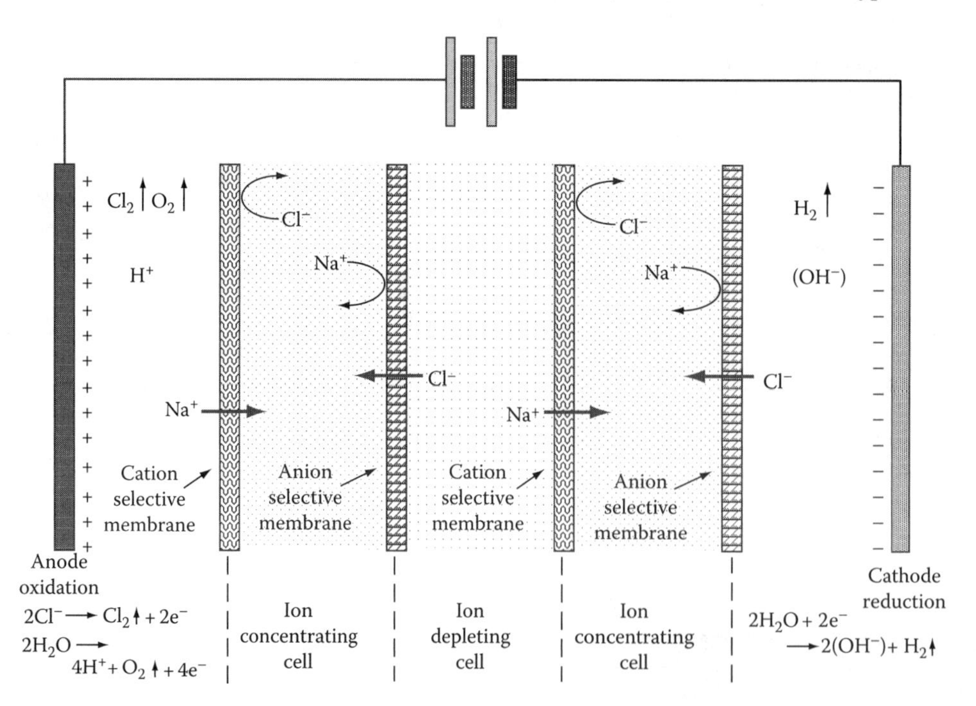

FIGURE 17.17 Electrodialysis stack of five cells. (Adapted from Lonsdale, H.K., *J. Membr. Sci.*, 10, 81, 1982.)

17.3.8.1 Reversible and Irreversible Fouling

Membranes invariably foul with time. Figure 17.18 illustrates four fouling cycles, in which flux declines with time but is restored partially by cleaning. The fouling cycle may be defined by three fouling terms, that is, (1) "total" fouling, seen as the overall loss of flux; (2) reversible fouling, which is that part of total fouling that may be restored by cleaning; and (3) irreversible fouling, which is that part of total fouling that is not restored by cleaning. For spiral-wound membranes, the time between the cleaning cycles depends on the feed-water quality, which depends, in turn, on the ambient water quality and the pretreatment. For spiral-wound membranes, the time between cleaning events may vary from weeks to months. Over time, however, irreversible fouling increases to such extent that replacement becomes economical. In the case of the RO membranes at Brighton, Colorado, the membranes were replaced only after several years of operation. In general, cleaning is scheduled after about 10% increase in pressure (increasing pressure is necessary to maintain constant flux density).

17.3.8.2 Natural Organic Matter

NOM may be a significant cause of fouling, both reversible and irreversible. Causes may be either adsorption (Champlin, 1998) or concentration polarization.

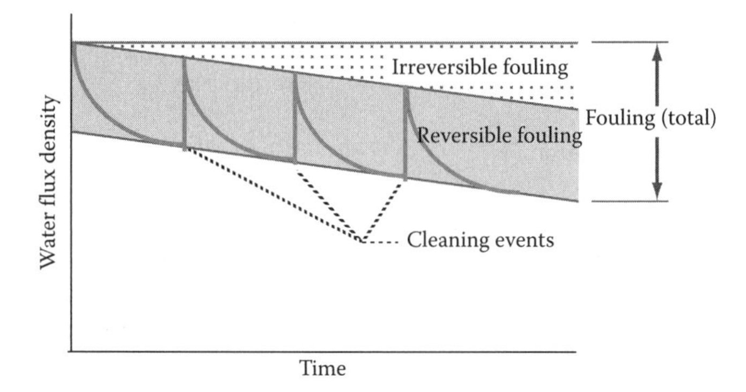

FIGURE 17.18 Fouling cycles for spiral-wound membrane.

17.3.8.3 Particle Fouling

Particles may deposit on membrane surfaces or collect within membrane pores. Deposited particles may remain near the surface (e.g., gel-polarization layer or pore blockage) or they may be transported back into the bulk flow by diffusion after the removal by shear. At steady state, the net transport of particles toward the membrane is in balance with the back-transport mechanisms (i.e., Brownian diffusion and turbulent diffusion if the advective flow is in the turbulent regime). Calculations performed by Wiesner and Chellam (1992) describing flow through thin plates showed that particles near 0.1 μm in radius would preferentially accumulate on the surface of the membrane. Theoretically, particles less than 0.1 μm in radius are transported primarily by Brownian diffusion, while particles greater than 0.1 μm in radius are transported primarily by turbulent diffusion.

Figure 17.19a and b shows scanning electron microscope images of two NF membrane surfaces. Both illustrate different types of foulants, for example, particle fouling and NOM fouling, respectively.

17.3.8.4 Inorganics

Inorganics such as carbonate, sulfate, fluoride, and phosphate salts, metal hydroxides, sulfides, and silicates may affect fouling by precipitating on membrane surfaces. Examples of precipitants include $CaSO_4$, CaF_2, $BaSO_4$, and $CaCO_3$. Such deposits increase the hydraulic resistance, that is, high transmembrane pressures, and therefore constitute fouling. Certain kinds of cleaning may remove a portion of this kind of fouling.

17.3.8.5 Concentration Polarization

Concentration polarization (also called, "gel" polarization) describes the increase in solute concentrations near the surface of the membrane. This phenomenon can lower water flux through a membrane due to either increased hydraulic resistance or due to higher local osmotic pressures. Which is dominant is debated in the literature (Rodgers and Sparks, 1992, p. 150).

(a)

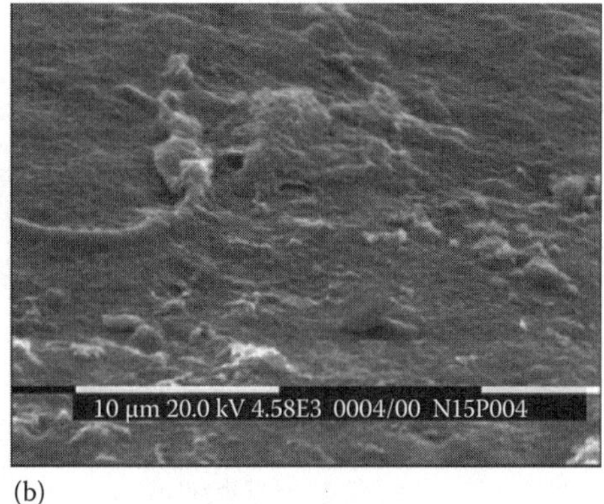

(b)

FIGURE 17.19 Nanofiltration membrane surface showing kinds of fouling. (Electron photomicrographs by Champlin, T.L., Natural organic matter and particle fouling of spiral-wound nanofiltration membrane elements, Doctoral thesis, Colorado State University, Fort Collins, CO, 1998. With permission.) (a) Particle fouling and (b) NOM fouling.

17.3.8.5.1 Depiction of Gel-Layer Development

During membrane filtration, solutes are transported from the bulk "cross-flow" to the surface of the membrane by advective transport as a part of the permeate flow. Solutes that are rejected accumulate near the surface forming either deposits or a gelatinous-type layer. These solutes either stay near the surface or diffuse back into the bulk flow by Brownian motion or shear-induced motion (i.e., turbulent diffusion). Figure 17.20 illustrates these processes. The thickness of deposition or the "gel-polarization layer" is established at steady state when the advective transport rate and back-transport rate (by diffusion) are equal. Equation 17.17 describes the mass balance between these two transport mechanisms,

$$-D_{\text{bl}}\frac{\text{d}C}{\text{d}x} + uC = 0 \tag{17.17}$$

where
- D_{bl} is the diffusion coefficient (m^2/s)
- $\text{d}C/\text{d}x$ is the concentration gradient over the differential element (kg/m^4)
- u is the permeation velocity (m/s)

Integration of Equation 17.22 yields:

$$u = \left(\frac{D_{\text{bl}}}{\delta_{\text{bl}}}\right)\ln\left(\frac{C_{\text{m}}}{C_{\text{b}}}\right) \tag{17.18}$$

where
- δ_{bl} is the thickness of the concentration polarization boundary layer (m)
- C_{m} is the concentration of the solute at the surface of the membrane (kg/m^3)
- C_{b} is the concentration of the solute in the bulk flow (kg/m^3)

The wall concentration, C_{m}, increases rapidly with increases in permeation velocity, u, and at some point, reaches the "gel-concentration," C_{g}, where the solution is no longer a fluid. At this point, u reaches a limiting value, u(min), defined as,

$$u(\min) = \left(\frac{D_{\text{bl}}}{\delta_{\text{bl}}}\right)\ln\left(\frac{C_{\text{g}}}{C_{\text{b}}}\right) \tag{17.19}$$

As observed by Equation 17.22, fluxes in and out of the boundary layer are independent of pressure. The equation is valid only when operating in the pressure-independent diffusion-controlled region, shown in Figure 17.20. Because C_{g} and C_{b} are controlled primarily by physiochemical properties, flux through the boundary layer can be improved only by increasing the ratio ($D_{\text{bl}}/\delta_{\text{bl}}$), for example, by reducing the thickness of the boundary layer, δ_{bl}, by higher cross-flow velocity. Attempts to increase flux (such as increasing pressure), however, is self-defeating as long as mechanisms are not provided to increase the rate of back diffusion into the bulk flow (Cheryan, 1986).

17.3.8.5.2 Factors That Affect Concentration Polarization

Variables that affect the concentration polarization layer include pressure, temperature, solute feed concentration, and cross-flow velocity. Figure 17.20 provides a basis for understanding how these variables relate to membrane fouling, with special focus on the concentration polarization layer.

17.3.8.5.3 Effect of Concentration Polarization Layer on Pressure–Flux Relationship

Water flux through the membrane is directly proportional to the applied pressure in the "pressure-controlled region," illustrated in Figure 17.21. In cases where pressures and feed concentrations are relatively low and cross-flow velocities

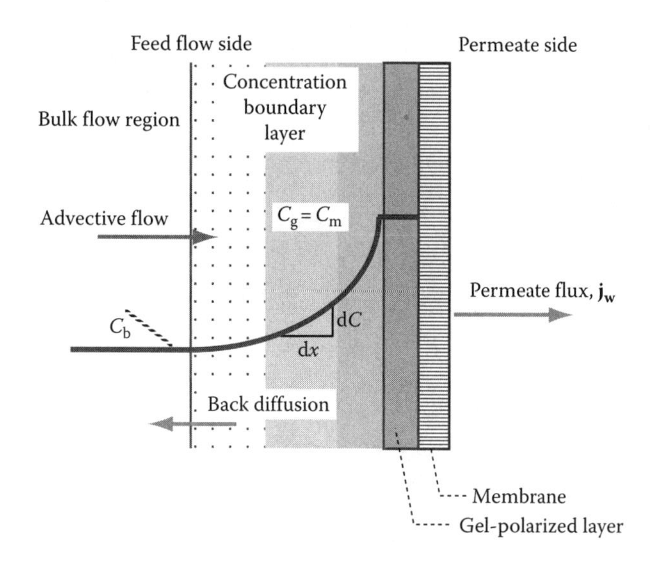

FIGURE 17.20 Depiction of how gel-layer is formed. (Adapted from Cheryan, M., *Ultrafiltration Handbook*, Technomic Publishing Company, Inc., Lancaster, PA, p. 83, 1986.)

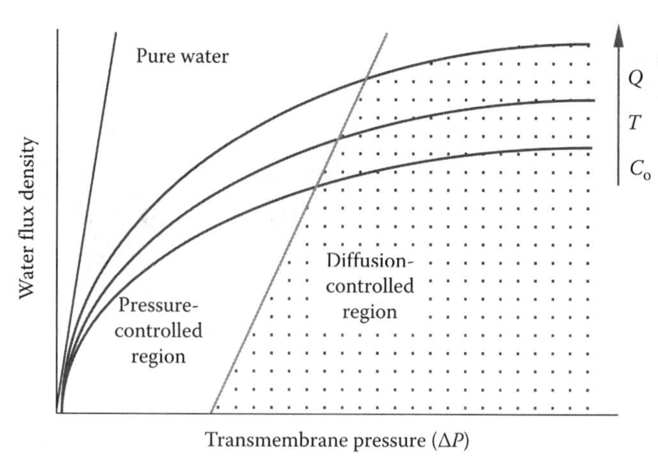

FIGURE 17.21 Pressure-controlled and diffusion-controlled regions in membrane operation (Adapted from Wiesner, M.R., An Overview of ΔP Membrane Processes, Presented to Association of Environmental Engineering Professors, San Antonio, TX, p. 37, June 7, 1993.)

are high, the relationship between water flux and pressure is valid. If, however, the process deviates from these conditions substantially, water flux may become independent of pressure. The asymptotic portion of the curves that depict the water flux vs. pressure relationship, as illustrated in Figure 17.21 in the "diffusion-controlled region," is considered a result of concentration polarization (Cheryan, 1986).

17.3.8.5.4 Effect of Temperature

In general, increasing the temperature of the feed water will increase flux in both the pressure-controlled and the diffusion-controlled regions of operation, as indicated in Figure 17.21. In the pressure-controlled region, temperature affects water flux due to its effect on viscosity. In the diffusion-controlled region, temperature affects the diffusivity of the feed solutes, thereby affecting concentration polarization effects (Cheryan, 1986).

17.3.8.5.5 Effect of Feed-Water Solute Concentration

A higher solute concentration in the feed flow can significantly decrease permeate flux and increase fouling by decreasing solute diffusivity, and thus increase concentration polarization effects. At the same time, increasing feed concentration may increase fouling by exceeding solute solubility's within the concentration polarization boundary, thus precipitating solutes on the membrane surface. Concentrated organics near the surface of the membrane can form a gelatinous layer or what is known as a gel layer (Cheryan, 1986). The gel layer can enhance the collection of particles and the growth of microorganisms on the surface of the membrane.

17.3.8.5.6 Cross-Flow Velocity

The fluid shear at the membrane surface is proportional to the cross-flow velocity. Deposited materials at the membrane surface are resuspended in proportion to this shear stress, being resisted by the adhesion between the foulant and the membrane and the cohesive forces within the foulant material. Higher cross-flow velocities decrease the effects of concentration polarization by reducing the boundary layer thickness at the surface of the membrane. A cross-flow velocity of 1–4 m/s will produce sufficient shear to resuspend most deposited materials (Cheryan, 1986). In the case of spiral-wound membrane elements, mesh-like materials are used for feed-flow spacers to promote turbulence. These spacers, however, create stagnant flow zones behind the spacers and may enhance deposition of particle matter (Champlin, 1998).

17.4 DESIGN

Only two aspects of process design are covered here: pretreatment and layout. Design of a plant includes, however, many ancillary aspects to make it work. A membrane skid is common and may be installed in a building sized to accommodate as many skids as needed for present and future flows. Some of the ancillary components may include header pipes, tubes, gages, etc. The support facilities may include pumps to pressurize the system, cleaning tanks with containment, pretreatment system, chemicals for pH adjustment, gas-stripping tower, storage of permeate water, blending tank, pumping to distribution system, sensors, actuated valves, pumps for backwash, laboratory, offices, computer control system, etc.

17.4.1 Pretreatment

Most systems require some kind of pretreatment to remove or reduce levels of whatever "foulants" may exist in the feed water. As noted, foulants are in four categories, that is, (1) particles, (2) organics, usually NOM, (3) bacteria, and (4) mineral substances such as silica. In addition, substances could be present, for example, chlorine and other chemicals, that could damage some membranes.

17.4.1.1 Cartridge Filters

For particles, cartridge filters may be the most economical choice to reduce particles to tolerable levels for a given membrane. The cartridge filter assembly for a plant may consist of a bank of such cartridges placed (in parallel) in a pressurized stainless steel vessel. For redundancy, at least two such vessels should be used. The main issue is the frequency of replacing the cartridge elements. For example, some surface waters may render cartridge filters "blinded" after only 1–2 days of use. In other cases, for example, well waters, where their use provides a margin of safety for expensive membranes, the change frequency may be several months.

17.4.1.2 Microfilter

For higher raw-water particle concentrations, MF may be integrated into the treatment train ahead of NF/RO. Again, pretreatment may be required ahead of MF.

17.4.1.3 Conventional Treatment

As noted, pretreatment may be required ahead of MF, and especially ahead of RO. For the case of Westminster, CO, a 57,000 m^3/day (15 mgd) MF plant was constructed (in 2002) to handle turbidities >20 NTU by coagulation, flocculation, and plate settling (without filtration).

17.4.1.4 Other Pretreatment

Other foulants, for example, NOM, bacteria, minerals, etc., each require treatment specific to the situation at hand. For example, NF or RO may be constructed following a conventional drinking water treatment plant to remove excess total organic carbon (TOC), or say, arsenic. In some cases, special pilot plant studies, or even research projects, may be necessary to determine the means of handling a given issue. Seawater has an abundance of biotic matter that will readily foul membranes; subocean floor intakes have been reported as a means to condition the feed water (Furukawa, 2006).

17.4.2 Membrane Layouts

NF/RO membrane modules are laid out commonly in a "tree" arrangement. In most circumstances, the fraction of feed water that may be filtered by an element is $0.10 \leq R < 0.30$.

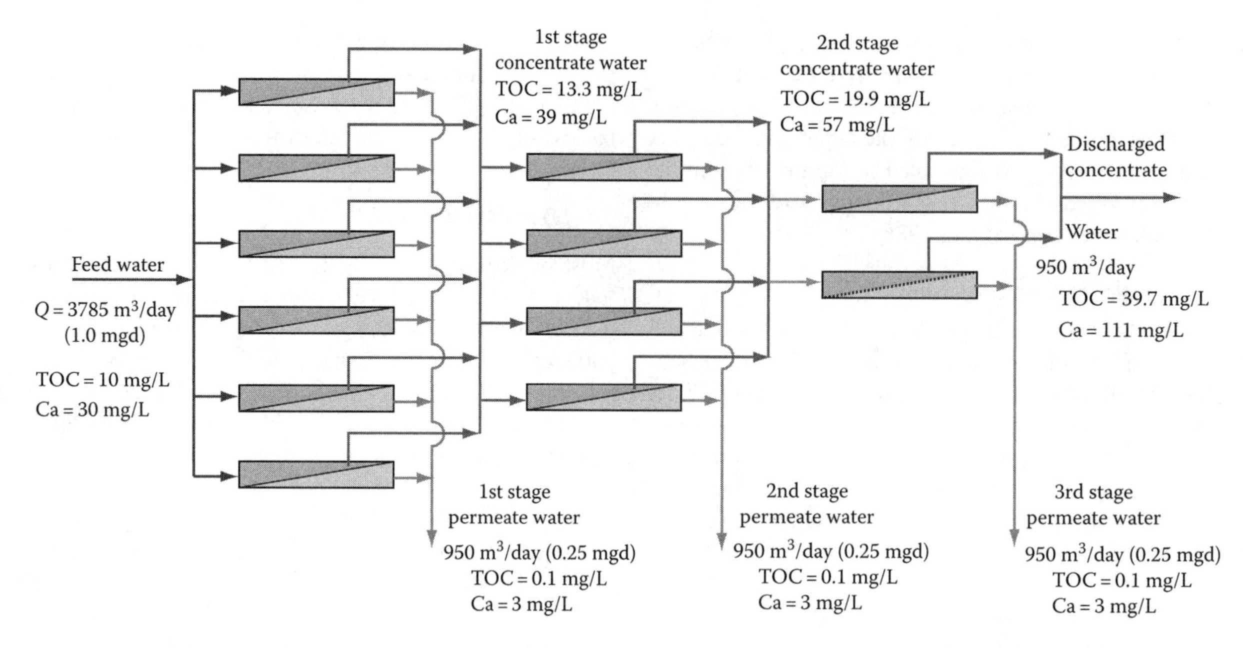

FIGURE 17.22 Illustrative "tree" layout of membrane filtration facility.

To increase the amount of feed water filtered, the concentrate from the first elements, say three, is treated further, as a rule, by additional membrane elements, for example, two and then one. The "tree" arrangement shown in Figure 17.22 illustrates this point where the concentrate flows are treated by successive membranes.

17.4.2.1 First Stage

In the example shown in Figure 17.22, $Q_{F1} = 3785\ m^3/day$ (1,000,000 gal/day), $C(TOC)_{F1} = 10$ mg/L, and $C(Ca)_{F1} = 30$ mg/L. For the first stage of six-membrane pressure vessels, $Q_{P1} = 950\ m^3/day$ (250,000 gal/day), $C(TOC)_{P1} = 0.1$ mg/L, $C(Ca)_{P1} = 3$ mg/L, and $Q_{C1} = 2800\ m^3/day$ (750,000 gal/day). The concentrate flow from the first stage, Q_{C1}, becomes the feed flow to the second stage, that is, $Q_{C1} = Q_{F2}$. At the same time, the concentrate has become more concentrated (hence the name) in both TOC and Ca, in which $C(TOC)_{C1} = 13.3$ mg/L and $C(Ca)_{C1} = 39$ mg/L.

17.4.2.2 Second Stage

After the second stage of four membranes (a lower flow requires fewer membranes), the permeate water has a $Q_{P2} = 950\ m^3/day$ (250,000 gal/day), a $C(TOC)_{P2} = 0.1$ mg/L, and $C(Ca)_{P2} = 3$ mg/L. As with the first stage, the concentrate flow from the second stage, Q_{C2}, becomes the feed flow to the third stage, that is, $Q_{C2} = Q_{F3} = 1900\ m^3/day$ (500,000 gal/day). As before, the concentrate becomes more concentrated with $C(TOC)_{C2} = 19.9$ mg/L and $C(Ca)_{C2} = 57$ mg/L.

17.4.2.3 Third Stage

After the third and final stage, $Q_{P3} = 950\ m^3/day$ (250,000 gal/day), $C(TOC)_{P3} = 0.1$ mg/L, and $C(Ca)_{P3} = 3$ mg/L. Also, $Q_{C3} = 950\ m^3/day$ (250,000 gal/day), $C(TOC)_{C3} = 39.7$ mg/L, and $C(Ca)_{C3} = 111$ mg/L. This concentrate flow from the third stage is discharged as waste.

TABLE 17.7
Typical System Recoveries for Different Membrane Types

Membrane Type	System Recovery (Fraction)
Reverse osmosis	0.20–0.80
Nanofiltration	0.60–0.90
Ultrafiltration	0.80–0.95+
Microfiltration	0.80–0.95+

Source: Wiesner, M. R., An Overview of ΔP Membrane Processes, Presented to Association of Environmental Engineering Professors, San Antonio, TX, p. 14, June 7, 1993.

17.4.2.4 Concentrate

For most installations, the system recovery, R, is $Q_P/Q_F \leq 0.80$–0.90. This fraction depends, however, on the membrane type as listed in Table 17.7. Higher recoveries are not practical since the high concentrations of substances removed after the third stage may cause major fouling problems. The remaining 0.10–0.20 fraction of concentrate water is usually disposed to the sea if near a salt water; this issue is more difficult for inland locations.

17.4.2.5 Recoveries

Table 17.7 lists typical recovery ranges for different types of membranes. For the low ranges, the water source is likely to be more abundant.

17.5 OPERATION

In general, membrane systems have low labor requirements. There are, however, certain operating concerns that are unique to membranes. One of the most important is to ensure that breaches may be detected if there is such an occurrence. Cleaning is another issue; to determine when replacement is needed is yet another. Cleaning chemicals must be ordered, stored, and utilized with special regard to safety and regulations. Pumps must be maintained, sensors calibrated, computer outputs monitored, and samples obtained and sent to a laboratory; also, reporting is required along with archiving data.

17.5.1 Integrity Testing

Integrity monitoring may be continuous, periodic, both, or variable; some form was required by 15 states (in 2001). Assurance of membrane capabilities is tied to integrity testing and monitoring. Most of this section was adapted from Hugaboom et al. (2003). Their work was referenced to six full-scale hollow-fiber MF and UF systems; some were dead end and two had cross-flow capability. Three had permeate backwash and three used permeate and air. Five were enclosed and one was immersed.

17.5.1.1 Breaches

A breach is a "leak" in the system, that is, a small portion of the feed water may reach permeate flow without passing through the membrane. Types of breaches include pinpoint, complete break, chemical or biological degradation, cross-connection via macropores, mechanical failures (o-rings, gaskets, glued fittings), etc. In hollow-fiber bundles, individual fibers may break, for example, one fiber out of thousands.

17.5.1.2 Testing

Two methods of integrity monitoring are direct and indirect evaluations. The direct methods include pressure decay, diffusive air flow, bubble point testing, and sonic sensor testing. Indirect methods include particle counting, particle monitoring, turbidity monitoring, laser turbidity monitoring, microbial challenge testing, spiked integrity monitoring, and routine microbial testing.

17.5.2 Cleaning

When water flux decline becomes significant, membrane modules are removed from service and washed using stored permeate water and chemical solutions. The chemicals used depend on the cause of the fouling. Most selections are based on trial and error, that is, what works. Chemical solutions include alkaline (basic) soap solutions and acid solutions. Alkaline soap solutions are used to remove organic and particle fouling. Acid solutions are used to remove precipitated inorganics.

Initially, membranes are rinsed with stored permeate water. Following this, an alkaline solution and then an acid solution are recycled through membrane modules. Permeate water is used to rinse the membrane between solution washes and before placing the system online.

Membranes are typically forward washed as shown in Figure 17.23a. Some modular designs, however, allow for reverse washing (opposite in direction to the cross-flow of the feed water) as shown in Figure 17.23b. Back-flushing, that is, causing the clean water to flow from the permeate side of the membrane to the feed-flow side, may help in dislodging deposited materials (Wiesner, 1993, p. 42).

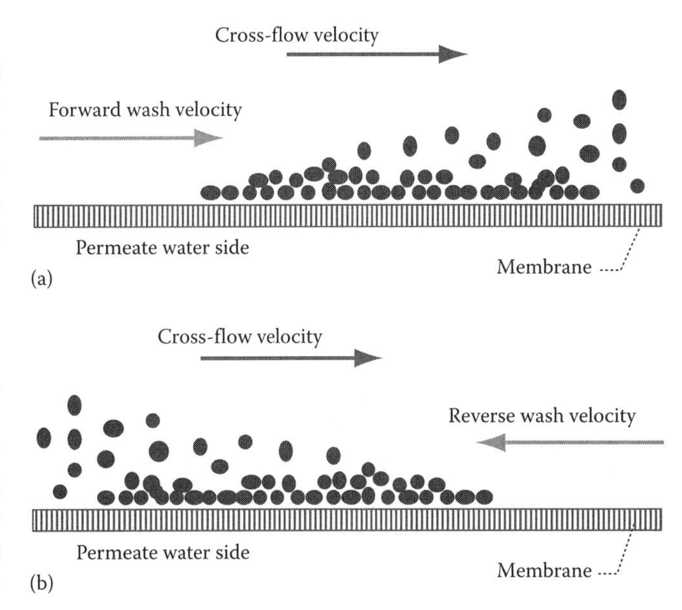

FIGURE 17.23 Depiction of membrane surface washing. (a) Forward washing (typical) and (b) reverse washing (not typical).

17.6 PILOT PLANTS

A pilot plant study will help to resolve many questions of design and operation. Such studies are used almost routinely as a basis for design and to establish operating protocol.

17.6.1 Utility of Pilot Plants

Questions for a pilot plant study may include the rate of reversible and irreversible fouling, pretreatment method, cleaning chemicals, cleaning protocol, removals of constituents, type of membrane, concentrate character and disposal options, membrane integrity testing, monitoring parameters, and frequency. After a given study is completed, the pilot plant may have utility in resolving questions of operation as they develop, in operator training, and for public tours.

17.6.1.1 Pilot Plant Design

A pilot plant may range from a single 51 mm (2 in.) diameter module to a larger size, for example, 102 mm (4 in.) diameter with several in parallel. Usually, at least two modules in parallel are set up that permit comparison of two different membranes, or two conditions, for example, two source waters, two pretreatment modes, two cleaning solutions, two operating pressures, etc. The more modules that are set up in parallel, the more that may be accomplished (and the higher the cost). Often, pilot plant skids are available from manufacturer's plant (perhaps for rent), or a pilot plant may be constructed without too much effort, depending on the situation.

17.6.1.2 Pilot Plant Operation

The parameters of operation may be measured manually, for example, by pressure gages, sampling taps, flow, etc. or by sensors placed at the needed locations, or both. For long-term operation, computer operation is most practical. For sensors, a "board" must be set up to take the signals (i.e., 4–20 ma) that is then interpreted by the software to give readings and provide for archiving and plots. Valves and pumps are operated by microrelays.

17.7 CASE

The wide varieties of membrane plants in all categories, that is, MF, UF, NF, and RO, have been built and are in operation (most are MF and RO). One case, Brighton, Colorado, is reviewed that illustrates some aspects of a design for an RO plant.

17.7.1 City of Brighton Reverse Osmosis Water Treatment Plant

Disposal of brine is a major consideration in application of RO. For that reason, interior locations, such as Brighton, Colorado, are not common. The Brighton case also illustrates the many and varied aspects of an RO design.

17.7.1.1 Background

The city of Brighton, Colorado, with a population of 14,000 in 1990, is located about 35 km (22 mi) north of the north-east side of Denver and about 2 km east of the South Platte River. Demand data are given in Table 17.8.

The source of the water supply was four shallow wells that tapped an alluvial aquifer that was contiguous with the South Platte River but about 2 km from the stream, and downstream about 40 km (25 mi) from the discharge of the Denver Waste Water Reclamation Plant (a regional facility). In addition, three wells were located near Barr Lake. Nitrate concentrations ranged 13–23 mg/L as N from summer to winter, respectively. The nitrate issue was a continuing concern from about 1970; the source was believed to be of fertilizer use on agricultural lands south-east of the city. TDS concentrations ranged 800–1140 mg/L, with hardness 370–480 mg/L as $CaCO_3$. About 0.60 fraction of homes had ion-exchange units installed to reduce hardness levels. Seven-day THMFPs have ranged 10–1000 mg/L, which added impetus to the search for adequate treatment.

17.7.1.2 Brighton Pilot Plant

A pilot plant study was initiated in October 1991 and continued through May 1992. Objectives were as following (Cevaal et al., 1993): (1) determine THM precursor rejection values, (2) confirm feed-water pressures and differential pressures required, (3) assess fouling rates for each of the membrane elements tested, (4) obtain data on concentrate water quality relative to a discharge permit, and (5) increase public awareness and acceptance of RO. Membranes evaluated included Hydranautics 4040 LSA NCM1®, Fluid Systems TFCL 4821 LP®, Filmtec BW30 4040®, and a later version of Hydranautics 4040 LSA NCM1®. Each membrane was a spiral-wound polyamide thin-film composite membrane. The polyamide thin membranes were selected over CA because of their lower feed pressures and higher nitrate rejections (Cevaal et al., 1993). Rejection of nitrates was about 97% for the second and third membranes listed. Also ≥ 0.92 rejection of THMFP was achieved. Based on the pilot plant study, the Fluid Systems TFCL membrane was selected.

Biological fouling was found to occur, caused by growth of *Pseudomonas*. The well water showed heterotrophic plate counts, $500 \leq HPC \leq 5000$ cfu/100 mL, TOC ≈ 3 mg/L, and DO ≈ 6 mg/L. Biological fouling was addressed in design of the full-scale RO plant by well rehabilitation consisting of strong acid/shock chlorine and hydrogen peroxide. Alternatives considered included membrane cleaning, well rehabilitation, deoxygenation of feed water, ultraviolet radiation of feed water, use of CA membranes instead of polyamide, chloramines in feed water, and chlorination/dechlorination. For the concentrate, TDS $\approx 880/3200$ mg/L and nitrate as $N \approx 16/56$ mg/L (raw water/concentrate water). The results of the pilot plant study were the basis for a permit being granted to discharge the concentrate into the South Platte River.

17.7.1.3 Design Parameters

The aggregate capacity of three treatment trains was Q(permeate) = 15,000 m^3/day (4.0 mgd). Pretreatment was

TABLE 17.8
Municipal Water Demand for Brighton, Colorado

		Q(avg.)		Q(max)		Q(peak-hour)	
Year	Population	(m^3/day)	(mgd)	(m^3/day)	(mgd)	(m^3/day)	(mgd)
1990	14,000	11,000	2.9	30,700	8.1	63,200	16.7
2000[a]	34,500	23,500	6.2	66,000	17.4	135,000	35.6
2010[a]	55,000	33,300	8.8	93,500	24.7	191,500	50.6

Source: Adapted from Cevaal, J.N. et al., Design of a reverse osmosis treatment system for nitrate removal for Brighton, CO, in: *AWWA Annual Conference*, San Antonio, TX, June 6–10, 1993.

[a] Projected data.

the addition of acid and anti-scalant followed by cartridge filtration. The purpose of the acid was to lower the pH sufficiently (from 7.2 to 6.8) to control $CaCO_3$ deposits on the membranes. Polyacrylic acid anti-scalant was used. Cartridge filtration was to remove silt and grit. Two pressure vessels of 5 μm carridge filters were used; a 70 kPa (10 psi) pressure drop was assumed. Table 17.9 provides selected design data.

17.7.1.3.1 Pumps

The required pressure was achieved by a 150 kW (200 hp) centrifugal pump (vertical turbine) for each RO skid to achieve a total head of 1350 kPa (450 ft). The vertical turbine pumps were selected because of higher efficiencies (i.e., 0.80–0.85 for high heads required), longer operating life, less space, and greater flexibility in adding additional stages. The usual total dynamic head required was 1050 kPa (353 ft). The difference between the design TDH and the usual operating TDH was substantial and could be wasted or matched to the conditions by a variable speed control. The latter was selected based on savings in energy costs.

TABLE 17.9
Data for Membrane System at Brighton, Colorado

Parameter	Magnitude SI	Magnitude US
Parameter		
RO permeate capacity per train	5050 m^3/day	1.333 mgd
RO concentrate capacity per train	1269 m^3/day	0.3326 mgd
RO recovery	80%	
RO permeate TDS	45 mg/L	
Chemical additions		
Sulfuric acid in feed (93% conc.)	23 L/h	6 gal/h
Anti-scalant-hypersperse 100 (as shipped)	1290 mL/h	0.34 gal/h
RO system design parameters		
RO trains	3 (1995); 5 (2000)	
Permeate capacity	3505 L/min/train	926 gpm/train
Pressure tube array	36 (stage 1):18 (stage 2)	
Membrane element per tube	6 M	
Membrane elements	324 per skid	
Pressure vessels	54 tubes/skid	
Water quality		
Constituent	Feed water (pretreat)	
NO_3^- as NO_3^-	66 mg/L	
SO_4^{2-} sulfate	296 mg/L	
Si (as SiO_2)	25 mg/L	
Temperature	13.4°C	
pH	6.8	
SDI	0.4	
TDS	1070	
THMFP (7 days)	100 μg/L	
Feed-water critical parameter specifications		
Pressure	≤2760 kPa	≤400 psig
pH	$3.0 \leq pH \leq 10$	
Turbidity	≤1.0 NTU	
Total suspended solids	≤1 mg/L	

Source: Adapted from Cevaal, J.N. et al., Design of a reverse osmosis treatment system for nitrate removal for Brighton, Colorado, in *AWWA Annual Conference*, San Antonio, TX, June 6–10, 1993.

17.7.1.3.2 Posttreatment

Posttreatment of permeate consisted of stripping carbon dioxide and addition of base (caustic), zinc orthophosphate, and chlorine in the clear-well. The permeate was then blended and pumped to the distribution reservoirs (two were available for one pressure zone).

17.7.1.3.3 Flux Density

Higher flux densities foul membranes at a faster rate than lower flux densities (Cevaal et al., 1993). Therefore, the $\mathbf{j}_w = 0.58$ m/day (14.2 gpd/ft^2) was selected.

17.7.1.3.4 Stages and Array

An array is the number of membranes in the various stages of an RO rack. As noted in Section 17.4.2, each stage of an array produces permeate with its concentrate feeding the next stage of the array. To achieve 0.80 recovery at Brighton, a two-stage unit was necessary. An array of 32 pressure vessels in the first stage and 16 in the second stage with anticipated feed pressure of 1600 kPa (231 psi) was recommended. Each pressure vessel would contain six standard 203 mm (8 in.) diameter membranes.

17.7.1.3.5 Cost

The cost for the RO treatment facility, based on bids, was \$6,118,892 with an additional \$1,823,742 for pipelines. The annual cost of operation and maintenance were projected as \$734,000, which included electric energy, chemicals, labor, membrane replacement, equipment repairs and replacement, insurance, and laboratory fees. Assuming annual production rate of 10,500 m^3/day (2.77 mgd) of permeate and blend water, the unit cost was calculated as \$0.020/m^3 (\$0.74/1000 gal).

17.7.1.4 Plant Layout

Figure 17.24 shows the layout of the Brighton RO membrane plant. The flow schematics are shown for raw water, concentrate, and permeate. The raw water flow is filtered by assemblies of bag filters, each housed in a separate pressure vessel. From the bag filter assemblies, the raw water flow becomes feed flow and is pressurized by three high-pressure pumps. The discharge from each pump then entered a manifold that distributed the feed water to each of three racks, and then to each of the membrane tubes.

The concentrate is collected also by a manifold and leaves the system through a pipeline for discharge to the South Platte River. The permeate water from each rack is pH adjusted by acid or base and is chlorinated, degassed, and discharged to a clear well. From the clear well, the product water is pumped to the distribution system, that is, first to one or more service

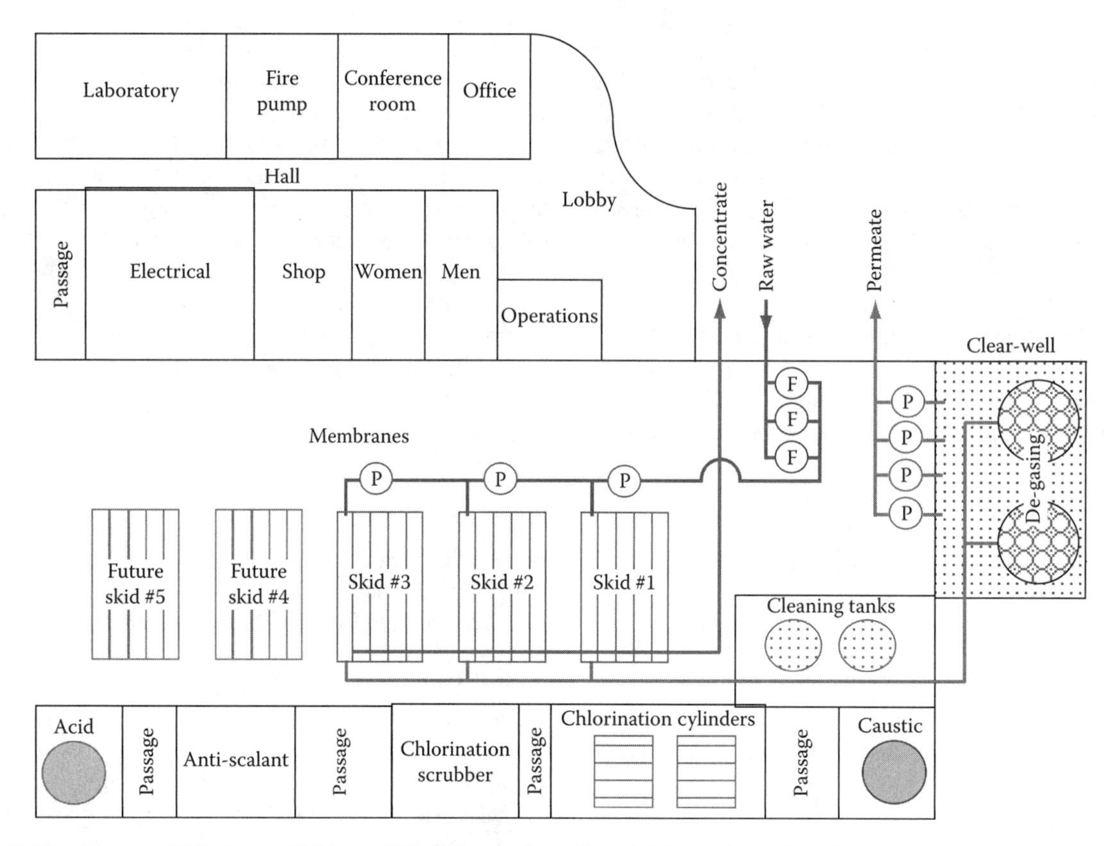

FIGURE 17.24 Layout of RO plant at Brighton, CO. (Adapted from Cevaal, J.N. et al., Design of a reverse osmosis treatment system for nitrate removal for Brighton, CO, in: *AWWA Annual Conference*, San Antonio, TX, June 6–10, 1993.)

reservoirs. In addition to the membrane process itself, a large amount of space is required for support, for example, laboratory, personnel needs, lobby for visitors, an operations room, shop, acid and base supply, chlorine room with associated scrubber room, anti-scalant chemical, and cleaning chemicals. In other words, as simple as a membrane process appears, the overall system still requires considerable support, as with any other water treatment process. Nevertheless, the building for the facility shown is simple and the floor is a slab in parts, albeit various sumps are needed for pumps, containment of chemical spills, and for the clear well. The design provided for modular expansion to add two additional membrane racks, which were added in 2003.

PROBLEMS

17.1 Flux Density Distribution Based on Pore-Size Distribution

Given

An MF membrane has pore-size distribution as follows: d(pore)0.33 fraction $= 1.5$ μm, d(pore)0.33 fraction $=$ 1.0 μm, and d(pore)0.33 fraction $= 0.5$ μm.

Required

(a) Determine the distribution of flux density for each pore size. (b) Show by a plot the distribution of pore sizes and the distribution of flux density.

17.2 Flux Density Distribution Model Based on Any Given Pore-Size Distribution

Given

An MF membrane has Gaussian pore-size distribution with d(pore)avg $= 1.2$ μm and $s = 0.22$ μm.

Required

(a) Determine the distribution of flux density. (b) Show by a plot the distribution of pore sizes and the distribution of flux density. (c) Set up a general mathematical model that would apply to any distribution.

17.3 Calculation of Flux Density

Given

An MF hollow-fiber membrane treatment plant has a total flux, $Q \equiv \mathbf{J_w} = 0.657\ \mathrm{m^3/s}$ (15 mgd). The plant is to be operated at about 2°C in the winter and 20°C in the summer. The pressure available is $\Delta p = 138$ kPa (20 psi).

Required

(a) Estimate the number of "modules" (tubes with bundles of fibers) required based on the information available from manufacturers. (b) If provided by the manufacturer's data, obtain the inside and outside diameters of the fibers, their length, the number of fibers per module, the total membrane surface area per module, whether the fiber is shell feed or bore feed, and any other information that may be of interest.

17.4 Effect of Temperature on Flux Density

Given

An MF membrane element has a given flux density, for example, $\mathbf{j_w}$ m^3 water/m^2 membrane/s for temperature 20°C.

Required

Calculate the relative flux density, $\mathbf{j_w}$, for 9°C.

17.5 Osmotic Pressure

Given

A brackish water has a TDS = 3500 mg/L.

Required

Calculate the osmotic pressure.

17.6 Calculation of Flux Density and Permeability for RO Membranes

Given

The problem is referenced to the RO membrane system at Brighton (see Section 17.7.1). The membranes are 8 in. diameter, 40 in. long with a reported area 33.9 m^2. The Brighton plant with three trains has 324 elements per train. A fourth train was being installed as we visited the plant on April 6, 2000. The influent pressure at start-up in 1994 was 154 psi; in December, 1996 when Tory Champlin and I visited the plant, the pressure was 164 psi. Let $T = 20$°C.

Required

(a) Calculate the design water flux density for the membranes at Brighton. (b) Estimate the intrinsic permeability coefficient, k, for the membranes.

ACKNOWLEDGMENTS

Ed Burke, director of utilities at Brighton, Colorado, and Rodney Evans, plant supervisor, provided ready access to their plant and provided whatever data were needed and gave permission to photograph the installation. Burke was helpful over a period of years in helping the author become acquainted with the membrane technology, beginning with the pilot plant phase and extending through operation. Photographs of the Brighton membrane facility were used with permission of the city of Brighton through Burke and Jodie Carroll.

This chapter has utilized sections from handouts and presentations by Dr. Tory L. Champlin, senior project engineer, Parsons, Tampa, Florida, while at Colorado State University during the period 1992–1998, including photomicrographs and findings from his doctoral thesis; all material was used with his permission. The chapter also utilized report material from sponsored research by the Environmental Protection Agency, Cincinnati, in which Dr. Champlin served as research assistant. In addition, Dr. Champlin has provided suggestions on parts of the manuscript concerning current practice, drawing on his experience in design and in pilot plant studies.

GLOSSARY

Acronyms for membrane materials and membranes: Acronyms are used commonly to designate different membrane materials. Some include polysulfone (PS), polyvinylidenefluoride (PVDF), polyacrylonitrile (PAN), and polyamines (PA). For membranes, microfiltration is MF, ultrafiltration is UF, nanofiltration is NF, and reverse osmosis is RO (RO has also been called "hyperfiltration").

ACS: Acronym for American Chemical Society.

Anisotropic: Contrast with: isotropic–exhibiting properties (as velocity of light transmission, conductivity of heat or electricity, and compressibility) with the same values when measured along axes in different directions (http://www.merriam-webster.com, 2003).

Anti-scalant: Usually an acid rinse used in the cleaning cycle to remove mineral scale that has deposited on the membrane surface as a foulant.

Array: See *membrane technologies*.

ASTM: American Society for Testing Materials.

Asymmetric: (1) Not symmetrical (http://www.merriam-webster.com, 2003). (2) As used with respect to membranes, a dense layer is on the separating surface, underlain by a support layer (Cheryan, 1986, p. 365). (3) Heating, that is, "annealing" certain membranes causes shrinkage of the pores on one side, but not on the other; discovered by Sidney Loeb and S. Sourirajan, c. 1960. The discovery was a "breakthrough" that paved the way toward development of commercial membranes since a homogeneous membrane had a flux density of only about 0.05 L/m^2/h, while with the asymmetric membrane, the flux density was about 14.5–0.05 L/m^2/h. This was due to the thin "skin" on one surface, about 0.1–0.2 μm thick, while the main body of the membrane was "sponge-like" with voids of high porosity. Previously, the homogeneous membranes had a thickness of about 100–200 μm (from Cheryan, 1986, pp. 10–13).

Clean-in-place: A more thorough cleaning of a membrane that occurs at an interval of perhaps every 30 days, depending on conditions and manufacturer's recommendations.

Cleaning: An action that removes a portion of the foulants from the membrane surface, that is, the "reversibly" fouling, such that the water flux density is restored as it was at the beginning of the cycle minus the "irreversible" fouling.

Composite membrane: A two-layer membrane consisting of a dense-film layer laminated to a microporous support, for example, a PS surface layer on top of a polyester support layer.

Concentrate: The portion of the feed flow that does not pass through the membrane but leaves the membrane element. The concentrate flow is sometimes called the "reject" flow.

Concentrate flow: The flow that is retained by the membrane, that is, does not pass through the membrane as permeate flow.

Concentration polarization: Deposition of a layer of solute on the membrane surface caused by the accumulation of salts due to the rate of advection of salts toward the membrane surface exceeding the back-diffusion of salts.

Cross-flow: In a spiral-wound membrane, the feed flow with velocity parallel to the membrane tube is called "cross-flow."

Dalton: (1) A unit of mass for expressing masses of atoms, molecules, or nuclear particles equal to 1/12 of the atomic mass of the most abundant carbon isotope, $6C_{12}$, which is about 1.66043×10^{-27} kg—called also mass unit (http://unabridged.merriam-webster.com). (2) Also synonymous atomic mass unit, which is a mass unit exactly equal to 1/12 the mass of a carbon atom-12 atom (Silberberg, 1996, p. G-2). The term is used commonly in membrane vernacular to give an idea of the sizes of molecules that may pass through the membrane, and perhaps an indication of a *pseudo*-"pore size." There are several measures of atomic size (Shriver and Atkins, 1999, p. 23), for example, atomic radii (metallic radius and covalent radii) and ionic radius. To give an idea of sizes, the atomic radius of carbon is given as 0.77 Å (p. 24), while for lead, it's 1.75 Å.

Depth filter: A filter in which removal of material occurs throughout its depth.

Dialysis: The transfer of solute molecules across a membrane by diffusion from a concentrated solution to a dilute solution.

Electrodialysis: The transfer of ions across membranes by an electric field (adapted from Helfferich, 1962, p. 397).

Element: The part of a membrane package as manufactured for the insertion into a pressure vessel (or open tank, in the case of a vacuum system). The "element" may be a single spiral-wound membrane, a "bundle" of hollow fibers, a ceramic tube, etc.

Equivalent: Molecular weight of an ion divided by its charge. For example, $Eq(Na^+) = MW(Na^+)/1 =$ 23.5 g/mol/1 Eq/mol = 23 g Na^+/Eq; $Eq(Ca^{2+}) =$ $MW(Ca^{2+})/1 =$ 40 g/mol/2 Eq/mol = 20 Ca^{2+}g/Eq.

Faraday: A Faraday, $F = 96{,}485$ C = 96,485 A-s.

Feed water: The influent flow of water to a membrane element.

***Flow balance*:** A key principle of analysis is that flow-in and flow-out of a membrane pressure vessel must be accounted for, that is, $Q_F = Q_P + Q_C$, in which Q_F = feed flow, Q_P = permeate flow, and Q_C = concentrate flow, also called "reject" flow (all in m^3/s). Figure 17.1a is a schematic depiction for a cross-flow membrane.

Flux: A flow of something, for example, water flux, salt flux.

Flux density: Flux per unit area.

Foulant: A material, e.g., bacteria, organic matter, particles, that deposit on the surface of a membrane which causes a reduction in permeate flux.

Fouling: An effect that causes a reduction in permeate flux.

Gel polarization: See *concentration polarization*.

Hollow fiber (HF): A type of membrane that is a dead-end tube with about 1 μm internal diameter and 2–3 μm outside diameter. Cheryan (1986, p. 87) mentions id ≈ 1.1 mm, length 635 mm, 660 fibers/tube, v(bore) ≈ 1.0 m/s. HF membranes been used since the early 1990s, mostly in MF. Two types of HF membranes are bore feed and shell feed. Bore feed has been used for UF (Cheryan, 1986, p. 64), with an interior "rejecting skin" and with bundles of 50–3000 fibers, and they have been manufactured for RO applications (Hydranautics, 2003, http://www.membranes.com). The shell-feed HF is through the pressure vessel, with the permeate passing through the bore and out the end of the fiber where the flow from a single bore is joined with the other flows, leaving the pressure vessel. Cleaning is by a hydraulic flush around the fibers to remove retained solids. To be consistent with the nomenclature for a spiral-wound membrane element, a bundle is also called here an "element."

Housing: See *pressure vessel*.

Hyperfiltration: A membrane with pore sizes mainly <5 Å. Sometimes, the term "hyperfiltration" is used instead of "reverse osmosis" (RO) since it is consistent with the series: MF, UF, and NF.

International Desalination Association (IDA): International organization started in 1985 focused on desalination technology, science, and practice with headquarters in Topsfield, Massachusetts; the association publishes the journal *Desalination*.

Intrinsic permeability: A characteristic of the inherent resistance of a membrane, k, defined in terms of Darcy's law that does not include viscosity (see Equation E.4, $K = k\rho g/\mu$). Alternatively, terms in the Poiseuille equation may be combined to give the same thing.

Mass flux: Flow of dissolved and/or suspended solids.

Membrane: (1) A thin soft pliable sheet or layer especially of animal or vegetable origin. (2) A limiting protoplasmic surface or interface (http://www.merriam-webster.com, 2003). (3) As used in this chapter, the reference is, in general, to synthetic membranes, for example, CA, PS, polyamide, etc., manufactured for the purpose of separating contaminants from water by causing the water to be transported through the membrane by means of an hydraulic gradient, that is, pressure differential, Δp. Contaminants are retained based on kind of membrane and associated pore size. For NF and RO, pore size may be a misconception since the water passes through the molecular structure of the membrane (actually, the retaining layer, as distinguished from the support layer).

Membrane element: A membrane package as manufactured for insertion into a pressure vessel that receives the feed flow. A spiral-wound membrane, as ready to be inserted into a pressure vessel, is an example. A hollow-fiber membrane element (also called a "module"; see also Mulder, 1996, p. 472) is the collection of fibers ready to be inserted into a pressure vessel.

***Membrane process*:** Retention of selected contaminants during the flow of water through a membrane material. For micro/ultra membranes, the retention is probably by "screening," that is, the membrane "pores" are smaller than the particles to be removed. For nano/RO membranes, the water passes through the molecular structure comprising the membrane material. The mechanism for the "rejection" of ions is not clear. It would seem that molecules are rejected based on size, but this is conjecture. In each case, the passage of water through the membrane material occurs under hydraulic gradient.

***Membrane technology*:** The membrane "package" provided by a manufacturer that is "plug-in" ready to operate in the context of an appropriate engineered system, that is, with pressure vessel, seals, connections for flows, taps for pressure measurement, etc. The "package" would include a manufactured membrane element, that is, a membrane sheet or a membrane fiber is the result of a manufacturing technology.

Membrane technology and hierarchy of components: The following is a nomenclature adopted for this chapter. It is not necessarily universally accepted but there was attempt both to adhere to accepted usage of terms and, at the same time, to provide a coherent set of definitions.

***Array*:** An array is the number of membrane pressure vessels that comprise a given stage of an RO rack. Each stage of an array produces permeate with its concentrate feeding the next stage of the array. To achieve 0.80 recovery at Brighton, a two stage unit was necessary: an array of 32 pressure vessels in the first stage and of 16 pressure vessels in the second stage. An anticipated feed pressure of 1600 kPa (231 psi) was recommended. Each pressure vessel contains six standard 203 mm (8 in.) diameter membranes.

***Element*:** A manufactured unit that "packages" the membrane material such that it is suitable for permeation of water across the membrane material. The term "membrane element" seems to be used mostly in relation to NF and RO.

***Module*:** A manufactured unit that "packages" the membrane material such that it is suitable for permeation of water across the membrane material. The term "module" seems to be used mostly in relation to MF and UF, especially in reference to hollow fiber membranes. The term is used also in a more general sense, for example, a "modular design," adding a "module," etc.

***Pass*:** A term used in desalination: (1) the permeate from one membrane serves as feed water to another perhaps different kind of membrane in series to remove a constituent not removed by the first membrane. (2) the reject water (also called "concentrate") from one membrane serves as feed water for another placed in series.

***Pressure vessel*:** A tube that accepts one or more membrane elements (e.g., spiral-wound membranes). Alternatively, the pressure vessel may be designed for a membrane module (e.g., a bundle of hollow fibers). Usually, the pressure vessel has end caps, gaskets, fittings for tubes, etc. such that the unit is suitable for hookup to feed-water tubes, permeate tubes, concentrate tubes, etc. Pressure vessels may have six to eight elements in series.

***Rack*:** A framework that holds a collection of membrane pressure vessels, arranged as a unit that may be added to with other racks in a "modular" fashion. The frame is fabricated on site.

***Skid*:** A framework that holds a collection of membrane pressure vessels, arranged as a unit that may be added to with other racks in a "modular" fashion. The framework is fabricated at a plant and can be delivered to the site and is ready to assemble, along with the membrane pressure vessels. Typically, a skid is six to eight pressure vessels high and four pressure vessels wide. The arrangements of how many elements in a pressure vessel and how many pressure vessels are related to cost (typically, elements cost $400–800 each and pressure vessels may cost $2500–3000). About seven to eight elements are the maximum for a pressure vessel because of pressure loss along the feed-water tube (which reduces the Δp (across the membrane) from the beginning to the end of the pressure vessel).

***Stage*:** If the concentrate from a first pressure vessel serves as the feed water to a second pressure vessel in series and then possibly to a third, each pressure vessel in the series is referred to as a "stage." The first pressure vessel is the first stage, the second pressure vessel is the second stage, and so forth. The first stage may be, actually, say, three pressure vessels in parallel; the second stage may be two in parallel, taking concentrate from the first stage of three vessels; and so forth. The first group may be called a "bank" of pressure vessels, which may go to a "bank" of stage 2 pressure vessels.

Microfiltration: Transport of water by means of a pressure gradient through a membrane of pore sizes of 1 μm in size.

Microporous: Membranes that may be isotropic or anisotropic. Microporous membranes are intended to retain all particles above its largest pore size, for example, 0.45 μm (this is an "absolute" size cutoff; in other words, particles that are smaller will be retained). Attempts to produce this type of filter with pore sizes in the UF range, for example, 1–10 nm, have not been successful. Microporous membranes are susceptible to plugging (Cheryan, 1986, p. 29).

Module: See *membrane technology*.

Molecular weight cutoff (MWCO): The pore sizes of a membrane may have a distribution. The MWCO is smallest molecular weight of a test molecule that will not pass the pores of the membrane to a significant extent (Wiesner, 1993, p. 5). Molecules having MW $\geq$ MWCO will not pass through the membrane. A portion of the molecules having MW $<$ MWCO will pass, depending on the size of the molecules and the pore-size distribution.

Nanofiltration: Membrane filtration that removes many organic molecules, and some larger ions, particularly ions that are complexes.

Osmosis: The diffusion of water from the pure water side of a semipermeable membrane to the saline side.

Osmotic pressure: If a semipermeable membrane separates a pure water and a saline solution, the water level in the saline solution side will rise to a level that is the osmotic pressure, π, of the saline solution. The osmotic pressure may be calculated by the van't Hoff equation for dilute solutions.

Package: The membrane element that includes whatever ancillary parts are necessary such that the system works as a technology (as used here for descriptive purposes; the term is not one that is used commonly in the industry).

Permeability: A characteristic of the flow resistance of a membrane, defined in terms of Darcy's law, Equation E.5, $v = -K(\mathrm{d}h/\mathrm{d}z)$, in which $K = k\rho g/\mu$, Equation E-4. Alternatively, the Poiseuille equation, Equation D.11, also describes the same, only it is in terms of viscous flow through a pipe, that is, $\mathrm{d}p/\mathrm{d}z = 32\mu v(\mathrm{avg})/d^2$; rearranged, $v(\mathrm{avg}) = (d^2/32\mu)(\mathrm{d}p/\mathrm{d}z) = (d^2\rho g/32\mu)(\mathrm{d}p/\mathrm{d}z)$. In the latter case, $K = d^2\rho g/32\mu$ and therefore, $k = d^2/32$. See also *intrinsic permeability*.

Permeate: The flow of water through the pores of the membrane.

Permeate flow: Flow of permeate, Q_P, that is, water that has been transported across the membrane, leaving contaminants behind, as illustrated in Figure 17.1.

Plate-and-frame: A configuration of membrane sheets in which the feed flow is between two membrane sheets. Spacers of some material are located on the other sides of the respective membrane sheets. The pattern is repeated such that perhaps several hundred sheet–spacer pairs comprise the membrane system.

Polarization: The point at which the current density (amps/m^2) is high enough to dissociate the water molecule, resulting in the formation of OH^- and H^+ ions (Meller, 1984).

Pressure vessel: Usually a tube that houses one or more membrane elements (see also *module*).

Product flow: See *permeate*.

Rack: See *membrane technology*.

Recovery: Ratio of product flow to feed flow. Typically, one pass through an RO membrane for seawater gives 40%–60% recovery (of permeate), that is, about 50% of the concentrate is brine. An RO membrane with brackish feed water gives typically 70%–80% recovery.

Reject flow: See *concentrate flow*.

Reverse osmosis: When pressure is applied to the saline water side of a semipermeable membrane that is in excess of the osmotic pressure that is due to the saline water, water will flow from the saline water side to the permeate side. While osmosis occurs by diffusion from the pure water side to the saline water side, "RO" causes water to flow in the reverse direction (see also "osmosis" and "hyperfiltration). Since the flow of water from the saline side to the permeate side is proportional to the pressure differential (after subtracting the osmotic pressure), one may infer that the Poiseuille law applies. (There may be some debate on this point.)

Screen filter: A filter in which suspended material larger than the openings is retained on the surface (Cheryan, 1986, p, 27).

Silt density index: An empirical test developed for membrane systems to measure the rate of fouling of a 0.45 μm filter pad by the suspended and colloidal particles in a feed water. This test involves the time required to filter a specified volume of feed at a constant 30 psi at time zero and then after 5, 10, and 15 min of continuous filtration. Typical RO element warranties list a maximum SDI of 4.0 at 15 min for the feed water. If the SDI test is limited to only 5 or 10 min readings due to plugging of the filter pad, the user can expect a high level of fouling for the RO. Deep wells typically have SDIs of three or less and turbidities less than one with little or no pretreatment. Surface sources typically require pretreatment for removal of colloidal and suspended solids to achieve acceptable SDI and turbidity values. Foregoing is described in, SDI–Silt Density Index Examined and Explained, Industrial Water Treatment Consulting, 2010, http://www.iwtc.com.au/resources/IWTC_SDI_examined_and_explained.pdf.

Solute: The dissolved solid portion of a solution. Most often, the term is used in reference to a particular species of dissolved solids, for example, Na^+, sucrose, etc.

Solute rejection: The fraction of solute that is "rejected" by the membrane and is carried out in the reject flow. The term has not been applied to hollow-fiber membranes (but could be defined based on the flow of water for hydraulic flushing over, say a 24 h period).

Spiral wound: A higher area of membrane surface area per unit volume is obtained by configuring an array of membrane sheets and spacers into a "roll," with the sheets and spacers wound around a hollow tube. The feed flow enters the end of the tube with velocity parallel to the tube as a "cross-flow." The permeate-flow velocity is normal to the cross-flow velocity, that is, the path is a spiral. The membrane sheets are attached to the inner tube with small holes between the sheets that receive the permeate flow.

Spiral-wound membrane element: A spiral-wound membrane element is a double layer of membrane sheets wound around a tube. Permeate spacers between the membrane sheets provide for permeate flow. Another kind of spacer, that is, feed-flow spacer (of a coarse mesh fabric), is between the membrane sheets. Several such sheet assemblies may be wound around the tube to comprise a membrane element. This whole element fits snugly in a cylindrical pressure vessel. An o-ring around the element fits against the pressure vessel and forces the flow to pass along the "feed-flow" spacer, parallel to the tube. The permeate flow enters a given spacer from the flow through the adjacent membranes and flows around the tube as a spiral and enters perforations along the tube; the tube collects water from each leaf assembly and the permeate then passes from the tube. Figure 17.2b shows a membrane element ready to insert into its pressure vessel.

Ultrafiltration: Characterized by membranes with pore size between MF and NF.

REFERENCES

Adham, S. Gramith, K., Chiu, K. P., Mysore, C., and Lainé, J. M., Development of a low pressure membrane knowledge base, in: *American Membrane Technology Association Annual Symposium*, Denver, CO, August 2–5, 2003.

Amjad, Z., Ed., *Reverse Osmosis—Membrane Technology, Water Chemistry, and Industrial Applications*, Van Nostrand Reinhold, New York, 1993.

Amy, G. L., Alleman, B. A., and Cluff, C. B., Removal of dissolved organic matter by nanofiltration, *Journal of Environmental Engineering, ASCE*, 116(1):200–205, 1990.

Anon., Advanced Waste Treatment Research Program Summary Report, Advanced Waste Treatment Branch, Robert A. Taft Sanitary Engineering Center, Federal Water Pollution Control Administration, Cincinnati, OH, 1968.

ASTM, ASTM D4194-85. Standard method for operating characteristics of reverse osmosis devices, in: *Annual Book of ASTM Standards, Water and Environmental Technology*, Vol. 11.02, *Water (II)*, Philadelphia, PA, 1987.

AWWA, AWWA Membrane Technology Research Committee, Committee Report: Membrane processes in potable water treatment, *Journal of the American Water Works Association*, 84(1):59–67, January 1992.

Baker, R. W., *Membrane Technology and Applications*, McGraw-Hill, New York, 2000.

Blau, T. J., Taylor, J. S., Morris, K. E., and Mulford, L. A., DBP control of nanofiltration: Cost and performance, *Journal of the American Water Works Association*, 84(12):104–116, December 1992.

Cevaal, J. N., Brunswick, R. J., Burke, J. E., and Suratt, W. B., Design of a reverse osmosis treatment system for nitrate removal for Brighton, Colorado, in: *AWWA Annual Conference*, San Antonio, Texas, June 6–10, 1993.

Champlin, T. L., Membrane Filtration (handout prepared for Operators Annual Short Course), University of Colorado, Boulder, CO, March 26, 1996.

Champlin, T. L., Natural organic matter and particle fouling of spiral-wound nanofiltration membrane elements, Doctoral thesis, Colorado State University, Fort Collins, CO, 1998.

Champlin, T. L. and Hendricks, D. W., Nanofiltration for treatment of low-turbidity waters, membrane fouling and removals of disinfection by-product precursors, Environmental Engineering Technical Report, Department of Civil Engineering, Colorado State University, Fort Collins, CO, December 1994.

Cheryan, M., *Ultrafiltration Handbook*, Technomic Publishing Company, Inc., Lancaster, PA, 1986.

Duranceau, S. J., Taylor, J. S., and Mulford, L. A., SOC removal in a membrane softening process, *Journal of the American Water Works Association*, 84(1):68–78, January 1992.

Furukawa, D. H., Desalination Technology, 2006, in: *AMBAG Conference*, Monterey, CA, September 27, 2006 (pdf file).

General Electric Company, *The Filtration Spectrum*, General Electric Company, Minnetonka, MN, 1993.

Global Water Intelligence (GWI), 19th International Desalination Association (IDA) Worldwide Desalting Plant Inventory, Media Analytic, Ltd., Oxford, U.K., 2006.

GWI, Desalination in 2008, Global Market Snapshot, 21st IDA Worldwide Desalting Plant Inventory, 2pp., 2008; http://www.globalwaterintel.com (this site has proprietary information on market conditions for water services and products, changing as new information is developed).

Hanft, S., Major reverse osmosis system components for water treatment: The global market, Report ID MST049B, Publication ID WA1176332, 297pp., BCC Research, September 1, 2005, http://www.bccresearch.com/report/MST049B (obtained June 2009).

Helferrich, F., *Ion Exchange*, McGraw-Hill, Inc., New York, 1962.

Hugaboom, D., Sethi, S., Crozes, G., Curl, J., and Marinas, B., Integrity testing for membrane filtration systems, in: *American Membrane Technology Association Annual Symposium*, Denver, CO, August 2–5, 2003.

Hydranautics, http://www.membranes.com, 2003.

Koros, W. J., Membranes: Learning a lesson from nature, *Chemical Engineering Progress*, 91(10):68–81, October 1995.

Laisure, D., Sung, L., and Taylor, J. S., Fundamental membrane phenomena, *Desalination and Water Reuse*, 3(3):10–13, 1993.

Loeb, S., The Loeb-Sourirajan Membrane: How it came about, in: Turbak, A. F., Ed., *Synthetic Membranes*, Volume 1, *Desalination*, ACS Symposium Series 153, American Chemical Society, Washington, DC, 1981.

Lonsdale, H. K., The growth of membrane technology, *Journal of Membrane Science*, 10:81–181, Elsevier Scientific Publishing Company, 1982.

Meller, F. H., Ed., *Electrodialysis (ED) & Electrodialysis Reversal (EDR) Technology*, Ionics Incorporated, Watertown, MA, 1984.

Mickley, M. and Hamilton, B., Disposal of membrane concentrate wastes: An AWWARF project status report, in: *Proceedings, Membrane Technologies in the Water Industry*, Orlando, FL, March 10–13, 1991 [American Water Works Association, Denver, CO, 1991].

Mulder, M., *Basic Principles of Membrane Technology*, Kluwer Academic Publishers, Dordrecht, the Netherlands, 1991.

Mulder, M., *Basic Principles of Membrane Technology*, 2nd edn., Kluwer Academic Publishers, Dordrecht, the Netherlands, 1996.

Pham, M. H., Pintauro, P. N., and Nobe, K., Application of a multicomponent membrane transport model to reverse-osmosis separation processes, in: *Reverse Osmosis and Ultrafiltration,* American Chemical Society Symposium Series, American Chemical Society, Washington, DC, 1985.

Porter, M. C., Ed., *Handbook of Industrial Membrane Technology,* Noyes Publications, Park Ridge, NJ. 1990.

Presswood, W. G., The membrane filter: Its history and characteristics, in: Dutha, B. J., Ed., *Membrane Filtration*, Marcel Dekker, Inc., New York, 1981.

Reahl, E. R., Half a century of desalination with electrodialysis—A short tour through fifty years, Technical Paper, TP1038EN 0603, GE Water & Process Technologies, General Electric Company, Trevose, PA, 2006.

Reid, C. W. and Breton, E. J., Water and ion flow across cellulosic membranes, *Journal Applied Polymer Science*, 1(2):133–143, 1959.

Rodgers, V. G. J. and Sparks, R. E., Effect of transmembrane pressure pulsing on concentration polarization, *Journal of Membrane Science*, 68:149–168, 1992.

Rouse, H., *Elementary Mechanics of Fluids*, John Wiley & Sons, New York, 1946.

Shriver, D. F. and Atkins, P. W., *Inorganic Chemistry*, 3rd edn., W. H. Freeman and Company, New York, 1999.

Silberberg, M., *Chemistry–The Molecular Nature of Matter and Change*, Mosby, St. Louis, MO, 1996.

Sourirajan, S., Ed., *Reverse Osmosis*, Academic Press, New York, 1970.

Sourirajan, S., Reverse osmosis: A new field of applied chemistry and chemical engineering, in: Turbak, A. F. (Ed.), *Synthetic Membranes*, Volume 1, *Desalination*, ACS Symposium Series 153, American Chemical Society, Washington, DC, 1981.

Thompson, M. A., The role of membrane filtration in drinking water treatment, Seminar S5—Engineering considerations for design of membrane filtration facilities for drinking water supply, in: *Proceedings of Annual Conference American Water Works Association,* Denver, CO, June 18, 1992.

Turbak, A. F., Ed., *Synthetic Membranes*, Volume 1, *Desalination*, ACS Symposium Series 153, American Chemical Society, Washington, DC, 1981a.

Turbak, A. F., Ed., *Synthetic Membranes*, Volume 2, *Hyper- and Ultrafiltration Uses*, ACS Symposium Series 154, American Chemical Society, Washington, DC, 1981b.

Van der Bruggen, B., Vandecasteele, C., Van Gestel, T., Doyen, W., and Leysen, R., A review of pressure-driven membrane processes in wastewater treatment and drinking water production, *Environmental Progress*, 22(1):46–56, April, 2003.

Weber, Jr., W. J., *Physicochemical Processes for Water Quality Control*, Wiley-Interscience, A Division of John Wiley & Sons, Inc., New York, 1972.

Wiesner, M. R. (1993). An Overview of ΔP Membrane Processes (Membrane Metaphysics...), Presented to Association of Environmental Engineering Professors, San Antonio, TX, June 7, 1993.

Wiesner, M. R. and Chellam, S., Mass transport considerations for pressure-driven membrane processes, *Journal of the American Water Works Association*, 84(1):88–95, January 1992.

Yun, T. I., Gabelich, C. J., Cox, M. R., Mofidi, A. A., and Lesan, R., Reducing costs for large-scale desalting plants using large-diameter, reverse osmosis membranes, *Desalination*, 189 (1–3):141–154, March 2006.

Zander, A. K., Chair, Committee on Advancing Desalination Technology, *Desalination—A National Perspective*, National Academies Press, Washington, DC, 2008.

18 Gas Transfer

The term "gas transfer," as used here, is a special case of "mass-transfer" limited to gases. A further limitation, for the purposes of this chapter, is that the focus is on the transfer of gases to and from water. The gases may be any, for example, oxygen, carbon dioxide, chlorine, chlorine dioxide, ozone, etc., and various volatile organic compounds (VOCs).

18.1 DESCRIPTION

The principles of mass transfer are founded on the three themes common to most reactor problems: (1) a materials balance, (2) equilibrium, and (3) kinetics. The approach of this chapter is to apply such fundamentals as the basis for operational equations, which differ by the characteristics of the particular application, for example, surface aerators, diffused aeration, packed towers, stream reaeration, etc. The differences are due essentially to the means in which interfacial surface area is created, for example, surface renewal or bubbles, and its rate of creation, for example, rate of pumping or airflow, and to the turbulence scale and intensity.

18.1.1 Gas Transfer In-a-Nutshell

Gas transfer involves (1) getting gases into solution, for example, oxygen, carbon dioxide, ozone, chlorine, chlorine dioxide, etc., and (2) getting gases out of solution, for example, ammonia, nitrogen, and a variety of VOCs. For each purpose, specialized industrial equipment has been developed, for example, surface aerators, diffusers for bubble aeration, packing for packed-towers, chlorinators, etc.

18.1.1.1 Comparison with Other Mass-Transfer Processes

Every natural and engineered reactor embodies mass-transfer processes, for example, advection, dispersion, turbulence, and molecular diffusion. In flotation, bubbles must be brought into contact with particles (by differential settling and/or turbulence); in filtration, particles must contact granular media (by advection, sedimentation, diffusion); in adsorption columns molecules must reach the particle–water interface, and then diffuse to the interior; in activated sludge, oxygen molecules, bacteria aggregates, and organic substrates are brought into contact by turbulence and then by diffusion to the point of reaction; etc. While the purposes are different in each case, the mass-transfer processes involve common principles.

18.1.1.2 Process Description

As stated, gas transfer involves transport either (1) from the gas phase to the dissolved aqueous phase or (2) from the dissolved aqueous phase to the gas phase. In the first case, the gas may be transported to the vicinity of the gas–water interface by advection and turbulence with diffusion being the final transport step to the interface. The water surface becomes "saturated" with respect to the gas concentration at the interface; the gas may be transported from the interface, that is, in the aqueous phase, by molecular diffusion, turbulence, or advection (or some combination). The process may be controlled by the rate of interfacial area created, which, in turn, is controlled by fluid transport processes, for example, the rate of pumping in the case of a turbine aerator. The same principles apply to natural processes, for example, reaeration in streams with oxygen, deoxygenation of streams of supersaturated oxygen concentration, carbon dioxide transport, etc. In each case, the system is striving for equilibrium, which is never attained in any "open" system.

18.1.2 Applications

Gas transfer to the aqueous phase occurs during aeration of water in activated sludge reactors, in trickling filters, in rotating biological contactors, in chlorination, in ozone uptake, etc. Gas transfer from the aqueous phase to the gas phase occurs in aeration to remove odors, such as hydrogen sulfide or radon (prevalent in some groundwaters); and in air stripping to remove VOCs, ammonia, etc. In nature, gas transfer occurs in stream reaeration; photosynthesis, for example, due to supersaturation with oxygen; anaerobic reactions in benthic muds with methane and carbon dioxide as products. Gas precipitation occurs in the latter situations. Table 18.1 gives some examples of each category of gas transfer and indicates the technology involved; the "notes" describe the context for each application.

18.1.3 History

The use of dissolved gases in water treatment goes back to 1902 when ozone disinfection was applied at Paderborn, Germany ($Q = 1.00\ m^3/min$) and then in 1906 at the Bon Voyage WTP at Nice, France ($Q = 13\ m^3/min$) (Hill and Rice, 1982, p. 11). The latter plant was considered the "birthplace" use of ozone in drinking water treatment. Chlorine was adopted for municipal drinking water starting in 1909 in Poughkeepsie, New York, in the form of hypochlorites (chloride of lime) and as chlorine gas in 1913 in Philadelphia (Baker, 1948, p. 340, 342, respectively). Removal of dissolved gases in drinking water treatment goes back to the 1920s for removal of taste and odors (Hale, 1932). Methods for the latter included the use of spray nozzles, perforated trays, diffused air, or the construction of cascades.

TABLE 18.1
Examples of Gas Transfer

Transfer Direction	Example	Technology	Notes
Gas to water	Aeration of activated sludge reactors	Diffused bubble aeration	Oxygen transfers from bubbles to water
		Turbine aeration	Oxygen transfers across interfacial surface area created by turbine pumping
		Brush aerator	Same as turbine aeration
	Ozonation	Diffused gas	Same as bubble aeration, except gas is ozone
	Chlorination	Diffused gas	Same as bubble aeration, except gas is chlorine
	Stream aeration	Passive occurrence	Oxygen transfer occurs across interfacial areas created continuously by stream turbulence
Water to gas	Odor removal	Spray aeration	Spray creates interfacial surface area
	VOC removal	Packed tower/cross flow tower/diffused bubbles	Air stripping, creating surface area
	Ammonia removal	Packed tower/cross flow tower	Air stripping, creating surface area, requires high pH
	Gas precipitation	Passive occurrence	Saturation of water with gas; for example, carbon dioxide, methane, oxygen, nitrogen
	Stream deaeration	Passive occurrence	Passive transfer of supersaturated DO to atmosphere

18.1.3.1 Theory

The modern theory of gas transfer is based on the classic two-film theory of W.K. Lewis and W.G. Whitman, described in their 1924 paper. The paper was a part of a symposium on gas absorption of the Division of Industrial and Engineering Chemistry of the American Chemical Society; the associated papers dealt with different gases, different rates of mixing, temperature, etc.; also, they recognized the role of Henry's constant. The symposium marked the beginning of current theory. Although Lewis and Whitman and the symposium recognized the role of mixing on film thickness, that is, a higher impeller speed decreases the thickness of a given liquid film, they did not link this to turbulence and the idea of the rate of surface renewal, which was to come. This idea was started by Higbie in 1935 and developed further by Danckwertz in 1951 and applied to stream aeration by O'Connor and Dobbins in 1958 (Eckenfelder and O'Connor, 1961, p. 82).

18.1.3.2 Stream Aeration

Stream reaeration was dealt with first by Streeter and Phelps in their classic 1924 paper on the topic. The context was the issue of BOD (biochemical oxygen demand) pollution of rivers and the associated concepts of "self-purification," assimilative capacity, and the rate of reaeration. Their differential equation compared the rate of reaeration and the rate of oxygen uptake due to BOD; its solution yielded a hypothetical BOD mass flux input limit in order to maintain aerobic conditions. The objective was to utilize the "assimilative-capacity" of streams, with treatment only to the extent needed to maintain aerobic conditions.

18.1.3.3 Oxygen Transfer in Activated Sludge

Airflow to an activated sludge reactor was based initially on an empirical criterion of so much airflow per unit volume of reactor. By the 1950s, the concept of oxygen utilization rate had evolved as a stoichiometric link to the various reactions which utilize oxygen, for example, BOD reaction, endogenous respiration of cells, nitrification, etc.

18.1.3.4 Spiral Flow Diffusers

Bewtra (1962) and Morgan and Bewtra (1959, 1960, 1963) related oxygen transfer rate to airflow for coarse and fine bubble diffusers, using a 4.6 m × 4.6 m × 1.33 m (15 ft × 15 ft × 4 in.) model section of an activated sludge reactor for measurement of oxygen transfer efficiency. An issue with fine-bubble diffusers was the maintenance, for example, frequent cleaning due to clogging, albeit oxygen transfer efficiencies were 8%–15%. The alternative of coarse-bubble diffusers was to accept low oxygen transfer efficiency, for example, about 6%–7%. The airflow was along one side, resulting in a spiral flow pattern, which was the practice until the 1980s.

18.1.3.5 Turbine Aeration

Turbine aerators came into vogue in the early 1960s; at that time, Kalinske (1963) related the rate of oxygen transfer to the rate of pumping by the impeller, depth of submergence, and other factors. The turbine aerators had appeal partly because they circumvented the issue of diffuser maintenance and they could be characterized as a "complete-mix," which was also in vogue at the time.

18.1.3.6 Grid Diffusers

In the 1980s, the issue of diffused (bubble) aeration was revisited and re-evaluated in terms of newer designs of diffusers and a grid layout (vis-à-vis spiral flow), and motivated by the significant energy reduction as compared with turbine aeration and coarse-bubble diffusers (see Boyle, 1985; Boyle et al., 1989).

18.1.3.7 Air Stripping

Another strand of gas transfer was the issue of air stripping, which has been around as an empirical practice in water treatment since the 1920s, to remove odors, and in wastewater treatment since the mid-1960s when the Lake Tahoe tertiary treatment plant was designed to remove ammonia. Then, in the 1980's, VOCs were an issue as hazardous waste sites were identified and the 1986 Safe Drinking Water Act had as one of its objectives the removal of certain VOCs (volatile organic compounds) from drinking water; air stripping was one of the treatment technologies used for their removal.

18.2 GAS TRANSFER THEORY

In-a-nutshell, gas transfer is limited by the rate of diffusion across a gas–water interface and the rate of surface renewal. The latter is related to the rate of advection to the gas–water interface and turbulence. Overall, the gas-transfer rate is proportional to the departure from equilibrium between a surface "film" and the interior of the reactor.

18.2.1 Equilibria

Most gas-transfer reactor systems are "open," which means that mass and/or energy flux may cross the boundaries. Any open system is striving toward equilibrium. The relative departure from equilibrium "drives" the reaction, that is, gas transfer, and controls the rate of the mass transfer.

18.2.1.1 Henry's Law

The equilibrium between a gas at a given pressure and its aqueous phase concentration is given by Henry's law, which has several forms (see Appendix H). The form favored here is the "solubility relation," which says merely that the aqueous-phase concentration of a gas, "A," is proportional to the partial pressure of "A" above the gas–water interface, that is,

$$C(A) = H^S(A) \cdot p(A) \tag{18.1}$$

where

$C(A)$ is the concentration of gas in aqueous phase in equilibrium with $p(A)$ (mg/L)
$H^S(A)$ is Henry's constant, a proportionality constant (mg/L/atm)
$p(A)$ is the partial pressure of gas "A" above the gas–water interface (atm)

Table H.5 gives values of H^S for different gases and for different temperatures. It is simple since for $p(A) = 1.0$ atm, $C(A) = H(A)$.

Equation H.28, that is, $p^*(A) = H(A) \cdot X^*(A)$, is the form used most commonly in physical chemistry and in chemical engineering; it says that the equilibrium pressure of the adjacent atmosphere is proportional to the mass fraction of "A" in the aqueous phase and is called here the "volatility relation."

Example 18.1 Illustration of Calculation of Dissolved Oxygen Concentration by Henry's Law

Given
Elevation = sea-level; $T = 20°C$; $p(O_2) = 0.209$ atm.

Required
Determine $C(O_2$, sea level, $20°C)$, that is, the aqueous concentration of oxygen at sea level at 20°C.

Solution
From Table H.5, $H^S(O_2, 20°C) = 43.39$ mg/L/atm. Application of Equation 18.1, the solubility form of Henry's law gives,

$$C(O_2) = (43.39\,\text{mg/L/atm}) \cdot (0.209\,\text{atm})$$
$$= 9.07\,\text{mg/L}$$

Discussion
$C(O_2) = 9.07$ mg/L is about the concentration measured in a water sample at sea level by the well-known Winkler titration.

18.2.2 Kinetics

The rate of a process is described by its "kinetics." As noted, for the case of gas transfer, such rate is dependent upon the gas diffusion flux in the gas phase and aqueous phase and the rate of creation of interfacial surface area. The diffusion is sequential, either gas-phase-then-aqueous-phase or aqueous-phase-then-gas-phase. For the sequence of gas-phase-then-aqueous-phase, as interfacial surface area is created, a gas-deficient surface film becomes instantaneously "saturated" with a given gas, for example, oxygen, which is then advected away from the gas–water interface. The opposite occurs in "gas stripping," that is, removal of dissolved gas from the aqueous phase.

18.2.2.1 Diffusion

The diffusion flux is due to the random thermal motion of molecules. Fick's law is the basis for the working equations.

18.2.2.1.1 Kinetic Energy of Gases

Molecules in a gas have a thermal kinetic energy,

$$\text{KE(thermal)} = \frac{1}{2}mv^2 = kT \tag{18.2}$$

where

KE(thermal) is the kinetic energy due to thermal motion (J/molecule)
m is the mass of molecule (g)
v is the velocity of molecule (cm/s)
k is the Boltzman constant ($1.38 \cdot 10^{-23}$ J/molecule/K)
T is the absolute temperature (K)

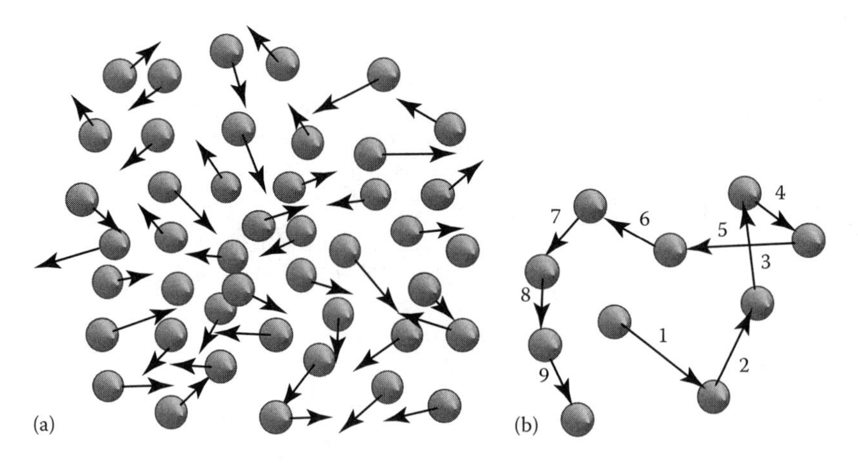

FIGURE 18.1 Random motion of molecules. (a) Collection of molecules, (b) single molecule.

Figure 18.1 illustrates the kind of motion involved in which the molecules travel from collision to collision, in random motion, as "billiard balls." Figure 18.1a represents a collection of molecules with motion frozen at a given time. The arrows represent random velocities. Figure 18.1b depicts the motion of a single molecule in a sequence of "steps" between collisions. As would be expected, the steps are random in length and orientation.

The kind of random motion depicted in Figure 18.1 causes a net transport of molecules from zones of higher concentration to zones of lower concentration. Mathematically, probability theory can describe this transport. Also, as indicated by Equation 18.2, the velocity of a given molecule changes with square root of absolute temperature. With increasing temperature, for example, the molecules have associated higher velocities, giving more steps per unit time. Therefore, as temperature increases or decreases, the transport of molecules across a given plane will change accordingly. The net transport of molecules will be in the direction of negative concentration gradient.

18.2.2.1.2 Diffusion in the Liquid Phase

The above theory applies to gases without qualification. The intermolecular forces between gas molecules are negligible relative to their kinetic energy, and the gas molecules move as if they were billiard balls. For liquids, however, intermolecular forces are substantial. Still, the molecules will move randomly, albeit at lower velocities. Thus the principles described above are applicable to molecules in liquids, as well as gases.

18.2.2.1.3 Fick's First Law

The net transport of molecules in the direction of their negative concentration gradient is expressed mathematically as Fick's first law, that is,

$$\mathbf{j} = -D\nabla C \tag{18.3}$$

where

$\mathbf{j}$ is the flux of molecules in the direction of negative concentration gradient ($kg/m^2/s$)

D is the diffusion constant (m^2/s)

C is the concentration of a given species of molecules (kg/m^3)

∇ is the del operator (partial differential in all coordinate directions)

Most often, Fick's law is expressed in one direction, that is,

$$\mathbf{j} = -D\frac{\partial C}{\partial x} \tag{18.4}$$

where x is the distance along x coordinate axis (m).

Fick's first law is the foundation for all mass transport equations based upon molecular diffusion (Box 18.1). Applications include carbon adsorption, ion exchange, and gas transfer. As related to the latter, Fick's first law is applicable to uptake of gases to the aqueous phase or removal by gas stripping.

BOX 18.1 ADOLF FICK

Adolf Eugen Fick was born September 3, 1829 and is one of the key persons, with Thomas Graham, to develop the modern ideas of diffusion (Cussler, 1984, p. 17). During his secondary schooling, Fick was enamored with mathematics, especially the work of Poisson. While he had intended to make his career in mathematics, his older brother, a professor of anatomy at the University of Marlburg, influenced him to pursue medicine instead. In the spring of 1847 Fick went to Marlburg, and was influenced by Carl Ludwig, who believed that medicine and all phenomena, must have a basis in mathematics, physics, and chemistry. In his first diffusion paper, about 1855, Fick codified Graham's experimental work and drew the analogies between diffusion and heat transfer and the flow of electricity. He developed his laws of diffusion through analogies with Fourier's work (c. 1822) on heat flow.

18.2.2.1.4 Fick's Second Law

A materials balance relation for the rate of accumulation of mass within an infinitesimal volume due to diffusion transport is called Fick's second law, and in one dimension is stated as

$$\frac{dC}{dt} = D\frac{\partial^2 C}{\partial X^2} \quad (18.5)$$

In three dimensions, Fick's second law is

$$\frac{dC}{dt} = D\nabla^2 C \quad (18.6)$$

Figure 18.2 illustrates how Equation 18.5 is derived. As seen, the difference between the mass flow to and from the infinitesimal element of thickness Δx and area A, is the rate of accumulation of mass within the element, and is expressed mathematically as Equation 18.7. By carrying out the subtraction indicated and dividing by ΔV gives $dC/dt = -\partial \mathbf{j}/\partial Z$. Now substituting Equation 18.4 into the foregoing yields Equation 18.5. Thus Fick's second law is only an extension of Fick's first law.

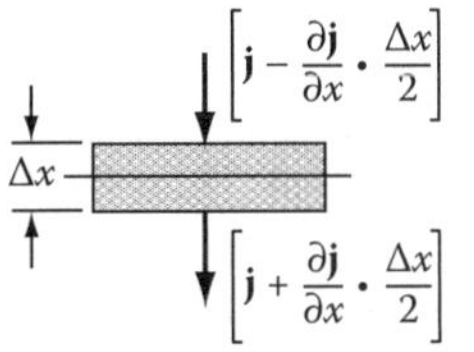

$$\left[j - \frac{\partial j}{\partial x}\cdot\frac{\Delta x}{2}\right] \quad \Delta x \quad \left[j + \frac{\partial j}{\partial x}\cdot\frac{\Delta x}{2}\right]$$

$$\frac{d(VC)}{dt} = \left[j - \frac{\partial j}{\partial x}\cdot\frac{\Delta x}{2}\right]\cdot A - \left[j + \frac{\partial j}{\partial x}\cdot\frac{\Delta x}{2}\right]\cdot A \quad (18.7)$$

FIGURE 18.2 Materials balance for diffusion transport to and from infinitesimal volume element and accumulation of mass within the element.

18.2.2.1.5 Diffusion Coefficients

Diffusivities of solutes in liquids are determined, most commonly, by the Wilke and Chang equation (Danckwerts, 1970), that is,

$$D = 7.4\cdot 10^{-8}\cdot\frac{T\cdot(x\cdot \mathrm{MW})^{0.5}}{\mu\cdot V^{0.6}} \quad (18.8)$$

where

- D is the diffusion coefficient of dissolved gas in solvent, for example, water (cm^2/s)
- T is the temperature of solvent (°K)
- MW is the molecular weight of the solvent, for example, water
- μ is the dynamic viscosity of solvent, for example, water (cP)
- V is the molecular volume of the solute at the normal boiling point ($cm^3/gmol$)
- x is the association factor (2.6 for water, as given by Danckwerts, 1970, p. 16)

Equation 18.8 predicts the diffusivity within about 10% (Danckwerts, 1970, p. 16). For water as the solvent, Equation 18.8 is

TABLE 18.2
Molecular Volumes of Simple Substances

Substance	V(solute) ($cm^3/gmol$)	Substance	V(solute) ($cm^3/gmol$)
H_2	14.3	N_2O	36.4
O_2	25.6	NH_3	25.8
N_2	31.2	H_2O	18.9
CO	30.7	H_2S	32.9
CO_2	34.0	COS	51.5
SO_2	44.8	Cl_2	48.4
NO	23.6	Br_2	52.2
		F_2	71.5

Source: Adapted from Danckwerts, P.V., *Gas–Liquid Reactions*, McGraw-Hill Book Company, New York, 1970, p. 17.

$$D(\text{water as solvent}) = 7.4\cdot 10^{-8}\cdot\frac{T\cdot(2.6\cdot \mathrm{MW(water)})^{0.5}}{\mu\cdot V(\text{solute})^{0.6}} \quad (18.9)$$

The molecular volumes of some solutes are given in Tables 18.2 and 18.3 shows diffusion coefficients calculated by Equation 18.9. For comparison, selected diffusion coefficients from experimental data are given also, in the far right column. From Table 18.3, the diffusion coefficients for O_2, N_2, and Cl_2 are not widely different at corresponding temperatures; also, temperature is seen to have an effect and should not be neglected, but its influence is not significant. Finally, and most important, the predictions of the diffusion coefficient by the Wilke–Chang relation agrees well with the experimental data shown in the far-right column.

Table 18.4 is a more extensive list of diffusion coefficients for substances in both the gas phase at one atmosphere pressure and as solutes dissolved in water. As is evident from Table 18.4, the gas phase diffusion coefficients are in the range 0.1–1 cm/s^2 at one atmosphere pressure and near room temperature (Cussler, 1984, p. 105).

18.2.2.2 Adaptation of Fick's Law to Two-Film Theory

When random molecular motion, that is, diffusion, occurs in sequence with other processes, and if it is the slowest process, then it limits the overall rate of the process (Cussler, 1984, p. 1). The rate of diffusion can be increased, however, by increasing the diffusion gradient, that is, ∇C, per Equation 18.3, or in one dimension, $\partial C/\partial x$. The diffusion gradient may be increased by increasing the rate of surface renewal, for example, by advection and turbulence.

18.2.2.2.1 Film Theory

Concentration gradients for pure diffusion, as described by Fick's law, are smooth and continuous. In most engineered and natural situations, however, the gas–liquid interfaces occur in the context of turbulent mixing, causing the concentration of gas in the "bulk of solution" to be homogeneous. From the bulk of solution, with concentration, C_o, to the

TABLE 18.3
Diffusion Coefficients Calculated by Wilke–Chang Relation[b,c]

	Solvent			Solute		Diffusion Coefficient	
Solvent	MW(solvent) (g/mol)	Temperature (°C)	μ^a (N · s/m²)	Solute	V(solute)[b] (cm³/gmol)	D^c (cm²/s)	D^d (cm²/s)
Water	18	0	0.001787	O_2	25.6	1.03E−05	
	18	20	0.001000		25.6	1.98E−05	
	18	25	0.000899		25.6	2.24E−05	2.10E−05
	18	30	0.000808		25.6	2.53E−05	
Water	18	0	0.001787	N_2	31.2	9.15E−06	
	18	20	0.001000		31.2	1.75E−05	
	18	25	0.000899		31.2	1.99E−05	1.88E−05
	18	30	0.000808		31.2	2.25E−05	
Water	18	0	0.001787	Cl_2	48.4	7.03E−06	
	18	20	0.001000		48.4	1.35E−05	
	18	25	0.000899		48.4	1.53E−05	1.25E−05
	18	30	0.000808		48.4	1.73E−05	

Source: Danckwerts, P. V., *Gas–Liquid Reactions*, McGraw-Hill Book Company, New York, 1970, p. 17.
[a] Calculated by polynomial equation for μ as given in Table CD/QR.7 and following Table B.9/Figure B.1.
[b] Calculated by Wilke–Chang: $D(\text{water as solvent}) = 7.4 \cdot 10^{-8} \cdot \dfrac{T \cdot (2.26 \cdot \text{MW(water)})^{1/2}}{\mu \cdot V(\text{solute})^{0.6}}$ as taken from Danckwerts (1970, p. 15), Sherwood et al. (1975, p. 25), Cussler (1984, p. 121).
[c] Source for experimental data was Cussler (1984, p. 16).

TABLE 18.4
Diffusion Coefficients of Substances in Both Air and in Water Solution[a]

Air Solution			Water Solution	
Substance	Temperature (°C)	D (cm²/s)	Substance	D^a (cm²/s)
CH_4	0	0.196	CH_4	$1.49 \cdot 10^{-5}$
CO_2	3	0.142	CO_2	$1.92 \cdot 10^{-5}$
CO_2	44	0.177		
H_2	0	0.611	H_2	$4.50 \cdot 10^{-5}$
O_2	0	0.177	O_2	$2.10 \cdot 10^{-5}$
Benzene	25	0.096	Benzene	$1.02 \cdot 10^{-5}$
Toluene	26	0.086		
Chlorobenzene	26	0.074	Ammonia	$1.64 \cdot 10^{-5}$
Water	16	0.282	H_2S	$1.41 \cdot 10^{-5}$
	25	0.260	Chlorine	$1.25 \cdot 10^{-5}$
	39	0.277	Air	$2.00 \cdot 10^{-5}$
			N_2	$2.6 \cdot 10^{-5}$
			Water	$2.44 \cdot 10^{-5}$

Source: Cussler, E.L., Diffusion, in *Mass Transfer in Fluid Systems*, Cambridge University Press, Cambridge, U.K., pp. 106–108, 116–117, 1984.
[a] Diffusion coefficients for water solutions are at 25°C.

gas–liquid interface, with concentration C_i, the diffusion concentration gradient is steep and occurs across a small distance, called the *film*. Actually such a film is assumed to occur across both phases, that is, gas and liquid, respectively; hence the designation, *two-film theory* (Lewis and Whitman, 1924).

Figure 18.3 illustrates graphically the concept of the two-film theory for the two conditions of gas transfer: (a) from the gas phase to the liquid phase, and (b) from the liquid phase to the gas phase. The lower portions of the illustrations depict the concentration gradients in terms of densities of the gases for the two phases; the upper portions define terms and show the associated slopes. For (a), the gas to liquid case, the gas pressure in the bulk of the gas solution is p_o, and the concentration at the interface is p_i. This difference occurs across the

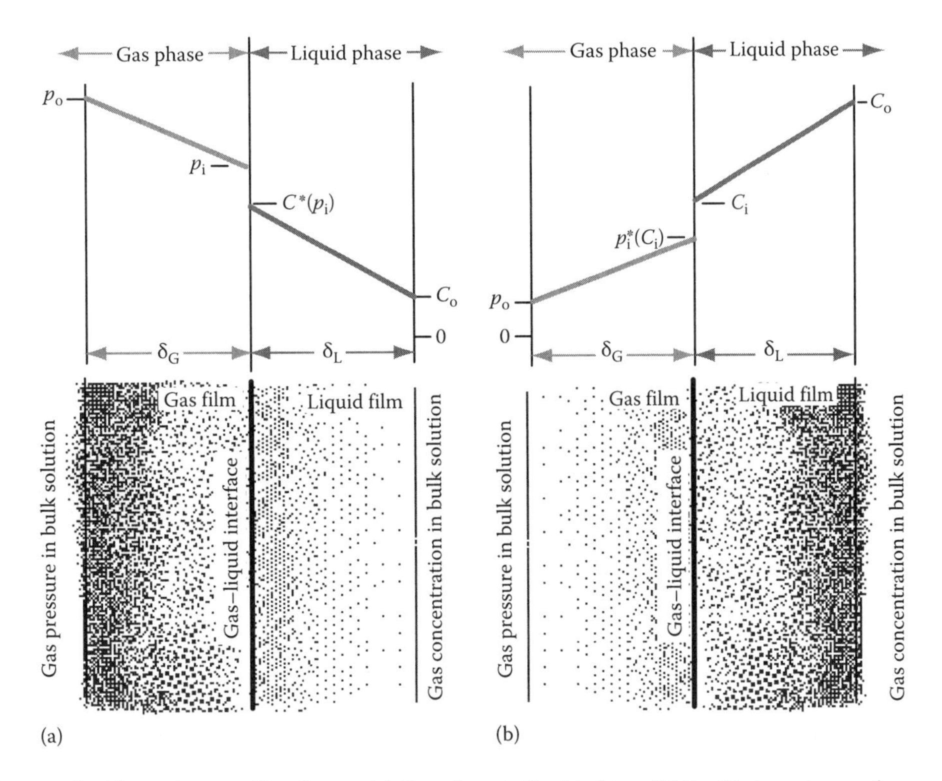

FIGURE 18.3 Gas transfer illustrating two-film theory. (a) Gas phase to liquid phase, (b) liquid phase to gas phase.

film of thickness, δ_G. The same is true for the liquid phase in which the interface concentration is $C^*(p_i)$, that is, saturation concentration in the aqueous phase at the interface, given the gas pressure at the interface, p_i. The assumption is that a monomolecular layer of liquid at the interface is "saturated" with respect to the gas phase interface pressure, p_i. Within the bulk of the aqueous solution, the concentration is C_o. An example of case (a) is an activated sludge reactor, in which C_o(oxygen) is constant at a fixed level, usually 2 mg/L, since oxygen is consumed at the rate supplied.

For (b), the aqueous phase to the gas phase, the gradients are reversed. An example of the (b) case is an "air stripping" system, for example, for ammonia, radon, a VOC, etc. The reactors may be either "batch," with C_o declining with time, or "continuous-flow" with C_o being constant with time. If the continuous-flow reactor is a column, C_o declines from entrance to exit; if it is a complete-mix, then C_o does not vary spatially.

18.2.2.2.2 Film Theory Mathematical Models

For convenience in deriving an expression for gas transfer across a "film," Fick's first law is restated, as a starting point, that is,

$$\mathbf{j} = -D\frac{\partial C}{\partial X} \tag{18.4}$$

$$\approx -D\frac{\Delta C}{\Delta X} \tag{18.10}$$

Equation 18.10 approximates the differentials for a film, gas or aqueous, such as illustrated in Figure 18.3. Applying the Equation 18.10 approximation to the liquid film of Figure 18.3a gives

$$\mathbf{j} = -D\frac{[C^*(p_i) - C_o]}{\delta_L} \tag{18.11}$$

Now assume that the gas film is negligible in gradient, such that: $p_i \approx p_o$, which means that

$$C^*(p_i) \approx C^*(p_o) \tag{18.12}$$

Since the interface pressure is not known, the approximation of Equation 18.12 may be substituted in Equation 18.11 to give

$$\mathbf{j} = -D\frac{[C^*(p_o) - C_o]}{\delta_L} \tag{18.13}$$

To simplify nomenclature, let $C^*(p_o) = C_s$ with C_s. Also, let $C_o = C$. Further, since the aqueous film is the main concern, let $\delta_L = \delta$. Using this simplified nomenclature, Equation 18.13 becomes

$$\mathbf{j} = -\left(\frac{D}{\delta}\right) \cdot (C_s - C) \tag{18.14}$$

where
- C_s is the saturation concentration of dissolved gas (with respect to gas pressure, p_o, in the bulk of the gas solution) in the aqueous phase at the gas–water interface (kg gas/m^3 aqueous solution)
- C is the concentration of dissolved gas in the bulk of solution (kg gas/m^3 aqueous solution)
- δ is the thickness of aqueous film across which diffusion is taking place (cm)

Multiplying the unit flux, that is, "flux-density," by the interface surface area, "**a**," gives

$$\mathbf{J} = \mathbf{j} \cdot \mathbf{a} = \frac{dm}{dt} = -\left(\frac{D}{\delta}\right) \cdot (C_s - C) \cdot \mathbf{a} \quad (18.15)$$

where
- $\mathbf{J}$ is the rate of mass transfer (g/s)
- m is the mass of substance being transferred (g)
- D is the diffusivity of gas through water (cm^2/s)
- $\mathbf{a}$ is the surface area across which diffusion is taking place (cm^2)
- C is the concentration of gas in water at any given time (g gas/mL water)

Equation 18.15 gives the rate of accumulation of mass by the volume to which the mass is being transferred in terms of the concentration difference across the liquid film (see also Eckenfelder and O'Connor, 1961, pp. 76–81). Now grouping the terms D/δ, and replacing these by k_L, gives Equation 18.16,

$$\mathbf{J} = \frac{dm}{dt} = -k_L \mathbf{a} \cdot (C_s - C) \quad (18.16)$$

in which k_L is the aqueous phase film transfer coefficient (cm/s).

A gas phase mass transport equation that corresponds to Equation 18.14 is

$$\mathbf{j}_g = -k_g(p_o - p_i) \quad (18.17)$$

in which k_g is the gas phase film transfer coefficient (cm/s).

The rate of mass accumulation, dm/dt, in a given reactor volume, V, is

$$\frac{dm}{dt} = V\frac{dC}{dt} \quad (18.18)$$

Substituting Equation 18.18 in Equation 18.16 and solving for dC/dt gives

$$\frac{dC}{dt} = k_L\left(\frac{\mathbf{a}}{V}\right)(C_s - C) \quad (18.19)$$

To aggregate the terms, k_L, "**a**," and V, a new coefficient is defined:

$$K_L a = k_L \frac{\mathbf{a}}{V} \quad (18.20)$$

Equation 18.19 then becomes

$$\frac{dC}{dt} = K_L a(C_s - C) \quad (18.21)$$

Equation 18.21 then is an operational equation for practical gas transfer situations and $K_L a$ is a well-known coefficient that characterizes a given system. The relation is used in design and in evaluation of aerators, that is, to determine $K_L a$. It is valid for both diffused and mechanical aeration and for any type of system.

As a word of caution, the $K_L a$ coefficient is unique for a particular system. As seen, the variables are K_L, "**a**," and V. Thus, the interfacial surface area, "**a**," under gas transfer at any given instant affects $K_L a$, as does the volume of the system.

18.2.2.2.3 Two-Film Theory

Across the gas–water interface, as illustrated in Figure 18.3, the flux density of gas "gas-A" in the gas phase must equal the flux density "gas-A" in the aqueous phase. Applying Equation 18.14 and using a corresponding equation for diffusion of "gas-A" in the gas phase, that is, $\mathbf{j}_g(A) = k_g(A) \cdot [p_o(A) - p_i(A)]$, gives for the equality

$$\mathbf{j}(A) = k_g(A)[p_o(A) - p_i(A)] = k_L(A)[C_i(A) - C(A)] \quad (18.22)$$

where
- $\mathbf{j}(A)$ is the mass flux density of solute, A across gas film or aqueous film ($kg/m^2/s$)
- $k_g(A)$ is the mass transfer coefficient for A in the gas phase ($kg/m^2/s/kPa$)
- $k_L(A)$ is the mass transfer coefficient for A in the liquid (aqueous) phase ($kg/m^2/s/kPa$)
- $p_o(A)$ is the partial pressure of species A in bulk of solution of gas mixture (kPa)
- $p_i(A)$ is the partial pressure of species A at gas–liquid interface (kPa)
- $C(A)$ is the concentration of species A in bulk of solution (kg A/m^3 solution)
- $C_i(A)$ is the concentration of species A at gas–liquid interface (kg A/m^3 solution)

Since the interface pressure, p_i, and the interface aqueous concentration, C_i, cannot be measured, two changes are imposed:

1. The partial pressure of species A in the gas phase at the gas–liquid interface is replaced by a pseudo interface partial pressure, $p^*(A)$. This is the partial pressure of "A" that would be in equilibrium with the liquid phase concentration taken to be that of the bulk of solution, $C(A)$.
2. The concentration of species A in the liquid phase at the gas–liquid interface is replaced by a pseudo interface concentration, $C^*(A)$. This is the interfacial aqueous phase concentration that would be in equilibrium with the gas phase concentration taken to be as the bulk of gas solution, $p_o(A)$.

With these surrogate measures, Equation 18.22 is modified as

$$\mathbf{j}(A) = K_g(A)[p_o(A) - p^*(A)] = K_L(A)[C^*(A) - C(A)] \quad (18.23)$$

where

$K_g(A)$ is the psuedo mass transfer coefficient for A in the gas phase, called the "overall" mass transfer coefficient in the literature (kg/m^2/s/kPa)

$K_L(A)$ is the psuedo mass transfer coefficient for A in the liquid phase, called the "overall" mass transfer coefficient in the literature (kg/m^2/s/kPa)

$p^*(A)$ is the pseudo partial pressure of A at gas–water interface at equilibrium with A in the aqueous phase at the interface (kPa)

$C^*(A)$ is the pseudo concentration of A at gas–water interface at equilibrium with A in the gas phase at the interface (kg A/m^3)

The pseudo mass transfer coefficients, K_g and K_L, are intended to compensate for the use of a pseudo interface equilibrium pressure, $p^*(A)$ and a pseudo interface aqueous concentration, $C^*(A)$. The variables of Equation 18.23, as defined, thus force a fit to the real mass flux density, $\mathbf{j}(A)$.

Again, to simplify nomenclature, the reference to A is omitted, to give

$$\mathbf{j} = K_g(p_o - p^*) = K_L(C^* - C) \quad (18.24)$$

where

K_g is the same as $K(A)_g$, that is, psuedo mass transfer coefficient for A in the gas phase, called the "overall" mass transfer coefficient in the literature (kg/m^2/s/kPa)

K_L is the same as $K(A)_L$, that is, psuedo mass transfer coefficient for A in the liquid phase, called the "overall" mass transfer coefficient in the literature (kg/m^2/s/kPa)

p^* is the same as $p^*(A)$, that is, pseudo partial pressure of A at gas–water interface at equilibrium with A in the aqueous phase at the interface (kPa)

C^* is the same as $C^*(A)$, that is, pseudo concentration of A at gas–water interface at equilibrium with A in the gas phase at the interface (kg A/m^3)

Since Equation 18.24 is an artifice, the coefficients, K_g and K_L must be determined to fit the variables. They were determined (derivations not shown) as

$$\frac{1}{K(A)_g} = \frac{1}{k(A)_g} + \frac{H(A)}{k(A)_L} \quad \text{or} \quad \frac{1}{K_g} = \frac{1}{k_g} + \frac{H}{k_L} \quad (18.25)$$

$$\frac{1}{K(A)_L} = \frac{1}{k(A)_L} + \frac{1}{H(A) \cdot k(A)_g} \quad \text{or} \quad \frac{1}{K_L} = \frac{1}{k_L} + \frac{1}{H \cdot k_g} \quad (18.26)$$

Combining Equations 18.25 and 18.26 gives

$$\frac{1}{K(A)_g} = \frac{H(A)}{K(A)_L} \quad \text{or} \quad \frac{1}{K_g} = \frac{H}{K_L} \quad (18.27)$$

Equations 18.25 through 18.27 provide a means to estimate whether K_g or K_L is rate controlling, summarized for large and small values of H (see also Dvorak et al., 1996, p. 946), that is,

		Gradient			
H	Solubility	Aqueous Phase	Gas Phase	K_g	K_L
Large	Low	Steep	Shallow	$\rightarrow 0$	$\approx k_L$
Small	High	Shallow	High	$\approx k_g$	$\rightarrow 0$

Interpreting values of H: From Equation 18.27, rearrangement shows that $K_L = H \cdot K_g$ and $K_g = K_L/H$. Thus K_L is proportional to H, that is, it is a linear function with H. At the same time, K_g is a hyperbolic function with H, that is, when $H \rightarrow 0$, K_g is very large and as H becomes large, $K_g \rightarrow 0$. The "volatility" definition of Henry's law, that is, Equation H.28 is used, that is, $p = HX$; the concentration may be any units desired since it is the form of the equation that is important.

18.2.2.3 Surface Renewal Models

The concept of the surface renewal model, or penetration model (Sherwood et al., 1975, p. 153), is that the surface interface is replaced by liquid from the interior that has the concentration of the bulk of the liquid. While an element of liquid is at the surface and is exposed to the gas, it absorbs or releases gas (Danckwerts, 1970, p. 100). The mass of a given gas that is transferred for a given element at the gas–water interface is its flux density, $\mathbf{j}$, times the element area, dA (element), times its time of exposure at the gas–water interface, dθ, that is, $\mathbf{j}$(element) $\cdot$ A(element) $\cdot$ dθ(element). The flux density of any given element is in accordance with the two-film theory. The rate of element replacement is proportional to the intensity of turbulence. At any given instant, the surface has a distribution of ages of elements exposed to the surface. Thus there is a mosaic of element ages. According to Danckwerts (1970, p. 101), the distribution function, ϕ, for the element ages is

$$\varphi = \mathbf{s} \cdot e^{-s\theta} \quad (18.28)$$

And the mass flux-density for any given element is

$$\mathbf{j} = (C^* - C_o)\left[\frac{D_A}{\pi\theta}\right]^{0.5} \int_0^\infty \mathbf{s}\, e^{-s\theta} d\theta \quad (18.29)$$

$$= (C^* - C_o)\mathbf{s}\left[\frac{D_A}{\pi}\right]^{0.5} \int_0^\infty \frac{e^{-s\theta}}{\theta^{0.5}} d\theta \quad (18.30)$$

$$= (C^* - C_o)[D_A \mathbf{s}]^{0.5} \quad (18.31)$$

and

$$K_L\varphi = [D \cdot \mathbf{s}]^{0.5} \tag{18.32}$$

where

ϕ is the probability that any element will be exposed to the time θ before being replaced by a fresh element

D_A is the gas-A diffusion coefficient, that is, as in accordance with Fick's law (m^2/s)

s is the fraction of area of surface which is replaced with fresh liquid per unit time (s^{-1})

θ is the time of exposure of element at surface (s)

j is the instantaneous rate of mass transfer ($kg/m^2/s$)

$K_L\varphi$ is the mass-transfer coefficient associate with surface renewal function (m)

Example 18.2 Surface Renewal Rate

Given

From Sherwood et al. (1975, p. 155), $K_L\varphi = 0.00147$ cm/s was measured for uptake of pure hydrogen in water in a small vessel stirred at 300 rpm at 25°C; also, $D(H_2) = 6.3 \times 10^{-5}$ cm²/s.

Required

Determine "**s**."

Solution

Determine D and then apply Equation 18.32

$$K_L\varphi = [D \cdot \mathbf{s}]^{0.5}$$

$$0.00147\ \frac{\text{cm}}{\text{s}} = \left[6.3 \cdot 10^{-5}\ \frac{\text{cm}}{\text{s}} \cdot \text{s}\right]^{0.5} \tag{Ex18.2.1}$$

$$s = 0.034\ \text{s}^{-1}$$

Discussion

The mass-transfer coefficient, $K_L\varphi$, is affected by the rate of surface renewal, which, in turn, is affected by the turbulence intensity.

18.2.2.4 K_La as a Design Parameter

The flux density, **j**, was given by Equations 18.16 through 18.18, 18.20, and 18.21, which are variations on the same theme, and Equation 18.31, which involves surface renewal. Selecting Equation 18.21 for the purpose at hand, both sides may be multiplied by V, the reactor volume, to give

$$\mathbf{J} = \left(\frac{dC}{dt}\right) \cdot V = K_La(C_s - C) \cdot V \tag{18.33}$$

For an activated sludge reactor, as an example, **J**(oxygen) = [$BOD_{in} - BOD_{out}$] (neglecting endogenous respiration and nitrification). For a given reactor volume, V, and for a flux demand, **J**, the K_La value is the design parameter that must be satisfied. Translated to variables that affect K_La, V may be adjusted in design; other factors are, for example, rate of surface renewal that is related to both design and operation. In the case of bubble aeration, Q'(airflow) or d(bubble diameter) may be changed (Section 18.2.2.5). For a turbine aerator, the rate of pumping may be changed by impeller diameter or impeller speed. For an air stripping tower, the packing and the ratio of airflow to water flow affect K_La.

18.2.2.5 Derivation of Working Equation

A variety of different gas-transfer technologies exist. For oxygen transfer, the two basic technologies are: (1) bubble aeration, and (2) turbine aeration.

18.2.2.5.1 Bubble Aeration Model

The term diffused aeration is the more common designation for what is called here, "bubble" aeration. The phase, "bubble" aeration is used to emphasize the point, that is, that the gas-transfer occurs across a gas–water interface of gas bubbles (Eckenfelder, 1959; McKeown and Okun, 1960; Eckenfelder and Ford, 1968). Gas-transfer by bubbles may be used for air stripping but is more commonly used for aeration, such as for activated sludge.

From the work of Bewtra (1962), oxygen transfer from bubbles occurs in three phases:

1. Formation of the bubble at the capillary opening
2. During bubble ascendency to the surface
3. Bursting of the bubble at the surface

For coarse bubble aeration, only the second phase is important. For fine bubble aeration, the first phase is a factor too, but is omitted here for brevity. The effect is assimilated, however, in the coefficient of Bewtra's mathematical model. The third phase was deemed not important.

Two kinds of diffuser layouts are: (1) spiral flow, in which the diffusers are along a line on one side of a plug-flow reactor, and (2) a grid, in which the diffusers cover the bottom. Figure 18.4 depicts each of these two types of layouts in Figure 18.4a and b, respectively.

Equation 18.19 is the starting point, in lieu of Equation 18.21, because the oxygen transfer surface area must be described. For convenience, Equation 18.19 is repeated here, letting $\mathbf{a} = a$(bubbles), that is,

$$\frac{dC}{dt} = k_L \frac{a(\text{bubbles})}{V}(C_s - C) \tag{18.34}$$

The area term, "**a**" and "a(bubbles)," is the term in Equation 18.19 that pertains to bubble aeration. Figure 18.5 illustrates the concept of a bubble column as developed from orifices from a header pipe at the bottom of an aeration tank. The depth of the water to the orifices is designated, "h_o." The a(bubbles) term is the aggregate amount of bubble area in the column at any given instant.

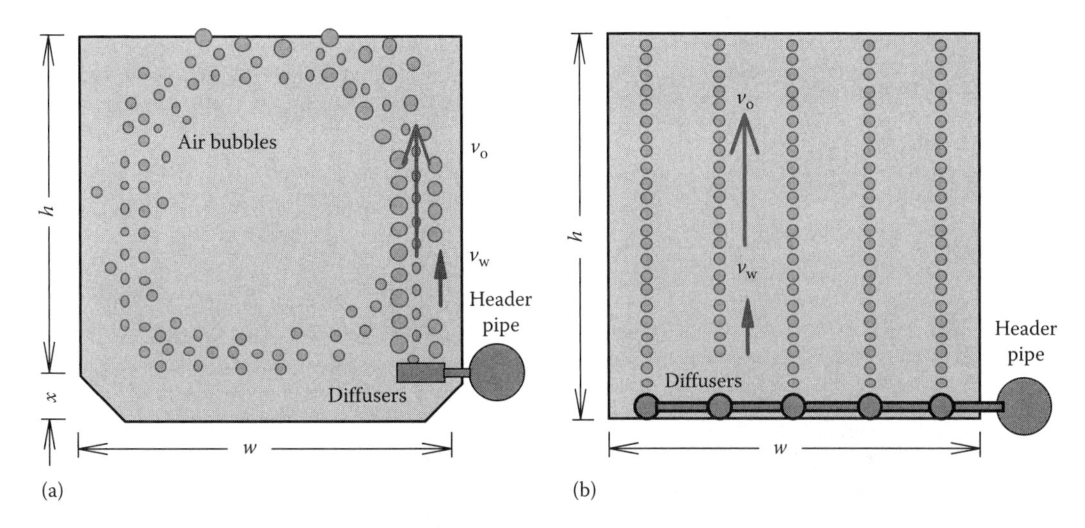

FIGURE 18.4 Cross sections illustrating two types of diffuser systems in plug-flow reactors. (a) Spiral flow reactor, (b) disk diffusers placed in grid.

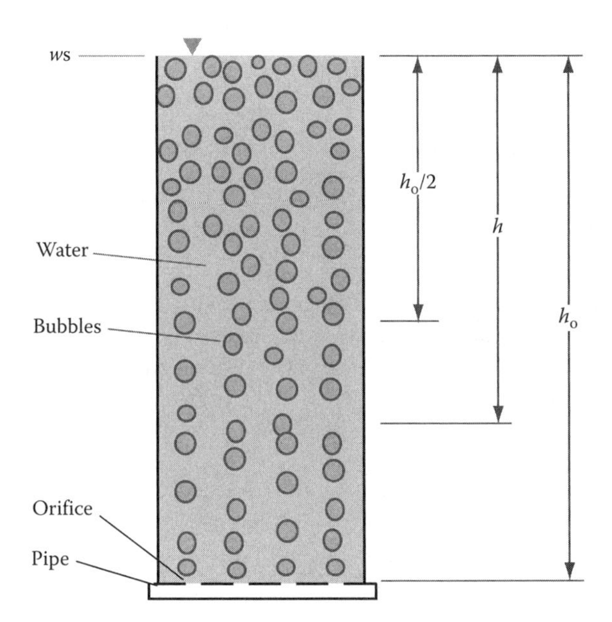

FIGURE 18.5 Bubble column.

For such a column, as illustrated in Figure 18.5, the area, a(bubbles), is

$$a(\text{bubbles}) = \pi d^2 \cdot N \qquad (18.35)$$

where

a(bubbles) is the collective surface area of all bubbles in ascent at a given instant (m^2)

d is the average diameter of bubbles (m)

N is the total number of bubbles in ascent

The number of air bubbles formed per unit time, dN/dt, is the airflow, Q', divided by the average bubble volume, that is,

$$\frac{\mathrm{d}N}{\mathrm{d}t} = \frac{Q'}{\frac{\pi}{6}d^3} \qquad (18.36)$$

where

t is the time (s)

Q' is the airflow from a given diffuser, corrected to temperature of water and pressure equal to atmospheric plus half the depth of submergence (m^3/s)

From Equation 18.36, the average total number of bubbles, rising at any given instant, N, is,

$$N = \frac{\mathrm{d}N}{\mathrm{d}t} \cdot T = \frac{Q'}{\frac{\pi}{6}d^3} \cdot T \qquad (18.37)$$

The term, T, can be calculated as the distance traveled by the bubble divided by the bubble velocity

$$T = \frac{\zeta h_o}{v} \qquad (18.38)$$

The absolute bubble velocity in the vertical direction v, is the sum of the velocity of the bubble relative to the water, v_o, and the vertical component of the average water velocity, v_w, that is,

$$v = v_o + v_w \qquad (18.39)$$

where

T is the average detention time of bubbles during ascent (s)

ζ is the ratio of average distance of bubble travel to depth of submergence of diffuser (relevant to spiral flow aeration systems)

h_o is the depth of submergence of diffuser (m)

v is the average absolute velocity of bubbles (m/s)

v_o is the average relative velocity of air bubbles with respect to water (m/s)

v_w is the velocity of circulating water mass entraining bulk of air bubbles (m/s)

Substituting Equation 18.38 into Equation 18.37 gives

$$N = \left[\frac{Q'}{\frac{\pi}{6}d^3}\right] \cdot \left[\frac{\zeta h}{v_o + v_w}\right] \quad (18.40)$$

The total area, a(bubbles) of all bubbles, collectively, in ascent at any given time is

$$a(\text{bubbles}) = \pi d^2 \cdot N \quad (18.41)$$

$$= \pi d^2 \cdot \frac{Q'}{\frac{\pi}{6}d^3} \cdot \frac{\zeta h_0}{v_o + v_w} \quad (18.42)$$

$$= \frac{6\zeta h_o}{v_0 + v_w} \cdot \frac{Q'}{d} \quad (18.43)$$

Equation 18.43 then is substituted in Equation 18.34, to give

$$\frac{dC}{dt} = \frac{k_L}{V}\left[\frac{\zeta h_o}{v_o + v_w} \cdot \frac{Q'}{d}\right] \cdot (C_s - C) \quad (18.44)$$

Comparing terms in Equations 18.19 and 18.21, K_La for bubble aeration is thus

$$K_La = \frac{k_L}{V}\left[\frac{\zeta h_o}{v_o + v_w} \cdot \frac{Q'}{d}\right] \quad (18.45)$$

Equation 18.45 shows that increasing Q', increasing h_o, having smaller bubbles, or having a smaller tank volume, will cause an increase in K_La. Further, v_w is dependent on Q' and d, and v_o is a function of d (Stokes law), so some terms are interdependent. Also, changing the basin geometry affects circulation, and thus ζ and v_w. Therefore, most terms are unique to the particular system of interest and to the operating conditions, for example, Q'. The effect of placing the diffusers on one side, vis-à-vis as a grid on the bottom, is that a spiral roll will occur for the former, and for the latter, $\zeta > 1.0$. Placing the diffusers in a grid configuration on the bottom of an aeration basin became more common since about 1980.

18.2.2.5.2 Application of Bubble Gas-Transfer Theory

In applying Equation 18.44 to a bubble aeration system, the conditions must be defined. The C_s value is defined by the partial pressure of the gas at the bubble location. The partial pressure of oxygen in a bubble as it emerges from an orifice is atmospheric pressure plus the depth of submergence, that is,

$$p(\text{bubble}, h) = p(\text{atm}) + \gamma_w \cdot h_o \quad (18.46)$$

where

p(bubble, h) is the pressure in bubble at depth of diffuser orifices (Pa)

p(atm) is the atmospheric pressure at elevation of aeration basin (Pa)

γ_w = specific weight of water (N/m^3)

$= \rho(\text{water}) \cdot g$

$g = 9.806\ 650$ m/s^2

ρ(water, 20°C) = 998.2 kg/m^3 (Table B.9 or polynomial equation, Appendix CDQR)

h_o is the depth of diffuser orifices (m)

The C_s value is calculated by Henry's law for solubility, Equation H.29, with p(bubble, h) as the argument, calculated by Equation 18.44. Such direct application of Equation 18.44 is not entirely accurate, however, for two reasons: (1) the pressure varies with depth and as the bubble rises the pressure declines proportionately, (2) as the bubble rises, oxygen is transferred from the bubble to the water and therefore the partial pressure of oxygen is reduced as the bubble rise. Thus, there are two effects on partial pressure: depth of water and loss of oxygen mass. In lieu of calculating mathematical relations to accurately predict these two effects, an approximate argument used for Henry's law is taken as the pressure of the bubble at half the depth, that is, $h/2$.

18.2.2.5.3 Bubble Gas-Transfer Theory Applied to Air Stripping

To apply Equation 18.42 to stripping of a gas from solution, assume the same gas, the same reactor volume, the same airflow, Q', the same orifice, etc. This means that all terms in the K_La coefficient are the same. The $(C_s - C)$ terms must be interpreted for the different conditions, that is, C is the concentration of gas to be reduced by stripping and C_s is a lower concentration at the interface, with "p_o" the argument in the bulk of gas solution. Since the air is the stripping gas, the gas "A" being removed, $p_o(A) \approx 0$ may be assumed, which means that $C_s(A) \approx 0$ at the interface. The aqueous-phase concentration gradient is thus, $(C_s - C) \approx (0 - C)$. The "gas-$A$" may be any gas, for example, radon, ammonia, a VOC, etc.

18.2.2.5.4 Mechanical Aeration Model

Mechanical aeration devices are pumps that circulate water and which create interface area for gas transfer. Two kinds of mechanical aerators are as follows: (1) turbine, and (2) brush. A turbine aerator is a "pump" impeller positioned at the water surface and in center of the area served, causing a rapid circulation of the water in tank volume. A brush aerator is a drum with blades projecting and is placed, as a rule, in a "racetrack" type of reactor, usually of shallow depth (and sometimes called a Pasveer ditch), and also is a pump of sorts (Berk, 1966). The axis of the drum is perpendicular to the velocity vector with the axis placed near the water surface. Two drums may be placed in the ditch so that the reactor does not become anoxic.

Figure 18.6a is a sketch of a turbine aerator as set up in a rectangular basin and illustrates the kind of circulation induced by the pumping. Figure 18.6b is a photograph of an installation which illustrates the violent agitation at the surface as the new interface area is created. The rate of oxygen transfer is proportional to the rate creation of interface area.

18.2.2.5.4.1 Theory A mechanical aerator transfers oxygen by both exposing new surfaces to the atmosphere and by entraining air in the jets created. Other notions of

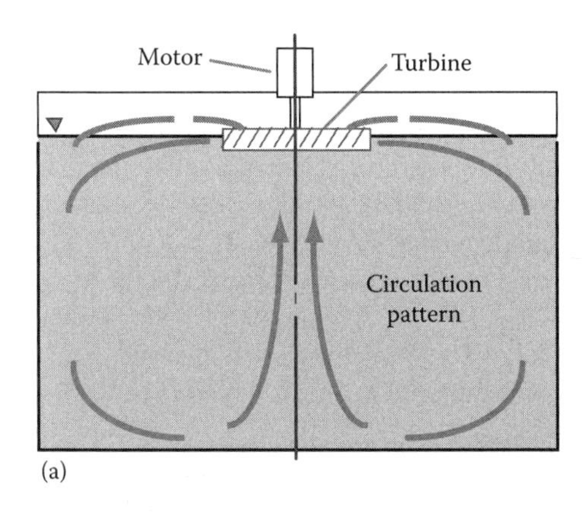

FIGURE 18.6 Turbine aerator illustrations. (a) Circulation pattern, (b) photograph. (Photo courtesy of S.K. Hendricks.)

oxygen transfer mechanisms were outlined by Kalinske (1963), an early proponent of mechanical aeration, who stated that the oxygen transfer rate was proportional to the pumping rate and the oxygen deficit. As a general relation, Equation 18.21 is applicable, but with K_La, being defined in terms of the surface renewal rate. The pumping rate is a surrogate; thus,

$$K_La = \frac{KQ_W}{V} \tag{18.47}$$

and Equation 18.21 becomes

$$\frac{dC}{dt} = \frac{KQ_W}{V}(C_s - C) \tag{18.48}$$

As seen, the K_La for a turbine aerator increases with pumping rate, Q_w, most likely increasing proportionately then approaching a limit asymptotically. Also, as seen, K_La is inversely proportional to V. The coefficient, K_La, incorporates a variety of influences, including diffusivity, turbulence intensity, circulation patterns as affected by geometry of the tank, etc. As for diffused aeration, the K_La term is unique for a particular aerator–aeration tank system. The pumping rate, Q_w depends on rotational velocity, ω, and impeller diameter, d(impeller).

18.2.3 Reactor Modeling

Reactor models for aeration may be batch, or continuous flow. A continuous-flow reactor may be complete mix, plug-flow, or a packed bed (which may be modeled as plug-flow).

18.2.3.1 Continuous-Flow Complete-Mix Reactor Modeling for Gas Transfer

A "complete-mix" reactor model was common for activated sludge in which turbine aeration was used from about 1962, but was used less after about 1980 when many reactors were retrofitted with diffused aeration with grid layout. The continuous-flow complete-mix model is reviewed here. The "plug-flow" model is used most often for diffused aeration, which occurs in a long narrow basin.

18.2.3.1.1 General Mass Balance: Kinetic Relations

Figure 18.7 depicts a continuous-flow complete-mix reactor, which shows the mass balance and kinetic terms. Bubbles rise from the bottom from a diffuser grid and transfer oxygen as they are dispersed by an impeller; some are always breaking the surface such that the number in suspension is always constant. Thus, the rate of oxygen uptake is constant.

As seen in Figure 18.7, the mass flux of gas into the reactor is QC_{in} and the mass flux out is QC_{out}, and the rate of uptake or depletion of gas is $[\partial C/\partial t]_r$. The "observed" rate of change of gas concentration in the reactor, $[\partial C/\partial t]_o$ is therefore

$$V\left[\frac{\partial C}{\partial t}\right]_o = QC_{in} - QC_{out} + V\left[\frac{\partial C}{\partial t}\right]_r \tag{18.49a}$$

where
- V is the volume of reactor (m^3)
- t is the time (s)
- Q_{in} is the flow into the reactor (m^3/s)
- Q_{out} is the flow out of the reactor (m^3/s)

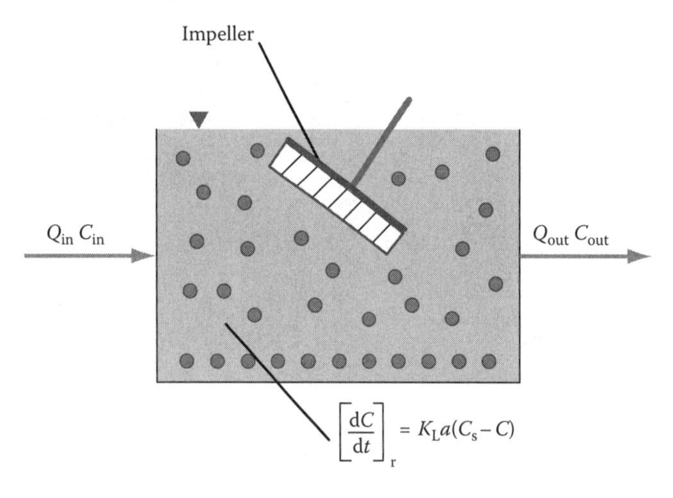

FIGURE 18.7 Continuous-flow aeration reactor.

C_{in} is the dissolved gas concentration of flow into the reactor (kg gas/m^3)
C_{out} is the dissolved gas concentration of flow out of the reactor (kg gas/m^3)
$[\partial C/\partial t]_o$ is the observed rate of change of dissolved gas concentration within the reactor (kg gas/m^3/s)
$[\partial C/\partial t]_r$ is the rate of dissolution of gas within the reactor (kg gas/m^3/s)

The sign for the kinetic term, $[\partial C/\partial t]_r$, in Equation 18.49a, is taken as positive since the proper sign will emerge when the details of the kinetic situation is known. For the case of gas transfer, the kinetic term is given by Equation 18.20, that is, $[\partial C/\partial t]_r = K_La(C_s - C)$. If $C < C_s$, then gas is transferred to the aqueous solution from the gas phase; thus the kinetic term is positive. If $C > C_s$, then gas must be transferred from the aqueous solution to the gas phase; thus the kinetic term is negative and the gas is "stripped" from solution. Recall that C_s is the gas concentration in the aqueous phase in equilibrium gas phase pressure.

18.2.3.1.2 Proprietary Equipment

The pressure in the gas bubbles (and thus the partial pressure of a given gas, "A") approaches zero gauge-pressure for a ShallowTray™ type of air stripping unit, since the depth of the unit is less than a meter. The gas transfer occurs largely across the large air–water interface area created and the evident high turbulence intensity. Figure 18.8a is a perspective drawing of a unit, which consists of a flow of water through open channels over a perforated bottom through which a flow of air bubbles emerge under pressure. The amount of turbulence and the shallow depth of the channels ensure that the bubbles do not have long residence time (which would accumulate the gas being stripped).

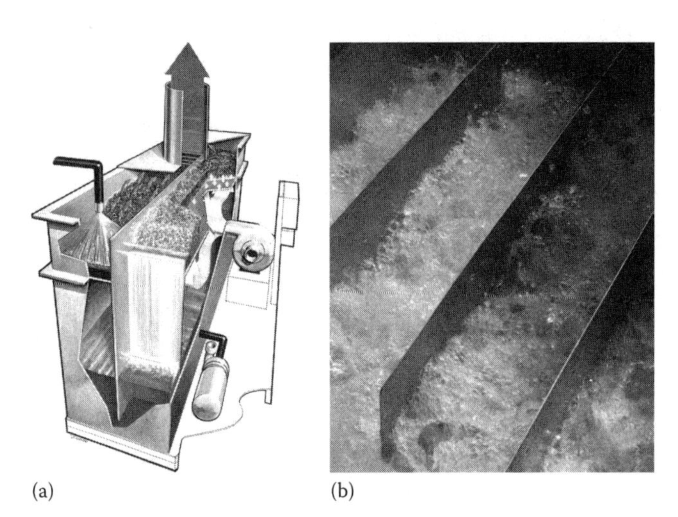

FIGURE 18.8 ShallowTray™ stripping unit for dissolved gases. (a) Perspective drawing of unit, (b) photograph illustrating turbulence. (Courtesy of BISCO Environmental NEEP Systems, Inc., Taunton, MA, 2010.)

Figure 18.8b is a photograph that illustrates the extremely high level of turbulence generated by the flow of diffused air through the perforated bottom. The value of K_La for such a unit is undoubtedly much higher than any traditional air stripping technology, due mostly to the high rate of air–water interface area created, and to a lesser extent due to the small film thickness created by the high turbulence intensity. An issue with applying the above analysis to the ShallowTray™ technology is that being a channel, it is not a compete-mix reactor. It is probably not a too inaccurate model, however, for understanding how the system works. Usually, a system such as the ShallowTray™ is selected based on experience, perhaps coupled with a pilot plant study (see also http://www.neepsystems.com).

18.2.3.2 Batch Reactor Aeration Modeling

Figure 18.9 depicts a batch aeration reactor, that is, $Q_{in} = 0$. $Q_{out} = 0$; therefore, the observed rate of change within the reactor is equal to the rate of dissolution of gas, for example, oxygen, within the reactor. Starting again with the general reactor statement,

$$V\left[\frac{\partial C}{\partial t}\right]_o = Q_{in}C_{in} - Q_{out}C_{out} + V\left[\frac{\partial C}{\partial t}\right]_r \tag{18.49b}$$

Since $Q_{in} = 0$ and $Q_{out} = 0$, and since Equation 18.21 applies to the $[\partial C/\partial t]_r$ term, Equation 18.49 becomes

$$V\left[\frac{\partial C}{\partial t}\right]_o = 0 - 0 + V \cdot K_La(C_s - C) \tag{18.50}$$

and

$$\left[\frac{\partial C}{\partial t}\right]_o = K_La(C_s - C) \tag{18.51}$$

In other words, the observed rate of increase in dissolved oxygen concentration equals the rate of oxygen uptake by the aeration system. That $[\partial C/\partial t]_o \neq 0$ means, by definition, that the reactor state is not "steady state." The foregoing

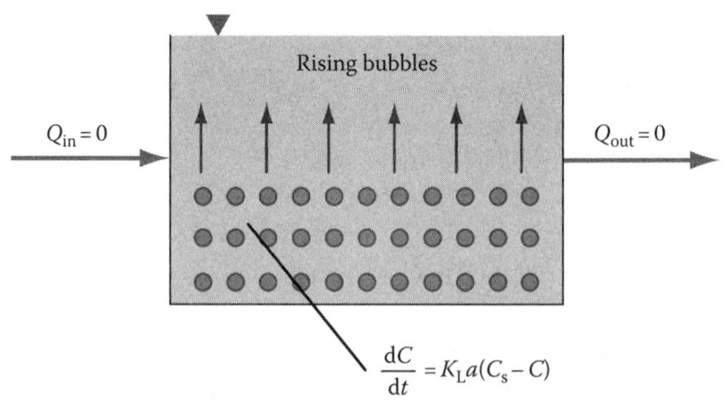

FIGURE 18.9 Batch aeration reactor.

equations apply to any batch of complete-mix system, whether bubble aeration is as depicted in Figure 18.9, or as in a turbine aeration system.

18.2.3.2.1 Aerator Testing

The two testing methods are (1) steady state, and (2) nonsteady state, that is, batch reactor test. The former is described by Mueller and Boyle (1988) and is necessary when evaluation is for process conditions (see also Eckenfelder and O'Connor, 1961). The nonsteady state method is outlined here.

18.2.3.2.1.1 Mathematics The kinetic model for any aeration system is Equation 18.21. For a batch reactor, Equation 18.51 may be integrated by separating variables as indicated below in Equation 18.52. The integrated form is Equation 18.53.

$$\int_{C_o}^{C} \frac{dC}{(C_s - C)} = K_L a \int_{0}^{t} dt \tag{18.52}$$

$$\log\left[\frac{C_s - C}{C_s - C_o}\right] = -\frac{K_L a}{2.3} \cdot t \tag{18.53}$$

where
- C_s is the saturation concentration of dissolved oxygen for ambient temperature and pressure (g/mL)
- C_o is the initial concentration of dissolved oxygen in aerator at time, $t = 0$ (g/mL)
- C is the concentration of dissolved oxygen in aerator at time, t (g/mL)
- t is the elapsed time from start of aeration test (s)

In graphical form Equation 18.53 has a negative sloped straight line on a semi-log plot. The anchor point at time $t = 0$ is $(C_s - C)/(C_s - C_o) = 1.0$. The slope of the straight line, measured as the time for a full log cycle, is $K_L a/2.3$. Example 18.3 gives results from an aerator test using diffused aeration data from a student laboratory at Colorado State University (CSU). Protocol used for a full-scale system can be followed to set up the apparatus for such a test, collection of data, and analysis of data.

Example 18.3 Processing and Analyzing Data from an Aerator Test

Given

1. Table 18.5 shows data obtained during a laboratory scale aerator test. The "aerator system" in this case was a 29 L (5 gal) carboy aerated with a small porous plug diffuser at the bottom.
2. The fixed conditions and the needed calculations to determine C_s also are shown in Table 18.5.

Required

Determine the $K_L a$ coefficient from aerator test data as given in Table 18.5.

Solution

From Table 18.5, Figure 18.10 is plotted. Figure 18.10 illustrates Equation 18.21 in graphical form. The slope of the straight line in the semi-log plot is $K_L a/2.3$, the slope turns out to be 1 log cycle/28 min. Then, from Figure 18.10, $K_L a/2.3 = 1/28$ min; therefore, $K_L a = 0.082\ \text{min}^{-1} = 4.9\ \text{h}^{-1}$.

Discussion

The $K_L a$ value obtained was unique for the system that was used for the test and for the conditions of the test. The result, $K_L a = 4.9\ \text{h}^{-1}$ was within the same range as found for some full-scale turbine aeration systems (which is mentioned only for reference to $K_L a$ magnitudes experienced).

18.2.3.2.1.2 Conducting an Aerator Test An unsteady state aerator test is accomplished by deaerating clean water in a test basin using sodium sulfite (about 0.12 kg/m^3 or 1 lb/1000 gal according to Huang, 1975) with a cobalt catalyst. Kalinske (1968) pointed out that only 0.5 mg/L cobalt, for example, cobalt chloride, is needed (if higher concentrations are used, say 2.0 mg/L, the cobalt will interfere with the Winkler dissolved oxygen test). When the aerator system is started, the test begins, and is maintained in continuous operation while the sodium sulfite is added and consumed in reaction with the oxygen. After all the sodium sulfite has reacted the concentration of dissolved oxygen begins to increase. This increase will be in accordance with Equation 18.21, but is plotted in terms of Equation 18.53 so that $K_L a$ can be determined, that is, as in Figure 8.13. Determining $K_L a$ and the corresponding brake horsepower of the turbine impellor or compressor, as the case may be, are the objectives of the test. One objective of such testing has been to determine, $W(O_2)$, the mass of oxygen transferred per unit of energy expended (e.g., kg O_2 transferred/kWh). The "wire" energy is easier to measure than the turbine impeller energy; the latter is measured by a "Prony-brake."

18.2.3.3 Column Reactor Modeling

Figure 18.11a depicts a column reactor for the case of gas flow from bottom diffusers with water flow in at the top having dissolved gas concentration, C_{in}; the dissolved gas concentration leaving the reactor is C_{out}. The dissolved gas concentration varies from top to bottom, that is, the reactor is not homogeneous. For such a reactor, as seen in Section 4.3.3.3, the materials balance analysis must be performed on an infinitesimal "slice" along the column length, of thickness, ΔZ. The reactor may be used either (1) to add gas to the aqueous solution, for example, oxygen, ozone, chlorine, ammonia, carbon dioxide, etc., as depicted in Figure 18.11b, or (2)to "strip" dissolved gas, for example, VOCs, from the incoming solution, with concentration profile depicted in Figure 18.11c. This particular reactor is not necessarily the most efficient for either case. The cases merely illustrate the modeling procedure, which is applicable to either gas addition or gas stripping. The reactor is "steady state" in that the

TABLE 18.5
Results of an Aeration Test—Laboratory Scale

Fixed Conditions

$V = 18.0$ L	$Q' = 42.47$ L/h	$P_{diffuser} = 80$ mmHg (gauge)
$T = 23°C$	$P_a = 623$ mmHg	Depth of diffuser $= 30.5$ cm

Variable Data

Elapsed Time (min)	C (mg/L)	$(C_s - C)$ (mg/L)	$\left[\frac{C_s - C}{C_s - C_o}\right]$
0	2.0	5.2	1.0
1	2.0	5.2	1.0
2	2.1	5.1	0.98
3	2.5	4.7	0.90
4	2.6	4.6	0.88
5	3.1	4.1	0.79
7	3.4	3.8	0.73
9	4.2	3.0	0.58
11	4.4	2.8	0.54
13	5.1	2.1	0.40
15	5.4	1.8	0.35
17	5.7	1.5	0.29
19	5.8	1.4	0.27
21	6.2	1.0	0.19

Calculations

Saturation concentration is obtained for the pressure at one half the depth of the diffuser submergence which is:

$$p(\text{bubbles}) = p_a + \gamma_w \cdot h_o$$
$$= \left[\frac{623\,\text{mmHg}}{760\,\text{mmHg}}\right] \cdot 101.325\,\text{kPa} + \left[\frac{1.00\,\text{kg water}}{\text{m}^3\ \text{water}}\right] \times \left[\frac{9.91\,\text{m}}{s^2}\right] \cdot (0.152\,\text{m water depth})$$
$$= 83.060\ \text{kPa} + 1.49\ \text{kPa}$$
$$= 84.55\ \text{kPa}$$
$$= 634\ \text{mmHg}$$

From Table H.5, $C_s(O_2, P = 760\ \text{mmHg}, T = 23°C) = 40.87$ mg O_2/L/atm O_2 which is actually a form of Henry's constant, that is, H_A^s, Equation H.29

For air at any elevation, the mole fraction of oxygen, $X(O_2) = 0.209476$ (Table H.1)

$$\text{The partial pressure of } O_2 \text{ is } p(O_2, 632\ \text{mmHg}) = 0.209476 \cdot 634\ \text{mmHg}$$
$$= 132.4\ \text{mmHg} = 0.175\ \text{atm } O_2$$

Applying Henry's law, Equation H.29, that is, $C_s(O_2) = H_A^s \cdot p(O_2)$, gives,

$$C_s(O_2) = (40.87\ \text{mg } O_2/\text{L solution/atm } O_2) \cdot (0.175\ \text{atm } O_2)$$
$$= 7.15\ \text{mg } O_2/\text{L solution}$$

Discussion

Although a laboratory exercise, the procedure is the same as has been used for full-scale aerator testing (diffused or turbine) in a large basin. In such large-scale testing, the oxygen concentration is measured at several points in the aerator basin and the concentration is averaged for use in Table 18.5.

dissolved gas concentrations at each ΔZ slice do not change with time.

The materials balance for an individual column "slice," as depicted in Figure 18.11a, is the same as for a granular activated carbon (GAC) column, as given in Equation 4.10 and in reduced form as Equation 4.16, repeated here as Equation 18.54, that is,

$$\left[\frac{\partial C}{\partial t}\right]_o = -v\frac{\partial C}{\partial Z} + D\frac{\partial^2 C}{\partial Z^2} + \left[\frac{\partial C}{\partial t}\right]_r \qquad (18.54)$$

To apply Equation 18.54 to the either the case of gas uptake or gas stripping, the observed rate of concentration change in any slice, $[\partial C/\partial t]_o = 0$, that is, its "steady-state" and the kinetic term is as given by Equation 18.21. Further, for simplicity, neglect the dispersion term to give

$$0 = -v\frac{\partial C}{\partial Z} + 0 + K_L a(C_s - C) \qquad (18.55)$$

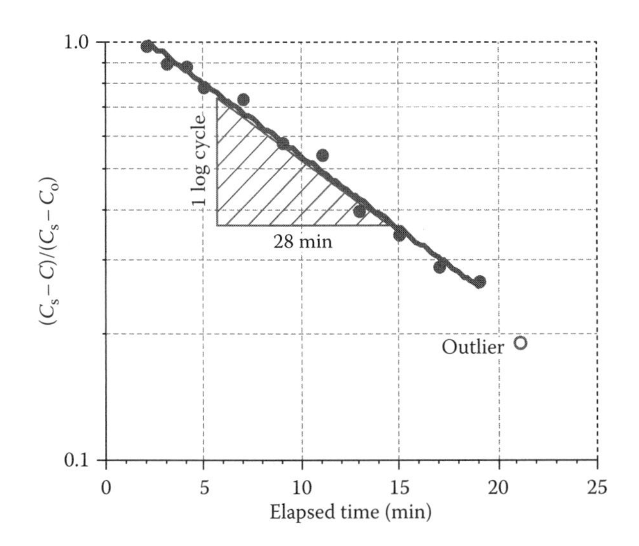

FIGURE 18.10 Results of an aeration test—laboratory scale (Table 18.5 data).

or

$$v\frac{\partial C}{\partial Z} = K_{L}a(C_s - C) \quad (18.56)$$

Separating variables and integrating gives

$$\int_{C_{in}}^{C_{pit}} \frac{dC}{(C_s - C)} = -\int_{0}^{Z_0} \frac{K_{L}a}{v} dZ \quad (18.57)$$

$$\ln\left[\frac{C_s - C}{C_s - C_{in}}\right] = -\frac{K_{L}a}{v}Z \quad (18.58)$$

or

$$\log\left[\frac{C_s - C}{C_s - C_{in}}\right] = -\frac{K_{L}a}{2.3 \cdot v}Z \quad (18.59)$$

Equation 18.59 plots as a semi-log relationship; if C data are obtained along the column, the slope is $-K_{L}a/2.3v$. As noted previously, the model is steady state, that is, $[\partial C/\partial t]_o = 0$.

18.2.3.3.1 Gas Uptake

Figure 18.11b illustrates the increase in dissolved gas concentration, C, as distance from the top of the column, Z, increases; $(C_s - C)$ decreases, as seen, as C increases.

18.2.3.3.2 Gas Stripping

Figure 18.11c illustrates the decrease in dissolved gas concentration, C, with distance due to gas stripping; usually, the stripping gas is air, although steam is mentioned in the literature. The assumption adopted here, for a simplistic mathematical depiction, is that the concentration of gas "A" being stripped is zero in the bubbles; therefore, $C_s = 0$. The result is that Equation 18.59 reduces to

$$\log\left[\frac{C}{C_{in}}\right] = -\frac{K_{L}a}{2.3 \cdot v}Z \quad (18.60)$$

Notes:

- The main assumption applied to Equation 18.54 is that dispersion is negligible. If the dispersion coefficient is measured, for example, by salt or dye tests (applying Equation 18.54 for the condition that there is no reaction), then Equation 18.54 must be solved numerically.

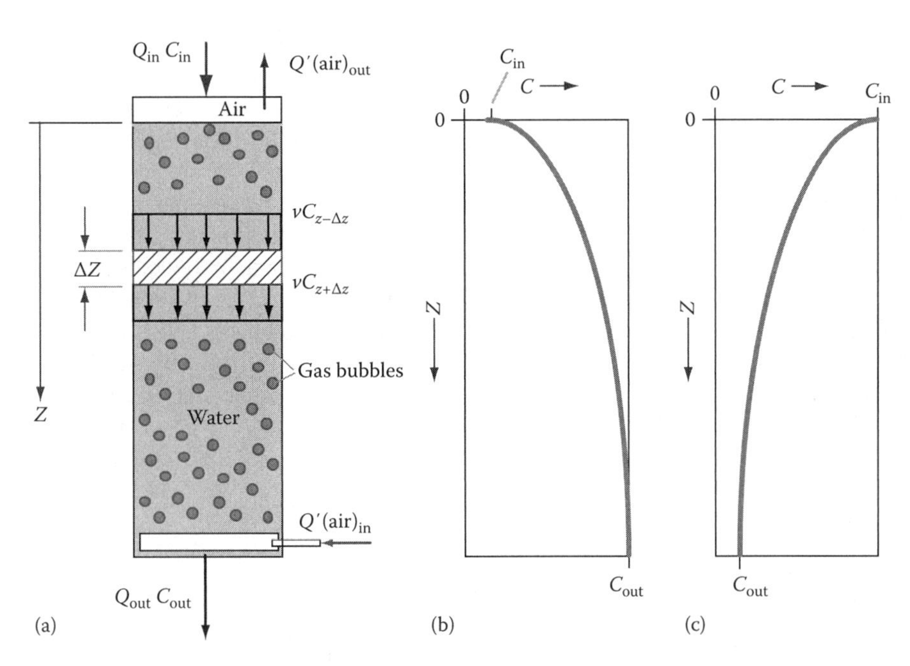

FIGURE 18.11 Depiction of steady state column reactor. (a) Terms in modeling, (b) adding gas, (c) stripping gas.

- Within a stripping bubble, the partial pressure of gas "*A*" (the gas being stripped from solution) will increase as the bubble rises, that is, it is not zero as assumed in Equation 18.60.
- The stream of air bubbles used for stripping will increase in concentration of the gas "*A*" being stripped such that the mass flux of "*A*" lost from solution equals the mass flux increase to the air stream.
- Theoretically, the mass-transfer coefficient, K_La, should be the same for different gases, differing by their respective diffusion coefficients.
- For moderately soluble gases, K_La once determined for a given gas, for example, oxygen, may be corrected to use for another gas (used in the same system and for the same conditions) by the ratio of diffusion coefficients for the two gases. This is based on substituting in Equation 18.20 the relation, $K_L = D/\delta$, to show the diffusion coefficient

$$K_La = \frac{D}{\delta}\frac{A}{V} \tag{18.61}$$

18.2.3.4 Column Reactor Modeling: Packed Beds

Figure 18.12a depicts a column reactor for the case of a packed bed for air stripping. The column is packed with relatively large media sizes, for example, 20–50 mm. Water flow, laden with dissolved gases to be removed, enters from the top and air leaves at the bottom with lower concentrations, being "stripped" of some of the dissolved gases. The water flow "percolates" down through the packing under gravity conditions, with each small stream bifurcating and falling to the next surface, coalescing with another stream and bifurcating again.

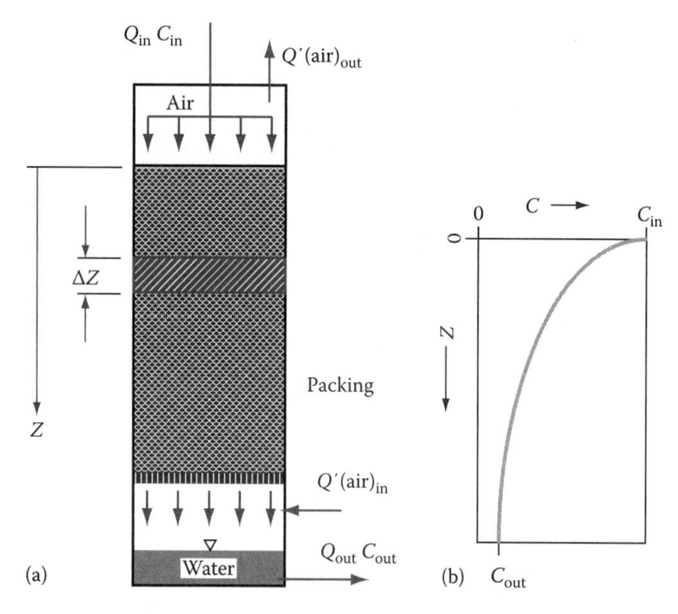

FIGURE 18.12 Depiction of packed bed reactor model. (a) Packed bed schematic, (b)aqueous-phase concentration profile.

The idea is to create a large air–water interface for uptake of dissolved gases from the water by the airflow. The water flow must be less than some critical value, above which will cause flooding. The water flow will pool at the bottom of the reactor and flow out under a hydraulic gradient. The airflow enters presumably without any of the gases to be stripped. As the airflow comes in contact with water surfaces, gas transfer occurs from the dissolved gases in aqueous solution to an air solution. Presumably, the gradients across the films are steep, due to these new air–water interfaces being continuously created as the water falls and as the air stream passes across these surfaces. For stripping a volatile gas, for example, benzene, the higher the airflow the higher the rate of gas transfer. Generally, packed towers are used with countercurrent flow, that is, water trickles down and air flows up. The air-to-water flow ratios range $10{:}1 \leq \mathbf{j}(\text{air})/\mathbf{j}(\text{water}) \leq 300{:}1$.

18.2.3.4.1 Mathematical Modeling

As seen by the depictions in Figure 18.12b, the dissolved gas concentration decreases with distance along the column. By definition, the reactor is not homogeneous and must be modeled by selecting an infinitesimal "slice" of the column, as indicated. Thus, Equation 18.54, the general materials balance relation is valid, as is Equation 18.56, that is, for the steady state.

These equations, per se, are not applied directly, however, in favor of presenting the "Onda correlation" (see Staudinger et al., 1990; Dvorak et al., 1996) for predicting the column height, *h*, needed to attain a specified level of dissolved gas concentration leaving the reactor.

18.2.3.4.2 Onda Correlations

A 1968 paper in Japan by Onda, Takeuchi, and Okumuto has been cited in the literature for its utility in predicting the column height needed to strip VOCs in a packed bed. A column is packed usually with manufactured objects designed to promote a large air–water interface per unit volume of packed bed. The objects are given such names as Pall rings, Rashig rings, saddles, etc., and may be made of polypropylene or ceramic material. The paper by Onda et al. provided empirical equations that predicted both liquid and gas phase mass-transfer coefficients, and the specific interfacial surface area (i.e., area of packing per unit volume of packed bed), that is, k_L, k_g, and **a**, respectively. The equations are called in the literature, the "Onda correlations." Their accuracy has been evaluated by Staudinger et al. (1990) and Dvorak et al. (1996). The Staudinger study found that the Onda correlations predicted rate constants with standard deviation ±17% (437 data points) corresponding to ±30% accuracy based on 90% confidence limits. Dvorak et al. (1996) found about 16% overprediction and 34% underprediction (at high gas flows). Appendix A reviews the Onda equations.

18.2.3.5 Effect of Gas on K_La and Uptake/Stripping Effects

Based on Equation 18.45, K_La should be directly proportional to Q' (but not exactly due to increased v_w due to fluid drag from the rising bubbles, especially for air flows along the side

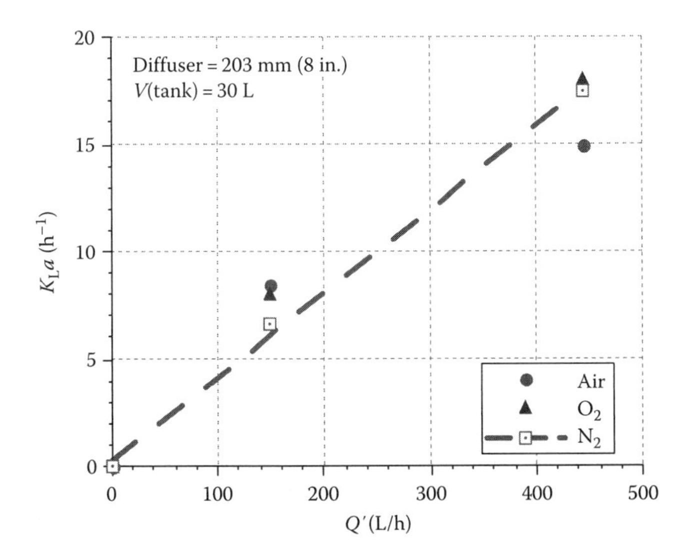

FIGURE 18.13 Effect of Q' and gases on K_La for lab-scale tests. (Adapted from Clunie, W. and Wu, Y., *Gas Transfer Coefficients for Bubble Diffusers*, CE 541 Unit Processes of Environmental Engineering Laboratory, Spring, Department of Civil Engineering, Colorado State University, Fort Collins, CO, 1995. With permission from William F. Clunie, 2010.)

of the basin). Also, the K_La for oxygen uptake should be the same for pure oxygen as for air and for stripping dissolved oxygen with nitrogen gas. Results of experiments from CSU laboratory classes confirm that these anticipated relations are approximately true. Figure 18.13 is a plot of K_Lav. Q' for three gases, air, pure oxygen, and nitrogen (nitrogen was used for stripping dissolved oxygen from solution). Oxygen concentration was measured in each case by a dissolved oxygen probe. For the several oxygen concentration versus time plots (semi-log) and one for each Q' and for each gas species used, $r^2 \geq 0.99$. Each plot yielded a single K_La point for a given Q' which was the basis for the plot of Figure 18.13. The gas flows were measured by a rotometer and calibrated by water displacement of an inverted cylinder. The same tank and the same diffuser were used in each case; the diffuser was a membrane type with slots. The gases in each case were from pressurized cylinders.

The data support the theory, that is,

- $K_La \propto Q'$ (approximately), which is in accordance with Equation 18.45, except that the bubble–water boundary layer is reduced as Q' increases and the velocity of the bubble increases as Q' increases, each of which will have offsetting effects (the magnitude of each is not known).
- K_La is approximately the same for aeration by either air or pure oxygen; whether the gas is oxygen in air or pure oxygen, the diffusion coefficient is about the same, except as affected by the oxygen gas pressure.
- K_La is approximately the same for stripping of dissolved oxygen as for uptake, since the diffusion coefficient and the film thickness should be the same, all other factors being the same. Also, the gas phase will not be the rate limiting part of the process for stripping a low solubility gas such as oxygen. If a highly soluble gas, for example, ammonia, is stripped, the control would shift to the gas phase.

18.3 DESIGN

As seen in "theory," the mass-transfer coefficient, K_La, is a single design parameter that integrates several fundamental variables. The value of K_La, along with the gradients within the gas film and the aqueous film, determine the gas transfer rate.

As a rule, K_La is proportional to the rate of creation of interface area, which for a given design, depends upon operating conditions, for example, higher rotational velocity for an turbine aerator, high airflow for an air stripping tower, etc. The intent is not necessarily to maximize K_La, but to find an overall design that is economical.

18.3.1 Aerator Design

Aeration to provide oxygen to a biological process is common for every wastewater treatment plant. An algorithm for sizing an aerator for an activated sludge reactor is outlined. The primary reference was Eckenfelder and O'Connor (1961), one of the first full elucidations of aeration theory and practice.

18.3.1.1 Algorithm for Aerator Sizing

The sequence of steps for sizing an aerator is enumerated for the two main classes of aerators, diffused (i.e., bubbles) and mechanical (i.e., turbine impeller). The output of the algorithm is the K_La required to meet a specified rate of oxygen uptake.

18.3.1.1.1 Determine Required Oxygen Uptake Rate

For an aerobic biological reaction, the oxygen utilization rate is proportional to the concentration of biochemically oxidizable organic substrate (BOD), and the concentration of cells (MLVSS), simultaneously undergoing respiration. Thus we can say (neglecting, for the present, the nitrification oxygen demand)

$$\frac{d[O_2]}{dt} = a'[BOD]_R + b'[MLVSS]_R \qquad (18.62)$$

where

$[O_2]$ is the concentration of dissolved oxygen in reactor solution (kg/m^3)
[BOD] is the concentration of biochemical oxygen demand in solution and suspension in reactor (kg/m^3)
[MLVSS] is the concentration of mixed liquor volatile suspended solids in reactor suspension (kg/m^3)
a' is the kinetic coefficient, first order (s^1)
b' is the kinetic coefficient (s^{-1})
t is the time (s)

Eckenfelder (1970) gave: $a' = 0.70$ and $b' = 0.12$. The equation may be expanded to include nitrification (i.e., oxidation of ammonia to nitrite/nitrate, which is omitted here for brevity). The total oxygen transfer rate is

$$J(O_2) = \frac{d[O_2]}{dt} \cdot V(\text{reactor}) \tag{18.63}$$

where

$J(O_2)$ is the mass flux of dissolved oxygen from gas phase to solution phase (i.e., within the reactor) due to aeration (kg O_2/s)

V(reactor) is the volume of reactor (m^3)

A "quick-and-dirty" way to estimate oxygen demand is to calculate the difference between mass flow of BOD into and out of the reactor on a daily basis:

$$J[O_2] = Q \cdot \{[BOD]_{in} - [BOD]_{out}\} \tag{18.64}$$

where

Q is the average daily plant flow (m^3/s)

$[BOD]_{in}$ is the average daily BOD concentration of water flowing into the reactor (kg/ m^3)

$[BOD]_{out}$ is the average daily BOD concentration of water leaving the reactor (kg/m^3)

18.3.1.1.2 Determine the $K_La(T)$ Needed: General

From Equation 18.21, and after adding nomenclature, for example, (W, T, p) to indicate ambient conditions

$$J(W, T, p) = \alpha \cdot K_La(T) \cdot [\beta C^*(T, p) - C] \cdot V \tag{18.65}$$

where

$J(W, T, p)$ is the flux of oxygen that must be generated by the aerator to meet the oxygen demand at temperature, T, of system, gas pressure, $p(O_2)$, at the air–water interface, and water type, W (kg O_2/s)

$K_La(T)$ is the aeration coefficient for a given aerator, basin, temperature, and clean water (s^{-1})

α = ratio of K_La for the process water divided by the K_La for clean water (dimensionless)

$\equiv K_La(\text{process water})/K_La(\text{clean water})$

β = ratio of C^* for process water to C^* for clean water (dimensionless)

$\equiv C^*(\text{process water})/C^*(\text{clean water})$

$C^*(T, p)$ is the saturation concentration (the same as C_s) for clean water at temperature, T, and oxygen partial pressure, $p(O_2)$, at gas–water interface (kg O_2/m^3)

C is the concentration of oxygen in reactor (kg O_2/m^3)

T is the temperature of reactor water (°C)

p is the pressure of gas at gas–water interface, for example, $p(O_2)$ (kPa)

W is the indication that water is any water, for example, wastewater

V is the volume of reactor (m^3)

The $K_La(T)$ term is the required mass-transfer coefficient for the oxygen demand as determined, for example, by Equation 18.65. In other words, whether the system has the capacity to provide the oxygen is determined by the $K_La(T)$ for the system. The $K_La(T)$ available may be determined by aerator testing under clean water conditions as illustrated in Example 18.3. The α and β terms in Equation 18.99 have been added to acknowledge that a difference may exist between clean water and wastewater. To account the possible effects of diffuser changes over time, for example, pores may clog on the air side due to particles in the air stream or the character of deposits on the water side may change due to precipitates or to biofilms that force the air through secondary passages defined by the film structure and its points of adhesion a term, F, was added to Equation 18.65 (Boyle et al., 1989). The F term is more likely to be lumped with the α term rather than to attempt to evaluate it.

Determination of terms for Equation 18.65 is described as follows:

- $J(T, p, W)$ may be calculated as outlined in Section 18.3.1.1.1; let $\alpha \approx 0.8$, and let $\beta \approx 1.0$, which are common assumptions for domestic wastewater.
- $C^*(T, p)$ is determined by calculation applying Henry's law, Equation H.29 solubility version, taking H from Table H.5 for the temperature of the reactor water. The saturation concentration, $C^*(T, p)$, is for whatever local atmosphere is in contact with the gas–water interface, for example, a bubble from a diffuser (with bubble pressure equal to atmospheric pressure plus the pressure due to the water depth) or the atmospheric pressure for a turbine aerator. The vapor pressure of water (see Table B.9) should be subtracted from the saturation pressure, but is ignored here since its magnitude is only 2.3 Pa at 20°C. As an alternative to Table H.5, Henry's constant may be calculated by Equation H.60, with coefficients, A and B from Table H.5. Recall also that the nomenclature for saturation pressure, $C^*(T, p)$, is the same as $C_s(T, p)$; the two forms are used interchangeably.
- For activated sludge reactors, $C = 2.0$ mg/L is assumed traditionally.

18.3.1.1.3 Determine the $K_La(T)$ Needed: Diffused Aeration

The pressure used for calculation of $C^*(T, p)$ in bubble aeration is the pressure of the bubble at the depth of submergence, that is,

$$p(\text{bubble}) = p_a + \gamma_w \cdot d(\text{water}) \tag{18.66}$$

where

p(bubble) is the pressure of gas in bubble at depth of bubble (kPa)

p_a is the pressure of atmosphere at measured barometric pressure or as estimated by elevation per Equation H.24

γ_w is the specific weight of water, that is, $\gamma_w = \rho_w g$ (N/m^3)
d(water) is the depth of water used for calculation of bubble pressure (m)

Since the pressure in the bubble decreases as the bubble rises, the average depth is used for the pressure argument in Henry's law, Equation H.60. In principle, the calculation of K_La for bubble aeration is the same as for turbine aeration or air stripping. The gas concentration in the aqueous phase is calculated by Henry's law, Equation 18.1.

18.3.1.1.4 Convert $K_La(T)$ to $K_La(ST)$

A traditional equation for temperature conversion (Eckenfelder and O'Connor, 1961; Boyle, 1989) is

$$\theta^{T-20} = \frac{K_La(T)}{K_La(\text{ST})} \tag{18.67}$$

where
$K_La(T)$ is the gas transfer coefficient at any temperature, T (s^{-1})
K_La(ST) is the gas transfer coefficient for standard temperature, that is, 20°C (s^{-1})
T is the temperature (°C)
θ = temperature conversion coefficient (dimensionless) = 1.024 is a common assumption (Boyle, 1989)

As to the source of Equation 18.67, Eckenfelder and O'Connor (1961, p. 68) show a derivation based on the van't Hoff–Arrhenius relations, Equation H.54. Looking at the definition of K_La, however, as seen in Equation 18.20, that is, $K_La = DA/\delta V$, and by the Wilke–Chang relation, Equation 18.9, K_La should be directly proportional to T. This has not been ascertained, however, and so Equation 18.67 remains as the means for taking temperature into account.

18.3.1.1.5 Calculate Equivalent J(ST)

In oxygen uptake, the mass flux of oxygen for a given reactor and given conditions is commonly converted to standard conditions, that is, W = clean water, T = 20°C, sea level pressure, and C(reactor) = 0 kg/m^3. The key is that K_La(ST) is determined as outlined in the preceding steps. Thus, J(ST) is calculated as

$$J(O_2, \text{ST}) = K_La(\text{ST}) \cdot C^*(O_2, \text{ST}) \cdot V(\text{reactor}) \tag{18.68}$$

where
$J(O_2$, ST) is the mass flux of dissolved oxygen from gas phase to solution phase due to aeration for clean water at 20°C (kg O_2/s)
C^*(ST) is the dissolved oxygen saturation concentration for clean water at temperature, T = 20°C, and sea level atmospheric pressure, that is, p(sea level) = 101.325 kPa, and at partial pressure of oxygen in atmosphere, that is, $X(O_2)$ = 0.0209

Calculation of C^*(ST) is in accordance with Henry's law, that is, Equation H.60, that is,

$$C^*(O_2, \text{ST}) = H(O_2, 20°\text{C}) \cdot X(O_2) \cdot p(\text{sea-level}) \tag{18.69}$$

where
$H(O_2$, 20°C) is Henry's constant for oxygen at 20°C (mg O_2/L/atm O_2)
$X(O_2)$ is the mole fraction of oxygen in atmosphere (0.2095 per Table B.7)
p(sea-level) is the atmospheric pressure at sea-level (1.00 atm or 101.325 kPa)

The reactor volume, V, is a part of the activated sludge reactor design. For the purpose of this illustration, however, V can be taken as the volume for a 6 h detention time using average daily flow at the end of the design period (which is a common assumption).

18.3.1.2 Oxygen Transferred per Unit of Energy Expenditure

To compare aeration systems, a parameter that has been used (during the 1960s) is mass of oxygen transferred per unit of energy required under standard conditions, for example kg O_2 transferred/kwh (lb O_2 transferred/hp-h). To evaluate a system, the test is done under whatever conditions exist to determine $K_La(T, p, W)$ and is then converted to K_La(clean water, sea level, 20°C), that is, K_La(ST), which permits calculation of the equivalent J(ST), that is, Equation 18.68. Measurement of the power consumed permits calculation of the mass of oxygen transferred per unit of energy expended, that is,

$$W(O_2, \text{ST}) = \frac{J(O_2, \text{ST})}{E_n} \tag{18.70}$$

where
$W(O_2$, ST) is the mass of oxygen transferred per unit of energy expended under equivalent standard conditions, that is, 20°C and sea level pressure (kg O_2 transferred/kwh or lb O_2 transferred/hp-h).
E_n is the energy expended (specified wire or impeller) over a defined time period (kwh or hph)

As seen in Equation (18.68), $J(O_2, \text{ST})$ depends on the volume of the reactor. Therefore, $W(O_2, \text{ST})$ is not a valid indicator of the impeller efficiency in transferring oxygen. The term is referred to only because of its past use, mostly in the 1960s.

18.3.2 Equipment

A variety of aeration equipment and gas stripping equipment has been manufactured over the years, for example, surface aerators, submerged diffusers, packed towers, etc. Activated sludge aeration equipment has included surface units, for example, turbines and brush rotors. The latter are used in shallow ditches <2m (6 ft) depth with a race-track shape. Submerged aerators include diffusers of various materials and shapes, jet aerators (pressurized air introduced into a water jet), static mixers, and a combination coarse bubble diffuser combined with a submerged turbine to shear and disperse the rising bubbles (Huang, 1975). Air stripping equipment is also varied, with packed towers being most common.

18.3.2.1 Reactor Types

Table 18.6 shows a sampling of reactor types used in gas transfer, along with K_La values and notes on characteristics of each. Within each reactor category, a variety of proprietary equipment is available, for example, bubble columns may be configured with coarse bubble or fine bubble diffusers; surface aerators include turbines, brush rotors, etc.; stripping towers may be packed with any of a range of shapes and materials (Lee and Tsui, 1999, p. 25).

18.3.2.2 Turbine Aerators

Turbine aerators have been used since the 1960s when they seemed to find favor over diffused bubble aeration. About 1980, fine bubble aeration technology had improved and evaluations established its economy as compared with surface aeration. Turbine aeration is reviewed here partly because it is

TABLE 18.6
Types of Reactors for Gas Transfer

Type	K_La (s^{-1})	Notes
Bubble columns	0.005–0.01	Low to moderate mixing intensity
		Mixing is by drag of gas bubbles
		High θ(water)
		Gas bubbles move as plug flow
		Simple construction, low capital cost
		No moving parts, low maintenance
		Gas bubbles may coalesce for height/diameter > 12
Turbine mixed tanks	0.02–0.2	Impeller type, size, speed give control of mixing intensity
		Large range for θ(water)
		Back-mixing
		Solids remain in suspension
Packed columns	0.005–0.02	Low Δp(gas)
		Packing is basis for interfacial area
		0.4–0.6 fraction of column is packing
		Not suitable for suspensions
		Flooding criterion must not be exceeded
Plate/tray columns		Higher θ(water) than packed columns
		Large number of transfer units possible
		Complete mixing for each tray; that is, similar to CFSTR's in series
		Can tolerate suspensions
		Higher capital costs
Spray tower	0.0007–0.015	Water is the dispersed phase; gas is the continuous phase
		Suitable for uptake of highly soluble gases
		High Δp for atomizing water
		Rapid coalescence of water droplets away from nozzle
Jet-loop	0.01–2.2	Jet of water or gas/water mixtures are injected into a column
		Jet causes mixing
		Higher K_La than bubble column alone
		Variable degrees of gas induction
		High power required
Venturi ejector	0.1–3	Unit is a Venturi section in a pipe
		Gas injection is in the high velocity (i.e., low pressure) section
		Very high mixing intensity
		High K_La
		Low θ(water); therefore, V(reactor) is low

Source: Adapted from Lee, S.Y. and Tsui, Y.P., *Chem. Eng. Prog.*, 95(7), 24, July 1999.
Note: CSTR, continuous flow stirred tank reactor.

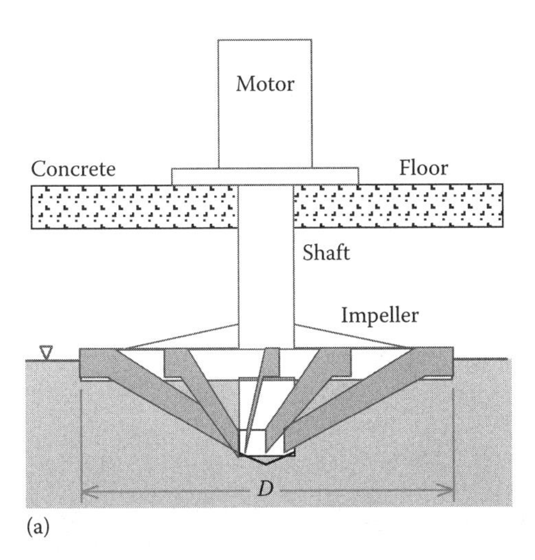

FIGURE 18.14 Turbine aeration. (a) Sketch of impeller, (b) photograph of turbine aerator basin, 1988. (Photo courtesy of S.K. Hendricks.)

still a technology that is "on-the-shelf," so to speak, and is still used in many situations, and partly because reviewing the topic helps to illustrate how basic principles may be applied to yet another kind of aeration device.

Figure 18.14a is a drawing of a turbine aerator with the impeller submersed in the water of a basin. Figure 18.14b is a photograph of an activated sludge basin with four such impellers spaced around the basin, with two shown. Figure 18.6 shows the same scene but taken as a close-up. The basin shown was later retrofitted, c. 1992, with fine bubble diffusers.

18.3.2.2.1 General Design Requirements

According to Kalinske (1969a,b) the basin size must be such that all portions of it are under the pumping influence of the aerator. Sizes of surface aerators have ranged from 3 to 150 hp and 1–3.7 m (3–12 ft) diameter, with rotational velocities 30–60 rpm (Huang, 1975, p. 65). In general, the oxygen transfer rate is proportional to the rate of creation of new air–water interface area, which is proportional to the pumping rate for a given turbine design. The pumping rate is (Kalinske, 1963):

$$Q(\text{pumping}) = k'D(\text{impeller}) \cdot h \cdot v_p \tag{18.71}$$

$$= k'D(\text{impeller})^2 \cdot h \cdot \omega \tag{18.72}$$

where

- Q(pumping) is the rate of flow of water through the impeller (m^3/s)
- k' is the proportionality coefficient, unique for a given impeller design
- D(impeller) is the diameter of impeller (m)
- h is the height of rotor blades (m)
- v_p is the peripheral velocity of impeller (m/s)
- ω is the rotational velocity of impeller (radians/s)

Figure 18.15 shows how oxygen uptake mass per unit of energy expended, $W(O_2)$, varies with impeller rotational speed for two impeller diameters. As seen, $W(O_2)$ increases with rpm to some peak and then declines, which is similar to a pump efficiency curve. As seen also, for a given impeller speed, a larger impeller diameter results in higher $W(O_2)$. As noted in Section 18.3.1.2, such a curve is unique for the basin at hand.

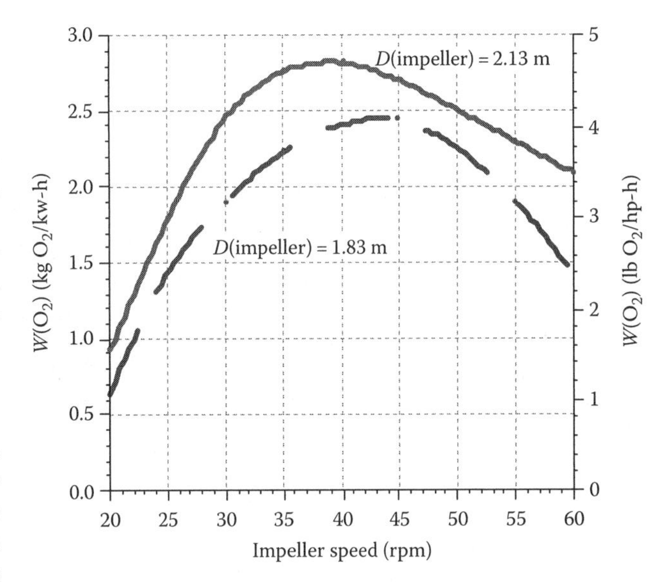

FIGURE 18.15 Effect of impeller speed on $W(O_2)$ for two impellers. (From Kalinske, A. A., Power consumption for oxygenation and mixing, in W.W. Eckenfelder and J. McCabe (Eds.), *Advances in Biological Waste Treatment*, Pergamon Press, New York, pp. 157–168, 1963. With permission.)

Example 18.4 Application of $W(O_2)$ versus rpm(Impeller) Characteristic Curve for System

Given

Assume $Q = 8000\ m^3/day$ ($333\ m^3/h$) (or 2.1 mgd) and $[BOD_{in} - BOD_{out}] = 200$ mg/L ($0.200\ kg/m^3$), which results in $J(O_2) = 67$ kg O_2/h. Assume also that the detention time for the reactor is $\theta = 6$ h and thus, $V(\text{reactor}) = (333\ \text{m}^3/\text{h}) \cdot 6\ \text{h} = 2000\ \text{m}^3$. Assume further that a single 2.13 m (7.0 ft) impeller is used and that placed in the basin described results in a characteristic curve as given by Figure 18.15. Let the basin be at sea level and let $T = 20°\text{C}$. Assume that a direct current motor is used (for control of rpm) and the power expended is proportional to the rotational velocity with 60 kW expended at 60 rpm.

Required

(a) Determine the K_La required

Solution

(a) The K_La(required) is given by the relation

$$J(O_2, \text{required}) = \alpha K_La(\beta C_s - C)V(\text{reactor})$$

which with numerical data is

$$67\ \text{kg O}_2/\text{h} = 1.0 \cdot K_La(1.0 \cdot 0.008 - 0.002)\ \text{kg O}_2/\text{m}^3 \cdot 2000\ \text{m}^3$$

which gives

$$K_La(\text{required}) = 5.6\ \text{h}^{-1}.$$

Discussion

An aerator test, as outlined in Example 18.3, is necessary to verify whether $K_La(\text{actual}) \geq K_La(\text{required})$ for a given system. For comparison, a surface aerator test value of $K_La = 6.5\ \text{h}^{-1}$ was obtained by Conway and Kumke (1966) for a 1703 m^3 (450,000 gal) basin and a 56 kW (75 hp) turbine aerator.

18.3.2.3 Diffused Aeration

Diffused aeration is the generation of gas bubbles emanating from pores or orifice openings submerged in a basin of water. Since the advent of activated sludge in 1914, a variety of diffusers have been used in practice to provide oxygen to aeration basins, for example, porous plates, porous tubes, saran-wrap tubes, pipes with orifices, etc. Clogging of the porous diffusers prompted the use of orifice-type materials in the 1920s. Later, air filters were developed to protect the fine-pore media. Surface aerators came on the scene primarily to circumvent the clogging problem. Foregoing is from Boyle et al. (1989, pp. 1, 2).

Evaluation of a diffused air system is commonly in terms of its oxygen transfer efficiency, defined as

$$E(O_2, \text{ST}) \equiv \frac{J(O_2\ \text{transferred, ST})}{J(O_2\ \text{delivered, ST})} \tag{18.73}$$

where

$E(O_2, \text{ST})$ is the standard oxygen transfer efficiency at $T = 20°\text{C}$, $p = 1.00$ atm, W = clean water (dimensionless); the term, $E(O_2, \text{ST})$ is the same as SOTE, the latter being used by Boyle et al. (1989)

$J(O_2$ transferred, ST) is the oxygen transfer rate for standard conditions (kg O_2 transferred/h)

$J(O_2$ delivered, ST) is the mass rate of oxygen delivered for standard conditions (kg O_2 delivered/h)

The mass fluxes of oxygen are calculated:

$$J(O_2\ \text{transferred, ST}) = K_La(20) \cdot C^*(p = 1.00\ \text{atm}, 20°\text{C}) \cdot V(\text{reactor}) \tag{18.74}$$

$$J(O_2\ \text{delivered, ST}) = Q' \cdot \rho(O_2\ \text{air at}\ 1.00\ \text{atm}, T = 20°\text{C}) \tag{18.75}$$

$$= Q' \cdot \rho(O_2, \text{ST}) \tag{18.76}$$

where

$K_La(20°\text{C})$ is the mass transfer coefficient obtained in aeration test for actual system, usually in clean water, converted to 20°C (h^{-1})

$C^*(p = 1.00$ atm, 20°C) is the concentration of dissolved oxygen in aeration basin for a hypothetical air pressure, $p(\text{atm}) = 1.00$ atm, $T = 20°\text{C}$, W = clean water (kg O_2/m^3)

V(reactor) is the volume of aeration basin (m^3)

Q'(ABasin) is the flow of air into diffuser system in aeration basin (m^3/s)

$\rho(O_2$ air at 1.00 atm, $T = 20°\text{C})$ is the density of oxygen based upon its partial pressure in air at $p(\text{air}) = 1.00$ atm, $T = 20°\text{C}$ (kg O_2/m^3)

$\rho(O_2, \text{ST})$ is the same as previous definition, that is, density of oxygen gas for standard conditions, only shorter notation (kg O_2/m^3)

The density of oxygen, $\rho(O_2)$, for any pressure and temperature may be calculated from the ideal gas law (see Table B.7),

$$\rho(O_2) = \frac{p(O_2)}{RT} \cdot \frac{\text{MW}(O_2)}{1000} \tag{18.77}$$

where

$\rho(O_2)$ is the mass density of oxygen (kg O_2/m^3)

$p(O_2)$ is the partial pressure of oxygen in atmosphere (N/m^2)

R is the gas constant (8.314 N m/mol K)

T is the absolute temperature (K)

$\text{MW}(O_2)$ is the molecular weight of oxygen (31.9988 g/mol)

Example 18.5 Calculation of Gas Density

Given

Air at standard temperature, $T = 20°\text{C}$, and standard pressure, $p(\text{sea-level}) = 101.325$ kPa.

Required

Calculate the density of air, ρ(air, ST) and of oxygen in air, $\rho(O_2, ST)$ for standard conditions

Solution

Substituting numerical data in Equation 18.75 yields ρ(air, ST) and $\rho(O_2, ST)$, respectively. For reference, the equivalent MW(air) = 29.964 g/mole (Table B.7) and the mole fraction of oxygen in air, $X(O_2/\text{air}) = 0.209476$ (also Table B.7)

$$\rho(\text{air, ST}) = \frac{101{,}325 \, \frac{\text{N}}{\text{m}^2}}{8.314 \, \frac{\text{N m}}{\text{mol K}} \cdot (273.15 + 20)\text{K}} \cdot \frac{28.964\,\text{g air}}{\text{mol air}} \cdot \frac{\text{kg}}{1000\,\text{g}}$$
$$= 1.204\,\text{kg air/m}^3$$

$$\rho(O_2, \text{ST}) = \frac{0.209476 \cdot 101{,}325 \, \frac{\text{N}}{\text{m}^2}}{8.314 \, \frac{\text{N m}}{\text{mol K}} \cdot (273.15 + 20)\text{K}} \cdot \frac{31.9988\,\text{g O}_2}{\text{mol O}_2} \cdot \frac{\text{kg}}{1000\,\text{g}}$$
$$= 0.2787\,\text{kg O}_2/\text{m}^3$$

Discussion

Application of the ideal gas law, as applied above is straightforward. The key points relate to (1) using the partial pressure of the gas, that is, oxygen in air; (2) applying the molecular weight of the gas to convert from molar density to mass density. Such calculations, as shown, are useful in determining the "mass flow of oxygen delivered," that is, $J(O_2, \text{delivered, ST})$, as in Equation (18.76), and in doing calculations that may involve the flow of air under any conditions, for example, as related to compressors. As with all calculations, any final numbers, for example, for mass flow, **J**, or power required, should be rounded to the number of significant figures applicable (commonly three for engineering purposes).

Combining Equations 18.71 through 18.73 gives the oxygen transfer efficiency yields

$$E(O_2, \text{ST}) \equiv \frac{K_L a(\text{ST}) \cdot C^*(O_2 \text{ in air, ST}) \cdot V(\text{reactor})}{Q'(\text{air}) \cdot \rho(O_2, \text{ST})} \quad \text{(Ex18.5.1)}$$

Equation 18.76 provides a relation for conversion between the oxygen transfer efficiency, $E(O_2, \text{ST})$, and $K_L a$, the mass-transfer coefficient for oxygen. Once $K_L a$ is determined, the oxygen transfer efficiency may be determined for any other condition, which includes calculating $Q'(\text{air})$ and compressor power required, etc.

Example 18.6 $K_L a$ from Plant Data (Data Taken from Boyle et al., 1989)

Given

(a) **Background**: The Plymouth, WI activated sludge plant operated between 1978 and 1985 using three aeration basins, each 12.2 m × 12.2 m × 4.6 m (40 ft × 40 ft × 15 ft), and each with one 30 kW (40 hp) surface aerator. Annual energy costs for aeration were estimated at $37,500 based on 45 A/aerator; annual operating days were 191 days with two aerators in operation and 174 days with three aerators; power cost was $0.05/kWh.

(b) **Retrofit—Diffused aeration**: In 1985, the aeration basins were retrofitted with a fine-pore diffused aeration system consisting of 1350 230 mm (9 in.) diameter diffusers with ratio diffuser surface area to tank floor area = 0.116. Each aeration basin had installed 450 ceramic disk diffusers, equally spaced at 660 mm cc (26 in.). Equipment was installed for in situ gas cleaning and for pressure monitoring.

(c) **Retrofit data**: Three positive displacement blowers were installed, each 37 kW (50 hp) with Q'(air per blower) = 0.400 m^3/s/blower (850 scfm/blower). Each blower was expandable to Q'(air per blower) = 0.802 m^3/s/blower (1700 scfm/blower) by retrofitting with a higher wattage electric motor. The design criteria were: $Q = 0.072$ m^3/s (1.65 mgd); $J(\text{BOD}) = 614$ kg BOD/day (1,356 lb/day); $J(NH_3\text{-N}) = 136$ kg NH_3-N/day (300 lb/day); $J(O_2 \text{ required}) = 1.15$ kg O_2/kg BOD applied + 4.6 kg O_2/kg NH_3-N applied. Also, $Q'(\text{air})/\text{diffuser} =$ 0.063 m^3 air/s/diffuser (1.33 scfm/diffuser).

Required

Estimate $K_L a$ for an aeration basin based upon the given data. Assume that $E(O_2, \text{ST}) = 0.15$ for the purpose of the calculation.

Solution

Apply Equation 18.76. First determine the input data:

(1) From Table H.5, $C^*(O_2/\text{air, ST}) = 0.209476 \cdot 43.39$ mg O_2/L = 9.07 mg O_2/L = 0.00907 kg O_2/m^3
(2) $V(\text{reactor}) = 12.2 \text{ m} \times 12.2 \text{ m} \times 4.6 \text{ m} = 684.66 \text{ m}^3$
(3) $Q'(\text{air})/\text{blower} = 0.400$ m^3/s/blower
(4) $\rho(O_2, \text{ST}) = 0.2787$ kg O_2/m^3 (Example 18.5)

Now apply Equation (18.78):

$$E(O_2, \text{ST}) \equiv \frac{K_L a(\text{ST}) \cdot C^*(O_2 \text{in air, ST}) \cdot V(\text{reactor})}{Q'(\text{air}) \cdot \rho(O_2, \text{ST})}$$

$$0.15 = \frac{K_L a(\text{ST}) \cdot 0.00907 \, \frac{\text{kg O}_2}{\text{m}^3} \cdot 684.66 \text{ m}^3}{0.400 \, \frac{\text{m}^3}{\text{s}} \cdot 0.2787 \, \frac{\text{kg O}_2}{\text{m}^3}}$$

$$K_L a(\text{ST}) = 0.027\,\text{s}^{-1}$$
$$= 9.7\,\text{h}^{-1}$$

Discussion

Once $K_L a$ is determined for the conditions described, a variety of what-if questions may be addressed (also called "scenarios"). For example, suppose pure oxygen is used instead of air. What happens if the airflow is increased (for air, not pure oxygen)? What if the elevation of the aeration basin is 1610 m (5280 ft)? At high elevations is there merit in using a deeper basin? Note that the round off was applied at the last step for $K_L a$.

18.3.2.3.1 Coarse-Bubble Aeration

Bubbles sizes in the range 6–10 mm are called "coarse-bubbles" and are produced by orifices in plates or pipes. In general, oxygen transfer efficiencies are about 6%–7% (Morgan and Bewtra, 1959, 1960, 1963). The coarse-bubble diffusers came into vogue mostly to avoid the need for air filters and to avoid clogging, both associated with fine-bubble diffusers.

18.3.2.3.2 Fine-Pore Aeration

Bubbles sizes in the range 2–5 mm are called "fine bubbles" and are produced by diffusers of materials such as ceramic, nonrigid plastic, rigid plastic, etc. These materials may be in the form of porous ceramic plates, disks, domes, and tubes. Air may flow through an ill-defined porous structure in ceramic or plastic materials or through perforated membranes comprising disk or tube surfaces (Boyle et al., 1989, p. 2). The pressure loss through the passages is due mostly to overcoming the surface energy to create a bubble; very little is due to frictional loss. In general the headloss due to bubble formation is $\geq$5 cm water (0.49 Pa), with range 8–51 cm water (0.78–5.00 kPa). The membrane slots are flexible and enlarge as air pressure increases and close if the airflow is zero, preventing backflow of water and are resistant to fouling. The air side fouling is considered less important than the water side fouling (Boyle et al., 1989, p. 4).

Figure 18.16 shows examples of diffusers in "plug-flow" aeration basins at two municipal wastewater treatment plants, illustrating coarse bubble aeration and fine-bubble aeration, respectively. Figure 18.16a shows a 6.1 m wide $\times$ 4.6 m deep (20 ft $\times$ 15 ft) basin with water flow; the turbulence at the surface is due to the breaking bubbles. Figure 18.16b shows an empty basin of the same plant showing "coarse-bubble" tube diffusers attached to two header pipes. Figure 18.16c shows a similar basin, but with fine-bubble diffusers. The "fine-bubble" disc diffusers were laid out in a grid, as seen in the empty basin, Figure 18.16d. When operation

FIGURE 18.16 Diffused aeration systems in two plants. (a) San Jose/Santa Clara Water Pollution Control Plant activated sludge reactor, (b) Tube diffusers (coarse-bubble) across bottom at San Jose/Santa Clara plant, (c) Metro wastewaters Reclamation Plant (Denver) Colorado activated sludge reactor, (d) Disc diffusers (fine-bubble) placed in grid at Denver Metro plant. ((a) and (b) Photo courtesy of San Jose WWRWWTP, Metro Wastewater Reclamation District, Denver, CO.)

TABLE 18.7
Efficiency Data for Different Conditions at Full-Scale Plants

Factor	Description	Diffuser	Place	αF	$\alpha F \cdot E(STp)$
Fine bubble	2 months	Perforated membrane tube	Eugene, Oregon		10.9
Coarse bubble	2 months				4.9
Fine bubble	14 months	Perforated membrane tube			8.7
Coarse bubble	14 months				6.6
Age	0 year	Flexible membrane	Durham, North Carolina	0.55	15.8
	3.5 years			0.45	12.8
Layout	Grid	Perforated membrane tube	Renton, Washington		6.5
	Spiral				4.9
Layout	Grid	Porous plastic disk	Renton, Washington		8.4
	Spiral				5.3

Source: Powell-Groves, K. et al., *Water Environ. Res.*, 64(5), 691, July/August, 1992.

began in 1966, the basins had saran-wrap fine-bubble diffusers on one side (resulting in a spiral flow). Bag air filters were provided to minimize clogging; the saran-wrap tubes were considered high maintenance, requiring removal for cleaning on a regular cycle.

18.3.3 Operation

About 50% of the energy used in an activated sludge wastewater treatment plant is due to the aeration requirement (Boyle et al., 1989). Therefore, the factors that affect the oxygen transfer efficiency are of interest in design and operation. In a study of 21 plants, Powell-Groves et al. (1992) addressed some of these factors, which included diffuser type, diffuser layout, diffuser age, solids retention time, and level of nitrification. Table 18.7 shows some of their data indicative of their field-testing. As noted by Boyle et al. (1989), it is not feasible, as a rule, to isolate α and F; hence the two variables are "lumped."

18.4 CASE STUDIES

Several case studies are reviewed to provide reference for actual designs. Fine bubble diffuser designs are summarized in tabular form and several air stripping cases are reviewed.

18.4.1 Fine-Bubble Diffusers

Table 18.8 shows data from six plants representative of a survey of municipal activated sludge plants (Houck, 1988). Each of the plants was fitted with either ceramic dome or membrane disk fine-pore diffusers. All of these aeration systems were installed in 1978 or later and before or during 1982. The plant sizes and the dimensions covered a wide range, as indicated by Q(avg). The design and operating data summarized include the diffuser densities, the airflow per diffuser, the total airflow to the system, the "wire" power used, and the oxygen transferred per unit of wire-power expended, $W(O_2)$.

18.4.2 Air Stripping

The following cases were taken from an EPA report (Rawe, 1991). The cases are indicative of the variation found in air stripping situations.

18.4.2.1 Sydney Mine at Valrico, Florida

A groundwater treatment system at Sydney Mine Site, Florida, consisted of air stripping followed by carbon adsorption for polishing. For the system, $Q = 570$ L/min (150 gpm); HLR $= 4.9$ m/h (12 gpm/ft^2); air/water ratio $= 200{:}1$. The air stripping tower was 1.219 m diameter $\times$ 12.20 m high with a 7.32 m depth of 90 mm polyethylene packing (4 ft $\times$ 42 ft/24 ftw/3.5 in. packing). Influent concentrations varied $25 \leq C(\text{TOC, influent}) \leq 700$ μg/L. For individual constituents, influent/effluent concentrations for the overall air stripping/GAC were (in μg/L): benzene 11/nd; chlorobenzene 1/nd; 1,1-dichloroethane 39/nd; trans-1,2-dichloropropane 1/nd; ethylbenzene 5/nd; methylene chloride 503/nd; toluene 10/nd; trichlorofluoromethane 71/nd; meta-xylene 3/nd; *ortho*-xylene 2/nd; 3-(1,1-dimethylethyl) phenol 32/nd; 2,4-D pesticide 4/nd; 2,4,5-TP 1/nd (nd means "not detected" at method detection limit of 1 μg/L).

18.4.2.2 Well 12A: City of Tacoma, Washington

Air stripping was used at Well 12A contaminated with chlorinated hydrocarbons, with $C(\text{VOC's}) \approx 100$ μg/L. For the well, $Q(\text{water}) = 221$ m^3/s (3500 gpm). Five air stripping towers were installed and began operation on July 15, 1983. Each tower was 3.66 m (12 ft) diameter and was packed with 25 mm (1 in.) polypropylene saddles to a depth of 6.10 m (20 ft). For each tower, $Q(\text{water, tower}) = 44.16$ m^3/s (700 gpm) with air/water ratio $= 310{:}1$. The towers removed 0.94–0.98 fraction of 1,1,2,2-tetrachloroethane; for other contaminants, removals were ≥ 0.98 fraction.

TABLE 18.8
Data from Fine Bubble Diffused Air Operating Systems

Place	*Q*(avg)		*L*		Width		Depth		Diffuser Density		*Q*′(air)/Diffuser		*Q*′(Air)		*P*(Used)	$W(O_2, T, W, p)$	
	(m^3/s)	(mgd)	(m)	(ft)	(m)	(ft)	(m)	(ft)	(#/m^2)	(#/ft^2)	(L/min/Diff)	(cfm/Diff)	(L/s)	(cfm)	(kW)	(kg O_2/kWh)	(lb O_2/hph)
Riverside	0.394	9.0	76.20	250	12.19	40	5.36	17.6	5.81	0.54	10.2	0.36	3,540	7,500	203	1.15	1.89
Whittier N. #1	0.548	12.5	91.44	300	9.14	30	4.39	14.4	2.80	0.26	32.3	1.14	3,288	6,966	207	1.18	1.94
Berlin, NH	0.074	1.7	30.48	100	7.62	25	4.57	15	2.91	0.27	20.1	0.71	160	340	8.3	2.27	3.74
Village Creek	2.401	54.8	72.85	239	31.70	104	4.21	13.8	5.38	0.50	15.9	0.56	13,082	27,720	812	2.41	3.97
Meriden	0.311	7.1	30.48	100	17.07	56	5.49	18	1.08	0.10	49.0	1.73	1,321	2,800	102	2.31	3.80
Ridgewood	0.131	3.0	35.36	116	7.32	24	4.57	15	2.80	0.26	17.6	0.62	316	670	19.6	2.52	4.15

Source: Adapted from Houck, D.H., *Aeration Survey and Evaluation of Fine Bubble Dome and Disc Diffuser Aeration Systems in North America*, Project Summary, EPA/600/S2–88/001, U.S. Environmental Protection Agency, Cincinnati, OH, September, 1988.

18.4.2.3 Wurtsmith AFB: Oscoda, Miami

The contamination at the site was trichloroethylene (TCE), caused by a leaking underground storage tank near a maintenance facility. Two packed-tower air strippers in series were installed, each 1.52 m (5 ft) diameter, 9.14 m (30 ft) high, with 5.49 m (18 ft) 16 mm pall-ring packing, with air/water ratio = 25:1 and Q(water) = 0.0378 m^3/s (600 gpm). The TCE removal was 0.86–0.98 fraction for a single tower and 0.999 fraction overall. Biological growths were a maintenance problem and required repeated removal and cleaning.

18.4.2.4 Hyde Park Superfund Site, New York

The site treated Q = 3.50 L/s (80,000 gpd) landfill leachate. Total organic carbon (TOC) (influent) = 4000 mg/L; the air stripper removed about 0.90 fraction TOC.

PROBLEMS

18.1 Applications of Henry's Law

Given/Required

Determine the concentration of atmospheric gases (O_2, N_2, CO_2) using Henry's law, Table H.5.

(a) At sea level at 25°C.
(b) At the Engineering Research Center (ERC) at Colorado State University (CSU) elevation, elevation is 1615 m (5300 ft) at 25°C based upon tabular data or formula for the effect of elevation on atmospheric pressure.
(c) Read the barometric pressure from the mercury barometer. Compare the reading with the estimate in (b).
(d) Determine the concentration of the three atmospheric gases in equilibrium with the atmosphere in distilled water based upon observed barometric pressure (or elevation of site).

18.2 Application of Henry's Law to Oxygen

Given/Required

Determine the concentration of oxygen using Henry's law:

(a) At sea level at 25°C.
(b) At the ERC elevation, elevation is 1615 m (5300 ft) at 25°C based upon tabular data for the effect of elevation on atmospheric pressure, Table B.6, and using Henry's law coefficients, Table H.5.

18.3 Application of Henry's Law to Carbon Dioxide

Given/Required

(a) Determine the concentration of carbon dioxide in equilibrium with the atmosphere in distilled water at your location.
(b) Bubbles are seen emerging from a primary clarifier. The bubbles are determined to be about two-thirds carbon dioxide and one-third methane. Assume that you have a Kemmerer water bottle (device to take a water sample at some depth) and you bring up the sample to measure the carbon dioxide concentration (by titration). From what you know of Henry's law, estimate the concentration that you would measure assuming no experimental error of analysis.
(c) Assume that you obtain a measurement of carbon dioxide in (b) by titration. Describe the major experimental error.

18.4 Gas Transfer Coefficient—Is Determination for Oxygen Applicable to Different Gases Using the Same System?

Given/Required

For a lab experiment, K_La was determined to be 0.029 min^{-1} for stripping or adding oxygen. Assume the reactor was "complete mix."

(a) Estimate K_La(benzene)
(b) Suppose, using the same reactor in the batch mode, you had to strip benzene at a concentration of 20 mg/L to a concentration of 0.5 mg/L. Estimate the time required.
(c) Suppose the same reactor is used as a continuous-flow reactor, with flow coming in at the top and out at the bottom. Does this affect K_La? What should be the detention time?

18.5 Application of $W(O_2)$ versus rpm(Impeller) Characteristic Curve for a System

Given

- Assume Q = 9000 m^3/d (375 m^3/h) (or 2.4 mgd).
- $[BOD_{in} - BOD_{out}]$ = 200 mg/L (0.200 kg/m^3).
- Assume also that the detention time for the reactor is $\theta = 6$ h.
- V(reactor) = (375 m^3/h) · 6 h = 2350 m^3.
- A single 2.13 m (7.0 ft) impeller is used.
- A characteristic curve is given by Figure 18.15.
- Elevation = 1610 m
- T = 15°C.
- A direct current motor is used.
- Power expended is proportional to the rotational velocity with 60 kW expended at 60 rpm.

Required

(a) Determine the K_La required.
(b) Calculate rpm(impeller) for the system to provide the required K_La.

Hint: Example 18.4 provides a pattern for solution. Note that C_s will reflect the elevation and temperature, which in turn will affect K_La.

18.6 Explanations of Test Results and Gas Transfer Variables

Given/Required

(a) Measurements of dissolved oxygen levels in full-scale aeration tests, at various sampling points, often show dissolved oxygen concentrations greater than saturation with respect to the atmosphere. Explain.
(b) Delineate in a mathematical format, the system variables which may influence K_La.
(c) What important principles must be operative in surface aerator performance with respect to turbulence (i.e., scale of eddies, intensity of turbulence).

18.7 Design of Aeration Systems

Given

Municipal waste from a city of population 20,000; BOD = 300 mg/L raw influent. Assume 35% BOD removal in primary sedimentation; 90% overall removal.

Required

Design the aeration system for an activated sludge design for a diffused aeration system. Consider for the diffused aeration system:

(1) Size of tanks and number
(2) Number, spacing, type and location of diffusers
(3) Blower capacity
(4) Power cost per day
(5) Estimate capital cost of complete system

18.8 Design of Aeration Systems

Set up a small laboratory vessel, about 10–100 L in size, for a diffused aeration system rating, and determine $K_{L}a(ST)$ for each of the following situations:

(a) For the system, set up in a way which will be the standard of comparison
(b) For double the air flow rate
(c) For half the volume (do this by partitioning down the middle)
(d) For half the volume (do this by reducing depth of water)
(e) For 10°C higher or lower temperatures
(f) For pure oxygen
(g) For sewage instead of pure water

18.9 Rate Limiting Mechanism of Gas Transfer—Ammonia Stripping

Given

Ammonia is to be stripped from a wastewater.

Required

(a) Indicate the gas concentration profiles in the gas phase and in the aqueous phase.
(b) Recommend a pH for most complete removal.

Hint: Refer to Equations 18.24 and 18.27, Figure 18.5, and Appendix H.

18.10 Rate Limiting Mechanism of Gas Transfer—Stripping and Uptake

Given

The following situations:

(a) Carbon dioxide is to be stripped from a water that has accrued an excess
(b) Carbon dioxide is to be added to a water (for pH control in water treatment)
(c) Chlorine gas is to be added to a water (as a disinfectant)
(d) Ozone is to be added to a water (as an oxidant to split organic molecules)
(e) Excess, that is, supersaturated, dissolved air is to be removed from a water coming through a pipeline from a high mountain source (the air has been entrained in the headworks and then dissolved in the pipeline under high pressure)
(f) Excess nitrogen gas is to be removed from water (e.g., after nitrification in biological treatment). [But is this necessary? Will the nitrogen essentially remove itself? How?]

Required

Indicate the gas concentration profiles in the gas phase and in the aqueous phase for each case. Recommend a pH for most complete removal or addition (as the case may be).

Hint: Refer to Equation 18.24 and 18.27, Figure 18.5, and Appendix H.

18.11 Estimating $K_{L}a$ from First Principles

Given

Assume conditions as given in Example 18.3, that is, an 18 L carboy with a carborundum diffuser at the bottom, which gives off a small stream of air bubbles. To repeat further, $Q'(\text{air}) = 42.5$ L/h, $p(\text{atmosphere}) = 623$ mmHg, $\Delta p(\text{diffuser}) =$ 80 mmHg (gauge), $h(\text{diffuser}) = 304$ mmHg, $T(\text{water}) = 23°\text{C}$. Assume $d(\text{bubbles}) \approx 2$ mm and $v_{w} \approx 0.5 \cdot v(\text{bubbles})$.

Required

Calculate $K_{L}a$ from "first principles," for example, by Equation 18.45; let $K_{L} = D/\delta$. Compare with the value calculated from experimental data in Example 18.3. Determine a value for the film thickness, d, such that the $K_{L}a$ value calculated from first principles is the same as obtained from experimental data (Example 18.3).

Hint: Set up a spreadsheet for solving for $K_{L}a$ as defined in Equation 18.45.

18.12 Onda Correlations

Given

A VOC, for example, chloroform, $CHCl_3$ (MW = 119.39, density = 1.489 g/mL at 20°C) is to be stripped from a water. The influent concentration is 190 CCl_4 μg/L. The effluent concentration desired is 6 μg/L. Assume the packing material is Pall rings with specific surface area, $a_{p} = 206.6$ m^2/m^3 packed volume; void fraction = 0.90; nominal diameter, $d_{p} = 25$ mm; $\text{HLR(water)} = Q/A = 0.0175$ m^3 water/m^2 section area/s; $\text{LR(air)} = Q'/A = 3.5$ m^3 air/m^2 section area/s (most of the foregoing data was from Dvorak et al. 1996, p. 947). Assume $T = 20°\text{C}$. (HLR is the "hydraulic loading rate" defined as the flow in m^3/s divided by the cross-section area (m^2) of a vessel; it's a "flux-density" of water. (2) LR(air) is the "loading-rate". Similar to HLR; its a "flux-density" of air.

Required

Determine the column height required. Set up a spreadsheet so that other scenarios may be imposed.

Determine also the K_La coefficient and compare with values given from aeration installations.

Hint: Obtain properties from Appendices and calculate diffusivities by the Wilke–Chang equation.

18.13 Aerator Test Data

Given

Aerator tests were performed using a basin at the Los Angeles County Sanitation District's Joint Treatment Facility (Yunt and Hancuff, 1988, p. 5). Data obtained were as follows:

Date	Depth (m)	Depth (ft)	P/V (hp/1000 ft^3)	Q'(Air) (scfm)	Q'(Air) (m^3/m)	$K_La(20)$ (h^{-1})	C^* (mg O_2/L)	E(ST)
03/24/78	7.62	25	0.28	73.8	2.08	5.34	11.42	0.495

The reactor data are as follows: V(reactor) = 6.1 m × 6.1 m × 6.1 m (20 ft × 20 ft × 25 ft deep) = 283.17 m^3.

[The foregoing data were obtained using a Norton® fine-bubble diffuser.]

Required

Calculate E(ST) and compare with that obtained in the foregoing table.

18.14 Calculation of $E(WTp)$ and K_La from Survey Data of Houck (1988)

Given

Table 18.8 is a compilation of various data, including aeration, obtained from some six activated sludge plants throughout the United States with two from Canada (19 of ours given in the original data).

Required

(a) Estimate the efficiency of oxygen transfer, that is, $E(W, T, p)$ for the conditions at hand of the plants, for example, wastewater, temperature, and atmospheric pressure at hand (assume sea level).

(b) Also calculate the K_La values for each plant. Set up a spreadsheet to calculate these parameters for each plant for which data are given. Discuss the results and the variation found.

18.15 Chlorine Gas Uptake to Aqueous Phase

Given

Chlorine gas is used commonly at municipal water treatment plants for disinfection.

Required

At the gas–water interface, is the gas film or the liquid film likely to have the greatest impedance to gas transfer? Sketch what you think should be the gas concentration gradients in the gas phase and aqueous phases, respectively. Provide the mathematical rationale for your responses.

18.16 Effects of Independent Variables on K_La for Bubble Aeration

Given

- Bubble aeration
- K_La(measured) = 2.5 h^{-1}.
- h(diffuser)$_1$ = 5.0 m (16.4 ft)

Required

Estimate the effects of changing one variable at a time in Equation 18.44, for example,

(a) Suppose the depth is decreased by one half
(b) Suppose the volume is decreased by one half
(c) Suppose the airflow is increased by a factor of two
(d) Suppose the bubble diameter is decreased by a factor of two
(e) As a second part of the equation, consider the effects of each variable changed on other variables, that is, discuss the effects

Hint: The effects of each of the variables indicated, one at a time, is seen by examining Equation 18.44. The interdependencies are with respect to the effect of V on ζ, the effect of h on ζ, the effect of Q' on ζ. Also, v_w may be affected directly by Q', and then begin to taper asymptotically up to some limit.

ACKNOWLEDGMENTS

The author is indebted to Tony Kalinske, director of research, and Gerry Shell, research engineer, for involvement in full-scale turbine aerator testing, c. 1968, while doing summer work at Eimco Corporation, Salt Lake City, Utah. Similar procedures were used by students in a laboratory course at Colorado State University, c. 1970–2000, to generate data for K_La determinations for air, pure oxygen, and oxygen stripping with nitrogen gas. Some of these data were used for this chapter.

William F. Clunie, technical manager, Water, AECOM, Wakefield, Massachusetts, gave permission (2010) to use Figure 18.13, generated at CSU, c. 1996, in conjunction with a lab partner, Y. Wu (who could not be located).

Dale W. Ihrke, PE, plant manager, San Jose/Santa Clara Water Pollution Control Plant, San Jose, California, kindly gave permission (2010) to use photographs of the aeration basins at the plant, Figure 18.16a and b, taken by the author with permission, c. 1991.

Steve Frank, public information officer, Metro Wastewater Reclamation District (Denver), provided photographs of the aeration basins, Figure 18.16c and d, at the Robert W. Hite Treatment Facility, North Plant, and granted permission for their use. The district administers some 370 km (232 mi) interceptor sewers as well as the 530 mL/day (140 mgd) Robert W. Hite Treatment Facility serving 59 local governmental entities, e.g., cities and sanitation districts, and some 1.6 million persons. The District and Frank have provided hospitality to the author along with many graduate classes over a period of three decades.

Frank D. Edwards, president, Bisco Environmental, Inc., Taunton, Massachusetts, kindly gave permission (2010) to use the two images of their product used in Figure 18.8. As a note, Northeast Environmental Products was acquired by BISCO Environmental in 2007. The product lines of both companies were merged to include a variety of products having to do separations (e.g., gas–water, oil–water, compressors), compressors, pumps, etc., as seen on their web site, (see BISCO, 2010), including the Shallow Tray™ product line. Their web site shows a variety of photographs of Shallow Tray™ air strippers in different situations, along with many other products.

APPENDIX 18.A: ONDA COEFFICIENTS

18.A.1 ONDA CORRELATIONS

A 1968 paper in Japan by Onda, Takeuchi, and Okumuto has been cited in the literature for its utility in predicting the column height needed to strip VOCs in a packed bed. A column is packed usually with manufactured objects designed to promote a large air–water interface per unit volume of packed bed. The objects are given such names as Pall rings, Rashig rings, saddles, etc., and may be made of polypropylene or ceramic material. The paper by Onda et al. provided empirical equations that predicted both liquid and gas phase mass-transfer coefficients, and the specific interfacial surface area (per unit volume of packed bed), that is, k_L, k_g, and **a**, respectively. The equations given are called in the literature, the "Onda correlations." Their accuracy has been evaluated by Staudinger et al. (1990) and Dvorak et al. (1996). The Staudinger study found that the Onda correlations predicted rate constants with standard deviation ±17% (437 data points) corresponding to ±30% accuracy based on 90% confidence limits. Dvorak et al. (1996) found about 16% overprediction and 34% underprediction (at high gas flows).

The application of a protocol that utilizes the Onda correlations to estimate the height of a gas stripping column was outlined by Dvorak et al. (1996) and is reviewed here. The protocol is to attain a specified effluent concentration for a given influent concentration. The sequence of equations is enumerated.

18.A.2 ONDA EQUATIONS

1. *Mass balance*: The mass balance for a control volume results in the relation (Dvorak et al., 1996)

$$Z = \frac{Q'}{K_L a \cdot A} \cdot \left(\frac{S}{S-1}\right) \ln\left\{\frac{1}{S} + \left(\frac{S-1}{S}\right) \cdot \left(\frac{C_{in}}{C_{out}}\right)\right\} \quad (18.A.1)$$

where
Z is the height of packed bed (m)
A is the cross section area of packed bed (m^2)

2. *Stripping factor*: A "stripping factor," S, is defined,

$$S = \frac{Q'H}{Q_w} \quad (18.A.2)$$

where
S is the stripping factor (dimensionless)
Q' is the air flow (m^3/s)
Q_w is the water flow (m^3/s)
H is Henry's constant (m^3 water/m^3 air)

3. *Mass-transfer coefficients*: Based on the two-film theory, the overall mass-transfer coefficient, $K_L a$, is

$$\frac{1}{K_L a} = \frac{1}{k_L \mathbf{a}} + \frac{1}{H k_g \mathbf{a}} \quad (18.A.3)$$

where
$k_L\mathbf{a}$ is the aqueous side mass-transfer coefficient (s^{-1})
$k_g\mathbf{a}$ is the gas side mass-transfer coefficient (s^{-1})
a is the interfacial area per unit volume of reactor (m^2 interfacial area/m^3 reactor)

Equation 18.87 is the same as Equation 18.26 except that in Equation 18.A.3, the specific interfacial area, **a**, is "lumped" in with the k_L and k_g terms, as for $K_L a$ (Dvorak et al., 1996). As a rule, either the liquid film or the gas film gas transfer dominates and so, most of the time, $K_L a$ is one or the other of the terms given. For example, if H is large, then, $1/(Hk_g\mathbf{a}) \to 0$ and $1/K_L a \approx 1/k_L a$; thus, $K_L a \approx k_L a$.

4. *Onda correlations*: The Onda correlations predict k_L, k_g, and "**a**," that is,

$$k_L = 0.0051\left(\frac{Q \cdot \rho(\text{water})}{A \cdot \mathbf{a} \cdot \mu(\text{water})}\right)^{\frac{2}{3}} \times \left(\frac{\mu(\text{water})}{\rho(\text{water}) \cdot D(\text{solute/water})}\right)^{-\frac{1}{2}} \times (a_p d_p)^{0.4}\left(\frac{\rho(\text{water})}{\mu(\text{water})g}\right)^{-\frac{1}{3}} \quad (18.A.4)$$

$$k_g = 5.23\left(\frac{Q' \cdot \rho(\text{gas})}{A \cdot a_p \cdot \mu(\text{gas})}\right)^{0.7} \times \left(\frac{\mu(\text{gas})}{\rho(\text{gas}) \cdot D(\text{solute/gas})}\right)^{\frac{1}{3}} \times (a_p d_p)^{-2}[a_p \cdot D(\text{solute/gas})] \quad (18.A.5)$$

$$\mathbf{a} = a_p\left\{1 - \exp\left[-1.45\left(\frac{\sigma_C}{\sigma_W}\right)^{0.75} \boldsymbol{R}^{0.1}\boldsymbol{F}^{-0.05}\boldsymbol{W}^{0.2}\right]\right\} \quad (18.A.6)$$

5. *Reynolds, Froude, and Weber numbers*: The Reynolds number, $\boldsymbol{R}$, the Froude number, $\boldsymbol{F}$, and the Weber number $\boldsymbol{W}$, are defined as follows:

$$\boldsymbol{R} = \frac{Q\rho(\text{water})}{A \cdot a_p \cdot \mu(\text{water})} \quad (18.A.7)$$

$$\boldsymbol{F} = \frac{Q^2 a_p}{A^2 g} \quad (18.A.8)$$

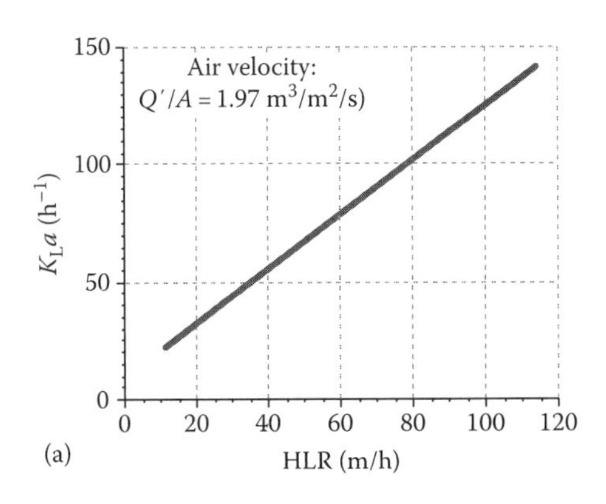

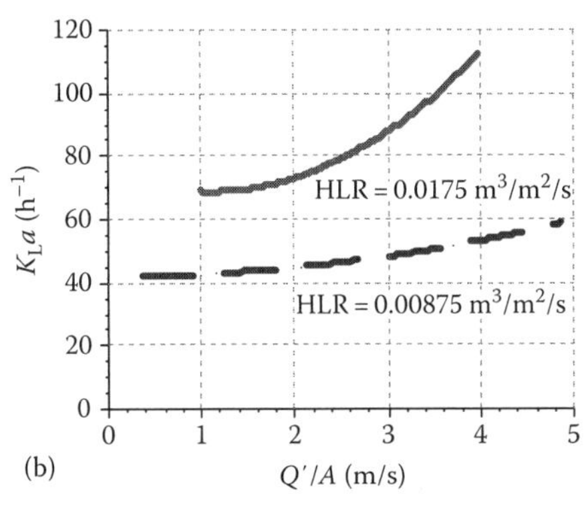

FIGURE 18.A.1 K_La as affected by HLR and air loading. (a) HLR effect, (b) air loading effect. (From Dvorak, B.I. et al., *Environ. Sci. Technol.*, 30(3), 951, March 1996.)

$$\mathbf{W} = \frac{Q^2\rho(\text{water})}{A^2 a_p \sigma(\text{water})} \quad (18.A.9)$$

6. *Terms*: Terms in the foregoing equations are defined as follows:
   a = interfacial area per unit volume of packing (m² air–water interface/m³ packing)
   a_p = specific surface area of packing material (m² packing material/m³ packing volume)
   A = cross sectional area of packing, that is, normal to mean velocity (m²)
   d_p = diameter of a packing unit (m)
   D(solute/water) = diffusion coefficient for dissolved gas in water (m²/s)
   D(solute/gas) = diffusion coefficient for dissolved gas in gas carrier, for example, in air (m²/s)
   g = gravitational constant (9.806 650 m/s²)
   Q = flow of water in column (m³/s)
   Q' = flow of gas in column at standard temperature and pressure (m³/s)
   μ(water) = viscosity of water, for example, air (N s/m²)
   μ(gas) = viscosity of carrier gas, for example, air (N s/m²)
   ρ(water) = density of water (kg/m³)
   ρ(gas) = density of gas, for example, air (kg/m³)
   σ(water) = surface tension of water (N/m)
   σ_c = critical surface tension of packing material (N/m)

As seen, the correlations incorporate measurable data, for example, Q, Q', and properties of the media, that is, a particular gas and water, and include diffusivity, viscosities, densities, and surface tension. The correlations may be used in lieu of a pilot plant study, albeit such a study would be preferred to supplement the foregoing calculations.

Figure 18.A.1 shows best fit curves of data obtained by Dvorak et al. (1996, p. 951), replotted without their data points shown. As seen in Figure 18.A.1a, at fixed $Q'A$, K_La varies directly with HLR; at some point, however, flooding of the media will occur and K_La will begin to decline. Figure 18.A.1b shows that at HLR = 31 m/h, K_La increases modestly with Q'/A but for HLR = 63 m/h, the increase is exponential.

7. *Protocol for application to determine column height*: To predict the column height, the Onda correlations provide empirical relations to predict k_L, k_g, and "**a**." The Onda correlations were found to give overpredictions of about 16% and underpredictions of about 34% (Dvorak et al., 1996). The protocol to estimate "h" then is to calculate K_La by Equation 18.87 from the computed k_L, k_g, and "**a**" values. The K_La value may then be plugged into Equation 18.85 to estimate Z (also designated "h" in some places) for specified C_{in} and C_{out} values, along with other design and operating values and looking up H (Henry's constant) for the gas in question.

GLOSSARY

Air stripping: The process of transferring contaminants from aqueous solution to air.

Air-to-water ratio: The ratio of air flux to water flux in a packed bed.

Arrhenius equation: Effect of temperature on kinetic constant, k, of a chemical reaction, that is, $k = Ae^\wedge (E_a/RT)$, where, k is the rate constant (s⁻¹); A is the constant (s⁻¹); E_a is the activation energy (J/mol); R is the gas constant (8.314 510 J/K/mol); T is the temperature (K). Proposed as an empirical equation by Svante Arrhenius in 1889 (Alberty and Silbey, 1992, p. 635). The constants, A and E_a may be evaluated by a plot of log k versus $1/T$; such plots describe the temperature effect of many unit processes. Integration of the equation gives the relation between rate constants for two respective temperatures, that is, $\ln(k_2/k_1) = (E_a/R)\cdot[(T_2 - T_1)/T_1T_2]$, which may be useful for reference.

BRV: Bubble-release vacuum; a test result which measures the negative pressure in cm of water required to form and release bubbles in a vacuum from a localized point on the surface of a wetted porous medium diffuser, for example, ceramic, porous disk, or for slotted orifices in membranes (Boyle et al., 1989, p. 21).

CERCLA: The federal legislation culminated in the form of the Comprehensive Environmental Response, Compensation, and Liability Act (Superfund) was enacted in 1980 by PL 96-510, which spawned an industry on cleanup of hazardous waste sites.

CFSTR: Continuous flow stirred tank reactor. Same as a "complete-mix" reactor.

Closed system: Mass transfer does not occur across system boundaries.

Coarse-bubble aeration: A diffused aeration system that creates gas bubbles, for example, air bubbles, by passage of a gas stream through orifices from a pressurized plenum or pipe. In general, the bubbles are 6–10 mm diameter (Boyle et al., 1989, p. 3).

Diffused aeration: Diffused aeration is the generation of gas bubbles emanating from pores or orifice openings submerged in a basin of water. In general, the bubbles are 2–5 mm diameter (Boyle et al., 1989, p. 3).

Fine-bubble aeration: A diffused aeration system that creates gas bubbles, for example, air bubbles, by passage of a gas stream through orifices from a pressurized plenum or pipe.

Fouling: Refers to diffuser airflow, in diffused aeration, being impaired by bacterial film, chemical precipitants, or air side clogging due to matter (particles, oil, etc.) covering or penetrating the pores (Boyle et al., 1989).

Gas transfer: A special case of mass transfer for gases.

Impeller: A rotor attached to a shaft and motor with shape designed to impart energy to water, for example, pressure in the case of a pump, and velocity in the case of placing the impeller in a basin open to the atmosphere.

K_La: Mass-transfer coefficient for uptake of a gas by an aqueous solution or stripping of gases from solution, used in the equation, $dC/dt = K_La(C^* - C)$ and defined as an aggregation of several variables, that is, $K_La = (D/\delta)\cdot[A(\text{interfacial surface area})/V(\text{basin})]$.

Lewis–Whitman two film theory: A model for transport of gases across the gas–liquid interface that attributes diffusion from the bulk of liquid solution across a pseudo liquid film to an interface and then across a pseudo gas film to the bulk of gas solution.

Mass transfer: A phrase that refers to the transport of molecules by advection, hydraulic dispersion, or diffusion.

Mechanical aeration: See *surface aerator*.

Onda correlations: A set of empirical equations that permit prediction of gas transfer coefficients and air–water interfacial area, with subsequent calculation of K_La. Once K_La is calculated, the height of a packed column may be calculated, given influent and effluent concentrations of VOCs (or any other kind of gas) from a materials balance relation.

Open system: Mass transfer occurs across system boundaries.

OTR: Oxygen transfer efficiency, a term from the report by Boyle et al. (1989).

Prony brake: Device that measures the "brake-power" of a rotating shaft by means of two brake pads applied to the shaft that create an equal and opposite torque. The brake power is the measured opposing torque times the rotational velocity of the shaft. The torque is the force measured on a lever arm times the length of the lever arm.

SOTE: Standard oxygen transfer efficiency, a term from the report by Boyle et al. (1989).

Surface aerator: An impeller located at the water surface, partially submerged, that causes mixing by large basin-size eddies with smaller eddies "cascading" from the large ones. The rate of oxygen uptake is proportional to the pumping rate by the rotating impeller, that is, by the rate of surface renewal and by rate of interface area created by the spray from the impeller. A surface aerator is known also as a "mechanical" aerator and sometimes as a "turbine" aerator.

Turbine aerator: A variety of impeller with vanes that impart a flow to the water being pumped. A radial-flow impeller has primarily radial vanes and imparts a primarily radial flow to the water.

van't Hoff relation: Effect of temperature on the equilibrium constant, K, for a chemical reaction, that is, $[\partial \ln K/\partial(1/T)]_P = -\Delta H^0/R$ (Alberty and Silbey, 1992, p. 159), where, K is the equilibrium constant (dimensionless); ΔH^0 is the standard state enthalpy of reaction (J/mol); R is the gas constant (8.314 510 J/K/mol); T is the temperature (K). Integration of the equation gives the relation between rate constants for two respective temperatures, that is, $\ln(K_2/K_1) = (\Delta H^0/R)\cdot[1/T_2 - 1/T_1]$, which may be useful for reference. The relation plots the same as the Arrhenius equation, that is, $\ln K$ versus $(1/T)$ is a straight line with slope, $\Delta H^0/R$. The intercept is $\Delta S^0/R$, where ΔS^0 is the standard state entropy of the reaction.

VOC: Volatile organic carbon.

REFERENCES

Alberty, R. A. and Silbey, R. J., *Physical Chemistry*, John Wiley & Sons, Inc., New York, 1992.

Baker, M. N., *The Quest for Pure Water*, The American Water Works Association, New York, 1948.

Berk, W. L., The oxidation ditch, *The American City*, September 1966.

Bewtra, J. K., Effect of diffuser arrangement of oxygen absorption in aeration tanks, Doctoral dissertation, University of Iowa, Iowa City, IA, 1962.

BISCO Environmental/Northeast Environmental Products, Inc., www.neepsystems.com, 2010; www.biscoenv.com, 2010.

Boyle, W. C. (Ed.), *ASCE Committee on Oxygen Transfer, Fine Pore (Fine Bubble) Aeration Systems*, Project Summary, EPA/625/8–85/010, U.S. Environmental Protection Agency, Cincinnati, OH, October 1985.

Boyle, W. C. et al. (Ed.), *Fine Pore Aeration Systems—Design Manual*, EPA/625/1–89/023, U.S. Environmental Protection Agency, Cincinnati, OH, September 1989.

Clunie, W. and Wu, Y., *Gas Transfer Coefficients for Bubble Diffusers*, CE 541 Unit Processes of Environmental Engineering Laboratory, Spring, Department of Civil Engineering, Colorado State University, Fort Collins, CL, 1995.

Conway, R. A. and Kumke, G. W., Field techniques for evaluating aerators, *Journal Sanitary Engineering Division*, ASCE, 92 (SA2):21, April 1966.

Cussler, E. L., Diffusion, in *Mass Transfer in Fluid Systems*, Cambridge University Press, Cambridge, U.K., 1984.

Danckwerts, P. V., Significance of liquid–film coefficients in gas absorption, *Industrial and Engineering Chemistry*, 43(6):1460–1467, June 1951.

Danckwerts, P. V., *Gas–Liquid Reactions*, McGraw-Hill Book Company, New York, 1970.

Dvorak, B. I., Lawler, D. F., Fair, J. R., and Handler, N. E., Evaluation of the Onda correlations for mass transfer with large random packings, *Environmental Science and Technology*, 30(3):945–853, March 1996.

Eckenfelder, W. W. Jr., Absorption of oxygen from air bubbles in water, *Journal Sanitary Engineering Division*, ASCE, 85:89–99, 1959.

Eckenfelder, W. W. Jr., *Water Quality for Practicing Engineers*, Barnes & Noble, Inc., New York, 1970.

Eckenfelder, W. W. Jr. and O'Connor, D. J., *Biological Waste Treatment*, Pergamon Press/Macmillan, New York, 1961.

Eckenfelder, W. W. Jr. and Ford, D. L., New concepts in oxygen transfer and aeration, in Gloyna, E. F. and Eckenfelder, W. W. Jr. (Eds.), *Advances in Water Quality Improvement*, University of Texas Press, Austin, TX, 1968.

Hale, F. E., Present status of aeration, *Journal American Water Works Association*, 24 (9):1401–1415, September 1932.

Higbie, R., The rate of adsorption of a pure gas into a still liquid during short periods of exposure, *Transactions American Institute of Chemical Engineers*, 31(1):365–380, 1935.

Hill, A. G. and Rice, R. G., Historical background, properties and applications, in Rice, R. G. and Netzer, A. (Eds.), *Handbook of Ozone Technology and Applications*, Vol. 1, Ann Arbor Science Publishers, Ann Arbor, MI, 1982.

Houck, D. H., *Aeration Survey and Evaluation of Fine Bubble Dome and Disc Diffuser Aeration Systems in North America*, Project Summary, EPA/600/S2–88/001, U.S. Environmental Protection Agency, Cincinnati, OH, September 1988.

Huang, J. Y. C., Selecting aerators for wastewater treatment, *Plant Engineering*, 29:6569, December 24, 1975.

Kalinske, A. A., Power consumption for oxygenation and mixing, in Eckenfelder, W. W. and McCabe, J. (Eds.), *Advances in Biological Waste Treatment*, Pergamon Press, New York, pp. 157–168, 1963.

Kalinske, A. A., Economics of aeration in waste treatment, in *Proceedings of the 23rd Purdue Industrial Waste Conference*, West Lafayette, IN, pp. 338–397, 1968.

Kalinske, A. A., Economic evaluation of aerator systems, *Environmental Science and Technology*, 3(3):229–234, March 1969a.

Kalinske, A. A., Field testing of aerators in waste treatment plants, in *ASME—AIChE Joint Conference on Stream Pollution and Abatement*, New Brunswick, New Jersey, June 1969b. ASME Reprint 69-PID-7.

Lee, S. Y. and Tsui, Y. P., Succeed at gas/liquid contacting, *Chemical Engineering Progress*, 95(7):23–49, July 1999.

Lewis, W. K. and Whitman, W. G., Principles of gas absorption, *Absorption Symposium, Industrial and Engineering Chemistry*, 16(12):1215–1239, December 1, 1924.

McKeown, J. J. and Okun, D. A., Absorption of oxygen from air bubbles in water, *Journal Sanitary Engineering Division*, ASCE, 86:37–42, 1960.

Mueller, J. A. and Boyle, W. C., Oxygen transfer under process conditions, *Journal Water Pollution Control Federation*, 60(3):332–341, March 1988.

Morgan, P. F. and Bewtra, J. K., Oxygen efficiency in full–scale controlled aeration tank, mm paper, presented at the *15th Annual Meeting of the Federation of Sewage and Industrial Wastes Associations*, Dallas, TX, October 13, 1959.

Morgan, P. F. and Bewtra, J. K., Air diffuser efficiencies, *Journal Water Pollution Control Federation*, 32(10), 1047, 1960.

Morgan, P. F. and Bewtra, J. K., Diffused air oxygen transfer efficiencies, in Eckenfelder, W. W. and McCabe, J. (Eds.), *Advances in Biological Waste Treatment*, Pergamon Press, New York, pp. 181–193, 1963.

Powell-Groves, K., Daigger, G. T., Simpkin, T. J., Redmon, D. T., and Ewing, L., Evaluation of oxygen transfer efficiency and alpha–factor on a variety of diffused aeration systems, *Water Environment Research*, 64(5):691–698, July/August 1992.

Rawe, J., Air stripping of aqueous solutions, *Engineering Bulletin*, U.S. Environmental EPA/540/2–91/022, Protection Agency, Cincinnati, OH, October 1991.

Sherwood, T. K., Pigford, R. L., and Wilke, C. R., *Mass Transfer*, McGraw-Hill, New York, 1975.

Staudinger, J., Knocke, W. R., and Randall, C. W., Evaluating the Onda mass transfer correlation for the design of packed–column air stripping, *Journal of the American Water Works Association*, 82(1):73–79, January 1990.

Yunt, F. W. and Hancuff, T. O., *Aeration Equipment Evaluation: Phase I—Clean Water Test Results*, Project Summary, EPA/600/S2–88/022, U.S. Environmental Protection Agency, Cincinnati, OH, May 1988.

19 Disinfection

The term, *disinfection*, refers to the inactivation of microorganisms. By inference, the inactivation is not 100%, but could be 1-log, 2-log, 3-log, or even 6-logs depending on the initial concentration of organisms, detection limit, and other factors. The idea is explained further in Box 19.1 and in the glossary.

This chapter reviews the fundamentals of disinfection, the characteristics of each major disinfectant, and the highlights of disinfection practice. Each of the disinfectants has its own body of literature and adherents.

19.1 FUNDAMENTALS

Basic notions of microorganisms, diseases, and disinfectants are reviewed here. As seen, the idea of disinfection of water has its roots in the science of microbiology, founded about 1880.

19.1.1 Microorganisms and Diseases

Table 19.1 lists three groups of microorganisms, that is, viruses, bacteria, and protozoa that are causes of waterborne diseases. Most disease-causing organisms occur in ambient waters through fecal contamination, for example, from runoff of fecal matter, from land, or from sewage discharges. Cross-connections are a possible source of contamination of treated water.

As another variation of microbial contamination, some bacteria, under conditions of stress, may form "endospores" (spores) that may be able to resist heat, chemicals, radiation, etc., and may survive for years, decades, or even centuries in some cases. The endospores of *Bacillus anthracis* may remain viable for years (Prescott et al., 1993, p. 552). Similarly, some protozoa, for example, aquatic, freeliving, or parasitic forms, may "encyst," which gives protection against environmental stresses (nutrient deficiency, desiccation, adverse pH, low oxygen partial pressure), and provide as a means to transfer between hosts. Excystation is triggered by a return to favorable environmental conditions, for example, ingestion by a host (Prescott et al., 1993, p. 550).

19.1.2 Disinfectants

An array of disinfectant chemicals are available and include chlorine, ozone, chlorine dioxide, iodine, bromine, and UV254 radiation. Others that have been mentioned (Chick, 1908) include mercuric chloride, silver nitrate, phenol, peroxide, etc. Also included are proprietary disinfectants, for example, MIOX®, which generates disinfectants onsite using sodium chloride and water as raw material reactants, to form Cl_2, HOCl, ClO_2, and O_3 as products at the anode of a cell that takes up the electrons produced (MIOX, 2003). The disinfectants that produce a residual include chlorine, chlorine dioxide, and chloramines. Ozone and UV do not hold residuals.

19.2 HISTORY

The practice of disinfection started about 1900, some 30 years after the discovery of microorganisms. This was almost 50 years after Dr. John Snow and coworkers, who in 1854 linked cholera and sewage contamination to the Broad Street Well in London, establishing, at the same time, a methodology for epidemiology investigations. In 1886, Koch conducted the first experiments on disinfection by using pure cultures of bacteria. His work was reviewed by Harriette Chick in her 1908 paper on the law of disinfection, that is, that $dN/dt = -kt$, in which k was found to vary with the disinfectant and with organism, which was the beginning of modern theory.

19.2.1 Chlorine

Chlorine was discovered in 1774 by a Swedish chemist, Scheele (Tiernan, 1948, p. 1042). The Greek word for green, "chloros" was adopted as being descriptive of the color of the gas. The first continuous application of chlorine was in 1908 as sodium hypochlorite, NaOCl at Jersey City, New Jersey (Doull, 1980, p. 18). Its first use for wastewater disinfection was in Hamburg in relation to the 1893 cholera epidemic (Isaac, 1996, p. 68). By 1910, of the 619 POTW's (POTW is an acronym from the 1972 Clean Water Act, PL92-500, meaning "publically operated treatment works"), 22% or 3.6% used chlorine; the percentage grew to 94.9% of 15,000 POTW's by 1979, but declined to 78% in 1990, with 19% using UV (Isaac, 1996, p. 68).

19.2.1.1 Story of Chlorine

As described by Tiernan (1948, p. 1042), the story of chlorine begins with the Electro Bleaching Gas Co., which was founded by E.D. Kingsley (who, just prior, in 1906, was superintendent of a department store). Kingsley had invested $1000 in a chlorine gas–conversion enterprise in Philadelphia; when the promoter disappeared, he took over the company and went to Germany to learn about the method of producing chlorine gas. He returned with a German chemical engineer, imported chlorine-compressing equipment from Germany, and set up the Electro Bleaching Gas Co. next to the Niagara Alkali Co. at

BOX 19.1 DISINFECTION DEFINITIONS

The term "disinfection" is used here to mean the "inactivation" of microorganisms. Two kinds of definitions, mathematical and mechanistic, are given in order to clarify their repeated use in this chapter.

Mathematical: A mathematical expression of inactivation is "log removal," that is, log R, defined, log $R = \log(C)/C_o)$, which is the same as used in filtration. Alternatively, inactivation may be expressed as "percent removal" or "percent inactivation" or "fraction inactivation," the latter being defined, $fR = (C_o - C)/C_o$, in which, C_o = initial concentration of microorganism s at time $t = 0$ (#/mL), and C = concentration of microorganisms at any time $t > 0$ (#/mL). If, for example, $C_o = 10{,}000$ /mL and $C = 1$ #/mL, log $R = \log [1/10{,}000] = \log(10^{-4}) = -4$, by the same token, fraction inactivation = (10,000 − 1)/10,000 = 0.9999 (99.99%). Both log R and fraction (or percent) removal, that is, inactivation, are used commonly.

Inactivation: The term "inactivation" means here that the microorganism is rendered not capable of reproduction, that is, its DNA is damaged or other parts of the cell are damaged, to the extent that it cannot replicate itself (Malley, 2002a, p. 12). Another definition is that an inactivated microorganism cannot metabolize, that is, process nutrients; which may be due to damaged enzymes (which goes back to Fair et al., 1948, p. 1056) and may be associated with disinfectant penetration of the cell wall.

Niagara Falls, New York. Chlorine was obtained from the Niagara Alkali Co., which was making chloride-of-lime.

Tiernan continued his account of how chlorine gas technology developed to give a colorful story describing also the start of their company, Wallace and Tiernan. The following is an abstract of his 1948 article. The story describes how they developed, by trial and error and other means, the technology for chlorine metering, control, dissolution, and concentrate feed.

In 1909, Martin F. Tiernan and C.F. Wallace began their association. Wallace was an electrical engineer and worked with the Gerard Ozone Process Co. where Martin F. Tiernan had been hired as a chemist. The ozone business folded and in 1911, they were also let go from their next job; with $1800 in savings they started their own business and in September 1911 they were involved in installing hypochlorite for Torrington, Connecticut to respond to a typhoid outbreak. After working with other hypochlorite delivery systems, Tiernan visited (in 1912) Dr. Carl Darnall, MD, U.S. Army Medical Corps, who suggested chlorine gas *in lieu* of hypochlorite for the Jersey City Water Company. After visiting the Electrochemical Bleaching Gas Co. to see if chlorine gas was available, he also visited with George Ornstein, a German chemist, who showed a method of making a chlorine solution from chlorine gas that was then fed through an orifice to the main flow. They confirmed that the bacterial reductions were comparable to that achieved by hypochlorite. After this investigation, Wallace and Tiernan proposed to install a chlorine feed system at Jersey City for $150 with performance guarantees (he mentioned that this was without having designed the apparatus prior to the proposal and ignorant of the properties of compressed chlorine gas). This first apparatus used hard rubber parts and solder for joints of metal tubing. When subject to chlorine under pressure the apparatus literally blew up in our faces. And, after many tries, we produced a controlling mechanism,... Further, to effect the solution of the gas in the stream, we put an inverted trough across the stream bottom, weighing it down with stones, and introduced the gas at one end of the trough. During inspection by the New Jersey State Board of Health, when the gas was turned on a considerable part came to the surface, but a good strong breeze diluted it so Randolph (the official) got just a very slight odor. During the night, the apparatus had sprung a leak and all the tools in the shop were coated with rust, and they had promptly thrown the device out the window. We redesigned the apparatus and made diffusers out of small alundum grinding wheels, cemented into the saucer of a flower pot. Silver tubing was later used and the installation was completed on February 22, 1913. The meter was a volumetric inverted siphon type, which we are still using.

The second installation was at Ford's Pond where they raised the price to $200. For the Stamford Water Works, which was considering hypochlorite, the consultant wanted to feed the disinfectant at a rate proportional to the main flow. Wallace and Tiernan offered to install a Venturi-operated automatic chlorinator for $500 before we had even made the design. As Tiernan had stated, This surely was the mark of our confidence and ignorance. It was only a few days, however, until Wallace had figured out a way of handling the problem. Three balanced diaphragms, functioning as two, were operated by the differential pressure across the Venturi throat in the water main, to maintain a drop in pressure across the gas control valve in proportion to the drop across the Venturi throat. This would give a proportional flow of chlorine to the water through a variable orifice (the control valve). A simple, low voltage dc toy motor that was operated by six dry cells and costing some six or eight dollars was cut in and out by contact operated by connections to the diaphragms. This motor operated the control valve in the gas line until the drop in pressure across the valve was proportioned to that across the Venturi throat. The installation was made at Stamford on September 3, 1913, and functioned perfectly from the very beginning. Chlorine was introduced into the intake well through a silver tube and a diffuser submerged to a depth of 25 ft (p. 1048). Later that year we installed automatics at New Haven, Torrington and Hartford, Connecticut. The automatic feature of these machines was modified to eliminate the electric motor. In June 1913 Herman Rosentretter of the Newark, New Jersey Water Dept., hearing of our installations for Jersey City, gave us an order for a direct-feed manual-control machine for the entire Newark supply. The installation was made at the Macopin intake, using diffusers which

TABLE 19.1
Waterborne Microorganisms and Associated Diseases[a]

Microorganism: Phylum/Class/Group	Microorganism: Genus/Species	Disease
Viruses		
	Adenovirus	Conjunctivitis
	Astrovirus	Gastroenteritis
	Calicivirus	Gastroenteritis
	Echovirus	Enteritis, meningitis
	Enteric adenovirus	Gastroenteritis
	Hepatitis A	Infectious hepatitis
	Norwalk virus	Gastroenteritis
	Poliovirus	Polio
	Rotavirus	Gastroenteritis
Bacteria		
	Escherichia coli—1885	Diarrhea
	Campylobacter jejuni	Diarrhea
	Salmonella serovars[b,c]	Salmonellosis
	Salmonella typhi—1880	Typhoid
	Shigella dysenteriae[b]—1898	Shigellosis, that is, dysentery
	Shigella paratyphi[b]	Paratyphoid
	Vibrio cholerae—1883	Cholera
Protozoa		
Apicomplexa (phylum)		
Coccidia (group)	*Cryptosporidium parvum*	Cryptosporidiosis
Cyclospora cayetanensis[d]		
Mastigophora (phylum)		
Ciliates		
Zoomastigophorea (class)	*Giardia lamblia*	Giardiasis
Sarcodina (subphylum)		
Rhizopoda (superclass)		
Amoebae (group)	*Entamoeba*	Amebiasis
	Entamoeba histolytica	Amebiasis, amebic dysentery

[a] *Source:* Prescott et al. 1993: viruses pp. 735–737; bacteria, p. 764; protozoa, pp. 552–558.
[b] White (1999, p. 343).
[c] *Salmonella* gastroenteritis is caused by over 2,000 *Salmonella* serovars (a serovar is a strain); the one reported most frequently is *Salmonella* serovar *typhimurium* (Prescott et al. 1993, p. 766).
[d] Wright (2000, p. 28).

fed into an open well. Then, In July 1913 we installed a direct-feed machine at the pumping station of the Bernardsville, New Jersey Water Co. The apparatus fed directly into the intake line of a pump. I remember quite clearly going by train to Bernardsville, carrying the machine under my arm, than taking a bus to the south, down toward the pumping station, and walking at least two miles with the machine on my shoulder. Wallace and I spent the July 4 holiday cutting a tap into the suction line in the engine room. It was a pretty hot job. We hung the apparatus high up on the wall, above the hydraulic gradient, figuring that the water would not get back into the machine. Of course, as soon as the apparatus was turned off, when the pump was shut down, the chlorine in the line from the apparatus was quickly absorbed by the water, and before morning, the machine was flooded. After the development of a satisfactory check valve, the installation was completed.

Another type of apparatus was a solution-feed machine. For Trenton, New Jersey, they developed an enclosed glass jar arrangement with an impinging jet, which for its size, had a very large dissolving capacity. Being a closed system, the danger of gas escaping was eliminated.

On April 4, 1916, we installed nine of our direct feed units to treat the water of the new and old Croton, N.Y., aqueducts, with a capacity of 340 mgd. This installation functioned very well until the water became cold, when chlorine hydrate forming in the diffusers caused difficulty. Later on, this installation was changed to the solution-feed type... By 1916, Wallace and Tiernan were well established in the chlorination field.

(a)

(b)

FIGURE 19.1 Chlorination control and metering system, c. 1930s; salvaged from Fort Collins Water Treatment Plant #1, decommissioned 1986. (a) Overall system showing tubing at bottom. (b) Closeup showing rotometer, control valve knob on right, pressure gages at top. (Courtesy of Kevin Gertig, Fort Collins, CO.)

Figure 19.1 shows a Wallace and Tiernan chlorinator system as salvaged from the Fort Collins Water Treatment Plant #1(decommissioned in 1986). The system is believed to have been a 1930s vintage system (see also Anon., 1968). The basic elements are the same as that found in a modern system, for example, rotometer for chlorine gas-flow measurement, control valve, pressure gages, etc.

19.2.1.2 Disinfection Byproducts Issue

While chlorine has several problems (mostly associated with safety and environmental effects) the one that emerged first as the most visible was the formation of tri-halomethanes, suspected to be a carcinogen (discovered in 1973 in New Orleans). The ensuing standard for THMs was set at 100 μg/L (per regulations promulgated by the USEPA in 1978). In the 1980s, however, with the prospect of more stringent standards in the future and of the inclusion of a broad array of "disinfection by-products" (DBPs), some water utilities were prompted to consider strategies to reduce the DBP's in their drinking water. These strategies were, in general, to (1) remove DBP precursors, that is, NOM (see Appendix A), (2) to use an alternative disinfectant that did not form DBP's, and (3) to remove the DBP's, once formed. The first and second strategies have been the prevailing direction of practice, that is, "enhanced coagulation" (see Chapter 9), and alternative disinfectants, respectively. The alternative disinfectants have included chlorine dioxide and ultraviolet radiation.

19.2.2 Ozone

Ozone was discovered in 1840 by a German chemist, Christian Schönbein, after observing a peculiar odor during electrolysis-and-sparking experiments. He concluded that the odor was due to a new substance that he named "ozone" from the Greek, "ozein," to smell. Not long after, that is, by 1848, ozone was found to be produced by subjecting pure, dry oxygen to an electric spark and that it was tri-atomic oxygen (Hill and Rice, 1982, p. 5). In 1857, Werner von Siemens developed ozone foil-coated generating tubes, which were two annular glass tubes, with dry oxygen being passed between. The disinfection capability of ozone was recognized in 1886 (Langlais, 1991, p. 2) and in 1897, an ozone water disinfection process was patented that produced about 10 g ozone/h, which was compressed and fed to the base of a tall bubble column equipped with perforated plates placed at intervals. The first full scale plant application was in 1893 at Oudshoorn, the Netherlands, followed by a plant in Paris in 1898 (Langlais, 1991, p. 2), then one at Paderborn, Germany in 1902, $Q = 60$ m^3/h (0.38 mgd), and in 1903 at Wiesbaden, $Q = 250$ m^3/h (1.6 mgd); the latter two plants were designed by the German firm, Siemens and Halske (Hill and Rice, 1982). A plate-type ozone generator was tested in 1898, which was the basis for a full-scale plant at Nice, France completed in 1906, $Q = 13$ m^3/min (4.9 mgd), named Bon Voyage, which operated until 1970, when it was replaced by the Super Rimiez Plant (Hill and Rice, 1982, p. 11). By 1906, 49 plants used ozone, with 26 being in France, and by 1977, 1036 plants had been identified, with about half in France and 150 in Switzerland (Hill and Rice, 1982).

In 1933, the Siemens Company introduced dual silica gel dryers, with one being thermally regenerated while the other was in operation. The company also introduced high frequency, that is, 10,000 Hz, into ozone generators of the type having tubular design with two concentric dielectrics. The model could generate 8 mg ozone/L at an energy efficiency of 35 g O_3/kWh. Also in 1933, the St. Maur WTP in Paris was expanded to treat $Q = 208$ m^3/min of water. Because of difficulties with a new refrigerated drying system, and the war, the plant did not start operation until 1953. The average ozone dose was $C(O_3) = 1.1$ mg/L. Energy consumption was 39 W h/g dissolved ozone, which included 12 W h for compressors, and 6–8 W h for refrigeration. The plant included 16 contact columns with a cross-sectional area of 14 m^2 and an effective contact depth = 6.3 m. Both water and ozone movement was in an upward direction. At the design flow 12 contactors were to be in operation at Q(contactor) = 17.4 m^3/contactor, with 4 contactors in reserve. Two types of ozone generators, each used for half the flow, were used: (1) plate generators operated at 18,000–20,000 V, producing 1,600 g ozone/h while consuming 35 kW power giving an ozone production rate of 45.7 g O_3/kWh; (2) tubular generators operated at 10,000 V, producing 1,600 g ozone/h while consuming 25 kW power giving an ozone production rate of 64 g O_3/kWh, with $C(O_3) = 2$–3 mg O_3/L. The St. Maur WTP produced about one-third of the Paris potable water supply. Absorption of ozone was 0.6–0.8 fraction of ozone applied. For the counter-current operation, the absorption efficiency was about 0.9–0.95 fraction (Hill and Rice, 1982).

The four largest water treatment plants at Paris are listed in Table 19.2 and have a combined ozone generating capacity of 8.7 metric tons/day. The plants and the date of being put online are listed along with the flows for each plant and the ozone concentrations used. At each plant, ozone is added: (1) to the raw water to give $C(O_3) < 1$ mg/L before storage for 2–3 days; (2) then again as before coagulant chemicals are added; (3) after sand filtration to give $C(O_3) = 0.4$ mg/L and $t = 10$ min. The treatment included GAC adsorption, followed by chlorination and dechlorination. The $C(O_3) = 0.4$ mg and $t = 10$ min, that is, $Ct = 4$, was predicated on inactivation of polio types 1, 2, 3 at log $R = 3$ (i.e., 99.9% reduction). The intent of the multistage ozonation was to react with organics in the first stage, the purpose of the second stage was to aid in coagulation, and the third stage of ozonation was for primary disinfection and further oxidization of the residual organics (Hill and Rice, 1982).

TABLE 19.2
Ozone Generation at Four Largest Plants for Paris, c. 1972

Plant	Date	Q (m^3/min)	$C(O_3)$ (mg/L)
Méry-sur-Oise	1965	208	3.6
Orly	1966	208	4.0
Choisy-le-Roi	1968	626	3.0
Neuilly-sur-Marne	1972	415	4.8

Source: Adapted from Hill, A.G. and Rice, R.G., Historical background, properties and applications, Chapter 1 in Rice, R.G. and Netzer, A. (Eds.), *Handbook of Ozone Technology and Applications*, Volume 1, Ann Arbor Science Publishers, Ann Arbor, MI, 1982, p. 17.

As an update to Table 19.2, 12 water treatment plants serving water to Paris and environs were stated by Langlais et al. (1991, p. 6) to have an ozone generating capacity of >500 kg O_3/h (12 metric tons/day) using air as the feed gas for treating Q(12 plants) = 3,000,000 m^3/day water for >10,000,000 people. Among the facilities was the Choisy-le-Roi, the second largest ozone plant worldwide with an output of 160 kg O_3/h. The Neuilly-sur-Marne WTP with a generating capacity of 140 kg O_3/h had generators with outputs of 30 kg O_3/h/generator, and were the largest ozone generators. As of 1990, over 700 water treatment plants in France used ozone. To provide a more specialized forum for the exchange of knowledge on ozone, the International Ozone Institute, later known as the International Ozone Association, was formed in 1973 at the *First International Symposium on Ozone for Water and Wastewater Treatment* (Loeb, 2002).

19.2.3 Chlorine Dioxide

Although Sir Humphrey Davy (Masschelein, 1992, p. 170) discovered chlorine dioxide, ClO_2, in 1811, its first use in water treatment was in 1944 to control phenolic tastes and odors at the Niagara Falls WTP. A 1956 survey indicated that of the 56 plants using ClO_2, most uses were for taste and odor control; the other uses were for algal control, 7 plants; iron and manganese removal, 3 plants; and disinfection, 15 plants. In 1977, about 100 plants were using chlorine dioxide, but mostly for taste and odor control. By 1986, however, 300–400 plants were using it for disinfection (Anon., 1986, p. 33), and by 1997, about 500 plants in the United States were using ClO_2. In Europe, several thousand utilities used ClO_2, mostly to maintain a disinfectant residual in the distribution system (Anon., 2001, pp. 1–4). Attributes of ClO_2 included (Aieta and Berg, 1986, p. 62): its effectiveness being comparable to HOCl; effective for $6.0 < pH < 8$; oxidizes NOM but without the formation of disinfection by-products; does not react with ammonia; maintains residual in the distribution system, it is generated onsite.

19.2.4 Ultraviolet Radiation

The source of UV radiation is mercury vaporization; this phenomenon was discovered in 1835 and in 1901 the mercury vapor lamp was developed (Snicer et al., 2000, p. 6). The first application of UV for disinfection of drinking water was in Marseilles, France in 1910; the system was not reliable due to its complicated technology and was not used since ozone came on the scene in Europe, along with chlorine in the United States. The emergence of DBP's in the 1970s stimulated a more concerted search for disinfectants other than chlorine.

In the 1980s, Professor Günther Schenck of the Max Planck Institute for Radiation Chemistry and Professor Heinz Bernhardt found that radiant exposures $\geq$400 J/m^2 could inactivate viruses and bacteria (Hoyer, 2000a, p. 22). At the same time, that is, in the 1980s, regulatory factors started to influence the development rate of the UV technology. First was the Innovative and Alternative Technologies program of the USEPA, which provided for replacement if an experimental technology failed to achieve compliance with a National Pollution Discharge Elimination System (NPDES) permit. Then, in 1988, the Uniform Fire Code began to impact facilities undergoing expansion by requiring chlorine scrubbers as protection against the accidental release of chlorine gas. The foregoing helped to expand interest in UV, particularly for the disinfection of municipal wastewaters. In 1984, only a few installations existed, but by 1999, the number exceeded 500. The technology developed during the 1990s had horizontal as well as vertical alignment of lamps and were equipped with self-cleaning mechanisms. Pulsed systems were developed also, giving radiation of high flux density. The foregoing is from Hunter (2000, p. 5).

In the United States, DeMers and Renner in 1992 determined that there were over 2000 drinking water installations in Europe using UV and by 1994 Montgomery–Watson Engineers determined that about 700 wastewater plants in the United States had UV (Snicer et al., 2000, p. 7). Masschelein (2002, p. 59) estimated the number of drinking water systems using UV disinfection at 3000–5000.

In 1998, UV was found to be effective in inactivating *Cryptosporidium*, which gave a final impetus to its widespread adoption, especially with EPA regulations pending (Malley, 2000, p. 9). The only alternatives for *Cryptosporidium* inactivation were ozone (with long detention time) or microfiltration (MF). The mechanism of UV inactivation (see also Section 19.3.9.4) was found to be damaging to the nucleic acids by absorption in the 200–300 nm range (Snicer et al., 2000, p. 32).

19.2.5 Other Disinfectants

A variety of disinfectants have been used since the early days of the science of microbiology, which was underway by about 1880 as a science. A few that have emerged that are important in water treatment are reviewed here.

19.2.5.1 Iodine

Iodine is the only common halogen that is solid at room temperature (Doull, 1980, p. 62). Free iodine, I_2, is considered an effective disinfectant, and a residual of 0.5–1.0 mg/L can be maintained in a distribution system, and it does not react with ammonia. The basic chemical reactions with water are

$$I_2 + H_2O \Leftrightarrow HOI + H^+ + I^- \tag{19.1}$$

$$HOI \Leftrightarrow H^+ + OI^- \tag{19.2}$$

The most effective form is I_2, which is predominant at lower pH values. For pH = 5, 0.99 fraction of the species is I_2 with about 0.01 fraction being HOI; at pH = 7, 0.5 fraction is I_2 and about 0.5 fraction is HOI; at pH = 8, 0.12 fraction is I_2 and about 0.88 fraction is HOI.

Iodine concentrations at 5–10 mg/L were effective against bacterial pathogens but were not effective as a cysticide at high pH; consequently, tablets (called "globaline") developed for the military included an acid buffer to lower the pH of the water and released 8 mg/L iodine. Some of the Ct values reported for 0.99 fraction inactivation at pH = 7.0, $23°C \leq T \leq 30°C$, were: Ct(coliforms) = 0.4; Ct(poliovirus 1) = 30 (Doull, 1980, p. 68). For cultured *E. histolytica* cysts and for pH = 5, Ct = 200, 130, and 65 at 3°C, 13°C, and 23°C, respectively.

19.2.5.2 Bromine

Bromine is another halide effective in inactivating microorganisms. Its chemistry in reacting with water is similar to chlorine and iodine, that is,

$$Br_2 + H_2O \Leftrightarrow HOBr + H^+ + Br^- \tag{19.3}$$

$$HOBr \Leftrightarrow H^+ + OBr^-. \tag{19.4}$$

Bromine may also be added to water as bromine chloride gas, that is, BrCl, which reacts with water to give HOBr as a product (along with H^+ and OBr^-). Bromine reacts with ammonia analogous to the reactions of chlorine with ammonia. The HOBr form is most effective as a germicide.

Bromine was first applied to water in the form of a liquid (as cited in a 1935 article by Doull, 1980, p. 72); it condenses as a liquid at $p = 1.00$ atm, $T = 5°C$. Its application is usually as $BrCl_2$ (due to the corrosive nature of Br_2) that is shipped and handled in a manner similar to chlorine, but is more reactive than chlorine with PVC plastics. A more usable form for small-scale application is potassium bromide, KBr.

Although bromine is considered to be an effective disinfectant, that is, similar to chlorine, data were sparse in the review by Doull (1980, pp. 72–82). For 2-log inactivation of *Bacillus subtilis* spores, for example, $Ct = 280$ (pH = 7, $T = 25°C$). But for 2-log inactivation of poliovirus 1, $Ct = 0.06$ (pH = 7.0, $T = 20°C$). Bromine as HOBr, remains the predominant form at pH < 8.5 (as contrasted with pH < 7.5 for chlorine).

19.2.5.3 Silver

Silver has been applied as a disinfectant in both metallic form and as silver nitrate. Usually dosages are <50 μg/L because of low solubility (for reference, the USEPA 1975 MCL = 0.050 mg/L. Values for Ct for 0.999 inactivation of *E. coli* were at $6.3 \leq pH \leq 8.7$, $5 \leq T \leq 25$, $0.01 \leq C$(Ag as $AgNO_3$) ≤ 0.27 mg/L, then $4.6 < Ct < 14$. The city of Zurich WTP has produced packaged water (1 L plastic bags) with silver used (1.0 mg Ag^+/L) to minimize possible growth during storage for use in emergencies (1982 visit).

19.3 THEORY

The main contribution of theory to practice is the Ct parameter, established by Chick (1908) and Watson (1908). Their combined theory evolved as the basis for the design of reactors and for the operation and development of extensive Ct databases.

19.3.1 Inactivation

The inactivation of microorganisms may be due to any one of several mechanisms. The classical explanation for chlorine was that HOCl reacts with the enzyme system of the cell through its ability to penetrate the cell wall because of its small size and its electroneutrality (Fair et al., 1948, p. 1056). Since the 1990s, opinion has converged toward the DNA being damaged, rendering the organism not capable of reproduction.

19.3.1.1 Factors

The factors that affect organism inactivation include time of contact, t; concentration of organisms, N; concentration of disinfectant, C; temperature, T; the disinfectant, for example, chlorine, ozone, etc.; and the specific organism, for example, *E. coli*, *Cryptosporidium parvum* oocysts, polio virus, MS-2 virus, etc. (Fair et al., 1948, p. 1057). In addition, mixing is essential in order to ensure that the reactants come into contact with the disinfectant (Doull, 1980, p. 11). Mixing is required also if radiant energy, for example, ultraviolet, is the basis for the inactivation so that all organisms suspended in the water are exposed to the radiation for the minimum dose, that is, the product of radiation intensity and time.

19.3.1.2 Mathematics

The rate of inactivation of microorganisms is expressed as a first order reaction, called Chick's law (reviewed by Smith et al., 1995, p. 204; White, 1999, p. 432; and presented originally by Chick, 1908), that is,

$$\frac{dN}{dt} = -kN \tag{19.5}$$

Integrating gives

$$\ln\frac{N}{N_0} = -kt \tag{19.6}$$

which, in exponential form, is

$$\frac{N}{N_0} = e^{-kt} \tag{19.7}$$

or alternatively,

$$\frac{N}{N_0} = 10^{-\frac{k}{2.3}t} \tag{19.8}$$

where

N is the concentration of organisms at any time after $t-0$ (org/mL)
N_0 is the concentration of organisms at time $t=0$ (org/mL)
k is the rate constant (min^{-1})
t is the time from the beginning of measurements (min)

As a semi-log plot, Equation 19.8 is a straight line. The data may deviate from a straight-line plot, however, due to microorganism aggregation, protection by extraneous matter, or changing disinfection concentration (Langlais, 1991, p. 219). These factors led to the use of an empirical relation by Watson (1908). Herbert Edmeston Watson in 1908, using Harriett Chick's data, refined her equation to produce an empirical relation that included changes in disinfectant concentration (Smith et al., 1995, p. 204),

$$\ln\frac{N}{N_0} = rC^n t \tag{19.9}$$

where

C is the concentration of disinfectant $[(\text{mg/L})^{1/n}]$
r is the coefficient of specific lethality ($\text{L/mg} \cdot \text{min}$)
n is the coefficient of dilution ($\text{L/mg} \cdot \text{min}$)

Moving the term, r, to the left side gives,

$$\frac{1}{r}\ln\frac{N}{N_0} = C^n t = K \tag{19.10}$$

For a given level of survival fraction, such as $N/N_0 = 0.01$ (2-log reduction that is 99%), the left side of Equation 19.10 is a constant, K, or,

$$K = C^n t \tag{19.11}$$

As pointed out by Fair et al. (1948, p. 1059), no theoretical significance can be attached to the value of n; the equation is without theoretical foundation and is just an expression for the correlation of data. The exponent, n is called a concentration exponent by Fair et al. (1948, p. 1059); more frequently it is called a coefficient of dilution. High values of n indicate that the disinfectant effectiveness decreases rapidly as the concentration is decreased. With low values of n, the time of contact becomes more important than the dosage. The values of n are determined by plotting log C versus log t and measuring the slope of the line.

The value of K depends upon the level of inactivation. If $n=1$, then Equation 19.7 becomes,

$$K = Ct \tag{19.12}$$

and a given level of inactivation is characterized by a specific Ct. To illustrate, consider the 0.99 fraction inactivation of coliforms by chlorine, which is characterized by $Ct = 0.1$, that is,

C (mg/L)	t (min)	Ct	Log R
1.0	0.1	0.1	2.0
0.1	1.0	0.1	2.0

As seen, for different levels of C a different t is required such that $Ct = 0.1$. As another illustration, consider chlorine disinfection of coxsackievirus by chlorine; for $\log R = 2$,

Ct(coxsackievirus) = 10, for example, if *C* = 5 mg chlorine/L, then *t* = 2 min.

For a given log *R*, *K* is a function of the organism, the disinfectant, pH, and temperature. Thus, *K* is different for different microorganisms, for example, *E. coli*, coxsackievirus, *Cryptosporidium parvum* oocysts, etc. *K* varies also with the disinfectant, for example, chlorine, ozone, chloramines, chlorine dioxide, ultraviolet radiation, etc. Also, *K* may vary with pH if the dissociation species is affected. Temperature influences kinetics in accordance with the Arrhenius equation.

19.3.1.3 *Ct*'s Compiled

Table 19.3 is a compilation of empirical *Ct* data for various organisms, for example, bacteria, viruses, and cysts, under conditions stated, for example, pH, temperature, and disinfectant. As seen, the *Ct*'s have a wide range depending on the organism and conditions. The *Ct* values are useful for design. The disinfectant concentration is limited, as a rule, based on practice or regulations. As an example, from Table 19.3, if *Ct* (*G. lamblia*, chlorine, pH = 7.0, *T* = 1°C) = 289, and if, *C*(chlorine) = 2 mg/L, then, $t \approx 150$ min (2.5 h).

19.3.1.4 *Ct*(chlorine) for *Giardia lamblia* Cysts

Equation 19.13 calculates *Ct*(*Giardia*) for chlorine as a function of conditions, that is, *C*(chlorine), pH, and *T*, and for log *R* = 4. The equation was obtained from regression analysis by Clark et al. (1989) of 167 data points from Hibler et al. (1987) for use in the EPA draft "Surface Water Treatment Rule." The Hibler et al. data were based on gerbil infectivity for a dose of 50,000 cysts per gerbil; experimental conditions were $0.4 \leq C(\text{chlorine}) \leq 4.2$ mg/L; $7.0 \leq \text{pH} \leq 9.0$; $0.5°\text{C} \leq T \leq 5°\text{C}$; the regression coefficient for the fit of Equation 19.13 to the data was, $r^2 = 0.80$. The *Ct*'s for other log *R* values may be calculated, for example, *Ct*(1 − log) = *Ct*(4 − log)/4. Similar regression analyses were done for *Cryptosporidium* and ozone (Clark et al., 2002) and *Cryptosporidium* and chlorine dioxide (Clark et al., 2003). Table CD19.4 (excerpt in text) calculates *Giardia Ct* values from Equation 19.13 for various log *R* values.

$$Ct(\text{4-log}) = 0.985C^{0.176}\text{pH}^{2.752}T^{0.147} \quad r^2 = 0.80 \quad (19.13)$$

where

Ct(4-log) is the Value of *Ct* for a 4-log inactivation of *Giardia lamblia* cysts by chlorine (dimensionless)
C is the concentration of chlorine (mg/L)
t is the time of contact (min)
pH is the pH of solution
T is the temperature of reacting vessel (°C)
r^2 is the regression coefficient for fit of equation to Hibler et al. (1987) data

19.3.1.5 Inactivation by Ozone

Some of the salient points regarding disinfection are encapsulated in a plot of *Ct* versus temperature data for ozone inactivation of *Giardia lamblia* cysts. Figure 19.2 shows *Ct* trends with temperature for log *R* = 1, 2, 3, respectively. The top scale shows the temperature for each point in °C and the lower scale is the inverse of absolute temperature. The plot shows (1) the log *Ct* versus $1/T$ (K) plots are linear and fit the Arrhenius relation; (2) the log *R* plots are displaced by a linear factor from each other (for example, log *R* = 2 values are twice the log *R* = 1 values; log *R* = 3 values are 3 times the log *R* = 1 values); (3) the *Ct* values for *Giardia* using ozone as a disinfectant are markedly lower than for chlorine.

19.3.2 Application of Chick–Watson Relation

Example 19.1 illustrates how Chick's law can be applied to utilize *Ct* data, per Equation 19.12, to estimate the percent inactivation for any time of contact. Example 19.2 illustrates how to convert *Ct*(99% kill) to *Ct*(99.9% kill) applying Equation 19.6. This is done by first calculating *t* for 99% inactivation, per Equation 19.12, for a given *Ct* and a given disinfectant concentration, and then determining the kinetic coefficient, *k*, per Equation 19.6, and finally, with *k* determined, the *N* can be determined for any *t* specified.

Example 19.1 Application of Chick's Law to a Practical Problem

Given

A city has a clear well with 60 min detention time and uses chloramines to disinfect. The chloramine concentration is maintained at 2.5 mg/L and is mostly in the form of monochloramine. Assume the concentration of heterotrophic plate count (HPC) bacteria for the flow entering the clear well is 10,000 organisms/mL (which is very high but is used for illustrative purposes).

Required

Estimate the average concentration of HPC bacteria leaving the clear well.

Solution

The approach is to define terms, then determine the time for 99% kill by the Watson relation, then obtain the kinetic coefficient for the disinfectant concentration specified by Chick's law. Knowing the kinetic coefficient, we can calculate the kill for any different contact time (without changing the disinfectant concentration).

Step 1: Define terms as given,

$$C(\text{monochloramine}) = 2.5 \text{ mg/L}$$
$$t = 60 \text{ min}$$
$$N_0 = 10{,}000 \text{ org/mL}$$
$$N = (\text{to be determined})$$

Step 2: Apply Equation 19.12 to determine *t*(99% kill). From Table 19.3, *K*(monochloramine, HPC) = 102

$$K = Ct(0.99 \text{ kill})$$

Applying Equation 19.12,

$$102 = 2.5 \text{ mg/L} \cdot t(99\% \text{ kill})$$
$$t(99\% \text{ kill}) = 40.8 \text{ min}$$

TABLE 19.3
***Ct* Data for Various Microorganisms and Disinfectants for 2-log (99%) Kill**

	Conditions		Disinfectant			
Microorganism	**pH**	***T* (°C)**	**Chlorine**	**Chloramine**	**Ozone**	**ClO_2**
Bacteria						
HPC			2[a]	102[b]		
E. coli			1[a]/0.2[a]	50[c]		
	6.0	5	0.04[d]			
	10.0	5	0.92 (OCl^-)[d]			
	4.5	15		5.5 ($NHCl_2$)[e]		
	9.0	5		175 (NH_2Cl)[e]		
	9.0	15		64 (NH_2Cl)[e]		
	9.0	25		40 (NH_2Cl)[e]		
	7.0	12			0.001[f]	
	7.2	1			0.02[f]	
	6.5	5				0.41[g]
Bacillus anthracis	7.2	0–5	400[a]			
	7.2	20–29	100[a]			
S. dysenteriae	7	20–29	0.5[a]			
Viruses						
Adenovirus 3			0.01[h]			
Polio			1.0[h]		0.9[i]	
Polio 1	6.0	5	1.0[d]			
	7.0	5				6.0[g]
	9.0	5		1420[c]		
	9.0	15		900 (NH_2Cl)[e]		
	7.2	5			<0.04[f]	
	7.0	25			0.42[f]	
Coxsackie A9	6.0	5	0.14[d]			
	7.0	15				0.4[g]
Coxsackie B5	6.0	5	1.7[d]			
	7.8	5	2.2[d]			
	7.0	25			0.53[f]	
Echo 1	6.0	5	0.24[d]			
	10.2	5	48 (OCl^-)[d]			
	7.0	25			0.44[f]	
Hepatitis A virus	6.0	5				0.50[j]
	7.0	5	0.70[j]		0.2[j]	
	8.0	5		296[c]		0.50[j]
Rotavirus						
MS-2 bacterioph.	6–9	<1	6[j]	2535[j]	0.9[j]	8.4[j]
Cysts						
G. lamblia cysts	7.0	<1	289[k]/140[j]	1243[j]	1.9[j]	42[j]
C. parvum			est. 7000[l]			
	6–11	1				606[m]
		5				442[m]
		13				230[m]
		22				111[m]
	6–8	10–27			5.0[n]	

(continued)

TABLE 19.3 (continued)
***Ct* Data for Various Microorganisms and Disinfectants for 2-log (99%) Kill**

	Conditions		Disinfectant			
Microorganism	pH	*T* (°C)	Chlorine	Chloramine	Ozone	ClO_2
E. histolytica	7.0	0–5	400[a]			
		20–29	60[a]			
	8.0	0–5	600[a]			
		20–29	200[a]			

[a] White (1999, pp. 223, 430) at pH 8.2, HOCL = 80%, OCl^- = 20%.
[b] Calculated from ratios using data in upper left *Ct* data cells.
[c] Sobsey (1989, p. 187), *Ct*(4-log)/2 = *Ct*(2-log).
[d] Doull, 1980, Table B.1, p. 33; *Ct* calculated from single concentration and time datum for each organism and each pH condition; all data at *T* = 5°C. *Ct* calculations corroborated by *Ct* data from Table II-2 p. 34.
[e] Doull, 1980, Table II-3, p. 40.
[f] Doull, 1980, Table II-5 for ozone, p. 50; compiled from various references.
[g] Doull, 1980, Table II-6 for chlorine dioxide, p. 58.
[h] Wastewater Treatment Plant Design, WPCF Manual of Practice No. 8 (1977, p. 380).
[i] Lev and Regli (1992).
[j] Malcolm Pirnie, Inc., and HDR Engineering, Inc. (1991).
[k] Hibler et al. (1987).
[l] Estimate of order of magnitude.
[m] Li et al. (2001, p. 601).
[n] Langlais (1991, p. 228).

TABLE CD19.4
Calculation of *Ct*'s for *Giardia* Cysts for Different *C*, pH, *T* Conditions (Excerpt)[a]

C (mg/L)	pH	*T* (°C)	*Ct*/log	*Ct*/2-log	*Ct*/3-log	*Ct*/4-log[b]
1.00	6.00	0.5	38	76	113	151
1.00	8.00	5.0	59	119	178	238

[a] From regression equation of Hibler's data by Clark et al. (1989, p. 85).
[b] $Ct(4\text{-log}) = 0.985C^{0.176}pH^{2.752}T^{-0.147}$.

Step 3: Determine the kinetic coefficient in Chick's law, Equation 19.6, based upon the time determined for 99% kill, that is, $C/C_0 = 0.01$, at $C = 2.5$ mg/L.

$$\ln\frac{N}{N_0} = -k(2.5\ \text{mg/L monochloramine})\cdot t(99\%\ \text{kill})$$

$$\ln\frac{1}{100} = -k\cdot 40.8\ \text{min}$$

$$\ln\frac{1}{100} = -k(2.5\ \text{mg/L monochloramine})\cdot 40.8\ \text{min}$$

$$k(2.5\ \text{mg/L monochloramine}) = 0.1129\ \text{min}^{-1}$$

Step 4: Apply Chick's law, Equation 19.6, again for system conditions, that is, *t*(clear well) = 60 min, *C*(clear well) = 2.5 mg/L.

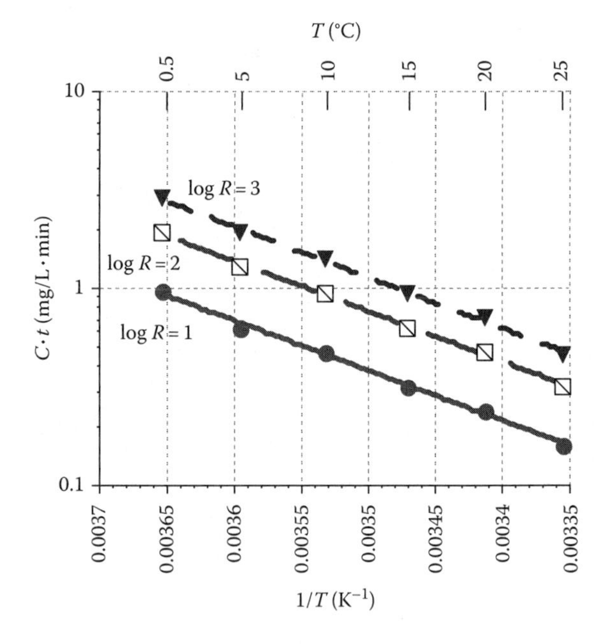

FIGURE 19.2 Ozone *Ct* values for the inactivation of *Giardia* cysts as affected by temperature and for different log *R*'s ($6 \le pH \le 9$). (From Langlais, B. et al., *Ozone in Water Treatment*, Lewis Publishers, Chelsea, MI, 1991, p. 220.)

$$\ln\frac{N}{N_0} = -k(2.5\ \text{mg/L monochloramine})\cdot t(\text{clear well})$$

$$\ln\frac{N}{10{,}000} = -0.1129\cdot 60\ \text{min}$$

$$N(t = 60\ \text{min}, \text{monochloramine at } 2.5\ \text{mg/L}) = 11.4\ \text{org/mL} \approx 11\ \text{org/mL}$$

Example 19.2 Apply the Watson–Chick Relations to Determine $C \cdot t$ for 99% Kill, That Is, $C \cdot t$(99% Kill), of *Giardia lamblia* Cysts, Given Data in Table CD19.4 for 99.9% Kill, That Is, $C \cdot t$(99.9% Kill)

Given

$C \cdot t$(99.9% kill) = 289 (Table CD19.4). Conditions are, $T = 0.5°C$, pH = 7.0.

Required

Calculate $C \cdot t$(99% kill) for $T = 0.5°C$, pH = 7.0.

Solution

The approach is to apply the Watson relation for the two conditions of interest.

Step 1: Apply the Watson relation, Equation 19.10.

$$\frac{1}{r}\ln\frac{N}{N_0} = C^n t$$

Step 2: Now apply the Watson relation for the two conditions of interest, that is, 99.9% kill and 99% kill, respectively. Let the exponent, $n = 1$; the $1/r$ term cancels.

$$Ct(99\%\,\text{kill}, G.\,\text{cysts}) = \frac{(1/r)\ln(1/100)}{(1/r)\ln(1/1000)} Ct(99.9\%\,\text{kill}, G.\,\text{cysts})$$

$$= \frac{-4.605}{-6.908} \cdot 289$$

$$= 193$$

Discussion

1. For reference, the EPA Guidance Manual gives $C \cdot t = 140$ for 99% inactivation.
2. From this exercise, one can see that Ct(2-log kill) = (log(1/100)/log(1/1000)] $\cdot$ Ct(3-log kill) = (−2/−3) $\cdot$ Ct(3-log kill), which can be generalized.

19.3.2.1 Examples of $C \cdot t$ Relation

Figure 19.3a and b are plots of concentration versus time data for four microorganisms and for two disinfectants, that is, HOCl and dichloramine, respectively. As noted, the product of concentration versus time, that is, Ct, is constant along the plot line for any microorganism. For example, Ct (*E. coli*, HOCl) ≈ 0.2 mg/L · min; Ct (*E. coli*, dichloramine) ≈ 6 mg/L · min. For coxsackievirus, Ct (coxsackie, HOCl) ≈ 5 mg/L · min; Ct (coxsackie, dichloramine) ≈ 300 mg/L · min. The plots illustrate further that Ct is approximately constant along a given plot line that dichloramine requires a much higher Ct than HOCl, and that different microorganisms have different Cts.

19.3.3 Chlorine Chemistry

The traditional disinfectant in the United States has been chlorine, which is effective and cheap. Chlorine is applied both as a primary disinfectant and to maintain a residual in the distribution system (usually, $0.2 \le C \le 0.5$ mg/L). The concentration in the contact basin may be 2–3 mg/L in order to achieve the ≥0.2 mg/L minimum residual.

19.3.3.1 Chlorine Properties

Chlorine is a greenish-yellow poisonous gas, with pungent odor, easily compressed to a liquid, but will volatilize upon release of pressure. Chlorine gas is highly corrosive if moisture is present and noncorrosive if dry. Some of the properties are MW(Cl_2) = 70.914, which is about 2.5 times heavier than air; at p(Cl_2) = 1.00 atm, the liquefying temperature is −34.5°C (−30.1°F). The specific gravity of liquid chlorine is SG(Cl_2, liquid) = 1.41. Liquid chlorine will attack and destroy PVC and rubber. Moist chlorine gas will attack all

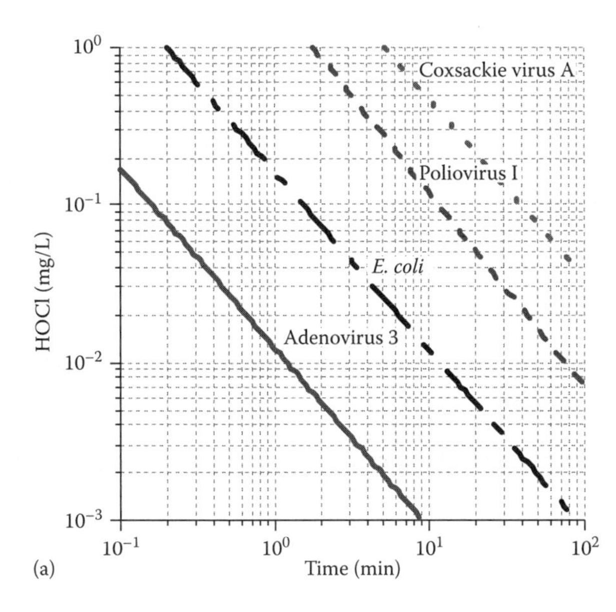

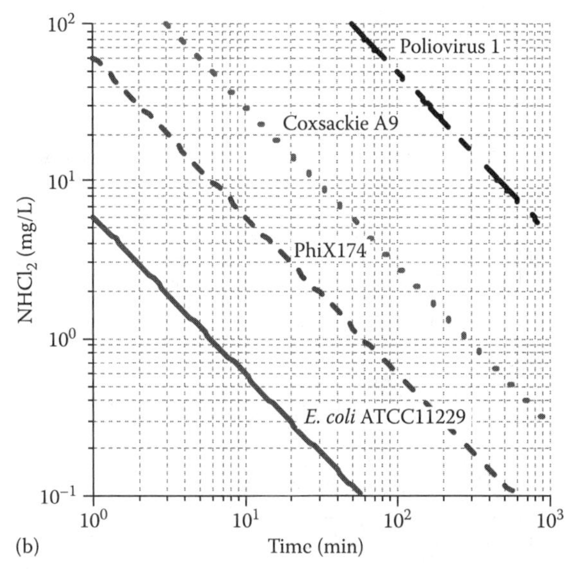

FIGURE 19.3 Plots of concentration time data for 99% inactivation of microorganisms shown to illustrate Ct concept. (a) HOCl at 0°C–6°C. (Adapted from ASCE-WPCF, Disinfection, in *Wastewater Treatment Plant Design*, Manual of Practice No. 8, ASCE, New York, 1977, Chapter 20.) (b) Dichloramine: pH = 4.5 and $T = 15°C$. (From Doull, J., *Drinking Water and Health*, Vol. 2, National Academy Press, Washington, DC, 1980, p. 28.)

ferrous metals, stainless steel, and copper, but not silver (since the silver chloride formed on contact is inert). Aqueous solutions are also corrosive. Materials used commonly for moist gas and solutions include PVC, fiberglass, Kynar, polyethylene, certain types of rubber, Saran, Kel-F, Viton, and Teflon (White, 1999).

19.3.3.2 Chlorine Demand

Since HOCl is a strong oxidizing agent, it reacts with a range of reducing substances, for example, NH_3, Fe^{2+}, Mn^{2+}, NO_3^-, H_2S, NOM, etc. (Fair et al., 1948, p. 1055). The aggregate of such reactions is termed, "chlorine-demand." The inorganic reactions are rapid while the NOM reactions are, in general, slow.

19.3.3.2.1 Breakpoint Chlorination

As seen by Equations 19.25 through 19.27, in subsequent Section 19.3.6.1, chlorine reacts with ammonia to form chloramines. Figure 19.4 shows a typical breakpoint chlorination relationship caused by the reaction with ammonia. As the chlorine concentration increases the "free available chlorine" (the solid line) increases but at a rate less than that of "applied chlorine" (solid line). The difference between the "applied chlorine" and the "free available chlorine" is the "chlorine demand." As the applied chlorine concentration continues to increase, the free available chlorine residual declines to near-zero, which is the "breakpoint." After the breakpoint, the free chlorine increases linearly with the applied chlorine.

19.3.3.2.2 Chlorine Reactions

Chlorine in water involves a sequence of reactions. Reactions include hydrolysis and acid–base. In addition, chlorine acid–base equilibrium is depicted in graphical form. Finally, the formation of hypochlorites is reviewed.

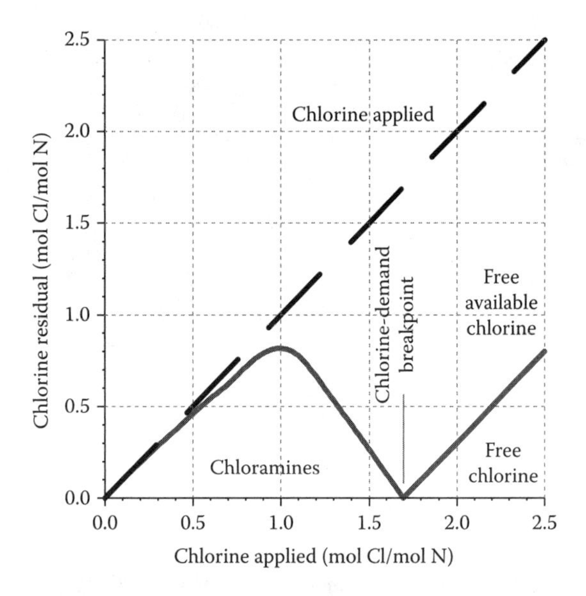

FIGURE 19.4 Illustration of breakpoint chlorination. (Adapted from Doull, J., *Drinking Water and Health*, Vol. 2, National Academy Press, Washington, DC, 1980, p. 22).

1. *Hydrolysis*: When chlorine gas is introduced in water, dissolution occurs readily, as seen by the Henry's law coefficient, that is, $H(Cl_2, 20°C) = 7283$ mg/L/atm (Table H.5). Upon dissolution, hydrolysis occurs with the reaction given as (Doull, 1980, p. 18)

$$Cl_2 + H_2O \Leftrightarrow HOCl + H^+ + Cl^- \quad (19.14)$$

The HOCl further dissociates to give

$$HOCl \Leftrightarrow OCl^- + H^+ \quad (19.15)$$

The two species, HOCl and OCl^- together are designated as "free chlorine," and also as "free available chlorine." As evident in Equations 19.14 and 19.15, the distribution of the species is pH dependent, with the HOCl predominating at pH ≤ 7.3; for reference, Cl_2 is the dominant species for pH ≤ 3.3. Also, for reference, at pH $= 6.0$, $T = 20°C$, HOCl $\approx$ 0.97 fraction; and at pH $= 7.0$, $T = 20°C$, HOCl $\approx$ 0.79 fraction (Fair et al., 1948, p. 1052; White, 1999, p. 218). The *Ct*'s for HOCl are generally 5–20 times higher than for OCl^- (see Table 19.3); in other words, chlorine is the most effective at pH ≤ 7.0.

2. *Acid–base reaction equilibria for chlorine*: The species present when chlorine gas reacts with water are seen in Equations 19.14 and 19.15, that is, Cl_2(gas), Cl_2(aq), HOCl, OCl^-, Cl^-, H^+, OH^-. Thus, the first step in the analysis of acid–base equilibria is to write the reaction equations, which permit the reactant and product species to be identified. Table 19.5 summarizes the reactions and the associated equilibrium equations, the latter being the second step. The reactions, of course, are linked. The third step is to write the mass balance relation, that is,

$$C = Cl_2(aq) = [HOCl] + [OCl^-] + [Cl^-] \quad (19.16)$$

and the charge balance may be stated,

$$[H^+] = [OCl^-] + [OH^-] + [Cl^-] \quad (19.17)$$

With the mass balance, charge balance, and five equilibrium equations, there are seven equations, which means that the seven unknowns may be solved. Often, the mass balance and the charge balance are combined; for example, the $[Cl^-]$ may be eliminated by subtracting Equation 19.17 from Equation 19.16.

3. *Graphical depiction of equilibria for chlorine*: Table CD19.6 shows the concentrations of the aqueous chlorine species, that is, Cl_2(aq), HOCl, OCl^-, along with H^+, OH^-; fractions of the first three are given also, for example, $\alpha_0 = [Cl_2]/C(\text{total})$, $\alpha_1 = [HOCl]/C(\text{total})$, $\alpha_2 = [OCl^-]/C(\text{total})$, respectively. The ancillary tables show the formulae for the "α" calculations, which are the basis for the calculations of the concentrations of $[Cl_2(aq)]$,

TABLE 19.5
Reactions and Equilibrium Statements for Chlorine Gas Dissolution

Reaction	Equilibrium Statement	
1. Equilibrium between gas and aqueous Cl_2 $Cl_2(g) \rightarrow Cl_2(aq)$	$K_H = \frac{Cl_2(aq)}{Cl_2(g)} = 7283\ mg/L/atm$	(19.18)
2. Aqueous chlorine reaction with water $Cl_2(aq) + H_2O \rightarrow HOCl + H^+ + Cl^-$	$K_1 = \frac{[HOCl][H^+][Cl^-]}{[Cl_2(aq)]} = 10^{-3.3}$	(19.19)
3. Hypochlorous acid dissociation $HOCl \rightarrow H^+ + OCl^-$	$K_{x2} = \frac{[H^+][OCl^-]}{[HOCl]} = 10^{-7.5}$	(19.20)
4. HCl dissociation $HCl \rightarrow H^+ + Cl^-$	$K_{x3} = \frac{[H^+][Cl^-]}{[HCl]} = 10^{+3}$	(19.21)
5. Water dissociation $H_2O \rightarrow H^+ + OH^-$	$K_w = [H^+][OH^-] = 10^{-14}$	(19.22)

K_H from Table H.5; full units are mg dissolved chlorine/L aqueous solution/atm Cl_2 gas.
K_1 and K_2 were obtained from Pankow (1991, pp. 435, 442); for comparison, White (1999, p. 217) gives $pK_2(20°C) = 7.7$.

TABLE CD19.6
Concentrations of $[Cl_2]$, [HOCL], and $[OCl^-]$ as a Function of pH for a Given $[Cl_2]$ Concentration

(a) Conditions for calculations within table

C = 5.7E−05 mol/L	4.0416 mg/L	0.0004	percent Cl
$pK_1 = 3.3$	$pK_2 = 7.3$		
$K_1 = 5.01E{-}04$	$K_2 = 5.01E{-}08$		
$K_w = 1.00E{-}14$	$K_2K_2 = 2.51E{-}11$		

(b) Calculations

pH	p $[H^+]$	$[H^+]$ (mol/L)	p $[OH^-]$	α_o	$[Cl_2]$ (mol/L)	$p[Cl_2]$	α_1	[HOCL] (mol/L)	p [HOCL]	α_2	[OCl−] (mol/L)	p [OCl−]
0	0.00	1.00E+00	14	9.995E−01	5.697E−05	4.244	5.007E−04	2.854E−08	7.545	2.509E−11	1.430E−15	14.845
1	1.00	1.00E−01	13	9.950E−01	5.672E−05	4.246	4.985E−03	2.841E−07	6.546	2.497E−09	1.424E−13	12.847
2	2.00	1.00E−02	12	9.523E−01	5.428E−05	4.265	4.771E−02	2.719E−06	5.566	2.390E−07	1.362E−11	10.866
3	3.00	1.00E−03	11	6.662E−01	3.797E−05	4.421	3.338E−01	1.902E−05	4.721	1.672E−05	9.532E−10	9.021
4	4.00	1.00E−04	10	1.663E−01	9.480E−06	5.023	8.333E−01	4.750E−05	4.323	4.175E−04	2.380E−08	7.624
5	5.00	1.00E−05	9	1.947E−02	1.110E−06	5.955	9.756E−01	5.561E−05	4.255	4.888E−03	2.786E−07	6.555
6	6.00	1.00E−06	8	1.897E−03	1.081E−07	6.966	9.505E−01	5.418E−05	4.266	4.762E−02	2.714E−06	5.566
7	7.00	1.00E−07	7	1.330E−04	7.579E−09	8.120	6.661E−01	3.797E−05	4.421	3.337E−01	1.902E−05	4.721
8	8.00	1.00E−08	6	3.321E−06	1.893E−10	9.723	1.664E−01	9.484E−06	5.023	8.336E−01	4.752E−05	4.323
9	9.00	1.00E−09	5	3.906E−08	2.226E−12	11.652	1.957E−02	1.115E−06	5.953	9.804E−01	5.588E−05	4.253
10	10.00	1.00E−10	4	3.976E−10	2.266E−14	13.645	1.992E−03	1.135E−07	6.945	9.980E−01	5.689E−05	4.245
11	11.00	1.00E−11	3	3.983E−12	2.270E−16	15.644	1.996E−04	1.137E−08	7.944	9.998E−01	5.699E−05	4.244
12	12.00	1.00E−12	2	3.984E−14	2.271E−18	17.644	1.996E−05	1.138E−09	8.944	1.000E+00	5.700E−05	4.244
13	13.00	1.00E−13	1	3.984E−16	2.271E−20	19.644	1.996E−06	1.138E−10	9.944	1.000E+00	5.700E−05	4.244
14	14.00	1.00E−14	0	3.984E−18	2.271E−22	21.644	1.996E−07	1.138E−11	10.944	1.000E+00	5.700E−05	4.244

[HOCl], $[OCl^-]$. The effect of input concentration on the species distribution and the pH levels may be seen merely by changing the concentration, C, in the second row. Table CD19.6 has associated plots for pC versus pH and α versus pH. Figure CD19.5a and b show examples of these two plots for C(total chlorine) ≈ 4 mg/L $= 5.7 \cdot 10^{-5}$ mol/L, that is, pC (total chlorine) ≈ 4.2, for which pH ≈ 5.8.

The plots are important in that they show the distributions of $[Cl_2(aq)]$, [HOCl], $[OCl^-]$; they also permit the pH to be determined for a given pC. The plot shows that the $Cl_2(aq)$ is the dominant species at pH < 3 and that HOCl is dominant at $3.3 < pH < 7.3$, while OCl^- is dominant at pH > 7.3. As noted, when chlorine gas is the source, pH ≈ 5.8, which is determined by electroneutrality, that is, where $[H^+] = [OCl^-]$. If sodium

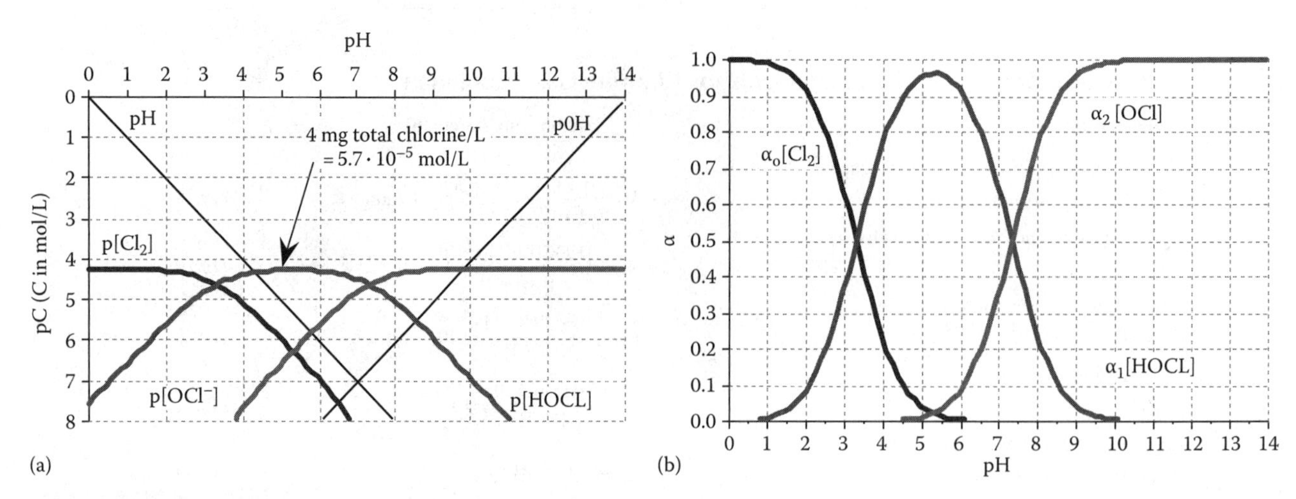

FIGURE CD19.5 Calculations based on 4 mg/L ($5.7 \cdot 10^{-5}$ molar) solution of Cl_2 added to solution (from mass balance, charge balance, equilibrium relations) as function of pH. (a) Concentrations of chlorine species. (b) Fraction, α, of different chlorine species.

hypochlorate, NaOCl, is the source, then the mass balance equation and the electroneutrality equation are combined to eliminate $[Na^+]$, such that the equality occurs where [HOCl] line crosses the $[OH^-]$ line, resulting in $pH \approx 9.1$.

Table CD19.7 (not excerpted in text but available as a download) shows the concentrations of the aqueous chlorine species, i.e., HOCl, OCl^-, along with H^+, OH^-, for a sodium hypochlorite solution, i.e., NaOCl, demonstrating the associated mass balance and electro-neutrality conditions (see also Example 19.3). In Table CD19.7, the mass balance/electroneutrality occurs along the point where $[HOCl] = [OH^-]$, which is the same as where $p[HOCl] = p[OH^-]$. Keep in mind that, by convention, the brackets indicate aqueous concentration in mol/L.

Generally, if Cl_2(gas) is bubbled into a side-stream of water, $p[Cl_2(gas)] = 1.0$ atm Cl_2(gas), which "drives" the system, that is, it determines the concentration of the other species. The concentrated side-stream is diluted in the main flow, and so the diluted concentration of the main flow is the "target" concentration, for example, for the clear well, let $[HOCl] = 3.0\ mg/L \cdot (1\ mol/52{,}500\ mg) = 5.7 \cdot 10^{-5}$ mol/L (Box 19.2).

4. *Hypochlorites*: The hypochlorites include calcium hypochlorite, $Ca(OCl)_2$ solid, and sodium hypochlorite, NaOCl solution (commercial bleach). When added to water, $Ca(OCl)_2$ dissolves and ionizes to give

$$Ca(OCl)_2 \rightarrow Ca^{2+} + 2OCl^- \tag{19.23}$$

The hypochlorite ions then combine with hydrogen ions from water, that is,

$$H_2O + OCl^- \Leftrightarrow HOCl + OH^- \tag{19.24}$$

As evident in Equation 19.24, by increasing the pH, that is, the OH^- concentration, drives the reaction to the left, increasing the proportion of OCl^-, which is less effective as a disinfectant than is HOCl. Thus, the pH should be adjusted to pH $\leq$ 7.0 for the chlorination reactor if one wishes to obtain the maximum inactivation effectiveness.

Bleach solutions of NaOCl undergo the same reactions as for $Ca(OCl)_2$. Calcium hypochlorite may form sludge with hard waters, a disadvantage of this form (ASCE-WPCF, 1977, p. 396). To summarize, the hypochlorites raise the pH, resulting in the less effective OCl^- form of chlorine. Gaseous chlorine, on the other hand, lowers the pH with the more effective HOCl resulting.

BOX 19.2 EQUILOGRAPHS FOR ACID–BASE REACTIONS

Graphical depiction of acid–base equilibria encapsulates the mass balance, electroneutrality, and equilibrium equations and the analysis permits the calculation of pH for different concentrations of chlorine. While the spreadsheet approach with the associated graphs, illustrated in Table CD19.6, calculate the equilibrium lines, the depiction may be shown by overlaying the equilibrium lines on top of a pC–pH grid, with $[H^+]$ and $[OH^-]$ lines at 45°. If the equilibrium lines for $[Cl_2(aq)]$, [HOCl], and $[OCl^-]$ are placed on a transparency, the array may be moved up and down with respect to the grid and the effect of pC on pH may be seen readily. In effect, the pC is the mass balance and is the horizontal line; the electroneutrality is along either the $[H^+]$ line or the $[OH^-]$ line. The equilibrium lines are given by the respective equilibrium equations. The foregoing graphical approach was developed by Professor Henry Freiser (Freiser, 1964), who later provided a spreadsheet approach (Freiser, 1996). The graphical approach for acid–base and other kinds of equilibria have been used widely in aqueous chemistry (see, for example, Snoeyink and Jenkins, 1980; Pankow, 1991; Benjamin, 2002).

Example 19.3 Sodium Hypochlorite Equilibria

Given
The label for household "bleach," for example, Clorox™, reads "6% sodium hypochlorite."

Required
Calculate the pH of the bleach solution.

Solution
The steps are enumerated.

1. Convert to concentration as given to molar concentration
   0.06 fraction NaOCl means literally, 60 g NaOCl per 1000 g H_2O, which is the same as (60 g NaOCl)/(L water). Since MW(NaOCl) = 74.5 g/mol, then 60 g NaOCl = 60/74.5 = 0.80 mol NaOCl, to give a molar concentration of 0.80 mol NaOCl/L solution. The associated pC(NaOCl) = −(−log[NaOCl]) = −(−log[0.805]) = (−0.094)−0.094 ≈ 0.10.
2. Mass balance, as in Equation 19.16,

$$C(\text{total}) = [Na^+] = [HOCl] + [OCl^-] + [Cl^-]$$

3. Charge balance, as in Equation 19.17,

$$[Na^+] + [H^+] = [OCl^-] + [Cl^-] + [OH^-]$$

4. Subtract charge balance from mass balance,

$$[OH^-] = [HOCl] + [H^+]$$

5. Solution pH
   5.1 As seen in the net equation, the charge balance occurs along the $p[OH^-]$ line in Figure CD19.5(a); the only place to satisfy the equation is where $p[OH^-] = p[HOCl]$ (since $[H^+]$ is very small)
   5.2 From Figure 19.5a for pC ≈ 0.10, pH(NaOCl solution) ≈ 10.7.
6. Solution pH for 6 mg/L NaOCl
   6.1 For NaOCl concentration, 6 mg/L, the molar concentration is 0.00008 mol/L, which gives pC ≈ 4.1.
   6.2 Entering Figure 19.5a for pC ≈ 4.1, gives pH ≈ 8.7.
   6.3 The concentration of OCl^- in mg/L is (51.5/74.5) · 6 = 4.1 mg/L OCl^-, which is just on the high side of practice for a clear well concentration.
   6.4 Note that the mol/L concentrations are the same for NaOCl and OCl^-.

Discussion

1. *pH of Clorox*: The problem illustrates that in the case of NaOCl, because of the Na^+, the electroneutrality occurs along the $p[OH^-]$ line where it is crossed by the [HOCl] line. This is seen in Figure CD19.5(a), which gives, for undiluted Clorox, C(NaOCl) = 60 g/L, pH(solution) ≈ 10.7. This compares with pH(Clorox from bottle, measured) = 11.5, using Oakton pH meter with calibration using pH = 4.7 and pH = 10 buffers. Also, for reference, pH(Diet Coke) = 3.2 and pH(Coke) = 2.6 (measurements by Bridgette Hendricks on September 3, 2009).
2. *pH of C(NaOCl) = 6.0 mg/L*: For the second case, that is, C(NaOCl) = 6.0 mg/L, pH(solution) ≈ 8.7, which is from the pC–pH diagram, Figure CD19.5a, for [NaOCl] = 0.00008 mol/L, which gives pC ≈ 4.1 as the point of entry to the diagram.
3. *Comparison with chlorine gas*: On the other hand, if chlorine gas is used, the equilibrium is along the pH line where it crosses the $[OCl^-]$ line. The implication for operation is that the use of NaOCl results in pH ≫ 7, such that OCl^- is the predominant form (which has much less "killing power" than HOCl.

Example 19.4 Chlorine Metering Requirement

Given
Calcium hypochlorite, $Ca(OCl)_2$ is to be dissolved in a feeder to result in a solution concentration of 100,000 mg $Ca(OCl)_2$/L (200,000 mg/L is feasible according to ASCE-WPCF, 1977, p. 396). The resultant chlorine concentration desired in the main flow is 12 mg HOCl/L. Let the main flow be $Q = 0.044\ m^3/s$ (1.0 mgd).

Required
Calculate the flow of concentrate, $Q[Ca(OCl)_2]$, required.

Solution
The steps are enumerated.

1. Determine the equivalent HOCl in a unit mass of $Ca(OCl)_2$. From Equation 19.23, apply the basic chemistry approach, that is,

$$\underset{\text{MW}=143}{\overset{100\text{ g}}{Ca(OCl)_2}} \rightarrow Ca^{2+} + \underset{\text{MW}=103}{\overset{x\text{ g}}{2OCl_2^-}}$$

$$x = 72.0\ \text{g/L}$$

2. If $C(OCl^-)$ = 72,000 mg/L, calculate the $Q(Ca(OCl)_2)$, by mass balance, that is,

$$Q \cdot C(\text{main flow}) = Q(\text{concentrate}) \cdot C(\text{concentrate})$$

$$44\ \text{L/s} \cdot 12\ \text{mg}\ OCl^-/\text{L} = Q(\text{concentrate}) \cdot 72{,}000\ \text{mg}\ OCl^-/\text{L}$$

$$Q(\text{concentrate}) = 7.3\ \text{mL/s}$$

$$= 631\ \text{L/day}$$

Discussion
In other words, for a 72 g OCl^-/L concentrate solution, 100 g $Ca(OCl)_2$ solid must be dissolved. The daily requirement is $J(Ca(OCl)_2)$ = 12 mg HOCl/L · 44 L/s · [100 g $Ca(OCl)_2$/72 g OCl^-] = 733 mg/s = 63 kg/day. Also, pH ≫ 7, which means that the form is mostly OCl^- that is less effective than HOCl. A diaphragm metering pump may be used to meter the hypochlorite flow into the main flow. Commercial systems are available that accomplish the same thing using a tablet feeder, for example, the Hammond® tablet feeder.

19.3.4 Chloramines

Some municipalities use chloramines for primary disinfection. Reasons include (1) taste and odor effects are essentially nonexistent, (2) the chloramines are persistent in the distribution system, and (3) disinfection by-products are not formed. On the negative side, the *Ct* requirement for 2-log inactivation of a given organism is much higher for the chloramines than for chlorine.

19.3.4.1 Chlorine–Ammonia Reactions

When chlorine reacts with ammonia, chloramines are formed. The sequence of reactions related to the formation of chloramines is enumerated.

1. The free ammonia, on the left side, reacts with HOCl to give a *monochloramine*,

$$NH_3 + HOCl \Leftrightarrow NH_2Cl + H_2O \tag{19.25}$$

2. The monochloramine reacts also with HOCl, to give a *dichloramine*,

$$NH_2Cl + HOCl \Leftrightarrow NHCl_2 + H_2O \tag{19.26}$$

3. The dichloramine is further oxidized to give *trichloramine*,

$$NHCl_2 + HOCl \Leftrightarrow NCl_3 + H_2O \tag{19.27}$$

4. The equilibrium distributions between monochloramine and dichloramine are at $pH = 5.0$, 0.84 fraction is $NHCl_2$; at $pH = 6.0$, 0.62 fraction is $NHCl_2$; at $pH = 7.0$, 0.35 fraction is $NHCl_2$; at $pH = 8.0$, 0.15 fraction is $NHCl_2$ (Fair et al., 1948, p. 1054).

19.3.4.2 Chloramine Disinfection

The effective form of chloramines is the dichloramine ($NHCl_2$) form. The relative bactericidal effectiveness of dichloramine ($NHCl_2$) to monochloramine (NH_2Cl) is about 35:1 but varies with the organism (Fair et al., 1948, p. 1054). *Ct* values for monochloramines are given in Table 19.3, along with chlorine and other disinfectants. The data show (1) dichloramine is confirmed as the most effective form of chloramines; (2) for 2-log inactivation of various organisms, very high *Ct* values are required.

19.3.5 Ozone Chemistry

Ozone gas has the formula, O_3, $MW = 48$ g/mol, $\rho(O_3) = 2.154$ g/L at $T = 0°C$, $p = 1.00$ atm (Hill and Rice, 1982, p. 42). The half-life of ozone in distilled water is about 25 min at 20°C. In natural waters, the half-life is much less, for example, about 10 min for filtered water from the Bodensee (Hill and Rice, 1982, p. 4), which is a large lake near Zurich. The reaction rate of ozone with most compounds is fast, particularly with organics. From Table H.5, Henry's constant is, $\mathbf{H}(O_3, 20°C) = 482$ mg O_3/L water/atm O_3. By comparison, $\mathbf{H}(Cl_2, 20°C) = 7{,}283$ mg Cl_3/L water/atm Cl_3; and from formula in Lide (1996, p. 6-5), $\mathbf{H}(ClO_2, 20°C) \approx 82{,}524$ mg ClO_2/L water/atm ClO_2.

The power requirement to generate ozone is about 13–22 kWh/kg ozone, when dry air is the source of oxygen and about half when the source is pure oxygen (Doull, 1980, p. 42). The reactor may be set up with diffuser in a counter-current mode, or some other method, such as described in Chapter 18. Commercial equipment is available to supply ozone in the range $0.002 < J(O_3) < 40$ kg/day (Doull, 1980, p. 43) at a voltage of up to 20,000 V is applied (ASCE-WPCF, 1977, p. 400). Ozone is produced at about 1% by weight when air is the oxygen source and about 2% when pure oxygen is the source gas. Its effectiveness is independent of pH but the range $6 < pH < 7$ appears most favorable. Table 19.3 shows that ozone is effective in the inactivation of *Giardia*, and *Cryptosporidium*, that is, Ct(*Giardia*, 2-log) ≈ 1.9 and Ct (*Cryptosporidium*, 2-log) ≈ 5.

The species of ozone in water, in addition to O_3, include hydroxyl radicals, $\cdot OH$; hydroperoxyl radicals, $HO_2\cdot$; oxide radicals, $O\cdot$; ozonide radicals, $O_3\cdot$; and possibly free oxygen, $\cdot O\cdot$ (Doull, 1980, p. 44). The hydroxyl radical is considered the most reactive.

19.3.6 Chlorine Dioxide

This section reviews some of the salient as aspects of chlorine dioxide, that is, its effectiveness, characteristics, and methods of synthesis.

19.3.6.1 Effectiveness of Chlorine Dioxide as a Disinfectant

On bactericidal effectiveness *Ct* values from Doull (1980, p. 59) were,

Disinfectant	Organism	pH	T (°C)	log R	Ct
ClO_2	*E. coli*	6	5	2	0.4
			10		0.25
			20		0.18
	Polio virus	7.0	5		5.5
		7.0	15		1.3
	Coxsackievirus	7.0	15		0.3
	Giardia lamblia cysts[a]	7.0	5	3	26
Chlorine	Giardia lamblia cysts[b]		5	3	149
Ozone	Giardia lamblia cysts[b]		5	3	1.9

In other words, comparing *Ct*'s, chlorine dioxide is similar to or stronger than chlorine in the inactivation of microorganisms. The maximum residual ClO_2 concentrations permitted for drinking water were Belgium, 0.25 mg/L; Germany, 0.2 mg/L; Switzerland, 0.15 mg/L (Masschelein, 1992, p. 191). For the United States, the maximum concentrations permitted by the EPA regulations for the distribution system (*Federal Register*, December 16, 1998) were 0.8 mg/L. For these maximum concentrations, the $Ct(\log R \approx 2)$ values can be achieved with $t \le 20$ min.

19.3.6.2 Characteristics of ClO_2

At concentrations $>10\%$ in air, ClO_2 may be explosive; therefore, it is generated onsite (Doull, 1980, p. 52). Based on the calculation by Henry's law, the concentration in the aqueous solution would have to be ≥ 8 g/L (at 20°C) in order to reach the $>10\%$ concentration level in air, and would require an air–water interface (Masschelein, 1992, p. 172). In fact, 4 g/L for a storage concentrate solution is the industry standard for reference 3000 mg/L stabilized solution is produced for commercial use by CDG, LLC (Gregory, 2009, 2010).

Since chlorine dioxide gas is highly soluble, that is, **H** (ClO_2, 20°C) = 1.0 mol/L/atm = 84,500 mg/L/atm (calculation from Note 8, Table H.5), the gas dissolves as quickly as it is produced. By comparison, chlorine gas solubility is **H**(Cl_2, 20°C) = 7283 mg/L/atm; in other words, by comparison, chlorine dioxide gas is more soluble than chlorine gas (which is highly soluble).

As to other characteristics, ClO_2 decomposes upon exposure to UV light to produce ClO_3, which in turn decomposes to chlorine and oxygen. In a weak acid solution, ClO_2 is stable at concentrations <10 g/L. In a basic solution, however, that is, pH > 8, ClO_2 hydrolyzes to form ClO_2^- and ClO_3^-. Solid sodium chlorite is explosive on heating or on contact with organic matter and is best stored in solution, for example, 300–400 g/L (Masschelein, 1992, p. 172).

19.3.6.3 Reaction Alternatives

Chlorine dioxide may be generated from the dissolution of chlorine gas and sodium chlorite and then combining the two to form ClO_2. Another approach, that requires special caution, is to bring solid sodium chlorite into contact with chlorine gas. Sodium chlorite is a white crystal with strong oxidizing capacity even in the solid form. Another reactant, less common, is sodium chlorate, $NaClO_3$.

1. *Chlorine gas reacting directly with sodium chlorite solution*: According to Masschelein (1992, p. 173), if chlorine gas is used, that is, added to the reactor, that is, with ClO_2^- fed into the reactor at the same time, the reaction occurs directly, that is,

$$Cl_2 + 2ClO_2^- \rightarrow 2ClO_2 + 2Cl^- \qquad (19.28)$$

   Equation 19.29 occurs because the reaction between the chlorine and the dissolved chlorite is faster than the hydrolysis of chlorine, that is, Equation 19.29, $Cl_2 + H_2O \rightarrow HOCl + H^+ + Cl^-$. This direct method, that is, Cl_2 with ClO_2^-, is the most common in practice and is done with a slight excess of chlorine, and is about 0.95-fraction complete.

2. *Chlorine gas forming hypochlorite solution to react with sodium chlorite solution*: Another approach, albeit not common, for the generation of chlorine dioxide is to react chlorine with water to form a hypochlorous acid solution, that is, HOCl. A sodium chlorite solution is brought into contact with the HOCl solution to form a ClO_2 solution (Aieta and Berg, 1986, p. 62). First, the reaction of chlorine gas with water gives HOCl as a product, that is,

$$Cl_2 + H_2O \rightarrow HOCl + H^+ + Cl^- \qquad (19.29)$$

BOX 19.3 THE DISINFECTANT RESIDUAL ISSUE

American disinfection practice for drinking water has utilized chlorine, since its inception about 1910. Chlorine also maintains a residual in the distribution system. Thus, if say 2–3 mg HOCl is applied at the clear well, about 0.5 mg HOCl may be measured in the distribution system. Usually, the system is monitored at various points and booster chlorine injections are added, if needed. Chlorine reacts, however, with organic substances in the finished water so that the residual becomes diminished.

The premise of maintaining residual chlorine is that the distribution system is protected from cross-connection contaminations, that is, contaminants entering the system by means of a negative hydraulic gradient (this may occur because of many kinds of activities, some of which may not have been anticipated by those who administer a system). Municipal plumbing codes, coupled with inspections and unwavering enforcement, are intended to minimize the occurrence of cross-connections. The frequency of cross-connection occurrences depends upon the code and the diligence of the inspections and enforcement. Despite all efforts, cross-connections seem to occur every so many years; either the code is violated or people find ingenious ways, albeit unintentional, to cause a cross-connection. The efforts on the distribution side are "slogging" and unglamorous, but are an essential part of any potable water system. In addition to reducing cross-connections, there are many other aspects to managing the distribution system, for example, having a program of flushing the mains and dead ends, providing for interchangeability of parts, controlling corrosion, detecting and fixing leaks, etc. Regardless of the sophistication and effectiveness of treatment, the "game may be lost" in the distribution system if there is not in place an equally effective program to minimize cross-connections.

The premise of the disinfectant residual has been questioned from time to time in that the dosage is probably not enough to eliminate the public health risk. Further, the disinfectant could attenuate the concentrations of indicator microorganisms, such as coliforms while it is less effective against say viral pathogens, that is, masking the occurrence of a cross-connection while not reducing the risk (suggested, c. 1984 by Henry Ongerth, retired from California Health Department). Along this line, some have pointed to European practice. For example, in Berlin, since 1979, treated water has not been post-chlorinated (Masschelein, 2002, p. 59).

The solutions from the two tanks, that is, HOCl and Na^+/ClO_2^-, are fed to a "reactor," which facilitates the following reaction (Doull, 1980, p. 190):

$$HOCl + 2ClO_2^- + H^+ \rightarrow 2ClO_2 + Cl^- + H_2O \quad (19.30)$$

3. *Chlorine gas reacting directly with solid sodium chlorite*: According to White (1999, p. 1160) a specially processed solid sodium chlorite (Saf-T-chlor®, CDG Environmental, LLC) reacts with chlorine gas to produce a high purity chlorine-free chlorine dioxide gas (quoting a patent by Rosenblatt et al. 1992). chlorine dioxide gas (quoting a patent by Rosenblatt et al. 1992), according to the following reaction,

$$Cl_2(g) + 2NaClO_2 \rightarrow 2ClO_2(g) + 2NaCl \quad (19.30')$$

On a mass basis, 1.97 kg ClO_2 reaction product is formed per kg Cl_2 (1.97 Lb ClO_2/Lb Cl_2). The chlorine gas is diluted approximately 90% (by volume) with air to maintain a safe partial pressure of ClO_2 in the product stream. The system is operated under vacuum produced by chlorine sidestream injector, similar to vacuum chlorination systems. One potential advantage of this process for water treatment facilities is that, because the reaction product is high-purity ClO_2 gas, excess chlorite ion, which does not exist as a gas, cannot be added to the main transmission line. This feature can be significant given the USEPA MCL of 1.0 mg/L for chlorite ion (Gregory, personal communication, January, 2010).

19.3.7 Ultraviolet Radiation

The effectiveness of ultraviolet radiation has been known since about 1910 (see Section 19.2.3), with the technology being largely dormant until about the 1980s. An upsurge of interest started during the 1980s, mostly in disinfection of wastewater effluents, with more growth during the 1990s. The impetus for its use in drinking water disinfection came in 1998 with the finding by Clancy et al. (1998, 2000) that low doses of UV inactivated *Cryptosporidium parvum* oocysts (Box 19.4).

BOX 19.4 THE INTERNATIONAL ULTRAVIOLET ASSOCIATION (IUVA)

The International UltraViolet Association became visible at the *AWWA Annual Conference* in Chicago in 1999. Also, at the trade show, UV was evident at a level not seen in previous years. It became evident later that the high level of activity in 1999 was the 1998 finding of Dr. Jennifer Clancy and associates that UV could inactivate *Cryptosporidium parvum* oocysts at much lower dosages than found by previous investigators. After *Giardia*, which emerged as an issue in the 1970s, *Cryptosporidium* came on the scene in the late 1986 and became one of the major issues in the drinking water industry. Until UV, ozone was the only known effective disinfectant.

Cryptosporidium was also the regulatory issue (i.e., the EPA "surface water treatment rule") that would eventually force municipalities with unfiltered water supplies to implement filtration. The UV finding was a "way out" for these utilities. Also, the finding gave other utilities that had filtration a second positive barrier to the oocysts, that is, an alternative to ozone or microfiltration. In addition, UV was found to be effective in inactivating other organisms, for example, viruses, bacteria, bacterial spores, and cysts. Finally, UV did not cause the formation of disinfection by-products. Thus, the interest in UV picked up from relatively "casual" to what might be termed, "high-profile."

The IUVA was formed in the spring of 1999, largely in the context of the foregoing circumstances. The organization started with a quarterly newsletter, *IUVA News*, and has held a biennial congress, the fifth being in Amsterdam in 2009. Dr. James P. Malley, University of New Hampshire, was the first president and Dr. James R. Bolton was named Executive Director, while Dr. Rip G. Rice became Editor of the newsletter (Malley, 1999).

19.3.7.1 Disinfection Rate by UV

In general, inactivation of organisms by UV follows the mathematical model of Chick, that is,

$$\frac{dN}{dt} = -k \cdot I \cdot N \quad (19.31)$$

where

N is the # microorganism/m^3
t is the elapsed time of contact (s)
k is the kinetic constant, a function of transmittance, microorganism, etc. (m^2/Js)
I is the intensity of UV radiation (W/m^2)

Integration gives the relation (Masschelein, 2002, p. 68),

$$\ln\left(\frac{N}{N_0}\right) = -kIt \quad (19.32)$$

The product, It, is defined as the "dose," that is,

$$\text{Dose} = It \quad (19.33)$$

where Dose is the energy quantity per unit area (J/m^2).

The "dose" is given usually as $mW \cdot s/cm^2$; the conversion is: 1 kJ/m^2 = 100 $mW \cdot s/cm^2$. Several notes relevant to UV in practice are

- UV radiation in the wavelength range $200 \leq \lambda \leq 300$ nm causes damage to the DNA, which prevents replication and hence inactivates the organism. Prescott et al. (1993) mention that $\lambda \leq 260$ nm is the most lethal to microorganisms. The predominant wavelength emitted by "low-pressure" UV lamps is $\lambda \leq 254$ nm, which is highly effective, though slightly less so than $\lambda \leq 260$ nm. Also, the dose should be high enough such that "photoreactivation," that is, repair in the presence of light, or "dark repair" repair in the absence of light, is not significant (Mackey, 2001, pp. 14, 15).

- For a given reactor, the distribution of contact times may be estimated by tracer studies.
- In general, the order of susceptibility of organisms to UV radiation is,

bacteria > viruses > bacterial spores > protozoan cysts.

- As a reference for the effect of "dose," $\log N/N_0 \approx -4$ was achieved for *Bacillus subtilis* spores for a UV dose ≈ 31 mJ/cm^2.

19.3.7.2 Log *R*'s by UV

Table 19.7 shows 1-log inactivation data for organisms in three groups, viruses, bacteria, and protozoan cysts. Original data were given for different log inactivations and were normalized to 1-log. For a 2-log inactivation rate, the values in Table 19.7 would be multiplied by "2"; for a 3-log rate, multiplied by "3," etc. Of the microorganisms listed, the Adenovirus, a double-stranded DNA virus, may be the most resistant to UV inactivation (Malley, 2000, p. 9; Malley, 2002a, p. 31). UV inactivation of different strains of *Cryptosporidium parvum*, that is, Iowa, Moredun, Glasgow, Maine, TAMU (Texas A&M University) were compared in a study by Clancy et al. (2002). The study showed that 4-log inactivation resulted from ≤ 10 mJ/cm^2 for the five strains. In a pilot plant, inactivation increased in proportion to detention time (for a given configuration of lamps). By comparison, for demineralized water, rotavirus, poliovirus, and hepatitis A, inactivation's were 12, 5, and 2 mJ/cm^2/log inactivation, respectively (Box 19.5).

19.3.7.3 Radiation Fundamentals

The starting point for understanding UV inactivation of microorganisms is the idea of radiant energy. Electromagnetic energy occurs in discrete bundles called photons or quanta; the energy of the radiation is proportional to its frequency (Willard et al., 1958, p. 2; Bolton, 2001, p. 5; Snicer et al., 2000, p. 9), that is,

TABLE 19.7
UV Dosages for 1-log Inactivation of Various Organisms[a]

Group	Organism	UV Dose (mJ/cm^2)
Viruses	Poliovirus Type I	4–6
	Coxsackievirus	6.9
	Hepatitis A	4–5
	Rotavirus strain SA 11	7–9
	Adenovirus Strain 40	30
	Adenovirus Strain 41	25
	Adenovirus Strain 41	30
Bacteria	*E. coli* ATCC 11229	2.5–5
	E. coli O157:H7 ATCC 43894	1.5
	Legionella pneumophila ATCC 43660	3
	Salmonella typhi ATCC 19430	2
	Shigella dysenteriae ATCC 290287	0.5
	Staphylococcus aureus ATCC 25923	4
	Vibrio cholerae ATCC 25872	1
Algae	*Cyanobacteria*	720–1200
Protozoa/cysts	*Microsporidia*	
	Enterocytozoon	No data
	Encephalitozoon	No data
	Giardia lamblia	1.3
	Cryptosporidium parvum	7

Source: Adapted from Mackey, E.D. et al., *Practical Aspects of UV Disinfection*, AWWA Research Foundation and American Water Works Association, Denver, CO, 2001, pp. 106–109.

[a] Compiled by Mackey (2001) from the work of about 10 investigators during period 1985–2000. The data given were rounded in some cases. For the most part, the log inactivations were linear with UV dose, but not in every case. Higher log inactivations may be approximated by assuming a linear extrapolation. For example, a 4-log inactivation for *Giardia lamblia* may be estimated by multiplying the value given for 1-log inactivation by "4," for example, 4-log inactivation (*Giardia lamblia*) $\approx 4 \cdot 1.3 = 5.2$.

BOX 19.5 CRYPTOSPORIDIUM AND ITS INACTIVATION BY UV

To review some of the pertinent background, *Cryptosporidium* became an issue in 1986 when a disease outbreak in Carrollton, Georgia, was traced to its filtered water supply. The issue was considered acute as a water treatment issue after an April 1993 outbreak of some 400,000 cases of cryptosporidiosis occurred in Milwaukee. At the same time, studies showed that *Cryptosporidium parvum* oocysts were pervasive in ambient waters and that chlorination was not effective in their inactivation. Thus, in traditional water treatment, filtration was the only effective barrier. Microfiltration was effective but expensive. Ozone was effective as a disinfectant but also expensive. Consequently, when UV was found to be effective in inactivation at low dosages, it was felt to be the answer.

Investigations of UV inactivation of *Cryptosporidium parvum* oocysts started about 1991. Clancy et al. (1998) followed work of these earlier researchers, confirming that high doses of low-pressure UV, for example, 8748 mJ/cm^2, achieved 4-log inactivation of *Cryptosporidium parvum* oocysts in trials at full-scale. Then, follow-up studies by Bukhari et al. (1999) found that medium-pressure UV at dosages $\approx$ 19 mJ/cm^2 resulted in log $R \approx 3.9$; the findings were corroborated by Clancy et al. (2000), who found that UV doses of 40 mJ/cm^2 achieved 3-log inactivation. Subsequent studies during Year-2000 showed that doses $\leq$19 mJ/cm^2 resulted in 3.9-log inactivation, and that 6–9 mJ/cm^2 resulted in $\geq$3.5-log inactivation. These studies, in aggregate, established that low doses of UV could inactivate *Cryptosporidium parvum* oocysts. Further studies established that UV dose $\approx$ 3 mJ/cm^2 could result in 4.5-log oocyst inactivation for filter backwash waters (in drinking water treatment) at turbidities $\leq$11 NTU. A detailed account of the events that led to the discovery that low-dosage UV could inactivate *Cryptosporidium parvum* oocysts at log $R \approx 4$ was summarized by Clancy (1999) and later by Clancy (2003).

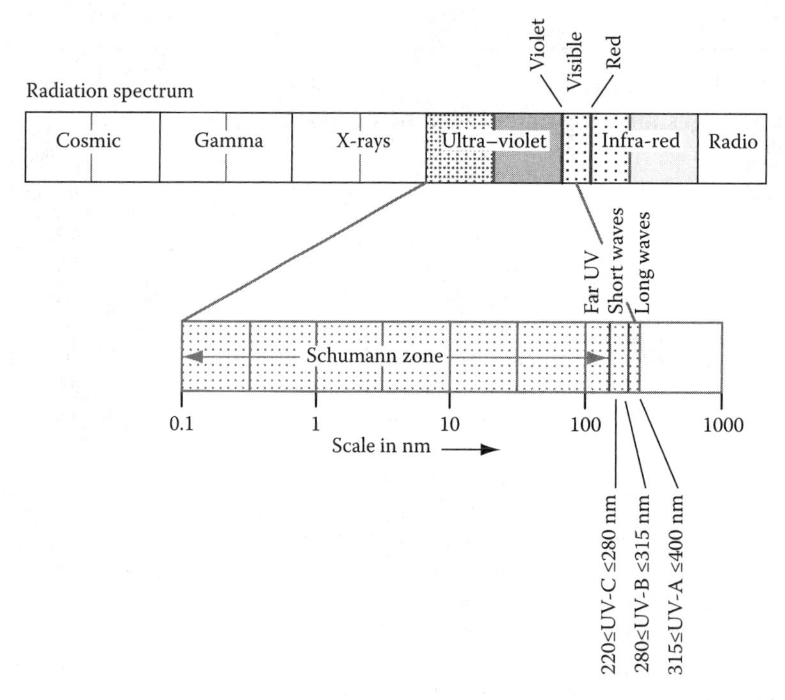

FIGURE 19.6 UV wavelengths within radiant energy spectrum. (Adapted from Masschelein, W.J., *Ultraviolet Light in Water and Wastewater Sanitation*, Lewis Publishers, Boca Raton, FL, 2002, p. 5.)

$$E = h\nu = h\left(\frac{c}{\lambda}\right) = hc\bar{\nu} \tag{19.34}$$

where

E is the energy of one photon (J)
h is the Planck's constant ($6.624 \cdot 10^{-34}$ J · s)
ν is the frequency (cycles/s or Hz)
c is the velocity of radiant energy in a vacuum, that is, $2.9976 \cdot 10^8$ m/s
λ is the wavelength (m)
$\bar{\nu}$ is the wave number, that is, $\bar{\nu} = 1/\lambda$ (1/m)

19.3.7.3.1 Radiant Energy Spectrum

Figure 19.6 shows the radiant energy spectrum. The designations for the various bands (Silberberg, 1996, p. 257) are $10^{-2} \leq \lambda(\text{x-rays}) \leq 10$; $10 \leq \lambda(\text{UV}) \leq 400$; $400 \leq \lambda(\text{visible}) \leq 750$; $750 \leq \lambda(\text{infrared}) \leq 600{,}000$; $600{,}000 \leq \lambda(\text{microwave}) \leq 10^8$; $10^8 \leq \lambda(\text{radio}) < 10^{12}$. The UV bandwidth includes the range from about $0.1 < \lambda \leq 400$ nm; within the UV band are the designations UV-A, UV-B, and UV-C, which are related to the sensitivity of the human skin. The UV-A range, $315 < \lambda(\text{UV-A}) < 400$, is the sun tanning range; the UV-B range $280 < \lambda(\text{UV-B}) < 315$ causes burning and eventually may induce skin cancer; the UV-C range $220 < \lambda(\text{UV-C}) < 280$ is absorbed by proteins, RNA and DNA, and can lead to cell mutations and cancer. The UV-C range is called the "germicidal" range (Bolton, 2001, p. 6); the portion of the UV-C range which is most germicidal is $250 < \lambda(\text{germicidal}) \leq 280$ nm with $\lambda(\text{optimum germicidal}) \approx 260$ nm. By comparison, the wavelength emitted by "low-pressure" lamps is a single wavelength, $\lambda = 253.7$ nm. The medium-pressure lamps emit radiation at several peaks in the range $240 \leq \lambda(\text{medium pressure}) \leq 580$ nm. The foregoing is from Bolton (2001) and Masschelein (2002).

19.3.7.3.2 *Mechanisms of Microorganism Inactivation*

The mechanism of microorganism inactivation was described by Bolton (2001, p. 32). For a given microorganism, its proteins, RNA, and DNA absorb UV radiation. Absorption by proteins in membranes at high dosages ultimately causes the disruption of the membrane and hence the death of the cell. At much lower dosages, however, the disruption of the DNA will inactivate the ability of the microorganism to replicate. As pointed out by Malley (2000, p. 8) the DNA damage caused by UV is the key to inactivation (rather than "killing" the cell). Hoyer (2000b, p. 22) described the distinction between chemical inactivation of a cell and UV inactivation; the former is due to the cell wall being destroyed by an oxidant and the latter is due to damage to the DNA.

19.3.7.3.3 *Mercury Lamps*

The activation of mercury atoms by electrical discharges causes the electrons to be elevated to an excited state, which then returns to the "ground" state, releasing radiation energy. Mercury has lower vapor pressure and is activated more easily than other metals, and therefore is the metal of choice in generating UV radiation (Masschelein, 2002, p. 7).

19.3.7.3.4 *Beer–Lambert Law*

Beer's law states that the intensity of a beam of monochromatic radiation decreases exponentially as the concentration of the absorbing medium increases, while Lambert's law states that the same kind of decline occurs in proportion to the thickness of the medium. The two may be combined to give,

$$\ln\left(\frac{I_o}{I}\right) = a_i bc \tag{19.35}$$

or,

$$I = I_o \cdot e^{-a_i bc} \tag{19.36}$$

where

I_o is the incident radiation intensity to a sample (kW/m^2)
I is the radiation intensity at a given distance from the incident radiation, I_o (kW/m^2)
a_i is the constant depending on the wavelength of incident radiation (kg/m^2)
b is the thickness of the cell of the medium, for example, water (m)
c is the concentration of the substance in the medium (kg/m^3)

In other words, the radiant energy, I, declines exponentially with distance, b, from the plane in which I_o is incident. At the same time, I, declines exponentially with increasing concentration, C, of whatever substance is in the water that absorbs the radiation of a given wavelength (which is the basis for determining the concentration by a spectrophotometric method). With respect to disinfection, this means that the radiation at a given point may be substantially reduced by the effect of turbidity or a chemical that absorbs the radiation of the wavelength that is incident.

For reference to the literature, the terms "absorbance," A, and "transmittance," T, are defined as $A = \ln(I_o/I)$, and $T = I/I_o$, respectively. As seen in Equation 19.35, for a given medium and radiation wavelength, the absorbance, A, increases with distance through the medium. And, as seen in Equation 19.36, the transmittance, T, declines exponentially with distance through the medium. In other words, an incident UV radiation intensity, I_o, loses its killing power in accordance with the foregoing relations, that is, either by increasing absorbance, A, or by decreasing transmittance, T (the "flip-side" of absorbance).

19.3.7.3.5 *Power of UV Radiation at Lamp*

Factors affecting UV disinfection include (Snicer et al., 2000, p. 14) the electric power applied to a tubular lamp and the distance from the axis of the lamp. The power intensity diminishes with the radial distance from the lamp simply because of the area of the surrounding concentric pseudo-cylinders increases, that is,

$$I = \frac{P(\text{lamp})/L(\text{lamp})}{4\pi r^2} \tag{19.37}$$

where

$P(\text{lamp})$ is the power emitted from lamp as radiant energy (kW)
$L(\text{lamp})$ is the length of lamp that emits radiant energy (m)
r is the radial distance from the lamp (m)

The radiation intensity at any of the concentric cylinders surrounding a tubular lamp (as modified from Snicer et al., 2000, p. 15) is the product of that given by Equation 19.37 and the Beer–Lambert law, Equation 19.46 ($I = I_o e^{-a_i bc}$), that is,

$$I = \frac{P(\text{lamp})/L(\text{lamp})}{4\pi r^2} \cdot e^{-a_i cr} \tag{19.38}$$

19.3.7.3.6 Radiant Power Efficiency

The radiant power efficiency of a lamp is defined (Bolton, 2001, p. 9) as

$$E(\text{lamp}) = \frac{P(\text{lamp})}{P(\text{wire})} \quad (19.39)$$

where

E(lamp) is the radiant power efficiency of lamp (dimensionless)

P(wire) is the power input to lamp from electric energy source (W)

In other words, if the efficiency, E(lamp), is known, for example, as a published value, one can calculate the power that must be supplied to a given lamp. As another point, if the wire power, P(wire), is increased, P(lamp) increases correspondingly.

19.3.7.3.7 Aggregate Effect of Multiple Tubes UV Radiation Intensity at Any Given Point

The intensity of radiation at any given coordinate point within a bank of UV tubes is the aggregate effect of the radiation "field" resulting from a collection of tubes. A computation such as Equation 19.38, done for each tube, would yield the UV field strength, $I(r, \theta)$ at any point in the cross section, as done by Snicer et al. (2000, p. 15).

19.3.7.3.8 Factors Affecting UV

The target microorganism must be exposed to the required flux density of the UV for such time duration that $I_o t \gg$ (dose-for-required-log R). Thus, if the lamps are fouled, or if radiation blocking particles are in the water, or if absorbing solutes are in the water, the UV emitted by a given lamp, that is, I_o, will be attenuated (Snicer et al., 2000, p. 2). A dose criterion may be met by adding sufficient banks of UV lamps in series, that is, by a longer reactor (preferable to parallel). Alternatively, the needed exposure, that is, It, may be achieved by a higher density of lamps (giving a higher probability that the needed It will be obtained), or by a higher emission strength of the lamps, that is, increasing I by increasing "P(wire)."

19.3.7.4 Reactor Design

Reactor design for UV is the same in principle as any other, but with a "target microorganism" as the basis for the mass balance reaction and kinetics.

19.3.7.4.1 Dose and Exposure

Bolton (2001) defines the UV inactivation "dose" as the UV radiation that is assimilated by the microorganism and that could cause damage to the cell. The UV dose received by a given microorganism depends on its specific path of exposure within the reactor. Of 100 microorganisms entering a reactor, a certain fraction will receive a dose that equals or exceeds the critical inactivating dose. Another fraction will receive a damaging, but repairable dose, and another fraction will have a dose that is less than what may cause damage. The distribution of doses among 100 microorganisms is due solely to the reactor hydraulics, given that other factors are the same.

Suppose, for a given reactor, that log R(inactivation) = 1.0 (i.e., 0.90 fraction inactivation for a sample collected over a period of time). This means that 90 cells were inactivated and 10 cells remained viable (and could reproduce). To achieve 4-log inactivation, that is, log R(inactivation) = 4, then four modules, of the type used for the test, must be placed in series (in the pipeline or in the channel).

19.3.7.4.2 Mathematical Modeling of a UV Reactor

Equation 19.40 gives the materials balance/kinetics modeling relation for a plug flow reactor, which for microorganisms becomes

$$\left[\frac{\partial N}{\partial t}\right]_o = -\bar{v}\frac{\partial N}{\partial Z} + D\frac{\partial^2 N}{\partial Z^2} - \left[\frac{\partial N}{\partial t}\right]_r \quad (19.40)$$

where

N is the concentration of the microorganisms ($\#/m^3$)

t is the elapsed time (min)

Z is the distance from the entrance to the reactor (m)

$\bar{v}$ is the mean velocity in reactor, that is, Q/A (m/s)

$[\partial N/\partial t]_o$ is the rate of change of microorganism concentration as "observed" in an infinitesimal reactor "slice," ΔZ, that is normal to the velocity vector, $\bar{v}$ ($\#/m^3/s$)

$[\partial N/\partial t]_r$ is the rate of change of microorganism concentration due to the reaction of microorganisms with the disinfectant, and as described by a kinetic equation ($\#/m^3/s$)

For a "steady-state" condition, the left side of Equation 19.40 is zero, that is, $[\partial N/\partial t]_o = 0$, which makes the equation more amenable to solution (see Chapter 4).

19.3.7.4.3 Kinetic Constant Determination

How to determine the kinetic constant exactly in Equation 19.5 is not apparent. Following the pattern of Example 19.1, however, one may calculate k for UV by a similar protocol:

1. For a given organism Table 19.7 gives the "dose," It, for $N/N_o = 0.1$, that is, log $R = -1$.
   For example, It(adenovirus strain 40) = 30 mJ/cm^2 for log $R = -1$.
2. From Equation 19.32, $\ln N/N_o = -k \cdot (It)$, solve for k.
   For example, $\ln(0.1) = -2.325851 = -k \cdot (30\ mJ/cm^2)$; solving for k, $k = 0.078\ cm^2/mJ\text{-}s$

The kinetic constant, k, may then be used in the kinetic equation for a quantitative solution to Equation 19.40, for example, by finite difference.

19.4 DESIGN

The heart of any design is the "reactor," which brings into contact the disinfectant and the organism. The reactor and its appurtenances comprise the "system." The system, for example, chlorine, hypochlorite, ozone, chlorine dioxide, UV, may be "packaged," in part or whole, as proprietary equipment.

19.4.1 Chlorine

The main parts of a chlorine disinfection system include (1) chlorine feed, and (2) the reactor. The chlorine feed system includes (1) providing for the logistics of delivery and the use of liquid chlorine, for example, taking delivery, storage of cylinders, specifying a proprietary feed system, monitoring the rate of consumption, etc.; (2) providing for safety; (3) sizing the reactor; and (4) monitoring chlorine throughout the system. The reactor is a volume in which contacts occur between the disinfectant and the organisms, such that the associated *Ct* results in a log *R* sufficient to meet the effluent standards.

19.4.1.1 Chlorine Feed

A chlorine feed system has several components that involves sizing and selection of equipment. The gas feed system is a proprietary "packaged" system that generally contains all of the components necessary to meter and control the gas flow and with the needed check valves and other safety devices.

19.4.1.1.1 Chlorine Demand

The term, chlorine demand, as used here means the sum of all consumption of chlorine by the various reactants, for example, ammonia, organics, biofilms, etc. For relatively "clean" product water, for example, for municipal drinking water that has a "low" chlorine demand; the chlorine concentration added to the finished water flow into the clear well may be only 2–3 mg/L as HOCl. The chlorine added to a municipal wastewater, by contrast, would be higher, for example, 6–9 mg HOCl/L (50–75 lb/million gallons, White, 1999, p. 704).

19.4.1.1.2 Chlorine Storage

Determining how often deliveries of chlorine ton-cylinders are scheduled is based on the daily use of chlorine and the number of cylinders that may be stored. Deliveries that are too frequent, for example, more than every few days, may be considered as causing an increased transport hazard and may be perceived as a nuisance to neighbors (depending on circumstances). On the other hand, a lot of storage, for example, more than several months, could be more costly than desired since the metering and storage requires special design to reduce the risks associated with chlorine use, and for security.

19.4.1.1.3 Flow Schematic

Figure 19.7 is a flow schematic of a chlorination system, which consists of (1) metering flow from the chlorine gas cylinder based on the signals from water flow, *Q*, and of the chlorine residual required as compared with the measured value; (2) mixing of the chlorine gas with a side-stream of water in a Venturi section; (3) mixing of chlorine-laden side-stream flow, *Q*(side-stream), that is, a concentrated chlorine solution, with the main flow, *Q*, by means of a diffuser, for example, a pipe with multiple orifices, a nozzle, or some other means to distribute the solution, or merely a pipe of concentrated flow in the main flow (if the main flow pipe turbulence is adequate). Regarding the third point, a multiple-orifice manifold is used, generally, for larger pipes and either an orifice manifold or nozzles are used for open channels (AWWA, 1973; ASCE-WPCF, 1977, p. 391). If a pump is required for the *Q*(concentrate) flow, it should be placed just before the point of injection into the main flow, so that a negative pressure can be maintained in the Venturi section. Figure 19.8 shows a flow controller, which is actuated by a signal from a computer that compares the chlorine flow to that required to maintain a "set-point" in the main flow.

19.4.1.1.4 Injector System

The side-stream for chlorine concentrate is designed to develop a negative pressure at the point at which the gas stream enters the side-stream flow, and mixes and dissolves in the water. This is the "injector" system, which is usually a Venturi section (other names given include "ejector," or "eductor;" see also Wolf, 1991). The negative pressure causes the gas to flow from the ton cylinder to the mixing chamber under a pressure gradient in the tubing that connects the two points. The rule-of-thumb for the side-stream flow, *Q*(side-stream), is $Q(\text{side-stream})/J(Cl_2) \approx 330$ L water/day/kg Cl_2/day, such that $C(Cl_2$ in side-stream$) \leq 3500$ mg Cl_2/L water (ASCE-WPCF, 1977, p. 389). So that the *Q*(side-stream), that is, the chlorine solution, may enter the

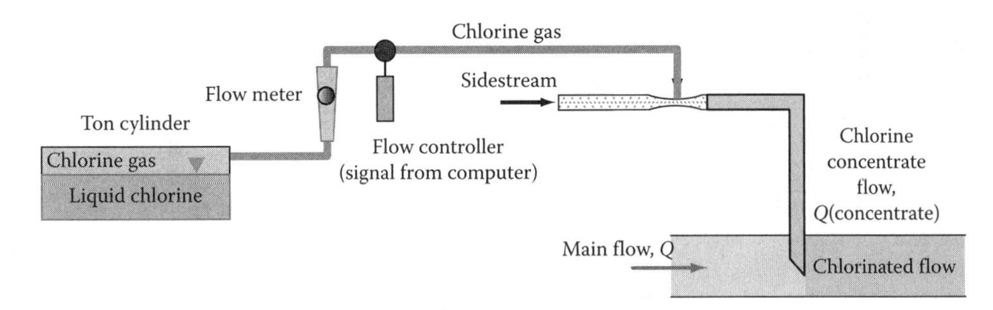

FIGURE 19.7 Schematic of chlorine gas feed system.

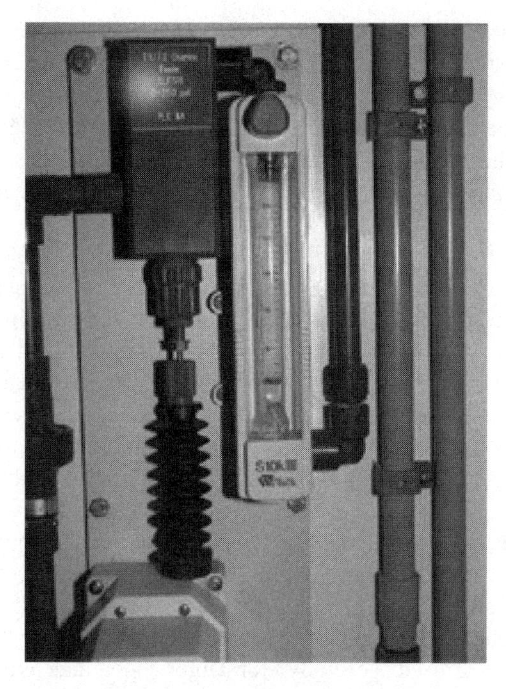

FIGURE 19.8 Rotometer for metering of chlorine gas flow with actuated control valve on left on exit side.

main flow, and have adequate pressure to both exceed the pressure in the main pipe (which should be low if the flow is entering a clear well) and overcome the pressure loss in the diffuser, a booster pump may be necessary.

19.4.1.1.5 Chlorine Metering and Control

Chlorine residual monitoring and feedback control are necessary components of a system. Either the flow signal is used to control the chlorine-metering orifice, which may be done pneumatically, for example, 20–100 kPa (3–15 psi), or electrically, for example, 4–20 mA, such that the chlorine mass flow is proportional to the water flow. Chlorination control may be by either a "compound-loop" system, or a chlorine residual signal. A compound-loop system uses two separate and independent signals to the chlorination device: (1) a flow-proportional signal to the chlorine metering orifice and (2) chlorine dosage signal to the vacuum regulating valve (or the dosage control device). The chlorination mechanism compounds these two signals to achieve a wide range of operation, for example, >100:1. On the other hand, a chlorination facility may be operated solely from a chlorine residual signal; the associated signal for chlorine dosage may be sent to the vacuum regulator on the chlorination device or to the chlorine orifice positioner. Chlorination systems using a vacuum signal are equipped with a vacuum gage calibrated to 0–2500 mm (0–100 in.).

19.4.1.1.6 Guidelines for Design

Post-chlorination, that is, chlorination following filtration, should be controlled as a proportion of flow as measured by an effluent flow meter. With respect to monthly use rate, if the plant capacity is 0.44 m^3/s (10 mgd), and if the chlorine demand is a maximum of 0.018 kg/m^3 or 18 mg/L (150 lb/mgal), which is very high, then the chlorination feed capacity must be 680 kg Cl_2/day (1500 lb/day), that is, $J(Cl_2) = Q \cdot C(Cl_2) = 0.44\ m^3/s \cdot 0.018\ kg\ Cl_2/m^3 = 0.00792$ kg Cl_2/s = 684 kg Cl_2/day. To provide for the logistics, that is, storage and delivery frequency, this is about 20,520 kg Cl_2/month, or 22 ton-cylinders/month. At 20°C and a vacuum backpressure, $J(Cl_2)max \leq 230$ kg Cl_2/day (500 lb/day) from a ton-cylinder; thus about three cylinders should be online in parallel (connected to a manifold).

19.4.1.2 Reactor Design

The chlorine reactor should provide for turbulent mixing at the diffuser, followed by plug flow to achieve the desired detention time. Long narrow channels should be used to minimize the extent of short-circuiting; perfect plug flow, is of course, not achievable. Tracer tests may be conducted to confirm dispersion curves for different flows, or computational fluid mechanics (CFD) modeling may be conducted to examine alternative designs. In most cases, retrofits are installed, for example, partitions in rectangular or circular basins to give a "serpentine" flow pattern. A reactor with about 15–60 min contact time, depending on chlorine dosage, is recommended for the disinfection of wastewaters (ASCE-WPCF, 1977, p. 394).

19.4.2 Hypochlorite

Calcium hypochlorite or sodium hypochlorite has been used increasingly as an alternative to chlorine gas as a disinfectant. Calcium hypochlorite, $Ca(OCl)_2$, is a solid and is favored. In addition, package feed and metering systems are available that feed $Ca(OCl)_2$ tablets into a concentrate solution, which then meters the solution into the main flow (Anon., 1999). The special operational problems of usage relate to scaling, corrosion, and gas coming out of the solution (Baur, 2001).

19.4.3 Ozone

Figure 19.9 is a schematic of an ozone system as set up for a 76 L/min (20 gpm) pilot plant (at the Engineering Research Center, Colorado State University, i.e., CSU). The ozone generator was a three-element unit (Model GS2-35, American Ozone Systems, Chicago, Illinois) that generated 35 g O_3/h (0.84 kg O_3/day). The categories of the system components included (1) the air preparation system, (2) the ozone generator, and (3) the reactor. The air preparation system consisted of a compressor, a water jacket air cooler, and a desiccators. The reactor was a PVC sewer pipe (oriented vertically), $d = 686$ mm (27 in.), $L = 4267$ mm (14 ft), $\theta = 21$ min, with two levels of three, 203 mm (8 in.) diameter diffusers. As seen, the reactor was a counter-current "plug flow" reactor (the plug flow being hypothetical).

Ozone generators are of two categories: (1) tube type, and (2) plate-type units. The tube-type is used most extensively and is composed of a number of tubular units; the outer tubes are stainless steel enclosed in a cooling water jacket. The inner tubes are glass dielectrics with a coated inner surface

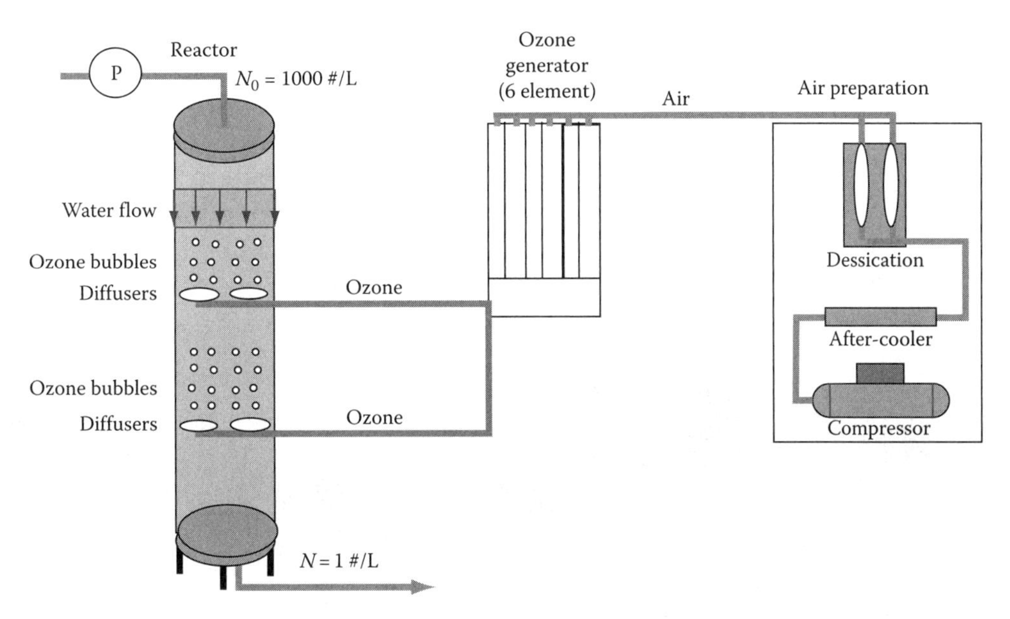

FIGURE 19.9 Schematic drawing of an ozone system, 0.84 kg O_3/day, θ(reactor) = 21 min, CSU pilot plant.

that acts as a second electrode. Feed gas flows within the space between the outer and inner tube, that is, the "discharge gap." The power required depends upon the oxygen content of the feed gas and the ozone concentration of the product gas. For a 1% ozone concentration, about 7.5–8 kWh are required to generate 0.5 kg O_3 (1 lb), for example, about 16 kWh/kg O_3.

Once the ozone is produced, it must be transferred to the water in the reactor, which requires a diffuser (see Chapter 18). Although a variety of diffusers have been used, a fine bubble diffuser set up in the counter-current mode in a pressure vessel reactor is most efficient.

The Henry's constant for ozone is $\mathbf{H}(O_3, 20°C) = 482$ mg O_3/L water/atm O_3, which is about 10 times higher solubility than oxygen. Once dissolved in water, the next task is to facilitate the contact between the ozone and other reactants, for example, microorganisms in the case of disinfection, and to provide sufficient detention time, θ, such that the needed *Ct*(target microorganism) is satisfied. Due to hydraulic dispersion, the organisms in the effluent flow will have a range of *Ct*'s.

19.4.4 Chlorine Dioxides

Chlorine dioxide is generated onsite also (i.e., as is ozone) using proprietary equipment (see, for example CDG Environmental, Bethlehem, Pennsylvania). In water treatment practice, chlorine-dioxide concentrate solutions are usually ≤4000 mg/L (Aieta and Berg, 1986, p. 62). The concentrate must be diffused throughout the reactor, for example, by a jet-mixer. The reactor should be of the "plug flow" type, with the detention time based on the *Ct* parameter. Because of hydraulic dispersion, the average detention time, θ, should be such that the calculated, $C\theta \gg Ct$(target organism), that is, due to the variation in residence times in the reactor (as seen by a tracer).

19.4.5 UV Reactors

Factors in the UV reactor design include (Malley, 2000, p. 8; Malley, 2002b, p. 2): number and type of UV lamps used; UV reactor hydraulics; level of inactivation required; location of UV in the treatment train; degree of redundancy desired; water quality characteristics that result in UV attenuation; minimum temperature for lamp operation; degree of lamp fouling by inorganic constituents such as hardness and iron; fouling by organics; number and type of UV sensors; reactor geometry; reactor material; provision for cleaning lamps; instrumentation; and controls. Low-pressure UV lamps, with operating temperature range, $40°C < T < 60°C$, may have unstable output at low water temperatures, for example, 0.5°C. Medium-pressure lamps operate at $400°C < T < 600°C$; low water temperature is not an issue (Malley, 2000, p. 11).

19.4.5.1 Hydraulics

UV reactors are designed as "plug flow" (*vis a vis* complete mix). As noted in Section 4.3.3.3, the hypothetical "plug flow" may be approximated most closely to a reactor that is long relative to its cross section. At the same time, turbulence is necessary in order to provide lateral dispersion; the idea is to expose a very large fraction of the organisms to "killing" levels of UV doses during their irregular, that is, random, longitudinal flow paths.

19.4.5.2 UV Reactors Volume

A UV reactor is a specified volume that can accommodate one or more "banks" of UV lamps and which results in the needed *Ct*'s for a specified log *R*. Due to the variation in the "residence times," that is, from hydraulic dispersion," the log *R* should be that achieved for a particular fraction of the samples, for example, log *R*(0.99 fraction of samples) ≥ 3. A UV reactor may be a section of a pipe, or an open channel. The former is found more commonly in drinking water treatment

(a)

(b)

FIGURE 19.10 UV reactor for drinking water treatment. (a) Pipe segment showing electric connections. (b) UV tubes oriented parallel to flow in pipe segment. (Courtesy of Kevin Gertig, Fort Collins, CO.)

installations and the latter in wastewater treatment. Figure 19.10 shows a UV reactor as installed in 2003 in the City of Fort Collins Water Treatment Plant. The reactor is a retrofit, installed in the 305 mm (12 in.) diameter return-flow pipeline from the solids settling basin, in a segment about 450 mm (18 in.) long.

In UV reactors for drinking water, those with low-pressure lamps are oriented with the lamp axis parallel to the flow. UV reactors for wastewater are typically in open channels and are oriented in horizontal or vertical banks. Reactor designs for medium-pressure reactors vary in the same fashion (Snicer et al., 2000, p. 20).

19.4.5.3 UV Lamps

The generation of UV by lamps involves passing an electric arc through mercury vapor. Three kinds are "low-pressure," medium-pressure," "pressure–pressure"; low-pressure lamps are the most common.

19.4.5.3.1 Low-Pressure UV Lamps

The most common lamps used for UV disinfection are "low-pressure" having a mercury vapor pressure, $10^{-3} \leq p(\text{Hg}) \leq 10^{-2}$ mm Hg with the surface operating temperature of 40°C–50°C; and as noted by DeMers and Renner in 1992, they mimic the behavior of conventional fluorescent lamps (Snicer et al., 2000, p. 18). About 0.85 fraction of the energy emitted is at wavelength, $\lambda = 253.7$ nm. The life span of a UV lamp is about 7500–8800 h, with the intensity declining with time, for example, to 0.75 fraction of its output at 0 h. Low-pressure mercury lamps are usually cylindrical, with $0.9 \leq \text{diameter} \leq 4$ cm, and $10 \leq \text{length} \leq 160$ cm (Masschelein, 2002, p. 15).

19.4.5.3.2 Medium-Pressure UV Lamps

The "medium-pressure" mercury lamps operate at pressure, $1 \leq p \leq 10$ atm, and temperature, $500°\text{C} \leq T \leq 800°\text{C}$ (Snicer et al., 2000, p. 18). These lamps emit radiation at several peaks in the range $240 \leq \lambda(\text{medium pressure}) \leq 580$ nm (from Figure 19.6, $100 < \lambda(\text{UV}) < 400$ nm). The bulb life ranges, $2000 \leq t(\text{life-span}) \leq 5000$ h. Although the power required to deliver a given UV dose is higher than that for a low-pressure lamp, there are fewer lamps, and the cost of lamps, lamp replacement, and cleaning are less.

19.4.5.3.3 High-Pressure UV Lamps

High-pressure lamps operate at $p(\text{lamp gas}) \approx 1000$ kPa, that is, 10 atm (Masschelein, 2002, p. 14). They emit continuous spectra of radiation in a range not appropriate for disinfection and thus they are not used to any extent in water treatment.

19.4.5.4 Lamp Components

A UV system has three main components: lamps, sleeves, and ballasts. In addition, the lamps must have the provision for cleaning either *in situ* or removed from the reactor. The cleaning may be done either by chemicals or mechanically by using brushes.

19.4.5.4.1 Sleeves

The mercury vapor in lamps are encased in clear, fused quartz envelopes, that is, sleeves that pass 0.85–0.90 fraction of the 253.7 nm energy generated by the lamp. The air gap between the bulb and the sleeve insulates the gas, thermally, from the flowing water. The water pressure limit is, $p(\text{water}) \leq 1034$ kPa (150 psi).

19.4.5.4.2 *Ballasts*

The lamp ballasts are transformers that control the lamp power. Two types are electromagnetic and electronic; the latter generate less heat and have a longer life than the former.

19.4.5.4.3 *UV Lamp Controls*

Control systems should include UV intensity sensors, calibrated to UV intensity; flow meters for each reactor; pressure gages for pressurized systems; hour meters to record lamp operation time. The control system should be able to shut down the UV lamps and flows in the event of high reactor temperature, low UV dose, or a higher flow rate, that causes the dose to fall below that specified.

19.4.5.5 UV Design Guidelines

UV design criteria for wastewater reclamation in California specified that under the worst operating conditions, the system should deliver a minimum UV flux of 140 $\mu W \cdot s/cm^2$ for the maximum week flow and 100 $\mu W \cdot s/cm^2$ at the peak hour for the maximum day flow. The minimum dose must be based on the following conditions (White, 1999, p. 1388):

- UV lamp output $\geq 0.70 \cdot$ new lamp output
- Transmittance through quartz sleeves ≥ 0.70
- UV flux density calculation method = point source summation
- UV dose to be achieved with a minimum of three UV banks in series

For $100 \leq \text{dose} \leq 120$ $\mu W \cdot s/cm^2$, pilot testing with a horizontal lamp configuration showed that 4-log inactivation of poliovirus was achieved, that is, $\log(N/N_0) = -4$. Backup power should be provided for continuous and reliable operation and the UV reactors should be designed for plug flow (White, 1999, p. 1390).

19.4.6 Costs

Malley (2000, p. 8) gave capital and operating/maintenance costs:

Cost per Unit of Installed Capacity				Cost per Unit of Water Treated			
$/m³		$/gal		$/m³		$/1000 gal	
Low	High	Low	High	Low	High	Low	High
12.2	21.1	0.05	0.08	0.0013	0.008	0.005	0.03

These costs, he stated, were as low as one-fifth the cost of ozone and one-tenth the cost of membrane filtration (MF) and would require only a small fraction of the space required for the other processes.

19.4.7 Case

While UV has been gaining acceptance in both drinking water and wastewater disinfection since the early 1990s, a major impetus came about in 1998 after the discovery that UV inactivated *Cryptosporidium parvum* at moderate dosage, for example, dose(UV) $\approx$ 6 mJ/cm^2. This discovery further postponed the implementation of filtration for many cities that had avoided filtration through waiver provision in regulations (Surface Water Treatment Rule, *Federal Register*, Vol. 54 (124):27486, June 29, 1989). The foregoing has led to the planning for one of the largest UV installations, that is, the New York City water supply. The Consultants Corner, *JAWWA*, Vol. 95(12):47 provided details. The contract was $35.7 million to Hazen and Sawyer, and CDM for permitting, design, and design services during construction for an 8 million m^3/day (2 bgd) facility for 9 million customers, with the completion scheduled for 2009 and the construction cost estimated to be about $500 million. The source was the Catskill and Delaware reservoir system, which provided about 90% of the water to the New York City system. The project allowed New York City to maintain its non-filtration status that precludes the need to a massive filtration facility, which could cost about $2 billion.

19.4.8 Summary

UV has been favored over chlorine for primary disinfection since about the late 1990s for several reasons: (1) the technology is effective in inactivating most organisms, including *Cryptosporidium parvum* at costs that are competitive with other technologies; (2) disinfection by-products are not formed; (3) transport and storage of hazardous chemicals is avoided (except for a much smaller rate of use of chlorine in order to provide a residual for the distribution system.)

19.5 OPERATION

Operation of the several kinds of disinfection systems involves monitoring, inspection, maintenance, and reporting. The topic is extensive and only some key issues are reviewed here.

19.5.1 Chlorine Operation

Aspects of chlorine operation include pacing chlorine feed rate to water flow with fine adjustment based on a "setpoint" for chlorine residual, maintaining a negative pressure in the chlorine storage room, providing a response procedure for chlorine leaks, maintaining a sodium hydroxide purge vat for chlorine gas, etc. In wastewater chlorination, de-chlorination has been common since the 1980s, using sulfur dioxide gas (Bernhardt, 1991). Other reducing agents such as sodium thiosulfate, sodium bisulfate, or sodium sulfite are possible but not common.

19.5.2 Ozone Operation

At the Neuilly-sur-Marne WTP, 1 of 12 plants that provides drinking water to Paris, ozone is generated by 4 ozonators, each with a capacity of 30 kg/h. Ozone is applied after rapid filtration to give an average dose, $C(O_3) \approx 1.5$ mg/L. Disinfection takes place in four reactors where ozone is diffused by porous plates, with $\theta(\text{avg}) \approx 12$ min. At the end of the reactors, $C(O_3) \approx 0.4$ mg/L, which is the setpoint for control; if a higher level is detected, the ozone dosage is reduced and *vice versa*. After ozone, chlorine is added for both secondary disinfection and maintaining a residual in the distribution system, $C(Cl_2) \approx 0.4$ mg/L (Langlais, 1991, p. 229). As to the effectiveness of ozone disinfection, data provided by Langlais (1991, p. 230) showed significant reductions of bacteria that were monitored, for example,

Microorganism	Units	Post-Filtration	Post-Ozonation
Heterotrophic Plate Count	(#/mL)	1580	3
Total coliform	(#/100 mL)	3580	0.2
Fecal coliform	(#/100 mL)	198	0.2
Fecal streptococci	(#/100 mL)	32	0

19.5.3 Ultraviolet Lamps

The issue in operation of UV lamps is to ensure that the radiation intensity is maintained. Malley (2000, p. 8) listed some of the UV operation and maintenance issues as: calibration and/or replacement of UV sensors; lamp and sleeve cleaning; lamp replacement determination; day-to-day performance monitoring protocol. The latter includes continuous monitoring of equipment performance such as sensors, lamp-out indicators, lamp hours, ballast temperatures. The monitoring data should be maintained in an easy-to-read tabular format, using plots where feasible. Data obtained by sensors may be shown in real-time plotted form directly on the computer screen. Other data may be plotted easily if the recording includes date, time, and monitored values as columns within a spreadsheet. The ensuing discussion is a brief overview of a few of the issues.

Fouling, for example, biofilms or chemical precipitants may occur on the quartz surfaces of lamps causing a decrease in the radiation intensity. Control may be by physical cleaning or by chemical bath. Sensors that detect UV intensity, I, at specified index locations will confirm that operation is continuing as intended. The sensors must be cleaned also at frequent intervals, and calibrated.

PROBLEMS

19.1 Wallace and Tiernan's Chlorination Apparatus

Given

A solution-feed apparatus is to be used to chlorinate a 1.00 m^3/s (22.8 mgd) water supply. The chlorine concentration is to be 3.00 mg HOCl/L, dissolved into a solution-feeder by a diffuser; the resulting solution is then to be fed into a pipeline between the filters and the clear well.

Required

(a) Calculate the rate of chlorine feed into the solution feeder. (b) Calculate the maximum concentration of HOCl in the solution feeder. (c) Assuming the maximum concentration of chlorine in the solution-feeder, determine the associated solution feeder flow, Q(solution).

19.2 Estimating Hypochlorite Concentration of *Giardia* Cyst Inactivation

Given

$Ca(OCl)_2$ is used as the primary disinfectant following filtration for drinking water. The average detention time in the clear well basin is 20 min and is 10 min in the pipeline prior to the first customer. The water temperature is as low as $T = 0.5°C$ in the winter. Let the required $\log R(\text{disinfection}) = 2$.

Required

Based on the spreadsheet, Table CD19.4, calculate the required Ct and then the needed $Ca(OCl)_2$ concentration.

19.3 Contact Time for Chlorine Disinfection

Given

A drinking water treatment plant applies chlorine to the flow into a clear well. The mean detention time in the clear well, a "plug flow" reactor, is $t(\text{clear well}) =$ 30 min. Conditions are: pH = 7.8, $T = 5°C$.

Required

Determine feasible combinations of °C and pH that will result in 2-log inactivation of *Giardia lamblia* cysts.

19.4 UV Reactor Design

Given

UV is to be installed in an existing WTP. $Q = 0.44$ m^3/s (10 mgd). The reactor is to be installed in a 1220 mm (48″) pipeline section; a length of 4.0 m (13.1 ft) may be used for the reactor, if needed. *Cryptosporidium parvum* is the target organism and log R(specified) = 3. Lamps are to be "low-pressure."

Required

(a) Determine the average UV radiation intensity, I, in mJ/cm^2 (i.e., $mW \cdot s/cm^2$) that must be provided by the UV lamps. Estimate the power input required for the lamps. (b) Set up a spreadsheet to explore the trade-off between θ and I(avg) in order to provide design that you may feel is more reasonable and for redundancy. All of this may be done most easily by means of a spreadsheet algorithm.

19.5 Mathematical Model for UV Reactor

Given

A UV reactor of length, Z_o, and area, A, has a detention time, θ, for flow Q.

Required

Set up a mathematical model for a steady-state UV reactor for a spreadsheet solution.

Hint: Sections 4.3.3 and 4.3.4.5, describes how this is done. Also, Equation 15.36 is a finite difference form of

Equation 19.40, which should be modified to the steady-state form, that is, $\Delta C/\Delta t = 0$ for left side. Also, the kinetic term, $\partial\bar{X}/\partial t$, is replaced by the UV kinetic term, Equation 19.34 in which I is constant along the length of the reactor and k is as determined in Section 19.3.7.4.3. Table 15.10 illustrates the pattern of spreadsheet set-up.

19.6 Application of a Mathematical Model for UV Reactor

Given

Let *Cryptosporidium parvum* oocysts be the target microorganism for a UV reactor for a municipal wastewater. Assume that 4-log inactivation is required. For calculation purposes, assume $N_o = 10{,}000$ oocysts/L (a typical concentrations of *Cryptosporidium* oocysts is $N_o \approx 200$ oocysts/L). Let $Q = 0.88$ m^3/s (10 mgd), $\bar{v} = 0.5$ m/s, and assume the reactor length, $Z_o = 3.0$ m, and assume $D/\bar{v} = 1.4$ cm (or 0.014 m). A weir or a Parshall flume is used to maintain a minimum depth (so that the UV lamps are always covered with water).

Required

Determine the concentration profile based on Equation 19.56 in finite difference form, using a spreadsheet for the profile calculations. Determine the required average in UV intensity in the reactor such that the 4-log inactivation occurs by the end of the reactor.

19.7 Estimating Effect of Dispersion

Given

Let *Cryptosporidium parvum* oocysts be the target microorganism for a UV reactor for a municipal wastewater. For calculation purposes, assume $N_o = 10{,}000$ oocysts/L (a typical concentrations of *Cryptosporidium* oocysts is $N_o \approx 200$ oocysts/L). Let $Q = 0.88$ m^3/s (10 mgd), $\bar{v} = 0.5$ m/s, and assume the reactor length, $Z_o = 3.0$ m, and assume $D/\bar{v} = 1.4$ cm (or 0.014 m). A weir or a Parshall flume is used to maintain a minimum depth (so that the UV lamps are always covered with water).

Required

(a) Calculate the dispersion profile for a frontal wave of salt moving through the reactor (b) Calculate the dispersion profile for a pulse wave of salt moving through the reactor. (c) Estimate the *Ct* for the 10% of flow, $\theta(10)$ that has moved through the reactor first. Estimate the associated concentration of oocysts, N that remain viable as they leave the reactor in the first 10% of flow. Compare the finite difference solution with the mathematical solution.

Hint: Let $[\partial N/\partial t]_r = 0$, such that Equation 19.56 becomes the same as Equation 4.5, that is, $\left[\frac{\partial N}{\partial t}\right]_o = -\bar{v}\frac{\partial N}{\partial Z} + D\frac{\partial^2 N}{\partial Z^2}$. This expression can be set up in a finite difference form and solved by a spreadsheet algorithm. The mathematical solution is seen in Equation 4.6.

ACKNOWLEDGMENTS

Kevin Gertig, water resources and treatment operations manager, City of Fort Collins, Colorado, has helped extensively with this chapter, as with others, for example, in providing photographs and reference items from the FCWTP library, and in various discussions and site visits related to the functioning of facilities.

Dr. James Malley, University of New Hampshire, graciously provided his powerpoint presentation for a 2002 AWWA (Rocky Mountain Section)-sponsored UV Workshop in Denver, Colorado, which was consulted and cited to clarify and amplify various issues on UV disinfection.

Dr. Dean Gregory, director of environmental technologies, CDG Environmental, LLC, Denver office, reviewed Section 19.3.8 on chlorine dioxide and added considerable knowledge on the topic pertinent to practice.

GLOSSARY

Absorbance: Defined as, $A = \ln(I_o/I)$, where I_o is the incident radiation, and I is the radiation at any given distance from the occurrence of the incident radiation. See also Transmittance.

Actinometry: A photochemical reaction, for which the quantum yield, is known; thus, the measurement of the chemical yield after exposure to light allows the determination of the photon flow (Bolton, 2001, p. 19).

Available chlorine: White (1999, pp. 221–223), considers this term a misnomer and that it "has no place in the field of water and waste treatment." Its origin according to White had to do with the idea of comparing the bleaching or disinfecting power of different chlorine compounds as measured by the starch-iodide test (the iodometric method) in which iodine is liberated.

Breakpoint chlorination: Satisfaction of the reaction demand for HOCl as an oxidant by reducing substances, for example, HS−, SO_3^{2-}, NO^{2-}, Fe^{2+}, ammonia, and organic compounds. Upon satisfaction of this demand, the "free chlorine" is available for disinfection.

Bromine: (1) A halogen element that reacts with water to form HOBr and OBr^-. Discovered in 1827 and named after the French word brome. (2) Nonmetallic halogen element that is isolated as a deep red corrosive toxic volatile liquid of disagreeable odor (http://www.merriam-webster.com/). MW = 79.904 g/mol. Disinfection properties are similar to chlorine except that bromine is effective as HOBr at pH < 8.5 (as compared with pH < 7.5 for chlorine); bromine also reacts with ammonia in a sequence of reactions similar to chlorine, that is, forming bromamines.

Calcium hypochlorite: $Ca(OCl)_2$ is a white granular powder with 70% "available chlorine." "Available chlorine" is calculated as 102 g HOCl/143 g Ca$(OCl)_2$ = 0.72 g HOCl/g $Ca(OCl)_2$. (ASCE-WPCF, 1977, p. 378). Sodium hypochlorite, NaOCl, is another non-gaseous alternative to chlorine gas. The latter was used by the San Jose Water Co. after a 1993 county toxic gas ordinance caused the switch from chlorine gas. After the switch, however, the 12.5% NaOCl in solution reacted with calcium carbonate (280 mg/L as $CaCO_3$) to form a concrete-type scale inside the pipes at the point of injection (Victorine, 1999, p. 1). The problem was solved by adding CO_2 gas that lowered the pH and eliminated the scaling.

CFR: The Code of Federal Regulations (CFR) is the codification of the general and permanent rules published in the *Federal Register* by the executive departments and agencies of the federal government. It is divided into 50 titles that represent broad areas subject to federal regulation. Each volume of the CFR is updated once each calendar year and is issued on a quarterly basis. Title 40 is Protection of the Environment (from/www.gpoaccess.gov/CFR).

cfu, colony forming unit: A measure of bacteria concentration.

Chlorine: Halogen element that is isolated as a heavy greenish yellow gas of pungent odor and is used especially as a bleach, oxidizing agent, and disinfectant in water purification. Atomic weight = 35.4527 g/mol; boiling point = −34°C (−29.3°F), which is the temperature at which chlorine liquid vaporizes. Chlorine hydrate, that is, $Cl_2 \cdot 8H_2O$, may crystallize at < 9.6°C (49.3 F) at atmospheric pressure. **H**(Cl_2, 20°C) = 7283 mg/L/atm. The density of liquid chlorine is 1.468 kg/L at 0°C. The pressure of liquid chlorine at 15.6°C (60°F) is 590 kPa (5.82 atm or 85.61 psi). Specific gravity of dry gas = 2.482 at STP. Viscosity of gas is about the same as saturated steam at $1 \leq p \leq 10$ atm; viscosity of liquid is about one-third that of water at the same temperature in the range $0°C < T < 66°C$. (Foregoing is from Chlorine Institute, 1969.) [The Chlorine Institute (1969, p. 17) stated that chlorine is only "slightly soluble" in water. However, Table H.5 shows that the chlorine solubility (20°C) at 1.00 atm Cl_2 is 7283 mg/L. While not as high as SO_2, that is, 112,800 mg/L at 20°C, p(SO_2) = 1.00 atm, chlorine solubility is much higher than O_2, that is, 43 mg/L at 20°C, p(O_2) = 1.00 atm. The comparisons are worth mentioning because the characterization as "slightly soluble" by the Chlorine Institute could be misinterpreted.]

Chlorine demand: The chlorine fed into the water that reacts with oxidizable impurities and, therefore, may not be available for disinfection, reported in units of mg/L.

Chlorine dioxide, ClO_2: An orange–yellow gas, liquid at 9.7°C at $p = 1.0$ atm. It is explosive in either its gaseous or pure liquid state (Doull, 1980, p. 190). For disinfection of drinking water, chlorine dioxide is usually generated onsite. MW = 35.4527 + 2 · 15.9994 = 67.4515 g/mol.

$$\mathbf{H}(ClO_2, 25°C) = 1.0\ \text{mol/L/atm} = 67451\ \text{mg/L/atm}.$$

Chlorine residual: The concentration of chlorine, in mg/L, remaining in the water after the chlorine demand has been met.

Cholera: Waterborne disease caused by *Vibrio cholerae*.

Clear well: A facility for the storage of treated water; usually the clear well provides detention time after a disinfectant chemical is applied such that the required minimum *Ct* for a specified log inactivation is achieved.

Coliform: A gram-negative, non-sporing, facultative rod that ferments lactose with gas formation within 48 h at 35°C (from Prescott et al., 1993, pG6). See also "enteric bacteria." The coliform group includes *E. coli*, *Enterobacter aerogenes*, and *Klebsiella pneumoniae* (p. 839).

Coliform removal, coliform removal efficiency: The extent of retention of coliform bacteria usually associated with removals by a filter.

Cross connection: A hydraulic connection between a given distribution system and an external source of contamination. The connection occurs by means of a negative hydraulic gradient from the source to the system. An example would be a sink of contaminated water on the third floor of a building that has a tube from a faucet extending to it. During periods of high demand, the water pressure in the main pipe may have lower pressure, which would then cause the sink contents to flow into the main.

***Cryptosporidium parvum*:** A pathogenic protozoan that causes enteritis and/or severe diarrhea; the oocyst is resistant to chlorine disinfection.

***Ct*:** The product of residual disinfectant concentration (C) in mg/L and the corresponding disinfectant contact time (t) in min.

Cylinder (for chlorine gas): Cylinder of ≤150 lb (68 kg) capacity used to transport and store chlorine gas in liquefied form (Chlorine Institute, 1969, p. 5). The cylinder in the weight class, "heavy," is about 10.5 in. (267 mm) diameter × 55 in. (1400 mm) with empty weight about 130 lb (59 kg). See also "ton container."

DBP: Disinfection by-product (see Appendix 2.A). DBP's may include a host of reaction products between a disinfectant chemical, for example, HOCl, and organic compounds; most occur with chlorine but ozone was found to produce a few. Chlorine dioxide does not produce disinfection by-products (Anon., 1986, p. 33). UV does not cause DBP's.

Detritus: Non-living organic matter that occurs as particles in natural waters (the definition used here).

Disinfectant: A substance, for example, chlorine as HOCl that may inactivate some or many microorganisms.

Disinfection: Inactivation of microorganism, that is, such that they cannot reproduce.

Dose: (1) The product of radiant energy incident on a unit area within a radiant energy field, *I*, and the time of exposure, θ, that is, dose $= I \cdot \theta$ (J/m^2). This is the more commonly used definition (see also, "fluence"). (2) Bolton (2001, p. 13) defines "dose" as that portion of the incident radiation that is absorbed by the organism, which is perhaps 1%.

Eductor: A term used in relation to the injection of chlorine gas into a Venturi section. Merriam–Webster (http://www.merriam-webster.com/) gave, "late Latin, one that leads out, from Latin educere." Sometimes either, "ejector" or "injector" is the term used. [These terms illustrate the common use of nebulous vernacular in the trade. Some terms used in the water and wastewater treatment industry are traditional and go back to the early part of the twentieth century when the prose was perhaps "classical" and more descriptive of "what works," rather than to indicate principles.]

Endospore: Dormant and resistant form of bacteria, particularly gram-positive bacteria, for example, *Bacillus*, *Clostridium*, etc. The spore structures are resistant to environmental stresses such as heat, ultraviolet radiation, chemical disinfectants, and desiccation. Their structure may include a spore coat, cortex, and inner spore membrane surrounding a protoplast; although distributed widely, they are largely soil inhabitants (p. 459). Some endospores have remained viable for 500 years. Some may survive boiling for an hour and so autoclaving is necessary for sterilization. The formation of an endospore is a complex process and results in a cell wall structure of several kinds of layers. Germination is also a complex process. [Foregoing from Prescott et al. 1993, p. 62.]

Enteric bacteria: (1) Members of the family *Enterobacteriaceae* (gram negative, straight rods, etc.,) (2) Bacteria that live in the intestinal tract. (both definitions from Prescott, et al., 1993, p. 435, pG9). Included are the genera, *Escherichia*, *Enterobacter*, *Proteus*, *Salmonella*, and *Shigella* (pp. 435–438). *Salmonella typhi* causes typhoid fever and gastroenteritis; *Shigella*, bacillary dysentery; some strains of *E. coli* cause gastroenteritis (p. 438).

Enterovirus: Genus of viruses of the human GI tract (of the family *Picornaviridae*), polio is an example (see Prescott et al., 1993, pA29).

Fecal coliform: Coliforms derived from the intestinal tract of warm-blooded animals that can grow at 44.5°C.

Field: (1) A region or space in which a given effect (as magnetism) exists (http://www.meriam-webster.com). (2) As applied to UV in disinfection of water, there would be a UV radiation field generated by multiple UV tubes. The field would vary in strength, that is, incident radiation, depending on the spatial position. The field strength, that is, energy flux density, would be reduced with radial distance from the tube, that is, $I = I_o/r$, due to the increasing cylindrical area, in which I_o = UV energy flux density (W/m^2) at the "sleeve" of the UV tube, and I = UV energy flux density (W/m^2) at any radial distance (in a vacuum in which there is no attenuation by absorbance), *r*, from the center of the tube. In reality, the radiant energy is reduced, not only by the increasing area covered by a given total energy flux, but by attenuation due to absorbance, in accordance with the Beer–Lambert law, Equation 19.35. Within a UV reactor, the particles (e.g., microorganisms) suspended in the flowing water are transported by advection through the reactor and by turbulence to various points laterally from one streamline to another. Thus, the particle will pass through a UV field that has varying strength spatially and it thus will be exposed to varying flux densities, depending upon its position within the UV energy field.

Fluence: Total radiant energy (flux density) over a given time period incident on a small sphere (J/m^2), as given by Bolton (2001, p. 12). Technically, according to Bolton, "dose" is the absorbed radiant energy, for example, by an organism, which is a small fraction of the incident radiation.

Fluence rate: Total radiant energy (flux density) incident on a small sphere (W/m^2), as given by Hoyer (2000a, p. 22).

Flux: (1) The rate of transfer of fluid, particles, or energy across a given surface (http://www.meriam-webster.com). (2) Radiant flux—the rate of emission or transmission of radiant energy (http://www.meriam-webster.com). (3) Flux may be derived from a "field," once the latter is defined, for example, as a magnetic force field, a velocity field, a pressure field. The gradient of a scalar that defines a field is a flux.

Flux density: The rate of transfer of fluid, particles, or energy across a given surface divided by the area of that surface.

Free available chlorine: Sum of [HOCl]+[OCl$^-$] in mol/L (ASCE-WPCF, 1977, p. 379; White, 1999, p. 223). Also called "free-chlorine."

Giardiasis: Disease caused by protozoan *Giardia lamblia*.

***Giardia*:** Name often used in conversation for the pathogenic organism, *Giardia lamblia*. There are several species of the genus *Giardia*. But only the *Giardia lamblia* species is felt to cause disease in humans.

***Giardia lamblia*:** A protozoan pathogenic to humans that causes severe diarrhea. The organism is ingested in the cyst form. Excystation occurs in the stomach.

***Giardia muris*:** A protozoan that infects mice and thought to be not pathogenic to humans.

Hydrolysis: Cleaving a molecule by reaction with water in which one part of the molecule bonds with to the water OH and the other to the water H (Silberberg, 1996, pG-9, p. 14.4).

Hypochlorite: Hypochlorite ion is OCl^-. Common forms are sodium hypochlorite, NaOCl, solution, called "bleach" and calcium hypochlorite, $Ca(OCl)_2$. As household bleach, the NaOCl solution strength varies 3%–6%. In water treatment, 12%–15% NaOCl solutions are used; NaOCl occurs only in the solution form. More commonly, however, $Ca(OCl)_2$, that is, "bleaching powder," is used for larger scale uses since it is cheaper than NaOCl solution and is produced in a solid form. Bleaching powder is actually a mixture of calcium hypochlorite $Ca(ClO)_2$ and the basic chloride $CaCl_2$, $Ca(OH)_2$, H_2O with some slaked lime, $Ca(OH)_2$.

Inactivation: (1) The rendering of an organism not capable for reproduction. (2) Killing of an organism, that is, rendering the organism not capable of metabolism (or of reproduction). [See also, *disinfection*.]

Iodine: A nonmetallic halogen element obtained usually as heavy shining blackish gray crystals and used especially in medicine, photography, and analysis. Discovered in 1814 and named from the Greek "ioeides." MW = 126.90447 g/mol. Iodine is as effective as chlorine as a germicide and does not react with ammonia and is effective at pH = 7 as about 50% I_2 and 50% HOI.

Irradiance: (1) Radiance. (2) The density of radiation incident on a given surface usually expressed in watts per square centimeter or square meter (http://www.merriam-webster.com).

Log removal: Log removal, that is, log **R**, as used in disinfection, is the same as that used in filtration; it is a means to compare the influent concentration (of a microorganism in the case of disinfection) entering a reactor to the effluent concentration leaving the reactor, that is,

$$\log \mathbf{R} \equiv \log C_o - \log C = \log \frac{C_o}{C}$$

where log **R** is the common log of the removal ratio. The relationship between the %**R** and log **R** is,

$$\%\mathbf{R} = 100 - 10^{-\log \mathbf{R}}$$

The definition for log **R** and the associated relations may be used for any reactor, for example, for adsorption, oxidation, filtration, etc.

Low-Pressure UV Lamps: The most common lamps used for disinfection are "low-pressure" having a mercury vapor pressure, $10^{-3} \leq p(\text{Hg}) \leq 10^{-2}$ mm Hg with surface operating temperature 40°C–50°C; as noted by DeMers and Renner in 1992, they mimic the behavior of conventional fluorescent lamps (Snicer et al., 2000, p. 18). About 0.85 fraction of the energy emitted is at $\lambda = 253.7$ nm; the conversion of electric energy applied to light is about 0.35–0.40. The life span is about 7500–8800 h (based on one restart every 8 h. The operating intensity declines with time, for example, to 0.75 fraction of its output at 0 h by the end of its life span, for example, t(life span) ≈ 8000 h.

Low-pressure high output (LPHO) lamp: The lamp has the same monochromatic output of $\lambda = 253.7$ nm as the low-pressure lamp, but the output power per unit length is about 2–3 times that of the low-pressure lamp (Bolton, 2001, p. 17).

MCL: Maximum contaminant level (*Federal Register*, December 16, 1998).

Medium-pressure UV lamps: The medium-pressure mercury lamps operate at pressure, $1 \leq p(\text{Hg}) \leq 10$ atm, and temperature 500°C–800°C. Because of its higher emission output, a medium-pressure lamp may replace 6–16 low-pressure lamps. The spectrum of light energy emitted is more broadband than low-pressure lamps. The lamp can carry up to 30,000 W. At maximum output, the bulb life ranges, $2000 \leq t(\text{life-span}) \leq 5000$ h. Although the power required to deliver a given UV dose is higher than for a low-pressure lamp, the present value analysis shows that the cost is less than low-pressure lamps (because there are fewer lamps, the cost of lamps, lamp replacement, and cleaning are less). The emission spectra cover $200 < \lambda < 400$ nm with several broad peaks, most of which are outside the germicidal range. Those within the germicidal range seem to have appreciably more energy than the low-pressure lamps (see Bolton, 2001, p. 18).

MRDL: Maximum residual disinfectant level (*Federal Register*, December 16, 1998).

MRDLG: Maximum residual disinfectant level goal (*Federal Register*, December 16, 1998).

Log removal, log R: Defined: $\log \mathbf{R} \equiv \log N_{in} - \log N_{out}$ in which N_{in} is the concentration in the influent and N_{out} is the concentration in the effluent of whatever constituent species is being measured. Conversion to %**R** is: $\%\mathbf{R} = 1 - 10^{-\log \mathbf{R}}$.

Monochromatic: (1) Radiation of a single wavelength. (2) From the Greek, "one color" (Silberberg, 1996, p. 258).

NPDES: National Pollution Discharge Elimination System (an acronym introduced with PL-500).

Ozone: A highly reactive gas, produced by passing a stream of dry air or oxygen between the electrodes. To be effective as a water disinfectant the ozone gas must be dissolved, usually by a diffuser located in the reactor. MW = 3 · 15.9994 = 47.9982 g/mol. $\mathbf{H}(O_3, 20°C) = 482$ mg/L/atm.

pC: Negative logarithm of the concentration of an ion or molecule. For example, for a Clorox solution as packaged, the label reads, "6% sodium hypochlorite," which is 60 g NaOCl/L solution, or 60/74.5 or 0.80 mol NaOCl/L solution. pC = −log(0.80) = −(−0.097) ≈ 0.10. If the dilution is ten times, pC ≈ 1.1; if diluted by a factor of "100," then pC ≈ 2.1.

pH: Negative logarithm of the hydrogen ion concentration. For example, if $[H^+] = 10^{-7}$ mol/L, then pH = –log $[10^{-7}] = 7.0$.

p*K*: Negative logarithm of the equilibrium constant for a given acid–base reaction. To illustrate, for the reaction, $HOCl \rightarrow H^+ + OCl^-$, the equilibrium statement is $K_2 = [H^+][OCl^-]/[HOCl] = 10^{-7.5}$. Therefore, $pK_2 = -\log(10^{-7.5}) = -(-7.5) = 7.5$.

If pH < 7.5, then HOCl predominates, if pH > 7.5, then OCl^- predominates.

Percent removal, %R: Defined: $\frac{N_{in} - N_{out}}{N_{in}} \cdot 100$, see also log removal.

Photochemistry: Breaking or rearrangement of chemical bonds within a molecule (Bolton, 2001, p. 23).

Pilot plant: A small-scale replica of a proposed or existing full-scale facility, useful in determining, at relatively low expense, the feasibility of the full-scale plant in achieving the desired finished water quality given the raw water characteristics.

Polio virus: The polio virus is of the family *Picornaviridae* ($MW \approx 2.5 \cdot 10^6$, diameter 22–30 nm), genera *Enterovirus* (viruses of the gastrointestinal tract p. A29). The shape of the polio virus is spherical, (p. 356) with a protein shell surrounding its nucleic acid called a "capsid", p. 355. (Foregoing from Prescott et al., 1993, pp. 354–356, p. A29).

Polychromatic: (1) Radiation of a many wavelengths. (2) From the Greek, "many colors" (Silberberg, 1996, p. 258).

Pre-ozonation: Oxidation of the raw water prior to filtration.

Pulsed UV: Delivers UV radiation in high intensity pulses that may be repeated many times per second. The pulses deliver intensities thousands of times stronger than conventional UV. [Foregoing by Anne Stobaugh, LightStream Technologies, as taken from the web site, http://www.wateronline.com, reported in *UVA News*, Vol. 2(2):8, 2000.]

Radiance: The flux density of radiant energy per unit projected area of radiating surface (http://www.merriam-webster.com).

Radiant energy: Energy transmitted through space as "waves" (or alternatively, photons). The waves have a frequency, ν, and wavelength, λ; the product is the speed of light. Alternatively, the energy form is called electromagnetic energy or electromagnetic radiation. The radiation covers a spectrum of different wavelengths (and frequencies), given designations such as gamma rays, x-rays, ultraviolet, visible, infrared, microwave, and radio (from short-wave to long-wave, respectively).

Rotometer: A rotometer is a tapered tube, oriented vertically with the narrow part at the bottom, with the cross section increasing with distance. A ball, usually plastic for gases, is smaller than the narrowest section. The height of the ball in the tube increases with flow and so the flow can be calibrated in accordance with the flow. A rotometer is called a "flow meter."

Silver: Metal that has been used to control bacterial growths, especially for small volumes of stored water. MW = 107.8682 g/mol.

Sodium hypochlorite: NaOCl solution; light-yellow liquid; 12%–15% available chlorine in commercial bulk quantities. 12% "available chlorine" = 12 g HOCl/100 g solution = 12 g HOCl/(12 g HOCl + 88 g H_2O) [ASCE- WPCF, 1977, p. 378].

Sterilization: 100% "kill" of organisms within a defined volume; the term "kill" means here that the organisms cannot metabolize nor, by corollary, reproduce.

SWTR: Surface water treatment rule (from *Federal Register*, Vol. 54, No. 124:27486, June 29, 1989).

THMFP: Trihalomethane formation potential, which is a measure of the quantity of chlorine that may be depleted from the solution over standard time duration, for example, 24 or 96 h.

Ton container: Large cylinder, diameter 30 in. (762 mm), length 80–82 in. (2050 mm) of 2000 lb (909 kg) capacity used to transport and store chlorine gas in liquefied form (Chlorine Institute, 1969, pp. 5, 6). Empty weight is about 1500 lb (680 kg).

Transmittance: (1) Defined as $T = I/I_o$, where I_o is the incident radiation and I is the radiation at any given distance from the occurrence of the incident radiation. See also *Absorbance*. (2) A measure of the intensity of radiation that passes through a given length of water (or other medium). Two kinds of transmittances are (1) internal transmittance, which is the energy loss due to absorption within the solution, and (2) total transmittance, which is due to absorption, reflection, scattering, etc. The transmittance is always for a given path length, for example, $l = 10$, 50, 100 mm. Potable water transmittance is typically 0.90, while the transmittance of wastewaters is typically 0.65. Over a distance of 20 cm, these waters attenuate UV(254) by 0.88 and 0.999, respectively. [Foregoing adapted from Bolton (2001, p. 24).]

Ultraviolet: (1) Radiation in the range 0.1–400 nm. For reference, $400 < \lambda(\text{visible light}) < 750$ nm and $\lambda(\text{x-rays}) < 10$ nm and $\lambda(\gamma\text{-rays}) < 0.1$ nm. (2) The UV ranges of greatest interest are designated:

UV-A: 315 < UV-A < 400 nm

UV-B: 280 < UV-B < 315 nm

UV-C: 220 < UV-C < 280 nm

Both UV-A and UV-B reach the earth's surface (foregoing from Masschelein, 2002). The germicidal range is the portion of UV-C in the range, $250 < \lambda(\text{UV-C/germicidal}) < 280$ nm.

UVA: Ultraviolet absorbance. Term relates to the instrumental method to measure dissolved organic carbon.

UV bank: An array of UV modules arranged in parallel.

UV dose: The product of the average UV intensity in milliwatts per square centimeter (mW/cm^2) and the average exposure time of the water treated in seconds, that is, $mW \cdot s/cm^2$. or mJ/cm^2 (White, 1999, p. 1388).

UV lamp—broadband pulsed: Alternating current is stored in a capacitor and energy is discharged through a high-speed switch to form a pulse of intense emission of light within about 100 μs (Masschelein, 2002, p. 36). The emission is similar in wavelength to solar light.

UV lamp—high pressure: A UV lamp with mercury gas pressure $p < 1000$ kPa (10 atm). The emission is a continuous spectra less appropriate for specific applications such as water disinfection (foregoing from Masschelein, 2002, p. 14).

UV lamp—low-pressure: A UV lamp with mercury gas pressure $0.1 < p < 1$ kPa. The wavelength of the resulting radiation is monochromatic at $\lambda = 253.7$ nm (Snicer et al., 2000, p. 16). [As a note, this wavelength is absorbed by organic compounds; in fact, the absorbance may be used as a surrogate measure for certain organics.]

UV lamp—medium pressure: A UV lamp with mercury gas pressure of $100 < p < 300$ kPa at a plasma temperature of $5000 < T < 7000$ K. The wavelength is polychromatic in the range $240 < \lambda < 580$ nm; the UV output is proportional to the voltage input (foregoing from Masschelein, 2002).

UV module: An array of UV lamps arranged parallel or perpendicular to flow.

Venturi: A pipe with a narrowing section, a "throat," and then an expanding section. The pressure at the throat is less than in the pipe before the constriction starts (or in the full pipe section after the expanding section) and for this reason is favored as a point for the injection of another fluid, for example, a gas. If the pressure is not too high in the main pipe flow, the pressure at the throat may be less than the atmospheric pressure, which provides a measure of safety in that controls are designed to shut down the gas flow if the pressure at the point of injection becomes zero or positive with respect to the atmosphere. Traditionally, Venturi sections are used for metering, particularly if the geometry of the section is in accordance with the accepted proportions (if not the Venturi section should be calibrated if used for metering).

Viability: For a parasite, viability is the ability of the organism to complete its life cycle in a suitable host (Finch et al., 2001, p. 260).

REFERENCES

Aieta, E. M. and Berg, J. D., A review of chlorine dioxide in drinking water treatment, *Journal of the American Water Works Association*, 78(6):62–72, June 1986.

Anon., The Wallace & Tiernan Series A-334 Chlorinator (Specifications), Industrial Products Division, Wallace & Tiernan, Inc., Belleville, NJ, 1968.

Anon., Chlorine dioxide—Theme introduction, *Journal of the American Water Works Association*, 78(6): 33, June 1986.

Anon., Cholera in the Americas—1991, *AIDIS—USA Newsletter*, 13(4): 10, December 1991.

Anon., Accu-Tab™ Tablet Chlorination System, PPG Industries, Evansville, IN, Inc., Hammonds Technical Services, Houston, TX, 1999.

Anon., The Chlorite Manual, Water Technologies, Sterling Pulp Chemicals, Toronto, Canada, 2001.

ASCE-WPCF, Chapter 20, Disinfection, in *Wastewater Treatment Plant Design*, Manual of Practice No. 8, ASCE, New York, 1977.

AWWA, Water Chlorination Principles and Practices, AWWA Manual M20, American Water Works Association, Denver, CO, 1973.

Baur, R., Puzzle me this, puzzle me that—Troubleshoot your hypochlorite system piece by piece, *Water, Environment & Technology*, 13(3): 59–62, March 2001.

Benjamin, M. M., *Water Chemistry*, McGraw-Hill, New York, 2002.

Bernhardt, S., Sulfur dioxide dechlorination of wastewater, *43rd Annual Convention of the Western Canada Water and Wastewater Association*, Winnipeg, Manitoba, Canada, September 25–27, 1991.

Bolton, J. R., *Ultraviolet Applications Handbook*, 2nd edn., Bolton Photosciences, Inc., Ayr, Ontario, Canada, 2001 (40 pages).

Bukhari, Z., Hargy, T., Bolton, J., Dussert, B., and Clancy, J., Medium pressure UV light for oocyst inactivation, *Journal of the American Water Works Association*, 91(3): 86–94, March 1999.

Chick, H., An investigation of the laws of disinfection, *Journal of Hygiene*, 8: 92–158, 1908.

Chlorine Institute, Chlorine Manual, Chlorine Institute, New York, 1969 (5th printing, April 1976).

Clancy, J. L., Ultraviolet light—A solution to the *Cryptosporidium* threat, *UV News,* International UltraViolet Association, 1(1): 18–22, 1999.

Clancy, J. L., Hargy, T. M., Marshall, M. M., and Dyksen, J. E., UV light inactivation of *Cryptosporidium parvum* oocycsts, *Journal of the American Water Works Association*, 90(9): 92–102, 1998.

Clancy, J. L., Bukhari, Z., Hargy, T. M., Bolton, B., Dussert, B., and Marshall, M. M., Using UV to inactivate *Cryptosporidium*, *Journal of the American Water Works Association*, 92(9): 97–104, 2000.

Clancy, J. L., Hargy, T. M., and Battigelli, D. A., Susceptibility of multiple strains of *C. Parvum* to UV light, AWWA Research Foundation and American Water Works Association, Denver, CO, 2002.

Clancy, J. L., Following the light: How UV research laid the groundwork for *Cryptosporidium* solutions, *Drinking Water Research*, 13(2): 2–6, 2003.

Clark, R. M., Read, E. J., and Hoff, J. C., Analysis of Inactivation of *Giardia lamblia* by chlorine, *Journal of Environmental Engineering*, American Society of Civil Engineers, 115(1): 80–90, February 1989.

Clark, R. M., Sivaganesan, M., Rice, E.W., and Chen, J., Development of a *Ct* equation for the inactivation of *Cryptosporidium* oocysts with ozone, *Water Research*, 36(12): 3141–3149, July 2002.

Clark, R. M., Sivaganesan, M., Rice, E.W., and Chen, J., Development of a *Ct* equation for the inactivation of *Cryptosporidium* oocysts with chlorine dioxide, *Water Research*, 37(11): 2773–2783, June 2003.

Doull, J., (Chair of Safe Drinking Water Committee), *Drinking Water and Health*, Volume 2, National Academy Press, Washington, DC, 1980.

Environmental Protection Agency (EPA), National Primary Drinking Water Regulations on Disinfection and Disinfection Byproducts, Final Rule, *Federal Register*, 63(242): 69389–69476, December 16, 1998.

Fair, G. M., Morris, J. C., Chang, S. L., Weil, I., and Burden, R. P., The behavior of chlorine as a water disinfectant, *Journal of the American Water Works Association*, 40(10): 1051–1061, October 1948.

Federal Register, Surface Water Treatment Rule, Vol. 54 (124):27486, June 29, 1989.

Finch, G. R., Haas, C. N., Oppenheimer, J. A., Gordon, G., and Trussell, R. R., Design criteria for inactivation of *Cryptosporidium* by ozone in drinking water, *Ozone: Science & Engineering, Journal of the International Ozone Association*, 23(4): 259–284, 2001.

Freiser, H., *Freiser Equilograph*, Freiser Equilograph Co., Tucson, Arizona, 1964 (the Fresier Equiligraphs are a set of overhead transparencies found in pocket of *Lecture Notes in Analytical Chemistry*).

Freiser, H., *Lecture Notes in Analytical Chemistry*, Department of Chemistry, University of Arizona, Freiser Equilograph Co., 1996.

Gregory, D., Personal Communication, November 2009.

Gregory, D., Personal Communication, January 2010.

Hibler, C., Hancock, C. M., Perger, L. M., Wegrzn, J. G., and Swabby, K. D., Inactivation of *Giardia* Cysts With Chlorine at 0.5 C to 5.0 C, American Water Works Association Research Foundation, 1987 (all data for 3-log inactivation).

Hill, A. G. and Rice, R. G., Historical background, properties and applications, Chapter 1 in Rice, R.G. and Netzer, A. (Eds.), *Handbook of Ozone Technology and Applications*, Volume 1, Ann Arbor Science Publishers, Ann Arbor, MI, 1982.

Hoyer, O., The status of UV technology in Europe, *IUVA News*, International Ultraviolet Association, 2(1): 22–27, 2000a.

Hoyer, O., UV disinfection of drinking water supply, *IUVA News*, International Ultraviolet Association, 2(5): 22–25, 2000b.

Hunter, G., The history of UV disinfection in the last 20 years, *UV News,* International UltraViolet Association, 2(3): 5–7, 2000.

Isaac, R. A., Disinfection dialogue—The evolving perception of need, benefits, and determents, *Water, Environment & Technology*, 8(5): 67–72, May 1996.

Langlais, B., Reckhow, D. A., and Brink, D. R., *Ozone in Water Treatment*, Lewis Publishers, Chelsea, MI, 1991.

Lev, O. and Regli, S., Evaluation of ozone disinfection systems: Characteristic time *T*, *Journal of Environmental Engineering*, 118(2):268–285, March/April 1992 (data for 0.5°C, pH 6–9).

Li, H., Finch, G. R., Smith, D. W., and Belosevic, M., Chlorine dioxide inactivation of *Cryptosporidium parvum* in oxidant demand-free phosphate buffer, *Journal of Environmental Engineering, American Society of Civil Engineers*, 127(7): 594–603, July 2001.

Lide, D. R., *Handbook of Chemistry and Physics*, 77th edn., CRC Press, Boca Raton, FL, 1996.

Loeb, B. L., Ozone: Science & engineering, the first 23 years, *Ozone: Science & Engineering, Journal of the International Ozone Association*, 24(6): 399–412, 2002.

Mackey, E. D., Cushing, R. S., and Crozes, G. F., *Practical Aspects of UV Disinfection*, AWWA Research Foundation and American Water Works Association, Denver, CO, 2001.

Malcolm Pirnie, Inc. and HDR Engineering, Inc., Guidance Manual for Compliance With the Filtration and Disinfection Requirements for Public Water Systems Using Surface Water Sources, American Water Works Association, Denver, CO, 1991.

Malley, J. P. Jr., U.S. EPA's workshop—Where do we go from here?, *UV News, International UltraViolet Association*, 1(1): 14–16, 1999.

Malley, J. P. Jr., Engineering of UV disinfection systems for drinking waters, *UV News*, International UltraViolet Association, 2(3):8–12, 2000.

Malley, J. P. Jr., Fundamentals and regulatory approval, Ultra-violet Disinfection Workshop, AWWA (Rocky Mountain Section), Denver, CO, May 2002a.

Malley, J. P. Jr., Water quality effects and other practical aspects of drinking water UV disinfection, Ultra-violet Disinfection Workshop, AWWA (Rocky Mountain Section), Denver, Colorado, May 2002b.

Masschelein, W. J., Chlorine dioxide, in Eckenfelder, W. W., Bowers, A. R., and Roth, J. A., (Eds.), Chemical oxidation—technologies for the nineties, *Proceedings of the First International Symposium*, Technomic Publishing, Lancaster, PA, 1992.

Masschelein, W. J. (translated to English by Rip G. Rice), *Ultraviolet Light in Water and Wastewater Sanitation*, Lewis Publishers, Boca Raton, FL, 2002. [International Standard Book Number 1-56670-603-3; Library of Congress Card Number 20002016078.]

MIOX Corporation, Mixed Oxidants Product Guide, in Engineering Catalog, MIOX Corporation, Albuquerque, NM, 2003. [See also http://www.miox.com]

Pankow, J. F., *Aquatic Chemistry Concepts*, CRC Press/Lewis Publishers, Boca Raton, FL, 1991. ISBN 0-87371-150-5.

Prescott, L. M., Harley, J. P., and Klein, D. A., *Microbiology*, 2nd edn., Wm. C. Brown Publishers, Dubuque, IA, 1993.

Rosenblatt, A. A., Rosenblatt, D. H., Feldman, D., Knapp, J. E., Battisti, D., and Morsi, B., Method and apparatus for chlorine dioxide manufacture, US Patent No. 5,110,580, May 5, 1992.

Silberberg, M., *Chemistry—The Molecular Nature of Matter and Change*, Mosby, St. Louis, MO, 1996.

Smith, D. B., Clark, R. M., Pierce, B. K., and Regli, S., An empirical model for interpolating C · t values for chlorine inactivation of *Giardia lamblia*, *Aqua, Journal of Water Supply Research and Technology*, 44(5):203–211, October 1995.

Snicer, G. A., Malley, J. P., Jr., Margolin, A. B., and Hogan, S. P., *UV Inactivation of Viruses in Natural Waters*, AWWA Research Foundation and American Water Works Association, Denver, CO, 2000.

Snoeyink, V. L. and Jenkins, D., *Water Chemistry*, John Wiley & Sons, Inc., New York, 1980 (ISBN 0-471-05196-9).

Sobsey, M. D., Inactivation of health-related microorganisms in water by disinfection processes, *Water Science and Technology*, 21(3):179–195, 1989.

Tiernan, M. F., Controlling the green Godess, *Journal of the American Water Works Association*, 40(10):1042–1050, October 1948.

Victorine, T., Hypochlorite Scaling Solved With CO_2, *Opflow*, American Water Works Association, 25(4):1–7, April 1999.

Watson, H. E., A note on the variation of the rate of disinfection with changes in the concentration of the disinfectant, *Journal of Hygiene*, 8: 536–542, 1908.

White, G. C., *Handbook of Chlorination and Alternative Disinfectants, Fourth Edition*, John Wiley & Sons, Inc., New York, 1999.

Willard, H. H., Merritt, L. L., Jr., and Dean, J. A., *Instrumental Methods of Analysis*, 3rd edn., D. Van Nostrand, Princeton, NJ, 1958.

Wolf, J., How to prolong the life of your gas chlorinator, *Opflow*, American Water Works Association, 17(7):1–5, July 1991.

Wright, H. B., Dose requirements for UV disinfection, *UV News, International UltraViolet Association*, 2(3):28–34, 2000.

20 Oxidation

Chemical oxidation is one of the pervasive reactions that occurs throughout the natural environment. It is also one of the unit processes in water treatment, albeit not a main one.

As applied to organic compounds, such as the substances classified as hazardous wastes or organic sludges, the goal is, in general, to "mineralize" the compound, that is, with carbon dioxide, water, and ions (e.g., Cl^-), as products. More often, however, the products are compounds that are of lesser molecular weight and possibly more biologically reactive. As an example, ozone is used sometimes as pretreatment for activated sludge treatment. Also, ozone may be used as pretreatment for granular activated carbon (GAC) adsorption, since the resulting lower molecular weight organic molecules may adsorb better. Inorganics, for example, Mn^{2+}, may be oxidized for removal as oxide particles, for example, MnO_2; the intent is to cause the reaction to occur within the treatment plant so that removal takes place, as opposed to within the distribution system.

20.1 DESCRIPTION

Oxidation is a means to transform an element to a different "oxidation state" or a compound from one species to another. In a reactor, an oxidizing agent is added in order to "reduce" the element or compound in question. Oxidation and reduction, by definition, occur simultaneously, that is, the oxidized element or compound looses electrons to the reduced element or compound which gains electrons. This simultaneous oxidation–reduction is termed a "redox" reaction. Redox reactions include the formation of a compound from its elements, all combustion reactions, reactions in batteries that generate an electric current, and the generation of biochemical energy. For synthetic organic compounds (SOCs), of which thousands have been synthesized, the "ultimate" reaction has been to form water and carbon dioxide as products, an elusive quest.

20.1.1 Applications of Oxidation Technology

Applications of the oxidation process include taste-and-odor control in drinking water, cleavage of organic molecules so that they may be more amenable to biodegradation, as a coagulation aid, etc. Also, in water treatment, ferrous iron, Fe^{2+}, may be converted to ferric iron, Fe^{3+}, and Mn^{2+} may be converted to Mn^{3+}. The respective hydroxide and oxide of these trivalent ions are insoluble in water and so may be removed by settling or filtration. The divalent ions, Fe^{2+}, and Mn^{2+}, are found in natural waters having low "redox potential," that is, with a reducing character, such as some groundwaters and some reservoir or lake waters, perhaps at deeper levels that may be devoid of oxygen. Without proper treatment, that is, oxidation and removal of the resulting insoluble precipitate, such waters are likely to cause "red-water" and "black-water" problems in the distribution system.

The use of oxidation for cleavage of organic molecules, for example, natural organic matter (NOM), came on the scene in the 1970s as a means to cause more effective adsorption or microbial degradation. As hazardous wastes became an issue, beginning from about 1980, stimulated in the USA by the 1980 Comprehensive Environmental Response, Compensation, and Liability Act (CERCLA), oxidation was one of the alternative approaches explored to cause a molecular transformation of synthetic organic molecules.

20.1.2 History of Oxidation Technology

An early application of oxidation was for taste and odor control. Oxidants used included ozone, chlorine dioxide, and permanganate. About 1960, faced with the growing problem of biosolids, that is, sewage sludge, wet-oxidation was patented but never saw wide application in practice. The successor technology, supercritical water oxidation (SCWO), came about in the early 1980s as the issue of hazardous wastes became prominent, but has not been implemented widely.

20.1.2.1 Oxidation Based on Electromotive Potential

The "Advanced Waste Treatment Research" program of the United States Public Health Service (USPHS) (active 1961–1968) explored extending the use of oxidants from such traditional areas as taste and odor control to "refractory" organics, that is, organic molecules that were not removed by biological treatment, and industrial wastes constituents such as sulfite or cyanide (Anon., 1965). The program recognized the important role of the hydroxyl radical, that is, OH•, and examined different approaches to generate the radical. The upshot, however, was that oxidation was expensive compared to other approaches, for example, adsorption by GAC.

20.1.2.1.1 Ozone

The first ozone plant in the United States was at Whittier, Indiana, constructed in 1940. In 1980, there were only five ozone plants; all were for taste and odor control. By 1996, some 130 water treatment plants used ozone for drinking water (Kaminski and Prendiville, 1996, p. 62). The largest was the (600 mgd) Los Angeles plant, which was for use as a coagulant aid. The Milwaukee drinking water treatment plants had ozone installed at the headworks of the plant, and was completed in 1998; the purpose was to reduce the probability of having viable cysts or oocysts in the drinking water (motivated by the occurrence of a cryptosporidiosis outbreak in April 1993 in which some 500,000 persons became infected).

In Europe, ozone was common from about 1906. By the 1990s, ozone had been installed in over 1000 municipal water treatment plants, mostly in France and Switzerland, with a large number in Germany. Many installations were for taste and odor control and color removal, and were backed up with chlorination. A consequence of the breakup of organic molecules by ozone is that the products can serve as more readily available substrate for microbes, which may then thrive in the form of biofilms, for example, in the distribution system. Higher doses of chlorine may be required to control such growths (White, 1999, p. 1203).

20.1.2.1.2 Permanganate

Permanganate, in the form of potassium permanganate, $KMnO_4$, has been used in drinking water treatment for taste and odor control since 1932 and for manganese removal since 1935 (these were the earliest references in Babbitt and Doland (1949, pp. 547, 549), an authoritative text of that period)).

20.1.2.1.3 Chlorine Dioxide

Chlorine dioxide had been in use for taste and odor control by 1944 (Babbitt and Doland, 1949, pp. 536, 550). This oxidant did not come into use, to any great extent, however, until about the 1990s due mostly to the search for alternatives to chlorine and due to its effectiveness both as a disinfectant and as an oxidant, for example, for oxidation of Mn^{2+}. In addition, chlorine dioxide has a variety of uses in wastewater treatment (Noack and Iacoviello, 1992).

20.1.2.1.4 Other Oxidants

During the early years of water treatment, that is, c. 1920–1950, aeration was a common unit process, with dissolved oxygen being the oxidant for manganese. Chlorine was another common oxidant for both iron and manganese control and for taste and odors.

20.1.2.2 Wet-Oxidation

A 1958 paper by F. J. Zimmerman in *Chemical Engineering* introduced "wet-oxidation" as an alternative to incineration. As a note, "wet-oxidation" means "wet air oxidation," that is, using air as a source of oxygen. The process was patented in 1958 by the Sterling Drug Co. and later was called the "Zimpro®" process. The Zimmerman process operated at $500°C < T < 700°C$, $82 < p(\text{reactor}) < 122$ atm (Anon., 1965, p. 91). For reference, $T_c = 374°C$, $p_c = 218$ atm (T_c and p_c are critical temperature and pressure, respectively). Thus, while the temperature of wet-oxidation, as described, is greater than critical, the pressure is less. Wet-oxidation was considered as a means of volume reduction, with carbon dioxide, water, and ash being the products of the reaction. The motivation for its adoption was the high mass fluxes of sludge from municipal wastewater treatment plants, such as Chicago, New York, Los Angeles, etc., and the limitations on disposal (e.g., ocean outfalls). Chicago tried the Zimpro process for several years but discontinued it in 1970 due to high costs of operation an instead transported its sludge to agricultural land in Indiana.

20.1.2.3 Supercritical Water Oxidation

The phenomenon of supercritical fluid (SCF) was discovered in 1821 by Charles Cagniard de la Tour, a French scientist (Jain, 1993, p. 806). The process was applied originally, about 1915, by the pulp and paper industry to oxidize waste lignin to carbon dioxide and water (Hochleitner, 1996, p. 48).

A state-of-the-art review of SCWO by Groves et al. (1985) listed few citations earlier than 1980. A patent by Professor M. Modell (Box 10.2) in 1982 and another in 1985 (Thornton and Savage, 1992, p. 321) stimulated the 1980s interest in SCWO. Applications of SCWO included hazardous wastes, for example, polychlorinated biphenols (PCBs) and obsolete explosives, which are insoluble in water and are difficult to treat; these compounds dissolve readily in supercritical water (SCW) because of its extraordinary solvating powers. Treatment of organic sludges from municipal wastewater treatment plants was another application.

In 1987, Professor Earnest F. Gloyna, University of Texas, Austin (Box 10.3), initiated a comprehensive and concerted research program with broad sponsorship and which was intended to bring out the potential of the process. To give an idea of the magnitude of the problem of sludge generation and hazardous wastes, Gloyna and Li (1995) stated that in 1982 the total U.S. sludge market consisted of 20 million tons of dry municipal sludge and 19 million tons of dry industrial sludge. In 1989, Texas industries alone either released or stored 893 million tons of toxic wastes. They stated further that in the United States there were about 127,000 hazardous waste generators and that at least 41,000 of these waste streams were amenable to SCWO treatment.

In 1995, the first SCWO plant was built at Austin, Texas, to treat wastes from the Texaco Chemical Co. (Svensson, 1995, p. 16). At the same time, progress was made toward overcoming some of the problems associated with reactor operation, such as sticky solids and corrosion from acids produced. A cost estimate to process municipal wastewater sludge using the MODEC® process was about \$220–\$275/metric ton dry solids (\$200/\$250/U.S. ton).

20.2 OXIDATION THEORY

Oxidation theory has a great deal of depth that is explored more fully in the body of literature and various texts (examples include: Langlais et al., 1991; Rice and Netzer, 1982; Eckenfelder et al. 1992; Snoeyink and Jenkins, 1980; Benjamin, 2002). The intent here is to provide both a "road map" for understanding the process and an introduction to supercritical oxidation.

20.2.1 Fundamentals

By definition, a redox reaction involves electron transfer. This is understood in terms of balancing an equation, for example, by the method of half-reactions, or alternatively by the method of oxidation numbers. The thermodynamics of a redox reaction is the same as for any other, except electrons are transferred.

20.2.1.1 Definitions

Some of the definitions used commonly in redox reactions are illustrated in the reaction between magnesium and oxygen, seen in flashbulbs that were common in photography before about 1970 (Silberberg, 1996, p. 153), that is,

Electrons:	Loses electrons		Gains electrons			
Role	Reducing agent		Oxidizing agent			
Reaction	**2Mg**	+	**O_2**	→	**2MgO**	(20.1)
Fate	Oxidized		Reduced			
Oxidation number	Increases		Decreases			

The reaction is shown in bold and effects are shown above and below, with the overall descriptions on the left side. As seen, the magnesium donates electrons and the oxygen accepts them. The magnesium is also the "reductant" and is "oxidized," while the oxygen is the "oxidant" and is "reduced.". Thus, it is clear that oxidation and reduction occur simultaneously. Oxidation of organic compounds is more complex but the redox principle still governs.

20.2.1.2 Enumeration of Reaction

Three ways to depict a reaction involving ions are

1. A "molecular equation," in which the compounds involved are shown in their un-dissociated state, for example,

$$Zn + 2HCl \rightarrow ZnCl_2 + H_2 \quad (20.2)$$

2. The "total ionic equation," which shows all ions that are in solution, for example,

$$Zn^o + 2H^+ + 2Cl^- \rightarrow Zn^{2+} + H_2 + 2Cl^- \quad (20.3)$$

3. The "net ionic equation" in which the Cl^- spectator ions are not shown, for example,

$$Zn^o + 2H^+ \rightarrow Zn^{2+} + H_2 \quad (20.4)$$

The first is not too useful for depicting redox reactions, and the second shows superfluous information, for example, the "spectator" ions, Cl^-, while the third shows only the participating species and is favored for the purpose of depicting a redox reaction.

20.2.1.3 Half Reactions

Another method of balancing a redox reaction is the "half-reaction" method since the overall redox reaction is the sum of an oxidation half-reaction and a reduction "half-reaction," each of which involves either the loss of electrons or the gain of electrons (Silberberg, 1996, p. 156); *standard half-cell potential* in glossary. The half-reaction method is illustrated in Example 20.1. Selected half-cell reactions are given in Table 20.1. The half-cell reactions are, by convention, shown with the electrons on the left side; in other words, the reduction half-reaction is always shown, along with the corresponding potential (referenced to the hydrogen half-cell). The "net ionic equation" is the sum of two half-cell reactions. The second half-cell reaction is the "oxidation" half with electrons on the right, that is, the half-cell reaction is reversed. The sum of potentials of the two half-cell reactions is the net potential and indicates whether the reactions will proceed spontaneously, that is, to $E^o_{reaction} > 0$. For example, in Example 20.1, the sum of the half-cell potentials is positive, that is, +0.76 V, and so the reaction is spontaneous.

TABLE 20.1
Standard Electrode (Half-Cell Potentials, 298 K)[a]

Half Reaction	$E^o_{half\text{-}cell}$ (V)
$F_2(g) + 2e^- \Rightarrow 2F^-(aq)$	+3.06
$\bullet OH + H^+ + e^- \Rightarrow H_2O(l)$	+2.87
$\bullet OH + H^+ + e^- \Rightarrow H_2O(l)$	+2.80[b]
$O_3(g) + 2H^+ + 2e^- \Rightarrow O_2(g) + H_2O(l)$	+2.07
$H_2O_2(aq) + 2H^+(aq) + 2e^- \Rightarrow 2H_2O(l)$	+1.77
$MnO_4^-(aq) + 8H^+(aq) + 5e^- \Rightarrow Mn^{2+}(aq) + 4H_2O$	+1.51
$2HOCl + 2H^+ + 2e^- \Rightarrow Cl_2(aq) + 2H_2O$	+1.60[c]
$2HOBr + 2H^+ + 2e^- \Rightarrow Br_2(l) + 2H_2O$	+1.59[c]
$Cl_2(g) + 2e^- \Rightarrow 2Cl^-(aq)$	+1.36
$Cr_2O_7^{2-}(aq) + 14H^+(aq) + 6e^- \Rightarrow 2Cr^{3+} + 7H_2O(l)$	+1.33
$MnO_2(s) + 4H^+(aq) + 2e^- \Rightarrow Mn^{2+}(aq) + 2H_2O(l)$	+1.23
$O_2(g) + 4H^+(aq) + 4e^- \Rightarrow 2H_2O(l)$	+1.23
$2NO_3^- + 12H^+ + 10e^- \Rightarrow N_2(g) + 6H_2O$	+1.24[c]
$ClO_2(aq) + e^- \Rightarrow ClO_2^-$	+1.15[c]
$2HOI + 2H^+ + 2e^- \Rightarrow I_2(s) + 2H_2O$	+1.09[c]
$Br_2(l) + 2e^- \Rightarrow 2Br^-(aq)$	+1.07
$CHCl_3 + H^+ + 2e^- \Rightarrow CH_2Cl_2 + Cl^-$	+1.02[d]
$Hg_2^{2+}(aq) + 2e^- \Rightarrow 2Hg(l)$	+0.85
$Ag^+(aq) + e^- \Rightarrow Ag(s)$	+0.80
$NO_2^- + 7H^+ + 6e^- \Rightarrow NH_3 + 2H_2O$	+0.78[d]
$Fe^{3+}(aq) + e^- \Rightarrow Fe^{2+}(aq)$	+0.77
$O_2(g) + 2H^+(aq) + 2e^- \Rightarrow H_2O_2(l)$	+0.68
$CH_3OH + 2H^+ + 2e^- \Rightarrow CH_4(g) + H_2O$	+0.63[c]
$MnO_4^-(aq) + 2H_2O(l) + 3e^- \Rightarrow MnO_2(s) + 4OH^-(aq)$	+0.59
$I_2(s) + 2e^- \Rightarrow 2I^-(aq)$	+0.53
$Cu^{2+}(aq) + 2e^- \Rightarrow Cu(s)$	+0.34
$2H^+(aq) + 2e^- \Rightarrow H_2(g)$	+0.00
$Pb^{2+}(aq) + 2e^- \Rightarrow Pb(s)$	−0.13
$Cd^{2+}(aq) + 2e^- \Rightarrow Cd(s)$	−0.40
$Fe^{2+}(aq) + 2e^- \Rightarrow Fe(s)$	−0.44
$Cr^{3+}(aq) + 3e^- \Rightarrow Cr(s)$	−0.74
$Zn^{2+}(aq) + 2e^- \Rightarrow Zn(s)$	−0.76
$Mn^{2+}(aq) + 2e^- \Rightarrow Mn(s)$	−1.18
$Al^{3+}(aq) + 3e^- \Rightarrow Al(s)$	−1.66
$Mg^{2+}(aq) + 2e^- \Rightarrow Mg(s)$	−2.37

[a] Silberberg (1996, p. 891), except as noted.
[b] Bull and Zeff (1992, p. 28).
[c] Snoeyink and Jenkins (1980, p. 450).
[d] Valsaraj (1995, p. 451).

By the same token, $\Delta G^o < 0$ for a spontaneous reaction, then this corresponds to $E^o_{reaction} > 0$. Thus, two half-reactions in which $E^o_{reaction} > 0$ will proceed spontaneously. Table 20.1 is a more extensive list of half-reactions, compiled from sources listed, that permits one to determine full reactions (illustrated in Example 20.1).

Example 20.1 Balancing by Means of Half-Reactions (Silberberg, 1996, p. 159)

Given

Reaction between zinc metal (Zn) and hydrochloric acid (HCl).

Required

Enumerate the redox reaction in terms of the two half-reactions.

Solution

The reaction is depicted first in terms of its "net ionic equation" and then as two half-reactions:

1. The "net ionic equation" is (removing the Cl^- spectator ions)

$$Zn^o + 2H^+ \rightarrow Zn^{2+} + H_2 \qquad \text{(Ex20.1.1)}$$

2. The two half-reactions are

$$Zn \rightarrow Zn^{2+} + 2e^- \quad E^o_{half\text{-}cell} = +0.76\ \text{V (oxidation half-reaction)} \qquad \text{(Ex20.1.2)}$$

$$2e^- + 2H^+ \rightarrow H_2 \quad E^o_{half\text{-}cell} = +0.00\ \text{V (reduction half-reaction)} \qquad \text{(Ex20.1.3)}$$

3. Next add the half-reactions, to give

$$Zn + 2H^+ \rightarrow Zn^{2+} + H_2 \quad E^o_{cell} = +0.76\ \text{V} \qquad \text{(Ex20.1.4)}$$

Discussion

1. As seen in Step 2, the reducing agent loses electrons, that is, Zn; the oxidizing agent, that is, H^+, gains electrons.
2. As seen in Step 3, the electrons shown cancel.
3. The simultaneous electron loss and electron gain is demonstrated.
4. The reaction will proceed spontaneously since E^o_{cell} is positive.

20.2.1.4 Oxidation Numbers

Chemists have devised a bookkeeping system to monitor which atom loses an electron and which gains it. Each atom is assigned an "oxidation number" (O.N.) or "oxidation state." Silberberg (1996, p. 155) has given rules for analyzing and balancing a reaction in terms of electrons and molar quantities, which relies on the use of oxidation numbers for each element or compound, with guidelines for assigning oxidation numbers. Latimer (1952) has published oxidation numbers for an extensive list of inorganic compounds (and potentials for many reactions). To summarize, oxidation, that is, loss of electrons, is represented by an increase in oxidation number. Reduction, that is, gain of electrons, is represented by a decrease in oxidation number (Silberberg, 1996, p. 156).

20.2.1.5 Thermodynamic Relations

The Nernst equation relates the actual concentrations of reduced species to the oxidized species for a half-reaction, that is,

$$E = E^o - \frac{0.5909}{n_e}\log\frac{[\text{Red}]}{[\text{Ox}]} \qquad (20.5)$$

in which

E is the electromotive force between electrodes of a cell, comprised of two cells, one where oxidation occurs and the other where reduction occurs (V)
E° is the electromotive force between electrodes of a cell, comprised of two cells, one where oxidation occurs and the other where reduction occurs, in which reactants and products are at unit activities, that is, concentrations for dilute solutions (V)
n_e is the number of electron involved in reaction
[Red] is the concentration of reduced species in product of half-cell (mol/L)
[Ox] is the concentration of oxidized species in reactant for half-cell (mol/L)

The Gibbs free energy equation corresponding to Equation 20.5 is

$$\Delta G = \Delta G^o - RT\ln Q \qquad (20.6)$$

and

$$\Delta G^o = -RT\ln K \qquad (20.7)$$

in which

ΔG is the free energy of reaction (J/mol)
ΔG^o is the standard state free energy of reaction (J/mol)
K is the equilibrium constant for reaction at hand (dimensionless)
Q is the reaction quotient; for the hypothetical reaction, $aA + bB \rightarrow cC + dD$, the reaction quotient is defined as

$$Q \equiv \frac{[C]^c[D]^d}{[A]^a[B]^b} \quad \text{(Snoeyink and Jenkins, 1980, p. 67).}$$

The Nernst equation for two half-reactions may be combined to obtain the potential for a cell (see for example, Alberty and Silbey (1992, pp. 236–243), Snoeyink and Jenkins (1980, pp. 322–343), and Benjamin (2002, pp. 472–482)).

The relationship between ΔG and E is $\Delta G = -zFE$, where z = equivalents per mole and F = coulombs per equivalent.

For reference, in 1887, Walther Nernst started as the principal assistant to Wilhelm Ostwald at Leipzig, one of the three founders of physical chemistry, along with Svante Arrhenius and Jacobus Henricus van't Hoff (see Box 2.3). Nernst received the 1909 Nobel Prize for developing a qualitative theory of catalysis (Servos, 1990, p. 48).

20.2.2 Oxidants

As seen in the previous section, the higher the half-cell potential, $E^{o}_{\text{half-cell}}$, the stronger the affinity of an oxidant for electrons. The strongest of the oxidants listed in Table 20.1 is thus •OH, the hydroxide radical. Other oxidants used in practice include ozone, peroxide, chlorine, chlorine dioxide, and permanganate. Combinations of oxidants include ozone or peroxide in conjunction with UV, which are effective because hydroxyl radicals are produced.

20.2.2.1 Chlorine

In addition to being a common disinfectant, chlorine has been used as an oxidizing agent in drinking water treatment for taste and odor control and for color removal. Examples of other uses of chlorine have included oxidation of Fe^{2+} and Mn^{2+} in groundwater and cyanide oxidation in industrial wastes. In domestic wastewater treatment, its uses have included odor control, sulfide oxidation, and ammonia removal. After about 1976, trihalomethanes (THMs) became an issue (and then disinfection by-products in general), and thus chlorine use was reduced. Subsequently, the search for alternative disinfectants and oxidants began.

20.2.2.2 Ozone

Ozone is generated most commonly by the "cold-plasma" discharge, in which ozone is formed by decomposition of diatomic oxygen (Glaze, 1987, p. 225):

$$O_2(\text{corona discharge}) \Rightarrow \bullet O + O\bullet \quad (20.8)$$

$$O\bullet + O_2 \Rightarrow O_3 \quad (20.9)$$

As described by Rakeness et al. (1996), the oxygen molecule is split to form oxygen ions that combine with oxygen molecules to form ozone. Ozone then reverts to oxygen in a matter of hours; the "half-life" depends upon conditions.

Large-scale generators may have as many as 400 double tubes and may generate about 600 kg O_3/day when dry oxygen is used as the feed gas. With air, the ozone yield is about 0.5 fraction of that generated using oxygen. Concentrations of O_3 in the gas are about 1%–3% for air feed and about 3%–7% for oxygen feed. Ozone may be generated also by UV light, $\lambda < 200$ nm (Glaze, 1987, p. 226). The ozone half-reaction is (Glaze, 1987, p. 227)

$$O_3 + 2H^+ + 2e^- \rightarrow H_2O + O_2 \quad \Delta G^o = -400 \text{ kJ/mol} \quad (20.10)$$

The high negative free energy means that the reaction is spontaneous. However, the reaction may be less strong than what appears from the ΔG^o value, since the reaction rate, that is, kinetics, may vary over eight orders of magnitude, depending on the substance reacting.

Ozone decomposition is complex and involves generation of hydroxyl radicals and other radicals. Figure 20.1 is a schematic depiction of the cycle of ozone reactions showing various intermediate reactants/products that occur sequentially when ozone is dissolved in water (see also von Sonntag et al., 1993, p. 207). The diagram illustrates the complexity of the ozone dissociation reactions. As seen, the •OH radical is one of the intermediate products that may be taken out of the cycle by reacting with organic compounds and inorganic "scavengers" that may be present. Also, ozone is unstable at high pH because the decomposition process is initiated by OH^- ions (Glaze, 1987, p. 227). The overall stoichiometry, that is, the "bottom-line" from Figure 20.1, is (Glaze, 1987, p. 227)

$$2O_3 \rightarrow 3O_2 \quad (20.11)$$

An ozone-induced reaction may occur directly with O_3, with the •OH radical, or with other intermediate radicals, with kinetic behavior depending on the particular intermediate (Carlson, 1992, p. 1). If the supply of O_3 and OH^- is steady, the chain process continues with molecular oxygen, O_2 being a final end product. As seen, all of the species in the middle of the chain are unstable, highly reactive free radicals. A direct reaction with O_3 is highly selective in terms of the functional groups and sites. As an example, O_3 oxidation of phenol occurs on the ortho- or para-positions on the aromatic ring. Free radicals are stronger oxidants and less selective with respect to where they will react and with what functional group (Carlson, 1992, p. 2).

20.2.2.2.1 Ozone Generation

As noted previously, ozone is generated in a corona discharge generator. The latter consists of two electrodes, one with a

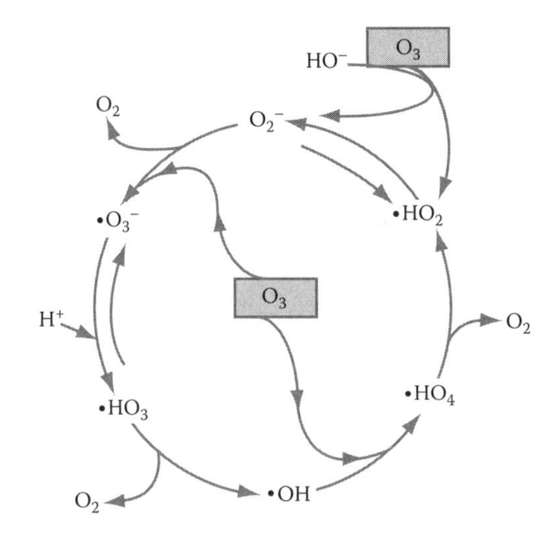

FIGURE 20.1 Schematic of ozone reactions. (Adapted from Glaze, W.H., Environ. Sci. Tech. 21(3), 224, 1987.)

dielectric, with space (gap) between, charged by an alternating current source; electrons flow across the gap in proportion to the voltage difference. The dielectric is commonly glass or ceramic material. Air or oxygen is passed between the plates and through the corona discharge; the oxygen reacts to become ozone, O_3. The electrodes may be in the shape of two plates in parallel, or two concentric cylinders. The process is not efficient, since only about 0.15 fraction of the energy is used for the O_2 to O_3 conversion; therefore about 0.85 fraction is heat and must be removed. The ozone production rate is a function of the peak voltage, frequency, and the geometry of the cell. Typical voltages are 10,000–19,000 V and most generators are in the medium frequency range of 60–1000 Hz. For every 100 kg/h of gas flowing, 1–10 kg/h ozone is produced, depending on applied voltage, whether the gas is air or oxygen, flow of feed gas, and cooling water temperature (Rakeness et al., 1996, p. 4).

20.2.2.2.2 Ozone Kinetics

Ozone has been found to react readily with some organic compounds but not with others. Also, in some cases, the reactions go to mineralization while with others a sequence of reactions is involved and in many cases the products may be less complex and perhaps amenable to biodegradation. Kinetics has everything to do with whether to use ozone or to modify the conditions such as with UV light or peroxide, or to search for some other oxidant, for example, ClO_2. The latter, for example, was found to have a much higher reaction rate than $KMnO_4$ or ozone in oxidation of Mn^{2+}, at low concentrations, to MnO_2 (Gregory, 1996, 1997).

In part, explanations of why ozone behavior varies have been explained in terms of the reaction kinetics between ozone and various compounds. The benchmark work in this area was a set of three papers: Hoigné and Bader (1983a,b) and Hoigné et al. (1985). As they point out, determination of ozone kinetics is complicated by the loss of ozone at any gas–air interfaces that may be present and by the rapid decomposition of ozone to oxygen, albeit through a complex pathway (as illustrated in Figure 20.1).

The rate law for ozone in aqueous solution is first order with respect to both ozone and solute concentration, that is,

$$\frac{d[O_3]}{dt} = -k(O_3)[O_3][M] \quad (20.12)$$

in which

$[O_3]$ is the concentration of ozone (mol/L)
[M] is the concentration of reactant, M (mol/L)
t is the time (s)
$k(O_3)$ is the rate constant (L/mol · s)

Figure 20.2 illustrates the variation in kinetic constants for five organic compounds, showing also the variation with temperature, which is in accordance with the Arrhenius relation; note that pH ≈ 2 for all tests. Activation energies were, in general, 35–50 kJ/mol (Hoigné and Bader, 1983a, p. 181).

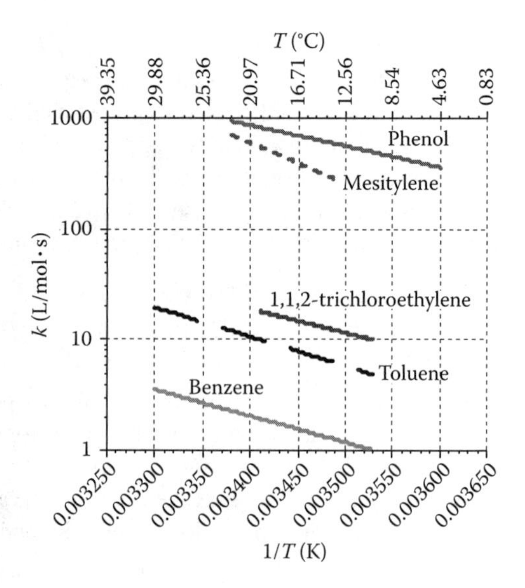

FIGURE 20.2 Arrhenius plot of kinetic constants for five organic compounds, pH ≈ 2. (From Hoigné, J. and Bader, H., *Water Res.*, 17(2), 178, 1983a.)

Kinetic constants were determined for some 67 compounds, in four groups, shown in columns below with representative compounds. The kinetic constants varied over wide order-of-magnitude ranges, for example, 10^{-3}–10^5 L/(mol · s), in each group, but with the third group being the lowest in absolute values, that is, <8 L/(mol · s).

Substituted Benzenes	Substituted Ethylene	Substituted Alkanes	Miscellaneous
Chlorobenzene	Tetrachloroethylene	1-Propylamine	Carbon tetrachloride
Benzene	Trichloroethylene	*tert*-Butanol	Chloroform
Toluene	Maleic acid	Ethanol	Bromoform
o-Xylene	Allylbenzene	1-Propanol	Urea
1,2,3-Trimethylbenzene	Styrene	1-Octanol	Saccharose
Phenol		Formaldehyde	Dioxane
Naphthalene		Propanal	Glucose

The half-life of a compound being oxidized is given (Hoigné and Bader, 1983a, p. 182),

$$t_{1/2} = \frac{0.69}{[O_3] \cdot k_{O_3}/\eta} \quad (20.13)$$

and if $\eta = 1$, and if $[O_3] = 10^{-5}$ mol/L (0.5 mg/L)

$$t_{1/2} = \frac{0.69}{10^{-5} \cdot k_{O_3}} \quad (20.14)$$

in which

$t_{1/2}$ is the time for depletion of 50% of the compound being treated (s)
η is the stoichiometric coefficient
$[O_3]$ is the concentration of ozone in reactor (mol/L)

Thus, only solutes that have a reaction rate constant, $k_{O_3} \geq 100$ L/mol/s, which gives $t_{1/2} = 11.5$ min, may be degraded within a reasonable time. For example, if θ(reactor) = 23 min, the degradation fraction is 0.75 of the target compound that will be degraded; if θ(reactor) = 35 min, the degradation fraction is 0.88. Only about 20 compounds meet this criterion among the 67 tested by Hoigné and Bader (1983a).

Conclusions from the foregoing are as follows: kinetic data from a reference list may be used to evaluate whether a particular compound may be oxidized by ozone, that is, within a feasible reaction time and to a significant degradation fraction; ozone degradation is not feasible for many organic compounds.

Ozone is also considered for oxidation of iron and manganese, that is,

$$2Fe^{2+} + O_3(aq) + 5H_2O \rightarrow 2Fe(OH)_3(s) + O_2(aq) + 4H^+ \quad (20.15)$$

$$Mn^{2+} + O_3(aq) + H_2O \rightarrow MnO_2(s) + O_2(aq) + 2H^+ \quad (20.16)$$

20.2.2.3 Hydroxyl Radical

As seen previously, the hydroxyl radical, •OH, is the most powerful oxidant among those listed in Table 20.1; the hydroxyl radical is the hydroxide ion minus one electron. The hydroxyl radical has been recognized as having a central role in redox reactions since the 1960s with the work of the Advanced Water Treatment Research (AWTR) program of the USPHS (Anon., 1965). As illustrated in Figure 20.1, the OH• radicals may be generated by ozone dissolution. Generation may be also by UV radiation of peroxide, H_2O_2, or other methods, for example, by electrical energy or titanium dioxide.

20.2.2.3.1 Radiation by UV

The UV photolysis of ozone in water yields hydrogen peroxide. The latter reacts in turn with UV radiation to form hydroxyl radicals. The sequence is (Topudurti et al., 1993)

$$O_3 + h\nu + H_2O \rightarrow H_2O_2 + O_2 \quad (20.17)$$

$$H_2O_2 + h\nu \rightarrow 2OH\bullet \quad (20.18)$$

20.2.2.4 Permanganate

A common oxidant in water treatment, most commonly used through trial and error, is potassium permanganate, $KMnO_4$ (Zawacki, 1992; Ma and Graham, 1996). As a case in point, the Collier County WTP, $Q(1985) \approx 0.17$ m^3/s (4 mgd), had water quality issues such as hardness of 280 mg/L, 7–15 mg H_2S/L, color units of 17–20, and high levels of NOM. The plant used lime softening to reduce hardness and chlorine oxidation mitigation of taste, odor, and algae. In 1989, the plant was expanded to $Q(1989) \approx 0.64$ m^3/s (15 mgd), at which time $KMnO_4$ was added as a pretreatment oxidant to address the foregoing problems. During plant trials, THM concentrations were reduced from about 80 to 10 mg/L; color was reduced from 20 to 7; the chlorine dose was reduced from 19 to 12 mg/L (Zawacki, 1992).

20.2.2.5 Chlorine Dioxide

Chlorine dioxide is a strong oxidant, on the same order as chlorine, with $E^\circ(ClO_2) = +1.15$ mV (Table 20.1). In addition to its use as a disinfectant, chlorine dioxide has found a place as one of the useful oxidants in water treatment for oxidation of Mn^{2+} or Fe^{3+} (Gregory, 1996, 1997) either of which may occur in groundwater sources or from surface water sources (the latter taken from water depths where a reducing atmosphere may be present). Without treatment, Mn^{2+} oxidizes readily when O_2 is present and forms MnO_2, a cause of "black water" in the distribution system, or in the case of Fe^{3+}, oxidation causes "red water" These are palatability and nuisance issues that may result in bad taste, stained fixtures, and stained laundry. The U.S. Environmental Protection Agency (USEPA) secondary maximum contaminant levels for iron and manganese are ≤ 300 and ≤ 50 μg/L, respectively.

Oxidation of Mn^{2+} or Fe^{3+} results in an insoluble precipitate, for example, MnO_2 or $Fe(OH)_3$ (or Fe_2O_3), as the case may be, with removal by coagulation-settling-filtration. As a case in point, Horsetooth reservoir in northern Colorado is a source water for two adjacent water treatment plants, Soldier Canyon WTP, Fort Collins (serving the urbanizing area surrounding the city) and the City of Fort Collins WTP, serving the city itself. Both plants draw water at about 65 m (200 ft) depth where anoxic conditions are prevalent during the late summer months and before the fall overturn. Consequently, from about June through November Mn^{2+} is common at concentrations ranging as high as 450 μg/L (Gregory, 1996, p. 1-1). Levels of manganese as low as 20 μg/L have caused problems and, therefore, concentrations ≤ 10 μg/L has been the goal in treatment. The City of Fort Collins WTP used $KMnO_4$ as an oxidant from 1991 to 1998 and had experienced less than satisfactory treatment results. At the same time, the Soldier Canyon WTP had used chlorine dioxide as an oxidant to remove Mn^{2+}, which was effective. Before switching to ClO_2, however, the City of Fort Collins embarked on an in-house research program consisting of bench- and pilot-scale experiments to investigate more formally the kinetics and other characteristics of three oxidants, that is, permanganate, $KMnO_4$, ozone, O_3, and chlorine dioxide, ClO_2, with respect to Mn^{2+} oxidation. Figure 20.3a shows the Mn^{2+} concentration time for each of the three oxidants. As seen, chlorine dioxide was the most effective oxidant, reducing $C(Mn^{2+}) \leq 2$ μg/L after 5 min. The initial concentrations of oxidants were $C(ClO_2) = 4 \cdot$ stoichiometric amount needed, that is, 0.5 mg/L; $C(KMnO_4) = 3 \cdot$ stoichiometric amount needed, that is, 0.36 mg/L; $C(O_3) = 1.5$ μg/L.

BOX 20.1 REACTIONS OF FIVE OXIDANTS WITH Fe^{2+} AND Mn^{2+}

A not uncommon need in oxidation is to remove Fe^{2+} or Mn^{2+} from raw water sources used for drinking water such that they do not cause nuisance effects in domestic water uses (due to their oxidation by dissolved oxygen), causing "red" water and "black" water, respectively. Both may cause palatability deterioration, that is, bad taste, stained fixtures, and stained laundry. The tabular summary of reactions that follows is for five oxidants, giving oxidation reactions for oxygen, ozone, HOCl, ClO_2, and $KMnO_4$ (Bablon et al., 1991, p. 139). The first, for oxygen, gives reactions of Fe^{2+} or Mn^{2+}; these reactions may occur during treatment, in the distribution system, or as the water is withdrawn at a household. The next four oxidants are those that may be used during treatment to remove Fe^{2+} or Mn^{2+} as $Fe(OH)_3(s)$ or $MnO_2(s)$, respectively. The second column shows the stoichiometric ratio of oxidant to target ion. Of the four oxidants, HOCl has not been used to any extent in practice. Ozone, ClO_2, or $KMnO_4$ have been shown to be effective. Of these three, ClO_2 was the method of choice for at least two utilities in Colorado for Mn^{2+} removal because of lower residual concentrations and about the same or faster kinetics (Gregory, 1996).

Reaction	Ratio
Oxygen	
$2Fe^{2+} + 1/2O_2(aq) + 5H_2O \rightarrow 2Fe(OH)_3(s) + 4H^+$	0.14 mg O_2/mg Fe
$2Mn^{2+} + 1/2O_2(aq) + H_2O \rightarrow MnO_2(s) + 2H^+$	0.29 mg O_2/mg Mn
Ozone	
$2Fe^{2+} + O_3(aq) + 5H_2O \rightarrow 2Fe(OH)_3(s) + O_2(aq) + 4H^+$	0.43 mg O_3/mg Fe
$Mn^{2+} + O_3(aq) + H_2O \rightarrow MnO_2(s) + O_2(aq) + 2H^+$	0.88 mg O_3/mg Mn
HOCl	
$2Fe^{2+} + HOCl + 5H_2O \rightarrow 2Fe(OH)_3(s) + Cl^- + 5H^+$	0.64 mg HOCL as O_2/mg Fe
$Mn^{2+} + HOCl + H_2O \rightarrow MnO_2(s) + Cl^- + 3H^+$	1.30 mg HOCL as O_2/mg Mn
ClO_2	
$Fe^{2+} + ClO_2(aq) + 3H_2O \rightarrow Fe(OH)_3(s) + ClO_2^- + 3H^+$	1.20 mg ClO_2/mg Fe
$Mn^{2+} + 2ClO_2(aq) + 2H_2O \rightarrow MnO_2(s) + 2ClO_2^- + 4H^+$	2.45 mg ClO_2/mg Mn
$KMnO_4$	
$3Fe^{2+} + MnO_4^- + 2H_2O \rightarrow 3Fe(OH)_3(s) + MnO_2(s) + 5H^+$	0.94 mg $KMnO_4$/mg Fe
$3Mn^{2+} + 2MnO_4^- + 2H_2O \rightarrow 5MnO_2(s) + 4H^+$	1.92 mg $KMnO_4$/mg Mn

Another concern was whether alum coagulation interfered with the oxidation process. Figure 20.3b shows the respective treatment train profiles for the three forms of Mn, that is, dissolved, colloidal, and particulate, respectively, for $C(ClO_2) = 0.85$ mg/L. As seen, the dissolved form is reduced to the $C(Mn^{2+}) \approx 4$ μg/L after reaction with ClO_2; the colloids and particulates are most likely assimilated into the alum floc after rapid mix. Filtration reduces all the colloid and particulate forms to negligible concentrations. As a negative point, about 0.90–0.95 fraction of chlorine dioxide reverts to chlorite ion, which, along with chlorate ion has been associated with health effects.

20.2.2.6 Titanium Dioxide

Titanium dioxide, TiO_2, has been speculated to be an effective catalyst for producing hydroxyl radicals in oxidation for water treatment. Its photo-efficiency, however, for degrading contaminants was found to be only 0.03–0.11 (i.e., in producing hydroxyl radicals). Professor J. R. Bolton (University of Western Ontario) considered these efficiencies so low as to make TiO_2-based water treatment systems economically unfeasible (Wilson, 1996, p. 29).

20.2.3 Supercritical Water Oxidation

SCF is a media in which chemical reactions, involving many difficult-to-treat substances, may be carried out toward achieving the "ideal" products of carbon dioxide and water. This kind of reaction is possible because of the extraordinary properties of SCF. The particular interest here is SCW.

20.2.3.1 Critical Point

The Figure 20.4 shows a phase diagram for water that indicates both the *triple point*, and the *critical point* (see glossary). The solid lines show the phase equilibria: the thin line from the triple point that slants slightly toward the left is the solid–liquid equilibria; the line from the triple-point to the right depicts the gas–liquid equilibria. Finally, at $T \geq 374°C$, $p \geq 218$ atm (22,089 kPa), "supercritical" water exists. At this point, the character of the water changes. At the critical point, the gas density and the liquid density are equal (Silberberg, 1996, p. 451). The shaded area (upper right) illustrates a portion of the supercritical region. Operating points in practice are higher than both the critical temperature and the critical pressure. For example, dissociation of urea to nitrogen gas, carbon dioxide, and water requires T(operating) > 650°C (Timberlake et al., 1982).

20.2.3.2 SCWO In-a-Nutshell

The density of SCW ranges 0.1–0.5 g/mL and is low enough and the temperature is high enough such that there is no hydrogen bonding (Timberlake et al., 1982, p. 1). The dielectric constant is reduced, for example, 3–10, and the water becomes nonpolar and an excellent solvent for organic compounds. At $T > 500°C$,

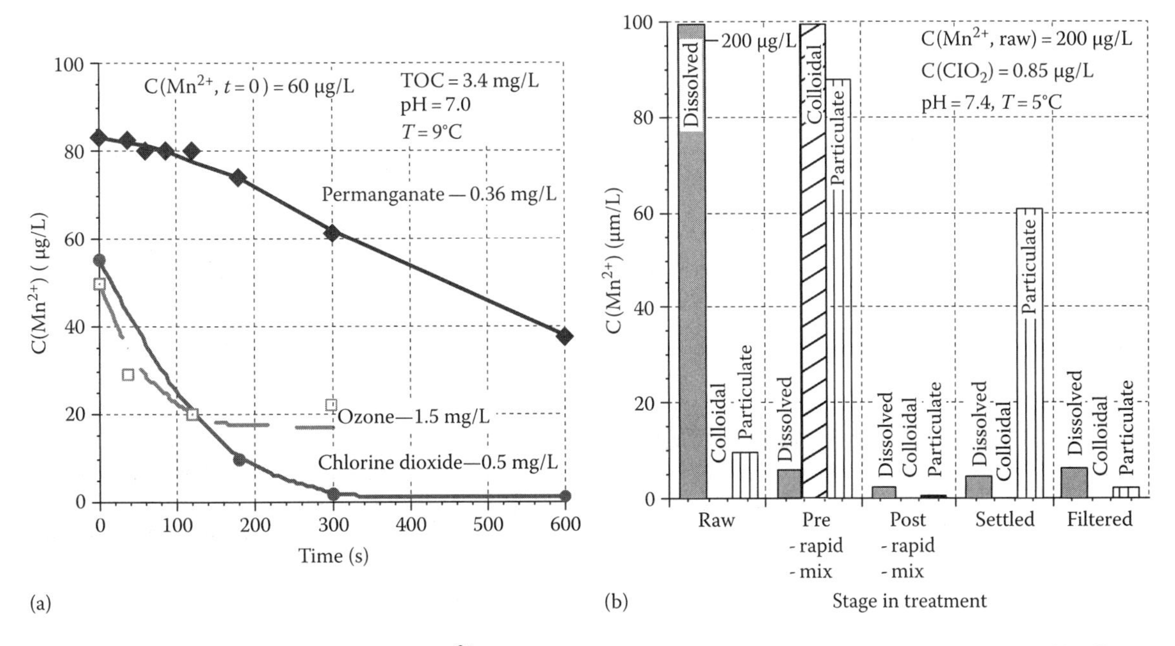

FIGURE 20.3 Results from experiments to reduce Mn^{2+} (from in-house research program at Fort Collins WTP reported by Gregory (1996, 1997). (a) Bench-scale kinetic results comparing ClO_2, $KMnO_4$, and ozone as oxidants, and (b) profile of Mn species across pilot plant treatment train following oxidation by ClO_2 (1.5 times stoichiometric amount).

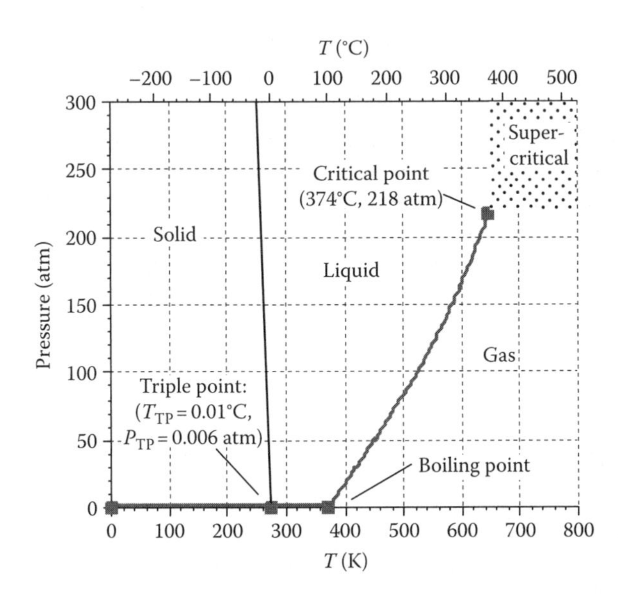

FIGURE 20.4 Phase diagram for water showing critical point. (Adapted from Silberberg, M., *Chemistry—The Molecular Nature of Matter and Change*, Mosby-Year Book, Inc., St. Louis, MO, 1996, p. 451.)

$0.05 < \rho(\text{water}, T > 500°C) < 0.1$ g/mL and the dielectric constant is less than 2 and inorganic salts are only sparingly soluble. Whereas many organic compounds tend to form a high molecular weight char at temperatures below 350°C, at supercritical conditions the same organics are reformed to gases, for example, CO_2, H_2, CH_4, CO_2, and volatile organic liquids, for example, alcohols, aldehydes, and furans without producing any char. Materials as complex as wood can be completely dissolved and reformed in SCW. These products then can be subjected to oxidation; for example, aqueous solutions of organics will undergo oxidation at temperatures, $200°C < T(\text{reactor}) < 300°C$, which is the basis for wet-oxidation (as exemplified by the Zimpro® process) operated under subcritical conditions but which requires residence times, $20 < \theta(\text{reactor}) < 50$ min, achieving destruction efficiencies of only 70%–95% in total organic carbon (TOC). Under supercritical conditions, however, $\theta(\text{reactor}) < 1$ min, and destruction efficiencies are >0.9999 fraction. In addition, oxygen is completely soluble in SCW and thus the oxidation can occur under homogeneous, that is, single phase, conditions (meaning that there is no mass-transfer across a gas–water interface). Thus, SCWO is highly efficient. The SCW functions as a carrier fluid, as a solvent for the feed material and reaction products, and as a reactant with organic materials (Modell, 1985, p. 97).

If the oxidation occurs in an adiabatic reactor, that is, no heat-transfer across the system boundaries, the heat of combustion is retained within the fluid, raising the temperature. If the concentration of carbon is higher than 2%–5%, the heat of oxidation is sufficient to bring the reactor temperature to $T(\text{reactor}) > 550°C$. The heat generated may be recovered for raising the temperature of the inflow stream. Except as noted, this section was based on an article by Timberlake et al. (1982, pp. 1, 2) and a comprehensive summary of the SCWO process by Modell (1988).

20.2.3.3 Characteristics of Supercritical Water Relevant to Engineering

Some of the characteristics of SCWO were given by Jain (1993) are as follows:

1. Supercritical water has high dissolving power, exhibiting the characteristics of a nonpolar organic solvent (Gloyna and Li, 1995, p. 183). Organic

compounds dissolve readily in SCW because of its extraordinary solvating powers. Examples of organic compounds that may dissolve and be oxidized include polychlorinated biphenols (PCBs), dioxin, benzene, DDT, urea, cyanide, explosives, chemical warfare agents, rocket propellants, chlorophenol, paint sludge, greases and lubricants, waste oil, mixed solvents, etc., and ammonia. Most such compounds are insoluble in ordinary water and are difficult to treat. Also, the solubility of oxygen is very high.

2. Supercritical water can sustain oxidation. In general, SCFs sustain oxidation reactions because they mix well with nonpolar organic compounds such as oxygen, carbon dioxide, methane, and other alkanes.
3. Solubilities of inorganic salts are reduced greatly (Li et al., 1993, p. 250) facilitating separation.
4. Reaction products of organics and oxygen include carbon dioxide and water, along with ammonia and a variety of low molecular weight acids, mostly acetic acid (Sawicki and Casas, 1993, p. 276).
5. The high specific heat capacity of SCW indicates that it is feasible to hold a high amount of heat within a SCWO reactor (Gloyna and Li, 1995, p. 183).
6. With respect to engineering design, (1) high diffusivities result in high mass-transfer rates, (2) low viscosity facilitates mixing, (3) high organic and oxygen miscibility result in more homogeneous reactors, and (4) low inorganic solubilities improve solids separation efficiencies (Gloyna and Li, 1995, p. 183).

To summarize, operating at supercritical conditions results in a homogeneous reaction mixture in which organics, water, and oxygen can exist in a single phase. Thus, interphase oxygen transport is not an issue, which was a limiting factor in wet-air oxidation (Thornton and Savage, 1992, p. 321). Companies that have emerged that utilize supercritical oxidation include Modar, Inc., Natick, Massachusetts; Modec, Inc., Framingham, Massachusetts; Eco-Waste, Inc., Austin, Texas; Air Products and Chemicals, Allentown, Pennyslvania. The first full-scale reactor was a Texaco SCWO unit in Austin, Texas, built in 1995, and which had a capacity $Q = 27.4\ m^3/day$ (5 gal/min) of organic wastes.

20.2.3.4 Supercritical Reactors

During SCWO, aqueous waste streams are first pressurized and heated until the water enters the supercritical phase. The organic components then react in an insulated reactor where the dissolved components break down further and readily combine with oxygen; the products approach the ideal sought, that is, water and carbon dioxide with about 0.9999 fraction conversion expected. With sufficient organic matter in the feed, that is, 2%–5%, either as the target reactant or as an ancillary reactant, the reaction is exothermic, that is, self-sustaining (Timberlake et al., 1982; Gloyna and Li, 1995; Modell et al., 1993). For reference, incinerators operate at $2000°C < T < 3000°C$ while SCWO reactors operate at $500°C < T < 600°C$ (Kruse and Schmieder, 1999, p. 234).

An engineered supercritical system involves subsystems to make the overall system function and to perform economically. Figure 20.5 depicts an SCWO system showing three subsystems: (1) SCWO, (2) steam generation, and (3) residuals separation. The first shows the organics feed (on the left) pressurized by a pump and then with temperature raised to supercritical through a heat exchanger. Oxygen is fed into the reactor input flow. Conversions of the organic reactants occur in the reactor, with $\theta \leq 10$ min, under supercritical conditions. In the case of compounds identified as hazardous wastes, $\theta \leq 10$ s for ≥ 0.999 fraction destroyed (Cansell et al., 1998, p. 240).

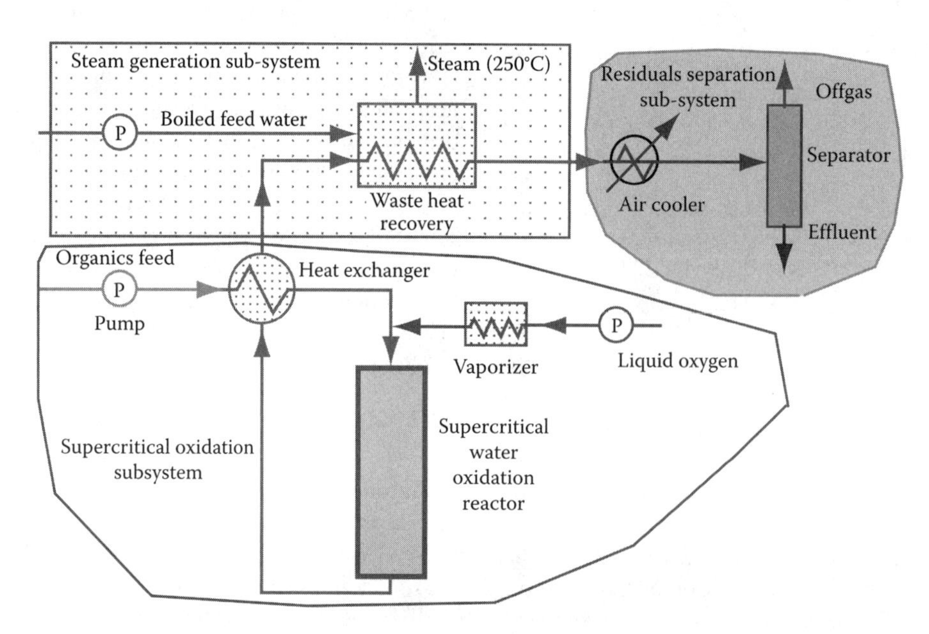

FIGURE 20.5 SCWO reactor and subsystems. (Adapted from Jain, V.K., *Environ. Sci. Technol.*, 27(5), 808, 1993.)

The products leave the reactor losing heat to the first heat exchanger and then to a second that is a part of the steam generation subsystem. Additional heat is lost by means of an air-cooler. The solids and off-gas are separated as the last step in the reactor effluent processing. Oxygen is preferred over air because the reactor volume may be less since nitrogen takes up volume and has no role in the reaction (Sawicki and Casas, 1993).

20.2.3.5 Research in the 1990s

Studies by Li et al. (1993) involving dinitrotoluene (DNT) wastewaters, also containing 2-nitrophenol, 4-nitrophenol, 4,6-dinitro-*o*-cresol, phenol, and DNT, found 99% TOC reduction at θ reactor) ≈ 1 min, T(reactor) ≥ 450°C, p(reactor) ≈ 306 atm (310 bar or 31,000 kPa). Both hydrogen peroxide and oxygen were added to the reactor in trials. Lower temperatures, that is, 200°C –300°C, were effective for H_2O_2, while oxygen produced better results at 200°C–500°C ($T_c = 374$°C). Biological sludges could provide the extra heat needed by the process. At p(reactor) > $p_c = 218$ atm, the higher pressures had no effect on removal efficiency (pressures were varied, 138 < p(reactor) < 306 atm). Removal efficiencies for municipal sludges were about 99.4% at T(reactor) = 450°C and p(reactor) = 27,600 kPa, which was for TOC(in) ≈ 14,200 and TOC(out) = 84 mg/L (Gloyna and Li, 1995, p. 183). Industrial wastewater TOC concentrations were reduced from about 1840 to about 4 mg/L at T(reactor) = 500°C.

20.2.3.6 Design Factors

Design considerations for SCWO include (Gloyna and Li, 1995, p. 2)

- Reactor detention time
- Reactor pressure and temperature
- Materials of construction for each unit operation
- Control and removal of solids either from the SCF or the treated effluent
- Operation and maintenance of the facility
- Design for safety (facilities, working conditions, regulatory compliance)
- Provision for analytical support and regulatory monitoring
- Provision for disposal of residuals

To give some idea of dimensions, the first commercial reactor at Austin, TX, built in 1995, was 4.0 m tall and 2.0 m diameter, with Q(reactor) = 25 L/min. The waste flow contained about 10% petrochemical organic compounds with fraction removed > 0.995. Previously, the waste was transported to a hazardous wastes incinerator at a cost of \$0.30–\$0.40 per liter, with total cost ≈ \$1 million/year.

Some of the limiting factors in reactor design have to do with the containment of high pressure and high temperature in the context of reactions that yield corrosive products. Thus, the reactor construction requires a cylinder designed for the stress caused by the high-pressure high-temperature, and material that may withstand corrosive conditions (Gloyna and Li, 1995, p. 189). No known material will handle all of these conditions simultaneously. A way around this was to construct an inner reaction vessel to contain the reaction and an outer vessel to handle the pressure (the stress within the walls of the inner vessel may be zero if the pressure is the same between the two sides). The inner vessel is also insulated from the outer vessel so that the temperature of the outer vessel may be maintained at lower temperature. A patent for the reactor was pending (in 1995) by Kimberly-Clark, Inc.

20.2.3.7 Case Study: SCWO of Pulp and Paper Mill Sludge

Blaney et al. (1995) described SCWO a pilot plant study of pulp and paper mill sludge that utilized the UT pilot facility with $Q = 150$ L/h (40 gal/h) capacity. About 64 million metric tons/year (70 million U.S. tons) of such sludge is generated worldwide. The sludge contains cellulosic fibrous fines and debris, inorganics such as fillers and clay, and small amounts of de-inking chemicals. Chlorinated organics present

BOX 20.2 SCWO: DEVELOPMENT OF A TECHNOLOGY

Professor Michael Modell, a chemical engineering professor at the Massachusetts Institute of Technology (MIT), began an odyssey of some 30 years in the field of SWCO; the odyssey began in 1975 when he patented a process for reforming organics in SCW. In 1980, he extended the work to the destruction of toxic and hazardous wastes and formed MODAR, Inc. to develop and commercialize the process, resigning his tenured academic appointment in 1982. The development activities at MODAR included proof of technical feasibility, kinetics, process design, destruction efficiency, economic evaluation, scale-up guidelines, inorganic precipitation, corrosion studies, and demonstration tests. In 1990, MODAR was sold to General Atomics, which applied the reactor technology to destruction of chemicals from weapons.

In 1987, Dr. Modell formed another company, MODEC, Inc., to develop a new reactor design, that is, a tubular reactor, to overcome the deposition of salts, to mitigate corrosion, and to collect the carbon dioxide product in liquefied form. This reactor was the basis for further work by Gloyna et al. at University of Texas (UT). Among Dr. Modell's many publications is a text, *Thermodynamics and Its Applications*, first edition, M. Modell and R.C. Reid, Prentice Hall, 1974, with a second edition in 1982, and a third in 1997 Tester and Modell (1997), the latter by J. Tester and M. Modell. Among his 12 patents, 5 have been in the field of SCWO. Applications of the SCWO technology have included work for the National Aeronautics and

(*continued*)

BOX 20.2 (continued) SCWO: DEVELOPMENT OF A TECHNOLOGY

Space Administration (NASA), which has continued since 1975 with the ultimate goal of recycling wastes for sustainable space travel. The NASA contracts were the genesis of his work in SCWO. Other applications have included degradation of hazardous wastes, urea, pulp mill sludges, various aqueous wastes, sludges from wastewater treatment, and retired chemical weapons. Applications to hazardous wastes remediation were stimulated by the "superfund" law, (PL 96-510, 1980), also known as Comprehensive Environmental Response, Compensation, and Liability Act (CERCLA). The ultimate goal of hazardous wastes treatment was "mineralization," which was achieved by SCWO. His pursuit of a SCWO technology has been continuous, from 1975 through the 1990s. A 1988 review paper (Modell, 1988) provided a comprehensive description of the SCWO process.

Michael Modell, c. 1990

BOX 20.3 EARNEST GLOYNA

What was lacking in the mid-1980s was an academic SCWO research program. This void was filled at the University of Texas (UT) in 1987 through the initiative of Professor Earnest F. Gloyna after he retired as Dean of Engineering at age 67, having served 17 years in that position (with UT recognized as one of the major engineering colleges in the United States). His expertise did not include SCWO when he embarked upon this new area (a point of interest that illustrates that basic education coupled with dedication are two of the requisites to launch a new research program). The SCWO program grew quickly in staff, facilities, sponsorship, and graduate students, with commensurate production of theses (10 PhD and 38 MS), papers, presentations, and collaborations with private start-ups. Research areas included reaction kinetics, reaction mechanisms, mass and heat transfer, catalyst development, by-product recovery, treatability, solubilities of salts, corrosion control, process design, model development and validation, control system design, and safety requirements. The intent was to develop SCWO as a technology, in the case of water treatment, a process that has the necessary appurtenances associated with it so that its implementation works. One of the interesting SCWO research areas was related to the destruction of chemical warfare agents (under contract with DARPA—see glossary). Professor Gloyna was joined in his SCWO research in 1988 by Dr. Lixiong Li who served with the program until 1998 (see references).

Professor Gloyna started his career at UT in 1947, after 4 years in the U.S. Army during WWII. His 50-year-plus career at UT has included a broad spectrum of teaching, research, administration, and consulting activities. Recognitions of his accomplishments have included essentially every major award and honor in the field of environmental engineering (see, e.g. Marquis *Who's Who in America*, 57th edition, 2003) including election to the National Academy of Engineering in 1970, service medals from several foreign governments, merit medals from various professional organizations, publication of several books, and over 250 papers. In addition, he has served as president of several major organizations, for example, Water Environment Federation, American Academy of Environmental Engineering, and American Association of Environmental Engineers and Scientists. He retired in 2001 as director, SWCO Center and from his endowed chair, but not from professional life. A biography of Professor Gloyna was prepared by Dr. Davis L. Ford (Ford, 2009) and is available from the Cockrell School of Engineering, University of Texas at Austin, Austin. Appendix K of the book is dedicated to SCWO and provides additional information on the role of Professor Gloyna in the development of this technology.

Earnest F. Gloyna, c. 1980

in trace amounts include: polychlorinated biphenyls (PCBs), polychlorinated dibenzo-*p*-dioxins (PCDDs), and polychlorinated dibenzofurans (PCDFs). Landfill disposal was practiced but was becoming less feasible because of both real and perceived toxic substances in the sludge. Incineration was thought to have been feasible, but the risk of toxic substances in gas emissions and ash was an issue that made it less attractive.

The waste stream from the pulp and paper plant was about 5%–7% solids (0.30 fraction inorganics of clays and fillers, 0.70 fraction organic cellulosic fiber fines, residual pulping, and de-inking chemicals), diluted to 2.5% solids for feed-flow to the pilot plant. Reactor temperature was $T(\text{reactor}) = 450°C$, but was increased to $T(\text{reactor}) = 500°C$ after adding methanol, 2% by weight; also, for $Q(\text{reactor}) = 76$ L/h (20 gph), $\theta(\text{reactor}) \approx 50$ s. At $T(\text{reactor}) = 500°C$, the reactor destroyed > 0.99 fraction of most dioxin-type compounds, 0.96 fraction of 2,3,7,8-TCDD and 2,3,7,8-TCDF, and >0.90 fraction PCBs. The destruction of organics was surmised to be mass-transfer limited, that is, in terms of reactor turbulence; the latter would be at a higher level in a full-scale reactor.

Cost estimates for alternative technologies for processing the wastes were

	Treatment Option	Cost (Dry Sludge) ($/metric ton) Low	High
1	Land fill—dewater to 0.50 fraction solids	113	353
2	Fluidized bed incineration, GAC off-gas treatment, ash to landfill	129	198
3	SCWO—0.10 fraction solids: pure O_2 oxidant; $p(\text{reactor}) = 245$ atm; $T = 500°C$; $\theta = 50$ s; ash to landfill	187	215
4	SCWO—0.12 fraction solids: pure O_2 oxidant; $p(\text{reactor}) = 135$ atm; $T = 500°C$; $\theta = 50$ s; recover salvage value of ash	120	140

As seen in the comparisons, the SCWO option (3) has higher costs, in general. Process modifications, for example, thickening the sludge to 0.12 fraction solids and obtaining $55/metric ton credit for calcium carbonate recovery as seen in (4), makes the cost competitive with the other processes. On the other hand, as landfill regulations have become more stringent and the cost of incineration has increased, 40%–45%, SCWO becomes more attractive.

20.3 PRACTICE

In most cases in drinking water treatment, oxidants are added to pipes or channels to suit the case at hand. For example, an oxidant for iron or manganese removal is usually added either before the rapid mix or in the rapid mix so that the products may be removed as a colloid or particle in flocculation/sedimentation or filtration. In taste and odor control, the oxidant is added at the end of the process so that only residual concentrations are removed. In most cases, the oxidation occurs as a part of an existing treatment train rather than in a designed reactor.

In oxidizing SOCs, as in groundwater remediation, oxidation is most likely one of the main unit processes, as opposed to being ancillary. Either a complete mix or a plug-flow reactor may be used, with the latter being favored. In mathematical modeling, kinetics is the underpinning. The principle of plug-flow reactor modeling is encompassed in Equation 4.18. As with other processes, a pilot plant is the surest way to reduce the uncertainty. Key questions have to do with stoichiometry, kinetics, mixing, dispersion, detention time, etc. For gases, for example, ozone, the method of introduction can be, for example, by fine-bubble diffusers, a Venturi intake, a proprietary mixer such as Water Champ® (U.S. Filter, Inc.).

PROBLEMS

20.1 Balancing a Redox Reaction by Half-Reaction Method

Suppose a redox reaction is between Fe^{3+} and Mg.

a. Write the half-reactions (always include the standard potential).
b. Write the overall reaction, balancing the equation (include the resultant standard potential).
c. Illustrate with a cell depiction.
d. Label the oxidants/oxidized species and reductants/reduced species.

20.2 Balancing a Redox Reaction by Half-Reaction Method

MnO_4^- is an oxidant considered for use to oxidize Mn^{2+} to MnO_2.

a. Write the half-reactions (always include the standard potential).
b. Write the overall reaction, balancing the equation (include the resultant standard potential).
c. Illustrate with a cell depiction.
d. Label the oxidants/oxidized species and reductants/reduced species.

20.3 Balancing a Redox Reaction by Oxidation Number Method

For the oxidation of Mn^{2+} to MnO_4^- by PbO_2, show the balancing by the oxidation number method. *Note*: Stumm and Morgan (1996, p. 428) do this by the half-reaction method.

20.4 Balancing a Redox Reaction by Oxidation Number Method

Balance the reaction in which Fe^{2+} is oxidized to Fe^{3+} by MnO_4^-; the latter is reduced to manganese dioxide $MnO_2(s)$. The reaction takes place in an alkaline solution.

ACKNOWLEDGMENTS

Kevin Gertig, Water Resources & Treatment Operations Manager, Fort Collins Utilities, and formerly Superintendent, Fort Collins Water Treatment Facility (WTF), helped in understanding, chlorine dioxide generation and application, and was available to clarify practices used at the plant.

Also, Dr. Dean Gregory, CDG Environmental, Bethlehem, Pennsylvania, was available to clarify issues related to chlorine dioxide generation. Dr. Gregory is based in Denver, Colorado, and conducted his master's thesis research on oxidation of Mn^{2+} while employed at the Fort Collins WTPF.

Dr. Earnest Gloyna gave permission (2010) to use his photograph as did Dr. Michael Modell (2010). Also, both Dr. Gloyna and Dr. Modell provided several of their published papers, at the request of the author, that were used in the development of the sections on SCWO. Dr. Lixiong Li provided leads on publications from the period that he was associated with Dr. Gloyna in research (1988–1998).

GLOSSARY

Advanced oxidation processes (AOP)

1. Oxidation processes based on generation of hydroxyl radical intermediates. Combinations based on H_2O_2, O_3, and UV have been investigated to a great extent (Bull and Zeff, 1992, p. 27).
2. Oxidative degradation reactions by HO• and HO_2• and other less important radicals in solution generated by various methods, for example, O_3/UV, H_2O_2/UV, and O_3/H_2O_2/UV photolysis and TiO_2 mediated photocatalysis in aqueous solution or photo-assisted Fenton processes (Kiwi et al., 2000, p. 2162). In all of the foregoing, the OH radical is thought to be the major reactive intermediate responsible for organic substrate oxidation (Glaze et al., 1993).

AWTR: Advanced Waste Treatment Research program of the Division of Water Supply and Pollution Control, U.S. Public Health Service. The Division of Water Supply and Pollution Control was the agency involved in wastewater treatment research before being transferred to the Department of the Interior in 1966 and before the formation of EPA in 1970. The AWTR program was active during the 1960s and was instrumental in the development of many wastewater treatment processes, for example, adsorption, electrodialysis, reverse osmosis, distillation, freezing, ion-exchange, oxidation, etc., that were not common before the advent of the program. The program was administered from the USPHS Taft Sanitary Engineering Center, Cincinnati, Ohio. The program functioned through both contracts and in-house research (Anon., 1965).

Corona

1. A faint glow adjacent to the surface of an electrical conductor at high voltage (www.merriam-webster.com, 2003).
2. An electrical discharge effect that causes ionization of oxygen and the formation of ozone, O_3 (Symons et al., 2000, p. 93).

Critical point: Point identified on p–T diagram where for $p > p_c$ and $T > T_c$, the liquid and gas phases no longer exist. For water, the critical point is: ($T_c = 374$°C, $p_c = 218$ atm) (Silberberg, 1996, p. 451).

DARPA: Defense Advance Research Project Agency (U.S. Department of Defense).

Dielectric: A nonconductor of direct electric current (http://www.merriam-webster.com, 2003).

EPA: U.S. Environmental Protection Agency. The agency was created by Executive Order of the President on December 2, 1970 to consolidate various environmental regulatory functions, for example, water pollution control, air pollution, pesticides, etc., into a single administrative agency.

Fenton oxidation: A sequence of redox reactions that form Fe complexes initially, with radical intermediates of HOO•, HO•, and O_2^{-}•.

Fenton's reagent: A ferrous salt, for example, ferrous sulfate, $Fe(SO_4)$, that acts as a catalyst for the production of hydroxyl radicals, •OH, from hydrogen peroxide, H_2O_2. During the reaction, the ferrous salt is oxidized to the ferric form and can participate in other chemical reactions, such as coagulation or metal complexation (Symons et al. 2000, p. 157).

Free radical: Atoms or groups of atoms with an unpaired valence electron, which causes high reactivity (Carlson, 1992, p. 2).

Geosmin: *trans*-1,10-Dimethyl-*trans*-9-decalol, a taste and odor causing compound that could be present in ambient waters (Metropolitan Water District of Southern California, and James M. Montgomery Consulting Engineers, Inc., 1991, pp. 3–5).

MIB: 2-methylisoborneol, a taste and odor causing compound that could be present in ambient waters (Metropolitan Water District of Southern California, and James M. Montgomery Consulting Engineers, Inc., 1991, pp. 3–5).

NTP: Normal temperature and pressure, that is, $T = 0$°C, $p = 1.00$ atm. NTP is common in European literature. See also STP, which is common to the United States.

Oxidation: A reaction in which a loss of electrons occurs, accompanied by an increase in oxidation number (Silberberg, 1996, p. G-13).

Oxidation number (O.N.): A number determined by a set of rules equal to the number of charges a bonded atom would have if electrons were held completely by the atom that attracts them more strongly

(Silberberg, 1996, p. G-13). (Same as oxidation state.)

Oxidation number method: A method for balancing redox reactions in which the change in oxidation numbers is used to determine balancing coefficients (Silberberg, 1996, p. G-13).

Oxidation–reduction reaction: A process in which electrons are transferred from one reactant, that is, the reducing agent, to another, that is, the oxidizing agent (Silberberg, 1996, p. G-13). See also *redox reaction*.

Oxidized: The element or compound in an oxidation reaction that loses electrons.

PCE: Acronym for tetrachloroethylene.

Photolysis: Incident radiant energy on a chemical species that causes a reaction resulting in one or more products.

Plasma: A collection of charged particles (as in the atmospheres of stars or in a metal) containing about equal numbers of positive ions and electrons and exhibiting some properties of a gas, but differing from a gas in being a good conductor of electricity and in being affected by a magnetic field (http://www.merriam-webster.com).

Reaction quotient: In the reaction, $a\text{A} + b\text{B} \rightarrow c\text{C} + d\text{D}$, the reaction quotient is defined as

$$Q \equiv \frac{[\text{C}]^c[\text{D}]^d}{[\text{A}]^a[\text{B}]^b} \text{ (Snoeyink and Jenkins, 1980, p. 67).}$$

Redox reaction: The coupling of two half-reactions in which one reactant is the oxidant and the other is the reductant.

Reduction: A reaction in which a gain of electrons occurs.

Scavenger: Chemical species other than the target compounds that consume oxidants. Scavengers may include anions such as bicarbonate, carbonate, sulfide, nitrite, bromide, and cyanide; metals in their reduced states such as trivalent chromium, ferrous iron, manganous ion, etc.; organics such as humic compounds (Topudurti et al., 1993).

SCWO: Abbreviation for supercritical water oxidation.

Standard cell potential: Potential, E^o_{cell}, measured at 25°C, with no current flow, and all components in their standard states, that is, 1 atm for gases, 1 M for solutions, pure solid for electrodes. For example, when $[Zn^{2+}] = [Cu^{2+}] = 1.00$ M, the cell is operating at standard conditions and produces 1.10 V at 298 K (Silberberg, 1996, p. 887), that is,

$$\text{Zn(s)} + \text{Cu}^{2+}\text{(aq)}\_ \Longrightarrow \geq \text{Zn}^{2+}\text{(aq)} + \text{Cu(s)} \quad E^o_{\text{cell}} = 1.10\text{ V}$$

Standard half-cell potential: Same as the *standard cell potential*, also designated, $E^o_{\text{half-cell}}$, measured at 25°C, with no current flow, and components in their standard states. By convention, a standard electrode potential refers to the half-reaction written as a reduction (Silberberg, 1996, p. 887), that is,

$$\text{oxidized form} + ne^- \rightarrow \text{reduced form} \quad E^o_{\text{half-cell}} > 0$$

When $E^o_{\text{half-cell}} > 0$ (i.e., +), which is the same as $\Delta F_o < 0$ (i.e., −), which means that the reaction is spontaneous. Changing the direction of the half-reaction reverses the sign, that is,

$$\text{reduced form} \rightarrow \text{oxidized form} + ne^- \quad E^o_{\text{half-cell}} < 0$$

Half-cell potentials are referenced to a "standard hydrogen electrode," which is a platinum electrode immersed in a 1 M strong acid through which H_2 gas is bubbled at 1 atm, for which $E^o_{\text{half-cell}} = 0$ V. Standard half-cell potentials for many half-cell reactions are given in standard reference works. For example, writing the half-reactions and subtracting the second from the first,

Reduction	$Cu^{2+}(aq) + 2e^- \Longrightarrow\geq Cu(s)$	$E^o_{\text{half-cell}} = +0.34$ V
Oxidation	$Zn^{2+}(aq) + 2e^- \Longrightarrow\geq Zn(s)$	$E^o_{\text{half-cell}} = -0.76$ V
Net reaction	$Zn(s) + Cu^{2+}(aq)\_ \Longrightarrow\geq Zn^{2+}(aq) + Cu(s)$	$E^o_{\text{cell}} = 1.10$ V

As seen, Cu^{2+} gains electrons and so is reduced. Reversing the direction of the second half-reaction results in a loss of electrons from Zn(s), which is oxidation. Therefore, both half-reactions are spontaneous, giving a net potential, $E^o_{\text{cell}} = 1.10$ V.

The "molecular equation" shows the reactants and products as if they were undissociated compounds, that is,

$$2\text{AgNO}_3\text{(aq)} + \text{Na}_2\text{CrO}_4\text{(aq)} \rightarrow \text{Ag}_2\text{CrO}_4\text{(s)} + 2\text{NaNO}_3\text{(aq)}$$

The more realistic depiction is the "total ionic equation," that is,

$$2\text{Ag}^+\text{(aq)} + 2\text{NO}_3^-\text{(aq)} + 2\text{Na}^+\text{(aq)} + \text{CrO}_4^{2-} \rightarrow \text{Ag}_2\text{CrO}_4\text{(s)} + 2\text{Na}^+\text{(aq)} + 2\text{NO}_3^-\text{(aq)}$$

The number of Na^+(aq) and NO_3^-(aq) ions are unchanged on both sides and are called "spectator ions," because they are not involved in chemical change, but are present as a part of the reactants.

The "net ionic equation" depicts only the actual change taking place, that is, with no spectator ions, that is,

$$2\text{Ag}^+\text{(aq)} + \text{CrO}_4^{2-} \rightarrow \text{Ag}_2\text{CrO}_4\text{(s)}$$

STP: Standard temperature and pressure, that is, $T = 20$°C, $p = 1.00$ atm. See also *NTP*.

Supercritical fluid: Supercritical fluids expand and contract like gases, but have the solvent properties of a liquid, which can be alerted by changing the density. Supercritical carbon dioxide has received the most attention by industry, extracting nonpolar ingredients from complex mixtures, such as caffeine from coffee beans, nicotine from tobacco, and fats from potato chips while leaving behind the taste and aroma. Supercritical water was also shown to dissolve nonpolar substances, for example, nonpolar organic toxins such as PCBs; adding O_2 gas oxidizes these substances to harmless molecules (Silberberg, 1996, p. 452).

Supercritical water oxidation (SCWO):

1. Oxidation that occurs in water at temperature and pressure that exceed the "critical point" for water, that is, $T_c = 374°C$, $p_c = 218$ atm.
2. Oxidation of a compound (organic in most cases) within a pressurized reactor in which water is at $T_c > 374°C$, $p_c > 218$ atm.

TCE: Acronym for trichloroethylene.

Triple point: The point on a p–T diagram where the three phase transition curves meet. For water, the triple point is ($T_{TP} = 0.01°C$, $p_{TP} = 0.006$ atm) (Silberberg, 1996, p. 451).

USPHS: United States Public Health Service. Its Division of Water Supply and Pollution Control, Cincinnati, OH administered research programs and conducted research. In 1966, the pollution control programs were transferred to the Department of the Interior and later to EPA when the agency was formed in 1970. Later (1974), the drinking water programs were transferred as well.

Wet-oxidation: Similar to SCWO except the temperature and pressure are both less than critical. (Thornton and Savage, 1992).

REFERENCES

Alberty, R. A. and Silbey, R. J., *Physical Chemistry*, 1st edn., John Wiley & Sons, New York, 1992.

Anon., Summary Report—Advanced Waste Treatment Research Program January 1962 through June 1964, AWTR-14, PHS Publication No. 999-WP-24, Environmental Health Series, Water Supply and Pollution Control, Advanced Waste Treatment Research Program, Robert A. Taft Sanitary Engineering Center, U.S. Public Health Service, April 1965.

Babbitt, H. E. and Doland, J. J., *Water Supply Engineering*, McGraw-Hill, New York, 1949.

Bablon, G. et al. Practical applications of ozone: Principles and case studies, in: Langlais, B., Reckhow, D. A., and Brink, D. R. (Eds.), *Ozone in Water Treatment*, Lewis Publishers, Chelsea, MI, 1991.

Benjamin, M. M., *Water Chemistry*, McGraw-Hill, New York, 2002.

Blaney, C. A., Li, L., Gloyna, E. F., and Hossain, S. U., Supercritical water oxidation of pulp and paper mill sludge as an alternative to incineration, in: Hutchenson, K. W., and Foster, N. R., (Eds.), *Innovations in Supercritical Fluids*, Chapter 30, pp. 444–455, *ACS Symposium Series* 608, American Chemical Society, Washington, DC, 1995.

Bull, R. A. and Zeff, J. D., Hydrogen peroxide in advance oxidation processes for treatment of industrial processes and contaminated groundwater, in: Eckenfelder, W. W., Bowers, A. R., and Roth, J. A. (Eds.), *Chemical Oxidation–Technologies for the Nineties, Proceedings of the First International Symposium*, Vanderbilt University, Nashville, TN, February 20–22, 1991, Technomic Publishing Co., Lancaster, PA, 1992.

Cansell, F., Beslin, P., and Berdeu, B., Hydrothermal oxidation of model molecules and industrial wastes, *Environmental Progress*, 17(4):240–245, Winter 1998.

Carlson, K., Ozone oxidation, unpublished paper given at *Rocky Mountain Section American Water Works Association—Water Environment Association Annual Conference*, Tamarron Resort, Durango, Durango, CO, September 1992.

Eckenfelder, W. W., Bowers, A. R., and Roth, J. A. (Eds.), *Chemical Oxidation—Technologies for the Nineties, Proceedings of the First International Symposium*, Vanderbilt University, Nashville, TN, February 20–22, 1991, Technomic Publishing Co., Lancaster, PA, 1992.

Ford, D. L., *Reflections of a Soldier and Scholar—The Life of Earnest F. Gloyna*, Cockrell School of Engineering, University of Texas at Austin, Austin, TX, October 2009.

Glaze, W. H., Drinking water treatment with ozone, Environmental Science & Technology, 21(3):224–230, March 1987.

Glaze, W. H., Kenneke, J. F., and Ferry, J. L., Chlorinated byproducts from the TiO_2-mediated photodegradation of trichloroethylene and tetrachloroethylene in water, *Environmental Science and Technology*, 27(1):177–184, January 1993.

Gloyna, E. F. and Li, L., Supercritical water oxidation research and development update, *Environmental Progress*, 14(3):182–192, August 1995.

Gregory, D., Oxidation and removal of dissolved manganese in water Treatment, MS Thesis, Department of Civil Engineering, Colorado State University, Fort Collins, CO, 1996.

Gregory, D., Determining the optimum chemical oxidant for Mn removal—The experience of the Fort Collins water utility, *Rumbles* (newsletter of the Rocky Mountain Section AWWA/WEA), May 1997.

Groves, F. R. Jr., Brady, B., and Knopf, F. R., State of the art on the supercritical extraction of organics from hazardous wastes, in: Straub, C.P. (Ed.), *Critical Reviews in Environmental Control*, Vol. 15(3), pp. 237–274, 1985, CRC Press, Boca Raton, FL, 1985.

Hochleitner, W. A., Analysis of non-isothermal, wet oxidation reactor data, *Environmental Progress*, 15(1):48–55, Spring 1996.

Hoigné, J. and Bader, H., Rate constants of reactions of ozone with organic and inorganic compounds in water—I, *Water Research*, 17(2):173–183, 1983a.

Hoigné, J. and Bader, H., Rate constants of reactions of ozone with organic and inorganic compounds in water—II, *Water Research*, 17(2):185–194, 1983b.

Hoigné, J., Bader, H., Haag, W. R., and Staehelin, J., Rate constants of reactions of ozone with organic and inorganic compounds in water—III, *Water Research*, 19(8):993–1004, 1985.

Jain, V. K., Supercritical fluids tackle hazardous wastes, *Environmental Science and Technology*, 27(5):806–808, 1993.

Kaminski, J. C. and Prendiville, P. W., Milwaukee's ozone upgrade, *Civil Engineering*, 66(9):62–64, September 1996.

Kiwi, J., Lopez, A., and Nadtochenko, V., Mechanism and kinetics of the OH-radical intervention during Fenton oxidation in the presence of a significant amount of radical scavenger, *Environmental Science and Technology*, 34(11):2162–2168, June 1, 2000.

Kruse, A. and Schmieder, H., Supercritical oxidation in water and carbon dioxide, *Environmental Progress*, 17(4):234–239, Winter 1999.

Langlais, B., Reckhow, D. A., and Brink, D. R., *Ozone in Water Treatment*, Lewis Publishers, Chelsea, MI, 1991.

Latimer, W. M., *The Oxidation States of the Elements and Their Potentials in Aqueous Solutions*, 2nd edn., Prentice-Hall, Inc., Englewood Cliffs, NJ, 1952.

Li, L., Gloyna, E. F., and Sawicki, J. E., Treatability of DNT process wastewater by supercritical water oxidation, *Water Environment Research*, 65(3):250–257, May/June 1993.

Ma, J. and Graham, N., Controlling the formation of chloroform by permanganate preoxidation—destruction of precursors, *Aqua*, 45(6):308–315, December 1996.

Metropolitan Water District of Southern California and James M. Montgomery Consulting Engineers, Inc., *Pilot-Scale Evaluation of Ozone and Peroxone*, AWWA Research Foundation, Denver, CO, 1991.

Modell, M., Gasification and liquefaction of forest products in supercritical water, in: Overend, R. P., Milne, T. A., and Mudge, L. K., (Eds.), *Fundamentals of Thermochemical Biomass Conversion*, Chapter 6, pp. 95–119, Elsevier Applied Science Publishers, New York, 1985.

Modell, M., Supercritical-water oxidation, in: Freeman, H. M. (Ed.), *Standard Handbook of Hazardous Waste Treatment and Disposal*, pp. 8.153–8.168, McGraw-Hill, New York, 1988.

Modell, M., Kuharich, E., and Rooney, M., Supercritical water oxidation process of organics with inorganics, U.S. Patent No. 5,252,224, issued October 12, 1993.

Noack, M. G. and Iacoviello, S. A., The chemistry of chlorine dioxide in industrial and wastewater treatment applications, in: Eckenfelder, W. W., Bowers, A. R., and Roth, J. A. (Eds.), *Chemical Oxidation—Technologies for the Nineties*, Volume 2, *Proceedings of the Second International Symposium*, Vanderbilt University, Nashville, TN, February 19–21, 1992, Technomic Publishing Co., Lancaster, PA, 1994.

Rakeness, K. L., DeMers, L. D., and Blank, B. D., Ozone system fundamentals for drinking water treatment, *Opflow* (American Water Works Association), 22(7):1–5, July 1996.

Rice, R. G. and Netzer, A. (Eds.), *Handbook of Ozone Technology and Applications*, Volume 1, Ann Arbor Science Publishers, Ann Arbor, MI, 1982.

Sawicki, J. E. and Casas, B., Wet oxidation system—process concept to design, *Environmental Progress*, 12(4):275–283, November 1993.

Servos, J. W., *Physical Chemistry from Ostwald to Pauling—The Making of A Science in America,* Princeton University Press, Princeton, NJ, 1990.

Silberberg, M., *Chemistry—The Molecular Nature of Matter and Change*, Mosby-Year Book, Inc., St. Louis, MO, 1996.

Snoeyink, V. L. and Jenkins, D., *Water Chemistry*, John Wiley & Sons, New York, 1980.

Stumm, W. and Morgan, J. J., *Aquatic Chemistry—Chemical Equilibria and Rates in Natural Waters*, John Wiley & Sons, Inc., New York, 1996.

Svensson, P., Look, no stack: Supercritical water destroys organic wastes, *Chemical Technology Europe*, 2(1):16–19, January/February 1995.

Symons, J. M., Bradley, L. C. Jr., and Cleveland, T. C., *The Drinking Water Dictionary*, American Water Works Association, Denver, CO, 2000.

Tester, J. and Modell, M., *Thermodynamics and Its Applications*, 3rd edn., Prentice Hall, Upper Saddle River, NJ, 1997.

Thornton, T. D. and Savage, P. E., Kinetics of phenol oxidation in supercritical water, *AIChE Journal,* 38(3), 321–327, March 1992.

Timberlake, S., Hong, G., Simson, M., and Modell, M., Supercritical water oxidation for wastewater treatment: Preliminary study of urea destruction, SAE Technical Paper Series, 820872, 1982.

Topudurti, K. V., Lewis, N. M., and Hirsh, S. R., The applicability of UV/oxidation technologies to treat contamination groundwater, *Environmental Progress*, 12(1):54–60, February 1993.

Valsaraj, K. T., *Elements of Environmental Engineering—Thermodynamics and Kinetics*, CRC-Lewis Publishers, Boca Raton, FL, 1995.

von Sonntag, C., Mark, G., Mertens, R., Schuchmann, M. N., and Schuchmann, H. P., UV radiation and/or oxidants in water pollution control, *Aqua*, 42(4):201–211, August 1993.

White, G. C., *Handbook of Chlorination and Alternative Disinfectants,* 4th edn., John Wiley & Sons, Inc., New York, 1999.

Wilson, E., TiO_2 appears inefficient for water treatment, *Chemical and Engineering News*, 74(27):29, July 1, 1996.

Zawacki, J., $KMnO_4$ contributes to least-cost treatment solution, *Water Engineering and Management*, 139:18–19, May, 1992.

FURTHER READING

Freeman, H. and Harris, E. F. (Eds.), Supercritical water oxidation, pp. 223–228, Chapter 23 in: *Hazardous Waste Remediation – Innovative Treatment Technologies*: (342pp), Technomic Publishing Company, Lancaster, PA, 1995. Also available as an e-book.

Gloyna, E. F. and Li, L., Waste treatment by supercritical water oxidation, waste, nuclear, reprocessing and treatment technologies to wastewater treatment, multilateral approach, in: McKetta, J. J. (Ed.) *Encyclopedia of Chemical Processing and Design*, pp. 272–303, Marcel Dekker, Inc., New York, 1998.

Gloyna, E. F. and Li, L., Supercritical water oxidation for wastewater and sludge remediation, in: Meyers, R.A. (Ed.), *Encyclopedia of Environmental Analysis*, pp. 4780–4794, John Wiley & Sons, New York, 1998. Contains 193 references.

Watts, R. J., *Hazardous Wastes: Sources, Pathways, Receptors*, John Wiley & Sons, New York, 1996.

Zou, L. Y., Li, Y., and Hung, Y., Wet air oxidation for waste treatment, in: Wang, L. K. and Hung, Y. H. (Eds.), *Advanced Physicochemical Treatment Technologies*, Volume 5, pp. 575–610, in Book Series, *Handbook of Environmental Engineering*, The Humana Press, Totowa, NJ, 2007. The chapter is on line at http://www.springerlink.com, 2007. ISBN 978-1-58829-860-7 (Print); 978-1-59745-173-4 (Online).

21 Precipitation

Chemical precipitation has been applied traditionally to softening of domestic water. Since removal of heavy metals became a major concern, during and after the 1970s, precipitation has been a "workhorse" process in the treatment of industrial wastes and mine drainage waters. This practice has evolved to a large extent by empirical designs and operation, with theory providing a rationale, that is, to explain rather than to lead. Nevertheless, theory provides the foundation for modern practice.

21.1 DESCRIPTION

In water treatment, precipitation is the addition of a chemical to cause an ion, usually a cation, to be removed as a solid. The "target" ions are those that may impair the use of water. Cations that are most likely to be an issue include calcium, magnesium, iron, manganese, and various heavy metals. Other cations may cause problems, depending on the situation and the concentration. Fewer anions are amenable to precipitation. Those that are amenable include phosphates, silica, sulfides, and cyanide. Nitrate, a "contaminant" in drinking water, is not amenable to removal by precipitation; neither are the innocuous anions such as chloride and sulfate. The exact nature of silica is complex. For example, SiO_2, SiO^-, $SiOH_2^+$ are included in the silica group by Letterman et al. (1999, p. 6.9). Stumm (1992, p. 175) gives a dissolution reaction as, $SiO_2\ (s) + H_2O \rightarrow H_4SiO_4(ag)$. The latter is orthosilicic acid, which may dissociate to $H_3SiO_4^-$ (and there are other derivatives).

21.1.1 Precipitation In-a-Nutshell

Chemical precipitation occurs when the solubility product is exceeded for the anions and cations of any given salt. The ions comprising the salt are thus removed from solution.

21.1.1.1 Definitions

Precipitation has not evolved to become a specialized field in which a unique vernacular, for example, acronyms and terms, has evolved. Many of the terms come from the field of chemistry, however, such that glossary may be useful for reference.

21.1.1.2 Comparison with Other Processes

Along with oxidation, precipitation involves a chemical reaction as the central issue of the process. It is characterized by the formation of a solid precipitate that is settled and/or filtered.

21.1.1.3 Process Description

The precipitation process has several steps between the ions in solution and the solid precipitate being settled and removed. The steps are as follows:

1. (a) Introduce an anion that will form a precipitate with the target cation (or that will cause a change in water chemistry, for example, to form carbonates or hydroxides that have an affinity for the target cation). (b) Introduce a cation that will remove the target anion.
2. Cause sufficient mixing, for example, by rapid mix, to cause a high rate of contacts between the anions and the target cations in order to effect formation of precipitant crystals.
3. Apply moderate turbulence, for example, by paddle-wheel flocculation, to cause floc formation and growth or, alternatively, cause flow through a floc blanket of previously formed floc to cause floc growth.
4. Permit settling by traditional continuous flow basin, by tube settlers, or by plate settlers with vacuum pickup of sludge or scraping into a pocket.
5. Apply filtration to remove residual floc particles.
6. Dewater sludge and transport to disposal site.
7. Solids disposal/recovery may be reclamation, disposal in a municipal landfill, or disposal in a secured landfill. The latter, used for heavy metals and toxic wastes, are usually constructed for a hypothetical 1000-year life (based upon EPA regulations).

Alternatively, the entire process may be done in a "reactor-clarifier," a single unit, that has a center-well chemical feed and mixing, a sludge blanket flocculation, an upflow settling. Units of this type are probably most common for small municipalities, industrial wastes, and mining.

21.1.2 Applications

Chemical precipitation is used commonly for softening of municipal or industrial water supplies, for removal of toxic metals (heavy metals) from mining wastes or industrial wastewaters, or for removal of certain anions, for example, phosphates, silica, or cyanide.

21.1.2.1 Softening

As described by Walker (1934, p. 77), the calcium and magnesium comprising "hard" water combines with soap, forming "insoluble sticky curds that tend to adhere to all things with which they come in contact." Hard water also causes scale, which reduces the effectiveness of pipes and boilers with ancillary economic damage. Hard water also has an economic loss in that larger quantities of soap are required for laundry purposes and clothes have shorter life.

21.1.2.2 Toxic Metals Removal

The toxic "heavy" metals include arsenic, cadmium, chromium, lead, mercury, and a few others. They are found in waste flows from various industries and in drainage from mines. In addition, zinc and copper are considered contaminants when found in such waste streams.

Some examples of heavy metal sources include metal processing and refining, metal plating, petroleum refining, chloroalkali production, battery manufacturing, steel production, pigment manufacturing, tanning, anodizing, photographic film manufacturing, automotive production, etc. (Banerjee, 2002, p. 181). Removals of metals are mostly by precipitation and ion exchange; other processes include oxidation/precipitation, reduction/precipitation, coagulation/coprecipitation (Patterson, 1990, p. 28). To remove metal ions from solution requires anions that when combined with a particular heavy metal cation, have a low solubility product. The most common anions for chemical precipitation are hydroxides, mostly because lime, the source, is cheaper than other chemicals, and because the solubility products, with most metals, are low.

21.1.3 History

The traditional application of chemical precipitation as a unit process has been the softening of municipal water supplies, with removal of heavy metals emerging as a concern during the 1970s. An application, not too well known after about 1900, was that, in England, precipitation was considered as a promising technology for treatment of municipal wastewaters. This can be seen by adding alum to raw sewage; a visible heavy precipitate forms readily and settles leaving a clarified supernatant.

21.1.3.1 Softening

The need for softening of "hard" water because of excessive soap consumption was known in the 1730s in England. The terms, "soft" water and "hard" water were common and the "alkaline salts" were used to remove hardness (Baker, 1948, pp. 415–420). The use of lime, that is, CaO, in softening, was described in 1841 by Thomas Clark, Professor of Chemistry, Aberdeen University, Scotland. Clark patented a method to remove hardness, which involved the use of lime, followed by "subsidence."

The first attempted use of lime softening for a municipal water supply was in 1854 by the Plumstead, Woolrich & Charlton Consumer's Water Co., to compete with the Kent Waterworks Co., each operating in the London metropolitan region. Both companies used wells as a source (Baker, 1948, p. 422). Although endorsed by the Royal Commission on Water Supply in an 1869 report, softening was not adopted. By 1900, only a few softening plants had been built in England. About that time the residual calcium carbonate precipitant, that is, after softening, was found to cause incrustation of filter sand, and clogging of service pipes and meters. This was remedied by "re-carbonation," that is, dissolving carbon dioxide to convert such residual $CaCO_3$ to Ca^{2+} and HCO_3^-, which became common practice in the United States by about 1921.

A 30 mgd plant was put on line in 1908 in Columbus, Ohio, which was the largest softening plant in the United States. The plant was designed by George A. Johnson, then with Hering & Fuller (Rudolph Hering and George W. Fuller). Just before rapid filtration, carbon dioxide gas was added to neutralize excess lime, that is, converting carbonates to bicarbonates. Other cities that employed lime softening included New Orleans in 1909, Cincinnati in 1938, Minneapolis in 1939, and St. Paul in 1940 (Baker, 1948, p. 432). By the 1930s the softening process employed rapid mix (baffled mixing was common), paddle-wheel mixing, sedimentation, re-carbonation, and filtration. A similar plant was described by Gelston (1934) for Quincy, Illinois to reduce hardness of water withdrawn from the Mississippi River (typically 150 mg/L as $CaCO_3$ total hardness). The clarifier was proprietary, that is, Dorr®, circular type, 24 m (80 ft) diameter, with center feed and sludge removal at the bottom, later called a "reactor-clarifier." Iron removal is a by-product of softening and may be achieved by first oxidizing ferrous ion (common in ground water) to the ferric state (Walker, 1934, p. 79).

21.1.3.2 Sewage Treatment

The beginning of sewage treatment was in England after about the year 1850. Prior to this, the streams of England were regarded as public dumps for ashes, cinders, demolition wastes, dead animals, etc. The Second Royal Commission of River Pollution, appointed in 1868, in its report on methods of treatment in 1870, reviewed three principal methods: irrigation, filtration, and chemical precipitation. Its fifth report in 1873 was on river pollution by mining and metal industries. The commission formulated standards of purity for British rivers by forbidding discharge of various substances. A Second Royal Commission of Sewage Disposal was appointed in 1898 (the first was 1857). Its fifth report in 1908 summarized the state-of-the-art in sewage treatment, which included sedimentation, chemical precipitation, contact beds, trickling filters, and land treatment (Metcalf and Eddy, 1916).

The chemical precipitation method was well established by 1870. Lime was used in most cases but $CaCl_2$, $MgCl_2$, alum, and other chemicals were used also. Patented processes were numerous. The effluent was reported as "clear and of good quality." The idea of chemical precipitation of treating sewage after primary settling was explored again in the late 1960s without reference to the early history of the process.

21.1.3.3 Heavy Metals

Until about 1970, the main emphasis of wastewater treatment was to reduce the biochemical oxygen demand (BOD) load on ambient waters. Along with the traditional concern with BOD, however, other issues began to emerge, as stream standards became more stringent to include a wide variety of contaminants, with restoration and preservation of stream ecology emerging as a principle goal. These issues and a host of others were addressed in the 1972 Clean Water Act, for example, PL92500 and its subsequent amendments. Control of toxic pollutants, which included metals, was included. As noted, precipitation is a favored method of metals removal, for example, by hydroxide, carbonate, sulfide, etc.

Biosolids (sludge) have always been a major issue in wastewater treatment. Ocean disposal had been common for coastal cities, through the 1960s, which was always controversial. With the Clean Water Act this changed and land disposal became favored. Because crops are usually grown or are a potential for any land, heavy metals in the biosolids had to be limited. Since industrial wastes have been the sources of most heavy metals, pretreatment was mandated. All of this evolved through the 1980s and into the 1990s, which resulted in biosolids with a very low fraction of metals, suitable for application to land which could grow selected crops.

As a note, removal of metals involves management strategies, such as reduction in metals flux, substitutions in industrial processes, separations within industrial processes, recovery and reuse, final disposal in secured landfills, etc. (Patterson, 1990, pp. 27–42). This addresses only "end-of-the-pipe treatment."

21.2 PRECIPITATION THEORY

Key variables in the precipitation process are (Kemmer, 1979, pp. 10–12) (1) solubility product, (2) temperature, (3) particle charge, and (4) time. The solubility product is temperature dependent, and the time has to do with the reaction kinetics and the rate of mixing. Particle size increases if the chemical reaction occurs on particles that have been formed previously (Kemmer, 1979, pp. 10–11).

The formation of precipitate occurs in three stages (Patterson et al. 1990, p. 95): nucleation, crystal growth, and aging. Most often, the nucleation occurs on preexisting particles of various types. Crystal growth occurs in two phases: (1) movement of solute to the crystal/water interface by advection and diffusion, and (2) adsorption of the solute onto the solid surface and incorporation into the crystal lattice.

21.2.1 EQUILIBRIA

The equilibrium between concentrations of particular ions in solution and the solid precipitate is expressed in terms of the solubility product. If a slightly soluble species is in equilibrium with a precipitate, for example, $PbCl_2$, there will be some undissociated $PbCl_2$ molecules and $PbCl^-$ ions; usually, however, these are ignored and only Pb^{2+} are considered (Silberberg, 1996, p. 803).

21.2.1.1 Solubility Law

Consider a particular reaction, that is, the dissolution of the solid, $Mg(OH)_2$:

$$Mg(OH)_2 \rightarrow Mg^{2+} + 2OH^- \quad (21.1)$$

The equilibrium statement is

$$K_{Mg(OH)_2} = [Mg^{2+}]\,[OH^-]^2 \quad (21.2)$$

$$= 5.6 \cdot 10^{-12} \quad (21.3)$$

in which

$K_{Mg(OH)_2}$ is the solubility product of magnesium hydroxide
$[Mg^{2+}]$ is the molar concentration of magnesium ion (mol/L)
$[OH^-]$ is the molar concentration of hydroxide ion (mol/L)
$5.6 \cdot 10^{-12}$ is the numerical value of solubility product of magnesium hydroxide (Table 21.1)

The solubility product is the same as any other equilibrium constant except that the activity of the solid precipitate, the unstated denominator, equals one. The concentrations of Mg^{2+} and OH^- are at "saturation" with respect to the solid. If the pH is changed, $[OH^-]$ changes and the $[Mg^{2+}]$ changes in accordance with Equation 21.2. If the concentrations of Mg^{2+} and OH^- exceed the solubility product, that is, if $[Mg^{2+}][OH^-]^2 > K_{Mg(OH)_2}$, then precipitation occurs; if less, that is, if $[Mg^{2+}][OH^-]^2 < K_{Mg(OH)_2}$, then the solid, $Mg(OH)_2$, dissolves. Precipitation requires nuclei, which are present in most waters.

21.2.1.2 Application of Solubility Law

To remove an undesired cation from solution, an anion may be added so that the solubility product of the precipitate product is exceeded. For example, if magnesium is to be removed, one may raise the pH; pH = 11 is common, such that the solubility product, $[Mg^{2+}][OH^-]^2 > K_{Mg(OH)_2} = 5.6 \cdot 10^{-12}$ is exceeded. Example 21.1 shows how to calculate the residual concentration of Mg^{2+} when the pH is raised to pH = 11.

Example 21.1 Mg^{2+} Concentration after Raising pH to 11

Given

Magnesium ion, Mg^{2+}, is a hardness component in a water and is to be removed by precipitation; its concentration in the ambient raw water is $C(Mg^{2+}) = 22$ mg/L as $CaCO_3$. The pH is to be raised to pH = 11 (in a reactor).

Required

Estimate the concentration of Mg^{2+} residual, that is, $[Mg^{2+}]$ solution.

TABLE 21.1
Solubility Products of Selected Ionic Compounds, 25°C[a,b]

Name	Formula	K_{sp}
Carbonates		
Barium carbonate	$BaCO_3$	$2.6 \cdot 10^{-9}$
Cadmium carbonate	$CdCO_3$	$1.8 \cdot 10^{-12}$
Calcium carbonate	$CaCO_3$	$3.4 \cdot 10^{-9}$
Cobalt(II) carbonate	$CoCO_3$	$1.0 \cdot 10^{-10}$
Copper(II) carbonate	$CuCO_3$	$3 \cdot 10^{-12}$
Iron(II) carbonate	$FeCO_3$	$3.1 \cdot 10^{-11}$
Lead(II) carbonate	$PbCO_3$	$7.4 \cdot 10^{-14}$
Magnesium carbonate	$MgCO_3$	$6.8 \cdot 10^{-6}$
Mercury(I) carbonate	Hg_2CO_3	$3.6 \cdot 10^{-17}$
Nickel(II) carbonate	$NiCO_3$	$1.4 \cdot 10^{-7}$
Strontium carbonate	$SrCO_3$	$5.6 \cdot 10^{-10}$
Zinc carbonate	$ZnCO_3$	$1.6 \cdot 10^{-10}$
Chromates		
Barium chromate	$BaCrO_4$	$1.2 \cdot 10^{-10}$
Calcium chromate	$PbCrO_4$	$1 \cdot 10^{-8}$
Lead(II) chromate	$PbCrO_4$	$2.3 \cdot 10^{-13}$
Silver chromate	Ag_2CrO_4	$1.1 \cdot 10^{-12}$
Cyanides		
Copper(I) cyanide	$CuCN$	$3.5 \cdot 10^{-20}$
Mercury(I) cyanide	Hg_2CN_2	$5 \cdot 10^{-40}$
Silver cyanide	$AgCN$	$6.0 \cdot 10^{-17}$
Halides		
Fluorides		
Barium fluoride	BaF_2	$1.8 \cdot 10^{-7}$
Calcium fluoride	BaF_2	$3.4 \cdot 10^{-11}$
Lead(II) fluoride	PbF_2	$3.3 \cdot 10^{-8}$
Magnesium fluoride	MgF_2	$5.2 \cdot 10^{-11}$
Strontium fluoride	SrF_2	$4.3 \cdot 10^{-9}$
Chlorides		
Copper(I) chloride	$CuCl$	$1.7 \cdot 10^{-7}$
Lead(II) chloride	$PbCl_2$	$1.7 \cdot 10^{-5}$
Silver chloride	$AgCl$	$1.8 \cdot 10^{-10}$
Bromides		
Copper(I) bromide	$CuBr^{\bullet}$	$6.3 \cdot 10^{-9}$
Silver bromide	$AgBr$	$5.0 \cdot 10^{-13}$
Iodides		
Copper(I) iodide	CuI_x	$1.3 \cdot 10^{-12}$
Lead(II) iodide	PbI_2	$9.8 \cdot 10^{-9}$
Mercury(I) Iodide	Hg_2I_2	$2.9 \cdot 10^{-29}$
Mercury(II) Iodide	HgI_2	$5.2 \cdot 10^{-29}$
Silver iodide	AgI	$8.5 \cdot 10^{-17}$
Hydroxides		
Aluminum hydroxide	$Al(OH)_3$	$7.2 \cdot 10^{-15}$
Cadmium hydroxide	$Ca(OH)_2$	$5.0 \cdot 10^{-6}$
Calcium hydroxide	$Ca(OH)_2$	$5.0 \cdot 10^{-6}$
Cobalt(II) hydroxide	$Co(OH)_2$	$5.9 \cdot 10^{-15}$
Copper(II) hydroxide	$Cu(OH)_2$	$2.2 \cdot 10^{-20}$
Iron(II) hydroxide	$Fe(OH)_2$	$4.9 \cdot 10^{-17}$
Iron(III) hydroxide	$Fe(OH)_3$	$2.8 \cdot 10^{-39}$
Lead(II) hydroxide	$Pb(OH)_2$	$1.4 \cdot 10^{-20}$
Magnesium hydroxide	$Mg(OH)_2$	$5.6 \cdot 10^{-12}$
Manganese(II) hydroxide	$Mn(OH)_2$	$1.6 \cdot 10^{-13}$
Nickel(II) hydroxide	$Ni(OH)_2$	$5.5 \cdot 10^{-16}$
Zinc hydroxide	$Zn(OH)_2$	$3 \cdot 10^{-17}$
Iodates		
Barium iodate	$Ba(IO_3)_2$	$4.0 \cdot 10^{-9}$
Calcium iodate	$Ca(IO_3)_2$	$6.5 \cdot 10^{-6}$
Lead(II) iodate	$Pb(IO_3)_2$	$3.7 \cdot 10^{-13}$
Silver iodate	$AgIO_3$	$3.2 \cdot 10^{-8}$
Strontium iodate	$Sr(IO_3)_2$	$1.1 \cdot 10^{-7}$
Zinc iodate dihydrate	$Zn(IO_3)_2 \cdot 2H_2O$	$4.1 \cdot 10^{-6}$
Oxalates		
Barium oxalate dihydrate	$BaCO_4 \cdot 2H_2O$	$1.1 \cdot 10^{-7}$
Calcium oxalate monohydrate	$CaC_2O_4 \cdot H_2O$	$2.3 \cdot 10^{-9}$
Strontium oxalate monohydrate	$SrCO_4 \cdot H_2O$	$5.6 \cdot 10^{-8}$
Phosphates		
Aluminum phosphate	$AlPO_4$	$9.8 \cdot 10^{-21}$
Cadmium phosphate	$Cd_3(PO_4)_2$	$2.5 \cdot 10^{-33}$
Calcium phosphate	$Ca_3(PO_4)_2$	$2.1 \cdot 10^{-33}$
Copper(II) phosphate	$Cu_3(PO_4)_2$	$1.4 \cdot 10^{-37}$
Magnesium phosphate	$Mg_3(PO_4)_2$	$1.0 \cdot 10^{-24}$
Nickel(II) phosphate	$Ni_3(PO_4)_2$	$4.7 \cdot 10^{-32}$
Silver phosphate	Ag_3PO_4	$8.9 \cdot 10^{-17}$
Sulfates		
Barium sulfate	$BaSO_4$	$1.1 \cdot 10^{-10}$
Calcium sulfate	$CaSO_4$	$4.9 \cdot 10^{-5}$
Lead(II) sulfate	$PbSO_4$	$2.5 \cdot 10^{-8}$
Magnesium sulfate	$MgSO_4$	$5.9 \cdot 10^{-3}$
Radium sulfate	$RaSO_4$	$3.7 \cdot 10^{-11}$
Silver sulfate	Ag_2SO_4	$1.2 \cdot 10^{-5}$
Strontium sulfate	$SrSO_4$	$3.4 \cdot 10^{-7}$
Sulfide[c]		
Cadmium sulfide	CdS	$1.0 \cdot 10^{-24}$
Cadmium sulfide (Lide)	CdS	$8 \cdot 10^{-7}$
Copper(II) sulfide	CuS	$8 \cdot 10^{-34}$
Iron(II) sulfide	FeS	$8 \cdot 10^{-16}$
Lead(II) sulfide	PbS	$3 \cdot 10^{-25}$
Manganese(II) sulfide	MnS	$3 \cdot 10^{-11}$
Mercury(II) sulfide	HgS	$2 \cdot 10^{-50}$
Silver sulfide	Ag_2S	$8 \cdot 10^{-48}$
Tin(II) sulfide	SnS	$1.3 \cdot 10^{-23}$
Zinc sulfide	ZnS	$2.0 \cdot 10^{-22}$

[a] The table format, categorized in terms of anions, was adapted from Silberberg (1996, p. 804).

[b] Except for sulfides, solubility product data were from Lide (1996, pp. 8–91, 8–93), since the reference is a standard source of data; some of the data differed from the Silberberg data but not greatly. In a few cases, i.e., $CoCO_3$, $CuCO_3$, Hg_2CN_2, SrF_2, AgBr, $Cu(OH)_2$, $Mn(OH)_2$, $BaCO_4 \cdot 2H_2O$, $SrCO_4 \cdot H_2O$, $MgSO_4$, the data were available only from Silberberg, p. 806. In other cases, i.e., $FeCO_3$, CuCN, $Al(OH)_3$, $AlPO_4$, $Cd_3(PO_4)_2$, $Cu_3(PO_4)_2$, $Ni_3(PO_4)_2$, data were available only from Lide.

[c] According to Silberberg (1996, p. 804), S^{2-} does not exist in water; rather, the reaction occurs, $S^{2-}(aq) + H_2O \rightarrow HS^-(aq) + OH^-$. Therefore, if an insoluble precipitate is shaken in water, the sulfide reacts immediately with water as indicated, such that the ion HS^- is present (rather than S^{2-}. Sulfide data are from Silberberg (1996, p. 806).

Solution

Equation 21.2, $K_{Mg(OH)_2} = [Mg^{2+}][OH^-]^2$, is the basis for the calculation. First, however, $[OH^-]$ must be calculated from the relation, $[H^+][OH^-] = 10^{-14}$, at pH = 11, $[H^+] = 10^{-11} \rightarrow [10^{-11}][OH^-] = 10^{-14}$; and $[OH^-] = 10^{-3}$ mol/L.

$$[Mg^{2+}][OH^-]^2 = 5.6 \cdot 10^{-12}$$

$$[Mg^{2+}][10^{-3}]^2 = 5.6 \cdot 10^{-12}$$

$$[Mg^{2+}] = 5.6 \cdot 10^{-6} \text{ mol/L}$$

$$C(Mg^{2+}) = \frac{5.6 \text{ mol} \cdot 10^{-6} \text{ Mg}^{2+}}{\text{L}} \cdot \frac{24{,}300 \text{ mg Mg}^{2+}}{\text{mol Mg}^{2+}}$$

$$= 0.14 \text{ mg Mg}^{2+}/\text{L}$$

Discussion

As seen, the residual concentration of Mg^{2+}, that is, that which is in equilibrium with $Mg(OH)_2$ particles, that is, "sludge," is very low.

21.2.1.3 Listing of Solubility Products

Table 21.1 lists compounds and solubility products, categorized by anions. From such a list, the precipitant compound may be selected.

For example, to remove lead, Table 21.1 shows that $K_{Pb(OH)_2} = 1.4 \cdot 10^{-20}$, which is the lowest listed. Removal involves merely raising the pH within the reactor, flocculation, and settling. The whole sequence is often done in the industry by means of a "reactor clarifier."

In other cases, anions, for example, CN^-, PO_4^{3-}, are targeted for removal. The same approach is applied, that is, finding a cation that when combined with the particular anion of interest, has a low solubility product. As seen in Table 21.1, phosphate is not difficult to remove because $K_{Ca_3(PO_4)_2} = 2.1 \cdot 10^{-33}$. Similarly, $K_{Cu(CN)} = 3.5 \cdot 10^{-20}$, which means that Cu(I) can remove CN^-.

21.2.1.4 Solubility pC–pH Diagrams

For precipitation reactions, as with acid–base reactions, a graphical solution facilitates understanding and adds a visualization component. As with an acid–base reaction, its construction requires an equilibrium statement, and equations for mass balance and electroneutrality. The graphical solution also has practical application in that whether precipitation occurs or not is evident for any pH–pC coordinate. Example 21.2 illustrates the method of constructing a pH–pC diagram for calcium hydroxide.

Example 21.2 Construction of pH-pC Diagram for $Ca(OH)_2$

Given

A solution has calcium chloride, $[CaCl_2] = 0.1$ mol/L and is titrated with sodium hydroxide in which, $[NaOH] = 1.0$ mol/L.

Required

Determine the pH versus $p[Ca^{2+}]$ curve as titration occurs.

Solution

State reactions:

$$CaCl_2 \rightarrow Ca^{2+} + 2Cl^-$$

$$H_2O \rightarrow H^+ + OH^-$$

$$NaOH \rightarrow Na^+ + OH^-$$

The net equation is, combining Equations 1 and 3 and excluding spectator ions, Na^+ and Cl^-:

$$Ca^{2+} + 2[OH^-] \rightarrow Ca(OH^-)_2$$

Species in solution:

$$Ca^{2+}, Cl^-, Na^+, OH^-, H^+$$

Mathematical relations:

Equilibrium equations

$$K_{Ca(OH)_2} = [Ca^{2+}][OH^-]^2 = 10^{-5.3}$$

$$K_w = [H^+][OH^-]$$

Electroneutrality

$$2[Ca^{2+}] + [Na^+] + [H^+] = [Cl^-] + [OH^-]$$

Mass balance

$$C = 2[Ca^{2+}] + 2[Ca(OH)_2] = [Cl^-]$$

Derivative relations:

Substitute (8) in (7) to eliminate $[Cl^-]$:

$$[Na^+] + [H^+] = [OH^-]$$

This means that as Na^+ is added, OH^- is added in the same amount.

Express (5) in log form:

$$\log [Ca^{2+}] + 2 \log [OH^-] = -5.3$$

or

$$p[Ca^{2+}] + 2p[OH^-] = 5.3$$

System point determination:

Substitute $[Ca^{2+}] = 0.1$ mol/L in (11), that is,

$$p[0.1] + 2p[OH^-] = 5.3$$

$$p[OH^-] = \frac{5.3 - 1.0}{2} = 2.15$$

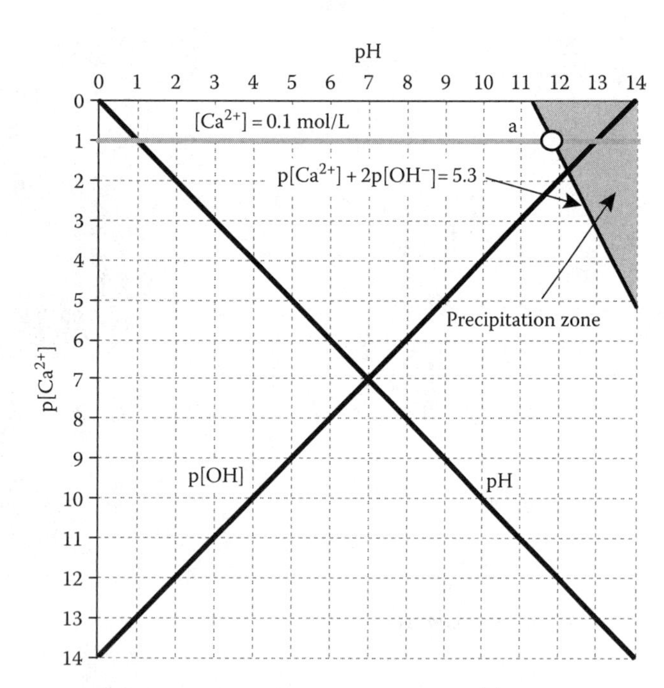

FIGURE CD21.1 pC versus pH diagram for Ca^{2+}.

Discussion

Figure CD21.1 (see also Figure 6-3, p. 255, Snoeyink and Jenkins) shows graphically the coordinate point where the solution occurs, that is, (pH, pC) = (pH = 11.85, C = 0.1 mol Ca^{2+}/L). As seen, Example "Express (5) in log form" second equation, that is, $p[Ca^{2+}] + 2p[OH^-] = 5.3$, expresses the equilibrium relation seen as the line passing through the point "a," the system point. The solution is "supersaturated" in the zone to the right of the equilibrium line. The area to the left represents undersaturation. In other words, precipitation may be induced by imposing high pH conditions; the lower the Ca^{2+} concentration, that is, as $p[Ca^{2+}]$ increases, the higher the pH required to cause precipitation. Table CD21.2 was the set up used for the calculations; Figure CD21.1 is linked to Table CD21.2 (using the spreadsheet). Table CD21.2 may be used as a template for other precipitates as well (modified to fit the chemical equations).

21.2.1.5 pε–pH Diagrams

For reactions in which both electron transfer and proton transfer occur, both control the species present at any given (pH, pε) coordinate. By plotting the equilibrium relations between different species in terms of pH and pε, the boundaries between species can be delineated (see Snoeyink and Jenkins, 1980, pp. 358–363). Figure 21.2 shows such a plot for iron. Such plots are called sometimes "predominance-area" diagrams or "Pourbaix" diagrams.

To explain the diagram, consider first, the diagonal line between H_2O and O_2; the area above the line represents the area where the potential is high enough that O_2 occurs. The diagonal between Fe^{2+} and $Fe(OH)_3(s)$ shows the equilibrium between the two species. For any coordinate point within the area above the line, $Fe(OH)_3(s)$ predominates; below the line, Fe^{2+} predominates.

TABLE CD21.2
Concentrations of $[Ca^{2+}]$ as a Function of pH

$C = 0.01$ mol/L
$K_a = 5.01E{-}06$
$pK_a = 5.3$

pH	p[H+]	[H+] (mol/L)	p[OH−]	log [OH−]	[OH−] (mol/L)	p[Ca2+]
0.00	0.00	1.00E+00	14	−14	1.00E−14	−22.7000
1.00	1.00	1.00E−01	13	−13	1.00E−13	−20.7000
2.00	2.00	1.00E−02	12	−12	1.00E−12	−18.7000
3.00	3.00	1.00E−03	11	−11	1.00E−11	−16.7000
4.00	4.00	1.00E−04	10	−10	1.00E−10	−14.7000
5.00	5.00	1.00E−05	9	−9	1.00E−09	−12.7000
6.00	6.00	1.00E−06	8	−8	1.00E−08	−10.7000
7.00	7.00	1.00E−07	7	−7	1.00E−07	−8.7000
8.00	8.00	1.00E−08	6	−6	1.00E−06	−6.7000
9.00	9.00	1.00E−09	5	−5	1.00E−05	−4.7000
10.00	10.00	1.00E−10	4	−4	1.00E−04	−2.7000
11.00	11.00	1.00E−11	3	−3	1.00E−03	−0.7000
11.20	11.20	6.31E−12	2.8	−2.8	1.58E−03	−0.3000
11.30	11.30	5.01E−12	2.7	−2.7	2.00E−03	−0.1000
11.40	11.40	3.98E−12	2.6	−2.6	2.51E−03	0.1000
11.60	11.60	2.51E−12	2.4	−2.4	3.98E−03	0.5000
11.80	11.80	1.58E−12	2.2	−2.2	6.31E−03	0.9000
12.00	12.00	1.00E−12	2	−2	1.00E−02	1.3000
13.00	13.00	1.00E−13	1	−1	1.00E−01	3.3000
14.00	14.00	1.00E−14	0	0	1.00E+00	5.3000

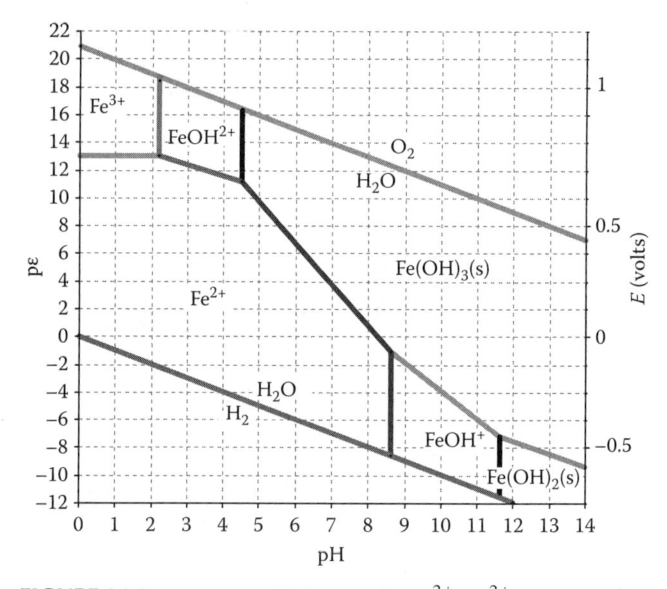

FIGURE 21.2 pε versus pH diagram for Fe^{2+}–Fe^{2+} system (25°C). (Adapted from Snoeyink, V.L. and Jenkins, D., *Water Chemistry*, John Wiley & Sons, New York, 1980, p. 362.)

To explain their utility of a pε vs. pH diagram in understanding the water quality issues of acid–mine drainage involving iron, we may consider the conditions within the underground environment. Here, acid conditions prevail with pH ≈ 4, and with reducing conditions, that is, low pε.

TABLE 21.3
Solubility Rules for Ionic Compounds in Water

Soluble Ionic Compounds	Insoluble Ionic Compounds
1. All common compounds of Group 1A(1) ions (Na^+, K^+, etc.) and ammonium ion (NH_4^+) are soluble	1. All common metal hydroxides are insoluble, except those of Group 1A(1) and the larger members of Group 2A(2) (beginning with Ca^{2+})
2. All common nitrates (NO_3^-), acetates (CH_3COO^-), and most perchlorates (ClO_4^-) are soluble	2. All common carbonates (CO_3^{2-}) and phosphates (PO_4^{3-}) are insoluble, except those of Group 1A(1) and NH_4^+
3. All common chlorides (Cl^-), bromides (Br^-), and iodides (I^-) are soluble, except those of Ag^+, Pb^{2+}, Cu^+, and Hg_2^{2+}	3. All common sulfides are soluble, except those of Group 1A(1), Group 2A(2), and NH_4^+
4. All common sulfates (SO_4^{2-}) are soluble, except those of Ca^{2+}, Sr^{2+}, Ba^{2+}, and Pb^{2+}	

Source: Adapted from Silberberg, M., *Chemistry—The Molecular Nature of Matter and Change*, Mosby-Year Book, Inc., St. Louis, MO, p. 147.
Group 1A(1) elements: H, Li, Na, K, Rb, Cs, and Fr.
Group 2A(2) elements: Be, Mg, Ca, Sr, Ba, Ra.

For such conditions, if iron is present, Fe^{2+} will be the species expected, as indeed occurs. As the water emerges from the mine and if exposed to the atmosphere, ambient oxygen will dissolve in solution, which will increase the redox potential (or pε). In treatment, if the pH is increased, the (pH, pε) coordinates will cause a shift toward $Fe(OH)_3(s)$ becoming the predominant species, which may be removed as a solid, for example, by settling and filtration.

21.2.1.6 General Rules of Solubility

Table 21.3 summarizes solubility rules for common ions; the left-hand columns show the rules for solubility, while the right-hand columns show the rules for insoluble compounds. The rules provide general guidelines for predicting solubilities.

21.2.2 Hardness

Hardness is defined as the soap consuming capacity of a water. Its cause is Ca^{2+}, Mg^{2+}, Fe^{2+}, Mn^{2+}, or any other ion or substance that may consume soap or cause deposits, for example, scale in boilers or precipitates in pipes. Most hardness that occurs in ambient waters is Ca^{2+}.

21.2.2.1 Occurrence of Hardness

Hardness is common in ambient waters, especially when groundwater is the source and in the lower reaches of river systems. Calcium is one of the most common anions and bicarbonate is one of the most common anions. Generally, <100 mg/L of hardness as $CaCO_3$ is not grounds for removal, 100–200 is a nuisance perhaps tolerable, and >300 is usually considered not tolerable. Usually, about 40–60 mg/L hardness as $CaCO_3$ is considered desirable, and is useful in order to have a "depositing" water, that is, water with a Langelier Index ≥ 0. Snowmelt waters, with total dissolved solids, TDS≤ 50 mg/L, $C(Ca^{2+}) \leq 30$ mg Ca^{2+}/L, and pH ≤ 7.0, may be corrosive in the distribution system with consequent leaks in water mains after decades of operation. The City of Fort Collins, in fact, adds lime, just following disinfection, to give $C(Ca^{2+}) \approx 60$ mg/L as $CaCO_3$, which results in a "slightly depositing" water, desired to protect the water mains.

21.2.2.2 Expressing Hardness as $CaCO_3$

By long-standing convention, hardness is nearly always expressed as calcium carbonate, that is, $CaCO_3$, MW $(CaCO_3) = 100$. Other units may be as mg/L of the particular ion or as mol/L of the ion (which should be the same as mol/L as $CaCO_3$). Conversion from any units to any other units may be in terms of mol/L as the intermediate expression. For example, to convert from mg Ca^{2+}/L to mg Ca^{2+} as $CaCO_3$/L, first convert to mol Ca^{2+}/L. The second step is to convert to mg Ca^{2+} as $CaCO_3$/L, that is, there are 100,000 mg $CaCO_3$/mol.

Consider a concentration of 10 mg HCO_3^-; the first step is to convert to molar concentration, that is,

$$\frac{10\ \text{mg}\ HCO_3^-}{\text{L}} \cdot \frac{\text{mol}\ HCO_3^-}{61{,}000\ \text{mg}\ HCO_3^-} = \frac{1.64 \cdot 10^{-4}\ \text{mol}\ HCO_3^-}{\text{L}}$$

The second step is to convert to mg as $CaCO_3$/L, that is,

$$\frac{1.64 \cdot 10^{-4}\ \text{mol}\ HCO_3^-}{\text{L}} \cdot \frac{100{,}000\ \text{mg}\ CaCO_3}{\text{mol}\ CaCO_3} = \frac{16.4\ \text{mg}\ CaCO_3}{\text{L}}$$

Examples 21.3 through 21.5 illustrate conversions based on the same principle, but applying "short cut" to each.

Example 21.3 Conversions of Carbonate Hardness

Given
$C(Ca^{2+}) = 62$ mg/L

Required
Hardness expressed as $CaCO_3$

Solution
62 mg Ca^{2+}/L · 100 mg $CaCO_3$/40 mg Ca^{2+} = 155 mg/L hardness as $CaCO_3$.

Example 21.4 Conversion of Ca^{2+} Concentration

Given
Calcium ion, Ca^{2+}, is a hardness component in a water; its concentration in the ambient raw water is $C(Ca^{2+}) =$ 33 mg/L as $CaCO_3$.

Required
Calculate the concentration of Ca^{2+} in: (a) mol Ca^{2+}/L, and (b) mg Ca^{2+}/L.

Solution

a. The concentration of Ca^{2+} in mol Ca^{2+}/L, that is, $[Ca^{2+}]$, is

$$[Ca^{2+}] = \frac{33 \text{ mg } Ca^{2+} \text{ as } CaCO_3}{L} \cdot \frac{\text{mol } CaCO_3}{100{,}000 \text{ mg } CaCO_3}$$

$$= \frac{3.3 \cdot 10^{-4} \text{ mol } Ca^{2+}}{L}$$

b. The concentration of Ca^{2+}, in mol Ca^{2+}/L, that is, $[Ca^{2+}]$, is

$$C(Ca^{2+}) = \frac{3.3 \cdot 10^{-4} \text{ mol } Ca^{2+}}{L} \cdot \frac{40{,}000 \text{ mg } Ca^{2+}}{\text{mol } Ca^{2+}}$$

$$= \frac{13.2 \text{ mg } Ca^{2+}}{L} \approx 13 \text{ mg } Ca^{2+}/L$$

Discussion
When expressing concentration as $CaCO_3$, conversion to the molar concentration is the basis for conversion to mg/L for the ion in question.

Example 21.5 Conversion of Mg^{2+} Concentration

Given
Magnesium ion, Mg^{2+}, is a hardness component in a water; its concentration in the ambient raw water is $C(Mg^{2+}) = 22$ mg/L as $CaCO_3$.

Required
Calculate the concentration of Mg^{2+} in: (a) mol Mg^{2+}/L, and (b) mg Mg^{2+}/L.

Solution

(a) The concentration of Mg^{2+} in mol Mg^{2+}/L, that is, $[Mg^{2+}]$, is

$$[Mg^{2+}] = \frac{22 \text{ mg } Mg^{2+} \text{ as } CaCO_3}{L} \cdot \frac{\text{mol } CaCO_3}{100{,}000 \text{ mg } CaCO_3}$$

$$= \frac{2.2 \cdot 10^{-4} \text{ mol } Mg^{2+}}{L}$$

(b) The concentration of Mg^{2+}, in mol Mg^{2+}/L, that is, $[Mg^{2+}]$, is

$$C(Mg^{2+}) = \frac{2.2 \cdot 10^{-4} \text{ mol } Mg^{2+}}{L} \cdot \frac{24{,}300 \text{ mg } Mg^{2+}}{\text{mol } Mg^{2+}}$$

$$= \frac{5.3 \text{ mg } Mg^{2+}}{L} \approx 5 \text{ mg } Mg^{2+}/L$$

Discussion
The conversion of Mg^{2+} concentration as $CaCO_3$ is the same, in principle, as for Ca^{2+}, or for bicarbonate or carbonate.

21.2.2.3 Other Definitions of Hardness

Several kinds of hardness definitions have been long used in water treatment practice. As a rule, all of the forms are expressed as $CaCO_3$ and include

- Carbonate hardness—that hardness that equals the equivalents of CO_3^{2-} and HCO_3^-.
- Total hardness—the sum of all hardness cations, expressed as $CaCO_3$.
- Permanent hardness—that portion of hardness not associated with carbonate or bicarbonate, but with other anions, for example, Cl^-, SO_4^-, etc.
- Temporary hardness—that portion of hardness associated with carbonate or bicarbonate.

21.2.2.4 Softening Reactions

When lime, CaO, is added to hard water, the following reactions are typical (Babbitt and Doland, 1949, p. 507):

$$\underset{\substack{\text{lime}\\\text{insoluble}}}{CaO} + \underset{\text{carbonic acid}}{2H_2CO_3} \rightarrow \underset{\substack{\text{calcium bicarbonate}\\\text{soluble causes hardness}}}{Ca(HCO_3)_2} + \underset{\text{water}}{H_2O} \quad (21.4)$$

Removal of temporary hardness, that is, the portion of Ca^{2+} associated with HCO_3^-:

$$\underset{\substack{\text{calcium bicarbonate}\\\text{temporary hardness}}}{Ca(HCO_3)_2} + \underset{\text{lime addition}}{Ca(OH)_2} \rightarrow \underset{\substack{\text{calcium carbonate}\\\text{insoluble}}}{2Ca(CO_3)} + \underset{\text{water}}{2H_2O} \quad (21.5)$$

Removal of permanent hardness, that is, the portion of Ca^{2+} associated with Cl^-, NO_3^-, SO_4^{2-}:

$$\underset{\substack{\text{calcium sulfate}\\\text{permanent hardness}}}{CaSO_4} + \underset{\substack{\text{sodium carbonate}\\\text{(soda ash)}}}{Na_2CO_3} \rightarrow \underset{\substack{\text{calcium carbonate}\\\text{insoluble}}}{CaCO_3} + \underset{\substack{\text{sodium sulfate}\\\text{soluble}}}{Na_2SO_4} \quad (21.6)$$

Explanations are

- Reaction (1), that is, Equation 21.4, expresses the reaction of lime with carbonate.
- Reaction (2), that is, Equation 21.5, expresses is the reaction that occurs when temporary hardness is removed by the addition of lime.
- Reaction (3), that is, Equation 21.6, occurs when permanent hardness is removed by the addition of soda ash. Normally, reaction (3) would be followed by reaction (2) in order to remove all hardness.

Magnesium, if it occurs, is removed as magnesium hydroxide, $Mg(OH)_2$, which is done usually at pH $\approx$ 11.0. Most softening reactions are done at pH $\approx$ 10; at this pH, the charge on $CaCO_3$ is negative and is positive on $Mg(OH)_2$. Therefore, coagulants are needed to charge neutralize the particles (Kemmer, 1979, p. 10–12).

Dosages of lime and soda ash, if needed, may be determined stoichiometrically, knowing the concentrations of Ca^{2+} and Mg^{2+} by analytical means. Usually, however, jar testing is done with different lime dosages, or lime and soda dosages, and with residual Ca^{2+} and Mg^{2+} concentrations measured as the dependent variables. The jar test results are plotted with Ca^{2+} and Mg^{2+} concentrations plotted on the ordinate against the lime or lime/soda dosages on the abscissa.

21.2.2.5 Lime-Soda Process

The portion of hardness that has no alkalinity counterions associated with it is "permanent" hardness. For example, if the counterions are Cl^-, SO_4^{2-}, NO_3^-, etc., the carbonate, that is, CO_3^{2-}, must be supplied from an external source; the most common is soda, that is, Na_2CO_3. Looking at the net reaction only, the associated reaction is

$$Ca^{2+} + CO_3^{2-} \rightarrow CaCO_3 \qquad (21.7)$$

The soda ash demand is the stoichiometric amount needed to satisfy the requirement for permanent hardness.

21.2.3 Chemistry of Metals

The chemistry of metal ions may be rather complex because of the different valence states and complexes that may form. For example, arsenic is stable in four oxidation states, that is, +5, +3, 0, and –3, depending on the redox potential (with values high to low, respectively). The aqueous forms are mostly arsenite, As(III), and arsenate, As(V); and in an oxidizing environment the arsenate predominates, and under moderately reducing conditions arsenite is the major species. Because of slow kinetics, however, both arsenite and arsenate may exist in either redox environment. Some forms of arsenic are removed more easily than others, with oxidation state being a major factor. To illustrate further the complexity of removal, the reaction between arsenate and iron (Fe^{3+}) forms an insoluble precipitate, ferric arsenate, $FeAsO_4$; on the other hand, ferric arsenite, $FeAsO_3$, is soluble in water and does not precipitate.

Many of the other metals, for example, Al, Cd, Cr, Cu, Fe, Hg, Mn, Pb, also have multiple oxidation states and/or may serve as the "central" atom with associated "ligands" to form a "complex" ion (or a "chelate"). Manganese, as another example, has several oxidation states, for example, Mn(s), which reacts with H^+ in the aqueous phase to form Mn^{2+}(aq) and H_2(g), $E° = 1.1.8$ V (Silberberg, 1996, pp. 982–984). Also, as stated by Silberberg, like chromium, manganese can use all its valence electrons in its compounds, exhibiting every possible oxidation state, with the +2, +4, and +7 states being most common. As the oxidation state of manganese increases, its "valence state electronegativity" also increases and its oxides change from basic to acidic. Mn(II) is basic, and manganese (III) oxide, Mn_2O_3, is amphoteric. Manganese (IV) oxide, MnO_2, is insoluble and thus shows no acid–base properties. Manganese (VII) oxide, Mn_2O_7, forms by reaction of Mn with pure O_2 and reacts with water to form permanganic acid, $HMnO_4$, a strong oxidizing agent. All manganese species with oxidation states >+2 act as oxidizing agents, but the permanganate ion is particularly powerful. As with chromium in its highest oxidation state, MnO_2^- is a much stronger oxidant in acidic solution than in a basic solution, that is,

$$MnO_4^-(aq) + 4H^+(aq) + 3e^- \rightarrow MnO_2(s) + 2H_2O \quad E° = 1.68 \text{ V} \qquad (21.8)$$

$$MnO_4^-(aq) + 2H_2O(aq) + 3e^- \rightarrow MnO_2(s) + 4OH^- \quad E° = 0.59 \text{ V} \qquad (21.9)$$

Table 21.4 summarizes the foregoing discussion, showing oxidation states of manganese, with examples on the line below, and the acid–base level associated with the respective oxidation state. The top line depicts the expected appearances of the respective oxidation states, with the second line describing color.

As related to natural cycles of manganese, Mn^{2+}, the reduced form, occurs in a reducing atmosphere, that is, anoxic. When withdrawn for drinking water, the Mn^{2+} will oxidize at some point in the system, with MnO_2 being a product; MnO_2 causes the black water noted by customers

TABLE 21.4
Examples of Oxidation States of Manganese

Depiction					
Color	Pale pink[a]	Black	Br-blake[a]	Green[a]	Purple[a]
Oxidation state	**Mn(II)**[b]	Mn(III)	**Mn(IV)**[b]	Mn(VI)	**Mn(VII)**[b]
Example	Mn^{2+}	Mn_2O_3	MnO_2	MnO_4^{2-}	MnO_4^-
Oxide acidity	Basic	→	→	→	Acidic

Source: Adapted from Silberberg, M., *Chemistry—The Molecular Nature of Matter and Change*, Mosby-Year Book, Inc., St. Louis, MO, 1996.

[a] Silberberg (1996, p. 984; Casale et al. 2002, p. 109).

[b] Most common species.

in water systems that have such a problem. As stated previously, MnO_4^- is a strong oxidant that in itself may oxidize free Mn^{2+} (in solution) to MnO_2.

21.3 PRACTICE

Removal of hardness and heavy metals involves first adding anions that are both cheap and have low solubility product when combined with the target ion. After the reaction has occurred, the task is to cause particle growth to form a settleable floc with polishing/removal by filtration.

21.3.1 LIME SOFTENING

Factors involved in the lime softening precipitation process include temperature, return of precipitated sludge, mixing turbulence, and time of mixing. Most often the process is carried out in a reactor clarifier, which involves rapid mix in a center feed well, removal of solids by means of a sludge pocket at the bottom center, upward passage of the precipitate suspension through a sludge blanket, overflow of the clarified water from peripheral weirs (or weirs placed in a radial configuration). Filtration is recommended as a final process in the treatment train.

Alternatively, for large installations involving hard groundwater, the same treatment train as alum coagulation may be used for precipitation softening, but dedicated to softening, that is, rapid-mix with lime added, flocculation, settling, and filtration. For hard surface water, softening may be carried out simultaneously with alum coagulation, with lime being added in the rapid-mix with alum.

21.3.2 PRECIPITATION OF HEAVY METALS

Mining sites are a source of heavy metals, for example, As, Cd, Hg, Pb, Se, and anions such as F^-, SO_4^{2-}, also Fe, Ni, Zn, Cu, are not uncommon. In most cases, the drainage from mines is acidic and is likely to come from a reducing environment, that is, low redox potential. For example, iron is likely to be in the ferrous state, that is, Fe^{2+}. If and when such water reaches the surface, oxidation to Fe^{3+} may occur. As a next step, $Fe(OH)_3$ will form, depending also on pH (see, e.g., the Pourbaix diagram for iron such as shown in Snoeyink and Jenkins, 1980, p. 362), causing deposits on the stream bed, with esthetic and ecological consequences (e.g., red deposits on rocks and clogging of insect niches in the gravel stream bottom, respectively). Similar reactions occur with Pb, Zn, Cd, and Hg, except that the elements are "heavy metals," and are toxic.

21.3.2.1 Common Chemical Reactions

The standard treatment method for removal of heavy metals from wastewater is chemical precipitation with hydroxide, sulfide, or carbonate anions. The most common anion used is hydroxide, with lime being the cheapest source and easiest to use in operation (Banerjee, 2002, p. 181). Treated effluent concentrations of metals such as copper, lead, and zinc are typically ≤ 0.5 mg/L by hydroxide precipitation after settling and filtration (Banerjee, 2002, p. 181). Sulfide is an effective alternative to hydroxide for removals of Cd, Cu, Zn, As, Se, etc. Removals to ≤ 0.1 mg/L may be expected. Sources of sulfide include sodium sulfide, hydrosulfide, or the slightly soluble ferrous sulfide introduced as slurry. On the negative side, hydrogen sulfide gas, and sulfide toxicity and odor are potential problems (Banerjee, 2002, p. 182).

Hydroxide treatment with lime after pH adjustment to $8.0 \leq \text{pH} \leq 11.0$ is a standard treatment to remove heavy metals (Robinson and Sum, 1980). The process includes rapid-mix, settling, and filtration. In experiments with five metals, that is, cadmium, copper, chromium, nickel, and zinc, the concentrations were reduced as indicated in Table 21.5. As seen, the sulfide polishing generally gives lower effluent concentrations, which is explained by the solubility products being lower than the hydroxides. Raw water concentrations are quite high in the case of chromium, nickel, and zinc.

Figure 21.3 shows the tanks for treatment of a mining waste, which is classed as acid mine drainage with pH < 4;

TABLE 21.5
Reductions in Metal Concentrations in Industrial Wastes[a]

Parameter	Concentrations in μg/L				
Metal	Cadmium	Copper	Chromium	Nickel	Zinc
Raw feed	4805	4136	26,000	18,617	259.865
Lime only[b]	103	673	58	1,561	6,629
Lime-sulfide[c]	8	70	86	751	422

Source: Robinson, A. K. and Sum, J. C., Sulfide precipitation of heavy metals, Report EPA-600/2-80-139, Industrial Environmental Research Laboratory, Office of Research and Development, U.S. Environmental Research Laboratory, Cincinnati, OH, 1980.

[a] Treatment train includes rapid-mix, settling, filtration.

[b] Lime only refers to the addition of calcium hydroxide after pH adjustment, i.e., $8.0 \leq \text{pH} \geq 11.0$.

[c] Lime-sulfide refers to adding soluble sulfide with the lime in the rapid mix.

FIGURE 21.3 Photograph of settling. (Courtesy of John H. Smith III, Sepco, Inc., Fort Collins, CO.)

heavy metals and iron must be removed. Lime is added in the first tank, called the precipitation tank, with $9 \leq pH \leq 10$. In the second tank, final clarification occurs, with overflow being discharged or in some cases filtered. Solids are generally dewatered and disposed of in an approved landfill.

In addition to direct precipitation reactions, iron coprecipitation/adsorption is common for removal of heavy metals. Oxides of iron, when fresh, are amorphous and have a high removal capacity in terms of moles removed per kg solid. The fresh precipitate may remove such cations as copper, lead, zinc, chromium, etc. with effluent concentrations being in the μg/L level. Depending on the pH, oxyanions of metals may be removed also, for example, arsenate, chromate, selenite, etc. Arsenic is often of a particular concern and may be removed by lime precipitation, coagulation, and coprecipitation/adsorption on metal hydroxides, precipitation softening, and sulfide precipitation (Banerjee, 2002, p. 182). Adsorption on alumina and ion exchange are other methods given.

21.3.2.2 Case: Mine Drainage

One of the oldest acid mine drainage sites in the intermountain region of the United States has been operating since the 1970s. The wastewaters were from three sources: (1) mine drainage, (2) surface water, and (3) the minerals processing plant; the waters contained Pb, Zn, Cd, Hg, Fe, F, and SO_4^{2-}, with suspended solids concentrations $\leq$100,000 mg/L. Later, the processing plant was closed and so water came from only the first two sources. Initial laboratory tests using lime resulted in a 2% (by weight) settled sludge. Mixing recirculated settled solids with the lime solution in a reactor, however, and then with the wastewater, which produced a 15% settled sludge. The precipitated solids settled quickly in the clarifier-thickener used, resulting in a lower volume rate of sludge accumulation stored in the temporary impounding area at the site. Because of the high concentration of iron, aeration was included as a part of the treatment train, which was considered to help in the flocculation and settling. The plant design was for $Q = 0.32$ m^3/s (5,000 gpm) to be treated in a clarifier-thickener, with area, A(settling) = 2,880 m^2 (31,000 ft^2), giving surface overflow rate (SOR) = 0.0065 m/min (0.16 gpm/ft^2); a polishing filter was not used.

Factors that are typical in design and operation in mine drainage situations include

1. Knowledge of water quality is most important in the design of a treatment system.
2. With respect to operation, two complete water analyses each year are recommended.
3. Variations may occur in flow, TDS, and total suspended solids (TSS) over the annual cycle.
4. In some cases there is potential for "blow-outs" of plugs or bulkheads.

The variations in water quality will affect solids settling, floc dosage, sludge concentration, life of filter media, and dewatering characteristics of settled sludge. Therefore, an equalization basin is recommended, which minimizes these variations, which in turn provides for more treatment consistency. The basin should have sufficient mixing, such as surface aerators, so that settling of solids does not occur or, alternatively, provision for removal of solids, such as by a front-end loader after draining the basin. As with all basins, provision should be made for drainage. The equalization basin also may mitigate the effect of a sudden flow surge, for example, due to a blowout. This section was adapted from the work of Smith (2000).

21.3.3 Precipitation of Anions

Certain anions may be targets for removal; some are amenable to removal by chemical precipitation. These include (1) phosphate, which may cause excessive algae growths if discharged into ambient waters; (2) cyanide, which is found in some industrial wastes and in some mining operations, particularly gold mining, which is toxic; and (3) chromate, CrO_4^{2-}, found in some industrial wastewaters.

21.3.3.1 Phosphate Precipitation

Municipal wastewater contains, in addition to orthophosphate, the polymeric phosphates, pyrophosphate, metaphosphate, and tripolyphosphate. These polyphosphates can be precipitated also by Al^{3+} with stoichiometric relationship about Al:P = 1; precipitates are not formed, however, until $Al/P \geq 1$, with control of the reaction being affected by pH. The most efficient pH range is $5 < pH < 6$.

Reduction of phosphate as a nutrient has been a traditional concern in treatment of municipal wastewaters. Precipitation with Al^{3+}, Fe^{3+}, or Ca^{2+} is feasible technically. In each case, removal of phosphate is by precipitation with the metal ion rather than by coagulation or adsorption. The stoichiometry is seen by the reaction equation

$$Al^{3+} + H_nPO_4^{-(3-n)} \rightarrow AlPO_4(s) + nH^+ \quad (21.10)$$

Equation 21.10 shows that 1 mol of Al^{3+} is necessary to precipitate 1 mol of phosphorous. With regard to competition with ion groups on colloids, the Al^{3+} has a stronger affinity for phosphorous than for such groups on mineral or biological colloids. Therefore, when the Al^{3+} is added to a suspension that includes phosphate and microorganisms, it will react first with the phosphate and will react with the colloids only after the phosphate has been precipitated.

At Al^{3+} concentration less than the stoichiometric amount, $AlPO_4$ will form colloids, which do not settle readily. At Al^{3+} concentration equal to the stoichiometric amount, however, a settleable precipitate will form. Removals are equally effective with Fe^{3+}. If Al^{3+} or Fe^{3+} is to be used, as in tertiary treatment, the phosphate demand must be satisfied first, that is, before other demands.

BOX 21.1 DESIGN

In reviewing the chemistry of water treatment, Moffett (1968) summarized the overall state of the art. Many of his comments are pertinent to understanding contemporary design and operation, with two of the most pertinent abstracted:

"For many years, both the design and the operation of water purification plants were considered to be arts. Little attention was given to the chemistry of water treatment and to the effects of chemical variables upon the engineering design or operation of a plant. Design had focused on steel and concrete. The sizing of plant facilities was based on millions of gallons of water to be treated per day and detention time.

Approximately ten years ago, there appeared to be an awakening to the fact that water treatment should be a scientific discipline,. . . . In the last 5 years, the number and competency of published articles has increased until now we can delineate the chemical and engineering variables in the process design and the operation of a water plant. Today, no water plant should be built without a complete process study prior to physical design" (Moffett, 1968, p. 1255).

21.3.3.2 Cyanide Precipitation

Removals of cyanides, free and complexed, at concentrations about 30 mg/L total cyanide were investigated with respect to mining wastes. In the first step, hydrogen peroxide and sodium thiosulfate are added at $7 < \text{pH} < 9$ in a 1:2 molar ratio to convert free and weakly complexed cyanide to nontoxic thiocyanate. Then steryldimethylbenzylammonium chloride is added to precipitate ferro-cyanide and, finally, ferric sulfate is added as a sweep floc and sequestrant for heavy metals. Final effluent concentrations were ≤ 0.02 mg/L (Schiller, 1983).

PROBLEMS

21.1 Precipitation Reactions

Given

Phosphate in wastewater is considered a nutrient pollutant and is removed by lime addition. Assume the concentration of phosphate as PO_4^{3+} is 30 mg/L, that is,

$$C_{PO_4^{3+}} = 30 \text{ mg/L}$$

Required

Determine the concentration of Ca^{2+} needed to precipitate the phosphate.

21.2 Equalization Basins

Given

The flow from a mine with $Q(\text{avg}) \approx 0.32$ m^3/s (5000 gpm) may vary from month to month with $Q(\text{max})/Q(\text{avg}) \approx 2.2$ and $Q(\text{min})/Q(\text{avg}) \approx 0.35$. Neglect variation in water quality.

Required

Determine the volume required for an equalization storage basin such that the average flow enters the treatment train; let $Q(\text{max})$ occur at $t \approx 9$ days; let $Q(\text{min})$ occur at $t \approx 21$ days, with the variation being a continuous function (appearing approximately sinusoidal).

Hint: Construct a mass flow diagram and solve for ΔV (equalization) graphically as the sum of the maximum differences between the cumulative actual flow above and below the cumulative average flow, that is, the sum of the maximum ordinate differences between the two.

21.3 Cooling Basin Salt Accumulation

Given

A clay-lined, that is, essentially zero seepage, cooling basin for a coal-fired power plant with a volume, $V = 12{,}300$ m^3 (10 acre-ft). To avoid applying for a national pollution discharge elimination system (NPDES) permit, the management elected to have zero discharge. The evaporation rate is about 250 mm/year. The incoming water, mostly snowmelt, has total dissolved solids concentration, TDS ≈ 50 mg/L.

Required

Estimate the number of years before TDS ≈ 1000 mg/L.

Hint: Do an annual mass balance, that is, $Q \cdot C = V \cdot dC/dt$ and $t(\text{years}) = 1000/(dC/dt)$.

ACKNOWLEDGMENT

John H. Smith III, president, Sepco, Inc., Fort Collins, Colorado, provided the photograph of the settling tank for heavy metals removal (Figure 21.3) and the material on the case study.

GLOSSARY

Amorphous solid: A solid that lacks molecular level order (Silberberg, 1996, p. 428).

Anion: Ion with one or more negative charges, for example, Cl^-, SO_4^{2-}.

Cation: Ion with one or more positive charges, for example, Na^+, Ca^{2+}.

Colligative properties: Of the four important solution properties as affected by solutes, it is the number of solute particles that makes the difference, not their chemical identity. The properties are vapor pressure lowering, boiling point elevation, freezing point depression, and osmotic pressure.

Complex: A central metal cation bonded to molecules and/or anions called ligands. To maintain electroneutrality, the complex ion is typically associated with simple ions, called counterions, for example, for the coordination compound, $[Co(NH_3)_6]Cl_3$, the complex ion is $[Co(NH_3)_6]^{3+}$, the three Cl^- ions are counterions, and the six NH_3 molecules bonded to the central Co^{3+} are ligands. The coordination number is the number of ligand atoms bonded directly to the central metal ion; thus, the coordination number for $[Co(NH_3)_6]^{3+}$ is 6 because 6 ligand atoms, that is, NH_3, are bonded to the Co^{3+} ion. Common ligands in coordination compounds include NH_3, F^-, Cl^- $[:C \equiv N:]^-$ etc.

See also chelate from Greek chela, crab's claw (Silberberg, 1996, pp. 988–990).

Coordination compound: One or more central atoms or central ions, usually metals, with a number of ligands attached.

Coprecipitation: In the formation of an amorphous precipitate, for example, $Fe_2O_3 \cdot H_2O$, trace elements (both dissolved and suspended) may be adsorbed onto and trapped within the precipitate (Banerjee, 2002, p. 182).

Covalent bond: Interatomic forces between two atoms are due to the attraction of the two nuclei to the mutually shared electrons. The latter forces are stronger than the repulsive forces between the nuclei, giving a net attractive force between the atoms, for example, H_2. Such bonding may occur also within polyatomic ions, for example, carbonate, CO_3^{2-} (Silberberg, 1996, p. 62). The shared electron pair, or bonding pair, is represented by a pair of dots or a line, for example, H:H or H–H (Silberberg, 1996, p. 62). The bond energy depends on the compound and the number of bonds, for example, H–H bond energy is 432 kJ/mol; C=C bond energy is 614 kJ/mol (Silberberg, 1996, p. 336).

Covalent compound: Covalent compounds form when elements share electrons, which usually occurs between nonmetals.

Crystal: A solid that is "ordered" at the molecular level; an ordered appearance is also visible since the order extends to the external appearance of the solid (Silberberg, 1996, p. 428).

Crystal lattice: (1) Three-dimensional framework of particles that form a crystal. (2) An array of points that forms a regular pattern that exists throughout the crystal; within the array, a "unit-cell" is the simplest arrangement of points that, when repeated, gives the lattice. The "particles" may be atoms, molecules, or ions (Silberberg, 1996, pp. 430–435).

Electron activity: The idea of electron activity, $\{e^-\}$, is analogous to proton activity, $\{H^+\}$. While pH is defined as $pH \equiv -\log\{H^+\}$, $p\varepsilon$ is defined as $p\varepsilon \equiv -\log\{e^-\}$ (Snoeyink and Jenkins, 1980, p. 339). The $p\varepsilon$ term is a measure of the availability of electrons in solution, although no free electrons exist in solution. For a half-reaction, $ox + ne^- \rightarrow red$, $p\varepsilon = p\varepsilon^\circ - (1/n)\log[(red)/(ox)]$ and $p\varepsilon^\circ = (1/n)\log K$, in which K is the equilibrium constant for the half reaction. The $p\varepsilon$ is related to a half-cell potential, E_H, as, $p\varepsilon = 16.9E_H$ (Snoeyink and Jenkins, 1980, p. 339).

Equivalent: Molecular weight of a given compound divided by its electric charge; for example, the equivalent weight of calcium as Ca^{2+} is 40 g/mol/2 = 20 g/mol.

Hardness: Cations that consume soap and which may cause precipitates that may deposit as scale in boilers or pipes, etc.

Heavy metals: Some metals that are toxic also have a high atomic weight; common ones in this category include As, Pb, Cd, and Hg, which are also toxic. Some authorities (e.g., SenGupta, 2002, p. 1) have suggested that the term "heavy metal" is a misnomer and the designation should be "toxic metals" as a more accurate description of the particular elements of concern.

Hydrolysis:

1. A ligand exchange reaction of hydrated metal ions with an acid or base. Example of stepwise hydrolysis of aquoaluminum(III) is

$$Al(H_2O)_6{}^{3+} + H_2O \rightarrow Al(H_2O)_5OH^{2+} + H_3O^+ \qquad (21.G.1)$$

$$Al(H_2O)_5OH^{2+} + H_2O \rightarrow Al(H_2O)_4(OH)_2{}^+ + H_3O^+ \qquad (21.G.2)$$

$$Al(H_2O)_4(OH)_2{}^+ + H_2O \rightarrow Al(H_2O)_3(OH)_3(s) + H_3O^+ \qquad (21.G.3)$$

$$Al(H_2O)_3(OH)_3(s) + H_2O \rightarrow Al(H_2O)_2(OH)_4{}^- + H_3O^+ \qquad (21.G.4)$$

   The foregoing reactions illustrates that hydrolysis of metal ions is a stepwise replacement of coordinated molecules of "water or hydration" by hydroxyl ions. In the reactions shown, this occurs by transfer of protons from waters of hydration to free water molecules to from hydronium ion. The species concentrations are functions of pH since they are proton transfers.

2. Another definition of hydrolysis is the reaction of a salt constituent, cation, and anion, with water. Within the framework of the Bronsted theory, the term hydrolysis is no longer necessary say Stumm and Morgan (1996, p. 91), since in principle there is no difference in the proteolysis of a molecule and that of a cation or anion to water.

Ionic bond: The central theme of ionic bonding is the transfer of electrons from metal to nonmetal to form ions that come together into a solid ionic compound (Silberberg, 1996, p. 328).

Langelier index: Professor W. F. Langelier devised an expression from the acid–base equilibrium statement, which provides an "index" as to whether a water is depositing or corrosive, which is defined (Babbitt and Doland, 1949, p. 558) as

$$pH_s = (pK_2' - pK_s') + pCa + pAlk \quad (21.G.5)$$

in which

pH_s is the pH that the water should have in order to be in equilibrium with $CaCO_3$.

pK_s' is the negative logarithm of the second dissociation constant for carbonic acid, that is, $K_s' = 6.35$

pK_2' is the negative logarithm of the activity product of $CaCO_3$, that is, $K_2' = 10.33$

pCa is the negative logarithm of molal concentration of Ca

pAlk is the negative logarithm of equivalent concentration of titratable base

The saturation index is obtained by subtracting $pH_s =$ from the pH of the water in question. A negative index indicates that the water is undersaturated with respect to $CaCO_3$. A positive index indicates that the water is supersaturated with respect to $CaCO_3$ and is depositing. The water analysis includes data on total solids, temperature, calcium, and alkalinity. With these data, the Langelier index may be determined from tables, a nomograph, or a computation wheel.

Ligand: Ions or molecules that are attached by a covalent bond to a central atom or ion as a part of a complex.

Lime: Calcium oxide, CaO. Slaked lime is CaO that has reacted with water to form $Ca(OH)_2$. The former is cheaper and if used must be slaked at the plant.

Lime-soda process: Softening by lime, $Ca(OH)_2$, to remove temporary hardness, followed by soda, $NaCO_3$, to remove permanent hardness.

Mass percent: Mass of solute per 100 g solution (Silberberg, 1996, p. 481), that is,

$$\text{Mass percent} \equiv \frac{\text{mass of solute}}{\text{mass of solute} + \text{mass of solvent}} \cdot 100 \quad (21.G.6)$$

$$\equiv \frac{\text{mass of solute}}{\text{mass of solution}} \cdot 100 \quad (21.G.7)$$

Molality: Moles of solute dissolved in 1000 g of solvent, that is, $m \equiv$ mol solute/kg solvent (Silberberg, 1996, p. 480).

Molarity: Moles of solute dissolved in 1 L of solution, that is, M $\equiv$ mol solute/L solution (Silberberg, 1996, p. 480).

Mole fraction: Ratio of solute moles to total moles of solute plus solvent (Silberberg, 1996, p. 482), that is,

$$X \equiv \frac{\text{moles of solute}}{\text{moles of solute} + \text{moles of solvent}} \quad (21.G.8)$$

Neutralization: A reaction in which water is a product, as when an acid reacts with a base. Soluble metal oxides act as strong bases in solution because when the oxide ion dissociates, it reacts with water to form hydroxide ion (Silberberg, 1996, p. 148), that is,

$$O^{2-}(aq) + H_2O(l) \rightarrow 2OH^-(aq) \quad (21.G.9)$$

Ammonia gas when dissolved in water also acts as a weak base, that is,

$$NH_3(g) + H_2O(l) \rightarrow NH_4^+(aq) + OH^-(aq) \quad (21.G.10)$$

Normality: Equivalents of solute dissolved in 1 L of solution, that is, M $\equiv$ equivalents of solute/L solution

Precipitation: Two soluble ionic compounds react to form an insoluble product, that is, a precipitate. For example, when silver nitrate and sodium chromate are mixed, a brick-red precipitate, Ag_2CrO_4 forms, which can be represented by any of three types of balanced equations, that is, molecular, total ionic, and net ionic (Silberberg, 1996, p. 145).

The "molecular equation" shows the reactants and products as if they were undissociated compounds:

$$2AgNO_3(aq) + Na_2CrO_4(aq) \rightarrow Ag_2CrO_4(s) + 2NaNO_3(aq) \quad (21.G.11)$$

The more realistic depiction is the "total ionic equation," that is,

$$2Ag^+(aq) + 2NO_3^-(aq) + 2Na^+(aq) + CrO_4^{2-} \rightarrow Ag_2CrO_4(s) + 2Na^+(aq) + 2NO_3^-(aq) \quad (21.G.12)$$

The number of $Na^+(aq)$ ions and $NO_3^-(aq)$ are unchanged on both sides and are called "spectator ions," because they are not involved in chemical change, but are present as a part of the reactants.

The "net ionic equation" depicts only the actual change taking place, that is, with no spectator ions:

$$2Ag^+(aq) + CrO_4^{2-} \rightarrow Ag_2CrO_4(s) \quad (21.G.13)$$

Re-carbonation: The addition of lime increases the tendency of a water to deposit calcium carbonate on filter sand, pipes, boiler tubes, etc. The carbonate balance may be partially or completely restored by re-carbonation. The process involves diffusing carbon dioxide gas through water. Reaction time should be $\theta \geq 20$ min. Equipment for production of carbon dioxide includes a burner, a scrubber, a compressor, and a diffuser.

Salt: An ionic compound that results from the reaction between an acid and a base (Silberber, 1996, p. 148).

Soda: Sodium carbonate, that is, $NaCO_3$, which is used to remove "permanent" hardness.

Softening: Removal of hardness.

Solubility: The maximum amount of a substance that can dissolve in a fixed amount of solvent to form a stable solution at a given temperature (Silberberg, 1996, pG-16).

Solubility product, K_{sp}: The equilibrium constant for the dissolving of a slightly soluble ionic compound in water (Silberberg, 1996, pG-16).

Solute: A substance that dissolves in a solvent (Silberberg, 1996, pG-16).

Solution: A mixture of solute, in molecular or ionic form as the case may be, with solvent.

Solvent: The substance, for example, water, in which a solute dissolves (Silberberg, 1996, pG-16).

Substitution reaction: Ligand replaces the coordinated water molecules, for example,

$$M(H_2O)_n + L \rightarrow ML(H_2O)_{n-1} + H_2O \quad (21.G.14)$$

Volume percent: Volume of solute per 100 volumes of solution (Silberberg, 1996, p. 481), that is,

$$\text{volume percent} \equiv \frac{\text{volume of solute}}{\text{volume of solution}} \cdot 100 \quad (21.G.15)$$

REFERENCES

Babbitt, H. E. and Doland, J. J., *Water Supply Engineering*, 3rd edn., McGraw-Hill, Inc., New York, 1949.

Baker, M. N., *The Quest for Pure Water*, The American Water Works Association, New York, 1948.

Banerjee, K., Case studies for immobilizing toxic metals with iron co-precipitation and adsorption, in: Gupta, A. K. (Ed.), *Environmental Separation of Heavy Metals—Engineering Processes*, Lewis Publishers, Boca Raton, FL, 2002.

Benjamin, M. M., *Water Chemistry*, McGraw-Hill, New York, 2002.

Campion, H. T., Lime sludge and its disposal, *Journal of the American Water Works Association*, 26(4):488–494, 1934.

Casale, R. J., LeChevallier, M. J., and Pontius, F. W., *Review of Manganese Control and Related Manganese Issues*, AWWA Research Foundation and American Water Works Association, Denver, CO, 2002.

Daily, C. M., The Howard Bend Plant of the St. Louis Water Works, *Journal of the American Water Works Association*, 26(4):495–500, 1934.

Fergusson, J. E., *The Heavy Elements—Chemistry, Environmental Impact, and Health Effects*, Pergamon Press, Oxford, U.K., 1990.

Freiser, H., *Lecture Notes in Analytical Chemistry*, Freiser Equilograph Co., Tucson, AZ, 1996.

Gelston, W. R. Jr., Water softening at Quincy, Illinois, *Journal of the American Water Works Association*, 26(1):70–76, January 1934.

Kemmer, F. N. (Ed.), *The Nalco Water Handbook*, McGraw-Hill, New York, 1979.

Letterman, R. D., Amirtharajah, A., and O'Melia, C. R., Chapter 6 Coagulation and flocculation, in: Letterman, R. D. (Tech. Ed.), *Water Quality and Treatment*, 5th edn., McGraw-Hill, New York, 1999.

Lide, D. R., *CRC Handbook of Chemistry and Physics, 1996–97*, 77th edn., CRC Press, Boca Raton, FL, 1996.

Metcalf, L. and Eddy, H. P., *American Sewerage Practice, Volume III, Disposal of Sewage*, McGraw-Hill, New York, 1916.

Moffett, J. W., The chemistry of high-rate water treatment, *Journal of the American Water Works Association*, 60(11):1255–1270, November 1968.

Patterson, J. W., Metals control technology; Past, present and future, in: Patterson, J. W. and Passino, R. (Eds.), *Metals Speciation, Separation, and Recovery, Volume II*, Lewis Publishers, Chelsea, MI, 1990. *Proceedings of the Second International Symposium on Metals Speciation, Separation, and Recovery*, Rome, Italy, May 14–19, 1989.

Patterson, J. W. and Passino, R. (Eds.), *Metals Speciation, Separation, and Recovery, Volume II*, Lewis Publishers, Chelsea, MI, 1990. *Proceedings of the Second International Symposium on Metals Speciation, Separation, and Recovery*, Rome, Italy, May 14–19, 1989.

Patterson, J. W., Luo, B., Marani, D., and Passino, R., Nucleation and crystal growth studies on precipitation of cadmium hydroxide from aqueous solution, in: Patterson, J. W. and Passino, R. (Eds.), *Metals Speciation, Separation, and Recovery, Volume II*, Lewis Publishers, Chelsea, MI, 1990. *Proceedings of the Second International Symposium on Metals Speciation, Separation, and Recovery*, Rome, Italy, May 14–19, 1989.

Roalson, S. R., Kweon, J., Lawler, D. F., and Speitel, G. E. Jr., Enhanced softening—effects of lime dose and chemical additions, *Journal of the American Water Works Association*, 95(11): 97–109, 2003.

Robinson, A. K. and Sum, J. C., Sulfide precipitation of heavy metals, Report EPA-600/2-80-139, Industrial Environmental Research Laboratory, Office of Research and Development, U.S. Environmental Research Laboratory, Cincinnati, OH, 1980.

Schiller, J. E., Removal of cyanide and metals from mineral processing waste waters, Report of Investigation 8836, U.S. Bureau of Mines, U.S. Department of the Interior, Washington, DC, 1983.

SenGupta, A. K., Principles of heavy metals separation: An introduction, in: SenGupta, A. K. (Ed.), *Environmental Separation of Heavy Metals—Engineering Processes*, Lewis Publishers, Boca Raton, FL, 2002.

Silberberg, M., *Chemistry—The Molecular Nature of Matter and Change*, Mosby-Year Book, Inc., St. Louis, MO, 1996.

Smith, J. H. III, AMD treatment, it works but are we using the right equipment, in: *Proceedings of the Seventh International Conference on Tailings and Mine Wastes'00*, Fort Collins, CO, January 23–26, 2000, A. A. Balkema, Rotterdam, the Netherlands, 2000, pp. 419–427.

Snoeyink, V. L. and Jenkins, D., *Water Chemistry*, John Wiley & Sons, New York, 1980.

Stumm, W., *Chemistry of the Solid-Water Interface*, John Wiley & Sons, New York, 1992.

Stumm, W. and Morgan, J. J., *Aquatic Chemistry*, 3rd edn., John Wiley & Sons, Inc., New York, 1996.

Walker, W. H., Municipal water softening, *Journal of the American Water Works Association*, 26(1):77–98, 1934.

Part V

Biological Treatment

Categories of microbiological reactions are (1) aerobic, (2) anaerobic, (3) anoxic, and (4) photosynthesis. To promote these reactions, a variety of reactor systems have been devised, such as septic tanks, Imhoff tanks, intermittent filtration, trickling filters, activated sludge, extended aeration, aerated lagoons, nitrification, denitrification, rotating biological contactors, bio-filtration, sludge digesters, anaerobic reactors, anaerobic ponds, facultative ponds, land treatment, and constructed wetlands. The foregoing is not an inclusive list. In other words, what we have is a spectrum of processes and associated technologies.

Part V consists only of Chapters 22 and 23, which deal with principles of biological treatment and reactor theory and practice, respectively. As with other topics, a body of knowledge is involved. The two chapters attempt to distill principles and lore that are applicable both to understand the theory and practice of biological treatment and to gain further knowledge on the literature. A few of the books on the topic were cited extensively, first because they have distilled from the literature, and second to provide a common basis for the use of terms, definitions, criteria, etc.

In biological treatment, supporting sciences include microbiology, organic chemistry, inorganic chemistry, biochemistry, physical chemistry, and analytical chemistry. Reactor engineering involves knowledge of the lore of past practices, hydraulics, kinetics, mathematical modeling, and equipment.

As a note, the references and glossary for this chapter serve all of Part V. The notation used is consistent to a large extent with that recommended by Grau et al. (1982, pp. 1501–1505) as a part of an IWA (International Water Association) Working Group.

REFERENCE

Grau, P., Sutton, P. M., Henze, M., Elmaleh, S., Grady, C. P. L., Gujer, W., and Koller, J., Recommended notation for use of in the description of biological wastewater treatment processes, *Water Research*, 16:1501–1505, 1982.

22 Biological Reactions and Kinetics

The purpose of a biological-treatment system is to convert "substrates," through the use of microbes, to benign reaction "products." The substrates are usually organic compounds; in the case of municipal wastewaters, examples include fats, sugars, proteins as general categories. The "products," in the case of an aerobic reaction, include carbon dioxide and water, and new cells. If the reaction is anaerobic, methane and carbon dioxide, and new cells, are products. Biological treatment is favored in most cases because, as a rule, it is usually "cheap" relative to alternatives.

Traditionally, biological treatment has been applied to wastewaters but during the 1980s its use for the removal of natural organic matter (NOM) was explored. Biological treatment may be applied to any situation in which a microbe can react with a substrate.

In understanding biological processes, reactions and kinetics are the underlying themes. When combined with the mathematical characteristics of a particular kind of reactor, reactions/kinetics are the bases for rational theory.

22.1 BACKGROUND

Biological treatment started experimentally about 1890 at the Lawrence Experiment Station in Massachusetts. This was some 10 years after the field of microbiology, the scientific underpinning, had developed an identity that evolved from Pasteur's discoveries, c. 1857 (Prescott et al., 1993, pp. 6–12). Sanitary engineering, as a field, evolved during the 1880s, and had an identity by about 1890, stimulated by an academic program at the Massachusetts Institute of Technology (MIT). The development of the MIT program was related to faculty and student involvement with the Lawrence Experiment Station.

22.1.1 1880–1980

The Lawrence Experiment Station was established by the Massachusetts Legislature in 1887 in response to some of the great issues of the day, including safe drinking water, sewerage, wastewater treatment, and stream pollution. While chemical treatment by lime precipitation and other methods was experimented with in England about 1875, the focus soon shifted to biological treatment. By the first decade of the twentieth century, trickling filters had developed, and by 1920 activated sludge was a fledgling technology. The purpose was to remove organic matter from wastewaters sufficiently so that the receiving stream did not become septic, nor develop sludge banks. The idea was to utilize the "assimilative capacity" of the receiving water, which became the modus operandi for treatment. The assimilative capacity concept was dislodged by the Water Quality Act of 1965 (PL 89-234), which required each state to establish stream standards.

About 1980, the perspective on biological treatment was expanded to drinking water. Biofilms on packed beds of granular media were used in research to explore the feasibility of the technology applied to NOM in drinking water treatment.

22.1.2 Practice

The Lawrence Experiment Station was important in launching the technologies of biological treatment, c. 1890, and used pilot plant studies to compare the effects of different variables. The practice that emerged by 1900, however, was largely empirical and has remained so. The idea of 6 h detention time was the basis for sizing activated-sludge reactors, along with "loading rate" with units, "lb BOD/ft^3 reactor volume/day." The same was true for trickling filters with loading-rate units, "lb BOD/ft^2 surface area/day" and "mgd/ft^2 surface area."

22.1.3 Theory

The research started on a more scientific track from after 1950. Mathematical modeling came into vogue (McCabe and Eckenfelder, 1956; Stewart et al., 1959; McKinney, 1962), and the role of microbiology became assimilated quantitatively with the introduction of such terms as "substrate," "cell-yield," "specific growth rate," "food-to-microorganism ratio" (see, for example, McKinney, 1962). With the modeling came a need to understand the reactions and kinetics (e.g., Gram, 1956), which, of course were microbiological.

BOX 22.1 CARBON DIOXIDE AS A GREENHOUSE GAS

The carbon dioxide from a biological reaction does not add fossil carbon, e.g., as in oil or coal, to the atmosphere. Rather, biological treatment merely makes the carbon dioxide available for photosynthesis as a part of the carbon cycle. It speeds up what would otherwise be a part of the natural process of microbial degradation in the ambient environment. While this explanation may not put the matter to rest, photosynthesis will occur at some point in the carbon cycle and convert the carbon dioxide again to organic matter, a sequestered form, as cells. Human activity may increase the rate of processes, such that they may occur in hours, days, or years; in natural cycles the processes may occur in years or decades, or perhaps centuries. Reduction of the photosynthesis capacity of the land reduces the rate of sequestering and if a carbon-cycle model were available the effect on accumulation in the atmosphere could be assessed.

Two other gases, methane, CH_4, and nitrous oxide, N_2O, are also emitted from wastewater treatment plants and are considered potent greenhouse gases. The captured methane from anaerobic digesters may be combusted to CO_2 and water and yield energy and thus not add to the greenhouse effect to the same extent. Anaerobic reaction also occurs in sludge stored at the bottom of primary clarifiers, however, and methane/carbon dioxide gas bubbles may be seen breaking the water surface and escaping the system. The same phenomenon may be observed in natural waters where the benthic debris may have a significant organic fraction. Nitrous oxide has a 300-fold stronger effect than CO_2 as a greenhouse gas and emerged as a concern of wastewater treatment probably after 1995. Kampschreur et al. (2009) gave 1990 global emissions from wastewater treatment as 0.22 TgN/year (220,000 metric tons/year), which is about 0.03 fraction of the total from the anthropogenic sources. They determined that N_2O occurs as a reaction product of both nitrification and denitrification and have recommended operational procedures to minimize N_2O production.

BOX 22.2 LAWRENCE EXPERIMENT STATION

In 1886 a new law reconstituted the Massachusetts State Board of Health to find practical methods for sewage disposal. Hiram Francis Mills (1836–1921), an eminent Boston consulting hydraulic engineer, was appointed to the Board. From this, the Lawrence Experiment Station took form in 1887, with Hiram Mills as the administrator, a position he held until 1915 when the Board was abolished and the Massachusetts State Department of Public Health was created. Mr. Mills developed a relationship with MIT which was symbiotic in that the MIT sanitary engineering program took form in 1891. The academics engaged included Professor William Thomson Sedgwick (a biologist), Professor William Ripley Nichols, Ellen Swallow Richards (1842–1911), the first female graduate of MIT, and Professor Thomas Messinger Drown. Allen Hazen (1869–1929) was at the Station from 1890 to 1892 where he developed the effective size and uniformity coefficient characterizations of filter sand. George Warren Fuller (1868–1934) was at the Station 1890–1895, after graduation from MIT. In 1895 he went on to Louisville, Kentucky where he conducted his famous work on rapid filtration for the purpose of filtering water from the Ohio River. Except for Mr. Mills, the aforementioned persons were either biologists or chemists. In 1904, Allen Hazen formed the consulting engineering firm of Hazen and Whipple and was recognized as one of the foremost of sanitary engineers (http://www.libraries.mit.edu/archives/exhibits/hazen). In 1911, Malcolm Pirnie (1889–1967) joined the firm and in 1929 the firm became the present Malcolm Pirnie.

At the time the Lawrence Experiment Station was organized, microbiology had just emerged as a scientific discipline and analytical chemistry was not a mature field. Mills was aware of the work of two European scientists who had concluded that the oxidation of organic matter in filtration through sand was most probably due to the action of microscopic organisms. Mills was determined to ferret out how to apply such knowledge to sewage treatment on a practical scale. He started with 10 tanks, each with under-drains and filled with gravel or sand and with different kinds of materials on the top surface, including garden soil, dust, sand, peat, etc. In 1890, aware of the role of water in transmitting the typhoid bacillus, he initiated studies of drinking water filtration. In 1893, the Station started studies on the biological treatment of industrial wastes, including such familiar examples as dairy, carpet mill, wool, textile, paper mill, etc., and acid neutralization. In 1993, the station was renamed, the Senator William X. Wall Experiment Station (www.mass.gov/dep/about/ organization/wespost.htm, February 13, 2007). The station remains a viable organization for research in environmental science and engineering.

22.1.4 Definitions

Biological treatment involves a body of knowledge with many definitions; some are given in the glossary rather than in the text. Some terms in the glossary are explained in considerable detail in order to complement the text but in some cases they would be difficult to place.

22.1.4.1 Reaction Classifications

Reaction classifications are reviewed here, however, so that they may be seen in one place. The definitions were taken from several sources, including Benefield and Randall (1980, pp. 25, 26); Orhon and Artan (1994, p. 45); and Rittman and McCarty (2001, p. 22). The definitions are based, in general, on their respective energy source. The following definitions are common:

1. Phototrophs: Organisms that use light as their energy source, e.g., algae, aquatic plants.
2. Chemotrophs: Organisms that use oxidation–reduction reactions as their energy sources:
   a. Heterotrophs (also called, chemoorganotrophs) use organic molecules as their electron donor; examples include carbohydrates, alcohols, and almost any organic molecule.
   b. Autotrophs (also called, chemoautotrophs, and sometimes, chemolithotrophs) use simple inorganic molecules as their electron donor with ammonia being the most prominent in wastewater treatment and carbon dioxide in photosynthesis.

For the purposes of this text, the reaction terms are categorized simply as heterotrophic and autotrophic. An example of a heterotrophic reaction is degradation of a municipal wastewater by activated sludge. An autotrophic reaction is nitrification where *Nitrosomonas* oxidizes NH_4^+ to NO_2^-. Another is conversion of NO_2^- to NO_3^- by *Nitrobacter*.

22.1.4.2 BOD Nomenclature

Regarding BOD nomenclature, the "five-day" BOD is the common measure of degradable organic carbon, and is designated in the literature as BOD_5; for this text the subscript is dropped since the five-day is used almost exclusively. The designation is BOD_u, refers to the *ultimate* BOD, a theoretical value. If the laboratory measurement of BOD is extended to 21 days, which may be done occasionally, the oxygen uptake continues, as a rule, but it is called a "second-stage" and usually involves nitrification, i.e., oxygen utilized by ammonia oxidation, and therefore only some fraction of the continued oxygen uptake is due to organic carbon. As a rule, $BOD_{21} \approx 0.99BOD_u$. Regarding industrial wastewaters, it is possible that the BOD test may not work because seeding organisms may not be appropriate for the substrate at hand.

22.1.4.3 Surrogates for Active Biomass Concentration, *X*

The term, *X* represents the viable cell mass; those cells that may reproduce and is pertinent to biological treatment. The quantity, *X*, is not measurable directly, however, and so surrogates have been used. The most common have been mixed liquor suspended solids (MLSS), and mixed liquor volatile suspended solids (MLVSS). The measure, MLSS, includes all suspended solids, determined by filtering the suspension and oven-drying. The MLVSS is only volatile solids, determined by the combustion of organic matter in lab furnace after oven drying. Since MLVSS does not include inert matter, it is considered a closer approximation to viable cell mass than is MLSS.

The chemical oxygen demand, COD, of the suspended solids is another surrogate, suggested by Orhon and Artan (1994, p. 85) as perhaps a closer approximation to *X*. The COD provides an electron equivalence between organic substrate, cells, and oxygen utilized (Orhon and Artan, 1994, p. 538).

22.1.5 Wastewaters

Municipal sewage comprises a variety of organic compounds in the general categories of sugars, fats, and proteins. In addition, metals are found, and since the 1990s, antibiotics and endocrine disruptors have become major concerns in both ecology and public health. Industrial wastes include a wide variety of substances but those of particular concern are the ones that may (1) "pass-through" municipal treatment, (2) cause problems with treatment, or (3) impose excessive loading on the receiving plant. The foregoing is indicative of the issues that must be dealt with in treatment.

22.1.5.1 Municipal Wastewaters

Traditionally, municipal wastewaters have been characterized in terms of organic matter. Nitrogen and phosphorous have been measured, since the 1950s (most likely earlier), and have been of increasing concern, decade by decade since the 1950s.

22.1.5.1.1 Traditional Measures

The aggregate measures of organic matter include volatile solids fraction (VS); total organic carbon (TOC); chemical oxygen demand (COD); biochemical oxygen demand (BOD). Table 22.1 gives typical values for raw and treated municipal wastewater for four size ranges: soluble, colloidal, supracolloid, and settleable. These values may vary from one place to another, depending on the domestic water and are given here for general reference. Settleable solids are another common parameter and have a general range of 200–300 mg/L. Since the 1970s aggregate parameters have been supplemented by specific contaminant measures which have been enabled by improvements in analysis technology for organic compounds, metals, and biological constituents (the associated instrumentation involves more than one specialty fields).

22.1.5.1.2 Nutrients

Table 22.2 provides data on nitrogen and phosphorous concentrations in raw municipal wastewater, after primary treatment, and after secondary treatment. The data are for reference.

22.1.5.2 Industrial Wastes

Virtually every industry generates wastes of some kind. Examples of industries include traditional ones such as sugar beet, meatpacking, brewery, dairy, textile, pulp and paper, to name a few. Others may include electronics, metal plating, refinery, oil field, photographic, fish hatchery, cattle feedlot, soft drink bottling, fertilizer, hospital, coal-fired power plants, cannery, refinery, etc. (see, for example, Nemerow

TABLE 22.1
Organic Fractions in Sewage and Secondary Effluent

			VS		TOC		COD			
Stage	**Fraction**	**Size (μm)**	**(mg/L)**	**(f)**	**(mg/L)**	**(f)**	**(mg/L)**	**(f)**	**BOD (mg/L)**	**T (JTU)[a]**
Raw sewage	Soluble	$<10^{-3}$	116	0.48	46	0.42	168	40		
	Colloidal	10^{-3}–1	23	0.10	12	0.11	43	10		
	Supracolloid	1–100	43	0.17	22	0.20	87	21		
	Settleable	>100	59	0.25	29	0.27	120	29		
	Total		241		109		418		200–250	50
Secondary effluent	Soluble		62	0.67	16.5	0.69	46	0.74		
	Colloidal		6	0.7	1.5	0.06	3	0.05		
	Supracolloid		24	0.26	6	0.25	13	0.21		
	Settleable		0	0.0	0	0.0	0	0.0		
	Total		92		24		62		30[b]	8

Source: Adapted from Rickert and Hunter, *J. Water Pollution Control Federation*, 44(1), 135, 1972.

[a] T is the turbidity in Jackson candle units (JTU).

[b] The value BOD ≈ 30 mg/L is a typical value activated sludge after secondary settling and was adopted commonly by states as an "effluent discharge standard," pursuant to the 1972 "Clean Water Act" (PL92-500).

TABLE 22.2
Nitrogen and Phosphorous in Raw Waste, Primary Effluent, and Secondary Effluent

Substance	**Form**	**Raw Waste (mg/L)**	**Eff. PS[a] (mg/L)**	**Eff. FS[b] (mg/L)**
Nitrogen as N	Organic	10–25	7–20	3–6
	Dissolved	4–15	4–15	1–3
	Suspended	4–15	2–9	1–5
	Ammonia	19–30	10–30	10–30
	Nitrite	0–0.1	0–0.1	0–0.1
	Nitrate	0–0.5	0–0.5	0–0.5
	Total	20–50	20–40	15–40
Phosphorous as P	Organic	1–3	0.5–2	0.5–1
	Ortho	2–8	1–7	1–8
	Condensed	2–8	2–8	1–3
	Total	4–14	3–12	3–11

Source: Adapted from McCarty, P.L., Phosphorous and nitrogen removal by biological systems, *Proceedings of the Wastewater Reclamation and Reuse Workshop at Lake Tahoe California* (226 pages), Sanitary Engineering Research Laboratory, University of California, p. 229, Berkeley, CA, 1970.

[a] Effluent from primary settling from municipal wastewater treatment.

[b] Effluent from secondary settling following biological treatment.

and Agardy, 1998). An axiom in dealing with industrial wastes has been to reduce the mass flows of different waste constituents based on a review of practices within a given industry or plant, a traditional approach codified by the 1972 Clean Water Act (PL92-500).

Another approach, especially when dealing with industries generating hazardous wastes, has been to find substitutes for process chemicals. The final step, after all other means have been implemented, is treatment, i.e., pretreatment for discharge into POTWs (public-owned treatment works as defined by PL92-500), or on-site treatment if discharge is into a stream or other water body. Each industry is unique and generates wastes that are characteristic and may range from a wide array of organic substances to various metals. The main issues in industrial waste discharge into a POTW include (1) whether it contains substances that impair treatment due to effect on organisms; (2) whether it contaminates sludge, such as by heavy metals; and (3) whether it passes through the plant untreated, perhaps causing a violation of an effluent discharge permit or a stream standard.

Whether or not a compound is amenable to treatment may be determined by a "treatability study." As an example, if a particular waste seems not amenable to treatment, the study is usually started at the laboratory scale with a diluted sample of the waste, which is brought into contact with activated sludge. The activated-sludge sample is usually obtained from a local municipal wastewater treatment plant (WWTP). The sample is aerated continuously, perhaps for several weeks or even months. During this time period the waste concentrations are increased week by week until full strength is reached. At the same time, the microbial culture "selects" toward the organisms that may metabolize the wastes, probably resulting in an approximate monoculture. The process is called "acclimatization."

Regarding references dedicated to industrial wastes, the texts are few with perhaps a dozen appearing since 1970; the most recent are Nemerow and Agardy (1998) and Eckenfelder et al. (2009). In addition, the *Purdue Industrial Waste Conference* has been the main forum for industrial wastes management starting with its inception in 1944 by Professor Don Bloodgood; as of the 1997 conference some

4000 papers had been published (with continuation by Professor James Alleman after the retirement of Professor Bloodgood). The principles of the various unit processes apply while management practices are most likely to be specific to the industry.

22.1.5.3 Contaminants

Traditionally, before 1965, municipal wastewater treatment plants were designed to degrade the organic matter and reduce suspended solids and pathogens. The goal was to maintain aerobic conditions in receiving waters, avoid depositions of solids, and reduce the infectious disease hazard. The increasing stringent legislation after 1965 first placed limits on degradable organic matter and suspended solids and then started to include ammonia, phosphorus, and toxic substances.

22.1.5.3.1 Organisms

In raw sewage, organisms include such groups as helminthes, rotifers, protozoa, cysts, bacteria, and viruses. The bacterium, *E. coli* is predominant, with concentrations on the order of $10^9/100$ L, a calculation based on an estimate of per capita coliform discharge of $100–400 \cdot 10^9$/c/d (Tchobanoglous and Burton, 1991, p. 93); dividing by 300 L/c/d wastewater discharge gives the 10^9/coliforms s/L. Also, to illustrate the diversity of organisms, *Giardia lamblia* cysts and *Cryptosporidium parvum* oocysts have been measured in samples at 200–300#/L. Municipal wastewaters are always presumed to contain pathogenic organisms. Most are removed during traditional biological treatment because they are not likely to survive long outside their niche, most likely in the intestinal track of warm-blooded animals. Some organisms "encyst," however, which equips them for longer survival in a hostile environment. In biological treatment an ecosystem develops that includes rotifers, protozoa, and bacteria in a hierarchal structure that facilitates the degradation of organic matter.

22.1.5.3.2 Nutrients

Any organism requires a variety of nutrients for growth, including sugars, fats, proteins for energy; phosphates for building ATP; nitrogen for protein synthesis; and sulfur, certain metals at low concentrations, etc. These substances are necessary for biological treatment and are present, as a rule in municipal wastewaters. In treating industrial wastewaters, it is often necessary to add nutrients in order to compensate for a deficiency. At the same time, discharges of these nutrients into ambient waters (even at the reduced concentrations after traditional biological treatment) may cause undesired organisms to thrive. This issue came into the forefront in the early 1960s when a focal point of the environmental movement was Lake Erie, said to have been "dead," being choked with algae, which was attributed to phosphates from various municipalities and industrial discharges, a condition known as "eutrophication." Because of its role as a nutrient, phosphorous, along with nitrogen, has been considered a major pollutant.

22.1.5.3.3 Nitrogen

Two sources of nitrogen in municipal wastewater are urea and proteins; ammonia is an end product of the breakdown of these compounds. Usually concentrations of ammonia are highest in the influent to a municipal wastewater treatment plant, e.g., in the range of 10–30 mg NH_3 as N/L (Benefield and Randall, 1980, p. 88). During the course of treatment, "nitrification" may occur (as stated by the conversion of ammonia to nitrite, NO_2^- by the organism, *Nitrosomonas*). The second stage is the conversion of nitrite to nitrate, NO_3^-, by *Nitrobacter*, and is rapid. A final step in nitrogen removal is "denitrification," i.e., the conversion of NO_3^- to nitrogen gas, N_2; usually, the organism is *Pseudomonas*, and the rate of reaction is "fast."

22.1.5.3.4 Ammonia

A problem of ecology recognized during the 1970s was that fish species are sensitive to low un-ionized ammonia, i.e., NH_3, concentrations (NH_3 is favored over NH_4^+ as pH increases where $pH > pK_a = 9.27$). Under most conditions natural water bodies have pH ranges of 6.5–7.5, which means that the NH_3 is concentration would be commensurately low. The pH usually increases with algae growths, but usually $pH < 8$. Stream standards have been established such that the level of un-ionized ammonia does not harm aquatic life (e.g., ≤ 0.06 mg/L as N). Therefore, "nitrification," i.e., conversion of ammonia to nitrate, has been required since the early 1980s, which provides latitude for uncertainty in both the possible formation of NH_3 and for the sensitivity of fish species. In cell synthesis, the bacteria prefer ammonia as a source of nitrogen for building bacterial proteins.

22.1.5.3.5 Synthetic Organics

New synthetic chemicals come to the forefront frequently and have included antibiotics, personal-care products, endocrine disruptors, etc. In a survey of pharmaceuticals and organic contaminants in ambient waters, the USGS (http://www.toxics.usgs.gov/pubs/FS-027-02, 2009) found steroids, insect repellents, plasticizers, insecticides, nonprescription drugs, prescription drugs, etc. Some compounds are persistent and resistant to biological treatment. For reference, the number of chemical substances cataloged in 2009 by the Chemical Abstract Service (CAS) registry exceeded 50 million. Many synthetic organics are regulated (see, for example, Chapter 2) but the number is small compared to the total.

22.2 CELL METABOLISM

Biological reactions are better understood in the context of a few fundamental notions from biochemistry. According to Stryer (1981, p. 235), over a thousand chemical reactions occur in even a simple organism such as *E. coli*. The sum of these reactions constitute metabolism, which has two parts: (1) catabolism, which yields energy; and (2) anabolism, which is cell synthesis. These ideas are reviewed here, along with a brief section on energy (thermodynamics). The term, "cell" as used in this text refers to a bacterium, which is a limited view.

BOX 22.3 ROLE OF LAW IN WATER TREATMENT

Prior to 1965, the objectives of wastewater treatment were (1) to maintain aerobic conditions, defined traditionally as ≥2 mg dissolved oxygen per liter in the receiving waters; (2) to not cause solids deposition; and (3) to reduce concentrations of pathogens. If a downstream water user was impacted by an upstream sewage discharge, the injured party had only the courts and "common law" (i.e., cases not covered by statute) for possible remedy. The process was long, arduous, and expensive, with outcome not certain. During the 1950s, however, public perceptions of water quality issues evolved and in 1965 the first federal law was enacted that had "teeth," which was PL 89-234, the Water Quality Act of 1965. The legislation required a "paradigm shift," i.e., a change from the idea of "assimilative capacity" to the requirement that states establish water-quality standards for interstate waters. The culmination was the 1972 Clean Water Act, PL 92-500, which had a "goal" of zero discharge of contaminants. The act included provisions for "effluent" standards, which were easier to enforce than stream standards. The term "contaminants" has been broadly interpreted and have included a host of chemicals, along with "nutrients," including nitrogen and phosphorous. The role of biological treatment has expanded commensurately. In Europe, the European Union (EU) has developed increasingly more stringent regulations for its member countries. In some cases, other countries have looked to the United States or the EU as possible models.

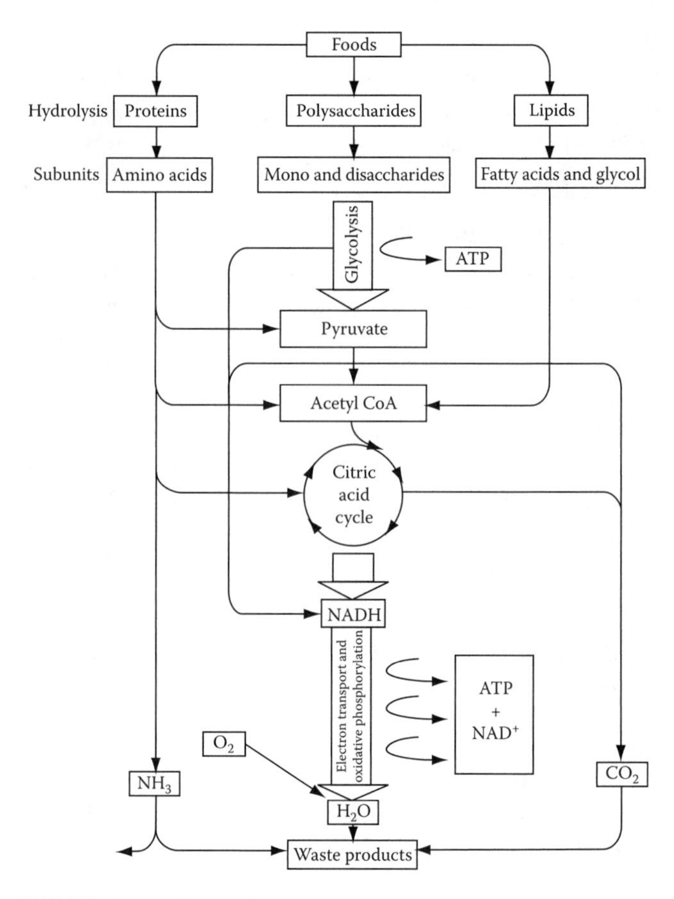

FIGURE 22.1 Catabolic biochemical pathways for ATP production. (Adapted from Rawn, J.D., *Proteins, Energy, and Metabolism*, Neil Patterson Publishers, Burlington, NC, 1989, p. 243, Figure 10-11.)

22.2.1 Metabolic Reactions

The term "substrate" is any substance that may be metabolized by an organism. Organic carbon is the substrate for a heterotrophic organism, also called a "chemoorganotroph." An inorganic molecule, e.g., ammonia, may serve as the substrate for an autotroph, also called a "chemolithotroph." In metabolism, substrates provide energy for catabolism in which energy is stored in the compound, ATP. In anabolism, the conversion of ATP to ADP provides the energy for cell synthesis. It is not an overstatement to say that the reactions involved are immensely complex.

22.2.1.1 Catabolism

Catabolism is a sequence of reactions that yield energy, with each reaction facilitated by a specific enzyme. The particular pathway depends upon the substrate and the microbe. The reaction sequence is "exothermic," meaning that the standard free energy of reaction is negative, i.e., $\Delta G^\circ < 0$, meaning that the reaction is "spontaneous."

Figure 22.1 depicts the phases of degradation of substrate, i.e., "food," in three forms (proteins, polysaccharides, lipids) to end products. The first phase, to acetyl-CoA, is anaerobic and produces two ATP molecules. The acetyl-CoA is formed from pyruvate if an electron acceptor is present, e.g., oxygen or nitrate (Orhon and Artan, 1994, p. 57). The acetyl-CoA then enters the citric acid cycle, which is aerobic. Water and carbon dioxide are end products, along with about 36 mol of ATP, produced by "oxidative phosphorylation."

Three points are relevant in further understanding: (1) if the starting point is glucose, the degradation is called glycolysis and may occur via "Embden–Myerhoff" pathway; (2) the reactions in the pathway approximate thermodynamic "reversibility" which means that the ATP energy yield is near the theoretical maximum; and (3) the NADH, which converts to NAD^+, functions to transport electrons along the pathway.

22.2.1.1.1 Electron Transfer (Abstracted from Orhon and Artan (1994, pp. 44–64)

In catabolism, a chain sequence of electron transfers enables a stepwise release of energy. As is well known, an *oxidized* compound loses electrons. For an organic compound, its oxidation involves the removal of two electrons and two protons (called dehydrogenation). To illustrate, Equation 22.1 depicts the oxidation of lactic acid to pyruvate,

$$\begin{array}{ccc} CH_3 & & CH_3 \\ | & & | \\ CHOH & \leftrightarrow & CH{=}O + 2H^+ + 2e^- \\ | & & | \\ COOH & & COOH \\ \text{lactic acid} & & \text{pyruvic acid} \end{array} \quad (22.1)$$

The electrons released are captured by nucleotide derivatives, especially nicotinamide adenine dinucleotide (NAD), which is an electron carrier in all cells. Other electron carriers include NADP (the same as NAD but with an extra phosphate group) and FAD (flavin adenine dinucleotide). Regarding NAD, which is illustrative, the nicotinamide portion of NAD accepts the electrons and is *reduced*, e.g., as NADH. The NADH then loses its electrons, via enzymes, to again form NAD. The reversible reaction is depicted,

$$\underset{\text{reduced form}}{NADH_2} \leftrightarrow \underset{\text{oxidized form}}{NAD} + 2H^+ + 2e^- \quad (22.2)$$

In aerobic metabolism, the NAD carries the electrons to the final electron acceptor, molecular oxygen, or nitrate. In the case of molecular oxygen, the reaction is, after a sequence of successive such reactions,

$$NADH_2 + \frac{1}{2}O_2 \rightarrow NAD + H_2O \quad (22.3)$$

The electron transfers occur in a sequence of reactions called the *respiratory chain*, through which ATP is formed. The NAD and FAC molecules donate their electrons to a series of carriers called *cytochromes*. At each step in the electron transfer free energy is released, which permits the formation of ATP. The final electron acceptor in an aerobic reaction is molecular oxygen. In an "anoxic" reaction, nitrate may serve but with a lower yield of ATP.

22.2.1.1.2 ATP Energy Yields

Table 22.3 compares reactions, moles ATP formed, ΔG°(reaction), and ΔG°(captured in ATP) for combustion, aerobic metabolism of glucose, and anoxic metabolism of glucose. In aerobic respiration with oxygen as the electron acceptor, 38 mol of ATP are generated, i.e., two in the Embden–Meyerhof pathway, two in the citric acid cycle, and 34 in the respiratory chain. The energy capture by ATP is thus, 38 mol ATP · 7.0 kcal/mol ATP = 1113 kJ (266 kcal). In anoxic respiration, only 26 mol ATP are formed per mol of glucose oxidized, i.e., 26 mol ATP 7.0 k cal/mol ATP = 761 kJ (182 kcal) (Orhon and Artan, 1994, p. 69).

22.2.1.1.3 Fermentation (Anaerobic Reaction)

If molecular oxygen is absent and if an inorganic oxidant is not present to act as an electron acceptor, an organic molecule may serve as both an electron donor and as an electron acceptor; such a reaction is termed, "fermentation," which is anaerobic (Section 22.3.3.4). For a fermentation reaction, the free energy of reaction is less than for the anoxic reaction, with correspondingly lower values of n(ATP) and cell yield.

22.2.1.2 Anabolism

Anabolism is also a sequence of reactions, with each reaction facilitated by a specific enzyme. As with degradation, the particular pathway depends upon the substrate and the microbe. The reaction sequence is, however, "endothermic," meaning that the standard free energy of reaction is positive, which means $\Delta G^\circ > 0$, meaning also that the reaction is not

TABLE 22.3
Aerobic and Anoxic Reactions and Energy Captured ATP

Reaction	ATP (mol ATP)	ΔG^o (Reaction) (kcal)	ΔG^o (Reaction) (kJ)	ΔG^o (Captured in ATP) (kcal)	ΔG^o (Captured in ATP) (kJ)
1. Combustion					
Oxidation of glucose (combustion) $C_6H_{12}O_6 + 6O_2 \rightarrow 6CO_2 + 6H_2O$ (glucose, oxygen)	0	−686	−2870	0	0
2. Aerobic Reaction					
Metabolism of glucose (EMP/citric acid cycle/respiratory chain/oxygen as electron acceptor) $C_6H_{12}O_6 + 38Pi + 38ADP + 6O_2 \rightarrow 6CO_2 + 38ATP + 44H_2O$ (glucose, oxygen)	38	−420	−1757	266	1113
3. Anoxic Reaction					
Metabolism of glucose (EMP/citric acid cycle/respiratory chain/nitrate as electron acceptor) $C_6H_{12}O_6 + 26Pi + 26ADP + 12NO_3^- \rightarrow 6CO_2 + 26ATP + 32H_2O + 12NO_2^-$ (glucose, nitrate)	26	−469	−1962	182	761

Source: Adapted from Orhon, D. and Artan, N., *Modeling of Activated Sludge Systems*, Technomic Publishing Co., Lancaster, PA, 1994, pp. 64–69.

Note: Pi is the symbol for phosphate.

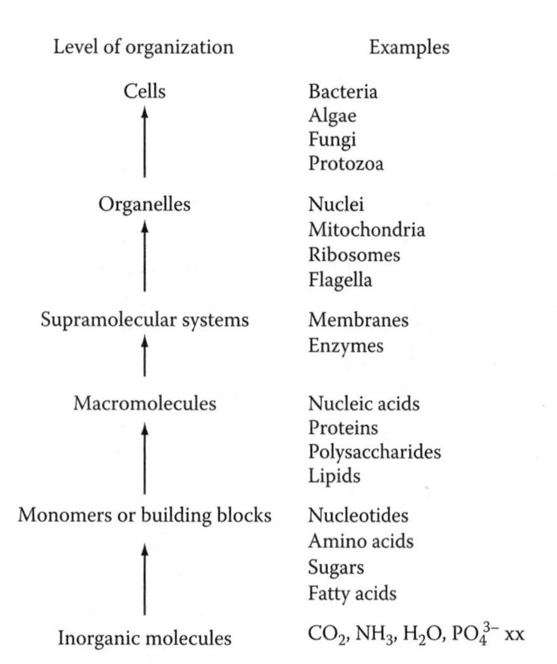

FIGURE 22.2 Biosynthesis—from inorganic molecules to cells. (Adapted from Prescott, L.M. et al., *Microbiology,* 6th edn., McGraw-Hill, Dubuque, IA, 2005, p. 200.)

"spontaneous" but depends on an input of energy. The ATP yield of the catabolism pathway provides the energy for the cell synthesis, in anabolism, i.e., the ATP-to-ADP conversion, as well as for cell maintenance.

Biosynthetic pathways are divergent, as contrasted to convergent catabolic pathways. For example, the compound "aspartate" is a precursor to the synthesized products, lysine, methionine, threonine, and isoleucine. The precursors of biological macromolecules in all cells are saccharides, fatty acids, amino acids, purines, and pyrimidines (Rawn, 1989, p. 244).

The microorganism begins with simple precursors and constructs evermore complex molecules until new organelles and cells arise, illustrated in Figure 22.2 (Prescott et al., 2005, p. 200). Most of the ATP available for biosynthesis is applied in protein synthesis. Proteins, of whatever composition, are made of only 20 amino acids, joined by peptide bonds; different proteins have different amino acid sequences (see Appendix 22.A).

22.2.1.3 Cell Division

The sequence of events from the formation of a new cell through the next division is called the "cell cycle." For example, a new *E. coli* cell grows in length with constant diameter and at some threshold mass (after perhaps 10 min) it will trigger its DNA replication and then at another threshold it will divide length-wise by binary fission. Each daughter cell receives at least one copy of the genetic material. The time sequence for the cell cycle is about 40 min for DNA replication and 20 min for cell division with a new cell created. After the separation of chromosomes a cross-wall or septum forms between the old cell and the new cell (abstracted from Prescott et al., 2005, pp. 280–281).

22.2.1.4 Photosynthesis

The empirical cell formula for an algae biomass is $C_{7.6}H_{8.1}O_{2.5}N$. An associated photosynthesis reaction is (Benefield and Randall, 1980, p. 326)

$$7.6CO_2 + 2.5H_2O + NH_3 \rightarrow C_{7.6}H_{8.1}O_{2.5}N + 7.6O_2 \tag{22.4}$$

The reaction is proportional to the radiant energy and varies diurnally. In monitoring dissolved oxygen (DO) concentrations over 24 h cycles, Kartchner et al. (1969) found an approximate sinusoidal variation during daylight hours. The oxygen produced has been considered a source for the aerobic treatment of wastewaters, such as in ponds. At the same time, the algae biomass "respires" and has an oxygen demand, which reduces the DO concentration after sunset with a steady decline until sunrise. The DO concentrations may reach levels in the range of 30–40 mg/L in some stagnant waters and as high as 16 mg/L in fast-moving mountain streams. At night the stagnant waters are likely to go "anoxic." Observations in a waste-stabilization pond at Logan, Utah (Hendricks and Pote, 1974) showed that while the biomass was highly concentrated throughout the summer, the pond water turned almost clear in the fall, with the only explanation being that the algae die and sink to the bottom with subsequent anaerobic decay.

22.2.1.5 Energy Principles

Thermodynamic principles apply to biological reactions, as for any chemical reaction, and provide another tool for analysis and description. The basic concepts are seen in texts on physical chemistry (see, for example, Daniels and Alberty, 1955) or in texts on aqueous chemistry (see, for example, Sawyer and McCarty, 1967). The few lines are not intended to replace a course on chemical thermodynamics, but to indicate some of the ideas that are relevant to biological treatment. If thermodynamics is familiar, then the outline here merely highlights a few principles that are common in biological treatment.

Free energy is a fundamental parameter and is usually measured by changes from one state to another. Any compound has a free energy of formation, ΔG_f. The free energy of a reaction, ΔG_R, is the sum of the free energies of formation of product compounds times the respective moles minus the free energies of formation of reactant compounds times the respective moles. If the reaction occurs under "standard" conditions, which is the common reference, the values can be calculated from handbook data and the reaction energy is designated, ΔG°_R. If it occurs under other conditions, such as the physiological concentrations of a cell, the free energy of a reaction can be calculated by the familiar relation, $\Delta G = \Delta G^\circ + RT\ln Q$, where Q represents concentration ratios with stoichiometric numbers as exponents. The free energy is a parameter invented to account for the two disparate aspects of any process: energy change as given by the enthalpy change, ΔH, and the entropy change, ΔS. For a reaction, $\Delta G = \Delta H - T\Delta S$. In a process, defined here as a "state" change, the ΔG wants to follow a negative gradient, and if such a path is available, the process is spontaneous.

This means that the enthalpy (defined as the internal energy plus PV energy) wants to go from higher to lower values. The entropy wants to increase, which results in a more negative value of ΔG. For example, a gas confined to a bottle will diffuse once the cap is off; its entropy is increasing giving more disorder. To get the gas back in the bottle (more order) requires energy. Essentially all water treatment involves trying to get more order from disorder. In biological treatment where we want to remove a dissolved substrate from water the cells have more order. Also, we can settle the cells and further process them (such as vacuum filtration), creating still more order.

The foregoing summary is intended to introduce a few of the thermodynamic terms that are mentioned in this text. The review is not intended to explain a topic that warrants a focused study. The idea of treating microbial cell synthesis as a chemical reaction and relating the free-energy of reaction to the yield of cells and applied to biological treatment was reviewed during the early to mid 1960s by Servizi and Bogan (1963, p. 64) and McCarty (1964) and are what might be termed, "classic" papers on the topic.

22.3 BIOLOGICAL TREATMENT OVERVIEW

The two aspects of biological treatment are (1) the reaction, and (2) kinetics. The reaction is best understood as a balanced chemical equation and the kinetics is usually expressed by the classic Monod equation. In addition to understanding principles, certain reference material is useful, including chemical formulae on substrates and cells; and coefficients such as cell yield, and Monod. The material reviewed has been building mostly since the 1950s; texts from that period through the recent decade have summarized available knowledge and were cited extensively. Of the reputed 1000 papers on activated sludge to the early 1970s, several have been "landmark," a few of which have been identified and, in some cases, cited.

22.3.1 Composition of Substrates

Formulae for substrate and cells are required as the first step in constructing a balanced chemical equation. Categories of substrates include organic compounds such as glucose, alcohols, proteins, etc; and inorganic compounds such as ammonia (occurring as NH_3 or NH_4^+, depending on pH), nitrates, and sulfates.

22.3.1.1 Domestic Wastewater and Organic Compounds

Formulae for representative organic substances of interest in wastewater treatment are given in Table 22.4. Domestic wastewater is first, and was derived from laboratory analysis, as was the formula for bacterial cells. A general formula for carbohydrates is given as CH_2O, which has a multitude of variations, a few of which are shown, e.g., glucose, sucrose, and lactose. Other substances listed are indicative of the variety found in raw wastewaters or may occur at some stage of treatment. Urea, in the bottom row, is a source of ammonia and comes from the breakdown of protein to amino acids, which breaks down further to give urea as a product, and is a physiological process (Rawn, 1989, p. 468).

TABLE 22.4
Formulae for Sampling of Organic Compounds

Substance	Composition	Formula	Reference
Domestic wastewater	Fats, protein, sugars	$C_{10}H_{19}O_3N$	Orhon and Artan (1994, p. 95)
Bacteria cells	Protein, fats, lipids, etc.	$C_5H_7O_2N$	Porges et al. (1956, p. 43)
Carbohydrates	Cellulose, starch, sugars	CH_2O	Eckenfelder and Weston (1956, p. 21)
Glucose		$C_6H_{12}O_6$	Eckenfelder and Weston (1956, p. 21)
Sucrose		$C_{12}H_{22}O_{11}$	Eckenfelder and Weston (1956, p. 21)
Lactose		$C_{12}H_{22}O_{11} \cdot H_2O$	Eckenfelder and Weston (1956, p. 21)
Protein	Amino acids, nitrogenous organics	$C_{16}H_{24}O_5N_4$	Orhon and Artan (1994, p. 95)
Grease	Fats and oils	$C_8H_{16}O$	Orhon and Artan (1994, p. 95)
Methyl alcohol		CH_3OH	Eckenfelder and Weston (1956, p. 21); Orhon and Artan (1994, p. 95)
Ethyl alcohol		CH_3CH_2OH	Eckenfelder and Weston (1956, p. 21); Orhon and Artan (1994, p. 95)
Acetic acid		CH_3COOH	Eckenfelder and Weston (1956, p. 21); Orhon and Artan (1994, p. 95)
Propionic acid		CH_3CH_2COOH	Orhon and Artan (1994, p. 95)
Urea		H_2NCONH_2	Stryer (1981, p. 33)

Sources: Orhon, D. and Artan, N., *Modeling of Activated Sludge Systems*, Technomic Publishing Co., Lancaster, PA, 1994; Eckenfelder, W.W. Jr. and Weston, R.F., in McCabe, J. and Eckenfelder, W.W. Jr., (Eds.), *Biological Treatment of Sewage and Industrial Wastes*, Volume 1, Aerobic Oxidation, Reinhold Publishing, New York, 1956.

TABLE 22.5
BOD and COD for Sampling of Industrial Wastewaters

Wastewater	BOD (mg/L)	COD (mg/L)	
Pharmaceutical	3290	5780	Eckenfelder et al. (2009, p. 27)
Cellulose	1250	3455	nd
Tannery	1160	4360	nd
Paper mill	380	686	nd
Domestic	135		Eckenfelder and Weston (1956, p. 30)
Tomato	450		nd
Packing house	1020		nd
Pharmaceutical	1684	3198	nd
Pharmaceutical	2532	3980	nd
Refinery	79	262	nd

22.3.1.2 Industrial Wastewaters

Industries that generate wastewaters number in the thousands. A sampling is shown in Table 22.5, which shows also the variation found in BOD and COD of the raw wastewaters. The treatment of industrial wastewaters is often specialized due to high concentrations, toxic substances, or limited nutrients.

22.3.2 Composition of Cells

Microbial cells (heterotrophic and autotrophic) are composed of elements, C, H, O, N, P, S, and trace metals. Empirical formulae can depict the element ratios in simple terms, but actually, the molecular composition and structure is complex.

22.3.2.1 Empirical Formulae for Cells

Porges et al. (1956, p. 42) gave an $N = 1$ based formula as $C_5H_7O_2N$ (MW = 113), which has been adopted most extensively and remains current (e.g., Grady et al., 1999; Orhan and Artan, 1994; Rittman and McCarty, 2002) and has been used for all categories of cells. With the inclusion of inert matter, they obtained the molecular weight, $MW(C_5H_7O_2N + inerts) = 124$.

Stumm (1964, pp. 216–230), gave a P = 1 based formula as, $C_{106}H_{180}O_{45}N_{16}P$. From the formula, the mole ratios are C:N:P = 106:16:1, which is useful in estimating nutrient additions, such as for an industrial waste. Also, the formula gives a basis for estimating the uptake of N and P, such as from a municipal wastewater. For reference, Stumm (1964, p. 219) gave a balanced equation for the oxidation of the cell matter as

$$C_{106}H_{180}O_{45}N_{16}P + 154O_2 = 106CO_2 + 90H_2O + 16NO_3^- + PO_4^{3-} \quad (22.5)$$

For anaerobic organisms, McCarty (1970, p. 236) gave a P = 1 based formula as $C_{60}H_{87}O_{23}N_{12}P$ (see also Sherrard, 1977, p. 1971), for which the molecular ratios were approximately the same as given by Porges et al. (1956). An N = 1 based formula for anaerobic cells was given as $C_5H_9O_3N$ (Sherrard, 1977, p. 1971).

Another formula by Orhon and Artan (1994, p. 81), based on the composition of total and organic solids from two municipal wastewater treatment plants, was $C_{66}H_{124}O_{26}N_{13}P$, with MW = 1545 g/mol. The formula differs little from the Porges et al. (1956, p. 42) formula for an N = 1 basis.

Example 22.1 Ash Content of Cells

Given
The atomic weight for bacterial cells was given by Porges et al. (1956, p. 42) as $MW(C_5H_7O_2N) = 113$ g/mol. When ash is included they gave $MW(C_5H_7NO_2 + ash) = 124$ g/mol.

Required
Determine the ash content of cells in (g ash/100 g cells with ash).

Solution
Assume the cell wall, protoplasm, etc. has an overall MW = 124 g/mol. In other words, for 1 mol of proteins (i.e., cells) their mass is 124 g, and the mass of ash is 11 g. Thus, for 100 g cell mass (which includes ash), the mass of ash is (11 g ash/124 g cells with ash) · (100 g cells with ash) = 8.87 g ash.

22.3.3 Biological Reactions

As noted, biological reactions are complex and involve biochemical cycles, energy transfers, and reaction stoichiometry. A reaction is depicted first on a conceptual basis, building to stoichiometric equations. The idea is to highlight the aspects of a reaction that are important for engineering purposes. Section 22.3 deals with the biochemistry, to add understanding to the reactions.

22.3.3.1 Substrate to Cells

A chemoorganotrophic, i.e., heterotrophic, biological reaction, may be depicted,

$$\text{substrate} + E(\textit{bacteria}) + \text{reactants} \rightarrow \text{new-cells}(\textit{bacteria}) + E(\textit{bacteria}) + \text{products} \quad (22.6)$$

where

substrate is any substance that may serve as an "electron-donor" for a biochemical reaction, e.g., degradable organic matter, ammonia, CO_2 (kg/m^3)
E(bacteria) is an enzyme contained in microbial cells that serves to degrade the organic matter
reactant is any chemical substance that serves as a reactant along with the other substances listed on the left side, e.g., oxygen (mg/L)
new-cells(*bacteria*) are the microbial cells that have been synthesized as a result of the reaction (kg cells/m^3)
products are any chemical substances that result due to the reaction, e.g., carbon dioxide, water (mg/L)

The enzyme, *E*, is provided by the "old" microbial cells, and is a necessary part of any biological reaction, and remains after the synthesis of new cells. A fraction of the "old" cells are subject to decay, i.e., endogenous respiration, so it is only the viable cells that provide the enzymes.

The quantity of "new" cells as a product per unit of substrate reacted is called the "cell yield," *Y*, and is a stoichiometric value. In other words, the cell synthesis is tied to the substrate degradation. The value of *Y* may be obtained from a balanced equation that shows substrate and cells. Not shown is the loss of cells by "endogenous-respiration"; empirical determinations of cell yield actually give the "net" amount of cells produced after loss by endogenous respiration, i.e., *Y*(net).

22.3.3.2 Heterotrophic

One of the first balanced equations for the oxidation of an organic substrate to yield microbial cells was by Porges et al. (1956, p. 43) and was cited by Rittman and McCarty (2001, p. 127), i.e.,

$$\underset{\text{(casein)}}{C_8H_{12}O_3N_2} + 3O_2 \rightarrow \underset{\text{(cells)}}{C_5H_7O_2N} + NH_3 + 3CO_2 + H_2O \qquad (22.7)$$

The stoichiometry of the cell-synthesis reaction depends, of course, on the substrate. Using sugar as the substrate, Porges et al. (1956, p. 43) depicted the synthesis reaction,

$$\underset{\text{(sugar)}}{8CH_2O} + 3O_2 + NH_3 \rightarrow \underset{\text{(cells)}}{C_5H_7O_2N} + 3CO_2 + 6H_2O \qquad (22.8)$$

As noted, the "ash" content of cells includes P, S, Fe, and various trace elements, which result in MW(124) or about 0.09 fraction by weight. In other words, the "ash" is the content of a cell that remains after the combustion at 550°C in a muffle furnace.

22.3.3.3 Autotrophic Involving Nitrogen

Organic nitrogen, in the form of amino groups, forms ammonia; this process is called "ammonification" (Grady et al., 1999, p. 20) and occurs usually in the sewers. Another source of ammonia is urea, which comes from the amino group, arginine, converted to urea in the body (Section 22.2.1.1).

The conversion of ammonia to nitrites, and then immediately after to nitrates, is called "nitrification." The organisms responsible are the "nitrifiers," *Nitrosomonas* and *Nitrobacter*, respectively. The further conversion of nitrates to nitrogen gas is "denitrification," which occurs commonly by the facultative bacteria, *Pseudomonas*, under anoxic conditions.

In biological treatment, nitrification occurs in an aerobic reactor, such as activated sludge or a biofilter, but toward the end of a treatment train. The anoxic condition for denitrification may be induced by cutting off the oxygen, such as in a final stage of a "plug-flow" activated-sludge reactor.

22.3.3.3.1 Nitrification

The overall aerobic reaction stoichiometry from ammonia ion to cells and nitrates is (Christensen and McCarty, 1975, p. 2656)

$$0.128NH_4^+ + 0.235O_2 + 0.0123CO_2 + 0.00304HCO_3^- \rightarrow 0.00304C_5H_7O_2N + 0.125NO_3^- + 0.25H^+ + 0.122H_2O \qquad (22.9)$$

From Equation 22.9, the following relations are obtained (Grady et al., 1999, p. 76):

- 6.71 mg HCO_3^- required/mg NH_4^+ reacted (8.62 mg HCO_3^-/mg NH_4^+-N)
- 3.30 mg O_2 required/mg NH_4^+ reacted (4.33 mg O_2/mg NH_4^+-N)
    - 3.22 mg O_2/mg NH_4^+ as N reacted used by *Nitrosomonas*
    - 1.11 mg O_2 /mg NH_4^+ as N reacted used by *Nitrobacter*
- 0.129 mg $C_5H_7O_2N$ synthesized/mg NH_4^+ reacted (0.166 mg $C_5H_7O_2N$/mg NH_4^+-as-N)
    - 0.146 mg *Nitrosomonas* is synthesized/mg NH_4^+-N (experimental: 0.04–0.13)
    - 0.020 mg *Nitrobacter* is synthesized/mg NH_4^+-N (experimental: 0.02–0.07)

As stated by Grady et al. (1999, p. 77) the cell production, i.e., nitrifying bacteria, is not large but the oxygen demand and the alkalinity requirement may be significant, depending on the ammonia concentration. The decimal fractions for the coefficients relate to the fact that the equation was "normalized" for one electron equivalent (e-eq). The half-reactions used as the basis for constructing the overall equation were done by canceling a single electron on each side (Section 22.3.3.5), described by Christensen and McCarty (1975, p. 2656). Because of the alkalinity requirement, the reaction requires pH $\geq$ 7.0 (Parker et al./EPA 1975, pp. 3–15).

22.3.3.3.2 Denitrification

The conversion of nitrate to nitrogen gas, i.e., "denitrification," occurs in a sequence, $NO_3^- \rightarrow NO_2^- \rightarrow NO \rightarrow N_2O \rightarrow N_2$. The reaction requires an electron donor, e.g., a carbon source. Using glucose, the overall balanced equation is (Rittman and McCarty, 2001, p. 133)

$$5C_6H_{12}O_6 + 24NO_3^- + 24H^+ \rightarrow 30CO_2 + 42H_2O + 12N_2 \qquad (22.10)$$

If a carbon source is limited, McCarty (1970, p. 241) suggested methanol as a source, depicted by the reaction,

$$\frac{5}{6}CH_3OH + NO_3^- \rightarrow \frac{1}{2}N_2 + \frac{5}{6}CO_2 + \frac{7}{6}H_2O + OH^- \qquad (22.11)$$

As noted by Parker et al./EPA (1975, pp. 3–29), *Pseudomonas*, *Micrococcus*, *Achromobacter*, and *Bacillus* are among the bacteria than can denitrify and can use either nitrate of oxygen as electron acceptors. They are called facultative heterotrophs and can shift easily from aerobic to anoxic conditions since the electron transport pathways are almost identical. They give the overall nitrate removal reaction as

$$1.08CH_3OH + NO_3^- + 0.24H_2CO_3 \rightarrow 0.056C_5H_7O_2N + 0.47N_2 + 1.68H_2O + HCO_3^- \quad (22.12)$$

As seen, about 2.5 kg methanol are required per kg NO_3^- as N. Methanol is a more readily available electron donor than wastewater and the denitrification rate for the latter is about one-third the rate for methanol, which requires a larger reactor volume (Parker et al./EPA 1975, pp. 3–35).

22.3.3.4 Anaerobic

As described by Stewart et al. (1959), two major metabolic groups of bacteria occur in anaerobic systems, *acid-producing* and *methane-forming*. The acid-producing bacteria are common facultative bacteria which exist in soil, activated-sludge floc, and trickling-filter slimes and have the ability to oxidize organic materials to simple organic acids, alcohols, and aldehydes. The metabolic activities of methane-forming bacteria result in the production of methane and carbon dioxide from the organic acids, alcohols, and aldehydes.

McCarty (1964a, p. 109) depicted the general reaction,

$$\text{complex organics} \xrightarrow{\text{acid formers}} \text{organic acids, } H_2 \xrightarrow{\text{methane formers}} CH_4 + CO_2 \quad (22.13)$$

First-stage waste conversion — Second-stage waste stabilization

In other words, the first stage reacts with the varied organic substance, often as found in primary sludge from a municipal wastewater treatment plant such as cellulose, proteins, sugars, lipids, cells, etc., to form organic acids such as those listed in Table 22.6. The second stage is rate limiting since the fermentation reactions in the first stage have a greater energy yield, ΔG(reaction), than in the second stage, i.e., the reactions involved in methane formation (Rittman and McCarty, 2001, p. 583). On the other hand, stage one may be slower for substances such as lignocellulose, grasses, crop residues, newsprint. The topic of anaerobic treatment was reviewed comprehensively in a series of four articles by McCarty (1964a,b,c,d) which explained a complex subject is a fashion useful for persons in practice.

The anaerobic reaction is called fermentation in that the electron donor and the electron acceptor are the same organic molecule. Using glucose as an example Rittman and McCarty (2001, p. 587) give the reaction stoichiometry as

$$C_6H_{12}O_6 + 0.24NH_4^+ + 0.24HCO_3^- \rightarrow 0.24C_5H_7O_2N + 2.4CH_4 + 2.64CO_2 + 0.96H_2O \quad (22.14)$$

TABLE 22.6
Common Volatile Acid Intermediates

Acid	Formula
Formic	$HCOOH$
Acetic	CH_3COOH
Propionic	CH_3CH_2COOH
Butyric	$CH_3(CH_2)_2COOH$
Valeric	$CH_3(CH_2)_3COOH$
Isovaleric	$(CH_3)_2CHCH_2COOH$
Caproic	$CH_3(CH_2)_4COOH$

Source: Adapted from McCarty, P.L., Anaerobic waste treatment fundamentals, Part one—Chemistry and microbiology, *Public Works*, September 1964a, p. 109.

From Equation 22.14 the relationships are seen by the coefficients, i.e., for each mole of glucose reacted, 0.24 mol of NH_4^+ is required, with yields of 0.24 mol of cells, 2.4 moles of CH_4, and 2.64 mol of CO_2.

Table 22.7 shows representative substrates that may be subject to the anaerobic reaction sequence, organic acid formation and methane formation, showing also the respective cell yields and endogenous coefficients. The Y coefficients (the mass of cells synthesized per unit of substrate reacted) were as calculated by Rittman and McCarty (2001, p. 586). In aerobic treatment, Y(aerobic) ≈ 0.5 g VSS/g BOD_L, which is on the order of 5–10 times the values shown in Table 22.7. This means that, in anaerobic treatment, the cell-disposal problem is much less. Also, in favor of anaerobic process, methane gas is formed, which has an energy value (p. 570) of 35.8 kJ/L (STP). A typical volumetric loading may be 5–10 kg COD/day/m^3 reactor volume as contrasted with <1 kg COD/day/m^3 for an aerobic system (p. 571); anaerobic treatment is especially suited for high-strength wastes. On the other hand, the minimum mean cell residence time,

TABLE 22.7
Substrate Formulae and Coefficients for Anaerobic Process to Final Products

Substrate	Formula	Y (g VSS/g BOD_L)	b (day^{-1})
Carbohydrates	$C_5H_{10}O_5$	0.20	0.05
Protein	$C_{16}H_{24}O_5N_4$	0.056	0.02
Fatty acids	$C_{16}H_{32}O_2$	0.042	0.03
Municipal sludge	$C_{10}H_{19}O_3N$	0.077	0.05
Ethanol	CH_3CH_2OH	0.077	0.05
Methanol	CH_3OH	0.11	0.05
Benzoic acid	C_6H_5COOH	0.077	0.05

Source: Adapted from Rittman and McCarty, Environmental Biotechnology: Principles and Applications McGraw-Hill, New York, 2001, p. 587.

Notes: Y is the yield coefficient for cell synthesis as calculated from stoichiometric equations. b is the endogenous decay coefficient.

$\theta_c^m(35°C) \approx 4$ day, while for design, the suggested mean cell residence time for a reactor is $\theta_c(35°C) \approx 10$ day (Tchobanoglous and Burton, 1991, p. 818), which apply to "complete-mix" digesters. In practice, $20 < \theta_c(35°C) < 30$ day, which allows for uncertainty in the process, such as the high sensitivity of the methane-former bacteria to environmental conditions. The most common application of anaerobic treatment is for municipal *sludge stabilization*, which is a traditional term that refers to the completion of an anaerobic reaction to the extent that the substrate is no longer noxious. In fact, the sludge from the underflow of a primary settler, a highly objectionable substance that does not dewater significantly, becomes after anaerobic digestion, a different material that has a musty, earthy odor and that dewaters easily in "sludge-drying beds." Also, it is easily worked into soil for agricultural disposal (subject to regulations).

22.3.3.5 Balancing Equations by Half-Reactions

The stoichiometry of any redox reaction can be determined by the technique of adding "half-reactions" (see Section 20.2.1.3). Tables of half-reactions, as in Table 20.1, are given in handbooks such as Latimer (1952), and Lide (1996, pp. 8-20–8-29). The technique is described in chemistry texts (e.g., Silberberg, 1996, p. 159) and is based on the principle that the electron balance is zero when oxidation and reduction half-reactions are added to obtain an overall reaction. The method is addressed here as an introduction and does not provide sufficient detail for an operational capability.

The half-reaction method applied to biochemical reactions was started by Professor Perry L. McCarty (e.g., McCarty, 1965, 1975; Christensen and McCarty, 1975) and is summarized in texts, e.g., Orhon and Artan (1994, pp. 86–107) and Rittman and McCarty (2001, pp. 132–161). The summary that follows was abstracted mostly from Rittman and McCarty (2001, pp. 132–161), which also provides representative half-reactions, R_a for the electron acceptors, R_c for cell synthesis, R_d for the electron donors. An overall reaction is given as (Rittman and McCarty, 2001, p. 143),

$$R = (1 - f_s)R_a + f_sR_c - R_d \qquad (22.15)$$

where

R is the overall reaction, written on an electron equivalent basis (e-eq.)

R_a is the half-reaction for electron acceptor, e.g., oxygen, nitrate, etc.

R_c is the half-reaction for cell synthesis, e.g., $C_5H_7O_2N$

R_d is the half-reaction for electron donor (the "substrate"), e.g., glucose

f_e is the fraction of electron equivalents from electron donor used for energy,

for which $f_e = (1 - f_s)$

f_s is the fraction of electron equivalents from electron donor used for cell synthesis

To illustrate the application of Equation 22.15, the reduction of nitrate to nitrogen gas is shown with benzoate, $C_6H_5COO^-$, as a carbon source, as abstracted from Rittman and McCarty (2001, p. 143), using half-reactions as given in their tables,

$$f_eR_a\text{: } 0.45 \cdot [1/5NO_3^- + 6/5H^+ + e^- \rightarrow 1/10N_2 + 3/5H_2O]$$

$$f_sR_c\text{: } 0.55 \cdot [1/28NO_3^- + 5/28CO_2 + 29/28H^+ + e^- \rightarrow 1/28C_5H_7O_2N + 11/28H_2O]$$

$$\underline{-R_d\text{: } 1/30C_6H_5COO^- + 13/30H_2O \rightarrow 1/5CO_2 + 1/30HCO_3^- + H^+ + e^-.}$$

$$R\text{: } 0.0333C_6H_5COO^- + 0.1096NO_3^- + 0.1096H^+ \rightarrow 0.0196C_5H_7O_2N + 0.045N_2 + 0.0333HCO_3^- + 0.1018CO_2 + 0.0528H_2O$$

As seen, the half-reactions are each for one electron-equivalent (e-eq), and the addition of the three half-reactions must result in zero electrons for the final equation, R. The coefficients in the final equation, R, are mol of each reactant and product for an electron-equivalent (since the overall equation balance is based upon zero resultant electrons). The final equation can be "normalized" about any reactant or any product by dividing by the respective coefficient. For example, to normalize about benzoate, divide the whole equation by 0.0333. The resultant equation also gives the stoichiometric requirements for the reaction (2.63 kg benzoate/kg NO_3^- as N),

$X/(0.0333 \text{ mol benzoate} \cdot 121 \text{ g benzoate/mol benzoate}) = 1 \text{ g } NO_3^- \text{ as N}/(0.1096 \text{ mol N} \cdot 14 \text{ } NO_3^- \text{ as N/mol N})$

$$X = 2.63 \text{ kg benzoate}/(\text{kg } NO_3^- \text{ as N})$$

22.3.3.5.1 Calculation of f_s and f_e: Synopsis

In the foregoing example, the fraction split between f_s and f_e was assumed (keeping in mind that $f_s + f_e = 1$). Actually, however, the fractions used for synthesis and energy, f_s and f_e, are unknown, but can be determined as described by Rittman and McCarty (2001, pp. 154–161). In their method, the free energy of the electron donor (e.g., glucose) oxidation, ΔG_r, is determined from tabular data. The value of ΔG_s is the ΔG from the electron donor half-reaction (benzoate to pyruvate) and the ΔG of the cell synthesis reaction (obtained from available data and an empirical relation, respectively); each of the latter is divided by a factor, ε, which is an efficiency factor for the respective energy conversions. From this a factor, A, is calculated, which is the fraction of electron donor that must be oxidized to provide the free energy to synthesize one equivalent of cells, ΔG_s. The two factors, ε, which is assumed and A, which is calculated, provide a basis for energy accounting, which is given by Rittman and McCarty (2001, pp. 156) as,

$$\varepsilon A \Delta G_r + \Delta G_s = 0 \qquad (22.16)$$

where

ε is an energy transfer efficiency factor where $0.55 \le \varepsilon \le 0.70$; a typical value is $\varepsilon \approx 0.6$ (dimensionless)

A is the fraction of electron donor that must be oxidized to provide the free energy to synthesize one equivalent of cells (dimensionless)

ΔG_r is the free energy released per equivalent of electron donor oxidized for energy generation under physiological conditions, which is within about 1% of the standard free energy of reaction, (kJ/e-eq); to illustrate, for the oxidation of ethanol, CH_3CH_2OH, $\Delta G_r \approx -111$ kJ/e-eq (p. 154) (Rittman and McCarty, 2001, p. 154).

ΔG_s is the free energy required to synthesize one equivalent of cells under physiological conditions (kJ/e-eq)

In other words, the product, εA, represents the fraction of energy of the reaction, ΔG_r, converted to the energy content of the cells, ΔG_s. To sum up, the energy expended, ΔG_r, results in the synthesis of ATP; the ATP energy when converted to ADP provides the energy for cell synthesis. As with any chemical reaction, a portion of the ΔG_r is lost as heat and entropy increase. Also, as seen, although A is obtained by calculation per Equation 22.16, the "ε" factor must be assumed, which makes it an uncertainty factor.

As a final step, the factors f_s and f_e, which permit an electron balance when adding the R_c and R_a equations, respectively, are calculated, for standard conditions, as, $f_s^{\circ} = 1/(1+A)$ and since $f_s^{\circ} + f_e^{\circ} = 1$, $f_e^{\circ} = A/(1+A)$. All of the foregoing is intended for a conceptual understanding of how the two equation multipliers, i.e., f_s and f_e, may be determined, but does not necessarily provide a computational capability, which is given by Rittman and McCarty (2001) and Orhon and Artan (1994). The ability to obtain a stoichiometrically valid overall equation for cell synthesis for any substrate (for which tabular ΔG° data are available) means that cell yield, Y, can be calculated along with substrate COD. Thus, Y units may be determined as (1) mol cells/mol substrate; (2) g cells/mol substrate; (3) g cells/g substrate; (4) g cells/g substrate COD (see Rittman and McCarty, 2001, p. 158).

22.4 CELL YIELD

The idea of cell yield, Y, has been around since the mid-1950s. It has been generated empirically from laboratory data and can be obtained from plant-operating data. The theoretical basis is the balanced reaction. It is defined as the cell mass synthesized per unit of substrate reacted, which is the *true* cell yield, Y. Another definition is, as the *observed* cell mass generated per unit of substrate degraded, which is the net yield, Y(net).

Although the concept of cell yield is simple, its use is complicated artificially by the fact that the mass of cells and the mass of substrate each may be assessed by a number of measures, e.g., moles, mass in metric units, VS, BOD, COD, TOC. In activated sludge, the cell mass is usually measured by the surrogates, MLSS and MLVSS, or by a spin test. Therefore, a variety of combinations are possible and used.

22.4.1 CELL-YIELD CALCULATION

Any balanced reaction equation may be the basis for the calculation of cell yield and other relationships. To put the issue in perspective, the relationships are already given in terms of moles. To convert from moles to mass involves obtaining the molecular weights (MW) of the substances involved. To illustrate, Orhon and Artan (1994, p. 87) give the aerobic reaction,

$$\begin{array}{llll} 1.0\ \text{g} & & x\ \text{g} & \\ 8CH_2O + 3O_2 + NH_3 & \rightarrow & C_5H_7O_2N + 3CO_2 + 6H_2O & \\ 8 \cdot 30\ \text{g/mol}\ \ 3 \cdot 32\ \text{g/mol} & & 113\ \text{g/mol} & (22.17) \end{array}$$

Associated calculated quantities include

$Y(\text{VSS}) = 113/(8 \cdot 30) = 0.47$ g cell VSS synthesized/g substrate utilized

$Y(O_2) = (3 \cdot 32)/(8 \cdot 30) = 0.40$ g O_2 utilized/g substrate reacted

$Y(O_2) = (3 \cdot 32)/(113) = 0.84$ g O_2 utilized/g cell VSS synthesized

For the oxidation of substrate, the reaction is

$$\begin{array}{ll} 1.0\ \text{g} & \\ 8CH_2O + 8O_2 \rightarrow 8CO_2 + 8H_2O & (22.18) \\ 8 \cdot 30\ \text{g/mol}\ \ 3 \cdot 32\ \text{g/mol} & \end{array}$$

which gives the capability to calculate quantities in terms of COD,

$f(\text{substrate COD}) = (8 \cdot 32)/(8 \cdot 30) = 1.06$ g COD/g substrate

$Y(\text{VSS/COD}) = (0.47$ g cell VSS/g substrate$)/(1.06$ g cell COD/g substrate$) = 0.44$ g cell VSS/g substrate COD

$Y(\text{VSS/BOD}) = (0.44$ g cell VSS/g substrate COD$)/(0.7$ g BOD/g COD$) = 0.63$ g cell VSS/g substrate BOD

The term, f(substrate COD) is the fraction of substrate that is COD in a unit mass of substrate. Relevant points are the following: (1) the stoichiometric equation provides the basis for a number of calculations; (2) the calculated quantities may be configured in units that may be useful; (3) the variation in the configuration of units may be confounding; and (4) recommended units are, "g cell COD/g substrate COD" (Orhon and Artan, 1994, p. 87). The principles involved in making calculations based on stoichiometric equation apply to any reactions, e.g., aerobic, anaerobic, anoxic, and nitrification. Total organic carbon (TOC) has also been suggested. For reference, BOD/COD ≈ 0.7 is a "rule of thumb" and varies from one substrate to another. Tchobanoglous and Burton (1991, p. 83) give $0.4 \leq$ BOD/COD ≤ 0.8, and $1.0 \leq$ BOD/TOC ≤ 1.6.

A final point implicit in Equation 22.18 is that the synthesis quantity of new cells is exactly tied to the degradation of substrate. The relation is depicted in a very simple stoichiometric relation, i.e.,

$$\frac{\Delta X}{\Delta S} = Y \qquad (22.19)$$

where

X is the cells synthesized by the biochemical reaction (e.g., mol cells, kg cells, COD cells, etc.)
S is the substrate, which is degradable organic matter, ammonia, etc., utilized by the bacterial enzymes that enable a reaction (mol substrate, kg substrate, COD substrate, etc.)
Y is the stoichiometric coefficient, usually called the "true yield" (e.g., mol cells synthesized/mol substrate degraded)

Equation 22.19 is utilized often in the literature as an empirical relation. As noted, the units for X and S may be selected based on their utility for a given problem.

Example 22.2 Convert Cell-Yield to Y(COD Cells/COD Substrate); Orhon and Artan (1994, p. 88)

Given
Reaction equations as listed in sequence.

Required
Determine cell yield with units (g cell COD)/(g substrate COD)

Solution
The procedure is given by Orhon and Artan (1994, p. 66). The units give identity to each term and following their cancellations facilitates explanation.

1. *Cell synthesis*

$$8CH_2O + 3O_2 + NH_3 \rightarrow C_5H_7O_2N + 3CO_2 + 6H_2O$$
$$8 \cdot 30\ g/mol \qquad\qquad 113\ g/mol \qquad (22.8)$$

Associated calculated quantities include
$Y = 113/(8 \cdot 30) = 0.47$ g cell VSS synthesized/g substrate utilized

2. *Oxidation of substrate*
For carbohydrate substrate, i.e., CH_2O, the oxidation equation is (Table 22.3)

$$8CH_2O + 8O_2 \rightarrow 8CO_2 + 8H_2O$$
$$8 \cdot 30 \qquad 8 \cdot 32$$

$$f(\text{substrate COD}) = (8 \cdot 32)/(8 \cdot 30) = 1.06\ \text{g COD/g substrate}$$

a. Cell yield mass per unit of substrate COD

$$Y(\text{cells/substrate COD}) = Y/f(\text{substrate COD})$$
$$= (0.47\ \text{g cell synthesized/g substrate utilized})/(1.06\ \text{g COD/g substrate})$$
$$= 0.44\ \text{g cells synthesized/g substrate COD}$$

3. *Oxidation of cells*
For biomass COD, the oxidation equation is

$$C_5H_7O_2N + 5O_2 \rightarrow 5CO_2 + NH_3 + 2H_2O \qquad (22.20)$$
$$113 \qquad 5 \cdot 32$$
$$f(\text{cell COD}) = (5 \cdot 32)/(113) = 1.42\ \text{g COD/g cells}$$

a. Cell yield as COD of cells per unit of substrate COD

$$Y(\text{COD cells/COD substrate})$$
$$= Y(\text{cells/substrate COD}) \cdot f(\text{cell COD})$$
$$= 0.44\ \text{g cells synthesized/g substrate COD} \cdot 1.42\ \text{g cell COD/g cells}$$

Discussion
The COD equivalent of the substrate, i.e., CH_2O, expended for energy and biomass is $8O_2$, with $3O_2$ used in the oxidation of CH_2O for energy; the rest is converted to cells, i.e., $C_5H_7O_2N$, which has a COD equivalent of $3O_2$ (Orhon and Artan, 1994, p. 89). Eckenfelder and Weston (1956, p. 19) also calculated, based on stoichiometric equations, 1.42 g cell COD/g cells based on cell material, $C_5H_7O_2N$. Further, they plotted experimental data (1956, p. 20) for sludge COD versus sludge VS for domestic sewage, pulp and paper waste, and pharmaceutical wastes; the plot showed 1.42 g sludge COD/g VS (the 19 data points showed very little scatter about the about the best fit line).

22.4.2 Cell Maintenance and Endogenous Respiration

Simultaneously with cell synthesis a fraction of the cells are consumed by "endogenous-respiration." This occurs especially when the cells are in a "starved" condition, which may be defined as a substrate concentration that results in a cell-division rate significantly lower than the "enzyme capacity." On the other hand, "cell maintenance" seems to be a necessary function of a cell, as opposed to a cell death. The two aspects are discussed in separate paragraphs to allow for the possible distinction.

22.4.2.1 Cell Maintenance

The maintenance function includes cell motility, rebuilding of proteins, transfer of solutes across the cell wall, etc. (Orhon and Artan, 1994, p. 74; Rittman and McCarty, 2001, p. 131). In this reaction the cells in suspension may consume a portion of the ATP energy available for cell synthesis (keep in mind that the ATP is present within the cell as a part of the biochemical cycles). Or the cells could oxidize a fraction of their own cell matter. It is not clear whether the energy comes from a fraction of the ATP energy available for synthesis or from the oxidation of the cell's own matter (Benefield and Randall, 1980, p. 54).

22.4.2.2 Endogenous Respiration

As described by Porges et al. (1956, p. 44), "After the cells are formed, there is a slow continuous oxygen requirement.

The organisms oxidize their own tissue to supply the energy of maintenance." The equation was given as

$$C_5H_7O_2N + 5O_2 \rightarrow 5CO_2 + NH_3 + 2H_2O \quad (22.20)$$

Based upon oxygen utilization rate studies with a "Warburg" apparatus, they determined that $d(O_2)/dt = 14.3$ mg O_2/g cell/h, which converts to $dX/dt = -10.2$ mg cell/h, or about 0.01 fraction of cell mass consumed per hour. Although endogenous respiration has been a continuing consideration in biological treatment, the foregoing description by Porges et al. (1956) remains valid.

In a later article, McKinney and Symons (1964, p. 441) refer to maintenance energy as "endogenous respiration," indicating that they are the same. To confirm this point they cite research based on C-14 labeled organisms that endogenous respiration proceeds uninterrupted at the same rate in the presence of an external source of substrate as in its absence. They further state that they require "a basic amount of energy per unit of active mass per unit time during growth as during starvation," which is sometimes called "basal metabolism."

Another depiction of endogenous respiration that is complementary to the foregoing was given by Grady et al. (1999, p. 98); the following summary is excerpted:

$$b \cdot X(\text{viable cells}) \rightarrow X(\text{nonviable cells})$$

$$X(\text{nonviable cells}) \rightarrow X(\text{lysed cells}) + X(\text{debris cells})$$

$$X(\text{lysed cells}) \rightarrow X(\text{lysed-cell substrate}) + X(\text{lysed-cell debris})$$

$$X(\text{lysed-cell substrate}) + E(\text{bacteria}) + O_2 \rightarrow X(\text{new cells}) + E(\text{bacteria}) + CO_2$$

To summarize, a fraction of the cells lyse, with a portion of the matter becoming substrate, and a portion becoming inert "debris." Also, some fraction of the viable cells is consumed by predation (e.g., by protozoa or rotifers). To illustrate another facet of endogenous respiration, if the substrate level declines to zero, the cell protoplasm is the only source of energy and the net rate of cell growth is zero; it then declines in proportion to the remaining viable cell concentration (see also, Sherrard, 1977, p. 1969).

22.4.2.3 Microbial Growth Curve and Debris Accumulation

In the context of reactors, if the system is a batch reactor with zero inflow and zero outflow, and if the system starts "fresh," i.e., with adequate substrate, then a typical "growth curve" results, which includes a lag phase, exponential growth, declining growth and stationary growth. Both the concentrations, X(debris cells) + X(lysed-cell-debris), increase along the growth curve. If the system is a steady-state continuous-flow reactor the concentration sum, X(debris cells) + X(lysed-cell-debris), is constant with value depending on the θ_c, the cell retention time in the system, as is X(net), which is the observed cell mass concentration. In other words, the cell debris accumulates to some equilibrium value.

22.4.3 Net Cell Yield, Y(net)

The mathematical relations for net cell yield, Y(net) are given, first, in terms of a mass relation and, second, in terms of a rate relation. Regarding nomenclature, the ΔX(net) and ΔX(obs) are the same, i.e., the net change in cell mass is the observed change.

22.4.3.1 Cell Mass Relations

The idea of net cell synthesis that was tied to cell synthesis and endogenous respiration was given by Eckenfelder and Weston (1956, p. 21), i.e.,

$$\Delta X(\text{obs}) = Y \cdot S - b' \cdot X(\text{obs}) \quad (22.21)$$

where
- X(obs) is a viable cell concentration as observed (mg cells that can reproduce/L)
- S is the substrate concentration (mg cells/L)
- Y is the cell yield (mg cells synthesized/mg substrate degraded)
- b' is the fraction of cells rendered nonviable (mg cells rendered nonviable/mg viable cells)

Dividing both sides of Equation 22.21 by ΔS gives

$$\frac{\Delta X(\text{obs})}{\Delta S} = Y - b' \cdot \frac{X(\text{obs})}{\Delta S} \quad (22.22)$$

to give an expression for Y(obs), i.e.,

$$Y(\text{obs}) = Y - b' \cdot \frac{X(\text{obs})}{\Delta S} \quad (22.23)$$

where $Y(\text{obs}) = \Delta X(\text{obs})/\Delta S$ (observed mg cells synthesized/mg substrate degraded).

In other words, the "observed" yield, Y(obs), is the true yield of viable cells, Y, minus the cells lost to endogenous decay. As seen in Equation 22.22, if $\Delta X(\text{obs}) = 0$, then $Y = b' \cdot X(\text{obs})/\Delta S$.

22.4.3.2 Cell Mass Rate Relations

The actual *observed* cell growth per unit of substrate, also called the "net" growth, X(net), was expressed in terms of rate by Rittman and McCarty (2001, p. 131),

$$\frac{dX(\text{net})}{dt} = Y \cdot \left[-\frac{dS}{dt}\right] - b \cdot X(\text{net}) \quad (22.24)$$

Dividing each term by $(-dS/dt)$,

$$Y(\text{net}) = \frac{dX(\text{net})/dt}{-dS/dt} \quad (22.25)$$

$$= Y - b \cdot \frac{X(\text{net})}{-dS/dt} \quad (22.26)$$

where

- $X(\text{net})$ is the viable cell concentration actually observed in a suspension (mg viable cells/L suspension)
- $Y(\text{net})$ is the observed yield coefficient, usually called the "net yield" (mol cells synthesized that are observed in suspension/mol substrate degraded)
- b is the decay rate of cells, an empirical coefficient (s^{-1}); for reference, the complete units are (mg cells rendered nonviable/s/mg viable cells)

Equation 22.25 defines $Y(\text{net})$ as the observed rate of cell growth divided by the rate of substrate utilization. If $[dX(\text{net})/dt] = 0$, meaning also that $Y(\text{net}) = 0$, the right side of Equation 22.26 may be rearranged, i.e.,

$$\frac{-dS/dt}{X(\text{net})} = \frac{b}{Y} = m \qquad (22.27)$$

where m is the "maintenance" level of substrate utilization rate per unit of cells in suspension (mg substrate utilized/L suspension/s)/(mg cells observed/L suspension).

If $[(-dS/dt)/X(\text{net})] < m$, then the cells cannot maintain their metabolic functions and the cells "starve" (Rittman and McCarty, 2001, p. 131), utilizing their own protoplasm, which is "endogenous respiration." Empirical values for b for domestic wastewater range from 0.01 to 0.14 (BOD basis) and from 0.016 to 0.068 (COD basis); see also Orhon and Artan (1994, p. 122).

22.4.4 Decline in Degradable VSS

As "sludge age," θ_c, increases, there is more time for decay of the cell mass and $Y(\text{net})$ decreases, as given by Tchobanoglous and Burton (1991, p. 388),

$$Y(\text{net}) = \frac{Y}{1 + b\theta_c} \qquad (22.28)$$

Based on mass balance and kinetics, the degradable fraction, f_d, of the viable cell-mass concentration was calculated for a food-processing wastewater (Eckenfelder, 2009, p. 240),

22.4.5 Cell-Yield Data

Table 22.8 provides data on cell yield, Y, as obtained from a stoichiometric equation for the oxidation of domestic wastewater to obtain microbial cells (based on methods given by Orhon and Artan, 1995).

22.5 KINETICS OF BIOLOGICAL REACTIONS

The rate of reaction may be expressed as the rate of conversion of substrate, dS/dt, or the rate of synthesis of cells, dX/dt. The current kinetic model, the Monod equation, known in the field since about 1952, was adopted by the late 1960s, and remains current. Its application has been limited by the uncertainty regarding coefficient values, Y, k_d, $\hat{\mu}$, and K_s.

22.5.1 Monod Description of Biological Reactions

The rate of substrate conversion is, as seen in Equation 22.28, related stoichiometrically to the rate of cell synthesis. Therefore, the rate of substrate utilization is related to the rate of cell synthesis by the yield coefficient, Y,

$$\left[\frac{dX}{dt}\right]_g = Y \cdot \left[\frac{dS}{dt}\right] \qquad (22.29)$$

The rate of cell synthesis, $[dX/dt]_g$, is a "first-order" reaction proportional to the cell concentration, i.e.,

$$\left[\frac{dX}{dt}\right]_g = \mu X \qquad (22.30)$$

where

- X is the cell concentration (mg cells/L suspension)
- t is the time (s)
- μ is the specific cell growth rate (s^{-1}); more common units are (h^{-1}) or (day^{-1}); a more complete designation for the units would be, for example kg COD substrate utilized/s/kg COD of cells in suspension; the units for the mass of substrate and the mass of cells may be other combinations (kg BOD of substrate utilized per second/kg MLVSS of suspension, both measured with reference to a given volume of suspension, e.g., 1.0 m^3 in SI units)

In 1942, Jaques Monod (1949) published a paper that described the coefficient, μ, as

$$\mu = \hat{\mu}\frac{S}{K_s + S} \qquad (22.31)$$

where

- $\hat{\mu}$ is the maximum specific cell growth rate (s^{-1})
- K_s is the substrate concentration at one-half maximum growth rate (mg substrate/L suspension)

Combining Equations 22.29 through 22.31 gives

$$\frac{dS}{dt} = \frac{1}{Y}\hat{\mu}\frac{S}{K_s + S}X \qquad (22.32)$$

Equation 22.31 was applied for a carbon substrate, with concentration, S. It has general applicability, however, with respect to any essential nutrient, ammonia, phosphorus, or any other substance that could be rate limiting for microbial growth.

22.5.2 K_s as the Half-Saturation Constant

The K_s constant in the Monod equation is called the "half-saturation" constant. It is the value of the substrate concentration for the condition when $\mu = \hat{\mu}/2$. For edification, the relation is illustrated graphically in Figure 22.3. It occurs when the bacterial enzymes present are "half-saturated" with substrate. At full saturation of the enzymes, $\mu \rightarrow \hat{\mu}$, in accordance with the Langmuir isotherm (see Appendix 22.B).

TABLE 22.8
Examples of Calculation of Cell-Yield and Other Values for Stoichiometric Equation for Oxidations of Substrates to Obtain Microbial Cells

Substrate **Domestic Wastewater, Formula: $C_{10}H_{19}O_3N$**

Reaction equation obtained from Orhon and Artan (1994, p. 101) based on half-reactions compiled in their Table 2.12

Assumption: $Y = 0.67$ e-eq. cells/e-eq. substrate

$1/50C_{10}H_{19}O_3N + 1/12O_2 + 1/75NH_4^+ + 1/75HCO_3^- \rightarrow 1/30C_5H_7O_2N + 7/150CO_2 + 25/150H_2O$(equation for one e^- eq)

$1/50 \cdot 201$ $\quad 1/12 \cdot 32$ $\quad 1/30 \cdot 113$

Cell yield and other values

(1) Y($C_{10}H_{19}O_3N$-stoichiometric) = 0.64 g cell COD/g substrate COD (Orhon and Artan, 1994, pp. 98)

(2) Y($C_{10}H_{19}O_3N$-experimental) = 0.66 g cell COD/g substrate COD (Orhon and Artan, 1994, pp. 98)

(3) Y($C_{10}H_{19}O_3N$-experimental) = 0.46 g cell VSS/g substrate COD (Orhon and Artan, 1994, pp. 98)

(4) f(conversion) = (1.42 g cell COD/g cell VSS) (Grady et al., 1999, p. 70)

(5) Y($C_{10}H_{19}O_3N$-stoichiometric) = 0.94 g cells synthesized/g substrate degraded (calculated)

(6) Y(O_2-stoichiometric) = 0.66 g O_2 used/g substrate degraded (calculated)

(7) Calculation of substrate COD based on one e-eq electron transfer in reaction:

$$(1/50 \cdot 201 \text{ g substrate/e-eq})/(8 \text{ g COD/e-eq}) = 0.5025 \text{ g substrate/g substrate COD}$$
$$= 1.99 \text{ g substrate COD/g substrate}$$

Substrate **Pentose, Formula: $C_5H_{10}O_5$**

1. ***Cell synthesis reaction equation***

Obtained from Orhon and Artan (1994, p. 103) based on half-reactions compiled in their Table 2.12.

Assumption: $Y = 0.67$ e-eq. cells/e-eq. substrate.

$1/20C_5H_{10}O_5 + 1/12O_2 + 2/60NH_4^+ + 2/60HCO_3^- \rightarrow 2/60C_5H_7O_2N + 7/60CO_2 + 13/60H_2O$

$1/20 \cdot 150$ $\quad 1/12 \cdot 32$ $\quad 2/60 \cdot 113$

2. ***Cell yield and other values***

(1) Y($C_5H_7O_2N$-stoichiometric) = 0.50 g cells synthesized/g substrate degraded

(2) Y(O_2-stoichiometric) = 0.36 g O_2 used/g substrate degraded

3. ***Oxidation of cells to get f (cell COD/g cells)***

Oxidation reaction for cells to get COD equivalent. Overall equation from half-reactions by Orhon and Artan (1994, p. 88); equation coefficients are for transfer of one e-eq needed to balance the two half-reactions.

$$1/20C_5H_7O_2N + 5/20O_2 \rightarrow 5/20CO_2 + 1/20NH_3 + 2/20H_2O$$

$1/20 \cdot 113$ $\quad 5/20 \cdot 32$

f(cell COD/g cells) = (5/20 · 32)/(1/20 · 113) = 1.42 g cell COD/g cells

4. ***Oxidation of pentose substrate to get f (substrate COD/g substrate)***—(see also Benefield and Randall, 1980, p. 73)

$$1/20\ C_5H_{10}O_5 + 1/4O_2 \rightarrow 1/4\ CO_2 + 1/4\ H_2O$$

$1/20 \cdot 150$ $\quad 1/4 \cdot 32$

f(substrate COD/g substrate) = (1/4 · 32)/(1/20 · 150) = 1.07 g substrate COD/g substrate

5. ***Calculation of Y (g cell COD/g substrate COD)***

Y(g cell COD/g substrate COD) = f(g cell COD/g cells) · [1/f(substrate COD/g substrate)] · Y(g cells/g substrate)
= (1.42 g cell COD/g cells) · [1/(1.07 g substrate COD/g substrate)] · (0.50 g cells/g substrate)
= 0.66 g cell COD/g substrate COD

22.5.3 Net Specific Growth Rate, μ(net)

The net increase $[dX/dt]_{net}$, in viable cell mass is the difference between growth rate and depletion rate:

$$\left[\frac{dX}{dt}\right]_{net} = \left[\frac{dX}{dt}\right]_g - \left[\frac{dX}{dt}\right]_d \quad (22.33)$$

$$= \mu X - bX \quad (22.34)$$

$$= (\mu - b)X \quad (22.35)$$

Dividing both sides of Equation 22.35 by X gives (Lawrence, 1975, p. 223)

$$\frac{[dX/dt]_{net}}{X} = \mu - b \quad (22.36)$$

And if we define,

$$\mu(\text{net}) \equiv \frac{[dX/dt]_{net}}{X} \quad (22.37)$$

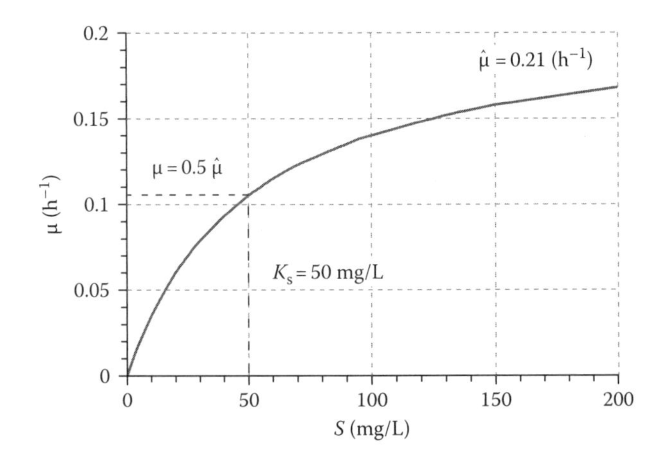

FIGURE 22.3 Illustration of K_s as the "half-saturation" constant.

then

$$\mu(\text{net}) = \mu - b \tag{22.38}$$

where μ(net) is the net kinetic coefficient, i.e., net specific growth rate, or the "observed" rate of change of viable-cells concentration per unit mass concentration of cells (s^{-1}); or (kg X observed change/m^3/s/)/(kg cells/m^3).

22.5.4 Temperature Effect

The effect of temperature on the reaction rate constant for a first-order reaction is (Daniels and Alberty, 1955, p. 341)

$$k = A \exp\left[-\frac{\Delta H_a}{RT}\right] \tag{22.39}$$

or

$$\log k = -\frac{\Delta H_a}{RT} + \log A \tag{22.40}$$

where
- k is the rate of reaction in Arrhenius model, which is the same as $\hat{\mu}$ (s^{-1})
- A is the intercept in plot of log k vs. $1/T$ (s^{-1})
- ΔH_a is the enthalpy of activation, which is the "energy hump" (J/mol)
- T is the absolute temperature (K)

The Arrhenius equation (see also, Glossary) serves as a model for the Monod constant, $\hat{\mu}$; in other words for a biological reaction, $\hat{\mu}$ may be used instead of k, which is applicable within the temperature limits for organism functioning. The interchange between k and $\hat{\mu}$ is seen as justified within the Michaelis–Menten description of enzyme kinetics, Appendix 22.B. The Arrhenius model, $\log \hat{\mu} = [\Delta H_a/2/303R] \cdot [1/T] + \log A$, may be used as a basis for plotting the temperature effect. The slope of the plot is $[\Delta H_a/2/303R]$, which is negative, and the intercept is "log A." The activation energy, ΔH_a, depends on the substrate and the organism; therefore, the slope is a practical fit and is treated usually as an empirical constant.

The Arrhenius equation is also given in the form, $k_2/k_1 = \exp(K(T_2 - T_1))$, where k_1 and k_2 are kinetic constants and K is a coefficient (Benefield and Randall, 1980, p. 12). The actual mathematical solution actually has a denominator, T_1T_2, which is incorporated into the constant, K. Also, for a biological reaction, $\hat{\mu}$'s may be used instead of k's. This form is mentioned because it is common in the literature.

22.5.5 Evaluation of Kinetic Constants

To make the kinetic equations operational it is necessary to determine the constants, $\hat{\mu}$, K_s, k_d, and Y. These data are not easy to obtain. Most of the available data were generated from laboratory experiments using a pure substrate, such as glucose, lactose, maltose, etc. In addition, the units for different data sets may be not compatible.

One approach to generate kinetic constants is to run a pilot plant under different conditions of X and S. The pilot plant should be run long enough to establish steady-state conditions. The materials balance equation, combined with the kinetic equation, permits the determination of needed constants. The experiments should systematically vary [S] so that a number of points are available. Schulze (1965) has illustrated the procedure in handling such data. Figure 22.4a shows some of his data plotted in accordance with Equation 22.31. To evaluate K_s and $\hat{\mu}$, however, Equation 22.31 can be restated in a linear form:

$$\frac{[S]}{\mu} = \frac{K_s}{\hat{\mu}} + \frac{[S]}{\hat{\mu}} \tag{22.41}$$

The corresponding plot of the Schulze data in terms of Equation 22.41 is given as Figure 22.4b. The K_s and $\hat{\mu}$ values determined from Figure 22.4b are 95.4 mg/L and 1.09 h^{-1} respectively. Schulze used a laboratory-scale system with glucose as the substrate and an *E. coli* culture.

22.5.5.1 Data on Kinetic Constants

Kinetic data are difficult and laborious to generate and are not abundant in the literature. Further, the measures are not uniform, which makes the data not easily comparable. Table 22.9 shows data obtained from literature and are provided for reference and may be useful as a starting point in various calculations. Examples may include estimating biological solids generated and minimum cell residence time.

22.5.6 Andrews/Haldane Model of Substrate Inhibition

A modified Monod model, the Haldane equation was identified by Andrews in 1968 for use in situations involving inhibitory substrate (Suidan, 1988; Grady et al., 1999, p. 81),

$$\mu = \hat{\mu}\frac{S}{K_s + S + S^2/K_i} \tag{22.42}$$

where K_i is the inhibition coefficient (mg substrate/L).

BOX 22.4 KINETICS IN ACTIVATED SLUDGE

One of the first inklings that reaction stoichiometry and Monod kinetics had entered activated-sludge reactor theory was a paper by Garrett and Sawyer (1952, p. 51), who referred to the relationship between cell growth and the removal of substrate and, in words (p. 75), described a loading parameter, $S_o/\Delta t)/X$, as kg substrate added per day divided by the mass of viable cells under aeration, which is the F/M ratio, i.e., $F/M \equiv S_o/\theta X$ (Tchobanoglous and Burton, 1991, p. 389). They also described the Monod equation but did not utilize it in analysis of their data. The idea was extended by Eckenfelder and Weston (1956, pp. 21–24), who showed a relation between net cell production and substrate utilization, $\Delta X = Y\Delta S$, Equation 22.19, a key part of a kinetic model.

In his doctoral thesis Andrew Gram III integrated the complete-mix concept (pp. 11, 12) and the Monod kinetics equation (p. 20) into a reactor mass balance model (Gram, 1956). Gram confirmed the validity of the Monod equation (p. 122) for his model, but in describing results, he deferred (p. 121) mostly to the "loading velocity" $(S_o - S)/(\theta \cdot X)$, which is defined currently as the "substrate utilization rate," U (Tchobanoglous and Burton, 1991, p. 388), which is the same as μ/Y.

Gram did not publish his work, but the Monod equation was applied by Stewart et al. (1959) as related to his doctoral research at Berkeley completed in 1958 under Pearson. But neither the reactor theory nor Monod kinetics were integrated into a complete model until done by Professor Erman Pearson in a paper given in April 1966 for a "special lecture series" held at the University of Texas at Austin (Pearson, 1968) and published as a book in 1968 (Gloyna and Eckenfelder, 1968). Pearson delineated a reactor model that was based on the mass balance principle and Michaelis–Menten kinetics. He also included the derivative parameters, substrate-removal velocity, U (derived from substrate mass balance and defined as $U \equiv (S_o - S)/\theta \cdot X = \mu/Y$); and sludge age, θ_c (derived from cell mass balance and defined as $\theta_c = VX/WX_r = \mu - b$). These emerged as parameters for practice; a major point is that they have their theoretical origins in substrate and cell mass balances, respectively, and kinetics. The parameters were delineated further with respect to design and operation by Lawrence and McCarty (1970), which has been the primary reference on the topic. Their paper was selected as the 1985 AEESP Outstanding Paper defined as "having stood the test of time."

As K_i for a particular substrate increases, implying a high tolerance for the substrate, the S^2/K_i term approaches zero and the Andrews/Haldane equation approximates the Monod equation. The Andrews/Haldane equation is most useful when describing the degradation of xenobiotic (synthetic organic) compounds under aerobic and/or anaerobic conditions. The inhibition coefficient K_i is, however, difficult to determine. In most cases, the Monod equation is sufficient to approximate the observed growth curve in the absence of an inhibitor; then the effect of the inhibitor can be seen by its incremental effect on $\hat{\mu}$ and K_s (Grady et al., 1999, p. 82).

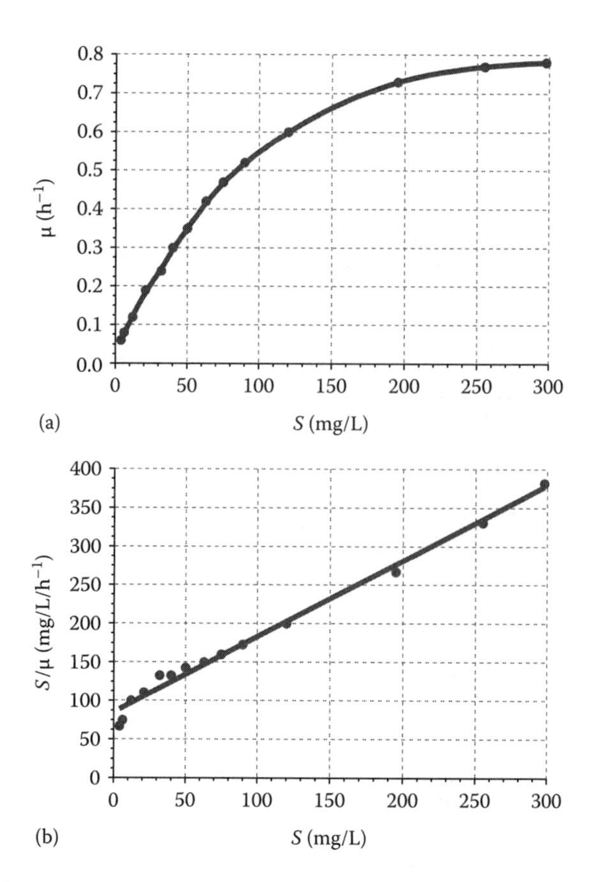

FIGURE 22.4 Relationship between substrate concentration and growth rate, μ. (a) Monod equation plot and (b) Linearized Monod plot (Adapted from Schulze, K.L., The activated sludge process as a continuous flow culture, Part I, Theory, 111(12), 526, December 1964; Part II, Application, *Water and Sewage Works*, 112(1), 11, January 1965.)

22.5.7 Kinetic Parameters

Several well-known parameters U, F/M, θ_c, θ_c^m, are used in activated sludge. The parameters are defined here since they are related to the "native" kinetic constants, μ and Y. Showing the relationships may help in relating to theory. The topic is considered further in Chapter 23 where reactor mass balances and kinetics are reviewed together.

22.5.7.1 Specific Substrate Utilization Rate, *U*

A design parameter is the specific substrate utilization rate, U, defined as (Tchobanoglous and Burton, 1991, p. 388)

$$U \equiv \frac{S_o - S}{\theta X} \tag{22.43}$$

TABLE 22.9
Kinetic Data

				Coefficients			
Source	**Basis**	**Substrate**	**Culture**	**Y (mg Cells/mg Substrate)**	**K_s (mg/L)**	**$\hat{\mu}$ (day^{-1})**	**b (day^{-1})**
Schulze (1965)		Concentration	Glucose	*E. coli*	0.44	95	26
Heukelekian (1951)	BOD	Domestic waste		0.50			0.055
Gram (1956)	BOD	Skim milk		0.48	100	2.5	0.045
Pearson-Haas (1965)	COD	Sewage		0.45			0.05
Stack and Conway (1959)	BOD	Glucose		0.42	355	1.26	0.09
Lawrence (1975, p. 252)	COD	Domestic wastewater	Activated sludge	0.35–0.45 mg VSS/mg COD	25–100 mg/L COD	2–4	0.05–0.10
Lawrence (1975, p. 254)	NH_3 oxidation	NH_3	Activated sludge	0.05 mg/mg N	1.0 mg N/L	0.33	ND
Gaudy and Gaudy (1980)	COD	Glucose	Sewage	0.62	390	7.7	ND
		Glucose	Sewage		100	9.1	ND
		Lactose		0.47			ND
		Sucrose		0.53			ND
	COD	Sewage	Sewage		63	12	ND
Benefield and Randall (1980)	NH_4^+ as N	NH_4^+		0.05 mg VSS/mg NH_4^+	0.5–2.0	0.3–0.5	ND
Hao et al. (2009)	MLSS	Synthetic wastewater with COD = 400 mg/L	Seeding from Beijing WWTP; 30 day operation				0.05
Hao et al. (2009)	NH_4^+ as N	200 mg NH_4^+ as N/L; to determine b, zero NH_4^+ as N/L	Seeding from Beijing WWTP; 30 day operation; then nitrifiers were predominant				0.08
Parker et al./EPA (1975)	NH_4^+ as N	NH_4^+	*Nitrosomonas* in suspended growth reactor	Y(net) ≈ 0.15 kg cells/NH_4^+ as N removed	0.6 mg NH_4^+ as N/L at 20°C	0.18 exp 0.116 (T-15) day^{-1} ≈ 0.32 day^{-1} at 20°C	ND
Parker et al./EPA (1975)	NO_3^- as N	NO_3^-/methanol	*Nitrobacter* in suspended growth reactor	$Y(\text{net}) = Y/(1 + b\theta_c)$; $Y = 0.6$–1.2 kg VSS/kg NO_3^--N removed;	0.08 mg NO_3^--N/L	1.04 (20°C)	0.04
Tchobanoglous and Burton (1991, p. 394)	Units as given	Domestic wastewater		0.6 mg VSS/mg BOD_5	60 mgBOD_5/L 40mgCOD/L	5	0.06

where

- U is the specific substrate utilization rate (kg substrate degraded/m^3/s)/(kg viable cells/m^3)
- S_o is the substrate concentration entering the reactor (mg substrate/m^3)
- S is the substrate concentration leaving the reactor (mg substrate/m^3)
- θ is the hydraulic detention time in reactor, V/Q (s)
- X is the viable cell concentration in reactor (kg cells/m^3)

If U is given and if S_o, S, and X are specified, then the reactor may be sized based on θ (since $V(\text{reactor}) = Q \cdot \theta$).

As may be seen by the substrate mass balance relation for a reactor, Chapter 23, Section 23.2.2.1, U is related to μ/Y; the mass reactor mass balance is, $Q(S_o - S) = [dS/dt]V$; and since $[dS/dt] = (1/Y)[\mu X]$, then rearranging gives $(\mu/Y) = (S_o - S)/(\theta \cdot X)$, or

$$U = \frac{\mu}{Y} \tag{22.44}$$

In other words, although U has been used as an empirical parameter, it is related to fundamental kinetic constants (see also Tchobanoglous and Burton, 1991, p. 392), The relation may be seen also by going back to the definition, $dX/dt = \mu X$, and since $dX/dt = Y(-dS/dt)$, then $(\mu/Y) = (-dS/dt)/X$, which is the differential from of Equation 22.43.

Example 22.3 illustrates the calculation of U from typical design and operating data for an activated sludge reactor. As seen, the criteria cover a wide range, indicating that experience and judgment are also involved. To obtain a lower U, the options are not many and include a larger aeration basin, e.g., let $\theta \approx 8$ h, or a higher MLVSS, which may not be feasible due to practical limits based on recycle rates, and the fact that the active mass declines as sludge age increases. Another approach is to try to improve the settling in the final clarifier to increase X_r, and thus reduce the recycle ratio. But, as is evident, several factors are interrelated, which points to the use of a spreadsheet.

Example 22.3 Calculate *U* from Design and Operating Data for an Activated-Sludge Reactor

Given

$Q = 1.0$ m^3/s (22.8 mgd)
$S_o = 280$ mg BOD/L (0.280 kg/m^3)
$S = 20$ mg BOD/L (0.020 kg/m^3)
$\theta = 6$ h
$X \approx 2000$ mg MLVSS/L (2.00 kg/m^3)

Required

Estimate U.

Solution

$$U \equiv \frac{S_o - S}{\theta \cdot X} \qquad (22.43)$$

$$= \left[(0.280 - 0.020)\ \text{kg/m}^3\right]/\left[6\ \text{h} \cdot (2.00\ \text{kg/m}^3)\right]$$

$$\approx 0.5\ \text{(kg substate degraded/d)/(kg MSVSS in reactor)}$$

Discussion

1. Estimate of μ/Y.
From Table 22.9, select $K_s = 60$ mg/L, $\hat{\mu} = 5$ day^{-1}, gives, $\mu = 5 \cdot [20/(60+20)] = 1.25$ day^{-1}; and $Y \approx 0.5$ kg VSS synthesized/kg BOD degraded; $\mu/Y = 0.62$ kg BOD degraded/day/kg VSS synthesized.
2. As seen, $U < \mu/Y$, or in numerical form, $0.5 < 0.6$.
3. Note that X, which theoretically is in terms of cell mass, is given as MLVSS. The units for the Monod constants and Y given are in terms of BOD and MLVSS, which are consistent with the data given.
4. For comparison, Tchobanoglous and Burton (1991, p. 534) give $0.05 < F/M < 1.0$.
5. As another comparison, Rittman and McCarty (2001, p. 330) give $0.2 < F/M < 0.5$ kg BOD degraded/day/kg MLSS in reactor.
6. The reactor sizing, based on θ would be $V(\text{reactor}) = Q \cdot \theta$. Another issue is that the Q varies, usually in a sinusoidal pattern, over a 24 h cycle, and is subject to seasonal variation, and an overall increase year by year (depending on location).
7. For such an apparently simple problem, a number of uncertainties are pertinent.

22.5.7.2 *F/M* Ratio

A well-known parameter, F/M, is defined as "mass substrate added per day divided by the mass of viable cells under aeration." It has been around since the Garrett and Sawyer (1952) article, Box 22.4, but became generally well known after a paper by McKinney (1962), which, arguably, "modernized" thinking about treatment, i.e., in terms of reactors, mass balance, kinetics. In mathematical terms the F/M ratio is defined (Tchobanoglous and Burton, 1991, p. 389),

$$\frac{F}{M} \equiv \frac{S_o}{\theta X} \qquad (22.45)$$

where F/M is the food-to-microorganism ratio (kg substrate/m^3 entering reactor/day)/(kg of cells in reactor volume).

Example 22.4 Relating *F/M* Word Definition to Mathematical Definition

Given

F/M ratio is defined, in words, as the "mass substrate added per day divided by the mass of viable cells under aeration," and mathematically as $F/M \equiv S_o/\theta X$.

Required

Relate the word definition of F/M to the terms in the equation.

Solution

$F/M \equiv$ "mass substrate added per day divided by the mass of viable cells under aeration"
$= [Q\ (\text{m}^3/\text{day}) \cdot S_o\ (\text{kg substrate/m}^3)]/V(\text{reactor})\ (\text{m}^3) \cdot X(\text{kg cells/m}^3)]$
$\equiv S_o/\theta X$

Discussion

The exercise may seem trivial but it may be confounding to try to relate the daily mass flux of substrate to the equation which is in terms of the entering substrate concentration S_o; cell concentration in the reactor, X; and hydraulic detention time, θ (where $\theta \approx 6$ h). The important thing is to express Q as m^3/day.

22.5.7.3 Conversion *F/M* to *U*

The conversion to U is given (Tchobanoglous and Burton, 1991, p. 389) as

$$\frac{F}{M} \equiv \frac{S_o}{\theta X} \cdot E \qquad (22.46)$$

$$U = \left[\frac{S_o}{\theta X}\right] \cdot \left[\frac{S_o - S}{S_o}\right] \qquad (22.47)$$

where E is the process efficiency, defined $E \equiv (S_o - S)/S_o$.

The reason for showing the conversion is so that the terms are seen as not necessarily independent conceptually. In practice, ideas continue because of tradition or utility. In engineering it is necessary to understand the science and the practice and, as feasible, to see the relationships.

22.5.7.4 Relating Monod Kinetics to U

Going back to the stoichiometric relationship between cell synthesis and substrate reacted, i.e., $dX/dt = Y(dS/dt)$, Equation 22.29, and since $dX/dt = \mu X$, Equation 22.30, then if we equate $\mu X = Y(dS/dt)$, we may rearrange to give

$$\frac{\mu}{Y} = \frac{dS/dt}{X} \tag{22.48}$$

In other words, $(dS/dt)/X$ is a micro-version of the "specific substrate utilization rate," and is an identity with U the macro-version. Importantly, it shows that

$$\frac{\mu}{Y} = U \tag{22.49}$$

It means that the plant loading equals the capacity of the microorganisms, μ/Y, to process the substrate. See also Tchobanoglous and Burton (1991, pp. 371, 388).

22.5.7.5 Sludge Age, θ_c

Another parameter common in practice is "sludge age," or "mean cell residence time," or "solids retention time," defined as (Lawrence, 1975, p. 224)

$$\theta_c = \frac{\text{total-active-mass-in-system}}{\text{total-active-mass-leaving-system-per-day}} \tag{22.50}$$

$$\theta_c = \frac{V(\text{reactor}) \cdot X}{[WX_r + (Q - W)X_e]} \tag{22.51}$$

As a rule, for conventional activated sludge, $X_e \ll X_r$. and so Equation 22.51 may be approximated as

$$\theta_c = \frac{V(\text{reactor}) \cdot X}{WX_r} \tag{22.52}$$

where

θ_c is the mass of viable cells in system/rate of cell synthesis (kg viable cells in system/kg cell synthesized/day)
W is the waste flow (removed from recycle flow in an activated-sludge reactor) (m^3/s)
X_r is the concentration of viable cells in recycle flow (kg cells/m^3)
X_e is the concentration of viable cells in flow leaving final clarifier, i.e., $(Q - W)$ (kg cells/m^3)

For reference, and as may be seen from Section 22.5.3 Net Growth, $X/[dX/dt]_{net} = 1/(\mu - b)$,

$$\theta_c \equiv \frac{1}{\mu - b} \tag{22.53}$$

22.5.7.6 Minimum Cell Regeneration Time, θ_c^m

The time for cell generation is designated, θ_c^m and is defined further as

$$\theta_c^m \equiv \frac{1}{\hat{\mu} - b} \tag{22.54}$$

where θ_c^m is the minimum cell residence time.

Notes on the foregoing are:

- θ_c is an operating parameter and is controlled by the W, the rate of wasting, which is based approximately on a routine. Usually the operator monitors the MLSS, a surrogate for X, and fine tunes W, based on whether the MLSS trend is increasing or decreasing. W may be an intermittent flow, rather than continuous. A centrifuge test is used, as a rule, in practice as a surrogate for MLSS.
- In operation, $\theta_c \gg \theta_c^m$; otherwise the cells will leave the system at a rate that exceeds their generation rate.
- The waste flow, W, is taken from the recycle flow, R, with concentration, X_r.
- As a guideline, $3 < \theta_c < 15$ day (Tchobanoglous and Burton, 1991, p. 534). From Example 22.3, for $\hat{\mu} = 5\ \text{day}^{-1}$, and $\mu = 1.25\ \text{day}^{-1}$, and if $b \approx 0.1$, $\theta_c^m \approx 0.2$ day and $\theta_c \approx 0.9$ day, respectively. In operation, they recommend that $\theta_c \gg \theta_c^m$.
- As $b \to 0$, then $\mu \to \hat{\mu}$, and $1/\hat{\mu} = \theta_c^m$.

22.5.7.7 Relationship between U and θ_c

The relationships between θ_c, F/M and U are (Tchobanoglous and Burton, 1991, p. 534)

$$\frac{1}{\theta_c} = Y \cdot \frac{F}{M} \cdot E - b \tag{22.55}$$

By definition, $F/M \equiv S_o/\theta X$ and $E \equiv (S_o - S)/S_o$; with the substitutions,

$$\frac{1}{\theta_c} = Y \cdot \left(\frac{S_o}{\theta X}\right) \cdot \left(\frac{S_o - S}{S_o}\right) - b \tag{22.56}$$

$$= Y \cdot \left(\frac{S_o - S}{\theta X}\right) - b \tag{22.57}$$

$$= YU - b \tag{22.58}$$

22.5.8 Nitrification/Denitrification

The Monod equation is applicable to any substrate in a biochemical reaction. In nitrification, the substrate is NH_3 as N; in denitrification, the substrate is NO_3^- as N. Values for the yield coefficient, Y, and the kinetic constants $\hat{\mu}$, K_s, and b, are given in Table 22.9, as obtained from the literature.

22.5.8.1 Nitrification: NH_4^+ to NO_3^-

The nitrification rate depends upon several factors including ammonia concentration; cell concentration, X(nitrifiers);

temperature; dissolved-oxygen concentration; and pH (Parker et al./EPA 1975, pp. 3-6–3-29). The kinetic rate constant, μ, for ammonia follows the Monod equation, i.e., with $C(NH_4^+)$ as the S. The cell concentration, X(nitrifiers), depends on the fraction, f(nitrifiers), of X (the MLSS in the reactor), which depends on the BOD/TKN ratio (Parker et al./EPA 1975, pp. 3–22). Oxygen is a reactant, per the stoichiometric equation, Equation 22.9, and affects the specific growth rate as defined by a "secondary" Monod equation, i.e., by the factor $[C(O_2)/(K_s(O_2)+C(O_2)]$, or $\mu \approx 0.19 \cdot [C(O_2)/(2.0+C(O_2)]\ \text{day}^{-1}$; the curve starts to level at $C(O_2) > 4$ mg O_2/L, with about 0.8 fraction of $\hat{\mu}$ at 4 mg/L. The effect of pH is due to the alkalinity requirement, per Equation 22.9, in which HCO_3^- is available at $7.0 < pH < 8.4$.

Nitrification is done most effectively in a separate nitrification activated-sludge reactor, then the cell-separator stage that follows the organic carbon (BOD) reactor and cell separator. In the design of the system, from $\theta_c \equiv 1/(\mu - b)$, then $\theta_c^m \equiv 1/(\hat{\mu} - b) \approx 1/(0.32\text{–}0.05)\ \text{day}^{-1} \approx 3.7$ day (note that $b \approx 0.05\ \text{day}^{-1}$ was assumed). The SRT(design) $> \theta_c^m$; thus, SRT(design) $>$ 3.7 day. Parker et al./EPA (1975, pp. 3-19–3-20) recommended SRT(design) $\geq$ 5 day, with an associated reduction in ammonia concentration >0.95 fraction.

A major factor in successful nitrification is that nitrifiers must comprise a significant fraction of the biomass, i.e., X, in the reactor. This requires a low BOD/TKN ratio in the reactor, calculated as BOD removed in mg/L and TKN oxidized in mg/L; to illustrate, if the numbers are 20 and 25, respectively, BOD/TKN $\approx$ 0.8. Further, if Y(net nitrifiers) $\approx$ 0.15 kg nitrifiers synthesized/kg NH_4^+ as N oxidized and if Y(net-cell-mass) $\approx$ 0.55 kg cells synthesized/kg BOD oxidized, then (f nitrifiers) $\approx (0.15 \cdot 25)/[(0.15 \cdot 25)+(0.55 \cdot 20)] = 0.25$. The low BOD/TKN ratio is achieved by adding a nitrifying reactor with sludge recycle, after the carbon reactor. In other words, to emphasize, the treatment train adds a nitrification reactor with a cell separator and recycle. For most systems, $1 \leq \text{BOD/TKN} \leq 3$, for which $0.21 > (f\text{-nitrifiers}) > 0.08$. Parker et al./EPA (1975, pp. 3–25) show that at 20°C, the nitrification rate, $U(N) = f(\text{nitrifiers}) \cdot [(\mu(\text{nitrifiers})/Y(\text{nitrifiers})] \approx$ (0.25 kg nitrifiers/kg all cells) $\cdot$ [0.32 day^{-1}/0.15 kg cells synthesize/kg NH_4^+-N oxidized] $\approx$ 0.53 kg NH_4^+-N oxidized/kg all cells/day. The "all cells" is measured as MLVSS. Thus, the rate of substrate utilization rate, U, depends to a large extent on the fraction of nitrifiers.

22.5.8.2 Denitrification: NO_3^- to N_2 Gas

Concerning denitrification, i.e., conversion of NO_3^- to N_2 gas, kinetic constants are given in Table 22.9. Parker et al./EPA (1975, pp. 3–40) use a secondary Monod relationship to calculate the effect of other variables on the maximum specific growth rate. For the effect of methanol as carbon source the relationship given was

$$\mu(\text{nitrifiers}) = \hat{\mu}(\text{nitrifiers})\frac{M}{K_M + M} \tag{22.59}$$

where

M is the methanol concentration (mg methanol/L)
K_M is the half-saturation constant for methanol (mg methanol/L)

They give $K_M = 0.1$ mg methanol/L; in other words an excess of ethanol above the stoichiometric requirement is not necessary as about 1 mg methanol/L in the reactor effluent is sufficient for the kinetic coefficient to approach the maximum. Concerning pH, the highest rate is in the range $7.0 \leq pH \leq .5$. As evident from Equation 22.10 the reaction is anoxic, which means that oxygen has a negative effect since it has a stronger affinity for electrons than does NO_3^-.

Since the reaction must occur without oxygen, an anoxic reactor must be added to follow the nitrification reactor. A carbon supplement is usually required.

22.5.9 Phosphorous Uptake

22.5.9.1 Occurrence in Wastewaters

The three forms of phosphorous in wastewater are orthophosphate, polyphosphate, and organically bound phosphate (Orhan and Artan, 1994, p. 471). The orthophosphates are easily assimilated by microorganisms. McCarty (1970, 227) gives the formulae of condensed phosphates as tripolyphosphate, $P_3O_{10}^{-5}$; and pyrophosphate as $P_2O_7^{-4}$, which must be hydrolyzed to the PO_4^{-3} form for assimilation; about 0.5–0.8 fraction are hydrolyzed during biological treatment (p. 228). The total P in wastewater ranges 6–20 mg/L and the orthophosphate, PO_4^{-3}-as-P ranges 4–15 mg/L (Orhan and Artan, 1994, p. 472). Table 22.2 provides additional resolution on concentrations and forms. Black and Veatch/EPA (1971, p. 2-1) gave a yearly mass P output per capita in municipal wastewater as 1.6 kg (3.5 lb) with about 0.5 kg (1.2 lb) from human excretions and about 1.1 kg (2.3 lb) from detergents with phosphates and gave an average total P in domestic raw wastewater as 10 mg P/L.

22.5.9.2 Uptake to Cells

Phosphorous is an integral part of cell matter such as DNA, RNA, ADP/ATP, etc., and therefore is an essential nutrient. From the P-based cell formula by Stumm, $C_{106}H_{180}O_{45}N_{16}P$, Equation 22.5, MW = 2429 g/mol and MW(P) $\approx$ 31, giving fraction(P) $\approx$ 0.013. Bowker and Stensel (EPA, 1987, p. 15) gave $0.015 < \text{fraction(P)} < 0.02$. They gave an estimated $0.10 < \text{fraction/removal(P)} < 0.30$. Their historical review referred to a 1955 finding that P uptake could exceed the cell fractions normally expected, and 1965 observations at a full-scale plant of $P \leq 1$ mg/L with adequate aeration, which was referred to as "luxury uptake." For such sludges, $0.02 < \text{fraction(P)} < 0.07$.

22.5.9.3 Theory

In the early 1970s the additional uptake was thought by some to have been due to chemical precipitation, which was not the case; the P was "stored" in the cells. Also, P release was observed in sludges under the anaerobic conditions in the final clarifier, with subsequent P increase in the reactor, i.e., higher than found in the raw wastewater. The anaerobically conditioned cells, in addition to the net release of P, were also characterized by having a higher rate of P uptake (as compared with cells not subject to an anaerobic environment). The foregoing observations led to the general approach in P removal: (1) promote "starved" microorganisms entering the reactor, and a "selection" of an organism such as *Acinetobacter*, that were amenable to the luxury uptake of P and at high rate; (2) get rid of the P either in the supernatant of sludge after being subjected to an anaerobic environment or in the sludge. To expand further, as abstracted from Bowker and Stensel (EPA, 1987, p. 17), the P removal was also observed in anoxic zones coincident with nitrate reduction (with N_2 as a product). As to mechanism of the luxury uptake of P, the bacteria showing this characteristic, stored the P as polyphosphates within "volutin" granules and is associated with the chemical, polyhydroxybutyrate (PHB).

22.5.9.4 Technologies

Based on the foregoing research, several approaches emerged to become technologies in biological treatment. Three of these processes, in order of development, are (1) the Phostrip process, (2) the modified Bardenpho process, and (3) the A/O process. As indicated they are proprietary. In the Phostrip process, a sidestream of settled sludge from the final clarifier is passed through an "anaerobic phosphorous stripper," where the supernatant is subjected to lime precipitation and leaves the system. The sludge is returned to the aeration basin. The other portion of the final clarifier underflow is handled normally with a portion wasted (WAS) and the other portion returned to the aeration basin (RAS) along with the sludge from the anaerobic P stripper. The modified Bardenpho process (p. 21) is both a nitrogen and P removal system. The raw-water inflow and the RAS are contacted in an anaerobic tank to promote fermentation and P release. Following this the flow passes through the four stage Barenpho process, i.e., anoxic, aerobic, anoxic, aerobic. In the first stage, NO_3^- from internal recycle from nitrification in the second stage, is reduced to N_2 gas utilizing the carbon in the raw wastewater (in lieu of an external carbon source), which removes about 0.7 fraction of the NO_3^- nitrogen. In the second stage, an aerobic stage, NH_4^+ oxidation, BOD removal, and P uptake occur. In the third stage, which is anoxic, additional denitrification occurs. In the fourth stage, which is aerobic, the aeration conditions the MLSS to minimize anaerobic condition in the final clarifier that may result in P release. The underflow from the final clarifier is handled as with conventional activated sludge, with a portion returned to the raw-water inflow (RAS) and a portion wasted (WAS). The A/O process has a sequence of four anaerobic stages followed by four aerobic stages, with RAS and WAS handled normally.

22.6 SUMMARY

Biological treatment processes, as stated, may be summarized in terms of reactions and kinetics. The three kinds of reactions include (1) heterotrophic, (2) autotrophic including nitrification and denitrification, and (3) anaerobic. A fourth reaction is endogenous respiration in which cell matter is oxidized. A key principle is that the cells synthesized are related stoichiometrically to the substrate degraded.

The Monod equation is the kinetic model, and has been adopted, universally since the late 1960s. A limitation is that the constants, $\hat{\mu}$, and K_s, are unique to the situation at hand. Data available are variable and depend upon the substrate and organisms and are not reported in consistent units.

Empirical parameters have been adopted for practice and include the substrate utilization rate, U, and the sludge age, θ_c. Each parameter has a rational basis.

Nitrification, i.e., conversion of ammonia to nitrate, follows the same principles as for carbon-based substrate degradation. The nitrifiers have a slower growth rate than heterotrophic organisms, and do not compete well. Therefore, a separate reactor is recommended to follow the carbon-substrate reactor, operated with sludge recycle. The nitrification reactions may be carbon limited and benefit by a supplemental carbon source such as methanol. Denitrification requires an anaerobic environment and a separate reactor. Biological phosphorous removal has evolved as a technology from observations in the 1950s to a rationale for practice by the early 1970s, with several proprietary technologies resulting.

PROBLEMS

22.1 Mass Balance for a Chemical Equation

Given

Equation 22.8 for oxidation of sugar and synthesis of cells.

$$8(CH_2) + 3O_2 + NH_3 \rightarrow C_5H_7NO_2 + 3CO_2 + 6H_2O$$

Required

Calculate the mass balance for each element for each side of Equation 22.4 for both a molar basis and for a mass basis.

22.2 Calculate Cell Yield, *Y* for Sugar as the Substrate

Given

Equation 22.8 gives an equation for cell synthesis, i.e.,

$$8(CH_2) + 3O_2 + NH_3 \rightarrow C_5H_7NO_2 + 3CO_2 + 6H_2O \quad (22.8)$$

MW 240,113

Required

Calculate the cell yield, Y.

22.3 COD of Cells

Given

The endogenous oxidation of a bacterial cell is given as Equation 22.30,

$$C_5H_7NO_2 + 5O_2 \rightarrow 5CO_2 + NH_3 + 2H_2O \quad (22.30)$$

Required

Calculate the chemical oxygen demand (COD) of bacterial cells.

22.4 Calculate Cell Yield, *Y*, and Oxygen Demand from Equation 22.4

Given

Sugar with the general formula, $8CH_2O$, reacts with oxygen and ammonia to form new cells and carbon dioxide and water, per Equation 22.4.

Required

Calculate the cell yield, the oxygen demand, and the ammonia metabolized.

22.5 Calculate COD and TOC of Microbial Cells from Equation 22.20

Given

Microbial cells with the empirical formula, $C_5H_7NO_2$ reacts with oxygen to form carbon dioxide, ammonia, and water, per Equation 22.20.

Required

Calculate the chemical oxygen demand (COD), total organic carbon (TOC), and nitrogen produced.

22.6 Calculation of P

Given

Activated sludge reactor with MLVSS ≈ 2000 mg/L

PO_4^{3-} as P = 15 mg/L

Required

Determine the uptake potential of P for a "normal" reaction, i.e., without "luxury" uptake.

ACKNOWLEDGMENTS

The author appreciates the detailed editorial review of Rachel E. Hanson, environmental process engineer, Golder Associates, Lakewood, Colorado. The manuscript was improved by a large measure due to her suggestions. The author is responsible for the manuscript.

APPENDIX 22.A: PROTEINS

22.A.1 Protein Molecules

A protein molecule, one of the components of municipal wastewater, comprises amino acids, which are the *building blocks*. An amino acid *skeleton* consists of an amino group, a carboxyl group, a hydrogen atom, and a distinctive R group bonded to a carbon atom, called the α-carbon. Figure 22.A.1 illustrates, showing the basic skeleton mentioned, common to all amino acids. For reference in reading the literature, two

COO^-
|
$^+HN_3$---C---H
|
R

FIGURE 22.A.1 Amino acid skeleton, comprising amino, carboxyl, hydrogen, R groups. (From Stryer, L., *Biochemistry*, W. H. Freeman, New York, 1981, p. 13.)

amino acids are joined by a peptide bond, also called an amide bond; the synthesis requires an input of energy. A sequence of amino acids, each joined by a peptide bond, forms a polypeptide chain with the amino group, $^+NH_3$, on one end and the carboxyl group, COO^-, on the other. The regularly repeating part is called the main chain, called the backbone, and the variable part has distinctive side chains. The synthesis of a protein molecule requires an input of energy since its Δ*G*(reaction) is positive.

The basic amino acids number about 20 (Stryer, 1981), with names such as alamine, arginine, asparagine, aspartic acid, cysteine, glutamine, glutamic acid, glycine, histidine, isoleucine, leucine, lysine, methionine, phenylalanine, proline, serine, threonine, tryptophan, tyrosine, and valine. From these building blocks, through the almost limitless combinations, the multitude of proteins is constructed. Figure 22.A.2 shows examples of four R groups, which illustrates the variety. These R groups, when combined with the skeleton structure shown in Figure 22.A.2, comprise the particular amino acids named above. As noted, the amino group of the skeleton is a source of ammonia, but in addition, some of the R groups have ammonia, as seen in Figure 22.A.2 for the R group of lysine. Note that the R group (the "side chain") for glycine is simply a hydrogen atom and that the R group for alanine is methane. A characteristic of proteins is that they have a well-defined three-dimensional structure, which is responsible for their respective functions.

The biosynthesis of amino acids, the building blocks of proteins, requires $[NH_4^+]$ as a reactant, with the reaction facilitated by the intermediates, glutamate and glutamine (Stryer, 1981). Of the basic set of 20 amino acids, 10 are synthesized by the citric acid cycle.

The surplus amino acids are used as metabolic fuel and converted to urea and other end products (Stryer, 1981, p. 407). One of the possible amino acid degradation reactions is

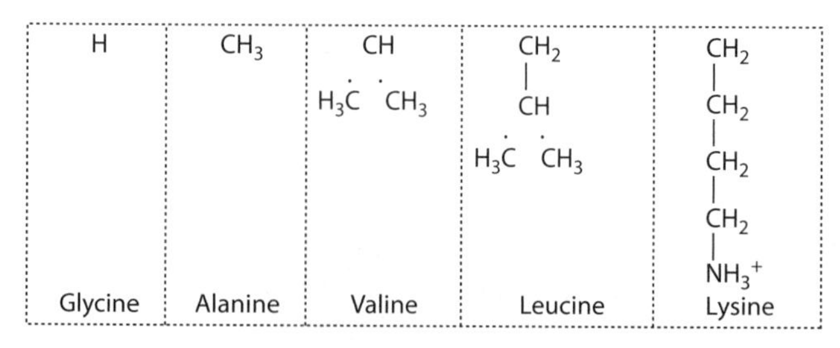

FIGURE 22.A.2 Examples of amino acid R groups.

$$\begin{array}{ccccc} & \text{H} & & \text{H} & \\ & | & & | & \\ {}^{+}\text{H}_3\text{N}\text{---} & \text{C} & \text{---R} \longrightarrow {}^{+}\text{H}_3\text{N}\text{---} & \text{C} & \text{---CH}_2\text{---COO}^- \longrightarrow \text{NH}_4{}^+ \\ & | & & | & \\ & \text{COO}^- & & \text{COO}^- & \\ & \text{amino acid} & & \text{glutamate} & \end{array} \quad (22.A.1)$$

Such degradation of amino acids, from protein molecules, occurs through the treatment process and within the activated-sludge reactor with ammonia as an end product. The strategy of amino acid degradation is to form major metabolic intermediates that can be converted to glucose or be oxidized by the Krebs cycle (Stryer, 1981, p. 415).

22.A.2 Urea

Another source of $[NH_4{}^+]$, in addition to the breakdown of proteins, depicted in Equation 22.A.1, is urea. Some of the $[NH_4{}^+]$ formed in the breakdown of amino acids is used in the biosynthesis of nitrogen compounds. In most terrestrial vertebrates, the excess $[NH_4{}^+]$ is converted into urea and then excreted. Humans excrete about 30 g/day of urea, an amount representing about 80%–90% of the total urine N (Bohinski, 1987, p. 679). The conversion is by the *urea cycle*, a metabolic pathway through several intermediate amino acids, which is also linked to the Krebs cycle. The structure of urea is

$$\begin{array}{c} \text{O} \\ \| \\ \text{H}_2\text{N---C---NH}_2 \end{array}$$

Urea

Bacteria which possess the enzyme *urease* may convert urea to ammonia and carbon dioxide as shown by Equation 22.A.2:

$$\begin{array}{c} \text{O} \\ \| \\ \text{H}_2\text{N---C---NH}_2 \end{array} + \text{HOH} \longrightarrow 2\text{NH}_3 + \text{CO}_2 \quad (22.A.2)$$

22.A.3 ATP

In driving synthesis, the conversion of ATP to ADP, the energy released to drive the biosynthesis is about $\Delta G_R{}^\circ$ (ATP $\rightarrow$ ADP) $\approx$ 7 kJ/mol. The respective structures are depicted (Stryer, 1981, p. 240),

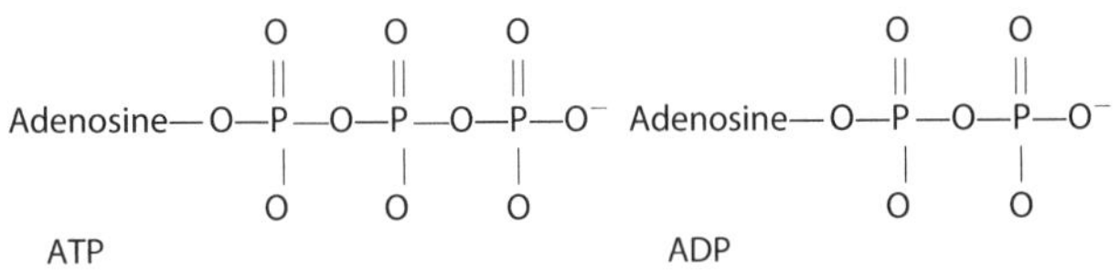

$$\underset{\text{ATP}}{\text{Adenosine—O—}\underset{\text{O}}{\overset{\text{O}}{\overset{\|}{\underset{|}{\text{P}}}}}\text{—O—}\underset{\text{O}}{\overset{\text{O}}{\overset{\|}{\underset{|}{\text{P}}}}}\text{—O—}\underset{\text{O}}{\overset{\text{O}}{\overset{\|}{\underset{|}{\text{P}}}}}\text{—O}^-} \qquad \underset{\text{ADP}}{\text{Adenosine—O—}\underset{\text{O}}{\overset{\text{O}}{\overset{\|}{\underset{|}{\text{P}}}}}\text{—O—}\underset{\text{O}}{\overset{\text{O}}{\overset{\|}{\underset{|}{\text{P}}}}}\text{—O}^-}$$

APPENDIX 22.B: MICHAELIS–MENTEN EQUATION

As noted, the Monod equation, $\mu = \hat{\mu}[S/(K_s + S)]$, Equation 22.31, was empirical, based on the observations of bacteria cultures. The Monod equation has been referenced in the literature more frequently than the earlier discovery in 1913 by Leonor Michaelis and Maud Menten, the latter being given in terms of enzyme kinetics (Stryer, 1981). Its derivation, given in many elementary texts, such as Stryer (1981) and Bohinski (1987), is worth reviewing in order to better understand its rationale.

22.B.1 Enzyme Kinetics

As with any kinetic formulation, the reaction is the starting point. The substrate-to-cell conversion reaction can be expressed as

$$\text{substrate} + \text{other reactants} + \text{enzymes} \leftrightarrow \text{substrate} \cdot \text{enzyme}$$
$$\rightarrow \text{cells} + \text{end products} + \text{enzymes} \quad (22.B.1)$$

where substrate · enzyme is the intermediate substrate–enzyme complex.

BOX 22.B.1 SEARCH FOR A SURROGATE FOR VIABLE CELLS

The total enzyme concentration, E_t, represents the *total catalytic activity* of the system, which is not a measurable quantity. Since viable cells provide the catalytic activity, then a measure of viable cell concentration would work as a good surrogate—or index.

In activated-sludge systems, mixed liquor volatile suspended solids, MLVSS, is an index of catalytic activity, easily measurable on a routine basis. The term does not, however, distinguish between viable and nonviable cells. An index more likely to be used is mixed liquor suspended solids, MLSS. The MLSS is easier to measure than MLVSS but it includes inert material as well, and so is another step removed from the ideal, a direct measure of catalytic activity. In making the transition from this theoretical ideal to operational kinetics, it is important to bear in mind these tacitly assumed proportionalities, i.e., [MLSS] $\approx$ a[MLVSS]; [MLVSS] $\approx a[E_t]$. The term X then may represent MLSS or MLVSS—whichever is the designation chosen as a system of measurement. The constants worked out, however, are associated with one or the other index and cannot be interchanged. The idea of using DNA as a surrogate has also been explored, starting in the 1960s, but has not supplanted the traditional measures. The use of COD was suggested by Orhan and Artan (1994).

Expressed symbolically, Equation 22.B.1 is

$$\mathrm{S} + \mathrm{R} + \mathrm{E} \underset{k_{-1}}{\overset{k_1}{\leftrightarrow}} \mathrm{E} \cdot \mathrm{S} \xrightarrow{k_2} \mathrm{X} + \mathrm{P} + \mathrm{E} \qquad (22.B.2)$$

where
R is the reactants; for example, O_2
E is the enzyme concentration
E · S is enzyme·substrate complex
X is the cells synthesized
P is the end products; for example, CO_2
k_1, k_{-1}, k_2 are the kinetic constants

Since the reactions depicted are hypothetical, there is no need to consider balancing them stoichiometrically. The following steps are based upon Equation 22.B.2 in terms of the associated kinetic expressions:

1. The rate of appearance of [ES] is

$$\frac{\mathrm{d[ES]}}{\mathrm{d}t} = k_1[\mathrm{E}]\,[\mathrm{S}] - k_{-1}[\mathrm{ES}] - k_2[\mathrm{ES}] \qquad (22.B.3)$$

2. The total enzyme in the system E_t is the sum of both combined and free enzyme:

$$[\mathrm{E_t}] = [\mathrm{ES}] + [\mathrm{E}] \qquad (22.B.4)$$

The total enzyme concentration, E_t, represents the *total catalytic activity* of the system, which is not a measurable quantity. Since viable cells provide the catalytic activity, then E_t should be proportional to the concentration of viable cells *X*.

3. Replace [E] in Equation 22.B.3 by Equation 22.B.4:

$$\frac{\mathrm{d[ES]}}{\mathrm{dt}} = k_1\{[\mathrm{E_t}] - [\mathrm{ES}]\}[\mathrm{S}] - k_{-1}[\mathrm{ES}] - k_2[\mathrm{ES}] \qquad (22.B.5)$$

4. Apply the "steady-state approximation" and get

$$0 = k_1\{[\mathrm{E_t}] - [\mathrm{ES}]\}[\mathrm{S}] - k_{-1}[\mathrm{ES}] - k_2[\mathrm{ES}] \qquad (22.B.6)$$

5. Solve for [ES] to give

$$[\mathrm{ES}] = \frac{[\mathrm{E_t}]\,[\mathrm{S}]}{\{(k_{-1} + k_2)/k_1\} + [\mathrm{S}]} \qquad (22.B.7)$$

6. Let

$$K_s = \frac{k_1 + k_2}{k_1} \qquad (22.B.8)$$

In other words, K_s is an equilibrium constant, and expresses the net effect of forward rates and backward rates of reaction.

7. Substitute Equation 22.B.8 in Equation 22.B.7:

$$[\mathrm{ES}] = \frac{[\mathrm{E_t}]\,[\mathrm{S}]}{K_s + [\mathrm{S}]} \qquad (22.B.9)$$

8. Express d*X*/d*t*, based upon [ES] Equation 22.B.2, as

$$\frac{\mathrm{d}X}{\mathrm{d}t} = k_2[\mathrm{ES}] \qquad (22.B.10)$$

9. Substituting Equation 22.B.9 in Equation 22.B.10 gives

$$\frac{\mathrm{d}X}{\mathrm{d}t} = k_2\frac{[\mathrm{E_t}]\,[\mathrm{S}]}{K_m + [\mathrm{S}]} \qquad (22.B.11)$$

10. Divide both sides of Equation 22.B.11 by $[E_t]$:

$$\frac{\mathrm{d}X}{\mathrm{d}t}/[\mathrm{E_t}] \equiv \mu \qquad (22.B.12)$$

11. Now also in Equation 22.B.11, let $[S] \gg K_s$, which gives

$$\left|\frac{\mathrm{d}X}{\mathrm{d}t} = k_2[\mathrm{E_t}]\right|_{[\mathrm{S}]>>K_m} \equiv V_{max} \qquad (22.B.13)$$

12. Substituting (22.B.13) in Equation 22.B.11 gives

$$\frac{\mathrm{d}X}{\mathrm{d}t} = V_{max}\frac{S}{K_m + S} \qquad (22.B.14)$$

13. Divide both sides of Equation 22.B.14 by $[E_t]$, and let

$$\hat{\mu} = \frac{V_{max}}{[\mathrm{E_t}]} \qquad (22.B.15)$$

14. The result is the Michaelis–Menten equation, which is the same as the Monod equation,

$$\mu = \hat{\mu}\frac{[\mathrm{S}]}{K_m + [\mathrm{S}]} \qquad (22.B.16)$$

It is easy to see, plotting Equation 22.B.9 as Figure 22.B.1, and Equation 22.B.14 as Figure 22.B.2, that the enzyme becomes

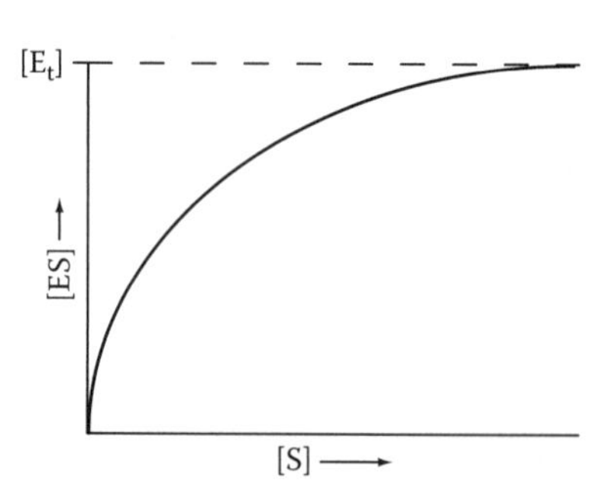

FIGURE 22.B.1 Enzyme isotherm plotted by Equation 22.B.9.

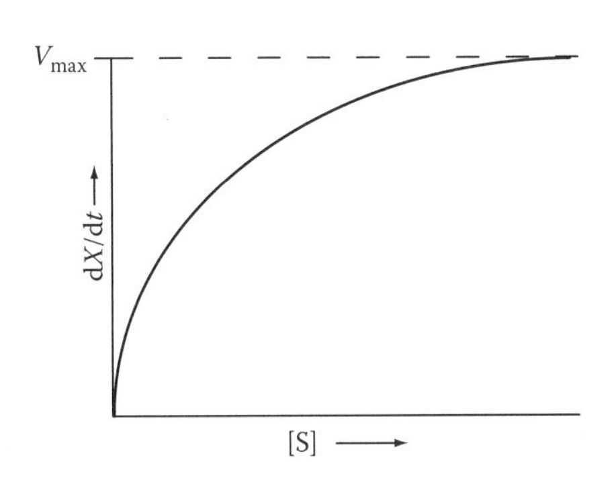

FIGURE 22.B.2 Enzyme reaction velocity plotted by Equation 22.B.14.

saturated with substrate as [ES] → [E_t], the reaction velocity, $dX/dt \rightarrow V$max. That is, when the enzyme becomes saturated the upper limit is reached in reaction velocity. In terms of specific reaction rate, this is $\hat{\mu}$. The form of Equation 22.B.16 is similar to the Langmuir isotherm equation and analogous to it in terms of adsorption sites available and sites occupied. In other words, as the enzyme becomes saturated with substrate, the reaction velocity, dX/dt approaches V_{max}.

GLOSSARY

Acetyl coenzyme A (acetyl CoA): A coenzyme A complex of acetic acid (Rittman and McCarty, 2001, p. 56).

Activation energy: (1) The energy required to overcome the "energy barrier" in an exergonic reaction, facilitated by enzymes. (2) See *Arrhenius*.

Activated sludge: The activated-sludge process offers a good example of controls on substrate supply and cell retention. Wastewater usually contains organic material, which serves as an electron donor, and adequate amounts of nitrogen and phosphorus. The substrate supply is controlled by aeration, which supplies the electron acceptor, oxygen. When enough oxygen is supplied, aerobic bacteria that oxidize organic matter grow and accumulate. Cell retention is controlled through the use of a clarifier that follows the aeration tank. The quiescent settling conditions of the clarifier favor retention of bacteria that aggregate into large flocs that settle out of the effluent flow. The settled flocs are recycled to the aeration tank, resulting in the buildup of the increasing amount of rapidly settling bacteria that oxidize the organic matter aerobically (Rittman, 1987). See also, *suspended-growth reactor*.

Active site: The part of an enzyme that binds the substrate to form an enzyme·substrate complex and catalyzes the reaction (Prescott et al., 1993, p. G1).

ADP: (1) Adenosine 5′ di-phosphate. The nucleoside diphosphate usually formed upon the breakdown of ATP when it provides energy for work (Prescott et al., 1993, p. G2). (2) The structure of ADP as adapted from Rawn (1989, p. 239) is

```
          O(−)     O(−)
          |        |
   (−)O—P—O—P—O—CH2
          ||       ||     |
          O        O      R
```

The R represents the adenosine part of the molecule consisting of three five-sided carbon rings.

Advanced wastewater treatment: Generally means any process, technique, or system not in common use (Anon., 1968, p. 1).

Aerobic bacteria: Bacteria that can use only oxygen acts as the electron acceptor; such bacteria are called "obligate aerobes."

Aerobic reaction: Oxygen acts as the electron acceptor.

Alga: Photosynthetic microorganisms lacking multicellular sex organs and conducting vessels (Prescott et al., 1993, p. G1). Algae react with dissolved carbon dioxide (based on equilibria between CO_2, H_2CO_3, and HCO_3^-) in the presence of sunlight to produce oxygen and new cells. In converting radiant energy from the sun to protoplasm, they are the base of the food chain.

Algae: Plural of alga.

Amino acid: A general formula for an amino acid is

```
        H  O
        |  ||
   R--αC--C--OH
        |
        NH2
```

The amino acids differ in the structure of the side chain, "R"; they contain an amino group on the α-carbon, i.e., the carbon adjacent to the carboxyl group. The "R" group may be any of about 20 structures, e.g., glycine, alanine, valine, leucine, isoleucine, serine, theonine, aspartic acid, glutamic acid, asparagines, glutamine, lysine, arginine, cysteine, cystine, methionine, phenylalanine, tyrosine, tryptophan, histidine, proline, hydroxyproline (Gaudy and Gaudy, 1980, p. 100–103).

Amino group: A basic amino group is $-NH_2$ (Rawn, 1989, p. 52).

Ammonification: The release of ammonia by the biodegradation of amino acids or other nitrogen-containing organic compounds (Grady et al., 1999, p. 49).

Anabolism: Biochemical pathways that result in production of cells. See *biosynthesis*.

Anaerobic: Organic compound serves both as electron acceptor and as electron donor. The biological system operates in the absence of free oxygen and the associated bacteria are referred to as "obligate anaerobes."

Anaerobic: The methanogenic fluidized-bed process offers a second illustration. Again, the wastewater contains an electron donor and sufficient N and P. In the methanogenic process, the supply of standard electron acceptors, i.e., O_2, NO_3^-, and SO_4^{2-}, is minimized. Without these electron acceptors, only fermenting and CO_2-reducing organisms can thrive. Included among those microorganisms are the methane-producing archaebacteria, the methanogens. Retention control is achieved by providing a very large surface area on the fluidized particles of materials such as sand, coal, or granular activated carbon. Cells able to attach to the small fluidized particles are retained and accumulated.

The controls on substrate supply and cell retention are a form of applied ecology. The composition and capabilities of the mixed population of microorganisms are regulated by controlling the availability of the substrate and the retention of cells in the process. Cells with the proper attributes fill the niche defined by the controls (Rittman, 1987).

Anaerobic reaction: (1) The organic molecule being metabolized serves as both the electron donor and electron acceptor (Prescott et al., 1993, p. 155).

Anoxic: (1) Electron acceptor is a mineral anion; NO_3^-, NO_2^-. SO_4^{2+}, are most common, with affinity in order stated. (2) Electron acceptor is a mineral anion, e.g., NO_3^-, NO_2^-. SO_4^{2+} with electron affinity in that order. This category of reaction is also called "anaerobic respiration" (Prescott et al., 1993, p. 158).

Archaea: Includes organisms that convert hydrogen and acetate to methane (methanogens); carbon dioxide is the electron acceptor for hydrogen oxidation (Rittman and McCarty, 2001, p. 21). For reference other main groupings are prokaryotic and eukaryotic.

Archaeobacteria: Bacteria that lack muramic acid in their cell walls, have membrane lipids with ether-linked branched chain fatty acids, and differ in many other ways from eubacteria (Prescott et al., 1993, p. G-3).

Aromatic compound: See Glossary in Appendix 2.A.

Arrhenius: (1) Svente Arrhenius (1859–1928), Professor of Chemistry, University of Uppsala, who in 1887 proposed in his doctoral dissertation the ion theory of solutions; the theory was not accepted until years later (Ehret, 1947). (2) The well-known "Arrhenius equation" is used commonly to account for the effect of temperature on the kinetic constant, i.e., $\ln k = -\Delta H_a/RT + \ln A$, where k is the reaction rate constant (s^{-1}); ΔH_a is the activation energy (J/mol); R is the universal gas constant (J/mol/K); T is the absolute temperature (K); A is the frequency factor (s^{-1}). The activation energy, ΔH_a, is the "energy hump" (Daniels and Alberty, 1955, p. 341). The equation form is also given as $k = A \exp(-\Delta H_a/RT)$; integration gives $\log(k_2/k_1) = (\Delta H_a/2.303RT) \cdot [(T_2 - T_1)/T_2T_1]$.

Attached growth reactor: (1) The bulk of the biomass is retained as a biofilm on fixed media such as rocks, lath, plastic cross-sheets. The media are characterized by high surface-area-to-volume ratio and large void spaces that facilitate the transport of "chunks" of sloughed-off biomass and airflow, to provide an aerobic environment within the reactor. See also *trickling filter*. (2) A "rotating biological contactor" (RBC) consists of an array of disks, perhaps 3 m (10 ft) diameter with spacing about 3 cm (1 in.), but is variable depending on the surface-area volume density. The disks are rotated by a shaft through a trough that contains the settled raw-water flow; as the submerged part of the disk emerges from the water the biofilm substrate is exposed to oxygen, which may then diffuse to a reaction site.

ATP: (1) Adenosine 5′-triphosphate; the triphosphate of the nucleoside adenosine, which is the high-energy molecule and serves as the cell's major form of energy currency (Prescott et al., 1993, p. G2). The breakdown of ATP to ADP and P_i, is given (Prescott et al., 1993, p. 136) as

$$\mathrm{ATP} + \mathrm{H_2O} \rightarrow \mathrm{ADP} + P_i \Delta G_R^\circ$$
$$= -30.5\ \mathrm{kJ/mol}\ (-7.3\ \mathrm{kcal/mol})$$

In other words, ATP gives up −30.5 kJ/mol in the reaction, which "drives" the cell-synthesis reaction during which energy losses occur (i.e., enthalpy and entropy).

(2) The structure of ATP as adapted from Rawn (1989, p. 239) is

```
        O(−)   O(−)   O(−)
         |      |      |
 (−)O—P—O—P—O—P—O—CH2
         ||     ||     ||      |
         O      O      O       R
```

The R represents the adenosine part of the molecule consisting of three five-sided carbon rings.

Autotroph: (1) Organisms that use carbon dioxide for cell synthesis (Rittman and McCarty, 2001, p. 16). (2) Organism that uses inorganic matter as substrate. (3) Organism that obtains its energy from inorganic compounds, such as ammonia, nitrite, nitrate. (4) Bacteria that use inorganic compounds as their electron donor and carbon dioxide as their source of carbon are called chemautotrophic bacteria, although most engineers call them autotrophs (Grady et al., 1999, p. 21). (5) An organism that uses carbon dioxide as its sole or principal source of carbon, such as an alga cell (Prescott et al., 1993, p. G2).

Bacterial cell composition: (1) The elemental cell composition for *E. coli* as an example is as follows: carbon, 0.50; oxygen, 0.20; nitrogen, 0.14; hydrogen, 0.08; phosphorous, 0.03; sulfur, 0.01; potassium, 0.01; sodium, 0.01; calcium, 0.05; magnesium, 0.05; chlorine, 0.05; iron, 0.02; others, 0.03. An empirical cell formula is $C_5H_7NO_2$. The oxidation of a cell is

$$C_5H_7O_2N + 5O_2 \rightarrow 5CO_2 + 2H_2O + NH_3$$
113 160
$$COD(cell) = 160/113 = 1.42 \text{ mg COD/mg cells}$$

Foregoing from Gaudy and Gaudy (1980, p. 59). (2) The empirical cell formula, $C_5H_7O_2N$, was given initially by Porges et al. (1956, p. 41), with MW ($C_5H_7O_2N$) = 113 g/mol; when ash was included they gave MW($C_5H_7O_2N$ + ash) = 124 g/mol. (3) A cell formula that includes N and P is $C_{60}H_{87}O_{23}N_{12}P$ (Sherrard, 1973, p. 1973).

Benzene ring: (1) A six-carbon ring that provides the structure for a multitude of compounds called "aromatic." (2) Benzene, C_6H_6, is the "parent" compound of the aromatic group (Sawyer and McCarty, 1967, p. 119).

Biochemical oxygen demand (BOD): The quantity of oxygen used in the biological oxidation of organic matter in a specified time, at a specified temperature, and under specified conditions (URS, 1973, p. K-1). The parameter is measured by a specified procedure. A sample is placed in several 300 mL BOD bottles, all previously aerated. The dissolved oxygen is measured in 1–3 bottles and the others are incubated at $T = 20°C$ for 5 days. After 5 days the dissolved-oxygen concentrations are measured in the incubated bottles. From the initial dissolved-oxygen concentration and the final, the 5 day BOD is calculated. For reference, the BOD test had been in use from some years when it was described by Metcalf and Eddy (1916, pp. 63–69).

Biochemical pathway: A complex sequence of reactions that degrade an organic molecule to certain end products. Examples include the citric acid cycle, also called the Krebs cycle or the TCA cycle.

Biological oxidation: The process whereby organisms in the presence of oxygen convert organic matter to carbon dioxide and new microorganisms (URS, 1973, p. K-1).

Biological process: (1) A biochemical reaction process that utilizes organisms to convert a given "substrate" to desired end products, such as new cells, carbon dioxide, water. If the reaction is anoxic, i.e., "without oxygen," the substrate may be ammonia, nitrite, or nitrate, with end products, nitrate, nitrite, and nitrogen, respectively. If the reaction is anaerobic, the substrate is an organic compound that is converted to carbon dioxide and methane.

Biological terms

Eukaryote: An organism consisting of a cell or cells in which the genetic material is DNA in the form of chromosomes contained within a distinct nucleus. Eukaryotes include all living organisms other than the eubacteria and archaebacteria. Compare with *prokaryote* (Apple Corporation, 2005).

Prokaryote: A microscopic single-celled organism, including the bacteria and cyanobacteria, which has neither a distinct nucleus with a membrane nor other specialized organelles. Compare with *eukaryote* (Apple Corporation, 2005).

Archaebacteria: Microorganisms that are similar to bacteria in size and simplicity of structure but radically different in molecular organization. They are now believed to constitute an ancient intermediate group between the bacteria and eukaryotes. Also called *archaea* (Apple Corporation, 2005).

Biomass: Cell mass.

Biosynthesis: The conversion of substrate to cells. Such conversion utilizes ATP, which is converted to ADP in the biosynthesis.

BNR: Acronym for biological nutrient removal, which refers to nitrogen and phosphorous. Although considered since the 1950s, the technologies for nitrogen and phosphorous removals emerged in practice during the 1980s.

BOD: Biological oxygen demand, which is understood, as a rule, to mean the 5-day BOD.

BOD_5: A 5-day biological oxygen demand based on a standard laboratory test in which the oxygen concentration in a BOD bottle is determined at $t = 0$ and again at $t = 5$ day after incubation at $T = 20°C$. The BOD_5 value is considered as being due to carbonaceous oxygen demand.

BOD(ultimate): Ultimate biological oxygen demand. The BOD test is extended to 20 days. The BOD(ultimate) value is considered as being due to the sum of carbonaceous oxygen demand and nitrogenous oxygen demand.

Bulking sludge: Sludge that is difficult to settle due to an excess of filamentous organisms or excess water.

Carbohydrate: (1) A class of compounds with the general formula, $C_n(H_2O)_n$ but which applies only to simple sugars, called monosaccharides (Campbell, 1991, p. 72). (2) Polymers built of repeating sugar molecules; glucose is the most common sugar (Campbell, 1991, p. 69). See also *glucose*.

Carboxyl group: An acidic carboxyl group is –COOH (Rawn, 1989, p. 52).

Catabolism: (1) The spontaneous conversion of reactants, e.g., proteins, polysaccharides, lipids to products. (2) An example of such conversion by aerobic means is given by Rawn (1989, p. 243), who shows three stages: (a) starting with proteins, polysaccharides, lipids, hydrolysis occurs to give amino acids, mono and disaccharides, and fatty acids and glycerol, respectively; (b) the foregoing products are converted by glycolysis to pyruvate and then to acetyl CoA giving off an ATP; (c) the acetyl CoA feeds into the citric acid cycle and is oxidized to give NADH, with products, ATP and NAD, and water. Other products, from Stage (a) are NH_3 (from amino acids), and carbon dioxide from Stage (b) and Stage (c). As contrasted to biosynthesis, catabolic pathways result in few end products (Rawn, 1989, p. 262). (3) A simple definition: the breakdown of nutrient molecules to provide energy (Campbell, 1991, p. 17).

Cell-yield: (1) Cell yield, Y, is a stoichiometric value; it is the mol of cells synthesized per mol of substrate reacted. Units are an issue in expressing Y since cells synthesized and substrate reacted may have units in terms of mass, volatile solids mass, COD, etc., and may have different combinations. (2) The preferred expression for Y is kg cell COD/kg substrate COD (Orhon and Artan, 1994, p. 87; Grady et al., 1999, p. 42). (3) Assuming a cell formula, $C_5H_7O_2N$, $Y = 1.42$ g cell COD/g cell VSS. If the ash content is assumed as 0.15 fraction, the value is $Y(\text{net}) = 1.20$ g cell COD/g SS of suspension (Grady et al., 1999, p. 42).

Chemical oxygen demand: See *COD*.

Chemoautotroph: Another term used for autotroph or chemotroph (see also *trophic*).

Chemostat: A continuous flow reactor used to attain steady-state equilibrium of microorganisms permitting the study of population changes and other factors (adapted from Jannasch and Egh, 1993, p. 214).

Citric acid cycle: (1) The citric acid cycle is considered the hub of metabolism and consists of eight enzyme-catalyzed reactions. It is the central pathway of aerobic metabolism. In most cells, the coupled reactions of the citric acid cycle and the respiratory electron-transport chain are responsible for the majority of energy production (Rawn, 1989, p. 329). (2) The substrate, the chemical entry point, for the cycle is acetyl-CoA which is oxidized after intermediate steps to carbon dioxide (Prescott et al., 2005, p. 178). (3) H. A. Krebs and W. A. Johnson completed the puzzle of the citric acid cycle in 1937; Krebs was awarded the Nobel Prize in 1953 for the discovery (Rawn, 1989, p. 330). (4) The citric acid cycle is also called the tri-carboxylic acid (TCA) cycle, or the Krebs cycle.

COD: (1) Chemical oxygen demand, which is the organic matter oxidized by di-chromate in accordance with Standard Methods. (2) The theoretical value of COD may be calculated if the balanced chemical equation is given. For example, consider the oxidation of ethanol,

$$C_2H_5OH + 3O_2 \rightarrow 2CO_2 + 3H_2O \Delta H^\circ = -295 \text{ kcal}$$

$$46 \quad\quad 96$$

$$COD = 96/46 = 2.09 \text{ mg } O_2/\text{mg } C_2H_5OH$$

By the same token,

$$TOC = 24/46 = 0.53 \text{ mg C/mg } C_2H_5OH$$

(Gaudy and Gaudy, 1980, p. 57). In other words, the COD approaches the theoretical oxygen demand as given by a balanced equation assuming that the substrate formula is known. (3) Some compounds are only partially oxidized, such as straight-chain acids, alcohols, amino acids; some are not oxidized, e.g., benzene, toluene, pyridine (Benefield and Randall, 1980, p. 73). (4) The COD is a measure of available electrons; the electron equivalent is 8 g COD/e-eq (where "e-eq" is an "electron equivalent" and is the basis for balancing equations using half-reactions), as seen in the half-reaction (Grady et al., 1999, p. 70), $\frac{1}{2}H_2O \rightarrow \frac{1}{4}O_2 + H^+ + e^-$. In other words, $\frac{1}{4} \cdot 32$ g O_2/mol O_2 per electron equivalent = 8 g COD/e-eq. Thus a Y value expressed per gram of COD can be converted to a Y value per available electron by multiplying by 8 (Grady et al., 1999, p. 42).

Coenzyme: Nonprotein substances that take part in enzyme reactions and are regenerated for further reaction. Two classes are (1) metal ions, and (2) organic molecules, such as vitamins. NAD^+ is a coenzyme in many redox reactions (Campbell, 1991, p. 248).

Constant: (1) A fundamental physical quantity; examples include the Boltzman constant, acceleration of gravity, and Planck's constant. (2) A nonchanging empirical value which constitutes a part of an equation, e.g., $y = ax + b$; the values, a and b are commonly termed "constants"; probably the term "coefficients" would be more appropriate so as to distinguish them from fundamental physical constants.

Contaminant: With reference to water, a contaminant is a substance in liquid water other than molecular water. As defined by PL93-523, the Safe Drinking Water Act of December 16, 1974, Section 1401: (3) The term *maximum contaminant level* (MCL) means the maximum permissible level of a contaminant in water, which is delivered to any user of a public water system. (6) The term "contaminant" means any physical, chemical, biological, or radiological substance or matter in water.

Cyst: An encapsulated form of protozoan that permits the organism's survival under adverse environmental conditions that do not permit metabolism. The organism is encapsulated in a shell. An example is the *Giardia lamblia* that forms a cyst after it is excreted and enters a nonsupportive environment.

Denaturization: (1) The disruption of an enzyme structure such that its activity is lost; it is caused by a temperature that is too high above the optimum or if the pH varies too much on either side of its optimum (Prescott et al., 2005, p. 159). (2) When a protein is exposed to conditions that deviate substantially from its normal environment, a structural change may occur that leaves the protein unable to function properly (Bailey and Ollis, 1977, p. 65).

Denitrification: (1) The conversion of nitrate, NO_3^-, to nitrogen gas, N_2. (2) Commonly, this conversion is by the bacterium, *Pseudomonas denitrificans*.

Detergent: An organic molecule other than soap, that serves as a wetting agent and emulsifier; it is normally used as a cleaner (Prescott et al., 1993, p. G8).

Diatoms: Photosynthetic microorganisms of the Chrysophyta group that contains silica.

DNA: (1) Deoxyribonucleic acid. (2) A linear polymer made up of deoxyribonucleotide repeating units (composed of the sugar 2-deoxyribose, phosphate, and a purine or pyrimidine base) linked by the phosphate groups joining at the 3′ position of one sugar and the 5′ position of the next sugar; DNA contains the genetic code (Rittman et al., 1990, p. 24). (3) Each molecule of DNA consists of two strands coiled around each other to form a double helix, a structure like a spiral ladder. Each rung of the ladder consists of a pair of chemical groups called bases (of which there are four types), which combine in specific pairs so that the sequence on one strand of the double helix is complementary to that on the other. It is the specific sequence of bases that constitutes the genetic information (Apple Corporation, 2005).

DNA: Double-stranded helix-shaped molecule that contains all the genetic information required for cell reproduction and contains in coded form all the genetic information needed to carry out cell functions. The unique genetic coding is in the particular sequence of four nitrogen bases, adenine, guanine, cytosine, and thymine. The information contained in the DNA is "read" and carried to the ribosomes for the production of proteins by RNA (Rittman and McCarty, 2001, p. 11).

DPN: Diphosphopyridine nucleotine, a coenzyme that facilitates electron transfer, or hydrogen transfer (McKinney, 1962, p. 54). DPNH is the same with a proton added.

Electron acceptor: (1) Matter reduced. (2) The final electron acceptor (Rittman and McCarty, 2001, p. 46) as a part of a complex biochemical pathway.

Electron carrier: Molecule that moves electrons from one compound to another (Rittman and McCarty, 2001, p. 46).

Electron donor: (1) Matter oxidized. (2) Primary electron donor. The initial compound that donates an electron (Rittman and McCarty, 2001, p. 46) in a complex biochemical pathway.

Embden–Meyerhoff pathway: (1) A biochemical pathway that degrades glucose to pyruvate (Prescott et al., 1993, p. G8). (2) Commonly called the glycolytic pathway, in which one molecule of glucose, a six-carbon compound, produces two molecules of pyruvate ion, a three-carbon compound, with a net gain of two molecules of ATP per molecule of glucose reacted. The pathway is anaerobic (Campbell, 1991, p. 294). See also *citric acid cycle*.

Endergonic: A reaction in which the standard free energy of reaction is positive, $\Delta G^\circ_R > 0$ and consequently, the equilibrium constant, K_{eq}, is less than one, i.e., $K_{eq} < 1$. The reaction is not spontaneous and requires an input of energy. See also, *exergonic*.

Endogenous respiration: (1) The digestion of an organism's own cellular matter by their own metabolic processes for energy production (Stewart et al., 1959, p. 312). The rate of autodigestion is a constant proportion of the cell matter present. (2) The basal energy requirement of bacteria, i.e., motion and enzyme activation that results in the metabolism of certain components of protoplasm. As endogenous metabolism proceeds, the bacteria reach a point at which they can no longer sustain life and die. A portion of the dead bacteria undergo lysing with a release of the nutrients that other bacteria can utilize. A residual organic fraction remains, which is insoluble and not metabolizable by other bacteria, and is believed to be polysaccharide (McKinney, 1963). (3) Those processes occurring when the organism is deprived of exogenous energy (Sherrard, 1977, p. 1969). (4) Thermodynamically, endogenous decay is the metabolism of the biomass itself to sustain life (Droste, 1998, p. 410). (5) The digestion of cellular tissue by the metabolic processes of an organism for energy production (p. 312); and most organisms, whether aerobic or anaerobic, maintain endogenous respiration during growth and that the rate of autodigestion is a constant proportion of the cell material present and is independent of substrate concentration (Stewart et al., 1959, p. 320).

Enthalpy: Defined as, $H = E + PV$, in which E = internal energy (kJ), P = pressure (kPa), V = volume (m^3). Enthalpy is the heat content of the substance.

Entropy: A measure of the "disorder" of a system in kJ. For example, a gas if released from a pressurized cylinder of volume $V(1)$ to one that is under vacuum that also has a volume $V(1)$ will expand spontaneously from $V(1)$ to $2V(1)$ with increase in entropy.

Enzyme: (1) A polymer of amino acids, i.e., a protein that catalyzes reactions (Rittman et al., 1990, p. 24). (2) Each protein enzyme exhibits a high degree of specificity, with specific sites to which the substrate molecule becomes bound, forming an enzyme · substrate complex, E · S. The enzyme allows an exothermic reaction to proceed by overcoming an energy barrier. The reaction velocity is affected by temperature in accordance with the Arrhenius equation, $\log k = \log A - E(\text{activation})/(2.303 \cdot RT)$, in which k = rate constant, A = constant (mols/L), R = gas constant 1.987 cal/degree/mol), T = temperature (K) (Gaudy and Gaudy, 1980, pp. 115–131). (3) Coupled biochemical reactions are catalyzed by enzymes; the enzyme provides a pathway that allows the reaction to occur (Rawn, 1989, p. 272). (4) The number of known enzymes is >1500 (Bailey and Ollis, 1977, p. 78).

Enzyme–substrate complex: (1) Enzymes bring substrates together at a special place on their surface called the "active site." The enzyme may interact with the substrate in two ways (in general): (a) by the "lock-and-key" model, in which the enzyme is rigid in shape and has a configuration that fits exactly a given substrate which binds to the surface of the enzyme; (b) the "induced-fit" model, in which the enzyme changes shape such that it surrounds the substrates. The enzyme may bring the substrates together at the active site, concentrating them, and lowering the activation energy of the reaction, and thus speeding up the rate of the reaction perhaps by a factor of $>10^5$ (Prescott et al., 2005, p. 158).

Equilibrium constant: See *Gibbs free energy*.

Eucarya: Fungi, algae, protozoa, rotifers, nematodes, and other zooplankton; all are characterized by a defined nucleus within the cell. The larger of this group may be defined as animals and plants but the smaller organisms cannot be so divided since the boundaries are blurred at the smaller scale (Rittman and McCarty, 2001, p. 22).

Eucaryotic: Cells that have a membrane-delimited nucleus. For reference, other groupings are archaeobacteria and procaryotic (Prescott et al., 1993, p. G-8; Prescott et al., 2005, p. 90). The eucarya and the procarya are similar in principle with respect to underlying metabolic processes and most of their most important biochemical pathways (Prescott et al., 2005, p. 91).

Exergonic: A reaction in which the standard free energy of reaction is negative, $\Delta G^\circ_R < 0$ and consequently, the equilibrium constant, K_{eq}, is greater than one, i.e., $K_{eq} > 1$. The reaction is spontaneous and releases energy to the "surroundings." See also, *endergonic*.

Facultative bacteria: Bacteria that can function in either aerobic or anaerobic conditions. The most significant are those that reduce nitrates to nitrogen gas, NO_3^- to N_2, called "denitrification," which occurs in an "anoxic" environment (see Grady et al., 1999, p. 22).

Fermentation: (1) An energy-yielding process in which organic molecules serve as both electron donors and acceptors (Prescott et al., 1993, p. G10). (2) A metabolic pathway in which sugars are broken down to simpler organic molecules and ATP is produced by the reaction of ADP with Pi. Fermentation is anaerobic, i.e., without oxygen as an electron acceptor, but loss of electrons occurs (Campbell, 1991, p. 18).

Fermentation reaction: An energy-yielding reaction in which organic molecules serve as both electron donors and electron acceptors (Prescott et al., 1993, p. 155).

Filamentous: Organisms that provide a "backbone" structure for activated-sludge flocs that may resist shear and permit settling in the final clarifier; when present in excess, filamentous organisms do allow compaction of the organisms during settling, called "bulking" (from Jenkins and Richard, 1982, p. 66, Richard et al., 1985a,b).

Food-to-microorganism ratio: Aeration tank loading parameter. Food may be expressed in kg suspended solids, COD, or BOD added per day to the aeration tank. Microorganisms may be expressed as mass (kg) of mixed liquor suspended solids (MLSS) or mixed liquor volatile suspended solids (MSVSS) in the aeration tank (URS, 1973, p. K-2).

F/M ratio: (1) Food-to-microorganism ratio; below a certain value, food is rate limiting (McKinney, 1963). (2) Also called a "loading factor" measured as substrate mass applied to reactor over a day divided by the mass of solids under aeration.

Free energy: See *Gibbs free energy*.

Fungi: The decomposers of such difficult-to-degrade matter as lignin, an aromatic polymer that binds cellulose in trees and grasses by means of peroxidase, the key enzyme for such breakdown, which also enables degradation of some resistant organic chemicals. Their slow rate of reaction makes the fungi not useful for reactor applications (Rittman and McCarty, 2001, p. 23).

Geosmin: A group of compounds produced by actinomycetes and cyanobacter that cause a characteristic "musty" odor of soils and often exhibited in natural waters (adapted from Prescott et al., 1993, p. G11).

Gibbs free energy: Defined as, $\Delta G = \Delta H - T \cdot \Delta S$, in which, G is the Gibbs free energy (kJ), H is the enthalpy (kJ), T is the temperature in Kelvin (K), S is the entropy (kJ). ΔH is the change in heat content of a given substance and ΔS is the change in entropy.

Gibbs free energy of formation: Free energy of the formation of a compound from its elements.

Gibbs free energy of reaction

1. For a given reaction (Rawn, 1989, pp. 265–272),

$$aA + bB \leftrightarrow cC + dD,$$

the "standard" values for the free energy of reaction, enthalpy of reaction, entropy of reaction, are respectively,

$$\Delta G^\circ_R(\text{reaction}) = \sum G^\circ_f(\text{products}) - \sum G^\circ_f(\text{reactants}). \text{ Also,}$$
$$\Delta H^\circ_R(\text{reaction}) = \sum H^\circ_f(\text{products}) - \sum H^\circ_f(\text{reactants}), \text{ and}$$
$$\Delta S^\circ_R(\text{reaction}) = \sum S^\circ_f(\text{products}) - \sum S^\circ_f(\text{reactants})$$

The subscripts "f" on the right sides of the respective equations refer to the "formation" values; $G_f^\circ(A)$ is the "standard free energy of formation" of the element or compound "A." The values of $G_f^\circ(A)$, $G_f^\circ(B)$, etc., may be obtained from handbooks for most elements and compounds, including *Lide's Handbook of Chemistry and Physics*, *Lange's Handbook of Chemistry*. For compounds of interest to water treatment, textbooks, e.g., Grady et al. (1999), Rittman and McCarty

(2001), or papers, for example, McCarty (1975), provide such data with examples on how to calculate ΔG°_R(reaction).

2. The designation, "standard" values mean that the values were obtained under "standard" conditions, meaning for "A," $T(A) = 298$ K; $p(A) = 1.0$ atm; or solution concentration of "A" is $[A] = 1.0$ molar, and pH(biological reactions) = 7.0.
3. The values for ΔG, ΔH, ΔS without the "°" superscript mean that they pertain to "physiological" conditions, i.e., whatever conditions occur in the cell or externally as may be the case,

$$\Delta G_R(\text{reaction}) = \sum G_f(\text{products}) - \sum G_f(\text{reactants}). \text{ Also,}$$
$$\Delta H_R(\text{reaction}) = \sum H_f(\text{products}) - \sum H_f(\text{reactants}), \text{ and}$$
$$\Delta S_R(\text{reaction}) = \sum S_f(\text{products}) - \sum S_f(\text{reactants})$$

An algebraic example of a calculation is (from Rawn, 1989, p. 271)

$$\Delta G_R(\text{reaction}) = [cG_f(C) + dG_f(D)] - [aG_f(A) + bG_f(B)]$$

where $G_f(A)$ is the Gibbs free energy of formation of "A" (kJ/mol A).

Also, for physiological conditions, i.e., any conditions,

$$\Delta G_R(\text{reaction}) = \Delta G^\circ_R(\text{reaction}) + 2.303RT \log\{[C]^c[D]^d\}/\{[A]^a[B]^b\}$$

where

$[A]^a$ is the molar concentration of "A" (mol A/L solution)

a are moles of "A" in stoichiometric, balanced equation

4. Gibbs free energy is defined, $\Delta G = \Delta H - T\Delta S$, where ΔG is the Gibbs free energy change, ΔH is the entropy change, T is the temperature in K, and ΔS is the entropy change, for any process, including a chemical reaction. Gibbs free energy, enthalpy, and entropy are "state" functions, meaning that the values are independent of the path taken between states, e.g., a biochemical pathway.
5. Nomenclature of thermodynamic functions: for standard state the designations are usually given as, ΔG°_R, ΔH°_R, ΔS°_R.

Glucose: The most common sugar, when assembled one way can form starch and assembled another way can form cellulose (Campbell, 1991, p. 69).

Glycosis: Fermentation of glucose to pyruvate and acetyl CoA, dominated by the Embden–Meyerhoff pathway (Bailey and Ollis, 1977, pp. 233, 243).

Growth curve: Bacteria multiply by binary fission. A few bacteria immersed in a beaker will thus multiply and decline in population in accordance with a classic growth curve. The stages of the growth curve includes the following phases: (1) lag phase, in which the organism "acclimatizes" to the media, synthesizing the needed enzymes to metabolize the substrate at hand; (2) the log phase, in which the organism growth rate is proportional to the concentration of organisms in suspension, i.e., $dX/dt = \mu(\max) \cdot X$; (3) the declining growth phase, in which the growth rate is limited by the substrate or some limiting nutrient, defined by the Monod equation; (4) the stationary growth phase, in which $dX/dt = 0$; (5) the death phase, in which the lack of nutrients causes a decline in organism concentration; (6) log death phase, in which the nutrients have been exhausted and the rate of decline in organism population is proportional to the concentration of organisms, i.e., $dX/dt = -k_d \cdot X$; (7) death phase, in which death takes over the culture and the growth cycle is complete (McKinney, 1962, p. 118).

Growth factor: An organic compound that must be supplied externally because it cannot be synthesized by the organism (adapted from Prescott et al., 1993, p. G11).

Heavy metal: Antimony, arsenic, beryllium, cadmium, chromium(III), chromium(IV), copper, lead, mercury, nickel, selenium, silver, thallium, zinc.

Heterotroph: (1) Microorganism that uses organic matter as a substrate. (2) Bacteria that use organic compounds as their electron donor and their source of carbon for cell synthesis (Grady et al., 1999, p. 21). (3) An organism that uses organic molecules as its principal source of carbon (Prescott et al., 1993, p. G12).

Krebs cycle: A name sometimes used for the citric acid cycle after Sir Hans Krebs, who first investigated the pathways and for which he received the Nobel Prize in 1953 (Campbell, 1991, p. 336).

Kjeldahl nitrogen: (1) Most of the nitrogen in domestic wastewaters occurs as proteins or their degradation products, such as polypeptides and amino acids. The concept behind the Kjeldahl method of measurement is to oxidize the carbon and hydrogen to carbon dioxide and water by sulfuric acid (called "digestion"). The amino group is released as ammonia and, due to the acid pH, in the form of NH_4^+. The solution is neutralized to give $pH \gg 8.0$. The ammonia as NH_3 may be measured by Nesslerization or back titration with a standard acid solution (foregoing from Sawyer and McCarty, 1967, p. 428). (2) The Kjeldahl method of protein analysis was invented by Johan Kjeldahl of Denmark in 1883 and remains common. The analysis has three major steps in the original procedure: (a) digestion in sulfuric acid at high temperature, which converts all nitrogen to ammonium acid sulfate; (b) distillation in which the digest is made alkaline and the liberated ammonia is steam distilled into a standard acid receiver; (c) titration in which the ammonia is back titrated of the excess standard acid with standard base (foregoing from Chen et al., 1988, p. 6081).

Liebig's law of the minimum: A microbial population will grow until some factor limits further growth (Prescott et al., 1990, p. G14).

Lipids: An organic molecule built mainly from long hydrocarbon chains of fatty acids, a rich source of stored energy (Campbell, 1991, p. 69).

Lumping: Defining a parameter collectively, rather than individually; examples include biochemical oxygen demand (BOD), chemical oxygen demand (COD), total organic carbon (TOC), total Kjeldahl nitrogen (TKN). Kinetic lumping refers to a kinetic study on a substrate mixture degraded by a mixture of organisms. Foregoing from Li et al. (1996, p. 841).

Macronutrient: Major fractions of biomass such as C, O, H, N, P, S (Eckenfelder and Grau, 1992, p. 2). Typically, the term is used for N, P.

Maintenance energy: The energy required by a microorganism in order to function, such as to rebuild proteins, provide motility, active transport of molecules. See also *endogenous respiration*.

Medium: The environment of the cell, e.g., temperature, pH, osmotic pressure, substrate, nutrients, and other components (Eckenfelder and Grau, 1992, p. 3).

Mesophilic: Refers to an organism with a temperature range optimum 20°C–45°C (Prescott et al., 1993, p. G15).

Metabolism: (1) The sum of anabolism (cell synthesis) and catabolism (substrate reaction that yields energy). (2) The aggregate of anabolism (cell synthesis) and catabolism (cell respiration). (3) The aggregate functioning of a cell, which includes taking up of nutrients, discarding waste products, reproduction.

Methanogenesis: Anaerobic reaction in which the electron equivalents of organic matter are used to reduce carbon to its most reduced oxidation state, −4, in methane, CH_4. Each mole of CH_4 contains 8 electron equivalents or 64 g of COD. Each mole of CH_4 has a volume of 22.4 L at STP; thus each gram of COD stabilized generates 0.35 L CH_4 gas at STP (Rittman and McCarty, 2001, p. 569).

Michaelis–Menten Equation: (1) Kinetic relation for cell growth based on enzyme reactions, published in 1913 by Leonor Michaelis and Maud Menten and was described by Sawyer and McCarty (1967, pp. 202–204) for waste treatment.

Microbe: A microorganism; examples include protozoa, algae, cyanobacteria, etc. (Prescott et al., 2005, p. 608). A specific definition seems to be lacking but the general idea is that it is an entity not visible to the eye and requires a microscope (as opposed to a "macroscopic" organism). The reason for using the term "entity" is that if the term "organism" were used, it may be that viruses would be excluded since some may not agree that a virus is a "living" entity.

Micronutrient: Elements found in biomass in small fractions such as Fe, Ca, Cu, K, Mn, Mg, Mo, Ni, Zn, Co (Eckenfelder and Grau, 1992, p. 2).

Mineralize: The conversion of an organic compound to carbon dioxide and water by means of a biological reaction.

Mitochondria: Organelles bounded by two membranes, with the inner membrane folded into cristae, and are responsible for energy generation by the tricarboxylic acid cycle, electron transport, and oxidative phosphorylation (Prescott et al., 2005, p. 92). Mitochondria are found in most eukaryotic cells and are the site of the TCA cycle and the generation of ATP and are bound by two membranes (Prescott et al., 2005, p. 82).

Mixed liquor suspended solids (MLSS): The concentration of solids in an activated-sludge basin, usually measured in mg/L. Mixed liquor volatile suspended solids (MLVSS) is also used.

Mole: The mass of a substance that contains $1.022 \cdot 10^{23}$ molecules, i.e., Avogadro's number. For example, a mole of oxygen gas, O_2, has a mass of 32.00 g. See also a chemistry text, e.g., Silberberg (1996, pp. 85–130).

Monod equation: Kinetic rate constant as defined by Jacques Monod in 1942, i.e., $\mu = \hat{\mu}[S/(K_s + S)]$.

NAD^+: Biological oxidizing agent that accepts electrons as a reactant in a redox reaction (Rawn, 1989, p. 241) and is the abbreviation for nicotinamide adenine dinucleotide.

NADH: Biological reducing agent that gives up electrons as a reactant in a redox reaction (Rawn, 1989, p. 241)

NADP: Biological oxidizing agent; it accepts electrons as a reactant in a redox reaction (Rawn, 1989, p. 241) and is the abbreviation for nicotinamide adenine dinucleotide phosphate.

NADPH: (1) Nicotinamide adenine dinucleotide phosphate. (2) Biological reducing agents; they give up electrons as a reactant in a redox reaction (Rawn, 1989, p. 241).

Nitrate: Nitrate (+5 oxidation state) or nitrite (+3 oxidation state).

Nitrification: (1) The oxidation of ammonia to nitrate; bacteria involved are of the family, *Nitrobacteriaceae* (Prescott et al., 1993, p. G17). (2) The conversion of ammonia N (NH_3) to nitrate N (NO_3^-), generally by autotrophic bacteria; *Nitrosomonas* oxidizes NH_3 to nitrite N (NO_2^-) and *Nitrobacter* oxidizes NO_2^- to NO_3^- (Grady et al., 1999, p. 26).

***Nitrosomonas*:** Bacteria that convert ammonia to nitrite (Anon, 1968, p. 7).

***Nitrobacter*:** Bacteria that convert nitrite to nitrate (Anon, 1968, p. 7).

***Nocardia*:** Bacteria that cause foams in activated sludge causing operating problems (Pagilla, et al., 1996, p. 235).

NUR: Nitrate uptake rate (kg $N_2/m^3/s$).

Nutrients: (1) Chemical elements required for cell synthesis. Nitrogen and phosphorous are the most common, but a variety of metal trace elements may be included, e.g., Cu, Zn, Cd, Mn, Co, Mb, Ni (Prescott et al., 1993, p. 97). (2) Substance that supports the reproduction and growth of bacteria (Prescott et al., 1993, p. G17).

Obligate anaerobe: An organism that utilizes CO_2 or carbonate as its terminal electron acceptor and reduces the CO_2 to methane (Prescott et al., 1993, p. 158). Oxygen is toxic to an obligate anaerobe.

Organelles: Intracellular structures that perform specific functions in cells analogous to organs in the body (Prescott et al., 2005. p. 75).

Organic compound: Compound containing carbon. Organic compounds number in the millions. Some categories include aromatic, aliphatic, heterocyclic aromatic.

ORP: Oxidation-reduction potential.

OUR: Oxygen uptake rate (kg $O_2/m^3/s$).

Oxidation: Compound that is oxidized loses electrons. When an organic compound is oxidized biochemically, it usually loses electrons in the form of hydrogen atoms; consequently, oxidation is synonymous with dehydrogenation (Bailey and Ollis, 1977, p. 230).

Oxidation potential: A ranking of the tendency of a substance to give up electrons, usually measured in terms of a half-reaction. Consider, for example, two coupled half-reactions (Latimer, 1952, p. 3): $Zn \rightarrow Zn^{2+} + 2e^-$, $E° = 0.763$ V; and $H_2 \rightarrow 2H^+ + 2e^-$, $E° = 0.0$ V (reference half-reaction). The coupled reaction is the addition of the two half-reactions; balancing electrons by multiplying the hydrogen half-reaction by "-1," and adding gives, $Zn + 2H^+ \rightarrow Zn^{2+} + H_2$, $E°(\text{reaction}) = +0.763$ V. By comparison, $Mg \rightarrow Mg^{2+} + 2e^-$, $E° = 2.37$ (Latimer, 1952, p. 340), which means that magnesium gives up its electrons more readily than zinc, i.e., it is more readily oxidized.

Oxidative phosphorylation: (1) NADH, produced by oxidation in catabolism, is reconverted to NAD^+ during respiration and with the production of ATP (Rawn, 1989, p. 242). (2) The production of ATP in a cell in which ADP is phosphorylated. (3) Process by which energy from electron transport, e.g., from NADH to oxygen, is used to make ATP (Prescott et al., 2005, p. 179).

Peptide bond: When amino acids are joined in a chain, the link between them is called a "peptide bond." The amide bond formed is between the carbonyl group, i.e., $C=O$, of one amino acid and the amino group, i.e., N–H, of the next. The "primary" structure of a protein is this linear sequence of amino acids; the chain length is indicated by the number of peptide bonds, e.g., dipeptide, tripeptide, polypeptide (Foregoing from Rawn, 1989, p. 61).

Phosphate: Inorganic ion, PO_4^{3-}, depicted in ADP–ATP reactions as P_i, meaning inorganic phosphate ion (Campbell, 1991, p. 285). On total phosphorous, Rohlich (1964, p. 207) give raw sewage per capita outputs of P for Minnesota communities as 1.5–3.7 g/person/day with a median value, 2.3 g/person/day, with a concentration level of about 7 mg total P/L.

Photosynthesis: The synthesis of new cells that utilize radiant energy as an energy source with carbon dioxide and the carbon source.

Plankton: Floating forms of aquatic life such as algae.

Pollutant: Usually considered an introduced contaminant and implies impaired utility of water.

Polysaccharide: A molecule comprising many monosaccharides that are bonded together (Campbell, 1991, p. 97).

Population: An assemblage of organisms of the same type (Prescott et al., 1993, p. G20).

POTW: Publicly owned treatment works, a term defined by PL92-500, the 1972 Clean Water Act of the U.S. Congress.

Primary treatment: Methods of wastewater treatment that remove a high fraction of suspended and floating solids (URS, 1973, p. A-1). The first settling basin in municipal wastewater treatment is considered essentially synonymous with primary treatment.

Priority pollutants: Antimony, arsenic, beryllium, cadmium, chromium(III), chromium(IV), copper, lead, mercury, nickel, selenium, silver, thallium, zinc, cyanide, asbestos, 2,3,7,8-TCDD (dioxin), acrolein, acrylonitrile, benzene, bromoform, carbon tetrachloride, chlorobenzene, chlorodibromomethane, chloroform, dichlorobromomethane, 1,2-dichloroethane, 1,1-dichloroethylene, 1,3-dichloropropylene, ethylbenzene, methyl bromide, methyl chloride, methylene chloride, 1,1,2,2-tetrachloroethane, tetrachloroethylene, toluene, 1,1,1-trichloroethane, 1,1,2-trichloroethane, trichloroethane, vinyl chloride, 2,4-dichlorophenol, 2-methyl-4,6-dinitrophenol, 2,4-dinitrophenol, pentachlorophenol, phenol, 2,4,6-trichlorophenol, acenaphthylene, anthracene, benzidine, benzo[*a*]anthracene, benzo[*a*]pyrene, benzo[*b*]fluoranthene, benzo[*g*,*h*,*i*]perylene, benzo[*k*]fluoranthene, bis(2-chloroethyl)ether, bis(2-chloroisopropyl)ether, bis(2-ethylhexyl)phthalate, Chrysene, dibenzo[*a*,*h*]anthracene, 1,2-dichlorobenzene, 1,3-dichlorobenzene, 1,4-dichlorobenzene, 3,3′-dichlorobenzidine, diethyl phthalate, dimethyl phthalate, di-*n*-butyl phthalate, 2,4-dinitrotoluene, 1,2-diphenylhydrazine, fluoranthene, fluorene, hexachlorobenzene, hexachlorobutadiene, hexachlorocyclopentadiene, hexachloroethane, indeno(1,2,3-cd)pyrene, isophorone, nitrobenzene, *N*-nitrosodimethylamine, *N*-nitrosodiphenylamine, phenanthrene, pyrene, aldrin, alpha-BHC, chlordane, 4,4′-DDT, 4,4′-DDE, 4,4′-DDD, dieldrin, alpha-endosulfan, beta-endosulfan, endosulfan sulfate, endrin, endrin aldehyde, heptachlor, heptachlor epoxide, PCB-1242, PCB-1254, PCB-1221, PCB-1232, PCB-1248, PCB-1260, PCB-1216, toxaphene (foregoing list is from Regulations

Review, *Pollution Engineering*, January 1, 1992, pp. 30, 32 and lists 105 water pollutants for which EPA had set numeric limits).

Prokaryotae: A kingdom comprising all the prokaryotic microorganisms (those lacking a true membrane-enclosed nuclei) (Prescott et al., 1993, p. G-20).

Prokaryotic cell: Cells that lack a true membrane-enclosed nucleus; bacteria are prokaryotic and have their genetic material located in a nucleoid. For reference, other groupings are archaeobacteria and eukaryotic (Prescott et al., 1993, p. G-20).

Prokaryotes: Kingdom that includes Bacteria and the Archaea (Rittman and McCarty, 2001, p. 6). They note that little is gained by the division into the two groups; the Bacteria convert complex organic matter into acetic acid and hydrogen and the Archaea convert acetic acid and hydrogen into methane gas. The latter group, in the past, has been called simply, "methane bacteria." As a matter of interest blue-green algae have no nucleus, a property of bacteria, not of plants, and are classified as *cyanobacteria*.

Prokaryotic cells: Single-celled organisms with nuclear material loosely organized; examples include bacteria and cyanobacteria (blue-green algae). Eukaryotes, on the other hand, have a true nucleus, set off from the rest of the cell by a membrane; examples include plant and animal cells (Campbell, 1991, p. 26).

Protein: (1) All proteins are polymers of α-amino acids joined by peptide bonds. The length of the polymer varies so that the molecular weight varies from a few thousand to several million. All proteins contain about 0.16 fraction nitrogen. Proteins comprise different combinations of the amino acids; the possible sequences of amino acids number over 10^{18} (assuming 20 different amino acids). Protein molecules have a primary structure, a "backbone," and also a secondary, tertiary, and quaternary structure, with a "conformal" three-dimensional structure; each protein has a specific spatial shape. Denaturing a protein causes a change in such molecular structure (Gaudy and Gaudy, 1980, p. 99–115). (2) All organisms, from bacteria to buffalo, have proteins made of the same set of 20 amino acids (Rawn, 1989, p. 52).

Protozoa: Single-celled organisms that can pursue and digest their food (Rittman and McCarty, 2001, p. 31). They form a part of the ecology of biological treatment systems.

Psychrophile: A microorganism that grows well at 0°C and has an optimum growth rate at ≤15°C with a maximum of about 20°C (Prescott et al., 1993, p. G21).

Pyruvate: (1) Pyruvate is a metabolic intermediate; under aerobic conditions, pyruvate is further oxidized by the citric acid cycle and NAD is again regenerated (Rawn, 1989, p. 314). (2) The carbon skeleton of pyruvate is incorporated into alanine, valine, and leucine, which are known as the pyruvate family of amino acids (Rawn, 1989, p. 603).

Recirculation rate: The recirculation of a fraction of settled activated sludge in the "final" settling tank; the other fraction is "wasted," also called, "return activated sludge," or "RAS," the latter having emerged since the 1990s.

Reduction: (1) Compound that is reduced and gains electrons; oxygen is the best-known example. (2) When an organic compound is reduced biochemically, it usually gains electrons in the form of hydrogen atoms; consequently, reduction is synonymous with hydrogenation (Bailey and Ollis, 1977, p. 230).

An example is the reduction of pyruvic acid to lactic acid when the former gains electrons, i.e., hydrogen atoms, i.e.,

$$\begin{array}{ccc} CH_3 & & CH_3 \\ | & & | \\ C{=}{=}O & + 2H \rightarrow & CHOH \\ | & & | \\ COOH & & COOH \\ \text{pyruvic acid} & & \text{lactic acid} \end{array}$$

Refractory: Refers to any substance that passes through a treatment process (Anon., 1965, p. iii). Chloride ion is an example. Synthetic organic chemicals are common.

Respiration: (1) The uptake of oxygen by cells. (2) An energy-yielding process in which an electron donor is oxidized using an inorganic electron acceptor; the electron acceptor may be either oxygen (aerobic respiration) or another inorganic acceptor (anaerobic respiration) (Prescott et al., 1993, p. G21). (3) An energy-producing process in which organic or reduced inorganic compounds are oxidized by inorganic compounds. When the oxidant is something other than oxygen, the process is termed *anaerobic respiration* (Bailey and Ollis, 1977, p. 241). It is also called an anoxic process.

Aerobic respiration: In the oxidation of an organic compound, oxygen is the final electron acceptor (Prescott et al., 1993, p. 134).

Ribosome: RNA protein particles that contain the enzymes for protein synthesis (Rittman and McCarty, 2001, p. 11).

Rotating biological contactor (RBC): See *attached growth reactor*. Descriptions and a design protocols are given by Tchobanoglous and Burton (1991, pp. 628–636) and by Grady et al. (1999, pp. 907–947).

Rotifer: (1) An organism of the class *Rotifera*, which are multicellular, strictly aerobic, and ingest small particulate matter, such as bacteria and algae, and other living or dead organic matter of similar size

(Rittman and McCarty, 2001, p. 35). A characteristic of the organism is its two wheel-like ciliates that cause an inflow of particles for feeding. (2) A single ciliate may ingest 60–70 bacteria per hour; in activated sludge, ciliates may remove suspended bacteria that have not settled (Prescott et al., 2005, p. 590).

RNA: (1) A linear polymer of ribonucleotide repeating units (composed of the sugar ribose, phosphate, and one of four heterocyclic bases); RNA is used in several ways to convert the genetic code to protein products (Rittman et al., 1990, p. 24). (2) Single-stranded helix-shaped molecule, similar to DNA, that serves as a "template" for the synthesis of protein molecules; uracil is a part of the molecule instead of thyamine. The three forms of RNA are messenger RNA (mRNA), which carries the information for protein synthesis from the DNA; transfer RNA, which carries amino acids to their proper site on the mRNA for the synthesis of a given protein; and ribosomal RNA (rRNA) which serves as the structural and catalytic components of the ribosomes, the sites where the proteins are synthesized.

Secondary treatment: The methods of wastewater treatment that remove a high fraction of fine suspended colloidal solids, dissolved solids, and organic matter by biological oxidation (URS, 1973, p. A-1).

Sequencing batch reactor: A sequence of batch reactors. Each reactor "reacts" for a finite time; the contents are then emptied to the next reactor; the previous reactor is filled again. See, for example, Orhon and Artan (1994, pp. 255–260).

Sludge: An accumulation of cells into a discernible mass.

Sludge age: (1) In activated sludge, the average number of days that the solids remain in the aeration system, as computed as the mass of solids in the aeration tank divided by the mass flux of solids leaving the system per day, i.e., $\theta_c \approx VX/RX_r$(d). (2) Mathematically, $\theta_c = [XV(\text{reactor}) + \text{additional mass in system}]/[RX_r + (Q - R)X_e]$, which is essentially the same as given by Lawrence and McCarty (1970, p. 769).

Sludge-density index: The reciprocal of SVI.

Sludge-volume index (SVI)

$$\text{SVI} = \frac{\text{volume of sludge after 30 min settling in1 L cylinder}}{\text{mass of MLSS in cylinder}}$$

(mL settled sludge/g oven-dried settled sludge).

Sludge wasting: Sludge wasted from final settling tank; a fraction of the sludge settled solids is "wasted" and the other fraction is returned to the aeration tank.

***Sphaerotilus*:** A strict aerobe, rod-shaped sheathed bacteria species that grows in chains of cells. The sheath may be surrounded by a polysaccharide slime layer. The organism is one of those commonly found in activated sludge and which is likely to cause poor settling sludge, if present in large numbers. The bacteria may attach to solid surfaces and may occur as massive growths if nutrients are available such as associated with paper mill wastes, dairies, etc. (Gaudy and Gaudy, 1980, p. 365). *Sphaerotilus* has been seen by the author growing, below a trout farm, as a stringy thick grey mass on the rocks of a fast moving mountain stream.

Spore: (1) An encapsulated form, with heavy cell wall, of a bacterium that permits the organism's survival under adverse environmental conditions that do not permit metabolism. The organism is encapsulated in a protein shell. An example is the *Salmonella typhosa* that forms a cyst after it is excreted and enters a nonsupportive environment. (2) Most bacteria do not form spores; those that do are mostly of two genera of gram-positive rods, *Bacillus* and *Clostridium.* A spore is dormant until induced to germinate and may remain viable for many years in the dormant state (Gaudy and Gaudy, 1980, p. 165).

Stoichiometry: The procedure of balancing a chemical relation such that the same molar quantities of elements are present on both sides of the equation, i.e., for the reactant side and for the product side.

Strain: A population of organisms that descends from a single organism or a pure culture (Prescott et al., 1993, p. G24).

Substrate: (1) Compound that serves as the source of energy for a microbiological reaction (Eckenfelder and Grau, 1992, p. 2). The compound may be organic or inorganic. (2) An organic molecule that reacts with an enzyme (adapted from Prescott et al., 1993, p. 140). (3) A substance that an enzyme acts on (Prescott et al., 1993, p. G24).

Suspended-growth reactor: A reactor in which microorganisms are suspended and mixed in water, making contact with the substrate by turbulent transport with the final transport step by diffusion. Activated-sludge reactors are the classic example. See *activated sludge*.

Suspended solids (SS): The suspended solids that are removed by filtration as measured in the laboratory from an aeration basin sample (mg/L).

***Sphaerotilus*:** A filamentous organism, *Sphaerotilus natans*, associated with sludge bulking.

***Sphaerotilus natans*:** A filamentous organism.

Stoichiometric: The coefficients in a balance chemical equation.

Strain: A population of genetically identical cells (Rittman et al., 1990, p. 24).

TCA cycle: Tricarboxylic acid cycle; a name sometimes used for the citric acid cycle, since some of the molecules involved are acids with three carboxyl groups (Campbell, 1991, p. 336). The Krebs cycle is another name sometimes used for the citric acid cycle.

Tertiary treatment: Methods of wastewater treatment that remove a high fraction of nutrients, residual organics, residual solids, and pathogens, by such methods as granular media filtration, chemical precipitation, carbon adsorption, ammonia stripping, electrodialysis, membrane treatment, etc. (URS, 1973, p. A-1). Generally, tertiary wastewater treatment follows secondary treatment.

Thermodynamic system: (1) Closed system: may exchange energy across a hypothetical or real boundary. (2) Open system: may exchange energy and mass across a hypothetical or real boundary (Bailey and Ollis, 1977, p. 226).

Thermophilic: Refers to an organism that can grow within the temperature range of 45°C–55°C (Prescott et al., 1993, p. G25).

TOC: (1) Total organic carbon. The measurement is done by TOC analyzer, an instrument that oxidizes organic carbon in a sample to carbon dioxide, the latter being measured by infrared absorbance. (2) Calculation is shown for ethanol for COD definition.

Total organic carbon: See *TOC*.

Treatability: Description as to whether a substance may be metabolized by a microbe and if so whether the rate of reaction may be reasonable.

Trickling filter: A traditional "attached growth reactor," usually filled with rocks 8–13 cm (3–5 in.) in size, and depth of about 2 m (6 ft), and a rotating arm that distributes the water flow over the plan area and permits airflow between passes. See also *attached growth reactor*.

Trophic classification: The method of categorizing microbial reactions that refer to energy source; some definitions are given below.

Chemotroph: (1) An organism that obtains its energy from the oxidation of a chemical compound (Prescott et al., 1993, p. G5). (2) Organisms that obtain energy from chemical reactions (Rittman and McCarty, 2001, p. 15). The two groups are

Chemoorganotroph: Organisms that obtain energy from organic chemical reactions and are commonly heterotrophic (Rittman and McCarty, 2001, p. 16). Most bacteria fit this category.

Chemolithotroph: Organisms that obtain energy from inorganic chemical reactions and are commonly autotrophic, using carbon dioxide for cell synthesis (Rittman and McCarty, 2001, p. 16). Alga cells fit this category.

Heterotroph: Organisms that use organic carbon for cell synthesis (Rittman and McCarty, 2001, p. 16).

Phototroph: Organisms that obtain energy from light (Rittman and McCarty, 2001, p. 15).

Turbidity: Light-scattering property of water due to suspended colloidal particles, measured in nephelometric turbidity units (NTU) since the advent of modern instruments in the 1960s. Before 1960 the common method was in terms of Jackson Candle Units (JCU), measured by a long tube mounted on a tripod above a candle; as a water sample was poured in, the level when the candle was no longer visible was the measure as determined by a standard scale etched in the tube.

Virus: A "nonliving" entity that can replicate only in association with a living cell and ranges in size from 15–300 nm. Viruses that infect bacteria are called bacteriophages (Rittman and McCarty, 2001, p. 36).

Volatile acid: Low molecular weight acids that can be volatilized at atmospheric pressure, such as acetic, propionic, and butyric (Gaudy and Gaudy, 1980, p. 75).

VS: Volatile solids; see *VSS*.

VSS: Volatile suspended solids as measured by laboratory procedure in which an oven-dried sample of sludge is placed in a muffle furnace for 24 h at about 550°C (in accordance with *Standard Methods*).

Warburg respirometer: A manufactured apparatus (Precision Scientific) fitted with 50 mL reactor flasks, each with a manometer to measure gas pressure, and heating and cooling elements to maintain constant temperature (Gram, 1956, p. 92). The apparatus was standard equipment in laboratories doing research on biological treatment, with particular reference to activated sludge.

Xenobiotic: (1) Resistant to biodegradation. (2) Relating to or denoting a substance, typically a synthetic chemical that is foreign to the body or to an ecological system (Apple Corporation, 2005). (3) Man-made organic compounds, mostly slowly degradable. Individual bacteria in activated sludge are "specialized" in degrading xenobiotic compounds (Chudoba et al., 1989).

Yield coefficient: Cell biomass related stoichiometrically to the substrate in a biochemical reaction, e.g., moles biomass synthesized per mole substrate reacted; units could be (kg COD biomass synthesized/kg COD substrate degraded).

REFERENCES

Anon., Summary Report—Advanced Waste Research Program, January 1962–June, 1964. The Advanced Waste Treatment Research Program, Environmental Health Series AWTR-14, Robert A. Taft Sanitary Engineering Center, Division of Water Supply and Pollution Control, U.S. Department of Health Education and Welfare, Cincinnati, OH, April 1965.

Anon., Summary Report—Advanced Waste Treatment, July 1964–July 1967. Water Pollution Control Research Series, Advanced Treatment Research, Publication WP-20-AWTR-19, Advanced Waste Treatment Branch, Division of Research, Robert A. Taft Sanitary Engineering Center Federal Water Pollution Control Administration, U.S. Department of the Interior, Cincinnati, OH, 1968.

Apple Corporation, *Aa Dictionary*, Apple Operating System 10.4, Cupertino, CA, 2005.

Bailey, J. E. and Ollis, D. F., *Biochemical Engineering Fundamentals*, McGraw-Hill, New York, 1977.

Benefield, L. D. and Randall, C. W., *Biological Process Design for Wastewater Treatment*, Prentice-Hall, Englewood Cliffs, NJ, 1980.

Black & Veatch, Consulting Engineers, Process design manual for phosphorous removal, Technology Transfer Program, Contract #14-12-936, U.S. Environmental Protection Agency, Washington, DC, 1971.

Bohinsky, R. C., *Modern Concepts in Biochemistry,* Allyn and Bacon, Inc., Boston, MA, 1987.

Bowker, R. P. G. and Stensel, H. D., Design manual-Phosphorous removal, Water Engineering Research Laboratory, Report EPA/625/1-87/001, U.S. Environmental Protection Agency, Cincinnati, OH, September, 1987.

Campbell, M. K., *Biochemistry*, Sanders College Publishing, Philadelphia, PA, 1991.

Chang, M. K., Voice, T. C., and Criddle, C. S., Kinetics of competitive inhibition and co-metabolism in biodegradation of benzene, Toulene, and p-Xylene by Two *Pseudomonas* Isolates. *Biotechnology and Bioengineering*, 41:1057–1065, 1993.

Chen, Y. S., Brayton, S. V., and Hach, C. C., Accuracy in Kjeldahl Protein Analysis, *American Laboratory*, pp. 6081–6084, June 1988.

Christensen, D. R. and McCarty, P. L., Multi-process biological treatment model, *Journal of Water Pollution Control Federation*, 47(11):2652–2664, November 1975.

Chudoba, J., Alboková, J., and Cech, J. S., Determination of kinetic constants of activated sludge microorganisms responsible for degradation of xenobiotics, *Water Research*, 22(11):1431–1438, 1989.

Dague, R. R., McKinney, R. E., and Pfeffer, J. T., Solids retention in anaerobic waste treatment systems, *Journal of Water Pollution Control Federation*, 42(2) (Part 2): R29-R46, February 1970.

Daniels, F. and Alberty, R. A., *Physical Chemistry*, John Wiley & Sons, New York, 1955.

Droste, R. L., Endogenous decay and bioenergetic theory for aerobic wastewater treatment, *Water Research*, 32(2):410–418, February 1998.

Eckenfelder, W. W. Jr. and Weston, R. F., Kinetics of biological oxidation, in McCabe, J. and Eckenfelder, W. W., Jr. (Eds.), *Biological Treatment of Sewage and Industrial Wastes, Volume 1 Aerobic Oxidation*, Reinhold Publishing, New York, 1956.

Eckenfelder, W. W. Jr. and Grau, P., *Activated Sludge Process Design and Control—Theory and Practice*, Technomic Publishing Co., Inc., Lancaster, U.K., 1992.

Eckenfelder, W. W. Jr., Ford, D. L., and Englande, A. J. Jr., *Industrial Water Quality*, 4th edn., McGraw-Hill, New York, 2009.

Ehret, W. F., *Smith's College Chemistry*, 6th edn., D. Appleton-Century, Inc., New York, 1947.

Garrett, M. T. Jr. and Sawyer, C. N., Kinetics of removal of soluble BOD by activated sludge, *Proceedings of the Seventh Industrial Waste Conference*, Purdue University, Lafayette, IN, pp. 51–77, May 7–9, 1952.

Gaudy, A. and Gaudy, E., *Microbiology for Environmental Scientists and Engineers*, McGraw-Hill, New York, 1980.

Gloyna, E. F. and Eckenfelder Jr., W. W. (Eds.), *Advances in Water Quality Improvement*, University of Texas Press, Austin, TX, 1968.

Grady, C. P. L. Jr., Daigger, G. T., and Lim, H. C., *Biological Wastewater Treatment, Theory and Applications*, 2nd edn., Marcel Dekker, Inc., New York, 1999.

Gram, A. L., III, Reaction kinetics of aerobic biological processes, PhD Dissertation, Department of Engineering, University of California, Berkeley, CA, May 15, 1956a.

Gram, A. L., III, Reaction kinetics of aerobic biological processes, Report No. 2, I.E.R. Series 90, Sanitary Engineering Research Laboratory, Department of Engineering, University of California, Berkeley, CA, May 15, 1956b.

Grau, P., Sutton, P. M., Henze, M., Elmaleh, S., Grady, C. P. L., Gujer, W., and Koller, J., Recommended notation for use of in the description of biological wastewater treatment processes, *Water Research*, 16:1501–1505, 1982.

Hao, X., Wang, Q., Zhang, X., Cao, Y., and van Mark Loosdrecht, C. M., Experimental evaluation of decrease in bacterial activity due to cell death and activity decay in activated sludge, *Water Research*, 43(14):3604–2612, August 2009.

Hendricks, D. W. and Pote, W. D. Thermodynamic analysis of a primary oxidation pond, *Journal of Water Pollution Control Federation,* 46(2): 333–351, February 1974.

Jannasch, H. W. and Egh, T., Microbial growth kinetics: A historical perspective, *Antonie van Leeuwenhoek*, 63(3/4):213–224, 1993.

Jenkins, D. and Richard, M. G., Factors affecting the selection of filamentous organisms in activated sludge, in: Kulpa, C. F., Irvine, R. L., and Sojka, S. A. (Eds.), *Proceedings of the Symposium on the Impact of Applied Genetics in Pollution Control*, University of Notre Dame, Notre Dame, IN, May 24–26, 1982.

Kampschreur, M. J., Temmink, H., Kleerebezem, R., Jetten, M. S. M., and van Loosdrecht, V. C. M., Nitrous oxide emission during wastewater treatment, *Water Research*, 43(17):4093–4103, September 2009.

Kartchner, A. D., Dixon, N. P., and Hendricks, D. W., Modeling diurnal fluctuations in stream temperature and dissolved oxygen, *Proceedings of the 23rd Purdue Industrial Waste Conference,* Purdue University, West Lafayette, IN, May 1969.

Latimer, W. M., *Oxidation Potentials*, 2nd edn., Prentice-Hall, Englewood Cliffs, NJ, 1952.

Lawrence, A. W., Modeling and simulation of slurry biological reactors, in: Keinath, T. M. and Wanielista, M. (Eds.), *Mathematical Modeling for Water Pollution Control Processes*, Ann Arbor Science Publishers, Ann Arbor, MI, 1975.

Li, L., Crain, N., and Gloyna, E. F., Kinetic lumping applied to wastewater treatment, *Water Environment Research*, 68(5): 841–854, July/August 1996.

Lide, D. R., *Handbook of Chemistry and Physics*, 77th edn., CRC Press, Boca Raton, FL, 1996.

McCabe, J. and Eckenfelder, W. W. Jr. (Eds.), *Biological Treatment of Sewage and Industrial Wastes,* Volume 1, *Aerobic Oxidation*, Reinhold Publishing, New York, 1956.

McCarty, P. L., Thermodynamics of biological synthesis and growth, *International Journal of Air and Water Pollution*, 9:621–639, 1965. The paper with discussions was also published (pp. 169–199) in *Proceedings of the Second International Water Pollution Research Conference*, Tokyo, Japan 1964, Pergamon Press, New York, Oxford, 1965 (a reprint was available).

McCarty, P. L., Phosphorous and nitrogen removal by biological systems, *Proceedings of the Wastewater Reclamation and Reuse Workshop at Lake Tahoe California* (226 pages), Sanitary Engineering Research Laboratory, University of California, Berkeley, CA, 1970.

McCarty, P. L., Stoichiometry of biological reactions, *Progress in Water Technology*, 7(1):157–172, 1975.

McKinney, R. E., *Microbiology for Sanitary Engineers*, McGraw-Hill, New York, 1962.

McKinney, R. E., Mathematics of complete mixing activated sludge (ASCE Paper 3516), *Transactions of ASCE*, 128(Part III): 497–534, 1963; Also discussions by Washington, Hetting, and Sathyanarayana.

McKinney, R. E. and Symons, J. E., Discussion (pp. 440–446) of McWhorter, T. R. and Heukelekian, H., Growth and endogenous phases in the oxidation of glucose (pp. 419–449 including discussion), in Eckenfelder, W. W. Jr. (Ed.), *Advances in Water Pollution Research* (Proceedings of the International Conference held in London, September 1962), Volume 2, Pergamon Press, The Macmillan Company, New York, 1964.

Metcalf, L. and Eddy, H. P., *American Sewerage Practice,* Volume III, *Disposal of Sewage*, McGraw-Hill, New York, 1916.

Metcalf & Eddy, Inc., *Wastewater Engineering—Treatment, Disposal, and Reuse*, 3rd edn., McGraw-Hill, Inc., New York, 1991.

Monod, J., *Annales de lInstitut Pasteur*, 68:444, 1942 (citation given by Monod, 1949).

Monod, J., The growth of bacterial cultures, *Annual Review of Microbiology*, Volume 3, Annual Reviews, Inc., Palo Alto, CA, 1949.

Morgan, P. F. and Bewtra, J. K., Air diffuser efficiencies, *Journal of the Water Pollution Control Federation*, 32(10):1047, 1960.

Nemerow, N. L. and Agardy, F. J., *Strategies of Industrial and Hazardous Waste Management*, Van Nostrand Reinhold, New York, 1998.

Orhon, D. and Artan, N., *Modeling of Activated Sludge Systems*, Technomic Publishing Co., Lancaster, PA, 1994.

Pagilla, K. R., Jenkins, D., and Kido, W. H., *Nocardia* control in activated sludge by classifying selectors, *Water Environment Research*, 68(2):235–239, March/April 1996.

Parker, D. S. *Process Design Manual for Nitrogen Control*, Technology Transfer, U.S. Environmental Protection Agency, Cincinnati, OH, 1975.

Pearson, E. A. and Haas, P., Unpublished field investigations, Sanitary Engineering Research Laboratory, University of California, Berkeley, CA, 1965. Cited in Pearson (1968).

Prescott, L. M., Harley, J. P., and Klein, D. A., *Microbiology*, 2nd edn., Wm. C. Brown Publishers, Dubuque, IA, 1993.

Porges, N., Jasewicz, L., and Hoover, S. R., Principles of biological oxidation, in: McCabe, J. and Eckenfelder, W. W. Jr. (Eds.), *Biological Treatment of Sewage and Industrial Wastes,* Volume 1, *Aerobic Oxidation*, Chapter 1–3, Reinhold Publishing, New York, 1956.

Prescott, L. M., Harley, J. P., and Klein, D. A., *Microbiology,* 6th edn., McGraw-Hill, Dubuque, IA, 2005.

Rawn, J. D., *Proteins, Energy, and Metabolism*, Neil Patterson Publishers, Burlington, NC, 1989.

Richard, M., Hao, O., and Jenkins, D., Growth kinetics of Sphaerotilus species and their significance in activated sludge bulking, *Journal of the Water Pollution Control Federation*, 57(1): 68–81, January 1985a.

Richard, M. G., Shimizu, G. P., and Jenkins, D., The growth and physiology of the filamentous organism Type 021N and its significance to activated sludge bulking, *Journal of the Water Pollution Control Federation*, 57(12): 1152–1162, 1985b.

Rickert, D. A. and Hunter, J. V., Colloidal matter in wastewaters and secondary effluents, *Journal of the Water Pollution Control Federation*, 44(1):134–139, January 1972.

Rittman, B. E., Aerobic biological treatment, *Environmental Science and Technology*, 21(2):128–136, 1987.

Rittman, B. E. and McCarty, P. L., *Environmental Biotechnology: Principles and Applications*, McGraw-Hill, New York, 2001.

Rittman, B. E., Smets, B. F., and Stahl, D. A., The role of genes in biological processes-Part I, *Environmental Science and Technology*, 24(1):23–29, 1990.

Rohlich, G. A., Methods for removal of phosphorous and nitrogen from sewage plant effluents, in: Eckenfelder, W. W. (Ed.), *Advances in Water Pollution Research* (Proceedings of the International Conference held in London, September, 1962), Volume 2, Pergamon Press, The Macmillan Company, New York, 1964.

Sawyer, C. N. and McCarty, P. L., *Chemistry for Sanitary Engineers*, McGraw-Hill, New York, 1967.

Schulze, K. L., The activated sludge process as a continuous flow culture-Part I, Theory, *Water and Sewage Works*, 111(12): 526–538, December 1964; Part II, Application, *Water and Sewage Works*, 112(1):11–17, January 1965.

Servizi, J. A. and Bogan, R. H., Free energy as a parameter in biological treatment, *Journal of Sanitary Engineering Division, American Society of Chemical Engineers*, 89(SA3): 17–40, June 1963.

Servizi, J. A. and Bogan, R. H., Thermodynamic aspects of biological oxidation and synthesis, *Journal of Water Pollution Control Federation*, 36(5):607–618, May 1964.

Sherrard, J. H., Kinetics and stoichiometry of completely mixed activated sludge, *Journal of the Water Pollution Control Federation*, 49(9):1968–1975, September 1977.

Silberberg, M., *Chemistry—The Molecular Nature of Matter and Change*, Mosby-Year Book, Inc., St. Louis, MO, 1996.

Stack, V. T. and Conway, R. A., Design data for completely mixed activated sludge treatment, *Sewage and Industrial Wastes*, 31(10):1180–1190, 1959.

Stewart, M. J., Pearson, E. A., and Hiramoto, E. M., Reaction kinetics of continuous flow anaerobic fermentation processes, *Proceedings of the 14th Industrial Waste Conference*, pp. 309–339, Purdue University, Lafayette, IN, May 5–7, 1959.

Stryer, L., *Biochemistry*, W. H. Freeman, New York, 1981.

Stumm, W., Discussion of paper, Rohlich, G. A., Methods for removal of phosphorous and nitrogen from sewage plant effluents, in: Eckenfelder, W. W. (Ed.), *Advances in Water Pollution Research* (*Proceedings of the International Conference* held in London, September, 1962), Volume 2, Pergamon Press, Macmillan Company, New York, 1964.

Suidan, M. T., Najm, I. N., Pfeffer, J. T., and Wang, Y. T., Anaerobic biodegradation of phenol: Inhibition kinetics and system stability, *Journal of the Environmental Engineering Division, ASCE*, 114(6):1359–1376, December 1988.

Tchobanoglous, G. and Burton, F. L. (Metalf & Eddy, Inc.), *Wastewater Engineering-Treatment, Disposal, Reuse*, 3rd edn., McGraw-Hill, New York, 1991.

URS Research Company, Procedures for Evaluating Performance of Wastewater Treatment Plants—A Manual, Under Contract No. 68-01-0107, Prepared for Office of Water Programs, Environmental Protection Agency, Washington, D.C., 1973.

Vagliasindi, F. and Hendricks, D. W., Wave front behavior in adsorption reactors, *Journal of the Environmental Engineering Division, ASCE*, 118(4):530–550, July/August 1992.

Washington, Hetting, and Sathyanarayana, discussion of paper by McKinney, R.E. Mathematics of complete mixing activated sludge (ASCE Paper 3516), *Transactions of ASCE*, 128(Part III):497–534, 1963.

Water Pollution Control Federation Manual of Practice No. 16, *Anaerobic Sludge Digestion*, 1968.

Yoshinaga, J. K., Hendricks, D. W., and Klein, D. A., Microbial growth kinetics for natural and synthetic organic compounds, Environmental Engineering Report, Department of Civil Engineering, Colorado State University, Fort Collins, CO, April, 1995.

23 Biological Reactors

The "idea" of a reactor is broad and may be defined as a volume of any size or shape in which a "change" occurs. In a biological reactor, microorganisms transform, or "change," substrate to desired reaction products.

The key themes in reactor theory are stoichiometry, kinetics, and mass balance. The stoichiometry and kinetics were reviewed in Chapter 22; in this chapter, the mass balance principle is added, which is the basis for modeling. Models are reviewed for three reactor categories: suspended growth, attached growth, and anaerobic. Practice in design and operation are summarized for each reactor category.

23.1 BIOLOGICAL REACTOR SPECTRUM

The idea of a reactor is to provide a favorable environment for a substrate to be transformed by bacterial enzymes to benign products. To accomplish this purpose a host of reactors have evolved to fit different purposes and conditions. All have the common themes: stoichiometry, kinetics, mass balance, which provide a framework for analysis and understanding, but are supplemented by empirical parameters.

To relate to the literature, Table 23.1 identifies some reactor names. Table 23.1 indicates that a classification of reactors has limited coherence; one may classify a plug-flow reactor, for example, as a continuous-flow reactor, but as a concept it applies to other kinds of reactors as well. Land application and aqua-culture are included to indicate that "passive" kinds of treatment are sometimes preferred. Some, e.g., granular activated carbon and "soil," are included so that the extent of the reactor concept is seen. As a note, the table is not intended to read across; the columns are intended to add some logic (or coherence) in categorization, but not the rows.

23.2 ACTIVATED SLUDGE

Reactor theory, i.e., materials balance combined with kinetics, is a major theme of most chapters. The theory was adapted to activated sludge from chemical engineering in the 1950s (see, for example, Gram, 1956), which provided a framework for understanding.

If the activated-sludge practice had started along the lines of reactor theory, the present state of the art would have a simple rationale as indicated above. Practice developed empirically, however, and the state of the art is compounded by the consequent lore which unfolded over the decades, and inconsistent units.

Over the decades since 1914 when Ardern and Lockett launched activated-sludge practice with their historic paper, a proverbial thousand papers, c. 1970, had been presented or published, along with a number of books. The review that follows abstracts key notions from the extensive lore and, at the same time, uses the reactor theory to explain the activated-sludge process and its variations. Reactor theory also gives a rationale for empirical parameters that were evolving during the 1950s.

23.2.1 History

The beginning was in 1914 when Ardern and Lockett published their paper, "Experiments on the oxidation of sewage. . . . " Actually, however, they stood on the shoulders of some of their predecessors. We can understand the development of the field by first looking at where it stood in 1914, and then the early development of practice and finally the milestones that enumerate how we arrived at current practice and knowledge.

23.2.1.1 Beginnings

The account given by Metcalf and Eddy (1930) gives insight into the evolution of the activated-sludge process. The nuances in their writing add flavor.

> The aeration of sewage in tanks, to hasten oxidation of organic matter, was investigated as early as 1882 by Dr. Angus Smith, who reported on it to the Local Government Board. It was studied subsequently by a number of investigators and in 1910 Black and Phelps reported that a considerable reduction in putrescibles could be secured by forcing air into sewage in basins (p. 636).

Following this, experiments by Clark and Gage at the Lawrence Experiment Station, conducted during 1912 and 1913 on sewage in bottles and in tanks partially filled with roofing slate spaced about 1 in. apart, showed that in aerated sewage, growths of organisms could be cultivated which would greatly increase the degree of purification obtained.

The results of the work at Lawrence were so striking that their knowledge led Fowler to suggest experiments along similar lines at Manchester, England, where Ardern and Lockett carried out valuable researches upon this subject. During the course of their experiments Ardern and Lockett found that the sludge played an important part in the results obtained by aeration, as announced in their paper of May 3, 1914, before the Manchester Section of the Society of Chemical Industry. At the outset it was necessary to aerate the sewage samples continuously for 5 weeks before complete nitrification was obtained. By repeatedly drawing off the clarified sewage and adding fresh raw sewage to the old sludge left in the experimental tank, the time for oxidation, however, was reduced to 24 h and eventually to a few hours

TABLE 23.1
Sampling of Biological Reactors

Suspended Growth Reactors	Attached Growth Reactors	Reaction Type
Activated sludge		
Conventional	Trickling filters	Aerobic
Plug-flow	Bio-filters	Anoxic
Complete-mix	Rotating biological contactors	Nitrification
Extended aeration	Soil, gravel	Anaerobic
Tapered aeration	Granular activated carbon	Sludge digesters
Hatfield process		Anaerobic reactors
Pasveer oxidation ditch		
Ponds	**Hybrid Reactors**	**Hydraulic Classification**
Aerobic	Imhoff tanks	Continuous-flow
Anaerobic	Septic tanks	Plug-flow
Facultative	Leach fields	Complete-mix
Wetlands	Land application	Batch
	Aquaculture	Sequencing batch
	Constructed wetlands	

only. This sludge accumulating in this manner and inducting such active nitrification was called "activated sludge." Hatton at Milwaukee, Rank at Baltimore, and others, finally showed that the process could be operated on a practical, continuous basis by running sewage through aeration tanks, the activated sludge being mingled with the entering sewage, and later separated from it after its passage through the tanks.

Metcalf and Eddy further reviewed the "state of the art" for the design of activated-sludge reactors. They report on the variations (p. 637) as follows:

1. *Diffused-air aeration or activated-sludge process proper*. In this method of treatment, air is blown into sewage as it flows through tanks, activated sludge being added to the incoming sewage and settled out in sedimentation basins from the tank effluent. A part of the sludge produced is fed into the influent as "return sludge," and a part is disposed of as "excess sludge."
2. *Mechanical aeration or bio-aeration*. Mechanical apparatus is employed in this method, to aerate the sewage and keep the tank contents in circulation. Absorption of air takes place from the atmosphere at the surface of the sewage.

Metcalf and Eddy go on to describe three distinct functions of diffused air: (1) to provide oxygen necessary to attain aerobic conditions and promote the growth of oxidizing organisms, chiefly bacteria; (2) to cause the activated sludge to move through the sewage and provide contact between the gelatinous surfaces of the sludge and the organic matter contained in the sewage; and (3) to prevent deposition of the sludge. Figure 23.1 shows two variations of tank sections.

Table 23.2 illustrates the extent of activated-sludge practice by 1930. The Houston North Side Plant was built in 1917, only 3 years after the Ardern and Lockett paper. Depths, widths, aeration times, etc., are essentially the same as current diffused aeration practice. Air consumption is on an "air volume per gallon of sewage" basis. Table 23.3 reports the performance of activated-sludge plants of the day. Parameters measured included suspended solids, BOD, ammonia, and nitrates. Plant performances were similar to contemporary plants of similar designs; actually, the designs have not changed much in concept. Reactors remain "plug-flow" and are long and narrow with a rectangular cross section, and the spiral flow was used until about 1980. Detention times were less than plants since the 1950s, however, and the surface loading has not been used, probably since the book was published. A volumetric loading was used by 1950.

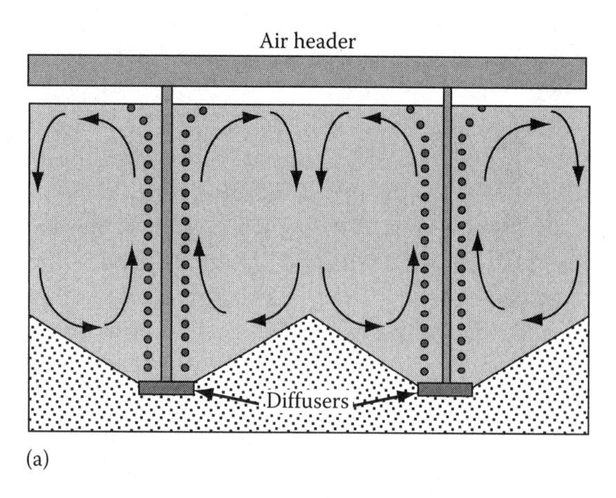

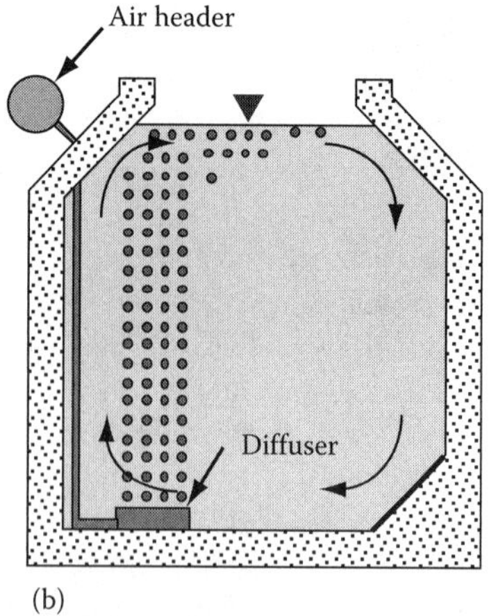

FIGURE 23.1 Two variations of conventional activated sludge using diffused air. (a) Ridge and furrow tank and (b) spiral flow tank. (Adapted from Metcalf, L. and Eddy, H.P., *Sewerage and Sewage Disposal*, McGraw-Hill, New York, 1930.)

TABLE 23.2
Some Diffused Air Activated-Sludge Plants in Operation in 1930 with Design Characteristics

	Milwaukee	Indianapolis	Chicago, North Side	Decatur	Manchester	Houston, North Side	Pasadena	Toronto
Date of construction	1925	1924	1926	1927	1923	1917	1924	1929
Population served	575,000	360,000	800,000	50,000	18,000	67,000	120,000	50,000
Preliminary treatment								
Grit chambers	Yes	Yes	Yes	Yes	None	Yes	None	Yes
Fine screens	Yes	Yes	None	None	None	None	Yes	None
Sedimentation (hours)	None	None	0.5	1.1	0.7	None	None	1.6
Aeration tanks								
Type	Ridge/fur	Spiral	Spiral	Spiral	Spiral	Ridge/fur	Ridge/fur	Spiral
Number	24	6	36	6	2	4	30	4
Length (m)	72	72.5	128	36.6	53.9	85.3	20.6	49.7
Width, one channel (m)	6.7	6.1	10.6	4.9	1.9	2.7	3.0	4.0
Depth (m)	4.6	4.6	4.6	4.4	2.1	3.0	4.6	3.2
Passes per tank	2	4	1	1	3	2	1	2
Aeration time (h)	6	4.7	6	2.5	7	2.3	4	4
Airflow (ft^3/gal)	1.5	1	0.8	1	0.9	1.4	1	1
Return sludge fraction	0.25	0.20	0.25	0.1	0.15	0.25	0.33	0.25
Area diffusers/area tank	1:4	1:13	1:9	1:9	1:17	1:7	1:6	1:10
Final sedimentation tanks								
Type	Dorr	Hopper bot.	Dorr	Dorr	Hopper bot.	Hopper bot.	Hardinge	Fidler
Number	11	12	30	2	1	40	5	2
Dimensions (m)	29.9 dia	23.8 × 12.8	23.5 × 23.5	23.6 × 23.6	11.0 × 11.0	5.8 × 3.0	15.2 × 15.2	19.8 × 19.8
Depth (m)	4.6	4.6	4.9	4.2	6.4	6.7	4.9	4.5
Detention time (h)	1.7	1.6	2	2.6	2	1.5	2	2
Surf. loading (gal/ft^2/d)	850	1380	980	830	1,000	1,330	880	710
Re-aeration tanks								
Number	None	None	None	None	1	4	5	None

Source: Adapted from Metcalf, L. and Eddy, H.P., *Sewerage and Sewage Disposal*, McGraw-Hill, New York, 1930.

TABLE 23.3
Performance of Activated-Sludge Plants

	Houston, North Side	Houston, South Side	Indianapolis	Milwaukee	Pasadena	San Marcos	Sherman
Year	1923	1923	1927	1927	1928–1929	1920	1920
Period covered	1 year	1 year	1 year	1 year	6 months	10 days	11 days
Suspended solids (mg/L)							
Influent	272	166	318	250–300	336	42	264
Effluent	34	9	27	25	15	3	73
Percent removal	87	94	92	90–92	96	92	72
BOD—5 day (mg/L)							
Influent	120		236	200–250	207	57	165
Effluent	8		32	≤ 6	9	16	33
Percent removal	93		86	≥ 97	96	72	80
Ammonia (mg/L)							
Influent	19	16			18	17	26
Effluent	7	5			8	3	23
Nitrates in effluent (mg/L)	4	11	0.7	4	4	9	1

Source: Metcalf, L. and Eddy, H.P., *Sewerage and Sewage Disposal*, McGraw-Hill, New York, 1930.

23.2.1.2 From Empiricism to Science

In sizing aeration tanks, Fair and Geyer (1954, p. 720) mention several sizing parameters including detention time, θ; volumetric loading (lb BOD/day/ft^3 tank volume); and BOD solids loading (lb BOD/day/1000 lb suspended solids per hour of aeration). The latter is called the F/M ratio (see Section 22.5.7.2). Detention time and F/M remain in practice for sizing the tank volume and substrate loading, respectively.

While reactor modeling was not underway until the mid-1950s, the biological character of activated sludge was recognized at inception and extended over the decades. Fair and Geyer (1954, p. 506), for example, describe activated sludge as an ecological system consisting of a variety of species of bacteria, fungi, protozoa, and rotifers. The 1950s at MIT saw the sciences, organic chemistry, biochemistry, and physical chemistry, integrated into graduate studies. Books with the modern science orientation had MIT roots, e.g., *Microbiology for Sanitary Engineers* (McKinney, 1962) and *Chemistry for Sanitary Engineers* (Sawyer and McCarty, 1967). Another milestone was the paper "Mathematics of complete mixing activated sludge" (McKinney, 1963). In 1962 MIT terminated its program in sanitary engineering; by that date, however, the transition was essentially complete. Its influence had spread, dating from perhaps 1948 and its graduates had disseminated to other universities.

23.2.1.3 Milestones

Activated sludge theory and practice have evolved from about 1880. Milestones are enumerated as follows, taken from references indicated.

1. *Lawrence*. In 1912, H.W. Clarke, at the Lawrence Experiment Station, Massachusetts, studied waste purification through its aeration in the presence of microorganisms. Dr. G.J. Fowler, consulting chemist for the Rivers Committee of the Manchester Corporation, observed some of the Lawrence experiments and suggested to Edward Ardern and William Lockett of the Davyhulme Sewage Works, Manchester Corporation that they carry out similar experiments (Goodman and Englande, 1974).
2. *Ardern and Lockett*. Ardern and Lockett (1914) found high removals of organic matter as measured by *oxygen absorption* after only several hours of aeration with *activated sludge*. The term *activated sludge* was given, for lack of another term, for the deposited solids resulting from the oxidation of sewage.
3. *Michaelis–Menten*. In 1913, Leonor Michaelis and Maud Menten (see Appendix 22.B) developed an expression for enzyme kinetics (Stryer, 1981). This was the same expression as the one later developed through empirical observations by Monod (Pearson, 1968).
4. *Monod*. In 1942, Monod (1949) published the results of his studies on continuous bacterial cultures, with an empirical equation describing results.
5. *Cell synthesis from substrate*. In 1951, Heukelekian et al. proposed the equation (Goodman and Englande, 1974),

$$\Delta X_v = a \cdot S_o - b \cdot X_v \tag{23.1}$$

where
ΔX_v is the volatile suspended solids (VSS) accumulation rate in aeration basin
(kg VSS synthesized/day)
S_o is the mass flow of BOD in influent flow (kg BOD flux-in./day)
X_v is the mixed liquor volatile suspended solids in suspension (kg VSS)
a is the VSS synthesis rate (kg VSS synthesized/day/kg BOD flux-in.)
b is the VSS oxidation rate kg VSS destroyed/day/kg VSS in aeration basin)

The Heukelekian equation was a first-order equation with respect to substrate, showing that cell synthesis is proportional to BOD, with endogenous respiration proportional to cell mass present (the units were changed to metric).

6. *Chemical formula for microorganisms*. In 1952, Hoover and Porges reported an empirical formula for the composition of activated-sludge microorganisms. The formula, $C_5H_7O_2N$, yields a molecular weight of 113, which is corrected to 124 to account for the ash content of the organisms (Goodman and Englande, 1974).
7. *Eckenfelder*. In 1954, Eckenfelder and O'Connor proposed a mathematical model for activated sludge which established a systematic mathematical approach to activated-sludge process design (Goodman and Englande, 1974).
8. *Kinetic model search at Berkeley*. In 1956, Mervin Stewart made a thorough appraisal of the kinetic descriptions of biological systems and from this study, selected the Michaelis–Menten kinetic model (Pearson, 1968). Stewart was doing doctoral research in methane fermentation kinetics under Erman Pearson. The research was continued by Frank J. Agardy and John F. Andrews. In addition, Andrew Gram (1956a,b) completed a doctoral dissertation incorporating Monod kinetics and a mass balance reactor model in an experimental study.
9. *Complete-mix model and F/M*. In 1962, McKinney (1962) proposed a complete-mixing model. Although the focus was on complete mixing, the approach was based upon a kinetic formulation of substrate utilization in the declining growth phase (of the growth curve) combined with a material-balance description of the reactor. McKinney used the food-to-microorganism ratio, i.e., F/M, and tied bacteria synthesis to substrate degradation. McKinney also

described extended aeration and aerated lagoon flow schematics. The paper recognized the need to break with the past, i.e.,

> The lack of understanding of the fundamental microbiology and the total dependence of engineers on the empirical design criteria are the causes of the problems concerning extended aeration systems.

and

> "....no longer can the design of biological waste treatment systems be based on crude empirical relationships."

10. *Michaelis–Menten verification.* In 1968, Pearson (1968) reported that after 8 years of laboratory research dealing with the growth kinetics of mixed cultures of anaerobic organisms over a wide range of growth rates, the results indicated that the Michaelis–Menten model fit the experimental data better than any other expression. In other words, although Pearson knew of the Michaelis–Menten/Monod model since the work of Andrew Gram, he let further experimental results confirm its validity before committing.
11. *Materials-balance model.* In 1968, Pearson (1968) described the materials-balance approach to reactor analysis, incorporating Michaelis–Menten kinetics. He proposed the model as applicable to either aerobic or anaerobic systems. His paper delineated substrate-balance, cell-balance, and oxygen-balance equations and established the basis for the modern approach to activated sludge. Pearson's paper reflected a chemical engineering influence which had been creeping into sanitary engineering education since the early 1960s (Pearson required his doctoral students to minor in chemical engineering).
12. *Pearson's critique of the state of the art.* Pearson in his 1968 paper analyzed the state of the art of knowledge, pointing out that most of the progress in wastewater treatment technology was the result of field experiments, done largely by trial and error. Further, he asserted that this past research did not accrue benefits because of the lack of an adequate theoretical basis upon which to design meaningful experiments and to obtain interpretable data. Much of the research, he stated, was undertaken within an empirical framework of analysis directed to specific system characteristics of interest or to practical problems to be resolved.
13. *Lawrence and McCarty.* In 1970, Lawrence and McCarty (1970) published a paper that established the concept of sludge age which became the major parameter for operation. They also formalized reactor theory as the theoretical basis for activated-sludge design. The paper became the defining modern characterization of activated sludge.
14. *Metcalf and Eddy.* In 1972, Metcalf & Eddy (1972) incorporated the reactor theory approach, which, in addition to the journal literature, accelerated the assimilation of science. The reactor theory culminated a 20 year quest for a rational basis for activated-sludge design theory. The 1972 book continued the classic series started by Leonard Metcalf and Harrison P. Eddy with their three-volume set, *American Sewerage Practice*, published in 1914, 1915, and 1916 with a textbook version in 1930.
15. *Nutrients.* Continuing with the story, during the 1970s, new demands were placed on the activated sludge, i.e., to nitrify, denitrify, and remove phosphorus as related to evolving standards for ambient waters. Activated-sludge processes were developed to deal with these new exigencies, resulting in a complex mix of associated technologies.
16. *IWA model.* To integrate the various facets of activated-sludge theory, the development of a comprehensive computer model was started in 1983 by an IWA Task Group, chaired by Professor Mogens Henze of the Technical University of Denmark (Henze, 1987). The model accommodated complexity, including process variations and nitrification, and was dynamic in that input variation could be included. The model development has continued with updates being provided by periodic publications by the Workgroup.

23.2.1.4 Modern History

The "modern" history of water pollution control is considered here to have begun about 1950. Educational institutions, consulting engineering firms, equipment manufacturers, governmental agencies, and professional organizations have had roles. A fervor in the field coincided essentially with the environmental movement; solving the problem of pollution was a focal point of the movement, and with it wastewater treatment.

Within the broad field of wastewater treatment, activated sludge was and is the "workhorse" of treatment processes. Issues have related to how to make the process work more effectively, more reliably, more cheaply, etc. This has involved improved understanding, better designs, operating guidelines, training operators, process modeling, etc.

To give a sense of one of the key persons who helped shape the foundation of our modern theory and knowledge of biological treatment, Box 23.1 reviews briefly, the career of Professor Erman Pearson. It is written in the first person since it is from personal interactions to a large extent.

23.2.2 Activated-Sludge Reactor Analysis

As noted, modern reactor theory is based on the mass balance of substrate and cells, combined with Monod kinetics. The empirical parameters, substrate utilization rate, U, and sludge age, θ_c, may be derived from reactor theory.

BOX 23.1 ERMAN PEARSON

Anaerobic treatment has had fewer adherents than has activated-sludge. While the latter had the proverbial thousand papers by the 1970s the anaerobic field just started to flower in the late 1950s. The anaerobic process is more complex and requires a great deal of persistence; everything is more difficult. One of a handful of academics who pursued this field, along with a number of able graduate students, was Professor Erman Pearson (1920–1985).

I knew him first in 1953 as an undergraduate student in two of his courses; he was always well prepared and articulate and transmitted volumes of basic and lasting knowledge on water supply and sewerage. As undergraduates we did not know of his other life as a researcher and as a leader in the field. I became familiar with these aspects in 1965 after getting my doctorate. The first thing I learned was that with Harvey Ludwig and about 20 professors, he was the instigator and first president of Association of Environmental Engineering and Science Professors (AEESP), formed in 1963. As with all of his compatriots, he had a presence when in a room, due to his size, his wit, and his incisive remarks. About the same time, 1960, he was helping to form what is now the International Water Association (IWA). The first biennial conference was in London in 1962, the second in Tokyo in 1964, the third in Munich in 1966, and the fourth in Prague in 1968. Professor Pearson followed Professor Eckenfelder as its third president, 1966–1968.

Professor Erman Pearson, c.1960. (Courtesy of Department of Civil and Environmental Engineering, University of California, Berkeley, CA.)

The first issue of *Water Research*, 1(1), January 1967 lists the governing board members, with an editorial by Pearson. Dr. Samuel H. Jenkins was one of the founders and the executive editor. Professor Eckenfelder and a number of academics from countries worldwide were also founders.

One of Professor Pearson's seminal contributions was a 1967 paper at a University of Texas lecture series, later published in a book (Pearson, 1968). His paper was a lucid exposition of activated-sludge reactor theory incorporating Michaelis–Menten kinetics with materials-balance, showing derivative parameters such as substrate removal velocity, U, sludge age, θ_c, net growth rate, and nutrient removal. I asked him how he got into activated-sludge theory. He remarked that when he was invited to lecture on the topic at Texas and started preparing he saw that "everyone had been flying in the wrong direction." His paper was to "try to set things straight." The statement illustrated both his sense of humor and, most likely, his frank assessment. The texas paper had outlined the basics of modern theory, and was a culmination of research that started with Andrew Gram in 1956, but with most of his doctroal students working on anaerobic topics. The transition to activated sludge theory was not an issue since the principles of biological growth and reactor theory are common to both aerobic and anaerobic treatment.

In addition to his research and leadership roles, he was a noted consultant and advisor to the State of California, which had advanced programs in pollution control. Although teaching itself has not too much recognition, he was a dedicated to this role, for both undergraduate and graduate areas. The field of environmental engineering has had, perhaps, more than its share of outstanding professionals, many who were and are known for their own unique character traits. Professor Pearson stood out as a professional and as a personality.

23.2.2.1 Materials Balance

A materials-balance statement for a reactor is merely a way to *account* for all the happenings within its volume (see also Chapter 4). The general statement is

$$\begin{aligned}&\text{accumulated net changes}\\&\quad = \text{materials inflow} - \text{materials outflow}\\&\qquad + \text{rate of production or depletion of material}\\&\qquad \text{within the reactor.}\end{aligned} \tag{23.2}$$

The material being accounted for may be anything of interest. For the case of activated sludge, the materials of interest are substrate and cells.

Procedure. The basic technique of analysis is to (1) sketch the system being analyzed, (2) draw a circle around the portion of the system of interest, and (3) write a materials-balance equation for the portion circumscribed.

Homogeneous volume. The most important concern in applying Equation 23.2 is that the contents of the volume being analyzed must be homogeneous. If the volume being considered is not homogeneous, then it must be broken down into homogeneous subunits. Or, for the purposes of an analysis, it is *assumed* to be homogeneous.

Reactors. Several kinds of reactors include complete-mix, plug-flow, extended aeration, aerated lagoon, batch, sequencing-batch, etc. The analysis of complete-mix and plug-flow is a basis for understanding variations.

1. *Complete-mix reactor*. A turbine aeration systems is "completely mixed" because of the high rate of pumping by the aerator. The assumption of a complete-mix reactor is that the entering fluid is instantly mixed throughout the reactor, i.e., the reactor is homogeneous. The substrate concentration, S, in the reactor effluent is the same within all parts of the reactor volume. The mass balance analysis for a "complete-mix" reactor is developed first, i.e., prior to the "plug-flow" reactor model, since it is simpler mathematically.
2. *Plug-flow reactor*. For a *plug-flow* activated-sludge reactor, such as for a diffused aeration system, the volume is not homogeneous, since the substrate concentration, S, declines along its length. Therefore, the reactor must be analyzed in terms of "slices," which are cross sections of finite thickness. It is not strictly true that a "slice" is a homogeneous volume but such an assumption simplifies the mass balance analysis. Another assumption is that longitudinal dispersion is negligible, not true, but it simplifies the mathematical depiction.
3. *Cell mass balance schemes*. The premise of activated sludge as established by Ardern and Lockett is that the reactor must contain previously synthesized viable cells. These cells convert the organic substrate (the waste material) to new cells as depicted in Section 22.3.3.1. To maintain the needed cell concentration, X (or the surrogates, MLVSS or MLSS), three basic schemes are used: (a) cell separation, recycle, and wasting; (b) cell separation, recycle, no wasting; and (c) no cell separation, no recycle, and no wasting. These three schemes are called conventional recycle activated sludge, extended aeration activated sludge, and aerated lagoon, respectively. The mathematical depictions follow these descriptions.

23.2.2.2 Conventional Activated Sludge

Figure 23.2 depicts *conventional* activated sludge, which by definition involves cell recycle and cell wasting. The system is also "complete-mix," since its shape is closer to square, whereas a "plug-flow" reactor is long and narrow in shape. The operator controls the process by maintaining a desired cell concentration, X, in the reactor, which is done by controlling the R, the return flow, and W, the flow of waste sludge.

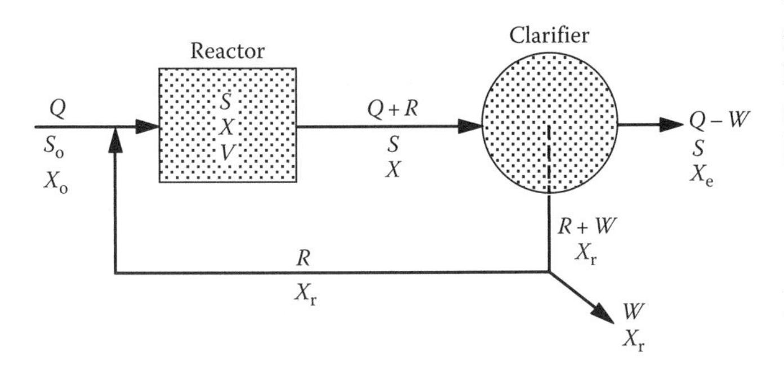

FIGURE 23.2 Schematic of a conventional activated-sludge reactor.

23.2.2.2.1 Substrate Balance

The substrate materials balance for the reactor and cell separator in Figure 23.2 is

$$\left[\frac{dS}{dt}\right]_o \cdot V = QS_o - (Q - W)S - WS - \left[\frac{dS}{dt}\right]_r \cdot V \tag{23.3}$$

where

$[dS/dt]_o$ is the observed rate of change of substrate concentration in the reactor (mg BOD/L/day)
V is the volume of reactor (L)
Q is the flow through the reactor (L/day)
S_o is the concentration of substrate in the influent flow; measured as BOD, COD, TOC, etc, as convenient as long as other units are consistent (mg/L)
S is the concentration of substrate in the reactor (mg/L)
$[dS/dt]_r$ is the rate of substrate conversion reaction (mg BOD/L/day)

Equation 23.3 is completely general; note that the "RS" into the reactor cancels the "RS" out (and to be strictly correct Equation 23.3 should have these terms). If carried through as a finite-difference equation it could be solved for transient conditions. The "steady-state" approximation is usually imposed, however, which requires that the flow, Q, and substrate inflow concentration, S_o, do not vary with time. Therefore, the observed rate of change of substrate concentration in the reactor is zero, i.e., $[dS/dt]_o = 0$. The resulting steady-state description is

$$0 = Q(S_o - S) - \left[\frac{dS}{dt}\right]_r \cdot V \tag{23.4}$$

Substituting the hydraulic detention time, $\theta = V/Q$,

$$(S_o - S) = \left[\frac{dS}{dt}\right]_r \cdot \theta \tag{23.5}$$

which says that the change in substrate concentration between the reactor influent and effluent is the rate of reaction times the hydraulic detention time. Incorporating the relation between substrate reaction and cell synthesis, i.e., $[dS/dt] = (1/Y) \cdot [dX/dt]_g = (1/Y) \cdot [\mu X]$, Equations 22.29 and 22.30,

$$(S_o - S) = \frac{\mu}{Y} X\theta \tag{23.6}$$

Rearranging, again gives, U, i.e.,

$$U \equiv \frac{(S_o - S)}{\theta X} = \frac{\mu}{Y} \tag{23.7}$$

In other words, the rate of substrate utilization, Equation 23.7, may be derived from the substrate mass balance, Equation 23.7. As seen, the difference $(S_o - S)$ is greater with higher cell concentration and longer detention time, resulting in lower effluent concentration, S. For reference, the F/M ratio is related to U by the relation, $U = F/M[(S_o - S)/S_o] = [S_o/\theta X] \cdot [(S_o - S)/S_o]$ as seen in Section 22.5.7.3.

The Monod equation, i.e., $[dS/dt]_r = (1/Y) \cdot \hat{\mu}[S/(K_s + S)] \cdot X$, Equation 22.31 may also be inserted into Equation 23.5, which gives a "final" equation that incorporates the basic variables, including the kinetic ones, i.e., Y, $\hat{\mu}$, K_s.

$$(S_o - S) = \frac{1}{Y}\hat{\mu}\frac{S}{K_s + S}X\theta \tag{23.8}$$

Equation 23.8 may be useful for a spreadsheet algorithm to explore the effect of X, and θ on performance, given assumptions on kinetic coefficients. The effect of uncertainties regarding $\hat{\mu}$ and K_s may also be explored as a "sensitivity-analysis." Keep in mind that "X" has an upper limit based on buildup of inert solids due to cell recycle.

23.2.2.2.2 *Cell Balance*

In terms of cells, and circumscribing the system, i.e., the reactor and the cell separator, the materials balance is

$$\left[\frac{dX}{dt}\right]_o \cdot V = QX_o - (Q - W)X_e - WX_r + \left[\frac{dX}{dt}\right]_g \cdot V - \left[\frac{dX}{dt}\right]_d \cdot V \tag{23.9}$$

where

$(dX/dt)_o$ is the observed rate of change of cells in the reactor (mg MLVSS/L/day)
W is the waste flow of cells of concentration, X_r, leaving system (L/day)
X_o is the concentration of cells entering reactor (mg MLVSS/L)
X_r is the concentration of cells in underflow from the cell separator (mg MLVSS/L)
X_e is the concentration of cells in effluent from the cell separator (mg MLVSS/L)
$[dX/dt]_g$ is the growth rate of cells in reactor (mg MLVSS/L/day)
$(dX/dt)_d$ is the death rate plus endogenous respiration rate of cells in reactor (mg MLVSS/L/day)

If the cell separator is performing well, $X_e \approx 0$, is assumed (actually, $X_e \approx 30$ mg/L, which compares with $X_r \approx 10{,}000$ mg/L). Also, let $X_o \approx 0$. Equation 23.45 then becomes

$$\left[\frac{dX}{dt}\right]_o \cdot V = -WX_r + \left[\frac{dX}{dt}\right]_g \cdot V - \left[\frac{dX}{dt}\right]_d \cdot V \tag{23.10}$$

Thus it is clear that for a steady-state system, where $[dX/dt]_o = 0$, then the rate of cell wastage equals the net rate of cell synthesis. Applying kinetic equations, and the steady-state approximation, Equation 23.10 becomes

$$WX_r = \mu XV - bXV \tag{23.11}$$

which can be rearranged to give

$$\frac{WX_r}{VX} = \mu - b \tag{23.12}$$

Also, by definition (Section 22.5.7.5),

$$\theta_c^m \equiv \frac{1}{(\hat{\mu} - b)} \tag{23.13}$$

where θ_c^m is the minimum mean cell residence time (day).

The term, θ_c^m is the minimum time required for cell regeneration, which also accounts for cell decay rate.

Therefore, from Equation 23.11, $VX/WX_r > \theta_c^m$. Actually, recalling Equation 22.52, the mean cell age, θ_c, is defined as

$$\theta_c = \frac{V(\text{reactor}) \cdot X}{WX_r} \tag{22.52}$$

At the same time, it must be true that $\theta_c > \theta_c^m$, which is necessary to avoid losing cells at a rate faster than they are regenerated. Tchobanoglous and Burton (1991, p. 393) recommend $2 < \theta_c/\theta_c^m < 20$. The mean cell age, θ_c, is a key operating parameter, adopted in practice almost universally. As seen, four variables, V, X, W, X_r, control θ_c.

The net growth rate, $(\mu - b)$, is also defined as the "net rates of cell production per unit mass of cells in the system," which in operation is the "mass flux of cells leaving per day per unit mass of cells in the system." Or, $(\mu - b) = (dX/dt)X$.

23.2.2.2.3 *Cell Recycle*

The mass balance for the cell separator, after neglecting the effluent cell flux $(Q - W)X_e$, is

$$(R + W)X_r = (Q + R)X \tag{23.14}$$

23.2.2.2.4 *Summary*

In summary, materials balance/kinetics is the starting point for reactor analysis. It is substrate mass balance and cell mass balance. The method: (1) for the substrate mass balance, circumscribe the reactor; (2) for the cell mass balance, circumscribe the system, i.e., both the reactor and cell separator; and (3) for the cell recycle ratio, circumscribe the reactor only. By the same token, from the substrate mass balance, the "substrate utilization rate," U, is derived (almost the same as the F/M ratio). From the cell mass balance, the sludge age, θ_c, is obtained. The cell mass balance around the reactor gives R/Q.

As to practical utility, Equation 23.7 can be used to determine a reactor volume, V, for a desired S (using kinetics from the previous sections). Equation 23.11 gives the required cell wastage rate, and Equation 23.54 gives the recirculation ratio in terms of needed X.

The needed X relates back to kinetic analysis; higher X results in higher reaction velocity. Higher values of X in the reactor are achievable by achieving a higher cell concentration, X_r, in the final clarifier underflow, or by increasing the R/Q ratio. Increasing the R/Q ratio is not a good approach, however, since it decreases the effective hydraulic detention time, θ, i.e., the value for θ should be calculated as $\theta = V/(Q+R)$, not $\theta = V/Q$. The latter is a simplifying assumption that permits illustration of relationships. A modification to the equation shown and solved on a spreadsheet provides a more accurate calculation of θ.

As a matter of interest, because of the uncertainty in the values of the kinetic constants, i.e., $\hat{\mu}$ and K_s, it is more expedient in practice to size the reactor, V(reactor), based on what is known to work through decades of experience, e.g., $\theta \approx 6$ h. When dealing with industrial wastes, the kinetics may warrant a selecting of different value for θ, usually based on pilot plant work or experience within the industry.

23.2.2.3 Extended Aeration

Figure 23.3 is the schematic for an extended aeration system. As noted previously, there is no cell wasting, which is the distinguishing feature of this reactor mode. The idea is that the cells remain in the system by continuous recycle with cell mass decreasing due to endogenous respiration.

23.2.2.3.1 Extended Aeration

The mass balance relations are the same as for conventional activated sludge, except that $W=0$. The associated resultant equations after applying assumptions are given in the same sequence, substrate and cells.

23.2.2.3.2 Substrate Mass Balance

Applying assumptions: for "steady-state," $dQ/dt=0$, and $dS_o/dt=0$, $\theta = V/Q$, and letting, $[dS/dt] = (1/Y)\cdot[dX/dt]_g = (\mu/Y)\cdot X$,

$$(S_o - S) = \frac{\mu}{Y} X\theta \qquad (23.6)$$

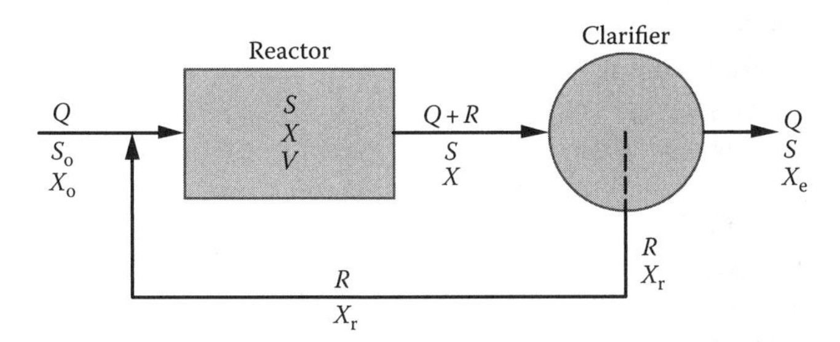

FIGURE 23.3 Schematic representation of an extended aeration system.

23.2.2.3.3 Cell Mass Balance

Circumscribing the system, i.e., the reactor and clarifier, $W=0$, assume steady state, $dQ/dt=0$, and let $X_o \to 0$, to give a resultant mass balance,

$$\left[\frac{dX}{dt}\right]_o \cdot V = -QX_e + (\mu - b)XV \qquad (23.15)$$

In other words, if the mass flow of viable cells leaving the clarifier is higher than the net cell production rate, the cell concentration in the reactor will decrease with time. By the same token, decreasing the mass flow, $Q\cdot X_e$, leaving the clarifier will cause an increase of cells in the reactor.

Next impose $[dX/dt]_o = 0$, which means that at steady state the observed rate of change in the reactor is zero, meaning that the mass flux of cells leaving the reactor equals the net rate of cell synthesis. Divide by Q, to give

$$X_e = (\mu - b)X\theta \qquad (23.16)$$

23.2.2.3.4 Cell Recycle

To determine, R, to maintain a required X, the materials balance is written about the reactor only, which gives, after rearrangement,

$$\frac{R}{Q} = \frac{X - X_e}{X_r - X} \qquad (23.17)$$

23.2.2.3.5 Summary

The two equations that tell the story are Equations 23.6 and 23.16 for substrate and cells, respectively, i.e.,

$$(S_o - S) = \frac{\mu}{Y} X\theta \qquad (23.6)$$

$$X_e = (\mu - b)X\theta \qquad (23.18)$$

Equation 23.6 shows that as the product, $X\theta$, increases, the effluent substrate concentration, S, declines; at the same time, X_e increases. Because the cells must leave in the clarifier effluent, the suspended solids limits will most likely exceed the regulatory limit. A rearrangement of Equation 23.18, is $\theta/\theta_c = X_e/X$. The idea of extended aeration is that b increases as θ_c increases; a larger fraction of the cells is "digested" aerobically.

23.2.2.4 Aerated Lagoon

The schematic representation of an aerated lagoon is shown in Figure 23.4. A lack of cell recirculation characterizes the system. The cell separation was omitted in the early years of use of this system, i.e., in the 1960s, but increasingly stringent regulations called for addition of this unit operation. Usually the clarifier is a pond in which the cell removal is by a batch process, such as bulldozer removal after draining, perhaps every several years. The cells may decay anaerobically on the bottom of the clarifier; since about 2000, methane

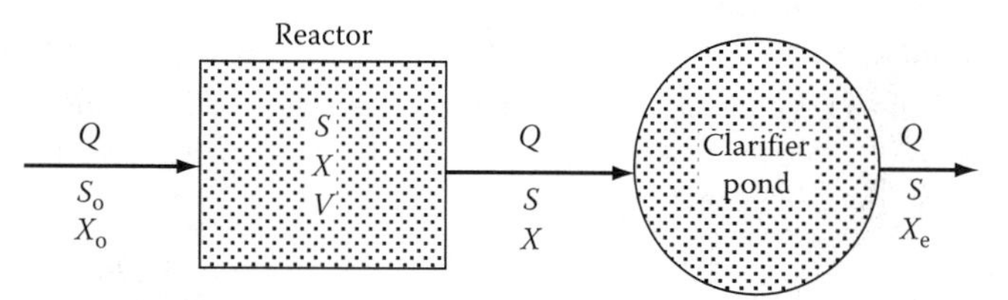

FIGURE 23.4 Aerated lagoon schematic.

emissions have been considered not acceptable. Separating the cells and sending to a digester would be a more contemporary solution.

The mass balance relations are the same as for conventional activated sludge, except that $W=0$. The associated resultant equations after applying assumptions are given in the same sequence, substrate, and cells.

23.2.2.4.1 Substrate Mass Balance

Applying assumptions: for "steady state," $dQ/dt=0$ and $dS_o/dt=0$, $\theta=V/Q$, and letting, $[dS/dt]=(1/Y)\cdot[dX/dt]_g=(\mu/Y)\cdot X$,

$$(S_o - S) = \frac{\mu}{Y}X\theta \tag{23.6}$$

23.2.2.4.2 Cells Mass Balance

Circumscribing the system, i.e., the reactor, assume steady state, $dQ/dt=0$, and let $X_o \rightarrow 0$, to give a resultant mass balance,

$$QX = (\mu - b)XV \tag{23.19}$$

or

$$\theta = \frac{1}{(\mu - b)} \equiv \theta_c \tag{23.20}$$

which means that the mass flow of cells in the reactor effluent equals the net mass rate of cell synthesis. As seen, the reactor cell concentration, X, cancels, and so, $1=(\mu-b)\theta$, or $\theta=\theta_c^m$. To prevent "washout," i.e., losing cells in the effluent faster than they are reproduced,

$$\theta \geq \theta_c^m \tag{23.21}$$

and in design, $\theta \gg \theta_c^m$. At the same time, from Equation 23.6, i.e., $(S_o-S)=(\mu/Y)\cdot X\theta$, then S decreases as θ increases.

23.2.2.4.3 Summary

The substrate balance, Equation 23.6, gives the basis for calculating the effluent substrate concentration, S. In design, $\theta \gg \theta_c^m$; if $\theta < \theta_c^m$, organism "washout" occurs. Typically $\theta \geq 24$ h. The mathematical model is the same as for an anaerobic digester except that the kinetic constants are different and the substrate is usually measured as VSS.

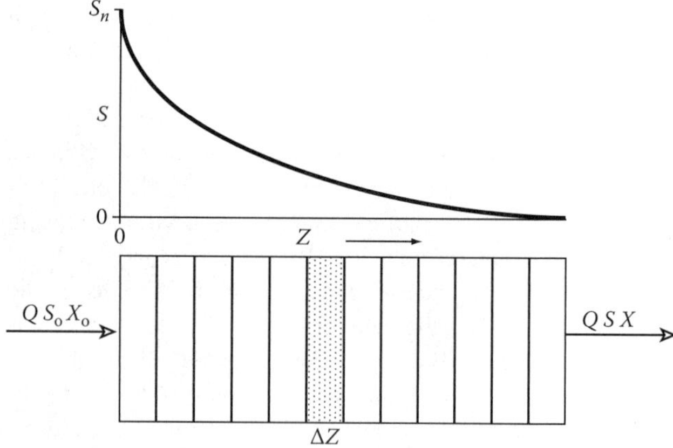

FIGURE 23.5 Schematic representation of a plug-flow reactor (clarifier not shown).

23.2.2.5 Plug-Flow Reactor

A schematic representation of a "plug-flow" reactor is shown in Figure 23.5; it is a long and narrow basin with a cross section of about 5 m × 5 m, with length perhaps 100 m. A characteristic of a "plug-flow" reactor is that the substrate concentration declines along the length, Z, of the reactor.

23.2.2.5.1 Substrate Mass Balance

The assumption, not true, is that a cross-sectional element of thickness, ΔZ, remains intact, i.e., without mass transfer across its pseudo boundaries; such an element is *assumed* homogeneous and therefore is amenable to mass balance modeling. The packed-bed reactor mathematics, Section, 4.3.3.3, applies to a plug-flow reactor and shows the mathematics more completely. A steady-state mass balance model, for substrate, cells, and recycle, respectively, is described in this section. A substrate mass balance for a plug-flow cross section is

$$\left[\frac{\partial(S\cdot\Delta Z\cdot A)}{\partial t}\right]_o = [\bar{v}S_{in}A - \bar{v}S_{out}A] + [j_{in}A - j_{out}A] + \left[\frac{\partial(S\cdot\Delta Z\cdot A)}{\partial t}\right]_r \tag{23.22}$$

where
- S is the substrate concentration within element, ΔZ (kg substrate/m^3)
- Z is the length of basin (m)
- ΔZ is the infinitesimal basin length, i.e., a "slice" (m)
- A is the cross-sectional area of basin (m^2)
- t is the time (s)
- subscript o denotes "observed" for infinitesimal basin volume, ΔZA
- v is the average velocity along length basin, i.e., Q/A (m/s)
- S_{in} is the substrate concentration at entrance to cross-sectional "slice" (kg substrate/m^3)

S_{out} is the substrate concentration at exit of cross-sectional "slice" (kg substrate/m^3)
j_{in} is the dispersion flux-density at entrance to cross-sectional "slice" (kg substrate/m^2/s)
j_{out} is the dispersion flux-density at exit from cross-sectional "slice" (kg substrate/m^2/s)
subscript r denotes "reaction" that occurs within an infinitesimal basin volume, ΔZA

The terms in Equation 23.22 are

advection-flux (in):

$$\bar{v}S_{in} = \bar{v}S - \left[\frac{\partial(\bar{v}S)}{\partial Z}\right]_r \cdot \frac{\Delta Z}{2} \text{ (kg substrate/m}^2\text{/s)}$$

advection-flux (out):

$$\bar{v}S_{out} = \bar{v}S - \left[\frac{\partial(\bar{v}S)}{\partial Z}\right]_r \cdot \frac{\Delta Z}{2} \text{ (kg substrate/m}^2\text{/s)}$$

dispersion (in) $= \boldsymbol{j}_{in}$

$$= \boldsymbol{j} - \left[\frac{\partial \boldsymbol{j}}{\partial Z}\right]_r \cdot \frac{\Delta Z}{2} \text{ (kg substrate/m}^2\text{/s)}$$

dispersion (out) $= \boldsymbol{j}_{out}$

$$= \boldsymbol{j} + \left[\frac{\partial \boldsymbol{j}}{\partial Z}\right]_r \cdot \frac{\Delta Z}{2} \text{ (kg substrate/m}^2\text{/s)}$$

where
S is the substrate concentration at center of the infinitesimal "slice" (kg substrate/m^3)
$\boldsymbol{j}$ is the dispersion flux density at center of the infinitesimal "slice" (kg substrate/m^2/s)

Also,

$$\boldsymbol{j} = \boldsymbol{D}\left[\frac{\partial S}{\partial Z}\right] \quad (23.23)$$

where $\boldsymbol{D}$ is dispersion coefficient for the turbulent flow within the reactor (m^2/s).

Substituting the appropriate terms in Equation 23.22 results in the differential equation,

$$\left[\frac{\partial S}{\partial t}\right]_o = -\bar{v}\frac{\partial S}{\partial Z} + \boldsymbol{D}\frac{\partial^2 S}{\partial Z^2} - \left[\frac{\partial S}{\partial t}\right]_r \quad (23.24)$$

The "steady-state" assumption is that $[\partial S/\partial t]_o = 0$. Also, although hydraulic dispersion is appreciable, the term is omitted to simplify further mathematical development, i.e.,

$$0 = -\bar{v}\frac{\partial S}{\partial Z} - \left[\frac{\partial S}{\partial t}\right]_r \quad (23.25)$$

or

$$\bar{v}\frac{dS}{dZ} = \frac{\hat{\mu}}{Y}\left[\frac{SX}{K_s + S}\right] \quad (23.26)$$

If $S \ll K_s$, then, $S(K_s + S) \approx S/K_s$, to give after grouping terms,

$$\frac{dS}{S} = \left[\frac{\hat{\mu}X}{\bar{v}YK_s}\right] \cdot dZ \quad (23.27)$$

Integrating between the limits, S_o and any S, and $Z = 0$ to any Z, gives

$$\ln\left[\frac{S}{S_o}\right] = \left[\frac{\hat{\mu}X}{\bar{v}YK_s}\right] \cdot Z \quad (23.28)$$

The exponential decline in S from S_o, is depicted in Figure 23.5. Another assumption in Equation 23.28 is that X is approximately unchanged with Z. As a rule, if at the entrance $X_r \approx 10{,}000$ mg/L, then $X > 2{,}000$ mg/L, which is assumed large, relative to ΔX during the reactor detention time. For more accuracy, however, the calculation is, $\Delta X(\text{reactor}) \approx \mu X\theta \approx [\hat{\mu}/Y][SX/(K_s + S)]\theta$. The calculation may be done by finite-difference approximation for Equation 23.28, i.e., $\Delta S/S_Z = [\hat{\mu}X_Z/\bar{v}YK_s] \cdot \Delta Z$, with numerical solution by spreadsheet.

As a note, the outcome of the integration along Z, as in Equation 23.27, is the same as considering a finite slice of reactor the same as a batch reactor, i.e., with zero transport across the pseudo boundaries, and letting the slice "float" downstream. The differential, dZ, in Equation 23.27 is d$Z = \bar{v}\,dt$, and the integration is between $t = 0$ and $t = \theta = Z/\bar{v}$.

23.2.2.5.2 Cells Mass Balance

The cell mass balance is, as a rule, dealt with the same as a complete-mix conventional system, i.e., with cell separator, settling, wasting a fraction of the underflow, with recycle of the other fraction, e.g., applying Equation 23.11 to the reactor as a whole. The model neglects the changes with distance along the length of the reactor.

23.2.2.5.3 Cell Recycle

The plug-flow reactor is dealt with the same as the complete-mix reactor and Equation 23.14 applies (since the cell-separator recycle relates only to the settling behavior of the suspension and its operation).

23.2.2.5.4 Summary

The plug-flow reactor is the other end of the reactor spectrum, i.e., as compared to the complete-mix reactor. With a plug-flow reactor, a salt-dispersion curve for a "pulse" injection shows a delay in first appearance with its peak approximately the same as the hydraulic detention time, θ, followed by a gradual decline. With a complete-mix reactor, on the other hand, the highest concentration appears immediately in the effluent, with exponential decline with time. The differences in the hydraulic models result in differences in reactor mathematical models. As noted, the plug-flow model is the same in concept as the "packed-bed" model in that both involve infinitesimal slices with cross sections normal to the respective flow velocities.

How we got here: Any discussion of activated sludge would not be complete without some knowledge of the persons who have shaped the current understandings. Many could be cited. The one person who would come to mind of virtually everyone in the field, however, is Professor Wes Eckenfelder. Box 23.2 gives a glimpse of his contributions since about 1952, and which have continued to the present, and a sense, perhaps, of his personality, which is serious, focused, and driven on the topic, but he also "has a life," of which others feel a part (see also Eckenfelder, 2009).

BOX 23.2 ECKENFELDER

Professor W. Wesley Eckenfelder, Jr. (1926–2010) has had a continuing presence in the field of biological treatment since early 1952 when the first of his papers appeared. His technical contributions have forged the foundation for much of current practice and have included such fundamental ideas as reaction stoichiometry, kinetic theory, oxygen transfer, reactor design protocol, biological solids handling, and the specialty area of industrial wastes. Equally important, he has given leadership to define the state of the art in biological treatment by convening numerous conferences, giving papers, writing books. In the early 1960s, he was one of the founders of what is now the International Water Association (IWA), its second president, and a charter member of AEESP.

Asked how he became a visible figure at such a young age, he noted that he graduated from high school at 16. Indeed, his biography shows a BSCE in 1946 from Manhattan College and an MS from Penn State in 1948. After developing an identity with Manhattan College, one of the focal points for sanitary engineering during the 1950s and beyond, he spent time at the University of Wisconsin with Professor Gerry Rohlich, moved to the University of Texas in 1964, invited by Professor Earnest Gloyna (Box 20.2), and later moved to Vanderbilt University, where he remained an emeritus professor. His career as a consulting engineer has included hundreds of experiences as both an individual and with his own firms, starting with Roy F. Weston in the 1950s. In the 1970s, Professor Eckenfelder and Professor Peter Krenkel formed AWARE, based in Nashville. In 1990, the 100-person firm was renamed Eckenfelder, Inc. by its employee-owners, which later became a part of Brown and Caldwell. Since 2004, he has been with AquAeTer, Brentwood, Tennessee.

Professor W.W. Eckenfelder, c. 2008

His sense of humor is legendary which explains, in part, why essentially all of his presentations have been to packed houses (from 100 to 1000). In addition, his explanations of complex topics are a model of distilling the essence of a concept, with the added touch of unorthodox slides coupled with humorous side comments. Of the thousands of people who know him, virtually all are known to him and would be considered friends. Over the years, any student could introduce himself or herself at a Purdue or WEFTEC conference and feel like they have gained from the experience. I ventured to ask him once at a Purdue conference, c.1968, in a spirit of trying to learn, why an equation in one of his books was at variance with established theory; he responded in good humor: "I do what works." The anecdotes from those who have experienced a presentation or personal interaction could fill more than one book. His autobiography, listed in References, Chapter 22, gives with light touch a good sense of his career and personality.

23.2.3 Numerical Modeling

The merit of a numerical (finite-difference) model is that assumptions do not have to be imposed and various kinds of functions may be incorporated. A numerical model may be as complex as desired in order to account for whatever "happenings" are known to occur and can be expressed mathematically. The limit is the storage capacity and speed of the computer available; the model must be compatible. For example, an activated carbon model with changes in distance and time has required 30 million iterations and 2–3 h of computer time using one of the largest computers (see, for example, Vagliasindi and Hendricks, 1992, reference in Chapter 15).

Two models are discussed here. The first is the "complete-mix" activated-sludge with recycle and is demonstrated for the case that the simplifying assumptions, $dQ_o/dt = 0$ and $dS_o/dt = 0$, are not imposed. In other words, if Q_o and S_o vary over a diurnal cycle, new values can be applied on the mass balance at the beginning of each new iteration. While complex compared to the "steady-state" differential equation, the model is simple as models go and can be set up on a spreadsheet. The model can show the effect of flow variation and substrate variation on the effluent substrate concentration, S, for example. The model is limited by the uncertainty of the kinetic constants. With the model, however, the "sensitivity" of S to the uncertainty may be tested. Other assumptions may be tested as well, such as other variations in Q_o and S_o, the effect of different recycle ratios, etc. Most models, even the complex ones, are approximations to real systems, and must be "calibrated" and "verified." In calibration, some of the coefficients are adjusted so that the model predictions match the real system. In verification, the predictions are tested over a range of input magnitudes. With a number of coefficients involved, however, in most models it is not likely that perfect calibration will be achieved.

The second model discussed is a more complex dynamic model that accounts for most of the important factors in the activated-sludge process and is the activated-sludge model (ASM) No. 1. This is the International Water Association (IWA) model and is described in Section 23.2.3.2.

23.2.3.1 Numerical Model Concept

A finite-difference model is demonstrated for "complete-mix" system with recycle. As seen, the finite-difference form of a differential equation involves merely, in the case shown, calculating $S_{t+\Delta t}$ and $X_{t+\Delta t}$ at the end of the time period, Δt. These values then become S_t and X_t for the start of a new iteration; the number iterations depend on the Δt selected and the time period to be simulated. An algorithm for a model may be set up for a spreadsheet solution.

An equation set showing the substrate and cell mass balances and recycle, is given here. The equations may be set up for a spreadsheet solution.

Three "native" mass balance equations apply to conventional activated sludge:

1. Substrate balance for whole system
2. Cell balance for whole system
3. Cell balance for final clarifier

These equations are given below in both differential and finite-difference forms.

1. Substrate balance for system in differential form and in equivalent finite-difference form,

$$\left[\frac{\mathrm{d}S}{\mathrm{d}t}\right]_{\mathrm{o}} \cdot V = Q_{\mathrm{o}}S_{\mathrm{o}} - (Q - W)\cdot S - \frac{1}{Y}\cdot\left[\hat{\mu}\frac{S}{K_{\mathrm{s}}+S}\right]\cdot X\cdot V \tag{23.29}$$

$$S_{t+\Delta t} = S_t + \left\{\frac{Q_{\mathrm{o}}S_{\mathrm{o}}}{V} - \left(\frac{Q-W}{V}\right)\cdot S_t - \frac{1}{Y}\cdot\left[\hat{\mu}\frac{S_t}{K_{\mathrm{s}}+S_t}\right]\cdot X\right\}\cdot\Delta t \tag{23.30}$$

2. Cell balances for the reactor and cell separator, in differential and finite-difference forms, and assuming for the finite-difference form, $X_{\mathrm{o}} \approx 0$ and $X_{\mathrm{e}} \approx 0$, gives

$$\left[\frac{\mathrm{d}X}{\mathrm{d}t}\right]_{\mathrm{o}} \cdot V = Q_{\mathrm{o}}X_{\mathrm{o}} - (Q - W)\cdot X_{\mathrm{e}} - WX_{\mathrm{r}} + \left[\hat{\mu}\frac{S}{K_{\mathrm{s}}+S} - b\right]\cdot X\cdot V \tag{23.9}$$

$$X_{t+\Delta t} = X_t + \left\{-\frac{W}{V}X_{\mathrm{r}} + \left[\hat{\mu}\frac{S_t}{K_{\mathrm{s}}+S_t} - b\right]\cdot X_t\right\}\cdot\Delta t \tag{23.31}$$

3. Cell balance for cell separator (final clarifier) in terms of averages (daily or any designated time period), after neglecting the effluent cell flux, $(Q - W)X_{\mathrm{e}}$, is

$$(R + W)\cdot X_{\mathrm{r}} = (Q + R)\cdot X \tag{23.32}$$

Such a model was set up for a spreadsheet, Table CD23.4, with an excerpt shown in the text. The spreadsheet has linked plots for the influent flow, influent BOD, computed effluent BOD, and computed MLSSS (not shown in text). Two of the earlier plots from the spreadsheet are shown as Figure CD23.6. The input data were obtained from the Fort Collins WWTP in 1990. The effluent BOD in Figure CD23.6b shows a steady increase (obtained from an early trial), but with adjustment to "force," a constant MLSS (as observed in the plant), the effluent BOD in the spreadsheet varied 0.6–25 mg/L over the 24 h cycle. The spreadsheet provides a simple dynamic model, which has a more realistic behavior than a steady-state model.

23.2.3.2 IWA Activated-Sludge Model

In 1983, the International Water Association (IWA) formed a task group to develop practical models for application to the design and operation of biological wastewater treatment, which resulted in the ASM No. 1 described by Henze et al. (1987, p. 515). The model was developed for application to carbon oxidation, nitrification, and denitrification in activated sludge and eight processes within. The model used COD for substrates and expressed nitrogen as N, which has helped to standardize units and gave typical parameter values, e.g., Y, $\hat{\mu}$, b, K_{s}, also emphasizing the need for experimentally determined values to fit the situation at hand. The model is adaptable to different flow configurations such as "step-feed," "contact stabilization," "denitrification," etc. The model development through ASM3 is described in the IWA Scientific and Technical Report No. 9 (www.iwapublishing.com), 2009. Many consulting firms have used the model in practical application, as intended.

23.2.4 Practice

A wide variety of empirical parameters have evolved for practice, mostly since the 1950s. Many are repeated from

TABLE CD23.4
Excerpt from Excel Spreadsheet for Dynamic Activated-Sludge Model

Variable Type	Group 1 Design			Group 2 Coefficients				Group 3 Operation		
Dependent	S_t =	Calc.	(mg/L)					R′ =	assumed	(L/h)
	X_t =	Calc.	(mg cells/L)					R =	calculated	(L/h)
					Original	Revised		X =	1,800	(mg cells/L)
Independent	Q_o =	31.56	(mil L/d)	μ(hat) =	0.205	0.25	(h^{-1})	X_r =	10,000	(mg cells/L)
	Q_e =	31.56	(mil L/d)	K_m =	5.8	300	(mg/L)			
	V =	11650000	(L)	Y =	0.31	0.6				
	Time, t	Calc.	(h)	k_d =	0.0042	0.0002	(h^{-1})			
	Incr., Δt	0.5	(h)							

	Inputs			Recirculation/Wasting			Substrate Degradation			Cell Synthesis	
Time (h)	**Q_o (L/h)**	**θ (h)**	**S_o (mg/L)**	**R (L/h)**	***R/Q***	**R′ (L/h)**	**μ (h^{-1})**	**S (mg/L)**	**S(t + Δt) (mg/L)**	**X (mg/L)**	**X(t + Δt) (mg/L)**
0.0	1,320,454	8.82	303	132,045	0.100	16,492	0.008	10.0	14.56	1800	1800
0.5	1,037,533	11.23	233	207,639	0.200	16,492	0.012	14.6	7.06	1800	1803
1.0	1,037,533	11.23	233	208,119	0.201	23,896	0.006	7.1	8.56	1803	1798

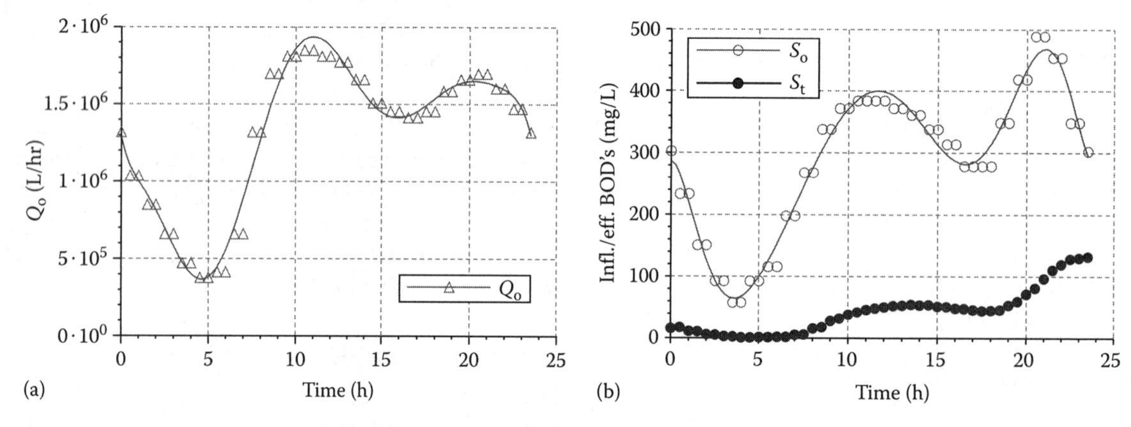

FIGURE CD23.6 Flow and BOD over 24 h and calculated effluent BOD, Fort Collins WWTP, 1990. (a) Influent flow. (b) Influent BOD and calculated effluent BOD.

Chapter 22 to indicate also their derivations from the mass-balance/kinetic theory.

23.2.4.1 Empirical Guidelines

Detention time, θ, for sizing tanks, and F/M ratio, along with sludge age, θ_c, for operation, are most common. As shown, these parameters are explained by theory. Their application to practice is addressed here along with their respective ranges. Values given are for conventional activated sludge.

23.2.4.1.1 Detention Time

More often than not, reactors are sized on the basis of detention time. The range is $4 \leq \theta \leq 8$ h (Tchobanoglous and Burton, 1991, p. 534), based on "average" flow, Q. Most state codes specify 6 h or greater. For conventional activated sludge, $\theta \approx 6$ h is a "rule of thumb." The detention time varies with flow, which usually has a 24 h cycle. The "average" annual flow increases with urban growth. In some cases, flow may vary over the annual cycle as well, e.g., due to the groundwater table increasing with summer lawn irrigation and the subsequent higher rate of infiltration to the sewerage system. A finite-difference model provides a more accurate "picture" of the effects of flow variation over the diurnal cycle.

23.2.4.1.2 F/M Ratio

The F/M ratio, the "food-to-microorganism ratio," is a common design parameter and is defined as

$$\frac{F}{M} \equiv \frac{S_o}{\theta X} \tag{22.45}$$

For conventional activated sludge, the F/M range is $0.2 \leq F/M \leq 0.4$ kg BOD applied/kg MLVSS/day (Tchobanoglous and Burton, 1991, p. 550). If S_o and X are expressed as concentrations, the volumes cancel. As seen by the definition, the lower the F/M ratio, the more the organisms are in a "starved" condition. A larger tank size gives lower values of F/M ratio.

23.2.4.1.3 Specific Substrate Utilization Rate

The "specific substrate utilization rate," U, is defined (Tchobanoglous and Burton, 1991, p. 533) as

$$U \equiv \frac{(S_o - S)}{\theta X} \tag{22.43}$$

where
$U = (F/M) \cdot E$
$E \equiv (S_o - S)/S_o$

The F/M parameter is used most often in practice while the U is seen more frequently in texts, since U is an identity with μ/Y (Equation 22.45). As seen by comparing F/M and U, if $S << S_o$, then $F/M \approx U$. Also for reference, $U \approx \mu/Y$; in other words, μ/Y is the "independent variable" and U is the "dependent variable"; U is determined by μ/Y. But this is not strictly true. To expand on this statement, for a given system, if θ is determined by Q and V, and if S_o depends on the influent flow value, and if X is determined by operation, S is a dependent variable. But actually, since $\mu = \hat{\mu}\ [S/(K_s + S)]$, a quadratic equation is involved to solve for S.

23.2.4.1.4 Sludge Age, θ_c

The sludge age is used in operation and is defined (Section 22.5.7.5) as

$$\theta_c = \frac{V(\text{reactor}) \cdot X}{WX_r} \tag{22.52}$$

where
- V(reactor) is the volume of aeration basin (m^3)
- X is the cell concentration, usually measured using the surrogate, MLVSS (kg/m^3)
- W is the waste sludge flow as taken as a sidestream from the return sludge flow (m^3/s)
- X_r is the cell concentration as "underflow" from the final clarifier, recycled (kg/m^3)

Also, keep in mind that $\theta_c^m = 1/(\mu - b)$, and for design, $\theta_c >> \theta_c^m$. From operating data in U.S. plants, θ_c ranges, $3 \leq \theta_c \leq 15$ day (Tchobanoglous and Burton, 1991, p. 534).

23.2.4.1.5 Volumetric Loading

The volumetric loading has been used for many decades; it is given here for reference, and is defined as

$$\text{VLF} = \frac{\text{kg BOD/day}}{\text{m}^3 \text{ reactor volume}}$$

The range is Tchobanoglous and Burton (1991, p. 550), $0.3 < \text{VL} < 0.6$ (kg BOD/day)/(m^3 reactor) or [$20 < \text{VL} < 40$ (lb BOD/day)/(1000 ft^3 reactor)].

23.2.4.1.6 Design Parameters

The F/M and θ_c are the recommended parameters for design and operation Tchobanoglous and Burton (1991, p. 534). The empirical parameters θ and VLF have been long used and may be useful for an approximate confirmation of results from F/M and θ_c.

23.2.4.1.7 Cell Production Rate

From Section 22.4.5, Table 22.8, the $Y \approx 0.94$ g cells/g substrate, or Y(COD/COD) ≈ 0.64 g cell COD/g substrate COD. In terms of BOD, the "rule of thumb" is Y(g cells/g BOD) ≈ 0.5 g cells synthesized/g BOD degraded (Eckenfelder, 2000, Figure 19). Example 23.1 illustrates the calculation of the mass flux of cells leaving the system.

Example 23.1 Cell-Wasting Rate

Given

$S_o \approx 280$ mg BOD/L or 0.280 kg BOD/m^3
$S \approx 30$ mg BOD/L
$Q = 18{,}925$ m^3/d (5.0 mgd)

Required

Rate of cell wasting, WX_r

Solution

Assume $X_r \approx 10{,}000$ mg/L $= 10.0$ kg cells/m^3

Y(g cells/g BOD) ≈ 0.5 g cells synthesized/g BOD degraded

Therefore, the mass rate of cell synthesis is

$$\begin{aligned} WX_r &= Q(S_o - S)Y \\ &= (18{,}925\ \text{m}^3/\text{day}) \cdot (0.280\text{–}0.030)\ \text{kg BOD/m}^3) \\ &\quad \cdot (0.5\ \text{g cells synthesized/g BOD degraded}) \\ &= 2{,}366\ \text{kg cells/day} \\ W &\approx (2{,}366\ \text{kg cells/day})/(10.0\ \text{kg cells/m}^3) \\ &= 237\ \text{m}^3/\text{day} \end{aligned}$$

Discussion

The cells wasted are usually mixed with the sludge underflow from the primary settling tank and sent to the anaerobic digester. A more accurate calculation would account for endogenous respiration, which means that about b fraction of the cells synthesized decay. The calculation is, ΔX(decay) $\approx bX$; let $b \approx 0.1$ day^{-1}; therefore, ΔX(decay) $\approx$ (0.1 kg cells decayed/day/kg viable cells/day^{-1}) $\cdot$ (2366 kg viable cells) $= 237$ kg cells decay/day. Net cell wasting rate ≈ 2366–$237 \approx 2129$ kg cells/day.

Oxygen demand:

The oxygen demand for a wastewater in a reactor has three parts: substrate oxidation, cell oxidation (as endogenous respiration), and ammonia oxidation.

Substrate oxidation: The oxygen demand for substrate oxidation may also be calculated from stoichiometry (Section 22.4.5, Table 22.8) based on the equation for the conversion of domestic wastes to cells. As seen, the conversion is

Y(O_2-stoichiometric)

$= 0.66$ g O_2 used/g substrate degraded

Endogenous respiration: In addition, the oxygen demand for cell oxidation may also be calculated from stoichiometry (Section 22.4.2.2; Example 22.2/3; Table 22.8) based on the equation for the conversion of cells to carbon dioxide and water. As seen, the conversion is

Y(O_2-stoichiometric/cell oxidation)

$= 1.42$ kg O_2 used/kg cells oxidized

Ammonia oxidation: If ammonia is oxidized to nitrite/nitrate, an additional oxygen demand is exerted. The oxygen demand for ammonia oxidation may be calculated from stoichiometry (Section 22.3.3.3/Nitrification, Equation 22.9) based on the equation for the conversion of ammonia to cells, nitrate, water. As seen, the conversion gives

Y(O_2-stoichiometric/ammonia)

$= 4.3$ g O_2 used/g ammonia oxidized

Example 23.2 Oxygen Required

Given

$S_o \approx 280$ mg BOD/L or 0.280 kg BOD/m^3
$S \approx 30$ mg BOD
$Q = 18{,}925$ m^3/day (5.0 mgd)
$b \approx 0.10$ day^{-1} x
C(NH_3-as-N) ≈ 22 mg/L (Section 22.1.5.3/Ammonia)

Required

Oxygen utilization rate (OUR) due to substrate, endogenous respiration of cells, and ammonia oxidation

Solution

Substrate oxidation: See Section 22.4.5, Table 22.8, for a stoichiometric equation for the oxidation of domestic wastewater, using the formula, $C_{10}H_{19}O_3N$ for domestic waste and $C_5H_7O_2N$ for cells. But since the BOD equivalents are not given, the BOD given is used as an index of substrate BOD demand (for this problem).

Since the substrate mass is given in terms of BOD, we can evaluate oxygen demand,

[Substrate O_2 demand] $\approx Q(S_o - S)$

$=$ (18,925 m^3/day) · [(0.280–0.030) kg BOD/m^3]

$=$ 4731 kg O_2/day

Endogenous respiration: See Section 22.4.5, Table 22.8, for stoichiometric equation for endogenous respiration (oxidation of cell matter).

From Example 23.1, ΔX(decay rate) ≈ 237 kg cells decay/day.

[O_2-demand/decay] $\approx Y$(O_2-stoichiometric/cell oxidation) · ΔX(decay rate)

$=$ [1.42 kg O_2 used/kg cells oxidized] · 237 kg cells decay/day

$= 336$ kg O_2 used/day

Ammonia oxidation: See Section 22.3.3.3 for stoichiometric equation for the oxidation of ammonia.

[O_2-demand/ammonia]

$\approx Y$(O_2-stoichiometric/ ammonia) · C(NH_3-as-N) · Q

$=$ [4.3 kg O_2 used/kg ammonia oxidized] · [0.022 kg NH_3-as-N/m^3] · (18,925 m^3/day)

$= 1790$ kg O_2 used/day

Total Oxygen Demand (Substrate, Endogenous Respiration, Ammonia)

The total oxygen required per day is

[Total O_2 demand rate] $=$ [Substrate O_2 demand] $+$ [O_2-demand/decay] $+$ [O_2-demand/ammonia]

$= 4731$ kg O_2 substrate/day $+ 336$ kg O_2 decay/day $+ 1790$ kg O_2 NH_3/day

$= 6857$ kg O_2/day

Discussion

In estimating oxygen demand based on BOD, the BOD model, i.e., measuring dissolved oxygen reduction in a 300 mL bottle (seeded and incubated at 20°C over a 5-day period) is not necessarily accurate; but it may be a reasonable estimate. Another approach in estimating oxygen demand would be to measure the TOC of the incoming substrate and apply the general substrate formula, $C_{10}H_{19}O_3N$, and use the carbon in the formula as the basis for an oxygen demand calculation as in Section 22.4.5, Table 22.8, for a stoichiometric equation for the oxidation of domestic wastewater.

The rate of oxygen delivered, J, is calculated by Equation 18.33, $J = (dC/dt) \cdot V = K_La(C_s - C) \cdot V$, Sections 18.2.2.5 and 18.3.2.3; the value of K_La depends on whether a surface aerator or a diffused aeration is used. For diffused aeration, Equation 18.45 may be used to calculate K_La, which is, $K_La = (k_L/V) \cdot [\zeta h_o Q'/(v_o + v_w)d]$. Another approach is to assume an oxygen transfer efficiency, e.g., E(oxygen to water) $= J$(oxygen transferred to water/J(oxygen delivered). With diffused air, the oxygen transfer efficiencies are about 0.06 for coarse-bubble aeration and 0.10–0.12 for fine-bubble aeration (from Morgan and Bewtra, 1960); see also Table 18.7. If we assume $E \approx 0.06$, then J(demand)/J(delivered) $=$ (6857 kg O_2/day)/J(delivered) ≈ 0.06, then J(O_2 delivered) $\approx 114{,}283$ kg O_2/day. From Table B.7, ρ(O_2, STP) $= 0.278$ kg O_2/m^3 and, ρ(air, STP) $= 1.204$ kg O_2/m^3, which gives J(air delivered) $= 494{,}952$ kg air/d $\approx$ 500,000 kg air/d. The airflow, Q(air) may be calculated by the ideal gas law, $PV = nRT$, Appendix H.

Sludge volume index

A routine measure of final clarifier underflow sludge concentration is the sludge volume index, SVI. This is the volume of 1 g of sludge after settling 30 min in a 1000 mL cylinder. It is measured by obtaining the oven-dry weight of the settled sludge and dividing this into the volume previously noted for the corresponding mass of settled sludge. Usually it is desired that

$$\text{SVI} < 100.$$

Larger values imply bulking sludge. Also it should be noted that

$$X_r = \frac{10^6}{\text{SVI}}$$

In other words, SVI ≈ 100 results in $X_r = 10^6/10^2 \approx 10{,}000$ mg MLVSS/L. A sample may be obtained by dipping a container attached to a pole, into the aeration basin.

Nutrients

A variety of trace metals and nutrients are necessary for activated-sludge microorganisms. Municipal sewage usually contains these at the necessary levels. Some industrial wastes may, however, need to be fortified.

Nutrients for industrial wastes: The nutrients of greatest concern are phosphorous and nitrogen. Eckenfelder and O'Connor (1961) give the following guidelines: $BOD/N \approx 32$; $BOD/P \approx 150$. It is not uncommon for nutrients to be added to many organic industrial wastewaters.

N and P in cells: As seen in Section 22.2.2.1, Stumm (1964) gave a $P = 1$ based formula for cells as, $C_{106}H_{180}O_{45}N_{16}P$, with mole ratios C:N:P = 106:16:1; the associated mass ratios are 1272:224:31, or 41:7:1. The purpose of his empirical formula was to estimate whether nitrogen or phosphorous may be rate limiting in cell synthesis. In other words, if HCO_3^- is a source of carbon in photosynthesis, then the C:N and C:P ratios may be calculated. For example, let $C(HCO_3^-) = 100$ mg/L, which gives, C(carbon) ≈ 20 mg/L; and from this, C(nitrogen) ≈ 3.4 mg/L; and C(phosphorous) ≈ 0.5 mg/L. In other words, nitrogen and phosphorous concentrations lower than the values indicated would be rate limiting. If carbon dioxide from the atmosphere is to be the source of carbon, then carbon would be rate limiting (if the concentrations of N and P were as calculated). More likely, the degradation of residual organics from a wastewater treatment plant would provide sufficient carbon dioxide for photosynthesis.

Composition of activated sludge

Activated sludge consists of a viable active mass, dead organisms, and inert suspended solids. Eckenfelder and O'Connor (1961) state active mass, X, ranges 0.50–0.70 fraction (measured as volatile suspended solids); and inert suspended solids range 0.30–0.40 fraction.

Sludge production

A rule of thumb generally accepted for municipal wastewater is that 0.5 kg VSS results from 1 kg BOD. Stoichiometric calculations provide a means to calculate Y based on the substrate.

23.2.4.2 Experience with Plants

Figure 23.7 is a plot of data from four sources compiled by Professor W. Wesley Eckenfelder (Eckenfelder, 1961, p. 78). The curve is explained as three operating phases, the first two showing exponential decline, each different. In the first phase, the soluble BOD was removed (and oxidized) rapidly and is followed by an exponential decline to $X\theta \approx 2000$. In the second phase, colloidal and suspended BOD was removed and required longer detention time to permit hydrolysis. In a third phase about 0.1 fraction remains as an inert residual. The third phase corresponds to long detention times (for example 20–30 h) with about 0.1 fraction residual biomass.

23.2.4.2.1 Process Variations

A number of activated-sludge process variations evolved by 1960. Table 23.5 is indicative of the spectrum of processes as

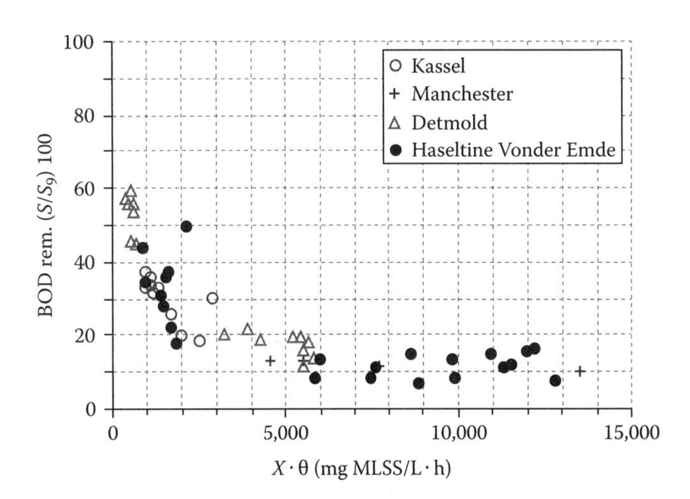

FIGURE 23.7 BOD removal relationship from domestic sewage. (Adapted from Eckenfelder, W.W. Jr., Activated sludge, in Eckenfelder, W.W. Jr. and T.P. Quirk (Eds.), Advances in Sewage Treatment Design, Proceedings of a Symposium at Manhattan College, Sanitary Engineering Division, Metropolitan Section, ASCE, Reston, VA, May 15, 1961, p. 78.)

reviewed by Stewart (1964). The processes are with reference to diffused aeration with "plug-flow" hydraulics characterized by a long, narrow basin; as a rule, the diffusers were placed along one side along the length of the basin, resulting in a "roll" and spiral flow.

The key feature of each configuration is stated in the "notes" column. Each has a rationale. For example, in "conventional" activated sludge (the first configuration depicted), the diffusers (for air) are spaced uniformly along the length of the basin; the rationale is that it has been tradition and is cheaper in capital cost. In "tapered-aeration" (the second configuration), the diffusers are spaced more closely at the head of the basin since demand for oxygen declines with distance. (As a note, the first and second depictions are the same since the diffuser layouts are not shown.) In step loading (the third depiction), substrate is injected along the basin length with the idea of providing a more even loading on the reactor. In contact aeration, the return activated sludge is given an exposure to oxygen with the idea of "conditioning" the cells prior to their main job, which is to react with substrate. The rationale for the Hatfield process is to add nutrients to the system from the anaerobic process in case some nutrients may be rate limiting. The idea of extended aeration is to reduce the sludge solids that must be further processed.

23.2.5 Operation

As with all unit processes, not to speak of the plant itself, operation is a field requiring both knowledge of principles and experience. An activated sludge reactor requires attention to control of the reactor suspended solids concentration which is controlled by effective settling and by the proper sludge wasting and recycle ratio (to give an idea). In settling, "bulking sludge" is a major issue, which involve both art and science. The extensive literature, seminars, training, research,

TABLE 23.5
Activated-Sludge Process Variations

Process Name	Notes	Loading: Volumetric (lb BOD/day/1000 ft^3)	Loading: F/M Ratio (lb BOD/day/lb VSS)	Sketch
Conventional	10%–30% return sludge	35	0.2–0.5	PS, Reactor, FS, Return sludge, Waste
Tapered aeration	Airflow is reduced along reactor length	35	0.2–0.5	PS, Reactor, FS, Return sludge, Waste
Step loading	Waste is injected along reactor length	50	0.2–0.5	Reactor, FS, Return sludge, Waste
Contact stabilization	Organisms assimilate and store in reactor and get hungry in re-aeration basin	70	0.2–0.5	Reactor, FS, Re-aeration, Return sludge, Waste
Hatfield process	Used for high organic wastes. Nutrients added from Anaerobic digester			PS, Reactor, FS, Re-aeration, Sludge, Waste, An-D, Primary sludges, Supernatent
Extended aeration	Sludge not wasted	20	0.2–0.5	Reactor, FS, Return sludge

Source: Adapted from Stewart, M.J., Activated sludge process variations, The complete spectrum, in *Water and Sewage Works*, Part I, April 1964; Part II, May 1964; Part III, June 1964.

and the involvement of many in the field, is an index of the importance of operation. As with other unit processes, design has a great deal to do with effective operation. Many "tradeoffs" can be seen between capital costs and operating costs ranging from effective operation of the processes to routing maintenance tasks. If the design lacks being "operator friendly," the capital investment is at risk. Since the 1980s construction of new plants is rare; plants are expanded often to take into account increasing population and industries and are called upon to comply with regulations that are increasingly more stringent. The outcome is usually increasing operating costs and at some point a new plant becomes "cost-effective."

23.2.5.1 Bulking Sludge

Bulking sludge is a sludge that settles poorly. The microorganisms do not coalesce into a dense floc. Usually, the floc is associated with an excessive growth of filamentous bacteria (van Niekerk et al., 1987). Causes include low dissolved oxygen concentration, nutrient deficiency, high sulfide concentration, and low organic loading (low food-to-microorganism ratio, F:M).

Bulking sludge is a not an uncommon problem and has been studied extensively since the 1960s by Professor David Jenkins at Berkeley and Dr. Michael Richard, Fort Collins, Colorado.

Rising sludge may be an issue due to gas precipitation. In primary clarifiers, methane and carbon dioxide are produced in the sludge-thickening zone at high enough concentrations to come out of solution as bubbles and attach to sludge agglomerations. In the final clarifier, the anoxic conditions may cause denitrification with the precipitation of nitrogen gas, which may cause rising sludge.

23.3 BIOFILM REACTORS

Biofilm reactors include traditional trickling filters, deep-bed trickling filters, rotating biological disks, and sand beds designed as biofilm reactors, also called "attached growth reactors" (Grady et al., 1999). The idea is that the biofilm that develops on surfaces is the site for a reaction. The media can be almost any material. Biofilms are found throughout the natural environment and occur spontaneously on almost any surface, e.g., rocks in streams, pipes in water distribution systems, in soils, membranes for water treatment, granular activated carbon, etc. Biofilms are important also in "passive" kinds of designs for water treatment, such as the leach field of a septic tank system where the water treatment occurs. The septic tank merely settles whatever will settle; the solids degrade anaerobically. Since about the late 1980s, research has been underway to explore biofilm reactors in the removal of natural organic matter in order to reduce disinfection by-product precursors.

The reactor design is a fixed bed in most cases, which means that the substrate concentration changes with bed depth. The qualification "most cases" is included because at least one proprietary innovation has involved fluidized beads with a biofilm. For a stationary-bed reactor, which includes virtually all cases, the associated reactor analysis must be for a finite element (since the bed as a whole is not homogeneous).

23.3.1 Biofilms

Not much was known about biofilms until about the 1970s. In recognition that biofilms were important but little understood, the Center for Biofilm Engineering was established in 1991 at Montana State University as a National Science Foundation Center of Excellence. Since then and from other sources, an understanding of biofilms has evolved along with knowledge on how to utilize them for practical purposes. In many cases biofilms are a problem and control is the objective. A book edited by Characklis and Marshall (1990) was one of the first to elucidate some of the characteristics of biofilms.

The stoichiometric relations and Monod kinetic equations as developed in Chapter 22 are applicable in general to biofilms. The reaction rate may be limited, however, by the diffusion-transport rate within the aqueous film and a *pseudo* solid biofilm. The biofilm reactors are mostly stationary and depend on advective transport to the microorganism (the reaction site). This is in contrast to the suspended growth reactors where the reactants come into contact with each other by fluid turbulence, with a diffusion-transport step.

23.3.1.1 Structure

Biofilms develop on virtually all surfaces in an aqueous environment. They are prone to form on surfaces adjacent to flowing water, where a steady nutrient supply facilitates organism growth. Along with cellular growth, extra-cellular organic polymers are excreted which are the slime layers on surfaces (Prescott et al., 2005, p. 897). Most bacteria excrete these polymers, which extend from the cell, forming a tangled matrix of fibers giving a structure to the associated biofilm (Characklis and Marshall, 1990, p. 4). After initial colonization by a single species, monolayer, layers of other species may be added; as the biofilm matures, the structure becomes complex with cell aggregates, interstitial pores, and channels. The biofilm may evolve into a "mat" with appreciable thickness.

23.3.1.2 Transport of Nutrients

The nutrients and dissolved oxygen are transported by advection to the vicinity of the biofilm, where the transport mechanism changes to diffusion across a liquid "film" and then into the complex macro-structure (see Chapter 18). As the biofilm develops macro-dimensions, the inner structure becomes anaerobic and may cause sloughing of discernible pieces.

23.3.2 Biofilm Reactors Model

Appendix 23.A shows the rationale for a biofilm reactor model starting with a mass balance equation for an infinitesimal slice of the reactor. From the mathematical development shown, a "final" equation, Equation 23.A.29, is the result, which has been used in practice since about 1960 when proposed by Eckenfelder (1961). Table CD23.6 is a spreadsheet model with finite elements; an excerpt is shown.

23.3.2.1 Empirical Equation

As seen by Equation 23.A the substrate concentration decreases exponentially with depth. The decline increases further with lower HLR. The model is characterized by a coefficient, k, which may be fitted to operating results or pilot plant experimental measurements.

$$S = S_o \exp\left\{-k\frac{Z}{\text{HLR}}\right\} \qquad (23.9')$$

where,

S is the substrate concentration at any depth (kg substrate/m^3)
S_o is the substrate concentration at entrance to the reactor (kg substrate/m^3)
k is the reactor coefficient (s^{-1})
Z is the depth within the reactor (m)
HLR is the hydraulic loading rate, HLR $= Q/A$ (m^3/m^2/day)

A simple stationary-bed model is developed to illustrate, again, the application of the materials-balance principle, combined with kinetics. The stationary bed could be a traditional trickling filter, about 2 m depth, or any deep-bed nitrification biofilm reactor. The approach is generally applicable.

TABLE CD23.6
Excerpt from Biofilm Trickling-Filter Model Spreadsheet

Variable Type	Group 1 Design			Group 2 Coefficients				Group 3 Operation
Dependent	S_t	= calculated	(mg/L)					
					Original	Revised		
Independent	Q_o	= 8.33	(mgd)	μ(hat) =	0.205	0.2	(h^{-1})	
		= 0.36491956	(m^3/s)			5.55556E−05	(s^{-1})	
	S_o	= 300	(mg/L)					
		= 0.300	(kg/m^3)					
	HLR	= 5.0	(gal/min/ft^2)	K_m =	5.8	5.8	(mg/L)	
		= 12.22	m/h			0.0058	(kg/m^3)	
		= 0.003395117	m/s	Y =	0.31	0.31	(kg cells synthesized/kg substrate degraded)	
	d(rocks)	= 0.1	m	k_d =	0.0042	0.0002	(h^{-1})	
	A(sphere)	= 0.031416	m^2/sphere			5.55556E−08	(s^{-1})	
	$\bar{A}$	= 31.416	m^2/m^3	$\bar{X}$ =		1.0	(kg cells/m^2)	
	f	= 1.0	No. rocks/no spheres rectangular packing				$S_{Z+\Delta Z} = S_Z - \frac{1}{Y}\left[\hat{\mu}\frac{S}{K_m + S}\right] \bar{X}\left[\frac{\pi f}{d_t}\right]\left(\frac{1}{HLR}\right) \cdot \Delta Z$	
	Time elapsed, t	Calculated	(h)	Factor,		0.018		
				F =	1.00			
	Incremental, ΔZ	0.1	(m)	F' =	1.00	0.01		
	Incremental, Δt	0.5	(h)					

Inputs

Elapsed Time (h)	Q_o (L/h)	S_o (mg/L)	Z (m)	S_Z (mg/L)	Z (m)	S_Z (kg/m^3)	$\mu(S/K_m + S)$	$X\pi f/d$ HLR Y	Product	Prod * ΔZ	ΔS	$S_{Z+\Delta Z}$ (kg/m^3)	$S_{Z+\Delta Z}$ (mg/L)
0.0	1320454	300	0.0	300.0	0.0	0.3000000	0.00005450	29834.121	1.62601486	0.16260149	0.002927	0.297073	297.07
0.5	1037533	233	0.1	297.1	0.1	0.2970732	0.00005449	29834.121	1.62571107	0.16257111	0.002926	0.294147	294.15
1.0	1037533	233	0.2	294.1	0.2	0.2941469	0.00005448	29834.121	1.62540142	0.16254014	0.002926	0.291221	291.22
1.5	848767	151	0.3	291.2	0.3	0.2912212	0.00005447	29834.121	1.62508572	0.16250857	0.002925	0.288296	288.30
2.0	848767	151	0.4	288.3	0.4	0.2882960	0.00005446	29834.121	1.62476381	0.16247638	0.002925	0.285371	285.37
2.5	660001	93	0.5	285.4	0.5	0.2853714	0.00005445	29834.121	1.62443549	0.16244355	0.002924	0.282447	282.45

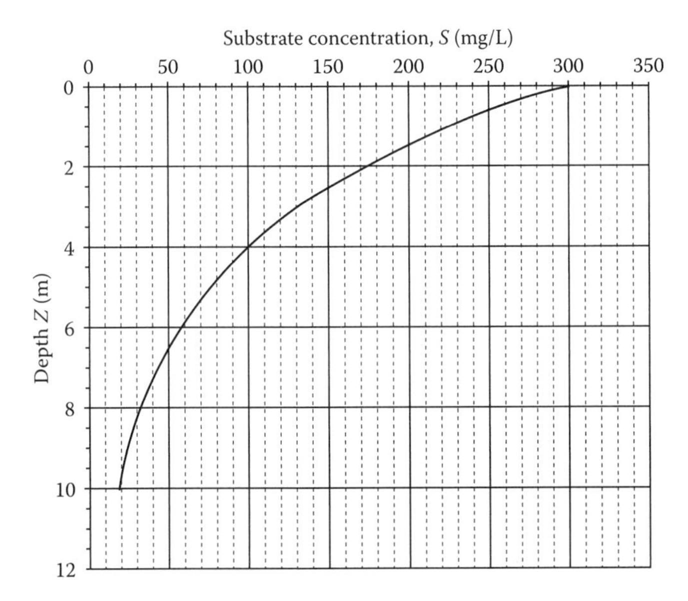

FIGURE CD23.8 Trickling-filter model output.

23.3.2.2 Trickling-Filter Spreadsheet Model

Table CD23.6 is an excerpt from a spreadsheet trickling-filter model. The spreadsheet is formatted to identify variables and to discern their respective effects on output through a linked plot. An example of a plot copied from the spreadsheet is shown as Figure CD23.8, which shows the characteristic decline of substrate concentration with depth. The finite-difference model, as opposed to a lumped-coefficient model, accounts for the effect of decreasing S on variables that affect S.

23.4 ANAEROBIC REACTORS

Anaerobic means "without oxygen." The anaerobic process is used most commonly in municipal wastewater treatment to "stabilize" settled organic solids from clarifiers. In this context it is called "anaerobic digestion." The solids react in a two-stage sequence first with "acid-former" bacteria; the reaction products from this first stage are organic acids. In the second stage, the organic acids react with methane bacteria to form methane and carbon dioxide gases. These same anaerobic reactions can occur with almost any organic substrate, including in the natural environment. Anaerobic treatment, as distinguished from "digestion" of municipal solids, is used normally only for high-strength wastewaters such as meatpacking wastes, but it has been advocated for broader applications.

The anaerobic process is the topic of this major section and is discussed mostly in the context of anaerobic sludge digestion. Its purpose is to transform noxious and infectious organic solids to a "stabilized" humus material, which can be easily dewatered and disposed to a landfill or to agricultural land.

Although anaerobic digestion—sludge drying—disposal remains the most common approach in handling organic solids, several other methods and combinations have emerged since the early 1960s, summarized in Table 23.7 (see also, McCarty, 1966). The permissible combinations of flow sequences are indicated; ocean disposal has not been permitted since about 1990. The processing and disposal of organic solids is a major issue and land disposal has emerged a frequent alternative, subject to stringent regulations (see for example, USEPA CFR Parts 257, 403, 503, 1993).

23.4.1 Evolution of Separate Sludge Digestion

The problem of sludge disposal was recognized early in the history of wastewater treatment. It was usual practice, prior to the advent and widespread practice of using separate sludge digestion tanks, to associate sludge digestion with primary settling. Thus the primary settling tanks during the early period of wastewater treatment, before and around the turn of the century, were designed also to hold large amounts of the accumulated organic solids for sludge digestion. These were called *single-story septic tanks*, as illustrated schematically in Figure 23.9.

TABLE 23.7
Permissible Flow Sequences of the Organic Solids Stream

		Initial Processes				Interm.		Final Disposal			
Origin	**Destination**	**Anaerobic Digestion**	**Vacuum Filtration**	**Heat Treatment**	**Wet Combustion**	**Grind, Heat, Bag**	**Sludge Drying Beds**	**Garden and Lawn**	**Agricultural Land**	**Land-fill**	**Ocean Dumping**
Raw material	Primary sludge	x	x	x	x						x
	Secondary sludge	x	x	x	x						
Initial processes	Anaerobic digestion					x	x				
	Vacuum filtration								x	x	
Intermediate processes	Heat treatment							x	x	x	
	Wet combustion										
	Grind heat bag									x	
	Sludge drying							x			
	beds							x	x	x	

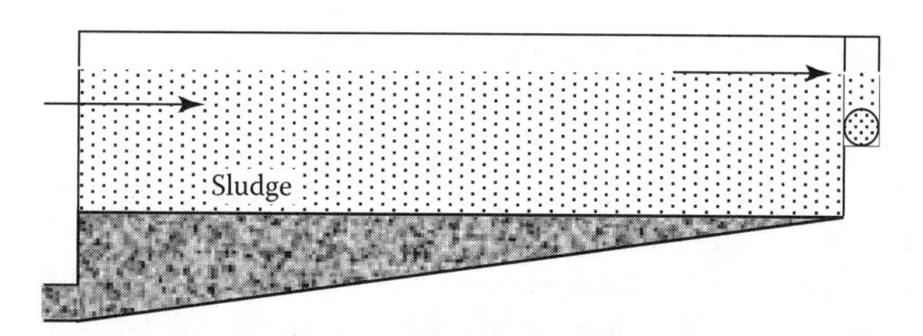

FIGURE 23.9 Settling and septic tank. (Adapted from Metcalf, L. and Eddy, H.P., *Sewerage and Sewage Disposal*, McGraw-Hill, New York, 1930.)

To circumvent some of the problems of single-story septic tanks (which included scum layers, rising sludge, etc.) the idea of a two-story tank was originated at the Lawrence Experiment Station of the Massachusetts Department of Public Health. This provided a distinct separation between the flowing wastewater and the digesting sludge. Karl Imhoff in Germany carried the idea further to an operational design about the turn of the century in a tank that provided for gas collection from the digesting sludge. Imhoff tanks were not uncommon up through the 1930s in the United States and have remained popular in Germany.

The next step was the recognition that the digestion process could be better controlled by a *separate sludge digestion* tank. The term *separate sludge digestion* then is a result of the sedimentation–digestion dual design conceptual background. The only design guidelines given by Metcalf and Eddy (1930) were as follows: (1) maximum of 6 months sludge storage, (2) 130 ft^3 of digester volume per million gallons of wastewater (0.00097 m^3 digester volume/m^3 wastewater), and (3) 2.4 ft^3 (0.065 m^3) per capita. Process design, being an innovation of the 1960s, was not relevant during this period of evolving and testing. The importance of temperature and other environmental conditions was well recognized, however; a standard operating temperature of 35°C (96°F) has been long used in practice.

23.4.2 Design Criteria

Until the late 1960s, most state codes specified digester capacity in terms of cu. ft. of digester volume per capita criterion (m^3 digester volume/capita). Other criteria include digester volume per unit mass of solids added daily (m^3 digester volume/kg solids/day), and digester volume per unit mass of volatile solids added daily (m^3 digester volume/kg VSS/day). For reference, Fair and Geyer (1954) gave the formula:

$$C = \left[V_f - \frac{2}{3}(V_f - V_d)\right]t \qquad (23.33)$$

where

C is the tank capacity in ft^3/capita

V_f is the volume of fresh sludge added per day per capita (ft^3/day/capita)

V_d is the volume of digested sludge leaving digester per day per capita (ft^3/day/capita)

t is the time for digestion (days)

TABLE 23.8
Digestion Time of Sludge at Different Temperatures

Temperature (°F)	50	60	70	80	90	100	110	120	130	140
Temperature (°C)	10	16	21	27	32	38	43	49	54	60
Reaction time (days)	75	56	42	30	25	24	26	16	14	18
Temperature range				Mesophillic					Thermophillic	

Source: Adapted from Fair, G.M. and Geyer, J.C., *Water Supply and Waste-Water Disposal*, John Wiley, New York, 1954.

Temperature has a marked effect on the rate of the anaerobic reactions. Table 23.8 gives guidelines for digestion times θ (reactor) as a function of temperature. The volume of the digester, V(digester) = θ(reactor) · Q(sludge flow).

The normal way a tank was operated was to add sludge, as produced from the primary clarifier, and then, after holding the sludge for usually 25–35 days, discharge a portion of the tank contents to sludge-drying beds. Single digesters operated in the fill-and-draw manner were later called *single-stage digesters*. This was after it was found that two digesters worked better than one; actually one was a reactor and one was a separator. Later yet, because it was also found that two-stage digesters were loaded at a greater rate than single-stage digesters; the latter were called *standard-rate* digesters. The loading criterion 0.05 lb VS/ft^3 of digester/day (0.81 kg VS/m^3 digester) was commonly used for the standard-rate digesters.

23.4.2.1 High-Rate Digestion

It was found in the mid-1950s that if the gas produced from the digester was circulated back into the tank and mixed with the contents, the reaction rate was greatly accelerated and the loading rate could be increased. The hypothesis was that the reaction velocity was accelerated by the presence of methane gas, which presumably was acting as a catalyst. Later it was found that the methane had nothing to do with the observed increased velocity of reaction. The mixing, induced by fluid drag from the methane gas bubbles, increased the reaction velocity by bringing together reactants by convective motion and by the dissipation of reaction-inhibiting end products. Loading rates in the range 0.1–0.4 lb VS/day/ft^3 (1.6–6.5 kg VS/m^3 digester) were found feasible. Hence the term *high-rate digestion* was used to designate a digester so operated. Since the contents of the digester are mixed, a second identical digester was used to separate the liquid and solids. Hence the designation *two-stage digestion* was adopted. Actually, the second stage functions to effect separation and any further digestion is only incidental to this main function. Figure 23.10 shows a digester cross section operated in accordance with the complete-mixing principle.

23.4.3 Process Design Principles

Again rational process design is based upon understanding the materials-balance concept coupled with reaction

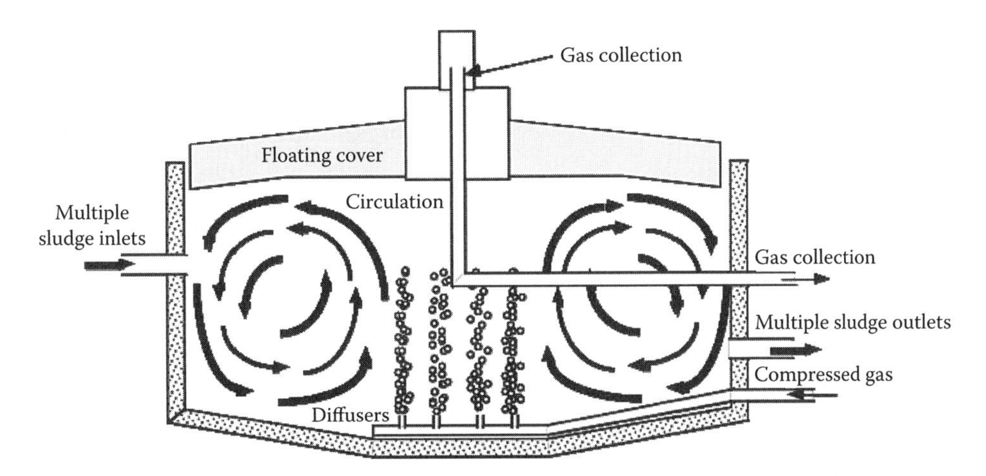

FIGURE 23.10 Complete-mixing ("high-rate") anaerobic digester.

kinetics. First, however, the reactions must be delineated (see also Section 22.2.3.4).

23.4.3.1 Reactions

Anaerobic digestion and anaerobic treatment are both fermentation reactions. Fermentation is by definition a reaction in which the reactant organic molecule acts as both the electron donor and electron receiver (in an aerobic reaction oxygen is the electron receiver). The organic molecules ordinarily found in wastewater sludge and high-strength wastes include fats, starches, sugars, proteins, etc. To convert these molecules to fermentation products requires enzymes. A good many bacteria have the necessary enzymes. Many in fact are facultative and need only a period of acclimatization. These bacteria break down organic molecules along complex biochemical pathways. For example, *Escherichia coli*, in fermenting 100 m moles of glucose will yield at pH 6.2: 1 and 2 macromoles glycerol, 423.4.8 ethanol, 2.43 formic acid, 36.5 acetic acid, 723.4.5 lactic acid, 10.7 succinic acid, 88.0 carbon dioxide, and 75.0 hydrogen (Andrews et al., 1964). Other bacteria will yield a different mix of these products and perhaps different compounds. The compounds formed are the final end products and represent what is usually called "first stage fermentation," or the "acid production phase." The reaction in simplified form is depicted,

$$\begin{aligned}&\text{organic molecule} + \text{enzymes (acid formers)}\\&\quad\rightarrow \text{enzymes (acid formers)} + \text{cells (acid formers)}\\&\quad+ \text{organic acids} + CO_2 + \text{other products}\end{aligned} \tag{23.34}$$

The organic acids are predominately those listed in Table 22.6, i.e., formic, acetic, propionic, butyric, valeric, isovaleric, and caproic (McCarty, 1964a). These acids, in turn go to methane. The product fractions along the different paths from the complex organics to organic acids, and in turn from the organic acids to methane are shown in Figure 23.11. This figure, published originally in 1964 in *Public Works* magazine in Part 1 of a four-part series by McCarty (1964a,b,c,d) has been probably the most frequently cited figure and article series on the topic of the anaerobic process. Figure 23.12 is a schematic adaptation from the original figure, which gives the numerical values.

Methane fermentation is the second stage in the overall process. It is characterized by low cell yield, a relatively slow reaction velocity, and rather fragile ecological requirements (i.e., pH should be about neutral, temperature should be about

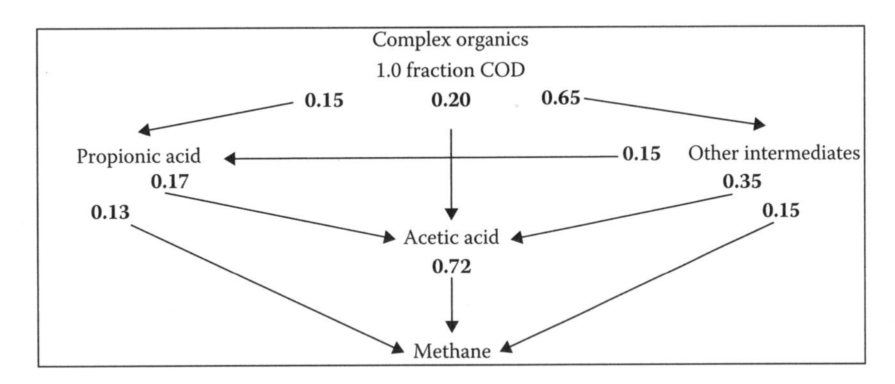

FIGURE 23.11 Pathways in methane fermentation. (Adapted from McCarty, P.L., Anaerobic waste treatment fundamentals, Part one—Chemistry and microbiology, *Public Works*, September 1964a, p. 111.)

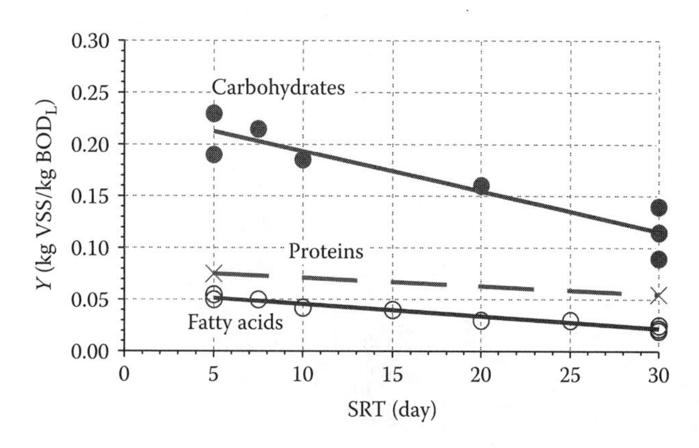

FIGURE 23.12 Cell yield from methane fermentation. (Adapted from McCarty, P.L., Anaerobic waste treatment fundamentals, Part one—Chemistry and microbiology, *Public Works*, September 1964a, p. 112.)

95°F, and various toxins should not be allowed to build up). The complex reaction is depicted, again, in simple terms:

$$\begin{aligned}&\text{organic acids} + \text{enzymes (methane bacteria)}\\&\quad\rightarrow \text{enzymes (methane bacteria)}\\&\quad+ \text{cells (methane bacteria)} + CO_2 + CH_4 \qquad (23.35)\end{aligned}$$

Specific methane-forming stoichiometric reactions for various organic acids previously listed were given by Andrews et al. (1964), along with nine species of methane formers. In some cases more than one species were shown to have participated in a reaction. A characteristic of methane bacteria is that they are obligate anaerobes and are sensitive to other environmental factors. If a digester goes "sour," the production of methane gas is reduced, which means the organic acids are increasing. Of the two reactions, the methane production is the slower.

Another characteristic of the anaerobic reaction is the lower cell yield, Y, compared with the aerobic reactions. Figure 23.12 shows the net cell yields, Y(net) for three substrates associated with methane fermentation. The respective net cell yields decline with the residence time in the reactor due to the endogenous decay of organisms.

23.4.3.2 Kinetics

As with aerobic reactions, Monod kinetics may apply to anaerobic reactions, as reviewed in Section 22.5. The Monod equation is restated,

$$\mu = \hat{\mu}\frac{S}{K_s + S} \qquad (22.31)$$

The kinetic equations apply to both stages of the anaerobic process, i.e., acid formation or methane formation, or empirically to the overall process, each with different constants. In sludge digestion the characterization of S in terms of VSS (volatile suspended solids) is hypothetical, i.e., not proven.

Table 23.9 shows kinetic data obtained by various sources. For comparison aerobic kinetic data are shown on the first line. The acid-former kinetic constants are seen to be similar to those of the aerobic constants (using COD as the substrate basis). Looking at the methane formers, the Y values are about 1/4th the aerobic Y's, but $\hat{\mu}$'s are not consistently different; the b's are about 1/3rd the aerobic values and the K_s values are markedly higher. The "aggregate" of acid fermentation and methane fermentation combined shows lower values of Y, $\hat{\mu}$, and b, but higher K_s. The data indicate generally that $Y(\text{anaerobic}) \approx 0.1 \cdot Y(\text{aerobic})$; that $\hat{\mu}(\text{anaerobic}) \approx 0.1 \cdot \hat{\mu}(\text{aerobic})$; that $b(\text{anaerobic}) \approx 0.3 \cdot b(\text{aerobic})$; that $K_s(\text{anaerobic}) >> K_s(\text{aerobic})$. Also, if we let $\hat{\mu}(\text{anaerobic}) \approx 0.26$, then $\theta_c^m \approx 4$ day. In practice, the SRT (solids retention time) ranges $10 < \text{SRT} < 30$ day. In other words, the kinetic values in Table 23.9 provide a sense of their respective magnitudes but would not warrant confidence for practice.

23.4.3.3 Influences on Reaction Velocity

As noted, in the anaerobic process, the methane formers are very sensitive to external influences. Temperature, mixing,

TABLE 23.9
Kinetic Constants for Anaerobic Reactions

Organism	Substrate	Basis	Y (mg VSS/mg Basis)	$\hat{\mu}$ (day^{-1})	b (day^{-1})	K_s (mg/L)	Reference
Aerobic		COD	0.67	3.8	0.07	22	Metcalf & Eddy (1972)
	Domestic wastewater	BOD	0.6	3.0	0.06	60	Tchobanoglous and Burton (1991)
Acid formers	Glucose	Glucose	0.22	30		22	Ghosh and Pohland (1974)
		COD	0.54	1.33	0.87		Andrews et al. (1964)
Methane form	Acetic acid	Acetic acid		3.37		600	Ghosh and Pohland (1974)
		COD	0.14	1.33	0.02		Andrews et al. (1964)
	Dextrose, tryptone	COD	0.15		0.03		Andrews (1963)
	Acetic acid 35°C			0.36		165	Grady et al. (1999, p. 630)
Aggregate	Acetic acid	Acetic acid	0.05	0.26	0.01	869	Lawrence and McCarty (1969)
	Municipal sludge	BOD	0.06		0.03		Tchobanoglous and Burton (1991)
	Protein	BOD	0.075		0.014		
	Dextrose, tryptone	COD	0.11	0.09	0.03	6700	Agardy et al. (1962)

pH, and the reaction chemical environment all influence reaction velocity. Consequently, the reactor must be designed and operated to maintain these conditions at "optimum" levels. In general 6.8 < pH(optimum) < 7.4 (Grady et al., 1999, p. 632). The problem may be exacerbated if the methane formers are affected and reduce their activity, which permits the volatile acid formation rate to exceed the methane production rate with a consequent accumulation of volatile acids resulting in a "sour" digester. Thus, monitoring volatile acids in the digester is done in practice. If the volatile acids start to accumulate, one remedy is to reduce the organic loading to permit two reactions to regain a balance. To mitigate large swings in pH, the alkalinity, present in the range of 1000–5000 mg/L as $CaCO_3$ (Grady et al., 1999, p. 634), may act as a buffer. Heavy metals and certain organic compounds may have an inhibiting effect.

23.4.3.4 Effect of Temperature

The Arrhenius equation is the model for temperature effect,

$$k = A\exp\left[-\frac{\Delta E_a}{RT}\right] \tag{23.36}$$

or

$$\log k = -\frac{\Delta E_a}{RT} + \log A \tag{23.37}$$

Figure 23.13 shows the effect of temperature on the rate of gas production formulated in an Arrhenius plot. The data were obtained from a 1000 mL anaerobic reactor operated using municipal primary sludge with 4% sludge concentration.

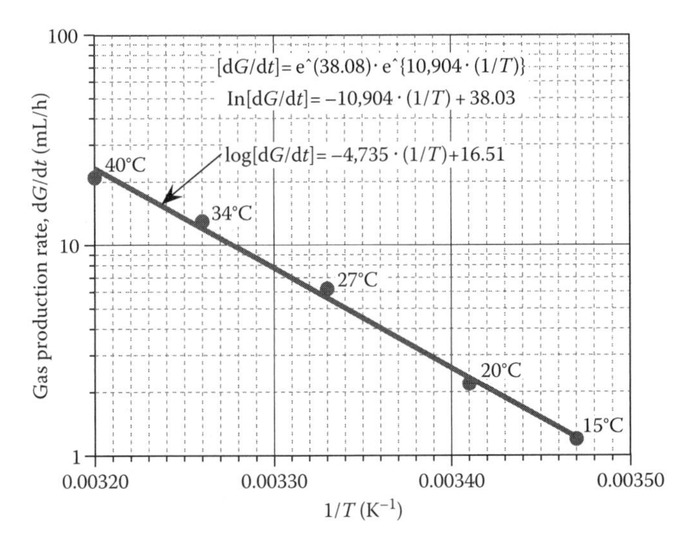

FIGURE 23.13 Gas production rate (STP) for municipal primary sludge (4% solids); measurements obtained using lab-scale anaerobic reactor. (From Pitkin, J., Effect of temperature on rate of gas generation from municipal sludge, MS Thesis, Department of Civil Engineering, Utah State University, Logan, UT, 1969.)

Measurement of reaction rate. Kinetics can be in terms of any of the reactants or products in a reaction equation. Once a rate is measured for any one of the reactants or products in a reaction, rates for any of the others may be calculated from the stoichiometric coefficients.

Measurement of the rate of gas production from municipal primary sludge was done by Pitkin (1969) at five temperatures, using a 1.00 L reactor in a water bath. Data are plotted in Figure 23.13 in accordance with the Arrhenius equation, i.e., log(gas production rate) vs. $(1/T)$. The solids concentration was 4% for all measurements; the reactor volume was 1.00 L. The best fit equation for Figure 23.13 is

$$\log\left[\frac{dG'}{dt}\right] = -4735\cdot\left(\frac{1}{T}\right) + 16.51 \tag{23.38}$$

where

- G' is the gas generated per unit of reactor volume (mL gas produced/L reactor volume)
- dG'/dt is the gas production rate (mL gas produced/L reactor volume/h)
- T is the temperature (K)

Equation 23.38 can be restated as

$$\left[\frac{dG}{dt}\right] = 10^{\left[-4735\cdot\left(\frac{1}{T}\right)+16.51\right]} \tag{23.39}$$

To convert the rate of gas production to m^3 gas produced/m^3 reactor volume/h requires multiplying (23.38) by 1000 L/m^3 (for reactor volume) and dividing by 10^6 mL/m^3 (for gas produced), to give

$$\left[\frac{dG}{dt}\right] = 10^{\left[-4735\cdot\left(\frac{1}{T}\right)+13.51\right]} \tag{23.40}$$

where G is the gas generated per unit of reactor volume (m^3 gas produced/h/m^3 reactor volume).

Keep in mind that the equation is applicable to a reactor with 4.0% sludge. If the gas flow from an operating digester is measured at two or three temperatures, the plot should follow the Arrhenius equation but with coefficients that are related to the local conditions. The gas flow from the digester can be measured if instrumented with an orifice meter with associated pressure gages or pressure transducers. To confirm the methane fraction, a gas sampling tube can be used to collect the gas from a sampling vent tube. Analysis can be with a gas-chromatograph (or GS/MS) with appurtenances for handling gas.

23.4.3.5 Mixing

It was discovered in the late 1950s that mixing, which brought the reactants in contact by convectively increasing the diffusion gradients in the transport process, markedly increased reaction rates. This led to the concept of "high-rate digestion" as outlined previously. There is no rational model which can predict the influence of mixing. Rather it is necessary to rely upon guidelines from experience, which are rather sparse. A CFD simulation by Meroney (2009) illustrated mixing patterns for various configurations of tank inflow points, which does provide an avenue for rational design.

Mixing may be accomplished by gas recirculation, by recirculating the sludge, or by a turbine mixer. Usually, the mixing requirement is satisfied as a part of the selection of proprietary equipment. This is selected also with the view toward not permitting grit to settle on the bottom of the digester, which requires about 2.0 fps local velocity.

23.4.3.6 Environmental Conditions

As stated, methane bacteria have a rather narrow ecological niche. First they are obligate anaerobes. While oxygen will not kill the cells it will make them dysfunctional. Heavy metals are toxic as they are to most organisms; toxic concentration levels have not been established. Proper pH is another highly important requirement. McCarty (1964b, p. 123) stated that the reaction may proceed in the pH range of $6.6 \leq pH \leq 7.6$, with an optimum range of $7.0 \leq pH \leq 7.2$. To maintain this pH range an alkalinity level of several thousand mg/L alkalinity as $CaCO_3$ is desirable, to act as a buffer in pH control (McCarty, 1964b, p. 125).

23.4.3.7 Materials Balance: Kinetic Model

The materials-balance concept is applicable also to the anaerobic reactor. Figure 23.14 is a flow scheme for an anaerobic system.

Mass balance for digester. The materials-balance equation, written for the reactor only, again merely says

observed rate of change of VSS in the reactor
= mass inflow of VSS − mass outflow of VSS
− rate of reaction − volume change (23.41)

In equation form this is

$$\left[\frac{d(VSS_o)}{dt}\right]_o \cdot V = Q \cdot VSS_o - Q \cdot VSS + VSS \cdot \frac{dV}{dt} - V\left[\frac{d(VSS)}{dt}\right]_r \tag{23.42}$$

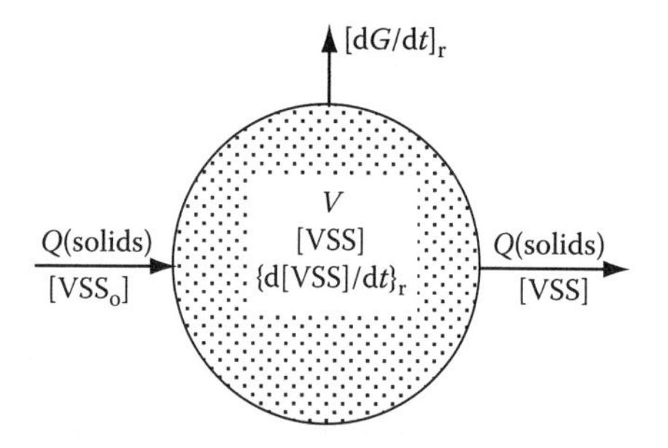

FIGURE 23.14 Anaerobic reactor showing terms in materials balance.

where

VSS_o is the volatile suspended solids concentration flowing into the reactor (mg/L)
VSS is the volatile suspended solids concentration in the reactor and flowing out (mg/L)
V is the volume of reactor (L)
Q is the flow into or out of the reactor (L/day)
$[d(VSS_o)/dt]_o$ is the observed rate of change of VSS concentration in the reactor (mg/L/day)
$[d(VSS)/dt]_r$ is the reaction rate of VSS in the reactor (mg VSS destroyed/L/day)

For the steady-state assumption $[d(VSS)dt]_o = 0$; also for simplicity in illustrating some other key ideas, $dV/dt = 0$, giving

$$0 = Q \cdot VSS_o - Q \cdot VSS - V\left[\frac{d(VSS)}{dt}\right]_r \tag{23.43}$$

Rearranging and substituting $V = Q\theta$,

$$\left[\frac{d(VSS)}{dt}\right]_{destroyed} = [VSS_o - VSS]/\theta \tag{23.44}$$

In other words, the rate of solids destroyed may be estimated by the difference between the inflow of VSS and the outflow of VSS divided by the detention time. A kinetic equation may be hypothesized as

$$\left[\frac{d(VSS)}{dt}\right]_r = \mu[VSS] \tag{23.45}$$

Substituting Equation 23.45 into Equation 23.44 gives

$$(VSS_o - VSS) = \theta\mu[VSS] \tag{23.46}$$

or

$$(VSS_o - VSS) = \left[\frac{\theta}{\theta_c}\right][VSS] \tag{23.47}$$

Equation 23.46 then relates volatile solids reduction to hydraulic residence time, $\vartheta = V/Q$, and cell growth rate μ, if desired, substitute, $\mu = 1/\vartheta_c$.

23.4.3.8 Practice

The process design of completely mixed digesters is based upon two criteria: (1) loading rate and (2) hydraulic detention time. The loading rate recommended is 0.1–0.4 lb VS/day/ft^3. Hydraulic detention times of 10–30 days are ordinarily recommended. Usually 15 days or greater is suggested because cell growth rate is usually 4–10 days (see Table 23.9).

Another way often advocated to increase digester performance is to have a longer solids-retention time, SRT. This is defined as

$$\text{SRT} \equiv \frac{\text{mass of solids in digestor}}{\text{mass of solids leaving per day}} \tag{23.48}$$

Ordinarily, $\theta = \text{SRT}$. However, if the mass of solids in the digester is increased by increasing the concentration of solids flowing into and maintained in the digester (through better thickening) then the SRT is increased correspondingly and $\text{SRT} \geq \theta$. Expressing Equation 23.48 mathematically,

$$\text{SRT} = \frac{V(\text{reactor}) \cdot [\text{VSS}]}{Q(\text{solids}) \cdot [\text{VSS}]} \tag{23.49}$$

or

$$\text{SRT} = \frac{V(\text{reactor})}{Q(\text{solids})} \tag{23.50}$$

Since, for a given waste loading on a plant, the mass-solids flux from the digester is constant, $Q(\text{solids}) \cdot [\text{VSS}] = \text{constant}$. Therefore, if Q(solids) is reduced due to improved thickening before the digester, and since V(reactor) does not change, then SRT is increased accordingly. Example 23.3 illustrates the application of the concept.

Example 23.3 Effect of Sludge Thickening on SRT

Given

$[\text{SS}]_o = 4.75\%$ $[= 47.5 \text{ kg/m}^3]$

$[\text{VSS}]_o = 0.80\,[\text{SS}]_o$ $[= 0.80 \cdot 47.5 \text{ kg SS/m}^3 = 38.0 \text{ kg VSS/m}^3]$

$Q(\text{solids}) = 40{,}000$ gal/day $[= 151.4 \text{ m}^3/\text{day} = 6.31 \text{ m}^3/\text{h}]$

$V(\text{reactor}) = 750{,}000 \text{ gal} = 2{,}839 \text{ m}^3$

Required

If the solids inflow is increased to $[\text{SS}]_o = 6.5\%$ $[= 65 \text{ kg/m}^3]$, calculate the new SRT.

Solution

1. SRT(1),

$$\begin{aligned}\text{SRT}_1 &= \frac{V(\text{reactor})}{Q(\text{solids})_1} \\ &= 2839 \text{ m}^3/151.4 \text{ m}^3/\text{day} \\ &= 19 \text{ day}\end{aligned} \tag{EX23.3.1}$$

2. Calculate $Q(\text{solids})_2$,

$$\begin{aligned}\text{Mass flow solids} &= Q(\text{solids})_1 \cdot V(\text{SS})_1 \\ &= [151.4 \text{ m}^3/\text{day}] \cdot [47.5 \text{ kg/m}^3] \\ &= 7192 \text{ kg solids/day}\end{aligned}$$

$$\text{Mass flow solids} = Q(\text{solids})_2 \cdot V(\text{SS})_2$$

$$7192 \text{ kg solids/d} = Q(\text{solids})_2\ [65 \text{ kg/m}^3]$$

$$Q(\text{solids})_2 = 110.6 \text{ m}^3/\text{day}$$

3. Calculate SRT(2),

$$\begin{aligned}\text{SRT}_1 &= \frac{V(\text{reactor})}{Q(\text{solids})_1} \\ &= 2839 \text{ m}^3/110.6 \text{ m}^3/\text{day} \\ &= 26 \text{ day}\end{aligned} \tag{Ex23.3.2}$$

Discussion

The longer SRT results in higher VSS reduction. The data are from the Fort Collins WWTP.

23.4.4 Operation and Monitoring

An issue that is often given only an afterthought level of attention in design is for the draining of all tanks with provisions for cleaning. This is true especially of anaerobic tanks where access to the bottom of the tank is, as a rule, not adequate. Another issue is safety, since explosions have been known around anaerobic digesters due to the methane gas. As a rule, the gas is collected and used for heating the digester. Methane is a greenhouse gas and should not be vented to the atmosphere; combustion (flaming) will convert the gas to carbon dioxide and water.

23.4.4.1 Process Upsets

Close operator control is necessary in operating the anaerobic process. The most likely problem is the acid production reaction getting ahead of the methane fermentation reaction and causing an acid condition, and an upset. Toxic inputs are another potential problem.

23.4.4.2 Indicators and Tests

Monitoring is essential if any impending upsets are to be discovered and avoided. Change is the key to detecting upsets. Some of the parameters monitored include pH, volatile acids, volatile suspended solids (in and out), and gas production. Since pH often changes after the fact, it has little use in operation. Volatile acids are recommended by some. However, gas production is the most sensitive indicator of an upset and is the one which is prominently mentioned in the list of indicators available. This is quite logical since CH_4 and CO_2 are the final end products of the methane reaction. The composition of the gas should also be measured periodically. Ordinarily this is about 70% methane and 30% carbon dioxide. Since carbon dioxide is also an end product of the acid-formation reaction, an increase in CO_2 levels also indicates an impending upset.

Measurements of volatile suspended solids in the incoming sludge, in the digester and leaving the digester, are essential so that the loading on the digester can be ascertained and the volatile solids reduction can be determined. Temperature and alkalinity should also be monitored.

23.4.4.3 Percent Reduction of Volatile Suspended Solids

A common measure of effectiveness of the anaerobic digestion process is "percent reduction" in volatile suspended solids (VSS). Figure 23.15 illustrates the method to calculate the percent reduction in VSS with calculations shown in Example 23.4.

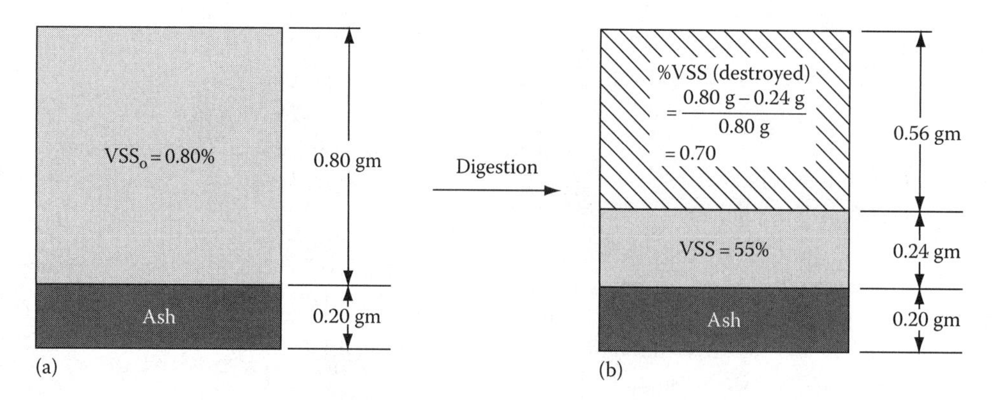

FIGURE 23.15 Illustration of method of calculation of percent reduction in volatile suspended solids. (a) Sludge to digester and (b) sludge from digester.

Example 23.4 Calculation of VSS Destroyed

Given

Figure 23.15a shows that $VSS_o \approx 0.80$ for a sludge entering a digester. For the sludge leaving the digester, let $VSS \approx 0.55$ in Figure 23.15b. Both were measured quantities at the plant.

Required

Calculate the fraction of VSS(destroyed) by the anaerobic process.

Solution

a. Ash

The ash mass remains constant and is 0.20 g.

b. VSS mass leaving digester

For the sludge leaving the digester,

$$f(\text{VS}) = \text{VSS}/(\text{ash} + \text{VSS})$$

$$0.55 = \text{VSS}/(0.20\ \text{g} + \text{VSS})$$

$$\text{VSS} = 0.24\ \text{g}$$

c. VSS destroyed

$$\begin{aligned}\text{VSS(destroyed)} &= \text{VSS}_o - \text{VSS}\\ &= 0.80\ \text{g} - 0.24\ \text{g}\\ &= 0.56\ \text{g}\end{aligned}$$

$$\begin{aligned}f(\text{VSS destroyed}) &= 0.56\ \text{g}/0.80\ \text{g}\\ &\approx 0.70\end{aligned}$$

Discussion

Without a "tip" a person new to the field is likely not to consider the ash. Also, a sketch helps to keep the calculations straight. Sampling to calculate the f(VS destroyed) is part of daily monitoring.

23.5 SUMMARY

A perspective on the state of the art and its evolution is outlined. In addition activated-sludge parameters are provided, both as a summary and as a convenience.

23.5.1 State of the Art

Biological treatment, as a technology, is still evolving; like most unit processes it has not reached closure. In the first 50 years of the twentieth century technologies such as septic tanks, Imhoff tanks for anaerobic treatment, two-stage digesters, activated sludge, trickling filters, etc. were developed. The approaches were empirical and experience was assimilated in converging on guidelines and practices that worked; theory had not taken hold.

From about 1950 to the present, theory has helped to complement practice. The idea of a tank as a "reactor" started to emerge in the 1950s. A "golden age" of research occurred, arguably, through the 1950s and into the 1970s, providing a knowledge base of principles. A defining break with the past was the 1970 paper of Lawrence and McCarty (1970) that delineated these principles with guidelines for design and operation. The 1972 Metcalf and Eddy book codified what had been learned in practice coupled with guidance from theory. Other books, e.g., by Orhan and Artan (1994), Grady et al. (1999), Rittman and McCarty (2001), added scientific rationale, essentially summarizing the state of the art gleaned from the last fifty years. In 1987 the IWA model provided a means to actually model the activated-sludge process and provided a flexible framework for modeling the anaerobic process.

Demands for a comprehensive control of pollutants during the 1970s led to sophisticated activated-sludge and bio-filter practices to remove nitrogen and phosphorous as well as organic carbon. The anaerobic process remains less glamorous but has been understood since the 1960s as a complex biochemical process which has been assimilated in design and operation. Operation is understood at a different level also than was the case through the 1960s. Plant operators run a complex chemical plant with monitoring and surveillance governed by restrictive regulations; the training and professionalism has been commensurate.

23.5.2 Parameters

The bases for most rational parameters are reaction stoichiometry, kinetics, and reactor mass balance. From these relations a number of parameters may be derived. Because they are often scattered in different sections of a book as related to their respective presentations, they are summarized in Table 23.10 as a matter of convenience for retrieval. Tchobanoglous and Burton/Metcalf and Eddy (1991) was the reference used for most of the equations. In some cases the nomenclature was altered. Also, this text used fewer defined coefficients than found in the literature, in order to keep a more fundamental orientation.

TABLE 23.10
Biological Treatment Equations and Parameters

Parameter

Monod kinetics

$$\mu = \hat{\mu}\frac{S}{K_s + S} \quad (22.31)$$

Substrate utilization rate and cell synthesis rate

$$\left[\frac{dX}{dt}\right]_g = Y \cdot \left[\frac{dS}{dt}\right] \quad (22.29)$$

$$\left[\frac{dX}{dt}\right]_g = \mu X \quad (22.30)$$

$$\frac{\mu}{Y} = \frac{dS/dt}{X} \quad (23.48)$$

Substrate degradation rate

$$\frac{dS}{dt} = \frac{1}{Y}\hat{\mu}\frac{S}{K_s + S}X \quad (22.32)$$

Net rate of cell synthesis

$$\frac{[dX/dt]_{net}}{X} = \mu - b \quad (22.36)$$

$$\mu(\text{net}) \equiv \frac{[dX/dt]_{net}}{X} \quad (22.37)$$

$$\mu(\text{net}) = \mu - b \quad (22.38)$$

Sludge age related to μ(net)

$$\theta_c = \frac{V(\text{reactor}) \cdot X}{WX_r} \quad (22.52)$$

$$\theta_c \equiv \frac{1}{\hat{\mu} - b} \quad (22.53)$$

$$\theta_c^m \equiv \frac{1}{\hat{\mu} - b} \quad (22.54)$$

Substrate utilization rate defined and related to μ/Y

$$U \equiv \frac{S_o - S}{\theta X} \quad (22.43)$$

$$U = \frac{\mu}{Y} \quad (22.44)$$

$$\frac{1}{\theta_c} = YU - b \quad (22.58)$$

F/M defined and related to U

$$\frac{F}{M} \equiv \frac{S_o}{\theta X} \quad (22.45)$$

$$U = \left[\frac{S_o}{\theta X}\right] \cdot \left[\frac{S_o - S}{S_o}\right] \quad (22.47)$$

$$E = \left[\frac{S_o - S}{S_o}\right]$$

Net cell yield as affected by sludge age

$$Y(\text{net}) = \frac{Y}{1 + b\theta_c} \quad (22.28)$$

PROBLEMS

Activated Sludge

23.1 Aerated Lagoon Mass Balance

Given

Figure P23.1 illustrates the conditions

Assume $S_o = 300$ mg BOD/L and that $\theta = 24$ h. Assume for analysis purposes that the aerated lagoon is "complete mix."

Required

Formulate a materials-balance–kinetics relation for an aerated lagoon, such as at Northglenn. From what you have done, estimate the effluent BOD at the Northglenn aerated lagoon.

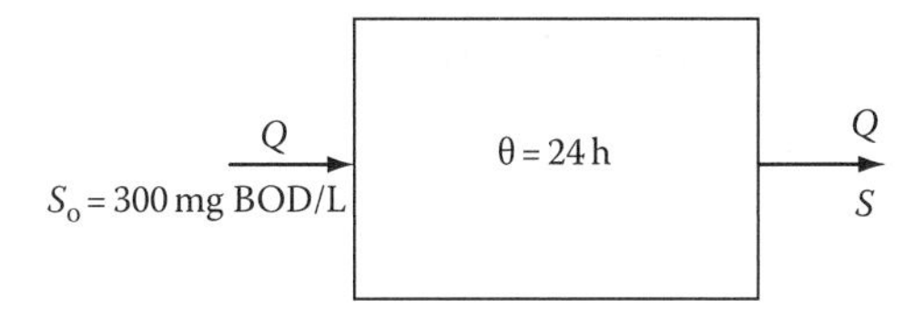

FIGURE P23.1 Aerated lagoon.

23.2 Conventional Activated-Sludge Wasting

Given

Assume the following data apply for complete-mix activated-sludge system that treats municipal wastewater

$Q = 0.066$ m^3/s (1.5 mgd)
V(reactor) = 1426 m^3
$S_o = 260$ mg/L.
$S = 30$ mg/L

Required

Calculate the sludge-wasting rate, R' (also called W).

23.3 Conventional Activated Sludge for New Conditions

Given

An activated-sludge system treats a municipal wastewater. Assume the system operates under steady-state conditions. The flow was $Q = 0.044$ m^3/s (1.0 mgd). The detention time in the reactor was 6 h. The effluent BOD, was measured as S(1.0 mgd) = 20 mg/L.

The flow is increased to $Q = 0.066$ m^3/s (1.5 mgd). Calculate S(1.5 mgd), the effluent BOD concentration. S_o does not change.

Required

1. Effluent BOD for $Q = 0.066$ m^3/s.
2. Calculate the sludge-wasting rate R'.

23.4 Conventional Activated-Sludge Treatment of Toluene

Given

An activated-sludge system treats toluene at 25°C with influent concentration, $S_o = 60$ mg/L. The flow was $Q = 0.044$ m^3/s (1.0 mgd). The effluent concentration, is required to be $S = 20$ mg/L. Assume the system operates under steady-state conditions. Kinetic coefficients from a study by Chang (1991) and reported in

Yoshinaga et al. (1995) using a laboratory chemostat were measured as

$$\hat{\mu} = 0.45\ \text{h}^{-1}$$

$$K_s = 1.88\ \frac{\text{mg substrate}}{\text{L}}$$

$$Y = 0.99\ \frac{\text{mg cells synthesized}}{\text{mg substrate degraded}}$$

and let

$$k_d = 0.05\ \frac{\text{mg cell mass lost}}{\text{mg cells in reactor h}}$$

Required

1. Determine the detention time, θ, for the reactor.
2. Calculate the sludge-wasting rate,, R', for the stated conditions.
3. The influent concentration increases to $S_o = 95$ mg/L. Determine S for this new condition.

BIO-FILTERS

23.5 Bio-Filter Performance

Given

A trickling filter for municipal wastewater treatment, after primary treatment, has been in place for some 30 years, The filter bed is 2000 mm in depth. The BOD_{in}, i.e., $S_o = 200$ mg/L and BOD_{out}, i.e., $S_e = 30$ mg/L; the corresponding HLR ≈ 5 gpm/ft^2. For the diurnal peak, HLR ≈ 8 gpm/ft^2, estimate the associated effluent BOD. Sketch the associated $S(Z)$ curves for each case.

Data summary

$D = 2.0$ m

HLR(normal) $= 5$ gal/min/ft^2
$= 0.2037$ m^3/m^2/min
$= 0.003395$ m^3/m^2/s

HLR(peak) $= 8$ gal/min/ft^2
$= 0.005432$ m^3/m^2/s

$S_o = 200$ mg BOD/L

$S_e = 30$ mg BOD/L

Required

S_e when HLR(peak) $= 8$ gal/min/ft^2.

23.6 Bio-Filter Depth

Given

Assume the following data apply for trickling filter that treats municipal wastewater:

$Q = 0.066$ m^3/s (1.5 mgd)
Depth of reactor, D(reactor) $= 1.8$ m
A(reactor) $= 194$ m^2
$S_o = 260$ mg/L
$S = 30$ mg/L
$T = 25$°C

Required

A similar installation using the same wastewater must be designed with a requirement that $S = 15$ mg/L. Determine the depth required for the same HLR.

23.7 Bio-Filter Depth for Loading Increase

Given

For a trickling-filter reactor, assume HLR $= 0.0034$ m^3/m^2/s (5.0 gpm/ft^2). Assume that S_o is increased by 50%.

Required

1. Estimate S for the new condition, i.e., $S(S_o = 1.5)$.
2. What new depth will result in the same S as before the 50% increase, i.e., $S(S_o = 1.0)$?

23.8 Bio-Filter Depth for Toluene and HLR Increase

Given

A biofilm filter reactor pilot plant ($D = 1.8$ m) was developed for the treatment of toluene at 25°C. The influent concentration is, $S_o = 60$ mg/L and the effluent concentration, S, was measured, $S = 15$ mg/L. Assume HLR $= 0.0034$ m^3/m^2/s (5.0 gpm/ft^2).

Required

1. Determine the depth needed to obtain an effluent concentration, $S = 4$ mg/L.
2. For whatever depth was worked out in "1", assume that it becomes necessary to handle a new flow, giving an HLR $= 0.0048$ m^3/m^2/s (7.0 gpm/ft^2). Determine the associated S.

ANAEROBIC

23.9 Effect of Sludge Thickening on SRT

Given

$[SS]_o = 4.75\%$ [$=47.5$ kg/m^3]
$[VSS]_o = 0.80 \cdot [SS]_o$ [$=0.80 \cdot 47.5$ kg SS/m$^3 = 38.0$ kg VSS/m^3]
Q(solids) $= 40{,}000$ gal/day [$=151.4$ m^3/d $= 6.31$ m^3/h]
V(reactor) $= 750{,}000$ gal $= 2839$ m^3

Required

If the solids inflow is increased to $[SS]_o = 6.5\%$ [$= 65$ kg/m^3], calculate the new SRT.

23.10 Anaerobic Reactor Volume and Gas Production

Given

An anaerobic reactor (a "digester") treats municipal primary sludge. The plant flow is $Q = 0.044$ m^3/s (1.0 mgd), and $T = 35$°C.

Required

Determine

1. Reactor volume
2. Rates of VSS(destroyed) and gas production

23.11 Anaerobic Reactor Volume with SS Increase and VSS Destroyed

Given

Although not needed, plant flow, $Q = 0.044$ m^3/s (1.0 mgd). For the problem, let the suspended solids concentration be 4%, i.e., $[SS]_o = 40$ kg/m^3, and let

Q(sludge) = 0.80 m^3/h. Also let $[VSS]_o = 32$ kg/m^3 (80% VSS) and assume $[VSS] = 1.5$ kg/m^3. (Notation: the o subscript is for inflow and values and the no subscript is for effluent values.)

Relevant data. From past Fort Collins reactor operation, which measured gas production rate at 35°C, the following was calculated:

$$\frac{dG}{dt} = 0.014 \frac{m^3 \text{ gas/h}}{m^3 \text{ reactor volume}}$$

Required

1. Determine *V*(digester).
2. For the your design in (1), the suspended solid concentration is increased to 5% $[SS]_o = 50$ kg/m^3, with no other changes. Determine the associated [VSS] and VSS(destroyed).

23.12 Digester Operating Data

Given

Visit a municipal WWTP in your area, subject to arrangements with the operator.

Required

1. Obtain digester data as listed from a municipal WWTP in your area.
   a. Average daily flow
   b. Suspended solids to plant, % volatile solids
   c. Wastage of activated sludge
   d. Pumpage of sludge to digester
   e. Fraction solids in primary clarifier underflow, fraction volatile solids
   f. Sludge leaving digester, fraction volatile solids
   g. Temperature in digester
   h. Digester size
   i. Daily gas production and gas analysis.
2. Examine the data and provide a discussion. Discuss the operation with the operator to learn from his or her experience.

23.13 Digester Operating Parameter Calculations

Given

Data from visit to WWTP digester in your area.

Required

From the data obtained, calculate the following operating parameters:

1. Measured solids loading to digester in (kg VS/day/m^3 digester volume)
2. Fraction of volatile solids destroyed by digester
3. Gas produced/kg of volatile solids destroyed
4. Solids retention time (SRT)
5. Hydraulic detention time (θ)

23.14 Digester Gas Production as Affected by Temperature

Given

The primary digester at Fort Collins Plant No. 2 produces 991 m^3/day (35,000 ft^3/day) of gas, operating at 90°F (32°C).

Required

What gas production might you expect if the temperature was maintained at 95°F (35°C)?

Hint: The Arrhenius equation is the theoretical basis for the determination.

23.15 Digester SRT to Deal with Increased Loading

Given

Community growth has caused an increase in average daily wastewater flow, from 0.219 m^3/s (5 mgd) to 0.350 m^3/s 8 mgd.

Required

How can the increased sludge production be processed by the existing digesters? Delineate mathematically.

Hint: Increase the solids concentration entering the digester.

23.16 Experiment with Primary Sludge and Denitrification

Given

Visit to a WWTP in your area.

Required

Obtain some primary sludge in a large carboy and place a portion in a 1 L beaker. Stir the sludge so that it settles to the bottom. Add some sodium nitrate (a few grams). Observe what happens. Explain what happens in terms of operative microorganisms, and the chemistry of the reaction.

23.17 Sour Digester

Given/Required

What should be done if a digester begins to go "sour"? What are the likely indicators?

ACKNOWLEDGMENTS

Professor Wesley W. Eckenfelder with cooperation from Christy Lewis, Aquaeter, Nashville, Tennessee, reviewed Box 23.2 for accuracy in November 2009; the narrative was the author's, based on friendship, interactions, and impressions over the time period since 1965. Professor Lisa Alvarez-Cohen, Chair, Department of Civil and Environmental Engineering, University of California, Berkeley, California, arranged for permission to use the photograph of Professor Pearson in Box 23.1.

APPENDIX 23.A: BIOFILM REACTOR MODEL

The biofilm model is an expansion of Section 23.2.1 to illustrate the rationale for a well-known empirical equation for the design of a biofilm reactor. The starting point here is the mass balance for an infinitesimal slice of a reactor. The "bottom line" is an empirical equation shown picked up in Section 23.2.1.

23.A.1 BIOFILM REACTORS MODEL

A simple stationary-bed model is developed to illustrate, again, the application of the materials-balance principle, combined with kinetics. The stationary bed could be a traditional trickling filter, about 2 m depth, or any deep-bed nitrification biofilm reactor. The approach is generally applicable.

23.A.1.1 Mathematics

The analysis of a biofilm reactor is described starting with the definition sketch, Figure 23.A.1. The approach is the same as the "plug-flow" model seen in Section 23.2.2.5 for activated sludge.

For the infinitesimal element shown in Figure 23.A.1 the materials balance is

$$\left[\frac{\partial S}{\partial t}\right]_{\mathrm{o}} P \cdot A \cdot \Delta Z = -\bar{v}\frac{\partial S}{\partial Z}\Delta Z \cdot P \cdot A - \left[\frac{\partial S}{\partial t}\right]_{\mathrm{r}} P \cdot A \cdot \Delta Z \qquad (23.A.1)$$

where
- S is the substrate concentration (kg/m^3 water volume)
- t is the elapsed time (s)
- P is the porosity
- A is the area of filter bed (m^2)
- Z is the depth of filter bed (m)
- ΔZ is the depth of finite element (m)
- X is the cell concentration (kg/m^3 filter volume)
- Y is the yield coefficient (kg cells synthesized/kg substrate degraded)

As seen, Equation 23.A.1 is common to any kind of fixed-bed reactor, such as used with granular activated carbon. The difference is with the kinetics equation. The Monod equation is used, but arguably it is an "artifice." The rate of substrate utilization, $[dS/dt]_r$, may be limited by diffusion transport (Chapter 18) of the carbon substrate or another reactant such as a nutrient to the microbial cell for reaction. As will be seen, the constants in the Monod equation will be "lumped" into an empirical coefficient so it does not make much difference as far as final outcome is concerned whether the rate is described actually by diffusion, i.e., Fick's first law, or the Monod equation. For reference, the Monod equation is

$$\left[\frac{\partial S}{\partial t}\right]_{\mathrm{r}} = \frac{1}{Y}\left[\hat{\mu}\frac{S}{K_{\mathrm{m}} + S}\right]X \qquad (23.A.2)$$

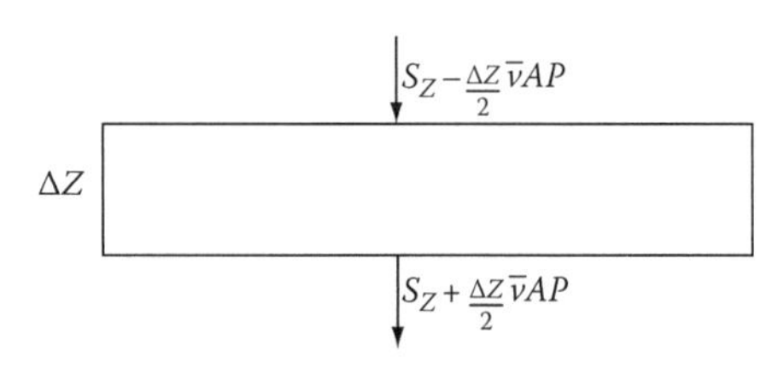

FIGURE 23.A.1 Definition sketch for finite element materials balance.

where
- X is the cell concentration (kg/m^3 filter volume)
- Y is the yield coefficient (kg cells synthesized/kg substrate degraded)
- $\hat{\mu}$ is the maximum specific rate of reaction (cells synthesized/cell mass/s)
- K_{m} is the half saturation coefficient (kg substrate/m^3)

The kinetics described by Williamson and McCarty (1976a,b) utilizes the Monod model but incorporates it into the diffusion process. This depiction does not consider diffusion and focuses on a macroscopic model.

Substituting (23.A.2) in Equation 23.A.1,

$$\left[\frac{\partial S}{\partial t}\right]_{\mathrm{o}} P \cdot A \cdot \Delta Z = -\bar{v}\frac{\partial S}{\partial Z}\Delta Z \cdot P \cdot A - \frac{1}{Y}\left[\hat{\mu}\frac{S}{K_{\mathrm{m}} + S}\right]X \cdot A \cdot \Delta Z \qquad (23.A.3)$$

The concentration of cells in the bulk volume is

$$X = \bar{X}\bar{A}(\text{rocks}) \qquad (23.A.4)$$

where
- $\bar{X}$ is the concentration of cells per unit of surface area (kg cells/m^2 rock surface)
- $\bar{A}$ is the surface area of rocks or other media per unit of bulk volume in filter bed (m^2 rock surface/m^3 bulk filter volume)

Substituting (23.A.4) in (23.A.3) gives

$$\left[\frac{\partial S}{\partial t}\right]_{\mathrm{o}} P \cdot A \cdot \Delta Z = -\bar{v}\frac{\partial S}{\partial Z}\Delta Z \cdot P \cdot A - \frac{1}{Y}\left[\hat{\mu}\frac{S}{K_{\mathrm{m}} + S}\right]\bar{X}\bar{A}(\text{rocks}) \cdot A \cdot \Delta Z \qquad (23.A.5)$$

The specific area, $\bar{A}$, can be estimated by assuming the rocks are spheres and that they have a rectangular packing.

$$\bar{A}(\text{rocks}) = A(\text{rock}) \cdot \bar{N}(\text{rocks}) \qquad (23.A.6)$$

$$= \pi d(\text{rock})^2 \cdot \bar{N}(\text{rocks}) \qquad (23.A.7)$$

where
- A(rock) is the surface area of one rock (m^2)
- $\bar{N}$(rocks) is the specific number of rocks (number of rocks/m^3 bulk filter volume)
- d(rock) is the diameter of one rock (m)

The minimum number of rocks per unit volume may be estimated by assuming that they have a rectangular packing. The number of rocks along a given side in a cube of 1.0 m is

$1.0/d(\text{rock})$. Therefore, the number of rocks in a unit volume, say 1.0 m, is

$$\bar{N}(\text{rocks}) = \left[\frac{1.0\ \text{m}}{d(\text{rock})}\right]^3 \tag{23.A.8}$$

Substituting (23.A.8) in (23.A.7),

$$\bar{A}(\text{rocks}) = \pi d(\text{rock})^2 \cdot \left[\frac{1.0\ \text{m}}{d(\text{rock})}\right]^3 \tag{23.A.9}$$

$$= \frac{\pi}{d(\text{rock})} \tag{23.A.10}$$

$$= \frac{\pi f}{d(\text{rock})} \tag{23.A.11}$$

where

f is the packing ratio (number of rocks actual/number of spherical rocks with rectangular arrangement)

$d_r \equiv d(\text{rock})$ (m)

Keeping in mind that the packing is not rectangular, a factor, f, is introduced in (23.A.10) to give (23.A.11); in which $f \geq 1.0$.

Substituting (23.A.11) in (23.A.5),

$$\left[\frac{\partial S}{\partial t}\right]_o P \cdot A \cdot \Delta Z = -\bar{v}\frac{\partial S}{\partial Z}\Delta Z \cdot P \cdot A - \frac{1}{Y}\left[\hat{\mu}\frac{S}{K_m + S}\right]\bar{X}\left(\frac{\pi f}{d_r}\right) \cdot A \cdot \Delta Z \tag{23.A.12}$$

Divide both sides of (23.A.12) by $PA\Delta Z$,

$$\left[\frac{\partial S}{\partial t}\right] = -\bar{v}\frac{\partial S}{\partial Z} - \frac{1}{Y}\left[\hat{\mu}\frac{S}{K_m + S}\right]\frac{\bar{X}}{P}\left(\frac{\pi f}{d_r}\right) \tag{23.A.13}$$

Recall that

$$\bar{v}PA = \text{HLR} \cdot A \tag{23.A.14}$$

or

$$\bar{v} = \frac{\text{HLR}}{P} \tag{23.A.15}$$

where HLR is the hydraulic loading rate (m^3 water/s/m^2 filter bed).

Substituting (23.A.15) in (23.A.13)

$$\left[\frac{\partial S}{\partial t}\right]_o = -\frac{\text{HLR}}{P}\frac{\partial S}{\partial Z} - \frac{1}{Y}\left[\hat{\mu}\frac{S}{K_m + S}\right]\frac{\bar{X}}{P}\left(\frac{\pi f}{d_r}\right) \tag{23.A.16}$$

Assuming steady state,

$$0 = -\text{HLR}\frac{\partial S}{\partial Z} - \frac{1}{Y}\left[\hat{\mu}\frac{S}{K_m + S}\right]\bar{X}\left[\frac{\pi f}{d_r}\right] \tag{23.A.17}$$

Rearranging (23.A.17) to solve for $\partial S/\partial Z$,

$$\frac{\partial S}{\partial Z} = -\frac{1}{Y}\left[\hat{\mu}\frac{S}{K_m + S}\right]\bar{X}\left[\frac{\pi f}{d_r}\right]\left(\frac{1}{\text{HLR}}\right) \tag{23.A.18}$$

which in finite-difference form is

$$\frac{\Delta S}{\Delta Z} = -\frac{1}{Y}\left[\hat{\mu}\frac{S}{K_m + S}\right]\bar{X}\left[\frac{\pi f}{d_r}\right]\left(\frac{1}{\text{HLR}}\right) \tag{23.A.19}$$

which in a form for numerical solution is

$$\frac{S_{Z+\Delta Z} - S_z}{\Delta Z} = -\frac{1}{Y}\left[\hat{\mu}\frac{S}{K_m + S}\right]\bar{X}\left[\frac{\pi f}{d_r}\right]\left(\frac{1}{\text{HLR}}\right) \tag{23.A.20}$$

or

$$S_{Z+\Delta Z} = S_z - \frac{1}{Y}\left[\hat{\mu}\frac{S}{K_m + S}\right]\bar{X}\left[\frac{\pi f}{d_r}\right]\left(\frac{1}{\text{HLR}}\right) \tag{23.A.21}$$

Any model must be "calibrated"; a calibration factor, F, is inserted in (23.A.21),

$$S_{Z+\Delta Z} = S_Z - \frac{1}{Y}\left[\hat{\mu}\frac{S}{K_m + S}\right]\bar{X}\left[\frac{\pi f}{d_r}\right]\left(\frac{1}{\text{HLR}}\right) \cdot F \cdot \Delta Z \tag{23.A.22}$$

The factor F is inserted in lieu of adjusting the kinetic coefficients. This model can be depicted by means of a spreadsheet.

23.A.1.2 Approximation Model by "Lumping" Coefficients

Another approach in modeling the packed-bed biofilm reactor is to "lump" the coefficients along with the S variable in the denominator in the Monod relation. This approach is used most often in other references. Repeating (23.A.18),

$$\frac{ds}{dz} = -\frac{1}{Y}\left[\hat{\mu}\frac{S}{K_m + S}\right]\bar{X}\left[\frac{\pi f}{d_r}\right]\left(\frac{1}{\text{HLR}}\right) \tag{23.A.18}$$

Regrouping terms in (23.A.18) and including F,

$$\frac{ds}{dz} = -\left[\hat{\mu}\frac{1}{K_m + S}\right]\left[\frac{\pi f}{d_r}\right]\left(\frac{\bar{X}}{Y}\right)F\frac{S}{\text{HLR}} \tag{23.A.23}$$

Separating the variables,

$$\frac{dS}{S} = -\left[\hat{\mu}\frac{1}{K_m + S}\right]\left[\frac{\pi f}{d_r}\right]\left(\frac{\bar{X}}{Y}\right)\frac{F}{\text{HLR}}dZ \tag{23.A.24}$$

$$\ln\frac{S}{S_o} = -\left[\hat{\mu}\frac{1}{K_m + S}\right]\left[\frac{\pi f}{d_r}\right]\left(\frac{\bar{X}}{Y}\right)F\frac{Z}{\text{HLR}} \tag{23.A.25}$$

$$\frac{S}{S_o} = \exp\left\{-\left[\hat{\mu}\frac{1}{K_m + S}\right]\left[\frac{\pi f}{d_r}\right]\left(\frac{\bar{X}}{Y}\right)F\frac{Z}{\text{HLR}}\right\} \quad (23.A.26)$$

$$S = S_o \exp\left\{-\left[\hat{\mu}\frac{1}{K_m + S}\right]\left[\frac{\pi f}{d_r}\right]\left(\frac{\bar{X}}{Y}\right)F\frac{Z}{\text{HLR}}\right\} \quad (23.A.27)$$

By assuming $K_m \gg [S]$ and letting,

$$k = \left[\hat{\mu}\frac{1}{K_m + S}\right]\left[\frac{\pi f}{d_r}\right]\left(\frac{\bar{X}}{Y}\right)F \quad (23.A.28)$$

Then, Equation 23.A.27 is

$$S = S_o \exp\left\{-k\frac{Z}{\text{HLR}}\right\} \quad (23.A.29)$$

Equation 23.A.29 was given by Eckenfelder (1961, p. 224) and has been used most widely in practice. As seen, the coefficient, k, must be calibrated. This can be done by "curve fitting" the model to trickling filter in operation. The k value obtained may be incorporated in the model for a new design at another site, assuming the wastewaters are similar. A reason for disaggregating k, as in Equation 23.A.27 is to better evaluate the role of operative variables and to utilize knowledge about their variation. For example, if the media density varies from one location to another or if one wishes to examine the effect of a variable by a spreadsheet model, the effects may be discernible.

GLOSSARY

Glossary is in Chapter 22.

REFERENCES

Agardy, F. J., Cole, R. D., and Pearson, E. A., Kinetic and activity parameters of anaerobic fermentation systems, Sanitary Engineering Research Laboratory Report 63-2, University of California, Berkeley, CA, 1963.

Andrews, J. F., Cole, R. D., and Pearson, E. A., Kinetics and characteristics of multistage methane fermentations, Sanitary Engineering Research Laboratory, Report No, 64-11, December 1, 1964.

Ardern, E. and Lockett, W. T., Experiments on the oxidation of sludge without the aid of filters, *Journal of the Chemical and Industrial Societies*, XXXII(10):523–539, 1914, London.

Characklis, W. G. and Marshall, K. C. (Eds.), *Biofilms*, John Wiley & Sons, New York, 1990. ISBN 0-471-82663-4.

Eckenfelder, W. W. Jr., Activated Sludge, in Eckenfelder, W. W. Jr. and T. P. Quirk, (Eds.), Advances in Sewage Treatment Design, *Proceedings of a Symposium at Manhattan College*, Sanitary Engineering Division, Metropolitan Section, ASCE, Reston, VA, May 15, 1961.

Eckenfelder, W. W. Jr. and O'Connor, D. J., *Biological Waste Treatment*, Pergamon Press (Macmillan), New York, 1961.

Eckenfelder, W. W. Jr., Activated sludge treatment of industrial wastewaters, Kappe Lecture, Colorado State University, American Academy of Environmental Engineers, handout, not-published (82 pages), Fort Collins, CO, October 27, 2000.

Eckenfelder, W. W., *W. Wesley Eckenfelder (Waste Water Extraordinaire)—The Life of a Pioneer*, Author House, Bloomington, IN, 2009.

Eckenfelder, W. W. Jr., Ford, D. L., and Englande, A. J. Jr., *Industrial Water Quality*, 4th edn., McGraw-Hill, New York, 2009.

Fair, G. M. and Geyer, J. C., *Water Supply and Waste-Water Disposal*, John Wiley, New York, 1954.

Ghosh, S. and Pohland, F. G., Kinetics of substrate assimilation and product formation in anaerobic digestion, *Journal of Water Pollution Control Federation*, 46(4):748–759, April 1974.

Goodwin, B. L. and Englande, A. J. Jr., A unified model of the activated sludge process, *Journal of the Water Pollution Control Federation*, 46(2):312–332, 1974.

Grady, C. P. L. Jr., Daigger, G. T., and Lim, H. C., *Biological Wastewater Treatment, Theory and Applications*, 2nd edn., Marcel Dekker, Inc., New York, 1999.

Gram, A. L., III, Reaction kinetics of aerobic biological processes, Report No. 2, I.E.R. Series 90, Sanitary Engineering Research Laboratory, Department of Engineering, University of California, Berkeley, CA, May 15, 1956.

Henze, M., Grady, C. P. Jr., Gujer, W., Marais, G. V. R., and Matsuo, T., A general model for single-sludge wastewater treatment systems, *Water Research*, 21(5):505–515, May 1987.

Lawrence, A. W. and P. L. McCarty, Kinetics of methane fermentation in anaerobic treatment, *Journal of Water Pollution Control Federation*, 41(2-Part 2):R1–R17, February 1969.

Lawrence, A. W. and McCarty, P. L., A unified basis for biological treatment design and operation, *Journal of the Sanitary Engineering Division, American Society of Chemical Engineers*, 96:757, 1970.

McCarty, P. L., Anaerobic waste treatment fundamentals, Part one—Chemistry and microbiology, *Public Works*, 95:107–112, September 1964a.

McCarty, P. L., Anaerobic waste treatment fundamentals, Part two—Environmental requirements and control, *Public Works*, 95:123–128, September 1964b.

McCarty, P. L., Anaerobic waste treatment fundamentals, Part three—Toxic materials and their control, *Public Works*, 95:91–94, November 1964c.

McCarty, P. L., Anaerobic waste treatment fundamentals, Part four—Process design, *Public Works*, 95:95–99, September 1964d.

McCarty, P. L., Sludge concentration – Needs, accomplishments, and future goals, *Journal of Water Pollution Control Federation*, 38(4):493–507, April 1966.

McKinney, R. E., *Microbiology for Sanitary Engineers*, McGraw-Hill, New York, 1962.

McKinney, R. E., Mathematics of complete mixing activated sludge (ASCE Paper 3516), *Transactions of ASCE*, 128(Part III): 497–534, 1963; Also discussions by Washington, Hetting, and Sathyanarayana.

Meroney, R. N., CFD simulation of mechanical draft tube mixing in anaerobic digester tanks, *Water Research*, 43(4):1040–1050, March 2009.

Metcalf, L. and Eddy, H. P., *Sewerage and Sewage Disposal*, McGraw-Hill, New York, 1930.

Metcalf & Eddy, Inc., *Wastewater Engineering—Collection, Treatment, Disposal*, McGraw-Hill, New York, 1972.

Monod, J., The growth of bacterial cultures, *Annual Review of Microbiology*, Volume 3, Annual Reviews, Inc., Palo Alto, CA, 1949.

Orhon, D. and Artan, N., *Modeling of Activated Sludge Systems*, Technomic Publishing Co., Lancaster, PA, 1994.

Pearson, E. A., Kinetics of biological treatment, in: Gloyna E. F. and Eckenfelder Jr., W. W. (Eds.), *Advances in Water Quality Improvement*, University of Texas Press, Austin, TX, 1968.

Pitkin, J., Effect of temperature on rate of gas generation from municipal sludge, MS Thesis, Department of Civil Engineering, Utah State University, Logan, UT, 1969.

Prescott, L. M., Harley, J. P., and Klein, D. A., *Microbiology,* 6th edn., McGraw-Hill, Dubuque, IA, 2005.

Rittman, B. E. and McCarty, P. L., *Environmental Biotechnology: Principles and Applications*, McGraw-Hill, New York, 2001.

Sawyer, C. N. and McCarty, P. L., *Chemistry for Sanitary Engineers*, McGraw-Hill, New York, 1967.

Stewart, M. J., Activated sludge process variations, The complete spectrum, *Water and Sewage Works*, Part I, April 1964; Part II, May 1964; Part III, June 1964.

Stryer, L., *Biochemistry*, W. H. Freeman, New York, 1981.

Stumm, W., Discussion of paper, Rohlich, G. A., Methods for removal of phosphorous and nitrogen from sewage plant effluents, in: Eckenfelder, W. W. (Ed.), *Advances in Water Pollution Research* (*Proceedings of the International Conference* held in London, September, 1962), Volume 2, Pergamon Press, Macmillan Company, New York, 1964.

Tchobanoglous, G. and Burton, F. L. (Metalf & Eddy, Inc.), *Wastewater Engineering-Treatment, Disposal, Reuse*, 3rd edn., McGraw-Hill, New York, 1991.

USEPA, Standards for the Use and Disposal of Sewage Sludge, Final Rule, *40 CFR Parts 257, 403, and 503*, Water Environment Federation, Stock #P0100, Alexandria, VA, 1993.

van Niekerk, A. M., Jenkins, D., and Richard, M. G., The competitive growth of Zoogloea ramigera and Type 021N in activated sludge and pure culture—A model for low F:M bulking, *Journal of the Water Pollution Control Federation*, 59 (5):262–273, May 1987.

Williamson, K. and McCarty, P. L., A model of substrate utilization by bacterial films, *Journal Water Pollution Control Federation*, Washington, D.C., 48(1):9–24, January 1976a.

Williamson, K. and McCarty, P. L., Verification studies of the biofilm model for bacterial substrate utilization, *Journal Water Pollution Control Federation*, Washington, D.C., 48(2):281–296, February 1976b.

Pre-Appendix Tables

TABLE CDQR.1
Constants—Quick Reference

Quantity	Symbol	Units	Numerical Value	Other Units	Values
Fundamental constants					
Atomic mass constant	m_u	kg	$1.660\ 540\ 2 \cdot 10^{-27}$	u	mass of $^{12}C/12$
Atmosphere, standard	p	Pa	101 325		
Avogadro constant	No	molecules/mol	$6.022 \cdot 10^{23}$/mol		
Boltzmann constant	k	J/K	$1.38 \cdot 10^{-23}$		
Elementary charge	e	Coulomb	$1.602\ 177\ 33 \cdot 10^{-19}$		
Electron mass	m_e	kg	$9.109\ 3897 \cdot 10^{-31}$		
Gas constant	R	J/K/mol	8.314 510		
Gravity acceleration	g	m/s^2	9.806 650		
Faraday constant	F	coulomb/mol	96 485	electric charge of one mole of electrons	
Permittivity of vacuum	ε_o	F/m	$8.854\ 188 \cdot 10^{-12}$	C/volt	$C^2/(mJ)$
Planck constant	h	J · s	$6.63 \cdot 10^{-34}$		
Velocity of light	c	m/s	$2.9979 \cdot 10^{8}$		
MW(air)		g/mol			28.964

TABLE CDQR.2
Units and Conversions—Quick Reference

Quantity	Symbol	From Units	Factor	To Units	Values
Base SI units					
Length	L	ft	0.3048	m	
		mil	25.4	μm	
		angstrom	$1 \cdot 10^{-10}$	m	
		angstrom	$1 \cdot 10^{-4}$	μm	
		angstrom	$1 \cdot 10^{-1}$	nm	
		dm	10	m	
Time	t	s			
Mass	M	lbm	0.44535 924	kg	
		slug	14.593 90	kg	
		slug	32.17	Lbm	
		ounces (avoir)	0.028 349 52	kg	
		ton (metric)	1000	kg	
Electric current	I	ampere (A)			
Thermodynamic temperature	T	kelvin (K)			
Amount of matter	n	mole (mol)			
Luminous intensity		candela (cd)			
Derived Units					
Area	A	hectare	10 000	m^2	
		hectare	100	km^2	
		acre	4046.873	m^2	
		mi^2	$2.589\ 998 \cdot 10^6$		
		ft^2	0.092 90	m^2	
		ft^2	0.09290304	m^2	
		in^2	$6.451\ 600 \cdot 10^{-4}$	m^2	
Concentration					
Molarity	M	mols/L			
	[*]	mols/L			
Molality		mols/kg solvent			
Concentration		mg/L	1/1000MW	mols/L	
		g/gal	17.711806	mg/L	
Density	ρ	$slug/ft^3$	515.379	kg/m^3	
		$slug/ft^3$	32.174	Lb/ft^3	
		Lb/ft^3	16 021.59	mg/L	
		Lb/ft^3	16.02159	kg/m^3	
Energy		J	1.0	kgm^2/s^2	Nm
		J	1.0	volt-coulomb	Nm
		J	1.0	watt-s	Nm
		J	10^7	erg	
		J	$9.9 \cdot 10^{-3}$	L-atm	
		J	$1.0365 \cdot 10^{-5}$	volt-faraday	
		J	$6.242 \cdot 10^{18}$	eV	
		calorie (thermo)	4.184	J	
		ft · lbf	1.355 818	J	
		volt-coulomb	1.0	J	
		kW-h	3600 000	J	
		watt-h	3600	J	
		Btu (thermo)	1055.35	J	
		watt · s	$2.7778 \cdot 10^{-7}$	kWh	
		btu	1.055056	kJ	

TABLE CDQR.2 (continued)
Units and Conversions—Quick Reference

Quantity	Symbol	From Units	Factor	To Units	Values
		100 ft^3 natural	108 720 000	J	
		gas	108 720	KJ	
			0.10872	GJ	
			30.2	kWh	
			1.030713	therms US	
			25 073 529.55	g-cal.	
Electrical					
Electric charge	Q	coulomb	1.0	A s	
				A · s	
Electric potential	V	volt	1.0	J/C	
				= N m/(As)	
				= W/A	
Electrical conductance	G	siemens (S)	1.0	A/V	$A^2s^3/kg\ m^2$
Electrical capacitance	C	farad (F)	1.0	C/V	$A^2s^4/(kg\ m^2)$
				= A · s/V	$= C^2/J = A\ s/V$
Electrical resistance	Ω	ohm	1.0	= V/A	
		siemens	1.0	A/V	S = A/V
		mho	1.0	siemens	
		coulomb	$2.7778 \cdot 10^{-4}$	A-h	
		coulomb	$1.0363 \cdot 10^{-5}$	faraday	
		faraday	96 487	coulomb	
		electron volt	$1.6022 \cdot 10^{-19}$	J	
Frequency		hertz (Hz)	1.0	s^{-1}	
Flow	Q	mgd	0.043 8078	m^3/s	
		gpm	0.000 063 090 20	m^3/s	
		ft^3/min	0.000 4719	m^3/s	
		ft^3/s	0.028 316 85	m^3/s	
		m^3/s	35.314 667	ft^3/s	
Flow per unit length	Q/L	$ft^3/min/ft$ length	0.001 548	$m^3/s/m$ length	
		$ft^3/s/ft$ length	0.092 903	$m^3/s/m$ length	
		$m^3/s/m$ length	10.76385	$ft^3/s/ft$ length	
Force	F	N	1	kgm/s^2	
		$kg\ m/s^2$	1	N	Definition
		kgm	9.806 650	N	Force of 1.0 kg mass on earth
		N	0.2248	Lbf	
		kg-force	9.806 650	N	
		Lbf	4.448 222	N	
		Lbf	32.17	poundal	
		poundal	0.13826	N	
		ozf	0.278	N	
Kinematic viscosity	ν	stoke	1	$cm^2\ s^{-1}$	
		stoke	0.0001	m^2/s	
Mass		grain	$6.479\ 891 \cdot 10^{-5}$	kg	
			0.064 798	g	
			64.798 91	mg	
Power	P	watt	1	$kg\ m^2/s^3$	
		watt	1	Nm/s	
		horsepower	745.6999	W (J/s)	
		ft · lbf/s	1.355 818	W	
Power intensity	P/V	hp/mgd	0.197	kW/mL/d	
		hp/ft^3	26.33	kW/m^3	

(continued)

TABLE CDQR.2 (continued)
Units and Conversions—Quick Reference

Quantity	Symbol	From Units	Factor	To Units	Values	
Pressure	p	Pa	1	N m^{-2}		
		atm	101325	Pa		
		atm	1.013	bar		
		atm	10.3327	m water		
		bar	100 000	Pa		
		mm Hg 0°C	133.322	Pa		
		psi	6894.757	Pa		
		kgf/m^2	9.80665	Pa		
		N/m^2	1	Pa		
		m water	9.806 246	kPa		
		Pa	101 325	atm		
		Pa	100 000	bar		
		torr	1.00	atm		
Revolutions		revolution	360	degree		
		revolution	2π	radians		
		rpm	0.104 719 8	rad/s		
Surface tension	γ	N/m		J/m^2		
					(°F)	(°C)
Temperature		°F	5/9(°F-32)	°C	50	10.0
		°C	°C+273.15	K	60	15.6
		°C	9/5°C + 32	°F	70	21.1
					80	26.7
Velocity	v	mi/h	1.609 344	km/h	90	32.2
			0.447 0400	m/s	95	35.0
		ft/s	0.304 8	m/s	110	43.3
		knots	0.514 444	m/s	120	48.9
			1.85	km/h	130	54.4
					140	60.0
Viscosity, dynamic	μ	poise	1	g cm^{-1}s^{-1}		
		poise	0.1	Pa · s		
		poise	0.1	N · s/m^2		
		cP	0.001	Pa · s		
		Lbf · s/ft^2	47.880 26	Pa · s		
Viscosity, kinematic	n	stoke	1	cm^{2}s^{-1}		
		stoke	0.0001	m^2/s		
Volume	V	gal	0.003 785 412	m^3		
		acre-ft	1233.489	m^3		
		barrel 42 gal	0.158 9873	m^3		
		ft^3	0.028 316 85	m^3		
		yd^3	0.764 5549	m^3		
		ounce (US liq)	$2.957353 \cdot 10^{-5}$	m^3		
Parameters						
Hydraulic loading rate	HLR	gal min^{-1}ft^{-2}	0.000679023	m/s		
= Q/A(cross section)			0.0407417	m/min		
			2.444502	m/h		
			58.667 616	m/d		
		ft^3/min/ft^2	0.3048	m/min		
		mgad	0.038 970 574	m/h		
			0.015 9422	gal/min/ft^2		
		m/s	1472.7035	gal/min/ft^2		
Porosity	P				0.35–0.45	
Permeability	D	Darcy	$0.987 \cdot 10^{-12}$	m^2		
Overflow velocity	v_o	gpd/ft^2	$4.074\ 14 \cdot 10^{-2}$	m^3/m^2/d		

TABLE CDQR.2 (continued)
Units and Conversions—Quick Reference

Quantity	Symbol	From Units	Factor	To Units	Values
=Q/A(plan)			$1.697\ 558 \cdot 10^{-3}$	$m^3/m^2/h$	
			$2.829\ 26 \cdot 10^{-5}$	$m^3/m^2/min$	
			$4.715 \cdot 10^{-7}$	$m^3/m^2/s$	
		gpm/ft^2	58.667 616	$m^3/m^2/d$	
			2.444 484	$m^3/m^2/h$	
			$4.074\ 14 \cdot 10^{-2}$	$m^3/m^2/min$	
			$6.790\ 233 \cdot 10^{-4}$	$m^3/m^2/s$	
Solids loading rate '-to clarifier	j(solids)	$Lb/d/ft^2$	4.793 806	$kg/d/m^2$	
Weir loading rate		gpd/ft	$1.241\ 7979 \cdot 10^{-2}$	$m^3/m/d$	
			$5.174\ 1579 \cdot 10^{-4}$	$m^3/m/h$	
			$8.623\ 5965 \cdot 10^{-6}$	$m^3/m/min$	
			$1.437\ 266 \cdot 10^{-7}$	$m^3/m/s$	
		gpm/ft	44.704 44	$m^3/m/d$	
			0.745 074	$m^3/m/h$	
			$1.241\ 7979 \cdot 10^{-2}$	$m^3/m/min$	
			$2.069\ 663 \cdot 10^{-4}$	$m^3/m/s$	

Multiply "from units" by factor to obtain "to units."

TABLE CDQR.3
SI Prefixes—Quick Reference

SI System Prefix	Symbol	Factor
exa	E	10^{18}
peta	P	10^{15}
tera	T	10^{12}
giga	G	10^{9}
mega	M	10^{6}
kilo	k	10^{3}
hecto	h	10^{2}
deca	da	10^{1}
deci	da	10^{-1}
centi	c	10^{-2}
milli	m	10^{-3}
micro	μ	10^{-6}
nano	n	10^{-9}
pico	p	10^{-12}
femto	f	10^{-15}
atto	a	10^{-18}

TABLE CDQR.4
Coefficients for Calculation of Variables—Polynomial Best Fit Equations

y-Variable	x-Variable	Limits	Polynomial Coefficients M0	M1	M2	M3	M4	M5	M6	Trial Argument	Formula Calculation Result
Vapor Pressure of water (kPa)	T(°C)	0–100°C	0.61052	0.044905	0.0013613	########	1.9829E−07	3.5164E−09	#########	20	2.3385
μ(water) (N · s/m²)	T(°C)	0–40°C	0.001787	−5.61E−05	1.00315E−06	−7.54E−09				20	#######
ρ(water) (kg/m³)	T(°C)	0–40°C	999.84	0.068256	−0.0091438	0.000103	−1.188E−06	7.152E−09		20	998.2040

Calculation formula: y = M0 + M1 × T + M2 × T∩2 + M3 × T∩3 + M4 × T∩4 + M5 × T∩5 + M6 × T∩6

TABLE CDQR.5
Coefficients for Calculation of Variables—Exponential Best Fit Equations

y-Variable	x-Variable	Limits	Intercept	Slope	Trial Argument	Formula Calculation
P(atmospheric)[a] (Pa)	Z (m)	0–5000 m	101,325	#########	1000	89,363
ρ(atmospheric)[b] (kg/m^3)	Z (m)	0–5000 m	1.225	#########	0	1.23
ρ(atm)[c] (mols/m^3)	Z (m)	0–5000 m	42.292	#########	0	42.3
σ(water)[d] (N/m)	T(°C)	0°C–40°C	0.076426	#########	20	0.073

[a] Atmospheric pressure (Pa) as function of elevation (m) above sea level.
[b] Atmospheric density (kg/m^3) as function of elevation (m) above sea level.
[c] Atmospheric density (mols/m^3) as function of elevation (m) above sea level.
[d] Surface tension of water.
Calculation formula: $y = \text{intercept} \cdot 10^{(\text{slope} \cdot x)}$.

TABLE CDQR.6
Miscellaneous Notes

Multiplicity of Units with Same Name

Frequently, in looking at unit conversions, there are several choices, for some units.
The normal quantities used in the United States are, according to Lindeburg (1993):

Barrel	31.5 gal (U.S. liquid)
Barrel	42 gal (petroleum)
Btu	Traditional (thermochemical) value
Calories	Thermochemical
Gallons	US liquid
Ounces	Avoirdupois
Pounds	Avoirdupois
Tons	Short tons of 2000 lb

Force and mass conversions

Conversions between different forces often requires going back to basics. To help, in the conversion task, force definitions from Wandmacher and Johnson (1995) are,

- Newton is the force required to accelerate 1.0 kgm at the rate of 1.0 m/s^2
- Poundal is the force required to accelerate 1.0 Lbm at the rate of 1.0 ft/s^2
- kgf is the force required to accelerate 1.0 kgm at 9.80665 m/s^2
- Lbf is the force required to accelerate 1.0 Lbm at the rate of 32.1740 ft/s^2

The related definitions for the derived mass units are,

- The slug is that mass which when acted upon by one pound force will be accelerated at the rate of 1.0 ft^2/s
- The gravitational metric unit of mass is that unit of mass which when acted upon by 1.0 kgf will be accelerated at 1.0 m/s^2. The o inference is to the kgm.
- The kgm and the Lbf are base units, not derived units.

Rounding Off

The number of places in any conversion result should be the number of places in the source number. Rather than round the conversion factor, the result of the conversion should be rounded. This can be done in a spreadsheet by setting the number of places in the calculated cell to that desired (Heausler, 1994).

Significant Figures

The "significant figures" represents the accuracy of a calculation. Its misleading to designate a result with more significant figures than warranted. The number of significant digits should not be greater than the lowest number of significant digits in any one of the numbers in a calculation. On the other hand, including all the significant figures up to, but not including, the final result helps to trace a sequence of calculation (which was the *modus operandi* for this text).

Example 1 Convert 8 in. to mm, i.e., 8 in. $\cdot$ 25.4 mm/in. = 203.2 mm. Only one significant digit is warranted, giving a result, 200 mm.

References

Heausler, T. F., Metric Conversion—Rounding Off to the Appropriate Tolerances, Technical Briefs, Burns & McDonnel, Kansas City, No. 4, 1994.
Lindeburg, M. R., *Engineering Unit Conversions*, 3rd edn., Professional Publications, Inc., Belmont, CA, 1993.
Wandmacher, C. and Johnson, A. I., *Metric Units in Engineering—Going SI*, ASCE Press, New York, 1995.
Elias, H. G., An Introduction to Polymer Science, VCH, Weinheim, New York, 1997.
Two Web sites found useful for conversions are given below. The respective, years of the last contacts are shown also.
http://www.Imnoeng.com/units.htm, 2010.
http://OnlineConversion.com, 2010.

Appendix A: International System of Units

The system of units used for measurement throughout the world is the International System of Units, called the SI system. In using the SI system, a set of conventions has evolved. This appendix reviews the origin and status of the SI system, and describes the conventions for the use of the system. The American Society for Testing Materials (ASTM) report (1991) was the primary reference for this review and is recommended for further reading. After completing this appendix, a book by Cardarelli (1999) was found, which has listed almost every obscure unit imaginable, each converted to SI units and with interesting and useful notes. Fundamental constants are also given. Online conversions were available at several Web sites (http://www.Imnoeng.com/units.htm and http://onlineconversion.com/) provided by colleagues at Colorado State University.

A.1 DEVELOPMENT OF THE INTERNATIONAL SYSTEM OF UNITS

The modernized metric system is called the International System of Units. This name in French, Le Systѐme International d'Unités, and the abbreviation, SI, were adopted in 1960 by the 11th conference (ASTM, 1991). The system is maintained by the *General Conference on Weights and Measures*.

The decimal system of units, the genesis of the present SI system, was developed in the sixteenth century. In 1790, the French National Assembly requested the French Academy of Sciences to develop a system of units suitable for adoption worldwide. The problems of doing business and communicating with a myriad of local units were recognized by all nations, and by the scientific community in particular. The international standardization began with an 1870 meeting of 15 nations in Paris that led to the May 2, 1875, International Metric Convention, and the establishment of a permanent International Bureau of Weights and Measures near Paris. A *General Conference on Weights and Measures* (*Conférence Générale des Poids et Mesures*, CGPM) was also constituted to handle international matters concerning the metric system. The CGPM meets at least every 6 years in Paris and controls the International Bureau of Weights and Measures, which preserves the metric system (ASTM, 1991).

A.2 THE SI/METRIC SYSTEM IN THE UNITED STATES

Article 1, Section 8, of the United States Constitution gives Congress the power to "... fix the standard of weights and measures." Although Thomas Jefferson and John Quincy Adams were early advocates of the metric system, the prevailing notion was to bring the U.S. measures in closer harmony with those of the English. The metric system was legalized, however, with the Metric Act of 1866. In 1893, the international meter and kilogram became the fundamental standards of length and mass in the United States (ASTM, 1991). In 1902, Congressional legislation requiring the federal government to adopt the metric system was defeated by a single vote. Despite these tendencies toward metrication, and while the rest of the world was adopting the metric system, the United States maintained its adherence to the English system.

The federal government began further movement toward metrication in 1968 legislation to study its benefits and problems. This was followed by the 1975 Metric Conversion Act (Interagency Council on Metric Policy, 1991; Carver, 1992). From this act, the expectation was that there would be a 10 year long voluntary transition period. The government failed to maintain leadership, however, and so the momentum underway was dissipated by about 1980.

The next legislation was in the 1988 Omnibus Trade and Competitiveness Act that had "metric usage" provisions for federal agencies and was an amendment to the 1975 Act. Then on July 25, 1991, President Bush issued Executive Order 12770, "Metric Usage in Federal Government Programs," which gave authority to the Secretary of Commerce to provide the needed leadership and coordination for adopting the metric system. The office implementing the program was the Metric Program of the National Institute of Standards and Technology (NIST), which has had this kind of responsibility since 1901. Executive Order 12770 mandated that each federal agency make a transition to metric units in government publications, work with government and private groups on metric implementation, and formulate and implement a Metric Transition Plan by November 30, 1991. To coordinate the implementation among federal agencies, the Interagency Council on Metric Policy (ICMP) was established under the Department of Commerce (ICMP, 1991). The Metrication Operating Committee, with 10 subcommittees, was given the responsibility to implement the metrication policy. The Construction Subcommittee established the goal of instituting the use of metric in all federal facilities by January, 1994 (ICMP, 1991).

The United States is the only country that has not adopted the SI system. Nevertheless, the system, commonly used by the scientific community, is legal and has been adopted widely in engineering, and by most professional organizations in the water quality field. The SI system is not new to engineers, as its predecessor, the metric system, e.g., CGS, MKS, has been a part of engineering and scientific education for many decades. Metric conversion is continuing, and a growing consensus indicates that complete conversion is inevitable. When the conversion is completed depends upon such factors as international competitiveness, costs, education, and leadership from the federal government.

In its efforts to move toward metrication, the ICMP (1991) reported that the United States has been moving toward metrication, and that the costs were much less than anticipated. For example, General Motors converted fully to metric at about 1% of the estimated cost. In fact, various industries have found that savings result from carrying fewer parts and sizes. One of the most compelling arguments for the continuing move toward metrication has been the problem of international competitiveness. In working abroad, the construction industry has used routinely metric in foreign work and has made the transition for that segment of the industry. Most engineering journals have specified the use of metric for many years.

A.3 LOGIC OF THE INTERNATIONAL SYSTEM OF UNITS

The original metric system provided a set of units for the measurement of length, area, volume, capacity, and mass based on two fundamental units, the meter and the kilogram. About 1900, the meter-kilogram-second (MKS) system was instituted. Other base units were added over the years to accommodate other fields, such as the ampere as the unit of electric current, the degree Kelvin (later renamed the kelvin) as the unit of temperature, and the candela as the unit of luminous intensity. In 1971, the seventh base unit, the mole, was added. Also in 1971, the pascal (Pa) was approved as the special name for the SI unit of pressure, i.e., Newtons per square meter.

The modernized metric system, the SI system, is a *coherent* system with seven base units. The SI system has only one unit for each physical quantity, which is the advantage of implementing the system universally. In other words, there is only one recognized unit for each variable. A second advantage of the SI system is its *coherence*. Such a system requires no conversion factors. For example, equations between the units of a coherent system contain only the number 1 as a conversion factor. In a coherent system, the product or quotient of any two unit quantities is the unit of the resulting quantity. For example, the velocity unit is length divided by time, and the unit of force is the mass multiplied by the unit of acceleration.

A.3.1 Base Units

The SI units are in three classes: base units, supplementary units, and derived units. The base units give the system its *coherence*. The base units, seven in number, are independent, and are listed in Table A.1.

A.3.2 Supplementary Units

The supplementary units are the radian and the steradian. These are dimensionless units and are the ratios between two lengths and two areas, respectively.

TABLE A.1
Base SI Units

Quantity	Unit	Symbol
Length	Meter	m
Mass	Kilogram	kg
Time	Second	s
Electric current	Ampere	A
Thermodynamic temperature	Kelvin	K
Amount of substance	Mole[a]	mol
Luminous intensity	Candela	cd

Source: Abridged from ASTM, *Standard Practice for Use of the International System of Units (SI) (The Modernized Metric System)*, Designation E380-91a, PCN 03-543191-34, American Society for Testing Materials, Philadelphia, PA, 1991.

[a] The mole is the amount of substance of a system, which contains as many elementary entities as there are atoms in 0.012 kg of carbon-12, as adapted by the 14th CGPM in 1971.

A.3.3 Derived Units

The derived units are formed by combining base units, supplementary units, and derived units. Some of the derived units with special names are listed in Table A.2. Some of the common derived units are listed in Table A.3. Other derived units include the pascal, the joule, the watt. Of special note is the Newton that is used in place of the kilogram-force.

A.3.4 Prefixes

The prefixes for decimal multiples and submultiples, and associated symbols, are listed in Table A.4. These may be used as a means to make more convenient the use of the

TABLE A.2
Derived Units with Special Names

Quantity	Unit	Symbol	Formula
Frequency	Hertz	Hz	1/s
Force	Newton	N	$kg \cdot m/s^2$
Pressure, stress	Pascal	Pa	N/m^2
Energy, work, quantity of heat	Joule	J	$N \cdot m$
Power, radiant flux	Watt	W	J/s
Quantity of electricity electric charge	Coulomb	C	$A \cdot s$
Electric potential, potential difference, electromotive force	Volt	V	W/A
Celsius temperature	Degree	°C	K
Luminous flux	Lumen	lm	$cd \cdot sr$
Activity	Becquerel	Bq	1/s
Absorbed dose	Gray	Gy	J/kg
Dose equivalent	Sievert	Sv	J/kg

Source: Abridged from ASTM, *Standard Practice for Use of the International System of Units (SI) (The Modernized Metric System)*, Designation E380-91a, PCN 03-543191-34, American Society for Testing Materials, Philadelphia, PA, 1991.

TABLE A.3
Common Derived Units

Quantity	Unit	Symbol
Absorbed dose rate	Gray per second	Gy/s
Acceleration	Meter per second squared	m/s^2
Angular acceleration	Radian per second squared	rad/s^2
Area	Square meter	m^2
Concentration	Mole per cubic meter	mol/m^3
Density, mass	Kilogram per cubic meter	kg/m^3
Energy density	Joule per cubic meter	J/m^3
Entropy	Joule per kelvin	J/K
Exposure (x and gamma rays)	Coulomb per kilogram	C/kg
Heat capacity	Joule per kelvin	J/K
Heat flux density irradiance	Watt per square meter	$W \cdot m^2$
Luminance	Candela per square meter	$cd \cdot m^2$
Molar energy	Joule per mole	J/mol
Molar entropy	Joule per mole kelvin	$J/(mol \cdot K)$
Molar heat capacity	Joule per mole kelvin	$J(mol \cdot K)$
Moment of force	Newton · meter	$N \cdot m$
Power density	Watt per square meter	W/m^2
Radiant intensity	Watt per steradian	W/sr
Specific heat capacity	Joule per kilogram kelvin	$J/(kg \cdot K)$
Specific energy	Joule per kilogram	J/kg
Specific entropy	Joule per kilogram kelvin	$J/(kg \cdot K)$
Specific volume	Cubic meter per kilogram	m^3/kg
Surface tension	Newton per meter	N/m
Thermal conductivity	Watt per meter kelvin	$W/(m \cdot K)$
Velocity	Meter per second	m/s
Viscosity, dynamic	Pascal second	$Pa \cdot s$
Viscosity, kinematic	Square meter per second	m^2/s
Volume	Cubic meter	m^3
Wave number	1 per meter	1/m

Source: Abridged from ASTM, *Standard Practice for Use of the International System of Units (SI) (The Modernized Metric System)*, Designation E380-91a, PCN 03-543191-34, American Society for Testing Materials, Philadelphia, PA, 1991.

SI system over a wide range of circumstances. For example, the kilopascal (kPa) is convenient because one atmosphere of pressure is 101,325 pascals (Pa) and it is easier to say 101 kPa.

A.4 UNITS IN USE

Some units that have been used commonly are depreciated under the new system. Examples include the calorie, kilogram-force, langley, metric horsepower, millimeter of mercury, standard atmosphere. The CGS system is to be avoided, e.g., the erg, the dyne, the poise, etc.

A.5 CONVENTIONS

A number of rules were adopted by the CGPM, which help in facilitating communication, i.e., in providing a common language. These are taken from ASTM (1991) and are enumerated in the following paragraphs:

TABLE A.4
SI Prefixes

Multiplication Factor	Prefix	Symbol
1 000 000 000 000 000 000 = 10^{18}.	exa	E
1 000 000 000 000 000 = 10^{15}.	peta	P
1 000 000 000 000 = 10^{12}.	tera	T
1 000 000 000 = 10^{9}.	giga	G
1 000 000 = 10^{6}.	mega	M
1 000 = 10^{3}.	kilo	k
100 = 10^{2}.	hecto	h
10 = 10^{1}.	deka	da
0.1 = 10^{-1}.	deci	d
0.01 = 10^{-2}.	centi	c
0.001 = 10^{-3}.	milli	m
0.000 001 = 10^{-6}.	micro	μ
0.000 000 001 = 10^{-9}.	nano	n
0.000 000 000 001 = 10^{-12}.	pico	p
0.000 000 000 000 001 = 10^{-15}.	femto	f
0.000 000 000 000 000 001 = 10^{-18}.	atto	a

Source: Abridged from ASTM, *Standard Practice for Use of the International System of Units (SI) (The Modernized Metric System)*, Designation E380-91a, PCN 03-543191-34, American Society for Testing Materials, Philadelphia, PA, 1991.

1. *Application of prefixes*: First, the SI form of the metric system is preferred for all applications. In some cases, units outside SI are appropriate.
2. *Prefixes for multiplication*: Table A.4 lists the prefixes associated with multiplication factors.
3. *Elimination of zero digits*: In general, the SI prefixes should be used to indicate the orders of magnitude, eliminating nonsignificant digits and leading zeros in decimals and giving an alternative to the power of ten notation. Examples are
   a. 12,300 mm is 12.3 m
   b. $12.3 \cdot 10^3$ m is 12.3 km
   c. 0.00123 μm is 1.23 nm
4. *Prefixes to control decimal location*: When expressing a quantity by a numerical value and unit, a prefix should be chosen so that the numerical value is between 0.1 and 1000. In expressing area and volume, the prefixes hex-, deck-, deci-, and centi- may be required, e.g., square hectometer, cubic centimeter.
5. *Deviations from SI*: For certain applications, one particular multiple is used by custom. For example, the millimeter is used for linear dimensions in engineering drawings, even when the values are far outside the 0.1–1000 mm range. The centimeter is often used for body measurements and clothing sizes.
6. *Calculations*: Errors in calculations are minimized if the base and the coherent derived SI units are used and the resulting numerical values are expressed in powers of 10 notation instead of using prefixes.

7. *Time*: The SI unit of time is the second. This unit is preferred, especially when technical calculations are involved. In cases where time relates to life customs or calendar cycles, the minute, the hour, and the day may be advisable. For example, vehicle speed is expressed as kilometers per hour.
8. *Angle*: The SI unit for the plane angle is the radian. The degree and its decimal submultiples is permissible when the radian is not convenient.
9. *Area*: The SI unit of area is the square meter. The hectare is a special name for a square hectometer (hm^2). Large land or water areas are expressed in hectares or in square kilometers (km^2).
10. *Volume*: The SI unit of volume is the cubic meter (m^3). This unit or one of the common multiples, such as cubic centimeter, is preferred. The special name liter (L) has been approved for the cubic decimeter. The prefixes milli- and micro- may be used with the liter.
11. *Energy*: The SI unit of energy is the joule, with its multiples, and is preferred for all applications.
12. *Pressure*: The SI unit of pressure and stress is the pascal (Newton per square meter) and should be used, with the proper prefixes. Other units are discouraged.
13. *Other units*: The CGS units are to be avoided. The poise and the stoke should not be used, along with the calorie, kgf, mm mercury, standard atmosphere, etc.
14. *Rules for symbols*:
    a. *Orientation*: Unit symbols should be upright regardless of surrounding text.
    b. *Case of symbols*: Letter unit symbols are written in lower case unless the unit name comes from a proper name, in which case, the first letter of the symbol is capitalized, e.g., W, Pa. The exception is the symbol for liter, L.
    c. *Spaces*: When a quantity is expressed as a numerical value and a unit symbol, a space should be left between them, e.g., 35 mm, not 35mm. An exception is that no space is left between the numerical value and the symbols for degree, minute, and second of a plane angle and degree Celsius, e.g., use 45°, 20°C.
    d. *Prefix space*: There is no space between the prefix and the unit symbols.
    e. *Multiplication product*: Products are expressed with a dot, e.g., N·m for Newton meter.
    f. *Quotients*: Quotients are expressed as m/s, or $\mathrm{m \cdot s^{-1}}$, or $\frac{\mathrm{m}}{\mathrm{s}}$.
    g. *Digit separations*: Digits are separated into groups of three, counting from the decimal from the left and the right, e.g., 2 141 596 73 722 7372 0.1335.
    h. In numbers of four digits on either side of the decimal place, the space is not necessary.
    i. *Thousands symbol, M*: The use of M to indicate thousands is deprecated.
    j. *Adding to symbols*: Attachment of letters to a unit symbol to give information about the quantity is incorrect. For example, do not use psia or psig to distinguish absolute and gage pressure. Instead, state "... at a gage pressure of 13 kPa," or ".... at an absolute pressure of 13 kPa." Where space is limited, one may write, "13 kPa (gage)," or "13 kPa (absolute).
    k. *Conversions and rounding*: The number of significant digits in conversions should be neither diminished nor exaggerated. For example, 125 ft converts exactly to 38.1 m. If, however, the 125 ft length had been obtained by rounding to the nearest 5 ft, the conversion is 38 m.

A.6 MASS, FORCE, AND WEIGHT

The SI system uses distinct units for mass and force. The kilogram is the unit of mass and the newton is the unit of force. The term, kilogram-force should not be used.

In the SI system, the term *weight*, means that the force, if applied to body, would give it an acceleration of free fall. When non-SI units are used, a distinction should be made between *force* and *mass*.

A.6.1 Newton

The force required to accelerate one kilogram at one meter per second is defined as the *Newton* (N). The acceleration due to gravity at the earth's surface is 9.8066 $\mathrm{m/s^2}$. The gravity force exerted by (weight of) 1 kg mass at the earth's surface is therefore

$$F = M \cdot g \tag{A.1}$$

$$= 1.0\ \mathrm{kg} \cdot 9.81\ \mathrm{m/s^2} \tag{A.2}$$

$$= 9.81\ \mathrm{N}$$

A.6.2 Conversion Factor, g_c

The relationship between force and mass causes some consternation because the units are not homogeneous between the two sides of the $F = ma$ equation. To account for this problem, an "artifice," g_c, was invented. The term g_c is contrived to cancel units on the right side, to give

$$F = \frac{1}{g_c} ma \tag{A.3}$$

Example A.1 Show the Dimensions of g_c for SI Units

1. Applying SI dimensions gives

$$1\ \mathrm{N} = \frac{1}{g_c}\ \mathrm{kg}\ \frac{\mathrm{m}}{\mathrm{s^2}} \tag{ExA.1.1}$$

Therefore

$$g_c \equiv \frac{\text{kg}}{\text{N}}\frac{\text{m}}{\text{s}^2} \quad \text{(ExA.1.2)}$$

As stated by Felder and Rousseau (1978) the term, g_c, is a factor for converting from one force unit to another, e.g., from (kg · m/s^2) to N. The term, g_c, is a constant and is always as defined in Equation A.3. In U.S. Customary units, g_c is defined, $g_c = 32.174$ lb$_m$ · ft/s^2/lb$_f$. Thus, applying Equation A.3, 1 lb$_m$ = 1 lb$_f$.

Example A.2 Calculate the Force of 1 kg Mass

1. Applying Equation A.3, with $a = g$, gives

$$F = \frac{1}{g_c}\text{mg} \quad \text{(ExA.2.1)}$$

$$= \frac{1}{(\text{kg/N})\ (\text{m/s}^2)} 1\ \text{kg}\frac{9.81\ \text{m}}{\text{s}^2} \quad \text{(ExA.2.2)}$$

$$= 9.81\ \text{N}$$

Example A.3 Calculate the Power Number, P, for Conditions Stated

1. The power number is defined as

$$\mathbf{P} \equiv \frac{P}{(\omega/2\pi)^3 \cdot D^5 \cdot \rho} \quad \text{(ExA.3.1)}$$

Applying units gives

$$\mathbf{P} = \frac{(\text{N} \cdot \text{m/s})}{(\text{rad}^3/\text{s}) \cdot \text{m}^5 \cdot (\text{kg/m}^3)} \quad \text{(ExA.3.2)}$$

$$= \frac{\text{N}}{\text{kg} \cdot (\text{m/s}^2)} \quad \text{(ExA.3.3)}$$

$$= 1$$

But to make Equation ExA.3.1 dimensionless, the term g_c is added, i.e.,

$$\mathbf{P} \equiv \frac{P \cdot g_c}{(\omega/2\pi)^3 \cdot D^5 \cdot \rho} \quad \text{(ExA.3.4)}$$

$$= \frac{((\text{N} \cdot \text{m})/\text{s}) \cdot ((\text{kg/N})(\text{m/s}^2))}{(\text{rad}^3/\text{s}) \cdot \text{m}^5 \cdot (\text{kg/m}^3)}$$

$$= 1 \quad \text{(ExA.3.5)}$$

Discussion

By applying the g_c term in Equation ExA.3.4, the conversion from mass to force is incorporated. By contrast, Equation ExA.3.1 with g_c not included, the residual is (kg · m/s^2), which is in the denominator and is equivalent to 1 N, giving **P** = 1. In other words, Equations ExA.3.1 and ExA.3.4 are equivalent. Using Equation ExA.3.1, one is conscious of the need to convert kg to N (or vice versa). With Equation ExA.3.4, however, the conversion is incorporated. Equation ExA.3.1 is favored in practice. Note that the gravitational constant 9.81 m/s^2 is not relevant.

As a final note for this example, suppose in Equation ExA.3.1 the actual numerical value for the density of water, i.e., $\rho = 998$ kg/m^3, is used as the equation is expanded numerically to give Equation ExA.3.2. Recall that $\gamma = \rho g$, and so, $\gamma = 1000$ kg/m^3 · 9.81 m/s^2 = 9810 N/m^3. Substituting in Equation ExA.3.1, $\rho = \gamma / g$, gives, 9810 N/m^3/9.81 m/s^2. The point of interest is that the 9.81 factor cancels out. While this "final note" may seem to belabor the point, it should serve to bring to a closure one of the gnawing aspects of this conversion from mass to force. This means that we can use the g_c factor with confidence (as in Equation ExA.3.4), without further concern about whether the 9.81 is a part of g_c. On the other hand, however, note that in U.S. Customary units, the 32.174 is proper as a part of g_c because in this way, it forces the equality, 1 lb$_m$ = 1 lb$_f$ (see note at end of Example A.1. [As a historical note, we would have one less problem to consider if President John Quincy Adams in 1821 had prevailed in his leaning toward the metric system. This is an example in history how, when faced with a choice of one path or the other, the selection has long-term consequences. Most likely, if President Adams had been able to see ahead 150 years, his selection would have been different.]

Example A.4 Conversions between lb$_m$, lb$_f$ and N

1. Convert 1.0 lb$_m$ to N,

$$1.0\ \text{lb}_m \cdot \frac{\text{lb}_f}{\text{lb}_m} \cdot \frac{4.48\ \text{N}}{\text{lb}_f} = 4.48\ \text{N}$$

2. Convert 1.0 N to lb$_f$,

$$1.0\ \text{N} \cdot \frac{\text{lb}_f}{4.448\ \text{N}} = 0.2248\ \text{lb}_f$$

A.7 U.S. CUSTOMARY UNITS

The base units in the U.S. Customary system are the *foot* (ft), *second* (s), and *pound* (lb). The unit of mass that receives an acceleration of 1 ft/s^2 when a force of 1 lb is applied to it is called the *slug*, and is a derived unit. The derivation is from $F = ma$, giving, 1 lb = 1 slug · 1 ft/s^2, or 1 slug = 1 lb · s^2/ft. The *weight* of one slug on earth is, W = mg = 1 slug · 32.2 ft/s^2 = 32.2 lb.

A.7.1 Pound-Mass and Pound-Force

Since the slug is not used in daily life, the *pound-mass* (lb-m) is convenient and is defined as the mass that has a weight of 1 *pound-force* (lb$_f$). The pound-mass is the slug divided by the units conversion factor, g_c, in which $g_c = 32.1740$ lb$_m$-ft/lb$_f$-s^2 at sea level. In other words, 1.0 pound-mass is 1/32.2 slugs.

While the pound-mass can be understood easily, another unit, the *poundal* has been introduced, which compounds the issue. The distinctions between these units are delineated by examples.

Example A.5 Illustrate the Distinction between Pound-Force and Pound-Mass

1. Consider 1.0 slug under the influence of gravitational acceleration, g, i.e.

$$\begin{aligned} F &= M \cdot g \\ &= 1.0 \text{ slug} \cdot 32.2 \text{ ft/s}^2 \\ &= 32.2 \text{ lb force} \end{aligned}$$

By corollary, a force of 1 lb will accelerate a one slug mass at 1 ft/s^2. As a second corollary, the mass that causes 1.0 lb force may be called a pound-mass (which is 1/32.2 slugs). These are all equivalent statements.

2. Now consider 1.0 pound-mass accelerated at gravitational acceleration, i.e.

$$\begin{aligned} F &= M/32.2 \cdot g \\ F &= 1.0 \text{ slug}/32.2 \cdot 32.2 \text{ ft/s}^2 \\ &= 1.0 \text{ lb-mass} \cdot 32.2 \text{ ft/s}^2 \\ &= 1.0 \text{ lb force} \end{aligned}$$

3. An alternate form of the equation, $F = M \cdot g$, is

$$\begin{aligned} F &= M \cdot (g/g_c) \\ &= 1 \text{ lb}_m \cdot \frac{32.1740 \text{ ft/s}^2}{32.1740 \text{ (lb}_m \text{ ft)/(lb}_f \text{ s}^2)} \\ &= 1 \text{b}_f \end{aligned}$$

Discussion
Again, the application of the g_c term incorporates the conversion from lb$_m$ to lb$_f$. While the conversion is straightforward using the SI system, and g_c is superfluous, the English system requires a great deal of familiarity, and the use of g_c minimizes confusion in the conversion.

A.7.1.1 Further Notes on the Slug, Pound-Mass, and Pound-Force

Some further notes on the distinction between the pound-mass, the slug, and the pound-force are paraphrased from Hawkins (1951, p. 76): Consistency of units in engineering calculations is important. In many cases, mixed units are employed; this practice leads to serious errors unless careful consideration is given to the conversion of the units.

The main difficulties usually arise in the use of the pound-force and the corresponding unit of mass, the slug. The slug is defined as that mass, which when acted upon by a force of 1 lb, will be accelerated at 1 ft/s^2.

There are times, however, when the pound-mass will be used as a unit of mass. In order to convert from slugs to lb$_m$ or from lb$_m$ to slugs, it is only necessary to remember the relation between them, namely, that 32.16 lb$_m$ equals 1 slug.

Confusion often results in the use of force and mass units in dealing with density, ρ, and specific weight, γ. Since the numerical value of density when expressed in lb$_m$ units, lb$_m$/ft^3, is equal to that of the specific weight, lb$_f$/ft^3, they are often considered identical. This is not the case, since in the same system of units, they differ by the gravitational acceleration, g, in accordance with the relation, $\gamma = \rho g$. The unit of density in the foot-pound-hour system is the slug per cubic foot. According to the above equation, the specific weight in pounds per cubic foot units may be converted to the density unit of slug per cubic foot by dividing by 32.16 ft/s^2. Hence, the density of water at 68°F is 62.305/32.16 = 1.937 slug/ft^3 or 62.305 lb$_m$/ft^3.

A.7.2 Poundal

The combination of units in which one pound-mass is accelerated at 1 ft/s^2, then the result is one *poundal*, i.e.

$$\begin{aligned} F &= M \cdot g \\ &= 1.0 \text{ lb-mass} \cdot 1.0 \text{ ft/s}^2 \\ &= 1.0 \text{ poundal force} \end{aligned}$$

Stated another way, a force of one poundal will accelerate a mass of 1 lb mass at 1 ft/s^2.

Example A.6 Convert a Force of 1 lb$_f$ to Poundals

1. Consider 1.0 slug accelerated at gravitational acceleration, g, i.e.

$$F = M \cdot g$$

$$1.0 \text{ poundal force} \equiv 1.0 \text{ lb}_m \cdot 1.0 \text{ ft/s}^2$$

But 1.0 lbf is attained by 1.0 lb$_m$ accelerated at 32.2 ft/s^2, and therefore, 1.0 lb$_m$ = 1.0 lb$_f$/32.2 ft/s^2

$$1.0 \text{ poundal force} = 1.0 \text{ lb}_f/32.2 \text{ ft/s}^2 \cdot 1.0 \text{ ft/s}^2$$

Therefore, 1 lb force is 32.2 poundals of force.

A.8 TEMPERATURE

The SI unit of thermodynamic temperature is the kelvin (K). The degree Celsius (°C) is also used. The kelvin temperature, T(K), for the freezing point of water is 273.15 K. The Celsius temperature, T(°C), is

$$T(^\circ\text{C}) = T(\text{K}) - 273.15 \qquad \text{(A.4)}$$

A.9 CONVERSION TABLES

The conversions of various common units to SI units are given in Tables CDQR. More extensive tables are given on Web sites.

A.10 CONVERSIONS FOR UNCOMMON DERIVED UNITS

Many of the derived units in the environmental engineering field are not common to other fields, and so may not be listed in many conversion tables. Therefore, the conversions must be done using the conversion factors available. This can be done most easily by a process of canceling units that are to be omitted, leaving only the SI units. Using a systematic format for setting up the conversion will help to minimize mistakes. The procedure should be familiar to most persons in engineering or science. Example A.7 illustrates the process.

Example A.7 Convert 1.0 million gal/acre/day to SI Units

In slow sand filtration, the common units for hydraulic loading rate have been in million gal/acre/day. The unit dates back earlier than 1900. After the advent of rapid filtration, the U.S. Customary unit has been, gal/min/ft^2. The latter units do not, however, fit slow sand well, and SI units are more appropriate, that is m^3/m^2/s, which reduce to m/s (but m/h units are more convenient).

1. Express the 1.0 million gal/acre/day in fraction format, i.e.

$$\begin{aligned}\text{HLR} &= 1.0 \text{ million gal/acre/day}\\ &= \frac{1.0 \times 10^6 \text{ gal}}{\text{acre} \cdot \text{day}}\end{aligned}$$

2. Convert to SI units using conversions from Table CDQR:

$$\begin{aligned}\text{HLR} &= 1.0 \text{ million gal/acre/day}\\ &= \frac{1.0 \times 10^6 \text{ gal}}{\text{acre} \cdot \text{day}}\\ &= \frac{1.0 \times 10^6 \text{ gal}}{\text{acre} \cdot \text{day}} \cdot \frac{\text{acre}}{4.046\,873 \times 10^3 \text{ m}^2}\\ &\quad \cdot \frac{3.785\,412 \times 10^{-3} \cdot \text{m}^3}{\text{gal}} \cdot \frac{\text{day}}{8.640\,000 \times 10^4 \text{ s}}\\ &= 1.082\,629 \times 10^{-5} \text{ m/s}\\ &\approx 1.1 \times 10^{-5} \text{ m/s}\end{aligned}$$

Significant digits: Note that based upon the accuracy given in the starting figure, only about two places are justified in the final converted result. So that conversion factors can be traced to the sources, all of the places are used as illustrated in Example A.7. The final answer, however, should be rounded so that the number of significant digits retained should neither sacrifice nor exaggerate accuracy. Any digit necessary to define a numerical value is said to be *significant*. The rule for multiplication or division is that the product or quotient should contain no more significant digits than are contained in the number with the fewest significant digits (i.e., as used in the multiplication or division).

3. AWWA (1982) recognizes the units, mm/s as more convenient and so the conversion gives

$$\begin{aligned}\text{HLR} &= 1.0 \text{ million gal/acre/day}\\ &\approx 1.1 \cdot 10^{-5} \text{ m/s}\\ &= 0.011 \text{ mm/s}\end{aligned}$$

4. A more convenient unit is m/h:

$$\begin{aligned}\text{HLR} &= 0.011 \; \frac{\text{mm}}{\text{s}} \cdot \frac{3600}{\text{h}} \cdot \frac{1 \text{ m}}{1000 \text{ mm}}\\ &= 0.0396 \text{ m/h}\\ &\approx 0.04 \text{ m/h}\end{aligned}$$

A.11 DIMENSIONAL HOMOGENEITY

A requisite to a correct mathematical expression requires dimensional homogeneity. In other words, both sides of a given equation must be dimensionally equivalent (Rouse, 1946). The application of SI units ensures that the homogeneity condition is satisfied.

The problem arises mostly when force appears in one term of an equation and mass appears in another term. To handle the problem, either the force term must be converted to mass or the mass term must be converted to a force. Since SI units are expressed in terms of kg-m-s, the former is preferred if the choice is arbitrary.

As another example, if a pressure term is in terms of kPa, the conversion to N/m^2 will help to ensure dimensional homogeneity. If a viscosity term is in poises in the same equation, the conversion to N-s/m^2 will be helpful. Should a force term remain, its conversion from N to kg/s^2 is appropriate (vis-à-vis from kg/s^2. to N). The conversion is not necessary, of course, if there is a reason for the force unit in the final equation.

REFERENCES

ASTM, *Standard Practice for Use of the International System of Units (SI) (The Modernized Metric System)*, Designation E380-91a, PCN 03-543191-34, American Society for Testing Materials, Philadelphia, PA, 1991.

AWWA, AWWA Metrication Committee, Final report on metric units and sizes, *Journal of the American Water Works Association*, 74(1):27–33, January 1982.

Cardarelli, F., *Scientific Unit Conversion*, 2nd edn., Springer, Berlin, Germany, 1999.

Carver, G. P., A metric America: A decision whose time has come—For real, NISTIR 4858, Metric Program Technology Services, National Institute of Standards and Technology, U.S. Department of Commerce, Washington, DC, June, 1992.

Construction Subcommittee, *Interagency Council on Metric Policy, Metric Guide for Federal Construction*, 1st edn., National Institute of Building Sciences, Washington, DC, 1991.

Felder, R. M. and Rousseau, R. W., *Elementary Principles of Chemical Processes*, John Wiley & Sons, New York, 1978.

Hawkins, G. A., *Thermodynamics*, 2nd edn., John Wiley & Sons, New York, 1951.

ICMP, *Interagency Council on Metric Policy, Metric Guide for Federal Construction*, 1st edn., National Institute of Building Sciences, Washington, DC, 1991.

Rouse, H., *Elementary Mechanics of Fluids*, John Wiley & Sons, New York, 1946.

Appendix B: Physical Constants and Physical Data

TABLE B.1
Standard Atomic Weights

Name	Symbol	Atomic Number	Atomic Weight
Actinium	Ac	89	227[a]
Aluminum	Al	13	26.981539
Americium	Am	95	243[a]
Antimony	Sb	51	121.760
Argon	Ar	18	39.948
Arsenic	As	33	74.92159
Astatine	At	85	210[a]
Barium	Ba	56	137.327
Berkelium	Bk	97	247[a]
Beryllium	Be	4	9.012182
Bismuth	Bi	83	208.98037
Boron	B	5	10.811
Bromine	Br	35	79.904
Cadmium	Cd	48	112.411
Calcium	Ca	20	40.078
Californium	Cf	98	251[a]
Carbon	C	6	12.011
Cerium	Ce	58	140.115
Cesium	Cs	55	132.90543
Chlorine	Cl	17	35.4527
Chromium	Cr	24	51.9961
Cobalt	Co	27	58.93320
Copper	Cu	29	63.546
Curium	Cm	96	247[a]
Dysprosium	Dy	66	162.5
Einsteinium	Es	99	252[a]
Erbium	Er	68	167.26
Europium	Eu	63	151.965
Fermium	Fm	100	257[a]
Fluorine	F	9	18.9984032
Francium	Fr	87	223[a]
Gadolinium	Gd	64	157.25
Gallium	Ga	31	69.723
Germanium	Ge	32	73.61
Gold	Au	79	196.96654
Hafnium	Hf	72	178.49
Hahnium	Ha	105	262[a]
Helium	He	2	4.002602
Holmium	Ho	67	164.93032
Hydrogen	H	1	1.00794
Indium	In	49	114.818
Iodine	I	53	126.90447
Iridium	Ir	77	192.217
Mercury	Hg	80	200.59
Molybdenum	Mo	42	95.94
Neodymium	Nd	60	44.24
Neon	Ne	10	20.1797
Neptunium	Np	93	237[a]
Nickel	Ni	28	58.6934
Niobium	Nb	41	92.90638
Nitrogen	N	7	14.00674
Nobelium	No	102	259[a]
Osmium	Os	76	190.23
Oxygen	O	8	15.9994
Palladium	Pd	46	106.42
Phosphorus	P	15	30.973762
Platinum	Pt	78	195.08
Plutonium	Pu	94	244[a]
Polonium	Po	84	209[a]
Potassium	K	19	39.0983
Praseodymium	Pr	59	140.90765
Promethium	Pm	61	145[a]
Protactinium	Pa	91	231.03588
Radium	Ra	88	226[a]
Radon	Rn	86	222[a]
Rhenium	Re	75	186.207
Rhodium	Rh	45	102.90550
Rubidium	Rb	37	85.4678
Ruthenium	Ru	44	101.07
Rutherfordium	Rf	104	261[a]
Samarium	Sm	62	150.36
Scandium	Sc	21	44.955910
Selenium	Se	35	78.96
Silicon	Si	14	28.0855
Silver	Ag	47	107.8682
Sodium	Na	11	22.989768
Strontium	Sr	38	87.62
Sulfur	S	16	32.066
Tantalum	Ta	73	180.9479
Technetium	Tc	43	98[a]
Tellurium	Te	53	127.60
Terbium	Tb	65	158.92534
Thallium	Tl	81	204.3833
Thorium	Th	90	232.0381
Thulium	Tm	69	168.93421
Tin	Sn	50	118.710

(continued)

TABLE B.1 (continued)
Standard Atomic Weights

Name	Symbol	Atomic Number	Atomic Weight	Name	Symbol	Atomic Number	Atomic Weight
Iron	Fe	26	55.845	Titanium	Ti	22	47.867
Krypton	Kr	36	83.8	Tungsten	W	74	183.84
Lanthanum	La	57	138.9055	Uranium	U	92	238.0289
Lawrencium	Lr	103	262[a]	Vanadium	V	23	50.9415
Lead	Pb	82	207.2	Xenon	Xe	54	131.29
Lithium	Li	3	6.941	Ytterbium	Yb	70	173.04
Lutetium	Lu	71	174.967	Yttrium	Y	39	88.90585
Magnesium	Mg	12	24.3050	Zinc	Zn	30	65.39
Manganese	Mn	25	54.93805	Zirconium	Zr	40	91.224
Mendelevium	Md	101	258[a]				

Source: Lide, D.R. (Ed.), *Handbook of Chemistry and Physics*, 77th edn., 1996–1997, CRC Press, Boca Raton, FL, 1996.

[a] Indicates the mass number of the longest-lived isotope of an element that has no stable isotopes and for which standard atomic weight cannot be defined because of wide variability in isotope composition, or complete absence, in nature.

TABLE B.2
Physical Constants

Name	Symbol	Value	Units
Avogadro constant	N_A	$6.022\ 1367 \cdot 10^{23}$	mol^{-1}
Gas constant	R	8.314 510	$J\ K^{-1}\ mol^{-1}$
		1.987 216	$cal\ K^{-1}\ mol^{-1}$
		0.082 066 7	$L\ atm\ K^{-1}\ mol^{-1}$
		0.000 082 066 7	$m^3\ atm\ K^{-1}\ mol^{-1}$
		0.083 145 1	$bar\ K^{-1}\ mol^{-1}$
Boltzman constant, R/N_A	k	1.380 658	$10^{-23}\ J\ K^{-1}$
Molar volume		22.410 10	$L\ mol^{-1}$
Electron volt	eV	1.602177 33	10^{-19} J
Faraday's constant	F	96 485.309	C/mol

Source: Lide, D.R. (Ed.), *Handbook of Chemistry and Physics*, 79th edn., 1998–1999, CRC Press, Boca Raton, FL, 1998.

Expressions for R in units other than SI from Alberty and Silbey (1992).

TABLE B.3
Standard Values

Name	Symbol	Value	Units
Electron volt	eV	1.602177 33	10^{-19} J
Standard atmosphere	atm	101 325	Pa
Standard acceleration of gravity	g	9.806 65	$m\ s^{-2}$
Unified atomic mass unit	u	1.660 540 2	10^{-27} kg
$1\ u = m_u = 1/12\ m(^{12}C)$			

Source: Lide, D.R. (Ed.), *Handbook of Chemistry and Physics*, 79th edn., 1998–1999, CRC Press, Boca Raton, FL, 1998.

TABLE B.4
Gas Constant in Different Units

Units of *V*, *T*, *n*			Units of *P*				
V	*T*	*n*	kPa	atm	Bar	lb/ft^2	psi
ft^3	K	mol	0.2936241	0.00289785			0.0425866
		lb-mol	133.1857	1.31444		2779	19.3169
	°R	mol	0.1631245	0.00160991		3.407	0.0236592
		lb-mol	73.99204	0.730245		1543	10.7316
cm^3	K	mol	8314.510	82.0578			1205.92
		lb-mol	3771.398	37220.8			546995
	°R	mol	4619.172	45.5877			669.954
		lb-mol	2095221	20678.2			303886
L	K	mol	8.314510	0.0820578	0.083 145		1.20592
		lb-mol	3771.398	37.2208			546.995
	°R	mol	4.619172	0.0455877			0.669954
		lb-mol	2095.221	20.6782			303.886
m^3	K	mol	0.008314510	0.0000820578			0.00120592
		lb-mol	3.771398	0.0372208			0.546995
	°R	mol	0.004619172	0.0000455877			0.000669954
		lb-mol	2.095221	0.0206782			0.303886

Source: Lide, D.R. (Ed.), *Handbook of Chemistry and Physics*, 77th edn., 1996–1997, CRC Press, Boca Raton, FL, 1996, pp. 1–43.

TABLE B.5
Gas Data and Calculations for Standard Conditions

Gas	MW (g/mol)	ρ(molar) (mol/ms^3)	ρ (kg/m^3)	*k*
Air	28.9641[a]	41.5606[a]	1.2038[d]	1.4
Ammonia	17.0312[b]	40.8637[c]	0.7080[d]	
Carbon dioxide	44.00982[b]	40.8637[c]	1.8295[d]	1.4
Carbon monoxide	28.0102[b]	40.8637[c]	1.1644[d]	
Hydrogen	2.015882[b]	40.8637[c]	0.0838[d]	
Nitrogen	28.01342[b]	40.8637[c]	1.1645[d]	1.4
Oxygen	31.99882[b]	40.8637[c]	1.3302[d]	1.4
Methane, CH_4	16.015882[b]	40.8637[c]	0.6658[d]	1.31

Notes: The volume and density of a gas must be referenced to the temperature and pressure. The standard temperature and pressure (STP) are defined as 20°C (68°F) and 101.325 kPa (1 atm), respectively. Another standard is "normal temperature and pressure" (NTP), defined as 0°C and 101.325 kPa (1 atm), respectively. In summary: $P = 101\ 325$ Pa; $T = 293.15$ K; $R = 8.31451$ J/K mol.

[a] Calculated from composition of air, Table B.7.
[b] Lide (1998).
[c] Calculated from $\rho(\text{molar}) = n/V = P/RT$, in which $R = 8.314\ 510\ \text{J K}^{-1}\ \text{mol}^{-1}$.
[d] Calculated from $\rho = \rho(\text{molar}) \cdot \text{MW(gas)}/1000$.

TABLE B.6
Atmospheric Pressure and Density as Function of Elevation

Z^a (m)	T^b (K)	P^c(atm) (Pa)	P^d(atm) (kPa)	ρ^e(molar) (mol/m^3)	ρ^f (kg/m^3)
0	288.15	101,325	101.30	42.282	1.225
500	284.9	95,460	95.46	40.299	1.167
1,000	281.65	89,880	89.88	38.381	1.112
1,500	278.4	84,560	84.56	36.531	1.058
2,000	275.15	79,500	79.50	34.750	1.007
2,500	271.91	74,690	74.69	33.037	0.957
3,000	268.66	70,120	70.12	31.391	0.909
3,500	265.41	65,790	65.78	29.813	0.864
4,000	262.17	61,660	61.66	28.287	0.819
4,500	258.92	57,750	57.75	26.826	0.777
5,000	255.68	54,050	54.05	25.425	0.736

[a–c] Lide (1996, p. 14-17).
[d] $P'(\text{atm}) = MO + M1 \cdot Z + M2 \cdot Z^2 + M3 \cdot Z^3 + M4 \cdot Z^4 + M5 \cdot Z^5 + M6 \cdot Z^6 +$
[e] $\rho(\text{molar}) = n/V = P/RT \quad R = 8.31451$ J/K mol (Table QR).
[f] ρ (kg/m^3) $= \rho$(molar) $\cdot$ MW(air)/1000 MW(air) = 28.964 g/mol (Table B.7).

Polynomial coefficients
MO = 101.3 M3 = −1.3476E−11
M4 = 8.2464E−15
M1 = −0.011944 M5 = −2.3906E−18
M2 = 5.3142E−07 M6 = 2.0382E−22

TABLE B.7
Composition of Air at Pressure and Temperature Specified

Gas^a	MW^a (g/mol)	X(gas)^a	ρ(molar)^b (mol/m^3)	ρ^c (kg/m^3)	wt avg^d
N_2	28.0134	0.78084	32.45226	0.909098125	21.87398326
O_2	31.9988	0.209476	8.70597	0.278580588	6.702980629
Ar	39.948	0.00934	0.38818	0.015506894	0.37311432
CO_2	44.0098	0.000314	0.01305	0.000574331	0.013819077
Ne	20.1797	0.00001818	0.00076	1.52472E−05	0.000366867
He	4.0026	0.00000524	0.00022	8.71679E−07	2.09736E−05
Kr	83.80	0.00000114	0.00005	3.97038E−06	0.000095532
Xe	131.29	0.000000087	0.00000	4.74716E−07	1.14222E−05
CH_4	16.0428	0.000002	0.00008	1.3335E−06	3.20856E−05
H_2	2.01588	0.0000005	0.00002	4.18907E−08	1.00794E−06
Sum		0.999997147	41.5606	1.2038^e	28.964^f

[a–c] Lide (1996, p. 4-37:98).
[b] $\rho(\text{molar}) = n/V = P/RT$ Gas law data: P(atm) = 101325 Pa, $R = 8.31451$ J/K mol.
[c] ρ(molar) $\cdot$ MW(gas)/1000 $T = 293.15$ K.
[d] Wt Avg $= MW_i \cdot X(\text{gas})_i$.
[e] $\rho(\text{air}) = \Sigma\rho(\text{gas})_i$.
[f] MW(air) = sum[MW(gas)$_i \cdot X$(gas)$_i$].

TABLE B.8
Properties of Air at Atmospheric Pressure

Temperature (°C)	Temperature (K)	ρ(molar)[a] (mol/m^3)	ρ(air)[b] (kg/m^3)	μ(air)[c] (N-s/m^2)	ν(air)[d] (m^2/s)
−73	200.15	60.9326	1.7649	**13.3×10^{-6}**	
−31.6	304.75			15.4	—
−23	250.15	48.7461	1.4119		
−10	283.15				12.45×10^{-6}
0	273.15	44.6148	1.2922	17.1	11.52×10^{-6}
10	283.15				14.12×10^{-6}
18	291.15			18.3	
20	293.15	41.5710	1.2041		14.86×10^{-6}
27	300.15	40.6015	1.1760	**18.6×10^{-6}**	
30	303.15				16.07×10^{-6}
37	310.15	39.2924	1.1381		
40	313.15			19.0	16.91×10^{-6}
54	327.15	38.0651	1.1025	19.6	

[a] $\rho(\text{molar}) = P/RT \quad R = 8.31451$ g/mol.
[b] $\rho = \rho(\text{molar}) \cdot (\text{MW(air)}/1000)$.
[c] Dynamic viscosity data in bold from Lide (1996, p. 6-206); not in bold from Weast (1978, p. F-58).
[d] Typed data; no reference.

TABLE B.9
Properties of Water

Temperature (°C)	Density[a] (kg/m^3)	Vapor Pressure[a] (N/m^2)	Surface Tension[a] (N/m)	Dynamic Viscosity[b] (poises)[c]	Kinematic Viscosity[d] (stokes)[c]	Dynamic Viscosity[e] (N · s/m^2)	Kinematic Viscosity[f] (m^2/s)
0	999.8426	611.29	0.07564	$1.787 \cdot 10^{-2}$	$1.787 \cdot 10^{-2}$	$1.787 \cdot 10^{-3}$	$1.787 \cdot 10^{-6}$
1	999.9015	657.16		$1.728 \cdot 10^{-2}$	$1.728 \cdot 10^{-2}$	$1.728 \cdot 10^{-3}$	$1.728 \cdot 10^{-6}$
2	999.9429	706.05		$1.671 \cdot 10^{-2}$	$1.671 \cdot 10^{-2}$	$1.671 \cdot 10^{-3}$	$1.671 \cdot 10^{-6}$
3	999.9672	758.13		$1.618 \cdot 10^{-2}$	$1.618 \cdot 10^{-2}$	$1.618 \cdot 10^{-3}$	$1.618 \cdot 10^{-6}$
4	999.9750	813.59		$1.567 \cdot 10^{-2}$	$1.567 \cdot 10^{-2}$	$1.567 \cdot 10^{-3}$	$1.567 \cdot 10^{-6}$
5	999.9668	872.60		$1.519 \cdot 10^{-2}$	$1.519 \cdot 10^{-2}$	$1.519 \cdot 10^{-3}$	$1.519 \cdot 10^{-6}$
6	999.9460	935.37		$1.472 \cdot 10^{-2}$	$1.472 \cdot 10^{-2}$	$1.472 \cdot 10^{-3}$	$1.472 \cdot 10^{-6}$
7	999.9043	1,002.1		$1.428 \cdot 10^{-2}$	$1.428 \cdot 10^{-2}$	$1.428 \cdot 10^{-3}$	$1.428 \cdot 10^{-6}$
8	999.8509	1,073.0		$1.386 \cdot 10^{-2}$	$1.386 \cdot 10^{-2}$	$1.386 \cdot 10^{-3}$	$1.386 \cdot 10^{-6}$
9	999.7834	1,148.2		$1.346 \cdot 10^{-2}$	$1.346 \cdot 10^{-2}$	$1.346 \cdot 10^{-3}$	$1.346 \cdot 10^{-6}$
10	999.7021	1,228.1	0.07423	$1.307 \cdot 10^{-2}$	$1.307 \cdot 10^{-2}$	$1.307 \cdot 10^{-3}$	$1.307 \cdot 10^{-6}$
11	999.6074	1,312.9		$1.271 \cdot 10^{-2}$	$1.271 \cdot 10^{-2}$	$1.271 \cdot 10^{-3}$	$1.271 \cdot 10^{-6}$
12	999.4996	1,402.7		$1.235 \cdot 10^{-2}$	$1.235 \cdot 10^{-2}$	$1.235 \cdot 10^{-3}$	$1.235 \cdot 10^{-6}$
13	999.3792	1,497.9		$1.202 \cdot 10^{-2}$	$1.203 \cdot 10^{-2}$	$1.202 \cdot 10^{-3}$	$1.203 \cdot 10^{-6}$
14	999.2464	1,598.8		$1.169 \cdot 10^{-2}$	$1.170 \cdot 10^{-2}$	$1.169 \cdot 10^{-3}$	$1.170 \cdot 10^{-6}$
15	999.1016	1,705.6		$1.139 \cdot 10^{-2}$	$1.140 \cdot 10^{-2}$	$1.139 \cdot 10^{-3}$	$1.140 \cdot 10^{-6}$
16	998.9450	1,818.5		$1.109 \cdot 10^{-2}$	$1.110 \cdot 10^{-2}$	$1.109 \cdot 10^{-3}$	$1.110 \cdot 10^{-6}$
17	998.7769	1,938.0		$1.081 \cdot 10^{-2}$	$1.082 \cdot 10^{-2}$	$1.081 \cdot 10^{-3}$	$1.082 \cdot 10^{-6}$
18	998.5976	2,064.4		$1.053 \cdot 10^{-2}$	$1.054 \cdot 10^{-2}$	$1.053 \cdot 10^{-3}$	$1.054 \cdot 10^{-6}$
19	998.4073	2,197.8		$1.027 \cdot 10^{-2}$	$1.029 \cdot 10^{-2}$	$1.027 \cdot 10^{-3}$	$1.029 \cdot 10^{-6}$
20	998.2063	2,338.8	0.07275	$1.002 \cdot 10^{-2}$	$1.004 \cdot 10^{-2}$	$1.002 \cdot 10^{-3}$	$1.004 \cdot 10^{-6}$
21	997.9948	2,487.7		$0.9779 \cdot 10^{-2}$	$0.9799 \cdot 10^{-2}$	$0.978 \cdot 10^{-3}$	$0.980 \cdot 10^{-6}$
22	997.7730	2,644.7		$0.9548 \cdot 10^{-2}$	$0.9569 \cdot 10^{-2}$	$0.954 \cdot 10^{-3}$	$0.957 \cdot 10^{-6}$
23	997.5412	2,810.4		$0.9325 \cdot 10^{-2}$	$0.9348 \cdot 10^{-2}$	$0.932 \cdot 10^{-3}$	$0.935 \cdot 10^{-6}$
24	997.2994	2,985.0		$0.9111 \cdot 10^{-2}$	$0.9136 \cdot 10^{-2}$	$0.911 \cdot 10^{-3}$	$0.914 \cdot 10^{-6}$
25	997.0480	3,169.0		$0.8904 \cdot 10^{-2}$	$0.8930 \cdot 10^{-2}$	$0.890 \cdot 10^{-3}$	$0.893 \cdot 10^{-6}$
26	996.7870	3,362.9		$0.8705 \cdot 10^{-2}$	$0.8733 \cdot 10^{-2}$	$0.870 \cdot 10^{-3}$	$0.873 \cdot 10^{-6}$
27	996.5166	3,567.0		$0.8513 \cdot 10^{-2}$	$0.8543 \cdot 10^{-2}$	$0.851 \cdot 10^{-3}$	$0.854 \cdot 10^{-6}$
28	996.2371	3,781.8		$0.8327 \cdot 10^{-2}$	$0.8359 \cdot 10^{-2}$	$0.833 \cdot 10^{-3}$	$0.836 \cdot 10^{-6}$
29	995.9486	4,007.8		$0.8148 \cdot 10^{-2}$	$0.8181 \cdot 10^{-2}$	$0.815 \cdot 10^{-3}$	$0.818 \cdot 10^{-6}$
30	995.6511	4,245.5	0.07120	$0.7975 \cdot 10^{-2}$	$0.8010 \cdot 10^{-2}$	$0.798 \cdot 10^{-3}$	$0.801 \cdot 10^{-6}$
31	995.3450	4,495.3		$0.7808 \cdot 10^{-2}$	$0.7844 \cdot 10^{-2}$	$0.781 \cdot 10^{-3}$	$0.784 \cdot 10^{-6}$
32	995.0302	4,757.8		$0.7647 \cdot 10^{-2}$	$0.7685 \cdot 10^{-2}$	$0.765 \cdot 10^{-3}$	$0.768 \cdot 10^{-6}$
33	994.7071	5,033.5		$0.7491 \cdot 10^{-2}$	$0.7531 \cdot 10^{-2}$	$0.749 \cdot 10^{-3}$	$0.753 \cdot 10^{-6}$
34	994.3756	5,322.9		$0.7340 \cdot 10^{-2}$	$0.7381 \cdot 10^{-2}$	$0.734 \cdot 10^{-3}$	$0.738 \cdot 10^{-6}$
35	994.0359	5,626.7		$0.7194 \cdot 10^{-2}$	$0.7237 \cdot 10^{-2}$	$0.719 \cdot 10^{-3}$	$0.724 \cdot 10^{-6}$
36	993.6883	5,945.3		$0.7052 \cdot 10^{-2}$	$0.7097 \cdot 10^{-2}$	$0.705 \cdot 10^{-3}$	$0.710 \cdot 10^{-6}$
37	993.3328	6,279.5		$0.6915 \cdot 10^{-2}$	$0.6961 \cdot 10^{-2}$	$0.692 \cdot 10^{-3}$	$0.696 \cdot 10^{-6}$
38	992.9695	6,629.8		$0.6783 \cdot 10^{-2}$	$0.6831 \cdot 10^{-2}$	$0.678 \cdot 10^{-3}$	$0.683 \cdot 10^{-6}$
39	992.5987	6,996.9		$0.6654 \cdot 10^{-2}$	$0.6703 \cdot 10^{-2}$	$0.665 \cdot 10^{-3}$	$0.670 \cdot 10^{-6}$
40	992.2204	7,381.4	0.06960	$0.6529 \cdot 10^{-2}$	$0.6580 \cdot 10^{-2}$	$0.653 \cdot 10^{-3}$	$0.658 \cdot 10^{-6}$
41		7,784.0		$0.6408 \cdot 10^{-2}$	$0.6461 \cdot 10^{-2}$	$0.641 \cdot 10^{-3}$	$0.646 \cdot 10^{-6}$
42		8,205.4		$0.6291 \cdot 10^{-2}$	$0.6345 \cdot 10^{-2}$	$0.629 \cdot 10^{-3}$	$0.636 \cdot 10^{-6}$
43		8,646.3		$0.6178 \cdot 10^{-2}$	$0.6234 \cdot 10^{-2}$	$0.618 \cdot 10^{-3}$	$0.623 \cdot 10^{-6}$
44		9,107.5		$0.6067 \cdot 10^{-2}$	$0.6124 \cdot 10^{-2}$	$0.607 \cdot 10^{-3}$	$0.612 \cdot 10^{-6}$
45		9,589.8		$0.5960 \cdot 10^{-2}$	$0.6019 \cdot 10^{-2}$	$0.596 \cdot 10^{-3}$	$0.602 \cdot 10^{-6}$
46		10,094		$0.5856 \cdot 10^{-2}$	$0.5916 \cdot 10^{-2}$	$0.586 \cdot 10^{-3}$	$0.592 \cdot 10^{-6}$
47		10,620		$0.5755 \cdot 10^{-2}$	$0.5817 \cdot 10^{-2}$	$0.576 \cdot 10^{-3}$	$0.582 \cdot 10^{-6}$

TABLE B.9 (continued)
Properties of Water

Temperature (°C)	Density[a] (kg/m^3)	Vapor Pressure[a] (N/m^2)	Surface Tension[a] (N/m)	Dynamic Viscosity[b] (poises)[c]	Kinematic Viscosity[d] (stokes)[c]	Dynamic Viscosity[e] (N · s/m^2)	Kinematic Viscosity[f] (m^2/s)
48		11,171		$0.5656 \cdot 10^{-2}$	$0.5719 \cdot 10^{-2}$	$0.566 \cdot 10^{-3}$	$0.572 \cdot 10^{-6}$
49		11,745		$0.5561 \cdot 10^{-2}$	$0.5626 \cdot 10^{-2}$	$0.556 \cdot 10^{-3}$	$0.563 \cdot 10^{-6}$
50		12,344	0.06794	$0.5468 \cdot 10^{-2}$	$0.5534 \cdot 10^{-2}$	$0.547 \cdot 10^{-3}$	$0.553 \cdot 10^{-6}$

[a] Lide (1996, p. 6-10, 6-8, 6-13).
[b] Weast (1978, p. F-51).
[c] 1 poise = gm · s/cm; 1 stoke = cm^2/s.
[d] Kinematic viscosity, ν, calculated as

$$\nu = \mu/\rho$$

in which
ν is the kinematic viscosity (cm^2/s)
μ is the dynamic viscosity (gm · s^{-1}/cm) = dyne-s/cm^2
ρ is the density of water (gm/cm^3)
Dynamic viscosity, μ, taken from column 5; density, ρ, taken from column 1
[e] To convert dynamic viscosity in poises to SI units:

$$\mu\left(\frac{\text{g}}{\text{cm} \cdot \text{s}}\right) = \left(\frac{\text{g}}{\text{cm} \cdot \text{s}}\right) \cdot \left(\frac{\text{kg}}{1000\,\text{g}}\right) \cdot \left(\frac{100\,\text{cm}}{\text{m}}\right) \cdot \left(\frac{(\text{N} \cdot \text{s}^2/\text{m})}{\text{kg}}\right) = 0.1\,\frac{\text{N} \cdot \text{s}}{\text{m}^2}$$

Multiply "poises" by $2.089 \cdot 10^{-3}$ to convert to $\text{lb}_F \cdot \text{s/ft}^2$.
[f] To convert kinematic viscosity in stokes (cm^2/s) to SI units:

$$\nu\left(\frac{\text{cm}^2}{\text{s}}\right) = \left(\frac{\text{cm}^2}{\text{s}}\right) \cdot \left(\frac{\text{m}^2}{10^4\,\text{cm}^2}\right) = 10^{-4}\,\text{m}^2/\text{s}$$

Multiply "stokes" by $1.0761 \cdot 10^{-3}$ to convert to ft^2/s.

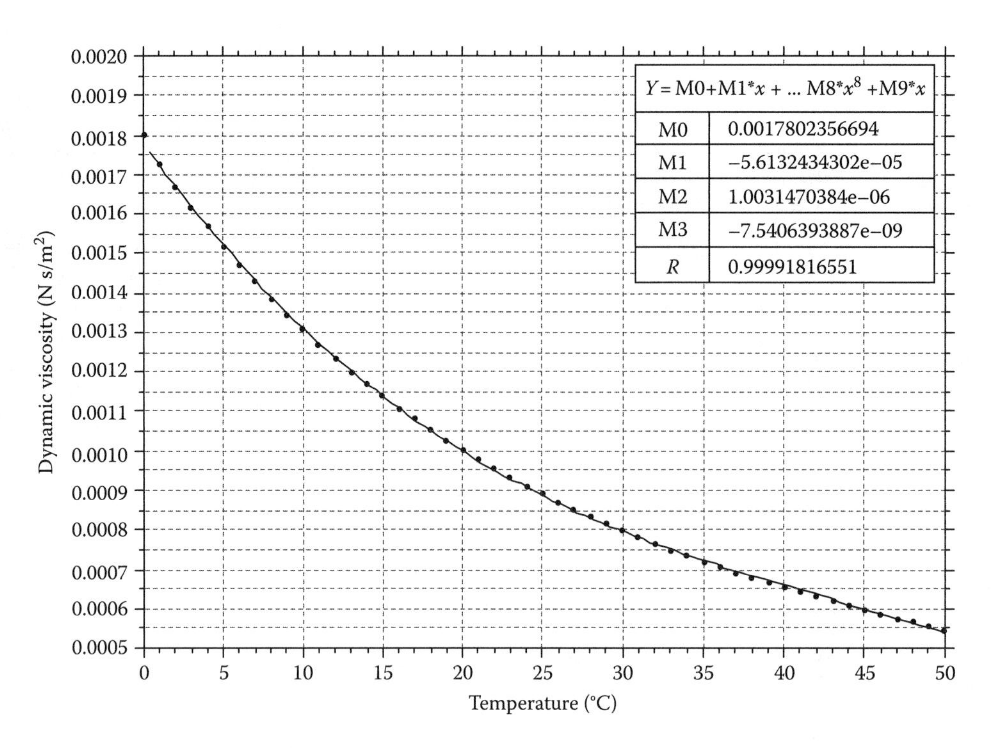

FIGURE B.1 Dynamic viscosity of water as a function of temperature. (From Weast, R.C., *Handbook of Chemistry and Physics*, 59th edn., CRC Press, Inc., Boca Raton, FL, 1978, p. F51.)

Polynomial best-fit equation ($R^2 = 0.9999$):

$$\mu\left(\frac{\mathrm{N\cdot s}}{\mathrm{m}^2}\right) = 0.0017802356694 - 5.6132434302\cdot 10^{-5}\cdot T(^\circ\mathrm{C}) + 1.0031470384\cdot 10^{-6}\cdot T(^\circ\mathrm{C})^2 - 7.5406393887\cdot 10^{-9}\cdot T(^\circ\mathrm{C})^3$$

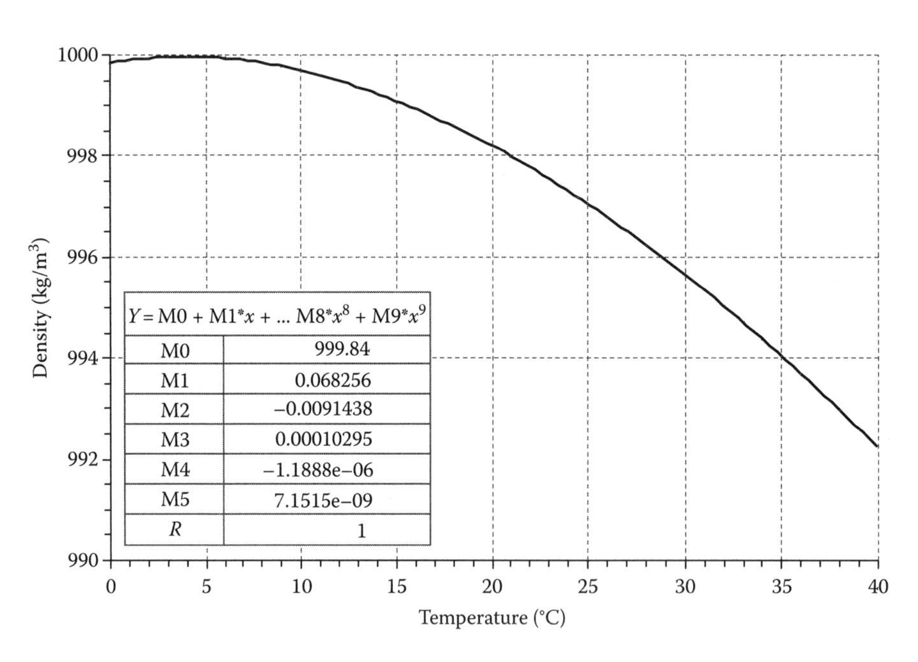

FIGURE B.2 Density of water as a function of temperature. (From Lide, D.R. (Ed.), *Handbook of Chemistry and Physics*, 77th edn., 1996–1997, CRC Press, Boca Raton, FL, 1996, p. 6-10.)

Polynomial best-fit equation ($R^2 = 1$):

$$\rho\left(\frac{\mathrm{kg}}{\mathrm{m}^3}\right) = 999.84 + 0.068256\cdot T(^\circ\mathrm{C}) - 0.0091438\cdot T(^\circ\mathrm{C})^2 + 0.00010295\cdot T(^\circ\mathrm{C})^3 - 1.1888\times 10^{-6}\cdot T(^\circ\mathrm{C})^4 + 7.1515\times 10^{-9}\cdot T(^\circ\mathrm{C})^5$$

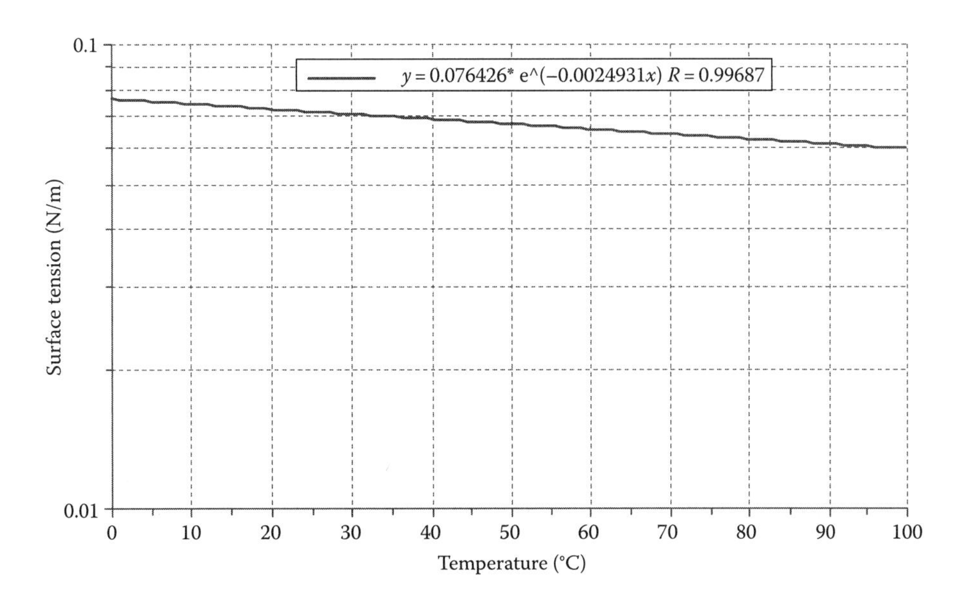

FIGURE B.3 Surface tension of water as a function of temperature. (From Lide, D.R. (Ed.), *Handbook of Chemistry and Physics*, 77th edn., 1996–1997, CRC Press, Boca Raton, FL, 1996, p. 6-10.)

Exponential best-fit equation ($R^2 = 10.99687$):

$$\sigma\left(\frac{\text{N}}{\text{m}}\right) = 0.076426 \cdot e^{-0.0024931 \cdot T(°\text{C})}$$

REFERENCES

Alberty, R. A. and Silbey, R. J., *Physical Chemistry*, 1st edn., John Wiley & Sons, Inc., New York, 1992.

Lide, D. R. (Ed.), *Handbook of Chemistry and Physics*, 77th edn., 1996–1997, CRC Press, Boca Raton, FL, 1996.

Lide, D. R. (Ed.), *Handbook of Chemistry and Physics*, 77th edn., 1998–1999, CRC Press, Boca Raton, FL, 1998.

Weast, R. C., *Handbook of Chemistry and Physics*, 59th edn., CRC Press, Inc., Boca Raton, FL, 1978.

Appendix C: Miscellaneous Relations

Often, a formula or a method must be retrieved from any one of several texts or reference books. This appendix is intended to provide for some of these needs, especially those that related to the topics in this text. The ideal gas law is covered briefly along with the ideas of gage pressure and absolute pressure. In addition, some geometric formulae are given, along with conversion rationale between common logs and natural logs. A few selections from statistics are included for convenience and also summaries of two ENR cost indexes are included, also for convenience.

C.1 IDEAL GAS LAW

The ideal gas law is expressed as

$$pV = nRT \tag{C.1}$$

in which

p is the absolute pressure (N/m^2)
V is the volume occupied by gas (m^3)
N is the moles of gas of species A (mol)
R is the universal gas constant (8.313 N · m/g-mol K)
T is the temperature (K)

The ideal gas law has a great deal of utility in that it can be applied in a number of ways. Example C.1 illustrates how the units are applied. The example also illustrates the coherency of the SI system of units.

Example C.1 Calculate the Density of Air at Sea Level at 20°C

1. Apply the ideal gas law for the conditions stated, i.e.,

$$pV = N(\text{air})RT \tag{ExC.1.1}$$

$$101\,325\ \frac{\text{N}}{\text{m}^2} \cdot V = n(\text{air}) \cdot 8.314510 \frac{\text{N} \cdot \text{m}}{\text{K mol}} \cdot 293.15\ \text{K} \tag{ExC.1.2}$$

$$\frac{n(\text{air})}{V} = 41.57\ \frac{\text{mol}}{\text{m}^3} \tag{ExC.1.3}$$

2. Convert density in mol/m^3 to kg/m^3, i.e.,

$$\rho(\text{air}) = \frac{n(\text{air})}{V} \cdot \text{molar density (air)} \tag{ExC.1.4}$$

$$= 41.57\ \frac{\text{mol}}{\text{m}^3} \cdot 0.029\ \frac{\text{kg}}{\text{mol}} \tag{ExC.1.5}$$

$$= 1.20\ \frac{\text{kg}}{\text{m}^3} \tag{ExC.1.6}$$

C.2 PRESSURE

Both absolute pressure and gage pressure are used. The relation between the two is understood most easily by graphical depiction, Figure C.1.

Example C.2 Calculate the Absolute Pressure at Point A if the Gage Pressure is 50 kPa and the Atmospheric Pressure is 101.3 kPa

1. Referring to Figure C.1a, the algebraic relation is

$$P(A)_{\text{abs}} = P_{\text{atm}} + P(A)_{\text{gage}} \tag{ExC.2.1}$$

$$= 101.3\ \text{kPa} + 50\ \text{kPa} \tag{ExC.2.2}$$

$$= 151\ \text{kPa} \tag{ExC.2.3}$$

C.3 MATHEMATICS

Quite often in handling data, it is necessary to retrieve certain formulae or methods that have been instilled over the years, but that require a referral to a text that one must keep on hand. Often, these texts are not readily available. This section is intended to provide some that are thought to be needed for use of this text.

C.3.1 Roots of a Quadratic Equation

The determination of the roots of a quadratic equation, e.g.,

$$ax^2 + bx + c = 0 \tag{C.2}$$

can be solved by the equation

$$x = \frac{-b \pm \sqrt{b^2 - 4ac}}{2a} \tag{C.3}$$

C.3.2 Geometry

1. Volume of sphere

$$V = \frac{\pi d^3}{6} \tag{C.4}$$

2. Area of sphere

$$A = \pi d^2 \tag{C.5}$$

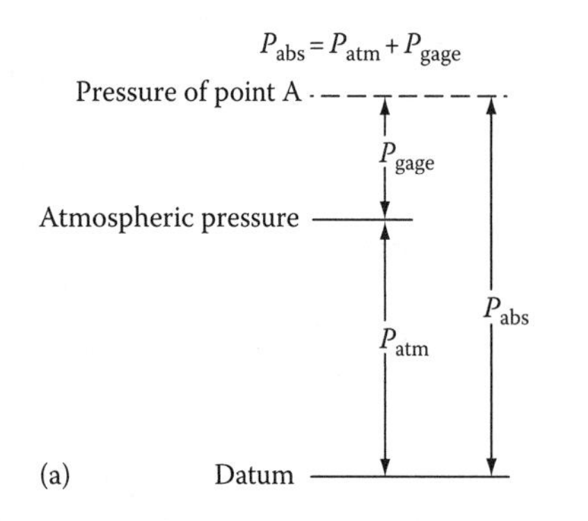

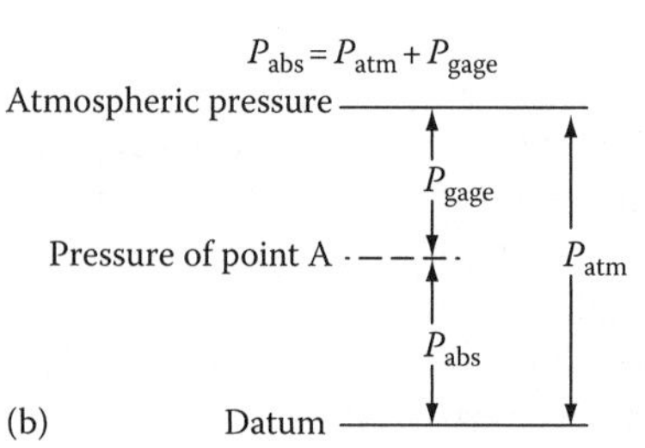

FIGURE C.1 Graphical and algebraic pressure relations for conversions between absolute pressure and gage pressure. Pressure at level A is the focus. (a) Positive gage pressure. (b) Negative gage pressure (vacuum).

C.3.3 Natural Logarithms

In converting between common log and natural log values, the rationale must be reconstructed (unless this task is done often). This section provides a review of the basics and can also be used more expediently to glean only the immediate needs of a conversion.

C.3.3.1 Compound Interest

We may look at say the way money is compounded as an introduction as to how some things happen in nature, say growth of a bacterial culture. Consider the money problem first. The rate at which a sum of money grows depends on the rate of interest, r. Suppose $r = 0.06$/year (i.e., 6% per annum) and the principal, $P = \$1.00$. The amount, A, accumulated at the end of 1 year is \$1.06, i.e., $A = P(1 + r)$. After 2 years, i.e., $n = 2$, $A(2\ \text{year}) = A(1\ \text{year}) \cdot (1 + r) = [P(1 + r)] \cdot (1 + r) = P(1 + r)^2$. After n years

$$A(n) = P(1 + r)^n \tag{C.6}$$

Now suppose the interest is compounded every half year. For this case, $r(0.5\ \text{year}) = 0.03$, or $r/2$, and the formula is, $A(n) = P(+r/2)^{2n}$. To generalize, if we let the value be compounded k times per year, then

$$A(n) = P\left(\frac{1 + r}{k}\right)^{kn} \tag{C.7}$$

in which

- $A(n)$ is the compounded amount at the end of n years
- P is the principal amount at $t = 0$
- r is the annual rate
- n is the number of years
- k is the number of times interest is compounded each year

C.3.3.2 Continuous Growth

Now suppose we compound the interest very often, or continuously. The rate of growth is then proportional to the amount at any instant. To illustrate, let, $P = 1.00$, $n = 1$, $r = 1.00$, and let $k = 1, 10, 100, 1000$. Table C.1 shows the amounts, A, for each k.

The limiting value of A is seen to be 2.718.

TABLE C.1
Amount after k Periods of Compounding

P	n	r	k	A
1.00	1	1.00	1	2.00000000
			10	2.59374246
			100	2.70481383
			1,000	2.71692393
			10,000	2.71814593
			100,000	2.71826824
			1,000,000	2.71828047

C.3.3.3 The Number e

The limit approached by the quantity, $(1 + 1/k)^k$ is denoted by e:

$$e = \underset{k\to\infty}{\text{L}} \left(1 + \frac{1}{k}\right)^k \tag{C.8}$$

$$= 2.71828 \tag{C.9}$$

and

$$\log \text{e} = 0.43429426 \tag{C.10}$$

Therefore, \$1 with 100% interest, compounded continuously for 1 year would result in e dollars. The limiting value of A in (C.6) is

$$A = Pe^{rn} \tag{C.11}$$

which is the amount of any principal, P, after n years with interest compounded continuously at any rate, r.

C.3.3.4 Kinetics

For any physical quantity that increases or decreases at a rate proportional to the concentration, the expression is

$$\frac{\text{d}A}{\text{d}t} = rA \tag{C.12}$$

Separating variables,

$$\frac{dA}{A} = r\,dt \tag{C.13}$$

Integrating between the limits of A_o and A and 0 and t, respectively,

$$\frac{\ln A}{A_o} = rt \tag{C.14}$$

or,

$$\frac{A}{A_o} = e^{rt} \tag{C.15}$$

which is the same as (C.11).

C.3.3.5 Laws of Exponents

The four laws of exponent are

1. Multiplying

$$10^x \cdot 10^y = 10^{(x+y)} \tag{C.16}$$

2. Dividing

$$\frac{10^x}{10^y} = 10^{(x-y)} \tag{C.17}$$

3. Powers

$$(10^x)^n = 10^{nx} \tag{C.18}$$

4. Roots

$$(10^x)^{1/n} = 10^{x/n} \tag{C.19}$$

C.3.3.6 Natural Logarithms

Relationships between natural logs and common logs are based on following a few fundamentals.

1. Values of natural logs, ln,

$$10 = e^{2.3026} \tag{C.20}$$

   Taking natural logs, both sides,

$$\ln 10 = 2.3026 \tag{C.21}$$

2. Conversion from log X and ln X. Let $X = 10$. Therefore,

$$C \log 10 = \ln 10 \tag{C.22}$$

$$C \cdot 1.0 = 2.3026 \tag{C.23}$$

$$C = 2.3026 \tag{C.24}$$

   Therefore,

$$2.3026 \log X = \ln X \tag{C.25}$$

   Example for $X = 5$:

$$2.3026 \log 5 = \ln 5 \tag{C.26}$$

$$2.3026 \cdot 0.69897 = 1.6094 \tag{C.27}$$

   A logarithm to the base e, designated ln, may be converted to a common log, or vice versa, by the relationship

$$X = e^Y = 10^{kY} \tag{C.28}$$

$$= 10^{Y/2.3026} \tag{C.29}$$

   Suppose $X = 5$, then

$$5 = e^{1.6094} = 10^{1.6094/2.3026} = 10^{0.69897} \tag{C.30}$$

C.3.3.7 Application to Problems

One of the common problems has to do with a semi-log plot, with the log scale based on common logs. The coefficients, i.e., slope and intercept must yield data in terms of natural logs.

1. *Best-fit equation for semi-log plot.* Suppose the best-fit equation of data from an experimental plot is

$$\log Y = \text{slope} \cdot X + \log B \tag{C.31}$$

   Multiply both side by 2.303:

$$2.303 \log Y = 2.303 \cdot \text{slope} \cdot X + 2.303 \cdot \log B \tag{C.32}$$

$$\ln Y = 2.303 \cdot \text{slope} \cdot X + \ln B \tag{C.33}$$

   The "slope" is from the log–log plot and is in log-cycles of Y per unit of X.

2. *Illustration in terms of van't Hoff equation.* To illustrate in terms of a common equation, i.e., the van't Hoff equation, for the effect of temperature on the equilibrium constant, i.e.,

$$K = \frac{C_{vh} e^{\wedge} \Delta H^\circ}{RT} \left(\text{or, } \ln K = -\Delta H^\circ/RT + \ln C_{vh}\right)$$

   If the experimental data are plotted as a semi-log relationship, i.e., log K vs. $1/T$, the slope is, $\Delta H^\circ/2.303R$ (ΔH° is the standard state enthalpy of reaction and R is the universal gas constant). Therefore

$$2.303 \cdot \text{slope} = \frac{-\Delta H^\circ}{R} \tag{C.34}$$

Substituting in (C.33):

$$\ln Y = \frac{-\Delta H^\circ}{R \cdot X + \ln B} \tag{C.35}$$

which answers the often nettlesome question as to whether to multiply or divide by 2.303. In terms of the van't Hoff equation, Equation C.35 is

$$\ln K = \frac{-\Delta H^\circ}{RT + \ln C_{vh}} \tag{C.36}$$

3. *Intercept*. To determine B, when $Y = 1.0$, $\log Y = \log 1.0 = 0$ and the intercept, i.e., where the relationship crosses $Y = 1.0$, is

$$0 = \frac{-\Delta H^\circ}{R \cdot X + \ln B} \tag{C.37}$$

For the intercept, log B, when

$$\frac{\Delta H^\circ}{R \cdot X} = 0, \quad \ln Y = \ln B \tag{C.38}$$

TABLE C.2
Time and Concentration Data (Hypothetical Lab Study)

t (s)	C (mg/L)	C/C_o
0.0	800.0	1.00000000
1.0	485.2	0.60653262
2.0	294.3	0.36788182
3.0	178.5	0.22313233
4.0	108.3	0.13533704
5.0	65.7	0.08208633
6.0	39.8	0.04978804
7.0	24.2	0.03019807
8.0	14.7	0.01831611
9.0	8.9	0.01110932
10.0	5.4	0.00673817

4. *Mathematical relation*. The functional relationship, once $\Delta H^\circ/R$ and B are found, is

$$Y = B \cdot e^{\frac{\Delta H_o}{R}X} \tag{C.39}$$

$$= B \cdot 10^{\frac{\Delta H_o}{2.303 R}X} \tag{C.40}$$

Example C.3 Analysis of Data that Exhibit an Exponential Decline with Time

Kinetic data, for a batch reactor are given in Table C.2, i.e., time in the left-hand column, concentration of a reacting solute in the middle column, and the calculated C/C_o in the right-hand column. Figure C.2 is a plot of C/C_o vs. t, plotted on arithmetic scales. The same data are re-plotted in Figure C.3, a semi-log plot. The slope is seen as one log cycle over 4.6 s, i.e., slope $= -0.2174$ cycles/s. (The kinetic constant, always in terms of natural log or e, is $k = \text{slope} \cdot 2.3026 = 0.2174$ cycles/s $\cdot$ $2.3026 = 0.50\ \text{s}^{-1}$.)

The slope in the semi-log plot, Figure C.2b, is 0.2174 cycles/s. From Figure C.2b

$$\text{Slope} = \frac{k}{2.3026} \tag{ExC.3.1}$$

$$-0.2174\ \text{cycles/s} = \frac{k}{2.3026} \tag{ExC.3.2}$$

$$k = -0.50\ \text{s}^{-1} \tag{ExC.3.3}$$

From Equation C.31:

$$\log\left(\frac{C}{C_o}\right) = \left(\frac{k}{2.3026}\right) \cdot t + \log B \tag{ExC.3.4}$$

$$\log(1.0) = -0.2174\ \text{cycles/s} \cdot 0 + \log B \tag{ExC.3.5}$$

$$0 = 0 + \log B \tag{ExC.3.6}$$

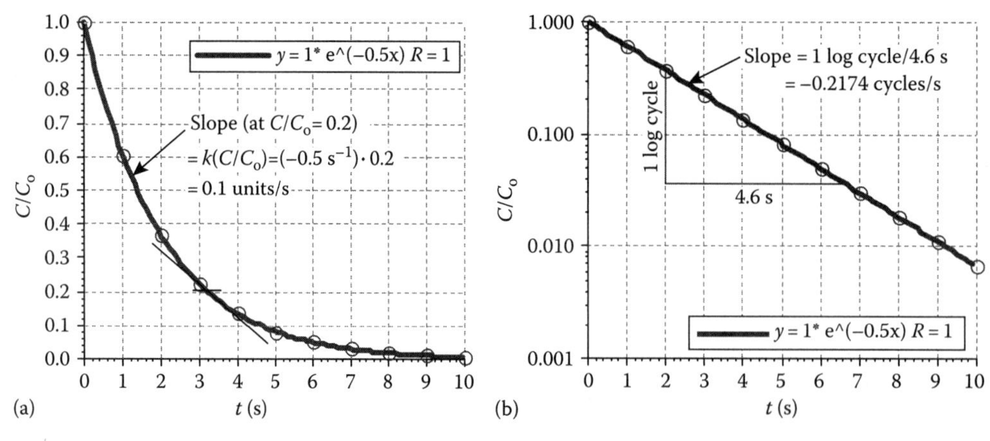

FIGURE C.2 Plot of Table C.2 data, showing best-fit curves, illustration of slopes, and method of determining slope and exponent. (a) Arithmetic plot; (b) semi-log plot.

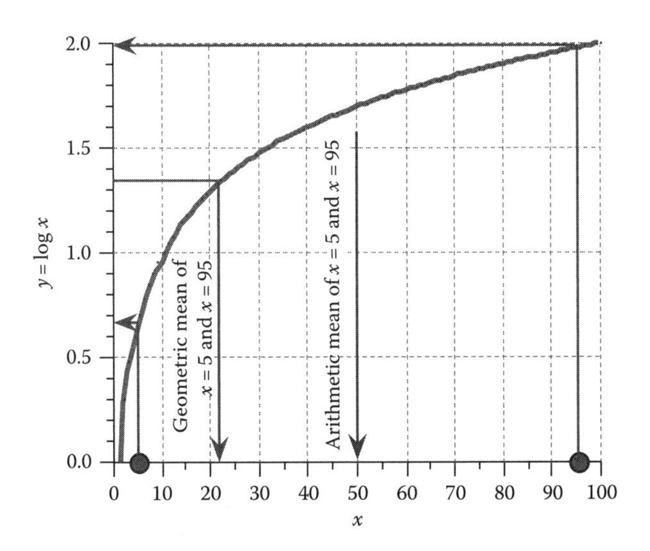

FIGURE C.3 Illustration of how geometric mean biases result toward lower result than given by data. (From Parkhurst, D.F., *Environmental Science and Technology*, 32(3), 92A, 1998.)

C.4 STATISTICS

C.4.1 Geometric Mean

Frequently, data are reported in terms of their "geometric means." Often this parameter is referred to in regulations. Geometric mean is defined as the antilog of the mean of the logarithms of a set of numbers (Parkhurst, 1998).

Parkhurst (1998) makes the case that the geometric mean has no rationale for its use and the arithmetic mean is a more accurate estimate of the population mean of any sample. The basic problem with the geometric mean is that it biases the mean toward the lower data values in a set as the larger values contribute less toward the geometric mean. This is illustrated in Figure C.3 that shows $0 < x < 100$ with the corresponding y values equal to log x.

C.4.2 Log Normal Distribution

A log normal distribution is defined (Parkhurst, 1998) as occurring when the logarithms of a set of numbers have a normal distribution.

C.4.3 Correlation

Suppose we have two tabulated values of x and y. And suppose that numerous values of y have been found associated with any one value of x. We cannot simply write $y = f(x)$. But the mean, $\bar{y}$, of the y values associated with any x, may vary with x in a fairly definite manner. Then, although individual predictions as to the y value to be expected with a given x will be subject to considerable uncertainty, we can determine whether, on the average, large or small values of y tend to go with large values of x. We can, in fact, make individual predictions with smaller average errors than if we guessed at random. The idea is expressed by saying that there is some "correlation" between x and y (Griffin, 1936).

C.5 SIEVE ANALYSIS

C.5.1 U.S. Standard Sieve Sizes

Sieves for the size distribution analysis of granular media are of two kinds, generally: (1) U.S. Standard sieve, and (2) Tyler series. Often, plotting paper is not available. Figure C.4 is a plot layout for plotting size distribution data based on U.S. Standard sieves. The data plot is most commonly as a straight line. In the analysis of granular media for filters, the d_{60} an d_{10} sizes are of greatest interest.

C.6 COST INDEXES

In places where cost data are used, the date of construction is given also (as a rule). This permits cost updating by the use of cost indexes. Using the ENR Construction Cost Index (CCI), for example

$$\text{Cost(present year)} = \text{Cost(year of construction)} \cdot \frac{\text{CCI(present year)}}{\text{CCI(year of construction)}} \quad \text{(C.41)}$$

in which
Cost(present year) = constructed cost of a given facility in current year
Cost(year of construction) = constructed cost of a given facility in year construction was completed (dollars)
CCI(present year) = ENR Construction Cost Index for present year (no units)
CCI(year of construction) = ENR Construction Cost Index for year construction was completed for the given facility (no units)

To provide a means for such updating is the purpose of reviewing the topic of cost indexes here. The application of a simple ratio may be the basis for getting an idea of the current cost of a given facility, it is also simplistic and its use should be limited to obtaining an initial estimate. Engineering cost estimating requires more depth of knowledge than given here, and experience in the use and interpretation of various kinds of cost indexes.

C.6.1 Caveats on Cost Indexes

As a caveat (complementing statements in the previous paragraph), one should be aware that in water treatment, factors comprising costs may change in ways that are different than some of the indexes. Membranes, e.g., were a new technology in 1970; the technology has evolved since that date, demand has increased, and prices have come down. The building that houses a primarily membrane plant may be

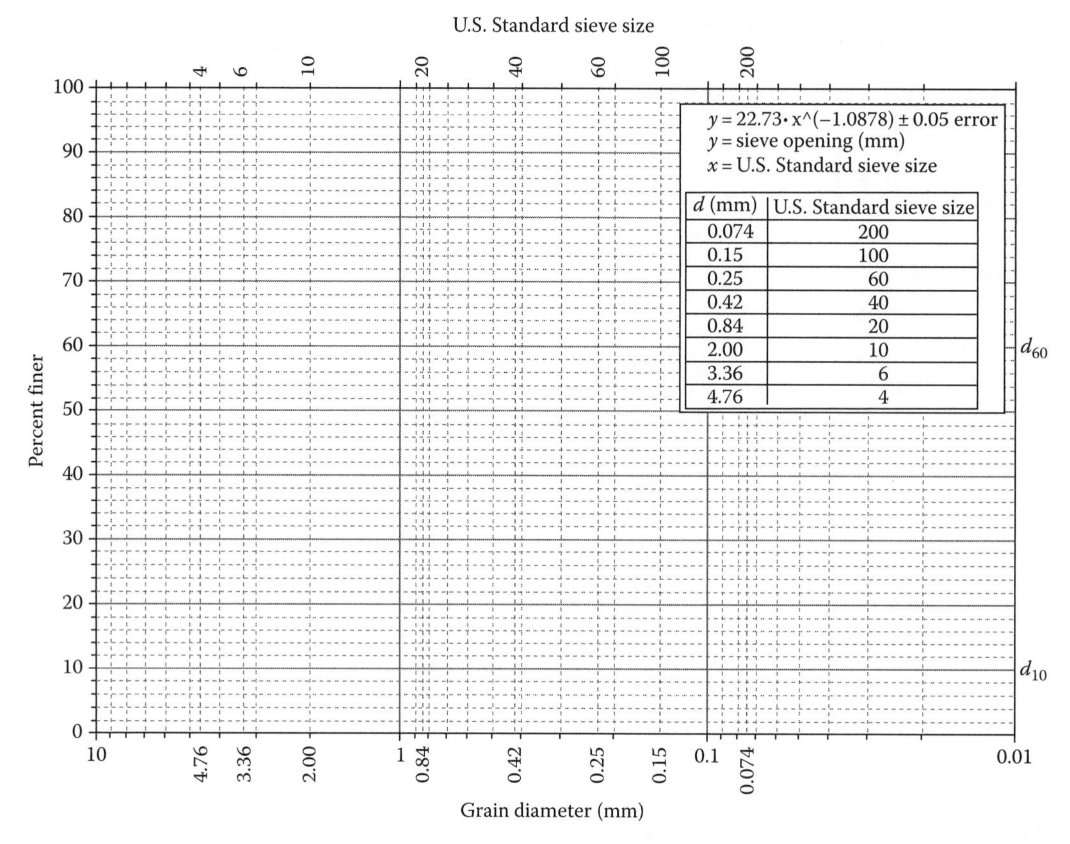

FIGURE C.4 Plotting form for sieve analysis data based on U.S. Standard sieves. (Sowers, G.B. and Sowers, G.F., Introductory Soil Mechanics and Foundations, Macmillan, New York, 1951.)

mostly a large "box," albeit pipe galleries, storage reservoirs, pumps, partitioned areas, etc., are still needed. For small installations, "package" plants that come directly from a manufacturer may be used. For larger plants, manufactured products may still comprise a significant part of the total cost. Thus, more specialized cost data are probably more appropriate for treatment plants, albeit they are less available that those published by ENR.

C.6.2 Kinds of Cost Indexes

Several sources of cost indexes are available and are compiled on a regular basis. In some cases, cost data for water treatment have been compiled for different regions or cities and as a function of size of plant (EPA has had such data compiled under contract and published from time to time). The ENR provides several kinds of cost data, e.g., the "common labor index," the "skilled labor index," the "building cost index" (BCI), and the "construction cost index" (CCI). The ENR also provides cost data averaged for 20 U.S. cities for various building materials (see, e.g., ENR, 2002, p. 35), for sewer, water, and drain pipe (see, e.g., ENR, 2003, p.19), etc.

The CCI was established in 1921 as a general-purpose tool to chart basic cost trends. The base year was 1913 for which the initial CCI was set at \$100 and was the value of a hypothetical package of goods with proportions: structural steel (0.11), portland cement (0.01), lumber (0.08), and common labor (0.80).

The BCI was introduced in 1938 to weigh the effect of skilled labor trades on construction costs. The base year is also 1913 and was set at \$100 and was the value of a hypothetical package of goods of structural steel (0.19), portland cement (0.03), lumber (0.14), and common labor (0.64).

C.6.3 ENR 20-City CCI and 20-City BCI

The ENR 20-city CCI and the ENR 20-city BCI are updated in the first issue of ENR each month (ENR is a weekly publication). The historical summaries are given in the quarterly cost report, published usually in the last issue of March, June, September, and December (ENR, 2002, p. 79). Table C.3 provides a summary of the CCI and BCI data obtained from ENR (2002). The footnotes for Table C.3 give the basis for both the CCI and the BCI. As noted, the CCI and the BCI differ in the labor

TABLE C.3
Historical ENR 20-City CCI and 20-City BCI

Year	CCI[a]	BCI[b]	Year	CCI[a]	BCI[b]
1910			1960	824	559
1911			1961	847	568
1912			1962	872	580
1913	100		1963	901	594
1914	89		1964	936	612
1915	93		1965	971	627
1916	130		1966	1019	650
1917	181		1967	1074	676
1918	189		1968	1155	721
1919	198	159	1969	1269	790
1920	251	207	1970	1381	836
1921	202	166	1971	1581	048
1922	174	155	1972	1753	1048
1923	214	186	1973	1895	1138
1924	215	186	1974	2020	1205
1925	207	183	1975	2212	1306
1926	208	185	1976	2401	1425
1927	206	186	1977	2576	1545
1928	207	188	1978	2776	1674
1929	207	191	1979	3003	1819
1930	203	185	1980	3237	1941
1931	181	168	1981	3535	2097
1932	157	131	1982	3825	2234
1933	170	148	1983	4066	2384
1934	198	167	1984	4146	2417
1935	196	166	1985	4195	2428
1936	206	172	1986	4295	2483
1937	235	196	1987	4406	2541
1938	236	197	1988	4519	2598
1939	236	197	1989	4615	2634
1940	242	203	1980	4732	2702
1941	258	211	1991	4835	2751
1942	276	222	1992	4985	2834
1943	290	229	1993	5210	2996
1944	299	235	1994	5408	3111
1945	308	239	1995	5471	3111
1946	346	262	1996	5620	3203
1947	413	313	1997	5826	3364
1948	461	341	1998	5920	3391
1949	477	352	1999	6059	3456
1950	510	375	2000	6221	3539
1951	543	401	2001	6334	3574
1952	569	416	2002	6538	3623
1953	600	431	2003	6695	3694
1954	628	446	2004	7115	3984
1955	660	469	2005	7446	4205
1956	692	491	2006	7751	4369
1957	724	509	2007	7967	4486
1958	759	525	2008	8310	4691
1959	797	548	2009	8570	4769

Source: Grogan, T., Building Cost Index History, 1923–2010, Engineering News Record, 264(10): 60, March 29, 2010.

[a] The CCI comprised 200 h of common labor at the 20-city average, plus 25 cwt of standard structural steel shapes, plus 1128 ton of portland cement at the 20-city price, plus 1088 board-ft of 2 × 4 lumber at the 20-city price.

[b] The CCI comprised 68.38 h of skilled labor (bricklayers, carpenters, and structural iron workers) at the 20-city average, plus 25 cwt of standard structural steel shapes, plus 1128 ton of portland cement at the 20-city price, plus 1088 board-ft of 2 × 4 lumber at the 20-city price.

components; the CCI incorporates "common labor" and the BCI incorporates "skilled labor." The materials components are the same for each index.

The 20 cities, with respective March 2002 CCI data, include: Atlanta (2943), Baltimore (3101), Birmingham (2825), Boston (4038), Chicago (4256), Cincinnati (6173), Cleveland (3624), Dallas (2683), Denver (3029), Detroit (4026), Kansas City (6482), Los Angeles (3695), Minneapolis (3758), New Orleans (2670), New York (5299), Philadelphia (4353), Pittsburgh (3583), St. Louis (3541), San Francisco (4031), Seattle (3738). As seen, the CCI vary from a low of 2671 for New Orleans to a high of 6173 for Cincinnati.

As stated in Equation C.41, cost may be updated by taking the ratio of the CCI for the current year and a given past year (or alternatively, the BCI may be used). Such methodology is subject to the caveats noted in C.41.

ACKNOWLEDGMENT

Dr. John Clark, Sear-Brown Consulting Engineers, Fort Collins, Colorado, provided the discussion of engineering economics and advice on using the ENR index.

REFERENCES

ENR, First Quarterly Cost Report, *ENR*, 248(10):64–83, March 18, 2002.

ENR, Construction Economics, Cost Indexes, *ENR*, 250(14):19–20, April 14, 2003.

Griffin, F. L., *An Introduction to Mathematical Analysis*, revised edition, Houghton Mifflin Company, New York, 1936.

Parkhurst, D. F., Arithmetic versus geometric means for environmental concentration data, *Environmental Science and Technology*, 32(3):92A–98A, February 1, 1998.

Sowers, G. B. and Sowers, G. F., *Introductory Soil Mechanics and Foundations*, Macmillan, New York, 1951, p. 19.

Appendix D: Fluid Mechanics—Reviews of Selected Topics

Fluid mechanics is pertinent in the design of nearly all unit processes. Topics of particular interest include drag, pipe flow (incompressible and compressible), flow distribution by manifolds, porous media flow, flow measurement, turbulence, and dispersion. These areas are reviewed here to provide a ready reference for key equations and associated coefficients. In the case of compressible flow, the logic is outlined as well as the final equations. The review is intended to provide reference for frequently sought equations, but may not be readily accessible. The equations may also be useful for spreadsheet applications.

D.1 FLUID DRAG

Drag is a fundamental notion of fluid mechanics and it permeates many kinds of problems. Examples include sedimentation, sediment transport, headloss in pipes, pump design, airfoil design, design of mixers, energy dissipation, etc. Problems of particular interest in unit operations include fall velocity of particles, paddle wheel design in flocculation, mixing, etc.

D.1.1 Drag Equations

Drag occurs when a fluid in motion contacts a surface of a "body," e.g., an impeller, a pipe wall, an airfoil, an automobile, a solid particle settling in water, a gas bubble rising in water, ad infinitum. The two types of drag are friction drag and form drag. Both occur in most situations and the total drag on a "body" is the sum of the two.

The *friction* drag is due to boundary layer shear. Headloss in a pipe, calculated by the Darcy–Weisbach equation is caused by friction drag. The resistance of an airfoil is due mostly to friction drag. The other kind of drag is *form* drag. The "lift" on an airfoil is due to form drag. A section of an axial flow impeller is really an "airfoil" and the associated pumping is due to lift. The pressure profile caused by flow around a body causes form drag (a "streamline" shape of an object reduces form drag). The shape of the body determines the pressure profile.

A flat rectangular plate provides a means to illustrate the two types of drag. If the plate is oriented with its surface parallel to the flow, the drag on the plate is caused by friction. If the plate is oriented with its surface perpendicular to the flow, the drag is caused by separation effects, called form drag. A pressure profile around the plate would show a nearly uniform positive pressure on the upstream face and a nearly uniform negative pressure on the downstream face. The additive effect of this pressure profile, i.e., the integration of the infinitesimal forces over the whole surface, causes a resultant force opposite to the plate velocity vector.

The calculation of the drag of a body is given by

$$F_{\mathrm{D}} = C_{\mathrm{D}}\rho\frac{v^2}{2}A \qquad \text{(D.1)}$$

in which

F_{D} is the drag force on body (N)
C_{D} is the coefficient of drag (dimensionless)
A is the area of body projection perpendicular to the direction of motion (m^2)
v_{o} is the velocity of body relative to fluid (m/s)
ρ is the density of fluid ($\mathrm{kg/m}^2$)

Equation D.1 is applicable to both friction drag and form drag. The coefficient, C_{D}, is quitc different in each case.

D.1.2 Flow Regime

The drag coefficient depends upon the shape of the body and the Reynolds number, i.e., **R**, defined,

$$\mathbf{R} = \frac{\rho v D}{\mu} \qquad \text{(D.2)}$$

in which

R is the Reynolds number (dimensionless)
v is the velocity of body relative to fluid (m/s)
D is the characteristic length (m)
μ is the dynamic viscosity ($\mathrm{N\ s/m}^2$)

The length term, D, in Equation D.2 is selected to characterize the system, e.g., pipe diameter, impeller diameter, etc. At low Reynolds numbers, i.e., $\mathbf{R} < 10$, the flow regime is laminar. As **R** increases, the flow becomes transitional, and then as **R** increases further, the regime is fully turbulent. The specific Reynolds number at each transition depends upon the kind of shape.

D.1.3 Drag Coefficients

As noted in Section D.1.2, drag coefficients depend on (1) the form (or shape) of the body and (2) the Reynolds number, **R**, associated with the velocity of the fluid. The drag coefficient, C_{D}, must be determined by experiment, using a wind tunnel, a towing tank, channel, or any other means to measure the drag force and fluid velocity. The fluid may be air, water, oil, alcohol, etc. The kind of fluid used does not affect the C_D versus **R** relationship.

Figure D.1 shows the drag coefficients for several bodies over a range, $0.1 \leq \mathbf{R} \leq 10^6$. As seen, at $\mathbf{R} < 10$, the flow regime is laminar with the drag coefficient calculated,

$$C_D = \frac{24}{\mathbf{R}} \tag{D.3}$$

Equation D.3 is applicable to all shapes, i.e., spheres, ellipse shaped rods, cylinder rods, ellipsoids, etc. As **R** increases, e.g., toward 10, the deviation from the 24/**R** line increases; the transition to the turbulent regime occurs from $10 < \mathbf{R} < 10^3$. At $\mathbf{R} > 10^3$, the regime is fully turbulent and the C_D's do not change with **R**, with exceptions noted at $\mathbf{R} > 10^5$ for certain shapes, as seen in Figure D.1. Thus, Figure D.1 shows that at $\mathbf{R} > 10^3$, the C_D's do not change with **R**, except as seen, and depend on shape.

D.1.4 Friction Coefficient

The Darcy–Weisbach equation was introduced by Julius Weisbach in about 1850 (Rouse and Ince, 1957, p. 163) for determining headloss due to fluid friction,

$$h_L = f\frac{L}{D}\frac{v^2}{2g} \tag{D.4}$$

in which

h_L is the headloss pipe due to friction (m)
L is the length of pipe (m)
D is the diameter of pipe (m)
v is the velocity of fluid in pipe (m/s)
g is the acceleration of gravity (m/s^2)
f is the friction coefficient from Moody diagram (dimensionless)

A major breakthrough in the use of the Darcy–Weisbach equation was that the friction factor, **f**, was related to the Reynolds number, **R**, by Blasius in 1913, which was adopted into engineering practice about the mid-1930s (Rouse, 1944). J. Nikuradse, one of Prandtl's students, later (in a 1933 paper) related **f** to relative pipe roughness, ε/D, and **R** by gluing sand grains of different uniform sizes in pipes of different diameters and performing experiments. Then in 1937 in Great Britain, C.F. Colebrook and C.M. White adapted the diagram of Nikuradse to commercial pipe (Rouse, 1943, 1946). Rouse (1943, p. 100) combined variables, patterned after Colebrook and White in a fashion that retained the fundamental character and resulted in a diagram that plotted $1/\sqrt{\mathbf{f}}$ versus $\mathbf{R}/\sqrt{\mathbf{f}}$ and ε/D, with ε values given for commercial pipe. The supplementary scales on the right side and at the top of the diagram were **f** and **R**, respectively. Professor Lewis F. Moody was in the audience when Rouse presented his paper in 1942 (at the Second Iowa Hydraulics Conference) and suggested that the results would be more useful for the practicing engineer by using only **f** and **R** for the plot. Rouse felt that to do so would be a retrogression as this would deviate from the spirit of maintaining an adherence to fundamentals (Rouse, 1976, p. 161). Professor Moody subsequently reworked the diagram and published it in 1944 in the *Journal of the American Association of Mechanical Engineers*. Rouse (1944) wrote a discussion of

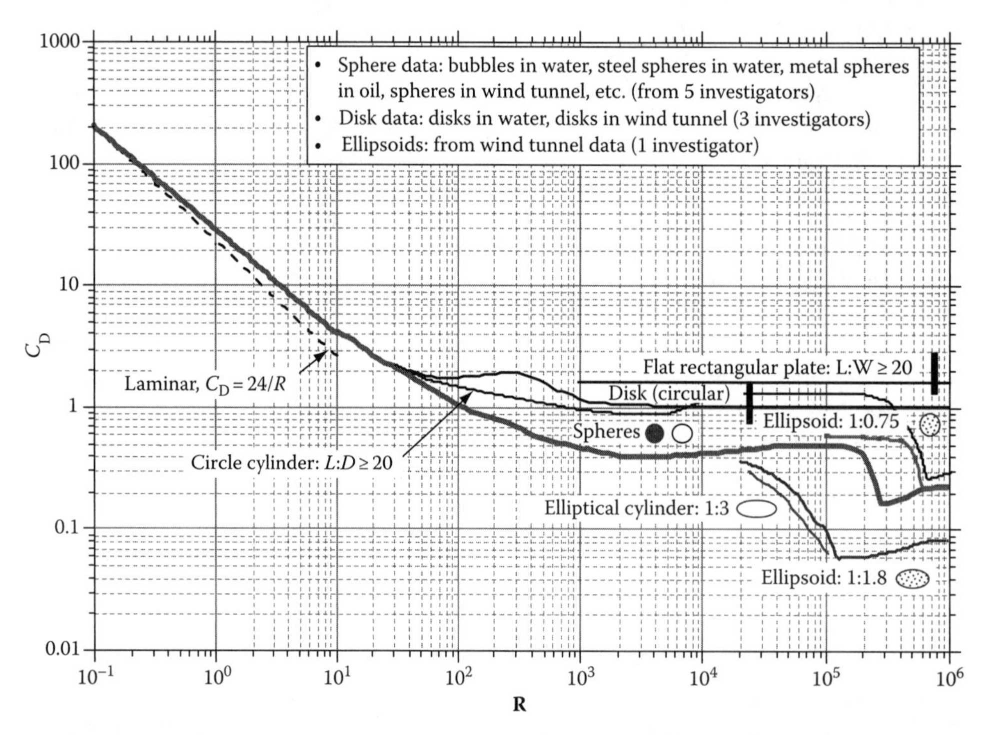

FIGURE D.1 Drag coefficients for different shapes as a function of Reynolds number, **R**. (Adapted from Rouse, H., *Elementary Mechanics of Fluids*, John Wiley & Sons, New York, p. 245, 1946).

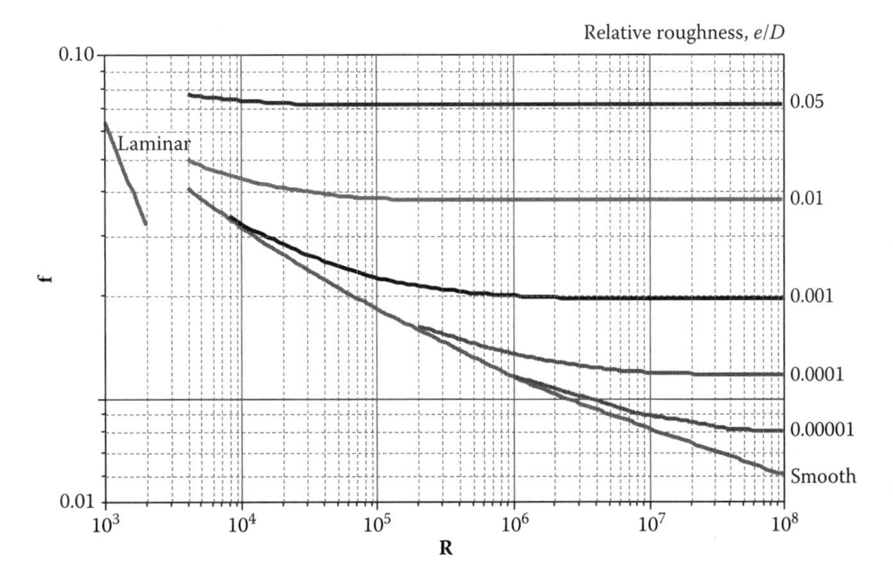

FIGURE D.2 Moody diagram **f** or friction factor as function of **R** for different values of relative roughness. (From Albertson, M.L. et al., *Fluid Mechanics for Engineers*, Prentice-Hall, Englewood Cliffs, NJ, p. 269, 1960).

Moody's paper, indicating his objections. The pipe friction diagram as adapted by Moody became well known worldwide and is known today as the "Moody diagram."

Figure D.2 is the Moody diagram for pipe friction, which shows how pipe friction varies with **R** and relative roughness, ε/D. Table D.1 gives "ε" values for various materials. Thus, to determine the friction factor, first determine the kind of material used and then select an "ε." For a selected **R** and a known (or assumed) pipe diameter, ε/D is calculated. From these "arguments," the friction factor, "**f**" is obtained from Figure D.2. After some years of use a pipe for water flow will contain mineral or bacterial deposits, which will alter "ε" (and pipe diameter). This change in "ε" should be considered in pipe sizing. Alternatives to the Darcy–Weisbach equation have included the Chezy equation and Hazen and Williams equation. The *C* coefficients for these equations are available in various handbooks. The Darcy–Weisbach is the preferred relation, albeit practice seems to employ, to a large extent, the latter two empirical equations (Box D.1).

TABLE D.1
Roughness Values for Different Conduit Materials

Material	ε[a] (ft)	ε[a] (mm)
Glass, drawn brass, copper, lead	Smooth	Smooth
Commercial steel, wrought iron	0.0001–0.0003	0.03–0.09
Asphalted cast iron	0.0002–0.0006	0.06–0.18
Galvanized iron	0.0002–0.0008	0.06–0.24
Cast iron	0.0004–0.002	0.12–0.61
Wood stave	0.0006–0.003	0.18–0.91
Concrete	0.001–0.01	0.30–3.0
Riveted steel	0.003–0.03	0.91–9.1
Corrugated metal pipe	0.1–0.2	30–61
Blasted rock tunnels	1.0	300

Source: Data from Albertson, M. L. et al., *Fluid Mechanics for Engineers*, Prentice-Hall, Englewood Cliffs, NY, 1960, p. 268.

[a] The "ε" is a roughness parameter for different pipe materials.

A mathematical relation that incorporates e/D and **R** to predict **f** is the Colebrook function (Daugherty and Ingersoll, 1954, p. 181; Munson et al. 1998, p. 494),

$$\frac{1}{\sqrt{\mathbf{f}}} = -2\log\left(\frac{e/D}{3.7} + \frac{2.51}{\mathbf{R}\sqrt{\mathbf{f}}}\right) \quad (D.5)$$

with a simplified version derived by Moody, which agrees with the former to within 5%,

$$f = 0.0055\left[1 + \left(20{,}000\frac{e}{D} + \frac{10^6}{\mathbf{R}}\right)\right] \quad (D.6)$$

Munson et al. (1998, p. 541) gave another form for Equation D.6,

$$\mathbf{f} = \frac{1.325}{\left\{\ln\left[(\varepsilon/3.7D) + \left(5.74/\mathbf{R}^{0.9}\right)\right]\right\}^2} \quad (D.7)$$

D.2 FLUID FLOW IN PIPES

The basis for pipe flow analysis is two equations: (1) Bernoulli and (2) Darcy–Weisbach. These two equations along with the continuity principle are the basis for solving practical problems. The idea of fluid shear is established first, however, since friction loss is caused by shear and the resulting Poiseuille equation is useful subsequently in helping to understand Darcy's law in porous media flow.

BOX D.1 REMINISCENCES ON PROFESSOR HUNTER ROUSE

Professor Hunter Rouse (1906–1996) was one of the traditional academics (defined here as those in the generations that preceded the author's). The author was a student in Rouse's 1961–1962 first-year fluid mechanics graduate course. Rouse followed his book closely, and, consistent with his book, there was little, if any, mention of the Moody diagram. Most probably, this was because his view was to advance knowledge in terms of fundamentals rather than by what seemed expedient (consistent with the discussion in the previous paragraph). He was Director of the Iowa Institute of Hydraulic Research (now IIHR-Hydroscience and Engineering) at the University of Iowa, known worldwide as the "Mecca" for attracting engineers and scholars in the field. As to persona, Rouse was always dignified and, to a student, seemed Teutonic. Maybe his time at Karlsruhe (Technische Hochschule, doctorate in 1932) had an influence; but whatever, Rouse seemed a natural German. His lectures were usually one or two simple questions that related to the topic of the day as described in his book and that confounded most of the students; usually, however, one or two students had the correct approach. The students did most of the talking, usually interspersed with an incisive question by Professor Rouse. Sometimes, he introduced a topic or gave a brief discourse, but seldom a full lecture. He expected the participation of all of the graduate students, writing on a pad the performance of each student during the session (one could only speculate on what he had written). A student could not do well, however, merely by reading his book. Each of his class sessions was an experience. Later, after he retired from Iowa (as Dean of Engineering), he came to CSU for the summers to give a shorter version of his course. Perhaps this was due to the influence of Professor Albertson, who was one of Rouse's doctoral students.

Hunter Rouse, c. 1960 (from the Archives IIHR—Hydroscience and Engineering, University of Iowa; used with permission)

Students who took his course seemed to sweat as they did at Iowa; I always advised my students to take the course, first to have the "experience" of Rouse, and second, because as an environmental engineer, one cannot know too much in the field of fluid mechanics. Rouse's writings were concise, eloquent, insightful, and substantive, i.e., what one should strive to emulate. This was true of his spoken word as well, i.e., articulate and always the model of decorum, never flamboyant. As a note, the formal style of academics in Rouse's generation was a contrast with the more relaxed approach of the subsequent generation in the United States (but probably not in Europe, i.e., from what the author has seen).

A more complete review of Professor Rouse's contributions in teaching, research, and writing was provided by Professor Robert Ettema (2006) and by Kennedy and Macagno (1971). Professor Ettema's paper (Ettema, 2006) brought to mind that Rouse was a perfect person for the "golden age" of hydraulic research, when the foundation for the modern state of the art was being established, starting with Prandtl, c. 1901, which was largely in place by the 1960s. A large proportion of the academics that continued hydraulic research and teaching in the United States and worldwide, e.g., from perhaps 1940, were from Iowa as were many of the hydraulic engineers who were doing hydraulic modeling and designing hydraulic structures. Rouse's contributions in teaching, research, writing, and leadership have influenced our understanding of the modern state of the art of fluid mechanics and hydraulics.

The foundation laid at Iowa in experimental hydraulics was the basis for a continuation in the form of computer modeling, started c. 1960, at IIHR using Fortran and mainframe computers and known since the early 1990s as computational fluid dynamics (CFD). The main task was to solve the classical Navier–Stokes equation by finite difference, which opened a new epoch in fluids modeling and, at the same time, became a tool of practice. The Navier–Stokes relation is a differential equation, intimidating at first glance, but was less so after Rouse's brief explanation, i.e., that the equation was merely Newton's second law, $F = ma$, in an expanded form. This characterized Rouse's style, which was to reduce a complex topic to its simple essence.

D.2.1 Fluid Shear

The equation for fluid shear in viscous flow is

$$\tau = \mu \frac{dv}{dy} \tag{D.8}$$

in which

τ is the shear stress (N/m^2)
μ is the dynamic viscosity ($N\ s/m^2$)

v is the velocity at any y (m/s)
y is the distance from a boundary along the y-coordinate and perpendicular to flow (m)

The shear is best seen on a cubic element in two dimensions as the rate of deformation. By means of a free body diagram shown in Figure D.3 for an infinitesimal fluid element, cubic but drawn in two dimensions, the relationship between pressure gradient in the x-direction and shear gradient in the y-direction is (Rouse, 1946, p. 154)

$$\frac{\mathrm{d}p}{\mathrm{d}x} = \frac{\mathrm{d}\tau}{\mathrm{d}y} \tag{D.9}$$

in which
p is the pressure (N/m^2)
x is the distance along the x-coordinate parallel to flow direction (N s/m^2)

From Equations D.8 and D.9, the velocity profile between two parallel boundaries can be shown to be parabolic. Applied to pipe flow, the Poiseuille equation is obtained, i.e.,

$$\frac{\mathrm{d}p}{\mathrm{d}x} = \frac{32\mu\bar{v}}{d^2} \tag{D.10}$$

or

$$p_1 - p_2 = \frac{32\mu v L}{d^2} \tag{D.11}$$

in which
$\bar{v}$ is the average velocity in pipe (m/s)
d is the diameter of pipe (m)
p_1 is the pressure in pipe at position "1" (N/m^2)
p_2 is the pressure in pipe at position "2" downstream in the x-direction (N/m^2)
L is the length of pipe between position "1" and position "2" (m)

For two dimensions, Shapiro (1958, p. 9) provides a definition sketch, Figure D.4, which shows that the velocity in the x-direction is u and changes with respect to y, while the velocity in the y-direction is v and changes with respect to x. Also, $\tau_{yx} = \tau_{xy}$ to give

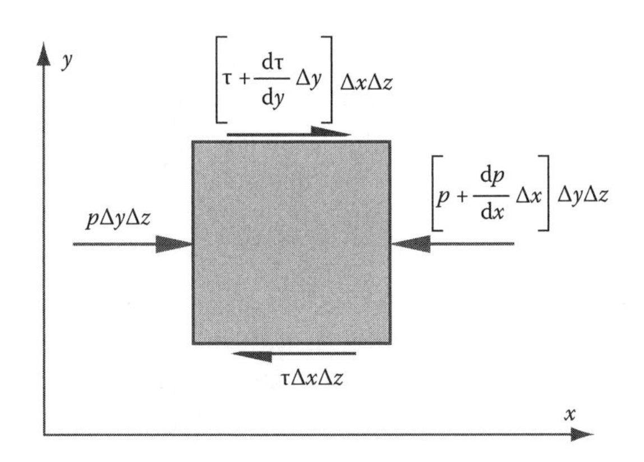

FIGURE D.3 Definition sketch.

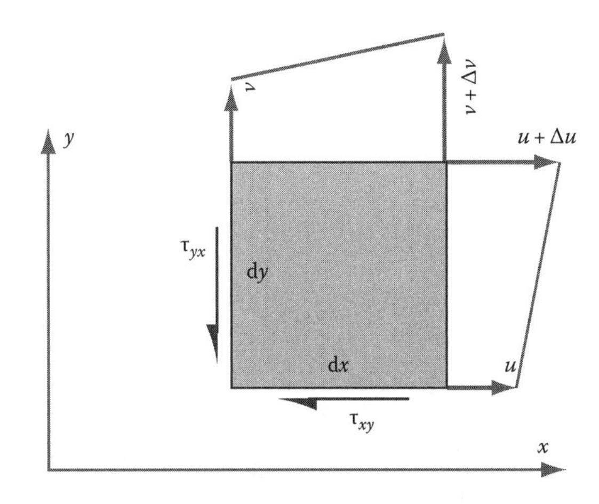

FIGURE D.4 Shear deformation.

$$\tau_{yx} = \tau_{xy} = \mu\left(\frac{\partial u}{\partial y} + \frac{\partial v}{\partial x}\right) \tag{D.12}$$

Camp and Stein (1943, p. 210 Civil Engineering Classics) show how these relations are expanded to three dimensions, i.e.,

$$\Phi = \mu\left[\left(\frac{\partial u}{\partial y} + \frac{\partial v}{\partial x}\right)^2 + \left(\frac{\partial u}{\partial z} + \frac{\partial w}{\partial x}\right)^2 + \left(\frac{\partial v}{\partial z} + \frac{\partial w}{\partial y}\right)^2\right] \tag{D.13}$$

in which
Φ is the work of shear per unit volume per unit of time at a point (J/s/m^3)
u, v, w are velocity components in x, y, z directions, respectively (m/s)

The three-dimensional relation is referred to by Camp as the work of shear per unit volume at a point, i.e., power per unit volume of fluid. The function is used in the development of mixing theory as outlined by Camp, i.e., $(\Phi/\mu)^{0.5}$ = velocity gradient, or G, as seen in mixing theory. Equation D.10 can be seen better in one dimension, in which $\Phi = \tau(\mathrm{d}v/\mathrm{d}y) = [\mu(\mathrm{d}v/\mathrm{d}y)](\mathrm{d}v/\mathrm{d}y) = \mu(\mathrm{d}v/\mathrm{d}y)^2$.

D.2.2 Materials Balance

A second major principle for any flow is that there must be an accounting of material fluxes, which is the conservation of mass. There are two conditions are for the general case of unsteady flow, i.e., flow changes with time and, the usual case of steady flow, i.e., no change with time. For the latter, we can say that the mass flux at any point in the pipeline equals the mass flux at any other point. Further, if the fluid is incompressible, i.e., density is constant, then the volumetric flow is constant.

D.2.2.1 Incompressible Flow

The conservation of mass principle for the steady state condition is

$$Q = Av\rho = \text{constant} \tag{D.14}$$

in which

Q is the flow in pipe (m^3/s)
A is the cross-sectional area of pipe (m^2)
v is the mean velocity within pipe (m/s)
ρ is the density of fluid within pipe (kg/m^3)

For any conduit, Q is constant, and if A changes, for example, from one pipe size to another, then $v\rho$ changes correspondingly. For an incompressible fluid, e.g., water, ρ is constant and the continuity equation is simply $Q = Av = \text{constant}$. In other words, $Av\rho$ is the same from one section to another for flow in a pipeline. For example, if A changes, v must change correspondingly, assuming ρ does not change.

D.2.2.2 Compressible Flow

Compressible flow must take into account the possible change of density with respect to time within a control volume, i.e.,

$$\frac{\partial \rho}{\partial t} dV = vA\rho_{in} - vA\rho_{out} \tag{D.15}$$

which is the materials balance for the control volume. For steady state conditions, the left side is zero and Equation D.15 reduces to the same expression as that for incompressible flow.

D.2.3 Conservation of Energy

The second of the two principles of pipe flow analysis is the conservation of energy, expressed as the Bernoulli relation, applied between two points, "1" and "2,"

$$z_1 + \frac{p_1}{\gamma_1} + \frac{v_1^2}{2g} = z_2 + \frac{p_2}{\gamma_2} + \frac{v_2^2}{2g} + h_{L(1-2)} \tag{D.16}$$

in which

z_1 is the elevation of point 1 in the pipe (m, ft)
p_1 is the pressure at point in the pipe (N/m^2, Lb/ft^2)
γ_1 is the specific weight of fluid at point 1 (N/m^3, Lb/ft^3)
v_1 is the velocity in pipe at point 1 (m/s, ft/s)
z_2 is the elevation of point 2 in the pipe (m, ft)
p_2 is the pressure at point in the pipe (N/m^2, Lb/ft^2)
γ_2 is the specific weight of fluid at point 1 (N/m^3, Lb/ft^3)
v_2 is the velocity in pipe at point 1 (m/s, ft/s)
$h_{L(1-2)}$ is the headloss between points 1 and 2, respectively, as calculated by (D.4) (m, ft)

Note that the z term is the elevation head and is the vertical distance from a selected "datum" to the pipe. The p/γ term is the pressure head and is in terms of the fluid being transported, e.g., meters (feet) of water. The equation is "dimensionally homogeneous" as can be seen by writing the dimensions of each term. For water, an "incompressible" fluid, γ does not change, i.e., $\gamma_1 = \gamma_2$.

D.2.3.1 Alternative Dimensions of Bernoulli Relation

In applying the Bernoulli relation to water, an incompressible fluid, the dimension of each term is length, i.e., with units "meters of water" in SI units (ft of head in U.S. Customary units). Really, however, the length dimension is an abbreviation for energy per unit of fluid, e.g., N m/N or J/N, which makes more sense in U.S. Customary units, e.g., ft-lb_f/lb_m. This term is not so convenient for a gas, however, as we usually do not express gas energy as say "ft of air," which would be a very large number in most cases. But we could express the *pressure* due to a gas as an equivalent "feet of water" or "mm of mercury," just by using manometers filled with such fluids for direct measure or by calculation. This can be observed, in fact, by tapping a pipe with gas flow for a manometer and observing the pressure by means of water or mercury liquid (keeping in mind that an atmosphere of pressure is 10.33 m (33.9 ft) of water and only 760 mmHg).

Alternatively, if we multiply both sides of Equation D.14 by γ, the specific weight of the fluid, the dimensions of each term is pressure, or F/L^2. Multiplying the equation by L/L gives $F \cdot L/L^3$, or energy per unit volume, e.g., J/m^3.

On the other hand, we can multiply Equation D.14 by g/g, which gives L^2/T^2 for each term, e.g., m^2/s^2 in SI units. Now, multiply by N/N to give $J/(Ns^2/m)$, which is J/kg, or energy per unit of fluid mass. This form is used often for compressible fluids, e.g.,

$$z_1 g + \frac{p_1}{\rho_1} + \frac{v_1^2}{2} = z_2 g + \frac{p_2}{\rho_2} + \frac{v_2^2}{2} + h_{L(1-2)} g \tag{D.17}$$

D.2.3.2 Conservation of Energy with Pump

The head (energy of water per unit mass of water) that must be supplied by a pump is the difference between head terms on the right side of Equation D.16 or Equation D.17 and those on the left side, i.e.,

$$z_1 + \frac{p_1}{\gamma} + \frac{v_1^2}{2g} + \Delta H(\text{pump}) = z_2 + \frac{p_2}{\gamma} + \frac{v_2^2}{2g} + h_{L(1-2)} \tag{D.18}$$

in which ΔH(pump) is the head developed by pump as installed in pipeline (m).

The relation can be visualized by sketching a pipeline with the pump and then sketching the hydraulic grade line (HGL) and then the energy grade line (EGL). As a practical note, the two are almost coincident if drawn to scale and so, for most purposes, only the HGL is drawn. The HGL rises by an amount, ΔH(pump), at the position of the pump.

D.2.4 Pump Power

The power supplied to the fluid is

$$P(\text{fluid}) = Q\gamma\Delta H(\text{pump}) \tag{D.19}$$

The power required by the shaft to the pump is

$$P(\text{shaft}) = \frac{Q\gamma \cdot \Delta H(\text{pump})}{\eta(\text{pump})} \tag{D.20}$$

in which
- P is the power supplied to fluid (Nm/s or W)
- P(shaft) is the power supplied to shaft (W)
- η(pump) is the pump efficiency defined as ratio P(fluid)/P(shaft)

The power required for the motor is another issue and is defined as

$$\eta(\text{motor}) = \frac{P(\text{shaft})}{VA} \tag{D.21}$$

in which
- P(motor) is the power from electric energy supplied to motor (watts)
- V is the voltage across motor (volts)
- A is the amperage of motor (amperes)

D.2.5 Manifolds

A problem that appears frequently in plant design is manifold sizing. A manifold is a means to distribute a fluid to other pipes or through openings within a given pipe. A pipe that serves as a manifold to other pipes, called "laterals," may be called a "header" pipe.

D.2.5.1 Basics of Manifold Hydraulics

Usually, the problem is to achieve uniform distribution of the flow from the header pipe to the lateral pipes or from a lateral pipe through orifices or diffusers in the side, illustrated in Figure D.5a and b, respectively. The quest of a manifold design is to achieve nearly equal flows to each lateral from the header pipe and then to each orifice. The ideal would be that $Q(\text{orifice})_{(1,1)} \approx Q(\text{orifice})_{(n,n)}$. A means to distribute the flow approximately uniformly is to oversize each manifold pipe such that the friction headloss is little in the header pipe, ensuring Δh is small between any two node points of a manifold.

Examples of manifold design include a filtration underdrain/backwash system in which uniform flow distribution is essential, distribution of air from a pipe gallery header to laterals in aeration basins and then further distribution from the laterals through diffusers. In any pipe gallery in either a water treatment plant or in a wastewater treatment plant, various pipes are evident as flows must be distributed to or taken from the adjacent basins. A manifold pipe (or header pipe) is the means to distribute such flows.

D.2.5.2 Effect of Headloss on Manifold Flows

In general, the purpose of a manifold is to distribute flows uniformly to each lateral and then to some kind of outlets such as orifices or diffusers (also with uniform flow to each). As stated in the foregoing paragraph, Figure D.5a and b illustrate this principle.

Figure D.6 is but a perspective drawing of a manifold system, e.g., as illustrated in Figure D.5. It depicts a system with a header pipe showing only four laterals, with orifices spaced along the length of each lateral. Lateral 2 ($i=2$) is highlighted for illustration. The hydraulic grade line (HGL) for the header pipe, *abc*, is shown with friction headloss indicated to the end of the pipe. The HGL for lateral 2 is shown from *b* with slope in the direction of the flow. Finally, the HGL must end at the water surface and the HGL drops precipitously at *e* to *i*, the water surface in the filter bed (also the terminus of the HGL). The HGL for the header, as sketched, shows a straight line, i.e., the slope is uniform. In reality, the slope changes at each lateral as the header loses flow, since the slope is $h_L/L = (f/D)(v^2/2g)$ and $v = Q/A$. The same is true for the HGL for each lateral; the HGL slope is reduced incrementally at each orifice (or diffuser). This means that in reality, because there is a pressure loss in any pipe, the flow through a given lateral, i.e., Q(lateral) is less than the previous one (after the first lateral). By the same token, along a given lateral, the orifice flows are reduced with distance along the lateral.

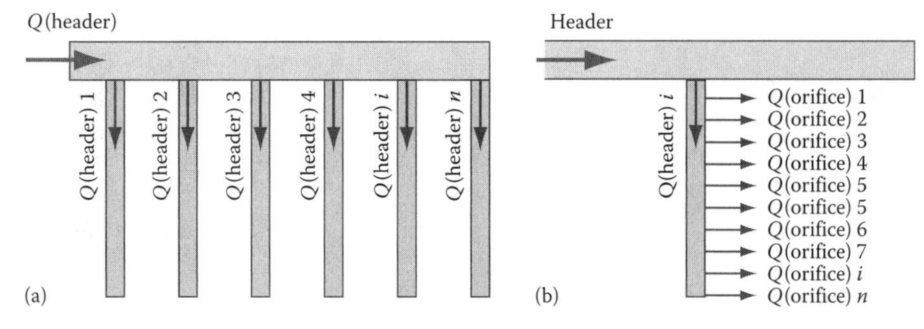

FIGURE D.5 Manifold system to illustrate flow distribution. (a) Header to laterals fix to Q(lateral); (b) lateral pipe to orifices.

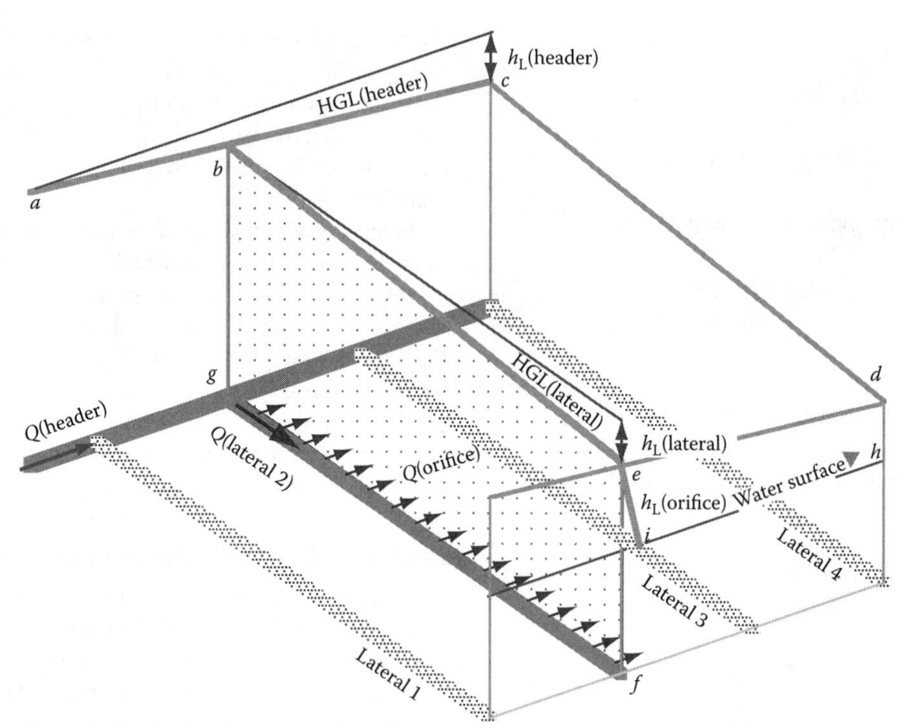

FIGURE D.6 Hydraulic grade line for header and lateral 2 of under-drain system; water surface "h–i" and is terminus for the HGL for any orifice.

D.2.5.3 Calculation Algorithm

As noted in the previous paragraph, the role of a manifold is to distribute flow and to do so uniformly (to the extent feasible). With a spreadsheet model, any changes can be incorporated and detailed calculations made so that pressure changes can be shown along the header pipe and each lateral pipe. The equations applicable are delineated in the following:

First, the flow through any orifice is

$$Q(\text{orifice}) = A(\text{orifice})C_D[2g\Delta h]^{0.5} \tag{D.22}$$

in which

Q(orifice) is the flow through any orifice (m^3/s)
A(orifice) is the area of any orifice (m^2)
C_D is the orifice coefficient—dimensionless
Δh is the headloss between two sides of orifice (m)

For equal flow from each orifice, the mathematical statement is

$$Q(\text{orifice})_{i,j=1} = Q(\text{orifice})_{i,j=2} = Q(\text{orifice})_{i,j=3} = \cdots = Q(\text{orifice})m,\ n \tag{D.23}$$

in which

$Q(\text{orifice})_{i,j}$ = flow through any orifice designated by lateral i and orifice j (m^3/s)

and to achieve this equality,

$$Q(\text{orifice})_{i,j=1} = \Delta h(\text{orifice})_{i,j=2} = \Delta h(\text{orifice})_{i,j=3} = \cdots = \Delta h(\text{orifice})m,\ n \tag{D.24}$$

in which

$\Delta h(\text{orifice})_{i,j}$ = headloss across any orifice designated by lateral i and orifice j (m)

The only means to achieve this uniform headloss for each orifice, i.e., Equation D.21, is by changing the size of the orifice along the lateral to compensate for the declining $\Delta h(\text{orifice})_i$ along the length of the lateral. This is not a practical solution. A "quick and dirty" means to approximate the conditions of Equations D.23 and D.24 is to use a large header pipe and then smaller but still large lateral pipes. The idea is that if the headlosses in these pipes are negligible, then $\Delta h(\text{orifice})_1 > \Delta h(\text{orifice})_n$, but only slightly. A spreadsheet solution to the problem, as seen in Figure D.6, provides a means to refine the solution (i.e., with regard to the problem of pipe sizing and orifice sizing). The needed equations for a computer solution are enumerated as follows, which comprise a solution "algorithm."

1. Continuity applies to the header–lateral relation,

$$Q(\text{header}) = \sum_{1}^{n} Q(\text{lateral})_i \tag{D.25}$$

in which
Q(header) is the flow through header pipe (m^3/s)
$Q(\text{lateral})_i$ is the flow through any lateral "i" (m^3/s)

2. Also, continuity applies to the lateral–orifice relation,

$$Q(\text{lateral})_i = \sum_j^n Q(\text{orifice})_j \quad \text{(D.26)}$$

in which

$Q(\text{orifice})_j$ = flow through any orifice j $\left(\text{m}^3/\text{s}\right)$

3. The water surface elevation, i.e., in the filter bed when backwash is involved, is the terminus of the HGL, which is illustrated as elevation "h" in Figure D.6. The summation of all headlosses through any HGL path from the water surface equals the HGL elevation at point "a" in the backwash header pipe, also as illustrated in Figure D.6. The mathematical statements is,

$$\text{HGL(elev. point a)} = \text{HGL(elev. at h)} + h_{\text{L}}(\text{orifice})_{i,j} + \sum_{i,j-1}^{m,m-1} h_{\text{L}}(\text{lateral})_{j,j-1} + \sum_{i,i-1}^{n,n-1} h_{\text{L}}(\text{manifold})_{i,i-1} \quad \text{(D.27)}$$

in which
HGL is the hydraulic grade line, i.e., the sum of pressure head and elevation head (m)
HGL(elev. point a) is the elevation of the HGL at point "a" in the header pipe (m)
HGL(elev. at h) is the elevation, "h," at the HGL terminus, e.g., the water surface of a rapid filter during backwash (m)
$h_{\text{L}}(\text{orifice})_{i,j}$ is the headloss across any orifice, i, j (m)
$h_{\text{L}}(\text{lateral})_{j,j+1}$ is the headloss in lateral between orifices j and $j+1$ (m)
$h_{\text{L}}(\text{manifold})_{i,i+1}$ is the headloss in manifold between laterals i and $i+1$ (m)

The "path" may be through any orifice, located on any lateral. The equality is the same regardless of the path. For example, referring to Figure D.6, the path could be from the last orifice in lateral 3, along lateral 3 to the manifold, and then to point "a." Or the path could be from any intermediate orifice of lateral 1 to point "a," or from any orifice of lateral 4. Note that the headloss terms will be different along each path, but that HGL(point a) and the HGL at the water surface in the filter bed are "anchor" points for the HGL.

4. The overall flow to the system is determined by the hydraulic loading rate (Q/A) times the area of the filter. Backwash is the most critical hydraulic condition, which should be the basis for design. The same principles apply to air flow, e.g., to an aerated grit chamber.
5. After an algorithm is set up on a spreadsheet, the variables to be explored are the size of the orifices, along with the diameters of the laterals and the manifold. These are the design questions to be explored. The measure of performance is the difference in $Q(\text{orifice})_{i=1}$ and $Q(\text{orifice})_{i-n}$.

D.2.5.4 Spreadsheet Algorithm

Table CDD.2(a) is a spreadsheet that incorporates the above guidelines to calculate orifice flows, lateral flows, velocity in a given lateral as affected by flow, losses through orifices, velocity head at a given point in the lateral, headloss in a given lateral between orifices, total head in lateral at a given distance, pressure head in a given lateral at a given distance from the header. Figure CDD.2, embedded in the spreadsheet and not shown in the text, is a definition sketch showing the layout of the header and laterals, incorporated in the spreadsheet so that it is self-contained.

Table CDD.2(b) is derived from Table CDD.2(a), but uses a larger y-increment (i.e., distance along the lateral) and has fewer columns. Table CDD.2(b) is linked to Figure CDD.7 (b), which shows the pressure surface as a function of x(orifice number) and y(distance along lateral) and is embedded in Table CDD.2(b) and not shown in the text.

In using the spreadsheet for design of a traditional under-drain system, the headings in Table CDD.2(a) are for the input data. The categories include "Data-Header," "Data-Laterals," and "Data-Orifices." The input variables are Q(header), L(header), d(header), n(laterals), L(laterals), d(laterals), d(orifices), $n(\text{orifices})/L(\text{lateral})$. The foregoing are all design variables and each selection is a unique combination that results, in turn, in a unique pressure surface. Thus, the design is, in general, a trial-and-error procedure. The spreadsheet, with linked pressure surface plot, provides an algorithm to facilitate the detailed calculations.

D.3 COMPRESSIBLE FLUID FLOW IN PIPES

In water treatment, air flow is required for a number of processes, e.g., aerated grit chambers, diffused aeration for activated sludge, air stripping (if bubble aeration should be used), carbon dioxide gas to adjust pH, etc. A compressor is required, as a rule, to provide the needed pressure to distribute the air flow through a manifold system and then to diffusers and orifices as depicted in Figure D.6 (which may apply to either compressible flow or incompressible flow).

Adaptation of the Bernoulli equation, friction losses, orifice flow, and calculation of compressor power are reviewed here. A main point is that, for the relatively low pressures involved in water and waste water applications, compressible flow can be handled in the same fashion as incompressible flow. For compressors, however, an adiabatic compression is involved.

TABLE CDD.2(a)
Spreadsheet Calculating Pressure Surface for Under-Drain System (Excerpt)

Formulae

$Q(\text{orifice})j{-}j{-}1 = A(\text{orif}) \cdot C \cdot (2g)0.5 \cdot (\Delta Hi, j)0.5$

$Q(\text{loss})j{-}j{-}1 = A(\text{orif}) \cdot C \cdot (2g)0.5 \cdot (\Delta Hi, j)0.5^{*} n(\text{orifices})/L^{*}\Delta L$

$Q(\text{lat})j{-}j{-}1 = Q(\text{lat})j{-}1 - Q(\text{loss})j{-}j{-}1$

$v(\text{lat})j{-}j{-}1 = Q(\text{lat})j{-}j{-}1/A(\text{lat})$

$\Delta hLj{-}j{-}1 = f(yj{-}j{-}1/d(\text{lat})^{*}(v_j^2/2g)$

Scenario

(1) Data entries are to develop model (they are not realistic)
(2) Under-drain system is generic with plastic pipe assumed for header and laterals
(3) Lateral missing for $x = 0$ m; also add laterals such that $x \approx y$

Data Inputs

Data-Header			Data—Laterals			Data—Orifices		
$f =$	0.012		$f =$	0.012		$C =$	0.620	
$Q(\text{header}) =$	1.00	m^3/s	$Q(\text{lateral}) =$	0.20	m^3/s	$d(\text{orifice}) =$	0.005	m
$L(\text{header}) =$	10	m	$L(\text{lateral}) =$	15	m	$n(\text{orifices})/L =$	55	orifices/m
$d(\text{header}) =$	0.50	m	$d(\text{lateral}) =$	0.25	m	$n(\text{orifices}) =$	825	orifices
$g =$	9.808	m/s^2	$n(\text{laterals}) =$	5		$Q(\text{orifice}) =$	$A(\text{orif}) \cdot C \cdot (2g)^{0.5} \cdot (\Delta Hi, j)^{0.5}$	
$H =$	20.000	m						

Preliminary Calculations

Lateral	#	A	1	2	3	4	5
x(header)	m	0.00	2.00	4.00	6.00	8.00	10.00
Q(header)	m^3/s	1.00	0.8	0.6	0.4	0.2	0
v(header)	m/s	5.093	4.074	3.056	2.037	1.019	0.000
Δh_L(header)	m	0.063	0.041	0.023	0.010	0.003	0.000
H	m	20.000	19.937	19.896	19.873	19.863	19.860

TABLE CDD.2(b)
Abstract of Head Levels and Flows in Header Pipes and Laterals for a Filter Bed[a]

	1.0			2.0				5.0
x (m) **Y (m)**	**$Q(\text{lat})_{j-j-1}$ (m^3/s)**	**H(i, j) (m)**	**p(i, j) (m)**	**$Q(\text{lat})_{j-j-1}$ (m^3/s)**	**H(i, j) (m)**	**p(i, j) (m)**		**p(i, j) (m)**
0.0	0.200	19.937	13.090	0.200	19.896	13.050	Columns 3.0 m	13.014
1.0	0.187	19.899	13.161	0.187	19.858	13.120	and 4.0 m	13.084
2.0	0.174	19.866	13.229	0.174	19.825	13.188	omitted	13.152
3.0	0.160	19.838	13.294	0.160	19.797	13.253		13.217
				Rows 4.0 m to 12.0 m not shown				
13.0	0.028	19.734	13.717	0.029	19.693	13.676		13.640
14.0	0.015	19.734	13.729	0.016	19.693	13.688		13.652
15.0	0.002	19.734	13.733	0.002	19.693	13.693		13.657

[a] Coarse grid—abstracted from Table CDD-2(a).

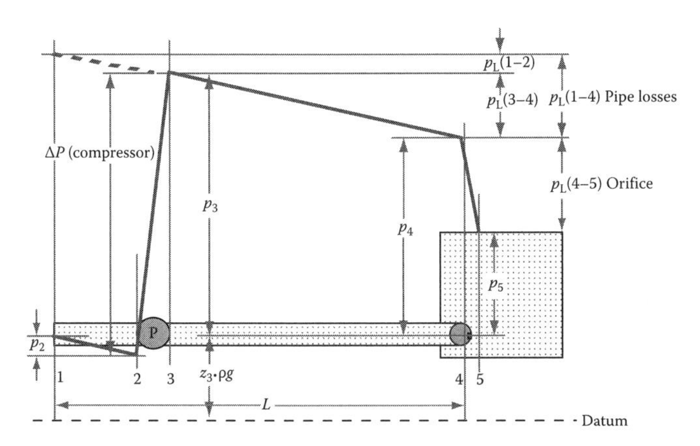

FIGURE CDD.7 Pneumatic grade line with delineation of changes for flow to submerged diffuser (figure is also embedded in Table CDD.3).

D.3.1 Friction Loss

For compressible fluid flow, the pipe friction relation, Equation D.4, and the Bernoulli relation, Equation D.17, are applicable, provided that the pressure changes are not great and that the velocities are sub-sonic (not an issue in the cases at hand). The equations are applied most conveniently, however, with a modification to express the energy dissipation either as energy per unit mass, or as pressure loss (energy per unit volume), such as shown in D.2.3.1. This change is done first for the friction loss relation by multiplying both sides of Equation D.4 by γ, the specific weight of the fluid, i.e.,

$$\Delta h_L \gamma = f \frac{L}{D} \frac{v^2}{2g} \gamma \tag{D.28}$$

In the next step, recall

$$\Delta p = h_L \cdot \gamma \tag{D.29}$$

and

$$\rho = \frac{\gamma}{g} \tag{D.30}$$

in which
- Δp is the pressure change between two points (N/m^2, lb/ft^2)
- γ is the specific weight of fluid (N/m^3, lb/ft^3)

Substituting (D.29) and (D.30) in (D.28) gives

$$\Delta p(\text{friction}) = f \frac{L}{D} \rho \frac{V^2}{2} \tag{D.31}$$

in which Δp(friction) is the friction loss between two points in the pipeline expressed as an equivalent pressure loss (N/m^2, lb/ft^2).

D.3.2 Bernoulli Equation—Modified Units

For gases, the Bernoulli equation may be modified to a form with pressure units by following the same procedure in converting Equation D.4 through D.31, i.e., multiplying both sides of Equation D.16 by γ, i.e.,

$$z_1\gamma_1 + \frac{p_1}{\gamma}\gamma_1 + \frac{v_1^2}{2g}\gamma_1 = z_2\gamma_2 + \frac{p_2}{\gamma}\gamma_2 + \frac{v_2^2}{2g}\gamma_2 + h_{L(1-2)}\gamma_2 \tag{D.32}$$

Following through for the three energy terms, given as Equation D.30(a), (b), (c),

$$\text{(a) } z\gamma = z\rho g \quad \text{(b) } p = \frac{p}{\gamma}\gamma \quad \text{(c) } \frac{v^2}{2g}\gamma = \rho\frac{v^2}{2} \tag{D.33}$$

gives for (D.29), with the pump term added,

$$z_1\rho_1 g + p_1 + \rho_1\frac{v_1^2}{2} + \Delta P(\text{pump}) = z_2\rho_2 g + p_2 + \rho_2\frac{v_2^2}{2} + \Delta p_{L(1-2)} \tag{D.34}$$

in which
- $z_1\rho_1 g$ is the pressure equivalent of the elevation of a given fluid with respect to a reference datum [also $z_1\rho_1 g = p_1$(elev)] (N/m^2, lb/ft^2)
- $\Delta p_{L(1-2)}$ is the energy loss between 1 and 2 in terms of pressure energy per unit volume of fluid (N $\cdot$ m/m^3, lb $\cdot$ ft/ft^3)
- ρ_1 is the density of fluid at point 1 (kg/m^3)

Note that the expression, "z(elev)ρg," is preferred in Equation D.31 to express the specific energy of a fluid due to elevation. In most cases of compressible pipe flow in a plant design, the elevation difference for a gas is not a major factor. Also, note that the density term is enumerated with subscripts, indicating that the density does change in accordance with the change in state conditions (p, T), but usually may be neglected. An analysis by Rouse (1946, pp. 338–342) of compressible flow showed that for either isothermal flow or adiabatic flow in a pipe, the assumption of constant density of the gas causes only about 2.5% discrepancy in pressure calculation, such as indicated by Equation D.31. The discrepancy of 2.5% applies to adiabatic flow if the velocity change is only 15–100 m/s (50–350 ft/s). A conservation of energy equation would also include internal energy and heat lost to or added from the surroundings, as outlined subsequently.

D.3.3 Operational Bernoulli Equation for Spreadsheet

Figure D.8 depicts a system that could be of interest for several kinds of design situations, e.g., aerated grit chamber, diffused aeration in activated sludge, or any situation involving bubbling of gas through orifices or diffusers. The system shown comprises an air intake pipe, a compressor, a header pipe, a submerged lateral pipe, and orifices within each lateral

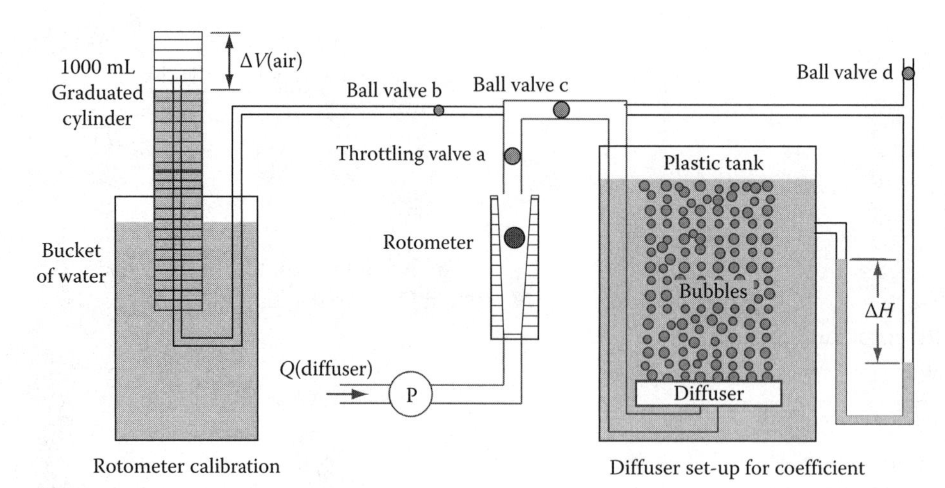

FIGURE D.8 Laboratory set-up for determination of diffuser coefficient.

pipe. The problem is to determine the pressure that the compressor must develop. The design may involve "scenarios" and so a spreadsheet is most useful. Several steps are involved in developing a spreadsheet solution. The spreadsheet algorithm is outlined.

D.3.3.1 Orifice Flow

The orifice equation is

$$Q(\text{orifice}) = A(\text{orifice})C\left[\frac{2\Delta p(\text{orifice})}{\rho(\text{gas})}\right]^{0.5} \quad \text{(D.35)}$$

in which

Q(orifice) is the flow through a single orifice (m^3/s)
A(orifice) is the area of a single orifice (m^2)
C is the orifice coefficient (dimensionless)
Δp(orifice) is the pressure difference across an orifice plate (Pa)
ρ is the density of gas (kg gas/m^3 gas)

The same form of equation applies to a diffuser, i.e.,

$$Q(\text{diffuser}) = K\left[\frac{\Delta p(\text{diffuser})}{\rho(\text{gas})}\right]^{0.5} \quad \text{(D.36)}$$

in which

Q(diffuser) is the flow through a single diffuser (m^3/s)
K is the "lumped" coefficient for diffuser (m^2)

D.3.3.1.1 Determination of K for a Diffuser

Diffusers are proprietary with K being unique for a particular design. The K must be either supplied by the manufacturer (usually not available) or determined by a laboratory test. The laboratory test involves measuring Q(diffuser) *v*. $\Delta p/\rho$ with enough points to define a straight line.

Figure D.8 shows the experimental set-up for the determination of K. To measure Q(diffuser), a rotometer should be selected for the range of air flow expected. Calibration of Q(air) *v*. ball position reading may be done by volumetric displacement of water. A graduated cylinder is filled with air-saturated water and inverted in the volume of water. An air tube is placed in the cylinder and at time $t = 0$, the ball valve b is opened, and the rotometer ball reading is taken. After a reasonable displacement, the volume V is measured and t is recorded, with $Q(\text{air}) = \Delta V/\Delta t$. The throttling valve "a" is changed so that Q(air) is different and the process is repeated. To determine K, the ball valves "b" and "d" are closed and ball valve "c" is opened. For a given setting of the throttling valve "a," the air flow is then measured by the rotometer and the headloss across the diffuser, Δh(diffuser), is measured at the same time. The air temperature should be measured also probably near valve "c." The throttling valve "a" is adjusted and the process is repeated. After enough points are obtained, K can be determined. The density ρ(air) is calculated by the gas law, i.e., $PV = nRT$, to give $\rho(\text{air}) = P \cdot \text{MW(air)}/RT$, with $P = \Delta h(\text{diffuser}) \cdot \gamma_w + p(\text{atm})$.

D.3.3.2 Submerged Flow

When an air bubble emerges from an orifice or a diffuser or appears spontaneously by gas precipitation, the absolute pressure inside the bubble equals the pressure due to the water depth plus the atmospheric pressure on the water surface, i.e.,

$$p_5 = \gamma_w D(\text{water}) + p(\text{atm}) \quad \text{(D.37)}$$

in which

D(water) is the depth of orifice below water surface in tank (m)
p(atm) is the atmospheric pressure under local conditions dependent on elevation, or more accurately, barometric pressure (Pa)

Figure D.8 shows the pneumatic grade line terminating at the surface of the water, which is the case if gage pressure is the basis for the diagram. If absolute pressure is used as the basis, all points of the pneumatic grade line have added one atmosphere of local pressure. Note that the pressure term in the ideal gas law is absolute pressure.

D.3.3.3 Density of Gas

According to the analysis of compressible pipe flow (Rouse, 1946, pp. 336–342), the change in density can be neglected for most engineering problems (where velocities are much less than the speed of sound). This, of course, greatly simplifies the application of Equations D.11 and D.15.

To determine the density at any point, the ideal gas law, $PV = nRT$, is the basis, the molar density being n/V and the mass density, $\rho(\text{mass density}) = (n/V)\text{MW}) = (p/RT)\text{MW}$. In other words, pressure and temperature must be known to calculate density (either molar density or mass density).

D.3.4 Pressure Increase Required by a Compressor

Figure D.8 illustrates a typical situation in which a compressor is installed in a pipeline to provide air or another gas, such as oxygen, to an installation. The question is to determine the magnitude of the pressure increase that the compressor must provide to achieve the required gas flow. This is done by an analysis of the pneumatic grade line, as shown in Figure D.8.

D.3.4.1 Bernoulli Relation between Two Points

Key features of Figure D.8 include (1) the compressor causes an increase in pipeline pressure from p_2 to p_3, i.e., $\Delta P(\text{compressor})$; (2) the pressure losses in air flow are similar to headlosses in water flow; (3) the pneumatic grade line ends at the water surface in the tank; and (4) the pressure outside the diffuser, p_5, equals the depth of the diffuser times the specific weight, γ_w, of the water. An analysis of the pneumatic grade line can be seen best by applying the Bernoulli relation between points 3 and 5. The equality is

$$p_3(\text{elev}) + p_3 + \rho\frac{v_3^2}{2} = p_5(\text{elev}) + p_5 + \rho\frac{v_5^2}{2} + \Delta p(\text{orifice}) + \Delta p(\text{friction})_{3-4} \quad \text{(D.38)}$$

D.3.4.2 Operational Form of Bernoulli Relation

Other equations may be substituted into Equation D.20 to permit determination of the needed Δp for the compressor, i.e., $\Delta p(\text{compressor})$. These are as follows:

1. Apply the Bernoulli relation between 1 and 3, and so the specific energy per unit volume of gas (J/m^3) at point "3" is

$$[p(\text{elev})_3 + p_3] = z_1\rho_1 g + p_1 - \Delta p(\text{friction})_{1-2} + \Delta p(\text{compressor}) \quad \text{(D.39)}$$

2. The gage pressure within the gas bubbles as they emerge through the orifice and into the tank at water depth, $D(\text{water})$,

$$p_5 = \gamma_w D(\text{water}) \quad \text{(D.40)}$$

3. The specific energy of the gas, due just to its position above some datum, is

$$z_3\rho_3 g \quad \text{(D.41)}$$

4. The pressure loss due to friction loss between two points, say "1" and "3," is

$$\Delta p(\text{friction}) = f\frac{L}{D}\rho(\text{gas})\frac{v^2}{2} \quad \text{(D.42)}$$

5. The flow through a single orifice is

$$Q(\text{orifice}) = A(\text{orifice})C\left[\frac{2\Delta p(\text{orifice})}{\rho(\text{gas})}\right]^{0.5} \quad \text{(D.43)}$$

in which

- $z(\text{elev})$ is the elevation of gas at point "1" (m)
- $D(\text{water})$ is the depth of orifice below water surface in tank (m)
- p_1 is the pressure of gas at point "1" (Pa)
- $z(\text{elev})_1$ is the elevation of gas at point "1" (m)
- $\rho(\text{gas})$ is the density of gas at a specified temperature and pressure (kg gas/m^3)
- g is the acceleration of gravity (m/s^2)
- v_1 is the velocity of gas in pipe (m/s)
- $\Delta p(\text{friction})_{1-2}$ is the pressure loss between points "1" and "2" due to friction in pipe (Pa)
- $Q(\text{orifice})$ is the flow through a single orifice (m^3/s)
- $A(\text{orifice})$ is the area of a single orifice (m^2)
- C_D is the orifice coefficient (about 0.62 for a sharp edge orifice plate) dimensionless
- $\Delta p(\text{orifice})$ is the pressure difference across an orifice plate (Pa)
- f is the friction factor from Moody diagram (about 0.012 for a smooth pipe)

Substituting Equation D.39 into Equation D.38 for the terms, $[p(\text{elev})_3 + p_3]$, gives

$$[z_1\rho g + p_1 - \Delta p(\text{friction})_{1-3} + \Delta p(\text{compressor})] + \rho\frac{v_3^2}{2} = z_5\rho g + \rho\frac{v_5^2}{2} + \Delta p(\text{friction})_{3-4} + \Delta p(\text{orifice}) + p_5 \quad \text{(D.44)}$$

Next, substituting Equations D.40 through D.43 in Equation D.42 gives

$$\left[z_1\rho g + p_1 - \left[f\frac{L}{D}\rho(\text{gas})\frac{v^2}{2}\right]_{1-3} + \Delta p(\text{compressor})\right] + \rho\frac{v_3^2}{2} = z_5\rho g + \rho\frac{v_5^2}{2} + \left[f\frac{L}{D}\rho(\text{gas})\frac{v^2}{2}\right]_{3-4} + \frac{Q(\text{orifice})^2\rho(\text{gas})}{2C^2A(\text{orifice})^2} + \gamma_w D(\text{water}) \quad \text{(D.45)}$$

Equation D.46 is an operational form of the Bernoulli relation applicable to Figure D.8, which has utility in solving for Δp(compressor). A spreadsheet may be set up to do this since the variables are design inputs or can be calculated, e.g.,

- Q(orifice) is the flow required for a given situation (such as from an empirical equation to give air flow per lineal meter of manifold as in an aerated grit chamber).
- A(orifice) is the area of the orifice from a given manifold; ρ(gas) is for air as a specified temperature and pressure and is determined by the ideal gas law; L is the length of pipe of a given diameter between specified points.
- z_i is the elevation of the flow at any point, i.
- D(water) is the depth of submergence of the orifices or diffusers.

The pipe friction losses may include several lengths of different diameters and so the terms may be consolidated as a summation, which should be taken into account in the spreadsheet depiction,

$$z_1\rho g + p_1 + \Delta p(\text{compressor}) + \rho\frac{v_3^2}{2} = z_5\rho g + \rho\frac{v_5^2}{2} + \sum_i^n \left[f\frac{L}{D}\rho(\text{gas})\frac{v^2}{2}\right]_{i-(i+1)} + \frac{Q(\text{orifice})^2\rho(\text{gas})}{2C^2A(\text{orifice})^2} + \gamma_w D(\text{water}) \qquad \text{(D.46)}$$

The sketch of Figure D.8 was done in terms of gage pressure, i.e., relative to atmospheric pressure. To show the same thing in terms of absolute pressure, the entire pneumatic grade line would be elevated by p(atm), the atmospheric pressure. Thus, in terms of absolute pressure, $p_1(\text{absolute}) = p(\text{atm})$ and $p_5(\text{absolute}) = \rho_w g D(\text{water}) + p(\text{atm})$. To be complete, the left side of Equation D.46 should also include minor losses, such as bends, constrictions, pipe irregularities, etc. Thus, the term $\sum[K_i\rho(\text{gas})v^2/2]$ should be included, in which K_i is the coefficient for any given pipe irregularity, i, a pipe elbow being most common example.

D.3.4.3 Spreadsheet Solution for Bernoulli Relation for Compressible Flow

Table CDD.3 is a spreadsheet that utilizes the algorithm as outlined above to calculate pressure energy at points "1" to "5" of Figure D.8. The spreadsheet requires certain inputs, e.g., air flow, temperature, ambient pressure, etc., and then "walks through" a calculation protocol that results in the determination of a Δp(compressor) required for the pipe system and manifold to deliver a required air flow.

D.3.5 Compressors

A compressor is a machine that increases the pressure of a gas and is essentially a pump for gases (see Cheremisinoff and Cheremisinoff, 1989). The ratio of final pressure, p_2, to the suction pressure, p_1, is called the compression ratio. The compressors are classified according to the compression ratio. Table D.4 indicates the three basic types of compressors with characteristics.

The compressor subcategories include centrifugal, axial, rotary centrifugal and rotary positive displacement, and reciprocating positive displacement; the definitions are not always precise. The rotary types are positive displacement, similar to gear pumps, while the reciprocating types are piston or diaphragm. Compressors are rated, ordinarily, in terms of flow in m^3/min (ft^3/min) at standard temperature and pressure; standards differ, however, among industries, as noted by McCabe et al. (1993).

D.3.5.1 Thermodynamics

Types of compressible fluid flow are (1) adiabatic, which means that heat exchange with the surroundings does not occur (i.e., there is no heat transfer across the pipe walls), (2) isothermal, which means that temperature does not change (with distance along the pipe in the case of pipe flow), and (3) poly-tropic, which means that volumes, pressures, and temperatures change.

D.3.5.1.1 Ideal Gas Equation

An integral part of most thermodynamic relations is the ideal gas law, which is applicable for all pressures likely to be encountered in pipe flow:

$$pV = nRT \qquad \text{(D.47)}$$

in which

p is the pressure of gas (N/m^2)
V is the volume of gas (m^3)
n is the moles of gas
R is the universal gas constant (8.314 510 J K^{-1} mol^{-1})
T is the temperature (K)

The ideal gas law may be manipulated to a number of different forms. One form used frequently is to state the law in terms of density. First, rearrange

$$\frac{n}{V} = \frac{p}{RT} \qquad \text{(D.48)}$$

and thus,

$$\rho(\text{molar}) = \frac{P}{RT} \qquad \text{(D.49)}$$

in which ρ(molar) is the molar density (mol gas i/m^3).

Then, to convert to mass density,

TABLE CDD.3
Pneumatic Analysis of Air Flow to Aerated Grit Chamber (by Bernoulli Relation) to Obtain Pressure Inputs to Compressor

1 Gas Constants

Constant	Value	Units	Reference
R =	8.314510	J $K^{-1}mol^{-1}$	Table B.2
MW(air) =	0.0289641	kg/mol	Table B.7
k =	1.395		
g =	9.81		

2 Hydraulic Constants

f = 0.012
ρ = 1000 kg/m^3
C = 0.61 orifice coefficient

3 Ambient Conditions						4 Intake Conditions at "1"			
z(elev) (ft)	z(elev) (m)	p(atm) (Pa)	T (atm) (C)	z_1 (m)	ρ(atm) (kg/m^3)	$z_1\rho g$ (Pa)	p_1 (Pa)	T_1 (C)	ρ_1 (kg/m^3)
0	0	101325	0	10	1.292	127	101325	0	1.2922
5500	1676	82083	0	10	1.047	103	82083	0	1.047
5500	1676	82083	0	10	1.047	103	82083	0	1.047
5500	1676	82083	0	10	1.047	103	82083	0	1.047
5500	1676	82083	0	10	1.047	103	82083	0	1.047
5500	1676	82083	0	10	1.047	103	82083	0	1.047
5500	1676	82083	0	10	1.047	103	82083	0	1.047
Elevation above sea level		Computed pressure from Lide data	Measured (or assumed) atmospheric temperature	Elevation of intake pipe above a datum	Density of air for ambient conditions	Energy of air at intake due to elev.	Pressure at intake	Temperature at "1"	Density of

Other categories calculated
5 Pipe design "1–2"
6 Calculation of HGL level at "2"
7 Conditions at "5"
8 Orifice flow/sizing/Δp determinations to get p_4
9 Pressure loss between "3" and "4" to get HGL_3
10 Summary—for input data to compressor

TABLE D.4
Characteristics of Compressors

Type	Description	p_2/p_1 Min	p_2/p_1 Max	Discharge (kPa)	Pressure[a] (atm)
Fan	Low speed, high flow[a]		<1.1	1–15	0.01–0.15
Blower	High speed rotary, using either positive displacement or centrifugal force[a]	1.1	<3.0	≤200	≤2
Compressor		>3.0		200–$n \cdot 10^5$	2–1000's

Source: Cheremisinoff, N. P. and Cheremisinoff, P. N., *Pumps/Compressors/Fans – Pocket Handbook*, Technomic Publishing Company, Lancaster, PA, p. 78, 1989.

[a] McCabe et al. (1993, p. 204).

$$\rho(\text{mass}) = \rho(\text{molar})\text{MW} = \left(\frac{n}{V}\right)\text{MW} = \frac{p}{RT}\text{MW} \quad \text{(D.50)}$$

in which

ρ(molar) is the molar density (mol gas i/m^3)

MW is the molecular weight (kg i/mol i)

D.3.5.1.2 First Law

The first law of thermodynamics is a means of energy accounting Pitzer and Brewer, 1961, p. 36) and is stated

$$\Delta E = \delta Q + \delta W \quad \text{(D.51)}$$

in which

ΔE is the change in internal energy of a given substance (J)

δQ is the increment of heat absorbed from the surroundings (J)

δW is the increment of work absorbed from the surroundings (J)

Also, Q and W are "path" dependent, meaning that the work done by the surroundings on the system may be, for example, $p_{\text{ext}}dV$, in which case, p_{ext} may be quite a bit different from p, the internal pressure. So, by definition, the work done is not reversible unless $(p_{\text{ext}} - p) = 0$. Regarding sign, the convention used by Pitzer and Brewer (1961, p. 34) is adopted here, i.e., heat that a system gains from the surroundings is positive (+), and work that is done on a system by the surroundings is positive (+). This may differ among authors and so it is always necessary to state the convention used. As another note, a "system" is defined as some part of the universe around which boundaries are drawn. The "surroundings" is everything outside those boundaries.

If the work of expansion is carried out such that no heat enters or leaves the system, then $Q = 0$ and the process is, by definition, adiabatic. An isothermal expansion means that the temperature is constant for the process.

D.3.5.1.3 Isothermal Compression

In a reversible expansion (i.e., $p_{\text{ext}} = p$), along a constant temperature line in a p-V diagram, if the internal energy remains constant through heat being absorbed from or assimilated by the surroundings, i.e., $-\delta W = \delta Q$ in the first law, we can say

$$Q = -W(\text{isothermal}) = nRT \ln\frac{V_2}{V_1} \quad \text{(D.52)}$$

$$= nRT \ln\frac{p_1}{p_2} \quad \text{(D.53)}$$

in which W(isothermal) is the work of an isothermal expansion (+) or compression (−) (J).

D.3.5.1.4 Adiabatic Compression

By definition, an adiabatic compression (or expansion) occurs when the heat transfer, Q, equals zero. Accordingly, the temperature of the fluid must rise.

D.3.5.1.5 Work of Adiabatic Compression

The work of adiabatic compression or an ideal frictionless gas (McCabe et al., 1993, p. 209) is

$$w = \int_{p_1}^{p_2} \frac{dp}{\rho} \quad \text{(D.54)}$$

in which w is the work of compression per unit mass of gas (+) (J/kg).

Subscripts "1" and "2" refer to positions in space (e.g., distance along a pipeline) or times t_1 and t_2. The path followed by the fluid in the compressor must be defined, e.g., whether it is a single sudden change in state or the change is in increments (as a note, when such increments are small the path approaches an irreversible path). The compressor may be any type, e.g., rotary positive displacement, reciprocating positive displacement, or centrifugal. If not cooled, the compressor follows an isentropic (i.e., frictionless or reversible) path in which w is

$$w = \frac{p_1}{\rho_1}\frac{\kappa}{\kappa - 1}\left[\left(\frac{p_2}{p_1}\right)^{(\kappa-1)/\kappa} - 1\right] \quad \text{(D.55)}$$

in which κ is the ratio of specific heats, i.e., $\kappa = c_p/c_v$ where c_p(constant pressure) and c_v (constant volume) = 1.395 and

$$R = c_p - c_v \quad \text{(D.56)}$$

For reference, c_v, c_p values for a monatomic ideal gas are $c_v = (3/2)R = 12.47$ J K^{-1} mol^{-1} and $c_p = (5/2)R = 20.77$ J K^{-1} mol^{-1} (Alberty and Silbey, 1992, p. 52). For other gases, $c_p(O_2) = 29.36$ J K^{-1} mol^{-1}, $c_p(H_2) = 28.82$ J K^{-1} mol^{-1}, $c_p(H_2Og) = 33.58$ J K^{-1} mol^{-1}, $c_p(Cl_2) = 33.91$ J K^{-1} mol^{-1}, $c_p(N_2) = 29.12$ J K^{-1} mol^{-1}, $c_p(CO_2) = 29.12$ J K^{-1} mol^{-1} (Alberty and Silbey, 1992, p. 850).

D.3.5.1.6 Power of Adiabatic Compression

The work of compression per unit mass times the mass flow, i.e., $Q\rho$, is the power required for the compression. Therefore,

$$P = Q\rho w \quad \text{(D.57)}$$

in which

P is the power for adiabatic compression for air flow, Q, and density, ρ (W)

Q is the flow of gas at given temperature and pressure (m^3/s)

ρ is the density of gas at given temperature and pressure (kg/m^3)

After substitution of Equation D.55 in D.57,

$$P = Q\rho \frac{p_1}{\rho_1} \frac{\kappa}{\kappa - 1} \left[\left(\frac{p_2}{p_1} \right)^{(\kappa-1)/\kappa} - 1 \right] \quad \text{(D.58)}$$

in which
- Q is the flow of gas at given temperature and pressure (m^3/s)
- ρ is the density of gas at given temperature and pressure (kg/m^3)

Now, if Q and ρ are for state "1," defined here as the compressor intake condition, the ρ's cancel, giving, after applying the subscript "1" for Q,

$$P = Q_1 p_1 \frac{\kappa}{\kappa - 1} \left[\left(\frac{p_2}{p_1} \right)^{(\kappa-1)/\kappa} - 1 \right] \quad \text{(D.59)}$$

This form is workable and is the actual power of the compressions under the conditions stated, i.e., whatever the conditions are that define states "1" and "2."

D.3.5.2 Power for Compressible Fluid Flow—In General

The power expended for any kind of fluid compression is, in general,

$$P = Q(\text{gas}) \cdot \rho(\text{gas}) \cdot w \quad \text{(D.60)}$$

in which w is the work of compression or expansion per unit mass, whether adiabatic, isothermal, or polytropic (J/kg)

Equation D.60 applies to any kind of compression, whether adiabatic, isothermal, or polytropic in which w is known. It applies also to irreversible processes, assuming w is known.

D.3.5.3 Power for Standard Temperature and Pressure

In some cases, conversion to standard temperature and pressure (stp) is needed. [As a note, "standard" conditions may be defined in accordance with the conventions of a particular group. There is not a universally adopted STP.]

A key principle for compressible fluid flow is that the mass flow is constant from section to section (for steady state conditions). Therefore, the mass flow at any section, "1," equals the mass flow at any other section, "2." If "2" is for "standard" conditions, then, the continuity relation is

$$Q_1\rho_1 = Q(\text{stp})\rho(\text{stp}) \quad \text{(D.61)}$$

in which
- $Q(\text{stp})$ is the flow of gas at standard temperature and pressure (m^3/s)
- $\rho(\text{stp})$ is the density of gas at standard temperature and pressure (kg/m^3)

with STP defined,
- $p(\text{stp})$ is the standard pressure, 101,325 Pa, or 1.00 atm
- $T(\text{stp})$ is the standard temperature of gas, 0°C or 273.15 K
- $V(\text{stp})$ is the volume of gas at STP, which is 0.0224 m^3 for mass of 1.00 mol

Substituting Equation D.61 in Equation D.59 gives

$$P(\text{stp}) = \left[Q(\text{stp}) \frac{\rho(\text{stp})}{\rho_1} \right] p_1 \frac{k}{k - 1} \left[\left(\frac{p_2}{p_1} \right)^{(k-1)/k} - 1 \right] \quad \text{(D.62)}$$

Equation D.62 is one method to obtain $P(\text{stp})$. An equivalent form is obtained by the sequence of steps that follow. First, multiply $\rho(\text{stp})/\rho_1$ by MW/MW to give $\rho(\text{molar, stp})/\rho(\text{molar})_1$. Then, substitute from the gas law,

$$p_1 = \rho(\text{molar})_1 RT_1 \quad \text{(D.63)}$$

to give

$$P(\text{stp}) = \left[Q(\text{stp}) \frac{\rho(\text{molar, stp})}{\rho(\text{molar})_1} \right] \rho(\text{molar})_1 RT_1 \frac{\kappa}{\kappa - 1} \times \left[\left(\frac{p_2}{p_1} \right)^{(\kappa-1)/\kappa} - 1 \right] \quad \text{(D.64)}$$

The $\rho(\text{molar})_1$ terms cancel to give

$$P(\text{stp}) = [Q(\text{stp})\rho(\text{molar, stp})] RT_1 \frac{\kappa}{\kappa - 1} \times \left[\left(\frac{p_2}{p_1} \right)^{(\kappa-1)/\kappa} - 1 \right] \quad \text{(D.65)}$$

and since

$$\rho(\text{molar, stp}) = \frac{n}{V} = \frac{p(\text{stp})}{RT(\text{stp})} \quad \text{(D.66)}$$

substituting Equation D.66 in D.65 gives

$$P(\text{stp}) = Q(\text{stp}) \frac{n(\text{stp})}{V(\text{stp})} RT_1 \frac{\kappa}{\kappa - 1} \left[\left(\frac{p_2}{p_1} \right)^{(\kappa-1)/\kappa} - 1 \right] \quad \text{(D.67)}$$

Now, since at STP, $n = 1.00$ mol and $V = 0.0224\ m^3$,

$$P(\text{stp}) = Q(\text{stp}) \frac{1.00\ \text{mol}}{0.0224\ \text{m}^3} \frac{8.321\ \text{J}}{\text{mol K}} T_1 \frac{\kappa}{\kappa - 1} \left[\left(\frac{p_2}{p_1} \right)^{(\kappa-1)/\kappa} - 1 \right] \quad \text{(D.68)}$$

$$= 321 Q(\text{stp}) T_1 \frac{\kappa}{\kappa - 1} \left[\left(\frac{p_2}{p_1} \right)^{(\kappa-1)/\kappa} - 1 \right] \quad \text{(D.69)}$$

Equation D.69 is the form given by McCabe et al. (1993, p. 211). It was derived here since it a common reference.

In conversion to U.S. Customary units, the equation is the same but the terms are in U.S. Customary units. Equation D.67 is repeated, i.e.,

$$P(\text{stp}) = Q(\text{stp})\frac{n(\text{stp})}{V(\text{stp})}RT_1\frac{\kappa}{\kappa-1}\left[\left(\frac{p_2}{p_1}\right)^{(\kappa-1/\kappa)}-1\right] \quad (D.70)$$

$$P(\text{stp}) = Q(\text{stp})\cdot\frac{1.0\ \text{lb-mol}}{359\ \text{ft}^3}\cdot\frac{1544\ \text{ft-lb}_\text{f}}{\text{lb-mol}\ ^\circ R}\cdot T_1\cdot\frac{\kappa}{\kappa-1} \times\left[\left(\frac{p_2}{p_1}\right)^{(\kappa-1)/\kappa}-1\right] \quad (D.71)$$

$$P(\text{stp}) = 4.30\cdot Q(\text{stp})\cdot T_1\cdot\frac{\kappa}{\kappa-1}\left[\left(\frac{p_2}{p_1}\right)^{(\kappa-1)/\kappa}-1\right] \quad (D.72)$$

Note that the adiabatic equations for compression apply to a reversible, i.e., isentropic, path and that the compressor efficiency term, η, should be added to the denominator (the fact that the "path" is "irreversible" is included in η along *with* the effects of mechanical friction). The pressure p_1 is the intake pressure in this case and p_2 is the pressure on the exit side of the compressor. Also, if the compression is truly adiabatic, i.e., no heat transfer, then the work of compression goes to increasing the internal energy (as seen by the first law, $\Delta E = q - w$, in which $q = 0$ for an adiabatic process. Thus, since $\Delta E = c_v \cdot \Delta T$, the temperature must increase. In a real compressor, the temperature rise may be a serious problem (causing bearing wear and other mechanical effects) and so the compressor may require cooling. If cooling is done, the compression is no longer adiabatic, by definition. Probably, such compression would be "polytropic," thus requiring analysis that is beyond the scope of this review.

D.3.5.4 Spreadsheet Calculation of Compressor Power

The power required by a compressor is given by the Equation D.59 for an adiabatic compression. If the compression is isothermal, then, the associated work function, Equation D.53, applies along with Equation D.60.

Table CDD.5 is a spreadsheet calculation of power for an adiabatic compression where the air flow is that required for an aerated grit chamber, which is used for the illustration. The spreadsheet was developed for both SI units and U.S. Customary units. The associated definitions in both systems of units are given below the calculation portion of the table in the respective columns. Guidelines for use of the spreadsheet are, briefly:

- Spreadsheet calculates air flow using an empirical relation.
- The pressure p(atmos) is the atmospheric pressure and is calculated by entering the elevation.
- A pressure, p_2, on the discharge of the compressor was assumed for the purpose of demonstrating the working of the spreadsheet.
- Temperature must be assumed.
- Spreadsheet then calculates power for adiabatic compression by the equations described previously.

D.3.5.5 Spreadsheet Calculation of Compressor Power Combined with Pipe Flow

A sequel to the two spreadsheets, Tables CDD.3 and CDD.5, is a combined spreadsheet that involves determining pipe diameters for a pipe system that provides a flow of air to an array of submerged diffusers and calculation of the pressures required. From these pressures, the compressor power can then be determined. Table CDD.6 combines the two spreadsheets, Tables CDD.3 and CDD.5, that do these tasks individually, i.e., pipe sizing/compressor pressure and compressor power, respectively.

D.4 FLOW MEASUREMENT

Flow measurement is a basic need in plant operation. Conventional technologies to measure flow are based upon the Bernoulli principle combined with the continuity equation. These include the traditional technologies, i.e., orifice plates, Venturi meters, weirs, and the Parshall flume. In recent years, new technologies have also become common, such as magnetic flow meters and electrical conductance. Propeller meters have been used for many years in pipes, both to give instantaneous flows and to integrate the flows over a given time to give a volume that has passed a given point. Figure D.9a and b are photographs of an orifice plate and a Venturi meter, respectively.

D.4.1 DEFINITIONS

Flow refers to the rate of fluid volume per unit time that passes a given point or section. The term "flow rate" is used commonly, but some have asserted that the term "flow" implies a rate and so the word "rate" is redundant. The term "rate of flow" is used by Rouse (1946).

For further reference, the term "flux" is defined as, "the rate of flow of fluid, particles, or energy through a given surface." The term, "flux density" is the flux per unit of area.

The measurement of a volume that has passed a point or a section of a pipe or open channel is sometimes called a "volumetric flow" measurement, a contradictory phrase, but one that is apropos. Volumetric measurement of a flow over a given time is necessary to calibrate a flow meter. Usually, a tank or basin is used for this purpose, with depth measurement at the beginning and the end of the time measurement.

D.4.2 CHARACTERISTICS OF FLOW MEASUREMENT

Flow meters should need low maintenance and be accurate, easy to use, reliable, and inexpensive. Such criteria

TABLE CDD.5
Air Flow Calculation for Compressor for an Aerated Grit Chamber (Example)

(a) Metric Units

1 Constants

Constant	Value	Units	Reference
R =	8.314510	J $K^{-1}mol^{-1}$	Table B.2
MW(air) =	0.0289641	kg/mol	Table B.7
k =	1.395		

Metcalf & Eddy (1991) gives

Q(air) =	2.0	5.0	ft^3/min/ft length
=	0.03333	0.08333	ft^3/s/ft length
=	0.00310	0.00774	m^3/s/m length

2 Air Flow			**3 Ambient and Operating Conditions**							
q(air) (m^3/min/m)	**L(Gr Ch) (m)**	**Q(air) (m^3/s)**	**Elev. (m)**	**p_1(atm) (Pa)**	**T_1 °C**	**ρ_1(air) (mol/m^3)**	**ρ_1(air) (kg/m^3)**	**p_2 (Pa)**	**w(op) (kw/kg)**	**P (kw)**
0.0031	10	0.031	1600	82,874	0	36.49	1.057	303,975	123.2	4.03
0.0031	10	0.031	0	101,325	0	44.61	1.292	303,975	101.0	4.04
0.0031	10	0.031	0	101,325	0	44.61	1.292	303,975	101.0	4.04
0.0031	10	0.031	0	101,325	0	44.61	1.292	303,975	101.0	4.04
0.0031	10	0.031	0	101,325	0	44.61	1.292	303,975	101.0	4.04
0.0031	10	0.031	4000	61,302	20	25.15	0.728	303,975	170.5	3.85
0.0031	10	0.031	0	101,325	20	41.57	1.204	303,975	108.4	4.04
0.0031	10	0.031	0	101,325	20	41.57	1.204	303,975	108.4	4.04
0.0031	10	0.031	0	101,325	20	41.57	1.204	303,975	108.4	4.04
0.0031	10	0.031	4000	61,302	40	23.54	0.682	303,975	182.1	3.85
0.0031	10	0.031	0	101,325	40	38.92	1.127	303,975	115.8	4.04
0.0031	10	0.031	0	101,325	40	38.92	1.127	303,975	115.8	4.04

q(air) = empirical guideline from Metcalf & Eddy
Q(air) = q(air) · L(Gr Ch)
Assumed length
Assumed elevation
Diffuser pressure + Δp(losses)
$w = p_1/\rho_1 \cdot (k/k - 1) \cdot [(p_2/p_1)\wedge((k - 1)/k) - 1]$
' = compression
Use barometric pressure or default value which has formula for elevation
i.e., p(atm) = 101,325 * 10^(−0.00005456*Z)

Other categories calculated
4 Standard temperature and pressure conditions and power

(b) U.S. Customary Units

are met by orifice plates for pipe installations and weirs, if an overflow situation is a part of the design. Venturi meters are favored by many, as headloss is low and they are accurate, but they are expensive relative to orifice plates, e.g., an order of magnitude higher. Installation of flow meters on both the influent and the effluent sides of a filter is most desirable. Influent side installation permits adjustment of flow, as desired by the operator. On the filter effluent side, a total flow meter is required, permitting measurements of total flow volume delivered.

The selection process described for orifice plates and weirs is based upon the procedures outlined by Roberson and Crowe (1985). Installation of any meter in a pipeline should be done downstream of bends or any other kind of disturbance. Although rules of thumb recommend locating several pipe diameters downstream from a disturbance, a better rule is "the longer the better." Fluctuations in manometers may be observed even, for example, after a meter is located 50–100 pipe diameters downstream of bends. Also, provision should be made for easy cleaning of the pressure taps, which may become clogged over time, and for inspection of the flow meter for deposits or erosion. Taps on the side of the pipe will minimize clogging due to sediment and will avoid the problem of gases, which could occur for taps at the top of a pipe. A flow meter is accurate only if the conditions of its calibration are duplicated in the field.

D.4.3 Orifice Plate Meter

Orifice meters are recommended for flow measurement because they are simple, cheap, and accurate. An orifice meter is a flat plate, with a hole in the center, placed between two flanges in the pipeline. The edge around the hole should be sharp, so that standard coefficients can be used, as given in Table D.7. The flow is proportional to the square root of

TABLE CDD.6
Combined Pipe Flow and Compression Spreadsheet

Flow is assumed to be isothermal in pipe and adiabatic across compressor

1 Gas Constants

Constant	Value	Units	Ref.
R =	8.314510	$J\ K^{-1}mol^{-1}$	Table B.2
MW(air) =	0.02896	kg/mol	Table B.7
k =	1.395		
g =	9.81		

2 Hydraulic Constants

Const.	Value	Units
f =	0.012	Pipe friction
ρ_w =	998	kg/m^3
C =	0.61	Orifice coefficient

4 Ambient Conditions						**5 Intake Conditions at "1"**			
z(elev)		**p(atm)**	**T(atm)**	z_1	**ρ(atm)**	$z_1\rho g$	P_1	T_1	ρ_1
(ft)	**(m)**	**(Pa)**	**(C)**	**(m)**	(kg/m^3)	**(Pa)**	**(Pa)**	**(C)**	(kg/m^3)
0	0	101,325	0	10	1.292	127	101,325	0	1.2922
5500	1676	82,083	0	10	1.047	103	82,083	0	1.047
5500	1676	82,083	0	10	1.047	103	82,083	0	1.047
5500	1676	82,083	0	10	1.047	103	82,083	0	1.047
5500	1676	82,083	0	10	1.047	103	82,083	0	1.047
5500	1676	82,083	0	10	1.047	103	82,083	0	1.047
5500	1676	82,083	0	10	1.047	103	82,083	0	1.047

Elevation above sea level — Density of air for ambient conditions — Temperature at "1"

Measured (or assumed) atmospheric temperature — Pressure at intake

Elevation of intake pipe above a datum — Density of

Computed pressure from Lide data — Energy of air at intake due to elev.

Other categories calculated in spreadsheet

3 Air flow for aerated grit chamber (Metcalf & Eddy, 1991),
6 Air flow to grit chamber
7 Pipe design "1–2"
8 Calculation of HGL level at "2"
9 Conditions at "5"
10 Orifice flow/sizing/Δp determinations to get p_4
11 Pressure loss between "3" and "4" to get HGL_3
12 Summary - for input data to compressor
13 Ambient and operating conditions
14 Standard temperature and pressure conditions and power
15 McCabe et al. (1993)

(a)

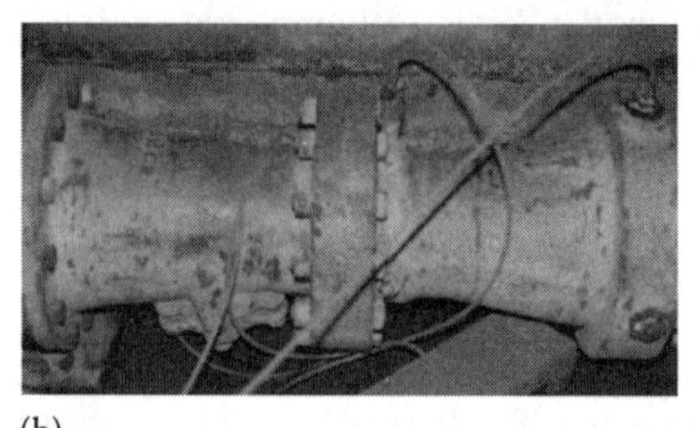

(b)

FIGURE D.9 (a) Orifice plates (b) Venturi meter (3 and 4 in. for 8 in. pipe).

TABLE D.7
Meter Coefficients for Flow Measurement

Meter Type	Metering Formula	d/D	C_D
Orifice	$Q = C_D A\,(2g\Delta H)^{0.5}$	0.10	0.60
		0.20	0.60
		0.30	0.60
		0.40	0.62
		0.50	0.63
		0.60	0.65
		0.70	0.70
		0.80	0.77
Venturi	$Q = C_D A(\text{throat}) \cdot [2g\Delta H]^{0.5}$	0.50	1.00
		0.60	1.04

Notes: (1) D is diameter of pipe and d is diameter of orifice or Venturi throat. (2) Coefficients are for $\mathbf{R} \geq 10^5$; for most flows, $\mathbf{R} \geq 10^5$ can be assumed. (3) For $\mathbf{R} < 10^5$ can, the coefficient increases for orifices and decreases for Venturi meters.

the pressure differential between upstream and downstream pressure taps, as indicated by Equation D.73. Figure D.10a is a sketch showing the main features of an orifice meter and terms used in Equation D.73.

The standard orifice equation is

$$Q = C_d A(\text{orifice}) \cdot [2g\Delta h]^{1/2} \tag{D.73}$$

and

$$C_D = f(d/D) \text{ per Table D.7} \tag{D.74}$$

in which
- Q is the flow of fluid in (m^3/s) or (ft^3/s)
- C_D is the orifice discharge coefficient, given in Table D.8 (dimensionless)
- A(orifice) is the cross-sectional area of orifice opening (m^2) or (ft^2)
- g is the acceleration of gravity (9.8 m/s^2) or (32. ft/s^2)
- Δh is the pressure differential across the plate (m) or (ft)
- d is the diameter of orifice in orifice plate (m) or (ft)
- D is the diameter of pipe (m) or (ft)

TABLE D.8
Weir Coefficients for Flow Measurement

Meter Type	Metering Formula	H/P	C_w
Rectangular weir	$Q = C_w A\,(2g)^{0.5} w \cdot \Delta H^{0.5}$	0.00	0.40
		0.10	0.405
		0.20	0.41
		0.30	0.415
		0.40	0.42
		0.50	0.43
		0.60	0.43
		0.80	0.44
Meter Type	**Metering Formula**	**θ**	CV_H
V-notch weir	$Q = (8/15) C_{VH} \cdot [2g]^{0.5} H^{2.5}$	60°	0.58
Metric	$Q = 0.79 H^{2.5}$	60°	(H in m)
U.S. Customary	$Q = 1.44 H^{2.5}$	60°	(H in ft)

Terms Q, flow in m^3/s (ft^3/s); C_w, weir coefficient; g, acceleration of gravity (9.81 m/s^2).

Notes: (1) For a rectangular weir, $C_w = 0.40 + 0.05H/P$. (2) For a V-notch weir, $C_{VN} = 0.58$ when $\theta = 60°$. (3) General formulae work for SI or U.S. Customary units.

The coefficient, C_D, is a function of the ratio, d/D, and is given in Table D.7. Over a period of time, as the average daily flow increases, a new orifice plate having a larger diameter may have to be installed. These plates should be on hand at plant start-up so that the operator has easy access to such replacements.

To measure pressure differential, pressure gauges are recommended. Mercury manometers should not be used because of the hazard of "blowing" the mercury, an acute health hazard. An orifice plate should be installed in a flanged section of the pipe so that removal is easy for cleaning of the pressure taps and to remove any debris trapped at the plate edge. The orifice plate should be located in a long straight length of pipe to minimize disturbances caused by eddies. In the event that continuous recording is desired, pressure transducers may be installed.

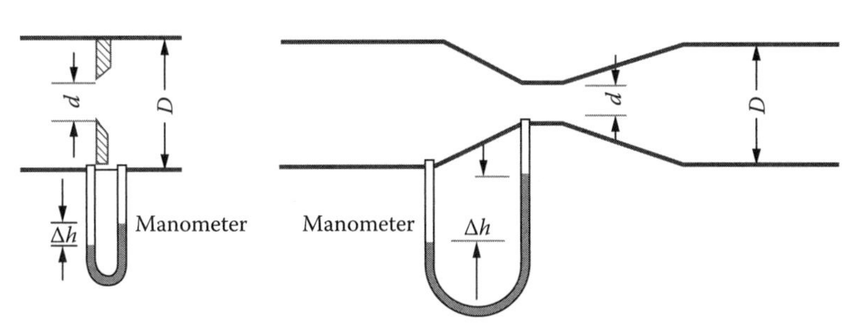

FIGURE D.10 Definition sketches for Table D.7.

An alternate form of Equation D.73 that is used more commonly for gas flow is

$$Q = C_d A(\text{orifice}) \cdot [\Delta p/\rho]^{1/2} \quad (D.75)$$

in which
Δp is the pressure loss across orifice (N/m^2)
ρ is the density of fluid (kg/m^3)

Note that Equations D.73 and D.75 are valid for any fluid, liquid or gas.

Example D.1 Orifice Plate Size

Calculate the size of orifice plate for Empire. Assume the pipe is 30.54 cm (12 in.). Assume also that permissible headloss across the orifice plate is 60 cm.

Given data for slow sand filter plant serving small town in Colorado

1. Q(max, 1,000 persons) = 3028 L/person/day × 1,000 persons = 3,028,000 L/day (= 800 gpcd · 1,000 persons = 800,000 gal/day)

Calculation

1. First trial.
   a. From Table D.7, select $C_D = 0.62$
   b. Calculate using Equation (D.73)

$$Q = 0.62\ A(\text{orifice}) \cdot (2gh)^{1/2}$$

$$3028/24/3600 = 0.62[\pi(d)2/4](2 \times 9.8 \times 0.60)1/2$$

$$d = 0.14 \text{ m } (5.7 \text{ in.})$$

2. Check C_D.
   a. $d/D = 0.14/0.35 = 0.4$
   b. From Table D.7, $C_D = 0.615$, which is close enough to the original assumption.
3. Determine headloss at the low flow during the season of minimum demand and with 500 persons, i.e., 278 Lcd (100 gpcd), giving $Q = 189{,}250$ L/day (50,000 gpd).

$$189/24/3600 = 0.62[\pi(0.14)2/4](2 \times 9.8 \times h)1/2$$

$$\Delta h = 0.003 \text{ m} \quad \text{or} \quad 3 \text{ mm } (0.12 \text{ in.})$$

Discussion

This amount of headloss for the low flow condition is too little to be measured accurately, so we must use a smaller orifice for the early years and substitute a larger orifice plate as the population increases. If we select say 50 mm as a measurable headloss for the low flow condition, the corresponding diameter is 70 mm (2.75 in.). Thus, we will use this smaller size for the early years.

For both orifice plates, headloss-flow graphs should be drawn permitting a determination of desired flow range for each of the two, based upon desired headloss range.

Figure D.9a is a photograph of an orifice plate, designed to fit into a flanged pipe. The graphical relationship with Δh for the abscissa and Q for the ordinate should be provided both as a plot and with proper coefficients for the SCADA system used in most plants since the early 1990s. The operator should have the means to convert immediately the headloss, as read by pressure gauges or piezometers, to flow.

D.4.4 Venturi Meter

The Venturi meter requires less headloss than an orifice plate and so may be more desirable when head is a major concern. Figure D.9b is a sketch of a Venturi meter showing terms used in the flow equation and the configuration of the manometer installation. *Note*: The manometer used in the drawing to illustrate pressure differential, pressure gauges, or transducers are recommended because manometer fluids may be toxic and should be avoided as they are subject to being "blown" into the flow.

The cost of a Venturi meter to fit into a 30.5 cm (12 in.) pipeline is expensive relative to an orifice plate. Equation D.76 is the flow equation, which is similar to the orifice equation. The coefficients for the Venturi meter are given in Table D.7. As noted, the coefficient is about 1.0, contrasted with about 0.6 for an orifice plate.

$$Q = C_v A_t (2g\Delta h)^{1/2} \quad (D.76)$$

and

$$C_v = f(d/D) \quad (D.77)$$

in which
Q is the flow of water in (m^3/s) or (ft^3/s)
C_v is the Venturi meter discharge coefficient = 1.0, Table D.8 (dimensionless)
A_t is the cross-sectional area of Venturi throat (m^2)
g is the acceleration of gravity (9.81/s^2)
Δh is the pressure head difference between upstream pipe section and throat (m)
d is the diameter of orifice in orifice plate (m)
D is the diameter of pipe (m)

D.4.5 Rectangular Weir

A rectangular weir can control water levels and measure flow. For the overflow within the filter box, a simple rectangular weir with contracted ends will control water level so that excess flow (i.e., in excess of what the filter can handle) will not over-top the filter box. Flow measurement is, of course, a requirement. To calculate the length of weir, the standard equation is applicable:

$$Q = C_W \sqrt{2g}\mathbf{w}H^{1.5} \quad (D.78)$$

in which

$$C_W = 0.40 + 0.05\frac{H}{P} \quad (D.79)$$

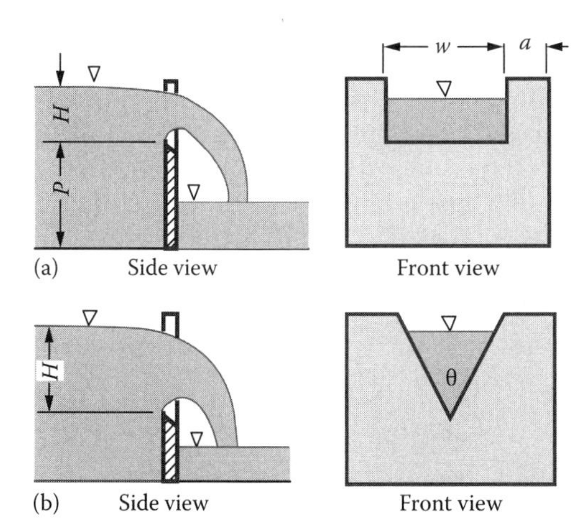

FIGURE D.11 Definition sketches for Table D.8. (a) Rectangular weir-end contractions, (b) triangular weir.

and

Q is the flow (m^3/s) or (ft^3/s)
C_w is the weir coefficient (dimensionless)
g is the acceleration of gravity (9.81 m/s^2) or (32.2 ft/s^2)
w is the length of weir crest (m) or (ft)
H is the height of water level above weir crest upstream from effect of drawdown (m) or (ft)
P is the distance from floor of channel to weir crest (m) or (ft)

Figure D.11a is a sketch of a rectangular weir with "end contractions," i.e., the crest does not extend across the channel. A "broad-crested" weir is one in which the weir plate extends across the channel, with the context being an open channel.

Example D.2 Rectangular Weir Design for Tailwater

Calculate the length of rectangular weir, **w**, for the tailwater overflow measurement, taking the entire flow, at Empire. Assume that the permissible depth of water above the weir crest, H, is 10 cm (0.328 ft) and that $P = 2.0$ m (6.6 ft).

Given data for Empire

1. Q(max, 1000 persons) = 3,028 L/person/day · 1,000 persons = 3,028,000 L/day
(= 800 gpcd · 1,000 persons = 800,000 gal/day)

Calculation

1. Substitute numerical data in Equation D.78:

$$Q = C_W\sqrt{2g}\,\mathbf{w}H^{1.5}$$

$$3028/24/3600 = \left[0.40 + 0.05\frac{0.10}{2.0}\right]\sqrt{2 \cdot 9.81}\,\mathbf{w}(0.10)^{\frac{3}{2}}\mathbf{w}$$

$$= 0.62 \text{ m } (2.0 \text{ ft})$$

Discussion

With a head range of only 10 cm, which corresponds to a flow of 3 million L/day, the accuracy of flow measurement will be acceptable. At the same time, the rectangular weir will control the tailwater elevation with only 10 cm change from zero flow to the maximum flow.

Example D.3 Tailwater Weir Design for 100 Mile House, B.C.

Show the required head, **H**, for the tailwater overflow weir used in the design at 100 Mile House, British Columbia.

Given data for 100 Mile House, British Columbia,
The design peak flow for 100 Mile House, British Columbia, for the slow sand filter plant of three cells, was 7.26 mil L/day (1.92 mgd).

1. Flow per Cell

$$Q[\text{cell}] = Q[\text{plant}]/\text{No. Cells}$$
$$= 7.260 \text{ m}^3/\text{day}/3 \text{ cells}$$
$$= 0.028 \text{ m}^3/\text{s/cell}$$

2. Assume that weir is circular, and has a diameter 0.40 m, and $P = 2.0$ m. Calculate the head on the weir.

Substitute numerical data in Equation D.78:

$$Q = C_W\sqrt{2g}\,\mathbf{bH}^{3/2}$$

$$0.028 \text{ m}^3/\text{s} = \left[0.40 + 0.05\frac{H}{2.0}\right]\sqrt{2 \cdot 9.81} \cdot [\pi(0.40)] \cdot \mathbf{H}^{3/2}$$

$$\mathbf{H} = 0.25 \text{ m } (0.82 \text{ ft})$$

Discussion

Overflow weir: The tailwater overflow weir for one of the cells at the 100 Mile House plant is circular. The overflow weir crest is the level of the maximum height of the sand bed. At the maximum flow, the head on the weir will be about 0.25 m (0.82 ft). The flow from the weir is captured by a circular vessel with pipe outflow to the chlorine contact basin. A flow meter is located after the exit from the filter under-drains.

Valve to control headwater elevation: Shown also is a valve on the downstream side of the flow meter. Such a valve may be used to raise the headwater elevation, after scraping, to alleviate sand bed erosion. After about 0.5 m of headloss, the valve may be opened fully. A valve may be used in lieu of a vertically movable tailwater weir, but requires more operator attention for a few days after the bed is scraped and before the initial 0.5 m of headloss.

Backfilling filter bed: A pipe should lead from the chlorine contact basin to the under-drain effluent pipe to backfill the filter after scraping. Such an ancillary component is easy to overlook, but is needed to permit start-up after scraping so that air-binding can be avoided and that the raw water flow can be re-introduced without eroding the sand bed.

D.4.6 Triangular Weir

A triangular weir, often called a "V-notch" weir, is accurate as a metering device and is cheap. Since the elevation difference is larger over a given flow range, it is less satisfactory than the rectangular weir for water surface elevation control. The flow-head relation is

$$Q = \frac{8}{15} C_{VH} \sqrt{2g} \tan\left(\frac{\theta}{2}\right) H^{2.5} \quad \text{(D.80)}$$

in which
- Q is the flow (m^3/s) or (ft^3/s)
- C_{VN} is the weir coefficient $= 0.58$ when $\theta = 60°$, given in Table D.8 (dimensionless)
- g is the acceleration of gravity (9.81 m/s^2) or (32.2 ft/s^2)
- b is the length of weir crest (m) or (ft)
- H is the height of water level above weir crest upstream from effect of drawdown (m) or (ft)
- θ is the angle of notch in triangular weir (degrees)

For a weir having $\theta = 60°$, $C_{VN} = 0.58$, the discharge equation is for metric units,

$$Q = 0.79H^{5/2} \quad \text{(D.81)}$$

and for U.S. Customary units,

$$Q = 1.44H^{5/2} \quad \text{(D.82)}$$

Terms are illustrated in Figure D.11b. Usually, a weir with a 60° notch is used since the coefficients are usually readily available from references.

Example D.4 Triangular Weir Head Calculation

Calculate the head on a 60° triangular weir at Empire for maximum and minimum flows.

1. For maximum flow,

Q(max, 1,000 persons)
$= 3{,}028$ L/person/day · 1,000 persons
$= 3{,}028{,}000$ L/day
($= 800$ gpcd · 1,000 persons $= 800{,}000$ gal/day
$= 106{,}952$ ft^3/day)

Metric	U.S. Customary
$Q = 0.79H^{5/2}$	$Q = 1.44H^{5/2}$
$3{,}028/24/3{,}600 = 0.79H^{5/2}$	$106{,}952/24/3{,}600 = 1.44H^{5/2}$
$H = 0.29$ m (29 cm)	$H = 0.94$ ft (11.3 in.)

2 For minimum flow,

Q(min, 500 persons)
$= 278$ L/person/day × 500 persons $= 189{,}250$ L/day
$= 100$ gpcd × 500 persons $= 50{,}000$ gal/day
$= 6{,}685$ ft^3/day)

Metric	U.S. Customary
$Q = 0.79H^{5/2}$	$Q = 1.44H^{5/2}$
$189/24/3600 = 0.79H^{5/2}$	$6685/24/3600 = 1.44H^{5/2}$
$H = 0.095$ m (9.5 cm)	$H = 0.31$ ft (3.7 in.)

Discussion

The tailwater level range from 9.5 to 29 cm should be acceptable, and so the triangular weir would be the choice for flow measurement or for the dual role of tailwater elevation control and flow measurement. The rectangular weir is the choice if tailwater control is the only function.

D.4.7 Propeller Meters

A propeller meter is simply a propeller placed in a matched pipe section that is coupled to a flow volume indicator (register). Such meters come in a range of sizes from 50 mm (2 in.) to 3 m (120 in.) and are used widely in all kinds of situations (Huth, 1990). Such meters are accurate to within ±2% throughout a 10:1 or 20:1 operating range.

The propeller has three or six blades and rotates in proportion to the velocity in the pipe section. The volume of flow that has passed the propeller is proportional to the number of revolutions, i.e.,

$$V = k(\text{m}) \cdot n(\text{rev}) \quad \text{(D.83)}$$

in which
- V is the volume of flow that has passed the propeller (m^3) or (ft^3)
- k(meter) is the coefficient of proportionality to calibrate flow meter (m^3/rev) or (ft^3/rev)
- n(rev) is the number of revolutions of propeller associated with volume V (rev)

Example D.5 Show the Derivation of Equation D.83

The volume of flow which has passed the pipe section, e.g.,

$$dV = Q \cdot dt$$
$$= v(\text{pipe}) \cdot A \cdot dt$$
$$= [k(\text{meter}) \cdot \text{rpm}] \cdot A \cdot dt$$
$$= [k(\text{meter}) \cdot dN/dt] \cdot A \cdot dt$$
$$V = k(\text{meter}) \cdot N$$

in which

V is the volume of flow that has passed the propeller (m^3) or (ft^3)
Q is the flow in pipe (m^3/s) or (ft^3/s)
t is the time for flow of volume, V (s)
v(pipe) is the velocity of water flowing within pipe (m/s) or (ft/s)
A(pipe) is the cross-sectional area of pipe (m^2) or (ft^2)
k(meter) is the coefficient of proportionality between velocity in pipe and rotational speed of propeller (m/s/rev) or (ft/s/rev)
k(meter) is the coefficient of proportionality to calibrate flow meter (m^3/rev)
N is the number of revolutions of propeller associated with volume V (rev)

From Equation D.83, the number of revolutions is converted to the flow volume reading on the register. Calibration of each flow meter must be done before installation to determine k'. The flow reading is by a digital register located at the top of the meter. Inside the meter, a coupling connects the propeller to the internal mechanism and may be direct drive or magnetic drive, as described by Huth. The magnetic drive eliminates the need for packing seals and thus water is not likely to enter the register.

The bearings of the propeller meter may be water lubricated ceramic sleeve bearings or stainless steel ball bearings. The ceramic sleeve type requires less maintenance and has longer life than the stainless steel bearings (Huth, 1990). Also, according to Huth (1990), the meters can be equipped to measure flow as well as volume. Clogging due to debris is a possible problem and so installing propeller meters on the upstream side of the sand bed is not advised if a debris hazard is present.

ACKNOWLEDGMENTS

On thermodynamics and discussion of compressors, Dr. Paul Wilbur, professor of mechanical engineering, provided orientation on practical aspects of compressor selection and helped in reviewing the thermodynamics of compressors. He is a faculty member at Colorado State University.

Dr. Larry Weber, director, and Cornelia Mutel, archivist and historian, IIHR-Hydroscience & Engineering at the University of Iowa, answered the question of exactly where and when Dr. Hunter Rouse received his doctorate in Germany. It turns out that his doctoral studies were under Professor Theodor Rehbock, a well-known hydraulic engineer and academic. Mutel also provided the photograph of Dr. Rouse, obtained from the Archives of IIHR-Hydroscience & Engineering.

GLOSSARY

Absolute temperature: Defined: $T(K) = 273.15 + °C$; $T(R) = 459.6 + °F$.

Adiabatic: Zero heat transfer during a change of state.

Bernoulli: The Bernoulli equation accounts for the various forms of energy in fluid flow; Daniel Bernoulli, 1700–1782 is credited with formulating the Bernoulli theorem, but the fact is that he did not quite achieve this as we know it today, according to Rouse and Ince (1957, pp. 91–100).

Cycle: A process in which the initial and final states are identical (Shapiro, 1958, p. 24).

Darcy–Weisbach: (1) Refers to equation for headloss in pipe flow, i.e., $h_L = f(L/D)(v^2/2g)$. (2) Julius Weisbach (1806–1871) was a professor of mathematics at the Freiberg School of Mines; his main interests were in hydraulics and geodesy. He published in 1877 a comprehensive treatise on hydraulics that Rouse and Ince (1957) credit as modernizing the topic and as being the model for twentieth century texts on fluid mechanics. He was the first to write the foregoing headloss equation and to understand the friction factor. (The foregoing from Rouse and Ince, 1957). The connection between Darcy and Weisbach was not indicated. (3) See also Brown (2000) for comprehensive history of the Darcy-Weisbach equation.

Entropy: Defined mathematically as $\Delta S = (q/T)_{rev}$; conceptually entropy increase is associated with increased "disorder."

Extensive property: Quantity depends upon mass, e.g., V, E, H, G, etc.

Friction factor: The Darcy–Weisbach friction factor, f. The Moody diagram is the source for f data currently.

Gas constant: The universal gas constant is the constant in the ideal gas equation; in SI units, $R = 8.314\ 510$ $N\ m \cdot K^{-1}\ mol^{-1}$. Also, $R = k$ (Boltzmann constant) $\cdot$ N (Avogadro's number).

Head: Specific energy of terms in the Bernoulli relation, i.e., $N\ m/(kg \cdot g)$

Header: A pressurized conduit that serves a collection of smaller conduits that are connected, usually in a "T" format, to the larger conduit. The header conduit is the main feed for a manifold.

Heat capacity: Two kinds of heat capacity are

c_P: defined as the rate of change of enthalpy with temperature at constant volume, i.e., $(\partial H/\partial T)_V$

c_V: defined as the rate of change of internal energy with temperature at constant pressure, i.e., $(\partial E/\partial T)_p$

Hydraulic grade line: Locus of points defined by $(z + p/\gamma)$ along a pipeline.

Hydraulic loading rate: Defined as flow divided by area of cross section, i.e., $v = Q/A$; same as superficial velocity.

Hydraulics: That branch of engineering describing the flow of liquids in terms of the associated dependent and independent variables.

Intensive property: Quantity that is independent of mass, e.g., temperature, pressure, partial molar free energy, density, etc.

Interstitial velocity: Defined as flow divided by pore area of cross section, i.e., $v = Q/(AP)$.

Irreversible work: If the system and the surroundings cannot be restored to their initial states, once a process is started, the process is irreversible. All real processes are irreversible. Three causes of irreversibility are viscosity, heat conduction, and diffusion and are molecular level phenomena (Shapiro, 1958, p. 30). For an irreversible process, it must always be true that $dS \geq dq/T$.

Isentropic: Adiabatic and frictionless change of state, meaning a reversible path is followed in the compression or expansion.

Isothermal: Constant temperature process.

κ: Ratio of specific heats, i.e., $c_p/c_v \approx 1.4$.

Manifold: A pipe or chamber having multiple apertures for making connections (Oxford American Dictionary, 1980).

Manometer: An instrument filled with a liquid, e.g., glycerin, mercury, that measures pressure by means of a U-tube. If, for example, the pressure is to be measured within a tank or pipe, one end to the tube is attached to the respective wall. The other end is exposed to the atmosphere, as a rule. The height of rise of the liquid, e.g., mercury, in the manometer is the basis for calculating the pressure in the pipe or tank. The specific weight of the fluid on each side of the measuring fluid, e.g., mercury, must be included in the calculation of pressure.

MW(air): Determined to be 28.9 g/mol = 0.028964 kg/mol.

Nozzle: A device shaped to accelerate a fluid (Munson et al., 1998, p. 121). Usually a nozzle is a short pipe length that has varying diameter with length, generally reducing to a "throat" section, intended to cause a high velocity at its outlet.

Orifice: A circular opening in a plate or pipe for controlled flow of a fluid.

Piezometer: A tube attached to a point at which pressure is to be measured. The height of rise of the liquid (usually, water is the liquid) in the tube is the basis for calculating the pressure at the point of attachment of the tube. For example, if the water rises 11.3 m (33.9 ft) above the point of attachment, the pressure is 11.3 m water head, which is 101.325 kPa (14.7 psi), or 1.00 atm.

Pneumatic grade line: (1) Locus of points defined by $(z\gamma + p)$ along a pipeline that flows with a gas. The pneumatic grade line is to the flow of gases as the hydraulic grade line is to the flow of liquids. But in lieu of using hydraulic head $(z + p/\gamma)$, pressure is used instead. The reason is that the "head" for a gas would be inordinately large, since its density, ρ, is much smaller than that of a liquid. (2) The idea of a pneumatic-grade-line was not found in the literature, but was based on discussions with Professor Robert Meroney, Colorado State University (Meroney, 1998) and may or may not be unique to this text. As a graphical depiction of pressures in a pipeline, however, its use has utility similar to that of the hydraulic-grade-line.

Pneumatic: A term that relates to gas, either static or flowing.

Pneumatics: That branch of engineering describing the flow of gases in terms of the associated dependent and independent variables or that describes the state relationships of a static gas, such as in a pressurized tank.

Polytropic: Change of state in which both temperature and pressure change.

Pressure head: Pressure due to a column of fluid above a given elevation, expressed as "meters of water" with dimension of length, as abbreviation of energy per unit of fluid mass; expressed mathematically as p/γ.

Process: A change of "state" of a system and may be described by the series of states passed through by the system (Shapiro, 1958, p. 24).

Reversible work: In a reversible expansion of a gas, the classic example of a reversible process, the internal pressure is just infinitesimally higher than the external pressure, and in the limit, $\Delta p \rightarrow dp$ (see Pitzer and Brewer, 1961, p. 35). In reversible work, there is no entropy production (p. 81). A reversible process is a standard against which real processes may be evaluated (Shapiro, 1958, p. 30). For a "cycle," the entropy change for a reversible process is zero.

State: The "state" of a gas is defined by T, P, V, which are "state" variables. In general, a state is the configuration of a system described in sufficient detail such that one state may be distinguished from another (the latter is from Shapiro, 1958, p. 24).

STP: Acronym for "standard temperature and pressure," sometimes called, NTP for "normal temperature and pressure." Standard values may vary, depending upon the application. For chemists, STP means 0°C and 1.00 atm pressure (Silberberg, 1996, p. 186). In engineering, a standard temperature of 20°C is common. A standard temperature of 25°C may be used for some purposes.

Venturi: Refers to the Venturi meter. Giovanni Batista Venturi (1746–1822) was an Italian physicist who conducted experiments and showed that a reduction of eddies was brought about by a gradual transition of a boundary rather than an abrupt transition and that the throat pressure was the minimum. The Venturi meter was, however, an invention of Clemens Herschel (1842–1930), described in an 1899 paper, "The Venturi Water Meter," with the name coming from the practice in his laboratory of calling the throat pressure the "Venturi." Hershel combined the already existing ideas into a meter for flow measurement (foregoing from Rouse and Ince, 1957).

Work: The effect produced by a system on its surroundings or by the surroundings on the system.

REFERENCES

Albertson, M. L., Barton, J. R., and Simons, D. B., *Fluid Mechanics for Engineers*, Prentice-Hall, Englewood Cliffs, NJ, 1960.

Brown, G. O., The history of the Darcy–Weisbach equation for pipe flow resistance, paper given at meeting commemorating the *150th Anniversary of the American Society of Civil Engineers*, November 2000. [Professor Brown's e-mail address is gbrown@okstate.edu; the paper was made available through e-mail as MS Word® and MS Power Point® files by Professor Neil Grigg, Colorado State University, Fort Collins, CO.]

Camp, T. R. and Stein, P. C., Velocity gradients and internal work in fluid motion, *Journal of the Boston Society of Civil Engineers*, 30:219–237, October, 1943 (reprinted in *Civil Engineering Classics—Outstanding Papers of Thomas R. Camp*, American Society of Civil Engineers, New York, 1973).

Cheremisinoff, N. P. and Cheremisinoff, P. N., *Pumps/Compressors/Fans—Pocket Handbook*, Technomic Publishing Company, Lancaster, PA, 1989.

Daugherty, R. L. and Ingersoll, A. C., *Fluid Mechanics*, McGraw-Hill, New York, 1954.

Ettema, R., Hunter Rouse—His work in retrospect, *Journal of Hydraulic Engineering*, 132(12):1248–1258, December 2006.

Huth, E., Propellor meters prove economical and durable, *Opflow*, 16(6):1, 1990.

Kennedy, J. F. and Macagno, E. O., *Selected Writings of Hunter Rouse*, Dover Publications, Inc., New York, 1971.

McCabe, W. L., Smith, J. C., and Harriott, P., *Unit Operations of Chemical Engineering*, 5th edn., McGraw-Hill, Inc., New York, 1993.

Meroney, R. N., personal communication (discussion of application of Bernoulli relation to gas flow), August 1998.

Munson, B. R., Young, D. F., and Okiishi, T. H., *Fundamentals of Fluid Mechanics*, 3rd edn., John Wiley & Sons, New York, 1998.

Pitzer, K. S. and Brewer, L. (revised by), *Thermodynamics—By Gilbert Lewis and Merle Randall*, 2nd edn., McGraw-Hill, New York, 1961.

Roberson, J. A. and Crowe, C. T., *Engineering Fluid Mechanics*, 3rd edn., Houghton Mifflin Co., Boston, MA, 1985.

Rouse, H., *Elementary Mechanics of Fluids*, John Wiley & Sons, New York, 1946.

Rouse, H., Evaluation of boundary roughness, in *Proceedings of the Second Hydraulics Conference*, University of Iowa Studies in Engineering, Bulletin No. 27, 1943, in Kennedy, J. F. and Macagno, E. O., *Selected Writings of Hunter Rouse*, Dover Publications, Inc., New York, 1971.

Rouse, H., Discussion of "Friction Factors for Pipe Flow," by Lewis F. Moody (Rouse's discussion published in *Transactions ASME*, 66(8), 1944, in Kennedy, J. F. and Macagno, E. O., *Selected Writings of Hunter Rouse*, Dover Publications, Inc., New York, 1971.

Rouse, H., *Hydraulics in the United States, 1776–1976*, Institute of Hydraulic Research, The University of Iowa, Iowa City, IA, 1976.

Rouse, H. and Ince, S., *History of Hydraulics*, Iowa Institute of Hydraulic Research, State University of Iowa, Iowa City, IA, 1957.

Shapiro, A. H., *The Dynamics and Thermodynamics of Compressible Fluid Flow*, Parts I and II from Volume I, The Ronald Press Co., New York, 1958.

Silberberg, M., *Chemistry*, Mosby—A Times Mirror Company, St. Louis, MO, 1996.

Appendix E: Porous Media Hydraulics

Porous media for this discussion includes any granular material that comprises a bed for treatment (but may include various kinds of materials with a pore structure that permits a flow under a pressure gradient). The hydraulics of porous media flow is reviewed (Box E.1); the idea is to be able to estimate headloss for a given hydraulic loading rate.

E.1 APPLICATIONS

Treatment processes that utilize granular media include adsorption by activated carbon, ion-exchange, and deep-bed filtration. Also diatomaceous earth, smaller in size and only 5–50 mm in depth of cake, is a porous medium. The characteristic sizes of media are about 0.5 mm for granular activated carbon (GAC), 0.5 mm for sand in deep-bed filtration, with about 0.9 mm for anthracite. In deep beds, such as say 3 m depth, a coarse medium such as 2 mm anthracite may be likely. In groundwater flow, the media may range from clays to sand and gravel, with a size range from microns to centimeters. In filter beds, the granular media is characterized by d_{10} and d_{50}; in other applications, the mean diameter is used commonly. Figure C.4 provides a graph that may be used for plotting the results of a sieve analysis to obtain these parameters. In addition, membrane filtration (Chapter 17) occurs under a hydraulic gradient in which Darcy's law is applicable.

E.2 HYDRAULIC FLOW REGIMES

Hydraulic flow regimes for porous media range from laminar to turbulent and are characterized by the Reynolds number. The starting point is to define the Reynolds number. Then, Darcy's law, the cornerstone for hydraulic theory (Freeze, 1994), can be defined. Box E.2 is a slight diversion from the main theme, which is intended to clarify terms, some of which have emerged since the 1990's.

E.2.1 Reynolds Number

The Reynolds number for flow through porous media, (Box E.2) is defined as

$$\mathbf{R}(\text{porous media}) \equiv \frac{\rho v d}{\mu} \tag{E.1}$$

in which

- $\mathbf{R}$(porous media) is the Reynolds number for porous media flow
- v is the superficial velocity defined as Q/A(cross section) (m/s)
- d_{50} is the mean diameter of granular media (m)
- μ is the dynamic viscosity of fluid at a given temperature (Ns/m^2)
- ρ is the dynamic viscosity of water at a given temperature (kg/m^3)

In the groundwater field, the characteristic length is d_{50}, but d_{10} is used here since it is used in the water treatment field to characterize media, along with the uniformity coefficient (UC). Conversion from d_{10} to d_{50} can be estimated by the relation, $d_{50} \approx d_{10}[1 + 0.8(\text{UC} - 1)]$. If a sieve analysis is done, the d_{50} can be obtained directly.

E.2.2 Four Flow Regimes

Table E.1 shows the four flow regimes—laminar, inertial, transition, and turbulent, based upon $\mathbf{R}$(porous media) criteria as given by Trussell and Chang (1999). The characteristics of each regime are described in the following enumeration:

1. *Laminar*: The "laminar" flow regime, i.e., $\mathbf{R}$(porous media) ≤ 1, is characterized by viscous effects governing flow and is described mathematically by Darcy's law (Box E.3), i.e.,

$$v = -K\frac{dh}{dz} \tag{E.2}$$

in which

- v is the superficial velocity, i.e., hydraulic load ing rate, Q/A (m/s)
- K is the hydraulic conductivity of porous medium (m/s)
- h is the hydraulic head, i.e., $(p/\gamma + z)$ (m)
- z is the coordinate is direction of velocity vector, generally vertical in filtration (m)

In the laminar regime, a dye injected at some point can be observed to follow a streamline path.

2. *Inertial*: As $\mathbf{R}$ increases slightly, the laminar condition will prevail, i.e., a dye will continue to follow the streamline path but the v versus dh/dz starts to deviate from linearity and is described mathematically by the Forchheimer equation,

$$\frac{dh_L}{dz} = \alpha_F v + \beta_F v^2 \tag{E.3}$$

in which

- α_F is the coefficient related to linear head loss (s/m)
- β_F is the coefficient related to nonlinear head loss (m^2/s^2)

BOX E.1 NOMENCLATURE ON PRESSURE

The idea of pressure is simple, but its not uncommon to see different terms used to mean the same thing, which can make one wonder what is meant. The explanation here is done to excess if one is familiar with the teachings of a first course in fluid mechanics. If not, the discussion may serve as a refresher and also may provide a common understanding for the terms used.

Pressure is a common measurement, done traditionally by a "piezometer," a "manometer," or a pressure gage, or more recently, a pressure transducer. A piezometer is merely a tube inserted at a point "A" where the measurement is to be taken, such as in a pipeline or a bed of porous medium. The liquid rises to a level, "h_A," above point "A," where the pressure, p_A, at "A" equals $h_A \cdot \gamma_w$, or $p_A = h_A \cdot \gamma_w$, where γ_w is the specific weight of water. A calibrated pressure gage measures p_A at directly as does a pressure sensor. The measurement, h_A, is the "hydraulic head at "A" (p_A/γ_w). A "manometer" measures the difference in pressure between two points (such as in a pipeline).

The commonly used term, "hydraulic-head" refers the sum of the elevation head, "z_A," at the point of measurement, "A," in a given "flow-field," plus the pressure head, "p_A/γ_w," where p_A is the pressure in (N/m^2 or Pa) at point "A" and γ_w is the specific weight of water in N/m^3. In other words, $h_A = p_A/\gamma_w + z_A$; velocity head is negligible. For reference, keep in mind that $\gamma_w(20°C) = \rho_w(20°C) \cdot g = (998.2\ \text{kg/m}^3) \cdot (9.807\ \text{m/s}^2) \approx 9789\ \text{N/m}^3$. The elevation head, "$z_A$," is the vertical distance (m) with respect to any arbitrary datum (a convenient reference elevation).

In any flow, pipeline or ground water, the hydraulic head gradient is usually called, simply, the "hydraulic gradient." The "velocity-head" "$v_A/2g$" is also included the definition for h_A, which is the kinetic energy of the water at "A"; usually its negligible in flow through porous media.

In ground water flow, h_A is also termed hydraulic potential with the symbol ϕ. In a "flow-net," the locus of points of constant potential is called a "potential" line. In ground water flow, the "flow-field," as depicted by a "flow-net" may have curvature, depending on the "boundary-conditions." In a "clean-bed" (homogeneous) rapid filter, the "flow-field" is characterized by an hydraulic-head gradient that is linear with depth and streamlines that are vertical. The associated "flow-net" is a rectangular grid. As the media "clogs" with floc, the hydraulic head gradient becomes steeper at the top where the floc deposits have a higher concentration. In a slow sand filter most of the headloss occurs at the surface, where a "schmutzdecke" usually forms.

The terms used in hydraulic gradient for a column are: Δh, which is the headloss across the sand bed of depth, ΔZ, and is the difference in elevation between the headwater and the tail-water in the column, or $\Delta h = h(\text{tailwater}) - h(\text{headwater})$. The gradient, $\Delta h/\Delta Z$, is the "hydraulic-gradient," which may also be called the "hydraulic-head-gradient," or the "potential-gradient." As noted, in a rapid filter, the hydraulic-gradient starts out uniform with depth in the filter bed but as clogging occurs, the gradient becomes steepest at the top and declines to a clean-bed gradient. After some hours of operation, there are no net deposits of floc in the upper few centimeters of the filter bed and the hydraulic gradient is constant (in the upper few centimeters).

BOX E.2 VELOCITIES

Two velocities in porous media flow are: (1) superficial velocity, v, and (2) interstitial velocity, v(pores). In addition, the term, "specific-discharge," $\boldsymbol{q}$, emerged during the 1990s, which has the same mathematical definition as "superficial velocity," v, but it has a different connotation, important in the field of groundwater hydraulics.

Superficial velocity, "v": Consider a vertically oriented empty column of cross-sectional area, A. Suppose the top half of the column is empty and the bottom half is packed with sand. Let a flow, Q, enter the top of the column and exit at the bottom. The calculated average velocity of water in the top half of the column is, $v = Q/A$. In the lower part of the column, this same calculation is used to characterize the velocity within the sand-bed, i.e., v(superficial) = Q/A(column). To emphasize the point, the "superficial-velocity is a "pseudo-velocity." To further emphasize the point, the "superficial-velocity," v, is also called the "approach-velocity," or the "face-velocity."

Interstitial velocity, "v(pores)": The real average velocity in the sand bed described is the "interstitial" velocity and is termed here, v(pores), calculated as, $v(\text{pores}) = Q/(A \cdot P)$, where P is the average porosity of the sand bed. Porosity is the volume of voids divided by the total volume of voids plus solids. Of course the pores of any packed-bed are not uniform and so, at the microscopic level, v(pores) varies within the pore space, which is random.

Specific discharge, "q": The term, "specific-discharge," $\boldsymbol{q}$, is used commonly in the ground water field i.e., $\boldsymbol{q} = Q/\text{A}$. Mathematically $\boldsymbol{q}$ is the same as the superficial velocity, v. But its conceptually different in that it's unequivocally "true." Also, explication of $\boldsymbol{q}$ is not necessary.

BOX E.2 (continued) VELOCITIES

Convention uses superficial velocity, "v": The porous media Reynolds number, ***R(porous media)***, uses "v," the superficial velocity, as is the velocity, "v," in Darcy's law, i.e., $v = K \cdot [\Delta h/\Delta Z]$. The use of superficial velocity, "v," as opposed to "v(pore)," is by convention.

TABLE E.1
Flow Regimes for Porous Media Flow with Governing Equations

Flow Regime	R(porous media)	Equation	Equation Statement
Laminar	≤ 1	Darcy	$v = K(dh/dZ)$
Inertial	1–100	Forchheimer	$dh/dZ = \alpha_F v + \beta_F v^2$
Transition	100–800	Forchheimer	$dh/dZ = \alpha_F v + \beta_F v^2$
Turbulent	>800	Forchheimer	$dh/dZ = \alpha_F v + \beta_F v^2$

BOX E.3 DARCY'S LAW

Equation E.2 is the cornerstone of porous media hydraulics. It's an understatement merely to state the equation without some discussion.

As seen in Equation E.2, Darcy's law is a simple statement. It's an empirical relationship, discovered by Henry Darcy in 1856, that relates flow (m^3/s) per unit of gross cross-section area (m^2) through a column of porous medium to the hydraulic gradient across the column. Mathematically, its $Q/A = K \cdot \Delta h/\Delta Z$. Darcy's law applies to any "boundary conditions," such as found in geologic formations, as well as to the simple ones of a column of filter sand or anthracite, or an ion-exchanger, or an activated carbon column, or any other "packed-bed" reactor. As noted by Trussell and Chang (1999), applications have taken two parallel but independent tracks by such groups as hydro-geologists and civil engineers, respectively. The first group has focused on situations found in nature while the second has been concerned with engineered systems.

At low **R** the dh/dz is linearly related to v with small v^2 dependence but as **R** increases, v^2 becomes dominant in the relationship. The first appearance of true turbulence, i.e., inertial effects, occurs at $40 < \mathbf{R} < 140$, based upon visual studies. The inertial effects are due to the changes in the velocity vector, i.e., (1) changes in magnitude due to expansions and contractions as the flow enters and exits from various cells, and (2) changes in direction due to curvilinear flow around media particles.

3. *Transition*: The transition regime is characterized by the transition from inertial flow to full turbulence. At the lower end of the regime, turbulence is just beginning to appear in some of the cells; at the upper end, turbulence is present in most cells. The upper limit of this regime is not well defined but is above $\mathbf{R} \approx 300$ and is likely in the range $600 < \mathbf{R} < 800$. The Forchheimer equation form remains, but the constants α and β change to α_T and β_T.
4. *Turbulent*: Full turbulence is present with random fluctuating micro-velocities about the mean throughout the media. The Forchheimer type equation applies.

Designs in water filtration are in the Reynolds number range, $0.5 < \mathbf{R} < 50$, meaning that they are in either the "laminar" or "inertial" ranges (Trussell and Chang, 1999). The Darcy equation applies in the laminar regime while the Forchheimer equation applies throughout both the laminar and the inertial regimes. As examples of **R** values that occur at the extremes of rapid rate filtration practice

1. Let $v = 6.1$ m/h (2.5 gpm/ft^2) with $d_{10} = 0.5$ mm and $T = 20°C$, then $\mathbf{R} \approx 0.9$.
2. Let $v = 37$ m/h (15 gpm/ft^2) with $d_{10} = 2.0$ mm and $T = 20°C$, then $\mathbf{R} \approx 20$.

For the upper limit for most conventional designs, e.g., say $v = 24$ m/h (10 gpm/ft^2) with $d_{10} = 1.0$ mm and $T = 20°C$, then $\mathbf{R} \approx 7$ and Darcy's law can be applied with little deviation from measured data. The Forchheimer equation, on the other hand, extends to the laminar range of **R**, since the v^2 term has little effect at small **R**. The latter two points are discussed subsequently. First, however, consider the equations applicable to laminar flow, i.e., the forms of the Darcy equation.

E.2.3 Experimental Demonstration of Darcy's Law for Filter Media

Data that illustrate the range of the applicability of Darcy's law for filter media were obtained by Chang et al. (1999) and are shown in Figure E.1a and b for 6 of 30 tests (the 30 tests involved 3 sizes of sand, 3 sizes of anthracite and glass beads, with tests for each media conducted for 3 or more porosity values). Figure E.1a shows that the $h_L/\Delta z$ versus v for 0.47 mm sand is linear throughout the range of data, i.e., $v \leq 0.01$ m/s, or 36 m/h (15 gpm/ft^2). Also of interest, $\mathbf{R} \approx 4.7$ at $v = 0.01$ m/s; thus Darcy's law is applicable as a means to predict headloss, for $\mathbf{R} < 5$ (the highest **R** for the data available).

Figure E.1b shows the same kind of $h_L/\Delta z$ versus v plot for 1.47 mm anthracite. The three curves start to deviate from linearity at about $v \approx 0.005$ m/s or 18 m/h (7.4 gpm/ft^2) at which $\mathbf{R} \approx 7$. Therefore, at $\mathbf{R} > 7$, the Forchheimer equation would be increasingly important as a means to describe the $h_L/\Delta z$ versus v relationship. But at $\mathbf{R} \leq 7$, Darcy's equation may be applied, which greatly simplifies the calculations of headloss.

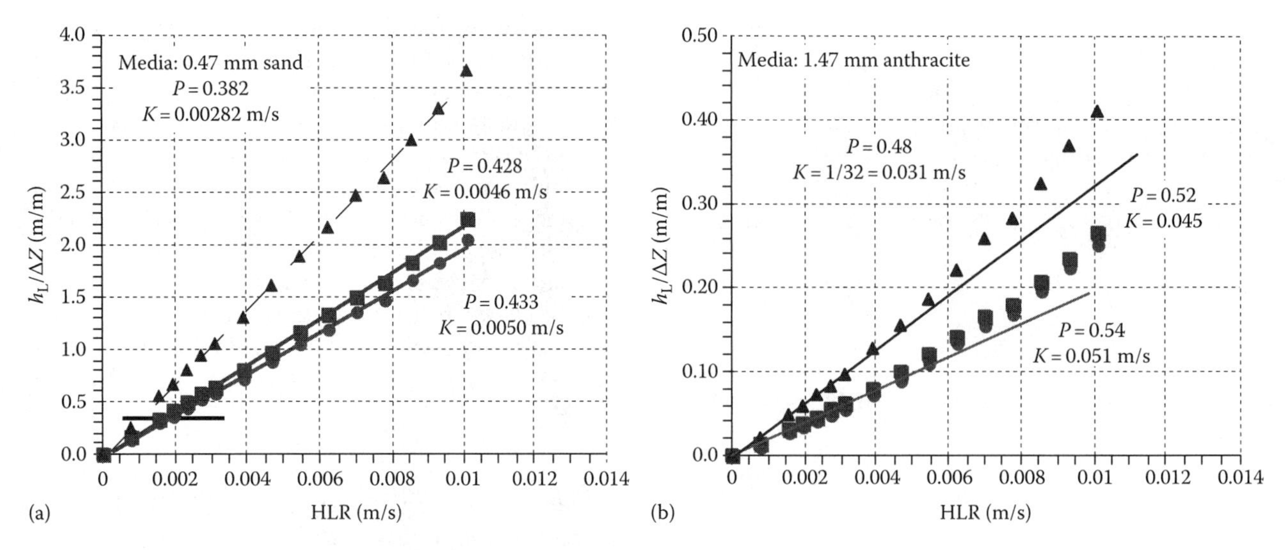

FIGURE E.1 Hydraulic gradient versus HLR for two media showing deviation from linearity and effect of porosity. (a) Sand with d_{10} = 0.47 mm (b) Anthracite with d_{10} = 1.47 mm. (Plots from data in Chang, M. et al., *Aqua*, 48, pp. 141&142, 1999.)

Figure E.1b characterizes the general shape of the other 24 plots with the point of deviation from linearity starting at $7 < \mathbf{R} < 8$. From these plots, we may generalize that Darcy's law may remain applicable for say $\mathbf{R} \leq 7$ for the tests conducted by Chang et al. (1999).

E.2.4 Headloss as a Parameter

In practice, filter beds are characterized often in terms of initial headloss, or "clean-bed" headloss, which does not incorporate the variables of Equation E.5, i.e., Δh is the only variable considered. Therefore the Δh is an index but is not as useful as K. Neither is the K term as useful as k.

E.3 INTRINSIC PERMEABILITY

The magnitude of K in Darcy's law is the function of both porous media and fluid properties. Therefore, the Darcy relation has broadest utility in terms of "intrinsic permeability," a property of the porous media per se.

E.3.1 Definition

The intrinsic permeability is related to hydraulic conductivity by the relation,

$$K = k\frac{\rho g}{\mu} \tag{E.4}$$

in which k is the intrinsic permeability of porous media (m^2).

The coefficient, k, is a property of the porous media as inferred by its dimensions, i.e., m^2, and is called "intrinsic permeability," and is distinguished from "hydraulic conductivity," which includes fluid properties.

E.3.2 Modified Darcy's Law

The fluid properties, μ and ρ_w, are the functions of temperature and so Darcy's law has greater utility if expressed,

$$v = -k\frac{\rho_w g}{\mu}\cdot\frac{dh}{dz} \tag{E.5}$$

The intrinsic permeability, k, is a function of the mean grain size, the statistical pore size distribution, and the pore structure; the latter has to do with the placement of the individual particles. The k values for slow sand media may vary significantly, since the uniformity coefficient may be perhaps as much as 3–5, as contrasted with a recommended 1.5 (since the 1.5 specification may be too expensive). On the other hand, the specifications for rapid rate filters are usually adhered to rather strictly and so the variation in k is probably not large from one installation to another. Data have not been collected, however, that permit a statistical view of k for different installations.

E.3.3 Conversions between Hydraulic Conductivity and Permeability

Equation E.4 provides the means to convert between hydraulic conductivity, K, and intrinsic hydraulic conductivity, k. The conversion may be done conveniently by means of a spreadsheet as illustrated in Table CDE.2(a) and (b), respectively. The μ and ρ_w values were calculated by polynomial formulae as given in Table CD/QR.4 (in this text, QR means "Quick Reference" and is an appendix table). Values of K may be determined from a hydraulic conductivity test as described in Section E.4.1.

Alternatively, if the intrinsic permeability k is given from empirical data, then K may be calculated for any given

TABLE CDE.2
Conversion between K and k Including Headloss Calculation from k

(a) K to k

	$g=9.807$				Enter K to Calculate k					
Media Name	d_{10} (mm)	d_{60} (mm)	d_{50} (mm)	UC	K (m/d)	K (m/s)	T (°C)	μ (Ns/m^2)	ρ_w (kg/m^3)	k (m^2)
Sand	0.50			1.5	2.42E+02	2.80E−03	3	0.00162	999.965	4.622E−10
Anthracite	0.91			1.5	1.26E+03	1.46E−02	3	0.00162	999.965	2.419E−09
Flatiron masonry	0.24			2.7	3.77E+01	4.37E−04	3	0.00162	999.965	7.215E−11
Flatiron masonry	0.24			2.7	4.08E+01	4.72E−04	3	0.00162	999.965	7.804E−11

(b) k to K

	$g=9.807$				Enter k to Calculate K					
Media Name	d_{10} (mm)	d_{60} (mm)	d_{50} (mm)	UC	k (m^2)	T (°C)	μ (Ns/m^2)	ρ_w (kg/m^3)	K (m/s)	K (m/d)
Sand	0.50			1.5	4.62E−10	3	0.00162	999.965	2.80E−03	2.4162E+02
Anthracite	0.91			1.5	2.42E−09	3	0.00162	999.965	1.46E−02	1.2644E+03
Flatiron masonry	0.24			2.7	7.21E−11	3	0.00162	999.965	4.37E−04	3.7717E+01
Flatiron masonry	0.24			2.7	7.80E−11	3	0.00162	999.965	4.72E−04	4.0800E+01

Notes:
$\mu(\text{water}) = 0.00178024 - 5.61324 \cdot 10^{-05} \cdot T + 1.003 \cdot 10^{-06}\, T^2 - 7.541 \cdot 10^{-09} \cdot T^3$.
$\rho(\text{water}) = 999.84 + 0.068256 \cdot T - 0.009144 \cdot T^2 + 0.00010295 \cdot T^3 - 1.1888 \cdot 10^{-06} \cdot T^4 + 7.1515 \cdot 10^{-09} \cdot T^5$.

conditions, e.g., as in Table CDE.2(b). From K, headloss may be calculated for a given HLR value and column length.

E.3.4 Permeability Data for Filter Media

As noted, the $h_L/\Delta z$ versus v data of Chang et al. (1999) for 30 tests were for 3 sand sizes, 3 anthracite sizes and 1.5 mm glass beads, and with 3 or more porosity levels for each media. Porosity was controlled, as feasible, by tapping on the 101 mm (4 in.) columns or by varying the rate of backwash termination. As noted, Figure E.1b characterized the trends found in each plot. The linear portion of a given plot was used to estimate the hydraulic conductivity, i.e., $1/\text{slope} = K$. The data also stated the temperature of each test which permitted the calculation of intrinsic permeability, k, by Equation E.5. As noted, the k is a characteristic of the media and so it has more utility than K (since temperature affects the latter).

Figures E.2 and E.3 show plots of k versus P and k versus d_{10}, respectively, derived from the Chang, et al. (1999) $h_L/\Delta z$ versus v data. The "groups" seen in Figure E.2 are for tests with different sands as characterized by their d_{10} values. A linear trend of k versus P is seen for each group.

Figure E.3 shows the same data but plotted as k versus d_{10}, and grouped by porosity; the two CSU data points are from Hendricks et al. (1991). An approximate envelope is shown for the data by the two lines (upper and lower).

From the k data, taken from either Figure E.2 or E.3, head loss may be calculated based on for any assumed set of design conditions, e.g., T, ΔZ, HLR. Note that any selection of k involves uncertainty, as suggested by the envelope of data seen in Figure E.3.

Example E.1 Calculation of Headloss for Given Conditions of Filter Media

Given
$d_{10} = 1.5$ mm anthracite, $P = 0.45$, $v = 18$ m/h (7.4 gpm/ft^2), media depth $= 2.0$ m, $T = 20$°C

Required
Headloss, i.e., clean bed

Solution

1. *Determine* k: From Figure E.3, for $d_{10} = 1.5$ mm and $P = 0.45$, $k \approx 4.0 \cdot 10^{-9}$ m^2. Comparing this with the k versus P plot of Figure E.2, adjust k upward slightly to give, $k \approx 4.5 \cdot 10^{-9}$ m^2.
2. *Determine* **R**: $\mathbf{R} = \rho v d_{10}/\mu = 998 \cdot 0.005 \cdot (1.5/1000)/0.001 = 7.5$, which is at the approximate upper limit for the application of the Darcy equation.
3. *Apply Darcy's equation*: Equation E.5 is the form applied and with numerical data substituted,

$$0.005 \text{ m/s} = 4.5 \cdot 10^{-9} \cdot (998 \cdot 9.81/0.001) \cdot (h_L/2.0)$$
$$= 0.0220 \cdot h_L$$
$$h_L = 0.227 \text{ m}$$

Discussion
First, the estimate of clean-bed headloss is reasonable, based upon experience. Second, although the plots, i.e., Figures E.2 and E.3 are not definitive, the trend in Figure E.3, i.e., k versus d_{10} for the loci of constant's, seems consistent. Third, there should be some estimate of uncertainty. If the porosity, P, was not stated, we would most likely state that $2 \cdot 10^{-9} < k < 4 \cdot 10^{-9}$ m^2 which would give, $100 < h_L < 200$ mm.

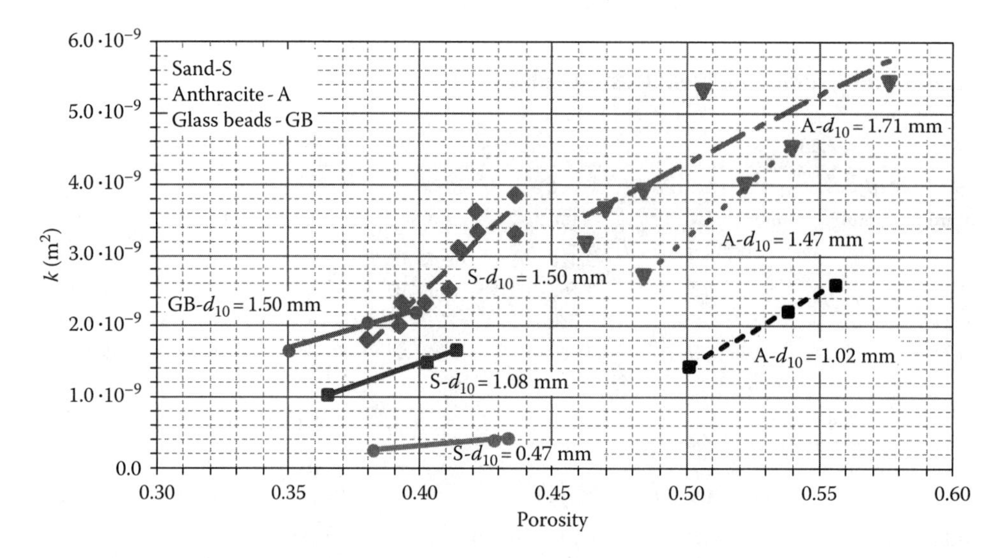

FIGURE E.2 Permeability k versus porosity for different d_{10} sizes. (Plots calculated by equation 5.5 from data in Chang, M. et al., *Aqua*, 48, pp. 141 & 142, 1999.)

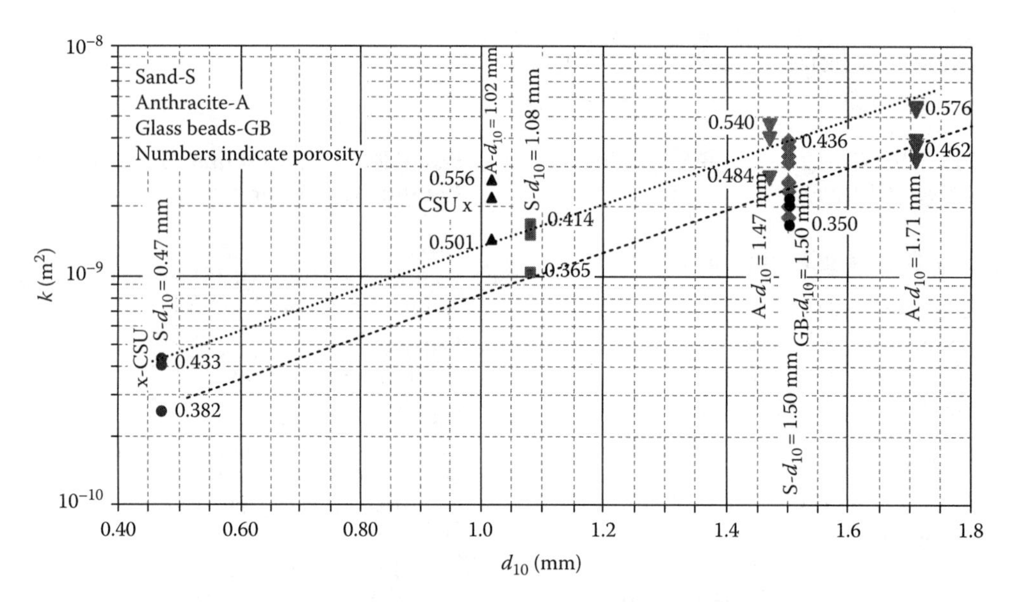

FIGURE E.3 Permeability k versus d_{10} with different porosities. (Plots calculated by Equation 5.5 from data in Chang, M. et al., *Aqua*, 48, pp. 141 & 142, 1999.)

E.3.5 Permeability Data for Aquifer Media

The intrinsic permeability, k, depends upon the average grain size, the pore size distribution, and the packing of the granular media and so is likely to vary even for media of the same average grain size. Therefore, in natural media these variables occur in various combinations and some, e.g., porosity and packing, may be different in situ compared with a laboratory test. This contrasts with granular media used in water treatment in which such variables are controlled by specifications.

Typical ranges in k for different categories of natural media are shown in Figure E.4. Media grain sizes are shown on the abscissa with ordinate values showing maximum and minimum values for k. The lighter vertical lines show the separation between categories of natural media, e.g., "clay and silt," "fine sand," etc. The logarithmic scales for both axes indicate the wide variation found in grain size and intrinsic permeability of natural media.

The shaded area at the top shows the envelope of Figure E.4 superimposed and indicates the contrast with natural media. The higher intrinsic permeability values for filter media may relate to the generally lower UC, specified usually as about 1.5. A datum for the Empire, Colorado, slow sand filter is shown with $d_{10} = 0.24$ mm and UC = 2.7, with $k = 7.2 \cdot 10^{-11}$ m^2. The other dots are for CSU pilot filters (calculations were from Mosher and Hendricks data, 1986) for sand with $d_{10} = 0.50$ mm and UC = 1.5 and $k = 4.6 \cdot 10^{-10}$ m^2; and for anthracite with $d_{10} = 0.91$ mm and UC = 1.5,

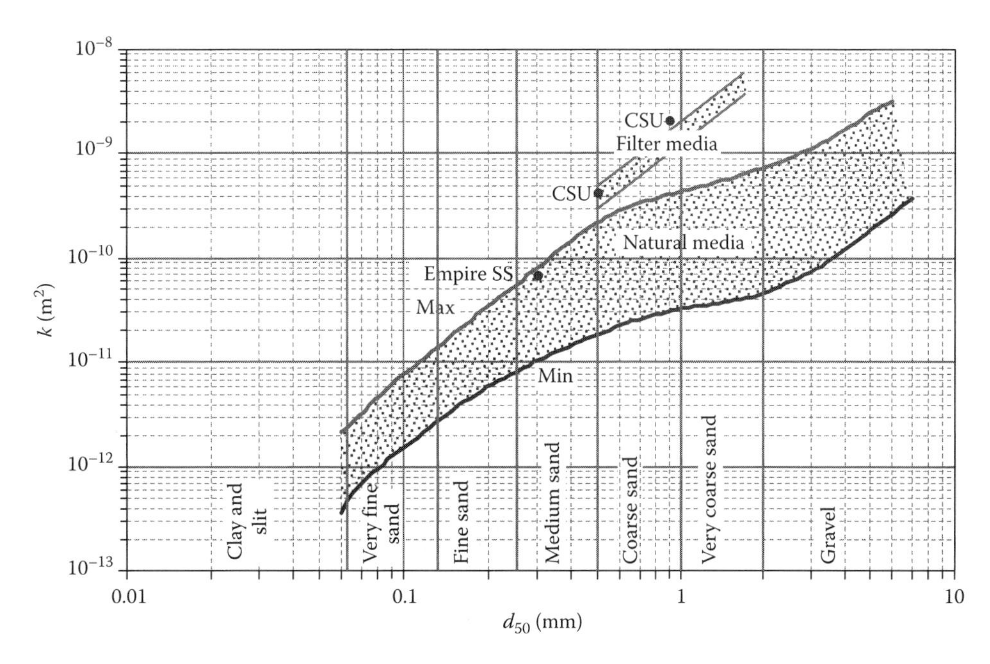

FIGURE E.4 Intrinsic permeabilities, k, for different average grain sizes and categories of granular media. (Adapted from Boulding, J.R., *Practical Handbook of Soil, Vadose Zone, and Ground-Water Contamination—Assessment Prevention, and Remediation*, Lewis Publishers, Ann Arbor, MI, 1995, p. 869; filter data calculated from Chang, M. et al., *Aqua*, 48, pp. 141 & 142, 1999.)

and $k = 2.4 \cdot 10^{-9}$ m^2. The headloss data were obtained from piezometers at 10 cm spacing. The slow sand data were for the whole bed (two beds in the filter) and the flow was not measured but judged based on the design flow capacity of the filter.

E.4 TESTS

Intrinsic permeability and porosity are two important variables in porous media flow. They are empirical and must be evaluated by laboratory tests.

E.4.1 Permeability Test

Always, for critical work, a laboratory test is preferred to determine the intrinsic permeability of a given media. Such a test involves a careful protocol which is summarized here conceptually. The basic idea of the test is to measure the hydraulic gradient through the media along with the HLR, i.e., v or 204, and then calculate K and then k (temperature being measured also).

Figure E.5 shows a bed of porous media, with length, $\Delta L = 18$ units, placed on its side with flow from "headwater"

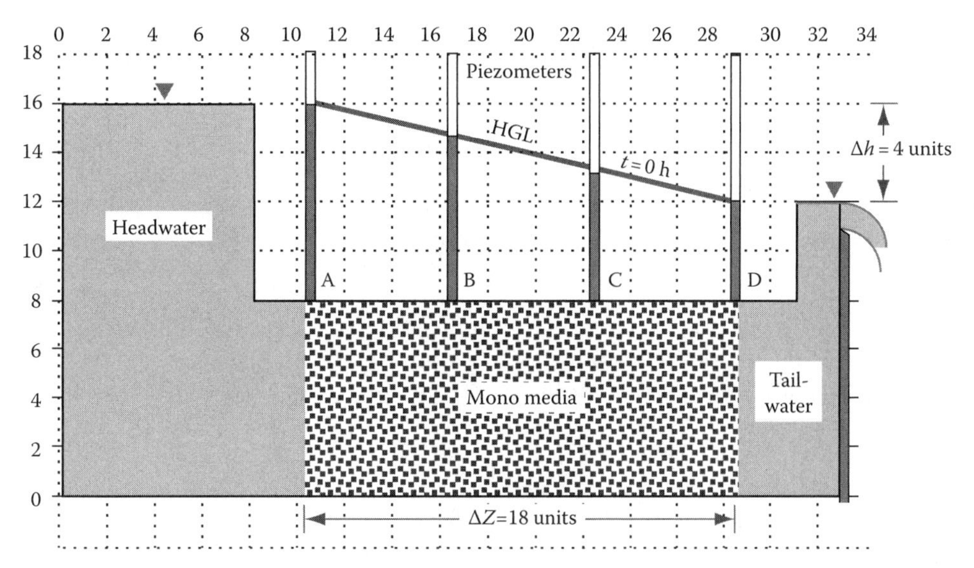

FIGURE E.5 Illustration of Darcy's law for porous media test bed with horizontal flow.

to "tailwater." Piezometers are placed at positions A, B, C, D. For the flow condition, the headloss, $\Delta h = 4$ units. Therefore, the hydraulic gradient, $dh/dz = \Delta h/\Delta z = 4/18 = 0.22$ units of head/unit of length. In addition to measuring hydraulic gradient, the flow, Q, must be measured, and the cross-section area must be measured ($v = Q/A$). Also the water temperature must be measured so that the fluid properties, ρ_w and μ may be calculated. From these data, k is calculated by Equation E.5.

The column may be oriented vertically or horizontally and for most situations Δh can be based on only the headwater and tailwater elevation difference (rather than a series of piezometers), which assumes a uniform hydraulic gradient. In some cases if a mat has developed on the surface of the bed, the piezometers give a more accurate hydraulic gradient. A caution is to remove the air from the column by using a slow displacement by air-free water rising from the bottom. The size of the column may be whatever is convenient but a size 10–15 cm (4–6 in.) diameter and perhaps 61 cm (24 in.) long would be sufficient to provide a "packing" that is statistically representative and to minimize wall effects. In addition, several tests, perhaps 10–12, should be conducted such that enough k values are obtained for calculating the statistics of the variation, e.g., average k and the standard deviation of the measurements. Repacking between tests would be preferred as opposed to a repetition with the same packing. As mentioned by Chang et al. (1999) the porosity of the media will affect the results of the test. For a given media, the porosity may vary, as indicated in Figure E.2, depending upon the rate of backwash termination or the amount of tapping on the side of the column. Thus, a technique is involved in packing the media. Porosity should be measured also to provide a more complete picture.

E.4.2 Porosity

Porosity is defined as the ratio of the volume of voids to the bulk volume. As shown by Chang et al. (1999) in Figure E.2, porosity for a given media has a significant effect on its intrinsic permeability. The porosity values, from Figure E.2 for media in pilot filters, show groupings as follows:

Media	UC	Porosity Range
Potter's beads	1.00	0.35–0.40
Sand	1.23–1.31	0.35–0.44
Anthracite	1.24–1.33	0.46–0.58

Trussell et al. (1999) estimated the porosity for a full-scale anthracite filter as $P \approx 0.48$, slightly less than the porosities measured for pilot scale filters of anthracite. The full-scale filter was located at the Aqueduct Filtration Plant at Sylmar, California, operated by the Los Angeles Department of Water and Power. The design capacity was 33.0 h/h (13.5 gpm/ft^2) and the bed was mono-media of anthracite, 1.8 m (6.0 ft) deep, with $d_{10} = 1.5$ mm, UC = 1.33. This single measurement, involving a considerable effort, constitutes available data for a full-scale filter. Even for pilot scale, the data provided by Chang et al. (1999) constitute available data.

Porosity data by Hsu (1994) were obtained for different media as follows:

Material	d_{10} (mm)	UC	Porosity
Coarse garnet	3.00	1.22	0.31
Anthracite	1.08	1.48	0.34
Garnet 8–12	1.43	1.60	0.33
Garnet 30–40	0.37	1.41	0.33
Fine garnet	0.11	3.44	0.36
Dowex 50 resin	0.53	1.31	0.37

The measurements by Hsu were by salt displacement. The columns were used for dispersion tests and had been tapped lightly to consolidate the media (i.e., to minimize arching). As seen, the data are on the low side of other measurements, such as by Chang et al. (1999) and by Boulding (1995).

As a matter of interest, the minimum possible porosity of a porous media of uniform spheres with rhombohedral packing is 0.259 (Scheidegger, 1960, p. 19; Muskat, 1946, p. 12). The maximum possible porosity for "face.centered" or cubic packing of spheres is 0.4764 (Muskat, 1946, p. 12). Muskat (1946, p. 13) states that the most stable array of spheres is rhombohedral since it has sufficient points of contact to provide support from any direction, while cubic is stable only to forces normal to the cell faces. He states further (p. 13) that:

> ...in natural assemblages, even when agitated to induce close packing, one should anticipate groups of spheres packed in orderly arrays separated by boundaries in which no orderly arrays are present and where the porosity is even higher than that of cubic packing. Such zones can be maintained because of the "bridging" of groups of particles under pressures less than the crushing strength of the particles. ...Moreover, it is found experimentally that assemblages of spheres, or even sand particles, will have porosities averaging about 40 percent in spite of careful efforts to induce closer packing, and even though the predominant array in the assemblage is rhombohedral with a porosity of only 26 percent. Theoretically, the actual size of the spheres has no influence on the porosity, but in the assemblages of natural materials this does not prove true.

Material	Porosity
Coarse sand	0.39–0.41
Medium sand	0.41–0.48
Fine sand	0.44–0.49
Fine sandy loam	0.50–0.54

Boulding (1995, p. 856) gave porosity ranges for different media: fine gravel, 0.20–0.40; coarse sand, 0.25–0.45; medium sand, 0.25–0.45; fine sand, 0.25–0.55; dune sand,

0.35–0.45; silt, 0.35–0.50; etc. Regarding filtration practice, Chang et al. (1999) noted that the rate at which backwash is terminated has a major effect on the porosity of the filter bed. A sudden stop of backwash, they noted resulted in higher bed porosity than a gradual termination.

E.4.3 Porosity Measurement

Two methods to determine porosity are (1) determining the volume of the media, and (2) determining the volume of the pores. The media volume method requires measuring the bulk volume of the media in place, *V*(bulk), and the oven-dry mass of the media, *M*(media), placed in the column and knowledge of its specific gravity, SG(media). The relation is basically that the bulk volume equals the volume of solids plus the volume of pores,

$$P = 1 - \frac{M(\text{media})}{\text{SG}(\text{media})\gamma_w V(\text{bulk})} \tag{E.6}$$

in which
P is the porosity of media
M(media) is the mass of media (kg)
SG(media) is the specific gravity of media, e.g., about 2.65 for sand and 1.4 for anthracite
ρ_w is the specific mass of water (kg/m^3)
V(bulk) is the bulk volume of media (m^3)

A second method is to determine the pore volume by first filling the media from the bottom with a concentrated salt solution. Briefly, the procedure is to fill the column from the bottom with a solution of NaCl at known concentration, *C*(voids), say 2000 mg/L, after first purging the media of residual water. The solution is brought exactly to the surface of the column. The volume of the column is calculated from its dimensions and the void space due to the media support and tubing is determined by water displacement. The salt solution is then purged with distilled water by displacing several volumes of void space and the volume is collected and measured to give *V*(purge). The chloride concentration of the dilute solution is measured by titration to give *C*(purge). Since the mass of salt is constant, the void volume can be calculated,

$$V(\text{voids}) \cdot C(\text{voids}) + V(\text{tubes}) \cdot C(\text{voids}) = V(\text{purge}) \cdot C(\text{purge}) \tag{E.7}$$

in which
V(voids) is the volume of voids in column of media (m^3)
C(voids) is the concentration of Cl^- in voids measured (kg/m^3)
V(purge) is the volume of solution collected after purging column with distilled water (m^3)
C(purge) is the concentration of Cl^- in V(purge) measured (kg/m^3)
V(tubes) is the volume of tubes and other support space under the media (m^3)

The chloride ion is an excellent tracer since it is largely nonreactive. Also, the larger the column, the less is the error of measurement of *V*(voids).

E.5 APPLICATION OF DARCY'S LAW

In flow of water through a bed of porous media the rate of headloss with respect to bed depth is constant if the media is uniform and is "clean." Figure E.6 illustrates such condition for $t = 0$. In rapid filtration, however, floc particles enter the bed and attach to the grains with the highest attachment density at the top, declining exponentially with depth. Such clogging causes k to decline with depth in the same fashion. Since Q must be constant if the flow does not change, then v must increase and so $\mathrm{d}h/\mathrm{d}z$ increases, which is depicted in the hydraulic profiles of Figure E.6.

Figure E.6 shows a filter bed oriented vertically, as in practice, with the velocity vector down. Piezometer taps are located at A, B, C, D. A series of hydraulic grade lines (HGL) are shown, for $t = 0$, 2, 4, 6 h. Note also that the bed is oriented vertically and so the hydraulic gradient is not seen as clearly is in Figure E.1, where the bed is on its side. At $t = 0$ h, the clean bed headloss is 4 units with the length of the column 12 units; therefore, $\Delta h/\Delta L = 4/12 = 0.33$ units of head/unit of length. A valve is located at E and the excess headloss between the headwater and tailwater is taken up by the valve (as shown on the right side of the HGL). The HGL at times 2, 4, 6 h shows the advance of the clogging front. Finally at 6 h, the valve at E is 100% open and the entire headloss is taken up by the media. As seen, most of the headloss is taken up by the clogged part of the media and the slope, $\Delta h/\Delta L$, is highest at the top of the bed. At the bottom of the bed the slope, $\Delta h/\Delta L$, remains as it was at $t = 0$ h. A valve at point E in Figure E.6 is common in filter design; the valve opens as the bed clogs, based upon maintaining constant flow. A simple design would be to omit the valve and let the water level rise as the filter bed clogs. The bed would have to be designed, however, with a higher weir crest so that the bed would have sufficient water depth above the media to avoid hydraulic scour.

E.6 MODELS OF PERMEABILITY

A quest of porous media modeling has been to calculate k from first principles. Such a quest is like seeking the Grail and in the case of porous media, like other modeling efforts, there is always one coefficient remaining that must be determined empirically. Then to determine that coefficient, one may as well have conducted the basic laboratory testing to get k in the first place.

Like many mathematical models, however, a more basic understanding is the fulfillment. Also, we may examine trends with mathematical models and do sensitivity analysis even if that one last coefficient is not determined (we can assume a number such as "1" for the purpose of exploration of trends).

The starting point for most models of flow through porous media is the Hagen–Poiseuille equation in which the pores of

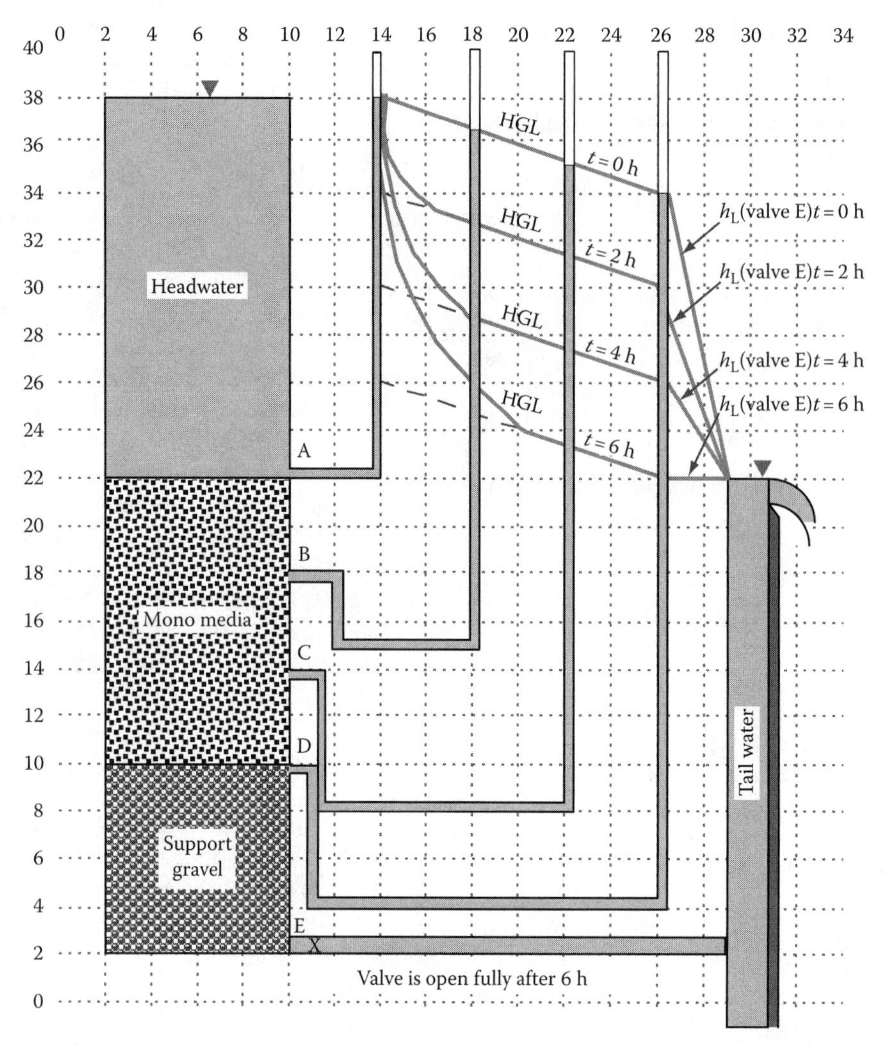

FIGURE E.6 Illustration of hydraulic gradient change with depth as filter bed clogs with time.

the media are considered a bundle of capillary tubes. Comparing this equation, i.e., Equation E.5, with the Darcy equation, K for a capillary tube is

$$K = \frac{d^2 \rho g}{32\mu} \tag{E.8}$$

and thus, $k = d^2/32$, for a single tube. The idea is that we can see that k is proportional to the cross-section area of a single tube and therefore its dimension, L^2 is clarified (see also Muskat, 1946, p. 12). This rationale leads to the Kozeny theory of 1927 which represents the porous medium as an assemblage of channels of various cross sections but definite length (Scheidegger, 1960, p. 125). The equation derived (Scheidegger, 1960, p. 128) was

$$k = \frac{cP^3}{S^2} \tag{E.9}$$

in which

c is the Kozeny constant which depends upon pore shape, i.e., 0.50 for a circle; 0.60 for an equilateral triangle
P is the porosity of media, i.e., ratio volume voids to volume media expressed here as a decimal fraction
S is the pore surface area per unit volume of porous media (m^2/m^3)

Trussell and Chang (1999) traced the historical development of porous media theory and showed how the Kozeny theory was extended to give a form

$$k = \frac{P^3}{(1-P)^2} \frac{1}{2\xi S^2} \tag{E.10}$$

in which ξ is the tortuosity coefficient.

If the media is uniform spheres, $S = (\pi d^2)/(\pi d^3/6)$ and letting $2\xi = 5$ as proposed by other experimenters as reviewed by Chang et al. (1999), Equation E.61 becomes

$$k(\text{spheres}) = \frac{1}{180} \frac{P^3}{(1-P)^2} d(\text{sphere})^2 \tag{E.11}$$

in which d(sphere) is the diameter of uniform spheres (m).

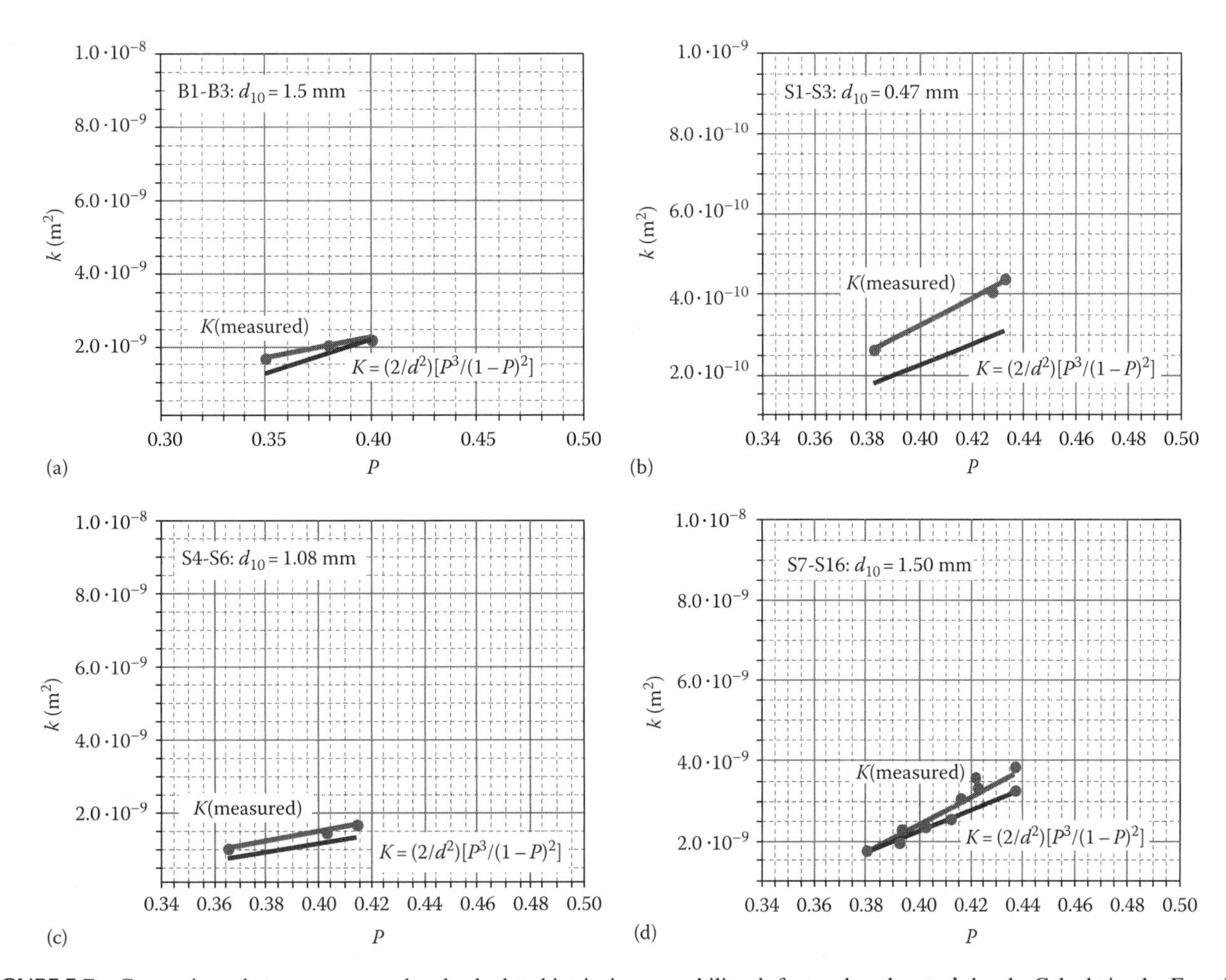

FIGURE E.7 Comparisons between measured and calculated intrinsic permeability, k, for sand, and potter's beads. Calculation by Equation E.6 based on d_{10}. (a) Beads $d_{10} = 1.50$ mm, UC $= 1.00$. (b) Sand $d_{10} = 0.47$ mm, UC $= 1.31$. (c) Sand $d_{10} = 1.08$ mm, UC $= 1.25$. (d) Sand $d_{10} = 1.50$ mm, UC $= 1.25$.

Figure E.7a shows the comparison between K(measured) and K(calculated) for 1.50 mm beads with UC = 1.00. The two curves are quite close in both trend and agreement, with difference ranging from 0.4% to 24%.

The dilemma in extending the calculation to nonuniform media was in selecting a surrogate that would characterize d(sphere) for the purposes of the calculation. Figure E.7b through d for the filter sand of Chang et al. (1999) was calculated using d_{10} as a trial and because it was convenient. The three comparisons show about the same trends, and with differences approximately 30%, 20%, and 15% for (b), (c), and (d), respectively.

Using an estimated d_{50} for the calculation resulted in slightly lower differences overall for the three sands. Using the same Equation E.6 for anthracite, with d_{10} as the basis, resulted in differences of 60%–200%, with calculated k being higher than the measured k for each of the three sizes.

The comparisons of Figure E.7 demonstrate that (1) the form of Equation E.10 predicts the trends, and (2) the accuracy is remarkably high considering the task. In other words, Equation E.10 is probably a valid model and may be applied as a reasonable approximation to sand, as well as potter's beads. On the other hand, the same kinds of comparisons for the anthracite k(measured) and k(calculated) showed that k(calculated) was appreciably higher than k(measured), e.g., by a factor of perhaps two and Equation E.9 would be improved with refined values for ξ and S.

E.7 FORCHHEIMER FLOW REGIME

As noted by Trussell and Chang (1999), the inertial, i.e., Forchheimer, flow regime is applicable to many instances of porous media flow in practice. This would apply especially to designs that started perhaps in the late 1980s that use a deep-bed mono media of anthracite, e.g., perhaps 2–3 m, with higher HLRs such as say 24 m/h (10 gpm/ft²); for such a design with $d_{10} \approx 1.5$ mm, $\mathbf{R} \approx 10$. They reviewed the historical development of porous media flow equations and have recommended "bottom-line" equations for the Forchheimer flow regime. The Forchheimer flow regime also applies to the laminar flow regime since the v^2 term becomes small at the low velocities of laminar flow and the

Forchheimer equation becomes the Darcy equation. The Forchheimer equation, proposed in 1901 (as reviewed by Chang et al. 1999) is

$$\frac{dh_L}{dZ} = \alpha_F v + \beta_F v^2 \tag{E.3}$$

in which

- α_F is the coefficient related to linear headloss (s/m)
- β_F is the coefficient related to nonlinear headloss (m^2/s^2)

The α_F term was defined in Equation E.9 and in a similar fashion β_F was determined by Trussell and Chang (1999) to give

$$\frac{dh_L}{dZ} = \left[\frac{\mu}{\rho g}\right] 2\xi \left[\frac{(1-P)^2}{P^3}\right]\left[\frac{S'}{d}\right]^2 v + \left[\frac{2\xi}{c}\right]\left[\frac{1}{g}\right]\left[\frac{(1-P)}{P^3}\right]\left[\frac{S'}{d}\right] v^2 \tag{E.12}$$

in which

- S' is the area-volume shape factor (m^2/m^3)
- c is the constant reflecting geometric properties of the porous media (unitless)

and

$$S' = Sd \tag{E.13}$$

in which d is the characteristic dimension of porous media.

The constants limit the application of Equation E.12 as there seems to be only "suggestions" as to values and then only for spheres. Trussell and Chang (1999) give $c = 49$ from the work of Ergun in 1952 and 3.3 was given by J. Ward in 1964. The magnitudes of these constants reflect the differences between the laminar and the inertial effects. Consequently, Trussell and Chang suggest "lumping" the constants in Equation E.6 such that

$$a = 2\xi S'^2 \tag{E.14}$$

TABLE E.3
Hydraulic Gradient Calculated by Forchheimer Equation

Medium	a	b	Typical Porosities
Crushed anthracite	210–245	3.5–5.3	0.47–0.52
Crushed sand	110–115	2.0–2.5	0.40–0.43
Glass beads	130–150	1.3–1.8	0.38–0.40

and

$$b = \sqrt{\frac{2\xi}{c}S'} \tag{E.15}$$

to give

$$\frac{dh_L}{dZ} = a\left[\frac{\mu}{\rho g}\right]\left[\frac{(1-P)^2}{P^3}\right]\left[\frac{1}{d}\right]^2 v + b\left[\frac{1}{g}\right]\left[\frac{(1-P)}{P^3}\right]\left[\frac{1}{d}\right] v^2 \tag{E.16}$$

Trussell and Chang give values for a and b in Table E.3, pointing out their "preliminary" nature.

Again, a laboratory column test will yield a and b coefficients that fit the actual media at hand. Once determined, the coefficients may be used with Equation E.16 to explore the effects of different depths, hydraulic loading rates, and media sizes on headloss. Example E.2 illustrates the application of Equation E.16.

Table CDE.4 is a spreadsheet solution for Equation E.16. For a given media, with conditions specified or assumed, the hydraulic gradient can be calculated and with bed depth stated the headloss can be estimated, as illustrated both in Table CDE.4 and Example E.2.

Example E.2 Headloss in the Inertial Flow Regime (From Trussell and Chang, 1999)

Given

Media is uniform crushed anthracite with $d = 1.55$ mm and $\Delta z = 2.54$ m (100 in.). HLR = 0.010185 m/s = 36.6 m/h (15 gpm/ft^2). $T = 20°C$.

TABLE CDE.4
Hydraulic Gradient Calculated by Forchheimer Equation

HLR			d	ΔZ				T	μ	ρ_w			h_L(vis)/ΔZ	h_L(tur)/ΔZ	h_L(tot)/ΔZ	h_L
(gpm/ft^2)	(m/h)	(m/s)	(mm)	(m)	a	b	P	(°C)	(Ns/m^2)	(kg/m^3)	α_F	β_F	(m/m)	(m/m)	(m/m)	(m)
2	4.88	0.0014	1.55	2.54	215	3.5	0.47	20	0.000998	998.371	24.68	1175	0.033	0.002	0.036	0.090
5	12.2	0.0034	1.55	2.54	215	3.5	0.47	20	0.000998	998.371	24.68	1175	0.084	0.013	0.097	0.247
8	19.52	0.0054	1.55	2.54	215	3.5	0.47	20	0.000998	998.371	24.68	1175	0.134	0.035	0.168	0.428
10	24.4	0.0068	1.55	2.54	215	3.5	0.47	20	0.000998	998.371	24.68	1175	0.167	0.054	0.221	0.562
15	36.6	0.0102	1.55	2.54	215	3.5	0.47	20	0.000998	998.371	24.68	1175	0.251	0.121	0.372	0.946
20	48.8	0.0136	1.55	2.54	215	3.5	0.47	20	0.000998	998.371	24.68	1175	0.335	0.216	0.551	1.398

Required
Clean-bed headloss, h_L

Solution

1. Constants Let $a = 215$ and $b = 3.5$ and $P = 0.47$
2. At 20°C, $\mu = 0.001002$ N s/m^2 and $\rho_w = 998.2$ kg/m^3
3. Compute Δh_L from Equation E.6,

$$\frac{\Delta h_L}{2.54} = 215\left[\frac{0.001}{998.2 \cdot 9.81}\right]\left[\frac{(1-0.47)^2}{0.47^3}\right]\left[\frac{1}{0.00155}\right]^2 0.010185$$

$$+3.5\left[\frac{1}{9.81}\right]\left[\frac{(1-0.47)}{0.47^3}\right]\left[\frac{1}{0.00155}\right]0.010185^2 \quad \text{(ExE.2.1)}$$

$$= 0.372 \text{ m/m} \quad \text{(ExE.2.2)}$$

$$\Delta h_L = 0.946 \text{ m} \quad \text{(ExE.2.3)}$$

Discussion
Table CDE.4 is a spreadsheet that provides a means to calculate Δh_L for any conditions. For the conditions stated, the distribution of headloss is 0.251 m/m laminar and 0.121 m/m turbulent. At HLR = 24.4 m/h (10 gpm/ft^2) the distribution is 0.167 m/m laminar and 0.054 m/m turbulent.

E.8 HYDRODYNAMICS

Steady flow through homogeneous isotropic porous media can be described mathematically (see Muskat, p. 129) by the hydrodynamic relation (the Laplace equation),

$$\nabla^2\Phi = 0 \quad \text{(E.17)}$$

in which Φ is the hydraulic potential (m).

The velocity at any point is proportional to the negative potential gradient. The "solution" to Equation E.6 can be seen graphically as a "flow net" which is characterized by all potential lines and all streamlines crossing normal to one another with the "stream tubes" conveying the same increment of flow and the $\Delta\Phi$ for adjacent potential lines being equal. In a column, such as rapid rate filter or a pilot plant filter, the flow net, looking at a side view in two dimensions, is simply a rectangular or square grid. Equation E.6 applies for the laminar flow regime, and as noted, probably could be extended into the inertial regime as long as the linear relationship between v and dh/dz is a reasonable approximation.

ACKNOWLEDGMENTS

Dr. Deanna Durnford, professor of civil and environmental engineering (Emeritus), Colorado State University, helped to set straight some of the nomenclature in Darcy's law, suggesting clarifications, and provided key references on hydraulic conductivity, on Darcy, and on the formulation of his well-known law on flow through porous media. The author is responsible for the interpretation of her advise.

GLOSSARY

Absolute temperature: Defined: T(K) = 273.15 + °C; T(R) = 459.6 + °F.

Darcy: Refers to Darcy's law stating that flow through sand is proportional to the hydraulic gradient. The results of Henry Darcy's (1803–1858) experiments, using a 2.50 m column 0.35 m diameter fitted with two manometer near the top and bottom, respectively, were published in 1856 in Paris, buried in a report of 647 pages that he had prepared that dealt with the development of a water supply for the City of Dijon. His work in pipe flow developed conclusive evidence that resistance to flow depends on the type and condition of the pipe and is usually linked with Weisbach. He was a part of the Corps des Pont et Chauseés, an elite fraternity of engineers and a government agency that gave engineers considerable status as intellectuals and professionals. Darcy was many things as a professional: the designer of the water supply for Dijon which was started in 1830 with water delivery in 1840, the administrator a large regional engineering office, a leader of the community, and a researcher. (The foregoing from Freeze, 1994; see also Brown, 2002)

Darcy: A unit of intrinsic permeability used sometimes by persons in the ground water field. The equivalent is: 1 Darcy = $0.987 \cdot 10^{-12}$ m^2. In other words, multiply a value in Darcys by the factor $0.987 \cdot 10^{-12}$ m^2 to obtain, k, the intrinsic permeability. For example Table 14.1 gives the permeability of Filter Cel as 0.07 Darcys; then $k = 0.07$ Darcy $\cdot 0.987 \cdot 10^{-12}$ m^2/Darcy = $0.07 \cdot 10^{-12}$ m^2.

d_{10}: In a sieve analysis this is the particle size in which 10% of the particles are smaller; the d_{10} size is called also the "effective size." The numbers d_{10}, d_{60}, and UC are used to characterize media size distribution in granular media filters used in water treatment.

d_{60}: In a sieve analysis this is the particle size in which 60% of the particles are smaller.

Dispersion: Super-position of random motion at the micro-level on the general advective transport of a fluid. The random motion is due to fluid turbulence in pipE.flow or open-channel flow, or atmospheric advection of air masses. Although molecular motion is also random and has the same effect, its effect is small except in laminar flow. The random motion results in a "normal" (i.e., Gaussian) distribution about the mean flow. The standard deviation of the normal distribution increases with the number of "steps," of which elapsed time is a surrogate measure. The mathematics may be described by probability theory, with each step a result of "coin-flipping." In flow through porous media, dispersion occurs due to the random distribution of pore velocities, which are larger and smaller than the mean velocity.

Forchheimer: German researcher who published in 1901 the nonlinear relationship between hydraulic gradient and velocity that occurs at higher Reynolds numbers (see Trussell and Chang, 1999).

Hydraulic conductivity: The constant K in Darcy's law, i.e., $v = K(\mathrm{d}h/\mathrm{d}z)$. The term K incorporates fluid properties ρ and μ which are the functions of temperature.

Intrinsic permeability: The constant k in Darcy's law, i.e., $v = (k\rho g/\mu)(\mathrm{d}h/\mathrm{d}z)$. The k term is a property of the porous medium with dimensions, L^2 and is preferred because the fluid properties, ρ and μ are isolated.

Permeability: A qualitative description of a porous medium, e.g., "this soil is highly permeable," meaning the soil has a relatively high intrinsic permeability. Hydraulic conductivity is also loosely referred to as permeability.

Superficial velocity: Defined as flow divided by cross-sectional area, i.e., $v = Q/A$.

Uniformity coefficient (UC): Defined, $UC = d_{60}/d_{10}$.

REFERENCES

Ahmed, N. and Sunada, D., Nonlinear flow in porous media, *J. Hydraulic Engineering Division, ASCE*, 95(6):1847–1857, June 1969.

Boulding, J. R., *Practical Handbook of Soil, Vadose Zone, and Ground-Water Contamination—Assessment Prevention, and Remediation*, Lewis Publishers, Ann Arbor, MI, 1995.

Brown, G. O., Henry Darcy and the making of a law, *Water Resources Research*, 38(7):11-1–11-12, July 2002.

Chang, M., Trussell, R. R., Guzman, V., Martinez, J., and Delany, C. K., Laboratory studies on the clean bed headloss of filter media, *Aqua*, 48:137–145, 1999.

Freeze, R. A., Translation of: Darcy, H., Determination of the laws of the flow of water through sand, from pp. 590–594 of *Les Fontaines Publiques de la Ville de Dijon*, Victor Dalmont, Paris, 647 pp. 1856. Translation reprinted in *Ground Water*, 32(1):260–261, 1994.

Freeze, R. A., Henry Darcy and the fountains of Dijon, *Ground Water*, 32(1):23–30, January–February 1994.

Hendricks, D. W., Barrett, J. M., Bryck, J., Collins, M. R., Janonis, B. A., and Logsdon, G. S., Manual of Design for Slow Sand Filtration, AWWA Research Foundation and American Water Association, Denver 1991.

Hsu, S., Dispersion and wave fronts, MS Thesis, Department of Civil Engineering, Colorado State University, Fort Collins, CO, 1994.

Kasenow, M., *Applied Ground-Water Hydrology and Well Hydraulics*, Water Resources Publications, Fort Collins, CO, 1997.

McWhorter, D. B. and Sunada, D. K., *Ground-Water Hydrology and Hydraulics*, Water Resources Publications, Denver, CO, 1977.

Mosher, R. R. and Hendricks, D. W., *Filtration of Giardia Cysts and Other Particles under Treatment Plant Conditions, Volume 2: Rapid Rate Filtration Using Field Scale Pilot Filters on the Cache La Poudre River*, AWWA Research Foundation Report for Contract 80-84, May, 1986.

Muskat, M., *The Flow of Homogeneous Fluids through Porous Media*, J. W. Edwards, Inc., Ann Arbor, MI, 1946

Seelaus, T., Hendricks, D. W., and Janonis, B., *Filtration of Giardia Cysts and Other Particles Under Treatment Plant Conditions, Volume 1: Slow Sand Filtration at Empire Colorado*, AWWA Research Foundation Report for Contract 80-84, May 1986.

Sheidegger, A. E., *The Physics of Flow through Porous Media*, Revised Edition, The Macmillan Co., New York, 1960.

Trussell, R. R. and Chang, M., Review of flow through porous media as applied to headloss in water filters, *Journal of Environmental Engineering Division, ASCE*, 125(11):998–1005, November 1999.

Trussell, R. R., Chang, M., Lang, J. S., and Hodges, W. E. Jr., Estimating the porosity of full-scale anthracite filter, *Journal of AWWA*, 91(12):54–63, November 1999.

Appendix F: Alum Data and Conversions

Aluminum ion, Al^{3+}, and the ferric ion, Fe^{3+}, are the common metal coagulants used in water treatment, with Al^{3+} being used most frequently. The Al^{3+} ion is provided commercially in the form of hydrated aluminum sulfate, $Al_2(SO_4)_3 \cdot 14H_2O$, called alum, and Fe^{3+} is provided as either $Fe_2(SO_4)_3 \cdot 7H_2O$ or $Fe_2Cl_3 \cdot 6H_2O$.

In using alum, the principles are straightforward. A given mass of alum added to a given volume of water results in a known concentration. To implement this ostensibly simple idea, however, requires knowledge of the manufacture of alum, and its resulting characteristics and conventions in expressing concentration. Both solid and liquid forms are considered here.

F.1 MANUFACTURER OF ALUM

The principal aluminum ore is bauxite, a mixture of hydrous aluminum oxides, varying physically according to their deposits. In general, commercial deposits have about 52% hydrated aluminum oxide. Most of the bauxite mined is refined into alumina, which has several forms of which aluminum oxide is the principal component (Britannica, 1974). The commercial chemical grade bauxite, from which alum is produced, is 59% Al, expressed as Al_2O_3. Metallurgical grade bauxite is lower assay and higher in contamination.

Aluminum sulfate is produced by the reaction between sulfuric acid and hydrated aluminum oxide with product $Al_2(SO_4)_3 \cdot 14H_2O$. The reaction, sans the waters of hydration is

$$Al_2O_3 + 3H_2SO_4 \rightarrow Al_2(SO_4)_3 + 3H_2O \qquad (F.1)$$

According to Harringer (1984), alum is manufactured by "digesting" an aluminum-bearing ore with sulfuric acid. Figure F.1 depicts the manufacture, showing the sequence of processes and operations. The waters of hydration, i.e., n in $Al_2(SO_4)_3 \cdot nH_2O$ is determined by the point at which the crystallization is arrested in the evaporation process with $n \leq 18$. The 14 waters of hydration is a stable form, losing water slowly. Also, with slightly lower mass than for $n = 18$, the cost to transport is less (the foregoing statements were adapted from Harringer, 1984). The hydrated state of commercial alum is approximately $14H_2O$ as a mixed hydrate, and is expressed normally as $14H_2O$. For reference, $14.3H_2O$ is used commonly in the literature, but 14 was suggested by General Chemical, a major manufacturer of alum in the United States.

The product formed called "dry" alum is a crystal, i.e., $Al_2(SO_4)_3 \cdot 14H_2O$, and must be ground to be used in water treatment. As seen in Figure F.1, grinding and screening produces powdered, ground, and granular, which are the common commercial sizes, with lump being available in some cases in other countries. The dry alum may be shipped in bulk and is also packaged in 50 or 100 lb multiwalled paper bags and, in some cases, plastic bags.

F.2 SOLID ALUM

The alum product manufactured is chemically, $Al_2(SO_4)_3 \cdot 14H_2O$, which has a solid crystal structure and is called "dry alum." Since this hydrated form is the "added mass" in water treatment, concentrations are expressed as mg $Al_2(SO_4)_3 \cdot 14H_2O$/L solution and mass flows are expressed as mass of $Al_2(SO_4)_3 \cdot 14H_2O$ per unit time.

F.2.1 DESCRIPTION

Table F.1 describes some of the properties of the "standard ground" dry alum. The data may be useful in designing storage for bulk alum, in sizing a reactor for dissolution (for standard ground alum), and in providing a general background concerning alum characteristics.

F.2.2 MASS FLOW OF ALUM

The dry chemical is free flowing and feeds well from bulk storage hoppers (Harringer, 1984). Feeders are of two types: volumetric and gravimetric. The accuracy of the volumetric feeder is within 1%–2% of the amount set and is common for smaller plants. The gravimetric feeders are more accurate and are favored by the larger plants. The designated rate of alum feed is dropped into a dissolving chamber which is agitated by a mixer and from here fed to the rapid mix.

Figure F.2 shows schematically the feed of granular alum from palettes to a hopper for storage and metering. The metered granular alum drops to a belt and then drops into a dissolution reactor. The dissolved alum concentrate is then metered by a positive displacement pump into the rapid mix (or into the flow to the rapid mix).

The mass flow of granular alum has three stages: (1) from the alum hopper to the dissolution reactor, (2) from the dissolution reactor to the rapid mix as a concentrated dissolved alum solution, and (3) as dissolved alum in the main

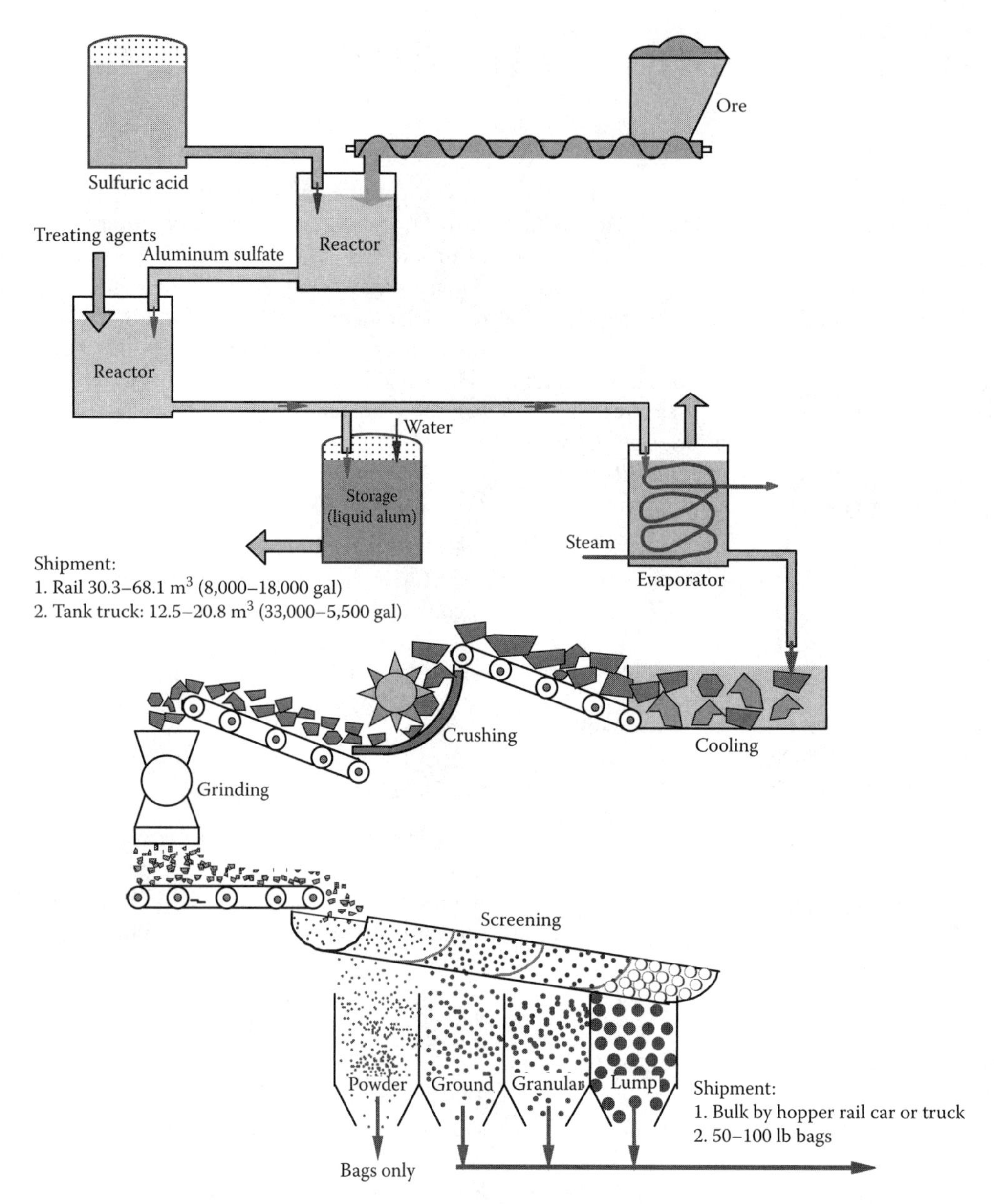

FIGURE F.1 Alum manufacture—schematic. (Adapted from Harringer, R.D., Aluminum sulfate coagulation low temperature water, unpublished paper presented at *CH2M-Hill Cold Water Coagulation Seminar*, Denver, CO, July 13, 1984; General Chemical, Aluminum Sulfate (Alum)—Technical Data, Brochure CHEM-M5-23, General Chemical Corporation, 1995.)

plant flow. These three flows are represented as one equation, having three parts, i.e.,

$$\dot{M}(\text{alum}) = Q(\text{alum feed}) \cdot C(\text{alum feed}) = Q(\text{plant}) \cdot C(\text{alum-in-plant-flow}) \quad \text{(F.2)}$$

in which

$\dot{M}$(alum) is the mass flow of alum
Q(alum feed) is the alum feed flow to rapid mix (m^3/s)
C(alum feed) is the concentration of alum in alum feed flow to rapid mix (kg/m^3)
Q(plant) is the raw water flow through plant (m^3/s)
C(alum in plant flow) is the concentration of alum as found in the rapid mix (kg/m^3)

As stated, Equation F.1 encompasses the mass flow of solid alum through the alum feeder, i.e., $\dot{M}$(alum), the mass flow of alum from the dissolution reactor (to the rapid mix), and the mass flow of alum in the feed flow through the rapid mix. The concentration of alum in the plant flow, i.e., C(alum in plant flow), is the "driver" variable and is determined by jar tests and/or by pilot plant tests and may vary with changing raw-water quality. Example F.1 illustrates calculations to determine the mass flow of alum, $\dot{M}$, and flow of alum concentrate feed, Q(alum feed). As a criterion, the alum concentration, C(alum feed) should be $>1\%$, i.e., 10 g $Al_2(SO_4)_3 \cdot 14H_2O$/L solution, since weaker solutions will hydrolyze and eventually plug lines.

TABLE F.1
Data for Dry Aluminum Sulfate[a,b]

Property	Condition	Data	
Formula	14 waters of hydration	$Al_2(SO_4)_3 \cdot 14H_2O$	
Common name	Name for $Al_2(SO_4)_3 \cdot 14H_2O$	dry alum	
Molecular weight	Approximate-for commercial	594	
Percent Al_2O_3		17	
Solubility	g/L	1050[b]	
pH	pH versus percent	%	**pH**
	$Al_2(SO_4)_3 \cdot 14H_2O$ by weight	1	3.5[a]
		10	3.2
		20	2.9
		30	2.7
		40	2.6
		50	2.4
Screen size[a]	Standard ground alum	**E-Mesh size**	**% Retained**
		8	2
		60	78
		100	7
	Powdered alum	100	3
		Ground	**Powdered**
Bulk density-ground	lb/ft^3	63–71[b]	38–45[a]
	g/L	1010–1138	609–722
	kg/m^3	1010–1138	609–722°
Angle of repose[a]		43°	
Rate of solution[a]	10 g added to 1000 mL water at 20°C, high-speed stirrer	**Time (s)**	**% dissolved**
		15	92
		30	98
		45	100

[a] General Chemical (1995).
[b] Harringer (1984) states solubility is 105 parts in 100 parts water.

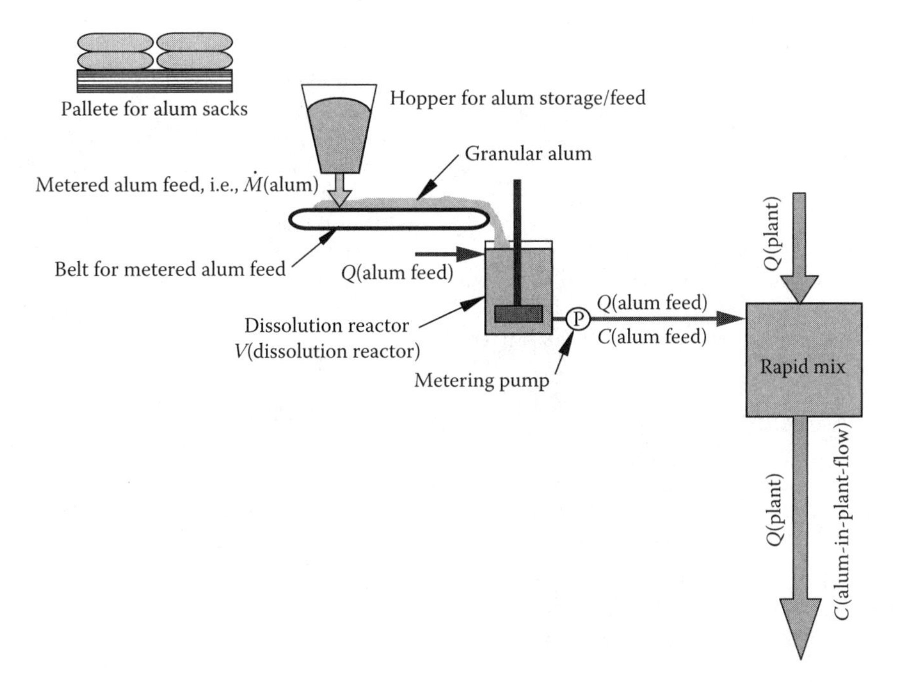

FIGURE F.2 Metered feed of granular alum—schematic drawing.

Example F.1 Determine Alum Feed Flow

Given

$$Q(\text{plant}) = 1.00\ \text{m}^3/\text{s}\ (22.83\ \text{mgd})$$

$$C(\text{in-plant-flow}) = 20\ \text{mg}\ Al_2(SO_4)_3 \cdot 14H_2O/L$$
$$= 0.020\ \text{kg/m}^3\ \text{(from jar tests or pilot testing)}$$

$$C(\text{alumfeed}) = 100\ \text{g}\ Al_2(SO_4)_3 \cdot 14H_2O/L$$
$$= 100\ \text{kg}\ Al_2(SO_4)_3 \cdot 14H_2O/\text{m}^3.$$

Required

(a) Q(alum feed)
(b) Mass feed rate of alum

Solution

(a) Calculate Q(alum feed) from (F.2),

$$Q(\text{plant}) \cdot C(\text{alum-in-plant-flow}) = Q(\text{alum feed}) \cdot C(\text{alum feed}) \quad \text{(F.2)}$$

$$1.00\ \text{m}^3/\text{s} \cdot 0.020\ \text{kg/m}^3 = Q(\text{alum feed}) \cdot 100\ \text{kg/m}^3$$

$$Q(\text{alum feed}) = 0.00020\ \text{m}^3/\text{s} = 200\ \text{mL/s}$$

(b) The mass feed rate of alum to the dissolution reactor (for feeder selection) is

$$M(\text{alum}) = Q(\text{plant}) \cdot C(\text{alum plant})$$
$$= 1.00\ \text{m}^3/\text{s} \cdot 0.020\ \text{kg/m}^3$$
$$= 0.020\ \text{kg/s}$$
$$= 1.2\ \text{kg}\ Al_2(SO_4)_3 \cdot 14H_2O/\text{min}$$
$$= 1728\ \text{kg}\ Al_2(SO_4)_3 \cdot 14H_2O/\text{day}$$

(c) Let the bulk density of granular alum be 1074 kg/m^3, i.e., from Table F.1, (1010 + 1038)/2, to give, the volumetric feed rate,

$$V(\text{feed rate}) = (1728\ \text{kg}\ Al_2(SO_4)_3 \cdot 14H_2O/\text{day})/(1074\ \text{kg/m}^3)$$
$$\approx 1.6\ \text{m}^3/\text{day}$$

Comments

To provide for say a 10-day supply of alum, about 16 m^3 would be required. This would be a silo that could have the dimensions, 2 m diameter × 6 m high. Most probably 3–4 such silos would be installed, i.e., for a 30-day supply plus one for standby.

F.2.3 Storage of Alum

Alum crystals will store well and therefore a long-term supply can be provided, such as 1–2 years for smaller plants and shorter durations for large plants in which such an amount of storage could be excessive (Harringer, 1984). For the large plants, alum would be provided in bulk rather than in sacks. In designing space for alum storage, the feed rate can be determined alum use on a monthly basis. Example F.2 illustrates the calculation, taking into account the monthly variation over the annual cycle.

In terms of 100 lb sacks, the number from Example F.2 would be about 1,140 per month (i.e., 51,849 kg/month · 2.2 lb/kg/100 lb/sack). For a plant of this size, handling the sacks would average about 1.6 per hour. As a matter of practice, plants using bagged alum will size a storage silo for less than a 10-day supply, e.g., normally 1–2 days. If, on the other hand, the alum is purchased in bulk and stored in bulk, the silos will be significantly larger, e.g., 30-day supply each. To illustrate, the volume required for 1-month supply in Example F.2 would be 47 m^3 (51,848 kg/1,100 kg/m^3). As an example of dimensions, a tank 3.5 m diameter · 4.0 m deep has such volume. Bulk storage should be designed for continuous mechanical feed with conical bottom. Several such tanks, such as 2–6 would help to reduce the number of deliveries. The sizing would depend also on the capacity of the trucks.

Example F.2 Determine Monthly Rate of Alum Use

Given

Result from Example F.1, i.e., $\dot{M} = 0.020$ kg/s

Required

Monthly use of alum

Solution

Monthly use of alum:

$$\dot{M}(\text{alum-monthly}) = 0.020\ \text{kg/s} \cdot 3600 \cdot 24\ \text{s/day} \cdot 30\ \text{day/month}$$
$$= 51{,}848\ \text{kg}\ Al_2(SO_4)_3 \cdot 14H_2O/\text{month}$$

Comments

The bulk storage volume required is 51,848 kg $Al_2(SO_4)_3 \cdot 14H_2O$/month/1,100 kg/m^3 = 47 m^3. In other words a storage volume of dimensions, 3.5 m · 3.5 m · 4.0 m would store a 1-month supply of alum at the rate of use calculated. Two or three such silos should be provided, depending on preferences in operation. With three silos, two could be active while the third would provide a supply while the alum is ordered for the other two. Also, the first two with metering could be cleaned and maintained while the third is placed on-line.

F.2.4 Cost of Solid Alum

The cost of alum depends on the amount ordered and whether the form is bagged or bulk, and the shipping distance. All of these factors and others will be reflected in a bid price for alum (if the bid method of purchase is used). Example F.3 illustrates the calculation of monthly cost. To compare alum cost when there are different forms, such as solid alum and liquid alum, the cost per unit of Al^{3+} provides a means.

Example F.3 Determine Monthly Cost of Alum

Given
Result from Example F.2, i.e., $\dot{M}$(alum monthly) = 51,848 kg $Al_2(SO_4)_3 \cdot 14H_2O$ /month

Required
Monthly cost of alum

Solution
Monthly cost of alum:

$$\text{Cost per 100 lb bag } Al_2(SO_4)_3 \cdot 14H_2O = \$11\text{–}15(\text{pick } \$15)$$

$$\text{Cost per kg } Al_2(SO_4)_3 \cdot 14H_2O = (\$15/100 \text{ lb}) \cdot (2.205 \text{ lb/kg}) = \$0.33/\text{kg } Al_2(SO_4)_3 \cdot 14H_2O$$

$$\begin{aligned} \text{Cost(monthly)} &= M(\text{monthly}) \cdot \text{Cost} \\ &= 51{,}840 \text{ kg/m} \cdot \$0.33/\text{kg} \\ &= \$17{,}146/\text{mo} \end{aligned}$$

(Cost is about \$11–15 per 100 lb (45 kg) bag, if purchased in large quantities.)

Example F.4 Determine the Cost of Alum Based upon Price per Unit of Al^{3+}

Given
Cost of alum is \$0.33/kg $Al_2(SO_4)_3 \cdot 14H_2O$

Required
Cost per unit of Al^{3+}

Solution
To convert cost from \$/kg $Al_2(SO_4)_3 \cdot 14H_2O$ to \$/kg Al^{3+} (solid granular form),

$$\begin{aligned} \$/\text{kg}Al^{3+} &= \$/\text{kg } Al_2(SO_4)_3 \cdot 14H_2O \\ &\quad \cdot \text{MW}\left(Al_2(SO_4)_3 \cdot 14H_2O/\text{MW}(Al^{3+})\right) \\ &= \$0.33/\text{kg } Al_2(SO_4)_3 \cdot 14H_2O \cdot (594/54) \\ &= \$3.63/\text{kg } Al^{3+} \end{aligned}$$

F.3 DISSOLUTION

Factors that affect the rate of dissolution of a substance in a stirred reactor include the propensity of the solid to dissolve, the target concentration after dissolution, the size of particles, water temperature, intensity of mixing, and the kind of mixer/basin configuration. The residual fraction (solids not dissolved) depends upon the product of the dissolution rate times the detention time, θ. Since the objective is not to have a residual, this product should be adjusted until the residual is near zero. This can be done for a given θ, by adjusting n, the mixing speed. Actually, the mass rate of dissolution decreases as the mass of substance in the reactor declines.

F.3.1 Criteria for Mixing

Complete mixing (defined as 99% blending) occurs if the contents of the volume mixed are circulated five times (McCabe et al., 1993). For a "standard" tank/mixer design (i.e., H(tank)/D(tank) ≈ 1, D(impeller)/D(tank) = 0.33, and for a six-blade flat-surface impeller), this amount of circulation occurs if the factor, $nt_T = 39$ at $\mathbf{R} \geq 2000$, in which n = rotational velocity of impeller in rev/s, t_r = time for five passes through the impeller, and $\mathbf{R}$ = Reynolds number, or $\mathbf{R} = nD_a^2\rho/\mu$.

Regarding alum dissolution, the 99% completion of mixing is merely for the alum to be dispersed uniformly throughout the solution which occurs by five circulation passes (McCabe et al., 1993); it does not ensure that the alum will have dissolved. If we wish that the fluid have 10 passes through the impeller (as a further hedge of our bets that the alum will dissolve, then we would have 99.99% completion of the mixing, and $nt_T = 78$.

As an index of the propensity of alum to dissolve, from Table F.1, 10 g $Al_2(SO_4)_3 \cdot 14H_2O$ at 20°C (68°F) will dissolve 100% within 45 s with high-speed mixing in 1000 mL water. Therefore, it would seem that if we choose $nt_T = 78$ (to give about 10 circulation passes), the probability would be high of say $\gg$99% dissolution of the granular alum being fed to the reactor.

F.3.2 Dispersion

A portion of the incoming solution will pass immediately through the reactor (see, for example, Figure 4.11) and will not be around for the 10 passes through the impeller. From Figure 4.11, about 0.37 fraction of the solution remains in the reactor when $t/\theta = 1.0$, about 0.14 fraction when $t/\theta = 2$, and about 0.05 fraction when $t/\theta = 3$. This means that if the time for 10 passes, t_T, is such that $t_T/\theta \leq 0.2$, then 0.90 fraction of the contents will have passed through the impeller 10 times, i.e., 0.90 fraction will have remained in the reactor.

F.3.3 Calculations

From the discussions above, two criteria must be satisfied: (1) $nt_T \geq K$, in which K is specified for the reactor and is for five passes through the impeller, and (2) $t_T/\theta \leq 0.2$, for 90% of the contents of the reactor to be retained. Example F.5 illustrates how these criteria may be applied.

Example F.5 Illustration of Method to Determine θ, n

Given
Standard design reactor

Required
a. Determine θ for 10 passes through the impeller
b. Determine n

Solution
1. Apply criterion

$$\frac{t_T}{\theta} \leq 0.2,$$

which if satisfied, 90% of the contents of the reactor remains to be circulated by the impeller. Now if we let $\theta = 90$ s, then,

$$t_T/90 \text{ s} \leq 0.2,$$
$$t_T \geq 18 \text{ s}$$

2. Apply criterion for 10 circulation passes (McCabe et al., 1993),

$$nt_T = 78$$
$$n \cdot 18 \text{ s} = 78$$
$$n = 4.3 \text{ rev/s}$$

(for 10 passes through the impeller).

Discussion

The mixing speed, $n = 4.3$ rev/s may be low and there will be no harm if n is much higher, say $n = 10$ rev/s. If we let $n = 10$ rev/s, then $t_T = 7.8$ s, a considerable reduction. This means that $t_T/\theta = 7.8$ s/90 s $= 0.086$. This can be interpreted in two ways: (1) about 95% of the contents of the reactor will have the 10 passes through the impeller in only 7.8 s, or (2) in 16 s, the number of passes for 90% of the contents through the impeller ≈20. This hedges the bet considerably more with respect to the fraction of alum dissolved. On the other hand, the detention time can be reduced considerably if we increase n further, let $n = 20$ rev/s, for example.

F.3.4 Sizing the Reactor and Mixer

The volume of the reactor may be calculated as

$$\theta = \frac{V(\text{reactor})}{Q(\text{alum feed})} \tag{F.3}$$

The mixing power may be determined by the power number relation,

$$\mathbf{P} = \frac{P}{\rho n^3 D^5} \tag{F.4}$$

in which

P is the power number (dimensionless)
P is the power applied to the impeller (Watts)
ρ is the density of solution (kg/m^3)
n is the rotational velocity of impeller (rev/s)
D is the diameter of impeller (m)

As a check, the criterion, P/V may be used, i.e., for "intense" mixing (McCabe et al., 1993),

$$\frac{P}{V} \approx 0.8 - 2.0 \text{ kW/m}^3 (4\text{–}10 \text{ hp/1000 gal}) \tag{F.5}$$

Example F.6 illustrates the application.

Example F.6 Size an Alum Dissolution Reactor

Given

Data from F.5
Q(alum feed) = 200 mL/s
C(alum feed) = 100 g $Al_2(SO_4)_3 \cdot 14H_2O$/L solution

Required

a. V(dissolution reactor)
b. Mixer speed and power

Solution

a. Calculate reactor volume, i.e.,

$$\theta = \frac{V}{Q}$$
$$90 \text{ s} = \frac{V}{(0.200 \text{ L/s})}$$
$$V = 18.0 \text{ L}$$

Dimensions

Dimensions for a cube are $D = H = 0.26$ m

And,

$$D(\text{impeller}) = 0.333 \cdot D = 0.086 \text{ m}$$

b. Mixer power

The reactor/impeller should be a "back-mix" type, i.e., with the impeller at the bottom causing a pumping effect to recirculate the flow. A "standard" design as described by McCabe et al. (1993) provides a known dimensionless power number, i.e., $\mathbf{P} = 6.0$

By definition, the power number, **P**, is

$$\mathbf{P} = \frac{P}{\rho n^3 D^5}$$

And the power number for a "standard" mixer is $\mathbf{P} = 6.0$, to give

$$6.0 = \frac{P}{\frac{998 \text{ kg}}{\text{m}^3} \cdot \left(\frac{5 \text{ rev}}{\text{s}}\right)^3 \cdot (0.0865 \text{ m})^5}$$

$$P = 3.5 \text{ W}$$

Discussion

- This is too small, so let $n = 10$ rps, giving $P = 28$ W (0.0013 hp)
- Try, $n = 20$ rps, giving $P = 224$ W (0.30 hp)
- Check $P/V = 224$ W/0.018 m$^3 = 12.4$ kW/m^3, which is $\gg P/V \approx 0.8$–2.0 kW/m^3 (for intense mixing)
- As a check, if we use $P/V = 2.0$ kW/m^3, then $P/0.018$ m$^3 = 2.0$ kW/m^3, then $P = 0.036$ kW = 36 W (0.05 hp). Thus, if $n = 10$ rps, the mixing would be "intense," which requires only 36 W. The second estimate, $P = 28$ W looks reasonable, but to hedge our bets, let $P = 36$ W (or the next larger mixer available in a catalog). Actually, to select an even larger size such as 224 W would

not be unreasonable. Allowing for losses, such a motor size would be reasonable.

- The power number approach, while rational, gives unreasonable results. The empirical approach, i.e., using the *P*/*V* criterion for intense mixing, remains the favored method (other than CFD) to estimate the power requirement for mixing. A slightly higher mixing power is preferred over undersizing.

F.4 LIQUID ALUM

Liquid alum came on the scene in water treatment in the about the 1950s (based on word of mouth recalled from the 1960s). In the United States, practice has shifted during the period from 1960 to 1990 from dry alum to mostly liquid alum. By 1990, for example, the dry alum equivalent of the liquid alum used was 500,000,000 kg (1.1 million tons) annually while the dry alum used (i.e., solid alum as $Al_2(SO_4)_3 \cdot 14H_2O$ crystals) was 33,000,000 kg (70,000 U.S. tons) annually. One reason for the shift has been that the production centers have been dispersed geographically, reducing the transport distances with consequent reduction in cost. The appeal of liquid alum is that it is more convenient to use than the solid form, labor costs are less than handling solid alum, and quality control is more easily assured. Liquid alum may be delivered in tank trucks or by rail and pumped or fed by gravity to storage tanks at the plant site.

F.4.1 Definition

The term "liquid alum" refers to the specifications of a manufactured product, i.e., a solution of dissolved $Al_2(SO_4)_3 \cdot 14H_2O$ in water that has a specific gravity of 1.335 (±0.002). The concentration of the solution is expressed as mass of $Al_2(SO_4)_3 \cdot 14H_2O$ per unit volume of solution. For SG = 1.335, *C*(alum) = 647 g $Al_2(SO_4)_3 \cdot 14H_2O$/L solution. This particular solution strength is specified merely because the freezing point is lowest.

F.4.2 Production

Liquid alum is manufactured as indicated by the schematic diagram, Figure F.1. The chemical reaction in the "reactor" in Figure F.1 is given by Equation F.1.

The intended specific gravity is variable, depending on the specifications of the finished product. Alum is manufactured to conform to the American Water Works Association Standard B403-98 for aluminum sulfate, which specifies ranges of strength for Al and Al_2O_3 of alum products. The industry controls on strength is based upon the specific gravity test. The SG standard stated above does not seem to be a common target industry wide.

F.4.3 Description

Selected properties of liquid alum are listed in Table F.2. Among the data given, some are for reference and others are requisite to calculations. Data include the molecular weight, specific gravity, concentration in g $Al_2(SO_4)_3 \cdot 14H_2O$/L solution. The latter is most important in metering the neat solution to the flow of a water treatment plant. Also important are the capacities of trucks, common sizes of storage tanks, pH, and freezing temperature. Note that the Baumé is a hydrometer reading; the scale is calibrated to give an equivalent specific gravity of the solution being tested.

Other properties of interest include pH, given in Figure F.3a as a function of mass concentration. As seen, pH ≈ 2.8 for a mass concentration of about 25%, which is equivalent to a 1:1 dilution. With pH ≤ 3.5, there should be no hydrolysis products which could reduce the effectiveness of the liquid alum.

The viscosity of liquid alum is given in Figure F.3b as a function of temperature for different mass concentrations of alum. The relationships are useful in hydraulic calculations.

F.4.4 Expressions for Alum Mass

Concentrations of alum have been expressed in five equivalent forms. They are enumerated in the paragraphs that follow.

$Al_2(SO_4)_3 \cdot 14H_2O$: As a rule, alum concentration is expressed as g/L of $Al_2(SO_4)_3 \cdot 14H_2O$, called "dry alum" in the industry. This is the commercial grade and the form manufactured for water treatment. The rationale for expressing concentration in this form is that the mass added to a given volume of water is as $Al_2(SO_4)_3 \cdot 14H_2O$.

$Al_2(SO_4)_3 \cdot 18H_2O$: If pilot plant experiments, or laboratory experiments, are conducted using reagent grade alum, i.e., $Al_2(SO_4)_3 \cdot 18H_2O$, then this is the expression usually adopted. Alternatively, such concentration may be converted to $Al_2(SO_4)_3 \cdot 14H_2O$ (using the ratio of molecular weights).

$Al_2(SO_4)_3$: Alum concentration may be expressed also as $Al_2(SO_4)_3$ with the rationale that this is the only solid that exists.

Al^{3+}: Increasingly, Al^{3+} is the favored expression since this is the only constituent from the alum that participates in the coagulation reaction. Expressing unit cost as dollars/kg Al^{3+} is a means to "normalize," i.e., finding a basis for comparison, alum costs.

Al_2O_3: Finally, the Al_2O_3 equivalent is used because the expression is a holdover from the days when chemicals were ignited to create their equivalent oxides and with gravimetric determination of the product. Thus, one sees iron salts as ferric oxide, i.e., Fe_2O_3, caustic soda and soda ash as Na_2O, etc. The industry is moving away from the oxide form of expression and toward the active metals form, as seen by AWWA standards for all chemicals used in water treatment.

F.4.5 Alum Conversions

Alum concentrations are expressed in a number of different forms. Guidance on how to convert between forms is given here.

TABLE F.2
Data for Liquid Alum (Dissolved Aluminum Sulfate)[a,b]

Property	Notes	Data	
Formula	14 waters of hydration	$Al_2(SO_4)_3 \cdot 14H_2O$	
Common name	Name for $Al_2(SO_4)_3 \cdot 14H_2O$	Liquid alum	
Molecular weight	Approximate-for commercial	594	
Specific gravity	g solution/g water at 60°F	1.335(± 0.002)	
Baumé, 60°F	Hydrometer reading	36.4	
	Also: quick measure of strength through use of manufacturer's tables (see also AWWA Standard B403-98 and Spreadsheet CDF.3 for quick conversions)		
Density	g solution/L solution	1.330–1.342[c]	
	lb solution/gal solution	11.1–11.2	
Concentration (SG = 1.335)	g $Al_2(SO_4)_3 \cdot 14H_2O$/L solution	647[c]	
	lb $Al_2(SO_4)_3 \cdot 14H_2O$/gal solution	5.4	
Al_2O_3	Percent by mass	8.1–8.4	
Solubility	g $Al_2(SO_4)_3 \cdot 14H_2O$/L solution	1,050[d]	
Shipping—tank truck	m^3	E-12.5-20.8[c]	
	gal	3,300–5,500[e]	
Shipping—rail car	m^3	E-30.3–68.1[c]	
	gal	8,000–18,000	
Storage tanks	Capacity-minimum	28.4–56.8[c]	
		7,500–15,000	
	Material (indicative list)	Fiberglass/epoxy	
		Polyester	
Appurtances	Liquid level	Manometer	
	Flush connections	Piping/pumps/metering	
pH	pH versus % $Al_2(SO_4)_3 \cdot 14H_2O$	**Mass%**	**pH[f]**
	(pH is given as a property)	1	3.5
		50	2.4
		Temp. (°F)	(%dry alum)
Freezing temperatures	Percent $Al_2(SO_4)_3 \cdot 14H_2O$ with corresponding temperatures at which freezing will occur	32	0
		30	10
		25	25
		20	33
		14	40
		4	48

[a] General Chemical (1995).
[b] General Chemical Product Data Sheet, General Chemical Corporation, Parsippany, NY, (1997b).
[c] Calculated from data in line below as given by General Chemical in U.S. Customary units.
[d] Harringer (1984) states solubility is 105 parts in 100 parts water.
[e] Tank truck maximum weight is 80,000 lb. Weight of truck plus fuel $\approx$32,000 lb, permitting a cargo weight $\approx$48,000 lb (Jones, 1994). Therefore, the calculated cargo weight/g(alum) = 48,000 lb/11.1 lb/gal = 4,324 gal.
[f] pH is of interest relative to dilutions that may be on interest in pilot plant work. The idea is to maintain the pH at levels such that the predominant species is Al^{3+} and hydrolysis products do not form. Generally pH $\ll$ 4.0 is safe for this purpose.

F.4.5.1 Equivalent Concentration Expressions

Equivalent alum concentrations for the five expressions are shown in Figure F.4a for SG(alum solution) = 1.335 with corresponding concentration of 647 g $Al_2(SO_4)_3 \cdot 14H_2O$/L solution. Using the latter as a basis, equivalent alum concentrations, seen in Figure F.4a, are 725 g $Al_2(SO_4)_3 \cdot 18H_2O$/L solution, 373 g $Al_2(SO_4)_3$/L solution, 111 g Al_2O_3/L solution, and 58.5 g Al_2/L solution. The conversions are based on molecular weight ratios, i.e., 666.132/594.136, 342.15/594.136, 101.961/594.136, 53.964/594.136, respectively. Figure F.4a is useful for quick reference if conversions are needed.

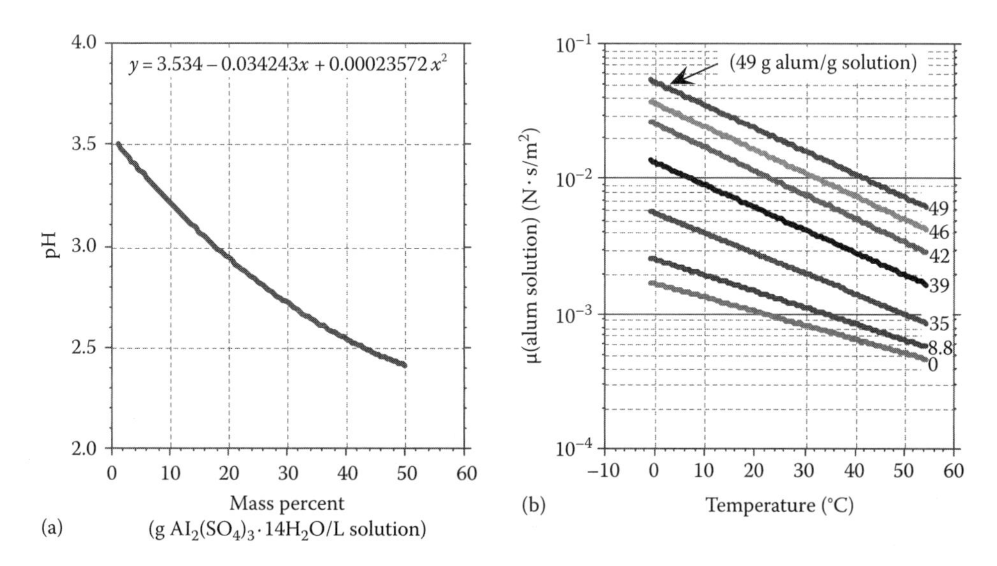

FIGURE F.3 pH and viscosity of liquid alum. (a) pH of alum versus mass percent solids. (b) Viscosity of alum versus temperature.

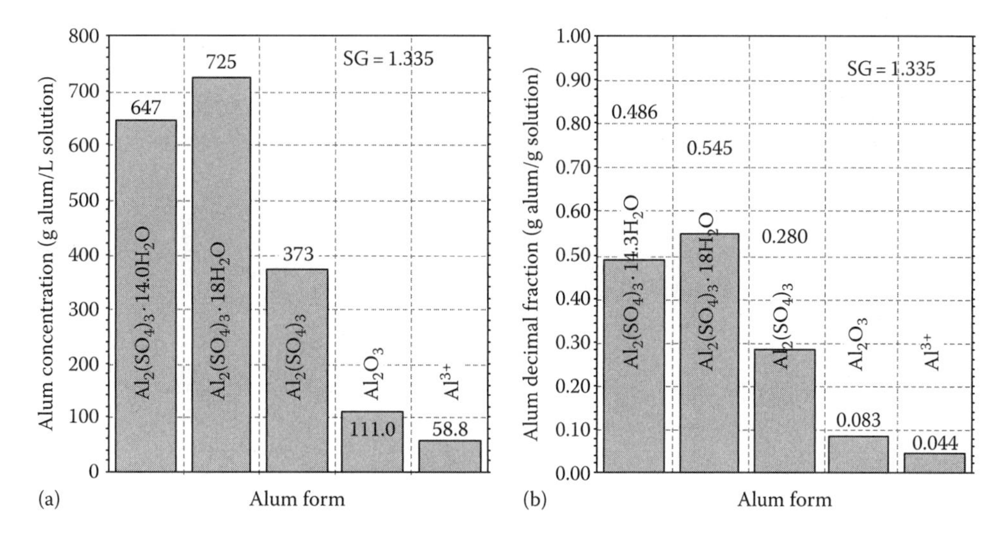

FIGURE F.4 Equivalent alum expressions for liquid alum, SG = 1.335. (a) Concentrations per unit volume. (b) Mass concentrations.

F.4.5.2 Standard Expressions

Alum concentration should be stated in terms of the form used. For example, in research the reagent grade alum, i.e., $Al_2(SO_4)_3 \cdot 18H_2O$/L is used frequently; concentration should be stated, e.g., 20 mg $Al_2(SO_4)_3 \cdot 18H_2O$/L, or 20 mg/L as $Al_2(SO_4)_3 \cdot 18H_2O$. The full expression is stated, e.g., "20 mg $Al_2(SO_4)_3 \cdot 14H_2O$/L" and not 20 mg/L alum; the latter alone is ambiguous. An ideal would be to use a common standard, i.e., as mg Al^{3+}/L.

Concentration as Al: While a standard expression for alum concentration has not been adopted by AWWA, the expression being favored by the industry is "g Al/L solution," which is codified by the AWWA Standard B403-98. This form is rational and has appeal as a simple unencumbered form without need for interpretation.

Concentration as mg $Al_2(SO_4)_3 \cdot 14H_2O$/L solution: If there is need to convert several kinds of concentrations to a single expression, a quasi-standard that seems to have emerged is "mg $Al_2(SO_4)_3 \cdot 14H_2O$/L solution." The full expression, albeit cumbersome, should be used. Not to do so leaves uncertainty.

F.4.5.3 Mass Percent

Another form of expression for alum concentration is as mass percent, which may be stated as "kg $Al_2(SO_4)_3 \cdot 14H_2O$/kg solution." The conversion from mass concentration is,

$$C(\text{mass percent}) = \frac{C(\text{mass concentration})}{\rho(\text{alum solution})} \quad (F.6)$$

$$= \frac{C(\text{mass concentration})}{SG(\text{alum solution}) \cdot \rho(\text{water})} \quad (F.7)$$

Equation F.7 was the basis for obtaining the values in Figure F.4b. The terms in the denominator for the conversions were, SG(alum solution) = 1.335, ρ(water) = 998.2 g/L (20°C).

F.4.6 Specific Gravity–Concentration Relation

The specific gravity (SG) of an alum solution depends on its concentration, i.e., mg $Al_2(SO_4)_3 \cdot 14H_2O$/L solution. From a set of experimental data such as given in Table CDF.3 (adapted from Table 1, General Chemical, 1995) one can use SG as a surrogate for concentration. Table CDF.3 has several sub-tables as follows:

Tables CDF.3(a) through CDF.3(d) are encapsulated in Figures F.5 and F.6. Table CDF.3(e) expands on Section F.4.7.

Figure F.5 shows a plot of the data from Table CDF.3(d). The polynomial coefficients, also shown in Figure F.5, are a

TABLE CDF.3
Alum Conversions

Table CDF.3(a) Molecular weight of alum forms
Table CDF.3(b) Polynomial coefficients for estimating alum concentrations from SG
Table CDF.3(c) Polynomial coefficients for estimating alum mass percent solids from SG
Table CDF.3(d) Alum concentration as function of "dry alum" added
Table CDF.3(e) Cost calculation for alum as Al^{3+} and sensitivity to SG of liquid alum and C(Al^{3+})
[Double click on "Table CDF." below to bring up Table CDF.3]

TABLE CDF.3
Alum Conversions (Excerpt Showing only Tables CDF.3(a) and (b))

Table CDF.3(a) Molecular Weights of Alum Forms				Table CDF.3(b) Polynomial Coefficients[a]			
Element	at. wt.	Compound	MW	Compound	M0	M1	M2
Al	26.982	$Al_2(SO_4)_3 \cdot 14H_2O$	594.136	$Al_2(SO_4)_3 \cdot 14H_2O$	−1011.9	312.25	699.2
S	32.066	$Al_2(SO_4)_3 \cdot 18H_2O$	666.132	$Al_2(SO_4)_3 \cdot 18H_2O$	−1134.3	350.01	783.01
O	15.999	$Al_2(SO_4)_3$	342.15	$Al_2(SO_4)_3$	−577.37	178	399.11
H	1.000	Al_2O_3	101.961	Al_2O_3	−172.09	53.103	118.91
O	15.999	Al_2	53.964	Al_2	−91.909	28.361	63.506

Note: ρ(water) = 998 kg/m^3.

[a] For alum concentrations from SG alum, i.e., $y = M0 + M1 \cdot x + M2 \cdot x^2$

TABLE F.3
Cost of Liquid Alum[a]

(1) The cost of liquid alum in metric units is

$$\frac{\text{dollars}}{\text{kg alum solution}} = \frac{\$181}{\text{ton alum solution}} \cdot \frac{\text{ton}}{2000\ \text{lb}} \cdot \frac{2.205\ \text{lb}}{\text{kg}} = \frac{\$0.20}{\text{kg alum solution}}$$

(2) In terms of Al, the cost is (using mass percent Al from Table F.3 for SG = 1.34)

$$\frac{\text{dollars}}{\text{kg Al}} = \frac{\$0.20}{\text{kg alum solution}} \cdot \frac{\text{kg alum solution}}{0.0445\ \text{kg Al}} = \frac{\$4.48}{\text{kg Al}}$$

Sources: Adapted from Letterman, R. D., *Overview of Operational Control of Coagulation and Filtration Processes*, Department of Civil Engineering, Syracuse University, Syracuse, NY, 2000, paper presented at *Seminar on Water Treatment*, New York Section, American Water Works Association Seminar, Rochester, NY, November 1985.

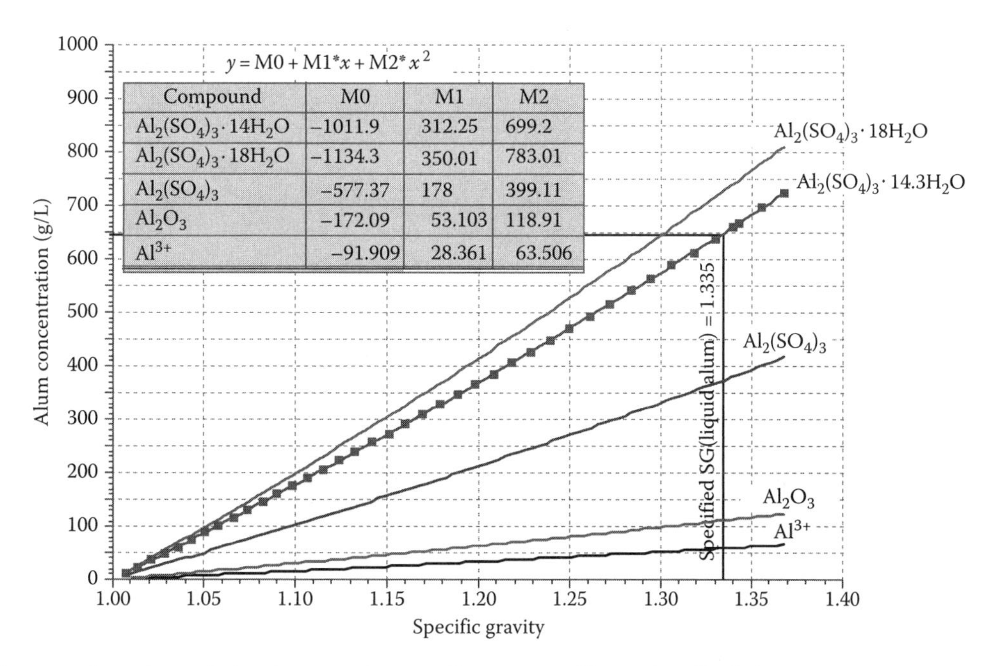

FIGURE F.5 Concentrations alum solution (expressed as stated) versus specific gravity of solution.

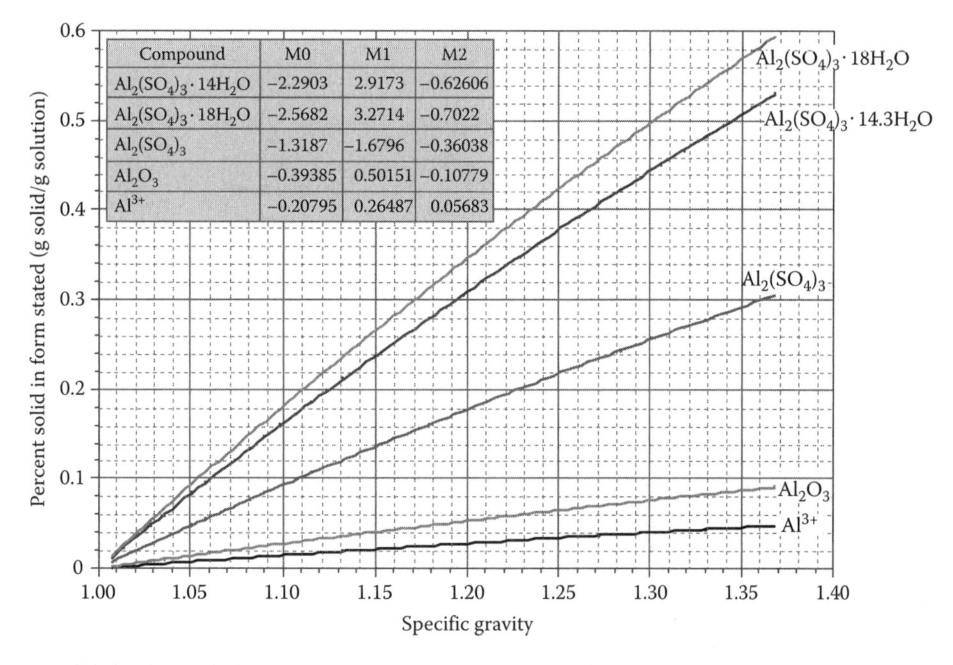

FIGURE F.6 Percent solids in alum solution (expressed as stated) versus specific gravity of solution.

means to calculate concentrations within 0.3%. The other curves in Figure F.5 were calculated by molecular weight ratios, using the data in Table CDF.3(d) for $Al_2(SO_4)_3 \cdot 14.3H_2O$ as the basis.

Figure F.6 shows mass percent of alum vs. SG(alum) for different expressions of alum. These curves were all derivatives, i.e., not original data, with calculations by Equation F.7. The full delineation of these curves is from original data (from General Chemical, 1995) given also in Table CDF.3(d).

F.4.7 Cost Calculations

The cost of liquid alum may be calculated in terms of Al^{3+} or as mass of $Al_2(SO_4)_3 \cdot 14H_2O$ crystals. Example F.7 illustrates the calculation. In addition, the density of alum is important and will vary a little from one shipment to the next and is measured at some plants (see F.2). The cost sensitivity to density (or SG) can be determined by changing this term in the calculations shown. The calculations can be set up in a spreadsheet form which can permit exploration of the effect of SG (see Table CDF.3(e))

Example F.7 Calculate Cost of Al in Liquid Alum Solution

Given
The cost of liquid alum at Denver, Colorado in 1999 was \$181/ton liquid alum.

Required
Determine the cost of Al in liquid alum in \$/kg Al.

Solution
Table F.3 outlines the solution.

Rationale: Again, the approach is to use the chain of conversions to achieve the result desired.

F.5 USING ALUM IN WATER TREATMENT

Some concerns in using alum relate to dilution, storage, mass flow, calibration of metering pumps, and cleaning lines. The issues of dilution and storage come up frequently in the operation of pilot plants. The others issues are included to emphasize their importance and to provide guidelines for operation.

F.5.1 DILUTING ALUM

In practice, liquid alum is metered into the rapid mix as a "neat" solution. In pilot plants, however, the required alum flow as a neat solution may be less than that of the minimum flow of the metering pumps available. In addition, the distribution of the alum flow throughout the raw water may be more difficult with only a small point flow of neat alum.

To alleviate both problems, the neat alum solution may be diluted. According to persons knowledgeable in the field, there is no limit to the dilution ratio as long as the pH of the diluted solution is less than 4.0, keeping in mind that the pH of the neat alum solution is about 2.4. The idea is that the pH should be low enough such that the equilibrium is toward Al^{3+} as the predominant alum species vis-à-vis hydrolysis products. Nevertheless, in pilot plant and full-scale operations, most persons the author knows use neat alum as long as it is feasible to do so with the metering equipment, in order to avoid any uncertainty.

Another aspect of dilution is that the terminology should be clear. Dilution by volume should state, for example, that 10 mL $Al_2(SO_4)_3 \cdot 14H_2O$ solution is diluted to result in 100 mL solution. The terminology 1:10 dilution may cause confusion as to whether the dilution is 10–100 mL solution or 10 mL added to 100 mL water. Example F.8 illustrates the rationale for dilution calculations.

Example F.8 Calculate Quantities of Liquid Alum to Obtain a Dilution of 10%

Given
Suppose that a neat liquid alum solution is to be diluted, say about 1:10, in order to increase the effectiveness of alum mixing in the raw water.

Required
Determine the quantities of liquid alum and water, respectively, to formulate the dilute solution.

Solution
Rationale: Keep in mind the 10% specification, or 1:10 dilution, are expressions that should be defined if they are used. Therefore, the term, 1:10 dilution is defined as adding x g of alum contained in 1.000 g of liquid alum, to 9.000 L of water, to give a solution of $x/10$ g/L concentration. In other words, express the new solution as a $0.1x$ g/L solution (i.e., as a concentration rather than as a "10%" alum solution). The latter is nebulous, having no agreed upon meaning.

Sequence of Steps

1. Conditions specified for the dilute solution.
   Suppose that the concentration of the diluted alum solution is to be

$$\frac{64.7 \text{ g } Al_2(SO_4)_3 \cdot 14.0H_2O}{L} \quad \text{and that 10.0 L is needed.}$$

2. Mass of alum needed for the 10.0 L is

$$\frac{64.7 \text{ g } Al_2(SO_4)_3 \cdot 14H_2O}{L} \cdot 10.000 \text{ L}$$
$$= 647 \text{ g } Al_2(SO_4)_3 \cdot 14H_2O$$

3. Volume of neat liquid alum that contains 647 g $Al_2(SO_4)_3 \cdot 14H_2O$ is 1.000 L.
4. Dilution protocol.
   Measure 1000 mL of neat liquid alum solution.
   Measure, for example, 5000 mL of water, with pH adjustment to pH $\approx$ 3.
   Add water to result in about 9.500 L of solution
   Make sure that pH $\ll$ 4.0
   Mix neat liquid alum and water
   Add water to result in 10.000 L
   Mix
5. pH check.
   After mixing, check the pH of the dilute alum solution to make sure that it is much less than 4.
6. Resulting concentration.
   The resulting alum solution concentration is

$$\frac{647 \text{ g } Al_2(SO_4)_3 \cdot 14H_2O}{10.000 \text{ L alum solution}} = \frac{64.7 \text{ g } Al_2(SO_4)_3 \cdot 14H_2O}{1.000 \text{ L alum solution}}$$

Discussion
The resulting alum solution should be referred to as a "64.7 g $Al_2(SO_4)_3 \cdot 14H_2O$/L alum solution." If one wishes to state that it is a 10% alum solution, to indicate the amount of dilution, then this may be understood as an approximation of what was done. But it has no precise meaning upon which calculations can be based, which should be conveyed in communication.

F.5.2 Storage of Liquid Alum

Under favorable storage conditions, i.e., no evaporation and temperatures above −10°C (10°F) to prevent freezing, liquid alum will remain chemically intact and usable. In other words, the "shelf life" of liquid alum is indefinite under such conditions. While temperature may be controlled, the storage tanks should be vented and so there is loss of water by evaporation. According to one person experienced in the field, the storage should be not so long then that the density of the alum changes appreciably.

Except for very small plants, the alum storage volume should be sufficient to handle a full load of alum from a truck or for several days' supply, whichever is the larger volume. As with most facilities, storage and feed facilities should be redundant, i.e., with duplicate facilities to ensure reliability and to provide for cleaning and maintenance.

F.5.3 Mass Flow Calculations

Alum flow into the raw water is determined by the materials balance principle, i.e.,

$$Q(\text{alum solution}) \cdot C(\text{alum}) = Q(\text{raw water}) \cdot C(\text{required alum in raw water}) \quad \text{(F.8)}$$

Example F.9 illustrates the calculation.

Example F.9 Calculate Metering Rate for Dilute Liquid Alum Solution to a Raw-Water Flow

Given

The rapid mix for a 20 gpm pilot plant is to receive an alum (as $Al_2(SO_4)_3 \cdot 14H_2O$) dosage of C(alum in raw water) = 18 mg/L. The alum-feed solution has a concentration, C(alum feed) = 647 g/L.

Required

Determine the flow of alum-feed solution Q(alum) to the rapid mix.

Solution

Convert all U.S. Customary Units to SI/Metric. The SI units are always preferred, but not always convenient, so apply appropriate metric units for this problem. The pilot plant flow is the only conversion needed, i.e.,

$$20\ \frac{\text{gal}}{\text{min}} \cdot \frac{3.785\ \text{L}}{\text{gal}} = 75.7\ \frac{\text{L}}{\text{min}}$$

Apply Materials Balance Principle

$$Q(\text{alum solution}) \cdot C(\text{alum}) = Q(\text{raw water}) \cdot C(\text{required alum in raw water})$$

$$Q(\text{alum solution}) \cdot 647{,}000\ \frac{\text{mg}}{\text{L}} = 75.7\ \frac{\text{L}}{\text{min}} \cdot 18\ \frac{\text{mg}}{\text{L}}$$

$$Q(\text{alum solution}) = 0.002106\ \frac{\text{L}}{\text{min}} = 2.106\ \frac{\text{mL}}{\text{min}}$$

Discussion

Some metering pumps (syringe or peristaltic) can achieve such low flows. Another problem is in the distribution of the alum throughout the raw-water flow. Therefore, dilution should be considered.

F.5.4 Metering and Calibration

In any plant, e.g. full-scale or pilot plant, the metering flow should be confirmed by volumetric measurement. Figure F.7 shows the setup for volumetric calibration of a metering pump. A graduated cylinder is placed in the line from the main alum storage to the rapid mix. For pump calibration valve b is opened and valve a is closed. At time t_1 the volume of alum in the graduated cylinder is measured, i.e., $V(\text{alum})_1$. At time t_2 the volume of alum in the graduated cylinder is measured again, i.e., $V(\text{alum})_2$. Therefore,

$$Q(\text{metering pump}) = \frac{[V(\text{alum})_2 - V(\text{alum})_1]}{(t_2 - t_1)} \quad \text{(F.9)}$$

In a pilot plant, such as a 75.7 L/min (20 gpm) size, the graduated cylinder may be a 100 mL burette. For a full-scale plant, the size would depend upon the rate of alum flow; the cylinder should be large enough to minimize the error of measurement. Note that after valve b, a waste line should be inserted with needed valves in order to clean the cylinder and lines between uses.

F.5.5 Cleaning Lines

In maintaining the alum-flow system the storage and feed should be in duplicate, such that one system can be rotated in service with another. Cleaning should be facilitated by inclusion of valves and tees so that the lines can be flushed easily with suitable provision for handling the waste flow.

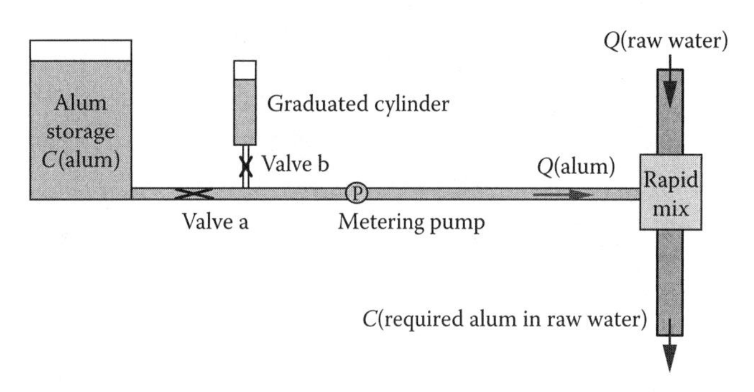

FIGURE F.7 Volumetric measurement of alum flow—calibration of metering.

F.6 ALUM POLYMER BLENDS AND FORMULATIONS

In the late 1970s, alum polymer blends and formulations were introduced, becoming increasingly popular during the 1980s. They provide alternatives to the pure metal coagulants, i.e., alum and iron salts and are being adopted worldwide.

F.7 ALUM POLYMERS

Polyaluminum chloride, aluminum chlorohydrate, and polyaluminum sulfate are different than alum polymer blends. They are distinct inorganic coagulants. Polyaluminum chloride was used first in Japan in 1967; by 1999 it was the third in demand in the United States after alum and iron salts.

ACKNOWLEDGMENTS

Kevin Gertig and Grant Williamson-Jones, Fort Collins Water Treatment Plant, were available to answer many questions on practical issues of using alum. Christopher Lind, General Chemical Co. Syracuse, New York, Parsippinay, New Jersey, responded to many questions about the use of alum and practices within the industry.

REFERENCES

AWWA, AWWA Standard for Aluminum Sulfate—Liquid, Ground, or Lump, American Water Works Association, ANSI/AWWA B403-88, Denver, CO, 1988.

Britannica, *The New Encyclopedia Britannica*, 15th edn., Encyclopedia Britannica, Inc., Chicago, IL, 1974.

General Chemical, Aluminum Sulfate (Alum)—Technical Data, Brochure CHEM-M5-23, General Chemical Corporation, 1995.

General Chemical, Aluminum Sulfate—Liquid, Product Data Sheet, General Chemical Corporation, 1997a.

General Chemical, Aluminum Sulfate—Dry, Product Data Sheet, General Chemical Corporation, 1997b.

Harringer, R. D., Aluminum sulfate coagulation low temperature water, unpublished paper presented at *CH2M-Hill Cold Water Coagulation Seminar*, Denver, CO, July 13, 1984.

Jones, G. W., personal communication by telephone, January 28, 1994.

McCabe, W. L., Smith, J. C., and Harriott, P., *Unit Operations of Chemical Engineering*, Fifth Edition, McGraw-Hill, 1993.

Appendix G: Dimensionless Numbers

Dimensionless groupings of variables have been derived using the Buckingham pi theorem in both fluid mechanics and chemical engineering. The former has included the Euler number, the Froude number, the Reynolds number, the Weber number, etc. Chemical engineers have adopted these numbers and have added the Schmidt number, the Sherwood number, the Power number, the Peclet number, etc. The nomenclature adopted, which is common, was to use bold fonts to indicate a given dimensionless number, i.e., the Euler number is **E**, the Froude number is **F**, the Reynolds number is **R**, and so on.

Dimensionless numbers are used frequently in the literature and often without definition or explanation. The intent of this appendix was to provide definitions for the commonly used dimensionless numbers and to review the topic in general.

G.1 THE WORLD OF DIMENSIONLESS NUMBERS

The world of dimensionless numbers is actually much more extensive than might be imagined, i.e., based on the usual exposures in textbooks. This is illustrated by the Land Chart of Dimensionless Numbers (Omega Engineering, Inc., 1997; http://www.omega.com/literature/posters, 2009), which lists some 154 dimensionless numbers. Table G.1 also lists these numbers; as seen, most are not recognizable from the literature. The purpose of Table G.1 is merely to give an indication of the extent to which dimensionless numbers have been proposed.

Table G.2 lists dimensionless numbers that could be applicable to treatment processes as compiled from the given references. Table G.2 gives the name, grouping of variables, definitions of variables, and nature of the ratios involved.

Table CDG.3 is a matrix of physical phenomena and associated dimensionless numbers (Omega Engineering, Inc., 1997). Table CDG.3 shows the 45 phenomena as columns and 154 dimensionless numbers as rows. The dimensionless numbers applicable to a given phenomena could be indicated by "x" in the appropriate columns (not done as given). The table is in the form of a spreadsheet on a CD disk.

G.2 UNDERSTANDING DIMENSIONLESS NUMBERS

Dimensionless numbers provide a means to group variables such that a large amount of data may be condensed into a single set of plots. For example, Reynolds number, **R**, defined as $\mathbf{R} = \rho v D/\mu$, can be applied to a wide range of combinations of fluid densities, velocities, diameters, and viscosity. Instead of conducting laboratory tests for an unknown condition such as a different temperature, the viscosity can be obtained from a handbook for the temperature in question, the Reynolds number may then be calculated, and for a given pipe material, the friction factor may be obtained from a Moody chart.

Dimensionless numbers have a rational basis as well. Consider those that describe bulk fluid flow, such as **E**, **F**, **R**. These dimensionless numbers, the Euler number, the Froude (pronounced "fru-d") number, and the Reynolds number are, respectively, the ratios of inertia forces to pressure forces, inertia forces to gravity forces, and inertia forces to viscous forces. They provide an empirical means to characterize fluid flow phenomena.

The Navier–Stokes equation has long been recognized as the most comprehensive mathematical description of bulk fluid flow. But as a "second-order non-linear partial differential equation" (White, 1979), it has been considered too complex for traditional mathematical solutions. The task of computational fluid mechanics is to solve the Navier–Stokes equation numerically for the particular boundary conditions of interest. The Navier–Stokes equation can be understood more easily if looked at as merely an expansion of Newton's second law, $F = ma$ (see, for example, Einstein, 1963). The force term on the left side incorporates terms for pressure, gravity, and viscous forces. The right side is the inertia term. If one force term is dominant, such as gravity, then the Froude number can serve as means to characterize the dynamics of the system, which is the ratio of gravity forces to the inertia forces.

Two kinds of forces may be important, however, especially in certain ranges. Consider, for example, pressure forces and viscous forces, as characterized by **E**, are a function of **R**.

An example is shown in Figure G.1 in which **E** declines rapidly with increasing **R** and then levels. **E** is influenced strongly by viscous forces at low **R**. Then as **R** increases, the inertia forces predominate over viscous forces and **E** is no longer affected by viscosity.

Such a curve as shown in Figure G.1 must be generated empirically by means of a laboratory setup. Discharge of a fluid through an orifice is a case in which the Euler number is important.

In this case, the discharge coefficient, C_d, is a mathematical identity with the Euler number and Figure G.1 is seen more commonly as C_d versus **R**. At high **R**, the Euler number (or C_d) is a function of the geometry only. As another example, the Euler number may be a function of the Weber number at low values of **W**.

In the same fashion, the Froude number, **F**, is a function of **R** and a plot would be similar to Figure G.1. As a matter of practical interest, the Froude number is an identity mathematically with the discharge coefficient for a weir.

TABLE G.1
List of 154 Dimensionless Numbers from Land Chart of Dimensionless Numbers

Dimensionless Numbers	Dimensionless Numbers	Dimensionless Numbers	Dimensionless Numbers
Acceleration	Deryagin	Karman-1	Prandtl heat transfer
Aeroelastic	Dulong	Kirpichev heat transfer	Prandtl mass transfer
Alfven	Ekman	Kirpichev mass transfer	Prandtl velocity ratio
Archimedes	Elasticity-1	Kirpitcheff	Predvodetlev
Arrhenius	Elasticity-2	Knudsen	Radiation pressure
Bagnold	Elasticity-3	Kossovich	Rayleigh
Bansen	Electric Reynolds	Lagrange-1	Regier
Bingham	Electroviscous	Lagrange-2	Reynolds
Biot heat transfer	Ellis	Leverett	Richardson
Biot mass transfer	Elsasser	Lewis	Rossby
Blake	Euler	Lundquist	Russell
Bodenstein	Evaporation-1	Lykoudis	Sachs
Boltzman	Evaporation-2	Mach	Schiller
Bond	Evaporation elasticity	Magnetic dynamic	Slosh time
Bouguer	Explosion	Magnetic force	Sommerfeld
Boussinesq	Fanning	Magnetic interaction	Specific heat ratio
Brinkman	Federov	Magnetic Prandtl	Specific speed
Bubble Nusselt	Fliegner	Magnetic pressure	Squeeze
Bubble Reynolds	Flow	Magnetic Reynolds	Stanton
Buoyancy	Fourier heat transfer	Marangoni	Stefan
Capillarity-1	Fourier heat transfer	Mass ratio	Stokes
Capillarity-2	Froude	McAdams	Strouhal
Capillarity-3	Frueh	Merkel	Structural merit
Capillarity-bouyancy	Galileo	Momentum	Suratman
Capillary	Goucher	Morton	Surface viscosity
Carnot	Graetz	Nusselt heat transfer	Taylor
Cavitation	Grashof	Nusselt mass transfer	Thoma
Centrifuge	Gravity	Nusselt film thickness	Toms
Clausius	Gukhman	Ocvirk	Truncation
Condensation-1	Hall	Ohnesorge	Two-phase flow
Condensation-2	Hartmann	Particle	Two-phase porous flow
Crispation	Heat transfer	Peclet heat transfer	Viscoelastic
Crocco	Hedstrom-1	Peclet mass transfer	Weber
Damköhler's first	Hersey	Pipeline	Weissenberg
Damköhler's second	Hodgson	Poiseuille	
Damköhler's third	*J*-Factor heat transfer	Poisson	
Damköhler's fourth	*J*-Factor mass transfer	Pomerantsev	
Darcy	Jacob	Porous flow	
Dean	Jakob	Posnov	
Debye	Joule	Power	

Source: Omega Engineering, Inc., The land chart of dimensionless numbers, U.S. Patent No. 5,465,838, Omega Engineering, Inc., Stamford, CT, 1997.

The dimensionless numbers, **E**, **R**, **F**, **W**, **P**, are related to the dynamics of fluid flow. On the other hand, the other dimensionless numbers have to do heat transfer and/or mass transfer and have interpretations along such lines. Some of the numbers are simply derivatives of others that are more basic. The derivative numbers may have special uses. As an example, the power number, **P**, is a derivative of the Euler number, **E**. Also, some combine the effects of two kinds of phenomena. For example, the Peclet number is the ratio of convective mass transfer to diffusion mass transfer and may be used in situations that compare the two kinds of mass transfer. As indicated in Table G.1, the world of dimensionless numbers is very large.

G.3 UTILITY OF DIMENSIONLESS NUMBERS

In practice, dimensionless numbers are used widely to display numerous kinds of empirical relationships. Examples include friction factor versus Reynolds number in pipe flow, Euler number versus Reynolds number in fluid flow

TABLE G.2
Selection of Dimensionless Numbers Applicable to Water Treatment Process Design

Name	Grouping	Variables	Ratio and Description
Euler (McCabe et al., 1993)	$\mathbf{E} = \dfrac{2F_D g_c}{\rho v_o^2 A}$	F_D = drag force on object (N) ρ = density of fluid (kg/m^3) v_o = velocity in core of jet (m/s) A = area of object (m^3) g_c = conversion factor (kg m/N s)	$\dfrac{\text{inertia forces}}{\text{pressure forces}}$
Friction factor (McCabe et al., 1993)	$\mathbf{f} = \dfrac{\Delta p D g_c}{2\rho \bar{v}^2 L}$	Δp = pressure loss in pipe due to friction in length, L (N/m^2) D = diameter of pipe (m) g_c = conversion factor (kg m/N s) ρ = density of fluid (kg/m^3) $\bar{v}$ = average velocity in pipe (m/s) L = length of pipe (m)	$\dfrac{\text{inertia forces}}{\text{friction shear forces}}$
Froude (Rouse, 1946; McCabe et al., 1993)	$\mathbf{F} = \dfrac{v^2}{gL}$	v = velocity (m/s) g = gravitational constant (9.18 m/s^2) L = characteristic length (m)	$\dfrac{\text{inertia forces}}{\text{gravity forces}}$
Mach (Rouse, 1946, p. 328)	$\mathbf{M} = \dfrac{v}{\sqrt{E/\rho}} = \dfrac{v}{c}$	v = velocity of object or fluid (m/s) E = bulk modulus of elasticity (N/m^2) ρ = density of fluid (kg/m^3) c = velocity of sound (m/s)	$\dfrac{\text{inertia forces}}{\text{elastic forces}}$
Peclet (McCabe et al., 1993)	$\mathbf{P_e} = \dfrac{vL}{D_v} = \mathbf{R_c^a S_c^b}$	v = velocity of fluid (m/s) L = characteristic length (m) D_v = molecular diffusivity in liquid (m^2/s) a = empirical exponent b = empirical exponent	$\dfrac{\text{convection mass transfer rate}}{\text{diffusion mass transfer rate}}$ (Weber and DiGiano, 1996, Omega, 1997)
Power (McCabe et al., 1993)	$\mathbf{P} = \dfrac{P g_c}{\rho n^3 D^5}$	P = power dissipated by turbulence (N m/s) g_c = conversion factor (kg m/N s) ρ = density of fluid (kg/m^3) n = rotational velocity of impeller (rev/s) D = diameter of impeller (m)	$\dfrac{\text{inertia forces}}{\text{drag forces}}$
Prandtl	$\mathbf{P_r} = \nu_v \left(\dfrac{\rho Q_H^\circ}{k_c}\right)$	ν_v = kinematic viscosity (m^2/s) ρ = density of fluid (kg/m^3) Q_H° = specific heat (kg m^2/s^2 K) or (mol m^2/s^2 K) k_c = thermal conductivity (kg m/s^3 K)	$\dfrac{\text{momentum dispersion}}{\text{heat diffusion}}$ (Weber and DiGiano, 1996, p. 212)
Reynolds (Rouse, 1946; McCabe et al., 1993)	$\mathbf{R} = \dfrac{\rho v D}{\mu}$	ρ = density of fluid (kg/m^3) v = velocity (m/s) D = characteristic length (m) μ = dynamic viscosity (kg/m s) or (N s/m^2)	$\dfrac{\text{inertia forces}}{\text{viscous forces}}$
Schmidt	$\mathbf{S_c} = \dfrac{\mu}{D_v \rho}$	μ = dynamic viscosity (kg/m s) or (N s/m^2) D_v = molecular diffusivity in liquid (m^2/s) ρ = density of fluid (kg/m^3	$\dfrac{\text{momentum dispersion}}{\text{mass diffusion}}$ (Weber and DiGiano, 1996, p. 212)
Sherwood	$\mathbf{S_h} = \dfrac{k_f \delta}{D_v}$	k_f = mass transfer coefficient (m/s) δ = characteristic length, which is the boundary layer thickness (m) D_v = molecular diffusivity in liquid (m^2/s)	$\dfrac{\text{interfacial mass transfer impedance}}{\text{molecular diffusion impedance}}$ (Weber and DiGiano, 1996, p. 212)

(continued)

TABLE G.2 (continued)
Selection of Dimensionless Numbers Applicable to Water Treatment Process Design

Name	Grouping	Variables	Ratio and Description
Specific speed (Rouse, 1946, p. 304, 305)	$N_s = \frac{nQ^{0.5}}{(g\Delta H)^{0.75}}$	N_s = specific speed n = rotational velocity (rev/s) Q = flow (m^3/s) g = gravitational constant (9.18 m/s^2) ΔH = head developed by pump (m)	$N_s \approx 0.05$, radial flow pump $N_s \approx 0.1$, radial mixed flow $N_s \approx 0.2$, mixed flow $N_s \approx 0.4$, axial mixed flow $N_s \approx 0.8$, axial flow
Weber (Rouse, 1946; McCabe et al., 1993)	$\mathbf{W} = \frac{v}{(\sigma/\rho D)}$	D = length of contact (m) ρ = density of fluid (kg/m^3) v = velocity (m/s) σ = coefficient of surface energy (N/m) or (kg/s^2)	$\frac{\text{inertia forces}}{\text{surface tension forces}}$

TABLE CDG.3
Matrix of Physical Phenomena and Associated Dimensionless Numbers

Source: Omega Engineering, Inc., The land chart of dimensionless numbers, U.S. Patent No. 5,465,838, Omega Engineering, Inc., Stamford, CT, 1997.)

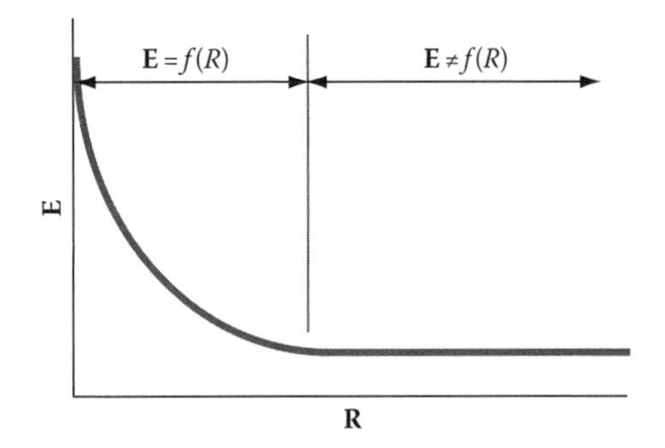

FIGURE G.1 Euler number versus Reynolds.

involving an immersed object (form drag), Sherwood number versus Reynolds number, etc.

Dimensionless numbers may be used to express experimental results in which many variables may be involved. The consolidation of variables into dimensionless numbers means that a given set of experimental results may be applicable to a wide range of conditions not investigated. The selection of the proper dimensionless numbers also identifies the key variables involved in an experiment, e.g., viscosity, diffusion, and gravity.

REFERENCES

Einstein, H. A., Engineering derivation of the Navier–Stokes equations, *Journal of the Engineering Mechanics Division, Proceedings of the American Society of Civil Engineers*, Proceedings Paper 3533, Vol. 89(EM3):1–7, June 1963.

McCabe, W. L., Smith, J. C., and Harriot, P., *Unit Operations of Chemical Engineering*, 5th edn., McGraw-Hill, New York, 1993, ISBN 0-07-112721-6.

Omega Engineering, Inc., The land chart of dimensionless numbers, U.S. Patent No. 5,465,838, Omega Engineering, Inc., Stamford, CT, 1997.

Rouse, H., *Elementary Mechanics of Fluids*, John Wiley & Sons, New York, 1946.

Weber, W. J. Jr. and DiGiano, G. A., *Process Dynamics in Environmental Systems*, John Wiley & Sons, New York, 1996, ISBN 0-471-01711-6.

White, F. M., *Fluid Mechanics*, McGraw-Hill, New York, 1979.

Appendix H: Dissolved Gases

The issue of dissolved gases comes up in a variety of situations both in unit processes and in the natural environment. The issues include (1) air stripping of dissolved gases as a unit process, (2) transfer of gas into solution as a unit process, (3) gas precipitation as a spontaneous occurrence. Gas transfer also occurs in the environment, e.g., oxygen uptake, carbon dioxide uptake, precipitation of various gases in a saturated local environment, e.g., oxygen, methane, carbon dioxide, nitrogen, etc. The equilibrium between the gas phase and the dissolved state for a given gas is expressed by Henry's law. Its application requires understanding the ideal gas law and Dalton's law of partial pressures. Other kinds of fundamental notions help to establish the background for understanding Henry's law, an ostensibly simple equation, e.g., the effect of elevation on atmospheric pressure, partial pressure of water vapor, molar composition of air, etc.

H.1 FUNDAMENTALS OF GAS BEHAVIOR

Dealing with gases requires a few notions of gas behavior and also include conventions in stating the pressure, the ideal gas law, Dalton's law, the effect of elevation on atmospheric pressure, composition of ambient air, and the partial pressure of water vapor. These fundamentals are reviewed here.

H.1.1 Conventions for Stating Pressure

Both the gage and absolute pressure are used in mathematical expressions, depending on the circumstances. Gage pressure is used commonly in practice while absolute pressure is necessary in calculations based on scientific principles (such as those involving the ideal gas law and Henry's law). The relation between the two is understood most easily by a graphical depiction, Figure H.1. Gage pressure is always relative to the atmosphere and is the difference between pressures, such as the pressure between the inside and outside of a pressure gage. Referring to Figure H.1, the relationship between absolute pressure and gage pressure is,

$$p_{\text{abs}} = p_{\text{atm}} + p_{\text{gauge}} \tag{H.1}$$

Example H.1 Conversions between Gage Pressure and Absolute Pressure

Calculate the absolute pressure at point A if the gage pressure is 50 kPa and the atmospheric pressure is 101.3 kPa.

Referring to Figure H.1a, and applying Equation H.1, the algebraic relation is, with subsequent substitution of numerical data,

$$p(A)_{\text{abs}} = p_{\text{atm}} + p(A)_{\text{gage}} \tag{H.1}$$

$$= 101.3 \text{ kPa} + 50 \text{ kPa}$$

$$= 151 \text{ kPa}$$

H.1.2 Ideal Gas Law

The ideal gas law is,

$$p_{\text{A}}V_{\text{A}} = n_{\text{A}}RT \tag{H.2}$$

where

p_{A} is the absolute pressure of gas "A" (Pa or N/m^2)
V_{A} is the volume occupied by gas "A" (m^3)
n_{A} is the moles of gas of species A (mol)
R is the universal gas constant (8.314 510 N · m/g-mol K)
T is the temperature of gas "A" (K)

The ideal gas equation is satisfactory for most engineering situations but is not accurate at very high pressures, which includes any gas near the condensation point. Van der Waal's equation or the virial equation (Alberty and Silbey, 1992) will more closely approximate high pressure conditions.

Example H.2 Application of Ideal Gas Law to Determine Density at NTP

Statement

Very often, gas densities must be determined. The basis for the calculation of gas densities is the ideal gas law. In such calculations, the SI system should be used. Conversion can be done to any other form desired after the basic calculation.

(a) Calculate the density of pure oxygen at NTP (normal temperature and pressure).

1. Apply the ideal gas law,

$$pV = nRT \tag{ExH.2.1}$$

2. Rearrange Equation ExH.2.1,

$$\frac{n}{V} = \frac{p}{RT}$$

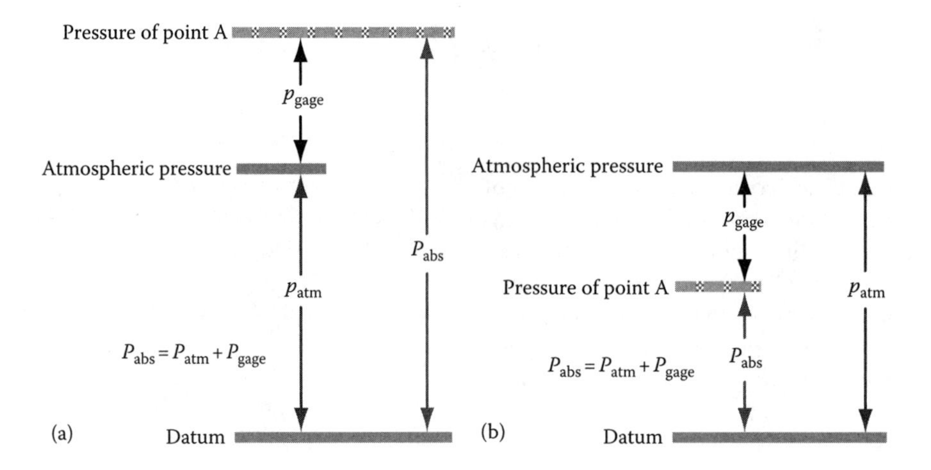

FIGURE H.1 Graphical and algebraic pressure relations for conversions between absolute pressure and gage pressure; pressure at level "A" is the focus. (a) Positive gage pressure and (b) negative gage pressure (vacuum).

3. Substitute NTP values.
   For NTP, substitute, $T = 0°C = 273.15$ K, $p = 1.013250 \cdot 10^5$ Pa, and let $V = 1.0$ m^3, with $R = 8.314510$ J K^{-1} mol^{-1}, to give

$$\left[\frac{n}{1.00\ \text{m}^3}\right] = \frac{1.013\,250 \cdot 10^5\ \text{Pa}}{(8.314510\ \text{J K}^{-1}) \cdot (273.15\ \text{K})}$$

then since n/V is molar density, ρ(molar, NTP), the result is,

$$\rho(\text{molar, NTP}) = 0.000446\ \frac{\text{mol O}_2}{\text{m}^3}$$

where ρ(molar, NTP) is the molar density of gas at NTP (mol O_2/m^3).

(b) Convert molar density, ρ(molar, NTP), to mass density ρ(mass, NTP).
   1. Apply chain of conversion equivalents,

$$\rho(\text{mass, NTP}) = 0.0004446\ \frac{\text{mol O}_2}{\text{m}^2} \cdot \frac{32.00\ \text{g O}_2}{\text{mol O}_2}$$

$$= 0.014227\ \frac{\text{g O}_2}{\text{m}^3}$$

$$= 14.23\ \frac{\text{g O}_2}{\text{L}}$$

Comments

The above calculations illustrate the methodology to calculate densities of gases.

Example H.3 Calculate the Density of Air at Sea Level at 20°C

1. Apply the ideal gas law for the conditions stated, i.e.,

$$pV = n(\text{air})RT \qquad \text{(H.2)}$$

$$101{,}325\ \frac{\text{N}}{\text{m}^2} \cdot V = n(\text{air}) \cdot 8.314\ \frac{\text{N} \cdot \text{m}}{\text{K mol}} \cdot 293.15\ \text{K}$$

$$\frac{n(\text{air})}{V} = 41.57\ \frac{\text{mol air}}{\text{m}^3}$$

2. Convert density in mol/m^3 to kg/m^3, i.e.,

$$\rho(\text{air}) = \frac{n(\text{air})}{V} \cdot \text{molar density(air)}$$

$$= 41.57\ \frac{\text{mol air}}{\text{m}^3} \cdot \frac{0.028964\ \text{kg air}}{\text{mol air}}$$

$$= 1.20\ \frac{\text{kg air}}{\text{m}^3}$$

H.1.3 Dalton's Law

In a mixture of gases, the total pressure in that mixture is the sum of the partial pressures of each of the species, i, i.e.,

$$p = \sum_n p_i \qquad \text{(H.3)}$$

where p_i is the partial pressure of gas i (Pa).

The partial pressure is also proportional to the mole fraction of gas, i,

$$p_i = \frac{n_i}{n} p \qquad \text{(H.4)}$$

Since the volume, V, is common for all gases in the mixture, it follows that the sum of the mole fractions is 1, i.e.,

$$1 = \frac{n_1}{n} + \frac{n_2}{n} + \cdots = \sum_i^{\infty} \frac{n_i}{n} \qquad \text{(H.5)}$$

Understanding how to apply Dalton's law and its variations is useful in dealing with mixtures of gases. Example H.4 illustrates how to apply Dalton's law to determine the partial pressure of oxygen.

Example H.4 Determine Partial Pressure of Oxygen in Air

The mole fraction of oxygen in air is 0.2095 (Table H.1).

a. Determine the partial pressure of oxygen at sea level.
   1. Apply Dalton's law for p(air, sea level),

$$p(O_2) = \frac{n(O_2)}{n(\text{air})} p(\text{air, sea level})$$
$$= \frac{0.2095 \text{ mol } O_2}{\text{mol gas mixture(air)}} 1.0 \text{ atm air}$$
$$= 0.2095 \text{ atm } O_2$$

b. Determine the partial pressure of oxygen at elevation, 1800 m,
   1. Apply Dalton's law for p(air, 1800 m) = 81.49 kPa (Equation H.24 or Figure H.2), which is 0.804 atm,

$$p(O_2) = \frac{n(O_2)}{n(\text{air})} p(\text{air, 1800 m})$$
$$= \frac{0.2095 \text{ mol } O_2}{\text{mol gas mixture (air)}} \cdot 0.804 \text{ atm}$$
$$= 0.168 \text{ atm } O_2 \text{ (17.0 kPa } O_2)$$

Comments

In applying Henry's law, the partial pressure of the gas species of interest must be determined first.

H.1.4 Atmospheric Pressure versus Elevation

Figure H.2 gives atmospheric pressure as a function of elevation, with plotting data obtained from Lide (1996, pp. 14–17). Equation H.6, a best fit polynomial that accurately depicts the plot of Figure H.2, was from the Kladiograph® software, which was used to develop the plot from the data provided. The actual pressure at any elevation may vary depending on the local conditions. For example, a mercury barometer located at the Engineering Research Center, Colorado State University, Fort Collins, Colorado, at a ground elevation of 1585 m (5200 ft) reads 634 mm Hg (which varies a few mm Hg from day-to-day with weather conditions); when compared, for this same elevation, Equation H.6 calculated 83.7 kPa, or 628 mm Hg (a 0.9% discrepancy). For any given elevation, the pressure will vary about a mean as weather conditions change.

$$P(\text{atm}) = \text{M0} + \text{M1} \cdot Z + \text{M2} \cdot Z^2 + \text{M3} \cdot Z^3 + \text{M4} \cdot Z^4 + \text{M5} \cdot Z^5 + \text{M6} \cdot Z^6 \qquad \text{(H.6)}$$

where
- P(atm) is the atmospheric pressure (kPa)
- Z is the elevation (m)
- M0, M1, M2, M3, M4, M5, M6 are polynomial coefficients
- M0 = 101.325
- M1 = −0.011944
- M2 = $5.3142 \cdot 10^{-07}$

TABLE H.1
Composition of Air and Calculation of Molecular Weight

Gas Law Data	P(atm) = 101,300 Pa		R = 8.31451 N m/K mol		T = 20°C = 293.15 K
Gas[a]	MW[a] (g/mol)	X(gas)[b]	ρ(molar)[c] (mol/m^3)	ρ[d] (kg/m^3)	MW Fraction[e] (g gas/mol air)
N_2	28.0134	0.78084	32.45226	0.9091	21.8740
O_2	31.9988	0.209476	8.70597	0.2786	6.7030
Ar	39.948	0.00934	0.38818	0.0155	0.3731
CO_2	44.0098	0.000314	0.01305	0.00057	0.0138
Ne	20.1797	0.00001818	0.00076	1.5247E−05	0.00037
He	4.0026	0.00000524	0.00022	8.7168E−07	2.0974E−05
Kr	83.80	0.00000114	0.00005	3.9703E−06	0.000095
Xe	131.29	0.000000087	0.00000	4.7472E−07	1.1422E−05
CH_4	16.0428	0.000002	0.00008	1.3335E−06	3.2086E−05
H_2	2.01588	0.0000005	0.00002	4.1891E−08	1.0079E−06
O_3[b]	47.999	$1.0 \cdot 10^{-8}$			
Rn[b]	222	$6.0 \cdot 10^{-20}$			
Sum		0.999997147	41.5606	1.2038	28.964[f]

[a] Lide, D. R., *Handbook of Chemistry and Physics*, 77th edn., CRC Press, Inc., Boca Raton, FL, 1996, pp. 4–37:98.
[b] Weast, R. C. (Ed.), *Handbook of Chemistry and Physics*, 59th edn., CRC Press, Inc., Boca Raton, FL, 1978, p. F205.
[c] ρ(molar) = $n/V = P(\text{gas})/RT = X(\text{gas}) \cdot P(\text{atm})/RT$.
[d] ρ = (molar) · MW(gas)/1000.
[e] MW fraction = MW(gas) · X(gas).
[f] MW(air) = sum[MW(gas)$_i$ · X(gas)$_i$].

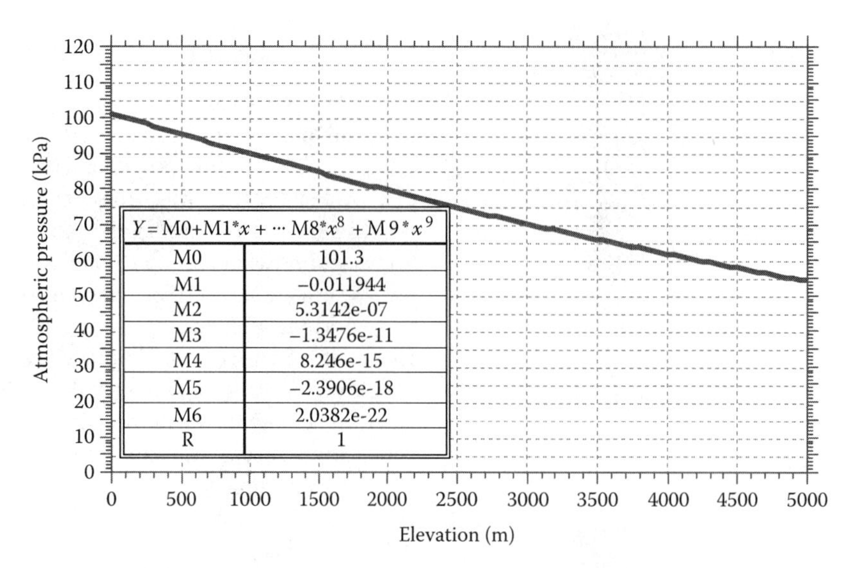

FIGURE H.2 Atmospheric pressure as function of elevation above sea level. (Figure plotted and regression equation from data as obtained in Lide, D. R. (Ed.), *Handbook of Chemistry and Physics*, 77th edn., CRC Press, Inc., Boca Raton, FL, pp. 14-17, 1996.)

$M3 = -1.3476 \cdot 10^{-11}$
$M4 = 8.2464 \cdot 10^{-15}$
$M5 = -2.3906 \cdot 10^{-18}$
$M6 = 2.0382 \cdot 10^{-22}$

The same data as given by Lide for pressure also provided temperature and density for different elevations. The temperature data showed a decline in elevation as given by Equation H.7, which was also a best fit of the data by the Kladiograph software,

$$T(\text{K}) = 1.2105 - 9.7673 \cdot 10^{-04} \cdot Z \qquad \text{(H.7)}$$

where T(K) is the temperature (K).

The associated density is depicted accurately by Equation H.8, also showing a decline with elevation, i.e.,

$$\rho(\text{air}) = 1.2105 - 9.7673 \cdot 10^{-04} \cdot Z \qquad \text{(H.8)}$$

where ρ(air) is the density of air (kg/m^3).

As a matter of interest, to cross reference with the utility of the ideal gas law, the density of a gas is a function of temperature and pressure and can be calculated by the pressure and temperature data, i.e., $\rho(\text{molar}) = n/V = P/RT$ and $\rho(\text{kg/m}^3) = \rho(\text{molar}) \cdot \text{MW(air)}/1000$. The value for MW(air) is given in Table H.1.

H.1.5 Composition of Ambient Air

Another interest is to know the composition of ambient air. Table H.1 gives the sea level composition of a dry atmosphere. Such data are required when applying Henry's law to problems involving atmospheric gases.

H.1.6 Water Vapor

Figure H.3 gives the vapor pressure of water at temperatures from 0°C to 100°C. The plot is given as reference for calculations requiring vapor pressure data. Equation H.9, also a polynomial describes the relationship, i.e.,

$$P(\text{vapor}) = M0 + M1 \cdot Z + M2 \cdot Z^2 + M3 \cdot Z^3 + M4 \cdot Z^4 + M5 \cdot Z^5 + M6 \cdot Z^6 \qquad \text{(H.9)}$$

where

P(vapor) is the pressure of water vapor in equilibrium with water surface (kPa)

M0, M1, M2, M3, M4, M5, M6 are polynomial coefficients for vapor pressure versus Z

M0	M1	M2	M3	M4	M5	M6
0.61052	0.044905	$1.3613 \cdot 10^{-03}$	$3.0315 \cdot 10^{-05}$	$1.9829 \cdot 10^{-07}$	$3.5164 \cdot 10^{-09}$	$-2.7009 \cdot 10^{-12}$

H.2 GAS SOLUBILITY IN WATER: HENRY'S LAW

The solubility of a gas in water is given by Henry's law, which has utility for innumerable situations. Although simple and clear, implementation of Henry's law may be complicated artificially. Reasons are (1) Henry's law has two forms, and (2) a variety of units for Henry's constant are in use, (3) Henry's constant data are scattered in the literature.

In this section, Henry's law is defined, applications are illustrated, and the issues that complicate its use are addressed. Hopefully, its pure simplicity is not obscured by the complicating issues. First, however, Henry's law has some interesting background that later was tied to thermodynamic theory.

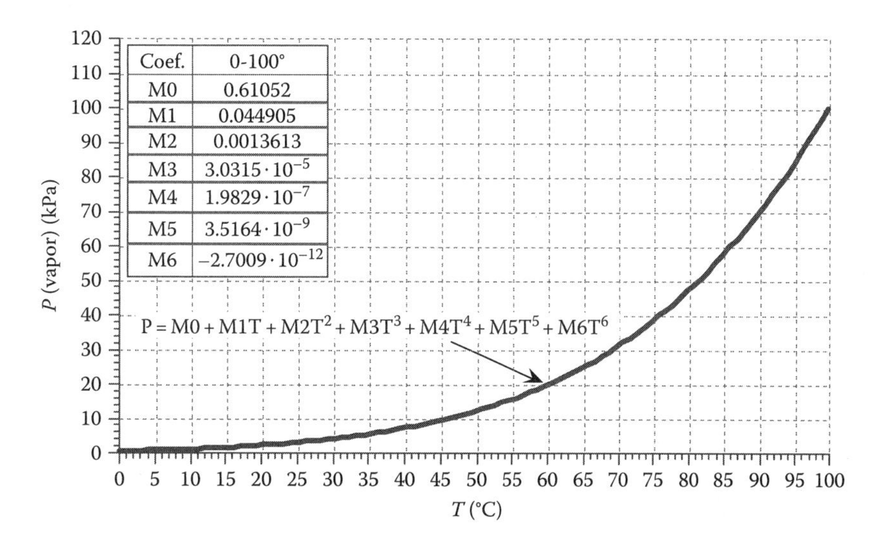

FIGURE H.3 Pressure of water vapor in equilibrium with water surface, $0 < T < 100$. (Figure plotted and regression equation from data as obtained in Lide, D. R. (Ed.), *Handbook of Chemistry and Physics*, 77th edn., CRC Press, Inc., Boca Raton, FL, pp. 6-13, 1996.)

H.2.1 Henry's Law: Discovery and Evolution

At the turn of the nineteenth century, having recently discarded the phlogiston theory, the science of chemistry, as formulated by Antoine Lavoisier in France (who, in 1789, published *Traité d'Elémentaire de Chemie*) and others across Europe, was just beginning to take shape in the modern sense. It was in this context that William Henry in 1802 presented to the Royal Society in London his observations that related pressure to the solubility for a few gases in water. His conclusion was,

> ... water takes up, of gas condensed by one, two or more additional atmospheres, a quantity which, ordinarily compressed, would be equal to twice, thrice, &c, the volume absorbed under the common pressure of the atmosphere.

William Henry was born in Manchester, England and started his career as a physician and drifted to chemistry taking over a chemical works established by his father. He was a member of the Manchester Literary and Philosophical Society and was a close friend to John Dalton, a Quaker schoolmaster, also a member. Dalton's principles were established over a 10-year period beginning in 1802. In 1805, Dalton presented his paper that established that the solubility of individual components in a gaseous mixture depended on their respective partial pressures, based upon the reasoning that each component in a gaseous mixture was independent of the other components. Dalton's law of partial pressures was assimilated into Henry's law.

Other scientific activity that related (Carroll, 1993) included the work of J. William Gibbs who in 1875 formulated the theory that supported phase equilibrium, a facet of his landmark work that applied thermodynamics to chemistry. Then, in 1887 François-Marie Raoult published his findings on vapor pressures of solutions, i.e., Raoult's law. Raoults law applies to high mole fractions of a solute, i, while Henry's law applies to low mole fractions. At the turn of the twentieth century, Gilbert Lewis postulated a new thermodynamic quantity, "fugacity," which facilitated the application of Gibbs' principles to real systems (as opposed to ideal).

H.2.2 Forms of Henry's Law

Two alternative expressions of Henry's law are (1) volatility of a dissolved gas and (2) solubility of a gas in the gas phase. One is the reciprocal of the other.

H.2.2.1 Volatility Expression

For any gas, "i," its partial pressure in the gas phase at equilibrium is proportional to its mole fraction in the liquid phase, which is a statement of Henry's law, as given by Alberty and Silbey (1992, p. 206), i.e.,

$$p_i^* = H_i^{\Delta} X_i^* \qquad \text{(H.10)}$$

where

p_i* is the partial pressure of gas i in equilibrium with aqueous phase (Pa)
X_i* is the mole fraction of gas i in dissolved state in equilibrium with gas phase (mol solute i · mol water^{-1})
H_i^{Δ} is the "volatility" form of Henry's law constant (Pa · mol water · mol solute i^{-1})

Some explanatory comments may help to further understand Equation H.10, i.e., beyond its face value, and its application:

- *Concept of Henry's law*. The choice of units for X_i expresses well a rationale for Henry's law, i.e., that the gas phase partial pressure of "i" is proportional to the interfacial surface area of "i." This is true only for the pseudo condition that there is no interaction between the molecules of "i" and the water molecules.

- *Units*. The "dimensions" of H_i^Δ are pressure divided by concentration. The "units" may be any combination of pressure (e.g., Pa, kPa, bars, atmospheres, etc.) and concentration (e.g., kg/m^3, mg/L, mol/m^3, mol/L, mol fraction, etc.). Henry's constant has been expressed in most of these combinations. This has been a major problem in the application of Henry's law.
- *Identifying units*. Certain units in Henry's constant may be considered (by some) as "dimensionless." This notion is false and adds confusion, especially when trying to accomplish conversions between units. For example, in using mole fraction as concentration unit, i.e., mol i/mol water the numerator is for moles of solute "i" and the denominator is moles of water. It is improper to omit the species in statements of units. The designation "mol/mol," often seen, is incorrect.
- *Identifying units—again*. A common unit is "Pa gas i/kg solute i/m^3 H_2O." Note that the species associated with each unit are delineated. Often the expression is Pa/kg/m^3, which is not sufficient.
- *Volatility*. The Henry's constant, H_i^Δ, is an expression of the "volatility" of "i," i.e., its tendency to favor the gas state (Sander, 1999). Therefore, the higher the value of H_i° is the higher is the tendency for i to be in the gas state.
- *The "volatility" Henry's constant*. The nomenclature adopted here, i.e., that H_i^Δ is a "volatility" Henry's constant, which is not in general use; the common designation is simply, Henry's constant. The term "volatility" is conceptually correct as a descriptor (see, e.g., Sander, 1999) but the main motivation was simply to distinguish it from its reciprocal, called H_i. Most often, one form or the other is noted, but not both, in a single writing.
- *Equilibrium constant*. Henry's constant is a form of equilibrium constant, i.e., the ratio of products to reactants (expressed as molar concentrations and with the stoichiometric coefficients as exponents). The notions of thermodynamic equilibrium are applicable.
- *Applicability*. The wide variety of applications of Henry's law includes dissolved oxygen levels in streams and lakes as affected by photosynthesis, carbon dioxide levels in natural waters and treatment plants, design of air stripping towers, oxygen transfer in secondary wastewater treatment, air binding in rapid rate filters, removal of excess air from waters before entering treatment, chlorine gas dissolution, etc.

H.2.2.2 Solubility Expression

The reciprocal form of Equation H.10 is,

$$C_i^* = H_1^S \cdot p_i^* \tag{H.11}$$

where

C_i^* is the concentration of gas "i" in dissolved state in equilibrium with gas phase (mg i $\cdot$ L^{-1})
H_1^S is the "solubility" form of Henry's law constant (mg i $\cdot$ L^{-1} $\cdot$ atm^{-1})
p_i^* is the partial pressure of gas "i" in equilibrium with aqueous phase (atm)

Equation H.11 simply says that the concentration of gas i in the aqueous phase is proportional to its partial pressure in the gas phase. Thus, the higher the magnitude of H_1^S, the higher is the gas solubility (and conversely, the lower its volatility). Comparing (H.10) and (H.11), we observe that

$$H_i^S = \frac{1}{H_i^\Delta} \tag{H.12}$$

The dimensions of one are the reciprocal of the other. The form of Equation H.11 was given by Silberberg (1996) and Sander (1999), indicating that the usage has not centered exclusively on Equation H.10. Explanatory comments regarding Equation H.11 help to understand its characteristics and utility:

- *Solubility*. The Henry's constant, H_1^S, is an expression of the "solubility" of "i," i.e., its tendency to favor the aqueous state (see Sander, 1999). Therefore, the higher the values of H_1^S, the higher is the tendency for "i" to be in the dissolved state (and H_i^Δ is correspondingly lower).
- *Solubility data*. In certain literature, solubility data, along with pressure, are given (see, for example, Battino, 1991, Fogg and Gerrard, 1981). In some cases, the data are given with concentration as mg/L and pressure at 1.0 atm. For such cases, H_1^S, may be taken directly from the solubility data, i.e., $H_1^S = (x$ mg i/L)/(1.0 atm i).
- *Units*. The "dimensions" of H_1^S are concentration divided by pressure. As with H_i^Δ the "units" may be any combination of pressure (e.g., Pa, kPa, bars, atmospheres, etc.) and concentration (e.g., kg/m^3, mg/L, mol/m^3, mol/L, mol fraction, etc.). The units selected for Equation H.29 were intended to relate to engineering use, i.e., (mg i/L water) = (mg i/L water/atm i) $\cdot$ (atm i).
- *Units conversions*. The equality of (H.12) assumes dimensional homogeneity. Therefore, if H_i is calculated from H_i^Δ then H_1^S will be in the inverse of those same units. A units conversion must be done if another form is desired, i.e., for H_i^Δ or H_1^S.

H.2.3 Units for Henry's Constant

Some of the units found in the literature for Henry's constants are given in Table H.2. The repertoire of units is not limited to those shown. The units shown are most common for being either the source or a target of a conversion.

TABLE H.2
Common Units for Henry's Constant, Volatility and Solubility Forms

		A	B	C
Volatile form, H_i^{Δ}	1	$\frac{\text{atm i(g)}}{\text{mol i(aq)/mol } H_2O}$ [a,b]	$\frac{\text{mol i(g)/L(g)}}{\text{mol i(aq)/L } H_2O}$ [b]	$\frac{\text{mol i(g)/mol(g)}}{\text{mol i(aq)/mol } H_2O}$ [b]
	2	$\frac{\text{atm i(g)}}{\text{mol i(aq)/m}^3\ H_2O}$ [b,c,d]		
Solubility form, H_i^S	3	$\frac{\text{mol i(aq)/L } H_2O}{\text{atm i(g)}}$ [e]	$\frac{\text{mol i(aq)/m}^3\ H_2O}{\text{Pa}}$ [e]	$\frac{\text{mg i(aq)/L } H_2O}{\text{atm i(g)}}$ [f]

[a] Alberty and Silbey (1992).
[b] Brennan et al. (1998).
[c] Ashworth et al. (1988).
[d] Yaws (1999).
[e] Sander (1999) [note that the second citation (3,B) is the official SI unit].
[f] Units adopted for this text.

For engineering purposes, the Equation H.11 form of Henry's law, i.e., 3C, is recommended here, with units as given, i.e., $H_i^S = (\text{mg i/L } H_2O/\text{atm i})$. Reasons include (1) a large number of problems are in terms of concentrations in mg/L and pressure in atmospheres (or kPa), and (2) tables of solubility data in mg/L and usually for 1.0 atmosphere partial pressure of gas gives an associated Henry's constant directly.

H.2.4 Conversions of Units for Henry's Constant

To convert from a given set of units for Henry's constant to another set, the chain of conversions approach will always work. As a caution, because conversions may be tedious, verification should be a part of the process. This can be done most easily by testing the conversion for a gas that is found in published literature in both the source units and the target units. Carbon dioxide and dissolved oxygen are examples of gas species that are likely to be found in several forms of units. Another is chloroform. Table H.3 gives the numerical values for each compound in the various units from Table H.2.

H.2.4.1 Procedure for Conversion of Units for Henry's Constant

To convert from one set of units to another, the "chain-of-conversions" principle always works. The procedure is illustrated in Examples H.5 and H.6, with the target units being H_i^S in mg i (aq)/L H_2O/atm i (g).

Example H.5 Conversion of Henry's constant as H_i^{Δ} to H_i^S

1. Obtain a Henry's constant from a literature source. Consider, for example, the Henry's constant for chloroform, $CHCl_3$, MW ($CHCl_3$) = 119.377, at 25°C from Yaws (1999, p. 407),

$$H^{\Delta}_{CHCl_3,25^\circ C} = 0.0041011\ \frac{\text{atm m}^3}{\text{mol}}$$

2. Expand the units to distinguish the solute (dissolved gas) from the solvent (water), i.e., place a label on each unit, and round off value given,

$$H^{\Delta}_{CHCl_3,25^\circ C} = 0.0041\ \frac{\text{atm } CHCl_3 \cdot \text{m}^3\ H_2O}{\text{mol } CHCl_3}$$

3. Apply chain-of-conversions

$$H^{\Delta}_{CHCl_3,25^\circ C} = 0.0041\ \frac{\text{atm } CHCl_3 \cdot \text{m}^3\ H_2O}{\text{mol } CHCl_3} \cdot \frac{\text{mol } CHCl_3}{119.377\ \text{g } CHCl_3} \cdot \frac{\text{g } CHCl_3}{10^3\ \text{mg } CHCl_3} \cdot \frac{10^3\ \text{L } H_2O}{\text{m}^3\ H_2O}$$

4. After canceling the terms,

$$H^{\Delta}_{CHCl_3,25^\circ C} = 0.0041\ \text{atm } CHCl_3 \cdot \frac{\text{L } H_2O}{119.377 \cdot \text{mg } CHCl_3} = 3.43 \cdot 10^{-5}\ \frac{\text{atm } CHCl_3 \cdot \text{L } H_2O}{\text{mg } CHCl_3}$$

5. Converting to $H^S_{CHCl_3}$

$$H^S_{CHCl_3,25} = \frac{1}{H^{\Delta}_{CHCl_3,25^\circ C}} = \frac{1}{3.43 \cdot 10^{-5}\ \frac{\text{atm } CHCl_3 \cdot \text{L } H_2O}{\text{mg } CHCl_3}} = 2.9 \cdot 10^4\ \frac{\text{mg } CHCl_3}{\text{L } H_2O} \Big/ \text{atm } CHCl_3$$

TABLE H.3
Henry's Constant for Three Cases in Different Units

Compound		Carbon Dioxide	Oxygen	Chloroform
Formula		CO_2	O_2	$CHCl_3$
MW		44.0098	31.998	119.377
Form	**Units**			
Volatility—H_i^{Δ}	$\frac{\text{Pa i(g)}}{\text{mol i(aq)/mol } H_2O}$ [a]	$0.167 \cdot 10^9$–25°[a]	$4.40 \cdot 10^9$–25°[a]	
	$\frac{\text{atm i(g)}}{\text{mol i(aq)/mol } H_2O}$ [b]	1510–20°[c] 1212.2 25°[d]	43000–20°[c]	170–20°[c] 227.84 25°[d]
	$\frac{\text{mol i(g)/L(g)}}{\text{mol i(aq)/L } H_2O}$ [b]			0.1905–20°[b]
	$\frac{\text{mol i(g)/mol(g)}}{\text{mol i(aq)/mol } H_2O}$ [b]			
	$\frac{\text{atm i(g)}}{\text{mol i(aq)/m}^3\ H_2O}$ [b,d,e]			0.004101–25°[d] 0.00332 20°[e] 0.00421 25°[e]
Solubility—H_i^S	$\frac{\text{mol i(aq)/L } H_2O}{\text{atm i(g)}}$ [f]	0.034 25°[f]	0.0013 25°[f]	0.27 25°[f]
	$\frac{\text{mol i(aq)/m}^3\ H_2O}{\text{Pa}}$ [f,g]			
	$\frac{\text{mg i(aq)/L } H_2O}{\text{atm i(g)}}$ [h]	1688–20°[h]	43.39–20°[h]	32231[f] 29080[d]

[a] Alberty and Silbey (1992).
[b] Brennan et al. (1998).
[c] Kavanaugh and Trussell (1981).
[d] Yaws (1999).
[e] Ashworth et al. (1988).
[f] Sander (1999).
[g] Official SI unit according to Sander (1999).
[h] This text; numerical values calculated from references indicated by footnotes.

Example H.6 Conversion of Henry's Constant as H_i^{Δ} in atm/mol Fraction to H_i^S in mg/L/atm

1. Consider again, chloroform, $CHCl_3$ with H_i^{Δ} given by Yaws (1999, p. 407).

$$H^{\Delta}_{CHCl_3,25°C} = 2.2784 \cdot 10^2 \frac{\text{atm}}{\text{mol fraction}}$$

2. Apply "labeling" of units and rounding off,

$$H^{\Delta}_{CHCl_3,25°C} = 2.2784 \cdot 10^2 \frac{\text{atm } CHCl_3}{\text{mol } CHCl_3/\text{mol } H_2O}$$

3. Now convert, by a chain of conversions,

$$H^{\Delta}_{CHCl_3,20°C} = 2.2784 \cdot 10^2 \frac{\text{atm } CHCl_3 \cdot \text{mol } H_2O}{\text{mol } CHCl_3} \cdot \frac{\text{mol } CHCl_3}{\text{MW(CHCl) g}} \cdot \frac{18.01528 \text{ g } H_2O}{\text{mol } H_2O} \cdot \frac{\text{L } H_2O}{998.21 \text{ g } H_2O(20°C)} \cdot \frac{\text{g}}{10^3 \text{ mg}}$$

$$= 2.2784 \cdot 10^2 \text{ atm } CHCl_3 \cdot \frac{18.01528 \cdot \text{L } H_2O}{\text{MW(CHCl)} \cdot 998.21 \cdot 10^3 \text{ mg}}$$

$$= 2.2784 \cdot 10^2 \text{ atm } CHCl_3 \cdot \frac{\text{L } H_2O}{\text{MW(CHCl)} \cdot 55.51 \cdot 10^3 \text{ mg}}$$

$$= 2.2784 \cdot 10^2 \text{ atm } CHCl_3 \cdot \frac{\text{L } H_2O}{119.377 \cdot 55.51 \cdot 10^3 \text{ mg}}$$

$$= 3.43 \cdot 10^{-5} \frac{\text{atm } CHCl_3 \cdot \text{L } H_2O}{\text{mg } CHCl_3 \text{ dissolved}}$$

4. Convert to H_i^S, with rounding off in the final step,

$$H^S_{CHCl_3,25} = \frac{1}{H^{\Delta}_{CHCl_3,25°C}}$$

$$= \frac{1}{3.43 \cdot 10^{-5} \frac{\text{atm } CHCl_3 \cdot \text{L } H_2O}{\text{mg } CHCl_3 \text{ dissolved}}}$$

$$= 2.9 \cdot 10^4 \frac{\text{mg } CHCl_3 \text{ dissolved}}{\text{L } H_2O} \Big/ \text{atm } CHCl_3$$

H.2.5 Effect of Temperature on Henry's Law Constant

A single Henry's constant at some given temperature is, as a rule, not adequate knowledge, since temperature has a strong effect. The temperature dependence of Henry's constant is given by the van't Hoff relation,

$$\ln H_i^{\Delta} = \ln A_i - \frac{\Delta H_i^{\circ}}{RT} \tag{H.13}$$

where

ΔH_i° is the standard state enthalpy change due to dissolution of component i in water (J/mol i)
R is the universal gas constant (8.314 510 cal mol^{-1} K^{-1})
T is the absolute temperature (K)
A_i is the constant for gas i (dimensionless)

In other words,

$$H_i^{\Delta} = A_i e^{\left(-\frac{\Delta H_i^{\circ}/R}{T}\right)} \tag{H.14}$$

H.2.5.1 Illustration of Temperature Effect

Figure H.4 shows a plot of experimental data given by Ashworth et al. (1988) for chloroform, $CHCl_3$. From the slope and intercept, obtained by Kladiograph plotting software, Henry's constant can be calculated, as indicated in Equation H.53. Ashworth et al. (1988) have provided such data for some 45 organic compounds given here as the ln A_i and $\Delta H_i^{\circ}/R$ data in Table H.4. The units for H_i in the temperature regression equation, Equation H.53 are [atm gas $i \cdot m^3$ H_2O/mol dissolved gas i].

Example H.7 Calculation of ΔH_i^{Δ} from Table H.4

Consider again, chloroform, i.e., $CHCl_3$, which shows ln $A_{CHCl3} = 11.41$, and $\Delta H_i^{\circ}/R = 5030$.

1. Substituting data in a modification of Equation H.14,

$$H_i^{\Delta} = e^{\left(\ln A_i - \frac{\Delta H_i^{\circ}/R}{T}\right)}$$

$$H_{CHCl_3}^{\Delta} = e^{\left(11.41 - \frac{5030}{298}\right)}$$

$$= 0.00451 \frac{\text{atm } CHCl_3 \cdot m^3 H_2O}{\text{mol } CHCl_3}$$

Comments

This compares with 0.0041 in Yaws (1999, p. 407), and with 0.0042, 0.0038 from Ashworth et al. (1988). Converting to $H_{CHCl_3}^{S}$ gives,

$$H_{CHCl_3}^{S} = 26\,469 \frac{\text{mg } CHCl_3/\text{L } H_2O}{\text{atm } CHCl_3}$$

The van't Hoff relation is a rational basis for determining the effect of temperature on Henry's constant from empirical data, as seen in Figure H.4. The van't Hoff relation is consistent with theory and is confirmed by most experimental data depiciting Henry's constant versus temperature.

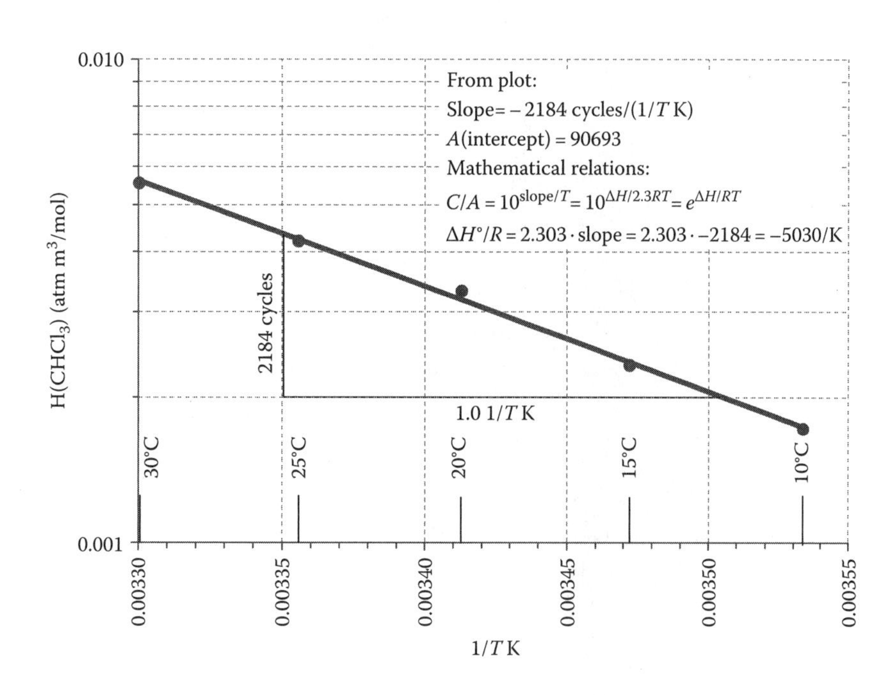

FIGURE H.4 Henry's constants from experimental data for chloroform plotted against 1/T K. (Figure plotted and regression equation from data as obtained in Ashworth, R.A. et al., *J. Hazard. Mater.*, 18, 25, 1988.)

TABLE H.4
Henry's Law Temperature Coefficients for Organic Compounds of Interest at U.S. Air Force Bases[a]

Compound	ln A_i	$\Delta H_i^\circ/R$
Nonane	−0.1847	202.1
n-Hexane	25.25	7530
2-Methylpentane	2.959	957.2
Cyclohexane	9.141	3238
Chlorobenzene	3.469	2689
1,2-Dichlorobenzene	−1.518	1422
1,3-Dichlorobenzene	2.882	2564
1,4-Dichlorobenzene	3.373	2720
o-Xylene	5.541	3220
p-Xylene	6.931	3520
m-Xylene	6.280	3337
Propylbenzene	7.835	3681
Ethylbenzene	11.92	4994
Toluene	5.133	3024
Benzene	5.534	3194
Methylethylbenzene	5.557	3179
1,1-Dichloroethane	5.484	3137
1,2-Dichloroethane	−1.371	1522
1,1,1-Trichloroethane	7.351	3399
1,1,2-Trichloroethane	9.320	4843
Cis-1,2-dichloroethylene	5.164	3143
Trans-1,2-dichloroethylene	5.333	2964
Tetrachloroethylene	10.65	4368
Trichloroethylene	7.845	3702
Tetralin	11.83	5392
Decalin	11.85	4125
Vinyl chloride	6.138	2931
Chloroethane	4.265	2580
Hexachloroethane	3.744	2550
Carbon tetrachloride	9.739	3951
1,3,5-Trimethylbenzene	7.241	3628
Ethylene dibromide	5.703	3876
1,1-Dichloroethylene	6.123	2907
Methylene chloride	8.483	4268
Chloroform	11.41	5030
1,1,2,2-Tetrachloroethane	1.726	2810
1,2-Dichloropropane	9.843	4708
Dibromochloromethane	14.62	6373
1,2,4-Trichlorobenzene	7.361	4028
2,4-Dimethylphenol	−16.34	−3307
1,1,2-Trichlorotrifluoroethane	9.649	3243
Methyl ethyl ketone	−26.32	−5214
Methyl isobutyl ketone	−7.157	160.6
Methyl cellosolve	−6.050	−873.8
Trichlorofluoromethane	9.480	3513

Source: Ashworth, R.A. et al., *J. Hazard. Mater.*, 18, 25, 1988.

[a] The Henry's coefficient is calculated from Equation H.53, i.e., $\ln H_i^{\Lambda} = \ln A_i - \Delta H_i^\circ/RT$ (note that H_i° is the standard state enthalpy of reaction).

H.2.6 Variability of Henry's Constant Data

As noted, Henry's constant data are often given to several significant places, e.g., 3–5. Examining the data from different sources, however, shows variability that indicates standard deviations of perhaps 10%–20% about a mean. Therefore, any final calculations should be rounded to about two decimal places, or perhaps three decimal places, depending upon the data provided.

H.2.7 Data Sources

Sources of data for Henry's constants (or solubility data) have not been compiled into a single document. Moreover, ferreting-out from different sources may be required. The work by Yaws (1999) approaches a comprehensive compilation and is close to a single source reference compared to the work presented by Sander (1999).

Prior to about 1980, solubility data and Henry's constant data were developed mostly for inorganic gases, such as in Table H.5. Solubility data from various sources for such gases were compiled in a comprehensive series such as the volume by Battino (1981). Brennan et al. (1998) summarized the state of knowledge, indicating that in 1981 data for only 35 chemicals were obtained from the literature, out of 70,000 compounds in current use. A problem they recognized was that Henry's constants have been reported in various forms and units, as noted here. Compilations for organic compounds have been developed mostly since the early 1980s stimulated by legislation relating to hazardous wastes. Gosset et al. (1984) included gas solubility in studies of air stripping, motivated by the problems faced by the U.S. Air Force. Table H.4 from Ashworth et al. (1988) includes compounds considered contaminants in air force bases. The most comprehensive compilations of data for organic compounds have been by Yaws (1999) and Sander (1999). Water solubility data with temperature coefficients for 151 paraffin hydrocarbons were given by Yaws et al. (1993) as related to the design of air stripping of water. Later, Yaws (1997) provided solubility data on disks with temperature coefficients for 217 compounds and included Henry's constants at representative temperatures (e.g., 20°C, 25°C) for 692 compounds. Similar data were published by Yaws (1999), which included solubility data with temperature coefficients for the same 217 compounds and Henry's constant data for 1360 compounds without temperature coefficients. The Henry's constant data were given in two kinds of units, i.e., atm/mol f, and atm gas i · m^3 H_2O/mol dissolved gas i. The data by Sander (1999) are comprehensive in that not only are a large number of compounds included (900 species), but the data for each compound from all of the various source (2200 data entries) were compiled (from 250 references) and presented in uniform units (atm m^3/mol) and temperature coefficients, i.e., $\Delta H_i^\circ/R$.

As another approach, since the Henry's constant is merely an equilibrium constant, it may be calculated from

TABLE H.5

Solubility of Gases in Water (mg Gas i/L Water) for Interfacial Pressure of Gas "i," $p_i = 1.00$ atm—and Temperature Coefficients; Solubility of Gas "i" is Same as Henry's Constant, i.e., mg Gas i Dissolved/L Water/atm Gas i

T (°C)	H_2	O_2	N_2	CO_2	H_2S	CH_4	Cl_2	SO_2	O_3	NH_3	CO	Rn
A	2.4543	64.75	27.593	3129.9	6659.5	36.396	1,1402	231,540	843.32	928,020	41.449	3633.5
B	−0.011708	−0.01862	−0.01710	−0.02955	−0.026105	−0.02063	−0.021520	−0.038103	−0.02719	−0.028579	−0.01747	−0.02156
A′	0.057647	0.13166	0.091101	0.2179	1.1525	0.037757	6.4933	3.4932	0.12786	60.481	0.12413	3.4488
$\Delta H_i^\circ/R$	1027.7	1702.3	1569.1	2625.6	2378.6	1887.3	2062.1	3037.7	2413.6	2,648	1595.9	1908.6
0	1.922	69.45	29.42	3,346	7,066	39.59		228,300	883	895,000	44.0	3,673
1	1.901	67.56	28.69	3,213	6,839	38.42		220,900	856		42.9	3,590
2	1.881	65.74	27.98	3,091	6,619	37.28		213,700	829		41.9	3,509
3	1.862	64.00	27.30	2,978	6,407	36.19		206,600	803		40.9	3,430
4	1.843	62.32	26.63	2,871	6,201	35.13		199,800	778	796,000	40.0	3,353
5	1.824	60.72	26.00	2,774	6,001	34.10		193,100	758		39.0	3,289
6	1.806	59.18	25.37	2,681	5,809	33.12		186,500	731		38.1	3,204
7	1.789	57.73	24.77	2,589	5,624	32.17		180,200	709		37.2	3,133
8	1.772	56.32	24.19	2,492	5,446	31.27		174,000	687	720,000	36.4	3,063
9	1.756	54.98	23.65	2,403	5,276	30.39		168,000	666		35.6	2,995
10	1.740	53.68	23.12	2,318	5,112	29.55	9,972	162,100	649	684,000	34.8	2,938
11	1.725	52.46	22.63	2,239	4,960	28.79	9,654	156,400	627		34.0	2,864
12	1.710	51.28	22.16	2,165	4,814	28.05	9,346	150,900	611	651,000	33.3	2,810
13	1.696	50.14	21.70	2,098	4,674	27.33	9,050	145,600	591	636,000	32.6	2,739
14	1.682	49.06	21.26	2,032	4,540	26.65	8,768	140,400	574		31.9	2,679
15	1.668	48.02	20.85	1,970	4,411	25.99	8,495	135,400	557		31.3	2,620
16	1.654	47.03	20.45	1,903	4,287	25.38	8,232	130,500	541	587,000	30.7	2,563
17	1.641	46.06	20.06	1,845	4,169	24.78	7,979	125,900	526		30.0	2,507
18	1.628	45.14	19.70	1,789	4,056	24.22	7,738	121,400	511		29.5	2,452
19	1.616	44.26	19.35	1,737	3,948	23.69	7,510	117,000	496		28.9	2,399
20	1.603	43.39	19.01	1,688	3,846	23.18	7,283	112,800	482	529,000	28.4	2,347
21	1.588	42.52	18.69	1,640	3,745	22.70	7,100	108,800	469		27.9	2,296
22	1.575	41.69	18.38	1,590	3,648	22.22	6,918	105,000	456		27.4	2,246
23	1.561	40.87	18.09	1,540	3,554	21.77	6,739	101,200	444		26.9	2,198
24	1.548	40.07	17.80	1,493	3,464	21.33	6,572	97,600	432	482,000	26.5	2,151
25	1.535	39.31	17.51	1,449	3,375	20.91	6,413	94,100	422		26.0	2,111
26	1.522	38.57	17.24	1,406	3,290	20.50	6,259	90,600	409		25.6	2,060
27	1.509	37.87	16.98	1,366	3,208	20.11	6,112	87,300	398		25.2	2,016
28	1.496	37.18	16.72	1,327	3,130	19.74	5,975	84,200	387	440,000	24.8	1,973
29	1.484	36.51	16.47	1,292	3,055	19.38	5,847	81,000	377		24.4	1,931
30	1.474	35.88	16.24	1,257	2,983	19.04	5,724	78,000	369	410,000	24.0	1,896
35	1.425	33.15	15.01	1,105	2,648	17.33	5,104	64,700	324		22.3	1,704
40	1.384	30.82	13.91	973	2,361	15.86	4,590	54,100	284	316,000	20.8	1,526
45	1.341	28.58	13.00	860	2,110	14.66	4,228		253		19.3	1,381
50	1.287	26.57	12.16	761	1,883	13.59	3,925		230	235,000	18.0	1,271
60	1.178	22.74	10.52	576	1,480	11.44	3,295		178	168,000	15.2	1,010
70	1.020	18.56	8.51		1,101	9.26	2,793			111,000	12.8	
80	0.79	13.81	6.60		765	6.95	2,227			65,000	9.8	
90	0.46	7.9	3.6		410	4.0	1,270			30,000	5.7	
100	0	0	0		0	0	0			0	0	

Notes: (1) The solubility form of Henry's constant may be calculated as: H_i^S (mg i/L H_2O)/atm i) $= Ae^{B\cdot(T\,°C)}$ (H.60) taking A and B from the table for the gas of interest.

(2) All columns except O_3 and Rn were from Dean, J. A. (Ed.), *Lange's Handbook of Chemistry*, 13th edn., McGraw-Hill, New York, 1985.

(*continued*)

TABLE H.5 (continued)

Solubility of Gases in Water (mg Gas i/L Water) for Interfacial Pressure of Gas "i," $p_i = 1.00$ atm—and Temperature Coefficients; Solubility of Gas "i" is Same as Henry's Constant, i.e., mg Gas i Dissolved/L Water/atm Gas i

(3) Ozone data were from Battino (1981, pp. 474–483) who reviewed most of the experimental data generated on ozone solubility. A problem in developing ozone solubility data was that ozone decomposes to oxygen shortly after introduction. The data recommended were those of Sullivan and Roth who provided a "smoothing" equation for Henry's constant, i.e., $H(O_3) = 38\,420\,000 \cdot e^{(-2428/T_{abs})} \cdot [OH^-]^{0.035}$ in which $H(O_3)$ is as defined in Equation H.10, i.e., atm/mol f, and [OH] is in mol/L. The data in this table were calculated (by Excel® spreadsheet) for pH = 7.0. The conversion to solubility in mg O_3/L water was:

$$X(O_3) = \frac{1.0 \text{ atm } O_3}{H(O_3)} \quad \text{and,}$$

$$C(O_3) = X(O_3)\frac{\text{mol } O_3}{\text{mol } H_2O} \cdot MW(O_3)\frac{48 \text{ g } O_3}{\text{mol } O_3} \cdot \frac{55.55 \text{ mol } H_2O}{\text{L } H_2O} \cdot \frac{1000 \text{ mg}}{\text{g}}$$

A sample calculation for $T = 10°C$ gives: $H(O_3) = 4107$ atm O_3/mols O_3/mol H_2O, and $X(O_3) = 0.0002435$ mol O_3/mol H_2O at $P(O_3) = 1.0$ atm O_3. Then, $C(O_3) = 649$ mg O_3/L water.

(4) Radon data are few and different experimental data sets give results that vary perhaps 20%. The radon data entered in this table were calculated from a best fit of experimental results generated by Lewis et al. (1987), represented by the equation,

$$\ln X = -2.01 + \frac{0.23}{(T/100)} - 3.88\ln(T/100) - 0.84(T/100)$$

The conversion to solubility in mg Rn/L water was by,

$$C(Rn) = X(Rn)\frac{\text{mol Rn}}{\text{mol } H_2O} \cdot MW(Rn)\frac{222 \text{ g Rn}}{\text{mol Rn}} \cdot \frac{55.55 \text{ mol } H_2O}{\text{L } H_2O} \cdot \frac{1000 \text{ mg}}{\text{g}}$$

A sample calculation for $T = 10°C$, gives, $\ln X(Rn) = -8.3422016$, $X(Rn) = 0.00023825$ mol Rn/mol H_2O at $P(Rn) = 1.0$ atm Rn. Then, $C(Rn) =$ 2938 mg Rn/L water.

(5) Also, it should be noted that, from Equation H.11 and for this table, there is an arithmetic identity that, numerically (not in units): $C_i^*(\text{mg/L}) = H_i^S$, since the data for this table are for pressure, $p_i^* = 1.00$ atm.

(6) The coefficients A and B are for best fit equations of the data in this table for a given gas, plotted in accordance with the form, $C(\text{mg/L}) = Ae^{B \cdot T°C}$, with $R^2 \geq 0.99$, in general and with deviations from data generally within 2%–4% (see also Note 1).

(7) The coefficients A' and $\Delta H_i^\circ/R$ are for best fit equations of the data plotted in accordance with the van't Hoff type relation, i.e., $C(\text{mg/L}) = A'e^{\frac{\Delta H_i^\circ/R}{T(K)}}$, with $R^2 \geq 0.99$, in general and with deviations from data generally within 2%–4%.

(8) For chlorine dioxide, $\mathbf{H}(ClO_2, 25°C) = 1.0$ mol/L/atm = 67,451 mg/L/atm. Lide (1996, pp. 6–5) gives, $\ln X(ClO_2) = A(ClO_2) + B(ClO_2)/T^* + C(ClO_2) \cdot \ln T^*$, in which, $X(ClO_2)$ = mol fraction of gas in solution, $A(ClO_2) = 7.9163$, $B(ClO_2) = 0.4791$, $C(ClO_2) = -11.0593$, $T^* = T(K)/100$; equation valid for $283.15 \leq T \leq 333.15$ K for $p(ClO_2) = 101.325$ kPa (1.00 atm) of pure gas.

the thermodynamic data. Again, this requires search, but data are found, to a limited extent, in standard handbooks (see, for example, Lide, 1996 or Dean, 1985) and sometimes in specialized publications.

H.2.8 Gas Solubility

Table H.5 gives solubility for 12 gases of frequent interest at temperatures ranging 0–100 K. The concentrations given are for equilibrium conditions at 1.00 atmosphere of pure gas above the water surface at each of the temperatures (stated in left column). Figure H.5 is a plot of the data of Table H.5, i.e., solubility of gas vs. temperature, and provides a sense of how the gases differ in solubility; and also the temperature effect on each gas.

The major utility of Table H.5 is that solubility at known pressures, i.e., 1.00 atm, is an "identity" with Henry's constant, H_i^S, with units (mg gas i in aqueous phase/L water/atm gas i). The coefficients, i.e., A_i and B_i, seen in the top two rows, are the intercept and slope of the best fit exponential equation, i.e.,

$$C_i \text{ (mg i/L } H_2O) \equiv H_i^S \text{ (mg i/L } H_2O)/\text{atm i} = Ae^{B \cdot (T°C)}. \quad \text{(H.15)}$$

where

A_i is the intercept for semi-log plot of Figure H.5 (mg i aq/L H_2O/atm i g)

B_i is the slope · 2.303 of Figure H.5 plot for a given gas, i

$\ln X(ClO_2) = A(ClO_2) + B(ClO_2)/T^* + C(ClO_2) \cdot \ln T^*$, in which, $X(ClO_2)$ = mol fraction of gas in solution, $A(ClO_2) = 7.9163$, $B(ClO_2) = 0.4791$, $C(ClO_2) = -11.0593$, $T^* = T(K)/100$; equation valid for $283.15 \leq T \leq 333.15$ K for $p(ClO_2) = 101.325$ kPa (1.00 atm) of pure gas.

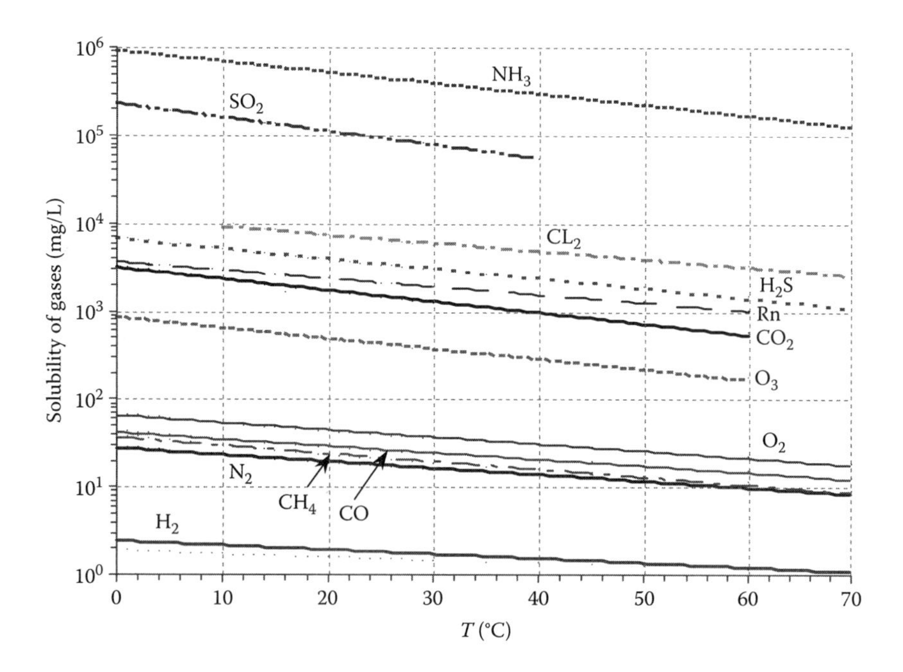

FIGURE H.5 Solubility of gases of Table H.4 as affected by temperature.

H.2.9 Application of Henry's Law

Examples H.8 through H.10 illustrate the application of Henry's law and Table H.5 provides the representations for several situations. As seen, Dalton's law is applied at the same time.

Example H.8 Oxygen Concentration

Determine oxygen concentration in water at 1585 m (5200 ft), at 10°C. Assume the water is in equilibrium with the atmosphere at that elevation.

1. Determine Henry's law coefficient for oxygen at 10°C.
   1.1 From Table H.5,

$$C[O_2,\ 1.0\ \text{atm},\ 10°C] = 53.68\ \text{mg}\ O_2/\text{L}\ H_2O$$

and therefore,

$$H^S[O_2,\ 1.0\ \text{atm},\ 10°C] = 53.68\ \text{mg}\ O_2/\text{L}\ H_2O/\text{atm}\ O_2$$

2. Calculate the partial pressure of oxygen at sea level, 1585 and 3048 m.
   2.1 Obtain from Equation H.34, seen in Figure H.2, the atmospheric pressure at 1585 m (elevation of ERC/CSU).

$$p(\text{atm},\ 1585\ \text{m}) = 628\ \text{mm Hg}$$

   2.2 Determine the partial pressure of oxygen in ambient air.

From Table H.1, mole fraction of oxygen in ambient air is:

$$nO_2/n = 0.209\ 476$$

From Dalton's law,

$$p(O_2\}/p = 0.2095$$

   2.3 Calculate the partial pressures of O_2 at 10°C for 1585 m.

$$\begin{aligned} p[O_2, 1585\ \text{m}] &= \frac{n[O_2, \text{sea level}]}{n} \cdot p[\text{atm}, 1585\ \text{m}] \\ &= 0.2095 \cdot 628\ \text{mm} \\ &= 131\ \text{mm} \\ &= 0.173\ \text{atm} \end{aligned}$$

3. Calculate dissolved oxygen concentration by applying Henry's law.
   3.1 General equation is:

$$C[O_2,\ 10°C] = H^S(O_2,\ 10°C) \cdot p(O_2)$$

   3.2 Apply for elevations 1585 m:

$$\begin{aligned} C[O_2, 10°C, 1585\ \text{m}] &= H(O_2, 10°C) \cdot p(O_2, 1585\ \text{m}) \\ &= \frac{53.68\ \text{mg}\ O_2/\text{L}\ H_2O}{\text{atm}\ O_2} \cdot 0.173\ \text{atm}\ O_2 \\ &= 9.3\ \text{mg/L} \end{aligned}$$

Example H.9 Carbon Dioxide Concentration at Sea Level at 20°C

1. Determine the partial pressure of carbon dioxide at sea level.

   From Table H.1, the mole fraction of carbon dioxide in the atmosphere at sea level is 0.000314, i.e., $n(CO2)/n = 0.000314$ mol carbon dioxide/mol air. Apply Dalton's law, i.e., Equation H.4,

$$\begin{aligned} p(CO_2) &= \frac{n(CO_2)}{n(\text{air})} p(\text{air, sea level}) \\ &= 0.000314 \cdot 1.00 \text{ atm} \\ &= 0.000314 \text{ atm} \end{aligned}$$

2. Look up Henry's constant

   From Table H.5,

$$H^S(CO_2, 20°C) = \frac{1688 \text{ mg } CO_2/\text{L } H_2O}{\text{atm } CO_2}$$

3. Apply Henry's law

   Knowing Henry's constant and the partial pressure of CO_2, the calculation is,

$$\begin{aligned} C(CO_2, \text{sea level}) &= H^S_{CO_2} \cdot p(CO_2) \\ &= \frac{1688 \text{ mg } CO_2/\text{L } H_2O}{\text{atm } CO_2} \cdot 0.000314 \text{ atm} \\ &= 0.53 \frac{\text{mg } CO_2}{\text{L } H_2O} \end{aligned}$$

Comments

Using Table H.5 as the source for Henry's constant data, the calculation is straight-forward.

Example H.10 Application of Henry's Law

Calculate the concentration of carbon dioxide in water at 5°C in mol/L at a pressure of 4.0 atm CO_2 (reported pressure of bottling by Silberberg, 1996, p. 479).

Solution

1. Apply Henry's law

$$\begin{aligned} C(CO_2, \text{bottle}) &= H^S_{CO_2} \cdot p(CO_2) \\ &= \frac{1688 \text{ mg } CO_2/\text{L } H_2O}{\text{atm } CO_2} \cdot 4.0 \text{ atm} \\ &= 6752 \frac{\text{mg } CO_2}{\text{L } H_2O} \\ &= 0.15 \frac{\text{mol } CO_2}{\text{L } H_2O} \end{aligned} \quad \text{(ExH.10.1)}$$

H.2.9.1 Equilibrium Constants from Thermodynamic Data

When equilibrium exists between the gas state and the aqueous state, as it must for Henry's law to be valid, the free energy of the reaction is zero and, thus, the general relation between standard-state free energy of reaction and the equilibrium constant (Henry's constant in this case) holds good, i.e.,

$$\Delta G^\circ_R = -RT \ln H^\Delta_i \quad \text{(H.16)}$$

where

ΔG°_R is the standard-state free energy of reaction (J/mol)
R is the gas constant, i.e., 8.314510 J/mol
T is the absolute temperature (K)

and,

$$\Delta G^\circ_R = \sum \Delta G^\circ_f(\text{products}) - \sum \Delta G^\circ_f(\text{reactants}) \quad \text{(H.17)}$$

where

ΔG°_f(product i) is the standard-state free energy of formation for product i (J/mol)
ΔG°_f(reactant i) is the standard-state free energy of formation for reactant i (J/mol)

In practice, while ΔG°_f data for the gas state are available in Lide (1996) for many substances, only a few data are given for the aqueous state. Some data have been compiled, however, by Pankow (1991) and by Snoeyink and Jenkins (1980). Therefore, if thermodynamic data are available, i.e., ΔG°_f for both the gas state and the aqueous state, H^Δ_i can be calculated. How to do this is illustrated in Example H.11 for carbon dioxide.

Example H.11 Thermodynamics of Carbon Dioxide Equilibrium (Modified from Sawyer and McCarty, 1967)

Determine for carbon dioxide the equilibrium constant between the gas and aqueous phases at 25°C.

1. Tabulate thermodynamic data,

Variable	CO_2(aq) -->	CO_2(g)	Reaction at 298.15 K
ΔG°_f (298.15 K)	−386.02[a]	−394.373 kJ/mol[b]	$\Delta G^\circ_R = -8.35$ kJ/mol
ΔH°_f(298.15 K)	−413.26[b]	−393.51 kJ/mol[b]	$\Delta H^\circ_R = 19.75$ kJ/mol
S°(298.15 K)	119.36[b]	213.785 kJ/mol[b]	$\Delta S^\circ_R = 94.43$ kJ/mol K

[a] Weast (1978, p. D-78).
[b] Lide (1996, p. 5–64).

2. Write the equation for the reaction,

$$CO_2(aq) \rightarrow CO_2(g) \quad \Delta G^\circ_R = -8.35 \text{ J/mol}$$

3. Calculate $H^{\Delta}_{CO_2}$ from the statement of thermodynamic equilibrium.

$$\Delta G^{\circ}_{R} = -RT \ln H^{\Delta}_{i}$$

$$-8350 \text{ J/mol} = -(8.314510 \text{ J/mol K}) \cdot (298.15 \text{ K}) \cdot \ln H^{\Delta}_{CO_2}$$

$$3.368 = \ln H^{\Delta}_{CO_2}$$

$$H^{\Delta}_{CO_2}(298 \text{ K}) = 29.02 \frac{\text{atm } CO_2 \cdot \text{L } H_2O}{\text{mol } CO_2} = 0.029 \frac{\text{atm } CO_2 \cdot \text{m}^3 H_2O}{\text{mol } CO_2}$$

This compares with 0.022 atm $CO_2 \cdot m^3$ H_2O/mol CO_2 in Yaws (1999, p. 407). Converted to $H^{S}_{CO_2}$,

$$H^{S}_{CO_2}(298 \text{ K}) = \frac{1}{29.02 \frac{\text{atm } CO_2 \cdot \text{L } H_2O}{\text{mol } CO_2} \cdot \frac{\text{mol } CO_2}{44{,}000 \text{ mg}}} = 1{,}516 \frac{\text{mg } CO_2}{\text{L } H_2O \cdot \text{atm } CO_2}$$

Comments

This value for $H^{S}_{CO_2}$ compares with 1449 mg CO_2/L H_2O/atm CO_2 in Table H.5. Comparing with Yaws (1999, p. 407), the 0.22 value converts to 2000 mg CO_2/L H_2O/atm CO_2 (which is on the high end of values found in the literature).

H.3 GAS PRECIPITATION

In many situations, a dissolved gas will occur in a "supersaturated" state with respect to the local pressure. When such condition occurs, the dissolved gas will "precipitate" forming bubbles of the pure gas. The local pressure is whatever occurs in the water (at any given elevation and at any given depth of water) irrespective of whether a gas–water interface is present. An everyday example of gas precipitation is observed when a bottle of carbonated beverage is opened; the pressure is released and bubbles appear spontaneously. Another example is boiling water, which is characterized by the spontaneous appearance of water vapor bubbles; boiling occurs when the vapor pressure of water equals atmospheric pressure. This occurs at lower temperatures as elevation increases, since atmospheric pressure declines with elevation.

Examples of gas precipitation include when (1) a bottle of soda is opened, carbon dioxide bubbles appear spontaneously within the bottle, (2) dissolved air flotation is due to a sudden reduction in pressure after supersaturated water reaches the flotation tank at which time the dissolved gas precipitates and forms bubbles, (3) oxygen dissolves continuously by photosynthesis up to a limit at which gas bubbles may be observed, (4) carbon dioxide and methane are produced in anaerobic environments and each form bubbles when saturation levels is reached, (5) air binding occurs in filters due to supersaturation, negative pressures, or both. Thus, in some cases gas precipitation is desired and is engineered to occur (as in dissolved-air-flotation), in other cases the effect is disruptive (as in filters), and in some cases the effect is expected (as in opening a bottle of soda). Other examples include floating sludge in a primary settling basin due to carbon dioxide and methane precipitating as bubbles; in an anaerobic lagoon, gas bubbles are an index that methane and carbon dioxide are being produced, a desired result; the "bends" in divers who rise too quickly; the "bends" in migrating salmon, swimming below a dam where nitrogen gas may be "supersaturated" due to a plunging nappe that entrains air bubbles.

H.3.1 Criterion for Gas Precipitation

In searching for an established criterion for the occurrence of gas precipitation, the literature provides little direct guidance. A probable explanation would be that the problem has not come to the attention of the physical chemists, who deal mostly with fundamentals as opposed to applied problems. Neither has it been articulated well for engineers and operators. To explain gas precipitation, theory provides a means for a coherent explanation. To interpret with a common-sense rationale then it can follow a theoretical understanding.

H.3.1.1 Nutshell Explanation for Gas Precipitation

In-a-nutshell, the gas precipitation may be explained first by a dissolved gas occurring at a "supersaturated" concentration in a given local environment. The gas may be transferred from a higher pressure region or could be generated. If the dissolved gas concentration exceeds that which could exist in equilibrium at the pseudo pressure of the pure gas at the pressure of the local environment, then the gas will come out of the solution as bubbles. For example, one may observe gas bubbles around a bloom of algae in stagnant water. From Table H.5, $C(O_2, 20C) = 43.39$ mg O_2/L water, which will occur if $p(O_2) = 1.00$ atm O_2. If oxygen is generated by the algae through photosynthesis at sea level at zero depth, when dissolved oxygen concentration exceeds 43.39 mg O_2/L water, then bubbles of pure oxygen will form. This can be confirmed by taking a water sample; usually about 30–35 mg O_2/L can be measured by a Winkler titration.

H.3.1.2 Chemical Potential Criterion for Equilibrium

The chemical potential (see, for example, Eisenberg and Crothers, 1979, pp. 271–290), can be defined for the dissolved state as

$$\mu_i(aq) = \mu^{\circ}_i(aq) + RT \ln [i] \quad (H.18)$$

and for the gas state as,

$$\mu_i(g) = \mu^{\circ}_i(g) + RT \ln p_i \quad (H.19)$$

where

$\mu_i(aq)$ is the chemical potential of species i in dissolved aqueous state (J/mol)

$\mu^{\circ}_i(aq)$ is the standard-state chemical potential of species i in aqueous state (J/mol)

$\mu_i(g)$ is the chemical potential of species i in gas state (J/mol)
$\mu_i^{\circ}(g)$ is the standard-state chemical potential of species i in gas state (J/mol)
[i] is the mole fraction of species i in aqueous state [mols i/(L water)]
p_i is the partial pressure of species i in gas state (atm)

Subtracting (product minus reactant),

$$[\mu_i(g) - \mu_i(aq)] = [\mu_i^{\circ}(g) - \mu_i^{\circ}(aq)] + RT\ln\frac{p_i}{[i]} \quad (H.20)$$

At equilibrium, $[\mu_i(g) - \mu_i(aq)] = 0$ and Equation H.20 becomes,

$$0 = [\mu_i^{\circ}(g) - \mu_i^{\circ}(aq)] + RT\ln\frac{p_i}{[i]} \quad (H.21)$$

From Henry's law, $p_i = H_i^{\Delta} \cdot [i]$, and when substituted in Equation H.21,

$$0 = [\mu_i^{\circ}(g) - \mu_i^{\circ}(aq)] + RT\ln H_i^{\Delta} \quad (H.22)$$

Also, since chemical potential and free energy of reaction per mole are identities,

$$\Delta G_R^{\circ} = -RT\ln H_i^{\Delta} \quad (H.16)$$

In this derivation, it is important to note that we have chosen, Equation H.10, for Henry's law definition, i.e.,

$$H_i^{\Delta} \equiv \frac{p_i}{[i]} \quad (H.23)$$

Equation H.23 is consistent with the literature definition for H_i^{Δ} as found in Equation H.16 and if one determines ΔG_R° the H_i^{Δ} calculated matches published values.

H.3.1.3 Chemical Potential Criterion for Gas Precipitation

Consider developing a criterion for gas precipitation in terms of "chemical-potential," i.e., "μ," i.e., Equations H.20 and H.22, repeated below,

$$[\mu_i(g) - \mu_i(aq)] = [\mu_i^{\circ}(g) - \mu_i^{\circ}(aq)] + RT\ln\frac{p_i}{[i]} \quad (H.20)$$

$$0 = [\mu_i^{\circ}(g) - \mu_i^{\circ}(aq)] + RT\ln H_i^{\Delta} \quad (H.22)$$

Now to replace $[\mu_i^{\circ}(g) - \mu_i^{\circ}(aq)]$ in Equation H.20, substitute Equation H.22, i.e.,

$$[\mu_i(g) - \mu_i(aq)] = -RT\ln H_i^{\Delta} + RT\ln\frac{p_i}{[i]} \quad (H.24)$$

$$= RT\ln\frac{[p_i/[i]]}{H_i^{\Delta}} \quad (H.25)$$

Equation H.24 is the key to developing a criterion for gas precipitation. We may assert that when, $\mu_i^{\circ}(aq) > \mu_i^{\circ}(g)$, then gas precipitation will occur. Mathematically,

$$[\mu_i^{\circ}(g) - \mu_i^{\circ}(aq)] < 0 \quad (H.26)$$

At the same time, the criterion of Equation H.26 can occur only when,

$$\frac{[p_i/[i]]}{H_i^{\Delta}} < 1 \quad (H.27)$$

or when,

$$[p_i/[i]] < H_i^{\Delta} \quad (H.28)$$

Still, another rearrangement expresses the relationship in more intuitive terms,

$$H_i^{\Delta} \cdot [i] > p_i \quad (H.29)$$

Equation H.29 says that when the dissolved gas concentration in high enough that the product, $H_i^{\Delta} \cdot [i]$ exceeds the local pressure, then gas precipitation will occur. At the time of gas precipitation, then,

$$H_i^{\Delta} \cdot [i]^* = p_i(\text{bubbles}) \quad (H.30)$$

where
[i]* is the dissolved gas concentration at equilibrium with bubbles (mol i/m^3 H_2O)
p_i(bubbles) is the partial pressure of gas i in bubbles formed by gas precipitation (kPa i)

When the bubbles form then an equilibrium has established itself, i.e., $\mu_i^{\circ}(aq) = \mu_i^{\circ}(g)$. The gas concentration, [i], can go no higher than [i]*. The pressure in the bubble is the "local" pressure. This is what occurs when a bottle of carbonated beverage is opened or when the dissolved gas in a flotation basin moves to the lower pressure zone, gas bubbles will form spontaneously as the system strives for a new equilibrium. In this case of a pressure release, $\mu_i^{\circ}(aq) \ll \mu_i^{\circ}(g)$ dissolved gas will come out of the solution as bubbles until the condition of Equation H.29 is met, i.e., $\mu_i^{\circ}(aq) = \mu_i^{\circ}(g)$.

H.3.1.4 Alternative Criterion for Gas Precipitation

In the development of a criterion for gas precipitation, the form of Henry's law expressed in Equation H.10 was used because it was compatible with the established thermodynamic relations. But subsequent to the thermodynamic development, the form expressed in Equation H.11 may be used as an alternative. This results in Equations H.30 through H.32, that correspond to H.28, H.29, and H.107, respectively, i.e.,

$$H_i^S < \frac{C_i}{p_i} \tag{H.30}$$

$$H_i^S p_i < C_i \tag{H.31}$$

$$H_i^S p_i[\text{bubbles}] = C_i^* \tag{H.32}$$

where C_i^* is the concentration of dissolved gas i in equilibrium with gas bubbles at pressure, p_i(bubbles) (mg i/L H_2O/atm i)

Equations H.30 through H.32, which are really the variations of a single equation, may be easier to use than any of the others because the units are common and H_i^S is found directly in Table H.5. Equation H.31 says that when the aqueous gas concentration of i exceeds the product, $H_i^S p_i$, or C_i^*, then gas precipitation will occur.

Examples will help to illustrate the utility of the criterion of Equations H.31 or H.32.

Example H.12 Gas Precipitation in Benthic Mud's

A lake at elevation 1524 m (5000 ft) has accumulated organic matter in its benthic zone and during the summer months, gas bubbles are observed breaking the surface of the lake. The lake is 5.0 m (16.4 ft) deep and the temperature is 30°C. Explain the situation with respect to dissolved gases.

Analysis

The benthic zone is most probably anaerobic, which means that methane and carbon dioxide are the products of the decomposition of the organic matter. These reaction products will be generated and accumulate in the dissolved state until the criterion of Equation H.31 is satisfied at which time gas precipitation will occur.

Solution

1. Apply Equation H.32 for methane first and then carbon dioxide, i.e.,

$$H_{CH_4}^S p_{CH_4}[\text{bubbles}] = C_{CH_4}^* \tag{H.32}$$

2. First $H_{CH_4}^S$(30°C) and p_{CH4}(bubbles) must be determined,

$$H_{CH_4}^S(30°C) = 19.04\ \frac{\text{mg CH}_4/\text{L H}_2\text{O}}{\text{atm CH}_4} \quad \text{(Table H.5)}$$

and

$$\begin{aligned} p_{CH_4}(\text{bubbles}) &= p(\text{atm}, 1524\,\text{m}) + \gamma_w h \\ &= 84.31\ \text{kPa} + 996\ \frac{\text{kg}}{\text{m}^3} \cdot 9.806\,65\ \frac{\text{m}}{\text{s}^2} \cdot 5.0\,\text{m} \\ &= 84.31\ \text{kPa} + 48.84\ \text{kPa} \\ &= 133.15\ \text{kPa} \\ &= 1.32\ \text{atm} \end{aligned}$$

Note that p(atm, 1524 m) was from Equation H.6; ρ_w was obtained from Figure H.2 and Equation H.6; g was from Table QR.1.

3. Substituting, the preceding calculated values for $H_{CH_4}^3$(30°C) and for P_{CH_4}(bubbles) in (H.32).

$$\begin{aligned} C_{CH_4}^* &= 19.04\ \frac{\text{mg CH}_4/\text{L H}_2\text{O}}{\text{atm CH}_4} \cdot 1.32\ \text{atm CH}_4 \\ &= 25.1\ \frac{\text{mg CH}_4}{\text{L H}_2\text{O}} \end{aligned}$$

4. For carbon dioxide the procedure is the same and the data are the same except that

$$H_{CO_2}^S(30°C) = 1257\ \frac{\text{mg CO}_2/\text{L H}_2\text{O}}{\text{atm CO}_2}$$

which results in,

$$C_{CO_2}^* = 1659\ \frac{\text{mg CO}_2}{\text{L H}_2\text{O}}$$

Comments

The concentrations of dissolved gases in the benthic zone of the lake will not exceed the levels given by $C_{CH_3}^*$, and $C_{CO_2}^*$. Note that methane has a much lower solubility than carbon dioxide. The calculations assume that the gases precipitate independently. It is likely that some of the bubbles will coalesce before reaching the water surface.

PROBLEMS

H.1 Bubbles in Water

When a glass of cold water is permitted to warm to room temperature, bubbles are observed. Explain.

H.2 Boiling Water

Explain why water boils as its temperature is elevated.

Solution

An everyday illustration of gas precipitation is seen in boiling water. For water, $[H_2O] = 1000$ mg/L. Now, as the temperature rises, the Henry's law coefficient rises also, which is the ratio of vapor pressure to concentration of water, which is 1000 mg/L. Finally, as the temperature reaches 100°C, the vapor pressure of water is 1.0 atm, and so we can say,

$$H^S[\text{H}_2\text{O}, 100°\text{C}] = 1000\ \text{mg/L}/1.0\ \text{atm}$$

The $H^S \cdot P$ product is,

$$\begin{aligned} &H^S[\text{H}_2\text{O}, 100°\text{C}] \cdot P(\text{local pressure} = 1\ \text{atm}) \\ &= 1000\ \text{mg/L/atm} \cdot 1.0\ \text{atm} \\ &= 1000\ \text{mg/L} \end{aligned}$$

Thus, since $[H_2O]_{actual} = 1000$ mg/L, the criterion for gas precipitation is satisfied and gas bubbles form. While boiling water is explained merely by the fact that boiling occurs when the vapor pressure of the water increases to the local atmospheric pressure, the Henry's law explanation shows the parallel with precipitation of any gas species.

H.3 Air Binding in Filter Media—General
A rapid filter in water treatment experiences air binding. Provide an analysis of how this can occur.

H.4 Air Binding in Filter Media—WTP
The Betasso Water Treatment Plant that serves Boulder, Colorado obtains is source water from Silver Lake at a high elevation. The water drops to a treatment plant more that 300 m lower elevation by means of a pipeline to the plant. Air binding in filters has been a chronic problem. Provide an analysis: (1) how the air gets into the water and (2) the point where the air will precipitate.

H.5 Remedies for Air Binding
How would you remedy the air binding filter media?

H.6 Algae as Possible Cause of Gas Binding
Algae occur in the summer months in Lake Whatcom, the source water for the Bellingham Water Treatment Plant, Washington. Air boils have been observed during backwash. Provide an analysis of the situation.

H.7 Quantification of Air Binding
Convert the dissolved gas in a water source to volume of air that may accumulate in a filter bed after gas precipitation.

H.8 Gas Production in Benthic Muds
Gas bubbles are observed breaking at the surface of a lake in Iowa. Explain.

H.9 Algae and Dissolved Gas
Gas bubbles are observed within an algae mass floating on the surface of a pond. Explain.

H.10 Gas Bubbles in Primary Clarifier
A water sample is obtained from the sludge zone of a primary clarifier in a wastewater treatment plant, using a Kemmerer water sampler. A portion of the sample is released to a 100 mL graduated cylinder and then poured into an evaporating dish where a carbon dioxide titration is carried out. The result was $C_{CO_2} \approx 1500$ mg/L. Gas bubbles were observed breaking the water surface of the clarifier. Explain.

H.11 Dissolved Gas Concentration from Diffused Aeration
A diffused aeration system is located at the bottom of a pond at elevation 1524 m (5000 ft). The pond is 10.33 m deep. Determine the dissolved oxygen concentration at the bottom of the pond.

H.12 The "Bends" in Salmon
In the Columbia River migrating salmon have been killed by the "bends" when swimming below a dam in the deep water below a dam (in the vicinity of a plunging nappe). Explain.

GLOSSARY

Bunsen coefficient: Volume (corrected to 0°C and 1.0 atm) of gas dissolved per unit volume of solvent at system temperature *T* when the partial pressure of the solute is 1.0 atm (Reid et al. 1977, p. 357, Fogg and Gerrard, 1991, p. 6).

Mass percent: Mass of solute divided by (mass of solute + mass of solvent)—as given by Silberberg (1996, p. 480).

Molality: Moles of solute dissolved in 1000 g solvent—as given by Silberberg (1996, p. 480).

Molarity: Moles of solute dissolved in 1 L of solution—as given by Silberberg (1996, p. 480).

Mole: A mole is defined (Alberty and Silbey, 1992, p. 9) as the amount of substance that has as many atoms or molecules as exactly 0.012 kg of ^{12}C. A gram-mole is the mass in grams of $6.022 \cdot 10^{23}$ molecules of a substance; for example, a mole of carbon has a mass of 12.011 g (Table B.1).

Mole fraction: Moles of solute dissolved divided by (moles of solute + mole of solvent)—as given by Silberberg (1996, p. 480).

Ostwald coefficient: Volume of gas at system temperature *T* and partial pressure *p* dissolved per unit volume of solvent. If the solubility is small and the gas phase is ideal, the Ostwald coefficient is independent of p and these two coefficients are simply related by

$$\text{Ostwald coefficient} = (T/273) \cdot \text{Bunsen coefficient}$$

REFERENCES

Alberty, R. A. and Silbey, R. J., *Physical Chemistry*, 1st edn., John Wiley & Sons, Inc., New York, 1992.

Ashworth, R. A., Howe, G. B., Mullins, M. E., and Rogers, T. N., Air-water partitioning coefficients of organics in dilute aqueous solutions, *Journal of Hazardous Materials*, 18:25–36, 1988.

Battino, R., *Solubility Data Series*, Volume 7, *Oxygen and Ozone*, International Union of Pure and Applied Chemistry, Pergamon Press, Oxford, 1981.

Brennan, R. A., Nirmalakhandan, N., and Speece, R. E., Comparison of predictive methods for Henry's law coefficients of organic chemicals, *Water Research*, 32(6):1901–1911, June 1998.

Carroll, J. J., Use of Henry's law for multicomponent mixtures, *Chemical Engineering Progress*, 88(8):53–58, August, 1992.

Carroll, J. J., Henry's law—a historical view, *Journal of Chemical Education*, 70(2):91–92, February 1993.

Carroll, J. J., Henry's law revisited, *Chemical Engineering Progress*, 95(1):49–56, January 1999.

Dean, J. A., *Lange's Handbook of Chemistry*, 13th edn., McGraw-Hill, New York, 1985.

Eisenberg, D. and Crothers, D., *Physical Chemistry—with Applications to the Life Sciences*, Benjamin/Cummings Publishing Co., Menlo Park, CA, 1979.

Fogg, P. G. T. and Gerrard, W., *Solubility of Gases in Liquids*, John Wiley & Sons, New York, 1991.

Gosset, J. M., Cameron, C. E., Eckstrom, B. P., Goodman, C., and Lincoff, A. H., Mass Transfer Coefficients and Henry's Constants for Packed-Tower Air Stripping of Volatile Organics: Measurement and Correlation, Final Report December 1981–May 1984, AD-A158 811 Engineering and Services Laboratory, Air Force Engineering and Services Center, Tyndall Air Force Base, FL, 1984.

Kavanaugh, M. C. and Trussell, R. R., Design of aeration towers to strip volatile contaminants from drinking water, *Journal American Water Works Association*, 71(12):684–692, 1980.

Lewis, C., Hopke, P. K., and Stukel, J. J., Solubility of radon in selected perfluorocarbon compounds and water, *Industrial and Engineering Chemistry Research*, 26:356–359, 1987.

Lide, D. R. (Ed.), *Handbook of Chemistry and Physics*, 77th edn., 1996–97, CRC Press, Inc., Boca Raton, FL, 1996.

Pankow, J. F., *Aquatic Chemistry Concepts*, CRC Press/Lewis Publishers, Boca Raton, FL, 1991.

Reid, R. C., Prausnitz, J. M., and Sherwood, T. K., *The Properties of Gases and Liquids*, 3rd edn., McGraw-Hill, New York, 1977.

Sander, R., Compilation of Henry's Law Constants for Inorganic and Organic Species of Potential Importance in Environmental Chemistry (Version 3), [http://www.mpcH.mainz.mpg.de/~sander/res/henry.html] Air Chemistry Department, Max-Planck Institute of Chemistry, Mainz, Germany, July 5, 1999.

Sawyer, C. N. and McCarty, P. L., *Chemistry for Sanitary Engineers*, McGraw-Hill, New York, 1967.

Silberberg, M., *Chemistry – The Molecular Nature of Matter and Change*, Mosby—Year Book, Inc., St. Louis, MO, 1996.

Snoeyink, V. L. and Jenkins, D., *Water Chemistry*, John Wiley & Sons, Inc., New York, 1980.

Weast, R. C. (Ed.), *Handbook of Chemistry and Physics*, 59th edn., 1978–79, CRC Press, Inc., Boca Raton, FL, 1978.

Yaws, C. L., *Property Data for Aqueous Systems* (software providing access to tabular data):

1. Solubility in Water (900 compounds), SOLUB4
2. Solubility in Salt Water (217 compounds), SOLUB3
3. Solubility in Water – Variation With Temperature (217 compounds), SOLUB2
4. Henry's Law Constant for Compounds in Water (692 compounds), HENRY
5. Diffusion Coefficient in Water (1359 compounds), DLIQ

Chemical Engineering Department, Lamar University, Beaumont, TX 77710, 1997.

Yaws, C. L., *Chemical Properties Handbook*, McGraw-Hill, New York, 1999.

Yaws, C. L., Pan, X., and Lin, X., Water solubility data for 151 hydrocarbons, *Chemical Engineering*, 100:108–111, February 1993.

Index

X

Z